Forms Containing $\sqrt{u^2 \pm a^2}$

37. $\int \sqrt{u^2 \pm a^2}\, du = \frac{u}{2}\sqrt{u^2 \pm a^2} \pm \frac{a^2}{2}\ln|u +$ …

38. … $a^2\, du = \frac{1}{3}(u^2 \pm a^2)^{3/2} + C$

39. $\int u^2\sqrt{u^2 \pm a^2}\, du = \frac{u}{8}(2u^2 \pm a^2)\sqrt{u^2 \pm a^2} - \frac{a^4}{8}\ln\left|u + \sqrt{u^2 \pm a^2}\right| + C$

40. $\int \frac{\sqrt{u^2 + a^2}}{u}\, du = \sqrt{u^2 + a^2} - a\ln\left|\frac{a + \sqrt{u^2 + a^2}}{u}\right| + C$

41. $\int \frac{\sqrt{u^2 - a^2}}{u}\, du = \sqrt{u^2 - a^2} - a\sec^{-1}\left|\frac{u}{a}\right| + C$

42. $\int \frac{\sqrt{u^2 \pm a^2}}{u^2}\, du = -\frac{\sqrt{u^2 \pm a^2}}{u} + \ln\left|u + \sqrt{u^2 \pm a^2}\right| + C$

43. $\int \frac{du}{\sqrt{u^2 \pm a^2}} = \ln\left|u + \sqrt{u^2 \pm a^2}\right| + C$

44. $\int \frac{u^2\, du}{\sqrt{u^2 \pm a^2}} = \frac{u}{2}\sqrt{u^2 \pm a^2} \mp \frac{a^2}{2}\ln\left|u + \sqrt{u^2 \pm a^2}\right| + C$

45. $\int \frac{du}{u\sqrt{u^2 + a^2}} = -\frac{1}{a}\ln\left|\frac{a + \sqrt{u^2 + a^2}}{u}\right| + C$

46. $\int \frac{du}{u\sqrt{u^2 - a^2}} = \frac{1}{a}\sec^{-1}\left|\frac{u}{a}\right| + C$

47. $\int \frac{du}{u^2\sqrt{u^2 \pm a^2}} = \mp\frac{\sqrt{u^2 \pm a^2}}{a^2 u} + C$

48. $\int (u^2 \pm a^2)^{3/2}\, du = \frac{u}{8}(2u^2 \pm 5a^2)\sqrt{u^2 \pm a^2} + \frac{3a^4}{8}\ln\left|u + \sqrt{u^2 \pm a^2}\right| + C$

49. $\int \frac{du}{(u^2 \pm a^2)^{3/2}} = \pm\frac{u}{a^2\sqrt{u^2 \pm a^2}} + C$

Forms Containing $\sqrt{a^2 - u^2}$

50. $\int \sqrt{a^2 - u^2}\, du = \frac{u}{2}\sqrt{a^2 - u^2} + \frac{a^2}{2}\sin^{-1}\frac{u}{a} + C$

51. $\int u^2\sqrt{a^2 - u^2}\, du = \frac{u}{8}(2u^2 - a^2)\sqrt{a^2 - u^2} + \frac{a^4}{8}\sin^{-1}\left(\frac{u}{a}\right) + C$

52. $\int \frac{\sqrt{a^2 - u^2}}{u}\, du = \sqrt{a^2 - u^2} - a\ln\left|\frac{a + \sqrt{a^2 - u^2}}{u}\right| + C$

53. $\int \frac{\sqrt{a^2 - u^2}}{u^2}\, du = -\frac{\sqrt{a^2 - u^2}}{u} - \sin^{-1}\frac{u}{a} + C$

54. $\int \frac{u^2}{\sqrt{a^2 - u^2}}\, du = -\frac{u}{2}\sqrt{a^2 - u^2} + \frac{a^2}{2}\sin^{-1}\frac{u}{a} + C$

55. $\int \frac{du}{u\sqrt{a^2 - u^2}} = -\frac{1}{a}\ln\left|\frac{a + \sqrt{a^2 - u^2}}{u}\right| + C$

56. $\int \frac{du}{u^2\sqrt{a^2 - u^2}} = -\frac{\sqrt{a^2 - u^2}}{a^2 u} + C$

57. $\int (a^2 - u^2)^{3/2}\, du = \frac{u}{4}(a^2 - u^2)^{3/2} + \frac{3a^2 u}{8}\sqrt{a^2 - u^2} + \frac{3a^4}{8}\sin^{-1}\frac{u}{a} + C$

58. $\int \frac{du}{(a^2 - u^2)^{3/2}} = \frac{u}{a^2\sqrt{a^2 - u^2}} + C$

Forms Involving $\sqrt{2au - u^2}$

59. $\int \sqrt{2au - u^2}\, du = \frac{u - a}{2}\sqrt{2au - u^2} + \frac{a^2}{2}\cos^{-1}\left(\frac{a - u}{a}\right) + C$

60. $\int u\sqrt{2au - u^2}\, du = \frac{2u^2 - au - 3a^2}{6}\sqrt{2au - u^2} + \frac{a^3}{2}\cos^{-1}\left(\frac{a - u}{a}\right) + C$

61. $\int \frac{\sqrt{2au - u^2}}{u}\, du = \sqrt{2au - u^2} + a\cos^{-1}\left(\frac{a - u}{a}\right) + C$

62. $\int \frac{\sqrt{2au - u^2}}{u^2}\, du = -\frac{2\sqrt{2au - u^2}}{u} - \cos^{-1}\left(\frac{a - u}{a}\right) + C$

63. $\int \frac{du}{\sqrt{2au - u^2}} = \cos^{-1}\left(\frac{a - u}{a}\right) + C$

64. $\int \frac{u\, du}{\sqrt{2au - u^2}} = -\sqrt{2au - u^2} + a\cos^{-1}\left(\frac{a - u}{a}\right) + C$

65. $\int \frac{u^2\, du}{\sqrt{2au - u^2}} = -\frac{(u + 3a)}{2}\sqrt{2au - u^2} + \frac{3a^2}{2}\cos^{-1}\left(\frac{a - u}{a}\right) + C$

66. $\int \frac{du}{u\sqrt{2au - u^2}} = -\frac{\sqrt{2au - u^2}}{au} + C$

(continued inside back cover)

CALCULUS
WITH ANALYTIC GEOMETRY

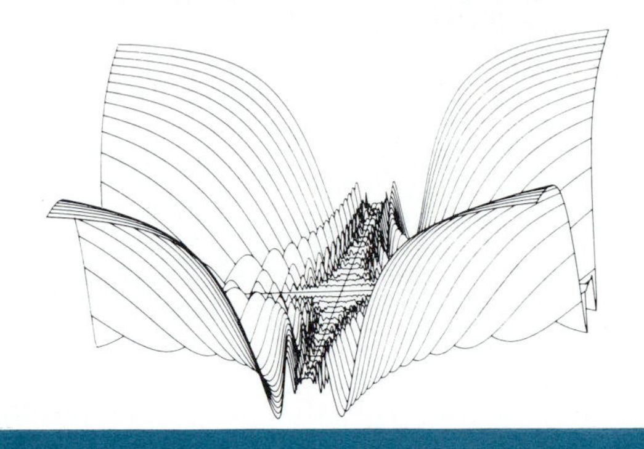

CALCULUS
WITH ANALYTIC GEOMETRY

LEONARD I. HOLDER
GETTYSBURG COLLEGE

WADSWORTH PUBLISHING COMPANY
BELMONT, CALIFORNIA
A DIVISION OF WADSWORTH, INC.

Mathematics Publisher: Kevin J. Howat
Developmental Editor: Anne Scanlan-Rohrer
Assistant Editor: Barbara Holland
Editorial Assistant: Sally S. Uchizono
Production: Cece Munson, The Cooper Company
Manuscript Editor: Carol Reitz
Interior and Cover Designer: Janet Bollow

Chapter Opening Art: Douglas Dunham
Chapter opening illustrations have been generated by computer. See page A-28 for the formulas.
Technical Illustrator: Scientific Illustrators
Compositor: Syntax International

Printed in the United States of America

1 2 3 4 5 6 7 8 9 10—92 91 90 89 88

ISBN 0-534-08202-5

LIBRARY OF CONGRESS CATALOGING-IN-PUBLICATION DATA

Holder, Leonard Irvin, 1923–
Calculus: with analytic geometry.

Includes index.
1. Calculus. 2. Geometry, Analytic. I. Title.
QA303.H576 1987 515 87-15969
ISBN 0-534-08202-5

TO JEAN

PREFACE

This text is designed for a standard three-semester, or four-quarter, course, primarily for students who are planning to major in mathematics, engineering, or one of the physical sciences. It is appropriate also for students in the social or life sciences, especially those in economics, whose interests tend toward the quantitative side of their field. The prerequisite is a good background in algebra and trigonometry, such as from a precalculus course at the high school or college level.

I have found that many students, even with the prerequisite background, can profit from a review of certain concepts. So I have provided such a review in appendixes and introductory material in the body of the text itself. One appendix contains a review of essential concepts from trigonometry and the other a review of mathematical induction and the binomial theorem. I believe it is advantageous to include trigonometric functions at an early stage in calculus, since they enlarge substantially the storehouse of functions available for use in both the theory and applications that follow. In particular, applications of the chain rule are much broader than is the case when only algebraic functions are available. The precalculus material in the text emphasizes the order properties of real numbers. I include here the completeness property (which will be something new for most students) because it fits in logically at this stage and will be needed later in the study of sequences and series. Completing the review material is a section on the coordinate plane, emphasizing the straight line and the circle, along with a brief treatment of the parabola (the more detailed analysis occurs in Chapter 12 on conic sections). My objective is to give a *brief* review of topics that for the most part will be put to immediate use in the study of calculus. I think it is important in motivating students to get to the calculus itself as soon as possible.

I have written the text in a style that I believe students can read and understand, but I have not "written down" to them. Through the use of cautionary notes and marginal comments I have tried to help students avoid some common pitfalls. I have tried to motivate new ideas and have generally proceeded intuitively before giving more abstract formulations. As an illustration, my use of the notion of a neighborhood of a point should help students gain an intuitive understanding of the concept of the limit of a function, and this should be a bridge to the abstract definition that follows. I have provided many examples, which should help students understand the theory and prepare them to do problems on their own.

Applications are important, both in their own right and in helping students to see the power of the calculus they are learning. So I have included a large number and variety of applications in the examples and exercises throughout the text. These range from the purely geometrical to those illustrating the use of calculus in physical and social sciences. Among the latter are applications involving such fields as physics, chemistry, engineering, biology, and economics. These should help motivate students by indicating the broad applicability of calculus.

In many respects problem sets are the heart of a mathematics textbook. So I have constructed these with care. All exercise sets except the supplementary exercises at the ends of chapters are divided into A and B categories (some also have C categories, consisting of problems to be done on a computer). The A exercises are chiefly for drill to fix ideas and techniques in students' minds, whereas the B exercises tend to be more challenging, requiring greater ingenuity on the part of students. The computer exercises (C) are entirely optional and are included for those instructors who wish to use the computer in the calculus course. No instruction in the use of the computer is provided in the text.

The appropriate level of rigor in a calculus course is a subject of continuing debate. Although I rely heavily on intuition, I do give detailed proofs of theorems that I feel are appropriate. In most cases proofs occur at the ends of sections or in a concluding section to a chapter. In this way instructors can include as many of these as they wish. In particular, proofs of theorems on limits and continuity are contained in the final section of Chapter 1 and can be partially included, wholly included, or omitted entirely.

Although the topics covered and the order in which they are taken do not differ significantly from those of other comparable texts, some things do set this one apart. Among these are the following:

1. I introduce the notion of a *neighborhood* of a point early and use it throughout. The first such use is in explaining the limit of a function. The neighborhood concept seems to me to have a simple geometric connotation, and it often simplifies language and symbolism. It is also a link between single and multivariate calculus.
2. I have pulled together several numerical techniques into one chapter (Chapter 9) called "Numerical Methods." I believe that by doing this the material is given greater emphasis than if it were buried here and there in other chapters.
3. As an important application of integration I have included a section on probability density functions (Section 5.8).
4. I have delayed introducing moments of continuously distributed masses until multiple integration is studied, since by doing so the extension from discrete mass systems to continuous ones is easier to explain and to understand than when only single integration is available.
5. I have divided the material on multivariate differential calculus into two chapters: Chapter 16 dealing with functions of two and three variables in general, along with partial derivatives, and Chapter 17 dealing primarily

with differentiability and its consequences. I believe that separating this material into two chapters gives greater emphasis to the crucial notion of differentiability, which is the cornerstone for the differential calculus of functions of more than one variable.

6. I have included chapters on both vector field theory (Chapter 19) and differential equations (Chapter 20). It is unlikely that both will be studied in their entirety in most courses, but having them available provides flexibility. It should be noted that the idea of a differential equation, along with the technique of separation of variables, is introduced in Chapter 6 in the study of exponential growth and decay. So even if the later chapter on differential equations is omitted, students will still have a brief introduction to the subject.

The usefulness of the text is further extended by the availability of the following supplements:

Student Solutions Manual containing solutions to every other even-numbered A and B and supplementary exercise. It also contains complete program listings and output in True BASIC for all C exercises.

Instructor's Solutions Manual containing solutions to all exercises.

Test Item Booklet and Computerized Test Bank giving instructors access to a large number of test questions designed for this text.

The Calculus Tutor, a study guide for students that emphasizes the precalculus concepts with which calculus students often have difficulty. A list of sections from the text that are studied in *The Calculus Tutor* can be found on page A-29 of the text.

Software packages from which instructors may choose, designed to enhance students' understanding of important concepts.

I wish to express my appreciation to all of the persons at Wadsworth Publishing Company who contributed to the production of this text. In particular, I want to thank Kevin Howat, publisher of the mathematics program, and former mathematics editors, Richard Jones and Jim Harrison, for their invaluable assistance and encouragement at various stages of this project. Special thanks also go to Barbara Holland, assistant mathematics editor, and to Anne Scanlan-Rohrer, developmental editor, for their creative approaches to various aspects of the production, including the supplements. Their editorial assistant, Sally Uchizono, handled a multitude of behind-the-scenes tasks with competence.

The production editor, Cece Munson of The Cooper Company, was a delight to work with, and she did an outstanding job. Finally, my typist, Donna Cullison, deserves much credit for her skill and perseverence in transforming my handwritten manuscript into a useable form.

ACKNOWLEDGMENTS

I wish to acknowledge the contributions made by many colleagues to the final form of this text. In its various drafts the manuscript was reviewed by the following persons, whose suggestions and criticisms led to many improvements:

James K. Baker, Jefferson Community College
James D. Blackburn, Tulsa Junior College
Nancy J. Boynton, SUNY College at Fredonia
Caspar R. Curjel, University of Washington
Ray Edwards, Chabot College
William Fuller, Purdue University
Herbert Gindler, San Diego State University
Stuart Goldenberg, California Polytechnic State University, San Luis Obispo
Vera R. Granlund, University of Virginia
John A. Hildebrant, Louisiana State University
Michael Laidacker, Lamar University
Paul McDougle, University of Miami
Jerry Metzger, University of North Dakota
Michael H. Ruge, Louisiana State University
Eugene P. Schlereth, University of Tennessee at Chattanooga
R. D. Semmler, Northern Virginia Community College
Hari Shankar, Ohio University
Marc P. Thomas, California State University at Bakersfield
A. Torchinsky, Indiana University
David V. V. Wend, Montana State University
J. D. Wilson, Murray State University

Having an error-free text is something all authors strive for but almost never achieve without the help of many others. Jerry Metzger of the University of North Dakota, Stuart Goldenberg of the California Polytechnic State University at San Luis Obispo, and Raymond Southworth of the College of William and Mary in Virginia made important contributions to the accuracy of this text by proofreading the galleys, independently from me. Professor Southworth also checked portions of the answer section for the text, and he collaborated with Gloria Langer of the University of Colorado and me on the solutions manuals. Hugh Edgar of San Jose State University and Karl Seydel of Stanford also worked on solutions and checked my answers to exercises. All did excellent work, and I am deeply indebted to them.

The test bank was prepared by George Feissner of SUNY, Cortland, and *The Calculus Tutor* by Ken Seydel of Skyline College. The computer programs for the C exercises were done by my colleague Carl Leinbach of Gettysburg College, who also offered many valuable suggestions for improvement. Students who make use of his solutions will appreciate his carefully written programs and the many accompanying comments explaining all of the crucial steps. The excellent computer graphics in the text were done by Douglas Dunham of the University of Minnesota at Duluth. George and Brian

Morris of Scientific Illustrators in Champaign, Illinois, did an outstanding job on the other art work.

TO THE STUDENT

A few words of advice may help you in your understanding of this text and in your mastery of its subject matter. First of all, *read* the text *before* attempting to do the exercises. Read it with pencil and paper at hand, working through examples and proofs, trying to supply any missing steps. It is counterproductive to try to read a mathematics text passively as you would a novel. You must be actively involved as you read. This will take some time, but it will be worth it. Not only will you be better able to do the exercises that follow, you will have a better understanding of the techniques involved and why they work.

Second, *do* the assigned homework and do it *when it is assigned.* Letting homework accumulate can be fatal. As important as reading the textbook is, it is not enough. To learn calculus you must do calculus problems—lots of them. In this text the A exercises are designed to fix ideas in your mind and so contain a number of ''drill type'' problems. The B exercises tend to be more challenging and often contain extensions of the theory. Do not be afraid of the B exercises though. Many of the most interesting problems are found there, and not all are difficult. However, some exercises are hard and you should struggle with them. Do not give up too quickly. Eventually obtaining the solution to such a challenging exercise can give you a real sense of satisfaction. Answers to odd-numbered exercises are given in the back of the book. You should use these only as a check on your work, after you have found a solution. Do not always assume your answer is wrong if it differs from the one in the text. Sometimes different-appearing answers are equivalent. When you get a different answer, a good practice is to check your work and, if you do not find a mistake, try to show your answer is equivalent to the one in the text. Look especially for combining of fractions or logarithms, factoring, and using trigonometric identities. In doing the exercises, if you find that you are having difficulty with algebraic or other precalculus techniques that are not addressed in the text or the appendixes, you may wish to obtain a copy of the supplement called *The Calculus Tutor.* Among other things this takes up the precalculus aspects of certain sections in the text. For reference, a list of these sections can be found on page A-29.

Applications in a wide variety of fields are given in the examples and exercises in the text. These should help you get a feel for the broad applicability of calculus. Applications of the theory cannot be properly understood, however, without an understanding of the theory itself. So theorems and their proofs are an important part of this textbook. Some proofs are omitted, since they require concepts usually studied only at the advanced calculus level. But those that are given help you to understand why certain results are true. They also help you see how to construct proofs on your own, and in some of the exercises you will have an opportunity to test that ability. Mathematical proofs are examples of deductive reasoning in its purest form, each result following logically from axioms, definitions, and propositions that have already been proved. Since much of the essence of mathematics is to be found in the theory, I encourage you to approach this aspect of calculus with an open and inquisitive mind. The rewards can be great.

Calculus is to me a fascinating subject. I hope I have been able to convey some of my enthusiasm for it in this text.

Leonard I. Holder

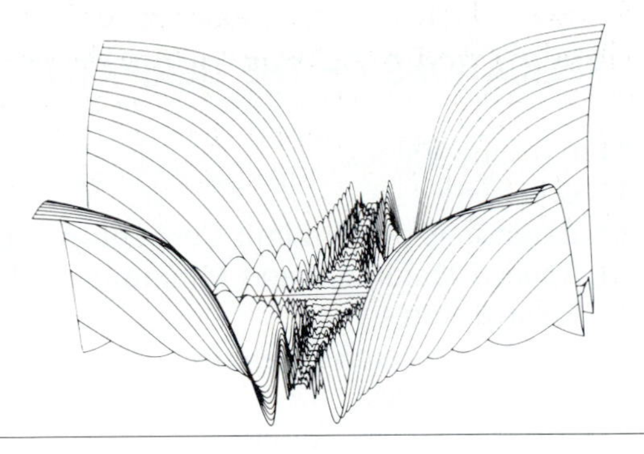

CONTENTS

CHAPTER 3 APPLICATIONS OF THE DERIVATIVE 95

CHAPTER 4 THE INTEGRAL 156

CHAPTER 5 APPLICATIONS OF THE INTEGRAL 190

CHAPTER 6 LOGARITHMIC AND EXPONENTIAL FUNCTIONS 241

CALCULUS
WITH ANALYTIC
GEOMETRY

CHAPTER 1

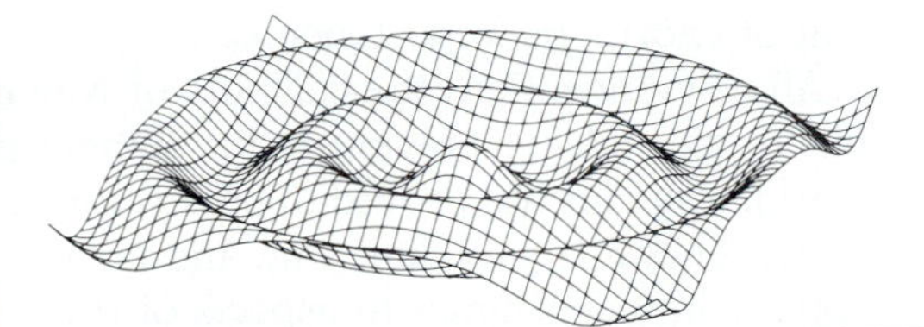

FUNCTIONS, LIMITS, AND CONTINUITY

1.1 HISTORICAL INTRODUCTION

The invention of calculus is one of the landmark achievements in the history of human thought. Its impact on scientific development, and so on society in general, has been profound. Major credit for the introduction of calculus as an organized body of knowledge goes to two geniuses of the 17th century, Gottfried Wilhelm Leibniz (1646–1716) and Sir Isaac Newton (1642–1727). Mathematics was but one area in which Leibniz made significant contributions. He has been called a "universal genius" for his mastery of virtually all branches of knowledge at the time, something that would be impossible today. Because Newton concentrated his efforts in science and mathematics, his contributions were less broad in scope, but his achievements were of monumental proportion. Probably no other single individual's work has had so great an impact on the future development of science. As a mathematician he is considered one of the greatest the world has yet seen.

In their formulation of calculus, Newton and Leibniz worked independently, and although the published work of Leibniz preceded that of Newton, it is now known that Newton obtained his results before Leibniz. (At the time the followers of the two men got into a bitter argument over whose work came first, and this unfortunate affair significantly slowed mathematical progress, especially in England.) The question of whose work came first is of little importance, however. Later refinements of the subject made use of the work of both men. An especially important contribution of Leibniz was a powerful notational system, still in use today.

The ideas of calculus did not occur full blown in the minds of Newton and Leibniz. Newton acknowledged this when he said, "If I have seen a little farther than others it is because I have stood on the shoulders of giants." Although rudiments of the subject can be traced back to the Greek mathematicians, Archimedes in particular, the primary impetus came in the 17th century with such men as Cavalieri in Italy, Fermat and Descartes in France, and Wallis and Barrow in England. The invention of analytical geometry by Fermat and Descartes was especially important to the development of calculus. But it was Newton and Leibniz who found the crucial relationships and provided the notational tools needed to bring the ideas together into a coherent calculational system—that is, into a *calculus*—and this remains a feat of enormous importance.

After its invention, the calculus of Newton and Leibniz was immediately put to use solving previously intractable problems relating to the physical world. Calculus has proved to be an indispensable tool in such areas as astronomy, chemistry, engineering, and physics, and it has also been extensively applied in recent times to aspects of the social and life sciences and to economics. It is rather surprising, then, that calculus evolved from the consideration of two seemingly simple geometric problems: (1) *finding the tangent line to a curve* in the plane and (2) *finding the area of a region* in the plane. The search for a solution to the first problem led to what is now called **differential calculus,** and the second led to **integral calculus.** Our approach to calculus will also be based on these two problems. We will quickly see, though, how the applications far transcend the problems themselves.

1.2 THE REAL LINE

Calculus is based on the **real number system,** and so it is appropriate that we begin by reviewing some of the properties of this most important number system. We denote the set of all real numbers by **R**. An important subset of **R** is the set **Q** of all **rational** numbers. These are numbers that can be expressed as the ratio of two integers—that is, in the form $\frac{m}{n}$, where m and n are integers with $n \neq 0$. This includes the integers themselves, since an integer m can be written as $\frac{m}{1}$. The real numbers that are not rational are called **irrational.** Some examples are $\sqrt{2}$, π, and $\sqrt[3]{7}$. In terms of their decimal representations, rational numbers are either *terminating,* such as $\frac{5}{4} = 1.25$, or *repeating,* such as $\frac{5}{3} = 1.666. . .$. All other decimal quantities represent irrational numbers. For example, 1.010010001. . . is neither terminating nor repeating and so is an irrational number.

One convenient way to visualize real numbers is by associating them with points on a line. By selecting points that correspond to 0 and 1, both a scale and a direction on the line are established. The point corresponding to 0 is called the **origin.** Points that correspond to other integers can be determined as shown in Figure 1.1. All other real numbers can be made to correspond to points on this line. The remarkable thing about this correspondence is that not only does every real number correspond to a point on the line, but also every point on the line corresponds to one and only one real number. We say there is a *one-to-one correspondence* between the points and the real numbers. Because of this identification, we frequently do not distinguish between a real number and the point that corresponds to it. So we might say, for example, "the point 2" rather than "the point corresponding to the number 2." When a line has been *coordinatized* in the manner indicated, we call it the **real number line,** or simply the **real line.**

We assume familiarity with the usual arithmetic properties of real numbers having to do with addition and multiplication. Recall that subtraction

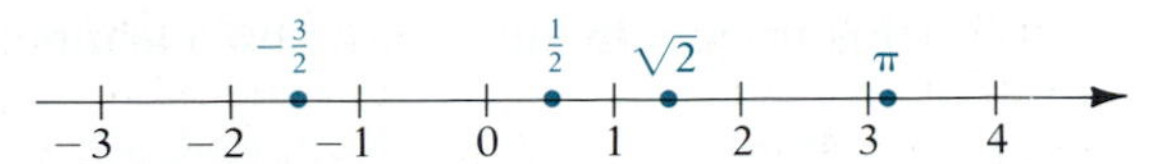

FIGURE 1.1 The real line

and division are defined in terms of addition and multiplication by

$$a - b = a + (-b) \quad \text{and} \quad \tfrac{a}{b} = a \cdot b^{-1} \qquad \text{if } b \neq 0$$

where $-b$ is the *additive inverse* of b and b^{-1} is the *multiplicative inverse* of b, provided $b \neq 0$. This means that $b + (-b) = 0$ and $b \cdot b^{-1} = 1$. Note carefully that the quotient $\frac{a}{b}$ is not defined if $b = 0$, so **division by 0 is excluded.**

We concentrate now on the *order* properties of **R**. If a and b are real numbers such that $b - a$ is positive, then we say **a is less than b** and write $a < b$. Thus,

$$a < b \text{ means } b - a \text{ is positive.}$$

When $a < b$, we may also say **b is greater than a** and write $b > a$. The symbol $a \leq b$ is shorthand for $a < b$ *or* $a = b$, and similarly for $b \geq a$. On the real line with positive direction to the right, if $a < b$, then a is to the *left* of b on the line. The following properties of inequality can be proved by using the definition of inequality and other properties of **R**.

PROPERTIES OF INEQUALITY

1. If $a < b$, then $a + c < b + c$.
2. If $a < b$, then $\begin{cases} ac < bc & \text{if } c > 0 \\ ac > bc & \text{if } c < 0 \end{cases}$.
3. If $a < b$ and $b < c$, then $a < c$.
4. If $a < b$ and $ab > 0$, then $\frac{1}{a} > \frac{1}{b}$.

Particular attention should be paid to Property 2 when $c < 0$. In effect, this says that multiplying both sides of an equality by a negative number reverses the sense of the inequality.

In the two examples that follow we illustrate ways to "solve" linear and quadratic inequalities. To solve an inequality means to find the set of all real numbers x for which the inequality is satisfied.

EXAMPLE 1.1 Solve the inequality

$$\frac{3 - x}{4} < \frac{x}{3} - 1$$

Solution To eliminate fractions we multiply both sides by the lowest common denominator, 12. Then we proceed by using the properties shown:

$$\begin{aligned} 9 - 3x &< 4x - 12 && \text{(Property 2)} \\ -7x &< -21 && \text{(Property 1)} \\ x &> 3 && \text{(Property 2)} \end{aligned}$$

Note that in the last step we multiplied both sides by the negative number $-\frac{1}{7}$ and so had to reverse the sense of the inequality. ■

A *set* is frequently designated by a symbol of the form

$$\{x: \underline{\qquad\qquad\qquad}\}$$

where a description of properties possessed by x follows the colon. As an illustration, in Example 1.1 we could write the solution set as $\{x: x > 3\}$. This is read as "the set of all x such that x is greater than 3." In such designations we will understand x to be a real number unless otherwise specified.

EXAMPLE 1.2 Solve the inequality $x(x-1) \leq 2$.

Solution We begin just as if this were a quadratic equation, using Property 1 to bring all terms to one side and then factoring:

$$x^2 - x - 2 \leq 0$$
$$(x-2)(x+1) \leq 0$$

The product on the left *equals* 0 when $x = 2$ or when $x = -1$. It is *less than* 0 when the factors $(x-2)$ and $(x+1)$ are of opposite sign. A convenient way to see where this occurs is to mark the points $x = 2$ and $x = -1$ on a real number line as in Figure 1.2. Then we test the product $(x-2)(x+1)$ for its sign in each of the three regions determined by these points. Because the product is of constant sign in each region, it is sufficient to test the sign at just one point in each region. For example, we could use $x = -2$, $x = 0$, and $x = 3$. The signs are readily seen to be those shown above the regions in the figure. We can now write the answer as

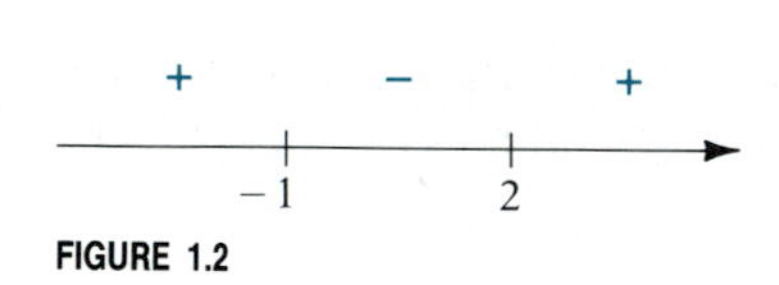

FIGURE 1.2

$$\{x: -1 \leq x \leq 2\}$$ ■

Figure 1.2 is an example of what we refer to as a **sign graph.** Here is another example illustrating the use of such a sign graph to solve a nonlinear inequality.

EXAMPLE 1.3 Solve the inequality

$$\frac{x(x+5)}{x+1} > 3$$

Solution First we add -3 to both sides and then combine in a single fraction, getting

$$\frac{x^2 + 2x - 3}{x+1} > 0 \qquad \text{(Verify.)}$$

Now we factor the numerator and make a sign graph:

$$\frac{(x+3)(x-1)}{x+1} > 0$$

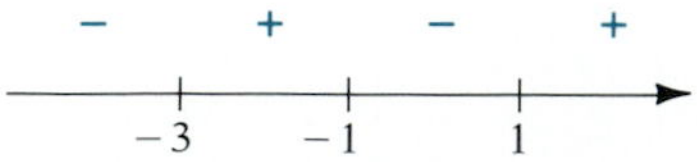

The points of division on the sign graph are points where either numerator *or denominator* is 0. The sign in each region is determined using a test value, with the results as shown in Figure 1.3. The solution is seen to be the points in either of the sets

– + – +
–3 –1 1
FIGURE 1.3

$$\{x: -3 < x < -1\} \quad \text{or} \quad \{x: x > 1\}$$ ■

Note carefully that in this example we did *not* clear the fractions, since this would have involved multiplying by the factor $x + 1$, whose sign is sometimes positive and sometimes negative, so it would not be clear which part of Property 2 to use.

The solution given in Example 1.3 can be indicated more briefly by using the following symbolism. Let A and B denote two sets. Then the **union of A and B,** written $A \cup B$, is defined as

$$A \cup B = \{\text{all elements in } \textit{either } A \textit{ or } B\}$$

So the solution in Example 1.3 can be written

$$\{x: -3 < x < -1\} \cup \{x: x > 1\}$$

It is also sometimes convenient to use the notion of the **intersection** of two sets A and B, written $A \cap B$, and defined as

$$A \cap B = \{\text{all elements in } \textit{both } A \textit{ and } B\}$$

The symbol $\varnothing$ is used to denote the **empty set.** If $A \cap B = \varnothing$, we say A and B are **disjoint.**

Sets like those in the solutions to the preceding inequalities occur in many other contexts as well; we give them the following special names and symbols:

Set	Name	Symbol
$\{x: a \le x \le b\}$	Closed interval	$[a, b]$
$\{x: a < x < b\}$	Open interval	(a, b)

A set of the form $\{x: a < x \le b\}$ is designated $(a, b]$ and is said to be a **half-open** (or **half-closed**) interval. The same is true for $\{x: a \le x < b\} = [a, b)$. It is also convenient to introduce the symbol ∞, read *infinity* and interpreted roughly as meaning "beyond all bound." Although this is *not* a number, we use it in interval notation as follows:

$$\{x: x \ge a\} = [a, \infty)$$
$$\{x: x > a\} = (a, \infty)$$
$$\{x: x \le a\} = (-\infty, a]$$
$$\{x: x < a\} = (-\infty, a)$$
$$\{\text{all real } x\} = (-\infty, \infty)$$

We will frequently need to use the **absolute value** of a number, which means the "magnitude" of the number without regard to its sign. More precisely, the absolute value of a real number a, denoted by $|a|$, is defined as

$$|a| = \begin{cases} a & \text{if } a \ge 0 \\ -a & \text{if } a < 0 \end{cases}$$

So $|a|$ is always nonnegative (that is, either positive or zero). For example, $|2| = 2$ and $|-2| = -(-2) = 2$. Geometrically, $|a|$ is the distance between a and 0 on the real line. Similarly, for any two real numbers a and b, $|a - b|$ is the distance between them on the real line:

$$\text{Distance between } a \text{ and } b = |a - b|.$$

The basic properties of absolute value are given here.

PROPERTIES OF ABSOLUTE VALUE

1. $|a| \geq 0$ and $|a| = 0$ if and only if $a = 0$
2. $|-a| = |a|$
3. $|ab| = |a|\,|b|$
4. $|a + b| \leq |a| + |b|$

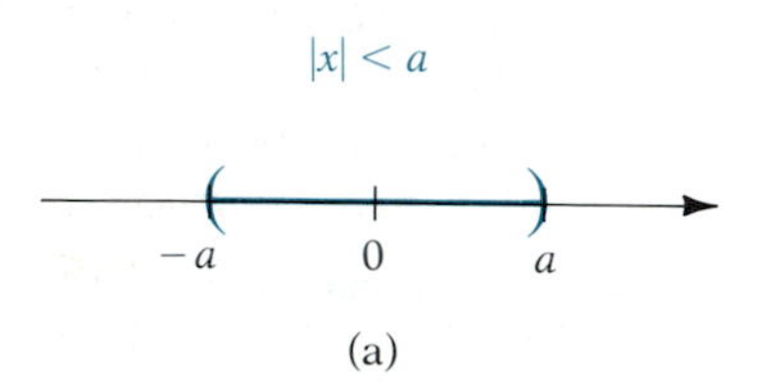

The first three properties can be seen to be true directly from the definition. Property 4 is called the **triangle inequality** and can be proved as follows. Since the absolute value of a number equals either the number or its negative, we have $-|a| \leq a \leq |a|$ and similarly $-|b| \leq b \leq |b|$. Adding corresponding members gives

$$-(|a| + |b|) \leq a + b \leq |a| + |b|$$

From this we get $(a + b) \leq |a| + |b|$ and $-(a + b) \leq |a| + |b|$. It follows that $|a + b| \leq |a| + |b|$, since $|a + b|$ is either $(a + b)$ or its negative.

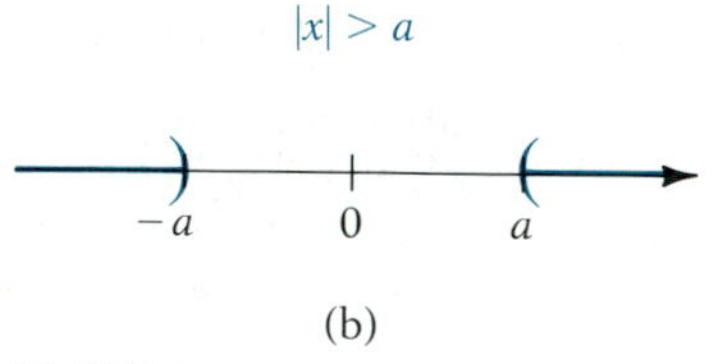

FIGURE 1.4

From the geometric interpretation of absolute value as the distance from a to 0, it is readily seen that for $a > 0$,

$$|x| < a \quad \text{if and only if} \quad -a < x < a$$

and

$$|x| > a \quad \text{if and only if} \quad x > a \quad \text{or} \quad x < -a$$

Figure 1.4 illustrates these relationships. They can also be verified by using the definition of absolute value. In interval notation, we see that the set of real numbers x that satisfy $|x| < a$ is the open interval $(-a, a)$, and the x values that satisfy $|x| > a$ are those in one or the other of the intervals $(-\infty, -a)$ or (a, ∞)—that is, in $(-\infty, -a) \cup (a, \infty)$.

EXAMPLE 1.4 Solve the inequality $2|3 - 4x| - 1 > 7$.

Solution Simplifying by using inequality Properties 1 and 2, we get $|3 - 4x| > 4$, which is satisfied if and only if

$$3 - 4x > 4 \quad \text{or} \quad 3 - 4x < -4$$

Solving each of these linear inequalities, we get

$$\begin{array}{ccc} -4x > 1 & & -4x < -7 \\ & \text{or} & \\ x < -\frac{1}{4} & & x > \frac{7}{4} \end{array}$$

The solution set can thus be written as

$$(-\infty, -\tfrac{1}{4}) \cup (\tfrac{7}{4}, \infty)$$

■

By a **neighborhood** of point a we mean an open interval centered on a. Thus a neighborhood of a is of the form $(a - r, a + r)$ for some positive number r called the *radius* of the neighborhood. This is illustrated in Figure 1.5. For example, (1.8, 2.2) is a neighborhood of 2 of radius 0.2. The Greek letters ε and δ are frequently used in calculus to designate the radii of neighborhoods. So $(2 - \delta, 2 + \delta)$ is a δ-neighborhood (a neighborhood of radius δ) of 2, and $(L - \varepsilon, L + \varepsilon)$ is an ε-neighborhood of L. From our discussion we see that an r-neighborhood of a is identical to the solution set of the inequality $|x - a| < r$.

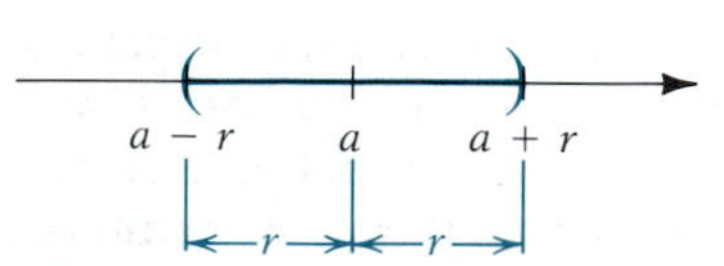

FIGURE 1.5 A neighborhood of a of radius r

If we delete the point $x = a$ from a neighborhood centered at a, we call the resulting set a **deleted neighborhood** of a. If the radius is r, then such a deleted neighborhood is the solution set of the inequality $0 < |x - a| < r$. Here the requirement that $0 < |x - a|$ ensures that $x \neq a$. A deleted neighborhood is illustrated in Figure 1.6. The open circle at a indicates that a is not included.

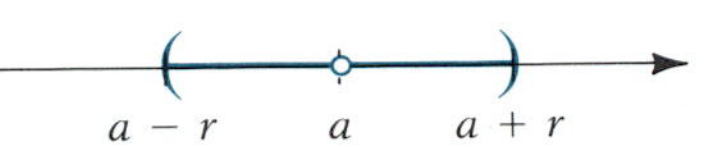

FIGURE 1.6 A deleted neighborhood of a

We conclude this section with a brief discussion of one of the fundamental properties of the real number system that has to do with the order relation. A subset S of $\mathbf{R}$ is said to be **bounded above** if there exists a real number M such that $x \leq M$ for all x in S. When this is true, M is said to be an **upper bound** of S. Thus M is an upper bound of S if, on the real line, no element of S is to the right of M. Clearly, if M is an upper bound of S, any number to the right of M is also an upper bound of S. The more critical question is whether there is an upper bound of S smaller than all other upper bounds. If so, this is called the **least upper bound** of S. To illustrate, the set

$$\{1, 1\tfrac{1}{2}, 1\tfrac{3}{4}, 1\tfrac{7}{8}, 1\tfrac{15}{16}, \ldots\}$$

is bounded above by 3, for example, but the *least* upper bound is 2, since 2 is an upper bound and nothing less than 2 will do.

A fundamental property of $\mathbf{R}$ is that for every nonempty subset of $\mathbf{R}$ that is bounded above, such a least upper bound does always exist. This is known as the *completeness axiom*, and we state it here for future reference.

THE COMPLETENESS AXIOM Every nonempty set of real numbers that is bounded above has a least upper bound in $\mathbf{R}$. ■

EXAMPLE 1.5 Determine the least upper bound of each of the following sets:

(a) $\{x: -1 \leq x \leq 3\}$ (b) $\{x: -1 < x < 3\}$

(c) $\{1 - \frac{1}{n}: n \text{ a positive integer}\}$ (d) $\{x: x^2 < 2\}$

Solution (a) An upper bound is 3, and since no number less than 3 is an upper bound, we conclude that 3 is the least upper bound.

(b) Again, 3 is an upper bound, and although 3 is not in the set, elements of the set come arbitrarily close to 3. So nothing less than 3 is an upper bound. For example, 2.99 is not an upper bound, since 2.999 is in the set and exceeds 2.99. Thus 3 is the least upper bound.

(c) Some elements of the set are

$$1 - \tfrac{1}{1} = 0 \qquad 1 - \tfrac{1}{2} = \tfrac{1}{2} \qquad 1 - \tfrac{1}{3} = \tfrac{2}{3} \qquad 1 - \tfrac{1}{4} = \tfrac{3}{4}$$

found by putting $n = 1, 2, 3, 4$, respectively. Since every member of the set is less than 1, it follows that 1 is an upper bound. In fact, it is the least upper bound, since by taking n large enough, we can make $1 - \frac{1}{n}$ as close to 1 as we wish. Nothing less than 1 is an upper bound to this set.

(d) Some upper bounds to the set are 3, 2, 1.5, 1.42, 1.415, since the square of each of these numbers exceeds 2. (Check this.) In fact, the least upper bound is $\sqrt{2}$, whose decimal expansion is neither repeating nor terminating. This example shows, incidentally, that the rational numbers do not possess the completeness property because there is no smallest *rational* number that bounds the set. ■

EXERCISE SET 1.2

A

1. Express each of the following rational numbers as either a terminating or a repeating decimal.
(a) $\frac{5}{8}$ (b) $\frac{8}{5}$ (c) $\frac{10}{27}$ (d) $-\frac{9}{37}$ (e) $\frac{10}{7}$
2. State which of the following are rational and which are not.
(a) $\frac{\sqrt{81}}{4}$ (b) -2 (c) $\sqrt{5}$
(d) $\sqrt[3]{-8}$ (e) $\frac{1}{2}+\frac{2}{3}$
3. Let $x = 1.242424\ldots$. By considering $100x - x$, express x in the form $\frac{m}{n}$ in lowest terms.
4. Use the idea of Exercise 3 to write $x = 0.0243243243\ldots$ in the form $\frac{m}{n}$ in lowest terms.
5. Replace the question mark with the correct inequality symbol.
(a) $-10 \,?\, 2$ (b) $-7 \,?\, -9$ (c) $\frac{2}{3} \,?\, \frac{5}{9}$
(d) $3.6 \,?\, \frac{15}{4}$ (e) $0 \,?\, -100$
6. Write each of the following sets in interval notation and show it on the real line.
(a) $\{x: -1 < x < 2\}$ (b) $\{x: x \leq 2\}$
(c) $\{x: 2 < x \leq 5\}$ (d) $\{x: 0 \leq x \leq 1\}$
7. Write the meaning of each of the following intervals using set notation, and show it on the real line.
(a) $[2, 5]$ (b) $(-2, 3]$ (c) $(3, 5)$
(d) $(-\infty, 2)$ (e) $[0, \infty)$
8. Give the value of each of the following.
(a) $|-6|$ (b) $|0|$
(c) $|3 - \pi|$ (d) $|a - b|$ if $b > a$
(e) $|-x|$ if $x < 0$

In Exercises 9–26 solve the inequality.

9. $4 - 2x < 6 - 3x$
10. $\frac{x}{2} - \frac{3}{4} > \frac{5x}{6} - \frac{1}{3}$
11. $0 < \frac{3x-2}{4} \leq 2$
12. $-3 \leq \frac{5-3x}{2} < 6$
13. $|3 - 2x| \geq 2$
14. $|3x - 5| \leq 2$
15. $2|2x - 1| - 3 < 5$
16. $\frac{|1-x|}{3} < 2$
17. $x^2 - 3x - 4 \leq 0$
18. $2x^2 - 9x + 4 \geq 0$
19. $x(2x - 3) < 5$
20. $3x^2 \geq 4(1 - x)$
21. $\frac{3-x}{x^3+2x^2} < 0$
22. $\frac{x}{x+4} > 0$
23. $(2x - 1)(x + 4)(x - 3) \geq 0$
24. $(x - 1)(x + 2)(x - 4) < 0$
25. $\frac{3x+1}{x-2} > 2$
26. $\frac{2x-3}{x-1} \leq 4$

In Exercises 27–32 express the given neighborhood in terms of an inequality and show it on the real line.

27. A neighborhood of 3 of radius 1
28. A neighborhood of -1 of radius $\frac{1}{2}$
29. A deleted neighborhood of 5 of radius 2
30. A deleted neighborhood of -2 of radius 0.1
31. An ε-neighborhood of 4
32. A deleted δ-neighborhood of 3

In Exercises 33 and 34 find a and r for which the given interval is an r-neighborhood of a.

33. (a) $(2, 12)$ (b) $(0, 4)$
34. (a) $(-5, 1)$ (b) $(-8, -3)$

In Exercises 35–37 show the set on the real line and give its least upper bound.

35. (a) $\{x: x < 3\}$ (b) $\{x: |x - 2| < 1\}$
36. (a) $[-3, 7]$ (b) $(-5, 0)$
37. (a) $\left\{\frac{2n-1}{n} : n \text{ a positive integer}\right\}$
(b) $\{-1, -\frac{1}{2}, -\frac{1}{3}, \ldots\}$

B

38. Prove inequality Property 1.
39. Prove inequality Property 2. (*Hint:* Use the fact that the product of two positive numbers is positive. Also, if $c < 0$, then $-c > 0$.)
40. Prove inequality Property 3. (*Hint:* Use the fact that the sum of two positive numbers is positive.)
41. Show that if $a > 0$, then $\frac{1}{a} > 0$. (*Hint:* Show $\frac{1}{a}$ cannot be negative or 0.)
42. Prove inequality Property 4. (*Hint:* Use Exercise 41.)
43. Prove absolute value Property 2.
44. Prove absolute value Property 3.

45. Prove that $|a - b| \leq |a| + |b|$.

46. Prove that $||a| - |b|| \leq |a - b|$. (*Hint:* Write the triangle inequality as $|x + y| \leq |x| + |y|$. First substitute $x = a$ and $y = b - a$. Starting again with the triangle inequality, substitute $x = b$ and $y = a - b$.)

47. Prove that if S is a set of real numbers that is bounded above, then M is the least upper bound of S provided the following two conditions are satisfied: (i) $x \leq M$ for all $x \in S$ and (ii) if $L < M$, there is an x in S such that $x > L$.

48. Formulate a definition of a **lower bound** and the **greatest lower bound** of a set S that is bounded below. Give conditions analogous to those in Exercise 47 that ensure that m is the greatest lower bound of S.

1.3 THE CARTESIAN PLANE

Consider two real number lines perpendicular to each other so that their origins coincide, as in Figure 1.7. We name the horizontal line the **x-axis** and the vertical line the **y-axis**, and we call their point of intersection the **origin.** These axes divide the plane into four **quadrants,** which we number as in the figure. Through any point in the plane we pass vertical and horizontal lines. Their intersections with the axes determine the **x-coordinate,** or **abscissa,** and the **y-coordinate,** or **ordinate,** of the point. If the abscissa is a and the ordinate is b, we represent the point by the ordered pair (a, b).* The numbers a and b are called the **coordinates** of the point. For brevity, we often say the point has coordinates (a, b). For example, in Figure 1.7 the point P has coordinates $(3, 2)$ and the point Q has coordinates $(-4, -3)$.

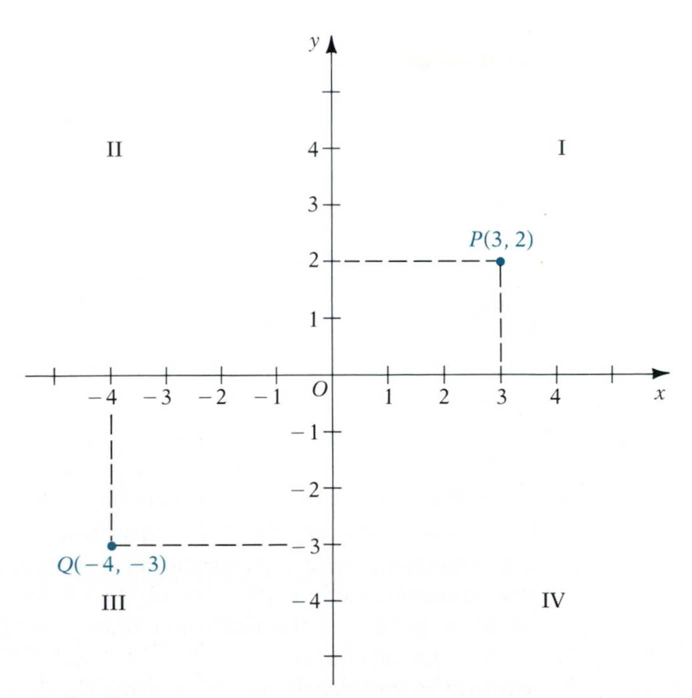

FIGURE 1.7

* The context will make clear whether (a, b) refers to a point in the plane or to an open interval on the line.

Conversely, if we are given an ordered pair of real numbers (a, b), we can locate the unique point that has these coordinates from the intersection of a vertical line through a on the x-axis and a horizontal line through b on the y-axis. In this way we obtain a one-to-one correspondence between points in the plane and ordered pairs of numbers. Frequently we do not distinguish between a point and its coordinates, saying, for example, "the point (3, 2)" rather than "the point with coordinates (3, 2)."

What we have described is referred to as a **rectangular,** or **Cartesian,** coordinate system. The latter name is for the French mathematician and philosopher René Descartes (1596–1650), who is usually given credit for originating analytic geometry, which is based on representing ordered pairs as points in the way we have described. Actually, many of the ideas were first introduced by another French mathematician, Pierre de Fermat (1601–1665). The plane, when coordinatized as we have shown, is often referred to as the **Cartesian plane.**

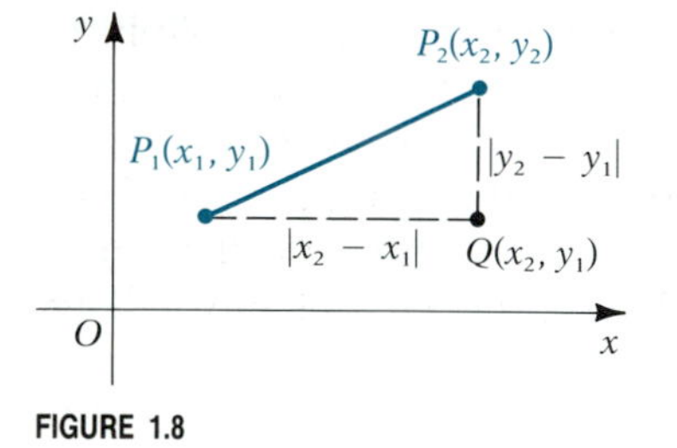

FIGURE 1.8

If we are given two points P_1 and P_2 in the plane with coordinates (x_1, y_1) and (x_2, y_2), as shown in Figure 1.8, we can find the distance d between them by using the Pythagorean theorem, which says that in a right triangle the square of the length of the hypotenuse equals the sum of the squares of the lengths of the legs. We introduce the point Q as shown, with coordinates (x_2, y_1). Then P_1QP_2 is a right triangle, and since P_1Q is horizontal, its length is $|x_2 - x_1|$. Similarly, QP_2 is vertical and its length is $|y_2 - y_1|$. Thus

$$d^2 = |x_2 - x_1|^2 + |y_2 - y_1|^2$$

or, equivalently, since $|a|^2 = a^2$, we have the following formula.

THE DISTANCE FORMULA

$$d = \sqrt{(x_2 - x_1)^2 + (y_2 - y_1)^2}$$

For example, the distance between $(1, -2)$ and $(-4, 7)$ is

$$d = \sqrt{(1 + 4)^2 + (-2 - 7)^2} = \sqrt{25 + 81} = \sqrt{106}$$

Notice that it does not matter which point is labeled (x_1, y_1) and which is labeled (x_2, y_2).

The real power of the contribution of Descartes and Fermat is in representing equations, which are *algebraic* objects, by means of collections of points in the plane, which are *geometric* objects. If the equation involves only two variables x and y, then the set of points in the Cartesian plane corresponding to all ordered pairs (x, y) for which x and y satisfy the equation is called the **graph of the equation.** For example, if we wish to graph the equation $y = 2x - 1$, we can find several ordered pairs (x, y) that satisfy the equation, such as $(0, -1)$, $(1, 1)$, $(2, 3)$, and so on. Then we can locate these as points in the Cartesian plane and (going on faith) connect them with a smooth curve (which in this case is a straight line). This method of plotting points clearly has its drawbacks, even if we plot a very large number of points. How can we be sure, for example, that they should be connected with a single unbroken curve? Later on, when we learn some calculus, we will have some aids to graphing that enable us to become less dependent on plotting points. Even without calculus, by analyzing certain curves we can

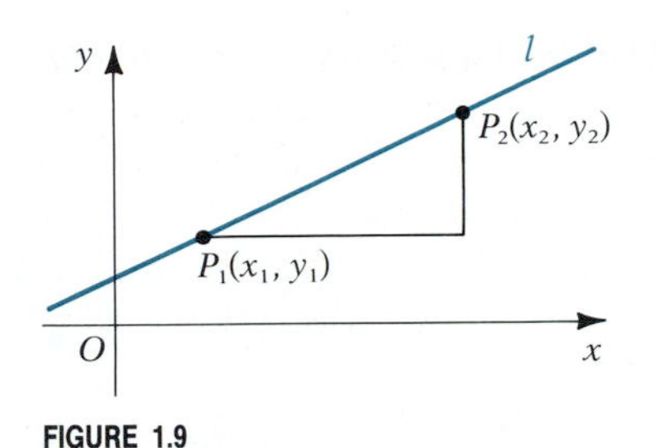

FIGURE 1.9

determine their equations, or conversely recognize the graph from the equation. For now, we do this for only three particularly simple, but very useful curves: the straight line, the circle, and the parabola.

The Straight Line

Consider a nonvertical line l as shown in Figure 1.9, and let P_1 and P_2 be any two distinct points on l with coordinates (x_1, y_1) and (x_2, y_2), respectively. Then we define the **slope** m of l by

$$m = \frac{y_2 - y_1}{x_2 - x_1}$$

You should convince yourself, using similar triangles, that the value of the slope is independent of which two points on l are used.

Lines that go upward to the right have positive slopes, and those that go downward to the right have negative slopes. This follows from the definition. Also, the slope of a horizontal line is 0, and the slope of a vertical line is not defined. Figure 1.10 illustrates these facts.

Suppose we know one point, say $P_1(x_1, y_1)$, on a nonvertical line l and that its slope m is given (or can be found). A point $P(x, y)$ distinct from P_1 will lie on l if and only if the slope as calculated using P_1 and P equals m:

$$\frac{y - y_1}{x - x_1} = m$$

If we clear this equation of fractions, we can write it as

$$y - y_1 = m(x - x_1) \tag{1.1}$$

Note that in this form, the equation is satisfied even if $P = P_1$. Thus, the point $P(x, y)$ lies on l if and only if its coordinates satisfy equation (1.1). The graph of equation (1.1) is therefore the straight line l. Equation (1.1) is called the **point-slope form** of the equation of a line.

EXAMPLE 1.6 Find the equation of the line through the points $(-1, 2)$ and $(4, 5)$.

Solution First we find the slope:

$$m = \frac{5 - 2}{4 - (-1)} = \frac{3}{5}$$

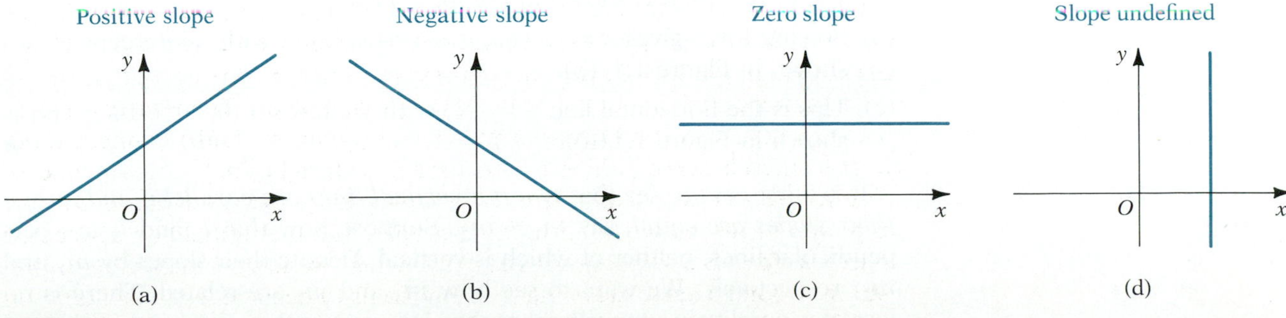

FIGURE 1.10

We may use either point as (x_1, y_1) in the point-slope equation (1.1). Choosing $(-1, 2)$, we get

$$y - 2 = \tfrac{3}{5}(x + 1)$$

which on simplifying becomes

$$3x - 5y + 13 = 0$$

■

The point at which a line crosses the y-axis is called its *y-intercept.* Let us designate this by $(0, b)$. If we use this as (x_1, y_1) in the point-slope equation (1.1), we get

$$y = mx + b \tag{1.2}$$

and this is called the **slope-intercept form** of the equation of a line.

Either equation (1.1) or (1.2) results in an equation of the form

$$Ax + By + C = 0 \tag{1.3}$$

when the terms are rearranged. Note that the final answer in Example 1.6 was written in this form. So we know that every nonvertical line has an equation that can be written in the form of equation (1.3). Conversely, suppose we are given such an equation, where A and B are not both 0. If $B \neq 0$, we can solve for y to get

$$y = -\tfrac{A}{B}\, x + \left(-\tfrac{C}{B}\right)$$

This is in the form of equation (1.2) and so represents a line with slope $m = -\frac{A}{B}$ and y-intercept $b = -\frac{C}{B}$. In particular, if $A = 0$, it is the horizontal line $y = b$.

If $B = 0$, then equation (1.3) has the form $Ax + C = 0$. This can be written as $x = a$ where $a = -\frac{C}{A}$. It is satisfied by all pairs (a, y) for y arbitrary, and hence represents a vertical line through $(a, 0)$.

We conclude that equation (1.3) always represents a line. We call it the **general equation** of a line.

EXAMPLE 1.7 Identify and sketch the graph of each of the following.

(a) $3x - 4y + 12 = 0$ (b) $2x - 3 = 0$ (c) $3y + 4 = 0$

Solution

(a) Solving for y, we get $y = \frac{3}{4}x + 3$. By the slope-intercept form $y = mx + b$, we see that this is a line with slope $m = \frac{3}{4}$ and y-intercept $(0, 3)$. One other point is sufficient to enable us to draw the line. Any point will do, but a particularly easy one to use is the x-intercept, $(-4, 0)$, found by putting $y = 0$. Its graph is shown in Figure 1.11(a).

(b) Solving for x gives $x = \frac{3}{2}$. This is a vertical line with x-intercept $(\frac{3}{2}, 0)$, shown in Figure 1.11(b).

(c) This is the horizontal line $y = -\frac{4}{3}$, with y-intercept $(0, -\frac{4}{3})$. Its graph is shown in Figure 1.11(c).

■

It is fairly easy to see that *two nonvertical lines are parallel if and only if their slopes are equal,* say $m_1 = m_2$. Suppose now that l_1 and l_2 are perpendicular lines, neither of which is vertical. Denote their slopes by m_1 and m_2, respectively. We want to see how m_1 and m_2 are related. There is no loss of generality in supposing that they intersect at the origin; if they do not,

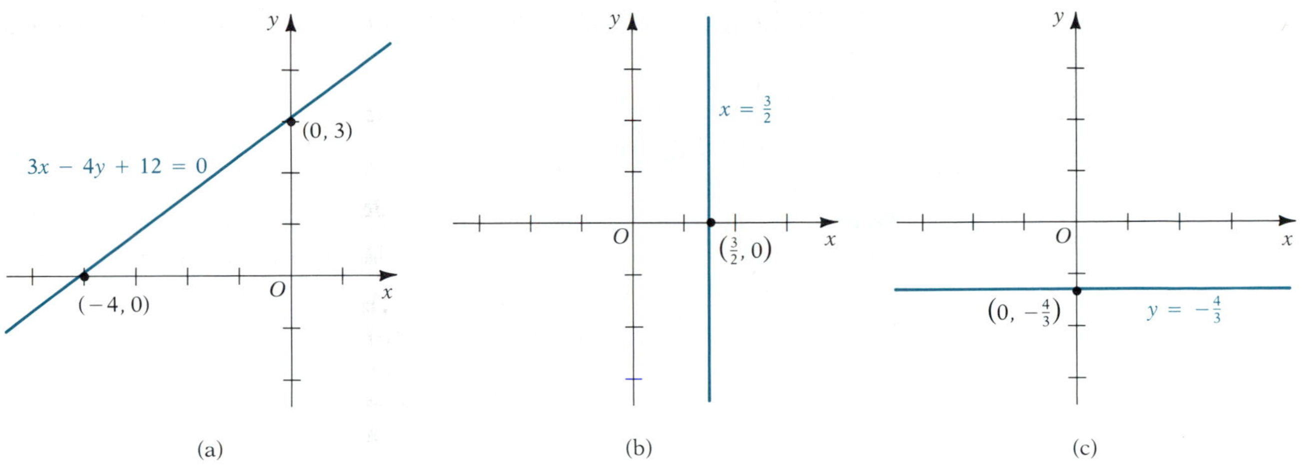

FIGURE 1.11

then we can consider lines parallel to l_1 and l_2 that do intersect at the origin, and these again have slopes m_1 and m_2. So we assume l_1 and l_2 intersect at 0, as in Figure 1.12. By the definition of slope, the point $P_1(1, m_1)$ lies on l_1, and $P_2(1, m_2)$ lies on l_2. Then, since P_1OP_2 is a right triangle, we have by the Pythagorean theorem, $\overline{OP}_1^2 + \overline{OP}_2^2 = \overline{P_1P}_2^2$, or using the distance formula,

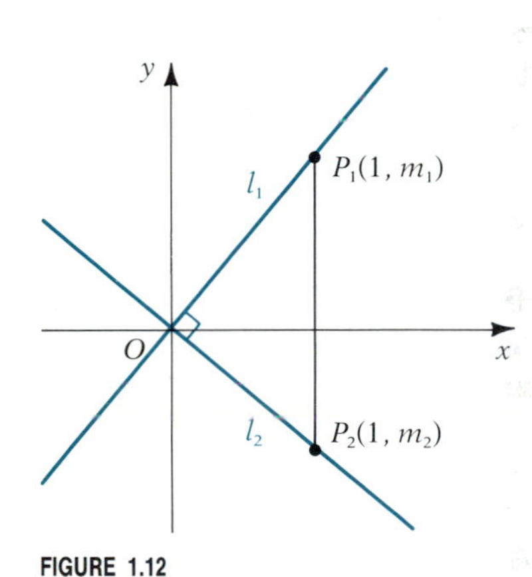

FIGURE 1.12

$$(1 + m_1)^2 + (1 + m_2)^2 = (m_1 - m_2)^2$$

Simplifying this, we get $m_1m_2 = -1$ or, equivalently,

$$m_2 = -\frac{1}{m_1} \tag{1.4}$$

We say that the slope of l_2 is the *negative reciprocal* of the slope of l_1 (and vice versa). When l_1 and l_2 are perpendicular, then equation (1.4) holds. Conversely, if equation (1.4) is true, by reversing steps we can see that l_1 and l_2 are perpendicular. We summarize our results here.

PARALLEL AND PERPENDICULAR LINES

Let l_1 and l_2 be nonvertical lines with slopes m_1 and m_2, respectively. Then l_1 and l_2 are **parallel** if and only if $m_1 = m_2$, and they are **perpendicular** if and only if $m_2 = -1/m_1$.

EXAMPLE 1.8 Find the equation of the line passing through $(2, -3)$ that is (a) parallel to and (b) perpendicular to the line $2x + 3y = 5$.

Solution By solving for y we get the given line in the slope-intercept form:

$$y = -\tfrac{2}{3}x + \tfrac{5}{3}$$

and so we see that its slope is $-\frac{2}{3}$.

(a) A line parallel to the given line also has slope $-\frac{2}{3}$. So by the point-slope equation (1.1) we get

$$y + 3 = -\tfrac{2}{3}(x - 2)$$

or $2x + 3y + 5 = 0$.

(b) A line perpendicular to the given line has slope $\frac{3}{2}$. So the desired equation is

$$y + 3 = \tfrac{3}{2}(x - 2)$$

or $3x - 2y - 12 = 0$. ■

Remark An alternate approach in the preceding example is to note that all lines parallel to $2x + 3y - 5 = 0$ have equations of the form $2x + 3y + C_1 = 0$, and all lines perpendicular to it have equations of the form $3x - 2y + C_2 = 0$. (Why?) These are called *families* of lines. By substituting the given point, the particular constant C_1 or C_2 can be found in each case.

The Circle

A circle is the set of all points at a fixed distance (the radius) from a fixed point (the center). Suppose the radius is r and the center is (h, k). Then a point $P(x, y)$ lies on the circle if and only if the distance from (h, k) to (x, y) is r or, equivalently, the square of the distance is r^2 (see Figure 1.13). This gives

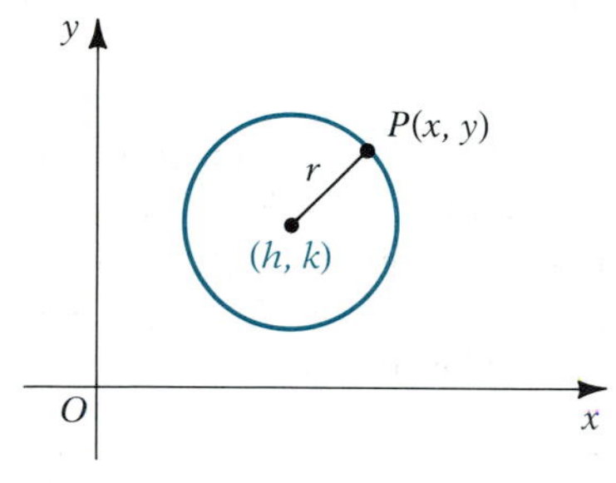

FIGURE 1.13

$$(x - h)^2 + (y - k)^2 = r^2 \tag{1.5}$$

We call this the **standard form** of the equation of a circle. If the center is at the origin, the equation assumes the particularly simple form

$$x^2 + y^2 = r^2$$

If equation (1.5) is expanded and terms are rearranged, it can be put in the form

$$x^2 + y^2 + ax + by + c = 0$$

The question now arises as to whether every equation of this type is the equation of a circle. To answer this, we can *complete the squares* on x and y to see whether it can be put in the form of equation (1.5). Recall that to complete the square on an expression of the form $x^2 + ax$, we add $(\frac{a}{2})^2$. Of course, we must then balance this by adding the same quantity to the right side. Completing the square on both x and y therefore gives

$$\left(x^2 + ax + \frac{a^2}{4}\right) + \left(y^2 + by + \frac{b^2}{4}\right) = -c + \frac{a^2}{4} + \frac{b^2}{4}$$

or

$$\left(x + \frac{a}{2}\right)^2 + \left(y + \frac{b}{2}\right)^2 = \frac{a^2 + b^2 - 4c}{4}$$

If $(a^2 + b^2 - 4c)/4$ is positive, call it r^2. Then we see by comparison with equation (1.5) that this is a circle with center $(-\frac{a}{2}, -\frac{b}{2})$ and radius r. If $(a^2 + b^2 - 4c)/4 = 0$, the equation is satisfied only at $(-\frac{a}{2}, -\frac{b}{2})$. This is called a *degenerate circle*. If $(a^2 + b^2 - 4c)/4 < 0$, there is no graph, since the left side is nonnegative.

EXAMPLE 1.9 Determine the nature of the graph of the equation

$$x^2 + y^2 - 2x + 4y - 4 = 0$$

and sketch the graph.

Solution Completing the squares in x and y gives

$$(x^2 - 2x + 1) + (y^2 + 4y + 4) = 4 + 1 + 4$$
$$(x - 1)^2 + (y + 2)^2 = 9$$

So this is a circle of radius 3 with center at $(1, -2)$. Its graph is shown in Figure 1.14. ■

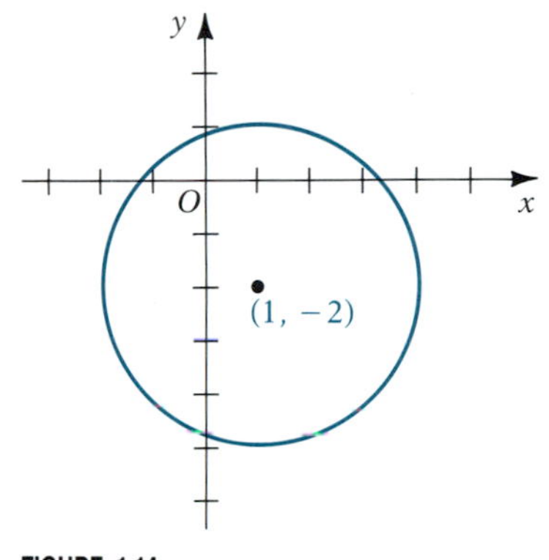

FIGURE 1.14

The Parabola

We will study parabolas in general in Chapter 12 along with the other so-called conic sections. For now we limit consideration to certain equations whose graphs can be shown to be parabolas.

We begin with the equation

$$y = ax^2$$

If $a > 0$, then $y \geq 0$ for all x, and if $a < 0$, then $y \leq 0$ for all x. A given value of x and its negative both produce the same value of y; that is, if (x, y) is on the graph, so is $(-x, y)$. Thus the graph is *symmetric* to the y-axis. Typical graphs for $a > 0$ and $a < 0$ are shown in Figure 1.15. Although we have not yet proved it, both graphs are parabolas. The origin is the **vertex,** and the **axis** of the parabola is the y-axis.

Consider next an equation of the form

$$y = ax^2 + k$$

Adding k has the effect of raising the parabola $y = ax^2$ by k units in the y direction if $k > 0$ or lowering it by $|k|$ units if $k < 0$. Thus the vertex is $(0, k)$.

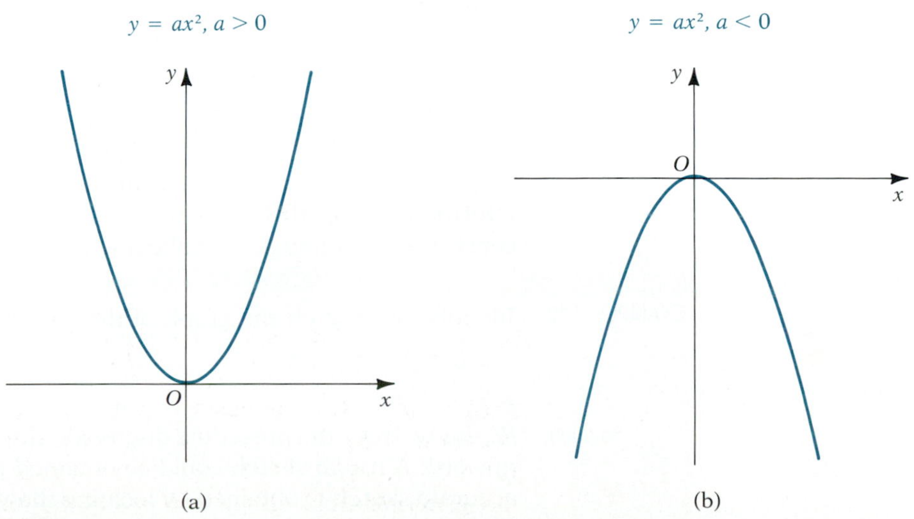

FIGURE 1.15

FIGURE 1.16

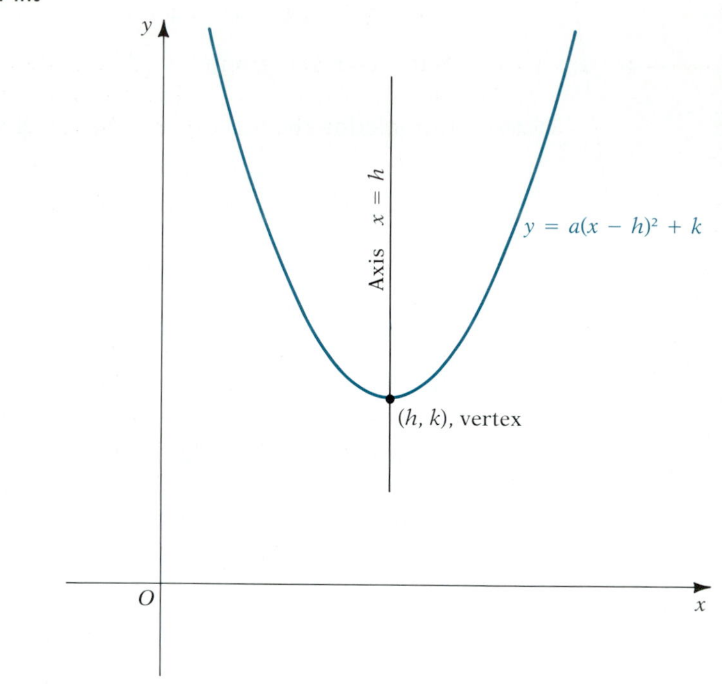

Finally, consider

$$y = a(x - h)^2 + k \tag{1.6}$$

Replacing x by $x - h$ has the effect of shifting the entire graph h units to the right if $h > 0$ or $|h|$ units to the left if $h < 0$. Thus the axis of the parabola is the line $x = h$. Its vertex is (h, k). In Figure 1.16 we show a typical case in which a, h, and k are all positive.

If equation (1.6) is expanded, it can be written in the form

$$y = ax^2 + bx + c \tag{1.7}$$

Conversely, if we are given an equation in this form, we can change it to the form of equation (1.6) by completing the square. So equation (1.7) has a graph that is a parabola opening upward if $a > 0$ and downward if $a < 0$. This information, together with two or three well-chosen points, such as intercepts, is often enough to make a sketch.

EXAMPLE 1.10 Identify and sketch the graph of the equation

$$y = 2x^2 - 5x - 3$$

Solution We know from the preceding discussion that the graph is a parabola opening upward. A rough sketch could be obtained by plotting a few points. A more accurate sketch is obtained by locating the vertex and axis. To complete the square on x, we first factor out the coefficient of x^2 from the terms involving

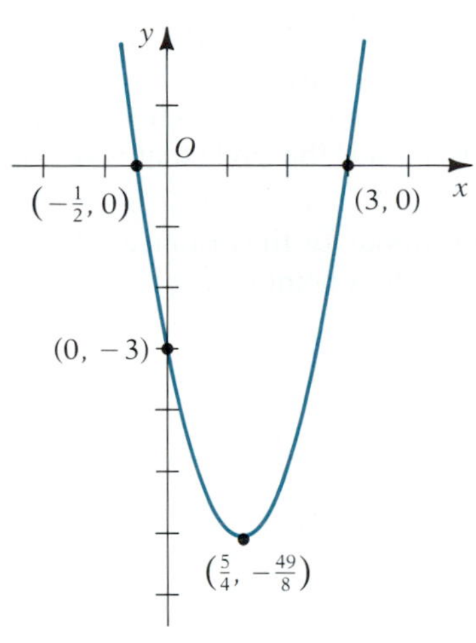

FIGURE 1.17

x. This gives

$$y = 2(x^2 - \tfrac{5}{2}x \qquad) - 3$$

Now we add the square of half the coefficient of x inside the parentheses and compensate for this by subtracting twice this amount, because of the factor 2, outside the parentheses. Thus we add $\frac{25}{16}$ inside and subtract $\frac{25}{8}$ outside:

$$y = 2(x^2 - \tfrac{5}{2}x + \tfrac{25}{16}) - 3 - \tfrac{25}{8}$$

or

$$y = 2(x - \tfrac{5}{4})^2 - \tfrac{49}{8}$$

This is in the form of equation (1.6). So we see the vertex is $(\frac{5}{4}, -\frac{49}{8})$ and the axis is the line $x = \frac{5}{4}$. It is still useful to plot a few points. In this case it is not too difficult to find the x- and y-intercepts, $(-\frac{1}{2}, 0)$, $(3, 0)$, and $(0, -3)$. The graph is shown in Figure 1.17. ■

EXERCISE SET 1.3

A

In Exercises 1–9 find the equation of the line that satisfies the given conditions.

1. Slope $\frac{2}{5}$, passing through $(-2, -5)$
2. Slope -2, passing through $(3, -4)$
3. Passing through $(3, 4)$ and $(-1, -2)$
4. Passing through $(-1, 3)$ and $(2, -5)$
5. Passing through $(5, -3)$, parallel to $3x - 4y = 12$
6. Passing through $(0, 4)$, parallel to $x - y = 4$
7. Passing through $(-1, 2)$, perpendicular to $y = 3x - 5$
8. Passing through $(3, -1)$, perpendicular to $2x - 3y - 4 = 0$
9. Passing through $(3, 5)$ and perpendicular to the line through $(-1, 3)$ and $(3, -5)$
10. Using slopes, show that the points $(3, -1)$, $(5, 4)$, $(-5, 8)$, and $(-7, 3)$ are vertices of a rectangle.
11. Using slopes, show that the points $(2, 2)$, $(0, -1)$, $(-4, 1)$, and $(-2, 4)$ are vertices of a parallelogram.
12. Show that the points $(2, 1)$, $(6, 9)$, and $(-2, 3)$ are vertices of a right triangle using (a) distances and (b) slopes.

In Exercises 13–18 determine whether the graph is a circle or a degenerate circle, or if there is no graph. If it is a circle, give its center and radius.

13. $x^2 + y^2 - 2x + 6y + 6 = 0$
14. $x^2 + y^2 - 2y = 0$
15. $x^2 + y^2 + 4x + 1 = 0$
16. $x^2 + y^2 - 10x + 4y + 29 = 0$
17. $x^2 + y^2 + 3x - 5y + 9 = 0$
18. $2(x^2 + y^2) + 3x - 4y - 3 = 0$

In Exercises 19–22 find the vertex and axis of the parabola, and draw the graph.

19. $y = x^2 + 4x + 5$
20. $y = 3x - x^2$
21. $y = 4 - 5x - 2x^2$
22. $y = 3x^2 - 2x + 4$

B

23. Prove that the midpoint of the segment from $P_1(x_1, y_1)$ to $P_2(x_2, y_2)$ is:

$$\text{midpoint of } P_1P_2 = \left(\frac{x_1 + x_2}{2}, \frac{y_1 + y_2}{2}\right)$$

24. A *median* of a triangle is a line segment from a vertex to the midpoint of the opposite side. Use Exercise 23 to find the equations of the medians of the triangle with vertices $(2, 5)$, $(0, -3)$, and $(-6, 1)$.

25. Prove that the line $3x - 4y + 15 = 0$ is tangent to the circle $x^2 + y^2 - 4x + 2y - 20 = 0$ at the point $(-1, 3)$.

26. Find the equation of the tangent line to the circle $x^2 + y^2 - 6x + 8y + 17 = 0$ at the point $(1, -2)$.

27. Find the point of intersection of the perpendicular bisectors of the sides of the triangle formed by the lines $4x - 3y = 21$, $x + 3y = 9$, and $2x + y = 3$. Show that this point is the center of the circumscribed circle, and find the equation of that circle.

28. Show that the circles $x^2 + y^2 - 6x + 8y = 0$ and $x^2 + y^2 + 4x - 16y + 4 = 0$ are tangent to each other. Find the equation of the line through their centers and the equation of their common tangent line.

29. Find the points of intersection of the parabola $x^2 - 2x - 3y = 5$ and the circle $x^2 + y^2 - 2x + 4y + 1 = 0$. Draw their graphs on the same set of axes and shade the area inside the circle that is above the parabola.

30. Derive the following formula for the distance d from the line $ax + by + c = 0$ to a point (x_1, y_1):

$$d = \frac{|ax_1 + by_1 + c|}{\sqrt{a^2 + b^2}}$$

[*Hint:* Find the point of intersection of the given line and the line through (x_1, y_1) and perpendicular to l.]

1.4 THE FUNCTION CONCEPT

The concept of a function is essential in calculus as well as in most branches of mathematics. To say that one quantity is a function of another means that the first quantity *depends on* the second in some specific way. For example, the cost of mailing a letter first class is a function of its weight. Similarly, the current value of an investment at a fixed interest rate is a function of the length of time the money has been invested. As a more explicit example, the circumference of a circle is a function of its radius; in fact, we know exactly the form of this functional relationship, $C = 2\pi r$. In every such functional relationship there is a rule that establishes a correspondence between two sets. This leads to the following definition.

DEFINITION 1.1 A function from the set A to the set B is a rule that associates with each element of A exactly one element of B. ■

The set A in this definition is called the **domain** of the function. If x is an element of A, the element y of B that the function assigns to x is called the **image** of x, and the totality of such images, as x varies over all of A, is called the **range** of the function. The range is therefore a subset of B, but it need not be all of B. When the range is all of B, we say that the function is from A **onto** B. Although the sets A and B can be any sets whatsoever, in our initial work in calculus they will always be subsets of the real numbers **R**. Later we will remove this restriction.

A function is usually designated by a letter, such as f. The image of an element x of A is then designated by $f(x)$, read "f of x." The image of x is also referred to as "the value of f at x." Figure 1.18 illustrates the idea of a function from A to B, with a typical element x of A and its image $f(x)$ in B.

Functions are often defined by means of an equation, such as

$$f(x) = \sqrt{x - 1}$$

and if the domain of f is not specified, we will understand it to be the largest subset of **R** for which the values of f are real—that is, for which the range of

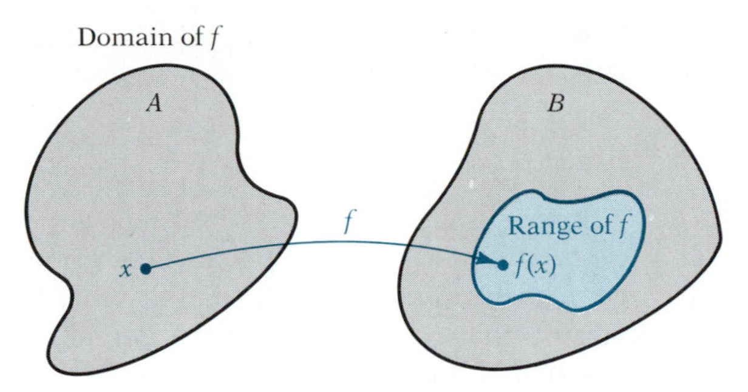

FIGURE 1.18

f is also in **R**. For this function, then, we understand the domain to be the set $\{x: x \geq 1\}$. Any other value of x would cause $f(x)$ to be imaginary. You should convince yourself that the range of this function is the set $\{y: y \geq 0\}$.

It is useful at times to refer to a function from A to B as a *mapping*. We can symbolize this with

$$f: A \to B$$

When the form of the function is known, we can indicate the mapping precisely; for example, in the function above,

$$f: x \to \sqrt{x - 1}$$

We interpret this as saying f *maps* a number x onto the number $\sqrt{x - 1}$.

A function establishes a pairing of elements in its domain with elements in its range. If we write $y = f(x)$, then the function f can be characterized by the totality of all ordered pairs (x, y), with x in A and $y = f(x)$. In fact, this is sometimes taken as the definition of f; that is,

$$f = \{(x, y): x \in A,\ y \in B, \text{ and each } x \text{ in } A \text{ is paired with a unique } y \text{ in } B\}$$

This characterization avoids the use of the term *rule* in Definition 1.1, but of course a rule is nevertheless implied by the determination of which y value goes with which x value. The pairing of values induced by the function leads in a natural way to the *graph of f*, which is the set of points in the plane corresponding to the ordered pairs $(x, f(x))$ with x in the domain of f.

For the function defined by $f(x) = \sqrt{x - 1}$, we let $y = f(x)$ and choose some convenient values of x as shown in the table. Connecting these with a smooth curve, we obtain the graph in Figure 1.19. It is the upper half of a parabola with a horizontal axis.

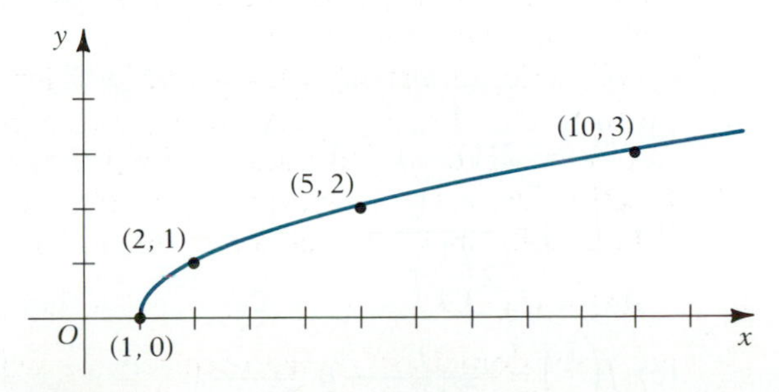

FIGURE 1.19

x	y
1	0
2	1
5	2
10	3

Not every curve in the plane is the graph of a function. There is a simple test to determine when this is the case. Since the definition of a function requires that each x value in the domain maps onto exactly one y value in the range, it follows that no two pairs (x, y_1) and (x, y_2), where $y_1 \neq y_2$, can belong to the function. Geometrically, this says that a vertical line drawn through a point $(x, 0)$, where x is in the domain of a function, will intersect the graph of f in only one point. So, if we are given a function, a vertical line will intersect its graph in at most one point. On the other hand, if we are given a graph in the plane that has this property, we can define a function f with this graph by the requirement that for each point (x, y) on the graph, $y = f(x)$. So we have the following test.

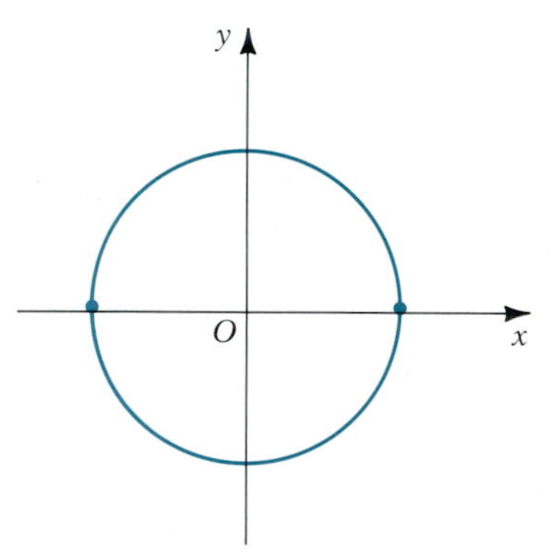

FIGURE 1.20

VERTICAL LINE TEST

A graph in the plane is the graph of a function if and only if each vertical line intersects the graph in at most one point.

We see, for example, that a circle is not the graph of a function (see Figure 1.20). However, we can consider the circle as composed of the upper semicircle and the lower semicircle, each of which does pass the vertical line test. In a similar way, many curves that are not graphs of functions can be divided into two or more parts, each of which is the graph of a function.

The functional notation $f(x)$ is particularly convenient when we want to indicate the value of f at certain specified values of x. For example, $f(2)$ is the value of f at 2 or, equivalently, the image of 2, and we find its value by applying the rule of f to 2. For example, if $f(x) = 3x^2 - 5$, then $f(2) = 3(2^2) - 5 = 7$. The following two examples illustrate this.

EXAMPLE 1.11 If

$$f(x) = \frac{1}{2x - 1}$$

find the following:

(a) $f(1)$ (b) $f\left(\frac{1}{4}\right)$ (c) $f\left(\frac{1}{x}\right)$ (d) $\frac{1}{f(x)}$ (e) $f(x + h)$

Solution (a) $f(1) = \dfrac{1}{2(1) - 1} = \dfrac{1}{1} = 1$

(b) $f\left(\dfrac{1}{4}\right) = \dfrac{1}{2\left(\dfrac{1}{4}\right) - 1} = \dfrac{1}{-\dfrac{1}{2}} = -2$

(c) $f\left(\dfrac{1}{x}\right) = \dfrac{1}{2\left(\dfrac{1}{x}\right) - 1} = \dfrac{1}{\dfrac{2}{x} - 1} = \dfrac{x}{2 - x}$ We substitute $\frac{1}{x}$ for x and simplify.

(d) $\dfrac{1}{f(x)} = \dfrac{1}{\dfrac{1}{2x-1}} = 2x - 1$

Note the difference in parts d and c.

(e) $f(x+h) = \dfrac{1}{2(x+h)-1} = \dfrac{1}{2x+2h-1}$

Again, we simply replace x by $x + h$. ■

EXAMPLE 1.12 If $f(x) = \sqrt{x}$, find

$$\frac{f(4+h) - f(4)}{h}, \qquad h \neq 0$$

and express the answer in a form with no radicals in the numerator.

Solution

$$\frac{f(4+h) - f(4)}{h} = \frac{\sqrt{4+h} - 2}{h}$$

To rationalize the numerator we multiply both numerator and denominator by $\sqrt{4+h} + 2$, and use the identity $(a-b)(a+b) = a^2 - b^2$:

$$\frac{f(4+h) - f(4)}{h} = \frac{\sqrt{4+h} - 2}{h} \cdot \frac{\sqrt{4+h} + 2}{\sqrt{4+h} + 2} = \frac{(4+h) - 4}{h(\sqrt{4+h} + 2)}$$

$$= \frac{h}{h(\sqrt{4+h} + 2)} = \frac{1}{\sqrt{4+h} + 2}$$

Problems of this type occur frequently in calculus. ■

EXAMPLE 1.13 Determine the domain and range of the functions f and g:

$$f(x) = \frac{x+3}{2x-5} \qquad g(x) = \sqrt{2 - x - x^2}$$

Solution The only value of x for which $f(x)$ fails to be a real number is $x = \frac{5}{2}$. So the domain of f is the set $\{x : x \neq \frac{5}{2}\}$. For the range, we want to know what values $y = f(x)$ are assumed as x varies over the domain. If we solve this equation for x, we get

$$x = \frac{5y+3}{2y-1} \qquad \text{(Verify.)}$$

The only value of y that fails to correspond to an x in the domain is $y = \frac{1}{2}$; that is, the range of f is the set $\{y : y \neq \frac{1}{2}\}$.

The function g has real values if and only if x satisfies the inequality $2 - x - x^2 \geq 0$. To solve this inequality, we use the idea introduced in Example 1.2. Factoring, we have

$$2 - x - x^2 = (2 + x)(1 - x)$$

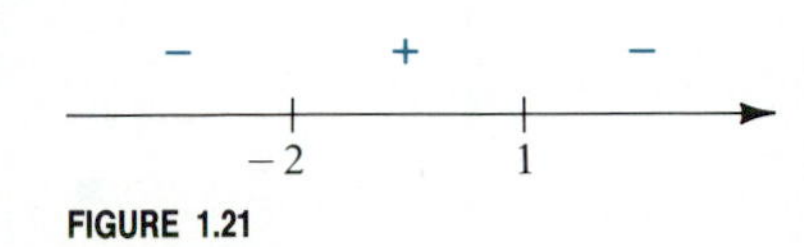

FIGURE 1.21

A sign graph for the product $(2 + x)(1 - x)$ is shown in Figure 1.21. Since we are interested in knowing where the product is nonnegative, we conclude that the domain of g is the set

$$\{x : -2 \leq x \leq 1\} = [-2, 1]$$

To find the range, we solve the equation $y = g(x)$ for x. Setting $y = g(x)$, we get

$$y = \sqrt{2 - x - x^2}$$

so that $y \geq 0$. Squaring and rearranging terms yields

$$x^2 + x + (y^2 - 2) = 0$$

The quadratic formula gives

$$x = \frac{-1 \pm \sqrt{9 - 4y^2}}{2}$$

We must have $y^2 \leq \frac{9}{4}$ so that x is real. Hence, $0 \leq y \leq \frac{3}{2}$. You can check to see that these y values correspond to x values in $-2 \leq x \leq 1$. Thus, the range of g is the set $\{y: 0 \leq y \leq \frac{3}{2}\}$ or $[0, \frac{3}{2}]$. ■

The particular letter used to designate an arbitrary element in the domain of a function is immaterial. The letter x is commonly used, but almost any other letter will do. All the following are equivalent:

$$f(x) = 2x + 3 \quad f(t) = 2t + 3 \quad f(u) = 2u + 3$$

Each equation states the defining rule of f. Similarly, we can use a letter such as y to designate an arbitrary element of the range of f. The letter used to designate an arbitrary domain element is sometimes called an **independent variable,** and the letter used to designate the corresponding range element is called the **dependent variable.** Thus, when we write $y = f(x)$, we are using x as the independent variable and y as the dependent variable. The independent variable is free to assume any value in the domain, but the dependent variable is then completely determined. In the next example we use the letter b, for base (of a triangle), as the independent variable, and A, for area, as the dependent variable.

EXAMPLE 1.14 Express the area A of an isosceles triangle that is inscribed in a circle of radius 2 as a function of its base b.

Solution A typical triangle is pictured in Figure 1.22. We denote the perpendicular distance from the center of the circle to the base of the triangle by y, as shown. Then the altitude of the triangle is $y + 2$ and so the area is

$$A = \tfrac{1}{2}b(y + 2)$$

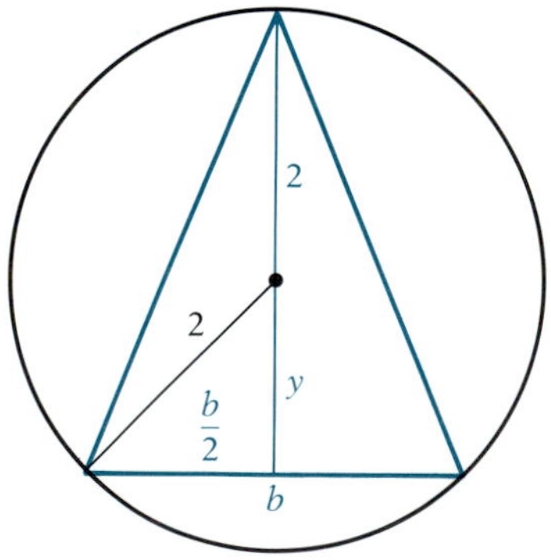

Now we must find y in terms of b. Using the Pythagorean theorem on the right triangle shown with legs $\frac{b}{2}$ and y and hypotenuse 2, we get

$$y^2 + \left(\frac{b}{2}\right)^2 = 4$$

$$y = \sqrt{4 - \frac{b^2}{4}} = \frac{\sqrt{16 - b^2}}{2}$$

FIGURE 1.22

Thus

$$A = \frac{b}{4}(\sqrt{16 - b^2} + 4)$$

■

A function is sometimes defined by different rules on different parts of its domain. We call this a *piecewise defined function.* The next example provides an illustration.

EXAMPLE 1.15 Draw the graph of the function F defined by

$$F(x) = \begin{cases} -1 & \text{if } x < 0 \\ 1 & \text{if } x = 0 \\ x + 2 & \text{if } x > 0 \end{cases}$$

Solution The graph is shown in Figure 1.23. Notice that the point (0, 1) is on the graph, as is indicated by the solid circle there, but (0, 2) and (0, −1) are not, as is indicated by the open circles.

Although F is defined in a piecewise fashion, it represents only one function. Its domain is all of **R**. ■

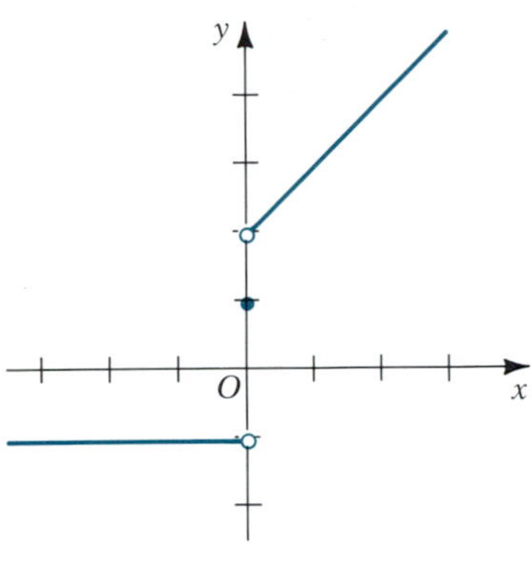

FIGURE 1.23

Suppose that for every x in the domain of a function f, $-x$ is also in its domain. Then f may fall into one of the following two categories: f is said to be **even** if $f(-x) = f(x)$, and f is said to be **odd** if $f(-x) = -f(x)$ for all x in the domain of f. For example, $f(x) = x^2 + 1$ is even and $f(x) = x^3$ is odd. On the other hand, $f(x) = x^3 + x^2 + 1$ is neither even nor odd.

For each point (x, y) on the graph of an even function, $(-x, y)$ is also on the graph, so that the graph is **symmetric to the *y*-axis.** For an odd function, when (x, y) is on the graph, so is $(-x, -y)$, and we say the graph is **symmetric to the origin.** Figure 1.24 illustrates both situations.

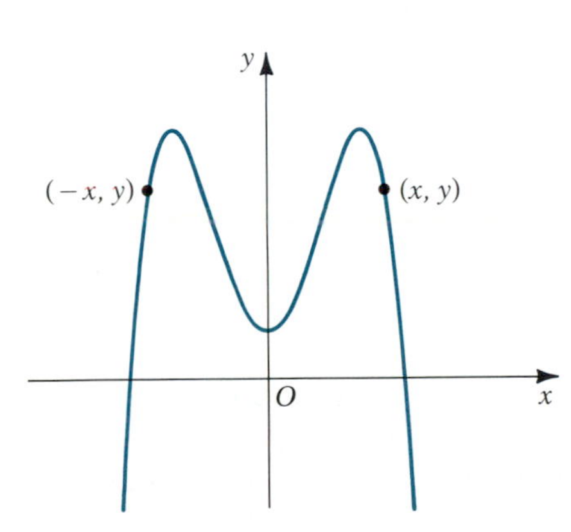

Graph of an even function

(a)

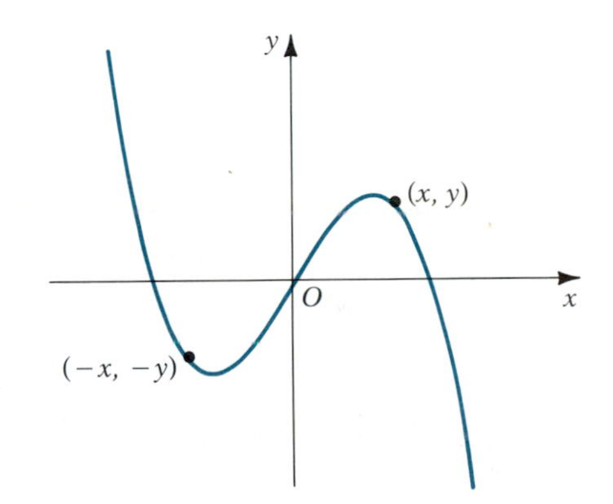

Graph of an odd function

(b)

FIGURE 1.24

EXERCISE SET 1.4

A

1. If $f(x) = \dfrac{x^2 - 1}{x + 2}$, find:

   (a) $f(1)$ (b) $f(-1)$ (c) $f(0)$
   (d) $f(a)$ (e) $f(x + 1)$

2. If $g(x) = \dfrac{2x - 3}{3x + 4}$, find:

   (a) $g(-2)$ (b) $g(\frac{5}{6})$ (c) $g(-\frac{2}{3})$
   (d) $g(2x)$ (e) $g(1 - x)$

3. If $F(t) = |t - 3|$, find:
(a) $F(4)$ (b) $F(1)$ (c) $F(0)$
(d) $F(t + 3)$ (e) $F(3 - x^2)$

In Exercises 4–9 find the domain and range of f.

4. $f(x) = 2x + 3$ **5.** $f(x) = x^2 - 4$

6. $f(x) = \dfrac{1}{x}$ **7.** $f(t) = \sqrt{1 - t}$

8. $f(u) = \dfrac{u + 1}{u - 2}$ **9.** $f(s) = s^3 + 1$

In Exercises 10–19 draw the graph of the given function.

10. $f(x) = 3x - 4$ **11.** $g(x) = x^2 + 1$

12. $h(x) = \frac{2}{x}$ **13.** $F(x) = 4x - x^2$

14. $f(x) = \sqrt{x + 9}$ **15.** $G(x) = |x + 1|$

16. $h(x) = \begin{cases} x^2 & \text{if } x \geq 0 \\ 1 - x & \text{if } x < 0 \end{cases}$

17. $f(x) = \begin{cases} 2x & \text{if } x \geq 1 \\ 2 & \text{if } 0 \leq x < 1 \\ x + 2 & \text{if } x < 0 \end{cases}$

18. $f(x) = \begin{cases} \sqrt{x} & \text{if } x > 0 \\ 1 & \text{if } x = 0 \\ -1 & \text{if } x < 0 \end{cases}$ **19.** $g(x) = |4 - x^2|$

20. If $f(x) = \dfrac{1}{x}$, find $\dfrac{f(x) - f(2)}{x - 2}$ for $x \neq 2$ or 0 and simplify the result.

21. If $g(x) = \sqrt{x}$, show that for $h \neq 0$,

$$\frac{g(x + h) - g(x)}{h} = \frac{1}{\sqrt{x + h} + \sqrt{x}}$$

22. Let $\phi(x) = \dfrac{x}{x - 3}$. Find

$$\frac{\phi(4 + h) - \phi(4)}{h} \qquad (h \neq 0)$$

and simplify the result.

23. Let $f(x) = \dfrac{x + 2}{2x + 1}$. Show that if $x \neq 0$,

$$f\left(\frac{1}{x}\right) = \frac{1}{f(x)}$$

24. Determine in each of the following whether the function is even, odd, or neither:
(a) $f(x) = \sqrt{x^2 + 1}$ (b) $f(x) = x^3 - 2x$
(c) $f(x) = x^2 + 2x$ (d) $g(x) = x - 1$
(e) $h(x) = 2x^2 - 3$

25. Follow the instructions for Exercise 24 for the following:

(a) $f(x) = \dfrac{x}{x^2 + 1}$ (b) $g(x) = \dfrac{1}{x^3 - 2x^5}$

(c) $h(t) = \dfrac{t + 2}{t - 1}$ (d) $\phi(x) = \dfrac{x^2 - 1}{x^2 + 4}$

(e) $F(t) = |t|$

26. A piece of equipment is purchased for \$200,000. For tax purposes the value is decreased by \$40,000 times the number of years x since the equipment was purchased. Let $f(x)$ = value of equipment x years after purchase.
(a) Write the rule for the function f.
(b) Find the domain of f if the value of the equipment cannot be negative.

27. If a price of x dollars per unit is charged for a certain product, the number of units sold will be $1000 - 25x$. Let $f(x)$ be the revenue when the price is x, where revenue is the price per unit times the number of units sold.
(a) Write the rule for the function f.
(b) Find the domain of f if the revenue is nonnegative.

28. The length of a rectangle is 3 more than twice its width. Express each of the following as a function of the width: (a) perimeter and (b) area.

29. A can is in the form of a right circular cylinder with its height equal to twice the radius of the base. Express each of the following as a function of the radius of the base: (a) volume, (b) lateral surface area, and (c) total surface area.

30. Express the area of an equilateral triangle as a function of one of its sides.

31. A rectangle is inscribed in an isosceles triangle of height 4 and base 6, as shown in the figure. Express the area of the rectangle as a function of its base b.

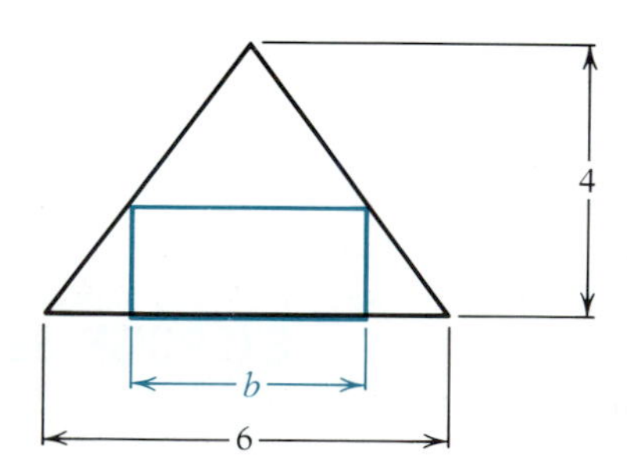

B

In Exercises 32–37 find the domain and range of the given function.

32. $f(x) = \sqrt{x^2 - 4}$

33. $f(x) = \sqrt{3 + 2x - x^2}$

34. $g(x) = \sqrt{\dfrac{x - 1}{x + 2}}$

35. $g(x) = \dfrac{2}{\sqrt{1 - x^2}}$

36. $h(x) = \dfrac{x - 1}{\sqrt{6 - x - x^2}}$

37. $h(x) = \dfrac{\sqrt{x^2 - 2x - 8}}{x + 5}$

38. A right circular cylinder is inscribed in a sphere of radius a. Express the volume of the cylinder as a function of its base radius r.

39. A right circular cone is inscribed in a sphere of radius a. Express the volume of the cone as a function of its base radius r.

40. A pyramid has a square base 4 meters on a side, and its height is 10 meters. Express the area of a cross section parallel to the base and h meters above the base as a function of h.

41. A storage tank is in the form of a cylinder x feet long and y feet in diameter, with hemispheres at both ends (see the figure).

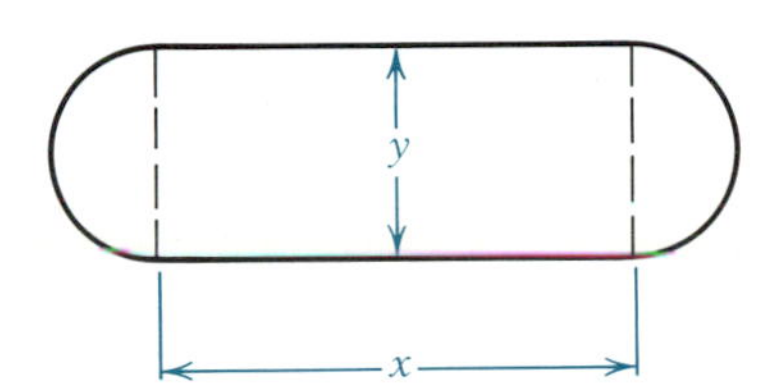

(a) Express the volume of the tank as a function of x and y.
(b) Express the surface area of the tank as a function of x and y.
(c) If $x = 4y$, express the volume and surface area as a function of y only.

42. An organization that issues credit cards charges cardholders interest of $1\frac{1}{2}\%$ per month for the first \$1000 of unpaid balance after 30 days and 1% per month on the unpaid balance above \$1000. Let $f(x)$ be the interest charged per month on an unpaid balance of x dollars, and find the rule for f.

43. A company deducts 2% of gross salary from each employee's pay as the employee's contribution to the pension fund. However, no employee pays more than \$400 into the fund in a single year. Let $f(x)$ be the contribution from an employee with a gross annual salary of x dollars. Write the rule for f.

44. Prove that the range of the function f for which $f(x) = 2x/(x^2 + 1)$ is the set $\{y: |y| \le 1\}$.

In Exercises 45–48 show that the given value of a is not in the domain of f. Then use a calculator to find $f(a + h)$ for each of the following values of h: ± 0.1, ± 0.01, ± 0.001, ± 0.0001. Conjecture how the function behaves as x comes arbitrarily close to a from each side.

45. $a = 3;\ f(x) = \dfrac{x^2 - 9}{x - 3}$

46. $a = -\dfrac{3}{2};\ f(x) = \dfrac{6x^2 + 17x + 12}{2x^2 + x - 3}$

47. $a = 2;\ f(x) = \dfrac{3x^3 - 4x^2 - 8}{3x^2 + x - 14}$

48. $a = 8;\ f(x) = \dfrac{x - 8}{2 - \sqrt[3]{x}}$

C

In Exercises 49–51 write a computer program that has as its output a table showing x values in one column and corresponding $f(x)$ values in the second. Use the results to draw on graph paper an accurate graph of f.

49. $f(x) = \dfrac{3x^3 + 2x^2 - 3}{x^4 + 1};\ x = 0, \pm 1, \pm 2, \pm 3, \pm 4, \pm 5$

50. $f(x) = \frac{1}{10}(x^5 + 4x^4 - 27x^3 - 38x^2 - 26x - 13);\ x = 0, \pm 1, \pm 2, \pm 3, \pm 4, \pm 5$. Locate the x values for which $f(x) = 0$ between consecutive integers.

51. $f(x) = x/\sqrt{1 - x^2}$ for x values 0.1 unit apart between $x = -0.9$ and 0.9

1.5 COMBINATIONS OF FUNCTIONS

Let f and g be any two functions, and let D denote the intersection of their domains. If D is nonempty, we can obtain new functions from f and g by the operations of addition, subtraction, multiplication, and division.

DEFINITION 1.2 Let D denote the intersection of the domain of f and the domain of g. We define the functions $f+g$, $f-g$, $f\cdot g$, and f/g as follows:

$$(f+g)(x) = f(x) + g(x) \qquad (f\cdot g)(x) = f(x)\cdot g(x)$$

$$(f-g)(x) = f(x) - g(x) \qquad \left(\frac{f}{g}\right)(x) = \frac{f(x)}{g(x)} \quad \text{if } g(x) \neq 0$$

The domain of $f+g$, $f-g$, and $f\cdot g$ is all of D. The domain of f/g is $\{x \in D: g(x) \neq 0\}$. ■

EXAMPLE 1.16 Let $f(x) = \sqrt{1-x}$ and $g(x) = \sqrt{1+x}$. Find $f+g$, $f-g$, $f\cdot g$, and f/g, and state the domain of each.

Solution The domain of f is the interval $(-\infty, 1]$, and the domain of g is the interval $[-1, +\infty)$. Their intersection is the interval $[-1, 1]$. So we have

$$(f+g)(x) = \sqrt{1-x} + \sqrt{1+x}, \qquad x \in [-1, 1]$$
$$(f-g)(x) = \sqrt{1-x} - \sqrt{1+x}, \qquad x \in [-1, 1]$$
$$(f\cdot g)(x) = \sqrt{1-x}\sqrt{1+x} = \sqrt{1-x^2}, \qquad x \in [-1, 1]$$
$$\left(\frac{f}{g}\right)(x) = \frac{\sqrt{1-x}}{\sqrt{1+x}} = \sqrt{\frac{1-x}{1+x}}, \qquad x \in (-1, 1]$$

Note that $x = -1$ is excluded from the domain of f/g. ■

The simplest functions are the **constant functions,** with values of the form $f(x) = c$, and the **identity function,** which maps each number onto itself and so has the form $f(x) = x$. By multiplying the identity function by itself repeatedly we obtain functions of the form $f(x) = x^n$, and when these are combined with constant functions using the operations of addition, subtraction, and multiplication, we obtain the important class of **polynomial functions.**

DEFINITION 1.3 A polynomial function P is a function of the form

$$P(x) = a_0 + a_1x + a_2x^2 + \cdots + a_nx^n \tag{1.8}$$

where each a_i is a constant. The domain of a polynomial function is the set of all real numbers. ■

When $a_n \neq 0$, the right side of equation (1.8) is called a **polynomial of degree n.** If the polynomial is of degree 0, we have $P(x) = a_0$, so that P is a constant function. The *zero polynomial* $P(x) = 0$ is not assigned a degree. For degrees 1 and 2 we call P a **linear function** and a **quadratic function,** respectively. These special polynomial functions are illustrated by the following:

$P(x) = 2$	P is a constant function.
$P(x) = 3x + 4$	P is a linear function.
$P(x) = 2x^2 - 3x + 5$	P is a quadratic function.

From Section 1.3 we know that the graph of a constant function is a horizontal line, that of a linear function is a line with nonzero slope, and that of a quadratic function is a parabola.

If P and Q are two polynomial functions, then the quotient function P/Q is called a **rational function.** For example, the function f defined by

$$f(x) = \frac{2x^2 - 3x + 4}{x^3 - 1}$$

is rational. The domain of a rational function P/Q is the set $\{x \in \mathbf{R}: Q(x) \neq 0\}$.

There is another very important way of combining functions, called **composition,** which we now define.

DEFINITION 1.4 The **composition of f and g,** denoted by $f \circ g$, is defined by

$$(f \circ g)(x) = f(g(x))$$

The domain of $f \circ g$ is the largest subset of the domain of g for which $g(x)$ is in the domain of f. ■

To illustrate, consider $f(x) = \sqrt{x - 4}$ and $g(x) = x^2$. Then

$$(f \circ g)(x) = f(g(x)) = \sqrt{x^2 - 4}$$

The domain of g is all of $\mathbf{R}$, and the domain of f is the set $\{x: x \geq 4\}$. So the domain of $f \circ g$ is the subset of $\mathbf{R}$ for which $g(x) \geq 4$—namely, $\{x: x^2 \geq 4\}$ or equivalently $\{x: |x| \geq 2\}$.

It is instructive to consider $g \circ f$ for these same two functions. We have

$$(g \circ f)(x) = g(f(x)) = (\sqrt{x - 4})^2 = x - 4$$

Although the result has meaning for each x in $\mathbf{R}$, the domain of $g \circ f$ is nevertheless restricted to those x values in the domain of f for which $f(x)$ is in the domain of g—namely, $\{x: x \geq 4\}$. This example shows, incidentally, that in general, two functions do not commute under the operation of composition; that is, $f \circ g \neq g \circ f$.

EXAMPLE 1.17 Let $f(x) = x/(x - 1)$ and $g(x) = 1/x^2$. Find $f \circ g$ and $g \circ f$, and give the domain of each.

Solution

$$(f \circ g)(x) = \frac{\frac{1}{x^2}}{\frac{1}{x^2} - 1} = \frac{1}{1 - x^2}$$

$$\text{domain of } f = \{x: x \neq 1\}$$
$$\text{domain of } g = \{x: x \neq 0\}$$

So

$$\text{domain of } f \circ g = \left\{x: x \neq 0 \text{ and } \frac{1}{x^2} \neq 1\right\} = \{x: x \neq 0, \pm 1\}$$

For $g \circ f$ we have

$$(g \circ f)(x) = \frac{1}{\left(\frac{x}{x - 1}\right)^2} = \left(\frac{x - 1}{x}\right)^2$$

$$\text{domain of } g \circ f = \left\{x: x \neq 1 \text{ and } \frac{x}{x - 1} \neq 0\right\} = \{x: x \neq 0, 1\}$$ ■

EXAMPLE 1.18 Let $F(x) = \sqrt{1 - x^2}$. Find two functions, f and g, for which $F = f \circ g$.

Solution There is more than one way to do this. For example, we can let $g(x) = 1 - x^2$ and $f(x) = \sqrt{x}$. Then

$$(f \circ g)(x) = f(g(x)) = \sqrt{1 - x^2}$$

Another possibility is to let $g(x) = x^2$ and $f(x) = \sqrt{1 - x}$. Then

$$(f \circ g)(x) = f(g(x)) = \sqrt{1 - x^2}$$

■

We have seen how polynomial and rational functions can be built up from constant functions and integral powers of x. If we also permit integral roots of x, say $\sqrt[n]{x}$ or equivalently $x^{1/n}$ (where $x \geq 0$ when n is even), and combine them using a finite number of the four arithmetic operations, together with composition, we generate a broad class known as **algebraic** functions.* So in addition to polynomials and other rational functions, this class includes such functions as

$$f(x) = (x^2 - 2x + 5)^{2/3} \quad \text{and} \quad g(x) = \frac{\sqrt{x - \sqrt[3]{x^2 - 7}}}{(x^3 - 1)^4}$$

Functions that are not algebraic are called **transcendental.** Probably the most familiar transcendental functions are the trigonometric functions. We will study others in Chapters 6 and 7. A review of the trigonometric functions is given in Appendix 2, and you should refer to this as needed. The next example involves the composition of an algebraic function and a transcendental function.

EXAMPLE 1.19 Let $f(x) = \sqrt{1 - x^2}$ and $g(x) = \sin x$. Find $(f \circ g)(x)$ and $(g \circ f)(x)$, and give the domain of each.

Solution Observe first the domain and range of each function.

	$f(x)$	$g(x)$
Domain:	$[-1, 1]$	All of $\mathbf{R}$
Range:	$[0, 1]$	$[-1, 1]$

Since the range of g is the same as the domain of f, the composition $f \circ g$ is valid for all x in the domain of g—namely, all of $\mathbf{R}$. So for all x,

$$(f \circ g)(x) = f(g(x)) = \sqrt{1 - \sin^2 x} = \sqrt{\cos^2 x} = |\cos x|$$

where we have made use of the identity $\sin^2 x + \cos^2 x = 1$.

For $g \circ f$, we have

$$(g \circ f)(x) = g(f(x)) = \sin \sqrt{1 - x^2}$$

and this is true for all x in the domain of f—namely, in $[-1, 1]$. ■

* Precisely, an algebraic function is any function f such that $y = f(x)$ satisfies an equation of the form

$$P_n(x)y^n + P_{n-1}(x)y^{n-1} + \cdots + P_1(x)y + P_0(x) = 0$$

where each $P_i(x)$ is a polynomial.

EXERCISE SET 1.5

A

In Exercises 1–12 find $f+g$, $f-g$, $f\cdot g$, and f/g, and give the domain of each.

1. $f(x) = 3x - 5$; $g(x) = 2x + 3$

2. $f(x) = 1 - x$; $g(x) = 1 + x$

3. $f(x) = \dfrac{x-1}{2}$; $g(x) = \dfrac{x+1}{4}$

4. $f(x) = \dfrac{1}{x-1}$; $g(x) = \dfrac{1}{x+1}$

5. $f(x) = \sqrt{x+4}$; $g(x) = \sqrt{4-x}$

6. $f(x) = x$; $g(x) = \sqrt{x-1}$

7. $f(x) = \dfrac{1}{x-1}$; $g(x) = \dfrac{x}{x+2}$

8. $f(x) = \dfrac{1}{x}$; $g(x) = \dfrac{x}{x-3}$

9. $f(x) = \begin{cases} -1 & \text{if } x < 0 \\ 1 & \text{if } x \geq 0 \end{cases}$; $g(x) = \begin{cases} 1 & \text{if } x < 0 \\ -1 & \text{if } x \geq 0 \end{cases}$

10. $f(x) = \begin{cases} x & \text{if } x \geq 0 \\ 0 & \text{if } x < 0 \end{cases}$; $g(x) = \begin{cases} 0 & \text{if } x \geq 0 \\ -x & \text{if } x < 0 \end{cases}$

11. $f(x) = \cos^2 x$; $g(x) = \sin^2 x$

12. $f(x) = \sec^2 x$; $g(x) = \tan^2 x$

13. Let $f(x) = \sqrt{x}$ and $g(x) = \sqrt{x-1}$. What is the domain of $f+g$ and $f-g$? Show on this domain:

$$(f-g)(x) = \frac{1}{(f+g)(x)}$$

14. For an arbitrary function f whose domain contains $-x$ whenever it contains x, define g and h as follows:

$$g(x) = \tfrac{1}{2}[f(x) + f(-x)] \qquad h(x) = \tfrac{1}{2}[f(x) - f(-x)]$$

(a) Show that g is even and h is odd.
(b) Show that $f = g + h$. (This proves that every such function can be expressed as the sum of an even function and an odd function.)

15. What conclusion can you draw about $f+g$, $f-g$, $f\cdot g$, and f/g with respect to being even or odd under the following conditions?
(a) f and g are both even.
(b) f and g are both odd.
(c) One is even and the other is odd.

16. In each of the following determine whether f is even, odd, or neither.

(a) $f(x) = \dfrac{\sin x}{x}$ (b) $f(x) = \dfrac{1 - \cos x}{x^2}$

(c) $f(x) = x + \tan x$ (d) $f(x) = \sin x + \cos x$

(e) $f(x) = x \cos x$

In Exercises 17 and 18 find $f \circ g$ and $g \circ f$, and give the domain of each. Use trigonometric identities to simplify the results where possible.

17. $f(x) = \cos x$; $g(x) = 2x^2 - 1$

18. $f(x) = \sqrt{x^2 - 1}$; $g(x) = \sec x$

19. Let $f(x) = x/(2x - 3)$. Find $(f \circ f)(x)$ and give its domain.

20. Let $f(x) = (2x - 3)/5$, $g(x) = x^2 - 3$, and $h(x) = (5x + 3)/2$. Find:
(a) $f(g(x))$ (b) $g(f(x))$ (c) $f(h(x))$
(d) $h(f(x))$ (e) $g(h(x))$

In Exercises 21–28 determine two functions f and g for which $F = f \circ g$.

21. $F(x) = \sqrt{2x + 3}$

22. $F(x) = (5 - 4x)^3$

23. $F(x) = \left(\dfrac{x}{x+2}\right)^{2/3}$

24. $F(x) = \dfrac{1}{(3x - 4)^2}$

25. $F(x) = \sqrt{(x^2 - 1)^3}$

26. $F(x) = 3(1 - x)^{-4}$

27. $F(x) = \sin \sqrt{x^2 + 1}$

28. $F(x) = 2 \cos^2 x - 3 \cos x + 2$

29. Let $f(x) = x^2 - 1$, $g(x) = 2x + 3$, and $h(x) = 1/(x + 1)$. Find $f \circ (g \circ h)$ and $(f \circ g) \circ h$. In general, what do you conjecture about $f \circ (g \circ h)$ and $(f \circ g) \circ h$ for arbitrary functions f, g, and h?

B

30. Prove that if P and Q are polynomial functions, then $P + Q$, $P - Q$, and $P \cdot Q$ are also polynomial functions. Discuss the degrees of each of these.

31. Prove that if f and g are rational functions, then $f+g$, $f-g$, $f\cdot g$, and f/g (if g is not the zero function) are also rational. Discuss the domains of each of these.

32. Prove that if f and g are polynomial functions, then $f \circ g$ is also a polynomial function. If $f(x)$ is of degree m and $g(x)$ is of degree n, what is the degree of $(f \circ g)(x)$?

33. Let $f(x) = |x|$ and $g(x) = |x - 1|$. Find $f + g$ and $f \cdot g$, and draw the graph of each.

34. Prove or disprove:
(a) $f \circ (g + h) = f \circ g + f \circ h$
(b) $(g + h) \circ f = g \circ f + h \circ f$

35. Let $f(x) = \frac{1}{x}$. Prove that for functions g and h,

$$f \circ \left(\frac{g}{h}\right) = \frac{f \circ g}{f \circ h}$$

What restrictions must be placed on the domains of g and h?

36. Let $f(x) = x^{2/3}$ and $g(x) = 4x^2 - 9$. Use a calculator to find each of the following correct to five significant figures.
(a) $(f + g)(x)$ for $x = 3.0257$
(b) $(f \cdot g)(x)$ for $x = 0.023768$
(c) $(\frac{f}{g})(x)$ for $x = 213.82$
(d) $(f \circ g)(x)$ for $x = 1.3956$
(e) $(g \circ f)(x)$ for $x = -17.248$

37. If a satellite of mass m is at a distance $s(t)$ from the center of the earth at time t, the force of gravity exerted by the earth on the satellite is $F(t) = (h \circ s)(t)$, where $h(x) = gR^2m/x^2$, g is the acceleration due to gravity at the earth's surface, and R is the radius of the earth.
(a) Find an explicit expression for $F(t)$ if

$$s(t) = -\tfrac{1}{2}gt^2 + v_0t + s_0$$

where v_0 is the initial velocity and s_0 the initial distance from the earth's center.
(b) Suppose the satellite has a mass of 2000 kilograms and is lifted from the earth's surface at an initial velocity of 4000 meters per second. Find the force of gravity on it after 10 seconds. Take $R = 6.37 \times 10^6$ meters and $g = 9.80$ meters/sec^2. The answer will be in **newtons,** the unit of force in the mks (meter–kilogram–second) system.

C

38. For the functions f and g below, write a computer program that will evaluate $f + g$, $f - g$, $f \cdot g$, f/g, $f \circ g$, and $g \circ f$ at a point x. Run the program for $x = 0, \pm 1, \pm 2, \pm 3, \pm 4$, and ± 5.

$$f(x) = (x^3 - 3x^2 + 4)^{3/5} \qquad g(x) = \sqrt{\frac{x^2 - 2x + 4}{x^2 + 9}}$$

Are there any restrictions on the domains of f or g for these combined functions to be defined? Prove your answer.

1.6 THE LIMIT OF A FUNCTION

The notion of **limit** is the single most important concept in calculus. To get an idea of the meaning of the limit of a function, consider the question: What number is approached by $f(x) = 2x - 3$ as x approaches 4? If you said 5, you would be correct. The values of $f(x)$ get closer and closer to 5 as x gets closer and closer to 4, so we say that the limit of $f(x)$ is 5 as x approaches 4 and write this more briefly as $\lim_{x \to 4}(2x - 3) = 5$. In this example you could find the limit by substituting 4 for x.

As you might expect, not all limits are this obvious. For example, consider the limit of $f(x) = (x^2 - 4)/(x - 2)$ as x approaches 2—that is,

$$\lim_{x \to 2} \frac{x^2 - 4}{x - 2}$$

As x gets closer and closer to 2, it is not immediately apparent what, if anything, this function approaches. We cannot simply substitute $x = 2$, since this would give 0/0, which has no meaning. Using a calculator we obtain the following table:

x	1.5	1.9	1.99	1.999	2.5	2.1	2.01	2.001
$f(x)$	3.5	3.9	3.99	3.999	4.5	4.1	4.01	4.001

This gives convincing evidence that as x approaches 2 from either side, $f(x)$ approaches 4. An even more convincing argument is as follows. When x is restricted to be different from 2, we have

$$\frac{x^2-4}{x-2} = \frac{(x+2)(x-2)}{x-2} = x+2$$ Do you see why we must not let x be 2?

Thus,

$$\lim_{x\to 2} \frac{x^2-4}{x-2} = \lim_{x\to 2}(x+2) = 4$$

Here is a similar example involving trigonometric functions.

EXAMPLE 1.20 Evaluate the limit

$$\lim_{x\to 0} \frac{1-\cos x}{\sin^2 x}$$

Solution Making use of the identity $\sin^2 x = 1 - \cos^2 x$, we have

$$\lim_{x\to 0} \frac{1-\cos x}{\sin^2 x} = \lim_{x\to 0} \frac{1-\cos x}{1-\cos^2 x} = \lim_{x\to 0} \frac{1-\cos x}{(1-\cos x)(1+\cos x)}$$

For x in a sufficiently small deleted neighborhood of 0, $1 - \cos x \neq 0$, so we can divide out the common factor and get

$$\lim_{x\to 0} \frac{1}{1+\cos x} = \frac{1}{2}$$

since $\cos x$ can be made arbitrarily close to 1 for x sufficiently close to 0. ■

We summarize these ideas in the informal definition below. This is satisfactory for our present purposes, but it must be more precisely stated before it can be used as a basis for proving theorems; we will give the formal definition in Section 1.8. The preceding examples should help you see why the x values in the definition are restricted to a *deleted* neighborhood of a.

INFORMAL DEFINITION OF LIMIT To say that **the limit of $f(x)$ is L as x approaches a** means that $f(x)$ will be arbitrarily close to L provided x is in a sufficiently small deleted neighborhood of a. When this is true, we write

$$\lim_{x\to a} f(x) = L$$ ■

EXAMPLE 1.21 Find

$$\lim_{x\to 2} \frac{x^2+5x-14}{x^2-x-2}$$

Solution We use the factoring technique illustrated earlier:

$$\lim_{x\to 2} \frac{x^2+5x-14}{x^2-x-2} = \lim_{x\to 2} \frac{\cancel{(x-2)}(x+7)}{\cancel{(x-2)}(x+1)} = \lim_{x\to 2} \frac{x+7}{x+1}$$

The cancellation of the common factor $x - 2$ is valid, since it is implicit in the limiting process that we are not permitting x to be 2, so that $x - 2 \neq 0$. In the last expression, as we take x closer and closer to 2, we get a number closer and closer to $\frac{9}{3} = 3$. In fact, it seems reasonable that we can make $(x + 7)/(x + 1)$ arbitrarily close to 3 by choosing x sufficiently close to 2. Later we will see how to make this reasonable assumption precise, but for now we are proceeding intuitively. So we have

$$\lim_{x \to 2} \frac{x^2 + 5x - 14}{x^2 - x - 2} = 3$$

■

The limit of a function can fail to exist in a variety of ways; three are illustrated in the next example.

EXAMPLE 1.22 Show that the limits fail to exist as $x \to 0$.

(a) $f(x) = \begin{cases} 1 & \text{if } x > 0 \\ -1 & \text{if } x < 0 \end{cases}$ (b) $g(x) = \dfrac{1}{x^2}$

(c) $h(x) = \sin \dfrac{1}{x}$

Solution The graphs of the functions are shown in Figure 1.25. The function f assumes both the values 1 and -1 in every deleted neighborhood of 0, so there is no single limit that is approached. (As we will show later, however, there is a *right-hand* and a *left-hand* limit.)

The function g becomes arbitrarily large in every deleted neighborhood of 0 and so does not approach a (finite) limit. In this case we can write $\lim_{x \to 0} g(x) = +\infty$, but this merely says that the limit fails to exist in a particular way by the function becoming arbitrarily large.

The function h oscillates wildly between 1 and -1 in every deleted neighborhood of 0, so no limit exists. ■

In our examples we have used certain properties of limits that we now state explicitly as a theorem. The proof will be given in Section 1.8.

THEOREM 1.1 If $\lim_{x \to a} f(x) = L$, $\lim_{x \to a} g(x) = M$, and c is any real number, then the following properties hold true.

1. $\lim_{x \to a} c = c$
2. $\lim_{x \to a} c \cdot f(x) = c \cdot L$
3. $\lim_{x \to a} (f + g)(x) = L + M$
4. $\lim_{x \to a} (f - g)(x) = L - M$
5. $\lim_{x \to a} (f \cdot g)(x) = L \cdot M$
6. $\lim_{x \to a} \left(\dfrac{f}{g}\right)(x) = \dfrac{L}{M}$, provided $M \neq 0$ ■

Note Property 1 means that the limit of the constant function with constant value c is the number c; that is, if $f(x) = c$ for all x, then $\lim_{x \to a} f(x) = c$ for all a in $\mathbf{R}$.

FIGURE 1.25

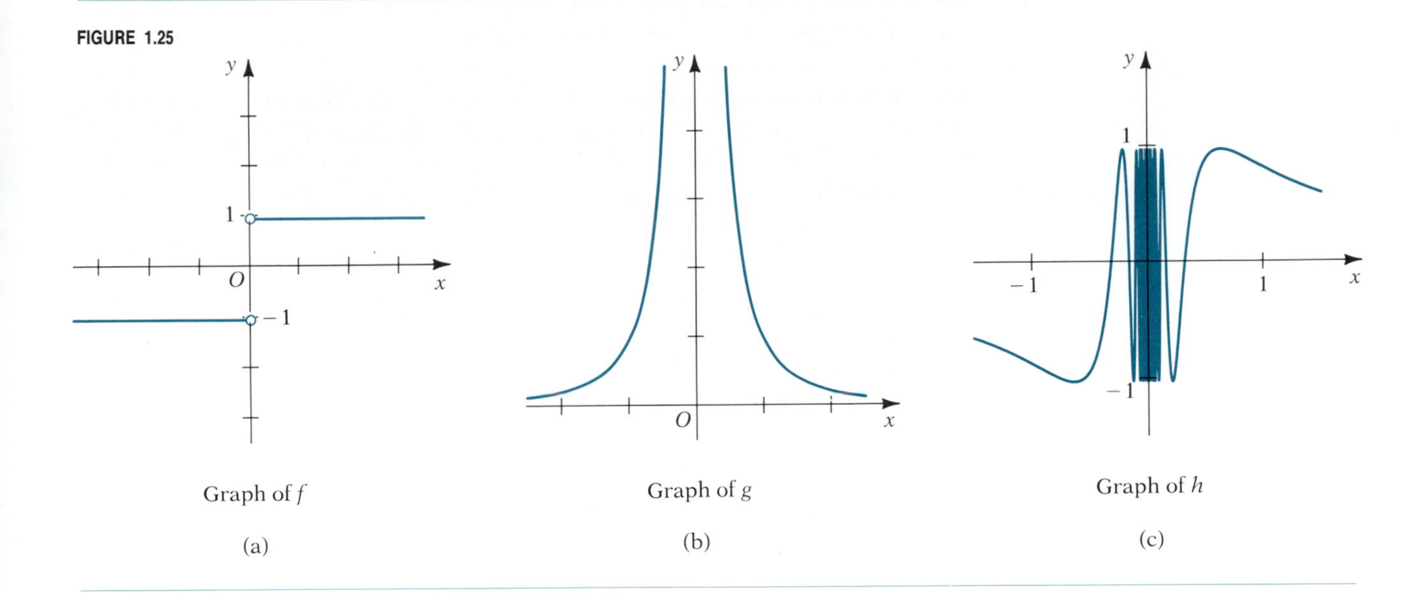

You should learn these properties in words as well as symbols. For example, Property 5 could be verbalized as "the limit of a product of functions is the product of their limits, if both limits exist."

In this as well as the other properties (except Property 1), the proviso that the limits of the individual functions exist (and are finite) is important, as can be seen by the example

$$f(x) = x \quad \text{and} \quad g(x) = \tfrac{1}{x}$$

The product function, $f \cdot g$, has the value

$$(f \cdot g)(x) = x \cdot \tfrac{1}{x} = 1$$

for all $x \neq 0$, and so by Property 1,

$$\lim_{x \to 0}(f \cdot g)(x) = 1$$

But $\lim_{x\to 0} g(x)$ does not exist, since $\frac{1}{x}$ is unbounded in every deleted neighborhood of 0. So it is meaningless to speak of the product of the limits in this case; that is,

$$\underbrace{\lim_{x\to 0} x \cdot \tfrac{1}{x}}_{=1} \neq \underbrace{\lim_{x\to 0} x}_{=0} \cdot \underbrace{\lim_{x\to 0} \tfrac{1}{x}}_{\text{does not exist}}$$

Properties 3 and 4 can be extended by mathematical induction (see Appendix 1) to any finite sum or product, and in the future we will assume these extended forms to be true.

If we begin with the obvious fact that $\lim_{x\to a} x = a$ (even this can be proved) and apply Property 5 repeatedly, we get $\lim_{x\to a} x^n = a^n$. Since polynomial functions are formed by the addition, subtraction, and multiplication of constant functions and powers of x, we conclude from Properties 1–4 that for any polynomial function P and any real number a,

$$\lim_{x \to a} P(x) = P(a)$$

By applying Property 5 we get a similar result for the limit of a rational function. The two results are summarized in the following theorem.

THEOREM 1.2 If P and Q are polynomial functions and a is any real number, then

$$\lim_{x \to a} P(x) = P(a)$$

and

$$\lim_{x \to a} \frac{P(x)}{Q(x)} = \frac{P(a)}{Q(a)} \qquad \text{if } Q(a) \neq 0$$

■

For the quotient $P(x)/Q(x)$, if $Q(a) = 0$, the limit may or may not exist. In Example 1.21 we considered

$$\lim_{x \to 2} \frac{x^2 + 5x - 14}{x^2 - x - 2}$$

Here the denominator is 0 when $x = 2$, so Theorem 1.2 is not applicable. Nevertheless, as we saw in Example 1.21 the limit exists and is 3, found by factoring and dividing out the common factor. Note that not only the denominator in this problem but also the numerator is 0 when $x = 2$.

As another example, let $P(x) = x^2 - 4$ and $Q(x) = (x - 2)^2$, and consider $\lim_{x \to 2} P(x)/Q(x)$. Again we see that not only $Q(2)$ but also $P(2)$ is 0. By factoring and dividing out the common factor, we get

$$\lim_{x \to 2} \frac{P(x)}{Q(x)} = \lim_{x \to 2} \frac{(x + 2)(x - 2)}{(x - 2)^2} = \lim_{x \to 2} \frac{x + 2}{x - 2}$$

Now we can see that no limit exists, since the numerator approaches 4 but the denominator approaches 0, so that the absolute value of the quotient becomes arbitrarily large.

We can summarize by saying that in considering $\lim_{x \to a} P(x)/Q(x)$:

1. If *$Q(a) = 0$ and $P(a) = 0$*, then the limit may or may not exist. This can be determined by factoring and dividing out the common factor.
2. If *$Q(a) = 0$ and $P(a) \neq 0$*, the limit does not exist.

The next theorem permits us to evaluate such limits as

$$\lim_{x \to 4} \sqrt{2x + 1}$$

in the natural way to get $\sqrt{9} = 3$. Its proof is given in Section 1.8.

THEOREM 1.3 If $\lim_{x \to a} f(x) = L > 0$, then

$$\lim_{x \to a} \sqrt{f(x)} = \sqrt{L}$$

■

Remark The more general result

$$\lim_{x \to a} \sqrt[n]{f(x)} = \sqrt[n]{L}$$

also holds true. When n is odd, the restriction that $L > 0$ can be removed.

EXAMPLE 1.23 Evaluate

$$\lim_{x \to 3} \frac{\sqrt{x+1} - 2}{x - 3}$$

Solution We obtain an equivalent fraction by rationalizing the numerator:

$$\lim_{x \to 3} \frac{\sqrt{x+1} - 2}{x - 3} = \lim_{x \to 3} \frac{\sqrt{x+1} - 2}{x - 3} \cdot \frac{\sqrt{x+1} + 2}{\sqrt{x+1} + 2}$$

$$= \lim_{x \to 3} \frac{x + 1 - 4}{(x-3)(\sqrt{x+1} + 2)}$$

Note that we leave the denominator in factored form.

$$= \lim_{x \to 3} \frac{\overset{1}{\cancel{x-3}}}{\underset{1}{\cancel{(x-3)}}(\sqrt{x+1} + 2)}$$

$$= \lim_{x \to 3} \frac{1}{\sqrt{x+1} + 2} = \frac{1}{\sqrt{4} + 2} = \frac{1}{4}$$

In the next to last step we used Property 5 of Theorem 1.1 and also Theorem 1.3. ■

The next theorem is often called the "squeezing theorem" (or "pinching theorem") for reasons that should be apparent. You will be asked to prove it in Exercise Set 1.8.

THEOREM 1.4 THE SQUEEZING THEOREM

If for all x in some deleted neighborhood of $x = a$, the functions f, g, and h satisfy

$$f(x) \le g(x) \le h(x)$$

and if f and h approach the common limit L as $x \to a$, then $g(x)$ must also approach L as $x \to a$. ■

For example, the inequality

$$-x \le x \sin \frac{1}{x} \le x$$

holds in every deleted neighborhood of 0, since $|\sin \frac{1}{x}| \le 1$. As $x \to 0$, the functions x and $-x$ both approach 0. So $x \sin \frac{1}{x}$ is squeezed toward 0 also; that is,

$$\lim_{x \to 0} x \sin \tfrac{1}{x} = 0$$

If, in the definition of the limit, we restrict x to be to the right of a, we get the **right-hand limit,** written

$$\lim_{x \to a^+} f(x) = L$$

This means that $f(x)$ will be arbitrarily close to L whenever $x > a$ and x is sufficiently close to a. The **left-hand limit** is defined in an analogous way,

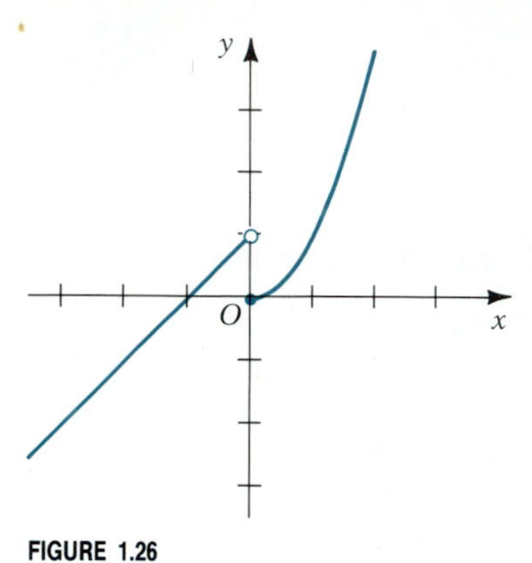

FIGURE 1.26

with $x < a$, and is written

$$\lim_{x \to a^-} f(x) = L$$

Consider, for example, the function defined by

$$f(x) = \begin{cases} x^2 & \text{if } x \geq 0 \\ 1 + x & \text{if } x < 0 \end{cases}$$

with the graph shown in Figure 1.26. Here we see that $\lim_{x \to 0^+} f(x) = 0$ and $\lim_{x \to 0^-} f(x) = 1$. The fact that the right- and left-hand limits are different in this case implies that no single limit exists, in the unrestricted sense. This is a general result, which can be stated as follows:

$\lim_{x \to a} f(x) = L$ if and only if $\lim_{x \to a^+} f(x) = L$ and $\lim_{x \to a^-} f(x) = L$.

EXERCISE SET 1.6

A

In Exercises 1–32 evaluate the limits.

1. $\lim_{x \to 1} (2x^2 - 3x + 1)$
2. $\lim_{x \to -2} \dfrac{2}{x + 1}$
3. $\lim_{x \to 4} \dfrac{2x + 1}{x + 3}$
4. $\lim_{x \to -2} \dfrac{x^2 - 4}{x + 1}$
5. $\lim_{x \to -3} \dfrac{2x^2 - 5x + 1}{x^2 - 2x}$
6. $\lim_{x \to 3} \dfrac{x - 2}{\sqrt{x^2 - x - 2}}$
7. $\lim_{x \to 0} \dfrac{1 - \sin x}{\cos x}$
8. $\lim_{x \to \pi/2} \sqrt{2 \sin x - 3 \cos x}$
9. $\lim_{x \to \pi/3} \dfrac{\tan x - 1}{\tan x + 1}$
10. $\lim_{x \to 3\pi/4} \dfrac{\tan x + \cot x}{\sin x \cos x}$
11. $\lim_{x \to -\pi/4} \dfrac{2 \sec x + \csc x}{\sqrt{1 - \sin 2x}}$
12. $\lim_{x \to 5\pi/6} \dfrac{\tan x - \sec x}{\cos x}$
13. $\lim_{x \to 5} \sqrt{\dfrac{x - 1}{x + 4}}$
14. $\lim_{x \to 8} \dfrac{2x - 5}{\sqrt{3x + 1}}$
15. $\lim_{x \to 6} (8 - x)\sqrt{3x - 2}$
16. $\lim_{x \to -4} \dfrac{x^2 - 16}{\sqrt{x^2 - 9}}$
17. $\lim_{x \to -2} \dfrac{x^2 - 4}{x + 2}$
18. $\lim_{x \to 3/2} \dfrac{4x^2 - 9}{2x - 3}$
19. $\lim_{x \to -5} \dfrac{x + 5}{x^2 - 25}$
20. $\lim_{x \to 1} \dfrac{x^2 - 2x + 1}{x^2 - 1}$
21. $\lim_{x \to 3} \dfrac{x^2 - x - 6}{x - 3}$
22. $\lim_{x \to 7} \dfrac{x^2 - 5x - 14}{x - 7}$
23. $\lim_{x \to 1} \dfrac{2x^2 + 3x - 5}{x^2 - 3x + 2}$
24. $\lim_{x \to -3} \dfrac{x^2 + 5x + 6}{x^2 + 6x + 9}$
25. $\lim_{x \to 1} \dfrac{\sqrt{x} - 1}{x - 1}$
26. $\lim_{x \to 2} \dfrac{\sqrt{2x - 3} - 1}{x - 2}$
27. $\lim_{x \to 5} \dfrac{\sqrt{2x - 1} - 3}{x - 5}$
28. $\lim_{x \to a} \dfrac{\sqrt{x} - \sqrt{a}}{x - a} \quad (a > 0)$
29. $\lim_{x \to \pi} \dfrac{1 + \cos x}{\sin^2 x}$
30. $\lim_{x \to \pi/2} \dfrac{\tan x - \cot x}{\csc 2x}$
31. $\lim_{x \to 0} \dfrac{\tan^2 x}{\sec x - 1}$
32. $\lim_{x \to -\pi/2} \dfrac{1 + \csc x}{\cot^2 x}$

In Exercises 33–40 evaluate

$$\lim_{x \to a} \frac{f(x) - f(a)}{x - a}$$

for the given values of a and f(x).

33. $a = 2$; $f(x) = x + 3$
34. $a = 3$; $f(x) = 2x + 1$
35. $a = 0$; $f(x) = x^2 + 3$
36. $a = -2$; $f(x) = 2x^2 - 3x + 1$
37. $a = -3$; $f(x) = 3 - 2x^2$
38. $a = 4$; $f(x) = 3x^2 - 5$

39. $a = 4;\ f(x) = \sqrt{x}$

40. $a = 4;\ f(x) = \sqrt{2x + 1}$

In Exercises 41–48 evaluate

$$\lim_{h \to 0} \frac{f(a + h) - f(a)}{h}$$

for the given values of a and f(x).

41. $a = 1;\ f(x) = 2 - x$

42. $a = -2;\ f(x) = 3x + 2$

43. $a = 3;\ f(x) = x^2$

44. $a = 2;\ f(x) = 2x^2 + 1$

45. $a = 5;\ f(x) = 3x^2 - 4x$

46. $a = -4;\ f(x) = 2 - 3x^2$

47. $a = 2;\ f(x) = \sqrt{x + 2}$

48. $a = 2;\ f(x) = \sqrt{3 - x}$

In Exercises 49–58 find $\lim_{x \to a^+} f(x)$ and $\lim_{x \to a^-} f(x)$ for the given values of a and f. State whether $\lim_{x \to a} f(x)$ exists and, if it does, give its value.

49. $a = 2;\ f(x) = \begin{cases} x^2 & \text{if } x \geq 2 \\ 6 - x & \text{if } x < 2 \end{cases}$

50. $a = 0;\ f(x) = \begin{cases} 2x - 3 & \text{if } x > 0 \\ x^2 + 1 & \text{if } x \leq 0 \end{cases}$

51. $a = 0;\ f(x) = \begin{cases} 1 & \text{if } x > 0 \\ 0 & \text{if } x = 0 \\ -1 & \text{if } x < 0 \end{cases}$

52. $a = 1;\ f(x) = \begin{cases} 2 - x & \text{if } x > 1 \\ -2 & \text{if } x = 1 \\ x^2 & \text{if } x < 1 \end{cases}$

53. $a = -1;\ f(x) = \begin{cases} x^3 & \text{if } x > -1 \\ 2 & \text{if } x = -1 \\ x^2 - 2 & \text{if } x < -1 \end{cases}$

54. $a = 1;\ f(x) = \begin{cases} 2x & \text{if } x > 1 \\ 0 & \text{if } x = 1 \\ 3 - x^2 & \text{if } x < 1 \end{cases}$

55. $a = 0;\ f(x) = |x|$

56. $a = 0;\ f(x) = \dfrac{|x|}{x}$

57. $a = 0;\ f(x) = x + |x| + 1$

58. $a = 1;\ f(x) = \dfrac{x^2 - x}{|x - 1|}$

In Exercises 59–62 find the indicated one-sided limits.

59. $\displaystyle\lim_{x \to 2^+} \frac{\sqrt{x - 2}}{x + 1}$

60. $\displaystyle\lim_{x \to 1^-} \frac{1 - x}{\sqrt{1 - x}}$

61. $\displaystyle\lim_{x \to -2^-} \sqrt{\frac{x + 2}{x + 3}}$

62. $\displaystyle\lim_{x \to 4^+} \frac{x - 4}{x^2 + x - 20}$

B

63. Let $g(t) = \dfrac{1}{\sqrt{t}}$. Find $\displaystyle\lim_{h \to 0} \frac{g(1 + h) - g(1)}{h}$.

64. Let $\phi(x) = x^3$. Find $\displaystyle\lim_{h \to 0} \frac{\phi(x + h) - \phi(x)}{h}$.

65. Find $\displaystyle\lim_{x \to 3} \frac{\dfrac{1}{\sqrt{x - 2}} - 1}{x - 3}$.

66. Evaluate $\displaystyle\lim_{x \to 2} \frac{f(x) - f(2)}{x - 2}$, where $f(x) = \dfrac{x}{x - 1}$.

67. Let $g(x) = x^{3/2}$. Find $\displaystyle\lim_{x \to 1} \frac{g(x) - g(1)}{x - 1}$.

68. Show that if either of the following limits exists, then so does the other, and they are equal.

$$\lim_{x \to a} \frac{f(x) - f(a)}{x - a} \qquad \lim_{h \to 0} \frac{f(a + h) - f(a)}{h}$$

(*Hint:* Let $x = a + h$.)

In Exercises 69–72 find the limits.

69. $\displaystyle\lim_{x \to 1^+} \frac{\sqrt{x} - 1}{\sqrt{x - 1}}$

70. $\displaystyle\lim_{x \to 1} \frac{\sqrt{x} - 1}{\sqrt[3]{x} - 1}$

71. $\displaystyle\lim_{x \to 0} \frac{1 - \cos x}{\sin x}$

72. $\displaystyle\lim_{x \to \pi} \frac{1 + \sec x}{\tan x}$

In Exercises 73 and 74 use mathematical induction (see Appendix 1) to verify the given formula. Assume that $\lim_{x \to a} f_k(x) = L_k$, where $k = 1, 2, \ldots, n$.

73. $\lim_{x \to a} (f_1 + f_2 + \cdots + f_n)(x) = L_1 + L_2 + \cdots + L_n$

74. $\lim_{x \to a} (f_1 \cdot f_2 \cdot \cdots \cdot f_n)(x) = L_1 \cdot L_2 \cdot \cdots \cdot L_n$

C

75. Write a computer program that will give the values of a function f at the points $a + h$ and $a - h$ for $h = 1, 0.1, 0.01, 0.001, 0.0001$, and 0.00001. Run the program for each of the following functions and values of a. On the basis of the results, what do you conjecture about $\lim_{x\to a^+} f(x)$, $\lim_{x\to a^-} f(x)$, and $\lim_{x\to a} f(x)$?

(a) $f(x) = \dfrac{x^3 - 2x - 4}{x^4 - 16}$; $a = 2$

(b) $f(x) = \dfrac{|x^2 + x - 6|}{x^3 - x^2 - 4}$; $a = 2$

(c) $f(x) = \dfrac{x^4 - x^2 + x + 1}{x^3 + x^2 - x - 1}$; $a = -1$

(d) $f(x) = \dfrac{2 - \sqrt[3]{x + 5}}{x - 3}$; $a = 3$

(e) $f(x) = \dfrac{\sin x}{x}$; $a = 0$

(f) $f(x) = \dfrac{1 - \cos x}{x}$; $a = 0$

1.7 CONTINUITY

Roughly speaking, a function is *continuous* if its graph has no breaks. This is not a suitable definition, however. How, in fact, can we tell whether it does or does not have breaks? Let us be more precise about what we mean by "no break." Suppose the domain of a function f includes a neighborhood of the point $x = a$. Suppose further that $\lim_{x\to a} f(x)$ exists. If this limiting value is equal to the function value $f(a)$, then we say that f is continuous at $x = a$. So continuity at $x = a$ means not only that the function *approaches* a limiting value as x approaches a, but also that this limit is *reached* when $x = a$ and is exactly the value of the function at that point.

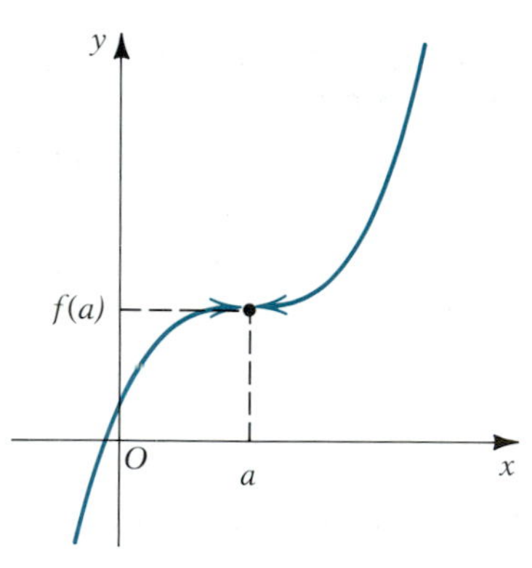

FIGURE 1.27

Figure 1.27 illustrates this situation. The limit approached by $f(x)$ as x approaches a is the same as the value of f at a. So there is no break in the graph; that is, f is continuous at $x = a$.

The following definition summarizes this discussion.

DEFINITION 1.5 CONTINUITY OF A FUNCTION

Let f be defined in a neighborhood of the point $x = a$. Then we say that ***f* is continuous at *a*,** provided that

$$\lim_{x\to a} f(x) = f(a)$$

If f is continuous at each point of an open interval I, then we say that ***f* is continuous on *I*.** ■

Continuity of f at a implies three conditions:

1. $\lim_{x\to a} f(x)$ exists.
2. $f(a)$ exists.
3. The values in conditions 1 and 2 are the same.

So a function can *fail* to be continuous—that is, it can be **discontinuous**—at $x = a$ by failing to satisfy one or more of these three conditions.

When a function is defined on a *closed* interval I, we modify the definition of continuity at the end points by using one-sided limits only. So if $I = [a, b]$, then f is **continuous from the right** at a if $\lim_{x\to a^+} f(x) = f(a)$ and **continuous from the left** at b if $\lim_{x\to b^-} f(x) = f(b)$.

It is instructive to look at various discontinuous functions. Here are some examples.

EXAMPLE 1.24 In each of the following, draw the graph of f and state where it is discontinuous. Explain in what ways Definition 1.5 fails.

(a) $f(x) = \dfrac{1}{x}$ (b) $f(x) = \begin{cases} x - 1 & \text{if } x \geq 1 \\ -1 & \text{if } x < 1 \end{cases}$

(c) $f(x) = \begin{cases} x + 2 & \text{if } x \neq 2 \\ 0 & \text{if } x = 2 \end{cases}$ (d) $f(x) = \dfrac{x^2 - 4}{x - 2}$

Solution

(a) The function is discontinuous at $x = 0$ (Figure 1.28). Not only does $f(0)$ fail to exist, but $\lim_{x\to 0} f(x)$ also fails to exist.

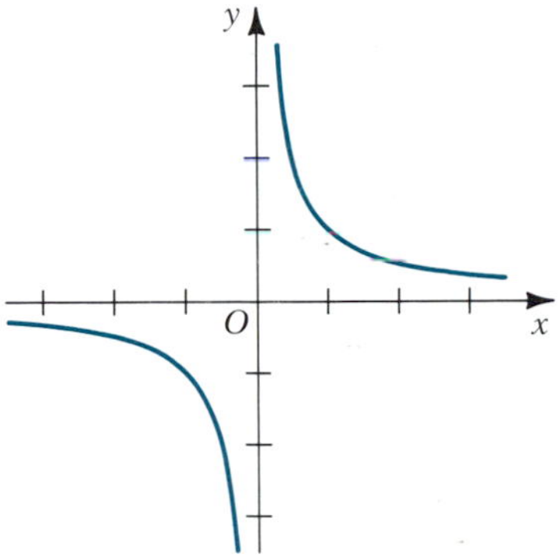

(b) In this case the point of discontinuity is $x = 1$ (Figure 1.29). Although $f(1)$ exists and equals 0, $\lim_{x\to 1} f(x)$ does not exist, since $\lim_{x\to 1^+} f(x) = 0$ and $\lim_{x\to 1^-} f(x) = -1$

(c) Here, $f(2)$ is given as 0, but $\lim_{x\to 2} f(x) = 4$ (Figure 1.30). So even though both values exist, they are unequal, and f is discontinuous at $x = 2$.

(d) Since $x = 2$ is not in the domain of f, the function cannot be continuous there (Figure 1.31). Note that since, for $x \neq 2$,

$$\frac{x^2 - 4}{x - 2} = \frac{(x + 2)\overset{1}{\cancel{(x - 2)}}}{\underset{1}{\cancel{x - 2}}} = x + 2$$

$\lim_{x\to 2} f(x) = 4$. So continuity fails because $f(2)$ does not exist. Observe that this function and the one in part c differ only by virtue of the fact that $f(2)$ is defined in part c. ■

FIGURE 1.28

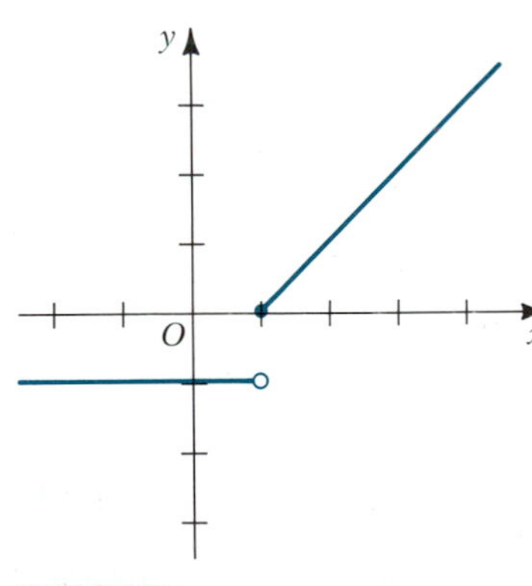

FIGURE 1.29

Remark The type of discontinuity illustrated in part a of the preceding example is called an **infinite** discontinuity, that in part b is called a **simple** discontinuity (sometimes referred to as a **jump** discontinuity), and that illustrated in parts c and d is called a **removable** discontinuity (by redefining the function at one point only, the discontinuity can be removed). There are still other types of discontinuities, some of which we will see later.

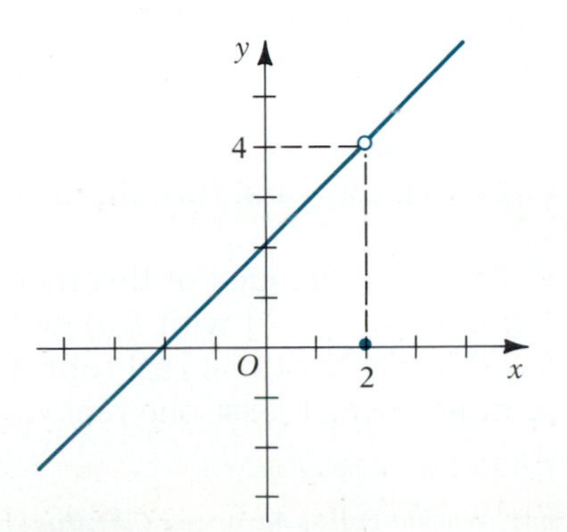

FIGURE 1.30

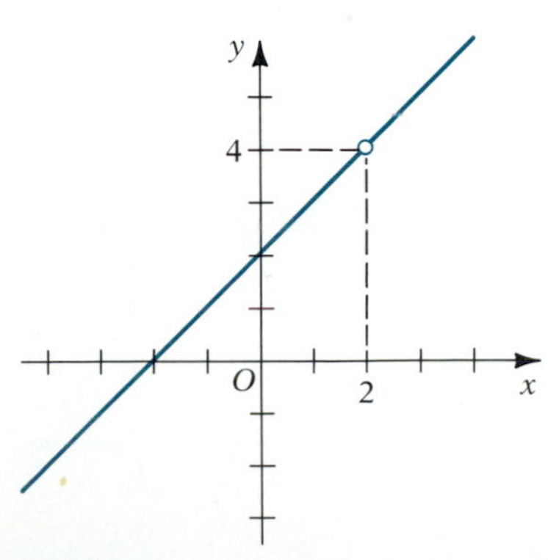

FIGURE 1.31

Suppose that f and g are continuous at $x = a$, so that $\lim_{x \to a} f(x) = f(a)$ and $\lim_{x \to a} g(x) = g(a)$. Then, by Theorem 1.1 we have

$$\lim_{x \to a}(f + g)(x) = f(a) + g(a) = (f + g)(a)$$

Thus $f + g$ is continuous at a. Similar calculations can be made with $f - g$, $f \cdot g$, and f/g. This, together with the results stated in Theorem 1.2, gives the following:

THEOREM 1.5 If f and g are continuous at $x = a$, then so are $f + g$, $f - g$, $f \cdot g$, and f/g provided $g(a) \neq 0$. In particular, every polynomial function is continuous on $\mathbf{R}$, and every rational function is continuous at all points of $\mathbf{R}$ except where the denominator is 0. ■

Thus we are able to recognize a broad class of continuous functions. This class can be extended further by the next theorem, which we will prove in Section 1.8. In essence it says that *a continuous function of a continuous function is continuous*. Recall that the domain of the composite function $f \circ g$ consists of all x in the domain of g for which $g(x)$ is in the domain of f. In the theorem we suppose $x = a$ to be such a point.

THEOREM 1.6 Suppose the composite function $f \circ g$ is defined in some neighborhood of the point $x = a$. If g is continuous at a and f is continuous at $g(a)$, then $f \circ g$ is continuous at a. ■

For example, let $f(x) = \sqrt{x}$ and $g(x) = x^2 + 1$. Since g is a polynomial, it is continuous on all of $\mathbf{R}$. Furthermore, $g(x) > 0$ for all x, so that $(f \circ g)(x) = \sqrt{x^2 + 1}$ is defined for all x. By Theorem 1.3 we see that for $a > 0$, $\lim_{x \to a} \sqrt{x} = \sqrt{a}$, so that f is continuous everywhere on its domain. Thus, by Theorem 1.5, $f \circ g$ is continuous on all of $\mathbf{R}$.

Functions that are continuous on an interval, especially on a closed interval, have many important properties. We will see some of these in later chapters. For now, we state one such property, whose truth seems intuitively obvious, but which is surprisingly difficult to prove. The proof can be found in textbooks on advanced calculus.*

THEOREM 1.7 THE INTERMEDIATE VALUE THEOREM

If f is continuous on the closed interval $[a, b]$ and $f(a) \neq f(b)$, then f takes on every value between $f(a)$ and $f(b)$. ■

Remark To say that f takes on every value between $f(a)$ and $f(b)$ means that if c is any number between $f(a)$ and $f(b)$, there is at least one number x_0 in (a, b) such that $f(x_0) = c$.

Figure 1.32 illustrates the idea of this theorem. One consequence of it is that if f is continuous on $[a, b]$ with $f(a)$ and $f(b)$ opposite in sign, then the equation $f(x) = 0$ has at least one real root. For example, every polynomial of odd degree must have at least one real zero. (See Exercise 33.)

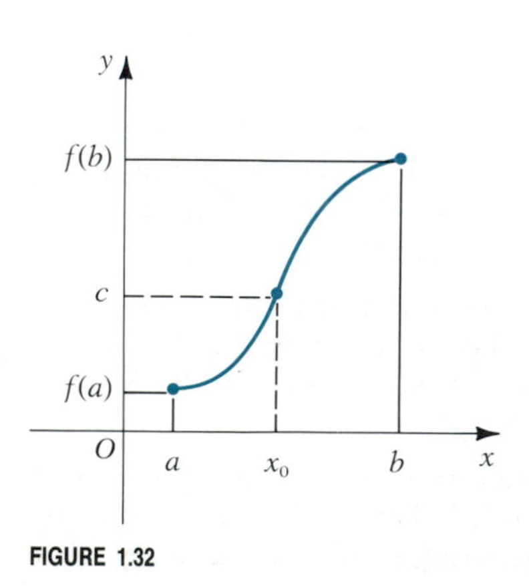

FIGURE 1.32

* See, for example, Watson Fulks, *Advanced Calculus* (John Wiley & Sons, New York, 1978), p. 67.

EXERCISE SET 1.7

A

In Exercises 1–14 determine whether the function f is continuous at x = a. If it is discontinuous, state in what way the definition fails.

1. $f(x) = x^2 - 3x + 4;\ a = 2$

2. $f(x) = 2 - 3x + 4x^3;\ a = -1$

3. $f(x) = \dfrac{x}{x-1};\ a = 3$

4. $f(x) = \dfrac{2x-3}{x^2-1};\ a = 4$

5. $f(x) = \dfrac{x^2-2x+1}{2x^2-x-3};\ a = -1$

6. $f(x) = \dfrac{x^3-1}{x^2+2x-8};\ a = -4$

7. $f(x) = |x-2|;\ a = 2$

8. $f(x) = \dfrac{|x|}{x};\ a = 0$

9. $f(x) = \dfrac{x-1}{x-2};\ a = 2$

10. $f(x) = \dfrac{x^2-4}{x-2};\ a = 2$

11. $f(x) = \dfrac{x+3}{x^2-9};\ a = -3$

12. $f(x) = \dfrac{x^2-2x-3}{x^2-5x+6};\ a = 3$

13. $f(x) = \begin{cases} \sin\frac{1}{x} & \text{if } x \neq 0 \\ 0 & \text{if } x = 0 \end{cases};\ a = 0$

14. $f(x) = \begin{cases} x\sin\frac{1}{x} & \text{if } x \neq 0 \\ 0 & \text{if } x = 0 \end{cases};\ a = 0$

In Exercises 15–24 graph the given function and state on what intervals it is continuous.

15. $f(x) = \begin{cases} x^2+1 & \text{if } x \geq 0 \\ 1-x & \text{if } x < 0 \end{cases}$

16. $g(x) = \begin{cases} x+2 & \text{if } x > 1 \\ 2x+1 & \text{if } x \leq 1 \end{cases}$

17. $h(x) = \begin{cases} \dfrac{1}{x} & \text{if } x > 0 \\ -x & \text{if } x \leq 0 \end{cases}$

18. $F(x) = \begin{cases} \dfrac{1}{x^2} & \text{if } x \neq 0 \\ 0 & \text{if } x = 0 \end{cases}$

19. $G(x) = \begin{cases} \dfrac{|x|}{x} & \text{if } x \neq 0 \\ 1 & \text{if } x = 0 \end{cases}$

20. $H(x) = \begin{cases} x^2-4 & \text{if } x > 2 \\ 1 & \text{if } x = 2 \\ 1-x & \text{if } x < 2 \end{cases}$

21. $f(x) = \begin{cases} \dfrac{x^2-9}{x-3} & \text{if } x \neq 3 \\ 6 & \text{if } x = 3 \end{cases}$

22. $g(x) = \begin{cases} \dfrac{4-x^2}{x-2} & \text{if } x \neq 2 \\ -4 & \text{if } x = 2 \end{cases}$

23. $h(x) = \begin{cases} 2x+1 & \text{if } x < 0 \\ 1-x & \text{if } 0 \leq x < 2 \\ x-2 & \text{if } x \geq 2 \end{cases}$

24. $\phi(x) = \begin{cases} \sqrt{x} & \text{if } x \geq 0 \\ -1 & \text{if } -2 \leq x < 0 \\ 2x+5 & \text{if } x < -2 \end{cases}$

B

25. Let $\phi(x) = \begin{cases} 1 & \text{if } x \text{ is rational} \\ 0 & \text{if } x \text{ is irrational} \end{cases}$. Discuss the continuity of ϕ.

26. Let $\psi(x) = \begin{cases} x & \text{if } x \text{ is rational} \\ 0 & \text{if } x \text{ is irrational} \end{cases}$. Discuss the continuity of ψ.

27. A function is said to be **bounded** if there is some positive number M such that the graph of f lies between the horizontal lines $y = M$ and $y = -M$. Give an example of a function that is continuous on an open interval but is not bounded there.

28. Give an example of a function that has a finite value at each point of a closed interval $[a, b]$ but is not bounded. (See Exercise 27.) (*Hint:* The function will have to be discontinuous.)

29. Does there exist a function that is discontinuous at every point? Does there exist a function that is continuous at one and only one point? (*Hint:* See Exercises 25 and 26.)

30. Suppose f is continuous on $\mathbf{R}$ and P is a polynomial function for which it is known that $f(x) = P(x)$ for all $x \in \mathbf{R}$ except the single point $x = a$. Show that

$f(a) = P(a)$; that is, $f(x)$ must be identical to $P(x)$ everywhere. What conclusion can you draw if it is only known that $f(x) = P(x)$ except for finitely many values of x? Explain.

31. Prove or disprove: If f and g are discontinuous at $x = a$, then (a) $f + g$ is discontinuous at $x = a$; (b) $f \cdot g$ is discontinuous at $x = a$.

32. Prove that if f is continuous on an interval I, then so is $|f|$. Is the converse of this also true? Prove or disprove.

33. Prove that every polynomial function P_n of *odd* degree has at least one real zero; that is, $P_n(x_0) = 0$ for some x_0.

C

34. Use the computer to determine whether the following function is continuous at $x = 3$:

$$f(x) = \begin{cases} \dfrac{x^3 + 18x - 81}{x^4 - 9x^2 - 43x + 129} & \text{if } x < 3 \\ \sqrt[4]{2x^5 - 7x^3 - 41} & \text{if } x \geq 3 \end{cases}$$

(*Note:* You should be able to obtain convincing evidence one way or the other on the computer, but recognize that it is not a proof.)

1.8 THE FORMAL LIMIT DEFINITION AND PROOFS OF LIMIT THEOREMS

So far we have been proceeding on our intuitive understanding of what is meant by the limit of a function. For proving theorems about limits and continuity, it is necessary to formulate the definition in a more precise way.

For the number L to be the limit of f as x approaches a, it must be true that if we choose any neighborhood of L (however small it may be), we can always find a deleted neighborhood of a contained in the domain of f such that if x is in this deleted neighborhood of a, then $f(x)$ will be in the given neighborhood of L. This is what we mean when we say that $f(x)$ can be made arbitrarily close to L [that is, $f(x)$ is in an arbitrarily small neighborhood of L] provided x is in a sufficiently small deleted neighborhood of a. In Figure 1.33 we illustrate these concepts. The neighborhood of L has as its radius an arbitrarily small positive number ε, and the radius of the corresponding deleted neighborhood of a is called δ. (These two Greek letters are customarily used for this purpose, but any two letters would do.) Note that if a smaller ε were chosen, δ would also be smaller. This is typically the case; that is, the size of δ is dependent on the size of ε. Observe also that the portion of the graph of f between the two vertical lines $x = a - \delta$ and $x = a + \delta$ must lie between the horizontal lines $y = L - \varepsilon$ and $y = L + \varepsilon$, except possibly the point $(a, f(a))$, if this exists.

For $f(x)$ to be in a neighborhood of L of radius ε, we must have

$$L - \varepsilon < f(x) < L + \varepsilon$$

or equivalently

$$-\varepsilon < f(x) - L < \varepsilon$$

That is, we must have

$$|f(x) - L| < \varepsilon$$

FIGURE 1.33

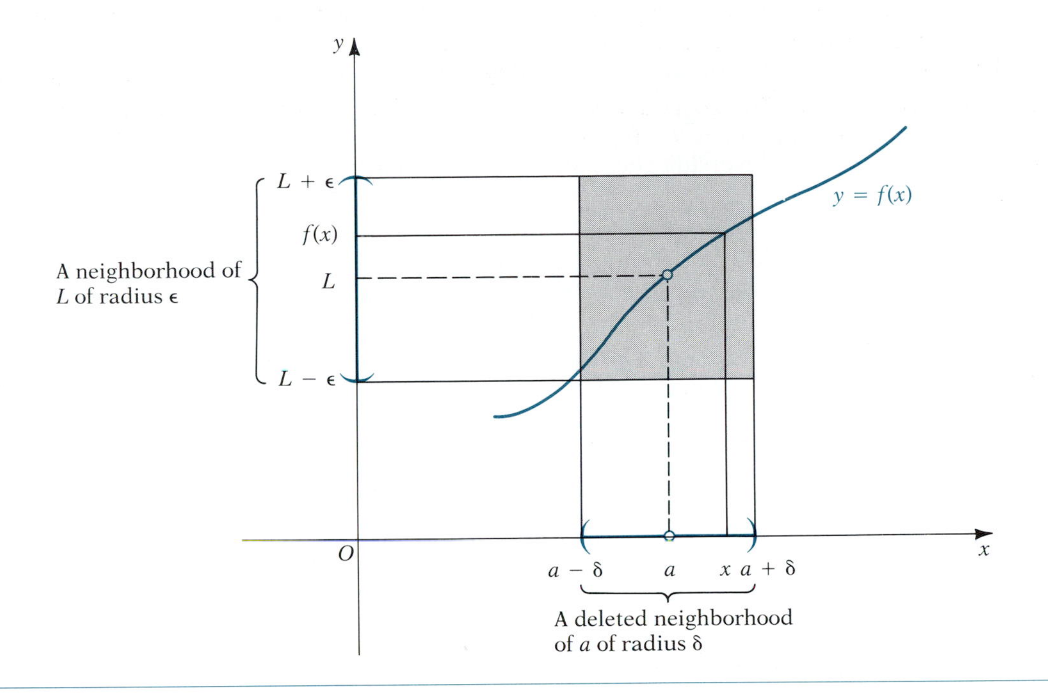

Similarly, for x to be in a deleted neighborhood of a of radius δ, we must have

$$|x - a| < \delta \quad \text{and} \quad x \neq a$$

The restriction that $x \neq a$ is accomplished by requiring that $0 < |x - a|$. Thus,

$$0 < |x - a| < \delta$$

All of these steps are reversible. We can therefore restate the limit definition in the following form:

DEFINITION 1.6 FORMAL LIMIT DEFINITION

Let f be defined in some deleted neighborhood of a. The limit of f is L as x approaches a provided that for each positive number ε, there exists a positive number δ such that

$$\text{if } 0 < |x - a| < \delta \text{ then } |f(x) - L| < \varepsilon.$$ ■

We illustrate this definition by proving that

$$\lim_{x \to 2}(3x + 4) = 10$$

First, we let ε denote an arbitrarily small, unspecified positive number. We must find a positive number δ such that $0 < |x - 2| < \delta$ implies $|(3x + 4) - 10| < \varepsilon$. We rewrite $|(3x + 4) - 10|$ in the following equivalent ways:

$$|(3x + 4) - 10| = |3x - 6| = 3|x - 2|$$

The inequality

$$3|x-2|<\varepsilon$$

will be satisfied provided

$$|x-2|<\tfrac{\varepsilon}{3}$$

Reversing these steps, we see that if $|x-2|<\frac{\varepsilon}{3}$, then

$$|(3x+4)-10|<\varepsilon$$

So the definition is satisfied with $\delta=\frac{\varepsilon}{3}$. Note that in this case we did not have to restrict x to be different from 2, since $f(2)$ exists and equals 10, and therefore the inequality $|(3x+4)-10|<\varepsilon$ is satisfied when $x=2$.

We now illustrate the power of this definition by proving the limit properties stated in Theorem 1.1.

PROPERTY 1 For any real number c, $\lim_{x\to a} c = c$.

Proof Write $f(x)=c$ and let $\varepsilon>0$ be given. Then for all x, $|f(x)-c|=|c-c|=0<\varepsilon$. So in this case, δ may be any positive number, and Definition 1.6 is satisfied. ■

PROPERTY 2 If $\lim_{x\to a} f(x)=L$ and c is any real number, then $\lim_{x\to a} cf(x)=cL$.

Proof Let $\varepsilon>0$ be given. If $c=0$, $|cf(x)-cL|=0<\varepsilon$ for all x. So the definition is satisfied in a trivial way in this case.

Suppose now that $c\neq 0$. The inequality $|cf(x)-cL|<\varepsilon$ will be satisfied if

$$|f(x)-L|<\frac{\varepsilon}{|c|}$$

Now $\varepsilon/|c|$ is a positive number, and since the limit of $f(x)$ as x approaches a is L, we know by Definition 1.6 that a positive number δ can be found such that

$$|f(x)-L|<\frac{\varepsilon}{|c|} \qquad \text{provided } 0<|x-a|<\delta$$

Thus,

$$|cf(x)-cL|<\varepsilon \qquad \text{provided } 0<|x-a|<\delta$$

This completes the proof. You might observe that Property 1 is a special case of Property 2 in which $f(x)=1$ for all x. ■

PROPERTY 3 If $\lim_{x\to a} f(x)=L$ and $\lim_{x\to a} g(x)=M$, then $\lim_{x\to a}(f+g)(x)=L+M$.

Proof Let ε denote any positive number. By Definition 1.6 there exist positive numbers δ_1 and δ_2 such that

$$\text{if } 0<|x-a|<\delta_1, \text{ then } |f(x)-L|<\tfrac{\varepsilon}{2}$$

and

$$\text{if } 0<|x-a|<\delta_2, \text{ then } |g(x)-M|<\tfrac{\varepsilon}{2}$$

Choose δ to be the smaller of δ_1 and δ_2. Then for $0 < |x - a| < \delta$, we have

$$\begin{aligned} |(f+g)(x) - (L+M)| &= |(f(x) + g(x) - (L+M)| \\ &= |(f(x) - L) + (g(x) - M)| \\ &\leq |(f(x) - L| + |g(x) - M)| \quad \text{By the triangle inequality} \\ &< \tfrac{\varepsilon}{2} + \tfrac{\varepsilon}{2} = \varepsilon \end{aligned}$$

and the proof is complete. ■

PROPERTY 4 If $\lim_{x \to a} f(x) = L$ and $\lim_{x \to a} g(x) = M$, then $\lim_{x \to a}(f - g)(x) = L - M$.

Proof Write $f(x) - g(x) = f(x) + (-1)g(x)$ and use Properties 2 and 3. ■

In the proofs of the next two properties we use the fact that if $\lim_{x \to a} f(x) = L$, then also $\lim_{x \to a}|f(x)| = |L|$, which follows from the inequality $||a| - |b|| \leq |a - b|$, a proof of which was called for in Exercise 46 in Exercise Set 1.2.

PROPERTY 5 If $\lim_{x \to a} f(x) = L$ and $\lim_{x \to a} g(x) = M$, then $\lim_{x \to a}(f \cdot g)(x) = L \cdot M$.

Proof We know that $\lim_{x \to a}|g(x)| = |M|$. Let h be the radius of a deleted neighborhood of a such that for all x in this deleted neighborhood, $|g(x)|$ is in the neighborhood $(|M| - 1, |M| + 1)$ of radius 1 about $|M|$. In particular, then, for these x values, $|g(x)| < |M| + 1$. Let K denote the larger of the numbers $|M| + 1$ and $|L| + 1$.

Now let $\varepsilon > 0$ be given. There exist positive numbers δ_1 and δ_2 such that

$$|f(x) - L| < \tfrac{\varepsilon}{2K} \qquad \text{if } 0 < |x - a| < \delta_1$$

and

$$|g(x) - M| < \tfrac{\varepsilon}{2K} \qquad \text{if } 0 < |x - a| < \delta_2$$

Choose $\delta = \min(\delta_1, \delta_2, h)$. Then for $0 < |x - a| < \delta$, we have

$$\begin{aligned} |(fg)(x) - LM| &= |f(x)g(x) - LM| \\ &= |f(x)g(x) - Lg(x) + Lg(x) - LM| \quad \text{We added and subtracted } Lg(x). \\ &\leq |f(x)g(x) - Lg(x)| + |Lg(x) - LM| \quad \text{By the triangle inequality} \\ &= |g(x)|\,|f(x) - L| + |L|\,|g(x) - M| \\ &< K \cdot \tfrac{\varepsilon}{2K} + K \cdot \tfrac{\varepsilon}{2K} = \varepsilon \end{aligned}$$

The conclusion follows. ■

PROPERTY 6 If $\lim_{x \to a} f(x) = L$ and $\lim_{x \to a} g(x) = M \neq 0$, then

$$\lim_{x \to a} \frac{f}{g}(x) = \frac{L}{M}$$

Proof First we show that

$$\lim_{x \to a} \frac{1}{g(x)} = \frac{1}{M}$$

Again, using the fact that $|g(x)|$ has the limit $|M|$ as $x \to a$, we bound $g(x)$ away from 0 by restricting x to be in a small enough deleted neighborhood of a, call its radius h, so that for x in this set, $|g(x)|$ is in the neighborhood $(|M|/2, 3|M|/2)$ of radius $|M|/2$ about $|M|$. In particular, then, for x so restricted, $|g(x)| > |M|/2$.

Now let $\varepsilon > 0$ be given. Choose $\delta \leq h$ such that when $0 < |x - a| < \delta$,

$$|g(x) - M| < \frac{M^2\varepsilon}{2}$$

Then, for all such x values, we have

$$\begin{aligned}\left|\frac{1}{g(x)} - \frac{1}{M}\right| = \left|\frac{M - g(x)}{g(x) \cdot M}\right| &= \frac{|g(x) - M|}{|g(x)|\,|M|}\\ &< \frac{|g(x) - M|}{\left|\frac{M}{2}\right||M|} \quad \text{(Why?)}\\ &= |g(x) - M| \cdot \frac{2}{M^2}\\ &< \frac{\varepsilon M^2}{2} \cdot \frac{2}{M^2} = \varepsilon\end{aligned}$$

Now we complete the proof that $(f/g)(x) \to L/M$ by noting that

$$\frac{f}{g}(x) = \frac{f(x)}{g(x)} = f(x) \cdot \frac{1}{g(x)} \to L \cdot \frac{1}{M} = \frac{L}{M}$$

as $x \to a$. ■

The definition of continuity is based on the notion of limit, and since we now have a precise definition of the limit of a function, we can reformulate the definition of continuity similarly.

DEFINITION 1.7 **THE $\varepsilon - \delta$ DEFINITION OF CONTINUITY**

Let the domain of f include some neighborhood of a. To say that f is continuous at a means that corresponding to each positive number ε there exists a positive number δ such that

if $|x - a| < \delta$, then $|f(x) - f(a)| < \varepsilon$. ■

Remark For continuity at the end points of a closed interval $[a, b]$, we use one-sided limits. So f is continuous at a **from the right** if, for each $\varepsilon > 0$, there exists a $\delta > 0$ such that

if $a \leq x < a + \delta$, then $|f(x) - f(a)| < \varepsilon$.

The analogous criterion for continuity **from the left** at b is:

if $b - \delta < x \leq b$, then $|f(x) - f(b)| < \varepsilon$.

EXAMPLE 1.25 Prove that $f(x) = \sqrt{x}$ is continuous at $x = a$, where a is any nonnegative number.

Solution Let ε denote any positive number. If $a = 0$, we consider continuity from the right. So we seek a number $\delta > 0$ such that when $0 \le x < \delta$, it will be true that $|\sqrt{x} - \sqrt{0}| < \varepsilon$. But $|\sqrt{x} - \sqrt{0}| = \sqrt{x}$, and $\sqrt{x} < \varepsilon$ if $0 \le x < \varepsilon^2$. So we take $\delta = \varepsilon^2$.

For $a > 0$, we have

$$|\sqrt{x} - \sqrt{a}| = \left|\frac{\sqrt{x} - \sqrt{a}}{1} \cdot \frac{\sqrt{x} + \sqrt{a}}{\sqrt{x} + \sqrt{a}}\right| = \frac{|x - a|}{\sqrt{x} + \sqrt{a}} \le \frac{|x - a|}{\sqrt{a}}$$

and this will be less than ε if $|x - a| < \varepsilon\sqrt{a}$. So we take $\delta = \varepsilon\sqrt{a}$. The proof is now complete. ■

EXAMPLE 1.26 Using Definition 1.7 prove that $f(x) = x^2$ is continuous on $\mathbf{R}$.

Solution Let $\varepsilon > 0$ be given, and let a denote any real number. Restrict x to be in a neighborhood of radius 1 about a, so that $|x - a| < 1$. Then

$$|x + a| = |x - a + 2a| \le |x - a| + 2|a| < 1 + 2|a|$$

Thus, for x so restricted, we have

$$|f(x) - f(a)| = |x^2 - a^2| = |x + a||x - a| < (1 + 2|a|)|x - a|$$

and this will be less than ε provided

$$|x - a| < \frac{\varepsilon}{1 + 2|a|}$$

Since we also want $|x - a| < 1$, we can satisfy both inequalities by requiring that $|x - a| < \delta$, where

$$\delta = \min\left(1, \frac{\varepsilon}{1 + 2|a|}\right)$$ ■

Finally, we prove Theorem 1.6 and an important corollary to it. We restate the theorem for completeness.

THEOREM 1.6 Suppose the composite function $f \circ g$ is defined in some neighborhood of the point $x = a$. If g is continuous at a and f is continuous at $g(a)$, then $f \circ g$ is continuous at a.

Proof Let ε denote any positive number. Write $y = g(x)$ and $b = g(a)$. Since f is continuous at b, there exists a positive number η such that if $|y - b| < \eta$, then $|f(y) - f(b)| < \varepsilon$. But $|y - b| = |g(x) - g(a)|$, and since g is continuous at a, there exists a $\delta > 0$ such that $|g(x) - g(a)| < \eta$ wherever $|x - a| < \delta$. Thus, for $|x - a| < \delta$, we have $|f(y) - f(b)| < \varepsilon$, or equivalently

$$|f(g(x)) - f(g(a))| < \varepsilon$$

So $f \circ g$ is continuous at a. ■

COROLLARY 1.6 If $\lim_{x \to a} g(x) = b$ and f is continuous at b, then

$$\lim_{x \to a} f(g(x)) = f\left(\lim_{x \to a} g(x)\right)$$

Proof Let g_1 be defined as follows:

$$g_1(x) = \begin{cases} g(x) & \text{if } x \neq a \\ b & \text{if } x = a \end{cases}$$

So g_1 and g are identical except possibly at $x = a$. Furthermore, g_1 is continuous at a. (Why?) So by Theorem 1.6 $f \circ g_1$ is continuous at a. Thus,

$$\lim_{x \to a} f(g_1(x)) = f(g_1(a)) = f\left(\lim_{x \to a} g_1(x)\right)$$

In the expressions $\lim_{x \to a} f(g_1(x))$ and $f(\lim_{x \to a} g_1(x))$ we restrict x to be different from a so that g can be substituted for g_1, giving the desired result:

$$\lim_{x \to a} f(g(x)) = f\left(\lim_{x \to a} g(x)\right)$$

■

Proof of Theorem 1.3 For convenience of notation we restate Theorem 1.3 in the following form:

If $\lim_{x \to a} g(x) = L > 0$, then $\lim_{x \to a} \sqrt{g(x)} = \sqrt{L}$.

In Example 1.25 we proved that $f(x) = \sqrt{x}$ is continuous for $x \geq 0$. So by the corollary to Theorem 1.6 we have

$$\lim_{x \to a} f(g(x)) = f\left(\lim_{x \to a} g(x)\right)$$

or

$$\lim_{x \to a} \sqrt{g(x)} = \sqrt{L}$$

It should be noted that using right-hand limits, this theorem remains true if $L = 0$. ■

EXERCISE SET 1.8

A

In Exercises 1–17 prove the statements by showing that Definition 1.6 is satisfied.

1. $\lim_{x \to 1}(2x + 3) = 5$
2. $\lim_{x \to 3}(4x - 5) = 7$
3. $\lim_{x \to 2}\left(\frac{x + 1}{3}\right) = 1$
4. $\lim_{x \to 4}\left(\frac{2x - 3}{5}\right) = 1$
5. $\lim_{x \to 2}(3 - 4x) = -5$
6. $\lim_{x \to 5}\left(\frac{1 - x}{2}\right) = -2$
7. $\lim_{x \to -1}(2x + 5) = 3$
8. $\lim_{x \to -2}(3 - 4x) = 11$
9. $\lim_{x \to 0}\frac{5 - 2x}{10} = \frac{1}{2}$
10. $\lim_{x \to -6}\frac{3 - x}{12} = \frac{3}{4}$
11. $\lim_{x \to 2}\frac{x^2 - 4}{x - 2} = 4$
12. $\lim_{x \to 3}\frac{9 - x^2}{x - 3} = -6$
13. $\lim_{x \to -1/2}\frac{1 - 4x^2}{2x + 1} = 2$
14. $\lim_{x \to 4/3}\frac{9x^2 - 16}{3x - 4} = 8$
15. $\lim_{x \to a} x = a$
16. If $\lim_{x \to a} f(x) = 0$ and $|g(x)| \leq M$, then $\lim_{x \to a} f(x) \cdot g(x) = 0$.
17. If $\lim_{x \to a} f(x) = 0$, then $\lim_{x \to a}[f(x)]^2 = 0$. [*Hint:* For x in a sufficiently small deleted neighborhood of a, show that $|f(x)| < 1$, and explain why it is true also that $[f(x)]^2 \leq |f(x)|$ for x in this deleted neighborhood.]
18. Prove that if $\lim_{x \to a} f(x) = L$, then $\lim_{x \to a}|f(x)| = |L|$.
19. Prove that $f(x) = |x|$ is continuous on $\mathbf{R}$.
20. Use the result of Exercise 19 and Theorem 1.5 to prove that if f is continuous at $x = a$, then so is $|f|$.

B

In Exercises 21–26 use Definition 1.6 to prove the given limit statement.

21. $\lim_{x\to a} x^3 = a^3$

22. $\lim_{x\to a} \frac{1}{\sqrt{x}} = \frac{1}{\sqrt{a}}$ if $a > 0$

23. $\lim_{x\to 2} \frac{x}{x-1} = 2$

24. $\lim_{x\to a} \frac{1}{x^2} = \frac{1}{a^2}$ if $a \neq 0$

25. $\lim_{x\to 0} f(x) = 1$ where $f(x) = \begin{cases} x^2+1 & \text{if } x > 0 \\ 3 & \text{if } x = 0 \\ 1-2x & \text{if } x < 0 \end{cases}$

26. $\lim_{x\to 1} g(x) = 0$ where $g(x) = \begin{cases} x^2-1 & \text{if } x > 1 \\ 1 & \text{if } x = 1 \\ \dfrac{x-1}{3} & \text{if } x < 1 \end{cases}$

27. Let $\phi(x) = \begin{cases} 0 & \text{if } x \text{ is irrational} \\ 1 & \text{if } x \text{ is rational} \end{cases}$. Prove that ϕ is discontinuous everywhere.

28. Let $f(x) = \begin{cases} 0 & \text{if } x \text{ is irrational} \\ x & \text{if } x \text{ is rational} \end{cases}$. Prove that f is continuous at $x = 0$ but discontinuous at all other points.

29. Prove that if f is continuous at $x = a$ and $f(a) > 0$, then $f(x) > 0$ in some neighborhood of a. State and prove an analogous result if $f(a) < 0$.

30. Prove the squeezing theorem (Theorem 1.4). [*Hint:* First show that for x in any sufficiently small deleted neighborhood of a, $|g(x) - L| \leq |f(x) - L| + |h(x) - L|$.]

1.9 SUPPLEMENTARY EXERCISES

In Exercises 1–4 solve the inequalities.

1. $2\left(\frac{3}{4} - 5x\right) - \frac{1}{3} \geq \frac{5x}{6} - 1$

2. $x(2x+5) > 12$

3. $2|4 - 3x| - 5 < 6$

4. $\frac{x^2 - 2x}{x+3} \leq 0$

5. (a) Express in set notation, making use of absolute values, the deleted neighborhood of -2 of radius 1.

(b) Express the following set using neighborhood terminology: $\{x : 0 < |x - 3| < \delta\}$.

6. Prove for real numbers a, b, and c, if $a < b$, then

$\frac{a}{c} < \frac{b}{c}$ if $c > 0$ and $\frac{a}{c} > \frac{b}{c}$ if $c < 0$.

7. Prove the converse of the statement in Exercise 47 of Exercise Set 1.2; that is, prove that if M is the least upper bound of a nonempty set S of real numbers, then conditions i and ii are satisfied.

8. Find the equations of (a) the medians and (b) the altitudes of the triangle with vertices $(-1, 2)$, $(3, 4)$, and $(5, -2)$.

9. Prove that the points $(-2, 9)$, $(-4, -2)$, $(1, -12)$, and $(3, -1)$ are vertices of a rhombus (a parallelogram all of whose sides have the same length). Prove that the diagonals are perpendicular.

10. Find the equation of the tangent line to the circle $x^2 + y^2 - 2x + 4y - 8 = 0$ at the point $(3, 1)$.

11. Find the equation of the circle that passes through the vertex and the two x-intercepts of the parabola $x^2 - 6x + 2y + 5 = 0$.

12. Find the points of intersection of the circle $x^2 + y^2 - 2x + 3y = 12$ and the parabola $2y = x^2 - 2x - 6$. Sketch both curves on the same set of axes and shade the area that is below the parabola and inside the circle.

13. Let $f(x) = x^2 - 3x + 5$. Identify and draw the graph of f. Find

$$\lim_{x\to 2} \frac{f(x) - f(2)}{x - 2}$$

14. Give the domain of each of the following functions:

(a) $f(x) = \sqrt{3x^2 - 8x - 35}$

(b) $g(x) = \sqrt{\dfrac{x^2 - 1}{x^2 - 2x - 3}}$

15. A circle is inscribed in an equilateral triangle, each side of which has length s. Express the area of the circle as a function of s.

16. At temperature x the thermal conductivity of a wall is given by $f(x) = k(1 + ax)$, where k and a are constants. Write the rule for the thermal resistance $g \circ f$, given that $g(x) = T/(Ax)$, where T is the thickness of the wall and A is its cross-sectional area.

17. Let $f(t) = t^2 - 3$ and $g(t) = \sqrt{1 - t}$. Find $f \circ g$ and $g \circ f$, and determine the domain of each.

18. Let

$$f(x) = \begin{cases} 1 & \text{if } x > 2 \\ \dfrac{1}{|x-2|} & \text{if } 0 \leq x < 2 \\ -x & \text{if } x < 0 \end{cases}$$

Draw the graph of f and find the following:

(a) $f(-3)$ (b) $f(0)$
(c) $f(1)$ (d) $\lim_{x\to 2^+} f(x)$
(e) $\lim_{x\to 2^-} f(x)$ (f) $\lim_{x\to 0^+} f(x)$
(g) $\lim_{x\to 0^-} f(x)$

19. Let $f(x) = x/(x+1)$. Find

$$\lim_{h\to 0}\frac{f(x+h)-f(x)}{h}$$

20. A manufacturer of solar energy cells finds that the cost, in dollars, of producing x cells per month is $30x + 1500$, and he sets the selling price per unit at $120 - 0.1x$ dollars. Find the profit function $P(x)$ and draw its graph. How many units should be sold in order for $P(x)$ to be greatest? What is this greatest profit?

21. Let $g(t) = 1/(t-1)$ and $h(t) = t/(t+2)$. Find $g \circ h$ and $h \circ g$, and give the domain of each.

22. Let $f(x) = x^2 - 2x - 3$, $g(x) = x^2 + 5x + 4$, and $h = \frac{f}{g}$. Find $\lim_{x\to -1} h(x)$.

23. Let $f(x) = 1 + x^2$ and $g(x) = \tan\frac{\pi x}{4}$. Find:
(a) $(f-g)(3)$ (b) $(\frac{f}{g})(-1)$ (c) $(f\circ g)(x)$

24. Find two functions f and g so that $F = f \circ g$, where

(a) $F(x) = 1 + \sqrt{(2x-3)^3}$ (b) $F(x) = \dfrac{3}{|x^2-4|}$

(c) $F(x) = \sin^2\left(\dfrac{1}{2x-1}\right)$

In Exercises 25–31 find the limits, or show that they fail to exist.

25. (a) $\displaystyle\lim_{x\to 2}\frac{x^2-4x+4}{x^2+2x-8}$ (b) $\displaystyle\lim_{x\to 3}\frac{\sqrt{x+1}-2}{x-3}$

26. (a) $\displaystyle\lim_{x\to 1^+}\frac{x^2-1}{\sqrt{x-1}}$ (b) $\displaystyle\lim_{x\to -1}\frac{x^{1/3}+1}{x^{2/3}-1}$

27. (a) $\displaystyle\lim_{x\to 2}\frac{x-2}{|x-2|}$ (b) $\displaystyle\lim_{x\to 0}\frac{x}{\sqrt{x^3+x^2}}$

28. (a) $\displaystyle\lim_{x\to -3^-}\frac{2x^2-1}{x^2+3x}$ (b) $\displaystyle\lim_{x\to 2^-}\frac{\sqrt{4-x^2}}{x-2}$

29. (a) $\displaystyle\lim_{x\to 2}\frac{\frac{1}{x}-\frac{1}{2}}{x-2}$ (b) $\displaystyle\lim_{x\to -1}\frac{|x-1|-2}{x+1}$

30. (a) $\displaystyle\lim_{x\to \pi/4}\frac{\sin x-\cos x}{1-\cot^2 x}$ (b) $\displaystyle\lim_{x\to \pi/2}\frac{\sec x-\tan x}{\sin x-1}$

31. (a) $\displaystyle\lim_{x\to \pi/2}(\sec x-\tan x)$ (b) $\displaystyle\lim_{x\to \pi}(\cot x+\csc x)$

32. Given that $\lim_{x\to 0}(\sin x)/x = 1$, find the following:

(a) $\displaystyle\lim_{x\to 0}\frac{\tan x}{x}$ (b) $\displaystyle\lim_{x\to 0}\frac{\sin 2x}{x}$

(c) $\displaystyle\lim_{x\to 0}\frac{\sin^2 x}{x}$ (d) $\displaystyle\lim_{x\to 0} x\cot x$

33. Let $f(x) = \sqrt{x}$, $g(x) = 2x + 3$, $F = f \circ g$, and $G = g \circ f$. Find

$$\lim_{h\to 0}\frac{F(x+h)-F(x)}{h} \quad\text{and}\quad \lim_{h\to 0}\frac{G(x+h)-G(x)}{h}$$

In Exercises 34–36 determine all points of discontinuity of the given function, and at each such point tell in what ways the definition of continuity fails to be satisfied.

34. (a) $f(x) = \dfrac{x+1}{x^2-3x-10}$ (b) $f(x) = \dfrac{1-x^2}{x+1}$

35. (a) $f(x) = \dfrac{|2-x|}{x-2}$, if $x \neq 2$; $f(2) = 1$

(b) $f(x) = \dfrac{|x|}{x^2-2}$

36. (a) $f(x) = \begin{cases} x^2+1 & \text{if } x > 0 \\ 0 & \text{if } x = 0 \\ 1-x & \text{if } x < 0 \end{cases}$

(b) $f(x) = \begin{cases} \frac{2}{x} & \text{if } x > 0 \\ \frac{x}{2} & \text{if } x \leq 0 \end{cases}$

In Exercises 37–40 prove the statements using Definition 1.6.

37. $\displaystyle\lim_{x\to 4}(2x-5) = 3$ **38.** $\displaystyle\lim_{x\to -1}\frac{1-x}{2} = 1$

39. $\displaystyle\lim_{x\to 4}|2-x| = 2$ **40.** $\displaystyle\lim_{x\to 0}\sqrt{2x+1} = 1$

41. Prove that the linear function $f(x) = mx + b$, where $m \neq 0$, is continuous everywhere, using Definition 1.7.

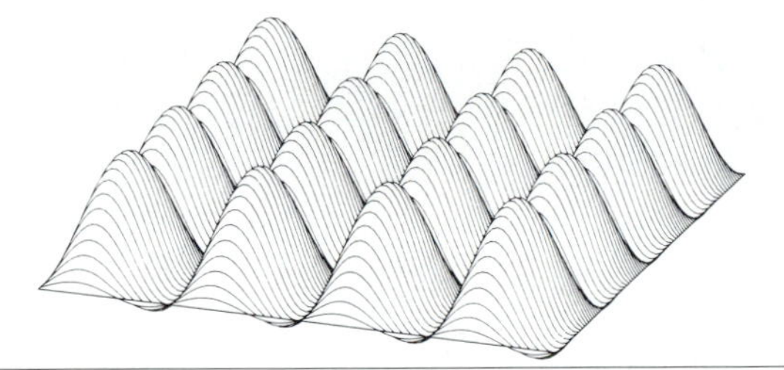

CHAPTER 2

THE DERIVATIVE

2.1 THE TANGENT PROBLEM AND THE VELOCITY PROBLEM

In this chapter we introduce the first of the two fundamental operations of calculus, called **differentiation.** As motivation we consider two related problems: (1) finding the tangent line to a curve and (2) finding the velocity of a moving particle.

We know how to find the tangent line to a circle, but how can we find tangent lines to other curves? More fundamentally, how is the tangent line defined? Students are often tempted to say something like "the line that touches the curve in only one point" or "the line that touches the curve but doesn't cross it." The curves in Figure 2.1 show, however, that neither of these descriptions suffices.

To see how to define the tangent line, consider the graph of $y = f(x)$, and let P be the point at which we want to draw the tangent line. As shown in Figure 2.2(a), let Q be another point on the graph, and consider the line

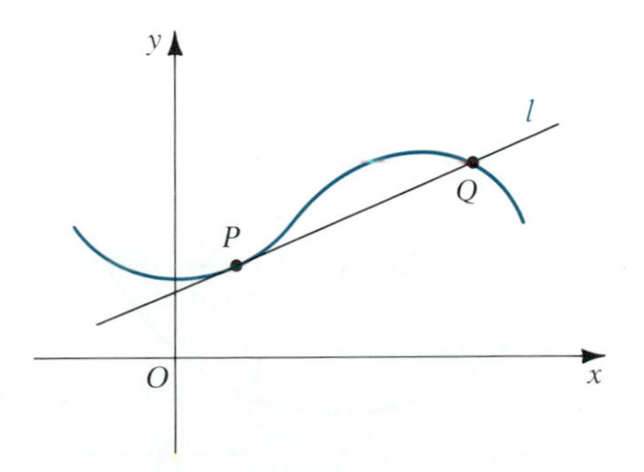

Line l is tangent to the curve at P but also crosses the curve at Q.

Line l is tangent to the curve at P and also crosses the curve there.

FIGURE 2.1

FIGURE 2.2

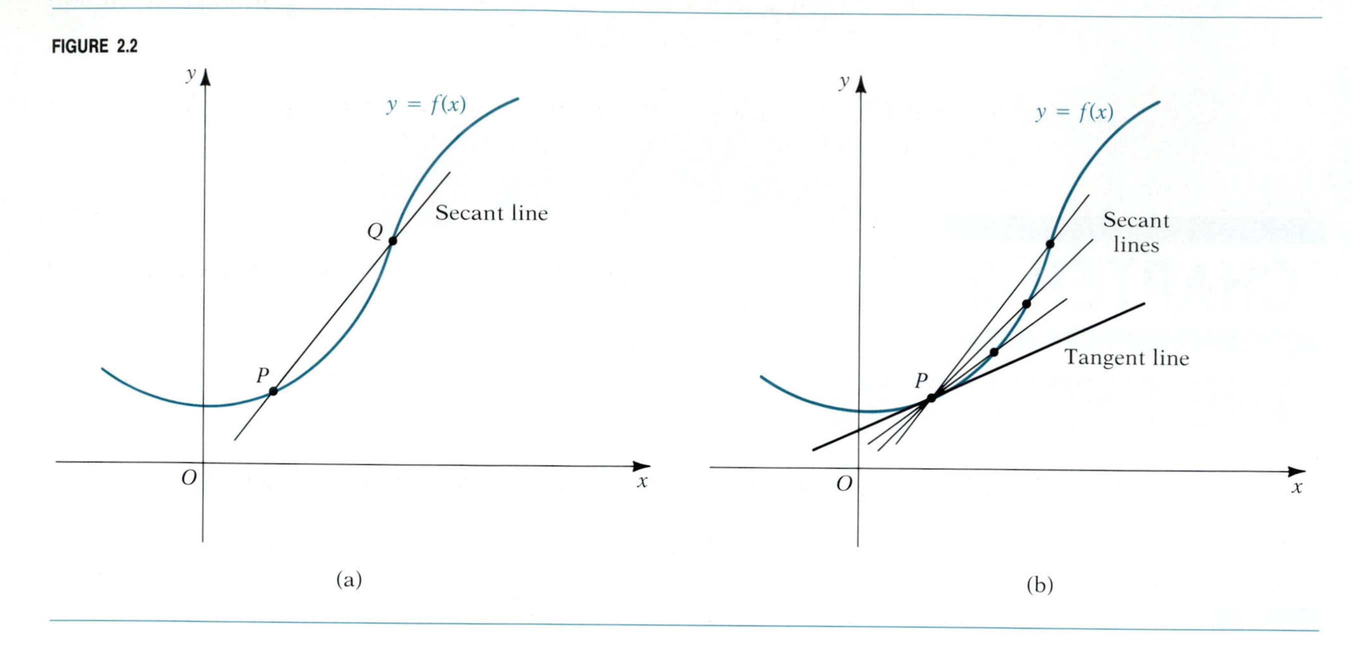

that joins P and Q. This line is called a *secant line*. Holding P fixed, we now let Q move along the curve toward P. In this way we get a collection of secant lines through P as in Figure 2.2(b). If these secant lines approach some limiting position as Q approaches P, then this limiting line is by definition the tangent line at P.

We can state the definition in a more useful way by considering slopes. Suppose P has coordinates $(x_1, f(x_1))$. The coordinates of Q can then be written in the form $(x_1 + h, f(x_1 + h))$ for some $h \neq 0$. The slope of the secant line PQ is then

$$m_{\text{sec}} = \frac{f(x_1 + h) - f(x_1)}{h}$$

FIGURE 2.3

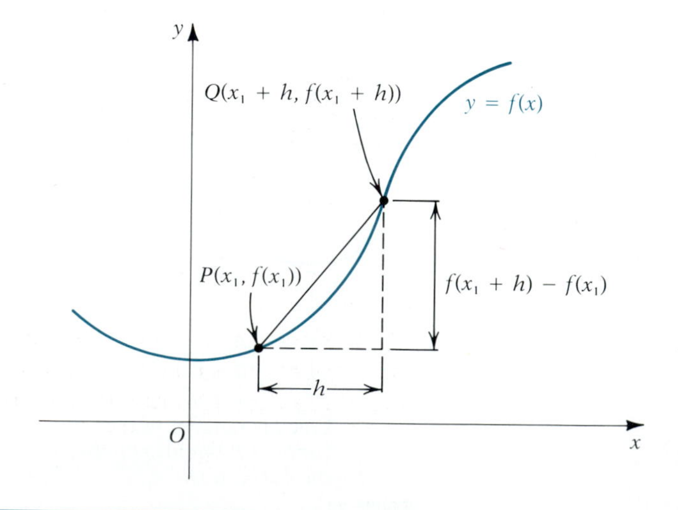

as seen by Figure 2.3. Now we can make Q approach P simply by letting $h \to 0$. This leads to the definition.

DEFINITION 2.1 Let $P(x_1, f(x_1))$ be a fixed point on the graph of the function f. A nonvertical tangent line at P exists if and only if the limit

$$m_{\tan} = \lim_{h \to 0} \frac{f(x_1 + h) - f(x_1)}{h} \tag{2.1}$$

exists as a finite number, and when this is true, the tangent line is the unique line through P with slope equal to $m_{\tan}$. ■

EXAMPLE 2.1 Show that the curve defined by $y = x^2 + 1$ has a tangent line at the point $(1, 2)$ and find its equation.

Solution Let $(x_1, f(x_1)) = (1, 2)$ and investigate the limit in equation (2.1):

$$\begin{aligned} \lim_{h \to 0} \frac{f(x_1 + h) - f(x_1)}{h} &= \lim_{h \to 0} \frac{[(1 + h)^2 + 1] - 2}{h} \\ &= \lim_{h \to 0} \frac{1 + 2h + h^2 + 1 - 2}{h} \\ &= \lim_{h \to 0} \frac{h^2 + 2h}{h} \\ &= \lim_{h \to 0} (h + 2) = 2 \end{aligned}$$

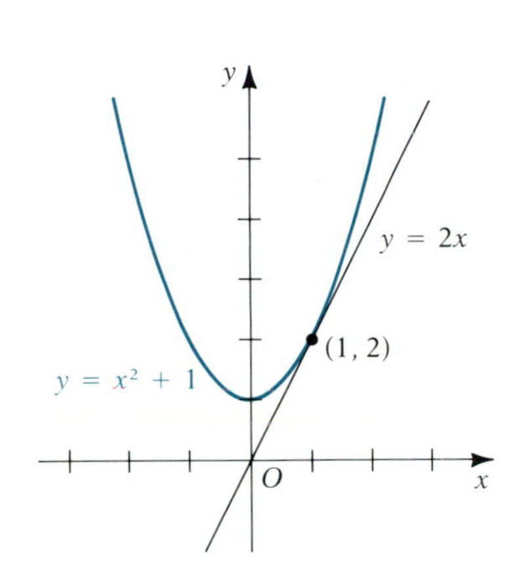

FIGURE 2.4

Since this limit exists, the tangent line exists and has slope $m_{\tan} = 2$. Thus, by equation (1.1) the equation of the tangent line is

$$y - 2 = 2(x - 1) \quad \text{or} \quad y = 2x$$

The situation is shown graphically in Figure 2.4. ■

Now let us consider how to find the instantaneous velocity of a particle moving along a straight path. Suppose we have set up a coordinate system on the path and know the location of the particle at time t. In particular, let $s(t)$ denote the position, measured from the origin, at time t. So that we can be specific, let t be measured in seconds and $s(t)$ in feet, but any time unit and distance unit will do. Suppose at time t_1 the particle is at P, as shown in Figure 2.5. We wish to find its velocity at the instant it passes through P. Toward this end, we consider its location Q at time $t_1 + h$, where $h \neq 0$. Now the *average velocity* in going from P to Q is the distance from P to Q divided by the time interval h:

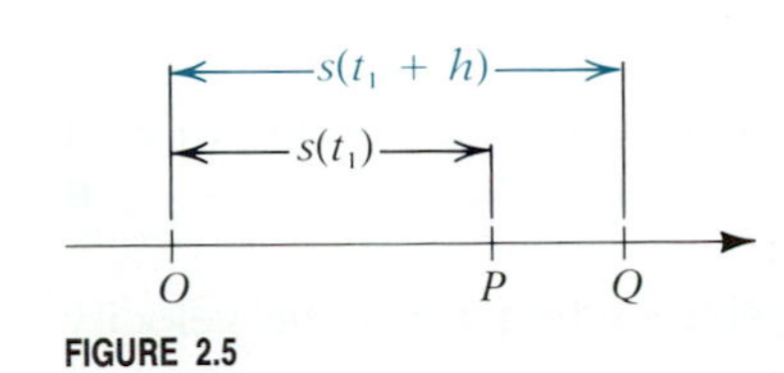

FIGURE 2.5

$$v_{\text{ave}} = \frac{s(t_1 + h) - s(t_1)}{h}$$

If we now consider smaller and smaller values of h, this average velocity may approach some limiting value. If it does, then this limiting value is by definition the **instantaneous velocity** at P:

$$v(t_1) = \lim_{h \to 0} \frac{s(t_1 + h) - s(t_1)}{h} \tag{2.2}$$

EXAMPLE 2.2 A particle moves in a straight line in such a way that its distance from the starting point, in meters, is given by

$$s(t) = 16\left(1 - \frac{1}{t+1}\right)$$

where t is nonnegative and is given in seconds.

(a) Find the average velocity of the particle over the time interval from $t = 1$ to $t = 3$.

(b) Find the instantaneous velocity of the particle when $t = 3$.

Solution (a) When $t = 1$, the particle is at a distance $s(1) = 16(1 - \frac{1}{2}) = 8$ m from the starting point, and when $t = 3$, it is $s(3) = 16(1 - \frac{1}{4}) = 12$ m from the starting point. The average velocity is the distance covered divided by the elapsed time:

$$v_{\text{ave}} = \frac{s(3) - s(1)}{3 - 1} = \frac{12 - 8}{2} = 2 \text{ m/sec}$$

(b) In equation (2.2) we let $t_1 = 3$ and get

$$\begin{aligned}
v(3) &= \lim_{h \to 0} \frac{s(3+h) - s(3)}{h} \\
&= \lim_{h \to 0} \frac{16\left(1 - \dfrac{1}{3+h+1}\right) - 12}{h} \\
&= \lim_{h \to 0} \frac{16 - \dfrac{16}{4+h} - 12}{h} \\
&= \lim_{h \to 0} \frac{4 - \dfrac{16}{4+h}}{h} \\
&= \lim_{h \to 0} \frac{16 + 4h - 16}{h(4+h)} \quad \text{We multiplied numerator and denominator by } 4 + h. \\
&= \lim_{h \to 0} \frac{4}{4+h} \\
&= 1
\end{aligned}$$

So at the instant when $t = 3$ sec, the particle is traveling at the rate of 1 m/sec. ■

From the preceding example we note that since when $t = 3$ the velocity is 1 m/sec, it follows that at that instant the particle is moving in the positive direction on its path. This can be seen intuitively by reasoning that if the velocity remained the same for 1 sec, then the distance would be 1 m greater after 1 sec had elapsed. More generally, since velocity is the rate of change of distance, we conclude that *when the velocity is positive, the distance is increasing*, and *when the velocity is negative, the distance is decreasing*. It should be emphasized that these are instantaneous conditions only. For example, at the instant when a particle changes direction, its velocity will be 0, but an observer would not be able to discern that it was not moving at that instant.

We will consider the motion of a particle in more detail after we have formalized the notion of a derivative.

EXERCISE SET 2.1

A

In Exercises 1–10 show that a tangent line to the graph of $y = f(x)$ exists at the point $P(x_1, f(x_1))$ and find its equation. Draw the graph of the function and show the tangent line at P.

1. $f(x) = \frac{x^2}{2}$; $x_1 = 2$
2. $f(x) = 1 - x^2$; $x_1 = 1$
3. $f(x) = \frac{4}{x}$; $x_1 = 2$
4. $f(x) = 4 - \frac{2}{x}$; $x_1 = 1$
5. $f(x) = \frac{2}{x-1}$; $x_1 = -1$
6. $f(x) = x^2 - x$; $x_1 = 1$
7. $f(x) = (x-2)^2$; $x_1 = 3$
8. $f(x) = 3x - x^2$; $x_1 = 2$
9. $f(x) = \frac{1}{2}x(x-2)$; $x_1 = -1$
10. $f(x) = \frac{3}{x+2}$; $x_1 = -4$

In Exercises 11–20 find the instantaneous velocity $v(t_1)$ for a particle moving on a straight-line path so that its distance from the origin at time t sec is $s(t)$ ft.

11. $s(t) = 2t + 3$; $t_1 = 2$
12. $s(t) = 3t - 4$; $t_1 = 3$
13. $s(t) = \frac{t+1}{2}$; $t_1 = 5$
14. $s(t) = t^2$; $t_1 = 1$
15. $s(t) = 2t^2 - 3$; $t_1 = 2$
16. $s(t) = t^2 - t$; $t_1 = 3$
17. $s(t) = 1 - \frac{1}{t}$; $t_1 = 2$
18. $s(t) = \frac{12}{t-1}$; $t_1 = 4$
19. $s(t) = \frac{t+4}{t}$; $t_1 = 3$
20. $s(t) = \frac{8}{t+2}$; $t_1 = 2$

B

In Exercises 21–24 find the slope of the tangent line to the graph of $y = f(x)$ at an arbitrary point $(x_1, f(x_1))$.

21. $f(x) = x^2 - 2x$
22. $f(x) = \frac{1}{x^2}$; $x_1 \neq 0$
23. $f(x) = \frac{x}{x+1}$; $x_1 \neq -1$
24. $f(x) = x - \frac{1}{x}$; $x_1 \neq 0$

25. Find the equation of the tangent line to the graph of $y = \sqrt{2x+1}$ at $x = 4$.
26. Find the average velocity during the time interval from $t = 0$ to $t = 3$ of a particle moving in a straight line according to the formula $s(t) = 4/\sqrt{t+1}$. What is the instantaneous velocity at $t = 3$? In what direction is the particle moving at the instant when $t = 3$?

C

In Exercises 27 and 28 write a computer program that will give the slopes of the secant lines joining $P(x_1, f(x_1))$ and $Q(x_2, f(x_2))$, where $x_2 = x_1 + h$ and h takes on the values $\pm\frac{1}{2}, \pm\frac{1}{4}, \pm\frac{1}{8}, \ldots, \pm\frac{1}{2^{10}}$. Run your program for the given value of x_1 and the given function f. Estimate the slope of the tangent line at P, if it is finite.

27. $f(x) = \frac{x^5 - 2x^3 + 4}{5x^4 + 3x^2 - 20}$; $x_1 = 2$
28. $f(x) = (x^3 - 8x - 3)^{2/3}$; $x_1 = 3$

29. Use the computer to find the average velocities of a particle moving on a line according to the formula

$$s(t) = \frac{5(t-1)}{\sqrt{t^2+9}}$$

over the time intervals $[4, 4+h]$, where h takes on the values 2, 0.2, 0.02, . . . , and 0.000002. Estimate the instantaneous velocity at $t = 4$.

2.2 THE DERIVATIVE

It should be evident by now that the tangent problem and the velocity problem involve exactly the same type of calculation. The interpretations are different, but mathematically they are the same. These are but two examples of a whole class of problems that involve this type of calculation. We will see some other examples later.

We now give the calculation a name and a symbol. Consider any function f, and let its domain contain the point x_1 as well as some neighborhood of x_1. Then we define the *derivative* of f at x_1 as follows.

DEFINITION 2.2 The **derivative** of f at x_1 is denoted by $f'(x_1)$ and is defined by

$$f'(x_1) = \lim_{h \to 0} \frac{f(x_1+h) - f(x_1)}{h} \tag{2.3}$$

provided this limit exists. ■

So if the graph of $y = f(x)$ has a tangent line at the point $(x_1, f(x_1))$, then its slope is $f'(x_1)$. Similarly, if $s(t)$ is the distance function for a particle moving along a line, then its velocity at time $t = t_1$ is $s'(t_1)$. We will mention other interpretations of the derivative shortly, but for now we concentrate on Definition 2.2 itself, without regard to any geometric or physical interpretation.

EXAMPLE 2.3 Find $f'(3)$ if $f(x) = \sqrt{x+1}$.

Solution

$$\begin{aligned} f'(3) &= \lim_{h \to 0} \frac{f(3+h) - f(3)}{h} \\ &= \lim_{h \to 0} \frac{\sqrt{(3+h)+1} - \sqrt{3+1}}{h} \\ &= \lim_{h \to 0} \frac{\sqrt{h+4} - 2}{h} \cdot \frac{\sqrt{h+4}+2}{\sqrt{h+4}+2} \qquad \text{Here we are rationalizing the numerator.} \\ &= \lim_{h \to 0} \frac{h+4-4}{h(\sqrt{h+4}+2)} = \lim_{h \to 0} \frac{h}{h(\sqrt{h+4}+2)} \\ &= \frac{1}{4} \end{aligned}$$

So the derivative of f at 3 is $\frac{1}{4}$. ■

There is an alternate way of writing the limit in Definition 2.2 that is sometimes useful. In it, to distinguish clearly between the fixed point x_1 and the variable point x that we introduce, we use the letter a instead of x_1. For

$f'(a)$, equation (2.3) gives

$$f'(a) = \lim_{h \to 0} \frac{f(a+h) - f(a)}{h}$$

Now we let $x = a + h$, so that $h = x - a$ and observe that $h \to 0$ if and only if $x \to a$. So we can write $f'(a)$ as

$$f'(a) = \lim_{x \to a} \frac{f(x) - f(a)}{x - a} \tag{2.4}$$

We can redo Example 2.3 using this alternate form as follows. Since $f(x) = \sqrt{x+1}$, $f(3) = \sqrt{4} = 2$. So, with $a = 3$, we get from equation (2.4),

$$f'(3) = \lim_{x \to 3} \frac{\sqrt{x+1} - 2}{x - 3} = \lim_{x \to 3} \frac{\sqrt{x+1} - 2}{x - 3} \cdot \frac{\sqrt{x+1} + 2}{\sqrt{x+1} + 2}$$

$$= \lim_{x \to 3} \frac{(x+1) - 4}{(x-3)(\sqrt{x+1} + 2)} = \lim_{x \to 3} \frac{\overset{1}{\cancel{x-3}}}{\cancel{(x-3)}(\sqrt{x+1} + 2)} = \frac{1}{4}$$

The process of calculating the derivative of a function is called **differentiation,** and to **differentiate** a function means to find its derivative. When the derivative of a function exists at a given point, we say the function is **differentiable** at that point. If a function is differentiable at all points of an interval I, we say it is **differentiable on *I*.** The quotient

$$\frac{f(x_1 + h) - f(x_1)}{h} \qquad (h \neq 0)$$

that occurs in the definition of the derivative of f at x_1 is called a **difference quotient** for f. So the derivative is a limit of difference quotients. If $I = [a, b]$ is a closed interval, then in calculating the derivative at a, we use the right-hand limit

$$\lim_{h \to 0^+} \frac{f(a+h) - f(a)}{h}$$

and call this the **right-hand derivative at *a*.** Similarly, at b we use the left-hand limit and call the result the **left-hand derivative at *b*.** In the future when we speak of differentiability on an interval, we will understand that if either end point is included, then the appropriate one-sided derivative must be used.

The values x_1 at which $f'(x_1)$ exists constitute a subset of the domain of f. To emphasize that x_1 can vary over this subset, we will in the future omit the subscript and denote by x any such value. In this way we see that we have actually defined a new function f', sometimes called the *derived function* for f. So we can indicate the nature of f' by

$$f' : x \to f'(x)$$

The domain of f' is the largest subset of the domain of f for which the limit in equation (2.3) exists. It is instructive to see how the graph of f' is related to the graph of f; we illustrate this in the next example.

FIGURE 2.6

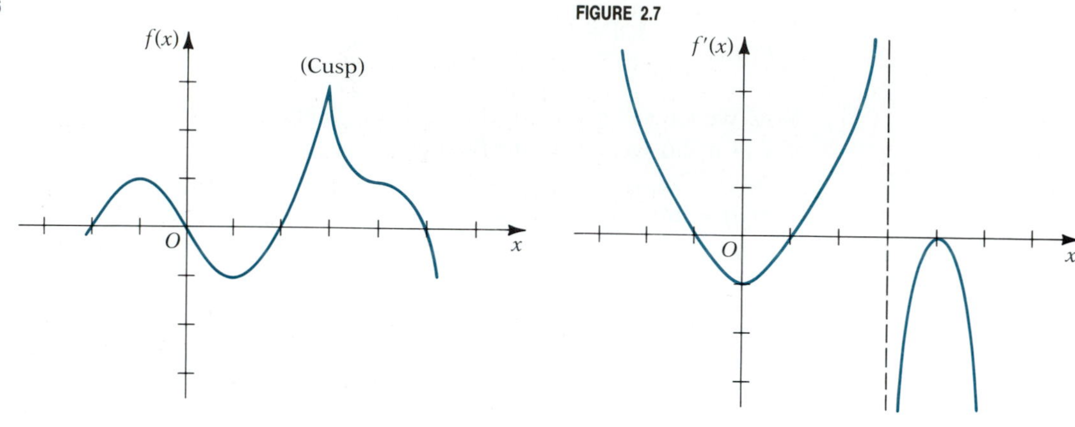

FIGURE 2.7

EXAMPLE 2.4 The graph of a function f is given in Figure 2.6. From it, obtain a sketch of the graph of f'.

Solution At selected values of x, we estimate the values of $f'(x)$ by means of the slope of the tangent line. This is a crude approximation only. At the integer points on the interval $[-2, 5]$, these values appear to be approximately those in the table. Notice that at $x = 3$ the tangent line is vertical, so there is no slope. The corresponding point on the graph is an example of a **cusp.** As 3 is approached from the left, the slope gets arbitrarily large, and as 3 is approached from the right, the slope is negative but arbitrarily large in absolute value. This results in a vertical asymptote for the graph of f'. We show the graph of f' in Figure 2.7.

x	-2	-1	0	1	2	3	4	5
$f'(x)$	2	0	-1	0	2	Undefined	0	-3

■

Remark on Notation There is nothing special about using the letter h in the difference quotient $[f(x+h) - f(x)]/h$. In fact, the symbol Δx, read "delta x," has been used historically where we have used h, and it still is in frequent use. The symbol Δx is understood to convey the idea of an *increment,* or *change,* in x—that is, the change from the point x to the point $x + \Delta x$. If we write $y = f(x)$, then the y increment that results from the increment Δx in x can be denoted by Δy; that is, $\Delta y = f(x + \Delta x) - f(x)$. In this notation, then,

$$f'(x) = \lim_{\Delta x \to 0} \frac{f(x + \Delta x) - f(x)}{\Delta x} = \lim_{\Delta x \to 0} \frac{\Delta y}{\Delta x}$$

Although in most cases we shall continue to use the h notation, at times the delta notation is more convenient.

Much of the importance of the derivative in applications of calculus lies in the fact that it represents the *instantaneous rate of change* of one quantity with respect to another. The simplest example of this is velocity, which is the rate of change of distance with respect to time. But the slope of the tangent line to a curve also can be interpreted as a rate of change—namely, the rate at which y changes with respect to x.

Here are some other examples that illustrate the derivative as an instantaneous rate of change.

1. Let $N(t)$ be the size of a bacteria culture at time t. Then $N'(t)$ is the rate of growth of the culture.
2. Let $Q(t)$ be the quantity of a radioactive substance on hand at time t. Then $Q'(t)$ is the rate of decomposition.
3. Let $X(t)$ be the amount present at time t of a compound being formed by the chemical reaction of two other compounds. Then $X'(t)$ is the reaction rate.
4. Let $C(x)$ be the cost of manufacturing x items of a certain type. Then $C'(x)$ is the rate of change of cost with respect to the number of items produced. This is called **marginal cost** in economics. Similarly, there is **marginal revenue,** $R'(x)$, and **marginal profit,** $P'(x)$.
5. Let $A(t)$ be the accumulated amount of money after t years that results from a certain principal invested at continuously compounded interest. Then $A'(t)$ is the rate of growth of the account.
6. Let $p(h)$ be the atmospheric pressure at height h feet above the earth's surface. Then $p'(h)$ is the rate at which pressure is changing with respect to altitude.
7. Let $L(t)$ be the current state of knowledge of a learner in an instructor/learner situation. Then $L'(t)$ is the rate at which the material is being learned.

We could give many other examples, but these suffice to give a feeling for the broad applications of the derivative as an instantaneous rate of change. In the next chapter we will return to applications.

EXERCISE SET 2.2

A

In Exercises 1–10 find $f'(x_1)$ using Definition 2.2.

1. $f(x) = 3x + 4;\ x_1 = 1$
2. $f(x) = \dfrac{2x - 1}{3};\ x_1 = 6$
3. $f(x) = x^2 - 4;\ x_1 = -2$
4. $f(x) = 4x - 3x^2;\ x_1 = 2$
5. $f(x) = 1 - \dfrac{1}{x};\ x_1 = 1$
6. $f(x) = \dfrac{1}{x - 1};\ x_1 = 2$
7. $f(x) = \sqrt{x};\ x_1 = 9$
8. $f(x) = \sqrt{x - 1};\ x_1 = 5$
9. $f(x) = \dfrac{x + 2}{x};\ x_1 = -1$
10. $f(x) = \dfrac{x}{x - 1};\ x_1 = 2$

In Exercises 11–20 redo the indicated problems, using equation (2.4), with $a = x_1$.

11. Exercise 1
12. Exercise 2
13. Exercise 3
14. Exercise 4
15. Exercise 5
16. Exercise 6
17. Exercise 7
18. Exercise 8
19. Exercise 9
20. Exercise 10

In Exercises 21–26 find $f'(x)$ using Definition 2.2.

21. $f(x) = 2x + 3$
22. $f(x) = 3x^2 - 5$
23. $f(x) = \dfrac{2}{x}$
24. $f(x) = \sqrt{x + 2}$
25. $f(x) = \dfrac{x}{x + 1}$
26. $f(x) = \dfrac{1}{\sqrt{x}}$

In Exercises 27–29 sketch the graph of f' using the method suggested in Example 2.4.

27. (a)

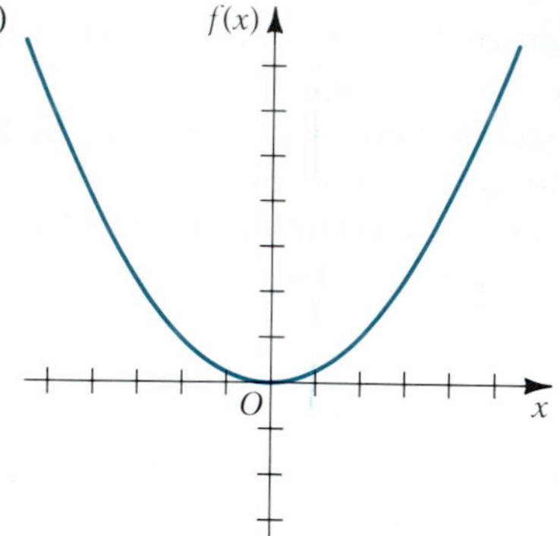

(b)

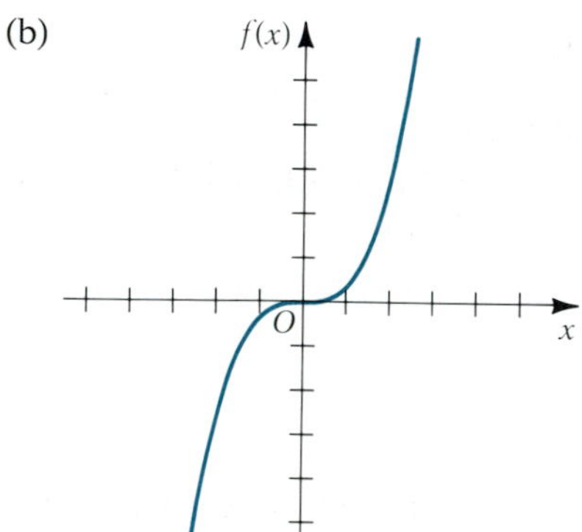

28. (a)

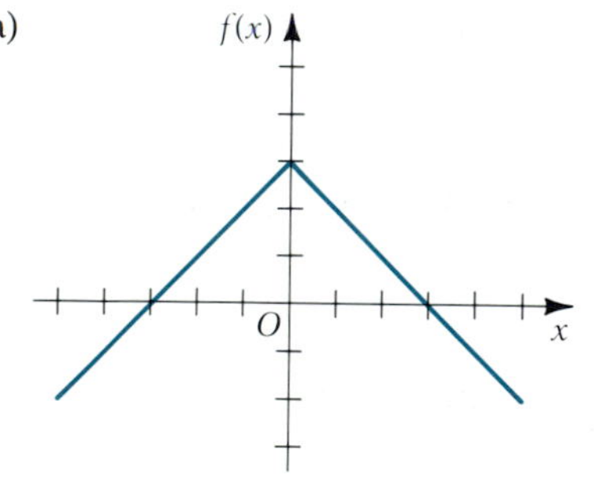

(b)

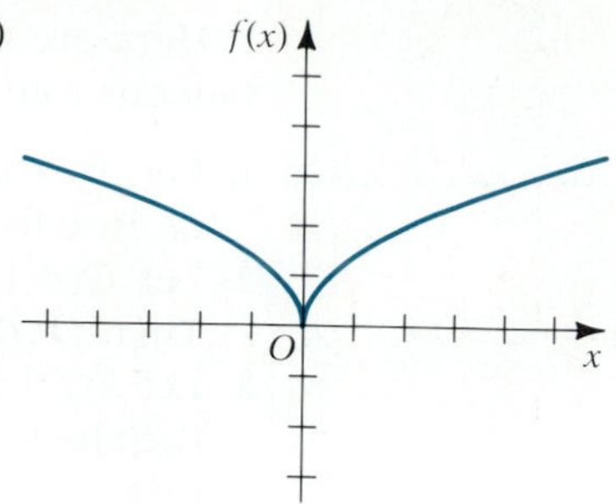

29. (a)

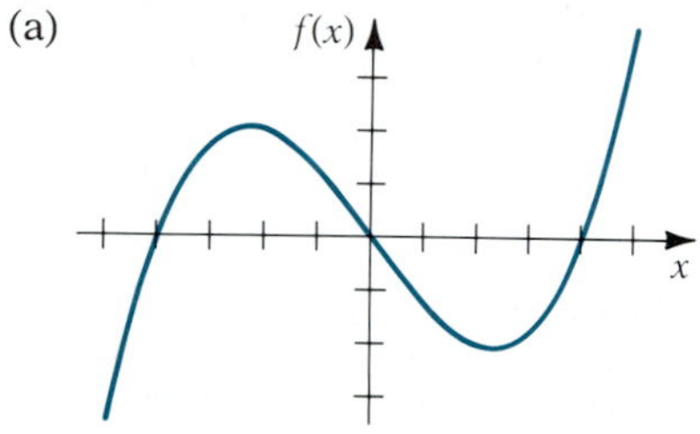

(b)

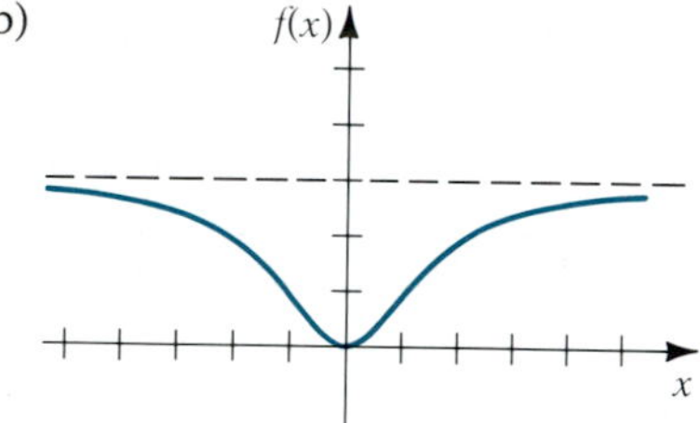

B

In Exercises 30–32 find $f'(x)$ using Definition 2.2.

30. $f(x) = \dfrac{1}{\sqrt{2x-3}}$; $x = 2$

31. $f(x) = x^3 - 2x$; $x = -1$

32. $f(x) = x^{3/2}$; $x \geq 0$

33. Let $f(x) = \sqrt{4 - x^2}$. Use equation (2.4) to find $f'(a)$. What restriction must be placed on a?

34. Show that

$$f'(x) = \lim_{t \to x} \frac{f(t) - f(x)}{t - x}$$

Use this to find $f'(x)$ for $f(x) = x/(x-1)$.

35. The accompanying figure shows the graph of the derivative of a function f. Sketch five possibilities for the graph of f. What do all of these have in common?

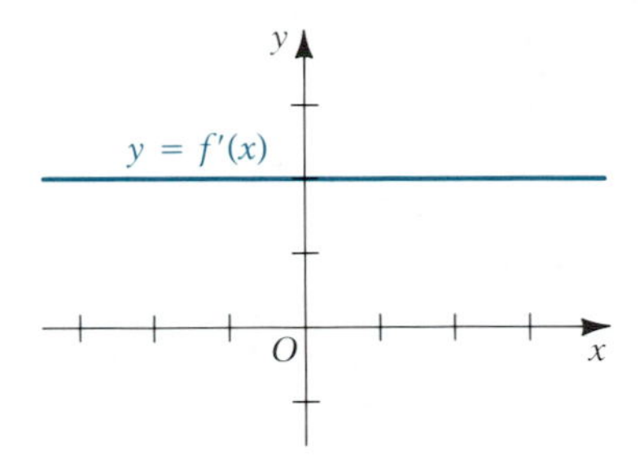

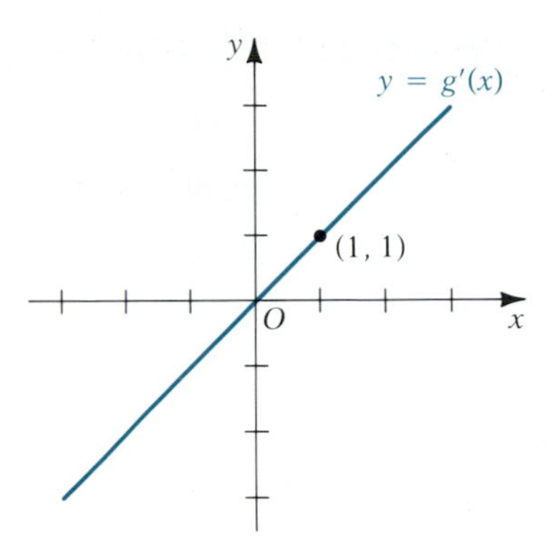

36. Repeat Exercise 35 for possible functions g, with g' as indicated in the figure.

C

In Exercises 37–39 use the computer to approximate $f'(x)$ for the given x values by finding the difference quotients for $h = \pm(0.1)^n$, with $n = 0, 1, 2, 3, 4,$ and 5.

37. $f(x) = \sqrt{x^3 - 2}$; $x = 3, 4, 5$

38. $f(x) = \dfrac{(x+5)^{2/3}}{x^2 - 4}$; $x = -1, 0, 3$. (Delete $n = 0$.)

39. $f(x) = x^a$; $x = 2, 4, 8$ and $a = 2, 3, 4, -1, -2, \frac{1}{2}, \frac{4}{3}, \sqrt{2}, \pi$. (Use 3.14159 for π.)

2.3 DERIVATIVES OF POLYNOMIAL FUNCTIONS

Calculating derivatives by means of the definition can become quite tedious (as you doubtless have discovered). Fortunately, in many cases the work can be greatly reduced by using results that we derive in the remainder of this chapter.

THEOREM 2.1 If $f(x) = c$, then $f'(x) = 0$ for all x; that is, *the derivative of a constant function is* 0.

Proof
$$f'(x) = \lim_{h \to 0} \frac{f(x+h) - f(x)}{h} = \lim_{h \to 0} \frac{c - c}{h} = \lim_{h \to 0} 0 = 0$$

Figure 2.8 shows this result geometrically. The tangent line coincides with the graph of f and has slope 0. ■

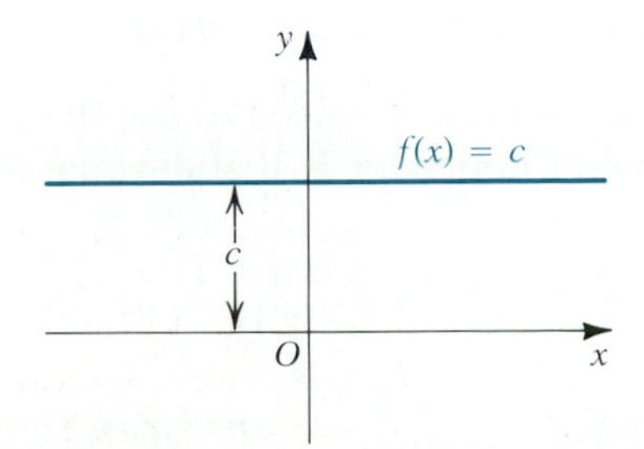

FIGURE 2.8

THEOREM 2.2 If c is any constant, then

$$(cf)'(x) = c \cdot f'(x)$$

for all x in the domain of f'.

Proof Since

$$\frac{(cf)(x+h) - (cf)(x)}{h} = \frac{cf(x+h) - cf(x)}{h} = c\,\frac{f(x+h) - f(x)}{h}$$

it follows that

$$(cf)'(x) = c \lim_{h \to 0} \frac{f(x+h) - f(x)}{h} = c \cdot f'(x) \qquad \text{By Theorem 1.1 (Property 2)}$$

■

In words, this theorem says that *the derivative of a constant times a function is the constant times the derivative of the function.*

THEOREM 2.3 For all x in the domain of f' and g',

$$(f \pm g)'(x) = f'(x) \pm g'(x)$$

Proof The difference quotient for $f + g$ can be written as

$$\begin{aligned}\frac{(f+g)(x+h) - (f+g)(x)}{h} &= \frac{[f(x+h) + g(x+h)] - [f(x) + g(x)]}{h} \\ &= \frac{f(x+h) - f(x)}{h} + \frac{g(x+h) - g(x)}{h}\end{aligned}$$

It now follows, on taking limits of both sides as $h \to 0$ and using Property 3 of Theorem 1.1, that

$$(f+g)'(x) = f'(x) + g'(x)$$

A similar proof can be given for $(f - g)'$, or alternately we may write $f - g = f + (-1)g$ and use Theorems 2.2 and 2.3. ■

Theorem 2.3 can be verbalized as: *the derivative of a sum (or difference) is the sum (or difference) of the derivatives.* This can be extended by associativity to any finite sum or difference. For example,

$$(f + g - h)'(x) = f'(x) + g'(x) - h'(x)$$

For the next theorem we need to use the *binomial theorem,* the proof of which is given in Appendix 1. It states that for any positive integer n,

$$\begin{aligned}(a+b)^n = a^n + na^{n-1}b &+ \frac{n(n-1)}{2!}a^{n-2}b^2 \\ &+ \frac{n(n-1)(n-2)}{3!}a^{n-3}b^3 + \cdots + b^n\end{aligned}$$

(Recall that $k!$ means $1 \cdot 2 \cdot 3 \cdot \cdots \cdot k$.)

THEOREM 2.4 THE POWER RULE

Let $f(x) = x^n$, where n is any positive integer. Then for all real numbers x,

$$f'(x) = nx^{n-1} \tag{2.5}$$

Proof

$$f'(x) = \lim_{h \to 0} \frac{f(x+h) - f(x)}{h} = \lim_{h \to 0} \frac{(x+h)^n - x^n}{h}$$

$$= \lim_{h \to 0} \frac{\left[x^n + nx^{n-1}h + \frac{n(n-1)}{2!}x^{n-2}h^2 + \cdots + h^n\right] - x^n}{h}$$

$$= \lim_{h \to 0} \left[nx^{n-1} + \frac{n(n-1)}{2!}x^{n-2}h + \cdots + h^{n-1}\right]$$

$$= nx^{n-1}$$

Note that in the next to last step all terms but the first contain h as a factor and so go to 0 in the limit. ■

EXAMPLE 2.5 Find $f'(x)$ if $f(x) = x^{10}$.

Solution By Theorem 2.4, $f'(x) = 10x^9$. ■

You might wish to compare this with the work of finding $f'(x)$ directly from the definition.

By combining Theorems 2.2 and 2.4 we see that if $f(x) = cx^n$, then

$$f'(x) = c(nx^{n-1}) = (cn)x^{n-1}$$

For example, if $f(x) = 3x^4$, then $f'(x) = 12x^3$.

We are now in a position to take the derivative of any polynomial function by using Theorems 2.1–2.4.

EXAMPLE 2.6 If $f(x) = 3x^4 + 5x^3 - 7x^2 + 2x - 9$, find $f'(x)$.

Solution

$$f'(x) = 12x^3 + 15x^2 - 14x + 2$$

You should check to see how we have made use of each of the four theorems. ■

EXAMPLE 2.7 If $f(x) = 2x^4 - 6x^2 + 7$, find $f'(2)$.

Solution First we find $f'(x)$ in general:

$$f'(x) = 8x^3 - 12x$$

Now we substitute $x = 2$:

$$f'(2) = 8(8) - 12(2) = 64 - 24 = 40$$ ■

A Word of Caution Do not infer that because you have now learned the "easy way" to differentiate a polynomial function, the definition of the derivative is no longer useful. There are many important functions that are not polynomials, and each time we encounter one of these, we will have to return to Definition 2.2 to develop a differentiation formula. So this definition is of fundamental importance and must be learned.

EXERCISE SET 2.3

A

In Exercises 1–20 find $f'(x)$, making use of Theorems 2.1–2.4.

1. $f(x) = 3$
2. $f(x) = x$
3. $f(x) = x^4$
4. $f(x) = 3x^5$
5. $f(x) = 1 - 2x$
6. $f(x) = (3x + 4)/5$
7. $f(x) = ax + b$
8. $f(x) = 2x^2 - 3x + 4$
9. $f(x) = 4x^3 - 7x^2 + 8x - 6$
10. $f(x) = 3 - x - 7x^2 + 2x^3$
11. $f(x) = x^5 - 6x^4$
12. $f(x) = \frac{2}{3}x^6 - 5x^4 + \frac{1}{2}x^2 - 8$
13. $f(x) = ax^2 + bx + c$
14. $f(x) = 2x - 3x^2 - 8x^3$
15. $f(x) = (2x - 1)(x + 3)$
16. $f(x) = x(1 - x)(2 + x)$
17. $f(x) = (3x + 4)^2$
18. $f(x) = (x + 2)^3$
19. $f(x) = 2x^3 - 3x(2 + x - x^2)$
20. $f(x) = x(2x - 3) - (3 - x)^2$

21. Find the equation of the tangent line to the graph of $y = 2x^2 - 3x + 4$ at the point $(2, 6)$.

22. Find the equation of the tangent line to the graph of $y = 3x^2 - x^4$ at the point for which $x = -2$.

23. Find k so that the slope of the tangent line to the graph of $y = 2x^2 - 3kx + 4$ is 4 at the point for which $x = 2$.

24. Find k so that the slope of the tangent line to the graph of $y = 2k^2x^2 - 2kx + 3$ is 2 at the point for which $x = 1$.

25. The line perpendicular to the tangent line to a curve is called the *normal line*. Find the equations of the tangent and normal lines to $y = x^3 - 3x^2 - 7x + 18$ at the point for which $x = 3$.

26. Find all points on the graph of $y = x^4 - 4x^3 + 4x^2 + 3$ at which the tangent line is horizontal.

In Exercises 27–30 a particle moves in a straight line so that its distance from the origin at time t is $s(t)$. Find the instantaneous velocity of the particle at time t_1.

27. $s(t) = 3t^2 - 4t + 5;\ t_1 = 4$
28. $s(t) = t(t^2 - 3t);\ t_1 = 3$
29. $s(t) = 1 - 2t(t + 1);\ t_1 = 2$
30. $s(t) = (2t - 3)(3t + 4);\ t_1 = 5$

B

31. Find a and b such that the graph of $y = ax^2 + bx$ passes through the point $(2, -3)$ and such that at this point the tangent line has slope 1.

32. Find all points on the graph of $y = 2x^3 - 3x^2 - 10x + 7$ at which the slope of the tangent line is 2. Find the equations of the tangent lines at these points.

33. Two tangent lines to the curve $y = x^2$ pass through the point $(0, -4)$. Find the points of tangency and the equations of the tangent lines.

34. Find the equations of all tangent lines to the graph of $f(x) = 3x - x^2$ drawn from the point $(4, 0)$. Draw the graph of f, showing these tangent lines along with their points of tangency.

35. If $s(t) = (2t^2 - 3)^3$ is the distance function for a particle moving in a straight line, find the velocity function $v(t)$. Also find the acceleration $a(t)$, defined as $v'(t)$.

2.4 THE PRODUCT AND QUOTIENT RULES

Our main objective in this section is to develop formulas for differentiating products and quotients of functions. These will prove to be indispensable in our continued study of derivatives. First we need a preliminary result, which we refer to as a *lemma*, although it turns out to be important in its own right.

LEMMA 2.1 If f is differentiable at x, then

$$\lim_{h \to 0} f(x+h) = f(x)$$

Proof We can write $f(x+h)$ in the following rather strange-looking way:

$$f(x+h) = \frac{f(x+h) - f(x)}{h} \cdot h + f(x), \qquad h \neq 0 \qquad \text{(Verify.)}$$

So, making use of the limit theorems and the definition of $f'(x)$, we get

$$\begin{aligned} \lim_{h \to 0} f(x+h) &= \lim_{h \to 0} \frac{f(x+h) - f(x)}{h} \cdot \lim_{h \to 0} h + \lim_{h \to 0} f(x) \qquad \text{(Why?)} \\ &= f'(x) \cdot 0 + f(x) \\ &= f(x) \end{aligned}$$

■

COROLLARY 2.1 If $f'(a)$ exists, then f is continuous at a.

Proof Continuity at a means (see Definition 1.5):

$$\lim_{x \to a} f(x) = f(a)$$

Write $h = x - a$. Then as $x \to a$, $h \to 0$. Also, since $x = a + h$, it follows that as $h \to 0$, $x \to a$; that is, $x \to a$ and $h \to 0$ are interchangeable. So replacing x by $a + h$ and $x \to a$ by $h \to 0$, we get

$$\lim_{x \to a} f(x) = \lim_{h \to 0} f(a+h)$$

The result now follows from the lemma. ■

It is important to observe that the converse of the result just proved is not true; that is, continuity at a does *not* imply differentiability there. To see this, consider the following example.

EXAMPLE 2.8 Show that the function f defined by $f(x) = |x|$ is continuous at $x = 0$ but that $f'(0)$ does not exist.

Solution If $x > 0$, $|x| = x$. So

$$\lim_{x \to 0^+} f(x) = \lim_{x \to 0^+} x = 0$$

If $x < 0$, $|x| = -x$, so

$$\lim_{x \to 0^-} f(x) = \lim_{x \to 0^-} (-x) = 0$$

Thus, $\lim_{x \to 0} f(x) = 0$. Since $f(0) = |0| = 0$, it follows that f is continuous at 0. Now consider

$$\frac{f(0+h) - f(0)}{h} = \frac{|0+h| - |0|}{h} = \frac{|h|}{h}$$

Again taking limits from the right and left, we have

$$\lim_{h\to 0^+}\frac{f(0+h)-f(0)}{h}=\lim_{h\to 0^+}\frac{|h|}{h}=\lim_{h\to 0^+}\frac{h}{h}=1$$

and

$$\lim_{h\to 0^-}\frac{f(0+h)-f(0)}{h}=\lim_{h\to 0^-}\frac{|h|}{h}=\lim_{h\to 0^-}\frac{(-h)}{h}=-1$$

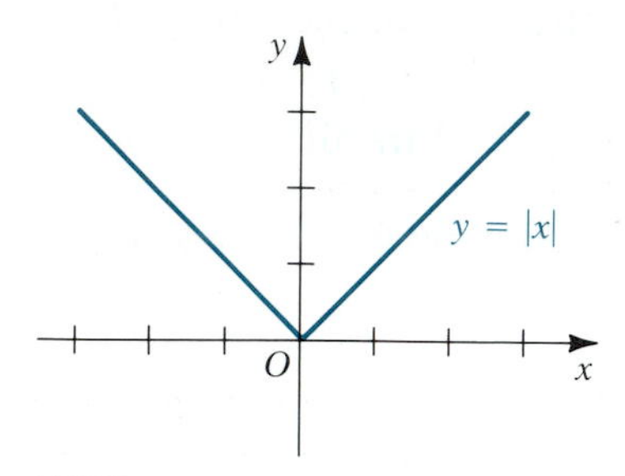

FIGURE 2.9

Since these one-sided limits are not equal, it follows that

$$\lim_{h\to 0}\frac{f(0+h)-f(0)}{h}$$

does not exist; that is, $f'(0)$ does not exist. The graph of f is shown in Figure 2.9. This clearly shows the nonexistence of a tangent line at $x = 0$. ■

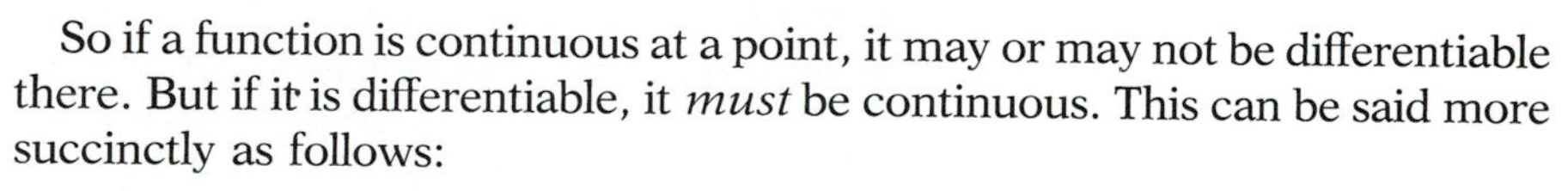

So if a function is continuous at a point, it may or may not be differentiable there. But if it is differentiable, it *must* be continuous. This can be said more succinctly as follows:

Continuity is *necessary* for differentiability but not *sufficient*.

We are ready now for the product rule.

THEOREM 2.5 THE PRODUCT RULE

If f and g are both differentiable at x, then so is fg, and

$$(fg)'(x) = f(x)\cdot g'(x) + g(x)\cdot f'(x) \tag{2.6}$$

Proof We write the difference quotient for fg in the following equivalent ways:

$$\frac{(fg)(x+h)-(fg)(x)}{h}$$

$$=\frac{f(x+h)g(x+h)-f(x)g(x)}{h} \quad \text{Definition of } fg$$

$$=\frac{f(x+h)g(x+h)-f(x+h)g(x)+f(x+h)g(x)-f(x)g(x)}{h} \quad \text{We added and subtracted } f(x+h)g(x).$$

$$=f(x+h)\frac{g(x+h)-g(x)}{h}+g(x)\frac{f(x+h)-f(x)}{h} \quad \text{Algebraic simplification}$$

Now we get the desired result by taking the limit of both sides as $h \to 0$, making use of limit properties (Theorem 1.1), Lemma 2.1, and the definition of $f'(x)$ and $g'(x)$. ■

It is useful to learn this result in words. We call f the first function and g the second.

The derivative of a product is the first function times the derivative of the second plus the second function times the derivative of the first.

EXAMPLE 2.9 Let $F(x) = (2x^3 - x + 3)(x^4 + 3x^2 - 8)$. Find $F'(x)$.

Solution We could, of course, multiply the factors to obtain a single polynomial. Instead, we will treat it as a product. So by equation (2.6),

$$F'(x) = \overbrace{(2x^3 - x + 3)}^{\text{first}} \cdot \overbrace{(4x^3 + 6x)}^{\substack{\text{derivative}\\\text{of}\\\text{the second}}} + \overbrace{(x^4 + 3x^2 - 8)}^{\text{second}} \cdot \overbrace{(6x^2 - 1)}^{\substack{\text{derivative}\\\text{of}\\\text{the first}}}$$

We will leave the answer in this form. ■

A Word of Caution In differentiating a product $f(x) \cdot g(x)$, it is tempting just to multiply the two derivatives and get $f'(x) \cdot g'(x)$, but you must not yield to this temptation, because *it is not correct*. The only correct result is given by equation (2.6).

We could prove the quotient rule in a similar way, but with the aid of the following lemma, which is also useful in its own right, we can obtain the result from the product rule.

LEMMA 2.2 If $g'(x)$ exists and $g(x) \neq 0$, then the derivative of $\frac{1}{g}$ at x exists and is given by

$$\left(\frac{1}{g}\right)'(x) = -\frac{g'(x)}{[g(x)]^2}$$

Proof Since $g'(x)$ exists, we know by Corollary 2.1 that g is continuous at x. Thus, since $g(x) \neq 0$, it follows that for all h sufficiently small in absolute value, $g(x + h) \neq 0$. (See Exercise 29 in Exercise Set 1.8.) For such h, with $h \neq 0$, the difference quotient for $\frac{1}{g}$ can be written in the following equivalent ways:

$$\frac{\dfrac{1}{g(x+h)} - \dfrac{1}{g(x)}}{h} = \frac{g(x) - g(x+h)}{hg(x)g(x+h)} = -\frac{g(x+h) - g(x)}{h} \cdot \frac{1}{g(x)g(x+h)}$$

The result now follows on letting $h \to 0$. (You should supply the reasons.) ■

THEOREM 2.6 **THE QUOTIENT RULE**

If f and g are both differentiable at x and $g(x) \neq 0$, then $\frac{f}{g}$ is differentiable at x, and

$$\left(\frac{f}{g}\right)'(x) = \frac{g(x)f'(x) - f(x)g'(x)}{[g(x)]^2}$$

Proof To simplify notation we write $g^2(x)$ for $[g(x)]^2$. From the definition of $\frac{f}{g}$, we have

$$\left(\frac{f}{g}\right)(x) = f(x) \cdot \frac{1}{g(x)}$$

Now we use the product rule together with Lemma 2.2:

$$\left(\frac{f}{g}\right)'(x) = f(x)\cdot\left(-\frac{g'(x)}{g^2(x)}\right) + \frac{1}{g(x)}\cdot f'(x)$$
$$= \frac{-f(x)g'(x) + g(x)f'(x)}{g^2(x)}$$
$$= \frac{g(x)f'(x) - f(x)g'(x)}{g^2(x)}$$ ■

We can state the quotient rule as follows:

> The derivative of a quotient is the denominator times the derivative of the numerator minus the numerator times the derivative of the denominator all divided by the square of the denominator.

EXAMPLE 2.10 Find $F'(x)$, where

$$F(x) = \frac{x^2 - 1}{2x^2 + 3}$$

Solution

$$F'(x) = \frac{\overbrace{(2x^2+3)}^{\text{denominator}} \cdot \overbrace{(2x)}^{\text{derivative of numerator}} - \overbrace{(x^2-1)}^{\text{numerator}} \cdot \overbrace{(4x)}^{\text{derivative of denominator}}}{\underbrace{(2x^2+3)^2}_{\text{square of denominator}}}$$

$$= \frac{4x^3 + 6x - 4x^3 + 4x}{(2x^2+3)^2} = \frac{10x}{(2x^2+3)^2}$$ ■

With the aid of Lemma 2.2 we can extend the power rule for derivatives (Theorem 2.4) to negative integers. Suppose $f(x) = x^n$, where n is a negative integer, and $x \neq 0$. Write $n = -m$, where m is a positive integer. Then

$$f(x) = x^{-m} = \frac{1}{x^m}$$

So if we take $g(x) = x^m$ in Lemma 2.2, we have

$$f'(x) = \left(\frac{1}{g}\right)'(x) = \frac{-g'(x)}{[g(x)]^2} = \frac{-mx^{m-1}}{x^{2m}} = -mx^{-m-1} = nx^{n-1}$$

For example, if $f(x) = x^{-3}$, then $f'(x) = -3x^{-4}$. So if n is any integer (even 0, for which the result is trivial) and $f(x) = x^n$, then

$$f'(x) = nx^{n-1}$$

Later on we will show that this result is in fact true for *all* real numbers n.

EXERCISE SET 2.4

A

In Exercises 1–20 $f(x)$ is given. Use the product rule or quotient rule, as appropriate, to find $f'(x)$.

1. $(x^2 - 2)(x^2 + 3)$
2. $(3x^2 + 1)(2x^2 - 4)$
3. $(1 - x^2)(2 + 5x^2)$
4. $(x^2 - 2x + 3)(4 - 3x^2)$
5. $(x^3 - 3x^2)(2x^2 + 1)$
6. $(4x - x^3)(x^2 + 2x + 3)$
7. $(x^4 - 2x^2 + 1)(x^3 - x + 3)$
8. $(3x^4 - 2x^2)(2x^4 + 3x^2 + 5)$
9. $(5x^4 + 6x^3 + 7)(2x^3 - 3x + 1)$
10. $(4 - x - x^3)(3x + 2x^3 - x^5)$
11. $\dfrac{x}{x-2}$
12. $\dfrac{3x+2}{4x-3}$
13. $\dfrac{x^2-1}{x^2+1}$
14. $\dfrac{2x^2-3x}{1-x^2}$
15. $\dfrac{2x-3}{x^2+2x}$
16. $\dfrac{3x-x^2}{2x+1}$
17. $\dfrac{x^2}{x^2-4}$
18. $\dfrac{4-2x+3x^2}{x^2+2}$
19. $\dfrac{x^3+8}{x^3-1}$
20. $\dfrac{4x^2+3x}{x^4-16}$

In Exercises 21–26 find $f'(x)$ using the power rule.

21. $f(x) = 2x^{-3}$
22. $f(x) = 3x^{-5}$
23. $f(x) = x^2 + x^{-2}$
24. $f(x) = 2 - 3x^{-1} + 4x^{-2}$
25. $f(x) = x + 3 + \dfrac{4}{x}$
26. $f(x) = 3 + 2x - \dfrac{4}{x^2}$
27. Let $g(x) = (x^2 - 4)/x^2$. Find $g'(x)$ in two ways: by the quotient rule and by writing $f(x)$ in the form $1 - 4x^{-2}$. Compare the two methods.
28. Let $h(x) = (3 - 4x + 5x^2)/2x$. Find $h'(x)$ without using the quotient rule. (See Exercise 27.)
29. Find the equation of the tangent line to the graph of
$$y = \frac{x+3}{x-1}$$
at the point (2, 5).
30. Find the equation of the tangent line to the graph of
$$y = \frac{x^2}{2x-3}$$
at the point for which $x = -3$.
31. A particle moves in a straight line so that its distance from the origin at time t is given by
$$s(t) = \frac{2t^2+3}{t+1}$$
Find its velocity at the instant when $t = 3$.

B

In Exercises 32 and 33 find $F'(x)$ using the quotient and product rules. Simplify the result.

32. $F(x) = \dfrac{(x^2-4)(x^2+1)}{x^2-9}$
33. $F(x) = \dfrac{2x-1}{(3x^2+5)(x^2-2x)}$
34. Give an alternate proof of the quotient rule without using Lemma 2.2. (*Hint:* Use the idea of the proof of the product rule.)
35. Let $f(x) = (x^2 - 4)/(x^2 + 4)$. Find $f'(x)$ without using the quotient rule or product rule. (*Hint:* Divide first and then make use of Lemma 2.2.)
36. Follow the instructions in Exercise 35 for
$$f(x) = (x^3 - x^2 + x)/(x^2 + 1)$$
37. Derive a formula for the derivative of the product fgh of three differentiable functions. [*Hint:* Write $(fgh)' = [(fg) \cdot h]'$ and apply the product rule twice.]
38. Using the result of Exercise 37, find $F'(x)$ if
$$F(x) = (2x + 1)(3x^2 - 4x)(x^2 + 4)$$
39. Conjecture from the result of Exercise 37 a formula for the derivative $(f_1 \cdot f_2 \cdot \cdots \cdot f_n)'$ of n differentiable functions. Prove your conjecture using mathematical induction.

C

Exercises 40–65 are to be done only if a symbolic differentiation package is available on your computer.

40–65. Redo Exercises 1–26 using the computer. Compare the answers with those you calculated by hand.

2.5 HIGHER ORDER DERIVATIVES AND OTHER NOTATIONS

Since for a differentiable function f, the derived function f' is a function in its own right, there is no reason why it may not also have a derivative. When it does, we denote its derivative by f''. So f'' is the derivative of f'. We also call f'' the **second derivative** of f. For example, suppose $f(x) = 3x^4 - 2x^2 + 7$. Then

$$f'(x) = 12x^3 - 4x$$

and so

$$f''(x) = 36x^2 - 4$$

There is no reason for stopping here, of course; we can take further derivatives, getting

$$f'''(x) = 72x$$
$$f''''(x) = 72$$

Note that all higher derivatives have the value 0 in this case.

It gets rather cumbersome to write so many primes, so we use the symbol

$$f^{(k)}(x)$$

to mean the *kth-order derivative.* Usually we go to this notation for $k > 3$. For example, $f^{(5)}(x)$ means the fifth derivative. The superscript k is put in parentheses to distinguish it from an exponent.

There are also other notations for derivatives. These are introduced not to make life more difficult for you, but rather because each notation has certain advantages and so it is useful to know them all. Moreover, in other books where calculus is used you might encounter these other notations.

Suppose that the rule for a function is given by an equation of the form $y = f(x)$. Then the derivative can be symbolized using any of the following forms:

$$y' \qquad \frac{dy}{dx} \qquad \frac{df}{dx} \qquad D_x y$$

Each means the same as $f'(x)$. The second notation, $\frac{dy}{dx}$, in particular, is widely used. For the present you should not think of this as a quotient but simply as a symbol for the derivative. Later we will give meaning to dy and dx separately. We can, however, give meaning to the symbol $\frac{d}{dx}$. We refer to this as a *derivative operator,* and when we follow it with an expression, as in

$$\frac{d}{dx}(\quad)$$

we mean to take the derivative with respect to x of what is inside the parentheses. For example,

$$\frac{d}{dx}(3x^2 + 2x) = 6x + 2$$

In this sense, then, $\frac{dy}{dx}$ can be thought of as $\frac{d}{dx}(y)$. Alternately, we can write $\frac{d}{dx}(f(x))$ or simply $\frac{df}{dx}$. The symbol D_x can be used in the same way as $\frac{d}{dx}$. Thus,

$$D_x(3x^2 + 2x) = 6x + 2$$

Suppose we are given that $f(x) = 3x^2 + 2x$. By setting $y = f(x)$, we can use any of the following for an indication of the derivative:

$$f'(x) \qquad y' \qquad \frac{dy}{dx} \qquad \frac{df}{dx} \qquad \frac{d}{dx}(3x^2 + 2x) \qquad D_x(3x^2 + 2x)$$

or even $(3x^2 + 2x)'$. In each case the value is $6x + 2$. If we want to evaluate the derivative at a specific point, say $x = 2$, the $f'(x)$ notation has the clear advantage; we simply write $f'(2)$. For the other notations we resort to the following device:

$$y'\Big|_{x=2} \qquad \frac{dy}{dx}\Big|_{x=2} \qquad \frac{d}{dx}(3x^2 + 2x)\Big|_{x=2}$$

and so on. We will see advantages to some of the other notations as we progress.

To indicate higher order derivatives with these new notations, we use the following:

$$y', y'', y''', y^{(4)}, \ldots$$
$$\frac{dy}{dx}, \frac{d^2y}{dx^2}, \frac{d^3y}{dx^3}, \frac{d^4y}{dx^4}, \ldots$$
$$D_xy, D_x^2y, D_x^3y, D_x^4y, \ldots$$

The notation d^2y/dx^2 can be explained by observing that the second derivative is the derivative of $\frac{dy}{dx}$ and so can be obtained from it by operating on $\frac{dy}{dx}$ with the derivative operator $\frac{d}{dx}$:

$$\frac{d}{dx}\left(\frac{dy}{dx}\right)$$

Since there are two d's on top and two dx's on the bottom, this gives rise to

$$\frac{d^2y}{(dx)^2}$$

but the parentheses on the bottom are not written. So we have

$$\frac{d^2y}{dx^2}$$

The higher derivatives in this notation can be explained in a similar way.

In the next chapter we will see some applications of second derivatives, but we can point out one application now. We know that if a particle moves in a straight line so that its distance from the origin at time t is $s(t)$, then its instantaneous velocity $v(t)$ is the derivative $s'(t)$. Since acceleration is the rate of change of velocity, the average acceleration over the time interval from t to $t + h$ is

$$\frac{v(t+h) - v(t)}{h}$$

and the limit of this as $h \to 0$ is the *instantaneous acceleration* at t. So we have

$$a(t) = v'(t) = s''(t)$$

EXERCISE SET 2.5

A

In Exercises 1–6 find y', y'', and y'''.

1. $y = 2x^4$
2. $y = 3x^{-2}$
3. $y = 4x^3 - 2x^2 + 1$
4. $y = x^5 - 2x^3 + 3$
5. $y = x^2 - x^{-2}$
6. $y = \frac{2}{x} - 3x^2$

In Exercises 7–12 find dy/dx and d^2y/dx^2.

7. $y = (2x + 3)(3x - 1)$
8. $y = (x^2 - 1)(2x^3 + 5)$
9. $y = x^2(x^2 - 3) + \dfrac{5}{x}$
10. $y = \dfrac{2}{x^3} - \dfrac{3}{2x^2}$
11. $y = \dfrac{x^2 - x + 3}{x}$
12. $y = \dfrac{x^2 - 1}{x^4}$

In Exercises 13–16 find the indicated derivatives.

13. $\dfrac{d}{dx}(x^3 - 3x^2 + 1)$
14. $\dfrac{d}{dx}(x^2 - 2x^{-3})$
15. $D_x\left(\dfrac{x-2}{x+3}\right)$
16. $D_x\left(\dfrac{3x-4}{2x+3}\right)$

In Exercises 17–20 find $f''(x)$.

17. $f(x) = x^{3/2} - x^{-1/2}$
18. $f(x) = \dfrac{3x - 2}{\sqrt{x}}$
19. $f(x) = \left(x - \dfrac{1}{2x}\right)^2$
20. $f(x) = \sqrt{x^2 + 1 + \dfrac{1}{4x^2}}$, $x > 0$
(*Hint:* Compare with Exercise 19.)

B

21. Let $P(x) = a_0 + a_1x + a_2x^2 + \cdots + a_nx^n$. Find $P^{(n)}(x)$. What is $P^{(k)}(x)$ for $k > n$?
22. Let $g(x) = \frac{1}{x}$. Find $g^{(n)}(x)$.
23. Let $f(x) = \sqrt{x}$. Find the first four derivatives of f and conjecture a formula for $f^{(n)}(x)$. Prove your formula by mathematical induction.
24. Let $y = f(x) \cdot g(x)$, where f and g have derivatives of order 3 or more. Find a formula for the following:
(a) $\dfrac{d^2y}{dx^2}$ (b) $\dfrac{d^3y}{dx^3}$
25. A particle moves in a straight line so that its distance from the origin at time t is given by
$$s(t) = 2t^2 - \frac{t^2 - 1}{t + 3}$$
Find formulas for its velocity $v(t)$ and its acceleration $a(t)$ at any time $t \geq 0$. Also find $v(1)$ and $a(1)$.

C

26. It can be shown that when $f''(x)$ exists, it can be approximated arbitrarily closely by the *second-difference quotient:*
$$\frac{f(x + 2h) - 2f(x + h) + f(x)}{h^2}$$
for h sufficiently small but different from 0. Use this and the computer to approximate $f''(x)$ for each of the functions given below at the points $x = -1$, 2, and 5. Use $h = (0.1)^n$, for $n = 0, 1, 2, 3, 4, 5$. Calculate the exact derivative by hand also in each case and compare answers.

(a) $f(x) = 2x^3 - 3x^2 + 4$

(b) $f(x) = \dfrac{x}{x^2 + 1}$

(c) $f(x) = \dfrac{2}{x^2} - \dfrac{x^2}{2}$

(d) $f(x) = (x^3 + 4)(3x^4 - 2x^2 + 5)$

27. Using the functions given in Exercise 26, approximate $f'''(x)$ at the same values of x, using the *third-difference quotient:*

$$\frac{f(x + 3h) - 3f(x + 2h) + 3f(x + h) - f(x)}{h^3}$$

Again, compare your results with the exact answers.

2.6 THE CHAIN RULE

In this section we will develop a formula, known as the **chain rule,** for the derivative of the composition of two functions. This is a powerful result that enables us to differentiate many otherwise intractable functions. We state the rule here but defer the proof to the end of this section because it is rather technical.

THEOREM 2.7 THE CHAIN RULE

Let g be differentiable at x and f be differentiable at $g(x)$. Then the composition $f \circ g$ is differentiable at x, and its derivative there is

$$(f \circ g)'(x) = f'(g(x)) \cdot g'(x) \tag{2.7}$$

■

Remark Since $(f \circ g)(x) = f(g(x))$, it is useful to refer to f as the *outer* function and g as the *inner* function. By the chain rule we first differentiate the outer function and evaluate it at the inner, getting $f'(g(x))$. Then we multiply this by the derivative of the inner function at x, $g'(x)$.

EXAMPLE 2.11 Use the chain rule to find $\frac{dy}{dx}$ if $y = (x^2 + 4)^{10}$.

Solution We can write y in the form $y = (f \circ g)(x)$, where $f(x) = x^{10}$ and $g(x) = x^2 + 4$. Since $f'(x) = 10x^9$, we see that $f'(g(x)) = 10(x^2 + 4)^9$. Thus by the chain rule,

$$\frac{dy}{dx} = f'(g(x)) \cdot g'(x) = \underbrace{10(x^2 + 4)^9}_{f'(g(x))} \cdot \underbrace{2x}_{g'(x)} = 20x(x^2 + 4)^9$$

■

If we write

$$y = f(u) \quad \text{and} \quad u = g(x)$$

so that $y = f(g(x))$, then the chain rule can be written in the alternate form

$$\frac{dy}{dx} = \frac{dy}{du} \cdot \frac{du}{dx} \tag{2.8}$$

In this form we could have done the problem in Example 2.11 as follows. Let $y = u^{10}$, where $u = x^2 + 4$. Then by equation (2.8),

$$\frac{dy}{dx} = \underbrace{10u^9}_{\frac{dy}{du}} \cdot \underbrace{2x}_{\frac{du}{dx}}$$

$$= 20x(x^2 + 4)^9 \quad \text{Substituting for } u \text{ and simplifying}$$

Derivatives of functions of the form $[g(x)]^n$ occur with sufficient frequency to warrant special consideration. This can be regarded as the composition $(f \circ g)(x)$, where $f(x) = x^n$. Assuming g is differentiable, we get from the chain rule,

$$\frac{d}{dx}[g(x)]^n = n[g(x)]^{n-1} \cdot g'(x) \tag{2.9}$$

This is sometimes called the **generalized power rule.** Our previous results show it to be valid for all integers n, and we will see later that it is true if n is any real number. By writing $u = g(x)$, the formula assumes the form

$$\frac{d}{dx}(u^n) = nu^{n-1}\frac{du}{dx}$$

Using this result on the function $y = (x^2 + 4)^{10}$ of Example 2.11, we get the answer immediately:

$$\frac{dy}{dx} = 10(x^2 + 4)^9 \cdot 2x = 20x(x^2 + 4)^9$$

The next two examples further illustrate this generalized power rule.

EXAMPLE 2.12 Find $\frac{dy}{dx}$ if

$$y = \frac{2}{(3x - x^2)^4}$$

Solution First write y in the form $y = 2(3x - x^2)^{-4}$. Then by equation (2.9),

$$\frac{dy}{dx} = -8(3x - x^2)^{-5}(3 - 2x)$$

■

EXAMPLE 2.13 Find $F'(x)$ if $F(x) = (x^2 - 1)^3(2x + 3)^4$.

Solution Using equation (2.9) along with the product rule, we get

$$F'(x) = (x^2 - 1)^3 \cdot 4(2x + 3)^3 \cdot 2 + (2x + 3)^4 \cdot 3(x^2 - 1)^2 \cdot 2x$$

A more useful form is obtained by factoring:

$$\begin{aligned} F'(x) &= 2(x^2 - 1)^2(2x + 3)^3[4(x^2 - 1) + 3x(2x + 3)] \\ &= 2(x^2 - 1)^2(2x + 3)^3(10x^2 + 9x - 4) \end{aligned}$$

■

We now give a proof of the chain rule.

Proof of the Chain Rule We want to prove that

$$(f \circ g)'(x) = f'(g(x))g'(x)$$

As usual, we work with the difference quotient to get it into an appropriate form before moving to the limit. To simplify notation, write $u = g(x)$. It is convenient in this proof to use the delta notation discussed in Section 2.2. In particular, we let $\Delta u = g(x + \Delta x) - g(x)$. Then $\Delta u \to 0$ as $\Delta x \to 0$ by Lemma 2.1. Also,

$$g(x + \Delta x) = g(x) + \Delta u = u + \Delta u$$

Thus the difference quotient for $f \circ g$ can be written as

$$\frac{f(g(x+\Delta x)) - f(g(x))}{\Delta x} = \frac{f(u+\Delta u) - f(u)}{\Delta x}, \qquad \Delta x \neq 0 \tag{2.10}$$

The result would be immediate if we could now multiply and divide by Δu and then let Δx (and hence also Δu) approach 0. The problem is that although we can require that $\Delta x \neq 0$, we have no control over Δu. It may equal 0. We get around this difficulty by defining a function η as follows:

$$\eta = \begin{cases} \dfrac{f(u+\Delta u) - f(u)}{\Delta u} - f'(u) & \text{if } \Delta u \neq 0 \\ 0 & \text{if } \Delta u = 0 \end{cases} \tag{2.11}$$

Then $\eta \to 0$ as $\Delta u \to 0$ (why?) and so also as $\Delta x \to 0$. From the top equation of (2.11), we get for $\Delta u \neq 0$

$$f(u+\Delta u) - f(u) = (f'(u) + \eta)\Delta u \tag{2.12}$$

But this also holds true when $\Delta u = 0$, since both sides equal 0. We can now substitute from equation (2.12) into equation (2.10) to get the difference quotient for $f \circ g$ in the form

$$[f'(u) + \eta] \cdot \frac{\Delta u}{\Delta x} = [f'(u) + \eta] \frac{g(x+\Delta x) - g(x)}{\Delta x}$$

As $\Delta x \to 0$, we get $f'(u)g'(x)$ for the limit; that is,

$$(f \circ g)'(x) = f'(g(x))g'(x)$$

■

EXERCISE SET 2.6

A

In Exercises 1–20 find $\frac{dy}{dx}$.

1. $y = (x^2 + 2)^4$
2. $y = (2x^3 - 3)^5$
3. $y = (1 - 2x^2)^5$
4. $y = (x^3 - x)^{-1}$
5. $y = (3x^4 - 2)^{10}$
6. $y = \left(\dfrac{x-1}{\sqrt{x}}\right)^{12}$
7. $y = (x^2 - 3x + 4)^3$
8. $y = 4(x^3 - 3x^2 + 5)^2$
9. $y = (3x^4 - 2x^2 + 1)^{-2}$
10. $y = (x^2 - 4x - 7)^{-3}$
11. $y = (2x + 1)^2(x^2 + 2)^3$
12. $y = x^5(x^3 - 2x^2 + 3)^4$
13. $y = \dfrac{2}{(x^2 + 1)^3}$
14. $y = \dfrac{3}{(2x^2 - 5)^4}$
15. $y = \dfrac{(3x - 4)^3}{x + 1}$
16. $y = \dfrac{(x^2 + 2)^2}{2x - 1}$
17. $y = \dfrac{x^2 + 1}{(x^2 - 1)^2}$
18. $y = \dfrac{2x^2 - x}{(3x - 4)^3}$
19. $y = \left(\dfrac{x-1}{x+1}\right)^3$
20. $y = \left(\dfrac{x^2}{x^2 - 1}\right)^{-2}$

In Exercises 21–26 find $(f \circ g)'(x)$ for the given values of f and g.

21. $f(x) = x^3$; $g(x) = 2x + 3$
22. $f(x) = 2x^4$; $g(x) = 1 - 3x$
23. $f(x) = x^{-2}$; $g(x) = x^2 + 4$
24. $f(x) = \frac{1}{x}$; $g(x) = x^2 - 1$
25. $f(x) = 3x^2$; $g(x) = x^3 - 8$
26. $f(x) = x^{10}$; $g(x) = 2x^2 + x$

In Exercises 27–32 use equation (2.8) to find $\frac{dy}{dx}$. Express the answer in terms of x.

27. $y = 2u^3$; $u = x^2 - 3$
28. $y = u^2 - 1$; $u = 2x + 5$
29. $y = 3/u^2$; $u = x^3 + 4$

30. $y = 4u^2 - 3$; $u = 2x^2 - 1$

31. $y = u^2 - 2u$; $u = 3x^2 - 4x$

32. $y = 1 - u^5$; $u = 1 - x^5$

In Exercises 33–36 find d^2y/dx^2.

33. $y = (x^2 + 1)^4$ **34.** $y = (2x^2 - 7)^5$

35. $y = x/(1 - x)$ **36.** $y = 1/(x^2 - 1)^2$

B

37. Let $h(t) = (t + 1)/(t - 1)$ and $g(t) = t^2$. Use the chain rule to find $(h \circ g)'(t)$ and $(g \circ h)'(t)$.

38. Let $\phi(u) = u^2 - 2u + 3$ and $\psi(u) = (2u - 1)/(3u + 1)$. Use the chain rule to find $(\phi \circ \psi)'(u)$ and $(\psi \circ \phi)'(u)$.

39. Let $z = t^3 - 3t^2 + 4$ and $t = y^2 + 3y - 1$. Use the chain rule to find $\frac{dz}{dy}$.

40. Find the equation of the tangent line and normal line to the graph of

$$y = \frac{(3x - 4)^2}{2(x^2 - 2)^2}$$

at the point for which $x = 2$.

41. Find $f''(x)$ if $f(x) = (2x - 1)/(x^2 + 1)^2$.

42. Derive a formula for $(f \circ g \circ h)'(x)$, assuming appropriate differentiability properties. [*Hint:* Write $f \circ g \circ h = (f \circ g) \circ h$ and apply the chain rule twice.]

43. Use the result of Exercise 42 to find y' if $y = [(2x^3 - 3)^2 + 4]^3$. What are f, g, and h in this case?

C

44. Use the computer to approximate $(f \circ g)'(x)$ for the functions f and g given below and for $x = -2, 1$, and 3. Follow these steps:

(1) Evaluate $u = g(x)$.
(2) Approximate $f'(u)$ using $[f(u + h) - f(u)]/h$.
(3) Approximate $g'(x)$ using $[g(x + h) - g(x)]/h$.
(4) Multiply the results in steps 2 and 3.

In steps 2 and 3 take $h = (10)^{-n}$, where $n = 1, 2, 3, 4, 5$. Also calculate $(f \circ g)'(x)$ by hand and compare the results.

$$f(x) = x^3 \qquad g(x) = \frac{x^2}{x + 1}$$

2.7 DERIVATIVES OF THE TRIGONOMETRIC FUNCTIONS

In the derivations and problems of this section we make use of definitions and properties of the trigonometric functions given in Appendix 2. You should refer to these as necessary.

The following two lemmas will be used in finding the derivative of the sine function.

LEMMA 2.3
$$\lim_{t \to 0} \frac{\sin t}{t} = 1$$

Proof Since both $\sin t$ and t are odd functions, $(\sin t)/t$ is even. (Why?) It is therefore sufficient to consider $t \to 0^+$. Then symmetry will show that the limit is the same as $t \to 0^-$.

In Figure 2.10 the circle has radius 1 (called a *unit circle*). The point A is t units of arc counterclockwise from C, where $0 < t < \frac{\pi}{2}$. By the definition of sine and cosine, A has coordinates $(\cos t, \sin t)$. Thus $\overline{OB} = \cos t$

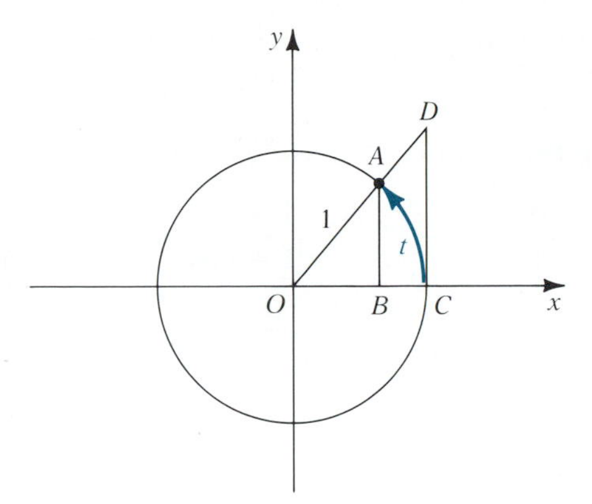

FIGURE 2.10

and $\overline{AB} = \sin t$. By similar triangles, we also have

$$\frac{\overline{AB}}{\overline{OA}} = \frac{\overline{CD}}{\overline{OC}} \quad \text{or} \quad \frac{\sin t}{\cos t} = \overline{CD}$$

since $\overline{OC} = 1$.

Using geometric reasoning, we see that

$$\text{area } \triangle OAB < \text{area sector } OCA < \text{area } \triangle OCD \tag{2.13}$$

We know how to find the areas of the two triangles. The area of the sector OCB is the fractional part of the area of the circle given by the ratio of the arc length t to the total circumference $2\pi r = 2\pi$. Since the area of the circle is $\pi r^2 = \pi$, we get

$$\text{area sector } OCA = \frac{t}{2\pi} \cdot \pi = \frac{t}{2}$$

For the triangles we have area $\triangle OAB = \frac{1}{2}(\overline{OA})(\overline{AB}) = \frac{1}{2} \cos t \sin t$ and area $\triangle OCD = \frac{1}{2}(\overline{OC})(\overline{CD}) = \frac{1}{2}(1)(\sin t/\cos t)$. So from equation (2.13),

$$\frac{1}{2} \sin t \cos t < \frac{1}{2} t < \frac{1}{2} \frac{\sin t}{\cos t}$$

On multiplying through by $2/\sin t$ (which is positive) and inverting all members (thus reversing the sense of each inequality), we get

$$\cos t < \frac{\sin t}{t} < \frac{1}{\cos t} \tag{2.14}$$

Now as $t \to 0^+$, $\cos t \to 1$. So both extremes of inequality (2.14) approach 1. Thus, by the squeezing theorem (Theorem 1.4),

$$\lim_{t \to 0^+} \frac{\sin t}{t} = 1$$

By our earlier remark, it follows that the limit from the left, and hence the limit itself, is also 1. ■

Note We used the letter t instead of x in this lemma because x was used as the abscissa in Figure 2.10. Now that we have the result, however, any letter, including x, can be used.

LEMMA 2.4 $\lim_{t\to 0} \frac{1 - \cos t}{t} = 0$

Proof Using the identity $1 - \cos^2 t = \sin^2 t$, we can write

$$\frac{1-\cos t}{t} = \frac{1-\cos t}{t} \cdot \frac{1+\cos t}{1+\cos t} = \frac{1-\cos^2 t}{t(1+\cos t)}$$
$$= \frac{\sin^2 t}{t(1+\cos t)} = \frac{\sin t}{t}\left(\frac{\sin t}{1+\cos t}\right)$$

Thus,

$$\lim_{t\to 0} \frac{1-\cos t}{t} = \left(\lim_{t\to 0} \frac{\sin t}{t}\right)\left(\lim_{t\to 0} \frac{\sin t}{1+\cos t}\right)$$

The first limit on the right is 1 by Lemma 2.3, and the second limit is $\frac{0}{2} = 0$. So

$$\lim_{t\to 0} \frac{1-\cos t}{t} = 0$$ ■

THEOREM 2.8 For all real numbers x,

$$\frac{d}{dx}(\sin x) = \cos x \tag{2.15}$$

Proof Let $f(x) = \sin x$. We write the difference quotient $[f(x+h) - f(x)]/h$ in the following equivalent forms:

$$\frac{\sin(x+h) - \sin x}{h} = \frac{(\sin x \cos h + \cos x \sin h) - \sin x}{h}$$
$$= \sin x\left(\frac{\cos h - 1}{h}\right) + \cos x\left(\frac{\sin h}{h}\right)$$

where we have made use of the identity

$$\sin(\alpha + \beta) = \sin\alpha\cos\beta + \cos\alpha\sin\beta$$

Using Lemmas 2.3 and 2.4, we get

$$f'(x) = \lim_{h\to 0} \frac{\sin(x+h) - \sin x}{h}$$
$$= (\sin x)(0) + (\cos x)(1)$$

That is,

$$\frac{d}{dx}(\sin x) = \cos x$$ ■

We can now obtain the derivative of the cosine, making use of the identities

$$\cos x = \sin(\tfrac{\pi}{2} - x) \quad \text{and} \quad \cos(\tfrac{\pi}{2} - x) = \sin x$$

$$\frac{d}{dx}\cos x = \frac{d}{dx}\sin\left(\frac{\pi}{2} - x\right) = \cos\left(\frac{\pi}{2} - x\right)\cdot(-1)$$ By equation (2.15) and the chain rule

$$= -\sin x$$

So we have

$$\frac{d}{dx}(\cos x) = -\sin x \tag{2.16}$$

Using the definitions

$$\tan x = \frac{\sin x}{\cos x} \qquad \cot x = \frac{\cos x}{\sin x} \qquad \sec x = \frac{1}{\cos x} \qquad \csc x = \frac{1}{\sin x}$$

we can get the derivatives of these other four trigonometric functions from those for the sine and cosine. We illustrate this for the tangent and leave the others as exercises:

$$\begin{aligned}
\frac{d}{dx}(\tan x) &= \frac{d}{dx}\left(\frac{\sin x}{\cos x}\right) \\
&= \frac{\cos x \dfrac{d}{dx}(\sin x) - \sin x \cdot \dfrac{d}{dx}(\cos x)}{\cos^2 x} \qquad \text{Quotient rule} \\
&= \frac{(\cos x)(\cos x) - (\sin x)(-\sin x)}{\cos^2 x} \\
&= \frac{\cos^2 x + \sin^2 x}{\cos^2 x} \\
&= \frac{1}{\cos^2 x} \\
&= \sec^2 x
\end{aligned}$$

We summarize here the derivatives of all six trigonometric functions.

DERIVATIVES OF THE TRIGONOMETRIC FUNCTIONS

$$\frac{d}{dx}(\sin x) = \cos x \qquad \frac{d}{dx}(\cot x) = -\csc^2 x$$

$$\frac{d}{dx}(\cos x) = -\sin x \qquad \frac{d}{dx}(\sec x) = \sec x \tan x$$

$$\frac{d}{dx}(\tan x) = \sec^2 x \qquad \frac{d}{dx}(\csc x) = -\csc x \cot x$$

The formulas for derivatives of the sine and cosine are valid for all real x, those for the tangent and secant for all $x \neq \frac{\pi}{2} + n\pi$, and those for the cotangent and cosecant for all $x \neq n\pi$, with $n = 0, \pm 1, \pm 2, \ldots$.

EXAMPLE 2.14 Find $\frac{dy}{dx}$ for each of the following:

(a) $y = \sin(x^2 + 2)$ (b) $y = \cos^3 2x$

(c) $y = \tan x \sec x$ (d) $y = \sin x/(1 + \cos x)$

Solution (a) Using the chain rule, we have

$$\frac{dy}{dx} = \underbrace{\cos(x^2 + 2)}_{\cos u} \cdot \underbrace{2x}_{\frac{du}{dx}} = 2x\cos(x^2 + 2)$$

(b) Remember that $\cos^3 2x$ means $(\cos 2x)^3$. So using the chain rule twice, we get

$$\frac{dy}{dx} = 3(\cos 2x)^2(-\sin 2x) \cdot 2 = -6 \sin 2x \cos^2 2x$$

(c) We use the product rule to get

$$\begin{aligned}\frac{dy}{dx} &= (\tan x)(\sec x \tan x) + (\sec x)(\sec^2 x) \\ &= \sec x \tan^2 x + \sec^3 x\end{aligned}$$

(d) By the quotient rule, we have

$$\begin{aligned}\frac{dy}{dx} &= \frac{(1 + \cos x)(\cos x) - \sin x(-\sin x)}{(1 + \cos x)^2} \\ &= \frac{1 + \cos^2 x + \sin^2 x}{(1 + \cos x)^2} \\ &= \frac{2}{(1 + \cos x)^2} \quad \text{Since } \sin^2 x + \cos^2 x = 1\end{aligned}$$

■

A Word of Caution One of the most frequent mistakes students make in differentiation is failure to apply the last step of the chain rule. For example, consider $\frac{d}{dx}(\sin 2x)$. All too frequently students give the answer as $\cos 2x$ and forget the final $\frac{du}{dx}$, in this case 2. The correct answer is $2 \cos 2x$. So be alert to this danger. *Do not forget the final* $\frac{du}{dx}$.

The next example mixes algebraic and trigonometric functions.

EXAMPLE 2.15 Find $f'(x)$ for each of the following:

(a) $f(x) = x^2 \sin 2x$ (b) $f(x) = (\tan x - x)^2$

Solution (a) This is a product, so we have

$$\begin{aligned}f'(x) &= x^2(2 \cos 2x) + (\sin 2x)(2x) \\ &= 2x(x \cos 2x + \sin 2x)\end{aligned}$$

(b)
$$\begin{aligned}f'(x) &= 2(\tan x - x)(\sec^2 x - 1) \\ &= 2(\tan x - x)(\tan^2 x) \quad \text{By the identity } 1 + \tan^2 x = \sec^2 x \\ &= 2 \tan^2 x(\tan x - x)\end{aligned}$$

■

EXERCISE SET 2.7

A

In Exercises 1–26, find $\frac{dy}{dx}$.

1. $y = \sin 3x$
2. $y = \sin(2x + 3)$
3. $y = \cos(1 - 2x)$
4. $y = 2 \cos(x^2 + 1)$
5. $y = \tan(3x^2)$
6. $y = \sec x + \tan x$
7. $y = 2 \csc 3x$
8. $y = 3 \cot(1 - x)$
9. $y = 5 \sin^2 x$
10. $y = 1 - 2 \cos^2 x$
11. $y = \tan^2(2x + 3)$
12. $y = \cos^2 x - \sin^2 x$
13. $y = \sec\left(\frac{1}{x}\right)$
14. $y = \csc 2x - \cot 2x$

15. $y = \tan^3 x^2$

16. $y = \dfrac{1 - \cos x}{\sin x}$

17. $y = \dfrac{\sec x + 1}{\sec x - 1}$

18. $y = \dfrac{1 - \cos x}{1 + \cos x}$

19. $y = (1 + \sin 2x)^2$

20. $y = (\cos x - \sin x)^2$

21. $y = \dfrac{1 + \tan x}{\sec x}$

22. $y = x^2 \cos 2x$

23. $y = (3x + 2) \csc 3x$

24. $y = \dfrac{x}{\sin x}$

25. $y = \sin(\cos x)$

26. $y = \tan(\sec^2 x)$

In Exercises 27–34 find $f''(x)$.

27. $f(x) = x \sin x$

28. $f(x) = x^2 \cos 2x$

29. $f(x) = \tan x^2$

30. $f(x) = \sec^3 2x$

31. $f(x) = \dfrac{1 - \cos x}{\sin x}$

32. $f(x) = \dfrac{\tan x}{1 - \tan x}$

33. $f(x) = \sin x - x \cos x$

34. $f(x) = \cos x^3$

35. Show that for all values of A and B, $y = A \cos x + B \sin x$ satisfies $y''' + y' = 0$.

B

36. Derive the differentiation formula for the following: (a) $\frac{d}{dx}(\cot x)$ (b) $\frac{d}{dx}(\sec x)$ (c) $\frac{d}{dx}(\csc x)$

37. Find $g'(t)$ and $g''(t)$ if $g(t) = \sec(\frac{1}{t})$.

38. If $y = (1 - \cos x)/(\sin x)$, show that $y'' - [y^2/(\sin x)] = 0$.

39. Show that $y = \cos 2x + \frac{1}{2}(1 - x \sin 2x)$ satisfies the equation $y'' + 4y = 4 \sin^2 x$.

40. A spring with a weight attached is set in motion and its position $x(t)$ in feet from the equilibrium position at time t sec is given by

$$x(t) = \tfrac{1}{4}(1 - 8t) \cos 16t + \tfrac{1}{8} \sin 16t$$

Find its velocity and acceleration when $t = \frac{5\pi}{3}$ sec.

C

41. Use the computer to verify that inequality (2.14) holds true for the following values of t: $\pm 1, \pm\frac{1}{2}, \pm\frac{1}{4}, \pm\frac{1}{8}, \ldots, \pm\frac{1}{2^{10}}$.

In Exercises 42–48 use the computer to estimate the given limit. Verify your result using trigonometric identities and the fact that $(\sin x)/x \to 1$ as $x \to 0$.

42. $\lim\limits_{x \to 0} \dfrac{1 - \cos x}{x}$

43. $\lim\limits_{x \to 0} \dfrac{\tan x}{x}$

44. $\lim\limits_{x \to 0} \dfrac{\sin 2x}{x}$

45. $\lim\limits_{x \to 0} \dfrac{\cos 2x - 1}{x^2}$

46. $\lim\limits_{x \to 0} \dfrac{x \sin x}{1 - \cos 2x}$

47. $\lim\limits_{x \to 0} \dfrac{\sqrt{1 - \cos x}}{|x|}$

48. $\lim\limits_{x \to 0} \dfrac{1}{x^2(1 + \cot^2 x)}$

49. For $h = 0.0001$ have the computer print two columns, one giving values of the difference quotient $[\sin(x + h) - \sin h]/h$ and the other giving values of $\cos x$ for values of x starting with 0 and increasing at increments of $\frac{\pi}{12}$ up to 2π. Compare the results.

2.8 IMPLICIT DIFFERENTIATION

Often we are presented with an equation that relates two variables, x and y say, but does not express y explicitly in terms of x. For example, consider the equation

$$x^2 + y^2 = 4$$

whose graph is a circle of radius 2 with center at the origin. The variables x and y are certainly related by this equation, but the equation does not indicate explicitly what y is for a given value of x. We could solve for y and get $y = \pm\sqrt{4 - x^2}$, and then we would have an explicit relationship, two of

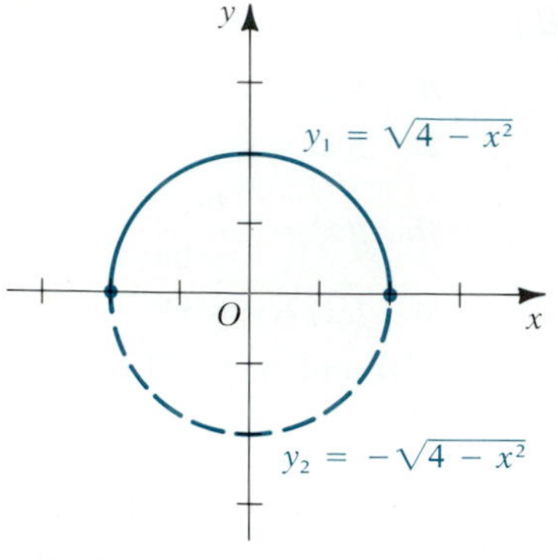

FIGURE 2.11

them, in fact. Let us denote the two explicit values of y by y_1 and y_2:

$$y_1 = \sqrt{4 - x^2} \qquad y_2 = -\sqrt{4 - x^2}$$

If we take the domain in each case to be $[-2, 2]$, these represent the upper and lower semicircles, as indicated in Figure 2.11. Each equation is now an explicit functional relationship, and we could, for example, find the derivatives y_1' and y_2'. It is an interesting fact that we can also find both of these derivatives at once, using the original equation $x^2 + y^2 = 4$.

To illustrate how this is done, we reason that if the equation were solved for y, we would get two equations of the form $y = f(x)$. [We know two possibilities for $f(x)$ in this case—namely, $\sqrt{4 - x^2}$ and $-\sqrt{4 - x^2}$—but that is not important for the present.] Since y must satisfy the equation $x^2 + y^2 = 4$, it follows that whatever $f(x)$ is, it satisfies

$$x^2 + [f(x)]^2 = 4 \tag{2.17}$$

for all values of x in the appropriate domain ($|x| \leq 2$ in this case). Equation (2.17) is an *identity* in x. You can convince yourself of this by substituting either of the values we found for $f(x)$ into this equation. Since the left-hand side of equation (2.17) is the same as the right-hand side for all admissible values of x, it follows that the derivative of the left-hand side is the same as the derivative of the right-hand side, provided f is a differentiable function. So we equate derivatives and get

$$2x + 2[f(x)] \cdot f'(x) = 0$$

Note how we used the chain rule in differentiating $[f(x)]^2$. Since we want to know $f'(x)$, we solve this last equation for $f'(x)$ and get

$$f'(x) = -\frac{x}{f(x)} \qquad [f(x) \neq 0]$$

Since $y = f(x)$, we can also write the answer as

$$\frac{dy}{dx} = -\frac{x}{y} \qquad (y \neq 0)$$

As stated earlier, this formula is correct regardless of which of the two values of y we choose.

In actual practice we usually shortcut the preceding procedure as follows:

$$x^2 + y^2 = 4$$

We differentiate both sides, under the assumption that y is a differentiable function of x:

$$2x + 2y \cdot y' = 0$$

Now solve for y':

$$y' = -\frac{x}{y} \qquad (y \neq 0)$$

It should be emphasized that when differentiating y^2, we must think of y as a function of x, so the chain rule has to be applied, just as if y were a known function of x. For example, when you differentiate $(x^3 + 1)^2$, you get

$$\frac{d}{dx}\underbrace{(x^3+1)}_{y}{}^2 = 2\underbrace{(x^3+1)}_{y}\cdot\underbrace{3x^2}_{y'}$$

In exactly the same way,

$$\frac{d}{dx}(y)^2 = 2y \cdot y'$$

We will not attempt at this stage to state conditions under which this procedure is valid, but we will *assume* in all of the problems that the given equation does define y as one or more differentiable functions of x on some suitable domain. The procedure we have described is called **implicit differentiation,** and the given equation in x and y is said to *define y implicitly* as one or more functions of x. The importance of implicit differentiation lies in the fact that derivatives can be found in situations where it is difficult or even *impossible* to solve for y in an explicit form. In our example of the equation of the circle, $x^2 + y^2 = 4$, we had a choice, either to solve for y and then differentiate or to differentiate implicitly. In the equation $x^3 - 2x^2y + y^4 - 3y^5 = 0$, for example, we have no such choice; we cannot solve explicitly for y, but we can still find y' using implicit differentiation.

EXAMPLE 2.16 Find y' in each of the following:

(a) $x^3 - 2xy + y^3 = 4$ (b) $y = x \sin y$

Solution (a) Proceeding as above, we have

$$3x^2 - 2(xy' + y) + 3y^2 \cdot y' = 0$$ The product rule is used on the xy term.

Now we rearrange terms and solve for y':

$$y'(3y^2 - 2x) = 2y - 3x^2$$
$$y' = \frac{2y - 3x^2}{3y^2 - 2x}$$

Of course, any points (x, y) for which the denominator is 0 must be excluded.

(b) At first, this looks like an explicit function, since it is of the form $y =$ something. But the "something" involves y, so it must be treated implicitly. Differentiating both sides, we get

$$y' = x(\cos y)y' + \sin y$$
$$y'(1 - x\cos y) = \sin y$$
$$y' = \frac{\sin y}{1 - x\cos y} \qquad (x\cos y \neq 1)$$

Note that the derivative of the term $x \sin y$ involved both the product rule and the chain rule. ■

EXAMPLE 2.17 Find y' and y'' if $x^3 - y^3 = 8$.

Solution We differentiate both sides with respect to x:

$$3x^2 - 3y^2 \cdot y' = 0$$

$$y' = \frac{x^2}{y^2} \qquad (y \neq 0)$$

Now we differentiate again, using the quotient rule on the right:

$$y'' = \frac{y^2(2x) - x^2(2y \cdot y')}{y^4}$$

We next divide out the common factor y on the numerator and denominator, replace y' by its value we have just found, and simplify:

$$y'' = \frac{2xy - 2x^2\left(\frac{x^2}{y^2}\right)}{y^3} = \frac{2x(y^3 - x^3)}{y^5}$$

Finally, observe that since the original equation was $x^3 - y^3 = 8$, it follows that $y^3 - x^3 = -8$, and the final answer can be written in the form

$$y'' = \frac{-16x}{y^5} \qquad (y \neq 0)$$

You should be alert to the possibility of simplifying final answers in this way. It frequently happens in finding second derivatives. ■

Implicit differentiation allows us to extend the power rule for differentiation to rational exponents, as the next theorem shows.

THEOREM 2.9 **POWER RULE FOR RATIONAL EXPONENTS**

For any rational number r, if $y = x^r$, then

$$\frac{dy}{dx} = rx^{r-1}$$

Proof Since r is rational, it can be written in the form $r = \frac{m}{n}$, where m and n are integers. So $y = x^{m/n}$. We raise both sides to the nth power and differentiate implicitly:

$$y^n = x^m$$

$$ny^{n-1} \cdot y' = mx^{m-1} \qquad \text{Since we know the power rule holds for integers}$$

$$y' = \frac{m}{n}\frac{x^{m-1}}{y^{n-1}}$$

Replacing y by $x^{m/n}$ and simplifying, we get

$$y' = \frac{m}{n}\frac{x^{m-1}}{x^{m(n-1)/n}} = \frac{m}{n}x^{m/n-1}$$

which is the desired result, since $r = \frac{m}{n}$. ■

EXAMPLE 2.18 Find $f'(x)$:

(a) $f(x) = \dfrac{\sqrt{x^2+1}}{x}$ (b) $f(x) = \sqrt{\cos x}$

Solution (a) We write this in the form

$$f(x) = \frac{(x^2+1)^{1/2}}{x}$$

and use the quotient rule, along with Theorem 2.9 and the chain rule:

$$f'(x) = \frac{x \cdot \frac{1}{2}(x^2+1)^{-1/2} \cdot 2x - (x^2+1)^{1/2}}{x^2}$$

This can be greatly simplified by multiplying numerator and denominator by the factor $(x^2+1)^{1/2}$:

$$f'(x) = \frac{x^2 - (x^2+1)}{x^2(x^2+1)^{1/2}} = \frac{-1}{x^2\sqrt{x^2+1}}$$

(b) $f(x) = \sqrt{\cos x} = (\cos x)^{1/2}$

$$f'(x) = \frac{1}{2}(\cos x)^{-1/2}(-\sin x) = -\frac{\sin x}{2\sqrt{\cos x}}$$ ∎

EXAMPLE 2.19 Find y' and y'' if

$$x^{2/3} + y^{2/3} = a^{2/3} \qquad (a \text{ is a constant})$$

Solution Differentiating implicitly, we obtain

$$\tfrac{2}{3}x^{-1/3} + \tfrac{2}{3}y^{-1/3} \cdot y' = 0$$

$$y' = -\frac{y^{1/3}}{x^{1/3}} \qquad (x \neq 0)$$

We find y'' as in Example 2.17:

$$y'' = -\frac{x^{1/3} \cdot \frac{1}{3}y^{-2/3} \cdot y' - y^{1/3} \cdot \frac{1}{3}x^{-2/3}}{x^{2/3}}$$

$$= -\frac{x^{1/3}y^{-2/3}\left(-\dfrac{y^{1/3}}{x^{1/3}}\right) - y^{1/3}x^{-2/3}}{3x^{2/3}}$$

$$= -\frac{-y^{-1/3} - y^{1/3} \cdot x^{-2/3}}{3x^{2/3}} \cdot \frac{x^{2/3}y^{1/3}}{x^{2/3}y^{1/3}}$$ This eliminates negative exponents.

$$= \frac{x^{2/3} + y^{2/3}}{3x^{4/3}y^{1/3}}$$

$$= \frac{a^{2/3}}{3x^{4/3}y^{1/3}}$$ Since $x^{2/3} + y^{2/3} = a^{2/3}$

Observe that y'' does not exist if either $x = 0$ or $y = 0$. ∎

EXERCISE SET 2.8

A

Use implicit differentiation to find y′ in Exercises 1–16.

1. $x^2 - y^2 = 1$
2. $x^2 + 2xy - 3y^2 = 0$
3. $x^2y^3 - 2x + y = 3$
4. $y^2 - 3xy - x^3 = 8$
5. $y = x^3 - y^3$
6. $x(y^2 - 3x) = 5$
7. $y(x^2 - 2y) = 3$
8. $\sqrt{x} + \sqrt{y} = 1$
9. $y = \sqrt{x + y}$
10. $x + 2\sqrt{xy} + y = 3$
11. $x \sin y + y \sin x = 1$
12. $y = \tan(x + y)$
13. $\sec(xy) = 3x$
14. $x = \cos\sqrt{xy}$
15. $\sin x \cos y = x - y$
16. $\sin^2 x + \cos^2 y = x^2 + y^2$

In Exercises 17–34 find $\frac{dy}{dx}$.

17. $y = \sqrt{x^2 + 1}$
18. $y = x\sqrt{1 - x}$
19. $y = \dfrac{1}{\sqrt{4 - x^2}}$
20. $y = \dfrac{x}{\sqrt{x + 4}}$
21. $y = \sqrt[3]{(2x^3 + 5)^2}$
22. $y = 1 + (1 - 3x^3)^{2/3}$
23. $y = \dfrac{x}{\sqrt{1 - x^2}}$
24. $y = \sin\sqrt{x + 1}$
25. $y = \sqrt{\sin x}$
26. $y = (\sqrt{\cos x})^3$
27. $y = \tan\left(\dfrac{1}{\sqrt{x}}\right)$
28. $y = \sin^2 x^2$
29. $y = \dfrac{\sec(2x + 3)}{\tan(2x + 3)}$
30. $y = (1 - \sin^2 x)^{3/2}$
31. $y = \sqrt{2(1 - \cos 2x)}$ $(0 \le x \le \pi)$
32. $y = \sin^4 x + \cos^4 x$
33. $y = \dfrac{\sqrt{4 - x^2}}{x}$
34. $y = \sqrt{\dfrac{1 - x}{1 + x}}$

In Exercises 35–42 find y″.

35. $x^2 + y^2 = 1$
36. $x^3 + y^3 = 8$
37. $x^{3/2} - y^{3/2} = 8$
38. $\sqrt{x} - \sqrt{y} = 1$
39. $y = \sqrt{1 - x^2}$
40. $y = 1/\sqrt{x^2 + 1}$
41. $y = \sin\sqrt{x}$
42. $y = \sqrt{\cos x}$

B

43. Find dy/dx and d^2y/dx^2 if $x^{4/3} - 2y^{4/3} = a^{4/3}$ where a is a constant.

44. Find $f'(x)$ and $f''(x)$ if $y = f(x)$ is a twice differentiable solution of $x = \tan(x + y)$.

45. Assuming g is a differentiable function for which $x = g(t)$ is a solution of

$$x^2t^3 - 4x^3t^2 + 3x^2 - 2t^5 - 7 = 0$$

find $g'(t)$.

46. (a) Find two functions f_1 and f_2 differentiable on $(-2, 2)$ such that $y = f_1(x)$ and $y = f_2(x)$ satisfy the equation $y^2 - 2xy + 2x^2 - 4 = 0$. (*Hint:* Use the quadratic formula.)
(b) Find $f_1'(x)$ and $f_2'(x)$.
(c) Using the equation of part a, differentiate implicitly to find y'.
(d) Show that the answers in parts b and c are the same.

47. (a) Show that $y = x$ satisfies the equation $y^3 - xy^2 - x^4y + x^5 = 0$, and find two other solutions. Differentiate each solution.
(b) Use implicit differentiation to find y' for the equation in part a, and show that this agrees with the derivatives found in part a.

48. A particle moves in a straight line so that its distance from the origin at time t is $s(t) = (t^2 - 1)^{2/3}$. Find its velocity $v(t)$ and its acceleration $a(t) = v'(t)$ at time t. What are $v(3)$ and $a(3)$?

C

49. (a) By solving the equation $x^{2/3} + y^{2/3} = 1$ for y, use the computer to generate a table of values (x, y) for $x = -1 + h$, where $h = 0, 0.1, 0.2, \ldots, 2.0$. Draw the graph of the equation. Determine two parts of the graph so that each part represents a function that is differentiable on $[-1, 1]$ except at one point. What is the exceptional point?

(b) For each of the functions in part a, use the computer and the difference quotient to approximate the derivative at $x = -1, -0.5, 0.5$, and 1. Use $h = 2^{-n}$ for $n = 0, 1, \ldots, 10$. In each case compare your answer with the value of y' as found in Example 2.19.

2.9 THE DIFFERENTIAL AND LINEAR APPROXIMATION

The Leibniz notation $\frac{dy}{dx}$ has been used up to now as one of several ways to designate the derivative $f'(x)$, where $y = f(x)$, but no meaning has been given to the symbols dx and dy individually. We now propose to give them meanings that will be consistent with the derivative notation.

Suppose f is differentiable at x. Let dx denote an independent variable that can take on any real value except 0. Then the **differential** of y, designated dy, is defined by

$$dy = f'(x)\,dx \tag{2.18}$$

Note that dy is a dependent variable, dependent on both x and dx. We may also write df in place of dy. We call dx the differential of x.

EXAMPLE 2.20 Find dy if $y = \sin^2 x$.

Solution Let $f(x) = \sin^2 x$. Then $f'(x) = 2 \sin x \cos x = \sin 2x$. So by equation (2.18),

$$dy = (\sin 2x)\,dx$$ ■

If we divide both sides of equation (2.18) by dx, we get

$$\frac{dy}{dx} = f'(x)$$

Now this seems to be just a restatement of a familiar fact. But there really is something new here, since now the left-hand side is a quotient of two differentials rather than simply a symbol for the derivative, as we have previously used it. The two interpretations give the same result, however.

All formulas for derivatives can now be phrased in terms of differentials as well. Suppose, for example, that $u = f(x)$ and $v = g(x)$, where f and g are differentiable in some common domain. Then we have for sums, differences, products, and quotients,

(i) $d(u \pm v) = du \pm dv$

(ii) $d(uv) = u\,dv + v\,du$

(iii) $d\left(\dfrac{u}{v}\right) = \dfrac{v\,du - u\,dv}{v^2} \qquad (v \neq 0)$

To illustrate how to prove these, consider equation (ii). Since $uv = (fg)(x)$, by definition of the differential,

$$\begin{aligned} d(uv) = (fg)'(x)\,dx &= [f(x)g'(x) + g(x)f'(x)]\,dx \\ &= \underbrace{f(x)}_{u}\underbrace{g'(x)\,dx}_{dv} + \underbrace{g(x)}_{v}\underbrace{f'(x)\,dx}_{du} \end{aligned}$$

These results provide an alternate means of finding derivatives implicitly, as the next example shows.

EXAMPLE 2.21 Use differentials to find $\frac{dy}{dx}$ if

$$x^2 + xy - 3y^2 = 5$$

Solution We take the differential of each term without regard to which variable is independent and which is dependent:

$$2x\,dx + (x\,dy + y\,dx) - 6y\,dy = 0$$

Now we collect the terms with dx on one side and those with dy on the other and then solve for the quotient $\frac{dy}{dx}$:

$$(2x + y)\,dx = (6y - x)\,dy$$

$$\frac{dy}{dx} = \frac{2x + y}{6y - x} \qquad (x \neq 6y)$$

■

Suppose now that $y = f(x)$ and $x = g(t)$, so that $y = f(g(t)) = (f \circ g)(t)$. Then, by equation (2.18) and the chain rule, we have

$$\begin{aligned} dy &= (f \circ g)'(t)\,dt \\ &= f'(g(t)) \cdot g'(t)\,dt \end{aligned}$$

But again by equation (2.18) applied to $x = g(t)$, we see that $dx = g'(t)\,dt$. So

$$dy = f'(\underbrace{g(t)}_{x}) \cdot \underbrace{g'(t)\,dt}_{dx} = f'(x)\,dx$$

This shows that when $y = f(x)$, the differential of y is given by equation (2.18) even when x is dependent on some other variable (assuming appropriate differentiability conditions).

The differentials dx and dy have a useful geometric interpretation as shown in Figure 2.12. Suppose $y = f(x)$. Let $P(x, f(x))$ be a point on the graph at which f' exists, and let l be the tangent line at P. Since l has slope $\frac{dy}{dx}$, it follows that the point $Q(x + dx, y + dy)$ always lies on l for all nonzero values of dx.

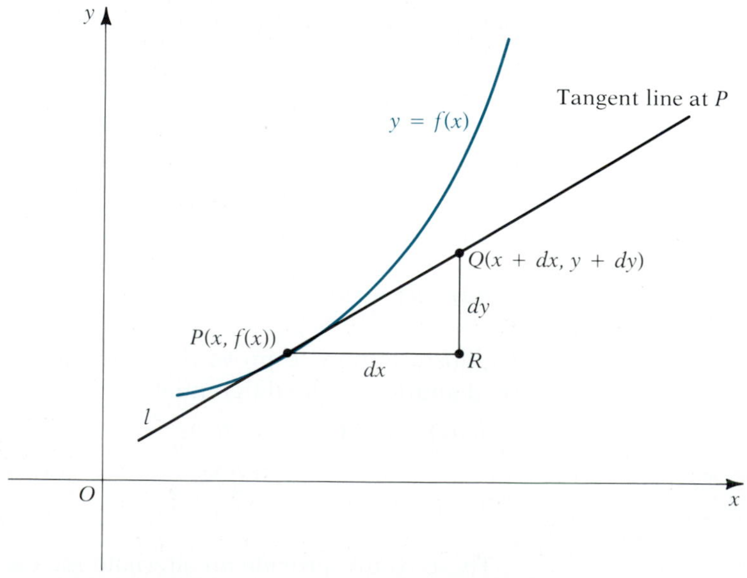

FIGURE 2.12

FIGURE 2.13

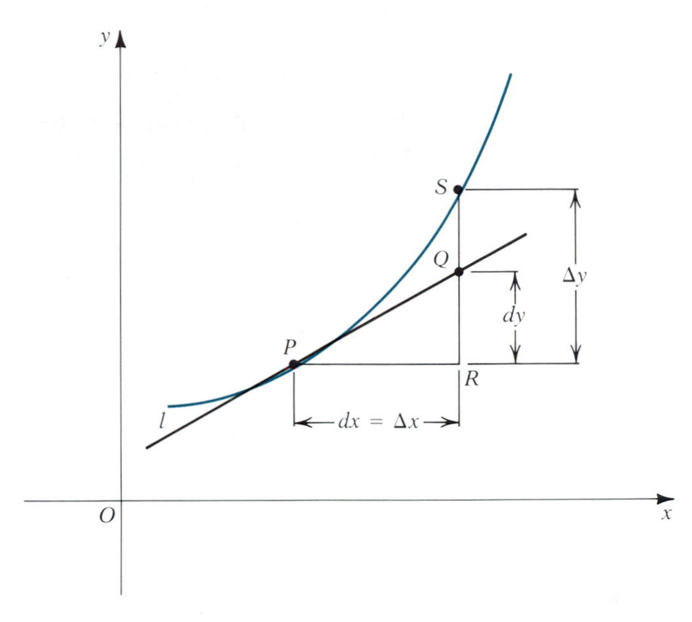

Thus, for any horizontal displacement dx from P, dy is the vertical displacement RQ to the tangent line l. Recall that we have previously used the symbol Δx to indicate an increment, or change, in x, and correspondingly we have used Δy to be the change in the function:

$$\Delta y = f(x + \Delta x) - f(x)$$

Suppose such an increment Δx is given. Since dx may be chosen arbitrarily, let us choose $dx = \Delta x$. Now we want to examine the relationship between dy and Δy. As we have seen, dy gives the vertical displacement from $P(x, f(x))$ *to the tangent line* drawn at P. The increment Δy, on the other hand, gives the vertical displacement from P *to the curve itself*. This is shown in Figure 2.13. Observe carefully that, although we are free to choose the differential dx of the *independent* variable equal to the increment Δx, the differential dy of the *dependent* variable will not, in general, equal the increment Δy. However, if $\Delta x\,(= dx)$ is sufficiently small, it seems reasonable to suppose that $\Delta y \approx dy$. To illustrate this, consider the following two examples.

EXAMPLE 2.22 Compare Δy and dy for $y = 2x^2 + 3x - 1$ when $x = 3$ and $\Delta x = 0.1$.

Solution Denote the given function by f, and let $dx = \Delta x = 0.1$. Then

$$\begin{aligned}\Delta y = f(x + \Delta x) - f(x) &= f(3.1) - f(3)\\ &= [2(3.1)^2 + 3(3.1) - 1] - [2(3)^2 + 3(3) - 1]\\ &= 27.52 - 26 = 1.52\end{aligned}$$

Since $f'(x) = 4x + 3$, we get for dy,

$$dy = f'(3)\,dx = (15)(0.1) = 1.5$$

So in this case the error in using dy to approximate Δy is 0.02. ■

EXAMPLE 2.23 A spherical tank to hold hydrogen is designed with a radius of 6 m. The allowable error in the actual radius is ± 2 cm. Find the approximate excess capacity the tank could have over its design volume. Compare this with the exact value of the maximum excess capacity.

Solution The volume $f(r)$ of a sphere is given by $V = \frac{4}{3}\pi r^3$. We take $r = 6$ and $\Delta r = dr = 0.02$, since 2 cm = 0.02 m. The approximate change in volume if the radius is 6.02 instead of 6 is

$$dV = f'(r)\,dr = 4\pi r^2\,dr = 4\pi(36)(0.2)$$
$$\approx 9.05 \text{ m}^3$$

The exact change is

$$\Delta V = f(r + \Delta r) - f(r) = f(6.02) - f(6)$$
$$= \tfrac{4}{3}\pi(6.02)^3 - \tfrac{4}{3}\pi(6)^3$$
$$\approx 9.0780 \text{ m}^3$$

to four decimal places. So dV approximates ΔV correct to one decimal place. ■

If we know the value of a function f at a given point x, then we can get its value at a nearby point $x + \Delta x$ by adding the change Δy to the original value; that is,

$$f(x + \Delta x) = f(x) + \Delta y$$

Furthermore, if $f'(x)$ exists, we can approximate Δy by dy if we take $dx = \Delta x$. So we have

$$f(x + \Delta x) \approx f(x) + dy$$

or equivalently

$$f(x + \Delta x) \approx f(x) + f'(x)\,\Delta x \tag{2.19}$$

This result is known as **linear approximation,** since we are, in effect, using values along the tangent line to approximate values on the curve.

EXAMPLE 2.24 Use equation (2.19) to approximate $\sqrt[3]{65}$.

Solution Let $f(x) = \sqrt[3]{x} = x^{1/3}$. The point nearest 65 at which we know the value of f is $x = 64$, since $\sqrt[3]{64} = 4$. So in equation (2.19) we take $x = 64$ and $\Delta x = 1$. Since $f'(x) = \frac{1}{3}x^{-2/3}$, we get

$$f(65) \approx f(64) + f'(64)\,\Delta x = 4 + \tfrac{1}{3}(64)^{-2/3}(1) = 4 + \tfrac{1}{3}(\tfrac{1}{16}) \approx 4.0208$$

By calculator we find $\sqrt[3]{65} = 4.02073$ (to five decimal places). So our approximation is quite good in this case. ■

We conclude this section by examining the size of the error in using the linear approximation given by equation (2.19)—that is, the error in using dy to approximate Δy. Assume $f'(x)$ exists and let $dx = \Delta x$. Then, as we know, $\frac{dy}{dx} = f'(x)$. Also,

$$\lim_{\Delta x \to 0} \frac{\Delta y}{\Delta x} = \lim_{\Delta x \to 0} \frac{f(x + \Delta x) - f(x)}{\Delta x} = f'(x)$$

Thus,

$$\lim_{\Delta x \to 0} \frac{\Delta y - dy}{\Delta x} = \lim_{\Delta x \to 0} \left(\frac{\Delta y}{\Delta x} - \frac{dy}{dx} \right) \quad \text{Since } dx = \Delta x$$
$$= f'(x) - f'(x) = 0$$

This tells us not only that the difference $\Delta y - dy$ approaches 0 as $\Delta x \to 0$, but also that it does so more rapidly than Δx itself. So the error $|\Delta y - dy|$ is small in comparison with the size of $|\Delta x|$. In case f has a bounded second derivative near x, a more explicit estimate of the error can be given. We state this below. Its proof depends on results we will obtain in Chapter 9.

THEOREM 2.10 If f has a bounded second derivative in some neighborhood $(x - r, x + r)$ of x, then for all Δx such that $|\Delta x| < r$,

$$|\Delta y - dy| \le \tfrac{M}{2}(\Delta x)^2 \tag{2.20}$$

where M is an upper bound on $|f''(t)|$ for t in $[x, x + \Delta x]$ (or $[x + \Delta x, x]$ if $\Delta x < 0$). ■

To illustrate this, consider again Example 2.24. There we had $f(x) = x^{1/3}$ and $f'(x) = \frac{1}{3}x^{-2/3}$, and so

$$f''(x) = -\frac{2}{9}x^{-5/3} = -\frac{2}{9x^{5/3}}$$

In that example we set $x = 64$ and $\Delta x = 1$, so that we want to know a bound on $|f''(t)|$ for t in the interval $[64, 65]$. On that interval,

$$|f''(t)| = \frac{2}{9t^{5/3}} \le \frac{2}{9(64)^{5/3}} = \frac{2}{9(4)^5} = \frac{1}{4608} \approx 0.000217$$

since the smallest value of t produces the largest possible fraction. This confirms that using dy to approximate Δy gives accuracy to at least three decimal places.

EXERCISE SET 2.9

A

In Exercises 1–16 find dy.

1. $y = x^2 + 2x - 3$
2. $y = 2x - x^3$
3. $y = \sqrt{4 - x^2}$
4. $y = 1/\sqrt{3x + 4}$
5. $y = x/\sqrt{1 - x}$
6. $y = x/\sqrt{1 - x^2}$
7. $y = x \sin x$
8. $y = \cot \sqrt{x}$
9. $y = \sec x + \tan x$
10. $y = \cos^2 3x$
11. $y = (x^6 + 1)^{2/3}$
12. $y = (x^{2/3} + a^{2/3})^{3/2}$
13. $y = (\tan x)/x$
14. $y = 2 \csc^3 2x$
15. $y = \sin^2 x^2$
16. $y = (\sin 2x)/(1 - \cos 2x)$

In Exercises 17–24 find $\frac{dy}{dx}$ implicitly using differentials.

17. $x^3 - 2xy^2 = 6$
18. $y^2 - x \sin y = \cos x$
19. $2x^2 - 3xy + 4y^2 - 10 = 0$
20. $\sqrt{x^2 + y^2} - 3 = 2xy$
21. $x^{2/3} + y^{2/3} = a^{2/3}$
22. $x^4 - 3x^2y^3 + y^5 = 10$
23. $x \sin y - y \sin x = 1$
24. $x^{4/3} - y^{4/3} = a^{4/3}$

In Exercises 25–28 use differentials for the approximations.

25. The side of a square is measured to be 18 cm. If there is a possible error in the measurement of 0.05 cm, find the approximate possible error in (a) the area and (b) the perimeter.

26. Find the approximate error in the calculated area of a circle with a diameter measured to 20 in. if in fact it is 20.04 in. What is the error in the circumference?

27. If $y = f(x)$ and x is in error by an amount Δx, then the corresponding error in y is Δy. The **relative error** in y is $\Delta y/y$, which is approximated by dy/y. Find the approximate relative error in each part of Exercise 25.

28. Find the approximate relative error in each part of Exercise 26. (See Exercise 27.)

In Exercises 29–32 estimate the value using linear approximation, and compare your result with the answer obtained from a calculator.

29. $\sqrt{10}$ **30.** $\sqrt[3]{7.5}$ **31.** $\sqrt[3]{-25}$ **32.** $1/\sqrt{102}$

33. A cardboard box has a square base, and its height is double the length of a side of the base. The box is supposed to be 2 ft × 2 ft × 4 ft, but it is subject to an error of $\pm\frac{1}{8}$ in. in each dimension. Find approximately the maximum and minimum values for the actual volume. Using a calculator, find the exact maximum and minimum values, and compare your results.

34. The radius of the orbit (assumed circular) of communications satellite B is 1 mi larger than that of satellite A. Approximately how much farther does satellite B travel in completing one orbit than satellite A? Show that linear approximation gives the exact value in this case.

35. From the top of a 75-feet-high lighthouse the angle of depression θ of a buoy that indicates the location of a sunken ship is measured to be 23°15′, and the distance x from the lighthouse to the buoy is calculated from this (see the figure). Find the approximate error in x if the measurement of θ is in error by 5′. (*Note:* Express θ and $d\theta$ in radians.)

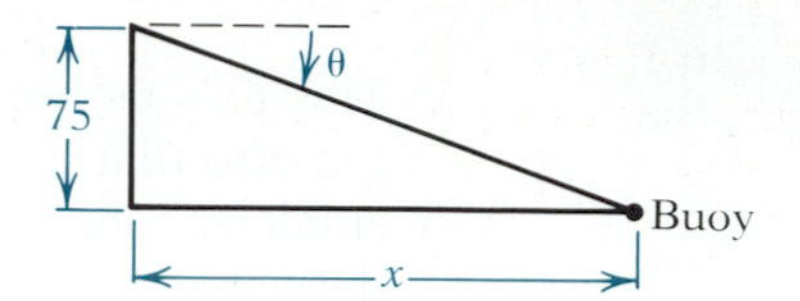

36. A projectile fired at an angle θ from the horizontal with an initial velocity of v_0 meters per second will rise to a maximum height of

$$y = \frac{v_0^2 \sin \theta}{2g}$$

where $g = 9.80$ m/sec² is the acceleration due to gravity. Approximately how much higher will a projectile fired at an angle of 31° go than one that is fired at an angle of 30°, if the initial velocity in each case is 80 m/sec?

37. The cost $C(x)$ in dollars of manufacturing x items of a certain type is given by

$$C(x) = 4000 - 15x + 0.015x^2$$

If 600 items are typically manufactured each day, find the approximate cost of manufacturing one more item. (*Note:* Although x actually takes on only integer values, for purposes of calculation we can treat it as though it is a continuous variable that takes on all positive real values.)

38. The profit function $P(x)$ for manufacturing and selling x items of a certain type is given by

$$P(x) = 100\sqrt{x} - 0.75x - 600$$

Approximately what additional profit would result from manufacturing and selling 405 items instead of 400 items?

B

39. Suppose $y = x\sqrt{4 - x^2}$ and $x = 2 \sin t$. Calculate dy in the following two ways and show that they are the same: (a) by replacing x in the first equation by its value from the second equation, and (b) by applying equation (2.18).

40. Verify the differential formula

$$d\left(\frac{u}{v}\right) = \frac{v\,du - u\,dv}{v^2}$$

where u and v are differentiable functions of x, with $v \neq 0$.

41. Find du if

$$u = \frac{z \tan z}{\sqrt{1 + z^2}}$$

42. Justify the linear approximation formula

$$f(x) \approx f(a) + f'(a)(x - a)$$

where x and a are two points in close proximity, both in an interval contained in the domain of f and such that f is differentiable at a.

43. The inner radius of a section of steel pipe for an oil pipeline is 2 ft, and the thickness of the metal is $\frac{1}{4}$ in. If the section is 32 ft long, use differentials to approximate the number of cubic feet of metal in the pipe.

44. The height of a mountain is calculated by measuring the angles of elevation α and β at points A and B, d meters apart, and using trigonometry. (See the figure.) Show that

$$h = \frac{d}{\cos \alpha - \cot \beta}$$

Suppose that B is chosen so that $\beta = 2\alpha$. If $d = 2000$ (assumed exact for purposes of this problem) and $\alpha = 20^\circ$, with a possible error of $\pm 0.5^\circ$, find the approximate possible error in the calculation of h.

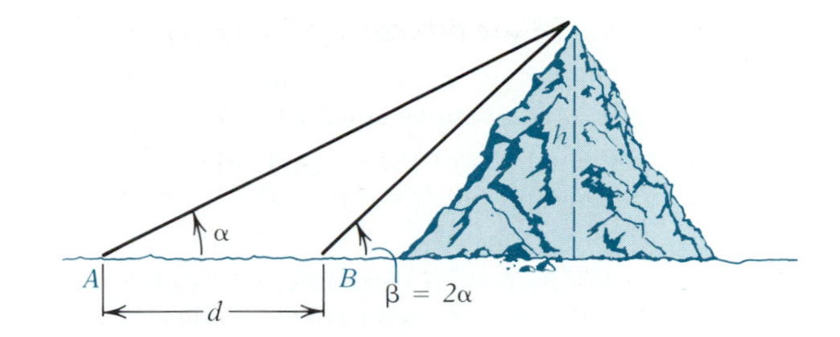

45. Use linear approximation to estimate each of the following, and find an upper bound on the error using Theorem 2.10:
(a) $\sqrt{4.2}$ (b) $(30)^{2/3}$

C

46. According to the *Verhulst* model for population growth, if $P(t)$ is the size of the population at time t (measured from some base point), the rate of change of the population, $P'(t)$, is approximated by

$$P'(t) = AP(t) - BP^2(t)$$

where A and B are constants. Using census data for the United States, it is possible to determine the values $A = 0.0384397$ and $B = 0.00013281$. If we take 1950 as time t_0, then $P(t_0) = 150.697$, measured in millions. Use the computer to approximate the population at yearly intervals from 1951 to 2000. Do this using the linear approximation *recursion formula:*

$$P(t_{n+1}) \approx P(t_n) + P'(t_n)\Delta t, \qquad n = 0, 1, \ldots, 50$$

where $\Delta t = t_{n+1} - t_n = 1$. Explain the basis for this formula. The 1980 census gave the population as 226.546 million. Compare this with your result.

2.10 SUPPLEMENTARY EXERCISES

In Exercises 1–4 find $f'(x_1)$ using Definition 2.2.

1. $f(x) = 2x^2 - 3x$; $x_1 = 3$
2. $f(x) = 2/\sqrt{x}$; $x_1 = 1$
3. $f(x) = x/(x + 1)$; $x_1 = 2$
4. $f(x) = \sqrt{2x + 3}$; $x_1 = -1$

In Exercises 5–14 find $f'(x)$.

5. $f(x) = (3x^2 - 2)^{10}$
6. $f(x) = \dfrac{2x^2 - 3x}{x^2 + 1}$
7. $f(x) = \dfrac{x}{\sqrt{1 - 2x}}$
8. $f(x) = x^2 \sin x^3$
9. $f(x) = \dfrac{\sin x}{1 + \cos x}$
10. $f(x) = \sqrt{\dfrac{1 + x}{1 - x}}$
11. $f(x) = (a^{2/3} - x^{2/3})^{3/2}$
12. $f(x) = \sec^3 x + \tan^3 x$
13. $f(x) = \dfrac{\sqrt{1 - \cos 2x}}{\cos x}$, $0 \le x < \dfrac{\pi}{2}$
14. $f(x) = \dfrac{x\sqrt{1 + x^2}}{1 - x^2}$

In Exercises 15 and 16 find the equation of the tangent line and normal line (line perpendicular to the tangent line) to the graph of f at the point $(x_1, f(x_1))$.

15. $f(x) = x^2/(x - 4)$; $x_1 = 2$
16. $f(x) = (\sqrt{x} - 2)^3$; $x_1 = 1$

In Exercises 17 and 18 the position function $s(t)$ of a particle moving on a straight line is given. Find its instantaneous velocity and acceleration when $t = t_1$.

17. $s(t) = 100\left(\dfrac{t^2 - 1}{t^2 + 1}\right)$; $t_1 = 3$
18. $s(t) = 16\sqrt[3]{t^3 - 3t + 6}$; $t_1 = 2$

In Exercises 19–22 find y' using implicit differentiation.

19. $2xy^2 + x^4 = y^3$
20. $y^3 + (x - y)^2 = 3$

21. $y \tan x - x \sec y = 1$ **22.** $\sqrt{x^2 + y^2}/(x + y) = 4$

23. Find y' and y'' if $x^{1/2} + y^{1/2} = a^{1/2}$, where $x > 0$, $y > 0$, and a is a positive constant.

24. Find the equations of the tangent line and normal line to the graph of $x^3 + x^2y^2 - y^3 = 1$ at the point $(-2, 3)$.

25. The height of a cone is three times the radius of the base. The base is supposed to have a radius of 12 cm but may be in error by as much as 1 mm. Use differentials to approximate the possible error in (a) the volume and (b) the lateral surface area.

26. Make use of differentials to approximate the area of the shaded circular segment in the figure. Leave your answer in exact form.

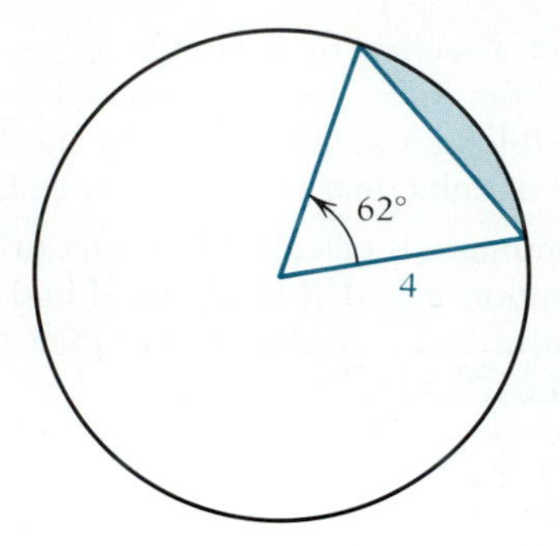

27. Verify that $y = x^{-1/2}(1 + \sin x)$ is a solution of the *differential equation*

$$x^2y'' + xy' + (x^2 - \tfrac{1}{4})y = x^{3/2}$$

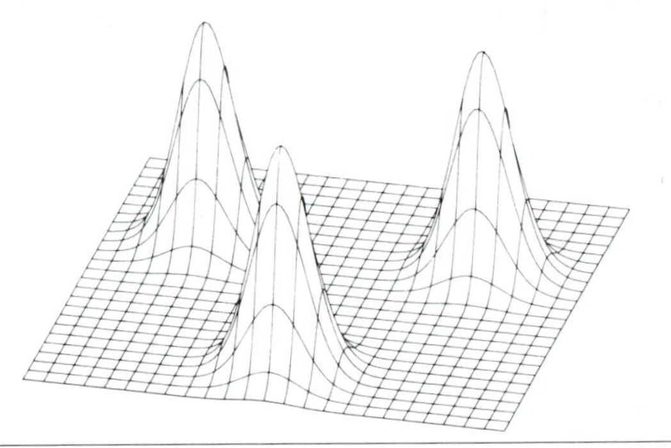

CHAPTER 3

APPLICATIONS OF THE DERIVATIVE

3.1 EXTREME VALUES

One of the most important applications of the derivative is in finding the largest or smallest value of a function. For example, we might wish to know the dimensions of a can that can hold a fixed volume and requires the least amount of material, or we might wish to know what price a theater owner should charge to obtain maximum revenue. As another example, an orchardist may be interested in knowing how many apple trees should be planted per acre to maximize the yield of apples. These problems and many others, when properly formulated, can be solved by making use of differentiation. Because graphs of functions exhibit much about their nature, we begin by considering how the derivative of a function can be used to determine certain information about its graph.

We know already that if $y = f(x)$, where f is a differentiable function, then $f'(x)$ gives the slope of the tangent line to the graph of the function. When this slope is positive, we would expect the curve to be rising as we move from left to right, and when the slope is negative, we expect the curve to be falling. Figure 3.1 illustrates this. These observations can be made more precise in the following definition.

DEFINITION 3.1 Let f be defined on an interval I. Then f is said to be **increasing** on I if, for any two numbers x_1 and x_2 in I with $x_1 < x_2$, $f(x_1) < f(x_2)$. Similarly, f is said to be **decreasing** on I if, for any two numbers x_1 and x_2 in I with $x_1 < x_2$, $f(x_1) > f(x_2)$. A function that is either increasing on I or decreasing on I is said to be a **monotone** function. ■

For example, if $f(x) = x^3$, then f is an increasing function on all of **R**, since if $x_1 < x_2$, then $x_1^3 < x_2^3$. The function g for which $g(x) = x^2$ is decreasing on $(-\infty, 0)$ and increasing on $(0, +\infty)$. The graphs of these two functions are shown in Figure 3.2.

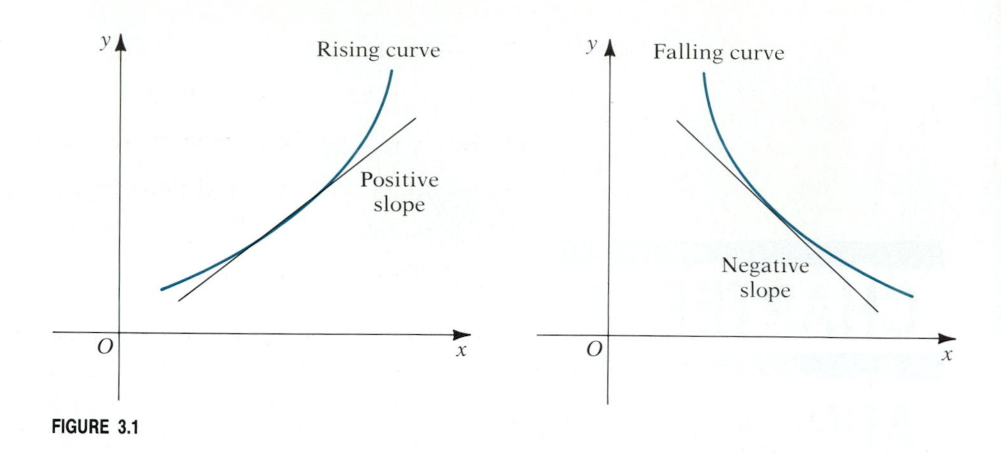

FIGURE 3.1

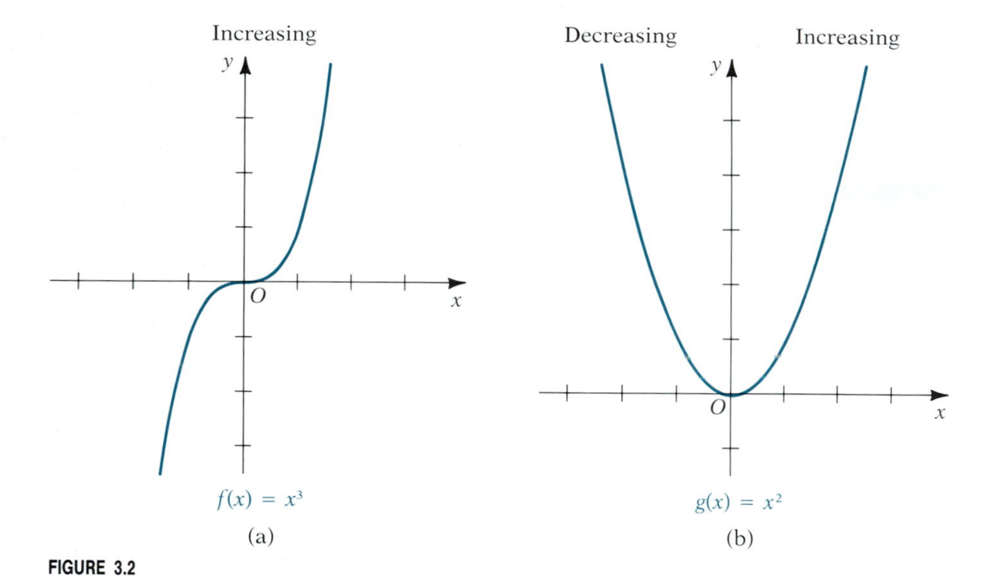

FIGURE 3.2

Our observation about the relation between derivatives and monotone functions is formalized in the following theorem.

THEOREM 3.1 If $f'(x) > 0$ for all x in an open interval I, then f is increasing on I, and if $f'(x) < 0$ for all x in I, then f is decreasing on I. ■

As we have seen, Theorem 3.1 is a plausible result, since $f'(x) > 0$ in I means that the tangent line is always sloping upward to the right, implying that the function is increasing. Similar remarks apply when $f'(x) < 0$. We will give a formal proof in Section 3.6 after we have developed more background. For now, we will proceed intuitively.

EXAMPLE 3.1 Determine the intervals on which the function f is increasing and the intervals on which it is decreasing:

$$f(x) = 2x^3 - 3x^2 - 12x + 4$$

Solution First calculate $f'(x)$:

$$f'(x) = 6x^2 - 6x - 12 = 6(x + 1)(x - 2)$$

From the sign graph for f', we conclude that

f is increasing on $(-\infty, -1)$ and on $(2, +\infty)$
f is decreasing on $(-1, 2)$

Figure 3.3 illustrates these results. ■

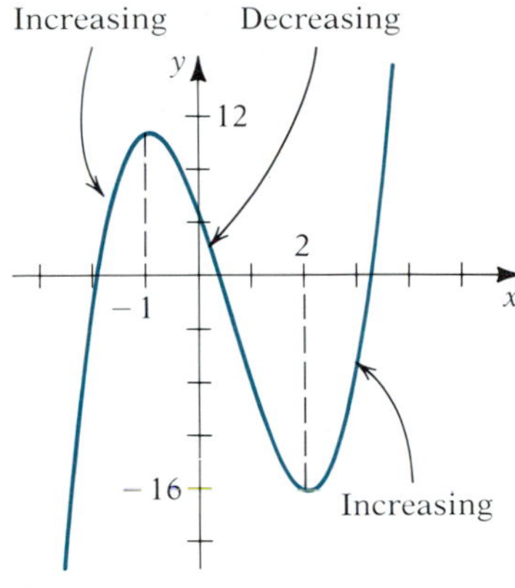

FIGURE 3.3

High points and low points on the graph of a continuous function occur when the function changes from increasing to decreasing, or vice versa. They might also occur at end points of intervals. Figure 3.4 illustrates some possibilities. The points P_2, P_4, and P_7 are all high points, in the sense that they are higher than other points in their immediate vicinity. Similarly, P_1, P_3, P_5, and P_8 are low points. If we ask for *the highest point,* it is P_7, and *the lowest point* is P_8. Note that except for P_1 and P_8, which are at the end points of the interval, the tangent lines at these high and low points appear to be either horizontal or vertical or else there is no tangent line. In particular, at P_2 there is no tangent line, since from the left the slopes approach one value and from the right they approach another. At P_3, P_4, and P_7 the tangent line is horizontal, and at P_5 it is vertical. Note also, however, that at P_6 there is a horizontal tangent line, even though P_6 is neither a high point nor a low point. The next definition makes these ideas more precise.

DEFINITION 3.2 A function f is said to have a **local maximum** at the point a of its domain if there exists a neighborhood N of a contained in the domain of f such that $f(a) \geq f(x)$ for all $x \in N$. Similarly, f has a **local minimum** at a if there exists a neighborhood N of a contained in the domain of f such that $f(a) \leq f(x)$ for all $x \in N$. ■

If f has a local maximum at a, then we say the point $(a, f(a))$ is a **local maximum point** on the graph of f, and $f(a)$ is called a **local maximum**

FIGURE 3.4

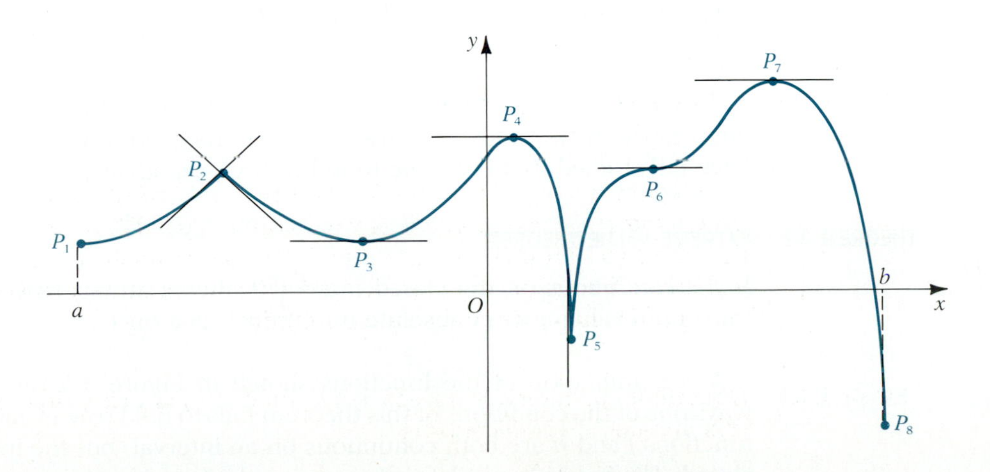

value of f. Similar terminology is used for a local minimum. The terms **maxima** and **minima** are often used to mean maximum and minimum values, respectively. When we wish to speak of maxima and minima collectively, we say either **extreme values** or **extrema.** Thus, if we ask for all local extrema, we mean the collection of all local maximum values and all local minimum values.

When the domain of f includes one or both of the end points of an interval, then the value of f at such a point may be greater or less than values at nearby points in the domain. It is convenient to refer to such values as **end point extrema** and to investigate them separately from local extrema that occur at interior points. For example, the function pictured in Figure 3.4 has end point minima at both $x = a$ and $x = b$.

Frequently we are interested in the largest or smallest value of a function over its entire domain, as opposed to local extrema. The next definition introduces terminology for such overall extrema.

DEFINITION 3.3 A function f is said to have an **absolute maximum** (or **global** maximum) at the point a of its domain if $f(a) \geq f(x)$ for all x in its domain, and it is said to have an **absolute minimum** (or **global** minimum) at a if $f(a) \leq f(x)$ for all x in its domain. ■

If f has an absolute maximum at a, then $f(a)$ is called the **absolute maximum value of *f*** (or simply the maximum value of f), and the point $(a, f(a))$ is called the absolute maximum point on the graph of f. Analogous terminology is used for the absolute minimum.

If a function has an absolute maximum value, it is the largest among all local maximum values and end point values, but as the graphs in Figure 3.5 show, not all functions have absolute maximum and minimum values. There may not even be local maxima or minima.

The function f in Figure 3.5(a), being defined only on the open interval $(0, 1)$, attains neither a maximum nor a minimum value on that interval. It comes arbitrarily close to the values 0 and 1, respectively, but does not reach either value. The function g in Figure 3.5(b) has no maximum but has end point minima at 0 and 1. The absolute minimum value is 0, attained at $x = 1$. For the function h in Figure 3.5(c), P_1 is a local minimum point and P_2 is a local maximum point, but there are no absolute maximum or minimum points. (We will see how to graph functions like h later on.)

Conditions that guarantee that a function will have an absolute maximum and minimum are given in the following theorem, which is of paramount importance in the study of extreme values. Its proof requires concepts we have not studied, but it can be found in texts on advanced calculus.

THEOREM 3.2 **EXTREME VALUE THEOREM**

If f is continuous on the closed interval I, then f attains both an absolute maximum value and an absolute minimum value on I. ■

A reexamination of the functions shown in Figure 3.5 reveals in what ways one of the conditions of this theorem fails to hold true in each case. The functions f and h are both continuous on an interval, but the interval is not closed. The function g is defined on a closed interval but g is not continuous.

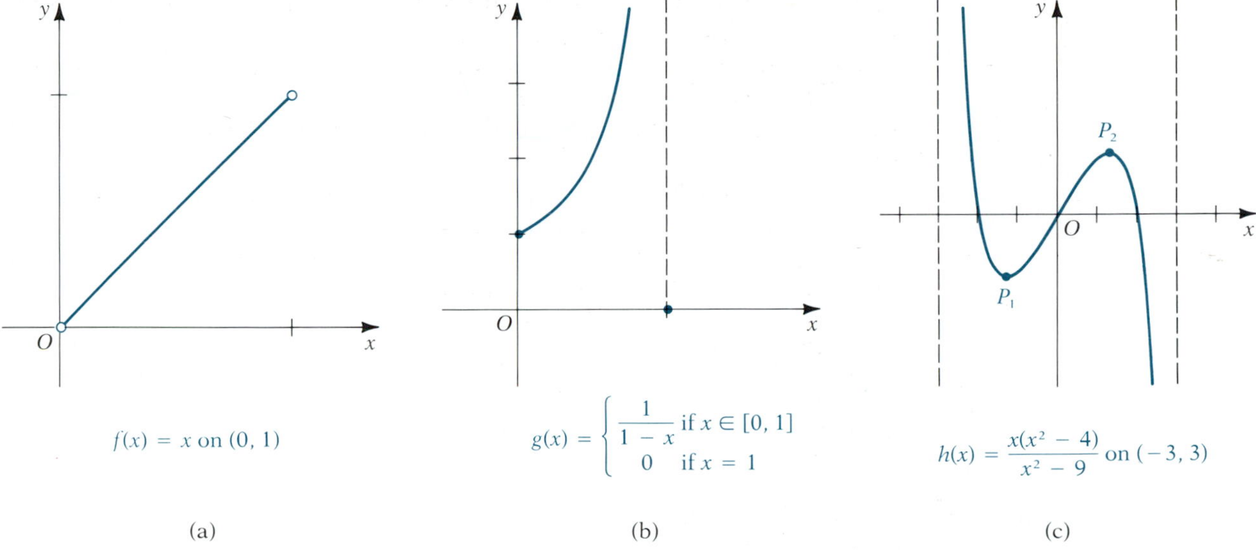

FIGURE 3.5

Suppose now that we are given a function and we want to find its maximum and minimum values. How do we go about this? If we could draw the graph with sufficient accuracy, it might suffice, but in most cases we would not be able to locate the maximum and minimum points exactly. As a matter of fact, usually we first find these extreme points and then use this information to help draw the graph, rather than the reverse. The key to locating these extreme values is given in the following theorem.

THEOREM 3.3 Let f be defined on an interval I. If f has a local maximum or minimum at an interior point a of I, and if $f'(a)$ exists, then $f'(a) = 0$.

Proof We prove the theorem for a local maximum only. By the definition of the derivative, we know that

$$f'(a) = \lim_{h \to 0} \frac{f(a + h) - f(a)}{h}$$

and since $f'(a)$ exists, we will get the same value by taking right-hand and left-hand limits. We consider these separately. First let $h \to 0$ from the right. Since $f(a)$ is a local maximum value, $f(a) \geq f(a + h)$ for all h in some deleted neighborhood of 0. Thus, $f(a + h) - f(a) \leq 0$ in this neighborhood, so that

$$\lim_{h \to 0^+} \frac{f(a + h) - f(a)}{h} \leq 0 \quad \text{Numerator negative, denominator positive}$$

(You will be asked to justify this in the exercises.) Similarly,

$$\lim_{h \to 0^-} \frac{f(a + h) - f(a)}{h} \geq 0 \quad \text{Numerator negative, denominator negative}$$

But the right-hand and left-hand limits are equal, and the only number that is both greater than or equal to 0 and less than or equal to 0 is 0 itself. Thus $f'(a) = 0$. ■

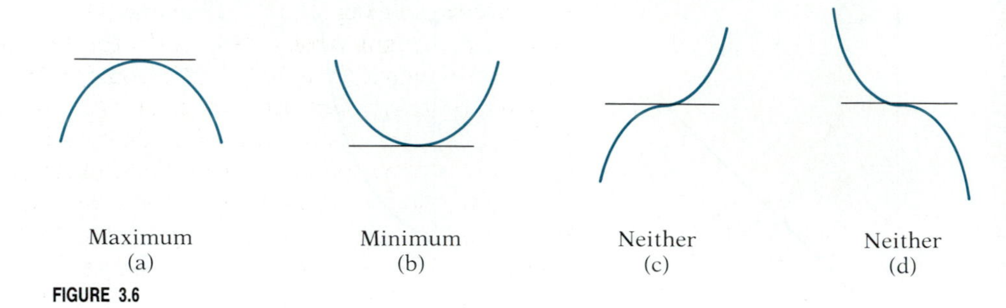

FIGURE 3.6

A consequence of this theorem is that local and absolute maximum and minimum values for a function defined on an interval can occur only at points of the interval of the following type:

1. An interior point where $f'(x) = 0$
2. An interior point where $f'(x)$ does not exist
3. An end point of the interval

Types 1 and 2 are called **critical points** of f. Critical points are not themselves maximum or minimum points but are *abscissas* of points on the graph that *may* be maximum or minimum points. So if a is a critical point, we substitute into the given function, f, to get $f(a)$, which is then a candidate for a maximum or minimum value. The end point values, if they are in the domain, must also be checked.

It should be emphasized that Theorem 3.3 gives a necessary condition but not a sufficient one to guarantee an interior local maximum or minimum. The fact that $f'(a) = 0$ does not guarantee that $(a, f(a))$ is a maximum or minimum point. The various possibilities are illustrated in Figure 3.6.

A critical point of type 2, for which f' does not exist, also may or may not be the abscissa of a maximum or minimum point. Some of the possibilities

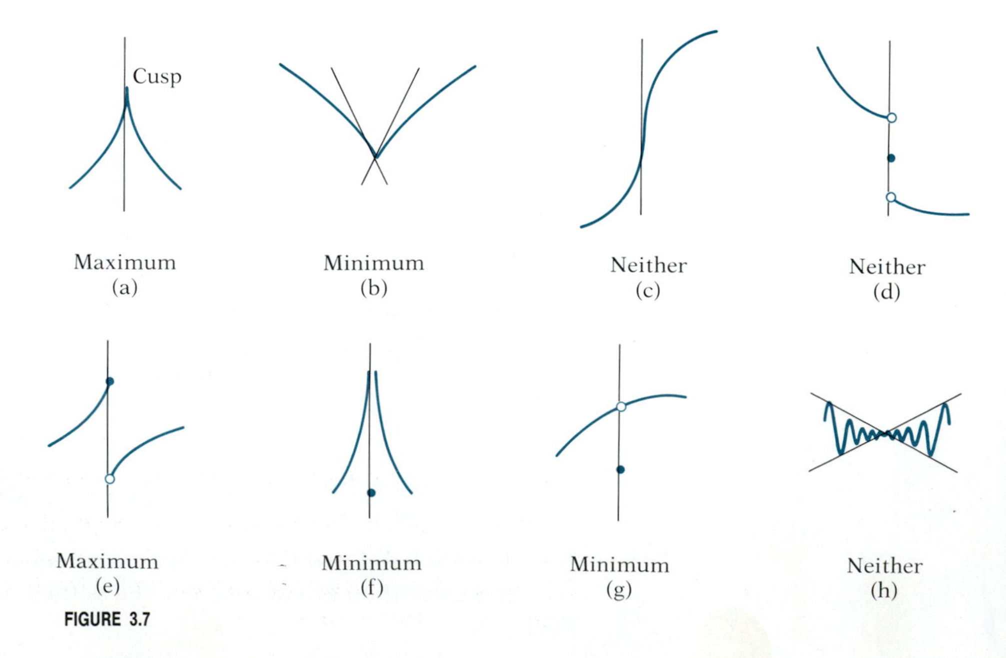

FIGURE 3.7

are illustrated in Figure 3.7. In parts (a) and (c) the tangent line is vertical and so the slope is undefined. In (b) the slopes from the left and from the right are unequal. In (d), (e), (f), and (g) the functions are discontinuous and so the derivatives cannot exist. The function in part (h) is given by $f(x) = x \sin \frac{1}{x}$ for $x \neq 0$, with $f(0)$ defined as 0. You will be asked in the exercises to show that $f'(0)$ does not exist in this latter case.

After finding the critical points, we must test in some way to determine the nature of the corresponding points on the graph. For a continuous function f there are two ways of doing this, one using f' and one using f''. The latter will be discussed in the next section. When f is discontinuous at a critical value, the possibilities are so numerous, as Figure 3.7 (d)–(g) show, that the only feasible test is to apply Definition 3.2 directly, comparing $f(a)$ with $f(x)$ for x near a.

The basis for the first derivative test is that for a continuous function, a maximum occurs when the function is increasing to the left of the critical value and decreasing to the right, and the reverse is true at a minimum. Thus, by Theorem 3.1 we have the following test.

FIRST DERIVATIVE TEST FOR LOCAL MAXIMA AND MINIMA

Let f be continuous at the critical point $x = a$. If there is a deleted neighborhood N of a in which f' exists and such that for $x \in N$:

1. $f'(x) > 0$ when $x < a$ and $f'(x) < 0$ when $x > a$, then f has a local maximum at a
2. $f'(x) < 0$ when $x < a$ and $f'(x) > 0$ when $x > a$, then f has a local minimum at a
3. $f'(x)$ does not change sign, then f has neither a local maximum nor a local minimum at a.

Figure 3.8 illustrates possibilities.

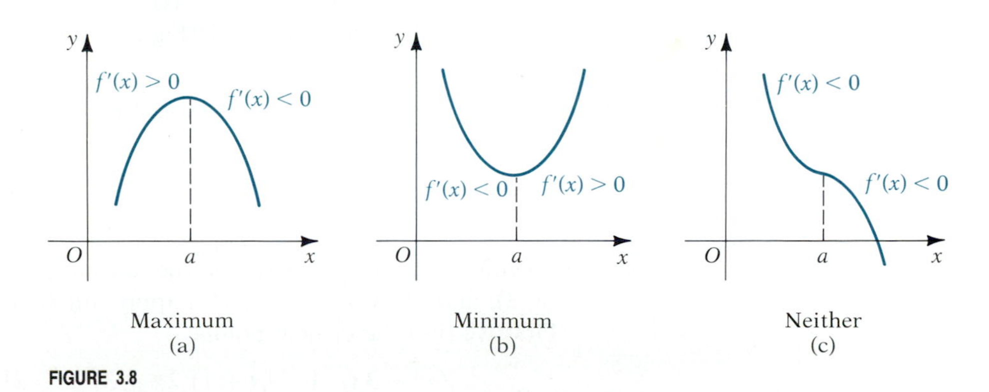

FIGURE 3.8

Remark We can state these results more succinctly:

1. f has a local maximum at a if f' changes from $+$ to $-$.
2. f has a local minimum at a if f' changes from $-$ to $+$.
3. f has neither a local maximum nor a local minimum if f' goes from $-$ to $-$ or from $+$ to $+$.

In applying this, we must understand that the signs indicated occur as we move from left to right through a.

We illustrate this test in the next three examples.

EXAMPLE 3.2 Find all local maxima and minima for the function f for which

$$f(x) = 2x^3 - 3x^2 - 12x + 4 \qquad (x \in \mathbf{R})$$

Solution To find the critical values we first calculate $f'(x)$:

$$f'(x) = 6x^2 - 6x - 12 = 6(x + 1)(x - 2)$$

Since $f'(x) = 0$ when $x = -1$ or $x = 2$, and $f'(x)$ exists everywhere, it follows that -1 and 2 are the only critical points. To test these we use the first derivative test. A useful device for doing this is the following sign graph for f':

+ − +

−1 2

Since f' changes from + to − as we pass through -1 and from − to + as we pass through 2, we conclude that f has a local maximum at -1 and a local minimum at 2. To determine these maximum and minimum values, we substitute into the original function:

$f(-1) = 11$ is a local maximum value.

$f(2) = -16$ is a local minimum value.

Alternately, we can say $(-1, 11)$ is a local maximum point on the graph of f, and $(2, 16)$ is a local minimum point on the graph.

By taking x large enough, $f(x)$ can be made arbitrarily large, and by taking negative x values large in absolute value, $f(x)$ can be made arbitrarily large in absolute value and negative. So since the domain of f is unlimited, it follows that there is no absolute maximum or minimum value of f. We showed the graph of f in Example 3.1 (Figure 3.3). ■

EXAMPLE 3.3 Find all local maxima and minima for the function

$$f(x) = \frac{x + 1}{x^2 - 3x} \qquad (x \neq 0, 3)$$

Solution The domain of f consists of the union of the three intervals $(-\infty, 0)$, $(0, 3)$, and $(3, +\infty)$. So we can apply our theory to each of these intervals. First we find the critical points:

$$f'(x) = \frac{(x^2 - 3x) \cdot 1 - (x + 1)(2x - 3)}{(x^2 - 3x)^2} = -\frac{x^2 + 2x - 3}{(x^2 - 3x)^2}$$
$$= -\frac{(x + 3)(x - 1)}{(x^2 - 3x)^2}$$

Since $f'(x) = 0$ when $x = -3$ and when $x = 1$, these are critical points. There are no values of x in the domain of f for which $f'(x)$ is undefined, so there are no other critical points. [Although $f'(x)$ is undefined at $x = 0$ and $x = 3$, these points are not in the domain of f.]

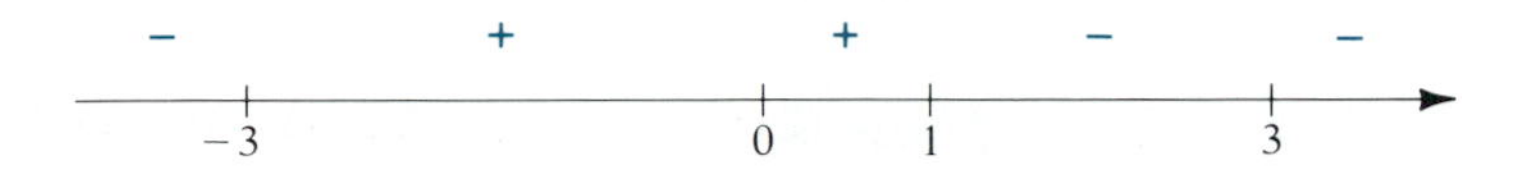

To test $x = -3$ and $x = 1$, we again look at a sign graph for f'. Notice that we show the points 0 and 3 on the line, since they are end points of intervals on which f is defined.

Since f' changes from $-$ to $+$ as we pass through -3 and from $+$ to $-$ as we pass through 1, we conclude that f has a local minimum at -3 and a local maximum at 1. The corresponding points on the graph are

$(-3, -\frac{1}{9})$, local minimum point

$(1, -1)$, local maximum point ■

EXAMPLE 3.4 Find all local and absolute extrema of the function f defined on the interval $[0, 2\pi]$ by

$$f(x) = \sin 2x + 2 \sin x$$

Solution $$f'(x) = 2 \cos 2x + 2 \cos x$$

Using the identity $\cos 2x = 2 \cos^2 x - 1$, this can be written as

$$f'(x) = 2(2 \cos^2 x + \cos x - 1) = 2(\cos x + 1)(2 \cos x - 1)$$

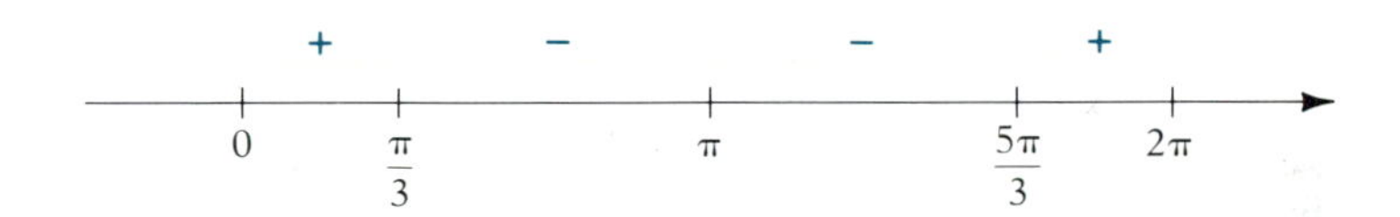

The critical points are those for which $\cos x = -1$ or $\cos x = \frac{1}{2}$—namely, $x = \pi$, $\frac{\pi}{3}$, and $\frac{5\pi}{3}$. The sign graph for f' shows that f has a local maximum at $\frac{\pi}{3}$ and a local minimum at $\frac{5\pi}{3}$. At $x = \pi$ there is neither a maximum nor a minimum. Substituting into the formula for $f(x)$, we find:

$$f\left(\frac{\pi}{3}\right) = \frac{3\sqrt{3}}{2}, \quad \text{local maximum value}$$

$$f\left(\frac{5\pi}{3}\right) = -\frac{3\sqrt{3}}{2}, \quad \text{local minimum value}$$

The end point values are $f(0) = 0$ and $f(2\pi) = 0$. Thus, the absolute maximum value of f is $3\sqrt{3}/2$ and the absolute minimum value is $-3\sqrt{3}/2$. ■

If f is continuous on the closed interval $[a, b]$ and the objective is to find absolute extrema, then the following procedure is all that is necessary:

1. Find all critical points.
2. Evaluate f at each critical point and at each end point.
3. Select the largest and smallest of the values found in step 2.

The next example illustrates this method.

EXAMPLE 3.5 Find the absolute extrema of f:

$$f(x) = (2x - 5)(x - 5)^{2/3} \quad \text{on } [3, 6]$$

Solution

$$\begin{aligned} f'(x) &= (2x - 5) \cdot \frac{2}{3}(x - 5)^{-1/3} + (x - 5)^{2/3} \cdot 2 \\ &= \frac{2}{3}(x - 5)^{-1/3}(2x - 5 + 3x - 15) \\ &= \frac{10(x - 4)}{3(x - 5)^{1/3}} \end{aligned}$$

The critical points are $x = 4$ and $x = 5$.

$$\begin{aligned} f(4) &= (3)(-1)^{2/3} = 3 \\ f(5) &= (5)(0) = 0 \\ f(3) &= (1)(-2)^{2/3} = \sqrt[3]{4} \\ f(6) &= (7)(1)^{2/3} = 7 \end{aligned}$$

Conclusions:

Absolute maximum value $= 7$ when $x = 6$.

Absolute minimum value $= 0$ when $x = 5$. ■

EXERCISE SET 3.1

A

In Exercises 1–12 find the intervals on which the function f is increasing and those on which it is decreasing.

1. $f(x) = 3 - 2x - x^2$
2. $f(x) = 2x^2 - x - 3$
3. $f(x) = x^3 - x^2 - 8x$
4. $f(x) = 2x^3 + 3x^2 - 36x + 5$
5. $f(x) = x^3 - 2x^2 - 4x + 7$
6. $f(x) = 12 + 40x + 7x^2 - 4x^3$
7. $f(x) = 3x^4 - 4x^3$
8. $f(x) = x^4 - 2x^3 - 5x^2$
9. $f(x) = \dfrac{1}{(x - 2)^2}$
10. $f(x) = \dfrac{x}{x + 3}$
11. $f(x) = \dfrac{1 - x}{3 + x^2}$
12. $f(x) = \dfrac{x + 3}{(x - 1)^2}$

In Exercises 13–24 find all local maximum and minimum points.

13. $f(x) = x^2 - 8x + 7$
14. $f(x) = 3 - x - 2x^2$
15. $f(x) = x^3 - 3x + 6$
16. $f(x) = 2x^3 + 3x^2 - 12x$
17. $f(x) = x^3 - 3x^2 - 24x + 4$
18. $f(x) = x^3 + 5x^2 - 8x + 3$
19. $f(x) = x^4 - 8x^2 + 12$
20. $f(x) = x^4 - 4x^3 - 8x^2 + 12$
21. $f(x) = x^4 - 4x^3 + 8$
22. $f(x) = 2x^5 + 5x^4 + 10$
23. $f(x) = 1 + 2(x - 3)^{2/3}$
24. $f(x) = (x^2 - 1)^{2/3}$

In Exercises 25–36 find local as well as absolute maximum and minimum values on the given interval.

25. $f(x) = x^2 - 4x + 3$ on $[0, 5]$
26. $f(x) = 5 - 3x - 2x^2$ on $[-6, 0]$
27. $f(x) = x^3 - 2x^2 - 4x + 8$ on $[-2, 3]$
28. $f(x) = 12x - x^3 - 9$ on $[-1, 3]$
29. $f(x) = x^4 - 2x^3 - 9x^2 + 27$ on $[-2, 4]$
30. $f(x) = 3x^4 - 16x^3 + 24x^2 - 21$ on $[-1, 3]$
31. $f(x) = \sin x + \cos x$ on $[0, \pi]$

32. $f(x) = \sin 2x - x$ on $[-\pi, \pi]$

33. $f(x) = \tan x - 2x + 1$ on $[-\frac{\pi}{3}, \frac{\pi}{3}]$

34. $f(x) = 2x + \sin 2x - 4 \sin x$ on $[-\pi, \pi]$

35. $f(x) = \sin 2x + 4 \cos x + x - 1$ on $[0, 2\pi]$

36. $f(x) = x \sin x + \cos x - \pi \sin x$ on $[0, 2\pi]$

Exercises 37–46 refer to earlier problems in this exercise set. In each case find the absolute maximum and absolute minimum values of f on the indicated interval. Make use of previously obtained results.

37. Exercise 15, on $[-3, 3]$

38. Exercise 16, on $[-3, 2]$

39. Exercise 17, on $[-3, 6]$

40. Exercise 18, on $[-6, 1]$

41. Exercise 19, on $[-3, 3]$

42. Exercise 20, on $[-2, 3]$

43. Exercise 21, on $[-4, 4]$

44. Exercise 22, on $[-3, 1]$

45. Exercise 23, on $[-5, 4]$

46. Exercise 24, on $[-3, 3]$

B

In Exercises 47–58 find the local and absolute extrema and state where they occur.

47. $f(x) = x\sqrt{x^2 - 1}$ on $[1, +\infty)$

48. $f(x) = (x - 1)^{2/3}(6 - x)$ on $[0, 6]$

49. $f(x) = x/(x - 1)^2$ on $(-\infty, \frac{1}{2}]$. Also discuss absolute extrema on $(-\infty, 1)$ and on $(1, +\infty)$.

50. $f(x) = x\sqrt[3]{x - 4}$ on $[0, 5]$

51. $f(x) = (x - 4)^{2/3}/x$ on $[3, 12]$

52. $f(x) = \sin^2 x - 2 \cos x + \sin x + x$ on $[\frac{\pi}{2}, \frac{3\pi}{2}]$

53. $f(x) = x^4 - 4x^3 - 18x^2 + 108x + 120$ on $[-5, 4]$

54. $f(x) = 2x^4 + 4x^3 - 9x^2 - 27x + 5$ on $[-2, 2]$

55. $f(x) = x^4 - 4x^3 + 16x - 8$ on $[-2, 3]$

56. $f(x) = x^4 - 2x^3 - 23x^2 + 24x + 64$ on $[-4, 5]$

57. $f(x) = 2.13x^3 + 3.02x^2 - 11.9x + 32.4$ on $[-3, 2]$. (Use a calculator.)

58. $f(x) = x^3 - 3.256x^2 + 1.432x + 5.875$ on $[-1.256, 2.435]$. (Use a calculator.)

59. Prove that if $f(x) \geq 0$ for all x in some deleted neighborhood of a and $\lim_{x \to a} f(x)$ exists, then $\lim_{x \to a} f(x) \geq 0$. Give an example to show that this limit may equal 0 even if $f(x) > 0$ for all x in the neighborhood.

60. Complete the proof of Theorem 3.3 by showing that if f' exists at an interior point a at which f attains a local minimum, then $f'(a) = 0$.

61. Use the definition of the derivative to show that $f'(0)$ does not exist for the function defined by

$$f(x) = \begin{cases} x \sin \frac{1}{x} & \text{if } x \neq 0 \\ 0 & \text{if } x = 0 \end{cases}$$

62. Use the definition of the derivative to show that for the function defined by

$$f(x) = \begin{cases} x^2 \sin \frac{1}{x} & \text{if } x \neq 0 \\ 0 & \text{if } x = 0 \end{cases}$$

$f'(0)$ exists, and find its value.

63. A function is **bounded** on its domain D if there exists a positive real number M such that $|f(x)| \leq M$ for all $x \in D$. Give an example of a function that is bounded on a closed interval I but does not attain a maximum value on I.

64. Is it possible for a continuous function on a closed interval to be unbounded? Explain. (See Exercise 63.)

65. Suppose both f and g are defined on an interval I and each has an absolute maximum there. Let $\max_{(I)} f$ and $\max_{(I)} g$ denote these maximum values. State everything you can about $\max_{(I)}(f + g)$ and $\max_{(I)}(fg)$.

C

A crude way to use the computer to estimate extreme values of a function on an interval is to have the computer evaluate the function at small increments across the interval; that is, pick a number n, let $h = (b - a)/n$, and evaluate $f(x)$ at $a, a + h, a + 2h, \ldots, a + nh = b$. In Exercises 66–69 use this procedure to estimate local and absolute maxima and minima and tell where each occurs, using $n = 100$.

66. $f(x) = 2x^4 - 3x^3 - 4x^2 + 5$ on $[-1, 2]$

67. $f(x) = x^4 + x^3 - 13x^2 - x + 12$ on $[-4, 3]$

68. $f(x) = x^5 - x^4 - 15x^3 + x^2 + 38x + 24$ on $[-3, 4]$

69. $f(x) = 2x^5 - 5x^4 - 11x^3 + 23x^2 + 8x - 15$ on $[-2, 3]$

70–73. Redo Exercises 66–69, finding the absolute maximum and absolute minimum by writing and running a program that selects the largest and smallest values among the finite set of numbers $\{f(a), f(a+h), f(a+2h), \ldots, f(b)\}$. Also, include in your program the x-coordinates of the points where the maximum and minimum values occur.

3.2 CONCAVITY

In the previous section we saw how the first derivative of a function gives certain critical information about its nature. In this section we investigate what can be learned from the second derivative. Consider first the curves shown in Figure 3.9. The curves in parts (a) and (b) are bending upward

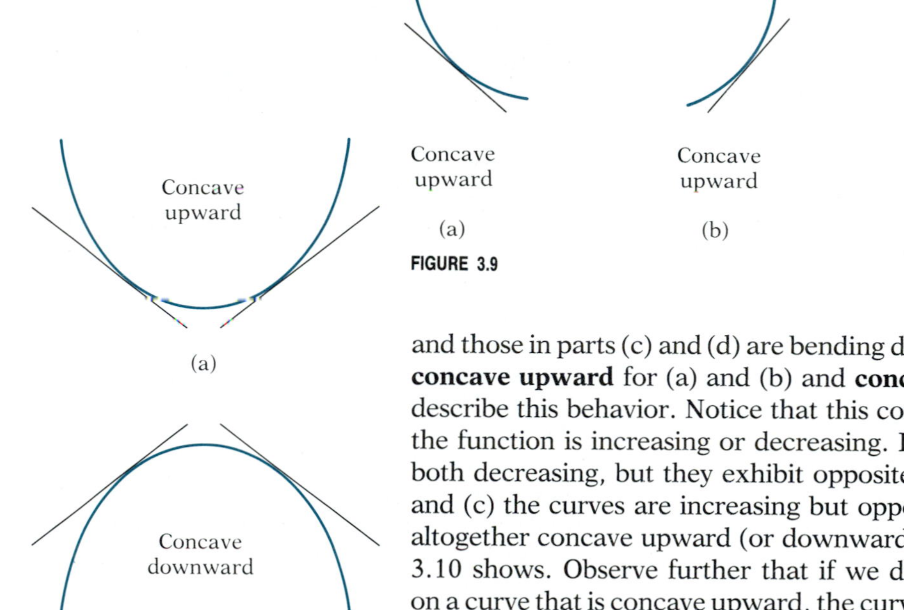

FIGURE 3.9

FIGURE 3.10

and those in parts (c) and (d) are bending downward. We use the terminology **concave upward** for (a) and (b) and **concave downward** for (c) and (d) to describe this behavior. Notice that this concavity is independent of whether the function is increasing or decreasing. In parts (a) and (d) the curves are both decreasing, but they exhibit opposite concavity. Similarly, in parts (b) and (c) the curves are increasing but opposite in concavity. A curve can be altogether concave upward (or downward) and not be monotone, as Figure 3.10 shows. Observe further that if we draw the tangent line at any point on a curve that is concave upward, the curve lies altogether above the tangent line in a deleted neighborhood of this point. For a curve that is concave downward, the curve is below the tangent line. It is this property, in fact, that is used in the following formal definition of concavity.

DEFINITION 3.4 Let f be differentiable at a and denote by L the tangent line to the graph of f drawn at $(a, f(a))$. If there exists a deleted neighborhood N of a such that for $x \in N$:

1. $f(x)$ is above L, then the graph of f is said to be **concave upward** *at a*
2. $f(x)$ is below L, then the graph of f is said to be **concave downward** *at a*.

The graph is said to be concave upward (downward) on an interval I if it is concave upward (downward) at each point in I. ■

Let us now examine more closely the slopes of tangent lines on curves that are concave upward. As Figure 3.11 shows, these slopes increase as we

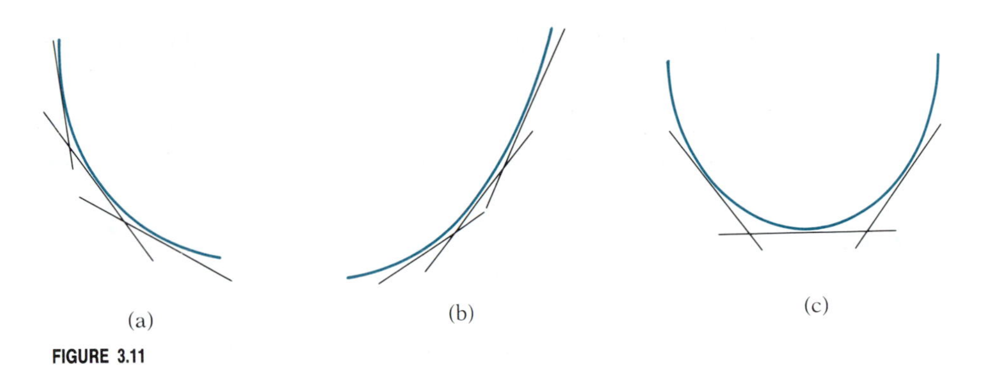

FIGURE 3.11

move from left to right. In part (a) the slopes are all negative, but as we move from left to right the absolute values of these slopes decrease, so that the slopes themselves are increasing. In part (b) the slopes are all positive and go from smaller to larger. In part (c) they go from negative, to zero, to positive. In all cases, the slopes increase. Exactly the opposite occurs for curves that are concave downward; that is, the slopes decrease as we move from left to right.

Since the slope of the tangent line to the graph of a function f is given by f', the preceding discussion implies that f' is an increasing function on a curve that is concave upward and a decreasing function on a curve that is concave downward. Since a function is increasing where its derivative is positive and decreasing where its derivative is negative, and since the derivative of f' is f'', we have the following theorem.

THEOREM 3.4 If $f''(a)$ exists, then the graph of f is concave upward at a if $f''(a) > 0$ and concave downward at a if $f''(a) < 0$. ■

Although we have given a geometric argument that suggests the validity of this theorem, its formal proof based on Definition 3.4 requires more analysis and will be given in Section 3.6. Meanwhile, we will use the results.

EXAMPLE 3.6 Determine where the graph of the function f defined by $f(x) = x^3 - 6x^2 + 9x + 2$ is concave upward and where it is concave downward.

Solution We need to determine where $f''(x) > 0$ and where $f''(x) < 0$. So we first calculate $f''(x)$:

$$f'(x) = 3x^2 - 12x + 9$$
$$f''(x) = 6x - 12 = 6(x - 2)$$

Thus $f''(x) > 0$ for all $x > 2$ and $f''(x) < 0$ for all $x < 2$. By Theorem 3.4 we conclude that the graph of f is concave upward on $(2, +\infty)$ and concave downward on $(-\infty, 2)$. ■

A point on the graph of a continuous function f at which the concavity changes is called an **inflection point;** that is, the point $(a, f(a))$ is an inflection point on the graph of f if there is a deleted neighborhood N of a

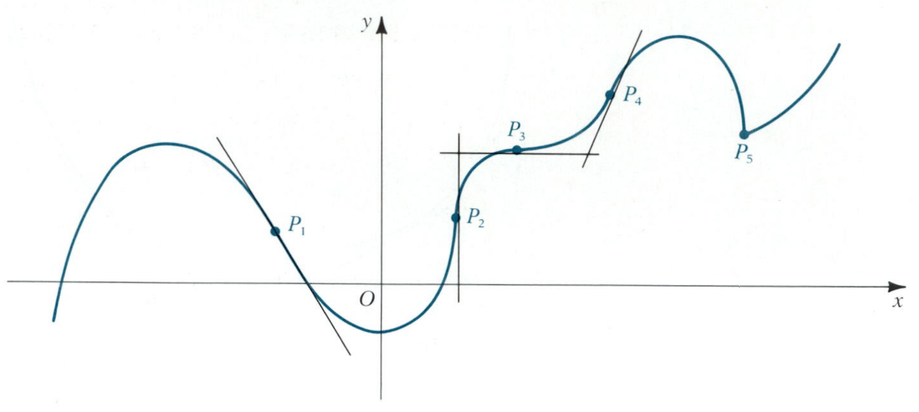

FIGURE 3.12

such that for x in N either of the following holds true:

1. The graph is concave downward for $x < a$ and concave upward for $x > a$.
2. The graph is concave upward for $x < a$ and concave downward for $x > a$.

In Example 3.6 the graph has an inflection point when $x = 2$, and the inflection point is therefore $(2, f(2)) = (2, 4)$. Figure 3.12 illustrates a graph with several inflection points. Observe that there is a change in concavity at each of the points P_1 through P_5, and so each of these is an inflection point. The points P_1 and P_4 are the most common types. At each of these both f' and f'' exist, with $f' \neq 0$ and $f'' = 0$. The points P_2, P_3, and P_5 all occur at critical points for x. At P_2 and P_5, f' is not defined, and at P_3, $f' = 0$. The tangent line is vertical at P_2 (and so its slope is not defined), and no tangent line exists at P_5. At P_3 both f' and f'' equal 0. This type of inflection point always has an appearance similar to one of those in Figure 3.13.

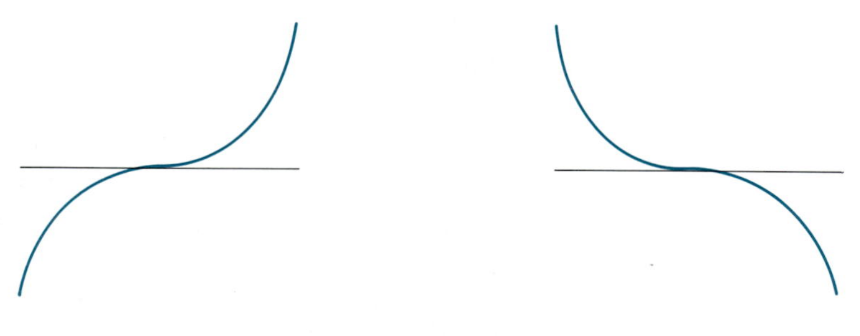

Inflection points at which both $f' = 0$ and $f'' = 0$

FIGURE 3.13

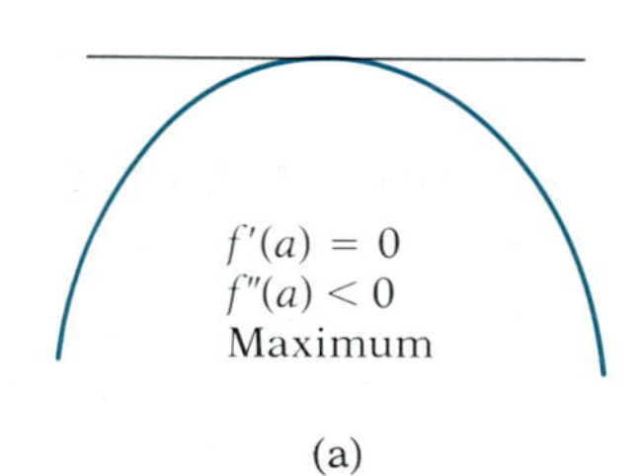

(a)

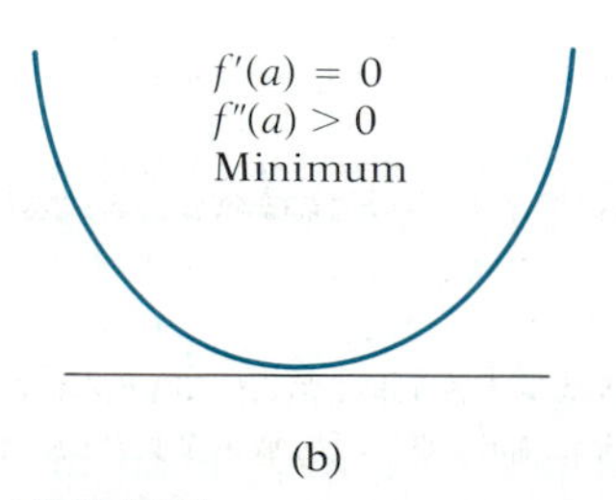

(b)

FIGURE 3.14

We can now see a relationship between concavity and maximum and minimum points. Suppose that $x = a$ is a critical point where $f'(a) = 0$ and $f''(a) < 0$. Then we know two things: the tangent line is horizontal and the graph is concave downward at $(a, f(a))$, as shown in Figure 3.14(a). If $f'(a) = 0$ and $f''(a) > 0$, then the situation is as pictured in Figure 3.14(b). We therefore have the following test for local maxima and minima.

SECOND DERIVATIVE TEST FOR MAXIMA AND MINIMA

Let $x = a$ be a critical point of f for which $f'(a) = 0$ and $f''(a)$ exists.

1. If $f''(a) > 0$, the point $(a, f(a))$ is a local minimum point.
2. If $f''(a) < 0$, the point $(a, f(a))$ is a local maximum point.

Comment If $f''(a) = 0$, this test is inconclusive. The point $(a, f(a))$ may be an inflection point of the type shown in Figure 3.13, or it may be a local maximum or minimum. Although a test involving higher derivatives can be derived, it is probably best to use the first derivative test in this case.

EXAMPLE 3.7 Use the second derivative test to find local maxima and minima for the function given by

$$f(x) = x^3 + 3x^2 - 24x - 20$$

Solution

$$\begin{aligned} f'(x) &= 3x^2 + 6x - 24 \\ &= 3(x^2 + 2x - 8) \\ &= 3(x + 4)(x - 2) \end{aligned}$$

The critical points are $x = -4$ and $x = 2$.

$$\begin{aligned} f''(x) &= 6x + 6 = 6(x + 1) \\ f''(-4) &= -18 < 0 \Rightarrow \text{local maximum at } -4 \\ f''(2) &= 18 > 0 \Rightarrow \text{local minimum at } 2 \end{aligned}$$

We substitute -4 and 2, respectively, into the original function to find the points on the graph:

$(-4, 60)$, local maximum point
$(2, -48)$, local minimum point ■

Whether to use the first derivative test or the second derivative test depends on several factors. First, we have no choice if a is a critical point for which either $f''(a) = 0$ or $f'(a)$ is not defined; the second derivative test will not work. When the function is a polynomial, the second derivative test often is simpler to apply. For other functions, it depends on the difficulty of calculating f''. If you need f'' anyway (for example, in determining concavity and inflection points), then use the second derivative test. If the only objective is to determine maxima and minima, and if it appears that the calculation of f'' would be quite laborious, then it is probably easier to use the first derivative test.

From the first two derivatives of a function we now know how to determine where it is increasing and where decreasing, the local maximum and minimum points on its graph, the intervals where it is concave upward and where concave downward, and the points of inflection. This is a wealth of information and is especially helpful in drawing the graph. The next two examples illustrate this.

EXAMPLE 3.8 For the function f, given by

$$f(x) = 2x^3 - 3x^2 - 12x + 8 \qquad (x \in \mathbf{R})$$

find the intervals on which f is increasing and on which f is decreasing, the maximum and minimum points, the intervals where f is concave upward and where concave downward, and the points of inflection. Use the information obtained to draw the graph of f.

Solution

$$f(x) = 2x^3 - 3x^2 - 12x + 8$$
$$f'(x) = 6x^2 - 6x - 12 = 6(x^2 - x - 2) = 6(x - 2)(x + 1)$$
$$f''(x) = 12x - 6 = 6(2x - 1)$$

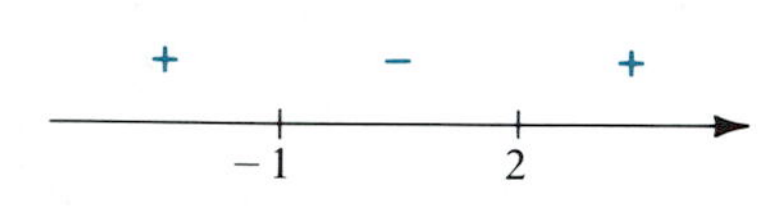

The critical points are $x = 2$ and $x = -1$, where $f'(x) = 0$. From the sign graph for f', we see that f is increasing on $(-\infty, -1)$ and on $(2, \infty)$, and f is decreasing on $(-1, 2)$.

Since $f''(-1) = -18$ and $f''(2) = 18$, we see by the second derivative test that f has a local maximum at -1 and a local minimum at 2. We calculate $f(-1) = 15$ and $f(2) = -12$. Thus $(-1, 15)$ is a local maximum point and $(2, -12)$ is a local minimum point.

For concavity, we consider $f''(x)$. When $x > \frac{1}{2}$, $f''(x) > 0$ and when $x < \frac{1}{2}$, $f''(x) < 0$. So the graph is concave upward on $(\frac{1}{2}, \infty)$ and concave downward on $(-\infty, \frac{1}{2})$. Furthermore, the point $(\frac{1}{2}, f(\frac{1}{2})) = (\frac{1}{2}, \frac{3}{2})$ is a point of inflection.

This information enables us to draw the graph of f with reasonable accuracy (Figure 3.15). Plotting a few additional points increases the accuracy. ■

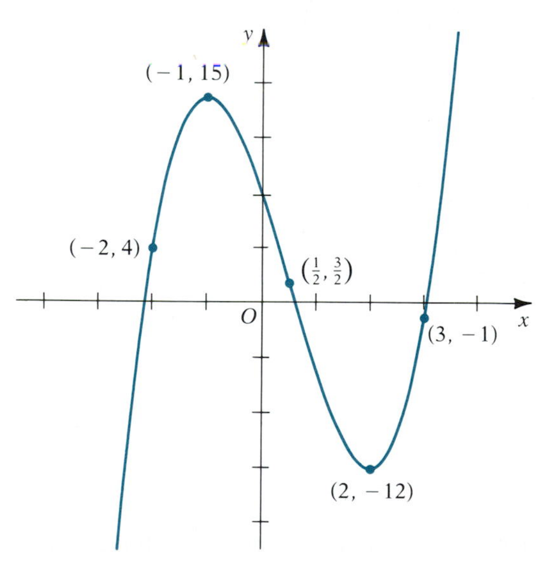

FIGURE 3.15

EXAMPLE 3.9 Draw the graph of the function g given by $g(x) = x^4 - 4x^3 + 16$ making use of the information obtained from g' and g''.

Solution

$$g(x) = x^4 - 4x^3 + 16$$
$$g'(x) = 4x^3 - 12x^2 = 4x^2(x - 3)$$
$$g''(x) = 12x^2 - 24x = 12x(x - 2)$$

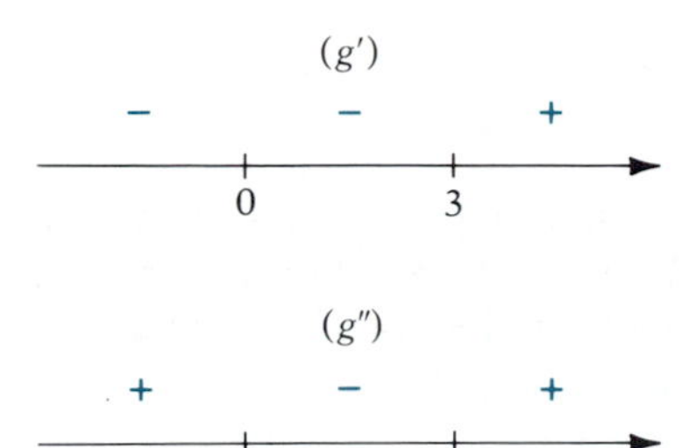

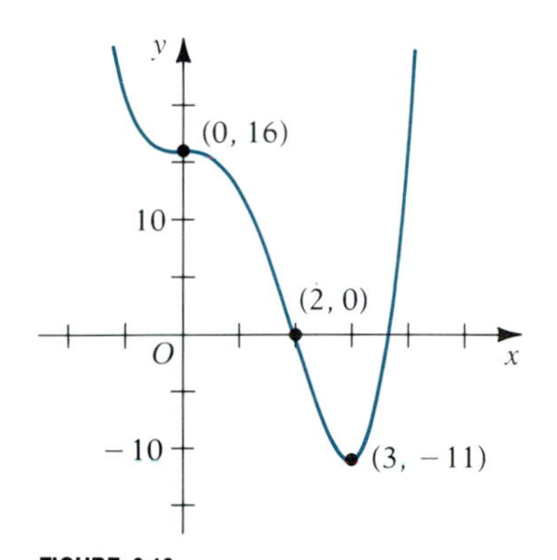

FIGURE 3.16

The critical points are $x = 0$ and $x = 3$. Since $g''(0) = 0$ and $g''(3) > 0$, we see that g has a minimum at $x = 3$, but we must examine $x = 0$ by the first derivative test. The sign graph of g' tells us that g does not have a maximum or a minimum at $x = 0$, but is decreasing both to the left and to the right of $x = 0$.

Next look at the sign graph for g''. We see that g has points of inflection at $x = 0$ and $x = 2$.

We calculate $g(x)$ at the points in question and summarize our findings:

g is increasing on $(3, \infty)$.

g is decreasing on $(-\infty, 0)$ and on $(0, 3)$.

g is concave upward on $(-\infty, 0)$ and on $(2, \infty)$.

g is concave downward on $(0, 2)$.

$(3, -11)$ is a minimum point.

$(0, 16)$ is a point of inflection where the tangent line is horizontal.

$(2, 0)$ is a point of inflection.

The graph is shown in Figure 3.16. ■

EXERCISE SET 3.2

A

In Exercises 1–10 determine the intervals on which the graph of the given function is concave upward and the intervals on which it is concave downward.

1. $f(x) = 2x^2 - 3x + 4$ **2.** $g(x) = 5 - 6x - x^2$

3. $f(x) = ax^2 + bx + c$ **4.** $f(x) = x^3 - 3x^2 + 4x$

5. $h(x) = x^3 + 6x^2 - 5x + 1$

6. $F(x) = 2x^3 - 3x^2 - x + 5$

7. $G(x) = x^4 - 2x^3$

8. $g(x) = x^4 + 2x^3 - 12x^2 + 10$

9. $f(x) = x^4 - 6x^2 + 8$

10. $F(x) = x^4 - 3x^3 - 15x^2 - 12x + 9$

In Exercises 11–20 find all local maximum and minimum points, using the second derivative test when applicable and the first derivative test otherwise.

11. $f(x) = 2x^3 - 6x + 4$

12. $g(x) = x^3 + 2x^2 - 4x + 5$

13. $h(x) = 2x^3 - 3x^2 + 7$

14. $\phi(x) = 2 - 3x + 5x^2 - x^3$

15. $F(x) = x^4 + 4x^3 - 8x^2 + 4$

16. $G(x) = 3x^5 - 25x^3 + 60x$

17. $f(x) = x + \frac{1}{x}$

18. $f(x) = 1/x^2 + 2/x + 3$

19. $g(x) = 3x^4 - 8x^3 + 6x^2$

20. $h(x) = 3x^5 - 20x^3 + 7$

In Exercises 21–40 determine intervals where the function is increasing and where decreasing, locate local maxima and minima, discuss concavity, and find all points of inflection. Use the information obtained to draw the graph of $y = f(x)$.

21. $f(x) = x^3 - 3x^2$

22. $f(x) = x^3 - 12x$

23. $f(x) = x^3 + 3x^2 - 9x - 15$

24. $f(x) = x^3 - x^2 - x + 1$

25. $f(x) = 4 + 3x - x^3$

26. $f(x) = 2x^3 + 4x^2 - 8x + 2$

27. $f(x) = x^4 - 4x^3 + 5$

28. $f(x) = x^3 - 6x^2 + 12x - 3$

29. $f(x) = x^4 - 6x^2 + 10$

30. $f(x) = 3 + 4x - x^4$

31. $f(x) = \frac{x^5}{5} - \frac{10x^3}{3} + 9x$

32. $f(x) = (x^2 - 4)(x - 2)$

33. $f(x) = (x - 3)^2(x + 3)$

34. $f(x) = (x - 2)^2(2x + 5)$

35. $f(x) = 2 \sin \frac{x}{2} + 1$ on $[0, 4\pi]$

36. $f(x) = \sin x + \cos x$ on $[0, 2\pi]$

37. $f(x) = x + \sin x$ on $[0, 4\pi]$

38. $f(x) = \cos x - \sqrt{3} \sin x$ on $[-\frac{\pi}{3}, \frac{5\pi}{3}]$

39. $f(x) = \cos 2x - 4 \cos x$ on $[-\pi, \pi]$

40. $f(x) = \sin 2x - 2 \sin^2 x$ on $[-\pi, \pi]$

B

In Exercises 41–46 find all local extrema, using the second derivative test when applicable and the first derivative test otherwise.

41. $f(x) = 3 + 4x - 2x^2 - \frac{x^3}{3} + \frac{x^4}{4}$

42. $f(x) = \frac{x^2 - x + 2}{x + 1}$

43. $f(x) = \frac{x^2}{2} - \frac{1}{x} + 2$

44. $f(x) = x^4 - 6x^2 + 8x - 3$

45. $f(x) = 3x + 2(4 - x)^{3/2}$

46. $f(x) = x^{2/3}(2x - 5)$

In Exercises 47–55 sketch the graph, using all of the information that can be obtained from f' and f''.

47. $f(x) = \frac{1}{16}(3x^4 - 8x^3 - 24x^2 + 96x)$

48. $f(x) = x^4/2 - x^3/3 - 4x^2 + 4x + 7$

49. $f(x) = x^5 - 5x^3 + 10x - 3$

50. $f(x) = 2 - 8x + 6x^2 - x^4$

51. $f(x) = x^4 - 4x^3 + 16x - 4$

52. $f(x) = x^4 - 2x^3 + 2x + 3$

53. $f(x) = \frac{3 \cos x}{2 - \sin x}$ on $[-\pi, \pi]$

54. $f(x) = \frac{\tan x}{2 \sec x - 1}$ on $[0, 2\pi]$

55. $f(x) = \cos^4 x - 4 \cos x + 3$ on $[-\frac{3\pi}{2}, \frac{3\pi}{2}]$

3.3 EXTENSIONS OF THE LIMIT CONCEPT; ASYMPTOTES

The nature of the graph of $y = f(x)$ for certain functions f can be further understood by investigating what are known as **infinite limits** and **limits at infinity.** In our considerations of the limit operation $\lim_{x \to a} f(x) = L$ up to this point we have assumed that both a and L are finite. In a sense that we will make precise later, we now are going to allow these to become infinite. Before giving the precise definition, however, let us consider the example

$$f(x) = \frac{1}{x - 2}$$

If we allow x to come closer and closer to 2, we see that the denominator comes arbitrarily close to 0, so that $f(x)$ has no finite limit. But a closer examination reveals some useful information about the graph of $y = f(x)$. Consider the following table of values:

x	1	1.5	1.9	1.99	1.999	3	2.5	2.1	2.01	2.001
y	−1	−2	−10	−100	−1000	1	2	10	100	1000

It appears that as x approaches 2 from the left, y becomes arbitrarily large in absolute value but is negative, and as x approaches 2 from the right, y becomes arbitrarily large through positive values. We will denote this situation symbolically by writing

$$\lim_{x \to 2^-} f(x) = -\infty \quad \text{and} \quad \lim_{x \to 2^+} f(x) = +\infty$$

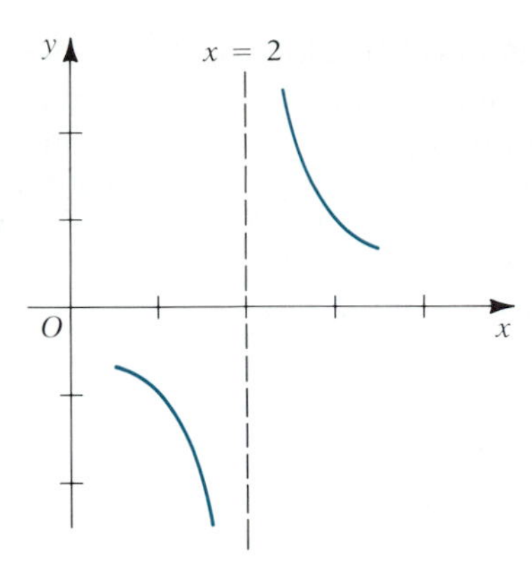

FIGURE 3.17

Graphically, what we have found is that this function behaves as shown in Figure 3.17 in the vicinity of $x = 2$. The line $x = 2$ that the curve approaches is called an **asymptote,** and the curve is said to be **asymptotic** to this line.

In this example the left-hand and right-hand limits are both infinite but are opposite in sign. It may happen for some functions that the one-sided limits are both $+\infty$ or both $-\infty$, in which case we write

$$\lim_{x\to a} f(x) = +\infty \quad \text{or} \quad \lim_{x\to a} f(x) = -\infty$$

For example, we can see readily that

$$\lim_{x\to 0} \frac{1}{x^2} = +\infty$$

for regardless of the sign of x, $1/x^2$ is unbounded positively in any deleted neighborhood of 0.

The following definitions summarize these ideas.

DEFINITION 3.5 The symbol

$$\lim_{x\to a} f(x) = +\infty$$

means that given any positive number M, however large, there exists a deleted neighborhood of a throughout which $f(x) > M$. Similarly, the symbol

$$\lim_{x\to a} f(x) = -\infty$$

means that given any positive number M, however large, there exists a deleted neighborhood of a throughout which $f(x) < -M$. ■

The one-sided infinite limits

$$\lim_{x\to a^+} f(x) = +\infty \quad \lim_{x\to a^-} f(x) = +\infty \quad \lim_{x\to a^+} f(x) = -\infty \quad \lim_{x\to a^-} f(x) = -\infty$$

are defined similarly, where x is restricted to lie on one side of a.

Although it is convenient to use such phrases as "$f(x)$ becomes infinite" or "$f(x)$ approaches infinity," it should be emphasized that this is just a short way of describing what is stated in Definition 3.5. It is a more accurate description to say "$f(x)$ increases beyond all bound." It is important to understand that $+\infty$ and $-\infty$ do not stand for real numbers.

DEFINITION 3.6 The line $x = a$ is called a **vertical asymptote** to the graph of f if, as x approaches a from either side, $f(x)$ becomes positively or negatively infinite—that is, provided

$$\lim_{x\to a^-} f(x) = \pm\infty \quad \text{or} \quad \lim_{x\to a^+} f(x) = \pm\infty$$

■

EXAMPLE 3.10 Find all vertical asymptotes to the graph of f, where

$$f(x) = \frac{x + 4}{x^2 - 2x - 3}$$

Solution On factoring the denominator we get

$$f(x) = \frac{x + 4}{(x + 1)(x - 3)}$$

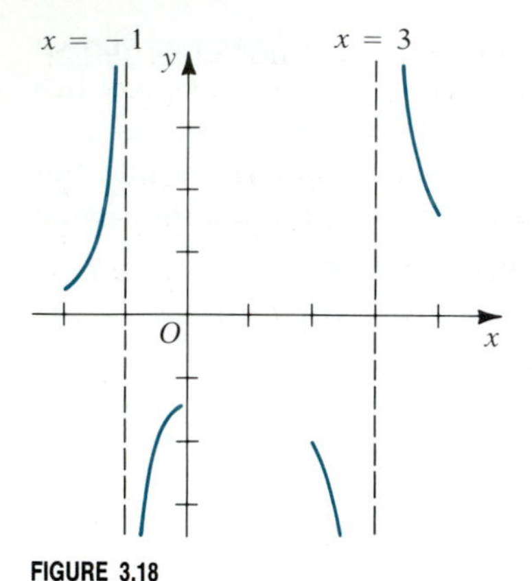

FIGURE 3.18

When x is close to -1 or to 3, the denominator will be close to 0, and we would expect $f(x)$ to be large in absolute value. In particular, when x is slightly greater than 3, all factors are positive, and $x - 3$ is close to 0. Thus when x is close enough to 3 from the right, the value of $f(x)$ can be made as large as we please. So we have

$$\lim_{x \to 3^+} f(x) = +\infty$$

Similar arguments can be used to see that

$$\lim_{x \to 3^-} f(x) = -\infty \qquad \lim_{x \to -1^+} f(x) = -\infty \qquad \lim_{x \to -1^-} f(x) = +\infty$$

It follows that the lines $x = -1$ and $x = 3$ are vertical asymptotes. In fact, in the vicinity of these lines the graph behaves as shown in Figure 3.18. ■

We can generalize the ideas of the preceding example to conclude that if f is a rational function

$$f(x) = \frac{P(x)}{Q(x)}$$

where P and Q are polynomials that have no linear factor in common, then the graph of f has a vertical asymptote at each number a for which $Q(a) = 0$ [since $P(a) \neq 0$, by hypothesis]—that is, at the **zeros** of the denominator. So the procedure for finding them is to factor the denominator and set each linear factor equal to 0. In Example 3.10, then, once we had written $f(x)$ in the form

$$f(x) = \frac{x + 4}{(x + 1)(x - 3)}$$

we could have said immediately that $x = -1$ and $x = 3$ are vertical asymptotes.

A Word of Caution For rational functions, setting the denominator equal to 0 yields vertical asymptotes *provided* the numerator and denominator have no linear factor in common. If a common factor does exist, it should first be divided out. For example, if

$$f(x) = \frac{x^2 - 4}{x - 2}$$

$x = 2$ is *not* a vertical asymptote, for we can write

$$f(x) = \frac{(x + 2)(x - 2)}{x - 2} = x + 2 \qquad (x \neq 2)$$

so that no value of x causes $f(x)$ to become infinite. On the other hand, if

$$g(x) = \frac{x^2 - 4}{(x - 2)^2}$$

then, on factoring and reducing, we get

$$g(x) = \frac{(x + 2)(x - 2)}{(x - 2)^2} = \frac{x + 2}{x - 2}$$

and so $x = 2$ *is* a vertical asymptote. The point is that you cannot tell whether a vertical asymptote exists at a zero of the denominator until the fraction is in lowest terms.

When we write $y = f(x)$, we can say that if y becomes infinite when x approaches a number a (from at least one side), then $x = a$ is a vertical asymptote. Now we are going to consider whether allowing x to become infinite causes y to approach some number b. If so, we will see that $y = b$ is a **horizontal asymptote.** To illustrate, consider

$$f(x) = \frac{2x^2 - 1}{x^2 + 3}$$

We want to allow x to become arbitrarily large and ask whether a limit exists. We write this as

$$\lim_{x \to +\infty} \frac{2x^2 - 1}{x^2 + 3}$$

We can see whether this limit exists by dividing the numerator and denominator by x^2 (which is permissible, since we are letting x be large and so can restrict it to be nonzero). We get

$$\lim_{x \to +\infty} \frac{2x^2 - 1}{x^2 + 3} = \lim_{x \to +\infty} \frac{2 - \dfrac{1}{x^2}}{1 + \dfrac{3}{x^2}}$$

The properties of limits stated in Theorem 1.2 continue to hold true as $x \to \pm\infty$ (called limits at infinity) provided each limit is finite. In particular, we can take the limit of a quotient by taking the limit of the numerator and denominator individually, provided the latter limit is nonzero. Using the limit properties also for sums and differences, we get

$$\lim_{x \to +\infty} \frac{2 - \dfrac{1}{x^2}}{1 + \dfrac{3}{x^2}} = \frac{2 - \lim\limits_{x \to +\infty} \dfrac{1}{x^2}}{1 + \lim\limits_{x \to +\infty} \dfrac{3}{x^2}}$$

Now it should be intuitively evident that the two limits on the right both equal 0, since when x gets larger and larger, both $1/x^2$ and $3/x^2$ get arbitrarily close to 0. (See Theorem 3.5.) Thus, we have

$$\lim_{x \to +\infty} \frac{2x^2 - 1}{x^2 + 3} = 2$$

So for large values of x, y is close to 2, and in fact, the graph comes arbitrarily close to the line $y = 2$. This line is a horizontal asymptote.

Following the same procedure, we could also show that

$$\lim_{x \to -\infty} \frac{2x^2 - 1}{x^2 + 3} = 2$$

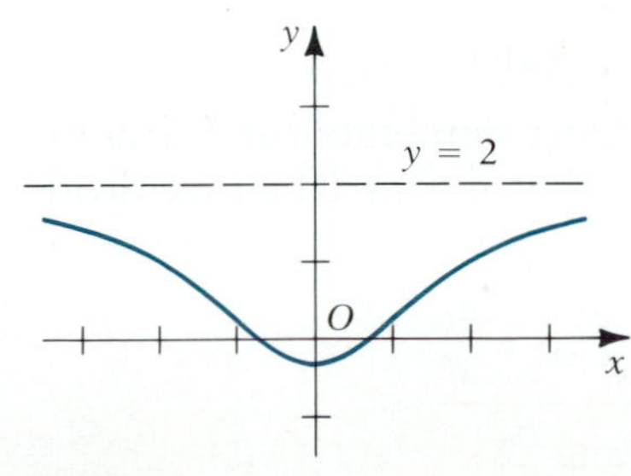

FIGURE 3.19

which indicates that the graph is asymptotic to $y = 2$ on the left side as well as on the right. The graph of this function is given in Figure 3.19, illustrating the horizontal asymptote.

The following definitions and theorems relate to the preceding discussion.

DEFINITION 3.7 Assume the domain of f includes an interval of the form (a, ∞). Then the symbol

$$\lim_{x \to +\infty} f(x) = L$$

means that $f(x)$ can be made arbitrarily close to L by taking x sufficiently large. More precisely, given any positive number ε, there exists a positive number N such that whenever $x > N$,

$$|f(x) - L| < \varepsilon$$

If the domain of f includes an interval of the form $(-\infty, b)$, then the symbol

$$\lim_{x \to -\infty} f(x) = L$$

is defined in an analogous fashion. ■

DEFINITION 3.8 The line $y = b$ is called a **horizontal asymptote** to the graph of f provided

$$\lim_{x \to +\infty} f(x) = b \quad \text{or} \quad \lim_{x \to -\infty} f(x) = b$$ ■

THEOREM 3.5 For any positive number p,

$$\lim_{x \to +\infty} \frac{1}{x^p} = 0$$

and if x^p is defined for all $x < 0$, then also

$$\lim_{x \to -\infty} \frac{1}{x^p} = 0$$ ■

You will be asked to give a proof of this theorem in Exercise 61 of the next exercise set.

THEOREM 3.6 Let f be the rational function

$$f(x) = \frac{a_m x^m + a_{m-1} x^{m-1} + \cdots + a_1 x + a_0}{b_n x^n + b_{n-1} x^{n-1} + \cdots + b_1 x + b_0} \qquad (a_m, b_n \neq 0)$$

Then the graph of f has the horizontal asymptote $y = b$, where

$$\begin{cases} b = 0 & \text{if } n > m \\ b = \dfrac{a_m}{b_n} & \text{if } n = m \end{cases}$$

If $m > n$, there is no horizontal asymptote.

Proof The proof makes use of the procedure of dividing numerator and denominator by the highest power of x. Suppose first that $n > m$. Then we divide by x^n to get

$$\lim_{x \to \pm\infty} f(x) = \lim_{x \to \pm\infty} \frac{\dfrac{a_m}{x^{n-m}} + \dfrac{a_{m-1}}{x^{n-m+1}} + \cdots + \dfrac{a_1}{x^{n-1}} + \dfrac{a_0}{x^n}}{b_n + \dfrac{b_{n-1}}{x} + \cdots + \dfrac{b_1}{x^{n-1}} + \dfrac{b_0}{x^n}}$$

By Theorem 3.5 all terms that have x in the denominator approach 0, so we get

$$\lim_{x \to \pm\infty} f(x) = \frac{0}{b_n} = 0$$

Thus $y = 0$ is a horizontal asymptote.

If $n = m$, then division by x^n yields

$$\lim_{x \to \pm\infty} \frac{a_m + \dfrac{a_{m-1}}{x} + \cdots + \dfrac{a_1}{x^{n-1}} + \dfrac{a_0}{x^n}}{b_n + \dfrac{b_{n-1}}{x} + \cdots + \dfrac{b_1}{x^{n-1}} + \dfrac{b_0}{x^n}} = \frac{a_m}{b_n}$$

If $m > n$, we divide by x^m to get

$$\lim_{x \to \pm\infty} \frac{a_m + \dfrac{a_{m-1}}{x} + \cdots + \dfrac{a_1}{x^{m-1}} + \dfrac{a_0}{x^m}}{\dfrac{b_n}{x^{m-n}} + \dfrac{b_{n-1}}{x^{m-n+1}} + \cdots + \dfrac{b_1}{x^{m-1}} + \dfrac{b_0}{x^m}}$$

Since by Theorem 3.5 each term in the denominator approaches 0, the limit does not exist, and so there is no horizontal asymptote. ■

Remark Put in a simpler way, this theorem states that as $x \to \pm\infty$, then when the higher power is in the denominator, the limit is 0; when the numerator and denominator are of the same degree, the limit is the ratio of the leading coefficients; and when the higher power is in the numerator, no limit exists.

EXAMPLE 3.11 Find the horizontal asymptotes, if they exist.

(a) $f(x) = \dfrac{x}{x^2 + 1}$ (b) $f(x) = \dfrac{2x^3 - 3x^2 - 5}{3x^3 + 5x - 7}$ (c) $f(x) = \dfrac{x^2 + 2x - 3}{x + 4}$

Solution (a) Since the denominator has the higher degree, the horizontal asymptote is $y = 0$.

(b) The numerator and denominator are of the same degree, so $y = \frac{2}{3}$ is the horizontal asymptote.

(c) There is no horizontal asymptote, since the numerator is of higher degree than the denominator. ■

Asymptotes, along with maximum, minimum, and inflection points, are valuable aids in drawing the graph of a function $y = f(x)$. Two other useful aids are **intercepts** and **symmetry.** The x-intercepts, if any exist, are found by setting $y = 0$, and the y-intercept, if one exists, is found by setting $x = 0$. Note that there may be more than one x-intercept but at most one y-intercept. (Why?) The graph is **symmetric to the y-axis** if $f(-x) = f(x)$ (that is, when f is even) and is **symmetric to the origin** if $f(-x) = -f(x)$ (that is, when f is odd). These two types of symmetry are illustrated in Figure 3.20.

The next example shows how these aids are used in graphing.

FIGURE 3.20

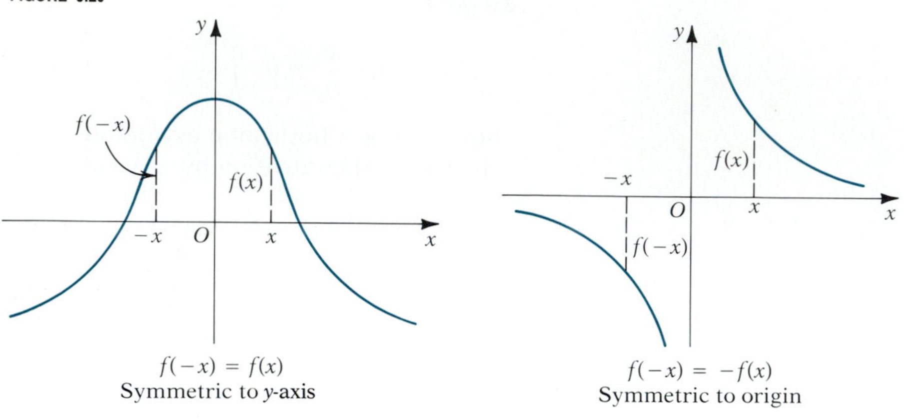

$f(-x) = f(x)$
Symmetric to y-axis

$f(-x) = -f(x)$
Symmetric to origin

EXAMPLE 3.12 Discuss and sketch the graph of

$$f(x) = \frac{x^2 - 1}{x^2 - 4}$$

Solution Let $y = f(x)$.

(a) *Intercepts.* Putting $y = 0$, we get $x = \pm 1$.
Putting $x = 0$, we get $y = \frac{1}{4}$.

(b) *Symmetry.* Since $f(x) = f(-x)$, the graph is symmetric to the y-axis.

(c) *Asymptotes.* The vertical asymptotes are $x = 2$ and $x = -2$, and the horizontal asymptote is $y = 1$.

(d) *Maxima and minima.* The derivative is found to be

$$f'(x) = \frac{-6x}{(x^2 - 4)^2}$$

so that $x = 0$ is the only critical point. Since $f'(x)$ changes from $+$ to $-$ as x goes from left to right through 0, we conclude that $(0, \frac{1}{4})$ is a local maximum point.

(e) *Inflection points.* Calculating $f''(x)$, we get, after simplifying,

$$f''(x) = \frac{6(3x^2 + 4)}{(x^2 - 4)^3}$$

For $|x| > 2$, $f''(x) > 0$, so that the graph is concave upward when $x > 2$ and when $x < -2$. For $|x| < 2$, $f''(x) < 0$, and the graph is therefore concave downward for $-2 < x < 2$. Since $x = 2$ and $x = -2$ are not in the domain of f, there are no points on the graph where the concavity changes, so there are no inflection points.

By plotting a few additional points, we can now draw the graph, as shown in Figure 3.21. ■

FIGURE 3.21

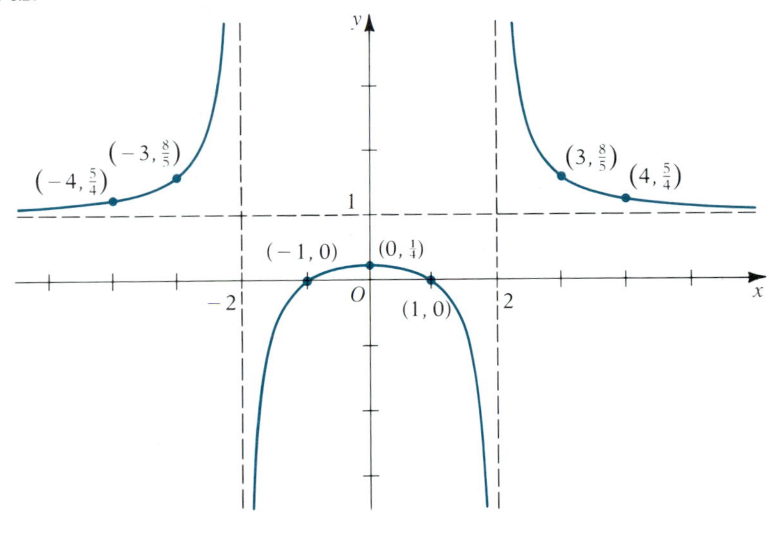

Infinite limits for f' can also provide useful information about the graph. Suppose that f is continuous at a and $\lim_{x \to a} f'(x) = +\infty$. Since $f'(x)$ gives the slope of the tangent line, we know that the slope is positive near a, and the tangent lines become steeper and steeper as x approaches a. At a itself, the tangent line is vertical. So the graph must be similar to Figure 3.22(a) in the vicinity of a. The other possibilities are shown in parts (b), (c), and (d).

In parts (a) and (b) of Figure 3.22 an inflection point occurs at $(a, f(a))$. The maxima and minima in parts (c) and (d) occur at cusps. The four possibilities shown assume that $\lim_{x \to a^+} f'(x)$ and $\lim_{x \to a^-} f'(x)$ are both infinite. If only one of these one-sided limits is infinite, or if f is discontinuous, various possibilities exist, some of which are shown in Figure 3.23.

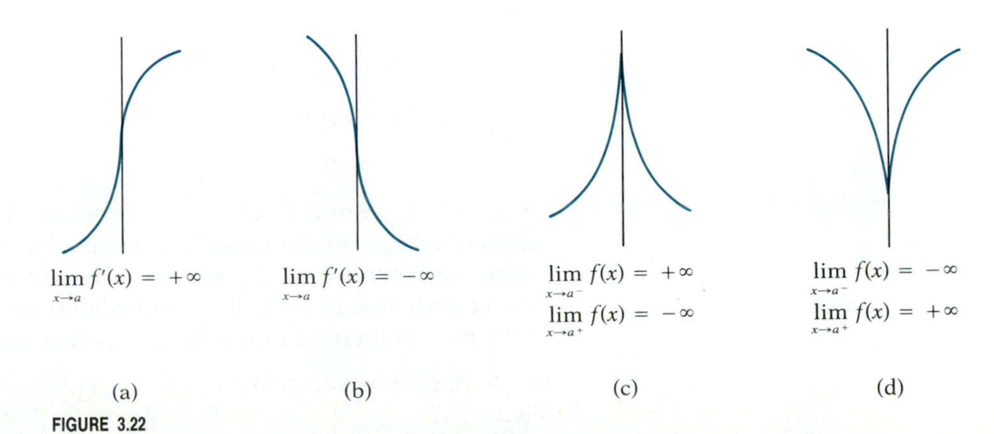

FIGURE 3.22

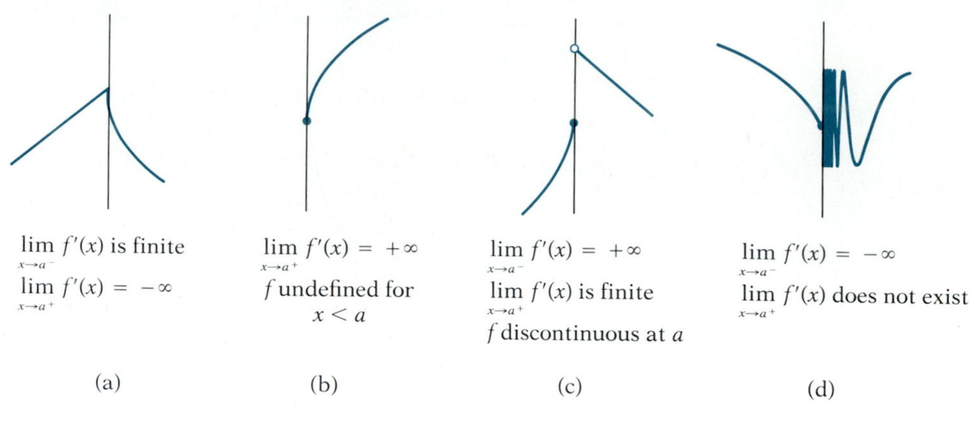

FIGURE 3.23

EXAMPLE 3.13 Discuss and sketch the graph of

$$f(x) = x + 3(1 - x)^{1/3}$$

Solution (a) *Intercepts.* $f(0) = 3$, so the y-intercept is 3. The x-intercepts are not easily obtained, but with the aid of a calculator it can be shown that one x-intercept is between -5 and -6 and one is between 4 and 5.

(b) *Symmetry.* There is no symmetry, since $f(-x) = -x + 3(1 + x)^{1/3}$, so that $f(-x)$ is not equal to $f(x)$ or $-f(x)$.

(c) *Asymptotes.* There are none.

(d) *Maxima and minima.*

$$f'(x) = 1 - (1 - x)^{-2/3}$$

The critical points are those for which $f'(x)$ is 0 or undefined. Since $f'(1)$ does not exist and $f(1)$ does exist, it follows that $x = 1$ is a critical point. Setting $f'(x) = 0$, we get

$$1 = \frac{1}{(1 - x)^{2/3}}$$

$$(1 - x)^{2/3} = 1 \qquad \text{Cube both sides.}$$

$$(1 - x)^2 = 1 \qquad \text{Take square roots.}$$

$$1 - x = \pm 1$$

$$x = 0 \text{ or } 2$$

We calculate $f''(x)$ to test 0 and 2:

$$f''(x) = -\tfrac{2}{3}(1 - x)^{-5/3}$$

Since $f''(0) < 0$ and $f''(2) > 0$, we conclude that f has a local maximum at 0 and a local minimum at 2. To determine the nature of f at the critical point 1, we cannot use the second derivative test. (Why?) Since $f'(x)$ does not change sign in a small deleted neighborhood of 1, f does not have a local maximum or minimum at $x = 1$. As x approaches 1 from each side, we find that

$$\lim_{x\to 1^-} f'(x) = \lim_{x\to 1^-}\left[1 - \frac{1}{(1 - x)^{2/3}}\right] = -\infty$$

and

$$\lim_{x \to 1^+} f'(x) = \lim_{x \to 1^+} \left[1 - \frac{1}{(1-x)^{2/3}} \right] = -\infty$$

So the graph has a vertical tangent line at $x = 1$.

(e) *Inflection points.* Since $f''(x) < 0$ for $x < 1$ and $f''(x) > 0$ for $x > 1$, it follows that (1, 1) is an inflection point.

In summary, the point (0, 3) is a local maximum point, (2, −1) is a local minimum point, and (1, 1) is an inflection point at which the tangent line is vertical. Furthermore, the graph is concave downward to the left of 1 and concave upward to the right of 1. We get a few additional points and draw the graph as shown in Figure 3.24.

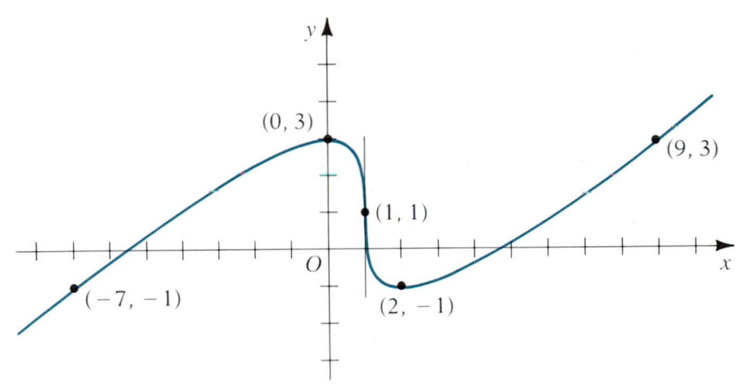

FIGURE 3.24

■

EXERCISE SET 3.3

A

In Exercises 1–6 determine whether the functions are unbounded positively [$\lim f(x) = +\infty$] *or negatively* [$\lim f(x) = -\infty$].

1. (a) $\displaystyle\lim_{x \to 2^-} \frac{1}{x-2}$ (b) $\displaystyle\lim_{x \to 2^+} \frac{1}{x-2}$

2. (a) $\displaystyle\lim_{x \to 3^-} \frac{x+1}{x-3}$ (b) $\displaystyle\lim_{x \to 3^+} \frac{x+1}{x-3}$

3. (a) $\displaystyle\lim_{x \to -1^-} \frac{2x-3}{x+1}$ (b) $\displaystyle\lim_{x \to -1^+} \frac{2x-3}{x+1}$

4. (a) $\displaystyle\lim_{x \to -2^-} \frac{1-x}{x^2-4}$ (b) $\displaystyle\lim_{x \to -2^+} \frac{1-x}{x^2-4}$

5. (a) $\displaystyle\lim_{x \to 5^-} \frac{3-4x}{(x-5)^2}$ (b) $\displaystyle\lim_{x \to 5^+} \frac{3-4x}{(x-5)^2}$

6. (a) $\displaystyle\lim_{x \to 4^-} \frac{x^2+x-1}{x^2-2x-8}$ (b) $\displaystyle\lim_{x \to 4^+} \frac{x^2+x-1}{x^2-2x-8}$

In Exercises 7–14 find the indicated limits or show that they fail to exist.

7. $\displaystyle\lim_{x \to +\infty} \frac{1}{2x+3}$

8. $\displaystyle\lim_{x \to -\infty} \frac{x}{x^2-4}$

9. $\displaystyle\lim_{x \to -\infty} \frac{x+1}{x-2}$

10. $\displaystyle\lim_{x \to +\infty} \frac{x^2+3}{2x^2-5}$

11. $\displaystyle\lim_{x \to +\infty} \frac{1-x+2x^2}{x^2-4}$

12. $\displaystyle\lim_{x \to -\infty} \frac{3x^3-2x-7}{2+4x^2-5x^3}$

13. $\displaystyle\lim_{x \to -\infty} \frac{x^2-1}{x+2}$

14. $\displaystyle\lim_{x \to +\infty} \frac{4-x-2x^2}{3x-7}$

In Exercises 15–30 find all vertical and horizontal asymptotes.

15. $f(x) = \dfrac{x}{x-3}$

16. $f(x) = \dfrac{x-1}{x^2-4}$

17. $f(x) = \dfrac{x^2 - 1}{x^2 - 9}$

18. $f(x) = \dfrac{2x^2 + x - 1}{x^2 - x - 6}$

19. $f(x) = \dfrac{x - 3}{6x^2 + x - 2}$

20. $f(x) = \dfrac{x^2 - 4}{x + 1}$

21. $f(x) = \dfrac{x^2 - 9}{x + 3}$

22. $f(x) = \dfrac{x + 3}{x^2 - 9}$

23. $f(x) = \dfrac{x^2 - 5x + 4}{3x^2 - x - 2}$

24. $f(x) = \dfrac{2x^2 + x - 10}{x^2 - x - 2}$

25. $f(x) = 1 - \cos \dfrac{2}{x}$

26. $f(x) = \dfrac{1}{x - 1} \sin \dfrac{1}{x}$

27. $f(x) = \dfrac{\cos x}{x(1 + \sin^2 x)}$

28. $f(x) = \dfrac{\sin^2 x}{\sqrt{x} + \cos^2 x}$

29. $f(x) = \dfrac{2 \cos^2 x - 3}{2 \sin^2 x - \sin x - 1}$

30. $f(x) = \dfrac{x^2 - 4}{4 \cos^2 x - 3}$

In Exercises 31–40 discuss intercepts, symmetry, vertical and horizontal asymptotes, local maxima and minima, and points of inflection, and sketch the graph.

31. $f(x) = \dfrac{x - 1}{x + 2}$

32. $f(x) = \dfrac{x}{x^2 - 1}$

33. $f(x) = \dfrac{2}{x^2 + 1}$

34. $f(x) = \dfrac{x}{x^2 + 1}$

35. $f(x) = \dfrac{x^2}{x^2 - 4}$

36. $f(x) = \dfrac{x^2 - 4}{x^2 - 1}$

37. $f(x) = \dfrac{x^2}{x^2 + 4}$

38. $f(x) = \dfrac{1 - x^2}{x^2 - 4}$

39. $f(x) = \dfrac{1}{x - 1} + \dfrac{1}{x + 1}$

40. $f(x) = \dfrac{1}{x} - \dfrac{1}{x + 1}$

In Exercises 41–48 show that the function has a vertical tangent line, and determine the nature of the graph at that point.

41. $f(x) = x^{2/3}$

42. $f(x) = 1 - x^{1/3}$

43. $f(x) = \sqrt[3]{1 - x}$

44. $f(x) = x^{3/5} - 2x^{2/5}$

45. $f(x) = \dfrac{x^{2/3}}{x - 1}$

46. $f(x) = \sqrt{1 - x^2}, \quad 0 \le x \le 1$

47. $f(x) = 1 - x^{1/2}, \quad x \ge 0$

48. $f(x) = \sqrt{\dfrac{x - 1}{x + 1}}, \quad x \ge 1$

B

In Exercises 49–53 discuss completely and sketch the graph.

49. $f(x) = 2x + 3(1 - x)^{2/3}$

50. $f(x) = \dfrac{3 - x^2}{x^3}$

51. $f(x) = \dfrac{x^2 - 1}{x^3 - 3x}$ (Use a calculator to estimate where points of inflection occur.)

52. $f(x) = \dfrac{3x^{1/3}}{2 - x}$

53. $f(x) = \dfrac{1 - \sin x}{\cos x - x}$

54. (a) Show that when $f(x)$ is a rational function of the form $P(x)/Q(x)$ in lowest terms with the degree of P *exactly one greater* than the degree of Q, then $f(x)$ can be written in the form

$$f(x) = mx + b + \frac{R(x)}{Q(x)}$$

where the degree of R is less than the degree of Q. (*Hint:* Use long division.)

(b) With the same hypotheses as in part a show that the line $y = mx + b$ is an asymptote to the graph of f by showing that

$$\lim_{x \to \pm\infty} |f(x) - (mx + b)| = 0$$

Such an asymptote is called an **oblique asymptote.**

In Exercises 55–58 use the procedure of Exercise 54 to find the oblique asymptote.

55. $f(x) = \dfrac{x^2 - 4}{x + 1}$

56. $f(x) = \dfrac{x^2 - 3x + 4}{x - 2}$

57. $f(x) = \dfrac{x^3 - x}{x^2 + 1}$

58. $f(x) = \dfrac{x^3 - 2x^2 + 3x - 4}{x^2 + 2x - 1}$

In Exercises 59 and 60 discuss completely, including oblique asymptotes (see Exercise 54), and sketch the graph.

59. $f(x) = \dfrac{x^3 + 1}{x^2}$

60. $f(x) = \dfrac{x^2 - 8}{x - 3}$

61. Prove Theorem 3.5.

C

Exercises 62–66 are for students who have studied computer graphics. Use the computer to graph the given function. From your graph, estimate where extreme values and points of inflection occur. Also from your graph, discuss intercepts, asymptotes (vertical, horizontal, and oblique), and concavity.

62. $f(x) = \dfrac{x^3 - 8}{x^3 + 2x^2}$

63. $f(x) = \dfrac{x - 1}{x^{2/3}}$

64. $f(x) = \dfrac{x^2 - x - 6}{x^3 - 1}$

65. $f(x) = \dfrac{x^4 - 16}{x^3 + 1}$

66. $f(x) = \sqrt{\dfrac{x^2 + 4}{x^2 - 2x}}$

3.4 APPLIED MAXIMUM AND MINIMUM PROBLEMS

As stated at the beginning of this chapter, one of the most important applications of the derivative is in finding maximum or minimum values of functions. Usually certain constraints are imposed on the variable, or variables, involved. Such problems are referred to collectively as **optimization problems,** since we are trying to find the best (optimal) result under the given constraints. Some optimization problems can be formulated in terms of one *continuous* variable (one that can assume any real value on an interval), and some require more than one such variable. We will concentrate here on the first type and study the second type later. It should also be mentioned that certain problems involve *discrete* variables (for example, variables that can assume only integer values). Such problems often can be solved by supposing the variable to be continuous and interpreting the final result appropriately. Others require totally different approaches studied in discrete mathematics courses.

We have the necessary background for finding extreme values, *provided we have the quantity to be maximized or minimized expressed as a function of one continuous variable.* It is this provision that we will concentrate on in this section. The main difficulty is that in real-life situations problems are usually presented in words, not as neat "textbook type" mathematical formulas. So the first step is to translate the word problem into an appropriate mathematical form. Then we solve the mathematical problem and finally interpret the answer in terms of the original problem. This process is known as **mathematical modeling,** and it is one of the most important aspects of the work of applied mathematicians and others who use mathematics in their work. Although there are no guaranteed paths to success in setting up a model of an optimization problem for which calculus is appropriate, the following steps can serve as a general guide.

1. Read the problem through, more than once if necessary, and identify the quantity to be maximized or minimized. We will refer to this quantity as the **objective function.** Give it a name (a suggestive letter such as C for cost, P for profit, V for volume, and so on, is a good idea).
2. Determine a variable on which the objective function depends and give it a name, such as x or t. This is called the *independent variable.* Note any limitations on the size of this variable. It sometimes appears that

there are two (or more) such variables, but in the present context enough information will be given to express one in terms of the other.

3. Express the objective function in terms of the independent variable. This is a key step and will be made easier if steps 1 and 2 have been carried out carefully. Frequently it is helpful to draw a figure. Sometimes well-known relationships such as "distance = rate × time" or "profit = revenue − cost" are used. Also mensuration formulas, such as for area or volume, may be useful.
4. Differentiate the objective function and find the critical points.
5. Test those critical points that lie in the admissible domain, and choose the one that gives the largest or smallest value (whichever is desired) of the objective function. Often there is only one critical point in the admissible domain, and the nature of the problem ensures that there is an extreme value interior to the interval in question, so that this must occur at the critical point. In such cases it is not necessary to test the critical point. However, it is a good idea to test it as a check.
6. Find the value of the objective function at the critical point found in step 5. If the end point values also are feasible solutions, find the value of the objective function at these points also. The desired extreme value can now be determined.
7. Mentally check to see that the answer found is reasonable. Sometimes it is possible to see that a mistake has been made because the answer obtained is clearly incorrect.

The following examples illustrate this procedure.

EXAMPLE 3.14 An open-top box is to have a square base and a volume of 10 cubic feet. If the bottom costs 15¢ per square foot and the sides 6¢ per square foot, find the dimensions of the most economical box, and find the minimum cost.

Solution The most economical box means the one with the minimum cost. So the objective function is the cost C, and we want to find the minimum value. We introduce as independent variable the side x of the square base (Figure 3.25). The height y is also a relevant variable, but it can be written in terms of x as follows. We know the volume of the box is 10 cu ft, and since the volume of a box is the area of the base times the height, we have

$$x^2y = 10$$
$$y = \frac{10}{x^2}$$

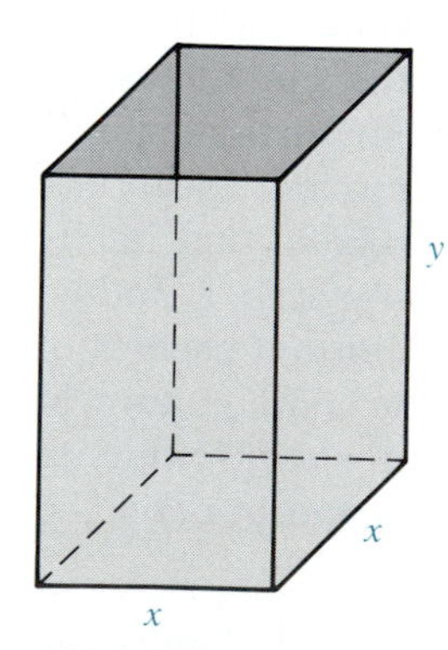

FIGURE 3.25

Clearly we must have $x > 0$.

Now we can find $C(x)$ from the given information. The base of the box has area x^2 sq ft, and each square foot costs 15¢. There are four vertical sides, each of area $xy = x(10/x^2) = 10/x$, and each costing 6¢ per square foot. Thus, the value $C(x)$ of the cost function is

$$C(x) = 15x^2 + 4(6)\left(\frac{10}{x}\right) = 15x^2 + \frac{240}{x}$$

We find the critical points from $C'(x)$:

$$C'(x) = 30x - \frac{240}{x^2}$$

Set $C'(x) = 0$:

$$30x - \frac{240}{x^2} = 0 \quad \text{Multiply by } x^2 \text{ and divide by 30.}$$

$$x^3 - 8 = 0$$

$$x^3 = 8$$

$$x = 2$$

It seems reasonable that this gives a minimum, since from $C(x)$ we see that for large x or x close to 0, the cost is large. However, we can check to see that this gives a minimum by substituting in $C''(x)$:

$$C''(x) = 30 + \frac{480}{x^3}$$

$$C''(2) > 0$$

So the minimum value of C occurs when $x = 2$. At this point we have

$$y = \frac{10}{2^2} = \frac{5}{2}$$

and

$$C(2) = 15(2^2) + \frac{240}{2} = 60 + 120 = 180$$

Thus, the most economical box has dimensions (in feet) of $2 \times 2 \times 2\frac{1}{2}$, and the minimum cost is 180¢ or \$1.80. ■

EXAMPLE 3.15 A rectangular flower bed is to be designed with one side adjacent to a building, and the other three sides are to be bordered by a walkway 3 ft wide. If the outer perimeter of the walkway is to be 172 ft, what should be the dimensions of the flower bed if it is to have maximum area? What is the maximum area?

Solution The objective function is the area A of the flower bed. As in Figure 3.26, let x be the width of the bed and y its length. Since the outer perimeter of the

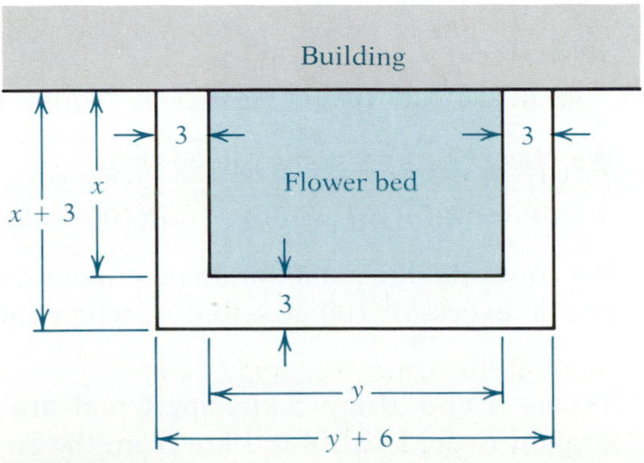

FIGURE 3.26

walkway is $2(x + 3) + (y + 6)$ (verify this), we must have

$$2x + 6 + y + 6 = 172$$
$$y = 160 - 2x$$

Since x and y are both positive, we must have $0 < x < 80$. The area of the bed is $xy = x(160 - 2x)$, so

$$A(x) = 160x - 2x^2$$
$$A'(x) = 160 - 4x$$
$$A'(x) = 0 \qquad \text{if } x = 40$$
$$A''(x) = -4$$

The only critical point is $x = 40$, and since $A''(40) < 0$, this produces a maximum value. Thus the dimensions that produce the maximum area are

$$x = 40 \quad \text{and} \quad y = 160 - 2(40) = 80$$

and the maximum area is

$$A(40) = (40)(80) = 3200 \text{ sq ft}$$

■

EXAMPLE 3.16 A boat is rented for an excursion for 100 passengers at \$10 per passenger. The price of each ticket will be reduced 5¢ per passenger for each person who goes in excess of 100. Find the number of passengers that will provide the owner of the boat the maximum revenue and find the maximum revenue. The boat will accommodate 200 passengers.

Solution Let R denote the revenue. The revenue is calculated by multiplying the number of passengers by the price of each ticket. Each of these quantities depends on the number of passengers *in excess* of 100. Call this number x. Then we have

$$100 + x = \text{total number of passengers}$$
$$10 - 0.05x = \text{price of each ticket}$$

Because of the capacity of the boat, we must have $0 \leq x \leq 100$. Thus we have

$$R(x) = (100 + x)(10 - 0.05x) = 1000 + 5x - 0.05x^2$$
$$R'(x) = 5 - 0.1x$$
$$R'(x) = 0 \quad \text{if } x = 50$$
$$R''(x) = -0.1$$
$$R(50) = (150)(10 - 2.50) = 150(7.50) = 1125$$

We check the end point values also:

$$R(0) = (100)(10) = 1000 \qquad R(100) = (200)(5) = 1000$$

We conclude that a maximum revenue of \$1125 is obtained when 50 passengers in excess of 100 go—that is, when 150 passengers go. ■

EXAMPLE 3.17 Towns A and B are 5 km apart and are located on the same side of a straight river. Town A is 1 km from the river and town B is 4 km from the river. A pumping station is to be built at the river's edge and pipes are to go

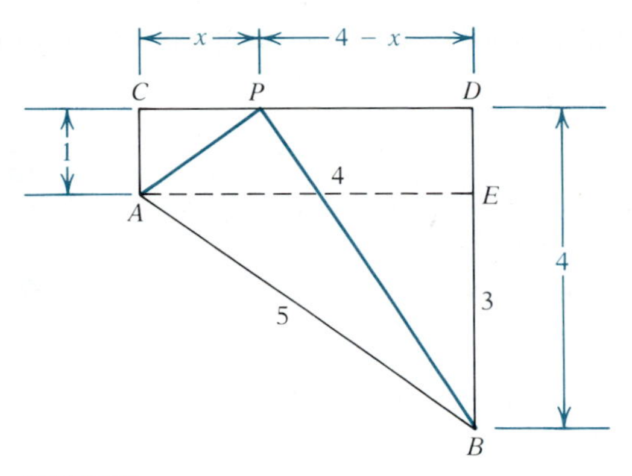

FIGURE 3.27

to the two towns. Where should the pumping station be located to minimize the amount of pipe?

Solution In Figure 3.27 we indicate by C the point on the shore nearest to A and by D the point on the shore nearest to B. Then the distance CD is the same as AE, which by the Pythagorean theorem is 4. Let P be the point at which the pumping station is to be located, and denote by x its distance from C. Then $4 - x$ is its distance from D. Clearly, $0 \le x \le 4$. If $L(x)$ denotes the total length of pipe, then $L(x) = \overline{PA} + \overline{PB}$. Using the Pythagorean theorem on triangles ACP and BDP, we get

$$L(x) = \sqrt{x^2 + 1} + \sqrt{(4 - x)^2 + 16}$$

$$L'(x) = \frac{x}{\sqrt{x^2 + 1}} - \frac{4 - x}{\sqrt{(4 - x)^2 + 16}}$$

Setting $L'(x) = 0$, we get

$$x\sqrt{(4 - x)^2 + 16} = (4 - x)\sqrt{x^2 + 1}$$

and on squaring and simplifying, this gives (you should check the details)

$$15x^2 + 8x - 16 = 0$$

$$(5x - 4)(3x + 4) = 0$$

$$x = \tfrac{4}{5} \,\big|\, x = -\tfrac{4}{3} \qquad \text{(Impossible)}$$

The answer $x = -\frac{4}{3}$ is ruled out on two counts: it is outside the admissible domain, and it fails to satisfy the equation that we squared. The solution $x = \frac{4}{5}$ does satisfy this equation (check this) and so it is a critical point. It can be shown that $L''(\frac{4}{5}) > 0$ so that a minimum value of L occurs at $x = \frac{4}{5}$. However, calculating L'' is a little messy, and we can avoid finding it (and also avoid using the first derivative test) by calculating $L(\frac{4}{5})$ and comparing it with the end point values $L(0)$ and $L(4)$. After a bit of arithmetic, we get

$$L(\tfrac{4}{5}) = \sqrt{41} \approx 6.40 \qquad L(0) = 1 + 4\sqrt{2} \approx 6.66 \qquad L(4) = 4 + \sqrt{17} \approx 8.12$$

Since L is continuous on $[0, 4]$, it reaches both a maximum and a minimum value there, and the only candidates for these extreme values are the three given here. So $L(\frac{4}{5})$ is the minimum value. Thus, the pumping station should be located $\frac{4}{5}$ km from C and $\frac{16}{5}$ km from D. ■

EXAMPLE 3.18 Prove that the isosceles triangle of maximum area that can be inscribed in a circle of fixed radius a is equilateral.

Solution Referring to Figure 3.28, we see that $y = \sqrt{a^2 - x^2}$, so that the area, $A(x)$, is

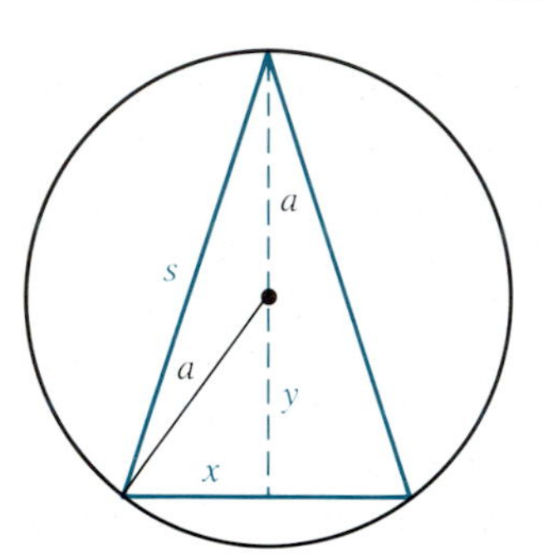

FIGURE 3.28

$$A(x) = x(\sqrt{a^2 - x^2} + a), \qquad 0 \le x \le a$$

We calculate $A'(x)$ and set it equal to 0. (We omit some of the algebra.)

$$A'(x) = \frac{-x^2}{\sqrt{a^2 - x^2}} + \sqrt{a^2 - x^2} + a = \frac{a^2 - 2x^2 + a\sqrt{a^2 - x^2}}{\sqrt{a^2 - x^2}}$$

$$A'(x) = 0 \qquad \text{if } 2x^2 - a^2 = a\sqrt{a^2 - x^2}$$

After squaring both sides and collecting terms, we get

$$x^2(4x^2 - 3a^2) = 0$$

so that $x = 0$ or $x = \pm a\sqrt{3}/2$. Since $x = 0$ gives an area of 0 and since x cannot be negative, the only critical point of interest is $a\sqrt{3}/2$. We calculate

$$A\left(\frac{a\sqrt{3}}{2}\right) = \frac{3a^2\sqrt{3}}{4}$$

which is greater than $A(a) = a^2$, so that $x = a\sqrt{3}/2$ does produce the maximum area.

To show that the triangle with this value of x is equilateral, we calculate the height,

$$y + a = \sqrt{a^2 - x^2} + a = \sqrt{a^2 - \frac{3a^2}{4}} + a = \frac{3a}{2}$$

The tangent of the base angle θ (Figure 3.29) is

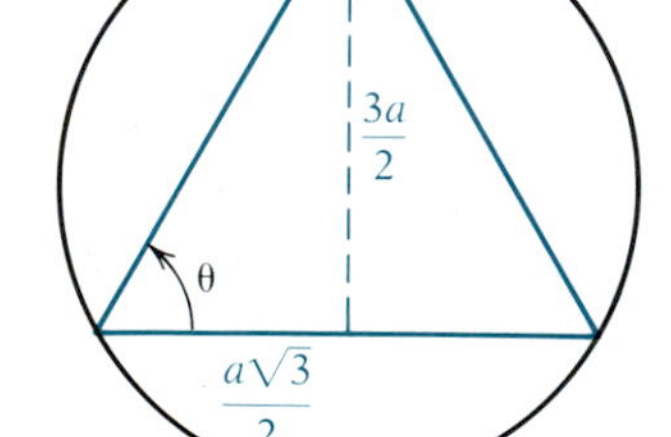

FIGURE 3.29

$$\tan\theta = \frac{\dfrac{3a}{2}}{\dfrac{a\sqrt{3}}{2}} = \sqrt{3}$$

so that $\theta = 60^\circ$. The base angles are equal, and since the three angles add to 180°, all three angles are equal to 60°. So the triangle is equiangular and hence equilateral. ■

EXAMPLE 3.19 Find the point on the parabola $y = (x - 3)^2$ that is nearest the origin.

Solution An arbitrary point P on the parabola has coordinates $(x, (x - 3)^2)$, and its distance $d(x)$ from the origin is

$$d(x) = \sqrt{x^2 + (x - 3)^4}$$

(Figure 3.30). This is the function we want to minimize. The problem can be simplified, however, by observing that the distance will be minimum if and only if the square of the distance is a minimum. We denote the squared distance by $s(x)$:

$$s(x) = x^2 + (x - 3)^4$$

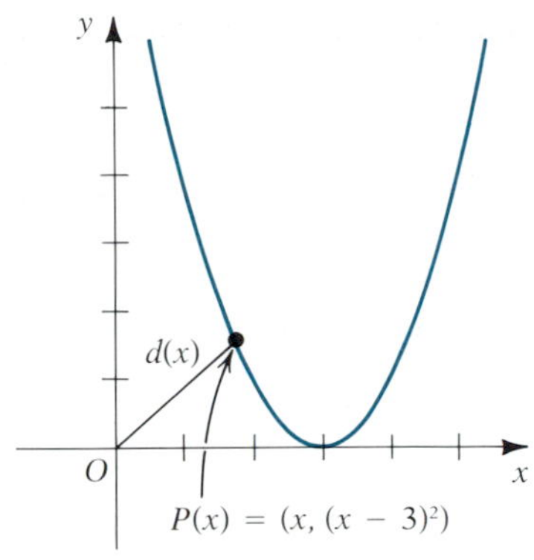

FIGURE 3.30

and proceed to minimize it:

$$s'(x) = 2x + 4(x - 3)^3 = 4x^3 - 36x^2 + 110x - 108$$

We set $s'(x) = 0$ and divide through by 2 to get

$$2x^3 - 18x^2 + 55x - 54 = 0$$

There are various techniques for solving equations of degree 3 or higher, some of which are usually studied in precalculus courses. If there are any rational roots, we can find them by trial and error, with the trials limited as follows. If $x = \frac{p}{q}$ is a root, then p must be a factor of the constant term, in this case 54, and q must be a factor of the leading coefficient, in this case 2. Starting with integer possibilities, we eliminate $x = 1$ but find that $x = 2$ works. (Synthetic division is useful here, but direct substitution can also be used.) After dividing out the factor $x - 2$, the other factor is found to be $2x^2 - 14x + 27$, which has no real zeros. So $x = 2$ is the only critical point. Since the geometry of the problem assures us that a minimum distance exists, we can conclude that $x = 2$ produces this minimum. We could also check this by observing that $s''(2) > 0$.

The point on the graph nearest the origin is therefore (2, 1), and its distance from the origin is $d(2) = \sqrt{5}$. ■

Some Concepts from Economics

Before giving the final example, we discuss briefly some basic concepts of economics. We will be concerned with maximizing the profit from the manufacture and sale of some commodity. If we let $C(x)$ denote the cost of producing x items and $R(x)$ denote the gross revenue obtained from selling them, then the profit, $P(x)$, is

$$P(x) = R(x) - C(x)$$

Although x in this situation can actually take on positive integer values only, we can treat it as if it could take on any positive real value. Then when we obtain the critical point x that maximizes the profit, if x is not an integer, we test the two integers closest to it to see which gives a maximum.

The cost function C usually consists of a **fixed cost** and a **variable cost.** The fixed cost is a constant, independent of the number of items produced. Its value is determined by such things as rent, mortgage payments, taxes, maintenance, basic utility bills, and salaries of persons unaffected by the number of items produced. The variable cost is that part of the cost that is dependent on the number of items produced. The formula for this may be obtained by analyzing the factors that affect the cost, or it may be obtained empirically. In the latter case records may be kept over a period of time and a graph plotted showing the cost versus the number of items produced. Then by a process known as "curve fitting," a function whose graph matches as closely as possible the one obtained experimentally is determined. The resulting function can take a variety of forms, but it is often a polynomial of first, second, or third degree.

To see how the variable cost might be obtained analytically, consider the following fairly typical situation. There is a certain cost c to produce each unit when some basic minimum number m of units is produced. Then for each additional unit produced, the cost per unit is reduced by some amount

a. Thus, if x units are produced, the variable cost $C_v(x)$ is

$$C_v(x) = \underbrace{\overset{\text{number produced}}{x}[c - a\overbrace{(x - m)}^{\text{number in excess of } m}]}_{\text{cost of producing } x \text{ items}}$$

$$= -ax^2 + (c + am)x$$
$$= -ax^2 + bx$$

where $b = c + am$. So in this case the variable cost is a quadratic polynomial. For this to be a reasonable model, we would expect some upper limit n on the number of units, so that $m \leq x \leq n$. In fact, if x exceeds n, it may be that the cost per unit begins to go up again, since more machines, more employees, more space, and so on, may be required. This would lead to the addition of a cubic term to the function.

The **revenue function** R is obtained by multiplying the number of items sold by the price p at which they are sold, so that

$$R(x) = xp(x)$$

We have indicated p as a function of x, since there is a definite functional relationship between price and number of items sold. The function $p(x)$ is sometimes called the **demand function,** since it gives the price per unit when there is demand for x units. Typically, when the price goes down, the number of items that can be sold at that price goes up.

The next example illustrates these ideas.

EXAMPLE 3.20 A company that produces cells for solar collectors finds through experience that on the average it can sell x cells per day when the price $p(x) = 100 - 0.05x$, where $250 \leq x \leq 800$. There is a fixed cost of \$4000 per day and a variable cost per unit of $60 - 0.01x$. Find the number of cells that should be produced each day to maximize profit, and find the maximum profit.

Solution The revenue is the number of units times the price per unit:

$$R(x) = xp(x) = x(100 - 0.05x) = 100x - 0.05x^2$$

The variable cost is the number of units times the cost per unit. So the total cost is

$$C(x) = 4000 + x(60 - 0.01x) = 4000 + 60x - 0.01x^2$$

Thus

$$\begin{aligned} P(x) &= R(x) - C(x) \\ &= 100x - 0.05x^2 - (4000 + 60x - 0.01x^2) \\ &= -0.04x^2 + 40x - 4000 \\ P'(x) &= -0.08x + 40 \\ P'(x) &= 0 \quad \text{if } x = 500 \end{aligned}$$

Since $P''(x) < 0$, the critical point $x = 500$ gives a maximum. We calculate $P(500) = 6000$ and the end point values $P(250) = 3500$ and $P(800) = 2400$.

So a maximum profit of $6000 per day is obtained when 500 units are produced. ■

Further Comments In general, as in the preceding example, the maximum profit will occur at a critical value where $P'(x) = 0$. Let x_0 be such a point. Then we have

$$P'(x_0) = R'(x_0) - C'(x_0) = 0$$

so that

$$R'(x_0) = C'(x_0)$$

That is, maximum profit occurs where R' and C' are equal. As we mentioned in Section 2.2, these quantities, $R'(x)$ and $C'(x)$, are called, respectively, **marginal revenue** and **marginal cost.** If we let x change by a small amount, say Δx, then the corresponding changes in $R(x)$ and $C(x)$ are approximated by their differentials:

$$\Delta R \approx dR = R'(x)\,\Delta x \quad \text{and} \quad \Delta C \approx dC = C'(x)\,\Delta x$$

In particular, if $\Delta x = 1$, we have $\Delta R \approx R'(x)$ and $\Delta C \approx C'(x)$. Thus, the marginal revenue is approximately the change in revenue caused by selling one additional item, and the marginal cost is approximately the change in cost caused by producing one additional item. Also, we have

$$\Delta P \approx dP = [R'(x) - C'(x)]\,\Delta x$$

So long as the marginal revenue $R'(x)$ exceeds the marginal cost $C'(x)$, there is an increase in profit by increasing the number of items. The value x_0 at which $R'(x_0) = C'(x_0)$ is the value where no further increase in profit would occur by increasing x, and so the profit is a maximum.

EXERCISE SET 3.4

A

1. A rectangular plot of ground is to be fenced on three sides to form a garden, with the fourth side bounded by a retainer wall. If there are 50 ft of fencing available, what should be the dimensions of the garden to produce the maximum area?
2. A rectangular lot adjacent to a road is to be enclosed by a fence. The fence along the road is to be reinforced and costs $7.00 a foot. Fencing that costs $5.00 a foot can be used for the other three sides. What is the largest area that can be enclosed for a cost of $3600, and what are the dimensions that give this maximum area?
3. A farmer wishes to fence a rectangular pasture with a total area of 12,000 sq m, and he wants to divide it into two parts with a fence across the middle. Fencing around the outside costs $7.50 per meter, but he can use less expensive fence at $3.00 per meter as the divider. What dimensions will result in the least cost?
4. A water trough is to be formed from a sheet of metal 2 ft wide by 10 ft long by bending up at a right angle equal amounts from each long side. (Pieces will then be welded on the ends.) How many inches on each side should be bent up to give the maximum volume?
5. A cardboard box is to be constructed having a volume of 9 cu ft and such that the bottom is a rectangle twice as long as it is wide. What should the dimensions be so that the least amount of cardboard is used?
6. An open-top box is to be made from a piece of cardboard 4 ft long and $2\frac{1}{2}$ ft wide by cutting out squares of equal size at each corner and bending up the flaps.

What should be the size of the squares to be removed in order to produce the maximum volume?

7. Find the dimensions of the rectangle of maximum area that can be inscribed in an isosceles triangle of height 6 and base 4 (see the figure). (*Hint:* Use similar triangles to find y in terms of x.)

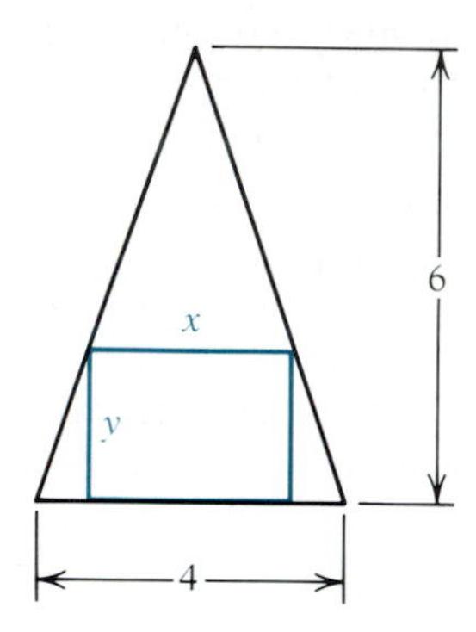

8. Find the dimensions of the rectangle of maximum area that can be inscribed as shown in a right triangle with legs of lengths 6 and 8, respectively.

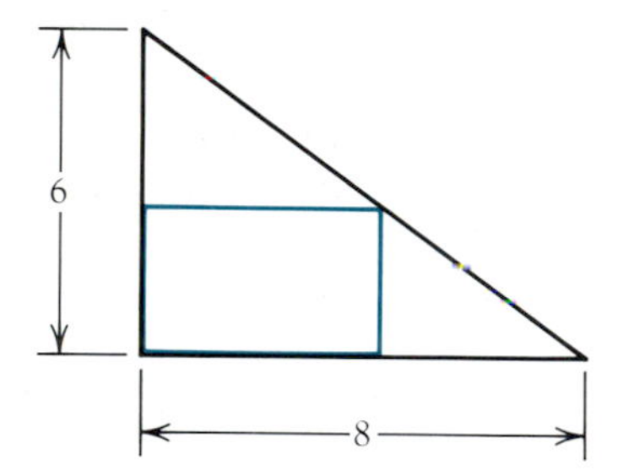

9. Show that the rectangle of maximum area that can be inscribed in a circle with fixed radius a is a square.

10. Find the dimensions of the rectangle of maximum area that can be inscribed in a semicircle of radius a as shown in the figure.

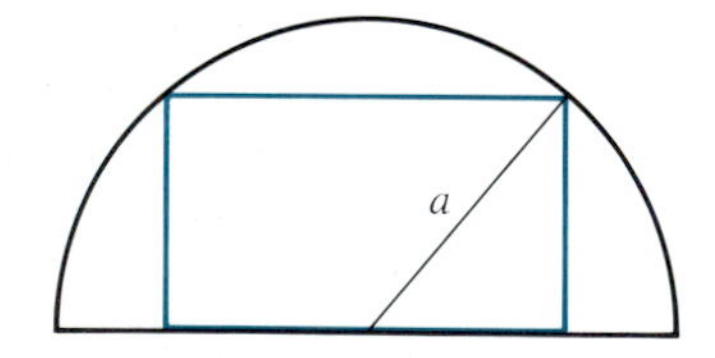

11. Find the dimensions of the right circular cylinder of maximum volume that can be inscribed in a sphere of fixed radius a. (*Hint:* Consider a cross section as shown in the figure at the top of the next column to find the altitude h of the cylinder in terms of its base radius r.)

12. In Exercise 11 maximize the lateral surface area instead of the volume.

13. The unit price of a manufactured item is constant at \$150 per item. The fixed cost is \$80, and the variable cost per item is found by experience to be approximately $x + 60$, when the number x of items is between 10 and 80. Assuming x is in this range and that all items that are manufactured are sold, find the number that should be manufactured to produce the maximum profit.

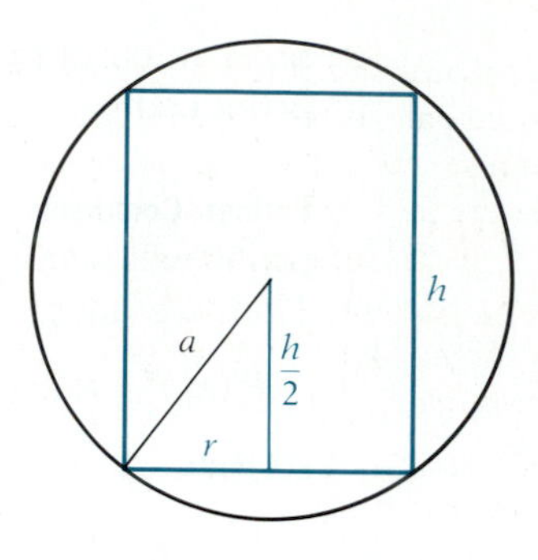

14. The total cost function for producing x tables in a certain furniture factory is determined to be $C(x) = 0.02x^2 + 20x + 100$ for $10 \leq x \leq 100$, and the unit price function $p(x) = 25 + \frac{120}{x} - 0.03x$ for x in this range. Find the number that should be produced to maximize profit. What is the maximum profit?

15. An orchardist has a 10-acre plot planted with 200 apple trees, and the average yield is 500 apples per tree. She plans to put apple trees of the same type on another 6-acre plot that has the same soil conditions. She knows that if she plants the trees closer together than on the first plot, the yield per tree will be less, but it is possible that with more trees the total yield per acre will be greater. Her estimate is that for each additional tree per acre she plants, the average yield per tree will be reduced by 10 apples. How many trees should she plant on the 6-acre plot to give the maximum total yield?

16. An excursion train is to be run to the Super Bowl. The railroad company sets the fare at \$10 per ticket if 200 people go but agrees to lower the cost of all tickets by 2¢ each for every passenger in excess of 200. How many passengers will produce the greatest revenue, and what is the maximum revenue? (The train has a capacity of 450 passengers.)

17. A yacht is rented for an excursion for 60 passengers at \$20 per passenger. For each person in excess of 60 that goes, up to an additional 40 passengers, the fare of each passenger will be reduced by 25¢. Find the number of passengers that will produce the maximum revenue.

18. A restaurant has a fixed price of \$12 for a complete dinner. The average number of customers per evening is 160. The owner estimates on the basis of prior experience that for each 50¢ increase in the cost of a dinner, there will be 5 fewer customers per evening on the average. What price should be charged to produce the maximum revenue, and what is the maximum revenue?

19. At a certain movie theater the price of admission is \$3.00, and the average daily attendance is 200. As an experiment the manager reduces the price by 5¢ and finds that the average attendance increases by 5 people per day. Assuming that for each further 5¢ reduction the average attendance would rise by 5, find the number of 5¢ reductions that would result in the maximum revenue.

20. A manufacturer makes aluminum cups in the form of right circular cylinders open at the top (no handle), having a volume of 16π cu in. If the cost of the material for the bottom is twice that for the sides, find the dimensions that will give the lowest cost.

21. The strength of a rectangular beam of fixed length is jointly proportional to the width and the square of the depth. Find the dimensions of the strongest beam that can be cut from a log 3 ft in diameter.

22. Find the point on the graph of $y = \sqrt{2x - 3}$ that is nearest the point (3, 0).

23. A picture with an area of 120 sq in. is rectangular and is framed so that there is a $2\frac{1}{2}$-in. matting on each side and a 3-in. matting at the top and bottom. Find the dimensions of the picture if the total area of the framed picture is as small as possible. (Test to see that you have found the minimum.)

24. Find the point on the upper half of the ellipse $4x^2 + 9y^2 = 36$ that is nearest the point (1, 0).

25. The printed matter on a page is to occupy an area of 80 sq in., and there are to be margins of 1 in. at the top, bottom, and right side, and $1\frac{1}{2}$ in. at the left. Find the overall dimensions of the page that requires the least amount of paper.

B

26. A rain gutter is to be formed as shown in the figure by bending up at equal angles sides of 10 cm from a long metal sheet 30 cm wide. Find the angle θ that will enable the gutter to handle the greatest flow of water.

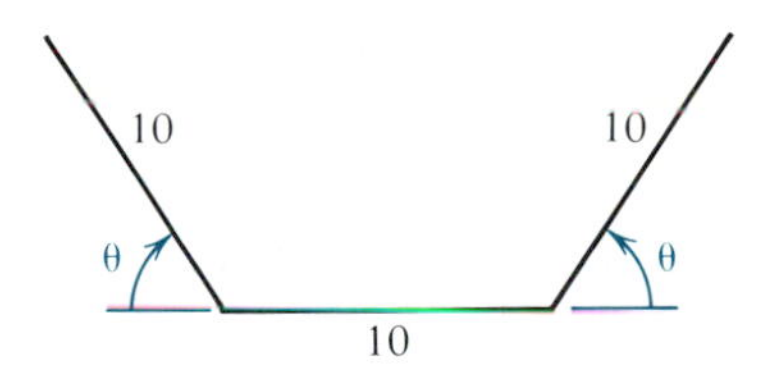

27. Find the dimensions of the right circular cone of maximum volume that can be inscribed in a sphere of fixed radius a.

28. Change Exercise 27 so that the cone is to have maximum lateral surface area.

29. A telephone cable is to be laid under a river from point A to point C (see the figure) and then underground to point B. The river is 20 m wide, and the distance from the point directly across the river from A to the point B is 40 m. If it costs $1\frac{1}{2}$ times as much to lay the cable under the river as it does to put it undergound, find where the point C should be located to minimize the cost.

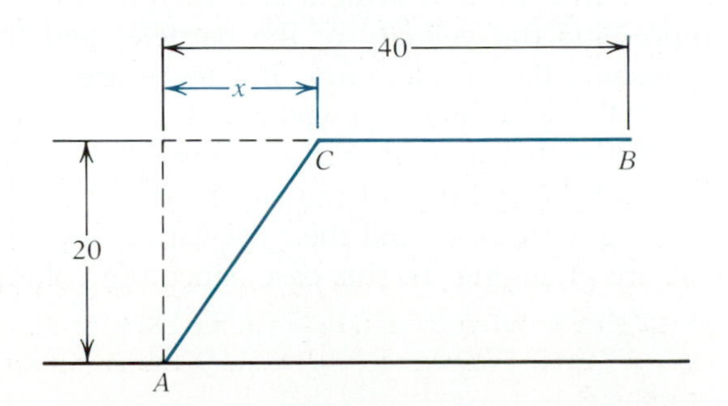

30. A window is to be designed in the shape of a rectangle surmounted by an equilateral triangle. The overall outside perimeter is to be 12 ft. Find the dimensions of the window if the area is to be a maximum.

31. In Exercise 30 suppose the triangle is replaced by a semicircle. Find the dimensions.

32. A silo is to be constructed in the form of a right circular cylinder surmounted by a hemisphere of the same radius as the cylinder. Find the dimensions that give the minimum cost if it is to have a volume of 600 cu m, and the unit cost of the material for the hemisphere is twice that for the cylinder.

33. A car leaves town A traveling at 80 km/hr due east on a straight road toward town B, 50 km away. At the same time a second car leaves town B traveling 60 km/hr due south on a straight road (see the figure). How long will it be from the time they left

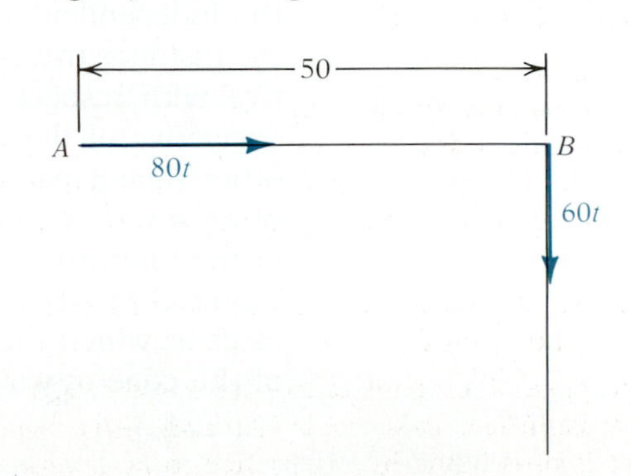

until the distance between them is minimum? At this instant where are the cars and how far apart are they?

34. The cost of operating a certain truck is determined on the basis of past experience to be $20 + \frac{x}{12}$ cents per kilometer when the truck is driven at a speed of x kilometers per hour. The truck driver earns $9.00 per hour of actual driving time. What is the most economical speed to operate the truck on a 600-km trip?

35. Find the length of the longest piece of straight pipe that can be carried horizontally around the corner from a 5-feet-wide hallway into a 3-feet-wide hallway. Assume the pipe has negligible thickness. (*Hint:* Choose as independent variable the angle θ as shown.)

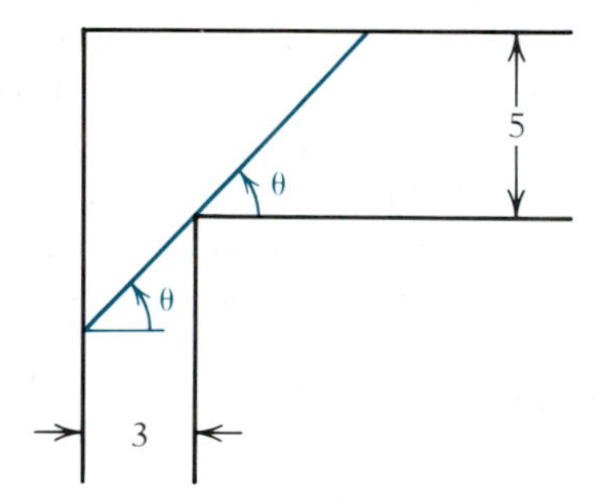

36. The illumination from a source of light varies directly as the intensity of the light source and inversely as the square of the distance from the source. One light is 1 m directly above one end of a table 3 m in length, and another light of twice the candlepower is directly opposite the first over the other end of the table and 2 m above it (see the figure). At what point on the table directly between the lights will the combined intensity of light be the least?

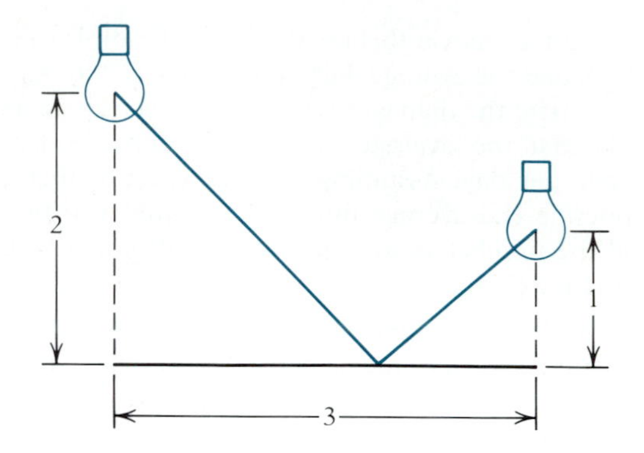

37. A long sheet of paper 6 in. wide is to be folded up as shown in the figure. Find where the point P of the fold should be in order that (a) the area of the folded part is a minimum and (b) the length PQ of the fold is a minimum.

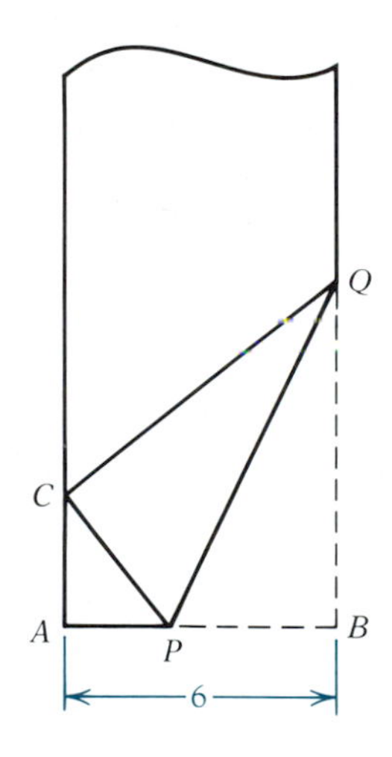

3.5 RELATED RATES

In this section we will apply the derivative to functions that have *time* as the independent variable. We already know that if the function represents the distance traveled by a particle moving in a straight line, then the derivative with respect to time represents the velocity of the particle, and the derivative of the velocity represents the acceleration. But there are many other time-dependent functions. For example, if a water tank in the shape of an inverted cone is being filled with water, then the volume V of water in the tank, the water level h, and the radius r of the circle formed by the top level of water are all changing with time, and their derivatives give the rates at which these variables are changing. In this case, since the volume of the cone of water is

$$V = \tfrac{1}{3}\pi r^2 h$$

we can differentiate implicitly with respect to time t to get

$$\frac{dV}{dt} = \frac{1}{3}\pi\left(r^2\frac{dh}{dt} + 2rh\frac{dr}{dt}\right)$$

and this equation relates the rates of change in question. It is because of such relationships that this section is entitled "related rates."

We now consider a particular case of the example just cited.

EXAMPLE 3.21 Suppose the conical water tank referred to has overall height 6 m and top radius 2 m and that water is being pumped into the tank at the constant rate of 500 L/min (1 L = 10^{-3} cu m). Find the rate at which the water level is rising (a) when the water level is 3 m and (b) when the tank is half full.

Solution As earlier, we denote by V, h, and r the volume of water in the tank, the height of the water, and the radius of the top circle of water, respectively, all at time t. It is important to note that all three of these variables are functions of time and that they are not constants. It would be incorrect at this stage, for example, in solving part a to put $h = 3$, because h is not always 3, and we must first find the relationship between the relevant rates of change before we consider what they are at a particular instant. *In short, before substituting a particular value for a variable quantity, we must differentiate.*

As noted, the volume of water at any instant is given by

$$V = \tfrac{1}{3}\pi r^2 h \tag{3.1}$$

In both parts a and b we want to know $\frac{dh}{dt}$ at a particular instant, so we will need to differentiate. First, however, it will simplify matters to replace r by its equivalent in terms of h, since h is the variable of interest in this problem. We can do this using similar triangles. Referring to Figure 3.31, we get

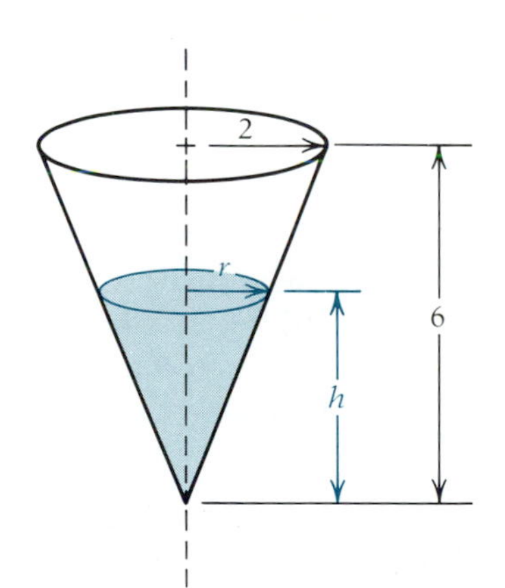

FIGURE 3.31

$$\frac{r}{2} = \frac{h}{6}$$

so that $r = \frac{1}{3}h$. Substituting this in equation (3.1) and simplifying give

$$V = \frac{1}{27}\pi h^3 \tag{3.2}$$

Now we differentiate both sides with respect to time t:

$$\frac{dV}{dt} = \frac{1}{9}\pi h^2 \frac{dh}{dt}$$

(Note the use of the chain rule.) Now $\frac{dV}{dt}$ is the rate of change of the volume of water, and since we are given that water is entering the tank at the rate of 500 L/min, or $\frac{1}{2}$ cu m/min, we have

$$\frac{dV}{dt} = \frac{1}{2}$$

So

$$\frac{1}{2} = \frac{1}{9}\pi h^2 \frac{dh}{dt}$$

or

$$\frac{dh}{dt} = \frac{9}{2\pi h^2} \tag{3.3}$$

Now we can find $\frac{dh}{dt}$ for any value of h. So for part a we have

$$\left.\frac{dh}{dt}\right|_{h=3} = \frac{9}{2\pi(3^2)} = \frac{1}{2\pi} \approx 0.159 \text{ m/min}$$

For part b we must find h when the tank is half full. The volume of the entire tank is $\frac{1}{3}\pi(2)^2 \cdot 6 = 8\pi$, so when it is half full the volume is 4π. Substituting this in equation (3.2) we get

$$4\pi = \tfrac{1}{27}\pi h^3$$
$$h^3 = 4(27)$$
$$h = 3\sqrt[3]{4}$$

Now we use equation (3.3) again:

$$\left.\frac{dh}{dt}\right|_{h=3\sqrt[3]{4}} = \frac{9}{2\pi(3\sqrt[3]{4})^2} = \frac{1}{4\pi\sqrt[3]{2}} \approx 0.0632 \text{ m/min}$$ ■

EXAMPLE 3.22 A car leaves from town A and travels due north at an average speed of 40 km/hr. At the same instant another car leaves from town B, 50 km due east of town A, and travels due south at an average speed of 80 km/hr. Find how fast the distance between the cars is increasing 1 hr later.

Solution As in Figure 3.32 let x denote the distance traveled by the first car, y the distance traveled by the second car, and z the distance between them, all after t hours. Note that x, y, and z are all variable quantities. The only constant is the distance 50 between A and B. From the given information we know that

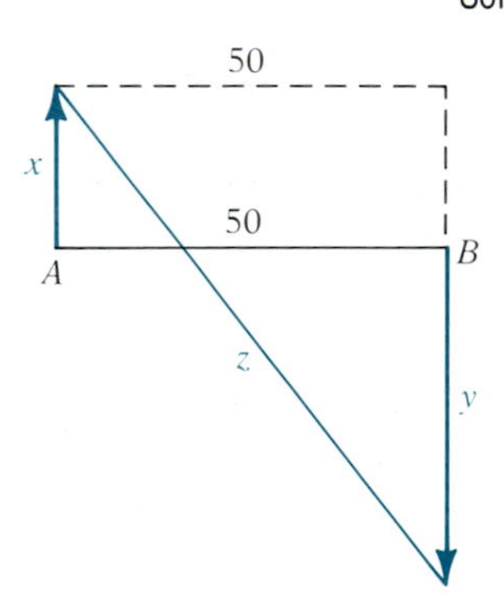

FIGURE 3.32

$$\frac{dx}{dt} = 40 \quad \text{and} \quad \frac{dy}{dt} = 80$$

and we want to find $\frac{dz}{dt}$ at a particular time. The first thing we must do is to find an equation relating the three variables. Then we will differentiate.

By means of the broken lines in Figure 3.32 we have constructed a right triangle with legs 50 and $x + y$, and hypotenuse z. By the Pythagorean theorem we have

$$z^2 = (50)^2 + (x + y)^2 \tag{3.4}$$

We differentiate both sides with respect to time:

$$2z\frac{dz}{dt} = 2(x + y)\left(\frac{dx}{dt} + \frac{dy}{dt}\right)$$

Substituting the known values of $\frac{dx}{dt}$ and $\frac{dy}{dt}$ and solving for $\frac{dz}{dt}$ give

$$\frac{dz}{dt} = 120\left(\frac{x + y}{z}\right) \tag{3.5}$$

which is the rate of change of the distance between the cars at any time. Now we wish to let $t = 1$. At that instant $x = 40$ and $y = 80$, and so from equation (3.4),

$$z = \sqrt{(50)^2 + (120)^2} = \sqrt{16{,}900} = 130$$

Thus, by equation (3.5),

$$\left.\frac{dz}{dt}\right|_{t=1} = (120)\left(\frac{40 + 80}{130}\right) = \frac{1440}{13} \approx 110.8 \text{ km/hr}$$ ■

These two examples are typical of related rate problems. Let us examine what they have in common. In both cases we are given one or more rates of change, and we wish to find another at a given instant. We were able to find a relationship between the variables whose rates of change we knew and the variable whose rate of change we were seeking. In Example 3.21 this was done by using the formula for the volume of a cone and then using similar triangles to get r in terms of h. In Example 3.22 it was done by using the Pythagorean theorem. In both cases drawing a figure was an important aid in finding the relationships. Next we differentiated with respect to time and substituted the known rates of change. Finally, we substituted the values of all variables at the particular instant in question and solved for the unknown rate of change.

We summarize these steps here.

1. If it is appropriate, draw a figure. On it label all variable parts with letters and all fixed parts with constants.
2. Find an equation relating the variables whose rates of change are known and the variable whose rate of change is desired. In doing this, look for such things as areas or volumes of geometric figures, similar triangles, the Pythagorean theorem, or trigonometric relationships.
3. Differentiate both sides of the equation found in step 2 with respect to time, and substitute in the known rates of change.
4. Calculate the values of all variables involved at the particular instant in question, and substitute them in the result of step 3.
5. Solve for the unknown rate of change.

A Word of Caution The importance of waiting until step 4 to substitute instantaneous values of variable quantities cannot be overemphasized. This is to be done only *after* differentiation. Substituting such values too soon is the most frequent mistake students make.

We will give one further example.

EXAMPLE 3.23 A television camera is 30 m directly opposite the finish line at an automobile racetrack and is trained on the lead car as it approaches the finish. If the car is traveling at the rate of 150 km/hr, find the rate in radians per second at which the camera is turning when the car is 40 m from the finish line.

Solution In Figure 3.33 we have labeled the relevant variables x and θ. We will measure x in meters and θ in radians. Note carefully that the distance from the car to the finish line is not shown as 40 because this distance is a *variable* quantity. We will put $x = 40$ only *after* we differentiate.

We are given that the car is approaching the finish line at 150 km/hr. Since other units are in meters and we want the answer to be in radians

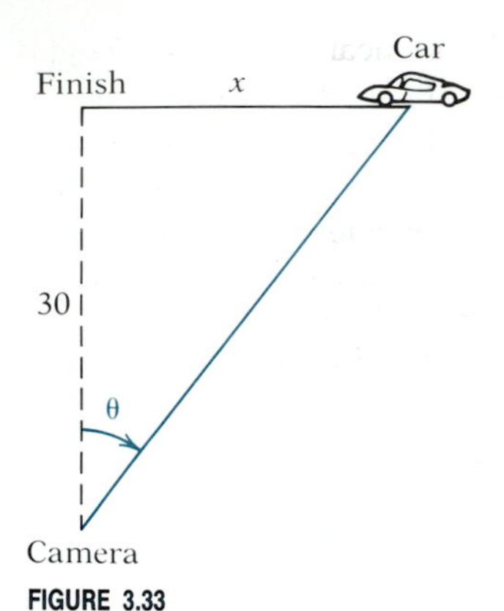

FIGURE 3.33

per second, we convert 150 km/hr to meters per second:

$$150 \text{ km/hr} = \frac{150{,}000}{3600} \text{ m/sec}$$
$$= \frac{125}{3} \text{ m/sec}$$

Now this is the rate at which the distance x is changing, but since x decreases with time, we know that $\frac{dx}{dt}$ must be negative. So

$$\frac{dx}{dt} = -\frac{125}{3}$$

An equation relating x and θ is

$$\tan \theta = \frac{x}{30}$$

We differentiate both sides with respect to time:

$$\sec^2 \theta \frac{d\theta}{dt} = \frac{1}{30}\frac{dx}{dt} = \frac{1}{30}\left(-\frac{125}{3}\right) = -\frac{25}{18}$$

So

$$\frac{d\theta}{dt} = \frac{1}{\sec^2 \theta}\left(-\frac{25}{18}\right) = -\frac{25}{18}\cos^2 \theta$$

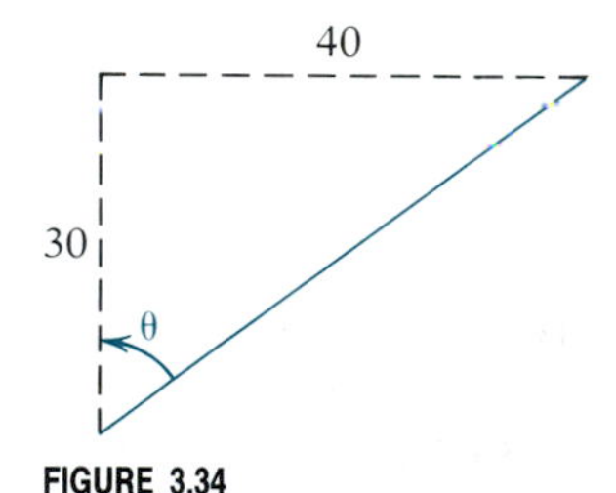

FIGURE 3.34

Now we are ready to consider the instantaneous situation when $x = 40$. At that instant, the right triangle is as shown in Figure 3.34 and so the hypotenuse is 50. Thus $\cos \theta = \frac{3}{5}$. Therefore, we have

$$\left.\frac{d\theta}{dt}\right|_{x=40} = -\frac{25}{18}\left(\frac{3}{5}\right)^2 = -\frac{25}{18}\cdot\frac{9}{25} = -\frac{1}{2} \text{ radian/sec}$$

The significance of the negative sign is that θ decreases with time. ■

EXERCISE SET 3.5

A

1. Two joggers leave simultaneously from the same point, one going east at 3 m/sec and the other going south at 4 m/sec. Find how fast the distance between them is increasing 30 min later.

2. The commander of a Coast Guard cutter is notified from an aircraft that a boat suspected of carrying illegal cargo is headed north in the Gulf of Mexico at the rate of 15 knots (1 knot is 1 nautical mile per hour). By heading east and traveling at 36 knots, the commander estimates that the cutter can intercept the boat in 50 min. Find how fast the two boats are approaching each other after 30 min.

3. A girl 5 ft tall is walking away from a lamppost at the rate of 3 ft/sec. If the height of the lamppost is 20 ft, find the rate at which her shadow is lengthening.

4. In Exercise 3, find the rate at which the tip of the girl's shadow is moving.

5. A child throws a rock into a pond, causing a circular ripple. If the radius of the circle increases at the constant rate of 1 m/sec, find how fast the area is increasing when the radius is 5 m.

6. A V-shaped trough has a cross section in the form of an inverted isosceles triangle with base 40 cm and

altitude 50 cm. If the trough is 3 m long and is being filled with water at the rate of 200 L/min (1 L = 10^{-3} cu m), find how fast the water level is rising when the water level is 30 cm.

7. A lighthouse is 500 m from a straight shore. If its light is revolving at the rate of 10 rpm, find how fast the beam of light is moving along the shoreline when the angle between the light beam and the shoreline is 30°.

8. Helium is being pumped into a spherical balloon at the rate of 600 cu cm/sec. At what rate is the radius increasing when the radius is 20 cm? At what rate is the surface area of the balloon increasing at this same instant?

9. Two runners are 7 m apart at the starting line, and they run on parallel paths, one averaging 8 m/sec and the other 7.6 m/sec. At what rate is the distance between the runners increasing after 1 min?

10. Corn in a grain elevator is being dropped through a chute to the floor, forming a conical pile whose height is always one-fourth the diameter of the base. If the corn is being dropped at the rate of 3 cu m/min, find how fast the height of the pile is increasing when it is 1 m high. How fast is the circumference of the base increasing at this same instant?

11. A watering trough 10 ft long has a cross section in the form of an isosceles trapezoid with dimensions as shown in the figure. If water is being pumped into the tank at the rate of 8 cu ft/min, find how fast the water level is rising when it is 1 ft deep.

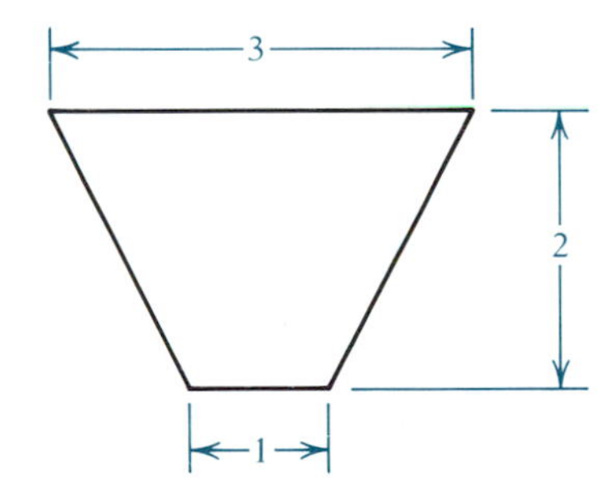

12. In Exercise 11 find how fast the water level is rising when the trough is half full.

13. A boat is being pulled in by means of a rope attached to a windlass on the dock. The windlass is 4.5 m above the level at which the rope is attached to the boat, and the rope is being pulled in at the rate of 2 m/sec. Find how fast the boat is approaching the base of the dock when it is 6 m from the base.

14. A helicopter is 500 m directly above an observer on the ground at a given instant and is flying horizontally at the rate of 50 m/sec. Find how fast the angle at the observer from the vertical to the line of sight of the helicopter is changing after 10 sec.

15. A projectile is fired from the ground and follows the path $y = 0.5x - 0.006x^2$, where x is in kilometers, and the origin is taken as the point of firing, as shown in the figure. If the horizontal speed of the missile is constant at $\frac{80}{3}$ km/min, find the rate at which the angle θ is changing when $x = 50$ km.

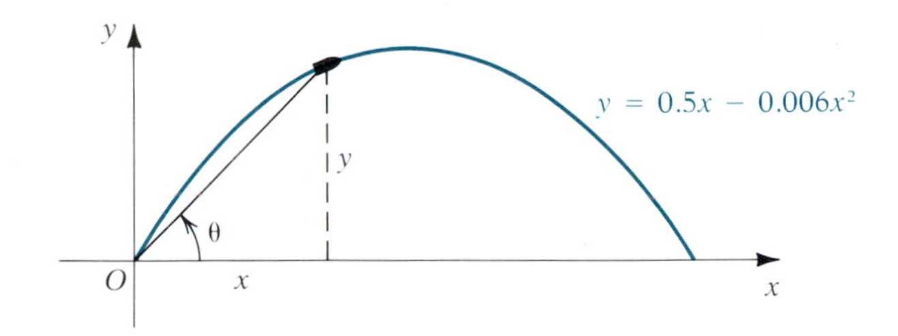

16. When a gas under pressure expands or contracts with no gain or loss of heat, the relationship between pressure p and volume v is given by $pv^{1.4} = C$, where C is a constant. This is known as the *adiabatic law*. Suppose for such a gas at a given instant the pressure is 200 dynes/sq cm, the volume is 64 cu cm, and the pressure is increasing at the rate of 10 dynes/sq cm each second. Find the rate of change of the volume at this instant.

17. The formula

$$\frac{1}{f} = \frac{1}{u} + \frac{1}{v}$$

expresses the relationship between the focal length f of a thin lens, the distance u of an object from the lens, and the distance v of its image from the lens. Suppose the focal length of a certain lens is 30 cm and an object is moving toward the lens at the rate of 3 cm/sec. Find the rate at which the image is receding from the lens when the object is 90 cm away.

18. An oil spill from a drilling platform in the Gulf of Mexico is approximately circular, and the diameter is observed to be increasing at the constant rate of 100 m per day. How fast is the area increasing when the diameter is 500 m?

19. Newton's universal law of gravitation states that two bodies of masses m_1 and m_2, at a distance r units apart, exert a force of attraction on each other given by the formula

$$F = G\frac{m_1 m_2}{r^2}$$

where G is the universal gravitational constant. Suppose that for two bodies of fixed masses m_1 and m_2, respectively, the force of attraction is 20 N (newtons) when they are 10 m apart, and that at that instant they are separating at the rate of 5 m/sec. How fast is the force of attraction decreasing?

20. A small airplane is flying 120 mi/hr at an altitude of 5000 ft. An observer on the ground is keeping a transit trained on the plane. Through how many radians per second must the observer rotate the scope of the transit when the plane is flying away from the observer and the angle of elevation of the plane is 60°. Neglect the height of the transit. (Be careful with units.)

21. Some children are rolling a ball of snow to make a snowman. If the radius of the ball is increasing at the rate of 20 cm/min, how fast is the volume of the ball increasing when it is 60 cm in diameter?

22. A searchlight located 30 m from a straight road is kept trained on a car traveling on the road at 50 km/hr. Through how many radians per second is the searchlight turning when the car is 20 m past the point on the road nearest the searchlight?

23. A girl flying a kite is letting string out at the rate of 2 m/sec, and the kite is 30 m high (from hand level) moving horizontally. Find the speed of the kite at the instant when 34 m of string are out.

24. The buildup of mineral deposits in a water pipe in an area of hard water gradually restricts the rate of flow. Assume that the buildup forms a uniform circular ring around the inside of a pipe (see the figure), and suppose that in a pipe of inside diameter 1.2 cm the area of the annular ring increases at the rate of 10π sq mm per year. At what rate is the effective diameter of the pipe decreasing when the thickness of the deposit is 2 mm?

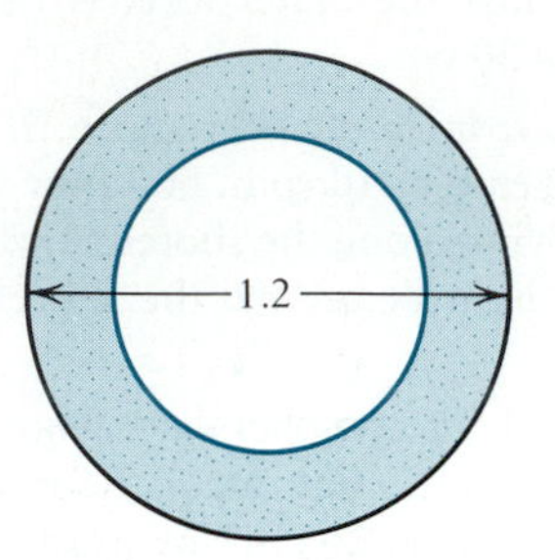

25. The larvae of a certain type of garden pest have the shape of a right circular cylinder with hemispheres at both ends and with length five times the diameter (see the figure). During the first stages of growth, the larvae increase in diameter at the rate of 0.2 mm per day. At what rate does the volume increase when the diameter is 3 mm?

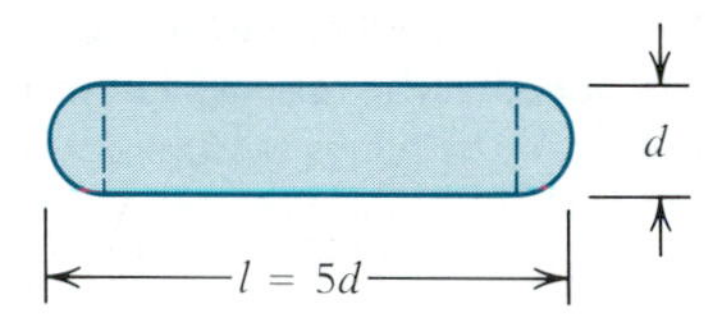

26. A water tank in the form of an inverted cone of height 6 m and base radius 3 m is being drained at the rate of 2 cu m/min. How fast is the water level falling when there are $\frac{2\pi}{3}$ cu m of water in the tank?

B

27. In a certain mechanical system parts A and B are connected by a cable 84 cm long that passes over a pulley at C 30 cm above the point D on a horizontal track on which parts A and B are constrained to move. The parts move back and forth in such a way that the cable is always taut. At a certain instant part B is 16 cm from D and is moving to the right at 17 cm/sec. How fast is part A moving at that instant?

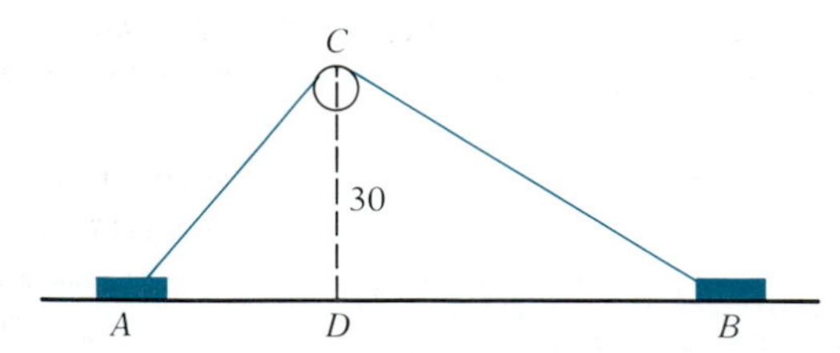

28. One end of an 85-ft rope is attached to a weight on the ground, and the rope goes over a pulley 45 ft above the ground, with the other end held by a man standing beside the weight. The end of the rope held by the man is 5 ft above the ground, and continuing to hold it at this height, he begins to walk away at the constant rate of 5 ft/sec, thereby raising the weight. How fast is the weight rising when the man has walked 30 ft?

29. An oil tank is in the form of a right circular cylinder with a horizontal axis, having length 3 m and diameter 1 m. If the tank is being filled at the rate of 60 L/min, find the rate at which the oil depth is increasing at the instant when the depth is 20 cm.

30. Cars A and B leave from the same point, traveling on roads that are 120° apart. Car A travels at an average speed of 60 km/hr. Car B leaves the intersection one-half hour after car A and travels at an average speed of 45 km/hr. How fast are the cars separating when car A has traveled 150 km?

31. A water tank is in the form of a frustum of a cone with upper radius 3 m, lower radius 1 m, and height 4 m (see the figure). Water is being pumped into the tank at the rate of 5 cu m/min and is being taken from the tank at the rate of 2 cu m/min. Find the rate at which the water level is rising when there are 30 cu m of water in the tank.

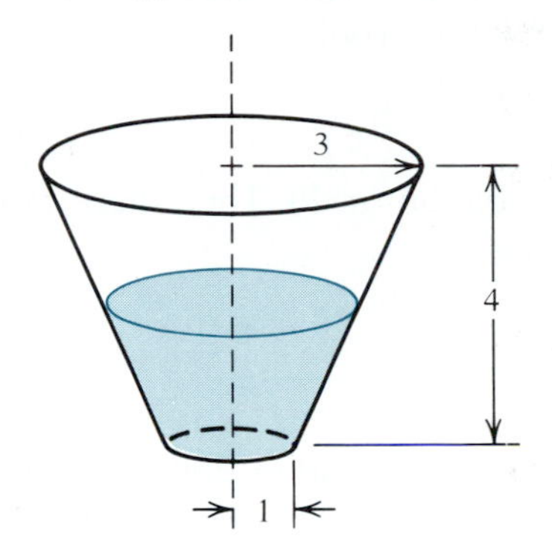

32. A bicyclist passes over the middle of a bridge at the same time a motorboat goes under the middle on a path at right angles to the bridge. The center span is 24 m above the water and is essentially level for a distance of 100 m. If the bicyclist is going at the rate of 10 m/sec and the boat is going at the rate of 15 m/sec, find the rate at which the distance between them is increasing 4 sec later.

3.6 THE MEAN VALUE THEOREM

In this section we prove a theorem that has far-reaching consequences, known as the **mean value theorem** for differential calculus, or the **law of the mean.** As one application of it, we will use it to prove some of the results of this chapter that we accepted earlier on intuitive grounds. We precede the main theorem with another result, which is used in the proof and so could be referred to as a lemma. This preliminary result is important in its own right, however, and it is known throughout the literature as Rolle's theorem, after the 17th-century French mathematician Michel Rolle.

THEOREM 3.7 ROLLE'S THEOREM

Let f be continuous on $[a, b]$ and differentiable on (a, b), with $f(a) = 0$ and $f(b) = 0$. Then there exists a number c in the open interval (a, b) such that $f'(c) = 0$.

Proof Figure 3.35 illustrates a typical situation. In this case there are three values of c for which $f'(c) = 0$. Since f is continuous on $[a, b]$, we know by Theorem 3.2 that it attains both an absolute maximum and an absolute minimum there. If $f(x) = 0$ for all x on $[a, b]$, then any number c between a and b satisfies the conclusion. If f is not constant, then either its maximum is

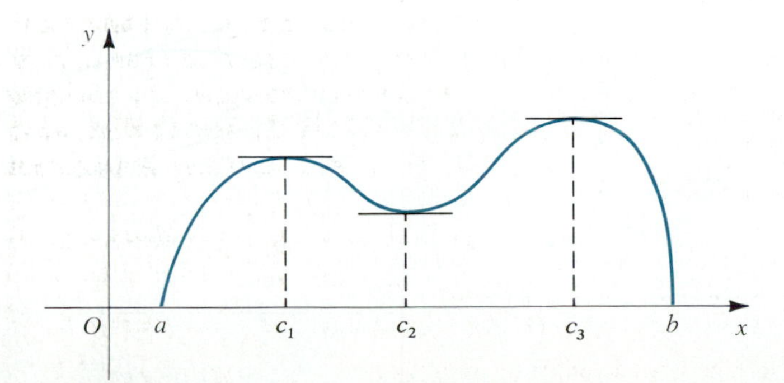

FIGURE 3.35

greater than 0 or its minimum is less than 0, or both (why?), and hence at least one of these must occur at an interior point c. But since $f'(c)$ exists, it must be 0 by Theorem 3.3. This completes the proof. ■

THEOREM 3.8 MEAN VALUE THEOREM

Let f be continuous on $[a, b]$ and differentiable on (a, b). Then there exists a point c in the open interval (a, b) such that

$$f'(c) = \frac{f(b) - f(a)}{b - a}$$

Proof Note that the only difference in the hypotheses here and in Rolle's theorem is the absence of the requirement that $f(a)$ and $f(b)$ be equal to 0. The quantity

$$\frac{f(b) - f(a)}{b - a}$$

is the slope of the secant line that joins the end points $(a, f(a))$ and $(b, f(b))$. So what we must show is the existence of some number c between a and b such that the slope $f'(c)$ of the tangent line to the curve at $(c, f(c))$ is equal to the slope of the secant line that joins the end points; that is, these lines are parallel. Figure 3.36 illustrates this situation.

For convenience, let

$$m = \frac{f(b) - f(a)}{b - a}$$

Then the equation of the secant line that joins the end points is

$$y = f(a) + m(x - a) \qquad \text{(Verify.)}$$

We are going to define a new function F whose value represents geometrically the vertical distance from this secant line to the curve. Thus, we define

$$F(x) = f(x) - [f(a) + m(x - a)] \tag{3.6}$$

FIGURE 3.36

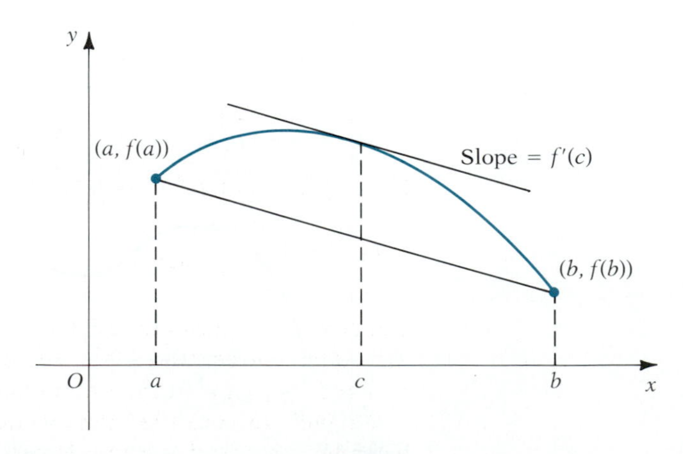

Now F is the difference of two functions, each of which is continuous on $[a, b]$ and differentiable on (a, b), and so F also has these properties. Furthermore, $F(a) = 0$ and $F(b) = 0$, as can be seen by direct substitution. (Verify.) Thus, F satisfies all hypotheses of Rolle's theorem and hence its conclusion; that is, there exists a number c in (a, b) such that $F'(c) = 0$. But from equation (3.6),

$$F'(x) = f'(x) - m$$

Now $F'(c) = f'(c) - m = 0$, from which we get that

$$f'(c) = m$$

and this is the desired conclusion. ■

EXAMPLE 3.24 Find a number c in the interval $(-1, 3)$ that satisfies the conclusion of the mean value theorem for the function

$$f(x) = x^3 - x^2 + 3$$

Solution Note first that the hypotheses of the theorem are satisfied, since f is continuous on $[-1, 3]$ and differentiable on $(-1, 3)$. In fact, f is differentiable and hence continuous everywhere, since it is a polynomial. We want to show the existence of a number c in $(-1, 3)$ for which

$$f'(c) = \frac{f(3) - f(-1)}{3 - (-1)} = \frac{21 - 1}{4} = 5$$

Since $f'(x) = 3x^2 - 2x$, we solve the equation

$$\begin{aligned} 3c^2 - 2c &= 5 \\ 3c^2 - 2c - 5 &= 0 \\ (3c - 5)(c + 1) &= 0 \\ c = \tfrac{5}{3} \quad &| \quad c = -1 \end{aligned}$$

But the requirement is for c to be in the *open* interval $(-1, 3)$, so we must have $-1 < c < 3$. The solution $c = \frac{5}{3}$ is therefore the only one that satisfies the theorem. ■

If f fails to satisfy the hypotheses of the mean value theorem, the conclusion may or may not be true. Consider, for example,

$$f(x) = 2 + (1 - x)^{2/3} \quad \text{on } [0, 2]$$

Here $f(0) = 3$ and $f(2) = 3$. So $[f(2) - f(0)]/2 = 0$. Since

$$f'(x) = -\frac{2}{3(1 - x)^{1/3}}$$

can never be 0, the conclusion of the mean value theorem cannot be satisfied. This does not contradict the theorem, however, because $f'(x)$ fails to exist at $x = 1$, which is in the open interval $(0, 2)$.

We will now use the mean value theorem to prove two of the results we accepted on intuitive grounds earlier.

Proof of Theorem 3.1

The theorem states that if $f'(x) > 0$ for all x in an open interval I, then f is increasing on I, and if $f'(x) < 0$ for all x in I, then f is decreasing on I. Suppose first that $f'(x) > 0$ on I. Choose x_1 and x_2 in I, with $x_1 < x_2$. Since $f'(x)$ exists on the interval I, f is differentiable and therefore continuous on the interval $[x_1, x_2]$, so that the hypotheses of the mean value theorem are more than satisfied. In that theorem we replace a by x_1 and b by x_2 and conclude that a number c exists in (x_1, x_2) such that

$$f'(c) = \frac{f(x_2) - f(x_1)}{x_2 - x_1}$$

Since $f'(c) > 0$ by hypothesis, and since $x_2 > x_1$ by choice, it follows that the numerator on the right must be positive; that is, $f(x_2) > f(x_1)$. In summary, we have shown that for $x_1 < x_2$, $f(x_1) < f(x_2)$, and this is what is meant by an increasing function. This completes the proof of the first part of the theorem. We leave the second part as an exercise. ■

Proof of Theorem 3.4

This theorem states that if $f''(a) > 0$, then f is concave upward at $(a, f(a))$, and if $f''(a) < 0$, then f is concave downward at $(a, f(a))$. Suppose first that $f''(a) > 0$. What we must show is that in some deleted neighborhood of a, the graph of f lies altogether above the tangent line drawn at a. The equation of this tangent line is

$$y = f(a) + f'(a)(x - a)$$

We define a new function F to be the difference in the ordinate of f and the ordinate to the tangent line (Figure 3.37):

$$\begin{aligned} F(x) &= f(x) - [f(a) + f'(a)(x - a)] \\ &= f(x) - f(a) - f'(a)(x - a) \end{aligned} \tag{3.7}$$

If we can show that this is positive throughout some deleted neighborhood of a, we will have proved the first part.

Using the alternate way of writing the derivative in equation (2.4), with f replaced by f', we have

$$f''(a) = \lim_{x \to a} \frac{f'(x) - f'(a)}{x - a}$$

It follows that there exists a deleted neighborhood N of a, throughout which

$$\frac{f'(x) - f'(a)}{x - a}$$

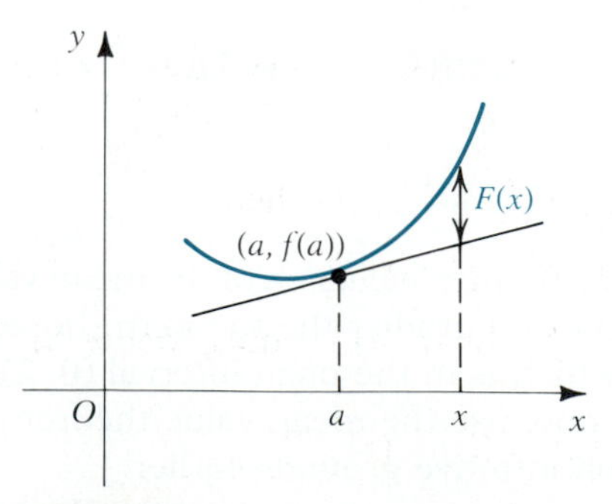

FIGURE 3.37

is positive, since otherwise the limit could not be positive. So when x is in N and $x > a$, we must have $f'(x) - f'(a) > 0$, and when $x < a$, we must have $f'(x) - f'(a) < 0$. Implicit in the existence of the limit defining $f''(a)$ also is the fact that $f'(x)$ exists throughout N.

Consider first $x > a$ and x in N. Then, since f is differentiable at a and throughout N, f satisfies the conditions of the mean value theorem on $[a, x]$. So there exists a number c between a and x such that

$$f'(c) = \frac{f(x) - f(a)}{x - a}$$

or

$$f(x) - f(a) = f'(c)(x - a)$$

We substitute this into equation (3.7) and get

$$F(x) = f'(c)(x - a) - f'(a)(x - a) = [f'(c) - f'(a)](x - a) \qquad (x \in N)$$

Since $a < c < x$, we know from the above discussion that $f'(c) - f'(a) > 0$, and since $x > a$, we see that $F(x) > 0$ throughout N; that is, for x in N and to the right of a, the curve lies above the tangent line. So f is concave upward.

The proof for x in N and $x < a$ is handled in a similar way and is left as an exercise. Also, the proof that the curve is concave downward when $f''(a) < 0$ is left as an exercise. ■

We conclude this section with two further consequences of the mean value theorem.

THEOREM 3.9 If $f'(x) = 0$ for all x in an interval I, then f is a constant function on I.

Proof Let a denote any fixed point in I, and let x denote any other point in I. We will show that $f(x) = f(a)$, and since $f(a)$ is constant, this will tell us that all functional values are equal to this constant. We can apply the mean value theorem to the interval $[a, x]$ (or to $[x, a]$ if $x < a$, but to simplify matters we will assume $x > a$). The existence of f' throughout I guarantees that the hypotheses of the theorem are satisfied. (Why?) Thus, there is a number c in (a, x) such that

$$f'(c) = \frac{f(x) - f(a)}{x - a}$$

But since c is in I, $f'(c) = 0$. Thus, $f(x) - f(a) = 0$, and $f(x) = f(a)$. Thus, $f(x)$ is constant for all x in I. ■

COROLLARY 3.9 If $f'(x) = g'(x)$ for all x in an interval I, then for all x in I, $f(x) = g(x) + C$ for some constant C.

Proof Define $h(x) = f(x) - g(x)$. Then

$$h'(x) = f'(x) - g'(x) = 0$$

for all x in I. So by the preceding theorem, $h(x)$ is a constant—say $h(x) = C$. Thus

$$f(x) - g(x) = C \qquad (x \in I)$$

which is equivalent to the desired conclusion. ■

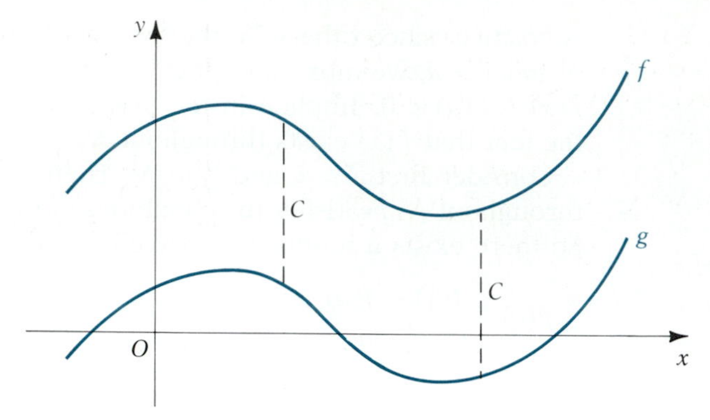

FIGURE 3.38

This corollary says that when two functions have the same slopes at all points, the curves are separated by a constant. In a certain sense the curves can be thought of as being parallel. Figure 3.38 illustrates this.

EXERCISE SET 3.6

A

In Exercises 1–8 show that the conditions of Rolle's theorem are satisfied by the function on the given interval, and find a number c that satisfies the conclusion of the theorem.

1. $f(x) = x^2 - 3x - 4$ on $[-1, 4]$
2. $f(x) = x^2 - 4x + 3$ on $[1, 3]$
3. $f(x) = 2x^2 - 3x - 2$ on $[-\frac{1}{2}, 2]$
4. $f(x) = x^3 + x^2 - x - 1$ on $[-1, 1]$
5. $f(x) = x^3 - 9x$ on $[0, 3]$
6. $f(x) = 2x^3 - x^2 - 8x + 4$ on $[-2, 2]$
7. $f(x) = (x^2 - 9)\sqrt{x+1}$ on $[-1, 3]$
8. $f(x) = (x - 1)\sqrt{3 - x^2}$ on $[1, \sqrt{3}]$

In Exercises 9–16 show that the conditions of the mean value theorem are satisfied by the function on the given interval, and find a number c that satisfies the conclusion of the theorem.

9. $f(x) = x^2 - 3x + 1$ on $[-1, 2]$
10. $f(x) = 4 - x - 3x^2$ on $[0, 3]$
11. $f(x) = x^3 - 2x^2 + 3$ on $[-1, 4]$
12. $f(x) = x^3 - 2x^2 + 5x + 8$ on $[-2, 2]$
13. $f(x) = x^3 + 5x^2 - 2x - 5$ on $[-1, 2]$
14. $f(x) = x - \frac{1}{x}$ on $[\frac{1}{2}, 2]$
15. $f(x) = \sqrt{2x+1}$ on $[0, 4]$
16. $f(x) = \dfrac{x-1}{x+1}$ on $[0, 5]$

In Exercises 17–20 show that the conclusion of Rolle's theorem is not satisfied by the function on the given interval, and explain why there is no contradiction with the theorem.

17. $f(x) = 1 - x^{2/3}$ on $[-1, 1]$
18. $f(x) = \dfrac{x^2 - 4}{x + 1}$ on $[-2, 2]$
19. $f(x) = \dfrac{2 \sin x - 1}{\cos x}$ on $\left[0, \dfrac{5\pi}{6}\right]$
20. $f(x) = \dfrac{1 - 2 \cos x}{\sin x}$ on $\left[\dfrac{-\pi}{3}, \dfrac{\pi}{3}\right]$

In Exercises 21–24 show that the conclusion of the mean value theorem is not satisfied by the function on the given interval, and explain why there is no contradiction with the theorem.

21. $f(x) = \dfrac{x^2 - 1}{x}$ on $[-1, 2]$
22. $f(x) = x - \dfrac{2}{x} + 3$ on $[-2, 2]$
23. $f(x) = (x - 1)^{2/3}$ on $[0, 2]$
24. $f(x) = x - 3x^{2/3}$ on $[-8, 8]$

25. Show that $f(x) = 1 - x + (3x - 1)^{1/3}$ fails to satisfy all of the hypotheses of Rolle's theorem on $[0, 3]$, yet the conclusion of the theorem is true. Does this represent a contradiction? Explain.

26. Show that $f(x) = x + 3(1 - x)^{1/3}$ fails to satisfy all of the hypotheses of the mean value theorem on $[0, 2]$, yet the conclusion of the theorem is true. Does this represent a contradiction? Explain.

27. Find two points in the interval $(-1, 3)$ for which the tangent line to the graph of

$$f(x) = 2x^3 - 5x^2 - 3x + 1$$

is parallel to the secant line that joins $(-1, f(-1))$ and $(3, f(3))$. Draw a figure showing the graph of f, the two tangent lines, and the secant line.

B

28. Let f satisfy the conditions of the mean value theorem on $[a, b]$, and let x be any number in $(a, b]$. Show that

$$f(x) = f(a) + f'(c)(x - a)$$

where c is some number that satisfies $a < c < x$.

29. Let f satisfy the conditions of the mean value theorem on $[a, b]$, and let h be any number that satisfies $0 < h \leq b - a$. Show that

$$f(a + h) = f(a) + h \cdot f'(a + \theta h)$$

for some number θ such that $0 < \theta < 1$.

30. Use the mean value theorem to show that for any two numbers x_1 and x_2,

$$|\sin x_1 - \sin x_2| \leq |x_1 - x_2|$$

31. Use the mean value theorem to show that $\tan x > x$ for all x that satisfy $0 < x < \frac{\pi}{2}$. (*Hint:* First pick an x in the given interval and then apply the mean value theorem to $f(x) = \tan x$ on $[0, x]$.)

32. Complete the proof of Theorem 3.1 by showing that if $f'(x) < 0$ for all x on an open interval I, then f is a decreasing function.

33. Complete the proof of the first part of Theorem 3.4 by showing that for x in N and $x < a$, the graph of f lies above the tangent line.

34. Prove the second half of Theorem 3.4 by showing that if $f''(a) < 0$, there is a deleted neighborhood of a throughout which the function lies below the tangent line drawn at $(a, f(a))$.

35. Prove that if f is differentiable on $[a, b]$ with $f'(a)$ and $f'(b)$ of opposite sign, then there is a number c between a and b such that $f'(c) = 0$. [*Hint:* Show that f must attain its maximum or its minimum (or both) somewhere between a and b. Then use Theorem 3.2.]

36. Prove the following **intermediate value theorem for derivatives:** if f is differentiable on $[a, b]$ and k is any number between $f'(a)$ and $f'(b)$, then there is a number c between a and b such that $f'(c) = k$. (Note that f' need not be a continuous function.) [*Hint:* Show that the result of Exercise 35 can be applied to $g(x) = f(x) - kx$.]

3.7 ANTIDERIVATIVES

Sometimes it is important to do the opposite of differentiating—that is, to find a function whose derivative we know. We call this process **antidifferentiation,** and a function arrived at in this way is called an **antiderivative** of the given function. For example, suppose we are given the function $3x^2$, and we are asked to find a function for which this is the derivative. We can do this mentally and arrive at x^3 as an answer. So we can say that x^3 is an antiderivative of $3x^2$. We are being very careful in using the article *an* here rather than *the* because there will always be many (infinitely many, in fact) antiderivatives of a function if there are any at all. In the present case, for example, we can see that each of the following is an antiderivative of $3x^2$: $x^3 + 2$, $x^3 - 4$, $x^3 + 100$. In fact, $x^3 + C$, where C is any real number, is an antiderivative of $3x^2$. Whether there can be any that are not of this form is answered by the next theorem.

THEOREM 3.10 Let $F(x)$ be an antiderivative of $f(x)$ on an interval I. Then every other antiderivative of $f(x)$ on I is of the form $F(x) + C$ for some constant C.

Proof We are given that $F'(x) = f(x)$ for all $x \in I$. Suppose G is any other antiderivative of $f(x)$ on I, so that $G'(x) = f(x)$ on I. Then for all $x \in I$, $G'(x) = F'(x)$. Therefore, by Corollary 3.9,

$$G(x) = F(x) + C$$

for some constant C. ■

If $F(x)$ is an antiderivative of $f(x)$, then we sometimes call $F(x) + C$ the **general antiderivative** of $f(x)$, so called because we leave the constant C unspecified. In the sense of the preceding theorem, $F(x) + C$ represents the *family* of *all* antiderivatives of $f(x)$. We will soon learn certain techniques for finding antiderivatives that are less obvious than the one we have considered so far. For the present we will concentrate on certain basic types.

We know that to differentiate a power of x, say x^n, we multiply by the exponent and then subtract 1 from the exponent, getting nx^{n-1}. To find an antiderivative of x^n we do the reverse; that is, we *add* 1 to the exponent and then *divide* by the resulting exponent, getting

$$\frac{x^{n+1}}{n+1}$$

For this to have meaning we must have $n \neq -1$. With this restriction, we can check the result by differentiating:

$$\frac{d}{dx}\left(\frac{x^{n+1}}{n+1}\right) = (n+1)\cdot\frac{x^n}{n+1} = x^n$$

For example, an antiderivative of x^3 is $x^4/4$, and an antiderivative of $x^{-1/2}$ is $2x^{1/2}$.

Because the derivative of a constant times a function is the constant times the derivative of the function, the same is true of antiderivatives. For example, an antiderivative of $5x^3$ is $5x^4/4$. Also, since the derivative of the sum or difference of functions is the sum or difference of the derivatives, the same is true for antiderivatives. This enables us to find antiderivatives of polynomials. For example, an antiderivative of $2x^2 - 3x + 4$ is

$$\frac{2x^3}{3} - \frac{3x^2}{2} + 4x$$

The following table summarizes some of the basic antiderivatives we already know.

Function	Antiderivative
x^n	$\dfrac{x^{n+1}}{n+1} + C \quad (n \neq -1)$
$\sin x$	$-\cos x + C$
$\cos x$	$\sin x + C$
$\sec^2 x$	$\tan x + C$
$\csc^2 x$	$-\cot x + C$
$\sec x \tan x$	$\sec x + C$
$\csc x \cot x$	$-\csc x + C$

All of these can be verified by differentiating the function on the right to get the function on the left. It should be clear that in the future, whenever we find a formula for the derivative of a function, we will also have found a formula for an antiderivative. If we denote by F any antiderivative of f, then we also have the following:

Function	Antiderivative
$cf(x)$	$cF(x)$
$f_1(x) \pm f_2(x)$	$F_1(x) \pm F_2(x)$

As we have seen, every function that has an antiderivative has infinitely many of them, differing from each other only by constants. Sometimes we are given enough information to determine one particular antiderivative, as the next example illustrates.

EXAMPLE 3.25 The slope of the tangent line at an arbitrary point on the graph of a certain function f is given by $3x^2 + 2x - 1$, and the point $(2, -1)$ is on the graph. Find $f(x)$.

Solution We know that the slope of the tangent line is given by $f'(x)$. So we have

$$f'(x) = 3x^2 + 2x - 1$$

and we want to know $f(x)$; that is, we want an antiderivative of $3x^2 + 2x - 1$. We know this is of the form

$$f(x) = x^3 + x^2 - x + C$$

This really gives a *family* of curves, all having the same slope. We want the particular member of this family that passes through $(2, -1)$, so we substitute $x = 2$ and $f(x) = -1$, getting

$$-1 = 8 + 4 - 2 + C$$
$$C = -11$$

Thus, the desired function is given by

$$f(x) = x^3 + x^2 - x - 11$$

You can verify that this satisfies the two requirements of the problem. ■

We know from Chapter 2 that when $s(t)$ represents the distance traveled by an object moving in a straight line at time t, then $s'(t)$ represents its velocity and $s''(t)$ its acceleration. If we begin with a known acceleration, we can work backward by antidifferentiation to get first the velocity and then the distance. The next two examples show how to do this.

EXAMPLE 3.26 From the top edge of a building 96 ft high a ball is thrown vertically upward with an initial velocity of 40 ft/sec. Find the position $s(t)$ of the ball above the ground t sec later. What is the velocity after 2 sec? Find the maximum height of the ball and also its terminal velocity.

Solution We take the distance as 0 at the ground level and choose the positive direction upward. The only force on the ball is assumed to be gravity, and we take the acceleration g caused by gravity to be 32 ft/sec^2 acting downward. Then, using $v(t)$ and $a(t)$, respectively, for velocity and acceleration, we are given the following:

$$a(t) = -32$$
$$s(0) = 96$$
$$v(0) = 40$$

Notice that the acceleration is negative, since we have chosen the positive direction to be upward. The two conditions $s(0) = 96$ and $v(0) = 40$ are called *initial conditions*. They will enable us to find the constants that occur in the antidifferentiation. To get $v(t)$, we find the general antiderivative of $a(t)$:

$$v(t) = -32t + C$$

To find C, we use the initial condition $v(0) = 40$:

$$40 = -32(0) + C$$
$$C = 40$$

Thus,

$$v(t) = -32t + 40$$

We take another antiderivative to get $s(t)$:

$$s(t) = -16t^2 + 40t + C_1$$

(We use C_1 for the constant, since it is not necessarily the same as the constant C we found above.) Since $s(0) = 96$, we have

$$96 = -16(0) + 40(0) + C_1$$
$$C_1 = 96$$

Thus,

$$s(t) = -16t^2 + 40t + 96$$

This is the position function at time t.

To get the velocity after 2 sec, we use the formula for $v(t)$ and substitute $t = 2$:

$$v(2) = -32(2) + 40 = -24$$

Since the sign is negative, we know the ball is moving downward at this instant, and its speed is 24 ft/sec.

The ball reaches its maximum height when its velocity changes from positive to negative—that is, when $v(t) = 0$. So we set $v(t) = 0$ and solve for t:

$$-32t + 40 = 0$$
$$t = \tfrac{40}{32} = \tfrac{5}{4}$$

Substituting this in the formula for $s(t)$, we get

$$s_{\text{max}} = s(\tfrac{5}{4}) = -16(\tfrac{25}{16}) + 40(\tfrac{5}{4}) + 96 = 121$$

So the ball attains a maximum height of 121 ft above the ground (25 ft above the building).

Finally, the terminal velocity means the velocity with which it hits the ground, and it hits the ground when $s(t) = 0$. So we set $s(t) = 0$ and solve for t:

$$\begin{aligned} -16t^2 + 40t + 96 &= 0 \\ 2t^2 - 5t - 12 &= 0 \\ (2t + 3)(t - 4) &= 0 \\ t = -\tfrac{3}{2} \quad &| \quad t = 4 \end{aligned}$$

We reject the negative solution as being meaningless in this case. So the ball strikes the ground when $t = 4$. Its velocity at that instant is

$$v_{\text{terminal}} = v(4) = -32(4) + 40 = -88$$

So it is traveling at a speed of 88 ft/sec when it strikes the ground. ■

EXAMPLE 3.27 A particle moves on a line so that its acceleration at time $t \geq 0$ is given by $a(t) = 6(1 - t)$. The initial conditions are $v(0) = 9$ and $s(0) = 0$.

(a) Find $v(t)$ and $s(t)$.

(b) Find when and where the particle reaches its greatest positive distance from the origin.

(c) Determine the time intervals and corresponding distance intervals in which the particle is moving to the right (positive direction) and those in which it is moving to the left.

(d) Find the time and distance intervals when it is slowing down and those in which it is speeding up.

Solution (a) We proceed as in the preceding example to get first $v(t)$ and then $s(t)$ by antidifferentiation:

$$\begin{aligned} v(t) &= 6t - 3t^2 + C && \text{Antiderivative of } a(t) \\ 9 &= 6(0) - 3(0) + C && \text{Since } v(0) = 9 \\ C &= 9 \end{aligned}$$

So $v(t) = -3t^2 + 6t + 9$.

$$\begin{aligned} s(t) &= -t^3 + 3t^2 + 9t + C_1 && \text{Antiderivative of } v(t) \\ 0 &= -0 + 3(0) + 9(0) + C_1 && \text{Since } s(0) = 0 \\ C_1 &= 0 \end{aligned}$$

Thus, for part a we have

$$\begin{aligned} v(t) &= -3t^2 + 6t + 9 \\ s(t) &= -t^3 + 3t^2 + 9t \end{aligned}$$

(b) To find the greatest positive distance from the origin, we look for the absolute maximum value of $s(t)$. First, we find critical points by setting $v(t) = s'(t) = 0$:

$$\begin{aligned} -3t^2 + 6t + 9 &= 0 \\ t^2 - 2t - 3 &= 0 \\ (t - 3)(t + 1) &= 0 \\ t = 3 \quad &| \quad t \neq -1 \quad \text{Since } t \geq 0 \end{aligned}$$

Since $s(3) = -27 + 3(9) + 9(3) = 27$, and $s''(3) = a(3) < 0$, it follows that 27 is a local maximum value of $s(t)$. The only end point value is $s(0) = 0$, so 27 is the absolute maximum value of $s(t)$.

(c) Since $v(t) = -3(t+1)(t-3)$, we see that for $t \geq 0$, the velocity is positive for $t < 3$ and negative for $t > 3$. So the particle moves to the right from $s(0) = 0$ to $s(3) = 27$. Then it turns around and moves to the left for all $t > 3$.

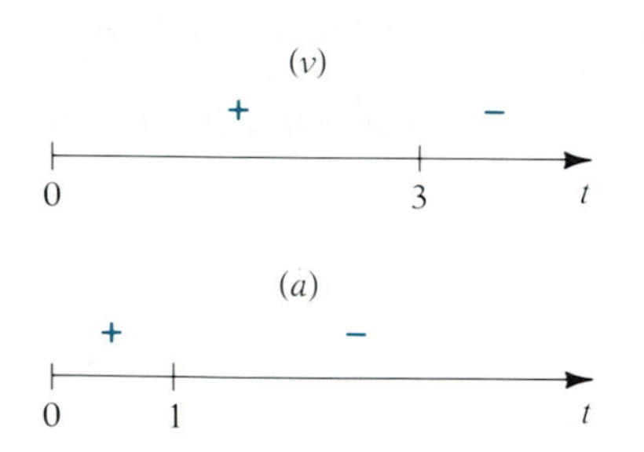

(d) Since acceleration measures the rate of change of velocity, it follows that when $v(t)$ and $a(t)$ are like in sign, the particle is speeding up, and when they are unlike in sign, it is slowing down. A comparison of sign graphs for v and a makes clear when each situation occurs. We see that from $t = 0$ to $t = 1$ it is speeding up to the right ($v > 0$, $a > 0$); from $t = 1$ to $t = 3$ it continues to the right but is slowing down ($v > 0$, $a < 0$), and for $t > 3$ it moves to the left faster and faster ($v < 0$, $a < 0$). Figure 3.39 summarizes our findings. The motion actually takes place on a line, but in this diagram we show it on two levels so as to see more clearly the nature of the motion in each direction.

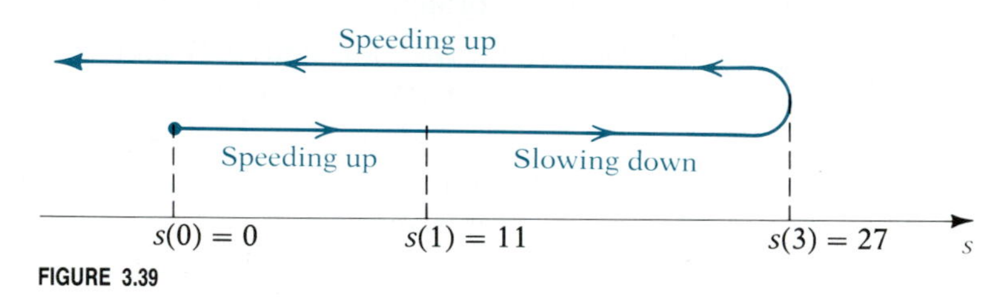

FIGURE 3.39

EXERCISE SET 3.7

A

In Exercises 1–20 find the general antiderivative of the given function.

1. $f(x) = x^5$
2. $f(x) = 3x^4$
3. $f(x) = 2x^2 - 4x + 5$
4. $f(x) = 6x^2 - 2x + 3$
5. $f(x) = 3x^3 - 5x^2 + 2x - 3$
6. $f(x) = 2 - x + 4x^2 - 6x^3$
7. $f(x) = 1/(2x^2) - x^2/2$
8. $f(x) = 2x^{-3} - 4x^{-2} + 1$
9. $f(x) = 2x^{1/2} - 3x^{-1/2}$
10. $f(x) = 1/\sqrt{x} - 2/x^2 + 1$
11. $f(x) = 3x - 4\sqrt{x} + 7$
12. $f(x) = x^{3/2} + 2x^{1/2}$
13. $f(x) = x(\sqrt{x} + 3)$
14. $f(x) = (x - 1)(1/\sqrt{x} + 1)$
15. $f(x) = 2 \sin x - 3 \cos x$
16. $f(x) = \sec x(\sec x + \tan x)$
17. $f(x) = x - 1/\sin^2 x$
18. $f(x) = \cos x + 1/\cos^2 x$
19. $f(x) = 1 - \cos x/\sin^2 x$
20. $f(x) = \tan x \cos x - \sin x \cot x$

In Exercises 21–28 find the function f that satisfies the given conditions.

21. $f'(x) = 2x - 3$; $f(0) = 5$
22. $f'(x) = 3x^2 - 2x + 1$; $f(-2) = 3$
23. $f'(x) = 1 - x - 4x^2$; $f(3) = 2$
24. $f'(x) = 3/x^2 - 2x + 3$; $f(1) = 5$
25. $f''(x) = 3x$; $f'(0) = 0$; $f(0) = 2$
26. $f''(x) = 1 - x$; $f'(1) = 2$; $f(1) = -1$

27. $f''(x) = x^2 - 1$; $f'(-1) = 0$; $f(0) = 2$

28. $f'''(x) = 2$; $f''(0) = 2$; $f'(0) = -1$; $f(0) = 4$

29. Find y as a function of x if $\frac{dy}{dx} = x + \sin x$ and the graph of the function goes through the point $(0, 3)$.

30. The graph of $y = f(x)$ goes through the point $(1, -2)$ and has a slope of $\frac{1}{2}$ at this point. If $y'' = 1 - x$, find $f(x)$.

In Exercises 31 and 32 assume that the acceleration due to gravity is 32 ft/sec², and assume there is no air resistance.

31. A ball is thrown vertically upward from the ground with an initial velocity of 64 ft/sec. Find expressions for its height $s(t)$ and its velocity $v(t)$ after t seconds. How high will the ball rise? When will it strike the ground, and what will its velocity be at that instant?

32. From a balloon 1600 ft above the ground a projectile is fired vertically downward at an initial velocity of 240 ft/sec. How long will it take to reach the ground, and what will be its terminal velocity?

33. From just over the top edge of a sheer cliff 58.8 m from the canyon floor below, a rock is thrown vertically upward at an initial velocity of 19.6 m/sec. How high will the rock rise, and when will it strike the canyon floor? (Take the acceleration due to gravity to be 9.8 m/sec².)

B

In Exercises 34–37 find the general antiderivative of the given function.

34. $f(x) = \dfrac{2 - x + x^4}{x^3}$

35. $f(x) = \dfrac{x^3 - 8x - 8}{x + 2}$

36. $f(x) = \sqrt{\dfrac{1}{4x^4} + 1 + x^4}$

37. $f(x) = \sqrt{x - 2 + \dfrac{1}{x}}$; $x \geq 1$

In Exercises 38–41 first change to an appropriate form using trigonometric identities and then find the general antiderivative.

38. $f(x) = \tan^2 x$

39. $f(x) = \dfrac{\sin 2x}{\sin^3 x}$

40. $f(x) = 2\cos^2 \dfrac{x}{2}$

41. $f(x) = \dfrac{\cos 2x}{\sin^2 x \cos^2 x}$

In Exercises 42–45 the acceleration of a particle moving on a line directed positively to the right is given, along with its initial position and velocity. Describe the motion completely for $t \geq 0$, using the pattern of Example 3.27.

42. $a(t) = t - 2$; $s(0) = 4$, $v(0) = 0$

43. $a(t) = 3t^2 - 6t - 9$; $s(0) = 0$, $v(0) = 27$

44. $a(t) = 15\sqrt{t}$; $s(0) = 12$, $v(0) = -10$

45. $a(t) = 3t^2 - 11t + 6$; $s(0) = 2$, $v(0) = 0$

3.8 SUPPLEMENTARY EXERCISES

In Exercises 1 and 2 find the intervals on which f is increasing and those on which it is decreasing.

1. (a) $f(x) = x^4 - 4x^3 - 2x^2 + 12x$
(b) $f(x) = (x + 2)^3(2x - 1)^2$

2. (a) $f(x) = (3x^2 - 8x)/(x^2 + 4)$
(b) $f(x) = (x - 1)^{2/3}(x + 7)^2$

In Exercises 3–10 find all local maxima and minima. Also find absolute maximum and minimum values if they exist.

3. $f(x) = 5 + 4x + 2x^2 - x^3$

4. $f(x) = 3x^5 - 25x^3 + 60x$

5. $f(x) = x^2/(x - 1)$

6. $f(x) = (x^2 + 3)/(x^3 - 5x)$ on $[-2, 2]$

7. $f(x) = x^4 - 6x^2 - 8x + 5$ on $[-2, 3]$

8. $f(x) = \sqrt{x - 1}\,(x - 6)^2$ on $[1, 10]$

9. $f(x) = \cos 2x + x$ on $[-\frac{\pi}{2}, \frac{\pi}{2}]$

10. $f(x) = 2\cos x + \sin 2x$ on $[-\pi, \frac{\pi}{2}]$

In Exercises 11 and 12 determine intervals on which the graph of f is concave upward and those on which it is concave downward.

11. (a) $f(x) = x^4 - x^3 - 3x^2 - 5x + 7$
(b) $f(x) = 3x^5 - 15x^4 - 10x^3 + 90x^2 - 100$

12. (a) $f(x) = (x - 2)/(x + 1)^2$
(b) $f(x) = (x - 2)^4(x + 1)^2$

In Exercises 13–15 find all vertical and horizontal asymptotes to the graph of f.

13. (a) $f(x) = \dfrac{3x-4}{x+1}$ (b) $f(x) = \dfrac{3x-4}{2x^2-3x-5}$

14. (a) $f(x) = x + \dfrac{2x-1}{x-3}$ (b) $f(x) = \dfrac{x^2-1}{x^3-4x}$

15. (a) $f(x) = \dfrac{\sin x + 1}{x(\cos x + 1)}$ (b) $f(x) = \dfrac{\cos x}{x \sin^2 x - x}$

In Exercises 16–32 discuss intercepts, symmetry, asymptotes, local maxima and minima, and points of inflection. Use this information to draw the graph of f.

16. $f(x) = 2x^3 - 3x^2 - 12x + 4$

17. $f(x) = x^4 - 18x^2 + 60$

18. $f(x) = \dfrac{x^2}{x^2-4}$

19. $f(x) = \dfrac{8x}{x^2+4}$

20. $f(x) = 3x^5 - 5x^3$

21. $f(x) = x^4 - 4x^3 + 15$

22. $f(x) = \frac{1}{3}(x-1)^3(x+3)$

23. $f(x) = x - \dfrac{2}{x-1}$

24. $f(x) = x^{1/3}(4+x)$

25. $f(x) = x - (1-3x)^{2/3}$

26. $f(x) = 3x^4 - 8x^3 + 6x^2 - 1$

27. $f(x) = \dfrac{4x}{(x-1)^2}$

28. $f(x) = \dfrac{4(1-x)}{x^2}$

29. $f(x) = 3x^{-1} - x^{-3}$

30. $f(x) = \dfrac{\sin x}{\cos x - 2}$

31. $f(x) = \dfrac{2\sin^2 x + 1}{2\sin^2 x - 1}$

32. $f(x) = \frac{1}{12}(3x^4 + 4x^3 - 30x^2 + 36x)$

33. A closed box with a square base is to be constructed so that its volume is 324 cu ft. The material for the top and bottom costs \$3 per square foot, and that for the sides costs \$2 per square foot. Find the dimensions so that the cost will be minimum.

34. Find the dimensions of the right circular cylinder of maximum volume that can be inscribed in a cone of base radius r and altitude h.

35. The largest box the United Parcel Service will accept is one for which the sum of the length and the girth (distance around) is 108 in. Find the dimensions of the box with square cross section having the greatest volume that can be sent by the United Parcel Service.

36. A car dealer sells an average of 100 cars of a certain type per month when the selling price is \$15,000. The cost to him is \$11,000 per car. He estimates that for every \$200 in rebate he offers, he will sell 10 more cars per month. What total rebate should he offer to obtain the maximum profit?

37. An open-top cylindrical boiler pan is to have a volume of 250π cu in. The lateral surface is to be made of stainless steel and the bottom of copper. If copper costs twice as much as stainless steel, what should the dimensions of the pan be for minimum cost?

38. An oil drilling platform in the Gulf of Mexico is 9 km from point A, the nearest point on shore. A second oil drilling platform is 3 km from the nearest point B to it on the shore. The distance from A to B is 5 km. A supply depot is located at a point C on the shore between A and B in such a way that the sum of the distances from C to the two platforms is a minimum. How far is it from A to C?

39. According to one model of population growth, the rate $\frac{dx}{dt}$ at which a population grows is jointly proportional to the current size x of the population and the remaining capacity to grow, $m - x$, where m is the theoretical maximum size of the population. Find how large the population is when it is growing most rapidly.

40. To produce a certain manufactured item there is for each unit a fixed cost of \$30 and variable cost of $20 - 0.1x$, where the number x of items does not exceed 100. The unit price function is

$$p(x) = 70 + \tfrac{100}{x} - 0.3x$$

Find how many units should be produced and sold to maximize profit.

41. An 8-ft-high fence stands 1 ft from a building. A ladder is to be placed against the building and over the fence (see the figure). What is the length of the shortest ladder that can be used?

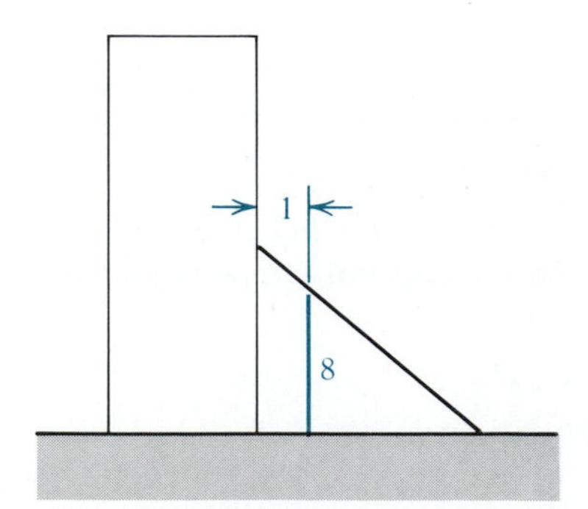

42. Car A and car B are approaching a town on roads that are at right angles to each other. Car A is traveling at 80 km/hr and car B at 55 km/hr. At the instant when car A is 5 km from the town, car B is 12 km from the town. How fast is the distance between the cars decreasing at that instant?

43. At the instant a hot air balloon begins rising a runner leaves from the takeoff point, running in a straight line. If the balloon rises at the constant rate of 22 ft/sec and the runner goes 8 mi/hr, find the rate at which the distance between the runner and the balloon is increasing after 10 min.

44. A yacht is cruising at 12 mi/hr. A person standing in the stern 15 ft above the water is pulling in a rope (assumed taut) attached to a rowboat at the rate of 50 ft/min. How fast is the rowboat moving through the water when there are 39 ft of rope out?

45. A tank in the form of an inverted cone is being filled with water at the rate of 4 cu ft/min, and at the same time water is being drained out the bottom at the rate of 2 cu ft/min. The top radius of the tank is 2 ft and its height is 8 ft. Find how fast the water level is rising when (a) the water level is 4 ft and (b) the tank is half full.

46. At a track meet a TV camerawoman is focusing her camera on the lead runner in a 100-yd dash. She is positioned 9 yd from the track, directly opposite the finish line. If the runner is going at the constant rate of 30 ft/sec, how many radians is the camera turning through per second at the instant the runner is 12 yd from the finish line?

In Exercises 47 and 48 show that Rolle's theorem is satisfied by f on the given interval, and find a number c that satisfies the conclusion of that theorem.

47. (a) $f(x) = x^3 - 3x - 2$ on $[-1, 2]$
(b) $f(x) = (x^2 - 1)/(x^2 + 2)$ on $[-1, 1]$

48. (a) $f(x) = x^{2/3}(1 - x)$ on $[0, 1]$
(b) $f(x) = \sqrt{1 - x}(x + 3)$ on $[-3, 1]$

In Exercises 49 and 50 show that the mean value theorem is satisfied by f on the given interval, and find a number c that satisfies the conclusion of that theorem.

49. (a) $f(x) = 2x^2 - 4x + 3$ on $[-1, 2]$
(b) $f(x) = x^3 - 2x^2 + 4x - 4$ on $[1, 3]$

50. (a) $f(x) = (x + 3)/(x - 2)$ on $[0, 1]$
(b) $f(x) = (3x - 4)^{4/3}$ on $[1, 4]$

51. (a) Suppose $f'(x) = 0$ for all x, and $f(2) = 7$. What is $f(5)$, and why?
(b) Suppose $f'(x) = g'(x)$ for all x, and $f(x) = x^2 + 4$. If $g(1) = 2$, what is $g(x)$, and why?

In Exercises 52–54 find the general antiderivative of f.

52. (a) $f(x) = 3x^2 - 2x + \dfrac{4}{x^2}$ (b) $f(x) = \dfrac{x^2 - 1}{\sqrt{x}}$

53. (a) $f(x) = 4 \sin x - 3 \cos x$
(b) $f(x) = \dfrac{\sec x + 2 \tan x}{\cos x}$

54. (a) $f(x) = \tan^2 x + \sec^2 x$
(b) $f(x) = \sin^2 x - \cos^2 x$

In Exercises 55–57 find the function f that satisfies the given conditions.

55. (a) $f'(x) = 3x^2 - 2/x^2$; $f(1) = 4$
(b) $f'(x) = 2 \sin x - x$; $f(0) = 5$

56. (a) $f''(x) = 2x$; $f(1) = -3$; $f'(1) = 2$
(b) $f''(x) = 4$; $f(0) = 7$; $f'(0) = 0$

57. (a) $f'''(x) = 1/x^2$; $f(-1) = 2$; $f'(-1) = 1$; $f''(-1) = 0$
(b) $f^{(4)}(x) = \cos x$; $f(0) = 2$; $f'(0) = -1$; $f''(0) = 1$; $f'''(0) = 0$

58. Find the function f such that $y = f(x)$ satisfies the equation

$$\frac{d^2y}{dx^2} + x = 2$$

and so that the graph of f passes through the point $(-1, 3)$ and has slope $\frac{1}{2}$ at that point.

59. A particle moves on a line directed positively to the right so that its acceleration at time t is $a(t) = 6t$. At time $t = 0$ it is at the origin and has initial velocity $v(0) = -12$. Consider $t \geq 0$.
(a) Find its velocity and position at time t.
(b) Find when and where it reaches its leftmost position.
(c) Find the intervals in which it is speeding up and those in which it is slowing down.
(d) Indicate its motion by means of a sketch.

60. From the edge of a building 128 ft high a rock is thrown vertically upward at a speed of 32 ft/sec. Find (a) how high the rock rises and (b) the speed with which it strikes the ground.

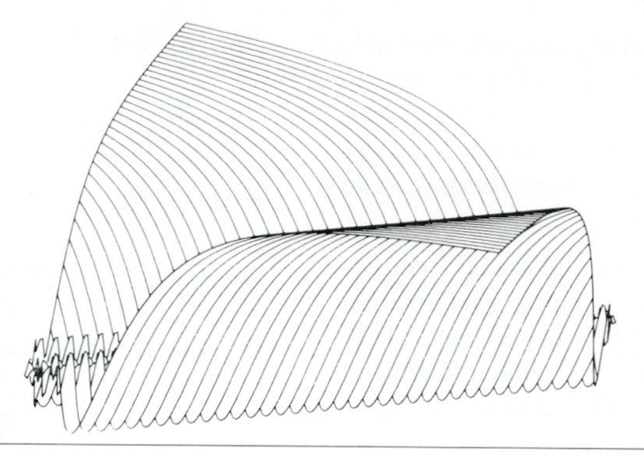

CHAPTER 4

THE INTEGRAL

4.1 THE PROBLEM OF AREAS

In this chapter we begin the study of the second major concept in calculus, called the **integral.** To motivate the definition we first consider how one might reasonably arrive at the area of a plane figure with one or more boundaries that are not necessarily straight lines. In particular, we consider how to find the area of the region between the x-axis and the graph of a nonnegative function f, bounded on the left by the line $x = a$ and on the right by $x = b$. Figure 4.1 illustrates a typical situation. We refer to this as the *area under the graph of f from a to b.* A prior question really should be whether the area exists, and the answer to this depends on how area is defined. So long as f is continuous on $[a, b]$, as in the figure, our intuition tells us that the area does exist, but if f is discontinuous, the question of area under the curve is not entirely obvious. We will get to these more basic questions later, but for the moment we will assume that f is continuous on $[a, b]$ and use geometric intuition about the meaning of area.

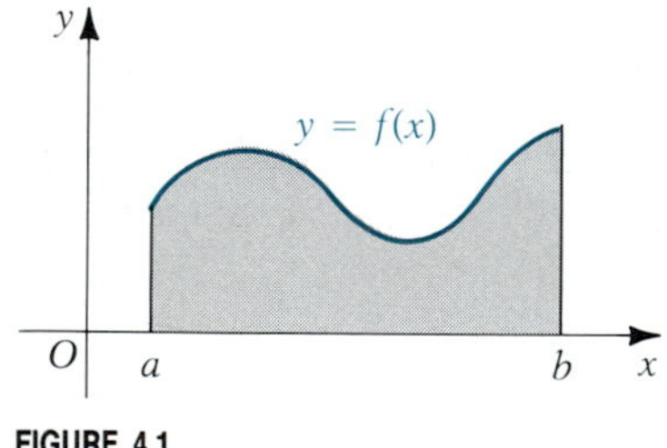

FIGURE 4.1

If you think about it, you can probably arrive at some method to estimate the area in Figure 4.1. There are several reasonable ways of doing this, but we are going to concentrate on just two. We call the first the method of *inscribed rectangles.* We divide up the interval $[a, b]$ into a number of subintervals of equal width. Next we construct rectangles that have these subintervals as bases and such that the height of each rectangle is the minimum value of $f(x)$ for x in the subinterval forming its base. We know by the theorem on extreme values for continuous functions (Theorem 3.2) that f does assume a minimum value (as well as a maximum value) on each of the closed subintervals. In Figure 4.2 we have shown two sets of inscribed rectangles; in the second set we have used a finer subdivision than in the first, causing the rectangles to be thinner and more numerous. We observe from this that the sum of the areas of the inscribed rectangles is an approximation to the true area, and this approximation always underestimates the true area. Furthermore, as we increase the number of subintervals, thereby making the

FIGURE 4.2

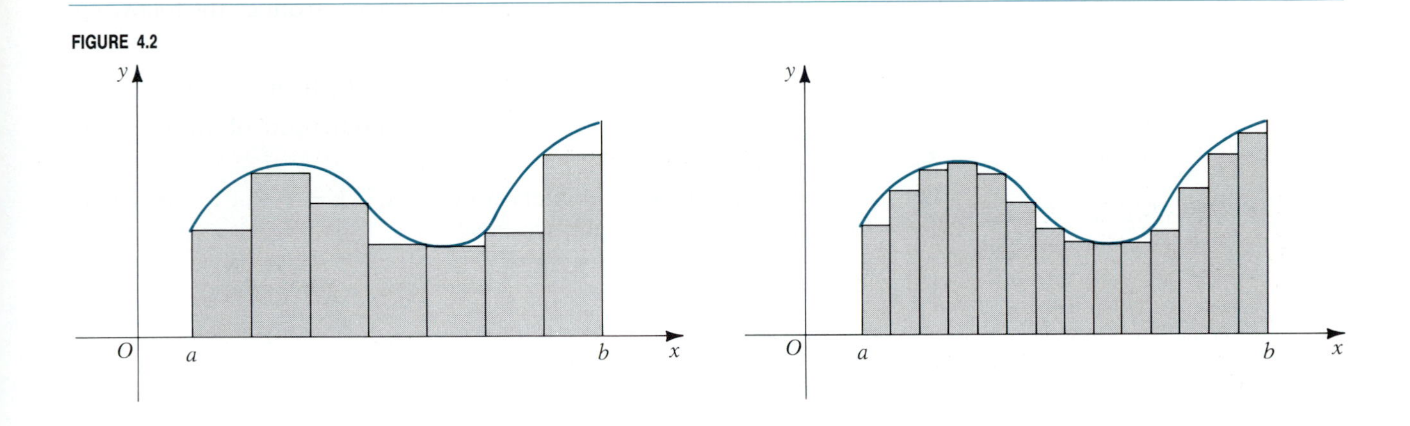

widths of the rectangles smaller, the error appears to get smaller. We refer to the sum of the areas of inscribed rectangles as a **lower sum.** So we have, for any subdivision of $[a, b]$,

lower sum $\leq$ true area

If instead of using the minimum value of f on a given subinterval, we use the maximum value, we arrive at what we call *circumscribed rectangles,* as in Figure 4.3, and the sum of their areas, called an **upper sum,** approximates the true area from above:

true area $\leq$ upper sum

Furthermore, as we take finer and finer subdivisions, the upper sums come closer and closer to the true area. So for a particular subdivision of $[a, b]$, we always have

lower sum $\leq$ true area $\leq$ upper sum

It appears that as we allow the number n of subintervals to increase indefinitely (that is, let $n \to \infty$), the lower and upper sums approach coincidence, and their common value is the area we are seeking.

FIGURE 4.3

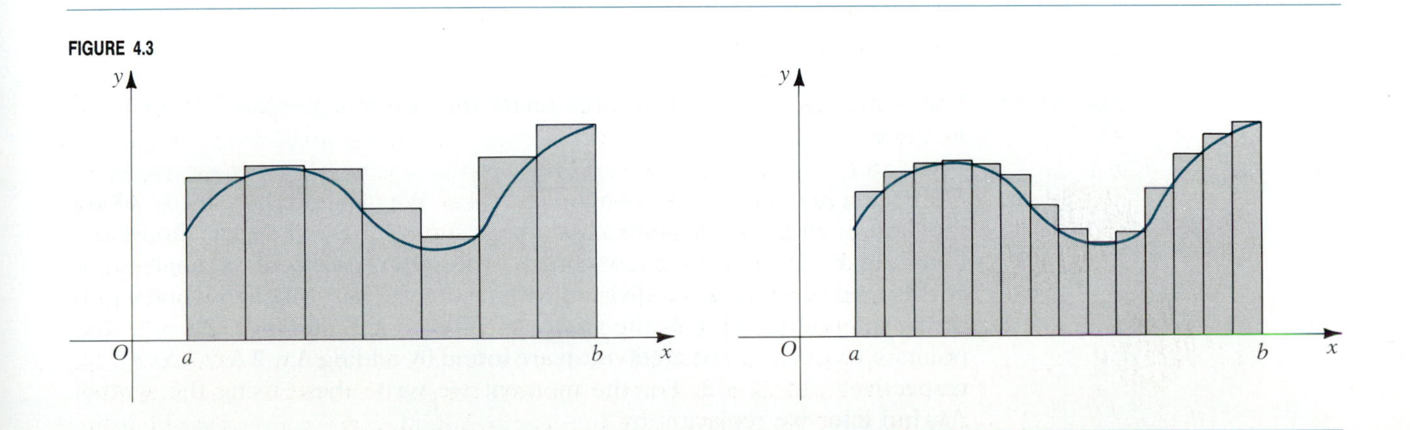

To make the foregoing discussion more precise, we introduce the following notation:

Let A = the area of the region under the graph of f from $x = a$ to $x = b$.

Let $x_0 < x_1 < x_2 < \cdots < x_n$ be the points of subdivision of $[a, b]$, with $x_0 = a$ and $x_n = b$.

Let Δx = the width of each subinterval $[x_{k-1}, x_k]$, $k = 1, 2, \ldots, n$. Then, since these widths are all equal, we have

$$\Delta x = \frac{b-a}{n}$$

Let c_k be the point in the subinterval $[x_{k-1}, x_k]$ for which $f(c_k)$ is the minimum value of f on this subinterval, and let d_k be the point in $[x_{k-1}, x_k]$ for which $f(d_k)$ is the maximum value of f there.

Let L_n = the lower sum and U_n = the upper sum. Then

$$\left.\begin{aligned} L_n &= [f(c_1) + f(c_2) + \cdots + f(c_n)]\,\Delta x \\ U_n &= [f(d_1) + f(d_2) + \cdots + f(d_n)]\,\Delta x \end{aligned}\right\} \tag{4.1}$$

From the foregoing discussion, we always have

$$L_n \leq A \leq U_n$$

$$L_n \leq L_{n+1} \quad \text{and} \quad U_{n+1} \leq U_n \qquad \text{for all } n$$

THEOREM 4.1 If f is continuous and nonnegative on $[a, b]$, then $\lim_{n\to\infty} L_n$ and $\lim_{n\to\infty} U_n$ both exist, and the two limits are the same. ■

DEFINITION 4.1 For f continuous and nonnegative on $[a, b]$, the area of the region under the graph of f and above the x-axis between $x = a$ and $x = b$ is the common value of $\lim_{n\to\infty} L_n$ and $\lim_{n\to\infty} U_n$. That is,

$$A = \lim_{n\to\infty} L_n \quad \text{or equivalently} \quad A = \lim_{n\to\infty} U_n$$ ■

As we have seen, Theorem 4.1 seems intuitively evident. Its proof involves concepts we have not studied, however, and so we omit it.

The precise meaning of $\lim_{n\to\infty} L_n = A$ is that, given any positive number ε, there is a positive integer N such that for all $n > N$, $|L_n - A| < \varepsilon$; that is, for all $n > N$, the numbers L_n lie in the ε-neighborhood of A. A similar definition holds for $\lim_{n\to\infty} U_n = A$.

EXAMPLE 4.1 Use Definition 4.1 to find the area under the graph of $y = x + 2$ from $x = 2$ to $x = 6$.

Solution The region in question is shown in Figure 4.4. We can find the area by taking the limit of either lower sums L_n or upper sums U_n. Suppose we choose the latter. In Figure 4.4 we show some of the circumscribed rectangles that correspond to a typical subdivision with n subintervals. Since the total width of the interval $[2, 6]$ is 4, the width Δx of each subinterval is $\Delta x = \frac{4}{n}$. The points $x_1, x_2, x_3, \ldots$ of subdivision are found by adding $\Delta x, 2\,\Delta x, 3\,\Delta x, \ldots,$ respectively, to $x_0 = 2$. For the moment we write these using the symbol Δx, but later we replace it by $\frac{4}{n}$.

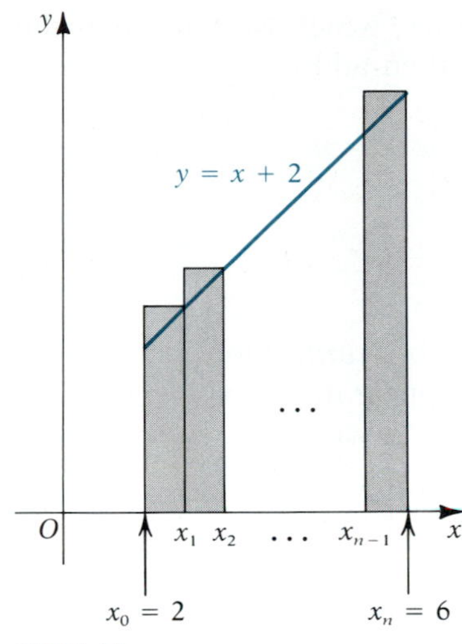

FIGURE 4.4

$$x_0 = 2$$
$$x_1 = 2 + \Delta x$$
$$x_2 = 2 + 2\,\Delta x$$
$$\vdots$$
$$x_n = 2 + n\,\Delta x$$

Now since f is an increasing function, its maximum value on each subinterval occurs at the right end point of that interval. So for the kth subinterval $[x_{k-1}, x_k]$, we have $d_k = x_k$. The maximum value is therefore

$$f(d_k) = f(x_k) = f(2 + k\,\Delta x) = (2 + k\,\Delta x) + 2 = 4 + k\,\Delta x$$

Using this for $k = 1, 2, \ldots, n$ and substituting into the formula for U_n from equation (4.1), we get

$$U_n = [(4 + \Delta x) + (4 + 2\,\Delta x) + (4 + 3\,\Delta x) + \cdots + (4 + n\,\Delta x)]\,\Delta x$$

On regrouping the terms inside the bracket, we get

$$\begin{aligned} U_n &= [(4 + 4 + \cdots + 4) + \Delta x + 2\,\Delta x + 3\,\Delta x + \cdots + n\,\Delta x]\,\Delta x \\ &= [4n + (1 + 2 + 3 + \cdots + n)\,\Delta x]\,\Delta x \end{aligned}$$

since there are n 4's. In the next section we will show that

$$1 + 2 + 3 + \cdots + n = \frac{n(n+1)}{2}$$

Assuming this is true (you might try to verify it yourself), we get

$$U_n = \left[4n + \frac{n(n+1)}{2}\,\Delta x\right]\Delta x = 4n\,\Delta x + \frac{n(n+1)}{2}(\Delta x)^2$$

Finally, we substitute $\Delta x = \frac{4}{n}$ and then pass to the limit as $n \to \infty$:

$$U_n = 4n\left(\frac{4}{n}\right) + \frac{n(n+1)}{2}\cdot\frac{16}{n^2} = 16 + \frac{8(n^2+n)}{n^2} = 16 + 8\left(1 + \frac{1}{n}\right)$$
$$A = \lim_{n\to\infty} U_n = 16 + 8 = 24$$

The region in question in this case is bounded by a trapezoid, and so you can easily find the area by elementary geometry in order to check the result. ■

We will return to problems of this sort after introducing some convenient notation in the next section.

4.2 SUMMATION NOTATION

As a shorthand way of writing sums such as

$$a_1 + a_2 + a_3 + \cdots + a_n$$

we use the capital Greek letter sigma as follows:

$$\sum_{k=1}^{n} a_k$$

This means precisely what we have written above, which we refer to as the *expanded form.* We first put $k = 1$ to get a_1, then add to this the term a_2 obtained by putting $k = 2$, then add the term a_3 for $k = 3$, and so on until we have added the final term a_n, obtained when $k = n$. As a specific case, we have

$$\sum_{k=1}^{5} \frac{1}{2^k} = \frac{1}{2^1} + \frac{1}{2^2} + \frac{1}{2^3} + \frac{1}{2^4} + \frac{1}{2^5}$$

The letter k as used here is called the *index of summation,* and it is a "dummy variable" in the sense that it does not appear in the expanded form, so that any other letter could be used and the result would be the same. For example,

$$\sum_{m=1}^{5} \frac{1}{2^m} = \frac{1}{2^1} + \frac{1}{2^2} + \frac{1}{2^3} + \frac{1}{2^4} + \frac{1}{2^5}$$

The only caution is that the letter used as the summation index should never be the same as a true variable appearing in the problem. For example, the sum

$$\sum_{k=1}^{n} \frac{1}{2^k}$$

could not be written

$$\sum_{n=1}^{n} \frac{1}{2^n}$$

Letters frequently used as indexes of summation are i, j, k, m, and n.

EXAMPLE 4.2 Write the expanded form of each of the following:

(a) $\displaystyle\sum_{j=1}^{4} \frac{1}{2j+1}$ (b) $\displaystyle\sum_{i=2}^{6} i^2$ (c) $\displaystyle\sum_{n=0}^{3} \frac{n+1}{n+2}$ (d) $\displaystyle\sum_{k=1}^{n} \left(\frac{1}{k} - \frac{1}{k+1}\right)$

Solution (a) $$\sum_{j=1}^{4} \frac{1}{2j+1} = \frac{1}{2(1)+1} + \frac{1}{2(2)+1} + \frac{1}{2(3)+1} + \frac{1}{2(4)+1} = \frac{1}{3} + \frac{1}{5} + \frac{1}{7} + \frac{1}{9}$$

(b) $$\sum_{i=2}^{6} i^2 = 2^2 + 3^2 + 4^2 + 5^2 + 6^2 = 4 + 9 + 16 + 25 + 36$$

(c) $$\sum_{n=0}^{3} \frac{n+1}{n+2} = \frac{0+1}{0+2} + \frac{1+1}{1+2} + \frac{2+1}{2+2} + \frac{3+1}{3+2} = \frac{1}{2} + \frac{2}{3} + \frac{3}{4} + \frac{4}{5}$$

(d) $$\sum_{k=1}^{n} \left(\frac{1}{k} - \frac{1}{k+1}\right) = \left(\frac{1}{1} - \frac{1}{2}\right) + \left(\frac{1}{2} - \frac{1}{3}\right) + \left(\frac{1}{3} - \frac{1}{4}\right) + \cdots + \left(\frac{1}{n} - \frac{1}{n+1}\right)$$

Observe that in part d the second term in parentheses except the last group cancels with the first term of the next parentheses. So the sum reduces to

$$1 - \frac{1}{n+1}$$

which is referred to as a *closed form* for the sum. This is an example of what is called a **telescoping sum.** ■

The following summation properties are easily established.

SUMMATION PROPERTIES

1. $\sum_{k=1}^{n} c = nc$
2. $\sum_{k=1}^{n} ca_k = c \sum_{k=1}^{n} a_k$
3. $\sum_{k=1}^{n} (a_k + b_k) = \sum_{k=1}^{n} a_k + \sum_{k=1}^{n} b_k$
4. $\sum_{k=1}^{n} (a_k - b_k) = \sum_{k=1}^{n} a_k - \sum_{k=1}^{n} b_k$

These can be verified by writing the expanded forms of the left-hand sides. For example,

$$\sum_{k=1}^{n} c = \underbrace{c + c + c + \cdots + c}_{n \text{ terms}} = nc$$

which proves Property 1. For Property 2 we have

$$\begin{aligned}\sum_{k=1}^{n} ca_k &= ca_1 + ca_2 + ca_3 + \cdots + ca_n = c(a_1 + a_2 + a_3 + \cdots + a_n) \\ &= c \sum_{k=1}^{n} a_k\end{aligned}$$

Similarly for Property 3,

$$\begin{aligned}\sum_{k=1}^{n} (a_k + b_k) &= (a_1 + b_1) + (a_2 + b_2) + (a_3 + b_3) + \cdots + (a_n + b_n) \\ &= (a_1 + a_2 + a_3 + \cdots + a_n) + (b_1 + b_2 + b_3 + \cdots + b_n) \\ &= \sum_{k=1}^{n} a_k + \sum_{k=1}^{n} b_k\end{aligned}$$

The rearrangement done in the second line is justified by the commutative and associative properties of real numbers. Property 4 is shown in a similar way.

Formulas giving closed forms for sums of powers of the first n natural numbers frequently occur in the calculation of lower and upper sums, and they have other uses as well. We used one of these formulas in Example 4.1. We now give the first three of these formulas but derive only the first one. They can all be proved by mathematical induction, but it may be more satisfying to see how the results can be derived.

$$\sum_{k=1}^{n} k = 1 + 2 + 3 + \cdots + n = \frac{n(n+1)}{2} \tag{4.2}$$

$$\sum_{k=1}^{n} k^2 = 1^2 + 2^2 + 3^2 + \cdots + n^2 = \frac{n(n+1)(2n+1)}{6} \tag{4.3}$$

$$\sum_{k=1}^{n} k^3 = 1^3 + 2^3 + 3^3 + \cdots + n^3 = \left[\frac{n(n+1)}{2}\right]^2 \tag{4.4}$$

Although equation (4.2) can be derived in a variety of ways, we choose one that suggests a procedure for deriving equations (4.3) and (4.4) also (and you will be asked to derive these in the exercises). We begin with the sum $\sum_{k=1}^{n} [k^2 - (k-1)^2]$ and write it in two ways. Then we equate the two results. First, the quantity inside the brackets can be simplified to give

$$k^2 - (k-1)^2 = k^2 - (k^2 - 2k + 1) = 2k - 1$$

So, using summation Properties 1, 2, and 4, we get

$$\sum_{k=1}^{n} [k^2 - (k-1)^2] = \sum_{k=1}^{n} (2k - 1) = \sum_{k=1}^{n} 2k - \sum_{k=1}^{n} 1 = 2\sum_{k=1}^{n} k - n \qquad (4.5)$$

Second, we use summation Property 4 on the original sum to get

$$\sum_{k=1}^{n} [k^2 - (k-1)^2] = \sum_{k=1}^{n} k^2 - \sum_{k=1}^{n} (k-1)^2 = n^2 \qquad (4.6)$$

since

$$\sum_{k=1}^{n} k^2 - \sum_{k=1}^{n} (k-1)^2 = [1^2 + 2^2 + \cdots + n^2] - [0^2 + 1^2 + 2^2 + \cdots + (n-1)^2]$$

and all terms cancel except n^2. Thus, equating the right-hand sides of equations (4.5) and (4.6), we get

$$n^2 = 2\sum_{k=1}^{n} k - n$$

Now we solve for $\sum_{k=1}^{n} k$:

$$\sum_{k=1}^{n} k = \frac{1}{2}(n^2 + n) = \frac{n(n+1)}{2}$$

which proves equation (4.2).

We conclude this section by finding another area, this time making use of summation notation. Note that the lower and upper sums L_n and U_n can be written compactly as

$$L_n = \sum_{k=1}^{n} f(c_k)\,\Delta x \quad \text{and} \quad U_n = \sum_{k=1}^{n} f(d_k)\,\Delta x \qquad (4.7)$$

where $f(c_k)$ is the minimum value of f on the kth subinterval, and $f(d_k)$ is the maximum value of f there.

EXAMPLE 4.3 Use lower sums to find the area under the graph of $f(x) = 16 - x^2$ from $x = 1$ to $x = 3$.

Solution The portion of the parabola $y = 16 - x^2$ between $x = 1$ and $x = 3$ is pictured in Figure 4.5, along with some typical inscribed rectangles. On this interval f is decreasing and so takes on its minimum value at the right end point of each subinterval. So for each k from 1 to n, $c_k = x_k$. Following the procedure used in Example 4.1, we have $\Delta x = \frac{2}{n}$, and for each k from 1 to n,

$$x_k = x_0 + k\,\Delta x = 1 + k\,\Delta x$$

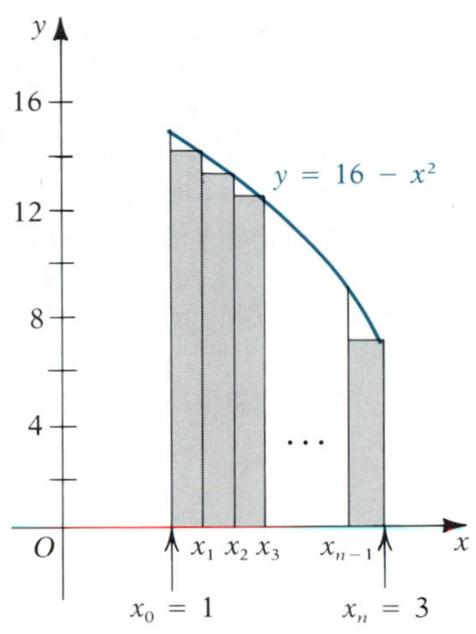

FIGURE 4.5

Thus, for the kth subinterval, $[x_{k-1}, x_k]$,

$$f(c_k) = f(x_k) = 16 - x_k^2 = 16 - (1 + k\,\Delta x)^2 = 15 - 2k\,\Delta x - k^2(\Delta x)^2$$

Substituting in equation (4.7) for L_n, we get

$$\begin{aligned} L_n = \sum_{k=1}^{n} f(c_k)\,\Delta x &= \sum_{k=1}^{n} [15 - 2k\,\Delta x - k^2(\Delta x)^2]\,\Delta x \\ &= \Delta x \sum_{k=1}^{n} 15 - 2(\Delta x)^2 \sum_{k=1}^{n} k - (\Delta x)^3 \sum_{k=1}^{n} k^2 \end{aligned}$$

Notice the use of the summation properties. Now we use equations (4.2), (4.3), and (4.4) to obtain L_n in the form

$$L_n = 15n\,\Delta x - 2(\Delta x)^2 \frac{n(n+1)}{2} - (\Delta x)^3 \frac{n(n+1)(2n+1)}{6}$$

On substituting $\Delta x = \frac{2}{n}$ and simplifying, this becomes

$$\begin{aligned} L_n &= 30 - \frac{4(n+1)}{n} - \frac{4(2n^2+3n+1)}{3n^2} \\ &= 30 - 4\left(1 + \frac{1}{n}\right) - \frac{4}{3}\left(2 + \frac{3}{n} + \frac{1}{n^2}\right) \end{aligned}$$

Finally, we let $n \to \infty$ to get the area A:

$$A = \lim_{n\to\infty} L_n = 30 - 4 - \tfrac{8}{3} = \tfrac{70}{3}$$ ■

You can imagine the work involved in computing the area by this method for higher degree polynomials or other types of functions. The method we have used differs little from a technique used by early Greek mathematicians, called the *method of exhaustion* (which seems appropriate!). You might reasonably hope that some shorter method exists and fortunately this is the case, as we shall soon see. But in order to arrive at the shorter method it is essential that the ideas presented so far be understood. So it is instructive to go through the work of actually calculating areas using upper or lower sums before proceeding.

EXERCISE SET 4.2

A

In Exercises 1–10 write out the expanded form of each sum. Simplify each term, but do not combine terms unless instructed to do so.

1. $\displaystyle\sum_{k=1}^{4} \frac{1}{k^2}$ **2.** $\displaystyle\sum_{n=1}^{5} \frac{1}{n(n+1)}$

3. $\displaystyle\sum_{i=0}^{3} \frac{i+1}{2i+1}$ **4.** $\displaystyle\sum_{j=3}^{6} \frac{j}{\sqrt{j+1}}$

5. $\displaystyle\sum_{k=1}^{n} (\sqrt{k} - \sqrt{k-1})$ Simplify by combining terms.

6. $\displaystyle\sum_{m=0}^{n-1} \frac{(-1)^m}{2^m}$ **7.** $\displaystyle\sum_{n=1}^{m} \frac{(-1)^{n-1}n}{n^2+1}$

8. $\displaystyle\sum_{k=m}^{n} \frac{1}{\sqrt{k}}$ **9.** $\displaystyle\sum_{k=1}^{n} \frac{\sin kx}{k}$

10. $\sum_{k=1}^{n}\left(\frac{1}{k}-\frac{1}{k+2}\right)$ Simplify by combining terms.

In Exercises 11–20 write the sum using the summation symbol $\sum$.

11. $1+\frac{1}{2}+\frac{1}{3}+\frac{1}{4}+\frac{1}{5}$

12. $1+\frac{1}{2}+\frac{1}{4}+\frac{1}{8}+\frac{1}{16}+\frac{1}{32}$

13. $\frac{1}{1\cdot 2}+\frac{1}{2\cdot 3}+\frac{1}{3\cdot 4}+\frac{1}{4\cdot 5}$

14. $\frac{1}{2}+\frac{2}{3}+\frac{3}{4}+\cdots+\frac{n}{n+1}$

15. $\frac{1}{3}+\frac{3}{9}+\frac{7}{27}+\cdots+\frac{2^n-1}{3^n}$

16. $\cos x+\frac{\cos 2x}{2}+\frac{\cos 3x}{3}+\cdots+\frac{\cos nx}{n}$

17. $\frac{4}{2^4}+\frac{5}{2^5}+\frac{6}{2^6}+\frac{7}{2^7}+\frac{8}{2^8}$

18. $1-\frac{1}{\sqrt{2}}+\frac{1}{\sqrt{3}}-\frac{1}{\sqrt{4}}+\frac{1}{\sqrt{5}}$

19. $\frac{2}{3}-\frac{4}{5}+\frac{6}{7}-\frac{8}{9}+\frac{10}{11}$

20. $2^{-m}+2^{-m-1}+2^{-m-2}+\cdots+2^{-n}$

Use the properties of summation together with equations (4.2)–(4.4) to write the sums in Exercises 21–24 in closed form.

21. $\sum_{k=1}^{n}(k^2-2k)$

22. $\sum_{k=1}^{n}(3k^2-4k+1)$

23. $\sum_{k=1}^{n}k(k^2-1)$

24. $\sum_{k=1}^{n}k(2k^2-3k+1)$

25. Let $S_n=1+2+3+\cdots+(n-2)+(n-1)+n$. Write an equivalent equation below this in which the order of terms on the right is reversed, and then add the two equations term by term. Deduce from this the result of equation (4.2).

26. Let $f(x)=x+1$ on $[0, 3]$. Find the lower sum L_6 and the upper sum U_6.

27. Let $f(x)=2x^2$ on $[0, 1]$. Find the lower sum L_5 and the upper sum U_5.

28. Let $f(x)=1-x^2$ on $[0, 1]$. Find the lower sum L_4 and the upper sum U_4.

29. Let $f(x)=\frac{2}{x}$ on $[1, 2]$. Find the lower sum L_5 and the upper sum U_5.

30. Let $f(x)=x$ on $[0, 2]$. Find L_n and U_n and show that they have the same limit as $n\to\infty$. Using geometry verify that this limit is the area of the region under the graph.

31. Follow the instructions given in Exercise 30 for the function $f(x)=2-x$ on $[0, 2]$.

32. Find the area of the region under the graph of $f(x)=x^2$ on $[0, 1]$ using upper sums.

33. Find the area of the region under the graph of $f(x)=1-x^2$ on $[0, 1]$ using lower sums.

B

34. Show each of the following by means of an example:
    (a) $\sum_{k=1}^{n}a_kb_k\neq\left(\sum_{k=1}^{n}a_k\right)\left(\sum_{k=1}^{n}b_k\right)$
    (b) $\left(\sum_{k=1}^{n}a_k\right)^2\neq\sum_{k=1}^{n}a_k^2$

35. (a) Prove that if the domain of f includes all nonnegative integers, then
$$\sum_{k=1}^{n}[f(k)-f(k-1)]=f(n)-f(0)$$
    (b) By letting $f(k)=k^p$, where p is a positive integer, use part a to show that
$$\sum_{k=1}^{n}[k^p-(k-1)^p]=n^p$$

36. (a) Show that $k^3-(k-1)^3=3k^2-3k+1$.
    (b) Following the pattern of the derivation of equation (4.2) derive equation (4.3). Make use of part a and Exercise 35, part b.

37. (a) Show that $k^4-(k-1)^4=4k^3-6k^2+4k-1$.
    (b) Derive equation (4.4) making use of part a and Exercise 35, part b.

38. Find the area of the region under the graph of $f(x)=x^3$ from $x=0$ to $x=1$ using upper sums.

39. Find the area bounded by the parabola $y=4-x^2$ and the x-axis. (*Hint:* Make use of symmetry.)

40. Find the area under the graph of $y=2x-x^2$ from $x=1$ to $x=2$.

41. Find the area under the graph of $f(x)=x^2-4x+6$ from $x=1$ to $x=4$. (*Hint:* First determine the intervals on which f is monotone.)

4.3 THE DEFINITE INTEGRAL

The concept of the integral of a function is abstracted from the area problem in much the same way that the derivative of a function is abstracted from the tangent problem. To arrive at the definition we first remove certain constraints placed on the function f, on the subdivision of the interval, and on the approximating sums for the area. Specifically we make the following changes:

1. We do not require that f be nonnegative or that it be continuous on $[a, b]$. The only requirements are that f be defined and **bounded** on $[a, b]$. To say that f is bounded on $[a, b]$ means that there are constants m and M such that $m \leq f(x) \leq M$ for all $x \in [a, b]$.
2. We do not require that the subintervals into which $[a, b]$ is divided be of equal width.
3. For the height of the kth rectangle, instead of using the minimum value $f(c_k)$ or the maximum value $f(d_k)$, we permit any other value of $f(x)$ for x in the kth subinterval. Thus, instead of using inscribed or circumscribed rectangles, we may use any rectangles between these extremes.

It will be convenient to introduce more notation to incorporate these changes. Let f be defined and bounded on $[a, b]$, and for any positive integer n, let $P = \{x_0, x_1, x_2, \ldots, x_n\}$ be a set of points with $x_0 = a$, $x_n = b$, and $x_0 < x_1 < x_2 < \cdots < x_n$. The set P is called a **partition** of $[a, b]$. The points of P divide $[a, b]$ into n subintervals of the form $[x_{k-1}, x_k]$, where k goes from 1 to n. Denote the width of the kth subinterval by Δx_k. Thus

$$\Delta x_k = x_k - x_{k-1}, \qquad k = 1, 2, \ldots, n$$

The largest of the widths Δx_k is called the **norm** (or *mesh*) of P and is denoted by $\|P\|$. (In Section 4.1 all subintervals were of the same width, so we had no need to speak of the largest one.) Let x_k^* denote any point in the kth subinterval; that is,

$$x_{k-1} \leq x_k^* \leq x_k, \qquad k = 1, 2, \ldots, n$$

Finally, form the sum

$$\sum_{k=1}^{n} f(x_k^*)\,\Delta x_k \tag{4.8}$$

This takes the place of the lower sum $L_n = \sum_{k=1}^{n} f(c_k)\,\Delta x$ or the upper sum $U_k = \sum_{k=1}^{n} f(d_k)\,\Delta x$ used in defining area. Note, however, that the sum (4.8) *may* be the same as L_n or U_n, depending on the choice of the points x_k^*.

In Section 4.1 we calculated the area under a graph by taking the limit of L_n or of U_n as $n \to \infty$. Allowing n to become infinite guaranteed that the width Δx of all rectangles approached 0, since $\Delta x = (b - a)/n$. We again want n to approach infinity, but this in itself no longer will guarantee that each of the widths Δx_k approaches 0, since these widths are no longer necessarily equal. To force the widths all to approach 0 we require that the *largest* of these widths—that is, the norm of P—approach 0. Then all smaller widths must necessarily also go to 0.

With these preliminaries we are ready for the definition of the integral.

DEFINITION 4.2 THE DEFINITE INTEGRAL

Let f be defined and bounded on $[a, b]$, and let P be a partition of $[a, b]$. If there exists a number I such that

$$I = \lim_{\|P\| \to 0} \sum_{k=1}^{n} f(x_k^*)\,\Delta x_k \tag{4.9}$$

where x_k^* is arbitrarily chosen in $[x_{k-1}, x_k]$, $k = 1, \ldots, n$, then I is called the **definite integral of f over $[a, b]$** and is denoted by

$$\int_a^b f(x)\,dx$$

■

Comments

1. Functions for which the integral over $[a, b]$ exists are said to be **integrable** on $[a, b]$. Continuity of f on $[a, b]$ is sufficient to guarantee integrability (see Theorem 4.2), but it is not necessary. Theorems 4.3 and 4.4 give other sufficient conditions.
2. The limit in equation (4.9) is a more complicated one than we have encountered in the past. Its precise meaning is as follows:

$$\lim_{\|P\| \to 0} \sum_{k=1}^{n} f(x_k^*)\,\Delta x_k = I$$

means that given any positive number ε, there exists a positive number δ such that

$$\left| \sum_{k=1}^{n} f(x_k^*)\,\Delta x_k - I \right| < \varepsilon$$

for all partitions P of $[a, b]$ with $\|P\| < \delta$, independently of how the points x_k^* are chosen in $[x_{k-1}, x_k]$, $k = 1, 2, \ldots, n$.
3. The symbol $\int$ in $\int_a^b f(x)\,dx$ is called an **integral sign.** It has the appearance of an elongated S, the first letter of the Latin word "summa." Leibniz introduced this. The end point values a and b that appear at the bottom and top of the integral sign are called **limits of integration,** and the function $f(x)$ is called the **integrand.** The dx that appears after $f(x)$ should simply be regarded as a part of the integral symbol. Later on we will see that it can be interpreted as a differential.
4. The symbol $\int_a^b f(x)\,dx$ is suggestive of the sum $\sum_{k=1}^{n} f(x_k^*)\,\Delta x_k$, with the summation symbol sigma being replaced with the integral sign, $f(x_k^*)$ by $f(x)$, and Δx_k by dx. The integral can be thought of as a sort of idealized summation, where by the limit process we pass from a discrete sum to a continuous one.
5. The variable x used in the symbol $\int_a^b f(x)\,dx$ is called the **variable of integration.** It is a dummy variable in the same sense that a summation index is a dummy variable. Any other letter could be used without changing the meaning. For example,

$$\int_a^b f(t)\,dt = \int_a^b f(x)\,dx$$

6. The adjective *definite* is often dropped. The integral of f over $[a, b]$ is understood to mean the definite integral. Later on, we will define an indefinite integral.

7. This definition is more general than that considered by Leibniz. Many mathematicians contributed to its eventual formulation, but the present form is primarily due to the German mathematician G. H. B. Riemann (1826–1866), and for this reason the integral is often called the **Riemann integral** and the sum (4.8) used in the definition is called a **Riemann sum.**

EXAMPLE 4.4 Find and evaluate the Riemann sum (4.8) for the function $f(x) = 1 + x$ on $[1, 2]$ with the partition $P = \{1, 1.2, 1.6, 1.7, 1.8, 2\}$, where $x_1^* = 1.1$, $x_2^* = 1.5$, $x_3^* = 1.7$, $x_4^* = 1.7$, and $x_5^* = 1.9$.

Solution It is easily verified that each x_k^* lies in the interval $[x_{k-1}, x_k]$ for $k = 1, 2, 3, 4, 5$. We calculate the lengths Δx_k of the intervals:

$$\begin{aligned}
\Delta x_1 &= x_1 - x_0 = 1.2 - 1 = 0.2 \\
\Delta x_2 &= x_2 - x_1 = 1.6 - 1.2 = 0.4 \\
\Delta x_3 &= x_3 - x_2 = 1.7 - 1.6 = 0.1 \\
\Delta x_4 &= x_4 - x_3 = 1.8 - 1.7 = 0.1 \\
\Delta x_5 &= x_5 - x_4 = 2 - 1.8 = 0.2
\end{aligned}$$

Thus, the Riemann sum is

$$\begin{aligned}
\sum_{k=1}^{5} f(x_k^*)\,\Delta x_k &= f(1.1)(0.2) + f(1.5)(0.4) + f(1.7)(0.1) + f(1.7)(0.1) \\
&\quad + f(1.9)(0.2) \\
&= (2.1)(0.2) + (2.5)(0.4) + (2.7)(0.1) + (2.7)(0.1) \\
&\quad + (2.9)(0.2) \\
&= 0.42 + 1.00 + 0.27 + 0.27 + 0.58 \\
&= 2.54
\end{aligned}$$

■

EXAMPLE 4.5 Evaluate the integral $\int_0^1 (4x + 3)\,dx$ using the definition.

Solution Since $f(x) = 4x + 3$ is continuous on $[0, 1]$, the integral exists (see comment 7 following Definition 4.2). We may therefore choose partition points in any manner; for simplicity we choose partitions with equally spaced points. If P is such a partition, with points $\{x_0, x_1, x_2, \ldots, x_n\}$, where $x_0 = 0$ and $x_n = 1$, then we have $\Delta x_k = \frac{1}{n}$ for all k. Note that $\|P\| = \frac{1}{n}$, so that as $n \to \infty$, $\|P\| \to 0$. We may also choose the points x_k^* in any manner, and we select the right end point of each interval. Thus,

$$\begin{aligned}
x_1^* &= 0 + \frac{1}{n} = \frac{1}{n} \\
x_2^* &= 0 + 2\left(\frac{1}{n}\right) = \frac{2}{n} \\
&\vdots \qquad \vdots \qquad \vdots \\
x_k^* &= 0 + k\left(\frac{1}{n}\right) = \frac{k}{n}
\end{aligned}$$

The corresponding Riemann sum is

$$\begin{aligned}
\sum_{k=1}^{n} f(x_k^*)\,\Delta x_k &= \sum_{k=1}^{n} f\left(\frac{k}{n}\right)\cdot\frac{1}{n} = \sum_{k=1}^{n}\left(4\cdot\frac{k}{n}+3\right)\cdot\frac{1}{n}\\
&= \frac{4}{n^2}\sum_{k=1}^{n} k + \frac{3}{n}\sum_{k=1}^{n} 1\\
&= \frac{4}{n^2}\cdot\frac{n(n+1)}{2} + \frac{3}{n}\cdot n\\
&= \frac{2}{n}(n+1) + 3\\
&= 2 + \frac{2}{n} + 3\\
&= 5 + \frac{2}{n}
\end{aligned}$$

Now letting $n \to \infty$ causes $\|P\| \to 0$, so we get

$$\int_0^1 (4x+3)\,dx = 5$$

Since $4x + 3 \geq 0$ on $[0, 1]$, we can interpret the result geometrically as the area under the graph of f. ■

We will see in the next chapter some applications of the integral. These go far beyond finding areas. It should be noted, however, that one application *is* finding areas under curves. Suppose f is nonnegative and continuous on $[a, b]$. Then, defining $f(c_k)$ and $f(d_k)$ as we have previously, as the minimum and maximum values of f on the kth subinterval of a partition of $[a, b]$, we must have

$$f(c_k) \leq f(x_k^*) \leq f(d_k) \tag{4.10}$$

for any x_k^* in $[x_{k-1}, x_k]$. This is illustrated in Figure 4.6. We are free to choose the spacing of the partition points in any way we wish, so in particular we can choose them equally spaced. The partition P is then said to be **regular.** With a regular partition all Δx_k values are the same and are equal to $(b - a)/n$. So as $n \to \infty$, $\|P\| \to 0$. We multiply each member of inequality (4.10) by Δx_k and sum from 1 to n:

$$\sum_{k=1}^{n} f(c_k)\,\Delta x_k \leq \sum_{k=1}^{n} f(x_k^*)\,\Delta x_k \leq \sum_{k=1}^{n} f(d_k)\,\Delta x_k$$

or

$$L_n \leq \sum_{k=1}^{n} f(x_k^*)\,\Delta x_k \leq U_n \tag{4.11}$$

Geometrically, the Riemann sum in the middle is the sum of areas of rectangles lying between inscribed and circumscribed rectangles. (See Figure 4.6.) If A denotes the area of the region under the graph of f from a to b, then we know that $\lim_{n\to\infty} L_n = A$ and $\lim_{n\to\infty} U_n = A$. So as $n \to \infty$ in inequality (4.11), the Riemann sum in the middle must also approach A (by an analog

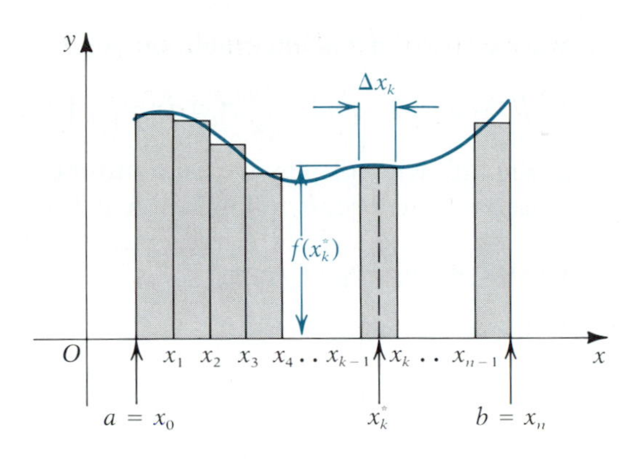

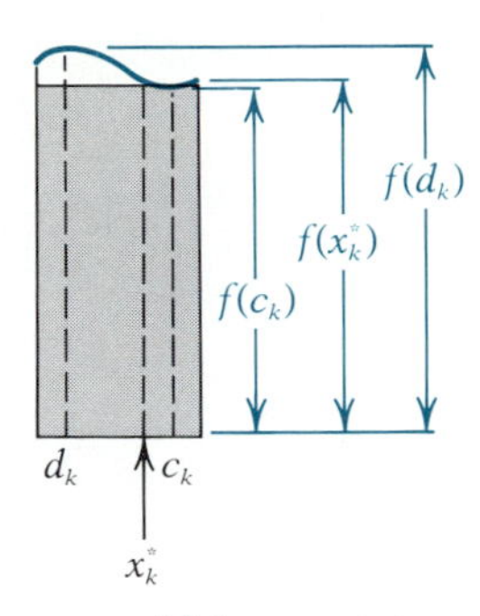

Enlargement of kth rectangle

FIGURE 4.6

of the squeezing theorem); that is,

$$\int_a^b f(x)\,dx = A \tag{4.12}$$

FIGURE 4.7

For equation (4.12) to be true, we have so far required f to be continuous and nonnegative. If f is continuous but negative for some values of x on $[a, b]$, the integral no longer gives the area bounded by the curve and the x-axis, but in a certain sense it can be interpreted as "net area." A typical situation is shown in Figure 4.7. The integral $\int_a^b f(x)\,dx$ in effect counts the areas of regions I and III as positive and the area of region II as negative. If we wanted the actual area, we would have to split the interval $[a, b]$ into the three parts $[a, c]$, $[c, d]$, and $[d, b]$. Then we would calculate the integral over each part and add their absolute values.

It is essential that f be bounded in order to be integrable; that is, *the Riemann integral of an unbounded function is not defined.* Not all bounded functions are integrable. The following three theorems, which we state without proof, give sufficient conditions to guarantee integrability.

THEOREM 4.2 If f is continuous on $[a, b]$, then f is integrable there. ■

THEOREM 4.3 If f is monotone on $[a, b]$, then f is integrable there. ■

THEOREM 4.4 If f is bounded on $[a, b]$ and discontinuous at only finitely many points, then f is integrable there. ■

In the definition of $\int_a^b f(x)\,dx$ it was implicit that $a < b$. We can extend the definition to cases where this is not true as follows.

DEFINITION 4.3 If $f(a)$ exists, then we define

$$\int_a^a f(x)\,dx = 0$$

and if f is integrable on $[a, b]$, we define

$$\int_b^a f(x)\,dx = -\int_a^b f(x)\,dx$$ ■

For example, if f is integrable on $[0, 3]$,

$$\int_3^3 f(x)\,dx = 0 \quad \text{and} \quad \int_3^0 f(x)\,dx = -\int_0^3 f(x)\,dx$$

In the following display we list a number of properties of the integral, and we discuss their proofs following their listing.

PROPERTIES OF THE INTEGRAL

1. If c is a constant,

$$\int_a^b c\,dx = c(b-a)$$

2. If f is integrable on $[a, b]$ and c is a constant,

$$\int_a^b cf(x)\,dx = c\int_a^b f(x)\,dx$$

3. If f is integrable on $[a, b]$ and c is between a and b,

$$\int_a^b f(x)\,dx = \int_a^c f(x)\,dx + \int_c^b f(x)\,dx$$

4. If f and g are integrable on $[a, b]$, then so are $f+g$ and $f-g$, and

$$\int_a^b (f(x) \pm g(x))\,dx = \int_a^b f(x)\,dx \pm \int_a^b g(x)\,dx$$

5. If f is integrable on $[a, b]$ and $f(x) \geq 0$, then

$$\int_a^b f(x)\,dx \geq 0$$

6. If f and g are integrable on $[a, b]$ and $f(x) \leq g(x)$, then

$$\int_a^b f(x)\,dx \leq \int_a^b g(x)\,dx$$

7. If f is integrable on $[a, b]$, then so is $|f|$, and

$$\left|\int_a^b f(x)\,dx\right| \leq \int_a^b |f(x)|\,dx$$

8. If f is integrable on $[a, b]$ and m and M are constants for which $m \leq f(x) \leq M$ for all x on $[a, b]$, then

$$m(b-a) \leq \int_a^b f(x)\,dx \leq M(b-a)$$

Partial Proofs of Properties

1. Property 1 is easily proved. Let $f(x) = c$ and consider any partition P of $[a, b]$. We form the Riemann sum

$$\sum_{k=1}^{n} f(x_k^*)\,\Delta x_k = \sum_{k=1}^{n} c\,\Delta x_k = c\sum_{k=1}^{n} \Delta x_k$$

But $\sum_{k=1}^{n} \Delta x_k$ is the sum of the lengths of the subintervals and so equals the length of the entire interval $[a, b]$—namely, $b-a$. Thus, for all partitions P of $[a, b]$,

$$\sum_{k=1}^{n} f(x_k^*)\,\Delta x_k = c(b-a)$$

Taking the limit of both sides as $\|P\| \to 0$, we get the desired result.

2. For Property 2 consider a Riemann sum for cf corresponding to a partition P. We have

$$\sum_{k=1}^{n} cf(x_k^*)\,\Delta x_k = c \sum_{k=1}^{n} f(x_k^*)\,\Delta x_k$$

Now we pass to the limit as $\|P\| \to 0$. Although we will not prove it, this type of limit is similar to limits of functions in that constants can be factored out. So we get

$$\lim_{\|P\|\to 0} \sum_{k=1}^{n} cf(x_k^*)\,\Delta x_k = c \cdot \lim_{\|P\|\to 0} \sum_{k=1}^{n} f(x_k^*)\,\Delta x_k$$

and this leads to the desired result.

3. The proof of Property 3 is somewhat tedious, and we will not even give a partial proof. We can point out its plausibility, however. If f is nonnegative and continuous as in Figure 4.8, we know that $\int_a^b f(x)\,dx$ gives the area under the curve from a to b. It is intuitively clear that this equals the sum of the area under f from a to c and the area from c to b. Thus in this case

$$\int_a^b f(x)\,dx = \int_a^c f(x)\,dx + \int_c^b f(x)\,dx$$

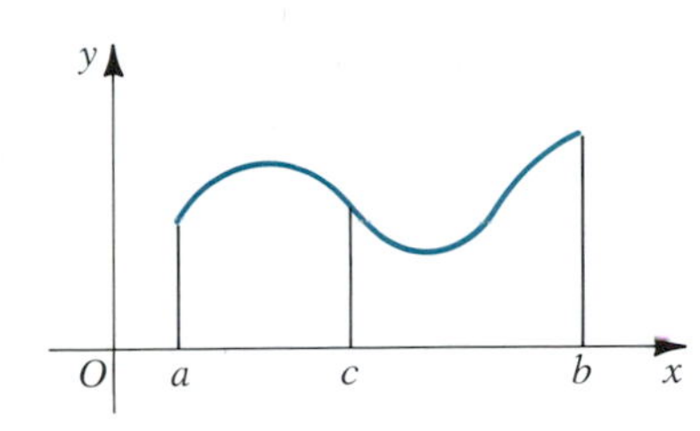

FIGURE 4.8

Even when f is not necessarily positive, a similar geometric argument strongly suggests the truth of the property in general.

4. Consider $f + g$ only, since $f - g$ is similar. Let P be a partition of $[a, b]$ and form the Riemann sum

$$\sum_{k=1}^{n} (f(x_k^*) + g(x_k^*))\,\Delta x_k = \sum_{k=1}^{n} f(x_k^*)\,\Delta x_k + \sum_{k=1}^{n} f(x_k^*)\,\Delta x_k$$

(by one of the summation properties). It can be shown that the limit of a sum is the sum of the limits with the limit process involved in the definition of the integral. Thus, letting $\|P\| \to 0$ on both sides of the preceding equation, we get the desired result.

5. If $f(x) \geq 0$ on $[a, b]$, then for all partitions P of $[a, b]$, the Riemann sum is nonnegative:

$$\sum_{k=1}^{n} f(x_k^*)\,\Delta x_k \geq 0$$

Thus, the limit must also be nonnegative as $\|P\| \to 0$, since otherwise we could find a partition that causes the Riemann sum corresponding to it to be so close to the limit that it also would be negative. So the limit must be nonnegative.

6. Property 6 can be deduced from Properties 4 and 5. We consider the function $g - f$. Since $f(x) \leq g(x)$, it follows that $g - f$ is nonnegative on $[a, b]$. So by Property 5,

$$\int_a^b (g(x) - f(x))\,dx \geq 0$$

or, using Property 4,

$$\int_a^b g(x)\,dx - \int_a^b f(x)\,dx \geq 0$$

and the conclusion now follows.

7. The proof that $|f|$ is integrable in $[a, b]$ whenever f is integrable there requires more background than we have introduced. If we assume this, however, we can prove the inequality. We have

$$-|f(x)| \leq f(x) \leq |f(x)| \qquad \text{(Why?)}$$

From the right-hand inequality and Property 6,

$$\int_a^b f(x)\,dx \leq \int_a^b |f(x)|\,dx$$

The left-hand inequality can be rewritten as $-f(x) \leq |f(x)|$, so again using Property 6 and also Property 2,

$$\int_a^b -f(x)\,dx = -\int_a^b f(x)\,dx \leq \int_a^b |f(x)|\,dx$$

Thus, both $\int_a^b f(x)\,dx$ and its negative are less than or equal to $\int_a^b |f(x)|\,dx$. Since the absolute value of a number is either the number itself or its negative, it follows that

$$\left|\int_a^b f(x)\,dx\right| \leq \int_a^b |f(x)|\,dx$$

8. Since $m \leq f(x) \leq M$, it follows by a double application of Property 6 that

$$\int_a^b m\,dx \leq \int_a^b f(x)\,dx \leq \int_a^b M\,dx$$

and by Property 1, the left-hand and right-hand integrals are $m(b-a)$ and $M(b-a)$, respectively. ■

EXAMPLE 4.6 Evaluate $\int_0^4 f(x)\,dx$, where f is defined as follows:

$$f(x) = \begin{cases} 1 - x^2 & \text{if } 0 \leq x \leq 2 \\ -3 & \text{if } 2 < x \leq 4 \end{cases}$$

Solution We show the graph of f in Figure 4.9. It looks as though it would be a good idea to divide the integral into two parts, and by Property 3 we can do this:

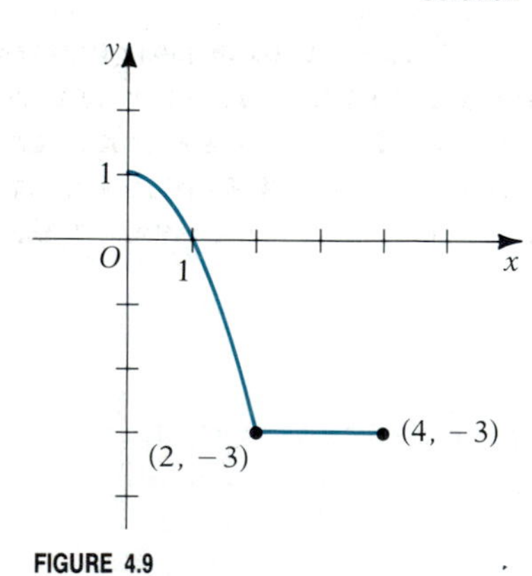

FIGURE 4.9

$$\int_0^4 f(x)\,dx = \int_0^2 f(x)\,dx + \int_2^4 f(x)\,dx$$

The second integral is easily evaluated, using Property 1:

$$\int_2^4 f(x)\,dx = \int_2^4 (-3)\,dx = -3(4-2) = -6$$

To evaluate the first integral we use the definition. Since f is continuous on $[0, 2]$ as well as being monotone, we know that the integral exists. We choose regular partitions only, and we choose x_k^* to be the right end point in the kth subinterval. (The left end point would have worked just as well.) So if P is any regular partition $\{x_0, x_1, x_2, \ldots, x_n\}$ of $[a, b]$, we have

$$\Delta x_k = \frac{b-a}{n} = \frac{2}{n} \quad \text{for all } k$$

and

$$x_k^* = k \cdot \frac{2}{n}, \qquad k = 1, 2, \ldots, n$$

(Verify that this gives the right end point.)

Now we form the Riemann sum

$$\begin{aligned}\sum_{k=1}^{n} f(x_k^*)\,\Delta x_k &= \sum_{k=1}^{n} (1 - x_k^{*2})\,\Delta x_k \\ &= \sum_{k=1}^{n}\left[1 - \left(\frac{2k}{n}\right)^2\right]\cdot\frac{2}{n} \\ &= \sum_{k=1}^{n}\frac{2}{n} - \frac{8}{n^3}\sum_{k=1}^{n} k^2 \\ &= n\cdot\frac{2}{n} - \frac{8}{n^3}\,\frac{n(n+1)(2n+1)}{6} \\ &= 2 - \frac{4}{3}\,\frac{2n^2+3n+1}{n^2} \\ &= 2 - \frac{4}{3}\left(2 - \frac{3}{n} + \frac{1}{n^2}\right)\end{aligned}$$

Since we are taking regular partitions only, $\|P\| \to 0$ if $n \to \infty$, and conversely. So we get

$$\int_0^2 f(x)\,dx = 2 - \tfrac{8}{3} = -\tfrac{2}{3}$$

Finally, then,

$$\int_0^4 f(x)\,dx = -\tfrac{2}{3} + (-6) = -\tfrac{20}{3}$$ ■

EXAMPLE 4.7 Verify each of the following inequalities:

(a) $\int_1^4 (-x^2 + 5x - 4)\,dx \geq 0$

(b) $\int_a^b x\,dx \leq \frac{1}{2}\int_a^b (1 + x^2)\,dx$ if $a < b$

(c) $16 \leq \int_0^4 (x^3 - 6x^2 + 9x + 4)\,dx \leq 32$

Solution (a) We make a sign graph for the integrand to determine where it is positive and where it is negative:

$$f(x) = -x^2 + 5x - 4 = -(x-1)(x-4)$$

So $f(x) \geq 0$ on $[1, 4]$. Therefore, by Property 5,

$$\int_1^4 (-x^2 + 5x - 4)\,dx \geq 0$$

(b) Let $f(x) = x$ and $g(x) = \frac{1}{2}(1 + x^2)$. Since

$$g(x) - f(x) = \tfrac{1}{2}(1 + x^2) - x = \tfrac{1}{2}(1 - 2x + x^2) = \tfrac{1}{2}(1 - x)^2 \geq 0$$

it follows that $f(x) \leq g(x)$ on every interval $[a, b]$. So by Properties 6 and 2 we have

$$\int_a^b x\,dx \leq \int_a^b \tfrac{1}{2}(1 + x^2)\,dx = \tfrac{1}{2}\int_a^b (1 + x^2)\,dx$$

(c) Let $f(x) = x^3 - 6x^2 + 9x + 4$. We will determine the absolute maximum M and the absolute minimum m of f on $[0, 4]$:

$$f'(x) = 3x^2 - 12x + 9 = 3(x^2 - 4x + 3) = 3(x - 1)(x - 3)$$
$$f''(x) = 6x - 12$$

The critical points are 1 and 3:

$$f(1) = 8$$
$$f(3) = 4$$

We must also check the end point values:

$$f(0) = 4$$
$$f(4) = 8$$

We conclude that the absolute maximum value of f on $[0, 4]$ is 8 and the absolute minimum is 4 (both values are assumed twice). Thus, for all x in $[0, 4]$, $4 \le f(x) \le 8$. Therefore, by Property 8,

$$4(4) \le \int_0^4 (x^3 - 6x^2 + 9x + 4)\,dx \le 8(4)$$

which is what we wanted to prove. ■

EXAMPLE 4.8 If it is given that

$$\int_{-1}^{2} x\,dx = \tfrac{3}{2} \quad \text{and} \quad \int_{-1}^{2} x^2\,dx = 3$$

find:

(a) $\int_{-1}^{2} (3x^2 + 2x)\,dx$ (b) $\int_{-1}^{2} (x - 1)^2\,dx$

Solution (a) By Property 2 we know that

$$\int_{-1}^{2} 3x^2\,dx = 3\int_{-1}^{2} x^2\,dx = 3(3) = 9$$

and

$$\int_{-1}^{2} 2x\,dx = 2\int_{-1}^{2} x\,dx = 2(\tfrac{3}{2}) = 3$$

So by Property 4,

$$\int_{-1}^{2} (3x^2 + 2x)\,dx = \int_{-1}^{2} 3x^2\,dx + \int_{-1}^{2} 2x\,dx = 9 + 3 = 12$$

(b) First we expand the integrand:

$$\int_{-1}^{2} (x^2 - 2x + 1)\,dx$$

and then proceed as in part a:

$$\int_{-1}^{2} (x^2 - 2x + 1)\,dx = \int_{-1}^{2} x^2\,dx - \int_{-1}^{2} 2x\,dx + \int_{-1}^{2} 1\,dx \quad \text{(Property 4)}$$
$$= 3 - 2(\tfrac{3}{2}) + 1 \cdot [2 - (-1)] \quad \text{(Properties 2 and 1)}$$
$$= 3 - 3 + 3 = 3$$

■

EXERCISE SET 4.3

A

In Exercises 1–4 find the Riemann sum corresponding to the given partition and the given choices of x_k^*. Simplify the result.

1. $f(x) = 2x - 3$ on $[-3, 2]$;
   $P = \{-3, -2, -1, 0, 1, 2\}$;
   $x_1^* = -\frac{5}{2},\ x_2^* = -1,\ x_3^* = -\frac{1}{4},\ x_4^* = 0,\ x_5^* = \frac{7}{4}$
2. $f(x) = 5 - 2x$ on $[0, 4]$; $P = \{0, 1, 3, 4\}$;
   $x_1^* = \frac{1}{2},\ x_2^* = \frac{5}{2},\ x_3^* = 4$
3. $f(x) = 10(x - 2)$ on $[1, 2]$;
   $P = \{1, 1.2, 1.3, 1.6, 1.9, 2\}$;
   $x_1^* = 1.1,\ x_2^* = 1.25,$
   $x_3^* = 1.4,\ x_4^* = 1.8,\ x_5^* = 1.95$
4. $f(x) = x(3 - x)$ on $[-1, 3]$; $P = \{-1, 0, \frac{3}{2}, 3\}$;
   $x_1^* = -\frac{1}{2},\ x_2^* = \frac{1}{2},\ x_3^* = 2$

In Exercises 5–10 evaluate the integral using the definition.

5. $\int_0^3 (2x - 1)\,dx$
6. $\int_0^2 (2 - 3x)\,dx$
7. $\int_1^2 (x - 3)\,dx$
8. $\int_{-1}^1 (3x - 4)\,dx$
9. $\int_0^3 3x^2\,dx$
10. $\int_1^3 (2 - x^2)\,dx$

In Exercises 11 and 12 use the following known values, $\int_{-2}^3 x^3\,dx = \frac{65}{4}$, $\int_{-2}^3 x^2\,dx = \frac{35}{3}$, and $\int_{-2}^3 x\,dx = \frac{5}{2}$, to evaluate the integrals. State which properties you use.

11. (a) $\int_{-2}^3 (4x^3 + 3x^2)\,dx$ (b) $\int_{-2}^3 (x^2 - 2x)\,dx$
12. (a) $\int_{-2}^3 (3x^2 - 2x + 4)\,dx$
    (b) $\int_{-2}^3 (x^3 + x^2 - x + 1)\,dx$

In Exercises 13–21 verify the given inequalities. State which properties you use.

13. $\int_0^2 (2x - x^2)\,dx \geq 0$
14. $\int_{-1}^2 (2 + x - x^2)\,dx \geq 0$
15. $\int_0^3 (6x + x^2 - x^3)\,dx \geq 0$
16. $\int_1^3 (-x^4 + 10x^2 - 9)\,dx \geq 0$
17. $\int_a^b 4x\,dx \leq \int_a^b (x^2 + 4)\,dx$
18. $\int_a^b (4x - 3)\,dx \leq \frac{4}{3}\int_a^b x^2\,dx$
19. $6 \leq \int_0^3 (x^2 - 2x + 3)\,dx \leq 18$
20. $-48 \leq \int_{-2}^4 \left(\frac{x^3}{3} - x^2 - 3x + 1\right) dx \leq 16$
21. $-2 \leq \int_{-2}^2 \frac{x}{x^2 + 1}\,dx \leq 2$
22. Justify each step of the following:
    $$\left|\int_0^2 \frac{\sin x}{3 + \cos x}\,dx\right| \leq \int_0^2 \left|\frac{\sin x}{3 + \cos x}\right| dx \leq \int_0^2 \frac{1}{2}\,dx = 1$$

B

In Exercises 23–26 use the definition and properties of the integral to evaluate.

23. $\int_{-1}^2 (2x^2 - 3x + 4)\,dx$
24. $\int_2^4 (3 - 4x - 2x^2)\,dx$
25. $\int_0^3 (x^3 - 2x^2)\,dx$
26. $\int_{-2}^0 (3 - 4x - x^3)\,dx$

In Exercises 27–29 justify the inequalities.

27. $\left|\int_{-1}^1 \frac{x - 2}{x + 2}\,dx\right| \leq 6$
28. $\left|\int_{-2\sqrt{2}}^{2\sqrt{2}} x\sqrt{8 - x^2}\,dx\right| \leq 16\sqrt{2}$
29. $\frac{3\pi}{2} \leq \int_{-\pi}^{\pi} (\sin x - \cos^2 x + 2)\,dx \leq 6\pi$

In Exercises 30 and 31 use the definition and the properties of the integral to evaluate the given integral.

30. $\int_{-2}^3 f(x)\,dx$, where $f(x) = \begin{cases} x + 1 & \text{if } -2 \leq x \leq 0 \\ 1 & \text{if } 0 < x \leq 1 \\ 2x - x^2 & \text{if } 1 < x \leq 3 \end{cases}$
31. $\int_{-2}^4 f(x)\,dx$, where $f(x) = \begin{cases} 2x^2 - 3 & \text{if } -2 \leq x < 0 \\ 3 & \text{if } 0 \leq x < 2 \\ 3x - 4 & \text{if } 2 \leq x \leq 4 \end{cases}$
32. Recall that an even function f is one for which $f(-x) = f(x)$, and an odd function is one for which $f(-x) = -f(x)$. Prove that if f is integrable on $[0, a]$,

then

$$\int_{-a}^{a} f(x)\,dx = \begin{cases} 2\int_0^a f(x)\,dx & \text{if } f \text{ is even} \\ 0 & \text{if } f \text{ is odd} \end{cases}$$

33. Let f be integrable on $[a, b]$, and for $x \in [a, b]$ let

$$f_+(x) = \begin{cases} f(x) & \text{if } f(x) \geq 0 \\ 0 & \text{if } f(x) < 0 \end{cases}$$

$$f_-(x) = \begin{cases} 0 & \text{if } f(x) \geq 0 \\ -f(x) & \text{if } f(x) < 0 \end{cases}$$

Show that $f(x) = f_+(x) - f_-(x)$ and $|f(x)| = f_+(x) + f_-(x)$. Assuming f_+ and f_- are integrable on $[a, b]$ (this can be proved), obtain an alternate proof of Property 7.

34. Prove the following extension of Property 3: if f is integrable on an interval I, and a, b, and c are any numbers in I, then

$$\int_a^b f(x)\,dx = \int_a^c f(x)\,dx + \int_c^b f(x)\,dx$$

35. Let $\phi(x) = \begin{cases} 1 & \text{if } x \text{ is rational} \\ 0 & \text{if } x \text{ is irrational} \end{cases}$ for $x \in [0, 1]$.
Prove that $\int_0^1 \phi(x)\,dx$ does not exist. (*Hint:* For any partition P, choose two Riemann sums, one with the x_k^* rational and the other with the x_k^* irrational.)

C

Write a computer program that will calculate Riemann sums for a function f on an interval [a, b] corresponding to a regular partition $P = \{x_0, x_1, x_2, \ldots, x_n\}$, *where* x_k^* *is chosen as the right end point of the subinterval* $[x_{k-1}, x_k]$. *In Exercises 36–41 run the program to approximate the given integral for* $n = 10$, 100, *and* 1000.

36. $\int_1^4 \frac{1}{\sqrt{x}}\,dx$

37. $\int_{-1}^3 \sqrt{1 + x^3}\,dx$

38. $\int_0^\pi \sin x\,dx$

39. $\int_0^{\pi/2} \frac{\sin x}{1 + \cos x}\,dx$

40. $\int_{-1}^3 \frac{x}{4 - x}\,dx$

41. $\int_0^\pi \frac{\sin x}{x}\,dx$

4.4 THE FUNDAMENTAL THEOREM OF CALCULUS

The concepts of the derivative and the integral are so completely different from each other that it would almost seem too much to hope to discover a relationship between them. That such a relationship does indeed exist is what we will show in this section. As a consequence of this relationship we will discover a much simpler way to evaluate certain integrals. We need two preliminary results before getting to the main theorem.

THEOREM 4.5 MEAN VALUE THEOREM OF INTEGRAL CALCULUS

If f is continuous on $[a, b]$, there exists a point c in $[a, b]$ such that

$$\int_a^b f(x)\,dx = f(c)(b - a)$$

Proof Since f is continuous on $[a, b]$, we know by the theorem on extreme values (Theorem 3.2) that f attains an absolute maximum value M and an absolute minimum value m on $[a, b]$.

By Property 8 for integrals, since $m \leq f(x) \leq M$, we have

$$m(b - a) \leq \int_a^b f(x)\,dx \leq M(b - a)$$

or, on dividing by $b - a$,

$$m \leq \frac{\int_a^b f(x)\,dx}{b - a} \leq M$$

Thus, the number

$$\frac{\int_a^b f(x)\,dx}{b-a}$$

is between the smallest and largest values of f on the interval. We can therefore apply the intermediate value theorem (Theorem 1.7) and conclude that there exists a number c in $[a, b]$ such that

$$f(c) = \frac{\int_a^b f(x)\,dx}{b-a}$$

Multiplying both sides by $b - a$ yields the desired result. ■

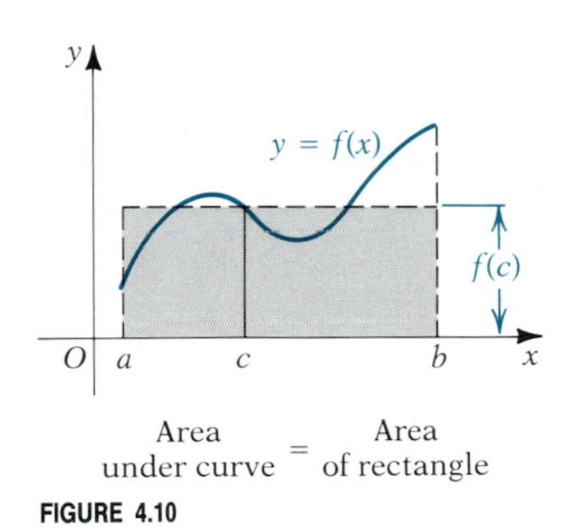

FIGURE 4.10

This theorem has an interesting geometric interpretation. Suppose f is continuous and $f(x) \geq 0$ on $[a, b]$. We know, then, that $\int_a^b f(x)\,dx$ gives the area under the graph of f. The mean value theorem for integrals says this area is equal to $f(c) \cdot (b - a)$ for some number c in $[a, b]$. But $f(c) \cdot (b - a)$ is just the area of the rectangle with height $f(c)$ and base equal to the interval $[a, b]$, as shown in Figure 4.10. The height $f(c)$ is therefore a sort of average value of the ordinates $f(x)$ of the graph; that is, if we replace the curve by the horizontal line $f(c)$ units high, the area under the line is the same as the area under the curve. Although in this discussion we have assumed f nonnegative and continuous on $[a, b]$, we extend this notion of the average value of f to any integrable function by means of the following definition.

DEFINITION 4.4 Let f be integrable on $[a, b]$. Then the **average value** of f on $[a, b]$ is defined as

$$f_{\text{ave}} = \frac{1}{b-a}\int_a^b f(x)\,dx$$ ■

The next theorem shows that the integral of a continuous function f over an interval of the form $[a, x]$ is an antiderivative of f. It thus provides one of the links between integration and differentiation. Recall that the variable of integration in a definite integral is a dummy variable, so that

$$\int_a^b f(x)\,dx = \int_a^b f(t)\,dt$$

for example. Because in the theorem we are going to use x as the upper limit of integration, we use t for the variable of integration to avoid using the same letter for two different purposes.

THEOREM 4.6 Let f be continuous on $[a, b]$. Then for x in $[a, b]$,

$$\frac{d}{dx}\int_a^x f(t)\,dt = f(x) \tag{4.13}$$

Proof We go back to the definition of the derivative. Let

$$F(x) = \int_a^x f(t)\,dt$$

Then

$$F'(x) = \lim_{h \to 0} \frac{F(x+h) - F(h)}{h} = \lim_{h \to 0} \frac{\int_a^{x+h} f(t)\,dt - \int_a^x f(t)\,dt}{h} \tag{4.14}$$

First consider the case in which $h > 0$. Then, by Property 3 for integrals,

$$\int_a^x f(t)\,dt + \int_x^{x+h} f(t)\,dt = \int_a^{x+h} f(t)\,dt$$

so that

$$\int_a^{x+h} f(t)\,dt - \int_a^x f(t)\,dt = \int_x^{x+h} f(t)\,dt$$

and by the mean value theorem for integrals there exists a number c between x and $x + h$ such that

$$\int_x^{x+h} f(t)\,dt = f(c) \cdot h \tag{4.15}$$

If $h < 0$, this same result holds true, and you will be asked to show it in the exercises. Thus the numerator in equation (4.14) can be replaced by the right-hand side of equation (4.15) to get

$$F'(x) = \lim_{h \to 0} \frac{f(c) \cdot h}{h} = \lim_{h \to 0} f(c)$$

Since c lies between x and $x + h$, it follows by the squeezing theorem that as $h \to 0$, $c \to x$, and so by continuity $f(c) \to f(x)$. Thus,

$$F'(x) = f(x)$$

which is what we wanted to prove. ■

One consequence of this theorem is that every function that is continuous on an interval $[a, b]$ has an antiderivative there—namely, $\int_a^x f(t)\,dt$. For example, an antiderivative of $\sqrt{1-x}$ on $[0, 1]$ is $\int_0^x \sqrt{1-t}\,dt$, since by the theorem,

$$\frac{d}{dx} \int_0^x \sqrt{1-t}\,dt = \sqrt{1-x}$$

for all x in $[0, 1]$.

This theorem also shows one aspect of what is essentially an inverse relationship between the operations of differentiation and integration; in a certain sense when we integrate a continuous function and then differentiate the result, we come back to the original function. The next theorem shows that except for a constant term, integrating the derivative also brings us back to the original function.

THEOREM 4.7 THE FUNDAMENTAL THEOREM OF CALCULUS

Let f be continuous on $[a, b]$ and let F be any antiderivative of f on $[a, b]$. Then

$$\int_a^b f(x)\,dx = F(b) - F(a)$$

Proof By Theorem 4.5 we know that $\int_a^x f(t)\,dt$ is an antiderivative of f on $[a, b]$, and we are given that F is also an antiderivative of f there. So by Corollary 3.9 these two antiderivatives differ by a constant; that is,

$$\int_a^x f(t)\,dt = F(x) + C$$

for some constant C. In fact, we can determine C. If we put $x = a$, we get

$$\int_a^a f(t)\,dt = F(a) + C$$

But the integral on the left is 0, so $C = -F(a)$. Thus,

$$\int_a^x f(t)\,dt = F(x) - F(a) \tag{4.16}$$

Finally, put $x = b$ to get

$$\int_a^b f(t)\,dt = F(b) - F(a)$$

This is what we wanted to prove. ■

The importance of this result can hardly be overemphasized, and that is why it is called the *fundamental theorem*. To give an idea of what it enables us to do, recall the work involved, for example, in evaluating an integral such as $\int_0^1 x^3\,dx$ making use of the limit of Riemann sums. Using the fundamental theorem, we can get the answer quickly and easily by first finding *any* antiderivative of x^3—for example, $F(x) = x^4/4$—and then calculating $F(1) - F(0)$:

$$\int_0^1 x^3\,dx = F(1) - F(0) = \tfrac{1}{4} - 0 = \tfrac{1}{4}$$

It is convenient to introduce the notation

$$F(x)\Big]_a^b$$

to mean $F(b) - F(a)$. Then we can write

$$\int_a^b f(x)\,dx = F(x)\Big]_a^b$$

for the result of the fundamental theorem. For example,

$$\int_0^1 x^3\,dx = \frac{x^4}{4}\Big]_0^1 = \frac{1}{4} - 0 = \frac{1}{4}$$

EXAMPLE 4.9 Evaluate each of the following using the fundamental theorem:

(a) $\int_{-1}^{2} (3x^2 - 2x + 4)\,dx$ (b) $\int_0^{\pi/2} \cos x\,dx$

(c) $\int_{-\pi/4}^{\pi/4} \sec^2 x\,dx$

Solution Note first that all of the integrands are continuous on the intervals involved, so the fundamental theorem applies.

(a) $$\int_{-1}^{2} (3x^2 - 2x + 4)\,dx = x^3 - x^2 + 4x\Big]_{-1}^{2} = (8 - 4 + 8) - (-1 - 1 - 4)$$
$$= 12 + 6 = 18$$

(b) $\int_0^{\pi/2} \cos x\,dx = \sin x\Big]_0^{\pi/2} = \sin\frac{\pi}{2} - \sin 0 = 1$

(c) $\int_{-\pi/4}^{\pi/4} \sec^2 x\,dx = \tan x\Big]_{-\pi/4}^{\pi/4} = \tan\frac{\pi}{4} - \tan(-\frac{\pi}{4}) = 1 - (-1) = 2$ ■

Remark Since in the fundamental theorem we are allowed to use *any* antiderivative of the integrand, we might as well choose the simplest one. Any other one would give the same result. For example,

$$\int_{-1}^{2} x^2\,dx = \frac{x^3}{3}\Big]_{-1}^{2} = \frac{8}{3} - \frac{(-1)}{3} = 3$$

If, instead of the simplest antiderivative, we had used some other, it would necessarily have been of the form $x^3/3 + C$, and so

$$\int_{-1}^{2} x^2\,dx = \frac{x^3}{3} + C\Big]_{-1}^{2} = \left(\frac{8}{3} + C\right) - \left(\frac{-1}{3} + C\right)$$
$$= \frac{8}{3} + \frac{1}{3} = 3$$

In other words, the constant C will cancel out.

A Word of Caution Do not infer from this that it is no longer necessary to add the constant when forming an antiderivative. We are speaking here of one particular use of the antiderivative where the constant can be deleted. In most cases we will need the constant.

In the fundamental theorem, F is an antiderivative of f; that is, $F'(x) = f(x)$. Therefore, we could write the result in the form

$$\int_a^b F'(x)\,dx = F(b) - F(a) \tag{4.17}$$

so long as F' is continuous on $[a, b]$. It can be proved, in fact, that equation (4.17) holds true even if F' is not continuous but is integrable on $[a, b]$. (See Exercise 55 in Exercise Set 4.4.) Going back to equation (4.16) in the proof of the theorem, if we also replace f by F', we get

$$\int_a^x F'(t)\,dt = F(x) - F(a)$$

Since this involves the function F only, we could just as well write it with another letter to designate the function. In particular, we can use our usual letter f and get

$$\int_a^x f'(t)\,dt = f(x) - f(a) \tag{4.18}$$

and this is true for all $x \in [a, b]$ provided f' is integrable there. Let us compare this with equation (4.13), which says if f is continuous on $[a, b]$, then

$$\frac{d}{dx}\int_a^x f(t)\,dt = f(x), \qquad x \in [a, b]$$

Here we differentiate the integral of f from a to x and get $f(x)$ as the result. In equation (4.18) we integrate from a to x the derivative of f and get $f(x)$ minus a constant, $f(a)$. It is in this sense that integration and differentiation can be considered inverse processes.

EXERCISE SET 4.4

A

Use the fundamental theorem to evaluate the integrals in Exercises 1–30.

1. $\int_1^3 (x^2 + 2x - 1)\, dx$

2. $\int_{-1}^1 (3x - x^2)\, dx$

3. $\int_{-2}^1 (x - 1)^2\, dx$

4. $\int_3^5 x(2x - 3)\, dx$

5. $\int_0^4 (x^3 - 3x^2 + 4)\, dx$

6. $\int_{-2}^0 (x + 2)^3\, dx$

7. $\int_0^1 (x - \sqrt{x} + 3)\, dx$

8. $\int_{-1}^1 (x^5 - 2x^3 + 4x)\, dx$

9. $\int_{-2}^2 (x^2 - 3)^2\, dx$

10. $\int_4^1 (x - 1)(x + 2)\, dx$

11. $\int_{-2}^{-1} \frac{x^3 - 1}{x^2}\, dx$

12. $\int_1^4 \frac{x + 1}{\sqrt{x}}\, dx$

13. $\int_0^\pi 2 \sin x\, dx$

14. $\int_{-\pi/2}^{\pi/2} 3 \cos x\, dx$

15. $\int_0^{\pi/3} \sec^2 x\, dx$

16. $\int_{-\pi/3}^{\pi/4} \sec x \tan x\, dx$

17. $\int_{\pi/6}^{\pi/2} 2 \csc x \cot x\, dx$

18. $\int_{\pi/4}^{3\pi/4} 4 \csc^2 x\, dx$

19. $\int_4^9 (t^{3/2} - 2t^{1/2} + 1)\, dt$

20. $\int_1^8 (t^{2/3} + 2t^{-1/3})\, dt$

21. $\int_1^8 \frac{u - 2}{\sqrt[3]{u}}\, du$

22. $\int_1^4 \frac{(v - 1)^3}{\sqrt{v}}\, dv$

23. $\int_{\pi/6}^{5\pi/6} (x + \sin x)\, dx$

24. $\int_{\pi/4}^{\pi/2} \frac{1 - \cos\theta}{\sin^2\theta}\, d\theta$

25. $\int_0^{\pi/4} (1 + \sin^2\phi \sec^2\phi)\, d\phi$

26. $\int_{\pi/6}^{\pi/3} \tan^2 x\, dx$

27. $\int_{2\pi/3}^{5\pi/6} \frac{\sec t - \csc t}{\sec t \csc t}\, dt$

28. $\int_{\pi/4}^{\pi/2} \cot^2 v\, dv$

29. $\int_1^3 \left(x - \frac{1}{x}\right)\left(2x - \frac{3}{x}\right) dx$

30. $\int_1^9 \frac{(\sqrt{t} - 2)(\sqrt{t} + 3)}{2\sqrt{t}}\, dt$

In Exercises 31–34 find the average value of the function on the given interval.

31. $f(x) = x^2 - 2x - 3$ on $[-2, 4]$

32. $f(x) = 1 + x - 2x^2$ on $[-1, 2]$

33. $f(x) = x^3 - 3x^2 - 4x + 2$ on $[-2, 3]$

34. $f(x) = 2x^3 - 3x - 4$ on $[-2, 2]$

In Exercises 35–38 find a number c in $[a, b]$ that satisfies $\int_a^b f(x)\, dx = f(c) \cdot (b - a)$.

35. $f(x) = \sqrt{x}$ on $[0, 9]$

36. $f(x) = 4/x^2$ on $[1, 4]$

37. $f(x) = x^2 - 1$ on $[0, 3]$

38. $f(x) = x^2 - 2x$ on $[-2, 1]$

In Exercises 39–42 use Theorem 4.6 to find the derivative. State on what intervals the result is valid.

39. $\frac{d}{dx} \int_1^x \frac{2}{\sqrt{1 + t^2}}\, dt$

40. $\frac{d}{dx} \int_2^x \frac{t}{t - 1}\, dt$

41. $F'(x)$ if $F(x) = \int_0^x \sqrt{t^3 + 8}\, dt$

42. $F'(x)$ if $F(x) = \int_{-3}^x \frac{\sin t}{2 + \cos t}\, dt$

43. Find $\frac{d}{dx} \int_{-1}^x (t^3 - 3t^2 + 1)\, dt$ by two methods: (a) Theorem 4.6 and (b) first performing the integration using the fundamental theorem and then differentiating.

B

44. If $f(x) = x/(x + 1)$, find $\int_0^2 f'(x)\, dx$.

45. Find $\int_0^2 \frac{d}{dt}(1/\sqrt{1 + t^3})\, dt$.

46. Derive a formula for

$$\frac{d}{dx} \int_x^b f(x)\, dt$$

where f is continuous on $[a, b]$ and $x \in [a, b]$.

47. Use the result of Exercise 46 to find $g'(x)$ if

$$g(x) = \int_x^4 \frac{\sin t}{t}\, dt$$

where $x > 0$.

In Exercises 48–53 simplify the integrands and then use the fundamental theorem to evaluate the integrals.

48. $\int_2^3 \sqrt{\frac{1}{x^4} - 1 + \frac{x^4}{4}}\, dx$

49. $\int_{-1}^2 \frac{2t^3 + 3t^2 + 4}{t + 2}\, dt$

50. $\int_{-\pi/6}^{\pi/3} \left(\frac{1}{1 + \sin\theta} + \frac{1}{1 - \sin\theta}\right) d\theta$

51. $\int_{\pi/2}^{5\pi/6} \frac{2}{1 - \cos 2x}\, dx$

52. $\int_{-\pi/4}^{\pi/3} \frac{1 - \sin\theta}{1 + \sin\theta}\, d\theta$ (*Hint:* Multiply numerator and denominator by $1 - \sin\theta$.)

53. $\int_{5\pi/6}^{\pi} \frac{1 - \sin 2x}{\sin x - \cos x}\, dx$

54. Complete the proof of Theorem 4.6 by considering $h < 0$.

55. Prove that equation (4.17) holds true if we assume only that F' is integrable on $[a, b]$. (*Hint:* Use the mean value theorem of differential calculus to show that for any partition P of $[a, b]$, there is a point $x_k^* \in (x_{k-1}, x_k)$, $k = 1, 2, \ldots, n$, such that

$$\sum_{k=1}^{n} [F(x_k) - F(x_{k-1})] = \sum_{k=1}^{n} F'(x_k^*)\,\Delta x_k$$

Show that the left-hand side reduces to $F(b) - F(a)$. Then let $\|P\| \to 0$.)

56. Let $f(x) = \begin{cases} x^2 \sin\frac{1}{x^2} & \text{if } x \neq 0 \\ 0 & \text{if } x = 0 \end{cases}$. Show that $f'(x)$ exists on $[0, 1]$ but that

$$\int_0^1 f'(x)\, dx \neq f(1) - f(0)$$

Why does this not contradict Exercise 55? (*Hint:* Show that f' is unbounded on $[0, 1]$.)

57. Prove that if f is integrable on $[a, b]$, the function $F(x) = \int_a^x f(t)\, dt$ is continuous on $[a, b]$. [*Hint:* Let M be such that $|f(x)| \leq M$ on $[a, b]$. (Justify.) Then use the $\varepsilon - \delta$ definition of continuity of F at x_0, making use of properties of the integral.]

58. Show that the mean value theorems of integral calculus and of differential calculus are essentially different formulations of the same result. [*Hint:* In $\int_a^b f(x)\, dx = f(c)(b - a)$, let F be an antiderivative of f so that $f = F'$.]

59. Prove that if f and g are defined on $[a, b]$, with $f(x) = g(x)$ on $[a, b)$ and g continuous on $[a, b]$, then $\int_a^b f(x)\, dx$ exists and equals $\int_a^b g(x)\, dx$ regardless of how $f(b)$ is defined. (*Hint:* Use Exercise 57.)

60. Prove that if f is integrable on $[a, b]$, then $F(x) = \int_x^b f(t)\, dt$ is continuous on $[a, b]$. (*Hint:* Use the hint for Exercise 57.)

61. Prove that if f and g are defined on $[a, b]$, with $f(x) = g(x)$ on $(a, b]$, and g is continuous on $[a, b]$, then $\int_a^b f(x)\, dx$ exists and equals $\int_a^b g(x)\, dx$ regardless of how $f(a)$ is defined. (*Hint:* Use Exercise 60.)

62. Suppose f is defined on $[a, b]$ and is continuous on (a, b), and $\lim_{x \to a^+} f(x)$ and $\lim_{x \to b^-} f(x)$ both exist. Prove that $\int_a^b f(x)\, dx$ exists, regardless of how $f(a)$ and $f(b)$ are defined. (*Hint:* Use the results of Exercises 59 and 61.)

4.5 MORE ON ANTIDERIVATIVES; THE TECHNIQUE OF SUBSTITUTION

Because of the fundamental theorem, antiderivatives take on a role of special importance in evaluating integrals. It will be useful therefore to learn techniques for finding antiderivatives beyond those in the list given in Section 3.7. We will see one such technique in this section. First we introduce a new notation.

DEFINITION 4.5 THE INDEFINITE INTEGRAL

The symbol $\int f(x)\, dx$ is called the **indefinite integral** of f, and its meaning is given by

$$\int f(x)\, dx = F(x) + C$$

where F is any function for which $F' = f$, and C is an arbitrary constant called the **constant of integration.** ■

In other words, the indefinite integral of f is just the familiar general antiderivative of f with a new name and a new symbol. Although definite and indefinite integrals differ only slightly in appearance, their meanings differ significantly. For example, consider the following:

Definite integral: $\int_{-1}^{2} x^2\,dx = \left.\frac{x^3}{3}\right]_{-1}^{2} = \frac{8}{3} - \left(-\frac{1}{3}\right) = 3$

Indefinite integral: $\int x^2\,dx = \frac{x^3}{3} + C$

In the first case the answer is a number, whereas in the second it is a family of functions. The definite integral and the indefinite integral are related by the fundamental theorem. In fact, we could write the conclusion of that theorem in the form

$$\int_a^b f(x)\,dx = \left.\int f(x)\,dx\right]_a^b$$

Just as with a definite integral, we refer to the $f(x)$ in $\int f(x)\,dx$ as the *integrand*, and the process by which we obtain the answer is called *integration*. The context will make clear whether the integration is being performed on a definite or an indefinite integral.

The following properties of the indefinite integral hold true on any interval I in which both f and g have antiderivatives.

PROPERTIES OF THE INDEFINITE INTEGRAL

1. $\frac{d}{dx}\int f(x)\,dx = f(x)$
2. $\int f'(x)\,dx = f(x) + C$, if f is differentiable on I
3. $\int kf(x)\,dx = k\int f(x)\,dx$; k is a constant
4. $\int [f(x) \pm g(x)]\,dx = \int f(x)\,dx \pm \int g(x)\,dx$

The proofs of these all follow from the definition of the indefinite integral and are left as exercises.

All the antidifferentiation formulas summarized in Section 3.7 can be written using the indefinite integral notation. For example, we can write

$$\int x^n\,dx = \frac{x^{n+1}}{n+1} + C \qquad (n \neq -1)$$

Similarly, $\int \sin x\,dx = -\cos x + C$ and $\int \cos x\,dx = \sin x + C$. The following theorem gives rise to an integration technique, called **integration by substitution,** that will enable us to integrate a much larger class of functions.

THEOREM 4.8 Suppose F and g are functions for which the composition $F \circ g$ is defined and differentiable on an interval I. Let $f = F'$. Then for $x \in I$,

$$\int f(g(x))g'(x)\,dx = F(g(x)) + C \tag{4.19}$$

Proof By the chain rule,

$$\frac{d}{dx}F(g(x)) = F'(g(x)) \cdot g'(x) = f(g(x))g'(x)$$

so that according to the definition of the indefinite integral, equation (4.19) is true. ■

If we make the substitution $u = g(x)$, then by the definition of the differential, we have $du = g'(x)\,dx$. Thus equation (4.19) takes the form

$$\int f(u)\,du = F(u) + C$$

where it is understood that in the answer u is to be replaced by $g(x)$.

Remark This substitution shows that the dx that appears in the indefinite integral really does play the role of a differential. The same is true in the definite integral, as we will see shortly.

EXAMPLE 4.10 Evaluate the following indefinite integrals:

$$\text{(a)} \int (x^2 - 1)^3 \cdot 2x\,dx \qquad \text{(b)} \int x\sqrt{1 - x^2}\,dx$$

Solution (a) The integrand is in the form that appears in equation (4.19), with $f(x) = x^3$ and $g(x) = x^2 - 1$. We use the substitution procedure outlined above. Let $u = x^2 - 1$; then $du = 2x\,dx$. Thus,

$$\int \underbrace{(x^2 - 1)^3}_{u^3} \cdot \underbrace{2x\,dx}_{du} = \int u^3\,du = \frac{u^4}{4} + C$$

All that remains is to replace u in the answer by $x^2 - 1$. So we have

$$\int (x^2 - 1)^3 \cdot 2x\,dx = \frac{(x^2 - 1)^4}{4} + C$$

(b) Since $\sqrt{1 - x^2}$ can be thought of as a composite function, we make a substitution for the inner function. Let $u = 1 - x^2$; then $du = -2x\,dx$. Now there is a slight complication. The integral can be rewritten as

$$\int (1 - x^2)^{1/2} x\,dx$$

and the first factor is $u^{1/2}$ by our substitution. However, the remaining $x\,dx$ is not du because the factor -2 is not present. There are two ways to proceed. First, by Property 2 for indefinite integrals, a constant factor can be taken inside or outside an integral without changing its value. So we can multiply by -2 inside the integral and compensate for this by multiplying by $-\frac{1}{2}$ outside, getting

$$\begin{aligned} -\frac{1}{2}\int \underbrace{(1 - x^2)^{1/2}}_{u^{1/2}}\underbrace{(-2x\,dx)}_{du} &= -\frac{1}{2}\int u^{1/2}\,du = -\frac{1}{2}\frac{u^{2/3}}{\frac{3}{2}} + C \\ &= -\frac{1}{2}\cdot\frac{2}{3}u^{3/2} + C \\ &= -\frac{1}{3}(1 - x^2)^{3/2} + C \end{aligned}$$

When we multiply by a constant on the inside and by its reciprocal on the outside in order to obtain du, we say we have "adjusted the differential."

Alternately, from $du = -2x\,dx$ we can solve for $x\,dx$, getting $x\,dx = -\frac{1}{2}\,du$, so that on substituting, we get

$$\int (1 - x^2)^{1/2} x\,dx = \int u^{1/2}\left(-\tfrac{1}{2}\,du\right) = -\tfrac{1}{2}\int u^{1/2}\,du$$

and from here on we proceed as we did before. ■

EXAMPLE 4.11 Evaluate the following indefinite integrals:

$$\text{(a)} \int \frac{\sin \sqrt{x}}{\sqrt{x}}\,dx \qquad \text{(b)} \int \cos^3 2x \sin 2x\,dx$$

Solution (a) Let $u = \sqrt{x}$; then $du = 1/(2\sqrt{x})\,dx$. We adjust the differential by multiplying on the inside by $\frac{1}{2}$ and compensate by multiplying on the outside by 2:

$$\begin{aligned}\int \frac{\sin \sqrt{x}}{\sqrt{x}}\,dx &= 2\int \sin \sqrt{x} \cdot \left(\frac{1}{2\sqrt{x}}\,dx\right)\\ &= 2\int \sin u\,du\\ &= -2\cos u + C\\ &= -2\cos \sqrt{x} + C\end{aligned}$$

(b) Let $u = \cos 2x$; then $du = -2\sin 2x\,dx$:

$$\begin{aligned}\int \cos^3 2x \sin 2x\,dx &= -\frac{1}{2}\int \cos^3 2x(-2\sin 2x\,dx)\\ &= -\frac{1}{2}\int u^3\,du\\ &= -\frac{1}{2}\frac{u^4}{4} + C\\ &= -\frac{1}{8}\cos^4 2x + C\end{aligned}$$

■

It is important that the substitution be chosen in such a way that everything inside the integral sign is in terms of u, including the differential. If the only thing preventing this is that a *constant* factor is missing, then you can adjust the differential by multiplying by that constant on the inside and by its reciprocal on the outside. But it should be stressed that you *cannot* multiply by a *variable* on the inside and by its reciprocal on the outside.

Another cautionary note is that in evaluating an indefinite integral by substitution, the final answer *must* be expressed in terms of the original variable of integration. The substitution of a new variable is a means to an end only, and it should not appear in the final answer. As the next theorem shows, however, when a substitution is made in a *definite* integral, if the limits of integration are appropriately changed, there is no need to revert to the original variable of integration.

THEOREM 4.9 Let F and g be differentiable functions such that $F \circ g$ is defined on $[a, b]$ and for which $\frac{d}{dx}(F \circ g)$ is continuous on $[a, b]$, and let $f = F'$. Then

$$\int_a^b f(g(x))g'(x)\,dx = \int_{g(a)}^{g(b)} f(u)\,du \tag{4.20}$$

Proof By Theorem 4.8, $F(g(x))$ is an antiderivative of $f(g(x))g'(x)$, so that

$$\int_a^b f(g(x))g'(x)\,dx = F(g(x))\Big]_a^b = F(g(b)) - F(g(a))$$

The right-hand side of equation (4.20) is

$$\int_{g(a)}^{g(b)} f(u)\,du = F(u)\Big]_{g(a)}^{g(b)} = F(g(b)) - F(g(a))$$

Thus, the left-hand and right-hand sides of equation (4.20) are the same. ■

The chief consequence of this theorem is that we can use the substitution method in definite integrals and, provided we also change the limits of integration in accordance with the substitution made, the integral can be evaluated using the new variable of integration. The next example illustrates this.

EXAMPLE 4.12 Evaluate the integrals:

(a) $\int_0^2 x^2\sqrt{1+x^3}\,dx$ (b) $\int_{1/\sqrt{\pi}}^{2/\sqrt{3\pi}} \frac{1}{x^3}\sec^2\frac{1}{x^2}\,dx$

Solution (a) Let $u = 1 + x^3$; then $du = 3x^2\,dx$. When $x = 0$, $u = 1$, and when $x = 2$, $u = 9$. We adjust the differential and change limits:

$$\begin{aligned}\int_0^2 x^2\sqrt{1+x^3}\,dx &= \tfrac{1}{3}\int_0^2 (1+x^3)^{1/2}(3x^2\,dx)\\ &= \tfrac{1}{3}\int_1^9 u^{1/2}\,du\\ &= \tfrac{1}{3}\cdot\tfrac{2}{3}u^{3/2}\Big]_1^9\\ &= \tfrac{2}{9}(9^{3/2} - 1^{3/2})\\ &= \tfrac{2}{9}(26) = \tfrac{52}{9}\end{aligned}$$

(b) Let $u = 1/x^2 = x^{-2}$:

$$du = -2x^{-3}\,dx = -\frac{2}{x^3}\,dx$$

When $x = 1/\sqrt{\pi}$, $u = (\sqrt{\pi})^2 = \pi$, and when $x = 2/\sqrt{3\pi}$, $u = (\sqrt{3\pi}/2)^2 = 3\pi/4$:

$$\begin{aligned}\int_{1/\sqrt{\pi}}^{2/\sqrt{3\pi}} \frac{1}{x^3}\sec^2\frac{1}{x^2}\,dx &= -\frac{1}{2}\int_{1/\sqrt{\pi}}^{2/\sqrt{3\pi}}\left(\sec^2\frac{1}{x^2}\right)\left(-\frac{2}{x^3}\,dx\right)\\ &= -\frac{1}{2}\int_{\pi}^{3\pi/4}\sec^2 u\,du\\ &= \frac{1}{2}\int_{3\pi/4}^{\pi}\sec^2 u\,du = \frac{1}{2}\tan u\Big]_{3\pi/4}^{\pi}\\ &= \frac{1}{2}\left[\tan\pi - \tan\frac{3\pi}{4}\right] = \frac{1}{2}[0-(-1)] = \frac{1}{2}\end{aligned}$$ ■

Remark In evaluating a definite integral by the substitution method, an alternate procedure is to revert back to the original variable of integration and use the original limits, but the method illustrated in the preceding example is usually easier.

It should be clear that there is nothing magic about using the letter u for a substitution, any more than there is in using the letter x as the original variable. For example, the following is perfectly valid. To evaluate $\int (t^2-1)^4 t\,dt$

let $v = t^2 - 1$; then $dv = 2t\,dt$:

$$\begin{aligned}\int (t^2 - 1)^4 t\,dt &= \frac{1}{2}\int (t^2 - 1)^4 \cdot 2t\,dt = \frac{1}{2}\int v^4\,dv \\ &= \frac{1}{2}\frac{v^5}{5} + C \\ &= \frac{1}{10}(t^2 - 1) + C\end{aligned}$$

With a little practice many substitutions can be done mentally, without ever actually introducing a new variable. Consider the following, for example:

$$\begin{aligned}\int \frac{x}{\sqrt{1 - x^2}}\,dx = -\frac{1}{2}\int \overbrace{(1 - x^2)^{-1/2}}^{u^{-1/2}}\overbrace{(-2x\,dx)}^{du} &= -\frac{1}{2}\cdot 2(1 - x^2)^{1/2} + C \\ &= -\sqrt{1 - x^2} + C\end{aligned}$$

Here, we *mentally* substituted u for $1 - x^2$ (sometimes writing u over what it is replacing is useful) and calculated its differential $-2x\,dx$, also mentally. Observing that we needed the factor -2 inside, we put it there and divided by it on the outside. Finally, we used the power rule on $\int u^{-1/2}\,du$. In the exercises that follow, after making several actual substitutions, you should try doing some of the others mentally.

EXERCISE SET 4.5

A

In Exercises 1–28 evaluate the indefinite integrals by making an appropriate substitution.

1. $\int x(1 + x^2)^3\,dx$

2. $\int x\sqrt{x^2 - 4}\,dx$

3. $\int x^2(x^3 - 1)^5\,dx$

4. $\int \frac{x}{\sqrt{9 - x^2}}\,dx$

5. $\int x\sqrt{3x^2 + 4}\,dx$

6. $\int \frac{(\sqrt{x} + 1)^5}{\sqrt{x}}\,dx$

7. $\int \sqrt{1 + 3x}\,dx$

8. $\int (4x - 3)^{2/3}\,dx$

9. $\int (x - 1)\sqrt{x^2 - 2x + 3}\,dx$

10. $\int \frac{2x + 1}{\sqrt{x^2 + x - 4}}\,dx$

11. $\int \sin 2x\,dx$

12. $\int \cos 3x\,dx$

13. $\int \sin kx\,dx$

14. $\int \cos kx\,dx$

15. $\int 2\sin x\cos x\,dx$

16. $\int \sin x\cos^2 x\,dx$

17. $\int x\sec^2 x^2\,dx$

18. $\int \sec 2x\tan 2x\,dx$

19. $\int \sqrt{\tan x}\sec^2 x\,dx$

20. $\int \sqrt{1 + \sin x}\cos x\,dx$

21. $\int \frac{\sin x}{(1 - \cos x)^2}\,dx$

22. $\int \frac{x + \cos 2x}{\sqrt{x^2 + \sin 2x}}\,dx$

23. $\int \frac{\csc^2\sqrt{x}}{\sqrt{x}}\,dx$

24. $\int \frac{1}{x^2}\cos\frac{1}{x}\,dx$

25. $\int \left(\frac{x - 1}{x}\right)^5 \cdot \frac{1}{x^2}\,dx$

26. $\int (x^2 - x)(2x^3 - 3x^2 + 1)^{3/2}\,dx$

27. $\int \sqrt{x}(x^{3/2} - 1)^{2/3}\,dx$

28. $\int \frac{(x^{2/3} + a^{2/3})^{3/2}}{\sqrt[3]{ax}}\,dx$

In Exercises 29–44 evaluate the definite integrals by making an appropriate substitution.

29. $\int_0^2 (1 + 4x)^{3/2}\,dx$

30. $\int_0^4 x\sqrt{x^2 + 9}\,dx$

31. $\int_{-1}^2 x(x^2 - 2)^4\,dx$

32. $\int_3^4 \frac{x}{\sqrt{25 - x^2}}\,dx$

33. $\int_4^9 \frac{(\sqrt{x} - 4)^6}{\sqrt{x}}\,dx$

34. $\int_0^2 x(3x^2 + 4)^{3/2}\,dx$

35. $\int_1^5 (2x + 1)(x^2 + x - 3)^{1/3}\,dx$

36. $\int_1^2 \left(\frac{5x^2 - 8}{3x^2}\right)^{2/3} \frac{dx}{x^3}$

37. $\int_0^{\pi/6} \cos 3x\,dx$

38. $\int_0^{\pi/2} \sin \frac{x}{2}\,dx$

39. $\int_{\pi/6}^{2\pi/3} \sin^3 x \cos x\,dx$

40. $\int_0^{\pi/2} \sqrt{\cos x} \sin x\,dx$

41. $\int_{-\pi/6}^{\pi/8} \tan^3 2x \sec^2 2x\,dx$

42. $\int_{-\pi/3}^{\pi/2} \frac{\sin x}{(1 + \cos x)^2}\,dx$

43. $\int_0^{\pi/3} \sec^3 x \tan x\,dx$

44. $\int_{\pi^2/4}^{\pi^2} \frac{\sin \sqrt{x}}{\sqrt{x}}\,dx$

B

45. Prove Properties 1–4 of the indefinite integral.

In Exercises 46–51 evaluate the indefinite integrals.

46. $\int x\sqrt{(x^4 + 12x^2 + 36)^3}\,dx$

47. $\int \frac{\sec^2 x}{\sec^2 x + 2\tan x}\,dx$

48. $\int \cos \frac{x}{2} \sin x\,dx$ on $[0, \pi]$

49. $\int \frac{dx}{\sqrt{x - x\sqrt{x}}}$

50. $\int (\sin 2x - \cos 2x)^2\,dx$

51. $\int \cos^2 2x \cos x\,dx$

Verify the equations in Exercises 52 and 53.

52. $\int x \sin 2x\,dx = -\frac{x\cos 2x}{2} + \frac{\sin 2x}{4} + C$

53. $\int \frac{1}{x^2} \cos \frac{1}{\sqrt{x}}\,dx = -2\left[\frac{1}{\sqrt{x}} \sin \frac{1}{\sqrt{x}} + \cos \frac{1}{\sqrt{x}}\right] + C$

54. Let $f(x) = \begin{cases} 1 & \text{if } x \geq 0 \\ -1 & \text{if } x < 0 \end{cases}$. Show that $\int_{-1}^1 f(x)\,dx$ exists, but $\int f(x)\,dx$ does not exist on $[-1, 1]$. [*Hint:* Consider the result of Exercise 35 in Exercise Set 3.6.]

4.6 SUPPLEMENTARY EXERCISES

1. Write each of the following sums in closed form:

(a) $\sum_{k=1}^{n} (3 + 2k - 4k^2)$ (b) $\sum_{k=1}^{n} k(1 - k^2)$

2. Let $f(x) = x^2 + 1$ on $[0, 2]$. Find U_n and L_n and show that they have the same limit as $n \to \infty$. Interpret the result geometrically.

3. Let $f(x) = x(1 + x)(3 - x)$ on $[-1, 4]$. Find the Riemann sum for f using the partition $P = \{-1, 0, 2, 3, 4\}$, with $x_1^* = -0.5$, $x_2^* = 0.6$, $x_3^* = 2.3$, and $x_4^* = 3.4$. Compare the result with the exact value of $\int_{-1}^4 f(x)\,dx$.

4. Find the exact value of the integral $\int_{-1}^3 (x^2 - 2x - 1)\,dx$ in two ways: (a) by taking the limit of the Riemann sum formed from a regular partition, with x_k^* the right end point of the kth subinterval, and (b) by using the fundamental theorem of calculus.

In Exercises 5 and 6 justify the inequalities.

5. (a) $\left|\int_{-1}^1 \frac{x + 1}{x - 2}\,dx\right| \leq 4$

(b) $\frac{\pi}{2} \leq \int_0^{\pi/4} \frac{\tan x + 2}{\sec x}\,dx \leq \frac{3\pi}{4}$

6. (a) $\int_a^b \sqrt{x^2 + 1}\,dx \leq \frac{1}{4}\int_a^b (x^2 + 5)\,dx$

(b) $-3 \leq \int_{-2}^4 \frac{x - 1}{x^2 + 3}\,dx \leq 1$

In Exercises 7–18 use the fundamental theorem to evaluate the integrals.

7. $\int_2^4 x(1-2x)\,dx$

8. $\int_{-3}^0 (x-\sqrt{1-x})\,dx$

9. $\int_1^4 \left(x-\dfrac{1}{\sqrt{x}}\right)^3 dx$

10. $\int_0^{\pi/3} \dfrac{1-\sin x}{\cos^2 x}\,dx$

11. $\int_{-8}^{-1} (u^{1/3}-u^{-2/3})\,du$

12. $\int_0^{\pi/4} \sin^3 t\cos t\,dt$

13. $\int_{-\pi/3}^{\pi/3} (\theta+\cos\theta)\,d\theta$

14. $\int_{-1}^2 (2x-1)(x+4)\,dx$

15. $\int_9^{16} \dfrac{(\sqrt{x}-2)^5}{\sqrt{x}}\,dx$

16. $\int_0^2 \dfrac{x^2}{\sqrt{1+x^3}}\,dx$

17. $\int_0^{\pi/4} \dfrac{\sin 2x}{(1+\cos 2x)^3}\,dx$

18. $\int_{27}^{125} t^{-1/3}(t^{2/3}-9)^{3/2}\,dt$

In Exercises 19–26 find the indicated antiderivatives.

19. $\int \dfrac{1}{x^2}\sec^2\dfrac{2}{x}\,dx$

20. $\int \dfrac{\tan^3\theta-\sin\theta}{\cos^2\theta}\,d\theta$

21. $\int \dfrac{x-1}{\sqrt{x^2-2x+4}}\,dx$

22. $\int x(x^4+4x^2+4)^{5/2}\,dx$

23. $\int \dfrac{\sqrt{1+\sqrt{x}}}{\sqrt{x}}\,dx$

24. $\int \dfrac{x(1+\cos^2 x^2)}{\sin^2 x^2}\,dx$

25. $\int \left[\dfrac{1}{1+\cos\theta}-\dfrac{1}{1-\cos\theta}\right]d\theta$

26. $\int \sin 2t(1+\cos^2 t)^3\,dt$

27. Verify that

$$\int x^2\cos 3x\,dx = \frac{1}{27}[(9x^2-2)\sin 3x+6x\cos 3x]+C$$

28. Evaluate the integral $\int_0^2 f'(t)\,dt$ if $f(x)=(3x^2-4)\sqrt{x^3+1}$.

29. Find $f'(t)$ if $f(t)=\int_0^t x/(\sin^2 x+1)\,dx$.

30. Evaluate:

(a) $\dfrac{d}{dx}\int_{\pi/6}^{x}\dfrac{\sin t}{t}\,dt$ at $x=\pi$

(b) $\int_{\pi/6}^{\pi}\dfrac{d}{dx}\dfrac{\sin x}{x}\,dx$

31. Find the average value of $f(x)=2x^3-4x^2-5x+2$ on $[-2, 2]$.

32. Find the average value of $g(x)=x/\sqrt{x^2+9}$ on $[0, 4]$.

33. Find c such that $\int_a^b f(x)\,dx=f(c)(b-a)$, where $f(x)=\sqrt{3x+4}$ and $[a, b]=[-1, 4]$.

34. Let f be continuous on $[a, b]$, and let g be a differentiable function on an interval I such that the range of g is contained in $[a, b]$. Derive a formula for

$$\frac{d}{dx}\int_a^{g(x)} f(t)\,dt \qquad (x\in I)$$

(*Hint:* Make use of the chain rule.)

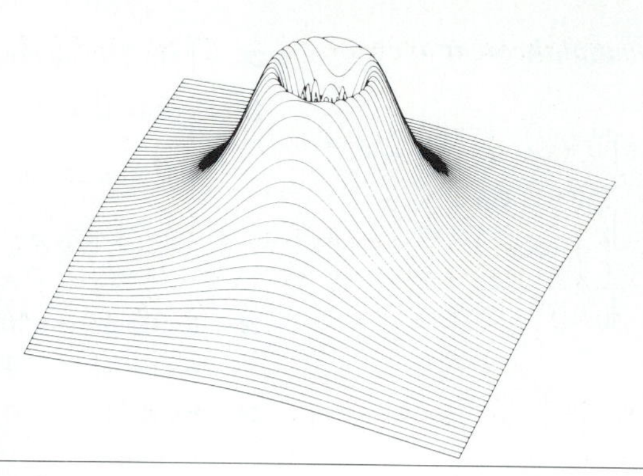

CHAPTER 5

APPLICATIONS OF THE INTEGRAL

5.1 AREAS

We know already that if f is a nonnegative continuous function on $[a, b]$, then the area of the region bounded by the graph of f, the x-axis, and the lines $x = a$ and $x = b$ is given by

$$A = \int_a^b f(x)\,dx$$

In this section we give two extensions of this.

First, we suppose f is not continuous but is nonnegative and bounded on $[a, b]$ and that it has only finitely many discontinuities, each of which is either a jump discontinuity or a removable discontinuity. This means that for each point of discontinuity x_k in (a, b), $\lim_{x \to x_k^+} f(x)$ and $\lim_{x \to x_k^-} f(x)$ both exist and that $\lim_{x \to a^+} f(x)$ and $\lim_{x \to b^-} f(x)$ exist. Such a function is said to be **piecewise continuous.** An example is shown in Figure 5.1.

Suppose f is such a piecewise continuous function and that the points of discontinuity in (a, b) are $x_1, x_2, \ldots, x_{n-1}$. Then letting $x_0 = a$ and $x_n = b$,

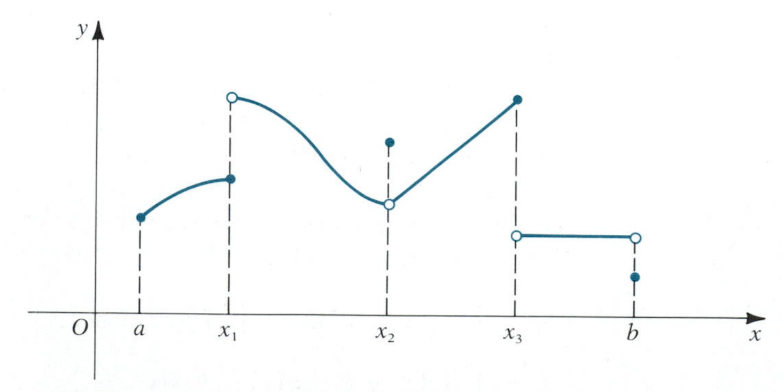

FIGURE 5.1 A piecewise continuous function

we define the area of the region under the graph of f between a and b to be

$$A = \sum_{k=1}^{n} \int_{x_{k-1}}^{x_k} f(x)\,dx$$

The fact that each of the integrals $\int_{x_{k-1}}^{x_k} f(x)\,dx$ exists follows from the continuity of f on (x_{k-1}, x_k) and the existence of the right- and left-hand limits, $\lim_{x\to x_{k-1}^+} f(x)$ and $\lim_{x\to x_k^-} f(x)$. (See Exercise 62 in Exercise Set 4.4.)

EXAMPLE 5.1 Find the area under the graph of

$$f(x) = \begin{cases} 1 - x^2 & \text{if } -1 \le x < 0 \\ 2 - x & \text{if } \ \ 0 \le x < 1 \\ 0 & \text{if } \ \ x = 1 \\ x^2 - 2x + 3 & \text{if } \ \ 1 < x \le 2 \end{cases}$$

from $x = -1$ to $x = 2$.

Solution The graph of f is shown in Figure 5.2. Since f is nonnegative and piecewise continuous, the area is given by

$$\begin{aligned} A &= \int_{-1}^{0} f(x)\,dx + \int_{0}^{1} f(x)\,dx + \int_{1}^{2} f(x)\,dx \\ &= \int_{-1}^{0} (1 - x^2)\,dx + \int_{0}^{1} (2 - x)\,dx + \int_{1}^{2} (x^2 - 2x + 3)\,dx \\ &= \left[x - \frac{x^3}{3}\right]_{-1}^{0} + \left[2x - \frac{x^2}{2}\right]_{0}^{1} + \left[\frac{x^3}{3} - x^2 + 3x\right]_{1}^{2} \\ &= \left[0 - \left(-1 + \frac{1}{3}\right)\right] + \left[\left(2 - \frac{1}{2}\right) - 0\right] + \left[\left(\frac{8}{3} - 4 + 6\right) - \left(\frac{1}{3} - 1 + 3\right)\right] \\ &= \frac{9}{2} \end{aligned}$$

■

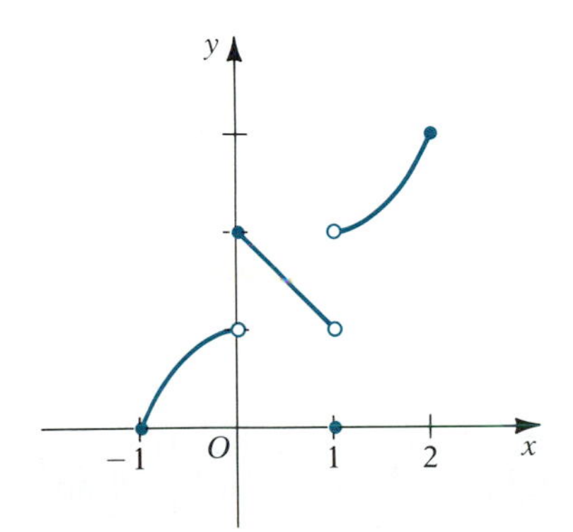

FIGURE 5.2

As a second extension of the use of integration to find areas, we consider the area of the region between the graphs of two continuous functions f and g from $x = a$ to $x = b$, where $f(x) \ge g(x)$ for all $x \in [a, b]$. Figure 5.3 illustrates a typical situation. Note that there is no requirement that either function be nonnegative. To motivate the procedure for finding the area, we are going to raise both curves by a constant amount k so that the entire region between them lies above the x-axis (Figure 5.4). This is possible because g has a minimum value on $[a, b]$. So if we take $k \ge |\min g(x)|$, then the functions $f + k$ and $g + k$ both are nonnegative on $[a, b]$. Furthermore, the region R' between the graphs of $f + k$ and $g + k$ is congruent to the region R between the graphs of f and g, and so the two regions have the same area. To get the area of R' we subtract the area under $g + k$ from the area under $f + k$ to get

$$\begin{aligned} A &= \int_a^b [f(x) + k]\,dx - \int_a^b [g(x) + k]\,dx \\ &= \int_a^b [f(x) + k - g(x) - k]\,dx \end{aligned}$$

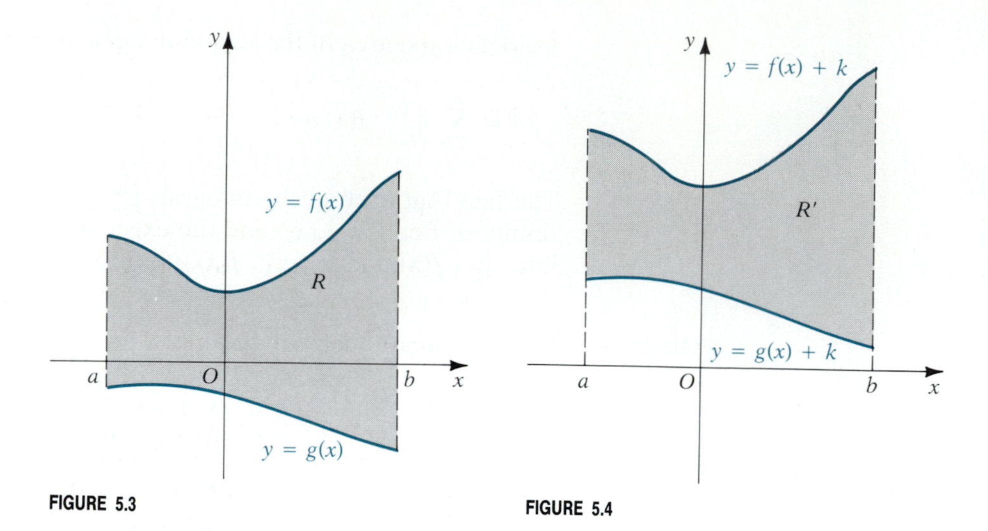

FIGURE 5.3

FIGURE 5.4

or

$$A = \int_a^b [\overbrace{f(x)}^{\text{upper curve}} - \overbrace{g(x)}^{\text{lower curve}}]\, dx \tag{5.1}$$

Observe that the constant k has disappeared, and we can obtain the integrand simply by subtracting the ordinate of the original lower curve from the ordinate of the original upper curve. This means that it is no longer necessary to raise the curves. We did this to arrive at equation (5.1), but now that we have the result, we can forget about the shift upward.

It is instructive to see how the result in equation (5.1) can be arrived at directly as a limit of Riemann sums, by partitioning the interval $[a, b]$ and

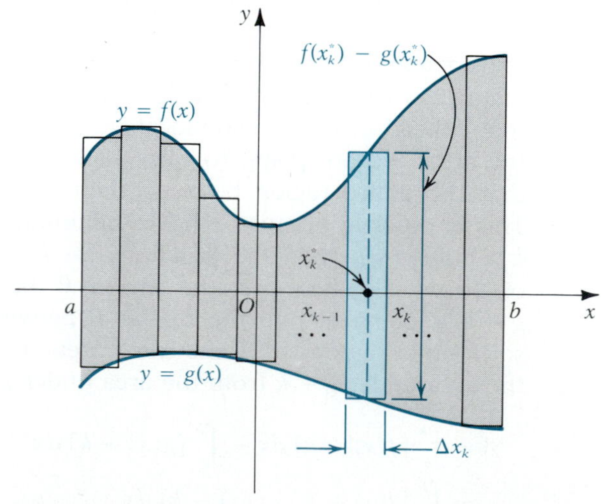

FIGURE 5.5

forming rectangles as shown in Figure 5.5. The kth rectangle has width $\Delta x_k = x_k - x_{k-1}$ and height $f(x_k^*) - g(x_k^*)$, where x_k^* is any point in the interval $[x_{k-1}, x_k]$, and so its area is $[f(x_k^*) - g(x_k^*)]\,\Delta x_k$. The sum of all such areas

$$\sum_{k=1}^{n} [f(x_k^*) - g(x_k^*)]\,\Delta x_k$$

is a Riemann sum for the continuous function $f - g$, and it approximates the area we want. Its limit, as $\|P\| \to 0$, is exactly this area; that is,

$$A = \lim_{\|P\|\to 0} \sum_{k=1}^{n} [f(x_k^*) - g(x_k^*)]\,\Delta x_k = \int_a^b [f(x) - g(x)]\,dx$$

Observe that once you have found the area of a typical rectangle, the form of the final integral is evident. In the examples and exercises that follow you may choose to use this way of obtaining the integral rather than to substitute into equation (5.1).

EXAMPLE 5.2 Find the area of the region bounded by the curves $y = 4 - x^2$ and $y = x^3 - x^2 - 3x$ between $x = -1$ and $x = 2$.

Solution Let $f(x) = 4 - x^2$ and $g(x) = x^3 - x^2 - 3x$. We show their graphs in Figure 5.6 together with a typical rectangular element of area. Since $f(x) \geq g(x)$ on $[-1, 2]$, equation (5.1) gives the area between them:

$$\begin{aligned}
A &= \int_{-1}^{2} [f(x) - g(x)]\,dx \\
&= \int_{-1}^{2} [(4 - x^2) - (x^3 - x^2 - 3x)]\,dx \\
&= \int_{-1}^{2} (-x^3 + 3x + 4)\,dx = -\frac{x^4}{4} + \frac{3x^2}{2} + 4x\Big]_{-1}^{2} \\
&= \left(-\frac{16}{4} + \frac{12}{2} + 8\right) - \left(-\frac{1}{4} + \frac{3}{2} - 4\right) = \frac{51}{4}
\end{aligned}$$

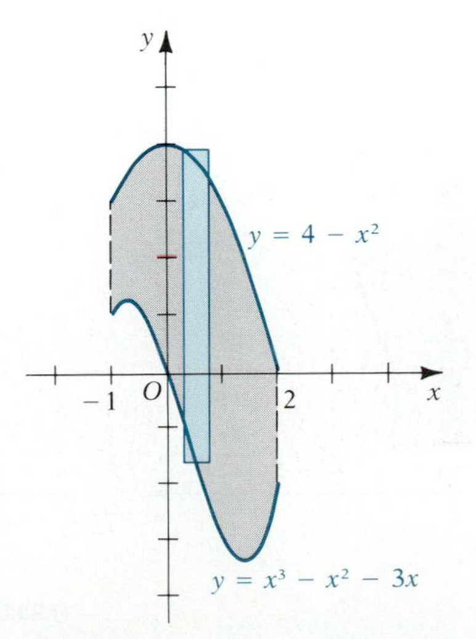

FIGURE 5.6

■

EXAMPLE 5.3 Find the area of the region bounded by the curves $y = x^2 - 2x$ and $x - y + 4 = 0$.

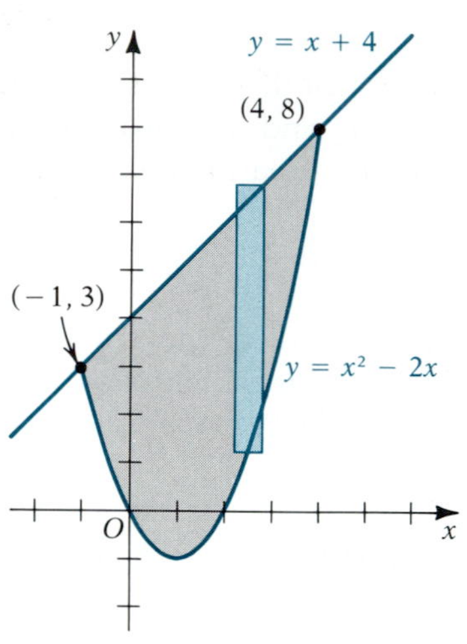

FIGURE 5.7

Solution The graphs are shown in Figure 5.7. The first step is to find the points of intersection, since the left and right limits are not specified. Equating the two y values gives

$$x^2 - 2x = x + 4$$
$$x^2 - 3x - 4 = 0$$
$$(x + 1)(x - 4) = 0$$
$$x = -1 \quad \Big| \quad x = 4$$
$$y = 3 \quad \Big| \quad y = 8$$

So the limits are from $x = -1$ to $x = 4$. Observe that the graph of $y = x + 4$ lies above that of $y = x^2 - 2x$ throughout this interval. Thus, the area is

$$A = \int_{-1}^{4} [\overbrace{(x+4)}^{\text{upper}} - \overbrace{(x^2 - 2x)}^{\text{lower}}]\, dx = \int_{-1}^{4} (-x^2 + 3x + 4)\, dx$$
$$= -\frac{x^3}{3} + \frac{3x^2}{2} + 4x\Big]_{-1}^{4} = \left(-\frac{64}{3} + \frac{48}{2} + 16\right) - \left(\frac{1}{3} + \frac{3}{2} - 4\right) = \frac{125}{6} \quad ■$$

Sometimes the roles of x and y are reversed; that is, the region whose area we want is between the y axis and a curve whose equation is of the form $x = g(y)$ from $y = c$ to $y = d$, where $g(y) \geq 0$, as in Figure 5.8. Or it may be between two curves, $x = g(y)$ and $x = h(y)$, with $g(y) \geq h(y)$, as in Figure 5.9. All of the theory we have had so far on areas can be modified in obvious ways to get the areas

$$A = \int_{c}^{d} g(y)\, dy \tag{5.2}$$

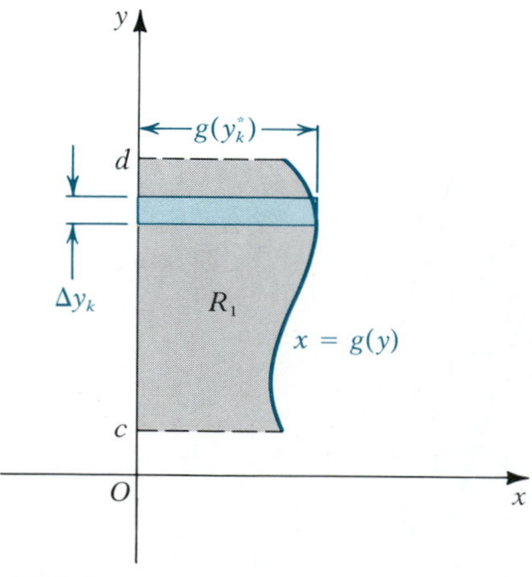

FIGURE 5.8

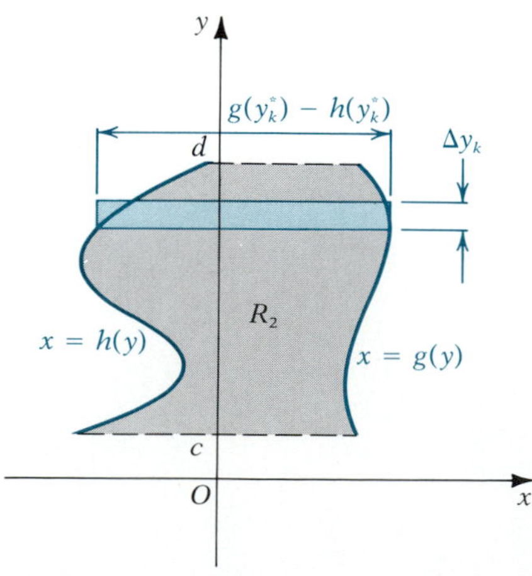

FIGURE 5.9

in the first case, and

$$A = \int_c^d [\overbrace{g(y)}^{\text{rightmost curve}} - \overbrace{h(y)}^{\text{leftmost curve}}]\, dy \tag{5.3}$$

in the second. We illustrate the use of equation (5.3) in the next example.

EXAMPLE 5.4 Find the area of the region bounded by the graphs of $x = \sqrt{3y + 4}$, $y = 3x$, $y = -1$, and $y = 4$.

Solution The region is pictured in Figure 5.10. Using equation (5.3), we have

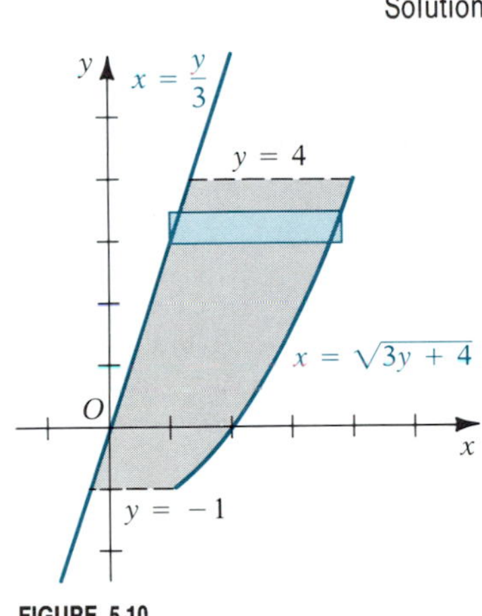

FIGURE 5.10

$$A = \int_{-1}^{4} \left(\sqrt{3y + 4} - \tfrac{y}{3}\right) dy$$

$$= \tfrac{1}{3}\int_{-1}^{4} \overbrace{(3y + 4)^{1/2}}^{u^{1/2}}\overbrace{3\,dy}^{du} - \tfrac{1}{3}\int_{-1}^{4} y\,dy$$

$$= \tfrac{1}{3}\cdot\left[\tfrac{2}{3}(3y + 4)^{3/2}\right]_{-1}^{4} - \left[\tfrac{1}{6}y^2\right]_{-1}^{4} = \tfrac{2}{9}(64 - 1) - \tfrac{1}{6}(16 - 1)$$

$$= 14 - \tfrac{5}{2} = \tfrac{23}{2}$$

Notice how in the integral of $(3y + 4)^{1/2}$ we multiplied and divided by 3 to adjust the differential and then carried out a mental substitution. ■

Finally, we note that if two curves cross each other, then finding the area between them requires that we split up the interval of integration at the crossing point and switch the upper and lower (or rightmost and leftmost) functions. For example, consider the region pictured in Figure 5.11. To get the area between f and g from $x = a$ to $x = b$, we use

$$A = \int_a^c [\overbrace{f(x)}^{\text{upper}} - \overbrace{g(x)}^{\text{lower}}]\, dx + \int_c^b [\overbrace{g(x)}^{\text{upper}} - \overbrace{f(x)}^{\text{lower}}]\, dx$$

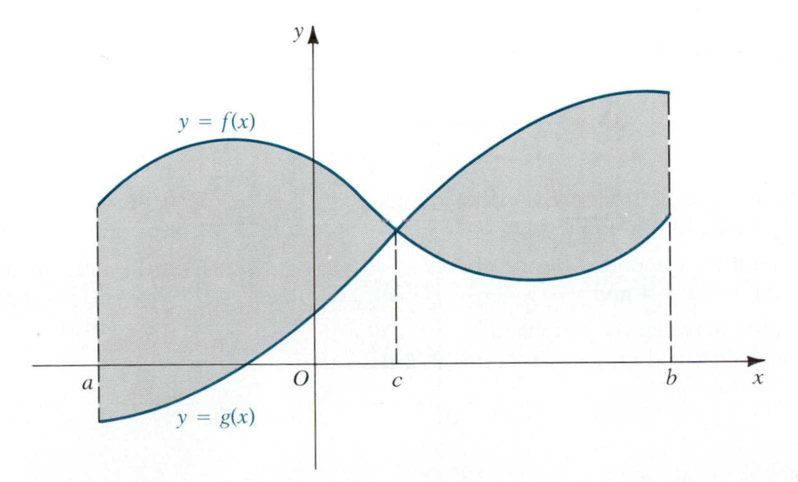

FIGURE 5.11

EXERCISE SET 5.1

A

In Exercises 1–10 find the area of the region between the x-axis and the graph of f from x = a to x = b, where [a, b] is the interval given. Sketch the graph in each case.

1. $f(x) = \begin{cases} x^2 & \text{if } 0 \le x < 2 \\ 4 - x & \text{if } 2 \le x \le 4 \end{cases}$ on $[0, 4]$

2. $f(x) = \begin{cases} -x & \text{if } -1 \le x < 0 \\ 1 & \text{if } x = 0 \\ 2 - x & \text{if } 0 < x \le 1 \end{cases}$ on $[-1, 1]$

3. $f(x) = \begin{cases} 4 - x^2 & \text{if } 0 \le x < 2 \\ x - 2 & \text{if } 2 \le x \le 4 \end{cases}$ on $[0, 4]$

4. $f(x) = \begin{cases} x^2 & \text{if } 0 \le x < 1 \\ 2x - x^2 & \text{if } 1 \le x \le 2 \end{cases}$ on $[0, 2]$

5. $f(x) = \begin{cases} \sqrt{x + 4} & \text{if } -4 \le x < 0 \\ 1 & \text{if } x = 0 \\ \dfrac{x^2}{2} & \text{if } 0 < x \le 2 \end{cases}$ on $[-4, 2]$

6. $f(x) = \begin{cases} x^3 + 1 & \text{if } -1 \le x < 1 \\ x^2 - 4x + 5 & \text{if } 1 \le x \le 3 \end{cases}$ on $[-1, 3]$

7. $f(x) = \begin{cases} \cos x & \text{if } -\frac{\pi}{2} \le x < 0 \\ \sin x & \text{if } 0 \le x \le \pi \end{cases}$ on $[-\frac{\pi}{2}, \pi]$

8. $f(x) = \begin{cases} \cos \frac{x}{2} & \text{if } -\pi \le x < \frac{\pi}{2} \\ \sin \frac{x}{2} & \text{if } \frac{\pi}{2} \le x \le 2\pi \end{cases}$ on $[-\pi, 2\pi]$

9. $f(x) = \begin{cases} 2 \sin \frac{3x}{2} & \text{if } 0 \le x < \frac{2\pi}{3} \\ 4 + 2 \sin \frac{3x}{2} & \text{if } \frac{2\pi}{3} \le x \le \frac{4\pi}{3} \end{cases}$ on $[0, \frac{4\pi}{3}]$

10. $f(x) = \begin{cases} x^2 + 2x + 2 & \text{if } -1 \le x < 0 \\ 1 + \sin \frac{\pi x}{2} & \text{if } 0 \le x \le 2 \end{cases}$ on $[-1, 2]$

In Exercises 11–30 find the area bounded by the graphs of the given equations. Sketch the graphs.

11. $y = x^2 + 1,\ y = x,\ x = 0,\ x = 1$
12. $y = 4 - x^2,\ y = x^2 - 1,\ x = -1,\ x = 1$
13. $y = x^3 - 4x,\ y = 8 - x^2,\ x = -2,\ x = 2$
14. $x^2 + y = 0,\ x - y + 2 = 0,\ x = -1,\ x = 1$
15. $y = x^2,\ y = x + 2$
16. $x + y = 4,\ y = -\sqrt{x + 5},\ x = -5,\ x = 4$
17. $x - y = 3,\ x^2 - 3x - y = 0$
18. $y = 8 - x^2,\ y = x^2$
19. $y = x^2,\ y^2 = x$
20. $y = x^2 - 3x + 4,\ y = -x^2 + 5x - 2$
21. $y = \sin \pi x,\ y = 4 + \cos 2\pi x,\ x = 0,\ x = 2$
22. $y = \sin \frac{\pi x}{2},\ y = x^2 - 2x$
23. $y = \sin 2x$, $y = \cos x$, between $x = -\frac{\pi}{2}$ and $x = \frac{\pi}{6}$
24. $x = 4 - y^2,\ x = 0$
25. $x = y^2 - 2,\ x = y$
26. $y = \sqrt{x},\ x - y = 2,\ y = 0$
27. $x = 2y + 1,\ x = y - 1,\ y = 2$
28. $x = y^3$, $y = x^2$ (Do this in two ways.)
29. $x = y^2 - 4y - 3,\ x - 1 = 2y^2$
30. $x = 2(y - 2)^2,\ x = y^2 - 4y + 8$
31. Find the area bounded by the graphs of the equations $y = 2x^2$, $x + y = 3$, and $y = 0$ (a) by using x as the independent variable and (b) by using y as the independent variable.
32. Follow the instructions for Exercise 31 for the area bounded by $x = \sqrt{y + 1}$, $x = 2 + \sqrt{3 - y}$, and $y = -1$.

B

33. Use integration to find the area of the triangle with vertices $(-1, 1)$, $(3, 2)$, and $(5, 4)$.
34. Find the total area bounded by the curves $y = x^3 - 3x^2 - 2x + 4$ and $y = 1 - x$.
35. Find the area between the graphs of $f(x) = 1 + 2 \sin x$ and $g(x) = \cos 2x$ on $[0, 2\pi]$.
36. Find the area between the graphs of $x = \sin \frac{\pi y}{2}$ and $y = x$.
37. Find the area of the region bounded on the left by the curves $x^2 - 2x - 4y + 9 = 0$ and $x + 2y = 5$ and on the right by $y^2 - 4y + x = 0$.

5.2 VOLUMES USING CROSS SECTIONS

In this section and the next we will see how the integral can be used to find volumes of certain solids. Consider the solid depicted in Figure 5.12. It is bounded on the left by a plane perpendicular to the x-axis at $x = a$, and it is bounded on the right by a plane perpendicular to the x-axis at $x = b$. If we pass a plane perpendicular to the x-axis at any point x between a and b, its intersection with the solid is a plane region called a **cross section** with respect to the x-axis. We denote the area of such a cross section, if the area exists, by $A(x)$. *Throughout the remainder of this section we will consider only solids for which $A(x)$ does exist and is a continuous function of x on $[a, b]$.*

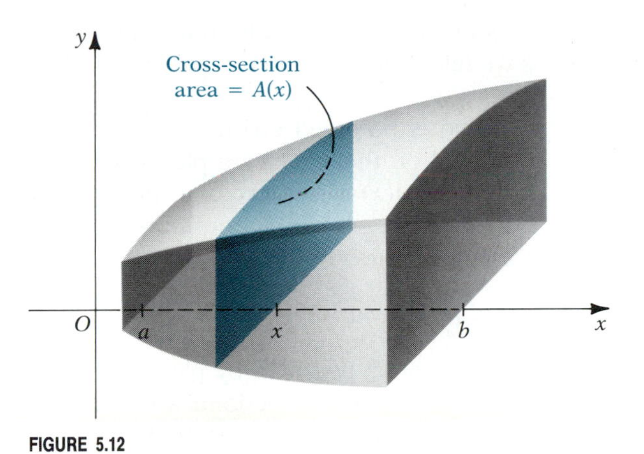

FIGURE 5.12

To find the volume of the solid we let $P = \{x_0, x_1, x_2, \ldots, x_n\}$ be a partition of $[a, b]$, where, as usual, we call $\Delta x_k = x_k - x_{k-1}$ $(k = 1, 2, \ldots, n)$, and in each of the subintervals $[x_{k-1}, x_k]$ we arbitrarily select a point x_k^*. Now if we pass planes perpendicular to the x-axis at each of the partition points, we divide the solid up into n slices. In Figure 5.13 we have pictured a typical

FIGURE 5.13

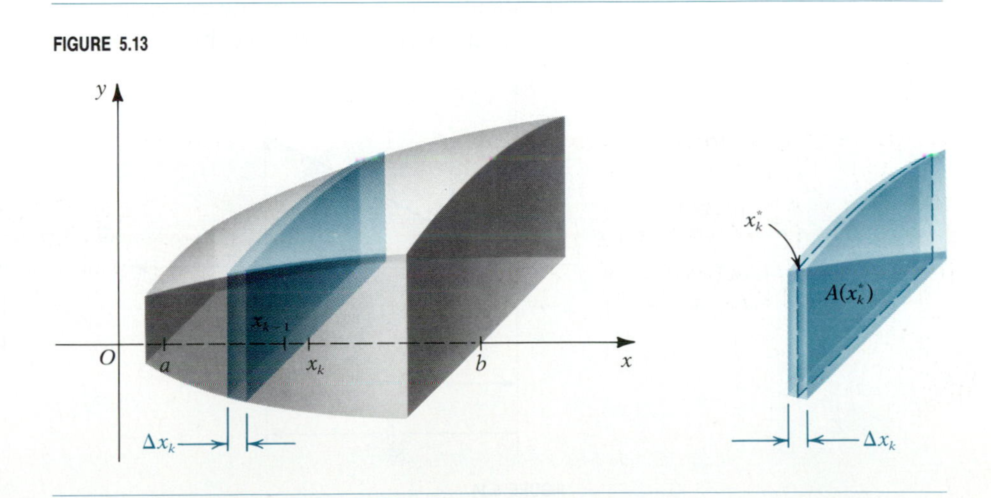

slice with faces that are the cross sections taken at x_{k-1} and x_k. The entire volume of the solid is the sum of the volumes of all the slices. In general, the cross sections will vary between x_{k-1} and x_k, but for Δx_k small, if we replace the kth slice with one that has constant cross-sectional area $A(x_k^*)$, as shown to the right in Figure 5.13, the volume will not differ much from that of the actual slice. We call this slice that has width Δx_k and cross-sectional area $A(x_k^*)$ a *typical element of volume* and denote its volume by ΔV_k. Its volume is defined as the *area of the face times the thickness:*

$$\Delta V_k = A(x_k^*)\,\Delta x_k$$

and the total volume is approximated by the sum of all of the ΔV_k values:

$$V \approx \sum_{k=1}^{n} \Delta V_k = \sum_{k=1}^{n} A(x_k^*)\,\Delta x_k$$

The sum on the right is a Riemann sum that approaches the integral $\int_a^b A(x)\,dx$ as we take finer and finer partitions, which leads to the following definition.

DEFINITION 5.1 If a solid is bounded on the left by a plane perpendicular to the x-axis at $x = a$ and on the right by a plane perpendicular to the x-axis at $x = b$, and if the area $A(x)$ of the cross section with respect to the x-axis of the solid at the point $x \in [a, b]$ is continuous for all x in $[a, b]$, then the volume V of the solid given by

$$V = \int_a^b A(x)\,dx \tag{5.4}$$ ■

If the solid is bounded by planes perpendicular to the y-axis at $y = c$ and $y = d$ and if the cross-sectional area with respect to the y-axis is a continuous function $A(y)$ on $[c, d]$, then similar reasoning leads to the equation

$$V = \int_c^d A(y)\,dy \tag{5.5}$$

EXAMPLE 5.5 Find the volume of a right circular cone of base radius r and altitude h.

Solution We orient the cone with respect to an xy-coordinate system as shown in Figure 5.14 so that x lies in the interval $[0, h]$. We need to find the cross-sectional area $A(x)$. Since the cross section is circular, it is sufficient to find its radius, which we denote by y. The area is then πy^2, but this must be

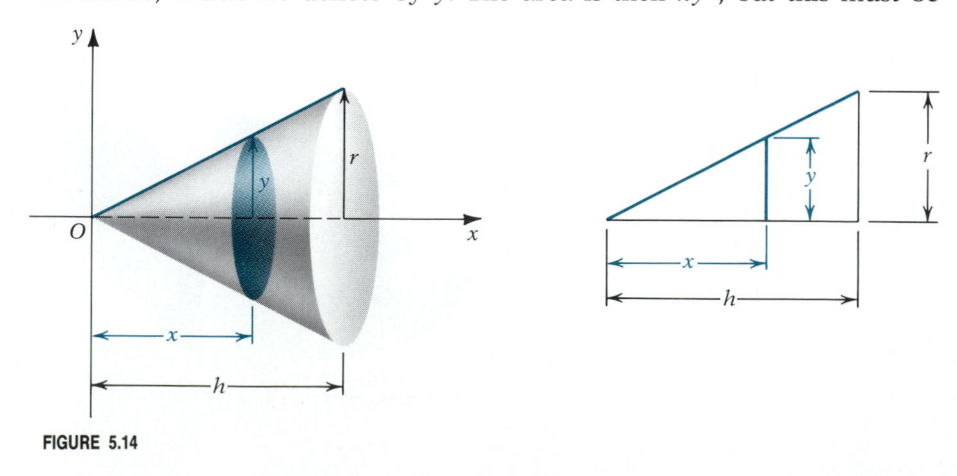

FIGURE 5.14

expressed as a function of x. To do so we use similar triangles (see Figure 5.14):

$$\frac{y}{r} = \frac{x}{h}$$

$$y = \frac{r}{h}x$$

Thus

$$A(x) = \pi y^2 = \pi \cdot \frac{r^2}{h^2}x^2$$

By equation (5.4) the volume is

$$V = \int_0^h \pi \frac{r^2}{h^2}x^2\,dx = \frac{\pi r^2}{h^2}\int_0^h x^2\,dx$$

$$= \frac{\pi r^2}{h^2}\left[\frac{x^3}{3}\right]_0^h = \frac{\pi r^2}{h^2} \cdot \frac{h^3}{3} = \frac{1}{3}\pi r^2 h$$

Since the area of the circular base of the cone is πr^2, we conclude that the volume of a right circular cone is *one-third the area of the base times the altitude.* ■

The cone is an example of a class of solids that can be formed by revolving an area about an axis. Such solids are known as **solids of revolution.** We could, for example, obtain the cone of Example 5.5 by revolving the triangular region in Figure 5.15 about the x-axis. We now look at such solids of revolution in general.

Consider a region R in the plane under the graph of f from $x = a$ to $x = b$, where f is continuous and nonnegative, as in Figure 5.16. The solid that results from rotating R about the x-axis has a circular cross section with

$$A(x) = \pi y^2 = \pi[f(x)]^2$$

So by equation (5.4) the volume is $\int_a^b \pi y^2\,dx$, or on replacing y by $f(x)$,

$$V = \pi \int_a^b [f(x)]^2\,dx \tag{5.6}$$

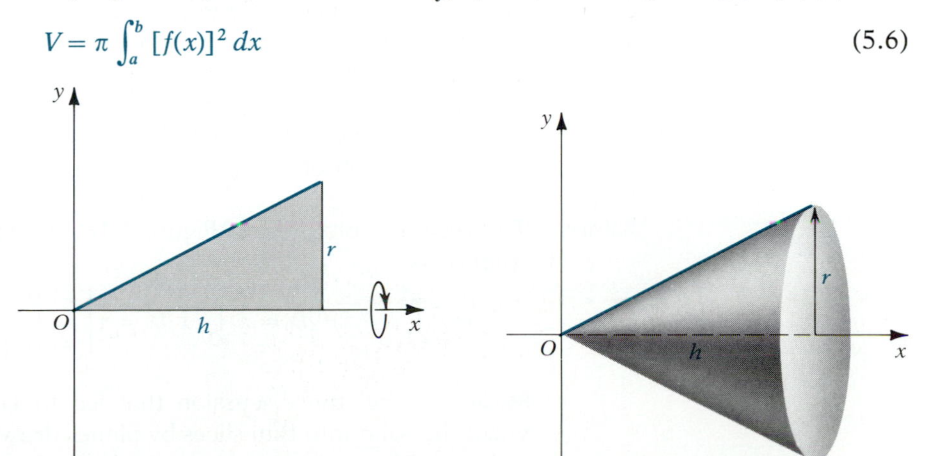

FIGURE 5.15

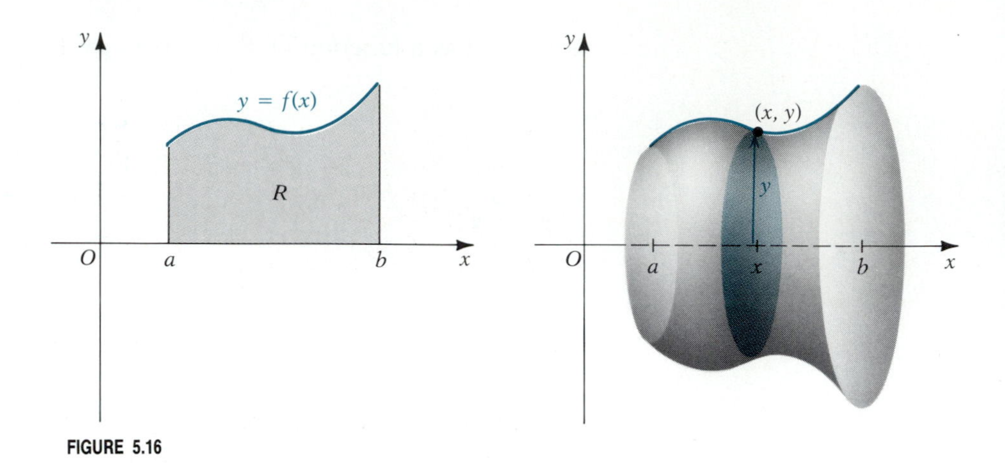

FIGURE 5.16

The analogous formula for rotating about the y-axis is

$$V = \pi \int_c^d [g(y)]^2\, dy \tag{5.7}$$

where the area to be rotated is between the y-axis and the graph of the non-negative continuous function g between $y = c$ and $y = d$.

Note When f is continuous, so is $f \cdot f = f^2$ and also πf^2. It follows that $A(x)$ is continuous, so that the requirements of Definition 5.1 are met for solids of revolution of the type we have described.

EXAMPLE 5.6 Redo Example 5.5 using equation (5.6).

Solution We obtain the volume of the cone by finding the volume of the solid of revolution (the cone) formed by revolving the triangular region R in Figure 5.17 about the x-axis. The line that forms the upper boundary of R has slope $\frac{r}{h}$, and since a line through the origin with slope m has the equation $y = mx$, this line has the equation $y = \frac{r}{h}x$. Thus, by equation (5.6),

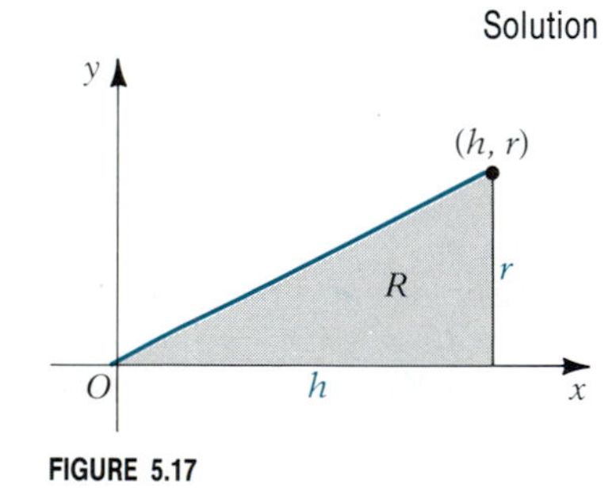

FIGURE 5.17

$$V = \pi \int_a^b \frac{r^2}{h^2} x^2\, dx = \frac{1}{3}\pi r^2 h$$

as we previously had. ■

EXAMPLE 5.7 Find the volume of the solid of revolution obtained by revolving the region under the graph of $y = \sqrt{x}$ from $x = 1$ to $x = 4$ about the x-axis.

Solution The region is pictured in Figure 5.18. By equation (5.6), the volume after rotation is

$$V = \pi \int_1^4 (\sqrt{x})^2\, dx = \pi \int_1^4 x\, dx = \pi \left[\frac{x^2}{2}\right]_1^4 = \pi\left[\frac{16}{2} - \frac{1}{2}\right] = \frac{15\pi}{2}$$ ■

Remark Recall the discussion that led to Definition 5.1, in which we divided the solid into thin slices by planes drawn at successive partition points of $[a, b]$. The typical element of volume that approximates the kth slice will have the appearance of a disk as shown in Figure 5.19 for solids of revolution of the type we described. For this reason, the method we have used to find the volume is often called the **disk method.**

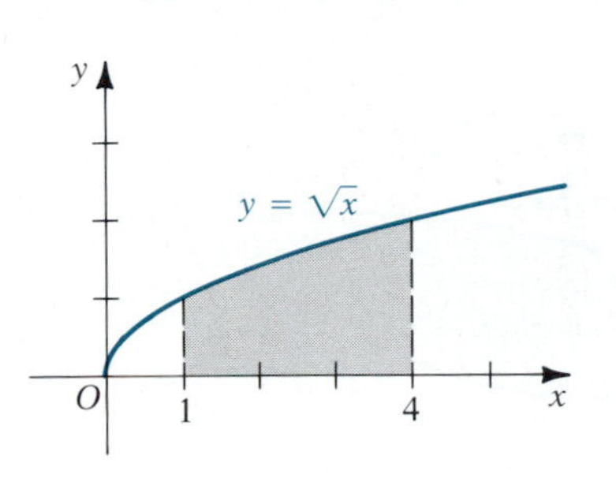

FIGURE 5.18

$f(x_k^*)$

Δx_k

FIGURE 5.19 Typical element of volume is a disk

Although we have required f (or g) to be nonnegative, this requirement is unnecessary, as the following reasoning shows. Suppose f is continuous on $[a, b]$. Then $|f|$ is also continuous on $[a, b]$ (see Exercise 20 in Exercise Set 1.8) and is nonnegative there. Figure 5.20 illustrates the graphs of f and $|f|$ for a typical function. When we rotate each of the shaded regions about the x-axis, we obtain identical solids, shown in Figure 5.21. But $|f|$ satisfies the conditions for equation (5.6) to hold. Thus

$$V = \pi \int_a^b |f(x)|^2 \, dx$$

and since $|f|^2 = f^2$, we get

$$V = \pi \int_a^b [f(x)]^2 \, dx$$

That is, equation (5.6) [and also (5.7)] holds true even when we drop the requirement that f be nonnegative.

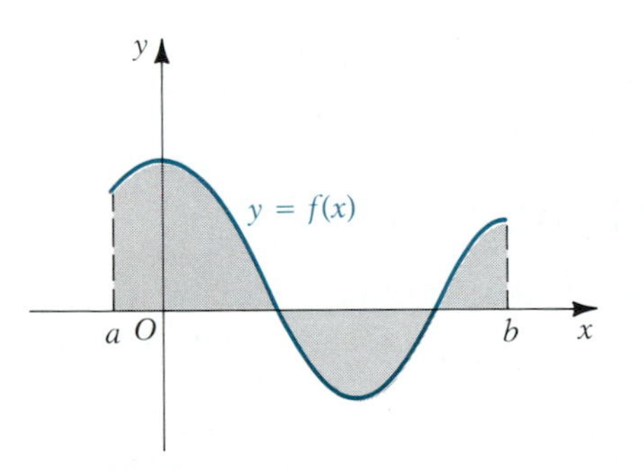

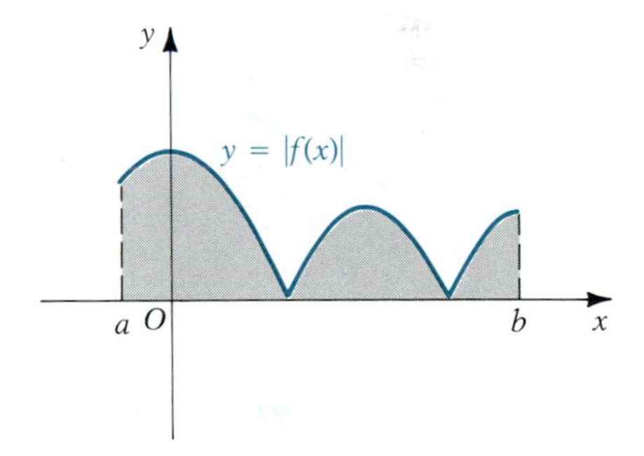

FIGURE 5.20

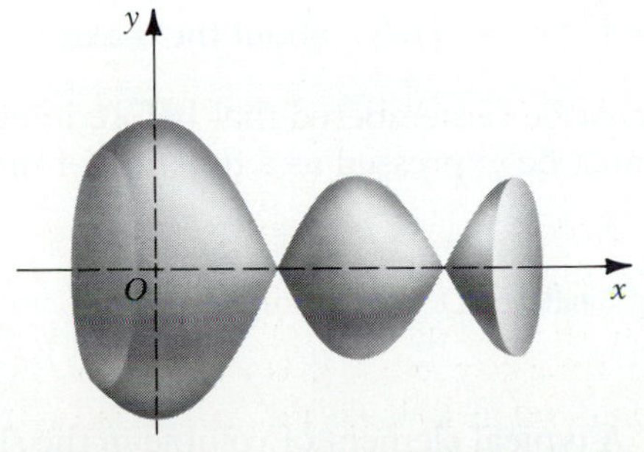

FIGURE 5.21 Same solid of revolution results from rotating either region

FIGURE 5.22

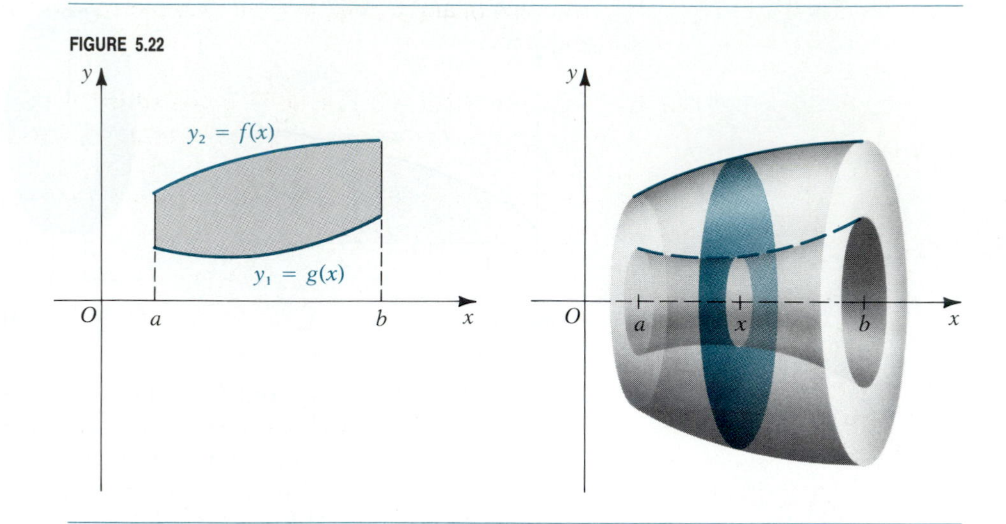

If the region to be rotated lies between the graphs of two nonnegative continuous functions $y_1 = g(x)$ and $y_2 = f(x)$, where $f(x) \geq g(x)$ on $[a, b]$, as shown in Figure 5.22, then the solid of revolution has a hole in it. A cross section is the region between two concentric circles of radii y_1 and y_2, respectively. Thus, its area is

$$A(x) = \pi y_2^2 - \pi y_1^2 = \pi(y_2^2 - y_1^2) = \pi\{[f(x)]^2 - [g(x)]^2\}$$

The volume, by equation (5.4), is therefore

$$V = \pi \int_a^b \{[f(x)]^2 - [g(x)^2]\} \, dx \tag{5.8}$$

The analogous formula for a region bounded by the graphs of $x_1 = h(y)$ and $x_2 = g(y)$, where $g(y) \geq h(y) \geq 0$, between $y = c$ and $y = d$, rotated about the y-axis, is

$$V = \pi \int_c^d \{[g(y)]^2 - [h(y)]^2\} \, dy \tag{5.9}$$

The use of the dependent variable notations $y_1 = g(x)$ and $y_2 = f(x)$ for equation (5.8) and $x_1 = h(y)$ and $x_2 = g(y)$ for equation (5.9) simplifies the appearance of these formulas to

$$V = \pi \int_a^b (y_2^2 - y_1^2) \, dx \quad \text{about the } x\text{-axis}$$

and

$$V = \pi \int_c^d (x_2^2 - x_1^2) \, dy \quad \text{about the } y\text{-axis}$$

but it must be remembered that before integration can take place, the integrand must be expressed as a function of the variable of integration in each case.

A Word of Caution Do not confuse $y_2^2 - y_1^2$ with $(y_2 - y_1)^2$. These are *not* the same.

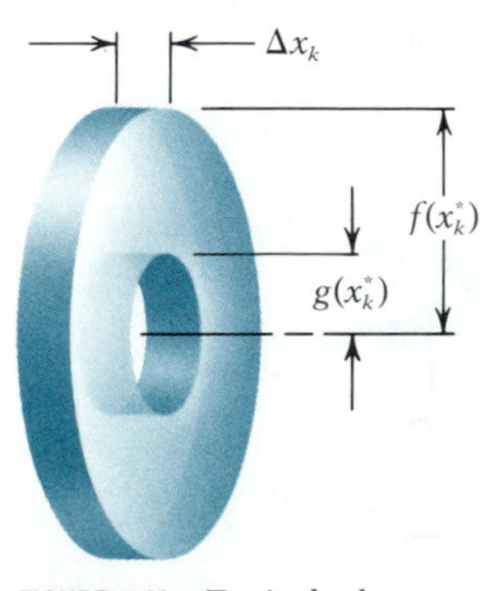

FIGURE 5.23 Typical element of volume is washer-shaped

Remark A typical element of volume in the slicing technique for solids of the type described is washer-shaped as shown in Figure 5.23 (for rotation about the x-axis), and for this reason the method is often referred to as the **washer method.**

EXAMPLE 5.8 Find the volume of the solid obtained by revolving the region between the graphs of $y = x^2 + 1$ and $y = \sqrt{x}$ from $x = 0$ to $x = 1$ about the x-axis.

Solution Let $f(x) = x^2 + 1$ and $g(x) = \sqrt{x}$. The region to be revolved is shown in Figure 5.24. By equation (5.8) the volume is

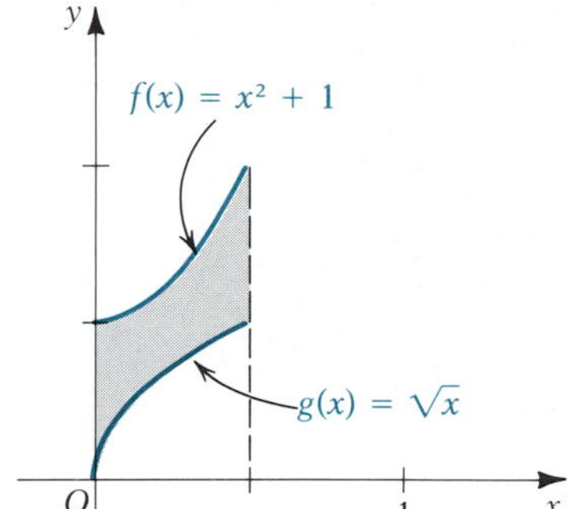

FIGURE 5.24

$$V = \pi \int_0^1 [(x^2 + 1)^2 - (\sqrt{x})^2]\, dx$$

$$= \pi \int_0^1 (x^4 + 2x^2 + 1 - x)\, dx = \pi \left[\frac{x^5}{5} + \frac{2x^3}{3} + x - \frac{x^2}{2}\right]_0^1$$

$$= \pi \left[\frac{1}{5} + \frac{2}{3} + 1 - \frac{1}{2}\right] = \frac{41\pi}{30}$$

As an aid in remembering equations (5.8) and (5.9), observe that the area of a cross section for the washer method is

$$\pi[(\text{outer radius})^2 - (\text{inner radius})^2] \tag{5.10}$$

and this is the form of the integrand in each case. As the next example shows, this formula for the cross-sectional area enables us to rotate areas about lines other than the x- and y-axes.

EXAMPLE 5.9 The region bounded by the graphs of $x = 2\sqrt{y}$ and $x = y$ is to be revolved about the line $x = 4$. Find the volume of the resulting solid of revolution.

Solution Figure 5.25 shows the region in question, together with a typical cross section of the rotated solid. Since the axis of rotation is vertical, we will use y as the variable of integration. By solving the two equations simultaneously we find their points of intersection to be (0, 0) and (4, 4). (Verify.) Thus the limits on y are 0 and 4. As shown in the figure, we let x_1 and x_2 denote the distances from the y-axis to the leftmost and rightmost boundaries, respectively. Thus, for each y in [0, 4],

$$x_1 = y \quad \text{and} \quad x_2 = 2\sqrt{y}$$

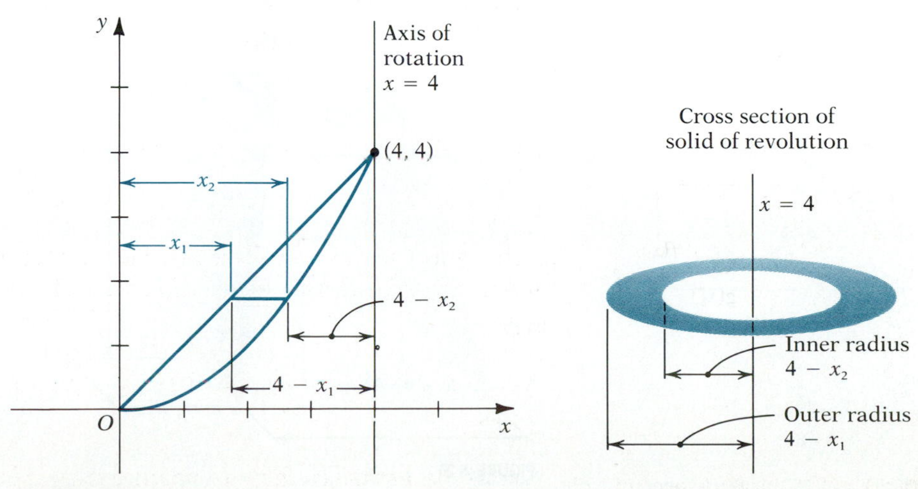

FIGURE 5.25

So the corresponding distances from the line $x = 4$ are $4 - x_1$ and $4 - x_2$. These are the outer and inner radii of the cross section of the solid of revolution taken at y:

$$\text{outer radius} = 4 - x_1 = 4 - y$$
$$\text{inner radius} = 4 - x_2 = 4 - 2\sqrt{y}$$

So the area $A(y)$ of the cross section, by equation (5.10), is

$$A(y) = \pi[(4 - y)^2 - (4 - 2\sqrt{y})^2]$$

and the volume of the solid is

$$\begin{aligned} V &= \pi \int_0^4 A(y)\,dy = \pi \int_0^4 [(4 - y)^2 - (4 - 2\sqrt{y})^2]\,dy \\ &= \pi \int_0^4 [16 - 8y + y^2 - 16 + 16\sqrt{y} - 4y]\,dy \\ &= \pi \int_0^4 (y^2 - 12y + 16y^{1/2})\,dy = 4\left[\frac{y^3}{3} - 6y^2 + \frac{32}{3}y^{3/2}\right]_0^4 \\ &= \pi\left[\frac{64}{3} - 96 + \frac{32}{3}(8)\right] = \frac{32\pi}{3} \end{aligned}$$

■

We conclude this section with two examples that show the application of Definition 5.1 where the solids are not solids of revolution.

EXAMPLE 5.10 Use integration to find the volume of a pyramid with an altitude of 10 m and a base that is a square 4 m on a side.

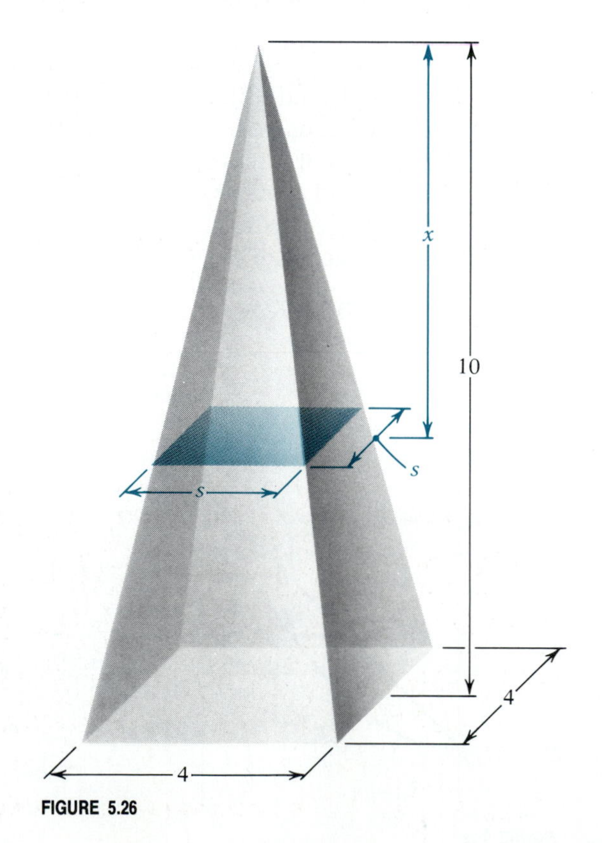

FIGURE 5.26

Solution The pyramid is shown in Figure 5.26. We take as an axis the line through the vertex and the center of the base, and let x be the distance along this axis, measured from the vertex. So the limits on x are 0 and 10. A typical cross section taken x units down is a square, as shown, whose side we designate as s. It is necessary to express s as a function of x, and we can do this using similar triangles, as shown in Figure 5.27:

$$\frac{s}{4} = \frac{x}{10}$$
$$s = \frac{4x}{10} = \frac{2x}{5}$$

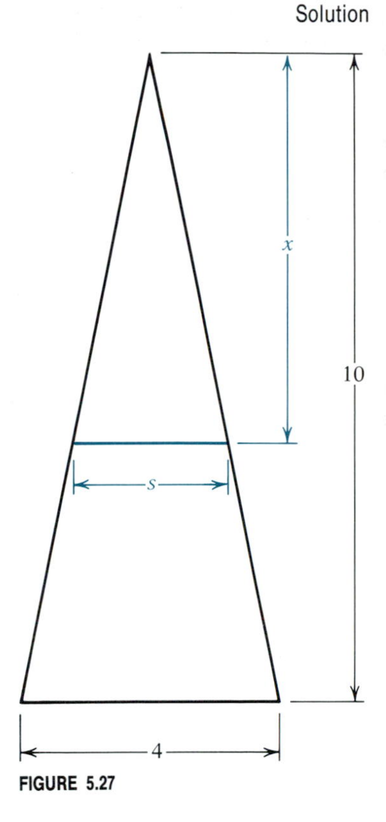

FIGURE 5.27

The area $A(x)$ of the cross section is therefore

$$A(x) = s^2 = \frac{4x^2}{25}$$

From equation (5.4) of Definition 5.1, the volume is

$$V = \int_0^{10} A(x)\,dx = \int_0^{10} \frac{4x^2}{25}\,dx = \frac{4}{25} \cdot \frac{x^3}{3}\Big]_0^{10}$$
$$= \frac{4}{25} \cdot \frac{1000}{3} = \frac{160}{3} \text{ cu m}$$

This completes the solution, but it is interesting to observe that this answer is one-third the area of the base times the altitude, just as was true of the cone, which shares essential features with a pyramid. ■

EXAMPLE 5.11 The base of a solid is the region in the xy-plane bounded by the graphs of $y = \sqrt{x}$, $y = 0$, and $x = 4$, and each cross section of the solid perpendicular to the x-axis is a square. Find its volume.

Solution The solid is depicted in Figure 5.28. The cross section at an arbitrary distance x between 0 and 4 is a square whose base is $y = \sqrt{x}$. Thus the area $A(x) = y^2 = (\sqrt{x})^2 = x$. So by equation (5.4) the volume is

$$V = \int_0^4 A(x)\,dx = \int_0^4 x\,dx = \frac{x^2}{2}\Big]_0^4 = 8$$ ■

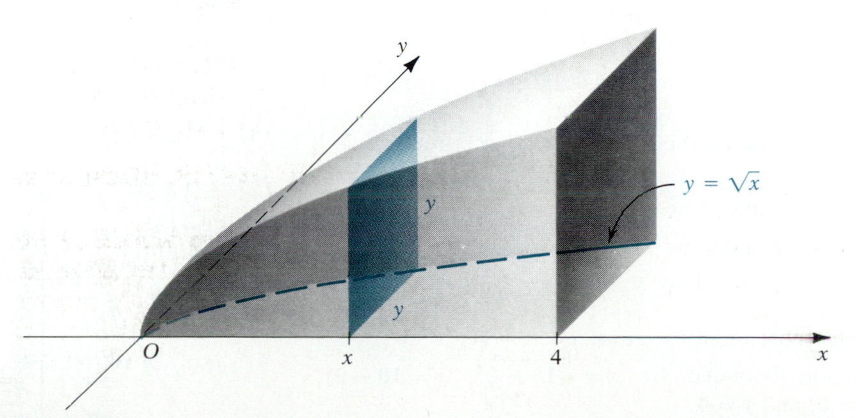

FIGURE 5.28

EXERCISE SET 5.2

A

In Exercises 1–20 find the volume of the solid obtained by rotating the given region about the x-axis.

1. Region under $y = (x + 3)/2$ from $x = 0$ to $x = 1$
2. Region under $y = x^2 + 1$ from $x = 0$ to $x = 2$
3. Region under $y = 4 - x^2$ from $x = 0$ to $x = 2$
4. Region under $y = 2x - x^2$ from $x = 0$ to $x = 2$
5. Region under $y = \sqrt{x + 1}$ from $x = -1$ to $x = 3$
6. Region under $y = \sqrt{2 - x}$ from $x = -2$ to $x = 2$
7. Region under $y = x^3 + 1$ from $x = -1$ to $x = 1$
8. Region under $y = \sqrt[3]{x}$ from $x = 1$ to $x = 8$
9. Region under $y = \sqrt{25 - x^2}$ from $x = -4$ to $x = 3$
10. Region under $y = 4x^2 - x^4$ from $x = -1$ to $x = 1$
11. Region between the x-axis and $y = x^2 - 1$ from $x = -2$ to $x = 2$
12. Region between the x-axis and $y = \sqrt[3]{3x + 1}$ from $x = -3$ to $x = 0$
13. Region between $y = x + 4$ and $y = 3 - \frac{x}{2}$ from $x = 1$ to $x = 4$
14. Region between $y = x$ and $y = \sqrt{x}$
15. Region between $y = x + 2$ and $y = x^2$
16. Region between $y = \sqrt{2x - 3}$ and $y = \frac{x}{2}$
17. Region between $y = x^2$ and $y = 2 - x^2$
18. Region between $y = \sqrt{x}$ and $y = \sqrt{3x + 4}$ from $x = 0$ to $x = 4$
19. Region between $y = 1$ and $y = \sqrt{5 - x^2}$
20. Region between $y = 2/\sqrt[3]{x}$ and $y = \sqrt[3]{9 - x}$

In Exercises 21–30 find the volume of the solid obtained by rotating the given region about the y-axis.

21. Region between the y-axis and $x + 2y = 4$ from $y = 0$ to $y = 2$
22. Region between the y-axis and $x = \sqrt{y}$ from $y = 4$ to $y = 9$
23. Region between the y-axis and $x = y^{1/3}$ from $y = -1$ to $y = 8$
24. Region bounded by the y-axis and $x = y^2 - y - 2$
25. Region bounded by $y = 2x$, $x = 2y$, and $y = 2$
26. Region bounded by $y = 2x$, $y = \sqrt{3 - 4x}$, and the x-axis
27. Region between $y = x$ and $x = 6 - y^2$
28. Region bounded by $x = \sqrt{16 - y^2}$, $x = 10 - 2y$, $y = 0$, and $y = 4$
29. Region bounded by $y^3 + x = 0$, $y + 2 = 0$, and $x - 1 = 0$
30. Region between $y^2 = x$ and $y^2 = 2x - 4$
31. Derive the formula for the volume of a sphere by rotating the region under the semicircle $y = \sqrt{a^2 - x^2}$ about the x-axis.
32. If the region under the semiellipse $y = \frac{b}{a}\sqrt{a^2 - x^2}$ is rotated about the x-axis, the resulting solid is called an *ellipsoid of revolution.* Find its volume.
33. The region bounded by $y = x$, $x = 2$, and $y = 0$ is rotated about the line $y = 3$. Find the volume of the resulting solid.
34. Find the volume of the solid of revolution obtained by rotating the region under $y = \sqrt{x}$ from $x = 1$ to $x = 4$ about the line $y + 1 = 0$.
35. Find the volume of the solid of revolution obtained by rotating the region bounded by $x = \sqrt{y}$, $y = 8 - 2x$, and the x-axis about (a) the y-axis and (b) the line $x = 4$.
36. Find the volume of the solid of revolution obtained by rotating the region in the first quadrant bounded by $x^3 = 2y$ and $y = 2x$ about (a) the line $x = 2$ and (b) the line $y = 4$.

In Exercises 37–40 find the volume of the solid of revolution obtained by rotating the region between the graph of f and the x-axis from x = a to x = b about the x-axis.

37. $f(x) = \sin x + \cos x$; $a = -\frac{\pi}{3}$, $b = \frac{\pi}{6}$
38. $f(x) = \sec x$; $a = 0$, $b = \frac{\pi}{4}$
39. $f(x) = 2 \csc x$; $a = \frac{\pi}{4}$, $b = \frac{3\pi}{4}$
40. $f(x) = \sqrt{2(1 - \sin x)}$; $a = -\frac{\pi}{2}$, $b = \frac{\pi}{6}$

In Exercises 41–44 find the volume using Definition 5.1.

41. The base of a solid is the triangular region bounded by $y = \frac{x}{2}$, $x = 4$, and $y = 0$, and each cross section perpendicular to the x-axis is a square. Find the volume.
42. Find the volume of the pyramid in the figure, whose

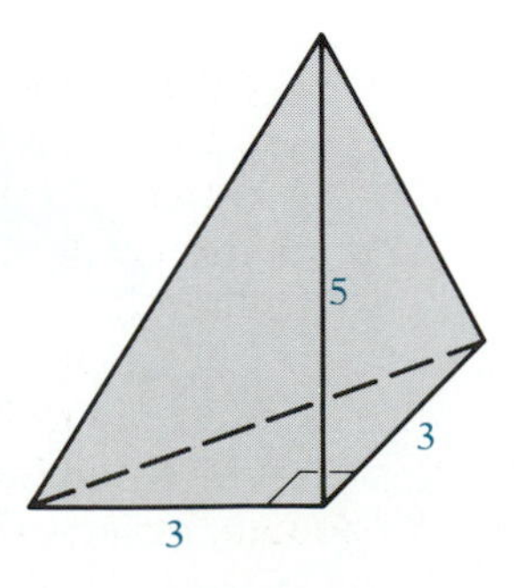

base is an isosceles right triangle with each leg 3 and altitude 5.

43. The base of a solid is the region enclosed by the parabola $y^2 = 2x$ and the line $x = 2$, and each cross section perpendicular to the x-axis is semicircular. Find the volume.

44. The base of a solid is the region enclosed by the graph of $y = 4 - x^2$ and the x-axis, and each cross section perpendicular to the y-axis is an isosceles triangle whose altitude is one-half its base. Find its volume.

B

45. Find the volume of the solid obtained by revolving the leftmost region bounded by $y = 3x$, $y = x^2/2$, and $x + y = 4$ about the x-axis. (*Hint:* Divide the region into two parts.)

46. Find the volume of the solid obtained by revolving the region bounded by the x-axis and the curve $y = x^3 - x^2 - 2x$ about the x-axis.

47. An oil tank is in the shape of a sphere with radius 2 m, and the oil in it is 1 m deep (see the figure). Find how many liters of oil are in the tank (1 cu m = 1000 L).

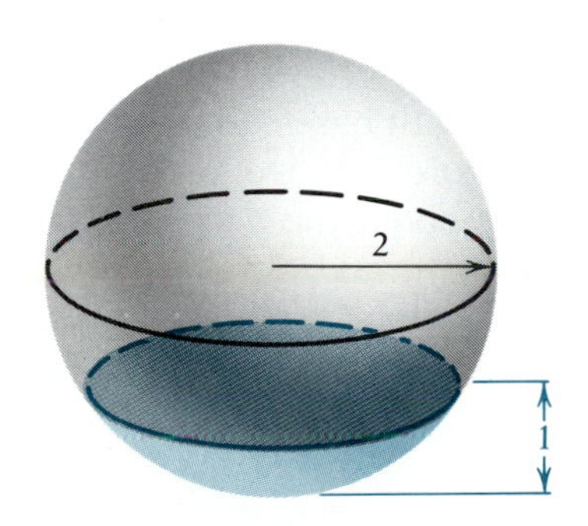

48. In a sphere of radius a, at a point b units above the center on a vertical axis ($b < a$) a horizontal plane is passed. The portion of the sphere above the plane is called a *spherical segment*. Find its volume.

49. A solid has as a base the region inside the ellipse $4x^2 + 9y^2 = 36$, and each cross section perpendicular to the x-axis is an equilateral triangle. Find the volume of the solid.

50. The upper half of the region inside the *four-cusp hypocycloid* $x^{2/3} + y^{2/3} = a^{2/3}$ (see the figure) is revolved around the x-axis. Find the volume generated.

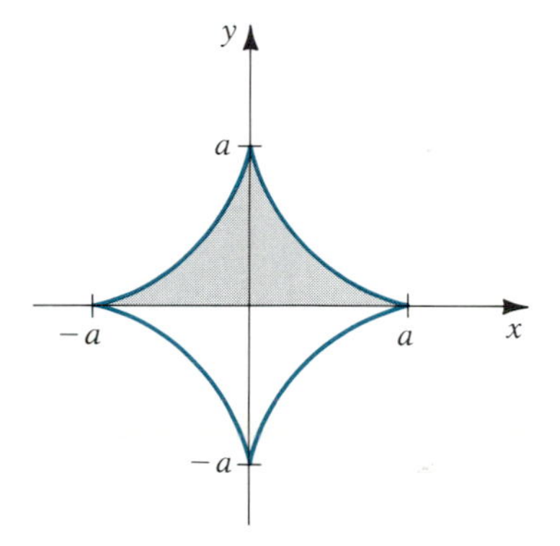

51. The region between the graphs of $y = \sqrt{x}$ and $y = 4 - \sqrt{x}$ from $x = 0$ to $x = 9$ is rotated about the line $y = -2$. Find the volume.

52. The region between the graphs of $y = \sin x$ and $y = \cos x$ from $x = \frac{\pi}{4}$ to $x = \pi$ is rotated about the line $y = 1$. Find the volume. (*Hint:* To perform the integration make use of a trigonometric identity.)

5.3 VOLUMES OF SOLIDS OF REVOLUTION BY CYLINDRICAL SHELLS

Using the method of disks or of washers to find the volume of a solid of revolution is sometimes difficult, or even impossible, because the resulting integral cannot easily be evaluated. In other cases the process is tedious because the shape of the region requires that it be divided into two or more parts in order to apply the method. These difficulties can sometimes be eliminated by an alternate procedure for finding volumes of solids of revolution that we present in this section.

The method involves the calculation of volumes of thin-walled hollow cylinders, called **cylindrical shells,** such as a piece of pipe. More precisely, a

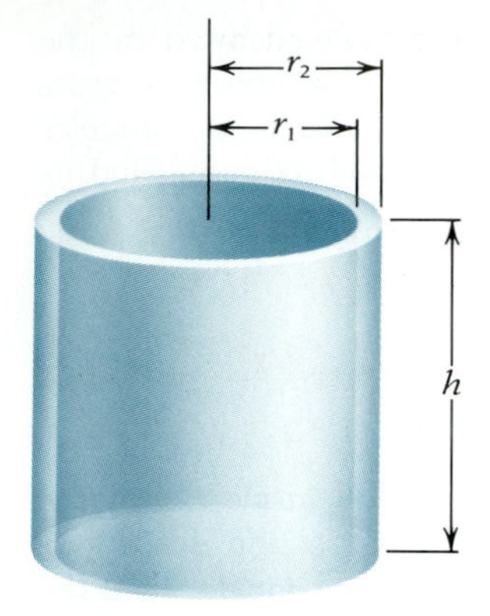

FIGURE 5.29

cylindrical shell is the solid between two concentric right circular cylinders of fixed height. We show such a cylindrical shell in Figure 5.29. If we denote the inner radius by r_1, the outer radius by r_2, and the height by h, then the volume V is the volume of the outer cylinder minus that of the inner cylinder:

$$V = \pi r_2^2 h - \pi r_1^2 h = \pi(r_2^2 - r_1^2)h$$

It is convenient to write this in the following equivalent way:

$$V = \pi(r_2 + r_1)(r_2 - r_1)h = 2\pi\left(\frac{r_2 + r_1}{2}\right)(r_2 - r_1)h$$

Since $\frac{1}{2}(r_1 + r_2)$ is the average radius, multiplying this by 2π gives the average circumference. Also, $r_2 - r_1$ is the thickness of the shell. Thus, we can state the formula for the volume in words:

$$\text{volume} = (\text{average circumference}) \times (\text{height}) \times (\text{thickness}) \tag{5.11}$$

Now let us consider again the region R between the x-axis and the graph of a continuous nonnegative function f between $x = a$ and $x = b$. Let $P = \{x_0, x_1, x_2, \ldots, x_n\}$ be a partition of $[a, b]$. Instead of choosing x_k^* arbitrarily in the kth subinterval, we let x_k^* be the midpoint of the subinterval:

$$x_k^* = \frac{x_{k-1} + x_k}{2}, \qquad k = 1, 2, \ldots, n$$

Next we construct a rectangle of height $f(x_k^*)$, with base equal to the subinterval $[x_{k-1}, x_k]$, for $k = 1, 2, 3, \ldots, n$. In Figure 5.30 we have shown a typical rectangle. If we rotate the kth rectangle about the y-axis, we get a cylindrical shell of average radius x_k^*, height $f(x_k^*)$, and thickness Δx_k, as shown in Figure 5.31. If this is done for $k = 1, 2, 3, \ldots, n$, we get n such concentric shells, which taken altogether approximate the solid of revolution that results from rotating the region R about the y-axis. By equation (5.11) the volume of the kth cylindrical shell is

$$\Delta V_k = \underbrace{2\pi x_k^*}_{\substack{\text{average} \\ \text{circumference}}} \cdot \underbrace{f(x_k^*)}_{\text{height}} \cdot \underbrace{\Delta x_k}_{\text{thickness}}$$

FIGURE 5.30

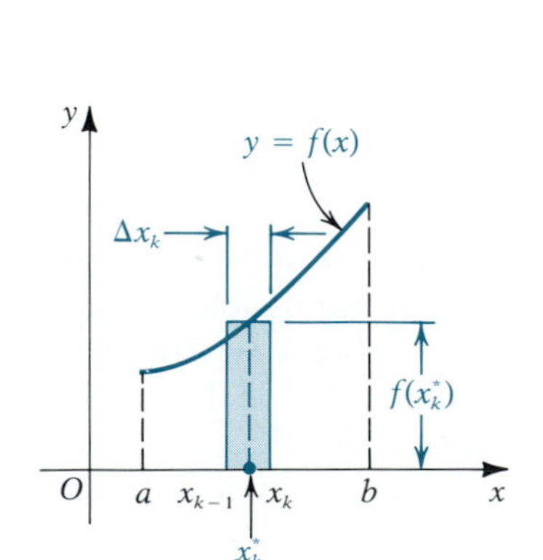

FIGURE 5.31

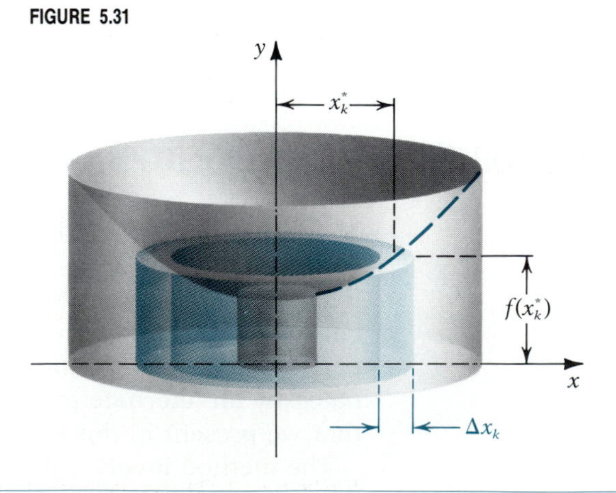

Thus, the volume V of the solid of revolution is approximated by:

$$V \approx \sum_{k=1}^{n} \Delta V_k = \sum_{k=1}^{n} 2\pi x_k^* f(x_k^*)\, \Delta x_k$$

The sum on the right is a Riemann sum for the continuous function $2\pi x f(x)$ on $[a, b]$. So if we let $\|P\| \to 0$, we come to the following definition.

DEFINITION 5.2 The volume of the solid of revolution obtained by rotating about the y-axis the area of the region under the graph of the continuous nonnegative function f from $x = a$ to $x = b$ is

$$V = 2\pi \int_a^b x f(x)\, dx \tag{5.12}$$ ■

Analogous formulas hold for rotation about the x-axis and for regions between two graphs. These are indicated along with the regions to be rotated in Figures 5.32, 5.33, and 5.34. All functions involved are assumed to be

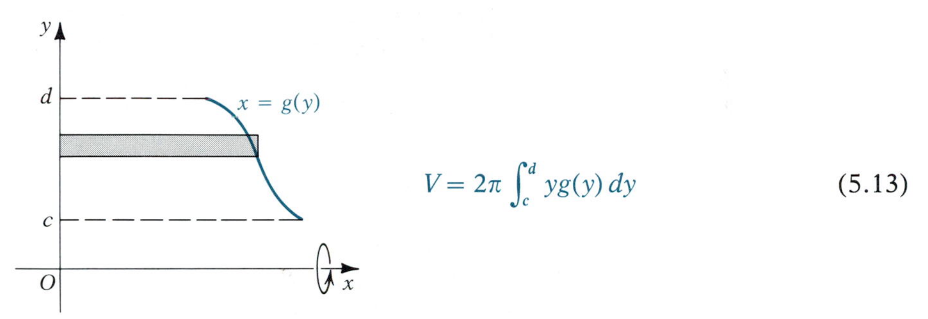

$$V = 2\pi \int_c^d y g(y)\, dy \tag{5.13}$$

FIGURE 5.32 Rotation about the x-axis

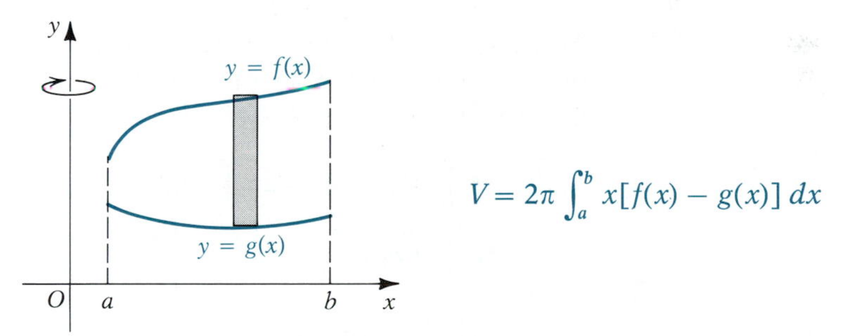

$$V = 2\pi \int_a^b x[f(x) - g(x)]\, dx \tag{5.14}$$

FIGURE 5.33 Rotation about the y-axis

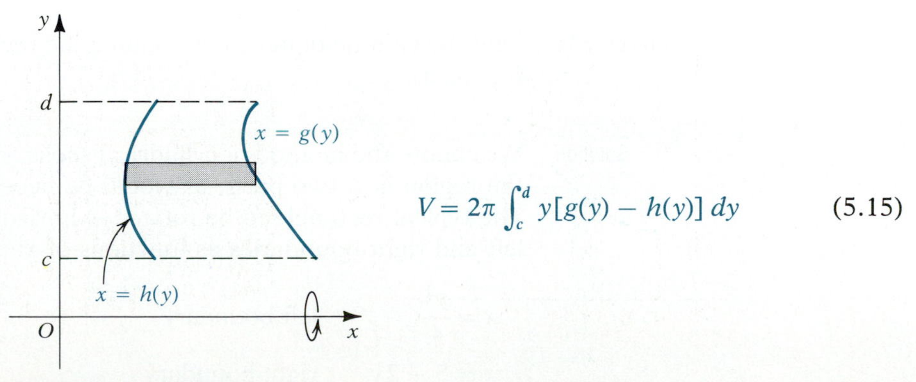

$$V = 2\pi \int_c^d y[g(y) - h(y)]\, dy \tag{5.15}$$

FIGURE 5.34 Rotation about the x-axis

continuous. You can convince yourself of their validity by finding the volume of a typical cylindrical shell.

Note Rather than to memorize equations (5.12)–(5.15), it is easier in each case to draw a typical rectangle and the resulting cylindrical shell. When its volume is calculated, this leads to the appropriate integrand.

For obvious reasons, calculating volumes in this way is called the **method of cylindrical shells.** We illustrate the technique with several examples.

EXAMPLE 5.12 Use the method of cylindrical shells to find the volume of the solid obtained by revolving the region under the graph of $y = x^2$ from $x = 0$ to $x = 1$ about the y-axis.

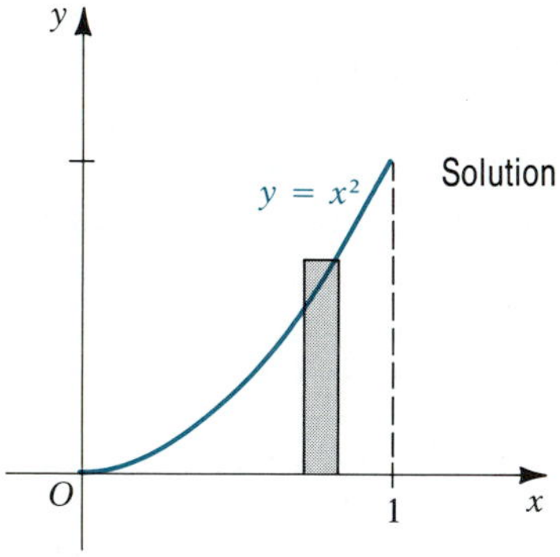

Solution We show the region to be rotated in Figure 5.35. Writing $f(x) = x^2$ and using equation (5.12), we get

$$V = 2\pi \int_0^1 xf(x)\,dx = 2\pi \int_0^1 x(x^2)\,dx = 2\pi \int_0^1 x^3\,dx$$

$$= 2\pi\left[\frac{x^4}{4}\right]_0^1 = \frac{\pi}{2}$$ ■

FIGURE 5.35

You might wish to compare this with the method of washers for calculating the same volume.

EXAMPLE 5.13 Find the volume of the solid obtained by revolving about the x-axis the region bounded by $g(y) = 2y - y^2$ and the y-axis.

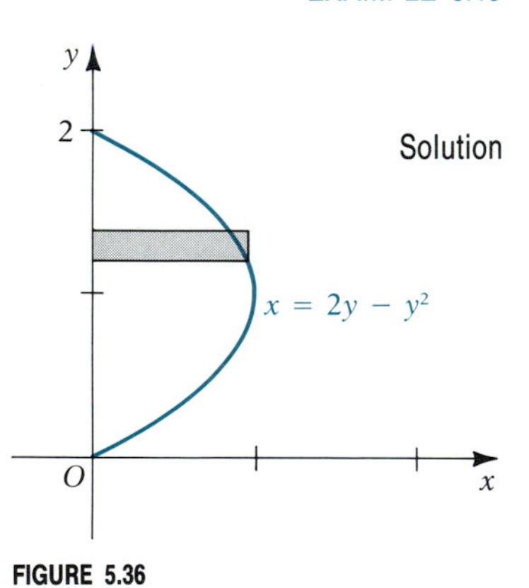

FIGURE 5.36

Solution The region is shown in Figure 5.36. We choose to use the method of cylindrical shells, since cross-sectional areas perpendicular to the x-axis are not easily calculated in this case. By equation (5.13) we obtain

$$V = 2\pi \int_0^2 yg(y)\,dy = 2\pi \int_0^2 y(2y - y^2)\,dy$$

$$= 2\pi \int_0^2 (2y^2 - y^3)\,dy = 2\pi\left[\frac{2y^3}{3} - \frac{y^4}{4}\right]_0^2$$

$$= 2\pi\left[\frac{16}{3} - 4\right] = \frac{8\pi}{3}$$ ■

EXAMPLE 5.14 Find the volume obtained by rotating the region shown in Figure 5.37 about the x-axis.

Solution We choose the method of cylindrical shells, since this avoids having to divide the region into two parts, as would be necessary with the washer method. The typical rectangle to be rotated is horizontal, so we need to express the left and right boundaries as functions of y:

$$x = \frac{y^2}{4} \qquad \text{left boundary}$$

$$x = 5 - 2y \qquad \text{right boundary}$$

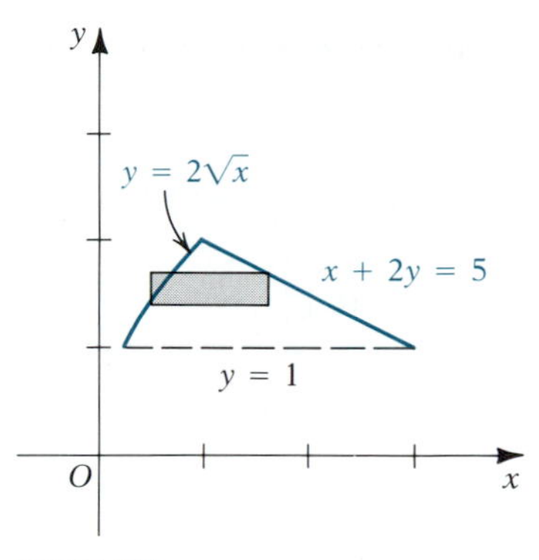

FIGURE 5.37

To get the upper limit on y we solve these two equations simultaneously by setting the x values equal to each other:

$$\frac{y^2}{4} = 5 - 2y$$

$$y^2 + 8y - 20 = 0$$

$$(y - 2)(y + 10) = 0$$

$$y = 2 \quad | \quad y \neq -10 \qquad \text{(outside domain)}$$

Using equation (5.15) with $g(y) = 5 - 2y$ and $h(y) = y^2/4$, we obtain the volume:

$$V = 2\pi \int_1^2 y\left(5 - 2y - \frac{y^2}{4}\right) dy = 2\pi \int_1^2 \left(5y - 2y^2 - \frac{y^3}{4}\right) dy$$

$$= 2\pi\left[\frac{5y^2}{2} - \frac{2y^3}{3} - \frac{y^4}{16}\right]_1^2$$

$$= 2\pi\left[\left(10 - \frac{16}{3} - 1\right) - \left(\frac{5}{2} - \frac{2}{3} - \frac{1}{16}\right)\right] = \frac{91\pi}{24}$$

■

EXAMPLE 5.15 Find the volume of the solid of revolution obtained by revolving the region bounded by the graphs of $y = x^2/2$ and $y = x + 4$ about the line $x = 6$.

Solution As Figure 5.38 shows, a typical area element when rotated produces a cylindrical shell with area

$$\Delta V_k = \underbrace{2\pi(6 - x_k^*)}_{\text{average circumference}} \underbrace{\left[(x_k^* + 4) - \frac{(x_k^*)^2}{2}\right]}_{\text{height}} \cdot \underbrace{\Delta x_k}_{\text{thickness}}$$

On summing and passing to the limit, we get

$$V = 2\pi \int_{-2}^{4} (6 - x)\left(x + 4 - \frac{x^2}{2}\right) dx$$

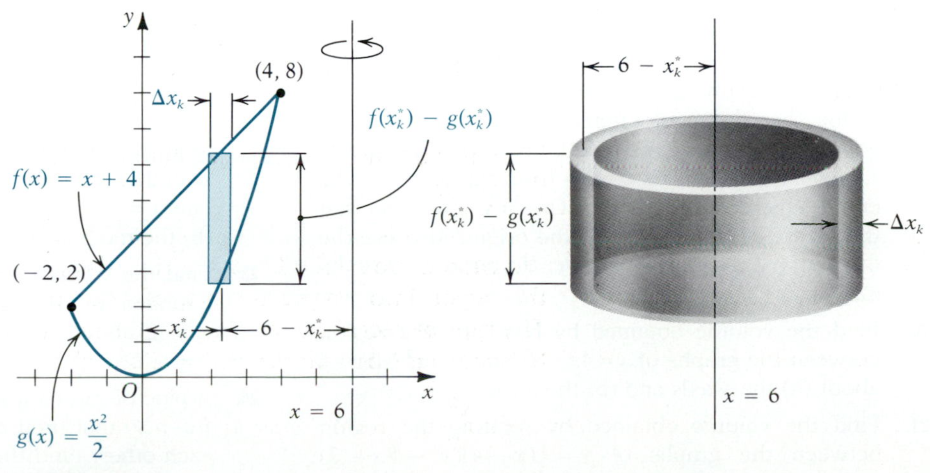

FIGURE 5.38

The limits of integration were found by solving the two equations simultaneously. We simplify the integrand and perform the integration:

$$\begin{aligned} V &= 2\pi \int_{-2}^{4} \left(\frac{x^3}{2} - 4x^2 + 2x + 24\right) dx = 2\pi\left[\frac{x^4}{8} - \frac{4x^3}{3} + x^2 + 24x\right]_{-2}^{4} \\ &= 2\pi\left[\left(32 - \frac{4\cdot 64}{3} + 16 + 96\right) - \left(2 + \frac{4\cdot 8}{3} + 4 - 48\right)\right] \\ &= 2\pi(90) = 180\pi \end{aligned}$$

■

EXERCISE SET 5.3

A

In Exercises 1–10 find the volume obtained by rotating the region bounded by the graphs of the given equations about the axis specified. Use the method of cylindrical shells.

1. $y = x,\ y = 0,\ x = 2$; y-axis
2. $y = \sqrt{x},\ y = 0,\ x = 4$; y-axis
3. $y = \sqrt{4 - x},\ x = 0,\ y = 0$; x-axis
4. $x = 1/\sqrt{y},\ y = 1,\ y = 4,\ x = 0$; x-axis
5. $x = \sqrt{y},\ x = 0,\ x + y = 6$; y-axis
6. $y = \sqrt{6 - x},\ y = x,\ y = 0$; x-axis
7. $y = (16 - x^2)/4,\ x - 2y + 6 = 0,\ x = 2,\ x = 4$; y-axis
8. $y = \sqrt{x + 5},\ y = x - 1,\ y = 1$; x-axis
9. $y = x^3/2,\ y = 1/2,\ y = 4,\ x = 0$; x-axis
10. $y^2 = x^3,\ x = 1$; y-axis

In Exercises 11–16 find by two methods the volume of revolution obtained by revolving the region bounded by the graphs of the given equations about the axis specified.

11. $y = x^2,\ y = 2x$; y-axis
12. $y = \sqrt{2x - x^2},\ y = x$; x-axis
13. $y^2 = 2x,\ x^2 = 2y$; x-axis
14. $x = 2\sqrt{4 - y^2},\ x + 2y = 4$; y-axis
15. $y^2 = x^2 - 1,\ 2x - y = 2$; x-axis
16. $y = \sqrt{8x},\ y = x$; y-axis
17. Find the volume of the solid obtained by revolving the region between the x-axis and the graph of $f(x) = 1 - x^2$ about the line $x = 2$.
18. Find the volume of the solid obtained by revolving the region between the graphs of $h(y) = y$ and $g(y) = 2 - y^2$ about the line $y = 1$.

B

19. A *torus* is a doughnut-shaped solid that can be obtained by revolving the region inside a circle about a line that does not intersect the circle. Find the volume of a torus generated by revolving a circle of radius a about an axis b units from the center of the circle, where $b > a$. (*Hint:* Orient the axes so that the center of the circle is at the origin. Also use the fact that $2\int_{-a}^{a} \sqrt{a^2 - x^2}\,dx$ gives the area of the circle, πa^2.)
20. Find the volume obtained by revolving the region between the graphs of $y = 4x - x^2$ and $y = x^2 - 2x + 4$ about (a) the y-axis and (b) the x-axis.
21. Find the volume obtained by rotating the region between the graphs of $y = \frac{1}{3}(x - 4)(x^2 - 8x - 3)$ and $x + y = 4$ about the y-axis.
22. Let
$$f(x) = \begin{cases} x^2 & \text{if } 0 \le x < 2 \\ 6 - x & \text{if } 2 \le x \le 6 \end{cases}$$
Find the volume obtained by revolving the region bounded by the graph of f and the x-axis, about the line $x = -1$ using (a) the method of washers and (b) the method of cylindrical shells.
23. Find the volume generated by revolving the region bounded by the graphs of $y = 3x - x^2$ and $y = x^2 - 2$ about (a) the line $x = -1$ and (b) the line $y = -2$.
24. A pipe of radius a is intersected by a pipe of radius $b > a$ so that their center lines are at right angles to each other. Find the volume inside the smaller pipe that is cut off by the larger pipe.

5.4 ARC LENGTH

Another application of integration is in finding the lengths of curves in the plane. If f is a function defined on $[a, b]$ for which f' is continuous at all points of $[a, b]$, then the graph of f is called a **smooth curve** and f is called a **smooth function.** Such a curve has a continuously turning tangent line and hence no sharp corners. The method we describe is limited to such curves.

Let f be a smooth function on $[a, b]$. As shown in Figure 5.39, we partition $[a, b]$ with the points $a = x_0 < x_1 < x_2 < \cdots < x_n = b$, and write $\Delta x_k = x_k - x_{k-1}$ for $k = 1, 2, \ldots, n$ in the usual way. This partitioning of the interval $[a, b]$ on the x-axis induces a partitioning of the curve $y = f(x)$ by means of the points $P_0, P_1, \ldots, P_n$, where for each $k = 1, 2, \ldots, n$, $P_k = (x_k, f(x_k))$. We can approximate the length of the curve by adding up the lengths of all the chords that join successive points P_{k-1} and P_k on the curve. If we denote the length of the chord that joins P_{k-1} and P_k by Δs_k, we have, by the distance formula,

$$\Delta s_k = \sqrt{(x_k - x_{k-1})^2 + (f(x_k) - f(x_{k-1}))^2} \tag{5.16}$$

Since f' exists on $[a, b]$, the conditions of the mean value theorem of differential calculus are satisfied on each of the intervals $[x_{k-1}, x_k]$. Thus, by that theorem there is a point x_k^* in (x_{k-1}, x_k) such that

$$\frac{f(x_k) - f(x_{k-1})}{x_k - x_{k-1}} = f'(x_k^*)$$

or

$$f(x_k) - f(x_{k-1}) = f'(x_k^*)(x_k - x_{k-1})$$

On substituting this in equation (5.16) and replacing $x_k - x_{k-1}$ by Δx_k, we get

$$\begin{aligned}\Delta s_k &= \sqrt{(\Delta x_k)^2 + [f'(x_k^*)]^2(\Delta x_k)^2} \\ &= \sqrt{1 + [f'(x_k^*)]^2}\,\Delta x_k\end{aligned}$$

FIGURE 5.39

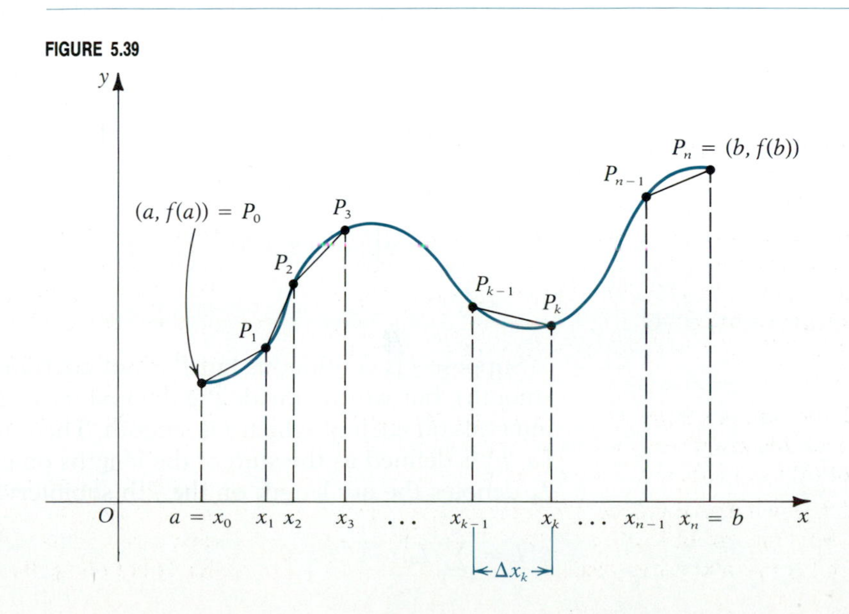

Thus, denoting the length of the graph by L:

$$L \approx \sum_{k=1}^{n} \sqrt{1 + [f'(x_k^*)]^2}\, \Delta x$$

As $\|P\| \to 0$ the Riemann sum on the right approaches the integral $\int_a^b \sqrt{1 + [f'(x)]^2}\, dx$. This leads to the following definition.

DEFINITION 5.3 If f is a smooth function on $[a, b]$, then the length of the graph of f, called **arc length,** from $(a, f(a))$ to $(b, f(b))$ is given by

$$L = \int_a^b \sqrt{1 + [f'(x)]^2}\, dx \tag{5.17}$$ ■

The analogous definition when g is a smooth function of y from $y = c$ to $y = d$ is

$$L = \int_c^d \sqrt{1 + [(g'(y)]^2}\, dy \tag{5.18}$$

EXAMPLE 5.16 Find the arc length of the curve $y = 2(x + 1)^{3/2}$ from $x = -1$ to $x = 6$.

Solution Let $f(x) = 2(x + 1)^{3/2}$. Then $f'(x) = 3(x + 1)^{1/2}$, so that by equation (5.17),

$$\begin{aligned} L &= \int_{-1}^{6} \sqrt{1 + 9(x + 1)}\, dx = \int_{-1}^{6} \sqrt{9x + 10}\, dx \\ &= \tfrac{1}{9} \int_{-1}^{6} (9x + 10)^{1/2} \cdot 9\, dx = \tfrac{1}{9} \cdot \tfrac{2}{3}(9x + 10)^{3/2}\Big]_{-1}^{6} \\ &= \tfrac{2}{27}[(64)^{3/2} - (1)^{3/2}] = \tfrac{2}{27}(8^3 - 1) = \tfrac{2}{27}(511) = \tfrac{1022}{27} \end{aligned}$$ ■

EXAMPLE 5.17 Find the length of arc on the graph of $x = \frac{2}{3}(y^2 - 1)^{3/2}$ from $y = 1$ to $y = 3$.

Solution The form in which the equation is written suggests using y as the independent variable, so that equation (5.18) is applicable:

$$g'(y) = \frac{dx}{dy} = \frac{2}{3} \cdot \frac{3}{2}(y^2 - 1)^{1/2} \cdot 2y = 2y(y^2 - 1)^{1/2}$$

Thus,

$$\begin{aligned} L &= \int_1^3 \sqrt{1 + [2y(y^2 - 1)^{1/2}]^2}\, dy = \int_1^3 \sqrt{1 + 4y^4 - 4y^2}\, dy \\ &= \int_1^3 \sqrt{(2y^2 - 1)^2}\, dy = \int_1^3 (2y^2 - 1)\, dy \\ &= \frac{2y^3}{3} - y\Big]_1^3 = (18 - 3) - (\tfrac{2}{3} - 1) \\ &= \tfrac{46}{3} \end{aligned}$$ ■

Suppose f is continuous but f' is not continuous on $[a, b]$ (so that f is not smooth), but we can divide the interval into a finite number, say m, of subintervals on each of which f is smooth. Then the length of the graph of f on $[a, b]$ is defined as the sum of the lengths on each of the subintervals. So if L_k denotes the arc length on the kth subinterval, then

$$L = \sum_{k=1}^{m} L_k$$

It is instructive to consider the *arc length function,* defined by

$$s(x) = \int_a^x \sqrt{1 + [f'(t)]^2}\, dt \qquad (x \in [a, b])$$

for the smooth function f on $[a, b]$. By Theorem 4.6,

$$s'(x) = \sqrt{1 + [f'(x)]^2}$$

and so the *differential of arc length,* $ds = s'(x)\,dx$, is given by

$$ds = \sqrt{1 + [f'(x)]^2}\, dx = \sqrt{1 + (\tfrac{dy}{dx})^2}\, dx \tag{5.19}$$

where in the last expression we have used the differential notation $\frac{dy}{dx}$ for $f'(x)$. If $x = g(y)$, where g is smooth on $[c, d]$, then ds is given by

$$ds = \sqrt{1 + [g'(y)]^2}\, dy = \sqrt{1 + (\tfrac{dx}{dy})^2}\, dy \tag{5.20}$$

We now can express both equations (5.17) and (5.18) in the briefer form

$$L = \int ds$$

where the appropriate limits of integration must be supplied, depending on whether ds is given by equation (5.19) or by equation (5.20).

In the equation

$$ds = \sqrt{1 + (\tfrac{dy}{dx})^2}\, dx$$

if we bring the final dx inside the radical and simplify, we get

$$ds = \sqrt{(dx)^2 + (dy)^2}$$

The same result can be obtained by bringing dy inside the radical in equation (5.20). Squaring both sides gives

$$(ds)^2 = (dx)^2 + (dy)^2 \tag{5.21}$$

This has an interesting geometric interpretation, as shown by Figure 5.40. The differential ds can be interpreted as the length of the hypotenuse of a right triangle with legs dx and dy. We let $dx = \Delta x$ be an arbitrary increment on x, which causes a corresponding increment on y, Δy. By Section 2.9 we know that the distance dy to the tangent line approximates the distance Δy to the curve when Δx is small. Calling Δs, the arc length from (x, y)

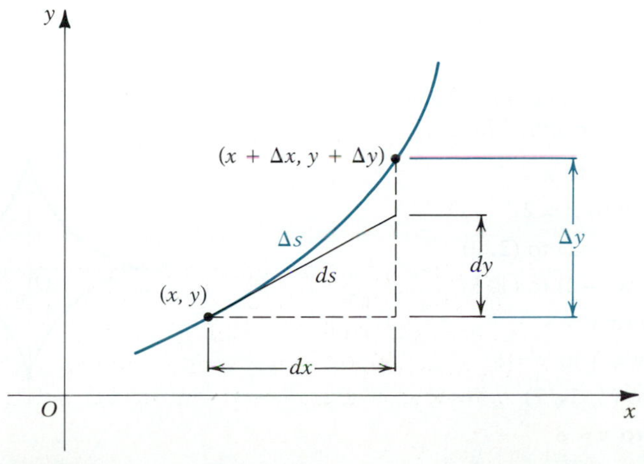

FIGURE 5.40

to $(x + \Delta x, y + \Delta y)$, we see then that $ds \approx \Delta s$. So we can think of the differential of arc as almost equal to a true section of arc when the change in x is small.

The use of this differential notation is illustrated in the following example.

EXAMPLE 5.18 Find the length of the curve defined by $x^4 - 6xy + 3 = 0$ from $(1, \frac{2}{3})$ to $(3, \frac{14}{3})$.

Solution We choose to solve for y, since it is much simpler than solving for x. We find that

$$y = \frac{x^3}{6} + \frac{1}{2x}$$

and for x in the interval $[1, 3]$ this is a smooth function of x. (Why?) So we use equation (5.19) to get ds:

$$\begin{aligned} ds &= \sqrt{1 + \left(\frac{dy}{dx}\right)^2}\, dx = \sqrt{1 + \left(\frac{x^2}{2} - \frac{1}{2x^2}\right)^2}\, dx \\ &= \sqrt{1 + \frac{x^4}{4} - \frac{1}{2} + \frac{1}{4x^4}}\, dx = \sqrt{\frac{x^4}{4} + \frac{1}{2} + \frac{1}{4x^4}}\, dx \\ &= \sqrt{\left(\frac{x^2}{2} + \frac{1}{2x^2}\right)^2}\, dx = \left(\frac{x^2}{2} + \frac{1}{2x^2}\right) dx \end{aligned}$$

Thus,

$$\begin{aligned} L &= \int_{x=1}^{x=3} ds = \int_1^3 \left(\frac{x^2}{2} + \frac{1}{2x^2}\right) dx \\ &= \frac{x^3}{6} - \frac{1}{2x}\Big]_1^3 = \left(\frac{27}{6} - \frac{1}{6}\right) - \left(\frac{1}{6} - \frac{1}{2}\right) \\ &= \frac{14}{3} \end{aligned}$$

■

EXERCISE SET 5.4

A

In Exercises 1–14 find the arc length on the graph of the given equation between the specified limits.

1. $y = 3x - 4$ from $x = -1$ to $x = 2$
2. $y = 3 - 2x$ from $x = 0$ to $x = 2$
3. $2x - 3y = 4$ from $(-1, -2)$ to $(2, 0)$
4. $4y - 5x = 7$ from $(-3, -2)$ to $(1, 3)$
5. $y = \frac{2}{3}x^{3/2}$ from $x = 0$ to $x = 3$
6. $x = (y - 1)^{3/2}$ from $y = 1$ to $y = 6$
7. $y^2 = x^3$ from $(0, 0)$ to $(5, 5\sqrt{5})$
8. $y = x^{2/3}$ from $x = 1$ to $x = 8$
9. $y = \frac{2}{3}(x^2 + 1)^{3/2}$ from $x = 0$ to $x = 2$

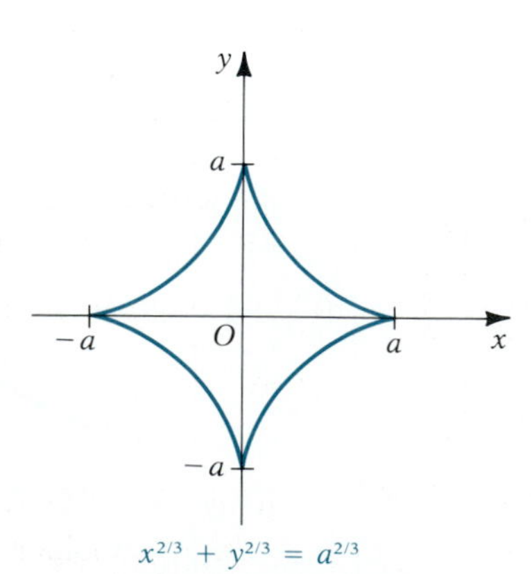

Figure for Exercise 15

10. $y = 1 + (3x/2)^{2/3}$ from $(0, 1)$ to $(\frac{2}{3}, 2)$

11. $y = \left(\dfrac{1-x}{2}\right)^{3/2} + 4$ from $x = -4$ to $x = 1$

12. $y = \dfrac{x^4}{8} + \dfrac{1}{4x^2}$ from $x = 1$ to $x = 2$

13. $4y^4 - 12xy + 3 = 0$ from $(\frac{7}{12}, 1)$ to $(\frac{109}{12}, 3)$

14. $9y^2 = (x^2 - 2)^3$ from $(\sqrt{2}, 0)$ to $(\sqrt{6}, \frac{8}{3})$

15. Find the total length of the four-cusp hypocycloid $x^{2/3} + y^{2/3} = a^{2/3}$ (see the figure on page 216).

B

16. Find the length of the arc on the graph of $x = \frac{1}{3}\sqrt{y}(y - 3)$ from $y = 1$ to $y = 4$.

17. Find the fallacy in the following attempt to find the length of the curve $2y = 3(x + 1)^{2/3}$ from $(-2, \frac{3}{2})$ to $(7, 6)$, and correct it.

$$x = \left(\frac{2y}{3}\right)^{3/2} - 1, \qquad \text{so } \frac{dx}{dy} = \left(\frac{2y}{3}\right)^{1/2}$$

$$L = \int_{3/2}^{6} \sqrt{1 + \frac{2y}{3}}\, dy = \left(1 + \frac{2y}{3}\right)^{3/2}\Bigg]_{3/2}^{6}$$
$$= 5\sqrt{5} - 2\sqrt{2}$$

18. Let

$$f(x) = \begin{cases} 1 - x & \text{if } 0 \le x < 1 \\ (x-1)^{3/2} & \text{if } 1 \le x < 2 \\ 2 - (x-1)^{2/3} & \text{if } 2 \le x \le 9 \end{cases}$$

(a) Draw the graph of f.

(b) Find the length of the graph of f from $x = 0$ to $x = 9$.

In Exercises 19 and 20 use ds to approximate the arc length in question.

19. Arc of $y = x^2$ from $x = 1.0$ to $x = 1.1$

20. Arc of $y = x/(x - 1)$ from $x = 2.05$ to $x = 2.07$

C

Write a computer program to estimate the length of arc on the graph of a smooth function $y = f(x)$ from $x = a$ to $x = b$, making use of the Riemann sum with 1000 equally spaced partition points taking $x_k^* = (x_{k-1} + x_k)/2$. Run the program for each of the following curves.

21. $f(x) = x^2$ on $[0, 2]$

22. $f(x) = \sin x$ on $[0, \pi]$

23. $f(x) = \sqrt{4 - x^2}$ on $[-2, 2]$

24. $f(x) = \frac{2}{3}\sqrt{9 - x^2}$ on $[-3, 3]$

5.5 AREAS OF SURFACES OF REVOLUTION

If an arc of a curve is revolved about an axis, a **surface of revolution** is generated. Our purpose in this section is to develop a means of finding the areas of such surfaces. First we must define precisely what we mean by the area in this situation.

As a starting point we need to know the area of a *frustum of a cone*. Such a frustum is pictured in Figure 5.41. An outline of how to find its area is presented in Exercise 20 at the end of this section. The result, which we take as known, is that the lateral surface area S is given by

$$S = (\text{average circumference}) \times (\text{slant height})$$

or

$$S = 2\pi r l \tag{5.22}$$

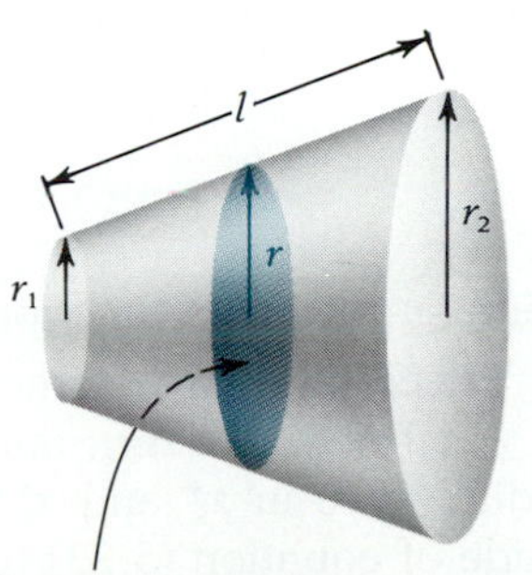

Average circumference $= 2\pi rl$
where $r = \frac{1}{2}(r_1 + r_2)$

FIGURE 5.41

where r is the average radius $= \frac{1}{2}(r_1 + r_2)$ and l is the slant height.

Now consider the arc of the graph of $y = f(x)$, where f is a smooth *non-negative* function, between $x = a$ and $x = b$. Just as in defining arc length,

FIGURE 5.42

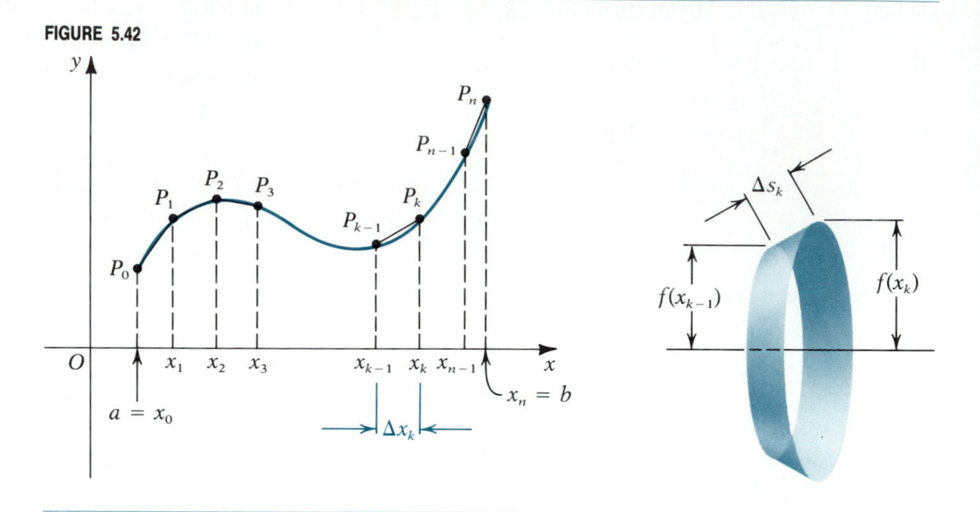

we partition $[a, b]$ and consider the chords that join successive points $P_k = (x_k, f(x_k))$ on the curve, as shown in Figure 5.42. For each $k = 1, 2, \ldots, n$, we rotate the chord $P_{k-1}P_k$ about the x-axis, generating a band, which is a frustum of a cone. In Figure 5.42 we show a typical band whose area we designate ΔS_k. The slant height Δs_k was found in Section 5.4 to be

$$\Delta s_k = \sqrt{1 + [f'(x_k^*)]^2}\,\Delta x_k$$

where x_k^* is between x_{k-1} and x_k. The average circumference is

$$2\pi \cdot \overbrace{\tfrac{1}{2}[f(x_{k-1}) + f(x_k)]}^{\text{average radius}}$$

and since $\frac{1}{2}[f(x_{k-1}) + f(x_k)]$ lies between $f(x_{k-1})$ and $f(x_k)$, we know by the intermediate value theorem (Theorem 1.7) that there is a number x_k^{**} in the interval $[x_{k-1}, x_k]$ such that

$$f(x_k^{**}) = \tfrac{1}{2}[f(x_{k-1}) + f(x_k)]$$

Thus, the surface area ΔS_k of the kth band is

$$\Delta S_k = 2\pi f(x_k^{**})\,\Delta s_k = 2\pi f(x_k^{**})\sqrt{1 + [f'(x_k^*)]^2}\,\Delta x_k$$

Summing all such areas, we arrive at what we would expect to be an approximation to the total surface area S:

$$S \approx \sum_{k=1}^{n} 2\pi f(x_k^{**})\sqrt{1 + [f'(x_k^*)]^2}\,\Delta x_k \tag{5.23}$$

The sum on the right looks almost like a Riemann sum for the function $2\pi f(x)\sqrt{1 + [f'(x)]^2}$, but it differs in that x_k^* and x_k^{**} need not be the same. However, they are both points in the interval $[x_{k-1}, x_k]$ of width Δx_k, and as this width shrinks to 0, we would expect the difference in x_k^* and x_k^{**} to become insignificant and the sum on the right side of equation (5.23) to differ so little from a true Riemann sum that their limits are the same. This is indeed the case and can be proved, but we will not do so here. Assuming the result, we are led to the following definition.

DEFINITION 5.4 Let f be a nonnegative smooth function on $[a, b]$. The area S of the surface of revolution generated by revolving the graph of f from $x = a$ to $x = b$ about the x-axis is given by

$$S = 2\pi \int_a^b f(x)\sqrt{1 + [f'(x)]^2}\, dx \tag{5.24}$$ ■

The corresponding result for revolving the graph of the nonnegative smooth function $g(y)$ from $y = c$ to $y = d$ about the y-axis is

$$S = 2\pi \int_c^d g(y)\sqrt{1 + [g'(y)]^2}\, dy \tag{5.25}$$

In Section 5.4 we showed that the differential of arc length ds can be written as either

$$ds = \sqrt{1 + [f'(x)]^2}\, dx \quad \text{or} \quad ds = \sqrt{1 + [g'(y)]^2}\, dy$$

Using these we can rewrite equations (5.24) and (5.25) in simpler forms:

Rotation about x-axis: $S = 2\pi \int y\, ds$ (5.26)

Rotation about y-axis: $S = 2\pi \int x\, ds$ (5.27)

It is understood that the form of ds used determines the variable of integration and that the limits of integration for that variable must be supplied.

Using the theory of inverse functions that we will develop in the next chapter, it is possible to prove that under certain conditions either of the forms for ds can be used in either of equations (5.26) and (5.27). In particular, if f is not only smooth and nonnegative but also *monotone* on $[a, b]$, then it is possible to solve the equation $y = f(x)$ for x, obtaining an equation of the form $x = g(y)$, with g a smooth monotone function of y on $[c, d] = [f(a), f(b)]$. Then equation (5.26) can be written as

$$S = 2\pi \int y\, ds = 2\pi \int_a^b f(x)\sqrt{1 + [f'(x)]^2}\, dx$$

or as

$$S = 2\pi \int y\, ds = 2\pi \int_c^d y\sqrt{1 + [g'(y)]^2}\, dy$$

Similar remarks apply to equation (5.27). We will illustrate this with the next example.

EXAMPLE 5.19 Find in two ways the area of the surface generated by revolving the arc of the curve $y = \sqrt{x}$ from $x = 1$ to $x = 4$ about the x-axis.

Solution Write $f(x) = \sqrt{x}$. Then on $[1, 4]$ f is smooth, nonnegative, and monotone increasing (Figure 5.43). So we have a choice as to how to write ds in equation (5.26).

In method 1, x is independent:

$$ds = \sqrt{1 + [f'(x)]^2}\, dx = \sqrt{1 + \left(\frac{1}{2\sqrt{x}}\right)^2}\, dx$$

$$= \sqrt{1 + \frac{1}{4x}}\, dx = \frac{\sqrt{4x + 1}}{2\sqrt{x}}\, dx$$

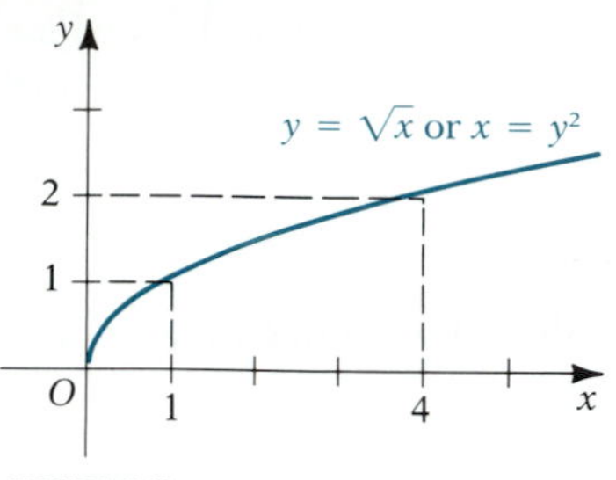

FIGURE 5.43

By equation (5.26),

$$S = 2\pi \int y\,ds = 2\pi \int_1^4 \sqrt{x}\,\frac{\sqrt{4x+1}}{2\sqrt{x}}\,dx = \frac{\pi}{4}\int_1^4 \sqrt{4x+1}(4\,dx)$$

$$= \frac{\pi}{4}\cdot\frac{2}{3}(4x+1)^{3/2}\Big]_1^4 = \frac{\pi}{6}\,[(17)^{3/2} - 5^{3/2}] \approx 30.85$$

In method 2, y is independent. We solve $y = \sqrt{x}$ for x to get $x = y^2$. When $x = 1$, $y = 1$, and when $x = 4$, $y = 2$, so the limits are from 1 to 2. Defining g by $g(y) = y^2$, we have

$$ds = \sqrt{1 + [g'(y)]^2}\,dy = \sqrt{1 + (2y)^2}\,dy = \sqrt{1 + 4y^2}\,dy$$

So by equation (5.26),

$$S = 2\pi \int y\,ds = \frac{2\pi}{8}\int_1^2 y\sqrt{1+4y^2}(8\,dy)$$

$$= \frac{\pi}{4}\cdot\frac{2}{3}(1+4y^2)^{3/2}\Big]_1^2 = \frac{\pi}{6}\,[(17)^{3/2} - 5^{3/2}] \approx 30.85$$ ■

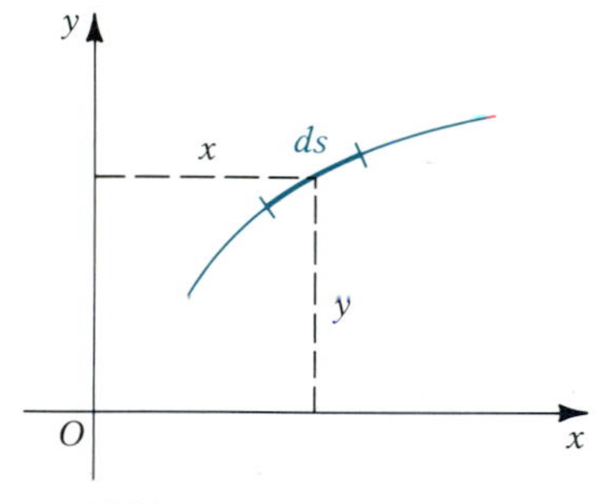

FIGURE 5.44

Another advantage of equations (5.26) and (5.27) is that they can be arrived at quite easily using the following device. Referring to Figure 5.44, we treat ds as if it were a small arc on the curve, so small that it can be considered straight. Let (x, y) be its center. Then, when this arc is rotated about the x-axis, we get a frustum of a cone (almost), whose surface area is

$$\text{(average circumference)} \times (\text{``slant height''}) = 2\pi y\,ds$$

The total surface area is then the "sum" $2\pi \int y\,ds$. The same sort of reasoning leads to $2\pi \int x\,ds$ for the area of the surface of rotation about the y-axis.

Remark It should be emphasized that what we have described is just a convenient device for arriving at a correct result. The reasoning is not mathematically rigorous.

The next two examples further illustrate the use of equations (5.26) and (5.27).

EXAMPLE 5.20 Find the area of the surface obtained by revolving the arc of $y = \frac{2}{3}x^2 + 1$ between $x = 0$ and $x = 1$ about the y-axis.

Solution Since f is monotone on $[0, 1]$ (Figure 5.45), we are free to choose either x or y as independent, and we choose x:

$$ds = \sqrt{1 + [f'(x)]^2}\,dx = \sqrt{1 + \left(\frac{dy}{dx}\right)^2}\,dx$$

$$= \sqrt{1 + \left(\frac{4}{3}x\right)^2}\,dx = \sqrt{1 + \frac{16x^2}{9}}\,dx$$

$$= \frac{\sqrt{9 + 16x^2}}{3}\,dx$$

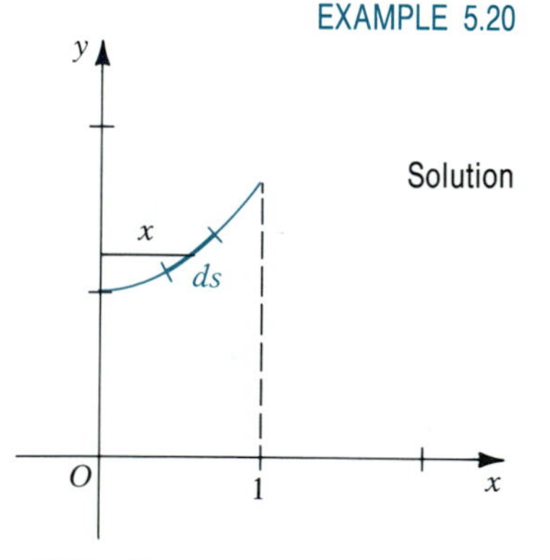

FIGURE 5.45

By equation (5.27),

$$S = 2\pi \int x\,ds = \frac{2\pi}{32}\int_0^1 x\frac{\sqrt{9+16x^2}}{3}(32\,dx)$$
$$= \frac{\pi}{48}\cdot\frac{2}{3}[(9+16x^2)^{3/2}]_0^1 = \frac{\pi}{72}[(25)^{3/2} - (9)^{3/2}]$$
$$= \frac{\pi}{72}[125-27] = \frac{49\pi}{36}$$

You may wish to do this problem using y as the independent variable to compare the difficulty of the two methods. ■

EXAMPLE 5.21 Find the surface area of the *segment* of a sphere of radius a cut off by a plane at a point b units from the center, where $0 < b < a$.

Solution In Figure 5.46 we show the sphere with center at the origin and the segment cut off by a plane through $(b, 0)$ perpendicular to the x-axis, and in Figure 5.47 we show an arc of the circle $x^2 + y^2 = a^2$ that generates the spherical segment when it is rotated around the x-axis.

We will show in this example how implicit differentiation can sometimes simplify calculations. Treating y as a function of x, we have

$$x^2 + y^2 = a^2$$
$$2x + 2yy' = 0$$
$$y' = -\frac{x}{y} \qquad (y \neq 0)$$

So

$$ds = \sqrt{1+(y')^2}\,dx = \sqrt{1+\frac{x^2}{y^2}}\,dx = \frac{\sqrt{x^2+y^2}}{y}\,dx = \frac{a}{y}\,dx$$

Here, we have used the fact that $\sqrt{x^2+y^2} = \sqrt{a^2} = a$ and also the fact that since $y > 0$ on the arc in question, $\sqrt{y^2} = y$. Thus,

$$S = 2\pi\int y\,ds = 2\pi\int_b^a y\left(\tfrac{a}{y}\right)dx = 2\pi a[x]_b^a$$
$$= 2\pi a(a-b)$$ ■

You may have observed that in calculating y' we required y to be nonzero, yet $y = 0$ at the end point $x = a$. This apparent contradiction can be overcome by observing that the integrand $y(\frac{a}{y})$ equals the constant a when x is in

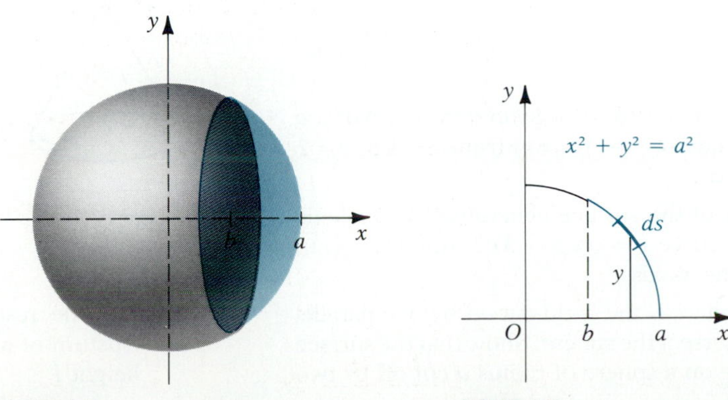

FIGURE 5.46

FIGURE 5.47

the interval $b \le x < a$, and since the left-hand limit as x approaches a exists ($\lim_{x \to a^-} a = a$), the integral on the closed interval exists and has the value shown. (See Exercise 59 in Exercise Set 4.4.)

EXERCISE SET 5.5

A

In Exercises 1–10 find the area of the surface generated by revolving the arc of the curve whose equation is given between the specified limits about the axis indicated.

1. $y = 2x + 1$ from $x = 0$ to $x = 3$; x-axis
2. $2x + 3y = 6$ from $(0, 2)$ to $(3, 0)$; y-axis
3. $y = \sqrt{2x}$ from $x = 0$ to $x = 2$; x-axis
4. $y = x^2$ from $(1, 1)$ to $(2, 4)$; y-axis
5. $y = \sqrt{1 - x^2}$ from $(0, 1)$ to $(1, 0)$; y-axis
6. $y = x^3$ from $x = 0$ to $x = 1$; x-axis
7. $y = x^{1/3} + 1$ from $(1, 2)$ to $(8, 3)$; y-axis
8. $x^2 + y^2 = 9$ from $(0, 3)$ to $(2, \sqrt{5})$; x-axis
9. $x = \sqrt{5 - y}$ from $y = 1$ to $y = 4$; y-axis
10. $y^2 - 2x + 3 = 0$ from $(2, 1)$ to $(6, 3)$; x-axis
11. Derive the formula for the surface area of a sphere of radius a.
12. Find the area of the surface that results from rotating the upper half of the four-cusp hypocycloid $x^{2/3} + y^{2/3} = a^{2/3}$ about the x-axis. (See Exercise 15 in Exercise Set 5.4.) Make use of implicit differentiation.
13. Find the area of the surface generated by revolving the segment of the line $2x + 3y = 6$ cut off by the x- and y-axes about (a) the line $x = -1$ and (b) the line $y = -2$.

B

14. Show that if f is smooth on $[a, b]$ but not necessarily nonnegative, the area of the surface generated by revolving the graph of f between $x = a$ and $x = b$ about the x-axis is

$$S = 2\pi \int_a^b |f(x)|\sqrt{1 + [f'(x)]^2}\, dx$$

15. Use the result of Exercise 14 to find the surface area generated by revolving the arc of the curve $y = x^3/6$ from $x = -1$ to $x = 2$ about the x-axis.
16. Find the area of the surface generated by revolving the arc of the curve

$$y = \frac{x}{2}\sqrt{\frac{1 - x^2}{2}}$$

from $x = -1$ to $x = 1$ about the x-axis. (See Exercise 14.)

17. Find the area of the surface generated by revolving the arc of the curve $y^4 + 3 = 6xy$ from $y = 1$ to $y = 2$ about the y-axis.
18. Find the area of the surface generated by revolving the arc of the curve $y = \sqrt{x}(x - 3)/3$ from $(1, -\frac{2}{3})$ to $(9, 6)$ about the x-axis.
19. A *zone* of a sphere is the band cut off by two parallel planes that intersect the sphere. Show that the surface area of a zone on a sphere of radius a cut off by two parallel planes h ($h \le 2a$) units apart is independent of the location of the planes (so long as both intersect the sphere). What is the area of such a zone?
20. (a) Find a formula for the lateral surface area of a cone as follows. Imagine cutting a cone of base radius r along a line AB as shown in part (a) of the figure. Then flatten out the cone so that it becomes a circular sector as in part (b). Calculate the area of this sector.

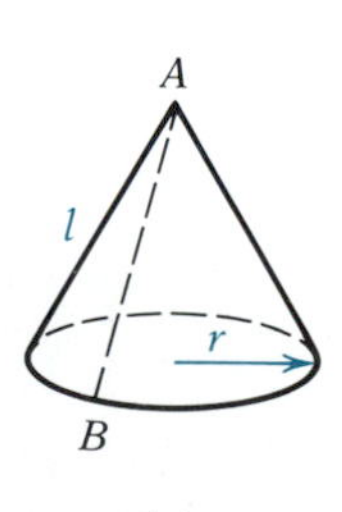

(a)

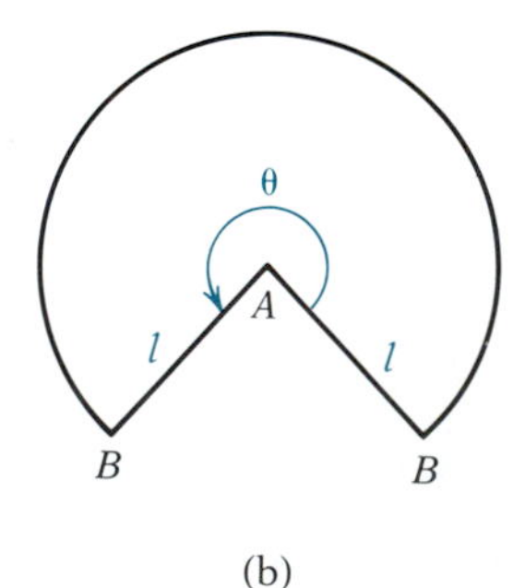

(b)

(b) Use the result of part a to find the area of a frustum of a cone with radii r_1 and r_2 and slant height l.

5.6 WORK

When a constant force F acts on an object causing it to move a distance d in the direction of the force, the **work** W done by the force is defined to be

$$W = Fd = (\text{force}) \times (\text{distance}) \tag{5.28}$$

In the English system force is typically measured in pounds and distance in feet or inches so that the unit of work is the *foot-pound* or the *inch-pound*. In the metric system two combinations of units are commonly used: (1) the centimeter, gram, second (cgs) system, in which force is measured in *dynes* and distance in centimeters, so that the unit of work is the *dyne-centimeter*, also called an *erg*, and (2) the meter, kilogram, second (mks) system, in which force is measured in *newtons*, with the corresponding unit of work a *newton-meter*, called a *joule*.

The application of equation (5.28) is limited to constant forces that act in the direction of motion. There are many situations in which a variable force acts, where both the magnitude and the direction of the force may change. For the present we will consider a variable force acting in a straight line. The case where the direction of the force changes will be taken up in a subsequent chapter. An example of a variable force is the force necessary to stretch a spring. The amount of force necessary varies with the distance through which the spring has been stretched. As another example, the force of repulsion between two like electrostatic charges changes as the distance between the charges changes.

To arrive at a satisfactory definition of work done by a variable force F acting in a straight line over a fixed distance, let us orient the x-axis so that the direction of force is along the axis, let $F(x)$ denote the amount of the force at x, and suppose F is a continuous function on the interval from $x = a$ to $x = b$. As in Figure 5.48 we partition the interval $[a, b]$ by means of the points $a = x_0 < x_1 < x_2 < \cdots < x_n = b$, and in the usual way let $\Delta x_k = x_k - x_{k-1} (k = 1, 2, \ldots, n)$ and let x_k^* denote an arbitrary point in $[x_{k-1}, x_k]$ for each $k = 1, 2, \ldots, n$. If the width Δx_k is sufficiently small, the error in replacing the variable force $F(x)$ by the constant force $F(x_k^*)$ on the kth subinterval will not be great. For this constant force, the work ΔW_k done across the kth subinterval is

$$\Delta W_k = \underbrace{F(x_k^*)}_{\text{force}} \; \underbrace{\Delta x_k}_{\text{distance}}$$

and it is reasonable to suppose that the total work W done by the force on the interval $[a, b]$ is approximated by the sum of the W_k values:

$$W \approx \sum_{k=1}^{n} \Delta W_k = \sum_{k=1}^{n} F(x_k^*) \Delta x_k$$

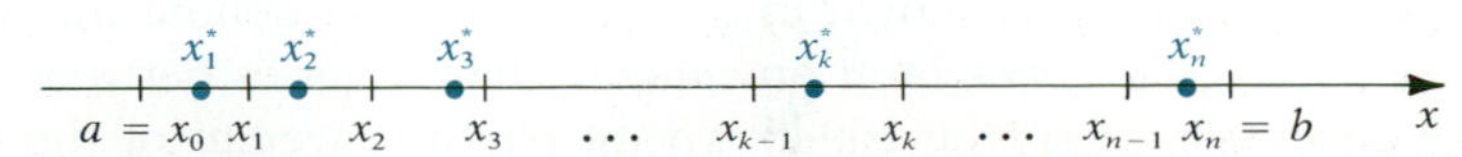

FIGURE 5.48

As the widths of the subintervals all approach 0 (that is, as $\|P\| \to 0$), the approximation becomes better and better, and since the Riemann sum on the right approaches the integral $\int_a^b F(x)\,dx$, we are led to the following definition.

DEFINITION 5.5 The **work** W done by a continuous force F acting along the x-axis from $x = a$ to $x = b$ is given by

$$W = \int_a^b F(x)\,dx \tag{5.29}$$ ■

To illustrate the use of this definition, let us consider the problem of stretching a spring. Within certain limits the force necessary to stretch a spring x units beyond its natural length is given by $F(x) = kx$ where k is a constant, called the *spring constant*. This result is known as **Hooke's law,** and it holds true also if the spring is compressed x units from its natural length.

EXAMPLE 5.22 A certain spring has a natural length of 12 in., and a 5-lb force is required to stretch it 3 in. Find the work done in stretching it from its natural length to a length of 16 in. Find the additional work done in stretching it 2 more inches (from 16 in. to 18 in.).

Solution By Hooke's law, $F(x) = kx$, and we know that $F(x) = 5$ when $x = 3$. So we can find k:

$$5 = k \cdot 3$$
$$k = \tfrac{5}{3}$$

Thus, for this spring

$$F(x) = \tfrac{5}{3}x$$

In stretching the spring from its natural length of 12 in. to a length of 16 in., x varies from 0 to 4 (remember that x is the distance *beyond its natural length* that it is stretched). So we have

$$W = \int_0^4 F(x)\,dx = \int_0^4 \frac{5}{3}x\,dx = \frac{5}{3}\left[\frac{x^2}{2}\right]_0^4 = \frac{40}{3}\text{ in.-lb}$$

For the spring to be stretched an additional 2 in., x will vary from 4 to 6, so the additional work done is

$$W = \int_4^6 \frac{5}{3}x\,dx = \frac{5}{3}\left[\frac{x^2}{2}\right]_4^6 = \frac{5}{3}[18 - 8] = \frac{50}{3}\text{ in.-lb}$$ ■

EXAMPLE 5.23 A 200-lb weight is to be hoisted from the ground to a point 20 ft high by a uniform chain passing over a pulley. If the chain weighs 2 lb/ft, find the work done in raising the weight.

Solution As shown in Figure 5.49, let x denote the distance from the pulley to the weight at any stage. Then x varies between 0 and 20. The total force $F(x)$ is the 200-lb weight plus the weight of the chain that is still to be raised.

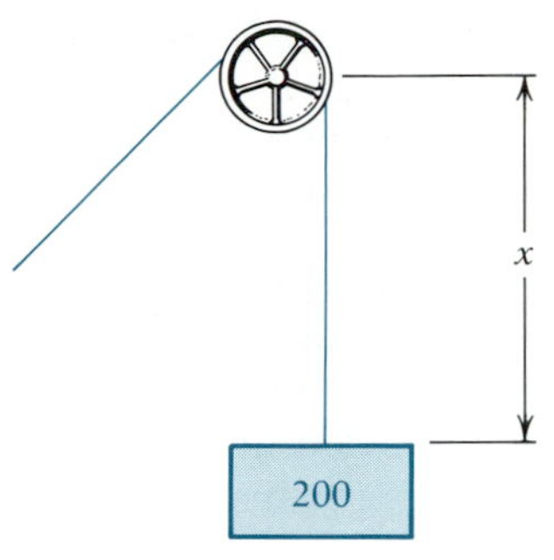

FIGURE 5.49

When x feet of chain are out, its weight is $2x$ pounds. So $F(x) = 200 + 2x$ is the variable force. Thus, by equation (5.29),

$$W = \int_0^{20} (200 + 2x)\,dx = 4400 \text{ ft-lb}$$ ■

As preparation for the next example, which is typical of a class of problems, consider a tank of height b units filled with a liquid to a depth a units from the top, as shown in Figure 5.50. Suppose at the distance x units down from the top of the tank, the cross-sectional area $A(x)$ is known and is a continuous function of x on $[a, b]$. We want to find the work done in pumping the liquid over the top rim of the tank.

Suppose the density of the liquid is ρ, defined as weight per unit volume (for example, pounds per cubic foot, lb/cu ft, or newtons per cubic meter, N/cu m). We can consider the liquid as being divided into thin layers by planes passed perpendicular to the vertical axis at points $\{x_0, x_1, x_2, \ldots, x_n\}$ that form a partition of $[a, b]$. As in our study of volumes by the cross-sectional method, the kth layer is approximated by a disk of constant cross-sectional area $A(x_k^*)$ and thickness $\Delta x_k = x_k - x_{k-1}$, where x_k^* is any point in $[x_{k-1}, x_k]$. We have shown a typical disk in Figure 5.50. We can imagine this liquid disk being lifted to the top, and the work required to do this is the weight of the disk (this is a force) multiplied by the distance x_k^* through which it is lifted. The weight of the kth disk is its volume times the density of the liquid, $\rho A(x_k^*)\,\Delta x_k$. So the work ΔW_k of lifting this weight to the top is

$$\Delta W_k = \overbrace{\rho A(x_k^*)\,\Delta x_k}^{\text{force (weight)}} \cdot \overbrace{x_k^*}^{\text{distance}}$$

and the total work is approximated by the sum

$$W \approx \sum_{k=1}^{n} \Delta W_k = \sum_{k=1}^{n} \rho x_k^* A(x_k^*)\,\Delta x_k$$

FIGURE 5.50

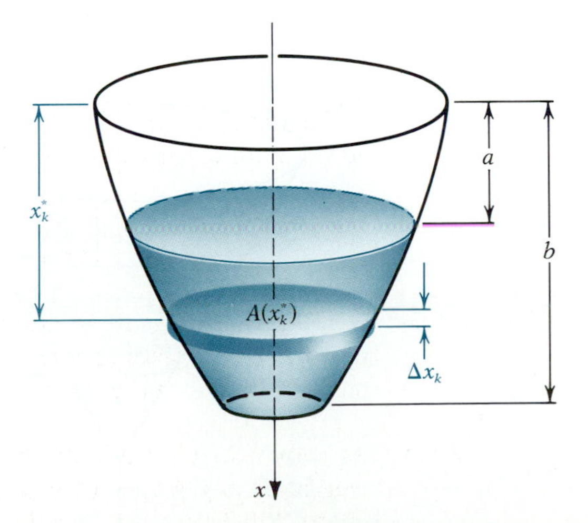

On passing to the limit as the thicknesses Δx_k all go to 0, we arrive at the definition

$$W = \int_a^b \rho x A(x)\, dx \tag{5.30}$$

Remark We have taken the density ρ to be *weight* per unit volume. Sometimes density is expressed as *mass* per unit volume. If we denote this by δ, then the relationship between δ and ρ is $\rho = \delta g$, where g is the acceleration due to gravity. For example, in the mks system the mass density of water is $\delta = 1000$ kg/cu m, and in this system $g = 9.81$ m/sec^2 approximately. So the weight density is $\rho = \delta g = 9810$ N/cu m.

EXAMPLE 5.24 A water tank in the form of an inverted right circular cone of height 20 ft and base radius 12 ft is filled with water to a depth of 15 ft. Find the work done in pumping all of the water over the top of the tank.

Solution As in the preceding discussion, we let x denote the distance from the top of the tank, measured positively downward. In order to apply equation (5.30) we need the cross-sectional area $A(x)$. Since the cross section is circular, it suffices to find the radius r. As can be seen in Figure 5.51, we have by similar triangles

$$\frac{r}{12} = \frac{20 - x}{20}$$

So $r = \frac{3}{5}(20 - x)$, and

$$A(x) = \pi r^2 = \frac{9\pi}{25}(20 - x)^2$$

The density of water is approximately 62.4 lb/cu ft. So by equation (5.30) we have

$$W = \int_a^b \rho x A(x)\, dx = \frac{(62.4)9\pi}{25} \int_5^{20} x(20 - x)^2\, dx$$

$$= 22.46\pi \int_5^{20} (400x - 40x^2 + x^3)\, dx \approx 695{,}000 \text{ ft-lb}$$

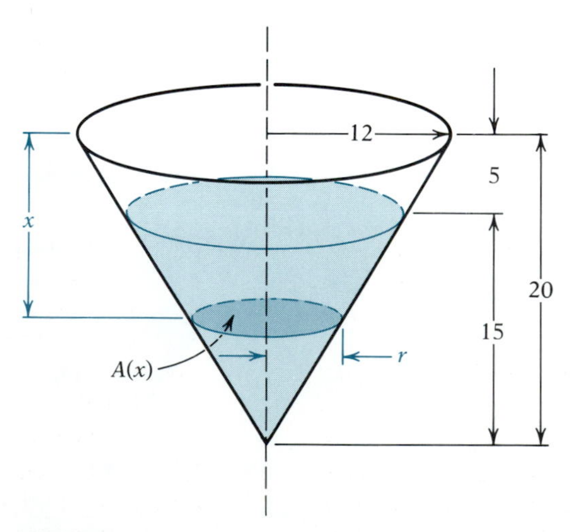

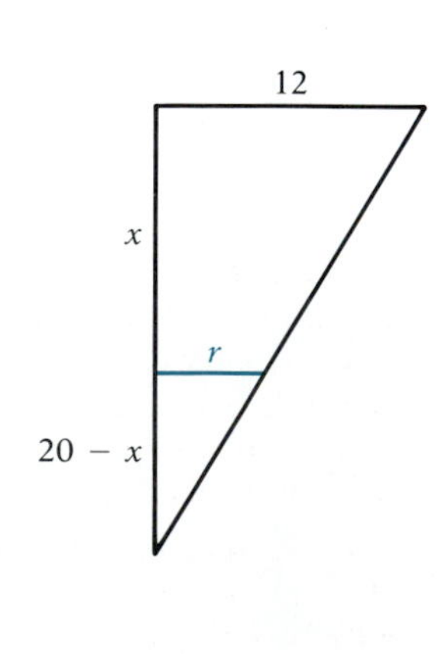

FIGURE 5.51

EXERCISE SET 5.6

A

1. A variable force of $2x - x^2$ N acts in the direction of the positive x-axis from $x = 0$ to $x = 2$ (in meters). Find the work done by the force.
2. A weight of 200 lb is to be pushed a distance of 12 ft up a smooth ramp that makes an angle of 30° with the horizontal. Neglecting any friction, find the work done.
3. A force of 4 lb is required to stretch a certain spring 3 in. beyond its natural length of 10 in. Find the work done in stretching it from its natural length to a length of 16 in.
4. For the spring of Exercise 3 find the work done in stretching it from a length of 15 in. to a length of 18 in.
5. A spring of natural length 20 cm is stretched to a length of 28 cm by a force of 5 dynes. Find the work done in compressing the spring 6 cm from its natural length. What is the work done in stretching the spring from a length of 22 cm to a length of 30 cm?
6. A heavy-duty spring with spring constant 10,000 N/m is used to cushion the shock of a subway car at the end of the line in the event it does not come to a complete stop before bumping the end. Find the work done by the force of the subway car if in bumping the end, the spring is compressed a distance of 1 m.
7. The work done in stretching a spring from its natural length 25 cm to a length of 30 cm is 10 dyne-cm. Find the spring constant.
8. A spring of natural length 14 in. requires 30 ft-lb of work to be compressed from a length of 12 in. to a length of 8 in. Find the spring constant. (Be careful with units.)
9. A uniform chain weighing 3 lb/ft is hanging over the edge of a building and is to be pulled up to the top. If 30 ft of chain are initially hanging down, find the work required to raise it to the top.
10. At a construction site a crane is to lift a bucket of cement from the ground to a point 30 ft above the ground. The distance from the end of the crane arm to the ground is 75 ft. If the bucket of cement weighs 500 lb and the cable weighs 5 lb/ft, find the work required to lift the cement.
11. Water is to be raised from a well 40 ft deep by means of a bucket attached to a rope. When the bucket is full of water it weighs 30 lb, but the bucket has a leak that causes it to lose water at the constant rate of $\frac{1}{4}$ lb for each foot that the bucket is raised. Neglecting the weight of the rope, find the work done in raising the bucket to the top.
12. A water tank is in the form of a vertical right circular cylinder with radius 6 ft and height 20 ft. If the tank is half full of water, find the work required to pump all of it over the top rim.
13. In Exercise 12 suppose the tank is initially full of water. Find the work required to pump all of it to a point 10 ft above the top of the tank.
14. A storage tank for liquid ammonia is in the form of a sphere with radius 3 m. If the tank is full, find the approximate work required to pump half of the ammonia to a point 5 m above the top of the tank. The *mass* density of ammonia is $\delta = 891$ kg/cu m. (The *weight* density is $\rho = \delta g$ N/cu m, where $g = 9.81$ m/sec^2.)
15. According to Coulomb's law the repulsive force between two like electrostatic charges Q_1 and Q_2 is given by

$$F = \frac{kq_1q_2}{x^2}$$

where q_1 and q_2 are the magnitudes of Q_1 and Q_2, respectively, x is the distance between them, and k is a constant of proportionality. Suppose $q_1 = 1$ C (coulomb) and $q_2 = 2$ C, and the charges are initially 6 cm apart. If Q_1 is stationary, find the work done in moving Q_2 on a line toward Q_1 to a point 3 cm from Q_1. (Leave the answer in terms of k.)
16. According to Newton's universal law of gravitation, the gravitational force exerted by the earth on a body of mass m at a distance x above the earth's surface is

$$F(x) = G\frac{mM}{(x + R)^2}$$

where R is the radius of the earth, M is the earth's mass, and G is a constant called the universal gravitational constant. It can be shown that $G = gR^2/M$. Taking $g = 9.81$ m/sec^2 and $R = 6.37 \times 10^6$ m, find the work done in lifting a satellite of mass 3.2×10^4 kg a distance of 300 km (300 km = 3×10^5 m) above the earth's surface.
17. Hooke's law can be extended to metal bars under tension or compression, in which case the force necessary to elongate it, or compress it, by x units

is given by

$$F(x) = \frac{EA}{L} x$$

where A is the cross-sectional area of the bar, L is its length, and E is a constant called the *modulus of elasticity* that is a property of the metal. Suppose a cylindrical aluminum bar of natural length 15 in. and cross-sectional radius 3 in. is stretched to a length of 15.5 in. Find the work done. For aluminum, $E = 10^7$ lb/sq in.

B

18. A water tank is in the form of a frustum of a cone that has bottom radius 5 ft, top radius 8 ft, and height 12 ft. If the tank is initially full of water, find the work done in pumping half of the water to a point 10 ft above the top. (Use $\rho = 62.4$ lb/cu ft.)

19. A swimming pool slopes from a depth of 1 m at the shallow end to a depth of 4 m at the deep end. It is 20 m long and 10 m wide. If it is full of water, find the work required to pump all of the water out over the top. (Use $\rho = \delta g$, where $\delta = 1000$ kg/cu m and $g = 9.81$ m/sec^2.)

20. A crane raises a bucket of sand weighing 500 lb from the ground to a height of 40 ft in 10 sec. Sand spills out of the bucket at the rate of 10 lb/sec as it is being raised. If the cable weighs 5 lb/ft, find the work done in raising the sand.

21. A tank has the shape of the surface generated by revolving the right half of the parabola $y = x^2/2$ from $x = 0$ to $x = 2$ about the y-axis. If x and y are measured in feet and the tank is initially half full of water, find the work done in pumping all of the water over the top.

22. The force on a piston caused by the expansion of a gas in a cylinder (see the figure) is $F = pA$, where A is the area of the face of the piston and p is the pressure (force per unit area) of the gas. The volume v occupied by the gas, when the distance from the piston to the end of the cylinder is x, is given by $v = Ax$. For an ideal gas under an adiabatic process (in which no heat is transferred into or out of the cylinder), the relationship between pressure and volume is $pv^{1.4} = c$, where c is a constant. If when $x = a$ the volume is v_1 and when $x = b$ the volume is v_2, show that the work done by the gas in expanding from v_1 to v_2 is

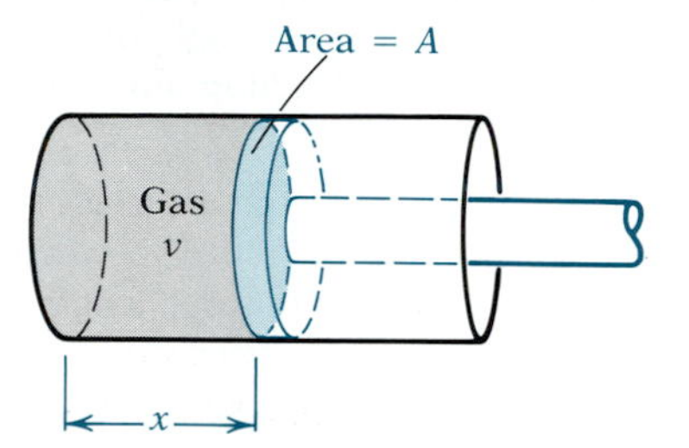

$$W = \int_a^b F(x)\,dx = \int_{v_1}^{v_2} cv^{-1.4}\,dv$$

If the radius of the piston is 3 in., find the work done by the gas in moving the piston from $x = 5$ in. to $x = 8$ in. Express the answer in terms of c.

5.7 FLUID PRESSURE

Our objective in this section is to show how integration can be used to calculate the force exerted by a liquid bearing against one side of a flat surface, such as the force of the water in a lake against the side of a dam. We will consider only static forces as opposed to dynamic ones. This means that we will suppose the liquid is not in motion.

Suppose the liquid in question has density ρ (weight per unit volume). If a flat plate of area A is submerged horizontally at a depth h, then the total force on the plate is just the weight of the liquid directly above the plate. The volume occupied by this liquid is the area times the height, Ah, and if this is multiplied by the weight per unit volume, ρ, we get the total weight of the liquid. So $F = \rho Ah$. For example, the force exerted by the oil in an oil storage tank on the bottom of the tank is the area of the bottom times the depth of the oil times its density.

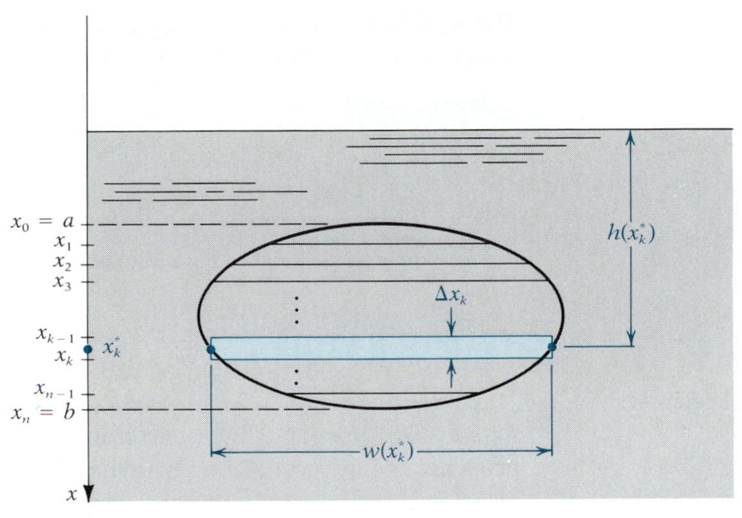

FIGURE 5.52

The **pressure** exerted by the liquid is defined as the force on one square unit of area. So taking $A = 1$, we get for the pressure p at depth h,

$$p = \rho h \tag{5.31}$$

It is an interesting fact, known as *Pascal's principle*, that the pressure p at a given depth is the same in all directions. Thus, if the plate is submerged horizontally, or vertically, or on a slant, or in any manner, the pressure at a fixed depth h units below the surface is given by equation (5.31).

When a plate whose area is known is submerged horizontally, calculating the force on it is straightforward, as we have seen. When it is submerged vertically, however, finding the force is more involved, since the pressure varies continuously from the top of the plate to the bottom. We introduce a vertical axis that we show in Figure 5.52 as being directed positively downward, but in certain applications it will be more convenient to choose its positive direction upward. The origin may be taken at any point, usually determined by the shape of the plate. In the examples we will see how certain choices can often simplify calculations. We assume that the vertical extremities of the plate are at $x = a$ and $x = b$, where $a < b$, and we assume that enough information is known about the shape of the plate that its width $w(x)$, corresponding to any x between a and b, can be determined and that w is a continuous function of x on $[a, b]$. We denote by $h(x)$ the depth at x.

In the usual way we partition the interval $[a, b]$ by the points $a = x_0 < x_1 < \cdots < x_n = b$, with $\Delta x_k = x_k - x_{k-1}$ for $k = 1, 2, \ldots, n$. Horizontal lines drawn at the partition points divide the plate into thin strips. If for each k we select a point x_k^* between x_{k-1} and x_k, then the kth strip can be approximated by a rectangle that has dimensions $w(x_k^*)$ by Δx_k. A typical such rectangle is shown in Figure 5.52. Although the pressure varies from the top to the bottom of this rectangle, if Δx_k is very small, the error induced by assuming the entire rectangle is at depth $h(x_k^*)$ will not be great. The pressure at this depth, by equation (5.31), is

$$p = \rho h(x_k^*)$$

The total force on the kth rectangle, which we designate by ΔF_k, is approximated by its area times this pressure:

$$\Delta F_k \approx \underbrace{\rho}_{\text{density}}\,\underbrace{h(x_k^*)}_{\text{depth}}\,\underbrace{w(x_k^*)\,\Delta x_k}_{\text{area}}$$

It is useful to learn this in words:

force on rectangular strip $\approx$ density $\times$ depth $\times$ area

We approximate the total force by summing all ΔF_k values:

$$F \approx \sum_{k=1}^{n} \Delta F_k \approx \sum_{k=1}^{n} \rho h(x_k^*) w(x_k^*)\,\Delta x_k$$

As we take finer and finer partitions, the Riemann sum on the right approaches the integral $\int_a^b \rho h(x) w(x)\,dx$. This leads to the following definition.

DEFINITION 5.6 The total **force** F on a flat plate submerged vertically in a liquid of density ρ is given by

$$F = \int_a^b \rho h(x) w(x)\,dx \tag{5.32}$$

where the vertical bounds of the plate are $x = a$ and $x = b$, respectively, $h(x)$ is the depth at x, and $w(x)$ is the width of the plate at x. ■

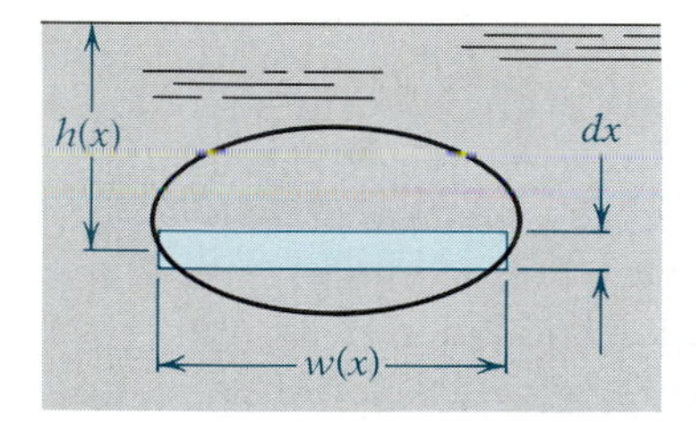

FIGURE 5.53

Remark As a memory device we can pretend that a typical rectangular element has width dx and length $w(x)$ as shown in Figure 5.53. So the force on this element is

density $\times$ depth $\times$ area $= \rho h(x) w(x)\,dx$

"Summing up" all such elements of force (in the sense of integration) we get the result of equation (5.32). Again, this is only a memory device, not a mathematically rigorous derivation.

EXAMPLE 5.25 A triangular plate with base 4 ft and altitude 2 ft is submerged vertically in a liquid of density ρ pounds per cubic foot. It is oriented as in Figure 5.54, with its upper vertex 3 ft below the surface of the liquid. Find the total force on one side of the plate.

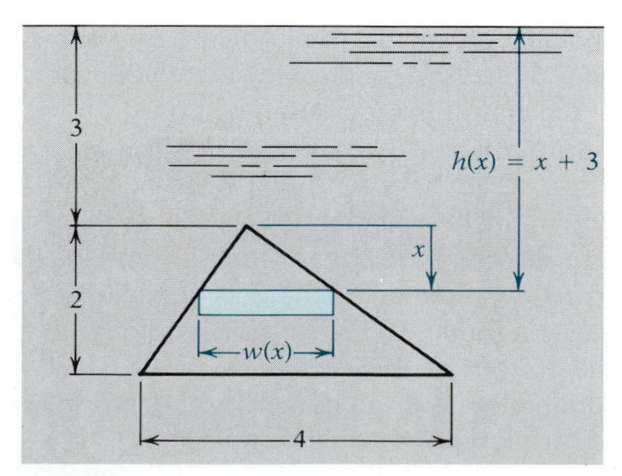

FIGURE 5.54

Solution We choose the origin for x as the upper vertex and direct x positively downward, as shown. The limits on x are then 0 and 2. To find the width $w(x)$, we use similar triangles:

$$\frac{w(x)}{4} = \frac{x}{2}$$

$$w(x) = 2x$$

The depth $h(x)$ is given by $h(x) = x + 3$. Thus, by equation (5.32),

$$F = \int_0^2 \underbrace{\rho}_{\text{density}}\underbrace{(x+3)}_{\text{depth}}\underbrace{(2x)\,dx}_{\text{"area"}}$$

$$= 2\rho \int_0^2 (x^2 + 3x)\,dx$$

$$= 2\rho \left[\frac{x^3}{3} + \frac{3x^2}{2}\right]_0^2$$

$$= \frac{52\rho}{3} \text{ lb}$$

In the exercises you will be asked to rework this example, choosing the origin for x as the level of the fluid so that you can see the advantages of the choice we made. ■

EXAMPLE 5.26 A dam across a river gorge is in the form of an inverted isosceles trapezoid, with upper base 100 ft, lower base 50 ft, and altitude 80 ft (Figure 5.55). Find the force of the water on the face of the dam (assumed vertical) when the water is within 10 ft of the top of the dam.

Solution Because the trapezoid is inverted, it is convenient to choose as origin for x the bottom of the dam and take the positive direction as upward. The limits on x are then 0 and 70, since it is only this portion of the dam that has water against it. The depth from the water level to a point with coordinate x is $h(x) = 70 - x$. To find the width $w(x)$, we again make use of similar triangles, as shown in Figure 5.56. From this figure we see that $w(x) = 50 + 2d$,

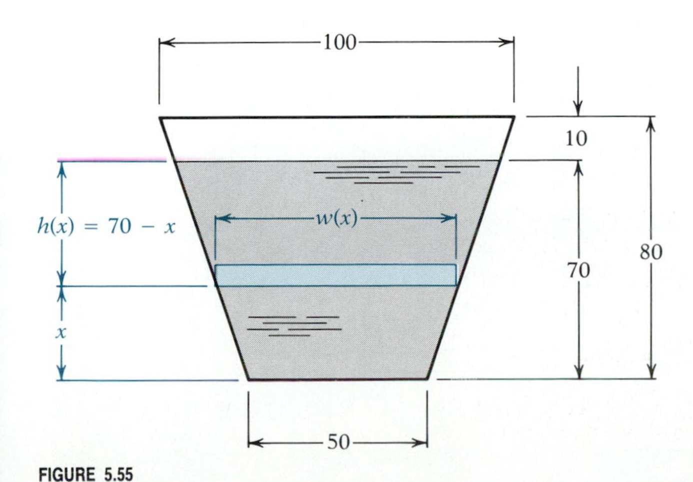

FIGURE 5.55

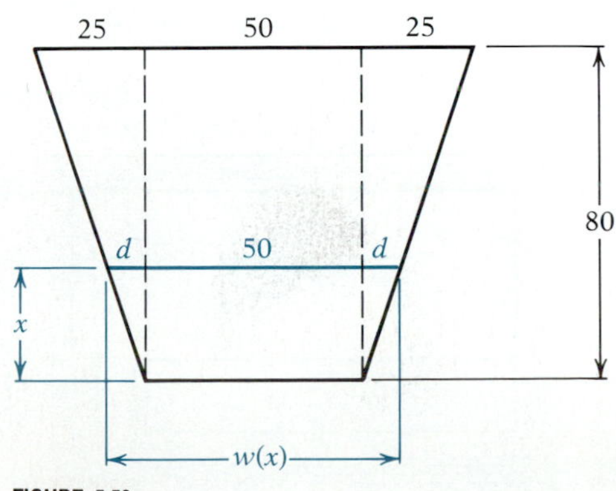

FIGURE 5.56

and using similar triangles, we have

$$\frac{d}{x} = \frac{25}{80}$$

$$d = \frac{5x}{16}$$

So

$$w(x) = 50 + 2\left(\frac{5x}{16}\right) = 50 + \frac{5x}{8} = \frac{5}{8}(80 + x)$$

Now, using equation (5.32), we have

$$F = \int_0^{70} \rho(70 - x) \cdot \frac{5}{8}(80 + x)\, dx = \frac{5\rho}{8} \int_0^{70} (5600 - 10x - x^2)\, dx$$

$$= \frac{5\rho}{8}\left[5600x - 5x^2 - \frac{x^3}{3}\right]_0^{70} = \frac{5\rho}{8}\left(\frac{759{,}500}{3}\right)$$

Taking $\rho = 62.4$ as the approximate density of water, we get finally $F =$ 9,873,500 lb. ■

EXERCISE SET 5.7

A

In Exercises 1–6 a plate is submerged vertically in a liquid of density ρ, as shown. Find the force of the liquid on one side of the plate.

1.

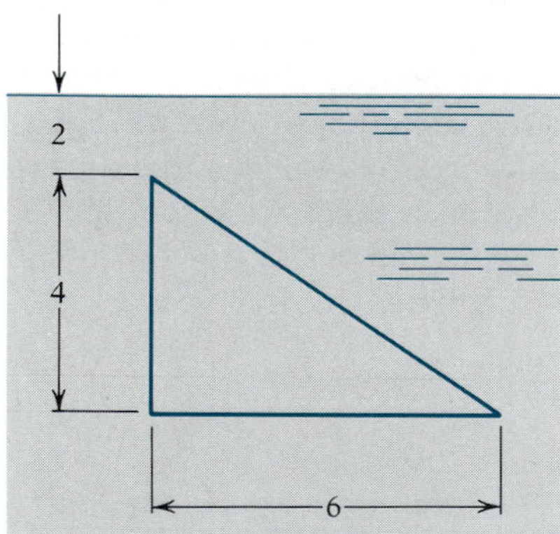

2.

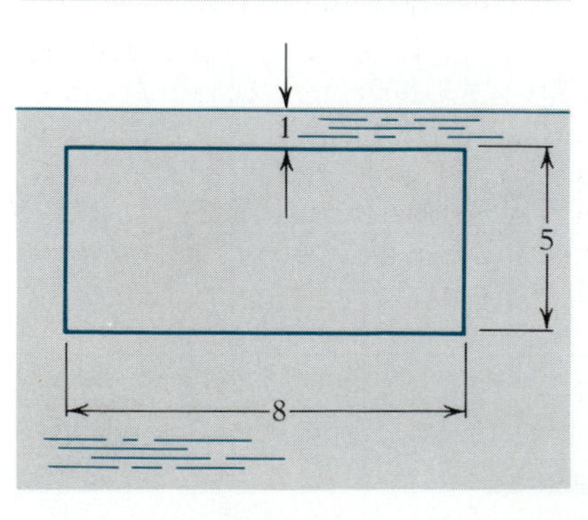

3.

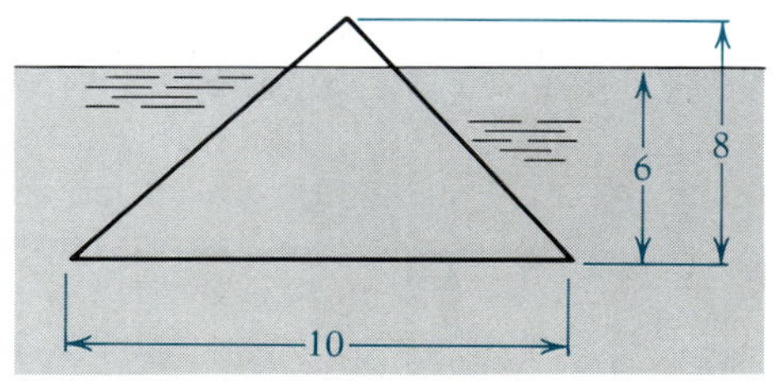

4.

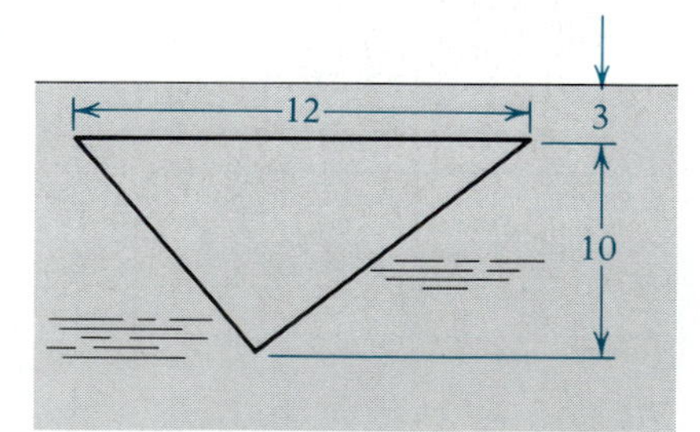

5.

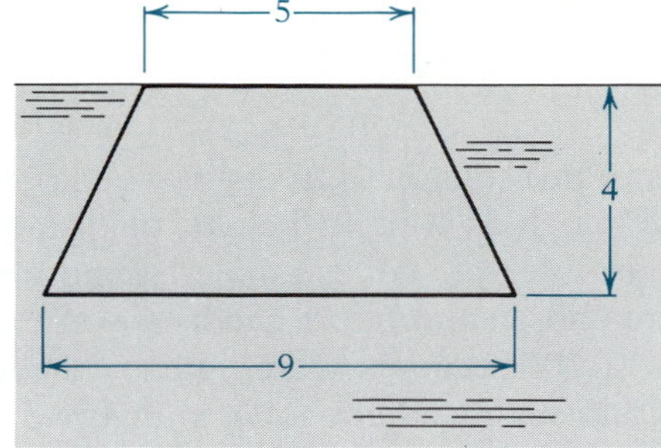

6.

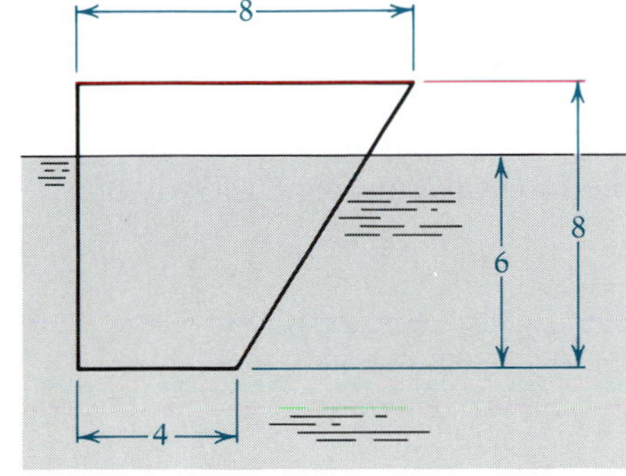

7. The ends of a watering trough are in the form of inverted isosceles triangles, having base 3 ft and altitude 2 ft. Find the force of the water in the trough on one of the ends when the trough is full of water. What is the force when the water level is 1 ft?

8. If each end of the watering trough in Exercise 7 forms the lower half of a circle of radius 2 ft, find the force on one of the ends when the trough is full of water.

9. A swimming pool is 10 m wide and 15 m long. At the shallow end it is 1 m deep, and at the other end it is 4 m deep, and the depth increases linearly from the shallow to the deep end. When the pool is full of water find the force on (a) each of the ends and (b) each of the sides.

10. An oil drum is cylindrical with diameter 3 ft. If it is half full of oil that has density 57.4 lb/cu ft, and it is lying on its side, find the force on each end.

11. Rework Example 5.25 choosing the level of the fluid as the origin for x.

12. Rework Example 5.26 choosing the water level as the origin for x.

B

13. The plate in the figure is parabolic and is submerged as shown in a fluid of density ρ. Find the force on one side of the plate. (*Hint:* If the origin is taken at the vertex of the parabola, its equation is of the form $y = ax^2$. Find a, and use y instead of x as the variable of integration.)

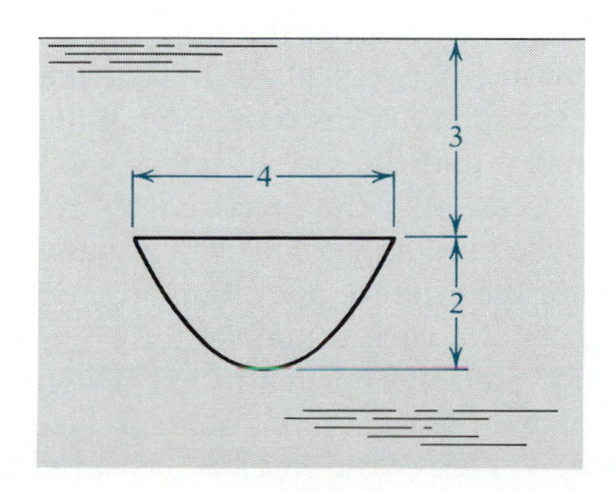

14. Each end of a gasoline tank is elliptical, with equation $x^2 + 4y^2 = 1$ when referred to a coordinate system with origin at the center of the ellipse and y-axis vertical, where x and y are in feet. Find the force on one end of the tank when it is half full. Take the density of gasoline as 45 lb/cu ft. (*Hint:* Use y as the variable of integration.)

15. Show that if a plate is submerged at an angle θ with the vertical, where $0 < \theta < \frac{\pi}{2}$, the force on it is given by

$$F = \int_a^b \rho h(x) w(x) \sec \theta \, dx$$

with $h(x)$ and $w(x)$ having the same meaning as in Definition 5.6, with the x-axis directed vertically downward.

16. Find the force on the bottom of the swimming pool in Exercise 9. (Use the result of Exercise 15.)

17. A porthole in an undersea observation laboratory is circular, with a 1 ft diameter. Assume the porthole lies in a vertical plane. If its center is at a depth of 1200 ft, find the force on the porthole. (The density of seawater is approximately 64.3 lb/cu ft.) (*Hint:* In carrying out the integration, use your knowledge of the area of the circle.)

5.8 PROBABILITY DENSITY FUNCTIONS

In applications of probability theory an investigator is frequently concerned with some characteristic or attribute, such as weight, height, or distance, that can assume any real value on an interval. For example, suppose the weight (in pounds) of a college student chosen at random can be any number between 80 and 300. If the weights of 100 students were taken, we could depict the results by means of a diagram such as that shown in Figure 5.57. In the diagram points stacked vertically indicate the number of students who have approximately the same weight. The *density* of weights is greatest in the 100–200 range and diminishes to the right and left of this interval. As another example, a researcher might be interested in the cholesterol levels in the blood of persons in a certain type of occupation. These values, too, can be any real numbers within some interval, say from 100 to 400.

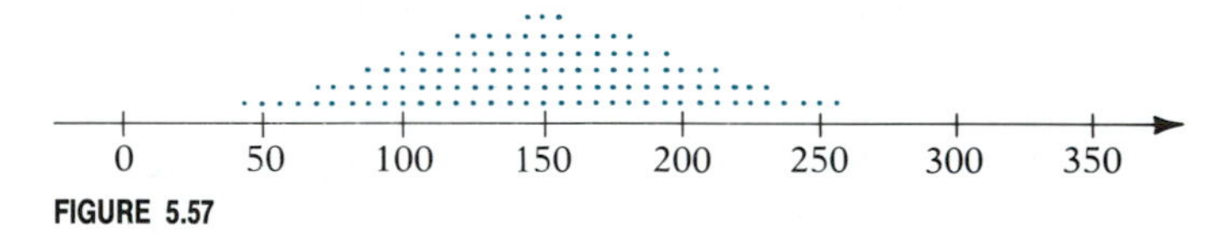

FIGURE 5.57

Weight and cholesterol level are examples of what are called **random variables.** It is often useful to be able to determine in advance the probability that the value of such a random variable will be in some specified interval. For example, we might want to know the probability that a college student chosen at random will weigh between 150 and 200 lb, or that the cholesterol level in a person in the occupation under study will exceed 250. In general, these probabilities cannot be determined with absolute certainty, but they can often be approximated by the use of integration.

Let f be a nonnegative integrable function on an interval I such that the area under the graph of f on I is 1. If a and b are two numbers in I, with $a < b$, then we know that the integral $\int_a^b f(x)\,dx$ gives the area under the graph from a to b, and this must be a number between 0 and 1, since the entire area under f is 1. We can think of the interval I as representing the set of all possible values of a random variable and the graph of f as showing how these values are distributed, in the sense that the *area* under f from a to b gives the probability that a value of the random variable will be between a and b. Figure 5.58 shows a function of the type we are discussing, which might indicate the distribution of weights of college students, for example. As in the diagram of Figure 5.57, this shows a heavy concentration of weights in the 100–200 range, so the probability that the weight of a student selected at random will be in this range is relatively high (close to 1) as measured by

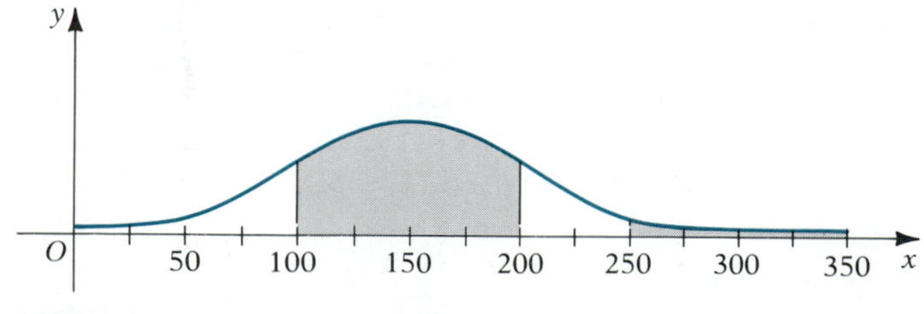

FIGURE 5.58

the area under the graph from 100 to 200. In contrast, the probability that the student weighs between 250 and 300 is low (close to 0).

We formalize this discussion in the following definition. We follow the standard practice of using a capital letter to designate a random variable, with the corresponding lowercase letter designating one of its values.

DEFINITION 5.7 Let f be a nonnegative integrable function on an interval I, such that the area under the graph of f on I is 1. If there exists a random variable X that has I as the set of all of its possible values, such that for any two numbers a and b in I with $a < b$, the probability that X takes on a value in $[a, b]$ is

$$P(a \leq X \leq b) = \int_a^b f(x)\,dx \tag{5.33}$$

then f is called a **probability density function** for X. ■

Note Since omitting one or both end points from the integral in equation (5.33) will not change its value,

$$P(a \leq X \leq B) = P(a < X \leq b) = P(a \leq X < b) = P(a < X < b)$$

In actual practice the probabilities given by equation (5.33) are usually only approximate, and the function f is an idealized mathematical model of the true probability density function for X. Many random variables have an apparent bell-shaped density function, similar to that shown in Figure 5.58. This type of graph is called a **normal curve** and the corresponding function a **normal density function.** We will come back to this type later. The choice of the particular density function to use depends on the nature of the random variable in question. Sometimes it can reasonably be predicted that the values of a random variable will be distributed in a certain way, and in other cases the choice of the model is based on experimental evidence.

A particularly simple probability density function is the **uniform** density, defined by

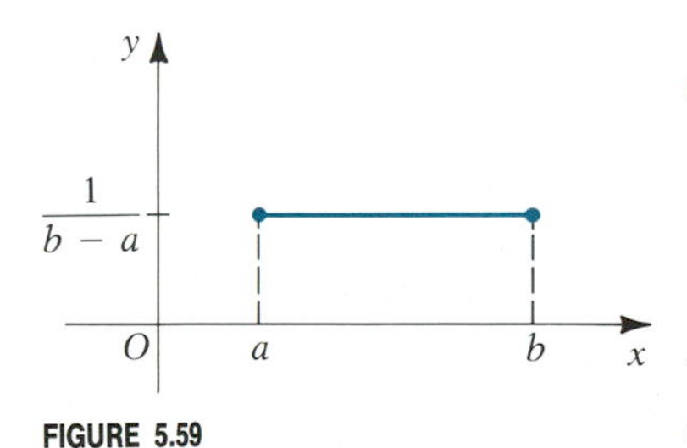

FIGURE 5.59

$$f(x) = \begin{cases} \dfrac{1}{b-a} & \text{if } a \leq x \leq b \\ 0 & \text{otherwise} \end{cases} \tag{5.34}$$

The graph is shown in Figure 5.59. The next example shows one case where such a density function is an appropriate model.

EXAMPLE 5.27 Suppose a bus is known to arrive at a certain bus stop between 9:00 and 9:10 each weekday morning and that from observation over a period of time, it can be assumed that its arrival time during the 10-min interval is uniformly distributed. Find the probability that a person who arrives at 9:00 will have to wait more than 7 min for a bus.

Solution The random variable X of interest in this case is the number of minutes after 9:00 that the bus arrives, and from the given information we know that the values have a uniform distribution on the interval [0, 10], as shown in Figure 5.60. Although we can find the area in this case without integration, we choose to illustrate equation (5.33). We want $P(7 < X \leq 10)$:

$$P = (7 < x \leq 10) = \int_7^{10} f(x)\,dx = \int_7^{10} \tfrac{1}{10}\,dx$$
$$= \tfrac{1}{10}[x]_7^{10} = \tfrac{3}{10}$$

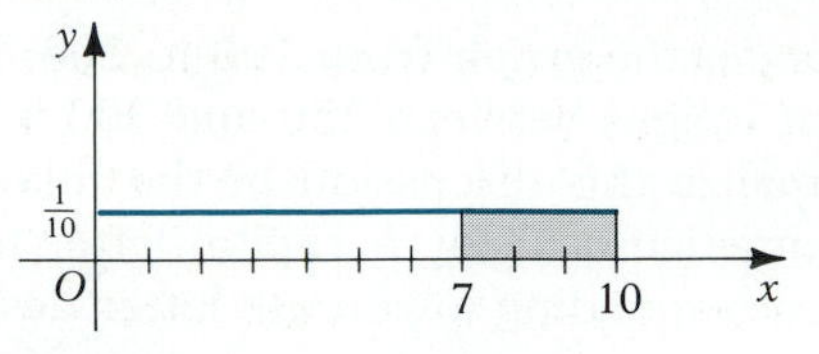

FIGURE 5.60

So if this person arrives at 9:00 each morning, we would expect that on average he or she would have to wait at least 7 min about 30% of the time. ■

The model for the next example is a special case of what is known as a **beta distribution**, with the general form

$$f(x) = Bx^{\alpha-1}(1-x)^{\beta-1}, \qquad 0 \le x \le 1 \tag{5.35}$$

where $\alpha > 0$ and $\beta > 0$, and the constant B is chosen so that the area under the graph is 1.

EXAMPLE 5.28 A certain grass seed mixture is advertised as containing at least 70% Kentucky blue grass, with the remainder composed of red fescue, rye, and other extraneous seeds. As a means of quality control, a sample of seeds is tested periodically, and if the proportion of non-Kentucky blue grass seeds exceeds 30%, the production is halted so that corrective measures can be taken. By experiment the probability density function for the proportion of undesirable seeds is found to be approximated by

$$f(x) = 10(1-x)^9, \qquad 0 \le x \le 1$$

Find the probability that on testing a random sample the production process will be halted.

Solution We define the random variable X to be the proportion of undesirable grass seeds in the sample. From the given information we see that this variable has a beta density function, equation (5.35), with $\alpha = 1$ and $\beta = 10$. (Verify this.) The production will be halted if the value of X exceeds 0.3. So the probability of halting production is

$$\begin{aligned} P(0.3 < X \le 1) &= \int_{0.3}^{1} 10(1-x)^9\, dx \\ &= -10\left[\frac{(1-x)^{10}}{10}\right]_{0.3}^{1} \\ &= -[0-(0.7)^{10}] \approx 0.028 \end{aligned}$$

So there is about a 3% chance of having to halt the production process. ■

Let X denote a random variable and f its probability density function, defined on the interval $I = [a, b]$.* We define the **expected value of X** as follows:

$$E[X] = \int_a^b xf(x)\, dx \tag{5.36}$$

* At this stage we consider only finite intervals. Later we will permit I to be infinite.

The expected value of X is also called the **mean** of the distribution and is denoted by μ; that is,

$$\mu = E[X] \tag{5.37}$$

It can be shown that if values of X are repeatedly sampled, there is a high probability they will be concentrated near μ. In a sense μ is the average value of X.

More generally, if g is any function such that fg is integrable on $[a, b]$, we define the expected value of $g(x)$ by

$$E[g(x)] = \int_a^b g(x)f(x)\,dx \tag{5.38}$$

An expected value of special interest is

$$E[(X-\mu)^2] = \int_a^b (x-\mu)^2 f(x)\,dx$$

and this is called the **variance** of X, denoted by σ^2. So

$$\sigma^2 = E[(X-\mu)^2] \tag{5.39}$$

The square root of the variance, denoted by σ, is called the **standard deviation** of X. Later on we will see the geometric significance of the mean and standard deviation for the normal curve.

EXAMPLE 5.29 A certain random variable X has probability density function $f(x) = kx^2$ on $[0, 2]$. Find the value of k and calculate μ, σ^2, and σ.

Solution The constant k must be chosen so that the total area under the graph of f is 1. Since

$$\int_0^2 kx^2\,dx = k\left[\frac{x^3}{3}\right]_0^2 = \frac{8k}{3}$$

we must have $8k/3 = 1$ or $k = 3/8$. Thus $f(x) = 3x^2/8$.

The mean μ is

$$\begin{aligned}\mu = E[X] &= \int_0^2 xf(x)\,dx = \int_0^2 x\left(\frac{3}{8}x^2\right)dx = \frac{3}{8}\int_0^2 x^3\,dx \\ &= \frac{3}{8}\left[\frac{x^4}{4}\right]_0^2 = \frac{3}{2}\end{aligned}$$

The variance σ^2 is

$$\begin{aligned}\sigma^2 = E[(X-\mu)^2] &= E\left[\left(X-\frac{3}{2}\right)^2\right] = \int_0^2\left(x-\frac{3}{2}\right)^2 f(x)\,dx \\ &= \int_0^2\left(x^2-3x+\frac{9}{4}\right)\cdot\frac{3}{8}x^2\,dx = \frac{3}{8}\int_0^2\left(x^4-3x^3+\frac{9x^2}{4}\right)dx \\ &= \frac{3}{8}\left[\frac{x^5}{5}-\frac{3x^4}{4}+\frac{3x^3}{4}\right]_0^2 = \frac{3}{8}\left[\frac{32}{5}-12+6\right] \\ &= \frac{3}{20}\end{aligned}$$

Thus, the standard deviation σ is

$$\sigma = \sqrt{\frac{3}{20}\cdot\frac{5}{5}} = \frac{\sqrt{15}}{10}$$

■

EXERCISE SET 5.8

A

In Exercises 1–10 find the constant k that will make f a probability density function on the given interval.

1. $f(x) = k(2x + 3)$ on $[0, 4]$
2. $f(x) = k(4 - 2x)$ on $[0, 2]$
3. $f(x) = k(x^2 + x)$ on $[1, 3]$
4. $f(x) = kx(1 - x)$ on $[0, 1]$
5. $f(x) = k/x^2$ on $[1, 4]$
6. $f(x) = k\sqrt{x - 1}$ on $[1, 5]$
7. $f(x) = kx^3(1 - x)^2$ on $[0, 1]$
8. $f(x) = kx/(1 + x^2)^2$ on $[0, 1]$
9. $f(x) = kx/\sqrt{1 + 2x^2}$ on $[0, 2]$
10. $f(x) = kx^2/\sqrt{1 + x^3}$ on $[0, 2]$

In Exercises 11–16 verify that the given function is a probability density function and find the probability specified for a random variable X that has f as its probability density function.

11. $f(x) = \frac{1}{2}x$ on $[0, 2]$; $P(\frac{1}{2} \leq X \leq \frac{3}{2})$
12. $f(x) = \frac{1}{6}(2x + 3)$ on $[-1, 1]$; $P(0 \leq X < 1)$
13. $f(x) = \frac{1}{24}(x^2 + 1)$ on $[-3, 3]$; $P(-1 < X \leq 2)$
14. $f(x) = 12x^2(1 - x)$ on $[0, 1]$; $P(\frac{1}{2} < X < 1)$
15. $f(x) = \frac{3}{125}x\sqrt{25 - x^2}$ on $[0, 5]$; $P(3 \leq X \leq 4)$
16. $f(x) = x/(2\sqrt{x^2 + 9})$ on $[0, 4]$; $P(0 < X \leq 1.25)$

In Exercises 17–22 find the mean and variance of the random variable that has the given probability density function.

17. $f(x) = \frac{2}{9}(x + 1)$ on $[-1, 2]$
18. $f(x) = \frac{1}{6}(3 - 2x)$ on $[-1, 1]$
19. $f(x) = 6x(1 - x)$ on $[0, 1]$
20. $f(x) = \frac{3}{16}\sqrt{x}$ on $[0, 4]$
21. $f(x) = 4x^3$ on $[0, 1]$
22. $f(x) = 3[1 - (x - 3)^2]/4$ on $[2, 4]$
23. Find the mean and variance of a random variable that has the uniform density function of equation (5.34).
24. The number X of parts per million of a certain pollutant emitted from the smokestack of a cement plant is known to range from a low of 1 to a high of 27. Measurements of samples collected over a period of time suggest that an appropriate probability density model for X is $f(x) = cx^{-2/3}$, where $1 \leq x \leq 27$.
    (a) Find the value of c.
    (b) Find the expected value of X.
    (c) Find the standard deviation of X.
25. The income level of persons in a certain city whose incomes exceed a certain amount $x_0 > 0$ is approximately modeled by a **Pareto** distribution:

$$f(x) = \begin{cases} \dfrac{\alpha x_0^\alpha}{x^{\alpha+1}} & \text{if } x \geq x_0 \\ 0 & \text{otherwise} \end{cases}$$

where α is a positive constant. If x is measured in thousands of dollars, $\alpha = 2$, and $x_0 = 10$, find the probability that a person with an income that exceeds \$10,000 makes less than \$20,000.

26. A **truncated Pareto** distribution (see Exercise 25) is of the form

$$f(x) = \begin{cases} c\left(\dfrac{x_0}{x}\right)^{\alpha+1} & \text{if } x_0 \leq x \leq x_1 \\ 0 & \text{otherwise} \end{cases}$$

where c and α are positive constants. Find the constant c in terms of x_0, x_1, and α so that f will be a probability density function.

27. The truncated Pareto function of Exercise 26 is sometimes used to model the distribution of the size of oil fields in a certain region. Suppose that x is measured in hundreds of millions of barrels, $\alpha = 2$, $x_0 = 2$, and $x_1 = 10$ for a certain region.
    (a) Find the probability that a randomly selected oil field in that region would produce more than 500 million barrels of oil.
    (b) Find the average (mean) number of barrels for oil fields in that region.
28. The proportion X of impurities in a product resulting from a certain chemical process has the beta distribution of equation (5.35) with $\alpha = 3$ and $\beta = 2$.
    (a) Find the constant B.
    (b) Find the expected value of X.
29. The lifetime, in hundreds of hours, of a certain type of lightbulb has been found empirically to have a probability density function approximated by

$$f(x) = \frac{5\sqrt{17}}{12(1 + x^2)^{3/2}}, \qquad 0 \leq x \leq 12$$

Find the expected lifetime of a bulb of this type chosen at random.

B

30. Let X be a random variable with probability density function f on $[a, b]$. Prove the following properties of expected value:
(a) $E[c] = c$, where c is a constant
(b) $E[cg(x)] = cE[g(x)]$, where c is a constant
(c) $E[g_1(x) + g_2(x)] = E[g_1(x)] + E[g_2(x)]$

31. Prove that $\sigma^2 = E[X^2] - \mu^2$. (*Hint:* Use the results of Exercise 30, along with the definitions of μ and σ^2.)

32. Use the result of Exercise 31 to find σ^2 for a random variable with the probability density function given by

$$f(x) = \frac{3}{20}\left(\frac{x+1}{\sqrt{x}}\right), \qquad 1 \le x \le 4$$

33. The random variable X with probability density function

$$f(x) = \frac{8}{\pi + 2}\frac{1}{(1 + x^2)^2} \quad \text{on } [0, 1]$$

is known to have variance

$$\sigma^2 = \frac{\pi^2 - 8}{(\pi + 2)^2}$$

Use this together with the result of Exercise 31 to evaluate the integral

$$\int_0^1 \frac{x^2\,dx}{(1 + x^2)^2}$$

34. Show that if X is a random variable with finite mean and t can assume any real value, then $E[(X - t)^2]$ takes on its minimum value when $t = \mu$.

5.9 SUPPLEMENTARY EXERCISES

In Exercises 1–10 find the area of the region indicated. Make a sketch of the region.

1. Under the graph of

$$f(x) = \begin{cases} 1 + \sqrt{2x} & \text{if } 0 \le x < 2 \\ x^2 - 4 & \text{if } 2 \le x \le 4 \end{cases} \quad \text{on } [0, 4]$$

2. Under the graph of

$$f(x) = \begin{cases} 1 - (x + 2)^{-2} & \text{if } -1 \le x < 0 \\ 0 & \text{if } x = 0 \\ 4 - 2x & \text{if } 0 < x \le 2 \end{cases} \quad \text{on } [-1, 2]$$

3. Bounded by $y = 3x + 4$, $y = 2 - x$, $x = 0$, and $x = 3$
4. Between $y = x^3 - x$ and $y = 1 - x^2$
5. Between $y = x^2 - 2x$ and $y = 4\sqrt{x}$
6. Bounded by $y = \sqrt{x + 5}$, $4y = 3x$, and $y = 0$
7. Between $y = x^2 - x$ and $y = \sin \pi x$
8. Between $y^2 = x + 2$ and $y^2 + x - 2y - 2 = 0$
9. One of the regions bounded above by $y = \sqrt{3}\cos x$ and below by $y = \sin x$
10. Bounded by $y = 1/x^2$, $y = 1$, $2x - 4y = 3$, and $x = 0$

In Exercises 11–20 find the volume obtained by revolving the region about the specified axis. Sketch the region.

11. Bounded by $y = \frac{1}{x}$, $x = 1$, $x = 2$, and the x-axis; x-axis
12. Bounded by $y = \sqrt{2x + 3}$, $x = -1$, $x = 3$, and the x-axis; x-axis
13. Bounded by $x = 8 - 2y$, $x = 2$, $x = 6$, and the x-axis; y-axis
14. Bounded by $y = x$, $y = 2x$, $x = 1$, and $x = 4$; y-axis
15. Between $y = x^2$ and $y = x + 6$; x-axis
16. Between $y = x^2 + 1$ and $y = 3x - x^2$; y-axis
17. Bounded by $y = \sqrt{2x + 1}$, $y = x - 1$, and $y = 1$; x-axis
18. Between $x + \sqrt{2y} = 0$ and $x = y - y^2$; x-axis
19. Region of Exercise 15 revolved about $x + 2 = 0$
20. Region of Exercise 18 revolved about $x = 1$
21. A wedge has as a base the region enclosed by the ellipse $x^2 + 2y^2 = 4$, and each cross section perpendicular to the x-axis is an isosceles right triangle with right angle on the upper half of the ellipse. Find the volume of the wedge.
22. In the accompanying figure ABC and DEF are equilateral triangles of sides 4 and 1, respectively, in planes perpendicular to edge AD, which is of length 6, as shown. Find the volume of the solid.

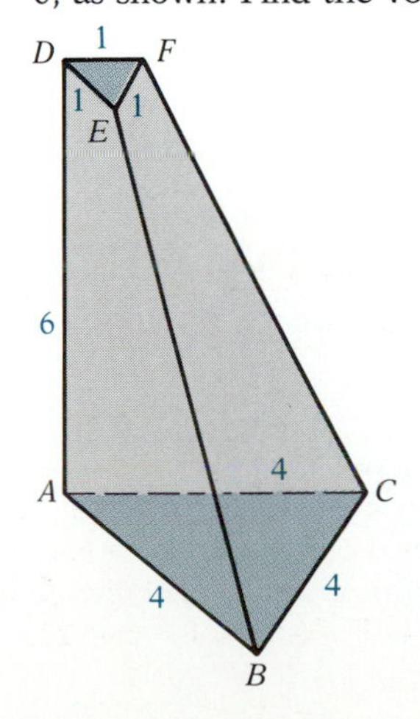

In Exercises 23–25 find the arc length of the curve on the specified interval.

23. $x = 2 + (y - 1)^{3/2}$ from $(2, 1)$ to $(10, 5)$

24. $y = \frac{1}{20}x^5 + \frac{1}{3}x^{-3}$ from $x = \frac{1}{2}$ to $x = 2$

25. $x = y^{3/2} - \frac{1}{3}y^{1/2}$ from $(\frac{2}{3}, 1)$ to $(\frac{14}{3}, 4)$

In Exercises 26–29 find the area of the surface formed by revolving the given arc about the specified axis.

26. $y = 2\sqrt{3x - 2}$ from $x = 3$ to $x = 6$, about x-axis

27. $x^2 - 3y = 3$ from $(0, -1)$ to $(2, \frac{1}{3})$, about y-axis

28. $x = \left(\dfrac{y + 5}{4}\right)^3$ from $(1, -1)$ to $(8, 3)$, about y-axis

29. The arc of Exercise 25 about the (a) x-axis and (b) y-axis.

30. A watering trough 20 ft long has ends in the shape of inverted isosceles trapezoids 4 ft across the top. 2 ft across the bottom, and 3 ft high. If the trough has water in it to a depth of 2 ft, find the work required to pump all of the water over the top of the trough.

31. A ship's anchor, weighing 1000 lb, is being raised. The anchor chain weighs 10 lb/ft, and it is being hoisted over a pulley that is 40 ft above the water. How much work is required to raise the anchor from the water level to a position 10 ft below the pulley?

32. A force of 12 lb is required to stretch a certain spring from its natural length of 20 in. to a length of 24 in. Find the work required to (a) stretch it from a length of 24 in. to a length of 30 in. and (b) compress it 4 in. from its natural length.

33. The ends of a long trough are parabolic segments with dimensions as shown in the figure (in feet). Find the total force on one end of the trough when it (a) is full of water and (b) has water in it to a depth of 1 ft.

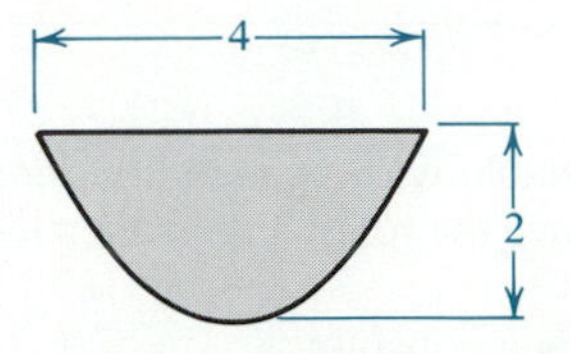

34. A plate as shown in the figure is submerged vertically in water so that the top edge of the plate is at a depth of 9 ft. Find the total force on one side of the plate.

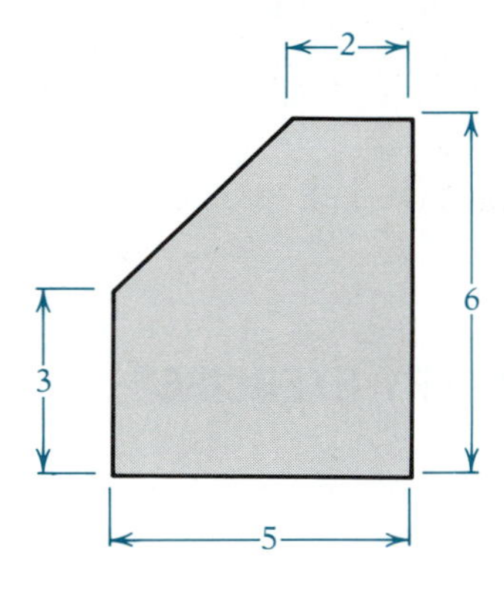

35. The probability density function for the random variable X, denoting the time for students in a certain class to complete a 3-hr final exam, is of the form $f(x) = \frac{x}{9}(1 + kx)$, where $0 \leq x \leq 3$. Find the following:
(a) The constant k (b) The mean μ
(c) The variance σ^2 (d) $P(X < 1)$
(e) $P(2 \leq X \leq 3)$

36. The proportion X of defective items in a large batch has a probability density function that is approximated by a beta function with $\alpha = 1$ and $\beta = 8$. Find (a) the constant B and (b) the probability that more than 20% are defective.

CHAPTER 6

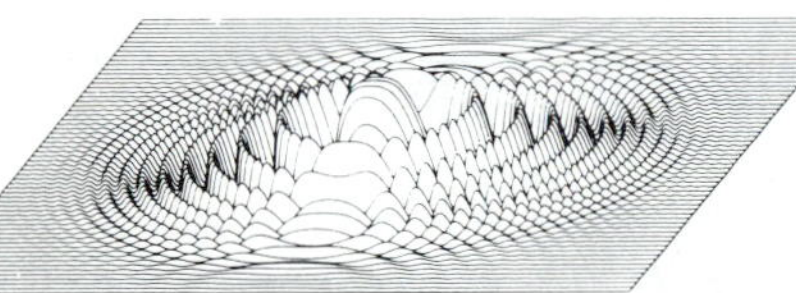

LOGARITHMIC AND EXPONENTIAL FUNCTIONS

6.1 INTRODUCTION

In your study of precalculus you may have learned a definition of the **logarithm** of a number something like the following:

$$\log_a x = y \quad \text{if and only if} \quad a^y = x \tag{6.1}$$

where $x > 0$, $a > 0$, and $a \neq 1$. The expression $\log_a x$ is read "the logarithm to the base a of x." The usefulness of logarithms stems in large measure from the following fundamental properties:

$$\left.\begin{aligned} \log_a xy &= \log_a x + \log_a y \\ \log_a \tfrac{x}{y} &= \log_a x - \log_a y \\ \log_a x^p &= p \log_a x \end{aligned}\right\} \tag{6.2}$$

To illustrate definition (6.1) we can say, for example, that $\log_2 8 = 3$, since $2^3 = 8$. However, to find y such that $\log_2 7 = y$ would require y to satisfy $2^y = 7$, and it is not at all clear how to go about finding y in this case, nor is it entirely clear that such a y exists. A problem also arises when one attempts to prove the properties stated in equations (6.2), especially the third one, when p is an irrational number.

The problem lies in the fact that up to now our knowledge of exponents has been limited to those that are rational. What, for example, is meant by $2^{\sqrt{2}}$ or, even worse, $\pi^{\sqrt{2}}$? Although it is possible to provide satisfactory answers based on the consideration of limits of exponential expressions involving rational exponents, it turns out to be simpler to arrive at such answers by first studying logarithms from a new point of view, based on calculus.

What we propose to do, then, is to begin with a new function, which we will call a logarithmic function, and then to define an exponential function based on the logarithm. We will eventually see that equations (6.1) and (6.2) are true, but an advantage of our approach is that we will have found a natural way to give meaning to a^y where y is irrational. This approach also provides an excellent opportunity to apply some of the fundamental principles of calculus that we have studied.

6.2 THE NATURAL LOGARITHM FUNCTION

Recall that if f is a continuous function on $[a, b]$, then the function F, defined by $F(x) = \int_a^x f(t)\,dt$ exists for all x in $[a, b]$, and its derivative (see Theorem 4.6) is $F'(x) = f(x)$. Since F' exists, it follows that F is continuous on $[a, b]$.

As a particular application we let $f(t) = \frac{1}{t}$ and $a = 1$, and we name the resulting function L (suggesting "logarithm"); that is, we let

$$L(x) = \int_1^x \frac{1}{t}\,dt, \qquad x > 0 \tag{6.3}$$

When $x > 1$, $L(x)$ can be interpreted geometrically as the area under the graph of $f(t) = \frac{1}{t}$ from 1 to x. If $0 < x < 1$, we have

$$L(x) = \int_1^x \frac{1}{t}\,dt = -\int_x^1 \frac{1}{t}\,dt$$

and this is the *negative* of the area under the graph of f from x to 1. Figure 6.1 shows these interpretations. We cannot allow x to be negative or 0, since $f(t) = \frac{1}{t}$ is unbounded in a neighborhood of 0 and the integral does not exist. Thus, L is defined for $x > 0$ only. Note also that when $x = 1$, $L(x) = 0$, since

$$L(1) = \int_1^1 \frac{1}{t}\,dt = 0$$

Since $L'(x) = \frac{1}{x}$, we can use the chain rule to get

$$\frac{d}{dx}L(g(x)) = L'(g(x)) \cdot g'(x) = \frac{1}{g(x)}\,g'(x) \tag{6.4}$$

FIGURE 6.1

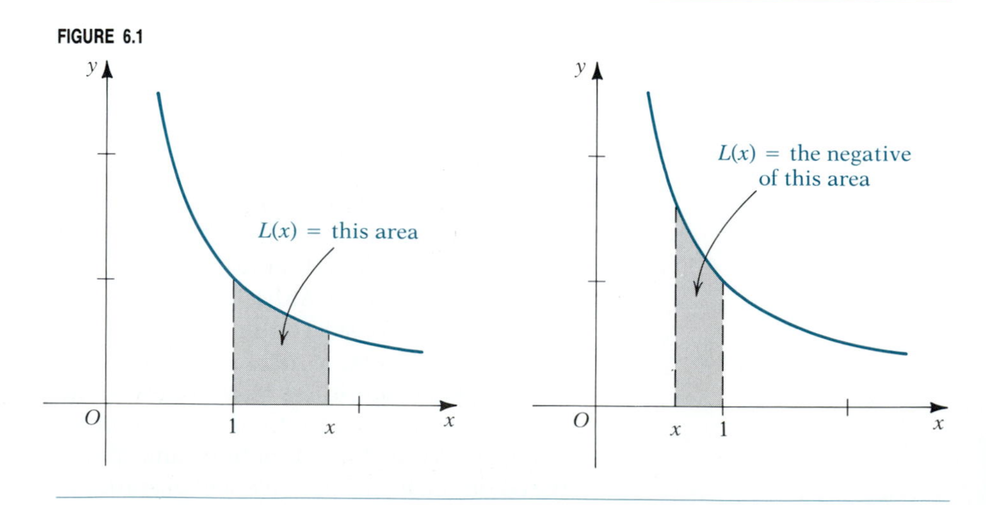

where $g(x) > 0$ and g is a differentiable function. Taking $g(x) = ax$, where $a > 0$, equation (6.4) gives

$$\frac{d}{dx}L(ax) = \frac{1}{ax} \cdot a = \frac{1}{x} = L'(x)$$

The fact that the derivatives $\frac{d}{dx}L(ax)$ and $L'(x)$ are equal implies that $L(ax)$ and $L(x)$ differ by a constant (Corollary 3.9); that is,

$$L(ax) - L(x) = C$$

Putting $x = 1$, and using the fact that $L(1) = 0$ we get $L(a) = C$. So

$$L(ax) - L(x) = L(a)$$

If we let $x = b$, where b is any positive number, we have, on rearranging,

$$L(ab) = L(a) + L(b) \tag{6.5}$$

Thus, L has the interesting property that it changes a product into a sum. (Does this sound familiar?)

We can use equation (6.5) also to find $L(\frac{a}{b})$ where $a > 0$ and $b > 0$ as follows:

$$L(a) = L\left(b \cdot \frac{a}{b}\right) = L(b) + L\left(\frac{a}{b}\right)$$

Thus,

$$L\left(\frac{a}{b}\right) = L(a) - L(b) \tag{6.6}$$

So L changes division into subtraction.

Let us apply equation (6.4) again with $g(x) = x^r$, where r is rational. We obtain

$$\frac{d}{dx} L(x^r) = \frac{1}{x^r} \cdot rx^{r-1} = r \cdot \frac{1}{x} = \frac{d}{dx} [rL(x)]$$

The equality of the derivatives implies again that the functions differ by a constant: $L(x^r) - rL(x) = C$, and on putting $x = 1$, we get $L(1^r) - rL(1) = C$, so $C = 0$; that is,

$$L(x^r) = rL(x), \qquad r \text{ rational} \tag{6.7}$$

So L changes exponentiation (raising to a power) to multiplication, so long as the exponent is rational.

We can get further insight into the nature of L by considering its graph. We know that for $x > 0$,

$$L'(x) = \frac{1}{x} > 0$$

Thus, L is an increasing function on $(0, +\infty)$. Also

$$L''(x) = -\frac{1}{x^2} < 0$$

so that L is always concave downward. Next we consider the limits $\lim_{x \to +\infty} L(x)$ and $\lim_{x \to 0^+} L(x)$. Since by equation (6.7) $L(2^n) = nL(2)$ and $L(2) > 0$, if we are given any positive number M, however large it may be, we can always find a positive integer n such that $L(2^n) > M$. Thus, if $x > 2^n$,

$$L(x) > L(2^n) > M$$

since L is an increasing function. So by choosing x large enough, we can make $L(x)$ arbitrarily large. This shows that

$$\lim_{x \to +\infty} L(x) = +\infty$$

Similarly, since $L(2^{-n}) = -nL(2)$, we can find an integer n such that $L(2^{-n}) < -M$. Thus, if $0 < x < 2^{-n}$,

$$L(x) < L(2^{-n}) < -M$$

This shows that

$$\lim_{x \to 0^+} L(x) = -\infty$$

so that the graph of L is asymptotic to the negative y-axis. The information we have obtained enables us to sketch the graph of L shown in Figure 6.2.

Note that as x increases, the rate at which $L(x)$ rises decreases. However, the graph does not become asymptotic to any horizontal line. As we have shown, in fact, $L(x)$ becomes arbitrarily large. Since L is continuous and it takes on values that are arbitrarily large (for x sufficiently large) and also values that are arbitrarily small (for x close enough to 0), it follows by the intermediate value theorem (Theorem 1.7) that L takes on all real values. Thus, the range of L is $(-\infty, +\infty)$.

We now give a new name to the function L we have been studying. It is called the **natural logarithm function,** and the value $L(x)$ is called the **natural logarithm of x,** designated *ln x*. The relationship of $\ln x$ to the logarithms you studied in precalculus mathematics will soon become clear. We now summarize what we know about the natural logarithm function.

FIGURE 6.2

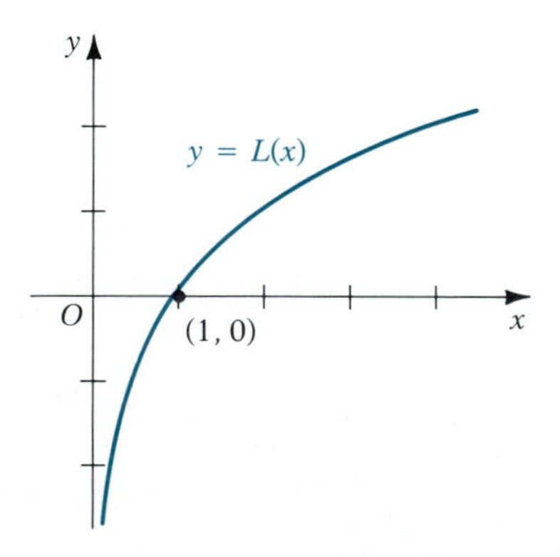

PROPERTIES OF THE NATURAL LOGARITHM FUNCTION

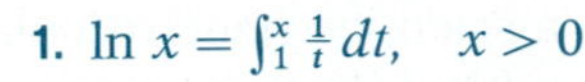

1. $\ln x = \int_1^x \frac{1}{t}\,dt, \quad x > 0$
2. Domain $= (0, +\infty)$. Range $= (-\infty, +\infty)$.
3. The natural logarithm function is differentiable, and hence continuous, on its domain, and

$$\frac{d}{dx}\ln x = \frac{1}{x}$$

4. The natural logarithm function is increasing, and its graph is concave downward.
5. $\ln x > 0$ when $x > 1$
   $\ln x < 0$ when $0 < x < 1$
   $\ln 1 = 0$
6. $\lim_{x\to+\infty} \ln x = +\infty$ and $\lim_{x\to 0^+} \ln x = -\infty$
7. For $a > 0$, $b > 0$, and r rational,
   (a) $\ln(ab) = \ln a + \ln b$
   (b) $\ln \frac{a}{b} = \ln a - \ln b$
   (c) $\ln a^r = r \ln a$

Although we know a good bit about the natural logarithm function, we do not as yet know any of its values except ln 1. We can estimate other values of $\ln x$ by Riemann sums based on the integral that defines $\ln x$. For example, to estimate ln 2, we use a Riemann sum to approximate $\int_1^2 \frac{1}{t}\,dt$. If we use a regular partition with $n = 10$, for example, and take the upper sum (see Figure 6.3), we get

$$\ln 2 \approx \sum_{k=1}^{n} \frac{0.1}{1 + (k-1)(0.1)} = 0.1\left(\frac{1}{1} + \frac{1}{1.1} + \frac{1}{1.2} + \cdots + \frac{1}{1.9}\right) \approx 0.72$$

(Verify this.) In the exercises you will be asked to show similarly that $\ln 3 \approx 1.1$ by using lower sums. Later, we will have occasion to use the fact that $\ln 2 < 1$ and $\ln 3 > 1$. Most values of $\ln x$ are irrational numbers and can therefore only be approximated. Later we will see better ways to obtain

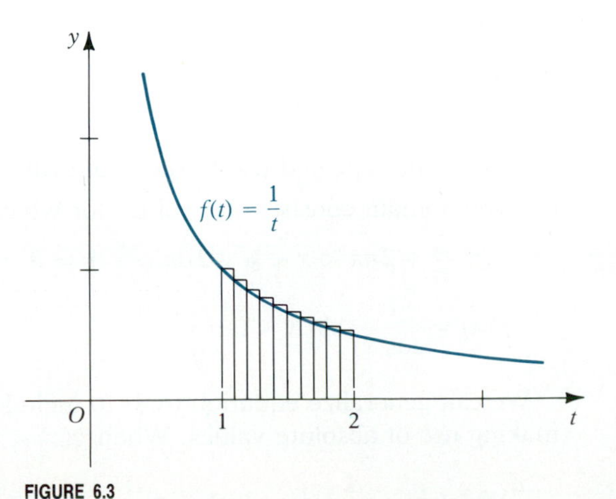

FIGURE 6.3

such approximations. Fortunately, tables have been worked out (such as Table 1 in the back of this book). Even better, highly accurate values can be obtained from a scientific hand calculator.

Equation (6.4) with $L(x)$ replaced by $\ln x$ states that for any positive differentiable function g,

$$\frac{d}{dx}\ln(g(x)) = \frac{1}{g(x)}\,g'(x)$$

Writing $u = g(x)$, we can write in the alternate notation

$$\frac{d}{dx}\ln u = \frac{1}{u}\frac{du}{dx}, \qquad u > 0 \tag{6.8}$$

EXAMPLE 6.1 State the domain of each of the following and find $f'(x)$.

(a) $f(x) = (\ln x\sqrt{1 + x^2})$ (b) $f(x) = (\ln x)/x$

(c) $f(x) = (\ln\sqrt{2x - 1})^3$ (d) $f(x) = \ln\cos x$

Solution (a) The domain consists of x values for which $x\sqrt{1 + x^2} > 0$—namely, $\{x : x > 0\}$. To find $f'(x)$ we first simplify, using Property 7:

$$\begin{aligned} f(x) = \ln(x\sqrt{1 + x^2}) &= \ln x + \ln(1 + x^2)^{1/2} \quad \text{(Property 7a)} \\ &= \ln x + \tfrac{1}{2}\ln(1 + x^2) \quad \text{(Property 7c)} \end{aligned}$$

So by equation (6.8),

$$f'(x) = \frac{1}{x} + \frac{1}{2}\cdot\frac{1}{1 + x^2}\cdot 2x = \frac{1}{x} + \frac{x}{1 + x^2}$$

(b) The domain is the set of all $x > 0$. By the quotient rule,

$$f'(x) = \frac{x\cdot\dfrac{1}{x} - \ln x}{x^2} = \frac{1 - \ln x}{x^2}$$

(c) The domain is the set of all $x > \frac{1}{2}$. By Property 7c we can write

$$f(x) = [\tfrac{1}{2}\ln(2x - 1)]^3 = \tfrac{1}{8}[\ln(2x - 1)]^3$$

So

$$f'(x) = \frac{1}{8}\cdot 3[\ln(2x - 1)]^2\cdot\frac{1}{2x - 1}\cdot 2 = \frac{3[\ln(2x - 1)]^2}{4(2x - 1)}$$

(Note the repeated use of the chain rule.)

(d) The domain consists of x values for which $\cos x > 0$—that is,

$$\{x : -\tfrac{\pi}{2} + 2n\pi < x < \tfrac{\pi}{2} + 2n\pi, \qquad n = 0, \pm 1, \pm 2, \ldots\}$$

$$f'(x) = \frac{1}{\cos x}\cdot\sin x = \tan x$$

■

We can generalize equation (6.8) to include cases where $u = g(x) < 0$ by making use of absolute values. When $g(x) < 0$, $|g(x)| = -g(x)$. So we have

$$\frac{d}{dx}\ln|g(x)| = \frac{d}{dx}\ln[-g(x)] = \frac{1}{-g(x)}\cdot -g'(x) = \frac{g'(x)}{g(x)}$$

In terms of u, this result is

$$\frac{d}{dx}\ln|u| = \frac{1}{u}\frac{du}{dx}, \qquad u \neq 0 \tag{6.9}$$

When $u > 0$, equation (6.9) is just a repetition of equation (6.8). So equation (6.9) holds true if $u > 0$ or $u < 0$. The next example shows one use of this result.

EXAMPLE 6.2 Find $f'(x)$ if

$$f(x) = \ln\frac{x(x-1)}{x+2}$$

Solution First we find the domain of f. A sign graph for

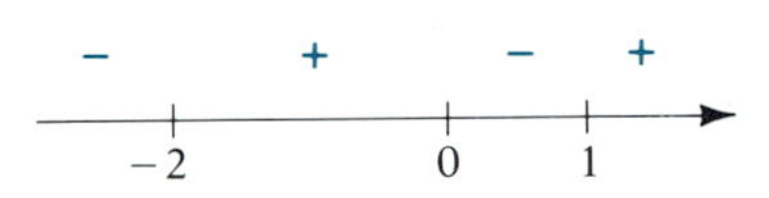

$$\frac{x(x-1)}{x+2}$$

shows where it is positive, so that the logarithm is defined.

The domain of f is therefore $(-2, 0) \cup (1, \infty)$. For x in this domain, we have

$$\begin{aligned} f(x) &= \ln\frac{x(x-1)}{x+2} = \ln\left|\frac{x(x-1)}{x+2}\right| = \ln\frac{|x|\,|x-1|}{|x+2|} \\ &= \ln|x| + \ln|x-1| - \ln|x+2| \quad \text{(Properties 7a and 7b)} \end{aligned}$$

Thus, by equation (6.9),

$$f'(x) = \frac{1}{x} + \frac{1}{x-1} - \frac{1}{x+2}$$

Observe that if we had simply written

$$f(x) = \ln\frac{x(x-1)}{x+2} = \ln x + \ln(x-1) - \ln(x+2)$$

we would be limiting the domain to the interval $(1, \infty)$ where each individual factor is positive. Nevertheless, the value of $f'(x)$ obtained from this is the same. ■

The next example shows how the calculation of derivatives of complicated expressions involving products, quotients, and powers can sometimes be simplified by first finding the logarithm of the expression. The technique illustrated is called **logarithmic differentiation.**

EXAMPLE 6.3 Find y' if

$$y = \frac{(x-1)^3}{(2x+3)\sqrt{1-2x}}$$

Solution We first find $\ln|y|$ and simplify:

$$\ln|y| = \ln\frac{|x-1|^3}{|2x+3|\,|1-2x|^{1/2}} = 3\ln|x-1| - \left[\ln|2x+3| + \frac{1}{2}\ln|1-2x|\right]$$

Then, by equation (6.9),

$$\frac{1}{y} \cdot y' = \frac{3}{x-1} - \left[\frac{1}{2x+3} \cdot 2 + \frac{1}{2} \cdot \frac{1}{1-2x}(-2)\right]$$
$$= \frac{3}{x-1} - \frac{2}{2x+3} + \frac{1}{1-2x}$$

Now we can get y' by multiplying both sides by y:

$$y' = \frac{(x-1)^3}{(2x+3)\sqrt{1-2x}}\left[\frac{3}{x-1} - \frac{2}{2x+3} + \frac{1}{1-2x}\right]$$

and this is valid so long as $y \neq 0$. Although the answer is rather complicated, it was not hard to obtain. You may wish to compare this method with the direct method using the product and quotient rules. ■

Remark Note that, like our observation in Example 6.2, we will always arrive at the correct answer in logarithmic differentiation by ignoring absolute values. This is because the derivatives of $\ln|y|$ and $\ln y$ are the same when both are defined. Strictly speaking, the absolute values should be used to account for values of x for which $y < 0$, but the answer will be valid for all $y \neq 0$ even if in the derivation y is assumed to be positive.

Each time we find a new derivative formula we also have a new antiderivative formula. In particular, from equation (6.9) we can deduce

$$\int \frac{1}{u}\,du = \ln|u| + C, \qquad u \neq 0 \tag{6.10}$$

You should observe that equation (6.10) fills a gap in the power rule for antiderivatives. Recall that in the formula

$$\int u^r\,du = \frac{u^{r+1}}{r+1} + C$$

we had to restrict r to be different from -1. Now we know the answer for $r = -1$, as given by equation (6.10).

EXAMPLE 6.4 Evaluate the indefinite integrals:

(a) $\int \frac{x}{x^2+1}\,dx$ (b) $\int \frac{\cos x\,dx}{1+2\sin x}$

(c) $\int \tan 2x\,dx$

Solution (a) Put $u = x^2 + 1$; then $du = 2x\,dx$. So we have

$$\int \frac{x}{x^2+1}\,dx = \frac{1}{2}\int \frac{2x\,dx}{x^2+1} = \frac{1}{2}\int \frac{du}{u} = \frac{1}{2}\ln|u| + C$$
$$= \frac{1}{2}\ln|x^2+1| + C$$
$$= \frac{1}{2}\ln(x^2+1) + C \quad \text{Since } x^2+1>0$$

We could also write the answer as $\ln\sqrt{x^2+1} + C$.

(b) $\displaystyle\int \frac{\cos x}{1 + 2\sin x}\,dx = \frac{1}{2}\int \frac{2\cos x}{1 + 2\sin x}\,dx = \frac{1}{2}\ln|1 + 2\sin x| + C$

(c) $$\int \tan 2x\,dx = \int \frac{\sin 2x}{\cos 2x}\,dx = -\frac{1}{2}\int \frac{-2\sin 2x\,dx}{\cos 2x}$$
$$= -\frac{1}{2}\ln|\cos 2x| + C$$ ■

EXAMPLE 6.5 Find a formula for the general antiderivative of $\sec x$.

Solution This involves a trick. We multiply numerator and denominator by $\sec x + \tan x$, getting

$$\int \sec\,dx = \int \frac{\sec^2 x + \sec x \tan x}{\sec x + \tan x}\,dx$$

The numerator is the derivative of the denominator. So we conclude that

$$\int \sec x\,dx = \ln|\sec x + \tan x| + C$$

We can now add this result to our basic list of trigonometric integrals. The integral of $\csc x$ can be similarly obtained. ■

We conclude this section with an example showing how to graph a function that involves a natural logarithm.

EXAMPLE 6.6 Discuss and sketch the graph of $y = \ln|x^2 - 1|$.

Solution Since $|x^2 - 1| > 0$ except when $x^2 = 1$, the domain consists of all real numbers except $x = \pm 1$.

(a) *Intercepts.* Setting $x = 0$, we obtain $y = \ln|-1| = \ln 1 = 0$. Setting $y = 0$, we obtain $\ln|x^2 - 1| = 0$, which implies $|x^2 - 1| = 1$. (Why?) So $x^2 - 1 = \pm 1$. If $x^2 - 1 = 1$, $x^2 = 2$ and $x = \pm\sqrt{2}$. If $x^2 - 1 = -1$, $x^2 = 0$, so that $x = 0$. Thus, we have

x-intercepts: $x = 0, \pm\sqrt{2}$

y-intercepts: $y = 0$

(b) *Symmetry.* x can be replaced by $-x$ to give the same value of y, so the graph is symmetric with respect to the y-axis.

(c) *Maxima and minima.* By equation (6.9),

$$y' = \frac{1}{x^2 - 1}\cdot 2x = \frac{2x}{x^2 - 1}$$

The only critical value is therefore $x = 0$ (since $x = \pm 1$ are outside the domain). The accompanying sign graph for y' shows that $(0, 0)$ is a local maximum point. It also shows that the graph is decreasing on $(-\infty, -1)$ and $(0, 1)$ and is increasing on $(-1, 0)$ and $(1, +\infty)$.

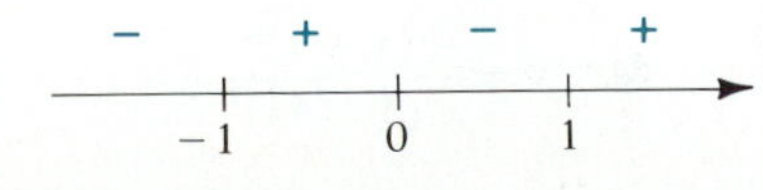

(d) *Concavity and inflection points.*

$$y'' = \frac{(x^2 - 1)\cdot 2 - 2x(2x)}{(x^2 - 1)^2} = \frac{-2(x^2 + 1)}{(1 - x^2)^2}$$

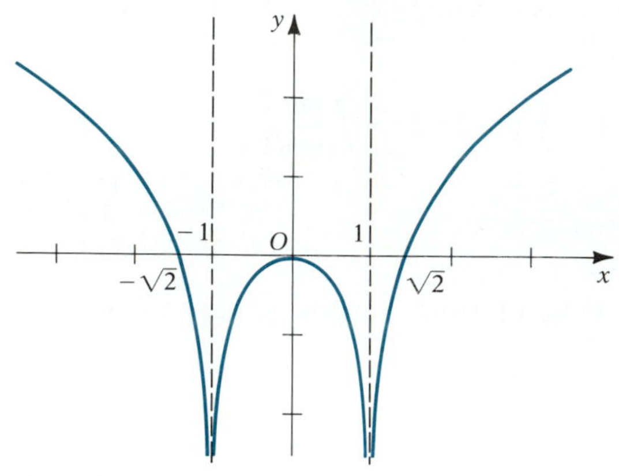

FIGURE 6.4

For all $x \neq \pm 1$, $y'' < 0$. So the graph is concave downward at all points of the domain. There are no inflection points.

(e) *Asymptotes.* As $x \to 1$, $|x^2 - 1| \to 0^+$, so that $\ln|x^2 - 1| \to -\infty$. Thus $x = 1$ is a vertical asymptote and, by symmetry, so is $x = -1$. As $x \to \pm\infty$, $|x^2 - 1| \to +\infty$, so that $\ln|x^2 - 1| \to +\infty$. Thus, there is no horizontal asymptote.

The graph is shown in Figure 6.4 ■

EXERCISE SET 6.2

A

In Exercises 1–20 give the domain of f and find $f'(x)$.

1. $f(x) = \ln \dfrac{1}{x}$
2. $f(x) = \ln \sqrt{x}$
3. $f(x) = \ln x^3$
4. $f(x) = (\ln x)^3$
5. $f(x) = \ln(x\sqrt{1 - x})$
6. $f(x) = \ln\left(\dfrac{x^2}{x - 1}\right)$
7. $f(x) = x \ln x$
8. $f(x) = \ln x^2$
9. $f(x) = \ln \sqrt{\dfrac{1 - x}{1 + x}}$
10. $f(x) = \sqrt{\ln x}$
11. $f(x) = \ln(\ln x)$
12. $f(x) = \ln \sin x$
13. $f(x) = \ln \sec x$
14. $f(x) = \sin(\ln x)$
15. $f(x) = \dfrac{1 - \ln x}{x}$
16. $f(x) = \ln\left(\dfrac{x - 2}{x + 1}\right)$
17. $f(x) = \ln \dfrac{\sqrt{x^2 - 1}}{2 - x}$
18. $f(x) = \ln \sqrt[3]{2x - x^2}$
19. $f(x) = \ln\left(\dfrac{x + 3}{x^2 - 2x}\right)$
20. $f(x) = \ln\left(\dfrac{x\sqrt{1 - x}}{1 + x}\right)$

In Exercises 21–30 find the antiderivatives.

21. $\displaystyle\int \frac{2x\,dx}{x^2 + 4}$
22. $\displaystyle\int \frac{\sin x\,dx}{1 - \cos x}$
23. $\displaystyle\int \frac{x\,dx}{2 - 3x^2}$
24. $\displaystyle\int \frac{x^2\,dx}{x^3 + 1}$
25. $\displaystyle\int \frac{(x - 1)\,dx}{x^2 - 2x + 3}$
26. $\displaystyle\int \cot x\,dx$
27. $\displaystyle\int \frac{dx}{\sqrt{x}(\sqrt{x} + 1)}$
28. $\displaystyle\int \frac{dx}{x \ln x}$
29. $\displaystyle\int \frac{x \cos x}{x \sin x + \cos x}\,dx$
30. $\displaystyle\int \frac{\sqrt{x} - 1}{x(2\sqrt{x} - 3)}\,dx$

In Exercises 31–34 use logarithmic differentiation to find y'.

31. $y = \dfrac{(x - 1)^2\sqrt{3x + 4}}{x + 2}$
32. $y = \dfrac{x\sqrt{1 - x}}{(x + 2)(x - 3)}$
33. $y = \sqrt{\dfrac{(x - 1)(x + 2)}{(x - 3)^3}}$
34. $y = \dfrac{\sqrt[3]{(x + 1)^2}}{x^4(3 - 4x)^2}$

In Exercises 35–40 find all local maxima and minima.

35. $y = x - \ln x$ **36.** $y = \ln\sqrt{x} - x^2$

37. $y = x^3 - \ln x^3$ **38.** $y = x + \ln\dfrac{x}{2x-1}$

39. $y = \ln\dfrac{\sqrt{1-x}}{3-2x}$ **40.** $y = \ln\dfrac{x(2-x)}{(x-3)^2}$

In Exercises 41–48 discuss domain, intercepts, symmetry, asymptotes, maxima and minima, concavity, and points of inflection and sketch the graph.

41. $y = \ln 2x$ **42.** $y = \ln x^2$

43. $y = \ln\frac{1}{x}$ **44.** $y = \ln(1 - x^2)$

45. $y = \ln\sqrt{x^2 - 1}$ **46.** $y = x - \ln x$

47. $y = \ln|x|$ **48.** $y = \ln(1 + x^2)$

49. Using the approximations $\ln 2 \approx 0.70$ and $\ln 3 \approx 1.1$, together with Property 7, find approximations for the following:
(a) $\ln 6$ (b) $\ln 8$ (c) $\ln 81$
(d) $\ln 1.5$ (e) $\ln 36$

B

In Exercises 50–53 find $f'(x)$ and $f''(x)$.

50. $f(x) = \dfrac{\ln(x+1)}{x+1}$ **51.** $f(x) = x(\ln x)^2$

52. $f(x) = \dfrac{x}{\ln x}$ **53.** $f(x) = \sqrt{\ln x^2}$

In Exercises 54 and 55 use implicit differentiation to find y'.

54. $x \ln y - y \ln x = 1$ **55.** $\ln x^2 y = (\ln y)/x$

In Exercises 56–59 find the antiderivatives.

56. $\displaystyle\int \frac{\tan 2x}{\ln \cos 2x}\,dx$

57. $\displaystyle\int \frac{\sin 2x}{1 + \cos^2 x}\,dx$

58. $\displaystyle\int \frac{dx}{x + x^{1/3}}$ (*Hint:* Factor the denominator.)

59. $\displaystyle\int \left(\frac{x-2}{x-1}\right)^2 dx$ (*Hint:* First divide.)

60. Derive the formula

$$\int \csc x\,dx = \ln|\csc x - \cot x| + C$$

In Exercises 61–63 discuss completely and sketch.

61. $y = \ln\left|\dfrac{x-1}{x+1}\right|$

62. $y = \ln\dfrac{x^2}{|x^2 - 2|}$

63. $y = \ln(x^2 - 3x) - \ln(x^2 + 3)$

C

64. Use a computer program for calculating Riemann sums to estimate $\ln 2$ and $\ln 3$. Take $N = 10, 20, 30, \ldots$, and stop when two consecutive answers agree to the first five decimal places.

6.3 INVERSE FUNCTIONS

Before proceeding to exponential functions we pause to introduce the notion of the **inverse** of a function. The ideas introduced here will be applied in defining the exponential function. Inverses of functions are also important in a broader context, however, and we will have other occasions to refer to what we present here.

Let f be a function with domain A and range B. We know that f assigns to each x in A exactly one y in B, which we designate as $f(x)$. It is entirely possible, however, that a given y value in B is the image of more than one x value. For example, if $f(x) = x^2$, then $y = 4$ is the image of $x = 2$ and of $x = -2$; that is, $f(2) = f(-2)$. On the other hand, if $f(x) = x^3$, then two

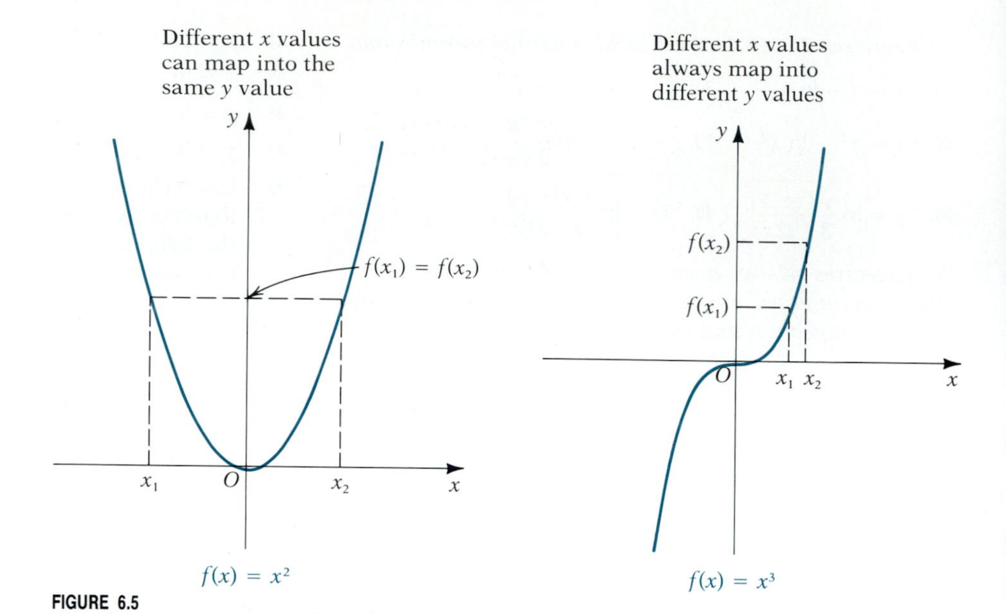

FIGURE 6.5

different x values will always determine two different y values. Figure 6.5 shows the graphs of these two functions. Functions such as $f(x) = x^3$ for which different x values always map into different y values are said to be **one-to-one,** written 1–1. The precise definition is as follows.

DEFINITION 6.1 A function f is said to be **one-to-one** on the domain A if for x_1 and x_2 in A,

$$x_1 \neq x_2 \text{ implies } f(x_1) \neq f(x_2)$$ ■

Remark An equivalent statement is that f is 1–1 if $f(x_1) = f(x_2)$ implies $x_1 = x_2$.

We know that the graph of every function intersects any *vertical* line in at most one point. From Definition 6.1 we can see that if f is 1–1, then its graph intersects each *horizontal* line in at most one point. In fact, this is a geometric condition that is both necessary and sufficient for f to be 1–1.

THE HORIZONTAL LINE TEST

A function f is 1–1 if and only if each horizontal line intersects the graph of f in at most one point.

Suppose now that f is 1–1 on the domain A and has range B. Then each y in B is the image of exactly one x in A. We define a new function g with domain B and range A as follows:

$$g(y) = x \quad \text{provided } y = f(x) \tag{6.11}$$

This definition is unambiguous, since there is one and only one value of g for each y in B. The relationship of f to g is shown schematically in Figure 6.6.

If we replace y by $f(x)$ in equation (6.11), the equation becomes

$$g(f(x)) = x$$

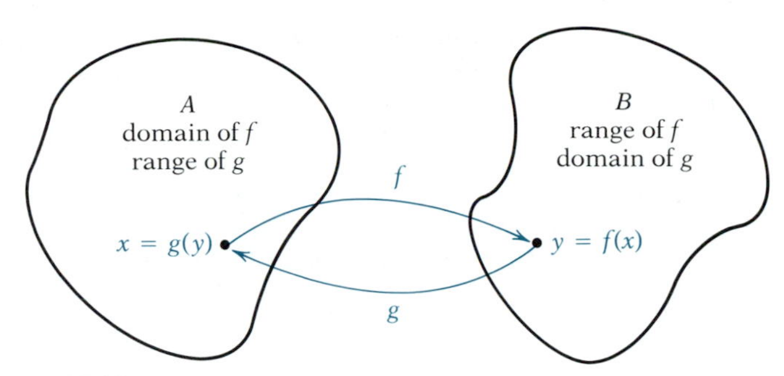

FIGURE 6.6

for every x in A. Again, from equation (6.11), if we apply the function f to both sides, we get $f(g(y)) = f(x)$, and since $y = f(x)$, this gives

$$f(g(y)) = y$$

for each y in B. These ideas are made formal in the following definition.

DEFINITION 6.2 Let f be a function with domain A and range B. If there exists a function g with domain B and range A such that

$$g(f(x)) = x \quad \text{for all } x \in A$$

and

$$f(g(y)) = y \quad \text{for all } y \in B$$

then g is called the **inverse** of f. To emphasize its relation to f, the inverse g is usually written f^{-1}. ■

Remark If g is the inverse of f, then f is also the inverse of g. Writing $g = f^{-1}$, this says that $(f^{-1})^{-1} = f$.

A Word of Caution The -1 in f^{-1} is *not* an exponent, so f^{-1} *does not mean* $\frac{1}{f}$.

We have used y to denote a typical element in the range of f, but since the range of f is the domain of f^{-1} and we often prefer to use the letter x as a typical domain element, we can express the relationships between f and f^{-1} in the following alternate way:

$$f^{-1}(f(x)) = x \quad \text{for all } x \in A \tag{6.12}$$

$$f(f^{-1}(x)) = x \quad \text{for all } x \in B \tag{6.13}$$

In Definition 6.2 we said g was *the* inverse of f. You will be asked to prove in the exercises that there can be only one inverse of a function; that is, if f has an inverse, then that inverse is unique.

As the preceding discussion shows, if f is 1–1, then f^{-1} always exists. If f were not 1–1, then any attempt to define f^{-1} would fail, since there would always be some y in B corresponding to two different x values in A, say x_1 and x_2. Then we would have to have $f^{-1}(y) = x_1$ and $f^{-1}(y) = x_2$, but this is impossible, since f^{-1} must assign a unique value to y. Thus, we have the following theorem.

THEOREM 6.1 The function f has an inverse if and only if it is one-to-one. ■

Suppose that f is 1–1. How do we go about finding f^{-1}? The key lies in equation (6.12). If we apply f^{-1} to both sides of the equation $y = f(x)$, we get

$$f^{-1}(y) = f^{-1}(f(x)) = x$$

In effect, this says that if we can solve the equation $y = f(x)$ for x, the result is $f^{-1}(y)$. Consider, for example, the function f defined by

$$f(x) = \frac{2x - 3}{4}$$

It is easily shown that f is 1–1, since $f(x_1) \neq f(x_2)$ if $x_1 \neq x_2$. We set $y = f(x)$ and solve for x:

$$y = \frac{2x - 3}{4}$$

$$x = \frac{4y + 3}{2}$$

By the preceding discussion, then, we have

$$f^{-1}(y) = \frac{4y + 3}{2}$$

This completely defines f^{-1}. It is customary, however, to use x as the independent variable, and so we could also write

$$f^{-1}(x) = \frac{4x + 3}{2}$$

We can summarize this procedure for finding the inverse of a 1–1 function f as follows:

1. Set $y = f(x)$.
2. Solve for x.
3. The result is $x = f^{-1}(y)$. The variables x and y may be interchanged to get $y = f^{-1}(x)$.

One difficulty that you may have thought about comes in step 2. Consider the function defined by

$$y = 2x^3 - 3x^2 + 18x + 30$$

We will see shortly how to prove that this function is 1–1 (see Exercise 33 in Exercise Set 6.3) and so it has an inverse. But solving the equation for x is no easy matter. In fact, there are 1–1 functions f for which it is *impossible* to solve $y = f(x)$ for x. Nevertheless, in these cases we know that the inverse exists, and we can determine much about it from our knowledge of f. In particular, if we know the graph of f, we can obtain the graph of f^{-1} whether or not we can find an explicit formula for f^{-1}, as the following analysis shows.

Suppose the point (a, b) is on the graph of f; that is, $b = f(a)$. Applying f^{-1} to both sides, we get $f^{-1}(b) = f^{-1}(f(a)) = a$. So $a = f^{-1}(b)$; that is, (b, a) is on the graph of f^{-1}. Figure 6.7 illustrates this. What this tells us is that the *graphs of f and f^{-1} are symmetric with respect to the line $y = x$*. So to get the graph of f^{-1} from the graph of f, we simply reflect the graph of f through the line $y = x$.

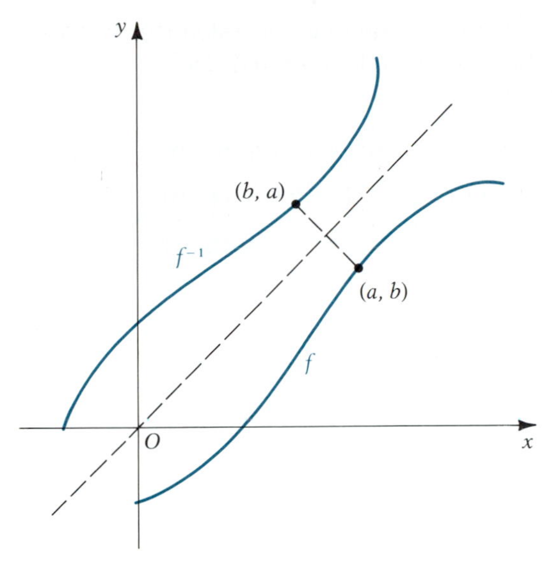

FIGURE 6.7

The next theorem shows that inverses always exist for monotone functions.

THEOREM 6.2 Let f be a continuous monotone function on an interval I, with range J. Then J is also an interval, and f is 1–1 from I onto J. Furthermore, f^{-1} is continuous on J and f^{-1} is also monotone, increasing if f is increasing and decreasing if f is decreasing.

Partial Proof To show that J is an interval, let y_1 and y_2 be any two points in J, and let x_1 and x_2 in I be such that $y_1 = f(x_1)$ and $y_2 = f(x_2)$. By the intermediate value theorem (Theorem 1.7), f assumes all values between y_1 and y_2. So every y between y_1 and y_2 is in the range J. This means that J is an interval.

Now suppose x_1 and x_2 are two numbers in I with $x_1 \neq x_2$. In particular, suppose $x_1 < x_2$. Then if f is increasing, $f(x_1) < f(x_2)$, and if f is decreasing, $f(x_1) > f(x_2)$. In either case, $f(x_1) \neq f(x_2)$. So f is 1–1 and hence has an inverse f^{-1}.

We will delete the proof that f^{-1} is continuous on J because it is rather technical. However, this certainly seems plausible, since the graph of f^{-1} is the same as the graph of f reflected through the line $y = x$. So the continuity of f would seem to imply that of f^{-1}.

We will show that f^{-1} is increasing when f is increasing. The other case is handled in a similar way. Let y_1 and y_2 be any two points in J with $y_1 < y_2$. We must show that $f^{-1}(y_1) < f^{-1}(y_2)$. Let x_1 and x_2 in I be such that

$$y_1 = f(x_1) \quad \text{so that} \quad x_1 = f^{-1}(y_1)$$
$$y_2 = f(x_2) \quad \text{so that} \quad x_2 = f^{-1}(y_2)$$

Suppose that $f^{-1}(y_1)$ is *not* less than $f^{-1}(y_2)$; that is, suppose $f^{-1}(y_1) \geq f^{-1}(y_2)$ or, equivalently, $x_1 \geq x_2$. Then by the fact that f is increasing, we would have $f(x_1) \geq f(x_2)$. But this says that $y_1 \geq y_2$, contrary to our choice of y_1 and y_2. So when $y_1 < y_2$, we *cannot* have $f^{-1}(y_1) \geq f^{-1}(y_2)$. Thus, $f^{-1}(y_1) < f^{-1}(y_2)$, which proves that f^{-1} is increasing. ■

EXAMPLE 6.7 Show that the function f defined by $f(x) = x^2 - 2x$ is monotone increasing on the interval $I = (1, \infty)$. Find f^{-1}, give its domain, and show that it is also increasing.

Solution We show that f is increasing by showing $f'(x) > 0$ on I:

$$f'(x) = 2x - 2 = 2(x - 1) > 0 \quad \text{for } x \in I$$

To find f^{-1} we follow the three-step process outlined earlier, using the quadratic formula to solve $y = f(x)$ for x:

$$y = x^2 - 2x$$
$$x^2 - 2x - y = 0$$
$$x = \frac{2 \pm \sqrt{4 + 4y}}{2} = 1 \pm \sqrt{1 + y}$$

Since $x > 1$, we must discard the negative sign. Thus,

$$f^{-1}(y) = 1 + \sqrt{1 + y}$$

The domain of f^{-1} is the same as the range of f, and we determine this by the requirement that $x > 1$, implying that $1 + \sqrt{1 + y} > 1$ or $\sqrt{1 + y} > 0$, which is true only when $y > -1$. Thus, the domain of f^{-1} is the interval $J = (-1, \infty)$.

We know by Theorem 6.2 that f^{-1} is increasing. However, we can also verify this by calculating its derivative:

$$(f^{-1})'(y) = \frac{1}{2\sqrt{1 + y}} > 0 \quad \text{for } y \in J$$

So f^{-1} is increasing. ■

In Example 6.7 we found that

$$f'(x) = 2(x - 1) \quad \text{and} \quad (f^{-1})'(y) = \frac{1}{2\sqrt{1 + y}}$$

and since we found also that $x = 1 + \sqrt{1 + y}$, it follows that

$$(f^{-1})'(y) = \frac{1}{2(x - 1)} = \frac{1}{f'(x)}$$

As you might expect, this is no accident but always holds true if $f'(x) \neq 0$, as the next theorem shows.

THEOREM 6.3 Let f be a differentiable function on an open interval I, with range J, and suppose $f'(x) \neq 0$ for all $x \in I$. Then f^{-1} exists and is differentiable on J. Furthermore, if $y \in J$ and $f^{-1}(y) = x$, then

$$(f^{-1})'(y) = \frac{1}{f'(x)} \tag{6.14}$$

Proof We show that f^{-1} exists by showing f is 1–1. Let x_1 and x_2 be distinct points in I. By the mean value theorem there is a number c between x_1 and x_2 such that

$$f(x_2) - f(x_1) = f'(c)(x_2 - x_1)$$

The right-hand side is nonzero, since $f'(c) \neq 0$ and $x_1 \neq x_2$. Thus $f(x_1) \neq f(x_2)$. So f is 1–1. It is also continuous, since f' exists, and as in the proof of Theorem 6.2, it follows that the range J is an interval.

We prove the existence of $(f^{-1})'$ and the validity of equation (6.14), making use of the definition of the derivative. Let y_0 be any fixed point in J and let $x_0 = f^{-1}(y_0)$, so that $y_0 = f(x_0)$. Since f^{-1} is continuous on J (by Theorem 6.2) it follows that as $y \to y_0$, $f^{-1}(y) \to f^{-1}(y_0)$ or equivalently $x \to x_0$, and if $y \neq y_0$, then $x \neq x_0$. (Why?) Also, since

$$\lim_{x \to x_0} \frac{f(x) - f(x_0)}{x - x_0} = f'(x_0) \neq 0$$

it follows that for all x in a sufficiently small deleted neighborhood of x_0,

$$\frac{f(x) - f(x_0)}{x - x_0} \neq 0$$

This fact is used when we take the reciprocal of this quotient in the following calculation:

$$\begin{aligned}(f^{-1})'(y_0) &= \lim_{y \to y_0} \frac{f^{-1}(y) - f^{-1}(y_0)}{y - y_0} = \lim_{x \to x_0} \frac{x - x_0}{f(x) - f(x_0)} \\ &= \lim_{x \to x_0} \frac{1}{\dfrac{f(x) - f(x_0)}{x - x_0}} = \frac{1}{\displaystyle\lim_{x \to x_0} \frac{f(x) - f(x_0)}{x - x_0}} = \frac{1}{f'(x_0)}\end{aligned}$$

Since y_0 is an arbitrary point in J, this says that for all $y \in J$,

$$(f^{-1})'(y) = \frac{1}{f'(x)}$$

■

Remark In the differential notation equation (6.14) takes on a particularly simple appearance. If $y = f(x)$ and $x = f^{-1}(y)$, then $(f^{-1})'(y)$ can be written $\frac{dx}{dy}$, so that we have

$$\frac{dx}{dy} = \frac{1}{\dfrac{dy}{dx}} \quad \text{if} \quad \frac{dy}{dx} \neq 0$$

EXAMPLE 6.8 Show that the function f defined by

$$f(x) = 3x^5 + 10x^3 + 3x + 2$$

is increasing on $(-\infty, \infty)$, and find $(f^{-1})'(2)$.

Solution We first calculate the derivative of f:

$$f'(x) = 15x^4 + 30x^2 + 3$$

This is always positive, so f is increasing. We cannot find an explicit form of f^{-1}, since solving a fifth-degree polynomial equation in general is not possible. However, by equation (6.14) we know that

$$(f^{-1})'(y) = \frac{1}{f'(x)} = \frac{1}{15x^4 + 30x^2 + 3}$$

and we want to substitute $y = 2$. Although the expression for $(f^{-1})'(y)$ is in terms of x, we observe that $f(0) = 2$, and by the monotonicity of f, no other x value gives $f(x) = 2$. So when $y = 2$, $x = 0$. Thus,

$$(f^{-1})'(2) = \frac{1}{f'(0)} = \frac{1}{3}$$

There will be times when we want to use x for the independent variable for f^{-1}, and then equation (6.14) can be written as

$$(f^{-1})'(x) = \frac{1}{f'(y)} = \frac{1}{f'(f^{-1}(x))}$$

since $y = f^{-1}(x)$. In differential form this becomes

$$\frac{dy}{dx} = \frac{1}{\frac{dx}{dy}}$$

EXERCISE SET 6.3

A

In Exercises 1–6 show that f is 1–1 on **R** *and find* f^{-1}.

1. $f(x) = 1 - 2x$
2. $f(x) = 3x - 5$
3. $f(x) = \frac{2x + 3}{4}$
4. $f(x) = \frac{4 - 3x}{2}$
5. $f(x) = x^3 - 1$
6. $f(x) = 3 - 2x^3$

In Exercises 7–14 show that f is 1–1 on the specified domain. Find f^{-1} *and give its domain.*

7. $f(x) = x^2$ on $[0, \infty)$
8. $f(x) = \sqrt{1 - x^2}$ on $[0, 1]$
9. $f(x) = 2(x - 1)^2 + 3$ on $[1, \infty)$
10. $f(x) = 1 - (x + 2)^2$ on $[-2, \infty)$
11. $f(x) = \frac{1}{x}$ on $(0, \infty)$
12. $f(x) = \frac{x}{1 - x}$ on $(1, \infty)$
13. $f(x) = \frac{x - 1}{x + 1}$ on $(-1, \infty)$
14. $f(x) = \frac{x^2 - 1}{x^2 - 4}$ on $(2, \infty)$

In Exercises 15–20 draw the graphs of f and of f^{-1} *on the same set of axes.*

15. $f(x) = (2 - 3x)/4$ on $(-\infty, \infty)$
16. $f(x) = \sqrt{x}$ on $[0, \infty)$
17. $f(x) = x^2 + 1$ on $(-\infty, 0]$
18. $f(x) = x/(x - 1)$ on $(1, \infty)$
19. $f(x) = x^{2/3}$ on $(0, \infty)$
20. $f(x) = (1 - x)^{3/2}$ on $(-\infty, 1)$

B

In Exercises 21–26 find a suitable restriction on the domain of f so that on the restricted domain f is 1–1. Find f^{-1} *and give its domain. Draw the graphs of f and of* f^{-1} *on the same set of axes.*

21. $f(x) = x^2 - 2x + 3$
22. $f(x) = x^2 + 4x - 5$
23. $f(x) = 2x - x^2$
24. $f(x) = 2x^2 - 3x + 1$
25. $f(x) = \frac{1}{(x - 1)^2}$
26. $f(x) = \frac{x^2 - 1}{x^2}$

In Exercises 27–30 prove that f is monotone on the interval given, and find $(f^{-1})'(y_0)$ *for the given value of* y_0.

27. $f(x) = x^3 + 2x$ on **R**; $y_0 = f(-1)$
28. $f(x) = x^3 - 2x^2 + x + 3$ on $(1, \infty)$; $y_0 = f(2)$
29. $f(x) = x^4 + 2x^2 - 3$ on $(0, \infty)$; $y_0 = f(1)$
30. $f(x) = 2x^3 - 3x^2 - 36x + 30$ on $(-2, 3)$; $y_0 = f(-1)$

31. Prove that $f'(x) \neq 0$ is not a necessary condition for a differentiable function to be 1–1. [*Hint:* Consider $f(x) = x^3$.]

32. Prove that if f has an inverse, then that inverse is unique. (*Hint:* Suppose both g and h are inverses of f. Show that $g = h$.)

33. Prove that if f is differentiable on an open interval I and $f'(x) \neq 0$ for all $x \in I$, then f is either increasing on I or decreasing on I. [*Hint:* Use the intermediate value property of derivatives (Exercise 59 in Exercise Set 3.1).]

34. Let f be a twice differentiable function on an open interval I with $f'(x) \neq 0$ for all $x \in I$. Prove that $(f^{-1})''(y)$ exists for all y on the range J of f, and

$$(f^{-1})''(y) = -\frac{f''(f^{-1}(y))}{[f'(f^{-1}(y))]^3}$$

6.4 THE EXPONENTIAL FUNCTION

We know that the natural logarithm function $L(x) = \ln x$ has the derivative $L'(x) = \frac{1}{x}$ on the interval $(0, \infty)$. Since $L'(x) > 0$, L is 1–1 and therefore has an inverse. Temporarily, let us denote L^{-1} by E (suggestive of "exponential"). We can get the graph of E by reflecting the graph of L through the line $x = y$, as in Figure 6.8.

We can deduce the following facts based on the results of Section 6.3 and the known properties of L:

Domain of $E = (-\infty, +\infty)$

Range of $E = (0, +\infty)$

E is continuous, increasing, and differentiable

$E(0) = 1$

$\lim_{x \to -\infty} E(x) = 0$

$\lim_{x \to +\infty} E(x) = +\infty$

FIGURE 6.8

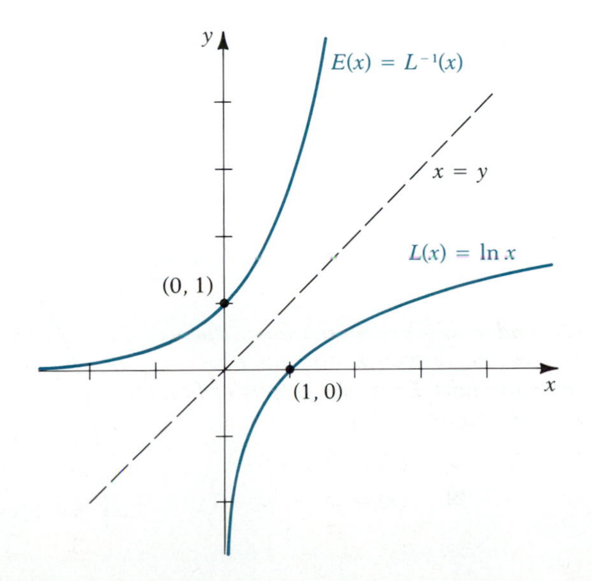

Before obtaining more information about E, we introduce a special number whose natural logarithm is 1. We know that such a number exists because the range of L is $(-\infty, +\infty)$. This number is designated by the letter e (after Leonhard Euler, the greatest mathematician of the 18th century and the most prolific mathematician of all time). Thus, e is defined by the equation

$$\ln e = 1$$

We show e graphically in Figure 6.9. Since, as we have seen, $\ln 2 < 1$ and $\ln 3 > 1$, it follows from the monotonicity of L that $2 < e < 3$. It can be shown that e is irrational and that its value to five decimal places is

$$e \approx 2.71828$$

(We will see a way to estimate e to any desired degree of accuracy in Section 6.5.)

Now if r is any rational number, we know that

$$\ln e^r = r \ln e = r$$

since $\ln e = 1$. Using the functional notation $L(x) = \ln(x)$, this can be written

$$L(e^r) = r$$

Applying $L^{-1} = E$ to both sides, we get

$$E(L(e^r)) = E(r)$$

or

$$e^r = E(r)$$

FIGURE 6.9

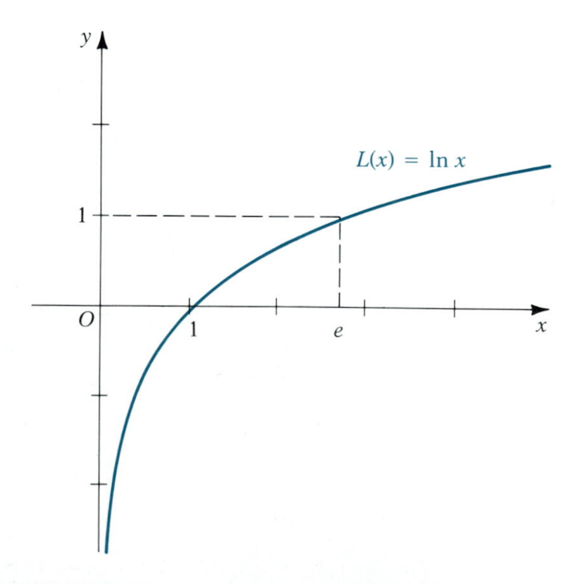

Up to this point no meaning has been given to e^x where x is an irrational number. By extending the result just obtained, we now give such a definition. If x is *any* real number, we *define* e^x to be

$$e^x = E(x) \tag{6.15}$$

Because E is the inverse of L, where $L(x) = \ln x$, we can state equation (6.15) in words as follows:

e^x is that number whose natural logarithm is x

Equivalently, we have

$$y = e^x \text{ if and only if } \ln y = x \tag{6.16}$$

Since $L(E(x)) = x$ for all x in the domain of E and $E(L(x)) = x$ for all x in the domain of L, we have

$$\begin{aligned} \ln e^x &= x \quad \text{for all } x \\ e^{\ln x} &= x \quad \text{for all } x > 0 \end{aligned} \tag{6.17}$$

The next theorem shows that the usual laws of exponents for powers of e continue to hold when the exponents involved are arbitrary real numbers.

THEOREM 6.4 If x and y are any real numbers and r is any rational number,

1. $e^{x+y} = e^x \cdot e^y$
2. $e^{x-y} = \dfrac{e^x}{e^y}$
3. $(e^x)^r = e^{rx}$

Proof

1. Using equation (6.16), we have

   $$\ln e^{x+y} = x + y$$

   and

   $$\ln(e^x \cdot e^y) = \ln e^x + \ln e^y = x + y$$

   Since the natural logarithm function is 1–1, the equality of $\ln e^{x+y}$ and $\ln(e^x \cdot e^y)$ guarantees the equality of e^{x+y} and $e^x \cdot e^y$.
2. This is proved in a similar way, and we leave it as an exercise.
3. Using the fact that $\ln a^r = r \ln a$, we have

   $$\ln(e^x)^r = r \ln e^x = rx$$

   Also,

   $$\ln e^{rx} = rx$$

   Again, the 1–1 nature of ln gives the desired result. ■

Note At this stage in part 3 we must limit r to be rational. We will soon remove this restriction.

Because of Theorem 6.4, the x in e^x behaves in a manner consistent with our previous understanding of an exponent, even when x is irrational. For this reason we name the function E the **exponential function** (also called the *natural* exponential function).

To find the derivative of the exponential function we write $y = e^x$ so that $x = \ln y$. Since $\frac{dx}{dy} = \frac{1}{y} > 0$, it follows from equation (6.14), with the roles of x and y reversed, that

$$\frac{dy}{dx} = \frac{1}{\frac{dx}{dy}} = \frac{1}{\frac{1}{y}} = y$$

and since $y = e^x$, this says that

$$\frac{d}{dx} e^x = e^x \tag{6.18}$$

This property of the derivative duplicating the function itself is quite remarkable. It holds true *only* for functions of the form $f(x) = Ce^x$, for constants C.

If $u = g(x)$ is a differentiable function, then equation (6.18) and the chain rule give

$$\frac{d}{dx} e^{g(x)} = e^{g(x)} \cdot g'(x)$$

or, in terms of u,

$$\frac{d}{dx} e^u = e^u \frac{du}{dx} \tag{6.19}$$

Consequently, we also have the antidifferentiation formula

$$\int e^u \, du = e^u + C \tag{6.20}$$

EXAMPLE 6.9 (a) Find $f'(x)$ if $f(x) = e^{\sin x}$.
(b) Evaluate the integral $\int xe^{x^2} \, dx$.

Solution (a) By equation (6.19) we have

$$f'(x) = e^{\sin x} \cos x$$

(b) Let $u = x^2$. Then $du = 2x \, dx$. So we have

$$\begin{aligned}\int xe^{x^2} \, dx &= \tfrac{1}{2} \int e^{x^2}(2x \, dx) = \tfrac{1}{2} \int e^u \, du \\ &= \tfrac{1}{2}e^u + C = \tfrac{1}{2}e^{x^2} + C\end{aligned}$$

■

We summarize now the main properties of the exponential function.

PROPERTIES OF THE EXPONENTIAL FUNCTION

1. $y = e^x$ if and only if $\ln y = x$.
2. Domain $= (-\infty, +\infty)$. Range $= (0, +\infty)$.
3. The exponential function is differentiable, and hence continuous, for all x, and

$$\frac{d}{dx} e^x = e^x$$

4. The exponential function is increasing and its graph is concave upward.
5. $\ln e^x = x$ and $e^{\ln x} = x$
6. $\lim_{x \to +\infty} e^x = +\infty$ and $\lim_{x \to -\infty} e^x = 0$
7. For all real numbers x and y, and for r rational,
   (a) $e^{x+y} = e^x \cdot e^y$
   (b) $e^{x-y} = e^x/e^y$
   (c) $(e^x)^r = e^{rx}$

The next two examples further illustrate the properties of the exponential function.

EXAMPLE 6.10 Evaluate the following integrals:

$$\text{(a)} \int \frac{e^x}{1 + e^x}\, dx \qquad \text{(b)} \int_0^{\ln 3} e^{-2x}\, dx$$

Solution (a) Let $u = 1 + e^x$; then $du = e^x\, dx$. So we have

$$\int \frac{e^x}{1 + e^x}\, dx = \int \frac{du}{u} = \ln|u| + C = \ln(1 + e^x) + C$$

Note that the absolute value symbol can be deleted, since $1 + e^x > 0$.

$$\text{(b)} \int_0^{\ln 3} e^{-2x}\, dx = -\frac{1}{2} \int_0^{\ln 3} \underbrace{e^{-2x}}_{e^u} \underbrace{(-2\, dx)}_{du} = -\frac{1}{2} [e^{-2x}]_0^{\ln 3}$$

$$= -\frac{1}{2}(e^{-2 \ln 3} - e^0) = -\frac{1}{2}\left(\frac{1}{e^{2 \ln 3}} - 1\right)$$

$$= -\frac{1}{2}\left(\frac{1}{e^{\ln 3^2}} - 1\right) = -\frac{1}{2}\left(\frac{1}{9} - 1\right) = \frac{4}{9}$$

Note that $e^{2 \ln 3} = e^{\ln 3^2} = e^{\ln 9} = 9$, since $e^{\ln x} = x$. ■

EXAMPLE 6.11 Discuss and sketch the graph of $y = xe^{-x}$.

Solution The only intercept is $x = 0$, $y = 0$. There is no symmetry.

In order to find extreme values and to discuss concavity, we calculate y' and y'':

$$y' = xe^{-x}(-1) + e^{-x} = e^{-x}(1 - x)$$
$$y'' = e^{-x}(-1) + (1 - x) \cdot e^{-x}(-1) = e^{-x}(x - 2)$$

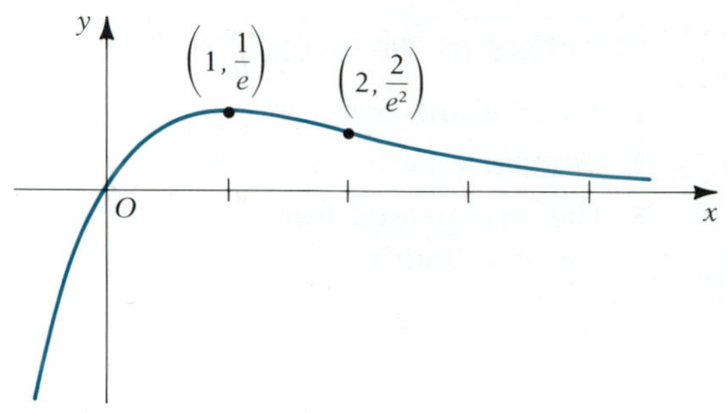

FIGURE 6.10

Since $e^{-x} = 1/e^x$ and $e^x > 0$ for all x, the only critical value is $x = 1$. Testing this in y'', we find

$$y''\Big|_{x=1} = e^{-1}(-1) = -\tfrac{1}{e} < 0$$

Thus, the function has a local maximum value at $x = 1$. Since

$$y\Big|_{x=1} = 1 \cdot e^{-1} = \tfrac{1}{e}$$

the maximum point is $(1, \frac{1}{e})$.

From $y'' = e^{-x}(x - 2)$, we see that $y'' > 0$ if $x > 2$ and $y'' < 0$ if $x < 2$. So the graph is concave upward on $(2, +\infty)$ and concave downward on $(-\infty, 2)$. The point $(2, 2/e^2)$ is a point of inflection.

We show the information obtained so far graphically in Figure 6.10.

Since $y > 0$ when $x > 0$, and the curve is concave upward but decreasing to the right of $x = 2$, it appears that the positive x-axis may be a horizontal asymptote. We will see how to prove this is true later by showing that

$$\lim_{x \to +\infty} \frac{x}{e^x} = 0$$

In fact, you can convince yourself of the plausibility of this by noting that e^x gets large much faster than x does as $x \to +\infty$. ■

EXERCISE SET 6.4

A

In Exercises 1–20 find $f'(x)$.

1. $f(x) = e^{x^2}$
2. $f(x) = e^{\cos x}$
3. $f(x) = e^{3 \ln x}$
4. $f(x) = \ln e^{2x}$
5. $f(x) = e^{\sqrt{x}}$
6. $f(x) = \sqrt{e^x}$
7. $f(x) = \dfrac{x}{e^{2x}}$
8. $f(x) = e^{1/x}$
9. $f(x) = e^{x + \ln x}$
10. $f(x) = e^{2x - 3 \ln x}$
11. $f(x) = e^{2x} \sin 3x$
12. $f(x) = e^{\sqrt{1 - x^2}}$
13. $f(x) = \dfrac{e^x - e^{-x}}{2}$
14. $f(x) = \dfrac{e^x + e^{-x}}{2}$
15. $f(x) = \dfrac{e^x - e^{-x}}{e^x + e^{-x}}$
16. $f(x) = e^{2x} \ln x$
17. $f(x) = \sin^2 e^x$
18. $f(x) = e^{-x} \cos 2x$
19. $f(x) = \dfrac{1}{1 - e^{-x}}$
20. $f(x) = e^{-x}(\sin x + \cos x)$

In Exercises 21–24 find y' using implicit differentiation.

21. $e^{x+y} = xy$
22. $e^{xy} = x + y$
23. $e^{-x} \sin y = e^{-y} \cos x$
24. $y = 1 - e^{-(x+y)}$

In Exercises 25–40 evaluate the integrals.

25. $\int xe^{(1-x^2)}\,dx$

26. $\int \frac{x\,dx}{e^{x^2+1}}$

27. $\int \frac{\sin x}{e^{\cos x}}\,dx$

28. $\int \frac{e^x}{1-e^x}\,dx$

29. $\int \frac{e^x - e^{-x}}{e^x + e^{-x}}\,dx$

30. $\int \frac{e^x}{\sqrt{e^x - 1}}\,dx$

31. $\int_0^{\ln 2} e^{3x}\,dx$

32. $\int_{\ln 2}^{\ln 5} \frac{e^x}{(1+e^x)^2}\,dx$

33. $\int_0^3 e^{2x}\sqrt[3]{1-e^{2x}}\,dx$

34. $\int_{2+1/e}^{3} \frac{dx}{x-2}$

35. $\int_0^{\ln 2} \frac{dx}{1+e^{-x}}$

36. $\int \frac{dx}{1+e^x}$ (*Hint:* Multiply numerator and denominator by e^{-x}.)

37. $\int_{1/\ln 3}^{2/\ln 3} \frac{e^{1/x}}{x^2}\,dx$

38. $\int_0^{\ln 2} \frac{e^{-x}}{4-3e^{-x}}$

39. $\int \frac{dx}{e^x - 2 + e^{-x}}$

40. $\int_{\ln \pi}^{\ln 4\pi/3} e^x \tan e^x\,dx$

In Exercises 41–46 discuss completely and draw the graph.

41. $y = e^{-x^2/4}$

42. $y = e^{1-|x|}$

43. $y = xe^{-x/2}$

44. $y = \frac{e^x + e^{-x}}{2}$

45. $y = \frac{1}{1+e^{-x}}$

46. $y = \frac{e^x}{1-e^x}$

47. Prove part 2 of Theorem 6.4.

48. Prove that $e^{-x} = 1/e^x$ for all real numbers x.

49. Prove the following, based on the definition of e^x:
(a) $e^0 = 1$ (b) $e^1 = e$

50. Find the area of the region under the graph of $y = e^{-x}$ from x_1 to x_2, where $e^{x_1} = 1$ and $e^{x_2} = 2$.

51. Find the volume obtained by revolving the region under the graph of $y = \sqrt{1+e^x}$ from $x = 0$ to $x = 1$ about the x-axis.

52. Find the length of the curve $y = (e^x + e^{-x})/2$ from $x = 0$ to $x = \ln 2$.

53. Find the volume obtained by revolving the region under the graph of $y = \sqrt{x}e^{(1-x^2)/2}$ from $x = 1$ to $x = 2$ about the x-axis.

B

54. Show that the function f defined by

$$f(x) = \frac{e^x - e^{-x}}{2}$$

is 1–1 on **R**, and find f^{-1}.

55. Find $(f^{-1})'(x)$ for the function in Exercise 54.

In Exercises 56–58 discuss completely and draw the graph.

56. $f(x) = e^{1/x}$

57. $f(x) = \frac{e^{2x}}{x^2}$

58. $f(x) = xe^{(1-x^2)/2}$

59. A number of types of oscillatory motion, such as that of an oscillating spring, when the oscillation is occurring in a resisting medium, can be described by an equation of the type

$$y(t) = e^{-kt}(C_1 \cos \omega t + C_2 \sin \omega t)$$

where $y(t)$ is the displacement from the equilibrium position at time t. Suppose $k = 1$, $C_1 = 0$, $C_2 = 1$, and $\omega = 2$. Find the velocity and acceleration when $t = \pi$. Draw the graph of $y(t)$.

60. When limitations on size are considered for the growth of a colony of bacteria (due to physical limitations or limitations on nutrients, for example), the number $Q(t)$ present at time t is given reasonably accurately by

$$Q(t) = \frac{m}{1 + \left(\frac{m - Q_0}{Q_0}\right)e^{-kmt}}$$

where $Q_0 = Q(0)$, m is the maximum size, and k is a positive constant. If $Q_0 = 200$, $m = 1000$, and $Q(\frac{1}{2}) = 400$, find the constant k. Using this value of k, draw the graph of $Q(t)$. When does the maximum rate of growth occur?

61. The normal probability density function with mean μ and standard deviation σ is given by

$$f(x) = \frac{1}{\sqrt{2\pi}\sigma} e^{-(x-\mu)^2/2\sigma^2}$$

(a) Draw the graph of f.
(b) Show that inflection points occur at $x = \mu \pm \sigma$.

6.5 EXPONENTIAL AND LOGARITHMIC FUNCTIONS WITH OTHER BASES

We know that for rational numbers r and any real number $a > 0$,

$$a^r = e^{\ln a^r} = e^{r \ln a} \tag{6.21}$$

We still have not assigned a meaning to a raised to an irrational power. We now do this. Let x denote any real number, rational or irrational. We define

$$a^x = e^{x \ln a}, \qquad a > 0 \tag{6.22}$$

This agrees with equation (6.21) when x is rational and extends the meaning in the natural way to irrational values. Note that this is a meaningful definition, since $\ln a$ is a well-defined real number for each $a > 0$, so that $x \ln a$ is also a real number, and since the exponential function is defined on all of **R**, $e^{x \ln a}$ is also well-defined. This does not mean it is easy to find a decimal approximation (without the aid of a calculator), but we know what the meaning is. For example,

$$\pi^{\sqrt{2}} = e^{\sqrt{2} \ln \pi}$$

and this can be approximated with the aid of tables or a calculator to be

$$\pi^{\sqrt{2}} \approx 5.0475$$

Observe that since $e^{g(x)} > 0$, it follows that $a^x > 0$ for all x.

Taking the natural logarithm of both sides of equation (6.22) gives an extension of Property 7c of the natural logarithm function:

$$\ln a^x = x \ln a, \qquad a > 0 \tag{6.23}$$

The next theorem shows that the laws of exponents for rational exponents extend to all real exponents.

THEOREM 6.5 LAWS OF EXPONENTS

Let a and b denote positive real numbers. If x and y are any real numbers, then

1. $a^{x+y} = a^x \cdot a^y$ **2.** $a^{x-y} = \dfrac{a^x}{a^y}$

3. $(a^x)^y = a^{xy}$ **4.** $(ab)^x = a^x b^x$

5. $\left(\dfrac{a}{b}\right)^x = \dfrac{a^x}{b^x}$

Proof We prove Properties 1 and 3 and leave the others as exercises.

1.
$$\begin{aligned} a^{x+y} &= e^{(x+y)\ln a} && \text{By equation (6.22)} \\ &= e^{x \ln a + y \ln a} && \text{Distributive law} \\ &= e^{x \ln a} \cdot e^{y \ln a} && \text{Property 7a for the exponential function} \\ &= a^x \cdot a^y && \text{By equation (6.22)} \end{aligned}$$

3. $(a^x)^y = e^{y \ln a^x}$ By equation (6.22)

$= e^{y(x \ln a)}$ By equation (6.23)

$= e^{(xy) \ln a}$ Associative and commutative laws

$= a^{xy}$ By equation (6.22) ■

We have often used the power rule

$$\frac{d}{dx} x^r = rx^{r-1} \qquad (x > 0)$$

for *rational* numbers r. Now we can extend this to all real exponents.

THEOREM 6.6 If $x > 0$ and α is any real number, then

$$\frac{d}{dx} x^\alpha = \alpha x^{\alpha - 1}$$

Proof By equation (6.22) and the derivatives of e^u and $\ln u$,

$$\frac{d}{dx} x^\alpha = \frac{d}{dx} e^{\alpha \ln x} = e^{\alpha \ln x}\left(\tfrac{\alpha}{x}\right)$$

$$= x^\alpha \cdot \tfrac{\alpha}{x} = \alpha x^{\alpha - 1}$$ ■

For example, if $y = x^\pi$, then

$$\frac{dy}{dx} = \pi x^{\pi - 1}$$

If we take a as a fixed positive real number, then for each real number, x, a^x is a well-defined real number, by equation (6.22). Thus, $f(x) = a^x$ is a function from **R** to **R** (but its range is not all of **R**). We call f the **exponential function with base *a*.** If the base $a = 1$, we have

$$1^x = e^{x \ln 1} = e^{x \cdot 0} = e^0 = 1$$

Thus, when $a = 1$, f is just the constant function $f(x) = 1$ for all x. Since the properties of constant functions are well known, *we will henceforth consider only those cases where $a > 0$ and $a \neq 1$.* Since e^u is differentiable when u is differentiable and

$$\frac{d}{dx} e^u = e^u \frac{du}{dx}$$

it follows from equation (6.22) that the derivative of f exists and is given by

$$f'(x) = \frac{d}{dx} e^{x \ln a} = e^{x \ln a} \cdot \ln a = a^x \ln a$$

Thus,

$$\frac{d}{dx} a^x = a^x \ln a \tag{6.24}$$

Note that when $a = e$, this reduces to

$$\frac{d}{dx} e^x = e^x$$

since $\ln e = 1$. This shows one reason why e is singled out as a special base. The derivative formula is simplified when e is used.

If $a > 1$, $\ln a > 0$, and since a^x is always positive,

$$\frac{d}{dx} a^x > 0 \qquad (a > 1)$$

If $0 < a < 1$, then $\ln a < 0$, so that

$$\frac{d}{dx} a^x < 0 \qquad (0 < a < 1)$$

Thus the graph of $f(x) = a^x$ is increasing if $a > 1$ and decreasing if $0 < a < 1$. For f'' we have

$$f''(x) = \frac{d}{dx}(a^x \ln a) = \left(\frac{d}{dx} a^x\right) \ln a = a^x(\ln a)^2$$

and this is always positive. So the graph is always concave upward. Using the properties of e^x, we can now graph $f(x) = a^x$ as in Figure 6.11.

Using the chain rule together with equation (6.24), we obtain

$$\frac{d}{dx} a^u = a^u \ln a \cdot \frac{du}{dx} \qquad (6.25)$$

where $u = g(x)$ is a differentiable function of x.

A Word of Caution It is important to distinguish between a^x and x^a, where a is fixed. We can contrast their derivatives as follows:

$y = a^x$	$y = x^a$
$\frac{dy}{dx} = a^x \ln a$	$\frac{dy}{dx} = ax^{a-1}$
($a > 0$, x any real number)	($x > 0$, a any real number)

FIGURE 6.11

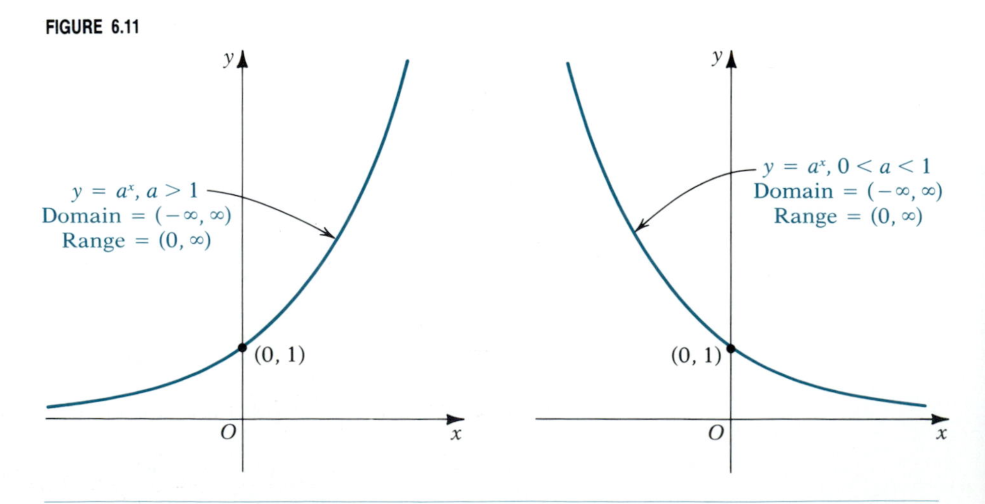

For example,

$$\frac{d}{dx}(3^x) = 3^x \ln 3 \quad \text{whereas} \quad \frac{d}{dx}x^3 = 3x^2$$

EXAMPLE 6.12 Find the derivative of each of the following:

(a) $y = 2^{x^2}$ (b) $y = x \cdot 3^{\sqrt{1-x}}$

Solution (a) $\dfrac{dy}{dx} = 2^{x^2} \ln 2 \cdot 2x = (2 \ln 2)x \cdot 2^{x^2}$

(b) From the product rule and equation (6.25),

$$\frac{dy}{dx} = x \cdot 3^{\sqrt{1-x}} \ln 3 \left(-\frac{1}{2\sqrt{1-x}}\right) + 3^{\sqrt{1-x}}$$

$$= 3^{\sqrt{1-x}}\left[1 - \frac{x \ln 3}{2\sqrt{1-x}}\right]$$ ■

From equation (6.25) we also have the antidifferentiation formula

$$\int a^u \, du = \frac{a^u}{\ln a} + C \tag{6.26}$$

where we are continuing to require $a > 0$ and $a \neq 1$.

EXAMPLE 6.13 Evaluate the integral

$$\int \frac{3^{\sqrt{x}}}{\sqrt{x}} \, dx$$

Solution Let $u = \sqrt{x}$. Then $du = dx/(2\sqrt{x})$. So we have

$$\int \frac{3^{\sqrt{x}}}{\sqrt{x}} \, dx = 2 \int \underbrace{3^{\sqrt{x}}}_{3^u} \underbrace{\left(\frac{1}{2\sqrt{x}} \, dx\right)}_{du} = 2\left(\frac{3^{\sqrt{x}}}{\ln 3}\right) + C$$ ■

The next example illustrates how the natural logarithm can be used to solve certain types of equations for a variable that appears as an exponent.

EXAMPLE 6.14 Solve for x:

(a) $2^x = 3^{1-x}$ (b) $3 \cdot 4^{2x} = \dfrac{2}{\sqrt{5^x}}$

Solution In each case we take the natural logarithm of both sides.

(a)

$$\begin{aligned} \ln 2^x &= \ln 3^{1-x} \\ x \ln 2 &= (1 - x) \ln 3 \\ x \ln 2 + x \ln 3 &= \ln 3 \\ x(\ln 2 + \ln 3) &= \ln 3 \\ x &= \frac{\ln 3}{\ln 2 + \ln 3} = \frac{\ln 3}{\ln 6} \approx 0.6131 \end{aligned}$$

(b) Rewrite the equation in the form

$$3 \cdot 4^{2x} = \frac{2}{5^{x/2}}$$

Then

$$\begin{aligned}
\ln(3 \cdot 4^{2x}) &= \ln\left(\frac{2}{5^{x/2}}\right) \\
\ln 3 + 2x \ln 4 &= \ln 2 - \frac{x}{2} \ln 5 \\
2 \ln 3 + 4x \ln 4 &= 2 \ln 2 - x \ln 5 \\
x(4 \ln 4 + \ln 5) &= 2 \ln 2 - 2 \ln 3 \\
x &= \frac{2 \ln 2 - 2 \ln 3}{4 \ln 4 + \ln 5} = \frac{\ln 4 - \ln 9}{4 \ln 4 + \ln 5} \approx -0.1133
\end{aligned}$$

Note that further attempts to simplify the answer before using the calculator are not worthwhile in this case. ■

We now know how to differentiate constants to variable powers (such as 3^x) and variables to constant powers (such as x^3). The next example shows how to differentiate variables to variable powers.

EXAMPLE 6.15 Find y' if $y = (\sin x)^x$, with $0 < x < \pi$.

Solution We will do this by two methods. In method 1 we use equation (6.22) to get

$$\begin{aligned}
y' = \frac{d}{dx} e^{x \ln \sin x} &= e^{x \ln \sin x}\left(x \cdot \frac{1}{\sin x} \cdot \cos x + \ln \sin x\right) \\
&= (\sin x)^x (x \cot x + \ln \sin x)
\end{aligned}$$

In method 2 we use logarithmic differentiation:

$$\begin{aligned}
\ln y &= x \ln \sin x \\
\frac{y'}{y} &= x \cdot \frac{1}{\sin x} \cdot \cos x + \ln \sin x \\
&= x \cot x + \ln \sin x \\
y' &= y(x \cot x + \ln \sin x) \\
&= (\sin x)^x (x \cot x + \ln \sin x)
\end{aligned}$$

■

Since for $a > 0$ and $a \neq 1$, $f(x) = a^x$ is monotone (increasing if $a < 1$ and decreasing if $0 < a < 1$), it is 1–1 and therefore has an inverse. We denote its inverse by

$$f^{-1}(x) = \log_a x \qquad (a > 0,\ a \neq 1,\ x > 0)$$

and call f^{-1} the **logarithm function with base *a*.** Similarly, we call $\log_a x$ the **logarithm to the base *a* of *x*.** Thus, for $a > 0$, $a \neq 1$, we have

$$y = \log_a x \text{ if and only if } a^y = x \qquad (x > 0) \tag{6.27}$$

and also

$$\begin{aligned} \log_a a^x &= x \quad \text{for all } x \\ a^{\log_a x} &= x \quad \text{for } x > 0 \end{aligned} \tag{6.28}$$

If $a = e$, then $\log_e x$ is the inverse of e^x. But $\ln x$ is the inverse of e^x, so we have

$$\log_e x = \ln x$$

That is, the natural logarithm of x is the same as the logarithm to the base e of x.

We obtain the graph of $y = \log_a x$ by reflecting the graph of $y = a^x$ through the line $y = x$, with the result shown in Figure 6.12.

We leave the proof of the next theorem for the exercises.

THEOREM 6.7 Assume x and y are in the domain of the logarithm function of base a, where $a > 0$ and $a \neq 1$. Then

1. $\log_a xy = \log_a x + \log_a y$
2. $\log_a \frac{x}{y} = \log_a x - \log_a y$
3. $\log_a x^\alpha = \alpha \log_a x$ for any real number α ■

Remark In particular, when $a = e$, Property 3 can be stated as

$$\ln x^\alpha = \alpha \ln x$$

This extends to all real numbers Property 7c for $\ln x$ that we previously had proved for only rational α.

Using equation (6.27) and the relationship

$$\frac{dy}{dx} = \frac{1}{\dfrac{dx}{dy}}$$

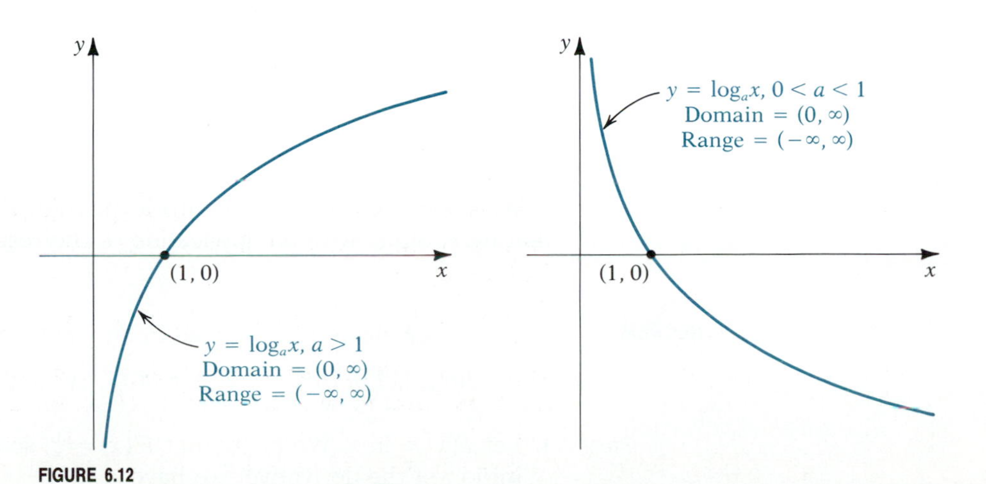

FIGURE 6.12

we obtain

$$\frac{d}{dx}\log_a x = \frac{1}{a^y \ln a} = \frac{1}{x \ln a} \tag{6.29}$$

since $x = a^y$. Here we see why e is the "natural" base for logarithms, for if $a = e$, the derivative becomes $\frac{1}{x}$. By the chain rule, from equation (6.28) we obtain for $u = g(x)$ a differentiable function with $u > 0$,

$$\frac{d}{dx}\log_a u = \frac{1}{u \ln a} \cdot \frac{du}{dx} \tag{6.30}$$

We could use equation (6.30) to obtain an antiderivative of $\frac{1}{u}$ in terms of $\log_a u$, but this is unnecessary, since we already know that an antiderivative is $\ln|u| + C$.

Sometimes it is useful to express a logarithm to one base in terms of a logarithm to another base. Suppose a and b are permissible bases (positive and different from 1). For $x > 0$, let

$$y = \log_b x \quad \text{so that } x = b^y$$

Now take the logarithm to the base a of both sides:

$$\log_a x = \log_a b^y = y \log_a b$$

Thus, since $y = \log_b x$, we have the *change of base formula*:

$$\log_b x = \frac{\log_a x}{\log_a b} \tag{6.31}$$

Since the most convenient base to work with is e, we usually want to use the following special case of equation (6.31):

$$\log_b x = \frac{\ln x}{\ln b} \tag{6.32}$$

For example, in Section 6.1 we posed the question about the value of $\log_2 7$. By equation (6.32) we have

$$\log_2 7 = \frac{\ln 7}{\ln 2} \approx 2.8074 \quad \text{By calculator}$$

We conclude this section with a theorem that provides a means of estimating e, along with an application of the result.

THEOREM 6.8 The number e can be expressed in either of the following ways:

1. $e = \lim_{h\to 0}(1 + h)^{1/h}$ **2.** $e = \lim_{n\to\infty}(1 + \frac{1}{n})^n$

Proof **1.** Let $F(x) = \ln x$. We know that $F'(x) = \frac{1}{x}$, so that $F'(1) = 1$. From the definition of the derivative, we have

$$\begin{aligned} F'(1) &= \lim_{h\to 0} \frac{F(1+h)-F(1)}{h} \\ &= \lim_{h\to 0} \frac{\ln(1+h)-\ln 1}{h} \\ &= \lim_{h\to 0} \frac{1}{h}\ln(1+h) && \text{Since } \ln 1 = 0 \\ &= \lim_{h\to 0} \ln(1+h)^{1/h} && \text{By Theorem 6.7} \\ &= \ln\left[\lim_{h\to 0}(1+h)^{1/h}\right] && \text{By continuity of ln} \end{aligned}$$

Thus,

$$\ln\left[\lim_{h\to 0}(1+h)^{1/h}\right] = 1$$

and since $\ln a = b$ implies $a = e^b$,

$$\lim_{h\to 0}(1+h)^{1/h} = e$$

2. Let $\frac{1}{n} = h$. As $n \to \infty$, $h \to 0$, so we have

$$\lim_{n\to\infty}(1+\tfrac{1}{n})^n = \lim_{h\to 0}(1+h)^{1/h} = e$$

by part 1. ■

With the aid of a calculator we get the following approximate values for $(1+\frac{1}{n})^n$:

n	$(1+\frac{1}{n})^n$
1	2.000000
10	2.59374
100	2.704814
1,000	2.716924
10,000	2.718146
100,000	2.718268
1,000,000	2.718280

This is convincing evidence that $e \approx 2.71828$, and it can be proved that this is correct to five decimal places.

The limit in part 1 of Theorem 6.8 arises naturally in compound interest problems. When an amount of money P (called the principal) is invested at an annual interest rate r (for example, if the interest rate is 8%, $r = 0.08$), and interest is compounded annually, the total amount A_1 in the account after t years is

$$A_1(t) = P(1+r)^t$$

(You will be asked to show this in the exercises.) If interest is compounded semiannually, the amount after t years becomes

$$A_2(t) = P(1+\tfrac{r}{2})^{2t}$$

and if it is compounded k times a year,

$$A_k(t) = P(1 + \tfrac{r}{k})^{kt} \tag{6.33}$$

Some banks compound interest daily, so that

$$A_{365}(t) = P(1 + \tfrac{r}{365})^{365t}$$

What we want to do now is to let $k \to \infty$ in equation (6.33). The result will give the amount that results from *continuous* compounding of interest. If we write $h = \frac{r}{k}$, then $h \to 0$ as $k \to \infty$, and we have

$$\begin{aligned} A(t) &= \lim_{k\to\infty} A_k(t) = \lim_{k\to\infty} P(1 + \tfrac{r}{k})^{kt} \\ &= \lim_{k\to\infty} P(1 + \tfrac{r}{k})^{k/r \cdot rt} \\ &= \lim_{h\to 0} P(1 + h)^{1/h \cdot rt} \\ &= P\left[\lim_{h\to 0}(1 + h)^{1/h}\right]^{rt} \quad \text{Using continuity of the exponential function} \\ &= Pe^{rt} \end{aligned}$$

So we have the formula for continuously compounded interest:

$$A(t) = Pe^{rt}$$

Comment The base e is by far the most commonly used base for logarithms and exponentials in calculus and its applications. The only other bases of any widespread use are 2 and 10. The base 2 is important in computer science. Logarithms with base 10 are called **common** logarithms. Before the advent of the hand calculator, common logarithms were used extensively in numerical calculations, but their use for this purpose is now virtually obsolete. In the exercises that follow we use a variety of bases to give practice in the theory of this section, but in the future we will concentrate on natural logarithms and exponentials.

EXERCISE SET 6.5

A

In Exercises 1 and 2 express each number in terms of the base e (natural exponentials or logarithms), and then approximate its value correct to five decimal places with the aid of a calculator.

1. (a) $\pi^{\sqrt{3}}$ (b) π^{π} (c) π^{e} (d) $(\sqrt{2})^{\sqrt{2}}$ (e) $(\sin 2)^{\pi}$

2. (a) $\log_3 4$ (b) $\log_{1/2} 5$ (c) $\log_6 4$ (d) $\log_7 3$ (e) $\log_2 100$

In Exercises 3–18 find the derivative.

3. $y = 2^x$

4. $y = 3^{1-x^2}$

5. $y = x^2(4^{-x})$

6. $y = \log_2 x$

7. $y = \log_3 \dfrac{x}{1 + x^2}$

8. $y = 3^x \cdot \log_3 x$

9. $y = \log_{10} \dfrac{\sqrt{x^2 - 1}}{x}$

10. $y = (\log_3 \sqrt{4 - x})^2$

11. $y = \log_2(\log_3 x)$

12. $y = (3^x)(x^3)$

13. $y = x^x$

14. $y = x^{\sin x}$

15. $y = x^{1/(1-x)}$

16. $y = (\cos x)^{\sin x}$

17. $y = (1 - x^2)^{1/x}$

18. $y = (\ln x)^x$

In Exercises 19–28 evaluate the integrals.

19. $\int 5^x \, dx$

20. $\int \dfrac{x}{2^{x^2}} \, dx$

21. $\int \frac{3^x}{1+3^x}\,dx$

22. $\int (\cos x)\cdot 2^{\sin x}\,dx$ **23.** $\int x^2 \cdot 2^{x^3-1}\,dx$

24. $\int \frac{dt}{1+2^{-t}}$ (*Hint:* Multiply by $2^t/2^t$.)

25. $\int_1^4 \frac{2^{\sqrt{x}}}{\sqrt{x}}\,dx$ **26.** $\int_1^3 \frac{3^t}{\sqrt{3^t-2}}\,dt$

27. $\int \frac{dx}{3^x(1-3^{-x})^2}$

28. $\int \frac{dt}{(2^t-2^{-t})^2}$ (*Hint:* Multiply by $2^{2t}/2^{2t}$.)

In Exercises 29 and 30 draw the graphs of each of the functions in parts a through d on the same set of axes.

29. (a) $y = 2^x$ (b) $y = e^x$
(c) $y = 10^x$ (d) $y = (\frac{1}{2})^x$

30. (a) $y = \log_2 x$ (b) $y = \log_e x$
(c) $y = \log_{10} x$ (d) $y = \log_{1/2} x$

31. Prove that for $x \neq 0$, $a > 0$, and $a \neq 1$,

$$\frac{d}{dx}\log_a|x| = \frac{1}{x}\ln a$$

32. Using the result of Exercise 31 along with the chain rule, differentiate each of the following:

(a) $y = \log_{10}\left|\frac{x-1}{x+2}\right|$ (b) $y = \log_2\left|\frac{x\sqrt{1-x}}{(2x-3)^3}\right|$

In Exercises 33 and 34 discuss completely and sketch.

33. $y = 2^{1-x^2}$ **34.** $y = \log_{10}\sqrt{x^2-1}$

In Exercises 35–38 solve for x in terms of natural logarithms and then approximate the answer using a calculator.

35. $3^{2x-1} = 4^x$ **36.** $2\cdot 3^{x+1} = 3\cdot 2^{2-x}$

37. $3e^{x+2} = 4\cdot 5^x$ **38.** $2^x\cdot 3^{1-2x} = 5\cdot 6^{-x}$

B

39. Complete the proof of Theorem 6.5 by proving Properties 2, 4, and 5.

40. Prove Theorem 6.7.

41. Prove the formula $A_1(t) = P(1+r)^t$ for the total amount of money $A_1(t)$ after t years when a principal P is invested at a rate of interest r, compounded annually.

42. Using the properties of logarithms, solve for x and check your answers:
(a) $\log_4(x+2) - \log_4(x-2) = \frac{1}{2}$
(b) $2\log_2|x-1| = 2 - \log_2(x+2)$

43. Find the volume obtained by revolving the region under the graph of $y = 2^{1-x^2}$ from $x = 0$ to $x = 1$ about the y-axis.

44. Find the volume obtained by revolving the region under the graph of $y = \log_3 x$ from $x = 1$ to $x = 3$ about the y-axis.

45. Obtain an alternate proof of part 2 of Theorem 6.8 by the following steps:

(a) For $t \in [1, 1+\frac{1}{n}]$ show that

$$\int_1^{1+1/n} \frac{1}{1+\frac{1}{n}}\,dt \le \int_1^{1+1/n} \frac{1}{t}\,dt \le \int_1^{1+1/n} 1\,dt$$

(b) From part a show that

$$\frac{1}{n+1} \le \ln\left(1+\frac{1}{n}\right) \le \frac{1}{n}$$

(c) In part b multiply through by n and show that the result can be put in the form

$$\frac{1}{1+\frac{1}{n}} \le \ln\left(1+\frac{1}{n}\right)^n \le 1$$

(d) In part c let $n \to \infty$ and show how it follows that

$$\lim_{n\to\infty}\left(1+\frac{1}{n}\right)^n = e$$

6.6 EXPONENTIAL GROWTH AND DECAY

An important application of the exponential function is in modeling certain phenomena of nature where the rate of change of some quantity at any given instant is approximately proportional to the amount of the quantity present at that instant. For example, under certain conditions a culture of bacteria will grow at a rate proportional to the number present: the more bacteria there are, the faster the rate of growth of the culture. Even in human populations a crude model is that the rate of population increase is proportional to the size of the population at any given time. This model was used by the English demographer, Thomas Malthus, in the late 18th century to predict future world populations and is referred to as the **Malthus model.** It is reasonably accurate for relatively short time intervals but because it does not take into account such limiting factors as war, famine, epidemics, and limits on available food supply and available space, its growth predictions are much too large for longer time intervals. We will indicate in the exercises (Exercise 24) a more realistic model.

Bacteria cultures and human (as well as other) populations are quantities that increase with time. In other quantities the rate of change is proportional to the amount present and the growth rate is negative; that is, the quantity decreases with time. An example of this is the rate of change of a radioactive substance, called *radioactive decay*. One interesting application of this is in determining the age of certain plant or animal remains. Archeologists in particular have used this technique. It is called *radiocarbon dating*, and we will describe how it works in one of the exercises (Exercise 25).

We could cite many other examples, but for now we want to show how the class of problems we are considering can be solved using calculus. We let $Q(t)$ denote the quantity in question at time t, where t is measured in appropriate units starting from a convenient time origin. Thus $Q(t)$ might denote the number of bacteria in a culture t days after the culture was prepared, or $Q(t)$ might be the population of the world t years after 1900, or $Q(t)$ might be the amount of strontium-90 (a radioactive substance) that will be left t years from now out of 200 g that are on hand. In some cases the actual quantity in question might change by only discrete amounts, such as population, which is always an integer. Nevertheless, in our model we will assume Q is a differentiable function, so that it varies in a continuous fashion over some interval. In cases where the answer must be an integer, we round our answer off to the nearest integer. We assume in all cases that $Q(t) > 0$.

The basic assumption is that the rate of change of Q is proportional to Q itself. Thus,

$$Q'(t) = kQ(t) \tag{6.34}$$

where k is the constant of proportionality, sometimes called the *growth constant*. We want to determine the explicit form of $Q(t)$. Since $Q(t) > 0$, we can write equation (6.34) in the equivalent form

$$\frac{Q'(t)}{Q(t)} = k$$

and since

$$\frac{d}{dt} \ln Q(t) = \frac{Q'(t)}{Q(t)}$$

we have

$$\frac{d}{dt} \ln Q(t) = k$$

Taking the antiderivative, we get

$$\ln Q(t) = kt + C$$

Hence,

$$Q(t) = e^{kt+C}$$

Since $e^{kt+C} = e^{kt} \cdot e^{C}$, we can write this result in the form

$$Q(t) = C_1 e^{kt}$$

where $C_1 = e^C$. Now suppose that when $t = 0$, $Q = Q_0$; that is, $Q(0) = Q_0$. Then we get, on putting $t = 0$,

$$Q(0) = C_1 e^{k \cdot 0} \quad \text{or} \quad Q_0 = C_1$$

Thus,

$$Q(t) = Q_0 e^{kt} \tag{6.35}$$

It is easy to show that this value of Q does actually satisfy equation (6.34) and has the correct value when $t = 0$. So this is the solution we were seeking. The growth constant k is determined in each individual instance from known data, usually based on observed values of Q at certain times. When $k > 0$ we say that Q **grows exponentially,** and when $k < 0$ we say that Q **decays exponentially.**

Equation (6.34) is an example of what is known as a **differential equation,** meaning an equation that involves derivatives (or differentials). We will consider differential equations in more detail in Chapter 20. The method we used to solve for $Q(t)$ is known as *separation of variables.* The reason for this terminology becomes clearer if we write Q for $Q(t)$ and $\frac{dQ}{dt}$ for $Q'(t)$. Then equation (6.34) becomes $\frac{dQ}{dt} = kQ$ or, in the equivalent differential form,

$$dQ = kQ\,dt$$

On dividing by Q, we get

$$\frac{dQ}{Q} = k\,dt$$

Now the variables Q and t are *separated* on opposite sides of the equation. Taking antiderivatives, we get, as before,

$$\ln Q = kt + C$$

and then we solve for $Q = Q(t)$ as we did earlier.

EXAMPLE 6.16 A culture of bacteria is observed to triple in size in 2 days. Assuming it grows exponentially, how large will the culture be in 5 days?

Solution We are not told the original size of the culture, so we let this be denoted by Q_0. We let $Q(t)$ be the size of the culture t days after the initial observation. When $t = 2$ we know that the size has tripled, so $Q(2) = 3Q_0$. By equation (6.35) we have

$$Q(t) = Q_0 e^{kt}$$

On putting $t = 2$, we get

$$3Q_0 = Q_0e^{k \cdot 2}$$
$$e^{2k} = 3$$

We write this in the equivalent logarithmic form to find k:

$$2k = \ln 3$$
$$k = \frac{\ln 3}{2}$$

Using a calculator or tables, we could get a decimal approximation for k. However, it is usually an advantage to leave k in an exact form at this stage. Substituting this value for k gives

$$Q(t) = Q_0e^{t(\ln 3)/2} = Q_0(e^{\ln 3})^{t/2}$$
$$Q(t) = Q_0(3)^{t/2}$$

(You can see now the advantage of leaving k in an exact form.) When $t = 5$ we have

$$Q(6) = Q_0(3)^{5/2} = 9\sqrt{3}Q_0 \approx 15.6Q_0$$

So after 5 days the culture will be approximately 15.6 times its original size. ■

A useful measure of the rate of decomposition of a radioactive substance is its **half-life,** defined to be the time required for half of the original amount to decompose. Thus if 100 g of a radioactive substance that has a half-life of 5 yr are on hand, then in 5 yr 50 g will remain; in 5 more years, 25 g will remain; and so on. The next example makes use of this concept.

EXAMPLE 6.17 Strontium-90 has a half-life of 25 yr. If 200 kg are on hand, how much will remain after 10 yr? How long will it be until 90% of the original amount has decomposed?

Solution Let $Q(t)$ be the number of kilograms that remain after t years. Then we know that $Q(0) = 200$, $Q(25) = 100$, and

$$Q(t) = Q_0e^{kt} = 200e^{kt}$$

Putting $t = 25$ gives

$$100 = 200e^{k \cdot 25}$$
$$e^{25k} = \tfrac{1}{2}$$
$$25k = \ln \tfrac{1}{2} = -\ln 2$$
$$k = -\tfrac{1}{25}\ln 2$$

Thus,

$$Q(t) = 200e^{(-\ln 2/25)t} = 200(e^{\ln 2})^{-t/25} = 200(2)^{-t/25}$$

So

$$Q(10) = 200(2)^{-10/25} = 200(2)^{-2/5} \approx 151.6 \text{ kg}$$

To determine how long it will be until 90% has decomposed, we first observe that at this time 10%, or 20 kg, will remain. Thus, we want to find t for which $Q(t) = 20$; that is

$$200(2)^{-t/25} = 20$$
$$2^{-t/25} = \tfrac{1}{10}$$

We take the natural logarithm of both sides to find t:

$$-\tfrac{t}{25}\ln 2 = \ln \tfrac{1}{10}$$

or

$$-\tfrac{t}{25}\ln 2 = -\ln 10$$
$$t = \frac{25 \ln 10}{\ln 2} \approx 83.05 \text{ yr}$$

■

EXAMPLE 6.18 In 1930 the world population was approximately 2 billion and in 1960 approximately 3 billion. Using the Malthus model, estimate the population in the year 2000.

Solution Let $Q(t)$ be the population in billions, with t in years measured from 1930. Then $Q(0) = 2$, $Q(30) = 3$, and, according to the Malthus model,

$$Q(t) = Q_0 e^{kt} = 2e^{kt}$$

Setting $t = 30$ and proceeding in the usual way, we get

$$3 = 2e^{30k}$$
$$e^{30k} = \tfrac{3}{2}$$
$$30k = \ln \tfrac{3}{2}$$
$$k = \tfrac{1}{30}\ln \tfrac{3}{2}$$

So

$$Q(t) = 2(e^{\ln 3/2})^{t/30} = 2 \cdot (\tfrac{3}{2})^{t/30}$$

In the year 2000, $t = 70$. So according to this model the population should be

$$Q(70) = 2 \cdot (\tfrac{3}{2})^{70/30} = 2 \cdot (\tfrac{3}{2})^{7/3} \approx 5.15 \text{ billion}$$

■

EXAMPLE 6.19 **Newton's law of cooling** states that when a body at initial temperature T_0 is introduced into a medium of temperature T_m, where $T_m < T_0$, the rate at which the body cools at any given instant is proportional to the difference between its temperature at that instant and the temperature T_m of the surrounding medium.

(a) Find a formula for the temperature of the body at time t units after being introduced into the medium.

(b) Suppose a body at 120°C is placed in air at 20°C and it has cooled to a temperature of 80°C after $\frac{1}{2}$ hr. Find its temperature after 1 hr.

Solution (a) Let $T(t)$ be the temperature of the body t units of time after it is introduced into the medium. Then by Newton's law of cooling,

$$T'(t) = k[T(t) - T_m]$$

Writing $T = T(t)$ and $\frac{dT}{dt} = T'(t)$, we obtain the equivalent differential equation

$$dT = k(T - T_m)\,dt$$

Now we separate variables and take antiderivatives:

$$\int \frac{dT}{T - T_m} = \int k\,dt$$

$$\ln(T - T_m) = kt + C \quad \text{Note that } T - T_m > 0.$$

$$T - T_m = e^{kt+C} = e^{kt} \cdot e^C = C_1 e^{kt}$$

So, on solving for $T = T(t)$, we get

$$T(t) = T_m + C_1 e^{kt}$$

To find the constant C_1 we put $t = 0$, observing that $T(0) = T_0$:

$$T_0 = T_m + C_1 e^0$$
$$C_1 = T_0 - T_m$$

Finally,

$$\begin{aligned} T(t) &= T_m + C_1 e^{kt} \\ &= T_m + (T_0 - T_m)e^{kt} \end{aligned} \tag{6.36}$$

(b) We are given that $T_0 = 120$, $T_m = 20$, and $T(\frac{1}{2}) = 80$. So from equation (6.36),

$$T(t) = 20 + 100e^{kt} \tag{6.37}$$

$$\begin{aligned} 80 &= 20 + 100e^{k(1/2)} \\ 60 &= 100e^{k/2} \\ e^{k/2} &= 0.6 \\ \tfrac{k}{2} &= \ln(0.6) \\ k &= 2\ln(0.6) = \ln(0.36) \end{aligned}$$

Substituting for k in equation (6.37) gives

$$\begin{aligned} T(t) &= 20 + 100e^{(\ln 0.36)t} \\ &= 20 + (100)(e^{\ln 0.36})^t \\ &= 20 + 100(0.36)^t \end{aligned}$$

Finally, we want $T(1)$:

$$T(1) = 20 + 100(0.36) = 56$$

So after 1 hr the temperature should be approximately 56°C. ■

Remark The result, equation (6.36), is valid also if the body is introduced into a warmer medium (so that $T_m > T_0$) and you will be asked to show this in the exercises.

EXERCISE SET 6.6

A

1. A culture of bacteria originally numbers 500. After 2 hr there are 1500 bacteria in the culture. Assuming exponential growth, find how many are present after 6 hr.

2. A radioactive substance has a half-life of 64 hr. If 200 g of the substance are initially present, how much will remain after 4 days?

3. A culture of 100 bacteria doubles after 2 hr. How long will it take for the number of bacteria to reach 3200?

4. How long will it take 100 g of a radioactive substance that has a half-life of 40 yr to be reduced to 12.5 g?

5. One hundred kilograms of a certain radioactive substance decay to 40 kg after 10 yr. Find how much will remain after 20 yr.

6. In the initial stages a bacteria culture grows approximately exponentially. If the number doubles after 3 hr, what proportion of the original number will be present after 6 hr? After 12 hr?

7. If a culture of bacteria doubles in size after 2 hr, how long will it take for it to triple in size? Assume exponential growth.

8. If 100 g of a radioactive material diminish to 80 g in 2 yr, find the half-life of the substance.

9. Find the half-life of a radioactive substance that is reduced by 30% in 20 hr.

10. A radioactive substance has a half-life of 6 yr. If 20 lb are present initially, how much will remain after 2 yr?

11. A culture of bacteria increases from 1000 to 5000 in 8 hr. Assuming the growth is exponential, find how much will be present after 20 hr.

12. The population of a certain Florida city increased by 40% between 1970 and 1978. If the population in 1978 was 210,000, what is the expected population in the year 2000 according to the Malthus model?

13. A certain radioactive substance has a half-life of 8 yr. If 200 g are present initially, how much will remain at the end of 12 yr? How long will it be until 90% of the original amount has decayed?

14. A body heated to 120°F is brought into a room in which the temperature is 70°F. After 15 min the temperature of the body is 100°F. According to Newton's law of cooling, how long will it take for the temperature of the body to drop to 80°F?

15. A thermometer registering 20°C is placed in a freezer in which the temperature is -10°C. After 10 min the thermometer registers 5°C. According to Newton's law of cooling, what will it register after 30 min?

16. According to Newton's law of cooling will the temperature of the body ever reach the temperature T_m of the surrounding medium? Show why or why not. Does this result agree with reality? If not, how do you explain the discrepancy?

17. Prove that the result in equation (6.36) remains valid if $T_m > T_0$. Use this to find how long it will take a thermometer that registers 5°C when on the outside to rise to 18°C after being brought to the inside where the temperature is 20°C, if it takes 10 min to rise to 12°C.

18. The rate of change of air pressure p with respect to altitude h is approximately proportional to the pressure. If at sea level the pressure is p_0, find a formula for p as a function of h. If $p_0 = 15$ lb/sq in., and at altitude $h = 10{,}000$ ft the pressure is 10 lb/sq in., find the approximate pressure at 20,000 ft.

19. The atmospheric pressure at sea level is approximately 15 lb/sq in., and at Denver, which is 1 mi high, the pressure is approximately 12 lb/sq in. What is the approximate pressure at the top of Vail Pass, which is 2 mi high? (See Exercise 18.)

20. The rate of change of the intensity I of light passing through a translucent material, with respect to the depth x of penetration of the light ray, is approximately proportional to the intensity at depth x. If the sunlight that strikes water is reduced to half of its original intensity at a depth of 10 ft, what will be the intensity (as a fractional part of the original intensity) at a depth of 30 ft?

21. Use the data of Exercise 20 to find at what depth the intensity of illumination will be 10% of the original.

22. In a certain type of chemical reaction a substance is converted in such a way that the rate of conversion is always proportional to the amount of unconverted substance. If there are 20 g of the substance originally, and half of it has been converted in 2 min, find how long it will be until only 4 g remain unconverted.

23. The rate at which the number of items of a certain type of merchandise are sold is proportional to the number still on hand. Let $N(t)$ be the number of items still on hand t days after the items were first introduced, and let $N_1 = N(1)$.

(a) Derive a formula for $N(t)$.

(b) If 20 items are sold on the first day and 10 on the fifth day, how many can be expected to be sold on the ninth day?

B

24. A more realistic model than the Malthus model for population growth is the **Verhulst model,** which asserts that the rate of growth of the population is jointly proportional to the size of the population and the further growth capacity. Thus, if $Q(t)$ denotes the size of the population at time t and m is the ultimate limit on population size,

$$\frac{dQ}{dt} = kQ(m - Q)$$

(a) Prove, by direct substitution into the differential equation, that

$$Q(t) = \frac{mQ_0}{Q_0 + (m - Q_0)e^{-mkt}}$$

is a solution of the Verhulst model, where $Q_0 = Q(0)$.

(b) Prove that for the solution in part a, $\lim_{t \to \infty} Q(t) = m$

(c) Redo Example 6.18 using the Verhulst model with the assumption that the maximum supportable world population is 10 billion.

25. A radioactive isotope of carbon, called carbon-14, is found in all living organisms. During the lifetime of an organism the amount that decays is replenished through the atmosphere, so that the amount in the organism remains constant, but after the death of the organism the decayed carbon-14 is no longer replenished. Because the half-life of carbon-14 is very long (approximately 5730 yr), the remains of organisms that died thousands of years ago may still contain measurable amounts of carbon-14. By comparing the amount remaining with the amount known to be present in the particular organism when it was alive, the age of the remains can be determined. Suppose that archeologists find a human skull and determine that 10% of the original amount of carbon-14 remains (the original amount being determined by measuring the amount in a similar present-day skull). Find the approximate age of the skull.

26. The radioactive substance einsteinium-253 decays at a rate given approximately by $\frac{dQ}{dt} = -0.0939Q$, where t is measured in days.

(a) Find the half-life.

(b) Determine the proportion of an original amount that will remain after 8 days.

(c) Find the number of days required for 80% of an original amount to decay.

27. Certain radioactive substances decay into a second substance. If $Q(t)$ denotes the amount of such a radioactive substance at time t and $R(t)$ denotes the amount of the second substance, then

$$Q(t) = Q_0 e^{kt}, \qquad Q_0 = Q(0)$$

and since the amount of substance that has decomposed is the original amount Q_0 minus the amount $Q(t)$ still present,

$$R(t) = Q_0 - Q(t) = Q_0(1 - e^{kt})$$

(a) Find a formula for the time required for the ratio of R to Q to equal a given value λ.

(b) Rubidium-87, with a half-life of 5×10^{11} yr, decays exponentially into strontium-87. Find the time required for there to be 1% as much strontium-87 as rubidium-87.

28. Refer to Exercise 27, part b. By measuring the ratio of strontium-87 to rubidium-87 in fossils, scientists can estimate the age of the fossil. Use this method to estimate the age of a fossil that contains 237 parts per million of rubidium and 1.85 parts per million of strontium.

6.7 SUPPLEMENTARY EXERCISES

In Exercises 1–7 find $f'(x)$.

1. (a) $f(x) = \ln(x\sqrt{x - 2})$ (b) $f(x) = \ln\left(\frac{x^2 - x}{x^2 - 2x - 8}\right)$

2. (a) $f(x) = \sqrt{\ln \sec^2 x}$ (b) $f(x) = x \ln \frac{1}{x^2}$

3. (a) $f(x) = x^2 e^{-x}$ (b) $f(x) = e^{\ln e^{-x}}$

4. (a) $f(x) = e^{\sin x + \ln x}$ (b) $f(x) = \ln\left(\frac{e^x - 1}{e^x + 1}\right)$

5. (a) $f(x) = \cos e^{-x} - \sin e^{-x}$
(b) $f(x) = (\ln \sqrt{e^x + e^{-x}})^2$

6. (a) $f(x) = x \log_3 x$ (b) $f(x) = \dfrac{2^x}{1 - 2^x}$

7. (a) $f(x) = x^{\ln x}$ (b) $f(x) = (\sec x)^{\tan x}$

8. Use logarithmic differentiation to find y':

(a) $y = \dfrac{x^3 \sqrt{2x+1}}{(x-3)^2}$ (b) $y = \dfrac{(x-1)^5(x^2+1)^{2/3}}{\sqrt{x^3 - 5x^2 + 6x}}$

9. Find y' using implicit differentiation:

(a) $e^y - \ln \dfrac{x}{y} = 1$ (b) $x = \ln\left(\dfrac{x+y}{x-y}\right)$

In Exercises 10–15 evaluate the given definite or indefinite integral.

10. (a) $\displaystyle\int \frac{\cos 2x}{1 + \sin 2x}\,dx$ (b) $\displaystyle\int \frac{x+1}{x^2 + 2x + 4}\,dx$

11. (a) $\displaystyle\int e^{-2 \ln x}\,dx$ (b) $\displaystyle\int \frac{e^x - 1}{e^{2x} - 2xe^x + x^2}\,dx$

12. (a) $\displaystyle\int \frac{\sqrt{x}}{\sqrt{x^3 + 4}}\,dx$ (b) $\displaystyle\int \frac{dx}{x\,(\ln 2x + 3)}$

13. (a) $\displaystyle\int_{\ln 2}^{\ln 3} \frac{e^{-2x}}{1 + e^{-2x}}\,dx$ (b) $\displaystyle\int_{e^{-2}}^{e^2} \frac{dx}{x + \sqrt{x}}$

14. (a) $\displaystyle\int_{(1/2)\ln 7}^{2 \ln 2} e^{2x}\sqrt{9 + e^{2x}}\,dx$

(b) $\displaystyle\int_{\ln 2}^{\ln 3} \frac{e^{2x} - 1}{e^{2x} + 1}\,dx$ (*Hint:* Multiply by e^{-x}/e^{-x}.)

15. (a) $\displaystyle\int \frac{2^{2x}}{4^x + 1}\,dx$ (b) $\displaystyle\int_0^2 \frac{\sqrt{1 - 2^{-t}}}{2^t}\,dt$

In Exercises 16–19 discuss completely and draw the graph.

16. $y = \dfrac{e^x - e^{-x}}{e^x + e^{-x}}$ **17.** $y = \ln|1 - x|$

18. $y = x \ln x^2$ **19.** $y = x^3 e^{-x^2/2}$

20. Show that f is 1–1 on the specified domain. Find f^{-1} and give its domain.

(a) $f(x) = \dfrac{3x - 1}{2x + 1}$; $x > -\dfrac{1}{2}$

(b) $f(x) = x^2 + 4x + 1$; $x \geq -2$

21. Show that f is monotone on the given interval, and find $(f^{-1})'(y_0)$.
(a) $f(x) = x^2 + 2x - 5$ on $(-1, \infty)$; $y_0 = f(1)$
(b) $f(x) = 2x^3 - 3x^2 - 12x + 5$ on $(-\infty, -1)$; $y_0 = 1$

22. Show that the function

$$f(x) = \frac{e^{2x} - 1}{e^{2x} + 1}$$

is 1–1 on **R**. Find f^{-1} and give its domain. Calculate $(f^{-1})'(x)$ in two ways: by using the explicit form of f^{-1} and by Theorem 6.3.

23. Find the volume obtained by revolving the region under the curve $y = e^{x^2/2}$ from $x = 0$ to $x = \sqrt{\ln 4}$ about the y-axis.

24. Find the surface area obtained by revolving the arc of the curve

$$y = e^{x/2} + e^{-x/2}$$

from $x = -\ln 3$ to $x = \ln 3$ about the x-axis.

25. Solve for x in terms of natural logarithms and approximate the answer using a calculator:
(a) $2 \cdot 3^{-x} = 5 \cdot 4^{2x}$
(b) $6e^{2x-1} - 3^x \cdot 2^{1-x} = 0$

26. Solve for x and check your answers:
(a) $\ln x + \ln(x - 1) = \ln 2$
(b) $2 \log_3 x - \log_3(2x - 3) = 1$

27. The radioactive isotope cobalt-60 has a half-life of approximately 5.26 yr. Of a given initial amount, how much will remain after 3 yr? How long will it take for 80% of the original amount to decay?

28. A culture of bacteria is observed to double in size in 13 hr. How long will it take to triple in size? Assume exponential growth.

29. A steel plate that has been heated to 40°C by exposure to direct sunlight is brought into an air-conditioned room where the temperature is 20°C and after 2 min it has cooled to 38°C. According to Newton's law of cooling, what will be the temperature of the plate after 5 min? How long will it take for the temperature to drop to 21°C?

30. A modification of the Malthus model for population growth is given by

$$Q'(t) = kQ \ln \frac{m}{Q}, \qquad Q < m$$

where k and m are positive constants. (This is known as the **Gompertz** model.) Find $Q(t)$ for this model. What is $\lim_{t\to\infty} Q(t)$?

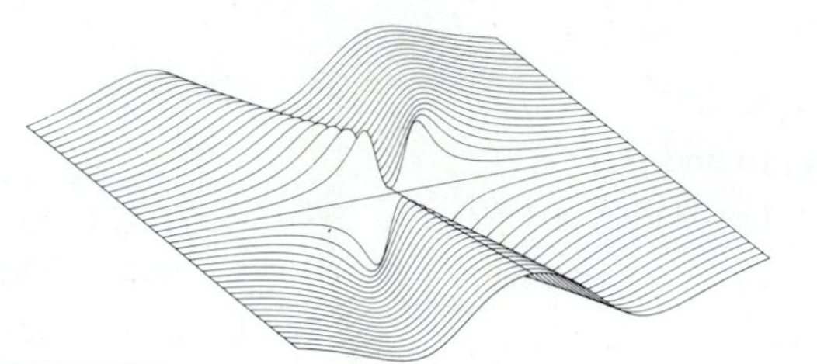

CHAPTER 7

INVERSE TRIGONOMETRIC FUNCTIONS AND HYPERBOLIC FUNCTIONS

7.1 INVERSE TRIGONOMETRIC FUNCTIONS

The graphs of the trigonometric functions clearly show that none of these functions is 1–1, so that inverses of them do not exist. For example, the graph of the sine function is shown in Figure 7.1, and it is clear that any horizontal line drawn between $y = -1$ and $y = 1$ intersects the graph in infinitely many points. We can, however, consider a restricted domain on which the sine function is 1–1. A convenient choice of such a restriction is $[-\frac{\pi}{2}, \frac{\pi}{2}]$. On that interval, the sine is increasing, and it assumes all values in the range $[-1, 1]$. We denote the inverse of the sine restricted in this way by $\mathbf{sin^{-1}}$ or by **arcsin.** We will normally use $\sin^{-1}$, since it is consistent with the notation f^{-1} that we have been using for inverses in general. However,

FIGURE 7.1

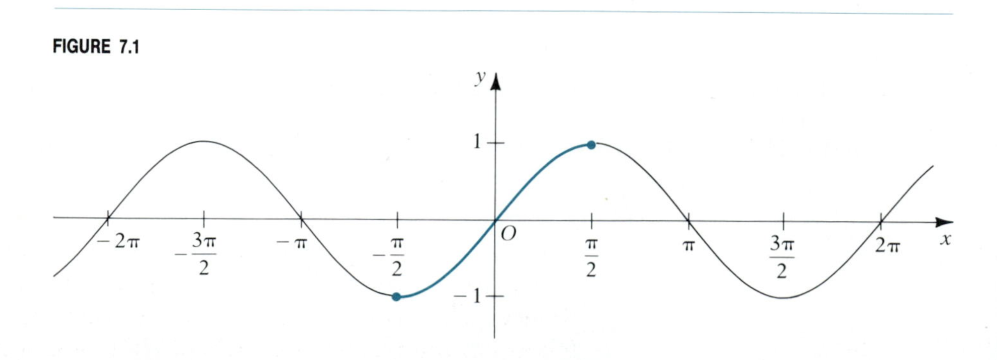

since both notations are in widespread use, you should be familiar with both. Relationships between the restricted sine function and its inverse are shown in Figure 7.2.

FIGURE 7.2

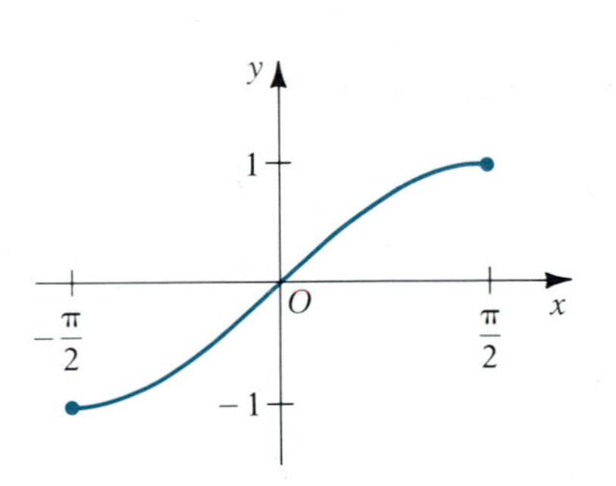

$y = \sin x$

Domain $= \left[-\frac{\pi}{2}, \frac{\pi}{2}\right]$

Range $= [-1, 1]$

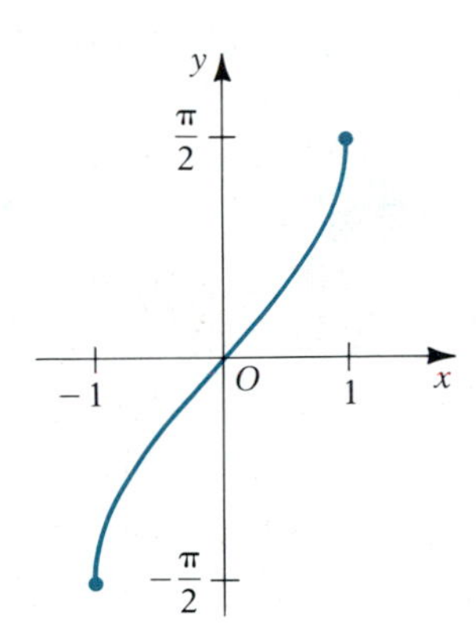

$y = \sin^{-1} x$

Domain $= [-1, 1]$

Range $= \left[-\frac{\pi}{2}, \frac{\pi}{2}\right]$

$y = \sin^{-1} x$ if and only if $\sin y = x$ and $-\frac{\pi}{2} \le y \le \frac{\pi}{2}$

$$\sin^{-1}(\sin x) = x \quad \text{if} \quad x \in [-\tfrac{\pi}{2}, \tfrac{\pi}{2}]$$
$$\sin(\sin^{-1} x) = x \quad \text{if} \quad x \in [-1, 1]$$

For example, we have

$$\sin^{-1} 1 = \tfrac{\pi}{2} \qquad \sin^{-1} \tfrac{1}{2} = \tfrac{\pi}{6}$$
$$\sin^{-1}(-1) = -\tfrac{\pi}{2} \qquad \sin^{-1}(-\tfrac{1}{2}) = -\tfrac{\pi}{6}$$

and so on. Note carefully that $\sin^{-1} x$ *must* be a number in the interval $[-\frac{\pi}{2}, \frac{\pi}{2}]$. It would be *incorrect*, for example, to say that $\sin^{-1}(-\frac{1}{2}) = \frac{11\pi}{6}$ even though $\sin \frac{11\pi}{6} = -\frac{1}{2}$, because $\frac{11\pi}{6}$ does not lie in the interval $[-\frac{\pi}{2}, \frac{\pi}{2}]$. The value of $\sin^{-1} x$, then, can be stated in words:

$\sin^{-1} x$ is that number in $[-\frac{\pi}{2}, \frac{\pi}{2}]$ whose sine is x

To find the derivative of the inverse sine, we let $y = \sin^{-1} x$, so that $x = \sin y$. Then by equation (6.14), with x and y interchanged, we get

$$\frac{dy}{dx} = \frac{1}{\frac{dx}{dy}} = \frac{1}{\cos y}, \quad \text{if } \cos y \neq 0$$

Since $\cos y > 0$ for $-\frac{\pi}{2} < y < \frac{\pi}{2}$, we conclude that the derivative of $\sin^{-1} x$ exists on the open interval $(-1, 1)$. To obtain the answer in terms of x, observe that $\sin^2 y + \cos^2 y = 1$, so that $\cos y = \sqrt{1 - \sin^2 y} = \sqrt{1 - x^2}$ (choosing the positive sign when taking the square root, since $\cos y > 0$).

Thus, we have

$$\frac{d}{dx}\sin^{-1} x = \frac{1}{\sqrt{1-x^2}} \quad \text{if } x \in (-1, 1) \tag{7.1}$$

Combining this with the chain rule in the usual way, we have for $u = g(x)$

$$\frac{d}{dx}\sin^{-1} u = \frac{1}{\sqrt{1-u^2}}\frac{du}{dx} \tag{7.2}$$

if g is a differentiable function such that $g(x) \in (-1, 1)$. This gives the antidifferentiation formula:

$$\int \frac{du}{\sqrt{1-u^2}} = \sin^{-1} u + C, \qquad u \in (-1, 1) \tag{7.3}$$

EXAMPLE 7.1 Find the derivative of each of the following:

(a) $y = \sin^{-1}\frac{1}{x}$ (b) $y = x\sin^{-1} 2x$

Solution (a) By equation (7.2), for $\frac{1}{x} \in (-1, 1)$,

$$\frac{dy}{dx} = \frac{1}{\sqrt{1-\frac{1}{x^2}}}\left(-\frac{1}{x^2}\right) = \frac{1}{\frac{1}{|x|}\sqrt{x^2-1}}\left(-\frac{1}{x^2}\right) = -\frac{|x|}{x^2}\cdot\frac{1}{\sqrt{x^2-1}}$$

$$= \begin{cases} -\dfrac{1}{x\sqrt{x^2-1}} & \text{if } x > 1 \\ \dfrac{1}{x\sqrt{x^2-1}} & \text{if } x < -1 \end{cases}$$

(b) Using the product rule and equation (7.2), we get

$$\frac{dy}{dx} = x\cdot\frac{1}{\sqrt{1-4x^2}}\cdot 2 + \sin^{-1} 2x = \frac{2x}{\sqrt{1-4x^2}} + \sin^{-1} 2x$$

and this is valid if $2x \in (-1, 1)$—that is, if $x \in (-\frac{1}{2}, \frac{1}{2})$. ■

For $u = g(x)$, a differentiable function such that for $a > 0$, $-a < u < a$, the following generalization of equation (7.3) can be proved (see Exercise 43 in Exercise Set 7.1):

$$\int \frac{du}{\sqrt{a^2-u^2}} = \sin^{-1}\frac{u}{a} + C, \qquad u \in (-a, a) \tag{7.4}$$

EXAMPLE 7.2 Evaluate the following integrals:

(a) $\displaystyle\int_0^1 \frac{dx}{\sqrt{4-x^2}}$ (b) $\displaystyle\int \frac{dx}{\sqrt{2x-x^2}}$

Solution (a) By equation (7.4),

$$\int_0^1 \frac{dx}{\sqrt{4-x^2}} = \sin^{-1}\frac{x}{2}\Big]_0^1 = \sin^{-1}\frac{1}{2} - \sin^{-1} 0 = \frac{\pi}{6} - 0 = \frac{\pi}{6}$$

(b) We first complete the square on x:

$$\int \frac{dx}{\sqrt{2x - x^2}} = \int \frac{dx}{\sqrt{1 - (x^2 - 2x + 1)}} = \int \frac{dx}{\sqrt{1 - (x - 1)^2}}$$

Now let $u = x - 1$; then $du = dx$. So we have

$$\int \frac{dx}{\sqrt{1 - (x - 1)^2}} = \int \frac{du}{\sqrt{1 - u^2}} = \sin^{-1} u + C = \sin^{-1}(x - 1) + C \qquad \blacksquare$$

Inverses of the other trigonometric functions suitably restricted can be obtained in a similar way to that of the sine. We outline these below and ask you to supply some of the details in the exercises.

$y = \cos x$

Domain $= [0, \pi]$
Range $= [-1, 1]$

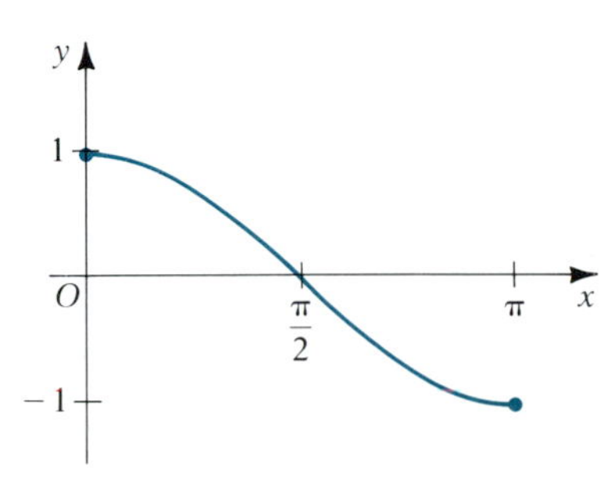

$y = \cos^{-1} x$

Domain $= [-1, 1]$
Range $= [0, \pi]$

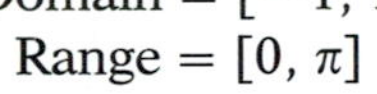

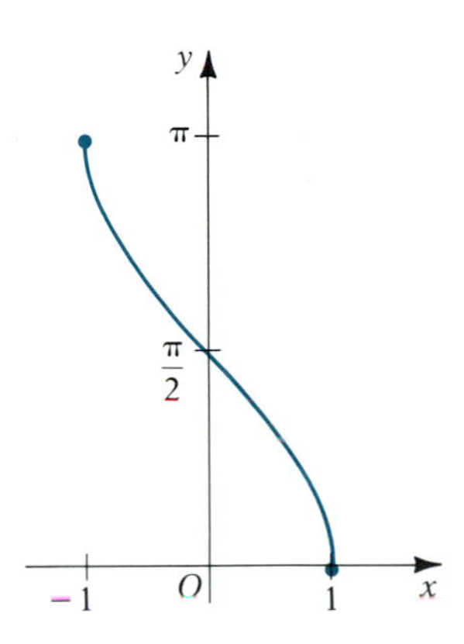

$y = \tan x$

Domain $= (-\frac{\pi}{2}, \frac{\pi}{2})$
Range $= (-\infty, \infty)$

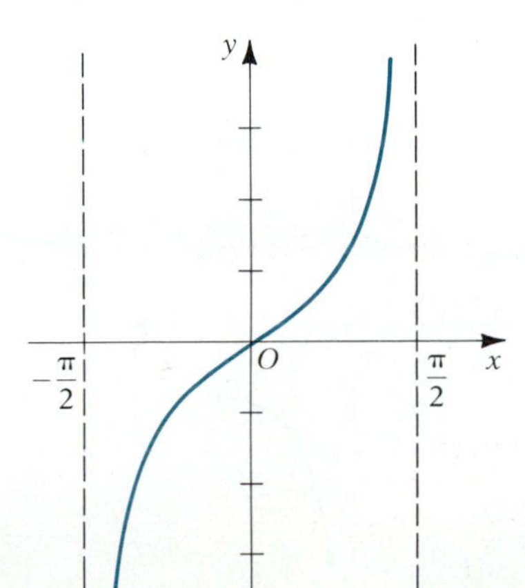

$y = \tan^{-1} x$

Domain $= (-\infty, \infty)$
Range $= (-\frac{\pi}{2}, \frac{\pi}{2})$

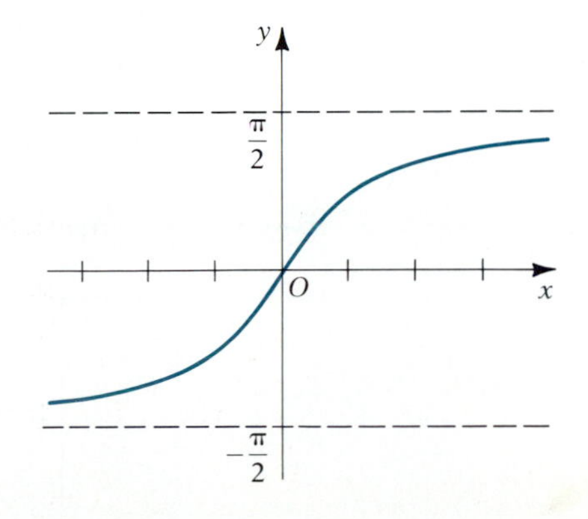

$y = \cot x$

Domain $= (0, \pi)$
Range $= (-\infty, \infty)$

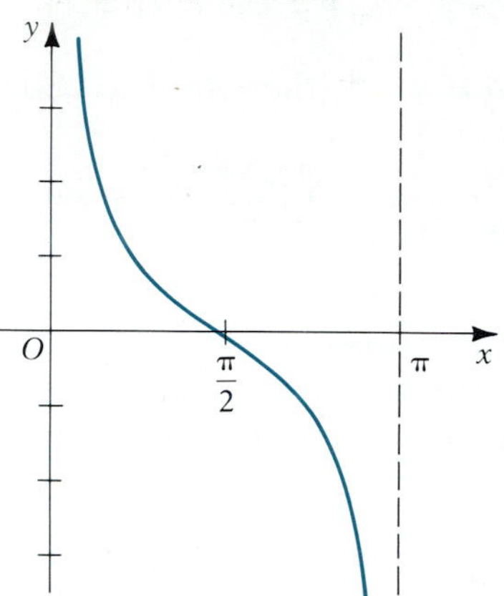

$y = \cot^{-1} x$

Domain $= (-\infty, \infty)$
Range $= (0, \pi)$

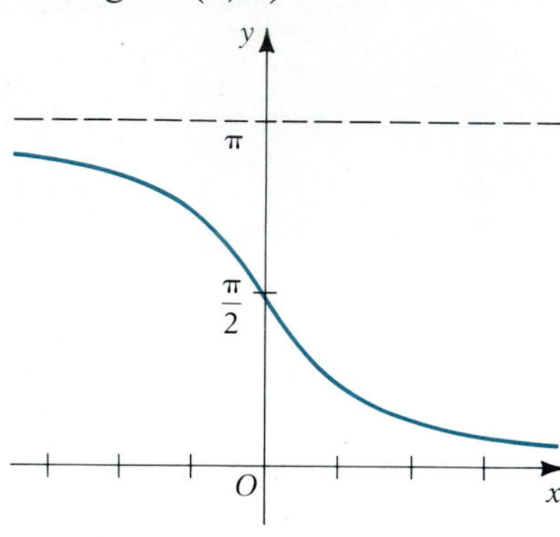

$y = \sec x$

Domain $= [0, \frac{\pi}{2}) \cup [\pi, \frac{3\pi}{2})$
Range $= [1, \infty) \cup (-\infty, -1]$

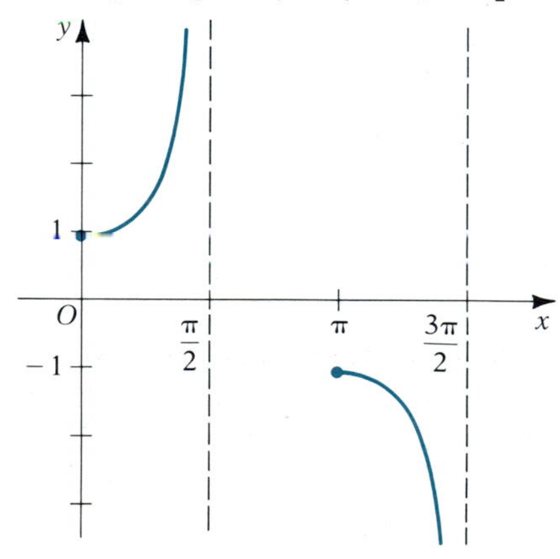

$y = \sec^{-1} x$

Domain $= [1, \infty) \cup (-\infty, -1]$
Range $= [0, \frac{\pi}{2}) \cup [\pi, \frac{3\pi}{2})$

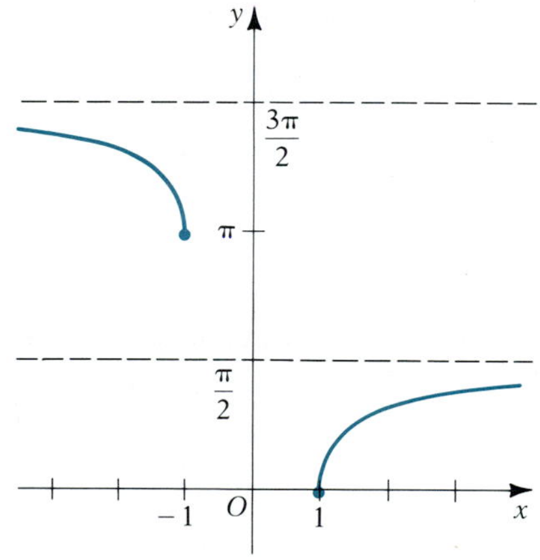

$y = \csc x$

Domain $= (0, \frac{\pi}{2}] \cup (\pi, \frac{3\pi}{2}]$
Range $= [1, \infty) \cup (-\infty, -1]$

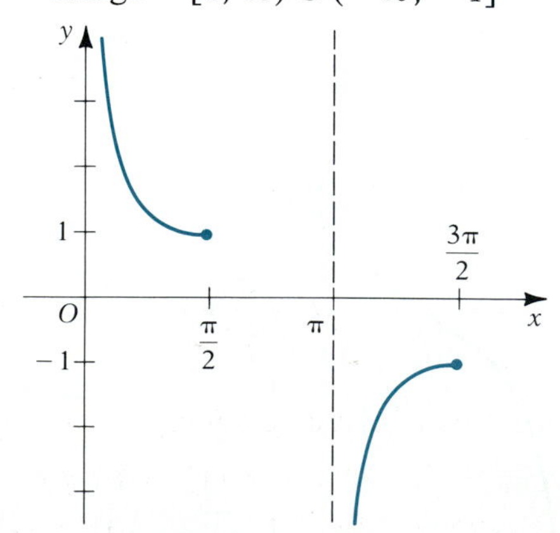

$y = \csc^{-1} x$

Domain $= [1, \infty) \cup (-\infty, -1]$
Range $= (0, \frac{\pi}{2}] \cup (\pi, \frac{3\pi}{2}]$

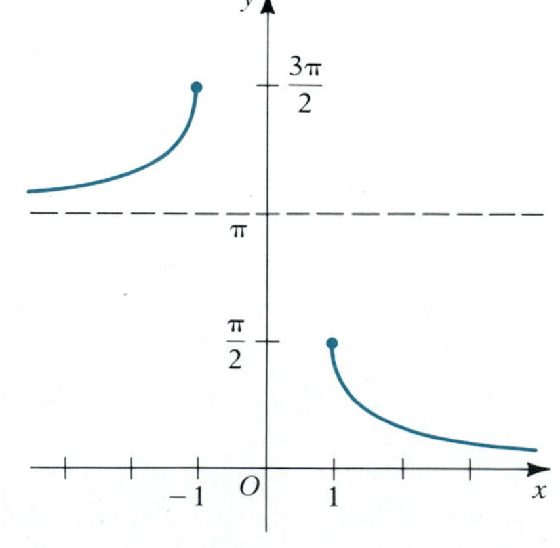

Remark There are other equally valid ways to restrict the domains of the secant and cosecant functions for forming the inverse. The choices we have made are motivated by the fact that the derivative formulas are simpler than with some of the other possible choices.

Derivatives

$$\frac{d}{dx}\cos^{-1} x = -\frac{1}{\sqrt{1-x^2}}, \qquad x \in (-1, 1) \tag{7.5}$$

$$\frac{d}{dx}\tan^{-1} x = \frac{1}{1+x^2}, \qquad x \in (-\infty, \infty) \tag{7.6}$$

$$\frac{d}{dx}\sec^{-1} x = \frac{1}{x\sqrt{x^2-1}}, \qquad |x| > 1 \tag{7.7}$$

$$\frac{d}{dx}\csc^{-1} x = -\frac{1}{x\sqrt{x^2-1}}, \qquad |x| > 1 \tag{7.8}$$

Formulas for the derivatives of $\cos^{-1} u$, $\tan^{-1} u$, $\sec^{-1} u$, and $\csc^{-1} u$, where u is a differentiable function of x that satisfies suitable restrictions, follow from these by the chain rule.

Integrals

$$\int \frac{du}{1+u^2} = \tan^{-1} u + C \tag{7.9}$$

$$\int \frac{du}{u\sqrt{u^2-1}} = \sec^{-1} u + C, \qquad |u| > 1 \tag{7.10}$$

Observe that there is never a need to write the integration formulas that arise from equations (7.5) and (7.8), since the negative sign can be factored out. For example, rather than to write

$$\int -\frac{dx}{\sqrt{1-x^2}} = \cos^{-1} x + C$$

(which is correct), we write

$$\int -\frac{dx}{\sqrt{1-x^2}} = -\int \frac{dx}{\sqrt{1-x^2}} = -\sin^{-1} x + C$$

The following generalizations of equations (7.9) and (7.10) are often useful. We assume $a > 0$.

$$\int \frac{du}{a^2+u^2} = \frac{1}{a}\tan^{-1}\frac{u}{a} + C \tag{7.11}$$

$$\int \frac{du}{u\sqrt{u^2-a^2}} = \frac{1}{a}\sec^{-1}\frac{u}{a} + C, \qquad |u| > a \tag{7.12}$$

EXAMPLE 7.3 Differentiate each of the following:

(a) $y = \cos^{-1}\sqrt{x}$ $(0 < x < 1)$ (b) $y = \tan^{-1} e^x$

(c) $y = x\sec^{-1} x$, $|x| > 1$ (d) $y = \cot^{-1}(\tan x)$

Solution (a) $\dfrac{dy}{dx} = \dfrac{-1}{\sqrt{1-x}} \cdot \dfrac{1}{2\sqrt{x}} = \dfrac{-1}{2\sqrt{x-x^2}}$

(b) $\dfrac{dy}{dx} = \dfrac{1}{1+e^{2x}} \cdot e^x = \dfrac{e^x}{1+e^{2x}}$

(c) $\dfrac{dy}{dx} = x \cdot \dfrac{1}{x\sqrt{x^2-1}} + \sec^{-1} x = \dfrac{1}{\sqrt{x^2-1}} + \sec^{-1} x$

(d) $\dfrac{dy}{dx} = -\dfrac{1}{1+\tan^2 x} \cdot \sec^2 x = -\dfrac{\sec^2 x}{\sec^2 x} = -1$ ■

EXAMPLE 7.4 Evaluate the following integrals:

(a) $\displaystyle\int_{-\sqrt{3}}^{3\sqrt{3}} \frac{dx}{9+x^2}$ (b) $\displaystyle\int \frac{dx}{x\sqrt{4x^2-9}}$

(c) $\displaystyle\int \frac{dx}{x^2+2x+5}$ (d) $\displaystyle\int_0^{(\ln 3)/2} \frac{e^x}{e^{2x}+1}\,dx$

Solution (a) By equation (7.11) with $a = 3$, we obtain

$$\int_{-\sqrt{3}}^{3\sqrt{3}} \frac{dx}{9+x^2} = \frac{1}{3}\tan^{-1}\frac{x}{3}\Big]_{-\sqrt{3}}^{3\sqrt{3}} = \frac{1}{3}\left[\tan^{-1}\sqrt{3} - \tan^{-1}\left(-\frac{1}{\sqrt{3}}\right)\right]$$

$$= \frac{1}{3}\left[\frac{\pi}{3} - \left(-\frac{\pi}{6}\right)\right] = \frac{1}{3}\left(\frac{\pi}{2}\right) = \frac{\pi}{6}$$

(b) To put this in the form of the integral in equation (7.12) we let $u = 2x$. Then $du = 2\,dx$. So on multiplying numerator and denominator by 2, we have

$$\int \frac{dx}{x\sqrt{4x^2-9}} = \int \frac{2\,dx}{2x\sqrt{4x^2-9}} = \int \frac{du}{u\sqrt{u^2-9}} = \frac{1}{3}\sec^{-1}\frac{u}{3} + C$$

$$= \frac{1}{3}\sec^{-1}\frac{2x}{3} + C$$

(c) First we complete the square on x:

$$\int \frac{dx}{x^2+2x+5} = \int \frac{dx}{(x^2+2x+1)+4} = \int \frac{dx}{(x+1)^2+4}$$

Now if we mentally substitute $u = x + 1$, we see that this is in the form that occurs in equation (7.11). So the answer is

$$\int \frac{dx}{x^2+2x+5} = \frac{1}{2}\tan^{-1}\frac{x+1}{2} + C$$

(d) This is in the form of the integral in equation (7.11) with $u = e^x$ and $a = 1$. So we have

$$\int_0^{(\ln 3)/2} \frac{e^x}{e^{2x}+1}\,dx = \tan^{-1} e^x\Big]_0^{(\ln 3)/2} = \tan^{-1}(e^{(\ln 3)/2}) - \tan^{-1} 0$$

$$= \tan^{-1}(e^{\ln\sqrt{3}}) - 0$$

$$= \tan^{-1}\sqrt{3} = \frac{\pi}{3}$$ ■

EXAMPLE 7.5 A billboard 10 ft high is at the edge of the top of a building 95 ft tall. How far should an observer on the ground stand from the base of the building to obtain the best view of the billboard, if the observer's eye is 5 ft above the ground? (The best view is assumed to be the one for which the angle at the observer's eye subtended by the billboard is a maximum.)

Solution The problem is to find x as shown in Figure 7.3 so as to maximize θ. We see that $\theta = \alpha - \beta$ and since $\cot \alpha = \frac{x}{100}$ and $\cot \beta = \frac{x}{90}$, we have

$$\alpha = \cot^{-1} \frac{x}{100} \quad \text{and} \quad \beta = \cot^{-1} \frac{x}{90}$$

(We could have used tangents instead, but the derivatives involved are somewhat simpler using cotangents.) So

$$\theta = \cot^{-1} \frac{x}{100} - \cot^{-1} \frac{x}{90} \qquad (x \geq 0)$$

Thus,

$$\begin{aligned}\frac{d\theta}{dx} &= -\frac{1}{1 + \left(\dfrac{x}{100}\right)^2} \cdot \frac{1}{100} + \frac{1}{1 + \left(\dfrac{x}{90}\right)^2} \cdot \frac{1}{90} \\ &= -\frac{100}{(100)^2 + x^2} + \frac{90}{(90)^2 + x^2}\end{aligned}$$

Setting this equal to 0 gives critical points. After simplification we get

$$\begin{aligned}10x^2 &= 90(100)^2 - 100(90)^2 \\ &= (90)(100)(100 - 90) \\ &= (9000)(10) \\ x^2 &= 9000 \\ x &= 30\sqrt{10} \approx 94.9 \text{ ft} \quad \text{Since } x > 0\end{aligned}$$

It is intuitively evident that as x varies from 0 to arbitrarily large values, θ attains a maximum value, and since we found only one critical point, this is the one we are seeking. We could confirm this by using either the first or second derivative test, but we will not do so.

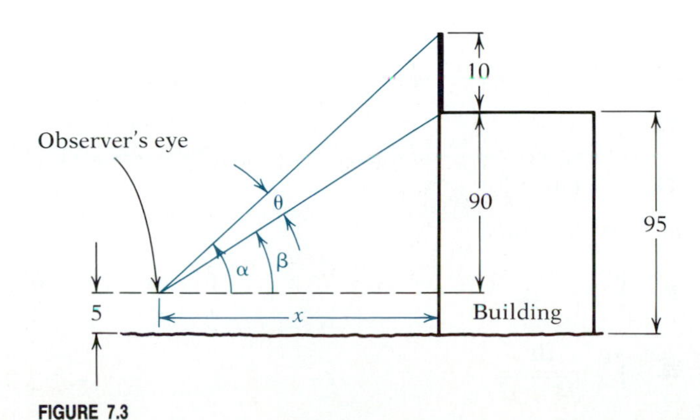

FIGURE 7.3

■

EXERCISE SET 7.1

A

In Exercises 1–12 give the exact value.

1. (a) $\sin^{-1}\frac{1}{2}$ (b) $\cos^{-1}\frac{1}{2}$ (c) arctan 1 (d) arcsec 2 (e) $\cot^{-1} 0$
2. (a) $\arccos(-\frac{1}{\sqrt{2}})$ (b) $\tan^{-1}(-1)$ (c) $\sin^{-1}(-\frac{\sqrt{3}}{2})$ (d) $\operatorname{arccsc}(-2)$ (e) $\sec^{-1}(-\frac{2}{\sqrt{3}})$
3. (a) $\arcsin(-1)$ (b) $\cos^{-1}(-1)$ (c) arctan 0 (d) $\cos^{-1}(-\frac{1}{2})$ (e) $\cot^{-1}(-1)$
4. (a) $\sin[\cos^{-1}(-\frac{1}{3})]$ (b) $\sec[\sin^{-1}(-\frac{1}{4})]$
5. (a) $\tan[\arccos\frac{3}{5}]$ (b) $\cos[\tan^{-1}(-\frac{4}{3})]$
6. (a) $\sin[\arctan\frac{5}{12}]$ (b) $\csc[\arccos(-\frac{2}{3})]$
7. (a) $\cos[2\sin^{-1}(-\frac{1}{3})]$ (b) $\sin[2\cos^{-1}(-\frac{3}{5})]$
8. (a) $\sin^{-1}(\cos\frac{\pi}{3})$ (b) $\cos^{-1}[\sin(-\frac{\pi}{6})]$
9. (a) $\cos^{-1}[\cos(-\frac{\pi}{5})]$ (b) $\sin^{-1}(\sin\frac{9\pi}{8})$
10. (a) $\tan^{-1}(\tan\frac{3\pi}{4})$ (b) $\cot^{-1}(\cot\frac{5\pi}{3})$
11. $\sin[\sin^{-1}\frac{3}{5} + \cos^{-1}(-\frac{5}{13})]$ [*Hint:* Use the identity for $\sin(\alpha + \beta)$.]
12. $\cos[\tan^{-1}(-\frac{1}{2}) - \sec^{-1}\frac{5}{3}]$ [*Hint:* Use the identity for $\cos(\alpha - \beta)$.]

In Exercises 13 and 14 use a calculator to find the approximate radian measure of θ, where $0 \le \theta < 2\pi$.

13. (a) $\sin\theta = \frac{2}{3}$, $\cos\theta < 0$ (b) $\cos\theta = \frac{1}{4}$, $\sin\theta < 0$
14. (a) $\tan\theta = 2$, $\sec\theta < 0$ (b) $\cos\theta = -\frac{2}{5}$, $\cot\theta > 0$

In Exercises 15–24 find y'.

15. $y = \tan^{-1}\frac{2x}{3}$
16. $y = \sin^{-1}(2x - 1)$
17. $y = \arccos\sqrt{1 - x}$
18. $y = \sec^{-1}\frac{1}{x}$ $(0 < x < 1)$
19. $y = (1 + x^2)\tan^{-1} x$
20. $y = \sin^{-1}\sqrt{1 - x^2}$ $(0 < x < 1)$
21. $y = \dfrac{\cos^{-1} x}{\sqrt{1 - x^2}}$
22. $y = x^2 \sec^{-1} x^2$
23. $y - \sin^{-1} y = x^2$
24. $y = \tan^{-1}\frac{y}{x}$

In Exercises 25–40 evaluate the integrals.

25. $\displaystyle\int^{\sqrt{3}/2} \frac{dx}{1 - x^2}$
26. $\displaystyle\int_{-1}^{\sqrt{3}} \frac{dx}{1 + x^2}$
27. $\displaystyle\int_{\sqrt{2}}^{2} \frac{dx}{x\sqrt{x^2 - 1}}$
28. $\displaystyle\int_{-1}^{1} \frac{dx}{\sqrt{4 - x^2}}$
29. $\displaystyle\int_0^4 \frac{dx}{16 + x^2}$
30. $\displaystyle\int_{4/\sqrt{3}}^{4} \frac{dx}{x\sqrt{x^2 - 4}}$
31. $\displaystyle\int \frac{dx}{\sqrt{16 - 9x^2}}$
32. $\displaystyle\int \frac{dx}{4x^2 + 25}$
33. $\displaystyle\int \frac{dx}{x\sqrt{16x^2 - 9}}$
34. $\displaystyle\int \frac{dx}{\sqrt{1 - (2x - 1)^2}}$
35. $\displaystyle\int \frac{dx}{1 + (3x - 2)^2}$
36. $\displaystyle\int_0^4 \frac{dx}{(x - 1)^2 + 3}$
37. $\displaystyle\int \frac{x\,dx}{\sqrt{1 - x^4}}$
38. $\displaystyle\int \frac{dx}{e^x + e^{-x}}$ [*Hint:* Multiply numerator and denominator by e^x.)
39. $\displaystyle\int_{-\sqrt{3}}^{-1} \frac{dx}{x\sqrt{4x^2 - 3}}$
40. $\displaystyle\int_1^2 \frac{dx}{\sqrt{4 - 3(x - 2)^2}}$
41. Find the area of the region under the graph of
$$y = \frac{1}{\sqrt{4 - x^2}}$$
from $x = -1$ to $x = 1$.
42. Find the area under the graph of
$$y = \frac{4}{4 + x^2}$$
from $x = -2$ to $x = 2$.
43. Find the volume generated by revolving the region under the graph of
$$y = \frac{1}{\sqrt{x^2 + 3}}$$
from $x = -1$ to $x = 3$ about the x-axis.
44. Find the volume generated by revolving the region under the graph of
$$y = \frac{1}{x^2\sqrt{x^2 - 9}}$$
from $x = 2\sqrt{3}$ to $x = 6$ about the y-axis.

B

In Exercises 45–53 verify the specified equation.

45. Equation (7.4)
46. Equation (7.5)
47. Equation (7.6)
48. Equation (7.7)
49. Equation (7.8)
50. Equation (7.9)
51. Equation (7.10)
52. Equation (7.11)
53. Equation (7.12)

54. Give the exact value of $\tan^{-1}\frac{1}{3} + \tan^{-1}(-2)$. [*Hint:* Use $\tan(\alpha + \beta)$.]

55. Prove that if $0 < x < 1$,

$$\sin^{-1} x + \sin^{-1}\sqrt{1 - x^2} = \frac{\pi}{2}$$

(*Hint:* Use differentiation to prove that the left-hand side is a constant, and then substitute a particular value of x to show that the constant is $\frac{\pi}{2}$.)

56. Prove that if $x > 0$,

$$\tan^{-1} x + \tan^{-1}\frac{1}{x} = \frac{\pi}{2}$$

(See the hint for Exercise 55.)

In Exercises 57–60 evaluate the integrals.

57. $\displaystyle\int_1^3 \frac{dx}{\sqrt{x}(1 + x)}$

58. $\displaystyle\int \frac{dx}{\sqrt{e^{2x} - 1}}$

59. $\displaystyle\int_{\pi/2}^{\pi} \frac{\sin x}{1 + \cos^2 x}\,dx$

60. $\displaystyle\int \frac{dx}{x\sqrt{x^6 - 1}}$

61. Find the volume generated by revolving the region under the graph of $y = (1 + x^4)^{-1}$ from $x = 0$ to $x = 1$ about the y-axis.

62. Discuss completely and draw the graph of $y = \tan^{-1}(x^2/4)$.

63. In an art gallery a painting 3 m high is to be hung, and a bench for seated viewing is to be placed in front of the painting. The gallery personnel hang the painting so that the bottom of it is 1 m above the eye level of an average observer seated on the bench. How far from the wall should the bench be placed so that the angle at an observer's eye subtended by the painting is a maximum?

64. A lighthouse is located on an island 2 km from the nearest point P on a straight shore. At a point on the shore 1 km from P the beam of light from the lighthouse beacon is observed to be moving at 40 km/min. Through how many radians per minute is the beacon turning? Through how many revolutions per minute is it turning?

65. A helicopter is initially 5000 ft above an observer on the ground, and it then moves horizontally at a speed of 80 ft/sec. What is the rate of change of the angle of elevation of the helicopter from the observer when the helicopter has gone 1000 ft?

7.2 THE HYPERBOLIC FUNCTIONS

The combinations of exponential functions

$$\frac{e^x + e^{-x}}{2} \quad \text{and} \quad \frac{e^x - e^{-x}}{2}$$

occur sufficiently often to warrant special consideration. In fact, they are given special names, as stated in the following definition.

DEFINITION 7.1 The **hyperbolic sine,** abbreviated **sinh,** and the **hyperbolic cosine,** abbreviated **cosh,** are defined by

$$\sinh x = \frac{e^x - e^{-x}}{2}$$

$$\cosh x = \frac{e^x + e^{-x}}{2}$$

For each function the domain is $(-\infty, \infty)$. ■

We will see later why these functions are called *hyperbolic*. Although the definition in no way suggests it, they have many properties similar to those of the trigonometric sine and cosine. For example, we can show the following.

SOME BASIC IDENTITIES

1. $\cosh^2 x - \sinh^2 x = 1$
2. $\sinh(x + y) = \sinh x \cosh y + \cosh x \sinh y$
3. $\cosh(x + y) = \cosh x \cosh y + \sinh x \sinh y$
4. $\sinh 2x = 2 \sinh x \cosh x$
5. $\cosh 2x = \cosh^2 x + \sinh^2 x$
   $= 2 \cosh^2 x - 1$
   $= 1 + 2 \sinh^2 x$
6. $\sinh \dfrac{x}{2} = \pm\sqrt{\dfrac{\cosh x - 1}{2}}$
7. $\cosh \dfrac{x}{2} = \sqrt{\dfrac{\cosh x + 1}{2}}$

The proofs are based on the definitions of $\sinh x$ and $\cosh x$. For example, to prove identity 1, we have

$$\cosh^2 x - \sinh^2 x = \left(\frac{e^x + e^{-x}}{2}\right)^2 - \left(\frac{e^x - e^{-x}}{2}\right)^2$$
$$= \frac{e^{2x} + 2 + e^{-2x}}{4} - \frac{e^{2x} - 2 + e^{-2x}}{4} = \frac{4}{4} = 1$$

For identity 2, it is easier to show that the right-hand side equals the left:

$$\sinh x \cosh y + \cosh x \sinh y = \frac{e^x - e^{-x}}{2} \cdot \frac{e^y + e^{-y}}{2} + \frac{e^x + e^{-x}}{2} \cdot \frac{e^y - e^{-y}}{2}$$

After multiplying terms and simplifying, using laws of exponents, we can write this as (verify)

$$\frac{e^{x+y} - e^{-(x+y)}}{2}$$

which by definition is $\sinh(x + y)$. Identity 3 is proved similarly, and the remaining ones are proved using identities 1, 2, and 3. We leave these as exercises.

By analogy with the trigonometric functions, we also define the hyperbolic tangent (tanh), hyperbolic cotangent (coth), hyperbolic secant (sech), and hyperbolic cosecant (csch) as follows.

DEFINITION 7.2

$$\tanh x = \frac{\sinh x}{\cosh x} \qquad \operatorname{sech} x = \frac{1}{\cosh x}$$
$$\coth x = \frac{\cosh x}{\sinh x}, \quad x \neq 0 \qquad \operatorname{csch} x = \frac{1}{\sinh x}, \quad x \neq 0$$

■

More identities can now be obtained, some of which we give here. You will be asked to prove these in the exercises.

MORE IDENTITIES

8. $1 - \tanh^2 x = \operatorname{sech}^2 x$

9. $\coth^2 x - 1 = \operatorname{csch}^2 x$

10. $\tanh 2x = \dfrac{2 \tanh x}{1 + \tanh^2 x}$

11. $\tanh \dfrac{x}{2} = \dfrac{\sinh x}{1 + \cosh x}$

$$= \frac{\cosh x - 1}{\sinh x}$$

You have probably observed the striking similarities between the identities we have listed and the analogous ones for the trigonometric functions. Some have exactly the same form; for example,

$$\sinh 2x = 2 \sinh x \cosh x \quad \text{and} \quad \sin 2x = 2 \sin x \cos x$$

but more often they differ in form on account of a sign, such as in

$$\cosh^2 x - \sinh^2 x = 1 \quad \text{and} \quad \cos^2 x + \sin^2 x = 1$$

The graphs of the hyperbolic functions can be obtained from our knowledge of the exponential function. Since

$$\sinh x = \frac{e^x - e^{-x}}{2} = \frac{e^x}{2} + \left(-\frac{e^{-x}}{2}\right)$$

we draw the graphs of $\frac{1}{2}e^x$ and $-\frac{1}{2}e^{-x}$ and add ordinates to get the graph of $\sinh x$. Similarly,

$$\cosh x = \frac{e^x + e^{-x}}{2} = \frac{e^x}{2} + \frac{e^{-x}}{2}$$

so we can add the ordinates of the graphs of $\frac{1}{2}e^x$ and $\frac{1}{2}e^{-x}$ to get the graph of $\cosh x$. The results are shown in Figures 7.4 and 7.5.

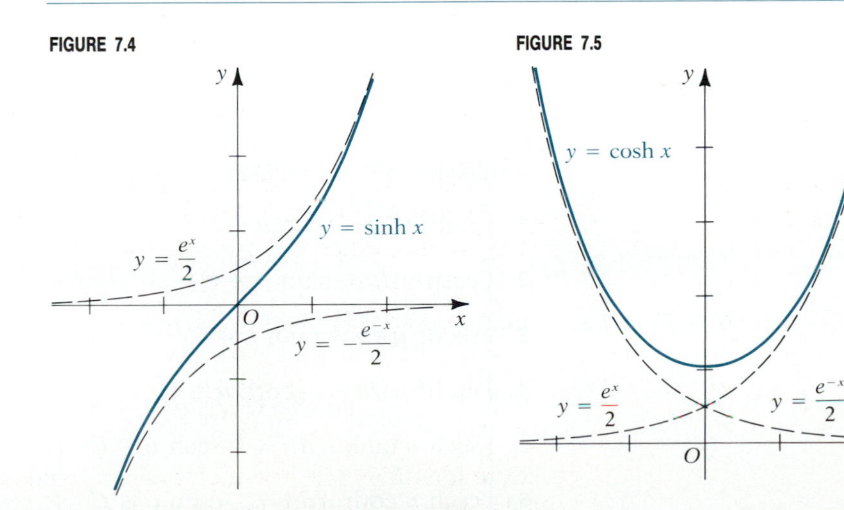

FIGURE 7.4

FIGURE 7.5

Note in particular that the hyperbolic sine is an odd function and the hyperbolic cosine is even; that is

$$\sinh(-x) = -\sinh x$$
$$\cosh(-x) = \cosh x$$

Also observe that $\cosh x \geq 1$ for all x and $\sinh 0 = 0$, $\cosh 0 = 1$.

The graphs of the other four hyperbolic functions can be obtained by first writing them in terms of exponentials and then analyzing their properties. You will be asked to do this in the exercises.

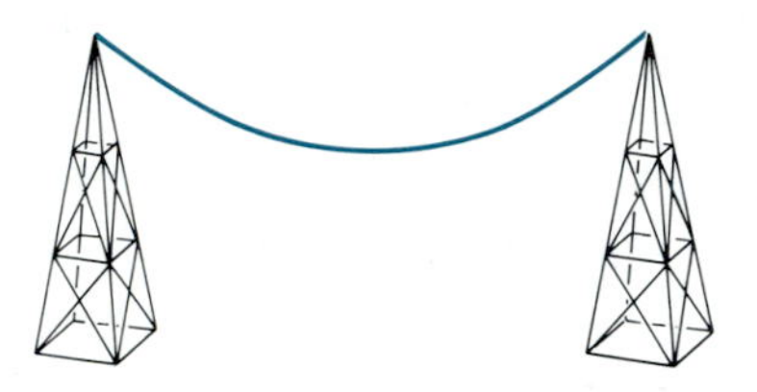

FIGURE 7.6

The hyperbolic cosine curve has an interesting application. When a uniform flexible cable hangs between two supports, as in Figure 7.6, if the only force acting on it is gravity, then it assumes a shape known as a **catenary.** If the axes are suitably chosen, the equation of the catenary is

$$y = a \cosh \frac{x}{a}$$

where a represents the height of the catenary at its lowest point.

Although graphs of the hyperbolic functions differ significantly from those of the trigonometric functions, the parallelism we saw between identities for the two types of functions extends to derivatives as well. The following can all be established on the basis of Definitions 7.1 and 7.2.

DERIVATIVES OF THE HYPERBOLIC FUNCTIONS

1. $\dfrac{d}{dx} \sinh x = \cosh x$
2. $\dfrac{d}{dx} \cosh x = \sinh x$
3. $\dfrac{d}{dx} \tanh x = \operatorname{sech}^2 x$
4. $\dfrac{d}{dx} \coth x = -\operatorname{csch}^2 x$
5. $\dfrac{d}{dx} \operatorname{sech} x = -\operatorname{sech} x \tanh x$
6. $\dfrac{d}{dx} \operatorname{csch} x = -\operatorname{csch} x \coth x$

These can all be generalized using the chain rule, replacing x by $u = g(x)$ for g differentiable and, in the case of formulas 4 and 6, having $g(x) \neq 0$. Then we obtain the following antiderivative formulas.

ANTIDERIVATIVES OF HYPERBOLIC FUNCTIONS

1. $\int \sinh u\, du = \cosh u + C$
2. $\int \cosh u\, du = \sinh u + C$
3. $\int \operatorname{sech}^2 u\, du = \tanh u + C$
4. $\int \operatorname{csch}^2 u\, du = -\coth u + C$
5. $\int \operatorname{sech} u \tanh u\, du = -\operatorname{sech} u + C$
6. $\int \operatorname{csch} u \coth u\, du = -\operatorname{csch} u + C$

EXAMPLE 7.6 Find the derivative of each of the following:

(a) $y = \sinh(\ln x)$ (b) $y = x - \tanh x$

Solution (a) By derivative formula 1 and the chain rule,

$$y' = \cosh(\ln x) \cdot \frac{1}{x} = \frac{\cosh(\ln x)}{x}$$

(b) By derivative formula 3 and identity 8,

$$y' = 1 - \operatorname{sech}^2 x = \tanh^2 x$$

■

EXAMPLE 7.7 Evaluate the following integrals:

(a) $\displaystyle\int \frac{\sinh x}{1 + \cosh x}\,dx$ (b) $\displaystyle\int_0^{\ln 2} \frac{\operatorname{sech}^2 x\,dx}{1 + \tanh^2 x}$ (c) $\displaystyle\int \cosh^2 x\,dx$

Solution (a) If we put $u = 1 + \cosh x$, then $du = \sinh x\,dx$. So we have

$$\int \frac{du}{u} = \ln|u| + C = \ln(1 + \cosh x) + C$$

Absolute values are not necessary, since $1 + \cosh x > 0$.

(b) Observe that this is in the form

$$\int \frac{du}{1 + u^2}$$

with $u = \tanh x$. Since this integrates to $\tan^{-1} u$, we have

$$\int_0^{\ln 2} \frac{\operatorname{sech}^2 x\,dx}{1 + \tanh^2 x} = \tan^{-1}(\tanh x)\Big]_0^{\ln 2}$$

Now

$$\tanh x = \frac{\sinh x}{\cosh x} = \frac{e^x - e^{-x}}{e^x + e^{-x}}$$

So $\tanh 0 = 0$, and

$$\tanh(\ln 2) = \frac{e^{\ln 2} - e^{-\ln 2}}{e^{\ln 2} + e^{-\ln 2}} = \frac{2 - \frac{1}{2}}{2 + \frac{1}{2}} = \frac{3}{5}$$

So we have

$$\int_0^{\ln 2} \frac{\operatorname{sech}^2 x\,dx}{1 + \tanh^2 x} = \tan^{-1}\frac{3}{5} \approx 0.5404$$

(c) From identity 5, solving the second form for $\cosh^2 x$ yields

$$\cosh^2 x = \frac{\cosh 2x + 1}{2}$$

So

$$\begin{aligned}\int \cosh^2 x\,dx &= \tfrac{1}{2}\int (\cosh 2x + 1)\,dx \\ &= \tfrac{1}{2}\cdot\tfrac{1}{2}\int \cosh 2x(2\,dx) + \tfrac{1}{2}\int dx \\ &= \tfrac{1}{4}\sinh 2x + \tfrac{1}{2}x + C\end{aligned}$$

■

EXAMPLE 7.8 A telephone cable hangs in the form of a catenary between two poles that are 20 m apart. With the axes chosen as in Figure 7.7 (x-axis on the ground), the equation of the catenary is $y = 10 \cosh \frac{x}{10}$.

(a) Find the height of the cable at its lowest and highest points.

(b) Find the length of the cable between the two poles.

Solution (a) The lowest point occurs at $x = 0$ and so is

$$y_{\text{min}} = 10 \cosh \tfrac{0}{10} = 10$$

The curve assumes its maximum height at each end. So

$$y_{\text{max}} = 10 \cosh \frac{10}{10} = 10 \cosh 1 = 10 \cdot \frac{e + e^{-1}}{2} = 5\left(e + \frac{1}{e}\right) \approx 15.43$$

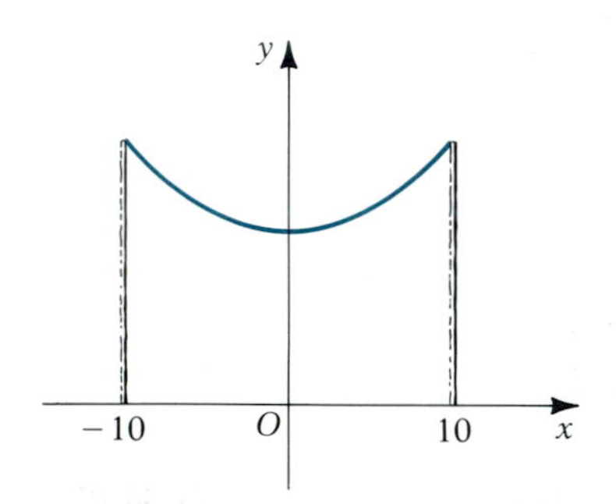

FIGURE 7.7

(b) The total length is twice the length from $x = 0$ to $x = 10$. So we have

$$\begin{aligned} s &= 2 \int_{x=0}^{x=10} ds = 2 \int_0^{10} \sqrt{1 + \left(\frac{dy}{dx}\right)^2}\, dx \\ &= 2 \int_0^{10} \sqrt{1 + \sinh^2 \frac{x}{10}}\, dx \\ &= 2 \int_0^{10} \cosh \frac{x}{10}\, dx \quad \text{By identity 1} \\ &= 20 \int_0^{10} \cosh \frac{x}{10} \left(\frac{1}{10}\, dx\right) \\ &= 20 \sinh \frac{x}{10}\Big]_0^{10} = 20(\sinh 1 - \sinh 0) \\ &= 20 \sinh 1 = 20\left(\frac{e - e^{-1}}{2}\right) = 10\left(e - \frac{1}{e}\right) \approx 23.50 \end{aligned}$$

■

We conclude this section by indicating why the name *hyperbolic* is applied to the functions we have been studying. Because of the trigonometric identity $\cos^2 t + \sin^2 t = 1$, we know that the point $(\cos t, \sin t)$ lies on the unit

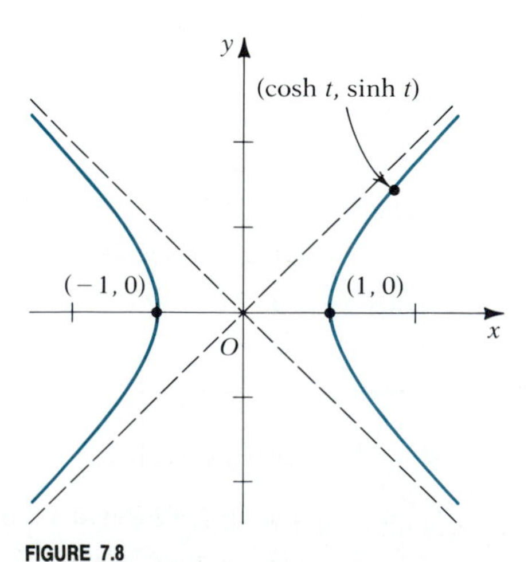

FIGURE 7.8

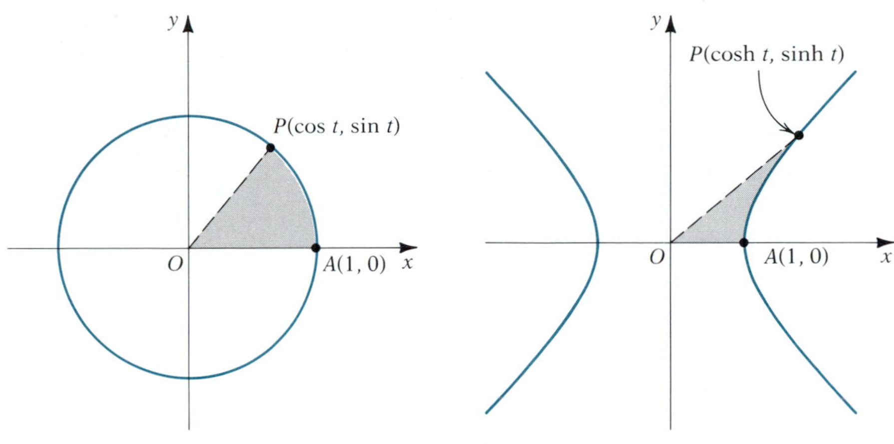

FIGURE 7.9

FIGURE 7.10

circle $x^2 + y^2 = 1$ for all real numbers t. The trigonometric functions are often called **circular functions** because of this relationship. Similarly, the hyperbolic identity $\cosh^2 t - \sinh^2 t = 1$ implies that for all real numbers t, the point $(\cosh t, \sinh t)$ lies on the "unit" hyperbola $x^2 - y^2 = 1$, whose graph is shown in Figure 7.8. The point is on the right-hand branch, since $\cosh t$ is always positive, and it is above the x-axis if $t > 0$ and below if $t < 0$.

The geometric analogy between the relationship of trigonometric functions to the unit circle and hyperbolic functions to the unit hyperbola goes even further. The shaded areas in Figures 7.9 and 7.10 are called a circular sector and a hyperbolic sector, respectively. For $0 < t < 2\pi$, the area of each sector can be shown to be $\frac{t}{2}$; that is, in each case

$$\text{area of sector } OAP = \tfrac{t}{2}$$

(See Exercises 52 and 53 in Exercise Set 7.2.)

EXERCISE SET 7.2

A

1. Give the domain and range of each of the hyperbolic functions.
2. Express $\tanh x$, $\coth x$, $\operatorname{sech} x$, and $\operatorname{csch} x$ in terms of exponential functions, and evaluate the following limits:
   (a) $\lim_{x\to+\infty} \tanh x$ (b) $\lim_{x\to0^+} \coth x$
   (c) $\lim_{x\to+\infty} \coth x$ (d) $\lim_{x\to+\infty} \operatorname{sech} x$
   (e) $\lim_{x\to0^+} \operatorname{csch} x$ (f) $\lim_{x\to+\infty} \operatorname{csch} x$
3. Use a calculator to approximate the value of each of the following for $x = 0.5$, 1, 2, and 4:
   (a) $\tanh x$ (b) $\coth x$
   (c) $\operatorname{sech} x$ (d) $\operatorname{csch} x$
4. Show that sech is an even function and that tanh, coth, and csch are odd functions.
5. Draw the graphs:
   (a) $y = \tanh x$ (b) $y = \coth x$
   (c) $y = \operatorname{sech} x$ (d) $y = \operatorname{csch} x$
   Make use of the results of Exercises 1–4.

In Exercises 6–13 prove the identities specified.

6. 3 **7.** 4 **8.** 5 **9.** 6
10. 7 **11.** 8 **12.** 9 **13.** 10

In Exercises 14–25 find $\frac{dy}{dx}$.

14. $y = \tanh^3 2x$ **15.** $y = \dfrac{\cosh x}{1 - \sinh x}$

16. $y = \ln(\sinh x)$ **17.** $y = \tan^{-1}(\sinh x)$

18. $y = \sec^{-1}(\cosh x)$ **19.** $y = x - \coth x$

20. $y = \sqrt{\operatorname{sech} x^2}$

21. $y = \dfrac{\tanh x}{1 + \operatorname{sech} x}$

22. $x \cosh y - y \sinh x = 1$

23. $y^2 = \ln(x \cosh y)$

24. $y = (\sinh x)^x$

25. $y = (\sinh x)^{\operatorname{sech} x}$

In Exercises 26–43 evaluate the integrals.

26. $\int \sinh x \cosh x\, dx$

27. $\int \dfrac{1 + \tanh x}{\cosh^2 x}\, dx$

28. $\int \dfrac{\sinh x\, dx}{\sqrt{1 + \sinh^2 x}}$

29. $\int \sinh^2 x\, dx$

30. $\int \dfrac{1 + \cosh x}{x + \sinh x}\, dx$

31. $\int e^{\sinh^2 x} \sinh 2x\, dx$

32. $\int \dfrac{\tanh x}{\ln \cosh x}\, dx$

33. $\int x \sinh x^2 \cosh^2 x^2\, dx$

34. $\int \sinh^3 x\, dx$

35. $\int \tanh^2 x\, dx$

36. $\int \cosh^3 x\, dx$

37. $\int \operatorname{sech}^3 x \tanh x\, dx$

38. $\int \operatorname{sech}^4 x \tanh x\, dx$

39. $\int \cosh^2 x \sinh^3 x\, dx$

40. $\int_{\ln 2}^{\ln 3} \coth x\, dx$

41. $\int_{-1}^{1} (\sinh x + \cosh x)^2\, dx$

42. $\int_0^{\ln 2} \dfrac{\cosh x}{\sqrt{9 - 4 \sinh^2 x}}\, dx$

43. $\int_{-\ln 3}^{\ln 3} \dfrac{\sinh x}{1 + \cosh^2 x}\, dx$

44. Find the volume obtained by rotating the region under the graph of $y = \operatorname{sech} x$ between $x = -1$ and $x = 1$ about the x-axis.

45. Find the volume obtained by rotating the region under the graph of $y = \tanh x$ between $x = 0$ and $x = 2$ about the x-axis.

46. Find the surface area obtained by rotating the arc of the curve $y = \cosh x$ from $x = 0$ to $x = 2$ about the x-axis.

In Exercises 47–50 verify that the equation is an identity for all admissible values of x by transforming the left-hand side into the right-hand side.

47. $\sinh x(\coth x - \tanh x) = \operatorname{sech} x$

48. $\dfrac{\sinh x}{\coth x - \operatorname{csch} x} = \cosh x + 1$

49. $\dfrac{2}{\coth x - \tanh x} = \sinh 2x$

50. $\dfrac{\coth x + \tanh x}{\coth x - \tanh x} = \cosh 2x$

B

51. Prove identity 11.

52. Prove that the area of the circular sector OAP shown in the figure is $\frac{t}{2}$, where $0 < t < \frac{\pi}{2}$. (*Hint:* Make use of your knowledge of the area of the circle.)

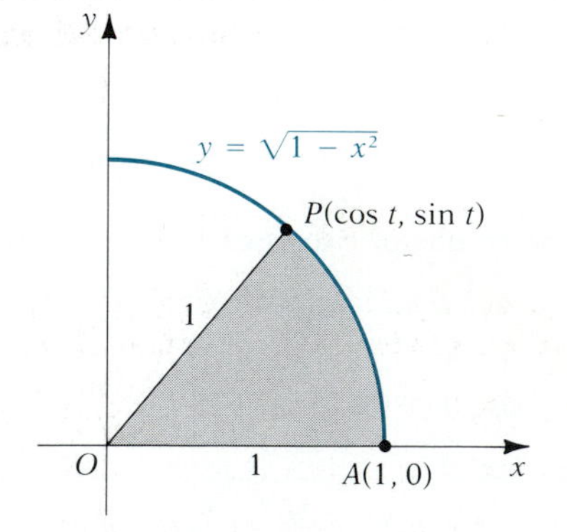

53. Prove that the area of the hyperbolic sector OAP shown in the figure is $\frac{t}{2}$, where $t > 0$, by completing the following steps.

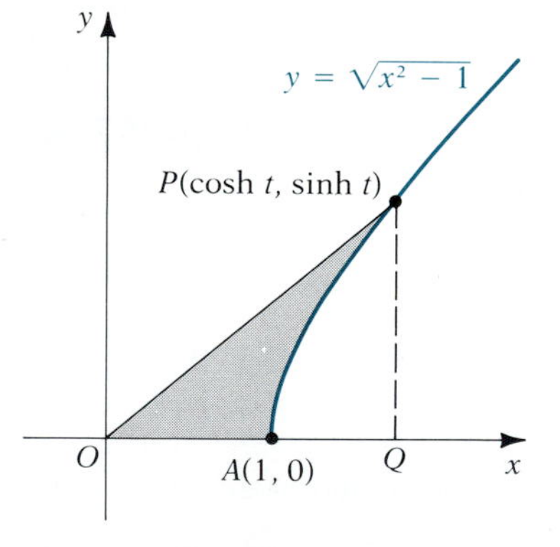

(a) Find the area of the triangle OPQ.

(b) Find the area under the arc AP by using area $= \int_1^{\cosh t} y\, dx$. To evaluate the integral, substitute $x = \cosh u$.

(c) Subtract the area in part b from the area in part a.

In Exercises 54–57 evaluate the integrals.

54. $\int \sinh^4 x\,dx$

55. $\int \tanh^5 x\,dx$

56. $\int \frac{dx}{\operatorname{sech}^2 x - \operatorname{sech}^4 dx}$

57. $\int \frac{dx}{\cosh x - 1}$ (*Hint:* Multiply numerator and denominator by $\cosh x + 1$.)

In Exercises 58 and 59 derive the formulas.

58. $\int \operatorname{sech} x\,dx = \tan^{-1}|\sinh u| + C$ (*Hint:* Write $\operatorname{sech} x = \cosh x/\cosh^2 x$.)

59. $\int \operatorname{csch} x\,dx = \ln|\tanh \frac{x}{2}| + C.$ (*Hint:* Write $\operatorname{csch} x = \sinh x/\sinh^2 x$.)

In Exercises 60–62 prove that the given equation is an identity for all admissible values of x.

60. $\cosh^4 x = \frac{3}{8} + \frac{\cosh 2x}{2} + \frac{\cosh 4x}{8}$

61. $\frac{\sinh 3x}{\sinh x} - \frac{\cosh 3x}{\cosh x} = 2$

62. $2\cosh^2 \frac{x}{2} \operatorname{csch} x = \coth \frac{x}{2}$

63. A uniform flexible cable suspended between two supporting towers 60 m apart hangs in the form of a catenary with the equation

$$y = 20 \cosh \tfrac{x}{20} + 10$$

where y is the distance above the ground.
(a) Find the height of the cable at its lowest point.
(b) Find the height at each end.
(c) Determine the length of the cable.

64. (a) Show that $y = A\cosh \omega t + B \sinh \omega t$ satisfies the differential equation

$$\frac{d^2y}{dt^2} - \omega^2 y = 0$$

(b) In part a find A and B such that when $t = 0$, $y = 3$ and $y' = 0$.

7.3 INVERSE HYPERBOLIC FUNCTIONS

The hyperbolic sine function is increasing for all x and so is 1–1. It therefore has an inverse, which we designate by $\sinh^{-1}$. Thus,

$$y = \sinh^{-1} x \quad \text{if and only if} \quad x = \sinh y$$

The graphs of both $y = \sinh x$ and $y = \sinh^{-1} x$ are shown in Figure 7.11. We know from Section 6.3 that $\sinh^{-1}$ also is increasing and is differentiable for all x, since the derivative of the hyperbolic sine exists everywhere and

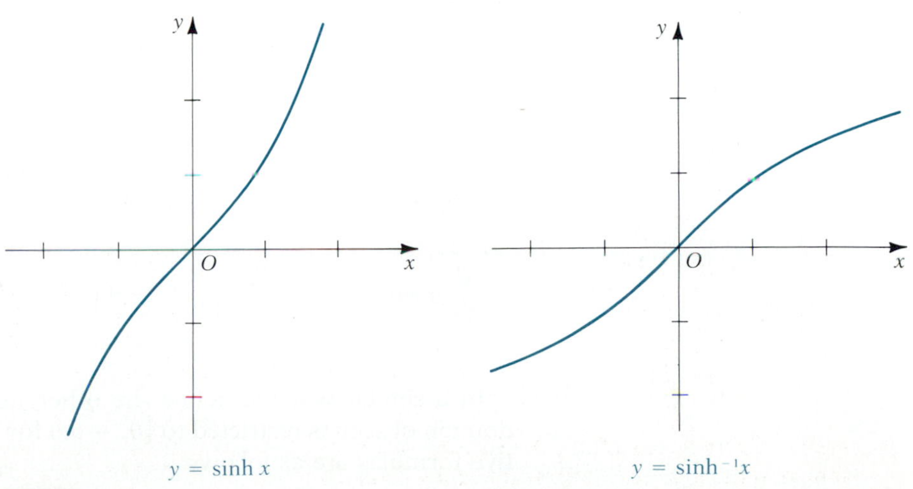

FIGURE 7.11

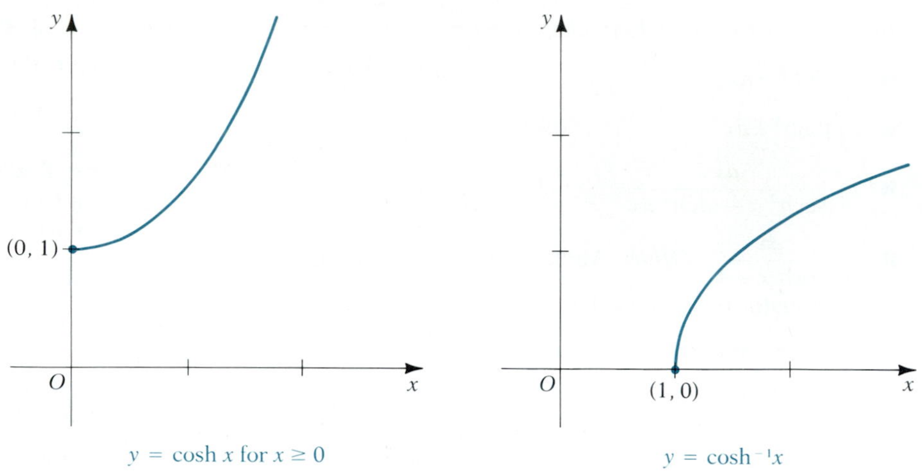

FIGURE 7.12

is never 0. Furthermore, by equation (6.14) with $y = \sinh^{-1} x$,

$$\frac{dy}{dx} = \frac{1}{\frac{dx}{dy}} = \frac{1}{\cosh y} = \frac{1}{\sqrt{1 + \sinh^2 y}} \quad \text{By identity 1}$$

But since $\sinh y = x$, this becomes

$$\frac{d}{dx} \sinh^{-1} x = \frac{1}{\sqrt{1 + x^2}} \tag{7.13}$$

The hyperbolic cosine is not 1–1, but by restricting its domain to the interval $[0, +\infty)$, it is 1–1 and has an inverse. So we define $\cosh^{-1}$ by

$$y = \cosh^{-1} x \quad \text{if and only if} \quad x = \cosh y \text{ and } y \geq 0$$

Graphs of $y = \cosh x$ for $x \geq 0$ and $y = \cosh^{-1} x$ are shown in Figure 7.12. On $[0, +\infty)$ the hyperbolic cosine is differentiable and its derivative is positive on $(0, +\infty)$. Thus, for $y = \cosh^{-1} x$,

$$\frac{dy}{dx} = \frac{1}{\frac{dx}{dy}} = \frac{1}{\sinh y} = \frac{1}{\sqrt{\cosh^2 y - 1}}, \qquad y > 0$$

Or, since $x = \cosh y$,

$$\frac{d}{dx} \cosh^{-1} x = \frac{1}{\sqrt{x^2 - 1}}, \qquad x > 1 \tag{7.14}$$

In a similar way we define the other inverse hyperbolic functions. The domain of sech is restricted to $[0, +\infty)$ for defining the inverse. The derivative formulas are as follows.

$$\frac{d}{dx}\tanh^{-1} x = \frac{1}{1-x^2}, \qquad |x| < 1 \tag{7.15}$$

$$\frac{d}{dx}\coth^{-1} x = \frac{1}{1-x^2}, \qquad |x| > 1 \tag{7.16}$$

$$\frac{d}{dx}\operatorname{sech}^{-1} x = -\frac{1}{x\sqrt{1-x^2}}, \qquad 0 < x < 1 \tag{7.17}$$

$$\frac{d}{dx}\operatorname{csch}^{-1} x = -\frac{1}{|x|\sqrt{1+x^2}}, \qquad x \neq 0 \tag{7.18}$$

One of the primary uses of inverse hyperbolic functions is in integration. From the derivative formulas (7.13)–(7.18) we have the following antiderivative formulas.

$$\int \frac{du}{\sqrt{1+u^2}} = \sinh^{-1} u + C \tag{7.19}$$

$$\int \frac{du}{\sqrt{u^2-1}} = \cosh^{-1} u + C, \qquad u > 1 \tag{7.20}$$

$$\int \frac{du}{1-u^2} = \begin{cases} \tanh^{-1} u + C & \text{if } |u| < 1 \\ \coth^{-1} u + C & \text{if } |u| > 1 \end{cases} \tag{7.21}$$

$$\int \frac{du}{u\sqrt{1-u^2}} = -\operatorname{sech}^{-1} u + C, \quad \text{if } 0 < u < 1 \tag{7.22}$$

$$\int \frac{du}{u\sqrt{1+u^2}} = -\operatorname{csch}^{-1} u + C, \qquad u > 0 \tag{7.23}$$

Since the hyperbolic functions are all expressible in terms of exponentials, it is not surprising that their inverses are expressible in terms of logarithms. The next theorem makes this explicit.

THEOREM 7.1 The inverse hyperbolic functions have the following equivalent forms.

$$\sinh^{-1} x = \ln(x + \sqrt{x^2+1}), \qquad -\infty < x < \infty \tag{7.24}$$

$$\cosh^{-1} x = \ln(x + \sqrt{x^2-1}), \qquad x \geq 1 \tag{7.25}$$

$$\tanh^{-1} x = \frac{1}{2}\ln\frac{1+x}{1-x}, \qquad |x| < 1 \tag{7.26}$$

$$\coth^{-1} x = \frac{1}{2}\ln\frac{x+1}{x-1}, \qquad |x| > 1 \tag{7.27}$$

$$\operatorname{sech}^{-1} x = \ln\left(\frac{1+\sqrt{1-x^2}}{x}\right), \qquad 0 < x \leq 1 \tag{7.28}$$

$$\operatorname{csch}^{-1} x = \ln\left(\frac{1}{x} + \frac{\sqrt{x^2+1}}{|x|}\right), \qquad x \neq 0 \tag{7.29}$$

■

We will prove equation (7.24) and leave the others as exercises.

Proof Let $y = \sinh^{-1} x$. Then $x = \sinh y$, so by the definition,

$$x = \frac{e^y - e^{-y}}{2}$$

Clearing of fractions and rearranging, we get

$$e^y - 2x - e^{-y} = 0$$

Now we multiply both sides by e^y:

$$e^{2y} - 2xe^y - 1 = 0$$

This is a quadratic equation in e^y, so by the quadratic formula

$$e^y = \frac{2x \pm \sqrt{4x^2 + 4}}{2} = x \pm \sqrt{x^2 + 1}$$

But $e^y > 0$, so that the negative sign cannot hold. Thus,

$$e^y = x + \sqrt{x^2 + 1}$$

or, in the equivalent logarithmic form,

$$y = \ln(x + \sqrt{x^2 + 1})$$

■

By Theorem 7.1 each of the antiderivative formulas (7.19)–(7.23) can be written in an equivalent form involving logarithms. We call attention in particular to the fact that the two parts of formula (7.21) can be written in the single form

$$\int \frac{du}{1 - u^2} = \frac{1}{2} \ln\left|\frac{1 + u}{1 - u}\right| + C, \qquad u \neq 1 \tag{7.30}$$

(Verify this.)

EXAMPLE 7.9 Find the derivative of each of the following:

(a) $y = \sinh^{-1}(\tan x), \qquad -\frac{\pi}{2} < x < \frac{\pi}{2}$

(b) $y = \tanh^{-1} \frac{1 + x}{1 - x}, \qquad x \neq 1$

Solution (a) By equation (7.13) and the chain rule,

$$\frac{dy}{dx} = \frac{1}{\sqrt{1 + \tan^2 x}} \cdot \sec^2 x = \frac{1}{\sec x} \cdot \sec^2 x = \sec x$$

Here we used the fact that $\sqrt{1 + \tan^2 x} = \sqrt{\sec^2 x}$ and, since $\sec x > 0$ for x in $(-\frac{\pi}{2}, \frac{\pi}{2})$, $\sqrt{\sec^2 x} = \sec x$.

(b) By equation (7.15) and the chain rule,

$$\frac{dy}{dx} = \frac{1}{1 - \left(\frac{1 + x}{1 - x}\right)^2} \cdot \frac{(1 - x) - (1 + x)(-1)}{(1 - x)^2} = \frac{2}{(1 - x)^2 - (1 + x)^2}$$

$$= \frac{2}{[(1 - x) + (1 + x)][(1 - x) - (1 + x)]} = -\frac{1}{2x}$$

Notice how we used the algebraic identity $a^2 - b^2 = (a + b)(a - b)$ to simplify the denominator in the third step. ■

EXAMPLE 7.10 Evaluate the following integrals:

(a) $\int \frac{x}{\sqrt{x^4 + 1}}\,dx$ (b) $\int_1^{7/2} \frac{dx}{\sqrt{4x^2 - 1}}$

Solution (a) Let $u = x^2$; then $du = 2x\,dx$, and we have

$$\int \frac{x}{\sqrt{x^4 + 1}}\,dx = \frac{1}{2}\int \frac{2x\,dx}{\sqrt{x^4 + 1}} = \frac{1}{2}\int \frac{du}{\sqrt{u^2 + 1}} = \frac{1}{2}\sinh^{-1} u + C$$

$$= \frac{1}{2}\sinh^{-1} x^2 + C$$

(b) Let $u = 2x$; then $du = 2\,dx$. When $x = 1$, $u = 2$, and when $x = \frac{7}{2}$, $u = 7$. So we have

$$\int_1^{7/2} \frac{dx}{\sqrt{4x^2 - 1}} = \frac{1}{2}\int_2^7 \frac{du}{\sqrt{u^2 - 1}} = \frac{1}{2}\cosh^{-1} u\Big]_2^7$$

$$= \frac{1}{2}[\cosh^{-1} 7 - \cosh^{-1} 2]$$

$$= \frac{1}{2}[\ln(7 + \sqrt{48}) - \ln(2 + \sqrt{3})]$$

$$= \frac{1}{2}[\ln(7 + 4\sqrt{3}) - \ln(2 + \sqrt{3})] = \frac{1}{2}\ln\frac{7 + 4\sqrt{3}}{2 + \sqrt{3}}$$

$$= \frac{1}{2}\ln\frac{7 + 4\sqrt{3}}{2 + \sqrt{3}} \cdot \frac{2 - \sqrt{3}}{2 - \sqrt{3}} = \frac{1}{2}\ln\frac{14 + \sqrt{3} - 12}{4 - 3}$$

$$= \frac{1}{2}\ln(2 + \sqrt{3})$$

Notice the use of equation (7.25). ■

EXERCISE SET 7.3

A

In Exercises 1–4 give the domain and range of f and draw its graph.

1. $f(x) = \tanh^{-1} x$
2. $f(x) = \coth^{-1} x$
3. $f(x) = \operatorname{sech}^{-1} x$
4. $f(x) = \operatorname{csch}^{-1} x$
5. (a) Prove $\operatorname{sech}^{-1} x = \cosh^{-1}\frac{1}{x}$, $x > 0$.
   (b) Prove $\operatorname{csch}^{-1} x = \sinh^{-1}\frac{1}{x}$, $x \neq 0$.

In Exercises 6–9 derive the specified equation.

6. Equation (7.15)
7. Equation (7.16)
8. Equation (7.17)
9. Equation (7.18)

In Exercises 10–19 find y′.

10. $y = \sinh^{-1}\sqrt{x}$
11. $y = \cosh^{-1} x^2$
12. $y = \tanh^{-1}\frac{1}{x}$
13. $y = \operatorname{sech}^{-1}(\cos x)$, $0 < x < \frac{\pi}{2}$
14. $y = \coth^{-1}\frac{x - 1}{x + 1}$
15. $y = x\operatorname{csch}^{-1} x$, $x > 0$
16. $y = \cosh^{-1}(\ln x)$

17. $y = \operatorname{sech}^{-1}\sqrt{x+1},\ -1 < x < 0$

18. $y = (\sinh^{-1} 2x)^3$

19. $y = \tanh^{-1}\dfrac{e^x}{1+e^x}$

In Exercises 20–29 evaluate the integrals.

20. $\displaystyle\int \frac{dx}{\sqrt{1+9x^2}}$ **21.** $\displaystyle\int \frac{dx}{\sqrt{4x^2-1}}$

22. $\displaystyle\int_{\ln 1/4}^{\ln 1/2} \frac{e^x}{1-e^{2x}}\,dx$ **23.** $\displaystyle\int \frac{dx}{x\sqrt{1-4x^2}}$

24. $\displaystyle\int \frac{\sin x\,dx}{\sqrt{1+\cos^2 x}}$

25. $\displaystyle\int \frac{\sec^2 x\,dx}{\sqrt{\tan^2 x-1}}\quad \left(\frac{\pi}{4} < x < \frac{\pi}{2}\right)$

26. $\displaystyle\int \frac{\sec^2 x}{\tan^2 x-1}\,dx\quad \left(-\frac{\pi}{4} < x < \frac{\pi}{4}\right)$

27. $\displaystyle\int_0^{1/2} \frac{x\,dx}{1-x^4}$

28. $\displaystyle\int_0^5 \frac{dx}{\sqrt{x^2+4x+3}}$

29. $\displaystyle\int \frac{dx}{\sqrt{x^2+2x+2}}$

In Exercises 30–34 verify the given antidifferentiation formula, where $a > 0$.

30. $\displaystyle\int \frac{dx}{\sqrt{a^2+x^2}} = \sinh^{-1}\frac{x}{a} + C$

31. $\displaystyle\int \frac{dx}{\sqrt{x^2-a^2}} = \cosh^{-1}\frac{x}{a} + C,\quad x > a$

32. $\displaystyle\int \frac{dx}{a^2-x^2} = \begin{cases} \dfrac{1}{a}\tanh^{-1}\dfrac{x}{a} + C & \text{if } |x| < a \\ \dfrac{1}{a}\coth^{-1}\dfrac{x}{a} + C & \text{if } |x| > a \end{cases}$

33. $\displaystyle\int \frac{dx}{x\sqrt{a^2-x^2}} = -\frac{1}{a}\operatorname{sech}^{-1}\frac{x}{a} + C,\quad 0 < x < a$

34. $\displaystyle\int \frac{dx}{x\sqrt{a^2+x^2}} = -\frac{1}{a}\operatorname{csch}^{-1}\frac{x}{a} + C,\quad x > 0$

In Exercises 35–40 use the results of Exercises 30–34 to evaluate the integrals.

35. $\displaystyle\int \frac{dx}{\sqrt{4+9x^2}}$ **36.** $\displaystyle\int \frac{dx}{\sqrt{4x^2-9}}$

37. $\displaystyle\int_0^1 \frac{dx}{9-4x^2}$ **38.** $\displaystyle\int_3^4 \frac{dx}{x\sqrt{25-x^2}}$

39. $\displaystyle\int \frac{dx}{\sqrt{4+e^{2x}}}$ **40.** $\displaystyle\int \frac{dx}{x^2-6x+5}$

B

41. Prove equations (7.25) and (7.26).

42. Prove equations (7.27) and (7.28).

43. Prove equation (7.29). Also show that the result can be put in the following equivalent form:

$$\operatorname{csch}^{-1} x = \begin{cases} \ln\left(\dfrac{1+\sqrt{x^2+1}}{x}\right) & \text{if } x > 0 \\ -\ln\left(\dfrac{1+\sqrt{x^2+1}}{|x|}\right) & \text{if } x < 0 \end{cases}$$

In Exercises 44–49 use Theorem 7.1 to arrive at an alternate proof of the specified differentiation formula.

44. Equation (7.13) **45.** Equation (7.14)

46. Equation (7.15) **47.** Equation (7.16)

48. Equation (7.17) **49.** Equation (7.18)

50. Prove that

$$\frac{d}{dx}\operatorname{sech}^{-1}|x| = -\frac{1}{x\sqrt{1-x^2}}\quad \text{if } |x| < 1 \text{ and } x \neq 0$$

Using this result show that the formula of Exercise 33 can be generalized to

$$\int \frac{du}{u\sqrt{a^2-u^2}} = -\frac{1}{a}\operatorname{sech}^{-1}\frac{|u|}{a} + C,\qquad 0 < |u| < a$$

51. Following the idea of Exercise 50, show that

$$\int \frac{du}{u\sqrt{a^2+u^2}} = -\frac{1}{a}\operatorname{csch}^{-1}\frac{|u|}{a} + C,\qquad u \neq 0$$

52. Make use of inverse hyperbolic functions to obtain the result

$$\int \sec u\,du = \ln|\sec u + \tan u| + C$$

[*Hint:* $\sec u = \dfrac{1}{\cos u} = \dfrac{\cos u}{\cos^2 u} = \dfrac{\cos u}{1-\sin^2 u}$. Now use equations (7.21) and (7.27).]

53. When a body falls under the influence of gravity and air resistance is taken into consideration, the downward force is the weight of the object (mg, where m is its mass) and the resisting force is known experimentally to be approximately proportional to the square of the velocity. According to Newton's second law of motion, the summation of forces acting on the body is its mass times its acceleration ($F = ma$). So we have

$$ma = mg - kv^2$$

Suppose that when $t = 0$, $v = 0$ and $s = 0$ (so that distance is measured positively downward from the initial position). Using the fact that $a = \frac{dv}{dt}$ and $v = \frac{ds}{dt}$, find (a) $v(t)$, (b) $s(t)$, and (c) $\lim_{t\to\infty} v(t)$. (*Note:* We assume $mg > kv^2$.)

7.4 SUPPLEMENTARY EXERCISES

In Exercises 1–4 give the exact value of the expression.

1. (a) $\sin^{-1} 1$ (b) $\cos^{-1}\left(-\frac{\sqrt{3}}{2}\right)$

(c) $\tan^{-1}\left(-\frac{1}{\sqrt{3}}\right)$ (d) $\operatorname{arcsec}(-\sqrt{2})$

(e) $\operatorname{arccot} 0$ (f) $\operatorname{arccsc}\left(-\frac{2}{\sqrt{3}}\right)$

2. (a) $\tan\left[\cos^{-1}\left(-\frac{3}{\sqrt{13}}\right)\right]$ (b) $\sin\left[2\tan^{-1}\left(-\frac{1}{2}\right)\right]$

3. (a) $\cos\left[2\sin^{-1}\left(-\frac{3}{5}\right)\right]$ (b) $\sin^{-1}\left(\sin\frac{8\pi}{9}\right)$

4. (a) $\sin\left[\tan^{-1}\frac{4}{3} + \sec^{-1}\left(-\frac{13}{5}\right)\right]$

(b) $\cos\left[\cos^{-1}\left(-\frac{1}{3}\right) - \tan^{-1}(-2)\right]$

5. Use a calculator to find the approximate radian measure of θ, if $0 \le \theta < 2\pi$.

(a) $\cos\theta = \frac{1}{3}$, $\tan\theta < 0$

(b) $\sin\theta = -\frac{3}{4}$, $\tan\theta > 0$.

In Exercises 6–12 find y'.

6. (a) $y = \frac{1}{x}\sin^{-1} x^2$, $0 < |x| < 1$
(b) $y = \tan^{-1}\sqrt{x^2 - 1}$, $|x| > 1$

7. (a) $y = \cos^{-1}(e^{-x})$, $x \ge 0$
(b) $y = x^2 \sec^{-1}\frac{1}{x} + \sin^{-1} x$, $0 < x < 1$

8. (a) $y = (\sin^{-1}\sqrt{x})^2 + (\cos^{-1}\sqrt{x})^2$
(b) $\tan^{-1}\frac{y}{x} + \ln\sqrt{x^2 + y^2} = 1$

9. (a) $y = \ln(\cosh^2 x)$
(b) $y = \cos^{-1}(\operatorname{sech} x)$, $x > 0$

10. (a) $y = \dfrac{\sinh x}{1 + \cosh x}$
(b) $\tan^{-1}(\sinh y) + \cot^{-1}(\sinh x) = 1$

11. (a) $y = \sinh^{-1}\sqrt{x^2 - 1}$, $x > 1$
(b) $y = \tanh^{-1}(\sin x)$, $|x| < \frac{\pi}{2}$

12. (a) $y = \coth^{-1}\dfrac{1 + x}{1 - x}$, $0 < x < 1$
(b) $y = \sqrt{x}\cosh^{-1}(x + 1)$, $x > 0$

In Exercises 13–28 evaluate the given definite or indefinite integrals.

13. $\displaystyle\int \frac{dx}{\sqrt{9 - 4x^2}}$ **14.** $\displaystyle\int_{-3/2}^{3/2} \frac{dx}{4x^2 + 3}$

15. $\displaystyle\int_{-2\sqrt{2}}^{-2} \frac{dx}{x\sqrt{x^2 - 2}}$ **16.** $\displaystyle\int_{-\sqrt{2}/4}^{0} \frac{dx}{\sqrt{1 - 4x^2}}$

17. $\displaystyle\int_1^e \frac{dx}{x[1 + (\ln x)^2]}$ **18.** $\displaystyle\int_{1/\sqrt{3}}^{1} \frac{dx}{x\sqrt{4x^2 - 1}}$

19. $\displaystyle\int \frac{dx}{\sqrt{e^{2x} + 1}}$ **20.** $\displaystyle\int \frac{x}{1 + x^4}\,dx$

21. $\displaystyle\int \frac{\sinh 2x}{1 + \cosh 2x}\,dx$ **22.** $\displaystyle\int \frac{\sinh 2x}{e^{\sinh^2 x}}\,dx$

23. $\displaystyle\int x \tanh x^2\,dx$ **24.** $\displaystyle\int_0^{\ln(1+\sqrt{2})} \frac{\cosh x}{1 + \sinh^2 x}\,dx$

25. $\displaystyle\int_0^{\ln 2} \frac{\tanh x}{\cosh^2 x}\,dx$ **26.** $\displaystyle\int_0^{\ln 2} \frac{e^x\,dx}{\sqrt{e^{2x} - 1}}$

27. $\displaystyle\int_{-1}^{1} \frac{dx}{4 - x^2}$ **28.** $\displaystyle\int_0^1 \frac{dx}{x^2 + 4x + 3}$

29. Find the volume obtained by revolving the region under the graph of

$$y = \frac{4}{x\sqrt{4 - x^2}}$$

from $x = 1$ to $x = \sqrt{3}$ about the y-axis.

30. Find the area between the graphs of

$$y = \frac{2e^x}{1 + e^x} \quad \text{and} \quad y = \frac{2e^x}{1 + e^{2x}}$$

from $x = 0$ to $x = \frac{1}{2}\ln 3$.

31. The bottom of a 15-ft-high billboard is 10 ft above an observer's eye level. At what distance from the billboard should the observer stand to get the best view of the billboard?

32. A helicopter is initially 200 ft directly above an observer on the ground. Then the helicopter begins to ascend at a speed of 50 ft/sec on a 45° path with the vertical. How fast is the angle of elevation of the helicopter changing, as measured from the observer, after 4 sec?

33. Derive a formula for $\coth(x + y)$ that involves $\coth x$ and $\coth y$ and no other hyperbolic functions.

In Exercises 34–37 prove the identities.

34. $$\frac{1}{1 - \tanh x} - \frac{1}{1 + \tanh x} = \sinh 2x$$

35. $$\frac{\coth^2 x + \cosh x \operatorname{csch}^2 x}{1 + \cosh x} = \operatorname{csch} x \coth x$$

36. $$\frac{\cosh x - \sinh x}{\cosh 2x - \sinh 2x} = \cosh x + \sinh x$$

37. $$\frac{1}{\operatorname{csch}^2 x + \operatorname{csch}^4 x} = \sinh^2 x - \tanh^2 x$$

38. Let $f(x) = 2 \cosh \frac{x}{2}$ on $[-1, 1]$.
(a) Find the area under the graph of f.
(b) Find the volume obtained by revolving the region under the graph of f about the x-axis.
(c) Determine the length of the graph of f.
(d) Find the surface area obtained by revolving the graph of f about the x-axis.

39. Show that the function

$$y = e^{-kt}(A \cosh \omega t + B \sinh \omega t)$$

satisfies the differential equation

$$\frac{d^2y}{dt^2} + 2k\frac{dy}{dt} + (k^2 - \omega^2)y = 0$$

where k and ω are positive constants. For $k = 2$ and $\omega = 1$, find A and B such that when $t = 0$, $y = -1$ and $\frac{dy}{dt} = 2$.

CHAPTER 8

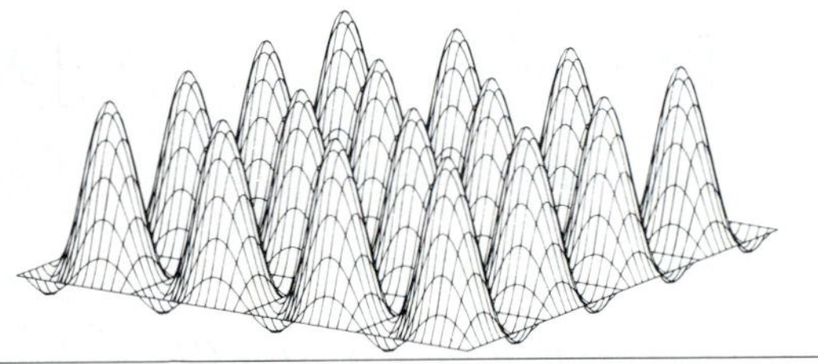

TECHNIQUES OF INTEGRATION

8.1 REVIEW OF BASIC FORMS

We now give a basic list of indefinite integrals that we have derived or have asked you to derive in the exercises. The main objective of this chapter is to introduce techniques that will enable you to change many integrals that do not conform to one in our basic list to one or more integrals that do. Unfortunately, there is no infallible step-by-step procedure (no *algorithm*) that will always lead to an integral that can be evaluated. It is easy, in fact, to give examples of continuous functions whose antiderivatives cannot be written in terms of familiar functions. The best we can do is to provide a variety of techniques from which you can choose in attempting to perform an integration. With a little practice you will usually be able to decide on the best technique.

Our basic list is by no means exhaustive. A more extensive list is given on the front and back end pages of this book, and there are tables that give hundreds of forms. We will discuss the uses of such tables later in this chapter, but it should be emphasized that these are not the answer to all of your problems. The techniques we will study are sometimes still necessary even when you use tables.

In our basic list we give the integrals in the most general form. We assume $u = g(x)$ is a differentiable function that satisfies the necessary restrictions in each case for the integral to exist. We do not state these restrictions explicitly. Whenever the constant a appears, we assume it to be positive.

BASIC LIST OF INTEGRALS

$$\int u^{\alpha}\,du = \frac{u^{\alpha+1}}{\alpha+1} + C, \quad \alpha \neq -1 \qquad \int \frac{du}{u} = \ln|u| + C$$

$$\int e^{u}\,du = e^{u} + C \qquad \int a^{u}\,du = \frac{a^{u}}{\ln a} + C, \quad a \neq 1$$

$$\int \sin u\,du = -\cos u + C \qquad \int \cos u\,du = \sin u + C$$

$$\int \sec^2 u\,du = \tan u + C \qquad \int \csc^2 u\,du = -\cot u + C$$

$$\int \sec u \tan u\,du = \sec u + C \qquad \int \csc u \cot u\,du = -\csc u + C$$

$$\int \tan u\,du = -\ln|\cos u| + C \qquad \int \cot u\,du = \ln|\sin u| + C$$

$$\int \sec u\,du = \ln|\sec u + \tan u| + C$$

$$\int \csc u\,du = \ln|\csc u - \cot u| + C$$

$$\int \frac{du}{\sqrt{a^2 - u^2}} = \sin^{-1}\frac{u}{a} + C \qquad \int \frac{du}{a^2 + u^2} = \frac{1}{a}\tan^{-1}\frac{u}{a} + C$$

$$\int \frac{du}{u\sqrt{u^2 - a^2}} = \frac{1}{a}\sec^{-1}\frac{u}{a} + C$$

$$\int \sinh u\,du = \cosh u + C \qquad \int \cosh u\,du = \sinh u + C$$

$$\int \operatorname{sech}^2 u\,du = \tanh u + C \qquad \int \operatorname{csch}^2 u\,du = -\coth u + C$$

$$\int \operatorname{sech} u \tanh u\,du = -\operatorname{sech} u + C$$

$$\int \operatorname{csch} u \coth u\,du = -\operatorname{csch} u + C$$

$$\int \operatorname{sech} u\,du = \tan^{-1}|\sinh u| + C$$

$$\int \operatorname{csch} u\,du = \ln\left|\tanh\frac{u}{2}\right| + C$$

$$\int \frac{du}{\sqrt{a^2 + u^2}} = \sinh^{-1}\frac{u}{a} + C = \ln\left(u + \sqrt{u^2 + a^2}\right) + C$$

$$\int \frac{du}{\sqrt{u^2 - a^2}} = \cosh^{-1}\frac{u}{a} + C = \ln\left(u + \sqrt{u^2 - a^2}\right) + C$$

$$\int \frac{du}{a^2 - u^2} = \left\{\begin{matrix} \dfrac{1}{a}\tanh^{-1}\dfrac{u}{a} + C, & |u| < a \\ \dfrac{1}{a}\coth^{-1}\dfrac{u}{a} + C, & |u| > a \end{matrix}\right\} = \frac{1}{2a}\ln\left|\frac{a+u}{a-u}\right| + C$$

$$\int \frac{du}{u\sqrt{a^2 - u^2}} = -\frac{1}{a}\operatorname{sech}^{-1}\frac{|u|}{a} + C = -\frac{1}{a}\ln\left(\frac{a + \sqrt{a^2 - u^2}}{|u|}\right) + C$$

$$\int \frac{du}{u\sqrt{a^2 + u^2}} = -\frac{1}{a}\operatorname{csch}^{-1}\frac{|u|}{a} + C = -\frac{1}{a}\ln\left(\frac{a + \sqrt{a^2 + u^2}}{|u|}\right) + C$$

8.2 INTEGRATION BY PARTS

One of the most important techniques for changing integrals that do not conform to basic forms into ones that do comes from the formula for the derivative of a product. If $u = f(x)$ and $v = g(x)$ are differentiable functions on some common domain, then we know that

$$\frac{d}{dx}[f(x)g(x)] = f(x)g'(x) + g(x)f'(x)$$

Taking antiderivatives of both sides, we get

$$f(x)g(x) = \int f(x)g'(x)\,dx + \int g(x)f'(x)\,dx$$

which can also be written in the form

$$\int f(x)g'(x)\,dx = f(x)g(x) - \int g(x)f'(x)\,dx \tag{8.1}$$

The use of this formula is called **integration by parts.** It is customary to write the result in terms of u and v. Since $du = f'(x)\,dx$ and $dv = g'(x)\,dx$, the formula becomes

$$\int u\,dv = uv - \int v\,du \tag{8.2}$$

We show how this is applied in the following examples.

EXAMPLE 8.1 Evaluate the integral $\int xe^x\,dx$.

Solution The secret lies in choosing the "parts" u and dv. Let us try letting $u = x$ and $dv = e^x\,dx$. (When a choice is made for u, the choice for dv is also determined.) So we have

$$\begin{array}{r|l} u = x & du = dx \\ dv = e^x\,dx & v = e^x \end{array}$$

Note that at this stage we do not use a constant of integration when we obtain v from dv. This is because we can combine all constants into one at the final step. (See Exercise 45 for justification of this.) Now we apply equation (8.2):

$$\int \overbrace{x}^{u}\ \overbrace{e^x\,dx}^{dv} = \overbrace{x}^{u}\ \overbrace{e^x}^{v} - \int \overbrace{e^x}^{v}\ \overbrace{dx}^{du}$$

The final integral is one we can integrate. So we have

$$\int xe^x\,dx = xe^x - e^x + C$$

■

EXAMPLE 8.2 Evaluate the integral $\int x^2 \cos x\,dx$.

Solution Let

$$\begin{array}{r|l} u = x^2 & du = 2x\,dx \\ dv = \cos x\,dx & v = \sin x \end{array}$$

Then by equation (8.2),

$$\begin{aligned}\int x^2 \cos x\,dx &= x^2 \sin x - \int \sin x(2x\,dx) \\ &= x^2 \sin x - 2\int x \sin x\,dx\end{aligned}$$

To evaluate the last integral we use integration by parts again, letting

$$\begin{array}{r|l} u = x & du = dx \\ dv = \sin x\,dx & v = -\cos x \end{array}$$

So

$$\begin{aligned}\int x \sin x\,dx &= -x \cos x + \int \cos x\,dx \\ &= -x\cos x + \sin x + C_1\end{aligned}$$

Thus,

$$\begin{aligned}\int x \cos x\,dx &= x^2 \sin x - 2\int x \sin x\,dx \\ &= x^2 \sin x - 2[-x\cos x + \sin x + C_1] \\ &= x^2 \sin x + 2x\cos x - 2\sin x - 2C_1\end{aligned}$$

But since C_1 is arbitrary, so is $-2C_1$. Calling this constant C, we get the final result:

$$\int x^2 \cos x\,dx = x^2 \sin x + 2x\cos x - 2\sin x + C$$

■

A useful formula can be derived to give the result of repeated integration by parts in problems such as Example 8.2. Suppose $u = x^n$ or, more generally, u is a polynomial of degree n, with $n \geq 2$. We assume that v and its successive antiderivatives can be obtained. Let u_0 and v_0 be the initial values of u and v, and for $k = 1, 2, \ldots, n$, let

$$u_k = u'_{k-1} \quad \text{and} \quad dv_k = v_{k-1}\,dx$$

Then the result of integrating by parts n times is (see Exercise 46)

$$\begin{aligned}\int u\,dv &= u_0v_0 - u_1v_1 + u_2v_2 - \cdots + (-1)^n u_nv_n + C \\ &= \sum_{k=0}^{n} (-1)^k u_kv_k + C \end{aligned} \tag{8.3}$$

To illustrate this, consider again the integral in Example 8.2, where we initially took $u = x^2$ and $dv = \cos x\,dx$. Thus, our first u and v are $u_0 = x^2$ and $v_0 = \sin x$. The accompanying table is a convenient way to organize the calculations. The first entries are u_0 and v_0. Each subsequent u_k is obtained by differentiating the preceding one, whereas each subsequent v_k is obtained by integrating the previous one. This is continued until a constant occurs in

u_k	v_k
x^2	$\sin x$
$2x$	$-\cos x$
2	$-\sin x$

the first column. (This is u_n.) We now multiply each u_k by the corresponding v_k, assign + signs and − signs alternately, and add the results:

$$\int x^2 \cos x\,dx = x^2 \sin x - (-2x\cos x) + (-2\sin x) + C$$
$$= x^2 \sin x + 2x\cos x - 2\sin x + C$$

EXAMPLE 8.3 Evaluate the integral $\int x^3 e^{2x}\,dx$ using equation (8.3).

Solution Let $u_0 = x^3$ and $dv_0 = e^{2x}\,dx$, so that $v_0 = e^{2x}/2$. From the table we get

$$\int x^3 e^{2x}\,dx = \frac{x^3 e^{2x}}{3} - \frac{3x^2 e^{2x}}{4} + \frac{6xe^{2x}}{8} - \frac{6e^{2x}}{16} + C$$
$$= \frac{e^{2x}}{24}(8x^3 - 18x^2 + 18x - 9) + C$$

u_x	v_k
x^3	$\frac{1}{2}e^{2x}$
$3x^2$	$\frac{1}{4}e^{2x}$
$6x$	$\frac{1}{8}e^{2x}$
6	$\frac{1}{16}e^{2x}$

■

EXAMPLE 8.4 Evaluate the integral $\int x \ln x\,dx$.

Solution In view of the choices of u and dv in the preceding examples, you might be tempted to try $u = x$ and $dv = \ln x\,dx$. But this will not help, since you do not know an antiderivative of $\ln x$ and so you cannot find v. So instead we take

$u = \ln x$	$du = \dfrac{1}{x}\,dx$
$dv = x\,dx$	$v = \dfrac{x^2}{2}$

Then we have

$$\int x \ln x\,dx = (\ln x)\left(\frac{x^2}{2}\right) - \int \frac{x^2}{2}\cdot\frac{1}{x}\,dx$$
$$= \frac{x^2}{2}\ln x - \frac{1}{2}\int x\,dx$$
$$= \frac{x^2}{2}\ln x - \frac{x^2}{4} + C$$

■

Integration by parts can be applied to a definite integral in two ways.

Method 1

$$\int_a^b u\,dv = uv\Big]_a^b - \int_a^b v\,du$$

where it is understood that a and b are limits for x and, as before, $u = f(x)$ and $v = g(x)$.

Method 2 First evaluate the indefinite integral

$$\int u\,dv = uv - \int v\,du$$

and after the result is obtained, substitute upper and lower limits.

We illustrate both methods in the next example.

EXAMPLE 8.5 Evaluate the integral $\int_0^1 \tan^{-1} x\,dx$.

Solution We use method 1. At first glance this integral does not seem suited to integration by parts, since there appears to be only one "part." But we can take dv as dx itself and so obtain

$$u = \tan^{-1} x \qquad du = \frac{1}{1+x^2}\,dx$$

$$dv = dx \qquad v = x$$

Thus,

$$\begin{aligned}\int_0^1 \tan^{-1} x\,dx &= x\tan^{-1} x\Big]_0^1 - \int_0^1 \frac{x}{1+x^2}\,dx \\ &= \tan^{-1} 1 - \frac{1}{2}\int_0^1 \frac{2x\,dx}{1+x^2} \\ &= \frac{\pi}{4} - \frac{1}{2}[\ln(1+x^2)]_0^1 \\ &= \frac{\pi}{4} - \frac{1}{2}\ln 2\end{aligned}$$

Using method 2 and taking u and dv as above, we have

$$\begin{aligned}\int \tan^{-1} x\,dx &= x\tan^{-1} x - \int \frac{x}{1+x^2}\,dx \\ &= x\tan^{-1} x - \frac{1}{2}\int \frac{2x}{1+x^2}\,dx \\ &= x\tan^{-1} x - \frac{1}{2}\ln(1+x^2) + C\end{aligned}$$

Thus, an antiderivative of $\tan^{-1} x$ is

$$x\tan^{-1} x - \frac{1}{2}\ln(1+x^2)$$

So by the fundamental theorem,

$$\begin{aligned}\int_0^1 \tan^{-1} x\,dx &= \left[x\tan^{-1} x - \frac{1}{2}\ln(1+x^2)\right]_0^1 \\ &= \tan^{-1} 1 - \frac{1}{2}\ln 2 \\ &= \frac{\pi}{4} - \frac{1}{2}\ln 2\end{aligned}$$

In general the first method is preferable, since limits are substituted in the uv term as the work progresses, and this sometimes simplifies the expression. ■

Remark Examples 8.4 and 8.5 suggest that when a problem involving a logarithm or an inverse trigonometric function (also an inverse hyperbolic function) is to be done by parts, u should be taken as the logarithm or the inverse function.

The next two examples show how repeated integration by parts can result in the recurrence of the original integral and how it can then be evaluated.

EXAMPLE 8.6 Evaluate the integral $\int e^{-x} \cos 2x\, dx$.

Solution Let

$$u = e^{-x} \qquad du = -e^{-x}\, dx$$
$$dv = \cos 2x\, dx \qquad v = \tfrac{1}{2} \sin 2x$$

Note that equation (8.3) for repeated integration by parts is not applicable, since u is not a polynomial. So we have

$$\int e^{-x} \cos 2x\, dx = \tfrac{1}{2}e^{-x} \sin 2x - \int (\tfrac{1}{2} \sin 2x)(-e^{-x}\, dx)$$
$$= \tfrac{1}{2}e^{-x} \sin 2x + \tfrac{1}{2} \int e^{-x} \sin 2x\, dx$$

To evaluate the last integral, we use integration by parts again with

$$u = e^{-x} \qquad du = -e^{-x}\, dx$$
$$dv = \sin 2x\, dx \qquad v = -\tfrac{1}{2} \cos 2x$$

Thus,

$$\int e^{-x} \sin 2x\, dx = -\tfrac{1}{2}e^{-x} \cos 2x - \int (-\tfrac{1}{2} \cos 2x)(-e^{-x}\, dx)$$
$$= -\tfrac{1}{2}e^{-x} \cos 2x - \tfrac{1}{2} \int e^{-x} \cos 2x\, dx$$

and substituting this in the equation that results from the first integration by parts gives

$$\int e^{-x} \cos 2x\, dx = \tfrac{1}{2}e^{-x} \sin 2x + \tfrac{1}{2}\left[-\tfrac{1}{2}e^{-x} \cos 2x - \tfrac{1}{2} \int e^{-x} \cos 2x\, dx \right]$$

or

$$\int e^{-x} \cos 2x\, dx = \tfrac{1}{2}e^{-x} \sin 2x - \tfrac{1}{4}e^{-x} \cos 2x - \tfrac{1}{4} \int e^{-x} \cos 2x\, dx$$

It might seem that we are going in circles, since the integral we are attempting to evaluate is expressed in terms of that integral itself. But the way out of this is simple—we just solve the equation algebraically for the unknown integral. Perhaps this is easier to see if we introduce a single letter, say I, for the unknown integral:

$$I = \int e^{-x} \cos 2x\, dx$$

Then we have arrived at

$$I = \tfrac{1}{2}e^{-x} \sin 2x - \tfrac{1}{4}e^{-x} \cos 2x - \tfrac{1}{4}I$$

To solve for I, we add $\tfrac{1}{4}I$ to both sides:

$$\frac{5I}{4} = \frac{e^{-x}}{4}(2 \sin 2x - \cos 2x)$$
$$I = \frac{e^{-x}}{5}(2 \sin 2x - \cos 2x)$$

In this process we did not introduce a constant of integration. To get the most general antiderivative, we must now do so:

$$\int e^{-x} \cos 2x\, dx = \frac{e^{-x}}{5}(2 \sin 2x - \cos 2x) + C$$ ■

Remark Example 8.6 is typical of integrals of the form

$$\int e^{ax} \cos bx\, dx \quad \text{or} \quad \int e^{ax} \sin bx\, dx$$

In these it is immaterial whether we take u as the exponential or the trigonometric function. We will ask you to redo Example 8.6 in the exercises, letting $u = \cos 2x$ and $dv = e^{-x}\, dx$ to show that this works just as well. It must be emphasized, however, that whatever choice is made initially, the same type of choice *must* be made on integrating by parts the second time. Otherwise, you really will go in circles, winding up with the true but unhelpful statement that the original integral equals itself!

EXAMPLE 8.7 Evaluate the integral $\int \sec^3 x\, dx$.

Solution We rewrite the integral in the form $\int \sec x(\sec^2 x\, dx)$ and choose

$$\begin{array}{r|l} u = \sec x & du = \sec x \tan x\, dx \\ dv = \sec^2 x\, dx & v = \tan x \end{array}$$

Then we have

$$\begin{aligned} \int \sec^3 x\, dx &= \sec x \tan x - \int \tan x(\sec x \tan x\, dx) \\ &= \sec x \tan x - \int \sec x \tan^2 x\, dx \end{aligned}$$

Now a fundamental trigonometric identity is that $\tan^2 x = \sec^2 x - 1$. Substituting this in the last integral gives

$$\begin{aligned} \int \sec^3 x\, dx &= \sec x \tan x - \int \sec x(\sec^2 x - 1)\, dx \\ &= \sec x \tan x - \int \sec^3 x\, dx + \int \sec x\, dx \end{aligned}$$

Here again, the unknown integral appears on both sides. Calling this I, we get

$$\begin{aligned} I &= \sec x \tan x - I + \ln|\sec x + \tan x| + C_1 \\ 2I &= \sec x \tan x + \ln|\sec x + \tan x| + C_1 \end{aligned}$$

Finally, divide by 2 and write $C = C_1/2$:

$$\int \sec^3 x\, dx = \tfrac{1}{2}(\sec x \tan x + \ln|\sec x + \tan x|) + C$$ ■

To illustrate what can happen if you make the wrong choice for u and dv, suppose in attempting to evaluate the integral $\int xe^x\, dx$ of Example 8.1 we had chosen $u = e^x$ and $dv = x\, dx$. Then we would have $du = e^x\, dx$ and $v = x^2/2$, so that

$$\int xe^x\, dx = \frac{x^2 e^x}{2} - \frac{1}{2}\int x^2 e^x\, dx$$

Now this is a true equation, but we would be worse off than when we started, since $\int x^2 e^x\,dx$ is more complicated than $\int xe^x\,dx$. Any attempt to integrate by parts again would just get us further into a quagmire. The lesson to be learned from this is that if you make a bad choice of parts, abandon that choice when it becomes evident it is not getting anywhere, and try another.

EXERCISE SET 8.2

A

Use integration by parts to evaluate all integrals.

1. $\int xe^{-x}\,dx$
2. $\int x\sin x\,dx$
3. $\int_0^{\pi/4} x\cos 2x\,dx$
4. $\int x\sec^2 x\,dx$
5. $\int x^4 e^x\,dx$
6. $\int x^2\cos^3 x\,dx$
7. $\int x\sec x\tan x\,dx$
8. $\int x\sinh 2x\,dx$
9. $\int_0^{\ln 2} x^2\cosh x\,dx$
10. $\int \ln x\,dx$
11. $\int x\csc^2 x\,dx$
12. $\int_0^{\sqrt{3}/2} \sin^{-1} x\,dx$
13. $\int x\ln\frac{1}{x}\,dx$
14. $\int_1^4 \frac{\ln x}{\sqrt{x}}\,dx$
15. $\int \cos x\ln\sin x\,dx$
16. $\int x^3(1-x)^9\,dx$
17. $\int_1^2 \frac{x}{(2x-1)^2}\,dx$
18. $\int xe^{1-2x}\,dx$
19. $\int x^3 e^{-2x}\,dx$
20. $\int x^3 e^{x^2}\,dx$ (*Hint:* Take $u = x^2$.)
21. $\int x\tan^2 x\,dx$
22. $\int \sqrt{x}\ln x^3\,dx$
23. Redo Example 8.6 taking $u = \cos 2x$ and $dv = e^{-x}\,dx$.
24. $\int e^x\sin x\,dx$
25. $\int e^x\cosh x\,dx$
26. $\int \csc^3 x\,dx$
27. $\int \operatorname{sech}^3 x\,dx$
28. $\int x\tan^{-1} x\,dx$ (*Hint:* In evaluating the integral that results from integration by parts, use long division.)
29. $\int x^2\cot^{-1} x\,dx$ (See the hint for Exercise 28.)
30. Find the volume obtained by revolving the region under the graph of $y = e^x$ from $x = 0$ to $x = \ln 3$ about the y-axis.
31. Find the volume obtained by revolving the region under the graph of $y = \cos\frac{x}{2}$ from $x = 0$ to $x = \pi$ about the y-axis.
32. Find the surface area generated by revolving the arc of $y = \cosh x$ from $(0, 1)$ to $(2, \cosh 2)$ about the y-axis.

B

33. $\int \sin\sqrt{x}\,dx$ (*Hint:* First make a substitution.)
34. $\int \cos(\ln x)\,dx$ (See the hint for Exercise 33.)
35. $\int x\sin^2 x\,dx$
36. $\int x\cos^2 x\,dx$
37. $\int \frac{x^3\,dx}{\sqrt{1-x^2}}$
38. $\int x^3(x^2-4)^{3/2}\,dx$
39. $\int e^{ax}\sin bx\,dx$
40. $\int e^{ax}\cos bx\,dx$
41. Prove the **reduction formulas**

(a) $$\int \sin^n x\,dx = -\frac{\sin^{n-1} x\cos x}{n} + \frac{n-1}{n}\int \sin^{n-2} x\,dx$$

(b) $$\int \cos^n x\,dx = \frac{\cos^{n-1} x\sin x}{n} + \frac{n-1}{n}\int \cos^{n-2} x\,dx$$

where n is a positive integer greater than 1.

42. Prove **Wallis' formula** for $n > 1$:

$$\int_0^{\pi/2} \sin^n x\,dx = \int_0^{\pi/2} \cos^n x\,dx = \begin{cases} \dfrac{2\cdot 4\cdot 6\cdot\cdots\cdot(n-1)}{1\cdot 3\cdot 5\cdot\cdots\cdot n} & \text{if } n \text{ is odd} \\[2ex] \dfrac{1\cdot 3\cdot 5\cdot\cdots\cdot(n-1)}{2\cdot 4\cdot 6\cdot\cdots\cdot n}\cdot\dfrac{\pi}{2} & \text{if } n \text{ is even} \end{cases}$$

(*Hint:* Use the result of Exercise 41.)

43. Use Wallis' formula from Exercise 42 to evaluate each of the following:

(a) $\int_0^{\pi/2} \sin^5 x\,dx$ (b) $\int_0^{\pi/2} \cos^6 x\,dx$

(c) $\int_0^{\pi/2} \sin^{10} x\,dx$ (d) $\int_0^{\pi/2} \cos^7 x\,dx$

44. Prove the reduction formulas

(a) $\int \sec^n x\,dx = \dfrac{\sec^{n-2} x \tan x}{n-1} + \dfrac{n-2}{n-1}\int \sec^{n-2} x\,dx$

(b) $\int \csc^n x\,dx = -\dfrac{\csc^{n-2} x \cot x}{n-1} + \dfrac{n-2}{n-1}\int \csc^{n-2} x\,dx$

where n is a positive integer greater than 1.

45. Prove that the result of integration by parts is unchanged if v is replaced by $v + C$. (This justifies omitting the constant in calculating v.)

46. Prove equation (8.3) for repeated integration by parts.

8.3 CERTAIN TRIGONOMETRIC FORMS

In this section we consider integrals of the following types:

I. $\int \sin^m x \cos^n x\,dx$

II. $\int \sec^m x \tan^n x\,dx$ and $\int \csc^m x \cot^n x\,dx$

III. $\int \sin mx \cos nx\,dx$, $\int \sin mx \sin nx$, and $\int \cos mx \cos nx\,dx$

We illustrate each type with several examples before giving general guidelines.

Type I

The simplest case is the one in which either m or n is 0. Consider the following example.

EXAMPLE 8.8 Evaluate the integrals:

(a) $\int \sin^3 x\,dx$ (b) $\int \cos^4 x\,dx$

Solution (a) We write the integral in the form

$$\int \sin^3 x\,dx = \int \sin^2 x(\sin x\,dx) = \int (1-\cos^2 x)\sin x\,dx$$
$$= \int \sin x\,dx - \int \cos^2 x(\sin x\,dx)$$

We know the integral of $\sin x$, and we observe that the other integral is almost in the form $\int u^2\,du$; all we need is a negative sign, which we can get by multiplying and dividing by -1. So we have

$$\int \sin^3 x\,dx = -\cos x + \frac{\cos^3 x}{3} + C$$

The important thing to observe here is that by splitting off the factor $\sin x$, the remaining factor ($\sin^2 x$) could be written in terms of the cosine, using $\sin^2 x = 1 - \cos^2 x$. Since $d(\cos x) = -\sin x\,dx$, we were able to bring the integral involving $\cos x$ into the form $\int u^n\,du$.

(b) This time we use the trigonometric identity

$$\cos^2 x = \frac{1 + \cos 2x}{2} \tag{8.4}$$

and get

$$\int \cos^4 dx = \int \left(\frac{1 + \cos 2x}{2}\right)^2 dx = \frac{1}{4}\int (1 + 2\cos 2x + \cos^2 2x)\, dx$$
$$= \frac{1}{4}\left[\int 1\, dx + \int 2\cos 2x\, dx + \int \cos^2 2x\, dx\right]$$

The first two integrals can be evaluated immediately, and in the third we use equation (8.4) again:

$$\int \cos^4 x\, dx = \frac{1}{4}\left[x + \sin 2x + \int \frac{1 + \cos 4x}{2}\, dx\right]$$
$$= \frac{x}{4} + \frac{\sin 2x}{4} + \frac{1}{8}\left[\int 1\, dx + \int \cos 4x\, dx\right]$$
$$= \frac{x}{4} + \frac{\sin 2x}{4} + \frac{x}{8} + \frac{\sin 4x}{32} + C$$
$$= \frac{3x}{8} + \frac{\sin 2x}{4} + \frac{\sin 4x}{32} + C$$

In evaluating $\int \cos 4x\, dx$ we mentally substituted $u = 4x$ and adjusted the differential by multiplying and dividing by 4, getting

$$\tfrac{1}{4}\int \overbrace{\cos 4x}^{\cos u}\overbrace{(4\, dx)}^{du} = \tfrac{1}{4}\sin 4x$$ ■

The essential difference between parts a and b of Example 8.8 is that in part a the exponent is odd and in part b the exponent is even. The two techniques illustrated carry over to situations where both the sine and cosine are present in the integrand, as the next example shows.

EXAMPLE 8.9 Evaluate the integrals:

(a) $\int \sin^2 x \cos^3 x\, dx$ (b) $\int \sin^4 x \cos^2 x\, dx$

Solution (a) We split off $\cos x$ to go with dx and write the rest of the integrand in terms of sine, using $\cos^2 x = 1 - \sin^2 x$:

$$\int \sin^2 x \cos^3 x\, dx = \int \sin^2 x \cos^2 x(\cos x\, dx)$$
$$= \int \sin^2 x(1 - \sin^2 x)\cos x\, dx$$
$$= \int (\sin^2 x - \sin^4 x)\cos x\, dx$$

Now we mentally substitute $u = \sin x$. Then $du = \cos x\, dx$. So we have two integrals in the form $\int u^n\, du$. Thus,

$$\int \sin^2 x \cos^3 x\, dx = \frac{\sin^3 x}{3} - \frac{\sin^5 x}{5} + C$$

(b) This time, since both exponents are even, we use equation (8.4) along with its counterpart for $\sin^2 x$:

$$\sin^2 x = \frac{1 - \cos 2x}{2} \tag{8.5}$$

$$\int \sin^4 x \cos^2 x \, dx = \int \left(\frac{1 - \cos 2x}{2}\right)^2 \left(\frac{1 + \cos 2x}{2}\right) dx$$

After a little algebraic simplification, we get

$$\begin{aligned}\int \sin^4 x \cos^2 x \, dx &= \tfrac{1}{8} \int (1 - \cos 2x - \cos^2 2x + \cos^3 2x) \, dx \\ &= \tfrac{1}{8}\left[\int 1 \, dx - \int \cos 2x \, dx - \int \cos^2 2x \, dx \right. \\ &\qquad \left. + \int \cos^3 2x \, dx\right]\end{aligned}$$

The first two integrals on the right are easily evaluated. In the last two we again use the even power–odd power approaches. We consider these individually:

$$\begin{aligned}\int \cos^2 2x \, dx &= \int \frac{1 + \cos 4x}{2} \, dx \quad \text{By equation (8.4)} \\ &= \frac{1}{2} \int (1 + \cos 4x) \, dx \\ &= \frac{x}{2} + \frac{\sin 4x}{8}\end{aligned}$$

For the last integral we split off $\cos 2x$ and get

$$\begin{aligned}\int \cos^3 2x \, dx &= \int \cos^2 2x (\cos 2x \, dx) \\ &= \int (1 - \sin^2 2x)(\cos 2x \, dx) \\ &= \int \cos 2x \, dx - \int \sin^2 2x \cos 2x \, dx \\ &= \frac{\sin 2x}{2} - \frac{\sin^3 2x}{6}\end{aligned}$$

(mentally adjusting differentials). Now we put everything together and get

$$\begin{aligned}\int \sin^4 x \cos^2 x \, dx &= \frac{1}{8}\left[x - \frac{\sin 2x}{2} - \left(\frac{x}{2} + \frac{\sin 4x}{8}\right)\right. \\ &\qquad \left. + \left(\frac{\sin 2x}{2} - \frac{\sin^3 2x}{6}\right)\right] + C \\ &= \frac{x}{16} - \frac{\sin^3 2x}{48} - \frac{\sin 4x}{64} + C\end{aligned}$$

■

From Examples 8.8 and 8.9 we can infer the following methods. *To integrate $\int \sin^m x \cos^n x \, dx$:*

1. If m is an odd positive integer, split off $\sin x$ to go with the differential, and use $\sin^2 x = 1 - \cos^2 x$ to write the rest of the integrand in terms of powers of $\cos x$. In this way one or more integrals of the form $\int u^k \, du$, with $u = \cos x$, are obtained.

2. If n is an odd positive integer, split off $\cos x$ to go with the differential, and use $\cos^2 x = 1 - \sin^2 x$ to write the rest of the integrand in terms of powers of $\sin x$. In this way one or more integrals of the form $\int u^k\,du$, with $u = \sin x$, are obtained.
3. If both m and n are nonnegative even integers, then use

$$\sin^2 x = \frac{1 - \cos 2x}{2} \qquad \cos^2 x = \frac{1 + \cos 2x}{2}$$

to reduce the powers. In the resulting integrals, if an odd power occurs, use technique 1 or 2, and if even powers only occur, use 3 again.

Type II

Again, we use examples to illustrate the techniques. The relevant trigonometric identities are

$$\sec^2 x - \tan^2 x = 1 \qquad \csc^2 x - \cot^2 x = 1$$

We use these along with

$$\frac{d}{dx}\tan x = \sec^2 x \qquad \frac{d}{dx}\sec x = \sec x \tan x$$

$$\frac{d}{dx}\cot x = -\csc^2 x \qquad \frac{d}{dx}\csc x = -\csc x \cot x$$

We use only integrals that involve secants and tangents in our examples because those with cosecants and cotangents are handled in a similar way.

EXAMPLE 8.10 Evaluate the integrals:

(a) $\int \sec^4 x \tan^3 x\,dx$ (b) $\int \sec^3 x \tan^3 x\,dx$ (c) $\int \sec^3 x \tan^2 x\,dx$

Solution (a) We split off $\sec^2 x$ to go with the differential and write the rest of the integrand in terms of $\tan x$:

$$\begin{aligned}\int \sec^4 x \tan^3 x\,dx &= \int \sec^2 x \tan^3 x(\sec^2 x\,dx)\\ &= \int (1 + \tan^2 x)\tan^3 x(\sec^2 x\,dx)\\ &= \int \tan^3 x(\sec^2 x\,dx) + \int \tan^5 x(\sec^2 x\,dx)\\ &= \frac{\tan^4 x}{4} + \frac{\tan^6 x}{6} + C\end{aligned}$$

The critical condition for this approach is that in $\int \sec^m x \tan^n x\,dx$, m is a positive *even* integer; that is, $\sec x$ is raised to an even power.

(b) This time we split off $\sec x \tan x$ to form the differential and write what is left in terms of $\sec x$:

$$\begin{aligned}\int \sec^3 x \tan^3 x\,dx &= \int \sec^2 x \tan^2 x(\sec x \tan x\,dx)\\ &= \int \sec^2 x(\sec^2 x - 1)(\sec x \tan x\,dx)\\ &= \int \sec^4 x(\sec x \tan x\,dx) - \int \sec^2 x(\sec x \tan x\,dx)\\ &= \frac{\sec^5 x}{5} - \frac{\sec^3 x}{3} + C\end{aligned}$$

Here the essential thing was for $\tan x$ to be to an *odd* power; that is, in the integral $\int \sec^m x \tan^n x\,dx$, n is a positive odd integer.

(c) This is an example in which the conditions for the methods used in parts a and b both fail; that is, in $\int \sec^m x \tan^n x\,dx$, m is odd and n is even. We can express the integrand in powers of $\sec x$ and then use integration by parts, as in Example 8.7:

$$\begin{aligned}\int \sec^3 x \tan^2 x\,dx &= \int \sec^3 x(\sec^2 x - 1)\,dx \\ &= \int \sec^5 x\,dx - \int \sec^3 x\,dx\end{aligned}$$

We obtained the result

$$\int \sec^3 x\,dx = \tfrac{1}{2}(\sec x \tan x + \ln|\sec x + \tan x|) + C$$

in Example 8.7. For $\int \sec^5 x\,dx$ we use the same technique. Let

$$\begin{array}{l|l} u = \sec^3 x & du = 3\sec^3 x \tan x \\ dv = \sec^2 x\,dx & v = \tan x \end{array}$$

$$\begin{aligned}\int \sec^5 x\,dx &= \sec^3 x \tan x - 3\int \sec^3 x \tan^2 x\,dx \\ &= \sec^3 x \tan x - 3\int \sec^3 x(\sec^2 x - 1)\,dx \\ &= \sec^3 x \tan x - 3\int \sec^5 x\,dx + 3\int \sec^3 x\,dx\end{aligned}$$

Solving for the unknown integral $\int \sec^5 x\,dx$, we get

$$\int \sec^5 x\,dx = \frac{1}{4}\left(\sec^3 x \tan x + 3\int \sec^3 x\,dx\right) + C$$

Now we can use the value of $\int \sec^3 x\,dx$ obtained earlier:

$$\int \sec^3 x \tan^2 x\,dx$$

$$= \frac{\sec^3 x \tan x}{4} - \frac{\sec x \tan x}{8} - \frac{1}{8}\ln|\sec x + \tan x| + C \quad \blacksquare$$

Comment In Exercise 44 of Exercise Set 8.2 we asked for a proof of reduction formulas for $\int \sec^n x\,dx$ and $\int \csc^n x\,dx$. You are not likely to commit the results to memory, but having these formulas available cuts down on the work involved in problems such as part c of the preceding example.

Here is a summary of type II problems. *To integrate* $\int \sec^m x \tan^n x\,dx$:

1. If m is an even positive integer, split off $\sec^2 x$ to go with the differential, and use $\sec^2 x = 1 + \tan^2 x$ to write the rest of the integrand in terms of powers of $\tan x$. In this way one or more integrals of the form $\int u^k\,du$, with $u = \tan x$, are obtained.
2. If n is an odd positive integer, split off $\sec x \tan x$ to go with the differential and use $\tan^2 x = \sec^2 x - 1$ to write the rest of the integrand in terms of powers of $\sec x$. In this way one or more integrals of the form $\int u^k\,du$, with $u = \sec x$, are obtained.
3. If m is an odd positive integer and n is an even positive integer, use $\tan^2 x = \sec^2 x - 1$ to obtain integrals of the form $\int \sec^k x\,dx$, where k is odd. Then use integration by parts (or the reduction formula of Exercise 44 in Exercise Set 8.2), along with the known result $\int \sec x\,dx = \ln|\sec x + \tan x| + C$.

Type III

For these problems we need the following trigonometric identities:

$$\sin A \cos B = \tfrac{1}{2}[\sin(A+B) + \sin(A-B)] \tag{8.6}$$

$$\sin A \sin B = \tfrac{1}{2}[\cos(A-B) - \cos(A+B)] \tag{8.7}$$

$$\cos A \cos B = \tfrac{1}{2}[\cos(A+B) + \cos(A-B)] \tag{8.8}$$

We illustrate the technique with the following example, which is typical of all type III problems.

EXAMPLE 8.11 Evaluate the integral $\int \sin 5x \cos 2x\, dx$.

Solution By equation (8.6) we have

$$\sin 5x \cos 2x\, dx = \tfrac{1}{2}\int (\sin 7x + \sin 3x)\, dx$$
$$= \tfrac{1}{2}\left[\int \sin 7x\, dx + \int \sin 3x\, dx\right]$$

In the first integral we mentally multiply and divide by 7 and in the second integral multiply and divide by 3, thus getting both integrals in the form $\int \sin u\, du$. This results in

$$\int \sin 5x \cos 2x\, dx = -\frac{\cos 7x}{14} - \frac{\cos 3x}{6} + C$$

We might note that an alternate approach to this problem is to integrate by parts twice and to solve the resulting equation for the original integral. You would obtain a different appearing result, but it could be shown to differ from our result by a constant. ■

Remark The techniques illustrated for types I, II, and III are also applicable to analogous integrals where hyperbolic functions are used instead of trigonometric functions. The relevant hyperbolic identities then must be used instead of trigonometric identities.

EXERCISE SET 8.3

A

In Exercises 1–40 evaluate the integrals.

1. $\int \cos^3 x\, dx$
2. $\int \sin^5 x\, dx$
3. $\int \sin^4 x\, dx$
4. $\int \cos^2 x\, dx$
5. $\int_0^{\pi/4} \sin^3 x \cos^2 x\, dx$
6. $\int \sin^2 x \cos^5 x\, dx$
7. $\int \sin^2 x \cos^2 x\, dx$
8. $\int \sin^2 x \cos^4 x\, dx$
9. $\int_0^{\pi/4} \sec^2 x \tan^3 x\, dx$
10. $\int \sec^4 x \tan^2 x\, dx$
11. $\int \sec x \tan^3 x\, dx$
12. $\int_0^{\pi/3} \sec^5 x \tan x\, dx$
13. $\int \sec x \tan^2 x\, dx$
14. $\int \sec^4 x \tan^4 x\, dx$
15. $\int_0^{\pi/3} \frac{\sin^3 x\, dx}{\cos x}$
16. $\int_{\pi/2}^{\pi} \sqrt{\sin x} \cos^3 x\, dx$
17. $\int \frac{\sec^4 x\, dx}{\tan^2 x}$
18. $\int \sec^{3/2} x \tan x\, dx$
19. $\int_0^{\pi/3} \tan^3 x\, dx$
20. $\int \tan^4 x\, dx$
21. $\int \sin^2 3x \cos^3 3x\, dx$
22. $\int_{\pi/12}^{\pi/4} \frac{\cos^5 2x}{\sin^4 2x}\, dx$

23. $\int_0^{\pi/8} \sqrt{\tan 2x}\, \sec^4 2x\, dx$

24. $\int \csc x \cot^2 x\, dx$

25. $\int x \sin^2(x^2) \cos^5(x^2)\, dx$

26. $\int \frac{\tan^3 \sqrt{x}\, \sec^4 \sqrt{x}\, dx}{\sqrt{x}}$

27. $\int \csc^2 \frac{x}{2} \cot^2 \frac{x}{2}\, dx$

28. $\int \sin^4 x \cos^4 x\, dx$

29. $\int \sinh x \cosh^3 x\, dx$

30. $\int \sinh^2 x \cosh^2 x\, dx$

31. $\int \tanh^3 x\, \text{sech}^4 x\, dx$

32. $\int \tanh^3 x\, \text{sech}^3 x\, dx$

33. $\int \sin 3x \cos 2x\, dx$

34. $\int \sin 3x \sin 2x\, dx$

35. $\int \cos 3x \cos 2x\, dx$

36. $\int \sin x \cos 2x\, dx$

B

37. $\int \sin^6 x\, dx$

38. $\int \sec^5 x \tan^4 x\, dx$

39. $\int \frac{\tan^5 x\, dx}{\sec x}$

40. $\int \sinh^3 x(1 + \sinh^2 x)^{3/2}\, dx$

41. For m and n positive integers, let

$$\delta_{mn} = \begin{cases} 1 & \text{if } m = n \\ 0 & \text{if } m \neq n \end{cases}$$

(δ_{mn} is known as the *Kronecker delta.*) Prove the following:

(a) $\int_{-\pi}^{\pi} \sin mx \sin nx\, dx = \int_{-\pi}^{\pi} \cos mx \cos nx\, dx$
$= \pi\delta_{mn}$

(b) $\int_{\pi}^{\pi} \sin mx \cos nx\, dx = 0$

42. Prove the following:

(a) $\int \sin^m x \cos^{2n+1} x\, dx = \int u^m(1 - u^2)^n\, du$, where $u = \sin x$ and n is a nonnegative integer.

(b) $\int \sin^{2m+1} x \cos^n x\, dx = -\int u^n(1 - u^2)^m\, du$, where $u = \cos x$ and m is a nonnegative integer.

(c) $\int \sec^{2m+2} x \tan^n x\, dx = \int u^n(1 + u^2)^m\, du$, where $u = \tan x$ and m is a nonnegative integer.

(d) $\int \sec^m x \tan^{2n+1} x\, dx = \int u^{m-1}(u^2 - 1)^n\, du$, where $u = \sec x$ and n is a nonnegative integer.

8.4 TRIGONOMETRIC SUBSTITUTIONS

Integrals involving one of the expressions

$$\sqrt{a^2 - u^2}, \quad \sqrt{a^2 + u^2}, \quad \text{or} \quad \sqrt{u^2 - a^2}$$

where $a > 0$, can often be evaluated making one of the following substitutions:

If integral contains	Make substitution
$\sqrt{a^2 - u^2}$	$u = a \sin \theta, \quad \frac{\pi}{2} \leq \theta \leq \frac{\pi}{2}$
$\sqrt{a^2 + u^2}$	$u = a \tan \theta, \quad \frac{\pi}{2} < \theta < \frac{\pi}{2}$
$\sqrt{u^2 - a^2}$	$u = a \sec \theta, \quad \begin{cases} 0 \leq \theta < \frac{\pi}{2} & \text{if } u \geq a \\ \pi \leq \theta < \frac{3\pi}{2} & \text{if } u \leq -a \end{cases}$

These choices are based on the trigonometric identities:

$$1 - \sin^2 \theta = \cos^2 \theta$$
$$1 + \tan^2 \theta = \sec^2 \theta$$
$$\sec^2 \theta - 1 = \tan^2 \theta$$

Using these and the substitutions shown, we eliminate the radicals:

$$\sqrt{a^2 - u^2} = \sqrt{a^2 - a^2 \sin^2 \theta} = \sqrt{a^2(1 - \sin^2 \theta)} = a\sqrt{\cos^2 \theta} = a \cos \theta$$
$$\sqrt{a^2 + u^2} = \sqrt{a^2 + a^2 \tan^2 \theta} = \sqrt{a^2(1 + \tan^2 \theta)} = a\sqrt{\sec^2 \theta} = a \sec \theta$$
$$\sqrt{u^2 - a^2} = \sqrt{a^2 \sec^2 \theta - a^2} = \sqrt{a^2(\sec^2 \theta - 1)} = a\sqrt{\tan^2 \theta} = a \tan \theta$$

Note In the first case, since $\cos \theta \geq 0$ for $\frac{\pi}{2} \leq \theta \leq \frac{\pi}{2}$, we have $\sqrt{\cos^2 \theta} = \cos \theta$. Similarly, $\sec \theta > 0$ for $-\frac{\pi}{2} < \theta < \frac{\pi}{2}$ so that $\sqrt{\sec^2 \theta} = \sec \theta$, and $\tan \theta > 0$ for θ satisfying either $0 \leq \theta \leq \frac{\pi}{2}$ or $\pi \leq \theta < \frac{3\pi}{2}$ so that $\sqrt{\tan^2 \theta} = \tan \theta$.

The restrictions placed on θ in each of the substitutions are precisely those we used in defining the corresponding inverse trigonometric function. It follows that in the first case $\theta = \sin^{-1} \frac{u}{a}$, in the second $\theta = \tan^{-1} \frac{u}{a}$, and in the third $\theta = \sec^{-1} \frac{u}{a}$. From now on in making these trigonometric substitutions, we will understand that the restrictions on θ are to be imposed, without explicitly stating them.

EXAMPLE 8.12 Evaluate the integral

$$\int \frac{x^3}{\sqrt{4 - x^2}}\, dx$$

Solution We make the substitution $x = 2 \sin \theta$. Then $dx = 2 \cos \theta \, d\theta$ and $\theta = \sin^{-1} \frac{x}{2}$. So we have

$$\int \frac{x^3}{\sqrt{4 - x^2}}\, dx = \int \frac{(2 \sin \theta)^3 (2 \cos \theta \, d\theta)}{\sqrt{4 - 4 \sin^2 \theta}} = \int \frac{8 \sin^3 \theta (2 \cos \theta \, d\theta)}{2\sqrt{1 - \sin^2 \theta}}$$
$$= 8 \int \frac{\sin^3 \theta \cos \theta \, d\theta}{\sqrt{\cos^2 \theta}} = 8 \int \sin^3 \theta \, d\theta$$

Now we use ideas of the preceding section:

$$8 \int \sin^3 \theta \, d\theta = 8 \int \sin^2 \theta (\sin \theta \, d\theta) = 8 \int (1 - \cos^2 \theta) \sin \theta \, d\theta$$
$$= 8\left(-\cos \theta + \frac{\cos^3 \theta}{3}\right) + C$$

But we are not through, since this must be expressed in terms of x. An easy way to do this is to draw a right triangle as in Figure 8.1, showing θ as an acute angle with its opposite side equal to x and hypotenuse 2. Then the adjacent side is $\sqrt{4 - x^2}$, so that $\cos \theta = \sqrt{4 - x^2}/2$. Thus,

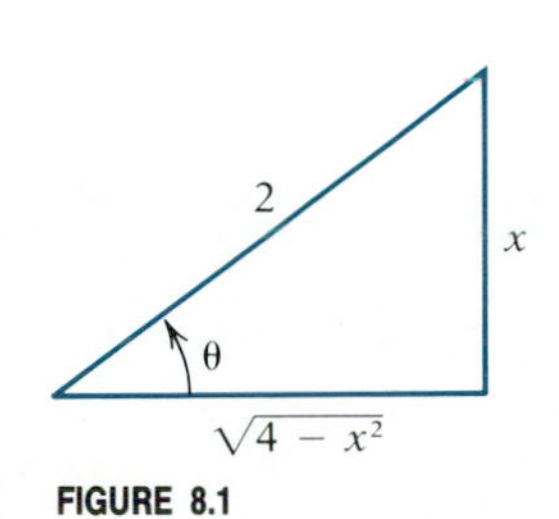

FIGURE 8.1

$$\int \frac{x^3}{\sqrt{4 - x^2}}\, dx = 8\left[-\frac{\sqrt{4 - x^2}}{2} + \frac{1}{3}\left(\frac{\sqrt{4 - x^2}}{2}\right)^3\right] + C$$
$$= -4\sqrt{4 - x^2} + \frac{(4 - x^2)^{3/2}}{3} + C$$ ■

A Word of Caution When making a substitution, do not forget to substitute for dx.

EXAMPLE 8.13 Evaluate the integral

$$\int_0^4 \frac{dx}{(9+x^2)^{3/2}}$$

Solution Since $(9+x^2)^{3/2} = (\sqrt{9+x^2})^3$, we let $x = 3\tan\theta$ so that $dx = 3\sec^2\theta\,d\theta$ and $\theta = \tan^{-1}\frac{x}{3}$. Because this is a definite integral, we also change limits. If $x = 0$, $\theta = \tan^{-1} 0 = 0$, and if $x = 4$, $\theta = \tan^{-1}\frac{4}{3}$. So we have

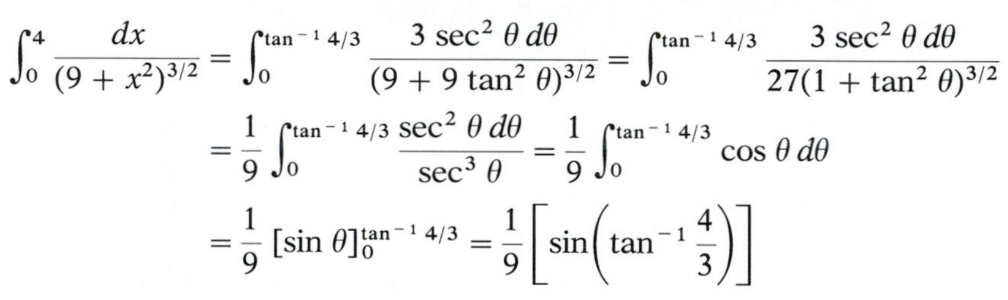

$$\int_0^4 \frac{dx}{(9+x^2)^{3/2}} = \int_0^{\tan^{-1} 4/3} \frac{3\sec^2\theta\,d\theta}{(9+9\tan^2\theta)^{3/2}} = \int_0^{\tan^{-1} 4/3} \frac{3\sec^2\theta\,d\theta}{27(1+\tan^2\theta)^{3/2}}$$

$$= \frac{1}{9}\int_0^{\tan^{-1} 4/3} \frac{\sec^2\theta\,d\theta}{\sec^3\theta} = \frac{1}{9}\int_0^{\tan^{-1} 4/3} \cos\theta\,d\theta$$

$$= \frac{1}{9}\left[\sin\theta\right]_0^{\tan^{-1} 4/3} = \frac{1}{9}\left[\sin\left(\tan^{-1}\frac{4}{3}\right)\right]$$

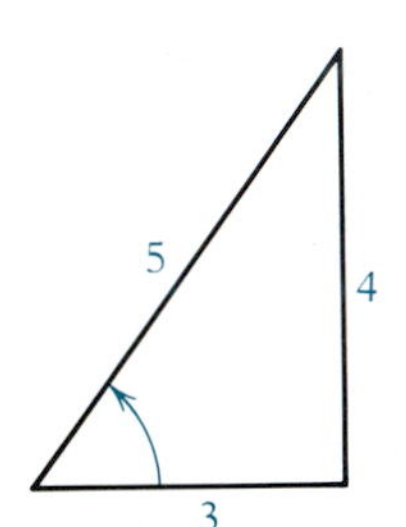

FIGURE 8.2

From Figure 8.2 we see that $\sin(\tan^{-1}\frac{4}{3}) = \frac{4}{5}$, so we have, finally,

$$\int_0^4 \frac{dx}{(9+x^2)^{3/2}} = \frac{1}{9}\left(\frac{4}{5}\right) = \frac{4}{45}$$

■

EXAMPLE 8.14 Evaluate the integral

$$\int \frac{\sqrt{4x^2-9}}{x}\,dx$$

Solution The radical is of the form $\sqrt{u^2-a^2}$ with $u = 2x$ and $a = 3$. Using $u = a\sec\theta$, we have $2x = 3\sec\theta$, or

$$x = \frac{3\sec\theta}{2}$$

Then $dx = \frac{3}{2}\sec\theta\tan\theta\,d\theta$ and $\theta = \sec^{-1}\frac{2x}{3}$. So the integral becomes

$$\int \frac{\sqrt{4x^2-9}}{x}\,dx = \int \frac{\sqrt{9\sec^2\theta-9}}{\frac{3}{2}\sec\theta}\cdot\left(\frac{3}{2}\sec\theta\tan\theta\,d\theta\right)$$

$$= \int 3\sqrt{\sec^2\theta-1}\,\tan\theta\,d\theta$$

$$= 3\int \tan^2\theta\,d\theta$$

$$= 3\int (\sec^2\theta-1)\,d\theta$$

$$= 3(\tan\theta-\theta)+C$$

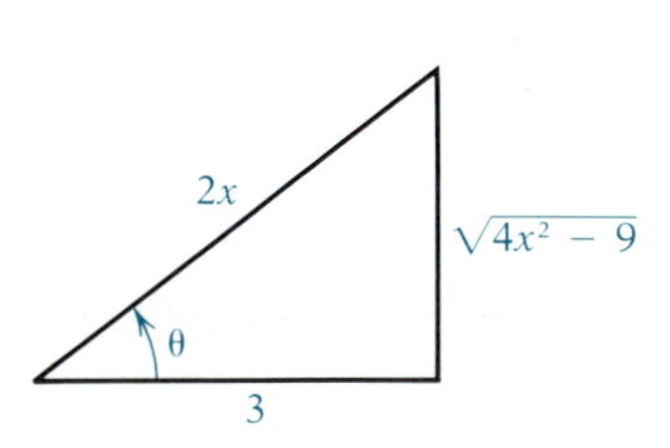

FIGURE 8.3

From the definition of θ, we get Figure 8.3, so that

$$\int \frac{\sqrt{4x^2-9}}{x}\,dx = 3\left(\frac{\sqrt{4x^2-9}}{3} - \sec^{-1}\frac{2x}{3}\right)+C$$

$$= \sqrt{4x^2-9} - 3\sec^{-1}\frac{2x}{3}+C$$

■

The next example shows that a trigonometric substitution is sometimes helpful even though there is no radical involved.

EXAMPLE 8.15 Evaluate the integral

$$\int \frac{x^3\,dx}{1 + 2x^2 + x^4}$$

Solution This can be written in the form

$$\int \frac{x^3\,dx}{(1 + x^2)^2}$$

Let $x = \tan\theta$, so that $dx = \sec^2\theta\,d\theta$ and $\theta = \tan^{-1} x$. This gives

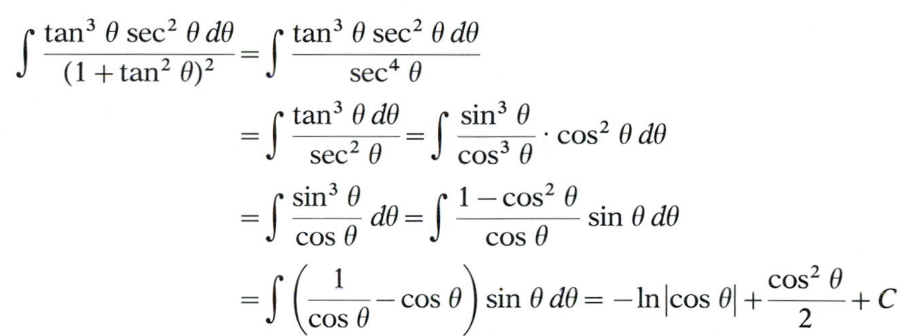

$$\int \frac{\tan^3\theta\sec^2\theta\,d\theta}{(1+\tan^2\theta)^2} = \int \frac{\tan^3\theta\sec^2\theta\,d\theta}{\sec^4\theta}$$

$$= \int \frac{\tan^3\theta\,d\theta}{\sec^2\theta} = \int \frac{\sin^3\theta}{\cos^3\theta}\cdot\cos^2\theta\,d\theta$$

$$= \int \frac{\sin^3\theta}{\cos\theta}\,d\theta = \int \frac{1-\cos^2\theta}{\cos\theta}\sin\theta\,d\theta$$

$$= \int\left(\frac{1}{\cos\theta} - \cos\theta\right)\sin\theta\,d\theta = -\ln|\cos\theta| + \frac{\cos^2\theta}{2} + C$$

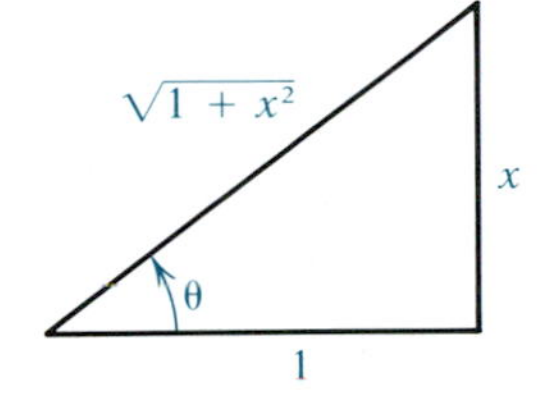

FIGURE 8.4

From Figure 8.4 we therefore have

$$\int \frac{x^3\,dx}{1 + 2x^2 + x^4} = -\ln\frac{1}{\sqrt{1 + x^2}} + \frac{1}{2}\left(\frac{1}{1 + x^2}\right) + C$$

$$= \frac{1}{2}\left[\ln(1 + x^2) + \frac{1}{1 + x^2}\right] + C \qquad \blacksquare$$

EXAMPLE 8.16 Derive the integration formula

$$\int \sqrt{a^2 - u^2}\,du = \frac{1}{2}\left(a^2\sin^{-1}\frac{u}{a} + u\sqrt{a^2 - u^2}\right) + C$$

Solution Let $u = a\sin\theta$; then $du = a\cos\theta\,d\theta$ and $\theta = \sin^{-1}\frac{u}{a}$. So we have

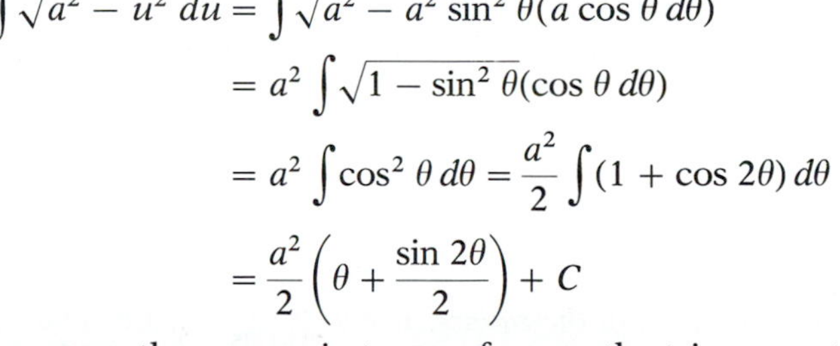

$$\int \sqrt{a^2 - u^2}\,du = \int \sqrt{a^2 - a^2\sin^2\theta}(a\cos\theta\,d\theta)$$

$$= a^2\int \sqrt{1 - \sin^2\theta}(\cos\theta\,d\theta)$$

$$= a^2\int \cos^2\theta\,d\theta = \frac{a^2}{2}\int (1 + \cos 2\theta)\,d\theta$$

$$= \frac{a^2}{2}\left(\theta + \frac{\sin 2\theta}{2}\right) + C$$

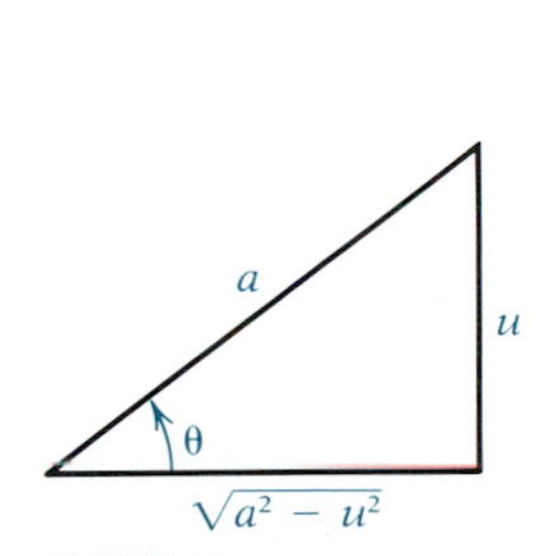

FIGURE 8.5

To express the answer in terms of x, use the trigonometric identity $\sin 2\theta = 2\sin\theta\cos\theta$ and refer to Figure 8.5. This gives

$$\int \sqrt{a^2 - u^2}\,du = \frac{a^2}{2}\left(\sin^{-1}\frac{u}{a} + \frac{u}{a}\cdot\frac{\sqrt{a^2 - u^2}}{a}\right) + C$$

$$= \frac{1}{2}\left(a^2\sin^{-1}\frac{u}{a} + u\sqrt{a^2 - u^2}\right) + C \qquad \blacksquare$$

Comment We could also use hyperbolic substitutions to rationalize the three types of expressions under study, but trigonometric substitutions are more commonly used.

EXERCISE SET 8.4

A

In Exercises 1–12 evaluate the integral by making a trigonometric substitution.

1. $\int x^3\sqrt{1-x^2}\,dx$
2. $\int \frac{x^3}{\sqrt{1+x^2}}\,dx$
3. $\int \frac{x^3\,dx}{\sqrt{x^2-1}}$
4. $\int \frac{dx}{\sqrt{x^2-4}}$
5. $\int \frac{\sqrt{9-x^2}}{x}\,dx$
6. $\int_{-1}^{1} \frac{dx}{(2-x^2)^{3/2}}$
7. $\int_1^3 \frac{\sqrt{x^2+3}\,dx}{x^4}$
8. $\int_{\sqrt{3}}^3 x^3\sqrt{4x^2-9}\,dx$
9. $\int_0^2 \frac{x^2\,dx}{(3x^2+4)^{3/2}}$
10. $\int_{-1/\sqrt{3}}^{1} \frac{x^3\,dx}{(4-3x^2)^{5/2}}$
11. $\int \frac{dx}{(x^2-1)^2}$
12. $\int \frac{x^5}{(1-x^2)^3}\,dx$

In Exercises 13–16 evaluate the integrals by two methods: (a) trigonometric substitution and (b) algebraic substitution.

13. $\int \frac{x\,dx}{\sqrt{9-4x^2}}$
14. $\int_0^1 x(3x^2+1)^{3/2}\,dx$
15. $\int_1^{\sqrt{3}} \frac{x}{\sqrt{4x^2-3}}\,dx$
16. $\int \frac{x}{1+2x^2+x^4}\,dx$

In Exercises 17–23 verify the given integration formula by using a trigonometric substitution to evaluate the integral.

17. $\int \frac{du}{\sqrt{a^2-u^2}} = \sin^{-1}\frac{u}{a} + C$
18. $\int \frac{du}{a^2+u^2} = \frac{1}{a}\tan^{-1}\frac{u}{a} + C$
19. $\int \frac{du}{u\sqrt{u^2-a^2}} = \frac{1}{a}\sec^{-1}\frac{u}{a} + C$
20. $\int \frac{du}{\sqrt{a^2+u^2}} = \ln\left|u+\sqrt{a^2+u^2}\right| + C$
21. $\int \frac{du}{u^2-a^2} = \frac{1}{2a}\ln\left|\frac{u-a}{u+a}\right| + C$
22. $\int \frac{du}{u\sqrt{a^2-u^2}} = -\frac{1}{a}\ln\left|\frac{a+\sqrt{a^2-u^2}}{u}\right| + C$
23. $\int \frac{du}{u\sqrt{a^2+u^2}} = -\frac{1}{a}\ln\left|\frac{a+\sqrt{a^2+u^2}}{u}\right| + C$
24. Determine an appropriate rationalizing hyperbolic substitution for integrands involving the following: (a) $\sqrt{a^2-u^2}$ (b) $\sqrt{a^2+u^2}$ (c) $\sqrt{u^2-a^2}$

In Exercises 25–27 redo the specified exercise using the appropriate hyperbolic substitution determined in Exercise 24.

25. Exercise 2
26. Exercise 3
27. Exercise 5
28. Find the length of the parabola $y = x^2/3$ from $x = 0$ to $x = 2$.
29. Find the area of the interior of the ellipse

$$\frac{x^2}{a^2} + \frac{y^2}{b^2} = 1$$

B

In Exercises 30–33, after completing the square, use a trigonometric substitution to evaluate the integral.

30. $\int \frac{dx}{\sqrt{x^2+2x+2}}$
31. $\int (2x-x^2)^{3/2}\,dx$
32. $\int \frac{dx}{x^2+4x-5}$
33. $\int \frac{(3x+2)^2\,dx}{(5-12x-9x^2)^{3/2}}$

Evaluate the integrals in Exercises 34–37.

34. $\int \frac{\cos t\,dt}{(1+\sin^2 t)^{3/2}}$ (*Hint:* First let $u = \sin t$.)
35. $\int \frac{\sec^2 t\tan^3 t\,dt}{\sqrt{\tan^2 t-1}}$
36. $\int \frac{\sin\theta\,d\theta}{(1+\cos^2\theta)^3}$
37. $\int \frac{e^{3x}\,dx}{\sqrt{1-e^{2x}}}$

8.5 PARTIAL FRACTIONS

We will now study techniques for integrating rational functions—that is, functions of the form $\frac{P(x)}{Q(x)}$, where $P(x)$ and $Q(x)$ are polynomials. Actually, what we study is an algebraic technique for expressing $\frac{P(x)}{Q(x)}$ as a sum of simpler rational functions when the degree of Q is 2 or greater. The simpler functions can then be integrated by means already at our disposal. As a way of introducing the ideas, consider the following addition of two fractions:

$$\frac{2}{x-1}+\frac{3}{x+2}=\frac{5x+1}{x^2+x-2}$$

Now suppose we are asked to perform the integration

$$\int \frac{5x+1}{x^2+x-2}\,dx$$

By replacing the integrand with the sum of the two simpler fractions we get

$$\begin{aligned}\int \frac{5x+1}{x^2+x-2}\,dx &= \int \left(\frac{2}{x-1}+\frac{3}{x+2}\right)dx\\ &= 2\ln|x-1|+3\ln|x+2|+C\\ &= \ln|(x-1)^2(x+2)^3|+C\end{aligned}$$

This is all well and good *provided* we know the two original fractions. The objective of this section is to introduce ways of determining the component fractions when we are given their sum; that is, we do the reverse of adding fractions: Given the sum, what are the simplest fractions that add together to give that sum? The process is called **decomposition into partial fractions.**

We begin with the assumption that $\frac{P(x)}{Q(x)}$ is in lowest terms and that the *degree of P is less than the degree of Q*—that is, that $\frac{P(x)}{Q(x)}$ is a *proper* fraction. If this is not the case, then by division we obtain

$$\frac{P(x)}{Q(x)}=P_1(x)+\frac{R(x)}{Q(x)}$$

where the degree of R is less than the degree of Q, and we would then apply our theory to $\frac{R(x)}{Q(x)}$.

It is proved in algebra that every polynomial with real coefficients can (in theory, at least) be factored into real linear and/or quadratic factors (such as x^2+1) that are **irreducible** over the reals. A quadratic factor ax^2+bx+c is irreducible over the reals if it cannot be factored into linear factors with real coefficients, and this is true when $b^2-4ac<0$. So we need to concern ourselves with only linear and irreducible quadratic factors for $Q(x)$. We illustrate in the next two examples how to handle situations in which $Q(x)$ has only linear factors.

EXAMPLE 8.17 Evaluate the integral

$$\int \frac{5x+1}{x^2+x-2}\,dx$$

Solution This is the example we considered earlier, and we know the answer, but we will do it again, this time showing how to decompose the integrand into its component fractions. First we factor the denominator:

$$\frac{5x+1}{x^2+x-2} = \frac{5x+1}{(x-1)(x+2)}$$

Now we would expect that the denominator $(x-1)(x+2)$ would arise from adding a fraction with $x-1$ as its denominator to one with $x+2$ as its denominator; that is, we expect that constants A and B exist such that

$$\frac{5x+1}{(x-1)(x+2)} = \frac{A}{x-1} + \frac{B}{x+2}$$

Assume for the moment that this is true, and clear the equation of fractions, obtaining

$$5x+1 = A(x+2) + B(x-1)$$

Now substitute, in turn, $x = 1$ and $x = -2$:

$$\underline{x=1}:\quad 5+1 = A(3)+B(0) \qquad \underline{x=-2}:\quad 5(-2)+1 = A(0)+B(-3)$$

$$3A = 6 \qquad\qquad -3B = -9$$

$$A = 2 \qquad\qquad B = 3$$

It is easy now to verify that these values of A and B do work—that is, that

$$\frac{5x+1}{(x-1)(x+2)} = \frac{2}{x-1} + \frac{3}{x+2}$$

The integration is straightforward and we get, as we have already seen,

$$\int \frac{5x+1}{x^2+x-2}\,dx = \ln\left|(x-1)^2(x+2)^3\right| + C$$ ■

This example is typical of those integrands $\frac{P(x)}{Q(x)}$ (where the fraction is proper) in which $Q(x)$ factors into distinct linear factors, each appearing to the first power. Each such linear factor gives rise to a fraction in the decomposition that has a constant in the numerator. The constants can be determined by clearing of fractions and substituting, in turn, values of x that cause each factor to be 0. This procedure is called the **method of substitution.**

EXAMPLE 8.18 Evaluate the integral

$$\int \frac{x^2-13x+20}{(x-1)(x-3)^2}\,dx$$

Solution The denominator again involves only linear factors, but $x-3$ is to the second power. We say that $x-3$ is a *repeated* linear factor. Now we have to ask what types of fractions have the common denominator $(x-1)(x-3)^2$.

The answer is

$$\frac{x^2 - 13x + 20}{(x-1)(x-3)^2} = \frac{A}{x-1} + \frac{B}{x-3} + \frac{C}{(x-3)^2}$$

Note carefully the presence of the second term on the right. If the factor $(x-3)^3$ had appeared, then we would need to allow for fractions with each of the denominators $x-3$, $(x-3)^2$, and $(x-3)^3$.

As before, we clear the equation of fractions:

$$x^2 - 13x + 20 = A(x-3)^2 + B(x-1)(x-3) + C(x-1) \tag{8.9}$$

Again we use the method of substitution, first putting $x = 1$ and then $x = 3$:

$$\begin{aligned} \underline{x=1}: \quad 8 &= A(-2)^2 \\ 4A &= 8 \\ A &= 2 \end{aligned} \qquad \begin{aligned} \underline{x=3}: \quad -10 &= 2C \\ C &= -5 \end{aligned}$$

But now we have exhausted the substitutions that cause factors to be 0 and we still do not know B. One procedure is to substitute any other value for x ($x = 0$ is an easy value to substitute) and make use of the known values of A and C. We prefer, however, to introduce a second method known as **comparison of coefficients** because we will find it useful in other situations as well.

Returning to equation (8.9), we rewrite the right-hand side, expanding and collecting together like powers of x. The result is (verify)

$$x^2 - 13x + 20 = (A + B)x^2 + (-6A - 4B + C)x + (9A + 3B - C)$$

Now this is to be an *identity* in x; that is, the polynomial on the right must be the same polynomial as the one on the left. So coefficients of corresponding powers of x must be equal:

$$\begin{aligned} &\text{coefficients of } x^2: & 1 &= A + B \\ &\text{coefficients of } x: & -13 &= -6A - 4B + C \\ &\text{constant terms:} & 20 &= 9A + 3B - C \end{aligned}$$

This system of three equations in three unknowns can be solved algebraically, but we already know A and C, so we can get B using any one of the equations. From the first one, we have

$$\begin{aligned} 1 &= 2 + B \\ B &= -1 \end{aligned}$$

We now know A, B, and C and so can write

$$\begin{aligned} \int \frac{x^2 - 13x + 20}{(x-1)(x-3)^2}\,dx &= \int \left[\frac{2}{x-1} - \frac{1}{x-3} - \frac{5}{(x-3)^2}\right] dx \\ &= 2\int \frac{dx}{x-1} - \int \frac{dx}{x-3} - 5\int \frac{dx}{(x-3)^2}\,dx \\ &= 2\ln|x-1| - \ln|x-3| - 5\,\frac{(x-3)^{-1}}{-1} + C \\ &= \ln\frac{(x-1)^2}{|x-3|} + \frac{5}{x-3} + C \end{aligned}$$

■

Remark The method of comparison of coefficients always works and can be used on all problems without first using substitution. However, it is usually easier to get as much information as possible by substitution and then go to comparison of coefficients, as we did in the preceding example.

The next two examples show how to handle irreducible quadratic factors in the denominator.

EXAMPLE 8.19 Evaluate the integral

$$\int \frac{x^2 + x - 7}{(x+2)(x^2+1)}\, dx$$

Solution In the decomposition in this case we must allow for the possibility that the numerator of the fraction with $x^2 + 1$ in the denominator is a first-degree polynomial rather than a constant. This is the worst case, in the sense that any higher degree numerator would result in an improper fraction. So we allow for the following type of decomposition:

$$\frac{x^2 + x - 7}{(x+2)(x^2+1)} = \frac{A}{x+2} + \frac{Bx + C}{x^2 + 1}$$

From here on we proceed much as in the preceding example:

$$x^2 + x - 7 = A(x^2 + 1) + (Bx + C)(x + 2)$$

$$\underline{x = -2}: \quad -5 = 5A$$
$$A = -1$$

Now we compare coefficients. We could collect like powers on the right, but we can do this mentally in this case and so will not bother to rewrite the equation. We take the coefficient of the highest and lowest powers only.

$$\begin{aligned} \text{coefficients of } x^2\text{:} \quad 1 &= A + B \\ 1 &= -1 + B \\ B &= 2 \\ \text{constant terms:} \quad -7 &= A + 2C \\ -7 &= -1 + 2C \\ 2C &= -6 \\ C &= -3 \end{aligned}$$

So we have

$$\begin{aligned} \int \frac{x^2 + x - 7}{(x+2)(x^2+1)}\, dx &= \int \left(\frac{-1}{x+2} + \frac{2x - 3}{x^2 + 1}\right) dx \\ &= -\int \frac{dx}{x+2} + \int \frac{2x\,dx}{x^2+1} - 3\int \frac{dx}{x^2+1} \\ &= -\ln|x+2| + \ln(x^2+1) - 3\tan^{-1} x + C \\ &= \ln \frac{x^2+1}{|x+2|} - 3\tan^{-1} x + C \end{aligned}$$

■

A Word of Caution Do not confuse a repeated linear factor with an irreducible quadratic factor; for example,

$x^2 - 2x + 1 = (x-1)^2$ is a repeated linear factor.

$x^2 - 2x + 2$ is an irreducible quadratic factor.

The first gives rise to fractions of the form

$$\frac{A}{x-1} + \frac{B}{(x-1)^2}$$

whereas the second gives rise to

$$\frac{Ax+B}{x^2-2x+2}$$

EXAMPLE 8.20 Evaluate the integral

$$\int \frac{2x-3}{x^5+2x^3+x}\,dx$$

Solution The integrand can be written as

$$\frac{2x-3}{x(x^4+2x^2+1)} = \frac{2x-3}{x(x^2+1)^2}$$

Here the denominator has the linear factor x and the repeated irreducible quadratic factor $x^2 + 1$. The decomposition follows a pattern similar to that for repeated linear factors:

$$\frac{2x-3}{x(x^2+1)^2} = \frac{A}{x} + \frac{Bx+C}{x^2+1} + \frac{Dx+E}{(x^2+1)^2}$$

As usual, we clear of fractions and, because we will rely extensively on comparison of coefficients, we group the right-hand side in powers of x. We leave to you to verify that the result is

$$2x - 3 = (A+B)x^4 + Cx^3 + (2A+B+D)x^2 + (C+E)x + A$$

Putting $x = 0$ gives $A = -3$. Comparing coefficients of like powers yields the following system. Note that powers of x on the left greater than 1 have coefficient 0.

$$\begin{aligned} A + B &= 0 \\ C &= 0 \\ 2A + B + D &= 0 \\ C + E &= 2 \end{aligned}$$

Since we have already found that $A = -3$, we get $B = 3$, $C = 0$, $D = 3$, and $E = 2$. Thus, the integral becomes

$$\begin{aligned} \int \frac{2x-3}{x^5+2x^3+x}\,dx &= \int \left[\frac{-3}{x} + \frac{3x}{x^2+1} + \frac{3x+2}{(x^2+1)^2}\right] dx \\ &= -3\int \frac{dx}{x} + \frac{3}{2}\int \frac{2x}{x^2+1}\,dx + \int \frac{3x+2}{(x^2+1)^2}\,dx \\ &= -3\ln x + \frac{3}{2}\ln(x^2+1) + \int \frac{3x+2}{(x^2+1)^2}\,dx \end{aligned}$$

To evaluate the last integral we make the substitution $x = \tan\theta$ and get

$$\begin{aligned}\int \frac{3x+2}{(x^2+1)^2}\,dx &= \int \frac{3\tan\theta + 2}{\sec^4\theta}\sec^2\theta\,d\theta \\ &= 3\int \frac{\tan\theta}{\sec^2\theta}\,d\theta + 2\int \frac{d\theta}{\sec^2\theta} \\ &= 3\int \frac{\sec\theta\tan\theta\,d\theta}{\sec^3\theta} + 2\int \cos^2\theta\,d\theta \\ &= -\frac{3}{2\sec^2\theta} + \int (1+\cos 2\theta)\,d\theta \\ &= -\frac{3}{2}\cos^2\theta + \theta + \frac{\sin 2\theta}{2} + C \\ &= -\frac{3}{2}\cos^2\theta + \theta + \sin\theta\cos\theta + C \\ &= -\frac{3}{2(1+x^2)} + \tan^{-1}x + \frac{x}{1+x^2} + C \\ &= \frac{2x-3}{2(1+x^2)} + \tan^{-1}x + C\end{aligned}$$

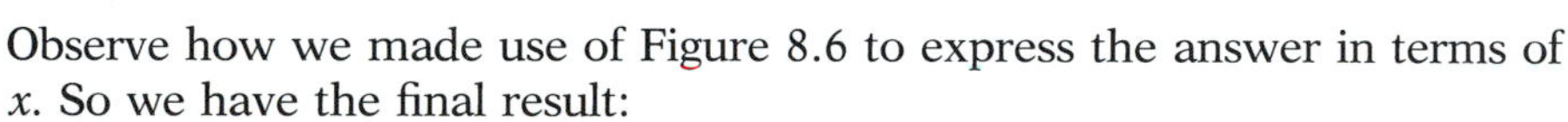

Observe how we made use of Figure 8.6 to express the answer in terms of x. So we have the final result:

$$\int \frac{2x-3}{x^5+2x^3+x}\,dx = \ln\frac{(x^2+1)^{3/2}}{|x|^3} + \frac{2x-3}{2(1+x^2)} + \tan^{-1}x + C \qquad \blacksquare$$

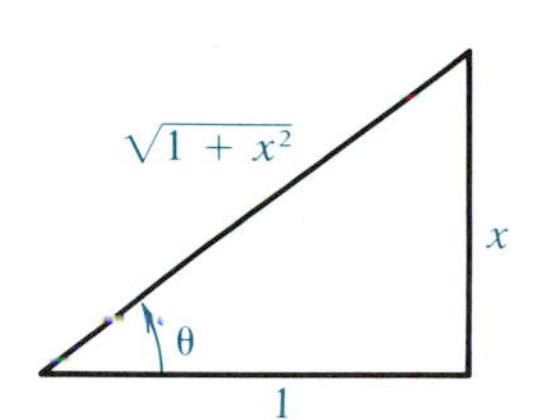

FIGURE 8.6

We now summarize the procedures we have illustrated.

1. If $\frac{P(x)}{Q(x)}$ is improper, divide to obtain a quotient plus a proper fraction. The next steps apply to proper fractions only.
2. Factor the denominator $Q(x)$.
3. For each factor of the form $(ax+b)^n$, write fractions

$$\frac{A_1}{ax+b} + \frac{A_2}{(ax+b)^2} + \cdots + \frac{A_n}{(ax+b)^n}$$

   (If $n = 1$, only one fraction is present.)
4. For each factor of the form $(ax^2+bx+c)^m$ where $b^2 - 4ac < 0$, write fractions of the form

$$\frac{B_1x+C_1}{ax^2+bx+c} + \frac{B_2x+C_2}{(ax^2+bx+c)^2} + \cdots + \frac{B_mx+C_m}{(ax^2+bx+c)^m}$$

5. Set $\frac{P(x)}{Q(x)}$ equal to the sum of all fractions obtained in steps 3 and 4 and clear of fractions.
6. Use substitution together with comparison of coefficients to find all unknown constants.

We will give one more example to illustrate how certain complications can be handled.

EXAMPLE 8.21 Evaluate the integral

$$\int_0^1 \frac{x^3 - x^2 - 11x + 10}{x^3 - 2x + 4}\,dx$$

Solution The first thing to observe is that the integrand is an improper fraction, so we divide:

$$\begin{array}{r|l} & 1 \\ \hline x^3 - 2x + 4 & x^3 - x^2 - 11x + 10 \\ & x^3 \qquad\ \ - 2x + 4 \\ \hline & \quad - x^2 - 9x + 6 \end{array}$$

We can therefore write

$$\frac{x^3 - x^2 - 11x + 10}{x^3 - 2x + 4} = 1 + \frac{-x^2 - 9x + 6}{x^3 - 2x + 4} = 1 - \frac{x^2 + 9x - 6}{x^3 - 2x + 4}$$

and we concentrate on the last fraction. The only possible rational zeros of the denominator are factors of 4, and by trial and error we find -2 to be a zero, so that $x + 2$ is a factor. By division, the other factor is found to be the irreducible quadratic $x^2 - 2x + 2$. So we have

$$\frac{x^2 + 9x - 6}{x^3 - 2x + 4} = \frac{x^2 + 9x - 6}{(x + 2)(x^2 - 2x + 2)} = \frac{A}{x + 2} + \frac{Bx + C}{x^2 - 2x + 2}$$

$$x^2 + 9x - 6 = A(x^2 - 2x + 2) + (Bx + C)(x + 2)$$

$$\underline{x = -2}: \quad 4 - 18 - 6 = A(4 + 4 + 2) + 0$$

$$10A = -20$$

$$A = -2$$

$$\text{coefficients of } x^2: \quad 1 = A + B$$

$$1 = -2 + B$$

$$B = 3$$

$$\text{constants:} \quad -6 = 2A + 2C$$

$$-6 = -4 + 2C$$

$$2C = -2$$

$$C = -1$$

So we have the decomposition

$$\frac{x^2 + 9x - 6}{x^3 - 2x + 4} = \frac{-2}{x + 2} + \frac{3x - 1}{x^2 - 2x + 2}$$

Returning to the original fraction and the result of division,

$$\begin{aligned} \int_0^1 \frac{x^3 - x^2 - 11x + 10}{x^3 - 2x + 4}\,dx &= \int_0^1 \left[1 - \frac{x^2 + 9x - 6}{x^3 - 2x + 4}\right] dx \\ &= \int_0^1 \left[1 - \left(\frac{-2}{x + 2} + \frac{3x - 1}{x^2 - 2x + 2}\right)\right] dx \\ &= \int_0^1 dx + 2\int_0^1 \frac{dx}{x + 2} - \int_0^1 \frac{3x - 1}{x^2 - 2x + 2}\,dx \\ &= x + 2\ln(x + 2)\Big]_0^1 - \int_0^1 \frac{3x - 1}{x^2 - 2x + 2}\,dx \\ &= 1 + 2\ln 3 - 2\ln 2 - \int_0^1 \frac{3x - 1}{x^2 - 2x + 2}\,dx \end{aligned}$$

To evaluate the last integral we force the derivative of the denominator into the numerator as follows:

$$\int \frac{3x-1}{x^2-2x+2}\,dx = \int \frac{\frac{3}{2}(2x-2)+2}{x^2-2x+2}\,dx$$
$$= \frac{3}{2}\int \frac{2x-2}{x^2-2x+2}\,dx + 2\int \frac{dx}{(x-1)^2+1}$$

The first integral is now in the form $\int du/u$ and the second $\int du/(u^2+1)$. So we have

$$\int_0^1 \frac{3x-1}{x^2-2x+2}\,dx = \left[\frac{3}{2}\ln(x^2-2x+2) + 2\tan^{-1}(x-1)\right]_0^1$$
$$= \frac{3}{2}\ln 1 + 2\tan^{-1} 0 - \frac{3}{2}\ln 2 - 2\tan^{-1}(-1)$$
$$= -\frac{3}{2}\ln 2 - 2\left(-\frac{\pi}{4}\right) = -\frac{3}{2}\ln 2 + \frac{\pi}{2}$$

Combining this with the results already obtained, we have finally

$$\int_0^1 \frac{x^3-x^2-11x+10}{x^3-2x+4}\,dx = 1 + 2\ln 3 - 2\ln 2 - \left(-\frac{3}{2}\ln 2 + \frac{\pi}{2}\right)$$
$$= 1 + 2\ln 3 - \frac{1}{2}\ln 2 - \frac{\pi}{4}$$ ■

EXERCISE SET 8.5

A

In Exercises 1–26 make use of partial fractions to evaluate the integrals.

1. $\int \frac{dx}{x^2+2x}$
2. $\int \frac{dx}{x^2+5x+6}$
3. $\int \frac{x\,dx}{x^2-x-6}$
4. $\int_{-2}^{0} \frac{5x+7}{x^2+2x-3}\,dx$
5. $\int_{-3}^{2} \frac{x-17}{x^2+x-12}\,dx$
6. $\int \frac{x+9}{2x^2+x-6}\,dx$
7. $\int \frac{3x+10}{x^2+5x+6}\,dx$
8. $\int \frac{2}{x^2-9x+20}\,dx$
9. $\int \frac{dx}{x^3-x}$
10. $\int \frac{10x+4}{4x-x^3}\,dx$
11. $\int \frac{x^2+4x-2}{x^2+x-2}\,dx$
12. $\int_0^2 \frac{2x^2-x-20}{x^2-x-6}\,dx$
13. $\int \frac{3x^2+32x+44}{(x+3)(x^2+2x-8)}\,dx$
14. $\int \frac{2x^2+19x-45}{(x-1)(x^2-x-6)}\,dx$
15. $\int \frac{2x^3-x^2-4x+5}{x^2-1}\,dx$
16. $\int \frac{5x+1}{(x+2)(x^2-2x+1)}\,dx$
17. $\int_{-1}^{1} \frac{x^2-6x+23}{(x+3)(x^2-4x+4)}\,dx$
18. $\int_1^2 \frac{4+4x-x^2}{x^3+2x^2}\,dx$
19. $\int \frac{2x^2-6x+12}{x^3-4x^2+4x}\,dx$
20. $\int \frac{6x-4}{(x-2)(x^2-4)}\,dx$
21. $\int \frac{5x+6}{(x-2)(x^2+4)}\,dx$
22. $\int_0^1 \frac{x^2+3}{(x+1)(x^2+1)}\,dx$
23. $\int_1^3 \frac{3x-6}{x^4+3x^2}\,dx$
24. $\int \frac{dx}{x^4-1}$
25. $\int \frac{dx}{x^3-1}$
26. $\int \frac{12}{x^3+8}\,dx$

27. (a) Find the area of the region under the graph of

$$y = \frac{1}{x^2 - x}$$

from $x = 2$ to $x = 3$.

(b) Find the volume obtained by rotating the region of part a about the x-axis.

28. Find the volume obtained by rotating the region under the graph of

$$y = \frac{1}{x\sqrt{x^2 + 4}}$$

from $x = 1$ to $x = 2$ about the x-axis.

B

In Exercises 29–33 derive the integration formulas for the given integrals, using partial fractions. For Exercises 29–32 compare your results with those given in the tables on the front and back end pages.

29. $\displaystyle\int \frac{du}{u(au + b)}$

30. $\displaystyle\int \frac{du}{u^2(au + b)}$

31. $\displaystyle\int \frac{du}{u(au + b)^2}$

32. $\displaystyle\int \frac{du}{u^2 - a^2}$

33. $\displaystyle\int \frac{du}{u^3 - a^3}$

In Exercises 34–38 use partial fractions to evaluate the integrals.

34. $\displaystyle\int \frac{x^4 + 5x^3 + 3x}{(x - 1)(x^2 + 2)^2}\,dx$

35. $\displaystyle\int \frac{4x}{x^3 - x^2 - x + 1}\,dx$

36. $\displaystyle\int \frac{x^2 + 3x + 6}{x^3 + x - 2}\,dx$

37. $\displaystyle\int \frac{x^3 - 3x^2 - 5x}{x^4 + 5x^2 + 4}\,dx$

38. $\displaystyle\int \frac{20 - 5x^2}{4x^4 + 9x^2 - 11x + 3}\,dx$

39. In Exercise 24 of Exercise Set 6.6 we gave the Verhulst model for population growth in the form of the differential equation

$$\frac{dQ}{dt} = kQ(m - Q)$$

where $Q(t)$ is the size of the population at time t, k and m are positive constants, and $m > Q$. Let $Q_0 = Q(0)$, and find Q as a function of t by separating variables and using partial fractions.

40. A second-order chemical reaction involves the interaction of molecules of a substance A with those of a substance B to form molecules of a new substance X. If the initial concentrations of substances A and B are a and b, respectively, and $x(t)$ is the concentration of X at time t, then the rate at which the reaction occurs is given by

$$\frac{dx}{dt} = k(a - x)(b - x)$$

where k is a positive constant. If $x(0) = 0$, find $x(t)$ at any time t by separating variables and using partial fractions.

8.6 MISCELLANEOUS SUBSTITUTIONS

In this section we introduce certain additional types of substitutions that work in integrals that occur with sufficient frequency to justify their inclusion here.

Integrals Involving $\sqrt[n]{ax + b}$

If only one such radical occurs, substitute $u = \sqrt[n]{ax + b}$.

EXAMPLE 8.22 Evaluate the integral

$$\int \frac{x\,dx}{\sqrt{1-x}}$$

Solution Let $u = \sqrt{1-x}$; then $u^2 = 1 - x$ and $x = 1 - u^2$. So $dx = -2u\,du$. Putting everything in terms of u, we get

$$\begin{aligned}
\int \frac{x\,dx}{\sqrt{1-x}} &= \int \frac{(1-u^2)(-2u\,du)}{u} = -2\int (1-u^2)\,du \\
&= -2\left(u - \frac{u^3}{3}\right) + C \\
&= -\frac{2u}{3}(3-u^2) + C \\
&= -\frac{2\sqrt{1-x}}{3}[3-(1-x)] + C \\
&= -\frac{2\sqrt{1-x}}{3}(x+2) + C
\end{aligned}$$

■

If two or more such radicals occur, put $u = \sqrt[m]{ax+b}$, where m is the least common multiple of the indexes of all the radicals that occur.

EXAMPLE 8.23 Evaluate the integral

$$\int \frac{dx}{\sqrt{x}(1+\sqrt[3]{x})}$$

Solution The indexes of the two radicals are 2 and 3 (square root and cube root) and the least common multiple is 6. So we let $u = \sqrt[6]{x}$. Then $x = u^6$ and $dx = 6u^5\,du$. Also

$$\sqrt{x} = x^{1/2} = (u^6)^{1/2} = u^3$$

and

$$\sqrt[3]{x} = x^{1/3} = (u^6)^{1/3} = u^2$$

(This shows why we choose the least common multiple of 2 and 3.) So we have

$$\int \frac{dx}{\sqrt{x}(1+\sqrt[3]{x})} = \int \frac{6u^5\,du}{u^3(1+u^2)} = 6\int \frac{u^2\,du}{1+u^2}$$

Since the integrand is an improper rational fraction, we divide and get

$$\begin{aligned}
6\int\left(1 - \frac{1}{1+u^2}\right)du &= 6(u - \tan^{-1} u) + C \\
&= 6(\sqrt[6]{x} - \tan^{-1}\sqrt[6]{x}) + C
\end{aligned}$$

■

Integrals Involving $\sqrt[n]{ax^2+b}$

The substitution $u = \sqrt[n]{ax^2+b}$ will rationalize the integrand provided x can be factored from the integrand, leaving *even powers* of x only.

EXAMPLE 8.24 Evaluate the integral

$$\int \frac{x^3\,dx}{\sqrt[3]{x^2+1}}$$

Solution Let $u = \sqrt[3]{x^2+1}$; then $u^3 = x^2 + 1$, so that

$$x^2 = u^3 - 1$$
$$2x\,dx = 3u^2\,du$$

Write the integral in the form

$$\begin{aligned}\frac{1}{2}\int \frac{x^2(2x\,dx)}{\sqrt[3]{x^2+1}} = \frac{1}{2}\int \frac{(u^3-1)(3u^2\,du)}{u} &= \frac{3}{2}\int (u^4 - u)\,du \\ &= \frac{3}{2}\left(\frac{u^5}{5} - \frac{u^2}{2}\right) + C \\ &= \frac{3u^2}{20}(2u^3 - 5) + C \\ &= \frac{3(x^2+1)^{2/3}}{20}[2(x^2+1) - 5] + C \\ &= \frac{3(x^2+1)^{2/3}}{20}(2x^2 - 3) + C\end{aligned}$$

Comment When $\sqrt[n]{ax+b}$ is the only radical in an integrand, the substitution $u = \sqrt[n]{ax+b}$ always leads to an integrand that is free of radicals; that is, this is a rationalizing substitution. When $\sqrt[n]{ax^2+b}$ is the only radical, the substitution $u = \sqrt{ax^2+b}$ rationalizes the integrand *provided* x can be factored out of the integrand leaving *even* powers of x only. If you forget this latter condition, you can always *try* the substitution to see whether it works.

Rational Functions of sin *x* and cos *x*

Make the substitution $u = \tan \frac{x}{2}$, where $-\pi < x < \pi$. Then $\frac{x}{2} = \tan^{-1} u$ and, using Figure 8.7, we get

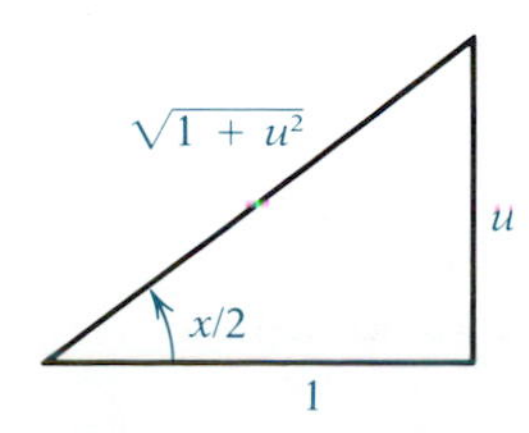

FIGURE 8.7

$$\begin{aligned}\sin x = 2\sin\frac{x}{2}\cos\frac{x}{2} &= 2\left(\frac{u}{\sqrt{1+u^2}}\right)\left(\frac{1}{\sqrt{1+u^2}}\right) \\ &= \frac{2u}{1+u^2}\end{aligned}$$

$$\cos x = 2\cos^2\frac{x}{2} - 1 = 2\left(\frac{1}{1+u^2}\right) - 1 = \frac{1-u^2}{1+u^2}$$

Also, $x = 2\tan^{-1} u$, so that

$$dx = \frac{2\,du}{1+u^2}$$

We summarize these substitutions.

$$u = \tan\frac{x}{2}, \qquad -\pi < x < \pi$$

$$dx = \frac{2\,du}{1+u^2}$$

$$\sin x = \frac{2u}{1+u^2} \qquad \cos x = \frac{1-u^2}{1+u^2}$$

EXAMPLE 8.25 Evaluate the integral

$$\int \frac{dx}{4\sin x - 3\cos x}$$

Solution Let $u = \tan\frac{x}{2}$. Then from what we have found,

$$\int \frac{dx}{4\sin x - 3\cos x} = \int \frac{\dfrac{2\,du}{1+u^2}}{4\left(\dfrac{2u}{1+u^2}\right) - 3\left(\dfrac{1-u^2}{1+u^2}\right)}$$

$$= \int \frac{2\,du}{3u^2+8u-3}$$

$$= \int \frac{2\,du}{(3u-1)(u+3)}$$

Using partial fractions, we find that this can be written in the form

$$\frac{1}{5}\int \left[\frac{3}{3u-1} - \frac{1}{u+3}\right] du = \frac{1}{5}\left[\ln|3u-1| - \ln|u+3|\right] + C$$

$$= \frac{1}{5}\ln\left|\frac{3u-1}{u+3}\right| + C = \frac{1}{5}\ln\left|\frac{3\tan\dfrac{x}{2} - 1}{\tan\dfrac{x}{2}+3}\right| + C \quad \blacksquare$$

Remark Before resorting to this substitution, which can lead to cumbersome integrands, check carefully to see that no simpler method can be used.

EXERCISE SET 8.6

A

Evaluate the integrals.

1. $\displaystyle\int \frac{x\,dx}{1+\sqrt{x}}$

2. $\displaystyle\int x\sqrt{2x+3}\,dx$

3. $\displaystyle\int \frac{\sqrt{x-1}}{x}\,dx$

4. $\displaystyle\int_{-1}^{1} \frac{\sqrt{1-x}}{x+3}\,dx$

5. $\displaystyle\int_0^1 \frac{x-2}{\sqrt{4-3x}}\,dx$

6. $\displaystyle\int_1^4 \frac{\sqrt{x}-1}{\sqrt{x}+1}\,dx$

7. $\displaystyle\int \frac{x^{1/4}+1}{x^{3/2}}\,dx$

8. $\displaystyle\int \frac{dx}{x^{2/3}-x^{1/2}}$

9. $\displaystyle\int \frac{\sqrt{x}\,dx}{1+\sqrt[3]{x}}$

10. $\displaystyle\int \frac{\sqrt[3]{x+1}}{\sqrt{x+1}}\,dx$

11. $\displaystyle\int \frac{x\,dx}{\sqrt{x^2+1}-1}$

12. $\displaystyle\int \frac{x^3\,dx}{\sqrt{2x^2-3}}$

13. $\int_{-1}^{2\sqrt{5}} \frac{x\,dx}{\sqrt[3]{x^2+7}}$
14. $\int (2x - x^3)\sqrt{1-x^2}\,dx$
15. $\int \frac{dx}{1+\sin x}$
16. $\int \frac{dx}{1+\cos x}$
17. $\int \frac{dx}{\sin x - \cos x - 1}$
18. $\int \frac{\cos x\,dx}{\sin x(1+\cos x)}$

B

19. $\int \frac{dx}{(\sqrt{x} - \sqrt[3]{x})^3}$
20. $\int_0^3 \frac{\sqrt[3]{1-3x}}{3x+7}\,dx$
21. $\int \frac{dx}{x\sqrt[3]{x^2+8}}$
22. $\int \frac{dx}{x\sqrt[4]{x^2+1}}$
23. $\int_{\pi/2}^{2\pi/3} \frac{\cot x\,dx}{\cot x + 2\csc x}$
24. $\int_0^{\pi/2} \frac{5\cos x\,dx}{3\sin x + 4\cos x}$
25. $\int \sqrt{1+e^x}\,dx$ (*Hint:* Let $u = \sqrt{1+e^x}$.)

8.7 USE OF INTEGRAL TABLES

Integral tables can be very useful, especially if you have several integrals to evaluate for which the techniques we have studied are time consuming. However, as mentioned earlier, they are not the answer to all your problems. To make a given integral conform to one in the table, it may first be necessary to make a substitution, or to integrate by parts, or to use partial fractions. Even when the integral can be found in the tables, it may take longer to look it up than to do it without using tables. A good approach to tables is to use them only when you know you *could* do the problem without tables, but use tables to save time.

We illustrate how to match up certain integrals with forms in the tables on the front and back end pages of this book with several examples. Far more extensive tables are available, such as in Burrington's *Handbook of Mathematical Tables and Formulas* (New York: McGraw-Hill) and the Chemical Rubber Company's *Standard Mathematical Tables* (Cleveland: Chemical Rubber Publishing Company).

EXAMPLE 8.26 Use the integral tables on the end pages of this book to evaluate the following integrals:

(a) $\int \sqrt{3-2x^2}\,dx$ (b) $\int \frac{x\,dx}{\sqrt{3x-4x^2}}$

(c) $\int \frac{x^2\,dx}{2x^2-7x+6}$ (d) $\int x\sqrt{x^2-4x+5}\,dx$

Solution (a) Formula 50 in the table for $\int \sqrt{a^2-u^2}\,du$ seems to be the appropriate one to use. We take $a = \sqrt{3}$ and $u = \sqrt{2}x$. In order to get du we must multiply and divide by $\sqrt{2}$. So we have

$$\int \sqrt{3-2x^2}\,dx = \frac{1}{\sqrt{2}}\int \sqrt{3-2x^2}\sqrt{2}\,dx$$
$$= \frac{1}{\sqrt{2}}\left[\frac{\sqrt{2}x}{2}\sqrt{3-2x^2} + \frac{3}{2}\sin^{-1}\frac{\sqrt{2}x}{\sqrt{3}}\right] + C$$

Note that to do this without using tables, we would have made the trigonometric substitution $x = \sqrt{\frac{3}{2}} \sin \theta$ (which is how the formula in the table was obtained).

(b) This can be made to match formula 64 by letting $2a = 3$ and $u = 2x$. So we adjust the differential by multiplying and dividing by 2 and get

$$\frac{1}{2}\int \frac{2x\,dx}{\sqrt{3x-4x^2}} = \frac{1}{2}\left[-\sqrt{3x-4x^2} + \frac{3}{2}\cos^{-1}\left(\frac{\frac{3}{2}-x}{\frac{3}{2}}\right)\right] + C$$

$$= -\frac{\sqrt{3x-4x^2}}{2} + \frac{3}{4}\cos^{-1}\left(\frac{3-2x}{3}\right) + C$$

Without tables we would have completed the square on the denominator and then either used a trigonometric substitution or split the integral into two, forcing one into the form $\int dv/v^{1/2}$ and the other into the form $\int dv/\sqrt{a^2-v^2}$. This is a case where the integral table does save some time and effort.

(c) There is no form in the table that works as the integral stands. Since the integrand is an improper rational fraction, we divide to get

$$\frac{x^2}{2x^2-7x+6} = \frac{1}{2} + \frac{\frac{7}{2}x-3}{2x^2-7x+6} = \frac{1}{2} + \frac{1}{2}\frac{7x-6}{(2x-3)(x-2)}$$

We split the integral into three parts:

$$\int \frac{x^2\,dx}{2x^2-7x+6} = \int \frac{1}{2}\,dx + \frac{7}{2}\int \frac{x\,dx}{(2x-3)(x-2)} - 3\int \frac{dx}{(2x-3)(x-2)}$$

The second integral matches formula 27 and the third formula 26, both with $a = -3$, $b = 2$, $c = -2$, and $d = 1$. So $ad - bc = -3 + 4 = 1 \neq 0$. Thus

$$\int \frac{x^2\,dx}{2x^2-7x+6}$$

$$= \frac{x}{2} + \frac{7}{2}\left[-\frac{3}{2}\ln|2x-3| + 2\ln|x-2|\right] - 3\ln\left|\frac{x-2}{2x-3}\right| + C$$

$$= \frac{x}{2} - \frac{21}{4}\ln|2x-3| + 7\ln|x-2| + 3\ln|2x-3| - 3\ln|x-2| + C$$

$$= \frac{x}{2} - \frac{9}{4}\ln|2x-3| + 4\ln|x-2| + C$$

In this case it would probably have been faster to use partial fractions and not bother with tables.

(d) Again, we do not find a form in the table that matches the integral, but by completing the square under the radical, we convert it to a standard form. We have

$$\int x\sqrt{x^2-4x+5}\,dx = \int x\sqrt{(x^2-4x+4)+1}\,dx$$

$$= \int x\sqrt{(x-2)^2+1}\,dx$$

Now if we put $u = x - 2$, we have $x = u + 2$ and $dx = du$. So our integral becomes

$$\int (u+2)\sqrt{u^2+1}\,du = \int u\sqrt{u^2+1}\,du + 2\int \sqrt{u^2+1}\,du$$

We certainly do not need a table for the first integral on the right (although it can be found there). For the second we use formula 37. Thus,

$$\begin{aligned}\int (u+2)\sqrt{u^2+1}\,du &\\ &= \tfrac{1}{2}(u^2+1)^{3/2}\cdot\tfrac{2}{3} + 2\,[\tfrac{u}{2}\sqrt{u^2+1} + \tfrac{1}{2}\ln|u+\sqrt{u^2+1}|] + C\\ &= \tfrac{1}{3}(u^2+1)^{3/2} + u\sqrt{u^2+1} + \ln|u+\sqrt{u^2+1}| + C\end{aligned}$$

and on replacing u by $x - 2$, we have

$$\begin{aligned}\int x\sqrt{x^2-4x+5}\,dx = \tfrac{1}{3}(x^2-4x+5)^{3/2} + (x-2)\sqrt{x^2-4x+5}\\ + \ln|x-2+\sqrt{x^2-4x+5}| + C\end{aligned}$$

■

EXERCISE SET 8.7

A

Use the tables on the front and back end pages to perform all integrations.

1. $\int \frac{dx}{x(2+3x)^2}$
2. $\int \frac{\sqrt{x^2-4}}{x}\,dx$
3. $\int x\tan^{-1} 2x\,dx$
4. $\int x^2 \sin 3x\,dx$
5. $\int \frac{dx}{x\sqrt{4x-x^2}}$
6. $\int \frac{x\,dx}{2x-3}$
7. $\int \frac{dx}{x^2(x-2)}$
8. $\int \frac{\sqrt{x^2-1}}{x^2}\,dx$
9. $\int \frac{dx}{(x^2+9)^{3/2}}$
10. $\int x^2\sqrt{9-4x^2}\,dx$
11. $\int \sqrt{2x-x^2}\,dx$
12. $\int (\ln x)^3\,dx$
13. $\int x^4 e^{3x}\,dx$
14. $\int \sin^6 x\,dx$
15. $\int \frac{dx}{x(3x+5)}$
16. $\int \frac{dx}{x^2\sqrt{3-2x^2}}$
17. $\int \frac{x\,dx}{(4x-3)^2}$
18. $\int \sec^5 x\,dx$
19. $\int e^{-2x}\cos 3x\,dx$
20. $\int \sqrt{x^2+2x+2}\,dx$

B

21. $\int \frac{x^2+1}{4+3x-x^2}\,dx$
22. $\int \frac{2x-3}{\sqrt{5+4x-x^2}}\,dx$
23. $\int x^2\sin^{-1} x\,dx$
24. $\int \frac{\sqrt{3x^2-6x+5}}{x-1}\,dx$

8.8 SUPPLEMENTARY EXERCISES

In Exercises 1–30 evaluate the given definite or indefinite integral.

1. $\int x^2 e^{-2x}\,dx$
2. $\int \sqrt{x}\ln\sqrt{x}\,dx$
3. $\int_0^1 x\tan^{-1} x^2\,dx$
4. $\int e^{-x}\cosh 2x\,dx$
5. $\int \sec^2\sqrt{x}\,dx$ (*Hint:* Let $t = \sqrt{x}$.)

6. $\int_{-\pi/6}^{\pi/3} \sin^2 x(\sin x - 1)\, dx$

7. $\int \sin^5 x \cos^3 x\, dx$

8. $\int \sec^3 x \tan^5 x\, dx$

9. $\int_{-\pi/3}^{\pi/4} \frac{\sin^2 x}{\cos^4 x}\, dx$

10. $\int \frac{\sin^2 x}{\cos^3 x}\, dx$

11. $\int x \sin x \cos^3 x\, dx$

12. $\int \sec^2 \sqrt{x} \tan \sqrt{x}\, dx$ (*Hint:* Let $t = \sqrt{x}$.)

13. $\int \sin 2x \cos 4x\, dx$

14. $\int \frac{\ln \cos x}{\sin^2 x}\, dx$ (*Hint:* Integrate by parts.)

15. $\int_2^{2\sqrt{3}} \frac{x^2}{(16 - x^2)^{3/2}}\, dx$

16. $\int_1^2 x^3 \sqrt{x^2 + 1}\, dx$ Do by two methods: integration by parts and trigonometric substitution.

17. $\int_{1/\sqrt{3}}^{1} \frac{\sqrt{4x^2 - 1}}{x^3}\, dx$

18. $\int \frac{x\, dx}{\sqrt{x^2 - 2x + 5}}$

19. $\int \frac{e^{4x}\, dx}{\sqrt{1 + e^{2x}}}$

20. $\int \frac{\sinh x \cosh^3 x}{\sqrt{\cosh^2 x - 4}}\, dx$

21. $\int \frac{x^2\, dx}{\sqrt{4x - x^2}}$

22. $\int \frac{4x - 9}{2x^2 + 5x - 3}\, dx$

23. $\int_1^2 \frac{x^2 - 3x + 2}{x^2 - 2x - 8}\, dx$

24. $\int \frac{5x + 1}{x^3 - x^2 - x + 1}\, dx$

25. $\int \frac{x^2 + x + 1}{x(x^2 + 1)^2}\, dx$

26. $\int \frac{x^4 + 20}{x^3 - 3x + 2}\, dx$

27. $\int_{-1}^{0} \frac{12\, dx}{x^3 - 8}$

28. $\int_0^4 \frac{dx}{\sqrt{x} + 1}$

29. $\int \frac{x^3}{\sqrt{x^2 - 4}}\, dx$ Do by two methods: trigonometric substitution and algebraic substitution.

30. $\int_0^{\pi/2} \frac{1 + \cos x}{1 + \sin x}\, dx$

31. Find the volume obtained by revolving the region under the graph of $y = \sin x$ from $x = 0$ to $x = \pi$ about (a) the x-axis and (b) the y-axis.

32. Find the length of the curve $y = x^2/4$ from $x = 0$ to $x = 2$.

33. Find the volume generated by revolving the region under the graph of

$$y = \frac{1}{\sqrt{5x - x^2}}$$

from $x = 1$ to $x = 4$ about (a) the x-axis and (b) the y-axis.

34. Find the area of the surface formed by revolving the arc of the curve $y = \cosh x$ from $x = 0$ to $x = \ln 3$ about (a) the x-axis and (b) the y-axis.

35. A model for a so-called *learning curve* in psychology is given by

$$\frac{dy}{dt} = k\sqrt{y^3(1 - y)^3}, \qquad 0 \le y \le 1$$

where k is a positive constant. Here $y = y(t)$ is the proportion of a task or of a body of knowledge learned by the subject at time t. For $0 < y < 1$, find y as a function of t by writing the equation in the form

$$\frac{dy}{y^{3/2}(1 - y)^{3/2}} = k\, dt$$

and integrating both sides. (*Hint:* To integrate the left-hand side, substitute $y = \sin^2 \theta$.)

CHAPTER 9

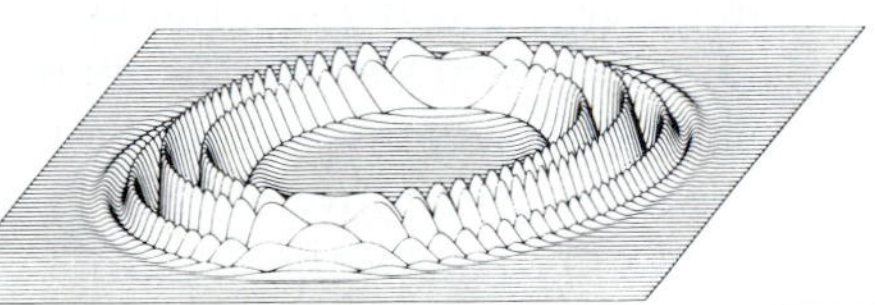

NUMERICAL METHODS

9.1 INTRODUCTION

In this chapter we will be concerned with approximations relating to calculus. This study introduces a part of mathematics called numerical analysis, the importance of which cannot be overemphasized in applications of mathematics. The integration techniques we have studied handle a large class of problems, but unfortunately in actual applications of mathematics, integrals are sometimes so complex that exact evaluations of them are prohibitively difficult or even impossible. In this case approximate methods are used. We will study two such methods. Similarly, equations that arise in practice are often impossible to solve in an exact way, and the techniques we will study provide a way to approximate the roots to any desired degree of accuracy. Finally, we will see how polynomial functions, which are the simplest functions of all, can be used to approximate more complicated functions.

A calculator is almost essential for the calculations in this chapter. In the actual application of the methods it is likely that a computer would be used, since there are many repetitive calculations, a job for which computers are especially well suited.

9.2 APPROXIMATING DEFINITE INTEGRALS

Despite the fact that we now have a substantial arsenal of techniques at our disposal for finding antiderivatives, which is expanded even further by tables of integrals, there are many continuous functions for which an antiderivative in the form of an elementary function either does not exist or is so hard to obtain as to make it impractical. Examples of integrals that cannot be expressed as elementary functions are

$$\int e^{x^2}\,dx \qquad \int \frac{\sin x}{x}\,dx \qquad \int \frac{dx}{\sqrt{1+x^3}}$$

In practice the primary reason for finding antiderivatives is to evaluate definite integrals by using the fundamental theorem. In this section we show two common means of approximating definite integrals, which may be used when an antiderivative cannot be found.

As a matter of fact, we already know one such approximation in the form of a Riemann sum for the integral in question, and we have used this in the past. In general, however, this method requires a large number of subintervals in the partition to obtain much accuracy. A better method is the so-called **trapezoidal rule,** which we now describe.

Let f be an integrable function on $[a, b]$ and let P be a regular partition of $[a, b]$ with n subintervals, so that

$$\Delta x = \frac{b-a}{n}$$

As usual, let the points of P be denoted by

$$a = x_0 < x_1 < x_2 < \cdots < x_n = b$$

and let $y_k = f(x_k)$ for each $k = 0, 1, 2, \ldots, n$. Note that since P is regular, we can say explicitly that

$$x_k = a + k\Delta x, \qquad k = 0, 1, 2, \ldots, n$$

In Figure 9.1 we have shown a typical situation where f is nonnegative and continuous, conditions that will aid in formulating the method but can be removed after we obtain the result.

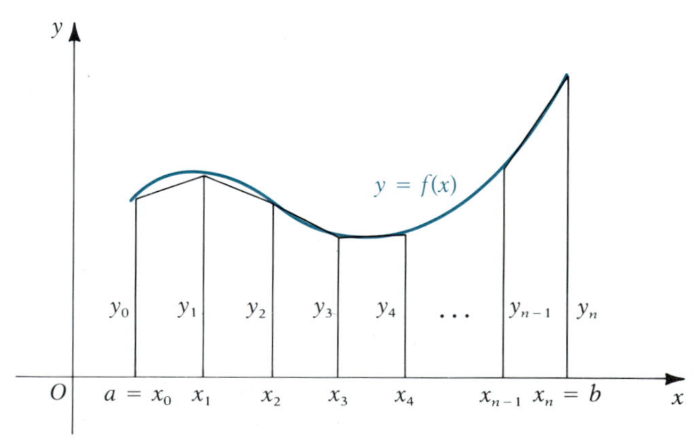

FIGURE 9.1

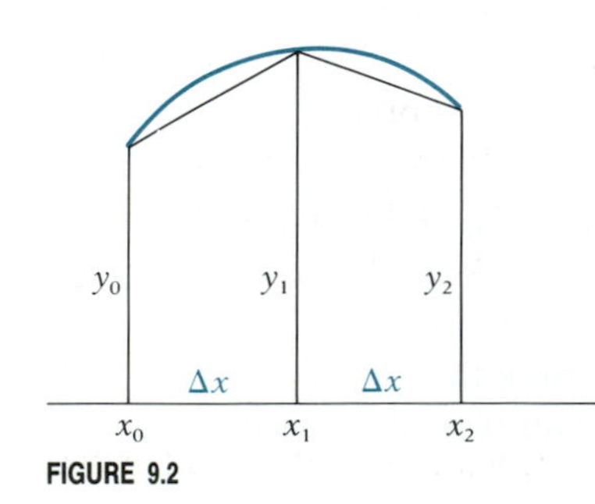

FIGURE 9.2

We draw the chords joining consecutive points (x_k, y_k) on the curve, thereby forming trapezoids. Figure 9.2 shows the first two such trapezoids enlarged. Now the area of a trapezoid is the average of the lengths of the parallel sides times the distance between them. So for the first two trapezoids the areas are

$$\tfrac{1}{2}(y_0 + y_1)\Delta x \quad \text{and} \quad \tfrac{1}{2}(y_1 + y_2)\Delta x$$

and when these are added, we get

$$\left(\tfrac{1}{2}y_0 + y_1 + \tfrac{1}{2}y_2\right)\Delta x = \tfrac{\Delta x}{2}(y_0 + 2y_1 + y_2)$$

Now we add the area of the third trapezoid, and the fourth, and so on until we get to the end. Note that each vertical distance y_k gets counted twice, except for y_0 and y_n. The final result for the total area T_n of the n trapezoids is

$$T_n = \tfrac{\Delta x}{2}(y_0 + 2y_1 + 2y_2 + 2y_3 + \cdots + 2y_{n-1} + y_n)$$

It is intuitively clear that T_n approximates $\int_a^b f(x)\,dx$ and that the approximation gets better as n increases. We can show this is true, in fact, for any integrable function f, whether or not it is nonnegative or continuous, as follows. Write T_n in the form

$$\begin{aligned} T_n &= \tfrac{1}{2}[(f(x_0) + f(x_1) + \cdots + f(x_{n-1}))\,\Delta x + (f(x_1) + f(x_2) + \cdots + f(x_n))\,\Delta x] \\ &= \tfrac{1}{2}\left[\sum_{k=1}^{n} f(x_{k-1})\,\Delta x + \sum_{k=1}^{n} f(x_k)\,\Delta x\right] \end{aligned}$$

Each of the last two sums is a Riemann sum of the form $\sum_{k=1}^{n} f(x_k^*)\,\Delta x_k$, where $x_k^* = x_{k-1}$ in the first case and $x_k^* = x_k$ in the second. Since all Riemann sums of f approach the integral $\int_a^b f(x)\,dx$ as the norms of partitions approach 0, and since with regular partitions the norm $\Delta x = (b - a)/n$ does approach 0 as $n \to \infty$, we conclude that

$$\lim_{n\to\infty} T_n = \tfrac{1}{2}\left[\int_a^b f(x)\,dx + \int_a^b f(x)\,dx\right] = \int_a^b f(x)\,dx$$

This result justifies approximating the integral by T_n, which is essentially the statement of the trapezoidal rule.

THE TRAPEZOIDAL RULE

If f is integrable on $[a, b]$ and $P = \{x_0, x_1, \ldots, x_n\}$ is a regular partition of $[a, b]$, with

$$\Delta x = \frac{b - a}{n}$$

then

$$\int_a^b f(x)\,dx \approx \frac{\Delta x}{2}\,[f(x_0) + 2f(x_1) + 2f(x_2) + \cdots + 2f(x_{n-1}) + f(x_n)]$$

An approximation is more meaningful if we have some way of knowing how close it is to the actual value. In general, of course, we do not know the actual value (or else we would not be using an approximation), but in certain cases we can find an upper bound on the error. It is proved in numerical analysis courses that if $f''(x)$ exists in $[a, b]$ and is bounded there, then the error E_n, defined by $E_n = \left|\int_a^b f(x)\,dx - T_n\right|$, satisfies

$$E_n \leq \frac{M(b - a)^3}{12n^2} \tag{9.1}$$

where M is any number for which $|f''(x)| \leq M$ for all x in $[a, b]$. We illustrate this in the following examples. In the first example f can be integrated exactly and is included only to compare the error as given by inequality (9.1) with the actual error.

EXAMPLE 9.1 Use the trapezoidal rule with $n = 10$ to approximate $\int_1^2 \frac{dx}{x}$. Find an upper bound on the error, using inequality (9.1), and compare this with the actual error.

Solution Since $a = 1$ and $b = 2$, we have

$$\Delta x = \frac{b-a}{n} = \frac{1}{10} = 0.1$$

Thus, the partition points are $x_0 = 1$, $x_1 = 1.1$, $x_2 = 1.2$, $x_3 = 1.3, \ldots,$ $x_{10} = 2$. So by the trapezoidal rule,

$$\int_1^2 \frac{dx}{x} \approx \left[\frac{0.1}{2}\left(\frac{1}{1}\right) + 2\left(\frac{1}{1.1}\right) + 2\left(\frac{1}{1.2}\right) + 2\left(\frac{1}{1.3}\right) + \cdots + 2\left(\frac{1}{1.9}\right) + \frac{1}{2}\right]$$

Using a calculator we get

$$\int_1^2 \frac{dx}{x} \approx 0.6938$$

To estimate the error we calculate f'', where $f(x) = \frac{1}{x}$:

$$f'(x) = -\frac{1}{x^2}$$

$$f''(x) = \frac{2}{x^3}$$

Since $f''(x)$ decreases for $x \geq 1$, its maximum value occurs when $x = 1$. So for all $x \in [1, 2]$, $|f''(x)| \leq 2$. Thus we can use $M = 2$ in inequality (9.1). This gives

$$E_{10} \leq \frac{2(b-a)^3}{12(10)^2} = \frac{2(1^3)}{12(10)^2} = \frac{2}{1200} \approx 0.00167$$

So our estimate of 0.6938 may be in error by at most 0.0017, which means that the true value of the integral lies between 0.6921 and 0.6955. So we should give our answer to two decimal places only and say that

$$\int_1^2 \frac{dx}{x} \approx 0.69$$

Since $\int_1^2 \frac{dx}{x} = \ln 2$, and from a calculator $\ln 2 = 0.6931472$, correct to seven places, our original estimate was actually correct to only two decimal places. ■

EXAMPLE 9.2 Use the trapezoidal rule to estimate

$$\int_0^1 \frac{dx}{\sqrt{1 + x^3}}$$

using $n = 10$ and obtain an upper bound on the error. How large should n be to ensure accuracy to four decimal places?

Solution For $n = 10$, $a = 0$, and $b = 1$, we have $\Delta x = 0.1$ and $x_0 = 0$, $x_1 = 0.1$, $x_2 = 0.2, \ldots, x_{10} = 1$. So

$$\int_0^1 \frac{dx}{\sqrt{1+x^3}} \approx \frac{0.1}{2}\left[\frac{1}{\sqrt{1+0^3}} + 2\frac{1}{\sqrt{1+(0.1)^3}} + 2\frac{1}{\sqrt{1+(0.2)^3}} + \cdots + 2\frac{1}{\sqrt{1+(0.9)^3}} + \frac{1}{\sqrt{1+1^3}}\right]$$

$$\approx 0.9092 \quad \text{By calculator}$$

To estimate the error we calculate (after some effort)

$$f''(x) = \frac{15x^4 - 12x}{4(1+x^3)^{5/2}}$$

where $f(x) = (1 + x^3)^{-1/2}$. Now it is not so easy to determine the maximum value of $|f''(x)|$ for x on $[0, 1]$, but we can at least find an upper bound by *maximizing the numerator* and *minimizing the denominator*. The absolute value of the numerator cannot exceed $15x^4 + 12|x|$ (why?), and this has its maximum value on $[0, 1]$ when $x = 1$. The minimum value of the denominator occurs when $x = 0$. So we have

$$|f''(x)| = \left|\frac{15x^4 - 12x}{(1+x^3)^{5/2}}\right| \le \frac{15x^4 + 12x}{4} \le \frac{27}{4} \quad \text{on } [0, 1]$$

So by inequality (9.1),

$$E_{10} \le \frac{27}{4}\,\frac{(b-a)^3}{12(10)^2} = \frac{27(1^3)}{4(12)100} = 0.005625$$

Thus, at best we can say that

$$\int_0^1 \frac{dx}{\sqrt{1+x^3}} \approx 0.91$$

and there is some uncertainty about the second decimal place.

To determine how large an n will guarantee accuracy to four decimal places, we want $E_n \le 0.00005$. So we solve

$$\frac{27}{4} \cdot \frac{1}{12n^2} \le 0.00005$$

for n and get

$$n \ge \sqrt{\frac{27}{48(0.00005)}} \approx 106.07$$

So to guarantee four-place accuracy, we would need to take $n = 107$, a formidable task with a calculator but no problem for a computer. ■

It should be emphasized that the trapezoidal rule is a means of estimating definite integrals, not just estimating areas. Whether the integral arises in calculating area, volume, work, pressure, the average value of a function, or any other application, the trapezoidal rule can be used to estimate the integral, provided values of the integrand are known at equally spaced intervals of the independent variable. It is not even necessary to know the form of the function being integrated, as the next example shows.

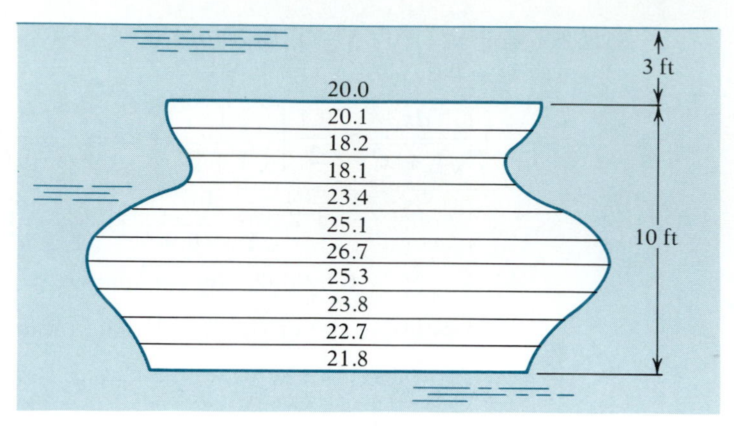

FIGURE 9.3

EXAMPLE 9.3 An irregularly shaped metal plate is submerged in water as shown in Figure 9.3. Widths are measured at 1-ft intervals as shown. Use the trapezoidal rule to estimate the total force on one face of the plate.

Solution Take x to be measured positively downward from the top of the plate. Then the depth at x is $h(x) = x + 3$, and if we denote the width at x by $w(x)$, we have the force F given by

$$F = \int_0^{10} \rho h(x)w(x)\,dx$$
$$= 62.4 \int_0^{10} (x + 3)w(x)\,dx$$

We use the trapezoidal rule to approximate the integral, taking $n = 10$, $a = 0$, $b = 10$, and $\Delta x = 1$, so that $x_0 = 0$, $x_1 = 1$, $x_2 = 2, \ldots, x_{10} = 10$. Thus,

$$\begin{aligned} F &\approx 62.4 \cdot \tfrac{1}{2}[(x_0 + 3)w(x_0) + 2(x_1 + 3)w(x_1) + 2(x_2 + 3)w(x_2) \\ &\quad + \cdots + 2(x_9 + 3)w(x_9) + (x_{10} + 3)w(x_{10})] \\ &= 31.2[3(20.0) + 2(4)(20.1) + 2(5)(18.2) + 2(6)(18.1) \\ &\quad + 2(7)(23.4) + 2(8)(25.1) + 2(9)(26.7) + 2(10)(25.3) \\ &\quad + 2(11)(23.8) + 2(12)(22.7) + (13)(21.8)] \\ &= 115{,}053 \end{aligned}$$

Of course, we have no way to estimate the error in this case because, not knowing $w(x)$ in general, we cannot compute the second derivative of the integrand. Since widths are given to three significant figures, however, we should not assume more than three-figure accuracy in the answer. Thus, we should say $F \approx 115{,}000$ lb. ■

A second method for estimating integrals is based on approximating the graph of f in segments by parabolic arcs rather than chords. In most cases this gives a better approximation than the trapezoidal rule. Again, we introduce the method by considering the area of the region under a continuous nonnegative function f, but we will later remove the continuity and nonnegativity requirements.

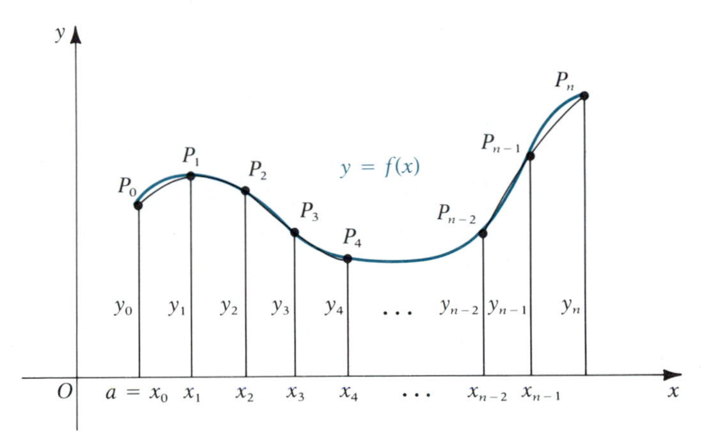

FIGURE 9.4

Let f be continuous and nonnegative on $[a, b]$, and partition $[a, b]$ into an *even number* of subintervals of equal width by means of the points

$$a = x_0 < x_1 < x_2 < \cdots < x_n = b \qquad (n \text{ even})$$

and let $\Delta x = (b - a)/n$ be the common width of the subintervals. As in Figure 9.4 let P_k be the point on the graph of f with coordinates (x_k, y_k), where $y_k = f(x_k)$, $k = 0, 1, 2, \ldots, n$. We are going to fit a second-degree polynomial (whose graph is a parabola) to the points P_0, P_1, and P_2, then fit another second-degree polynomial to P_2, P_3, and P_4, then to P_4, P_5, and P_6, and so on until we get to the last group of three P_{n-2}, P_{n-1}, and P_n.

Consider first the problem of fitting a second-degree polynomial to the points P_0, P_1, and P_2. What we really want is the integral, which is equal to the area under the graph, and this area is unchanged if we translate the origin to coincide with the point $(x_1, 0)$, as in Figure 9.5. With respect to the new axes, the coordinates of P_0, P_1, and P_2 are $(-\Delta x, y_0)$, $(0, y_1)$, and $(\Delta x, y_2)$. Let

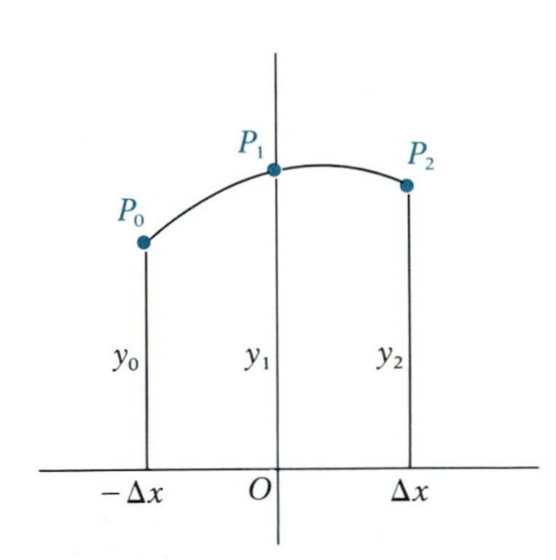

FIGURE 9.5

$$y = ax^2 + bx + c$$

be the equation of the parabola in the new coordinate system that passes through P_0, P_1, and P_2. Since the coordinates of P_0, P_1, and P_2 each satisfy this equation, we have

$$\begin{cases} y_0 = a(-\Delta x)^2 + b(-\Delta x) + c = a(\Delta x)^2 - b(\Delta x) + c \\ y_1 = a(0)^2 + b(0) + c = c \\ y_2 = a(\Delta x)^2 + b(\Delta x) + c \end{cases}$$

From these we can determine a, b, and c. First, however, since it is the area we really want, we calculate

$$\int_{-\Delta x}^{\Delta x} (ax^2 + bx + c)\,dx = \frac{ax^3}{3} + \frac{bx^2}{2} + cx\Big]_{-\Delta x}^{\Delta x} = \frac{\Delta x}{3}\,[2a(\Delta x)^2 + 6c]$$

We know from the system of equations above that $c = y_1$, and by adding the first and third equation, we get

$$y_0 + y_2 = 2a(\Delta x)^2 + 2c$$

or

$$2a(\Delta x)^2 = y_0 + y_2 - 2y_1$$

Thus,

$$\int_{-\Delta x}^{\Delta x} (ax^2 + bx + c)\,dx = \frac{\Delta x}{3}[(y_0 + y_2 - 2y_1) + 6y_1]$$

$$= \frac{\Delta x}{3}[(y_0 + 4y_1 + y_2)] \qquad (9.2)$$

This is the area under the parabolic arc from P_0 to P_2, passing through P_1. Notice that it depends only on the width Δx and the ordinates of the three points.

Exactly the same reasoning could be used to get the area under the parabolic arc that passes through the next group of three points, P_2, P_3, and P_4. But there is no need to go through the derivation again because we already know the result. We simply change the subscripts in equation (9.2), and similarly for the next group of three and so on. The sum of these areas, which we denote by S_n, is therefore

$$S_n = \frac{\Delta x}{3}(y_0 + 4y_1 + y_2) + \frac{\Delta x}{3}(y_2 + 4y_3 + y_4) + \cdots + \frac{\Delta x}{3}(y_{n-2} + 4y_{n-1} + y_n)$$

$$= \frac{\Delta x}{3}[y_0 + 4y_1 + 2y_2 + 4y_3 + \cdots + 2y_{n-2} + 4y_{n-1} + y_n]$$

Although it is more difficult, it is possible to show, just as for the trapezoidal rule, that

$$\lim_{n\to\infty} S_n = \int_a^b f(x)\,dx$$

where f is any integrable function on $[a, b]$. The use of S_n to approximate $\int_a^b f(x)\,dx$ is known as **Simpson's rule** (after the 18th-century English mathematician Thomas Simpson).

SIMPSON'S RULE

If f is integrable on $[a, b]$ and $P = \{x_0, x_1, \ldots, x_n\}$ is a regular partition of $[a, b]$ that has an even number n of subintervals, with

$$\Delta x = \frac{b-a}{n}$$

then

$$\int_a^b f(x)\,dx \approx \frac{\Delta x}{3}[y_0 + 4y_1 + 2y_2 + 4y_3 + \cdots + 2y_{n-2} + 4y_{n-1} + y_n]$$

As with the trapezoidal rule, we let E_n denote the error in the approximation. Thus $E_n = \left|\int_a^b f(x)\,dx - S_n\right|$. Although we cannot in general find E_n exactly, it is proved in numerical analysis courses that if $f^{(4)}(x)$ exists and is bounded in $[a, b]$, then

$$E_n \leq \frac{K(b-a)^5}{180n^4} \tag{9.3}$$

where K is any number for which $|f^{(4)}(x)| \leq K$ for all x in $[a, b]$.

Note Although we use the same symbol E_n to denote the error for the trapezoidal rule and for Simpson's rule, the context will make its meaning clear.

EXAMPLE 9.4 Redo Example 9.1 using Simpson's rule instead of the trapezoidal rule.

Solution The problem, recall, is to approximate $\int_1^2 \frac{dx}{x}$ using $n = 10$ (which satisfies the requirement that n be even). So $\Delta x = 0.1$ as before, and the partition points are the same. Only the formula used for the approximation is different. By Simpson's rule,

$$\begin{aligned}\int_1^2 \frac{dx}{x} &\approx \frac{0.1}{3}\left[\frac{1}{1} + 4\left(\frac{1}{1.1}\right) + 2\left(\frac{1}{1.2}\right) + 4\left(\frac{1}{1.3}\right) + 2\left(\frac{1}{1.4}\right) + 4\left(\frac{1}{1.5}\right)\right.\\ &\quad \left. + 2\left(\frac{1}{1.6}\right) + 4\left(\frac{1}{1.7}\right) + 2\left(\frac{1}{1.8}\right) + 4\left(\frac{1}{1.9}\right) + \frac{1}{2.0}\right]\\ &\approx 0.69315\end{aligned}$$

To use inequality (9.3) to estimate the error, we first calculate $f^{(4)}(x) = -6/x^4$. On the interval $[1, 2]$, therefore, $|f^{(4)}(x)| \leq 6$. So by inequality (9.3),

$$E_n \leq \frac{6(1)^5}{(180)10^4} = 0.0000033$$

which means that our answer is correct to five decimal places. A comparison with the value by calculator, $\ln 2 = 0.6931472$, shows the correctness of this result.

Note that with the trapezoidal rule we obtained accuracy to only two decimal places. So in this case Simpson's rule gives a far better result. ■

It stands to reason that Simpson's rule would in general give greater accuracy than the trapezoidal rule, since a parabolic arc in all likelihood comes closer to matching the actual curve over a short interval than does a straight line, as in the trapezoidal rule. Of course, if the graph of f happens to be a straight line or a collection of straight lines, then the trapezoidal rule will give the *exact* value of the integral when the partition points are properly chosen. (In fact, Simpson's rule does also.) This is an exceptional case, however. Let us compare the bounds on the errors for the trapezoidal and Simpson's rules.

Since $\Delta x = (b - a)/n$, the error estimates of inequalities (9.1) and (9.3) can be put in the following forms:

Trapezoidal Rule	**Simpson's Rule**
$E_n \leq \frac{M(b-a)}{12}(\Delta x)^2$	$E_n \leq \frac{K(b-a)}{180}(\Delta x)^4$

For small values of Δx, $(\Delta x)^4$ is much smaller than $(\Delta x)^2$. So unless K is substantially larger than M, it follows that the bound on the error for

Simpson's rule will be much less than the bound on the error for the trapezoidal rule.

Another interesting consequence of the bound on E_n for Simpson's rule is that the sum S_n gives the *exact* value of $\int_a^b f(x)\,dx$ if f is a polynomial function of degree less than or equal to 3. This is because the fourth derivative of such a polynomial is 0 (verify), so that K can be taken as 0. This is a curious result, since in Simpson's rule we use second-degree curves to approximate the given curve, and so it would seem that the exact value would be obtained only when f itself is of second degree. But, in fact, f can be of degree 0, 1, 2, or 3 (thus showing that Simpson's rule as well as the trapezoidal rule gives the exact value when f is linear). This result is of more theoretical than practical value, since if f is a polynomial we can easily get $\int_a^b f(x)\,dx$ using the fundamental theorem.

The next example shows how the error bounds on the trapezoidal rule and Simpson's rule can be used to determine how large an n will guarantee a specified degree of accuracy.

EXAMPLE 9.5 Determine n so that the error in approximating the integral $\int_1^4 x \ln x\,dx$ will not exceed 10^{-4} when using (a) the trapezoidal rule and (b) Simpson's rule.

Solution Let $f(x) = x \ln x$. Then

$$f'(x) = \frac{x}{x} + \ln x = 1 + \ln x$$

$$f''(x) = \frac{1}{x}$$

$$f'''(x) = -\frac{1}{x^2}$$

$$f^{(4)}(x) = \frac{2}{x^3}$$

(a) On $[1, 4]$ $|f''(x)| \leq 1$. So in inequality (9.1) we take $M = 1$, and we find the smallest n such that

$$\frac{M(b-a)^3}{12n^2} \leq 10^{-4}$$

That is,

$$\frac{1(3)^3}{12n^2} \leq 10^{-4}$$

or

$$n^2 \geq \frac{3^3(10^4)}{12} = 22{,}500$$

and so $n = 150$.

(b) On $[1, 4]$, $|f^{(4)}(x)| \leq 2$, so in inequality (9.3) we take $K = 2$ and find n such that

$$\frac{K(b-a)^5}{180n^4} \leq 10^{-4}$$

or

$$\frac{2(3)^5}{180n^4} \leq 10^{-4}$$

$$n^4 \geq \frac{2(3^5)10^4}{180}$$

By calculator we find $n \approx 12.8$. But n must be an even integer. So we take $n = 14$. Contrast this with the result in part a. ■

In the trapezoidal rule we used straight lines to approximate the curve, and in Simpson's rule we used parabolic arcs. An even better fit can usually be obtained if we use arcs of higher degree polynomials, such as cubic or quartic curves. Formulas for such approximations are developed in numerical analysis courses, and we will not take them up here. Simpson's rule is one of the most widely used methods. There are some considerations, however, that might cause one to choose the trapezoidal rule. For one thing, the idea behind it is intuitively clear and it can easily be derived. For another, it is usually easier to find the bound on the error than in Simpson's rule, since only the bound M on $|f''(x)|$ is needed. Computing fourth derivatives can become tedious!

EXERCISE SET 9.2

A

In Exercises 1–8 estimate the integral in two ways: (a) by the trapezoidal rule and (b) by Simpson's rule. Find bounds on the error in each case, and by performing the integration compare the exact value with the two approximations.

1. $\int_0^1 \frac{dx}{(x+1)^2}$; $n = 4$ **2.** $\int_0^4 \frac{dx}{\sqrt{2x+1}}$; $n = 10$

3. $\int_0^2 e^{-x}\,dx$; $n = 8$ **4.** $\int_1^3 \ln x\,dx$; $n = 10$

5. $\int_{-1}^1 \frac{2\,dx}{1+x^2}$; $n = 10$

6. $\int_1^4 (x \ln x - x)\,dx$; $n = 12$

7. $\int_{-1}^2 (1 + x^3)\,dx$; $n = 6$

8. $\int_2^4 \frac{dx}{x(x-1)}$; $n = 8$

In Exercises 9–14 determine a value of n for (a) the trapezoidal rule and (b) Simpson's rule that will ensure that the error in approximating the integral does not exceed 10^{-4}.

9. $\int_1^3 \frac{dx}{\sqrt{x+1}}$ **10.** $\int_{-1/2}^2 \sqrt[3]{x+1}\,dx$

11. $\int_1^2 \frac{\ln x}{x}\,dx$ **12.** $\int_{-1}^2 e^{-2x}\,dx$

13. $\int_3^5 \cos^2 x\,dx$ **14.** $\int_{-1}^1 \tan \frac{\pi x}{4}\,dx$

15. Use the trapezoidal rule to estimate

$$\int_{-1}^1 \frac{dx}{\sqrt{1+x^2}}$$

correct to two decimal places. Compare your result with the exact value.

16. Use Simpson's rule to approximate

$$\int_{-2}^0 \frac{x}{x-1}\,dx$$

correct to three decimal places. Compare your result with the exact value.

17. A variable force F acts along the x-axis from $x = 0$ to $x = 4$, where x is in feet. The formula for $F(x)$ is not known, but its value in pounds is measured at regular intervals as given in the table. Estimate the work done by F, using Simpson's rule.

x	0	0.4	0.8	1.2	1.6	2.0	2.4	2.8	3.2	3.6	4.0
$F(x)$	12.1	13.3	15.7	16.9	14.3	12.6	8.9	11.3	12.6	14.1	15.9

18. The average depth of a farm water tank is known to be approximately 2 m. The width of the tank at 2-m increments is measured and found to be as shown (in meters) in the accompanying figure. Use Simpson's rule to estimate the volume of water in the tank.

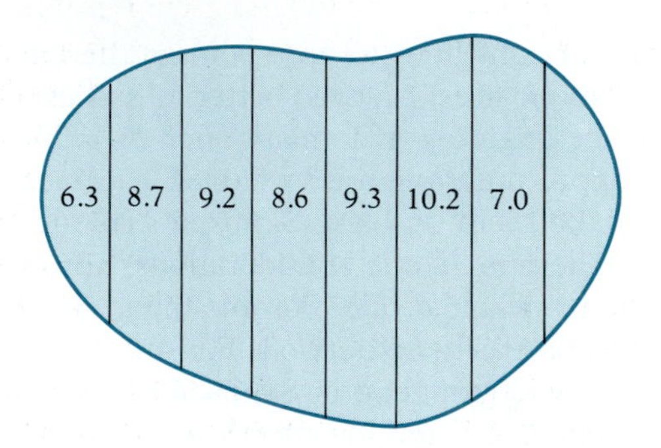

19. The widths of a dam constructed across a gorge are measured at regular 10-ft intervals from bottom to top as shown (in feet) in the accompanying figure. Use Simpson's rule to estimate the force on the dam when the water level is even with the top of the dam. (Assume the face of the dam is vertical.)

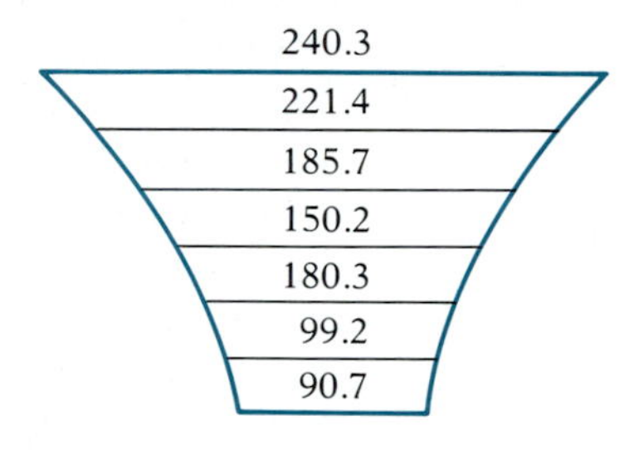

B

In Exercises 20–23 estimate the integral using (a) the trapezoidal rule and (b) Simpson's rule. In each case obtain an upper bound on the error. In each case take $n = 10$.

20. $\int_{-1}^{1} \sin x^2\, dx$ **21.** $\int_0^1 e^{-x^2}\, dx$

22. $\int_1^2 \frac{\sin x}{x}\, dx$ **23.** $\int_0^1 \frac{1 - e^x}{1 + e^x}\, dx$

24. Use the trapezoidal rule to approximate the length of the ellipse $4x^2 + y^2 = 8$ from $(0, 2\sqrt{2})$ to $(1, 2)$, using $n = 5$. Find a bound on the error.

C

25. Write a computer program to compute the sum T_n used in the trapezoidal rule and run the problem for each of the following.

(a) $\int_{-1}^{1} 3\, dx/\sqrt{4 - x^2}$; $n = 20$. Compute the actual value of the integral and determine the error in using the trapezoidal rule.

(b) $\int_1^{10} \frac{dx}{x}$ with a maximum allowable error of 0.00005. Have the computer calculate and print a suitable value of n after you have calculated by hand a constant M that is a bound for $|f''(x)|$.

(c) $\int_0^1 e^{-x^2}\, dx$ for $n = 6, 10, 14, 18, 22$. Discuss the accuracy of your results.

26. Repeat Exercise 25 using Simpson's rule (replacing T_n by S_n and M by K).

27. Use the program for Simpson's rule from Exercise 26 to do the following.

(a) Approximate $\int_a^b \sqrt{1 + x^3}\, dx$ with $n = 20$, where the interval $[a, b]$ is $[0, 1]$, $[1, 2]$, and $[2, 10]$.

(b) Approximate $\int_1^5 \sqrt{1 + \ln x}\, dx$ by using $n = 10, 12, 14, \ldots$ until two consecutive values of S_n agree in the first five decimal places.

9.3 APPROXIMATING ROOTS OF EQUATIONS

We mentioned that no formulas exist for finding exact solutions in terms of radicals of polynomial equations of degree greater than 4. And even though there are such formulas for cubic and quartic equations, they are hard to apply in many cases. When we consider nonpolynomial equations, the situation is likely to be much worse, as only very special types lend themselves to exact methods of solution. Even the simple looking equation $\tan x - x = 0$ cannot be solved in an exact form for any root other than the obvious one, $x = 0$. For these reasons methods of approximating roots of equations are important to learn. Many such methods exist, but we will concentrate on just one because it makes use of calculus. It is, in fact, one of the most efficient of all the methods.

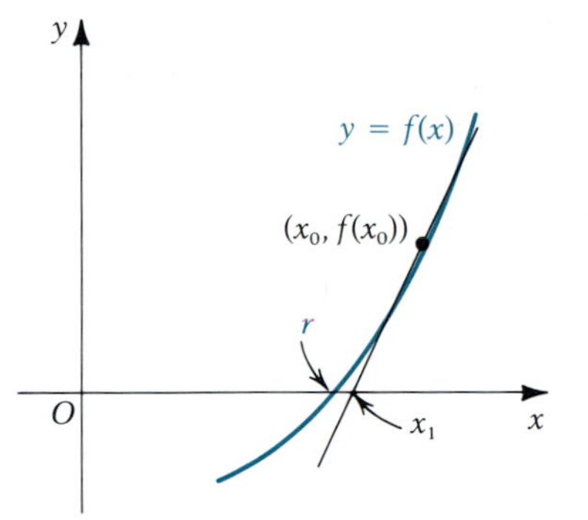

FIGURE 9.6

The basic idea of the method we describe originated with Newton, and it is generally referred to as **Newton's method.** However, the form in which it is presently used was given by Joseph Raphson (1648–1715), so it is also known as the **Newton–Raphson method.** The method is easy to describe geometrically. Suppose that by graphical means or otherwise we find that the equation $f(x) = 0$ has at least one real root. Let r be the true value of one such root. We make an initial guess at its value and denote this first guess by x_0. Sketching the graph of f or making a table of values is usually sufficient to provide a reasonable guess for x_0. Unless we are unusually lucky, x_0 will not equal r. So we locate the point $(x_0, f(x_0))$ and draw the tangent line to the graph of f there. As in Figure 9.6 we let x_1 be the x-intercept of this tangent line. This value, x_1, is our next approximation to r. In general, if x_0 was close enough to r, x_1 will be closer to r than x_0. (If this is not so, try another x_0 as the starting point.) Now we repeat the process. We draw the tangent line to the graph of f at $(x_1, f(x_1))$ and obtain the next approximation to r from its x-intercept, which we call x_2. This process is continued until we obtain the desired degree of accuracy.

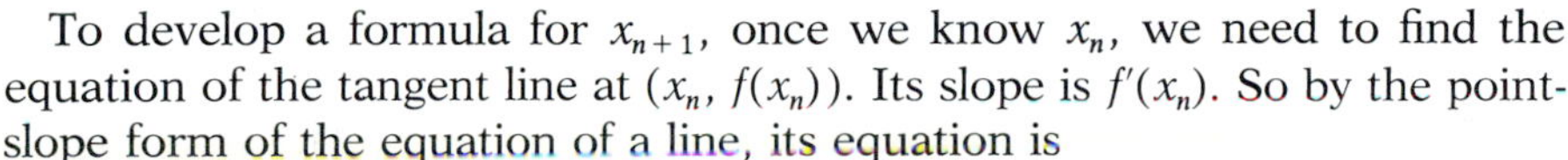

To develop a formula for x_{n+1}, once we know x_n, we need to find the equation of the tangent line at $(x_n, f(x_n))$. Its slope is $f'(x_n)$. So by the point-slope form of the equation of a line, its equation is

$$y - f(x_n) = f'(x_n)(x - x_n)$$

To get the x-intercept, we set $y = 0$:

$$-f(x_n) = f'(x_n)(x - x_n)$$

We solve for x and name the result x_{n+1}:

$$x_{n+1} = x_n - \frac{f(x_n)}{f'(x_n)} \tag{9.4}$$

Of course, $f'(x_n)$ must be different from 0. This result is called a *recursion formula,* since it gives the new value in terms of the old. We guess at x_0, then get x_1 from the recursion formula, then get x_2 by using the formula again, and so on. We now summarize the procedure.

THE NEWTON–RAPHSON METHOD

1. Make an initial guess x_0 for a root r of the equation $f(x) = 0$.
2. Find $x_1, x_2, x_3, \ldots$ using

$$x_{n+1} = x_n - \frac{f(x_n)}{f'(x_n)} \qquad (f'(x_n) \neq 0)$$

by taking $n = 0, 1, 2, 3, \ldots$.

If the successive approximations $x_0, x_1, x_2, \ldots$ come arbitrarily close to r as $n \to \infty$, we say the process *converges*. Before considering criteria that guarantee convergence, we consider two examples. In the first we use an equation for which an exact solution can be found, for comparison purposes.

EXAMPLE 9.6 Approximate the positive root of $x^2 - 2x - 2 = 0$ using Newton's method. Continue the process until two consecutive approximations agree in the first five decimal places and round the answer to four places. Compare this with the exact answer.

Solution We show the graph of $f(x) = x^2 - 2x - 2$ in Figure 9.7. Since $f(2) = -2$ and $f(3) = 1$, the root we are seeking is between 2 and 3, and from the graph it appears to be closer to 3. A good first approximation might therefore be $x_0 = 2.6$. To apply Newton's method we need f':

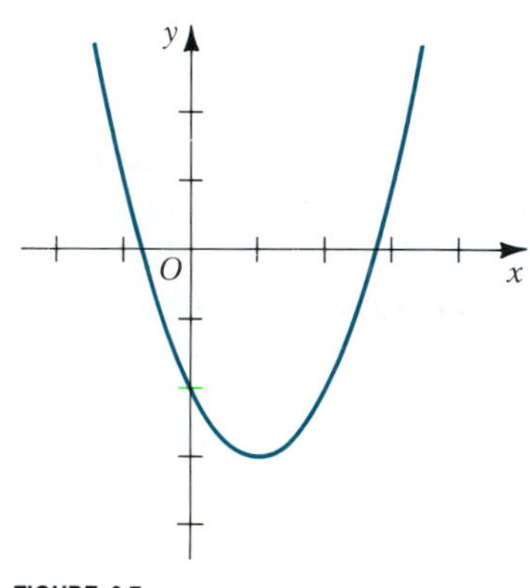

FIGURE 9.7

$$f'(x) = 2x - 2 = 2(x - 1)$$

Now we apply the recursion formula (9.4), using a calculator:

$$\underline{n = 0}: \quad x_1 = x_0 - \frac{f(x_0)}{f'(x_0)} = 2.6 - \frac{f(2.6)}{f'(2.6)} = 2.7375$$

$$\underline{n = 1}: \quad x_2 = x_1 - \frac{f(x_1)}{f'(x_1)} = 2.72918$$

$$\underline{n = 2}: \quad x_3 = x_2 - \frac{f(x_2)}{f'(x_2)} = 2.73205$$

$$\underline{n = 3}: \quad x_4 = x_3 - \frac{f(x_3)}{f'(x_3)} = 2.7320508$$

Since x_3 and x_4 agree in the first five decimal places, we stop the process here and say that $r \approx 2.7321$. The fact that x_3 and x_4 agree to five places does not guarantee that the answer is correct to four places, but it strongly suggests this is so.

To compare the exact answer, we solve for x by the quadratic formula, getting

$$x = \frac{2 \pm \sqrt{4 + 8}}{2} = 1 \pm \sqrt{3}$$

The positive root is $1 + \sqrt{3}$, which to seven places is 2.7320508. Note that x_4 agrees with this in all seven places, so four iterations of Newton's method yielded very good results. ■

EXAMPLE 9.7 Use Newton's method to approximate the smallest positive root of the equation $\tan x - x = 0$.

Solution Let $f(x) = \tan x - x$. We will need f':

$$f'(x) = \sec^2 x - 1 = \tan^2 x$$

Rather than to graph f, it is easier to graph $y_1 = \tan x$ and $y_2 = x$ and to estimate where $y_1 = y_2$, as we have done in Figure 9.8. It appears that the root r we are seeking is close to $\frac{3\pi}{2} \approx 4.71$. We try $x_0 = 4.5$ as a first approximation

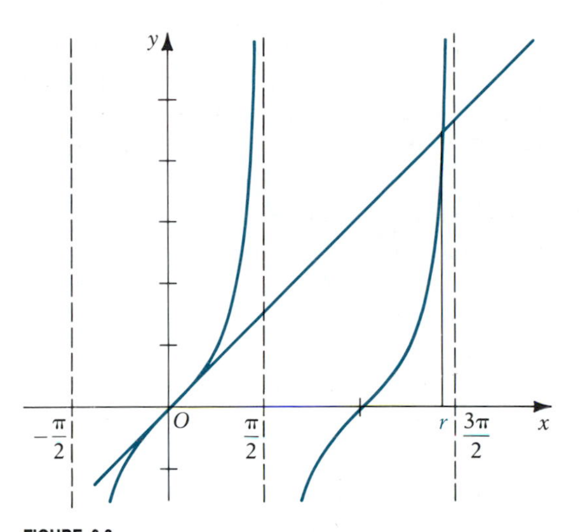

FIGURE 9.8

of r, and proceed with Newton's method. The recursion formula is

$$x_{n+1} = x_n - \frac{f(x_n)}{f'(x_n)} = x_n - \frac{\tan x_n - x_n}{\tan^2 x_n}$$

This time we use a table:

n	x_n	x_{n+1}
0	4.5	4.4936
1	4.4936	4.49341
2	4.49341	4.49341

So in just three iterations it appears that we have five-place accuracy and can say that $r \approx 4.49341$. ■

Remark It was essential in the preceding example that we have the calculator set in the *radian* mode.

We turn our attention now to ways in which Newton's method can fail. We know already that if $f'(x_n) = 0$ for any n, then the method fails. As shown in Figure 9.9, even if x_n is near a point where $f'(x) = 0$, then all subsequent approximations may be farther from r than x_n is because of the flatness of the slope of the tangent line. Some other ways in which the process may go

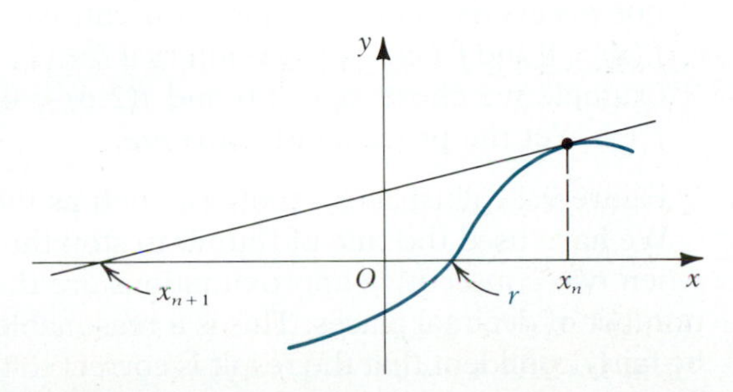

FIGURE 9.9

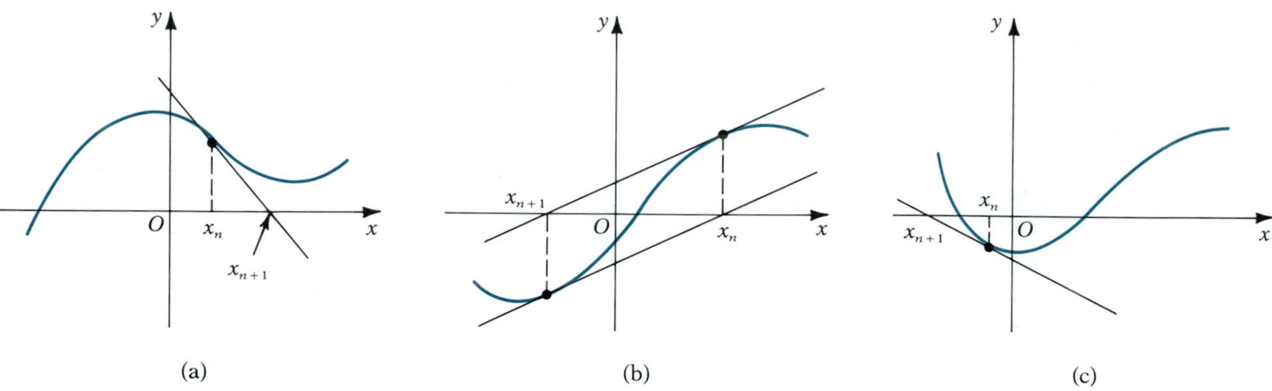

FIGURE 9.10

awry are shown in Figure 9.10. In part (a) the approximations again get farther and farther from r. In part (b) the process is in an endless loop, and in part (c) the x_n values are approaching the wrong root.

It is usually easy to tell when the process is not converging toward the root because successive approximations are not settling down. If this happens, it usually suffices to try a new starting point. Various criteria will assure convergence, and for the sake of completeness we state one of these, but we omit the proof.

THEOREM 9.1 Let I be an open interval containing a root r of the equation $f(x) = 0$ such that for all x in I both $f'(x)$ and $f''(x)$ exist and are of constant sign (either always positive or always negative). If x_0 is any point in I for which $f(x_0)$ agrees in sign with $f''(x)$, then the approximations x_n generated by Newton's method using x_0 as the starting point will converge to r as $n \to \infty$. ■

Comments

1. The requirement that $f'(x)$ be of one sign assures us that there is only one root in I, since f is monotone and hence 1–1 on I. It also assures us that f has no maxima or minima in I.
2. The requirement that $f''(x)$ be of one sign in I assures us that f has no point of inflection in I. The graph is always concave upward or else always concave downward.
3. When f is concave upward, we choose x_0 such that $f(x_0)$ is positive. When f is concave downward, we choose x_0 such that $f(x_0)$ is negative.
4. The conditions of the theorem are *sufficient* to guarantee convergence but not *necessary*. For example, you can easily verify that in Example 9.6 $f'(x) > 0$ and $f''(x) > 0$ in the interval $I = (2, 3)$ that contains a root r. In that example we chose $x_0 = 2.6$ and $f(2.6) < 0$, which is opposite in sign to $f''(x)$. Yet the process did converge.

Figure 9.11 illustrates situations such as the theorem describes.

We have used the rule of thumb to stop the iteration in Newton's method when two consecutive approximations are the same up to a predetermined number of decimal places. This is a reasonable way to proceed, and you can be fairly confident that the result is correct to that number of decimal places. The next theorem gives more precise information about errors.

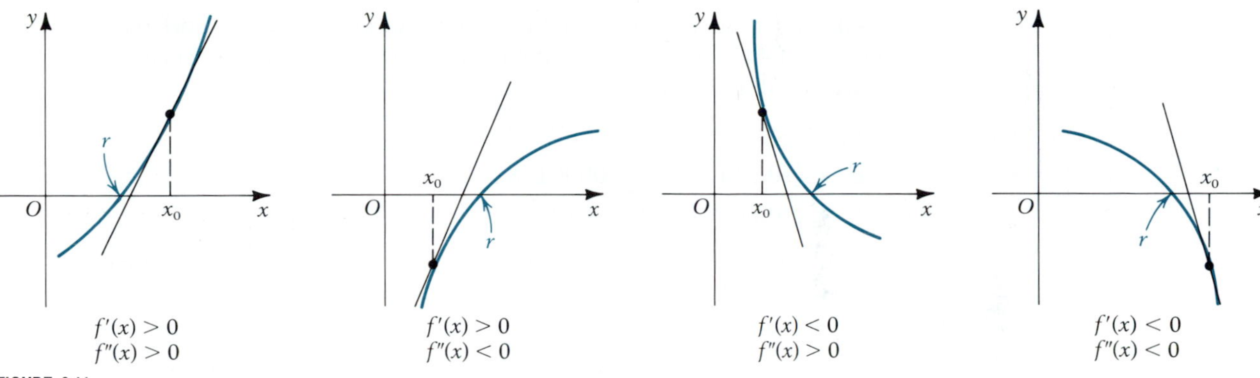

FIGURE 9.11

THEOREM 9.2 Suppose f satisfies the conditions of Theorem 9.1 in an interval I containing a root r of $f(x) = 0$. In addition, suppose there exist positive constants m and M such that for all x in I, $|f'(x)| \geq m$ and $|f''(x)| \leq M$. If the error in the nth approximation is denoted by E_n—that is, if $E_n = |x_n - r|$—then E_{n+1} and E_n are related by

$$E_{n+1} \leq \frac{M}{2m} E_n^2 \qquad (9.5)$$ ■

A proof of this theorem will be given in the next section. Meanwhile, we illustrate its use and show an interesting consequence.

EXAMPLE 9.8 Show that the function

$$f(x) = x^3 - 3x^2 - 24x + 48$$

satisfies the conditions of Theorem 9.2 on the interval $I = (1, 2)$. Find a starting point x_0 in I that will guarantee the convergence of Newton's method and determine a bound on the error in x_3.

Solution Since $f(1) = 22$ and $f(2) = -4$, there is a root r to the equation $f(x) = 0$ in I. We calculate f' and f'':

$$f'(x) = 3x^2 - 6x - 24 = 3(x + 2)(x - 4)$$
$$f''(x) = 6x - 6 = 6(x - 1)$$

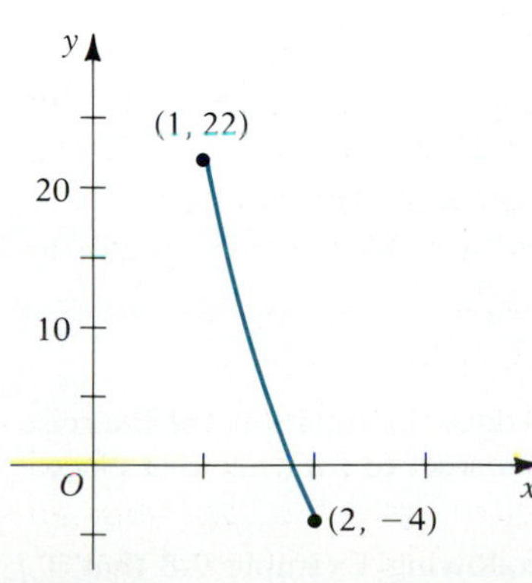

FIGURE 9.12

For $x \in (1, 2)$, $f'(x) < 0$ and $f''(x) > 0$, so both derivatives are of constant sign in I. We need a lower bound m on $|f'(x)|$ and an upper bound M on $|f''(x)|$. Since $f'(x) < 0$ on I, it follows that $-f'(x) > 0$, so that $|f'(x)| = -f'(x)$. Furthermore, since $f''(x) > 0$, f' is increasing, so that $-f'$ is decreasing. Thus,

$$|f'(x)| = -f'(x) = -3(x + 2)(x - 4) > -3(2 + 2)(2 - 4) = 24$$

Also

$$|f''(x)| = |6(x - 1)| = 6(x - 1) < 6(2 - 1) = 6$$

So we can take $m = 24$ and $M = 6$.

Figure 9.12 shows the nature of the graph of f in the interval in question. According to Theorem 9.1 we can be assured of convergence of Newton's method if we choose x_0 so that $f(x_0) > 0$ [since $f''(x) > 0$]. From the graph it appears that $x_0 = 1.7$ might be a reasonable choice, and we find that $f(1.7) = 3.443$, so this is a suitable value. Since the root is between 1.7 and 2, we see that $E_0 < 0.3$. By inequality (9.5), therefore,

$$E_1 < \tfrac{6}{48}(0.3)^2 = \tfrac{0.09}{8} = 0.01125$$
$$E_2 < \tfrac{1}{8}(0.01125)^2 = 0.0000158$$
$$E_3 < \tfrac{1}{8}(0.0000158)^2 < 3.13 \times 10^{-11}$$

So after just three iterations we have accuracy to ten decimal places. ■

A consequence of Theorem 9.2 is that when $M < 4m$, if x_n is correct to the first k decimal places, x_{n+1} is correct to the first $2k$ decimal places. A proof of this will be called for in the exercises. This shows how rapidly Newton's method converges. The condition $M < 4m$ is not very restrictive and is found to be true in most cases.

EXERCISE SET 9.3

A

In Exercises 1–9 approximate each root using Newton's method, continuing until two successive approximations agree in the first five decimal places. Round the answer to four places.

1. The positive root of $2x^2 - 4x - 1 = 0$ (Compare the exact answer.)
2. The real root of $x^3 - 4x^2 + 7x - 5 = 0$
3. The negative root of $x^4 - 2x^3 - 2 = 0$
4. The positive root of $2 \sin x - x = 0$
5. The real root of $e^x + x = 2$
6. The smallest positive root of $\ln x - x + 2 = 0$
7. Both real roots of $2x^4 - 3x^3 - 2x + 2 = 0$
8. The positive root of $x^2 - \cos x = 0$
9. The real root of $\sinh x - x + 2 = 0$

In Exercises 10–13 write an equation that has the given root as a solution, and use Newton's method to find the root to four decimal places.

10. $\sqrt{2}$ **11.** $\sqrt[3]{4}$ **12.** $\sqrt[3]{-9}$ **13.** $\sqrt[4]{6}$

In Exercises 14–17 find an open interval I that contains the root r for which the conditions of Theorem 9.1 are satisfied. Find a suitable starting point x_0 for which the theorem guarantees convergence of the successive approximations to r. Sketch the graph of f for x in I.

14. $f(x) = x^3 - 5x^2 - 8x - 3$; $r > 0$
15. $f(x) = x^4 - 8x^2 - 3$; $r > 0$
16. $f(x) = 3(x - 1)^{2/3} - x$; $1 < r < 9$
17. $f(x) = 2 \cos x + x - 2$; $0 < r < \pi$

In Exercises 18–21 show that I contains a root of the equation $f(x) = 0$, that f satisfies the conditions of Theorem 9.2 in I, and that x_0 is a starting point that will guarantee convergence of Newton's method. How many decimal places of accuracy can we be assured of in the approximation x_3?

18. $f(x) = 2x^3 - 3x^2 + 4x - 8$; $I = (1, 2)$, $x_0 = 1.8$
19. $f(x) = x^4 - 2x^3 - 3x - 8$; $I = (-2, -1)$, $x_0 = -1.2$
20. $f(x) = e^{-x} - x^2$; $I = (0, 1)$, $x_0 = 0.8$
21. $f(x) = \ln x^2 - (x - 4)^2$; $I = (2, 3)$, $x_0 = 2.7$

B

22. Find all local maximum and minimum points correct to four decimal places on the graph of the function f defined by

$$f(x) = 3x^4 - 8x^3 - 5x^2 + 4x - 20$$

23. How many real zeros does the function f of Exercise 22 have? Find them, correct to four decimal places. Draw the graph of f.
24. Prove the assertion following Example 9.8 that if f satisfies the conditions of Theorem 9.2 in I and $M <$

$4m$ there, then whenever x_n is correct to k decimal places, x_{n+1} is correct to $2k$ decimal places. [*Hint:* The approximation x_n is correct to k places if $E_n \leq 5 \times 10^{-(k+1)}$.]

25. Let f satisfy the conditions of Theorem 9.1 in I, and let x_0 be chosen as in that theorem. A method known as the **method of secants** for approximating the root r in I is as follows. Choose any point z_0 in I such that $f(z_0)$ is opposite in sign to $f(x_0)$. Let z_1 be the x-intercept of the line joining $(z_0, f(z_0))$ and $(x_0, f(x_0))$ (the secant line joining these points). Let z_2 be the x-intercept of the line joining $(z_1, f(z_1))$ and $x_0, f(x_0))$, and so on.

(a) Using Figure 9.11 give a geometric argument to show that the method of secants converges to r.

(b) Show that

$$z_{n+1} = z_n - \frac{f(z_n)}{m_n}$$

where m_n is the slope of the secant line joining $(z_n, f(z_n))$ and $(x_0, f(x_0))$.

(c) Show that if $x_0, x_1, x_2, \ldots$ are the approximations generated by Newton's method and $E_n = |x_n - r|$, then $E_n \leq |x_n - z_n|$.

C

For Exercises 26–30 do the following:

(a) Use the computer to locate all roots of $f(x) = 0$ between consecutive integers, for x in the prescribed range. For each root located, carry out steps b–d.

(b) If r is a root between n and $n + 1$, test f' and f'' for constancy of sign there. If either f' or f'' is not of constant sign, bisect the interval $(n, n + 1)$ and test the half that contains r for constancy of sign. Continue this process until an interval I containing r is found on which f' and f'' are of constant sign.

(c) Use the computer to determine a suitable starting point x_0 in I, as specified in Theorem 9.1.

(d) Use the computer to generate the sequence of approximations $x_1, x_2, x_3, \ldots$ obtained by Newton's method. Stop when x_{n+1} agrees with x_n to 11 decimal places. Round the answer to ten places.

26. $f(x) = x^3 - 6x^2 + 7x - 4;\ -10 \leq x \leq 10$

27. $f(x) = x^4 - 2x^3 + 2x^2 - 8x - 8;\ -5 \leq x \leq 5$

28. $f(x) = x^3 - 2x^2 - 6x + 8;\ -8 \leq x \leq 8$

29. $f(x) = x^2 - 2\sqrt{x} - 3;\ 0 \leq x \leq 10$

30. $f(x) = \ln x - \sin x;\ 1 \leq x \leq 5$

9.4 APPROXIMATION OF FUNCTIONS BY POLYNOMIALS; TAYLOR'S THEOREM

Polynomial functions are in many ways the simplest functions of all. First, they are the easiest to evaluate, requiring only addition and multiplication (since raising to positive integral powers is just repeated multiplication). Second, they possess derivatives of all orders on all of **R**. Third, they can be integrated with ease over any interval. On the other hand, nonpolynomial functions can be difficult or impossible to evaluate, differentiate, or integrate. Even the exponential function $f(x) = e^x$ does not lend itself to easy evaluation. For example, try finding $e^{0.1}$ without the aid of a calculator or tables. Similarly, ln 2, sin 0.5, and $\sqrt[3]{9.2}$ cannot readily be approximated as decimal quantities. We encounter the same problem in evaluating definite integrals. Although we can readily get the "answer":

$$\int_{\pi/6}^{\pi/2} \cot x\, dx = \ln(\sin x)\Big]_{\pi/6}^{\pi/2} = \ln(\sin \tfrac{\pi}{2}) - \ln(\sin \tfrac{\pi}{6})$$

$$= \ln 1 - \ln \tfrac{1}{2} = \ln 2$$

we are again confronted with giving this as a decimal quantity.

Of course, calculators and tables *are* available, and with these a very large number of functions can be evaluated to a high degree of accuracy. It is instructive, however, to learn how calculators might arrive at the answers and how tables are constructed. Also, it is important to be able to determine just how accurate the approximations are. Answers to these questions are provided in the procedures developed in this section.

To motivate the definition that follows consider again the exponential function $f(x) = e^x$. We want to find the polynomial function P_n that best approximates f in a neighborhood of $x = 0$. We must determine coefficients $a_0, a_1, a_2, \ldots, a_n$ such that

$$P_n(x) = a_0 + a_1x + a_2x^2 + \cdots + a_nx^n$$

comes "closer" in some sense than any other nth-degree polynomial to fitting $f(x) = e^x$ in the vicinity of $x = 0$. In particular we want the two functions to be identical when $x = 0$; that is, $P_n(0) = f(0)$. Also, a reasonable definition of "best fit" is one that requires that P_n have the same slope and same concavity as f at 0, so we want $P_n'(0) = f'(0)$ and $P_n''(0) = f''(0)$. Generalizing this, we will require that $P_n^{(k)}(0) = f^{(k)}(0)$ for all $k = 0, 1, 2, \ldots, n$, where for convenience of notation we define $f^{(0)}(x) = f(x)$. Now we write the successive derivatives of P_n and evaluate them at $x = 0$:

$$\begin{aligned}
P_n(x) &= a_0 + a_1x + a_2x^2 + a_3x^3 + \cdots + a_nx^n & P_n(0) &= a_0\\
P_n'(x) &= a_1 + 2a_2x + 3a_3x^2 + \cdots + na_nx^{n-1} & P_n'(0) &= a_1\\
P_n''(x) &= 2\cdot 1a_2 + 3\cdot 2a_3x + \cdots + n(n-1)a_nx^{n-2} & P_n''(0) &= 2\cdot 1a_2\\
P_n'''(x) &= 3\cdot 2\cdot 1a_3 + \cdots + n(n-1)(n-2)a_nx^{n-3} & P_n'''(0) &= 3\cdot 2\cdot 1a_3
\end{aligned}$$

Continuing in this way we find that for any k,

$$P_n^{(k)}(0) = k\cdot(k-1)\cdot(k-2)\cdot\cdots\cdot 2\cdot 1a_k$$

This can be simplified by using the *factorial* notation $k! = 1\cdot 2\cdot 3\cdot\cdots\cdot k$ and, again for notational convenience, we define $0! = 1$. So

$$P_n^{(k)}(0) = k!\,a_k$$

Since for $f(x) = e^x$, $f^{(k)}(x) = e^x$ for all k, we have $f^{(k)}(0) = 1$. When we equate $P^{(k)}(0)$ and $f^{(k)}(0)$, we therefore obtain

$$k!\,a_k = 1$$

$$a_k = \frac{1}{k!}$$

Thus

$$P_n(x) = 1 + x + \frac{x^2}{2!} + \frac{x^3}{3!} + \cdots + \frac{x^n}{n!} \tag{9.6}$$

This polynomial best approximates e^x in a neighborhood of $x = 0$ in the sense that e^x and $P_n(x)$ have equal derivatives up through the nth order at 0. For x near 0, then, we would expect the functions to be very close to one another. We will soon see how to measure the discrepancy.

In Figure 9.13 we show the graphs of e^x and of $P_n(x)$ for $n = 1, 2,$ and 3. Note that when x is close to 0, P_3 seems to be very close to e^x. Let us use P_3 to approximate $e^{0.1}$:

$$\begin{aligned}
P_3(0.1) &= 1 + 0.1 + \frac{(0.1)^2}{2!} + \frac{(0.1)^3}{3!} \approx 1 + 0.1 + 0.005 + 0.000167\\
&= 1.105167
\end{aligned}$$

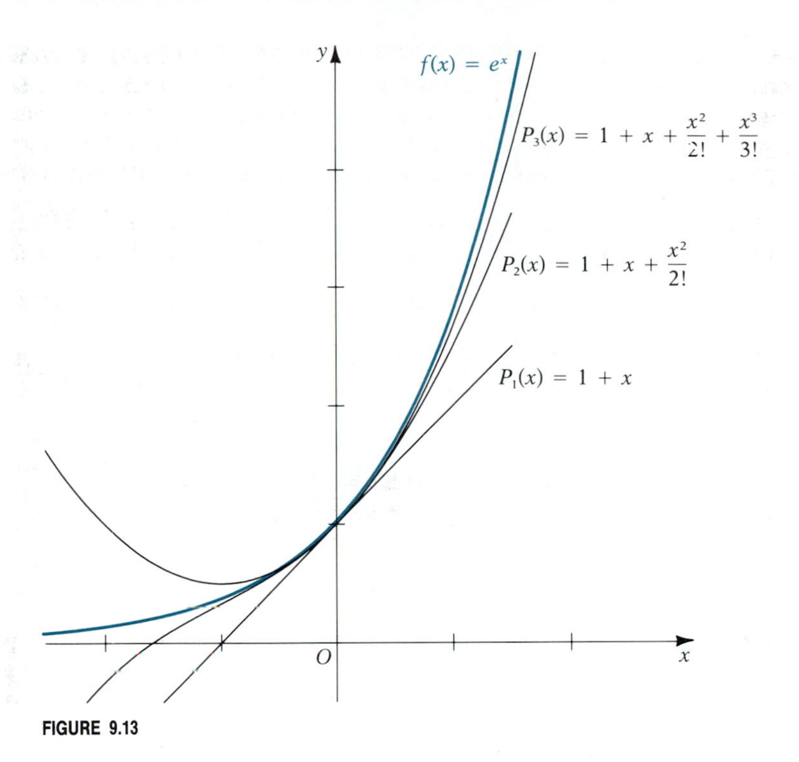

FIGURE 9.13

By calculator we find that $e^{0.1} = 1.1051709$. So P_3 is a very good approximation. For x values farther from 0, we would need to use a higher degree polynomial to get a good approximation. For example, to estimate $e^1 = e$, let us try P_5:

$$\begin{aligned} P_5(1) &= 1 + 1 + \frac{1}{2!} + \frac{1}{3!} + \frac{1}{4!} + \frac{1}{5!} \\ &= 1 + 1 + \frac{1}{2} + \frac{1}{6} + \frac{1}{24} + \frac{1}{120} = \frac{326}{120} \approx 2.71667 \end{aligned}$$

Since we know that $e \approx 2.71828$, the estimate given by P_5 is correct to only two decimal places. A higher value of n would result in P_n being a better approximation.

Let us now generalize the procedure we followed in determining the polynomial approximation for e^x near 0. Let f be any function for which $f^{(n)}(x)$ exists throughout some neighborhood N of 0. (The existence of $f^{(n)}$ implies the existence, and in fact the continuity, of all derivatives of order less than n.) Exactly as in calculating the coefficients of $P_n(x)$ for $f(x) = e^x$, we find that for $k = 0, 1, 2, \ldots, n$,

$$a_k = \frac{f^{(k)}(0)}{k!}$$

so that

$$P_n(x) = f(0) + f'(0)x + \frac{f''(0)}{2!}x^2 + \cdots + \frac{f^{(n)}(0)}{n!}x^n$$

is the polynomial that best fits $f(x)$ in N, in the sense that P_n and f have the same values at 0 for derivatives of all orders through the nth.

We can generalize this result still further by replacing 0 by any other number, say a, in a neighborhood of which $f^{(n)}(x)$ exists. The approximating polynomial is more conveniently written in this case as

$$P_n(x) = a_0 + a_1(x - a) + a_2(x - a)^2 + \cdots + a_n(x - a)^n$$

Just as when $a = 0$, the coefficients are calculated to be (see Exercise 33):

$$a_k = \frac{f^{(k)}(a)}{k!}, \qquad k = 0, 1, 2, \ldots, n$$

The polynomial $P_n(x)$ so defined is called the **nth Taylor polynomial about $x = a$ for the function f**, named after the English mathematician Brook Taylor (1685–1731). We make this formal in the following definition.

DEFINITION 9.1 If f is defined in a neighborhood N of $x = a$ and $f^{(n)}(x)$ exists in N, then the polynomial

$$\begin{aligned} P_n(x) &= f(a) + f'(a)(x - a) + \frac{f''(a)}{2!}(x - a)^2 + \cdots + \frac{f^{(n)}(a)}{n!}(x - a)^n \\ &= \sum_{k=0}^{n} \frac{f^{(k)}(a)}{k!}(x - a)^k \end{aligned} \tag{9.7}$$

is called the **nth Taylor polynomial about $x = a$ for f.** ■

We know that $P_n(x)$ and its first n derivatives agree with $f(x)$ at $x = a$, and that for x near a, $P_n(x) \approx f(x)$. But just how good is this approximation? The next theorem provides the answer. We defer its proof until the end of this section.

THEOREM 9.3 Let $f^{(n+1)}(x)$ exist in a neighborhood N of $x = a$, and let $P_n(x)$ be the nth Taylor polynomial about $x = a$ for f. Define

$$R_n(x) = f(x) - P_n(x)$$

for x in N. Then for each x in N there exists a number c in the open interval between a and x such that

$$R_n(x) = \frac{f^{(n+1)}(c)}{(n+1)!}(x - a)^{n+1} \tag{9.8}$$ ■

Comments

1. $R_n(x)$ is called the **Lagrange form of the remainder,** after Joseph Louis Lagrange (1736–1813), one of the greatest mathematicians of the 18th century. The term *remainder* is used here to mean what is left when the approximation $P_n(x)$ is subtracted from $f(x)$. It is the error in approximating $f(x)$ by $P_n(x)$.
2. The number c as stated in the theorem is dependent on x. If $x > a$, then c lies in the interval (a, x). If $x < a$, then c lies in the interval (x, a).

3. The neighborhood N may be the entire real line, $(-\infty, \infty)$, or it may be finite, in which case it will be of the form $N = (a - r, a + r)$ for some $r > 0$. The size of the neighborhood is determined by the point closest to a for which $f^{(n+1)}(x)$ fails to exist. If this point is r units from a, then $N = (a - r, a + r)$. If $f^{(n+1)}(x)$ exists everywhere, then $N = (-\infty, \infty)$.

Since $R_n(x) = f(x) - P_n(x)$, we can write

$$f(x) = P_n(x) + R_n(x), \qquad x \in N$$

This result is known as **Taylor's formula with remainder.** When we write the values of $P_n(x)$ and $R_n(x)$, it takes the following form.

TAYLOR'S FORMULA WITH REMAINDER

$$f(x) = f(a) + f'(a)(x - a) + \frac{f''(a)}{2!}(x - a)^2 + \cdots + \frac{f^{(n)}(a)}{n!}(x - a)^n + \frac{f^{(n+1)}(c)}{(n + 1)!}(x - a)^{n+1} \qquad (9.9)$$

where $x \in N$ and c is between a and x.

Note that the remainder term [the final term in equation (9.9)] follows the same pattern as all preceding terms *except* that the derivative is evaluated at the intermediate value c between a and x, rather than at a.

When $n = 0$ in Taylor's formula and we replace x by b, where $b \in N$ and $b \neq a$, the result is

$$f(b) = f(a) + f'(c)(b - a)$$

or equivalently,

$$f'(c) = \frac{f(b) - f(a)}{b - a}$$

where c is between a and b. But this is just the statement of the conclusion in the mean value theorem. For this reason Theorem 9.3 (which leads to Taylor's formula with remainder) is sometimes called the *extended mean value theorem.*

EXAMPLE 9.9 Find $R_n(x)$ about $x = 0$ for $f(x) = e^x$. Use the result to find a bound on the error in estimating $e^{0.1}$ using $P_3(0.1)$.

Solution Note first that $f(x) = e^x$ possesses derivatives of all orders everywhere, all equal to e^x, so that the neighborhood of validity of the result in Theorem 9.3 is $(-\infty, \infty)$. Taking $a = 0$ in Theorem 9.3, we have

$$R_n(x) = \frac{f^{(n+1)}(c)}{(n + 1)!}x^{n+1} = \frac{e^c}{(n + 1)!}x^{n+1}$$

where c is between 0 and x. Thus, the error in using $P_3(0.1)$ to estimate $e^{0.1}$ is

$$R_3(0.1) = \frac{e^c}{4!}(0.1)^4, \qquad 0 < c < 0.1$$

Although we do not know c, we can get an upper bound on e^c by the fact that

$$e^c < e^{0.1} < e^1 < 3$$

This is a very generous upper bound, since $e^{0.1}$ is much less than e, but since we presumably do not know $e^{0.1}$, we take something larger with a value that we do know. Thus,

$$R_3(0.1) < \frac{3}{4!}(0.1)^4 = \frac{3}{4 \cdot 3 \cdot 2 \cdot 1}(0.0001) = 0.0000125$$

Thus, even using the generous upper bound on e^c, we see that the error is small, resulting in at least four decimal place accuracy for $P_3(0.1)$ as the approximate value of $e^{0.1}$. Recall that we earlier found $P_3(0.1) \approx 1.105167$ actually was correct to five decimal places. ■

EXAMPLE 9.10 Find Taylor's formula with remainder for arbitrary n for the function $f(x) = \ln x$ about $x = 1$, and state the interval in which it is valid.

Solution In abbreviated form, Taylor's formula is $f(x) = P_n(x) + R_n(x)$. For P_n we need the first n derivatives of f, and for R_n we need $f^{(n+1)}$. So we begin by calculating the successive derivatives:

$$\begin{array}{ll}
f(x) = \ln x & f(1) = 0 \\
f'(x) = \frac{1}{x} = x^{-1} & f'(1) = 1 \\
f''(x) = -x^{-2} & f''(1) = -1 \\
f'''(x) = 2x^{-3} & f'''(1) = 2 \\
f^{(4)}(x) = -3 \cdot 2x^{-4} = -3!x^{-4} & f^{(4)}(1) = -3! \\
f^{(5)}(x) = 4 \cdot 3!x^{-5} = 4!x^{-5} & f^{(5)}(1) = 4! \\
\vdots \qquad \vdots & \vdots \\
f^{(n)}(x) = (-1)^{n-1}(n-1)!x^{-n} & f^{(n)}(1) = (-1)^{n-1}(n-1)! \\
\hline
f^{(n+1)}(x) = (-1)^n n!x^{-(n+1)} & f^{(n+1)}(c) = (-1)^n n!c^{-(n+1)}
\end{array}$$

Observe carefully how we handled the coefficients of the successive derivatives. First, they alternate in sign, being positive when the order is odd and negative when the order is even. The factor $(-1)^{n-1}$ in the nth derivative accomplishes this alternation. Second, we did not multiply out the factors as they accumulated but rather left the multiplication as indicated. This enabled us to recognize the general pattern. We also used the fact that $k! = k(k-1)!$ for $k \geq 1$. You should convince yourself this is true.

Since

$$f^{(n+1)}(x) = \frac{(-1)^n n!}{x^{n+1}}$$

is not defined when $x = 0$, the largest neighborhood of 1 for which Taylor's formula is valid is $(0, 2)$. (Later we will see that the formula remains valid at the right end point $x = 2$, but for now we will consider only the open interval.)

Now we can find $P_n(x)$ and $R_n(x)$. For $x \in (0, 2)$,

$$\begin{aligned} f(x) &= P_n(x) + R_n(x) \\ &= f(1) + f'(1)(x-1) + \frac{f''(1)}{2!}(x-1)^2 + \frac{f'''(1)}{3!}(x-1)^3 \\ &\quad + \frac{f^{(4)}(1)}{4!}(x-1)^4 + \cdots + \frac{f^{(n)}(1)}{n!}(x-1)^n + \frac{f^{(n+1)}(c)}{(n+1)!}(x-1)^{n+1} \\ &= 0 + 1(x-1) + \frac{-1}{2!}(x-1)^2 + \frac{2}{3!}(x-1)^3 + \frac{-3!}{4!}(x-1)^4 + \cdots \\ &\quad + \frac{(-1)^{n-1}(n-1)!}{n!}(x-1)^n + \frac{(-1)^n n!}{(n+1)!c^{n+1}}(x-1)^{n+1} \end{aligned}$$

or

$$\begin{aligned} f(x) &= (x-1) - \frac{(x-1)^2}{2} + \frac{(x-1)^3}{3} - \frac{(x-1)^4}{4} + \cdots \\ &\quad + \frac{(-1)^{n-1}(x-1)^n}{n} + R_n(x) \end{aligned}$$

where

$$R_n(x) = \frac{(-1)^n(x-1)^{n+1}}{(n+1)c^{n+1}}, \qquad c \text{ between } 1 \text{ and } x$$ ■

EXAMPLE 9.11 Use the results of Example 9.10 here.

(a) Find $P_5(1.2)$ and an upper bound on the error in using this to approximate ln 1.2.

(b) Find the smallest integer n that will guarantee that $P_n(1.2)$ estimates ln 1.2 with eight decimal place accuracy.

Solution (a) From Example 9.10,

$$P_5(x) = (x-1) - \frac{(x-1)^2}{2} + \frac{(x-1)^3}{3} - \frac{(x-1)^4}{4} + \frac{(x-1)^5}{5}$$

and

$$R_5(x) = \frac{(-1)^5(x-1)^6}{6c^6}, \qquad c \text{ between } 1 \text{ and } x$$

So

$$\begin{aligned} P_5(1.2) &= 0.2 - \frac{(0.2)^2}{2} + \frac{(0.2)^3}{3} - \frac{(0.2)^4}{4} + \frac{(0.2)^5}{5} \\ &\approx 0.182331 \end{aligned}$$

The error is given by $R_5(1.2)$. Since $1 < c < 1.2$, we have

$$|R_5(1.2)| = \left|\frac{(0.2)^6}{6c^6}\right| = \frac{0.00064}{6c^6} \le \frac{0.00064}{6} \approx 0.000017$$

So the answer is correct to at least four decimal places.

(b) From Example 9.10,

$$R_n(x) = \frac{(-1)^n(x-1)^{n+1}}{(n+1)c^{n+1}}, \qquad c \text{ between 1 and } x$$

With $x = 1.2$, an upper bound on $|R_n(1.2)|$ can be obtained by replacing c by 1:

$$|R_n(1.2)| \leq \frac{(0.2)^{n+1}}{n+1}$$

To obtain eight decimal places of accuracy, we find the smallest integer n such that

$$\frac{(0.2)^{n+1}}{n+1} \leq 5 \times 10^{-9}$$

By trial and error, using a calculator, we find that $n + 1 = 11$, or $n = 10$ is the smallest value that works. So $P_{10}(1.2)$ estimates ln 1.2 with at least eight decimal places of accuracy. ■

EXAMPLE 9.12 Find Taylor's formula with remainder for arbitrary n for the function $f(x) = \cos x$ about $x = 0$.

Solution We calculate the first n derivatives of $f(x)$ and evaluate them at $x = 0$:

$$\begin{aligned}
f(x) &= \cos x & f(0) &= 1 \\
f'(x) &= -\sin x & f'(0) &= 0 \\
f''(x) &= -\cos x & f''(0) &= -1 \\
f'''(x) &= \sin x & f'''(0) &= 0 \\
f^{(4)}(x) &= \cos x & f^{(4)}(0) &= 1 \\
&\vdots & &\vdots
\end{aligned}$$

The derivatives divide into two types, depending on whether n is even or odd. For n even, we can write $n = 2k$, and for n odd, $n = 2k + 1$. Thus,

$$f^{(n)}(x) = \begin{cases} f^{(2k)}(x) = (-1)^k \cos x & \text{if } n = 2k \\ f^{(2k+1)}(x) = (-1)^{k+1} \sin x & \text{if } n = 2k+1 \end{cases}$$

So

$$f^{(n)}(0) = \begin{cases} (-1)^k & \text{if } n = 2k \\ 0 & \text{if } n = 2k+1 \end{cases}$$

It follows that only even powers appear in the expansion of $P_n(x)$. For $n = 2k$, we have

$$\begin{aligned}
P_{2k}(x) &= f(0) + f'(0)x + \frac{f''(0)}{2!}x^2 + \cdots + \frac{f^{(2k)}(0)}{(2k)!}x^{2k} \\
&= 1 - \frac{x^2}{2!} + \frac{x^4}{4!} - \frac{x^6}{6!} + \cdots + \frac{(-1)^k}{(2k)!}x^{2k}
\end{aligned}$$

Observe that $P_{2k+1}(x) = P_{2k}(x)$, since the final term of $P_{2k+1}(x)$ is 0. It follows that $R_{2k+1}(x) = R_{2k}(x)$, and this common value is

$$R_{2k+1}(x) = \frac{f^{(2k+2)}(c)}{(2k+2)!}x^{2k+2} = \frac{(-1)^{k+1}\cos c}{(2k+2)!}x^{2k+2}, \qquad c \text{ between 0 and } x$$

Finally then we have Taylor's formula:

$$\cos x = 1 - \frac{x^2}{2!} + \frac{x^4}{4!} - \frac{x^6}{6!} + \cdots + \frac{(-1)^k}{(2k)!} x^{2k} + \frac{(-1)^{k+1} \cos c}{(2k+2)!} x^{2k+2}$$

for $k = 0, 1, 2, \ldots$. This is valid on $(-\infty, \infty)$ since derivatives of $\cos x$ of all orders exist everywhere. ■

We complete this section by providing two proofs we have deferred until now.

Proof of Theorem 9.3 Let N be a neighborhood of a throughout which $f^{(n+1)}(x)$ exists (where $n \geq 0$), and let x be any point in N distinct from a. Throughout the proof we will hold x fixed. For definiteness, we consider the case $x > a$. The proof for $x < a$ is almost identical. We define a new function $F(t)$ that satisfies Rolle's theorem on the interval $[a, x]$. This requires a bit of ingenuity, but as we shall show, the following will work:

$$F(t) = f(x) - \left[f(t) + f'(t)(x-t) + \frac{f''(t)}{2!}(x-t)^2 + \frac{f'''(t)}{3!}(x-t)^3 + \cdots + \frac{f^{(n)}(t)}{n!}(x-t)^n + K(x-t)^{n+1} \right] \tag{9.10}$$

where K is independent of t and is yet to be determined. Let us check the conditions of Rolle's theorem.

1. F is continuous on $[a, x]$ and differentiable on (a, x). This is true, since we are given that $f^{(n+1)}$ exists in N and hence on $[a, x]$, and the existence of $f^{(n+1)}$ implies the continuity of all lower ordered derivatives. Because F involves products and sums of differentiable functions, F itself is differentiable and hence also continuous on $[a, x]$.
2. $F(a) = F(x) = 0$. Direct substitution shows that $F(x) = 0$. We will force $F(a)$ to be 0 by choosing K properly. On substituting $t = a$, we get

$$\begin{aligned} F(a) &= f(x) - \left[f(a) + f'(a)(x-a) + \frac{f''(a)}{2!}(x-a)^2 + \frac{f'''(a)}{3!}(x-a)^3 + \cdots + \frac{f^{(n)}(a)}{n!}(x-a) + K(x-a)^{n+1} \right] \\ &= f(x) - [P_n(x) + K(x-a)^{n+1}] \end{aligned}$$

where $P_n(x)$ is the nth Taylor polynomial for f about a. Setting $F(a) = 0$ and solving for K, we get

$$K = \frac{f(x) - P_n(x)}{(x-a)^{n+1}} = \frac{R_n(x)}{(x-a)^{n+1}}$$

With this choice of K, the conditions of Rolle's theorem are all satisfied.

The conclusion of Rolle's theorem must hold; namely, there is a number c in (a, x) such that $F'(c) = 0$. When we substitute the value of K we found into equation (9.10) and differentiate, we get a telescoping effect, with many terms canceling out. For example, the first few terms of $F'(t)$ are

$$F'(t) = -f'(t) + f'(t) - f''(t)(x-t) + f''(t)(x-t) - \frac{f'''(t)}{2!}(x-t)^2 \cdots$$

(You will be asked to calculate F' completely in the exercises.) The final result is

$$F'(t) = -\frac{f^{(n+1)}(t)}{n!}(x-t)^n + (n+1)\frac{R_n(x)}{(x-a)^{n+1}}(x-t)^n$$

Rolle's theorem tells us that when $t = c$, the left-hand side is 0. Substituting $t = c$ and solving for $R_n(x)$ give

$$R_n(x) = \frac{f^{(n+1)}(c)}{n!(n+1)}(x-a)^{n+1} = \frac{f^{(n+1)}(c)}{(n+1)!}(x-a)^{n+1}$$

where c is between a and x. This completes the proof. ■

Proof of Theorem 9.2

The theorem states that if f' and f'' exist in an interval I containing a root r of the equation $f(x) = 0$ and each is of constant sign there, and if $|f'(x)| \geq m$ and $|f''(x)| \leq M$ in I, then

$$E_{n+1} \leq \frac{M}{2m} E_n^2$$

where $E_n = |r - x_n|$ and $E_{n+1} = |r - x_{n+1}|$. The relationship between x_{n+1} and x_n is

$$x_{n+1} = x_n - \frac{f(x_n)}{f'(x_n)}$$

We apply Taylor's formula with $n = 1$ for f about x_n, using t as the variable instead of x, to avoid confusion with x_n:

$$f(t) = f(x_n) + f'(x_n)(t - x_n) + R_2(t)$$

where

$$R_2(t) = \frac{f''(c)}{2!}(t - x_n)^2$$

and c is between t and x_n. Now put $t = r$. Since $f(r) = 0$, we get

$$0 = f(x_n) + f'(x_n)(r - x_n) + R_2(r)$$

Since $f'(x_n) \neq 0$, we can divide by this to get

$$-\frac{f(x_n)}{f'(x_n)} = r - x_n + \frac{R_2(r)}{f'(x_n)}$$

So we have

$$\begin{aligned} x_{n+1} - r &= \left[x_n - \frac{f(x_n)}{f'(x_n)}\right] - r = x_n - r - \frac{f(x_n)}{f'(x_n)} \\ &= x_n - r + r - x_n + \frac{R_2(r)}{f'(x_n)} \\ &= \frac{R_2(r)}{f'(x_n)} \\ &= \frac{f''(c)}{2f'(x_n)}(r - x_n)^2 \end{aligned}$$

Thus,

$$|x_{n+1} - r| = \frac{|f''(c)|}{2|f'(x_n)|}|r - x_n|^2 \leq \frac{M}{2m}|r - x_n|^2$$

or, equivalently,

$$E_{n+1} \leq \frac{M}{2m}E_n^2$$

This completes the proof. ■

EXERCISE SET 9.4

A

In Exercises 1–6 find the Taylor polynomial for f about $x = 0$ of the specified degree.

1. $f(x) = e^{-x}$; $n = 5$
2. $f(x) = \sin x$; $n = 7$
3. $f(x) = 1/(x + 1)$; $n = 6$
4. $f(x) = \sqrt{x + 1}$; $n = 5$
5. $f(x) = \cosh x$; $n = 8$
6. $f(x) = \tan x$; $n = 5$

In Exercises 7–12 find the Taylor polynomial of degree n for f about the specified point a.

7. $f(x) = 1/x$; $a = 1$
8. $f(x) = e^{x-2}$; $a = 2$
9. $f(x) = \sin x$; $a = 0$
10. $f(x) = 1/(1 - x)^2$; $a = -1$
11. $f(x) = \ln(2 - x)$; $a = 1$
12. $f(x) = \sinh x$; $a = 0$

Exercises 13–18 refer to Exercises 7–12. In each case find $R_n(x)$ and give the neighborhood of a throughout which $f(x) = P_n(x) + R_n(x)$.

13. Exercise 7
14. Exercise 8
15. Exercise 9
16. Exercise 10
17. Exercise 11
18. Exercise 12

In Exercises 19–32 estimate the specified function value with P_n for the given value of n and an appropriate value of a. Find an upper bound on the error, and from this determine the number of decimal places of accuracy. (When appropriate you may use the results of preceding problems or examples.)

19. e^{-1}; $n = 5$
20. $\sqrt{1.2}$; $n = 5$
21. $\sin 1.5$; $n = 7$
22. $\ln 1.5$; $n = 6$
23. $\cos 2$; $n = 8$
24. $\sinh 2$; $n = 7$
25. $\sqrt{5}$; $n = 5$ (*Hint:* Take $a = 4$.)
26. $\ln \frac{4}{5}$; $n = 8$
27. e^3; $n = 12$
28. $\sqrt{e}$; $n = 4$
29. $\cosh 1$; $n = 6$
30. $\sin 32°$; $n = 5$ (*Hint:* Change to radians and use $a = \frac{\pi}{6}$.)
31. $\cos 58°$; $n = 4$ (*Hint:* Take $a = \frac{\pi}{3}$.)
32. $\sqrt[3]{7}$; $n = 4$
33. Prove that if $f^{(n)}(x)$ exists in a neighborhood of $x = a$, and we require of the polynomial

$$P_n(x) = a_0 + a_1(x - a) + a_2(x - a)^2 + \cdots + a_n(x - a)^n$$

that for $k = 0, 1, 2, \ldots, n$, $P_n^{(k)}(a) = f^{(k)}(a)$, then P_n is the nth Taylor polynomial for f about $x = a$; that is,

$$a_k = \frac{f^{(k)}(a)}{k!}, \qquad k = 0, 1, 2, \ldots, n$$

B

34. Use the result of Example 9.12 to find the degree of the Taylor polynomial for $\cos x$ about $x = 0$ that approximates $\cos 72°$ to six decimal places of accuracy. Find this approximation and compare it with the value given by a calculator.
35. Find $\sin 80°$ correct to five decimal places of accuracy using the appropriate Taylor polynomial about $x = 0$. (See Exercises 9 and 15.)
36. Find $\sinh 2$ to five decimal places of accuracy using the appropriate Taylor polynomial about $x = 0$. (See Exercises 12 and 18.)
37. Find $P_n(x)$ and $R_n(x)$ for $f(x) = \sqrt{x}$ about $x = 4$. In what interval is the result valid? Use the result to find $\sqrt{3}$ correct to four decimal places.
38. Prove that if f is twice differentiable in a neighborhood $N = (x - r, x + r)$ and $|\Delta x| < r$, the error

$|\Delta y - dy|$ in estimating $\Delta y = f(x + \Delta x) - f(x)$ by $dy = f'(x)\,dx$ (where $dx = \Delta x$) satisfies

$$|\Delta y - dy| \le \tfrac{M}{2}(\Delta x)^2$$

provided $|f''(t)| \le M$ for all t in $[x, x + \Delta x]$ (or in $[x + \Delta x, x]$ if $\Delta x < 0$).

39. Let $f(x) = (1 + x)^\alpha$, where α is real.
(a) Find $P_n(x)$ and $R_n(x)$ for f about $x = 0$.
(b) Show that if $\alpha = m$, where m is a positive integer, then $f(x) = P_m(x)$ for all x.
(c) If α is not a positive integer, show that $f(x) = P_n(x) + R_n(x)$ for $-1 < x < 1$.
(d) Use the result of part a to estimate $\sqrt[3]{1.5}$ taking $n = 4$. Find an upper bound on the error.

40. (a) Show that if $f(x)$ is a polynomial of degree n, then the Taylor polynomial $P_n(x)$ for f about $x = a$ is identical to $f(x)$.
(b) Express the polynomial

$$f(x) = 2x^5 - 3x^4 + 7x^3 - 8x^2 + 2$$

in powers of $x - 1$.

41. For $-1 < x < 1$, let

$$f(x) = \ln\frac{1 + x}{1 - x}$$

(a) Show that if u is any positive number, there is an x in $(-1, 1)$ such that $f(x) = \ln u$.
(b) Find $P_n(x)$ and $R_n(x)$ for f about $x = 0$. (*Hint:* Simplify f before taking derivatives.)
(c) Use parts a and b to find ln 2 correct to three decimal places.

42. (a) In Exercise 41 part a let $u = (N + 1)/N$, where N is a positive integer. What is x so that $f(x) = \ln u$?
(b) Use the result of part a, together with Exercise 41, to show that

$$\ln(N + 1) = \ln N + 2\sum_{k=1}^{n} \frac{1}{(2k - 1)(2N + 1)^{2k-1}} + R_{2n}\left(\frac{1}{2N + 1}\right)$$

and that

$$\left|R_{2n}\left(\frac{1}{2N + 1}\right)\right| \le \frac{2}{(2n + 1)(2N)^{2n+1}}$$

(Note that $R_{2n-1}(x) = R_{2n}(x)$, since the Taylor polynomial consists of odd powers only.)
(c) Find ln 3 correct to three decimal places making use of part b and Exercise 41 part c.

43. For $F(t)$ defined by equation (9.9), prove that

$$F'(t) = \frac{-f^{(n+1)}(t)}{n!}(x - t)^n + (n + 1)\frac{R_n(x)}{(x - a)^{n+1}}(x - t)^n$$

C

44. Use the computer to find e correct to ten decimal places as follows. For the Taylor polynomial $P_n(x)$ of $f(x) = e^x$ about $x = 0$, find an upper bound on the error $|R_n(1)|$. Have the computer evaluate this upper bound for $n = 1, 2, \ldots$ until a value of n is found for which the bound is less than 5×10^{-11}. For this n, have the computer evaluate $P_n(1)$.

45. Follow a procedure similar to that in Exercise 44 to obtain cos 5° correct to ten decimal places.

46. Use the results of Exercise 42 to have the computer calculate ln N for $N = 2, 3, \ldots, 10$, correct to seven decimal places. Compare the results with those given by a calculator.

9.5 SUPPLEMENTARY EXERCISES

In Exercises 1–4 approximate the integral by both the trapezoidal rule and Simpson's rule for the given value of n, and find an upper bound on the error in each case. Compare with the exact value of the integral.

1. $\int_0^2 \sin^2 \frac{\pi x}{2}\,dx;\ n = 8$

2. $\int_0^1 \frac{2x}{e^x}\,dx;\ n = 10$

3. $\int_{-2}^0 \frac{x}{x + 4}\,dx;\ n = 4$

4. $\int_3^5 \frac{2\ln\sqrt{x - 1}}{x}\,dx;\ n = 8$

In Exercises 5 and 6 find n such that the error in estimating the integral by the trapezoidal rule does not exceed 10^{-4}. Repeat using Simpson's rule.

5. $\int_1^4 \frac{dx}{\sqrt{5-x}}$ **6.** $\int_0^3 e^{-x} \cos x\, dx$

In Exercises 7 and 8 approximate the integral using the trapezoidal rule and Simpson's rule with $n = 10$. Estimate the error in each case.

7. $\int_0^1 \frac{e^x}{x+1}\, dx$ **8.** $\int_0^2 \ln(\cosh x)\, dx$

9. Use the trapezoidal rule with $n = 10$ to estimate the volume of the solid formed by rotating the region under the graph of $y = \sqrt[3]{x^2+4}$ from $x = 0$ to $x = 2$ about the x-axis. Estimate the error.

10. The slope of a curve with an unknown equation is estimated at regular intervals as given in the table. Approximate the length of the curve from $x = 0$ to $x = 3$ using (a) the trapezoidal rule and (b) Simpson's rule.

x	0	0.5	1.0	1.5	2.0	2.5	3.0
$f'(x)$	−1.9	−1.4	0.5	1.2	2.1	2.4	1.6

In Exercises 11–14 approximate the specified roots of the equation using Newton's method, continuing until two successive approximations agree in the first five decimal places. Round the answer to four places.

11. The real root of $x^3 + x^2 + 2x - 3 = 0$

12. All real roots of $x^4 - 2x^2 + 5x - 8 = 0$

13. All real roots of $x^2 + e^{-x} = 4$

14. The nonzero real roots of $2 \tan^{-1} x = x$

15. Use Newton's method to find the maximum value of $f(x) = x \cos x$ on $(0, \frac{\pi}{2})$ correct to six decimal places.

16. Show that if $a \neq 0$, $\sqrt[3]{a}$ can be approximated by the sequence of iterations

$$x_{n+1} = \frac{2x_n^3 + a}{3x_n^2}$$

and this will converge so long as x_0 has the same sign as a.

17. What value of n will ensure that in using Newton's method x_n will approximate the root of $x^3 - 3x^2 + 2x - 4 = 0$ that lies in $[2, 3]$ to an accuracy of ten decimal places, if x_0 is taken as 2.8? (*Hint:* First show that the root lies between 2.7 and 2.8.)

In Exercises 18 and 19 show that a root r lies in the given interval and that the conditions of Theorem 9.1 are satisfied there. Then find a suitable x_0 that will ensure convergence.

18. $f(x) = \frac{3}{x-1} - x$ on $(2, 3)$

19. $f(x) = \cos x - \sqrt{x}$ on $(\frac{\pi}{6}, \frac{\pi}{3})$

20. Use Theorem 9.2 to find n such that the nth approximation by Newton's method to the solution of $\tan x = \sqrt{2x}$ in $(0, \frac{\pi}{2})$ is correct to eight decimal places. Find this value of x_n.

21. Find the Taylor polynomial about $x = 0$ of degree 5 for $f(x) = \ln(1 - x)$, and determine how accurately it approximates $f(x)$ for $|x| \leq 0.2$.

22. Compute cos 62° with six decimal place accuracy using Taylor's formula with an appropriate value of a.

23. Find the Taylor polynomial of degree 5 about $x = \frac{\pi}{4}$ for $f(x) = \tan x$.

24. Find the nth Taylor polynomial for $f(x) = \sin^2 x$ about $x = 0$. What is $R_n(x)$? (*Hint:* Use a trigonometric identity.)

25. Find the nth Taylor polynomial about $x = \frac{\pi}{4}$ for $f(x) = \sin x - \cos x$.
[*Hint:* Show that $f(x) = \sqrt{2} \sin(x - \frac{\pi}{4})$.]

26. Use a cubic polynomial to approximate $\sqrt[3]{25}$. Estimate the error.

27. Find $P_3(x)$ and $R_3(x)$ about $x = 1$ for $f(x) = \tan^{-1} x$.

28. Show that the approximation

$$\cos x \approx 1 - \frac{x^2}{2!} + \frac{x^4}{4!} - \frac{x^6}{6!} + \frac{x^8}{8!}$$

gives accuracy to at least seven decimal places for all x in $[-\frac{\pi}{4}, \frac{\pi}{4}]$.

29. Let $P_n(x)$ be the nth Taylor polynomial about $x = 0$ for $f(x) = e^x$. Make a table showing e^x (by calculator), $P_1(x)$, $P_2(x)$, $P_3(x)$, and $P_4(x)$ for each of the values $x = \pm 0.25, \pm 0.50, \pm 0.75$, and ± 1.

30. Write out Taylor's formula with remainder with $a = 0$ for $f(x) = xe^{-x}$.

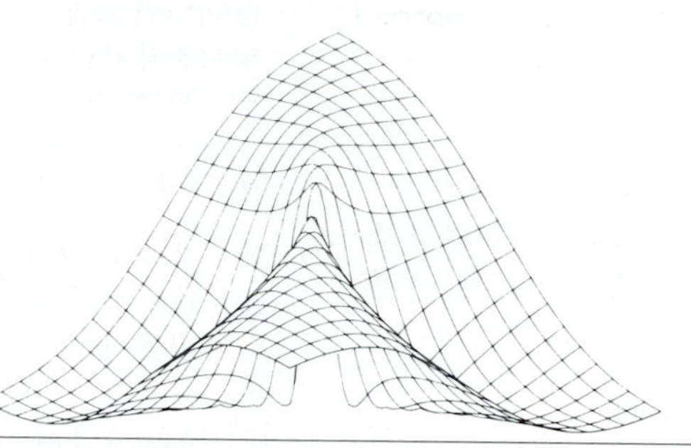

CHAPTER 10

INDETERMINATE FORMS AND IMPROPER INTEGRALS

10.1 L'HÔPITAL'S RULE

In our work with limits we have often encountered problems of the type $\lim_{x\to a}\frac{f(x)}{g(x)}$ in which $\lim_{x\to a} f(x) = 0$ and $\lim_{x\to a} g(x) = 0$. For example,

$$\lim_{x\to 2}\frac{x^2-4}{x-2} \qquad \lim_{x\to 4}\frac{\sqrt{x}-2}{x-4} \qquad \lim_{x\to 0}\frac{\sin x}{x}$$

are all of this type. Each of these is "indeterminate" initially in the sense that it cannot be determined from the limits of the numerator and denominator individually whether the limit exists or, if so, what its value is. In Chapter 2 we showed how to find these limits by using algebraic techniques on the first two (factoring and canceling a common factor on the first and rationalizing the numerator on the second) and a geometric procedure on the third. It should be noted that by its very definition, the derivative of a function at a is of the type we are discussing, since

$$f'(a) = \lim_{x\to a}\frac{f(x)-f(a)}{x-a}$$

and both numerator and denominator approach 0.

There are many other examples of this type for which we have no immediate algebraic or geometric technique for determining the limit. One such example is

$$\lim_{x\to 0}\frac{e^x-1}{x}$$

and we will soon see many more. A closely related type of limit problem is $\lim_{x\to a}\frac{f(x)}{g(x)}$, where both $f(x)$ and $g(x)$ become infinite as $x \to a$. For example,

$$\lim_{x\to +\infty}\frac{x^3-2x+3}{2x^3+4}$$

is of this type. Remember that we handled this by first dividing numerator and denominator by x^3 (getting the limit $\frac{1}{2}$). But this approach will not always work. Consider, for example,

$$\lim_{x \to 0^+} \frac{\ln x}{x^{-1}}$$

Here $\ln x \to -\infty$ and $x^{-1} \to +\infty$ as $x \to 0^+$, and so the limit is initially indeterminate (since $\frac{-\infty}{+\infty}$ is not defined). It is not clear how to proceed in this case to find the limit, if it exists.

The object of this section is to present a technique that will enable us to find limits of the two types we have described for a wide class of functions. The result is quite remarkable, and it is easy to apply. First, we make more explicit what we mean by the two types of indeterminate forms we are discussing.

DEFINITION 10.1 INDETERMINATE FORMS OF TYPES $\frac{0}{0}$ and $\frac{\infty}{\infty}$

1. If $\lim_{x \to a} f(x) = 0$ and $\lim_{x \to a} g(x) = 0$, then $\frac{f(x)}{g(x)}$ is said to be an indeterminate form of type $\frac{0}{0}$ at a.
2. If $\lim_{x \to a} f(x) = \pm\infty$ and $\lim_{x \to a} g(x) = \pm\infty$, then $\frac{f(x)}{g(x)}$ is said to be an indeterminate form of type $\frac{\infty}{\infty}$ at a.

In each of these, $x \to a$ can be replaced by any of the following: $x \to a^+$, $x \to a^-$, $x \to +\infty$, or $x \to -\infty$. ■

We are now ready to state the main theorem. It is attributed to G. F. L'Hôpital (1661–1704), who incidentally published the first calculus book. The result was actually discovered by his teacher, Johann Bernoulli (1667–1748), who communicated it to L'Hôpital in 1694. The result is universally known as *L'Hôpital's rule.*

THEOREM 10.1 L'HÔPITAL'S RULE

Let $\frac{f(x)}{g(x)}$ be an indeterminate form of type $\frac{0}{0}$ or $\frac{\infty}{\infty}$ at a. Suppose that f' and g' exist at all points of some open interval I containing a, except possibly at a itself, and that $g'(x) \neq 0$ for $x \neq a$ and x in I. If

$$\lim_{x \to a} \frac{f'(x)}{g'(x)} = L$$

where L is either finite or $+\infty$ or $-\infty$, then also

$$\lim_{x \to a} \frac{f(x)}{g(x)} = L$$

The conclusion continues to hold true if a is an end point of I, or if a is replaced by $+\infty$ or $-\infty$, in which case $x \to a$ is to be replaced by $x \to a^+$, $x \to a^-$, $x \to +\infty$, or $x \to -\infty$, as appropriate. ■

Comments If the hypotheses of the theorem are satisfied, then

$$\lim_{x \to a} \frac{f(x)}{g(x)} = \lim_{x \to a} \frac{f'(x)}{g'(x)}$$

Do not make the mistake of writing

$$\frac{f(x)}{g(x)} = \frac{f'(x)}{g'(x)}$$

The two expressions have the same *limit*, but they are not equal in general. Also, do not confuse $\frac{f'(x)}{g'(x)}$, which is the quotient of the derivatives, with the derivative of the quotient. In other words, *do not* use the quotient rule.

Before taking up the proof of L'Hôpital's rule, we consider several examples.

EXAMPLE 10.1 Use L'Hôpital's rule to evaluate the limit

$$\lim_{x\to 0} \frac{\sec x - 1}{\tan x}$$

Solution Since $(\sec x - 1)/\tan x$ is indeterminate at 0, of type $\frac{0}{0}$, and both numerator and denominator are differentiable near 0, with the derivative of the denominator nonzero, the hypotheses of L'Hôpital's rule are all satisfied. So we have,

$$\lim_{x\to 0} \frac{\sec x - 1}{\tan x} \lim_{x\to 0} \frac{\sec x \tan x}{\sec^2 x} = \lim_{x\to 0} \frac{\tan x}{\sec x} = \frac{0}{1} = 0$$ ■

Note In subsequent examples we will not explicitly show that all the hypotheses of L'Hôpital's rule are satisfied. However, in all cases, you should make a mental check that they are satisfied.

EXAMPLE 10.2 Find the following limits or show that they fail to exist.

(a) $\displaystyle\lim_{x\to 0} \frac{e^x - 1}{x}$ (b) $\displaystyle\lim_{x\to 0^+} \frac{\ln x}{x^{-1}}$

Solution (a) This is of type $\frac{0}{0}$, and so by L'Hôpital's rule,

$$\lim_{x\to 0} \frac{e^x - 1}{x} = \lim_{x\to 0} \frac{e^x}{1} = 1$$

(b) As noted earlier, $\ln x/x^{-1}$ is an indeterminate form of type $\frac{\infty}{\infty}$, so we apply L'Hôpital's rule to get

$$\lim_{x\to 0^+} \frac{\ln x}{x^{-1}} = \lim_{x\to 0^+} \frac{\frac{1}{x}}{-x^{-2}} = \lim_{x\to 0^+} \frac{\frac{1}{x}}{-\frac{1}{x^2}} = \lim_{x\to 0^+} (-x) = 0$$

Note carefully that after applying L'Hôpital's rule, we simplified the fraction before proceeding. ■

EXAMPLE 10.3 Evaluate the following limits:

(a) $\displaystyle\lim_{x\to +\infty} \frac{x^3}{e^x}$ (b) $\displaystyle\lim_{x\to 0^-} \frac{\cos x - 1}{x^3}$

Solution (a) This is indeterminate of type $\frac{\infty}{\infty}$. Applying L'Hôpital's rule, we get

$$\lim_{x \to +\infty} \frac{x^3}{e^x} = \lim_{x \to +\infty} \frac{3x^2}{e^x}$$

Now $3x^2/e^x$ is still of the form $\frac{\infty}{\infty}$, so we use L'Hôpital's rule again, and continue doing so until we no longer have an indeterminate form. The complete solution is

$$\lim_{x \to +\infty} \frac{x^3}{e^x} = \lim_{x \to +\infty} \frac{3x^2}{e^x} = \lim_{x \to +\infty} \frac{6x}{e^x} = \lim_{x \to +\infty} \frac{6}{e^x} = 0$$

Notice that $6/e^x$ is not indeterminate, so we can evaluate the limit.

(b) This is of the form $\frac{0}{0}$. Again, we apply L'Hôpital's rule more than once:

$$\lim_{x \to 0^-} \frac{\cos x - 1}{x^3} = \lim_{x \to 0^-} \frac{-\sin x}{3x^2} = \lim_{x \to 0^-} \frac{-\cos x}{6x} = +\infty$$

The quotient $-\cos x/6x$ is not indeterminate, since $-\cos x$ approaches -1 and $6x$ approaches 0 through negative values. Thus the fraction becomes arbitrarily large through positive values. ■

As this example illustrates, L'Hôpital's rule may be applied repeatedly so long as the quotient at each stage is of the form $\frac{0}{0}$ or $\frac{\infty}{\infty}$. When a stage is reached where the quotient no longer has one of these forms, the limit can be determined or be shown not to exist. *Do not* apply the rule again.*

A Word of Caution Do not make the mistake of writing either $\frac{1}{\infty} = 0$ or $\frac{1}{0} = \infty$. Both of these are meaningless, since ∞ is not a number and division by 0 is not permitted. For example, *do not write*

$$\lim_{x \to +\infty} \frac{1}{x^2} = \frac{1}{+\infty} = 0 \quad \text{or} \quad \lim_{x \to 0} \frac{1}{x^2} = \frac{1}{0} = +\infty$$

Instead, write these simply as

$$\lim_{x \to +\infty} \frac{1}{x^2} = 0 \quad \text{and} \quad \lim_{x \to 0} \frac{1}{x^2} = +\infty$$

Before considering the proof of L'Hôpital's rule, we need a preliminary lemma, known as *Cauchy's mean value theorem*, after the famous French mathematician Augustin Cauchy (1789–1857).

THEOREM 10.2 CAUCHY'S MEAN VALUE THEOREM

If f and g are continuous on $[a, b]$ and differentiable on (a, b), with $g'(x) \neq 0$ for all x in (a, b), then there exists a number c in (a, b) such that

$$\frac{f(b) - f(a)}{g(b) - g(a)} = \frac{f'(c)}{g'(c)}$$

Proof Observe that if $g(x) = x$, the theorem is just a statement of the familiar mean value theorem, so this is a generalization of the mean value theorem. Since

* Actually the rule will work when $\frac{f(x)}{g(x)}$ has the form $\frac{k}{\infty}$ at a, even if k is finite. But there is no advantage to using the rule in this case.

$g'(x) \neq 0$ on (a, b), it follows that $g(a) \neq g(b)$, for if $g(a) = g(b)$, Rolle's theorem would imply that $g'(x) = 0$ for some x in (a, b), contrary to our hypothesis. (See Exercise 44.)

We define the new function

$$F(x) = [f(b) - f(a)][g(x) - g(a)] - [g(b) - g(a)][f(x) - f(a)]$$

The hypotheses on f and g imply that F is continuous on $[a, b]$ and differentiable in (a, b). Furthermore, $F(a) = 0$ and $F(b) = 0$, as you can easily verify. Therefore, by Rolle's theorem there is a number c in (a, b) for which $F'(c) = 0$; that is,

$$[f(b) - f(a)]g'(c) - [g(b) - g(a)]f'(c) = 0$$

If we divide by $g'(c)[g(b) - g(a)]$, which is nonzero, we get

$$\frac{f(b) - f(a)}{g(b) - g(a)} = \frac{f'(c)}{g'(c)}$$

and the theorem is proved. ■

We will prove L'Hôpital's rule only for the $\frac{0}{0}$ type of indeterminate form in which a is interior to I and the limit of $\frac{f'(x)}{g'(x)}$ is finite. The other cases for $\frac{0}{0}$ are similar. The proof for $\frac{\infty}{\infty}$ is more difficult, and we omit it entirely.

Partial Proof of Theorem 9.1

We assume that $\frac{f(x)}{g(x)}$ is indeterminate of type $\frac{0}{0}$ at a where a is finite, and that $f'(x)$ and $g'(x)$ exist throughout some open interval I containing a, except possibly at $x = a$. Furthermore, we assume $g'(x) \neq 0$ for x in I and $x \neq a$, and that $\lim_{x\to a} \frac{f'(x)}{g'(x)} = L$ where L is finite. We want to prove that $\lim_{x\to a} \frac{f(x)}{g(x)} = L$.

The functions f and g need not be defined at a, and if they are they may not be continuous there. To ensure that the functions we are working with *are* continuous at $x = a$, we define, for x in I,

$$F(x) = \begin{cases} f(x) & \text{if } x \neq a \\ 0 & \text{if } x = a \end{cases} \quad \text{and} \quad G(x) = \begin{cases} g(x) & \text{if } x \neq a \\ 0 & \text{if } x = a \end{cases}$$

Then, for $x \neq a$, $F' = f'$ and $G' = g'$, and so F' and G' exist on all of I except possibly at a. Also, since $\lim_{x\to a} F(x) = \lim_{x\to a} f(x) = 0$ and $\lim_{x\to a} G(x) = \lim_{x\to a} g(x) = 0$, F and G are continuous at a, and therefore on all of I, since they are differentiable elsewhere on I. Furthermore, $G'(x) \neq 0$ if $x \neq a$ and x is in I.

Now let x be any point in I for which $x > a$. Then the hypotheses of Cauchy's mean value theorem are satisfied for F and G on $[a, x]$. Thus, there is a number c in (a, x) such that

$$\frac{F(x) - F(a)}{G(x) - G(a)} = \frac{F'(c)}{G'(c)}$$

But $F(a) = G(a) = 0$, and since both x and c are different from a, this can be written in the form

$$\frac{f(x)}{g(x)} = \frac{f'(c)}{g'(c)}, \qquad a < c < x$$

As $x \to a^+$, we must also have $c \to a^+$, since $a < c < x$. Thus,

$$\lim_{x \to a^+} \frac{f(x)}{g(x)} = \lim_{c \to a^+} \frac{f'(c)}{g'(c)} = L$$

This completes the proof for $x > a$. If $x < a$, then the Cauchy mean value theorem is satisfied by F and G on $[x, a]$, and the proof is essentially the same, so that we get

$$\lim_{x \to a^-} \frac{f(x)}{g(x)} = \lim_{c \to a^-} \frac{f'(c)}{g'(c)} = L$$

Since the right-hand and left-hand limits of $\frac{f(x)}{g(x)}$ are both equal to L as $x \to a$, the limit exists and equals L. ■

EXERCISE SET 10.1

A

In Exercises 1–34 verify that L'Hôpital's rule is applicable, and use it to find the limit.

1. $\lim_{x \to 1} \frac{x^2 - 1}{x - 1}$
2. $\lim_{x \to 2} \frac{x^2 + x - 6}{x^2 - 4}$
3. $\lim_{x \to 1} \frac{\sqrt{x} - 1}{x - 1}$
4. $\lim_{x \to +\infty} \frac{2x^4 - 3x}{1 - x - x^4}$
5. $\lim_{x \to 0} \frac{\tan x}{x}$
6. $\lim_{x \to 0} \frac{1 - \cos x}{x^2}$
7. $\lim_{x \to 1} \frac{x - 5x^5 + 4x^6}{(x - 1)^2}$
8. $\lim_{x \to 0^+} \frac{x^{-1/2}}{\ln x}$
9. $\lim_{x \to \pi/2^-} \frac{\cot x}{\sin 2x}$
10. $\lim_{x \to -1} \frac{\ln x^2}{x + 1}$
11. $\lim_{x \to 0} \frac{x^3}{x - \sin x}$
12. $\lim_{x \to 1} \frac{x^2 + x - 2}{\sin \pi x}$
13. $\lim_{x \to 1^+} \frac{\sqrt{x - 1}}{\ln \sqrt{x}}$
14. $\lim_{x \to 0} \frac{\tan^{-1} x}{x}$
15. $\lim_{x \to 2} \frac{x - 2}{\sqrt{x + 2} - 2}$
16. $\lim_{x \to 0} \frac{x^2}{e^x - x - 1}$
17. $\lim_{x \to 0} \frac{x - \sin x}{\cos 2x - 1}$
18. $\lim_{x \to 0} \frac{\cos x - 1}{\cos 2x - 1}$
19. $\lim_{x \to 0^+} \frac{\csc x}{\ln x}$
20. $\lim_{x \to 0^+} \frac{\cot x}{\ln x}$
21. $\lim_{x \to 1^-} \frac{\tanh^{-1} x}{\ln(1 - x)}$
22. $\lim_{x \to 0} \frac{\cosh x - x^2 - 1}{\sinh^2 x}$
23. $\lim_{x \to 0} \frac{\cot x + x}{\csc x}$
24. $\lim_{x \to +\infty} \frac{e^{1/x} - \frac{1}{x} - 1}{\frac{1}{x^2}}$
25. $\lim_{x \to +\infty} \frac{e^{-x}}{\cot^{-1} x}$
26. $\lim_{x \to +\infty} \frac{1 + (\ln x)^3}{x \ln x}$
27. $\lim_{x \to 1^+} \frac{\ln(\ln x)}{(x - 1)^{-2}}$
28. $\lim_{x \to -1^-} \frac{x \ln|x|}{(x + 1)^2}$
29. $\lim_{x \to 0} \frac{1 - \cos x}{\sin x^2}$
30. $\lim_{x \to 0} \frac{x \sin x}{\sec x - 1}$

B

31. $\lim_{x \to +\infty} \frac{e^{\sqrt{x}}}{x^2}$
32. $\lim_{x \to 1^-} \frac{e^{1/(1-x)}}{\frac{1}{(1 - x)^2}}$
33. $\lim_{x \to -\infty} \frac{2^x}{x^{-3}}$ (*Hint*: First rewrite the fraction using reciprocals.)
34. $\lim_{x \to 0^-} \frac{e^{1/x}}{x}$ (See hint for Exercise 33.)
35. Show that if $\alpha > 0$ is any real number, then
$$\lim_{x \to +\infty} \frac{x^\alpha}{e^x} = 0$$
36. Show that if $\alpha > 0$ is any real number, then
$$\lim_{x \to +\infty} \frac{(\ln x)^\alpha}{x} = 0$$

In Exercises 37–40 first rewrite the function in the form of a quotient that is indeterminate of type $\frac{0}{0}$ or $\frac{\infty}{\infty}$ and then apply L'Hôpital's rule.

37. $\lim_{x \to 0^+} x \ln x$

38. $\lim_{x \to \infty} x^2 e^{-x}$

39. $\lim_{x \to 0} (\frac{1}{x} - \cot x)$

40. $\lim_{x \to \pi/2} (\sec x - \tan x)$

41. Prove L'Hôpital's rule for the $\frac{0}{0}$ case in which $a = +\infty$ by making the change of variable $x = \frac{1}{t}$, so that $\lim_{x \to +\infty} \frac{f(x)}{g(x)}$ becomes

$$\lim_{t \to 0^+} \frac{f(\frac{1}{t})}{g(\frac{1}{t})}$$

Then apply L'Hôpital's rule for $a = 0^+$ to this limit, and change back to x at the end.

42. In each of the following, find the limit without using L'Hôpital's rule. Then apply L'Hôpital's rule. How do you explain the results?

(a) $\lim_{x \to 0} \dfrac{x^2 \sin \dfrac{1}{x}}{\sin x}$ (b) $\lim_{x \to +\infty} \dfrac{x + \cos x}{x}$

43. If all the other hypotheses of L'Hôpital's rule are satisfied but $\lim_{x \to a} \frac{f'(x)}{g'(x)}$ does not exist, is it still possible for $\lim_{x \to a} \frac{f(x)}{g(x)}$ to exist? (*Hint:* See Exercise 42.)

44. Verify the statement made in the proof of Cauchy's mean value theorem that if f is continuous on $[a, b]$ and f' exists in (a, b) and is different from 0 there, then $f(a) \neq f(b)$.

10.2 OTHER INDETERMINATE FORMS

Limiting procedures often lead to forms other than $\frac{0}{0}$ or $\frac{\infty}{\infty}$ that are initially indeterminate. In many of these cases we can still use L'Hôpital's rule after rewriting the limit problem in an appropriate way. The trick is to rewrite it so as to be in either the $\frac{0}{0}$ or $\frac{\infty}{\infty}$ form, for L'Hôpital's rule is applicable *only* to these two types of indeterminate forms.

Type $0 \cdot \infty$

A product $f(x) \cdot g(x)$ is said to be indeterminate of type $0 \cdot \infty$ at a if $f(x) \to 0$ and $g(x) \to +\infty$ or $-\infty$ as $x \to a$. Again a can be either finite or infinite, and the limit may be from one side only. By writing $f(x) \cdot g(x)$ in one of the forms

$$\frac{f(x)}{1/g(x)} \quad \text{or} \quad \frac{g(x)}{1/f(x)}$$

we change the product into a quotient that is either of the form $\frac{0}{0}$ or $\frac{\infty}{\infty}$. Sometimes it is obvious which quotient to use, and sometimes it is not. In the latter case, try one way, and if it does not seem to be getting anywhere, try the other. The next example illustrates this.

EXAMPLE 10.4 Evaluate the following limits:

(a) $\lim_{x \to +\infty} x^2 e^{-x}$ (b) $\lim_{x \to 0^+} \sqrt{x} \ln x$

Solution (a) This is indeterminate of type $\infty \cdot 0$, which is equivalent to $0 \cdot \infty$. Because of the negative exponent, it seems natural to write

$$x^2 e^{-x} = \frac{x^2}{e^x}$$

and this is indeterminate of the form $\frac{\infty}{\infty}$ at $+\infty$. So we apply L'Hôpital's rule:

$$\lim_{x\to+\infty} x^2e^{-x} = \lim_{x\to+\infty} \frac{x^2}{e^x} = \lim_{x\to+\infty} \frac{2x}{e^x} = \lim_{x\to+\infty} \frac{2}{e^x} = 0$$

(b) This is of type $0 \cdot \infty$, since $\ln x \to -\infty$ as $x \to 0^+$. Generally speaking, when changing a product to a quotient, if $\ln x$ is involved, leave it alone and invert the other factor. If we do this, we get

$$\sqrt{x}\ln x = \frac{\ln x}{x^{-1/2}}$$

and this is of the form $\frac{\infty}{\infty}$. So we get

$$\lim_{x\to0^+} \sqrt{x}\ln x = \lim_{x\to0^+} \frac{\ln x}{x^{-1/2}} = \lim_{x\to0^+} \frac{\frac{1}{x}}{-\frac{1}{2x^{3/2}}} = \lim_{x\to0^+}(-2\sqrt{x}) = 0$$ ■

Type $\infty - \infty$

By this we mean $f(x) - g(x)$ where both $f(x)$ and $g(x)$ become infinite *of the same sign* (both $+\infty$ or else both $-\infty$) as $x \to a$. Note that $(+\infty) - (-\infty)$ and $(-\infty) - (+\infty)$ are not indeterminate. The following example illustrates how to proceed.

EXAMPLE 10.5 Evaluate the following limits:

(a) $\lim_{x\to0}\left(\frac{1}{x} - \csc x\right)$ (b) $\lim_{x\to0} \frac{1}{\ln(x+1)} - \frac{1}{x}$

Solution (a) If $x \to 0^+$, both $\frac{1}{x}$ and $\csc x$ become positively infinite, and if $x \to 0^-$, they both become negatively infinite. So this is type $\infty - \infty$. We make the difference into a quotient as follows:

$$\frac{1}{x} - \csc x = \frac{1}{x} - \frac{1}{\sin x} = \frac{\sin x - x}{x \sin x}$$

This is indeterminate of type $\frac{0}{0}$. So we apply L'Hôpital's rule to get

$$\lim_{x\to0}\left(\frac{1}{x} - \csc x\right) = \lim_{x\to0} \frac{\sin x - x}{x\sin x} = \lim_{x\to0} \frac{\cos x - 1}{x\cos x + \sin x}$$

$$= \lim_{x\to0} \frac{-\sin x}{-x\sin x + \cos x + \cos x} = 0$$

(b) This also is type $\infty - \infty$, since both functions become positively infinite as $x \to 0^+$ and negatively infinite as $x \to 0^-$. We write the difference as a single fraction and test to see whether it is indeterminate:

$$\frac{1}{\ln(x+1)} - \frac{1}{x} = \frac{x - \ln(x+1)}{x\ln(x+1)}$$

As $x \to 0$, both numerator and denominator approach 0, so we can apply L'Hôpital's rule:

$$\lim_{x\to 0}\left(\frac{1}{\ln(x+1)} - \frac{1}{x}\right) = \lim_{x\to 0}\frac{x-\ln(x+1)}{x\ln(x+1)}$$

$$= \lim_{x\to 0}\frac{1-\dfrac{1}{x+1}}{\dfrac{x}{x+1}+\ln(x+1)}$$

Multiply numerator and denominator by $x + 1$.

$$= \lim_{x\to 0}\frac{x+1-1}{x+(x+1)\ln(x+1)}$$

$$= \lim_{x\to 0}\frac{x}{x+(x+1)\ln(x+1)}$$

$\frac{0}{0}$, so apply L'Hôpital's rule again.

$$= \lim_{x\to 0}\frac{1}{1+\dfrac{x+1}{x+1}+\ln(x+1)}$$

$$= \frac{1}{2}$$

■

Types 0^0, 1^∞, ∞^0

Each of these stands for an expression of the form $f(x)^{g(x)}$, where $f(x)$ and $g(x)$ approach the limits indicated as $x \to a$. For example, we say that $(1-x)^{1/x}$ is indeterminate of the form 1^∞ as $x \to 0^+$. Here $f(x) = 1 - x$ and $g(x) = \frac{1}{x}$. Note carefully that $f(x)$ is not identically 1 but *approaches* 1. In fact, if $f(x)$ were identically 1, the expression $f(x)^{g(x)}$ would also be identically 1 for all x in the domain of g and hence would not be indeterminate. The important thing to understand is that each of the forms 0^0, 1^∞, and ∞^0 is a shorthand way of describing limiting behaviors of $f(x)$ and $g(x)$. It should also be noted that 0^∞ is not an indeterminate form. (See Exercise 51.)

The procedure for determining the limit, if it exists, is the same in all three types and can be outlined as follows:

1. Set $y = f(x)^{g(x)}$.
2. Take the natural logarithm of both sides, getting $\ln y = g(x)\ln f(x)$.
3. Apply the procedure for products of type $0 \cdot \infty$ (or $\infty \cdot 0$).
4. If $\lim_{x\to a}\ln y = L$, then $\lim_{x\to a} y = e^L$.

The last step is justified because e^u is continuous, so by the Corollary to Theorem 1.5,

$$\lim_{x\to a}\ln y = L \quad \text{implies} \quad \lim_{x\to a} e^{\ln y} = e^L$$

and $e^{\ln y} = y$.

EXAMPLE 10.6 Evaluate the following limits:

(a) $\lim_{x\to 0^+}(\sin x)^x$ (b) $\lim_{x\to 0}(1-x)^{1/x}$

Solution (a) $(\sin x)^x$ is of type 0^0 at $x = 0$. We follow the procedure outlined above. Let

$$y = (\sin x)^x$$

$$\ln y = x \ln \sin x = \frac{\ln \sin x}{x^{-1}} \qquad \left(\tfrac{\infty}{\infty}\right)$$

$$\lim_{x \to 0^+} \frac{\ln \sin x}{x^{-1}} = \lim_{x \to 0^+} \frac{\frac{\cos x}{\sin x}}{-x^{-2}}$$

$$= \lim_{x \to 0^+} \frac{-x^2 \cos x}{\sin x} \qquad \left(\tfrac{0}{0}\right)$$

$$= \lim_{x \to 0^+} \frac{x^2 \sin x - 2x \cos x}{\cos x}$$

$$= 0$$

Thus, $\lim_{x \to 0^+} \ln y = 0$, so $\lim_{x \to 0^+} y = e^0 = 1$; that is, $\lim_{x \to 0^+} (\sin x)^x = 1$.

(b) This is of type 1^∞. We proceed as above. Let

$$y = (1 - x)^{1/x}$$

$$\ln y = \frac{1}{x} \ln(1 - x) = \frac{\ln(1 - x)}{x}$$

This is indeterminate of type $\frac{0}{0}$, so we apply L'Hôpital's rule:

$$\lim_{x \to 0} \ln y = \lim_{x \to 0} \frac{\ln(1 - x)}{x} = \lim_{x \to 0} \frac{\frac{-1}{1 - x}}{1} = -1$$

Thus,

$$\lim_{x \to 0} (1 - x)^{1/x} = \lim_{x \to 0} y = e^{-1} = \frac{1}{e}$$

■

A Word of Caution It is tempting to stop after finding $\lim_{x \to a} \ln y$, since the result of applying L'Hôpital's rule has been obtained. You must resist this temptation and take the last step to get $\lim_{x \to a} y$.

EXERCISE SET 10.2

A

In Exercises 1–50 evaluate the limits using L'Hôpital's rule.

1. $\lim_{x \to +\infty} x^2 e^{-x}$
2. $\lim_{x \to 0} x \ln x^2$
3. $\lim_{x \to 0^+} x \ln \sin x$
4. $\lim_{x \to +\infty} x(e^{1/x} - 1)$
5. $\lim_{x \to 0} \left(\cot x - \frac{1}{x} \right)$
6. $\lim_{x \to 0} (\csc x - \cot x)$
7. $\lim_{x \to 1^-} \left(\frac{x}{1 - x} + \frac{1}{\ln x} \right)$
8. $\lim_{x \to 0} \left(\frac{1}{e^x - 1} - \frac{1}{x} \right)$
9. $\lim_{x \to 0^+} x^x$
10. $\lim_{x \to 0^+} x^{\tan x}$
11. $\lim_{x \to 0} (1 + \sin x)^{1/x}$
12. $\lim_{x \to \pi/2^-} (\tan x)^{\cos x}$
13. $\lim_{x \to 0^+} x^2 e^{1/x}$
14. $\lim_{x \to \pi^-} \left(\tan \frac{x}{2} - \sec \frac{x}{2} \right)$

15. $\lim_{x\to 0}(\sec x)^{\cot x}$

16. $\lim_{x\to+\infty}(x^2)^{1/x}$

17. $\lim_{x\to+\infty}(\sqrt{x^2+x}-x)$ (*Hint:* Let $x = 1/t$.)

18. $\lim_{x\to\pi/2^-}(1-2\cos x)^{\tan x}$

19. $\lim_{x\to 0^-}\left(1-\dfrac{3}{x}\right)^x$

20. $\lim_{x\to 0} x\cot x$

21. $\lim_{x\to-\infty}\left(1-\dfrac{2}{x^2}\right)^{x^2}$

22. $\lim_{x\to 1}\left(\dfrac{3}{x^2-1}-\dfrac{\sqrt{4x+5}}{x^2-1}\right)$

23. $\lim_{x\to 1^+}(\ln x)^{1-x}$

24. $\lim_{x\to+\infty} x^2\,\mathrm{sech}\, x$

25. $\lim_{x\to 0}(\mathrm{csch}\, x-\coth x)$

26. $\lim_{x\to 1} x^{1/(1-x)}$

27. $\lim_{x\to+\infty}\left(\dfrac{x+1}{x}\right)^{2x}$

28. $\lim_{x\to 0}(\cos x-2\sin x)^{\csc x}$

29. $\lim_{x\to 0^+}\left(\dfrac{1}{\ln(x+1)}-\dfrac{x+1}{x}\right)$

30. $\lim_{x\to 0^+}(1-\sqrt{x})^{1/\sqrt{x}}$

31. $\lim_{x\to+\infty}\left(\dfrac{x+1}{x-1}\right)^x$

32. $\lim_{x\to+\infty}[\ln(2x+1)-\ln(x+2)]$

33. $\lim_{x\to 0}(\sec x)^{\cot^2 x}$

34. $\lim_{x\to 0^+}\sqrt{x}(\ln x)^2$

35. $\lim_{x\to 0} x^{-1}\tan^{-1}(2x)$

36. $\lim_{x\to 0}(e^x+x)^{1/x}$

B

37. $\lim_{x\to 0}(\csc^2 x-x^{-2})$

38. $\lim_{x\to 1^-}(\ln x)\ln(1-x)$

39. $\lim_{x\to 0^+}\left(\dfrac{1}{x}-e^{1/x}\right)$

40. $\lim_{x\to 0^+}\left(\dfrac{2}{x}+\ln x\right)$

41. $\lim_{x\to 0^+} x(\ln x)^n$ (n a positive integer)

42. $\lim_{x\to+\infty} x^n e^{-x}$ (n a positive integer)

43. $\lim_{x\to 0^+} x^{x^x}$

44. $\lim_{x\to 0}\left(\dfrac{x}{\sin x}\right)^{2/x}$

45. $\lim_{x\to 0}\left(\dfrac{2-2\cos x}{x^2}\right)^{1/x}$

46. $\lim_{x\to 0} x^{-3}e^{-1/x^2}$

47. $\lim_{x\to 0} x^{-1}\int_0^x e^{-t^2}\,dt$

48. $\lim_{x\to 0} x^{-2}\int_0^x \dfrac{t\,dt}{1+t^3}$

49. $\lim_{x\to 0^+} x^\alpha(\ln x)^\beta;\ \alpha>0,\ \beta>0$

50. $\lim_{x\to+\infty} x^\alpha e^{-\beta x};\ \alpha>0,\ \beta>0$

51. The form 0^∞ at $x=a$ means $f(x)^{g(x)}$, where $f(x)\geq 0$, and as $x\to a$, $f(x)\to 0$, $g(x)\to\pm\infty$. Show that this is not an indeterminate form, and that its value is 0 if $g(x)\to+\infty$ and is $+\infty$ if $g(x)\to-\infty$. (*Hint:* Use logarithms.)

52. Let

$$f(x)=\begin{cases} e^{-1/x^2} & \text{if } x\neq 0\\ 0 & \text{if } x=0\end{cases}$$

Show that $f'(0)=0$ and draw the graph of f.

53. Let

$$f(x)=\begin{cases} x^{1/x} & \text{if } x>0\\ 0 & \text{if } x\leq 0\end{cases}$$

Show that $f'(0)=0$. Find all local maxima and minima, find the horizontal asymptote, and draw the graph of f.

54. Discuss completely and draw the graph of each of the following. Pay particular attention to horizontal asymptotes.
(a) $y=x^3e^{-x}$ (b) $y=(1/x^2)\ln x$

10.3 IMPROPER INTEGRALS; UNBOUNDED INTERVAL OF INTEGRATION

In the definition of $\int_a^b f(x)\,dx$ it was implicit that a and b represented real numbers. In other words, the integral was defined only over finite intervals $[a, b]$. In this section we extend the definition to unbounded intervals of the types $[a,\infty)$, $(-\infty, a]$, and $(-\infty,\infty)$. (Here we are using ∞ to mean $+\infty$.)

Suppose, for example, we want to know whether the area of the region under the graph of $f(x)=1/x^2$ for $x\geq 1$ is finite, and if so, its value. For

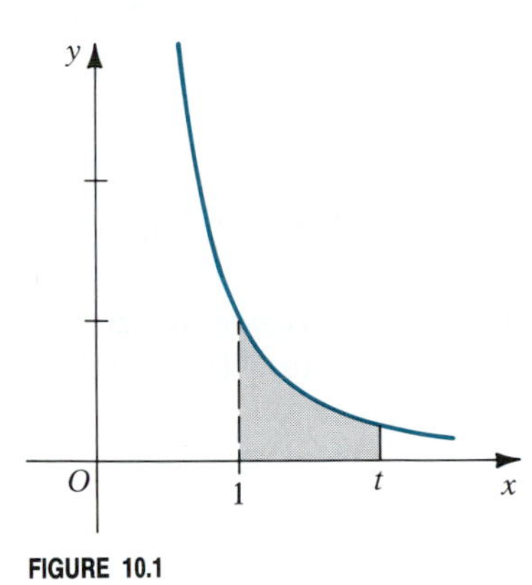

FIGURE 10.1

any real number $t > 1$, the area from 1 to t is finite, since f is continuous on $[1, t]$ (see Figure 10.1). Let us compute this area:

$$A_t = \int_1^t \frac{dx}{x^2} = -\frac{1}{x}\bigg]_1^t = -\frac{1}{t} + 1$$

Now we let $t \to \infty$ and see whether A_t approaches a finite limit:

$$\lim_{t\to\infty} A_t = \lim_{t\to\infty}\left(-\frac{1}{t} + 1\right) = 1$$

Thus, it seems reasonable to say that the area of the region under the graph of f over the unbounded interval $[1, \infty)$ is finite and equals 1. We denote the integral in question by

$$\int_1^\infty \frac{dx}{x^2}$$

What we have described by this example is really a definition of the meaning of $\int_a^\infty f(x)\,dx$. First we carry out the integration from a to t, and then we let $t \to +\infty$ and see whether the limit exists. The formal definition is now given.

DEFINITION 10.2 Let f be continuous on $[a, \infty)$. Then

$$\int_a^\infty f(x)\,dx = \lim_{t\to\infty} \int_a^t f(x)\,dx$$

provided the limit on the right exists. ■

The integral $\int_a^\infty f(x)\,dx$ is called an **improper integral.** When the limit in the definition exists, the improper integral is said to **converge.** Otherwise it is said to **diverge.**

We can also define improper integrals of the form $\int_{-\infty}^a f(x)\,dx$ and $\int_{-\infty}^\infty f(x)\,dx$ as follows.

DEFINITION 10.3 Let f be continuous on $(-\infty, a]$. Then

$$\int_{-\infty}^a f(x)\,dx = \lim_{t\to-\infty} \int_t^a f(x)\,dx$$

provided the limit on the right exists. ■

DEFINITION 10.4 Let f be continuous on $(-\infty, \infty)$. Then if a is any real number,

$$\int_{-\infty}^\infty f(x)\,dx = \int_{-\infty}^a f(x)\,dx + \int_a^\infty f(x)\,dx$$

provided each of the integrals on the right converges. ■

Comment It is important to note that in Definition 10.4 we require each integral on the right to converge. This means that each of the limits

$$\lim_{t_1\to-\infty} \int_{t_1}^a f(x)\,dx \quad \text{and} \quad \lim_{t_2\to\infty} \int_a^{t_2} f(x)\,dx$$

must exist (be a finite number) independent of the other. If either limit fails to exist, the improper integral $\int_{-\infty}^\infty f(x)\,dx$ diverges. This result should be contrasted with

$$\lim_{t \to \infty} \int_{-t}^{t} f(x)\,dx$$

where the limit may exist even though the two individual limits do not. (See Exercise 42.)

Note also that the convergence or divergence of an improper integral is unaffected if the integrand is multiplied by a nonzero constant (although if convergent, the value of the integral will be changed).

EXAMPLE 10.7 Determine the values of p for which

$$\int_{1}^{\infty} \frac{dx}{x^p}$$

converges and those for which it diverges.

Solution From Definition 10.2, if $p \neq 1$, then

$$\begin{aligned}\int_{1}^{\infty} \frac{dx}{x^p} = \lim_{t \to \infty} \int_{1}^{t} x^{-p}\,dp &= \lim_{t \to \infty} \left[\frac{x^{-p+1}}{-p+1}\right]_1^t \\ &= \lim_{t \to \infty} \frac{1}{1-p}\left[\frac{1}{t^{p-1}} - 1\right] \\ &= \frac{1}{1-p}\left[\lim_{t \to \infty} \frac{1}{t^{p-1}} - 1\right]\end{aligned}$$

Now the limit will be finite provided $p - 1 > 0$—that is, provided $p > 1$. It will be infinite if $p - 1 < 0$, or $p < 1$. When $p = 1$, we have

$$\int_{1}^{\infty} \frac{dx}{x} = \lim_{t \to \infty} \int_{1}^{t} \frac{dx}{x} = \lim_{t \to \infty} \ln x \Big]_1^t = \lim_{t \to \infty} \ln t = \infty$$

We can summarize the findings as follows:

$$\int_{1}^{\infty} \frac{dx}{x^p} \begin{cases} \text{converges} & \text{if } p > 1 \\ \text{diverges} & \text{if } p \leq 1 \end{cases}$$

■

Integrals of the type in the preceding example can often be used to determine the convergence or divergence of other integrals for which antiderivatives of the integrand are not readily available. The basis for this is given in the following theorem.

THEOREM 10.3 COMPARISON TEST FOR IMPROPER INTEGRALS

If, for all $x \geq a$, f and g are continuous and $0 \leq f(x) \leq g(x)$, then

$$\int_{a}^{\infty} f(x)\,dx \text{ converges if } \int_{a}^{\infty} g(x)\,dx \text{ converges}$$

and

$$\int_{a}^{\infty} g(x)\,dx \text{ diverges if } \int_{a}^{\infty} f(x)\,dx \text{ diverges}$$

■

A similar result holds for integrals on $(-\infty, a]$. We will not prove this theorem, but we prove an analogous result in the next chapter. Intuitively, the result seems plausible. From Figure 10.2, the area under f is smaller

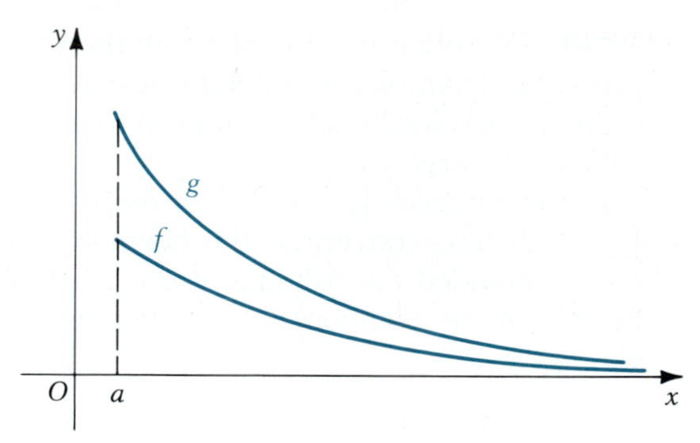

FIGURE 10.2

than the area under g. So if the area under g is finite, we would expect the area under f to be finite also. On the other hand, if the area under f is infinite, then the area under g, being greater, must also be infinite.

The next example illustrates how this theorem can be applied.

EXAMPLE 10.8 Test each of the following for convergence or divergence:

(a) $\displaystyle\int_1^\infty \frac{dx}{\sqrt{x^3+4}}$ (b) $\displaystyle\int_1^\infty \frac{dx}{\sqrt[3]{2x^2-1}}$

Solution (a) Since for all x on $[1, \infty)$,

$$0 \le \frac{1}{\sqrt{x^3+4}} < \frac{1}{\sqrt{x^3}} = \frac{1}{x^{3/2}}$$

and since by Example 10.7,

$$\int_1^\infty \frac{dx}{x^{3/2}} \qquad (p = \tfrac{3}{2} > 1)$$

converges, it follows by the comparison test (Theorem 10.3) that

$$\int_1^\infty \frac{dx}{\sqrt{x^3+4}}$$

also converges.

(b) We observe that for all $x \ge 1$,

$$\frac{1}{\sqrt[3]{2x^2-1}} > \frac{1}{\sqrt[3]{2x^2}} = \frac{1}{\sqrt[3]{2}} \cdot \frac{1}{x^{2/3}} > 0$$

Since by Example 10.7,

$$\int_1^\infty \frac{dx}{x^{2/3}}\, dx \qquad (p = \tfrac{2}{3} < 1)$$

diverges, and multiplying by the constant $1/\sqrt[3]{2}$ does not alter this divergence, it follows by the comparison test that

$$\int_1^\infty \frac{dx}{\sqrt[3]{2x^2-1}}$$

also diverges. ■

Remark Although we do not know the value of the integral in part a of the preceding example, it is helpful just to know that it converges. We could estimate its value by using Simpson's rule, say, on the interval $[1, t]$ for some sufficiently large t.

If the integral $\int_a^\infty f(x)\,dx$ converges, then for any $c > a$, the integral $\int_c^\infty f(x)\,dx$ also converges. Furthermore, this integral also converges when $c < a$, provided f is continuous on $[c, a]$. These facts enable us to be more flexible in the comparison test. For example, if in Example 10.8 we had wanted to test

$$\int_0^\infty \frac{dx}{\sqrt{x^3 + 4}}$$

we would have determined convergence on the interval $[1, \infty)$ as in the example. Then, since the integrand is continuous on $[0, 1]$, it would follow that the integral from 0 to ∞ converges.

The following examples further illustrate the ideas of this section.

EXAMPLE 10.9 Investigate the convergence or divergence of the following. If convergent, find the value.

(a) $\displaystyle\int_{-\infty}^\infty \frac{x}{1 + x^2}\,dx$ (b) $\displaystyle\int_0^\infty xe^{-x}\,dx$

Solution (a) According to Definition 10.4 we must split the integral into two separate integrals of the form $\int_{-\infty}^a + \int_a^\infty$, where a can be any real number. A convenient value is $a = 0$. So we have

$$\int_{-\infty}^\infty \frac{x}{1 + x^2}\,dx = \int_{-\infty}^0 \frac{x}{1 + x^2}\,dx + \int_0^\infty \frac{x}{1 + x^2}\,dx$$

and the integral on the left converges provided each integral on the right converges. To investigate the second integral, consider

$$\int_0^t \frac{x}{1 + x^2}\,dx = \frac{1}{2}\int_0^t \frac{2x\,dx}{1 + x^2} = \frac{1}{2}\ln(1 + x^2)\Big]_0^t$$

$$= \frac{1}{2}\ln(1 + t^2) = \ln\sqrt{1 + t^2}$$

Thus,

$$\int_0^\infty \frac{x}{1 + x^2}\,dx = \lim_{t\to\infty}\int_0^t \frac{x}{1 + x^2}\,dx = \lim_{t\to\infty}\ln\sqrt{1 + t^2} = \infty$$

Since this integral diverges, we conclude that

$$\int_{-\infty}^\infty \frac{x}{1 + x^2}\,dx$$

also diverges. (Note that if the integral from 0 to ∞ had been convergent, we would also have had to consider the integral from $-\infty$ to 0, but since the first was found to be divergent, we were finished.)

(b) First consider

$$\int_0^t xe^{-x}\,dx$$

and after finding its value, we will see whether it approaches a finite limit as $t \to \infty$. We integrate by parts, taking

$$\begin{array}{l|l} u = x & du = dx \\ dv = e^{-x}\,dx & v = -e^{-x} \end{array}$$

$$\int_0^t xe^{-x}\,dx = [-xe^{-x}]_0^t + \int_0^t e^{-x}\,dx = -te^{-t} - [e^{-x}]_0^t$$
$$= -te^{-t} - e^{-t} + 1$$

So

$$\int_0^\infty xe^{-x}\,dx = \lim_{t\to\infty}[-te^{-t} - e^{-t} + 1]$$
$$= \lim_{t\to\infty} te^{-t} - \lim_{t\to\infty} e^{-t} + 1$$

To get the first limit, we use L'Hôpital's rule, since te^{-t} is indeterminate:

$$\lim_{t\to\infty} te^{-t} = \lim_{t\to\infty} \frac{t}{e^t} = \lim_{t\to\infty} \frac{1}{e^t} = 0$$

Thus, $\int_0^\infty xe^{-x}\,dx$ converges to 1. ■

EXAMPLE 10.10 Find the constant k so that

$$f(x) = \frac{k}{1 + x^2}$$

will be a probability density function on $(-\infty, \infty)$. Then find $P(x \geq 1)$.

Solution Recall that the requirements for a probability density function are that it be nonnegative and that its integral over the interval in question be equal to 1. The nonnegativity requirement is satisfied if $k > 0$. We want to find k such that

$$\int_{-\infty}^\infty \frac{k\,dx}{1 + x^2} = 1$$

First we must show that the integral converges. We write

$$\int_{-\infty}^\infty \frac{k\,dx}{1 + x^2} = \int_{-\infty}^0 \frac{k\,dx}{1 + x^2} + \int_0^\infty \frac{k\,dx}{1 + x^2}$$

Because the integrand is an even function, it suffices to show that the integral from 0 to ∞ converges, since the two integrals are equal. (See Exercise 43.) We could show convergence by the comparison test, but since we also want its value, we write

$$\int_0^\infty \frac{k\,dx}{1 + x^2} = \lim_{t\to\infty} \int_0^t \frac{k\,dx}{1 + x^2} = \lim_{t\to\infty}[k \tan^{-1} x]_0^t$$
$$= \lim_{t\to\infty}(k \tan^{-1} t) = \frac{k\pi}{2}$$

Thus, the integral does converge, and its value is

$$\int_{-\infty}^\infty \frac{k\,dx}{1 + x^2} = k\pi$$

So we choose $k = \frac{1}{\pi}$.

For the probability that x is greater than or equal to 1, we have

$$P(x \geq 1) = \int_1^\infty f(x)\,dx = \frac{1}{\pi}\int_1^\infty \frac{dx}{1+x^2}$$

To evaluate the integral, we first integrate from 1 to t and then let $t \to \infty$:

$$\int_1^t \frac{dx}{1+x^2} = \tan^{-1} x\Big]_1^t = \tan^{-1} t - \frac{\pi}{4}$$

So

$$\frac{1}{\pi}\int_1^\infty \frac{dx}{1+x^2} = \frac{1}{\pi}\lim_{t\to\infty}\int_1^t \frac{dx}{1+x^2} = \frac{1}{\pi}\lim_{t\to\infty}\left[\tan^{-1} t - \frac{\pi}{4}\right]$$
$$= \frac{1}{\pi}\left[\frac{\pi}{2} - \frac{\pi}{4}\right] = \frac{1}{4}$$

■

EXAMPLE 10.11 Show that the area under the standardized normal probability density function

$$\phi(x) = \frac{1}{\sqrt{2\pi}}\,e^{-x^2/2}, \qquad -\infty < x < \infty$$

is finite. (*Note:* For this to be a probability density function, not only must the area be finite, but its value must be 1. Here we are concerned only with convergence. In Chapter 13 we will see how we can show that its value is 1.)

Solution Since ϕ is an even function, it suffices to show that the integral from 0 to ∞ is finite. We will use the comparison test, since $e^{-x^2/2}$ has no elementary antiderivative. For $x \geq 1$, we have

$$0 < e^{-x^2/2} = \frac{1}{e^{x^2/2}} \leq \frac{1}{e^{x/2}} = e^{-x/2}$$

since $x^2 \geq x$ and the exponential function is increasing. So if we can show that $\int_1^\infty e^{-x/2}\,dx$ converges, it will follow that $\int_1^\infty e^{-x^2/2}$ converges. For $t > 1$,

$$\int_1^t e^{-x/2}\,dx = -2\int_1^t e^{-x/2}\left(-\frac{1}{2}\,dt\right) = -2[e^{-x/2}]_1^t$$
$$= -2[e^{-t/2} - e^{-1/2}]$$

So

$$\int_1^\infty e^{-x/2}\,dx = \lim_{t\to\infty}[-2e^{-t/2} + 2e^{-1/2}] = \frac{2}{\sqrt{e}}$$

Since this is finite, it follows by the comparison test that $\int_1^\infty e^{-x^2/2}\,dx$ converges. Since $e^{-x^2/2}$ is continuous on $[0, 1]$, the integral over that interval exists, and so both

$$\int_0^\infty e^{-x^2/2}\,dx \quad \text{and} \quad \int_{-\infty}^\infty e^{-x^2/2}\,dx$$

exist. Finally, the convergence is unaltered if we multiply by the constant $1/\sqrt{2\pi}$. ■

EXAMPLE 10.12 Evaluate the integral

$$\int_1^\infty \frac{dx}{x^2 + 2x}$$

Solution We can see that the integral is convergent by comparison with

$$\int_1^\infty \frac{dx}{x^2} \qquad (p = 2 > 1)$$

which we have shown to be convergent. In order to find an antiderivative of $1/(x^2 + 2x)$ we first resolve this into partial fractions:

$$\frac{1}{x^2 + 2x} = \frac{1}{x(x+2)} = \frac{A}{x} + \frac{B}{x+2}$$

$$1 = A(x+2) + Bx$$

$$\underline{x = 0}: \quad 1 = 2A \qquad \underline{x = -2}: \quad 1 = -2B$$

$$A = \tfrac{1}{2} \qquad\qquad B = -\tfrac{1}{2}$$

So

$$\int \frac{dx}{x^2 + 2x} = \frac{1}{2}\int \frac{dx}{x} - \frac{1}{2}\int \frac{dx}{x+2} = \frac{1}{2}[\ln x - \ln(x+2)] = \frac{1}{2}\ln\frac{x}{x+2}$$

It is important to combine the two logarithms as we did, since otherwise we would obtain $\infty - \infty$ when we pass to the limit.

For $t > 1$ we therefore have

$$\int_1^t \frac{dx}{x^2 + 2x} = \frac{1}{2}\ln\frac{x}{x+2}\bigg]_1^t = \frac{1}{2}\ln\left[\frac{t}{t+2} - \ln\frac{1}{3}\right]$$

$$= \frac{1}{2}\ln\frac{t}{t+2} + \ln\sqrt{3}$$

Finally, we let $t \to \infty$:

$$\int_1^\infty \frac{dx}{x^2 + 2x} = \lim_{t\to\infty}\left[\frac{1}{2}\ln\frac{t}{t+2} + \ln\sqrt{3}\right]$$

Since

$$\lim_{t\to\infty}\frac{t}{t+2} = 1$$

and the logarithm function is continuous, it follows that

$$\lim_{t\to\infty}\frac{1}{2}\ln\frac{t}{t+2} = \frac{1}{2}\ln\left(\lim_{t\to\infty}\frac{t}{t+2}\right) = \frac{1}{2}\ln 1 = 0$$

Thus,

$$\int_1^\infty \frac{dx}{x^2 + 2x} = \ln\sqrt{3}$$ ■

EXAMPLE 10.13 Evaluate the integral

$$\int_1^{\infty} \frac{\sqrt{x^2+1}}{x^4}\,dx$$

or show that it diverges.

Solution The integral does converge, since on $[1, \infty)$,

$$\frac{\sqrt{x^2+1}}{x^4} \le \frac{\sqrt{x^2+x^2}}{x^4} = \frac{\sqrt{2}}{x^3}$$

and we know that

$$\int_1^{\infty} \frac{dx}{x^3}$$

converges. By definition,

$$\int_1^{\infty} \frac{\sqrt{x^2+1}}{x^4}\,dx = \lim_{t\to+\infty} \int_1^{t} \frac{\sqrt{x^2+1}}{x^4}\,dx$$

Now let $x = \tan\theta$ with $-\frac{\pi}{2} < \theta < \frac{\pi}{2}$. Then $dx = \sec^2\theta\,d\theta$. When $x = 1$, $\theta = \frac{\pi}{4}$, and when $x = t$, $\theta = \tan^{-1} t$. Denote $\tan^{-1} t$ by α. Then as $t \to +\infty$, $\alpha \to (\frac{\pi}{2})^-$. So

$$\begin{aligned}
\lim_{t\to+\infty} \int_1^{t} \frac{\sqrt{x^2+1}}{x^4}\,dx &= \lim_{\alpha\to\pi/2^-} \int_{\pi/4}^{\alpha} \frac{\sqrt{\tan^2\theta+1}\,\sec^2\theta\,d\theta}{\tan^4\theta} \\
&= \lim_{\alpha\to\pi/2^-} \int_{\pi/4}^{\alpha} \frac{\sec^3\theta\,d\theta}{\tan^4\theta} = \lim_{\alpha\to\pi/2^-} \int_{\pi/4}^{\alpha} \frac{\cos\theta}{\sin^4\theta}\,d\theta \\
&= \lim_{\alpha\to\pi/2^-} \left[-\frac{1}{3\sin^3\theta}\right]_{\pi/4}^{\alpha} = \lim_{\alpha\to\pi/2^-} \left[-\frac{1}{3}\left(\frac{1}{\sin^3\alpha} - 2\sqrt{2}\right)\right] \\
&= \frac{2\sqrt{2}-1}{3}
\end{aligned}$$

■

EXERCISE SET 10.3

A

In Exercises 1–26 evaluate the integral or show that it diverges.

1. $\int_1^{\infty} \frac{dx}{x^3}$

2. $\int_1^{\infty} \frac{dx}{\sqrt{x^3}}$

3. $\int_{-\infty}^{0} e^x\,dx$

4. $\int_0^{\infty} e^{-2x}\,dx$

5. $\int_{-\infty}^{\infty} \frac{dx}{x^2+4}$

6. $\int_2^{\infty} \frac{dx}{x\sqrt{x^2-1}}$

7. $\int_0^{\infty} \frac{e^{-x}\,dx}{1+e^{-x}}$

8. $\int_{-\infty}^{0} \frac{e^x\,dx}{e^x+1}$

9. $\int_0^{\infty} \frac{x\,dx}{(1+x^2)^2}$

10. $\int_2^{\infty} \frac{x\,dx}{\sqrt{x^2+1}}$

11. $\int_1^{\infty} x^2 e^{-x}\,dx$

12. $\int_1^{\infty} \frac{x+1}{(x^2+2x-2)^{3/2}}\,dx$

13. $\int_2^{\infty} \frac{dx}{x(\ln x)}$

14. $\int_2^{\infty} \frac{dx}{x(\ln x)^2}$

15. $\int_{-\infty}^{\infty} \frac{(x-1)\,dx}{(x^2-2x+4)^2}$

16. $\int_1^{\infty} \frac{dx}{x^2+x}$

17. $\int_2^{\infty} \frac{2\,dx}{x^2-1}$

18. $\int_2^{\infty} \frac{(x+1)\,dx}{x^3-x^2}$

19. $\int_3^{\infty} \frac{3}{x^2-x-2}\,dx$

20. $\int_0^{\infty} \tan^{-1} x\,dx$

21. $\int_1^{\infty} \frac{dx}{x\sqrt{1+x^2}}$

22. $\int_2^{\infty} \frac{\sqrt{x^2-4}}{x^3}\,dx$

23. $\int_{-\infty}^{-3} \frac{5x\,dx}{x^2 - x - 6}$

24. $\int_1^{\infty} \frac{dx}{\sqrt{x}(1+\sqrt{x})^2}$

25. $\int_3^{\infty} \frac{\sqrt{x^2 - 9}}{x^4}\,dx$

26. $\int_{-\infty}^{\infty} \frac{dx}{(x^2+1)^{3/2}}$

In Exercises 27–34 determine whether the integral is convergent or divergent, using the comparison test.

27. $\int_1^{\infty} \frac{dx}{x^3 + x + 1}$

28. $\int_2^{\infty} \frac{dx}{x^{2/3} - 1}$

29. $\int_3^{\infty} \frac{x\,dx}{\sqrt{x^3 - 1}}$

30. $\int_0^{\infty} \frac{x\,dx}{x^4 + 4}$

31. $\int_3^{\infty} \frac{dx}{x^2 \ln x}$

32. $\int_3^{\infty} \frac{(x+1)\,dx}{x^2 - x - 1}$

33. $\int_{-\infty}^{\infty} \frac{dx}{e^{x^2} + 1}$

34. $\int_1^{\infty} \frac{dx}{x^2 + \sin^2 x}$

In Exercises 35–38 find the value of k so that f will be a probability density function for a random variable X on the indicated interval, and then find the specified probability.

35. $f(x) = ke^{-x/3}$ on $[0, \infty)$; $P(X \geq 1)$

36. $f(x) = kxe^{-x/2}$ on $[0, \infty)$; $P(X > 2)$

37. $f(x) = kxe^{-x^2}$ on $[0, \infty)$; $P(1 < X \leq 2)$

38. $f(x) = k\,[e^x/(1+e^x)^2]$ on $[0, \infty)$; $P(X \geq 1)$

39. Find the area of the region under the graph of $y = 1/(1+x^2)$ on $[1, \infty)$.

40. Find the volume obtained by revolving the area of the region under the graph of $y = 1/(1+x)$ on $[0, \infty)$ about the x-axis.

41. Find the volume obtained by revolving the region under the graph of

$$y = \frac{1}{\sqrt{x}\ln x}, \qquad 2 \leq x < \infty$$

about the x-axis.

42. Show that

$$\lim_{t\to\infty} \int_{-t}^{t} \frac{x\,dx}{x^2+1} = 0$$

but that

$$\int_{-\infty}^{\infty} \frac{x\,dx}{x^2+1}$$

does not exist. Explain this result.

43. Prove each of the following.

(a) If f is even and $\int_0^{\infty} f(x)\,dx$ converges, then

$$\int_{-\infty}^{\infty} f(x)\,dx = 2\int_0^{\infty} f(x)\,dx$$

(b) If f is odd and $\int_0^{\infty} f(x)\,dx$ converges, then

$$\int_{-\infty}^{\infty} f(x)\,dx = 0$$

44. A random variable X has the **exponential distribution** if its probability density function is

$$f(x) = \frac{1}{\alpha}e^{-x/\alpha}, \qquad x \geq 0$$

where α is a positive constant. Find the mean and variance of X. (See Section 5.8.)

45. The lifetime X in hours of a certain type of lightbulb is a random variable with an exponential probability density function with $\alpha = 1000$. (See Exercise 44.) Find the probability that a lightbulb of this type selected at random will last at least 1500 hr.

46. In an experiment with mice in a maze it is found that a reasonable model of the probability density function of the time X in minutes for a mouse to go through the maze is given by

$$f(x) = \frac{a}{x^2}, \qquad x \geq a$$

where a is the minimum time required.

(a) Verify that f is a probability density function on $[a, \infty)$.

(b) Find the probability that a mouse selected at random will take longer than $2a$ minutes to go through the maze.

B

47. Determine the values of p for which the integral

$$\int_2^{\infty} \frac{dx}{x(\ln x)^p}$$

converges and those for which it diverges.

48. Evaluate

$$\int_0^{\infty} \frac{dx}{1+e^x}$$

49. Use the following outline to prove:

If $\int_a^{\infty} |f(x)|\,dx$ converges, then $\int_a^{\infty} f(x)\,dx$ converges.

(a) Explain why $0 \leq f(x) + |f(x)| \leq 2|f(x)|$.

(b) Why does it follow that $\int_a^{\infty} [f(x) + |f(x)|]\,dx$ converges?

(c) Write $f(x) = [f(x) + |f(x)|] - |f(x)|$, and explain why $\int_a^{\infty} f(x)\,dx$ must therefore converge.

50. Use the result to Exercise 49, together with the comparison test, to show that each of the following converges:

(a) $\int_0^\infty \frac{\sin x}{1+x^2}\,dx$ (b) $\int_0^\infty e^{-x}\cos x\,dx$

51. Let $f(t)$ be continuous on $0 \le t < \infty$. The **Laplace transform** of $f(t)$, denoted by $F(s)$, is defined by

$$F(s) = \int_0^\infty f(t)e^{-st}\,dt$$

provided the integral converges. Find the Laplace transform of each of the following:
(a) $f(t) = 1$ (b) $f(t) = t$
(c) $f(t) = e^t$ (d) $f(t) = \sin t$

52. Refer to Exercise 51. Prove that if there are positive constants M and k such that

$$|f(t)| \le Me^{kt} \quad \text{for } 0 \le t < \infty$$

then $F(s)$ exists for $s > k$.

53. Let $f(x) = \frac{1}{x}$ for $1 \le x < \infty$. Consider the surface S formed by revolving the graph of f about the x-axis. Show that (a) the volume enclosed by S is finite and (b) the area of the surface S is infinite. (Someone has described this phenomenon by saying that a vessel in the shape of S would hold a finite amount of paint, but it would take an infinite amount of paint to cover the surface of S!)

54. The normal probability density function is of the form

$$f(x) = \frac{1}{\beta\sqrt{2\pi}}\,e^{-(x-\alpha)^2/2\beta^2}, \qquad -\infty < x < \infty$$

Take it as given that $\int_{-\infty}^{\infty} f(x)\,dx = 1$, and prove that $\mu = \alpha$ and $\sigma = \beta$. (Refer to Section 5.8.)

55. The probability density function for the random variable X representing the length of life of certain plants and animals is often modeled by the **Weibull distribution:**

$$f(x) = \frac{m}{\alpha}\,x^{m-1}e^{-x^m/\alpha}, \qquad x > 0$$

Let $m = 2$ and $\alpha = 4$ and find the following:
(a) $P(X \ge 2)$ (b) $E[X^2]$

10.4 IMPROPER INTEGRALS; UNBOUNDED INTEGRAND

In the previous section we studied improper integrals in which the interval of integration was unbounded. Now we consider extending the definition of the integral to those in which the integrand becomes unbounded at one or more points in the interval of integration. In the definition of the Riemann integral it is an explicit requirement that the integrand be a bounded function. So if $f(x)$ is unbounded on $[a, b]$, then $\int_a^b f(x)\,dx$ does not exist in the ordinary sense. Whatever meaning we may assign to this will be something new, and so it is also called an improper integral.

FIGURE 10.3

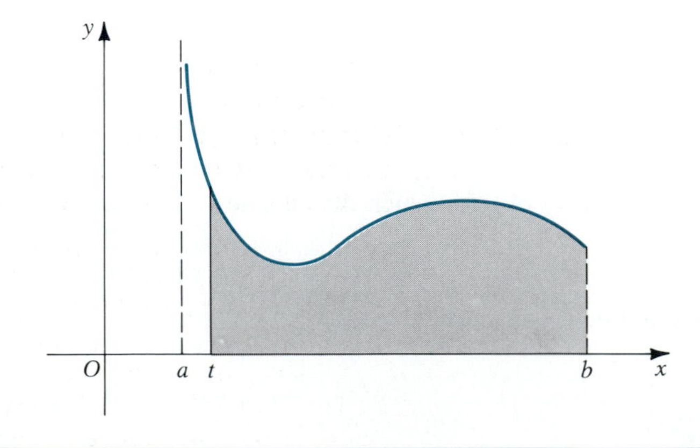

Consider first the case in which f is continuous on $(a, b]$ (where a and b are finite) and $\lim_{x\to a^+} f(x) = \pm\infty$. Figure 10.3 shows such a function. If t is in (a, b), the integral $\int_t^b f(x)\,dx$ is defined, since f is continuous on $[t, b]$. After evaluating this integral, we let $t \to a^+$ and see whether there is a finite limit. If so, this is the value we give to $\int_a^b f(x)\,dx$.

Consider, for example,

$$\int_0^1 \frac{dx}{\sqrt{x}}$$

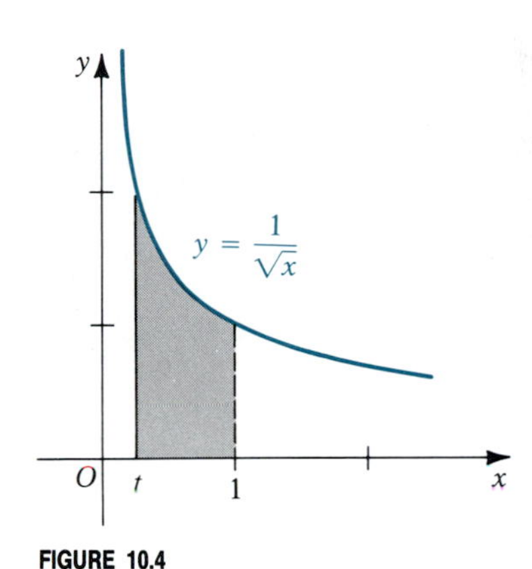

FIGURE 10.4

The graph of $f(x) = 1/\sqrt{x}$ is shown in Figure 10.4. As we said, we first evaluate the integral from t to 1 for $0 < t < 1$:

$$\int_t^1 \frac{dx}{\sqrt{x}} = 2\sqrt{x}\Big]_t^1 = 2 - 2\sqrt{t}$$

Now we let $t \to 0^+$:

$$\int_0^1 \frac{dx}{\sqrt{x}} = \lim_{t\to 0^+} (2 - 2\sqrt{t}) = 2$$

We can say then that the area under this curve equals 2.

If f is unbounded at the right end point, we integrate from a to t and then let $t \to b^-$ (see Figure 10.5). The formal definition of improper integrals of the type we are considering is given now.

FIGURE 10.5

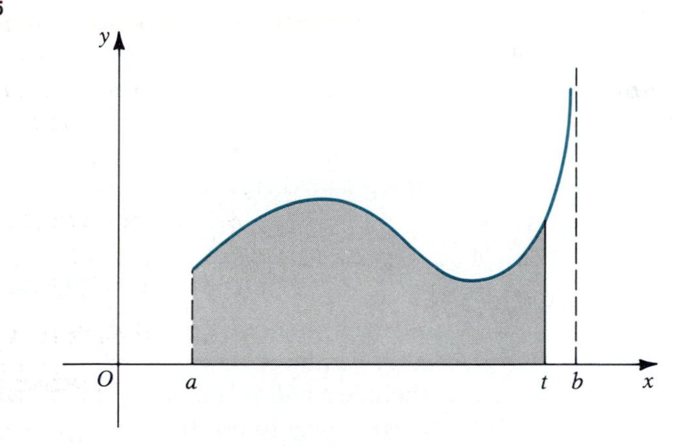

DEFINITION 10.5 **1.** If f is continuous on $(a, b]$ but is unbounded as $x \to a^+$, then

$$\int_a^b f(x)\,dx = \lim_{t\to a^+} \int_t^b f(x)\,dx$$

provided the limit exists.

2. If f is continuous on $[a, b)$ but is unbounded as $x \to b^-$, then

$$\int_a^b f(x)\,dx = \lim_{t\to b^-} \int_a^t f(x)\,dx$$

provided the limit exists. ■

As with the improper integrals with an unbounded interval of integration, when the limit in this definition does exist, we say the integral *converges*. Otherwise it *diverges*.

Note Definition 10.5 concerns integrals with unbounded integrands. If f is continuous and *bounded* on (a, b), then it can be shown that $\int_a^b f(x)\,dx$ always exists. For this reason we do not classify such integrals as improper. For example, each of the following exists:

$$\int_0^1 \frac{\sin x}{x}\,dx \qquad \int_0^1 \frac{x^2-1}{x-1}\,dx \qquad \int_0^1 \cos\frac{1}{x}\,dx$$

The point where the integrand becomes unbounded may occur interior to the interval $[a, b]$, and in this case we divide the integral into two improper integrals, as stated in the next definition.

DEFINITION 10.6 If f is continuous on $[a, c)$ and on $(c, b]$ but is unbounded on at least one of these intervals, then

$$\int_a^b f(x)\,dx = \int_a^c f(x)\,dx + \int_c^b f(x)\,dx$$

provided each of the integrals on the right converges. ■

Remark If there are finitely many points in $[a, b]$ in a neighborhood of which f is unbounded, then we divide the interval up so that each of these points is an end point of one of the subintervals.

The situation covered by Definition 10.6 can be the most troublesome of all, since failure to recognize the bad point c in the interval can lead to incorrect results. Consider, for example,

$$\int_0^2 \frac{dx}{(x-1)^2}$$

If we carelessly applied the fundamental theorem, we would get

$$\int_0^2 \frac{dx}{(x-1)^2} = -\frac{1}{x-1}\Big]_0^2 = -1 - 1 = -2$$

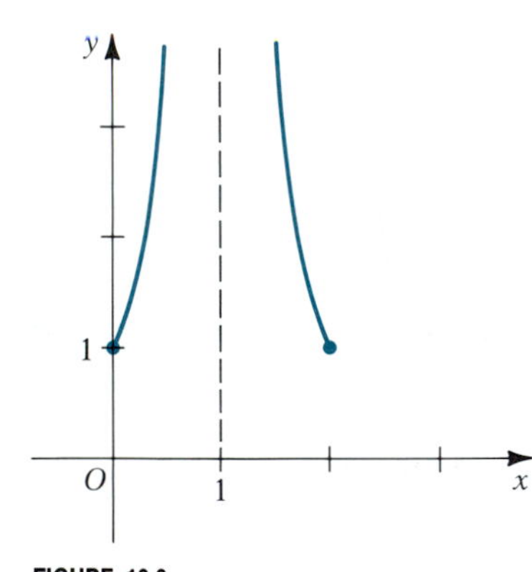

FIGURE 10.6

But this cannot be right, since the integrand is nonnegative, so that its integral, whatever its value, cannot be negative. The trouble, of course, is that the integrand is unbounded in a neighborhood of 1. (See Figure 10.6.) The correct way to do the problem, therefore, is to write

$$\int_0^2 \frac{dx}{(x-1)^2} = \int_0^1 \frac{dx}{(x-1)^2} + \int_1^2 \frac{dx}{(x-1)^2}$$

and then to apply Definition 10.5 to each integral on the right. For the first of these we integrate from 0 to t and then let $t \to 1^-$:

$$\int_0^t \frac{dx}{(x-1)^2} = -\frac{1}{x-1}\Big]_0^t = -\frac{1}{t-1} - 1$$

Thus,

$$\int_0^1 \frac{dx}{(x-1)^2} = \lim_{t\to 1^-}\left[-\frac{1}{t-1} - 1\right] = +\infty$$

Since this integral diverges, there is no need to go further. The original integral also diverges.

Remark Improper integrals with unbounded intervals of integration are easy to recognize, since $-\infty$ or ∞ appears as a limit of integration. The type we are considering here is more subtle. There is no clear warning that you are dealing with an unbounded integrand. You must be on the alert for this possibility. This is especially true when the bad point occurs interior to the interval, since you could get the wrong answer and never know it if you did not stop to think about its plausibility. When the bad point is an end point, if the integral diverges, you will see that something is wrong when you try to substitute limits. If it converges, you will normally get the right answer even if you failed to see that it was improper (but do not rely on this—always check to see whether it is improper).

Here are some more examples.

EXAMPLE 10.14 Evaluate the following integrals, or show that they are divergent:

(a) $\int_0^1 \frac{dx}{x}$ (b) $\int_0^1 \frac{dx}{\sqrt{1-x^2}}$

Solution (a) The integrand is unbounded as $x \to 0^+$, so by Definition 10.5 we have

$$\int_0^1 \frac{dx}{x} = \lim_{t\to 0^+} \int_t^1 \frac{dx}{x} = \lim_{t\to 0^+} [\ln x]_t^1 = \lim_{t\to 0^+} (-\ln t) = +\infty$$

So the integral diverges.

(b) The bad point is $x = 1$. So we have

$$\int_0^1 \frac{dx}{\sqrt{1-x^2}} = \lim_{t\to 1^-} \int_0^t \frac{dx}{\sqrt{1-x^2}} = \lim_{t\to 1^-} [\sin^{-1} x]_0^t$$
$$= \lim_{t\to 1^-} (\sin^{-1} t - 0) = \sin^{-1} 1 = \frac{\pi}{2}$$

■

EXAMPLE 10.15 Evaluate the following integrals, or show that they are divergent:

(a) $\int_1^e \frac{dx}{x\sqrt{\ln x}}$ (b) $\int_{-\pi}^{\pi} \frac{\sin x}{1-\cos x}\, dx$

Solution (a) Since $\ln 1 = 0$, the integrand is unbounded near 1. Since $d(\ln x) = \frac{1}{x}\, dx$, we have

$$\int_1^e \frac{dx}{x\sqrt{\ln x}} = \lim_{t\to 1^+} \int_t^e \frac{dx}{x\sqrt{\ln x}} = \lim_{t\to 1^+} [2\sqrt{\ln x}]_t^e$$
$$= \lim_{t\to 1^+} (2 - \sqrt{\ln t}) = 2$$

(b) We observe that the denominator is 0 at $x = 0$, and the numerator is 0 there also. So we use L'Hôpital's rule to find the limit:

$$\lim_{x\to 0} \frac{\sin x}{1-\cos x} = \lim_{x\to 0} \frac{\cos x}{\sin x}$$

When $x \to 0^+$ the limit is $+\infty$, and when $x \to 0^-$ the limit is $-\infty$. In either case, the integrand is unbounded. So we write

$$\int_{-\pi}^{\pi} \frac{\sin x}{1 - \cos x}\,dx = \int_{-\pi}^{0} \frac{\sin x}{1 - \cos x}\,dx + \int_{0}^{\pi} \frac{\sin x}{1 - \cos x}\,dx$$

For the last integral we have

$$\begin{aligned} \int_{0}^{\pi} \frac{\sin x}{1 - \cos x}\,dx &= \lim_{t\to 0^+} \int_{t}^{\pi} \frac{\sin x}{1 - \cos x}\,dx \\ &= \lim_{t\to 0^+} [\ln(1 - \cos x)]_t^{\pi} \quad \text{Note that } 1 - \cos x \geq 0. \\ &= \lim_{t\to 0^+} [\ln 2 - \ln(1 - \cos t)] = +\infty \end{aligned}$$

So this integral diverges, and hence the original integral also diverges. ■

Note If we calculate $\int_{-\pi}^{0} (\sin x)/(1 - \cos x)\,dx$, we will get $-\infty$, but the two integrals do not add to give 0. Each integral must be finite.

EXAMPLE 10.16 Determine the values of p for which the integral

$$\int_0^1 \frac{dx}{x^p}$$

converges and those for which it diverges.

Solution If $p \leq 0$ the integral is not improper, so we need only consider $p > 0$. By Definition 10.5 we have, if $p \neq 1$,

$$\begin{aligned} \int_0^1 \frac{dx}{x^p} &= \lim_{t\to 0^+} \int_t^1 \frac{dx}{x^p} = \lim_{t\to 0^+} \left[\frac{1}{1-p} x^{1-p}\right]_t^1 \\ &= \lim_{t\to 0^+} \left[\frac{1}{1-p}(1 - t^{1-p})\right] = \begin{cases} \dfrac{1}{1-p} & \text{if } p < 1 \\ +\infty & \text{if } p > 1 \end{cases} \end{aligned}$$

If $p = 1$ we have

$$\int_0^1 \frac{dx}{x} = \lim_{t\to 0^+} \int_t^1 \frac{dx}{x} = \lim_{t\to 0^+} [\ln x]_t^1 = \lim_{t\to 0^+} [-\ln t] = +\infty$$

In summary,

$$\int_0^1 \frac{dx}{x^p} \begin{cases} \text{converges} & \text{if } p < 1 \\ \text{diverges} & \text{if } p \geq 1 \end{cases}$$

■

EXAMPLE 10.17 Evaluate the integral

$$\int_{-2}^{2} \frac{dx}{x^2\sqrt{4 - x^2}}$$

or show that it diverges.

Solution Because the integrand is even, we need to investigate only the integral from 0 to 2. If it converges, then we will have

$$\int_{-2}^{2} \frac{dx}{x^2\sqrt{4 - x^2}} = 2\int_{0}^{2} \frac{dx}{x^2\sqrt{4 - x^2}}$$

The integral

$$\int_0^2 \frac{dx}{x^2\sqrt{4-x^2}}$$

is doubly improper, in the sense that the integrand becomes unbounded as $x \to 0^+$ and as $x \to 2^-$. To treat one of these problems at a time, we divide the interval $[0, 2]$ at some convenient point, say $x = 1$, and write the integral as the sum of two integrals:

$$\int_0^2 \frac{dx}{x^2\sqrt{4-x^2}} = \int_0^1 \frac{dx}{x^2\sqrt{4-x^2}} + \int_1^2 \frac{dx}{x^2\sqrt{4-x^2}}$$

Each integral on the right is improper because the integrand is unbounded in a neighborhood of just one of the end points. The integral on the left converges if and only if each integral on the right converges. Consider the last integral first. We will integrate from 1 to t and then let $t \to 2^-$. We make the trigonometric substitution $x = 2 \sin \theta$. When $x = 1$, $\theta = \frac{\pi}{6}$, and when $x = t$, $\theta = \sin^{-1}(\frac{t}{2})$. Thus,

$$\int_1^t \frac{dx}{x^2\sqrt{4-x^2}} = \int_{\pi/6}^{\sin^{-1}(t/2)} \frac{2\cos\theta\, d\theta}{4\sin^2\theta \cdot 2\cos\theta} = \frac{1}{4}\int_{\pi/6}^{\sin^{-1}(t/2)} \csc^2\theta\, d\theta$$

$$= -\frac{1}{4}\Big[\cot\theta\Big]_{\pi/6}^{\sin^{-1}(t/2)} = -\frac{1}{4}\left[\cot\left(\sin^{-1}\frac{t}{2}\right) - \sqrt{3}\right]$$

As $t \to 2^-$, $\sin^{-1}\frac{t}{2} \to \sin^{-1} 1 = \frac{\pi}{2}$. So we get

$$\int_1^2 \frac{dx}{x^2\sqrt{4-x^2}} = -\frac{1}{4}(0 - \sqrt{3}) = \frac{\sqrt{3}}{4}$$

Using the same substitution on the integral

$$\int_t^1 \frac{dx}{x^2\sqrt{4-x^2}}$$

we get

$$\int_t^1 \frac{dx}{x^2\sqrt{4-x^2}} = \frac{1}{4}\int_{\sin^{-1}(t/2)}^{\pi/6} \csc^2\theta = -\frac{1}{4}\Big[\cot\theta\Big]_{\sin^{-1}(t/2)}^{\pi/6}$$

$$= -\frac{1}{4}\left[\sqrt{3} - \cot\left(\sin^{-1}\frac{t}{2}\right)\right]$$

This time we let $t \to 0^+$, so that $\sin^{-1}\frac{t}{2} \to 0^+$ and $\cot(\sin^{-1}\frac{t}{2}) \to +\infty$. It follows that

$$\int_0^1 \frac{dx}{x^2\sqrt{4-x^2}}$$

diverges. Thus, the integral

$$\int_0^2 \frac{dx}{x^2\sqrt{4-x^2}}$$

and hence also the original integral both diverge. ■

EXAMPLE 10.18 Evaluate the integral

$$\int_0^\infty \frac{dx}{x^2 + x}$$

or show that it diverges.

Solution This integral is improper both because it has an unbounded interval of integration and because the integrand is unbounded in a neighborhood of $x = 0$. So that we can treat each problem separately, as in the preceding example, we write

$$\int_0^\infty \frac{dx}{x^2 + x} = \int_0^1 \frac{dx}{x^2 + x} + \int_1^\infty \frac{dx}{x^2 + x}$$

Without evaluating it, we can see by the comparison test that the second integral converges. (Do you see why?) So we consider the first integral. By partial fractions we find that

$$\frac{1}{x^2 + x} = \frac{1}{x(x + 1)} = \frac{1}{x} - \frac{1}{x + 1} \qquad \text{(Verify.)}$$

So

$$\int \frac{dx}{x^2 + x} = \int \left(\frac{1}{x} - \frac{1}{x + 1}\right) dx = \ln|x| - \ln|x + 1| = \ln\left|\frac{x}{x + 1}\right|$$

Thus,

$$\begin{aligned}\int_0^1 \frac{dx}{x^2 + x} = \lim_{t \to 0^+} \int_t^1 \frac{dx}{x^2 + x} &= \lim_{t \to 0^+} \left[\ln \frac{x}{x + 1}\right]_t^1 \\ &= \lim_{t \to 0^+} \left[\ln \frac{1}{2} - \ln \frac{t}{t + 1}\right] \\ &= +\infty\end{aligned}$$

since as $t \to 0^+$, $\ln t/(t + 1) \to -\infty$. Thus, this integral and hence also the original integral both diverge. ■

EXERCISE SET 10.4

A

In Exercises 1–38 evaluate the improper integral, or else show that it diverges.

1. $\displaystyle\int_0^1 \frac{dx}{1 - x}$

2. $\displaystyle\int_0^1 \frac{dx}{\sqrt{1 - x}}$

3. $\displaystyle\int_1^2 \frac{x\,dx}{\sqrt{x^2 - 1}}$

4. $\displaystyle\int_0^1 \frac{x\,dx}{\sqrt[3]{1 - x^2}}$

5. $\displaystyle\int_1^2 \frac{dx}{x\sqrt{x^2 - 1}}$

6. $\displaystyle\int_0^2 \frac{dx}{\sqrt{4 - x^2}}$

7. $\displaystyle\int_2^3 \frac{dx}{(x - 2)^{3/2}}$

8. $\displaystyle\int_2^3 \frac{dx}{(x - 2)^{2/3}}$

9. $\displaystyle\int_0^3 \frac{dx}{(x - 1)^2}$

10. $\displaystyle\int_{-2}^1 \frac{dx}{x + 1}$

11. $\displaystyle\int_1^2 \frac{dx}{x \ln x}$

12. $\displaystyle\int_{1/2}^2 \frac{dx}{x\sqrt[3]{\ln x}}$

13. $\displaystyle\int_{-2}^{-1} \frac{dx}{x\sqrt{\ln|x|}}$

14. $\displaystyle\int_0^{\ln 2} \frac{e^x\,dx}{\sqrt{e^x - 1}}$

15. $\displaystyle\int_0^{\ln 3} \frac{e^{-x}\,dx}{(e^{-x} - 1)^{2/3}}$

16. $\displaystyle\int_0^{\pi/2} \frac{\cos x}{1 - \sin x}\,dx$

17. $\displaystyle\int_0^\pi \frac{\sin x}{\sqrt[3]{\cos x}}\,dx$

18. $\displaystyle\int_0^\pi \tan^2 x \sec^2 x\,dx$

19. $\int_0^{\pi/3} \frac{\sec^2 x\,dx}{(\tan x - 1)^2}$
20. $\int_0^{\pi/3} \frac{\sec x \tan x\,dx}{2 - \sec x}$

21. $\int_{-2}^{2} \frac{x^3\,dx}{\sqrt{4 - x^2}}$
22. $\int_1^{\sqrt{3}} \frac{dx}{x^3\sqrt{x^2 - 1}}$

23. $\int_0^2 \frac{dx}{x\sqrt{x^2 + 1}}$
24. $\int_{3/2}^2 \frac{x^2\,dx}{(4x^2 - 9)^{5/2}}$

25. $\int_0^2 \frac{dx}{x^2 - 2x}$
26. $\int_1^3 \frac{dx}{x^2 - 4x + 3}$

27. $\int_{-1}^3 \frac{dx}{x^2 + x - 2}$
28. $\int_{-1}^1 \frac{dx}{x^3 + x^2}$

29. $\int_1^2 \frac{dx}{\sqrt{2x - 1} - 1}$
30. $\int_0^4 \frac{\sqrt{x} + 2}{\sqrt{x} - 2}\,dx$

31. $\int_{-1}^8 \frac{x^{1/3}\,dx}{x^{2/3} - 4}$
32. $\int_0^a \frac{\sqrt{x}\,dx}{(x^{3/2} - a^{3/2})^{2/3}}$

33. $\int_0^1 \frac{dx}{x^{2/3} + x}$
34. $\int_0^{\pi} \frac{dx}{1 - \sin x}$

35. $\int_0^2 \frac{dx}{\sqrt{2x - x^2}}$
36. $\int_a^{2a} \frac{dx}{\sqrt{x^2 - a^2}}$

37. $\int_0^{\ln 2} \frac{dx}{e^x\sqrt{1 - e^{-2x}}}$

38. $\int_{-1}^1 \sqrt{\frac{1 + x}{1 - x}}\,dx$ (*Hint:* Under the radical multiply numerator and denominator by $1 + x$.)

39. Find the area of the region under the graph of $y = 1/\sqrt{1 - x}$ from 0 to 1, or show that it fails to exist.

40. Find the volume (if it exists) generated by revolving the region above $y = x$ and below $y = x^{-3/2}$ between $x = 0$ and $x = 1$ about the y-axis.

B

In Exercises 41–48 evaluate the integral or show that it diverges.

41. $\int_0^{\infty} \frac{dx}{\sqrt{x}(1 + x)}$
42. $\int_1^{\infty} \frac{dx}{x\sqrt{x - 1}}$

43. $\int_0^{\infty} \frac{dx}{e^x - 1}$
44. $\int_0^{\infty} \frac{dx}{\sqrt{x + x^2}}$

45. $\int_{-\infty}^{\infty} \frac{dx}{x(\ln|x|)^2}$
46. $\int_{-\infty}^{\infty} \frac{dx}{x^{2/3} + x^{5/3}}$

47. $\int_0^{\pi} \frac{\cot x\,dx}{\ln \sin x}$
48. $\int_0^2 \frac{dx}{(x - 1)\sqrt{2x - x^2}}$

49. Show that $\int_0^1 \cos \frac{1}{x}\,dx$ exists without reference to the note following Definition 10.5. (*Hint:* Let $u = \frac{1}{x}$.)

50. Evaluate the integral

$$\int_0^e \frac{dx}{x \ln x(\ln|\ln x|)^2}$$

or show that it diverges. (*Hint:* Look for all zeros of the denominator.)

51. There are comparison tests for improper integrals with unbounded integrands analogous to the one for those that have an unbounded interval of integration. For $f(x)$ unbounded in a neighborhood of a, the test is as follows: If $0 \le f(x) \le g(x)$ for all x on $(a, b]$, and f and g are continuous on $(a, b]$, then

$\int_a^b f(x)\,dx$ converges if $\int_a^b g(x)\,dx$ converges

and

$\int_a^b g(x)\,dx$ diverges if $\int_a^b f(x)\,dx$ diverges

A similar test applies if f is unbounded near b. Use this test along with the result of Example 10.16 to determine the convergence or divergence of the following integrals.

(a) $\int_0^{\pi} \frac{\cos^2 x}{\sqrt{x}}\,dx$
(b) $\int_{-\pi/3}^{\pi/3} \frac{\sec x}{x^{3/2}}\,dx$

(c) $\int_0^{\pi/2} \frac{1 - \sin x}{x^{2/3} + x}\,dx$
(d) $\int_0^1 \frac{1 + e^{-x}}{x - x^2}\,dx$

52. Show that the formula for the length of an arc of the four-cusp hypocycloid $x^{2/3} + y^{2/3} = a^{2/3}$ (see the figure) results in an improper integral. Find the total length.

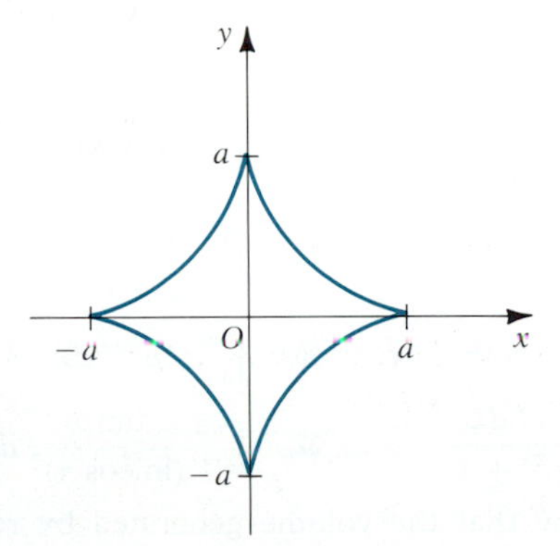

53. (a) Find the area of the region under the graph of $f(x) = 1/\sqrt{4 - x^2}$ from $x = 0$ to $x = 2$.
(b) Find the volume generated by revolving the region in part a about the line $x = 2$.
(c) Without finding its value, show that the surface obtained by revolving the graph of f in part a about the line $x = 2$ has a finite area.

10.5 SUPPLEMENTARY EXERCISES

In Exercises 1–16 evaluate the limits.

1. $\lim_{x\to 1} \dfrac{x^3 - 3x + 2}{2x^3 - 3x^2 + 1}$

2. $\lim_{x\to 4} \dfrac{\sqrt[3]{2x} - 2}{\sqrt{x} - 2}$

3. $\lim_{x\to 0} \dfrac{\sin x - \tan x}{x^3}$

4. $\lim_{x\to 1^+} \dfrac{\ln(\ln x)}{\tan \dfrac{\pi x}{2}}$

5. $\lim_{x\to 0} x \coth 2x$

6. $\lim_{x\to 0^-} e^{1/x} \ln|x|$

7. $\lim_{x\to 1}(1 - x) \tan \dfrac{\pi x}{2}$

8. $\lim_{x\to 1^+} \ln x[\ln(\ln x)]$

9. $\lim_{x\to 0^+} x \ln \tanh x$

10. $\lim_{x\to +\infty} [\ln x - \ln(x + 2)]$

11. $\lim_{x\to 0}\left[\dfrac{1}{e^x - 1} - \operatorname{csch} x\right]$

12. $\lim_{x\to 1}\left[\dfrac{1}{x - 1} - \dfrac{1}{\ln x}\right]$

13. $\lim_{x\to 1^+} (x - 1)^{\ln x}$

14. $\lim_{x\to 0^+} [\ln(x + 1)]^x$

15. $\lim_{x\to 0^+} \left(\dfrac{1}{e^x - 1}\right)^{\sqrt{x}}$

16. $\lim_{x\to 0}(\cosh x)^{\coth^2 x}$

In Exercises 17–32 evaluate the integral or show that it diverges.

17. $\displaystyle\int_1^{\infty} \frac{dx}{x^4 + x^2}$

18. $\displaystyle\int_1^{\infty} \frac{dx}{x^4 - x^2}$

19. $\displaystyle\int_2^{\infty} \frac{dx}{x\sqrt{x^2 - 4}}$

20. $\displaystyle\int_{-1}^{8} \frac{x + 1}{\sqrt[3]{x}}\, dx$

21. $\displaystyle\int_2^{4} \frac{x^3\, dx}{\sqrt{x^2 - 4}}$

22. $\displaystyle\int_1^{\infty} \frac{x\, dx}{\sqrt{x^4 + 1}}$

23. $\displaystyle\int_0^{1} \sqrt{\frac{1 - x}{x}}\, dx$

24. $\displaystyle\int_3^{\infty} \frac{3\, dx}{x^2 - x - 2}$

25. $\displaystyle\int_{-1}^{1} \frac{dx}{x^2\sqrt{1 - x^2}}$

26. $\displaystyle\int_0^{1} x \ln x\, dx$

27. $\displaystyle\int_{-\infty}^{\infty} \frac{dx}{1 + x^3}$

28. $\displaystyle\int_1^{e} \frac{dx}{x\sqrt{\ln x}}$

29. $\displaystyle\int_{-\ln 2}^{\ln 2} \coth x\, dx$

30. $\displaystyle\int_0^{\infty} x^3 e^{-x^2}\, dx$

31. $\displaystyle\int_0^{\infty} \frac{e^x\, dx}{e^{2x} + 1}$

32. $\displaystyle\int_{\pi/3}^{\pi/2} \frac{\tan x}{(\ln \cos x)^2}\, dx$

33. Show that the volume generated by revolving the region under the graph of $y = xe^{-x}$ on $[0, \infty)$ about the x-axis is finite, and find its value.

34. Show that $\int_0^1 \sin(1/x^\alpha)\, dx$ converges for $\alpha > 0$.

35. Let

$$f(x) = \frac{1}{x \ln x[\ln(\ln x)]^p}, \qquad x > e$$

Show that $\int_e^{e^2} f(x)\, dx$ converges if and only if $p < 1$, whereas $\int_{e^2}^{\infty} f(x)\, dx$ converges if and only if $p > 1$.

36. The **Pareto** probability density function is of the form

$$f(x) = \frac{k\theta^k}{x^{k+1}}$$

if $x \geq \theta$ and $f(x) = 0$ otherwise, where k and θ are positive constants.

(a) Show that f is a probability density function.

(b) For what values of k is the mean μ finite? Find μ for these values.

(c) For what values of k is the variance σ^2 finite? Find σ^2 for these values.

37. An important function, called the **gamma** function, is defined by

$$\Gamma(\alpha) = \int_0^{\infty} x^{\alpha - 1} e^{-x}\, dx, \qquad \alpha > 0.$$

Show the following:

(a) For $x \in (0, 1]$, $x^{\alpha-1}e^{-x} \leq x^{\alpha-1}$.

(b) For x sufficiently large, $x^{\alpha-1}e^{-x} \leq 1/x^2$.

(c) By parts a and b the integral that defines $\Gamma(\alpha)$ converges if $\alpha > 0$.

38. Refer to Exercise 37. Prove the following:

(a) $\Gamma(\alpha + 1) = \alpha\Gamma(\alpha)$

(b) For all natural numbers n,

$$\Gamma(\alpha + n) = (\alpha + n - 1)(\alpha + n - 2) \cdots (\alpha + 1)\Gamma(\alpha)$$

(c) For all natural numbers n, $\Gamma(n + 1) = n!$ Show why it is natural to define $0! = 1$.

39. The **gamma probability density function** is defined by

$$f(x) = \frac{1}{\beta^\alpha \Gamma(\alpha)} x^{\alpha-1} e^{-x/\beta} \quad \text{for } x > 0$$

and $f(x) = 0$ otherwise, where α and β are positive constants. (See Exercise 37.) Prove that f is a probability density function, and find μ and σ^2.

40. Analyze the function

$$f(x) = \begin{cases} e^{1-1/x} & \text{if } x \neq 0 \\ 0 & \text{if } x = 0 \end{cases}$$

and draw its graph. Pay particular attention to the right-hand derivative at 0.

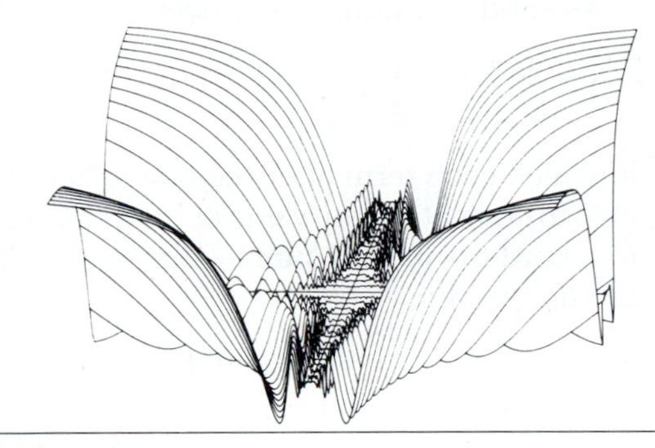

CHAPTER 11

INFINITE SERIES

11.1 INTRODUCTION

The representation of functions by infinite series has played a major role in both the theory and application of calculus from its inception, and infinite series continue to be of fundamental importance in present-day applications. As we shall see, functions that are difficult to evaluate, or to differentiate or integrate, are often more manageable when expressed as infinite series. Moreover, some functions occur naturally in the form of infinite series, and this may be the only way of representing them. This is especially true of functions obtained as solutions to certain differential equations of physics and engineering.

In mathematics the word *series* refers to a *sum*, and so we would expect an infinite series to be an "infinite sum." But how can we add infinitely many terms? Certainly some new definition of addition is required, for in ordinary algebra only finitely many terms can be added. To get at such a definition, let us begin with a familiar example involving an unending decimal. We all know that the fraction $\frac{1}{3}$ can be written as $0.3333\ldots$. This unending decimal can also be written as the infinite sum

$$0.3 + 0.03 + 0.003 + 0.0003 + \cdots$$

or equivalently,

$$\frac{3}{10} + \frac{3}{10^2} + \frac{3}{10^3} + \frac{3}{10^4} + \cdots$$

Now if we add any finite number of these terms, the result will be close to, but not equal to, $\frac{1}{3}$. The more terms we add on, the closer the result will be to $\frac{1}{3}$. In a sense yet to be made precise, the infinite sum is the limit

$$\lim_{n\to\infty}\left(\frac{3}{10} + \frac{3}{10^2} + \frac{3}{10^3} + \cdots + \frac{3}{10^n}\right)$$

and, as we know, the result is $\frac{1}{3}$.

As another example, consider the infinite series

$$1 + \frac{1}{2} + \frac{1}{4} + \frac{1}{8} + \cdots$$

in which each term after the first is half of the preceding term. You may be able to conjecture the value of the sum. A reasonable way to guess its value is to record the **partial sums**—that is, the successive finite sums, starting with the first term:

$$\begin{aligned} \text{first term} &= 1 \\ \text{sum of first two terms} &= 1\tfrac{1}{2} \\ \text{sum of first three terms} &= 1\tfrac{3}{4} \\ \text{sum of first four terms} &= 1\tfrac{7}{8} \end{aligned}$$

Now it does not take any great insight to see that these partial sums are getting closer and closer to 2. In fact, our definition will show that the infinite sum *is* 2, so that we can write

$$1 + \frac{1}{2} + \frac{1}{4} + \frac{1}{8} + \cdots = 2$$

In each of these cases the value of the infinite sum was obtained by finding the limit of finite (partial) sums. In the first case we knew the limit at the outset, and in the second we guessed the limit by observing the pattern of the first few partial sums.

Not all infinite series of numbers result in a finite limit, and even when they do, it is not always so easy to find that limit. Consider the following:

a. $1 + 2 + 3 + 4 + 5 + \cdots$

b. $1 + \dfrac{1}{2} + \dfrac{1}{3} + \dfrac{1}{4} + \dfrac{1}{5} + \cdots$

c. $1 + \dfrac{1}{2^2} + \dfrac{1}{3^2} + \dfrac{1}{4^2} + \dfrac{1}{5^2} + \cdots$

In a it is evident that as we add on more and more terms, we get larger numbers. We could say the sum is $+\infty$, or simply say the series diverges. The situation in b is less clear. You may wish to look at the first few partial sums and try to conjecture the value of the infinite sum. We will leave the question open for now and answer it in Section 11.3. The series in c has an interesting history. The Bernoulli brothers, Jakob and Johann, Swiss mathematicians who lived in the latter part of the 17th century and early 18th century, spent many years trying to find the sum.* It was finally discovered by Johann's student, Leonhard Euler, in 1736, after Jakob had died. On learning of this, Johann wrote: "And thus is fulfilled the earnest desire of my brother, who, recognizing the investigation of this sum to be more difficult than anyone had thought, frankly admitted his own work had been baffled. . . . I wish that you were present, Brother!" Incidentally, Euler

* The Bernoulli family was quite remarkable, producing eight mathematicians in three generations and a host of other descendants who distinguished themselves in many fields.

found that

$$1 + \frac{1}{2^2} + \frac{1}{3^2} + \frac{1}{4^2} + \cdots = \frac{\pi^2}{6}$$

It is not likely you would ever have guessed this result!

The process of listing the partial sums of an infinite series and then investigating whether these finite sums approach a limit is the key element in defining an infinite series. The list of partial sums is an example of what is called a **sequence.** Because sequences are fundamental in studying infinite series, we begin our formal study with them and return to infinite series in Section 11.3.

11.2 SEQUENCES

Roughly speaking, a sequence is a list of terms (numbers, functions, or any kind of mathematical entities). For example,

$$1, 1\tfrac{1}{2}, 1\tfrac{3}{4}, 1\tfrac{7}{8}, 1\tfrac{15}{16}, \ldots$$

is a sequence we looked at in the preceding section. The important idea in a sequence is that we can identify a first term, a second, a third, and so on; that is, corresponding to any natural number, there is a well-defined term of the sequence. This sounds like the definition of a function in which the domain is the set of natural numbers. In fact, this is precisely the meaning of a sequence.

DEFINITION 11.1 A **sequence** is a function whose domain is the set **N** of natural numbers. ■

If f is a sequence, its range is the set $\{f(1), f(2), f(3), \ldots\}$. It is customary to call this ordered listing of the range elements the sequence itself, although strictly speaking it should be referred to as the *values* of the sequence. It is also customary to designate the values with subscripts, writing, for example, a_n instead of $f(n)$. When we use the set notation $\{a_n\}$ we mean the entire sequence, $a_1, a_2, a_3, a_4, \ldots$. For each natural number n, a_n is called the ***n*th term,** or the **general term,** of the sequence $\{a_n\}$. Specifying the formula for a_n is an unambiguous way of defining the sequence. For example, if

$$a_n = \frac{(-1)^{n-1}n}{n+1}$$

then the sequence $\{a_n\}$ is completely defined. Its first few terms are

$$\frac{1}{2}, \frac{-2}{3}, \frac{3}{4}, \frac{-5}{6}, \ldots$$

Usually the first term of the sequence is obtained by putting $n = 1$ in the formula for a_n, but there are times when it is more convenient to begin with $n = 0$ or possibly some other value of n. *Unless otherwise specified, we will assume n begins with 1.*

Sometimes a partial listing of terms is given, and we must infer the general term. For example, if we are given the sequence

$$1, \frac{1}{3}, \frac{1}{5}, \frac{1}{7}, \ldots$$

we can safely infer that

$$a_n = \frac{1}{2n-1}$$

However, merely listing the first few terms does not necessarily make the pattern clear. For example, the listing

$$1, \frac{2}{3}, \frac{5}{8}, \frac{10}{17}, \ldots$$

is *not* a satisfactory way of designating a sequence, since it is unclear how to form further terms.

Here then are some ways of designating a sequence:

Partial listing, nth term clear: $1, \frac{1}{3}, \frac{1}{5}, \frac{1}{7}, \ldots$

Braces enclosing nth term: $\left\{\frac{1}{2n-1}\right\}$

Partial listing, showing nth term: $1, \frac{1}{3}, \frac{1}{5}, \frac{1}{7}, \ldots, \frac{1}{2n-1}, \ldots$

There is yet another way of designating a sequence, called a *recursive* definition. An example is the sequence $\{a_n\}$ for which

$$a_1 = 1 \quad \text{and} \quad a_n = \frac{a_{n-1}}{2}, \qquad n \geq 2$$

To obtain successive terms we begin with a_1 and then use the *recursion formula*, $a_n = a_{n-1}/2$. The result is

$$1, \frac{1}{2}, \frac{1}{4}, \frac{1}{8}, \frac{1}{16}, \ldots$$

Sometimes the recursion formula may involve the two preceding terms, as in

$$a_1 = 1, \quad a_2 = 1, \quad a_n = a_{n-1} + a_{n-2}, \qquad n \geq 3$$

The first few terms are

$$1, 1, 2, 3, 5, 8, 13, \ldots$$

This is a famous sequence, called the *Fibonacci sequence*, that occurs in nature in a variety of ways.

If $f(n) = a_n$ is a sequence of real numbers, we can plot the points $f(1)$, $f(2)$, $f(3)$, . . . in the usual way. For example, if

$$a_n = f(n) = \frac{n}{n+1}$$

then the graph of f is as shown in Figure 11.1. Notice that we do not connect the points with a smooth curve, since the domain of f consists of positive

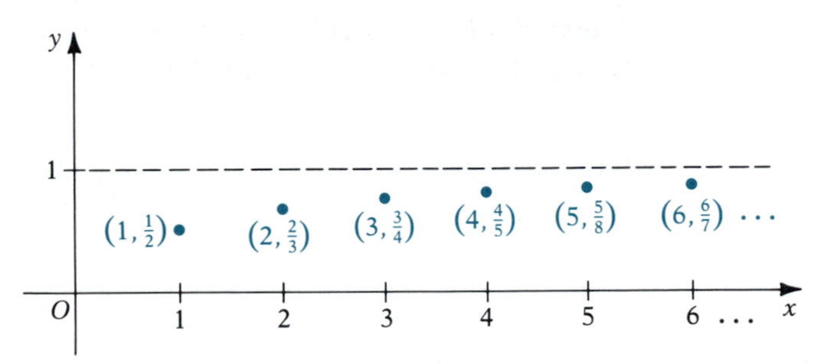

FIGURE 11.1

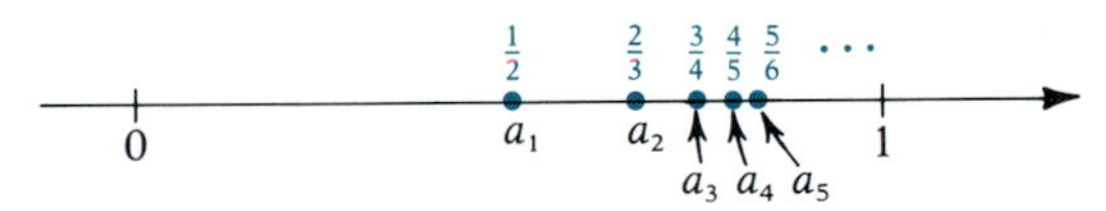

FIGURE 11.2

integers only. An alternate way of graphing a sequence is to show the values on a number line. The sequence above would then be shown as in Figure 11.2. This second method is often preferable.

The primary question we want to answer about a sequence $\{a_n\}$ is whether, as we take larger and larger values of n, the terms a_n approach some limit. The following definition shows precisely what this means.

DEFINITION 11.2 The sequence $\{a_n\}$ is said to **converge to the limit L** provided that, given any positive number ε, there exists a natural number N such that for all $n > N$, $|a_n - L| < \varepsilon$. If no such limit L exists, the sequence is said to **diverge.** ■

Another way of describing the limit L is to say that any given neighborhood of L will contain all terms of the sequence from some point onward. The number ε is the radius of such an arbitrary neighborhood $(L - \varepsilon, L + \varepsilon)$, and the natural number N specifies the term beyond which all higher numbered terms lie in the neighborhood. In general, N depends on ε—the smaller ε is, the larger N has to be. These ideas are shown graphically in Figure 11.3.

To illustrate the definition, consider the sequence $\{a_n\}$ for which

$$a_n = 1 + \frac{(-1)^n}{n}$$

FIGURE 11.3

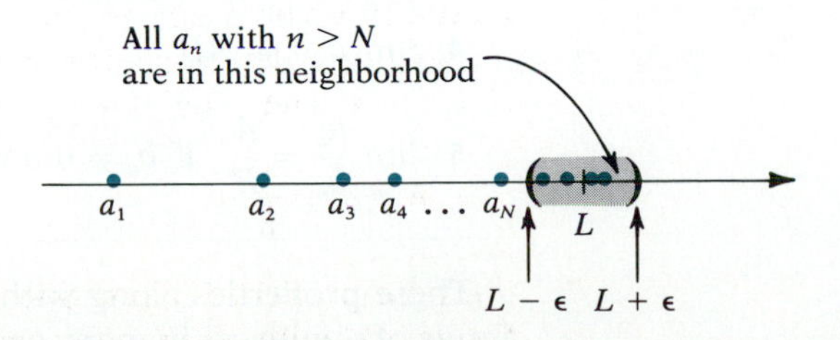

The first few terms are

$$0, \frac{3}{2}, \frac{2}{3}, \frac{5}{4}, \frac{4}{5}, \frac{7}{6}, \ldots$$

It appears that the terms are clustering about the number 1 as we go farther and farther out. So we try $L = 1$ and see whether it works:

$$|a_n - L| = \left|1 + \frac{(-1)^n}{n} - 1\right| = \left|\frac{(-1)^n}{n}\right| = \frac{1}{n}$$

Now let ε denote any positive number. Then $|a_n - L|$ will be less than ε if $\frac{1}{n} < \varepsilon$ or, equivalently, if $n > \frac{1}{\varepsilon}$. So we choose N as any natural number such that $N \geq \frac{1}{\varepsilon}$. Then for $n > N$, $|a_n - L| < \varepsilon$, and so the definition is satisfied; that is,

$$\lim_{n \to \infty} \left[1 + \frac{(-1)^n}{n}\right] = 1$$

As another example consider the sequence

$$\left\{\frac{(-1)^{n-1}n}{n+1}\right\}$$

that we looked at earlier. As Figure 11.4 shows, the even-numbered terms approach -1 and the odd-numbered terms $+1$. There is no single number L that is a candidate for the limit. You might be tempted to say there are two limits, but Definition 11.1 rules that out, since it is impossible for *all* terms beyond the Nth one to be in an arbitrarily small neighborhood of two different points. (See Exercise 47.) Thus, the sequence diverges. This example shows that a sequence can diverge without its terms becoming infinite.

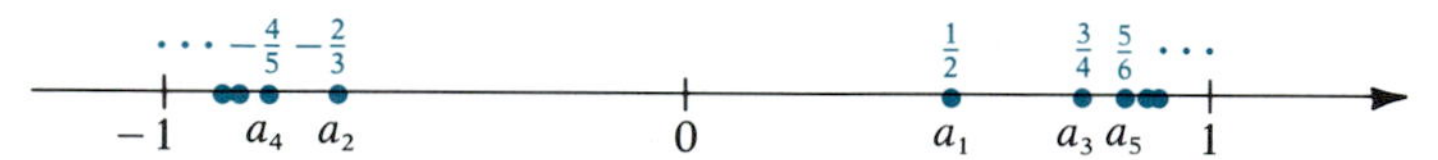

FIGURE 11.4

The following properties of limits of sequences are analogous to those in Section 1.6 for limits of functions and can be proved in a similar way.

PROPERTIES OF LIMITS OF SEQUENCES

Let $\{a_n\}$ and $\{b_n\}$ be convergent sequences with limits A and B, respectively.

1. $\lim_{n \to \infty} ca_n = cA$
2. $\lim_{n \to \infty} (a_n + b_n) = A + B$
3. $\lim_{n \to \infty} (a_n - b_n) = A - B$
4. $\lim_{n \to \infty} a_n b_n = AB$
5. $\lim_{n \to \infty} \frac{a_n}{b_n} = \frac{A}{B}$ if $b_n \neq 0$ and $B \neq 0$

These properties along with the following theorem enable us to calculate limits of sequences in many cases without having to go back to the definition.

THEOREM 11.1 Let $\{a_n\}$ be a sequence and f be a function defined on $[k, \infty)$ for some $k > 0$, and suppose that $a_n = f(n)$ for $n \geq k$. If $\lim_{x\to\infty} f(x) = L$, then also $\lim_{n\to\infty} a_n = L$. ■

This follows from the fact that $|f(x) - L|$ being arbitrarily small for x sufficiently large implies that $|f(n) - L|$ is also arbitrarily small for n sufficiently large. A detailed proof is called for in the exercises. This theorem enables us to use our knowledge of the behavior of functions of a continuous variable x as $x \to \infty$ to infer results about sequences. In particular, we can often make use of L'Hôpital's rule, as the next example shows.

EXAMPLE 11.1 Find the limit of the sequence $\{a_n\}$ where

$$a_n = \frac{(\ln n)^2}{n}$$

Solution The function f defined by

$$f(x) = \frac{(\ln x)^2}{x}$$

is defined on $[1, \infty)$ and for all $n \geq 1$, $f(n) = a_n$. So by Theorem 11.1, if we can find $\lim_{x\to\infty} f(x)$, this will also be the value of $\lim_{x\to\infty} a_n$. Now f is indeterminate of the form $\frac{\infty}{\infty}$, and L'Hôpital's rule applies. So we have

$$\lim_{x\to\infty} f(x) = \lim_{x\to\infty} \frac{\frac{2\ln x}{x}}{1} = \lim_{x\to\infty} \frac{2\ln x}{x}$$

$$= \lim_{x\to\infty} \frac{\frac{2}{x}}{1} = 0$$

Thus, by Theorem 11.1,

$$\lim_{n\to\infty} \frac{(\ln n)^2}{n} = 0$$ ■

EXAMPLE 11.2 Find the limit of each of the following:

(a) $\left\{\frac{n^2 - 1}{2n^2 + 1}\right\}$ (b) $\left\{\frac{n \sin n}{1 + n^2}\right\}$

Solution (a)

$$\lim_{n\to\infty} \frac{n^2 - 1}{2n^2 + 1} = \lim_{n\to\infty} \frac{1 - \frac{1}{n^2}}{2 + \frac{1}{n^2}}$$ We divided numerator and denominator by n^2.

$$= \frac{\lim_{n\to\infty}\left(1 - \frac{1}{n^2}\right)}{\lim_{n\to\infty}\left(2 + \frac{1}{n^2}\right)}$$ Property 5

$$= \frac{1}{2}$$ Properties 2 and 3

(b) Since $|\sin n| \leq 1$, we have

$$\left|\frac{n \sin n}{1 + n^2}\right| \leq \frac{n}{1 + n^2} < \frac{n}{n^2} = \frac{1}{n}$$

and this will be arbitrarily close to 0 if n is sufficiently large. Thus,

$$\lim_{n \to \infty} \frac{n \sin n}{1 + n^2} = 0$$

■

The next theorem will be particularly useful when we study infinite series.

THEOREM 11.2 If $|r| < 1$, then $\lim_{n \to \infty} r^n = 0$.

Proof If $r = 0$, the result is trivial, so we consider $r \neq 0$. It is also sufficient to show that $|r|^n \to 0$, since $|r^n - 0| = \big||r|^n - 0\big|$. Thus, we want to show that $|r|^n \to 0$, where $0 < |r| < 1$.

Since $|r| < 1$, we can write

$$|r| = \frac{1}{1 + p}$$

where $p > 0$, and so

$$|r|^n = \frac{1}{(1 + p)^n}$$

Since, by the binomial theorem (see Appendix 1),

$$(1 + p)^n = 1 + np + \frac{n(n - 1)}{2!} p^2 + \cdots + p^n$$

and all terms on the right are positive, it follows that

$$(1 + p)^n \geq 1 + np$$

Thus,

$$|r|^n = \frac{1}{(1 + p)^n} \leq \frac{1}{1 + np} < \frac{1}{np}$$

So if ε denotes any positive number, $\frac{1}{np} < \varepsilon$ if $np > \frac{1}{\varepsilon}$ or $n > \frac{1}{\varepsilon p}$. We therefore take N as any natural number such that $N \geq \frac{1}{\varepsilon p}$. Then for $n > N$, $\big||r|^n - 0\big| < \varepsilon$. Thus, $\lim_{n \to \infty} |r|^n = 0$ and hence also $\lim_{n \to \infty} r^n = 0$. ■

A sequence $\{a_n\}$ of real numbers is said to be **bounded** if there exists a number M such that $|a_n| \leq M$ for all n. The sequence is **bounded above** if for some M, $a_n \leq M$ for all n, and it is **bounded below** if for some m, $m \leq a_n$ for all n. Clearly, a bounded sequence is both bounded above and bounded below.

We saw by the example

$$\left\{\frac{(-1)^n n}{n + 1}\right\}$$

that boundedness alone does not guarantee convergence. However, if a bounded sequence is also *monotone,* it always converges. For sequences, monotonicity means essentially the same as for functions, either increasing or decreasing, except that we allow for the possibility of consecutive terms remaining the same. The next definition makes this explicit.

DEFINITION 11.3 A sequence $\{a_n\}$ is said to be

$$\begin{cases} \textbf{increasing} & \text{if } a_n \leq a_{n+1} \\ \textbf{decreasing} & \text{if } a_n \geq a_{n+1} \end{cases} \quad \text{for all } n$$

A sequence that is either increasing or decreasing is said to be **monotone.** ■

Note The terms *nondecreasing* and *nonincreasing* are often used where we have used increasing and decreasing, respectively.

If it is necessary to distinguish between a sequence for which $a_n < a_{n+1}$ for all n and one for which $a_n \leq a_{n+1}$, we will refer to the former as *strictly* increasing, and similarly for strictly decreasing, but it is seldom necessary to make this distinction. By our definition each of the following is an increasing sequence:

$$1, 2, 3, 4, 5, \ldots$$
$$1, 2, 2, 3, 4, 4, 5, \ldots$$

The first one is strictly increasing.

The next theorem is fundamental in the study of infinite series.

THEOREM 11.3 Every monotone bounded sequence in **R** converges.

Proof We will prove the theorem for increasing sequences only. The proof for decreasing sequences is similar.

Let $\{a_n\}$ denote an increasing, bounded sequence of real numbers. By the completeness of **R** (see Section 1.2) there exists a least upper bound to the values of the sequence. Denote this least upper bound by L, and let ε denote any positive number. Now consider the neighborhood $(L - \varepsilon, L + \varepsilon)$ of L, as shown in Figure 11.5. Since $L - \varepsilon$ is *not* an upper bound to the sequence (why?), there must be some member of the sequence, call it a_N, such that

$$L - \varepsilon < a_N \leq L$$

And since the sequence is increasing, all subsequent members of the sequence fall between a_N and L; that is, for all $n > N$,

$$L - \varepsilon < a_N \leq a_n \leq L < L + \varepsilon$$

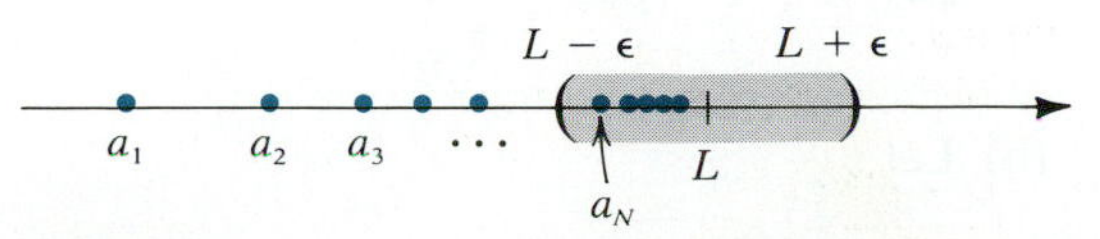

FIGURE 11.5

In particular, for all $n > N$,

$$L - \varepsilon < a_n < L + \varepsilon$$

or equivalently,

$$|a_n - L| < \varepsilon$$

This completes the proof for increasing sequences. ■

Note In applying Theorem 11.3 it is sufficient to show monotonicity *from some point onward,* since convergence or divergence depends only on the behavior of the sequence as n gets arbitrarily large. For example, the sequence

$$20, 33, -75, 1, \tfrac{1}{2}, \tfrac{1}{4}, \tfrac{1}{8}, \tfrac{1}{16}, \ldots$$

converges, since starting with the fourth term it is decreasing and the sequence is bounded below by -75. The limit is easily seen to be 0.

Sometimes it is evident that a sequence is monotone and bounded. For example, the sequence $\{\frac{1}{n}\}$ is clearly decreasing, and it is bounded below by 0. At times, however, some analysis is required to determine whether a sequence possesses one or both of these properties. We illustrate with the next example three possible ways of showing monotonicity.

EXAMPLE 11.3 Show that each of the following sequences is monotone and bounded, and hence convergent. If possible, find the limit.

(a) $\left\{\dfrac{n}{n+1}\right\}$ (b) $\left\{\dfrac{\sqrt{1+n^2}}{n}\right\}$ (c) $\left\{\dfrac{n!}{n^n}\right\}$

Solution (a) We consider the difference $a_{n+1} - a_n$. If we can show that this is greater than or equal to 0, then the sequence is increasing; if it is less than or equal to 0, the sequence is decreasing:

$$a_{n+1} - a_n = \frac{n+1}{n+2} - \frac{n}{n+1} = \frac{n^2 + 2n + 1 - n^2 - 2n}{(n+1)(n+2)}$$
$$= \frac{1}{(n+1)(n+2)} > 0$$

So $a_{n+1} > a_n$, and thus the sequence is increasing. It is also bounded, since

$$\frac{n}{n+1} < 1$$

for all n. The limit is

$$\lim_{n\to\infty} \frac{n}{n+1} = \lim_{n\to\infty} \frac{1}{1 + \dfrac{1}{n}} = 1$$

(b) Let

$$f(x) = \frac{\sqrt{1+x^2}}{x}$$

Then $f(n) = a_n$ and f is a differentiable function on $[1, \infty)$. If we can show $f'(x) \geq 0$, the function and hence the sequence are increasing, whereas if $f'(x) \leq 0$, both are decreasing:

$$f'(x) = \frac{x \cdot \dfrac{2x}{2\sqrt{1 + x^2}} - \sqrt{1 + x^2}}{x^2} = \frac{x^2 - 1 - x^2}{x^2\sqrt{1 + x^2}}$$

$$= \frac{-1}{x^2\sqrt{1 + x^2}} < 0$$

Thus, the sequence is decreasing. Since all terms are positive, 0 is a lower bound. So the sequence converges. To get the limit, we divide numerator and denominator by n (dividing under the radical by n^2):

$$\lim_{n \to \infty} \frac{\sqrt{1 + n^2}}{n} = \lim_{n \to \infty} \frac{\sqrt{\dfrac{1}{n^2} + 1}}{1} = 1$$

(c) We calculate the ratio a_{n+1}/a_n. If $a_{n+1}/a_n \geq 1$, then $a_{n+1} \geq a_n$ and $\{a_n\}$ is increasing. If $a_{n+1}/a_n \leq 1$, then $a_{n+1} \leq a_n$ and $\{a_n\}$ is decreasing:

$$\frac{a_{n+1}}{a_n} = \frac{(n + 1)!}{(n + 1)^{n+1}} \cdot \frac{n^n}{n!} = \frac{(n + 1)n^n}{(n + 1)^{n+1}} = \frac{n^n}{(n + 1)^n}$$

$$= \left(\frac{n}{n + 1}\right)^n < 1$$

since $n/(n + 1) < 1$. Thus, the sequence is decreasing. All terms are positive, so it is bounded below by 0 and hence converges. It is possible to show that the limit is 0, but it is not evident how to do this. (A hint will be given in Exercise 56.) ■

To summarize, we can show the monotonicity of $\{a_n\}$ by any one of the following methods:

1. Calculate the difference $a_{n+1} - a_n$. If

$$\begin{cases} a_{n+1} - a_n \geq 0, \text{ then } \{a_n\} \text{ is increasing} \\ a_{n+1} - a_n \leq 0, \text{ then } \{a_n\} \text{ is decreasing} \end{cases}$$

2. Calculate $f'(x)$, where f is a differentiable function on $[1, \infty)$ that has the same form as a_n; that is, $f(n) = a_n$. If

$$\begin{cases} f'(x) \geq 0, \text{ then } \{a_n\} \text{ is increasing} \\ f'(x) \leq 0, \text{ then } \{a_n\} \text{ is decreasing} \end{cases}$$

3. Calculate the ratio a_{n+1}/a_n. If

$$\begin{cases} \dfrac{a_{n+1}}{a_n} \geq 1, \text{ then } \{a_n\} \text{ is increasing} \\ \dfrac{a_{n+1}}{a_n} \leq 1, \text{ then } \{a_n\} \text{ is decreasing} \end{cases}$$

Again, in each instance it is sufficient to show that the condition holds from some point onward, say for $n \geq n_0$. Sometimes there is a choice of method, whereas at other times only one method will work. When products, powers, or factorials are involved, method 3 is probably the best one to use.

EXERCISE SET 11.2

A

In Exercises 1–3 write out the first five terms of the sequence $\{a_n\}$.

1. (a) $a_n = \dfrac{(-1)^{n-1}n}{2n-1}$ (b) $a_n = \dfrac{2^n}{n!}$

2. (a) $a_n = \dfrac{2 \cdot 4 \cdot 6 \cdot \cdots \cdot (2n)}{1 \cdot 3 \cdot 5 \cdot \cdots \cdot (2n-1)}$
 (b) $a_n = \dfrac{\sin[(2n+1)\pi/2]}{n(n+1)}$

3. (a) $a_1 = 1,\ a_n = -na_{n-1}$ if $n \geq 2$
 (b) $a_1 = 1,\ a_2 = 2,\ a_n = \dfrac{a_{n-1}}{a_{n-2}}$ if $n \geq 3$

In Exercises 4–6 determine a formula for the nth term of the sequence.

4. (a) $1, -\frac{1}{3}, \frac{1}{5}, -\frac{1}{7}, \frac{1}{9}, \ldots$
 (b) $\frac{1}{2}, \frac{2}{5}, \frac{3}{10}, \frac{4}{17}, \frac{5}{26}, \ldots$

5. (a) $\frac{1}{2}, \frac{3}{4}, \frac{7}{8}, \frac{15}{16}, \frac{31}{32}, \ldots$
 (b) $\frac{2}{5}, -\frac{4}{7}, \frac{6}{9}, -\frac{8}{11}, \frac{10}{13}, \ldots$

6. (a) $\frac{1}{7}, -\frac{3}{10}, \frac{5}{13}, -\frac{7}{16}, \frac{9}{19}, \ldots$
 (b) $-1, 3, \frac{5}{3}, \frac{7}{5}, \frac{9}{7}, \ldots$

In Exercises 7–30 find $\lim_{n\to\infty} a_n$ or show that the sequence diverges.

7. $a_n = \dfrac{n^2 - 2n}{3n^2 + n - 1}$
8. $a_n = \dfrac{n+1}{2n^2 - n + 1}$
9. $a_n = \dfrac{2n - n^3}{(n+1)(n+3)}$
10. $a_n = \dfrac{n}{\sqrt{1+n^2}}$
11. $a_n = \dfrac{\cos n\pi}{\sqrt{n}}$
12. $a_n = \dfrac{1 + \sin n}{\ln(n+1)}$
13. $a_n = \dfrac{1 - (-1)^n}{2}$
14. $a_n = n^3 e^{-n}$
15. $a_n = \tanh n$
16. $a_n = \dfrac{\ln n}{\sqrt{n}}$
17. $a_n = (0.9999)^n$
18. $a_n = \dfrac{1 - \cosh n}{\sinh n}$
19. $a_n = \ln\left(1 + \dfrac{1}{n}\right)$
20. $a_n = \dfrac{\ln n}{\ln(\ln n)}$
21. $a_n = \dfrac{2^n}{e^{n-1}}$
22. $a_n = \dfrac{(-1)^{n-1}(2\sin n + \cos n)}{n \sec n}$
23. $a_n = \sin n + \cos n$
24. $a_n = (1.00001)^n$
25. $a_n = \dfrac{n+1}{n}\left(\dfrac{2}{3}\right)^n$
26. $a_n = \dfrac{(-1)^n e^{2n}}{1 + 9^n}$
27. $a_n = \dfrac{n \ln n}{1 + n^2}$
28. $a_n = \dfrac{\sqrt{1+n^3}}{\sqrt[3]{1+n^4}}$
29. $a_n = n^2 \cdot 2^{-n}$
30. $a_n = n \sin \dfrac{1}{n}$

In Exercises 31–40 show that the sequence $\{a_n\}$ is monotone for $n \geq n_0$. If $n_0 \neq 1$, give the smallest value of n_0.

31. $a_n = \dfrac{2n-1}{n+1}$
32. $a_n = \dfrac{n}{2n-3}$
33. $a_n = n^2 e^{-n}$
34. $a_n = \dfrac{(\ln n)^2}{n}$
35. $a_n = \dfrac{2^n}{n!}$
36. $a_n = 2^{1/n}$
37. $a_n = \dfrac{n!}{1 \cdot 3 \cdot 5 \cdot \cdots \cdot (2n-1)}$
38. $a_n = \dfrac{1 \cdot 3 \cdot 5 \cdot \cdots \cdot (2n-1)}{2^n \cdot n!}$
39. $a_n = \ln n - \ln(n+1)$
40. $a_n = \tan^{-1} n$

In Exercises 41–46 prove that the sequence $\{a_n\}$ converges.

41. $a_n = \dfrac{e^n}{n!}$
42. $a_n = \dfrac{2^n n!}{(2n)!}$
43. $a_n = n^{1/n}$
44. $a_n = \dfrac{1 \cdot 3 \cdot 5 \cdot \cdots \cdot (2n-1)}{2 \cdot 4 \cdot 6 \cdot \cdots \cdot 2n}$
45. $a_1 = 1,\ a_n = 1 + \dfrac{a_{n-1}}{2}$ for $n \geq 2$
46. $a_1 = 2,\ a_n = \sqrt{a_{n-1}}$ for $n \geq 2$

B

47. Prove that a sequence of real numbers can have at most one limit.

In Exercises 48–52 prove the indicated limit property for sequences.

48. Property 1 **49.** Property 2

50. Property 3 **51.** Property 4

52. Property 5

53. Prove that $\lim_{n\to\infty} a_n = 0$ if and only if $\lim_{n\to\infty} |a_n| = 0$.

54. Complete the proof of Theorem 11.3 by showing that if $\{a_n\}$ is decreasing and bounded, then it converges.

55. Prove that if $r > 1$ or $r \le -1$, then $\{r^n\}$ diverges.

56. Prove that $\lim_{n\to\infty} (n!/n^n) = 0$. (*Hint:* Write

$$a_n = \frac{n!}{n^n} = \frac{1 \cdot 2 \cdot 3 \cdot \cdots \cdot n}{n \cdot n \cdot n \cdot \cdots \cdot n}$$

and show that $0 < a_n \le \frac{1}{n}$.)

57. Prove that the sequence $\{a_n\}$, where

$$a_n = \frac{1}{n+1} + \frac{1}{n+2} + \frac{1}{n+3} + \cdots + \frac{1}{2n}$$

is convergent. *Hint:* To show boundedness, observe that

$$a_n \le n\left(\frac{1}{n+1}\right)$$

58. Let $a_1 = \sqrt{3}$ and for $n \ge 2$, $a_n = \sqrt{3a_{n-1}}$. Prove that $\{a_n\}$ converges. What is its limit? [*Hint:* Use mathematical induction (see Appendix 1) to prove that for all n, $a_{n+1} > a_n$ and $a_n < 3$.]

11.3 INFINITE SERIES

We are now ready to return to the subject of infinite series. First we need some definitions. Let $\{a_n\}$ be a sequence of real numbers. The indicated sum

$$a_1 + a_2 + a_3 + \cdots + a_n + \cdots \tag{11.1}$$

is called an **infinite series,** or simply a series. In summation notation, the series can be written as

$$\sum_{n=1}^{\infty} a_n$$

Any other index of summation would do just as well. For example, we could also write $\sum_{k=1}^{\infty} a_k$. We will sometimes refer to the expression (11.1) as the expanded form of the series. For example, the expanded form of the series

$$\sum_{n=1}^{\infty} \frac{1}{n^2}$$

is

$$1 + \frac{1}{2^2} + \frac{1}{3^2} + \cdots + \frac{1}{n^2} + \cdots$$

The numbers $a_1, a_2, a_3, \ldots$ are called **terms** of the series, and the term a_n is called the ***n*th term,** or **general term,** of the series. Notice that in writing the expanded form of a series, we show the first few terms, three dots, the general term, and three more dots to indicate that the series continues indefinitely. When the pattern of the terms is obvious, it is permissible to delete the general term. In the example above, for instance, we could simply write

$$1 + \frac{1}{2^2} + \frac{1}{3^2} + \frac{1}{4^2} + \cdots$$

Be sure to show the three dots, though, since otherwise a finite sum is indicated.

Normally, the index of summation begins with 1, but sometimes it is more convenient to begin with another integer, such as 0. Notice that

$$\sum_{n=0}^{\infty} \frac{1}{2^n} \quad \text{and} \quad \sum_{n=1}^{\infty} \frac{1}{2^{n-1}}$$

are two ways of representing the same series. Sometimes we will simply write $\sum a_n$, and when we do so, we will understand that the index runs from 1 to ∞.

As we indicated in Section 11.1, to arrive at a meaning for the sum of an infinite series, $\sum a_n$, we consider the **sequence of partial sums,** $\{S_n\}$, defined as follows:

$$\begin{aligned} S_1 &= a_1 \\ S_2 &= a_1 + a_2 \\ S_3 &= a_1 + a_2 + a_3 \\ &\vdots \\ S_n &= a_1 + a_2 + a_3 + \cdots + a_n \end{aligned}$$

If this sequence converges, its limit is the number we assign as the sum of the series. This is made precise in the following definition.

DEFINITION 11.4 Let

$$a_1 + a_2 + a_3 + \cdots + a_n + \cdots$$

be an infinite series, and for each $n = 1, 2, 3, \ldots$, let

$$S_n = a_1 + a_2 + a_3 + \cdots + a_n$$

If the sequence $\{S_n\}$ converges to a finite limit S, then we say the series $\sum a_n$ **converges** and that its sum is S, and we write

$$\sum_{n=1}^{\infty} a_n = S$$

If the sequence $\{S_n\}$ does not converge, then we say that the series $\sum a_n$ **diverges.** ■

In brief,

$$\sum_{n=1}^{\infty} a_n = \lim_{n\to\infty}(a_1 + a_2 + a_3 + \cdots + a_n)$$

It is useful to observe that the convergence or divergence of a series is unaffected if we delete, or otherwise alter, any finite number of terms (but the sum of the series *is* affected, provided it converges). We state this result as a theorem.

THEOREM 11.4 If for $n \geq N$, $a_n = b_n$, where N is a fixed positive integer greater than 1, then the two series $\sum a_n$ and $\sum b_n$ either both converge or both diverge.

Proof Let the partial sums of $\sum a_n$ be denoted by A_n and those of $\sum b_n$ be denoted by B_n. Then for $n \geq N$,

$$a_N + a_{N+1} + \cdots + a_n = b_N + b_{N+1} + \cdots + b_n$$

or equivalently,

$$A_n - A_{N-1} = B_n - B_{N-1}$$

Since A_{N-1} and B_{N-1} are constants, it follows that if either A_n or B_n approaches a limit as $n \to \infty$, so does the other, and if either does not approach a limit, neither does the other. This completes the proof. ■

In most cases it is not feasible to apply Definition 11.4 directly to determine the convergence or divergence of a series. In subsequent sections we will develop means of testing for convergence or divergence without having to apply the definition directly. There are, however, certain series for which the definition can be used, and we will now show two of these. The first, called a **geometric series,** is of fundamental importance.

Let a and r denote any two real numbers, with $a \neq 0$. The series

$$\sum_{n=1}^{\infty} ar^{n-1} = a + ar + ar^2 + ar^3 + \cdots \tag{11.2}$$

is called a *geometric series, with first term a and common ratio r.* The nth partial sum, S_n, of this series is

$$S_n = a + ar + ar^2 + ar^3 + \cdots + ar^{n-1}$$

If $r = 1$, then $S_n = a + a + a + \cdots + a = na$, and as $n \to \infty$, $S_n \to +\infty$ if $a > 0$ and $S_n \to -\infty$ if $a < 0$. In either case, the sequence $\{S_n\}$ diverges, and hence the series (11.2) also diverges.

For $r \neq 1$, we write S_n in a closed form by subtracting rS_n from S_n and simplifying the result:

$$S_n - rS_n = (a + ar + ar^2 + ar^3 + \cdots + ar^{n-1}) - (ar + ar^2 + ar^3 + \cdots + ar^n)$$

All the terms on the right except a and ar^n cancel, so we have

$$S_n(1 - r) = a - ar^n$$

Solving for S_n, we get

$$S_n = \frac{a(1 - r^n)}{1 - r} \tag{11.3}$$

You may recognize this from your study of precalculus as the formula for the sum of the first n terms of a *geometric progression* (which is the same as the nth partial sum of a geometric series). If $|r| < 1$, we showed in Theorem 11.2 that $r^n \to 0$ as $n \to \infty$. Thus,

$$\lim_{n\to\infty} S_n = \frac{a}{1 - r} \quad \text{if } |r| < 1$$

If $r > 1$, it is easy to see that $r^n \to \infty$ as $n \to \infty$. (See Exercise 55 in Exercise Set 11.2.) If $r = -1$, then r^n alternates between -1 and $+1$, so that the sequence $\{S_n\}$ becomes $a, 0, a, 0, a, 0, \ldots$, and since $a \neq 0$, this sequence does not converge. Finally, if $r < -1$, then r^n oscillates from negative to positive through larger and larger absolute values. So r^n and hence S_n both diverge. (See Exercise 55 in Exercise Set 11.2.)

Because of its importance, we state the result just proved as a theorem.

THEOREM 11.5 The geometric series

$$a + ar + ar^2 + ar^3 + \cdots + ar^{n-1} + \cdots$$

converges to the sum S, where

$$S = \frac{a}{1-r}$$

provided $|r| < 1$. If $|r| \geq 1$, the series diverges. ■

EXAMPLE 11.4 Determine which of the following series converge and which diverge. For those that converge, find the sum.

(a) $1 - \frac{2}{3} + \frac{4}{9} - \frac{8}{27} + \cdots$ (b) $\sum_{n=0}^{\infty} e^{-n}$

(c) $3 + 4 + \frac{16}{3} + \frac{64}{9} + \cdots$ (d) $\sum_{k=1}^{\infty} \frac{(-1)^{k-1}}{2^k}$

Solution (a) This is a geometric series with $a = 1$ and $r = -\frac{2}{3}$. (Verify.) Since $|r| < 1$, the series converges to

$$\frac{a}{1-r} = \frac{1}{1 - \left(-\frac{2}{3}\right)} = \frac{3}{3+2} = \frac{3}{5}$$

So we can write

$$1 - \frac{2}{3} + \frac{4}{9} - \frac{8}{27} + \cdots = \frac{3}{5}$$

(b) Writing the series in expanded form,

$$\sum_{n=0}^{\infty} e^{-n} = 1 + \frac{1}{e} + \frac{1}{e^2} + \frac{1}{e^3} + \cdots$$

we recognize it as a geometric series with $a = 1$ and $r = \frac{1}{e}$. Since $e > 1$, $|r| < 1$. Thus,

$$\sum_{n=0}^{\infty} e^{-n} = \frac{a}{1-r} = \frac{1}{1 - \frac{1}{e}} = \frac{e}{e-1}$$

(c) This is geometric, with $a = 3$ and $r = \frac{4}{3}$. Since $|r| > 1$, the series diverges.

(d) In expanded form, the series is

$$\frac{1}{2} - \frac{1}{4} + \frac{1}{8} - \frac{1}{16} + \cdots$$

which is geometric, with $a = \frac{1}{2}$ and $r = -\frac{1}{2}$. So the series converges, and we have

$$\sum_{k=1}^{\infty} \frac{(-1)^{k-1}}{2^k} = \frac{a}{1-r} = \frac{\frac{1}{2}}{1 - \left(-\frac{1}{2}\right)} = \frac{1}{2+1} = \frac{1}{3}$$ ■

EXAMPLE 11.5 Make use of geometric series to write each of the following repeating decimals in the form $\frac{m}{n}$, where m and n are integers:

(a) 0.333 . . . (b) 1.272727 . . .

Solution (a) We can write

$$0.333\ldots = 0.3 + 0.03 + 0.003 + 0.0003 + \cdots$$

The series on the right is geometric, with $a = 0.3$ and $r = 0.1$. Thus,

$$0.333\ldots = \frac{a}{1-r} = \frac{0.3}{1-0.1} = \frac{0.3}{0.9} = \frac{3}{9} = \frac{1}{3}$$

(This confirms what we already knew.)

(b) Write the given decimal in the form

$$1 + (0.27 + 0.0027 + 0.000027 + \cdots)$$

The series in parentheses is geometric, with $a = 0.27$ and $r = 0.01$. Thus, its sum is

$$\frac{a}{1-r} = \frac{0.27}{1-0.01} = \frac{0.27}{0.99} = \frac{27}{99} = \frac{3}{11}$$

Therefore, the entire sum is

$$1.272727\ldots = 1 + \frac{3}{11} = \frac{14}{11}$$

■

EXAMPLE 11.6 Find all values of x for which the series

$$\sum_{n=1}^{\infty} \frac{(-1)^{n-1}x^n}{2^{n-1}}$$

converges, and express the sum in terms of x when it does converge.

Solution Writing the series in expanded form, we have

$$\sum_{n=1}^{\infty} \frac{(-1)^{n-1}x^n}{2^{n-1}} = x - \frac{x^2}{2} + \frac{x^3}{4} - \frac{x^4}{8} + \cdots$$

This is a geometric series, with $a = x$ and $r = -\frac{x}{2}$. So the series converges when $|r| < 1$—that is, when

$$\left|-\frac{x}{2}\right| = \frac{|x|}{2} < 1 \quad \text{or} \quad |x| < 2$$

It diverges for all other values of x. When $|x| < 2$, the sum is

$$\frac{a}{1-r} = \frac{x}{1-\left(-\frac{x}{2}\right)} = \frac{2x}{2+x}$$

So we have

$$\sum_{n=1}^{\infty} \frac{(-1)^{n-1}x^n}{2^{n-1}} = \frac{2x}{x+2} \quad \text{if } |x| < 2$$

■

EXAMPLE 11.7 A ball is dropped from a height of 50 ft, and on each bounce it goes three-fourths as high as before. Approximate the total distance traveled by the ball in coming to rest.

Solution Although the ball moves only along a vertical line, we can get a better picture by showing the path as in Figure 11.6. First it drops 50 ft, then it rises $37\frac{1}{2}$ ft and falls the same distance. Thereafter it rises and falls three-fourths of the previous up and down total. To make the first motion analogous to the others, we can pretend it started from the ground, rising 50 ft and then falling 50 ft. We will subtract this imaginary initial 50-ft rise after getting the total. With this understanding, the total distance covered is

$$\left[100 + \frac{3}{4}(100) + \left(\frac{3}{4}\right)^2(100) + \left(\frac{3}{4}\right)^3(100) + \cdots\right] - 50$$

The expression in brackets is a geometric series, with $a = 100$ and $r = \frac{3}{4}$. So its sum is

$$\frac{a}{1-r} = \frac{100}{1 - \frac{3}{4}} = 400$$

Thus, the (theoretical) total distance covered is 350 ft.

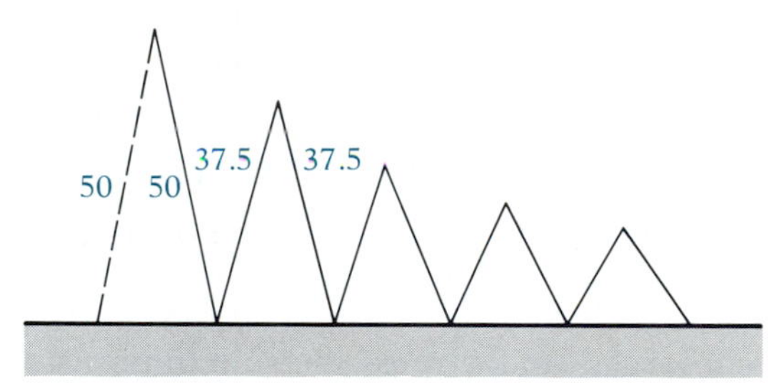

FIGURE 11.6

EXAMPLE 11.8 Write each of the following as an infinite series in powers of x, and give the domain of validity:

(a) $\dfrac{1}{1-2x}$ (b) $\dfrac{6}{3+x}$

Solution What we seek to do in each case is to write the fraction in the form $a/(1-r)$, which we know to be the sum of a geometric series with initial term a and common ratio r, provided $|r| < 1$.

(a) This is already in the proper form, with $a = 1$ and $r = 2x$. So if $|2x| < 1$ or, equivalently, $|x| < \frac{1}{2}$, we have

$$\begin{aligned}\frac{1}{1-2x} &= a + ar + ar^2 + ar^3 + \cdots \\ &= 1 + 2x + (2x)^2 + (2x)^3 + \cdots \\ &= 1 + 2x + 4x^2 + 8x^3 + \cdots\end{aligned}$$

(b) We force this into the form $a/(1 - r)$ by dividing numerator and denominator by 3 (to get 1 as the first term on the denominator) and using the fact that $\frac{x}{3} = -(-\frac{x}{3})$:

$$\frac{6}{3 + x} = \frac{2}{1 + \dfrac{x}{3}} = \frac{2}{1 - \left(-\dfrac{x}{3}\right)}$$

Thus $a = 2$ and $r = -\frac{x}{3}$. So if $\left|-\frac{x}{3}\right| < 1$ or, equivalently, $|x| < 3$, we have

$$\begin{aligned}\frac{6}{3 + x} &= a + ar + ar^2 + ar^3 + \cdots \\ &= 2 + 2\left(-\frac{x}{3}\right) + 2\left(-\frac{x}{3}\right)^2 + 2\left(-\frac{x}{3}\right)^3 + \cdots \\ &= 2\left[1 - \frac{x}{3} + \frac{x^2}{9} - \frac{x^3}{27} + \cdots\right]\end{aligned}$$

■

Another type of series for which a direct application of Definition 11.4 is possible is what is sometimes called a **telescoping series.** The next example illustrates how it works.

EXAMPLE 11.9 Show that the series

$$\sum_{k=1}^{\infty} \frac{1}{k^2 + k}$$

converges, and find its sum.

Solution Using partial fractions, we can obtain the general term in the form

$$\frac{1}{k^2 + k} = \frac{1}{k(k + 1)} = \frac{1}{k} - \frac{1}{k + 1}$$

Thus, the nth partial sum of the series is

$$\begin{aligned}S_n &= \sum_{k=1}^{n} \left(\frac{1}{k} - \frac{1}{k + 1}\right) \\ &= \left(\frac{1}{1} - \frac{1}{2}\right) + \left(\frac{1}{2} - \frac{1}{3}\right) + \left(\frac{1}{3} - \frac{1}{4}\right) + \cdots + \left(\frac{1}{n} - \frac{1}{n + 1}\right)\end{aligned}$$

All terms except the first and last cancel out (the expression "telescopes"), and so

$$S_n = 1 - \frac{1}{n + 1}$$

As $n \to \infty$, $S_n \to 1$. This shows that the original series converges, since the sequence $\{S_n\}$ converges, and also that its sum is 1; that is,

$$\sum_{k=1}^{\infty} \frac{1}{k^2 + k} = 1$$

■

Another very useful result can be obtained from Definition 11.4, which we state as a theorem.

THEOREM 11.6 If the series $\sum_{n=1}^{\infty} a_n$ converges, then $\lim_{n\to\infty} a_n = 0$.

Proof Let S_n be the nth partial sum of $\sum_{n=1}^{\infty} a_n$; that is,

$$S_n = a_1 + a_2 + a_3 + \cdots + a_n$$

Then, if $n > 1$, we also have

$$S_{n-1} = a_1 + a_2 + a_3 + \cdots + a_{n-1}$$

On subtracting S_{n-1} from S_n, all terms drop out except a_n:

$$a_n = S_n - S_{n-1}$$

Now since the series converges, we know that $\lim_{n\to\infty} S_n$ exists. Call its value S. But as $n \to \infty$, we also have $(n-1) \to \infty$. So both $\lim_{n\to\infty} S_n = S$ and $\lim_{n\to\infty} S_{n-1} = S$. Thus,

$$\begin{aligned}\lim_{n\to\infty} a_n &= \lim_{n\to\infty}(S_n - S_{n-1})\\ &= \lim_{n\to\infty} S_n - \lim_{n\to\infty} S_{n-1}\\ &= S - S = 0\end{aligned}$$

■

An equivalent way of stating Theorem 11.6 (called its *contrapositive*) is the following:

If $\lim_{n\to\infty} a_n \neq 0$, then $\sum_{n=1}^{\infty} a_n$ diverges.

Thus, if we can show that $\lim_{n\to\infty} a_n \neq 0$, we can conclude that the series $\sum a_n$ definitely diverges. For example, the series

$$\sum_{n=1}^{\infty} \frac{n}{2n+1}$$

diverges, since

$$\lim_{n\to\infty} a_n = \lim_{n\to\infty} \frac{n}{2n+1} = \frac{1}{2} \neq 0$$

A Word of Caution If you find that $\lim_{n\to\infty} a_n = 0$, you cannot draw any definite conclusion regarding the convergence or divergence of $\sum a_n$ except that the series at least has a chance to converge. It is a common mistake to assume the converse of Theorem 11.6 and suppose that $\lim_{n\to\infty} a_n = 0$ implies the convergence of $\sum a_n$. That this is *not true* can be seen by finding a divergent series whose terms approach 0, as we do now.

The best known example of a divergent series whose nth term tends to 0 is the so-called **harmonic series**—namely,

$$\sum_{n=1}^{\infty} \frac{1}{n} = 1 + \frac{1}{2} + \frac{1}{3} + \cdots + \frac{1}{n} + \cdots$$

Clearly, $\lim_{n\to\infty} \frac{1}{n} = 0$, yet the series diverges, as the following analysis of certain of its partial sums shows:

$$S_1 = 1$$

$$S_2 = 1 + \frac{1}{2}$$

$$S_4 = 1 + \frac{1}{2} + \frac{1}{3} + \frac{1}{4} > 1 + \frac{1}{2} + \left(\frac{1}{4} + \frac{1}{4}\right) = 1 + \frac{1}{2} + \frac{1}{2}$$

$$S_8 = 1 + \frac{1}{2} + \frac{1}{3} + \frac{1}{4} + \frac{1}{5} + \frac{1}{6} + \frac{1}{7} + \frac{1}{8}$$

$$> 1 + \frac{1}{2} + \left(\frac{1}{4} + \frac{1}{4}\right) + \left(\frac{1}{8} + \frac{1}{8} + \frac{1}{8} + \frac{1}{8}\right) = 1 + \frac{1}{2} + \frac{1}{2} + \frac{1}{2}$$

Continuing in this way, we show that

$$S_{2^n} \geq 1 + n\left(\frac{1}{2}\right), \qquad n = 0, 1, 2, \ldots$$

(mathematical induction can be used to give a formal proof). Thus, $\lim_{n\to\infty} S_{2^n} = +\infty$. It follows that $\lim_{n\to\infty} S_n$ does not exist. Suppose it did and was equal to the finite number S. Then, since $2^n \to \infty$ as $n \to \infty$, it would follow that $\lim_{n\to\infty} S_{2^n} = S$, contrary to what we have just shown. Thus, the series diverges. Because we will have frequent occasions to refer to this result, we set it off for emphasis.

The **harmonic series**

$$\sum_{n=1}^{\infty} \frac{1}{n} = 1 + \frac{1}{2} + \frac{1}{3} + \cdots + \frac{1}{n} + \cdots$$

diverges.

It is interesting to note that the sum of the first billion terms of the harmonic series is only about 21; that is,

$$S_{10^9} = 1 + \frac{1}{2} + \frac{1}{3} + \cdots + \frac{1}{10^9} \approx 21$$

and each succeeding term is *very* small. Yet if we continue adding terms indefinitely, the cumulative effect is to cause the sum to become infinite! Sometimes infinite processes defy one's intuition.

We conclude this section with a theorem that is a direct consequence of Definition 11.4 and corresponding properties of sequences.

THEOREM 11.7 If $\sum_{n=1}^{\infty} a_n$ and $\sum_{n=1}^{\infty} b_n$ are convergent series with sums A and B, respectively, and if c is any real number, then each of the series $\sum_{n=1}^{\infty} ca_n$, $\sum_{n=1}^{\infty}(a_n + b_n)$, and $\sum_{n=1}^{\infty}(a_n - b_n)$ converges, and

1. $\sum_{n=1}^{\infty} ca_n = cA$
2. $\sum_{n=1}^{\infty} (a_n + b_n) = A + B$
3. $\sum_{n=1}^{\infty} (a_n - b_n) = A - B$ ∎

EXAMPLE 11.10 Show that the series

$$\sum_{n=1}^{\infty}\left(\frac{1}{2^{n-1}}+\frac{1}{n^2+n}\right)$$

converges, and find its sum.

Solution The series

$$\sum_{n=1}^{\infty}\frac{1}{2^{n-1}}$$

is a geometric series, with $a = 1$ and $r = \frac{1}{2}$, and so it converges to

$$\frac{a}{1-r}=\frac{1}{1-\dfrac{1}{2}}=2$$

The series

$$\sum_{n=1}^{\infty}\frac{1}{n^2+n}$$

is the telescoping series we considered in Example 11.9, where we found that it converged to 1. It follows by property 2 of Theorem 11.7 that

$$\sum_{n=1}^{\infty}\left(\frac{1}{2^{n-1}}+\frac{1}{n^2+n}\right)=2+1=3$$

■

EXERCISE SET 11.3

A

In Exercises 1–4 write the expanded form of the series and give the first four partial sums.

1. $\sum_{k=1}^{\infty}\frac{k}{k^2+1}$
2. $\sum_{n=1}^{\infty}\frac{(-1)^{n-1}}{\sqrt{2n-1}}$
3. $\sum_{m=1}^{\infty}\frac{\ln(m+1)}{m!}$
4. $\sum_{n=0}^{\infty}\frac{(-1)^n(n+1)}{2^n}$

In Exercises 5–8 write the series using summation notation. Also, give the nth partial sum.

5. $1-\frac{1}{3}+\frac{1}{9}-\frac{1}{27}+\frac{1}{81}-\cdots$
6. $1-\frac{1}{2!}+\frac{1}{4!}-\frac{1}{6!}+\frac{1}{8!}-\cdots$
7. $\frac{1}{2\ln 2}+\frac{1}{3\ln 3}+\frac{1}{4\ln 4}+\cdots$
8. $1-\frac{2}{3^2}+\frac{3}{5^2}-\frac{4}{7^2}+\frac{5}{9^2}-\cdots$

In Exercises 9–12 the nth partial sum, S_n, of a series $\sum a_n$ is given. Determine whether the series converges and, if it does, give its sum.

9. $S_n=\frac{n}{n+1}$
10. $S_n=1-(-1)^n$
11. $S_n=2-\frac{\ln(n+1)}{n+1}$
12. $S_n=\frac{n^3-2n+1}{\sqrt{n^4+n^2+4}}$

In Exercises 13–20 show that the given series is geometric and determine whether it converges. If it converges, find its sum.

13. $2-1+\frac{1}{2}-\frac{1}{4}+\frac{1}{8}-\cdots$
14. $\sum_{k=1}^{\infty}\frac{2}{3^{k-1}}$
15. $\sum_{n=0}^{\infty}3\cdot 2^{-n}$
16. $\frac{1}{\ln 3}+\frac{1}{(\ln 3)^2}+\frac{1}{(\ln 3)^3}+\cdots$

17. $\frac{2}{e} - \frac{4}{e^2} + \frac{8}{e^3} - \frac{16}{e^4} + \cdots$

18. $\sum_{n=1}^{\infty} \left(\frac{5}{4}\right)^n$ **19.** $\sum_{n=0}^{\infty} (0.99)^n$

20. $\sum_{k=1}^{\infty} \frac{(-1)^{k-1}3^k}{4^{k-1}}$

In Exercises 21–24 express the repeating decimal in the form $\frac{m}{n}$, in lowest terms, where m and n are integers, making use of geometric series.

21. 0.151515 . . . **22.** 2.181818 . . .

23. 0.148148148 . . . **24.** 1.135135135 . . .

In Exercises 25–28 show that the series is telescoping, and find its sum.

25. $\sum_{k=1}^{\infty} \frac{1}{(k+1)(k+2)}$ **26.** $\sum_{n=1}^{\infty} \frac{2}{n(n+2)}$

27. $\sum_{n=1}^{\infty} \frac{2}{4n^2 - 1}$ **28.** $\sum_{k=1}^{\infty} \frac{1}{k^2 + 4k + 3}$

In Exercises 29–32 show that the series converges, and find its sum.

29. $\sum_{n=1}^{\infty} \left(\frac{1}{2^n} + \frac{2}{3^n}\right)$

30. $\sum_{k=1}^{\infty} \left[\left(\frac{2}{3}\right)^k - \left(\frac{3}{4}\right)^{k-1}\right]$

31. $\sum_{n=1}^{\infty} \left(\frac{1}{n^2 + n} - \frac{1}{3^{n-1}}\right)$

32. $\sum_{n=1}^{\infty} \left[\frac{(-1)^{n-1}5}{2^n} + \frac{2}{n^2 + 2n}\right]$

33. A ball is dropped from a height of 10 m, and on each successive bounce it rises two-thirds as high as on the preceding bounce. Find the total distance the ball travels (theoretically).

34. Determine whether the series $\sum_{k=100}^{\infty} \frac{1}{k}$ converges or diverges. Justify your answer.

In Exercises 35 and 36 show that each series diverges.

35. (a) $\sum_{n=1}^{\infty} \frac{n}{100n + 1}$ (b) $\sum_{n=2}^{\infty} \frac{n}{(\ln n)^2}$

36. (a) $\sum_{n=1}^{\infty} \frac{2n^2 - 3n + 4}{3n^2 + n + 5}$ (b) $\sum_{n=1}^{\infty} \frac{(-1)^{n-1}n}{\sqrt{1 + n^2}}$

37. Indicate which of the following statements are true and which are false.
(a) If $\sum a_n$ diverges, then $\lim_{n\to\infty} a_n = 0$.
(b) If $\lim_{n\to\infty} a_n = 0$, then $\sum a_n$ converges.
(c) If $a_1 + a_2 + \cdots + a_n = \frac{1}{n}$, then $\sum a_n$ converges to 0.
(d) If $\sum(a_n + b_n)$ converges, so do $\sum a_n$ and $\sum b_n$.
(e) If $c \neq 0$ and $\sum ca_n$ converges, then so does $\sum a_n$.

B

38. Prove that if $\sum a_n$ diverges, then $\sum ca_n$ also diverges for every constant $c \neq 0$.

39. Prove that if $\sum a_n$ converges and $\sum b_n$ diverges, then $\sum(a_n + b_n)$ diverges.

40. Show that the series

$$\sum_{n=1}^{\infty} \ln \frac{n}{n+1}$$

diverges. (*Hint:* Show that it is a telescoping series and find S_n.)

41. Find all values of x for which

$$\sum_{n=1}^{\infty} \frac{(-1)^{n-1}2^n x^{n-1}}{3^{n-1}}$$

converges, and find the sum as a function of x.

42. Prove that if $a_n \geq 0$ and $a_1 + a_2 + a_3 + \cdots + a_n \leq k$ for all n, where k is a constant, then $\sum a_n$ converges.

43. A pendulum 3 ft long is released from a position in which its angle with the vertical is 60°. On each swing after the first, it reaches a maximum angle with the vertical 0.9 times as large as the angle reached on the previous swing. Find the total distance covered by the bob of the pendulum in coming to rest.

C

Write a computer program to calculate the nth partial sum of a series, $\sum_{k=1}^{\infty} a_n$, and use it in Exercises 44–49.

44. Estimate $\sum_{k=1}^{\infty} 1/k^2$ using S_n for $n = 10, 20, 30, \ldots, 100$. Compare S_{100} with the actual sum of $\pi^2/6$.

45. Estimate $\sum_{k=1}^{\infty} 1/k^3$ using S_n for $n = 10, 20, 30, \ldots,$ and continuing until two partial sums agree in the first six decimal places. (*Note:* No one knows the exact value of this series. This is a famous unsolved problem.)

46. Estimate $\sum_{k=1}^{\infty}(-1)^{k-1}2^{-k}$ using S_{10}. Calculate the error in this approximation.

47. Estimate $\sum_{k=1}^{\infty}(-1)^{k-1}/k$ using S_n for $n = 100, 200, \ldots, 1000$. The exact value of this sum is ln 2. Compare S_{1000} with this value. What conclusion do you draw?

48. For the series $\sum_{k=1}^{\infty}\frac{1}{k}$ calculate S_n for $n = 100, 200, \ldots, 1000$. Based on these values, what would you conjecture about convergence or divergence?

49. Repeat Exercise 48 with $\sum_{k=1}^{\infty} 1/[(k+1)\ln(k+1)]$.

11.4 SERIES OF POSITIVE TERMS

Given an infinite series, there are two primary questions we usually want to answer: (1) Does the series converge? and (2) If it does converge, what is its sum? The second question in general is much harder to answer than the first (geometric series and telescoping series are exceptions). However, if we know that a series converges, we can at least approximate its sum by using a partial sum S_n for sufficiently large n. So it is very useful just to be able to answer the first question. In this section we give certain tests for convergence that are applicable when the terms are all positive.

The basis for our tests is the following theorem.

THEOREM 11.8 If $a_n \geq 0$ for all n, then the series $\sum a_n$ converges if and only if its sequence of partial sums is bounded.

Proof Since the terms of the series are nonnegative, the sequence $\{S_n\}$ is increasing, so if the sequence is also bounded, it converges (by Theorem 11.3). Thus, by Definition 11.4, $\sum a_n$ also converges.

Now suppose $\sum a_n$ converges. If $\{S_n\}$ were unbounded, we would have $\lim_{n\to\infty} S_n = +\infty$, since $\{S_n\}$ is increasing. But this would contradict the fact that $\sum a_n$ converges. Thus $\{S_n\}$ must be bounded. So the proof is complete. ■

We are ready now to give the first of our tests for convergence.

THEOREM 11.9 **THE COMPARISON TEST**

If $\sum a_n$ and $\sum b_n$ are series of nonnegative terms and for all n, $a_n \leq b_n$, then

$$\sum a_n \text{ converges if } \sum b_n \text{ converges}$$

and

$$\sum b_n \text{ diverges if } \sum a_n \text{ diverges}$$

Proof Let the nth partial sums of $\sum a_n$ and $\sum b_n$ be denoted by A_n and B_n, respectively. Now suppose $\sum b_n$ converges. Then by Theorem 11.8 $\{B_n\}$ is bounded; that is, there exists some positive constant M such that $B_n \leq M$ for all n. Since $a_n \leq b_n$, it follows that $A_n \leq B_n$ and hence $A_n \leq M$. Thus, $\{A_n\}$ is a bounded sequence, so by Theorem 11.8 $\sum a_n$ converges.

Now suppose $\sum a_n$ diverges. Then $\{A_n\}$ is unbounded (for if it were bounded, Theorem 11.8 would say that $\sum a_n$ is convergent). Thus, since $B_n \geq A_n$, the sequence $\{B_n\}$ is also unbounded and hence divergent. Thus $\sum b_n$ is also divergent. This completes the proof. ■

Remarks

1. It is sufficient that $a_n \leq b_n$ hold from some n onward, say for $n \geq N$, since convergence or divergence is unaffected by neglecting any finite number of terms (see Theorem 11.4).
2. When a_n and b_n are nonnegative and $a_n \leq b_n$, we say that $\sum b_n$ *dominates* $\sum a_n$ or that $\sum a_n$ *is dominated by* $\sum b_n$. So the theorem can be rephrased by saying that for series of nonnegative terms, a series that is dominated by a convergent series also converges, and a series that dominates a divergent series also diverges.

EXAMPLE 11.11 Use the comparison test to determine the convergence or divergence of each of the following series:

(a) $\sum_{n=1}^{\infty} \frac{1}{n+2^n}$ (b) $\sum_{n=1}^{\infty} \frac{1}{\sqrt{n}}$

Solution (a) For $n \geq 1$, we have

$$\frac{1}{n+2^n} < \frac{1}{2^n}$$

and since $\sum 1/2^n$ is a convergent geometric series, it follows from the comparison test that $\sum_{n=1}^{\infty} 1/(n+2^n)$ also converges.

(b) For $n \geq 1$,

$$\frac{1}{\sqrt{n}} \geq \frac{1}{n}$$

and $\sum 1/n$ is the divergent harmonic series. So by the comparison test, $\sum 1/\sqrt{n}$ also diverges. ■

The next test is a variation on the comparison test that often is easier to apply than the comparison test itself.

THEOREM 11.10 THE LIMIT COMPARISON TEST

If $a_n > 0$ and $b_n > 0$ for all n and

$$\lim_{n \to \infty} \frac{a_n}{b_n} = L$$

where $0 < L < +\infty$, then $\sum a_n$ and $\sum b_n$ either both converge or both diverge.

Proof Suppose the conditions of the theorem are satisfied. Since a_n/b_n approaches L as a limit, from some n onward, say for $n > N$, a_n/b_n lies in the neighborhood $(\frac{L}{2}, \frac{3L}{2})$. Thus, for all $n > N$,

$$\frac{a_n}{b_n} < \frac{3L}{2} \quad \text{or} \quad a_n < \frac{3L}{2} b_n$$

and

$$\frac{a_n}{b_n} > \frac{L}{2} \quad \text{or} \quad b_n > \frac{L}{2} a_n$$

Now suppose $\sum b_n$ converges. Then so does $\sum \frac{3L}{2} b_n$ (by Theorem 11.7), and since $a_n \le \frac{3L}{2} b_n$ for $n > N$, it follows by the comparison test that $\sum a_n$ converges.

If $\sum a_n$ diverges, then so does $\sum \frac{L}{2} a_n$. (See Exercise 38 in Exercise Set 11.3.) Thus, since for $n > N$, $b_n \ge \frac{L}{2} a_n$, we see by the comparison test that $\sum b_n$ also diverges. This completes the proof. ■

Remark When the conditions of Theorem 11.10 are met, we say that a_n and b_n are *of the same order of magnitude* as $n \to \infty$. So we can rephrase the limit comparison test by saying that if a_n and b_n are of the same order of magnitude, then $\sum a_n$ and $\sum b_n$ behave in the same way.

EXAMPLE 11.12 Test each of the following for convergence or divergence:

(a) $\displaystyle\sum_{n=1}^{\infty} \frac{n^2 - n}{2n^3 + 3n - 4}$ (b) $\displaystyle\sum_{n=1}^{\infty} \frac{1}{\sqrt{3n^4 - 2n}}$

Solution (a) By neglecting all but the highest degree terms in the numerator and denominator, we can see that the nth term is of the same order of magnitude as $\frac{1}{n}$. To confirm this, we consider the limit

$$\lim_{n\to\infty} \frac{a_n}{b_n} = \lim_{n\to\infty} \frac{n^2 - n}{2n^3 + 3n - 4} \div \frac{1}{n}$$
$$= \lim_{n\to\infty} \frac{n^2 - n}{2n^3 + 3n - 4} \cdot n$$
$$= \lim_{n\to\infty} \frac{n^3 - n^2}{2n^3 + 3n - 4} = \frac{1}{2}$$

Since this limit is a nonzero finite number, the nth term of the given series and the term $\frac{1}{n}$ *are* of the same order of magnitude. We know that $\sum \frac{1}{n}$ diverges. Thus, the given series also diverges.

(b) We have shown that the telescoping series

$$\sum \frac{1}{n^2 + n}$$

is convergent (see Example 11.9), and this appears to be a useful series to compare with the given series, since the denominator in the given series is effectively of second degree. Using the limit comparison test, we have

$$\lim_{n\to\infty} \frac{1}{\sqrt{3n^4 - 2n}} \div \frac{1}{n^2 + n} = \lim_{n\to\infty} \frac{1}{\sqrt{3n^4 - 2n}} \cdot (n^2 + n)$$
$$= \lim_{n\to\infty} \frac{1 + \frac{1}{n}}{\sqrt{3 - \frac{2}{n^3}}}$$

We divided numerator and denominator by n^2.

$$= \frac{1}{\sqrt{3}}$$

Thus the given series converges. ■

Note We need to expand our list of known convergent and divergent series to make the two comparison tests more broadly applicable. The next test will enable us to do this.

THEOREM 11.11 THE INTEGRAL TEST

If $\{a_n\}$ is a monotone decreasing sequence of positive numbers and f is a continuous, positive, monotone decreasing function on $[1, \infty)$ for which $f(n) = a_n$, then $\sum_{n=1}^{\infty} a_n$ and $\int_1^{\infty} f(x)\,dx$ either both converge or both diverge.

Remark Since deleting any finite number of terms of the series does not affect the convergence or divergence, it is sufficient that the conditions hold for $n \geq N$ and $x \in [N, \infty)$, where N is a positive integer greater than 1.

Proof Suppose first that the integral $\int_1^{\infty} f(x)\,dx$ converges. As Figure 11.7(a) shows, if we partition the interval $[1, n]$ at the integer points, the sum of the areas of the inscribed rectangles is

$$\begin{aligned} f(2)\cdot 1 + f(3)\cdot 1 + f(4)\cdot 1 + \cdots + f(n)\cdot 1 &= a_2 + a_3 + a_4 + \cdots + a_n \\ &= S_n - a_1 \end{aligned}$$

where S_n is the nth partial sum of $\sum a_n$. Since the sum of these areas cannot exceed the area under f, we have

$$\begin{aligned} S_n - a_1 &\leq \int_1^n f(x)\,dx \\ S_n &\leq a_1 + \int_1^n f(x)\,dx \leq a_1 + \int_1^{\infty} f(x)\,dx \end{aligned}$$

Since the integral converges, the right-hand side is a finite constant. Thus $\{S_n\}$ is bounded. Therefore, by Theorem 11.8, $\sum a_n$ converges.

Now suppose we know that $\sum a_n$ converges. We want to show that the integral also converges. As in Figure 11.7(b), we partition the interval $[1, n+1]$ at the integer points and consider the circumscribed rectangles.

FIGURE 11.7

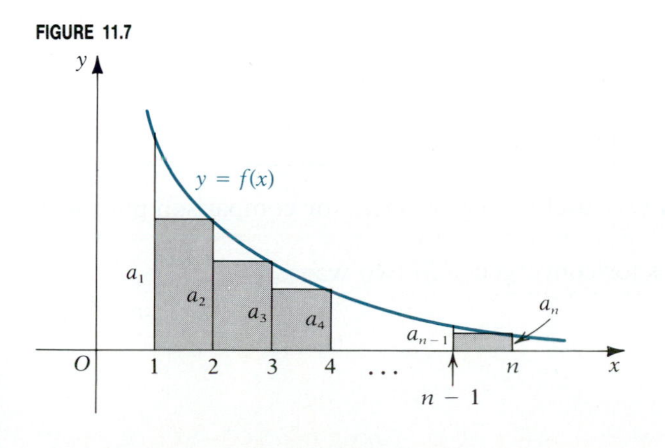

(a)

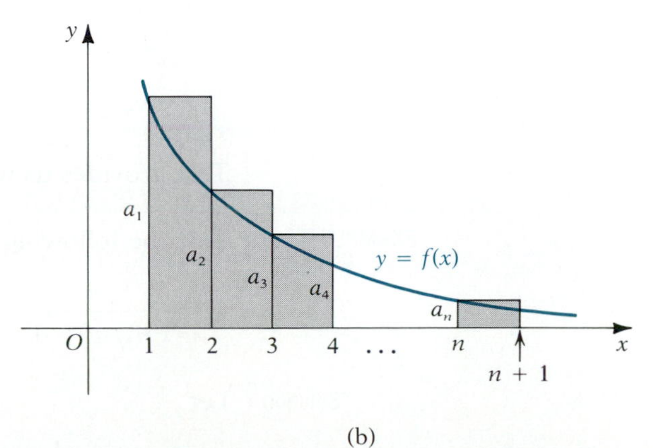

(b)

The area of the kth rectangle is $f(k) \cdot 1 = a_k \cdot 1 = a_k$. So we have

$$\int_1^{n+1} f(x)\,dx \le \sum_{k=1}^{n} a_k \le \sum_{k=1}^{\infty} a_k$$

As n increases, the integrals on the left form a monotone increasing sequence that is bounded by the finite constant $\sum_{k=1}^{\infty} a_k$. Thus, $\lim_{n\to\infty} \int_1^{n+1} f(x)\,dx$ exists. It is not difficult to show that this limit coincides with $\int_1^{\infty} f(x)\,dx$, as we have previously defined it.

This completes the proof, for if $\sum a_n$ diverges, $\int_1^{\infty} f(x)\,dx$ could not converge (otherwise we would contradict the first part of the proof). Similarly, if $\int_1^{\infty} f(x)$ diverges, $\sum a_n$ must also diverge. ■

EXAMPLE 11.13 Determine all values of p for which the series $\sum 1/n^p$ converges.

Solution Let $f(x) = 1/x^p$. Then for $p > 0$, the conditions of the integral test are satisfied. If $p \ne 1$, we have

$$\int_1^{\infty} \frac{1}{x^p}\,dx = \lim_{t\to\infty} \int_1^t x^{-p}\,dx = \lim_{t\to\infty}\left[\frac{x^{-p+1}}{-p+1}\right]_1^t$$
$$= \lim_{t\to\infty}\left[\frac{t^{-p+1}}{-p+1} - \frac{1}{1-p}\right]$$

This limit will be finite only if the exponent $-p+1$ is negative—that is, if $p > 1$. Thus $\sum 1/n^p$ converges if $p > 1$ and diverges if $p < 1$. If $p = 1$, we have

$$\int_1^{\infty} \frac{1}{x}\,dx = \lim_{t\to\infty} \int_1^t \frac{1}{x}\,dx = \lim_{t\to\infty} [\ln x]_1^t$$
$$= \lim_{t\to\infty} [\ln t] = \infty$$

Thus $\sum 1/n$ diverges. We knew this already, since $\sum 1/n$ is the harmonic series, but this provides an alternate proof of its divergence. ■

The series tested in the preceding example is usually referred to as the *p-series*. The result we just proved deserves to be emphasized.

The ***p*-series**

$$\sum_{n=1}^{\infty} \frac{1}{n^p}$$

converges if $p > 1$ and diverges if $p \le 1$.

This provides us with a very useful class of series for comparison purposes.

EXAMPLE 11.14 Test the following series for convergence in two ways:

$$\sum_{n=2}^{\infty} \frac{1}{n\sqrt{n^2-1}}$$

Solution Let

$$f(x) = \frac{1}{x\sqrt{x^2-1}} \quad \text{on } [2, \infty)$$

The conditions of the integral test are satisfied. (See the remark following Theorem 11.11.) We consider $\int_2^\infty f(x)\,dx$:

$$\int_2^\infty \frac{dx}{x\sqrt{x^2-1}} = \lim_{t\to\infty}\int_2^t \frac{dx}{x\sqrt{x^2-1}} = \lim_{t\to\infty}[\sec^{-1} x]_2^t$$

$$= \lim_{t\to\infty}\left[\sec^{-1} t - \frac{\pi}{3}\right] = \frac{\pi}{2} - \frac{\pi}{3} = \frac{\pi}{6}$$

Since this is finite, it follows by the integral test that

$$\sum_{n=2}^{\infty} \frac{1}{n\sqrt{n^2-1}}$$

converges.

As a second method we use the limit comparison test. The general term of the series appears to be of the same order of magnitude as $1/n^2$ (determined by considering only the highest power of n). To confirm this we consider the limit

$$\lim_{n\to\infty} \frac{1}{n\sqrt{n^2-1}} \div \frac{1}{n^2} = \lim_{n\to\infty} \frac{n}{\sqrt{n^2-1}} = \lim_{n\to\infty} \frac{1}{\sqrt{1-\dfrac{1}{n^2}}} = 1$$

We know that $\sum_{n=1}^{\infty} 1/n^2$ converges, since it is a p-series with $p = 2 > 1$. Thus, by the limit comparison test,

$$\sum_{n=2}^{\infty} \frac{1}{n\sqrt{n^2-1}}$$

also converges. ■

The next test is particularly well suited to series with general terms that involve powers or products, especially factorials.

THEOREM 11.12 **THE RATIO TEST**

If, for all n, $a_n > 0$ and

$$\lim_{n\to\infty} \frac{a_{n+1}}{a_n} = L$$

exists, then $\sum a_n$ converges if $L < 1$ and diverges if $L > 1$. If $L = 1$, the test is inconclusive.

Proof Suppose first that $L < 1$. As in Figure 11.8, we choose any number r such that $L < r < 1$. Because $\lim_{n\to\infty} a_{n+1}/a_n = L$, the quotients a_{n+1}/a_n all lie in a neighborhood of L of radius $r - L$ if n is sufficiently large, say $n \geq N$. In particular,

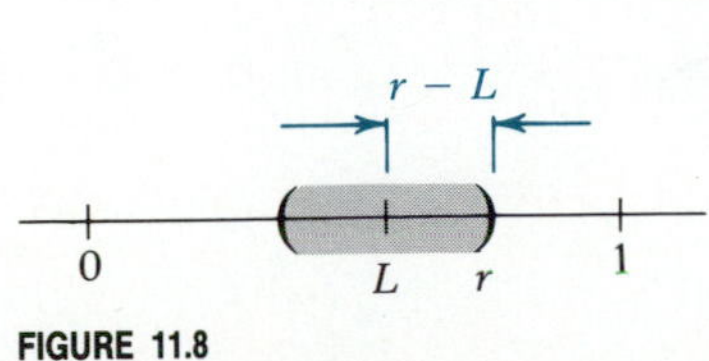

FIGURE 11.8

$$\frac{a_{n+1}}{a_n} < r \quad \text{if } n \geq N$$

or equivalently,

$$a_{n+1} < ra_n \quad \text{if } n \geq N$$

Thus,

$$\begin{aligned} a_{N+1} &< r a_N \\ a_{N+2} &< r a_{N+1} < r^2 a_N \\ a_{N+3} &< r a_{N+2} < r^3 a_N \\ &\vdots \end{aligned}$$

So the terms of the series $\sum_{k=1}^{\infty} a_{N+k}$ are dominated by the terms of $\sum_{k=1}^{\infty} a_N r^k$. This latter series is a geometric series with common ratio $r < 1$, and so is convergent. Therefore, by the comparison test, $\sum_{k=1}^{\infty} a_{N+k}$ also converges. Since

$$\sum_{k=1}^{\infty} a_{N+k} = \sum_{n=N+1}^{\infty} a_n \qquad \text{(Verify)}$$

it follows that $\sum_{n=1}^{\infty} a_n$ also converges, since adding the first N terms does not alter convergence (see Theorem 11.4).

If $L > 1$, we have, for all sufficiently large n,

$$\frac{a_{n+1}}{a_n} > 1 \quad \text{or} \quad a_{n+1} > a_n$$

so that the terms increase and thus cannot approach 0. So $\sum a_n$ diverges.

Finally, $\sum 1/n$ and $\sum 1/n^2$ both satisfy

$$\lim_{n\to\infty} \frac{a_{n+1}}{a_n} = 1 \qquad \text{(Verify)}$$

yet the first series diverges and the second converges. So when $L = 1$, we cannot draw any definite conclusion concerning convergence. We say in this case that the ratio test fails. ■

A Word of Caution In applying the ratio test to show convergence, it is not enough to show that $a_{n+1}/a_n < 1$ for all n. (This is true for the harmonic series.) It is the *limit* of this ratio that we must show is less than 1.

EXAMPLE 11.15 Test each of the following for convergence:

(a) $\displaystyle\sum_{n=1}^{\infty} \frac{2^n}{n!}$ (b) $\displaystyle\sum_{n=1}^{\infty} \frac{3^{n-1}}{n^2 \cdot 2^n}$

Solution In each case we apply the ratio test.

(a) $$\lim_{n\to\infty} \frac{a_{n+1}}{a_n} = \lim_{n\to\infty} \frac{2^{n+1}}{(n+1)!} \cdot \frac{n!}{2^n} = \lim_{n\to\infty} \frac{2}{n+1} = 0$$

Since $L = 0 < 1$, the given series converges.

(b) $$\begin{aligned} \lim_{n\to\infty} \frac{a_{n+1}}{a_n} &= \lim_{n\to\infty} \frac{3^n}{(n+1)^2 \cdot 2^{n+1}} \cdot \frac{n^2 \cdot 2^n}{3^{n-1}} \\ &= \lim_{n\to\infty} \left(\frac{n}{n+1}\right)^2 \cdot \frac{3}{2} = \frac{3}{2} \end{aligned}$$

Thus, $L = \frac{3}{2} > 1$, so the given series diverges. ■

We conclude our series of tests with the **root test.** Its proof is similar to that of the ratio test, and we leave it for the exercises.

THEOREM 11.13 THE ROOT TEST

If $a_n \geq 0$ and $\lim_{n\to\infty} \sqrt[n]{a_n} = L$ exists, then $\sum a_n$ converges if $L < 1$ and diverges if $L > 1$. If $L = 1$, the test is inconclusive. ■

The root test is more powerful than the ratio test in the sense that whenever the ratio test gives a definite conclusion concerning convergence or divergence, the same will be true of the root test. But there are series for which the ratio test is inconclusive and the root test gives definite information. However, the ratio test is usually easier to apply.

EXAMPLE 11.16 Test the series

$$\sum_{n=1}^{\infty} \left(\frac{n}{2n+1}\right)^n$$

Solution

$$\lim_{n\to\infty} \sqrt[n]{a_n} = \lim_{n\to\infty} \frac{n}{2n+1} = \frac{1}{2} < 1$$

So the series converges. ■

You may have wondered why there are so many tests for convergence (there are still more, but we have given the main ones). The answer is that no single test works on all positive-term series. Even when more than one test *could* be used, one may be much easier to apply than the others. With a little practice you will probably develop a feel for which test to apply in a given situation.

EXERCISE SET 11.4

A

In Exercises 1–6 use the comparison test to determine convergence or divergence.

1. $\sum_{n=1}^{\infty} \frac{1}{n^2+1}$ **2.** $\sum_{n=1}^{\infty} \frac{1}{\sqrt{n^3+2n}}$

3. $\sum_{n=1}^{\infty} \frac{2}{2n-1}$ **4.** $\sum_{n=1}^{\infty} \frac{n+1}{n^2}$

5. $\sum_{n=1}^{\infty} \sqrt{\frac{n+1}{n}}$ **6.** $\sum_{n=1}^{\infty} \frac{n}{2n^3+1}$

In Exercises 7–12 use the limit comparison test to test for convergence or divergence.

7. $\sum_{n=1}^{\infty} \frac{2n-1}{3n^2+4n-2}$ **8.** $\sum_{n=1}^{\infty} \frac{3n}{n^4-2}$

9. $\sum_{n=1}^{\infty} \frac{\sqrt{2n+3}}{n^3}$ **10.** $\sum_{n=1}^{\infty} \frac{n}{\sqrt{n^3-2n+4}}$

11. $\sum_{n=1}^{\infty} \frac{1+2n}{(n^2+1)^{3/2}}$ **12.** $\sum_{n=1}^{\infty} \frac{n^2-2}{\sqrt{3n^5+1}}$

In Exercises 13–18 use the integral test to determine convergence or divergence.

13. $\sum_{n=1}^{\infty} \frac{1}{\sqrt{2n-1}}$ **14.** $\sum_{n=2}^{\infty} \frac{1}{n(\ln n)^2}$

15. $\sum_{n=1}^{\infty} \frac{n}{e^{n^2}}$ **16.** $\sum_{n=1}^{\infty} \frac{1}{n^2+4}$

17. $\sum_{n=3}^{\infty} \frac{1}{n\sqrt{\ln n}}$ **18.** $\sum_{n=1}^{\infty} \frac{n}{(n^2+1)^{3/2}}$

In Exercises 19–24 use the ratio test to determine convergence or divergence.

19. $\sum_{n=1}^{\infty} ne^{-n}$ **20.** $\sum_{n=1}^{\infty} \frac{n^2 \cdot 2^n}{3^{n-1}}$

21. $\sum_{n=1}^{\infty} \frac{n!}{10^n}$ **22.** $\sum_{n=1}^{\infty} \frac{n!}{1 \cdot 3 \cdot 5 \cdot \cdots \cdot (2n-1)}$

23. $\sum_{n=1}^{\infty} \frac{1 \cdot 4 \cdot 7 \cdot \cdots \cdot (3n-2)}{3 \cdot 5 \cdot 7 \cdot \cdots \cdot (2n+1)}$ **24.** $\sum_{n=1}^{\infty} \frac{(n!)^2}{(2n)!}$

In Exercises 25–30 use the root test to determine convergence or divergence.

25. $\sum_{n=1}^{\infty} \frac{2^{3n}}{3^{2n}}$ **26.** $\sum_{n=1}^{\infty} \frac{n^n}{2^{n^2}}$

27. $\sum_{n=1}^{\infty} \frac{\sqrt{3^n}}{2^n}$ **28.** $\sum_{n=1}^{\infty} \frac{2^{3n+1}}{e^{2n}}$

29. $\sum_{n=1}^{\infty} \frac{n \cdot 2^n}{3^n}$ (*Hint:* $\sqrt[n]{n} \to 1$ as $n \to \infty$.)

30. $\sum_{n=1}^{\infty} \frac{(n+1)^{n/2}}{n \cdot 2^{n+1}}$

In Exercises 31–60 test for convergence or divergence by any appropriate means.

31. $\sum_{n=1}^{\infty} \frac{n \sin^2 n}{1+n^3}$ **32.** $\sum_{n=1}^{\infty} n^2 e^{-n}$

33. $\sum_{n=0}^{\infty} \frac{\sqrt{n+1}}{n^2+1}$ **34.** $\sum_{k=1}^{\infty} \frac{\ln k}{k}$

35. $\sum_{k=1}^{\infty} \frac{2k^2}{3k^2-2}$ **36.** $\sum_{m=1}^{\infty} \frac{m!}{2 \cdot 4 \cdot 6 \cdot \cdots \cdot (2m)}$

37. $\sum_{j=0}^{\infty} \frac{j+1}{j^3+2}$ **38.** $\sum_{n=0}^{\infty} \frac{3^n}{4^n+3}$

39. $\sum_{k=2}^{\infty} \left(\frac{2k^2+1}{3k^2-4}\right)^k$ **40.** $\sum_{n=0}^{\infty} \frac{1}{\cosh n}$

41. $\sum_{n=1}^{\infty} \frac{n^2}{2^n}$ **42.** $\sum_{k=0}^{\infty} \frac{e^k}{1+e^{2k}}$

43. $\sum_{n=1}^{\infty} \frac{n}{\sqrt{e^n}}$ **44.** $\sum_{n=0}^{\infty} \frac{2^{n^2}}{n!}$

45. $\sum_{m=1}^{\infty} \frac{(m+1)^{2m}}{3^{m^2}}$ **46.** $\sum_{n=1}^{\infty} n^3 \cdot 3^{-n}$

47. $\sum_{n=1}^{\infty} \frac{1+\cos^2 n}{n^{3/2}}$ **48.** $\sum_{k=2}^{\infty} \frac{2k^2-3k+1}{k^5+2k^2-1}$

49. $\sum_{n=1}^{\infty} n^{2n} e^{-n^2}$ **50.** $\sum_{m=1}^{\infty} \frac{m}{\sqrt{100m^2+1}}$

51. $\sum_{n=0}^{\infty} \frac{(100)^{2n+1}}{(2n+1)!}$ **52.** $\sum_{k=1}^{\infty} k^2 \left(\frac{2}{3}\right)^k$

53. $\sum_{n=1}^{\infty} \frac{3n-2}{(n+1)(n+2)}$ **54.** $\sum_{m=1}^{\infty} \frac{m \sec^2 m}{1+m^2}$

B

55. $\sum_{n=1}^{\infty} \left(1 - \frac{1}{n}\right)^{n^2}$ **56.** $\sum_{n=1}^{\infty} \frac{2^n \cdot n!}{1 \cdot 3 \cdot 5 \cdot \cdots \cdot (2n-1)}$

57. $\sum_{n=3}^{\infty} \frac{1}{n \ln n (\ln \ln n)}$ **58.** $\sum_{n=1}^{\infty} \frac{\sinh n}{\cosh 2n}$

59. $\sum_{n=1}^{\infty} \frac{n \ln n}{1+n^3}$ **60.** $\sum_{n=1}^{\infty} \frac{\sqrt{n+1}-\sqrt{n}}{n}$

61. Prove that if $a_n \geq 0$ and $\sum a_n$ converges, then so does $\sum a_n^2$.

62. Prove the root test (Theorem 11.13).

63. Determine all values of p for which the series

$$\sum_{n=2}^{\infty} \frac{1}{n(\ln n)^p}$$

converges.

64. Prove the following extensions of the limit comparison test for series $\sum a_n$ and $\sum b_n$ of positive terms.
 (a) If $\lim_{n\to\infty} a_n/b_n = 0$ and $\sum b_n$ converges, then $\sum a_n$ also converges.
 (b) If $\lim_{n\to\infty} a_n/b_n = +\infty$ and $\sum b_n$ diverges, then $\sum a_n$ also diverges.

65. Prove the following version of the ratio test for the series $\sum a_n$ with $a_n > 0$.
 (a) If there is a number r that satisfies $0 < r < 1$ such that $a_{n+1}/a_n \leq r$ for all $n \geq N$, where N is some positive integer, then $\sum a_n$ converges.
 (b) If $a_{n+1}/a_n \geq 1$ for all $n \geq N$, where N is some positive integer, then $\sum a_n$ diverges.

66. Prove that if $a_n \geq 0$ and $\lim_{n\to\infty} na_n$ exists and is positive, then $\sum a_n$ diverges.

11.5 SERIES WITH TERMS OF VARIABLE SIGNS; ABSOLUTE CONVERGENCE

In general when a series consists of both positive and negative terms, tests for convergence are more complex than those for positive-term series. There is one important test, however, that is relatively simple, applicable to so-called **alternating series,** in which the terms are alternately positive and negative. It is sufficient to consider such series in which the first term is positive, writing, for example,

$$\sum_{n=1}^{\infty} (-1)^{n-1} a_n = a_1 - a_2 + a_3 - a_4 + \cdots$$

(where for all n, $a_n \geq 0$). If the first term is negative, we could write

$$-a_1 + a_2 - a_3 + a_4 - \cdots = -(a_1 - a_2 + a_3 - a_4 + \cdots)$$

and convergence or divergence can be determined by studying the series in parentheses, beginning with a positive term. An example of such a series is

$$\sum_{n=1}^{\infty} \frac{(-1)^{n-1}}{n} = 1 - \frac{1}{2} + \frac{1}{3} - \frac{1}{4} + \cdots$$

This is called the **alternating harmonic series.**

THEOREM 11.14 ALTERNATING SERIES TEST

If the sequence $\{a_n\}$ is monotone decreasing and converges to 0, then the alternating series $\sum_{n=1}^{\infty} (-1)^{n-1} a_n$ converges.

Note As usual, it is sufficient that the conditions hold from some point onward, and so if from some term onward the signs alternate and $\{a_n\}$ decreases monotonically to 0, the series converges. If $\lim_{n\to\infty} a_n = 0$ but $\{a_n\}$ is not monotone, it is still possible that the series converges, but this would have to be tested by other means.

Proof First consider the partial sums of even order:

$$\begin{aligned} S_2 &= a_1 - a_2 \\ S_4 &= (a_1 - a_2) + (a_3 - a_4) \\ &\vdots \\ S_{2n} &= (a_1 - a_2) + (a_3 - a_4) + \cdots + (a_{2n-1} - a_{2n}) \end{aligned}$$

The indicated grouping shows that $S_{2n} \geq 0$, since the monotonicity guarantees that $a_k - a_{k+1} \geq 0$ for all k. Furthermore,

$$S_{2n+2} = S_{2n} + (a_{2n+1} - a_{2n+2}) \geq S_{2n}$$

So the even-ordered partial sums form a monotone increasing sequence. This sequence is also bounded, since

$$S_{2n} = a_1 - (a_2 - a_3) - (a_4 - a_5) - \cdots - (a_{2n-2} - a_{2n-1}) - a_{2n} \leq a_1$$

Thus $\{S_{2n}\}$ is both monotone and bounded, and so it converges. Call its limit S: $\lim_{n\to\infty} S_{2n} = S$.

Each odd-ordered partial sum after the first can be written in the form

$$S_{2n+1} = S_{2n} + a_{2n+1}$$

For example,

$$S_3 = (a_1 - a_2) + a_3 = S_2 + a_3$$

By hypothesis, $\lim_{n\to\infty} a_{2n+1} = 0$, so we have

$$\lim_{n\to\infty} S_{2n+1} = \lim_{n\to\infty} S_{2n} + \lim_{n\to\infty} a_{2n+1} = S + 0 = S$$

So the odd-ordered partial sums also approach S; that is, $\lim_{n\to\infty} S_n = S$ independently of whether n is even or odd. This completes the proof. ■

COROLLARY 11.14 Under the hypotheses of Theorem 11.14,

$$|S_n - S| < a_{n+1}$$

where S_n is the nth partial sum and S is the sum of the series.

Proof We showed in the proof of the theorem that the even partial sums are bounded above by a_1. It follows that $S \le a_1$, since $\lim_{n\to\infty} S_{2n} = S$. This says that the sum of an alternating series of terms that decrease monotonically to 0 can be no greater than the first term. But

$$\begin{aligned} |S_n - S| &= |a_{n+1} - a_{n+2} + a_{n+3} - \cdots| \\ &= a_{n+1} - a_{n+2} + a_{n+3} - \cdots \qquad \text{(Why?)} \end{aligned}$$

is itself an alternating series of terms that decrease monotonically to 0 and so the sum is no greater than its first term, a_{n+1}. So the corollary is proved. ■

EXAMPLE 11.17 Show that the alternating harmonic series

$$\sum_{n=1}^{\infty} \frac{(-1)^{n-1}}{n} = 1 - \frac{1}{2} + \frac{1}{3} - \frac{1}{4} + \cdots$$

converges, and determine an upper bound on the error in estimating the sum using S_{100}.

Solution The sequence $\{\frac{1}{n}\}$ is monotone decreasing and converges to 0. Since the signs of the series alternate, it follows by Theorem 11.14 that the series converges. Let S denote the sum of the series. Then the error incurred in using S_{100} to estimate S is $|S_{100} - S|$, and by Corollary 11.14 this is at most $a_{101} = \frac{1}{101}$; that is,

$$|S_{100} - S| \le \frac{1}{101} \approx 0.009$$

Thus, even by taking 100 terms of the series to estimate the sum, we cannot be assured of accuracy even to two decimal places. This is an example of a very slowly converging series. ■

EXAMPLE 11.18 Determine the convergence or divergence of each of the following:

$$\text{(a)} \sum_{n=1}^{\infty} (-1)^{n-1} \ln \frac{n+1}{n} \qquad \text{(b)} \sum_{n=1}^{\infty} (-1)^{n-1} \frac{n}{2n+1}$$

Solution (a) To show monotonicity, let

$$f(x) = \ln \frac{x+1}{x}$$

and consider $f'(x)$. First, simplify $f(x)$ by writing

$$f(x) = \ln(x+1) - \ln(x)$$

So

$$f'(x) = \frac{1}{x+1} - \frac{1}{x} = \frac{x-(x+1)}{x(x+1)} = \frac{-1}{x(x+1)} < 0$$

for positive values of x. Thus f is decreasing, and hence so is

$$a_n = \ln \frac{n+1}{n}$$

Furthermore,

$$\lim_{n\to\infty} a_n = \lim_{n\to\infty} \ln \frac{n+1}{n} = \ln\left(\lim_{n\to\infty} \frac{n+1}{n}\right)$$
$$= \ln 1 = 0$$

The conditions of the alternating series test are met, so the series converges.

(b) Since

$$\lim_{n\to\infty} \frac{2}{2n+1} = \frac{1}{2} \neq 0$$

the series diverges. ■

Some series with variable signs have the property that, when the signs are changed to be all positive, the resulting series converges. When this happens, we say the original series is *absolutely convergent*. This can be said more succinctly as follows.

DEFINITION 11.5 The series $\sum a_n$ is said to be **absolutely convergent** if $\sum |a_n|$ converges. ■

Remarks Here we are allowing a_n to be either positive or negative and we are not explicitly showing the signs, as we did with alternating series. If $a_n \geq 0$ for all n, then it is pointless to speak of absolute convergence. So this concept is meaningful only when the original series has variable signs (not necessarily alternating, however).

EXAMPLE 11.19 Test each of the following for absolute convergence:

(a) $\sum_{n=1}^{\infty} \frac{(-1)^{n-1}}{n}$ (b) $\sum_{n=1}^{\infty} \frac{(-1)^{n-1}}{n^2}$

Solution (a) This is the alternating harmonic series. To test for absolute convergence, we consider the series of absolute values—namely, $\sum_{n=1}^{\infty} 1/n$. But this is the harmonic series itself, which we know to be divergent. Thus, the alternating harmonic series is not absolutely convergent.

(b) To test for absolute convergence, we consider $\sum 1/n^2$. This is a p-series with $p = 2$ and hence is convergent. So the original series is absolutely convergent. ■

In Example 11.17 we showed that the alternating harmonic series converges, but as we have just seen, it is not absolutely convergent. This type of convergence is given a name, as follows.

DEFINITION 11.6 If $\sum a_n$ converges but $\sum |a_n|$ diverges, then $\sum a_n$ is said to be **conditionally convergent.** ■

Thus, of the two series in Example 11.19, the first is conditionally convergent and the second is absolutely convergent.

We now know that a series may be convergent without being absolutely convergent (in which case it is conditionally convergent). However, if a series is absolutely convergent, then it must also be convergent, as the next theorem shows.

THEOREM 11.15 If $\sum a_n$ is absolutely convergent, then it is convergent.

Proof Since $|a_n|$ is either a_n or $-a_n$, it follows that

$$0 \le a_n + |a_n| \le 2|a_n|$$

Thus, since $\sum |a_n|$ (and hence also $\sum 2|a_n|$) converges, it follows by the comparison test that $\sum (a_n + |a_n|)$ converges. Finally, $a_n = (a_n + |a_n|) - |a_n|$, so by property 3 of Theorem 11.7, $\sum a_n$ converges and

$$\sum a_n = \sum (a_n + |a_n|) - \sum |a_n|$$

since each series on the right converges. This completes the proof. ■

The next theorem, which we state without proof, shows that absolute convergence is a much stronger condition than conditional convergence. It deals with *rearrangements* of terms—that is, altering the order in which the terms occur. For example,

$$1 + \frac{1}{3} - \frac{1}{2} + \frac{1}{5} - \frac{1}{4} + \frac{1}{7} - \frac{1}{6} + \cdots$$

is a rearrangement of the alternating harmonic series.

THEOREM 11.16 If $\sum a_n$ is absolutely convergent, then every rearrangement of it converges to the same sum. If $\sum a_n$ is conditionally convergent only, then by a suitable rearrangement it can be made to converge to any real number or to diverge to $+\infty$ or $-\infty$. ■

Comment According to this theorem, no amount of rearranging of terms disturbs the convergence or the value of an absolutely convergent series. On the other hand, a conditionally convergent series is delicately balanced. Any change in the order may cause it to diverge or to converge to another sum. In fact, by a suitable rearrangement, the series can be made to converge to any sum we want. (See Exercise 40.) The sum is therefore *conditional* on the particular way the terms are written.

When one is confronted with a series of variable signs, a good way to proceed is to test first to see that $a_n \to 0$ (so that it has a chance to converge). If $a_n \to 0$, then test for absolute convergence, using one of the tests for positive-term series. If the given series is absolutely convergent, then we know by Theorem 11.15 that it is convergent. If it is not absolutely convergent and the signs alternate, then the alternating series test can be applied. If the signs do not alternate, there are more delicate tests that can be used, but we will not study them here.

EXERCISE SET 11.5

A

In Exercises 1–10 use the alternating series test to determine convergence or divergence.

1. $\sum_{n=1}^{\infty} \frac{(-1)^{n-1}}{\sqrt{n}}$

2. $\sum_{n=1}^{\infty} \frac{(-1)^{n-1}}{\ln(n+1)}$

3. $\sum_{n=1}^{\infty} \frac{(-1)^{n-1}\ln n}{n}$

4. $\sum_{n=1}^{\infty} \frac{(-1)^n n}{\sqrt{1+n^2}}$

5. $\sum_{n=2}^{\infty} \frac{(-1)^n}{n \ln n}$

6. $\sum_{n=1}^{\infty} (-1)^n \ln\left(1+\frac{1}{n}\right)$

7. $\sum_{n=1}^{\infty} \frac{(-1)^n n}{100n+1}$

8. $\sum_{n=1}^{\infty} \frac{\cos n\pi}{n}$

9. $\sum_{n=1}^{\infty} \frac{(-1)^n(n+1)}{n^2-2}$

10. $\sum_{n=1}^{\infty} (-1)^n \ln \frac{2n+3}{2n+1}$

In Exercises 11–14 show that the given series converges, and determine an upper bound on the error in using S_n to approximate the sum, for the specified value of n.

11. $\sum_{n=1}^{\infty} \frac{(-1)^{n-1}}{n^3}$; $n = 9$

12. $\sum_{n=1}^{\infty} \frac{(-1)^{n-1}}{n!}$; $n = 7$

13. $\sum_{n=1}^{\infty} \frac{(-1)^{n-1}}{n^n}$; $n = 5$

14. $\sum_{n=1}^{\infty} \frac{(-1)^n\sqrt{n}}{n+1}$; $n = 9999$

In Exercises 15–18 show that the series converges, and determine the smallest value of n for which the error in using S_n to approximate the sum does not exceed 0.005.

15. $\sum_{n=1}^{\infty} \frac{(-1)^n}{\sqrt{n+1}}$

16. $\sum_{n=2}^{\infty} \frac{(-1)^n}{n(\ln n)^2}$

17. $\sum_{n=1}^{\infty} \frac{(-1)^n(n+1)}{2n^2-3}$

18. $\sum_{n=1}^{\infty} \frac{(-1)^{n-1}(\ln n)^3}{n^2}$

In Exercises 19–30 test for absolute convergence, conditional convergence, or divergence.

19. $\sum_{n=1}^{\infty} \frac{(-1)^{n-1}}{\sqrt{n^3+1}}$

20. $\sum_{n=1}^{\infty} \frac{(-1)^{n-1}\sqrt{n+1}}{n}$

21. $\sum_{n=2}^{\infty} \frac{(-1)^n}{n \ln \sqrt{n}}$

22. $\sum_{n=1}^{\infty} \frac{(-1)^{n-1} n}{2^n}$

23. $\sum_{n=2}^{\infty} \frac{(-1)^n}{\ln n}$

24. $\sum_{n=0}^{\infty} (-1)^n \frac{e^n}{e^{2n}+1}$

25. $\sum_{n=1}^{\infty} \frac{(-1)^{n-1} n}{\sqrt{2n^2+3}}$

26. $\sum_{n=1}^{\infty} \frac{(-1)^{n-1}}{n(n+2)}$

27. $\sum_{n=1}^{\infty} \frac{\cos n}{n^2}$

28. $\sum_{n=1}^{\infty} \frac{n \sin n}{1+n^3}$

29. $\sum_{n=0}^{\infty} ne^{-n}\cos n$

30. $\sum_{n=0}^{\infty} \frac{(1-\pi)^n}{2^{n+1}}$

31. Determine the fallacy in the following "proof" that $2 = 1$. Let S denote the sum of the alternating harmonic series:

$$S = 1 - \frac{1}{2} + \frac{1}{3} - \frac{1}{4} + \frac{1}{5} - \frac{1}{6} + \frac{1}{7} - \frac{1}{8} \cdots$$

Then

$$\begin{aligned} 2S &= 2 - 1 + \frac{2}{3} - \frac{1}{2} + \frac{2}{5} - \frac{1}{3} + \frac{2}{7} - \frac{1}{4} + \cdots \\ &= 1 - \frac{1}{2} + \left(\frac{2}{3} - \frac{1}{3}\right) - \frac{1}{4} + \left(\frac{2}{5} - \frac{1}{5}\right) - \frac{1}{6} + \cdots \\ &= 1 - \frac{1}{2} + \frac{1}{3} - \frac{1}{4} + \frac{1}{5} - \frac{1}{6} + \cdots = S \end{aligned}$$

So $2S = S$, and since $S \neq 0$, $2 = 1$.

B

32. Let

$$a_n = \frac{2^n n!}{3 \cdot 5 \cdot 7 \cdot \cdots \cdot (2n+1)}$$

Prove that $\sum_{n=0}^{\infty} (-1)^n a_n$ is conditionally convergent. (*Hint:* Show that a_n can be written in the form

$$a_n = \frac{2 \cdot 4 \cdot 6 \cdot \cdots \cdot (2n)}{3 \cdot 5 \cdot 7 \cdot \cdots \cdot (2n+1)}$$

Compare this with $1/(2n+1)$. To show that $a_n \to 0$, verify the following:

$$a_n < \frac{3 \cdot 5 \cdot 7 \cdot \cdots \cdot (2n+1)}{4 \cdot 6 \cdot 8 \cdot \cdots \cdot (2n+2)} = \frac{1}{a_n(n+1)}$$

so that

$$a_n^2 < \frac{1}{n+1} \quad \text{or} \quad a_n < \frac{1}{\sqrt{n+1}}\Bigg)$$

33. Prove that

$$\sum_{n=1}^{\infty} (-1)^{n-1} \frac{1 \cdot 3 \cdot 5 \cdot \cdots \cdot (2n-1)}{2 \cdot 4 \cdot 6 \cdot \cdots \cdot (2n)}$$

is conditionally convergent. (*Hint:* Use the idea of the hint in Exercise 32.)

34. Under the hypotheses of Theorem 11.14 show that the partial sums S_{2n+1} of odd order form a monotone decreasing bounded sequence.

35. It can be shown that in any convergent series the associative law holds unrestrictedly; that is, we may group terms by introducing parentheses (but not rearranging the order of the terms). Use this fact to show that

$$\sum_{n=1}^{\infty} \frac{1}{(2n-1)2n} = \frac{1}{1 \cdot 2} + \frac{1}{3 \cdot 4} + \frac{1}{5 \cdot 6} + \cdots$$

converges to the same sum as the alternating harmonic series. (*Hint:* In the alternating harmonic series, group terms by pairs.)

36. Let $\sum a_n$ be conditionally convergent and write

$$p_n = \begin{cases} a_n & \text{if } a_n \geq 0 \\ 0 & \text{if } a_n < 0 \end{cases} \quad \text{and} \quad q_n = \begin{cases} 0 & \text{if } a_n \geq 0 \\ -a_n & \text{if } a_n < 0 \end{cases}$$

(a) Show that $\sum a_n = \sum (p_n - q_n)$ and $\sum |a_n| = \sum (p_n + q_n)$.

(b) Show that $\sum p_n$ and $\sum q_n$ both diverge.

37. Show that in an absolutely convergent series the series of positive terms and the series of negative terms both converge.

38. Let p_n and q_n be defined as in Exercise 36 for the alternating harmonic series. Use a calculator to do the following. Add terms of $\sum p_n$ until the sum first exceeds 2. Then add one or more terms of $\sum(-q_n)$ until the sum is less than 2. Then continue with terms of $\sum p_n$ until the sum again exceeds 2. Continue in this way until you have added a total of 50 nonzero terms. If this process were continued, what could you conclude about the convergence of this rearrangement of the alternating harmonic series? Justify your conclusion.

39. Use the idea of Exercise 38 with the roles of $\sum p_n$ and $\sum(-q_n)$ reversed in order to obtain a rearrangement of the alternating harmonic series with partial sums that can be made arbitrarily close to -1.

40. Use the result of Exercise 36 to explain how a suitable rearrangement of any conditionally convergent series can be made to converge to any sum S we choose.

C

In Exercises 41–44 use a computer to calculate S_n for the smallest n for which $|S_n - S| < \varepsilon$ for the given value of ε.

41. $\displaystyle\sum_{n=1}^{\infty} \frac{(-1)^{n-1}\sqrt[3]{n+1}}{n+2}$; $\varepsilon = 0.01$

42. $\displaystyle\sum_{n=1}^{\infty} \frac{(-1)^n 2^n}{n^n}$; $\varepsilon = 0.0001$

43. $\displaystyle\sum_{n=1}^{\infty} \frac{(-1)^{n+1}}{(n+1)\ln(n+1)}$; $\varepsilon = 0.005$

44. $\displaystyle\sum_{n=1}^{\infty} \frac{(-1)^{n-1} n}{n^3+1}$; $\varepsilon = 0.0005$

11.6 POWER SERIES

Up to this point we have concentrated on series of constants. In this section we consider series with variable terms. Some examples are:

$$\sum_{n=1}^{\infty} \frac{x^n}{n} \qquad \sum_{n=0}^{\infty} \frac{(-1)^n x^{2n}}{(2n)!} \qquad \sum_{n=0}^{\infty} \frac{(x-2)^n}{1+n^2}$$

These are called **power series**, since the variable is raised to powers. The first two are power series in x and the third is a power series in $x - 2$. More specifically, a **power series in x** is a series of the form

$$\sum_{n=0}^{\infty} a_n x^n = a_0 + a_1 x + a_2 x^2 + \cdots + a_n x^n + \cdots \tag{11.4}$$

where the coefficients a_n are constants. A **power series in $x - c$,** where c is a constant, is a series of the form

$$\sum_{n=0}^{\infty} a_n(x-c)^n = a_0 + a_1(x-c) + a_2(x-c)^2 + \cdots + a_n(x-c)^n + \cdots \quad (11.5)$$

Note We use the convention that the left-hand side of equation (11.4) equals a_0 when $x = 0$ and the left-hand side of equation (11.5) equals a_0 when $x = c$.

We will concentrate most of our attention on power series in x, since the theory we develop can be extended easily to power series in $x - c$. Since each partial sum of the series (11.4) or (11.5) is a polynomial, a power series can be thought of as a generalized polynomial.

Concerning convergence, the appropriate question to ask for a power series is not "Does the series converge?" but rather "For what value of x does the series converge?" This is because convergence is defined in terms of series of constants, and a power series becomes a constant series only when x is given a value. All power series in x converge in a trivial way for $x = 0$, as can be seen by putting $x = 0$ in equation (11.4). Similarly, the series (11.5) converges for $x = c$. Some series converge for all real values of x, and others converge on an interval and diverge outside that interval. The nature of the set of x values where convergence occurs can be determined from the following theorem.

THEOREM 11.17 If the power series $\sum a_n x^n$ converges at $x_0 \neq 0$, then it converges absolutely for all x such that $|x| < |x_0|$. If the series diverges at $x_1 \neq 0$, then it diverges for all x such that $|x| > |x_1|$.

Note We will frequently write $\sum a_n x^n$ to mean $\sum_{n=0}^{\infty} a_n x^n$.

Proof Since $\sum a_n x_0^n$ converges, we know by Theorem 11.6 that $\lim_{n\to\infty} a_n x_0^n = 0$, and so for n sufficiently large, say $n > N$, $|a_n x_0^n| < 1$. Now let x be any number for which $|x| < |x_0|$, and denote the ratio $|x/x_0|$ by r, so that $0 \leq r < 1$. Then we have

$$|a_n x^n| = |a_n x_0^n| \cdot \left|\frac{x}{x_0}\right|^n = |a_n x_0^n| \cdot r^n < r^n$$

if $n > N$. Since $\sum r^n$ is a convergent geometric series, it follows by the comparison test that $\sum |a_n x^n|$ converges; that is, $\sum a_n x^n$ converges absolutely.

For the second part, if $\sum a_n x_1^n$ diverges and $|x| > |x_1|$, the series could not converge at x, for if it did, it would converge absolutely at x_1, by what we have just proved. But this contradicts the given fact that the series diverges at x_1. So $\sum a_n x^n$ diverges when $|x| > |x_1|$. ■

As a consequence of Theorem 11.17 and the meaning of the least upper bound, the following result can be proved. (See Exercise 41.)

THEOREM 11.18 For the power series $\sum a_n x^n$, one and only one of the following holds true:

1. The series converges only for $x = 0$.
2. The series converges absolutely for all x.
3. The series converges absolutely for $-R < x < R$ and diverges for $|x| > R$, where R is the least upper bound of the set of x values for which $\sum a_n x^n$ is absolutely convergent ■

The number R in case 3 is called the **radius of convergence** of the series. For convenience, if case 1 holds, we also write $R = 0$, and if case 2 holds, $R = +\infty$. For case 3, in which $0 < R < +\infty$, nothing is said about convergence at $x = R$ or $x = -R$. The series may or may not converge at these points, and if it does converge, the convergence may be absolute or conditional. These end points of the interval $(-R, R)$ must be individually tested using means studied in the last two sections. Depending on convergence or divergence at these end point values, the series will converge in an interval of one of the types $(-R, R)$, $[-R, R]$, $[-R, R)$, or $(-R, R]$, and it will diverge elsewhere. The appropriate interval for a given series is called its **interval of convergence** (sometimes called the *complete* interval of convergence to emphasize that the end points $x = \pm R$ have been tested). When $R = 0$, the interval of convergence degenerates to the single point $x = 0$, and if $R = +\infty$, it is the entire real line $(-\infty, \infty)$.

Theorem 11.18 as well as the notions of radius of convergence and interval of convergence extend, with obvious modifications, to power series of the form $\sum a_n(x - c)^n$. If the radius of convergence is R, for example, where $0 < R < +\infty$, the interval of convergence is of the form $(c - R, c + R)$, or this interval together with one or both of its end points.

When $\lim_{n\to\infty} |a_{n+1}/a_n|$ exists, the radius of convergence can be found using the ratio test. We illustrate this in the next three examples.

EXAMPLE 11.20 Find the interval of convergence of the series

$$\sum_{n=0}^{\infty} \frac{x^n}{2n + 1}$$

Solution We use the ratio test on the series of absolute values. (Remember that the ratio test is applicable only to series of positive terms, and since x can be either positive or negative, we must consider absolute values.)

$$\lim_{n\to\infty} \left| \frac{x^{n+1}}{2(n + 1) + 1} \cdot \frac{2n + 1}{x^n} \right| = \lim_{n\to\infty} \frac{2n + 1}{2n + 3} |x| = |x|$$

The series therefore converges absolutely when $|x| < 1$ and diverges when $|x| > 1$. It follows that $R = 1$. Now we must test the end point values $x = \pm 1$. We do this by substituting in the original series:

$\underline{x = -1}$: $$\sum_{n=0}^{\infty} \frac{(-1)^n}{2n + 1} = 1 - \frac{1}{3} + \frac{1}{5} - \frac{1}{7} + \cdots$$

This is an alternating series, and $1/(2n + 1)$ decreases monotonically to 0. Thus, the series converges.

$\underline{x = 1}$: $$\sum_{n=0}^{\infty} \frac{1}{2n + 1} = 1 + \frac{1}{3} + \frac{1}{5} + \frac{1}{7} + \cdots$$

The term $1/(2n + 1)$ appears to be of the same order of magnitude as $1/n$, so we can try the limit comparison test:

$$\lim_{n\to\infty} \frac{1}{2n + 1} \div \frac{1}{n} = \lim_{n\to\infty} \frac{n}{2n + 1} = \frac{1}{2}$$

This confirms that our series and the harmonic series $\sum 1/n$ behave in the same way. Thus, at $x = 1$ our series diverges.

The complete interval of convergence of the original series is therefore $-1 \leq x < 1$. Note that the convergence is conditional at $x = -1$. ■

EXAMPLE 11.21 Find the interval of convergence of the series

$$\sum_{n=0}^{\infty} \frac{(-1)^n x^{2n}}{(2n)!}$$

Solution Again using the ratio test with the series of absolute values, we have

$$\lim_{n\to\infty} \frac{x^{2(n+1)}}{[2(n+1)]!} \div \frac{x^{2n}}{(2n)!} = \lim_{n\to\infty} \frac{x^{2n+2}}{(2n+2)!} \cdot \frac{(2n)!}{x^{2n}}$$
$$= \lim_{n\to\infty} \frac{x^2}{(2n+2)(2n+1)} = 0$$

for all x. Since $0 < 1$ always, we conclude that the series converges absolutely for all real x; that is, $R = +\infty$ and the interval of convergence is $(-\infty, \infty)$.

Note that in applying the ratio test, we discarded the factor $(-1)^n$, since this is always $+1$ in absolute value. ■

EXAMPLE 11.22 Find the interval of convergence of the series $\sum_{n=0}^{\infty} n!x^n$.

Solution By the ratio test we have

$$\lim_{n\to\infty} \left| \frac{(n+1)!x^{n+1}}{n!x^n} \right| = \lim_{n\to\infty} (n+1)|x|$$

If $x = 0$, the limit is 0, and we have convergence. If $x \neq 0$, the limit is $+\infty$, and so the series diverges. Thus, $R = 0$ and the interval of convergence degenerates to the single point $x = 0$. ■

The same technique can be applied to power series in $x - c$ to find the interval of convergence. The next example illustrates this.

EXAMPLE 11.23 Find the interval of convergence of the series

$$\sum_{n=0}^{\infty} \frac{(-1)^n (x-2)^n}{(n+1)^2 \cdot 3^n}$$

Solution By the ratio test we have

$$\lim_{n\to\infty} \left| \frac{(x-2)^{n+1}}{(n+2)^2 3^{n+1}} \cdot \frac{(n+1)^2 3^n}{(x-2)^n} \right| = \lim_{n\to\infty} \frac{1}{3} \left(\frac{n+1}{n+2} \right)^2 |x-2| = \frac{|x-2|}{3}$$

Thus, the series converges absolutely if $|x-2|/3 < 1$ or, equivalently, if $|x-2| < 3$, and it diverges if $|x-2| > 3$. Now we test the end point values $(x-2) = \pm 3$:

$$\underline{(x-2) = 3:} \quad \sum_{n=0}^{\infty} \frac{(-1)^n 3^n}{(n+1)^2 \cdot 3^n} = \sum_{n=0}^{\infty} \frac{(-1)^n}{(n+1)^2}$$

and this converges absolutely, since the series of absolute values is a p-series with $p = 2$.

$$\underline{(x-2) = -3:} \quad \sum_{n=0}^{\infty} \frac{(-1)^n(-3)^n}{(n+1)^2 \cdot 3^n} = \sum_{n=0}^{\infty} \frac{(-1)^n \cdot (-1)^n \cdot 3^n}{(n+1)^2 \cdot 3^n} = \sum_{n=0}^{\infty} \frac{1}{(n+1)^2}$$

and again this is a convergent p-series. So the complete interval of convergence is $|x-2| \leq 3$, and the convergence is absolute throughout the interval. It is customary to write the answer in interval notation as follows:

$$|x-2| \leq 3$$
$$-3 \leq x - 2 \leq 3$$
$$-1 \leq x \leq 5$$

Thus, the interval of convergence is $[-1, 5]$. Notice that the center of the interval is $x = 2$ and the radius is $R = 3$. ■

It is possible to give an explicit formula for R, based essentially on the method used in the preceding four examples. For the sake of completeness, we give this in the following theorem. As a practical matter, however, it is probably just as well to use the ratio test as we did in the examples.

THEOREM 11.19 Suppose the limit

$$\lim_{n\to\infty} \left| \frac{a_{n+1}}{a_n} \right|$$

is either a finite number L or $+\infty$. Then the radius of convergence R of the series $\sum a_n(x-c)^n$ is

$$\begin{cases} R = \dfrac{1}{L} & \text{if } L \neq 0 \\ R = +\infty & \text{if } L = 0 \\ R = 0 & \text{if the limit is } +\infty \end{cases}$$

Proof Note that the *radius* of convergence is independent of the number c, which is the center of the *interval* of convergence. So this theorem gives R for power series in x (by taking $c = 0$) as well as power series in $x - c$.

As in the preceding examples, we apply the ratio test:

$$\lim_{n\to\infty} \left| \frac{a_{n+1}(x-c)^{n+1}}{a_n(x-c)^n} \right| = \lim_{n\to\infty} \left| \frac{a_{n+1}}{a_n} \right| \cdot |x-c| = L \cdot |x-c|$$

if $\lim_{n\to\infty} |a_{n+1}/a_n| = L$ is finite. If $L = 0$, then $L \cdot |x-c| = 0 < 1$ for all x, so that $R = +\infty$. If $L \neq 0$, the series converges absolutely when $L \cdot |x-c| < 1$ and diverges when $L \cdot |x-c| > 1$; that is, it converges when $|x-c| < \frac{1}{L}$ and diverges when $x - c > \frac{1}{L}$. Thus, $R = \frac{1}{L}$. Finally, if $\lim_{n\to\infty} |a_{n+1}/a_n| = +\infty$, then

$$\lim_{n\to\infty} \frac{a_{n+1}}{a_n} \cdot |x-c| = +\infty$$

except when $x = c$. So the radius of convergence is 0. ■

Remark If $\lim_{n\to\infty}|a_{n+1}/a_n|$ does not exist (and is not $+\infty$), then more sophisticated means that are beyond the scope of this book must be used to find R.

EXERCISE SET 11.6

A

In Exercises 1–36 find the complete interval of convergence.

1. $\sum_{n=1}^{\infty} \frac{x^n}{n}$
2. $\sum_{n=0}^{\infty} \frac{x^n}{n!}$
3. $\sum_{n=0}^{\infty} \frac{(-1)^n n x^n}{n+1}$
4. $\sum_{n=0}^{\infty} \frac{(-1)^n x^n}{1+n^2}$
5. $\sum_{n=0}^{\infty} \frac{x^{2n+1}}{(2n+1)!}$
6. $\sum_{n=0}^{\infty} \frac{(x-1)^n}{\sqrt{n^2+1}}$
7. $\sum_{n=0}^{\infty} \frac{(x+1)^n}{2^n}$
8. $\sum_{n=0}^{\infty} \frac{(x+2)^n}{3^n(n+1)}$
9. $\sum_{n=0}^{\infty} \frac{(-1)^n 2^{n+1} x^n}{3^n}$
10. $\sum_{n=2}^{\infty} \frac{(-1)^n x^{n-2}}{n \ln n}$
11. $\sum_{n=2}^{\infty} \frac{\ln n}{n} x^{n-2}$
12. $\sum_{n=1}^{\infty} \frac{2^n x^{n-1}}{n^2}$
13. $\sum_{n=0}^{\infty} (-1)^n \sqrt{\frac{n+1}{n+2}}\, x^n$
14. $\sum_{n=0}^{\infty} \frac{n+1}{n^2+1} x^n$
15. $\sum_{n=0}^{\infty} \frac{n!(x-1)^n}{2^n}$
16. $\sum_{n=0}^{\infty} \frac{n!}{(2n)!} x^n$
17. $\sum_{n=0}^{\infty} n e^{-n} x^n$
18. $\sum_{n=1}^{\infty} \frac{(-1)^{n-1} 2^n (x-3)^{n-1}}{n^2}$
19. $\sum_{n=0}^{\infty} \frac{(-1)^n \sqrt{n}}{n+1} (x+2)^n$
20. $\sum_{n=2}^{\infty} \left(\frac{3}{4}\right)^{n-2} (x+1)^n$
21. $\sum_{n=2}^{\infty} \frac{(-2x)^{n-2}}{\ln n}$
22. $\sum_{n=0}^{\infty} \frac{n(-3x)^n}{2n+1}$
23. $\sum_{n=0}^{\infty} \frac{(-1)^n n^4 x^n}{e^n}$
24. $\sum_{n=0}^{\infty} \frac{\sqrt{n}(x-5)^n}{1+n^2}$
25. $\sum_{n=0}^{\infty} \frac{n^2-1}{n^2+1} x^{2n}$
26. $\sum_{n=0}^{\infty} \frac{(-1)^n x^{2n+1}}{(2n+1)!}$
27. $\sum_{n=1}^{\infty} \frac{(2x-1)^n}{n\sqrt{n+1}}$
28. $\sum_{n=0}^{\infty} \frac{(3x-2)^n}{(n+1)^{2/3}}$
29. $\sum_{n=0}^{\infty} \frac{(1-x)^n}{2^{n+1}}$
30. $\sum_{n=0}^{\infty} \frac{n!(x+4)^{2n}}{3^n}$

B

31. $\sum_{n=0}^{\infty} \frac{n! x^n}{1 \cdot 3 \cdot 5 \cdot \cdots \cdot (2n+1)}$ (See Exercise 32 in Exercise Set 11.5.)
32. $\sum_{n=1}^{\infty} \frac{1 \cdot 3 \cdot 5 \cdot \cdots \cdot (2n-1)}{2 \cdot 4 \cdot 6 \cdot \cdots \cdot (2n)} x^{n-1}$ (See Exercise 33 in Exercise Set 11.5.)
33. $\sum_{n=1}^{\infty} \frac{n^n x^n}{2^{n+1}}$
34. $\sum_{n=0}^{\infty} \frac{(n!)^2}{(2n)!} x^{2n}$
35. $\sum_{n=0}^{\infty} \frac{2^n+n}{3^n+2} x^n$
36. $\sum_{n=1}^{\infty} \frac{x^n}{(\sqrt{n})^n}$
37. Use the root test to derive a formula for the radius of convergence of $\sum a_n (x-c)^n$, assuming $\lim_{n\to\infty} \sqrt[n]{|a_n|}$ exists.
38. Use the result of Exercise 37 to find the interval of convergence of

$$\sum_{n=1}^{\infty} \left(\frac{n+1}{n}\right)^{n^2} x^n$$

39. If $\sum a_n x^n$ has radius of convergence R, where $0 < R < \infty$, prove that the radius of convergence of $\sum a_n x^{2n}$ is $\sqrt{R}$.
40. Find the radius of convergence of the series $\sum_{n=0}^{\infty} \binom{\alpha}{n} x^n$, where α is a fixed real number and

$$\binom{\alpha}{n} = \frac{\alpha(\alpha-1)(\alpha-2)\cdots(\alpha-n+1)}{n!}$$

41. Prove case 3 of Theorem 11.18 using the following as a guide.

(a) If $x_0 \in (-R, R)$, choose x_1 such that $|x_0| < |x_1| \leq R$ and such that $\sum a_n x_1^n$ is absolutely convergent. Explain why such a choice is possible based on the definition of R. (*Hint:* Otherwise $|x_0|$ would be an upper bound to the set of points where $\sum a_n x^n$ is absolutely convergent. Why is this a contradiction?)

(b) Now use Theorem 11.17. How can you conclude that $\sum a_n x^n$ converges absolutely for *all* x in $(-R, R)$?

(c) To prove divergence for $|x| > R$, assume to the contrary that $\sum a_n x^n$ converges for some x for which $|x| > R$ and arrive at a contradiction.

11.7 DIFFERENTIATION AND INTEGRATION OF POWER SERIES

Within its interval of convergence a power series defines a function. So it is appropriate to write, for example,

$$f(x) = \sum_{n=0}^{\infty} a_n x^n, \qquad x \in I$$

where I is the interval of convergence. For any x_0 in I, $f(x_0)$ is equal to the sum of the convergent series of constants, $\sum a_n x_0^n$. It is appropriate to ask, then, what properties such a function has. For example, we may want to know about continuity, differentiability, or integrability. The following rather remarkable theorem answers these questions. Its proof is quite technical, so we omit it. (A proof can be found in most textbooks on advanced calculus.)

THEOREM 11.20 Let $\sum a_n x^n$ have nonzero radius of convergence R, and for $-R < x < R$ write

$$f(x) = \sum_{n=0}^{\infty} a_n x^n$$

Then

1. f is continuous on the interval $(-R, R)$.
2. f is differentiable on $(-R, R)$ and

$$f'(x) = \sum_{n=0}^{\infty} \frac{d}{dx}(a_n x^n) = \sum_{n=1}^{\infty} n a_n x^{n-1}$$

The series on the right also has radius of convergence R.

3. f is integrable over any interval $[a, b]$ contained in $(-R, R)$, and

$$\int_a^b f(x)\,dx = \sum_{n=0}^{\infty} \int_a^b a_n x^n\,dx$$

Furthermore, f has an antiderivative in $(-R, R)$ given by

$$\int f(x)\,dx = \sum_{n=0}^{\infty} \int a_n x^n\,dx = \sum_{n=0}^{\infty} \frac{a_n x^{n+1}}{n+1} + C$$

The series on the right also has radius of convergence R. ■

Remarks

1. Property 2 says that a power series may be differentiated term by term, and the resulting power series has the same radius of convergence as the original series. Since the differentiated series is itself a power series with radius of convergence R, it too can be differentiated term by term to give $f''(x)$. Repeating this process, it follows that *a power series is infinitely differentiable within* $(-R, R)$. This is a very powerful result.
2. We can paraphrase properties 2 and 3 by saying that a power series can be differentiated or integrated term by term within $(-R, R)$. If the integral is a definite integral over $[a, b]$, then we must have $-R < a < b < R$. If it is an indefinite integral (an antiderivative), the integration is valid for any x in $(-R, R)$.
3. None of the properties can be assumed to be true at the end points $x = \pm R$, even if the original series converges at one or both of these points. Actually it can be proved that continuity *does* extend to an end point if the series converges there. So a power series is continuous on its entire interval of convergence.
4. The theorem extends in an obvious way to $f(x) = \sum a_n(x - c)^n$.

The following examples illustrate some of the consequences of Theorem 11.20.

EXAMPLE 11.24 Find the sum of the series

$$\sum_{n=1}^{\infty} (-1)^{n-1}nx^n = x - 2x^2 + 3x^3 - 4x^4 + \cdots$$

and state the domain of validity.

Solution On factoring out an x we get

$$x(1 - 2x + 3x^2 - 4x^3 + \cdots)$$

and we observe that the series in parentheses is the derivative of

$$x - x^2 + x^3 - x^4 + \cdots = \sum_{n=1}^{\infty} (-1)^{n-1}x^n$$

This latter series is geometric, with $a = x$ and $r = -x$. So if $|x| < 1$, it converges to

$$\frac{a}{1-r} = \frac{x}{1+x}$$

Thus, by property 2 of Theorem 11.20, we have

$$\begin{aligned}\sum_{n=1}^{\infty} (-1)^{n-1}nx^n &= x\frac{d}{dx}\left(\sum_{n=1}^{\infty} (-1)^{n-1}x^n\right)\\ &= x\left(\frac{d}{dx}\frac{x}{1+x}\right)\\ &= \frac{x}{(1+x)^2}\end{aligned}$$

and this is valid for $|x| < 1$. ■

EXAMPLE 11.25 Use the geometric series

$$\frac{1}{1+x} = \sum_{n=0}^{\infty} (-1)^n x^n, \quad |x| < 1$$

to find a power series whose sum is $\ln(1 + x)$.

Solution Since

$$\ln(1+x) = \int \frac{dx}{1+x}$$

we can obtain the desired result by integrating both sides of the given equation. Thus,

$$\int \frac{dx}{1+x} = \sum_{n=0}^{\infty} \int (-1)^n x^n \, dx$$

$$\ln(1+x) = \sum_{n=0}^{\infty} \frac{(-1)^n x^{n+1}}{n+1} + C, \quad |x| < 1$$

To find C we can substitute $x = 0$:

$$\ln 1 = 0 + C$$
$$C = 0$$

So we have

$$\ln(1+x) = \sum_{n=0}^{\infty} \frac{(-1)^n x^{n+1}}{n+1} = x - \frac{x^2}{2} + \frac{x^3}{3} - \frac{x^4}{4} + \cdots$$

and this is valid for $|x| < 1$. Note that when $x = 1$, the series on the right is convergent (verify), and as stated in remark 3 above, the continuity of the series thus extends to include the end point $x = 1$. This means, then, that

$$\lim_{x \to 1} \ln(1+x) = \lim_{x \to 1} \sum_{n=0}^{\infty} \frac{(-1)^n x^{n+1}}{n+1}$$

and since the natural logarithm function also is continuous, we get

$$\ln 2 = 1 - \frac{1}{2} + \frac{1}{3} - \frac{1}{4} + \cdots$$

So we now know that the sum of the alternating harmonic series is ln 2. ■

EXAMPLE 11.26 Find a power series representation of $\tan^{-1} x$ valid near $x = 0$.

Solution We use the idea of the previous example. Since

$$\tan^{-1} x = \int \frac{1}{1+x^2} \, dx$$

we can solve the problem provided we can determine a power series representation of $f(x) = 1/(1 + x^2)$. Writing this in the form

$$\frac{1}{1-(-x^2)}$$

we can interpret it as being the sum $a/(1 - r)$ of the geometric series with first term $a = 1$ and $r = -x^2$. The series will converge for $|-x^2| < 1$ or, equivalently, $|x| < 1$. Thus

$$\frac{1}{1+x^2} = 1 - x^2 + x^4 - x^6 + \cdots = \sum_{n=0}^{\infty} (-1)^n x^{2n}$$

and so

$$\begin{aligned} \tan^{-1} x &= \sum_{n=0}^{\infty} \int (-1)^n x^{2n}\, dx \\ &= \sum_{n=0}^{\infty} \frac{(-1)^n x^{2n+1}}{2n+1} + C \end{aligned}$$

For $x = 0$ we get

$$\begin{aligned} \tan^{-1} 0 &= 0 + C \\ C &= 0 \end{aligned}$$

So

$$\begin{aligned} \tan^{-1} x &= \sum_{n=0}^{\infty} \frac{(-1)^n x^{2n+1}}{2n+1} \\ &= x - \frac{x^3}{3} + \frac{x^5}{5} - \frac{x^7}{7} + \cdots \end{aligned}$$

valid so long as we have $|x| < 1$.

It is readily seen that the series on the right converges at both end points. (Check this.) So, as we have indicated, the series represents a continuous function on the closed interval $[-1, 1]$. Since $\tan^{-1} x$ is also continuous at $x = \pm 1$, it follows as in the preceding example that

$$\tan^{-1} 1 = 1 - \frac{1}{3} + \frac{1}{5} - \frac{1}{7} + \cdots$$

and

$$\tan^{-1}(-1) = -1 + \frac{1}{3} - \frac{1}{5} + \frac{1}{7} - \cdots$$

Using the first of these we find that

$$\frac{\pi}{4} = 1 - \frac{1}{3} + \frac{1}{5} - \frac{1}{7} + \cdots$$

or

$$\pi = 4\left[1 - \frac{1}{3} + \frac{1}{5} - \frac{1}{7} + \cdots\right]$$

We can view this result in two ways: first, it gives a means of calculating π to any degree of accuracy (although it is not efficient, since the series converges very slowly), and second, it is a formula for the sum of the series on the right. ■

EXAMPLE 11.27 Find a power series in x that converges to

$$f(x) = \frac{x^2}{(1-x)^2}$$

and determine the interval of convergence.

Solution As in the preceding example, we make use of the known geometric series

$$\frac{1}{1-x} = 1 + x + x^2 + x^3 + \cdots + x^n + \cdots = \sum_{n=0}^{\infty} x^n$$

valid for $|x| < 1$. By Theorem 11.20 we have, by differentiation,

$$\frac{1}{(1-x)^2} = 1 + 2x + 3x^2 + \cdots + nx^{n-1} + \cdots = \sum_{n=1}^{\infty} nx^{n-1}$$

and this continues to be valid for $|x| < 1$. For x in this interval, multiplying both sides by x^2 yields the desired result:

$$\frac{x^2}{(1-x)^2} = x^2 + 2x^3 + 3x^4 + \cdots + nx^{n+1} + \cdots = \sum_{n=1}^{\infty} nx^{n+1}$$

with interval of convergence $(-1, 1)$. ■

EXAMPLE 11.28 Show that the series

$$\sum_{n=0}^{\infty} \frac{x^n}{n!} = 1 + x + \frac{x^2}{2!} + \frac{x^3}{3!} + \cdots + \frac{x^n}{n!} + \cdots$$

converges to e^x for all real x.

Solution Let

$$f(x) = \sum_{n=0}^{\infty} \frac{x^n}{n!}$$

Applying the ratio test, we have

$$\lim_{n\to\infty} \left| \frac{x^{n+1}}{(n+1)!} \cdot \frac{n!}{x^n} \right| = \lim_{n\to\infty} \frac{|x|}{n+1} = 0$$

Since this limit is always less than 1, the given series converges for all values of x. Its derivative is

$$f'(x) = \sum_{n=1}^{\infty} \frac{nx^{n-1}}{n!} = \sum_{n=1}^{\infty} \frac{x^{n-1}}{(n-1)!}$$

$$= 1 + x + \frac{x^2}{2!} + \frac{x^3}{3!} + \cdots = f(x)$$

That is, $f'(x) = f(x)$ for all values of x. We know that e^x is a function with the property that its derivative equals itself. To show that our $f(x)$ and e^x are identical, consider the derivative

$$\frac{d}{dx}\left(\frac{f(x)}{e^x}\right) = \frac{e^x \cdot f'(x) - f(x) \cdot e^x}{e^{2x}}$$

Since $f'(x) = f(x)$, it follows that this derivative is 0, so that

$$\frac{f(x)}{e^x} = c$$

for some constant c. Putting $x = 0$, we get $c = 1$, since both $f(0)$ and e^0 equal 1. Thus, $f(x) = e^x$. ■

The result of the preceding example deserves special emphasis.

$$e^x = \sum_{n=0}^{\infty} \frac{x^n}{n!} = 1 + x + \frac{x^2}{2!} + \frac{x^3}{3!} + \cdots + \frac{x^n}{n!} + \cdots \tag{11.6}$$

valid for $-\infty < x < \infty$.

A consequence of the convergence of the series in equation (11.6) is that the nth term goes to 0 as $n \to \infty$, and this too is a result we will need later.

$$\lim_{n\to\infty} \frac{x^n}{n!} = 0 \quad \text{for all real } x \tag{11.7}$$

EXAMPLE 11.29 Make use of equation (11.6) to find

$$\int_0^1 \frac{1 - e^{-x}}{x}\, dx$$

in the form of an infinite series, and use this to give the answer correct to three decimal places.

Solution Replacing x by $-x$ in equation (11.6), we get

$$e^{-x} = 1 - x + \frac{x^2}{2!} - \frac{x^3}{3!} + \frac{x^4}{4!} - \cdots$$

Hence,

$$1 - e^{-x} = x - \frac{x^2}{2!} + \frac{x^3}{3!} - \frac{x^4}{4!} + \cdots$$

Thus,

$$\begin{aligned}\int_0^1 \frac{1 - e^{-x}}{x}\, dx &= \int_0^1 \left(1 - \frac{x}{2!} + \frac{x^2}{3!} - \frac{x^3}{4!} + \cdots\right) dx \\ &= x - \frac{x^2}{2 \cdot 2!} + \frac{x^3}{3 \cdot 3!} - \frac{x^4}{4 \cdot 4!} + \cdots \Big]_0^1 \\ &= 1 - \frac{1}{2 \cdot 2!} + \frac{1}{3 \cdot 3!} - \frac{1}{4 \cdot 4!} + \cdots\end{aligned}$$

Note that since this is an alternating series in which the terms $1/n \cdot n!$ go monotonically to 0, we can estimate the error in stopping after n terms by

$$\text{error} \le \frac{1}{(n + 1)(n + 1)!}$$

With $n = 5$, the error is at most $\frac{1}{6} \cdot 6! \approx 0.00023$. So we have

$$\int_0^1 \frac{1 - e^{-x}}{x}\, dx \approx 0.797$$

correct to three decimal places. ■

EXERCISE SET 11.7

A

In Exercises 1–6 differentiate the given series term by term, and determine in what interval the resulting series is the derivative of the function defined by the given series.

1. $1 + x + \dfrac{x^2}{2} + \dfrac{x^3}{3} + \cdots$

2. $x - \dfrac{x^3}{3} + \dfrac{x^5}{5} - \dfrac{x^7}{7} + \cdots$

3. $\sum_{n=0}^{\infty} \dfrac{(-1)^n x^{2n}}{(2n)!}$ **4.** $\sum_{n=0}^{\infty} \dfrac{(-1)^n x^{2n+1}}{(2n+1)!}$

5. $\sum_{n=0}^{\infty} \dfrac{x^n}{2^{n+1}}$ **6.** $\sum_{n=1}^{\infty} \dfrac{x^n}{n^2 + n}$

In Exercises 7–12 find the indicated antiderivatives and give the domain of validity.

7. $\int \sum_{n=0}^{\infty} \dfrac{x^n}{n+1}\, dx$

8. $\int \sum_{n=1}^{\infty} (-1)^{n-1} n x^n\, dx$

9. $\int \sum_{n=0}^{\infty} \dfrac{(-1)^n x^{2n}}{(2n)!}\, dx$

10. $\int \sum_{n=1}^{\infty} \dfrac{x^{2n}}{1 \cdot 3 \cdot 5 \cdot \cdots \cdot (2n-1)}\, dx$

11. $\int \sum_{n=0}^{\infty} \dfrac{(-1)^n (2n+1) x^{2n}}{2^n}\, dx$

12. $\int \sum_{n=1}^{\infty} \dfrac{(-1)^{n-1} x^{2n-1}}{(2n-1)!}\, dx$

In Exercises 13–18 show that term-by-term integration is valid over the given interval, and carry out the integration. Where possible, express the answer in closed form.

13. $\int_0^{1/2} \sum_{n=0}^{\infty} (n+1) x^n\, dx$

14. $\int_{-1}^{1} \sum_{n=0}^{\infty} \dfrac{x^n}{2^n}\, dx$

15. $\int_1^2 \sum_{n=0}^{\infty} \dfrac{x^n}{n!}\, dx$

16. $\int_{-1}^{2} \sum_{n=0}^{\infty} \dfrac{(-1)^n x^n}{3^{n+1}}\, dx$

17. $\int_{-1/2}^{1/2} \sum_{n=0}^{\infty} \dfrac{(-1)^n x^{2n}}{2n+1}\, dx$

18. $\int_{-1}^{3} \sum_{n=0}^{\infty} \dfrac{(-1)^n x^{2n}}{(2n)!}\, dx$

In Exercises 19–30 use differentiation or integration of an appropriate geometric series to find a series representation in powers of x of the given function, and give its radius of convergence.

19. $\dfrac{1}{(1+x)^2}$ **20.** $\dfrac{2}{(2-x)^2}$

21. $\ln(1-x)$ **22.** $\tanh^{-1} x$

23. $\dfrac{2x}{(1-x^2)^2}$ **24.** $\dfrac{2x}{(1-2x)^2}$

25. $\ln\sqrt{1+x}$ **26.** $\ln \dfrac{1+x}{1-x}$

27. $\tan^{-1} \dfrac{x}{2}$ **28.** $6\left(\dfrac{x}{3-2x}\right)^2$

29. $\dfrac{1}{(1-x)^3}$ **30.** $\ln(3+2x)$

31. Show by integrating the series in equation (11.6) that $\int e^x\, dx = e^x + C$.

32. Let $f(x) = e^{2x}$. In equation (11.6) replace x by $2x$ to find a series representation of f. Use this result to calculate $\int_0^1 f(x)\, dx$. What conclusion can you draw concerning the sum of the series

$$1 + \frac{2}{2!} + \frac{2^2}{3!} + \frac{2^3}{4!} + \cdots + \frac{2^n}{(n+1)!} + \cdots$$

B

33. Find the sum of the series $\sum_{n=1}^{\infty} nx^n$ for $|x| < 1$. (*Hint:* Differentiate an appropriate geometric series and then multiply by x.)

34. Find the sum of the series $\sum_{n=0}^{\infty} n^2x^n$. (*Hint:* Write $n^2 = n + n(n-1)$ and use the idea of Exercise 33.)

35. Let $f(x) = xe^x$. Make use of equation (11.6) and $f'(1)$ to find the sum of the series

$$\sum_{n=0}^{\infty} \frac{n+1}{n!}$$

36. Make use of equation (11.6) to find a series for $(e^x - 1)/x$. By differentiating the result, show that

$$\frac{1}{2!} + \frac{2}{3!} + \frac{3}{4!} + \cdots = 1$$

37. (a) Use partial fractions and geometric series to find a power series in x for the function

$$f(x) = \frac{1}{2 + x - x^2}$$

and give its radius of convergence.

(b) Using part a, find a power series representation of

$$\frac{2x-1}{(2+x-x^2)^2}$$

What is the radius of convergence?

38. Use equation (11.6) to evaluate

$$\int_0^1 e^{-x^2}\,dx$$

correct to three decimal places.

39. Define f and g on $(-\infty, \infty)$ by

$$f(x) = \sum_{n=0}^{\infty} \frac{(-1)^n x^{2n}}{(2n)!} \qquad g(x) = \sum_{n=0}^{\infty} \frac{(-1)^n x^{2n+1}}{(2n+1)!}$$

Show the following:
(a) $f''(x) + f(x) = 0$ and $g''(x) + g(x) = 0$
(b) $f(0) = 1$, $f'(0) = 0$, $g(0) = 0$, and $g'(0) = 1$
(c) $f'(x) = -g(x)$ and $g'(x) = f(x)$

11.8 TAYLOR SERIES

We now know that every power series with a positive radius of convergence defines a function that is infinitely differentiable and integrable in the interior of its interval of convergence. Sometimes we can write the function defined by the series in a familiar form, as in the case of the geometric series and other series obtained by differentiating or integrating the geometric series. Also, we found that $\sum x^n/n!$ converges to the exponential function e^x. In many cases, though, the series itself is the only way we have of representing the function it defines. For example,

$$\sum_{n=0}^{\infty} \frac{x^n}{\sqrt{n+1}}$$

defines a function on $(-1, 1)$, but we do not know any other way to write the function than the series itself.

In this section we want to take a different approach, essentially the opposite to that discussed above; that is, given a function f defined on some interval $(-R, R)$, we want to determine whether a power series $\sum a_n x^n$ exists for which $f(x) = \sum a_n x^n$. If the answer is yes, then we want to know the coefficients a_n. One requirement on f is immediately apparent: it must possess derivatives of all orders on $(-R, R)$, since the power series $\sum a_n x^n$ has this property. Let us *assume* for the moment that this differentiability requirement is satisfied and that $f(x) = \sum_{n=0}^{\infty} a_n x^n$. We say that f is *represented* by the series, or alternately that the series *represents* f, on $(-R, R)$, when this is true. The coefficients a_n can be determined by successive differentiation and then putting $x = 0$, as follows:

$$\begin{array}{ll}
f(x) = a_0 + a_1x + a_2x^2 + a_3x^3 + a_4x^4 + \cdots & f(0) = a_0 \\
f'(x) = a_1 + 2a_2x + 3a_3x^2 + 4a_4x^3 + \cdots & f'(0) = a_1 \\
f''(x) = 2a_2 + 3 \cdot 2a_3x + 4 \cdot 3a_4x^2 + \cdots & f''(0) = 2a_2 \\
f'''(x) = 3 \cdot 2a_3 + 4 \cdot 3 \cdot 2a_4x + \cdots & f'''(0) = 3 \cdot 2a_3 \\
\vdots & \vdots \\
f^{(n)}(x) = n!a_n + (n+1)n!a_{n+1}x + \cdots & f^{(n)}(0) = n!a_n \\
\vdots & \vdots
\end{array}$$

Solving for a_n, we get

$$a_n = \frac{f^{(n)}(0)}{n!}, \qquad n = 0, 1, 2, \ldots$$

[where $f^{(0)}(0)$ means $f(0)$, and $0! = 1$]. Thus,

$$f(x) = f(0) + f'(0)x + \frac{f''(0)}{2!}x^2 + \cdots + \frac{f^{(n)}(0)}{n!}x^n + \cdots \tag{11.8}$$

The series in equation (11.8) is called **the Taylor series for f about $x = 0$.** It is also called **the Maclaurin series for f,** after the Scottish mathematician Colin Maclaurin (1698–1746).

If f is infinitely differentiable in an interval of the form $(a - R, a + R)$ for some constant a, and if f can be represented by a power series in $x - a$, then by reasoning similar to that given earlier, the series must be of the form

$$f(x) = f(a) + f'(a)(x-a) + \frac{f''(a)}{2!}(x-a)^2 + \cdots + \frac{f^{(n)}(a)}{n!}(x-a)^n + \cdots \tag{11.9}$$

and this series is called **the Taylor series for f about $x = a$.** Note that the Maclaurin series for f is the special case of the Taylor series in which $a = 0$.

We state the preceding result as a theorem.

THEOREM 11.21 If f is infinitely differentiable in $(a - R, a + R)$ for some $R > 0$ and if f can be represented by a power series in $x - a$ in this interval, then that power series is the Taylor series for f about $x = a$; that is,

$$f(x) = \sum_{n=0}^{\infty} \frac{f^{(n)}(a)}{n!}(x-a)^n$$ ■

We have the following immediate consequence.

COROLLARY 11.21 If f has a power series representation in powers of $x - a$ in an interval $(a - R, a + R)$, then that series is unique. ■

This follows from the fact that whatever the series representation of f is, it must be its Taylor series. An important consequence of this corollary is that whatever appropriate means we might use to arrive at a power series representation for a function (such as by differentiation or integration of a known series), the result will be the same as if we had used equation (11.9).

Using equation (11.9) is the direct method and will always work, but it is not always the easiest method.

EXAMPLE 11.30 Find the Taylor series about $x = 0$ (that is, the Maclaurin series) for e^x.

Solution By Corollary 11.21 we should get the series given by equation (11.6). We will calculate the coefficients by equation (11.8) to verify that this is the case.

Since, for $f(x) = e^x$, we have

$$f(x) = f'(x) = f''(x) = \cdots = e^x$$

it follows that $f^{(n)}(0) = e^0 = 1$ for all n. Thus by equation (11.8) the Maclaurin series for e^x is

$$1 + x + \frac{x^2}{2!} + \frac{x^3}{3!} + \cdots + \frac{x^n}{n!} + \cdots = \sum_{n=0}^{\infty} \frac{x^n}{n!}$$

Note that in Example 11.28 we showed that this series does represent e^x for all real x. ■

EXAMPLE 11.31 Find the Maclaurin series for

$$f(x) = \frac{1}{(1-x)^2}$$

by two methods.

Solution We first use the direct method. To make use of equation (11.8) we need to calculate the successive derivatives and evaluate them at $x = 0$:

$$\begin{aligned}
f(x) &= \frac{1}{(1-x)^2} = (1-x)^{-2} && f(0) = 1 \\
f'(x) &= -2(1-x)^{-3}(-1) = 2(1-x)^{-3} && f'(0) = 2 \\
f''(x) &= -3 \cdot 2(1-x)^{-4}(-1) = 3!(1-x)^{-4} && f''(0) = 3! \\
f'''(x) &= -4 \cdot 3!(1-x)^{-5}(-1) = 4!(1-x)^{-5} && f'''(0) = 4! \\
&\vdots && \vdots \\
f^{(n)}(x) &= (n+1)!(1-x)^{-(n+2)} && f^{(n)}(0) = (n+1)! \\
&\vdots && \vdots
\end{aligned}$$

The general term in the Maclaurin series (11.8) is therefore

$$\frac{f^{(n)}(0)}{n!}x^n = \frac{(n+1)!}{n!}x^n = (n+1)x^n$$

Thus the Maclaurin series for f is

$$1 + 2x + 3x^2 + 4x^3 + \cdots = \sum_{n=0}^{\infty} (n+1)x^n$$

For the second method we use differentiation. Since the geometric series $\sum x^n$ converges to $1/(1-x)$ for $|x| < 1$, we have

$$\frac{1}{1-x} = 1 + x + x^2 + x^3 + x^4 + \cdots = \sum_{n=0}^{\infty} x^n, \qquad |x| < 1$$

So by Theorem 11.20 we can differentiate to get

$$\frac{1}{(1-x)^2} = 1 + 2x + 3x^2 + 4x^3 + \cdots = \sum_{n=1}^{\infty} nx^{n-1}, \qquad |x| < 1$$

and this is the same result we obtained by the first method. Actually, it is a stronger result, for we not only found the Maclaurin series but also showed it *does* represent the function in $(-1, 1)$. ■

EXAMPLE 11.32 Find the Taylor series about $x = 1$ for $f(x) = \ln x$, and find its interval of convergence.

Solution We use equation (11.9) with $a = 1$:

$$\begin{aligned}
f(x) &= \ln x & f(1) &= 0 \\
f'(x) &= \tfrac{1}{x} = x^{-1} & f'(1) &= 1 \\
f''(x) &= -x^{-2} & f''(1) &= -1 \\
f'''(x) &= 2x^{-3} & f'''(1) &= 2 \\
f^{(4)}(x) &= -3 \cdot 2x^{-4} & f^{(4)}(1) &= -3! \\
f^{(5)}(x) &= 4 \cdot 3 \cdot 2x^{-5} & f^{(5)}(1) &= 4! \\
&\vdots & &\vdots \\
f^{(n)}(x) &= (-1)^{n-1}(n-1)!x^{-n} & f^{(n)}(1) &= (-1)^{n-1}(n-1)!, \qquad n \geq 1 \\
&\vdots & &\vdots
\end{aligned}$$

The general term in equation (11.9) is therefore

$$\frac{f^{(n)}(1)}{n!}(x-1)^n = \frac{(-1)^{n-1}(n-1)!}{n!}(x-1)^n = \frac{(-1)^n}{n}(x-1)^n, \qquad n \geq 1$$

Thus, the Taylor series for $\ln x$ about $x = 1$ is

$$\sum_{n=1}^{\infty} \frac{(-1)^{n-1}}{n}(x-1)^n = (x-1) - \frac{(x-1)^2}{2} + \frac{(x-1)^3}{3} - \frac{(x-1)^4}{4} + \cdots$$

To find the interval of convergence, we use the ratio test:

$$\lim_{n\to\infty} \left| \frac{(x-1)^{n+1}}{n+1} \cdot \frac{n}{(x-1)^n} \right| = \lim_{n\to\infty} \frac{n}{n+1}|x-1| = |x-1|$$

So the series converges absolutely when $|x-1| < 1$ and diverges when $|x-1| > 1$. Thus the radius of convergence is 1. The inequality $|x-1| < 1$ is equivalent to $-1 < x - 1 < 1$, or $0 < x < 2$. We must test the end points:

$\underline{x = 0}$: $\displaystyle\sum_{n=1}^{\infty} \frac{(-1)^{n-1}(-1)^n}{n} = -\sum_{n=1}^{\infty} \frac{1}{n}$

This is the negative of the harmonic series and is therefore divergent.

$\underline{x = 2}$: $\displaystyle\sum_{n=1}^{\infty} \frac{(-1)^{n-1}}{n}$

This is the alternating harmonic series, which we have shown to be convergent.

The complete interval of convergence of the series is therefore $(0, 2]$. It should be emphasized, however, that we have *not* yet shown that the series represents the function on this interval. We know only that *if* $\ln x$ can be represented by a series in powers of $x - 1$, then the series is the one we have found. ■

Before giving other examples, we will investigate conditions under which the Taylor series of a function actually converges to the function. The answer can be found in Taylor's formula with remainder (Theorem 9.3), which states that if f has derivatives up through order $n + 1$ in a neighborhood of $x = a$, then

$$f(x) = P_n(x) + R_n(x)$$

for all x in that neighborhood. Recall that

$$P_n(x) = f(a) + f'(a)(x - a) + \frac{f''(a)}{2!}(x - a)^2 + \cdots + \frac{f^{(n)}(a)}{n!}(x - a)^n$$

and

$$R_n(x) = \frac{f^{(n+1)}(c)}{(n + 1)!}(x - a)^{n+1}$$

where c is some number between a and x. Comparing $P_n(x)$ with the Taylor series (11.9), we see that $P_n(x)$ is the nth partial sum (with n starting at 0) of the Taylor series. The Taylor series therefore converges to f if and only if the limit of the partial sums is f:

$$\lim_{n \to \infty} P_n(x) = f(x)$$

This is true if and only if

$$\lim_{n \to \infty} (f(x) - P_n(x)) = 0$$

Since $f(x) - P_n(x) = R_n(x)$, we have the following theorem.

THEOREM 11.22 Let f have derivatives of all orders in the interval $(a - R, a + R)$, where $R > 0$. Then the Taylor series for f converges to $f(x)$ in this interval if and only if $\lim_{n\to\infty} R_n(x) = 0$, where $R_n(x)$ is the remainder term in Taylor's formula:

$$R_n(x) = \frac{f^{(n+1)}(c)}{(n + 1)!}(x - a)^{n+1}, \quad c \text{ between } a \text{ and } x$$ ■

EXAMPLE 11.33 Show that the Maclaurin series for e^x converges to e^x for all real x.

Solution Since all derivatives of e^x are the same as the original function, we have

$$R_n(x) = \frac{f^{(n+1)}(c)}{(n + 1)!}x^{n+1} = \frac{e^c}{(n + 1)!}x^{n+1}$$

where c is between 0 and x. For any $x > 0$, then,

$$|R_n(x)| < e^x \cdot \frac{x^{n+1}}{(n + 1)!}$$

since $0 < c < x$ and the exponential function is increasing. From the limit (11.7) we know that $x^{n+1}/(n + 1)! \to 0$ as $n \to \infty$. Thus, for fixed $x > 0$,

$|R_n(x)|$ can be made arbitrarily small by taking n sufficiently large; that is, $\lim_{n\to\infty} R_n(x) = 0$.

If $x < 0$, then $x < c < 0$ and so $e^c < e^0 = 1$. Thus,

$$|R_n(x)| < 1 \cdot \left|\frac{x^{n+1}}{(n+1)!}\right|$$

and again by limit (11.7) the right-hand side can be made arbitrarily small by choosing n large enough. So $\lim_{n\to\infty} R_n(x) = 0$ in this case also.

When $x = 0$, the Maclaurin series for e^x reduces to the single number 1, which does equal e^0. Thus, making use of Theorem 11.22, we see that for any real number x, e^x is correctly represented by its Maclaurin series. ■

EXAMPLE 11.34 Find the Maclaurin series for $f(x) = \cos x$, and show that it represents $\cos x$ for all real x.

Solution First we find the series

$$\begin{aligned} f(x) &= \cos x & f(0) &= 1 \\ f'(x) &= -\sin x & f'(0) &= 0 \\ f''(x) &= -\cos x & f''(0) &= -1 \\ f'''(x) &= \sin x & f'''(0) &= 0 \\ f^{(4)}(x) &= \cos x & f^{(4)}(0) &= 1 \\ &\vdots & &\vdots \end{aligned}$$

Continuing in this way, we see that we should consider the cases n even and n odd separately. For n even, say $n = 2k$, we have

$$f^{(n)}(x) = f^{(2k)}(x) = (-1)^k \cos x; \qquad f^{(2k)}(0) = (-1)^k$$

For n odd, say $n = 2k + 1$, we have

$$f^{(n)}(x) = f^{(2k+1)}(x) = (-1)^{k-1} \sin x; \qquad f^{(2k+1)}(0) = 0$$

Thus, the Maclaurin series is

$$\sum_{k=0}^{\infty} \frac{(-1)^k}{(2k)!} x^{2k} = 1 - \frac{x^2}{2!} + \frac{x^4}{4!} - \frac{x^6}{6!} + \cdots$$

To show that this converges to $\cos x$ for all x, we need to consider the remainder term

$$R_n(x) = \frac{f^{(n+1)}(c)}{(n+1)!} x^{n+1}$$

Since $f^{(n+1)}(c)$ is either $\pm\cos c$ or $\pm\sin c$, we have, for $x \neq 0$,

$$|R_n(x)| \le 1 \cdot \left|\frac{x^{n+1}}{(n+1)!}\right|$$

and so by equation (11.7) $\lim_{n\to\infty} R_n(x) = 0$. If $x = 0$, the series is finite, reducing to the single term 1, which does equal $\cos 0$. Thus, by Theorem 11.22, for all real x,

$$\cos x = 1 - \frac{x^2}{2!} + \frac{x^4}{4!} - \frac{x^6}{6!} + \cdots$$ ■

EXAMPLE 11.35 Use infinite series to estimate

$$\int_0^1 \left(\frac{\sin x}{x}\right)^2 dx$$

correct to three decimal places.

Solution Since $\sin^2 x = (1 - \cos 2x)/2$, we can use the result of Example 11.34 to write

$$\begin{aligned}\sin^2 x &= \frac{1}{2}\left[1 - \left(1 - \frac{(2x)^2}{2!} + \frac{(2x)^4}{4!} - \frac{(2x)^6}{6!} + \cdots\right)\right]\\ &= \frac{2x^2}{2!} - \frac{2^3x^4}{4!} + \frac{2^5x^6}{6!} - \cdots, \qquad -\infty < x < \infty\end{aligned}$$

Thus,

$$\begin{aligned}\int_0^1 \frac{\sin^2 x}{x^2}\,dx &= \int_0^1 \left[1 - \frac{2^3x^2}{4!} + \frac{2^5x^4}{6!} - \frac{2^7x^6}{8!} + \cdots\right] dx\\ &= x - \frac{(2x)^3}{3\cdot 4!} + \frac{(2x)^5}{5\cdot 6!} - \frac{(2x)^7}{7\cdot 8!} + \cdots\Big]_0^1\\ &= 1 - \frac{2^3}{3\cdot 4!} + \frac{2^5}{5\cdot 6!} - \frac{2^7}{7\cdot 8!} + \cdots\end{aligned}$$

From the first three terms we get

$$\int_0^1 \frac{\sin^2 x}{x^2}\,dx \approx 0.898$$

with an error not exceeding the next term:

$$\text{error} \le \frac{2^7}{7\cdot 8!} \approx 0.000454$$

■

We wish to consider finally an important series known as the **binomial series**, which is the Maclaurin series for $f(x) = (1 + x)^\alpha$ for an arbitrary real number α. If $\alpha = 0$, then $f(x) = 1$, and if α is a positive integer, say $\alpha = n$, we know from the binomial formula that

$$(1 + x)^\alpha = (1 + x)^n = 1 + nx + \frac{n(n-1)}{2!}x^2 + \frac{n(n-1)(n-2)}{3!}x^3 + \cdots + x^n$$

which is a finite series—that is, a polynomial—so that there is no question of convergence. For all other values of α we proceed as usual to find the Maclaurin series:

$$\begin{aligned}
f(x) &= (1 + x)^\alpha & f(0) &= 1\\
f'(x) &= \alpha(1 + x)^{\alpha-1} & f'(0) &= \alpha\\
f''(x) &= \alpha(\alpha-1)(1 + x)^{\alpha-2} & f''(0) &= \alpha(\alpha-1)\\
f'''(x) &= \alpha(\alpha-1)(\alpha-2)(1 + x)^{\alpha-3} & f'''(0) &= \alpha(\alpha-1)(\alpha-2)\\
&\vdots & &\vdots\\
f^{(n)}(x) &= \alpha(\alpha-1)(\alpha-2)\cdots & f^{(n)}(0) &= \alpha(\alpha-1)(\alpha-2)\cdots\\
&\quad(\alpha-n+1)(1 + x)^{\alpha-n} & &\quad(\alpha-n+1)
\end{aligned}$$

So the binomial series is

$$1 + \alpha x + \frac{\alpha(\alpha-1)}{2!}x^2 + \frac{\alpha(\alpha-1)(\alpha-2)}{3!}x^3 + \cdots + \frac{\alpha(\alpha-1)(\alpha-2)\cdots(\alpha-n+1)}{n!}x^n + \cdots \tag{11.10}$$

It is convenient to introduce the notation

$$\binom{\alpha}{n} = \frac{\alpha(\alpha-1)(\alpha-2)\cdots(\alpha-n+1)}{n!}$$

with $\binom{\alpha}{0}$ defined as 1. Then we can write the series in the compact form

$$\sum_{n=0}^{\infty}\binom{\alpha}{n}x^n$$

It is not difficult to show that the radius of convergence for this series is $R = 1$ (see Exercise 27), so that the series converges in $(-1, 1)$, with convergence at the end points in question. This in itself, however, does not guarantee that the function represented by the series in this interval is what we hope it will be—namely, $(1+x)^\alpha$. We can show this by proving that $\lim_{n\to\infty} R_n(x) = 0$. The details of this are tedious, however. We simply state the result in the following theorem.

THEOREM 11.23 If α is any real number other than a nonnegative integer, the binomial series

$$\sum_{n=0}^{\infty}\binom{\alpha}{n}x^n = 1 + \alpha x + \frac{\alpha(\alpha-1)}{2!}x^2 + \frac{\alpha(\alpha-1)(\alpha-2)}{3!}x^3 + \cdots$$

converges to $(1+x)^\alpha$ in the interval $(-1, 1)$ and diverges if $|x| > 1$. If α is a nonnegative integer, the series is finite and represents $(1+x)^\alpha$ for all real values of x. ■

Note When end point values are taken into consideration, it can be shown that the series represents the function precisely in the following intervals, depending on the size of α:

$$\begin{cases} -1 \le x \le 1 & \text{if } \alpha > 0 \\ -1 < x < 1 & \text{if } \alpha \le -1 \\ -1 < x \le 1 & \text{if } -1 < \alpha < 0 \end{cases}$$

EXAMPLE 11.36 Find the Maclaurin series for $f(x) = 1/\sqrt{1-x^2}$.

Solution Since $f(x) = (1-x^2)^{-1/2}$, we can use the binomial series in which $\alpha = -\frac{1}{2}$ and x is replaced by $-x^2$. So we have

$$\begin{aligned}\frac{1}{\sqrt{1-x^2}} &= \sum_{n=0}^{\infty}\binom{-\frac{1}{2}}{n}(-x^2)^n = \sum_{n=0}^{\infty}(-1)^n\binom{-\frac{1}{2}}{n}x^{2n} \\ &= 1 - \left(-\frac{1}{2}\right)x^2 + \frac{\left(-\frac{1}{2}\right)\left(-\frac{3}{2}\right)}{2!}x^4 - \frac{\left(-\frac{1}{2}\right)\left(-\frac{3}{2}\right)\left(-\frac{5}{2}\right)}{3!}x^6 + \cdots \\ &= 1 + \frac{1}{2}x^2 + \frac{1\cdot 3}{2^2\cdot 2!}x^4 + \frac{1\cdot 3\cdot 5}{2^3\cdot 3!}x^6 + \cdots \\ &= 1 + \sum_{n=1}^{\infty}\frac{1\cdot 3\cdot 5\cdot\cdots\cdot(2n-1)}{2^n\cdot n!}x^{2n}\end{aligned}$$

The series converges to the given function for $|-x^2| < 1$ or, equivalently, $-1 < x < 1$. ■

EXAMPLE 11.37 Find the Maclaurin series for $\sin^{-1} x$ and give its radius of convergence.

Solution Since

$$\sin^{-1} x = \int \frac{dx}{\sqrt{1 - x^2}} + C$$

we can integrate the series obtained in the preceding example to get

$$\begin{aligned}
\sin^{-1} x &= \int \left[1 + \sum_{n=1}^{\infty} \frac{1 \cdot 3 \cdot 5 \cdot \cdots \cdot (2n-1)}{2^n \cdot n!} x^{2n}\right] dx + C \\
&= x + \sum_{n=1}^{\infty} \int \frac{1 \cdot 3 \cdot 5 \cdot \cdots \cdot (2n-1)}{2^n \cdot n!} x^{2n}\, dx + C \\
&= x + \sum_{n=1}^{\infty} \frac{1 \cdot 3 \cdot 5 \cdot \cdots \cdot (2n-1)}{2^n \cdot n!} \cdot \frac{x^{2n+1}}{2n+1} + C
\end{aligned}$$

Putting $x = 0$ on each side, we see that $C = 0$. So we can write

$$\sin^{-1} x = x + \frac{1}{2} \cdot \frac{x^3}{3} + \frac{1 \cdot 3}{2^2 \cdot 2!} \cdot \frac{x^5}{5} + \frac{1 \cdot 3 \cdot 5}{2^3 \cdot 3!} \cdot \frac{x^7}{7} + \cdots$$

Since the radius of convergence of the series for $(1 - x^2)^{-1/2}$ is $R = 1$, it follows from Theorem 11.20 that the series for $\sin^{-1} x$ has the same radius. ■

For reference, we list some of the more frequently used Maclaurin series. It would be well to commit the first three of these to memory.

FREQUENTLY USED MACLAURIN SERIES

$$e^x = \sum_{n=0}^{\infty} \frac{x^n}{n!} = 1 + x + \frac{x^2}{2!} + \frac{x^3}{3!} + \cdots, \qquad -\infty < x < \infty$$

$$\begin{aligned}
\sin x &= \sum_{n=0}^{\infty} \frac{(-1)^n x^{2n+1}}{(2n+1)!} \\
&= x - \frac{x^3}{3!} + \frac{x^5}{5!} - \frac{x^7}{7!} + \cdots, \qquad -\infty < x < \infty
\end{aligned}$$

$$\begin{aligned}
\cos x &= \sum_{n=0}^{\infty} \frac{(-1)^n x^{2n}}{(2n)!} \\
&= 1 - \frac{x^2}{2!} + \frac{x^4}{4!} - \frac{x^6}{6!} + \cdots, \qquad -\infty < x < \infty
\end{aligned}$$

$$\begin{aligned}
\ln(1 + x) &= \sum_{n=1}^{\infty} \frac{(-1)^{n-1} x^n}{n} \\
&= x - \frac{x^2}{2} + \frac{x^3}{3} - \frac{x^4}{4} + \cdots, \qquad -1 < x \le 1
\end{aligned}$$

$$\begin{aligned}
\tan^{-1} x &= \sum_{n=1}^{\infty} \frac{(-1)^{n-1} x^n}{2n-1} \\
&= x - \frac{x^3}{3} + \frac{x^5}{5} - \frac{x^7}{7} + \cdots, \qquad -1 \le x \le 1
\end{aligned}$$

$$\sinh x = \sum_{n=0}^{\infty} \frac{x^{2n+1}}{(2n+1)!}$$
$$= x + \frac{x^3}{3!} + \frac{x^5}{5!} + \frac{x^7}{7!} + \cdots, \qquad -\infty < x < \infty$$
$$\cosh x = \sum_{n=0}^{\infty} \frac{x^{2n}}{(2n)!}$$
$$= 1 + \frac{x^2}{2!} + \frac{x^4}{4!} + \frac{x^6}{6!} + \cdots, \qquad -\infty < x < \infty$$

EXERCISE SET 11.8

A

In Exercises 1–18 find the Maclaurin series for the given function and determine its interval of convergence.

1. $\sin x$ **2.** $\sinh x$

3. $\cosh x$ **4.** $\ln(1-x)$

5. $\tanh^{-1} x$ **6.** $\sqrt{1-x}$ (Do not test end points.)

7. $\sin^2 x$ (*Hint:* $\cos 2x = 1 - 2\sin^2 x$.)

8. $\cos^2 x$ (*Hint:* $\cos 2x = 2\cos^2 x - 1$.)

9. $\dfrac{1}{(1-x)^3}$

10. $\sin x \cos x$ (*Hint:* $\sin 2x = 2 \sin x \cos x$.)

11. 2^x **12.** xe^{-x}

13. $\dfrac{1-\cos x}{x^2}$ (defined as its limiting value at 0)

14. $\sin x^2$ **15.** $\sqrt[3]{8-x}$ (Do not test end points.)

16. $\dfrac{\sin x}{x}$ (defined as its limiting value at 0)

17. $\cos\sqrt{x}$ **18.** $x \ln\sqrt{1+x}$

In Exercises 19–24 find the Taylor series for the given function about $x = a$ for the specified value of a, and determine the interval of convergence of the series.

19. e^x; $a = 1$ **20.** $\cos x$; $a = \frac{\pi}{3}$

21. $\frac{1}{x}$; $a = -1$ **22.** $\sqrt{x+3}$; $a = 1$

23. $\sin \frac{\pi x}{6}$; $a = 1$ **24.** $1/(x-1)^2$; $a = 2$

25. For each of the following show that the Maclaurin series of the given function represents the function for all values of x.
(a) $\sin x$ (b) $\sinh x$ (c) $\cosh x$

26. Suppose that for all x in $(a - R, a + R)$, where $R > 0$,
$$\sum_{n=0}^{\infty} a_n(x-a)^n = \sum_{n=0}^{\infty} b_n(x-a)^n$$
Does it follow that $a_n = b_n$? Explain why or why not.

B

27. Show that the radius of convergence of the binomial series is 1.

28. Prove that if $f^{(n)}(x)$ exists for all n and x and there exists a constant M such that $|f^{(n)}(x)| \le M$ for all n and x, then the Taylor series of f about $x = a$ of f exists and converges to $f(x)$ everywhere.

29. For $x \neq 0$, define $f(x) = e^{-1/x^2}$ and let $f(0) = 0$. Prove that the Maclaurin series for f exists and converges everywhere but that it represents f only at $x = 0$.

In Exercises 30–33 approximate the integrals, making use of three terms of the Maclaurin series of the integrand. Estimate the error.

30. $\displaystyle\int_{-1}^{1} \frac{1-\cos x}{x^2}\,dx$ **31.** $\displaystyle\int_0^{0.5} \sqrt{1+x^3}\,dx$

32. $\displaystyle\int_0^{0.4} \frac{\tan^{-1} x}{x}\,dx$ **33.** $\displaystyle\int_{-0.2}^{0} e^{-x^2}\,dx$

34. Find the first three nonzero terms of the Maclaurin series for $\tan x$.

35. Find the first three nonzero terms of the Maclaurin series for $\sec^2 x$. (*Hint:* Use the result of Exercise 34.)

36. The error function is defined by

$$\operatorname{erf}(x) = \frac{2}{\sqrt{\pi}} \int_0^x e^{-t^2}\, dt, \qquad -\infty < x < \infty$$

Find the Maclaurin series for $\operatorname{erf}(x)$, and use it to approximate $\operatorname{erf}(1)$ correct to three decimal places.

11.9 SUPPLEMENTARY EXERCISES

In Exercises 1 and 2 find $\lim_{n\to\infty} a_n$, or show that the limit fails to exist.

1. (a) $a_n = \dfrac{2n^3 - 3n + 1}{n^3 + 2n + 3}$ (b) $a_n = \dfrac{\sqrt{n+10}}{n+2}$

2. (a) $a_n = \dfrac{\sqrt{n^5+1}}{10n^2 + 3n}$ (b) $a_n = \dfrac{n \cos n\pi}{n+1}$

In Exercises 3 and 4 show that the sequence is monotone. Then determine whether it converges by investigating boundedness.

3. (a) $\{e^{1-1/n}\}$ (b) $\left\{\dfrac{2n-3}{n+3}\right\}$

4. (a) $\left\{\dfrac{e^{2n} n!}{(2n-1)!}\right\}$ (b) $\{n \ln(n+1) - (n+1)\ln n\}$

5. Prove that if $\lim_{n\to\infty} a_n = L$, then $\lim_{n\to\infty} a_n^2 = L^2$. Is the converse also true? Prove or disprove.

6. Let

$$S_n = \frac{1}{2n+1} + \frac{1}{2n+2} + \frac{1}{2n+3} + \cdots + \frac{1}{3n}$$

Prove that $\{S_n\}$ converges.

7. A sequence $\{S_n\}$ is defined recursively by $S_1 = 1$, and for $n \geq 2$, $S_n = S_{n-1} + 2/(n^2 + n)$. Find an explicit formula for S_n. Then show that $\{S_n\}$ converges and find its limit.

8. What can you conclude about the convergence or divergence of the series $\sum_{k=1}^{\infty} a_k$ if:

(a) $a_n = \dfrac{2n^2 - 1}{n^2 + 1}$ (b) $a_1 + a_2 + \cdots + a_n = \dfrac{2n^2 - 1}{n^2 + 1}$

Give reasons.

9. Find the sum of each of the following series:

(a) $\displaystyle\sum_{n=0}^{\infty} \frac{(-1)^n 2^{n-1}}{3^n}$ (b) $\displaystyle\sum_{k=2}^{\infty} \frac{1}{k^2 - 1}$

10. Use geometric series to express each of the following repeating decimals as the ratio of two integers:
(a) 1.297297297 . . . (b) 3.2454545 . . .

In Exercises 11–34 test for convergence or divergence. If the series has variable signs, test also for absolute convergence.

11. $\displaystyle\sum_{n=1}^{\infty} \frac{n}{\sqrt{n^3+2}}$

12. $\displaystyle\sum_{n=1}^{\infty} \frac{\sin n + \cos n}{n\sqrt{n+1}}$

13. $\displaystyle\sum_{n=0}^{\infty} \frac{e^{-n}}{1+e^{-n}}$

14. $\displaystyle\sum_{n=2}^{\infty} \frac{\cos n\pi}{\ln n}$

15. $\displaystyle\sum_{n=1}^{\infty} \frac{\tanh n}{n}$

16. $\displaystyle\sum_{n=1}^{\infty} ne^{-2n}$

17. $\displaystyle\sum_{k=1}^{\infty} \frac{\ln k}{k^2}$

18. $\displaystyle\sum_{n=1}^{\infty} \left(\frac{n}{3n+4}\right)^n$

19. $\displaystyle\sum_{k=1}^{\infty} \frac{(-1)^{k-1} \ln k}{\sqrt{k}}$

20. $\displaystyle\sum_{k=1}^{\infty} \frac{3k^2 + 2k - 1}{k^5 + 2}$

21. $\displaystyle\sum_{k=2}^{\infty} \frac{\sin k}{k(\ln k)^2}$

22. $\displaystyle\sum_{n=1}^{\infty} \frac{n-1}{(n+1)^2}$

23. $\displaystyle\sum_{n=0}^{\infty} \frac{2^n n!}{(2n)!}$

24. $\displaystyle\sum_{k=0}^{\infty} (-1)^k k^2 e^{-k}$

25. $\displaystyle\sum_{k=1}^{\infty} \frac{k^k}{3^k k!}$

26. $\displaystyle\sum_{n=2}^{\infty} \frac{\ln n}{n^3 + 4}$

27. $\displaystyle\sum_{n=2}^{\infty} (-1)^n \ln\left(\frac{n+1}{n-1}\right)$

28. $\displaystyle\sum_{k=0}^{\infty} \frac{\sin(k+\frac{1}{2})\pi}{k+\frac{1}{2}}$

29. $\displaystyle\sum_{n=1}^{\infty} \frac{n^{2n}}{(n^3+1)^n}$

30. $\displaystyle\sum_{n=1}^{\infty} \left(\frac{1}{n} - \frac{1}{n+2}\right)$

31. $\displaystyle\sum_{k=1}^{\infty} \frac{2 + \cos n}{\sqrt{2n^2 - 1}}$

32. $\displaystyle\sum_{n=0}^{\infty} \frac{\operatorname{sech}^2 n}{1 + \tanh n}$

33. $\displaystyle\sum_{n=0}^{\infty} (-1)^n(\sqrt{n+1} - \sqrt{n})$

34. $\displaystyle\sum_{n=0}^{\infty} \frac{(-1)^n (n-1)}{2n+1}$

In Exercises 35–42 find the complete interval of convergence.

35. $\displaystyle\sum_{n=0}^{\infty} \frac{2^n x^n}{3^{n+1}}$

36. $\displaystyle\sum_{n=0}^{\infty} \frac{(-1)^n n x^n}{2n+1}$

37. $\displaystyle\sum_{k=0}^{\infty} \frac{(x-2)^k}{\sqrt{2k+1}}$

38. $\displaystyle\sum_{k=0}^{\infty} \frac{(-1)^k (x+1)^{2k}}{3^k}$

39. $\displaystyle\sum_{k=1}^{\infty} \frac{kx^k}{\ln(k+1)}$

40. $\displaystyle\sum_{n=1}^{\infty} \frac{(nx)^n}{n!e^n}$ (Do not check end points.)

41. $\displaystyle\sum_{n=0}^{\infty} \frac{(-1)^n(2x-1)^n n!}{1\cdot 3\cdot 5\cdot \cdots \cdot(2n+1)}$ (See Exercise 32 in Exercise Set 11.5.)

42. $\displaystyle\sum_{n=1}^{\infty} \frac{\cosh n}{n^2}(x-1)^n$

In Exercises 43–46 use integration or differentiation of an appropriate series to find the Maclaurin series of f. Give the radius of convergence.

43. $f(x) = \ln(1-x^2)$ **44.** $f(x) = x/(2+x)^2$

45. $f(x) = \sinh^{-1} x$ **46.** $f(x) = \frac{1}{a}\tan^{-1}\frac{x}{a}$

47. Use infinite series to verify the following limits:

(a) $\displaystyle\lim_{x\to 0} \frac{\sin x}{x} = 1$ (b) $\displaystyle\lim_{x\to 0} \frac{1-\cos x}{x^2} = \frac{1}{2}$

(c) $\displaystyle\lim_{x\to 0} \frac{e^x - 1}{x} = 1$

In Exercises 48–50 find the Taylor series about $x = a$.

48. $f(x) = 1/(2-x);\ a = 1$

49. $f(x) = \ln\sqrt{x+2};\ a = -1$

50. $f(x) = \sin x;\ a = \frac{\pi}{4}$. Show that the series represents the function for all values of x.

In Exercises 51 and 52 approximate the integral using the first four nonzero terms of the Maclaurin series of the integrand. Give an upper bound on the error.

51. $\displaystyle\int_0^{1/2} \frac{1-e^{-x}}{x}\,dx$ **52.** $\displaystyle\int_0^{1/2} \frac{dx}{\sqrt{1+x^3}}$

53. Find the Taylor series for $f(x) = \cos^2 x - \sin^2 x$ about $x = \frac{\pi}{4}$, and show that it converges to f for all x. (*Hint:* First use a trigonometric identity.)

54. Show that the Taylor series about $x = 0$ for $f(x) = \ln(1+x)$ represents f for all x in $[0, 1]$. [*Note:* It represents f for all x in $(-1, 0)$ also, but this is more difficult to show.]

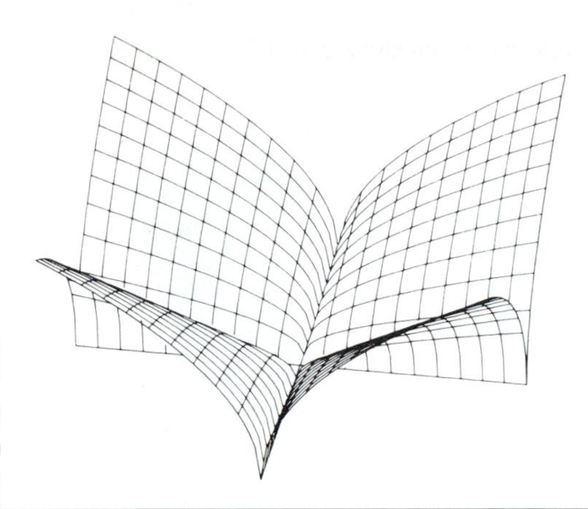

CHAPTER 12

THE CONIC SECTIONS

12.1 INTRODUCTION

The circle, parabola, ellipse, and hyperbola are curves that have played a special role in mathematics since the time of the ancient Greeks, and their importance today is undiminished. Although we have already encountered them in our study of calculus, we devote this chapter to a more detailed analysis of them.

These four curves are known as **conic sections,** or simply **conics,** because they are curves of intersection of a plane with a right circular cone of two nappes, as shown in Figure 12.1, where the plane does not pass through the vertex of the cone. When the intersecting plane does contain the vertex of the cone, the intersection is a single point, a single line, or two intersecting lines. Each of these is called a *degenerate conic.*

The definition of conic sections as curves resulting from planes passing through cones is primarily of historical interest. It was Apollonius of Perga, in the third century B.C., who first studied these curves extensively, writing eight books on the subject. Apollonius did not have the tools of analytical geometry at his disposal, which makes his discoveries all the more remarkable.

Present-day applications of conic sections are found in such diverse fields as art, architecture, astronomy, engineering, and physics. They play an important role in atomic and electromagnetic field theory, as well as in acoustics and optics. The orbits of planets around the sun and satellites around the earth are elliptical. Parabolic reflectors are used for radar, radio telescopes, searchlights, and solar energy devices. Certain atomic particles follow paths that are approximately hyperbolic.

Our approach to each conic will be to give a geometric definition and from this derive its algebraic equation. We have already done this for the circle in Chapter 1. To summarize, we defined the circle as the set of all points in a plane at a fixed distance (the radius) from a fixed point in the plane (the center). From this we obtained, using the distance formula, the

Plane perpendicular to axis of cone; intersection is a circle.

(a)

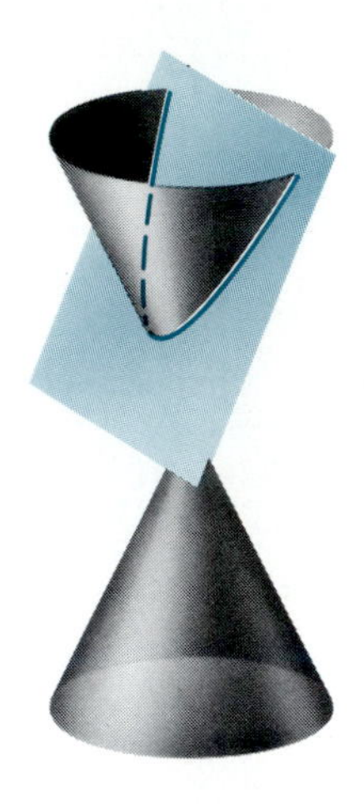

Plane parallel to element of cone; intersection is a parabola.

(b)

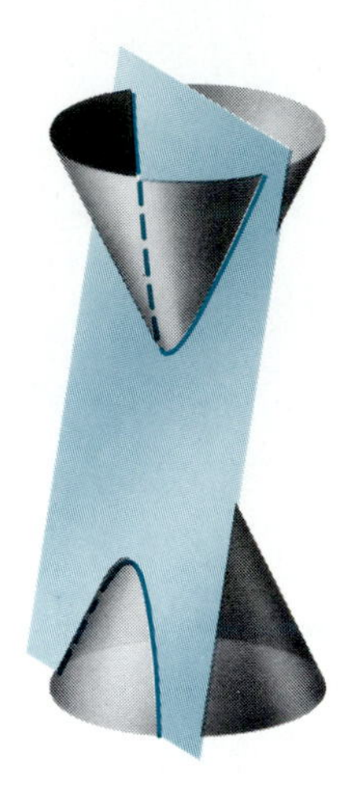

Plane intersecting both nappes of cone; intersection is a hyperbola.

(c)

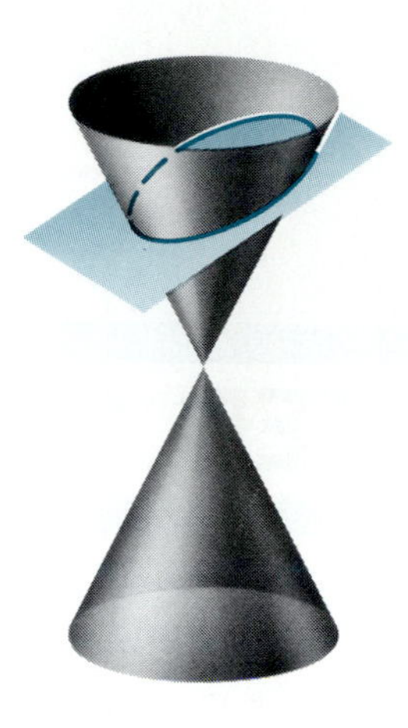

Any other plane intersecting one nappe; intersection is an ellipse.

(d)

FIGURE 12.1

standard form

$$(x - h)^2 + (y - k)^2 = r^2$$

where the center is at (h, k) and the radius is r. We showed also that every equation of the form

$$x^2 + y^2 + ax + by + c = 0$$

has for its graph a circle or a point (a degenerate circle), if it has a graph at all. The key to recognizing this is the combination of second-degree terms, $x^2 + y^2$. So when the second-degree terms in a quadratic equation in x and y occur only in the combination $x^2 + y^2$, you should immediately think of a circle.

Now we proceed with a similar analysis of the parabola, the ellipse, and the hyperbola.

12.2 THE PARABOLA

A parabola is defined as follows.

DEFINITION 12.1 A **parabola** is the set of all points in a plane equidistant from a fixed line and a fixed point not on the line. ■

The fixed line in this definition is called the **directrix** of the parabola, and the fixed point is called the **focus.** The line through the focus perpendicular to the directrix is called the **axis,** and the point where the parabola crosses its axis is called the **vertex.** From Definition 12.1 it follows that the vertex is

FIGURE 12.2

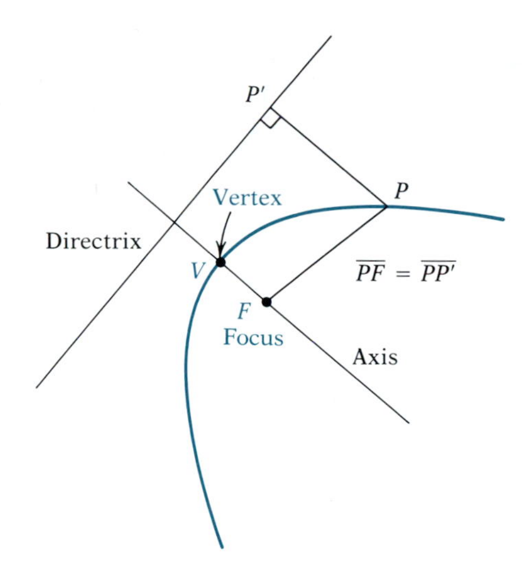

midway between the focus and the directrix. These ideas are illustrated in Figure 12.2.

The simplest equations for parabolas occur when the vertex is at the origin and the axis is along either the x-axis or the y-axis. We begin with the case shown in Figure 12.3, with the axis on the x-axis and the focus at a point p units to the right of the origin, where $p > 0$. It follows that the directrix is p units to the left of the origin and so has the equation $x = -p$. By Definition 12.1 a point $P(x, y)$ will be on the parabola if and only if $\overline{PF} = \overline{PP'}$, where F is the focus and P' is the horizontal projection of P onto the directrix and so has coordinates $(-p, y)$. Using the distance formula, this requirement becomes

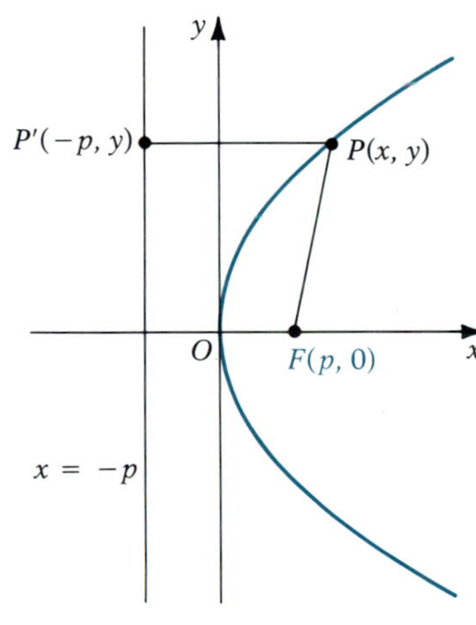

FIGURE 12.3

$$\sqrt{(x - p)^2 + y^2} = x + p$$

These two nonnegative numbers are equal if and only if their squares are equal:

$$x^2 - 2px + p^2 + y^2 = x^2 + 2px + p^2$$

or

$$y^2 = 4px \tag{12.1}$$

This is the equation of the parabola in question. When a parabola is in this position, we refer to it as being in **standard position I** with vertex at the origin. So standard position I means a horizontal axis opening to the right. Positions II, III, and IV are obtained by rotating position I counterclockwise by 90°, 180°, and 270°, respectively. It is easy to show that when the vertex is at the origin, the equations for the parabolas in these other three positions are those shown in Figure 12.4.

Summarizing, we can say that if a parabola with vertex at the origin has a horizontal axis, its equation is of the form $y^2 = \pm 4px$, opening to the right if the sign is plus and to the left if the sign is minus. If the axis is vertical, the equation is of the form $x^2 = \pm 4py$, opening upward if the sign is plus and downward if it is minus.

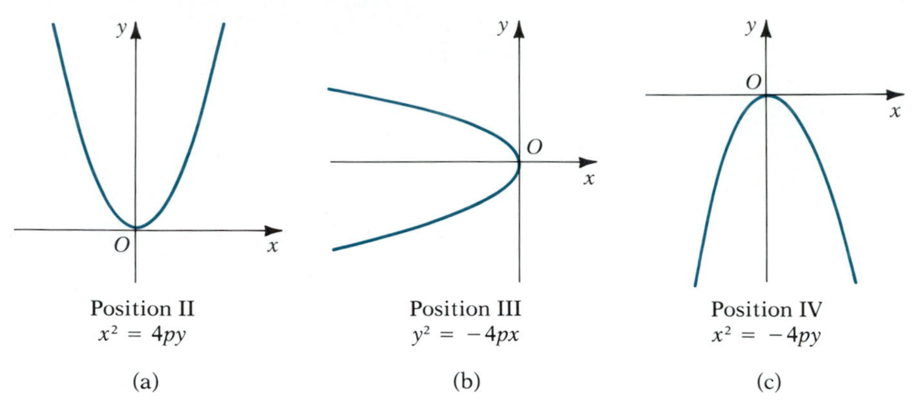

FIGURE 12.4

Next we want to determine equations of parabolas in one of the standard positions but with the vertex not at the origin. For this purpose we consider the general question of how equations are altered when the axes are moved. The results will be useful in a broader context, in particular when we study the other two conics.

Translation of Axes

Suppose that in a given xy-coordinate system we introduce new x'- and y'-axes parallel to the x- and y-axes, respectively, and with the same orientation, as shown in Figure 12.5. Then we say that the new system is a **translation** of the old. Let the origin of the new system have coordinates (h, k) with respect to the original system. A point P in the plane then has two sets of coordinates, one referred to the old xy system and one referred to the new $x'y'$ system. The relationships between the new and the old coordinates are seen from Figure 12.5 to be

$$\begin{cases} x = x' + h \\ y = y' + k \end{cases} \tag{12.2}$$

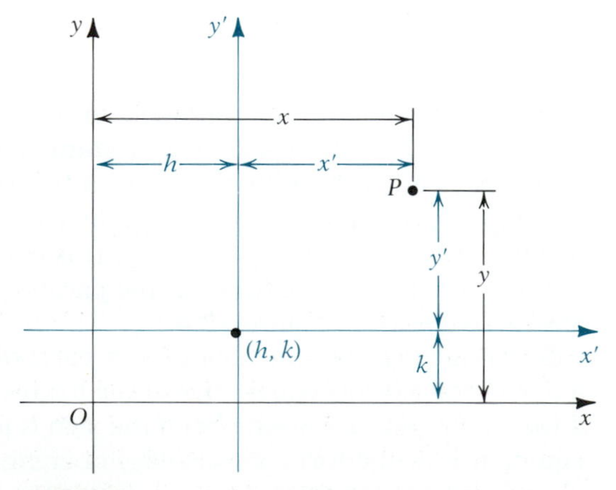

FIGURE 12.5

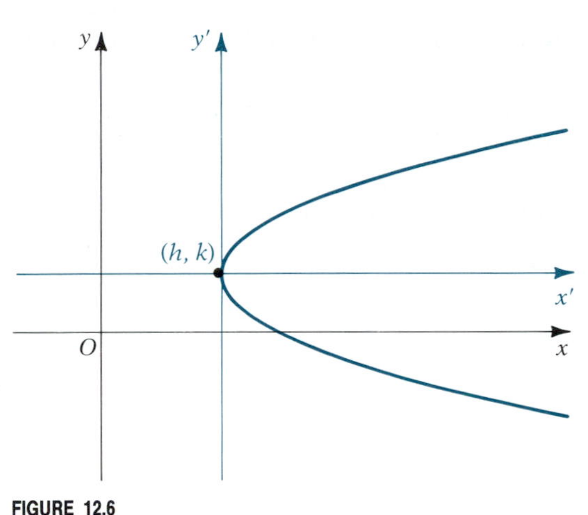

FIGURE 12.6

or, equivalently,

$$\begin{cases} x' = x - h \\ y' = y - k \end{cases} \tag{12.3}$$

If we know the equation of a curve in x and y, then we can get the new equation with respect to the translated system by replacing x by $x' + h$ and y by $y' + k$. Conversely, if we know the equation in x' and y', we can replace x' by $x - h$ and y' by $y - k$ to get the equation with respect to the original system.

Now suppose a parabola in standard position I has its vertex at the point (h, k). If we introduce new axes, translated so that the origin of the new system is at (h, k), as shown in Figure 12.6, then the equation of the parabola with respect to the new system is

$$y'^2 = 4px'$$

By the translation equations (12.3), it follows that when referred to the original axes, the equation is

$$(y - k)^2 = 4p(x + h)$$

Using a similar analysis for the other three positions, we get the following standard forms for the vertex at (h, k).

STANDARD FORMS OF EQUATIONS OF PARABOLAS

$(y - k)^2 = \pm 4p(x - h)$,	axis horizontal	(12.4)
$(x - h)^2 = \pm 4p(y - k)$,	axis vertical	(12.5)

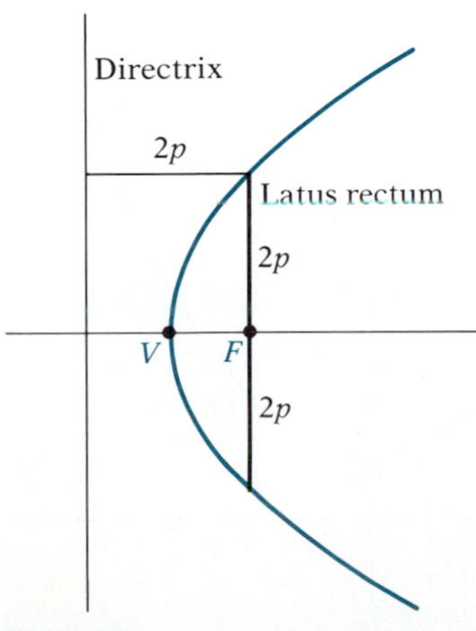

FIGURE 12.7

Recall that the number p is the distance from the vertex to the focus or equivalently, the distance from the vertex to the directrix.

The line segment through the focus of a parabola, perpendicular to its axis and terminating on the parabola, is called the **latus rectum** of the parabola. We show this in Figure 12.7. From the definition of the parabola, the length of the latus rectum must be $4p$, since the distance from one of its end points to the directrix is $2p$, so that this also is the distance from the end point to the focus. A rapid sketch of a parabola can be made in most cases by using the vertex and the ends of the latus rectum. We illustrate this in the examples that follow.

EXAMPLE 12.1 Find the equation in standard form of each of the following parabolas, and sketch their graphs:

(a) Focus $(-3, 2)$; directrix $x = 1$

(b) Vertex $(1, -4)$; end points of latus rectum $(-1, -3)$ and $(3, -3)$

Solution (a) Since the directrix is vertical, the axis is horizontal, and since the focus is to the left of the directrix, the parabola opens to the left (position III). So the equation is

$$(y - k)^2 = -4p(x - h)$$

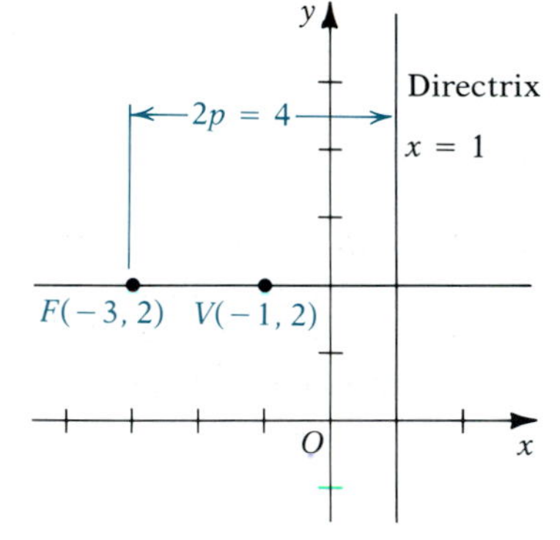

FIGURE 12.8

The distance from the focus to the directrix is $2p$, and so $2p = 4$ or $p = 2$. The vertex is midway between the focus and the directrix, at $(-1, 2)$. So $h = -1$ and $k = 2$. (See Figure 12.8.) We now have all the relevant information, and the equation is

$$(y - 2)^2 = -8(x + 1)$$

A sketch, making use of the end points $(-3, 6)$ and $(-3, -2)$ of the latus rectum, is shown in Figure 12.9.

(b) From the given information we can sketch the graph, as in Figure 12.10. The length of the latus rectum is $4p$, so we see that $4p = 4$ and $p = 1$. The focus is at $(1, -3)$. The equation is

$$(x - h)^2 = 4p(y - k)$$

and with $h = 1$, $k = -4$, and $p = 1$, this becomes

$$(x - 1)^2 = 4(y + 4)$$

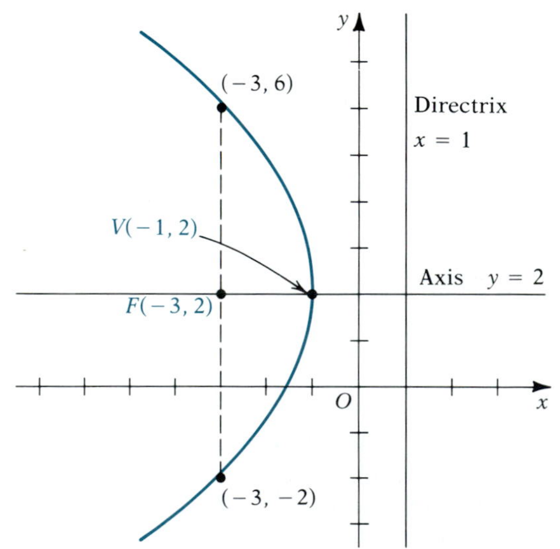

FIGURE 12.9

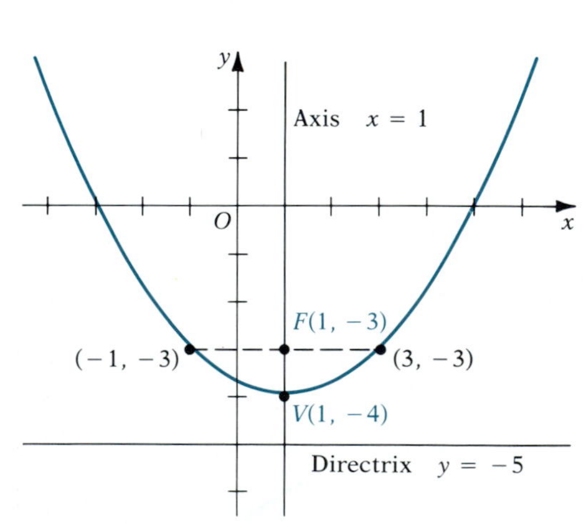

FIGURE 12.10

If either of equations (12.4) is multiplied out, we get an equation of the form

$$y^2 + ax + by + c = 0, \qquad \text{axis horizontal} \tag{12.6}$$

Similarly, equations (12.5) expand to the form

$$x^2 + ax + by + c = 0, \qquad \text{axis vertical} \tag{12.7}$$

If we are given an equation in the form of (12.6) or (12.7), just as with the circle we can put it in one of the standard forms by completing the square, as we show in the next example.

EXAMPLE 12.2 Discuss and sketch the graph of each of the following:

(a) $y^2 - 6x + 4y + 16 = 0$ (b) $3x^2 - 5x + 2y - 4 = 0$

Solution (a) Completing the square in y gives

$$y^2 + 4y + 4 = 6x - 16 + 4$$
$$(y + 2)^2 = 6x - 12$$
$$(y + 2)^2 = 6(x - 2)$$

So this is a parabola that opens to the right, with $4p = 6$ so that $p = \frac{3}{2}$. The vertex is $(2, -2)$, so that the focus is $(\frac{7}{2}, -2)$ and the directrix is the line $x = \frac{1}{2}$. In Figure 12.11 we show the graph, making use of the latus rectum, which extends $2p = 3$ units to either side of the axis.

(b) First we divide through by 3 and then complete the square:

$$x^2 - \tfrac{5}{3}x + \tfrac{2}{3}y - \tfrac{4}{3} = 0$$
$$x^2 - \tfrac{5}{3}x + \tfrac{25}{36} = -\tfrac{2}{3}y + \tfrac{4}{3} + \tfrac{25}{36}$$
$$(x - \tfrac{5}{6})^2 = -\tfrac{2}{3}y + \tfrac{73}{36}$$
$$(x - \tfrac{5}{6})^2 = -\tfrac{2}{3}(y - \tfrac{73}{24})$$

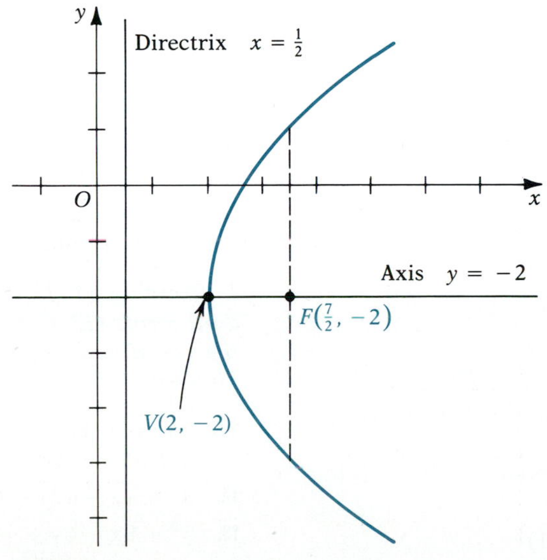

FIGURE 12.11

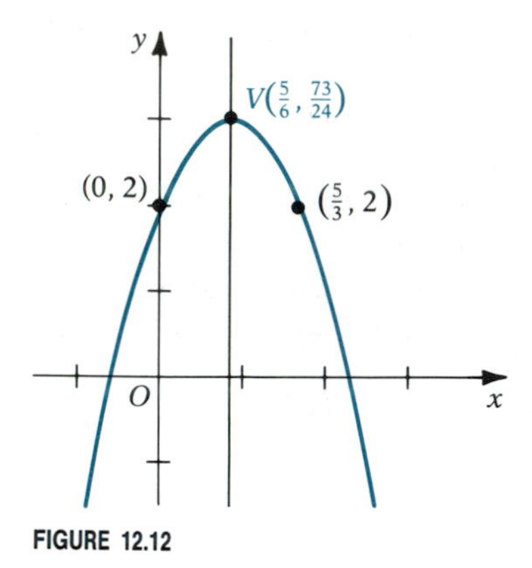

FIGURE 12.12

So this is a parabola with vertex $(\frac{5}{6}, \frac{73}{24})$, opening downward. Since $4p = \frac{2}{3}$, $p = \frac{1}{6}$. The focus is therefore $\frac{1}{6}$ unit down from the vertex, and the latus rectum extends $\frac{1}{3}$ unit to either side. Since the focus is so close to the vertex, the latus rectum is not of much help in the sketch this time. The y-intercept (0, 2) is readily obtained, however, and hence the point $(\frac{5}{3}, 2)$, symmetric to (0, 2) with respect to the axis of the parabola, is also on the curve. The graph is shown in Figure 12.12. ■

By the procedure of the preceding example we can show that so long as $a \neq 0$ in equation (12.6), the graph is a parabola with horizontal axis, and so long as $b \neq 0$ in equation (12.7), the graph is a parabola with vertical axis. The key to recognizing a second-degree polynomial equation in x and y (without an xy term) as being the equation of a parabola is the presence of *either* x^2 *or* y^2, *but not both.* If the equation involves only one variable [$a = 0$ in equation (12.6) or $b = 0$ in equation (12.7)], it will have a graph that consists of two parallel or coincident lines (a degenerate parabola) or else there will be no graph.

FIGURE 12.13

Parabolas have an interesting and useful focusing property. If a parabolic arc is rotated about its axis, a surface called a *paraboloid* is formed. When sound waves or light rays parallel to the axis strike such a surface, they are all reflected to the focus of the parabola (see Figure 12.13). Reflecting telescopes make use of this principle, for example, as do radio telescopes. In reverse, if light is emitted from the focus, then all rays are reflected off the surface in rays parallel to the axis of the parabola (see Exercise 29). For this reason headlights on a car are in the shape of paraboloids.

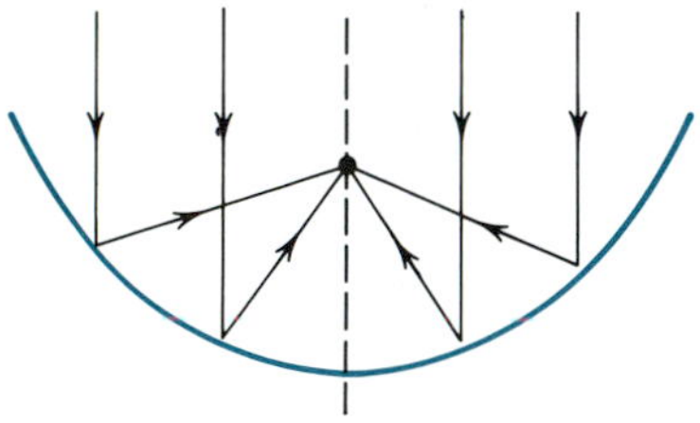

EXERCISE SET 12.2

A

In Exercises 1–10 find the equation of the parabola described.

1. (a) Focus (2, 0); directrix $x = -2$
(b) Focus (0, -3); directrix $y = 3$

2. (a) Vertical axis; vertex at the origin; passing through (-3, 2)
(b) Horizontal axis; vertex at the origin; passing through (-4, -3)

3. (a) Vertex (1, -2); directrix $y = -4$
(b) Vertex (-3, 1); focus (-3, -1)

4. (a) Focus (2, 4); directrix $x = 4$
(b) Focus (-4, -2); vertex (-5, -2)

5. (a) Vertex (2, -3); end points of latus rectum (-4, 0) and (8, 0)
(b) Directrix $x + 2 = 0$; end points of latus rectum (1, 2) and (1, -4)

6. (a) Vertex (2, 3); focus (2, -1)
(b) Focus (0, 4); directrix $x = 6$

7. Vertex (2, -4); axis vertical; passing through (-2, 0)

8. Vertex (3, -2); axis horizontal; passing through (-1, 2)

9. Axis horizontal; passing through (0, 3), (-2, 1), and (6, -3) [*Hint:* Use equation (12.6).]

10. Axis vertical; passing through (1, 1), (-1, 3), and (3, 5) [*Hint:* Use equation (12.7).]

In Exercises 11–17 write the equation in one of the standard forms (12.4) or (12.5) and sketch the graph. Give the coordinates of the focus and the equation of the directrix.

11. (a) $y^2 = 8x$ (b) $y^2 + 4x = 0$
(c) $x^2 = -6y$ (d) $x^2 - 8y = 0$

12. $x^2 - 4x - 12y - 8 = 0$

13. $y^2 + 8x - 6y + 41 = 0$

14. $y = x^2 - 2x$

15. $x = y^2 - 3y + 4$

16. $2y^2 - 5x + 3y - 4 = 0$

17. $4y^2 + 4y + 24x - 35 = 0$

In Exercises 18–21 draw both graphs on the same coordinate system, and find the indicated area.

18. Area between $y = 3x - x^2$ and $y = x - 3$

19. Area between $(x + 1)^2 = 2(y - 3)$ and $x + y = 2$

20. Area between $(y - 1)^2 = 4(x - 1)$ and $(y - 1)^2 = -(x - 6)$

21. Area between $x^2 - 6x + 4y - 11 = 0$ and $x^2 - 6x - 8y + 1 = 0$

22. Find the volume generated by revolving the area bounded by the upper half of the parabola $y^2 = x$ and the line $x - 2y = 0$ about (a) the x-axis and (b) the line $x = 4$.

23. (a) Find the length of the arc on the parabola $x^2 = 2y$ from (0, 0) to (2, 2).
(b) Find the surface area generated by revolving the arc of part a about the y-axis.

24. A searchlight is in the form of a parabola rotated about its axis (a paraboloid). If the depth of the light is 6 in. and the diameter across the face is 16 in., where should the bulb be placed to send rays out parallel to the axis?

25. The underneath side of a masonry bridge is in the form of a parabolic arc with a span of 60 m and a maximum height of 10 m. Find the height at 10-m intervals from one end to the other.

B

26. Find the area between the graphs of $y^2 = 8x$ and $8x^2 + 5y = 12$. Sketch the curves.

27. Find the area inside the circle

$$x^2 + y^2 - 2x + 4y + 1 = 0$$

and above the parabola $x^2 - 2x - 3y = 5$.

28. Find the equation of the parabola having directrix $x - 2y - 4 = 0$ and focus $(-2, 0)$. Sketch the parabola. (*Hint:* Use the result of Exercise 30 in Exercise Set 1.3.)

29. Prove that if P is any point on a parabola with focus F, the angle α from the tangent line drawn at P to the line PF equals the angle β from a line parallel to the axis through P to the tangent line. (See the figure.)

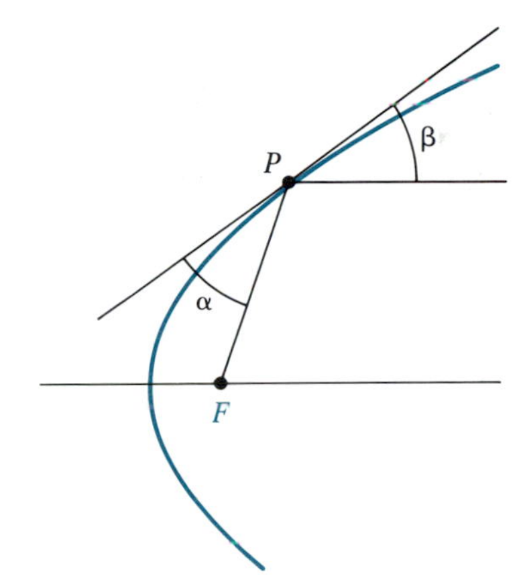

12.3 THE ELLIPSE

An ellipse is defined as follows.

DEFINITION 12.2 An **ellipse** is the set of points in a plane the sum of whose distances from two fixed points in the plane is a constant. ■

The two fixed points in this definition are called the **foci** of the ellipse, and the midpoint of the line segment joining the foci is called the **center.** Figure 12.14 illustrates the definition. For any point P on the ellipse, the sum $\overline{PF_1} + \overline{PF_2}$ is constant, where F_1 and F_2 are the foci. The points V_1 and V_2 where the ellipse crosses the line through the foci are called **vertices** of the ellipse. The chord of the ellipse joining the vertices is called the **major axis,** and the chord through the center perpendicular to the major axis is called the **minor axis.**

FIGURE 12.14

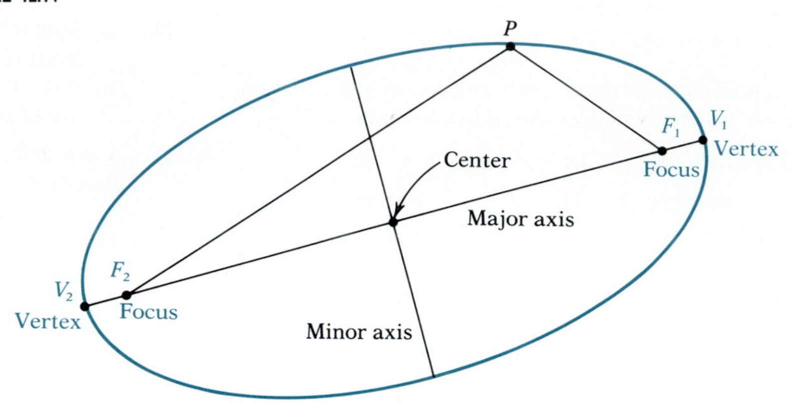

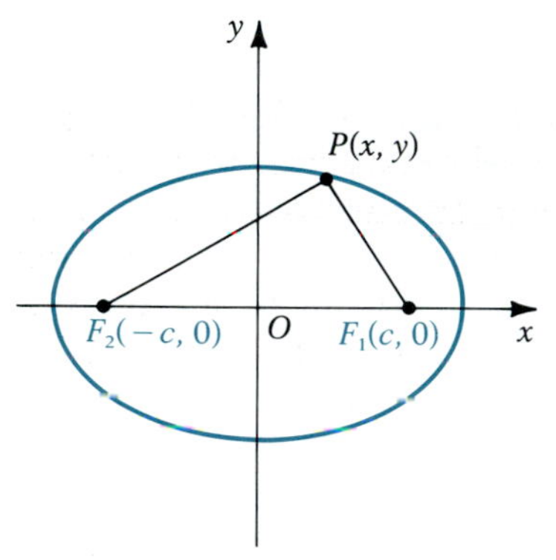

FIGURE 12.15

The simplest equations of ellipses result when the center is at the origin and the major axis lies on one of the coordinate axes. We derive the equation when the major axis lies on the x-axis, as shown in Figure 12.15, and leave the other case as an exercise.

As shown in Figure 12.15, we designate the foci by $F_1(c, 0)$ and $F_2(-c, 0)$. A point $P(x, y)$ lies on the ellipse if and only if the sum of the distances $\overline{PF_1}$ and $\overline{PF_2}$ is a constant. We designate this constant by $2a$, and observe that we must have $a > c$. (Why?) So the requirement for P to be on the ellipse is

$$\overline{PF_1} + \overline{PF_2} = 2a$$

or, from the distance formula,

$$\sqrt{(x-c)^2 + y^2} + \sqrt{(x+c)^2 + y^2} = 2a$$

To simplify, we isolate one radical and square, then simplify and isolate the remaining radical, and square again:

$$\begin{aligned}\sqrt{(x-c)^2 + y^2} &= 2a - \sqrt{(x+c)^2 + y^2}\\ x^2 - 2cx + c^2 + y^2 &= 4a^2 - 4a\sqrt{(x+c)^2 + y^2} + x^2 + 2cx + c^2 + y^2\\ a\sqrt{(x+c)^2 + y^2} &= a^2 + cx\\ a^2(x^2 + 2cx + c^2 + y^2) &= a^4 + 2a^2cx + c^2x^2\\ (a^2 - c^2)x^2 + a^2y^2 &= a^2(a^2 - c^2)\end{aligned}$$

Since $a > c$, we can simplify further by writing

$$b^2 = a^2 - c^2 \qquad (12.8)$$

This gives

$$b^2x^2 + a^2y^2 = a^2b^2$$

or, on dividing through by a^2b^2,

$$\frac{x^2}{a^2} + \frac{y^2}{b^2} = 1 \qquad (12.9)$$

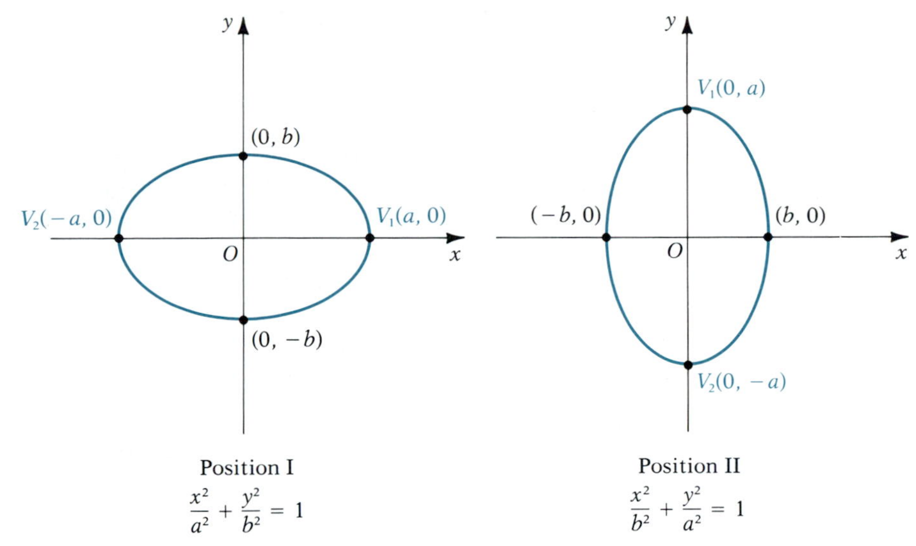

FIGURE 12.16

We refer to equation (12.9) as the **standard form** of the equation of an ellipse with center at the origin and horizontal major axis. From it, we see that the ellipse is symmetric to both axes. The fact that the squaring operations in this derivation led to equivalent equations can be demonstrated by showing that each time this was done, both sides of the equation represented positive numbers, and since two positive numbers are equal if and only if their squares are equal, the equivalence follows. When the major axis is on the y-axis, a similar analysis gives the standard form

$$\frac{x^2}{b^2} + \frac{y^2}{a^2} = 1 \tag{12.10}$$

where again $b^2 = a^2 - c^2$.

If in equation (12.9) we put $y = 0$, we get the two x-intercepts $(a, 0)$ and $(-a, 0)$. These are the vertices. Putting $x = 0$ gives the y-intercepts $(0, b)$ and $(0, -b)$. Thus we have,

$$\begin{cases} \text{length of major axis} = 2a \\ \text{length of minor axis} = 2b \end{cases} \tag{12.11}$$

In equation (12.10) the roles of x and y are reversed, so that the vertices are at $(0, a)$ and $(0, -a)$, whereas the end points of the minor axis are $(b, 0)$ and $(-b, 0)$ and equations (12.11) remain unchanged. Figure 12.16 illustrates the two cases. We will refer to an ellipse with a horizontal major axis as being in **position I** and one with a vertical major axis as being in **position II.**

Remark The position is determined by the *sizes* of the denominators when the equation is in standard form. Whichever denominator is larger is a^2. Thus,

$$\frac{x^2}{16} + \frac{y^2}{9} = 1$$

is in position I, whereas

$$\frac{x^2}{9} + \frac{y^2}{16} = 1$$

is in position II. In each case $a = 4$ and $b = 3$.

To get the standard forms for ellipses in position I or II with center at (h, k), we use the translation equations (12.3) to obtain the following equations.

STANDARD FORMS OF EQUATIONS OF ELLIPSES

$$\frac{(x-h)^2}{a^2} + \frac{(y-k)^2}{b^2} = 1, \quad \text{major axis horizontal} \tag{12.12}$$

$$\frac{(x-h)^2}{b^2} + \frac{(y-k)^2}{a^2} = 1, \quad \text{major axis vertical} \tag{12.13}$$

When the foci of an ellipse are close to the center, so that c is small, from $b^2 = a^2 - c^2$ we see that b is almost as large as a. The ellipse is nearly circular in this case. On the other hand, as the foci move farther apart, so that they are near the vertices, b is small in comparison to a and the ellipse is long and narrow. A convenient way to measure this effect is with the ratio $\frac{c}{a}$, which is called the **eccentricity**, designated by e:

$$e = \frac{c}{a} \tag{12.14}$$

If c is close to 0, e is small and the ellipse is nearly circular, whereas if c is close to a, e is close to 1 and the ellipse is narrow. Figure 12.17 illustrates these extremes. Note that in all cases $0 < e < 1$. If we permit the two foci to coincide at the center, then $e = 0$ and $a = b$. The ellipse then is a circle. For this reason a circle is sometimes classified as a special case of an ellipse.

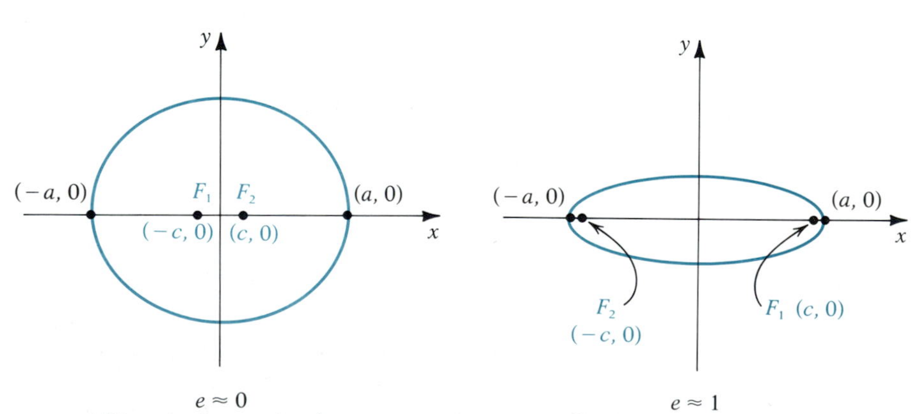

FIGURE 12.17

EXAMPLE 12.3 Find the equation in standard form of an ellipse with the following properties:

(a) Center at $(3, -4)$, one focus at $(0, -4)$, and one vertex at $(8, -4)$

(b) Vertices at $(1, 8)$ and $(1, -4)$, eccentricity $\frac{2}{3}$

Solution (a) Since the y-coordinates of the center and the given focus are the same, we conclude that the ellipse is in position I. The distance from the center to the given focus is 3, so $c = 3$. The distance from the center to the given vertex is 5, so $a = 5$. Thus

$$b^2 = a^2 - c^2 = 25 - 9 = 16$$

Using equation (12.12) we therefore have

$$\frac{(x-3)^2}{25} + \frac{(y+4)^2}{16} = 1$$

(b) The center is at the midpoint of the major axis—namely, (1, 2)—and a is half its length, so $a = 6$. From $e = \frac{c}{a}$ we get

$$\frac{c}{6} = \frac{2}{3}$$
$$c = 4$$

Now we can get b using

$$b^2 = a^2 - c^2 = 36 - 16 = 20$$

The ellipse is in position II, since the vertices are on a vertical line, so from equation (12.13) the equation is

$$\frac{(x-1)^2}{20} + \frac{(y-2)^2}{36} = 1$$

■

If either of the equations (12.12) or (12.13) is multiplied out and terms are rearranged, the result is an equation of the form

$$ax^2 + by^2 + cx + dy + e = 0 \tag{12.15}$$

where a and b are like in sign but are unequal. Reversing this, we want to know whether all equations of this type represent ellipses. Rather than to consider this question in general, we illustrate all the possibilities that can occur by means of examples.

EXAMPLE 12.4 Discuss the graph of the equation

$$4x^2 + 9y^2 + 16x - 18y - 11 = 0$$

Solution First we group the x terms and y terms separately and factor out the coefficient of x^2 from the x terms and the coefficient of y^2 from the y terms. Then we complete the squares:

$$4(x^2 + 4x \qquad) + 9(y^2 - 2y \qquad) = 11$$
$$4(x^2 + 4x + \textcircled{4}) + 9(y^2 - 2y + \textcircled{1}) = 11 + \textcircled{16} + \textcircled{9}$$

Notice that we added 4 and 1, respectively, inside the parentheses to complete the squares, and this had the effect of adding 16 and 9 because of the coefficients outside the parentheses. So we had to balance the equation by adding 16 and 9 on the right. Simplifying, we get

$$4(x+2)^2 + 9(y-1)^2 = 36$$
$$\frac{(x+2)^2}{9} + \frac{(y-1)^2}{4} = 1$$

We recognize this as being an ellipse in position I with center at $(-2, 1)$, with $a = 3$ and $b = 2$. Its graph is shown in Figure 12.18.

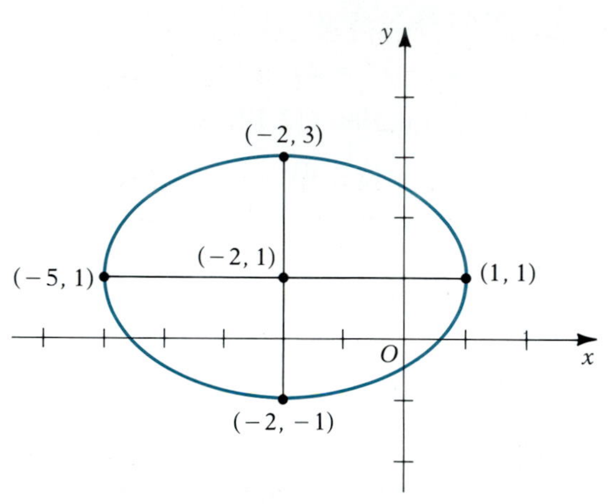

FIGURE 12.18

EXAMPLE 12.5 Discuss and sketch the graph of the equation

$$9x^2 + 4y^2 - 54x - 16y + 61 = 0$$

Solution Proceeding as in Example 12.4, we get

$$9(x^2 - 6x + 9) + 4(y^2 - 4y + 4) = -61 + 81 + 16$$
$$9(x-3)^2 + 4(y-2)^2 = 36$$
$$\frac{(x-3)^2}{4} + \frac{(y-2)^2}{9} = 1$$

This is in position II with center at $(3, 2)$, $a = 3$, and $b = 2$. The graph is shown in Figure 12.19.

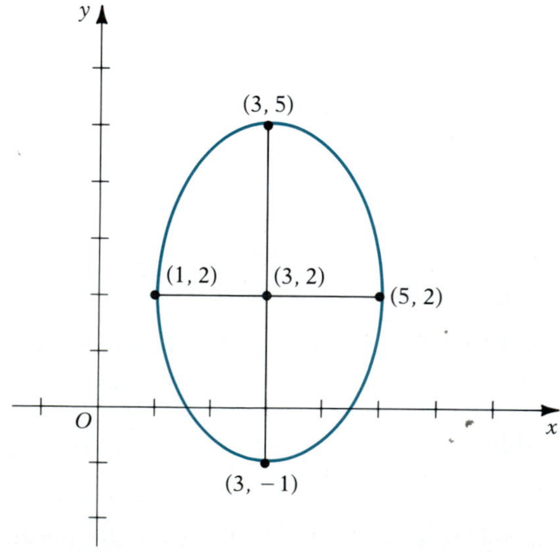

FIGURE 12.19

EXAMPLE 12.6 Discuss the graph of the equation

$$4x^2 + 5y^2 - 8x + 20y + 24 = 0$$

Solution Completing the squares we get

$$4(x^2 - 2x + 1) + 5(y^2 + 4y + 4) = -24 + 4 + 20$$
$$4(x - 1)^2 + 5(y + 2)^2 = 0$$

Since the right-hand side is 0, this cannot be put in standard form. In fact, the equation is satisfied only by the single point $(1, -2)$. We call this a degenerate ellipse. ■

EXAMPLE 12.7 Discuss the graph of the equation

$$4x^2 + 5y^2 - 8x + 20y + 25 = 0$$

Solution This is just like the equation of the previous example, except for the constant. The result of completing the square is

$$4(x - 1)^2 + 5(y + 2)^2 = -1$$

Since the left-hand side is always nonnegative, the equation is not satisfied by any point, and there is no graph. ■

These examples illustrate all the possibilities for equations of the form (12.15) with $ab > 0$. The graph is either an ellipse in standard position I or II, or a point (degenerate ellipse), or else there is no graph. When you see the combination $ax^2 + by^2$ in an equation like (12.15), *where a and b are unequal but like in sign* (that is, $ab > 0$), you should immediately conclude that if it has a nondegenerate graph, it is an ellipse.

EXERCISE SET 12.3

A

In Exercises 1–8 find the standard form of the equation of the ellipse with the given properties. Sketch the graph.

1. (a) x-intercepts ± 4; y-intercepts ± 2
   (b) x-intercepts ± 5; y-intercepts ± 8
2. Foci at $(-4, 3)$ and $(2, 3)$; one vertex at $(4, 3)$
3. Foci at $(-2, 0)$ and $(-2, 8)$; one vertex at $(-2, -1)$
4. Foci at $(2, 4)$ and $(2, -2)$; one end point of minor axis at $(4, 1)$
5. End points of minor axis at $(4, -3)$ and $(4, -7)$; one focus at $(8, -5)$
6. Major axis vertical; one vertex at $(-3, 4)$; one end point of minor axis at $(-5, 1)$
7. One vertex at $(1, 3)$; corresponding focus at $(1, 0)$; eccentricity $\frac{1}{2}$
8. End points of minor axis at $(-3, 3)$ and $(-3, -5)$; eccentricity $\frac{3}{5}$

In Exercises 9–15 write the equation in standard form, and sketch the graph.

9. (a) $25x^2 + 4y^2 = 100$ (b) $4x^2 + 5y^2 = 20$
10. (a) $x^2 = 4(1 - y^2)$ (b) $y^2 = 4(4 - x^2)$
11. (a) $x^2 + 9y^2 = 4$ (b) $5x^2 = 3(1 - 5y^2)$
12. $9x^2 + 4y^2 + 36x - 24y + 36 = 0$
13. $x^2 + 4y^2 - 2x + 16y + 13 = 0$
14. $4x^2 + 7y^2 - 48x - 70y + 319 = 0$
15. $2x^2 + 4y^2 - 5x + 6y - 4 = 0$
16. Find the equation of the parabola whose vertex and focus coincide with the upper vertex and

focus, respectively, of the ellipse $16x^2 + 12y^2 - 64x - 24y - 116 = 0$.

17. An arch is to be made in the shape of a semiellipse. It is to be 12 m from end to end and 4 m high at the center. Find how high the arch is at a point halfway from the center to one end.

18. Prove that the area of an ellipse with semiaxes of lengths a and b, respectively, is πab.

19. Find the equations of the tangent line and the normal line to the ellipse $4x^2 + y^2 - 8x + 4y = 0$ at the point $(2, -4)$. Draw the graph of the ellipse showing these tangent and normal lines.

20. The cross section of a horizontal tank is the ellipse with major axis 4 ft long (horizontal) and minor axis 2 ft long (vertical). The tank is 6 ft long and is half filled with a liquid of density ρ lb/cu ft. Find the total force on one end of the tank.

21. Find the dimensions of the rectangle of greatest area with sides parallel to the coordinate axes that can be inscribed in the ellipse

$$\frac{x^2}{a^2} + \frac{y^2}{b^2} = 1$$

22. Find the volume obtained by rotating the first-quadrant area below the ellipse $4x^2 + 9y^2 = 36$ and above the line $x - 2y = 0$ about (a) the x-axis and (b) the y-axis.

23. Find the equations of the inscribed and circumscribed circles to the ellipse

$$x^2 + 2y^2 + 4x + 12y - 3 = 0$$

Sketch all three curves.

24. Each chord of an ellipse through a focus and perpendicular to the major axis is called a latus rectum of the ellipse. If the lengths of the major and minor axes are $2a$ and $2b$, respectively, prove that the length of each latus rectum is $2b^2/a$.

25. Derive equation (12.10).

B

26. Find the area of the surface formed by revolving the ellipse $y^2 = 2(1 - x^2)$ about (a) the x-axis and (b) the y-axis.

27. Prove that at any point (x_0, y_0) on the ellipse

$$\frac{x^2}{a^2} + \frac{y^2}{b^2} = 1$$

except where $y_0 = 0$, the equation of the tangent line is

$$\frac{xx_0}{a^2} + \frac{yy_0}{b^2} = 1$$

28. Find the equation of the ellipse with foci at $(-1, 2)$ and $(1, -2)$ with major axis of length 10. (*Hint:* Use the definition of the ellipse.)

29. Find the smaller area between the curves $y^2 - 2x + 2y + 5 = 0$ and $4x^2 + 9y^2 - 16x + 18y - 11 = 0$. (*Hint:* The problem can be simplified by making a suitable translation of axes.)

30. Prove that the angles α and β in the accompanying figure are equal, where l is the tangent line to the ellipse at P and F_1 and F_2 are the foci. (*Note:* This explains the "whispering chamber" phenomenon in which a person standing at one focus in an elliptical room can whisper and be heard by a person at the other focus. It also shows how light emitted at a focus of an elliptical mirror is reflected to the other focus.)

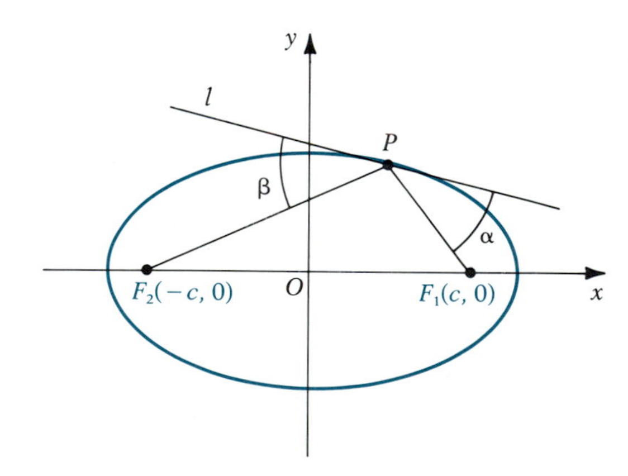

12.4 THE HYPERBOLA

If, instead of the requirement that the *sum* of the distances of a point from two foci is constant, as is true for the ellipse, we require the *difference* of these distances to be a constant, the resulting curve is a hyperbola.

FIGURE 12.20

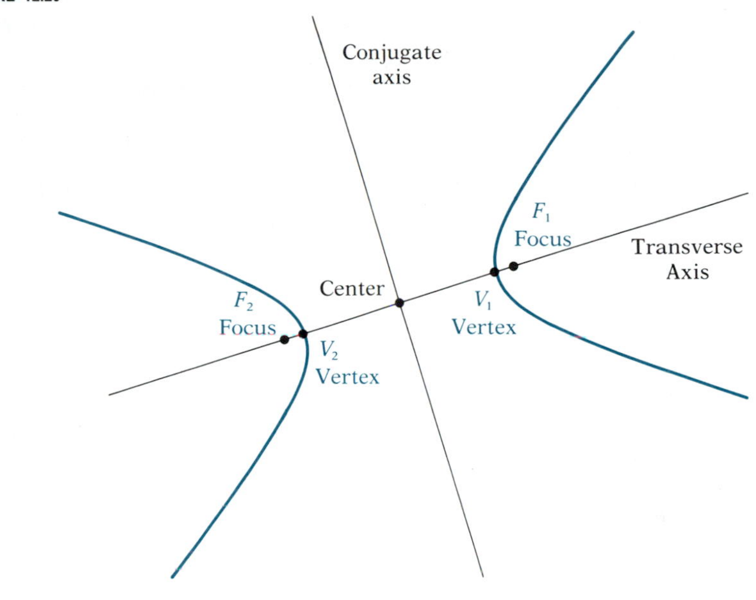

DEFINITION 12.3 A **hyperbola** is the set of points in a plane for which the absolute value of the difference in their distances from two fixed points in the plane is a constant. ■

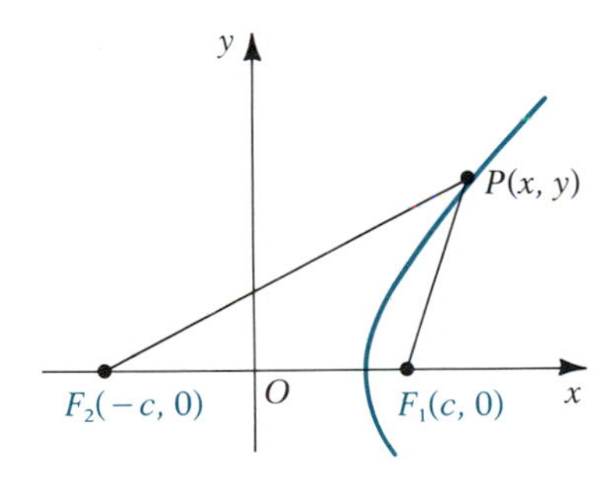

FIGURE 12.21

Just as with the ellipse, we call the fixed points the **foci,** and the midpoint of the line segment joining the foci is called the **center.** The line through the foci is called the **transverse axis,** and the line perpendicular to it through the center is called the **conjugate axis.** The **vertices** are the points where the hyperbola crosses its transverse axis. These definitions are illustrated in Figure 12.20.

A hyperbola consists of two separate branches, and it is symmetric to both its transverse axis and its conjugate axis.

Just as with the ellipse, we refer to a hyperbola that has a horizontal transverse axis as being in *position I* and one with a vertical transverse axis as being in *position II.* We consider first a hyperbola in position I with center at the origin. Figure 12.21 shows the two foci at $(c, 0)$ and $(-c, 0)$ and a typical point P on the curve. The definition asserts that P will lie on the curve if and only if $|\overline{PF_1} - \overline{PF_2}|$ is a constant, which we designate by $2a$. This condition can also be written as

$$\overline{PF_1} - \overline{PF_2} = \pm 2a$$

or, in terms of coordinates,

$$\sqrt{(x-c)^2 + y^2} - \sqrt{(x+c)^2 + y^2} = \pm 2a$$

To simplify this, we isolate one radical, square, isolate the remaining radical, and square again. The result (which you should check) is

$$(c^2 - a^2)x^2 - a^2y^2 = a^2(c^2 - a^2) \tag{12.16}$$

Since the sum of the lengths of two sides of a triangle exceeds the length of the third side, we see from Figure 12.21 that $\overline{F_1F_2} + \overline{PF_1} > \overline{PF_2}$ and similarly, $\overline{F_1F_2} + \overline{PF_2} > \overline{PF_1}$. Thus, $\overline{F_1F_2} > \overline{PF_2} - \overline{PF_1}$ and $\overline{F_1F_2} > \overline{PF_1} - \overline{PF_2}$, from which it follows that $\overline{F_1F_2} > |PF_1 - PF_2|$; that is, $2c > 2a$ and so $c > a$. Hence $c^2 - a^2$ is a positive number, which we designate by b^2:

$$b^2 = c^2 - a^2 \tag{12.17}$$

Substituting this in equation (12.16) yields

$$b^2x^2 - a^2y^2 = a^2b^2$$

and on dividing through by a^2b^2, we get

$$\frac{x^2}{a^2} - \frac{y^2}{b^2} = 1 \tag{12.18}$$

The corresponding result for a vertical transverse axis is

$$\frac{y^2}{a^2} - \frac{x^2}{b^2} = 1 \tag{12.19}$$

Equations (12.18) and (12.19) are the **standard forms** for positions I and II, respectively, when the center is at the origin.

Remark In the case of the ellipse, a^2 was the larger of the denominators when the equation was in standard form. In the case of the hyperbola, it is the *sign* of the term that determines a^2 and b^2, with a^2 being the denominator of the positive term. In determining position, the relative sizes of a and b are of no consequence. Thus,

$$\frac{x^2}{4} - \frac{y^2}{9} = 1$$

is in position I, and

$$\frac{y^2}{9} - \frac{x^2}{4} = 1$$

is in position II.

The x-intercepts for equation (12.18) are $(\pm a, 0)$ and the y-intercepts for equation (12.19) are $(0, \pm a)$. In each case these are the coordinates of the vertices. There are no intercepts on the conjugate axis in either case. We will see the geometric significance of b shortly.

In equation (12.18) if we solve for y we get

$$y = \pm\frac{b}{a}\sqrt{a^2 - x^2} = \pm\frac{bx}{a}\sqrt{1 - \frac{a^2}{x^2}}$$

Since

$$\lim_{x \to \pm\infty} \sqrt{1 - \frac{a^2}{x^2}} = 1$$

we would expect that for $|x|$ large, the hyperbola would approach the straight lines $y = \pm\frac{bx}{a}$. To make this more precise, consider the vertical distance $d(x)$ between the first-quadrant portion of the hyperbola, given by $y =$

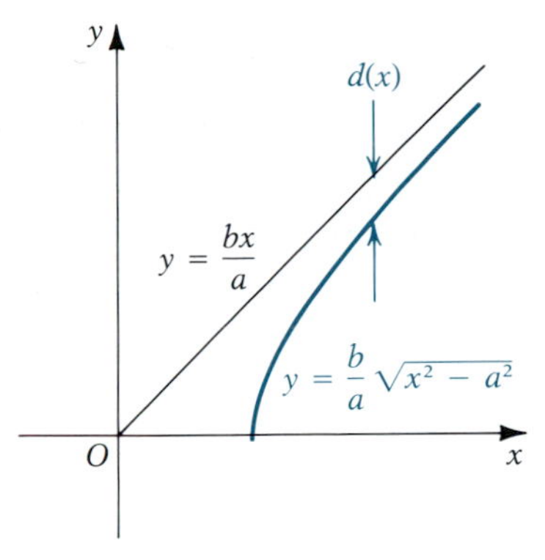

FIGURE 12.22

$b\sqrt{x^2 - a^2}/a$, and the line $y = \frac{bx}{a}$, as shown in Figure 12.22. Then

$$\begin{aligned} d(x) &= \frac{bx}{a} - \frac{b}{a}\sqrt{x^2 - a^2} \\ &= \frac{b}{a}(x - \sqrt{x^2 - a^2}) \\ &= \frac{b}{a}(x - \sqrt{x^2 - a^2}) \cdot \frac{x + \sqrt{x^2 - a^2}}{x + \sqrt{x^2 - a^2}} \\ &= \frac{b}{a}\left(\frac{a^2}{x + \sqrt{x^2 - a^2}}\right) \\ &= \frac{ab}{x + \sqrt{x^2 - a^2}} \end{aligned}$$

So $\lim_{x \to +\infty} d(x) = 0$. It follows that the hyperbola becomes asymptotic to the line $y = \frac{bx}{a}$ as $x \to +\infty$. By symmetry, we see that $y = -\frac{bx}{a}$ also is an asymptote. The situtation for equation (12.19) is similar, except the roles of x and y are reversed. Thus,

For $\frac{x^2}{a^2} - \frac{y^2}{b^2} = 1$, the asymptotes are $y = \pm\frac{bx}{a}$.

For $\frac{y^2}{a^2} - \frac{x^2}{b^2} = 1$, the asymptotes are $y = \pm\frac{ax}{b}$.

An easy way to sketch the asymptotes without memorizing these results is as follows. On the transverse axis mark the coordinates of the vertices, a units to each side of the center. Similarly, mark the points on the conjugate axis b units on each side of the center. Then draw the rectangle with sides parallel to the axes of the hyperbola through the four points obtained. We call this the **fundamental rectangle.** The diagonals of this rectangle are the asymptotes. This holds true for any position of the hyperbola. For the center at the origin and transverse axis horizontal, we show the fundamental rectangle and its diagonals in Figure 12.23(a). Once we have the fundamental

FIGURE 12.23

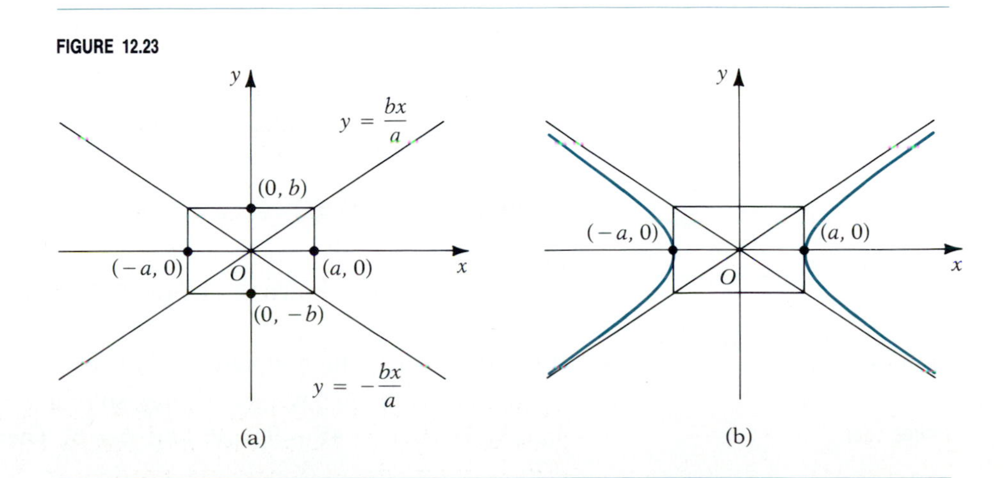

rectangle, it is simple to sketch in the hyperbola, as we have shown in Figure 12.23(b).

The foregoing discussion can be extended to hyperbolas in positions I and II, with centers at (h, k), by the equations of translation. The result is as follows.

STANDARD FORMS OF EQUATIONS OF HYPERBOLAS

$$\frac{(x-h)^2}{a^2} - \frac{(y-k)^2}{b^2} = 1, \quad \text{transverse axis horizontal} \tag{12.20}$$

$$\frac{(y-k)^2}{a^2} - \frac{(x-h)^2}{b^2} = 1, \quad \text{transverse axis vertical} \tag{12.21}$$

The asymptotes for the hyperbola (12.20) are $y - k = \pm b(x - h)/a$, and for the hyperbola (12.21), $y - k = \pm a(x - h)/b$. These are most easily obtained from the fundamental rectangle. They can also be gotten by replacing the 1 by 0 on the right-hand side of equation (12.20) or (12.21), respectively. You should verify that this is so.

Comment For both the ellipse and the hyperbola the number a is the distance from the center to each vertex, and the number c is the distance from the center to each focus. In the case of the ellipse $a > c$, and in the case of the hyperbola $c > a$. The number b is related to a and c by:

$$b^2 = a^2 - c^2, \quad \text{ellipse}$$
$$b^2 = c^2 - a^2, \quad \text{hyperbola}$$

It is important not to confuse these.

The **eccentricity** e of a hyperbola is defined just as for the ellipse, $e = \frac{c}{a}$, and since $c > a$, we always have $e > 1$.

EXAMPLE 12.8 Find the equation in standard form of the hyperbola that satisfies the following conditions. Sketch the graph.

(a) Center at $(-1, 3)$; one focus at $(4, 3)$; one vertex at $(-4, 3)$

(b) Vertices at $(2, -2)$ and $(2, 4)$; eccentricity $= \sqrt{5}$

Solution (a) A sketch of the given information in Figure 12.24 shows that the transverse axis is horizontal, $c = 5$, and $a = 3$. So $b^2 = c^2 - a^2 = 25 - 9 = 16$, and $b = 4$. The equation is

$$\frac{(x+1)^2}{9} - \frac{(y-3)^2}{16} = 1$$

and a sketch is shown in Figure 12.25. Note the use of the fundamental rectangle.

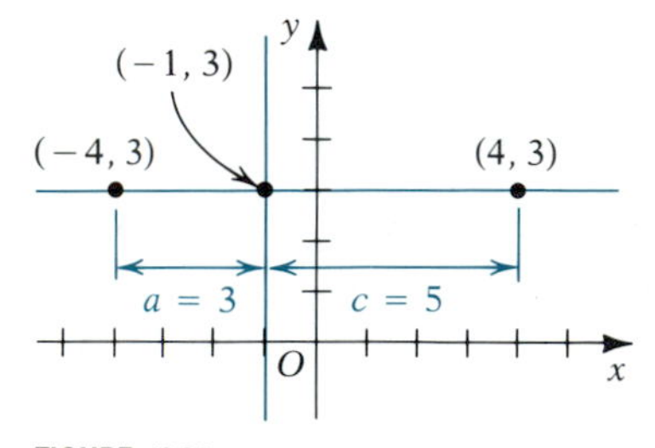

FIGURE 12.24

(b) The center is the midpoint $(2, 1)$ between the vertices. Also, $2a = 6$, so $a = 3$. From $e = \frac{c}{a}$ we get $\sqrt{5} = \frac{c}{3}$, so that $c = 3\sqrt{5}$. Thus $b^2 = c^2 - a^2 = (3\sqrt{5})^2 - 3^2 = 45 - 9 = 36$ and $b = 6$. The transverse axis is vertical,

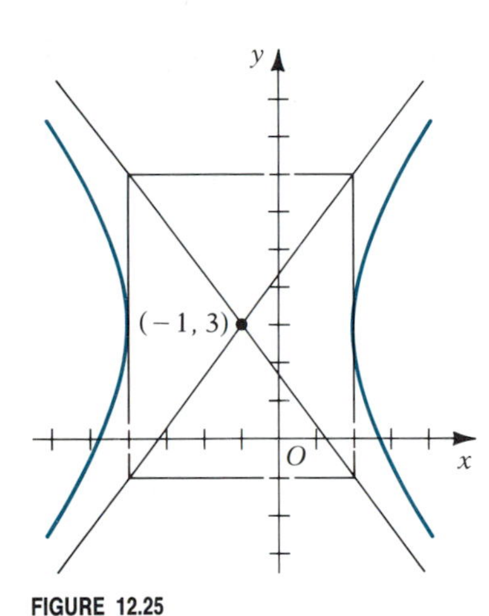

FIGURE 12.25

so that the equation has the form (12.21):

$$\frac{(y-1)^2}{9} - \frac{(x-2)^2}{36} = 1$$

The graph is shown in Figure 12.26.

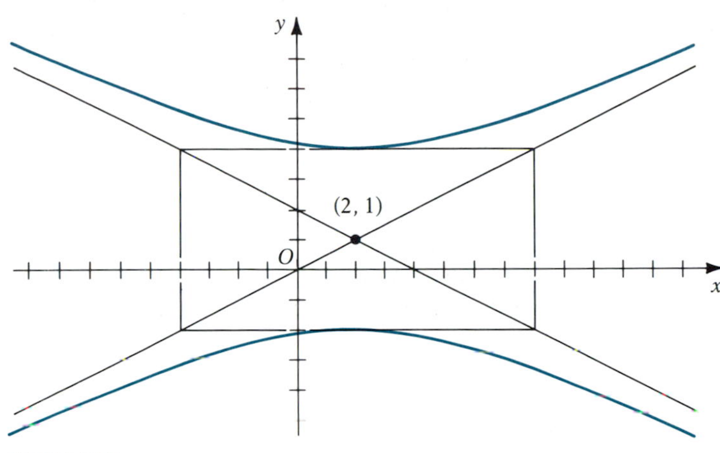

FIGURE 12.26

When either equation (12.20) or (12.21) is multiplied out, the result can be put in the form

$$ax^2 + by^2 + cx + dy + e = 0 \tag{12.22}$$

in which *a and b are opposite in sign.* Reversing this, we can show that such an equation always has a graph that is a hyperbola in one of the standard positions or a degenerate of one. The next three examples illustrate all the possibilities.

EXAMPLE 12.9 Discuss and sketch the graph of

$$4x^2 - 9y^2 - 16x - 18y - 29 = 0$$

Solution We complete the squares, as with the ellipse:

Be careful here.

$$4(x^2 - 4x + 4) - 9(y^2 \oplus 2y + 1) = 29 + 16 - 9$$
$$4(x-2)^2 - 9(y+1)^2 = 36$$
$$\frac{(x-2)^2}{9} - \frac{(y+1)^2}{4} = 1$$

So this is a hyperbola in position I, center at $(2, -1)$, $a = 3$, and $b = 2$. Thus $c^2 = a^2 + b^2 = 9 + 4 = 13$, and $e = \sqrt{13}/3$. The asymptotes are

$$y + 1 = \pm\tfrac{2}{3}(x - 2)$$

We show the graph in Figure 12.27.

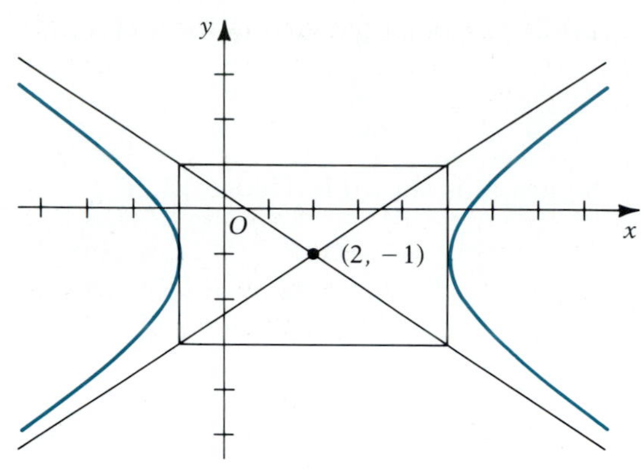

FIGURE 12.27

■

EXAMPLE 12.10 Discuss and sketch the graph of

$$x^2 - 4y^2 + 4x + 24y - 28 = 0$$

Solution

$$(x^2 + 4x + 4) - 4(y^2 - 6y + 9) = 28 + 4 - 36$$

$$(x + 2)^2 - 4(y - 3)^2 = -4 \quad \text{Divide through by } -4.$$

$$\frac{(y - 3)^2}{1} - \frac{(x + 2)^2}{4} = 1$$

This is a hyperbola in position II, center at $(-2, 3)$, $a = 1$, and $b = 2$. So $c^2 = a^2 + b^2 = 5$, and $e = \sqrt{5}$. The asymptotes are

$$y - 3 = \pm\tfrac{1}{2}(x + 2)$$

The graph is shown in Figure 12.28.

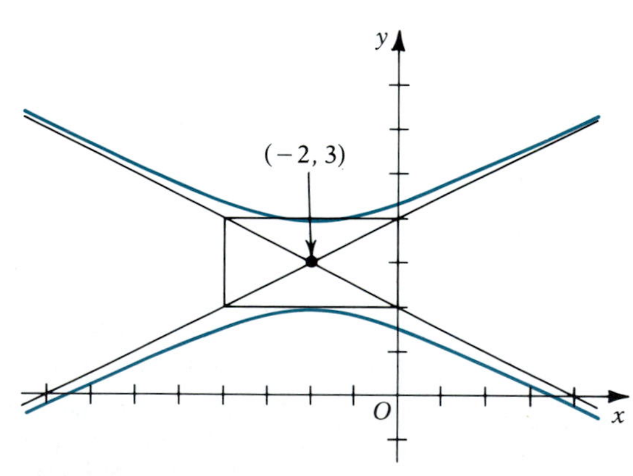

FIGURE 12.28

■

EXAMPLE 12.11 Discuss and sketch the graph of

$$16x^2 - 25y^2 - 64x + 200y - 336 = 0$$

Solution

$$16(x^2 - 4x + 4) - 25(y^2 - 8y + 16) = 336 + 64 - 400$$

$$16(x - 2)^2 - 25(y - 4)^2 = 0$$

This cannot be put in one of the standard forms, since the right-hand side is 0, so it does not represent a hyperbola. Its graph consists of the two straight lines

$$y - 4 = \pm\tfrac{4}{5}(x - 2)$$

These are, in fact, the asymptotes of all hyperbolas of the form $16(x-2)^2 - 25(y-4)^2 = K$, where $K \neq 0$. As $K \to 0$ the vertices of all such hyperbolas approach the center, and the limiting case is the graph of the two asymptotes. For this reason, the graph is called a degenerate hyperbola. ■

The key to recognizing a hyperbola (or a degenerate) in an equation of the form (12.22) is that *the coefficients of x^2 and y^2 are opposite in sign.* When you see a minus sign between the x^2 and y^2 terms, you know immediately the graph is a hyperbola.

EXERCISE SET 12.4

A

In Exercises 1–8 find the equation of the hyperbola with the given properties. Sketch the graph.

1. (a) Center at the origin; one vertex at $(2, 0)$; one focus at $(-3, 0)$
(b) Center at the origin; one vertex at $(0, -3)$; one focus at $(0, 4)$

2. Center at $(2, 3)$; one vertex at $(5, 3)$; one focus at $(-3, 3)$

3. Center at $(-2, -4)$; one vertex at $(-2, -7)$; one focus at $(-2, -9)$

4. Vertices at $(-1, 2)$ and $(-1, -4)$; slope of one asymptote $\frac{3}{4}$

5. Intersection of conjugate axis and fundamental rectangle $(2, -7)$ and $(2, 5)$; one vertex at $(0, -1)$

6. Foci at $(0, 2)$ and $(6, 2)$; eccentricity $\frac{3}{2}$

7. Eccentricity $\frac{5}{4}$; intersections of conjugate axis and fundamental rectangle $(3, 5)$ and $(3, -1)$

8. Center at $(4, -2)$; axis vertical; passing through $(5, -4)$ and $(-3, 8)$

In Exercises 9–15 write the equation in standard form and sketch the graph. Give the coordinate of the vertices, the eccentricity, and the equations of the asymptotes.

9. (a) $4x^2 - 9y^2 = 36$ (b) $4x^2 - 9y^2 + 36 = 0$

10. (a) $x^2 = 4(y^2 + 1)$ (b) $y^2 = x^2 + 9$

11. (a) $9x^2 - 4y^2 = -36$ (b) $25(x^2 - 9) = 9y^2$

12. $4x^2 - y^2 - 32x + 4y + 56 = 0$

13. $x^2 - 4y^2 - 2x - 16y - 19 = 0$

14. $9x^2 - 16y^2 + 36x + 96y - 108 = 0$

15. $2x^2 - y^2 + 4x - 4y = 0$

16. Find the equation of the ellipse that has the same foci as the hyperbola

$$\frac{x^2}{9} - \frac{y^2}{27} = 1$$

and with eccentricity the reciprocal of that of the hyperbola. Sketch both curves.

In Exercises 17–19 find the area bounded by the two curves. Draw the graphs, showing the area involved.

17. $y = \sqrt{9 + x^2}$ and $x - 2y + 6 = 0$

18. $x^2 - y^2 = 1$ and $2y = 4 - x^2$

19. $y = \sqrt{1 + x^2}$ and $y = \sqrt{4 - 2x^2}$

20. Find the equation of the parabola whose vertex and focus coincide with the upper vertex and focus, respectively, of the hyperbola $(y^2/100) - (x^2/44) = 1$. Sketch both curves carefully, and compare their behavior for large x. Do you think any parabola would exactly fit the upper half of this (or any other) hyperbola? Explain your reasoning.

21. Show that each latus rectum of a hyperbola has length $2b^2/a$, where a latus rectum is a line segment through a focus, perpendicular to the transverse axis, that is cut off by the hyperbola.

22. Find the volume obtained by rotating the region inside the right branch of the hyperbola $3x^2 - y^2 = 12$ between its vertex and latus rectum (see Exercise 21) about the x-axis.

23. Find the surface area obtained by rotating the arc of the curve $y = \sqrt{x^2 - 1}$ from $x = 1$ to $x = 2$ about the x-axis.

24. Find the equations of the tangent line and normal line to the hyperbola $3x^2 - y^2 + 6x + 4y + 8 = 0$ at the point $(2, -4)$.

25. A line with slope 2 is tangent to the hyperbola $3x^2 - y^2 = 3$ at a point in the first quadrant. Find the point of tangency.

B

26. Prove that the equation of the tangent line to the hyperbola

$$\frac{x^2}{a^2} - \frac{y^2}{b^2} = 1$$

at (x_0, y_0) can be written in the form

$$\frac{xx_0}{a^2} - \frac{yy_0}{b^2} = 1$$

27. Let d_1 and d_2 denote the perpendicular distances of a point on a hyperbola to its two respective asymptotes. Prove that

$$d_1 d_2 = \frac{a^2 b^2}{c^2}$$

where a, b, and c have their usual meanings. (*Hint:* Since the distances involved are independent of the location or orientation of the hyperbola, choose one of the simplest positions.)

28. A point moves in the xy-plane so that the absolute value of the difference of its distances from $(1, 1)$ and $(5, 3)$ is always 4. Find the equation of the path it traces out, and draw the graph.

29. Find the area bounded by the graphs of $4x^2 - y^2 - 8x + 4y - 12 = 0$ and $y^2 - 2x - 4y + 6 = 0$. (*Hint:* The problem can be simplified by making a suitable translation.)

30. Prove that the angles α and β in the accompanying figure are equal, where l is tangent to the hyperbola and F_1 and F_2 are its foci.

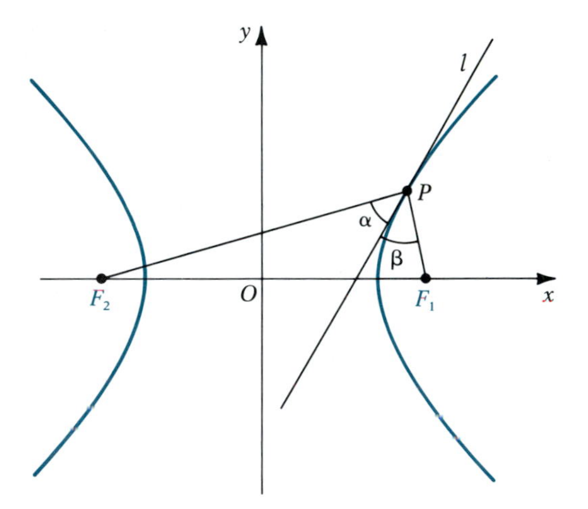

12.5 THE GENERAL QUADRATIC EQUATION IN TWO VARIABLES; ROTATION OF AXES

The purpose of this section is to study graphs of equations of the form

$$Ax^2 + Bxy + Cy^2 + Dx + Ey + F = 0 \tag{12.23}$$

This is called the **general quadratic equation in *x* and *y*.** If $B = 0$, we already know how to identify the curve and draw its graph. The equation is then

$$Ax^2 + Cy^2 + Dx + Ey + F = 0$$

and the nature of the graph depends on the coefficients A and C as follows:

1. If $A = C$, the graph is a circle or a degenerate of a circle, or there is no graph.
2. If $AC = 0$, A or C (but not both) is 0, so the curve is a parabola or a degenerate of a parabola. If $A \neq 0$ and $C = 0$, the axis is vertical, and if $A = 0$ and $C \neq 0$, its axis is horizontal.
3. If $AC > 0$ but $A \neq C$, A and C are like in sign, so the graph is an ellipse with major axis either vertical or horizontal, or it is a degenerate ellipse, or else there is no graph.

4. If $AC < 0$, A and C are opposite in sign, so the graph is a hyperbola with transverse axis either vertical or horizontal, or it is a degenerate hyperbola.

We focus our attention for the remainder of this section on equation (12.23) in which $B \neq 0$. As we shall see, the graph (if it exists) is still a conic, but with a rotated axis.

Consider a rectangular coordinate system with x- and y-axes, and let a second rectangular system with x'- and y'-axes be superimposed upon the first, so that the origins coincide and such that the angle from the positive x-axis to the positive x'-axis is θ, as shown in Figure 12.29. Then we say that the new system is rotated through an angle θ from the old. A point P now has two sets of coordinates, (x, y) and (x', y'). To see how they are related, let α be the angle as shown from the positive x'- axis to the line OP. Then from triangles OPQ and OPR, we have

$$\begin{cases} x = \overline{OP}\cos(\theta + \alpha) \\ y = \overline{OP}\sin(\theta + \alpha) \end{cases} \tag{12.24}$$

and

$$\begin{cases} x' = \overline{OP}\cos\alpha \\ y' = \overline{OP}\sin\alpha \end{cases} \tag{12.25}$$

Using the addition formulas for sine and cosine, we can rewrite equations (12.24) in the form

$$x = \overline{OP}(\cos\theta\cos\alpha - \sin\theta\sin\alpha)$$
$$y = \overline{OP}(\sin\theta\cos\alpha + \cos\theta\sin\alpha)$$

and substituting from equations (12.25), we get

$$\begin{cases} x = x'\cos\theta - y'\sin\theta \\ y = x'\sin\theta + y'\cos\theta \end{cases} \tag{12.26}$$

If we know the equation of a curve in the xy system, then replacing x and y by their equivalents in terms of x' and y', from equations (12.26), gives the equation of the same curve referred to the rotated system. We illustrate this with a frequently occurring type of equation.

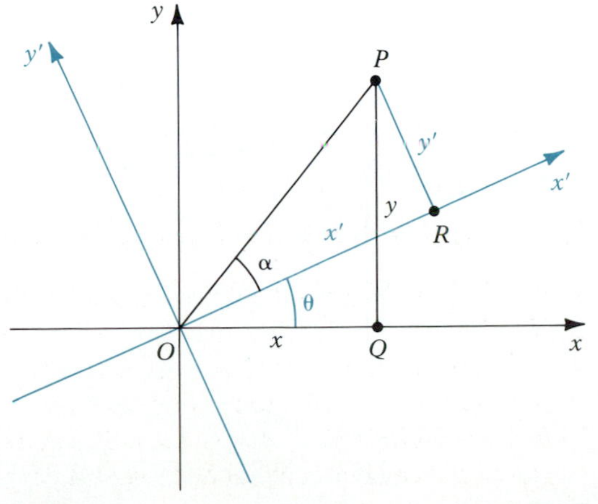

FIGURE 12.29

Consider an equation of the form

$$xy = K \tag{12.27}$$

where $K \neq 0$, and let us see what effect a 45° rotation has on it. By equations (12.26), we have

$$x = x' \cos 45° - y' \sin 45° = x'\left(\frac{1}{\sqrt{2}}\right) - y'\left(\frac{1}{\sqrt{2}}\right) = \frac{x' - y'}{\sqrt{2}}$$

and

$$y = x' \sin 45° + y' \cos 45° = \frac{x' + y'}{\sqrt{2}}$$

Substituting in equation (12.27) gives

$$\left(\frac{x' - y'}{\sqrt{2}}\right)\left(\frac{x' + y'}{\sqrt{2}}\right) = K$$

or

$$x'^2 - y'^2 = 2K$$

This we recognize as a hyperbola. In standard form, the equation is

$$\frac{x'^2}{2K} - \frac{y'^2}{2K} = 1$$

If $K > 0$, the transverse axis is along the x'-axis (position I), and if $K < 0$, the transverse axis is along the y'-axis (position II). Since $a = b = \sqrt{2|K|}$, the fundamental rectangle is a square. Whenever this is the case, the hyperbola is said to be **rectangular** or **equilateral.** Figure 12.30 shows typical graphs of the rectangular hyperbolas $xy = K$ for $K > 0$ and $K < 0$.

If the equation is in the form $(x - h)(y - k) = K$, then the graph is a rectangular hyperbola with center at (h, k).

When we are given an equation of the form (12.23) with $B \neq 0$, we want to find an angle θ through which to rotate the axes so that the equation in

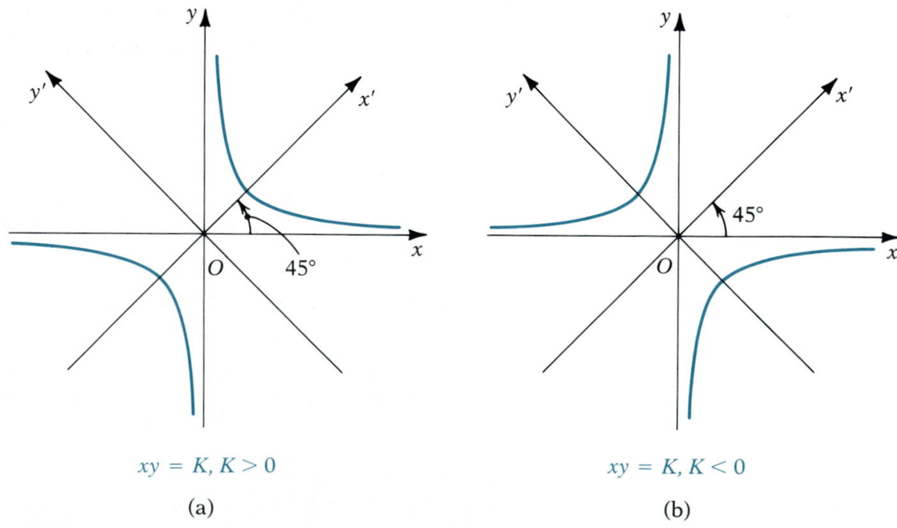

FIGURE 12.30

x' and y' has no $x'y'$ term. This will permit us to identify and draw the graph using the rotated axes. The result will be the graph of the original equation with respect to the x- and y-axes. The key, then, is to make an appropriate choice of θ. The process outlined below will lead to such a choice.

First we substitute for x and y from the rotation equations (12.26) into (12.23). The result is a new equation of the form

$$A'x'^2 + B'x'y' + C'y'^2 + D'x' + E'y' + F' = 0$$

where the coefficients are:

$$\begin{aligned}
A' &= A\cos^2\theta + B\sin\theta\cos\theta + C\sin^2\theta \\
B' &= B(\cos^2\theta - \sin^2\theta) - 2(A - C)\sin\theta\cos\theta \\
C' &= A\sin^2\theta - B\sin\theta\cos\theta + C\cos^2\theta \\
D' &= D\cos\theta + E\sin\theta \\
E' &= E\cos\theta - D\sin\theta \\
F' &= F
\end{aligned}$$

You will be asked in Exercise 20 to verify these results. Our object is to choose θ so that $B' = 0$. Using double angle formulas, we can rewrite the equation for B' in the form

$$B' = B\cos 2\theta - (A - C)\sin 2\theta$$

and setting $B' = 0$ yields

$$\cot 2\theta = \frac{A - C}{B} \tag{12.28}$$

(We choose to write the result in terms of the cotangent rather than the tangent to avoid division by 0 in case $A = C$.) Since it is always sufficient to rotate the axes through a positive acute angle, we choose θ using equation (12.28) so that $0 < \theta < \frac{\pi}{2}$. The next two examples illustrate this result.

EXAMPLE 12.12 Use a suitable rotation of axes to draw the graph of $8x^2 - 4xy + 5y^2 = 36$.

Solution From equation (12.28) we want to choose an acute angle θ so that

$$\cot 2\theta = \frac{A - C}{B} = \frac{8 - 5}{-4} = -\frac{3}{4}$$

Figure 12.31 depicts the angle 2θ. We can obtain $\sin\theta$ and $\cos\theta$ using the double angle formulas

$$\sin^2\theta = \frac{1 - \cos 2\theta}{2} \quad \text{and} \quad \cos^2\theta = \frac{1 + \cos 2\theta}{2}$$

Reading $\cos 2\theta = -\frac{3}{5}$ from Figure 12.31, we get

$$\sin^2\theta = \frac{1 - \left(-\frac{3}{5}\right)}{2} = \frac{4}{5}$$

$$\cos^2\theta = \frac{1 + \left(-\frac{3}{5}\right)}{2} = \frac{1}{5}$$

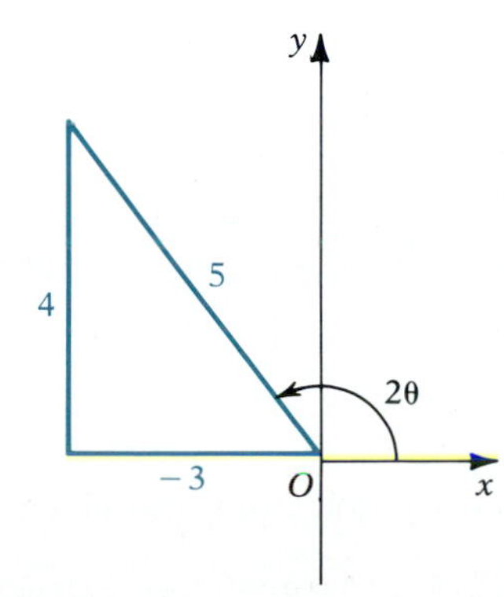

FIGURE 12.31

Thus,

$$\sin\theta = \frac{2}{\sqrt{5}} \quad \text{and} \quad \cos\theta = \frac{1}{\sqrt{5}}$$

The rotation equations now give

$$x = \frac{1}{\sqrt{5}}(x' - 2y') \quad \text{and} \quad y = \frac{1}{\sqrt{5}}(2x' + y')$$

We substitute these in the original equation:

$$\tfrac{8}{5}(x'^2 - 4x'y' + 4y'^2) - \tfrac{4}{5}(2x'^2 - 3x'y' - 2y'^2) + \tfrac{5}{5}(4x'^2 + 4x'y' + y'^2) = 36$$

which simplifies to $4x'^2 + 9y'^2 = 36$ or, in standard form,

$$\frac{x'^2}{9} + \frac{y'^2}{4} = 1$$

We recognize this as an ellipse with center at the origin in position I, $a = 3$, and $b = 2$. To draw the rotated axes, we note that since $\sin\theta = 2/\sqrt{5}$ and $\cos\theta = 1/\sqrt{5}$, the point (1, 2) lies on the x'-axis. The graph is shown in Figure 12.32.

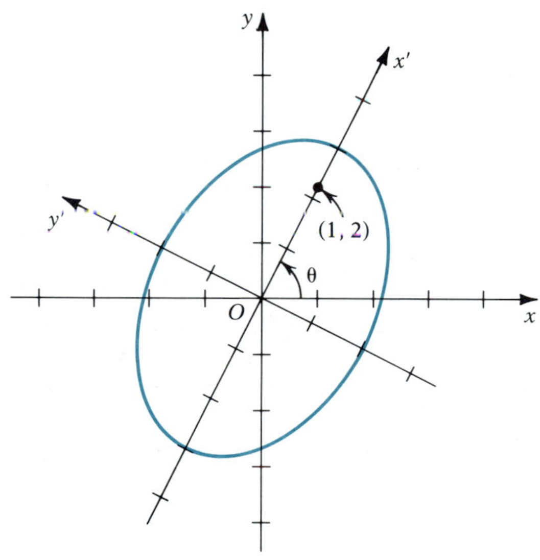

FIGURE 12.32

EXAMPLE 12.13 Use a suitable rotation of axes to draw the graph of $x^2 - 2xy + y^2 - 2x - 2y = 0$.

Solution

$$\cot 2\theta = \frac{A - C}{B} = \frac{0}{-2} = 0$$

Thus, $2\theta = 90°$ and $\theta = 45°$. So $\sin\theta = \cos\theta = 1/\sqrt{2}$. The rotation equations are therefore

$$x = \frac{1}{\sqrt{2}}(x' - y') \quad \text{and} \quad y = \frac{1}{\sqrt{2}}(x' + y')$$

On substituting these into the original equation, we get, after simplification (verify),

$$y'^2 = \sqrt{2}x'$$

So the graph is a parabola in position I with vertex at the origin, as shown in Figure 12.33.

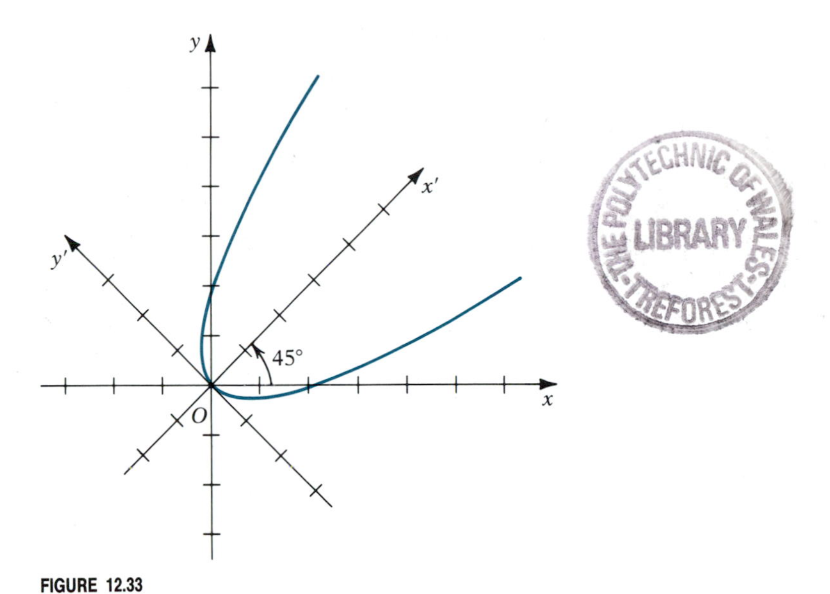

FIGURE 12.33

Comment It should be emphasized that introducing rotated x'- and y'-axes is a means to an end. The object is to graph the original equation in the xy-coordinate system. After we have drawn the graph, we could just as well erase the x'- and y'-axes.

The following theorem, whose proof is called for in Exercise 22, enables us to identify the type of conic we have even before rotating the axes. It refers to the **discriminant** of equation (12.23), defined as $B^2 - 4AC$. The discriminant of the equation after rotation, similarly, is $B'^2 - 4A'C'$.

THEOREM 12.1 The discriminant of a quadratic equation in two variables is invariant under rotation; that is,

$$B^2 - 4AC = B'^2 - 4A'C'$$

COROLLARY 12.1 If a quadratic equation in two variables has a graph and does not represent a degenerate conic, then the graph is one of the following:

1. A parabola if $B^2 - 4AC = 0$.
2. An ellipse if $B^2 - 4AC < 0$.
3. A hyperbola if $B^2 - 4AC > 0$.

Proof of Corollary 12.1 Let

$$Ax^2 + Bxy + Cy^2 + Dx + Ey + F = 0$$

be the given equation, and suppose $B \neq 0$. (If $B = 0$, the results can readily be seen to be true.) Rotate the axes to eliminate the $x'y'$ term. The resulting equation is of the form

$$A'x'^2 + C'y'^2 + D'x' + E'y' + F' = 0$$

We know the graph of this equation is one of the following:

1. A parabola if A' or C' is 0.
2. An ellipse if A' and C' have the same sign.
3. A hyperbola if A' and C' are opposite in sign.

Furthermore, we know from Theorem 12.1 that $B^2 - 4AC = -4A'C'$, since $B' = 0$. So in case 1 $B^2 - 4AC = 0$, in case 2 $B^2 - 4AC < 0$, and in case 3 $B^2 - 4AC > 0$. The result now follows. ■

It should be noted that the equation of a circle will never involve an xy term, so we have not considered circles in the discussion.

EXAMPLE 12.14 Identify each of the following, assuming a nondegenerate graph exists.

(a) $2x^2 - 3xy + 5y^2 = 7$
(b) $x^2 + 9xy + 3y^2 - 2x + 5y = 4$
(c) $8x^2 - xy - 2y^2 + x + 3y = 0$
(d) $9x^2 - 24xy + 16y^2 + 7x - 5y = 8$

Solution In each case we calculate $B^2 - 4AC$ and use Corollary 12.1.

(a) $B^2 - 4AC = 9 - 40 = -31 < 0$; ellipse
(b) $B^2 - 4AC = 81 - 12 = 69 > 0$; hyperbola
(c) $B^2 - 4AC = 1 + 64 = 65 > 0$; hyperbola
(d) $B^2 - 4AC = 576 - 576 = 0$; parabola ■

EXERCISE SET 12.5

A

In Exercises 1–6 draw the rectangular hyperbolas.

1. $xy = 2$
2. $xy + 4 = 0$
3. $(x - 3)(y + 2) = 8$
4. $(2x + 1)(3y - 4) = 12$
5. $xy - 2x + y = 0$
6. $xy - 3x - 2y = 2$

In Exercises 7–14 rotate the axes to remove the x′y′ term. Identify and sketch the graph.

7. $x^2 - 3xy + 5y^2 = 22$
8. $6x^2 - 24xy - y^2 = 150$
9. $2x^2 - 5xy + 2y^2 = 18$
10. $x^2 + 2xy + y^2 + 2\sqrt{2}(x - y) = 0$
11. $20x^2 - 8xy + 5y^2 = 84$
12. $4x^2 + 4xy + y^2 + 5x - 10y = 0$
13. $9x^2 - 12xy + 4y^2 + 8\sqrt{13}x + 12\sqrt{13}y = 52$
14. $x^2 + 3xy - 3y^2 = 42$

In Exercises 15 and 16 without rotating axes identify the graph, assuming it exists and is nondegenerate.

15. (a) $3x^2 + 2xy - y^2 + 4x - 3y = 7$
(b) $x^2 - 4xy + 4y^2 - 3x + 2y - 5 = 0$
(c) $2x^2 - 3xy + 5y^2 = 4$
(d) $x^2 + 3xy - 7x + 4y = 0$
(e) $5x^2 + 7xy - 2x = 8$
16. (a) $2xy - y^2 + 4x - 5y = 7 + x^2$
(b) $xy = x^2 - 2y^2 + 4$
(c) $y^2 = 3x - 10x^2 + 4xy - 9$
(d) $x^2 + y^2 = 4xy - 2x + 3y + 11$
(e) $x(3x - 2y) + y(x + 5y) = 3$

B

In Exercises 17–19 identify and sketch the graph after a suitable rotation of axes.

17. $2x^2 - \sqrt{3}xy + y^2 + 5\sqrt{3}x - 5y = 0$
18. $9x^2 + 24xy + 16y^2 - 110x + 20y + 125 = 0$
19. $x^2 - 4xy - 2y^2 + 22x + 4y = 5$
20. Verify the formulas given for A', B', C', D', E', and F'.
21. Prove that the quantity $A + C$ is invariant under rotation—that is, that $A + C = A' + C'$.
22. Prove Theorem 12.1.

12.6 SUPPLEMENTARY EXERCISES

In Exercises 1–8 write the equation in standard form and identify the nature of its graph. If there is a nondegenerate graph, draw it.

1. $x^2 - 4y^2 - 4x + 8y + 16 = 0$

2. $x^2 + y^2 + 6x - 8y + 21 = 0$

3. $y^2 + 2x + 6y + 7 = 0$

4. $4x^2 + 3y^2 - 12y = 36$

5. $3x^2 + 3y^2 - 18x + 42y + 174 = 0$

6. $3x^2 + 5y^2 + 12x - 10y + 18 = 0$

7. $4x^2 - 9y^2 + 8x + 36y - 32 = 0$

8. $4x(x + 6) = 3(y - 10)$

9. Each focus and each latus rectum of a hyperbola coincide with those of the ellipse $x^2 + 4y^2 = 16$. Find its equation. Sketch both curves.

In Exercises 10–13 find the equations of the conics that have the given properties. Identify and sketch the curves.

10. (a) Eccentricity $\frac{3}{2}$; center at $(1, -2)$; one focus at $(-5, -2)$
(b) Foci at $(-2, 3)$ and $(-2, -1)$; one vertex at $(-2, 5)$

11. (a) All points on the graph are equidistant from the point $(3, 2)$ and the line $y = 6$.
(b) All points on the graph are twice as far from $(2, 4)$ as from $(-1, -2)$.

12. (a) Eccentricity $\frac{3}{5}$; ends of minor axis at $(3, -1)$ and $(3, -5)$
(b) Center at $(3, 2)$; a focus at $(3, 7)$; a vertex at $(3, -1)$

13. (a) All points on the graph are twice as far from the line $x + 2 = 0$ as from the point $(1, 0)$.
(b) The absolute value of the difference of the distances from each point on the graph to the points $(2, -1)$ and $(-1, 0)$ is 2.

14. A parabola that opens upward has its vertex at the center of the circle $x^2 + y^2 - 4x + 6y - 7 = 0$, and its latus rectum is a chord of the circle. Find the equation of the parabola.

15. Find the equations of the inscribed and circumscribed circles to the triangle with vertices $(2, -1)$, $(-1, 5)$, and $(-6, -5)$.

16. Show that the circle $x^2 + y^2 + 2x - 6y + 5 = 0$ and the parabola $(x + 1)^2 = -2(y - 6)$ are tangent to each other at each of their points of intersection. Find these points and the equations of the common tangent lines. Sketch both curves.

17. Find the area of the region inside the ellipse
$$4x^2 + 3y^2 = 12$$
that lies above the parabola $4x^2 = 9y$. Sketch both curves.

18. Two tangent lines to the parabola
$$x^2 - 8x + y + 14 = 0$$
pass through the point $(1, 9)$. Find the points of tangency.

19. Find the volume obtained by revolving the region between the semihyperbola $2y = 3\sqrt{x^2 + 16}$ and the horizontal line through its focus about the x-axis.

20. Prove that the circle $x^2 + y^2 + 2x - 4y = 3$ is tangent internally to the circle $x^2 + y^2 - 4x + 2y = 45$. Find the equation of their common tangent line.

21. Find the equations of the tangent line and normal line to the parabola $x^2 + 2x - 8y + 25 = 0$ at the right end point of its latus rectum.

22. Determine whether the area between the hyperbola
$$\frac{x^2}{a^2} - \frac{y^2}{b^2} = 1$$
and its asymptotes is finite or infinite. If finite, find its value.

23. Find the equation of the tangent line to the hyperbola $9x^2 - 16y^2 + 18x + 64y = 199$ at one end point of a latus rectum. Find the angles this tangent line makes with the asymptotes to the hyperbola.

In Exercises 24–28 identify the conic. Then make a suitable rotation and sketch the graph.

24. $2x^2 + xy + 2y^2 = 30$

25. $x^2 - 4xy + y^2 - 3 = 0$

26. $xy - 4x + 2y - 6 = 0$

27. $2x^2 - 3xy + 6y^2 = 39$

28. $9x^2 - 24xy + 16y^2 + 4x + 3y = 0$

29. A point P in the plane moves so that the ratio of its distance from a fixed point F to its distance from a fixed line l is a positive constant e. Prove that the curve traced out by P is an ellipse if $e < 1$, a parabola if $e = 1$, or a hyperbola if $e > 1$. (*Hint:* Orient the axes so that F is on the x-axis and the y-axis is parallel to l.)

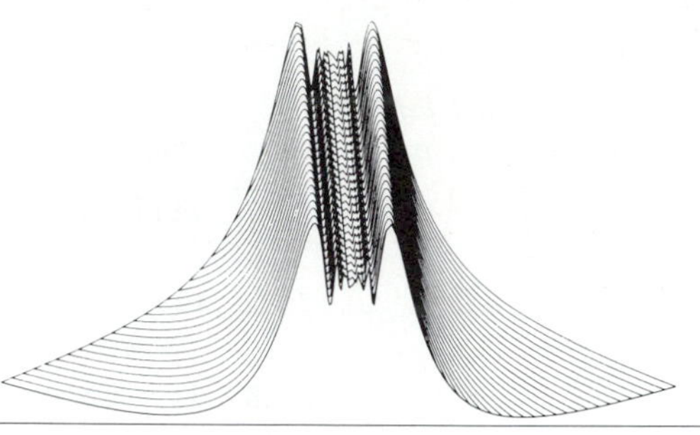

CHAPTER 13

PARAMETRIC EQUATIONS AND POLAR COORDINATES

13.1 CURVES DEFINED PARAMETRICALLY

Except for the conic sections, most of the curves we have considered so far have been graphs of functions—that is, graphs of equations of the form $y = f(x)$. One characteristic feature of such a curve is that a vertical line intersects it in at most one point. But there are many curves that do not have this property. For example, all the conic sections, with the exception of parabolas with vertical axes, fail the vertical line test. So does each of the graphs in Figure 13.1. Yet each of these is what we would call a curve.

To study these more general types of curves we introduce **parametric equations.** These have the form

$$\begin{cases} x = f(t) \\ y = g(t) \end{cases}$$

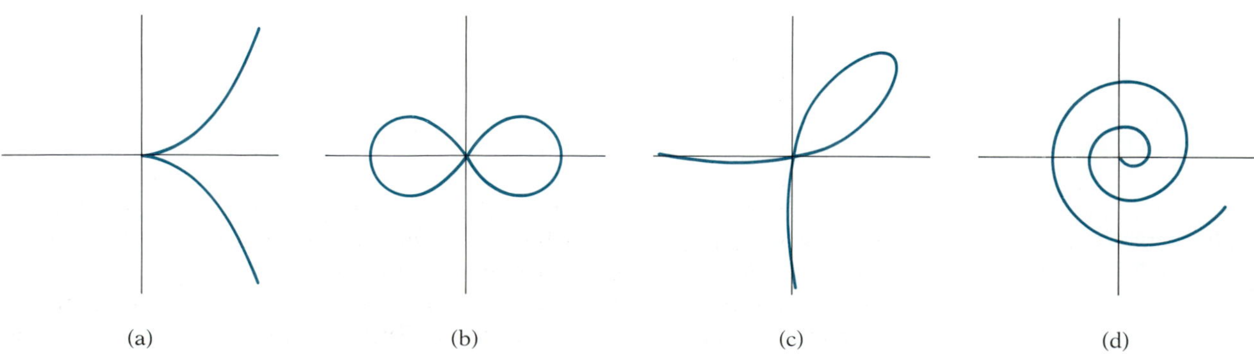

FIGURE 13.1

where t ranges over some interval I (which may be all of $\mathbf{R}$). The variable t is called a **parameter.** It provides a means of determining x and y but does not appear on the graph. As an example, consider the parametric equations

$$\begin{cases} x = t^2 \\ y = 2t - 1 \end{cases} \qquad -\infty < t < \infty$$

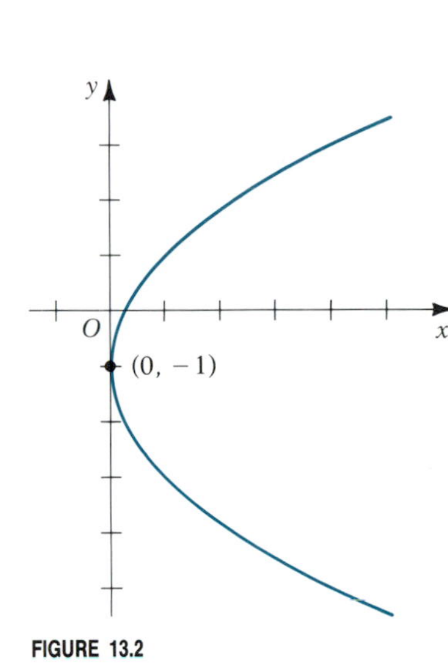

FIGURE 13.2

By assigning values of t, points (x, y) on the graph are determined as in the table.

t	0	1	2	3	4	−1	−2	−3	−4
x	0	1	4	9	16	1	4	9	16
y	−1	1	3	5	7	−3	−5	−7	−9

Now we can plot the points (x, y) and connect them with a smooth curve. (We are making an assumption by connecting them in this way, and we will discuss this assumption later.) The result is shown in Figure 13.2. As you see, this looks like a parabola, and we will find shortly that this is the case. (You might think about it now, however, and see whether you can discover how to show this.)

The letter t is frequently used for a parameter because often in physical applications t represents time, and the point (x, y) is the position of some particle at time t. However, other letters can also be used. For example, we can obtain parametric equations of a circle most conveniently using the angle θ shown in Figure 13.3 as a parameter. If the radius of the circle is r and its center is at the origin, as shown, then

$$\begin{cases} x = r\cos\theta \\ y = r\sin\theta \end{cases}$$

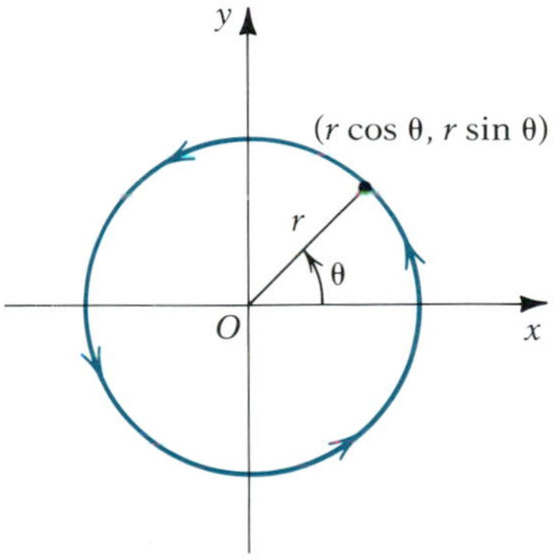

and as θ varies from 0 to 2π, the points (x, y) so obtained give the entire circle. In this way we say we have *parameterized* the circle in terms of the parameter θ. Notice that the circle is traced out in a counterclockwise direction as θ increases from 0 to 2π.

FIGURE 13.3

We need to be more precise about what we mean by a curve, and we do this in the following definition.

DEFINITION 13.1 A **curve** in the plane is the set $\{(x, y): x = f(t),\ y = g(t)\}$, where f and g are continuous for t in some interval I. ■

Returning to the earlier example

$$\begin{cases} x = t^2 \\ y = 2t - 1 \end{cases} \qquad -\infty < t < \infty$$

we see that this does define a curve, since for t in the interval $(-\infty, \infty)$, $f(t) = t^2$ and $g(t) = 2t - 1$ are continuous. The significance of the continuity requirement is to guarantee that small changes in t produce small changes in x and y, thus assuring that there are no breaks in the graph.

Sometimes we can eliminate the parameter between two parametric equations and find how x and y are related. We refer to the result as the **rectangular** equation. The next two examples illustrate this.

EXAMPLE 13.1 Find the rectangular equation of the curve defined by each of the following:

(a) $\begin{cases} x = t^2 \\ y = 2t - 1 \end{cases} \quad -\infty < t < \infty$ (b) $\begin{cases} x = 2\cos\theta \\ y = 3\sin\theta \end{cases} \quad 0 \leq \theta \leq 2\pi$

Solution (a) Solving for t from the second equation gives $t = (y + 1)/2$, and if we substitute this in the equation for x, we get

$$x = \frac{(y+1)^2}{4} \quad \text{or} \quad (y+1)^2 = 4x$$

Note that this is the equation of a parabola with horizontal axis, opening to the right, with vertex at $(0, -1)$. This confirms our earlier observation.

(b) To eliminate θ, we make use of the identity $\sin^2\theta + \cos^2\theta = 1$. Divide the first equation by 2, divide the second by 3, square, and add:

$$\left(\frac{x}{2}\right)^2 = \cos^2\theta$$
$$\left(\frac{y}{3}\right)^2 = \sin^2\theta$$
$$\overline{\frac{x^2}{4} + \frac{y^2}{9} = 1}$$

So this is an ellipse, with center at the origin and major axis along the y-axis. ■

EXAMPLE 13.2 Find the rectangular equation of each of the following, and draw the graph:

(a) $\begin{cases} x = \cosh t \\ y = \sinh t \end{cases} \quad -\infty < t < \infty$ (b) $\begin{cases} x = \sqrt{t} \\ y = t + 2 \end{cases} \quad t \geq 0$

Solution (a) By squaring both sides of each equation and subtracting, we can use the identity $\cosh^2 t - \sinh^2 t = 1$ to eliminate the parameter:

$$x^2 = \cosh^2 t$$
$$y^2 = \sinh^2 t$$
$$\overline{x^2 - y^2 = 1}$$

This is the equation of an equilateral hyperbola, but we have to exercise care in drawing the graph, since the original parametric equations impose a restriction not seen in the rectangular equation. In particular, since $\cosh t \geq 1$ for all t, it follows that $x \geq 1$. Thus, it is the right branch only of the hyperbola that is determined by the parametric equations. We show this in Figure 13.4. Note that as t goes from $-\infty$ to ∞, the curve is traced out as shown by the arrows.

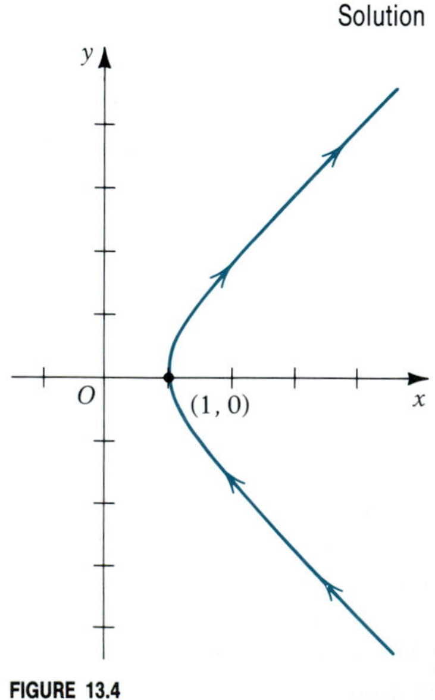

FIGURE 13.4

(b) One way to eliminate t is to square both sides of the first equation and substitute into the second. This gives $y = x^2 + 2$, which is the equation of a parabola. But again, x and y are restricted by the original parametric equations so that $x \geq 0$ and $y \geq 2$. We therefore have only the right half of the parabola, as shown in Figure 13.5. ■

A Word of Caution As the last example makes clear, when you eliminate the parameter between two parametric equations, it is essential to check the original equations for any restrictions imposed on x or y, and to take these into account in drawing the graph.

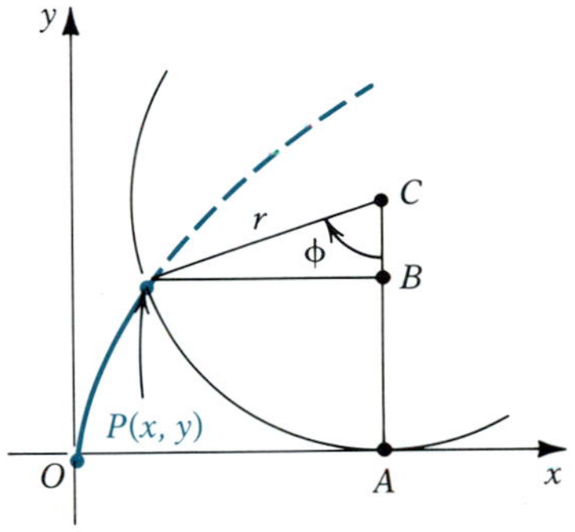

As we noted, when the parameter runs through its domain, with increasing values, a direction or *orientation* is induced on the curve. So a parameterized curve always has a direction, whereas in rectangular form no direction is implied.

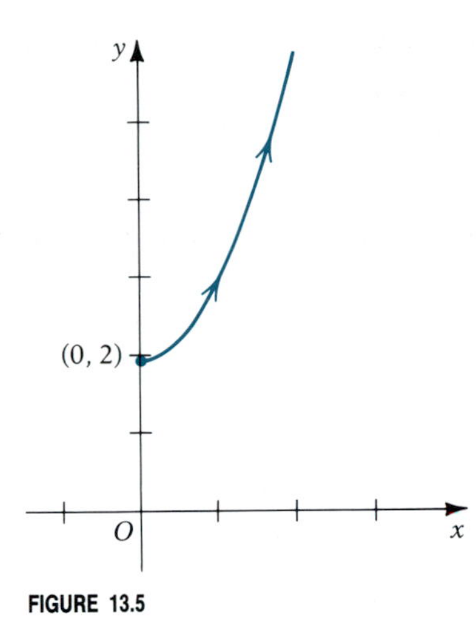

FIGURE 13.5

We conclude this section by deriving parametric equations of a curve called a **cycloid.** This is the path traced out by a point on the periphery of a wheel as the wheel moves along a straight line. For example, if a point is marked on a bicycle tire and the bicycle moves in a straight line, the point traces out a cycloid. The cycloid has many interesting properties, two of which we will mention after we derive the equations.

In Figure 13.6 we show the point, originally at the origin, after the wheel has moved to the right. The angle ϕ (in radians) shown will be used as a parameter. Our object is to write x and y in terms of ϕ. Let the radius of the wheel be r. From the figure we see that

FIGURE 13.6

$$x = \overline{OA} - \overline{PB}$$
$$y = \overline{AC} - \overline{BC}$$

From triangle PBC, we determine by trigonometry that $\overline{PB} = r \sin \phi$ and $\overline{BC} = r \cos \phi$. Also $\overline{AC} = r$, and $\overline{OA}$ is equal to the arc length $\widehat{PA}$. (Why?) But the length of arc $\widehat{PA} = r\phi$. So we have

$$\begin{cases} x = r\phi - r\sin\phi \\ y = r - r\cos\phi \end{cases}$$

or

$$\begin{cases} x = r(\phi - \sin\phi) \\ y = r(1 - \cos\phi) \end{cases} \qquad (13.1)$$

By allowing the wheel to roll to either the left or right, we see that ϕ can be any real number.

The graph of the cycloid can be obtained from equations (13.1) by substituting values of ϕ. The result is shown in Figure 13.7. Note that the graph goes through one complete cycle as ϕ goes from 0 to 2π, causing x to go from 0 to $2\pi r$. Then it repeats itself in the next x interval of length $2\pi r$, and so on; that is, the cycloid is periodic of period $2\pi r$.

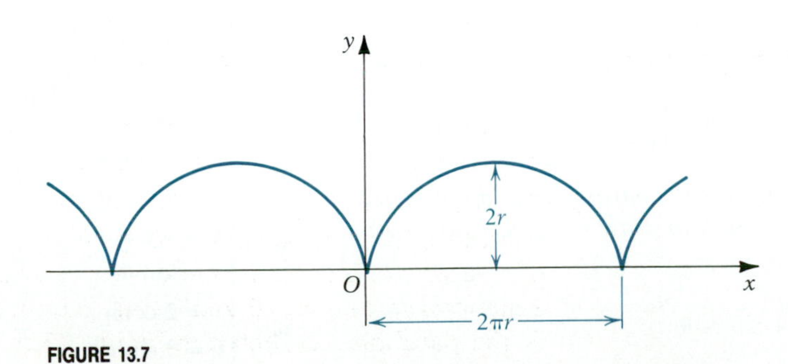

FIGURE 13.7

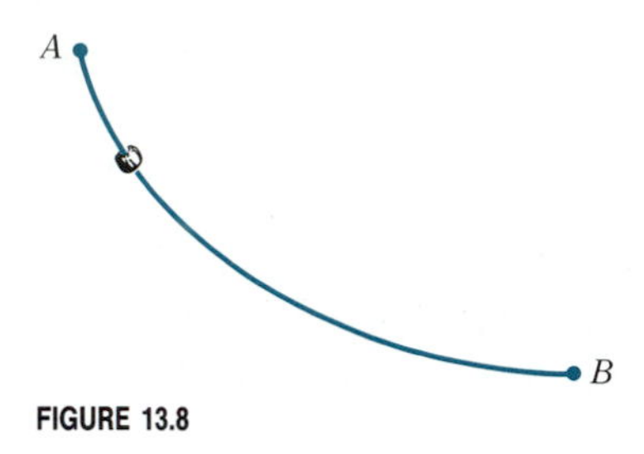

FIGURE 13.8

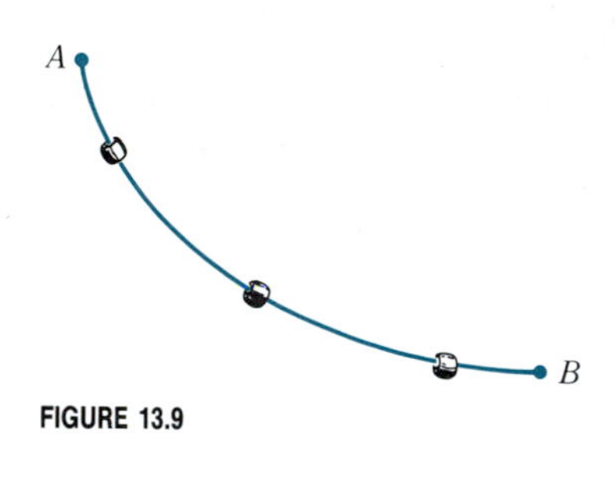

FIGURE 13.9

Although it is possible to eliminate the parameter ϕ between the equations (13.1) of the cycloid (see Exercise 29), it is not advantageous to do so, since the result is rather complex. The cycloid thus illustrates that it is sometimes easier to graph a curve from the parametric equations than from the rectangular equation.

Two famous problems in mechanics are known as the **brachistochrone** and the **tautochrone** problems. The brachistochrone is the "curve of quickest descent." Suppose, for example, a bead is allowed to slide (without friction) along a wire from a point A to a lower point B, as in Figure 13.8. In what shape should the wire be bent so that the time required for the bead to go from A to B is a minimum? Surprisingly, the answer is not a straight line (the shortest distance) but is an inverted cycloid in which A is on the x-axis at a cusp of the cycloid. This discovery was made independently by a number of 17th-century mathematicians, including Johann Bernoulli and Blaise Pascal.

The tautochrone is the "curve of equal time." Suppose beads are initially placed on a wire joining A and B, as in the brachistochrone problem, and then released. What should be the shape of the wire so that all of the beads arrive at B at the same time? (See Figure 13.9.) Again, the answer is the inverted cycloid. This is quite a remarkable result, for even if one of the beads starts very near to B and another starts at A, when the curve is a cycloid, they will arrive at the same instant. The Dutch physicist and astronomer Christian Huygens (1629–1695) made this discovery in his effort to construct a pendulum clock in which the period of the pendulum is independent of the amplitude of its swing.

EXERCISE SET 13.1

A

In Exercises 1–6 draw the curve defined by the parametric equations by substituting values for the parameter. Indicate the orientation on the graph.

1. $\begin{cases} x = 2t + 3 \\ y = 3t - 1 \end{cases} \quad -\infty < t < \infty$

2. $\begin{cases} x = t^2 - 1 \\ y = t^2 + 1 \end{cases} \quad -3 \le t \le 0$

3. $\begin{cases} x = t - 2 \\ y = t^2 \end{cases} \quad -2 \le t \le 3$

4. $\begin{cases} x = u^2 \\ y = u^3 \end{cases} \quad -\infty < u < \infty$

5. $\begin{cases} x = \frac{1}{v} \\ y = v \end{cases} \quad v > 0$

6. $\begin{cases} x = \sqrt{t} \\ y = t \end{cases} \quad t > 0$

In Exercises 7–20 eliminate the parameter and make use of the rectangular equation to draw the given curve. Indicate the orientation.

7. $\begin{cases} x = 3t - 2 \\ y = 1 - t \end{cases} \quad -1 \le t \le 4$

8. $\begin{cases} x = 4 - t^2 \\ y = t^2 + 2 \end{cases} \quad 0 \le t \le 3$

9. $\begin{cases} x = \cos t \\ y = -\sin t \end{cases} \quad 0 \le t \le \pi$

10. $\begin{cases} x = 1 - \cos\theta \\ y = \sin\theta \end{cases} \quad 0 \le \theta \le 2\pi$

11. $\begin{cases} x = e^t \\ y = e^{-t} \end{cases} \quad -\infty < t < \infty$

12. $\begin{cases} x = 2t - 1 \\ y = 1 - t^2 \end{cases} \quad t \ge -1$

13. $\begin{cases} x = \sin^2\theta + 1 \\ y = \cos\theta \end{cases} \quad 0 \le \theta \le \pi$

14. $\begin{cases} x = 3\sin\theta \\ y = 4\cos\theta \end{cases} \quad 0 \le \theta \le 2\pi$

15. $\begin{cases} x = \sec\theta \\ y = \tan\theta \end{cases} \quad -\frac{\pi}{2} < \theta < \frac{\pi}{2}$

16. $\begin{cases} x = 2\cosh t \\ y = 3\sinh t \end{cases} \quad -\infty < t < \infty$

17. $\begin{cases} x = 2 - \cos t \\ y = 1 + 3\sin t \end{cases}$ $\quad 0 \le t \le 2\pi$

18. $\begin{cases} x = \sqrt{4 - t} \\ y = \sqrt{t} \end{cases}$ $\quad 0 \le t \le 4$

19. $\begin{cases} x = \sin\theta \\ y = \cos 2\theta \end{cases}$ $\quad 0 \le \theta \le \frac{\pi}{2}$

20. $\begin{cases} x = 2\sec^2\theta - 3 \\ y = \tan\theta + 1 \end{cases}$ $\quad 0 \le \theta < \frac{\pi}{2}$

21. Compare the curves C_1, C_2, C_3, and C_4. Include a discussion of orientation.

C_1: $\begin{cases} x = 1 + 2t \\ y = -2 + 4t \end{cases}$ $\quad 0 \le t \le 1$

C_2: $\begin{cases} x = 3 - 2t \\ y = 2 - 4t \end{cases}$ $\quad 0 \le t \le 1$

C_3: $\begin{cases} x = t \\ y = 2t - 4 \end{cases}$ $\quad 1 \le t \le 3$

C_4: $\begin{cases} x = \tan^2 t \\ y = 2\sec^2 t - 6 \end{cases}$ $\quad \frac{\pi}{4} \le t \le \frac{\pi}{3}$

22. If a curve crosses itself, the point where this occurs is called a **double point.** Show that the curve C defined by

$$\begin{cases} x = t^3 - 3t + 1 \\ y = t^2 + t - 1 \end{cases} \quad -\infty < t < \infty$$

has a double point, and find this point. What are the two distinct values of t that produce the double point?

23. If a curve has no double points (see Exercise 22), it is said to be a **simple** curve. Show that the curve C defined by

$$\begin{cases} x = t^3 + 1 \\ y = t^2 - 1 \end{cases} \quad 0 \le t \le 2$$

is simple. Draw its graph.

24. A curve C defined by $x = f(t)$ and $y = g(t)$, for $a \le t \le b$, is said to be a **simple closed curve** if it is simple for $a \le t < b$ (see Exercise 23) and $(f(a), g(a))$ is the same point as $(f(b), g(b))$. Show that the curve defined by

$$\begin{cases} x = t^2 - 3 \\ y = t^3 - 4t - 1 \end{cases} \quad -2 \le t \le 2$$

is a simple closed curve.

B

In Exercises 25–28 draw the curves, showing the orientation.

25. $\begin{cases} x = t^2 - 1 \\ y = t^3 - t \end{cases}$ $\quad -\infty < t < \infty$

26. $\begin{cases} x = \cos^3\theta \\ y = \sin^3\theta \end{cases}$ $\quad 0 \le \theta \le 2\pi$

27. $\begin{cases} x = \cos^2\theta \\ y = \frac{1}{2}\sin 2\theta \end{cases}$ $\quad 0 \le \theta \le \pi$

28. $\begin{cases} x = (t^2 - 2)(t + 1) \\ y = (t^2 - 2)(t - 1) \end{cases}$ $\quad -\infty < t < \infty$

29. Find the rectangular equation of the cycloid.

30. Find the area under one arch of the cycloid.

31. A *hypocycloid* is a curve traced out by a fixed point on a circle rolling on the inside of a fixed larger circle. (See the figure.) If the radius of the larger circle is a and the radius of the smaller circle is b, derive the parametric equations

$$\begin{cases} x = (a - b)\cos\theta + b\cos\dfrac{a - b}{b}\theta \\ y = (a - b)\sin\theta - b\sin\dfrac{a - b}{b}\theta \end{cases} \quad 0 \le \theta \le 2\pi$$

(*Hint:* Observe that $\widehat{AB} = \widehat{BP}$, so that $a\theta = b\phi$.)

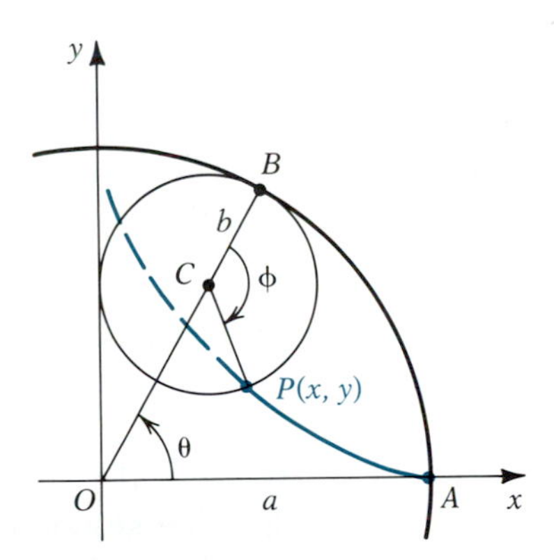

32. Refer to Exercise 31 and show that if $a = 4b$ the equations can be put in the form

$$\begin{cases} x = a\cos^3\theta \\ y = a\sin^3\theta \end{cases}$$

Also find the rectangular equation. (This is the four-cusp hypocycloid.)

33. If the smaller circle of Exercise 31 rolls on the outer circumference of the larger circle, the point traces out an **epicycloid.** (See the figure.) Find the parametric equations of the epicycloid.

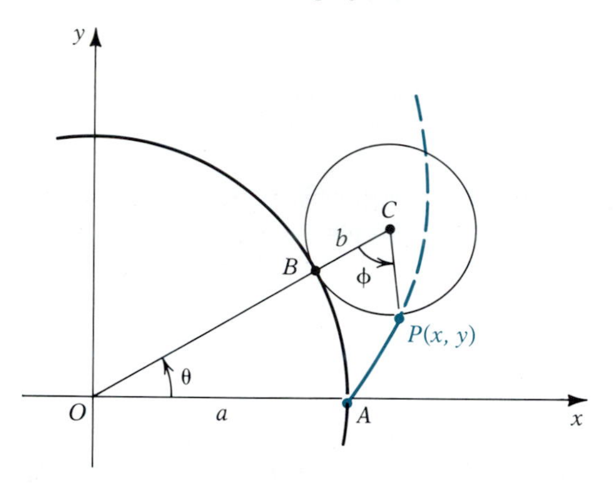

34. Suppose a point P is attached to an extended spoke of a wheel as in the figure and the wheel rolls along the x-axis. If the distance from the center of the wheel to P is b and the radius of the wheel is a (with $a < b$), find parametric equations of the path of P. This curve is called a **trochoid.**

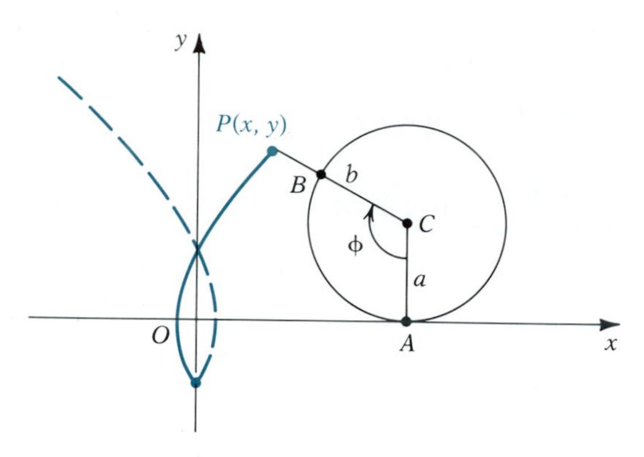

13.2 DERIVATIVES OF FUNCTIONS DEFINED PARAMETRICALLY

Without more restrictions on the curves given by Definition 13.1, they can behave in erratic ways. For example, they can have many cusps so that no tangent line exists, they can cross themselves at various points (double points), and they can have infinite length even when they are bounded. To avoid these problems, we restrict ourselves in this section and the next to curves of the type described by the following definition.

DEFINITION 13.2 A curve C defined parametrically by

$$\begin{cases} x = f(t) \\ y = g(t) \end{cases} \qquad t \in I$$

is said to be **smooth** provided f' and g' are continuous on I and are not simultaneously 0 there. ■

Remark If f' and g' are both 0 at $t_0 \in I$, then $(f(t_0), g(t_0))$ is called a **singular point** of the curve. So smooth curves have no singular points. As we shall see, the smoothness requirement guarantees that the curve has a continuously turning tangent line. There are no sharp corners.

If a smooth curve $\{(f(t), g(t)) : t \in I\}$ also has the property that $f'(t) \neq 0$ for all $t \in I$, then y can be expressed as a differentiable function of x, whose derivative has a particularly simple form. This is the result of the following theorem.

THEOREM 13.1 If C is a smooth curve defined parametrically by

$$\begin{cases} x = f(t) \\ y = g(t) \end{cases} \qquad a \le t \le b$$

and if $f'(t) \neq 0$ for all t in $[a, b]$, then y can be expressed as a differentiable function of x, $y = F(x)$, in the interval $[f(a), f(b)]$, and

$$F'(x) = \frac{g'(t)}{f'(t)}$$

for all x in this interval.

Proof Since f' is continuous on $[a, b]$ and f' is never 0 there, it follows that either $f'(t) > 0$ or $f'(t) < 0$ for all $t \in [a, b]$. (Why?) So f is strictly monotone and hence by Theorem 6.3 has a differentiable inverse f^{-1} whose derivative is given by

$$(f^{-1})'(x) = \frac{1}{f'(t)}$$

This is valid for x in the interval $[f(a), f(b)]$. For x in this interval we have $t = f^{-1}(x)$ and so $y = g(f^{-1}(x))$. Writing $F = g \circ f^{-1}$, we get $y = F(x)$ and, by the chain rule,

$$\begin{aligned} F'(x) &= g'(f^{-1}(x)) \cdot (f^{-1})'(x) \\ &= \frac{g'(t)}{f'(t)} \end{aligned}$$

where in the last expression it is understood that $t = f^{-1}(x)$. This completes the proof. ■

This result can be more easily remembered in the Leibniz notation:

$$\frac{dy}{dx} = \frac{\frac{dy}{dt}}{\frac{dx}{dt}} \quad \text{if } \frac{dx}{dt} \neq 0 \tag{13.2}$$

Remark If the curve is smooth and $\frac{dx}{dt} = 0$, then a vertical tangent line exists at that point.

EXAMPLE 13.3 Find $\frac{dy}{dx}$ when $t = 2$ for the curve C defined parametrically by

$$\begin{cases} x = 2t^2 - 1 \\ y = t^3 - 3t \end{cases} \quad 1 \leq t \leq 4$$

Solution It is easily verified that C is smooth and that $\frac{dx}{dt} \neq 0$ in $[1, 4]$. So we have

$$\frac{dy}{dx} = \frac{\frac{dy}{dt}}{\frac{dx}{dt}} = \frac{3t^2 - 3}{4t}$$

and at $t = 2$,

$$\left.\frac{dy}{dx}\right|_{t=2} = \frac{9}{8}$$

■

EXAMPLE 13.4 Find the equation of the tangent line to the curve C defined by

$$\begin{cases} x = \frac{4}{t} \\ y = \sqrt{t} \end{cases} \qquad 1 \le t \le 9$$

at the point (1, 2).

Solution The conditions of Theorem 13.1 are met, so we have

$$\frac{dy}{dx} = \frac{\frac{dy}{dt}}{\frac{dx}{dt}} = \frac{\frac{1}{2\sqrt{t}}}{-\frac{4}{t^2}} = -\frac{t^{3/2}}{8}$$

The value of t that produces the point (1, 2) is $t = 4$. Thus, at this point,

$$\left.\frac{dy}{dx}\right|_{t=4} = -\frac{(4)^{3/2}}{8} = -1$$

Using the point-slope form of the equation of a line, we therefore obtain the tangent line

$$y - 2 = -(x - 1)$$

or

$$x + y = 3$$

■

If the second derivative of y with respect to x exists, then the same technique as above can be used to find it. Letting $y' = dy/dx$ and $y'' = d^2y/dx^2$, we can get y'' by

$$y'' = \frac{dy'}{dx} = \frac{\frac{dy'}{dt}}{\frac{dx}{dt}}$$

Higher order derivatives, if they exist, can be found similarly. For example,

$$y''' = \frac{dy''}{dx} = \frac{\frac{dy''}{dt}}{\frac{dx}{dt}}$$

EXAMPLE 13.5 Find y'' for the function defined in Example 13.4.

Solution In Example 13.4 we found that

$$y' = \frac{dy}{dx} = -\frac{t^{3/2}}{8}$$

Then, since $\frac{dx}{dt} = -4t^{-2}$, we have

$$y'' = \frac{dy'}{dx} = \frac{\frac{dy'}{dt}}{\frac{dx}{dt}} = \frac{\frac{-3t^{1/2}}{16}}{-\frac{4}{t^2}} = \frac{3t^{5/2}}{64}$$

■

EXERCISE SET 13.2

A

In Exercises 1–10 find $\frac{dy}{dx}$ at the indicated point without eliminating the parameter.

1. $\begin{cases} x = t^2 - 1 \\ y = 2t + 3 \end{cases}$ at $t = 1$

2. $\begin{cases} x = \sqrt{t-1} \\ y = t^2 \end{cases}$ at $t = 2$

3. $\begin{cases} x = t^3 \\ y = t^2 \end{cases}$ at $t = -2$

4. $\begin{cases} x = 2 \sin t \\ y = 3 \cos t \end{cases}$ at $t = \frac{\pi}{3}$

5. $\begin{cases} x = 4 \sin^2 t - 3 \\ y = 2 \cos t \end{cases}$ at $t = \frac{\pi}{3}$

6. $\begin{cases} x = 2 \sinh t \\ y = \tanh t \end{cases}$ at $t = 0$

7. $\begin{cases} x = e^{-t} \\ y = 1 + e^t \end{cases}$ at $t = \ln 2$

8. $\begin{cases} x = \ln t \\ y = 2 - \ln \sqrt{t} \end{cases}$ at $t = e^2$

9. $\begin{cases} x = t^3 - 2t \\ y = \sqrt{t-1} \end{cases}$ at $(4, 1)$

10. $\begin{cases} x = \frac{1}{t} \\ y = 4(1 - t^2) \end{cases}$ at $(-2, 3)$

In Exercises 11–14 find the equation of the tangent line to the curve defined by the parametric equations at the given point.

11. $\begin{cases} x = 3t^2 - 2t \\ y = t^2 + t - 1 \end{cases}$ at $t = -1$

12. $\begin{cases} x = \dfrac{1}{t-1} \\ y = \dfrac{1}{t+1} \end{cases}$ at $t = \frac{1}{2}$

13. $\begin{cases} x = t^3 + 3t^2 \\ y = 2t^2 - 3 \end{cases}$ at $(2, -1)$

14. $\begin{cases} x = t - \frac{2}{t} \\ y = t^2 \end{cases}$ at $(1, 4)$

15. Find all points on the curve C defined by

$$\begin{cases} x = t^3 - 4t^2 - 3t \\ y = 2t^2 + 3t - 5 \end{cases} \quad -\infty < t < \infty$$

where the tangent line is vertical and all points where the tangent line is horizontal.

16. (a) Show that the curve C defined by

$$\begin{cases} x = t^2 - 2t - 8 \\ y = t^2 - t - 2 \end{cases} \quad -\infty < t < \infty$$

is smooth.

(b) In what intervals do the parametric equations in part a define y as a differentiable function of x? Find y' in terms of t for the intervals in which it exists.

17. Follow the instructions for Exercise 16 for the curve C defined by

$$\begin{cases} x = \dfrac{t^2 - 5}{t - 2} \\ y = \dfrac{1}{t - 2} \end{cases} \quad t \neq 2$$

In Exercises 18–21 find y'' in terms of t without eliminating the parameter.

18. $\begin{cases} x = t^2 \\ y = 2t^3 \end{cases}$ $t > 0$

19. $\begin{cases} x = 2 \cos t \\ y = 3 \sin t \end{cases}$ $0 < t < \pi$

20. $\begin{cases} x = \sinh t \\ y = \cosh^2 t \end{cases}$ $-\infty < t < \infty$

21. $\begin{cases} x = \sqrt{1 - t^2} \\ y = 1 + t^2 \end{cases}$ $0 < |t| < 1$

B

22. Show that the curve defined by

$$\begin{cases} x = t^2 + 1 \\ y = t^3 - t - 1 \end{cases} \quad -\infty < t < \infty$$

has a double point, and find equations of the two tangent lines at this point.

23. Use y' and y'' to show that the cycloid

$$\begin{cases} x = a(t - \sin t) \\ y = a(1 - \cos t) \end{cases} \quad -\infty < t < \infty$$

has local (and absolute) maxima at $t = (2n + 1)\pi$ and local (and absolute) minima at $t = 2n\pi$, and that the curve is concave downward in each interval of the form $(2n\pi, 2(n+1)\pi)$, for $n = 0, \pm 1, \pm 2, \ldots$.

24. When a smooth curve C is defined parametrically by $x = f(t)$ and $y = g(t)$ for t in some interval I, the notation $\dot{x} = dx/dt$, $\dot{y} = dy/dt$, $\ddot{x} = d^2x/dt^2$, and $\ddot{y} = dy^2/dt^2$ is sometimes used (it was introduced by Newton). Assuming the equations define y as a twice differentiable function of x, show that

$$\frac{d^2y}{dx^2} = \frac{\dot{x}\ddot{y} - \dot{y}\ddot{x}}{\dot{x}^3}$$

Similarly, find an expression for d^3y/dx^3.

25. Find the equation of the normal line to the cycloid at an arbitrary point $P(x, y)$ for which the parameter ϕ is in the open interval $(0, 2\pi)$. Show that the x-intercept of this line is the x-coordinate of the point of contact of the generating circle with the x-axis for the given value of ϕ.

13.3 ARC LENGTH AND SURFACE AREA

In our earlier study of arc length we were concerned only with curves that were graphs of functions, $y = f(x)$. We can now extend the result to curves defined parametrically. Let C be defined by

$$\begin{cases} x = f(t) \\ y = g(t) \end{cases} \qquad a \le t \le b$$

Partition the interval $[a, b]$ by means of the points

$$a = t_0 < t_1 < t_2 < \cdots < t_n = b$$

Each value t_k in this partition gives rise to a point P_k on C, having coordinates $(f(t_k), g(t_k))$. As shown in Figure 13.10, we connect consecutive points P_{k-1} and P_k $(k = 1, 2, \ldots, n)$ on the curve with straight line segments. By adding up the lengths $\overline{P_{k-1}P_k}$ of all such segments we obtain an approximation to the length of C. In fact, we *define* the length of C to be the limit of this sum as the norm $\|P\|$ of such partitions goes to 0, provided this limit exists.

DEFINITION 13.3 The curve C defined by $x = f(t)$ and $y = g(t)$ for t in $[a, b]$ is said to be **rectifiable** if

$$\lim_{\|P\| \to 0} \sum_{k=1}^{n} \overline{P_{k-1}P_k} \qquad (13.3)$$

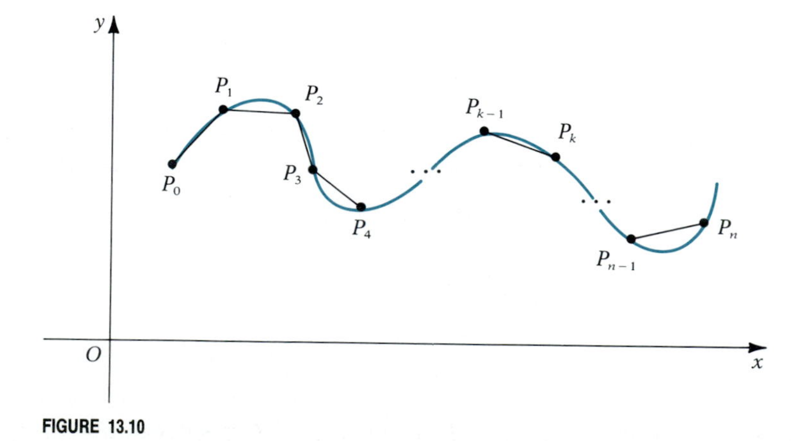

FIGURE 13.10

exists, where $P = \{t_0, t_1, \ldots, t_n\}$ is a partition of $[a, b]$ and $P_k = (f(t_k), g(t_k))$. If C is rectifiable, its **length** is defined as the limit in (13.3). ■

It is interesting to note that there are curves that are bounded, in the sense of being completely contained within some rectangle in the plane, that are infinite in length and so are not rectifiable. Curves can behave in strange ways!

It is beyond the scope of this course to study conditions that are both necessary and sufficient for rectifiability. However, we will now show that continuity of f' and g' on $[a, b]$ is sufficient to guarantee that C is rectifiable. Suppose this is the case. Then we have

$$\overline{P_{k-1}P_k} = \sqrt{(f(t_k) - f(t_{k-1}))^2 + (g(t_k) - g(t_{k-1}))^2}$$

and since f' and g' are continuous on $[t_{k-1}, t_k]$, f and g each satisfies the conditions of the mean value theorem there (Theorem 3.7). So there exists numbers c_k and d_k in (t_{k-1}, t_k) such that

$$f(t_k) - f(t_{k-1}) = f'(c_k)(t_k - t_{k-1})$$

and

$$g(t_k) - g(t_{k-1}) = g'(d_k)(t_k - t_{k-1})$$

Since we have defined Δt_k as $t_k - t_{k-1}$, we therefore have

$$\begin{aligned}\overline{P_{k-1}P_k} &= \sqrt{[f'(c_k)\,\Delta t_k]^2 + [g'(d_k)\,\Delta t_k]^2} \\ &= \sqrt{[f'(c_k)]^2 + [g'(d_k)]^2}\,\Delta t_k\end{aligned}$$

Thus,

$$\sum_{k=1}^{n} \overline{P_{k-1}P_k} = \sum_{k=1}^{n} \sqrt{[f'(c_k)]^2 + [g'(d_k)]^2}\,\Delta t_k$$

If c_k were the same as d_k, the sum on the right would be a Riemann sum. Even though they may differ, both are in the interval (t_{k-1}, t_k), whose length shrinks toward 0 as the norm $\|P\| \to 0$, and it can be proved that the sum behaves as if it were a Riemann sum, in that its limit is the integral

$$\int_a^b \sqrt{[f'(t)]^2 + [g'(t)]^2}\,dt$$

This is finite, since f' and g', and hence the integrand, are continuous on $[a, b]$. Thus C is rectifiable and its length is this integral. We state this result as a theorem.

THEOREM 13.2 If f' and g' are continuous on $[a, b]$, then the curve C defined by $x = f(t)$ and $y = g(t)$ for t in $[a, b]$ is rectifiable, and its length L is given by

$$L = \int_a^b \sqrt{[f'(t)]^2 + [g'(t)]^2}\,dt$$

or, equivalently,

$$L = \int_a^b \sqrt{\left(\frac{dx}{dt}\right)^2 + \left(\frac{dy}{dt}\right)^2}\,dt \tag{13.4}$$ ■

Remark In particular, if C is smooth on $[a, b]$, it is rectifiable.

EXAMPLE 13.6 Find the length of the curve C defined by

$$\begin{cases} x = t^3 - 1 \\ y = 2t^2 + 3 \end{cases} \qquad 0 \le t \le 1$$

Solution We have

$$\frac{dx}{dt} = 3t^2 \quad \text{and} \quad \frac{dy}{dt} = 4t$$

and these are continuous on $[0, 1]$. By equation (13.4),

$$\begin{aligned} L &= \int_0^1 \sqrt{(3t^2)^2 + (4t)^2}\, dt = \int_0^1 \sqrt{9t^4 + 16t^2}\, dt \\ &= \int_0^1 t\sqrt{9t^2 + 16}\, dt = \tfrac{1}{18}[(9t^2 + 16)^{3/2} \cdot \tfrac{2}{3}]_0^1 \\ &= \tfrac{1}{27}[125 - 64] = \tfrac{61}{27} \end{aligned}$$

■

To see that equation (13.4) for the arc length of a curve defined parametrically really is a generalization of equation (5.17) for arc length on the graph of $y = f(x)$ from $x = a$ to $x = b$, we can introduce the parameter $t = x$ and get

$$\begin{cases} x = t \\ y = f(t) \end{cases} \qquad a \le t \le b$$

Then, applying equation (13.4), we find the length of the arc:

$$L = \int_a^b \sqrt{\left(\frac{dx}{dt}\right)^2 + \left(\frac{dy}{dt}\right)^2}\, dt = \int_a^b \sqrt{1 + [f'(t)]^2}\, dt$$

which is equivalent to equation (5.17). Thus, when a curve is the graph of a smooth function, equation (13.4) gives the same length as equation (5.17). But equation (13.4) also applies to curves that are not graphs of functions. So it is a generalization of equation (5.17).

As in Section 5.4, defining the arc length function s by

$$s(t) = \int_a^t \sqrt{\left(\frac{dx}{dt}\right)^2 + \left(\frac{dy}{dt}\right)^2}\, dt$$

we obtain

$$\frac{ds}{dt} = \sqrt{\left(\frac{dx}{dt}\right)^2 + \left(\frac{dy}{dt}\right)^2}$$

so that the differential of arc length is

$$ds = \sqrt{\left(\frac{dx}{dt}\right)^2 + \left(\frac{dy}{dt}\right)^2}\, dt \qquad (13.5)$$

The discussion of the surface area of a surface generated by revolving a curve about an axis can be carried out in a manner analogous to that in Section 5.5. We will briefly indicate the main ideas.

Again we require of the curve C

$$\begin{cases} x = f(t) \\ y = g(t) \end{cases} \qquad a \le t \le b$$

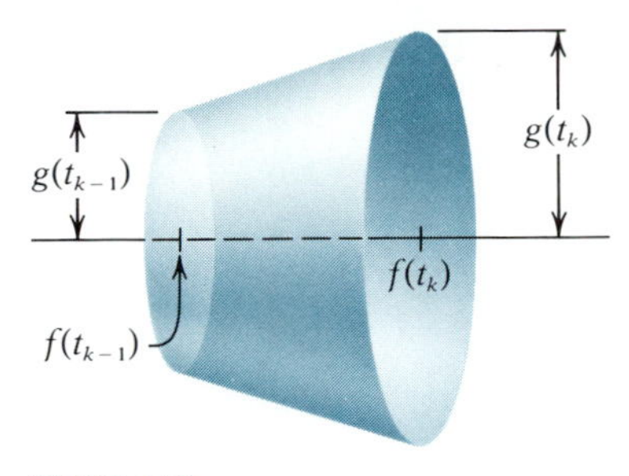

FIGURE 13.11

that f' and g' be continuous on $[a, b]$. For the surface formed by revolving about the x-axis we require further that $g(t) \geq 0$ for all t in $[a, b]$. This guarantees that the curve lies altogether above the x-axis, so that the surface generated is well defined. For revolving about the y-axis, we would require instead that $f(t) \geq 0$. For the remainder of this discussion, we assume C is to be revolved about the x-axis.

Partitioning $[a, b]$ as before, we consider the sum of the areas of the frustums of cones generated by revolving each line segment that joins consecutive points on the curve corresponding to the partition points of the t interval. (See Figure 13.10.) A typical frustum is shown in Figure 13.11. Its surface area ΔS_k is the average circumference times the slant height:

$$\Delta S_k = 2\pi\left[\frac{g(t_{k-1}) + g(t_k)}{2}\right]\sqrt{[f(t_k) - f(t_{k-1})]^2 + [g(t_k) - g(t_{k-1})]^2}$$
$$= 2\pi\left[\frac{g(t_{k-1}) + g(t_k)}{2}\right]\sqrt{[f'(c_k)]^2 + [g'(d_k)]^2}\,\Delta t_k$$

where we have applied the mean value theorem as in the development of the arc length formula. When we sum these areas and take the limit as the norm of the partition approaches 0, we get the total surface area:

$$S = \int_a^b 2\pi g(t)\sqrt{[f'(t)]^2 + [g'(t)]^2}\,dt$$

or, in differential notation, replacing $g(t)$ by y, we get

$$S = 2\pi \int_a^b y\sqrt{\left(\frac{dx}{dt}\right)^2 + \left(\frac{dy}{dt}\right)^2}\,dt \tag{13.6}$$

The analogous result for rotating C about the y-axis [under the assumption that $x = f(t) \geq 0$] is

$$S = 2\pi \int_a^b x\sqrt{\left(\frac{dx}{dt}\right)^2 + \left(\frac{dy}{dt}\right)^2}\,dt \tag{13.7}$$

Using equation (13.5) for the differential of arc length, we can write these results in the simplified form

$$S = 2\pi \int_{t=a}^{t=b} y\,ds, \qquad \text{rotation about } x\text{-axis}$$

and

$$S = 2\pi \int_{t=a}^{t=b} x\,ds, \qquad \text{rotation about } y\text{-axis}$$

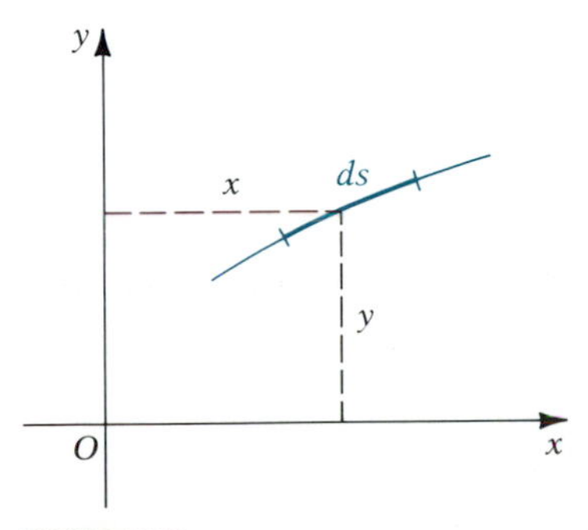

FIGURE 13.12

These simplified forms can be remembered by thinking of ds as a small element of arc that, when rotated about the x-axis, forms a "frustum of a cone" with surface area $2\pi y\,ds$, and when rotated about the y-axis a "frustum" with area $2\pi x\,ds$. (See Figure 13.12.) Then, "summing up" such areas by integration, we get the correct results. It should be emphasized that this is a memory device only and not a mathematically rigorous derivation.

EXAMPLE 13.7 Find the area of the surface generated by revolving the curve C defined by

$$\begin{cases} x = t^2 - 1 \\ y = 3t \end{cases} \qquad 0 \le t \le 2$$

about the x-axis.

Solution We have

$$\frac{dx}{dt} = 2t \quad \text{and} \quad \frac{dy}{dt} = 3$$

So

$$ds = \sqrt{\left(\frac{dx}{dt}\right)^2 + \left(\frac{dy}{dt}\right)^2}\, dt = \sqrt{4t^2 + 9}\, dt$$

and the surface area is

$$\begin{aligned} S &= 2\pi \int_{t=0}^{t=2} y\, ds = 2\pi \int_0^2 3t\sqrt{4t^2+9}\, dt \\ &= \frac{6\pi}{8} \int_0^2 8t\sqrt{4t^2+9}\, dt \\ &= \frac{3\pi}{4} (4t^2+9)^{3/2} \cdot \frac{2}{3}\Big]_0^2 \\ &= \frac{\pi}{2} [(25)^{3/2} - 9^{3/2}] \\ &= \frac{\pi}{2} [125 - 27] = \frac{\pi}{2}(98) = 49\pi \end{aligned}$$

■

EXERCISE SET 13.3

A

In Exercises 1–16 find the length of the curve defined by the given parametric equations over the specified interval.

1. $\begin{cases} x = 2t - 1 \\ y = 3t + 4 \end{cases} \quad 0 \le t \le 4$

2. $\begin{cases} x = 2 - t \\ y = 2t - 1 \end{cases} \quad -2 \le t \le 2$

3. $\begin{cases} x = t \\ y = t^{3/2} \end{cases} \quad 0 \le t \le 5$

4. $\begin{cases} x = \sin t \\ y = \cos t \end{cases} \quad 0 \le t \le \pi$

5. $\begin{cases} x = 4t^2 \\ y = t^3 \end{cases} \quad 0 \le t \le 2$

6. $\begin{cases} x = t + \frac{1}{t} \\ y = 2 \ln t \end{cases} \quad 1 \le t \le 3$

7. $\begin{cases} x = 3e^t \\ y = 2e^{3t/2} \end{cases} \quad \ln 3 \le t \le 2 \ln 2$

8. $\begin{cases} x = \sqrt{1 - t^2} \\ y = 1 - t \end{cases} \quad 0 \le t \le \frac{1}{2}$

9. $\begin{cases} x = e^t \sin t \\ y = e^t \cos t \end{cases} \quad 0 \le t \le \ln 3$

10. $\begin{cases} x = \cos^3 t \\ y = \sin^3 t \end{cases} \quad 0 \le t \le \frac{\pi}{2}$

11. $\begin{cases} x = \sin t - t\cos t \\ y = \cos t + t \sin t \end{cases} \quad 1 \le t \le 2$

12. $\begin{cases} x = \ln \cos t \\ y = t - \tan t \end{cases}$ $0 \le t \le \frac{\pi}{3}$

13. $\begin{cases} x = 4\sqrt{t} \\ y = t^2 + \frac{1}{2t} \end{cases}$ $\frac{1}{2} \le t \le 2$

14. $\begin{cases} x = \tanh t \\ y = \ln(\cosh^2 t) \end{cases}$ $-2 \le t \le 3$

15. $\begin{cases} x = 2 \sin t \\ y = \cos 2t \end{cases}$ $0 \le t \le \frac{\pi}{3}$

16. $\begin{cases} x = 3 \cos t \\ y = \cos 2t \end{cases}$ $0 \le t \le \frac{\pi}{2}$

In Exercises 17–24 find the area of the surface formed by revolving the given curve about the specified axis.

17. $\begin{cases} x = 3t^2 - 1 \\ y = 2t \end{cases}$ $0 \le t \le \sqrt{7}$; x-axis

18. $\begin{cases} x = 6t^2 \\ y = t^4 \end{cases}$ $0 \le t \le 2$; y-axis

19. $\begin{cases} x = \sin t + 2 \\ y = \cos t - 3 \end{cases}$ $0 \le t \le \frac{\pi}{2}$; y-axis

20. $\begin{cases} x = \cos 2t \\ y = 4 \sin t \end{cases}$ $0 \le t \le \frac{\pi}{2}$; x-axis

21. $\begin{cases} x = \tan^2 t \\ y = 2 \sec t \end{cases}$ $0 \le t \le \frac{\pi}{3}$; x-axis

22. $\begin{cases} x = 6t^2 \\ y = t^4 - 2 \end{cases}$ $0 \le t \le 2$; y-axis

23. $\begin{cases} x = 4e^t - 1 \\ y = 3e^t - 1 \end{cases}$ $0 \le t \le \ln 2$; about the line $x + 2 = 0$

24. $\begin{cases} x = \cos^3 t \\ y = \sin^3 t \end{cases}$ $0 \le t \le \frac{\pi}{2}$; about the line $y + 1 = 0$

B

25. (a) Find the length of one arch of the cycloid

$$\begin{cases} x = a(\theta - \sin \theta) \\ y = a(1 - \cos \theta) \end{cases}$$

(b) Find the surface area generated by revolving the arch of the cycloid of part a about the x-axis.

26. Find the length of the epicycloid

$$\begin{cases} x = 2a \cos t - a \cos 2t \\ y = 2a \sin t - a \sin 2t \end{cases} \quad 0 \le t \le 2\pi$$

27. For the curve defined by

$$\begin{cases} x = \frac{1}{t} \\ y = \ln t \end{cases} \quad 1 \le t \le \sqrt{3}$$

find (a) its length and (b) the surface area formed by revolving the curve about the y-axis.

28. Show that the graph of

$$\begin{cases} x = t^2 - 3 \\ y = \frac{1}{3}t^3 - t \end{cases} \quad -\infty < t < \infty$$

has a loop, and find the length of the loop. Draw the graph.

29. Approximate the length of the curve

$$\begin{cases} x = \sqrt{2}t \\ y = \frac{2}{5}t^{5/2} \end{cases} \quad 0 \le t \le 1$$

with the aid of a calculator, using four terms of a binomial series. Estimate the error.

C

Write a computer program that approximates the length of a curve

$$\begin{cases} x = f(t) \\ y = g(t) \end{cases} \quad a \le t \le b$$

by summing the lengths of chords $\overline{P_{i-1}P_i}$, *where* $P_i = (f(t_i), g(t_i))$, *and* $\{t_0, t_1, \ldots, t_n\}$ *is a regular partition of* $[a, b]$. *Use the program to approximate the lengths of the curves in Exercises 30–32. In each case show that the use of equation (13.4) results in an integral that cannot be evaluated by elementary means.*

30. $\begin{cases} x = 3 \sin t \\ y = 4 \cos t \end{cases}$ $0 \le t \le 2\pi$; $n = 50$

31. $\begin{cases} x = t \ln t \\ y = t + \dfrac{1}{t} \end{cases}$ $1 \le t \le 5$; $n = 10, 20, 30, 40, 50$

32. $\begin{cases} x = t^4 - 2t \\ y = 2t^3 + t^2 \end{cases}$ $0 \le t \le 2$; $n = 10, 20, 30 \ldots$

Stop when two consecutive approximations differ by less than 0.000001.

13.4 THE POLAR COORDINATE SYSTEM

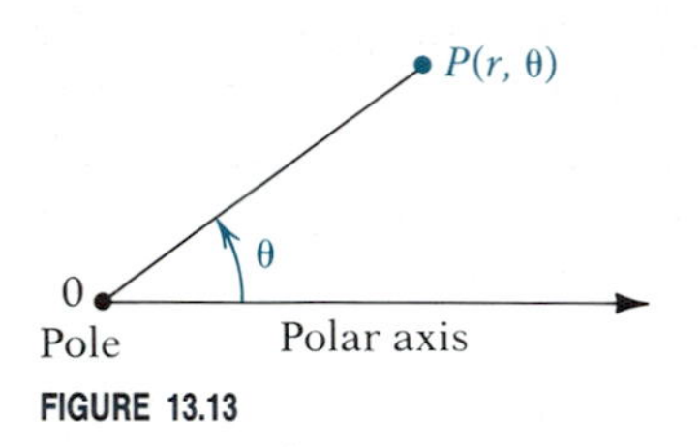

FIGURE 13.13

Although the rectangular system of coordinates we have used in all of our work so far is suitable for most graphing needs, other systems are sometimes more convenient to use. The most important alternative system is called the **polar coordinate system.** In it we take as references a fixed point O, called the **pole,** and a ray emanating from O, called the **polar axis.** It is customary to direct the polar axis horizontally to the right. If P is any point in the plane other than the pole, we indicate its location by means of the radial distance r from O to P and the measure θ of the angle from the polar axis to the ray from O through P. This is illustrated in Figure 13.13. The numbers r and θ are called the **polar coordinates** of P, and we indicate them as the ordered pair (r, θ). Although it is possible to take θ as either the degree or radian measure of the angle, we will always use radian measure. Coordinates of the pole are taken to be $(0, \theta)$ for any value of θ.

If a ray from the pole makes an angle α with the polar axis, we refer to it as the **α ray.** For example, the polar axis is the 0 ray and the $\frac{\pi}{2}$ ray is directed vertically upward. These correspond to the positive x- and y-axes, respectively, in a rectangular system. On a given ray distances are measured positively in the direction of the ray. In Figure 13.14 we have plotted the points $A(3, \frac{\pi}{4})$, $B(1, \frac{2\pi}{3})$, and $C(2, \frac{3\pi}{2})$. We also define the negative direction on a ray to be the opposite of the positive direction. More precisely, the negative direction of an α ray is in the direction of the $(\alpha + \pi)$ ray. With this understanding we can permit r to be negative, and to plot (r, θ) for r negative we measure $|r|$ units in the negative direction of the θ ray. In Figure 13.15 we have plotted $P(-2, 0)$ and $Q(-3, \frac{4\pi}{3})$.

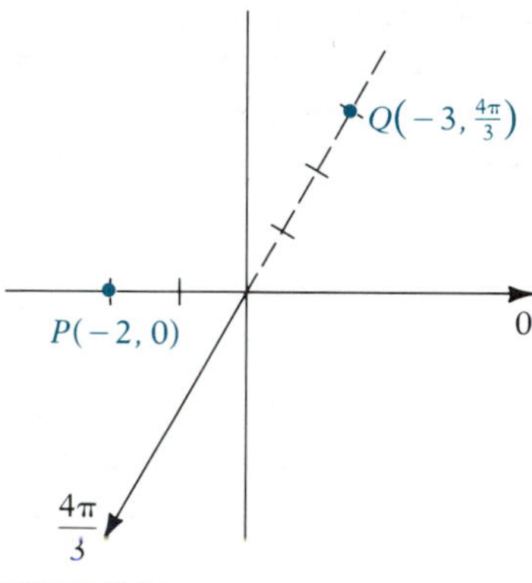

FIGURE 13.15

When we are given polar coordinates (r, θ), the point P represented by them is uniquely determined. Unfortunately the converse is not true. If we are given a point P, there are infinitely many sets of polar coordinates for it. This contrasts with the rectangular system, where there is a one-to-one

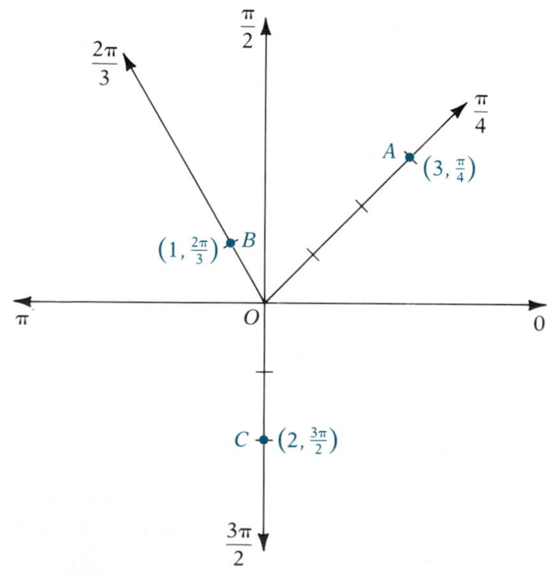

FIGURE 13.14

correspondence between points in the plane and the set of all ordered pairs (x, y). You should convince yourself that $(1, \frac{\pi}{4})$, $(1, \frac{9\pi}{4})$, $(1, -\frac{7\pi}{4})$, $(-1, \frac{5\pi}{4})$, and $(-1, -\frac{3\pi}{4})$are all polar coordinates of the same point. In fact, $(1, \frac{\pi}{4} + 2n\pi)$ and $(-1, \frac{5\pi}{4} + 2n\pi)$ for $n = 0, \pm 1, \pm 2, \ldots$ represent this point. This lack of uniqueness of representation is not a serious drawback to the polar system, but it does mean that special care must be exercised—in analyzing polar graphs, for example. Although two different ordered pairs may represent the same point, they are not equal as ordered pairs. For example, $(1, \frac{\pi}{4})$ and $(-1, \frac{5\pi}{4})$ are coordinates of the same point, but $(1, \frac{\pi}{4}) \neq (-1, \frac{5\pi}{4})$, since by equality of ordered pairs we mean that first coordinates are equal and second coordinates are equal.

It is frequently useful to change from the polar coordinate system to the rectangular system, or vice versa. To see how to do this, superimpose a rectangular system onto a polar system so that the origin coincides with the pole, and the positive x-axis coincides with the polar axis. Let P be any point other than the origin. Then P has unique rectangular coordinates (x, y). As in Figure 13.16(a), let (r, θ) be coordinates of P for which $r > 0$ and θ is between 0 and 2π. Then from the definitions of $\sin\theta$ and $\cos\theta$, we have $\cos\theta = \frac{x}{r}$ and $\sin\theta = \frac{y}{r}$ or, equivalently,

$$\begin{cases} x = r\cos\theta \\ y = r\sin\theta \end{cases} \tag{13.8}$$

The same relations hold true if P is represented by coordinates of the form $(r, \theta + 2n\pi)$, where $r > 0$, since $\cos(\theta + 2n\pi) = \cos\theta$ and $\sin(\theta + 2n\pi) = \sin\theta$. The fact that we have shown P in the first quadrant is not significant. Equations (13.8) are true wherever P is located. If $r < 0$, then $P(r, \theta)$ is on the negative extension of the θ ray; that is, P is on the $(\theta + \pi)$ ray at a distance $|r|$ from 0, as shown in Figure 13.16(b). An equivalent polar representation of P is therefore $(|r|, \theta + \pi)$. Since $|r| > 0$ we can use equations (13.8) to conclude that

$$x = |r|\cos(\theta + \pi) = (-r)(-\cos\theta) = r\cos\theta$$
$$y = |r|\sin(\theta + \pi) = (-r)(-\sin\theta) = r\sin\theta$$

since $|r| = -r$. So equations (13.8) also hold true if $r < 0$. Finally, if P is the origin, we take $r = 0$ so that equations (13.8) again hold true.

If we square both sides of equations (13.8) and add, we get $r^2 = x^2 + y^2$, and if $x \neq 0$, on dividing the second equation by the first, we get $\tan\theta = \frac{y}{x}$.

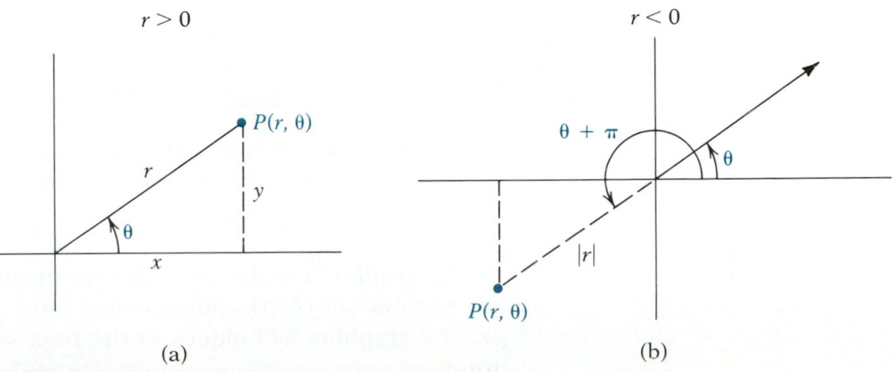

FIGURE 13.16

Thus, equations for r and θ in terms of x and y are

$$\begin{cases} r^2 = x^2 + y^2 \\ \tan\theta = \dfrac{y}{x}, \quad x \neq 0 \end{cases} \tag{13.9}$$

To change from polar coordinates to rectangular coordinates we use equations (13.8) and get an unambiguous answer. To change from rectangular coordinates to polar coordinates we use equations (13.9), taking account of the quadrant in which the point lies and choosing $r = \sqrt{x^2 + y^2}$ or $-\sqrt{x^2 + y^2}$, depending on which value of θ is chosen. It will always work to take $r > 0$ if we choose θ as a positive angle from the polar axis to the ray from O through P where P is the point with rectangular coordinates (x, y). There are times, however, when we might prefer some other polar representation of P, consistent with equations (13.9). We illustrate these ideas in the next example.

EXAMPLE 13.8 (a) Change the polar coordinates $(-3, \frac{4\pi}{3})$ to rectangular coordinates. Plot the point.

(b) Give three sets of polar coordinates for the point P that has rectangular coordinates $(\sqrt{3}, -1)$. For one set choose $r > 0$ and $\theta > 0$, and for the other two choose r and θ opposite in sign.

Solution (a) The point P with polar coordinates $(-3, \frac{4\pi}{3})$ is plotted in Figure 13.17. Note that a simpler set of coordinates for P is $(3, \frac{\pi}{3})$, but it is not necessary to make this change. By equations (13.8) we have

$$x = r\cos\theta = (-3)\cos\tfrac{4\pi}{3} = (-3)(-\tfrac{1}{2}) = \tfrac{3}{2}$$
$$y = r\sin\theta = (-3)\sin\tfrac{4\pi}{3} = (-3)(-\tfrac{\sqrt{3}}{2}) = \tfrac{3\sqrt{3}}{2}$$

So the rectangular coordinates of P are $(\frac{3}{2}, \frac{3\sqrt{3}}{2})$

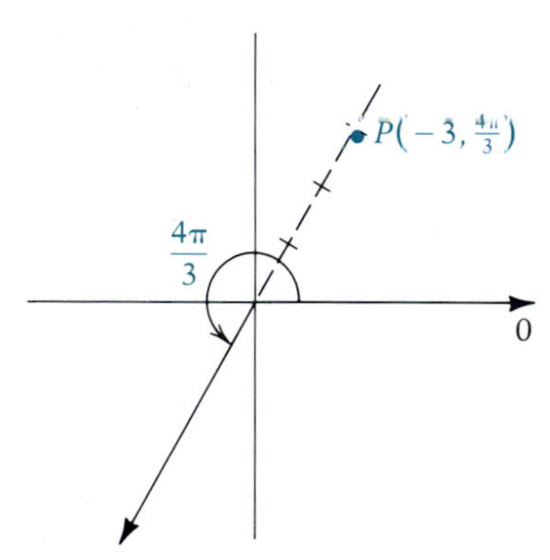

FIGURE 13.17

(b) The point P is plotted in Figure 13.18. From equations (13.9) we have

$$r^2 = x^2 + y^2 = 3 + 1 = 4$$

So $r = \pm 2$. Also

$$\tan\theta = \tfrac{y}{x} = \tfrac{-1}{\sqrt{3}}$$

From our knowledge of special angles in trigonometry, we see that θ has a reference angle of 30°, or $\frac{\pi}{6}$ radians. So three sets of polar coordinates for P can be taken as:

$(2, \frac{11\pi}{6})$	$(2, -\frac{\pi}{6})$	$(-2, \frac{5\pi}{6})$
$r > 0, \theta > 0$	$r > 0, \theta < 0$	$r < 0, \theta > 0$

The second of these is probably the simplest form in this case. ■

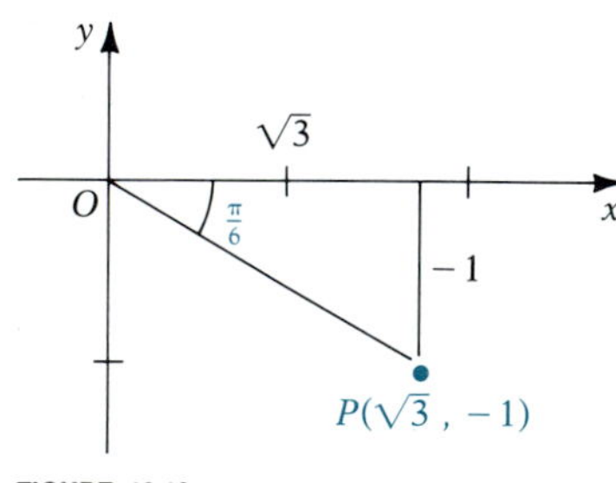

FIGURE 13.18

A **polar equation** is an equation whose only variables are the polar variables r and θ. For example, each of the following is a polar equation.

$$r = 1 - 2\cos\theta \qquad r^2 = \sin 2\theta \qquad r = 2 \qquad \theta = \tfrac{\pi}{4}$$

By the **graph** of a polar equation we mean the set of all points P that have *at least one* set of coordinates that satisfy the equation. We will consider general graphing techniques in the next section. For now we observe that graphs of polar equations can sometimes be more easily obtained by finding a rectangular equation that has the same graph. On the other hand, complicated rectangular equations can sometimes be graphed more easily using an equivalent polar equation. Thus, it is desirable to be able to change an equation in one system to an equation in the other system that has the same graph. The equations of transformation (13.8) and (13.9) enable us to do this.

In the examples and exercises that follow, when we ask for a polar equation to be transformed to an equivalent rectangular equation, or vice versa, we mean that the resulting equation will have the same graph as the given equation.

EXAMPLE 13.9 (a) Transform $r = 2 \sin \theta$ to an equivalent rectangular equation. Identify and draw the graph.

(b) Transform the equation $(x^2 + y^2)^2 = x^2 - y^2$ to an equivalent polar equation.

Solution (a) By multiplying both sides of the equation by r, we can bring it to a form in which the transformation equations can be more easily applied. This gives $r^2 = 2r \sin \theta$ or, since $r^2 = x^2 + y^2$ and $r \sin \theta = y$,

$$x^2 + y^2 = 2y$$

This is the equation of a circle, which in standard form is $x^2 + (y - 1)^2 = 1$. So the center is (0, 1) and the radius is 1. (See Figure 13.19.)

We should note that multiplication by r in this procedure did not change the graph, since the only possible new point it could have introduced would have been the pole, $r = 0$, and the pole is on the graph of the original equation, since when $\theta = 0$, $r = 0$.

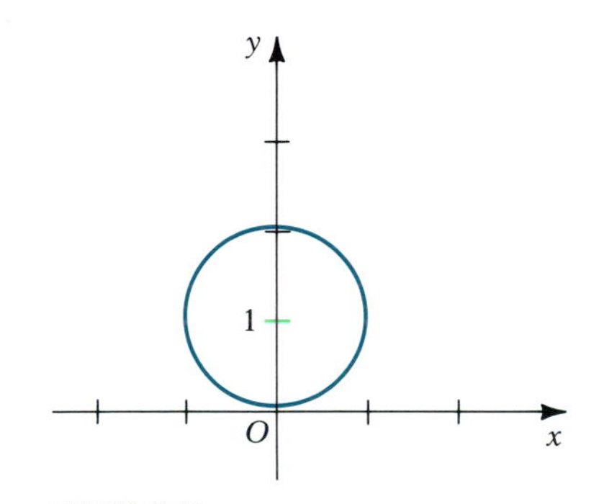

FIGURE 13.19

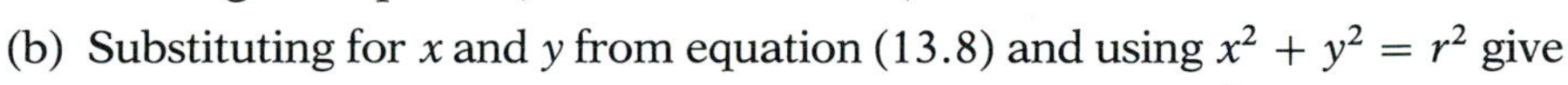

(b) Substituting for x and y from equation (13.8) and using $x^2 + y^2 = r^2$ give

$$r^4 = r^2 \cos^2 \theta - r^2 \sin^2 \theta$$

or, since $\cos 2\theta = \cos^2 \theta - \sin^2 \theta$,

$$r^4 = r^2 \cos 2\theta$$

Rewriting this as $r^4 - r^2 \sin 2\theta = 0$, we can factor to get $r^2(r^2 - \cos 2\theta) = 0$. So the graph consists of all points with coordinates that satisfy $r^2 = 0$ (the pole) or $r^2 = \cos 2\theta$. But the pole is on the graph of $r^2 = \cos 2\theta$, since when $\theta = \frac{\pi}{4}$, $r = 0$. Thus, the equation $r^2 = \cos 2\theta$ is equivalent to the original equation. We will see in the next section how to draw the graph using its polar equation. This is an example in which the graph is easier to get from the polar equation than from the rectangular equation. ■

EXERCISE SET 13.4

A

In Exercises 1 and 2 plot the points in a polar coordinate system.

1. (a) $(2, \pi)$ (b) $(3, -\frac{\pi}{2})$
   (c) $(-1, \frac{\pi}{4})$ (d) $(0, \frac{5\pi}{9})$

2. (a) $(-2, -\frac{\pi}{3})$ (b) $(4, \frac{7\pi}{2})$
   (c) $(2, -5\pi)$ (d) $(-3, -\frac{5\pi}{6})$

3. Show the approximate location of the points that have the following polar coordinates. Use the fact that 1 radian $\approx 57.3°$.
(a) $(1, 2)$ (b) $(2, 1)$
(c) $(-5, 8)$ (d) $(-3, -4)$

4. Let P be the point determined by the given polar coordinates. Determine two other pairs of polar coordinates for P with $r > 0$ and two with $r < 0$.
(a) $(2, \frac{\pi}{4})$ (b) $(-1, -\frac{\pi}{2})$
(c) $(-3, \frac{4\pi}{5})$ (d) $(4, -\frac{2\pi}{9})$

In Exercises 5 and 6 change the given polar coordinates to equivalent rectangular coordinates.

5. (a) $(2, \frac{5\pi}{6})$ (b) $(-3, \frac{3\pi}{2})$
(c) $(0, \frac{17\pi}{13})$ (d) $(-2, -\frac{3\pi}{4})$

6. (a) $(3, 0)$ (b) $(4, -\frac{11\pi}{6})$
(c) $(-2, 5\pi)$ (d) $(7, \frac{4\pi}{3})$

In Exercises 7 and 8 change the given rectangular coordinates to an equivalent pair of polar coordinates in which $r > 0$ and $0 \le \theta < 2\pi$.

7. (a) $(4, 0)$ (b) $(0, 4)$
(c) $(-4, 0)$ (d) $(0, -4)$

8. (a) $(1, -\sqrt{3})$ (b) $(-5, 5)$
(c) $(-4\sqrt{3}, 4)$ (d) $(-2\sqrt{2}, -2\sqrt{2})$

In Exercises 9–22 find an equivalent rectangular equation. Also identify the graph and draw it in Exercises 9–14.

9. (a) $r = 2$ (b) $\theta = \frac{\pi}{4}$

10. (a) $r = -2$ (b) $\theta = -\frac{\pi}{4}$

11. (a) $r \cos \theta = 3$ (b) $r \sin \theta = -1$

12. (a) $r = -2 \csc \theta$ (b) $r = 3 \sec \theta$

13. (a) $2r \cos \theta - 3r \sin \theta = 4$
(b) $r(3 \cos \theta + 2 \sin \theta) = 5$

14. (a) $r = 2 \cos \theta$ (b) $r = -2 \sin \theta$

15. $r = a(1 + \cos \theta)$ **16.** $r^2 = 2a \cos 2\theta$

17. $r = 1 - 2 \cos \theta$ **18.** $r = 2 + \sin \theta$

19. $r = a \cos 2\theta$ **20.** $r = a \sin 2\theta$

21. $r = \dfrac{2}{1 - \cos \theta}$ **22.** $r = \dfrac{1}{1 + 2 \sin \theta}$

In Exercises 23–31 find an equivalent polar equation.

23. (a) $x^2 + y^2 = a^2$ (b) $y = \sqrt{3}x$

24. (a) $x + y = 0$ (b) $2x - 3y = 4$

25. (a) $y + 1 = 0$ (b) $x - 3 = 0$

26. $x^2 - y^2 = 4$ **27.** $y = x^2$

28. $x^2 + y^2 = 2x$ **29.** $x^2 - y^2 - 4x = 0$

30. $x^2y^2 = 4(x^2 + y^2)^{3/2}$ **31.** $x^2 - y^2 = \dfrac{2xy}{x^2 + y^2}$

B

32. Prove that if P has rectangular coordinates (x, y) with $x \neq 0$, then as one set of coordinates (r, θ) for P we may take $\theta = \tan^{-1} \frac{y}{x}$ (so that $-\frac{\pi}{2} < \theta < \frac{\pi}{2}$) by taking $r = \sqrt{x^2 + y^2}$ if $x > 0$, but $r = -\sqrt{x^2 + y^2}$ if $x < 0$.

In Exercises 33–35 show that an equivalent polar equation is of the form $r = ep/(1 + e \cos \theta)$ for suitable values of p and e.

33. $y^2 = 1 - 2x$

34. $3x^2 + 4y^2 + 4x = 4$

35. $3x^2 - y^2 - 4x + 1 = 0$

36. By writing an equivalent rectangular equation, show that the graph of $r = ep/(1 + e \cos \theta)$ for $p \neq 0$ is a parabola if $e = 1$, an ellipse if $0 < e < 1$, and a hyperbola if $e > 1$.

37. By means of an example show that a point P can lie on the graph of a polar equation even though a particular set of coordinates for P does not satisfy the equation.

13.5 GRAPHING POLAR EQUATIONS

As we have seen, the graph of a polar equation can sometimes be obtained by changing to an equivalent rectangular equation. The graphs of most interest, however, generally have rather complicated rectangular equations,

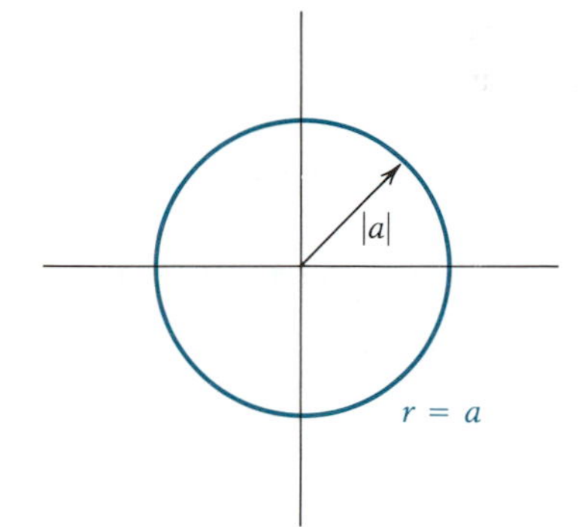

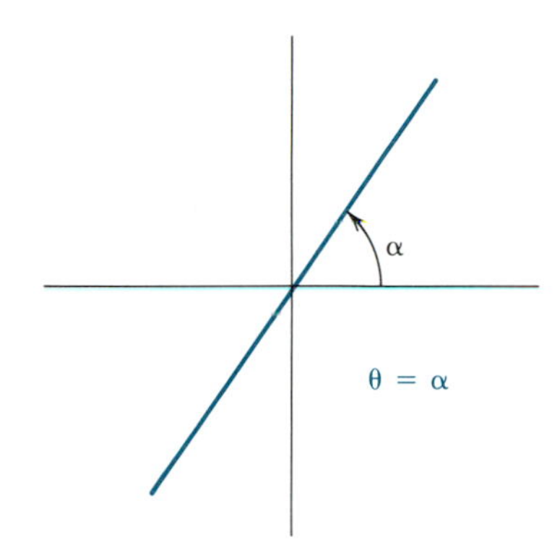

FIGURE 13.20

and so it is important to be able to draw the graph directly from the polar equation. By substituting convenient values of θ and solving the equation for r, a table of values can be obtained from which a reasonable approximation of the graph might be drawn. This "point plotting" method is not entirely satisfactory though, since it is not always clear how to connect the points. Some additional analysis is usually needed.

The simplest polar equations to graph are of the form

$$r = a \quad \text{and} \quad \theta = \alpha$$

where a and α are constants. The first equation puts no restriction on θ, and so its graph consists of points (a, θ) for all possible values of θ. Thus, its graph is a circle of radius $|a|$ if $a \neq 0$ and is the pole if $a = 0$. The equation $\theta = \alpha$ places no restriction on r, and so its graph consists of all points of the form (r, α) for all possible values of r. Thus, its graph is the line consisting of the α ray together with its extension in the opposite direction. Graphs of these equations are shown in Figure 13.20.

For more complicated equations we want to consider several aids to graphing that can reduce reliance on plotting a large number of points.

1. *Symmetry*. We consider three types of symmetry: (a) symmetry with respect to the line $\theta = 0$ (x-axis), (b) symmetry with respect to the line $\theta = \frac{\pi}{2}$ (y-axis), and (c) symmetry with respect to the pole (origin). As Figure 13.21 indicates, in each case symmetry will exist if, whenever $P(r, \theta)$ is on the graph, the point P' as indicated is also on the graph. Since each point P' has many equivalent representations, devising a satisfactory test for symmetry is complicated. The following is perhaps the simplest test, but it should be emphasized that it provides *sufficient* conditions only. In Exercise 35 you will be asked to give a more general test, providing conditions that are both necessary and sufficient. (This is accomplished by considering all possible polar representations of the points P' in Figure 13.21.)

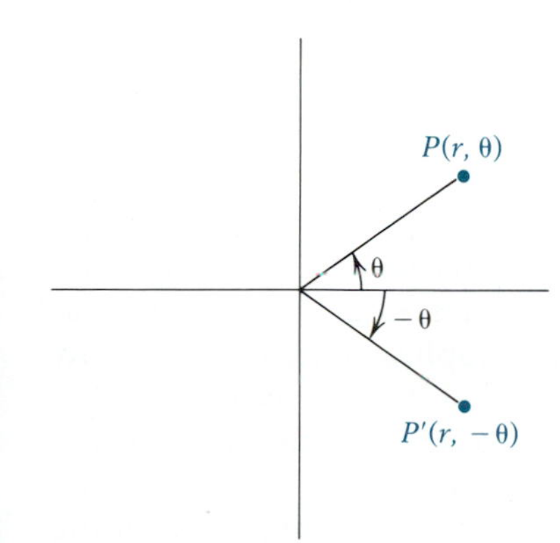

Symmetry to line $\theta = 0$

(a)

P′(r, π − θ)
P(r, θ)
π − θ
θ

Symmetry to line $\theta = \frac{\pi}{2}$

(b)

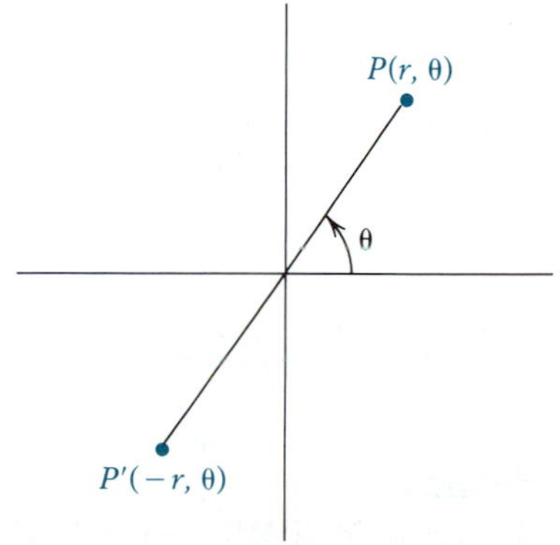

Symmetry to pole

(c)

FIGURE 13.21

A TEST FOR SYMMETRY

Consider the graph of a polar equation, $F(r, \theta) = 0$.

(a) If $F(r, -\theta) = \pm F(r, \theta)$, then the graph is symmetric with respect to the line $\theta = 0$ (the polar axis).

(b) If $F(r, \pi - \theta) = \pm F(r, \theta)$, then the graph is symmetric with respect to the line $\theta = \frac{\pi}{2}$.

(c) If $F(-r, \theta) = \pm F(r, \theta)$, then the graph is symmetric with respect to the pole.

Comment Since $F(r, \theta) = 0$ and $-F(r, \theta) = 0$ are equivalent equations, this test can be paraphrased by saying that if an equivalent equation results when:

(a) θ is replaced by $-\theta$, then the graph is symmetric to $\theta = 0$ (the x-axis).

(b) θ is replaced by $\pi - \theta$, then the graph is symmetric to $\theta = \frac{\pi}{2}$ (the y-axis).

(c) r is replaced by $-r$, then the graph is symmetric to the pole (the origin).

For example, the graph of $r = 1 + \cos\theta$ is symmetric to $\theta = 0$, since $\cos(-\theta) = \cos\theta$. Similarly, since $\sin(\pi - \theta) = \sin\theta$, the graph of $r = 2\sin\theta$ is symmetric to $\theta = \frac{\pi}{2}$. And the graph of $r^2 = e^\theta$ is symmetric to the pole, since replacing r by $-r$ results in an equivalent equation.

To illustrate the limitations of this test, consider the equation $r = \sqrt{\sin 2\theta}$. It fails all three tests, yet the graph is symmetric with respect to the pole. This can be seen by using the coordinates $(r, \pi + \theta)$ for P' in Figure 13.21(c). The test is to replace θ by $\pi + \theta$ and see whether an equivalent equation results, which it does in this case, since $\sin 2(\pi + \theta) = \sin(2\pi + 2\theta) = \sin 2\theta$.

2. *Intercepts.* The intercepts on what would correspond in a rectangular system to the x- and y-axes are almost always key points in drawing the graph. These are found by setting $\theta = 0, \frac{\pi}{2}, \pi$, and $\frac{3\pi}{2}$ and solving for r.

3. *Periodicity.* Since many polar equations involve $\sin\theta$ or $\cos\theta$, the fact that each of these has period 2π can limit the domain for θ that needs to be considered. For example, in graphing $r = 1 + \cos\theta$ it is sufficient to consider values of θ that satisfy $-\pi < \theta \le \pi$. In fact, since the graph is symmetric to the line $\theta = 0$, we need only consider θ in the range $0 \le \theta \le \pi$. The graph for $-\pi < \theta \le 0$ is then found by reflecting through the line $\theta = 0$.

4. *Tangents at the origin.* If the graph has one or more tangent lines at the origin, these can be found by setting $r = 0$ and solving for θ. We can reason as follows to see why this is so. In Figure 13.22 the line OP that joins the pole O and a point $P(r, \theta)$ on the graph is a secant line. Now suppose that when $\theta = \theta_0$, $r = 0$. If we let $\theta \to \theta_0$, the secant line OP approaches the tangent line at O. The slope of OP is $\tan\theta$, and this approaches $\tan\theta_0$ as $\theta \to \theta_0$. So the tangent line at O is the line with slope $\tan\theta_0$—namely, $\theta = \theta_0$. This reasoning will be made more precise in the next section, but we will use the result now, since it is often helpful in graphing.

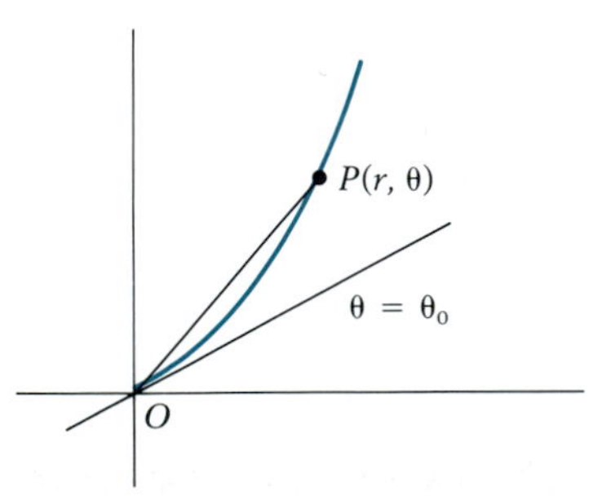

FIGURE 13.22

To illustrate, consider the equation $r = 2 - 4\sin\theta$. Setting $r = 0$, we get $\sin\theta = \frac{1}{2}$, so that the lines $\theta = \frac{\pi}{6}$ and $\theta = \frac{5\pi}{6}$ are tangent to the graph at the origin.

We now proceed to several examples that illustrate these graphing aids.

EXAMPLE 13.10 Draw the graph of the equation $r = 2\cos\theta$.

Solution This is similar to the equation in Example 13.9 part a, and we could proceed as in that example to find an equivalent rectangular equation. The result (you should check this) is

$$(x-1)^2 + y^2 = 1$$

which is the equation of a circle of radius 1 with center at $(1, 0)$, with respect to a rectangular coordinate system.

To illustrate how to work directly from the given polar equation, using the aids to graphing we have discussed, we proceed as follows. Since $\cos(-\theta) = \cos\theta$, the equation is unchanged when θ is replaced by $-\theta$. The graph is therefore symmetric with respect to the horizontal line $\theta = 0$. Because of this symmetry and the periodicity of $\cos\theta$, we need only consider θ in the interval $[0, \pi]$. The key intercepts in this interval are $(2, 0)$, $(0, \frac{\pi}{2})$, and $(-2, \pi)$. Setting $r = 0$ yields $\cos\theta = 0$, so that the curve is tangent to the line $\theta = \frac{\pi}{2}$ at the origin. Finally, we plot a few additional points for convenient values of θ.

θ	$\frac{\pi}{6}$	$\frac{\pi}{4}$	$\frac{\pi}{3}$	$\frac{2\pi}{3}$	$\frac{3\pi}{4}$	$\frac{5\pi}{6}$
r	$\sqrt{3}$	$\sqrt{2}$	1	-1	$-\sqrt{2}$	$-\sqrt{3}$

The graph is shown in Figure 13.23 and, as we already know, is a circle. Note that the entire circle is traced out as θ varies from 0 to π, with the upper semicircle corresponding to θ between 0 and $\frac{\pi}{2}$, and the lower semicircle corresponding to θ between $\frac{\pi}{2}$ and π.

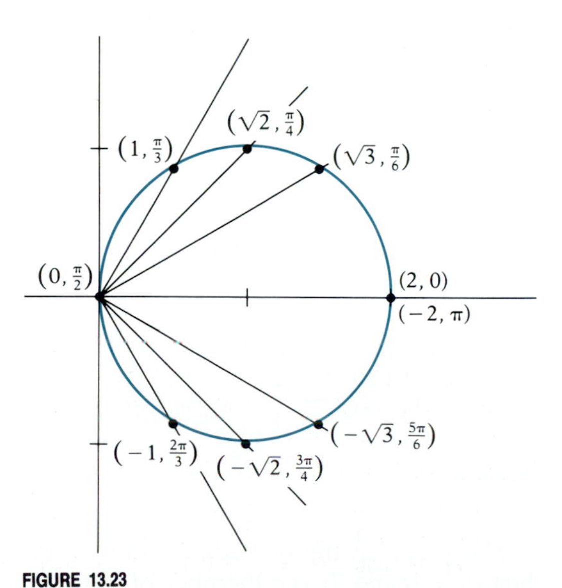

FIGURE 13.23 ■

The equations in Examples 13.9 part a and 13.10 are special cases of the circles illustrated in Figure 13.24.

FIGURE 13.24

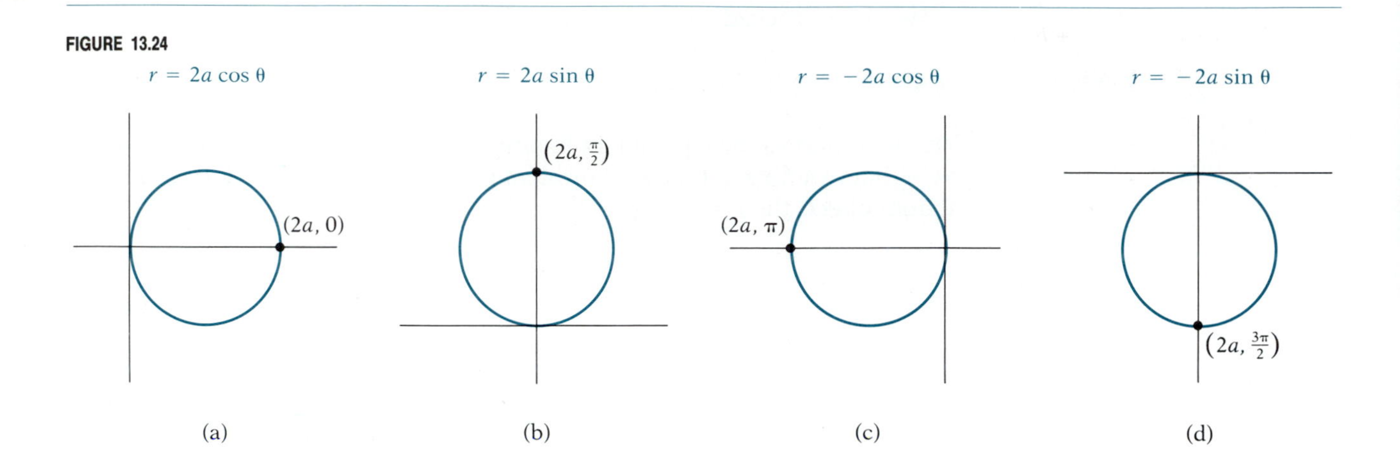

EXAMPLE 13.11 Draw the graph of the equation $r = 1 + \sin\theta$.

Solution Since $\sin(\pi - \theta) = \sin\theta$, the graph is symmetric to the vertical line $\theta = \frac{\pi}{2}$. Because of this and the periodicity of $\sin\theta$, it is sufficient to consider θ in the interval $[-\frac{\pi}{2}, \frac{\pi}{2}]$. The key intercepts in this interval are $(0, -\frac{\pi}{2})$, $(1, 0)$, and $(2, \frac{\pi}{2})$. Setting $r = 0$ yields $\sin\theta = -1$, so the tangent line at the origin is $\theta = -\frac{\pi}{2}$. Additional selected points are found as shown in the table.

θ	$-\frac{\pi}{3}$	$-\frac{\pi}{4}$	$-\frac{\pi}{6}$	$\frac{\pi}{6}$	$\frac{\pi}{4}$	$\frac{\pi}{3}$
r	$1 - \frac{\sqrt{3}}{2} \approx 0.13$	$1 - \frac{\sqrt{2}}{2} \approx 0.29$	$1 - \frac{1}{2} = 0.50$	$1 + \frac{1}{2} = 1.50$	$1 + \frac{\sqrt{2}}{2} \approx 1.71$	$1 + \frac{\sqrt{3}}{2} \approx 1.87$

The graph can now be drawn with reasonable accuracy, as in Figure 13.25.

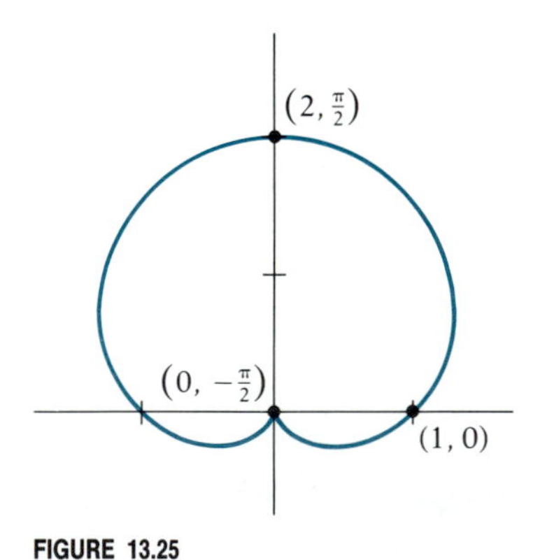

FIGURE 13.25

■

The graph in the preceding example is called a **cardioid** because of its heartlike shape. It is a member of a family of curves called **limaçons.** The three types of limaçons are shown in Figure 13.26, along with their standard equations when the axis of symmetry is horizontal. When the axis of symmetry is vertical, $\cos\theta$ is replaced by $\sin\theta$. The curves shown are with a

FIGURE 13.26 Limaçon: $r = a + b\cos\theta$

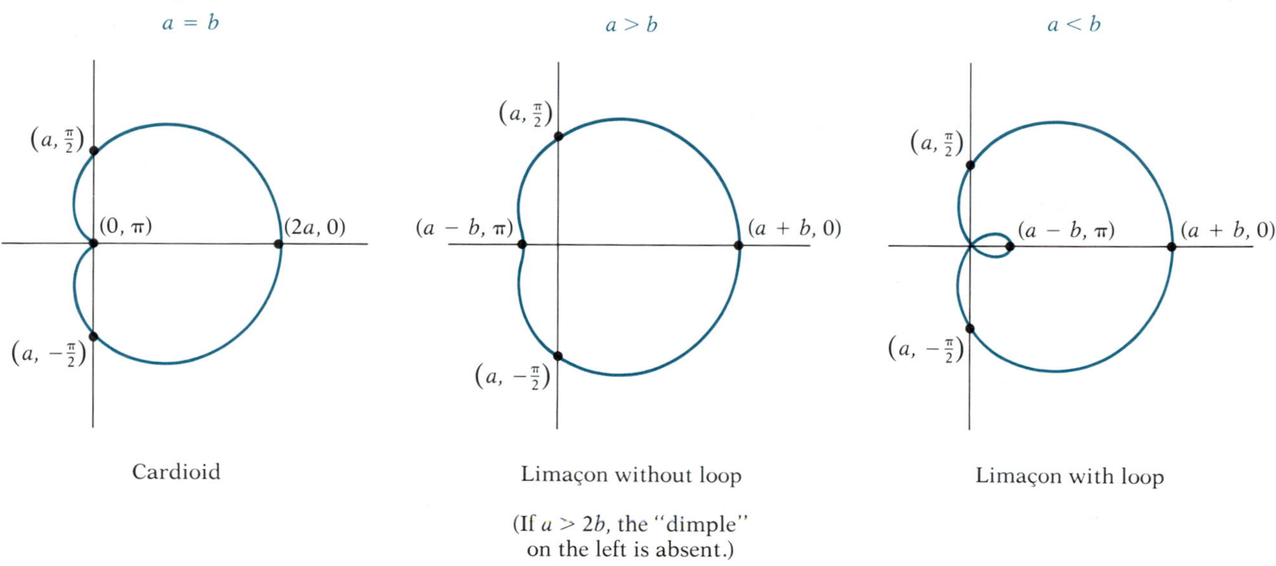

and b positive. If b is negative, the graph is rotated 180°; if a is negative and b positive, the graph is unchanged.

EXAMPLE 13.12 Draw the graph of the equation $r^2 = 4\cos 2\theta$.

Solution The graph is symmetric to the line $\theta = 0$ and to the pole, as can be seen by replacing, in turn, θ by $-\theta$ and r by $-r$. It follows that it is also symmetric to the vertical line $\theta = \frac{\pi}{2}$. Since the period of $\cos 2\theta$ is π, it is sufficient to consider θ in the interval $[0, \pi]$. Solving for r yields $r = \pm 2\sqrt{\cos 2\theta}$, and we see that θ is restricted to satisfy $\cos 2\theta \geq 0$. Thus, we must have $0 \leq 2\theta \leq \frac{\pi}{2}$, or $0 \leq \theta \leq \frac{\pi}{4}$. The tangent lines at the origin are solutions of $\cos 2\theta = 0$—namely, $\theta = \frac{\pi}{4}$ and $\theta = -\frac{\pi}{4}$. Using the intercepts $(2, 0)$ and $(0, \frac{\pi}{4})$, together with the one additional point $(\sqrt{2}, \frac{\pi}{6})$, we can now draw the portion of the graph shown in Figure 13.27(a). Then, by symmetry, we get the entire graph, shown in Figure 13.27(b).

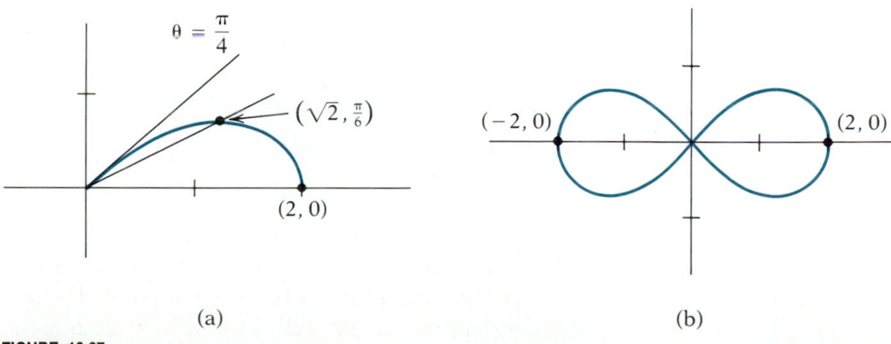

FIGURE 13.27

■

FIGURE 13.28 Lemniscates

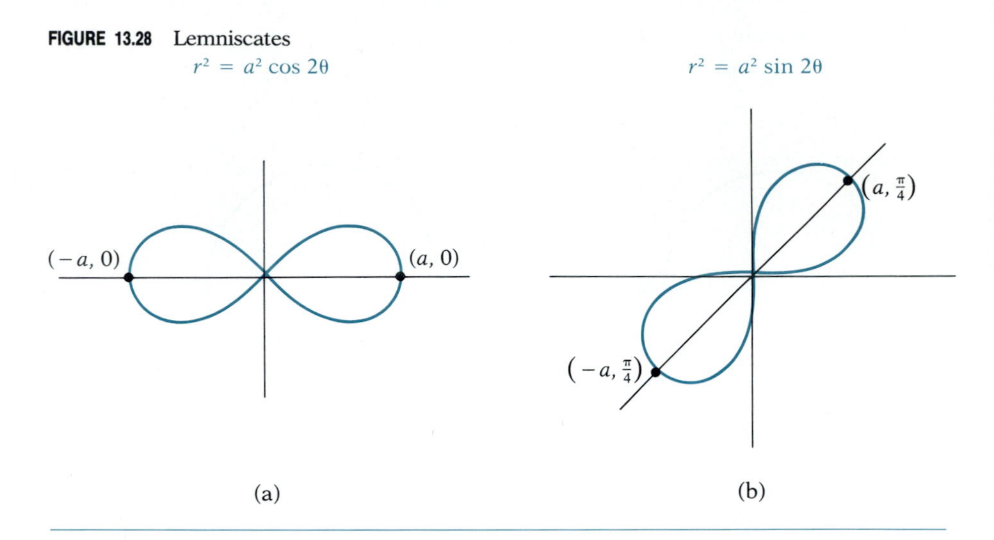

The curve in Example 13.12 is called a **lemniscate.** These curves typically have equations of the form $r^2 = a^2 \cos 2\theta$ or $r^2 = a^2 \sin 2\theta$. The graphs of these two basic forms are shown in Figure 13.28. Note that when $\cos 2\theta$ is replaced by $\sin 2\theta$, the graph is rotated through an angle of $\frac{\pi}{4}$ radians. In Exercise 40 you will be asked to show in general that if $\cos n\theta$ is replaced by $\sin n\theta$, where the original equation is of the form $r = f(\cos n\theta)$, then the graph is rotated through an angle of $\frac{\pi}{2n}$ radians.

EXAMPLE 13.13 Draw the graph of the equation $r = \sin 2\theta$.

Solution Since $\sin 2(\pi - \theta) = \sin(2\pi - 2\theta) = \sin 2\theta$, the graph is symmetric with respect to $\theta = \frac{\pi}{2}$. Also, $\sin 2(\pi + \theta) = \sin(2\pi + 2\theta) = \sin 2\theta$, and since the points $P(r, \theta)$ and $P'(r, \pi + \theta)$ are symmetrically placed with respect to the pole, it follows that the graph is symmetric with respect to the pole. Thus, it is also symmetric with respect to the line $\theta = 0$. The period of $\sin 2\theta$ is π

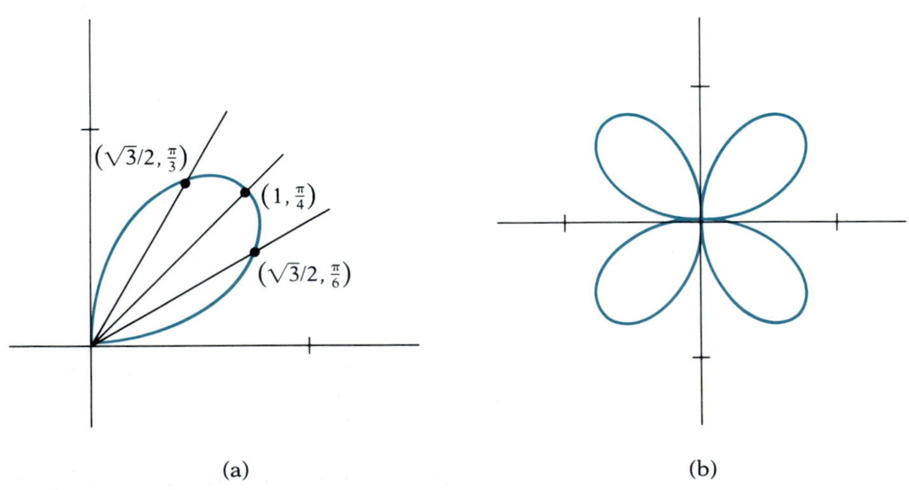

FIGURE 13.29

FIGURE 13.30 Rose curves

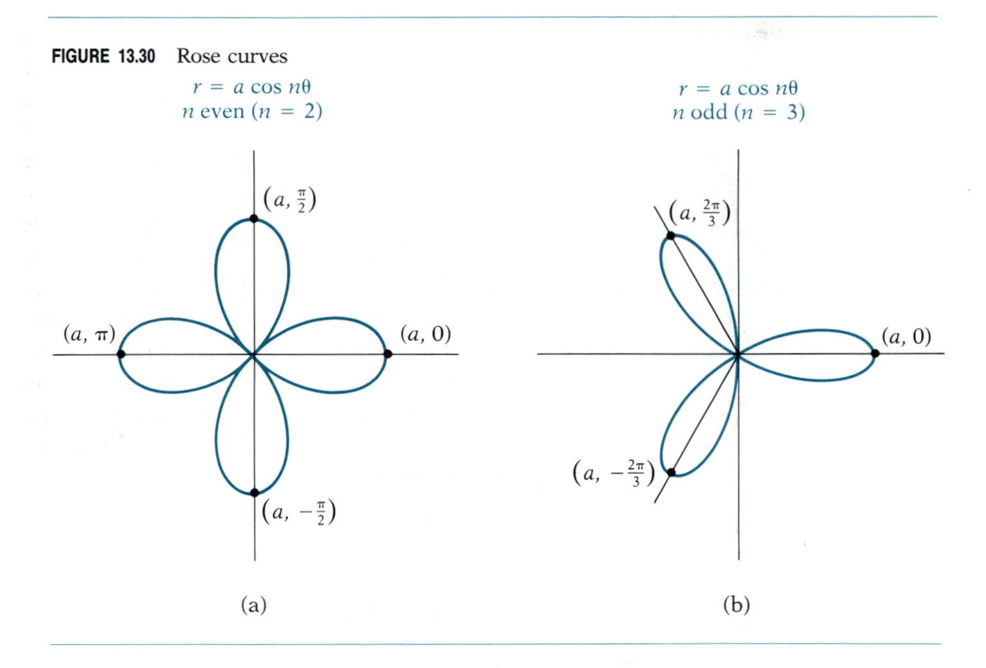

and so, because of symmetry, we can limit our consideration of θ to values between 0 and $\frac{\pi}{2}$. Tangent lines at the origin are $\theta = 0$ and $\theta = \frac{\pi}{2}$, found by solving $\sin 2\theta = 0$. Useful points are tabulated and the graph of the portion of the curve for θ in $[0, \frac{\pi}{2}]$ is drawn in Figure 13.29(a). Symmetry is then used to get the complete graph, shown in Figure 13.29(b).

θ	0	$\frac{\pi}{6}$	$\frac{\pi}{4}$	$\frac{\pi}{3}$	$\frac{\pi}{2}$
r	0	$\frac{\sqrt{3}}{2}$	1	$\frac{\sqrt{3}}{2}$	0

■

The graph of this example is called a **four-leaf rose.** Standard equations of rose curves are $r = a \cos n\theta$ and $r = a \sin n\theta$. When n is even, there are $2n$ petals, as in Example 13.13, whereas when n is odd, there are n petals. This is illustrated in Figure 13.30, using $r = a \cos 2\theta$ and $r = a \cos 3\theta$.

EXAMPLE 13.14 Draw the graph of the equation $r = e^{\theta/4}$.

Solution The graph has no symmetry and $e^{\theta/4}$ is not periodic. Nor can r ever be 0, so there are no tangent lines at the origin. We see, in fact, that $r > 0$ for all values of θ, and that r increases as θ increases. Furthermore,

$$\lim_{\theta \to -\infty} r = 0 \quad \text{and} \quad \lim_{\theta \to +\infty} r = +\infty$$

There are infinitely many intercepts on every ray. The graph is shown in Figure 13.31. Points are found using a calculator. ■

The graph in the preceding example is called a **logarithmic spiral.** You will be asked to draw other types of spirals in the exercises.

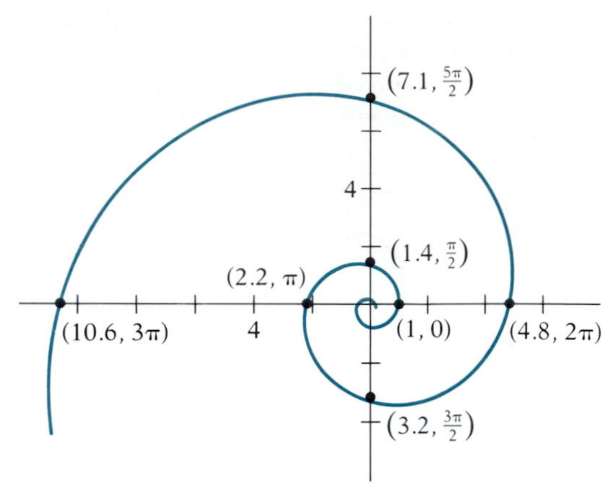

FIGURE 13.31

EXERCISE SET 13.5

A

In Exercises 1–24 identify the curve and draw its graph.

1. (a) $r = 3$ (b) $r = -3$

2. (a) $\theta = \frac{3\pi}{4}$ (b) $\theta = \frac{\pi}{6}$

3. (a) $r \cos \theta = 2$ (b) $r \sin \theta = -3$

4. (a) $r = -3 \csc \theta$ (b) $r = 2 \sec \theta$

5. $r = 3 \cos \theta$ **6.** $r = -2 \sin \theta$

7. $r = 2(1 - \cos \theta)$ **8.** $r = 1 - \sin \theta$

9. $r = -2(1 + \sin \theta)$ **10.** $r = \cos \theta - 1$

11. $r = 5 - 4 \cos \theta$ **12.** $r = 4 - 5 \sin \theta$

13. $r = 2 + 3 \sin \theta$ **14.** $r = 3 + 2 \cos \theta$

15. $r = 4 - 2 \sin \theta$ **16.** $r = -2 + 4 \cos \theta$

17. $r^2 = 4 \sin 2\theta$ **18.** $r^2 = -2 \cos 2\theta$

19. $r^2 = 6 \sin \theta \cos \theta$ **20.** $r^2 = \cos^2 \theta - \sin^2 \theta$

21. $r = 2 \cos 3\theta$ **22.** $r = 3 \cos 2\theta$

23. $r = -2 \sin 2\theta$ **24.** $r = \sin 3\theta$

In Exercises 25–28 draw the spirals.

25. $r = a\theta$, with $a = 1$ **(spiral of Archimedes)**

26. $r = \frac{a}{\theta}$, with $a = 2$ **(hyperbolic spiral)**

27. $r^2 = a\theta$, with $a = 1$ **(parabolic spiral)**

28. $r = e^{a\theta}$, with $a = \frac{1}{2}$ **(logarithmic spiral)**

In Exercises 29–34 draw the graph by first changing to rectangular coordinates.

29. $r = \dfrac{6}{2 \cos \theta - 3 \sin \theta}$

30. $r^2 - r(2 \cos \theta + 4 \sin \theta) + 1 = 0$

31. $r = \dfrac{2}{1 - \cos \theta}$ **32.** $r = \dfrac{1}{1 + 2 \sin \theta}$

33. $r = \dfrac{4}{\sqrt{2} - \sin \theta}$ **34.** $r = \dfrac{3}{1 + \cos \theta}$

B

35. By using the most general representations of the points P' in Figure 13.21 devise tests for symmetry that cover all possibilities; that is, devise tests that give conditions that are both necessary and sufficient.

36. Using the result of Exercise 35, show that the graph of $F(r, \theta) = 0$ is symmetric to:
(a) $\theta = 0$ if $F(-r, \pi - \theta) = \pm F(r, \theta)$
(b) $\theta = \frac{\pi}{2}$ if $F(-r, -\theta) = \pm F(r, \theta)$
(c) The pole if $F(r, \pi + \theta) = \pm F(r, \theta)$

37. Use the result of Exercise 36 to show the specified type of symmetry for the following. Also show that the test as stated in this section fails in each case.
(a) $r = \tan \theta$; symmetric to $\theta = 0$, $\theta = \frac{\pi}{2}$, and the pole
(b) $r = \sin 2\theta \cos \theta$; symmetric to $\theta = \frac{\pi}{2}$
(c) $r = 2 \csc 2\theta - 1$; symmetric to the pole
(d) $r = \sin 2\theta + \cos \theta$; symmetric to $\theta = 0$

38. Prove that if the polar axis is rotated through an angle α radians, a point P that has coordinates (r, θ)

with respect to the original system has coordinates (r', θ') with respect to the new system, where $r' = r$ and $\theta' = \theta - \alpha$. Conclude from this that the graph of $F(r, \theta - \alpha) = 0$ can be obtained by rotating the graph of $F(r, \theta) = 0$ through the angle α.

39. Use Exercise 38 to prove that if an equation is expressible in the form $F(r, \cos\theta) = 0$, then:
(a) When $\cos\theta$ is replaced by $\sin\theta$, the graph is rotated $\frac{\pi}{2}$ radians.
(b) When $\cos\theta$ is replaced by $-\cos\theta$, the graph is rotated π radians.
(c) When $\cos\theta$ is replaced by $-\sin\theta$, the graph is rotated $\frac{3\pi}{2}$ radians.

40. Prove that the graph of an equation of the form $F(r, \sin n\theta) = 0$ is the same as that of $F(r, \cos n\theta) = 0$ after rotating through an angle of $\frac{\pi}{2n}$ radians. (See Exercise 38.)

41. By multiplying and dividing the right-hand side of the equation $r = a\sin\theta + b\cos\theta$ by $\sqrt{a^2 + b^2}$, show that its graph is the circle $r = \sqrt{a^2 + b^2}\cos\theta$ rotated through the angle α for which $\sin\alpha = a/\sqrt{a^2 + b^2}$ and $\cos\alpha = b/\sqrt{a^2 + b^2}$. (See Exercise 38.)

42. Use the result of Exercise 40 to draw the graph of each of the following:
(a) $r = \sin\theta + \cos\theta$ (b) $r = 3\cos\theta - 4\sin\theta$

In Exercises 43–47 draw the graph.

43. $r = 2\cos\frac{\theta}{2}$

44. $r = \sqrt{4\sin\theta}$

45. $r = a\sin 2\theta\cos\theta$, for $a = 1$ **(bifolium)**

46. $r = a\sin\theta\tan\theta$, for $a = 2$ **(cissoid)**

47. $r = a\sec\theta + b$, for $a = 1$, $b = -2$ **(conchoid)**

13.6 AREAS IN POLAR COORDINATES

In this section we consider graphs of polar equations of the form $r = f(\theta)$, where f is continuous and nonnegative for θ in the closed interval $[\alpha, \beta]$, with $\beta < \alpha + 2\pi$. Our objective is to find the area of the region bounded by the graph and the α ray and β ray, as shown in Figure 13.32. We proceed in a familiar way. Let $P = \{\theta_0, \theta_1, \theta_2, \ldots, \theta_n\}$ be a partition of $[\alpha, \beta]$, where $\alpha = \theta_0 < \theta_1 < \theta_2 < \cdots < \theta_n = \beta$. For $k = 1, 2, \ldots, n$, write $\Delta\theta_k = \theta_k - \theta_{k-1}$, and let $\|P\|$ denote the norm of the partition (maximum $\Delta\theta_k$). Finally, let θ_k^* be any number in $[\theta_{k-1}, \theta_k]$ for $k = 1, 2, \ldots, n$.

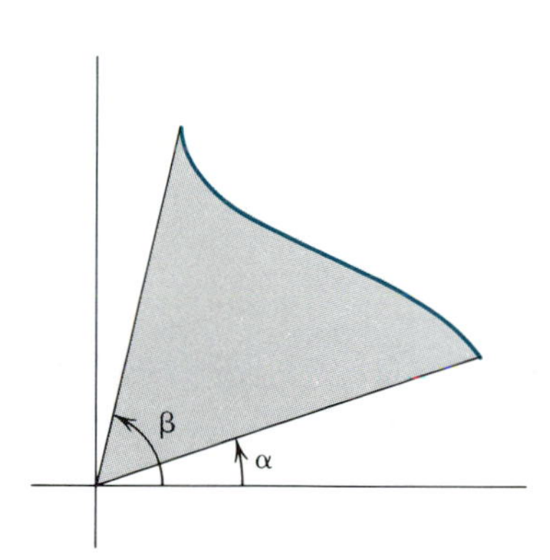

FIGURE 13.32

As we see in Figure 13.33(a), the θ_k rays divide the region in question into n subregions, each shaped somewhat like a sector of a circle. In Figure 13.33(b) we have shown an enlarged typical subregion, and an exact circular sector that has radius $r_k^* = f(\theta_k^*)$. The area of this sector only approximates the area of the subregion, since the outer boundary is not necessarily circular. Nevertheless, if $\Delta\theta_k$ is small, the error will be small. The area of this circular sector is

$$\Delta A_k = \tfrac{1}{2}(r_k^*)^2\,\Delta\theta_k = \tfrac{1}{2}[f(\theta_k^*)]^2\,\Delta\theta_k$$

If we sum these for $k = 1, 2, \ldots, n$, we obtain an approximation to the desired area:

$$A \approx \sum_{k=1}^{n} \Delta A_k$$

and the true area is the limit of this sum as we take finer and finer partitions; that is,

$$A = \lim_{\|P\|\to 0} \sum_{k=1}^{\infty} \Delta A_k = \lim_{\|P\|\to 0} \sum_{k=1}^{\infty} \tfrac{1}{2}[f(\theta_k^*)]^2\,\Delta\theta_k$$

The sum on the right is a Riemann sum and its limit is the integral

$$A = \int_{\alpha}^{\beta} \tfrac{1}{2}[f(\theta)]^2\,d\theta$$

FIGURE 13.33

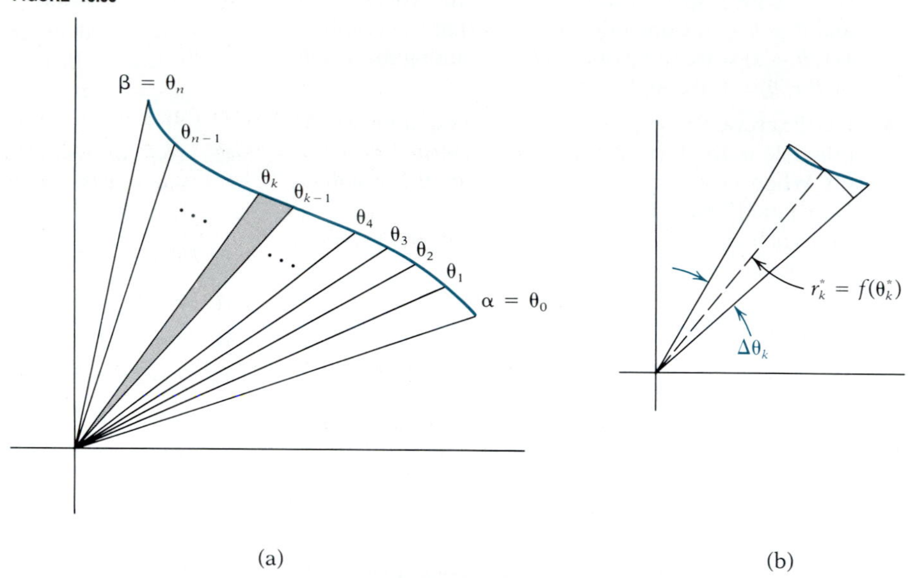

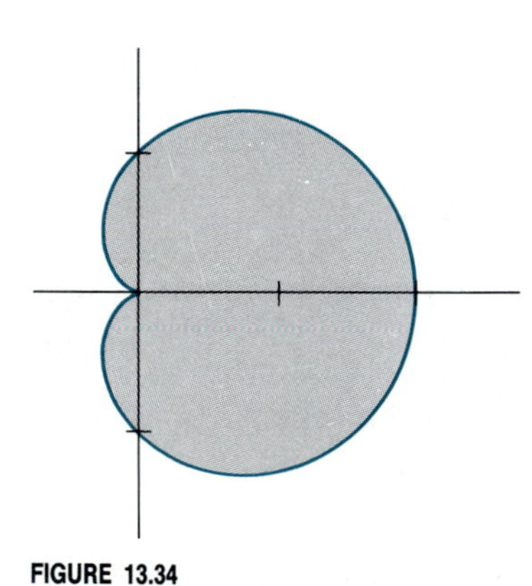

FIGURE 13.34

Since $r = f(\theta)$, we can write this in the simpler form

$$A = \tfrac{1}{2}\int_{\alpha}^{\beta} r^2\, d\theta \tag{13.10}$$

EXAMPLE 13.15 Find the area inside the cardioid $r = 1 + \cos\theta$.

Solution The graph is shown in Figure 13.34. By symmetry it is sufficient to find the area between $\theta = 0$ and $\theta = \pi$ and then double the result. So we have, by equation (13.10),

$$\begin{aligned}
A = 2\cdot\frac{1}{2}\int_0^{\pi} r^2\, d\theta &= \int_0^{\pi} (1+\cos\theta)^2\, d\theta \\
&= \int_0^{\pi} (1 + 2\cos\theta + \cos^2\theta)\, d\theta \\
&= \int_0^{\pi} \left(1 + 2\cos\theta + \frac{1+\cos 2\theta}{2}\right) d\theta \\
&= \int_0^{\pi} \left(\frac{3}{2} + 2\cos\theta + \frac{\cos 2\theta}{2}\right) d\theta \\
&= \frac{3\theta}{2} + 2\sin\theta + \frac{\sin 2\theta}{4}\Bigg]_0^{\pi} = \frac{3\pi}{2}
\end{aligned}$$

■

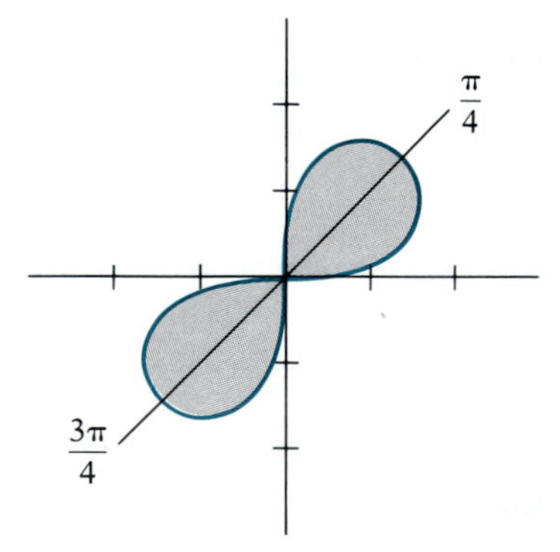

FIGURE 13.35

EXAMPLE 13.16 Find the area inside the lemniscate $r^2 = 4\sin 2\theta$.

Solution The curve is shown in Figure 13.35. We again use symmetry (to the pole) by finding the area between $\theta = 0$ and $\theta = \frac{\pi}{2}$ and doubling the result:

$$A = 2 \cdot \frac{1}{2} \int_0^{\pi/2} r^2 \, d\theta = \int_0^{\pi/2} 4 \sin 2\theta \, d\theta = -2 \cos 2\theta \Big]_0^{\pi/2}$$
$$= -2[-1 - 1] = 4$$

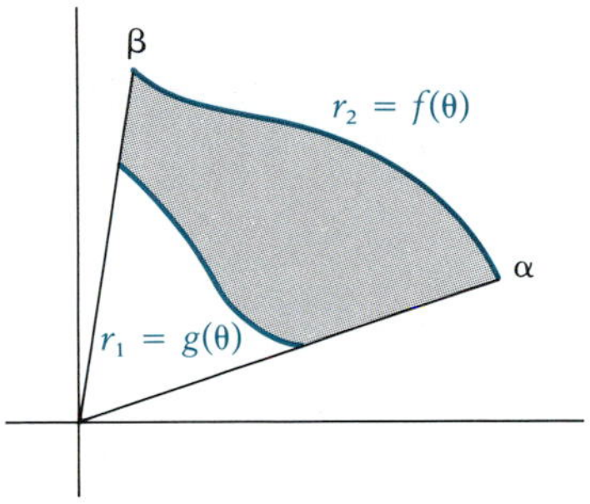

■

Suppose now we want the area between the graphs of two continuous functions f and g for which $0 \le g(\theta) \le f(\theta)$ for θ between α and β, where again we have $\beta < \alpha + 2\pi$. This is illustrated in Figure 13.36. Then we have

$$A = \tfrac{1}{2} \int_\alpha^\beta [f(\theta)]^2 \, d\theta - \tfrac{1}{2} \int_\alpha^\beta [g(\theta)]^2 \, d\theta$$
$$= \tfrac{1}{2} \int_\alpha^\beta \{[f(\theta)]^2 - [g(\theta)]^2\} \, d\theta$$

FIGURE 13.36

Writing $r_1 = g(\theta)$ and $r_2 = f(\theta)$, we can write the result in the simpler form

$$A = \tfrac{1}{2} \int_\alpha^\beta (r_2^2 - r_1^2) \, d\theta \qquad (13.11)$$

EXAMPLE 13.17 Find the area outside the circle $r = 1$ and inside the circle $r = 2 \cos \theta$.

Solution The two circles and the area in question are shown in Figure 13.37. Our first task is to find the points of intersection so that we will know the limits on θ. Solving the equations simultaneously, we get

$$1 = 2 \cos \theta$$
$$\cos \theta = \tfrac{1}{2}$$

So $\theta = \pm\frac{\pi}{3}$. Because of symmetry we will find the area from 0 to $\frac{\pi}{3}$ and double the result. By equation (13.11) we have

$$A = 2 \cdot \frac{1}{2} \int_0^{\pi/3} [(2 \cos \theta)^2 - 1^2] \, d\theta$$
$$= \int_0^{\pi/3} (4 \cos^2 \theta - 1) \, d\theta = \int_0^{\pi/3} [2(1 + \cos 2\theta) - 1] \, d\theta$$
$$= \int_0^{\pi/3} (2 \cos 2\theta + 1) \, d\theta = \sin 2\theta + \theta \Big]_0^{\pi/3} = \frac{\sqrt{3}}{2} + \frac{\pi}{3}$$

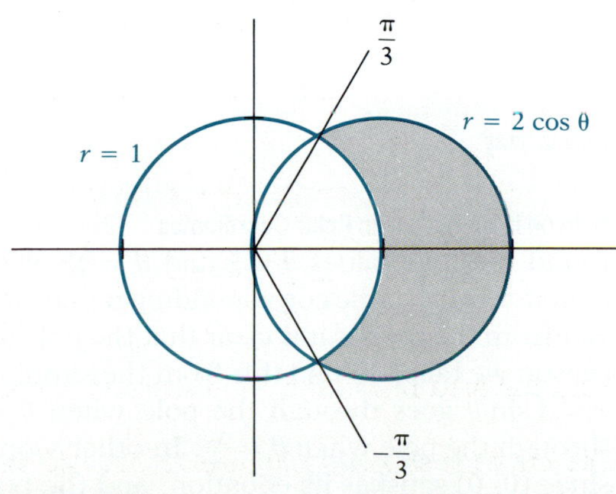

FIGURE 13.37

■

EXAMPLE 13.18 Find the area inside the circle $r = 3\sin\theta$ and outside the cardioid $r = 1 + \sin\theta$.

Solution The graphs are shown in Figure 13.38. Setting the two r values equal to each other gives

$$3\sin\theta = 1 + \sin\theta$$
$$2\sin\theta = 1$$
$$\sin\theta = \frac{1}{2}$$
$$\theta = \frac{\pi}{6}, \frac{5\pi}{6}$$

So we have, using symmetry,

$$\begin{aligned} A &= 2\cdot\tfrac{1}{2}\int_{\pi/6}^{\pi/2} [(3\sin\theta)^2 - (1+\sin\theta)^2]\, d\theta \\ &= \int_{\pi/6}^{\pi/2} (8\sin^2\theta - 2\sin\theta - 1)\, d\theta \\ &= \int_{\pi/6}^{\pi/2} [4(1-\cos 2\theta) - 2\sin\theta - 1]\, d\theta \\ &= \int_{\pi/6}^{\pi/2} (-4\cos 2\theta - 2\sin\theta + 3)\, d\theta \\ &= [-2\sin 2\theta + 2\cos\theta + 3\theta]_{\pi/6}^{\pi/2} = \tfrac{3\pi}{2} - (-\sqrt{3} + \sqrt{3} + \tfrac{\pi}{2}) = \pi \end{aligned}$$

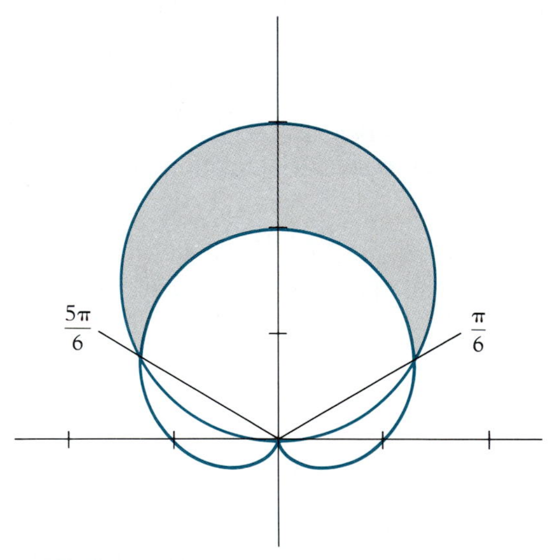

FIGURE 13.38 ■

Note on Intersections in Polar Coordinates Observe that in the preceding example we found the two values $\theta = \frac{\pi}{6}$ and $\theta = \frac{5\pi}{6}$ when we solved the two equations simultaneously. The corresponding points of intersection are $(\frac{3}{2}, \frac{\pi}{6})$ and $(\frac{3}{2}, \frac{5\pi}{6})$. But from the graphs it is clear that the pole also is a point of intersection. The reason we failed to find this from the simultaneous solution is that the curve $r = 3\sin\theta$ goes through the pole when $\theta = 0$, whereas $r = 1 + \sin\theta$ goes through the pole when $\theta = \frac{3\pi}{2}$. In other words, the pole is on the first graph, since $(0, 0)$ satisfies its equation, and the pole is on the second graph, since $(0, \frac{3\pi}{2})$ satisfies its equation. Although $(0, 0)$ and $(0, \frac{3\pi}{2})$ represent the same *point* (the pole), they are not the same as *ordered pairs.* This phenomenon occurs in polar coordinates because of the multiplicity of coordinates for each point. In order to find *all* points of intersection of two graphs, it is usually sufficient to solve their equations simultaneously and then draw the graphs to see whether there are any "nonsimultaneous" solutions. That is, from the graph see whether there are points, such as the pole, that are on both graphs but that did not show up when the equations were solved simultaneously. In Exercise 32 we give an analytical way of finding all points of intersection and ask you to verify it.

EXERCISE SET 13.6

A

In Exercises 1–10 find the area enclosed by the graph of the given equation. Sketch the graph.

1. $r = 2(1 - \cos\theta)$
2. $r = 3 + 2\sin\theta$
3. $r^2 = 9\cos 2\theta$
4. $r = \cos 2\theta$
5. $r = \sin\theta - 1$
6. $r = 4 + 2\cos\theta$
7. $r = 2\sin 3\theta$
8. $r^2 = 2\sin\theta\cos\theta$
9. $r = -4\cos\theta$
10. $r = \sin\theta + \cos\theta$

In Exercises 11–20 find the area outside the graph of the first equation and inside the graph of the second. Sketch the graphs, showing the area in question.

11. $r = 2$; $r = 2(1 + \cos\theta)$
12. $r = \sqrt{3}$; $r = 2\sin\theta$
13. $r = \sqrt{2} - \cos\theta$; $r = \cos\theta$
14. $r = \sqrt{2}$; $r^2 = 4\cos 2\theta$
15. $r = 2\cos\theta$; $r^2 = 2\sqrt{3}\sin 2\theta$
16. $r = 1$; $r = 2\cos 2\theta$
17. $r = 3 + 2\cos\theta$; $r = 1 - 2\cos\theta$
18. $r = 2 - \sin\theta$; $r = 2(1 + \sin\theta)$
19. $r = 3(1 + \cos\theta)$; $r = 1 - \cos\theta$
20. $r = \sin\theta$; $r = \sqrt{\sin 2\theta}$

21. Find the area enclosed by the small loop of the limaçon $r = 1 - 2\cos\theta$.
22. Find the area inside the limaçon $r = 1 + 2\cos\theta$ that is to the right of the line $r = \sec\theta$.
23. Find the area inside the limaçon $r = 3 - 2\sin\theta$ that lies below the line $r = -2\csc\theta$.
24. Find the area inside both the circle $r = 3\sin\theta$ and the cardioid $r = 1 + \sin\theta$.
25. Find the area inside the cardioid $r = 2(1 + \sin\theta)$ and also inside the circle $r = 2\cos\theta$.

B

26. Find the smallest area bounded by $r = \ln\theta$, $\theta = 1$, and $\theta = e$. Sketch the graph.
27. Find the area inside the limaçon $r = 1 - \sqrt{2}\sin\theta$ that lies outside its small loop.
28. Find the area inside the graph of $r = 2\cos 3\theta$ and outside the graph of $r = -2\cos\theta$.
29. Find the area inside the circle $r = -2\sin\theta$ that lies outside the small loop of the limaçon $r = 1 - 2\sin\theta$.
30. (a) Find all points of intersection of the graphs of $r = 2\cos\frac{\theta}{2}$ and $r = -2\cos\theta$. Draw the graphs.
(b) Find the area inside the graph of $r = -2\cos\theta$ that lies outside the two small loops of $r = 2\cos\frac{\theta}{2}$.
31. Find the shaded area shown in the accompanying figure.

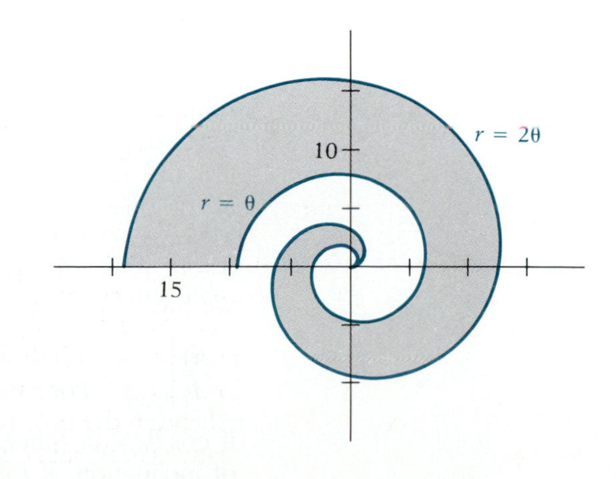

32. Show that a point P_0 is a point of intersection of the graphs of $r = f(\theta)$ and $r = g(\theta)$ if and only if (a) there exist numbers θ_1 and θ_2 such that $f(\theta_1) = 0$ and $g(\theta_2) = 0$, in which case P_0 is the pole, or (b) there exists a number θ_0 such that

$$f(\theta_0) = (-1)^k g(\theta_0 + k\pi)$$

for some integer k. In case (b) P_0 has coordinates $(f(\theta_0), \theta_0)$ or, equivalently, $(g(\theta_0 + k\pi), \theta_0 + k\pi)$. [*Hint:* If P_0 has coordinates (r_0, θ_0) for $r_0 \neq 0$, then all other coordinates of P_0 have one of the forms $(r_0, \theta_0 + 2n\pi)$ or $(-r_0, \theta_0 + (2n+1)\pi)$, where $n = 0, \pm 1, \pm 2, \ldots$.]

In Exercises 33 and 34 find all points of intersection, making use of Exercise 32.

33. $r = 2$, $r = 4 \sin 2\theta$

34. $r^2 = 25 \cos \theta$, $r = 3 + 2 \cos \theta$

13.7 TANGENT LINES TO POLAR GRAPHS; ARC LENGTH

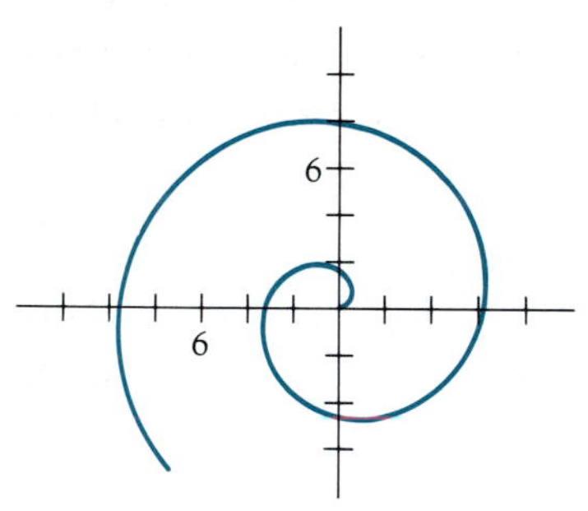

FIGURE 13.39

In rectangular coordinates if $y = f(x)$ and f is a differentiable function, then $f'(x)$ gives the slope of the tangent line to the graph. In polar coordinates if $r = f(\theta)$ and f is a differentiable function, it might be tempting to assume that $f'(\theta)$ also gives the slope of the tangent line to the graph. But this clearly is not so, as the example $r = \theta$ shows. Here, $f(\theta) = \theta$, and so $f'(\theta) = 1$. But the graph (Figure 13.39) is the spiral of Archimedes, which does *not* always have slope 1. So we have to approach the slope of a polar graph in a different way.

The secret of success lies in changing the equation of a polar curve to parametric equations, with θ as a parameter. Suppose $r = f(\theta)$, where f is differentiable on the interval $[\alpha, \beta]$. Then, on superimposing rectangular coordinates, we know that $x = r \cos \theta$ and $y = r \sin \theta$. Replacing r by $f(\theta)$ gives

$$\begin{cases} x = f(\theta) \cos \theta \\ y = f(\theta) \sin \theta \end{cases} \qquad \alpha \leq \theta \leq \beta \tag{13.12}$$

These are the parametric equations of the curve. Now by Theorem 13.1,

$$\frac{dy}{dx} = \frac{\dfrac{dy}{d\theta}}{\dfrac{dx}{d\theta}} = \frac{f'(\theta) \sin \theta + f(\theta) \cos \theta}{f'(\theta) \cos \theta - f(\theta) \sin \theta} \tag{13.13}$$

provided the denominator is nonzero. This is not a very satisfactory result, since the formula is rather complicated. There is a more suitable measure than slope for describing the behavior of the tangent line in polar coordinates that we discuss below. First, though, we can use the result, equation (13.13), to confirm our observation of the last section about tangent lines at the pole.

Suppose that $r = f(\theta)$ and f is differentiable at the pole. Set $r = 0$ and let $\theta = \theta_0$ be a solution. Assume further that $f'(\theta_0) \neq 0$. Then, since $f(\theta_0) = 0$, we have from equation (13.13) that the slope at the origin is

$$\left.\frac{dy}{dx}\right|_{(0,0)} = \frac{f'(\theta_0) \sin \theta_0}{f'(\theta_0) \cos \theta_0} = \tan \theta_0$$

if $\cos \theta_0 \neq 0$. Since the slope of the tangent line is the tangent of its angle of inclination, it follows that its angle of inclination is θ_0, or that it differs from this by a multiple of π. In any case, the line $\theta = \theta_0$ is tangent to the curve at the origin. If $\cos \theta_0 = 0$, then the tangent line is vertical, and again $\theta = \theta_0$ is this tangent line, since $\theta_0 = \frac{\pi}{2}$ or an odd multiple of $\frac{\pi}{2}$. Note that if $f'(\theta_0) = 0$, then both $\frac{dx}{d\theta}$ and $\frac{dy}{d\theta}$ are 0 at the origin, so that the origin is a singular point.

To find a way to describe the tangent line to a polar graph that is more convenient than its slope, consider the graph of the differentiable function $r = f(\theta)$. At an arbitrary point P on the curve, other than the pole, let ϕ be the angle of inclination of the tangent line at P. Then $\tan \phi$ is the slope, and so its value is given by equation (13.13), provided $\phi \neq \frac{\pi}{2}$:

$$\tan \phi = \frac{f'(\theta) \sin \theta + f(\theta) \cos \theta}{f'(\theta) \cos \theta - f(\theta) \sin \theta} \tag{13.14}$$

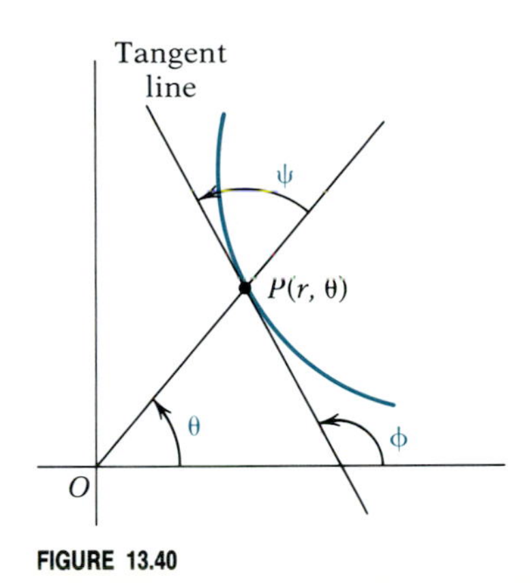

FIGURE 13.40

As shown in Figure 13.40 let ψ denote the smallest positive angle from the ray OP to the tangent line. From the figure, $\psi = \phi - \theta$. Although this figure illustrates the special case in which $\phi > \theta$ and $0 \leq \theta < 2\pi$, it can be shown that in all cases

$$\psi = \phi - \theta + n\pi \tag{13.15}$$

where n is an integer. Because the period of the tangent function is π, it follows that if $\psi \neq \frac{\pi}{2}$, $\tan \psi = \tan(\phi - \theta + n\pi) = \tan(\phi - \theta)$. By the trigonometric identity for $\tan(\phi - \theta)$, therefore,

$$\tan \psi = \frac{\tan \phi - \tan \theta}{1 + \tan \phi \tan \theta}$$

It is left as an exercise to show that when $\tan \phi$ is replaced by the right-hand side of equation (13.14) and $\tan \theta$ is replaced by $\sin \theta/\cos \theta$, we get the simple formula

$$\tan \psi = \frac{f(\theta)}{f'(\theta)} \tag{13.16}$$

provided both $f(\theta)$ and $f'(\theta)$ are nonzero. Since $r = f(\theta)$, this can also be written in the form

$$\tan \psi = \frac{r}{dr/d\theta}$$

EXAMPLE 13.19 Find the angle ψ at the point $(3, \frac{\pi}{3})$ on the graph of $r = 2(1 + \cos \theta)$. Show the result graphically.

Solution By equation (13.16) for $\theta = \frac{\pi}{3}$,

$$\tan \psi = \frac{f(\theta)}{f'(\theta)} = \frac{2(1 + \cos \theta)}{-2 \sin \theta} = -\frac{1 + \cos \theta}{\sin \theta} = -\frac{\frac{3}{2}}{\frac{\sqrt{3}}{2}} = -\sqrt{3}$$

So $\psi = \frac{2\pi}{3}$. The graph is shown in Figure 13.41. Note that at this point the tangent line is horizontal.

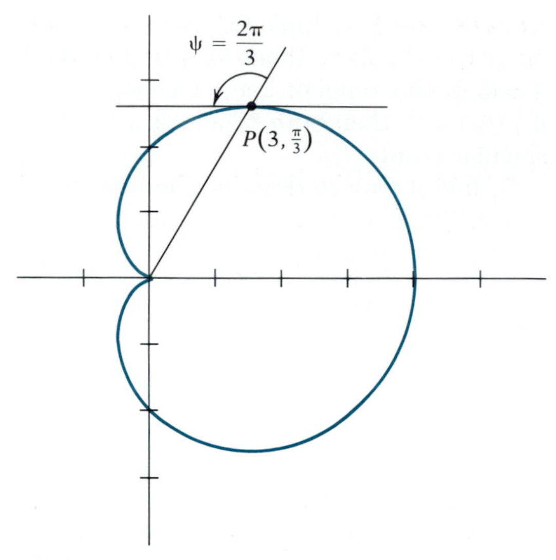

FIGURE 13.41

From equation (13.15) it follows that the angle of inclination ϕ of the tangent line is $\theta + \psi$ or differs from this by a multiple of π. In any case, if the tangent line is nonvertical,

$$\tan \phi = \tan(\theta + \psi) \tag{13.17}$$

This is an alternative to equation (13.13) for calculating the slope, should it be needed. For example, suppose we want the slope of $r = \cos^2 \theta$ at $\theta = \frac{\pi}{4}$. First we calculate $\tan \psi$:

$$\tan \psi = \frac{r}{dr/d\theta} = \frac{\cos^2 \theta}{-2 \sin \theta \cos \theta} = -\frac{1}{2} \cot \theta$$

At $\theta = \frac{\pi}{4}$, $\tan \psi = -\frac{1}{2}$ and $\tan \theta = 1$. So the slope is

$$\tan \phi = \tan(\theta + \psi) = \frac{\tan \theta + \tan \psi}{1 - \tan \theta \tan \psi} = \frac{1 + \left(-\frac{1}{2}\right)}{1 - (1)\left(-\frac{1}{2}\right)} = \frac{1}{3}$$

When two curves intersect, the angle between them at a point of intersection is defined as the angle between their tangent lines. If the equations are $r_1 = f(\theta)$ and $r_2 = g(\theta)$, then it is not difficult to show that the smallest positive angle α from the first curve to the second is $\psi_2 - \psi_1$, or differs from this by an integral multiple of π (see Exercise 47). So

$$\tan \alpha = \tan(\psi_2 - \psi_1) \tag{13.18}$$

This assumes that $\alpha \neq \frac{\pi}{2}$. When $\alpha = \frac{\pi}{2}$, the tangent lines are perpendicular, and the curves are said to be *orthogonal*. In this case $\tan \alpha$ is undefined, but $\cot \alpha = 0$. You will be asked to show in Exercise 31 that this leads to the condition

$$\tan \psi_1 \tan \psi_2 = -1 \tag{13.19}$$

for orthogonality, provided neither ψ_1 nor ψ_2 is $\frac{\pi}{2}$. Note the similarity to the perpendicularity condition $m_1 m_2 = -1$ for slopes m_1 and m_2; that is, $\tan \phi_1 \tan \phi_2 = -1$.

EXAMPLE 13.20 Find the angle from $r_1 = 2 - \cos\theta$ to $r_2 = 2(1 + \cos\theta)$ at each of the points of intersection.

Solution First we find the points of intersection. Setting $r_1 = r_2$ gives

$$\begin{aligned} 2 - \cos\theta &= 2 + 2\cos\theta \\ 3\cos\theta &= 0 \\ \cos\theta &= 0 \\ \theta &= \pm\tfrac{\pi}{2} \end{aligned}$$

So the curves intersect at $(2, \pm\frac{\pi}{2})$. An examination of the graphs (Figure 13.42) shows that there are no other points of intersection. By symmetry we need only consider the angle between the curves when $\theta = \frac{\pi}{2}$. In order to use equation (13.18) we need $\tan\psi_1$ and $\tan\psi_2$:

$$\tan\psi_1 = \frac{r_1}{dr_1/d\theta} = \frac{2 - \cos\theta}{\sin\theta}$$

and

$$\tan\psi_2 = \frac{r_2}{dr_2/d\theta} = \frac{2(1 + \cos\theta)}{-2\sin\theta} = \frac{1 + \cos\theta}{-\sin\theta}$$

At $\theta = \frac{\pi}{2}$, $\tan\psi_1 = 2$ and $\tan\psi_2 = -1$. Thus,

$$\begin{aligned} \tan\alpha = \tan(\psi_2 - \psi_1) &= \frac{\tan\psi_2 - \tan\psi_1}{1 + \tan\psi_1 \tan\psi_2} \\ &= \frac{2 - (-1)}{1 + 2(-1)} = -3 \end{aligned}$$

From a calculator we find $\alpha \approx 1.89$ radians, or approximately $108.43°$.

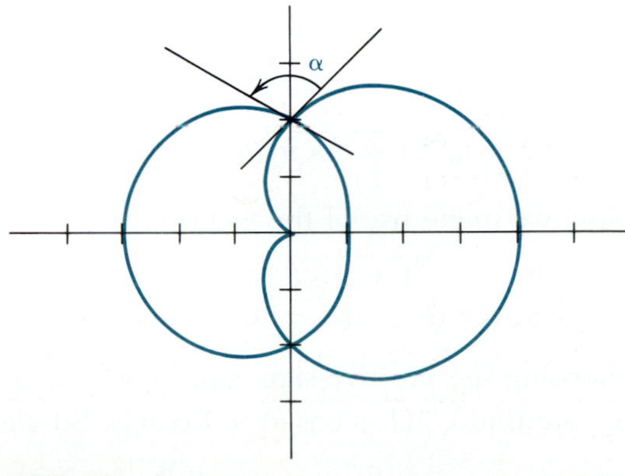

FIGURE 13.42 ■

The parametric equations (13.12) enable us to use the results of Section 13.3 to find the length of a curve $r = f(\theta)$, where f' is continuous on $[\alpha, \beta]$. Again superimposing a rectangular coordinate system onto the polar system in the usual way, we have from equation (13.12),

$$\frac{dx}{d\theta} = f'(\theta) \cos \theta - f(\theta) \sin \theta$$

and

$$\frac{dy}{d\theta} = f'(\theta) \sin \theta + f(\theta) \cos \theta$$

So by equation (13.5) with t replaced by θ,

$$\begin{aligned} ds &= \sqrt{\left(\frac{dx}{d\theta}\right)^2 + \left(\frac{dy}{d\theta}\right)^2}\, d\theta \\ &= \sqrt{[f'(\theta) \cos \theta - f(\theta) \sin \theta]^2 + [f'(\theta) \sin \theta + f(\theta) \cos \theta]^2}\, d\theta \end{aligned}$$

We leave it as an exercise (Exercise 48) to show that this reduces to

$$ds = \sqrt{[f(\theta)]^2 + [f'(\theta)]^2}\, d\theta \tag{13.20}$$

We also note that if $f'(\theta)$ is continuous for θ in $[\alpha, \beta]$, both $\frac{dx}{d\theta}$ and $\frac{dy}{d\theta}$ are continuous there. It therefore follows from Theorem 13.2 that arc length in polar coordinates is given by

$$L = \int_\alpha^\beta \sqrt{[f(\theta)]^2 + [f'(\theta)]^2}\, d\theta \tag{13.21}$$

provided f' is continuous in $[\alpha, \beta]$. Since $r = f(\theta)$, we can also write equation (13.21) in the form

$$L = \int_\alpha^\beta \sqrt{r^2 + \left(\frac{dr}{d\theta}\right)^2}\, d\theta$$

EXAMPLE 13.21 Find the length of the cardioid $r = 1 + \cos \theta$.

Solution Because of symmetry it is sufficient to find the length for θ in $[0, \pi]$ and double the result. Since $r' = -\sin \theta$, we have, by equation (13.21),

$$\begin{aligned} L &= 2 \int_0^\pi \sqrt{r^2 + \left(\frac{dr}{d\theta}\right)^2}\, d\theta \\ &= 2 \int_0^\pi \sqrt{1 + 2 \cos \theta + \cos^2 \theta + \sin^2 \theta}\, d\theta \\ &= 2 \int_0^\pi \sqrt{2(1 + \cos \theta)}\, d\theta \end{aligned}$$

Now we make use of the identity

$$\cos \frac{\theta}{2} = \pm \sqrt{\frac{1 + \cos \theta}{2}}$$

choosing the positive sign, since $\cos(\frac{\theta}{2})$ is positive for θ in $[\theta, \pi]$. From this we see that $\sqrt{2(1 + \cos \theta)} = 2 \cos(\frac{\theta}{2})$. So we have

$$L = 2 \int_0^\pi 2 \cos \frac{\theta}{2}\, d\theta = 8 \sin \frac{\theta}{2} \Big]_0^\pi = 8$$

■

EXERCISE SET 13.7

A

In Exercises 1–10 find $\tan\psi$ *at the indicated value of* θ.

1. $r = 2(1 - \sin\theta)$; $\theta = \frac{\pi}{6}$
2. $r = 3\cos 2\theta$; $\theta = \frac{\pi}{6}$
3. $r = \cos^2\theta$; $\theta = \frac{\pi}{4}$
4. $r = 4\sin\theta$; $\theta = \frac{\pi}{3}$
5. $r = 4 - 2\sin\theta$; $\theta = \frac{5\pi}{6}$
6. $r = \sqrt{\cos 2\theta}$; $\theta = -\frac{\pi}{6}$
7. $r = \dfrac{1}{1 - \cos\theta}$; $\theta = \dfrac{2\pi}{3}$
8. $r = \dfrac{2}{2 - \sin\theta}$; $\theta = -\dfrac{\pi}{6}$
9. $r = \sin\theta\cos\theta$; $\theta = \frac{\pi}{3}$
10. $r = 1 - 2\sin\frac{\theta}{2}$; $\theta = \frac{\pi}{2}$

In Exercises 11–20 make use of the fact that $\tan\phi = \tan(\theta + \psi)$, *where* ϕ *is the angle of inclination of the tangent line, to find the slope of the curve at the given value of* θ.

11. $r = 1 - \cos\theta$; $\theta = \frac{2\pi}{3}$
12. $r = 2 + \sin\theta$; $\theta = -\frac{\pi}{6}$
13. $r = \dfrac{1}{1 - \cos\theta}$; $\theta = \frac{\pi}{3}$
14. $r = 2\cos\theta$; $\theta = \frac{\pi}{4}$
15. $r = 2\cos\frac{\theta}{2} - 1$; $\theta = \pi$
16. $r = \sqrt{\sin 2\theta}$; $\theta = \frac{\pi}{3}$
17. $r = 1 - 2\sin\theta$; $\theta = \pi$
18. $r = \cos^2\theta$; $\theta = \frac{\pi}{4}$
19. $r = \dfrac{2}{3 - 2\sin\theta}$; $\theta = \pi$
20. $r = \sec\theta + \tan\theta$; $\theta = \frac{2\pi}{3}$

In Exercises 21 and 22 show that the curve has a horizontal tangent line at the indicated value of θ, *using* $\tan\phi = \tan(\theta + \psi)$.

21. $r = 4 + 4\cos\theta$; $\theta = -\frac{\pi}{3}$
22. $r = 2\sqrt{\cos 2\theta}$; $\theta = \frac{5\pi}{6}$

In Exercises 23 and 24 show that the curve has a vertical tangent line at the indicated point, using the fact that $\cot\phi = \cot(\theta + \psi)$.

23. $r = 1 - \cos\theta$; $\theta = \frac{\pi}{3}$
24. $r = 3 + \tan^2\theta$; $\theta = \frac{3\pi}{4}$
25. Show that for the logarithmic spiral $r = ae^{b\theta}$, the angle ψ is constant. What is ψ when $b = 1$?

In Exercises 26–30 find the tangent of the angle from the first curve to the second at each of their points of intersection.

26. $r_1 = 2 - \cos\theta$; $r_2 = 3\cos\theta$
27. $r_1 = \dfrac{1}{1 - \cos\theta}$; $r_2 = \dfrac{1}{2 + \cos\theta}$
28. $r_1 = \dfrac{3}{1 + \sin\theta}$; $r_2 = 4\sin\theta$
29. $r_1 = \sqrt{3}\cos\theta$; $r_2 = \sin 2\theta$
30. $r_1 = \sqrt{2} + \sin\theta$; $r_2 = 3\sin\theta$
31. Prove the orthogonality condition, equation (13.19).

In Exercises 32–34 show that the graphs are orthogonal at all points of intersection.

32. $r_1 = \sin\theta$; $r_2 = \cos\theta$
33. $r_1 = 3\cos\theta$; $r_2 = 2 - \cos\theta$
34. $r_1 = \dfrac{1}{1 - \sin\theta}$; $r_2 = \dfrac{1}{1 + \sin\theta}$

In Exercises 35–37 show that the curves intersect orthogonally for all nonzero values of a and b.

35. $r_1 = a\cos\theta$; $r_2 = b\sin\theta$
36. $r_1 = a\theta$; $r_2 = be^{-\theta^2/2}$
37. $r_1 = a(1 - \sin\theta)$; $r_2 = b(1 + \sin\theta)$, except at the pole

In Exercises 38–44 find the length of the curve on the given interval.

38. $r = 2a\cos\theta$; $0 \le \theta \le 2\pi$
39. $r = e^{3\theta}$; $0 \le \theta \le \ln 2$
40. $r = \cos\theta - 1$; $0 \le \theta \le \pi$
41. $r = 2\cos^2\frac{\theta}{2}$; $0 \le \theta \le \pi$
42. $r = \sin^3\frac{\theta}{3}$; $0 \le \theta \le 3\pi$
43. $r = \theta^2 - 1$; $0 \le \theta \le 3$
44. $r = \sin\theta - \cos\theta$; $-\frac{\pi}{2} \le \theta \le \pi$

B

45. Prove equation (13.16).
46. Prove that the condition for orthogonality of the graphs of $r = f(\theta)$ and $r = g(\theta)$ can be put in the form $f(\theta)g(\theta) + f'(\theta)g'(\theta) = 0$.
47. Verify equation (13.18). (*Hint:* Express $\phi_2 - \phi_1$ in terms of ψ_2 and ψ_1.)
48. Verify equation (13.20).
49. Find the length of the curve $r = \sin^2\theta$ from $\theta = 0$ to $\theta = \frac{\pi}{2}$.
50. Find the length of that portion of the curve $r = 2(1 - \sin\theta)$ that lies inside $r = -6\sin\theta$.

51. (a) Suppose that f' is continuous and that $f(\theta)\sin\theta \geq 0$ for $\alpha \leq \theta \leq \beta$. Derive the following formula for the surface area obtained by rotating the graph of $r = f(\theta)$ about the line $\theta = 0$:

$$S = 2\pi \int_\alpha^\beta f(\theta)\sin\theta\sqrt{[f(\theta)]^2 + [f'(\theta)]^2}\,d\theta$$

(b) State appropriate conditions and derive an analogous formula for the surface area formed by rotation about the line $\theta = \frac{\pi}{2}$.

In Exercises 52–55 use the results of Exercise 51 to find the surface area formed when the graph of $r = f(\theta)$ is rotated about the specified axis.

52. $r = 1 - \cos\theta$, $0 \leq \theta \leq \pi$; about $\theta = 0$

53. $r = \sqrt{\sin 2\theta}$, $0 \leq \theta \leq \frac{\pi}{2}$; about $\theta = \frac{\pi}{2}$

54. $r = e^{2\theta}$, $0 \leq \theta \leq \pi$; about $\theta = 0$

55. $r = 2\sin^2\frac{\theta}{2}$, $0 \leq \theta \leq \frac{\pi}{2}$; about (a) $\theta = 0$ and (b) $\theta = \frac{\pi}{2}$

13.8 SUPPLEMENTARY EXERCISES

In Exercises 1–6 eliminate the parameter and use the result to sketch the curve defined by the parametric equations.

1. $\begin{cases} x = 2t - 1 \\ y = \sqrt{1-t} \end{cases} \quad -\infty < t \leq 1$

2. $\begin{cases} x = \sin^2 t \\ y = \cos t \end{cases} \quad -\infty < t < \infty$

3. $\begin{cases} x = \ln t \\ y = 1 - \frac{1}{t} \end{cases} \quad 0 < t < \infty$

4. $\begin{cases} x = 1 + e^t \\ y = 1 - e^{-t} \end{cases} \quad -\infty < t < \infty$

5. $\begin{cases} x = \tan^2\theta - 1 \\ y = \sec\theta + 1 \end{cases} \quad 0 \leq \theta < \frac{\pi}{2}$

6. $\begin{cases} x = 2\sinh t - 1 \\ y = 3\cosh t + 2 \end{cases} \quad -\infty < t < \infty$

In Exercises 7–10 find dy/dx and d^2y/dx^2 without eliminating the parameter.

7. $\begin{cases} x = t^2 - 1 \\ y = 2t^3 \end{cases}$

8. $\begin{cases} x = \sqrt{t} \\ y = t^2 + 2 \end{cases}$

9. $\begin{cases} x = \sin^2 t \\ y = \ln\cos t \end{cases}$

10. $\begin{cases} x = \dfrac{t-1}{t} \\ y = \ln\sqrt{t} \end{cases}$

11. Find the equations of the tangent line and normal line to the curve defined by $x = \ln(t^2 - 3)$ and $y = t/\sqrt{t^2 - 3}$ at $(0, 2)$.

12. Without eliminating the parameter, find all maximum and minimum points on the curve defined by

$$\begin{cases} x = \sqrt{t^2 - 1} \\ y = 10t^3 - 3t^5 \end{cases} \quad t \geq 1$$

Use the second derivative test for testing critical values.

13. Find the length of the curve defined by

$$\begin{cases} x = \cosh^2 t \\ y = 2\sinh t \end{cases} \quad 0 \leq t \leq \ln 3$$

14. Find the area of the surface formed by revolving the curve of Exercise 13 about (a) the x-axis and (b) the y-axis.

15. For the curve defined by

$$\begin{cases} x = \ln t - 1 \\ y = 2\sqrt{t} \end{cases} \quad 1 \leq t \leq 3$$

find (a) its length and (b) the area of the surface formed by revolving the curve about the x-axis.

16. Let R denote the region enclosed by the x-axis and one arch of the cycloid

$$\begin{cases} x = a(t - \sin t) \\ y = a(1 - \cos t) \end{cases}$$

Find (a) the area of R and (b) the volume of the solid formed by revolving R about the x-axis.

In Exercises 17 and 18 change to an equivalent rectangular equation, identify the curve, and draw its graph.

17. (a) $r = 2(\sin\theta + \cos\theta)$ (b) $r(2 + \sin\theta) = 1$

18. (a) $r = \dfrac{\sec\theta}{1 - \tan\theta}$ (b) $r = \sec^2\frac{\theta}{2}$

In Exercises 19–21 discuss and graph the given equation.

19. (a) $r = 2\sin^2\frac{\theta}{2}$ (b) $r^2 = 2\sin\theta\cos\theta$

20. (a) $r = -1 - 2\cos\theta$ (b) $r = 5 + 3\sin\theta$

21. (a) $r = \sin^2\theta - \cos^2\theta$ (b) $r\theta = 1$, $\theta > 0$

In Exercises 22–24 find the area of the specified region.

22. Inside both $r = \sqrt{3}\cos\theta$ and $r = \sin\theta$

23. Inside both $r = 1 - \cos\theta$ and $r = 2 + \cos\theta$

24. Outside $r = 2\cos 2\theta$ and inside $r^2 = 6\sin 2\theta$

25. Let C_1 be the curve $r = 2$ and C_2 the curve $r = 4\cos 3\theta$.
Find the areas of the following regions:
(a) Outside C_1 and inside C_2
(b) Inside C_2 and inside C_1
(c) Outside C_2 and inside C_1

In Exercises 26 and 27 find the angle between the two curves at each of their points of intersection.

26. $r = 5\sin\theta$, $r = 2 + \sin\theta$

27. $r = \dfrac{1}{1 + \sin\theta}$, $r = 2(1 - \sin\theta)$

In Exercises 28 and 29 show that the curves intersect orthogonally.

28. $r = \tan\frac{\theta}{2}$, $r = e^{\cos\theta}$; $0 < \theta < \pi$

29. $r(1 + \cos\theta) = 3$, $r = 4\cos\theta$; $-\frac{\pi}{2} < \theta < \frac{\pi}{2}$

In Exercises 30 and 31 find the length of the given curve.

30. $r = \sqrt{1 - \sin 2\theta}$; $-\frac{\pi}{4} \le \theta \le \frac{\pi}{4}$

31. $r = \frac{7}{\theta}$; $1 \le \theta \le 7$

32. Find the area of the surface formed by revolving the arc of the curve $r = \sec^2\frac{\theta}{2}$ from $\theta = 0$ to $\theta = \frac{2\pi}{3}$ about the polar axis.

33. The **witch of Agnesi** is a curve generated by the point P (as in the figure) moving so that QPR is always a right triangle with right angle at P. The point Q is on the circle of radius a centered at $(0, a)$ and R is on the line $y = 2a$. Find parametric equations for this curve, using θ as a parameter, and draw its graph. (*Hint:* Make use of the polar equation of the circle.)

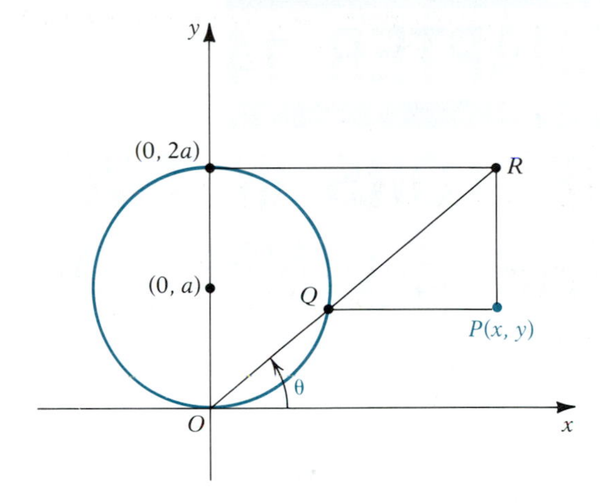

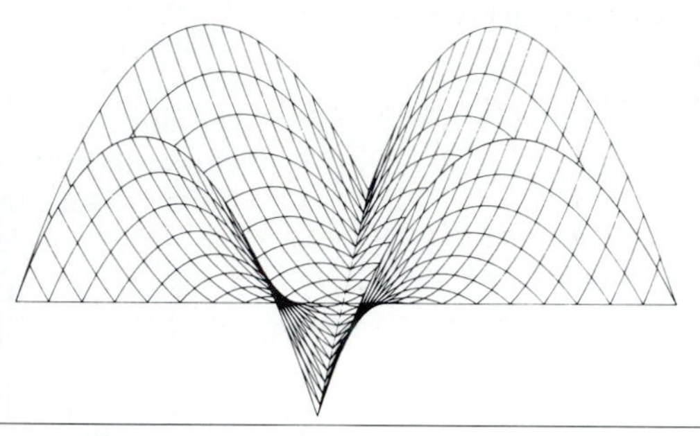

CHAPTER 14

VECTORS IN TWO AND THREE DIMENSIONS

14.1 INTRODUCTION

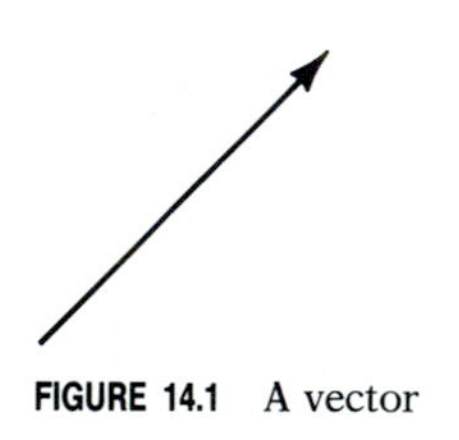

FIGURE 14.1 A vector

Certain physical quantities have both *magnitude* and *direction*; examples are force, velocity, acceleration, and displacement of a moving particle. A convenient way to represent such quantities is with a directed line segment, such as the one in Figure 14.1. The length of the segment, to some scale, represents the magnitude of the quantity in question, and the direction is indicated by the inclination of the segment and by the arrowhead. Such a directed line segment is called a **vector.** For example, the vector in Figure 14.1 might represent a wind velocity of 20 mi/hr blowing in a northeasterly direction. The length would then be taken as 20 units, to some convenient scale. Any quantity that has both magnitude and direction can be represented in this way and for this reason is called a *vector quantity*.

Vector quantities are in contrast with such things as area, mass, time, and distance, which can be adequately described by a single number. These are called *scalar quantities* (since they are measured according to some scale), and the numbers used to measure them are called **scalars.** For our purposes, then, scalars are just real numbers. It should be mentioned, however, that in certain more advanced treatments both vectors and scalars can be different from what we are discussing here.

The subject now called vector analysis was developed in the latter part of the 19th century by the American physicist and mathematician Josiah Willard Gibbs (1839–1903) and the English engineer Oliver Heavyside (1850–1925), working independently. Many of the ideas came earlier, however, especially from the Irish mathematician William Rowan Hamilton (1805–1865) in his work on *quaternions,* and the Scottish physicist James Clerk Maxwell (1831–1879), who used some of Hamilton's ideas in his study of electromagnetic field theory. So the subject is strongly grounded in the physical

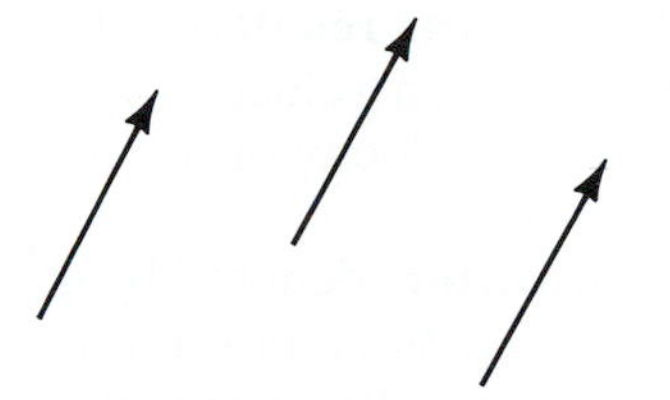

FIGURE 14.2 Equivalent vectors

sciences and engineering and is still an important tool in these fields. The applications have expanded greatly now, even to economics and some of the other social sciences. Moreover, vectors have contributed significantly to the continuing development of mathematics itself.

In the early work with vectors their geometric properties were dominant. As we shall see, though, the advantages of vectors can only be fully realized when their algebraic properties are used in conjunction with their geometric ones. In the next section we explore the geometric nature of vectors in the plane and then formulate vectors and their properties in algebraic terms.

14.2 VECTORS IN THE PLANE

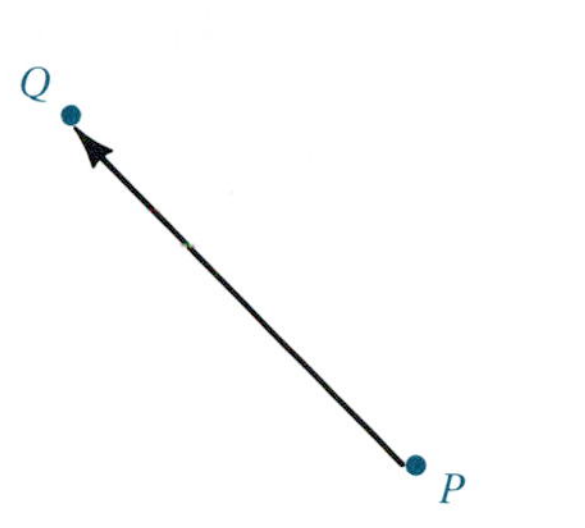

FIGURE 14.3

Typically boldface letters such as **u**, **v**, and **w** are used to designate vectors. Throughout this section we consider vectors that lie in a plane, although most of the results have natural extensions to three (or more) dimensions. Two vectors **u** and **v** are said to be **equivalent** if they have the same magnitude and direction, and in this case we write $\mathbf{u} = \mathbf{v}$. Three equivalent vectors are illustrated in Figure 14.2. We do not distinguish between equivalent vectors, so that in effect we can shift a vector from one location to another so long as its original magnitude and direction are retained. Because of this freedom of movement, we say that we are working within a system of *free* vectors.

Suppose, as in Figure 14.3, a vector extends from a point P to a point Q. When we wish to emphasize this we use the notation $\overrightarrow{PQ}$ to designate the vector. The point P is called the **initial point** and Q the **terminal point.** Sometimes we also use "tail" and "tip" instead of initial point and terminal point, respectively.

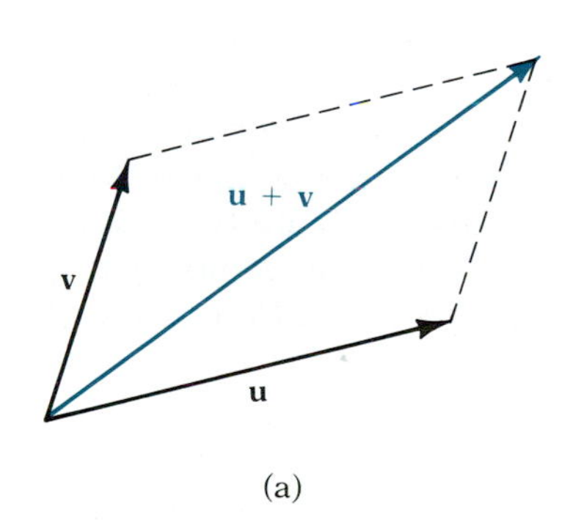

(a)

Two nonparallel vectors are **added** according to the **parallelogram law,** illustrated in Figure 14.4(a). The vectors are drawn with a common initial point, and a parallelogram is constructed with **u** and **v** as adjacent sides. The vector $\mathbf{u} + \mathbf{v}$ is then defined as the vector along the diagonal from the common initial point to the opposite vertex. An alternate method is to place the initial point of **v** at the terminal point of **u**. Then $\mathbf{u} + \mathbf{v}$ is the vector shown in Figure 14.4(b) drawn from the initial point of **u** to the terminal point of **v**. You should convince yourself that the triangle in part b is just the lower half of the parallelogram in part a. This second method is sometimes called the "tail to tip" method of adding. If **u** and **v** are parallel vectors, then the parallelogram of part a is degenerate. The tail to tip method still works, however.

From our definition of addition it is easy to see that

$$\mathbf{u} + \mathbf{v} = \mathbf{v} + \mathbf{u}$$

that is, vector addition is commutative. It is also associative; that is,

$$\mathbf{u} + (\mathbf{v} + \mathbf{w}) = (\mathbf{u} + \mathbf{v}) + \mathbf{w}$$

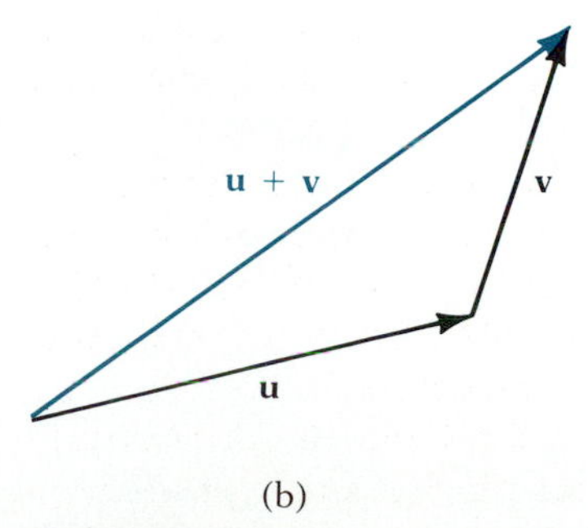

(b)

FIGURE 14.4

You will be asked in the exercises to give a geometric argument for this.

Vector addition is consistent with observed results. For example, if **u** and **v** represent forces acting on an object, then the net effect is $\mathbf{u} + \mathbf{v}$; that is, the two individual forces **u** and **v** could be replaced by the force $\mathbf{u} + \mathbf{v}$, and the effect would be the same. In this case we call $\mathbf{u} + \mathbf{v}$ the **resultant** of $\mathbf{u}$ and $\mathbf{v}$. Similarly, if $\mathbf{u}$ is a vector representing the indicated velocity of an airplane and $\mathbf{v}$ is the wind velocity vector, then the true velocity of the airplane relative to the ground is $\mathbf{u} + \mathbf{v}$.

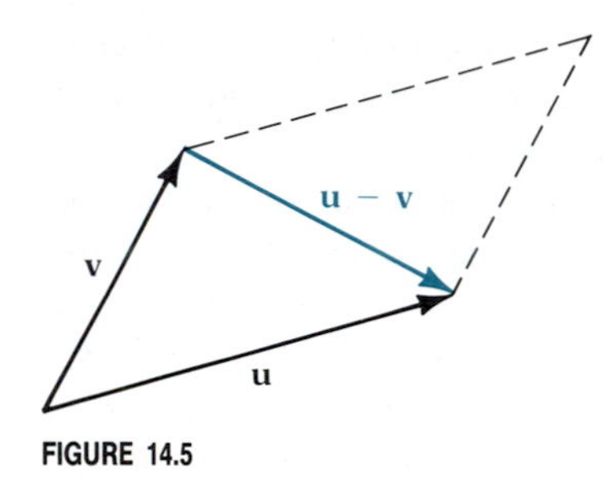

FIGURE 14.5

It is convenient to introduce the notion of the **zero vector,** denoted by $\mathbf{0}$, with a magnitude of 0 and assigned no direction. We may think of the zero vector as a single point. If $\mathbf{v}$ is a nonzero vector, then $-\mathbf{v}$ is the vector that has the same length as $\mathbf{v}$ but with a direction opposite to that of $\mathbf{v}$. We now define **subtraction** by

$$\mathbf{u} - \mathbf{v} = \mathbf{u} + (-\mathbf{v})$$

Thus, $\mathbf{u} - \mathbf{v}$ is the vector that when added to $\mathbf{v}$ gives $\mathbf{u}$. This is illustrated in Figure 14.5. Notice that when $\mathbf{u}$ and $\mathbf{v}$ are drawn with the same initial point, $\mathbf{u} - \mathbf{v}$ is the vector *from the tip of* $\mathbf{v}$ *to the tip of* $\mathbf{u}$. Notice also that when we construct the parallelogram with $\mathbf{u}$ and $\mathbf{v}$ as adjacent sides, $\mathbf{u} - \mathbf{v}$ is directed along the diagonal from the tip of $\mathbf{v}$ to the tip of $\mathbf{u}$, in contrast to $\mathbf{u} + \mathbf{v}$, which is directed along the other diagonal. From this definition we see that, as we would expect,

$$\mathbf{v} - \mathbf{v} = \mathbf{0}$$

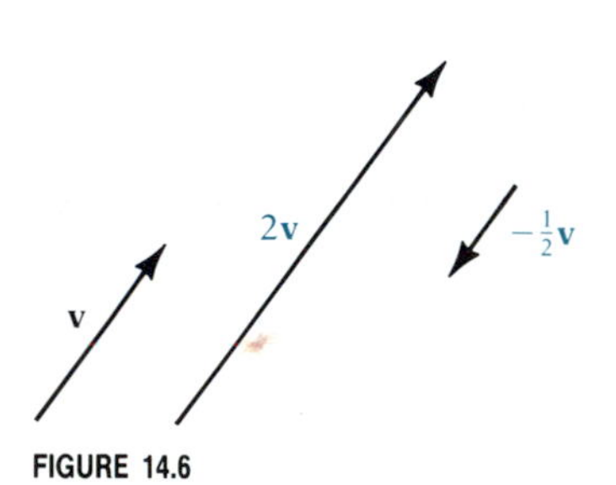

FIGURE 14.6

Vectors can be multiplied by scalars as follows. If $k > 0$ and $\mathbf{v}$ is a nonzero vector, then $k\mathbf{v}$ is a vector that has the same direction as $\mathbf{v}$ and magnitude k times the magnitude of $\mathbf{v}$. If $k < 0$, then $k\mathbf{v}$ has direction opposite to that of $\mathbf{v}$ and magnitude $|k|$ times the magnitude of $\mathbf{v}$. If $k = 0$, we define $k\mathbf{v}$ as the zero vector, and if $\mathbf{v} = \mathbf{0}$, then $k\mathbf{v} = \mathbf{0}$ for all scalars k. In Figure 14.6 we picture a vector $\mathbf{v}$, along with the vectors $2\mathbf{v}$ and $-\frac{1}{2}\mathbf{v}$.

Further insight into the properties of vectors can be gained by introducing a rectangular coordinate system. Suppose that a vector $\mathbf{v} = \overrightarrow{P_1P_2}$, where the coordinates of P_1 and P_2 are (x_1, y_1) and (x_2, y_2), respectively. As shown in Figure 14.7, the horizontal displacement from P_1 to P_2 is $x_2 - x_1$ and the vertical displacement is $y_2 - y_1$. We call $x_2 - x_1$ the **horizontal component** (or x component) and $y_2 - y_1$ the **vertical component** (or y component) of $\mathbf{v}$. For example, if the coordinates of P_1 are (3, 2) and of P_2 are (7, 5), then the horizontal component of $\mathbf{v}$ is $7 - 3 = 4$ and the vertical component is $5 - 2 = 3$. So every vector has a unique pair of components. Conversely, if

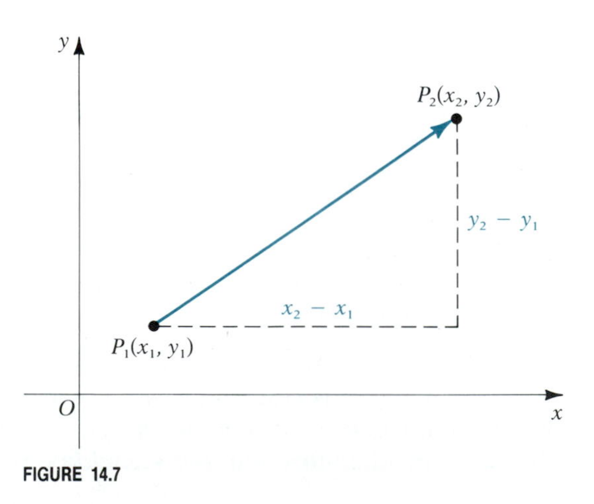

FIGURE 14.7

we are given a pair of components, then these uniquely determine the collection of equivalent vectors that have these components. For example, given the x component 3 and y component 2, we can determine all vectors that have these components. The simplest of these is the one with initial point at the origin and terminal point at (3, 2). Since we do not distinguish between equivalent vectors, we can in effect say that a vector is uniquely determined by its components.

This identification of a vector with its components enables us to look at vectors in a new way. We use the symbol $\langle a, b\rangle$ to indicate a vector with x component a and y component b, and we call this ordered pair of numbers itself a vector. When we wish to distinguish between vectors as directed line segments and vectors as ordered pairs, we say the former is a *geometric vector* and the latter is an *algebraic vector*. By the discussion above, given a geometric vector $\mathbf{v}$, we can determine the corresponding algebraic vector $\langle a, b\rangle$ and conversely, and we write $\mathbf{v} = \langle a, b\rangle$. Any geometric vector corresponding to $\langle a, b\rangle$ is called a **geometric representative** of $\langle a, b\rangle$.

EXAMPLE 14.1 (a) Express the vector $\overrightarrow{AB}$ in algebraic form, where $A = (-1, 2)$ and $B = (3, -4)$. Draw the geometric vector.

(b) Draw the geometric representative of the vector $\langle -2, 3\rangle$ whose initial point is $(4, -1)$. What is its terminal point?

Solution (a) $\overrightarrow{AB} = \langle 3 - (-1), -4 - 2\rangle = \langle 4, -6\rangle$

The vector is shown geometrically in Figure 14.8.

(b) Beginning at $(4, -1)$, we go 2 units to the left and 3 units up, giving the terminal point (2, 2), as shown in Figure 14.9.

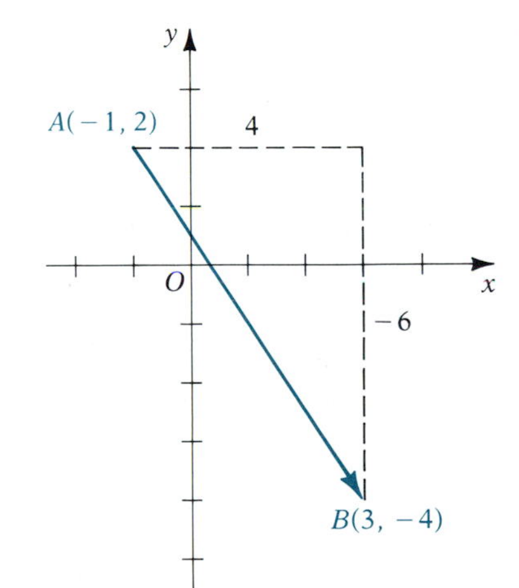

FIGURE 14.8

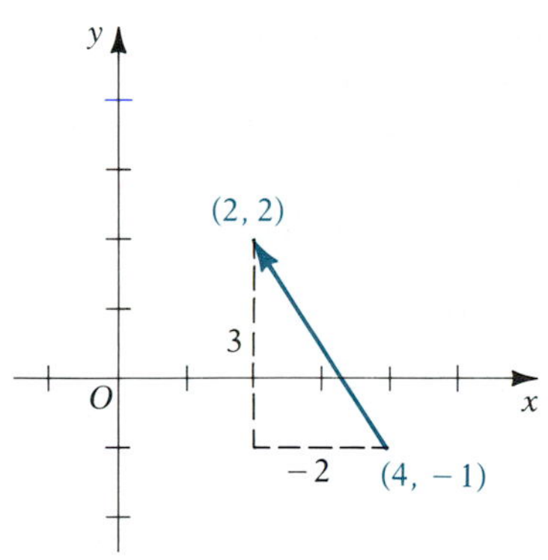

FIGURE 14.9 ■

We can now state properties of vectors in terms of their algebraic representations. First we give a definition of addition and multiplication by a scalar. You will be asked in the exercises to show that these are consistent with the corresponding geometric definitions.

DEFINITION 14.1 For ordered pairs $\langle a, b\rangle$ and $\langle c, d\rangle$ of real numbers,

1. $\langle a, b\rangle = \langle c, d\rangle$ if and only if $a = c$ and $b = d$
2. $\langle a, b\rangle + \langle c, d\rangle = \langle a + c, b + d\rangle$
3. $k\langle a, b\rangle = \langle ka, kb\rangle$ for any scalar k

The set of all such ordered pairs of real numbers with equality, addition, and multiplication by a scalar defined by equations 1, 2, and 3 is called a **vector space of dimension two,** designated V_2, and each element of V_2 is called a **two-dimensional vector.** ■

THEOREM 14.1 If $\mathbf{u} = \langle a, b \rangle$, $\mathbf{v} = \langle c, d \rangle$, and $\mathbf{w} = \langle e, f \rangle$ are arbitrary elements of V_2, then

1. $\mathbf{u} + \mathbf{v} = \mathbf{v} + \mathbf{u}$
2. $\mathbf{u} + (\mathbf{v} + \mathbf{w}) = (\mathbf{u} + \mathbf{v}) + \mathbf{w}$

and for any scalars k and l,

3. $k(\mathbf{u} + \mathbf{v}) = k\mathbf{u} + k\mathbf{v}$
4. $(k + l)\mathbf{u} = k\mathbf{u} + l\mathbf{u}$
5. $k(l\mathbf{u}) = (kl)\mathbf{u}$

(The proofs will be called for in the exercises.) ■

DEFINITION 14.2
1. The element $\langle 0, 0 \rangle$ of V_2 is called the **zero vector** and is denoted by $\mathbf{0}$.
2. If $\mathbf{u} = \langle a, b \rangle$ is any element of V_2, then $-\mathbf{u} = \langle -a, -b \rangle$.
3. If $\mathbf{u} = \langle a, b \rangle$ and $\mathbf{v} = \langle c, d \rangle$ are arbitrary elements of V_2, then

$$\mathbf{u} - \mathbf{v} = \mathbf{u} + (-\mathbf{v})$$ ■

THEOREM 14.2 For any element $\mathbf{u} = \langle a, b \rangle$ in V_2,

1. $\mathbf{u} + \mathbf{0} = \mathbf{u}$
2. $\mathbf{u} + (-\mathbf{u}) = \mathbf{0}$
3. $1\mathbf{u} = \mathbf{u}$
4. $(-1)\mathbf{u} = -\mathbf{u}$
5. $0\mathbf{u} = \mathbf{0}$
6. $k\mathbf{0} = \mathbf{0}$ for all scalars k

(The proofs are left as an exercise.) ■

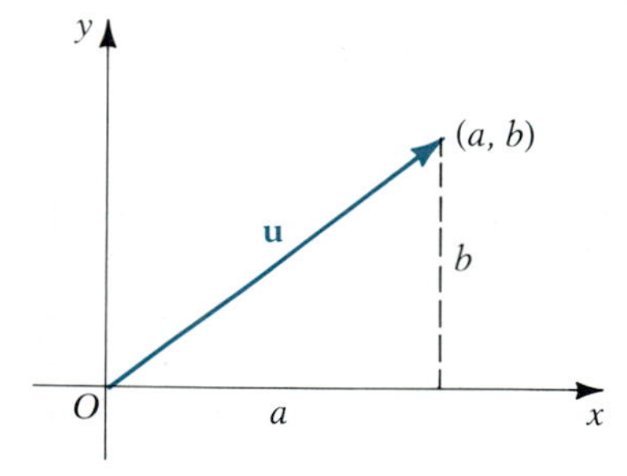

FIGURE 14.10

Remark Because the addition of $\mathbf{0}$ to a vector leaves that vector unchanged, $\mathbf{0}$ is the *additive identity* for V_2. Also, because $-\mathbf{u}$ adds to $\mathbf{u}$ to give $\mathbf{0}$, $-\mathbf{u}$ is the *additive inverse* of $\mathbf{u}$.

Consider a vector $\mathbf{u} = \langle a, b \rangle$. Its geometric counterpart can be represented with initial point at the origin and terminal point at (a, b) as in Figure 14.10. The magnitude of this geometric vector is its length, which by the Pythagorean theorem is $\sqrt{a^2 + b^2}$. This leads to the following definition.

DEFINITION 14.3 Let $\mathbf{u} = \langle a, b \rangle$ be any vector in V_2. The **magnitude** (or *length*) of $\mathbf{u}$, denoted by $|\mathbf{u}|$, is defined by

$$|\mathbf{u}| = \sqrt{a^2 + b^2} \tag{14.1}$$ ■

Comment It is reasonable to use the same symbol to designate the magnitude of a vector as we use to denote the absolute value of a real number. If x is a real number, then $|x|$ can be interpreted geometrically as the distance from 0 to x on a number line. Similarly, when a vector $\mathbf{u}$ is interpreted geometrically with initial point at the origin, $|\mathbf{u}|$ is the distance from the origin to the terminal point of $\mathbf{u}$.

THEOREM 14.3 Let $\mathbf{u} = \langle a, b \rangle$ and $\mathbf{v} = \langle c, d \rangle$ be any two vectors in V_2, and let k be any scalar. Then

1. $|\mathbf{u}| \geq 0$ and $|\mathbf{u}| = 0$ if and only if $\mathbf{u} = \mathbf{0}$
2. $|-\mathbf{u}| = |\mathbf{u}|$

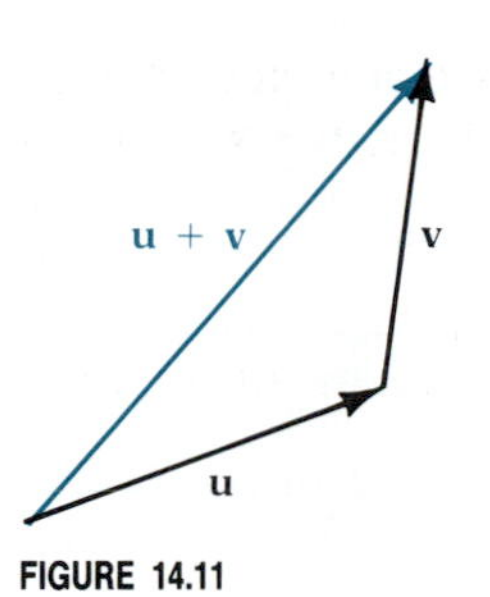

FIGURE 14.11

3. $|k\mathbf{u}| = |k|\,|\mathbf{u}|$
4. $|\mathbf{u} + \mathbf{v}| \le |\mathbf{u}| + |\mathbf{v}|$ (triangle inequality) ■

Proofs of properties 1, 2, and 3 are called for in the exercises. An algebraic proof of property 4 will be given in the next section, but it is evident geometrically, as an examination of Figure 14.11 shows. The inequality simply reflects the fact that the length of one side of a triangle is less than the sum of the lengths of the other two sides (which explains the name "triangle inequality"). You should reflect on the circumstances when equality occurs.

In keeping with the geometric relationship between **u** and $k\mathbf{u}$, we have the following definition.

DEFINITION 14.4 PARALLEL VECTORS

Two nonzero vectors **u** and **v** in V_2 are said to be **parallel** if there exists a nonzero scalar k such that $\mathbf{v} = k\mathbf{u}$. We also say that **0** is parallel to every vector in V_2. ■

EXERCISE SET 14.2

A

Exercises 1–12 refer to the vectors $\mathbf{u} = \langle -2, 3\rangle$, $\mathbf{v} = \langle 4, 2\rangle$, *and* $\mathbf{w} = \langle -1, -2\rangle$. *In each case give the result as an algebraic vector. Also give a geometric construction illustrating the given operations.*

1. $\mathbf{u} + \mathbf{v}$
2. $\mathbf{v} + \mathbf{w}$
3. $\mathbf{u} + \mathbf{w}$
4. $\mathbf{u} + \mathbf{v} + \mathbf{w}$
5. $\mathbf{u} - \mathbf{v}$
6. $\mathbf{v} - \mathbf{w}$
7. $\mathbf{w} - \mathbf{u}$
8. $2\mathbf{u} + \frac{1}{2}\mathbf{v}$
9. $\mathbf{u} - 2\mathbf{w}$
10. $\mathbf{u} + \frac{3}{2}\mathbf{v} - \mathbf{w}$
11. $-2\mathbf{u} + \mathbf{v} - 3\mathbf{w}$
12. $2\mathbf{u} - \frac{1}{2}\mathbf{v} + 3\mathbf{w}$

In Exercises 13–16 find the algebraic vector corresponding to $\overrightarrow{P_1P_2}$.

13. $P_1 = (3, 4)$, $P_2 = (-1, 2)$
14. $P_1 = (-4, -2)$, $P_2 = (3, -1)$
15. $P_1 = (0, 4)$, $P_2 = (-3, 0)$
16. $P_1 = (7, -3)$, $P_2 = (-1, -8)$

In Exercises 17–20 draw the vector $\overrightarrow{P_1P_2}$ *that corresponds to the given algebraic vector and the given initial point* P_1. *Determine the coordinates of* P_2.

17. $\langle 3, -2\rangle$; $P_1 = (0, 0)$
18. $\langle -2, 4\rangle$; $P_1 = (1, 2)$
19. $\langle 0, 3\rangle$; $P_1 = (-2, -3)$
20. $\langle -3, -4\rangle$; $P_1 = (4, 2)$

In Exercises 21–26 find $|\mathbf{v}|$.

21. $\mathbf{v} = \langle 3, 4\rangle$
22. $\mathbf{v} = \langle -8, 6\rangle$
23. $\mathbf{v} = \langle 8, 15\rangle$
24. $\mathbf{v} = \langle -12, -5\rangle$
25. $\mathbf{v} = \langle 2, 1\rangle$
26. $\mathbf{v} = \langle 4, -6\rangle$
27. If $\mathbf{u} = \langle -3, 1\rangle$ and $\mathbf{v} = \langle 2, -3\rangle$, find:
(a) $|\mathbf{u} + \mathbf{v}|$ (b) $|\mathbf{u} - \mathbf{v}|$
(c) $|2\mathbf{u} + 3\mathbf{v}|$ (d) $|3\mathbf{u} - 2\mathbf{v}|$
28. In the accompanying figure $\mathbf{F}_1$ and $\mathbf{F}_2$ are forces acting on an object as shown. If $|\mathbf{F}_1| = 80$ lb, $|\mathbf{F}_2| = 60$ lb, $\alpha = 25^\circ$, and $\beta = 115^\circ$, find the magnitude and direction of the resultant both geometrically and using trigonometry. (*Hint:* Use the Pythagorean theorem.)

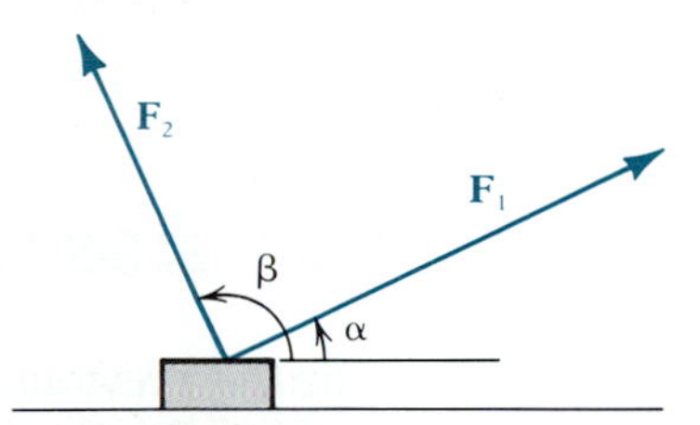

29. Repeat Exercise 28 if $|\mathbf{F}_1| = 20$ N, $|\mathbf{F}_2| = 30$ N, $\alpha = 10^\circ$, and $\beta = 70^\circ$. (*Hint:* For the trigonometric solution use the law of cosines.)
30. In air navigation direction is given by the angle measured clockwise from north (this angle is called the *heading*). If an airplane is flying at an indicated heading of 120° at a speed of 300 mi/hr and a wind of 50 mi/hr is blowing from 210°, find the actual speed and direction of the airplane (relative to the ground). Do this geometrically and also using trigonometry.

31. Repeat Exercise 30 for an airplane flying at an indicated heading of 230° at 260 mi/hr with a 60-mi/hr wind blowing from 110°.

32. Show by a geometric argument that vector addition is associative; that is, $\mathbf{u} + (\mathbf{v} + \mathbf{w}) = (\mathbf{u} + \mathbf{v}) + \mathbf{w}$.

B

33. Show that Definition 14.1 is consistent with the corresponding geometric definitions of equivalence of vectors, addition of vectors, and multiplication of a vector by a scalar.

34. Prove Theorem 14.1.

35. Prove Theorem 14.2.

36. Prove parts 1, 2, and 3 of Theorem 14.3.

37. Let $\mathbf{u} = \langle 1, 2\rangle$, $\mathbf{v} = \langle -2, 1\rangle$, and $\mathbf{w} = \langle 4, 3\rangle$. Find scalars a and b such that $\mathbf{w} = a\mathbf{u} + b\mathbf{v}$.

38. Prove that if $\mathbf{u} = \langle u_1, u_2\rangle$ and $\mathbf{v} = \langle v_1, v_2\rangle$ are nonzero vectors, then they are parallel if and only if $u_1v_2 - u_2v_1 = 0$.

39. Prove that if $\mathbf{u} = \langle u_1, u_2\rangle$ and $\mathbf{v} = \langle v_1, v_2\rangle$ are nonzero vectors with $\mathbf{u} \neq k\mathbf{v}$ for all scalars k, and $\mathbf{w} = \langle w_1, w_2\rangle$ is any other vector, then there exists scalars a and b such that $\mathbf{w} = a\mathbf{u} + b\mathbf{v}$.

40. Prove that if $\mathbf{u} = k\mathbf{v}$, with $k \geq 0$, then

$$|\mathbf{u} + \mathbf{v}| = |\mathbf{u}| + |\mathbf{v}|$$

Give a geometric argument to show that if $\mathbf{u}$ and $\mathbf{v}$ are not related in this way, $|\mathbf{u} + \mathbf{v}| < |\mathbf{u}| + |\mathbf{v}|$.

41. Use vectors to show that the diagonals of a parallelogram bisect each other.

42. Let P_1, P_2, P_3, and P_4 be any four points in the plane. Show both geometrically and algebraically that $\overrightarrow{P_1P_2} + \overrightarrow{P_2P_3} + \overrightarrow{P_3P_4} + \overrightarrow{P_4P_1} = \mathbf{0}$.

43. Use vectors to show that the line segments joining consecutive midpoints of the sides of an arbitrary quadrilateral form a parallelogram.

44. Three forces of magnitude $|\mathbf{F}_1| = 30$ lb, $|\mathbf{F}_2| = 45$ lb, and $|\mathbf{F}_3| = 56$ lb are acting on an object as shown in the figure. Find the magnitude of the resultant and its angle from $\mathbf{F}_1$ both geometrically and using trigonometry. (See the hint for Exercise 29.)

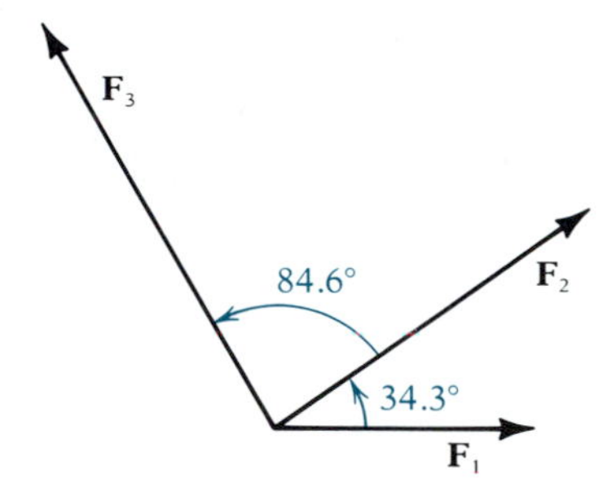

45. A pilot is to fly from A to B, 350 mi due north of A, and then return to A. There is a wind blowing from 310° at 55 mi/hr. If the average air speed of the plane is 210 mi/hr, find the heading (see Exercise 30) the pilot should take on each part of the trip. What will be the total flying time?

14.3 THE DOT PRODUCT AND BASIS VECTORS

In the previous section we considered addition and subtraction of vectors and multiplication by a scalar, but we have not yet considered the product of two vectors. We introduce one type of product now and later consider a second type, applicable to three-dimensional vectors only.

DEFINITION 14.5 Let $\mathbf{u} = \langle u_1, u_2\rangle$ and $\mathbf{v} = \langle v_1, v_2\rangle$ be any two vectors in V_2. Then the **dot product** of $\mathbf{u}$ and $\mathbf{v}$, written $\mathbf{u} \cdot \mathbf{v}$, is defined as

$$\mathbf{u} \cdot \mathbf{v} = u_1v_1 + u_2v_2 \tag{14.2}$$ ■

Remark Observe that the dot product of two vectors is not a vector but is a scalar. For this reason the dot product is sometimes called the **scalar product**.

EXAMPLE 14.2 Let $\mathbf{u} = \langle 3, -2\rangle$ and $\mathbf{v} = \langle -4, -5\rangle$. Find $\mathbf{u} \cdot \mathbf{v}$.

Solution

$$\mathbf{u} \cdot \mathbf{v} = 3(-4) + (-2)(-5) = -12 + 10 = -2$$

■

The dot product of vectors has a number of properties in common with products of real numbers, as the following theorem shows.

THEOREM 14.4 If $\mathbf{u}$, $\mathbf{v}$, and $\mathbf{w}$ are vectors in V_2 and k is a scalar, then

1. $\mathbf{u} \cdot \mathbf{v} = \mathbf{v} \cdot \mathbf{u}$ (commutative law)
2. $\mathbf{u} \cdot (\mathbf{v} + \mathbf{w}) = \mathbf{u} \cdot \mathbf{v} + \mathbf{u} \cdot \mathbf{w}$ (distributive law)
3. $k(\mathbf{u} \cdot \mathbf{v}) = (k\mathbf{u}) \cdot \mathbf{v} = \mathbf{u} \cdot (k\mathbf{v})$
4. $\mathbf{u} \cdot \mathbf{0} = 0$
5. $\mathbf{u} \cdot \mathbf{u} = |\mathbf{u}|^2$

■

We will prove part (2) and leave the other parts for the exercises. Let $\mathbf{u} = \langle u_1, u_2\rangle$, $\mathbf{v} = \langle v_1, v_2\rangle$, and $\mathbf{w} = \langle w_1, w_2\rangle$. We prove equation 2 by calculating each side independently and showing that we get the same result. For the left-hand side we have

$$\begin{aligned}
\mathbf{u} \cdot (\mathbf{v} + \mathbf{w}) - \langle u_1, u_2\rangle \cdot [\langle v_1, v_2\rangle + \langle w_1, w_2\rangle]& \\
= \langle u_1, u_2\rangle \cdot \langle v_1 + w_1, v_2 + w_2\rangle &\quad \text{By Definition 14.1} \\
= u_1(v_1 + w_1) + u_2(v_2 + w_2) &\quad \text{By Definition 14.5} \\
= (u_1v_1 + u_1w_1) + (u_2v_2 + u_2w_2) &\quad \text{By the distributive law of real numbers}
\end{aligned}$$

For the right-hand side we have

$$\begin{aligned}
\mathbf{u} \cdot \mathbf{v} + \mathbf{u} \cdot \mathbf{w} = \langle u_1, u_2\rangle \cdot \langle v_1, v_2\rangle + \langle u_1, u_2\rangle \cdot \langle w_1, w_2\rangle& \\
= (u_1v_1 + u_1v_2) + (u_1w_1 + u_2w_2) &\quad \text{By Definition 14.5} \\
= (u_1v_1 + u_1w_1) + (u_2v_2 + u_2w_2) &\quad \text{By commutativity and associativity of real numbers}
\end{aligned}$$

The equality of the results proves equation 2.

An important property of the dot product has to do with the angle between two vectors, which we now define.

DEFINITION 14.6 The **angle** between two nonzero vectors $\mathbf{u}$ and $\mathbf{v}$ in V_2 is the smallest positive angle between geometric representatives of $\mathbf{u}$ and $\mathbf{v}$ that have the same initial point.

■

If we denote the angle between $\mathbf{u}$ and $\mathbf{v}$ by θ, it follows that $0 \leq \theta \leq \pi$. The angle is 0 if $\mathbf{u}$ and $\mathbf{v}$ are in the same direction, and it is π if they are in opposite directions. Figure 14.12 illustrates various possibilities for θ.

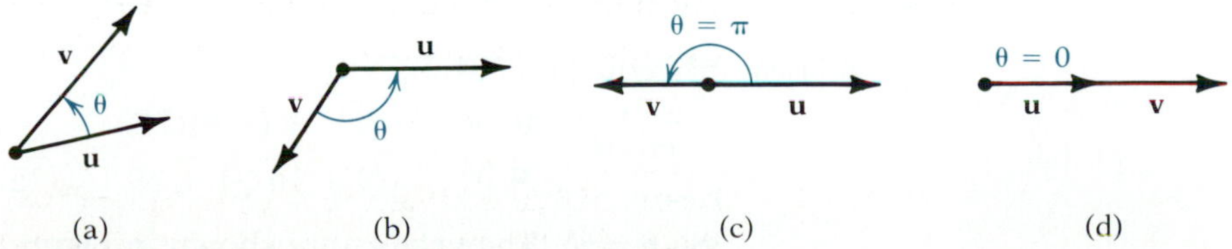

(a) (b) (c) (d)

FIGURE 14.12

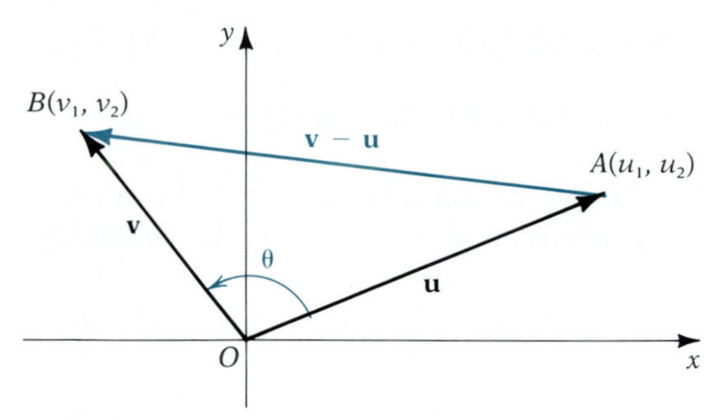

FIGURE 14.13

THEOREM 14.5 If θ is the angle between nonzero vectors **u** and **v**, then

$$\cos\theta = \frac{\mathbf{u}\cdot\mathbf{v}}{|\mathbf{u}|\,|\mathbf{v}|} \tag{14.3}$$

Proof Assume first that **u** and **v** are not parallel, and choose geometric representatives of them such that each has its initial point at the origin, as in Figure 14.13. If **u** and **v** have components given by $\langle u_1, u_2\rangle$ and $\langle v_1, v_2\rangle$, respectively, then it follows that the terminal point A of **u** is (u_1, u_2) and the terminal point B of **v** is (v_1, v_2). The vector $\overrightarrow{AB} = \mathbf{v} - \mathbf{u}$, and so by the law of cosines we have

$$|\mathbf{v}-\mathbf{u}|^2 = |\mathbf{u}|^2 + |\mathbf{v}|^2 - 2|\mathbf{u}|\,|\mathbf{v}|\cos\theta$$

or, in terms of components,

$$(v_1 - u_1)^2 + (v_2 - u_2)^2 = u_1^2 + u_2^2 + v_1^2 + v_2^2 - 2|\mathbf{u}|\,|\mathbf{v}|\cos\theta$$

so that

$$\begin{aligned}|\mathbf{u}|\,|\mathbf{v}|\cos\theta &= u_1v_1 + u_2v_2\\ &= \mathbf{u}\cdot\mathbf{v}\end{aligned}$$

The conclusion now follows.

If **u** and **v** are parallel, then $\mathbf{v} = k\mathbf{u}$, and $\theta = 0$ or $\theta = \pi$, according to whether $k > 0$ or $k < 0$.

For $k > 0$, we have $\cos\theta = \cos 0 = 1$, and

$$\frac{\mathbf{u}\cdot\mathbf{v}}{|\mathbf{u}|\,|\mathbf{v}|} = \frac{\mathbf{u}\cdot(k\mathbf{u})}{|\mathbf{u}|\,|k\mathbf{u}|} = \frac{k(\mathbf{u}\cdot\mathbf{u})}{|\mathbf{u}|\,k|\mathbf{u}|} = \frac{|\mathbf{u}|^2}{|\mathbf{u}|^2} = 1$$

so the result is true in this case. For $k < 0$, $\cos\theta = \cos\pi = -1$, and

$$\frac{\mathbf{u}\cdot\mathbf{v}}{|\mathbf{u}|\,|\mathbf{v}|} = \frac{\mathbf{u}\cdot(k\mathbf{u})}{|\mathbf{u}|\,|k\mathbf{u}|} = \frac{k(\mathbf{u}\cdot\mathbf{u})}{|k|\,|\mathbf{u}|\,|\mathbf{u}|} = \frac{k|\mathbf{u}|^2}{-k|\mathbf{u}|^2} = -1$$

Thus, the result is true in all cases. This completes the proof. ■

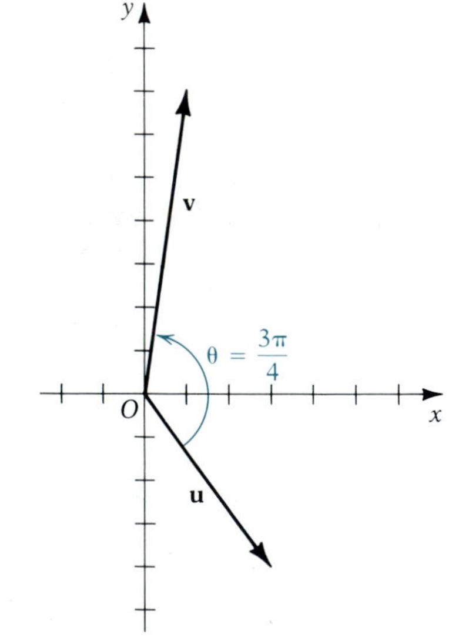

FIGURE 14.14

EXAMPLE 14.3 Find the angle between the vectors $\mathbf{u} = \langle 3, -4\rangle$ and $\mathbf{v} = \langle 1, 7\rangle$. Draw the vectors with initial point at the origin.

Solution By equation (14.3),

$$\cos\theta = \frac{\mathbf{u}\cdot\mathbf{v}}{|\mathbf{u}|\,|\mathbf{v}|} = \frac{3(1)+(-4)(7)}{\sqrt{9+16}\sqrt{1+49}} = \frac{-25}{\sqrt{25}\sqrt{50}} = \frac{-25}{5(5\sqrt{2})} = -\frac{1}{\sqrt{2}}$$

So $\theta = \frac{3\pi}{4}$. The vectors are shown in Figure 14.14. ■

The following corollary is an immediate consequence of equation (14.3).

COROLLARY 14.5a THE CAUCHY-SCHWARZ INEQUALITY

For any two vectors **u** and **v** in V_2,

$$|\mathbf{u} \cdot \mathbf{v}| \leq |\mathbf{u}|\,|\mathbf{v}| \tag{14.4}$$

Proof The result is trivial if either **u** or **v** is the zero vector, since both sides of equation (14.4) are 0. For nonzero vectors **u** and **v**, with angle θ between them, we have by equation (14.3)

$$\frac{|\mathbf{u} \cdot \mathbf{v}|}{|\mathbf{u}|\,|\mathbf{v}|} = |\cos\theta| \leq 1$$

Thus, $|\mathbf{u} \cdot \mathbf{v}| \leq |\mathbf{u}|\,|\mathbf{v}|$. ■

This corollary enables us to give an algebraic proof of the triangle inequality, as follows.

$$\begin{aligned}|\mathbf{u} + \mathbf{v}|^2 &= (\mathbf{u} + \mathbf{v}) \cdot (\mathbf{u} + \mathbf{v}) \\ &= \mathbf{u} \cdot \mathbf{u} + 2\mathbf{u} \cdot \mathbf{v} + \mathbf{v} \cdot \mathbf{v} = |\mathbf{u}|^2 + 2\mathbf{u} \cdot \mathbf{v} + |\mathbf{v}|^2 \\ &\leq |\mathbf{u}|^2 + 2|\mathbf{u} \cdot \mathbf{v}| + |\mathbf{v}|^2 \leq |\mathbf{u}|^2 + 2\,|\mathbf{u}|\,|\mathbf{v}| + |\mathbf{v}|^2 \\ &= (|\mathbf{u}| + |\mathbf{v}|)^2\end{aligned}$$

By the Cauchy-Schwarz inequality

Now we take square roots to get the result:

$$|\mathbf{u} + \mathbf{v}| \leq |\mathbf{u}| + |\mathbf{v}|$$

If the angle between two nonzero vectors is $\frac{\pi}{2}$, the vectors are said to be **orthogonal.** So geometric representatives of orthogonal vectors are perpendicular to each other. Since $\cos\frac{\pi}{2} = 0$, it follows from Theorem 14.5 that if **u** and **v** are orthogonal, then $\mathbf{u} \cdot \mathbf{v} = 0$. Conversely, if $\mathbf{u} \cdot \mathbf{v} = 0$, then $\cos\theta = 0$ by equation (14.3), and so $\theta = \frac{\pi}{2}$. Thus **u** and **v** are orthogonal. If either **u** or **v** is the zero vector, then $\mathbf{u} \cdot \mathbf{v} = 0$, and it is convenient in this case too to call **u** and **v** orthogonal; that is, we agree to say that **0** *is orthogonal to every vector.* We therefore have the following additional corollary to Theorem 14.5.

COROLLARY 14.5b Two vectors **u** and **v** in V_2 are orthogonal if and only if $\mathbf{u} \cdot \mathbf{v} = 0$. ■

EXAMPLE 14.4 Show that the vectors $\mathbf{u} = \langle 6, -4 \rangle$ and $\mathbf{v} = \langle -2, -3 \rangle$ are orthogonal.

Solution

$$\mathbf{u} \cdot \mathbf{v} = 6(-2) + (-4)(-3) = -12 + 12 = 0$$

So by Corollary 14.5b, **u** and **v** are orthogonal. ■

EXAMPLE 14.5 Find x so that the vectors $\langle x, 2 \rangle$ and $\langle 1 - x, 3 \rangle$ are orthogonal.

Solution Let $\mathbf{u} = \langle x, 2 \rangle$ and $\mathbf{v} = \langle 1 - x, 3 \rangle$. Then

$$\begin{aligned}\mathbf{u} \cdot \mathbf{v} = \langle x, 2 \rangle \cdot \langle 1 - x, 3 \rangle = x - x^2 + 6 &= 6 + x - x^2 \\ &= (3 - x)(2 + x)\end{aligned}$$

and so $\mathbf{u} \cdot \mathbf{v} = 0$ if $x = 3$ or $x = -2$. Either value of x causes **u** and **v** to be orthogonal. For $x = 3$ we get $\mathbf{u} = \langle 3, 2 \rangle$ and $\mathbf{v} = \langle -2, 3 \rangle$, and for $x = -2$ we get $\mathbf{u} = \langle -2, 2 \rangle$ and $\mathbf{v} = \langle 3, 3 \rangle$. ■

A **unit vector** is a vector with a magnitude of 1. It is always possible to find a unit vector in the direction of a given nonzero vector **v** by dividing **v** by its own magnitude, since

$$\left|\frac{\mathbf{v}}{|\mathbf{v}|}\right| = \frac{|\mathbf{v}|}{|\mathbf{v}|} = 1$$

EXAMPLE 14.6 Find a unit vector in the direction from $P_1(3, 2)$ toward $P_2(5, -2)$.

Solution Let $\mathbf{v} = \overrightarrow{P_1P_2}$. Then $\mathbf{v} = \langle 5-3, -2-2\rangle = \langle 2, -4\rangle$. So the desired unit vector is

$$\frac{\mathbf{v}}{|\mathbf{v}|} = \frac{\langle 2, -4\rangle}{\sqrt{20}} = \frac{\langle 2, -4\rangle}{2\sqrt{5}} = \left\langle \frac{1}{\sqrt{5}}, -\frac{2}{\sqrt{5}}\right\rangle$$ ■

If **v** is a nonzero vector, $\mathbf{v}/|\mathbf{v}|$ is a unit vector, and so for any scalar k, $k\mathbf{v}/|\mathbf{v}|$ is a vector of magnitude $|k|$ that has the same direction as **v** if $k > 0$ and the opposite direction if $k < 0$. This fact enables us to find a vector of a specified length in the direction of a given vector or in the opposite direction. The following example illustrates this.

EXAMPLE 14.7 Find a vector in the same direction as $\langle -1, 2\rangle$ that has magnitude 10.

Solution Let $\mathbf{v} = \langle -1, 2\rangle$. The desired vector is

$$10\frac{\mathbf{v}}{|\mathbf{v}|} = 10\left\langle \frac{-1}{\sqrt{5}}, \frac{2}{\sqrt{5}}\right\rangle = \langle -2\sqrt{5}, 4\sqrt{5}\rangle$$ ■

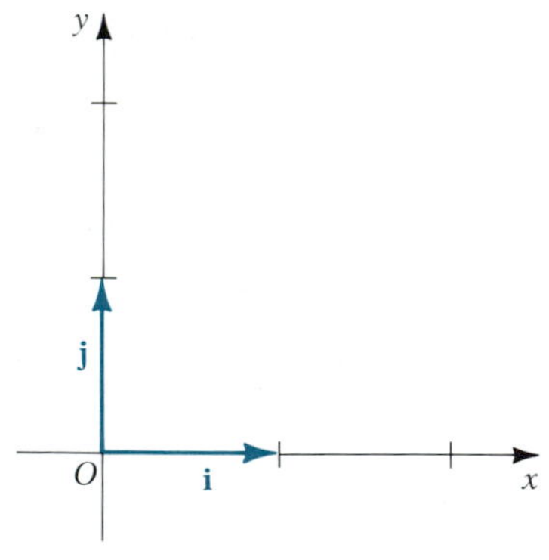

FIGURE 14.15

The two unit vectors $\langle 1, 0\rangle$ and $\langle 0, 1\rangle$ are of particular importance. They are given the special names

$$\mathbf{i} = \langle 1, 0\rangle \qquad \mathbf{j} = \langle 0, 1\rangle$$

So **i** is a unit vector in the positive x direction and **j** is a unit vector in the positive y direction. (See Figure 14.15.) Notice that **i** and **j** are orthogonal, since $\mathbf{i}\cdot\mathbf{j} = 0$.

Suppose $\mathbf{u} = \langle a, b\rangle$ is an arbitrary vector in V_2. Then we can write

$$\mathbf{u} = \langle a, 0\rangle + \langle 0, b\rangle = a\langle 1, 0\rangle + b\langle 0, 1\rangle$$

that is,

$$\langle a, b\rangle = a\mathbf{i} + b\mathbf{j} \tag{14.5}$$

The expression $a\mathbf{i} + b\mathbf{j}$ is called a **linear combination** of **i** and **j**. So every vector in V_2 is uniquely expressible as a linear combination of **i** and **j**. Because of this, the vectors **i** and **j** constitute what is called a **basis** for V_2. As Exercise 39 of Exercise Set 14.2 shows, any two nonzero vectors that are not parallel also form a basis for V_2, but **i** and **j** provide the simplest basis.

EXAMPLE 14.8 Let $P_1 = (7, -4)$ and $P_2 = (-3, 1)$. Express $\overrightarrow{P_1P_2}$ as a linear combination of **i** and **j**.

Solution $\overrightarrow{P_1P_2} = \langle -3-7, 1+4\rangle = \langle -10, 5\rangle = -10\mathbf{i} + 5\mathbf{j}$ ■

Because of the equality (14.5) we can use the representation $a\mathbf{i} + b\mathbf{j}$ as an alternative to $\langle a, b\rangle$. In the future both representations will be used. Thus, for example, we may speak of the vector $2\mathbf{i} - 3\mathbf{j}$, and we will understand this to mean the same as the vector $\langle 2, -3\rangle$. Using this alternative representation, we have, in particular,

$$|a\mathbf{i} + b\mathbf{j}| = \sqrt{a^2 + b^2}$$

and

$$(a\mathbf{i} + b\mathbf{j}) \cdot (c\mathbf{i} + d\mathbf{j}) = ac + bd$$

EXAMPLE 14.9 Find the angle between $2\mathbf{i} - 4\mathbf{j}$ and $\mathbf{i} + \mathbf{j}$.

Solution

$$\cos\theta = \frac{(2\mathbf{i} - 4\mathbf{j}) \cdot (\mathbf{i} + \mathbf{j})}{|2\mathbf{i} - 4\mathbf{j}|\,|\mathbf{i} + \mathbf{j}|} = \frac{2 - 4}{\sqrt{20}\sqrt{2}} = \frac{-2}{2\sqrt{10}} = \frac{-1}{\sqrt{10}}$$

From a calculator we find $\theta \approx 1.65$ radians. ■

Sometimes it is useful to find components of a vector in directions other than horizontal and vertical. To understand this, let $\mathbf{u}$ be any vector and suppose we want the displacement of $\mathbf{u}$ in the direction of some nonzero vector $\mathbf{v}$. Figure 14.16 illustrates this. We use geometric representatives $\overrightarrow{OP}$ and $\overrightarrow{OQ}$ of $\mathbf{u}$ and $\mathbf{v}$, respectively, and designate by P' the foot of the perpendicular from P to the line joining O and Q. Let θ be the angle between $\mathbf{u}$ and $\mathbf{v}$. Then we define the **component of u along v,** designated $\text{Comp}_\mathbf{v}\mathbf{u}$, by

$$\text{Comp}_\mathbf{v}\mathbf{u} = |\mathbf{u}| \cos\theta \tag{14.6}$$

If $0 \le \theta \le \frac{\pi}{2}$, then $\text{Comp}_\mathbf{v}\mathbf{u} \ge 0$, as in Figure 14.16(a), whereas if $\frac{\pi}{2} < \theta \le \pi$, $\text{Comp}_\mathbf{v}\mathbf{u} < 0$, as in Figure 14.16(b). If $\mathbf{u} \ne \mathbf{0}$, $\cos\theta = \mathbf{u} \cdot \mathbf{v}/|\mathbf{u}|\,|\mathbf{v}|$, so that equation (14.6) becomes

$$\text{Comp}_\mathbf{v}\mathbf{u} = \frac{\mathbf{u} \cdot \mathbf{v}}{|\mathbf{v}|} \tag{14.7}$$

This result is valid also for $\mathbf{u} = \mathbf{0}$. It is easy to show (see Exercise 32) that if $\mathbf{u} = \langle a, b\rangle$ and $\mathbf{v}$ is directed along the positive x-axis, then $\text{Comp}_\mathbf{v}\mathbf{u} = a$. Similarly, if $\mathbf{v}$ is directed along the positive y-axis, then $\text{Comp}_\mathbf{v}\mathbf{u} = b$. So our definition generalizes horizontal and vertical components.

FIGURE 14.16

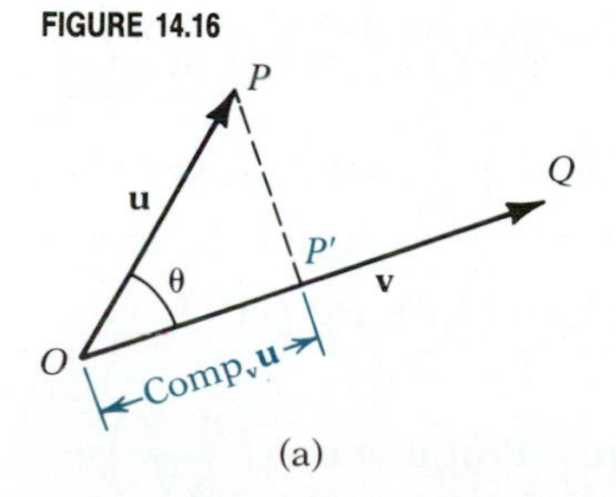

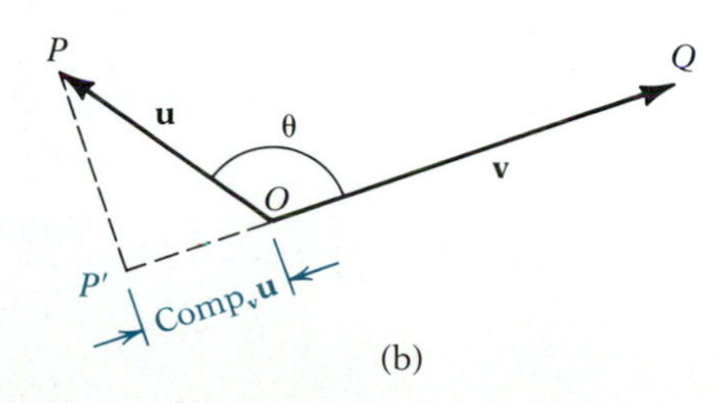

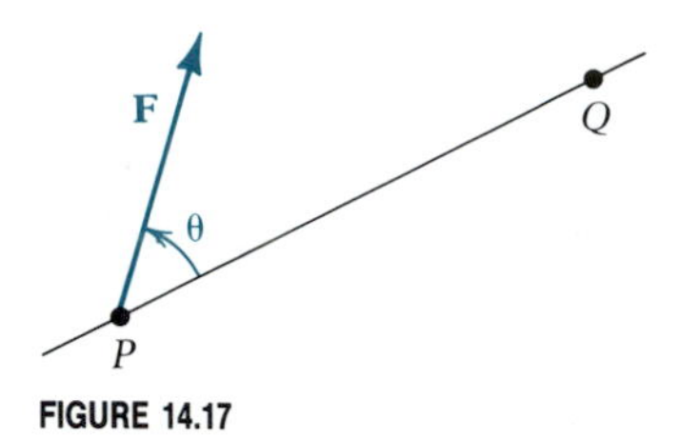

FIGURE 14.17

As an application of this concept, consider the work done by a constant force **F** in moving a particle along a straight line from P to Q. If **F** acts in the direction of $\overrightarrow{PQ}$, then we know that work $= |\mathbf{F}|\,|\overrightarrow{PQ}|$ (force × distance). Suppose, however, that **F** acts at some fixed angle θ with $\overrightarrow{PQ}$ as in Figure 14.17. Then it is natural to define work as the component of **F** along $\overrightarrow{PQ}$ times the distance; that is,

$$W = (\text{Comp}_{\overrightarrow{PQ}}\mathbf{F})|\overrightarrow{PQ}|$$

Substituting from equation (14.7), this becomes simply

$$W = \mathbf{F} \cdot \overrightarrow{PQ} \tag{14.8}$$

EXAMPLE 14.10 The force $\mathbf{F} = 3\mathbf{i} + 5\mathbf{j}$ moves an object along the line segment from $(1, -2)$ to $(3, 5)$. If the magnitude of **F** is in pounds and distance is measured in feet, find the work done by **F**.

Solution Let $P = (1, -2)$ and $Q = (3, 5)$. Then $\overrightarrow{PQ} = 2\mathbf{i} + 7\mathbf{j}$, and

$$\begin{aligned} W = \mathbf{F} \cdot \overrightarrow{PQ} &= (3\mathbf{i} + 5\mathbf{j}) \cdot (2\mathbf{i} + 7\mathbf{j}) \\ &= 6 + 35 = 41 \text{ ft-lb} \end{aligned}$$

■

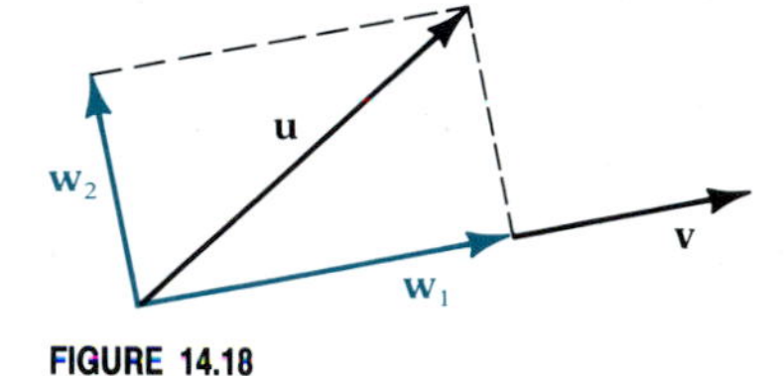

FIGURE 14.18

When we write a vector $\mathbf{u} = \langle a, b \rangle$ in the form $u = a\mathbf{i} + b\mathbf{j}$, we are in effect expressing u as the sum of two mutually perpendicular vectors, one acting horizontally and the other vertically. Sometimes it is desirable to express **u** as the sum of two mutually perpendicular vectors, one in a prescribed nonhorizontal direction and the other perpendicular to this direction. This is easily done geometrically, as Figure 14.18 shows. There we show the given vector **u** and a direction as determined by the vector **v**. We construct a rectangle with sides $\mathbf{w}_1$ and $\mathbf{w}_2$ having **u** as a diagonal and one side along **v**. Then $\mathbf{u} = \mathbf{w}_1 + \mathbf{w}_2$, as required. We call the vector $\mathbf{w}_1$ the **projection of u on v** and designate it by $\text{Proj}_{\mathbf{v}}\mathbf{u}$. The vector $\mathbf{w}_2$ is called the **projection of u orthogonal to v** and is designated by $\text{Proj}_{\mathbf{v}}^{\perp}\mathbf{u}$. So we always have, for any vector **u** and $\mathbf{v} \neq 0$,

$$\mathbf{u} = \text{Proj}_{\mathbf{v}}\mathbf{u} + \text{Proj}_{\mathbf{v}}^{\perp}\mathbf{u} \tag{14.9}$$

To find algebraic representations of these projections, observe that $\mathbf{w}_1$ can be obtained by multiplying the component of **u** along **v** by a unit vector in the direction of **v**. Thus,

$$\text{Proj}_{\mathbf{v}}\mathbf{u} = (\text{Comp}_{\mathbf{v}}\mathbf{u})\,\frac{\mathbf{v}}{|\mathbf{v}|} \tag{14.10}$$

If we replace $\text{Comp}_{\mathbf{v}}\mathbf{u}$ by its value from equation (14.7), we obtain

$$\text{Proj}_{\mathbf{v}}\mathbf{u} = \left(\frac{\mathbf{u} \cdot \mathbf{v}}{|\mathbf{v}|}\right)\frac{\mathbf{v}}{|\mathbf{v}|} = \left(\frac{\mathbf{u} \cdot \mathbf{v}}{|\mathbf{v}|^2}\right)\mathbf{v} \tag{14.11}$$

and from equation (14.8), we get

$$\text{Proj}_{\mathbf{v}}^{\perp}\mathbf{u} = \mathbf{u} - \text{Proj}_{\mathbf{v}}\mathbf{u} = \mathbf{u} - \left(\frac{\mathbf{u} \cdot \mathbf{v}}{|\mathbf{v}|^2}\right)\mathbf{v} \tag{14.12}$$

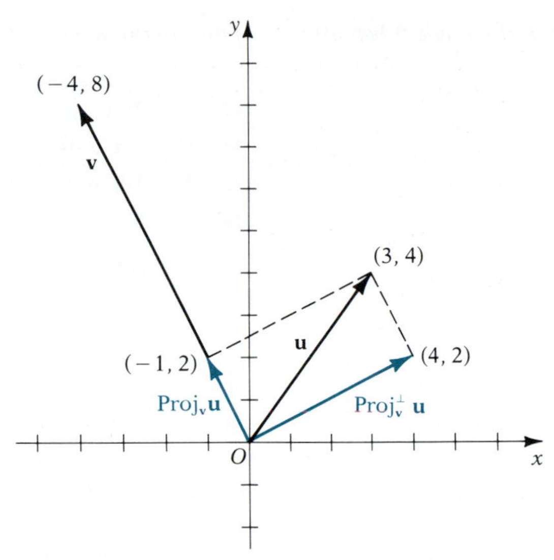

FIGURE 14.19

EXAMPLE 14.11 Let $\mathbf{u} = \langle 3, 4\rangle$ and $\mathbf{v} = \langle -4, 8\rangle$. Express $\mathbf{u}$ as the sum of two vectors, one parallel to $\mathbf{v}$ and the other perpendicular to $\mathbf{v}$. Show the results geometrically.

Solution Note that the desired vectors are $\text{Proj}_{\mathbf{v}}\mathbf{u}$ and $\text{Proj}_{\mathbf{v}}^{\perp}\mathbf{u}$. By equation (14.11),

$$\begin{aligned}\text{Proj}_{\mathbf{v}}\mathbf{u} &= \left(\frac{\mathbf{u}\cdot\mathbf{v}}{|\mathbf{v}|^2}\right)\mathbf{v} = \frac{\langle 3, 4\rangle\cdot\langle -4, 8\rangle}{|\langle -4, 8\rangle|^2}\langle -4, 8\rangle \\ &= \frac{-12+32}{80}\langle -4, 8\rangle = \frac{1}{4}\langle -4, 8\rangle \\ &= \langle -1, 2\rangle\end{aligned}$$

By equation (14.12),

$$\text{Proj}_{\mathbf{v}}^{\perp}\mathbf{u} = \mathbf{u} - \text{Proj}_{\mathbf{v}}\mathbf{u} = \langle 3, 4\rangle - \langle -1, 2\rangle = \langle 4, 2\rangle$$

The results are illustrated in Figure 14.19. ■

Note $\text{Proj}_{\mathbf{v}}\mathbf{u}$ and $\text{Proj}_{\mathbf{v}}^{\perp}\mathbf{u}$ are sometimes called **vector components** of $\mathbf{u}$ in the direction of $\mathbf{v}$ and orthogonal to $\mathbf{v}$, respectively, and when these are found, the vector $\mathbf{u}$ is said to have been *resolved into vector components* in these directions.

EXERCISE SET 14.3

A

In Exercises 1–4 find the dot product of **u** *and* **v**.

1. $\mathbf{u} = \langle 4, 7\rangle$, $\mathbf{v} = \langle -5, 2\rangle$
2. $\mathbf{u} = \langle -3, -6\rangle$, $\mathbf{v} = \langle 5, -2\rangle$
3. $\mathbf{u} = 2\mathbf{i} - 3\mathbf{j}$, $\mathbf{v} = \mathbf{i} + 2\mathbf{j}$
4. $\mathbf{u} = -\mathbf{i} + 4\mathbf{j}$, $\mathbf{v} = 3\mathbf{i} + \mathbf{j}$

In Exercises 5–8 show that **u** *and* **v** *are orthogonal.*

5. $\mathbf{u} = \langle 4, -8\rangle$, $\mathbf{v} = \langle -4, -2\rangle$
6. $\mathbf{u} = \langle -10, 6\rangle$, $\mathbf{v} = \langle 12, 20\rangle$
7. $\mathbf{u} = 2\mathbf{i} - 3\mathbf{j}$, $\mathbf{v} = 9\mathbf{i} + 6\mathbf{j}$
8. $\mathbf{u} = -3\mathbf{i} + 4\mathbf{j}$, $\mathbf{v} = -12\mathbf{i} - 9\mathbf{j}$

In Exercises 9–15 find the cosine of the angle θ between **u** *and* **v**.

9. $\mathbf{u} = \langle 4, -4 \rangle$, $\mathbf{v} = \langle 1, 7 \rangle$

10. $\mathbf{u} = \langle 1, -2 \rangle$, $\mathbf{v} = \langle -1, 1 \rangle$

11. $\mathbf{u} = -\mathbf{i} + 3\mathbf{j}$, $\mathbf{v} = -2\mathbf{i} - \mathbf{j}$

12. $\mathbf{u} = 4\mathbf{i} + 6\mathbf{j}$, $\mathbf{v} = 4\mathbf{i} - 2\mathbf{j}$

13. $\mathbf{u} = \langle 6, 8 \rangle$, $\mathbf{v} = \langle -3, 4 \rangle$

14. $\mathbf{u} = \langle 4, 8 \rangle$, $\mathbf{v} = \langle -1, 3 \rangle$; also find θ.

15. $\mathbf{u} = \mathbf{i} + \sqrt{3}\mathbf{j}$, $\mathbf{v} = 2\mathbf{i}$; also find θ.

In Exercises 16–19 find a unit vector in the direction of **v**.

16. $\mathbf{v} = 3\mathbf{i} - 4\mathbf{j}$ **17.** $\mathbf{v} = 5\mathbf{i} + 12\mathbf{j}$

18. $\mathbf{v} = \langle -4, 8 \rangle$ **19.** $\mathbf{v} = \langle -2, -3 \rangle$

In Exercises 20–25 find a vector **w** *that is in the direction of the given vector* **v**, *with the specified magnitude.*

20. $\mathbf{v} = \langle -4, 3 \rangle$; $|\mathbf{w}| = 10$

21. $\mathbf{v} = \langle 2, -4 \rangle$; $|\mathbf{w}| = 10$

22. $\mathbf{v} = \mathbf{i} + \mathbf{j}$; $|\mathbf{w}| = 2$

23. $\mathbf{v} = 6\mathbf{i} + 8\mathbf{j}$; $|\mathbf{w}| = 4$

24. $\mathbf{v} = 7\mathbf{i} - 24\mathbf{j}$; $|\mathbf{w}| = 5$

25. $\mathbf{v} = 3\mathbf{i} - 6\mathbf{j}$; $|\mathbf{w}| = 15$

26. Find all values of x so that $\mathbf{u} = \langle 3x, 1 - x \rangle$ and $\mathbf{v} = \langle x, -4 \rangle$ will be orthogonal.

27. Find all values of x so that the angle between $\mathbf{u} = \langle 4, -3 \rangle$ and $\mathbf{v} = \langle x, 1 \rangle$ will be $\frac{\pi}{4}$.

In Exercises 28–31 find $\text{Comp}_{\mathbf{v}}\mathbf{u}$.

28. $\mathbf{u} = \langle 7, -4 \rangle$, $\mathbf{v} = \langle -3, 4 \rangle$

29. $\mathbf{u} = \langle -2, -3 \rangle$, $\mathbf{v} = \langle 1, 1 \rangle$

30. $\mathbf{u} = \mathbf{i} - 2\mathbf{j}$, $\mathbf{v} = 2\mathbf{i} - \mathbf{j}$

31. $\mathbf{u} = 3\mathbf{i} - 4\mathbf{j}$, $\mathbf{v} = \mathbf{i} + 7\mathbf{j}$

32. Show that if $\mathbf{u} = \langle a, b \rangle$ and $\mathbf{v} = k\mathbf{i}$ for $k > 0$, then $\text{Comp}_{\mathbf{v}}\mathbf{u} = a$. Also show that if $\mathbf{v} = k\mathbf{j}$ for $k > 0$, then $\text{Comp}_{\mathbf{v}}\mathbf{u} = b$.

In Exercises 33–36 find the work done by the force **F** *acting on a particle along a line segment from the first point to the second. Assume* $|\mathbf{F}|$ *is in pounds and distance is in feet.*

33. $\mathbf{F} = 2\mathbf{i} + 3\mathbf{j}$; $(1, 2)$ to $(6, 8)$

34. $\mathbf{F} = -\mathbf{i} + 4\mathbf{j}$; $(-2, 3)$ to $(3, 5)$

35. $\mathbf{F} = 10\mathbf{i} + 20\mathbf{j}$; $(2, 3)$ to $(1, 5)$

36. $\mathbf{F} = 5\mathbf{i} - 7\mathbf{j}$; $(-4, -1)$ to $(6, -6)$

In Exercises 37–40 find $\text{Proj}_{\mathbf{v}}\mathbf{u}$ *and* $\text{Proj}_{\mathbf{v}}^{\perp}\mathbf{u}$.

37. $\mathbf{u} = \langle 3, -2 \rangle$, $\mathbf{v} = \langle 2, 4 \rangle$

38. $\mathbf{u} = \langle -2, -1 \rangle$, $\mathbf{v} = \langle -3, 4 \rangle$

39. $\mathbf{u} = 6\mathbf{i} + 2\mathbf{j}$, $\mathbf{v} = 3\mathbf{i} - 4\mathbf{j}$

40. $\mathbf{u} = -2\mathbf{i} + 3\mathbf{j}$, $\mathbf{v} = 7\mathbf{i} + \mathbf{j}$

41. A block that weighs 1000 lb is being held in place on an inclined plane that makes a 30° angle with the horizontal by a person pulling on a rope attached to the block and passing over a pulley as shown in the figure. Assuming no friction, what is the magnitude of the force **F** that must be exerted?

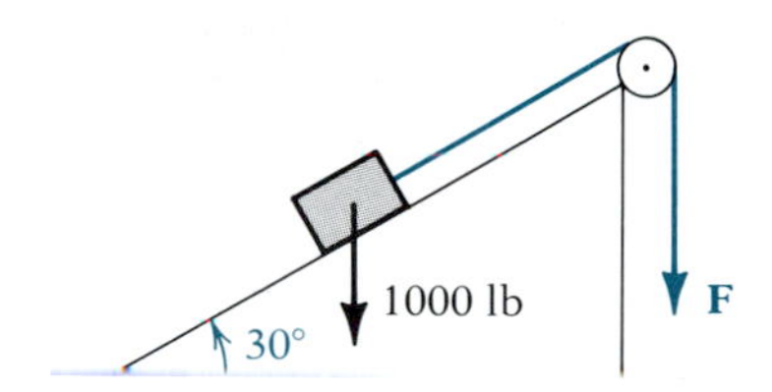

42. Resolve the vector $\mathbf{w} = 6\mathbf{i} - 4\mathbf{j}$ into vector components parallel and perpendicular, respectively, to the line that joins $(-1, -2)$ and $(2, 2)$.

43. Use vector methods to show that the points $(2, 1)$, $(6, 9)$, and $(-2, 3)$ are vertices of a right triangle. What is its area?

44. Use vector methods to show that the points $(3, -1)$, $(5, 4)$, $(-5, 8)$, and $(-7, 3)$ are vertices of a rectangle. What is its area?

45. Use vector methods to show that the points $(-5, 2)$, $(-3, -2)$, $(6, 1)$, and $(4, 5)$ are vertices of a parallelogram. Find the interior angles.

B

Prove the identities in Exercises 46 and 47.

46. (a) $(\mathbf{u} + \mathbf{v}) \cdot (\mathbf{u} - \mathbf{v}) = |\mathbf{u}|^2 - |\mathbf{v}|^2$
(b) $(\mathbf{u} + \mathbf{v}) \cdot (\mathbf{u} + \mathbf{v}) = |\mathbf{u}|^2 + 2\mathbf{u} \cdot \mathbf{v} + |\mathbf{v}|^2$

47. (a) $|\mathbf{u} + \mathbf{v}|^2 + |\mathbf{u} - \mathbf{v}|^2 = 2(|\mathbf{u}|^2 + |\mathbf{v}|^2)$
(b) $|\mathbf{u} + \mathbf{v}|^2 - |\mathbf{u} - \mathbf{v}|^2 = 4\mathbf{u} \cdot \mathbf{v}$

48. Prove that **u** and **v** are orthogonal if and only if

$$|\mathbf{u} + \mathbf{v}| = |\mathbf{u} - \mathbf{v}|$$

49. Give an algebraic proof that $\text{Proj}_{\mathbf{v}}^{\perp}\mathbf{u}$ is orthogonal to $\text{Proj}_{\mathbf{v}}\mathbf{u}$.

50. Prove that the vector $\mathbf{n} = a\mathbf{i} + b\mathbf{j}$ is perpendicular to the line $ax + by + c = 0$. (*Hint:* Consider two points on the line.)

51. Let $P_0(x_0, y_0)$ be any point on the line $ax + by + c = 0$ and let $P_1(x_1, y_1)$ be any point not on the line. Show that the distance d from the line to the point P_1 is $d = |\text{Comp}_{\mathbf{n}}\overrightarrow{P_0P_1}|$, where $\mathbf{n} = a\mathbf{i} + b\mathbf{j}$. From this, verify the distance formula

$$d = \frac{|ax_1 + by_1 + c|}{\sqrt{a^2 + b^2}}$$

(*Hint:* Use the result of Exercise 50.)

52. Use vector methods to prove that any triangle inscribed in a semicircle, with one side coinciding with the diameter, is a right triangle. (*Hint:* In the figure, find $\overrightarrow{AB}$ and $\overrightarrow{BC}$ in terms of $\mathbf{u}$ and $\mathbf{v}$, and use the fact that $|\mathbf{u}| = |\mathbf{v}|$ together with the result of Exercise 46a.)

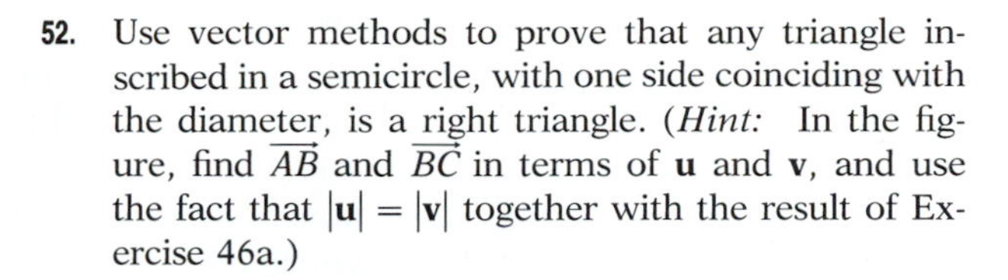

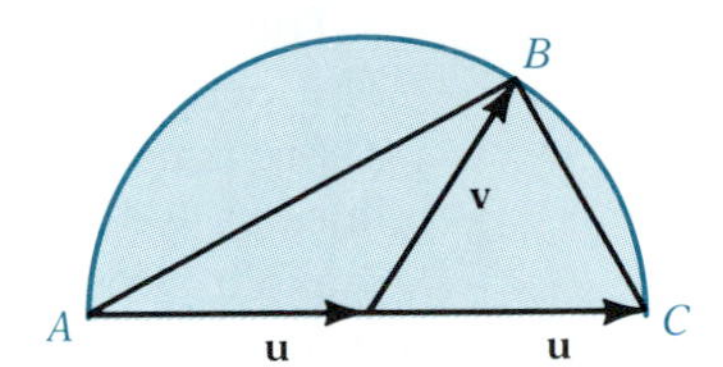

14.4 VECTORS IN SPACE

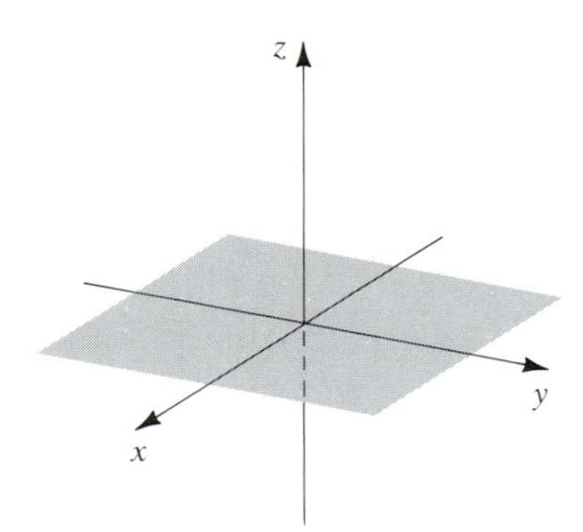

FIGURE 14.20

In this section we extend the vector concept to three-dimensional space. It is necessary first to introduce a rectangular coordinate system. To do this we begin with a horizontal plane that has a two-dimensional rectangular coordinate system with x- and y-axes in their usual orientation, the positive y-axis being 90° counterclockwise from the positive x-axis. Through the origin we introduce a vertical axis, called the z-axis, coordinatized to the same scale as the x- and y-axes, directed positively upward, with its origin coinciding with that of x and y. This is illustrated in Figure 14.20. We now have three mutually perpendicular axes oriented according to what is called the *right-hand rule*: if you point the index finger of your right hand in the positive x direction and the middle finger in the positive y direction, as in Figure 14.21, then the thumb will point in the positive z direction.

Each pair of axes determines a plane. We call these the **xy-plane,** the **xz-plane,** and the **yz-plane.** Frequently we will refer to the xy-plane as the **horizontal plane.** These three planes are the **coordinate planes.** Now let P denote any point in space. Through P pass planes parallel to each of the coordinate planes. If these cut the x-axis, y-axis, and z-axis at x_0, y_0, and z_0, respectively, then these three numbers are called the *coordinates* of P, and we write them as the ordered triple (x_0, y_0, z_0). This is illustrated in Figure 14.22. Reversing this, if we begin with the ordered triple (x_0, y_0, z_0), we locate P by proceeding x_0 units from the origin along the x-axis, then y_0 units

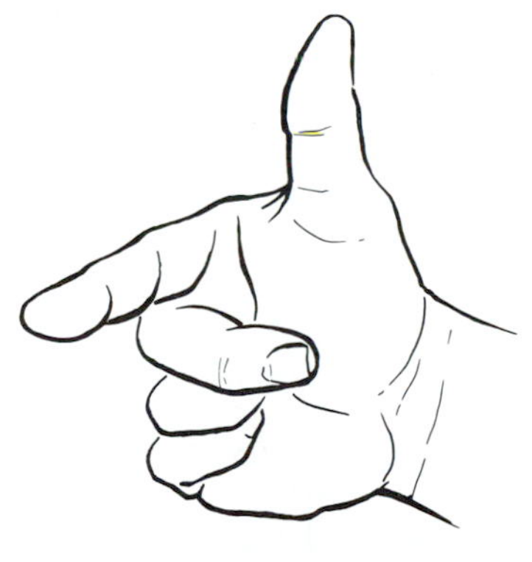

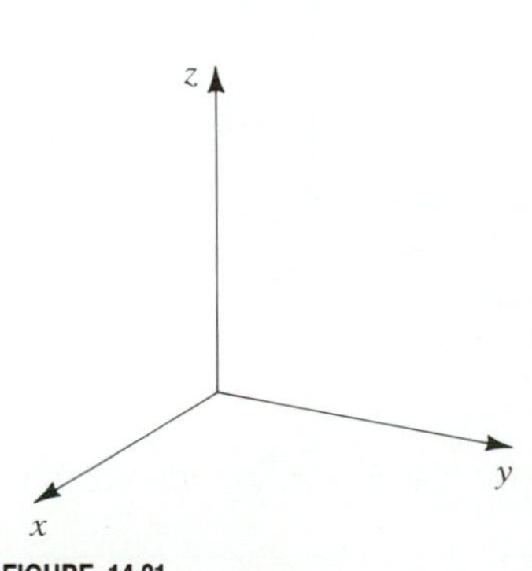

FIGURE 14.21

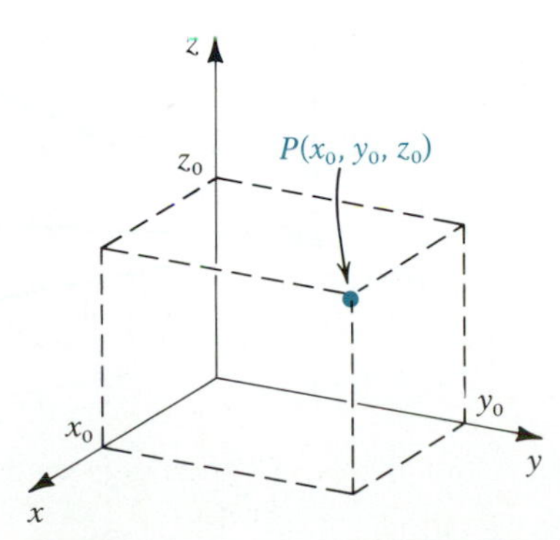

FIGURE 14.22

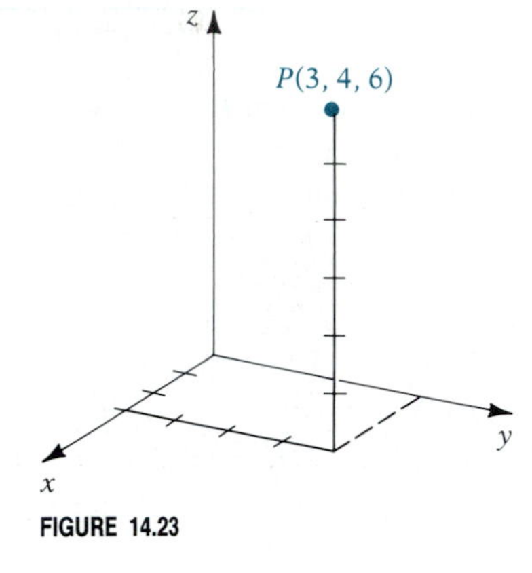

FIGURE 14.23

parallel to the y-axis, and then z_0 units parallel to the z-axis, in each case using directed distances. In this way we establish a one-to-one correspondence between all points in three-dimensional space and all ordered triples of real numbers. We often will not distinguish between a point and its coordinates, saying, for example, "the point (2, 3, −4)" rather than "the point whose coordinates are (2, 3, −4)."

The three coordinate planes divide space into eight regions, called **octants.** The octant in which all coordinates are positive is called the first octant. There is no need to number the others. In plotting points it is useful to show lines as we have done in plotting $P(3, 4, 6)$ in Figure 14.23. These help to make it appear that P is not in the plane of the paper. In this case we have shown the positive axes only, since the point is in the first octant.

To determine a formula for the length of a vector, we need to know the distance between two points in space. Let $P_1(x_1, y_1, z_1)$ and $P_2(x_2, y_2, z_2)$ be any two such points. Construct a rectangular box with sides parallel to the coordinate planes so that P_1 and P_2 are at opposite corners of the box, as in Figure 14.24. With vertices Q and R as shown, triangle P_1QR is a right triangle in a horizontal plane, and triangle P_1RP_2 is a right triangle in a vertical plane. Using $d(P_1, P_2)$ to mean the distance from P_1 to P_2, we have from the first triangle,

$$[d(P_1, R)]^2 = [d(P_1, Q)]^2 + [d(Q, R)]^2$$

and from the second,

$$[d(P_1, P_2)]^2 = [d(P_1, R)]^2 + [d(R, P_2)]^2$$

So

$$[d(P_1, P_2)]^2 = [d(P_1, Q)]^2 + [d(Q, R)]^2 + [d(R, P_2)]^2$$

But $d(P_1, Q) = |y_2 - y_1|$, $d(Q, R) = |x_2 - x_1|$, and $d(R, P_2) = |z_2 - z_1|$. Making these substitutions, we get the following.

FIGURE 14.24

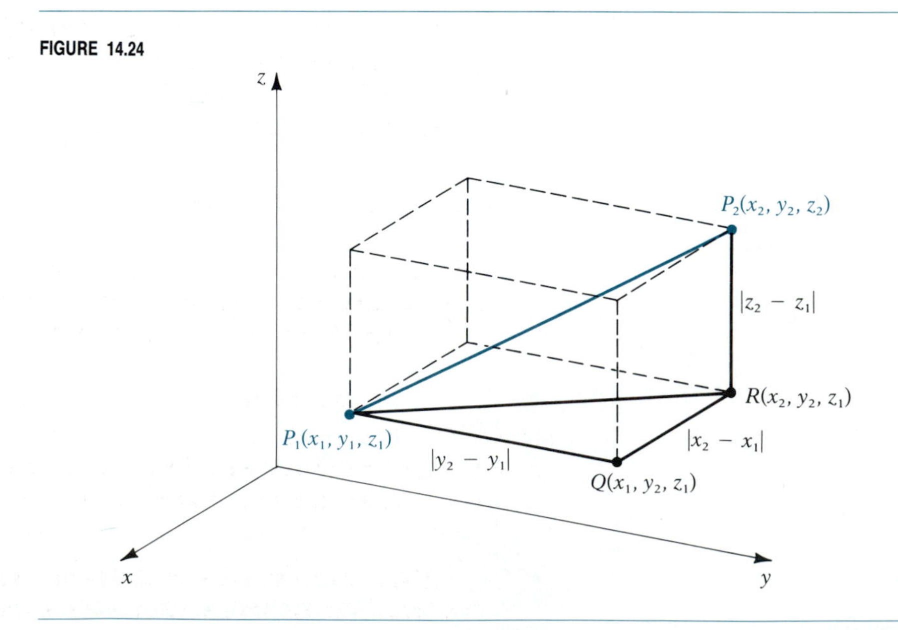

DISTANCE FORMULA IN THREE DIMENSIONS

$$d(P_1, P_2) = \sqrt{(x_2 - x_1)^2 + (y_2 - y_1)^2 + (z_2 - z_1)^2} \tag{14.13}$$

EXAMPLE 14.12 Plot the points $P(3, -4, 5)$ and $Q(-2, 3, 4)$ and find the distance between them.

Solution The points are shown in Figure 14.25. By equation (14.13),

$$\begin{aligned} d(P, Q) &= \sqrt{(3 + 2)^2 + (-4 - 3)^2 + (5 - 4)^2} \\ &= \sqrt{25 + 49 + 1} \\ &= \sqrt{75} \\ &= 5\sqrt{3} \end{aligned}$$

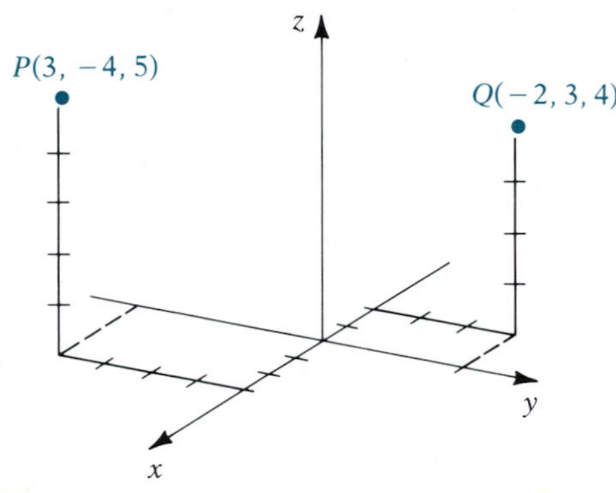

FIGURE 14.25

With this background we are ready to extend vectors to three dimensions. The notion of a geometric vector as a directed line segment is exactly as it was for two dimensions, with operations on vectors done in exactly the same way. Suppose a geometric vector has its initial point at $P_1(x_1, y_1, z_1)$ and its terminal point at $P_2(x_2, y_2, z_2)$. Then, analogous to the two-dimensional case, we identify the ordered triple $\langle x_2 - x_1, y_2 - y_1, z_2 - z_1\rangle$ with the vector $\overrightarrow{P_1P_2}$. Conversely, if we are given an ordered triple $\langle a, b, c\rangle$, we identify with this any geometric vector that has x displacement a, y displacement b, and z displacement c. The simplest such vector is the one with initial point at the origin and terminal point at (a, b, c). In keeping with the geometric definitions of equivalence, addition, and multiplication by a scalar, we define:

1. $\langle a_1, a_2, a_3\rangle = \langle b_1, b_2, b_3\rangle$ if and only if $a_1 = b_1$, $a_2 = b_2$, and $a_3 = b_3$
2. $\langle a_1, a_2, a_3\rangle + \langle b_1, b_2, b_3\rangle = \langle a_1 + b_1, a_2 + b_2, a_3 + b_3\rangle$
3. $k\langle a_1, a_2, a_3\rangle = \langle ka_1, ka_2, ka_3\rangle$ for any scalar k

With these definitions the set of all such ordered triples is called a **vector space of dimension 3,** denoted by V_3, and each element of V_3 is called a **three-dimensional vector.** If $\mathbf{u} = \langle u_1, u_2, u_3\rangle$, then u_1, u_2, and u_3 are called the **components** of $\mathbf{u}$. The negative of $\mathbf{u}$ is $-\mathbf{u} = \langle -u_1, -u_2, -u_3\rangle$, and the zero vector is $\langle 0, 0, 0\rangle$. Subtraction is defined by $\mathbf{u} - \mathbf{v} = \mathbf{u} + (-\mathbf{v})$.

The **magnitude** of a vector $\mathbf{u} = \langle u_1, u_2, u_3\rangle$ is defined by

$$|\mathbf{u}| = \sqrt{u_1^2 + u_2^2 + u_3^2} \tag{14.14}$$

Magnitude is by equation (14.13) the length of a geometric representative of **u**. The **dot product** of two vectors $\mathbf{u} = \langle u_1, u_2, u_3 \rangle$ and $\mathbf{v} = \langle v_1, v_2, v_3 \rangle$ is defined by

$$\mathbf{u} \cdot \mathbf{v} = u_1 v_1 + u_2 v_2 + u_3 v_3 \tag{14.15}$$

The angle θ between two nonzero vectors **u** and **v** is defined as for two-dimensional vectors, and $\cos\theta$ is again given by

$$\cos\theta = \frac{\mathbf{u} \cdot \mathbf{v}}{|\mathbf{u}|\,|\mathbf{v}|} \tag{14.16}$$

The proof is the same as before. With the agreement again that **0** is orthogonal to every vector, we have that **u** *and* **v** *are orthogonal if and only if* $\mathbf{u} \cdot \mathbf{v} = 0$.

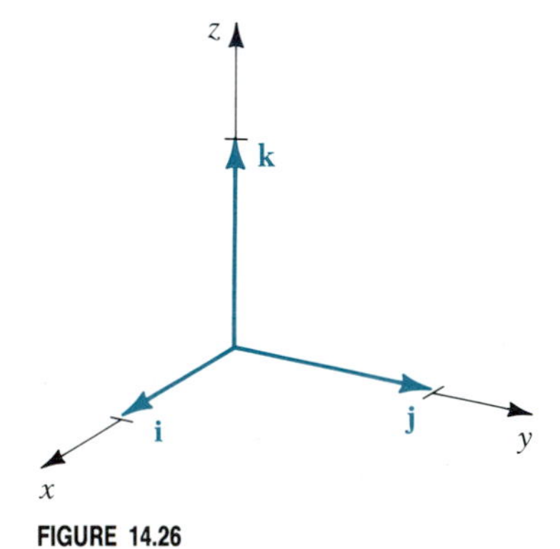

FIGURE 14.26

For any nonzero vector **u**, $\mathbf{u}/|\mathbf{u}|$ is a *unit* vector, since its magnitude is 1. The unit vectors $\mathbf{i} = \langle 1, 0, 0 \rangle$, $\mathbf{j} = \langle 0, 1, 0 \rangle$, and $\mathbf{k} = \langle 0, 0, 1 \rangle$ form a *basis* for V_3, since

$$\langle a, b, c \rangle = a\mathbf{i} + b\mathbf{j} + c\mathbf{k} \tag{14.17}$$

means that every vector in V_3 is expressible as a linear combination of **i**, **j**, and **k**. Note that $\mathbf{i} \cdot \mathbf{j} = \mathbf{i} \cdot \mathbf{k} = \mathbf{j} \cdot \mathbf{k} = 0$, so that **i**, **j**, and **k** are mutually orthogonal. Geometrically, when placed with initial points at the origin, they are unit vectors directed along the positive x-axis, y-axis, and z-axis, respectively, as shown in Figure 14.26.

All other definitions and theorems in Sections 14.2 and 14.3 have natural extensions to vectors in V_3, and we will not repeat them. Proofs of the theorems are in many cases identical in V_3 with proofs in V_2, and at most require obvious modifications. Some of the results are illustrated in the examples that follow.

EXAMPLE 14.13 Let $\mathbf{u} = \langle 1, -2, 2 \rangle$ and $\mathbf{v} = \langle -3, -4, 5 \rangle$. Find:

(a) $|3\mathbf{u} - 2\mathbf{v}|$ (b) The angle between **u** and **v**

Solution (a)
$$\begin{aligned} 3\mathbf{u} - 2\mathbf{v} &= 3\langle 1, -2, 2 \rangle - 2\langle -3, -4, 5 \rangle \\ &= \langle 3, -6, 6 \rangle - \langle -6, -8, 10 \rangle \\ &= \langle 3, -6, 6 \rangle + \langle 6, 8, -10 \rangle \\ &= \langle 9, 2, -4 \rangle \end{aligned}$$

So

$$|3\mathbf{u} - 2\mathbf{v}| = \sqrt{81 + 4 + 16} = \sqrt{101}$$

(b)
$$\begin{aligned} \cos\theta = \frac{\mathbf{u} \cdot \mathbf{v}}{|\mathbf{u}|\,|\mathbf{v}|} &= \frac{\langle 1, -2, 2 \rangle \cdot \langle -3, -4, 5 \rangle}{\sqrt{1 + 4 + 4}\sqrt{9 + 16 + 25}} \\ &= \frac{-3 + 8 + 10}{3\sqrt{50}} = \frac{15}{3(5\sqrt{2})} = \frac{1}{\sqrt{2}} \end{aligned}$$

So $\theta = \frac{\pi}{4}$. ■

EXAMPLE 14.14 Find the work done by the force $\mathbf{F} = 4\mathbf{i} + 5\mathbf{j} - 8\mathbf{k}$ in moving a particle from $P(-1, 2, 4)$ to $Q(3, 6, -8)$. Assume $|\mathbf{F}|$ is in dynes and distance is in centimeters.

Solution $\overrightarrow{PQ} = \langle 4, 4, -12 \rangle = 4\mathbf{i} + 4\mathbf{j} - 12\mathbf{k}$

so by equation (14.8),

$$W = \mathbf{F} \cdot (\overrightarrow{PQ}) = (4\mathbf{i} + 5\mathbf{j} - 8\mathbf{k}) \cdot (4\mathbf{i} + 4\mathbf{j} - 12\mathbf{k})$$
$$= 16 + 20 + 96 = 132 \text{ ergs}$$

■

EXAMPLE 14.15 Find $\text{Comp}_\mathbf{v}\mathbf{u}$ and $\text{Proj}_\mathbf{v}\mathbf{u}$ if $\mathbf{u} = 4\mathbf{i} - 6\mathbf{j} + \mathbf{k}$ and $\mathbf{v} = -3\mathbf{i} - 2\mathbf{j} + 5\mathbf{k}$.

Solution From equation (14.7),

$$\text{Comp}_\mathbf{v}\mathbf{u} = \frac{\mathbf{u} \cdot \mathbf{v}}{|\mathbf{v}|} = \frac{4(-3) + (-6)(-2) + (1)(5)}{\sqrt{9 + 4 + 25}} = \frac{5}{\sqrt{38}}$$

By equation (14.10),

$$\text{Proj}_\mathbf{v}\mathbf{u} = (\text{Comp}_\mathbf{v}\mathbf{u})\frac{\mathbf{v}}{|\mathbf{v}|} = \frac{5}{\sqrt{38}}\,\frac{-3\mathbf{i} - 2\mathbf{j} + 5\mathbf{k}}{\sqrt{38}}$$
$$= -\frac{15}{38}\mathbf{i} - \frac{5}{19}\mathbf{j} + \frac{25}{38}\mathbf{k}$$

■

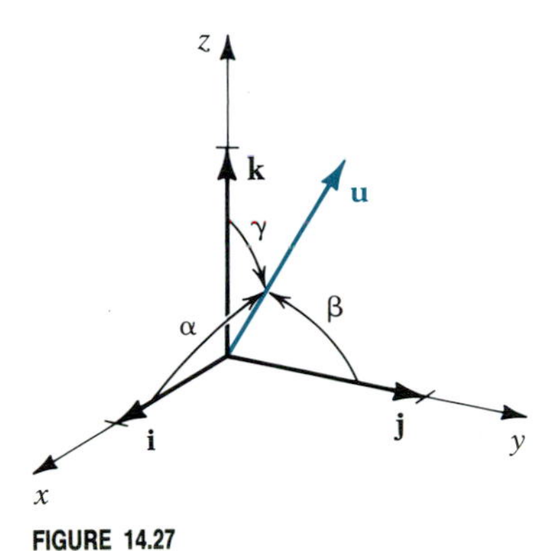

FIGURE 14.27

The angles a nonzero vector **u** makes with **i**, **j**, and **k** are called *direction angles* of **u** and are designated by α, β, and γ, respectively. These are illustrated in Figure 14.27. The cosines of these angles are called *direction cosines* of **u**. If $\mathbf{u} = \langle u_1, u_2, u_3 \rangle$, we have

$$\cos\alpha = \frac{\mathbf{u} \cdot \mathbf{i}}{|\mathbf{u}|\,|\mathbf{i}|} = \frac{u_1}{|\mathbf{u}|} \qquad \cos\beta = \frac{\mathbf{u} \cdot \mathbf{j}}{|\mathbf{u}|\,|\mathbf{j}|} = \frac{u_2}{|\mathbf{u}|} \qquad \cos\gamma = \frac{\mathbf{u} \cdot \mathbf{k}}{|\mathbf{u}|\,|\mathbf{k}|} = \frac{u_3}{|\mathbf{u}|} \tag{14.18}$$

If we square and add, we get

$$\cos^2\alpha + \cos^2\beta + \cos^2\gamma = \frac{u_1^2}{|\mathbf{u}|^2} + \frac{u_2^2}{|\mathbf{u}|^2} + \frac{u_3^3}{|\mathbf{u}|^2}$$

or

$$\cos^2\alpha + \cos^2\beta + \cos^2\gamma = 1 \tag{14.19}$$

If **u** is a unit vector, then by equation (14.18) its components are precisely its direction cosines:

$$\mathbf{u} = \langle \cos\alpha, \cos\beta, \cos\gamma \rangle \quad \text{if } |\mathbf{u}| = 1$$

EXAMPLE 14.16 Find the direction cosines of the vector with initial point $P(7, -2, 4)$ and terminal point $Q(5, 3, 0)$.

Solution Let $\mathbf{u} = \overrightarrow{PQ} = \langle -2, 5, -4 \rangle$. Then $|\mathbf{u}| = \sqrt{4 + 25 + 16} = \sqrt{45} = 3\sqrt{5}$. So by equation (14.18),

$$\cos\alpha = \frac{-2}{3\sqrt{5}} \qquad \cos\beta = \frac{5}{3\sqrt{5}} \qquad \cos\gamma = \frac{-4}{3\sqrt{5}}$$

■

EXAMPLE 14.17 A unit vector **u** makes an angle of 60° with the positive *x*-axis and with the positive *y*-axis. What angle does it make with the positive *z*-axis? What are the components of **u**?

Solution By the given information, $\alpha = \beta = \frac{\pi}{3}$, and by equation (14.19),

$$\cos^2\gamma = 1 - \cos^2\alpha - \cos^2\beta = 1 - \frac{1}{4} - \frac{1}{4} = \frac{1}{2}$$

So $\cos \gamma = \pm 1/\sqrt{2}$. Thus $\gamma = \frac{\pi}{4}$ or $\frac{3\pi}{4}$. There are therefore two possibilities for $\mathbf{u}$:

$$\mathbf{u} = \left\langle \frac{1}{2}, \frac{1}{2}, \frac{1}{\sqrt{2}} \right\rangle \quad \text{or} \quad \mathbf{u} = \left\langle \frac{1}{2}, \frac{1}{2}, -\frac{1}{\sqrt{2}} \right\rangle$$ ■

EXERCISE SET 14.4

A

1. Plot each of the following points.
   (a) $(3, 2, 4)$
   (b) $(4, -2, 1)$
   (c) $(-3, 2, 4)$
   (d) $(0, -5, -2)$
   (e) $(-4, -3, -6)$
2. Find the distance between P and Q.
   (a) $P(2, 0, -1)$, $Q(3, 5, 7)$
   (b) $P(-3, 5, 2)$, $Q(-1, -1, 4)$
   (c) $P(8, 2, 0)$, $Q(7, 6, -3)$
   (d) $P(4, -2, -3)$, $Q(-1, -3, 2)$
3. Identify each of the following three-dimensional point sets.
   (a) The set of all points for which $z = 0$
   (b) The set of all points for which $x = 0$
   (c) The set of all points for which $y = 0$
   (d) All points of the form $(x, 0, 0)$
   (e) All points for which $xyz = 0$
4. Let $P(x, y, z)$ be an arbitrary point in space and Q be the fixed point (h, k, l). Write an equation expressing the fact that $d(P, Q) = a$, where a is a positive constant. Clear the equation of radicals. How would you describe the set of all points P that satisfy this equation?
5. Express the vector $\overrightarrow{PQ}$ in terms of its components.
   (a) $P(7, 3, -1)$, $Q(5, -1, 2)$
   (b) $P(2, -3, -4)$, $Q(-1, -2, 0)$
   (c) $P(0, -2, 7)$, $Q(3, 0, 5)$
   (d) $P(-4, 6, 10)$, $Q(2, 4, 8)$
6. Find the magnitude of $\overrightarrow{PQ}$ for each part of Exercise 5.
7. Let $\mathbf{u} = \langle 3, 1, -2 \rangle$, $\mathbf{v} = \langle -1, 0, 4 \rangle$, and $\mathbf{w} = \langle 4, 1, 5 \rangle$. Find the following:
   (a) $\mathbf{u} \cdot \mathbf{w} - |\mathbf{v}|^2$ (b) $\mathbf{v} \cdot (\mathbf{u} - \mathbf{w})$
   (c) $|\mathbf{u} - 2\mathbf{v}|$ (d) $3\mathbf{u} + 2\mathbf{v} - \mathbf{w}$
8. Let $\mathbf{u} = 3\mathbf{i} - 2\mathbf{j} - \mathbf{k}$, $\mathbf{v} = 5\mathbf{i} - 4\mathbf{k}$, and $\mathbf{w} = -4\mathbf{i} + 6\mathbf{j} + 2\mathbf{k}$. Find the following:
   (a) $\mathbf{u} \cdot (\mathbf{v} - \mathbf{w})$ (b) $|3\mathbf{u} + 2\mathbf{w}|$
   (c) $|\mathbf{u}||\mathbf{w}| - |\mathbf{u} \cdot \mathbf{w}|$ (d) $(\mathbf{v} + \mathbf{w}) \cdot (\mathbf{v} - \mathbf{w})$
9. Find a unit vector in the direction of $\mathbf{u}$.
   (a) $\mathbf{u} = \langle 2, -1, 2 \rangle$ (b) $\mathbf{u} = \langle 4, 3, -5 \rangle$
   (c) $\mathbf{u} = \mathbf{i} - \mathbf{j} + \mathbf{k}$ (d) $\mathbf{u} = 2\mathbf{i} - 4\mathbf{j} + 5\mathbf{k}$
10. Find the cosine of the angle between $\mathbf{u}$ and $\mathbf{v}$.
    (a) $\mathbf{u} = \langle 3, -2, 6 \rangle$, $\mathbf{v} = \langle 1, 1, 1 \rangle$
    (b) $\mathbf{u} = 4\mathbf{i} + 2\mathbf{j} - 2\mathbf{j}$, $\mathbf{v} = -7\mathbf{i} + 4\mathbf{j} + 5\mathbf{k}$
11. (a) Show that $\mathbf{u} = 2\mathbf{i} - 3\mathbf{j} + \mathbf{k}$ and $\mathbf{v} = 4\mathbf{i} + 2\mathbf{j} - 2\mathbf{k}$ are orthogonal.
    (b) Find x so that $\mathbf{u} = \langle x, -1, 2 \rangle$ and $\mathbf{v} = \langle 6, 4, x \rangle$ will be orthogonal.
12. Use vector methods to show that the points $A(-1, 2, -3)$, $B(1, -1, 2)$, and $C(0, 5, 6)$ are vertices of a right triangle.
13. Find the direction cosines of $\mathbf{u}$.
    (a) $\mathbf{u} = 2\mathbf{i} - \mathbf{j} + 2\mathbf{k}$ (b) $\mathbf{u} = \langle 3, -5, 4 \rangle$
14. Find the vector $\mathbf{u}$ with magnitude 3 whose z component is positive if $\cos \alpha = \frac{1}{3}$ and $\cos \beta = -\sqrt{2}/3$.
15. Find the unit vector for which $\beta = \frac{\pi}{3}$ and $\gamma = \frac{\pi}{4}$ and whose x component is negative.
16. If a vector makes equal acute angles with $\mathbf{i}$, $\mathbf{j}$, and $\mathbf{k}$, what is this angle?
17. Show that if the direction angles α, β, and γ of a vector are all acute and $\alpha \geq \frac{\pi}{3}$ and $\beta \geq \frac{\pi}{3}$, then $\gamma \leq \frac{\pi}{4}$.

In Exercises 18–21 find the component of $\mathbf{u}$ *along* $\mathbf{v}$.

18. $\mathbf{u} = \langle -3, 1, 4 \rangle$, $\mathbf{v} = \langle 2, -1, -2 \rangle$
19. $\mathbf{u} = \langle 5, 4, -4 \rangle$, $\mathbf{v} = \langle 3, 0, -4 \rangle$
20. $\mathbf{u} = 7\mathbf{i} - 2\mathbf{k}$, $\mathbf{v} = 3\mathbf{i} - 5\mathbf{j} + 4\mathbf{k}$
21. $\mathbf{u} = \mathbf{i} + 2\mathbf{j} - \mathbf{k}$, $\mathbf{v} = \mathbf{i} - \mathbf{j} + \mathbf{k}$

In Exercises 22–24 find the work done by the force $\mathbf{F}$ *in moving a particle along the line segment from P to Q.*

22. $\mathbf{F} = 10\mathbf{i} + 12\mathbf{j} - 8\mathbf{k}$; $P(2, -1, 4)$, $Q(3, 5, 2)$; $|\mathbf{F}|$ in newtons, distance in meters
23. $\mathbf{F} = 20\mathbf{i} - 12\mathbf{j} + 6\mathbf{k}$; $P(3, 4, 6)$, $Q(8, -1, 10)$; $|\mathbf{F}|$ in dynes, distance in centimeters
24. $\mathbf{F} = 6\mathbf{i} + 2\mathbf{j} + 8\mathbf{k}$; $P(-1, 3, 5)$, $Q(4, -1, 9)$; $|\mathbf{F}|$ in pounds, distance in feet

In Exercises 25–28 find $\text{Proj}_{\mathbf{v}}\mathbf{u}$ *and* $\text{Proj}_{\mathbf{v}}^{\perp}\mathbf{u}$.

25. $\mathbf{u} = \langle 5, -1, 3\rangle$, $\mathbf{v} = \langle 2, 6, -4\rangle$

26. $\mathbf{u} = \langle 2, -3, 0\rangle$, $\mathbf{v} = \langle -5, 1, -2\rangle$

27. $\mathbf{u} = 2\mathbf{i} - 3\mathbf{j} - 5\mathbf{k}$, $\mathbf{v} = \mathbf{i} + 2\mathbf{j} - 3\mathbf{k}$

28. $\mathbf{u} = 4\mathbf{j} - 5\mathbf{k}$, $\mathbf{v} = 3\mathbf{i} - 5\mathbf{j} + 4\mathbf{k}$

B

In Exercises 29–33 prove that the indicated theorem continues to hold true for vectors in V_3.

29. Theorem 14.1 **30.** Theorem 14.2

31. Theorem 14.3 **32.** Theorem 14.4

33. Theorem 14.5

34. Find a nonzero vector $\mathbf{x} = \langle x_1, x_2, x_3\rangle$ that is perpendicular to each of the vectors $\mathbf{u} = \langle 2, 1, -3\rangle$ and $\mathbf{v} = \langle -1, 1, 2\rangle$. (*Hint:* Obtain two equations with three unknowns. Choose one of the unknowns arbitrarily.)

35. Find a unit vector orthogonal to each of the vectors $\mathbf{u} = \mathbf{i} - \mathbf{j} + 2\mathbf{k}$ and $\mathbf{v} = 3\mathbf{i} + 2\mathbf{j} - 2\mathbf{k}$. (See the hint in Exercise 34.)

36. Find scalars a and b such that $\mathbf{w} = a\mathbf{u} + b\mathbf{v}$, where $\mathbf{u} = \langle 3, -2, 4\rangle$, $\mathbf{v} = \langle 1, 1, -2\rangle$, and $\mathbf{w} = \langle 6, 1, -2\rangle$. Interpret the result geometrically.

37. Let $\mathbf{u}_1 = \langle 1, -1, 0\rangle$, $\mathbf{u}_2 = \langle 0, 1, -1\rangle$, and $\mathbf{u}_3 = \langle 1, 1, 1\rangle$. Find scalars a, b, and c such that $\mathbf{v} = a\mathbf{u}_1 + b\mathbf{u}_2 + c\mathbf{u}_3$, where $\mathbf{v} = \langle 3, 2, -1\rangle$.

38. With $\mathbf{u}_1$, $\mathbf{u}_2$, and $\mathbf{u}_3$ as in Exercise 37, show that *every* vector $\mathbf{v}$ in V_3 can be expressed as a linear combination of $\mathbf{u}_1$, $\mathbf{u}_2$, and $\mathbf{u}_3$. (*Note:* This shows that $\mathbf{u}_1$, $\mathbf{u}_2$, and $\mathbf{u}_3$ constitute a basis for V_3.)

39. Prove that the sum of any two of the three direction angles α, β, and γ must be greater than or equal to $\frac{\pi}{2}$. (*Hint:* Suppose, for example, that $\alpha + \beta < \frac{\pi}{2}$, so that $\alpha < \frac{\pi}{2} - \beta$. Show that $\cos^2\alpha + \cos^2\beta > 1$.)

14.5 THE CROSS PRODUCT

For vectors $\mathbf{u}$ and $\mathbf{v}$ in V_3 there is a second type of product, called the **cross product,** written $\mathbf{u} \times \mathbf{v}$, that results in another vector, rather than a scalar, as with the dot product. For this reason the cross product is sometimes called the **vector product.**

DEFINITION 14.7 The **cross product** of $\mathbf{u} = \langle u_1, u_2, u_3\rangle$ and $\mathbf{v} = \langle v_1, v_2, v_3\rangle$ is the vector

$$\mathbf{u} \times \mathbf{v} = \langle u_2v_3 - u_3v_2, u_3v_1 - u_1v_3, u_1v_2 - u_2v_1\rangle \tag{14.20}$$ ■

This definition can be more easily remembered using determinant notation, which we review briefly. A second-order determinant is defined by

$$\begin{vmatrix} a_1 & a_2 \\ b_1 & b_2 \end{vmatrix} = a_1b_2 - a_2b_1$$

For example,

$$\begin{vmatrix} 3 & 2 \\ -1 & 4 \end{vmatrix} = 3(4) - (2)(-1) = 12 + 2 = 14$$

A third-order determinant can be evaluated as follows:

$$\begin{vmatrix} a_1 & a_2 & a_3 \\ b_1 & b_2 & b_3 \\ c_1 & c_2 & c_3 \end{vmatrix} = a_1\begin{vmatrix} b_2 & b_3 \\ c_2 & c_3 \end{vmatrix} - a_2\begin{vmatrix} b_1 & b_3 \\ c_1 & c_3 \end{vmatrix} + a_3\begin{vmatrix} b_1 & b_2 \\ c_1 & c_2 \end{vmatrix}$$

Each second-order determinant is then evaluated as above. The formula we have given is sometimes referred to as *expansion by the first row*. It is possible to expand by any row or column, but for our purposes the first row is the most convenient. To illustrate, consider the following:

$$\begin{vmatrix} 2 & -1 & -3 \\ 4 & 2 & 1 \\ 0 & 5 & -4 \end{vmatrix} = 2\begin{vmatrix} 2 & 1 \\ 5 & -4 \end{vmatrix} - (-1)\begin{vmatrix} 4 & 1 \\ 0 & -4 \end{vmatrix} + (-3)\begin{vmatrix} 4 & 2 \\ 0 & 5 \end{vmatrix}$$
$$= 2(-8-5) + (-16) - 3(20) = -26 - 16 - 60 = -102$$

Now let $\mathbf{u} = \langle u_1, u_2, u_3 \rangle$ and $\mathbf{v} = \langle v_1, v_2, v_3 \rangle$, and form the third-order determinant

$$\begin{vmatrix} \mathbf{i} & \mathbf{j} & \mathbf{k} \\ u_1 & u_2 & u_3 \\ v_1 & v_2 & v_3 \end{vmatrix}$$

Since the first row consists of vectors instead of numbers, this is not a proper determinant. Nevertheless, if we formally expand it by the first row, we get (writing the scalars times the vectors instead of the reverse)

$$\begin{vmatrix} u_2 & u_3 \\ v_2 & v_3 \end{vmatrix}\mathbf{i} - \begin{vmatrix} u_1 & u_3 \\ v_1 & v_3 \end{vmatrix}\mathbf{j} + \begin{vmatrix} u_1 & u_2 \\ v_1 & v_2 \end{vmatrix}\mathbf{k} = (u_2v_3 - u_3v_2)\mathbf{i} + (u_3v_1 - u_1v_3)\mathbf{j}$$
$$+ (u_1v_2 - u_2v_1)\mathbf{k}$$

and the result is seen by comparison with equation (14.20) to be the **i**, **j**, **k** notation for $\mathbf{u} \times \mathbf{v}$. Thus, with this understanding of what is meant by the determinant with vector entries in the first row, we have

$$\mathbf{u} \times \mathbf{v} = \begin{vmatrix} \mathbf{i} & \mathbf{j} & \mathbf{k} \\ u_1 & u_2 & u_3 \\ v_1 & v_2 & v_3 \end{vmatrix} \tag{14.21}$$

EXAMPLE 14.18 Find $\mathbf{u} \times \mathbf{v}$ where $\mathbf{u} = \langle 3, -1, 4 \rangle$ and $\mathbf{v} = \langle -2, 2, 5 \rangle$.

Solution By equation (14.21),

$$\mathbf{u} \times \mathbf{v} = \begin{vmatrix} \mathbf{i} & \mathbf{j} & \mathbf{k} \\ 3 & -1 & 4 \\ -2 & 2 & 5 \end{vmatrix} = \begin{vmatrix} -1 & 4 \\ 2 & 5 \end{vmatrix}\mathbf{i} - \begin{vmatrix} 3 & 4 \\ -2 & 5 \end{vmatrix}\mathbf{j} + \begin{vmatrix} 3 & -1 \\ -2 & 2 \end{vmatrix}\mathbf{k}$$
$$= (-5-8)\mathbf{i} - (15+8)\mathbf{j} + (6-2)\mathbf{k}$$
$$= -13\mathbf{i} - 23\mathbf{j} + 4\mathbf{k}$$

Equivalently, $\mathbf{u} \times \mathbf{v} = \langle -13, -23, 4 \rangle$. ■

One of the most important properties of the cross product is given by the following theorem.

THEOREM 14.6 The vector $\mathbf{u} \times \mathbf{v}$ is orthogonal to both **u** and **v**.

Proof We will show that $\mathbf{u} \cdot (\mathbf{u} \times \mathbf{v}) = 0$ and leave it as an exercise to show that $\mathbf{v} \cdot (\mathbf{u} \times \mathbf{v}) = 0$. By equation (14.20),

$$\mathbf{u} \cdot (\mathbf{u} \times \mathbf{v}) = u_1(u_2v_3 - u_3v_2) + u_2(u_3v_1 - u_1v_3) + u_3(u_1v_2 - u_2v_1)$$
$$= u_1u_2v_3 - u_1u_3v_2 + u_2u_3v_1 - u_2u_1v_3 + u_3u_1v_2 - u_3u_2v_1$$
$$= 0$$

So **u** and $\mathbf{u} \times \mathbf{v}$ are orthogonal. ■

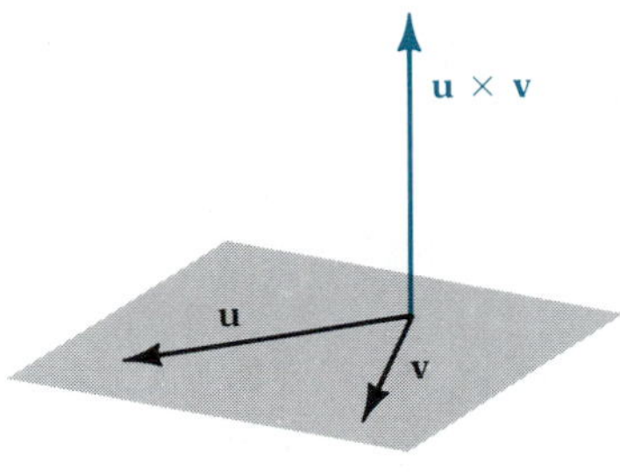

FIGURE 14.28

If **u** and **v** are nonzero and are not parallel, then we know from the preceding theorem that $\mathbf{u} \times \mathbf{v}$ is orthogonal to both **u** and **v**. Suppose geometric representatives of **u**, **v**, and $\mathbf{u} \times \mathbf{v}$ are drawn with the same initial point. Then $\mathbf{u} \times \mathbf{v}$ is perpendicular to the plane containing **u** and **v**, as shown in Figure 14.28. The direction of $\mathbf{u} \times \mathbf{v}$ is determined according to the following right-hand rule: if you curl the fingers of your right hand in the direction that would rotate **u** into **v** (through an angle of less than π), then your extended thumb will point in the direction of $\mathbf{u} \times \mathbf{v}$.

The magnitude of $\mathbf{u} \times \mathbf{v}$ is related to the magnitudes of **u** and **v** as given in the following theorem.

THEOREM 14.7 If θ is the angle between the nonzero vectors **u** and **v**, then

$$|\mathbf{u} \times \mathbf{v}| = |\mathbf{u}|\,|\mathbf{v}| \sin \theta \tag{14.22}$$

Proof By equation (14.20) we have

$$|\mathbf{u} \times \mathbf{v}|^2 = (u_2v_3 - u_3v_2)^2 + (u_3v_1 + u_1v_3)^2 + (u_1v_2 - u_2v_1)^2$$

We leave it as an exercise for you to show that if the right-hand side is expanded and terms are appropriately grouped, it can be written in the form

$$(u_1^2 + u_2^2 + u_3^2)(v_1^2 + v_2^2 + v_3^2) - (u_1v_1 + u_2v_2 + u_3v_3)^2$$

Thus,

$$|\mathbf{u} \times \mathbf{v}|^2 = |\mathbf{u}|^2|\mathbf{v}|^2 - (\mathbf{u} \cdot \mathbf{v})^2$$

Since $\mathbf{u} \cdot \mathbf{v} = |\mathbf{u}|\,|\mathbf{v}| \cos \theta$, this gives

$$\begin{aligned} |\mathbf{u} \times \mathbf{v}|^2 &= |\mathbf{u}|^2|\mathbf{v}|^2 - |\mathbf{u}|^2|\mathbf{v}|^2 \cos^2 \theta \\ &= |\mathbf{u}|^2|\mathbf{v}|^2(1 - \cos^2 \theta) \\ &= |\mathbf{u}|^2|\mathbf{v}|^2 \sin^2 \theta \end{aligned}$$

Taking square roots, we get the desired result. ■

COROLLARY 14.7 Two vectors **u** and **v** in V_3 are parallel if and only if $\mathbf{u} \times \mathbf{v} = \mathbf{0}$.

Proof If either **u** or **v** is **0**, the result is trivial. If they are both nonzero, they are parallel if and only if $\theta = 0$ or $\theta = \pi$ or, equivalently, $\sin \theta = 0$. Thus, by equation (14.22), they are parallel if and only if $|\mathbf{u} \times \mathbf{v}| = 0$, and hence if and only if $\mathbf{u} \times \mathbf{v} = \mathbf{0}$. ■

Equation (14.22) has an interesting geometric interpretation. Let **u** and **v** be nonzero, with angle θ between them. Choose geometric representatives of **u** and **v** that have the same initial point. Complete the parallelogram with **u** and **v** as adjacent sides, as in Figure 14.29. From that figure we see that the height h of the parallelogram from the base **u** is $h = |\mathbf{v}| \sin \theta$. Thus, its area is $|\mathbf{u}|h = |\mathbf{u}|\,|\mathbf{v}| \sin \theta$. So by Theorem 14.7,

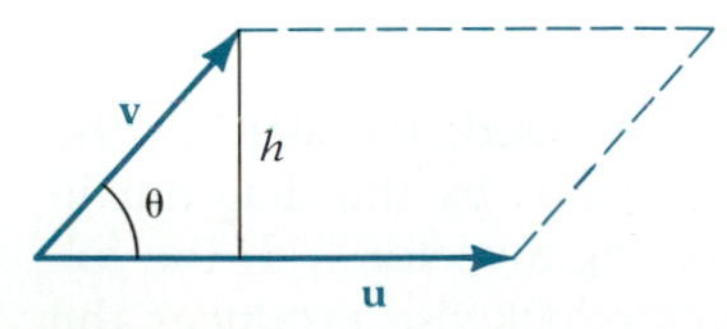

FIGURE 14.29

$|\mathbf{u} \times \mathbf{v}|$ = the area of the parallelogram with adjacent sides **u** and **v**.

EXAMPLE 14.19 Find the area of the triangle with vertices $A(2, 1, 4)$, $B(3, -1, 7)$, and $C(-1, 2, 5)$.

Solution Let $\mathbf{u} = \overrightarrow{AB}$ and $\mathbf{v} = \overrightarrow{AC}$. Then the area of the triangle is one-half of the area of the parallelogram determined by $\mathbf{u}$ and $\mathbf{v}$, or

$$\text{area} = \tfrac{1}{2}|\mathbf{u} \times \mathbf{v}|$$

We first find $\mathbf{u} \times \mathbf{v}$:

$$\mathbf{u} = \overrightarrow{AB} = \langle 1, -2, 3 \rangle$$
$$\mathbf{v} = \overrightarrow{AC} = \langle -3, 1, 1 \rangle$$

$$\mathbf{u} \times \mathbf{v} = \begin{vmatrix} \mathbf{i} & \mathbf{j} & \mathbf{k} \\ 1 & -2 & 3 \\ -3 & 1 & 1 \end{vmatrix} = \begin{vmatrix} -2 & 3 \\ 1 & 1 \end{vmatrix}\mathbf{i} - \begin{vmatrix} 1 & 3 \\ -3 & 1 \end{vmatrix}\mathbf{j} + \begin{vmatrix} 1 & -2 \\ -3 & 1 \end{vmatrix}\mathbf{k}$$
$$= -5\mathbf{i} - 10\mathbf{j} - 5\mathbf{k}$$

Thus, the area of the triangle is

$$\text{area} = \tfrac{1}{2}|\mathbf{u} \times \mathbf{v}| = \tfrac{1}{2}\sqrt{25 + 100 + 25} = \tfrac{1}{2}\sqrt{150} = \frac{5}{2}\sqrt{6}$$ ■

The next theorem provides some other properties of the cross product. The proof of each part can be shown by direct application of Definition 14.7. In some cases the proof can be facilitated, however, using the determinant equation (14.21) and the following two properties of determinants:

1. If two rows in a determinant are identical, the value of the determinant is 0.
2. If two rows in a determinant are interchanged, the result is the negative of the original determinant.

In the exercises you will be asked to verify these properties for third-order determinants, and you will also be asked to prove the theorem.

THEOREM 14.8 For vectors $\mathbf{u}$, $\mathbf{v}$, and $\mathbf{w}$ in V_3,

1. $\mathbf{u} \times \mathbf{v} = -\mathbf{v} \times \mathbf{u}$ (anticommutative property)
2. $k(\mathbf{u} \times \mathbf{v}) = (k\mathbf{u}) \times \mathbf{v} = \mathbf{u} \times (k\mathbf{v})$ for any scalar k
3. $\mathbf{u} \times \mathbf{0} = \mathbf{0}$
4. $\mathbf{u} \times \mathbf{u} = \mathbf{0}$
5. $\mathbf{u} \times (\mathbf{v} + \mathbf{w}) = \mathbf{u} \times \mathbf{v} + \mathbf{u} \times \mathbf{w}$ (left distributive property)
6. $(\mathbf{u} + \mathbf{v}) \times \mathbf{w} = \mathbf{u} \times \mathbf{w} + \mathbf{v} \times \mathbf{w}$ (right distributive property)
7. $\mathbf{u} \times (\mathbf{v} \times \mathbf{w}) = (\mathbf{u} \cdot \mathbf{w})\mathbf{v} - (\mathbf{u} \cdot \mathbf{v})\mathbf{w}$
8. $\mathbf{u} \cdot (\mathbf{v} \times \mathbf{w}) = (\mathbf{u} \times \mathbf{v}) \cdot \mathbf{w}$ ■

It is useful to learn the various cross products involving pairs of the basis vectors $\mathbf{i}$, $\mathbf{j}$, and $\mathbf{k}$. Direct application of Definition 14.7 gives:

$$\mathbf{i} \times \mathbf{j} = \mathbf{k} \qquad \mathbf{j} \times \mathbf{k} = \mathbf{i} \qquad \mathbf{k} \times \mathbf{i} = \mathbf{j}$$

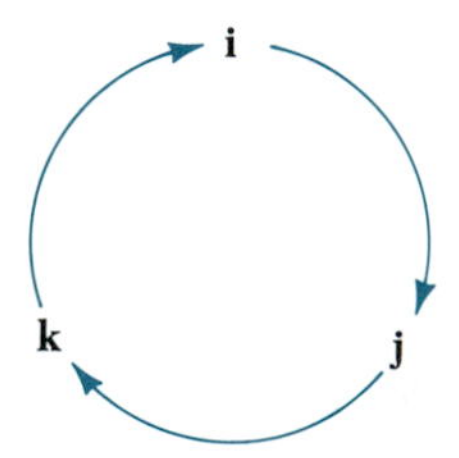

FIGURE 14.30

By Theorem 14.8, if the factors on the left are reversed, the sign on the right becomes negative. One way to remember this is by the diagram in Figure 14.30. We see that, going clockwise, crossing a vector with the following one produces the next one. Going counterclockwise produces the negative of the next one.

Vector multiplication is neither commutative nor associative in general. Noncommutativity follows from part 1 of Theorem 14.8, and nonassociativity can be seen, for example, by the following calculations:

$$(\mathbf{i} \times \mathbf{j}) \times \mathbf{j} = \mathbf{k} \times \mathbf{j} = -\mathbf{i} \quad \text{but} \quad \mathbf{i} \times (\mathbf{j} \times \mathbf{j}) = \mathbf{i} \times \mathbf{0} = \mathbf{0}$$

The product $\mathbf{u} \cdot (\mathbf{v} \times \mathbf{w})$ is called the **triple scalar product** of $\mathbf{u}$, $\mathbf{v}$, and $\mathbf{w}$. Applying Definition 14.7 to $\mathbf{v} \times \mathbf{w}$, we get

$$\begin{aligned}\mathbf{u} \cdot (\mathbf{v} \times \mathbf{w}) &= u_1(v_2w_3 - v_3w_2) + u_2(v_3w_1 - v_1w_3) + u_3(v_1w_2 - v_2w_1)\\ &= u_1\begin{vmatrix} v_2 & v_3 \\ w_2 & w_3 \end{vmatrix} - u_2\begin{vmatrix} v_1 & v_3 \\ w_1 & w_3 \end{vmatrix} + u_3\begin{vmatrix} v_1 & v_2 \\ w_1 & w_2 \end{vmatrix}\end{aligned}$$

The right-hand side is the result of expanding by the first row the determinant whose rows, in order, are the components of $\mathbf{u}$, $\mathbf{v}$, and $\mathbf{w}$, respectively. Thus,

$$\mathbf{u} \cdot (\mathbf{v} \times \mathbf{w}) = \begin{vmatrix} u_1 & u_2 & u_3 \\ v_1 & v_2 & v_3 \\ w_1 & w_2 & w_3 \end{vmatrix} \tag{14.23}$$

The triple scalar product has an interesting geometric interpretation. Let $\mathbf{u}$, $\mathbf{v}$, and $\mathbf{w}$ be nonzero vectors that do not lie in the same plane. Take geometric representatives of $\mathbf{u}$, $\mathbf{v}$, and $\mathbf{w}$ that have the same initial point, and construct a parallelepiped, with these vectors as edges, as in Figure 14.31. Using as a base the parallelogram determined by $\mathbf{v}$ and $\mathbf{w}$, the altitude h is the absolute value of the component of $\mathbf{u}$ perpendicular to this base; that is,

$$h = |\text{Comp}_{\mathbf{v} \times \mathbf{w}}\mathbf{u}|$$

As we have seen, the area of the base is $|\mathbf{v} \times \mathbf{w}|$, and the volume of the parallelepiped is the area of the base times the altitude. So we have

$$\begin{aligned}\text{vol} = h|\mathbf{v} \times \mathbf{w}| &= |\text{Comp}_{\mathbf{v} \times \mathbf{w}}\mathbf{u}|\,|\mathbf{v} \times \mathbf{w}|\\ &= \frac{|\mathbf{u} \cdot (\mathbf{v} \times \mathbf{w})|}{|\mathbf{v} \times \mathbf{w}|}|\mathbf{v} \times \mathbf{w}|\\ &= |\mathbf{u} \cdot (\mathbf{v} \times \mathbf{w})|\end{aligned}$$

FIGURE 14.31

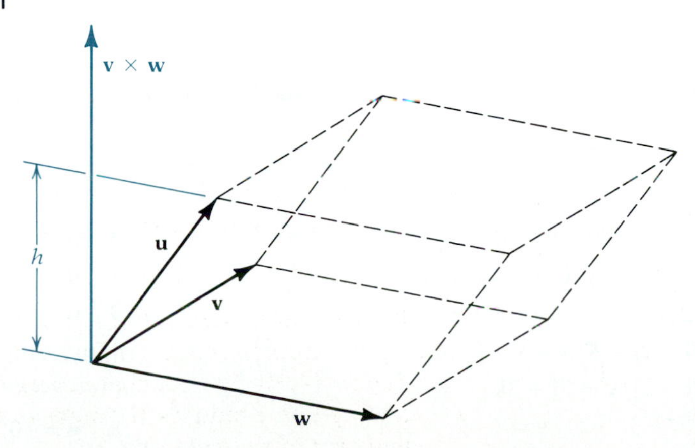

In words, this says

> $|\mathbf{u} \cdot (\mathbf{v} \times \mathbf{w})|$ = the volume of the parallelepiped with adjacent edges **u**, **v**, and **w**.

EXAMPLE 14.20 Given the points $A(3, -1, 1)$, $B(2, 3, -2)$, $C(0, 1, 3)$, and $D(-1, 2, 4)$, find the volume of the parallelepiped determined by the vectors $\overrightarrow{AB}$, $\overrightarrow{AC}$, and $\overrightarrow{AD}$.

Solution Let

$$\begin{aligned}\mathbf{u} &= \overrightarrow{AB} = \langle -1, 4, -3\rangle\\ \mathbf{v} &= \overrightarrow{AC} = \langle -3, 2, 2\rangle\\ \mathbf{w} &= \overrightarrow{AD} = \langle -4, 3, 3\rangle\end{aligned}$$

By equation (14.23),

$$\mathbf{u} \cdot (\mathbf{v} \times \mathbf{w}) = \begin{vmatrix} -1 & 4 & -3 \\ -3 & 2 & 2 \\ -4 & 3 & 3 \end{vmatrix} = -1\begin{vmatrix} 2 & 2 \\ 3 & 3 \end{vmatrix} - 4\begin{vmatrix} -3 & 2 \\ -4 & 3 \end{vmatrix} - 3\begin{vmatrix} -3 & 2 \\ -4 & 3 \end{vmatrix}$$

$$= 0 - 4(-1) - 3(-1) = 7$$

The volume is $|\mathbf{u} \cdot (\mathbf{v} \times \mathbf{w})| = |7| = 7$. ■

In arriving at the volume of the parallelepiped determined by **u**, **v**, and **w** as $|\mathbf{u} \cdot (\mathbf{v} \times \mathbf{w})|$, we assumed the vectors were not coplanar. However, an analysis of the computations we made will show that for any nonzero vectors **u**, **v**, and **w**,

$$|\mathbf{u} \cdot (\mathbf{v} \times \mathbf{w})| = |\text{Comp}_{\mathbf{v} \times \mathbf{w}}\mathbf{u}|\,|\mathbf{v} \times \mathbf{w}|$$

and if **u** is in the same plane with **v** and **w**, $\text{Comp}_{\mathbf{v} \times \mathbf{w}}\mathbf{u} = 0$, since $\mathbf{v} \times \mathbf{w}$ is orthogonal to **u**. Conversely, if $\text{Comp}_{\mathbf{v} \times \mathbf{w}}\mathbf{u} = 0$, **u** is orthogonal to $\mathbf{v} \times \mathbf{w}$ and hence in the plane of **v** and **w**. We therefore conclude that if **u**, **v**, and **w** have the same initial point,

> **u**, **v**, and **w** are coplanar if and only if $\mathbf{u} \cdot (\mathbf{v} \times \mathbf{w}) = 0$.

EXERCISE SET 14.5

A

In Exercises 1–8 find $\mathbf{u} \times \mathbf{v}$.

1. $\mathbf{u} = \langle 3, 1, -2\rangle$, $\mathbf{v} = \langle -1, 1, 1\rangle$
2. $\mathbf{u} = \langle 2, 0, -1\rangle$, $\mathbf{v} = \langle 0, 2, 1\rangle$
3. $\mathbf{u} = 4\mathbf{i} - 2\mathbf{j} + \mathbf{k}$, $\mathbf{v} = \mathbf{i} + \mathbf{j} - 2\mathbf{k}$
4. $\mathbf{u} = 3\mathbf{i} - 2\mathbf{j}$, $\mathbf{v} = 2\mathbf{i} + 3\mathbf{k}$
5. $\mathbf{u} = \langle 5, -3, -2\rangle$, $\mathbf{v} = \langle -2, -3, 1\rangle$
6. $\mathbf{u} = \langle 2, -1, 2\rangle$, $\mathbf{v} = \langle -3, 4, -1\rangle$
7. $\mathbf{u} = \mathbf{i} - 3\mathbf{j} + 4\mathbf{k}$, $\mathbf{v} = 2\mathbf{i} - \mathbf{j} - 5\mathbf{k}$
8. $\mathbf{u} = 6\mathbf{i} - 5\mathbf{j} + 4\mathbf{k}$, $\mathbf{v} = 4\mathbf{i} - 3\mathbf{j} - \mathbf{k}$

In Exercises 9–12 find a vector orthogonal to each of the given vectors.

9. $\mathbf{u} = \langle 0, 1, -3\rangle$, $\mathbf{v} = \langle 2, 4, -1\rangle$

10. $\mathbf{u} = \langle 3, 2, -3 \rangle$, $\mathbf{v} = \langle 2, 1, -4 \rangle$
11. $\mathbf{u} = 3\mathbf{i} - 2\mathbf{j} - 5\mathbf{k}$, $\mathbf{v} = \mathbf{i} + 4\mathbf{j} + 3\mathbf{k}$
12. $\mathbf{u} = 2\mathbf{i} - 3\mathbf{j} - \mathbf{k}$, $\mathbf{v} = 3\mathbf{i} - 2\mathbf{j} - \mathbf{k}$

In Exercises 13–19 $\mathbf{u} = 2\mathbf{i} - \mathbf{j} + \mathbf{k}$, $\mathbf{v} = \mathbf{i} + 2\mathbf{j} - 3\mathbf{k}$, *and* $\mathbf{w} = 3\mathbf{i} + 2\mathbf{j} - \mathbf{k}$. *Compute the value of the given expressions in Exercises 13–18.*

13. $\mathbf{u} \cdot (\mathbf{v} \times \mathbf{w})$
14. $(\mathbf{u} \times \mathbf{v}) \cdot \mathbf{w}$
15. $\mathbf{u} \times (\mathbf{v} \times \mathbf{w})$
16. $(\mathbf{u} \times \mathbf{v}) \times \mathbf{w}$
17. $(\mathbf{u} \times \mathbf{v}) \cdot (\mathbf{u} \times \mathbf{w})$
18. $(\mathbf{u} \times \mathbf{v}) \times (\mathbf{u} \times \mathbf{w})$
19. Show that $\mathbf{v}$ and $\mathbf{u} \times \mathbf{v}$ are orthogonal.

In Exercises 20–23 find the area of the parallelogram that has $\overrightarrow{AB}$ *and* $\overrightarrow{AC}$ *as adjacent sides.*

20. $A(3, 1, 0)$, $B(2, 2, -1)$, $C(4, 0, 2)$
21. $A(-1, 1, 3)$, $B(1, 3, 2)$, $C(-2, 2, -1)$
22. $A(4, -2, -7)$, $B(3, 1, -5)$, $C(-1, 2, 0)$
23. $A(0, 2, -1)$, $B(4, 0, 2)$, $C(3, -1, -4)$
24. Find the area of the triangle with vertices $A(4, -2, 3)$, $B(6, 1, -1)$, and $C(5, 2, 3)$.
25. Find the area of the triangle with vertices $A(1, 0, -2)$, $B(-3, 2, 1)$, and $C(4, -2, -3)$.
26. Find a vector perpendicular to the plane that contains the points $A(3, 4, 5)$, $B(-1, 2, 4)$, and $C(2, 3, 1)$.
27. Find a unit vector perpendicular to the plane that contains the points $P(0, -1, 3)$, $Q(1, 3, 2)$, and $R(2, -1, 4)$.

In Exercises 28 and 29 find the volume of the parallelepiped that has $\overrightarrow{AB}$, $\overrightarrow{AC}$, *and* $\overrightarrow{AD}$ *as edges.*

28. $A(3, 2, -5)$, $B(1, 4, -2)$, $C(-2, 3, 0)$, $D(4, 3, -8)$
29. $A(-2, 0, 4)$, $B(1, 1, 2)$, $C(0, 3, -1)$, $D(-3, -2, 4)$
30. Show that the vectors $\mathbf{u} = 2\mathbf{i} - 3\mathbf{j} + 4\mathbf{k}$, $\mathbf{v} = \mathbf{i} + 2\mathbf{j} - \mathbf{k}$, and $\mathbf{w} = 7\mathbf{i} + 5\mathbf{k}$ are coplanar.
31. Show that the points $A(1, -1, 2)$, $B(3, -4, 1)$, $C(0, 1, 2)$, and $D(1, 0, 1)$ all lie in the same plane.

B

32. Supply the missing steps in the proof of Theorem 14.7.
33. Prove that for a third-order determinant if two rows are identical, the value of the determinant is 0.
34. Prove that for a third-order determinant if two rows are interchanged, the resulting determinant is the negative of the original.
35. Prove parts 1, 2, and 3 of Theorem 14.8.
36. Prove parts 4, 5, and 6 of Theorem 14.8.
37. Prove parts 7 and 8 of Theorem 14.8.

In Exercises 38–42 prove the given identities based on the properties given in Theorem 14.8, where $\mathbf{u}$, $\mathbf{v}$, $\mathbf{w}$, *and* $\mathbf{z}$ *are vectors in* V_3.

38. $\mathbf{u} \cdot (\mathbf{u} \times \mathbf{v}) = 0$
39. $(\mathbf{u} + \mathbf{v}) \times (\mathbf{u} - \mathbf{v}) = 2(\mathbf{v} \times \mathbf{u})$
40. $(\mathbf{u} \times \mathbf{v}) \times \mathbf{w} = (\mathbf{u} \cdot \mathbf{w})\mathbf{v} - (\mathbf{v} \cdot \mathbf{w})\mathbf{u}$
41. $\mathbf{u} \times (\mathbf{v} + \mathbf{w}) + \mathbf{v} \times (\mathbf{w} + \mathbf{u}) + \mathbf{w} \times (\mathbf{u} + \mathbf{v}) = \mathbf{0}$
42. $(\mathbf{u} \times \mathbf{v}) \cdot (\mathbf{w} \times \mathbf{z}) = \begin{vmatrix} \mathbf{u} \cdot \mathbf{w} & \mathbf{u} \cdot \mathbf{z} \\ \mathbf{v} \cdot \mathbf{w} & \mathbf{v} \cdot \mathbf{z} \end{vmatrix}$ (*Hint:* First apply part 8 of Theorem 14.8 to the left-hand side, then part 7, and use properties of the dot product.)
43. (a) Let $P_1(x_1, y_1, z_1)$, $P_2(x_2, y_2, z_2)$, and $P_3(x_3, y_3, z_3)$ be any three noncollinear points in space. Show that the area of the triangle that has these points as vertices is

$$A = \tfrac{1}{2}|\overrightarrow{P_1P_2} \times \overrightarrow{P_1P_3}|$$

(b) By treating points in two dimensions as points in three dimensions with the z-coordinate equal to 0, use part a to show that the area of a triangle in a two-dimensional coordinate system that has $P_1(x_1, y_1)$, $P_2(x_2, y_2)$, and $P_3(x_3, y_3)$ as vertices can be put in the form

$$A = \pm\tfrac{1}{2}\begin{vmatrix} x_1 & y_1 & 1 \\ x_2 & y_2 & 1 \\ x_3 & y_3 & 1 \end{vmatrix}$$

where the sign is chosen so that the result is nonnegative.

14.6 LINES IN SPACE

A line in space can be described by a point on the line and a direction for the line. The direction is specified by means of a vector, called a **direction vector,** that is parallel to the line. This is similar to the two-dimensional case

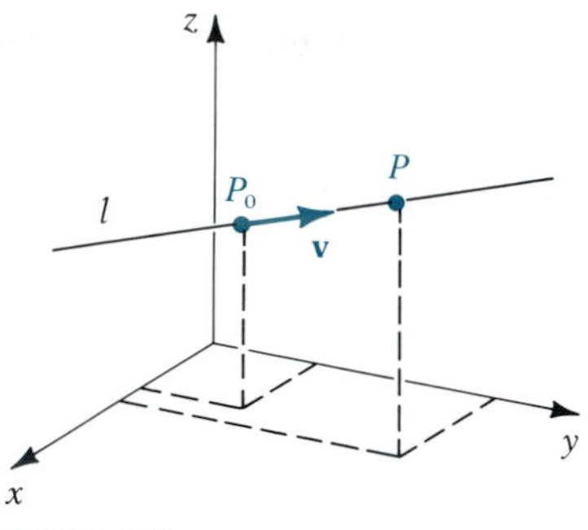

FIGURE 14.32

where a point and a slope are given. Suppose l is a line that passes through $P_0(x_0, y_0, z_0)$ and has direction vector $\mathbf{v} = \langle a, b, c \rangle$. Position $\mathbf{v}$ so that its initial point is at P_0, as in Figure 14.32. A point $P(x, y, z)$ will be on l if and only if

$$\overrightarrow{P_0P} = t\mathbf{v} \tag{14.24}$$

for some scalar t. As we allow t to range over all real numbers, P traces out the entire line.

Equation (14.24) can be put in another form using *position vectors*. If $P(x, y, z)$ is a point in space, then its **position vector** is the vector $\overrightarrow{OP}$ that has initial point at the origin and terminal point at P. The components of the position vector for P are precisely the coordinates of P—namely, $\langle x, y, z \rangle$. Now let $\mathbf{r}_0$ and $\mathbf{r}$ be the position vectors of P_0 and P, respectively, on the line l. Then, as illustrated in Figure 14.33, $\overrightarrow{P_0P} = \mathbf{r} - \mathbf{r}_0$. So equation (14.24) can be written in the form

$$\mathbf{r} = \mathbf{r}_0 + t\mathbf{v}, \qquad -\infty < t < \infty \tag{14.25}$$

and we call this **a vector equation for the line l.**

EXAMPLE 14.21 (a) Find a vector equation of the line l passing through the point $(1, 5, -4)$ and parallel to the vector $\langle 3, -2, 1 \rangle$.

(b) Given the equation

$$\mathbf{r} = (2 - t)\mathbf{i} + (4 + 3t)\mathbf{j} + (-2 + 5t)\mathbf{k}, \qquad -\infty < t < \infty$$

describe its graph.

Solution (a) Let $\mathbf{r}_0$ be the position vector $\langle 1, 5, -4 \rangle$ of the given point. The vector $\langle 3, -2, 1 \rangle$ is a direction vector for l. So by equation (14.25) the position vector $\mathbf{r}$ of any point on l is

$$\mathbf{r} = \langle 1, 5, -4 \rangle + t\langle 3, -2, 1 \rangle = \langle 1 + 3t, 5 - 2t, -4 + t \rangle$$

(b) We can rewrite the equation as

$$\mathbf{r} = (2\mathbf{i} + 4\mathbf{j} - 2\mathbf{k}) + t(-\mathbf{i} + 3\mathbf{j} + 5\mathbf{k})$$

By comparison with equation (14.25) we see that this is a vector equation for the line passing through $(2, 4, -2)$ and parallel to the vector $-\mathbf{i} + 3\mathbf{j} + 5\mathbf{k}$. ■

Comment Notice that we call equation (14.25) *a* vector equation for l rather than *the* vector equation. This is because neither $\mathbf{r}_0$ nor $\mathbf{v}$ is unique. We are free to use *any* point on l as P_0 and *any* nonzero vector that is parallel to l as its direction vector $\mathbf{v}$.

EXAMPLE 14.22 Find a vector equation of the line l passing through the points $P(2, -1, 4)$ and $Q(3, 2, -1)$.

Solution We can use either point as the fixed point of equation (14.25). Letting this be P, we have $\mathbf{r}_0 = \langle 2, -1, 4 \rangle$. The direction vector $\mathbf{v}$ can be taken as $\overrightarrow{PQ} = \langle 1, 3, -5 \rangle$. Thus, a vector equation for l is

$$\mathbf{r} = \langle 2, -1, 4 \rangle + t\langle 1, 3, -5 \rangle = \langle 2 + t, -1 + 3t, 4 - 5t \rangle$$

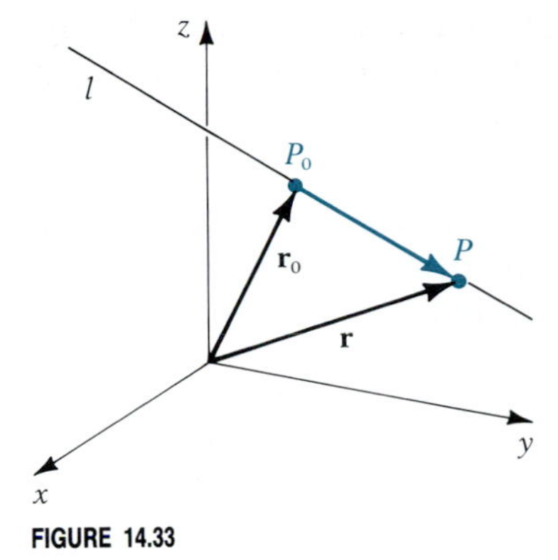

FIGURE 14.33

Notice that when $t = 0$, $\mathbf{r}$ is the position vector of P and when $t = 1$, $\mathbf{r}$ is the position vector of Q. ■

If we write equation (14.25) in component form, we get

$$\begin{aligned}\langle x, y, z\rangle &= \langle x_0, y_0, z_0\rangle + t\langle a, b, c\rangle \\ &= \langle x_0 + at, y_0 + bt, z_0 + ct\rangle\end{aligned}$$

and equating components yields

$$\begin{cases} x = x_0 + at \\ y = y_0 + bt \qquad -\infty < t < \infty \\ z = z_0 + ct \end{cases} \tag{14.26}$$

Equations (14.26) are **parametric equations for the line *l*.** If we know a vector equation for l, we can write parametric equations, and conversely.

EXAMPLE 14.23 Find parametric equations of the line in Example 14.22.

Solution In that example we found that

$$\mathbf{r} = \langle 2 + t, -1 + 3t, 4 - 5t\rangle$$

So parametric equations are

$$\begin{cases} x = 2 + t \\ y = -1 + 3t \qquad -\infty < t < \infty \\ z = 4 - 5t \end{cases}$$

■

EXAMPLE 14.24 Give two points and a direction vector for the line l that has parametric equations $x = 3 - 5t$, $y = -4 + 7t$, and $z = 10 - 8t$.

Solution One point on the line is $(3, -4, 10)$ corresponding to $t = 0$. We get another point using a different value of t, say $t = 1$. This gives $(-2, 3, 2)$. A direction vector is $\langle -5, 7, -8\rangle$. ■

The numbers a, b, and c in equations (14.26) that are the components of a direction vector $\mathbf{v}$ for l are also called **direction numbers** for l. Since $k\mathbf{v} = \langle ka, kb, kc\rangle$ is also a direction vector for l for any nonzero scalar k, it follows that ka, kb, and kc also are direction numbers for l. Knowing a set of direction numbers for l is equivalent to knowing a direction vector for l.

EXAMPLE 14.25 Find a set of direction numbers for the line

$$\mathbf{r} = \langle 3 - 6t, 2 + 4t, -5 - 8t\rangle$$

Solution One set of direction numbers is $-6, 4, -8$. Another set is $3, -2, 4$, obtained by multiplying the first set by $-\frac{1}{2}$. ■

By the *angle between two lines* l_1 and l_2, we mean the angle between a direction vector for l_1 and a direction vector for l_2, or its supplement, whichever does not exceed $\frac{\pi}{2}$. If $\mathbf{u}$ is any direction vector for l_1 and $\mathbf{v}$ is any direction vector for l_2, then by our definition the angle θ between l_1 and l_2 satisfies

$$\cos\theta = \frac{|\mathbf{u}\cdot\mathbf{v}|}{|\mathbf{u}|\,|\mathbf{v}|}, \qquad 0 \le \theta \le \tfrac{\pi}{2}$$

The absolute value of $\mathbf{u}\cdot\mathbf{v}$ is needed to ensure that $\cos\theta \ge 0$. The lines are *parallel* if $\mathbf{u}$ and $\mathbf{v}$ are parallel, and they are *orthogonal* if $\mathbf{u}$ and $\mathbf{v}$ are orthogonal. Lines that do not intersect and are not parallel are called **skew** lines.

EXAMPLE 14.26 Find the angle between the lines l_1 and l_2, defined by $\mathbf{r} = \langle 1 - 2t, 3 + t, -2 + 3t\rangle$ and $\mathbf{r} = \langle -2 + t, 4, 3 - t\rangle$, respectively.

Solution A direction vector for l_1 is $\mathbf{u} = \langle -2, 1, 3\rangle$, and a direction vector for l_2 is $\mathbf{v} = \langle 1, 0, -1\rangle$. So for the angle θ between l_1 and l_2,

$$\cos\theta = \frac{|\mathbf{u}\cdot\mathbf{v}|}{|\mathbf{u}|\,|\mathbf{v}|} = \frac{|\langle -2, 1, 3\rangle\cdot\langle 1, 0, -1\rangle|}{\sqrt{4+1+9}\sqrt{1+1}} = \frac{5}{2\sqrt{7}}$$

$$\theta = \cos^{-1}\frac{5}{2\sqrt{7}} \approx 0.3335 \text{ radian} \approx 19.11^\circ$$

■

EXAMPLE 14.27 Let l_1 be the line with vector equation $\mathbf{r} = \langle 3 + 3t, 5 - t, -1 + 2t\rangle$ and l_2 be the line with vector equation $\mathbf{r} = \langle -1 + 4t, 7 + 2t, 3 - 5t\rangle$.

(a) Show that l_1 and l_2 are orthogonal.

(b) Find a vector equation of a line passing through $(4, -2, 0)$ that is parallel to l_1.

Solution (a) Direction vectors for l_1 and l_2 are $\mathbf{u} = \langle 3, -1, 2\rangle$ and $\mathbf{v} = \langle 4, 2, -5\rangle$, respectively. Since $\mathbf{u}\cdot\mathbf{v} = 12 - 2 - 10 = 0$, it follows that $\mathbf{u}$ and $\mathbf{v}$, and hence l_1 and l_2, are orthogonal.

(b) We may use the same direction vector $\mathbf{u} = \langle 3, -1, 2\rangle$ as for l_1. Only the fixed point need be changed. So the equation is

$$\mathbf{r} = \langle 4, -2, 0\rangle + t\langle 3, -1, 2\rangle = \langle 4 + 3t, -2 - t, 2t\rangle$$

■

EXAMPLE 14.28 Find a vector equation of the line passing through $(7, -1, 2)$ that intersects the line $\mathbf{r} = \langle 2 - t, 4 + 3t, 5 - 2t\rangle$ orthogonally.

Solution First we find the point of intersection. Let l_1 be the given line, and let t_1 be the parameter value for l_1 of the point of intersection P_1 of the two lines. Let Q be the point $(7, -1, 2)$. Then, as seen in Figure 14.34, $\overrightarrow{P_1Q}$ is a direction vector for the desired orthogonal line l_2. Since P_1 has coordinates $(2 - t_1, 4 + 3t_1, 5 - 2t_1)$, we obtain for $\overrightarrow{P_1Q}$:

$$\overrightarrow{P_1Q} = \langle 5 + t_1, -5 - 3t_1, -3 + 2t_1\rangle$$

A direction vector for l_1 is $\mathbf{v} = \langle -1, 3, -2\rangle$. For orthogonality we must have $\mathbf{v}\cdot\overrightarrow{P_1Q} = 0$, or

$$\begin{aligned} -5 - t_1 - 15 - 9t_1 + 6 - 4t_1 &= 0 \\ -14t_1 &= 14 \\ t_1 &= -1 \end{aligned}$$

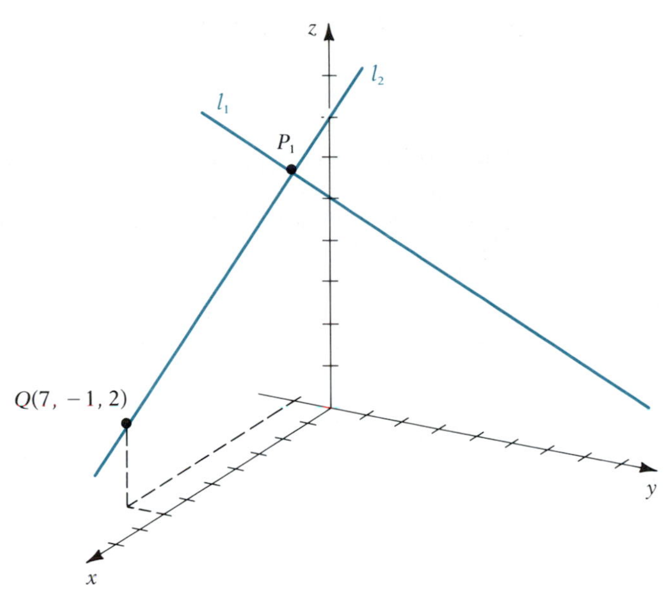

FIGURE 14.34

So the lines intersect at (3, 1, 7), and the direction vector $\overrightarrow{P_1Q}$ for l_2 is $\langle 4, -2, -5\rangle$. An equation for l_2 is therefore

$$\mathbf{r} = \langle 3, 1, 7\rangle + t\langle 4, -2, -5\rangle$$

or

$$\mathbf{r} = \langle 3 + 4t, 1 - 2t, 7 - 5t\rangle$$

■

EXAMPLE 14.29 Let l_1 and l_2 be defined parametrically by

$$l_1: \begin{cases} x = 3 - 2t \\ y = -1 + t \\ z = 2 + 3t \end{cases} \qquad l_2: \begin{cases} x = 1 - s \\ y = 2 + 3s \\ z = 7 + 4s \end{cases}$$

Determine whether l_1 and l_2 intersect, are parallel, or are skew. If they intersect, find the point of intersection.

Solution Notice that different letters are used to designate the parameters. This is important, since otherwise we would disguise the fact that, if the lines intersect, the point of intersection might occur for different values of the parameters for l_1 and l_2.

Direction vectors for l_1 and l_2 are $\mathbf{u} = \langle -2, 1, 3\rangle$ and $\mathbf{v} = \langle -1, 3, 4\rangle$, and since $\mathbf{u} \neq k\mathbf{v}$, the lines are not parallel. To see whether they intersect, our approach will be to determine values of s and t that give the same values of x and y and then test to see whether the z values also are the same.

Setting the x and y values equal to each other gives

$$\begin{cases} 3 - 2t = 1 - s \\ -1 + t = 2 + 3s \end{cases} \quad \text{or} \quad \begin{cases} -2t + s = -2 \\ t - 3s = 3 \end{cases}$$

The simultaneous solution is easily found to be $t = \frac{3}{5}$, $s = -\frac{4}{5}$. The corresponding values of x and y are $x = \frac{9}{5}$, $y = -\frac{2}{5}$. The critical test is for z. For

l_1 we have

$$z = 2 + 3t = 2 + 3(\tfrac{3}{5}) = \tfrac{19}{5}$$

and for l_2,

$$z = 7 + 4s = 7 + 4(-\tfrac{4}{5}) = \tfrac{19}{5}$$

The fact that we get the same value of z tells us that the lines do intersect. The point of intersection is $(\frac{9}{5}, -\frac{2}{5}, \frac{19}{5})$.

If the z values had been different, we would have concluded that the lines are skew. ■

EXERCISE SET 14.6

A

In Exercises 1–4 a point P_0 and a vector $\mathbf{v}$ *are given. Find (a) a vector equation and (b) parametric equations for the line through P_0 that has direction vector* $\mathbf{v}$.

1. $P_0(2, 5, -1)$, $\mathbf{v} = \langle -3, 1, 2\rangle$
2. $P_0(-1, -2, 4)$, $\mathbf{v} = \langle 2, 5, -7\rangle$
3. $P_0(5, 8, -6)$, $\mathbf{v} = 2\mathbf{i} - 3\mathbf{j} + 4\mathbf{k}$
4. $P_0(3, -9, 4)$, $\mathbf{v} = 3\mathbf{i} + 2\mathbf{j} - 5\mathbf{k}$

In Exercises 5–8 find a vector equation for the line through P and Q.

5. $P(4, -1, 8)$, $Q(3, 2, 5)$
6. $P(-1, 5, -6)$, $Q(2, 3, -1)$
7. $P(7, -2, -4)$, $Q(3, 1, -2)$
8. $P(4, 6, 9)$, $Q(1, -1, 5)$

In Exercises 9–12 find the angle between l_1 and l_2.

9. l_1: $\mathbf{r} = \langle 1 - 2t, 3 + t, 4 - 5t\rangle$
   l_2: $\mathbf{r} = \langle 2 - t, 1 - 2t, 3 + 2t\rangle$
10. l_1: $\mathbf{r} = (3 + 4t)\mathbf{i} + (2 - t)\mathbf{j} + (2 + 3t)\mathbf{k}$
    l_2: $\mathbf{r} = (1 - 3t)\mathbf{i} + (4 + t)\mathbf{j} + (7 - 2t)\mathbf{k}$
11. l_1: $x = 5 + 3t$, $y = 7 + 4t$, $z = 11 - 2t$
    l_2: $x = 4 - t$, $y = 5 + 2t$, $z = -1 + 3t$
12. l_1 has direction numbers $-1, 2, 3$ and l_2 has direction numbers $3, 5, -2$.

In Exercises 13 and 14 show that l_1 and l_2 are orthogonal.

13. l_1: $\mathbf{r} = \langle 8 + 3t, -6 - 2t, 7 + t\rangle$
    l_2: $\mathbf{r} = \langle 11 + 7t, 9 + 8t, 3 - 5t\rangle$
14. l_1: $x = 13 - 4t$, $y = -7 - 3t$, $z = 4 + 3t$
    l_2: $x = 6 + 6t$, $y = 8 - 5t$, $z = 12 + 3t$
15. Find parametric equations of the line through $(3, -1, 2)$ that is parallel to the line $\mathbf{r} = \langle 2 - 3t, 7 + t, 8 + 5t\rangle$.
16. Find a vector equation of the line through $P_1(4, -1, 3)$ that is parallel to the line through $P_2(-1, 0, 4)$ and $P_2(1, 3, 2)$.

In Exercises 17 and 18 find a vector equation of the line that passes through P_0 and has a direction vector that is orthogonal to both lines whose equations are given.

17. $P_0(5, 2, -3)$; $\mathbf{r} = \langle 2 + t, 3 - 2t, 4 - 5t\rangle$,
    $\mathbf{r} = \langle 1 - t, 2t, 3 + 4t\rangle$
18. $P_0(0, -1, 2)$; $\mathbf{r} = (3 + 2t)\mathbf{i} + (4 - 3t)\mathbf{j} + (-2 - t)\mathbf{k}$,
    $\mathbf{r} = (2 - 4t)\mathbf{i} + (-1 + t)\mathbf{j} + 2\mathbf{k}$
19. Show that l_1 and l_2 intersect, and find parametric equations of a line orthogonal to both l_1 and l_2 at their point of intersection.

$$l_1: \begin{cases} x = 2 - 3t \\ y = 1 + t \\ z = 5 - 4t \end{cases} \qquad l_2: \begin{cases} x = 5 + 3s \\ y = -2 - 2s \\ z = 3 + s \end{cases}$$

In Exercises 20–23 determine whether l_1 and l_2 are parallel, intersecting, or skew. If they intersect, find their point of intersection.

20. l_1: $\mathbf{r} = \langle 11 - t, 7 + 2t, 8 - 3t\rangle$
    l_2: $\mathbf{r} = \langle 4 + 3s, 2 - 6s, 5 + 9s\rangle$
21. l_1: $x = 4 - t$, $y = 2t$, $z = 3 + 4t$
    l_2: $x = 2 + 3s$, $y = 1 - s$, $z = 4 + s$
22. l_1: $\mathbf{r} = (3 - 4t)\mathbf{i} + (2 + 3t)\mathbf{j} + (1 - t)\mathbf{k}$
    l_2: $\mathbf{r} = (2 + 2s)\mathbf{i} + (5 - 3s)\mathbf{j} + s\mathbf{k}$
23. l_1: $\mathbf{r} = \langle 3 - 4t, 2 + t, 2t\rangle$
    l_2: $\mathbf{r} = \langle 3 + 2s, 1 - s, 8 + 3s\rangle$
24. Show that l_1 and l_2 are the same line.
    l_1: $\mathbf{r} = \langle 3 - 4t, 7 + 2t, -8 - 3t\rangle$
    l_2: $\mathbf{r} = \left\langle 11 + 2s, 3 - s, \dfrac{-4 + 3s}{2}\right\rangle$

B

25. If a, b, and c are all nonzero, show that the line $\mathbf{r} = \langle x_0 + at, y_0 + bt, z_0 + ct \rangle$ can be described by the equations

$$\frac{x - x_0}{a} = \frac{y - y_0}{b} = \frac{z - z_0}{c} \qquad (14.27)$$

These are called **symmetric equations** for the line. (*Hint:* Write parametric equations and eliminate the parameter.)

26. Referring to Exercise 25, suppose a line l has the symmetric equations

$$\frac{x - 2}{3} = \frac{y + 1}{2} = \frac{z - 3}{-5}$$

(a) Give two points on l.
(b) Find a unit direction vector for l.
(c) Give parametric equations for l.
(d) Give a vector equation for l.

27. A line l has direction numbers -2, 4, 3 and it contains the point $(3, -1, 4)$.
(a) Find symmetric equations for l. (See Exercise 25.)
(b) Find parametric equations for l.
(c) Determine a vector equation for l.
(d) Find the points where l pierces each of the coordinate planes.

28. Find a vector equation of the line passing through $(4, 0, -5)$ that intersects $\mathbf{r} = \langle 3 - t, 2t, 4 + 3t \rangle$ orthogonally.

29. Let l_1 be the line $\mathbf{r} = (3 + t)\mathbf{i} + (4 - 2t)\mathbf{j} + (5 + 2t)\mathbf{k}$ and let l_2 be a line passing through $(-1, 2, 8)$ that intersects l_1 so that the angle between l_1 and l_2 is $\frac{\pi}{4}$. Find a vector equation for l_2. (There are two solutions.)

30. Let l_1 and l_2 be the lines with vector equations $\mathbf{r} = \langle 3 - 2t, 4 + 3t, 1 - t \rangle$ and $\mathbf{r} = \langle 5 + 4t, 2 - 6t, 3 + 2t \rangle$ respectively.
(a) Show there is a line l_3 passing through $(2, 5, 0)$ that intersects l_1 and l_2 orthogonally, and find a vector equation for l_3.
(b) Find a vector equation of a line l_4 passing through the point $(5, -1, 7)$ that is perpendicular to the plane containing l_1 and l_2.

14.7 PLANES

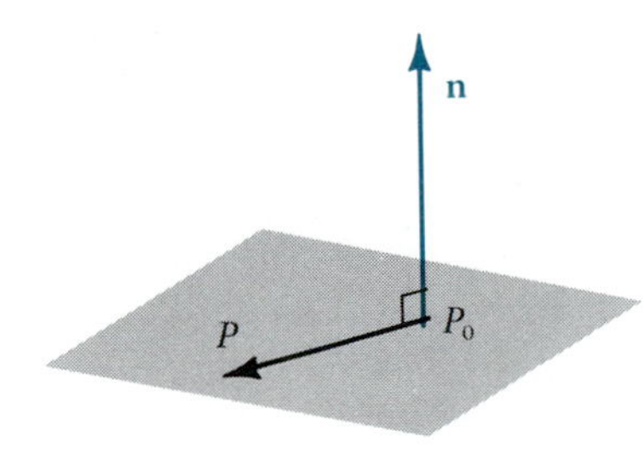

FIGURE 14.35

Given a point $P_0(x_0, y_0, z_0)$ and a nonzero vector $\mathbf{n} = \langle a, b, c \rangle$, the set of all points P such that $\overrightarrow{P_0P}$ and $\mathbf{n}$ are orthogonal is a plane (see Figure 14.35). The vector $\mathbf{n}$ is called a **normal vector** (or simply a **normal**) to the plane. The condition for orthogonality can be written as

$$\mathbf{n} \cdot \overrightarrow{P_0P} = 0 \qquad (14.28)$$

As with equations of lines, if we let $\mathbf{r}_0$ and $\mathbf{r}$ be the position vectors of P_0 and P, respectively, then $\overrightarrow{P_0P} = \mathbf{r} - \mathbf{r}_0$, so that equation (14.28) becomes

$$\mathbf{n} \cdot (\mathbf{r} - \mathbf{r}_0) = 0 \qquad (14.29)$$

We call this **a vector equation of the plane.** In terms of components, $\mathbf{r} - \mathbf{r}_0 = \langle x - x_0, y - y_0, z - z_0 \rangle$, so that equation (14.29) becomes

$$a(x - x_0) + b(y - y_0) + c(z - z_0) = 0 \qquad (14.30)$$

This is known as a **standard form** of the equation of the plane with normal $\langle a, b, c \rangle$ and containing the point (x_0, y_0, z_0). Neither the vector equation nor the standard form is unique, since we may use any point P_0 on the plane and any normal vector $\mathbf{n}$. Note, however, that all normal vectors to a given plane are parallel.

If we carry out the indicated multiplications in equation (14.30) and simplify, we get an equation of the form

$$ax + by + cz + d = 0 \qquad (14.31)$$

We can also reverse this procedure. Suppose we are given an equation in the form (14.30). We find a point (x_0, y_0, z_0) that satisfies the equation (for example, by choosing x_0 and y_0 arbitrarily and solving for z_0). Then, $ax_0 + by_0 + cz_0 + d = 0$, so that $d = -ax_0 - by_0 - cz_0$. Substituting this value of d into equation (14.31), we get

$$ax + by + cz + (-ax_0 - by_0 - cz_0) = 0$$

or

$$a(x - x_0) + b(y - y_0) + c(z - z_0) = 0$$

This is the standard form (14.30), and so we know it represents a plane. An equation of the form (14.31) is called **linear** (meaning first degree) in x, y, and z. So what we have shown is that *every linear equation in x, y, and z represents a plane in space.* Furthermore, the coefficients of x, y, and z, in order, are components of a normal vector to the plane. We also refer to equation (14.31) as a **general form** of the equation of a plane.

EXAMPLE 14.30 Find a general form of the equation of the plane passing through the point $(3, -1, 4)$ and having normal vector $\langle 2, 5, -3 \rangle$.

Solution An equation in the standard form (14.30) is

$$2(x - 3) + 5(y + 1) - 3(z - 4) = 0$$

Simplifying, we get

$$2x + 5y - 3z + 11 = 0$$

■

A normal vector is not always given directly but can be found from the given information. Frequently, this involves a cross product, as the next two examples show.

EXAMPLE 14.31 Find an equation for the plane that contains the point $P(1, 0, -3)$, $Q(2, -5, -6)$, and $R(6, 3, -4)$.

Solution As we show in Figure 14.36, the vectors $\overrightarrow{PQ}$ and $\overrightarrow{PR}$ lie in the plane, and so a normal can be found by taking their cross product:

$$\overrightarrow{PQ} = \langle 1, -5, -3 \rangle \quad \text{and} \quad \overrightarrow{PR} = \langle 5, 3, -1 \rangle$$

$$\overrightarrow{PQ} \times \overrightarrow{PR} = \begin{vmatrix} \mathbf{i} & \mathbf{j} & \mathbf{k} \\ 1 & -5 & -3 \\ 5 & 3 & -1 \end{vmatrix} = \begin{vmatrix} -5 & -3 \\ 3 & -1 \end{vmatrix} \mathbf{i} - \begin{vmatrix} 1 & -3 \\ 5 & -1 \end{vmatrix} \mathbf{j} + \begin{vmatrix} 1 & -5 \\ 5 & 3 \end{vmatrix} \mathbf{k}$$
$$= 14\mathbf{i} - 14\mathbf{j} + 28\mathbf{k}$$

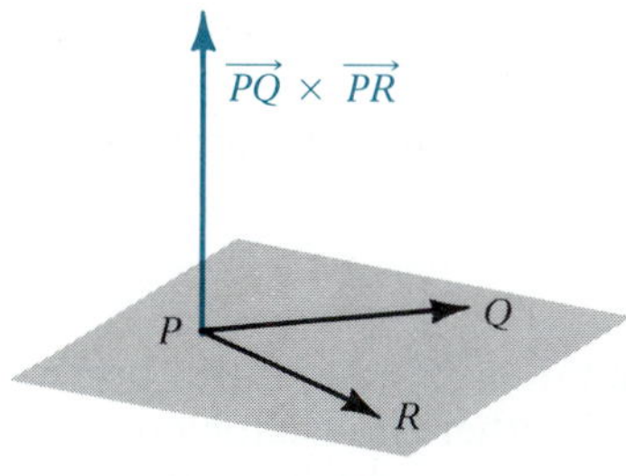

FIGURE 14.36

Since any nonzero vector perpendicular to the plane is a suitable normal, we use $\frac{1}{14}$ of this cross product, getting $\mathbf{n} = \langle 1, -1, 2 \rangle$. We can use any one of P, Q, or R as the point (x_0, y_0, z_0). Choosing P, we get for the equation in standard form

$$(x - 1) - (y - 0) + 2(z + 3) = 0$$

or in general form

$$x - y + 2z + 5 = 0$$

■

EXAMPLE 14.32 Find an equation of the plane that contains the line $\mathbf{r} = \langle 2 - t, 3 + 4t, -1 - 2t \rangle$ and the point $(5, -2, 7)$.

Solution A point P_0 on the line is $(2, 3, -1)$, and a direction vector for the line is $\mathbf{v} = \langle -1, 4, -2 \rangle$. Let Q be the given point $(5, -2, 7)$. Then $\overrightarrow{P_0Q}$ is a vector in the plane. So a normal can be found by taking the cross product $\mathbf{v} \times \overrightarrow{P_0Q}$. This is illustrated in Figure 14.37. Since $\overrightarrow{P_0Q} = \langle 3, -5, 8 \rangle$, we get

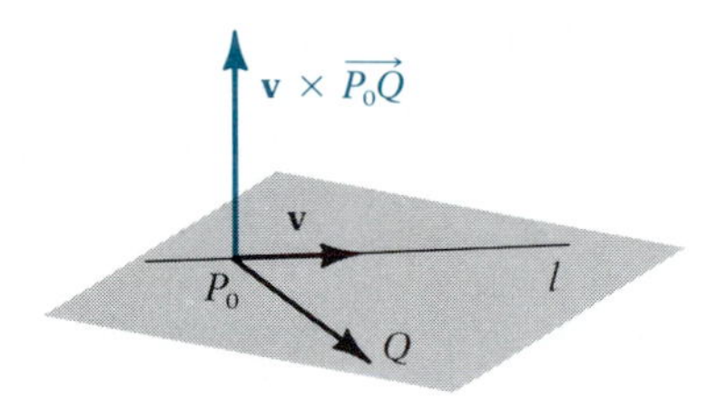

FIGURE 14.37

$$\mathbf{v} \times \overrightarrow{P_0Q} = \langle -1, 4, -2 \rangle \times \langle 3, -5, 8 \rangle = \langle 22, 2, -7 \rangle$$

(You should supply the missing steps.) We can write the equation of the plane as

$$22(x - 2) + 2(y - 3) - 7(z + 1) = 0$$

or, equivalently,

$$22x + 2y - 7z - 57 = 0$$

■

Two planes are parallel if their respective normals are parallel, and *two planes are perpendicular* if their respective normals are orthogonal. A line and a plane are parallel if a direction vector for the line and a normal to the plane are orthogonal. By the angle between two planes with normals $\mathbf{n}_1$ and $\mathbf{n}_2$, we mean the angle between $\mathbf{n}_1$ and $\mathbf{n}_2$ or its supplement, whichever does not exceed $\pi/2$. If we call this angle θ, then

$$\cos \theta = \frac{|\mathbf{n}_1 \cdot \mathbf{n}_2|}{|\mathbf{n}_1| \, |\mathbf{n}_2|}, \qquad 0 \le \theta \le \tfrac{\pi}{2}$$

EXAMPLE 14.33 Find an equation of the plane that contains the line

$$\mathbf{r} = (2 + t)\mathbf{i} + (-3 + 4t)\mathbf{j} + (1 - t)\mathbf{k}$$

and that is perpendicular to the plane $3x - 4y - 5z + 7 = 0$.

Solution A normal to the desired plane is orthogonal to the direction vector $\mathbf{i} + 4\mathbf{j} - \mathbf{k}$ of the line and to the normal $3\mathbf{i} - 4\mathbf{j} - 5\mathbf{k}$ of the perpendicular plane. We take the cross product to get

$$(\mathbf{i} + 4\mathbf{j} - \mathbf{k}) \times (3\mathbf{i} - 4\mathbf{j} + 5\mathbf{k}) = 16\mathbf{i} - 8\mathbf{j} + 8\mathbf{k}$$

We use for $\mathbf{n}$ the simpler normal $2\mathbf{i} - \mathbf{j} + \mathbf{k}$. A point in the plane can be taken as any point on the given line. The one corresponding to $t = 0$ is $(2, -3, 1)$. Thus, the desired equation is

$$2(x - 2) - (y + 3) + (z - 1) = 0$$

or, in general form,

$$2x - y + z - 8 = 0$$

■

We will find it helpful later to be able to show planes graphically. Since planes are unbounded, it is impossible to graph an entire plane, but we can give an indication of its graph by showing its **traces** on the coordinate planes. These are the lines of intersection of the given plane with the coordinate planes. Finding *intercepts* on the coordinate axes is helpful in getting the traces.

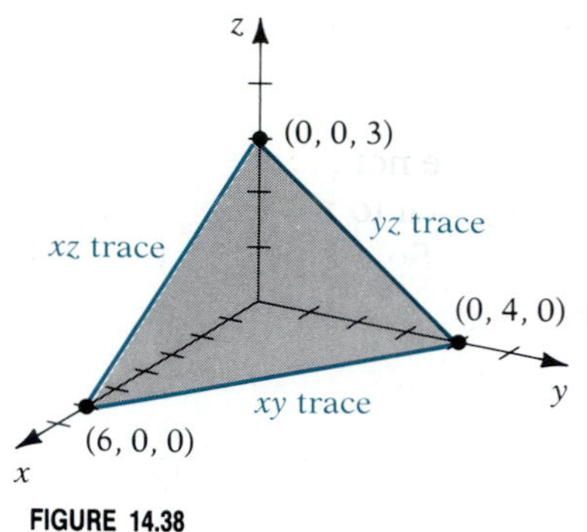

FIGURE 14.38

For example, the plane

$$2x + 3y + 4z - 12 = 0$$

has x-intercept 6, y-intercept 4, and z-intercept 3. Connecting these gives the traces, as shown in Figure 14.38. Two-dimensional equations of the traces are found by setting one of the variables equal to 0. For example, if we set $z = 0$ in the equation of this plane, we get

$$2x + 3y = 12$$

which is the equation of the trace in the xy-plane.

Not all planes have intercepts on each axis. In the next example we illustrate some of these.

EXAMPLE 14.34 Describe the planes that have the following equations, and sketch their graphs.

(a) $x = 3$ (b) $z = 4$ (c) $2x + 3y = 6$

Solution (a) A normal vector is $\langle 1, 0, 0 \rangle = \mathbf{i}$. So the plane is perpendicular to the x-axis, and hence parallel to the yz-plane. The x-intercept is 3. A portion of its graph is shown in Figure 14.39(a).

(b) A normal vector is $\langle 0, 0, 1 \rangle = \mathbf{k}$. So this plane is perpendicular to the z-axis, crossing it at $z = 4$. It is parallel to the xy-plane. A portion of its graph is shown in Figure 14.39(b).

(c) The x-intercept is 3, and the y-intercept is 2. The xy trace has the same equation as that of the plane, since putting $z = 0$ does not alter the equation. A normal vector is $\langle 2, 3, 0 \rangle$, which is a vector in a horizontal plane, so the plane itself is vertical—that is, parallel to the z-axis. A portion of its graph is shown in Figure 14.39(c). ■

Two nonparallel planes intersect in a line, and we can find an equation of the line from the equations of the planes by the procedure shown in the next example.

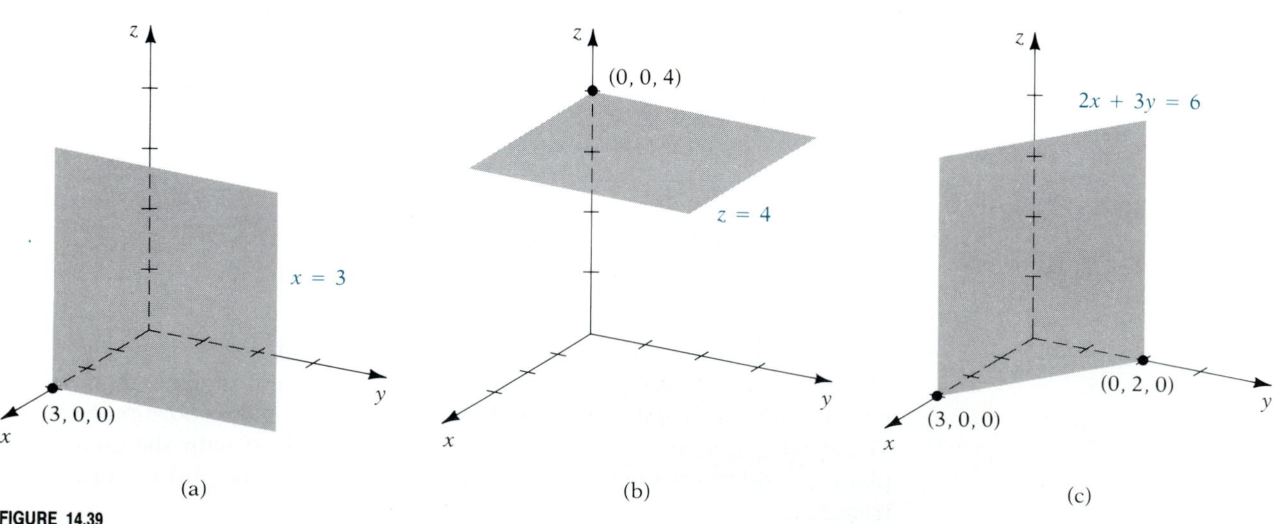

FIGURE 14.39

EXAMPLE 14.35 Find a vector equation of the line of intersection of the planes $3x - 4y + 2z = 7$ and $x + 2y - 3z = 4$.

Solution The planes are not parallel because their normals are not parallel. Since the line of intersection lies in both planes, it is orthogonal to the normals $\mathbf{n}_1 = \langle 3, -4, 2 \rangle$ and $\mathbf{n}_2 = \langle 1, 2, -3 \rangle$ of the two planes. So a direction vector $\mathbf{v}$ of the line is the cross product of these normals:

$$\mathbf{v} = \mathbf{n}_1 \times \mathbf{n}_2 = \langle 3, -4, 2 \rangle \times \langle 1, 2, -3 \rangle = \langle 8, 11, 10 \rangle \qquad \text{(Verify.)}$$

We can find a point on the line by solving simultaneously the equations of the planes. Since there are three variables and two equations, we can select one variable arbitrarily and solve for the other two. Letting $z = 0$, we have

$$\begin{cases} 3x - 4y = 7 \\ x + 2y = 4 \end{cases}$$

The solution is found to be $x = 3$, $y = \frac{1}{2}$. So a point on the line is $(3, \frac{1}{2}, 0)$. A vector equation of the line is therefore

$$\mathbf{r} = \langle 3, \tfrac{1}{2}, 0 \rangle + t\langle 8, 11, 10 \rangle$$

or

$$\mathbf{r} = \langle 3 + 8t, \tfrac{1}{2} + 11t, 10t \rangle$$

■

We conclude this section by deriving a formula for the distance between a plane and a point not on the plane. Let $ax + by + cz + d = 0$ be the equation of the plane, and let the point be $P_1(x_1, y_1, z_1)$. Let $P_0(x_0, y_0, z_0)$ be any point on the plane. Then, as Figure 14.40 shows, the distance D between the plane and the point P_1 is the length of the projection of $\overrightarrow{P_0P_1}$ along the normal $\mathbf{n} = \langle a, b, c \rangle$. The length of this projection is $|\text{Comp}_{\mathbf{n}} \overrightarrow{P_0P_1}|$. So we have

$$\begin{aligned} D = |\text{Comp}_{\mathbf{n}} \overrightarrow{P_0P_1}| &= \frac{|\mathbf{n} \cdot \overrightarrow{P_0P_1}|}{|\mathbf{n}|} = \frac{|\langle a, b, c \rangle \cdot \langle x_1 - x_0, y_1 - y_0, z_1 - z_0 \rangle|}{\sqrt{a^2 + b^2 + c^2}} \\ &= \frac{|a(x_1 - x_0) + b(y_1 - y_0) + c(z_1 - z_0)|}{\sqrt{a^2 + b^2 + c^2}} \\ &= \frac{|ax_1 + by_1 + cz_1 - (ax_0 + by_0 + cz_0)|}{\sqrt{a^2 + b^2 + c^2}} \end{aligned}$$

Since P_1 is on the plane, its coordinates satisfy the equation of the plane. So $ax_0 + by_0 + cz_0 = -d$. Thus, we obtain the following formula.

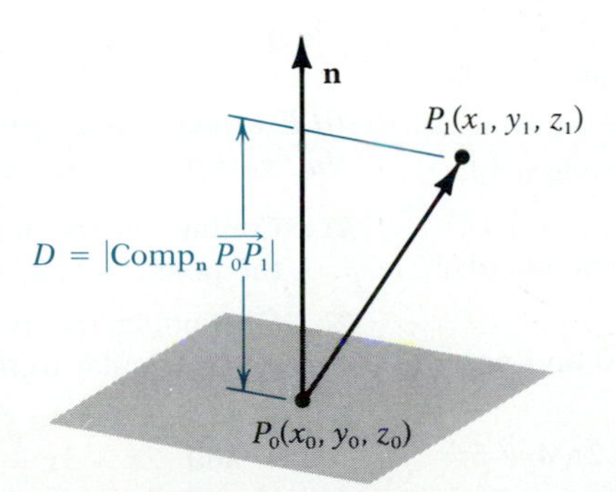

FIGURE 14.40

The distance D between the plane $ax + by + cz + d = 0$ and the point (x_1, y_1, z_1) is

$$D = \frac{|ax_1 + by_1 + cz_1 + d|}{\sqrt{a^2 + b^2 + c^2}} \tag{14.32}$$

Note Compare this result with the two-dimensional case of the distance between a line and a point in Exercise 51 of Exercise Set 14.3.

EXAMPLE 14.36 Find the distance between the plane $3x - 4y + 5z - 8 = 0$ and the point $(2, 1, -1)$.

Solution By equation (14.32),

$$D = \frac{|3(2) - 4(1) + 5(-1) - 8|}{\sqrt{9 + 16 + 25}} = \frac{11}{5\sqrt{2}}$$

■

EXERCISE SET 14.7

A

In Exercises 1–4, find an equation of the plane through the given point perpendicular to the given vector in (a) vector form, (b) standard form, and (c) general form.

1. $(4, 2, -6)$, $\langle 3, 2, -1 \rangle$

2. $(1, 0, -3)$, $\langle -1, 2, 4 \rangle$

3. $(5, -3, -4)$, $2\mathbf{i} + 3\mathbf{j} - 4\mathbf{k}$

4. $(6, 1, 2)$, $\mathbf{i} - 2\mathbf{j} + 2\mathbf{k}$

In Exercises 5–12 find a general form of the equation of the plane that satisfies the given conditions.

5. Perpendicular to the line $\mathbf{r} = \langle 2 - t, 3 + 2t, 1 + 4t \rangle$ at the point $(2, 3, 1)$

6. Perpendicular to the line $\mathbf{r} = (4 + 5t)\mathbf{i} + 3t\mathbf{j} + (1 - 2t)\mathbf{k}$ at the point $(4, 0, 1)$

7. Containing the point $(1, 4, -5)$ and parallel to the plane $3x - 2y + 4z = 7$

8. Containing the point $(-2, 3, 4)$ and parallel to the plane $2x - 3z = 4$

9. Containing the point $(3, -2, 4)$ and perpendicular to the z-axis

10. Containing the point $(-1, 8, 11)$ and perpendicular to the y-axis

11. Parallel to the plane $3x - 4y + 5z + 8 = 0$ and passing through the origin

12. Perpendicular to the line $\mathbf{r} = \langle 2 + t, 3 - 2t, 4 + 5t \rangle$ and passing through the origin

In Exercises 13–16 find a general equation of the plane that contains the three given points.

13. $(2, 4, -5)$, $(1, -3, 4)$, $(3, -1, 2)$

14. $(0, 3, -1)$, $(2, 4, 2)$, $(-1, 2, -3)$

15. $(1, 0, -1)$, $(2, 3, 1)$, $(4, -3, 2)$

16. $(5, 4, -3)$, $(2, -1, -2)$, $(4, 2, 3)$

In Exercises 17 and 18 find the angle between the two planes.

17. $2x - 3y - 4z = 8$, $3x + 2y - z = 4$

18. $3x + 4y - 2z = 3$, $2x - y - 3z = 5$

In Exercises 19 and 20 show that the two planes are perpendicular.

19. $3x - 4y + 2z = 5$, $2x + 3y + 3z = 7$

20. $5x + 3y - 4z = 8$, $2x - 6y - 2z = 15$

In Exercises 21–24 find a general equation of the plane that satisfies the given conditions.

21. Containing the line $\mathbf{r} = \langle 3 - 2t, 2 + t, 4 + 3t \rangle$ and the point $(-1, 2, 4)$

22. Containing the points $(4, -2, 1)$ and $(3, 1, 2)$ and perpendicular to the plane $3x + 2y - 4z = 5$

23. Perpendicular to each of the planes $3x + 5y - 4z = 4$ and $2x - 3y - z = 2$ and containing the point $(3, -3, 1)$

24. Containing the line $\mathbf{r} = 2\mathbf{i} + (3 - t)\mathbf{j} + (4 + 2t)\mathbf{k}$ and parallel to the line $\mathbf{r} = (3 + 2t)\mathbf{i} + (1 - t)\mathbf{j} + (-2 + 3t)\mathbf{k}$

In Exercises 25–34 use intercepts and traces on the coordinate planes to sketch the given plane.

25. $3x + 2y + z - 6 = 0$
26. $9x + 2y + 6z = 18$
27. $2x - y + z = 4$
28. $x + y - z + 4 = 0$
29. $3x + 4y = 12$
30. $2x + z = 8$
31. $y = 2z$
32. $x + 1 = 0$
33. $y = 4$
34. $x + y - 2z = 0$

In Exercises 35 and 36 find parametric equations of the line of intersection of the two planes.

35. $3x - y - 2z = 4$
    $5x + y + z = -2$
36. $x + 4y + 3z = 3$
    $2x - 7y + z = 11$

In Exercises 37–38 find the distance between the plane and the point.

37. $3x - 4y + 10z = 5$; $(1, -1, 2)$
38. $2x - y - 2z + 3 = 0$; $(6, -1, -4)$
39. Show that the planes $x - 2y + 2z = 3$ and $3x - 6y + 6z + 5 = 0$ are parallel, and find the distance between them. (*Hint:* Find the distance from one of the planes to a point on the other.)

B

40. Show that l_1 and l_2 are parallel, and find an equation of the plane that contains them.

    l_1: $\mathbf{r} = \langle 2 - 3t, 4 + 2t, -3 + t\rangle$
    l_2: $\mathbf{r} = \langle 6t, 3 - 4t, 5 - 2t\rangle$

41. Show that l_1 and l_2 intersect, and find an equation of the plane that contains them.

    l_1: $\mathbf{r} = \langle 1 + 2t, -1 + 3t, 2 - t\rangle$
    l_2: $\mathbf{r} = \langle 5 + 3t, 6 + 5t, -1 - 2t\rangle$

42. Find an equation of the plane perpendicular to the line of intersection of the planes $3x - y + 4z = 2$ and $x + 2y - z = 3$ at the point where this line pierces the xy-plane.
43. Find the point where the line $\mathbf{r} = (2 + t)\mathbf{i} + (3 - 2t)\mathbf{j} + (1 - t)\mathbf{k}$ pierces the plane $3x - 2y + 4z + 5 = 0$.
44. Find the minimum distance between the two skew lines

    l_1: $x = 3 + t, y = 2 - t, z = 4 + 3t$
    l_2: $x = -1 - 2t, y = 5 + 4t, z = -3t$

    (*Hint:* First find a plane that contains one of the lines and is parallel to the other.)

14.8 SUPPLEMENTARY EXERCISES

Exercises 1 and 2 refer to the vectors $\mathbf{u} = 3\mathbf{i} - 4\mathbf{j}$, $\mathbf{v} = \mathbf{i} + 2\mathbf{j}$, *and* $\mathbf{w} = 5\mathbf{i} - 7\mathbf{j}$. *Find the specified quantities.*

1. (a) $(2\mathbf{u} - 3\mathbf{v}) \cdot (\mathbf{v} + 2\mathbf{w})$
   (b) A vector in the direction of $\mathbf{v}$ with length $|3\mathbf{u} - 2\mathbf{w}|^2$
2. (a) $\text{Proj}_{\mathbf{v}}\mathbf{u}$ and $\text{Proj}_{\mathbf{v}}^{\perp}\mathbf{u}$
   (b) Scalars a and b such that $\mathbf{w} = a\mathbf{u} + b\mathbf{v}$

Exercises 3–6 refer to the vectors $\mathbf{u} = \langle 1, 1, -4\rangle$, $\mathbf{v} = \langle -2, 0, 3\rangle$, *and* $\mathbf{w} = \langle -2, 1, 2\rangle$. *Find the specified quantities.*

3. (a) $|\mathbf{u}|^2 - (\mathbf{v} \cdot \mathbf{w})^2$ (b) The angle between $\mathbf{u}$ and $\mathbf{w}$
4. (a) $(\mathbf{u} \times \mathbf{v}) \cdot \mathbf{w}$ (b) $\mathbf{u} \times (\mathbf{v} \times \mathbf{w})$
5. (a) $|2\mathbf{u} - 3\mathbf{v} + \mathbf{w}|$ (b) $\text{Proj}_{\mathbf{w}}\mathbf{v}$
6. (a) The component of $\mathbf{v}$ in the direction of $\mathbf{w} \times \mathbf{u}$
   (b) The volume of the parallelepiped with adjacent sides $\mathbf{u}$, $\mathbf{v}$, and $\mathbf{w}$

7. Using vector methods prove that the line segment joining the midpoints of two sides of a triangle is parallel to the third side and one-half its length.
8. Forces $\mathbf{F}_1$, $\mathbf{F}_2$, and $\mathbf{F}_3$ of magnitudes 25.3 N, 14.8 N, and 19.6 N, respectively, are acting on an object as shown in the figure. Find the magnitude and direction of the resultant force.

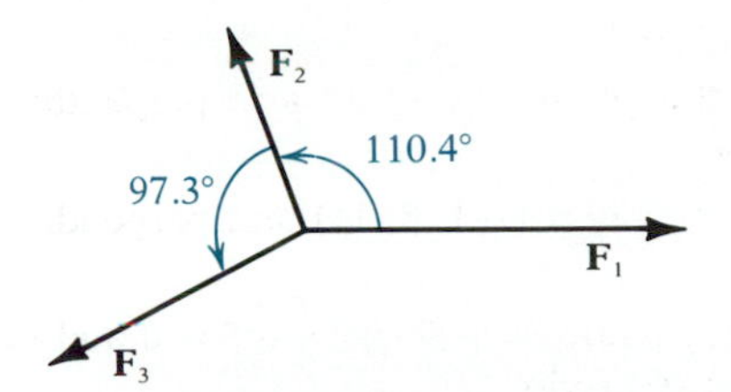

9. A pilot wants to fly from town A to town B, 400 mi due south of town A. A 60 mi/hr wind is blowing

from 210°. In order to make the trip in 2 hr, at what heading and at what average speed should she fly?

10. Prove that for arbitrary vectors $\mathbf{u}$, $\mathbf{v}$, and $\mathbf{w}$ in V_3, $(\mathbf{u} \times \mathbf{v}) \times \mathbf{w} + (\mathbf{v} \times \mathbf{w}) \times \mathbf{u} + (\mathbf{w} \times \mathbf{u}) \times \mathbf{v} = \mathbf{0}$. (This is known as *Jacobi's identity.*)

In Exercises 11–13 find both vector and parametric equations of the line that satisfies the given conditions.

11. Through $(3, 1, -2)$, parallel to $\mathbf{r} = \langle 2 - t, 4 + 2t, 3 + 5t \rangle$

12. Formed by the intersection of the planes $2x - 3y - z = 4$ and $x + y + 2z + 3 = 0$.

13. Through $(4, 2, -1)$, perpendicular to the plane $5x - 3y + 4z = 7$

14. Find the point where the line $\mathbf{r} = \langle 1 - 3t, 2t, 3 + t \rangle$ pierces the plane $3x - 5y + 4z = 5$. Also find the angle between the line and the plane (defined as the complement of the angle between the line and a normal to the plane).

15. Determine whether the lines $\mathbf{r}_1 = \langle 3 - t, 4 + 2t, -1 + t \rangle$ and $\mathbf{r}_2 = \langle 2s, 1 - s, 3 + 4s \rangle$ intersect. If so, find their point of intersection.

16. Show that the lines $\mathbf{r}_1 = \langle 2 - t, 4 + 2t, -3 + 4t \rangle$ and $\mathbf{r}_2 = \langle 1 + t, 5 - 3t, 3 - 2t \rangle$ intersect, and find an equation of the plane that contains them.

In Exercises 17 and 18 find an equation of the plane that satisfies the given conditions.

17. Containing the points $(2, 0, -1)$, $(3, 2, 0)$, and $(-4, -2, 3)$

18. Perpendicular to the line of intersection of the planes $3x - y + z + 3 = 0$ and $x + 2y - z = 9$ at the point where this line pierces the xy-plane

19. (a) Find the angles the vector $\mathbf{u} = -2\mathbf{i} + 4\mathbf{j} + 5\mathbf{k}$ makes with the coordinate axes.
    (b) For a certain vector $\mathbf{v}$ of length 12, $\cos \beta = \frac{7}{9}$ and $\cos \gamma = \frac{4}{9}$. Find $\mathbf{v}$, given that it has a negative x component.

20. A force $\mathbf{F}$ of magnitude 30 N acts in a direction perpendicular to the plane $3x - 4y + 5z = 10$, and its z component is positive. Find the work done by $\mathbf{F}$ in moving an object from $A(4, 1, -2)$ to $B(2, -5, -1)$.

21. (a) Find the distance from the line $3x - 4y = 7$ to the point $(5, -2)$.
    (b) Find the distance from the plane $x - 2y + 2z = 7$ to the point $(2, -1, 4)$.

22. Find x, y, and z so that the vectors $x\mathbf{i} + \mathbf{j} - 3\mathbf{k}$, $3\mathbf{i} + y\mathbf{j} - 7\mathbf{k}$, and $8\mathbf{i} + 5\mathbf{j} + z\mathbf{k}$ will be mutually orthogonal.

23. Prove that if $\mathbf{u}$, $\mathbf{v}$, and $\mathbf{w}$ are mutually orthogonal nonzero vectors, then they are *linearly independent;* that is, the equation

$$a\mathbf{u} + b\mathbf{v} + c\mathbf{w} = \mathbf{0}$$

is satisfied only if $a = b = c = 0$. (*Hint:* In turn, find the dot product of $\mathbf{u}$, $\mathbf{v}$, and $\mathbf{w}$ with both sides.)

24. Unit vectors that are mutually orthogonal are said to form an **orthonormal** set. Prove that if $\mathbf{u}$, $\mathbf{v}$, and $\mathbf{w}$ are any three noncoplanar vectors in V_3, then the vectors $\mathbf{e}_1$, $\mathbf{e}_2$, and $\mathbf{e}_3$, defined by

$$\mathbf{e}_1 = \frac{\mathbf{u}}{|\mathbf{u}|} \qquad \mathbf{e}_2 = \frac{\mathbf{v} - (\mathbf{v} \cdot \mathbf{e}_1)\mathbf{e}_1}{|\mathbf{v} - (\mathbf{v} \cdot \mathbf{e}_1)\mathbf{e}_1|}$$

$$\mathbf{e}_3 = \frac{\mathbf{w} - (\mathbf{w} \cdot \mathbf{e}_1)\mathbf{e}_1 - (\mathbf{w} \cdot \mathbf{e}_2)\mathbf{e}_2}{|\mathbf{w} - (\mathbf{w} \cdot \mathbf{e}_1)\mathbf{e}_1 - (\mathbf{w} \cdot \mathbf{e}_2)\mathbf{e}_2|}$$

form an orthonormal set.

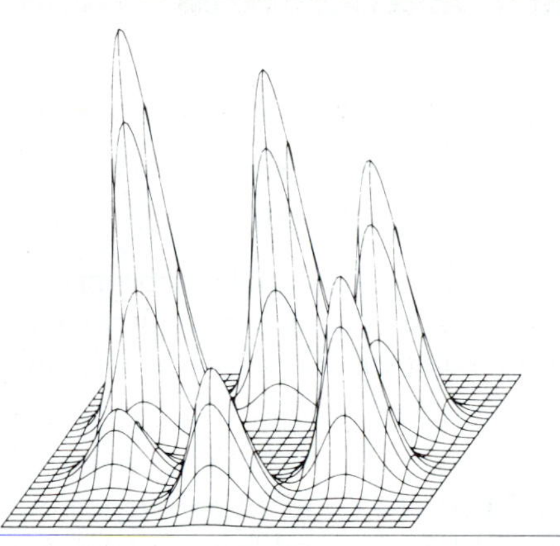

CHAPTER 15
VECTOR-VALUED FUNCTIONS

15.1 THE CALCULUS OF VECTOR FUNCTIONS

A function whose domain is a subset of **R** and whose range is a subset of V_2 or V_3 is called a *vector-valued function* or, more briefly, a *vector function.* Since vectors in V_2 can be identified with vectors in V_3 in which the third component is 0, we will concentrate on vector functions with values in V_3. If f_1, f_2, and f_3 are real-valued functions of a real variable, also called *scalar functions,* then the function **F** defined by

$$\mathbf{F}(t) = \langle f_1(t), f_2(t), f_3(t) \rangle$$

is a vector function. Alternately we may write $\mathbf{F}(t)$ in terms of the basis vectors **i**, **j**, and **k** as $\mathbf{F}(t) = f_1(t)\mathbf{i} + f_2(t)\mathbf{j} + f_3(t)\mathbf{k}$. The functions f_1, f_2, and f_3 are called the *component functions* of **F**. The domain of **F**, unless otherwise specified, will be taken to be the intersection of the domains of its component functions.

We encountered functions of this type in the study of lines. For example, consider

$$\mathbf{r}(t) = \langle 1 - t, 2 + 3t, 4t \rangle$$

For each t in **R**, $\mathbf{r}(t)$ is a vector, and if we interpret it geometrically as the position vector of a point in space, then as t varies over **R**, the tip of $\mathbf{r}(t)$ traces out a line that passes through the point (1, 2, 0), parallel to the vector $\langle -1, 3, 4 \rangle$.

In the next section we will consider geometric interpretations of vector functions other than lines, but for now we want to concentrate primarily on the calculus concepts of limit, continuity, derivative, and integral for these functions.

Vector functions can be combined, or combined with scalar functions, in ways that are natural extensions of combinations of real-valued functions. In particular, for vector functions **F** and **G** and scalar function α, we define

$\mathbf{F} \pm \mathbf{G}$, $\alpha\mathbf{F}$, $\mathbf{F} \cdot \mathbf{G}$, $\mathbf{F} \times \mathbf{G}$, and $\mathbf{F} \circ \alpha$ by

$$(\mathbf{F} \pm \mathbf{G})(t) = \mathbf{F}(t) \pm \mathbf{G}(t) \qquad (\alpha\mathbf{F})(t) = \alpha(t)\mathbf{F}(t) \qquad (\mathbf{F} \cdot \mathbf{G})(t) = \mathbf{F}(t) \cdot \mathbf{G}(t)$$
$$(\mathbf{F} \times \mathbf{G})(t) = \mathbf{F}(t) \times \mathbf{G}(t) \qquad (\mathbf{F} \circ \alpha)(t) = \mathbf{F}(\alpha(t))$$

In each case we must make appropriate assumptions about domains so that all functions are defined. Note that each combination results in a new vector function except $\mathbf{F} \cdot \mathbf{G}$, which is a scalar function.

DEFINITION 15.1 Let $\mathbf{F}(t) = \langle f_1(t), f_2(t), f_3(t) \rangle$.

1. The **limit** of $\mathbf{F}(t)$ as t approaches a exists if and only if $\lim_{t\to a} f_1(t) = l_1$, $\lim_{t\to a} f_2(t) = l_2$, and $\lim_{t\to a} f_3(t) = l_3$ exist, and in this case $\lim_{t\to a} \mathbf{F}(t) = \mathbf{L}$, where $\mathbf{L} = \langle l_1, l_2, l_3 \rangle$.
2. If $\lim_{t\to a} \mathbf{F}(t) = \mathbf{F}(a)$, then $\mathbf{F}$ is said to be **continuous** at $t = a$. ■

Remarks To say that $\lim_{t\to a} \mathbf{F}(t) = \mathbf{L}$ means that the vectors $\mathbf{F}(t)$ will be arbitrarily close to the vector $\mathbf{L}$ *in both magnitude and direction* for all values of t sufficiently close to a. From (1) and (2) it follows that $\mathbf{F}$ is continuous if and only if each of its component functions is continuous.

EXAMPLE 15.1 (a) Let $\mathbf{F}(t) = \langle t^2, \cos t, (\sin t)/t \rangle$. Find $\lim_{t\to 0} \mathbf{F}(t)$.
(b) Show that $\mathbf{F}(t) = \langle e^t, \ln t, \sinh t \rangle$ is continuous at $t = a$, where $a > 0$.

Solution (a) Since

$$\mathbf{L} = \left\langle \lim_{t\to 0} t^2, \lim_{t\to 0} \cos t, \lim_{t\to 0} \frac{\sin t}{t} \right\rangle = \langle 0, 1, 1 \rangle$$

exists, we have, $\lim_{t\to 0} \mathbf{F}(t) = \langle 0, 1, 1 \rangle$.

(b) Each of the component functions is continuous at $t = a$, where $a > 0$. Thus,

$$\lim_{t\to a} \mathbf{F}(t) = \left\langle \lim_{t\to a} e^t, \lim_{t\to a} \ln t, \lim_{t\to a} \sinh t \right\rangle = \langle e^a, \ln a, \sinh a \rangle = \mathbf{F}(a)$$

So $\mathbf{F}$ is continuous at $t = a$. ■

DEFINITION 15.2 The **derivative** of a vector function $\mathbf{F}$ at t is defined by

$$\mathbf{F}'(t) = \lim_{h\to 0} \frac{\mathbf{F}(t + h) - \mathbf{F}(t)}{h}$$

provided this limit exists. If $\mathbf{F}'(t)$ exists, we say $\mathbf{F}$ is **differentiable** at t. ■

THEOREM 15.1 If $\mathbf{F}(t) = \langle f_1(t), f_2(t), f_3(t) \rangle$, then $\mathbf{F}'(t)$ exists if and only if $f_1'(t)$, $f_2'(t)$, and $f_3'(t)$ exist, and in this case,

$$\mathbf{F}'(t) = \langle f_1'(t), f_2'(t), f_3'(t) \rangle$$

Proof

$$\frac{\mathbf{F}(t + h) - \mathbf{F}(t)}{h} = \left\langle \frac{f_1(t + h) - f_1(t)}{h}, \frac{f_2(t + h) - f_2(t)}{h}, \frac{f_3(t + h) - f_3(t)}{h} \right\rangle$$

The limit on the left exists if and only if the limits of the components on the right exist, and these are $f'(t)$, $g'(t)$, and $h'(t)$ by definition of the derivative of a real-valued function.

The conclusion now follows. ■

EXAMPLE 15.2 Find $\mathbf{F}'(\frac{\pi}{3})$ if $\mathbf{F}(t) = \langle \sin 2t, 2\cos t, \tan t \rangle$.

Solution By Theorem 15.1, $\mathbf{F}'(t) = \langle 2\cos 2t, -2\sin t, \sec^2 t \rangle$. So $\mathbf{F}'(\frac{\pi}{3}) = \langle -1, -\sqrt{3}, 4 \rangle$. ■

THEOREM 15.2 If $\mathbf{F}$ and $\mathbf{G}$ are differentiable vector functions and α is a differentiable scalar function, then

1. $(\mathbf{F} \pm \mathbf{G})'(t) = \mathbf{F}'(t) \pm \mathbf{G}'(t)$
2. $(\alpha\mathbf{F})'(t) = \alpha(t)\mathbf{F}'(t) + \alpha'(t)\mathbf{F}(t)$
3. $(\mathbf{F} \cdot \mathbf{G})'(t) = \mathbf{F}(t) \cdot \mathbf{G}'(t) + \mathbf{F}'(t) \cdot \mathbf{G}(t)$
4. $(\mathbf{F} \times \mathbf{G})'(t) = \mathbf{F}(t) \times \mathbf{G}'(t) + \mathbf{F}'(t) \times \mathbf{G}(t)$
5. $(\mathbf{F} \circ \alpha)'(t) = \mathbf{F}'(\alpha(t))\alpha'(t)$ (the chain rule) ■

The proof of each part follows from the analogous result for scalar functions, since derivatives of vector functions are computed component by component. Again, appropriate conditions on domains must be assumed.

Remarks

1. In equation 2 of Theorem 15.2, if α is a constant, say $\alpha(t) = k$, then we obtain $(k\mathbf{F})'(t) = k\mathbf{F}'(t)$.
2. It is essential in equation 4 that the order of the factors not be reversed, since the cross product is noncommutative.
3. In equation 5 the scalar $\alpha'(t)$ would normally precede the vector $\mathbf{F}'(\alpha(t))$. We wrote it as we did so that it would have the same appearance as the chain rule for scalar functions. In general, for a vector $\mathbf{v}$ and a scalar k, $\mathbf{v}k$ means the same as $k\mathbf{v}$.
4. The Leibniz notation, d/dt, for differentiation can be used throughout. For example, the chain rule can be written as

$$\frac{d\mathbf{F}}{dt} = \frac{d\mathbf{F}}{d\alpha}\frac{d\alpha}{dt}$$

EXAMPLE 15.3 Let $\mathbf{F}(t) = \langle t, 2 - t^3, 2t \rangle$, $\mathbf{G}(t) = \langle 1, t^2, t^3 \rangle$, and $\alpha(t) = 2/t^2$. Use Theorem 15.2 to find the following:

(a) $(\mathbf{F} - \mathbf{G})'(1)$ (b) $(\alpha\mathbf{F})'(1)$ (c) $(\mathbf{F} \cdot \mathbf{G})'(1)$

(d) $(\mathbf{F} \times \mathbf{G})'(1)$ (e) $(\mathbf{F} \circ \alpha)'(1)$

Solution First we calculate $\mathbf{F}'$, $\mathbf{G}'$, and α' and then evaluate each function and its derivative at $t = 1$:

$$\mathbf{F}'(t) = \langle 1, -3t^2, 2 \rangle \qquad \mathbf{G}'(t) = \langle 0, 2t, 3t^2 \rangle \qquad \alpha'(t) = -\frac{4}{t^3}$$

$$\mathbf{F}(1) = \langle 1, 1, 2 \rangle \qquad \mathbf{G}(1) = \langle 1, 1, 1 \rangle \qquad \alpha(1) = 2$$

$$\mathbf{F}'(1) = \langle 1, -3, 2 \rangle \qquad \mathbf{G}'(1) = \langle 0, 2, 3 \rangle \qquad \alpha'(1) = -4$$

For part e we will also need $\mathbf{F}'(\alpha(1)) = \mathbf{F}'(2) = \langle 1, -12, 2\rangle$. Now we apply Theorem 15.2:

(a) $(\mathbf{F} - \mathbf{G})'(1) = \mathbf{F}'(1) - \mathbf{G}'(1) = \langle 1, -3, 2\rangle - \langle 0, 2, 3\rangle = \langle 1, -5, -1\rangle$

(b) $$(\alpha\mathbf{F})'(1) = \alpha(1)\mathbf{F}'(1) + \alpha'(1)\mathbf{F}(1) = 2\langle 1, -3, 2\rangle + (-4)\langle 1, 1, 2\rangle = \langle 2, -6, 4\rangle + \langle -4, -4, -8\rangle = \langle -2, -10, -4\rangle$$

(c) $$(\mathbf{F}\cdot\mathbf{G})'(1) = \mathbf{F}(1)\cdot\mathbf{G}'(1) + \mathbf{F}'(1)\cdot\mathbf{G}(1) = \langle 1, 1, 2\rangle\cdot\langle 0, 2, 3\rangle + \langle 1, -3, 2\rangle\cdot\langle 1, 1, 1\rangle = \langle 0, 2, 6\rangle + \langle 1, -3, 2\rangle = \langle 1, -1, 8\rangle$$

(d) $$(\mathbf{F}\times\mathbf{G})'(1) = \mathbf{F}(1)\times\mathbf{G}'(1) + \mathbf{F}'(1)\times\mathbf{G}(1) = \langle 1, 1, 2\rangle\times\langle 0, 2, 3\rangle + \langle 1, -3, 2\rangle\times\langle 1, 1, 1\rangle = \langle -1, -3, 2\rangle + \langle -5, 1, 4\rangle = \langle -6, -2, 6\rangle$$

(e) $$(\mathbf{F}\circ\alpha)'(1) = \mathbf{F}'(\alpha(1))\alpha'(1) = \mathbf{F}'(2)\alpha'(1) = \langle 1, -12, 2\rangle(-4) = \langle -4, 48, -8\rangle$$ ■

The symbol $\int \mathbf{f}(t)\,dt$ is used to mean the general antiderivative of $\mathbf{f}$; that is, $\int \mathbf{f}(t)\,dt = \mathbf{F}(t) + \mathbf{C}$, where $\mathbf{F}'(t) = \mathbf{f}(t)$. Since differentiation is carried out by components, so too is antidifferentiation; that is, if $\mathbf{f}(t) = \langle f_1(t), f_2(t), f_3(t)\rangle$, then

$$\int \mathbf{f}(t)\,dt = \left\langle \int f_1(t)\,dt, \int f_2(t)\,dt, \int f_3(t)\,dt \right\rangle$$

The definite integral is defined similarly:

$$\int_a^b \mathbf{f}(t)\,dt = \left\langle \int_a^b f_1(t)\,dt, \int_a^b f_2(t)\,dt, \int_a^b f_3(t)\,dt \right\rangle$$

If $\int_a^b \mathbf{f}(t)\,dt$ exists, we say $\mathbf{f}$ is integrable on $[a, b]$.

The following theorem can be proved by considering components.

THEOREM 15.3 **1.** If $\mathbf{f}$ is continuous on $[a, b]$, then for t in $[a, b]$,

$$\frac{d}{dt}\int_a^t \mathbf{f}(u)\,du = \mathbf{f}(t)$$

2. If $\mathbf{F}$ is any antiderivative of $\mathbf{f}$ on $[a, b]$, then

$$\int_a^b \mathbf{f}(t)\,dt = \mathbf{F}(b) - \mathbf{F}(a) \qquad \text{(fundamental theorem of calculus)}$$ ■

EXAMPLE 15.4 Let $\mathbf{f}(t) = \langle 2t, 3, t^2\rangle$. Find $\int \mathbf{f}(t)\,dt$ and $\int_1^4 \mathbf{f}(t)\,dt$.

Solution

$$\int \mathbf{f}(t)\,dt = \left\langle \int 2t\,dt, \int 3\,dt, \int t^2\,dt \right\rangle = \left\langle t^2 + C_1, 3t + C_2, \frac{t^3}{3} + C_3 \right\rangle = \left\langle t^2, 3t, \frac{t^3}{3} \right\rangle + \langle C_1, C_2, C_3\rangle$$

We can write this as $\mathbf{F}(t) + \mathbf{C}$, where $\mathbf{F}(t) = \langle t^2, 3t, t^3/3 \rangle$ and $\mathbf{C} = \langle C_1, C_2, C_3 \rangle$. By the fundamental theorem,

$$\begin{aligned}\int_1^4 \mathbf{f}(t)\,dt = \mathbf{F}(t)\Big]_1^4 &= \left\langle t^2, 3t, \frac{t^3}{3} \right\rangle\Big]_1^4 \\ &= \left\langle 16, 12, \frac{64}{3} \right\rangle - \left\langle 1, 3, \frac{1}{3} \right\rangle \\ &= \langle 15, 9, 21 \rangle\end{aligned}$$

■

Additional properties of integrals of vector functions are given in the next three theorems. We omit the proof of the first of these, since it follows directly from properties of the component functions.

THEOREM 15.4 If $\mathbf{f}$ and $\mathbf{g}$ are integrable on $[a, b]$ and k is any scalar,

1. $\displaystyle\int_a^b [\mathbf{f}(t) \pm \mathbf{g}(t)]\,dt = \int_a^b \mathbf{f}(t)\,dt \pm \int_a^b \mathbf{g}(t)\,dt$
2. $\displaystyle k\int_a^b \mathbf{f}(t)\,dt = \int_a^b k\mathbf{f}(t)\,dt$
3. For any c in $[a, b]$,

$$\int_a^b \mathbf{f}(t)\,dt = \int_a^c \mathbf{f}(t)\,dt + \int_c^b \mathbf{f}(t)\,dt$$

■

THEOREM 15.5 If $\mathbf{f}(t) = \langle f_1(t), f_2(t), f_3(t) \rangle$ is integrable on $[a, b]$ and $\mathbf{k} = \langle k_1, k_2, k_3 \rangle$ is any constant vector, then

$$\mathbf{k} \cdot \int_a^b \mathbf{f}(t)\,dt = \int_a^b \mathbf{k} \cdot \mathbf{f}(t)\,dt$$

Proof

$$\begin{aligned}\mathbf{k} \cdot \int_a^b \mathbf{f}(t)\,dt &= k_1 \int_a^b f_1(t)\,dt + k_2 \int_a^b f_2(t)\,dt + k_3 \int_a^b f_3(t)\,dt \\ &= \int_a^b k_1 f_1(t)\,dt + \int_a^b k_2 f_2(t)\,dt + \int_a^b k_3 f_3(t)\,dt \\ &= \int_a^b [k_1 f_1(t) + k_2 f_2(t) + k_3 f_3(t)]\,dt = \int_a^b \mathbf{k} \cdot \mathbf{f}(t)\,dt\end{aligned}$$

where we have used properties of integrals of scalar functions. ■

If $\mathbf{f}$ is a vector function with component functions f_1, f_2, and f_3, by $|\mathbf{f}|$ we mean the scalar function defined by

$$|\mathbf{f}|(t) = |\mathbf{f}(t)| = \sqrt{(f_1(t))^2 + (f_2(t))^2 + (f_3(t))^2}$$

THEOREM 15.6 If both $\mathbf{f}$ and $|\mathbf{f}|$ are integrable on $[a, b]$, then

$$\left| \int_a^b \mathbf{f}(t)\,dt \right| \le \int_a^b |\mathbf{f}(t)|\,dt$$

Note Even though this has the same appearance as a similar property for scalar functions, it must be realized that the left-hand side is the magnitude of a vector, whereas the right-hand side is the integral of a nonnegative scalar function.

Proof Let $\mathbf{k} = \int_a^b \mathbf{f}(t)\,dt$. Then $\mathbf{k}$ is a constant vector. If $\mathbf{k} = \mathbf{0}$, the inequality is trivial, so assume $\mathbf{k} \neq \mathbf{0}$. Then we have

$$|\mathbf{k}|^2 = \mathbf{k} \cdot \mathbf{k} = \mathbf{k} \cdot \int_a^b \mathbf{f}(t)\,dt = \int_a^b \mathbf{k} \cdot \mathbf{f}(t)\,dt \quad \text{By Theorem 15.5}$$

Now $\mathbf{k} \cdot \mathbf{f}(t) \leq |\mathbf{k} \cdot \mathbf{f}(t)|$, and so

$$\int_a^b \mathbf{k} \cdot \mathbf{f}(t)\,dt \leq \int_a^b |\mathbf{k} \cdot \mathbf{f}(t)|\,dt \leq \int_a^b |\mathbf{k}|\,|\mathbf{f}(t)|\,dt \quad \text{By the Cauchy-Schwarz inequality}$$

Thus

$$|\mathbf{k}|^2 \leq |\mathbf{k}| \int_a^b |\mathbf{f}(t)|\,dt$$

and since $|\mathbf{k}| > 0$, we get the desired result by dividing both sides by $|\mathbf{k}|$. ■

EXERCISE SET 15.1

A

In Exercises 1–4 give the domain of each vector function.

1. $\mathbf{F}(t) = \langle 2t, \frac{1}{t}, \sqrt{1-t} \rangle$
2. $\mathbf{F}(t) = \langle \ln(t+2), e^{-t}, \ln(1-t) \rangle$
3. $\mathbf{F}(t) = \langle \tan t, \cot t, \sin \sqrt{t} \rangle$
4. $\mathbf{F}(t) = \langle \sqrt{1-t^2}, \ln t, e^{\sin t} \rangle$

Evaluate the limits in Exercises 5–8.

5. $\lim_{t \to 0} \langle e^{-t}, 2e^{3t}, 3e^t \rangle$
6. $\lim_{t \to 1} \left\langle \dfrac{t-1}{t^2-1}, \dfrac{1-t}{1+t} \right\rangle$
7. $\lim_{t \to 0} \left(\dfrac{\sin t}{t}\mathbf{i} + \dfrac{t^2}{1-\cos t}\mathbf{j} \right)$
8. $\lim_{t \to 0^+} \left[(t \ln t)\mathbf{i} + t^2\left(1 - \dfrac{1}{t}\right)\mathbf{j} + 3t\mathbf{k} \right]$

In Exercises 9–12 determine the set of t values for which **F** *is continuous.*

9. $\mathbf{F}(t) = \left\langle t, \dfrac{1}{t-1}, \sqrt{1-t} \right\rangle$
10. $\mathbf{F}(t) = \langle t^{3/2}, e^{-t}, \ln t \rangle$
11. $\mathbf{F}(t) = \sqrt{\dfrac{1-t}{1+t}}\,\mathbf{i} + \dfrac{1}{(t-1)^2}\mathbf{j}$
12. $\mathbf{F}(t) = (\ln \cosh t)\mathbf{i} + (\ln \sinh t)\mathbf{j}$

In Exercises 13–15 evaluate the expressions for the vector functions **F** *and* **G** *and scalar function* α *defined by* $\mathbf{F}(t) = \langle 2t, t, 3 \rangle$, $\mathbf{G}(t) = \langle 1-t, 2, t \rangle$, *and* $\alpha(t) = 1 - t$.

13. (a) $(\mathbf{F} - \mathbf{G})(t)$ (b) $(2\mathbf{F} + 3\mathbf{G})(t)$
14. (a) $(\alpha\mathbf{F})(t)$ (b) $(\mathbf{F} \cdot \mathbf{G})(t)$
15. (a) $(\mathbf{F} \times \mathbf{G})(t)$ (b) $\mathbf{G}(\alpha(t))$

In Exercises 16–18 repeat the given exercise for $\mathbf{F}(t) = \cos t\mathbf{i} + \sin t\mathbf{j} + \sin t\mathbf{k}$, $\mathbf{G}(t) = \sin t\mathbf{i} + \cos t\mathbf{j} + \mathbf{k}$, *and* $\alpha(t) = 2t$.

16. Exercise 13 17. Exercise 14 18. Exercise 15

In Exercises 19–26 find $\mathbf{F}'(t)$.

19. $\mathbf{F}(t) = \langle t^2, e^{2t}, \frac{1}{t} \rangle$
20. $\mathbf{F}(t) = \langle \ln t, \sin t, \cos 2t \rangle$
21. $\mathbf{F}(t) = \langle t \sin t, t \ln t \rangle$
22. $\mathbf{F}(t) = \langle \sin^{-1} t, \tan^{-1} t, \ln(1+t) \rangle$
23. $\mathbf{F}(t) = (1 - e^{-t})\mathbf{i} + te^{2t}\mathbf{j}$
24. $\mathbf{F}(t) = \sinh t\mathbf{i} + \cosh t\mathbf{j} + \tanh t\mathbf{k}$
25. $\mathbf{F}(t) = \dfrac{t}{\sqrt{1-t^2}}\mathbf{i} + \sin^{-1} t\mathbf{j} + \sqrt{1-t^2}\,\mathbf{k}$
26. $\mathbf{F}(t) = \dfrac{1-t}{1+t}\mathbf{i} + \dfrac{t}{1-t^2}\mathbf{j}$

In Exercises 27–34 evaluate the integrals.

27. $\int \langle \sin t, 1 - \cos t, t \rangle\,dt$
28. $\int \left\langle te^{-t^2}, \dfrac{\ln t}{t}, \dfrac{1}{t \ln t} \right\rangle dt$
29. $\int_0^1 \left\langle \dfrac{1}{1+t^2}, \sqrt{1-t}, t\sqrt{1-t^2} \right\rangle dt$
30. $\int_0^{\ln 2} \langle e^{-t}, e^{2t}, 6e^{-3t} \rangle\,dt$

31. $\int_0^{\pi/2} (\sin^2 t\mathbf{i} + \sin t \cos^2 t\mathbf{j})\, dt$

32. $\int_1^{\sqrt{3}} \left(\frac{1}{1+t^2}\mathbf{i} + \frac{1}{\sqrt{4-t^2}}\mathbf{j}\right) dt$

33. $\int_{2\pi/3}^{\pi} (\cos^3 t\mathbf{i} + \tan^2 t\mathbf{j} + \mathbf{k})\, dt$

34. $\int_0^1 \left(\frac{t-1}{t+1}\mathbf{i} + \frac{1}{\sqrt{t}(\sqrt{t}+1)}\mathbf{j} + \frac{t}{(t+1)^2}\mathbf{k}\right) dt$

Exercises 35–40 refer to the vector functions **F** *and* **G** *and the scalar function* α, *defined by* $\mathbf{F}(t) = t^2\mathbf{i} + (\frac{2}{t})\mathbf{j} + t^3\mathbf{k}$, $\mathbf{G}(t) = 2t\mathbf{i} + t^3\mathbf{j} - t^2\mathbf{k}$, *and* $\alpha(t) = \frac{4}{t}$. *Evaluate, using Theorem 15.2.*

35. $(\mathbf{F} + \mathbf{G})'(2)$

36. $(\mathbf{F} - \mathbf{G})'(2)$

37. $(\alpha\mathbf{F})'(2)$

38. $(\mathbf{F} \cdot \mathbf{G})'(2)$

39. $(\mathbf{F} \times \mathbf{G})'(2)$

40. $(\mathbf{F} \circ \alpha)'(2)$

41. Evaluate the integral $\int_1^{2\sqrt{2}} |\mathbf{f}(t)|\, dt$ if $\mathbf{f}(t) = \langle t^2 \sin t,\, t^2 \cos t,\, t^2\rangle$.

42. Prove that if $\mathbf{F}(t) = \mathbf{C}$, where $\mathbf{C} = \langle C_1, C_2, C_3\rangle$ is a constant vector, then $\mathbf{F}'(t) = \mathbf{0}$ for all t.

B

43. For $\mathbf{f}(t) = \langle \sin t, \cos t, 1\rangle$, verify the inequality in Theorem 15.6.

44. Prove equations 1 and 2 of Theorem 15.2.

45. Prove equations 3 and 4 of Theorem 15.2.

46. Prove equation 5 of Theorem 15.2.

47. Prove Theorem 15.3.

48. Prove Theorem 15.4.

49. State and prove a theorem analogous to Theorem 15.5 involving the cross product.

50. Prove that if $\mathbf{F}'(t) = \mathbf{G}'(t)$ for all t on an interval I, then $\mathbf{F}(t) = \mathbf{G}(t) + \mathbf{C}$ on I.

15.2 GEOMETRIC INTERPRETATION; TANGENT VECTORS AND ARC LENGTH

Each value of t in the domain of a vector function corresponds to a vector in V_3 (or in V_2), and this vector has many equivalent geometric representatives. We will concentrate now on the geometric representative that has its initial point at the origin—that is, the **position vector** of its terminal point. Throughout the remainder of this chapter we will use **r** to name a function when we are interpreting $\mathbf{r}(t)$ as a position vector.

Let $\mathbf{r}(t) = \langle f(t), g(t), h(t)\rangle$, and let the domain of **r** be D. By the **graph** of **r** we mean the set $\{(f(t), g(t), h(t)): t \in D\}$. The graph of **r** is thus the set of terminal points of the position vectors $\mathbf{r}(t)$ as t ranges over D. The graphs of most interest to us are those of continuous functions with intervals as domains. These are called curves.

DEFINITION 15.3 Let **r** be a continuous vector function on an interval I with values in V_3. Then the graph of **r** is called a **space curve**. If we designate this curve by C, then

$$\mathbf{r}(t) = \langle f(t), g(t), h(t)\rangle$$

is called a **vector equation** of C. The corresponding **parametric equations** of C are

$$\begin{cases} x = f(t) \\ y = g(t) \qquad t \in I \\ z = h(t) \end{cases}$$

The variable t is called a **parameter**. ■

Comment If the range of $\mathbf{r}$ is in V_2, then its graph is a **plane curve** and its parametric equations are the same as those we studied in Chapter 13. Generally, we will use *curve* to mean either a plane curve or a space curve. The context will make clear which one we mean.

Observe that parametric equations for a curve are immediately apparent from the vector equation, and vice versa. An advantage of the vector form is that all information is contained in a single equation.

We will sometimes say, for brevity, "the curve $\mathbf{r}(t) = \langle f(t), g(t), h(t)\rangle$" rather than "the curve with vector equation $\mathbf{r}(t) = \langle f(t), g(t), h(t)\rangle$."

We assign a **direction** to a curve C, with the positive direction corresponding to increasing values of the parameter.

EXAMPLE 15.5 Discuss and sketch the graph of $\mathbf{r}(t) = \langle \cos t, \sin t, t\rangle$ for $t \geq 0$.

Solution The parametric equations are $x = \cos t$, $y = \sin t$, and $z = t$. Since $x^2 + y^2 = \cos^2 t + \sin^2 t = 1$, the projections of points on the curve to the xy-plane lie on the unit circle $x^2 + y^2 = 1$. As t increases, the z-coordinate increases. The result is the climbing circular curve shown in Figure 15.1. It is as if the curve were wrapped around a right circular cylinder of radius 1. The direction is as indicated, with the initial point $(1, 0, 0)$. ∎

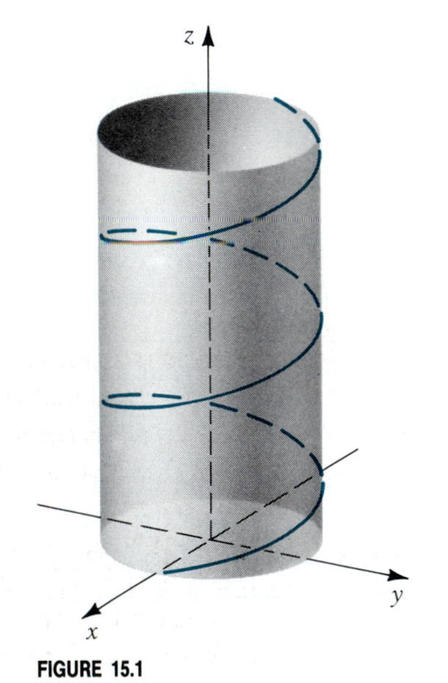

FIGURE 15.1

The curve in Example 15.5 is a **circular helix.** The general equation is $\mathbf{r}(t) = \langle a \cos \omega t, a \sin \omega t, bt\rangle$.

Let C be the graph of the continuous vector function $\mathbf{r}$ on $I = [a, b]$. If $\mathbf{r}(t_1) = \mathbf{r}(t_2)$ for $t_1 \neq t_2$ in I, then the corresponding point on C is called a **double point.** If C has no double point, it is called **simple.** If $\mathbf{r}(a) = \mathbf{r}(b)$, then C is **closed,** and C is said to be a **simple closed curve** if the only double point occurs at the end points of I. Finally, if $\mathbf{r}'$ is continuous on I and $\mathbf{r}'(t) \neq 0$ for all t in I, then C is said to be **smooth.** These definitions put in vector form corresponding definitions for plane curves defined parametrically in Chapter 13, and they extend the concepts to space curves as well.

At times it is useful to change the parameterization for a curve. Suppose, for example, that $\mathbf{r}$ is continuous on the interval I. Let C be the graph of $\mathbf{r}$, and suppose the parameter is t. Now let $t = \alpha(u)$, where α is a differentiable scalar function whose domain is an interval J for which the range of α is I. If $\alpha'(u) \neq 0$ for all u in J, then it can be proved that the graph of $\mathbf{r} \circ \alpha$ also is the curve C. The direction may, however, be changed. In particular, if $\alpha'(u) > 0$ on J, then the direction is unchanged, whereas if $\alpha'(u) < 0$ on J, the direction is reversed. We illustrate this in the next example.

EXAMPLE 15.6 Let C be the curve defined by $\mathbf{r}(t) = \langle t, 1 - e^{2t}, e^{-t}\rangle$ on $I = [0, 1]$. Express the equation of C in terms of the parameter u, where $u = e^t$.

Solution First we solve for t, getting $t = \ln u$. So in terms of the preceding discussion, $\alpha(u) = \ln u$. To have the range of α equal to I, we choose the domain $J = [1, e]$, since $\ln 1 = 0$ and $\ln e = 1$. Observe that $\alpha'(u) = \frac{1}{u} > 0$ for all u in J. Thus, the graph of $\mathbf{r} \circ \alpha$ is C. Letting $\mathbf{R} = \mathbf{r} \circ \alpha$, we obtain the vector equation for C parameterized by u as

$$\begin{aligned}\mathbf{R}(u) = \mathbf{r}(\ln u) &= \langle \ln u, 1 - e^{2 \ln u}, e^{-\ln u}\rangle \\ &= \langle \ln u, 1 - u^2, \tfrac{1}{u}\rangle, \quad u \in [1, e]\end{aligned}$$

The direction of C is preserved in this case. ∎

FIGURE 15.2

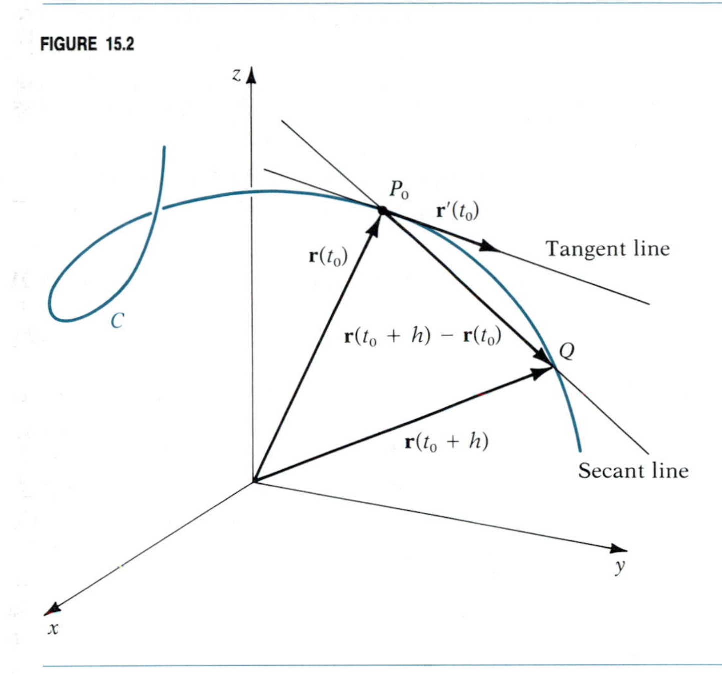

Now let us consider the geometric significance of the derivative $\mathbf{r}'(t) = \langle f'(t), g'(t), h'(t)\rangle$. It should come as no surprise that this has some relationship with the tangent line to the curve.

Let C be the curve $\mathbf{r}(t) = \langle f(t), g(t), h(t)\rangle$, let t_0 be a point in the domain I of $\mathbf{r}$ at which $\mathbf{r}'$ exists, and suppose that $\mathbf{r}'(t_0) \neq 0$. Choose any $h \neq 0$ so that $t_0 + h \in I$, and denote the tip of $\mathbf{r}(t_0)$ by P_0 and the tip of $\mathbf{r}(t_0 + h)$ by Q. Then the vector $\overrightarrow{P_0Q} = \mathbf{r}(t_0 + h) - \mathbf{r}(t_0)$ is a direction vector for the secant line through P_0 and Q, as shown in Figure 15.2. The vector

$$\frac{\mathbf{r}(t_0 + h) - \mathbf{r}(t_0)}{h} = \frac{1}{h}\,[\mathbf{r}(t_0 + h) - \mathbf{r}(t_0)]$$

being a scalar multiple of $\overrightarrow{P_0Q}$, is also a direction vector for this secant line. It is natural to define the tangent line at P_0 as the line with direction vector equal to the limit of this direction vector for the secant line as $h \to 0$—namely,

$$\lim_{h\to 0} \frac{\mathbf{r}(t_0 + h) - \mathbf{r}(t_0)}{h} = \mathbf{r}'(t_0)$$

DEFINITION 15.4 Let C be the curve with vector equation $\mathbf{r}(t) = \langle f(t), g(t), h(t)\rangle$, where $t \in I$, and suppose $\mathbf{r}'(t_0)$ exists and is not $\mathbf{0}$. Let P_0 be the point $(f(t_0), g(t_0), h(t_0))$. The vector $\mathbf{r}'(t_0)$ is called a **tangent vector** to C at P_0, and the **tangent line** to C at P_0 is the line through C with direction vector $\mathbf{r}'(t_0)$. ■

Since we are using $\mathbf{r}(t)$ to designate the position vector of a point on C, we will need some other notation for the position vector of a point on the tangent line. Calling this $\lambda(u)$, we have

$$\lambda(u) = \mathbf{r}(t_0) + u\mathbf{r}'(t_0), \qquad -\infty < u < \infty \tag{15.1}$$

as a vector equation of the tangent line to C at P_0. In component form this is

$$\lambda(u) = \langle f(t_0) + uf'(t_0),\ g(t_0) + ug'(t_0),\ h(t_0) + uh'(t_0)\rangle$$

EXAMPLE 15.7 Find a tangent vector to the circular helix $\mathbf{r}(t) = \cos t\mathbf{i} + \sin t\mathbf{j} + t\mathbf{k}$ at $t = \frac{\pi}{3}$. Also, find a vector equation of the tangent line.

Solution $\mathbf{r}'(t) = -\sin t\mathbf{i} + \cos t\mathbf{j} + \mathbf{k}$

So

$$\mathbf{r}'\left(\frac{\pi}{3}\right) = -\frac{\sqrt{3}}{2}\mathbf{i} + \frac{1}{2}\mathbf{j} + \mathbf{k}$$

is a tangent vector at $P_0(1/2, \sqrt{3}/2, \pi/3)$. The tangent line through P_0 has direction vector $\mathbf{r}'(\pi/3)$, and so can be written as

$$\begin{aligned}\lambda(u) &= \mathbf{r}\left(\frac{\pi}{3}\right) + u\mathbf{r}'\left(\frac{\pi}{3}\right)\\ &= \left(\frac{1}{2} - \frac{u\sqrt{3}}{2}\right)\mathbf{i} + \left(\frac{\sqrt{3}}{2} + \frac{u}{2}\right)\mathbf{j} + \left(\frac{\pi}{3} + u\right)\mathbf{k}\end{aligned}$$ ■

The *length* of a space curve C on an interval $I = [a, b]$ is defined analogously to the length of a plane curve given in Chapter 13 (Definition 13.3). When C has a finite length, it is said to be **rectifiable.** For a plane curve C, with parameterization $x = f(t)$ and $y = g(t)$, Theorem 13.2 stated that if f' and g' are continuous on $[a, b]$, then C is rectifiable and its length L is given by

$$L = \int_a^b \sqrt{[f'(t)]^2 + [g'(t)]^2}\, dt = \int_a^b \sqrt{\left(\frac{dx}{dt}\right)^2 + \left(\frac{dy}{dt}\right)^2}\, dt$$

Letting $\mathbf{r}(t) = \langle f(t), g(t)\rangle$, we can write L in the succinct form

$$L = \int_a^b |\mathbf{r}'(t)|\, dt$$

The following theorem generalizes this result.

THEOREM 15.7 Let C be the graph of the continuous vector function $\mathbf{r}$ on $[a, b]$, and suppose $\mathbf{r}'$ also is continuous on $[a, b]$. Then C is rectifiable, and its length L is given by

$$L = \int_a^b |\mathbf{r}'(t)|\, dt \tag{15.2}$$ ■

The proof follows the same lines as that of Theorem 13.2. When C is a plane curve, this result is merely a restatement of Theorem 13.2, but it holds true for space curves as well. If $\mathbf{r}(t) = \langle f(t), g(t), h(t)\rangle$, then equation (15.2) can be written in the form

$$L = \int_a^b \sqrt{[f'(t)]^2 + [g'(t)]^2 + [h'(t)]^2}\, dt = \int_a^b \sqrt{\left(\frac{dx}{dt}\right)^2 + \left(\frac{dy}{dt}\right)^2 + \left(\frac{dz}{dt}\right)^2}\, dt$$

EXAMPLE 15.8 Find the length of the circular helix $\mathbf{r}(t) = \langle \cos t, \sin t, t \rangle$ from $t = 0$ to $t = 2\pi$.

Solution $\mathbf{r}'(t) = \langle -\sin t, \cos t, 1 \rangle$

So by equation (15.2),

$$L = \int_0^{2\pi} |\mathbf{r}'(t)|\, dt = \int_0^{2\pi} \sqrt{\sin^2 t + \cos^2 t + 1}\, dt$$
$$= \int_0^{2\pi} \sqrt{1 + 1}\, dt = 2\sqrt{2}\pi$$

■

EXAMPLE 15.9 Find the length of the curve $\mathbf{r}(t) = 3t^2\mathbf{i} + (1 - 4t^2)\mathbf{j} + 2t^3\mathbf{k}$ from $t = 0$ to $t = 4$.

Solution We have $\mathbf{r}'(t) = 6t\mathbf{i} - 8t\mathbf{j} + 6t^2\mathbf{k}$, so that

$$|\mathbf{r}'(t)| = \sqrt{36t^2 + 64t^2 + 36t^4} = \sqrt{100t^2 + 36t^4} = 2t\sqrt{25 + 9t^2}$$

since $t \geq 0$. Thus, by equation (15.2),

$$L = \int_0^4 2t\sqrt{25 + 9t^2}\, dt = \tfrac{1}{9} \cdot \tfrac{2}{3}[(25 + 9t^2)^{3/2}]_0^4$$
$$= \tfrac{2}{27}[(169)^{3/2} - (25)^{3/2}] = \tfrac{4144}{27}$$

■

We would expect that the length of a curve would be unaffected by a change in parameter, and this is indeed the case, as you will be asked to show in Exercise 43. Specifically, suppose the length is L, calculated by equation (15.2), and we make the change of parameter $t = \alpha(u)$, where α' is continuous and nonzero on $[c, d]$, with the range of α equal to $[a, b]$. Then, writing $\mathbf{R} = \mathbf{r} \circ \alpha$, you will be asked to show that

$$L = \int_c^d |\mathbf{R}'(u)|\, du$$

If C has the vector equation $\mathbf{r}(t) = \langle f(t), g(t), h(t) \rangle$ for t in $[a, b]$ and $\mathbf{r}'$ is continuous on $[a, b]$, we know that C is rectifiable. For an arbitrary t in $[a, b]$ we designate the length of C on $[a, t]$ by $s(t)$; that is,

$$s(t) = \int_a^t |\mathbf{r}'(u)|\, du, \qquad a \leq t \leq b \tag{15.3}$$

(where u is used as the variable of integration to avoid confusion with the upper limit t). In this way we are defining a scalar function s, called the **arc length function** for C. Note that $s(a) = 0$ and $s(b) = L$, the total length of C on $[a, b]$. Since $|\mathbf{r}'|$ is continuous on $[a, b]$, it follows by Theorem 4.6 that

$$s'(t) = |\mathbf{r}'(t)| \tag{15.4}$$

Equivalently,

$$\frac{ds}{dt} = \sqrt{\left(\frac{dx}{dt}\right)^2 + \left(\frac{dy}{dt}\right)^2 + \left(\frac{dz}{dt}\right)^2} \tag{15.5}$$

or, in terms of differentials,

$$ds = \sqrt{\left(\frac{dx}{dt}\right)^2 + \left(\frac{dy}{dt}\right)^2 + \left(\frac{dz}{dt}\right)^2}\, dt \tag{15.6}$$

EXAMPLE 15.10 Find $s(t)$ for the curve C defined by

$$\mathbf{r}(t) = \frac{t^2 + 2t}{2}\mathbf{i} + \frac{(2t+2)^{3/2}}{3}\mathbf{j} + t\mathbf{k}, \qquad t \geq 0$$

Solution

$$\mathbf{r}'(t) = (t+1)\mathbf{i} + (2t+2)^{1/2}\mathbf{j} + \mathbf{k}$$
$$|\mathbf{r}'(t)| = \sqrt{(t^2 + 2t + 1) + (2t+2) + 1}$$
$$= \sqrt{t^2 + 4t + 4} = \sqrt{(t+2)^2} = t + 2$$

since $t + 2 > 0$. Thus, by equation (15.3),

$$s(t) = \int_0^t |\mathbf{r}'(u)|\, du = \int_0^t (u+2)\, du = \frac{(u+2)^2}{2}\Big]_0^t$$
$$= \frac{(t+2)^2}{2} - 2 = \frac{t^2 + 4t}{2}$$

■

It is sometimes useful to parameterize a curve by arc length s, measured along the curve. If C has length L, then its vector equation with s as a parameter has the form

$$\mathbf{r}(s) = \langle f(s), g(s), h(s) \rangle, \qquad s \in [0, L]$$

Equation (15.3) then gives $s = \int_0^s |r'(u)|\, du$, where we are assuming $\mathbf{r}'$ is continuous on $[0, L]$. If we differentiate both sides with respect to s, we obtain $1 = |\mathbf{r}'(s)|$ for all s in $[0, L]$. Thus, when C is parameterized by arc length, the tangent vector $\mathbf{r}'(s)$ is a unit vector at every point on C. We will make use of this fact in Section 15.4.

EXERCISE SET 15.2

A

In Exercises 1–8 discuss and sketch the curve with the given vector equation.

1. $\mathbf{r}(t) = \langle 2 - t, 3 + t, 5 + 2t \rangle, \quad -\infty < t < \infty$
2. $\mathbf{r}(t) = (3 + 2t)\mathbf{i} + (-2 + 4t)\mathbf{j} + (2 - t)\mathbf{k}, \quad -\infty < t < \infty$
3. $\mathbf{r}(t) = \langle 2\cos t, 3\sin t \rangle, \quad 0 \leq t \leq 2\pi$
4. $\mathbf{r}(t) = t\mathbf{i} + t^2\mathbf{j}, \quad -\infty < t < \infty$
5. $\mathbf{r}(t) = \langle 2\cos t, 2\sin t, \frac{3t}{2} \rangle, \quad t \geq 0$
6. $\mathbf{r}(t) = \langle 1 - t^2, t, 2 \rangle, \quad 1 \leq t \leq 3$
7. $\mathbf{r}(t) = 2\cos t\mathbf{i} + 3\sin t\mathbf{j} + t\mathbf{k}, \quad t \geq 0$
8. $\mathbf{r}(t) = \langle t, t^2, t \rangle, \quad t \geq 0$

In Exercises 9–14 find parametric equations of the tangent line to the curve C at t_0, where C has the given vector equation.

9. $\mathbf{r}(t) = \langle 2t, 3t^2, t^3 \rangle, \quad t_0 = 1$
10. $\mathbf{r}(t) = \langle 2\sin t, 3\cos t, 4t \rangle, \quad t_0 = \frac{\pi}{2}$
11. $\mathbf{r}(t) = \frac{1}{t}\mathbf{i} + \sqrt{t}\mathbf{j} + 4t^2\mathbf{k}, \quad t_0 = \frac{1}{4}$
12. $\mathbf{r}(t) = e^t\mathbf{i} + e^{-t}\mathbf{j} + e^{2t}\mathbf{k}, \quad t_0 = \ln 2$
13. $\mathbf{r}(t) = \left\langle \frac{t}{t-1}, t\sqrt{t+2}, \frac{1}{t-1} \right\rangle, \quad t_0 = 2$
14. $\mathbf{r}(t) = \left\langle t\ln t, 2\ln t, \frac{\ln t}{t} \right\rangle, \quad t_0 = 1$

In Exercises 15–18 find an equivalent vector equation of the curve under the specified change in parameter. What is the parameter interval for u? State whether the direction is unchanged or reversed.

15. $\mathbf{r}(t) = \langle t, 2t, \frac{1}{t} \rangle, \quad t > 0;\ u = t^2$
16. $\mathbf{r}(t) = \langle e^{-2t}, 1 + e^t, 2e^{2t} \rangle, \quad -1 \leq t \leq 1;\ u = e^t$

17. $\mathbf{r}(t) = \dfrac{t}{t+1}\mathbf{i} + \dfrac{1}{t+1}\mathbf{j} + \dfrac{t+2}{t+1}\mathbf{k}, \quad 0 \le t \le 1;$

$u = \dfrac{1}{t+1}$

18. $\mathbf{r}(t) = \sqrt{4-t^2}\mathbf{i} + \dfrac{1}{\sqrt{4-t^2}}\mathbf{j} + \dfrac{t}{\sqrt{4-t^2}}\mathbf{k}, \quad 0 \le t \le 1;$

$u = \sqrt{4-t^2}$

In Exercises 19–24 find a tangent vector to the given curve at the designated point.

19. $\mathbf{r}(t) = \langle t^2, 1-t^3, 3t\rangle, \quad -\infty < t < \infty; (1, 0, 3)$

20. $\mathbf{r}(t) = \left\langle \dfrac{1}{t}, \dfrac{1-t}{t}, \dfrac{1}{t-1}\right\rangle, \quad t > 0; (2, 1, -2)$

21. $\mathbf{r}(t) = 2 \sin t\mathbf{i} + 4 \cos t\mathbf{j} + \sin 2t\mathbf{k}, \quad 0 \le t < 2\pi;$ $(\sqrt{3}, -2, -\sqrt{3}/2)$

22. $\mathbf{r}(t) = 2e^{-t}\mathbf{i} + 3\mathbf{e}^{2t}\mathbf{j} - e^{-3t}\mathbf{k}, \quad -\infty < t < \infty;$ $(2, 3, -1)$

23. $x = 1 + 3t^2,\ y = 1 + 2t,\ z = 3 - t^3, \quad -\infty < t < \infty;$ $(4, -1, 4)$

24. $x = 2 + \ln t,\ y = 1 - t\ln t,\ z = \dfrac{2 \ln t}{t}, \quad t > 0;$

$(2, 1, 0)$

In Exercises 25–30 find the length of the curve on the specified interval

25. $\mathbf{r}(t) = \langle 2-t, 3+2t, 5-3t\rangle, \quad 1 \le t \le 3$

26. $\mathbf{r}(t) = \langle 3t, 2t^{3/2}, 4\rangle, \quad 0 \le t \le 8$

27. $\mathbf{r}(t) = (3t^2+1)\mathbf{i} + 3t^2\mathbf{j} + 2t^3\mathbf{k}, \quad 0 \le t \le 2$

28. $\mathbf{r}(t) = 2 \sin^2 t\mathbf{i} + \cos^3 t\mathbf{j} + \sin^3 t\mathbf{k}, \quad 0 \le t \le \frac{\pi}{2}$

29. $x = 3t,\ y = 2 \cos 3t,\ z = 2 \sin 3t, \quad 0 \le t < \frac{\pi}{3}$

30. $x = 2e^t,\ y = e^{-t},\ z = 2t, \quad -1 \le t \le 1$

In Exercises 31–36 find s(t).

31. $\mathbf{r}(t) = \langle 3-2t, 4+6t, 5t\rangle, \quad -1 \le t \le 10$

32. $\mathbf{r}(t) = \langle 2 \sin t, 4t, 2 \cos t\rangle, \quad t \ge 0$

33. $x = 3t \sin t,\ y = 3t \cos t,\ z = (2t)^{3/2}, \quad t \ge 0$

34. $x = 5e^{-t},\ y = (3e^{-1} + 1),\ z = -4e^{-t}, \quad t \le 0$

35. $\mathbf{r}(t) = 7\mathbf{i} + t^3\mathbf{j} - t^2\mathbf{k}, \quad t \ge 0$

36. $\mathbf{r}(t) = \ln t^2\mathbf{i} + \frac{1}{t}\mathbf{j} + 2t\mathbf{k}, \quad t \ge 1$

B

37. Prove that the tangent vector $\mathbf{r}'(t)$ always points in the direction of increasing values of t. [*Hint:* Consider the difference quotient $(\mathbf{r}(t+h) - \mathbf{r}(t))/h$ for $h > 0$ and for $h < 0$ separately.]

38. Let $\mathbf{r}(t) = (2t-1)\mathbf{i} + (2e^t + 3)\mathbf{j} + e^{-t}\mathbf{k}$ for t in the interval $0 \le t \le \ln 2$. Find the length of the curve C defined by $\mathbf{r}$. Now make the change of parameter $u = e^{-t}$, and let $\mathbf{R}(u)$ be the resulting vector function. What is the domain of $\mathbf{R}$ so that its graph is also C? Find the length of C using $\mathbf{R}(u)$, and verify that the answer is the same as that found using $\mathbf{r}(t)$.

39. Find the length of the curve $\mathbf{r}(t) = \langle t^2, t^3, -3t^2\rangle$ from $(1, -1, -3)$ to $(1, 1, -3)$.

40. Let C be the curve defined parametrically by $x = t$, $y = t^2/2$, and $z = 2t$.
(a) Find a unit vector tangent to C at the point $(2, 2, 4)$.
(b) Find the length of C from $(-2, 2, -4)$ to $(2, 2, 4)$.
(c) Sketch C over the range from $t = 0$ to $t = 4$, and show the unit tangent vector found in part a.

41. Let C be the curve with position vector $\mathbf{r}(t) = \langle 3t^2/2, 4+2t^2, 3\rangle$ on $[1, \infty)$. Make the change of parameter $u = s(t)$, where s is the arc length function, and let $\mathbf{R}(u)$ be the resulting position vector for C. Show that for all $u \ge 0$, $|\mathbf{R}'(u)| = 1$.

42. Let C be the graph of the differentiable vector function $\mathbf{r}$ on the interval I. Let α be a differentiable scalar function on an interval J that has range I, suppose $\alpha'(u) \ne 0$ for all u in J, and let $\mathbf{R} = \mathbf{r} \circ \alpha$. If $t_0 = \alpha(u_0)$, prove that $\mathbf{r}'(t_0) = k\mathbf{R}'(u_0)$ for some scalar k. What is the geometric significance of this result?

43. Under the hypotheses of Exercise 42, prove that the length of the graph of $\mathbf{r}$ over I equals the length of the graph of $\mathbf{R}$ over J. (This proves that the length of a curve is invariant under a parameter change of the type described.)

44. Let C be the graph of the continuously differentiable function $\mathbf{r}(t)$ for t in $[a, b]$, and let the length of C be L. Change the parameter to arc length $s = s(t)$ for s in $[0, L]$, and denote the result by $\mathbf{R}(s)$. Show that for every s in $[0, L]$, $|\mathbf{R}'(s)| = 1$. (This is an alternate proof to the one at the end of this section that when a smooth curve is parameterized by arc length, the tangent vector is always a unit vector.) [*Hint:* Let $t = t(s)$ be the inverse of $s = s(t)$, and use the chain rule, along with the fact that

$$\frac{dt}{ds} = \frac{1}{\dfrac{ds}{dt}}$$

15.3 MOTION ALONG A CURVE

In the preceding section we considered curves as sets of terminal points of position vectors $\mathbf{r}(t)$ for continuous functions $\mathbf{r}$. No concept of movement was involved. We were looking at curves from the *static* point of view. Now we want to look at them from the *dynamic* point of view; that is, we want to consider a curve as the path traced out by a particle moving in space (or in a plane).

We let t denote time, measured in whatever units we choose, from some convenient time origin, and we let $\mathbf{r}(t) = \langle f(t), g(t), h(t)\rangle$ be the position vector of the moving particle at time t. Since a moving particle always goes in a continuous path, it follows that $\mathbf{r}$ is continuous, so that the path of the particle is a curve in the sense of Definition 15.1. Suppose the time interval of interest is I. We assume further that $\mathbf{r}'$ and $\mathbf{r}''$ both exist for all t in I. We define the velocity, acceleration, and speed of the particle as follows.

DEFINITION 15.5 Let $\mathbf{r}(t)$ be the position vector of a moving particle for t in the interval I, and suppose $\mathbf{r}'$ and $\mathbf{r}''$ exist in I. Then the **velocity** $\mathbf{v}(t)$ and **acceleration** $\mathbf{a}(t)$ are defined as

$$\mathbf{v}(t) = \mathbf{r}'(t)$$
$$\mathbf{a}(t) = \mathbf{v}'(t) = \mathbf{r}''(t)$$

The **speed** of the particle is defined as the magnitude of the velocity vector:

$$\text{speed} = |\mathbf{v}(t)| = |\mathbf{r}'(t)|$$ ■

Remarks Observe that both velocity and acceleration are vectors, whereas speed is a scalar. Velocity is the instantaneous rate of change of position, and acceleration is the instantaneous rate of change of velocity. These agree with our intuitive understanding of these concepts. We know from equation (15.4) that $s'(t) = |\mathbf{r}'(t)|$; that is, speed $= ds/dt$. Thus, speed can be interpreted as the rate at which the arc length s is changing with time or, equivalently, the rate at which the distance along the path covered by the particle is changing with time. We observe further that the velocity vector is directed along the tangent line and points in the direction of motion.

EXAMPLE 15.11 A particle moves on the curve $\mathbf{r}(t) = \langle t, \frac{1}{t}, 2\sqrt{t}\rangle$ for $t > 0$. Find its velocity, acceleration, and speed when $t = 1$. Show the results graphically.

Solution

$$\mathbf{v}(t) = \mathbf{r}'(t) = \left\langle 1, -\frac{1}{t^2}, \frac{1}{\sqrt{t}}\right\rangle$$

$$\mathbf{a}(t) = \mathbf{v}'(t) = \left\langle 0, \frac{2}{t^3}, -\frac{1}{2t^{3/2}}\right\rangle$$

So when $t = 1$, $\mathbf{v}(1) = \langle 1, -1, 1\rangle$ and $\mathbf{a}(1) = \langle 0, 2, -\frac{1}{2}\rangle$. The speed at this instant is therefore $|\mathbf{v}(1)| = \sqrt{3}$. From the parametric equations $x = t$, $y = \frac{1}{t}$, and $z = 2\sqrt{t}$, we see that for $t > 0$, all three coordinates are positive. Also, since $xy = 1$, the curve lies above the first-quadrant branch of this equilateral hyperbola. By plotting points corresponding to $t = \frac{1}{9}, \frac{1}{4}, 1, 4$, and 9, we get a reasonably accurate sketch, as shown in Figure 15.3. ■

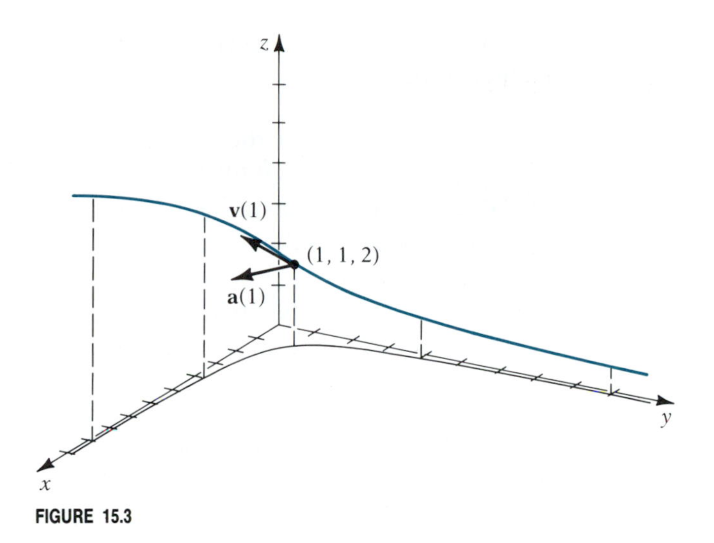

FIGURE 15.3

EXAMPLE 15.12 Find the force acting on a particle that moves in a circular path of radius r with constant speed v_0.

Solution Introduce a coordinate system so that the center of the circle is at the origin and when $t = 0$ the particle is at $(r, 0)$, and assume the motion is counterclockwise. Let $\theta(t)$ be the polar angle to the position vector $\mathbf{r}(t)$ at time t, as shown in Figure 15.4, and let $s(t)$ be the arc length covered by the particle, measured from $(r, 0)$. Then, $x = r\cos\theta(t)$ and $y = r\sin\theta(t)$ are the parametric equations of the circle. In vector form, the equation is

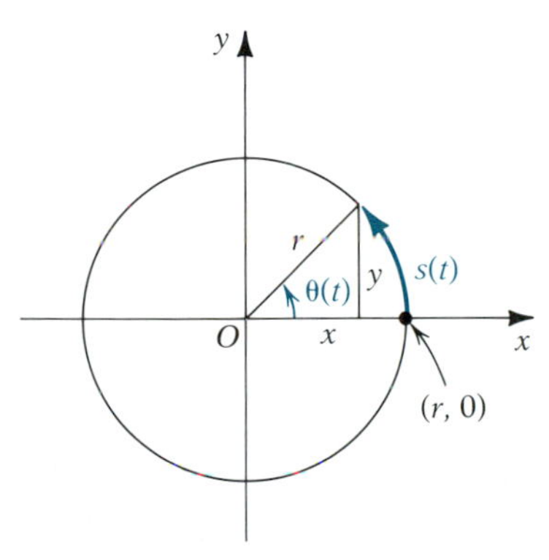

FIGURE 15.4

$$\mathbf{r}(t) = \langle r\cos\theta(t),\ r\sin\theta(t)\rangle$$

To find an explicit expression for $\theta(t)$, we use the relationship $s = r\theta$ to get $ds/dt = r\,d\theta/dt$. But ds/dt is the speed of the particle, which is equal to the constant v_0 by hypothesis. So $d\theta/dt = v_0/r$, which is also a constant. We designate this constant by ω and call it the **angular speed** because it is the rate of change of θ with respect to time. Thus, $d\theta/dt = \omega$. Integrating this, we get $\theta(t) = \omega t + C$. Since $\theta(0) = 0$, it follows that $C = 0$, and so $\theta(t) = \omega t$. The equations of motion now become

$$\begin{aligned}
\mathbf{r}(t) &= (r\cos\omega t)\mathbf{i} + (r\sin\omega t)\mathbf{j} \\
\mathbf{v}(t) &= (-r\omega\sin\omega t)\mathbf{i} + (r\omega\cos\omega t)\mathbf{j} \\
\mathbf{a}(t) &= (-r\omega^2\cos\omega t)\mathbf{i} + (-r\omega^2\sin\omega t)\mathbf{j} \\
&= -\omega^2[(r\cos\omega t)\mathbf{i} + (r\sin\omega t)\mathbf{j}] \\
&= -\omega^2\mathbf{r}(t)
\end{aligned}$$

We also have

$$|\mathbf{v}(t)| = \sqrt{r^2\omega^2\sin^2\omega t + r^2\omega^2\cos^2\omega t} = r\omega$$

which confirms that the constant speed v_0 is $r\omega$: $v_0 = r\omega$. Since $|\mathbf{r}(t)| = r$, $|\mathbf{a}(t)| = |-\omega^2||\mathbf{r}(t)| = r\omega^2$. So each of the vectors $\mathbf{r}(t)$, $\mathbf{v}(t)$, and $\mathbf{a}(t)$ has constant magnitude.

According to Newton's second law of motion, the force $\mathbf{F}$ acting on the body satisfies $\mathbf{F} = m\mathbf{a}$, where m is the mass of the body. Thus,

$$\mathbf{F} = -m\omega^2\mathbf{r}(t)$$

This force, which is of constant magnitude $mr\omega^2$ and is directed toward the center of the circle, is called **centripetal force.** ■

EXAMPLE 15.13 (a) Find a formula for the speed necessary to maintain a satellite of mass m in a fixed orbit h miles above the earth's surface.

(b) Find a formula for the time required for the satellite of part a to complete one revolution.

(c) Taking the radius of the earth as approximately 4000 mi, find the speed of a satellite in orbit 1000 mi above the earth's surface. Find the number of hours required to complete one revolution.

Solution We make the assumptions that the orbit is circular, that the acceleration due to gravity at the earth's surface is constant at $g \approx 32.2$ ft/sec^2, and that the gravitational attraction of other bodies is negligible.

(a) Let R denote the radius of the earth. Then the radius of the orbit of the satellite is $r = R + h$. We found in Example 15.12 that the centripetal force necessary to keep the satellite in orbit has magnitude $mr\omega^2$, and since $v = r\omega$, we get

$$|\mathbf{F}| = mr\left(\frac{v}{r}\right)^2 = \frac{mv^2}{r}$$

This force is produced by the earth's gravitational pull, which according to Newton's law of universal gravitation is given by

$$|\mathbf{F}| = \frac{GMm}{r^2}$$

where M is the mass of the earth and G is a constant, called the universal gravitational constant. Equating the two expressions for $|\mathbf{F}|$ gives

$$v^2 = \frac{GM}{r}$$

A more useful form is obtained by observing that when the satellite is on the earth's surface, so that $r = R$, the attractive force is just the weight, mg, of the satellite. So by Newton's gravitational law with $r = R$,

$$mg = \frac{GMm}{R^2}$$

and thus $GM = R^2g$. Making this substitution in the formula for v^2 gives $v^2 = R^2g/r = R^2g/(R + h)$. So

$$v = R\sqrt{\frac{g}{R+h}}$$

(b) The angular speed ω is the number of radians through which the position vector turns per unit of time. Since one revolution is equivalent to 2π radians, the time T required for the satellite to complete one revolution is

$$T = \frac{2\pi}{\omega} = \frac{2\pi r}{v} = \frac{2\pi(R+h)}{v}$$

From the result of part a, we can substitute for v to obtain T in the alternate form:

$$T = \frac{2\pi(R+h)^{3/2}}{R\sqrt{g}}$$

(c) We will use distances in miles and time in hours. So $R = 4000$ and $h = 1000$. To convert $g \approx 32.2 \text{ ft/sec}^2$ to mi/hr^2, we divide by 5280 and multiply by $(3600)^2$, getting $g \approx 79{,}036 \text{ mi/hr}^2$. From part a, the velocity is

$$v = R\sqrt{\frac{g}{R+h}} \approx 4000\sqrt{\frac{79{,}036}{5000}} \approx 15{,}903 \text{ mi/hr}$$

and from part b, the time T for one revolution is

$$T = \frac{2\pi(R+h)}{v} \approx \frac{2\pi(5000)}{15{,}903} \approx 1.975 \text{ hr}$$

■

As a further application of motion on a curve, consider a projectile fired from the ground at an angle θ with the horizontal with an initial velocity $\mathbf{v}_0$. This is essentially two-dimensional motion, and so we introduce x- and y-axes with the origin coinciding with the point from which the projectile is fired, as in Figure 15.5. We make the assumptions that the curvature of the earth is negligible in the interval in question, that air resistance and wind can be neglected, and that the acceleration due to gravity, g, is constant. (It should be noted that none of these assumptions is, strictly speaking, valid, but for moderate distances the errors introduced by making them are small.) For simplicity we also assume that the projectile is fired from ground level. The questions of interest are:

1. What is the path of the projectile?
2. How high does it rise?
3. What is its range; that is, how far does it go in a horizontal direction?
4. What is its terminal velocity?

We consider each of these.

1. Let $\mathbf{r}(t)$ be the position vector of the projectile (considered as a point mass) at time t. Then $\mathbf{r}(0) = \mathbf{0}$ and $\mathbf{v}(0) = \mathbf{v}_0$. By our assumptions, the

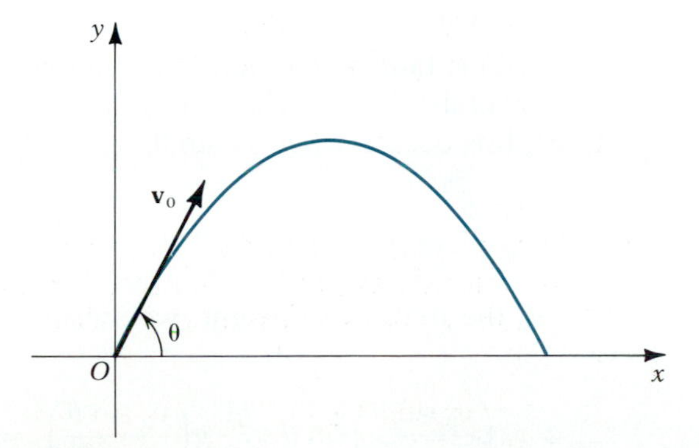

FIGURE 15.5

only force that acts on the projectile is the pull of gravity. The magnitude of this force is the weight of the projectile—namely, mg—where m is its mass. Since the force is downward, we have $\mathbf{F} = -mg\mathbf{j}$. From Newton's second law of motion we know that $\mathbf{F} = m\mathbf{a}$, where $\mathbf{a}$ is the acceleration of the projectile. Thus, $m\mathbf{a} = -mg\mathbf{j}$, or

$$\mathbf{a}(t) = -g\mathbf{j}$$

To find $\mathbf{v}(t)$, we integrate $\mathbf{a}(t)$, since $\mathbf{a}(t) = \mathbf{v}'(t)$:

$$\mathbf{v}(t) = \int \mathbf{a}(t)\,dt = -gt\mathbf{j} + \mathbf{C}_1$$

where $\mathbf{C}_1$ is a constant vector. Substituting $t = 0$, we get $\mathbf{v}(0) = \mathbf{C}_1$, and since $\mathbf{v}(0) = \mathbf{v}_0$,

$$\mathbf{v}(t) = -gt\mathbf{j} + \mathbf{v}_0$$

Now we integrate again to find $\mathbf{r}(t)$:

$$\mathbf{r}(t) = \int \mathbf{v}(t)\,dt = -\tfrac{1}{2}gt^2\mathbf{j} + \mathbf{v}_0 t + \mathbf{C}_2$$

Again letting $t = 0$, we find that $\mathbf{C}_2 = \mathbf{r}(0) = \mathbf{0}$. Thus,

$$\mathbf{r}(t) = -\tfrac{1}{2}gt^2\mathbf{j} + \mathbf{v}_0 t$$

If we write $v_0 = |\mathbf{v}_0|$, then $\mathbf{v}_0 = (v_0 \cos\theta)\mathbf{i} + (v_0 \sin\theta)\mathbf{j}$. So $\mathbf{r}(t)$ becomes

$$\mathbf{r}(t) = (v_0 t\cos\theta)\mathbf{i} + (v_0 t\sin\theta - \tfrac{1}{2}gt^2)\mathbf{j} \tag{15.7}$$

and $\mathbf{v}(t)$ becomes

$$\mathbf{v}(t) = (v_0\cos\theta)\mathbf{i} + (v_0\sin\theta - gt)\mathbf{j} \tag{15.8}$$

Equation (15.7) is a vector equation of the path of the projectile. To analyze it further, we use the parametric equations

$$\begin{cases} x = v_0 t\cos\theta \\ y = v_0 t\sin\theta - \tfrac{1}{2}gt^2 \end{cases} \tag{15.9}$$

By eliminating the parameter t, we obtain the rectangular equation (verify)

$$y = x\tan\theta - \frac{gx^2}{2v_0^2\cos^2\theta}, \qquad 0 \le \theta < \tfrac{\pi}{2}$$

which is the equation of a parabola.

2. Since $\mathbf{v}(t)$ is tangent to the path, the maximum height occurs when $\mathbf{v}(t)$ is horizontal—that is, when the y component of $\mathbf{v}(t)$ is 0. From equation (15.8), this occurs when $v_0\sin\theta - gt = 0$, or

$$t = \frac{v_0\sin\theta}{g}$$

To find the maximum height, we substitute this into equation (15.9) for y:

$$y_{max} = v_0\left(\frac{v_0\sin\theta}{g}\right)\sin\theta - \frac{1}{2}g\left(\frac{v_0\sin\theta}{g}\right)^2 = \frac{1}{2}\frac{v_0^2\sin^2\theta}{g}$$

3. To find the range, we set $y = 0$ from equation (15.9):

$$t(v_0 \sin \theta - \frac{1}{2}gt) = 0$$

$$t = 0 \quad \text{or} \quad t = \frac{2v_0 \sin \theta}{g}$$

Clearly, the second value is the one we want. The range is found by substituting this value of t in equation (15.9) for x:

$$\text{range} = x_{\max} = v_0\left(\frac{2v_0 \sin \theta}{g}\right)\cos \theta = \frac{v_0^2 \sin 2\theta}{g}$$

4. The terminal velocity is the velocity at impact with the ground. We already know this occurs when $t = (2v_0 \sin \theta)/g$. Putting this in equation (15.8) and writing $\mathbf{v}_T$ for terminal velocity, we get

$$\begin{aligned}\mathbf{v}_T &= (v_0 \cos \theta)\mathbf{i} + (v_0 \sin \theta - 2v_0 \sin \theta)\mathbf{j} \\ &= (v_0 \cos \theta)\mathbf{i} - (v_0 \sin \theta)\mathbf{j}\end{aligned}$$

Also, if $v_T = |\mathbf{v}_T|$, then

$$v_T = \sqrt{(v_0 \cos \theta)^2 + (-v_0 \sin \theta)^2} = v_0$$

So the terminal speed is the same as the initial speed.

EXERCISE SET 15.3

A

In Exercises 1–10 assume a particle moves so that its position vector at time t is $\mathbf{r}(t)$. Find its velocity, acceleration, and speed at t_0. Draw the graph of the curve followed by the particle, showing $\mathbf{v}(t_0)$ and $\mathbf{a}(t_0)$, drawn from the tip of $\mathbf{r}(t_0)$.

1. $\mathbf{r}(t) = \langle 2 \cos t, 3 \sin t\rangle$; $t_0 = \frac{\pi}{4}$

2. $\mathbf{r}(t) = t^2\mathbf{i} + t^3\mathbf{j}$; $t_0 = -1$

3. $\mathbf{r}(t) = 2e^t\mathbf{i} + 3e^{-t}\mathbf{j}$; $t_0 = 0$

4. $\mathbf{r}(t) = \langle \cosh t, \sinh t\rangle$; $t_0 = 0$

5. $\mathbf{r}(t) = \langle 2t - 1, t^2 + 3\rangle$; $t_0 = 2$

6. $\mathbf{r}(t) = 2 \cos t\mathbf{i} + 2 \sin t\mathbf{j} + 3t\mathbf{k}$; $t_0 = \frac{\pi}{3}$

7. $\mathbf{r}(t) = 2t\mathbf{i} + t^2\mathbf{j} + t^3\mathbf{k}$; $t_0 = 1$

8. $\mathbf{r}(t) = \langle t + 1, 3t, t^2\rangle$; $t_0 = 1$

9. $\mathbf{r}(t) = \langle e^t, 2t, e^{-t}\rangle$; $t_0 = 0$

10. $\mathbf{r}(t) = \cos^2 t\mathbf{i} + 2 \sin t\mathbf{j} + 2t\mathbf{k}$; $t_0 = \frac{\pi}{4}$

In Exercises 11–15 find $\mathbf{v}(t)$, $\mathbf{a}(t)$, and the speed at an arbitrary t in the given range.

11. $\mathbf{r}(t) = (5 - 2t)^{3/2}\mathbf{i} + \frac{1}{2}(t^2 + 4t)\mathbf{j}$; $t < \frac{5}{2}$

12. $\mathbf{r}(t) = (\ln t^2)\mathbf{i} + \frac{1}{t}\mathbf{j} + 2t\mathbf{k}$; $t > 0$

13. $\mathbf{r}(t) = \left\langle t \cos t, t \sin t, \frac{(2t)^{3/2}}{3}\right\rangle$; $t > 0$

14. $\mathbf{r}(t) = \langle \cos^3 t, \sin^3 t, \cos 2t\rangle$; $0 \le t \le \frac{\pi}{2}$

15. $\mathbf{r}(t) = \langle e^t \cos t, e^t \sin t, e^t\rangle$; $-\infty < t < \infty$

In Exercises 16–20 use the results of Examples 15.12 and 15.13.

16. A 2-lb weight attached to one end of a 10-ft rope is being whirled around horizontally in a circular path by a child holding the other end of the rope. If the speed of the weight is 12 ft/sec, find the force exerted by the child on the rope. Through how many revolutions per minute is the weight turning? (*Note:* $w = mg$, where w is weight and m is mass.)

17. In Exercise 16 find the effect on the force exerted by the child if (a) the speed is doubled and (b) the rope is half as long.

18. A satellite moves in a circular orbit 200 mi above the earth. What is its speed? How long does it take to complete one orbit?

19. A satellite is in a circular orbit h miles above the earth. If its speed is 17,600 mi/hr, find h.

20. If a satellite circles the earth once every 90 min, find its height above the earth and its velocity.

21. A projectile is fired from the earth's surface with an initial speed of 2000 ft/sec at an angle of 30° with the horizontal. Find the maximum height and range of the projectile.

22. At what angle θ should the projectile of Exercise 21 be fired for the range to be 30 mi?

23. If a projectile is fired from the ground at an angle of 42° and attains a maximum height of 5000 ft, what is its initial speed?

24. In Exercise 23 if, instead of the known height, we are given that the range is 36 km, what is the initial speed? (*Note:* $g \approx 9.80$ m/sec^2.)

B

25. Prove that if a particle moves along a curve with constant speed, then its velocity and acceleration vectors are always orthogonal. (*Hint:* Use the fact that $|\mathbf{v}|^2 = \mathbf{v} \cdot \mathbf{v} = C$, and differentiate.)

26. Prove that if the position vector of a particle moving in space is of the form

$$\mathbf{r}(t) = (\cos \omega t)\mathbf{A} + (\sin \omega t)\mathbf{B}$$

where $\mathbf{A}$ and $\mathbf{B}$ are arbitrary constant vectors, then $\mathbf{a}(t) = -\omega^2\mathbf{r}(t)$.

27. Prove that the position vector of a moving particle is of the form $\mathbf{r}(t) = t\mathbf{A} + \mathbf{B}$ where $\mathbf{A}$ and $\mathbf{B}$ are constant vectors if and only if $\mathbf{a}(t) = \mathbf{0}$ for all t. Describe the motion in this case.

28. A communications satellite is located above the equator, and its speed and altitude are such that it remains stationary relative to the earth. What are its speed and altitude?

29. A projectile is fired at an angle of 25° with the horizontal from the top of a hill 1000 m above the plain below. If the initial speed of the projectile is 500 m/sec, find the range and the speed at impact.

30. In order to feed the cattle in winter a rancher drops bales of hay from a light airplane. If a bale is dropped from a height of 600 ft while the airplane is flying horizontally at 150 mi/hr, how far is it from the point on the ground below the airplane when the bale is dropped to the point where it hits the ground? What is its speed at impact? (Neglect air resistance and assume the ground is flat.)

31. For a certain particle moving in space it is known that $\mathbf{a}(t) = -t\mathbf{k}$ and that $\mathbf{v}(0) = 2\mathbf{i} - 3\mathbf{j} + \mathbf{k}$ and $\mathbf{r}(0) = 4\mathbf{i} + 2\mathbf{j}$. Find $\mathbf{v}(t)$ and $\mathbf{r}(t)$.

32. Redo Exercise 31 if $\mathbf{a}(t) = e^{-t}\mathbf{i} + 2e^t\mathbf{j} + te^t\mathbf{k}$, $\mathbf{v}(0) = 2\mathbf{i} + 6\mathbf{j} - \mathbf{k}$, and $\mathbf{r}(0) = \mathbf{i} + 2\mathbf{j} - 2\mathbf{k}$.

33. In Example 15.12 imbed the problem in a three-dimensional coordinate system, with the circle lying in the xy-plane, and write $\mathbf{r}(t)$ as a vector in V_3 with third component 0. Let $\boldsymbol{\omega} = \omega\mathbf{k}$. Prove that $\mathbf{v}(t) = \boldsymbol{\omega} \times \mathbf{r}(t)$. Show that the same result holds true if the circle is in any plane parallel to the xy-plane, with the center on the z-axis.

15.4 UNIT TANGENT AND NORMAL VECTORS; CURVATURE

We saw in Section 15.2 that when a smooth curve C is parameterized by arc length, the tangent vector $\mathbf{r}'(s)$ always has length 1. We call this vector the **unit tangent vector** for C and designate it by $\mathbf{T}$. So

$$\mathbf{T} = \frac{d\mathbf{r}}{ds} \tag{15.10}$$

Since $|\mathbf{T}| = 1$, we have $\mathbf{T} \cdot \mathbf{T} = |\mathbf{T}|^2 = 1$, and if $d\mathbf{T}/ds$ exists, we can differentiate both sides of $\mathbf{T} \cdot \mathbf{T} = 1$ to get

$$\mathbf{T} \cdot \frac{d\mathbf{T}}{ds} + \frac{d\mathbf{T}}{ds} \cdot \mathbf{T} = 0$$

$$2\mathbf{T} \cdot \frac{d\mathbf{T}}{ds} = 0$$

Thus $\mathbf{T} \cdot d\mathbf{T}/ds = 0$, and so $\mathbf{T}$ and $d\mathbf{T}/ds$ are orthogonal. If $d\mathbf{T}/ds$ is nonzero, we make it into a unit vector by dividing by its length, and we designate the result by $\mathbf{N}$:

$$\mathbf{N} = \frac{\dfrac{d\mathbf{T}}{ds}}{\left|\dfrac{d\mathbf{T}}{ds}\right|} \tag{15.11}$$

We call $\mathbf{N}$ the **principal unit normal vector** for C.

Since $\mathbf{T}$ has constant length 1, the derivative $d\mathbf{T}/ds$ reflects the change in *direction* of $\mathbf{T}$ only. Its magnitude $d\mathbf{T}/ds$ thus provides a measure of how rapidly the unit tangent vector $\mathbf{T}$ is turning as arc length is traversed. We call this magnitude the **curvature** of C and designate it by the Greek letter kappa:

$$\kappa = \left|\frac{d\mathbf{T}}{ds}\right| \tag{15.12}$$

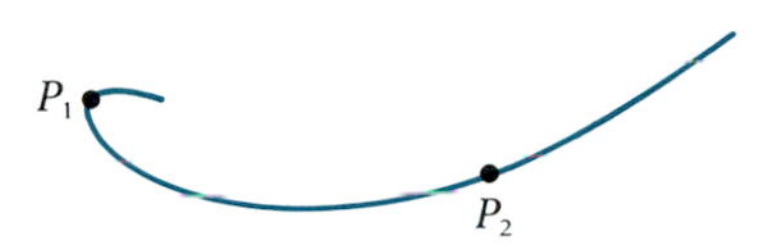

FIGURE 15.6 At P_1 κ is large and at P_2 κ is small

In Figure 15.6 we show two points on a curve C, one at which the curvature is large ($\mathbf{T}$ is turning rapidly) and one at which it is small ($\mathbf{T}$ is turning slowly). We can now write equation (15.11) in the form

$$\mathbf{N} = \frac{1}{\kappa}\frac{d\mathbf{T}}{ds}$$

so that

$$\frac{d\mathbf{T}}{ds} = \kappa\mathbf{N} \tag{15.13}$$

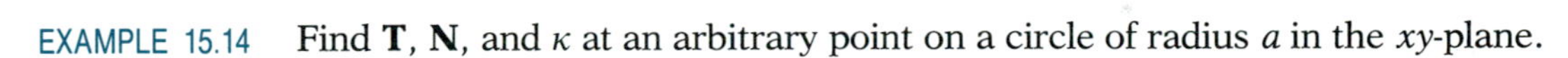

EXAMPLE 15.14 Find $\mathbf{T}$, $\mathbf{N}$, and κ at an arbitrary point on a circle of radius a in the xy-plane.

Solution For convenience we take the center at the origin. From Figure 15.7 we see that parametric equations for the circle are $x = a\cos\theta$ and $y = a\sin\theta$, where $0 \le \theta \le 2\pi$, and since $s = a\theta$, we have $\theta = s/a$. So we can write the vector equation with s as parameter in the form

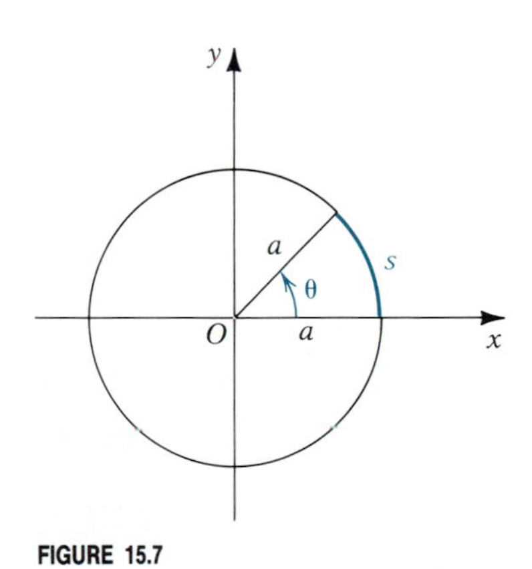

FIGURE 15.7

$$\mathbf{r}(s) = a\cos\left(\frac{s}{a}\right)\mathbf{i} + a\sin\left(\frac{s}{a}\right)\mathbf{j}, \qquad 0 \le s \le 2\pi a$$

Then, from equation (15.10),

$$\mathbf{T} = \frac{d\mathbf{r}}{ds} = -\sin\left(\frac{s}{a}\right)\mathbf{i} + \cos\left(\frac{s}{a}\right)\mathbf{j}$$

To find $\mathbf{N}$ and κ we calculate $\frac{d\mathbf{T}}{ds}$:

$$\frac{d\mathbf{T}}{ds} = -\frac{1}{a}\cos\left(\frac{s}{a}\right)\mathbf{i} - \frac{1}{a}\sin\left(\frac{s}{a}\right)\mathbf{j}$$

$$\left|\frac{d\mathbf{T}}{ds}\right| = \sqrt{\frac{1}{a^2}\cos^2\left(\frac{s}{a}\right) + \frac{1}{a^2}\sin^2\left(\frac{s}{a}\right)} = \frac{1}{a}$$

So

$$\mathbf{N} = -\left[\cos\left(\frac{s}{a}\right)\mathbf{i} + \sin\left(\frac{s}{a}\right)\mathbf{j}\right] = -\mathbf{r}(s)$$

and $\kappa = 1/a$. Observe that $\mathbf{N}$ is directed opposite to $\mathbf{r}(s)$ and so points toward the center of the circle. ∎

Remark That the curvature of a circle is the reciprocal of the radius makes sense on intuitive grounds. When the radius is small, the unit tangent vector changes direction rapidly as we progress around the circle and so the curvature is large, whereas for a circle with a large radius, this change is slow and the curvature is small.

It is easy to parameterize the circle in terms of arc length. Unfortunately, for most curves this is difficult. So we need computational formulas for $\mathbf{T}$, $\mathbf{N}$, and κ in terms of other parameters. If t is any other parameter and $\mathbf{r}(t)$ is the position vector for C, we know that in theory at least, we can introduce as the parameter $s = s(t)$ given by equation (15.3) provided $\mathbf{r}'$ is continuous on the t interval $[a, b]$ and $\mathbf{r}'(t) \neq 0$ there (that is, C is smooth). It follows that $s(t) = |\mathbf{r}'(t)| > 0$ so that s is an increasing function of t and hence has an inverse, say $t(s)$, that is also increasing. Furthermore by equation (6.14),

$$\frac{dt}{ds} = \frac{1}{\dfrac{ds}{dt}}$$

So, using the chain rule, we have

$$\mathbf{T} = \frac{d\mathbf{r}}{ds} = \frac{d\mathbf{r}}{dt}\frac{dt}{ds} = \frac{\dfrac{d\mathbf{r}}{dt}}{\dfrac{ds}{dt}}$$

and since $ds/dt = |\mathbf{r}'(t)|$, we can write

$$\mathbf{T} = \frac{\mathbf{r}'(t)}{|\mathbf{r}'(t)|} \tag{15.14}$$

To find $\mathbf{N}$ also as a function of t, we assume $\mathbf{r}''(t)$ exists and get, again by the chain rule,

$$\mathbf{N} = \frac{\dfrac{d\mathbf{T}}{ds}}{\left|\dfrac{d\mathbf{T}}{ds}\right|} = \frac{\dfrac{d\mathbf{T}}{dt}\dfrac{dt}{ds}}{\left|\dfrac{d\mathbf{T}}{dt}\dfrac{dt}{ds}\right|} = \frac{\dfrac{d\mathbf{T}}{dt}\dfrac{dt}{ds}}{\left|\dfrac{d\mathbf{T}}{dt}\right|\dfrac{dt}{ds}} = \frac{\dfrac{d\mathbf{T}}{dt}}{\left|\dfrac{d\mathbf{T}}{dt}\right|}$$

since $dt/ds > 0$. Thus,

$$\mathbf{N} = \frac{\dfrac{d\mathbf{T}}{dt}}{\left|\dfrac{d\mathbf{T}}{dt}\right|} \tag{15.15}$$

Equations (15.14) and (15.15) are valid for both plane curves and space curves. For space curves the cross product $\mathbf{T} \times \mathbf{N}$ is a vector orthogonal to both $\mathbf{T}$ and $\mathbf{N}$. It is called the **binormal** for C, and we designate it by $\mathbf{B}$:

$$\mathbf{B} = \mathbf{T} \times \mathbf{N} \tag{15.16}$$

The binormal is also a unit vector, as we see by

$$|\mathbf{B}| = |\mathbf{T} \times \mathbf{N}| = |\mathbf{T}|\,|\mathbf{N}| \sin \tfrac{\pi}{2} = 1$$

The triple **T**, **N**, **B** thus is a set of mutually orthogonal unit vectors, much like the triple **i**, **j**, **k**. But whereas the latter triple is fixed in direction, **T**, **N**, **B** can vary at different points on the curve. This triple is often called the *moving trihedral* for C.

In order to find a more useful computational formula for κ than equation (15.12), we derive some preliminary results. From equation (15.14) we can write $\mathbf{r}'(t)$ in the form

$$\frac{d\mathbf{r}}{dt} = \left|\frac{d\mathbf{r}}{dt}\right|\mathbf{T} = \frac{ds}{dt}\mathbf{T}$$

So

$$\frac{d^2\mathbf{r}}{dt^2} = \frac{d^2s}{dt^2}\mathbf{T} + \frac{ds}{dt}\frac{d\mathbf{T}}{dt}$$

By the chain rule and equation (15.13),

$$\frac{d\mathbf{T}}{dt} = \frac{d\mathbf{T}}{ds}\frac{ds}{dt} = \kappa\frac{ds}{dt}\mathbf{N}$$

Thus,

$$\frac{d^2\mathbf{r}}{dt^2} = \frac{d^2s}{dt^2}\mathbf{T} + \kappa\left(\frac{ds}{dt}\right)^2\mathbf{N} \tag{15.17}$$

We assume now that C is a space curve so that all vectors are in V_3. The calculations are valid as well, however, for curves in the xy-plane if we treat vectors in V_2 as being in V_3 with third component 0. We take the cross product of $\mathbf{r}'(t)$ and $\mathbf{r}''(t)$, getting

$$\begin{aligned}\frac{d\mathbf{r}}{dt} \times \frac{d^2\mathbf{r}}{dt^2} &= \frac{ds}{dt}\mathbf{T} \times \left[\frac{d^2s}{dt^2}\mathbf{T} + \kappa\left(\frac{ds}{dt}\right)^2\mathbf{N}\right]\\ &= \left(\frac{ds}{dt}\frac{d^2s}{dt^2}\right)(\mathbf{T}\times\mathbf{T}) + \kappa\left(\frac{ds}{dt}\right)^3(\mathbf{T}\times\mathbf{N})\\ &= \kappa\left(\frac{ds}{dt}\right)^3\mathbf{B}\end{aligned}$$

since $\mathbf{T}\times\mathbf{T} = \mathbf{0}$. Since $|\mathbf{B}| = 1$ and both κ and ds/dt are nonnegative,

$$\left|\frac{d\mathbf{r}}{dt} \times \frac{d^2\mathbf{r}}{dt^2}\right| = \kappa\left(\frac{ds}{dt}\right)^3$$

Finally, we solve for κ, writing $\mathbf{r}'$ and $\mathbf{r}''$ for $d\mathbf{r}/dt$ and $d^2\mathbf{r}/dt^2$, respectively, and replacing ds/dt by $|\mathbf{r}'|$:

$$\kappa = \frac{|\mathbf{r}' \times \mathbf{r}''|}{|\mathbf{r}'|^3} \tag{15.18}$$

EXAMPLE 15.15 Find **T**, **N**, and κ at an arbitrary point on the circular helix $\mathbf{r}(t) = \langle 2\cos 3t, 2\sin 3t, 8t\rangle$.

Solution We will need $\mathbf{r}'$, $\mathbf{r}''$, and $|\mathbf{r}'|$:

$$\begin{aligned}\mathbf{r}'(t) &= \langle -6\sin 3t, 6\cos 3t, 8\rangle\\ \mathbf{r}''(t) &= \langle -18\cos 3t, -18\sin 3t, 0\rangle\\ |\mathbf{r}'(t)| &= \sqrt{36\sin^2 3t + 36\cos^2 3t + 64} = \sqrt{36+64} = 10\end{aligned}$$

By equation (15.14),

$$\mathbf{T} = \frac{\mathbf{r}'(t)}{|\mathbf{r}'(t)|} = \frac{\langle -6 \sin 3t, 6 \cos 3t, 8\rangle}{10} = \left\langle -\frac{3}{5} \sin 3t, \frac{3}{5} \cos 3t, \frac{4}{5}\right\rangle$$

To get $\mathbf{N}$, we calculate $d\mathbf{T}/dt$:

$$\frac{d\mathbf{T}}{dt} = \left\langle -\frac{9}{5} \cos 3t, -\frac{9}{5} \sin 3t, 0\right\rangle$$

So by equation (15.15),

$$\mathbf{N} = \frac{\dfrac{d\mathbf{T}}{dt}}{\left|\dfrac{d\mathbf{T}}{dt}\right|} = \frac{\left\langle -\frac{9}{5} \cos 3t, -\frac{9}{5} \sin 3t, 0\right\rangle}{\sqrt{\frac{81}{25} \cos^2 3t + \frac{81}{25} \sin^2 3t}} = \frac{\left\langle -\frac{9}{5} \cos 3t, -\frac{9}{5} \sin 3t, 0\right\rangle}{\frac{9}{5}}$$
$$= \langle -\cos 3t, -\sin 3t, 0\rangle$$

Finally we calculate κ, using equation (15.18):

$$\mathbf{r}' \times \mathbf{r}'' = \begin{vmatrix} \mathbf{i} & \mathbf{j} & \mathbf{k} \\ -6 \sin 3t & 6 \cos 3t & 8 \\ -18 \cos 3t & -18 \sin 3t & 0 \end{vmatrix}$$
$$= 144 \sin 3t\mathbf{i} - 144 \cos 3t\mathbf{j} + (108 \sin^2 3t + 108 \cos^2 3t)\mathbf{k}$$
$$= 36(4 \sin 3t\mathbf{i} - 4 \cos 3t\mathbf{j} + 3\mathbf{k})$$
$$|\mathbf{r}' \times \mathbf{r}''| = 36\sqrt{16 \sin^2 3t + 16 \cos^2 3t + 9} = 36\sqrt{25} = 180$$

Thus,

$$\kappa = \frac{|\mathbf{r}' \times \mathbf{r}''|}{|\mathbf{r}'|^3} = \frac{180}{1000} = \frac{9}{50}$$

Notice that κ is constant in this case. ■

For a curve C in the xy-plane the unit tangent and normal vectors can be found from equations (15.14) and (15.15), respectively. In this case, however, there is no binormal vector. Equation (15.18) is valid, as remarked earlier, if we write two-dimensional vectors as three-dimensional vectors with the third component 0. When we do this, we arrive at a formula for κ that is applicable to plane curves only. To see this, suppose C has position vector $\mathbf{r}(t) = \langle f(t), g(t)\rangle$. Then we treat C as a space curve lying in the xy-plane, so that $\mathbf{r}(t) = \langle f(t), g(t), 0\rangle$. For brevity, we can write $x = f(t)$ and $y = g(t)$, so that $\mathbf{r}' = \langle x', y', 0\rangle$ and $\mathbf{r}'' = \langle x'', y'', 0\rangle$. Then $\mathbf{r}' \times \mathbf{r}'' = \langle 0, 0, x'y'' - x''y'\rangle$. (You should verify this.) So $|\mathbf{r}' \times \mathbf{r}''| = |x'y'' - x''y'|$, and

$$\kappa = \frac{|x'y'' - x''y'|}{(x'^2 + y'^2)^{3/2}} \tag{15.19}$$

In case the equation of C is in the form $y = f(x)$, we can use the parameterization $x = t$ and $y = f(t)$, and since $x' = 1$ and $x'' = 0$, equation (15.19) becomes

$$\kappa = \frac{|y''|}{[1 + (y')^2]^{3/2}} \tag{15.20}$$

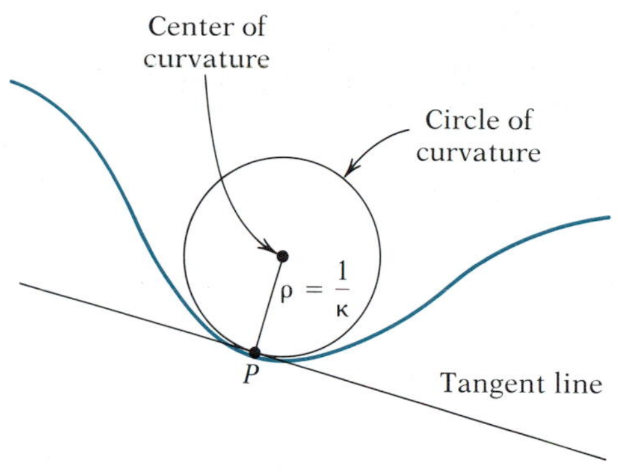

FIGURE 15.8

At a point P on a curve C where $\kappa \neq 0$, we define $\rho = 1/\kappa$ as the **radius of curvature** of C at P. If C is a plane curve, the circle of radius ρ that is tangent to C at P on its concave side (the direction of $\mathbf{N}$) is called the **circle of curvature,** or **osculating circle.** Its center is called the **center of curvature.** If C itself is a circle, Example 15.14 shows that its curvature is the reciprocal of its radius, and hence its radius is the radius of curvature. So C is its own circle of curvature. This shows that for any plane curve C at a point P where $\kappa \neq 0$, the circle of curvature has the same curvature as C at P. In the sense that C and its circle of curvature have the same tangent line and the same curvature at P, the circle of curvature is the circle that best "fits" C in a neighborhood of P. These ideas are illustrated in Figure 15.8.

EXAMPLE 15.16 Find the curvature and center of curvature at the point $P(2, 1)$ on the curve $\mathbf{r}(t) = \langle 2t^2, 2 - t^3 \rangle$, where $t > 0$.

Solution The parametric equations corresponding to $\mathbf{r}(t)$ are $x = 2t^2$ and $y = 2 - t^3$. So we have

$$x' = 4t \qquad y' = -3t^2$$
$$x'' = 4 \qquad y'' = -6t$$

The point P is given by $t = 1$. At this point $x' = 4$, $x'' = 4$, $y' = -3$, and $y'' = -6$. So by equation (15.19),

$$\kappa = \frac{|x'y'' - x''y'|}{(x'^2 + y'^2)^{3/2}} = \frac{|4(-6) - (4)(-3)|}{(16 + 9)^{3/2}} = \frac{12}{125}$$

The radius of curvature ρ is therefore $\frac{125}{12}$. The center of curvature is ρ units in the direction of the normal from P. So we need $\mathbf{N}$. Since $t > 0$, we have

$$\mathbf{T} = \frac{\mathbf{r}'(t)}{|\mathbf{r}'(t)|} = \frac{\langle 4t, -3t^2 \rangle}{\sqrt{16t^2 + 9t^4}}$$
$$= \frac{\langle 4t, -3t^2 \rangle}{t\sqrt{16 + 9t^2}} = \left\langle \frac{4}{\sqrt{16 + 9t^2}}, \frac{-3t}{\sqrt{16 + 9t^2}} \right\rangle$$

Thus, we find (omitting some details)

$$\frac{d\mathbf{T}}{dt} = \left\langle \frac{-36}{(16 + 9t^2)^{3/2}}, \frac{-48}{(16 + 9t^2)^{3/2}} \right\rangle$$

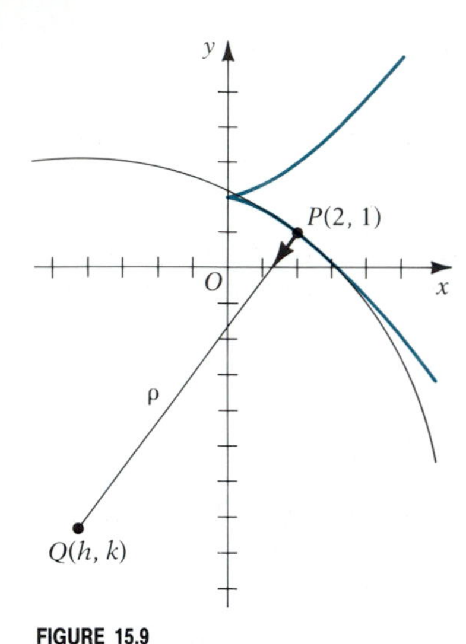

FIGURE 15.9

At $t = 1$,

$$\frac{d\mathbf{T}}{dt} = \left\langle \frac{-36}{125}, \frac{-48}{125} \right\rangle = \frac{-12}{125}\langle 3, 4\rangle$$

and

$$\left|\frac{d\mathbf{T}}{dt}\right| = \frac{12}{25}$$

Thus, by equation (15.15) the unit normal at P is

$$\mathbf{N} = \frac{\dfrac{d\mathbf{T}}{dt}}{\left|\dfrac{d\mathbf{T}}{dt}\right|} = \left\langle -\frac{3}{5}, -\frac{4}{5} \right\rangle$$

Now let $Q(h, k)$ denote the center of curvature. Then, as seen in Figure 15.9, $\overrightarrow{PQ} = \rho\mathbf{N}$, or

$$\langle h - 2, k - 1\rangle = \frac{125}{12}\left\langle -\frac{3}{5}, -\frac{4}{5} \right\rangle = \left\langle -\frac{25}{4}, -\frac{25}{3} \right\rangle$$

So $h - 2 = -\frac{25}{4}$ and $k - 1 = -\frac{25}{3}$, from which we get the center

$$(h, k) = \left(\frac{-17}{4}, \frac{-22}{3}\right)$$

■

EXERCISE SET 15.4

A

In Exercises 1–10 find **T** *and* **N** *at the prescribed point.*

1. $\mathbf{r}(t) = \langle t, t^2\rangle$; $t = 2$
2. $\mathbf{r}(t) = \langle 1 - t^2, 2 + 3t\rangle$; $t = 1$
3. $\mathbf{r}(t) = 2\cos t\mathbf{i} + 2\sin t\mathbf{j}$; $t = \frac{\pi}{4}$
4. $\mathbf{r}(t) = 3\cos t\mathbf{i} + 4\sin t\mathbf{j}$; $t = \frac{\pi}{2}$
5. $\mathbf{r}(t) = \langle 2\cos t, 2\sin t, 4t\rangle$; $t = \frac{3\pi}{4}$
6. $\mathbf{r}(t) = \langle t, t^2, 3 - t^2\rangle$; at $(1, 1, 2)$
7. $\mathbf{r}(t) = 2t^2\mathbf{i} + t^3\mathbf{j} + 3\mathbf{k}$; at $(2, 1, 3)$
8. $\mathbf{r}(t) = 2t^3\mathbf{i} + 3t^2\mathbf{j} + 3t\mathbf{k}$; $t = -1$
9. $\mathbf{r}(t) = \langle e^t, 2e^{-t}, 2t\rangle$; at $(1, 2, 0)$
10. $\mathbf{r}(t) = \langle 2t, \frac{1}{t}, 2\ln t\rangle$; at $(2, 1, 0)$

In Exercises 11–25 find κ *at the prescribed point.*

11. $\mathbf{r}(t) = \langle t^2, t^3\rangle$; $t = 1$
12. $\mathbf{r}(t) = \langle 1 + t, 1 - t^2\rangle$; $t = 0$
13. $\mathbf{r}(t) = (t\ln t)\mathbf{i} + \frac{1}{t}\mathbf{j}$; $t = 1$
14. $\mathbf{r}(t) = \sqrt{t^2 - 3}\,\mathbf{i} + \dfrac{t}{\sqrt{t^2 - 3}}\mathbf{j}$; $t = 2$
15. $x = \sin 2t$, $y = 4\cos 2t$; $t = \frac{\pi}{3}$
16. $x = t - \cos t$, $y = 1 - \sin t$; $t = \frac{\pi}{6}$
17. $y = 2x - x^2$; at $(0, 0)$
18. $y = x^3 - 2x^2 + 3$; at $(1, 2)$
19. $y = \frac{2}{x}$; at $(2, 1)$
20. $y = \sec x$; at $x = \frac{3\pi}{4}$
21. $\mathbf{r}(t) = \langle t^2, t^3, 1 - 2t\rangle$; $t = -1$
22. $\mathbf{r}(t) = \langle 4\cos t, 4\sin t, 3t\rangle$; $t = \frac{\pi}{3}$
23. $\mathbf{r}(t) = \langle 2e^t, e^{-t}, 2t\rangle$; at $(2, 1, 0)$
24. $\mathbf{r}(t) = t^2\mathbf{i} + t^3\mathbf{j} - 3t^2\mathbf{k}$; $t = \frac{1}{2}$
25. $\mathbf{r}(t) = (e^t\sin t)\mathbf{i} + (e^t\cos t)\mathbf{j} + e^t\mathbf{k}$; $t = 0$

In Exercises 26–30 find the center of the circle of curvature at P. Sketch the graph in a neighborhood of P, showing the circle of curvature.

26. $\mathbf{r}(t) = \langle t, 1 + t^2\rangle$; $P(0, 1)$
27. $\mathbf{r}(t) = \langle 4\cos t, 3\sin t\rangle$; $P(4, 0)$
28. $\mathbf{r}(t) = t\mathbf{i} + \sqrt{2t}\mathbf{j}$; $P(\frac{1}{2}, 1)$
29. $\mathbf{r}(t) = t\mathbf{i} + e^t\mathbf{j}$; $P(0, 1)$
30. $\mathbf{r}(t) = \langle t - \sin t, 1 - \cos t\rangle$; $t = \frac{\pi}{3}$

B

31. Let C be a plane curve that is the graph of $\mathbf{r}(s)$, where s is the arc length parameter, and suppose $\mathbf{r}''(s)$ exists on $[0, L]$, where L is the length of C. Let $\theta = \theta(s)$ be the angle between the vectors $\mathbf{i}$ and $\mathbf{T}$ at an arbitrary point on C. Show the following:
(a) $\mathbf{T} = \langle \cos\theta, \sin\theta \rangle$

(b) $\kappa = \left|\dfrac{d\theta}{ds}\right|$

(c) $\mathbf{N} = \langle -\sin\theta, \cos\theta \rangle$ if $d\theta/ds > 0$ and
$\mathbf{N} = \langle \sin\theta, -\cos\theta \rangle$ if $d\theta/ds < 0$

Use the result of part c to show that $\mathbf{N}$ always is directed toward the concave side of C.

32. Suppose $\mathbf{N} = \langle n_1, n_2 \rangle$ is the unit normal vector at a point $P(x, y)$ on the plane curve C. Prove that the center of curvature at P is the point $(x + \rho n_1, y + \rho n_2)$, where ρ is the radius of curvature.

33. Find $\mathbf{T}$, $\mathbf{N}$, $\mathbf{B}$, and κ at an arbitrary point on the circular helix $\mathbf{r}(t) = \langle a\cos bt, a\sin bt, ct \rangle$.

34. Find $\mathbf{T}$, $\mathbf{N}$, $\mathbf{B}$, and κ at an arbitrary point on the curve $\mathbf{r}(t) = \langle t, t^2, 2t^3/3 \rangle$.

35. Use equation (15.17) to show that $(\mathbf{r}' \times \mathbf{r}'') \times \mathbf{r}' = |\mathbf{r}'|^2(ds/dt)^2\kappa\mathbf{N}$, and hence obtain the formula

$$\mathbf{N} = \frac{(\mathbf{r}' \times \mathbf{r}'') \times \mathbf{r}'}{\kappa|\mathbf{r}'|^4}$$

In Exercises 36 and 37 assume C is a smooth curve parameterized by arc length and that $d\mathbf{T}/ds$, $d\mathbf{N}/ds$, and $d\mathbf{B}/ds$ all exist.

36. By differentiating both sides of $\mathbf{B} \cdot \mathbf{B} = 1$, show that $\mathbf{B}$ and $d\mathbf{B}/ds$ are orthogonal. Explain why it follows that

$$\frac{d\mathbf{B}}{ds} = \alpha\mathbf{T} + \beta\mathbf{N}$$

for some scalars α and β. By differentiating both sides of $\mathbf{B} \cdot \mathbf{T} = 0$, show that $\alpha = 0$.

37. (Continuation of Problem 36) Write $\tau = -\beta$, so that $d\mathbf{B}/ds = -\tau\mathbf{N}$. (The scalar τ is called the **torsion.**) Show that $\mathbf{N} = \mathbf{B} \times \mathbf{T}$. Prove that

$$\frac{d\mathbf{N}}{ds} = \tau\mathbf{B} - \kappa\mathbf{T}$$

(*Note:* The formulas

$$\frac{d\mathbf{T}}{ds} = \kappa\mathbf{N} \qquad \frac{d\mathbf{N}}{ds} = \tau\mathbf{B} - \kappa\mathbf{T} \qquad \frac{d\mathbf{B}}{ds} = -\tau\mathbf{N}$$

are called the **Frenet formulas.**)

38. Find the torsion τ as a function of t for the curve of Exercise 34. Verify Frenet's formulas for this curve. (See Exercise 37.)

15.5 TANGENTIAL AND NORMAL COMPONENTS OF ACCELERATION

From Section 15.3, when a curve C is the path of a moving particle whose position vector at time t is $\mathbf{r}(t)$, we know that its velocity $\mathbf{v}(t) = \mathbf{r}'(t)$, its acceleration $\mathbf{a}(t) = \mathbf{v}'(t) = \mathbf{r}''(t)$, and its speed $|\mathbf{v}(t)| = ds/dt$. Then equation (15.14) becomes

$$\mathbf{T} = \frac{\mathbf{v}(t)}{|\mathbf{v}(t)|} \tag{15.21}$$

Similarly, we can write equation (15.17) as

$$\kappa = \frac{|\mathbf{v} \times \mathbf{a}|}{|\mathbf{v}|^3} \tag{15.22}$$

where, for brevity, we are using $\mathbf{v}$ for $\mathbf{v}(t)$ and $\mathbf{a}$ for $\mathbf{a}(t)$. Some students may find this the easiest formula for curvature to learn.

Let us examine equation (15.17) more closely from the moving particle point of view. It can be written

$$\mathbf{a} = \frac{d^2s}{dt^2}\mathbf{T} + \kappa\left(\frac{ds}{dt}\right)^2\mathbf{N} \tag{15.23}$$

One immediate consequence of this is that the acceleration vector at a point on the path always lies in the plane of **T** and **N** at that point. If we take the dot product of equation (15.23) first with **T** and then with **N**, we get

$$\mathbf{a} \cdot \mathbf{T} = \frac{d^2 s}{dt^2}\mathbf{T} \cdot \mathbf{T} + \kappa\left(\frac{ds}{dt}\right)^2 \mathbf{N} \cdot \mathbf{T} = \frac{d^2 s}{dt^2}$$

$$\mathbf{a} \cdot \mathbf{N} = \frac{d^2 s}{dt^2}\mathbf{T} \cdot \mathbf{N} + \kappa\left(\frac{ds}{dt}\right)^2 \mathbf{N} \cdot \mathbf{N} = \kappa\left(\frac{ds}{dt}\right)^2$$

where we have used $\mathbf{T} \cdot \mathbf{T} = \mathbf{N} \cdot \mathbf{N} = 1$ and $\mathbf{T} \cdot \mathbf{N} = 0$. Now

$$\mathbf{a} \cdot \mathbf{T} = \frac{\mathbf{a} \cdot \mathbf{T}}{|\mathbf{T}|} = \text{Comp}_{\mathbf{T}}\mathbf{a}$$

and

$$\mathbf{a} \cdot \mathbf{N} = \frac{\mathbf{a} \cdot \mathbf{N}}{|\mathbf{N}|} = \text{Comp}_{\mathbf{N}}\mathbf{a}$$

So the components of **a** along **T** and **N**, respectively, are

$$\text{Comp}_{\mathbf{T}}\mathbf{a} = \frac{d^2 s}{dt^2}$$

$$\text{Comp}_{\mathbf{N}}\mathbf{a} = \kappa\left(\frac{ds}{dt}\right)^2$$

We designate these components by the symbols $a_{\mathbf{T}}$ and $a_{\mathbf{N}}$, respectively, and call them the **tangential** and **normal components of acceleration.** So

$$a_{\mathbf{T}} = \frac{d^2 s}{dt^2} \qquad a_{\mathbf{N}} = \kappa\left(\frac{ds}{dt}\right)^2 \tag{15.24}$$

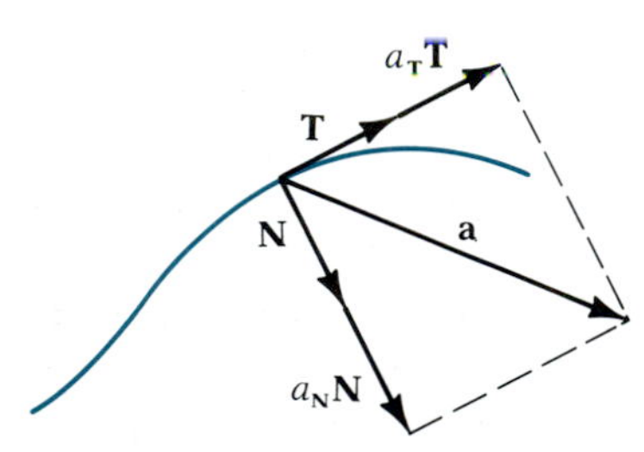

FIGURE 15.10

Equation (15.23) can now be written

$$\mathbf{a} = a_{\mathbf{T}}\mathbf{T} + a_{\mathbf{N}}\mathbf{N} \tag{15.25}$$

where $a_{\mathbf{T}}\mathbf{T} = \text{Proj}_{\mathbf{T}}\mathbf{a}$ and $a_{\mathbf{N}}\mathbf{N} = \text{Proj}_{\mathbf{N}}\mathbf{a}$. This is illustrated in Figure 15.10.

From equation (15.25) we also have

$$\begin{aligned}\mathbf{a} \cdot \mathbf{a} &= (a_{\mathbf{T}}\mathbf{T} + a_{\mathbf{N}}\mathbf{N}) \cdot (a_{\mathbf{T}}\mathbf{T} + a_{\mathbf{N}}\mathbf{N}) \\ &= a_{\mathbf{T}}^2\mathbf{T} \cdot \mathbf{T} + 2a_{\mathbf{T}}a_{\mathbf{N}}\mathbf{T} \cdot \mathbf{N} + a_{\mathbf{N}}^2\mathbf{N} \cdot \mathbf{N} \\ &= a_{\mathbf{T}}^2 + a_{\mathbf{N}}^2\end{aligned}$$

that is,

$$|\mathbf{a}|^2 = a_{\mathbf{T}}^2 + a_{\mathbf{N}}^2 \tag{15.26}$$

(This can also be seen from Figure 15.10, using the Pythagorean theorem.) In practice it is usually easier to find $a_{\mathbf{T}}$ than $a_{\mathbf{N}}$ from equations (15.24). Then equation (15.26) can be used to get $a_{\mathbf{N}} = \sqrt{|a|^2 - a_{\mathbf{T}}^2}$. The next example illustrates this.

EXAMPLE 15.17 Find the tangential and normal components of acceleration of a particle whose position vector is $\mathbf{r}(t) = 3t\mathbf{i} + 2t^3\mathbf{j} + 3t^2\mathbf{k}$.

Solution We have

$$\mathbf{v} = \mathbf{r}'(t) = 3\mathbf{i} + 6t^2\mathbf{j} + 6t\mathbf{k} = 3(\mathbf{i} + 2t^2\mathbf{j} + 2t\mathbf{k})$$
$$\mathbf{a} = \mathbf{r}''(t) = 12t\mathbf{j} + 6\mathbf{k} = 6(2t\mathbf{j} + \mathbf{k})$$
$$\frac{ds}{dt} = |\mathbf{v}| = 3\sqrt{1 + 4t^4 + 4t^2} = 3\sqrt{(2t^2 + 1)^2} = 3(2t^2 + 1)$$
$$\frac{d^2s}{dt^2} = 12t$$
$$|\mathbf{a}| = 6\sqrt{4t^2 + 1}$$

So by the first of equations (15.24), $a_\mathbf{T} = 12t$. Then, from equation (15.26),

$$a_\mathbf{N} = \sqrt{|\mathbf{a}|^2 - a_\mathbf{T}^2} = \sqrt{36(4t^2 + 1) - 144t^2} = 6$$

Notice that we can now obtain κ using the formula for a_N in equations (15.24):

$$\kappa = \frac{a_\mathbf{N}}{\left(\dfrac{ds}{dt}\right)^2} = \frac{6}{9(2t^2 + 1)^2} = \frac{2}{3(2t^2 + 1)^2}$$

You may wish to compare this "backdoor" method of finding κ with the direct method. ■

The tangential component of acceleration is $a_\mathbf{T} = d^2s/dt^2$, which is the rate of change of speed, and this agrees with our intuitive idea of acceleration. To help understand the normal component, $a_\mathbf{N} = \kappa(ds/dt)^2$, consider an automobile going around a curve. Since force is mass times acceleration, the normal component of the force necessary to hold the car on the road (the centripetal force) is $ma_\mathbf{N}$, where m is the mass of the car. This is the magnitude of the force of friction between the tires and the road. If the curve is sharp, so that κ is large, this frictional force has a large magnitude. Similarly, if the speed ds/dt is large, the magnitude of the force is large. In fact, it increases as the square of the speed. Of course, if the curve is sharp *and* the speed is great, it is unlikely the car will stay on the road.

We can obtain another representation of $a_\mathbf{T}$ and $a_\mathbf{N}$ from the products $\mathbf{v} \cdot \mathbf{a}$ and $\mathbf{v} \times \mathbf{a}$. Since $\mathbf{v} = |\mathbf{v}|\mathbf{T}$, we have from equation (15.25)

$$\mathbf{v} \cdot \mathbf{a} = |\mathbf{v}|\mathbf{T} \cdot (a_\mathbf{T}\mathbf{T} + a_\mathbf{N}\mathbf{N}) = |\mathbf{v}|a_\mathbf{T}$$

and

$$\mathbf{v} \times \mathbf{a} = |\mathbf{v}|\mathbf{T} \times (a_\mathbf{T}\mathbf{T} + a_\mathbf{N}\mathbf{N}) = |\mathbf{v}|a_\mathbf{N}\mathbf{B}$$

where we have used $\mathbf{T} \cdot \mathbf{T} = 1$, $\mathbf{T} \cdot \mathbf{N} = 0$, $\mathbf{T} \times \mathbf{T} = 0$, and $\mathbf{T} \times \mathbf{N} = \mathbf{B}$. Thus, since $|\mathbf{B}| = 1$ and $a_\mathbf{N} \geq 0$,

$$a_\mathbf{T} = \frac{\mathbf{v} \cdot \mathbf{a}}{|\mathbf{v}|} \qquad\qquad a_\mathbf{N} = \frac{|\mathbf{v} \times \mathbf{a}|}{|\mathbf{v}|} \tag{15.27}$$

EXAMPLE 15.18 Use equations (15.27) to find $a_{\mathbf{T}}$ and $a_{\mathbf{N}}$ at time $t = 1$ for a particle with position vector

$$\mathbf{r}(t) = \left\langle 4\sqrt{t},\ 1 - 2t^2,\ \frac{8(t-1)}{\sqrt{t+3}} \right\rangle$$

Solution First we calculate $\mathbf{v}(t)$ and $\mathbf{a}(t)$:

$$\mathbf{v}(t) = \left\langle \frac{2}{\sqrt{t}},\ -4t,\ \frac{4(t+7)}{(t+3)^{3/2}} \right\rangle$$

$$\mathbf{a}(t) = \left\langle -\frac{1}{t^{3/2}},\ -4,\ \frac{-2(t+15)}{(t+3)^{5/2}} \right\rangle$$

So

$$\mathbf{v}(1) = \langle 2, -4, 4\rangle, \quad |\mathbf{v}(1)| = \sqrt{4 + 16 + 16} = 6,$$

and

$$\mathbf{a}(1) = \langle -1, -4, -1\rangle$$

Then, by equations (15.27),

$$a_{\mathbf{T}} = \frac{|\mathbf{v} \cdot \mathbf{a}|}{|\mathbf{v}|} = \frac{\langle 2, -4, 4\rangle \cdot \langle -1, -4, -1\rangle}{6} = \frac{-2 + 16 - 4}{6} = \frac{5}{3}$$

$$a_{\mathbf{N}} = \frac{|\mathbf{v} \times \mathbf{a}|}{|\mathbf{v}|} = \frac{|\langle 2, -4, 4\rangle \times \langle -1, -4, -1\rangle|}{6} = \frac{|\langle 20, -2, -12\rangle|}{6}$$

$$= \frac{\sqrt{137}}{3}$$

■

EXERCISE SET 15.5

A

In Exercises 1–6 find the tangential and normal components of acceleration at the indicated time. Draw the graph of **r**, *showing the acceleration, together with its tangential and normal projections at the point in question.*

1. $\mathbf{r}(t) = 2t\mathbf{i} - t^2\mathbf{j};\ t = 2$
2. $\mathbf{r}(t) = \langle t^2, t^3\rangle;\ t = 1$
3. $\mathbf{r}(t) = \langle 2\cos t, 4\sin t\rangle;\ t = \frac{\pi}{3}$
4. $\mathbf{r}(t) = \cosh t\mathbf{i} + \sinh t\mathbf{j};\ t = \ln 2$
5. $\mathbf{r}(t) = (t-1)\mathbf{i} + 4\sqrt{t}\mathbf{j};\ t = 1$
6. $\mathbf{r}(t) = \langle e^{2t} - 1, e^t\rangle;\ t = 0$

In Exercises 7–12 use the results of the specified problem to find the curvature at the indicated point.

7. Exercise 1
8. Exercise 2
9. Exercise 3
10. Exercise 4
11. Exercise 5
12. Exercise 6

In Exercises 13–18 find $a_{\mathbf{T}}$ and $a_{\mathbf{N}}$ at an arbitrary value of t in the domain.

13. $\mathbf{r}(t) = \langle a\cos\omega t, a\sin\omega t, bt\rangle;\ t \geq 0,\ a > 0$
14. $\mathbf{r}(t) = \left\langle 2\ln t, \frac{t-1}{t}, t\right\rangle;\ t > 0$
15. $\mathbf{r}(t) = t\mathbf{i} + t^2\mathbf{j} + \frac{2}{3}t^3\mathbf{k};\ -\infty < t < \infty$
16. $\mathbf{r}(t) = (3t\sin t)\mathbf{i} + (3t\cos t)\mathbf{j} + (2t)^{3/2}\mathbf{k};\ t > 0$
17. $\mathbf{r}(t) = \langle 2t, 2e^t, e^{-t}\rangle;\ -\infty < t < \infty$
18. $\mathbf{r}(t) = \langle \sin t - t\cos t, \cos t + t\sin t, t^2\rangle;\ t \geq 0$
19. Show that if $a_{\mathbf{N}} \neq 0$,

$$\mathbf{N} = \frac{\mathbf{a} - a_{\mathbf{T}}\mathbf{T}}{a_{\mathbf{N}}}$$

In Exercises 20–24 find **T**, **N**, *and* κ *for the curve in the indicated problem, using* $a_{\mathbf{T}}$ *and* $a_{\mathbf{N}}$ *as previously found, together with the result of Exercise 19.*

20. Exercise 14 **21.** Exercise 15 **22.** Exercise 16

23. Exercise 17 **24.** Exercise 18

25. Let $v = |\mathbf{v}|$. Show that

$$\mathbf{a} = \frac{dv}{dt}\mathbf{T} + \frac{v^2}{\rho}\mathbf{N}$$

where ρ is the radius of curvature.

B

26. Prove that if the normal component of acceleration of a particle is constantly 0, the particle moves in a straight line.

27. Prove that if the force on a particle is always centripetal (directed along the normal to the path), its speed is constant.

Exercises 28–32 form a sequential unit.

28. Let a particle move in the xy-plane in which a polar coordinate system is superimposed, with the polar axis coinciding with the positive x-axis. Let $\mathbf{r} = \mathbf{r}(t)$ be its position vector at time t, and let $r = |\mathbf{r}|$. If $\theta = \theta(t)$ is the polar angle to the vector $\mathbf{r}$ at time t, define $\mathbf{u}_r = \langle \cos\theta, \sin\theta \rangle$ and $\mathbf{u}_\theta = \langle -\sin\theta, \cos\theta \rangle$. Show each of the following.

(a) $\mathbf{u}_r$ and $\mathbf{u}_\theta$ are orthogonal unit vectors, and $\mathbf{u}_\theta$ is rotated 90° counterclockwise from $\mathbf{u}_r$.

(b) $\mathbf{r} = r\mathbf{u}_r$

(c) $\dfrac{d\mathbf{u}_r}{dt} = \mathbf{u}_\theta \dfrac{d\theta}{dt}$ and $\dfrac{d\mathbf{u}_\theta}{dt} = -\mathbf{u}_r \dfrac{d\theta}{dt}$

(*Note:* $\mathbf{u}_r$ and $\mathbf{u}_\theta$ are called the **radial** and **transverse** unit vectors, respectively.)

29. (a) Show that $\mathbf{v} = \dfrac{dr}{dt}\mathbf{u}_r + r\dfrac{d\theta}{dt}\mathbf{u}_\theta$.

(b) Show that speed $= \sqrt{\left(\dfrac{dr}{dt}\right)^2 + r^2\left(\dfrac{d\theta}{dt}\right)^2}$.

30. Show that

$$\mathbf{a} = \left[\frac{d^2r}{dt^2} - r\left(\frac{d\theta}{dt}\right)^2\right]\mathbf{u}_r + \left[r\frac{d^2\theta}{dt^2} + 2\frac{dr}{dt}\frac{d\theta}{dt}\right]\mathbf{u}_\theta$$

31. Show the following:

(a) $\text{Comp}_{\mathbf{u}_r}\mathbf{a} = \dfrac{d^2r}{dt^2} - r\left(\dfrac{d\theta}{dt}\right)^2$

(b) $\text{Comp}_{\mathbf{u}_\theta}\mathbf{a} = r\dfrac{d^2\theta}{dt^2} + 2\dfrac{dr}{dt}\dfrac{d\theta}{dt}$

32. Let $\mathbf{a}_r = \text{Comp}_{\mathbf{u}_r}\mathbf{a}$ and $\mathbf{a}_\theta = \text{Comp}_{\mathbf{u}_\theta}\mathbf{a}$. Show that $|\mathbf{a}|^2 = \mathbf{a}_r^2 + \mathbf{a}_\theta^2$

33. A particle moves in the horizontal plane so that its polar coordinates at time t are $r = 1 + \cos t^2$ and $\theta = t^2$. Use the results of Exercises 29 and 30 to resolve **v** and **a** into radial and transverse vector components.

34. A particle moves in the xy-plane with position vector $\mathbf{r}(t) = (t^2 \cos t)\mathbf{i} + (t^2 \sin t)\mathbf{j}$ at time t. Find the polar coordinates of a point on the path at time t. Find the radial component $\mathbf{a}_r$ of acceleration and the transverse component $\mathbf{a}_\theta$, at time t. (See Exercise 31.)

15.6 KEPLER'S LAWS

In the early 17th century the German mathematician and astronomer Johannes Kepler (1571–1630) postulated the following three laws governing the orbits of planets around the sun.

KEPLER'S LAWS OF PLANETARY MOTION

1. The orbit of each planet is an ellipse with the sun at one focus.
2. The radius vector from the sun to a planet sweeps out area at a constant rate.
3. The square of the time for a planet to complete one revolution around its elliptical orbit is proportional to the cube of the length of the semimajor axis of the ellipse.

Kepler deduced these laws based on the astronomical observations of his friend and mentor Tycho Brahe. This required years of analyzing masses of data with laborious calculations. His discoveries rank as one of the outstanding achievements in the history of science. Based as they were on empirical evidence, however, they lacked a sound theoretical basis until Newton, some 50 years later, deduced Kepler's laws using his newly invented calculus. Newton based his proofs on the following two principles:

1. *Newton's second law of motion.* For a body of constant mass m moving under the action of a force $\mathbf{F}$,

$$\mathbf{F} = m\mathbf{a} \tag{15.28}$$

where $\mathbf{a}$ is the acceleration of the body.

2. *Newton's law of gravitation.* The force $\mathbf{F}$ of attraction between two bodies of masses M and m, respectively, is proportional to the product of the masses and inversely proportional to the square of the distance, r, between them:

$$\mathbf{F} = -\frac{GMm}{r^2}\mathbf{u}_r \tag{15.29}$$

where $\mathbf{u}_r$ is a unit vector directed from one mass toward the other.

The constant G in equation (15.29) is called the *universal gravitational constant* and has the approximate value

$$G = 6.673 \times 10^{-8} \frac{\text{cm}^3}{(\text{gm})(\text{sec})^2}$$

To demonstrate the power of the vector calculus we have studied, we will prove the first of Kepler's laws. In the exercises we outline proofs of the other two laws. You should follow the proof below using pencil and paper, verifying all steps. Note carefully the use of properties of the dot and cross products.

We take the origin as the center of mass of the sun and let M be its mass. Let $\mathbf{r}(t)$ to be the position vector at time t of the center of mass of a planet that has mass m. Let $r = |\mathbf{r}|$, and let $\mathbf{u}_r$ be the unit vector $\mathbf{r}/r$, so that $\mathbf{r} = r\mathbf{u}_r$. We assume that the forces of attraction between this planet and all other bodies are negligible compared with the gravitational attraction between it and the sun.

Equating the right-hand sides of equations (15.28) and (15.29), we get the acceleration in the form

$$\mathbf{a} = -\frac{GM}{r^2}\mathbf{u}_r \tag{15.30}$$

Then

$$\mathbf{r} \times \mathbf{a} = r\mathbf{u}_r \times \left(-\frac{GM}{r^2}\mathbf{u}_r\right) = -\frac{GM}{r}(\mathbf{u}_r \times \mathbf{u}_r) = \mathbf{0}$$

and since $\mathbf{v} \times \mathbf{v} = \mathbf{0}$, we also have

$$\frac{d}{dt}(\mathbf{r} \times \mathbf{v}) = \mathbf{r} \times \mathbf{a} + \mathbf{v} \times \mathbf{v} = \mathbf{0}$$

Thus,

$$\mathbf{r} \times \mathbf{v} = \mathbf{c} \tag{15.31}$$

where **c** is a constant vector. Since **r** is orthogonal to $\mathbf{r} \times \mathbf{v}$ and hence to **c**, this says that at all times, the vector $\mathbf{r}(t)$ is perpendicular to the fixed vector **c**. So the path of the planet lies in a plane. There is no loss of generality in assuming that this is the xy-plane and that **c** is directed along the positive z-axis.

Let $\theta = \theta(t)$ be the polar angle from the positive x-axis to $\mathbf{r}(t)$. Then, $\mathbf{u}_r = \langle \cos\theta, \sin\theta, 0\rangle$, and if we define $\mathbf{u}_\theta = \langle -\sin\theta, \cos\theta, 0\rangle$, $\mathbf{u}_\theta$ is a unit vector such that $\mathbf{u}_r \cdot \mathbf{u}_\theta = 0$ and $\mathbf{u}_r \times \mathbf{u}_\theta = \langle 0, 0, 1\rangle = \mathbf{k}$. As in Exercises 28 and 29 of Exercise Set 15.5, we obtain from differentiating both sides of $\mathbf{r} = r\mathbf{u}_r$,

$$\mathbf{v} = \frac{dr}{dt}\mathbf{u}_r + r\frac{d\theta}{dt}\mathbf{u}_\theta$$

Thus, using equation (15.31),

$$\begin{aligned}\mathbf{c} = \mathbf{r} \times \mathbf{v} &= (r\mathbf{u}_r) \times \left(\frac{dr}{dt}\mathbf{u}_r + r\frac{d\theta}{dt}\mathbf{u}_\theta\right)\\ &= r^2\frac{d\theta}{dt}(\mathbf{u}_r \times \mathbf{u}_\theta) = r^2\frac{d\theta}{dt}\mathbf{k}\end{aligned} \tag{15.32}$$

From equations (15.30) and (15.32) we get

$$\begin{aligned}\frac{d}{dt}(\mathbf{v} \times \mathbf{c}) = \mathbf{a} \times \mathbf{c} &= \left(-\frac{GM}{r^2}\mathbf{u}_r\right) \times \left[r^2\frac{d\theta}{dt}\mathbf{k}\right]\\ &= -GM\frac{d\theta}{dt}[\mathbf{u}_r \times \mathbf{k}]\\ &= GM\frac{d\theta}{dt}\mathbf{u}_\theta\\ &= \frac{d}{dt}(GM\mathbf{u}_r)\end{aligned}$$

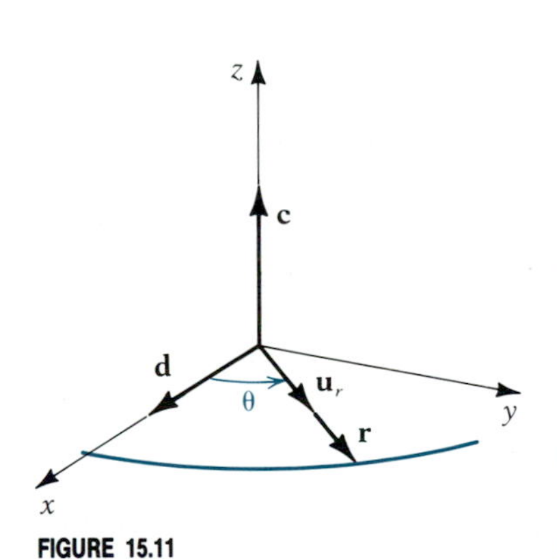

FIGURE 15.11

The equality of these two derivatives implies that

$$\mathbf{v} \times \mathbf{c} = GM\mathbf{u}_r + \mathbf{d} \tag{15.33}$$

The constant vector **d** is a linear combination of $\mathbf{u}_r$ and $\mathbf{v} \times \mathbf{c}$, each of which lies in the xy-plane, and so **d** also lies in the xy-plane. We may assume, again without loss of generality, that it is directed along the positive x-axis. The relationship among the vectors **c**, **d**, $\mathbf{u}_r$, and **r** is shown in Figure 15.11.

Let $c = |\mathbf{c}|$ and $d = |\mathbf{d}|$. Then from equations (15.31) and (15.33),

$$\begin{aligned}c^2 = \mathbf{c} \cdot \mathbf{c} &= (\mathbf{r} \times \mathbf{v}) \cdot \mathbf{c}\\ &= \mathbf{r} \cdot (\mathbf{v} \times \mathbf{c})\\ &= (r\mathbf{u}_r) \cdot (GM\mathbf{u}_r + \mathbf{d})\\ &= rGM + r\mathbf{u}_r \cdot \mathbf{d}\\ &= r(GM + d\cos\theta)\end{aligned}$$

Solving for r gives

$$r = \frac{c^2}{GM + d\cos\theta}$$

If we let $e = d/GM$ and $p = c^2/GMe$, this becomes, on dividing numerator and denominator by GM,

$$r = \frac{ep}{1 + e\cos\theta} \tag{15.34}$$

In Exercise 36 of Exercise Set 13.4, you were asked to show that an equation of the form (15.34) is the polar equation of an ellipse if $0 < e < 1$, a hyperbola if $e > 1$, and a parabola if $e = 1$. Since it is known that planets travel in closed orbits, it follows that the equation is that of an ellipse. In Exercise 2 you will be asked to show that one focus of the ellipse (15.34) is at the origin. This will complete the proof of Kepler's first law.

EXERCISE SET 15.6

A

1. Write the polar equation (15.34) for $0 < e < 1$ in the standard rectangular form

$$\frac{(x-h)^2}{a^2} + \frac{(y-k)^2}{b^2} = 1$$

2. From the result of Exercise 1 show the following:
   (a) $b = a\sqrt{1-e^2}$
   (b) The center of the ellipse is at $\left(\frac{-e^2p}{1-e^2}, 0\right)$.
   (c) The distance from the center to each focus is $\frac{e^2p}{1-e^2}$.
   (d) One focus is at $(0, 0)$ and the other is at $\left(\frac{-2e^2p}{1-e^2}, 0\right)$.

In Exercises 3 and 4 each equation was used in the proof of Kepler's first law. Verify each one.

3. (a) $\mathbf{u}_r \cdot \mathbf{u}_\theta = 0$ (b) $\mathbf{u}_r \times \mathbf{u}_\theta = \mathbf{k}$
   (c) $\mathbf{v} = \frac{dr}{dt}\mathbf{u}_r + r\frac{d\theta}{dt}\mathbf{u}_\theta$

4. (a) $(r\mathbf{u}_r) \times \left(\frac{dr}{dt}\mathbf{u}_r + r\frac{d\theta}{dt}\mathbf{u}_\theta\right) = r^2\frac{d\theta}{dt}(\mathbf{u}_r \times \mathbf{u}_\theta)$
   (b) $-GM\frac{d\theta}{dt}(\mathbf{u}_r \times \mathbf{k}) = GM\frac{d\theta}{dt}\mathbf{u}_\theta$
   (c) $GM\frac{d\theta}{dt}\mathbf{u}_\theta = \frac{d}{dt}(GM\mathbf{u}_r)$

5. Referring to the proof of Kepler's first law, explain fully the justification for the following assertions:
   (a) $\mathbf{v} \times \mathbf{c}$ lies in the xy-plane.
   (b) $\mathbf{d}$ lies in the xy-plane.

B

In Exercises 6 and 7 fill in the details of the outlines given of proofs of Kepler's second and third laws.

6. (a) Using the formula $A = \frac{1}{2}\int_\alpha^\beta r^2\,d\theta$ for area in polar coordinates, show that the area between any fixed angle $\theta_0 = \theta(t_0)$ and the angle $\theta = \theta(t)$, bounded by the ellipse of equation (15.34), is

$$A(t) = \int_{t_0}^{t} r^2\frac{d\theta}{dt}\,dt$$

   (b) By using part a together with equation (15.32), show that

$$\frac{dA}{dt} = \frac{c}{2}$$

   (c) Conclude that area is swept out by $r(t)$ at a constant rate. This proves Kepler's second law.

7. Let T be the time required for a planet to complete one revolution around the sun (the *period*).
   (a) Using Exercise 6, show that the total area enclosed by the ellipse is

$$A = \tfrac{1}{2}cT$$

   (b) Recall from Exercise 18 in Exercise Set 12.3 that the area also is given by $A = \pi ab$, where a and b are the lengths of the semimajor and semiminor axes of the ellipse. (If you have not previously shown this, show it now.) Combining this with part a and Exercise 2, part a, show that

$$T = \frac{2\pi a^2}{c}\sqrt{1-e^2}$$

(c) From the result of Exercise 1 show that $1 - e^2 = \frac{ep}{a}$ and hence obtain

$$T^2 = \frac{4\pi^2 ep}{c^2} a^3$$

This proves Kepler's third law. By replacing p and e with the values assigned to them, rewrite the result in the form

$$T^2 = \frac{4\pi^2}{GM} a^3 \qquad (15.35)$$

8. A reasonable approximation to the period of a planet is obtained by replacing the semimajor axis a in equation (15.35) by the mean distance of the planet from the sun.
(a) The mean distance of Mars from the sun is approximately $1\frac{1}{2}$ times that of the earth. Find the approximate time (in "earth days") it takes Mars to complete one revolution around the sun.
(b) It takes Mercury approximately 88 earth days to orbit the sun. If the mean distance of the earth from the sun is approximately 93 million miles, find the approximate mean distance of Mercury from the sun.

9. Let r_0 denote the minimum value of the distance r of a planet from the sun and let v_0 be its speed when $r = r_0$.
(a) Show that $r_0 = ep/(1 + e)$ and this occurs when $\theta = 0$.
(b) Use equation (15.33) to show that

$$v_0 = \frac{(GM + d)}{c}$$

(*Hint*: Use $|\mathbf{v} \times \mathbf{c}| = |\mathbf{v}|\,|\mathbf{c}|$. Why is this true?)
(c) Show that

$$v_0 = \frac{c}{r_0}$$

10. (a) Use the result of Exercise 7, part b, to express v_0 in terms of a, e, and T.
(b) The earth takes approximately 365.26 days to complete its orbit around the sun. Its semimajor axis a is approximately 9.2956×10^7 mi, and the eccentricity of the orbit is approximately 0.016732. Find r_0 and v_0 for the earth. (Express v_0 in miles per hour.)

15.7 SUPPLEMENTARY EXERCISES

1. Let $\mathbf{F}(t) = \langle t^2, t^3, 1 - t\rangle$, $\mathbf{G}(t) = \langle t \ln t, -2, t + 3\rangle$, and $\alpha(t) = e^t$. Find the following:
(a) $(\mathbf{F}\cdot\mathbf{G})(t)$ (b) $\mathbf{F}(\alpha(t))$ (c) $(\mathbf{F} - \mathbf{G})(t)$
(d) $(\mathbf{G} \circ \alpha)(0)$ (e) $(\mathbf{F}' \times \mathbf{G}')(1)$

2. Find $\mathbf{F}'(t)$ if:
(a) $\mathbf{F}(t) = \langle t \ln \sqrt{t}, t^2 e^{-t}\rangle$, $t > 0$
(b) $\mathbf{F}(t) = \left(\frac{t}{\sqrt{t^2 - 1}}\right)\mathbf{i} + \left(\sin^{-1}\frac{1}{t}\right)\mathbf{j} + (t\sqrt{t^2 - 1})\mathbf{k}$

3. Evaluate the integrals:
(a) $\int \left\langle \frac{1}{\sqrt{1 - t^2}}, \frac{t}{\sqrt{1 - t^2}}, \frac{1}{1 - t^2} \right\rangle dt$
(b) $\int_0^{\pi/3} [(\cos^2 t)\mathbf{i} - (\sin^3 t)\mathbf{j} + (\tan^2 t)\mathbf{k}]\, dt$

4. Let $\mathbf{F}(t) = t^2\mathbf{i} + (3t - 2)\mathbf{j} + (1 - t^2)\mathbf{k}$ and $\alpha(t) = \cos t$. Find:
(a) $\int (\alpha\mathbf{F})(t)\, dt$ (b) $\int (\mathbf{F} \circ \alpha)(t)\, dt$

5. Let $\mathbf{f}(t) = (\ln t^2)\mathbf{i} - \frac{2}{\sqrt{t}}\mathbf{j} + 4\sqrt{t}\mathbf{k}$. Find:
(a) $\int_1^4 \mathbf{f}'(t)\, dt$ (b) $\int_1^4 |\mathbf{f}'(t)|\, dt$

6. Verify the result of Theorem 15.6 for $\mathbf{f}(t) = \langle 20t, 9t^2, 12t^2\rangle$ on $[0, 1]$.

7. Let C be the graph of $\mathbf{r}(t) = (2t^3/3)\mathbf{i} + (1 - 2t^2)\mathbf{j} + 4t\mathbf{k}$ for $0 \le t \le 3$. Find the length of C.

8. Sketch the graph of $\mathbf{r}(t) = \langle \cos t, t, \sin t\rangle$ for $0 \le t \le 4\pi$. Find parametric equations of the tangent line to the graph at $t = \frac{4\pi}{3}$.

9. Let C be the graph of $\langle \ln(\cosh t), 2\tan^{-1} e^t, \sqrt{3}\,t\rangle$ on $[-1, 2]$.
(a) Find the length of C.
(b) Make the change of variables $u = e^{-t}$, and let $\mathbf{R}(u)$ be the new position vector for C. Find $\mathbf{R}(u)$ and the u interval. Is the orientation of C preserved or reversed?
(c) Find the length of C using $\mathbf{R}(u)$ and show it is the same as that found in part a.

10. Let C be the graph of $\mathbf{r}(t) = \langle 2e^t \sin t, 2e^t \cos t, e^t\rangle$ on $[0, \infty)$. Introduce arc length s as a parameter and let $\mathbf{R}(s)$ be the resulting position vector for C. Show that for all $s > 0$, $|\mathbf{R}'(s)| = 1$.

11. Use the arc length parameterization of C in Exercise 10 to find $\mathbf{T}$, $\mathbf{N}$, and κ at an arbitrary $s \ge 0$. Evaluate each of these at the point $(0, 2, 1)$.

12. A particle moves in the xy-plane so that its position vector at time t is $\mathbf{r}(t) = \langle 3 - 2\sqrt{t}, t + 1 \rangle$. Find its velocity, acceleration, and speed when $t = 4$. Identify the curve.

13. A particle moves so that its position vector at time t is $\mathbf{r}(t) = \langle e^{-t} \sin 2t, e^{-t} \cos 2t, 2e^{-t} \rangle$. Find $\mathbf{v}(t)$, $\mathbf{a}(t)$, and $\frac{ds}{dt}$ at an arbitrary t.

14. A projectile is fired from ground level with an initial velocity $\mathbf{v}_0 = 0.4\mathbf{i} + 0.3\mathbf{j}$, with the magnitude in kilometers per second. Find the range of the projectile ($x_{\max}$) and the maximum height it attains ($y_{\max}$).

15. (a) A satellite is in orbit 150 mi above the earth's surface. What is its speed?
(b) A satellite completes one orbit around the earth every 2 hr. Find its altitude and speed.

16. Let $\mathbf{r}(t) = t^3\mathbf{i} - 4t^2\mathbf{j}$ for $t \in [1, 4]$. At the point $(8, -16)$, find $\mathbf{T}$, $\mathbf{N}$, κ, and the center of curvature.

17. (a) Find the curvature of $y = \ln \csc x$ for any $x \neq n\pi$.
(b) Find the curvature of $x^3 + 3xy - y^3 = 3$ at the point $(2, -1)$.

18. Let C be defined by

$$\mathbf{r}(t) = \langle -1 + 5 \sin t, 3 \cos t, 1 - 4 \cos t \rangle$$

for t in $[0, 2\pi]$. Find $\mathbf{T}$, $\mathbf{N}$, and $\mathbf{B}$ at an arbitrary point on C.

19. A particle moves so that its position vector at time $t \geq 0$ is $\mathbf{r}(t) = \langle \ln(t + 1), 2t, t^2 + 2t \rangle$. Find $\mathbf{a}$, $a_{\mathbf{T}}$, and $a_{\mathbf{N}}$.

20. A particle moves in the horizontal plane so that its polar coordinates at time t are $r = 1 + 2 \sin e^t$ and $\theta = e^t$. Describe its path for $t \geq 0$. Find the radial and transverse components of acceleration. (See Exercises 29 and 30 in Exercise Set 15.5.)

21. If a planet has the elliptical orbit $r = ep/(1 + e \cos \theta)$, show that its speed v_m at the point on its orbit farthest from the sun is

$$v_m = \frac{2\pi a}{T} \sqrt{\frac{1 - e}{1 + e}}$$

where a is the length of the semimajor axis and e is the eccentricity of the ellipse and T is the period. Use the data given for the earth in Exercise 10, part b, of Exercise Set 15.6 to find v_m for the earth.

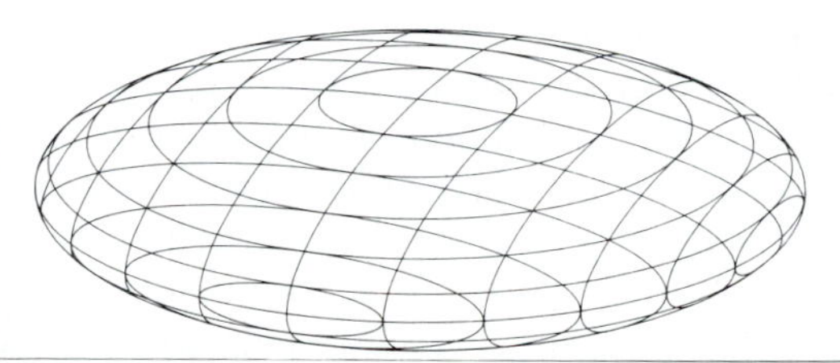

CHAPTER 16

FUNCTIONS OF SEVERAL VARIABLES

16.1 FUNCTIONS OF MORE THAN ONE VARIABLE

In Chapter 1 we defined a function from a set A to a set B as a rule that assigns to each element of A one and only one element of B. For the functions we have studied up to now, the elements of A (the domain) have been single real numbers. The set B also has consisted of real numbers, except for vector-valued functions, where B was a subset of V_2 or V_3.

Now we want to consider functions for which the set A is a collection of pairs (x, y) or triples (x, y, z) of real numbers. Geometrically, this amounts to saying that the domain is a subset of the xy-plane or of three-dimensional space. The ideas we present can be extended in natural ways to higher dimensions, but we will not pursue this. It is possible also to permit B to be multidimensional, but for now we will restrict attention to the case in which B is a subset of $\mathbf{R}$.

A real-valued function with a domain that is a set of ordered pairs of real numbers is said to be a *function of two (real) variables.* If f is such a function and (x, y) is a point in its domain, then we denote the image of this point by $f(x, y)$. Similarly, when the domain consists of ordered triples of real numbers, we call f a *function of three (real) variables* and denote the image of (x, y, z) by $f(x, y, z)$.

Actually, you are already familiar with certain functions of this type. For example, the area A of a rectangle of length x and width y is given by $A = xy$. So A is a function of two variables, x and y. Similarly, $A = P(1 + r)^t$ gives the amount of money A that results from an investment of P dollars at interest rate r, compounded annually for t years. So A is a function of the three variables P, r, and t. As one more example, if $\mathbf{r} = \langle x, y, z\rangle$ is the vector from a body of mass m to a body of mass M, then by Newton's law of gravitation, the magnitude of the force $\mathbf{F}$ of attraction between the bodies is the function of three variables given by

$$F(x, y, z) = \frac{GMm}{x^2 + y^2 + z^2}$$

The set of all ordered pairs of real numbers is frequently denoted by $\mathbf{R}^2$ and the set of all ordered triples of real numbers by $\mathbf{R}^3$. Geometrically, $\mathbf{R}^2$ is the entire xy-plane (two-dimensional space) and $\mathbf{R}^3$ is all of three-dimensional space. So the functions we are considering are mappings from $\mathbf{R}^m$ to $\mathbf{R}$, where $m = 2$ or 3. It is interesting to note that this is essentially the reverse of the situation with vector-valued functions. If we make points in $\mathbf{R}^2$ or $\mathbf{R}^3$ correspond to their position vectors, we see that the vector-valued functions we have studied are essentially mappings from $\mathbf{R}$ to $\mathbf{R}^n$, where $n = 2$ or 3. A natural question to ask at this point is whether there are functions from $\mathbf{R}^m$ to $\mathbf{R}^n$, where both m and n are 2 or more. The answer is that such functions do exist, and we will see an example of them in Chapter 19.

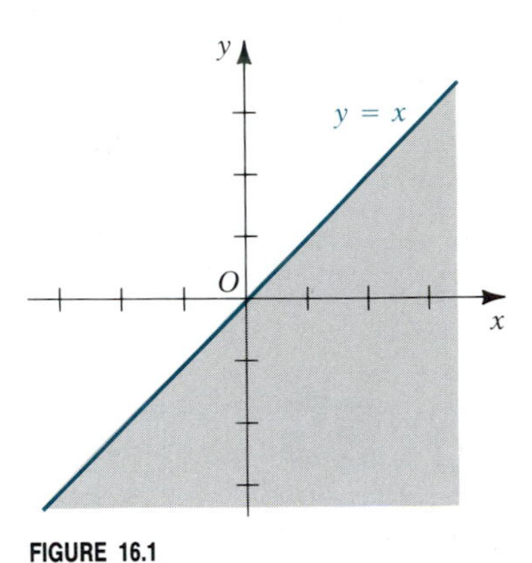

FIGURE 16.1

For the present we will concentrate on functions of two variables—that is, functions from $\mathbf{R}^2$ to $\mathbf{R}$. When the domain of such a function is not explicitly stated, we will assume it to be the largest subset of $\mathbf{R}^2$ for which all function values are real. The next two examples illustrate this.

EXAMPLE 16.1 Find the domain of f and describe it graphically.

(a) $f(x, y) = \dfrac{x + y}{x - y}$ (b) $f(x, y) = \sqrt{x - y}$

Solution (a) The values $f(x, y)$ will be real so long as $x - y \neq 0$. Thus, the domain of f is the set $\{(x, y): x \neq y\}$. Geometrically, this is the entire xy-plane with the exception of the line $y = x$.

(b) We must have $x - y \geq 0$. So the domain is the set $\{(x, y): y \leq x\}$. This is shown graphically in Figure 16.1. To get this, we draw the boundary line $y = x$ and then note that for a point to satisfy $y < x$, it must be *below* this line. Thus the line $y = x$ and all points below it make up the domain. ■

EXAMPLE 16.2 (a) Find the domain of f and show it graphically.

(a) $f(x, y) = \dfrac{1}{\sqrt{1 - x^2 - y^2}}$ (b) $f(x, y) = \ln(y - x^2)$

Solution (a) For the denominator to be nonzero *and* for the value of $f(x, y)$ to be real, we must have $1 - x^2 - y^2 > 0$. Thus, the domain is the set $\{(x, y): x^2 + y^2 < 1\}$. This is the set of points whose distance from the origin is less than 1—namely, the interior of the unit circle. We show this in Figure 16.2. The unit circle itself is shown by a broken line to indicate that it is not part of the domain.

(b) The logarithm is defined only when $y - x^2 > 0$. So the domain is the set $\{(x, y): y > x^2\}$. To sketch this, we first draw the bounding parabola $y = x^2$ and then note that points for which $y > x^2$ are *above* this parabola (Figure 16.3). Again we show the parabola as a broken line since it is not included. ■

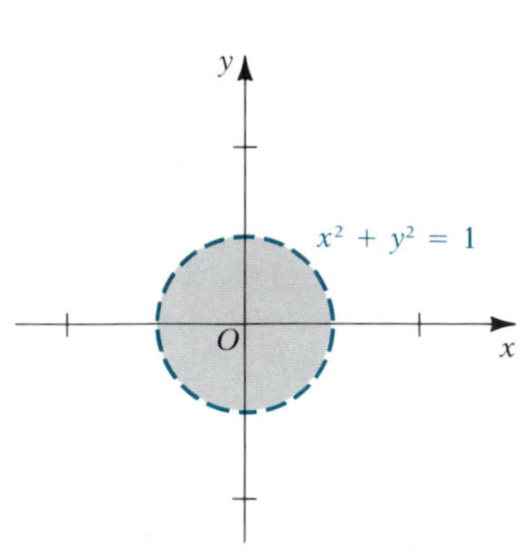

FIGURE 16.2

Comment As we have illustrated in these examples, one way to sketch the region defined by an inequality of type $y < g(x)$ or $y \leq g(x)$ is to draw the boundary curve $y = g(x)$ and then observe that points *below* this boundary satisfy the inequality. The boundary itself is included if $y \leq g(x)$ and is excluded if $y < g(x)$. Similar remarks apply to inequalities of the form $y >$

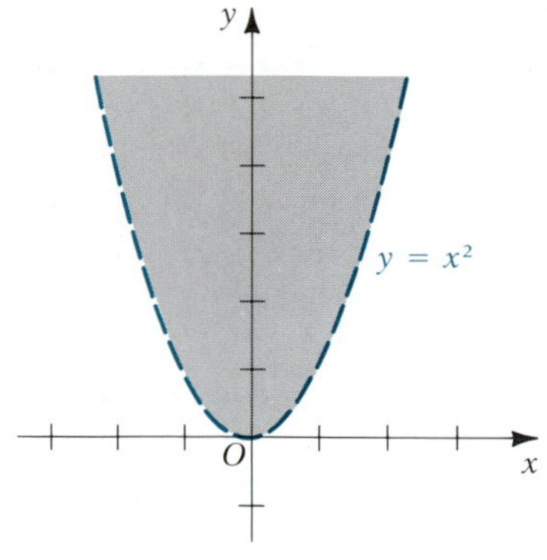

FIGURE 16.3

$g(x)$ and $y \geq g(x)$, where the region *above* the boundary satisfies the inequality. When the roles of x and y are reversed, so that $x < g(y)$ or $x \leq g(y)$ in the first instance, then the region to the *left* of the boundary curve $x = g(y)$ is the one in question, and when $x > g(y)$ or $x \geq g(y)$, the region is to the *right* of the boundary. For more complex inequalities, after drawing the boundary curve (or curves), you can test specific points in each region to see which satisfy the inequality.

If we set $z = f(x, y)$, we call x and y *independent variables* and z the *dependent variable.* By the *graph* of f we mean the set of all points (x, y, z) in $\mathbf{R}^3$ for which $z = f(x, y)$. Such a graph is called a **surface,** and although it is virtually impossible to draw most surfaces, certain important special cases can be sketched. We will study these, as well as other ways of gaining insight into graphs of functions, in the next two sections. It should be mentioned that, using the graphical capabilities of computers, many more surfaces can be illustrated than we ourselves are capable of drawing.

EXERCISE SET 16.1

A

1. If $f(x, y) = \dfrac{2xy - y^2}{x^2 + 3xy}$, find:

(a) $f(-1, 1)$ (b) $f(1, 0)$
(c) $f(2, 3)$ (d) $f(a, a)$; $a \neq 0$

2. If $g(x, y) = 2 \tan^{-1} \dfrac{y}{x} - \sin^{-1} \dfrac{x}{\sqrt{x^2 + y^2}}$, find:

(a) $g(-1, 0)$ (b) $g(1, -1)$ (c) $g(-1, \sqrt{3})$
(d) $g(-3, \sqrt{3})$

3. If $f(x, y, z) = \sqrt{12 - x^2 - y^2 - z^2}$, find:

(a) $f(1, -3, 1)$ (b) $f(-1, 1, -1)$
(c) $f(2, 0, -2)$ (d) $f(-3, 1, \sqrt{2})$

4. If $f(x, y) = \ln(2x + y)$, find:

(a) $f(-1, 3)$ (b) $f(0, e^2)$ (c) $f(x + h, y)$
(d) $f(x, y + k)$

5. If $g(x, y) = \dfrac{x^2 - y}{x - y^2}$, find:

(a) $g(-1, 1)$ (b) $g(a, \frac{1}{a})$; $a \neq 0$
(c) $g(x + \Delta x, y)$ (d) $g(x, y + \Delta y)$

6. If $f(x, y) = x^2 - 2xy + 3y^2$, find each of the following and simplify:

(a) $\dfrac{f(x + h, y) - f(x, y)}{h}$; $h \neq 0$

(b) $\dfrac{f(x, y + k) - f(x, y)}{k}$; $k \neq 0$

7. If $f(x, y) = \dfrac{x + y}{x - y}$, find each of the following and simplify:

(a) $\dfrac{f(-2 + h, 1) - f(-2, 1)}{h}$; $h \neq 0$

(b) $\dfrac{f(-2, 1 + k) - f(-2, 1)}{k}$; $k \neq 0$

8. Let $f(x, y) = e^{x-y} \sin(x + y)$, $g(t) = 3t$, and $h(t) = t$. Find $f(g(t), h(t))$.

9. Let $f(x, y) = \dfrac{2xy}{x^2 - y^2} + \dfrac{1}{x^2 + y^2}$, $g(t) = \cos t$, and $h(t) = \sin t$. Find $f(g(t), h(t))$.

10. Let $f(x, y) = xy - \frac{x}{y}$, $g(u, v) = uv$, and $h(u, v) = \frac{u}{v}$. Find $f(g(u, v), h(u, v))$.

In Exercises 11–21 find the domain of f and show it graphically.

11. (a) $f(x, y) = \dfrac{1}{x^2 - y}$ (b) $f(x, y) = \dfrac{1}{x^2 - y^2}$

12. $f(x, y) = \sqrt{x + y}$

13. $f(x, y) = \sqrt{xy}$

14. $f(x, y) = \ln(2x - y)$

15. $f(x, y) = \dfrac{xy}{x^2 + y^2}$

16. $f(x, y) = \sqrt{x + y} - \sqrt{x - y}$

17. $f(x, y) = \ln(x^2 - y^2)$

18. $f(x, y) = \dfrac{x + y}{\sqrt{x^2 + y^2 - 1}}$

19. $f(x, y) = \ln \sinh(x^2 - 2y)$

20. $f(x, y) = \ln\left(\dfrac{2x + y}{2x - y}\right)$

21. $f(x, y) = \ln \sqrt{xy}$

In Exercises 22–25 give the domain of f and describe it geometrically.

22. $f(x, y, z) = \dfrac{\sin xyz}{\sqrt{x^2 + y^2 + z^2}}$

23. $f(x, y, z) = \dfrac{2x - 3y + z}{(x-1)(y+2)(z-3)}$

24. $f(x, y, z) = \sqrt{xyz}$

25. $f(x, y, z) = e^{-(x^2+y^2)} \ln(x + y - z)$

B

26. Let $f(x, y) = \sqrt{4x + 3y}$. Show that if $h \neq 0$,

$$\frac{f(1+h, 4) - f(1, 4)}{h} = \frac{2}{2 + \sqrt{h+4}}$$

27. Let $f(x, y) = x^2 - 3xy + 2y^2$. Show that

$$f(x + \Delta x, y + \Delta y) = f(x, y) + (2x - 3y)\Delta x + (4y - 3x)\Delta y + \varepsilon_1 \Delta x + \varepsilon_2 \Delta y$$

where $\varepsilon_1 = g(\Delta x, \Delta y)$ and $\varepsilon_2 = h(\Delta x, \Delta y)$. Give the explicit forms of $g(x, y)$ and $h(x, y)$.

In Exercises 28 and 29 find the domain of f and show it graphically.

28. $f(x, y) = \sqrt{\dfrac{x - 3y}{x - y}}$

29. $f(x, y) = \ln(1 - |x| - |y|)$

30. (a) Express the volume of a right circular cone as a function of its base radius r and its altitude h.
(b) Express the surface area of an open-top box as a function of its length l, width w, and depth d.

31. A water tank is to be constructed in the form of a right circular cylinder of radius r and altitude h, with the top in the form of a hemisphere. The hemispherical top costs twice as much per unit area as the lateral surface and bottom. Express the total cost as a function of r, h, and the price p per square unit for the lateral surface and bottom.

32. A company manufactures two types of washing machines, the deluxe model A and the standard model B. When the price of each A model is p dollars and the price of each B model is q dollars, x model A machines and y model B machines can be sold each week. These price functions (called *demand functions*) are found by experience to be approximated by the equations

$$p = p(x, y) = 600 - 0.4x - 0.2y$$
$$q = q(x, y) = 400 - 0.3x - 0.5y$$

(a) Find the weekly revenue $R(x, y)$ from producing x A models and y B models.
(b) If the weekly cost of producing x A models and y B models is $C(x, y) = 120x + 90y + 600$, find the profit function $P(x, y)$ for this weekly production.

16.2 SKETCHING SURFACES

We learned in Chapter 14 that the graph of every linear equation $ax + by + cz = d$ in three variables is a plane. Recall that the lines of intersection of a plane with the coordinate planes are called its *traces*, and these are helpful in sketching a portion of the plane. The traces are found by setting one variable at a time equal to 0. For example, the traces of the plane $2x + 3y + 4z = 12$ are the lines $2x + 3y = 12$ (xy trace), $3y + 4z = 12$ (yz trace), and $x + 2z = 6$ (xz trace). Using intercepts, we can draw these traces easily, with the result as shown in Figure 16.4. Some planes are better depicted by showing selected traces with planes parallel to one of the coordinate planes. For example, the plane $x + y - z = 0$ that passes through the origin is shown in Figure 16.5 by drawing the xz trace $x = z$, the yz trace $y = z$, and the trace on a horizontal plane $z = k$ for an arbitrary positive number k, giving the line $x + y = k$.

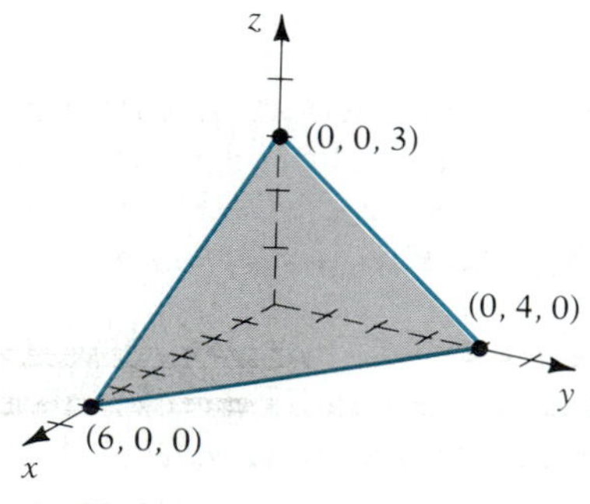

FIGURE 16.4

In general the graphs of nonlinear equations in three variables are *surfaces* in three dimensions. As remarked earlier, most of these are difficult, if not impossible, to draw, but there are some important exceptions that

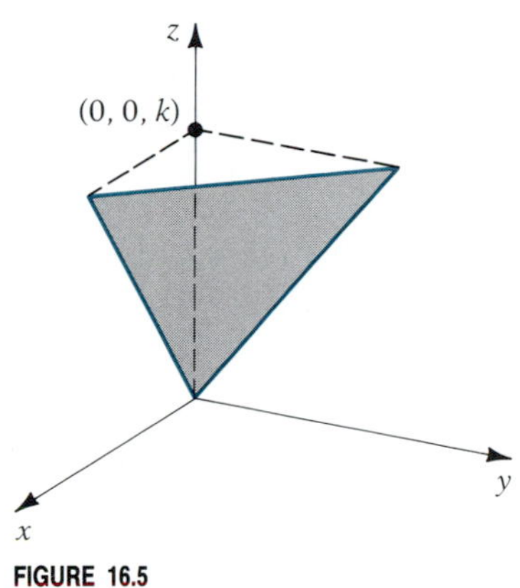

FIGURE 16.5

we discuss below. The principal aids to graphing, as with the plane, are intercepts, traces on the coordinate planes, and traces on planes parallel to one or more coordinate planes. These latter traces are also called *sections*. So a section is the curve of intersection of the surface with a plane perpendicular to one of the coordinate axes.

When you are sketching surfaces, it is probably easiest to show the yz-plane in the plane of the paper so that the yz trace can be shown full-size, as with any other two-dimensional drawing. The positive x-axis should be shown at an angle of about 135° from the positive y-axis, and a unit of distance on it should be about two-thirds of a unit in the y and z directions. This is to give the illusion of the x-axis projecting outward from the plane of the paper. We illustrate this in Figures 16.6–16.10. All other three-dimensional drawings in this book have been done with the aid of a computer, using a program that shows the surfaces from the most advantageous perspective. With this technique none of the coordinate planes coincides with the plane of the paper. Once you learn the shapes of the basic surfaces, you should be able to sketch them yourself, using the ideas discussed above.

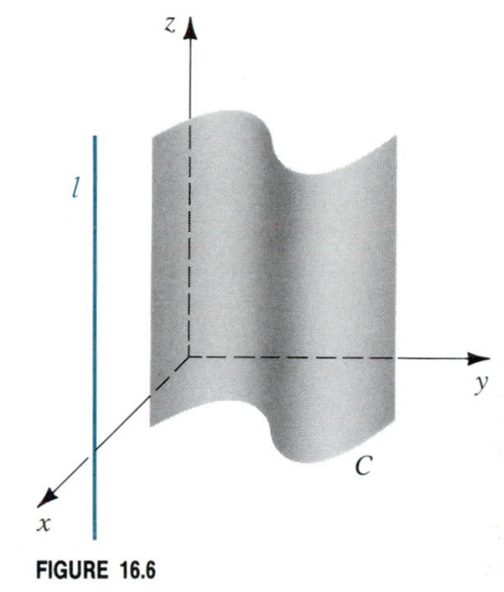

FIGURE 16.6

Cylinders

Let C be a plane curve and let l be a line not in the plane of C. The surface generated by a line moving along C so as to be always parallel to l is called a **cylinder,** or a **cylindrical surface.** Figure 16.6 illustrates such a surface, where C is in the xy-plane and l is perpendicular to this plane. The curve C is called the **directrix** of the cylinder, and the line that moves along C parallel to l is called a **generator.** The line l is sometimes called a **generatrix** for the cylinder. The most familiar cylinder is the right circular cylinder for which C is a circle and l is perpendicular to the plane of the circle.

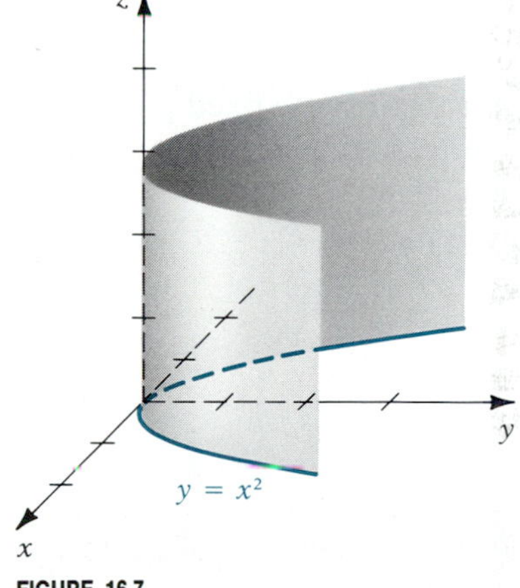

FIGURE 16.7

We will restrict our consideration of cylinders to those for which C is in one of the coordinate planes and l is perpendicular to that plane. Thus, l may be taken as one of the coordinate axes. For example, consider the cylinder whose directrix C is the parabola $y = x^2$ and that is generated by a line parallel to the z-axis. The graph is shown in Figure 16.7. For obvious reasons, we call this a **parabolic cylinder.** Notice how we have made use of sections perpendicular to the y-axis and z-axis, respectively, to aid in visualizing the cylinder, which is actually infinite in extent. All points on the

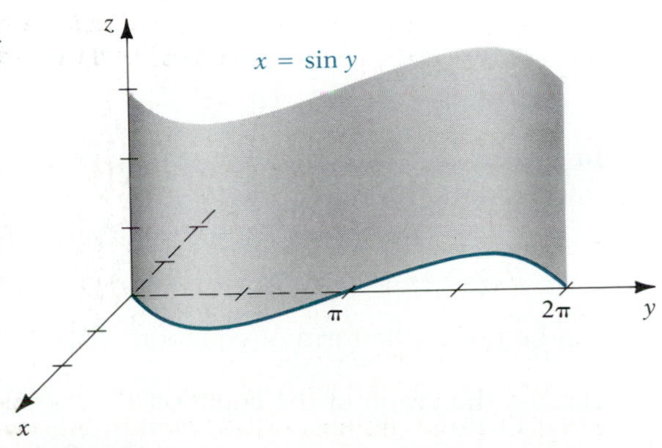

FIGURE 16.8

cylinder satisfy the equation $y = x^2$, and only these points satisfy the equation. Thus, $y = x^2$ is the equation of the cylinder. When viewed as a curve in $\mathbf{R}^2$, this equation represents a parabola, but when viewed as a surface in $\mathbf{R}^3$, it represents a parabolic cylinder.

More generally, in $\mathbf{R}^3$, the graph of an equation involving x and y only, say $f(x, y) = 0$, is a cylinder parallel to the z-axis. The two-dimensional graph of $f(x, y) = 0$ is its directrix. Analogous results hold for $f(x, z) = 0$ and $f(y, z) = 0$.

If an equation involves only two of the variables x, y, and z, its graph in $\mathbf{R}^3$ is a cylinder parallel to the axis of the missing variable. The directrix is the curve in $\mathbf{R}^2$ defined by the given equation.

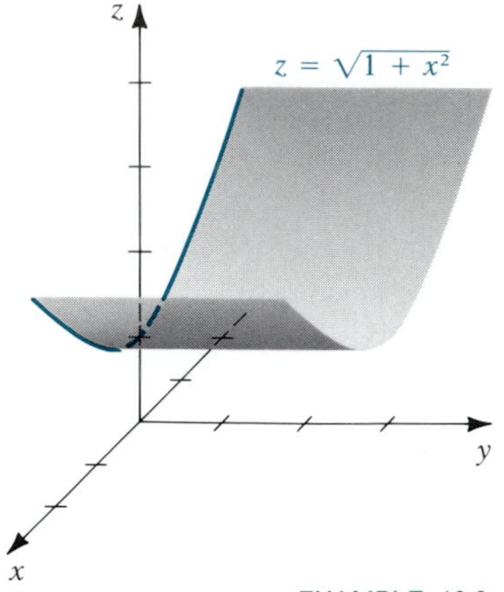

FIGURE 16.9

EXAMPLE 16.3 Sketch the graph of the following equations:

(a) $x = \sin y \quad (0 \le y \le 2\pi)$ (b) $z = \sqrt{1 + x^2}$ (c) $9y^2 + 4z^2 = 36$

Solution

(a) First we draw the curve $x = \sin y$ in the xy-plane and then extend it parallel to the z-axis, as in Figure 16.8.

(b) In the xz-plane this is the upper branch of the hyperbola $z^2 - x^2 = 1$. So in $\mathbf{R}^3$ it is the upper part of a hyperbolic cylinder parallel to the y-axis (Figure 16.9).

(c) In standard form the equation is

$$\frac{y^2}{4} + \frac{z^2}{9} = 1$$

This is an ellipse in the yz-plane, so in $\mathbf{R}^3$ it is an elliptical cylinder parallel to the x-axis, as shown in Figure 16.10. ■

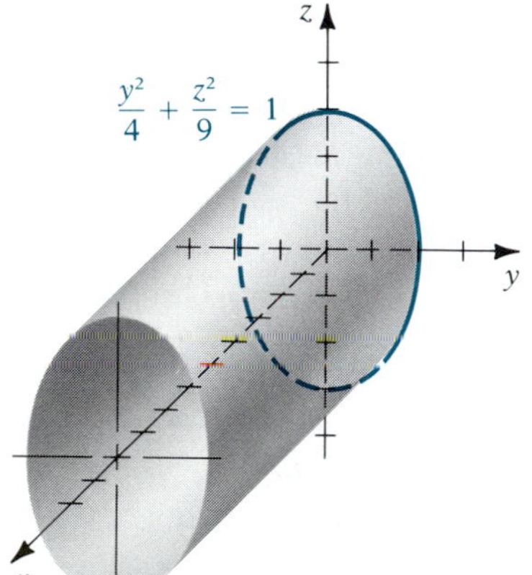
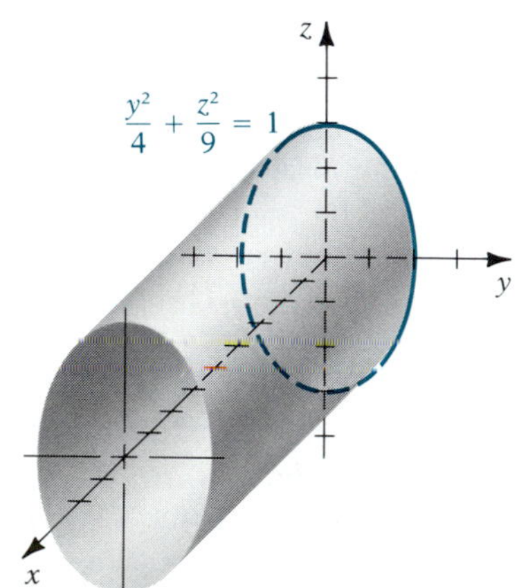

FIGURE 16.10

Spheres

A sphere is the set of all points in $\mathbf{R}^3$ equidistant from some fixed point. Let $P_0(h, k, l)$ be the fixed point (the center), and let a be the common distance of points on the sphere to P_0 (so a is the radius). Then a point $P(x, y, z)$ lies on the sphere if and only if $|P_0P| = a$ or, equivalently, $|\overrightarrow{P_0P}|^2 = a^2$. In terms of coordinates, this equation is

$$(x - h)^2 + (y - k)^2 + (z - l)^2 = a^2 \tag{16.1}$$

If P_0 is the origin, the equation simplifies to

$$x^2 + y^2 + z^2 = a^2$$

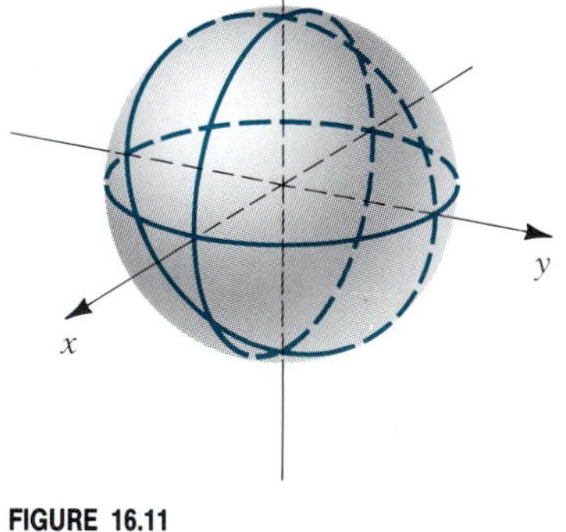

FIGURE 16.11

In Figure 16.11 we sketch a sphere centered at the origin by showing its traces on the coordinate planes.

The next example shows how an equation of the form

$$x^2 + y^2 + z^2 + ax + by + cz + d = 0$$

can be put in the form of equation (16.1).

EXAMPLE 16.4 Discuss the graph of the equation

$$x^2 + y^2 + z^2 - 4x - 10y - 6z + k = 0$$

where

(a) $k = 34$ (b) $k = 38$ (c) $k = 42$

Solution (a) We complete the square on x, y, and z, getting

$$(x^2 - 4x + 4) + (y^2 - 10y + 25) + (z^2 - 6z + 9) = -34 + 4 + 25 + 9$$

or

$$(x - 2)^2 + (y - 5)^2 + (z - 3)^2 = 4$$

So by equation (16.1) this is a sphere of radius 2 centered at (2, 5, 3). Its graph is shown in Figure 16.12.

(b) This is the same as part a except that we get 0 on the right-hand side:

$$(x - 2)^2 + (y - 5)^2 + (z - 3)^2 = 0$$

The equation is satisfied only by the single point (2, 5, 3). This is an example of a degenerate sphere.

(c) Again, the only change is on the right-hand side:

$$(x - 2)^2 + (y - 5)^2 + (z - 3)^2 = -4$$

Since the left-hand side is nonnegative for all points (x, y, z), there is no graph. ■

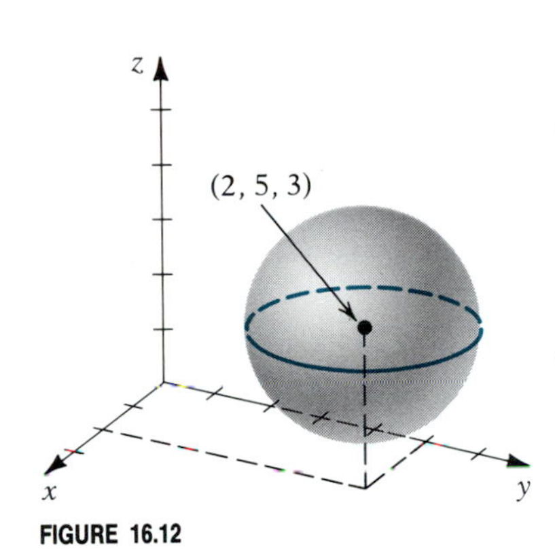

FIGURE 16.12

Quadric Surfaces

Equation (16.1) of the sphere, when expanded, is seen to be a special case of the *general quadratic equation in three variables:*

$$Ax^2 + By^2 + Cz^2 + Dxy + Exz + Fyz + Gx + Hy + Iz + K = 0 \qquad (16.2)$$

It can be proved that the graph of every such equation, if it has a graph at all, is one of nine types of surfaces, called **quadric surfaces,** or a degenerate of one of these. Quadric surfaces can be thought of as three-dimensional analogues of conic sections in the plane. Although the sphere is a quadric surface, it is such an important special case that we consider it separately.

Parabolic cylinders, elliptic cylinders, and hyperbolic cylinders are three of the nine types of quadric surfaces. For example, in equation (16.2) if the only nonzero coefficients are A and H, so that the equation is $Ax^2 + Hy = 0$, the graph is a parabolic cylinder parallel to the z-axis. Similarly if all coefficients are 0 except B, C, and K, leaving the equation $By^2 + Cz^2 + K = 0$, the graph will be an elliptic cylinder parallel to the x-axis if B and C are positive and K is negative. It will be a hyperbolic cylinder if B and C are opposite in sign. You can easily find other special cases of equation (16.2) that result in one of these types of cylinders.

We now give a brief description of the remaining six types of quadric surfaces. Each equation is given in *standard form,* which, if expanded, can be seen to be a special case of equation (16.2). None involves the mixed second-degree terms xy, xz, or yz; that is, in all the cases, D, E, and F in equation (16.2) are 0. It can be proved that if one or more of these coefficients is nonzero, by a suitable rotation of the axes an equation of one of the types we illustrate can always be obtained. The equations we give are for surfaces conveniently placed with respect to the origin and the coordinate axes. When interchanges are made among the variables x, y, and z, the resulting surface is of the same type as illustrated but the orientation is changed. If x, y, and z are replaced by $x - h$, $y - k$, and $z - l$, respectively, the given surface is translated h units in the x direction, k units in the y direction, and l units in the z direction.

Ellipsoid

$$\frac{x^2}{a^2} + \frac{y^2}{b^2} + \frac{z^2}{c^2} = 1$$

xy trace: ellipse $\dfrac{x^2}{a^2} + \dfrac{y^2}{b^2} = 1$

xz trace: ellipse $\dfrac{x^2}{a^2} + \dfrac{z^2}{c^2} = 1$

yz trace: ellipse $\dfrac{y^2}{b^2} + \dfrac{z^2}{c^2} = 1$

Sections perpendicular to each coordinate axis between intercepts are ellipses. A special case of the ellipsoid is the sphere, in which $a = b = c$. The graph is shown in Figure 16.13.

The key to recognizing an ellipsoid is that its standard equation involves the sum of the squares of all three variables.

Elliptic Paraboloid

$$\frac{x^2}{a^2} + \frac{y^2}{b^2} = cz$$

We illustrate the case $c > 0$.

xy trace: the origin

xz trace: parabola $\dfrac{x^2}{a^2} = cz$

yz trace: parabola $\dfrac{y^2}{b^2} = cz$

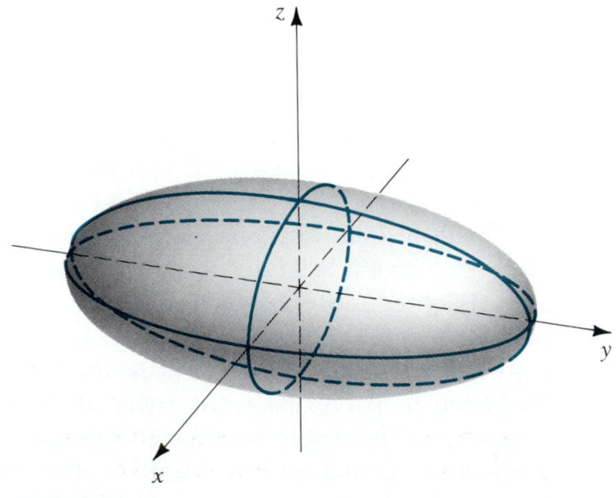

FIGURE 16.13

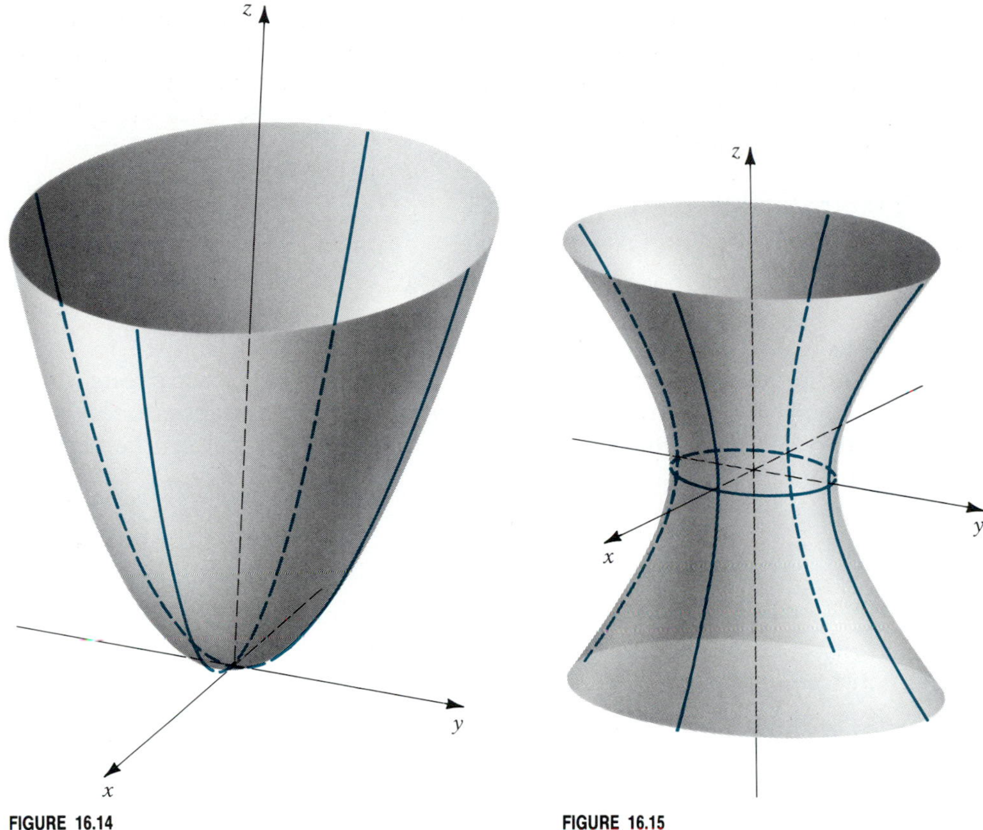

FIGURE 16.14

FIGURE 16.15

Sections perpendicular to the positive z-axis are ellipses. For the given standard form, the paraboloid opens upward as in Figure 16.14 if $c > 0$, and it opens downward if $c < 0$.

The key to recognizing a paraboloid is that its equation can be written so that one side involves the sum of the squares of two of the variables and the other side involves the first power of the third variable. The axis corresponds to the first-degree variable.

Elliptic Hyperboloid of One Sheet

$$\frac{x^2}{a^2} + \frac{y^2}{b^2} - \frac{z^2}{c^2} = 1$$

xy trace: ellipse $\dfrac{x^2}{a^2} + \dfrac{y^2}{b^2} = 1$

xz trace: hyperbola $\dfrac{x^2}{a^2} - \dfrac{z^2}{c^2} = 1$

yz trace: hyperbola $\dfrac{y^2}{b^2} - \dfrac{z^2}{c^2} = 1$

Sections perpendicular to the z-axis are ellipses. See Figure 16.15.

The key to recognizing the equation of a hyperboloid of one sheet is that the standard form involves the squares of all three variables, two with positive signs and one with a negative sign. The axis is that of the variable in the negative term.

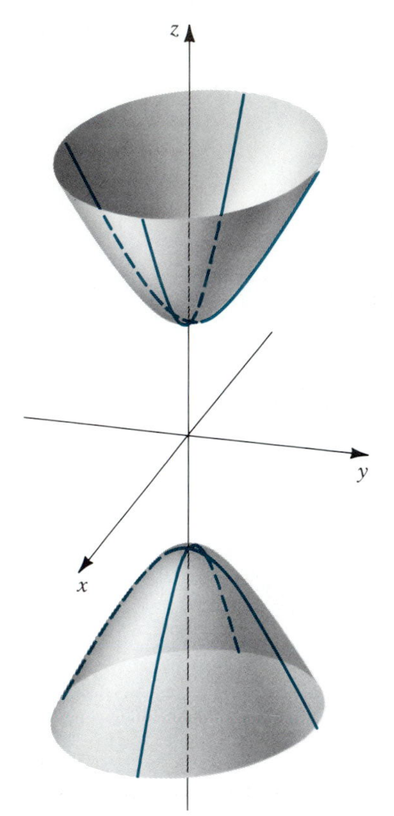

FIGURE 16.16

Elliptic Hyperboloid of Two Sheets

$$\frac{z^2}{c^2} - \frac{x^2}{a^2} - \frac{y^2}{b^2} = 1$$

xy trace: none

xz trace: hyperbola $\dfrac{z^2}{c^2} - \dfrac{x^2}{a^2} = 1$

yz trace: hyperbola $\dfrac{z^2}{c^2} - \dfrac{y^2}{b^2} = 1$

For $|k| > c$, sections on planes $z = k$ are ellipses. See Figure 16.16.

The key to recognizing the equation of a hyperboloid of two sheets is that the standard form involves the squares of all three variables, one with a positive sign and two with negative signs. The axis is that of the variable with a positive sign.

Elliptic Cone

$$\frac{x^2}{a^2} + \frac{y^2}{b^2} = \frac{z^2}{c^2}$$

xy trace: origin

xz trace: two lines $\dfrac{x}{a} = \pm\dfrac{z}{c}$

yz trace: two lines $\dfrac{y}{b} = \pm\dfrac{z}{c}$

Sections perpendicular to the z-axis are ellipses. See Figure 16.17.

For the cone the equation can be written so that one side involves the sum of the squares of two of the variables and the other sides involves the square of the third variable, with the axis that of the latter variable.

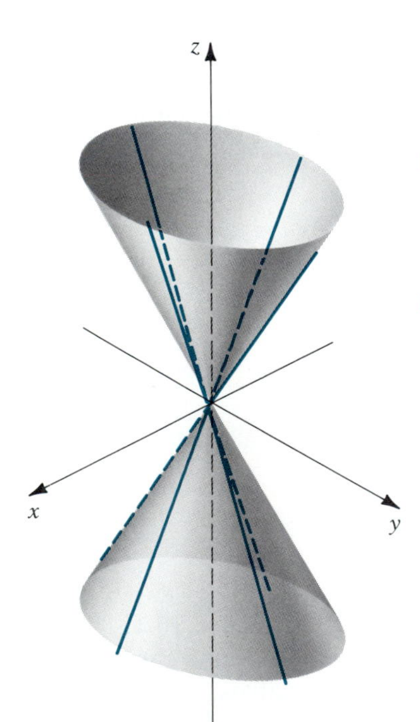

FIGURE 16.17

Note The upper and lower parts of the cone are called *nappes*. In customary usage when one refers to a cone (or to a right circular cone) only one nappe of the total cone is intended, and sections perpendicular to the axis are circular.

Hyperbolic Paraboloid

$$\frac{y^2}{b^2} - \frac{x^2}{a^2} = cz$$

xy trace: two lines $\dfrac{y}{b} = \pm\dfrac{x}{a}$

xz trace: parabola $-\dfrac{x^2}{a^2} = cz$

yz trace: parabola $\dfrac{y^2}{b^2} = cz$

Sections perpendicular to the z-axis are hyperbolas, and sections perpendicular to the x-axis or y-axis are parabolas. A hyperbolic paraboloid is pictured in Figure 16.18 with $c > 0$.

FIGURE 16.18

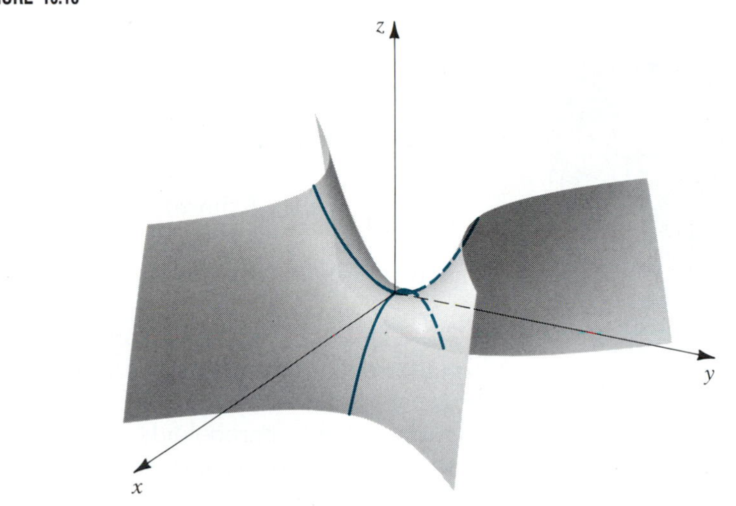

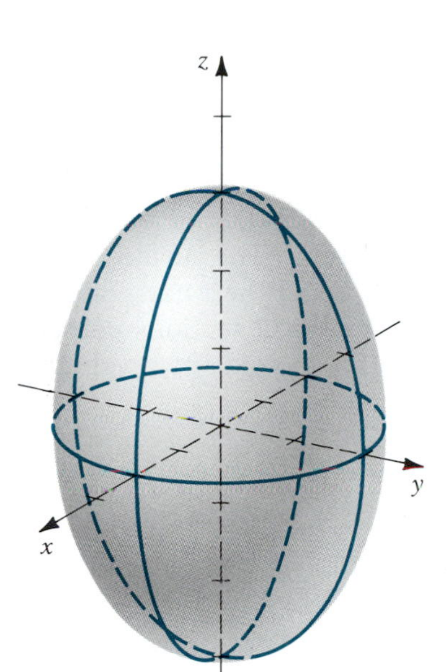

FIGURE 16.19

Note the similarity in the equations of the elliptic and hyperbolic paraboloids, the former involving the sums of squares and the latter the difference of squares of two of the variables. This difference of squares set equal to the first power of the third variable is the key to recognizing the hyperbolic paraboloid.

Because of its saddlelike shape in the vicinity of the origin, the origin is called a **saddle point** on the surface.

EXAMPLE 16.5 Identify and sketch each of the following surfaces:

(a) $9x^2 + 9y^2 + 4z^2 = 36$ (b) $4x^2 + z^2 = 4y$

Solution (a) The equation in standard form is

$$\frac{x^2}{4} + \frac{y^2}{4} + \frac{z^2}{9} = 1$$

This is an ellipsoid. The traces are as follows:

xy trace: $\frac{x^2}{4} + \frac{y^2}{4} = 1$ or $x^2 + y^2 = 4$

xz trace: $\frac{x^2}{4} + \frac{z^2}{9} = 1$

yz trace: $\frac{y^2}{4} + \frac{z^2}{9} = 1$

The graph is shown in Figure 16.19.

(b) The standard form is

$$\frac{x^2}{1} + \frac{z^2}{4} = y$$

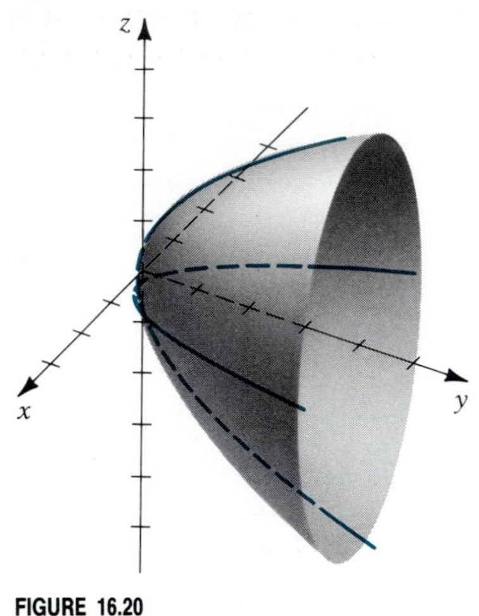

FIGURE 16.20

This is an elliptic paraboloid with axis along the y-axis.

xy trace: $x^2 = y$

xz trace: origin

yz trace: $z^2 = 4y$

Section at $y = k > 0$:

$$\frac{x^2}{1} + \frac{z^2}{4} = k \qquad \text{(ellipse)}$$

The graph is shown in Figure 16.20. ■

The ellipsoid in part a of Example 16.5 is called an *ellipsoid of revolution* because sections perpendicular to the z-axis are circular. The surface could be generated by revolving either the xz or yz trace about the z-axis. Similar remarks apply to paraboloids and hyperboloids. When all sections perpendicular to the axis are circular, these are called paraboloids of revolution and hyperboloids of revolution, respectively.

EXAMPLE 16.6 Identify and sketch the graph of each of the following:

(a) $3x^2 - y^2 + 2z^2 + 6 = 0$ (b) $z = \sqrt{x^2 + y^2}$

Solution (a) We first put the equation in standard form:

$$\frac{y^2}{6} - \frac{x^2}{2} - \frac{z^2}{3} = 1$$

This is a hyperboloid of two sheets with axis along the y-axis.

xy trace: $\dfrac{y^2}{6} - \dfrac{x^2}{2} = 1$

xz trace: none

yz trace: $\dfrac{y^2}{6} - \dfrac{z^2}{3} = 1$

For $|k| > \sqrt{6}$, the planes $y = k$ intersect the surface in an ellipse. The graph is shown in Figure 16.21.

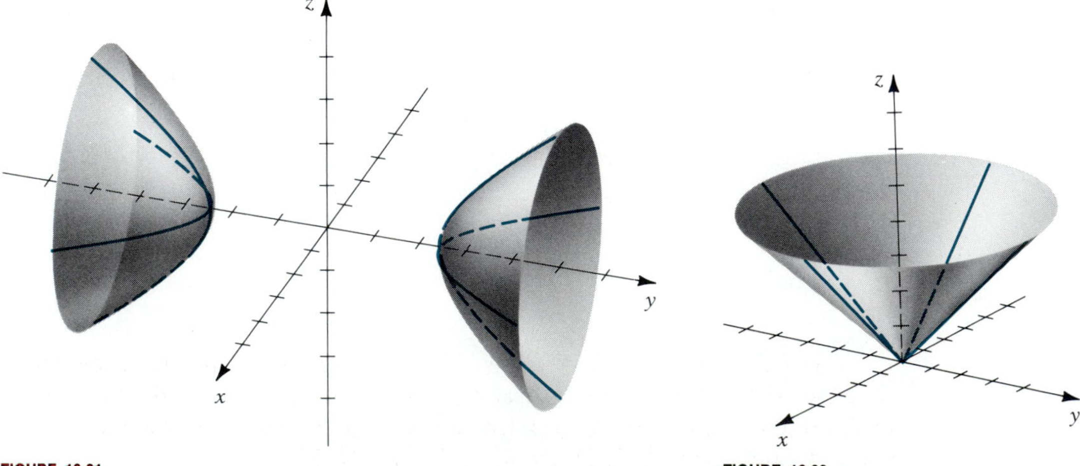

FIGURE 16.21

FIGURE 16.22

(b) Squaring both sides, we get $z^2 = x^2 + y^2$, which is the equation of a circular cone. The given equation corresponds to the upper nappe only, as shown in Figure 16.22.

xy trace: origin

xz trace: $z = x,\ z \geq 0$

yz trace: $z = y,\ z \geq 0$

■

EXERCISE SET 16.2

A

In Exercises 1–12 draw the graph of the given equation in three dimensions.

1. $y^2 = 2x$

2. $x^2 + y^2 = 4$

3. $y = 4 - x^2$

4. $z = x^2/4$

5. $z = \sqrt{4 - y^2}$

6. $4x^2 + 9y^2 = 36$

7. $x^2 - y^2 = 1$

8. $xy = 2$

9. $z = e^{-x}$

10. $z = \cos y,\ 0 \leq y \leq 2\pi$

11. $z = \ln y$

12. $y = \cosh x$

In Exercises 13–17 determine whether the graph is a sphere or a degenerate sphere, or if there is no graph. If the graph is a sphere, give its center and radius and draw the graph.

13. $x^2 + y^2 + z^2 - 2x - 6y - 8z + 10 = 0$

14. $x^2 + y^2 + z^2 - 8y - 4z + 11 = 0$

15. $x^2 + y^2 + z^2 + 10x - 2y + 6z + 35 = 0$

16. $x^2 + y^2 + z^2 - 10x + 4y + 8z + 47 = 0$

17. $x^2 + y^2 + z^2 - 6x + 8y - 8z + 33 = 0$

In Exercises 18–33 identify the quadric surface and draw its graph, showing traces and sections where useful.

18. $36x^2 + 9y^2 + 16z^2 = 144$

19. $36x^2 - 9y^2 + 16z^2 = 144$

20. $36x^2 - 9y^2 + 16z^2 + 144 = 0$

21. $36x^2 - 9y^2 + 16z^2 = 0$

22. $36z = 9x^2 + 4y^2$

23. $16x - 4y^2 - 9z^2 = 0$

24. $4x^2 - y^2 + 4z = 0$

25. $z = x^2 - y^2$

26. $x^2 = y^2 + z^2$

27. $z = \sqrt{4 - x^2 - y^2}$

28. $z = 4 - x^2 - y^2$

29. $x^2 + y^2 = 1 + z^2$

30. $x^2 - y^2 = 1 + z^2$

31. $4x^2 + 2y^2 = 8 - z^2$

32. $3x^2 + 2z^2 - 6y = 0$

33. $y = \sqrt{4x^2 + z^2}$

B

In Exercises 34–39 make an appropriate translation to identify and draw the graph.

34. $25x^2 + 9y^2 + 15z^2 - 100x - 54y - 60z + 16 = 0$

35. $4x^2 + y^2 - 2z^2 - 8x - 4y - 8z + 8 = 0$

36. $9x^2 - 4y^2 + 9z^2 - 54x - 16y - 18z + 38 = 0$

37. $4x^2 + y^2 - 24x - 4y - 4z + 20 = 0$

38. $4x^2 + 3y^2 - z^2 - 32x - 12y + 2z + 75 = 0$

39. $2x^2 - 3y^2 - 8x - 12y + 12z - 52 = 0$

In Exercises 40–45 show the volume in the first octant bounded by the given surfaces.

40. $x^2 + z^2 = 4,\ y = x,\ y = 0,\ z = 0$

41. $z = x^2 + y^2,\ x + y + z = 4$

42. $x^2 + z^2 = 4,\ y^2 + z^2 = 4,\ x = 0,\ y = 0,\ z = 0$

43. $z = 4 - y^2,\ x^2 = 2y,\ x = 0,\ z = 0$

44. $z = 4 - x^2 - y^2,\ z^2 = x^2 + y^2$

45. $4x^2 + 2y^2 + 3z^2 = 48,\ y = 2x,\ y = 4,\ x = 0,\ z = 0$

16.3 GRAPHS OF FUNCTIONS

Let f be a function of two variables. By the *graph of f*, we mean the graph of the equation $z = f(x, y)$. For example, if $f(x, y) = 2x^2 + y^2$, then the graph of f is the graph of $z = 2x^2 + y^2$, which we know from the previous section is an elliptic paraboloid, as shown in Figure 16.23.

The definition of a function requires that each point in the domain correspond to exactly one point in the range. Thus, the graph of $z = f(x, y)$ must satisfy the vertical line test: *each vertical line intersects the surface in at most one point.* This eliminates such quadric surfaces as the ellipsoid, cone, and hyperboloids as graphs of functions. However, a portion of one of these can be the graph of a function, as the next example shows.

EXAMPLE 16.7 Identify and sketch the graph of f. Give the domain and range.

(a) $f(x, y) = \sqrt{1 - x^2 - y^2}$ (b) $f(x, y) = \sqrt{1 + x^2 + y^2}$

Solution (a) Set $z = \sqrt{1 - x^2 - y^2}$. On squaring and rearranging, we get $x^2 + y^2 + z^2 = 1$, which is the equation of a sphere of radius 1 centered at the origin. The original equation requires that z be nonnegative, so its graph is the upper hemisphere pictured in Figure 16.24. The domain is the set of all points (x, y) such that $x^2 + y^2 \leq 1$, and the range is the set of all points z such that $0 \leq z \leq 1$.

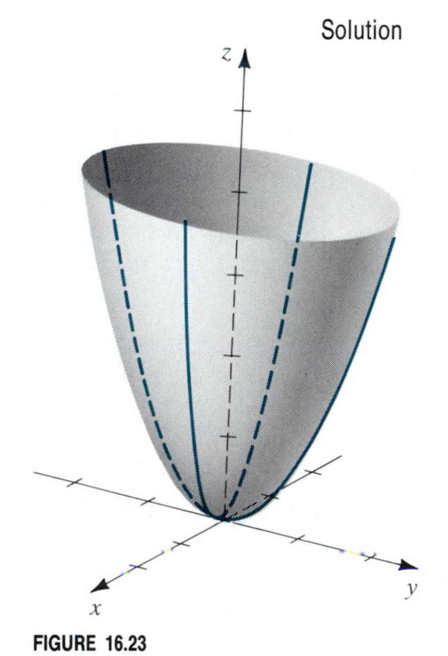

FIGURE 16.23

(b) Following the same procedure, we get $z = \sqrt{1 + x^2 + y^2}$ and $z^2 - x^2 - y^2 = 1$. The graph of the latter equation is a hyperboloid of revolution of two sheets, with axis along the z-axis. So the graph of the given function is the upper sheet as pictured in Figure 16.25. The domain of f is the entire xy-plane, and the range is the set of all points z such that $z \geq 1$.

■

Unless the graph of a function of two variables is one of the special surfaces we have studied, it is unlikely that we can draw it in any meaningful way. It may still be possible, however, to gain some insight into the nature of the graph by using **level curves.** For a function f a level curve is the two-dimensional graph of the equation $f(x, y) = c$, where c is some constant

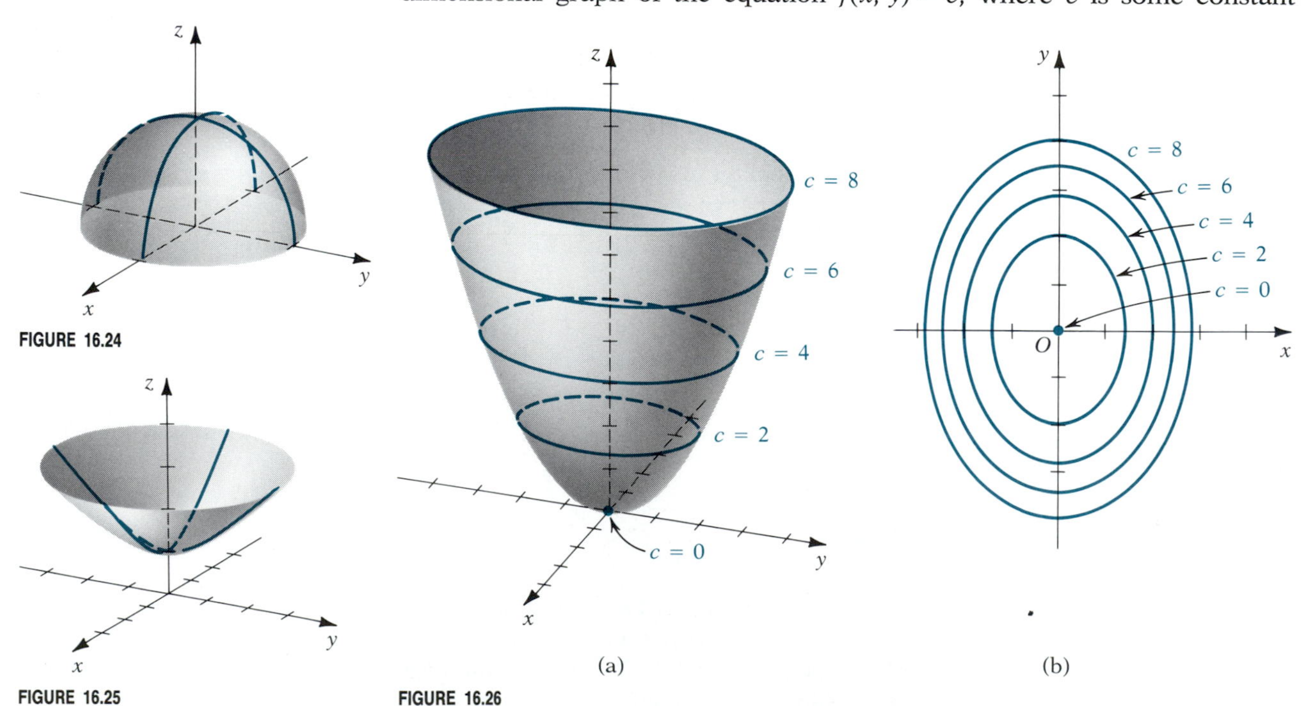

FIGURE 16.24

FIGURE 16.25

(a) (b)

FIGURE 16.26

value in the range of f. By choosing various constants c, we get a collection of level curves in the xy-plane, called a **contour map** of f. A level curve $f(x, y) = c$ can be thought of as the projection onto the xy-plane of the curve of intersection of the surface $z = f(x, y)$ and the horizontal plane $z = c$. It is useful to choose c values so that the different planes are equally spaced. Then when the corresponding level curves are close to one another, we will know the function is increasing or decreasing rapidly, whereas if level curves are far apart, the function is changing slowly.

These ideas are illustrated in Figure 16.26(a) where we have shown the graph of $f(x, y) = x^2 + y^2/2$, which we know to be an elliptic paraboloid. The planes $z = c$ for $c = 2, 4, 6$, and 8 intersect the surface in ellipses, which are level curves shown in Figure 16.26(b). The level curve corresponding to $c = 0$ is the single point at the origin. Notice that as c gets larger, the level curves are more closely spaced, telling us that the surface is rising more steeply. In this example, we could draw the graph of f, and so there was little need for the contour map. You can see from it, though, that the contour map alone tells much about the surface. In the next example, drawing the graph of f is quite difficult, and we analyze it directly from the contour map.

EXAMPLE 16.8 Draw a contour map for the function $f(x, y) = 2xy$, showing the level curves $f(x, y) = c$ for $c = 0, \pm 1, \pm 2, \pm 3$.

Solution For $c > 0$ the level curves $2xy = c$ are the equilateral hyperbolas with branches in quadrants I and III on the contour map shown in Figure 16.27, and for $c < 0$ they are the equilateral hyperbolas with branches in quadrants II and IV. If $c = 0$, the level curve consists of the two axes, $x = 0$ and $y = 0$. Trying to visualize features of the surface from this, we see that starting from the origin the surface rises for (x, y) in quadrants I and III and falls for (x, y) in quadrants II and IV. The most rapid ascent is along the line $y = x$, and the most rapid descent is along $y = -x$. It follows that $x = 0$ is a saddle point. It is possible to show, in fact, using methods from Chapter

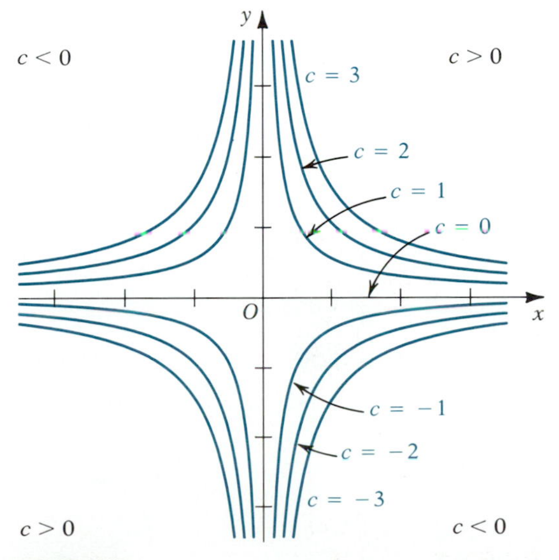

FIGURE 16.27

12, that the graph is the hyperbolic paraboloid $z = x^2 - y^2$ rotated horizontally through a 45° angle. ■

Computer graphics can be used to illustrate surfaces as well as contour maps. This technique is especially useful for figures that are difficult to draw by hand. Figure 16.28(a) shows a contour map and Figure 16.28(b) the corresponding surface, both of which were generated by a computer.

The idea of a contour map has many applications. In a topographic map, for example, level curves are used to show topographic features, such as hills and valleys in a given terrain. Figure 16.29 illustrates this. In part a a mountain is shown, along with curves of constant elevations of 300, 600, 900, and 1200 m. These curves are plotted in the xy-plane in part b and form the topographic map. An experienced map reader can form a mental picture of the terrain by studying the topographic map.

Weather maps are another example of a contour map. Typically, these show curves along which the temperature is constant—that is, level curves for the temperature function. These level curves are called **isothermals.** A weather map might also show curves of constant barometric pressure, called **isobars.** Still another example is the **equipotentials** in an electric field. These are the curves along which the electrostatic force is constant.

When we go from functions of two variables to functions of three variables, the notion of a graph becomes more complicated. Although we can say that the graph of $w = f(x, y, z)$ is the set of "points" (x, y, z, w) in four-dimensional space for which w satisfies the given equation, it is impossible to draw such a graph. The extension to three dimensions of the idea of a level curve can be drawn, however. These are the **level surfaces** $f(x, y, z) = c$ for constants c in the range of f. As an example, consider the function

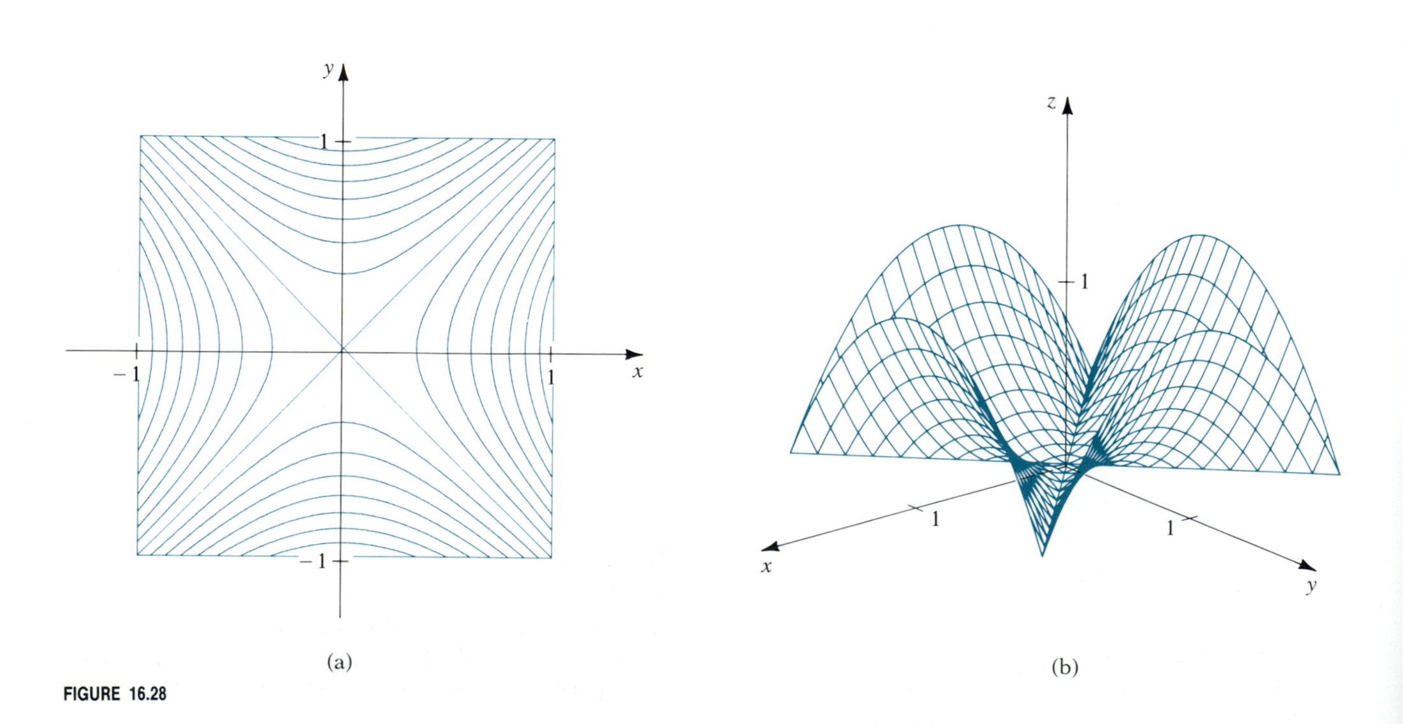

FIGURE 16.28

FIGURE 16.29

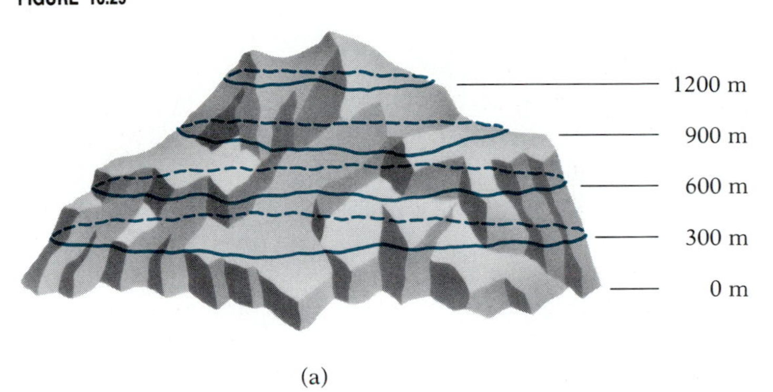

(a)

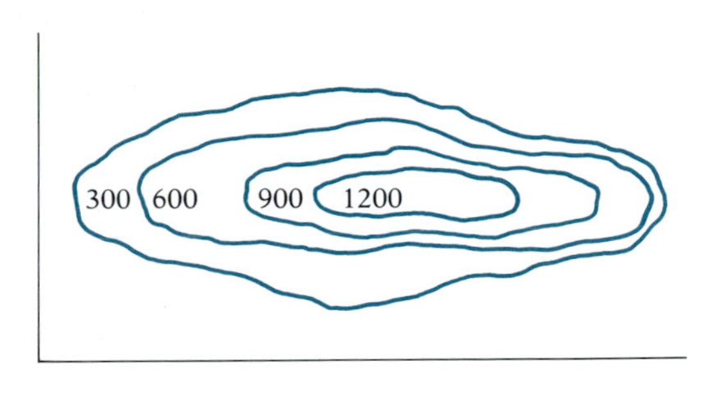

(b)

defined by

$$f(x, y, z) = \sqrt{16 - x^2 - y^2 - z^2}$$

The level surfaces $f(x, y, z) = c$ for $0 \le c \le 4$ are the spheres

$$x^2 + y^2 + z^2 = 16 - c^2$$

The function assumes its smallest value, 0, on the sphere of radius 4, and as c increases, the spheres on which $f(x, y, z) = c$ shrink in size, finally contracting to the single point $(0, 0, 0)$ when $c = 4$, which is the maximum value the function assumes.

EXERCISE SET 16.3

A

In Exercises 1–20 identify and draw the graph of f. State the domain and range.

1. $f(x, y) = 4 - x - y$
2. $f(x, y) = 4 - x$
3. $f(x, y) = x + 2y$
4. $f(x, y) = y^2$
5. $f(x, y) = \sqrt{x}$
6. $f(x, y) = \sqrt{4 - x^2}$
7. $f(x, y) = 9 - y^2$
8. $f(x, y) = \sqrt{1 + x^2}$
9. $f(x, y) = 2x^2 + y^2$
10. $f(x, y) = 4 - x^2 - y^2$
11. $f(x, y) = \sqrt{12 - 4x^2 + 3y^2}$
12. $f(x, y) = \sqrt{36 - 4x^2 - 9y^2}$
13. $f(x, y) = 1 + x^2 + y^2$
14. $f(x, y) = \sqrt{x^2 + y^2}$
15. $f(x, y) = \sqrt{x^2 + y^2 - 4}$
16. $f(x, y) = \sqrt{4 - x^2 + y^2}$
17. $f(x, y) = \sqrt{4y - x^2}$
18. $f(x, y) = \sqrt{4 + 4x^2 + y^2}$
19. $f(x, y) = y^2 - x^2$
20. $f(x, y) = 2x^2 - y^2$

In Exercises 21–31 draw a contour map with at least six level curves. Use both positive and negative values of c where appropriate.

21. $f(x, y) = 3x - 5y$
22. $f(x, y) = \frac{x}{y}$
23. $f(x, y) = x^2 - 2y$
24. $f(x, y) = ye^{-x}$
25. $f(x, y) = x^2y$
26. $f(x, y) = y - \cos x$
27. $f(x, y) = (x^2 - 1)/y$
28. $f(x, y) = \sqrt{4 - x^2 - y^2}$
29. $f(x, y) = x^2 - y^2$
30. $f(x, y) = y - \ln x$
31. $f(x, y) = y - \sin \pi x$
32. On pages 634 and 635 there are six contour maps labeled (a)–(f) and six surfaces labeled (1)–(6). Match up each contour map with the correct surface.

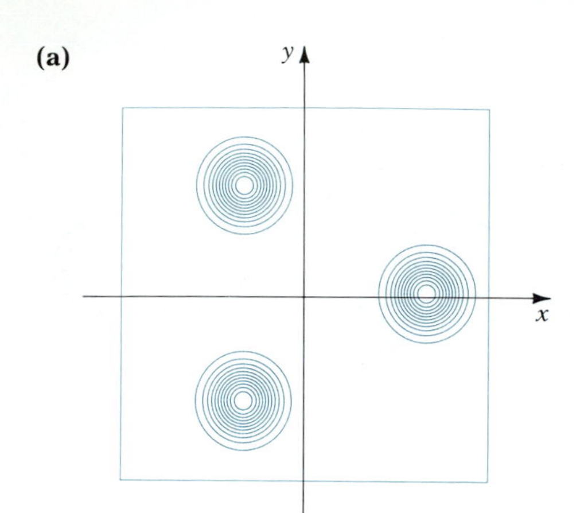
(a)
y
x

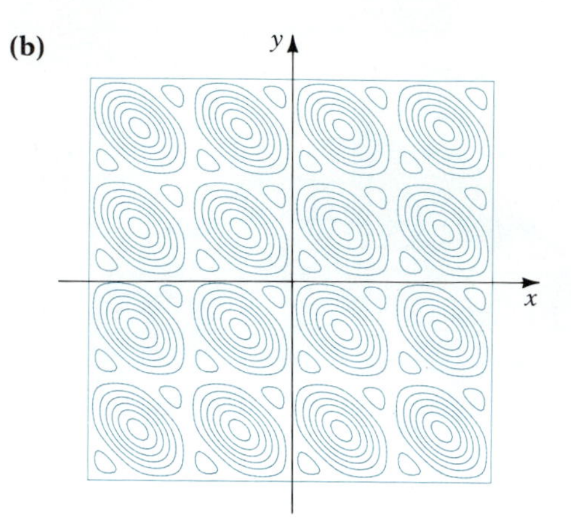
(b)
y
x

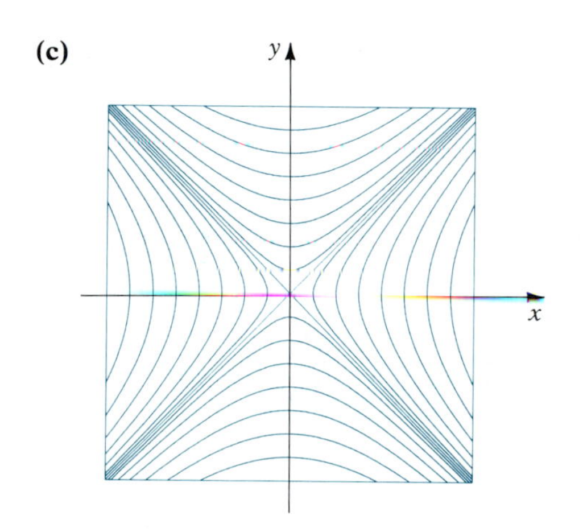
(c)
y
x

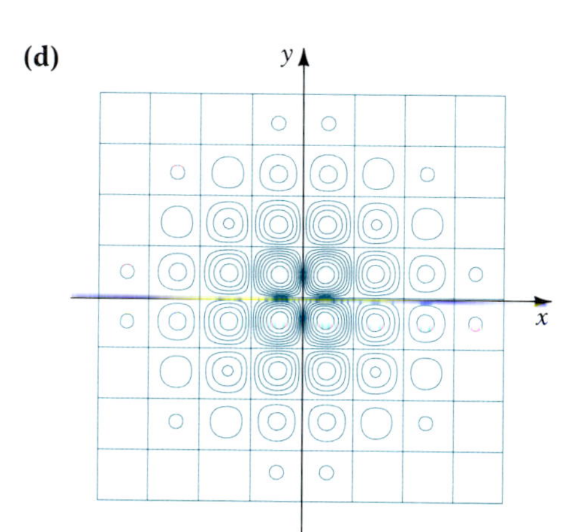
(d)
y
x

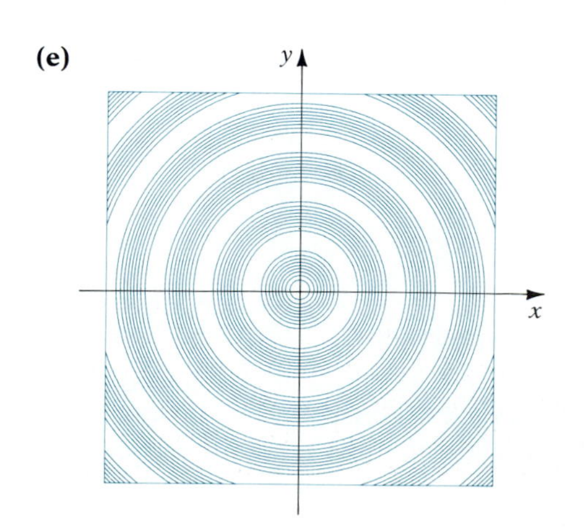
(e)
y
x

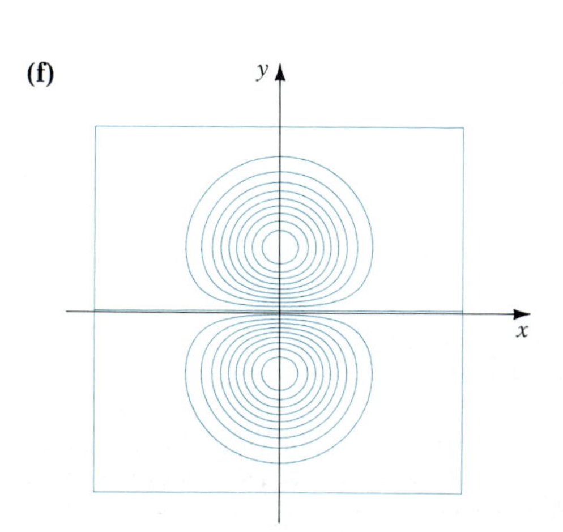
(f)
y
x

(1)

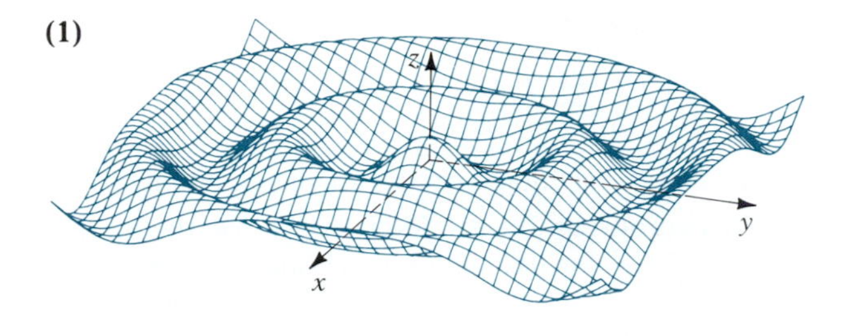

(2)

(3)

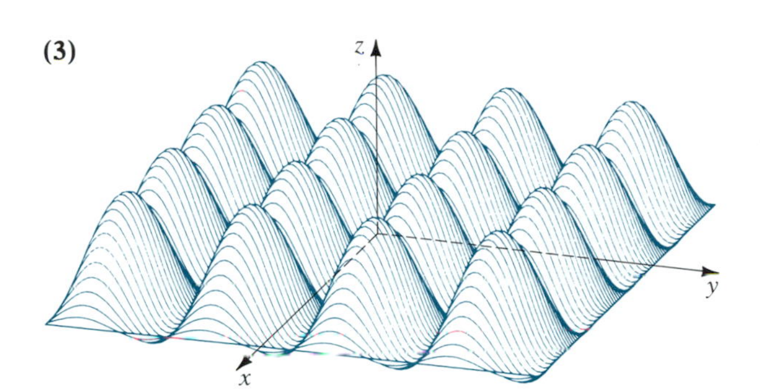

(4)

(5)

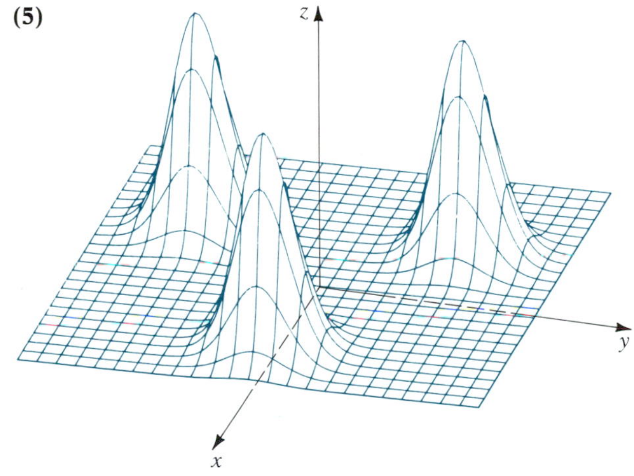

(6)

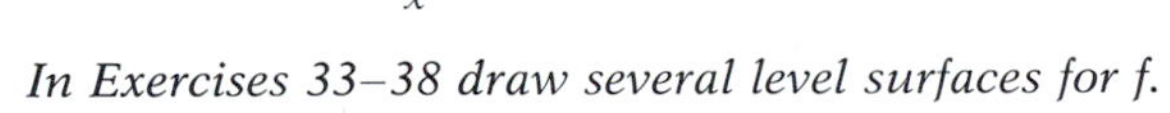

In Exercises 33–38 draw several level surfaces for f.

33. $f(x, y, z) = x + y + z$

34. $f(x, y, z) = z - x^2 - y^2$

35. $f(x, y, z) = 2x^2 + 4y^2 + z^2$

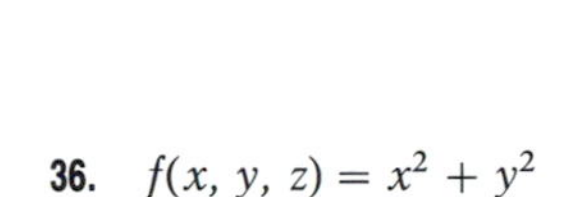

36. $f(x, y, z) = x^2 + y^2$

37. $f(x, y, z) = x - y^2$

38. $f(x, y, z) = z^2 - x^2 - y^2$

B

39. A flat plate in the xy-plane is heated from a point source at the origin. The temperature T in degrees Celsius at a point on the plate varies inversely as the square of its distance from the origin. Describe the isothermals. Suppose $T = 50$ at the point $(3, 4)$. Find all points at which $T = 30$.

40. The *ideal gas law* states that the temperature T of a gas, the volume V it occupies, and its pressure P are related by the equation $PV = kT$, where k is a constant. Draw several isothermal curves in the PV-plane and interpret the results. Express P as a function of T and V, and draw several isobars in the TV-plane. Interpret the results.

41. The speed of sound in an ideal gas is given by

$$v = \sqrt{\frac{kp}{d}}$$

where p is the pressure of the gas, d is its density, and k is a positive constant. Draw some level curves for v in the pd-plane. Solve the equation for p as a function of v and d, and draw some isobars in the vd-plane.

42. The accompanying figure shows a cross section of a circular cylinder lying on a horizontal plane. The cylinder and the plane are held at two different electric potentials. The electric potential in the shaded region is given by

$$V(x, y) = \frac{ky}{x^2 + y^2}$$

Draw several equipotentials for this function.

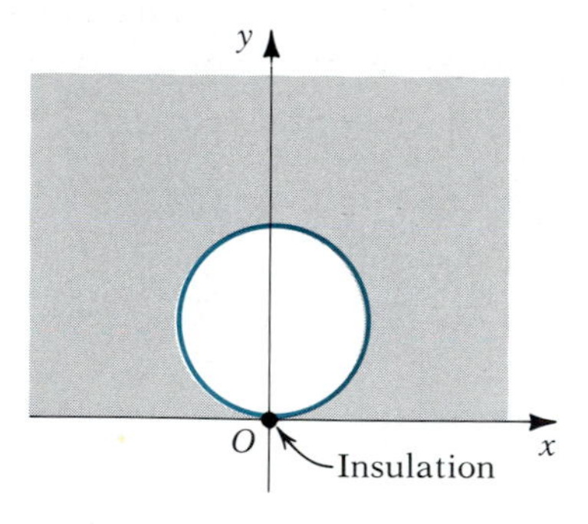

43. In hydrodynamics the level curves for the *stream function* ψ for two-dimensional fluid flow are called *streamlines.* A streamline is the path along which a given particle of the fluid moves. In the accompanying figure a fluid (such as water) flows from the negative x-axis toward the positive x-axis, with $y > 0$. There is a semicircular obstruction as shown centered at the origin. (You can think of the hump as a half-buried pipe at the bottom of a stream.) The stream function in this case is

$$\psi(x, y) = y - \frac{y}{x^2 + y^2}$$

Draw several streamlines. (*Hint:* Write the equations of the level curves in polar coordinates, and use a calculator to aid in plotting points.)

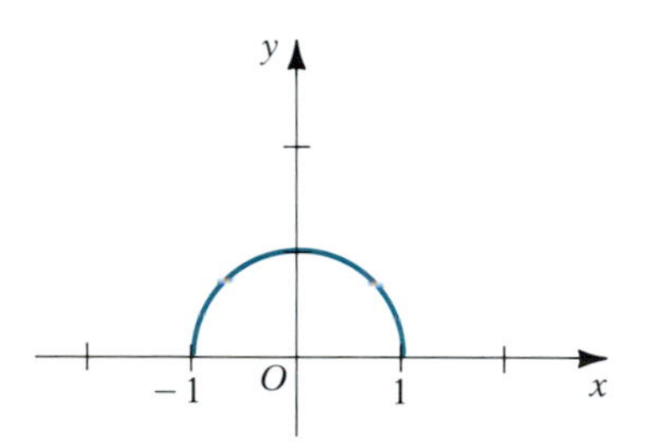

16.4 PARTIAL DERIVATIVES

Let f be a function of x and y. If we treat y as a constant and differentiate with respect to x, we obtain what is called the **partial derivative of f with respect to x,** denoted by $f_x(x, y)$. Similarly, if we treat x as a constant and differentiate with respect to y, we get the **partial derivative of f with respect to y,** denoted by $f_y(x, y)$. For example, if $f(x, y) = x^2 - 2xy - 3y^2$, we have

$f_x(x, y) = 2x - 2y$ We held y fixed and differentiated with respect to x.
$f_y(x, y) = -2x - 6y$ We held x fixed and differentiated with respect to y.

Recalling the definition of the derivative of a function of one variable, we can state the definitions of f_x and f_y in the following form.

DEFINITION 16.1 Let f be a function of the two variables x and y, with domain D. The functions f_x and f_y with values given by

$$f_x(x, y) = \lim_{h \to 0} \frac{f(x + h, y) - f(x, y)}{h} \tag{16.3}$$

$$f_y(x, y) = \lim_{k \to 0} \frac{f(x, y + k) - f(x, y)}{k} \tag{16.4}$$

at all points of D where these limits exist are called, respectively, the partial derivative of f with respect to x and the partial derivative of f with respect to y. ■

Remarks on Notation Just as with ordinary derivatives, different notations for partial derivatives are commonly used. Analogous to the Leibniz notation df/dx, the symbols

$$\frac{\partial f}{\partial x} \quad \text{and} \quad \frac{\partial f}{\partial y}$$

are alternate notations for f_x and f_y, respectively. The symbol ∂ (sometimes referred to as "curly d") replaces the d in ordinary differentiation to signify that more than one variable is involved. The symbols $\partial/\partial x$ and $\partial/\partial y$ can be regarded as *partial derivative operators,* which instruct you to take the partial derivative of whatever follows. For example,

$$\frac{\partial}{\partial x}(2x^2y + x^3) = 4xy + 3x^2 \quad \text{and} \quad \frac{\partial}{\partial y}(2x^2y + x^3) = 2x^2$$

When a dependent variable is introduced, say $z = f(x, y)$, the partial derivatives with respect to x and y, respectively, can be written as z_x or $\partial z/\partial x$ and z_y or $\partial z/\partial y$.

EXAMPLE 16.9 Let $f(x, y) = \tan^{-1} \frac{y}{x}$. Find $f_x(x, y)$; $f_y(x, y)$; $f_x(4, -3)$; and $f_y(4, -3)$.

Solution Holding y fixed, we get

$$f_x(x, y) = \frac{1}{1 + \left(\frac{y}{x}\right)^2}\left(-\frac{y}{x^2}\right) = \frac{-y}{x^2 + y^2}$$

Holding x fixed, we get

$$f_y(x, y) = \frac{1}{1 + \left(\frac{y}{x}\right)^2} \frac{1}{x} = \frac{x}{x^2 + y^2}$$

So we have

$$f_x(4, -3) = \frac{3}{25} \quad \text{and} \quad f_y(4, -3) = \frac{4}{25}$$ ■

EXAMPLE 16.10 Find $\frac{\partial}{\partial x}(e^{-xy} \cos y)$ and $\frac{\partial}{\partial y}(e^{-xy} \cos y)$.

Solution

$$\frac{\partial}{\partial x}(e^{-xy} \cos y) = -ye^{-xy} \cos y$$

$$\begin{aligned}\frac{\partial}{\partial y}(e^{-xy} \cos y) &= e^{-xy}(-\sin y) - xe^{-xy} \cos y \quad \text{Product rule}\\ &= -e^{-xy}(\sin y + x \cos y)\end{aligned}$$ ■

EXAMPLE 16.11 If $z = xy/(x + y)$, show that

$$x\frac{\partial z}{\partial x} + y\frac{\partial z}{\partial y} = z$$

Solution

$$\frac{\partial z}{\partial x} = \frac{(x + y)y - xy}{(x + y)^2} = \frac{y^2}{(x + y)^2} \quad \text{Quotient rule}$$

$$\frac{\partial z}{\partial y} = \frac{(x + y)x - xy}{(x + y)^2} = \frac{x^2}{(x + y)^2}$$

$$x\frac{\partial z}{\partial x} + y\frac{\partial z}{\partial y} = \frac{xy^2}{(x + y)^2} + \frac{x^2y}{(x + y)^2} = \frac{xy(y + x)}{(x + y)^2} = \frac{xy}{x + y} = z$$

■

Let us now consider the geometric significance of partial derivatives of a function f of x and y. Let (x_0, y_0) be a point in the domain of f at which both f_x and f_y exist. Write $z_0 = f(x_0, y_0)$ and let $P_0 = (x_0, y_0, z_0)$. Now consider the surface with equation $z = f(x, y)$. Its intersection with the plane $y = y_0$ is a plane curve that passes through P_0, having equation $z = f(x, y_0)$. The slope of its tangent line at P_0 is $f_x(x_0, y_0)$, since this is the derivative of z with respect to x, with y held fixed at y_0. This is illustrated in Figure 16.30(a). Similarly, the slope of the tangent line at P_0 on the curve of intersection of the surface and the plane $x = x_0$ is given by $f_y(x_0, y_0)$, as illustrated in Figure 16.30(b).

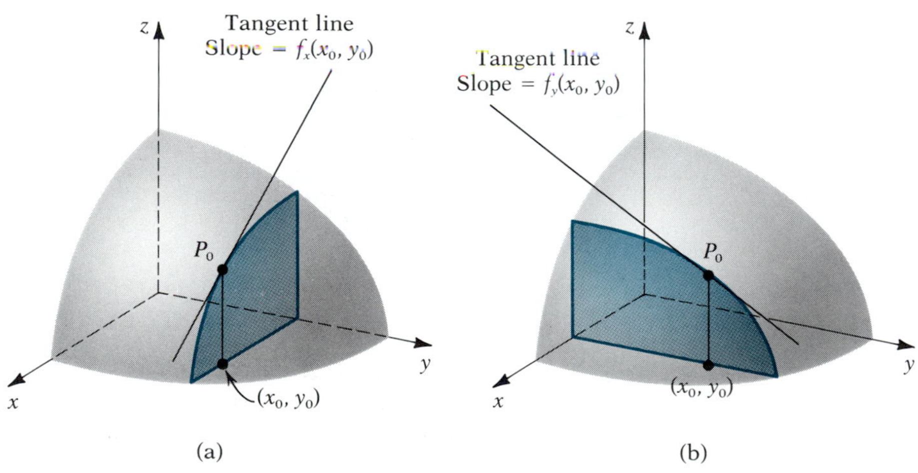

FIGURE 16.30

EXAMPLE 16.12 Find the slope of the tangent line to the curve of intersection of the surface $z = \sqrt{10 - 2x^2 - y^2}$ and the plane (a) $y = 1$ and (b) $x = 2$ at the point $(2, 1, 1)$. Show the results graphically.

Solution By the preceding discussion, the slope of the tangent line in part a is $\partial z/\partial x$ and in part b is $\partial z/\partial y$, each evaluated at $(2, 1)$:

$$\frac{\partial z}{\partial x} = \frac{-2x}{\sqrt{10 - 2x^2 - y^2}} \quad \text{and} \quad \frac{\partial z}{\partial y} = \frac{-y}{\sqrt{10 - 2x^2 - y^2}}$$

So for the curve of part a,

$$\text{slope} = \left.\frac{\partial z}{\partial x}\right|_{(2,1)} = \frac{-4}{1} = -4$$

and for the curve of part b,

$$\text{slope} = \left.\frac{\partial z}{\partial y}\right|_{(2,1)} = \frac{-1}{1} = -1$$

The results are illustrated in Figure 16.31. Notice that the given surface is the upper half of the ellipsoid

$$\frac{x^2}{5} + \frac{y^2}{10} + \frac{z^2}{10} = 1$$

We show the first-octant portion only.

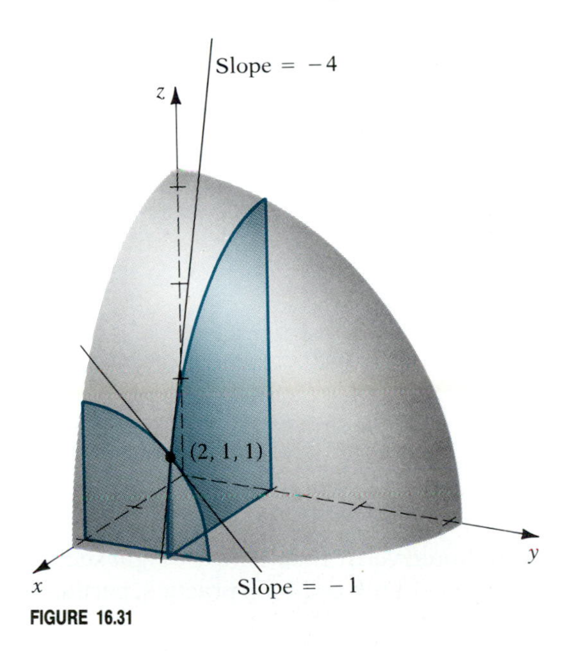

FIGURE 16.31 ■

Since f_x and f_y are themselves functions of the two variables x and y, it is possible that they too have partial derivatives, called *second-order partial derivatives of f* (or just "second partials" for short). These are denoted as follows:

f_{xx}, second partial of f with respect to x

f_{xy}, second mixed partial of f, first with respect to x and then with respect to y

f_{yy}, second partial of f with respect to y

f_{yx}, second mixed partial of f, first with respect to y and then with respect to x

Note carefully the order of differentiation in the mixed partials. For example, f_{xy} means $(f_x)_y$, so we first differentiate with respect to x and then differentiate the result with respect to y.

Using the alternate notation, we write

$$\frac{\partial^2 f}{\partial x^2} \quad \text{same as } f_{xx}$$

$$\frac{\partial^2 f}{\partial y \, \partial x} \quad \text{same as } f_{xy}$$

$$\frac{\partial^2 f}{\partial y^2} \quad \text{same as } f_{yy}$$

$$\frac{\partial^2 f}{\partial x \, \partial y} \quad \text{same as } f_{yx}$$

These notations are suggested by applying the partial differential operators twice—for example, as in

$$\frac{\partial}{\partial x}\left(\frac{\partial f}{\partial x}\right) = \frac{\partial^2 f}{\partial x^2}$$

and

$$\frac{\partial}{\partial y}\left(\frac{\partial f}{\partial x}\right) = \frac{\partial^2 f}{\partial y \, \partial x}$$

Again, observe carefully the order of differentiation. Compare the following, for example:

$$\frac{\partial^2 f}{\underset{\text{2nd}}{\partial y}\,\underset{\text{1st}}{\partial x}} = f_{\underset{\text{1st}}{x}\underset{\text{2nd}}{y}}$$

We could continue to higher order partials, using notations such as

$$f_{xxy} \quad \text{or} \quad \frac{\partial^3 f}{\partial y \, \partial x^2}$$

You should verify that there are eight such third-order partials, and in general 2^n partials of nth order. In practice, partials of orders higher than 2 are seldom used.

EXAMPLE 16.13 Let $f(x, y) = 2x^3y^2 - 3xy^4$. Find $f_{xx}(x, y)$; $f_{xy}(x, y)$; $f_{yy}(x, y)$; and $f_{yx}(x, y)$.

Solution

$$f_x(x, y) = 6x^2y^2 - 3y^4 \qquad f_y(x, y) = 4x^3y - 12xy^3$$

$$f_{xx}(x, y) = 12xy^2 \qquad f_{yy}(x, y) = 4x^3 - 36xy^2$$

$$f_{xy}(x, y) = 12x^2y - 12y^3 \qquad f_{yx}(x, y) = 12x^2y - 12y^3$$

■

Note that in this example $f_{xy} = f_{yx}$. As you might expect, this is no accident. Although it is not always true, it is true for "well-behaved" functions. We will give sufficient conditions to guarantee the equality of the second mixed partials in the next section.

If f is a function of three (or more) variables, we define partial derivatives in a similar way to those for functions of two variables. All variables except one are held fixed, and the derivative with respect to the remaining variable is taken. For example, if

$$f(x, y, z) = \ln(x^2 + y^2 + z^2)$$

we have

$$f_x(x, y, z) = \frac{2x}{x^2 + y^2 + z^2} \qquad f_y(x, y, z) = \frac{2y}{x^2 + y^2 + z^2}$$

$$f_z(x, y, z) = \frac{2z}{x^2 + y^2 + z^2}$$

EXERCISE SET 16.4

A

In Exercises 1–10 find $f_x(x, y)$ and $f_y(x, y)$.

1. $f(x, y) = x^2 + y^2$
2. $f(x, y) = \sqrt{x^2 + y^2}$
3. $f(x, y) = \sin xy$
4. $f(x, y) = e^x \cos y$
5. $f(x, y) = \dfrac{x + y}{x - y}$
6. $f(x, y) = \ln \dfrac{x}{y}$
7. $f(x, y) = x \ln y - y \ln x$
8. $f(x, y) = \tan^{-1} \dfrac{x}{y}$
9. $f(x, y) = \dfrac{xy}{\sqrt{x^2 - y^2}}$
10. $f(x, y) = e^{x-y} \sin(x - y)$

In Exercises 11–16 find $\partial z/\partial x$ and $\partial z/\partial y$.

11. $z = (1 - x^2 - y^2)^{-1/2}$
12. $z = \sin x \cosh y$
13. $z = \ln \cos(x - y)$
14. $z = \ln\left(\dfrac{x^2 - 2xy}{3xy - y^2}\right)$
15. $z = \sin^{-1} \sqrt{1 - x^2y^2}$, $x > 0$, $y > 0$, $xy < 1$
16. $z = \sqrt{\dfrac{x - 2y}{x + 2y}}$
17. Find $f_x(3, -2)$ and $f_y(3, -2)$ if $f(x, y) = x^2y - 2y^2$.
18. If $g(x, y) = 2xy/(x - y)$, find $g_x(-1, 1)$ and $g_y(-1, 1)$.
19. If $f(r, \theta) = e^r \cos \theta$, find $f_r(0, \frac{\pi}{3})$ and $f_\theta(0, \frac{\pi}{3})$.
20. If $w = \frac{u}{v} - \frac{v}{u}$, find $\partial w/\partial u$ and $\partial w/\partial v$ when $u = -2$ and $v = 2$.
21. If $w = 1/(2s - t^2)^2$, find $\partial w/\partial s$ and $\partial w/\partial t$ when $s = 3$ and $t = -2$.
22. Let $w = (u + v)/(u - v)$. Show that $v(\partial w/\partial u) + u(\partial w/\partial v) = 2w$.
23. Let $z = \tan^{-1} \frac{y}{x}$. Show that $x(\partial z/\partial y) - y(\partial z/\partial x) = 1$.
24. Find the slope of the curve of intersection of the cone $z = \sqrt{4x^2 + 3y^2}$ and (a) the plane $y = 4$ and (b) the plane $x = -2$ at the point $(-2, 4, 8)$.
25. Let $f(x, y) = 2x^2 - 3xy^2 + 3y^3$, and let C_1 and C_2 be the curves of intersection of the graph of f and the planes $y = 2$ and $x = 3$, respectively. Show that both C_1 and C_2 have horizontal tangent lines at the point $(3, 2, 6)$.

In Exercises 26–31 find all second-order partial derivatives of f.

26. $f(x, y) = \sqrt{x - 2y}$
27. $f(x, y) = \ln(x^2 + 3y^2)$
28. $f(x, y) = \sin xy$
29. $f(x, y) = e^{x^2y}$
30. $f(x, y) = \dfrac{x - y}{x + y}$
31. $f(x, y) = e^{-x} \sin y + e^{-y} \cos x$
32. Let $f(x, y, z) = \dfrac{x - y}{y - z}$. Find f_x, f_y, and f_z.

B

*A function f of two variables x and y is said to be **harmonic** if $f_{xx} + f_{yy} = 0$. In Exercises 33–36 show that f is harmonic at all points where f_{xx} and f_{yy} are defined.*

33. $f(x, y) = \cos x \cosh y$
34. $f(x, y) = \ln \sqrt{x^2 + y^2}$
35. $f(x, y) = \tan^{-1} \frac{y}{x}$
36. $f(x, y) = x/(x^2 + y^2)$

*If f is a harmonic function and g is a harmonic function such that $f_x = g_y$ and $f_y = -g_x$, then g is said to be a **harmonic conjugate** of f. In Exercises 37–40 show that g is a harmonic conjugate of f.*

37. $g(x, y) = -\sin x \sinh y$; $f(x, y)$ in Exercise 33
38. $g(x, y) = \tan^{-1} \frac{y}{x}$; $f(x, y)$ in Exercise 34
39. $g(x, y) = \ln [1/\sqrt{x^2 + y^2}]$; $f(x, y)$ in Exercise 35
40. $g(x, y) = -y/(x^2 + y^2)$; $f(x, y)$ in Exercise 36

41. Let $f(x, y) = x^2 - y^2 + 2x + 1$ and $g(x, y) = 2xy + 2y$.
(a) Show that f and g are harmonic.
(b) Show that g is a harmonic conjugate of f.
(c) Show that the level curves of f and the level curves of g intersect at right angles except for $f(x, y) = 0$ and $g(x, y) = 0$.
(d) Draw several level curves for f and g on the same graph.

42. Let $f(x, y, z) = xy \ln z + xz \ln y + yz \ln x$. Show the following:
(a) $f_{xxy} = f_{xyx} = f_{yxx}$
(b) $f_{xyy} = f_{yxy} = f_{yyx}$
(c) $f_{xzz} = f_{zxz} = f_{zzx}$
(d) The equality of all third-order partials of f with respect to x, y, and z, taken in any order

43. Let

$$f(x, y) = \begin{cases} \dfrac{x^3 - y^3}{x^2 + y^2} & \text{if } (x, y) \neq (0, 0) \\ 0 & \text{if } (x, y) = (0, 0) \end{cases}$$

Use Definition 16.1 to compute $f_x(0, 0)$ and $f_y(0, 0)$.

16.5 LIMITS AND CONTINUITY

To extend the notions of limit and continuity to functions of two and three variables, it is useful to introduce some concepts that help to describe regions in two- and three-dimensional space.

First, if $P_0 = (x_0, y_0)$ is any point in $\mathbf{R}^2$, by a **neighborhood** of P_0 we mean the set of all points inside some circle centered at P_0. If $P_0 = (x_0, y_0, z_0)$ is in $\mathbf{R}^3$, then a neighborhood of P_0 is the set of all points inside some sphere centered at P_0. We can describe both two- and three-dimensional neighborhoods of a point P_0 as the set of all points P such that $|\overrightarrow{P_0P}| < r$ for some positive number r, called the radius of the neighborhood. If we delete P_0 itself from the neighborhood, we call the resulting set a **deleted neighborhood** of P_0. Such a deleted neighborhood is described by the inequality $0 < |\overrightarrow{P_0P}| < r$. These definitions are analogous to those for the one-dimensional case.

Now let S denote any set of points in $\mathbf{R}^2$ or $\mathbf{R}^3$. By a **boundary point** of S is meant a point P_0 such that every neighborhood of P_0 contains at least one point in S and one point not in S. The **boundary** of S is the set of all its boundary points. A point in S that is not a boundary point is called an **interior point** of S.

A set S in $\mathbf{R}^2$ or $\mathbf{R}^3$ is said to be **open** if S contains none of its boundary points, and S is said to be **closed** if it contains all of its boundary points. Note that an open set consists of interior points only. For example, the set of points inside an ellipse is open, and this set together with the ellipse itself is closed. Observe also that a set may be neither open nor closed.

If P_1 and P_2 are two points in $\mathbf{R}^2$ or $\mathbf{R}^3$, by a **polygonal path** from P_1 to P_2 we mean a finite number of line segments joined end to end, extending from P_1 to P_2. An open subset of $\mathbf{R}^2$ or $\mathbf{R}^3$ is said to be **connected** if any two points in S can be joined by a polygonal path lying entirely in S. Figure 16.32(a) shows such an open connected subset of $\mathbf{R}^2$, and Figure 16.32(b) shows a subset that is **disconnected**—that is, one that is not connected.

Up to now we have used the term *region* in a general way. We now give it a precise definition.

A **region** in $\mathbf{R}^2$ or $\mathbf{R}^3$ is an open connected set together with some, none, or all of its boundary.

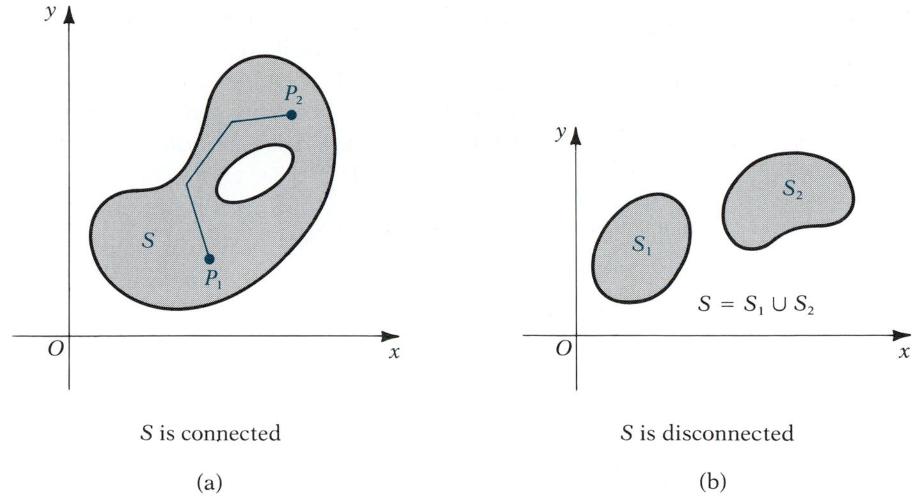

FIGURE 16.32

From now on this is what we will mean by a region. Note that in Figure 16.32(a) S is a region (whether or not the boundary is included), whereas in part b S is not a region but is the union of two disjoint regions.

EXAMPLE 16.14 Let S be the set of points (x, y) in $\mathbf{R}^2$ such that

(a) $\begin{cases} 1 < x < 2 \\ 1 < y < 2 \end{cases}$ (b) $\begin{cases} 1 \le x \le 2 \\ 1 \le y \le 2 \end{cases}$ (c) $\begin{cases} 1 \le x < 2 \\ 1 \le y < 2 \end{cases}$

Determine in each case the boundary of S and whether S is open, closed, or neither.

Solution Figure 16.33 illustrates S in each case. In all cases the boundary of S is the square formed by the lines $x = 1$, $x = 2$, $y = 1$, and $y = 2$. To see that all points on this square are boundary points, consider a typical point P_0 on the square as shown in each part of the figure. Any neighborhood of P_0 must

FIGURE 16.33

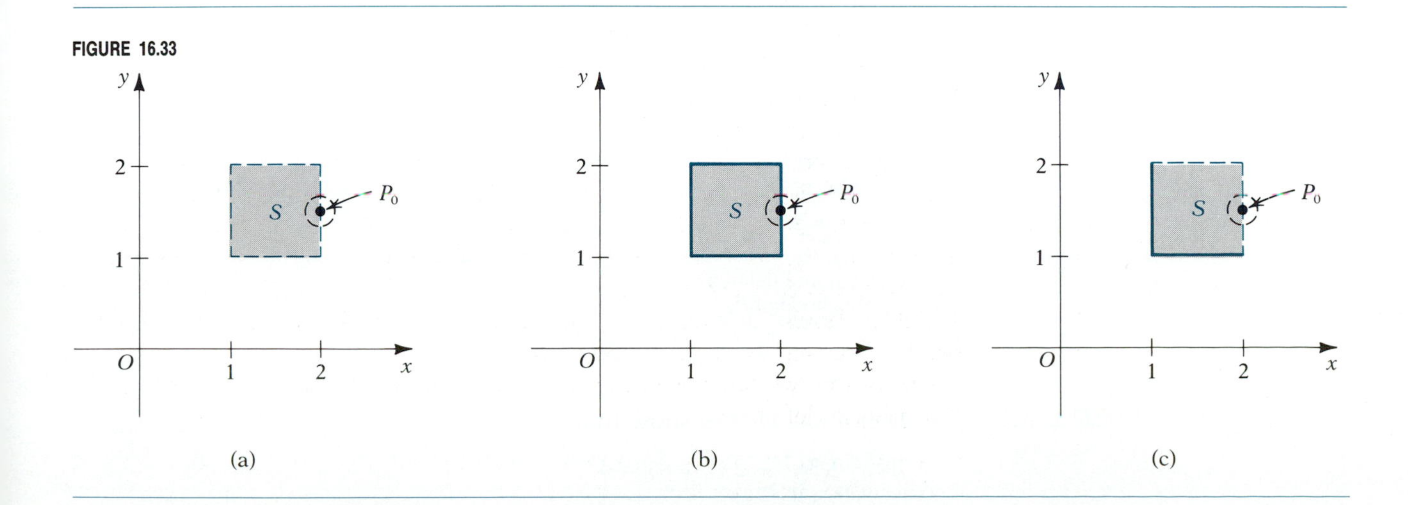

contain points in S and points not in S. So P_0 is a boundary point. In part a, S contains none of its boundary points and so is open. In part b, S contains all of its boundary points and so is closed. In part c, S is neither open nor closed, since it contains some, but not all, of its boundary points. Observe that in each case S is a region. ■

Suppose now that f is a function of either two or three variables. So that we can discuss both cases at once, we will write $f(P)$ to mean $f(x, y)$ or $f(x, y, z)$, depending on whether the domain of f is in $\mathbf{R}^2$ or $\mathbf{R}^3$. Roughly speaking, to say that the limit of $f(P)$ is the number L as P approaches P_0 means that the function values $f(P)$ all can be made arbitrarily close to L provided only that P is sufficiently close to P_0 but not equal to P_0. When this is the case we write $\lim_{P \to P_0} f(P) = L$. Of course, $f(P)$ must be defined for all points near P_0, except possibly at P_0 itself. Thus, the domain of f must include some deleted neighborhood of P_0. (It might also include P_0, but this is not relevant insofar as the limit concept is concerned.) The following definition makes the meaning of limit precise.

DEFINITION 16.2 Let f be a function of two or three variables defined in some deleted neighborhood of P_0. Then

$$\lim_{P \to P_0} f(P) = L$$

provided that, given any positive number ε, there exists a positive number δ such that

$$|f(P) - L| < \varepsilon \quad \text{whenever} \quad 0 < |P_0P| < \delta$$ ■

Remark That this definition is analogous to the one-dimensional case is made clearer by phrasing it in neighborhood terminology. It says that the function values $f(P)$ will be in an arbitrarily small neighborhood of L provided the points P lie within a sufficiently small deleted neighborhood of P_0. In the definition ε is the radius of the given neighborhood of L, and δ is the radius of the deleted neighborhood of P_0. This can be indicated schematically as in Figure 16.34, where the domain of f is taken as two dimensional.

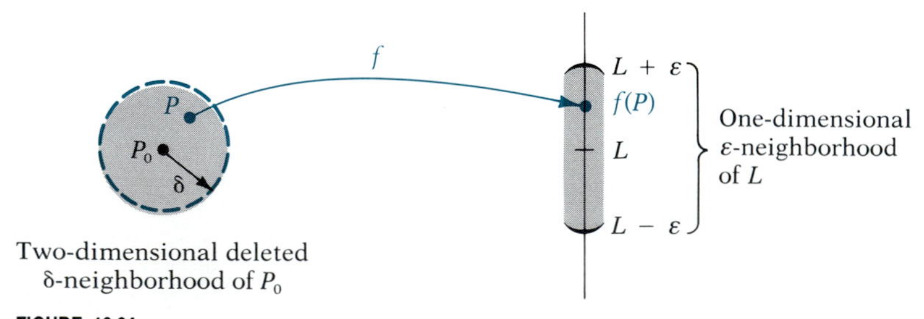

FIGURE 16.34

EXAMPLE 16.15 Use Definition 16.2 to show that

$$\lim_{(x,y) \to (0,0)} (3x^2 + 2xy + 4) = 4$$

Solution We begin by letting ε denote an arbitrary positive number. Our goal is to find a number δ that satisfies the conditions of the definition. To accomplish this, we try to determine a restriction on $|\overrightarrow{P_0P}|$ that will guarantee $|f(P) - L| < \varepsilon$. In this example, $P_0 = (0, 0)$ and $P = (x, y)$. So $|\overrightarrow{P_0P}| = \sqrt{x^2 + y^2}$. For $|f(P) - L|$ we have

$$\begin{aligned} |f(P) - L| &= |(3x^2 + 2xy + 4) - 4| \\ &= |3x^2 + 2xy| \leq 3x^2 + 2|x|\,|y| \quad \text{(Why?)} \end{aligned}$$

Now for all points (x, y), $x^2 \leq x^2 + y^2$, and since $|x| = \sqrt{x^2} \leq \sqrt{x^2 + y^2}$ and $|y| = \sqrt{y^2} \leq \sqrt{x^2 + y^2}$, we obtain

$$|f(P) - L| \leq 3(x^2 + y^2) + 2(\sqrt{x^2 + y^2})(\sqrt{x^2 + y^2}) = 5(x^2 + y^2)$$

Thus $|f(P) - L|$ will be less than ε provided $5(x^2 + y^2) < \varepsilon$ or, equivalently, $x^2 + y^2 < \frac{\varepsilon}{5}$. Since $|\overrightarrow{P_0P}| = \sqrt{x^2 + y^2}$, we conclude that

$$\text{if} \quad 0 < |\overrightarrow{P_0P}| < \sqrt{\frac{\varepsilon}{5}} \quad \text{then} \quad |f(P) - L| < \varepsilon$$

So $\delta = \sqrt{\varepsilon/5}$ satisfies the conditions of the definition. (*Note:* In this case it was not necessary to restrict P to be different from P_0.) ■

EXAMPLE 16.16 Use Definition 16.2 to show that

$$\lim_{(x,y)\to(2,1)} (3x - 4y + 9) = 11$$

Solution Let ε denote any positive number. As in the preceding example, we want to obtain an inequality that compares $|f(P) - L|$ with some expression involving $|\overrightarrow{P_0P}|$, which in this case equals $\sqrt{(x-2)^2 + (y-1)^2}$. We do this in the following way:

$$\begin{aligned} |f(P) - L| &= |(3x - 4y + 9) - 11| = |3x - 4y - 2| \\ &= |3(x - 2) - 4(y - 1)| \\ &\leq 3|x - 2| + 4|y - 1| \end{aligned}$$

Now we use the fact that

$$|x - 2| = \sqrt{(x-2)^2} \leq \sqrt{(x-2)^2 + (y-1)^2}$$

and

$$|y - 1| = \sqrt{(y-1)^2} \leq \sqrt{(x-2)^2 + (y-1)^2}$$

to get

$$\begin{aligned} |f(P) - L| &\leq 3\sqrt{(x-2)^2 + (y-1)^2} + 4\sqrt{(x-2)^2 + (y-1)^2} \\ &= 7\sqrt{(x-2)^2 + (y-1)^2} = 7|\overrightarrow{P_0P}| \end{aligned}$$

Thus,

$$|f(P) - L| < \varepsilon \quad \text{provided} \quad 0 < |\overrightarrow{P_0P}| < \frac{\varepsilon}{7}$$

So $\delta = \varepsilon/7$ satisfies the condition of the definition. ■

EXAMPLE 16.17 Prove that

$$\lim_{(x,y)\to(0,0)} \frac{x^2 - y^2}{x^2 + y^2}$$

does not exist.

Solution Let $P = (x, y)$ and $P_0 = (0, 0)$. If P lies on the x-axis with $x \neq 0$, then

$$f(P) = f(x, 0) = \frac{x^2}{x^2} = 1$$

whereas if P lies on the y-axis with $y \neq 0$, then

$$f(P) = f(0, y) = \frac{-y^2}{y^2} = -1$$

This shows that in every deleted neighborhood of P_0 there are points P for which $f(P) = 1$ (occurring when P is on the x-axis) and points P for which $f(P) = -1$ (occurring when P is on the y-axis). So there can be no single number L that $f(P)$ approaches as P approaches P_0. ■

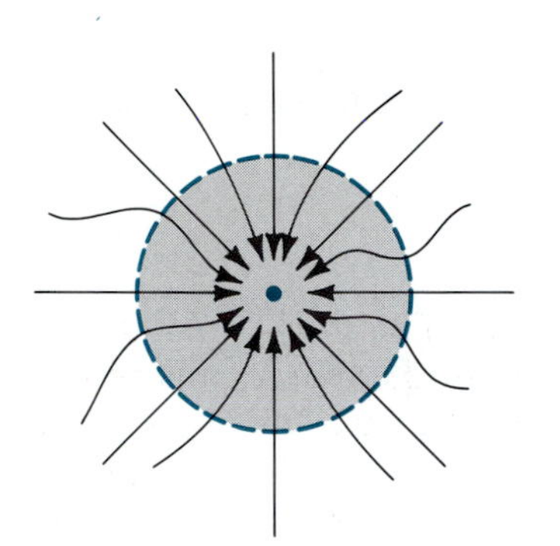

FIGURE 16.35 P can approach P_0 along infinitely many different paths

Example 16.17 illustrates a way to show that a function fails to approach a limit. The general procedure is to try to find two different paths along which P approaches P_0 such that $f(P)$ approaches different values along these paths. This is much like showing that left-hand and right-hand limits are different in single-variable functions. In the case of functions of two or three variables, P can approach P_0 in an infinite variety of ways. This is indicated for $\mathbf{R}^2$ in Figure 16.35. We summarize as follows:

If $f(P)$ approaches different limits as P approaches P_0 along two different paths, then $\lim_{P\to P_0} f(P)$ does not exist.

A Word of Caution If you get the *same* limit for f(P) as P approaches P_0 along two or more different paths, this proves nothing. Since it is impossible to test all paths of approach, this method cannot be used to show the existence of the limit.

The next example illustrates the nonexistence of the limit, where one of the paths of approach is along a parabola.

EXAMPLE 16.18 Prove that

$$\lim_{(x,y)\to(0,0)} \frac{x^2 y}{x^4 + y^2}$$

does not exist.

Solution If the point $P = (x, y)$ is on either the x-axis or y-axis but is not the origin, then $f(P) = 0$. So we do not get different limits along these two axes. In fact, the approach along any line $y = mx$ still gives the limit 0. However, if P is on the parabola $y = x^2$, then

$$f(P) = f(x, x^2) = \frac{x^2 x^2}{x^4 + x^4} = \frac{x^4}{2x^4} = \frac{1}{2}$$

so long as $x \neq 0$. So when P approaches the origin along this parabola, the limit is $\frac{1}{2}$. But, as we have seen, when P approaches the origin along a straight line, the limit is 0. Thus, there is no unique limit as P approaches the origin unrestrictedly. ■

Properties of limits analogous to those for single-variable functions can be proved from the definition, and you will be asked to prove some of them in the exercises. In particular, if f is a constant function, say $f(P) = c$, then $\lim_{P \to P_0} f(P) = c$ for all points P_0, and if $\lim_{P \to P_0} f(P) = L$, then $\lim_{P \to P_0} cf(P) = cL$ for all constants c. Also, if $\lim_{P \to P_0} g(P) = M$, then

$$\lim_{P \to P_0} (f \pm g)(P) = L \pm M \qquad \lim_{P \to P_0} (fg)(P) = LM$$

and

$$\lim_{P \to P_0} \frac{f}{g}(P) = \frac{L}{M} \quad \text{if } M \neq 0$$

We turn now to the concept of the continuity of a function of two or three variables. Let f be such a function and D be its domain.* If P_0 is an *interior* point of D, then f is continuous at P_0 provided

$$\lim_{P \to P_0} f(P) = f(P_0) \tag{16.5}$$

Let us recall the meaning of the limit equation (16.5). For any neighborhood of $f(P_0)$ we can find a δ-neighborhood of P_0 such that the images of all points in this δ-neighborhood lie in the given ε-neighborhood of $f(P_0)$. [Note that for continuity it is not necessary that the δ-neighborhood of P_0 be a deleted neighborhood, since when $P = P_0$, $f(P) = f(P_0)$.] The fact that P_0 is an interior point of D makes it possible to choose δ such that the δ-neighborhood of P_0 lies entirely in D. (Why?) Suppose now that P_0 is a *boundary* point of D. We will in this case say that f is continuous at P_0 under the same conditions as before except that we require only that the points *of* D lying in the δ-neighborhood of P_0 map into the ε-neighborhood of $f(P_0)$. Because P_0 is a boundary point of D, there will always be points outside of D in every neighborhood of P_0. We ignore these points. (See Figure 16.36.)

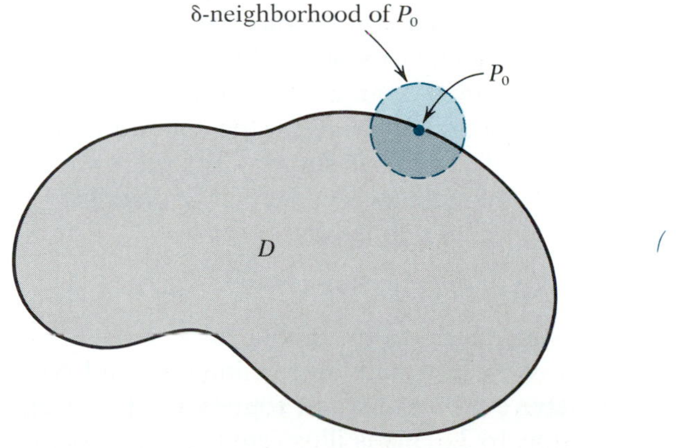

FIGURE 16.36

The definitions of continuity at an interior point and at a boundary point of the domain of f are both given in the following definition, which summarizes the preceding discussion.

* We will primarily consider functions whose domains are regions, but this is not essential. For example, D may be the union of two disjoint regions.

DEFINITION 16.3 Let f be a function of two or three variables with domain D, and let $P_0 \in D$. Then f is continuous at P_0 provided that, corresponding to each positive number ε, there is a positive number δ such that

$$|f(P) - f(P_0)| < \varepsilon \quad \text{whenever} \quad P \in D \text{ and } |\overrightarrow{P_0P}| < \delta$$ ■

If f is continuous at all points of some subset S of D, we say *f is continuous on S*. When f is continuous on all of D, we sometimes say simply that *f is continuous*.

In Example 16.15 we showed that

$$\lim_{(x,y)\to(0,0)} (3x^2 + 2xy + 4) = 4$$

Since $f(0, 0) = 4$ also, it follows from equation (16.5) that this function is continuous at $P_0 = (0, 0)$. Similarly, in Example 16.16 we showed that

$$\lim_{(x,y)\to(2,1)} (3x - 4y + 9) = 11$$

and by direct substitution we get $f(2, 1) = 11$. Thus, this function also is continuous at the point (2, 1). These are examples of *polynomial functions in two variables*, and it can be shown that all such functions are continuous everywhere. Similarly, it can be shown that quotients of polynomial functions—that is, *rational functions*—are continuous except where the denominator is 0. (See Exercise 56.) Thus, for example,

$$f(x, y) = \frac{3x^2 - 2xy + y^4}{x^2 - y}$$

is continuous except for points on the parabola $y = x^2$.

More generally, it can be shown, based on properties of limits, that the *sum, difference, product,* and *quotient* of two functions, each continuous at P_0, also are continuous at P_0, provided that in the case of the quotient, the denominator is not 0 at P_0. (See Exercise 54.) Furthermore, a *continuous function of a continuous function* is also continuous in the following sense. Suppose the function f of two or three variables is continuous at P_0 and that g is a single-valued function whose domain contains the range of f. Then if g is continuous at $f(P_0)$, the composite function $g \circ f$ is continuous at P_0. An example of this is $\sin(x^2 + y^2)$. Here we could take $f(x, y) = x^2 + y^2$, which is continuous everywhere in $\mathbf{R}^2$, and $g(u) = \sin u$, continuous for all u in $\mathbf{R}$. Then $g \circ f$, defined by

$$(g \circ f)(x, y) = \sin(x^2 + y^2)$$

is continuous throughout $\mathbf{R}^2$.

We can usually determine points of continuity of functions based on the above properties, but sometimes it is necessary to appeal directly to Definition 16.3. This is illustrated in the next example.

EXAMPLE 16.19 Let f be defined on $\mathbf{R}^2$ by

$$f(x, y) = \begin{cases} \dfrac{x^2y^2}{x^2 + y^2} & \text{if } (x, y) \neq (0, 0) \\ 0 & \text{if } (x, y) = (0, 0) \end{cases}$$

Determine whether f is continuous at the origin.

Solution We might note that at all points other than the origin f is continuous, since it is a rational function with nonzero denominator. But we have to handle the origin differently. Let $P = (x, y)$ and $P_0 = (0, 0)$. Then $f(P_0) = 0$ and for $P \neq P_0$, $f(P) = x^2y^2/(x^2 + y^2)$. So for $P \neq P_0$,

$$|f(P) - f(P_0)| = \left| \frac{x^2y^2}{x^2 + y^2} - 0 \right| = \frac{x^2y^2}{x^2 + y^2}$$

Now $x^2 \leq x^2 + y^2$ and $y^2 \leq x^2 + y^2$. So

$$|f(P) - f(P_0)| \leq \frac{(x^2 + y^2)^2}{x^2 + y^2} = x^2 + y^2$$

If $\varepsilon > 0$ is given, we will have

$$|f(P) - f(P_0)| < \varepsilon \quad \text{if } x^2 + y^2 < \varepsilon$$

or, equivalently, if $|\overrightarrow{P_0P}| = \sqrt{x^2 + y^2} < \sqrt{\varepsilon}$. So taking $\delta = \sqrt{\varepsilon}$ satisfies the condition for continuity in Definition 16.3 at the point $P_0 = (0, 0)$. ■

EXAMPLE 16.20 Let f be defined on $\mathbf{R}^2$ by

$$f(x, y) = \begin{cases} \dfrac{xy}{x^2 + y^2} & \text{if } (x, y) \neq (0, 0) \\ 0 & \text{if } (x, y) = (0, 0) \end{cases}$$

Show that f is discontinuous at the origin.

Solution Again, write $P_0 = (0, 0)$ and $P = (x, y)$. If $P \neq P_0$ and P lies on the x-axis, then

$$f(P) = f(x, 0) = \frac{0}{x^2} = 0$$

On the other hand, if $P \neq P_0$ and P lies on the line $y = x$, then

$$f(P) = f(x, x) = \frac{x^2}{x^2 + x^2} = \frac{1}{2}$$

Thus, as $P \to P_0$ along the x-axis, $f(P)$ approaches 0 (in fact, it is 0), and as $P \to P_0$ along the line $y = x$, $f(P)$ approaches $\frac{1}{2}$. It follows that $\lim_{P \to P_0} f(P)$ does not exist, and hence f is discontinuous at P_0. ■

We conclude this section by stating a theorem that provides a sufficient condition for the equality of mixed second-order partial derivatives f_{xy} and f_{yx}. Its proof can be found in advanced calculus textbooks.

THEOREM 16.1 Let f be a function of two variables, and suppose f_{xy} and f_{yx} both exist in a neighborhood of a point P_0. Then if f_{xy} and f_{yx} are continuous at P_0, they are equal there:

$$f_{xy}(P_0) = f_{yx}(P_0)$$

■

It should be emphasized that continuity of the mixed partials is a sufficient condition for their equality, but failure to be continuous does not guarantee they are unequal.

EXERCISE SET 16.5

A

In Exercises 1–10 determine the limits by inspection.

1. $\lim_{(x,y)\to(2,-1)} (3x - 2y + 5)$

2. $\lim_{(x,y)\to(-1,-2)} (2x^2y - 3xy^3)$

3. $\lim_{(x,y)\to(-2,3)} \dfrac{x - 2y}{x + y}$

4. $\lim_{(x,y)\to(1,3)} \dfrac{x^2 - y^2}{x^2 + y^2}$

5. $\lim_{(x,y,z)\to(1,-1,2)} (xz - yz + xy)$

6. $\lim_{(x,y,z)\to(-4,5,3)} \dfrac{x^2 - 2yz + z^2}{x^2 + y^2 + z^2}$

7. $\lim_{(x,y)\to(2,2)} \dfrac{x^2 - y^2}{x - y}$

8. $\lim_{(x,y)\to(2,-1)} \dfrac{x^2 - 4y^2}{x + 2y}$

9. $\lim_{(x,y)\to(0,0)} f(x, y)$, where
$$f(x, y) = \begin{cases} \sqrt{x^2 + y^2} & \text{if } (x, y) \neq (0, 0) \\ 1 & \text{if } (x, y) = (0, 0) \end{cases}$$

10. $\lim_{(x,y)\to(1,-1)} f(x, y)$, where
$$f(x, y) = \begin{cases} \dfrac{x^2 - 2xy}{2x + 3y} & \text{if } (x, y) \neq (1, -1) \\ 3 & \text{if } (x, y) = (1, -1) \end{cases}$$

In Exercises 11–16 determine $\lim_{(x,y)\to(0,0)} f(x, y)$ by inspection, and then use Definition 16.2 to verify your result.

11. $f(x, y) = x^2 + y^2 + 4$
12. $f(x, y) = \sqrt{x^2 - y^2}$
13. $f(x, y) = 2xy - 5$
14. $f(x, y) = 4x - 3y + 7$
15. $f(x, y) = x^2 - 3xy + y^2$
16. $f(x, y) = \dfrac{x^4 - y^4}{x^2 + y^2}$

In Exercises 17–24 prove that $\lim_{(x,y)\to(0,0)} f(x, y)$ does not exist. In each case the domain of f is all of $\mathbf{R}^2$ except the origin.

17. $f(x, y) = \dfrac{x^2 - y^2}{x^2 + y^2}$
18. $f(x, y) = \dfrac{2x^2 + 3y^2}{x^2 + y^2}$
19. $f(x, y) = \dfrac{xy}{x^2 + y^2}$
20. $f(x, y) = \dfrac{x + y}{x^2 + y^2}$
21. $f(x, y) = \dfrac{3x^2 - 4y^2}{4x^2 + y^2}$
22. $f(x, y) = \dfrac{x^2 - xy + y^2}{x^2 + y^2}$
23. $f(x, y) = \dfrac{xy^2}{x^2 + y^4}$
24. $f(x, y) = \dfrac{x^3y}{x^6 + y^2}$

In Exercises 25–36 determine the largest subset of $\mathbf{R}^2$ on which f is continuous.

25. $f(x, y) = 2x^2 - 3xy + 7y^3$
26. $f(x, y) = \dfrac{x^2 - y^2}{x^2 + y^2}$
27. $f(x, y) = \dfrac{2x - 3y}{x^2 + y^2 + 1}$
28. $f(x, y) = \dfrac{x + y}{x - y}$
29. $f(x, y) = \dfrac{2x - 3y}{x^2 + y^2 - 1}$
30. $f(x, y) = \dfrac{x^2 + y^2}{x^2 - y^2}$
31. $f(x, y) = \dfrac{2x^2 - 3xy + y^2}{x^2 - y + 1}$
32. $f(x, y) = e^{2xy}$
33. $f(x, y) = e^{-x/y}$
34. $f(x, y) = \ln \sqrt{x^2 + y^2}$
35. $f(x, y) = \cos(x^2 - 2xy)$
36. $f(x, y) = \tan^{-1} \dfrac{y}{x}$

In Exercises 37–44 use Definition 16.3 to prove that f is continuous at $P_0 = (0, 0)$.

37. $f(x, y) = 3x + 2y + 4$
38. $f(x, y) = x^2 + 2y^2 - 1$
39. $f(x, y) = \dfrac{x^2 + y^2}{\sqrt{x^2 + y^2} + 1}$
40. $f(x, y) = \dfrac{x^2 - y^2}{x^2 + 4}$
41. $f(x, y) = \begin{cases} \dfrac{x^4 - y^4}{x^2 + y^2} & \text{if } (x, y) \neq (0, 0) \\ 0 & \text{if } (x, y) = (0, 0) \end{cases}$
42. $f(x, y) = \begin{cases} \dfrac{xy}{\sqrt{x^2 + y^2}} & \text{if } (x, y) \neq (0, 0) \\ 0 & \text{if } (x, y) = (0, 0) \end{cases}$

43. $f(x, y) = \begin{cases} \dfrac{xy(x+y)}{x^2+y^2} & \text{if } (x, y) \neq (0, 0) \\ 0 & \text{if } (x, y) = (0, 0) \end{cases}$

44. $f(x, y) = \begin{cases} \dfrac{x^3 - y^3}{x^2 + xy + y^2} & \text{if } (x, y) \neq (0, 0) \\ 0 & \text{if } (x, y) = (0, 0) \end{cases}$

B

45. Prove that $\lim_{(x,y)\to(0,0)} \sqrt{x^2+y^2+1} = 1$.

In Exercises 46 and 47 determine whether f is continuous at (0, 0). Prove your result.

46. $f(x, y) = \begin{cases} \dfrac{2xy^2 - 3x^2y}{3x^2+2y^2} & \text{if } (x, y) \neq (0, 0) \\ 0 & \text{if } (x, y) = (0, 0) \end{cases}$

47. $f(x, y) = \begin{cases} \dfrac{x^2 + 2xy + y^2}{x^2+y^2} & \text{if } (x, y) \neq (0, 0) \\ 1 & \text{if } (x, y) = (0, 0) \end{cases}$

In Exercises 48–53 assume f and g are both functions of two variables or else both functions of three variables, c is a constant, $\lim_{P\to P_0} f(P) = L$ and $\lim_{P\to P_0} g(P) = M$. Use Definition 16.2 to prove the given property. (See Section 1.8 for ideas on proofs.)

48. $\lim_{P\to P_0} c = c$
49. $\lim_{P\to P_0} cf(P) = cL$
50. $\lim_{P\to P_0} (f+g) = L + M$
51. $\lim_{P\to P_0} (f-g)(P) = L - M$
52. $\lim_{P\to P_0} (fg)(P) = LM$
53. $\lim_{P\to P_0} \frac{f}{g}(P) = \frac{L}{M}$, if $M \neq 0$
54. Prove that if f and g are both functions of two variables or else both functions of three variables, each of which is continuous at P_0, then each of the following is also continuous at P_0.
(a) cf (b) $f \pm g$
(c) fg (d) $\frac{f}{g}$ if $g(P_0) \neq 0$
(*Hint:* Use the results of Exercises 48–53.)

55. (a) Prove that $f(x, y) = x$ and $g(x, y) = y$ are continuous at all points (x_0, y_0) in $\mathbf{R}^2$.
(b) Prove that $f(x, y) = cx^m y^n$ is continuous on $\mathbf{R}^2$, where c is a constant and m and n are nonnegative integers.

56. (a) Using the results of Exercises 54 and 55, prove that all polynomial functions in two variables are continuous on all of $\mathbf{R}^2$.
(b) Prove that every rational function of two variables is continuous at all points of $\mathbf{R}^2$ except where the denominator is 0.

57. Let $f(x, y) = \begin{cases} \dfrac{xy}{x^2+y^2} & \text{if } (x, y) \neq (0, 0) \\ 0 & \text{if } (x, y) = (0, 0) \end{cases}$

Prove that $f_x(0, 0)$ and $f_y(0, 0)$ both exist, but that f is discontinuous at (0, 0).

58. Let f have domain D in $\mathbf{R}^2$ or $\mathbf{R}^3$. A boundary point P_0 of D is said to be **isolated** if there is some neighborhood of P_0 that contains no point of D other than P_0 itself. On the basis of Definition 16.3 show that every function f is continuous at an isolated boundary point of its domain.

16.6 SUPPLEMENTARY EXERCISES

1. Give the domain of f and show it graphically.
(a) $f(x, y) = \ln(x^2 + y^2 - 1)$
(b) $f(x, y) = xy/\sqrt{x^2 - y^2}$

In Exercises 2–5 identify the surface and draw its graph.

2. (a) $4x^2 + y^2 + 4z = 8$ (b) $y = \sqrt{x^2 + z^2}$
3. (a) $y^2 - 2x - 4y + 4 = 0$
(b) $x^2 + y^2 + z^2 - 6x - 4y - 2z + 10 = 0$
4. (a) $x^2 + 4z^2 = 4$ (b) $9(x^2 + z^2) = 4(y^2 + 9)$
5. (a) $x^2 - 4y^2 + 4z = 0$ (b) $z = 1 + \tan^{-1} y$
6. Show the region in the first octant bounded by the given surfaces.
(a) $4x^2 + 3y^2 + 6z^2 = 48$, $x = 3$, $y = 2$, $z = 0$
(b) $4x^2 + y^2 = 4$, $z = 4 - y^2$, and the coordinate planes
7. Give the domain and range of f, identify its graph, and draw it.
(a) $f(x, y) = \sqrt{4 + x^2 + 4y^2}$
(b) $f(x, y) = 4 - x^2 - y^2$
8. Draw several level curves in part (a) and level surfaces in part (b) for both positive and negative values of the constant c.
(a) $f(x, y) = y^2 - x^2$ (b) $f(x, y, z) = \dfrac{x^2 + y^2}{z}$

9. A coordinate system is set up on a round metal plate with the origin at its center. A point source of heat is at the origin, and the temperature $T(x, y)$ in degrees Celsius at any point on the plate is given by

$$T(x, y) = 100e^{-(x^2+y^2)/2}$$

Draw the isothermals $T(x, y) = C$ for $C = 80, 60, 40$, and 20. How rapidly is the temperature changing in the direction of the positive y-axis at the point $(1, 3)$? Is it increasing or decreasing?

In Exercises 10 and 11 find f_x and f_y.

10. $f(x, y) = e^{\sin xy}$ (b) $f(x, y) = \ln \dfrac{x^2 + 2y^2}{\sqrt{3x + 4y}}$

11. (a) $f(x, y) = \dfrac{y}{x} \cosh^2 \dfrac{x}{y}$

(b) $f(x, y) = \sin^{-1}\left(\dfrac{\sqrt{y^2 - 2x}}{y}\right)$, $y > 0$

12. Let $w = (2u - v)/(u + 2v)$. Show that

$$u\frac{\partial w}{\partial u} + v\frac{\partial w}{\partial v} = 0$$

13. Show that $f(x, y) = \sin x \cosh y$ and $g(x, y) = \cos x \sinh y$ are harmonic conjugates. (See the definitions preceding Exercises 33 and 37 in Exercise Sct 16.4.)

14. Let $f(x, y) = xy \ln \frac{x}{y}$. What is the domain of f? Find all second-order partials and show that $f_{xy} = f_{yx}$.

15. Let $w = \sqrt{z^2 - x^2 - y^2}$. Find each of the following:

(a) $\dfrac{\partial^2 w}{\partial z^2}$ (b) $\dfrac{\partial^2 w}{\partial x \partial z}$ (c) $\dfrac{\partial^2 w}{\partial y \partial z}$ (d) $\dfrac{\partial^2 w}{\partial y \partial x}$

16. Show that

$$\lim_{(x,y)\to(0,0)} \frac{\sin xy}{\sqrt{x^2 + y^2}} = 0$$

using Definition 16.2. (*Hint:* $|\sin \theta| \le |\theta|$.)

17. Let $f(x, y) = (x^3 - 3x^2y)/(x^2 + 2y^2)$ if $(x, y) \ne (0, 0)$. What value should be assigned to $f(0, 0)$ to make f continuous at the origin? Prove your result.

18. For $(x, y) \ne (0, 0)$ let

$$f(x, y) = (x^2 + y^2)\left(\sin \frac{1}{x^2 + y^2} + \cos \frac{1}{x^2 + y^2}\right)$$

and let $f(0, 0) = 0$. Use Definition 16.1 to show that $f_x(0, 0)$ and $f_y(0, 0)$ both exist and equal 0.

19. Use Definition 16.2 to prove the following:

(a) $\lim_{(x,y)\to(0,0)} (3x - 2y + 4) = 4$

(b) $\lim_{(x,y)\to(0,0)} \dfrac{x^3 - y^3}{x - y} = 0$

20. Prove that $\lim_{(x,y)\to(0,0)} f(x, y)$ does not exist.

(a) $f(x, y) = \dfrac{x^3 - 2y^3}{3x^3 + 4y^3}$ (b) $f(x, y) = \dfrac{3xy^2}{2x^2 + 5y^4}$

21. Let

$$f(x, y) = \begin{cases} \dfrac{2x^3 - y^3}{x^2 + 2y^2} & \text{if } (x, y) \ne (0, 0) \\ 0 & \text{if } (x, y) = (0, 0) \end{cases}$$

(a) Use Definition 16.1 to find $f_x(0, 0)$.
(b) Find $f_x(x, y)$ for $(x, y) \ne (0, 0)$ using the quotient rule.
(c) Show that f_x is not continuous at the origin.
(d) Carry out steps a, b, and c for f_y.

22. Let

$$f(x, y) = \begin{cases} \dfrac{x^3y - xy^3}{x^2 + y^2} & \text{if } (x, y) \ne (0, 0) \\ 0 & \text{if } (x, y) = (0, 0) \end{cases}$$

(a) Prove that f is continuous at $(0, 0)$.
(b) Find $f_x(0, y)$ and $f_y(x, 0)$ using Definition 16.1.
(c) From part b find $f_{xy}(0, 0)$ and $f_{yx}(0, 0)$.
(d) What conclusion about the continuity of f_{xy} and f_{yx} can you draw from part c?

23. Prove that a nonempty set S in $\mathbf{R}^2$ or $\mathbf{R}^3$ is open if and only if, for every point P in S, some neighborhood of P lies altogether in S.

24. If S is a set, its **complement** is the set of all points not in S. Prove that a set S in $\mathbf{R}^2$ or $\mathbf{R}^3$ is closed if and only if its complement is open.

CHAPTER 17

MULTIVARIATE DIFFERENTIAL CALCULUS

17.1 DIFFERENTIABILITY

In the case of a function of one variable we define differentiability to mean the existence of the derivative. There is no simple analogue of this concept for functions of more than one variable, since there is no single derivative. The existence of the first partial derivatives turns out not to be a satisfactory criterion, since among other things we want differentiability to imply continuity, and as Exercise 57 in Exercise Set 16.5 shows, a discontinuous function can still have partial derivatives. To extend the notion of differentiability to a function of two or more variables, we first reformulate its meaning for a function of one variable.

If $f'(x)$ exists, then

$$\lim_{\Delta x \to 0} \frac{f(x + \Delta x) - f(x)}{\Delta x} = f'(x) \tag{17.1}$$

Let $\Delta f = f(x + \Delta x) - f(x)$. Then by equation (17.1), the difference $\frac{\Delta f}{\Delta x} - f'(x)$ approaches 0 as $\Delta x \to 0$. We denote this difference by ε, so that

$$\frac{\Delta f}{\Delta x} - f'(x) = \varepsilon$$

or, equivalently,

$$\Delta f = f'(x)\,\Delta x + \varepsilon\,\Delta x \tag{17.2}$$

where $\varepsilon \to 0$ as $\Delta x \to 0$. The differentiability of a function of one variable can thus be characterized by equation (17.2). In Section 2.9 we defined the differential df to be the first term on the right-hand side of equation (17.2) and showed that for Δx small, the true change in f, Δf, is approximated by df, since the error term $\varepsilon\,\Delta x$ approaches 0 more rapidly than Δx itself approaches 0. So we have

$$df = f'(x)\,\Delta x \tag{17.3}$$

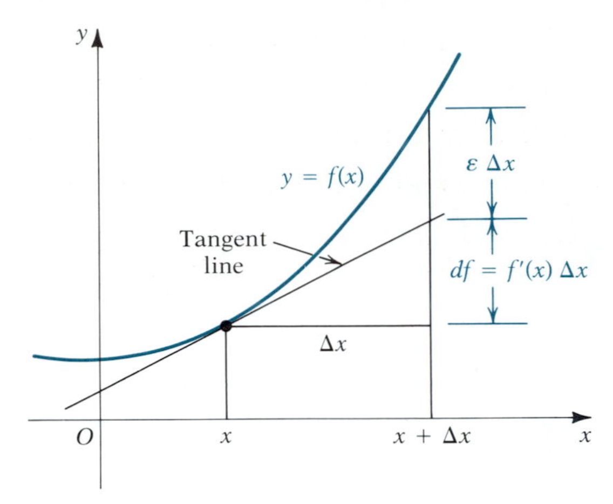

FIGURE 17.1

and

$$\Delta f = df + \varepsilon\,\Delta x \tag{17.4}$$

Recall also the geometric interpretation of equation (17.4) as shown in Figure 17.1.

We will use equation (17.2) as a basis for defining differentiability for a function of two variables, and we will define the differential by analogy with equation (17.3). In Section 17.4 we will see that a geometric interpretation similar to that shown in Figure 17.1 also holds, in which the tangent line to the curve $y = f(x)$ is replaced by the *tangent plane* to the surface $z = f(x, y)$.

DEFINITION 17.1 Let f be a function of two variables defined throughout some neighborhood N of the point $P = (x, y)$ and suppose f_x and f_y both exist in N. Let Δx and Δy be independent variables subject to the restriction that $(x + \Delta x, y + \Delta y)$ lies in N. Write

$$\Delta f = f(x + \Delta x, y + \Delta y) - f(x, y) \tag{17.5}$$

Then f is said to be **differentiable** at P provided there exist functions ε_1 and ε_2 of Δx and Δy such that

$$\Delta f = f_x(P)\,\Delta x + f_y(P)\,\Delta y + \varepsilon_1\,\Delta x + \varepsilon_2\,\Delta y \tag{17.6}$$

where $\varepsilon_1 \to 0$ and $\varepsilon_2 \to 0$ as $(\Delta x, \Delta y) \to (0, 0)$. ■

Before defining the differential, we will illustrate a differentiable function and also show an important consequence of differentiability.

EXAMPLE 17.1 Show that the function f defined by $f(x, y) = x^2 - 2xy + 3y^2$ is differentiable everywhere in $\mathbf{R}^2$.

Solution Let $P = (x, y)$ be any point in $\mathbf{R}^2$. Then for Δx and Δy arbitrary, we have by equation (17.5)

$$\begin{aligned}\Delta f &= f(x + \Delta x, y + \Delta y) - f(x, y)\\ &= (x + \Delta x)^2 - 2(x + \Delta x)(y + \Delta y) + 3(y + \Delta y)^2 - x^2 + 2xy - 3y^2\\ &= (2x - 2y)\,\Delta x + (-2x + 6y)\,\Delta y + (\Delta x - \Delta y)\,\Delta x + (3\,\Delta y - \Delta x)\,\Delta y\end{aligned}$$

(You should verify the last step.) Now

$$f_x(x, y) = 2x - 2y \quad \text{and} \quad f_y(x, y) = -2x + 6y$$

So if we define $\varepsilon_1 = \Delta x - \Delta y$ and $\varepsilon_2 = 3\,\Delta y - \Delta x$, we have

$$\Delta f = f_x(x, y)\,\Delta x + f_y(x, y)\,\Delta y + \varepsilon_1\,\Delta x + \varepsilon_2\,\Delta y$$

and furthermore, $\varepsilon_1 \to 0$ and $\varepsilon_2 \to 0$ as $(\Delta x, \Delta y) \to (0, 0)$. Thus, f is differentiable at P. Since P is any point in $\mathbf{R}^2$, the solution is complete. ■

THEOREM 17.1 If f is differentiable at $P_0 = (x_0, y_0)$, then f is continuous at P_0.

Proof We must show that $\lim_{P \to P_0} f(P) = f(P_0)$ or, equivalently,

$$\lim_{P \to P_0} [f(P) - f(P_0)] = 0$$

Let $P = (x, y)$ and let $\Delta x = x - x_0$ and $\Delta y = y - y_0$. Then we can write P in the form $(x_0 + \Delta x, y_0 + \Delta y)$. Furthermore, $P \to P_0$ if and only if $(\Delta x, \Delta y) \to (0, 0)$. Thus,

$$\begin{aligned}\lim_{P \to P_0} [f(P) - f(P_0)] &= \lim_{(\Delta x, \Delta y) \to (0,0)} [f(x_0 + \Delta x, y_0 + \Delta y) - f(x_0, y_0)] \\ &= \lim_{(\Delta x, \Delta y) \to (0,0)} \Delta f\end{aligned}$$

Since f is differentiable, this becomes, by equation (17.6),

$$\lim_{(\Delta x, \Delta y) \to (0,0)} [f_x(x_0, y_0)\,\Delta x + f_y(x_0, y_0)\,\Delta y + \varepsilon_1\,\Delta x + \varepsilon_2\,\Delta y] = 0$$

and this completes the proof. ■

Remark This theorem shows that differentiability implies continuity, just as with single-valued functions. The converse is not true because there are continuous functions that fail to be differentiable.

If f is differentiable at $P = (x, y)$, we define the **differential** df of f to be $f_x(x, y)\,\Delta x + f_y(x, y)\,\Delta y$. It is customary to replace Δx by dx and Δy by dy in this context, remembering that these are independent variables. So we have for the differentiable function f at (x, y):

$$df = f_x(x, y)\,dx + f_y(x, y)\,dy \tag{17.7}$$

Observe that when f is differentiable, equation (17.5) can be written

$$\Delta f = df + \varepsilon_1\,\Delta x + \varepsilon_2\,\Delta y \tag{17.8}$$

and since ε_1 and ε_2 both approach 0 as $(\Delta x, \Delta y) \to (0, 0)$, we see that $\Delta f \approx df$ for Δx and Δy small in absolute value. Since $\Delta f = f(x + \Delta x, y + \Delta y) - f(x, y)$, we may rewrite this approximation in the form

$$f(x + \Delta x, y + \Delta y) \approx f(x, y) + f_x(x, y)\,\Delta x + f_y(x, y)\,\Delta y$$

We refer to this as *linear approximation*, analogous to the one-variable case. We will see the geometric significance in Section 17.4.

When we introduce a dependent variable for $f(x, y)$, say $z = f(x, y)$, in equations (17.5)–(17.8) we can replace the symbols Δf and df by Δz and dz, respectively.

The following example illustrates one application of the differential as an approximation to the true change in a function.

EXAMPLE 17.2 A can is in the shape of a right circular cylinder. The base radius is measured to be 4.5 cm and the height is measured to be 10.2 cm. If the error in measuring the radius is at most 0.05 cm and the error in measuring the height is at most 0.02 cm, find the approximate maximum error in the total surface area, using differentials. Also compute the actual maximum error and compare results.

Solution The total surface area is given by

$$S = 2\pi r^2 + 2\pi rh \qquad \text{(Verify.)}$$

It can easily be verified that S is a differentiable function of r and h. So with appropriate modifications in variable names, we have from equation (17.7)

$$\begin{aligned} dS &= \frac{\partial S}{\partial r}\,dr + \frac{\partial S}{\partial h}\,dh \\ &= (4\pi r + 2\pi h)\,dr + 2\pi r\,dh \end{aligned}$$

We take r and h as the measured values and dr and dh as their maximum errors. So $r = 4.5$, $h = 10.2$, $dr = 0.05$, and $dh = 0.02$. Thus,

$$\begin{aligned} dS &= [(4\pi)(4.5) + (2\pi)(10.2)](0.05) + (2\pi)(4.5)(0.02) \\ &= 2.10\pi \approx 6.60 \end{aligned}$$

For the actual maximum error ΔS in surface area, we have by equation (17.5)

$$\begin{aligned} \Delta S &= S(r + \Delta r, h + \Delta h) - S(r, h) \\ &= [2\pi(4.55)^2 + 2\pi(4.55)(10.22)] - [2\pi(4.5)^2 + 2\pi(4.5)(10.2)] \\ &= 2\pi[20.7025 + 46.5010] - 2\pi[20.25 + 45.90] \\ &= 2.107\pi \approx 6.62 \end{aligned}$$

The approximation $dS \approx \Delta S$ in this case is quite good. ■

The existence of the first partial derivatives of a function in a neighborhood of a point does not guarantee its differentiability there. However, if these partials are also *continuous* at the point, the function is differentiable there, as the next theorem states.

THEOREM 17.2 Let f be a function of two variables with P_0 an interior point of its domain. If f_x and f_y exist throughout some neighborhood N of P_0 and are continuous at P_0, then f is differentiable at P_0.

Proof Let $P_0 = (x_0, y_0)$ and choose Δx and Δy such that $(x_0 + \Delta x, y_0 + \Delta y)$ lies in the neighborhood N. Then we can write

$$\begin{aligned} \Delta f &= f(x_0 + \Delta x, y_0 + \Delta y) - f(x_0, y_0) \\ &= [f(x_0 + \Delta x, y_0 + \Delta y) - f(x_0, y_0 + \Delta y)] + [f(x_0, y_0 + \Delta y) - f(x_0, y_0)] \end{aligned} \tag{17.9}$$

where we have added and subtracted the term $f(x_0, y_0 + \Delta y)$ on the right. In the first bracket, y is held fixed at $y_0 + \Delta y$, so that only x is changing. Since $f_x(x, y_0 + \Delta y)$ exists for all x on the closed interval between x_0 and $x_0 + \Delta x$, the mean value theorem is applicable there, and so a number c exists between

x_0 and $x_0 + \Delta x$ such that

$$f(x_0 + \Delta x, y_0 + \Delta y) - f(x_0, y_0 + \Delta y) = f_x(c, y_0 + \Delta y)\,\Delta x \tag{17.10}$$

Similarly, in the second bracket on the right-hand side of equation (17.9), x is held fixed at x_0, and $f_y(x_0, y)$ exists for all y in the closed interval from y_0 to $y_0 + \Delta y$. Thus, a number d exists between y_0 and $y_0 + \Delta y$ such that

$$f(x_0, y_0 + \Delta y) - f(x_0, y_0) = f_y(x_0, d)\,\Delta y \tag{17.11}$$

Now if we let Δx and Δy each approach 0, the points $(c, y_0 + \Delta y)$ and (x_0, d) both approach (x_0, y_0), and since f_x and f_y are continuous at (x_0, y_0), it follows that

$$\lim_{(\Delta x, \Delta y) \to (0,0)} f_x(c, y_0 + \Delta y) = f_x(x_0, y_0)$$

and

$$\lim_{(\Delta x, \Delta y) \to (0,0)} f_y(x_0, d) = f_y(x_0, y_0)$$

So if we write

$$\varepsilon_1 = f_x(c, y_0 + \Delta y) - f_x(x_0, y_0)$$

and

$$\varepsilon_2 = f_y(x_0, d) - f_y(x_0, y_0)$$

then $\varepsilon_1 \to 0$ and $\varepsilon_2 \to 0$ as $(\Delta x, \Delta y) \to (0, 0)$. Replacing $f_x(c, y_0 + \Delta y)$ in equation (17.10) by $f_x(x_0, y_0) + \varepsilon_1$ and $f_y(x_0, d)$ in equation (17.11) by $f_y(x_0, y_0) + \varepsilon_2$, we get, on substituting for the two bracketed terms in equation (17.9),

$$\Delta f = f_x(x_0, y_0)\,\Delta x + f_y(x_0, y_0)\,\Delta y + \varepsilon_1\,\Delta x + \varepsilon_2\,\Delta y$$

Thus, by definition, f is differentiable at $P_0 = (x_0, y_0)$. ■

EXAMPLE 17.3 Prove that if $f(x, y) = \tan^{-1}\frac{y}{x}$, then f is differentiable at all points (x, y) with $x \neq 0$.

Solution We note first that all points for which $x \neq 0$ lie in the domain of f. If we can show that f_x and f_y are continuous throughout this domain, then Theorem 17.2 will guarantee differentiability:

$$f_x(x, y) = \frac{1}{1 + \left(\dfrac{y}{x}\right)^2} \cdot \frac{-y}{x^2} = \frac{-y}{x^2 + y^2}$$

$$f_y(x, y) = \frac{1}{1 + \left(\dfrac{y}{x}\right)^2} \cdot \frac{1}{x} = \frac{x}{x^2 + y^2}$$

Since $x \neq 0$, the denominator $x^2 + y^2 \neq 0$ in each case. Both $f_x(x, y)$ and $f_y(x, y)$ are rational functions with nonzero denominators. Thus, each is continuous for all (x, y) with $x \neq 0$. It now follows that f is differentiable at all such points. ■

The notions of differentiability and the differential extend in natural ways to functions of more than two variables. For three variables, for example, we say that a function f is differentiable at $P = (x, y, z)$ if the domain of f includes a neighborhood of P; f_x, f_y, and f_z exist at P; and the change

$$\Delta f = f(x + \Delta x, y + \Delta y, z + \Delta z) - f(x, y, z)$$

can be written in the form

$$\Delta f = f_x(x, y, z)\,\Delta x + f_y(x, y, z)\,\Delta y + f_z(x, y, z)\,\Delta z + \varepsilon_1\,\Delta x + \varepsilon_2\,\Delta y + \varepsilon_3\,\Delta z$$

where ε_1, ε_2, and ε_3 are functions of $(\Delta x, \Delta y, \Delta z)$ that approach 0 as $(\Delta x, \Delta y, \Delta z) \to (0, 0, 0)$. If f is differentiable at P, then the differential of f is defined as

$$df = f_x(x, y, z)\,dx + f_y(x, y, z)\,dy + f_z(x, y, z)\,dz$$

where $dx = \Delta x$, $dy = \Delta y$, and $dz = \Delta z$ are independent variables.

EXERCISE SET 17.1

A

In Exercises 1–8 use Definition 17.1 to verify that f is differentiable at every point in $\mathbf{R}^2$.

1. $f(x, y) = 2x + 3y$
2. $f(x, y) = xy$
3. $f(x, y) = x^2 + 3y^2$
4. $f(x, y) = x^2 - y^2$
5. $f(x, y) = 2x^2y$
6. $f(x, y) = x^2 - 3xy + 2y^2$
7. $f(x, y) = (x - 3y)^2$
8. $f(x, y) = x^3 - y^3$

In Exercises 9–20 find df.

9. $f(x, y) = 2x^2 - 3xy + y^3$
10. $f(x, y) = x^4y^2 - 2xy^3$
11. $f(x, y) = e^{xy}$
12. $f(x, y) = \ln\sqrt{x^2 + y^2}$
13. $f(x, y) = \dfrac{x - y}{x + y}$
14. $f(x, y) = x \sin y + y \cos x$
15. $f(x, y) = \tan^{-1}\dfrac{x}{y}$
16. $f(x, y) = \ln\dfrac{2x^2y}{x + y}$
17. $f(x, y) = \ln\cos 2xy$
18. $f(x, y) = y \sin^{-1} x$
19. $f(x, y, z) = x^2 + y^2 + z^2$
20. $f(x, y, z) = x^2y^3z^4$

In Exercises 21–26 use differentials to approximate Δf.

21. $f(x, y) = 2x^3 - 3x^2y + y^2$; $(x, y) = (2, -1)$, $\Delta x = 0.02$, $\Delta y = 0.01$
22. $f(x, y) = \sqrt{\dfrac{2x - y}{x + 3y}}$; $(x, y) = (4, -1)$, $\Delta x = -0.03$, $\Delta y = 0.05$
23. $f(x, y) = \ln(x - y)^2$ from $(3, -2)$ to $(3.04, -1.99)$
24. $f(x, y) = \tan^{-1}\frac{y}{x}$ from $(-3, 4)$ to $(-3.02, 3.98)$
25. $f(s, t) = \ln\sqrt{s^2 + t^2}$ from $(-4, 3)$ to $(-3.98, 2.99)$
26. $f(x, y, z) = \dfrac{1}{x^2 + y^2 + z^2}$ from $(1, -1, 2)$ to $(0.97, -1.01, 2.05)$
27. The dimensions of a room are measured to be as follows: length 21 ft, width 12 ft, and height 8 ft. If each measurement is accurate only to the nearest tenth of a foot, find the approximate maximum error in volume calculated from the measured values, making use of differentials.
28. In Example 17.2 find the approximate maximum error in volume, making use of differentials, and compare this with the actual maximum error.
29. Suppose the temperature at a point in a thin square metal sheet is given by
$$T(x, y) = \frac{x^2y^2}{x^2 + y^2}$$
where the origin is at one corner. Using differentials, find the approximate difference in temperature at the points $(1.15, 2.05)$ and $(1, 2)$. Compare your answer with the actual change.
30. A company manufactures and sells two models of automatic ice cream makers, the standard model A and the deluxe model B. Through an analysis of sales over a period of time, it is found that the weekly profit from producing and selling x model A and y model B machines is approximately
$$P(x, y) = 200x + 300y - 0.2x^2 - 0.3y^2 - 0.4xy - 50$$

where P is in dollars. On average, 50 model A and 20 model B machines are sold each week. Use differentials to approximate the effect on profit of selling one fewer model A and two more model B machines per week than the average.

31. The cost of producing one unit of a certain manufactured item is given by

$$C(x, y) = 2x^2 + 3xy + y^2 + 15x + 6y + 50$$

where x is the hourly wage rate for the workers and y is the cost per pound of raw materials. Currently the hourly wage is \$9.50 and raw materials cost \$6.00 per pound. Use differentials to estimate the increase in cost if the labor cost goes up by \$0.50 per hour and the material cost increases by \$0.40 per pound.

In Exercises 32–35 use Theorem 17.2 to show that f is differentiable in the specified region.

32. $f(x, y) = \dfrac{x+y}{x-y}$; $x \neq y$

33. $f(x, y) = e^x \sin y$; all of $\mathbf{R}^2$

34. $f(x, y) = \sin^{-1} \dfrac{y}{x}$; $x^2 > y^2$

35. $f(x, y) = \ln \dfrac{xy}{x^2 + y^2}$; $xy > 0$

B

36. Use Definition 17.1 to verify that f is differentiable on all of $\mathbf{R}^2$, where $f(x, y) = (x + y)^3$.

37. Use the extension of Definition 17.1 to three variables to show that f is differentiable on all of $\mathbf{R}^3$, where $f(x, y, z) = xy - 2yz + 3xz - 4xyz$.

38. Let

$$f(x, y) = \begin{cases} \dfrac{xy}{x^2 + y^2} & \text{if } (x, y) \neq (0, 0) \\ 0 & \text{if } (x, y) = (0, 0) \end{cases}$$

Prove that $f_x(0, 0)$ and $f_y(0, 0)$ both exist but that f is not differentiable at the origin. What can you conclude about the continuity of f_x and f_y at the origin? (*Hint:* Make use of Theorems 17.1 and 17.2.)

39. Prove that $f(x, y) = \sqrt{x^2 + y^2}$ is not differentiable at the origin. (*Hint:* Show that neither f_x nor f_y exists there.)

40. Let

$$f(x, y) = \begin{cases} \dfrac{x^2y^2}{x^2 + y^2} & \text{if } (x, y) \neq (0, 0) \\ 0 & \text{if } (x, y) = (0, 0) \end{cases}$$

Prove that f is differentiable in all of $\mathbf{R}^2$. (*Hint:* Use Theorem 17.2, paying special attention to the origin.)

41. Let $f(x, y) = \sqrt{|x, y|}$. Prove the following:
(a) f is continuous at $(0, 0)$
(b) $f_x(0, 0) = f_y(0, 0) = 0$
(c) f is not differentiable at $(0, 0)$
[*Hint:* Show that equation (17.6) becomes $\sqrt{|\Delta x \Delta y|} = \varepsilon_1 \Delta x + \varepsilon_2 \Delta y$, and by taking $\Delta x = \Delta y$ conclude that ε_1 and ε_2 do not approach 0 as $(\Delta x, \Delta y) \to (0, 0)$.]

17.2 CHAIN RULES

For a function f of one variable x, if x in turn is a function of t, say $x = g(t)$, then the chain rule for differentiation can be written

$$(f \circ g)'(t) = f'(g(t))g'(t) \tag{17.12}$$

This holds true at all points t in the domain of g at which g' exists and for which $f'(g(t))$ exists. When we introduce the dependent variable $y = f(x)$, equation (17.12) becomes, in the Leibniz notation,

$$\frac{dy}{dt} = \frac{dy}{dx}\frac{dx}{dt} \tag{17.13}$$

This is the familiar form of the chain rule. Care should be taken to distinguish between $\frac{dy}{dt}$ and $\frac{dy}{dx}$. On the left, it is understood that x is first replaced by

$g(t)$ and then the derivative is taken, so that

$$\frac{dy}{dt} \quad \text{means} \quad \frac{d}{dt} f(g(t))$$

On the right, $\frac{dy}{dx}$ means first calculate $f'(x)$ and then replace x by $g(t)$, so that

$$\frac{dy}{dx} \quad \text{means} \quad f'(x)\Big|_{x=g(t)} \quad \text{or } f'(g(t))$$

It is convenient in this context to call x the *intermediate variable* and t the *final variable*.

The form of the extension of equation (17.12) or (17.13) to two or more variables depends on the number of intermediate variables and final variables. We begin with the simplest case, in which $z = f(x, y)$, $x = g(t)$, and $y = h(t)$. Here x and y are intermediate variables and t is the single final variable. If we first substitute for x and y, we get $z = f(g(t), h(t))$, so that z is finally a function of the single variable t. We want to find a formula for $\frac{dz}{dt}$. The secret lies in the definition of differentiability. We assume that $g'(t)$ and $h'(t)$ both exist and that f is differentiable at $P = (g(t), h(t))$. Let

$$\Delta x = g(t + \Delta t) - g(t)$$
$$\Delta y = h(t + \Delta t) - h(t)$$

Then because g and h are continuous at t (why?), Δx and Δy both approach 0 as $\Delta t \to 0$. Now, by Definition 17.1 we can write

$$\Delta z = f_x(P)\,\Delta x + f_y(P)\,\Delta y + \varepsilon_1\,\Delta x + \varepsilon_2\,\Delta y \tag{17.14}$$

where $\varepsilon_1 \to 0$ and $\varepsilon_2 \to 0$ as $(\Delta x, \Delta y) \to 0$ and hence also as $\Delta t \to 0$. Here Δz means

$$\begin{aligned} \Delta z = \Delta f &= f(x + \Delta x, y + \Delta y) - f(x, y) \\ &= f(g(t + \Delta t), h(t + \Delta t)) - f(g(t), h(t)) \end{aligned}$$

Dividing equation (17.14) on both sides by Δt, we get

$$\frac{\Delta z}{\Delta t} = f_x(P)\frac{\Delta x}{\Delta t} + f_y(P)\frac{\Delta y}{\Delta t} + \varepsilon_1\frac{\Delta x}{\Delta t} + \varepsilon_2\frac{\Delta y}{\Delta t}$$

Now we let $\Delta t \to 0$ to obtain the result

$$\frac{d}{dt} f(P) = f_x(P)g'(t) + f_y(P)h'(t)$$

or, in Leibniz notation, since $z = f(x, y)$,

$$\frac{dz}{dt} = \frac{\partial z}{\partial x}\frac{dx}{dt} + \frac{\partial z}{\partial y}\frac{dy}{dt} \tag{17.15}$$

It is understood that on the right-hand side, $\partial z/\partial x$ and $\partial z/\partial y$ are to be evaluated at the point $P = (g(t), h(t))$.

EXAMPLE 17.4 Let

$$z = \frac{x - 2y}{2x + 3y} \quad \text{and} \quad \begin{cases} x = 2t - 3 \\ y = t^2 + 1 \end{cases}$$

Use the chain rule to find $\frac{dz}{dt}$ when $t = -1$.

Solution By equation (17.15),

$$\frac{dz}{dt} = \frac{\partial z}{\partial x}\frac{dx}{dt} + \frac{\partial z}{\partial y}\frac{dy}{dt}$$
$$= \frac{7y}{(2x+3y)^2}(2) + \frac{-7x}{(2x+3y)^2}(2t) \qquad \text{(Verify.)}$$

When $t = -1$, $x = -5$, and $y = 2$,

$$\left.\frac{dz}{dt}\right|_{t=-1} = \frac{14}{(-4)^2}(2) + \frac{(35)}{(-4)^2}(-2) = -\frac{21}{8}$$

■

When the number of final variables is greater than one, the derivation goes much the same way, but the results are partial derivatives instead of total derivatives. For example, if

$$z = f(x, y) \quad \text{and} \quad \begin{cases} x = g(u, v) \\ y = h(u, v) \end{cases}$$

then z is a function of the two final variables u and v, and the chain rule provides a way of calculating $\partial z/\partial u$ and $\partial z/\partial v$. The results are:

$$\begin{aligned} \frac{\partial z}{\partial u} &= \frac{\partial z}{\partial x}\frac{\partial x}{\partial u} + \frac{\partial z}{\partial y}\frac{\partial y}{\partial u} \\ \frac{\partial z}{\partial v} &= \frac{\partial z}{\partial x}\frac{\partial x}{\partial v} + \frac{\partial z}{\partial y}\frac{\partial y}{\partial v} \end{aligned} \tag{17.16}$$

Each of these is obtained from equation (17.14) for Δz, where in the one case we divide by Δu throughout and then let $\Delta u \to 0$, and in the other case we do the same with Δv. We omit the details.

EXAMPLE 17.5 If $z = x^2 - 2xy + 3y^2$, $x = uv$, and $y = u^2 - v^2$, find $\partial z/\partial u$ and $\partial z/\partial v$.

Solution By equations (17.16),

$$\begin{aligned} \frac{\partial z}{\partial u} &= \frac{\partial z}{\partial x}\frac{\partial x}{\partial u} + \frac{\partial z}{\partial y}\frac{\partial y}{\partial u} = (2x - 2y)(v) + (-2x + 6y)(2u) \\ &= 2(uv - u^2 + v^2)v + 4(-uv + 3u^2 - 3v^2)u \\ &= 12u^3 - 6u^2v - 10uv^2 + 2v^3 \\ \frac{\partial z}{\partial v} &= \frac{\partial z}{\partial x}\frac{\partial x}{\partial v} + \frac{\partial z}{\partial y}\frac{\partial y}{\partial v} = (2x - 2y)(u) + (-2x + 6y)(-2v) \\ &= 2(uv - u^2 + v^2)u - 4(-uv + 3u^2 - 3v^2)v \\ &= -2u^3 - 10u^2v + 6uv^2 + 12v^3 \end{aligned}$$

■

Remark Note that we first obtained $\partial z/\partial u$ and $\partial z/\partial v$ as a mixed expression involving both intermediate and final variables. Then we replaced the intermediate variables x and y by their values in terms of the final variables u and v. In Example 17.4 we avoided the last step, since we were evaluating the derivative at a specific point, and we substituted the values of dx/dt and dy/dt at that point.

EXAMPLE 17.6 If $z = f(u/v, v/u)$, show that

$$u\frac{\partial z}{\partial u} + v\frac{\partial z}{\partial v} = 0$$

Solution We can view z as the composite of $z = f(x, y)$ and $x = u/v$, $y = v/u$. The chain rule (17.16) then gives

$$\frac{\partial z}{\partial u} = \frac{\partial z}{\partial x}\frac{1}{v} + \frac{\partial z}{\partial y}\left(-\frac{v}{u^2}\right)$$

$$\frac{\partial z}{\partial v} = \frac{\partial z}{\partial x}\left(-\frac{u}{v^2}\right) + \frac{\partial z}{\partial y}\left(\frac{1}{u}\right)$$

Even though we do not know the values of $\partial z/\partial x$ and $\partial z/\partial y$, we see that multiplying the first equation by u and the second by v and adding gives

$$u\frac{\partial z}{\partial u} + v\frac{\partial z}{\partial v} = 0$$

■

The following theorem generalizes the chain rule to any number of intermediate variables and any number of final variables.

THEOREM 17.3 GENERALIZED CHAIN RULE

Suppose $z = f(x_1, x_2, \ldots, x_n)$ and $x_i = g_i(u_1, u_2, \ldots, u_m)$ for $i = 1, 2, \ldots, n$, and the composition of f with $g_1, g_2, \ldots, g_n$ is defined in some domain D. If each g_i is differentiable at a point $P = (u_1, u_2, \ldots, u_m)$ of D and f is differentiable at $(g_1(P), g_2(P), \ldots, g_n(P))$, then at P,

$$\frac{\partial z}{\partial u_j} = \frac{\partial z}{\partial x_1}\frac{\partial x_1}{\partial u_j} + \frac{\partial z}{\partial x_2}\frac{\partial x_2}{\partial u_j} + \cdots + \frac{\partial z}{\partial x_n}\frac{\partial x_n}{\partial u_j} \tag{17.17}$$

for $j = 1, 2, \ldots, m$. ■

Here there are n intermediate variables $x_1, x_2, \ldots, x_n$, and m final variables $u_1, u_2, \ldots, u_m$. There are m equations of the form (17.17), each having n terms on the right.

There is nothing special, of course, about the particular letters used to designate intermediate and final variables. The next example illustrates this.

EXAMPLE 17.7 Let $w = r^2 + s^2 - 2t^2$ and $r = e^{-y}\cos z$, $s = e^{-y}\sin z$, and $t = e^{-y}$. Find $\partial w/\partial y$ and $\partial w/\partial z$.

Solution With appropriate changes in variable names, equation (17.17) gives

$$\begin{aligned}
\frac{\partial w}{\partial y} &= \frac{\partial w}{\partial r}\frac{\partial r}{\partial y} + \frac{\partial w}{\partial s}\frac{\partial s}{\partial y} + \frac{\partial w}{\partial t}\frac{\partial t}{\partial y} \\
&= (2r)(-e^{-y}\cos z) + (2s)(-e^{-y}\sin z) + (-4t)(-e^{-y}) \\
&= (2e^{-y}\cos z)(-e^{-y}\cos z) + (2e^{-y}\sin z)(-e^{-y}\sin z) \\
&\quad + (-4e^{-y})(-e^{-y}) \\
&= 2e^{-2y}(-\cos^2 z - \sin^2 z + 2) \\
&= 2e^{-2y}
\end{aligned}$$

$$\begin{aligned}\frac{\partial w}{\partial z} &= \frac{\partial w}{\partial r}\frac{\partial r}{\partial z} + \frac{\partial w}{\partial s}\frac{\partial s}{\partial z} + \frac{\partial w}{\partial t}\frac{\partial t}{\partial z}\\ &= (2r)(-e^{-y}\sin z) + (2s)(e^{-y}\cos z) + (-4t)(0)\\ &= (2e^{-y}\cos z)(-e^{-y}\sin z) + (2e^{-y}\sin z)(e^{-y}\cos z)\\ &= 0\end{aligned}$$

■

The next example shows how the chain rule (17.15) can be used to compute the second derivative d^2z/dt^2 when $z = f(x, y)$ and $x = g(t)$, $y = h(t)$. In the exercises you will be asked to derive similar formulas for second-order partial derivatives where there are two final variables.

EXAMPLE 17.8 Let $z = f(x, y)$ and $x = g(t)$, $y = h(t)$. Assuming suitable differentiability conditions, derive a formula for d^2z/dt^2.

Solution For the first derivative we have from equation (17.15)

$$\frac{dz}{dt} = \frac{\partial z}{\partial x}\frac{dx}{dt} + \frac{\partial z}{\partial y}\frac{dy}{dt}$$

Using the product rule, we get

$$\frac{d^2z}{dt^2} = \frac{\partial z}{\partial x}\frac{d^2x}{dt^2} + \frac{dx}{dt}\frac{d}{dt}\left(\frac{\partial z}{\partial x}\right) + \frac{\partial z}{\partial y}\frac{d^2y}{dt^2} + \frac{dy}{dt}\frac{d}{dt}\left(\frac{\partial z}{\partial y}\right) \tag{17.18}$$

Now $\partial z/\partial x$ and $\partial z/\partial y$ are initially functions of the intermediate variables x and y. So to find their derivatives with respect to the final variable t, we use the chain rule (17.15) again, with z replaced by $\partial z/\partial x$ in the one case and by $\partial z/\partial y$ in the other:

$$\frac{d}{dt}\left(\frac{\partial z}{\partial x}\right) = \frac{\partial^2 z}{\partial x^2}\frac{dx}{dt} + \frac{\partial^2 z}{\partial y \partial x}\frac{dy}{dt}$$
$$\frac{d}{dt}\left(\frac{\partial z}{\partial y}\right) = \frac{\partial^2 z}{\partial x \partial y}\frac{dx}{dt} + \frac{\partial^2 z}{\partial y^2}\frac{dy}{dt}$$

Now we substitute these into equation (17.18) and simplify to get

$$\frac{d^2z}{dt^2} = \frac{\partial z}{\partial x}\frac{d^2x}{dt^2} + \frac{\partial z}{\partial y}\frac{d^2y}{dt^2} + \frac{\partial^2 z}{\partial x^2}\left(\frac{dx}{dt}\right)^2 + 2\frac{\partial^2 z}{\partial y \partial x}\left(\frac{dx}{dt}\frac{dy}{dt}\right) + \frac{\partial^2 z}{\partial y^2}\left(\frac{dy}{dt}\right)^2$$

Note that we are assuming equality of the second-order mixed partials. ■

The chain rule can be used to derive a very convenient formula for the derivative of a function defined implicitly. As in Section 2.8, suppose that $F(x, y) = 0$ defines y as a differentiable function of x, say $y = f(x)$, on some domain D. We want a formula for $\frac{dy}{dx}$. We can think of this as a chain rule situation with intermediate variables x and y and final variable x. However, to distinguish the two roles played by x, we will use the letter t (temporarily) as the final variable. So we have

$$F(x, y) = 0 \quad \text{and} \quad \begin{cases} x = t \\ y = f(t) \end{cases}$$

and since $F(t, f(t)) = 0$ for all t in D, it follows that

$$\frac{d}{dt} F(t, f(t)) = 0$$

there also. By the chain rule (17.15),

$$\begin{aligned} \frac{d}{dt} F(t, f(t)) &= \frac{\partial F}{\partial x}\frac{dx}{dt} + \frac{\partial F}{\partial y}\frac{dy}{dt} \\ &= \frac{\partial F}{\partial x}(1) + \frac{\partial F}{\partial y}\frac{dy}{dt} = 0 \end{aligned}$$

Thus, if $\partial F/\partial y \neq 0$, we can solve for dy/dt to get

$$\frac{dy}{dt} = -\frac{\dfrac{\partial F}{\partial x}}{\dfrac{\partial F}{\partial y}}$$

Now we can replace t by x and write the answer in the alternate form:

$$\frac{dy}{dx} = -\frac{F_x}{F_y} \quad \text{if } F_y \neq 0 \tag{17.19}$$

For example, suppose we have the equation

$$x^3 - 2xy + y^3 - 4 = 0$$

Assuming this defines y as a differentiable function of x on some domain D, we obtain from equation (17.19)

$$\frac{dy}{dx} = -\frac{3x^2 - 2y}{-2x + 3y^2} = \frac{2y - 3x^2}{3y^2 - 2x}$$

at all points of D for which $3y^2 - 2x \neq 0$. This same problem was solved in Example 2.16, part a. You might be interested in comparing the new method with the one used in Chapter 2.

A similar method can be applied when an equation $F(x, y, z) = 0$ defines z implicitly as a differentiable function of x and y, say $z = f(x, y)$, on some domain D of $\mathbf{R}^2$. To find formulas for $\partial z/\partial x$ and $\partial z/\partial y$, we proceed as follows. Treat x, y, and z as intermediate variables and u and v as final variables, where

$$\begin{cases} x = u \\ y = v \\ z = f(u, v) \end{cases}$$

Then for (u, v) in D, $F(u, v, f(u, v)) = 0$. So $\partial F/\partial u = 0$ and $\partial F/\partial v = 0$. By equation (17.17),

$$\frac{\partial F}{\partial u} = \frac{\partial F}{\partial x}\frac{\partial x}{\partial u} + \frac{\partial F}{\partial y}\frac{\partial y}{\partial u} + \frac{\partial F}{\partial z}\frac{\partial z}{\partial u} = \frac{\partial F}{\partial x}(1) + \frac{\partial F}{\partial y}(0) + \frac{\partial F}{\partial z}\frac{\partial z}{\partial u}$$

and

$$\frac{\partial F}{\partial v} = \frac{\partial F}{\partial x}\frac{\partial x}{\partial v} + \frac{\partial F}{\partial y}\frac{\partial y}{\partial v} + \frac{\partial F}{\partial z}\frac{\partial z}{\partial v} = \frac{\partial F}{\partial x}(0) + \frac{\partial F}{\partial y}(1) + \frac{\partial F}{\partial z}\frac{\partial z}{\partial v}$$

Since each of these equals 0, we can solve for $\partial z/\partial u$ and $\partial z/\partial v$ provided $\partial F/\partial z \neq 0$, getting

$$\frac{\partial z}{\partial u} = -\frac{\dfrac{\partial F}{\partial x}}{\dfrac{\partial F}{\partial z}} \quad \text{and} \quad \frac{\partial z}{\partial v} = -\frac{\dfrac{\partial F}{\partial y}}{\dfrac{\partial F}{\partial z}}$$

Finally, since $u = x$ and $v = y$,

$$\frac{\partial z}{\partial x} = -\frac{F_x}{F_z} \quad \text{and} \quad \frac{\partial z}{\partial y} = -\frac{F_y}{F_z} \quad \text{if } F_z \neq 0 \tag{17.20}$$

Note the similarity between these equations and equation (17.19).

EXAMPLE 17.9 If z is defined implicitly as a differentiable function of x and y by the equation

$$z^3 - 2xz + y^2 - x^3 - 11 = 0$$

find $\partial z/\partial x$ and $\partial z/\partial y$ at the point $(-1, 3, 1)$.

Solution Here $F(x, y, z)$ is the left-hand side of the given equation. So we have, by equations (17.20),

$$\frac{\partial z}{\partial x} = -\frac{F_x}{F_z} = -\frac{-2z - 3x^2}{3z^2 - 2x}$$

$$\frac{\partial z}{\partial y} = -\frac{F_y}{F_z} = -\frac{2y}{3z^2 - 2x}$$

On substituting $(-1, 3, 1)$ and simplifying, we get

$$\left.\frac{\partial z}{\partial x}\right|_{(-1,3,1)} = 1 \quad \text{and} \quad \left.\frac{\partial z}{\partial y}\right|_{(-1,3,1)} = -\frac{6}{5}$$ ■

EXERCISE SET 17.2

A

In Exercises 1–16 make use of a chain rule. In Exercises 1–4 find dz/dt at the specified value of t.

1. $z = x^2 + y^2$, $x = t^2 + 1$, $y = 2 - t^2$; at $t = 2$
2. $z = 2x^2 - 3xy$, $x = e^t$, $y = te^t$; at $t = 0$
3. $z = xe^y - ye^x$, $x = \ln t$, $y = \ln \frac{1}{t}$; at $t = 1$
4. $z = x \cos xy$, $x = 2t^2$, $y = \frac{1}{t}$; at $t = \frac{\pi}{4}$

In Exercises 5–8 find dz/dt. Express answers in terms of t.

5. $z = \ln \frac{y}{x}$, $x = \sin t$, $y = \cos t$
6. $z = \tan^{-1} \frac{y}{x}$, $x = \sin 2t$, $y = \cos 2t$
7. $z = r^2(1 - \cos \theta)$, $r = \sqrt{t}$, $\theta = t^2$
8. $z = uv^2w^3$, $u = e^t$, $v = e^{-t}$, $w = te^t$

In Exercises 9–12 find $\partial z/\partial u$ and $\partial z/\partial v$ at the specified point.

9. $z = x^2 - 2xy$, $x = \frac{u}{v}$, $y = uv$; at $(u, v) = (2, -1)$
10. $z = \ln \sqrt{x^2 + y^2}$, $x = u + v$, $y = 2u - 3v$; at $(u, v) = (3, 1)$
11. $z = e^{xy}$, $x = \frac{u}{v}$, $y = 2v$; at $(u, v) = (1, \frac{1}{2})$
12. $z = \sin(x + y)$, $x = u^2 - v^2$, $y = 2uv$; at $(u, v) = (\sqrt{\pi}, \sqrt{\pi}/2)$

In Exercises 13–16 find the $\partial z/\partial u$ and $\partial z/\partial v$ at an arbitrary point (u, v).

13. $z = \ln \dfrac{x + y}{x - y}$; $x = \cos^2 uv$, $y = \sin^2 uv$

14. $z = x^2 + 2xy - y^2$; $x = \cosh u + \sinh v$, $y = \cosh u - \sinh v$

15. $z = r^2(3\sin^2\theta - 2\cos^2\theta)$; $r = \sqrt{u^2 + v^2}$, $\theta = \tan^{-1} v/u$

16. $z = \sqrt{rst}$; $r = u^2v$, $s = u/v$, $t = v^2/u \quad (uv > 0)$

In Exercises 17–22 assume the given equation defines y as a differentiable function of x on some domain, and use equation (17.19) to find dy/dx there.

17. $x^3 - 2xy^2 + y^4 = 0$

18. $2x^2y^2 - 3xy^3 + 4y - 5 = 0$

19. $x \sin y - y \sin x = 1$

20. $x \ln y + y \ln x = xy$

21. $e^{-xy}(x^2 - y^2) = 4$

22. $xy \tan xy + 1 = 0$

In Exercises 23–28 assume the given equation defines z as a differentiable function of x and y on some domain, and use equations (17.20) to find ∂z/∂x and ∂z/∂y there.

23. $x^2yz - 2y^2 + 3z^2 = 0$

24. $x^2 + y^2 + z^2 - 2xy + 3xz - 4yz = 1$

25. $\ln[(x + y)/\sqrt{z}] - 2xyz = 4$

26. $\tan^{-1}\frac{x}{y} - \tan^{-1}\frac{y}{z} = 2$

27. $\sin xz + \ln \cos yz = 0$

28. $z - xy \ln z = 1$

29. If $w = (u + v)/u$ and $u = \sqrt{x^2 + y^2 + z^2}$, $v = 2xyz$, find $\partial w/\partial x$, $\partial w/\partial y$, and $\partial w/\partial z$ as functions of x, y, and z.

30. Let $f(x, y, z) = xyz$ and $F(s, t) = f(s + t, s - t, st)$. Using an appropriate chain rule, find $F_s(2, -1)$ and $F_t(2, -1)$.

31. Let $z = f(x, y)$ and let the xy-coordinates be transformed to polar coordinates by the equations $x = r\cos\theta$, $y = r\sin\theta$. Show that

$$\left(\frac{\partial z}{\partial r}\right)^2 + \frac{1}{r^2}\left(\frac{\partial z}{\partial \theta}\right)^2 = \left(\frac{\partial z}{\partial x}\right)^2 + \left(\frac{\partial z}{\partial y}\right)^2$$

32. The relationship between pressure P, volume V, and temperature T of a certain gas is given by $PV = 12T$. If the pressure is decreasing at the constant rate of 3 psi/min and the temperature is constantly increasing at 4°K per minute, find the rate at which the volume is changing when $P = 10$ psi and $T = 298$°K.

33. An oil slick in the Gulf of Mexico from a ruptured oil tanker is approximately triangular. When the altitude is 2 km, the base is 3 km, and at that instant the altitude and base are increasing at the rates of 200 m/hr and 320 m/hr, respectively. Find the rate at which the area is increasing at that instant.

B

34. Let $z = f(x - y, y - x)$. Show that

$$\frac{\partial z}{\partial x} + \frac{\partial z}{\partial y} = 0$$

(*Hint:* Let $u = x - y$ and $v = y - x$.)

35. Let $z = f(\frac{x}{y})$. Show that

$$x\frac{\partial z}{\partial x} + y\frac{\partial z}{\partial y} = 0$$

36. The equation

$$\frac{\partial^2 w}{\partial t^2} = c^2\frac{\partial^2 w}{\partial x^2}$$

is known as the *wave equation*. Prove that for any twice differentiable functions f and g, $w = f(x + ct) + g(x - ct)$ is a solution of the wave equation.

37. Let $z = f(x, y)$, $x = g(u, v)$, and $y = h(u, v)$. Derive formulas for $\partial^2 z/\partial u^2$ and $\partial^2 z/\partial v^2$.

38. Assume that the equation $F(x, y) = 0$ defines y as a twice differentiable function, $y = f(x)$, on a domain D. Derive the formula.

$$f''(x) = -\frac{F_{xx}F_y^2 - 2F_xF_yF_{xy} + F_{yy}F_x^2}{F_y^3} \quad (F_y \neq 0)$$

39. If $z = f(x, y)$, the expression $\partial^2 z/\partial x^2 + \partial^2 z/\partial y^2$ is called the *Laplacian* of z. Show that on changing the xy-coordinates to polar coordinates by means of the transformation $x = r\cos\theta$, $y = r\sin\theta$, the Laplacian of z becomes

$$\frac{\partial^2 z}{\partial r^2} + \frac{1}{r}\frac{\partial z}{\partial r} + \frac{1}{r^2}\frac{\partial^2 z}{\partial \theta^2}$$

40. In a grain elevator grain deposited through a chute at the rate of 60 cu ft/min assumes the form of a cone as it accumulates on the floor. After 5 min the base radius of the cone is 10 ft, and at that instant it is increasing at the rate of 0.5 ft/min. How fast is the height of the cone changing at that instant?

41. Part of a certain hydraulic lift has a triangular shape that varies in size as the lift is actuated. At the instant when two adjacent sides of the triangle are 3.1 ft and 1.7 ft long and the angle between them is 30°, these sides and this angle are increasing at the respective rates of 0.5 ft/sec, 0.8 ft/sec, and 2°/sec. Find the rate at which the area of the triangle is changing at that instant.

FIGURE 17.2

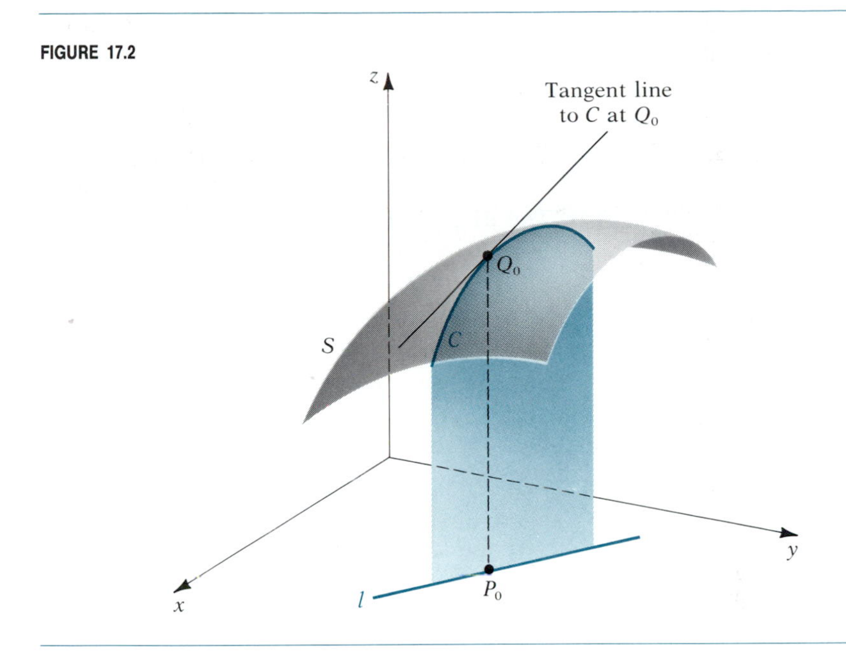

17.3 DIRECTIONAL DERIVATIVES

In this section we introduce a generalization of partial derivatives that will be useful in many applications. We motivate the idea geometrically.

Let $z = f(x, y)$ and assume that f is continuous in some region D that contains the point $P_0 = (x_0, y_0)$. Denote by S the surface that is the graph of f, and let Q_0 be the point on S on a vertical line through P_0. Then $Q_0 = (x_0, y_0, f(x_0, y_0))$. Now consider an arbitrary plane that contains the line segment P_0Q_0. We show a typical situation in Figure 17.2. This plane intersects S in some curve C. Our objective is to find the slope of the tangent line to C at Q_0, provided this exists. Observe that if the plane happens to be parallel to the xz-plane, then the slope in question is $f_x(x_0, y_0)$, and if it is parallel to the yz-plane, this slope is $f_y(x_0, y_0)$.

We can specify the direction of the plane by giving a direction vector of the line l of its intersection with the xy-plane, and we choose to give this as a unit vector, say $\mathbf{u} = \langle u_1, u_2 \rangle$. Then parametric equations for l are

$$\begin{cases} x = x_0 + u_1 t \\ y = y_0 + u_2 t \end{cases}$$

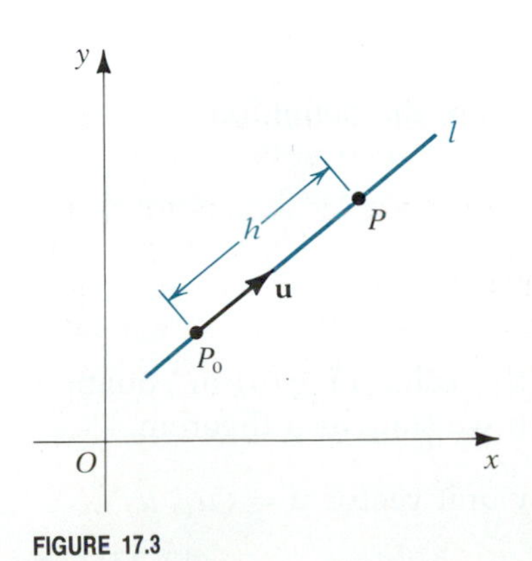

FIGURE 17.3

Let $P(t) = (x_0 + u_1 t, y_0 + u_2 t)$ be an arbitrary point on l. Then $P(0) = P_0$. As we have done, we will frequently write $f(P_0)$ to mean $f(x_0, y_0)$ and, more generally, $f(P(t))$ to mean $f(x_0 + u_1 t, y_0 + u_2 t)$. To arrive at an expression for the desired slope of the tangent line to the curve C, we consider a point P on l that is $|h|$ units from P_0, in the direction of $\mathbf{u}$ if $h > 0$ and in the opposite direction if $h < 0$. Figure 17.3 illustrates the case $h > 0$. Thus $P = P(h) = (x_0 + u_1 h, y_0 + u_2 h)$. Then the slope of C at Q_0, if it exists, is by definition

$$\lim_{h\to 0}\frac{f(P)-f(P_0)}{h}=\lim_{h\to 0}\frac{f(x_0+u_1h,\,y_0+u_2h)-f(x_0,\,y_0)}{h}$$

We give this slope a special name and symbol in the following definition.

DEFINITION 17.2 The **directional derivative** of f at $P_0=(x_0,\,y_0)$ in the direction of the unit vector $\mathbf{u}=\langle u_1,\,u_2\rangle$ is designated by $D_{\mathbf{u}}f(P_0)$ or $D_{\mathbf{u}}f(x_0,\,y_0)$ and has the value

$$D_{\mathbf{u}}f(P_0)=\lim_{h\to 0}\frac{f(x_0+u_1h,\,y_0+u_2h)-f(x_0,\,y_0)}{h} \tag{17.21}$$

provided this limit exists. ■

Remarks

1. If a nonzero direction vector $\mathbf{v}$ is given that is not a unit vector, then by the directional derivative of f in the direction of $\mathbf{v}$ we mean $D_{\mathbf{u}}f(P_0)$ where $\mathbf{u}=\mathbf{v}/|\mathbf{v}|$. So *it is essential that the direction vector be made into a unit vector if it is not one already.*
2. When we wish to speak of the directional derivative at an arbitrary point $P(x,\,y)$ of the domain, we can use either symbol $D_{\mathbf{u}}f(x,\,y)$ or $D_{\mathbf{u}}f(P)$. Sometimes, for brevity, we write simply $D_{\mathbf{u}}f$.
3. By comparison with Definition 16.1, it can be seen that when $\mathbf{u}=\langle 1,\,0\rangle$, $D_{\mathbf{u}}f(P_0)=f_x(P_0)$, and when $\mathbf{u}=\langle 0,\,1\rangle$, $D_{\mathbf{u}}f(P_0)=f_y(P_0)$. It is in this sense that the directional derivative generalizes the notion of a partial derivative.

As is true of other derivatives, calculation of the directional derivative directly from the definition can be quite tedious. If f is a differentiable function, we can develop a convenient computational formula for $D_{\mathbf{u}}f$ as follows. For points on the curve C described earlier, we have both $z=f(x,\,y)$ and $x=x_0+u_1t$, $y=y_0+u_2t$. We can therefore calculate dz/dt by the chain rule, where x and y are intermediate variables and t is the final variable. If we write $g(t)=f(P(t))=f(x_0+u_1t,\,y_0+u_2t)$, we obtain

$$g'(t)=f_x(P(t))\frac{dx}{dt}+f_y(P(t))\frac{dy}{dt}=f_x(P(t))u_1+f_y(P(t))u_2$$

In particular, at $t=0$,

$$g'(0)=f_x(P_0)u_1+f_y(P_0)u_2 \tag{17.22}$$

where again $P_0=P(0)$.

Alternately, we can calculate $g'(0)$ directly from the definition:

$$\begin{aligned}g'(0)&=\lim_{h\to 0}\frac{g(0+h)-g(0)}{h}\\&=\lim_{h\to 0}\frac{f(x_0+u_1h,\,y_0+u_2h)-f(x_0,\,y_0)}{h}\end{aligned}$$

But this is $D_{\mathbf{u}}f(P_0)$. If we equate this with the value of $g'(0)$ in equation (17.22), we obtain the following result, which we state as a theorem.

THEOREM 17.4 If f is differentiable at $P=(x,\,y)$, then for any unit vector $\mathbf{u}=\langle u_1,\,u_2\rangle$,

$$D_{\mathbf{u}}f(P)=f_x(P)u_1+f_y(P)u_2 \tag{17.23}$$

■

Note In the statement of the theorem we omitted the subscript on P, since it can be any point at which f is differentiable. We often write the result more briefly as $D_{\mathbf{u}}f = f_x u_1 + f_y u_2$.

EXAMPLE 17.10 Find the directional derivative of the function $f(x, y) = x^2 - 2xy^3$ at $(-2, 1)$ in the direction of the vector $\mathbf{v} = \langle 3, 4 \rangle$.

Solution First we obtain a unit direction vector:

$$\mathbf{u} = \frac{\mathbf{v}}{|\mathbf{v}|} = \left\langle \frac{3}{5}, \frac{4}{5} \right\rangle$$

Now we use equation (17.23):

$$D_{\mathbf{u}}f(P) = f_x(\tfrac{3}{5}) + f_y(\tfrac{4}{5}) = (2x - 2y^3)(\tfrac{3}{5}) + (-6xy^2)(\tfrac{4}{5})$$

Setting $P = (-2, 1)$, we get

$$D_{\mathbf{u}}f(-2, 1) = (-6)(\tfrac{3}{5}) + (12)(\tfrac{4}{5}) = 6$$

■

EXAMPLE 17.11 A coordinate system is established on a flat metal plate, and it is determined that at a point (x, y) on the plate, other than the origin, the temperature $T(x, y)$ in degrees Celsius is given by $T(x, y) = 100(x^2 + y^2)^{-1/2}$. Find the instantaneous rate of change of temperature at the point $P(2, 6)$ in the direction from P toward $Q(4, 2)$.

Solution The instantaneous rate of change of T in the direction of $\overrightarrow{PQ}$ is the directional derivative at P in the direction of the unit vector:

$$\mathbf{u} = \frac{\overrightarrow{PQ}}{|\overrightarrow{PQ}|} = \frac{\langle 2, -4 \rangle}{\sqrt{20}} = \left\langle \frac{1}{\sqrt{5}}, \frac{-2}{\sqrt{5}} \right\rangle$$

So we have, by equation (17.23),

$$D_{\mathbf{u}}T(x, y) = \left[\frac{-100x}{(x^2 + y^2)^{3/2}}\right]\left(\frac{1}{\sqrt{5}}\right) + \left[\frac{-100y}{(x^2 + y^2)^{3/2}}\right]\left(-\frac{2}{\sqrt{5}}\right)$$

and after simplification we obtain

$$D_{\mathbf{u}}T(2, 6) = \left(\frac{-200}{80\sqrt{10}}\right)\left(\frac{1}{\sqrt{5}}\right) + \left(\frac{-600}{80\sqrt{10}}\right)\left(-\frac{2}{\sqrt{5}}\right) = \frac{5\sqrt{2}}{4} \approx 1.768$$

Thus, from P, moving one unit on a line toward Q, one would expect an increase in temperature of approximately $1.768°$. (Remember, though, that this is an instantaneous rate of change, so that after moving away from P the rate changes.) ■

The right-hand side of equation (17.23) has the appearance of the dot product of two vectors. In fact, it is the dot product of the unit vector $\mathbf{u}$ and the vector $\langle f_x(P), f_y(P) \rangle$. We give this latter vector a name and symbol as follows.

DEFINITION 17.3 The **gradient of f** at a point $P(x, y)$ is denoted by $\nabla f(P)$ or $\nabla f(x, y)$ and is defined as the vector

$$\nabla f(P) = \langle f_x(P), f_y(P) \rangle \tag{17.24}$$

provided these partials exist. ■

Comment The symbol ∇ is read "del." So one says "del f" and writes ∇f. Sometimes the gradient of f is symbolized by **Grad** f. So ∇f and **Grad** f mean the same thing. We will use ∇f exclusively in this book.

In view of Definition 17.3 we can rewrite equation (17.23) in the form

$$D_{\mathbf{u}}f(P) = \nabla f(P) \cdot \mathbf{u} \tag{17.25}$$

EXAMPLE 17.12 Let $f(x, y) = \ln \frac{x}{y}$. Find ∇f and use this to find the directional derivative of f at (1, 2) in the direction $\mathbf{u} = \langle \frac{1}{2}, -\sqrt{3}/2 \rangle$.

Solution First we write $f(x, y) = \ln x - \ln y$. Thus,

$$\nabla f = \langle f_x, f_y \rangle = \left\langle \frac{1}{x}, -\frac{1}{y} \right\rangle$$

At (1, 2) this becomes

$$\nabla f(1, 2) = \left\langle 1, -\frac{1}{2} \right\rangle$$

By equation (17.25),

$$D_{\mathbf{u}}f(1, 2) = \left\langle 1, -\frac{1}{2} \right\rangle \cdot \left\langle \frac{1}{2}, \frac{-\sqrt{3}}{2} \right\rangle = \frac{1}{2} + \frac{\sqrt{3}}{4} = \frac{2 + \sqrt{3}}{4}$$ ■

Sometimes we want to know the direction in which a function changes most rapidly. The answer is provided in the following theorem.

THEOREM 17.5 Let f have continuous partial derivatives f_x and f_y in a neighborhood of a point P_0 at which $\nabla f \neq 0$. Then $D_{\mathbf{u}}f(P_0)$ is maximum when $\mathbf{u}$ is in the direction of $\nabla f(P_0)$. Furthermore, the maximum value of $D_{\mathbf{u}}f(P_0)$ is $|\nabla f(P_0)|$.

Proof As in Figure 17.4, let $\mathbf{u}$ be any unit vector, and let θ be the angle between $\nabla f(P_0)$ and $\mathbf{u}$. Then we have

$$D_{\mathbf{u}}f(P_0) = \nabla f(P_0) \cdot \mathbf{u} = |\nabla f(P_0)|\,|\mathbf{u}| \cos \theta = |\nabla f(P_0)| \cos \theta$$

since $|\mathbf{u}| = 1$. Now the maximum value that $\cos \theta$ can assume is 1, and this occurs when $\theta = 0$. Thus, to obtain the maximum value of $D_{\mathbf{u}}f(P_0)$, we take the angle θ between $\mathbf{u}$ and $\nabla f(P_0)$ to be 0; that is, we take $\mathbf{u}$ in the same direction as $\nabla f(P_0)$, so that $\mathbf{u} = \nabla f(P_0)/|\nabla f(P_0)|$. Furthermore, max $D_{\mathbf{u}}f(P_0) = |\nabla f(P_0)|$. This completes the proof. ■

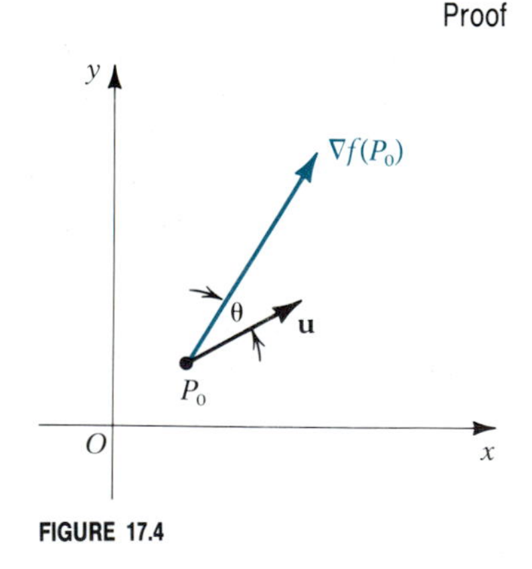

FIGURE 17.4

The *minimum* value of $D_{\mathbf{u}}f(P_0)$ occurs when $\cos \theta = -1$, namely, when $\theta = \pi$, and so when $\mathbf{u}$ is directed opposite to $\nabla f(P_0)$. The minimum value is $-|\nabla f(P_0)|$.

The relationship between $\nabla f(P_0)$ and the level curve to the surface defined by $z = f(x, y)$ that passes through P_0 is helpful in understanding the maximum value of $D_{\mathbf{u}}f(P_0)$. Let $c = f(x_0, y_0)$. Then the level curve $f(x, y) = c$ passes through P_0. Suppose this curve is parameterized by $x = x(t)$, $y = y(t)$ and that $P_0 = (x(t_0), y(t_0))$. Then since $f(x(t), y(t)) = c$, we have by the chain rule

$$f_x(P_0)x'(t_0) + f_y(P_0)y'(t_0) = 0$$

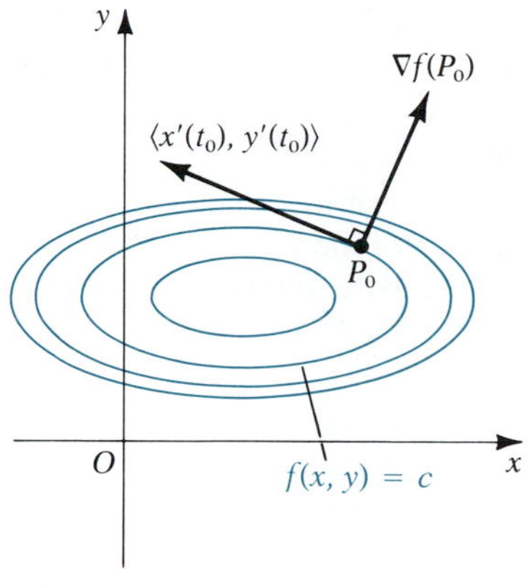

FIGURE 17.5

or, equivalently,

$$\nabla f(P_0) \cdot \langle x'(t_0), y'(t_0)\rangle = 0$$

It follows that $\nabla f(P_0)$ is orthogonal to the vector $\langle x'(t_0), y'(t_0)\rangle$. But this latter vector is a tangent vector to the level curve at P_0 (see Figure 17.5). We therefore have shown the following:

$\nabla f(P_0)$ is orthogonal to the level curve $f(x, y) = c$ passing through P_0.

Combining this result with Theorem 17.5, we conclude that from any point in its domain a function $f(x, y)$ increases most rapidly in a direction perpendicular to the level curve through that point. For example, on a weather map to move in the direction of the most rapid change in temperature, one would move in a direction perpendicular to the isotherms.

All the concepts of this section can be extended in natural ways to functions of three (or more) variables. In particular, if $\mathbf{u} = \langle u_1, u_2, u_3\rangle$ is a unit vector and $P_0 = (x_0, y_0, z_0)$, then for the differentiable function $f(x, y, z)$,

$$D_{\mathbf{u}}f(P_0) = f_x(P_0)u_1 + f_y(P_0)u_2 + f_z(P_0)u_3 = \nabla f(P_0) \cdot \mathbf{u}$$

where $\nabla f(P_0) = \langle f_x(P_0), f_y(P_0), f_z(P_0)\rangle$. The maximum value of the directional derivative occurs when $\mathbf{u} = \nabla f(P_0)/|\nabla f(P_0)|$ and is equal to $|\nabla f(P_0)|$. Furthermore, this direction of maximum change in f is orthogonal to the level surface $f(x, y, z) = c$ passing through P_0.

EXERCISE SET 17.3

A

In Exercises 1–14 find the directional derivative of f at the given point in the direction of the given vector.

1. $f(x, y) = 2x^2y - 3xy^2$ at $(3, 1)$; $\mathbf{u} = \langle \frac{3}{5}, \frac{4}{5}\rangle$
2. $f(x, y) = x^3 - 2xy + y^3$ at $(1, 0)$; $\mathbf{u} = \langle 1/\sqrt{10}, 3/\sqrt{10}\rangle$
3. $f(x, y) = \ln [xy/(x^2 + y^2)]$ at $(1, 2)$; $\mathbf{v} = \langle -2, 2\rangle$
4. $f(x, y) = xe^y - ye^x$ at $(0, 0)$; $\mathbf{v} = \langle 8, -1\rangle$
5. $f(x, y) = x(\cos y - \sin y)$ at $(-1, \frac{\pi}{2})$; $\mathbf{v} = 4\mathbf{i} - 3\mathbf{j}$
6. $f(x, y) = \tan^{-1} \frac{y}{x}$ at $(2, 1)$; $\mathbf{v} = \mathbf{i} + \mathbf{j}$
7. $f(x, y) = x \cosh y + y \cosh x$ at $(0, 0)$; $\mathbf{v} = -2\mathbf{i} + 4\mathbf{j}$
8. $f(x, y) = y^2 \ln \sqrt{x}$ at $(1, -2)$; $\mathbf{v} = 3\mathbf{i} + 2\mathbf{j}$
9. $f(x, y) = \sqrt{9 - x^2 - y^2}$ at $(2, -1)$; $\mathbf{v} = \langle 1, -1\rangle$
10. $f(x, y) = e^{(x-y)/(x+y)}$ at $(1, 1)$; $\mathbf{v} = \langle 5, 12\rangle$
11. $f(x, y, z) = x^2 - 2y^2 + 3z^2$ at $(2, 0, -1)$; $\mathbf{u} = \langle \frac{1}{3}, \frac{2}{3}, -\frac{2}{3}\rangle$
12. $f(x, y, z) = \dfrac{x + y}{x + z}$ at $(3, -2, -1)$; $\mathbf{u} = \left\langle \dfrac{1}{\sqrt{6}}, -\dfrac{1}{\sqrt{6}}, \dfrac{2}{\sqrt{6}}\right\rangle$
13. $f(x, y, z) = \ln [x^2/(yz^3)]$ at $(1, 1, 3)$; $\mathbf{v} = 3\mathbf{i} - 4\mathbf{j} + 5\mathbf{k}$
14. $f(x, y, z) = x \cosh(y + z)$ at $(3, 2, -1)$; $\mathbf{v} = \mathbf{i} + \mathbf{j} - \mathbf{k}$
15. Find the directional derivative of $f(x, y) = 3x^2 - 2xy^2$ at $(3, -2)$ in the direction from $(3, -2)$ toward $(5, 4)$.
16. Find the directional derivative of $f(x, y) = \ln \sqrt{x^2 - 2y^2}$ at $(2, 1)$ in the direction from $(2, 1)$ toward $(5, -3)$.

In Exercises 17–20 find the unit vector $\mathbf{u}$ *for which* $D_{\mathbf{u}}f(P_0)$ *is a maximum, and give this maximum value.*

17. $f(x, y) = \sqrt{\dfrac{x - y}{x + y}}$; $P_0 = (5, 4)$
18. $f(x, y) = \ln \cos(x + 2y)$; $P_0 = (\frac{\pi}{4}, \frac{\pi}{4})$
19. $f(x, y) = y^2 + e^{(\sin x)/y}$; $P_0 = (0, -1)$
20. $f(x, y, z) = \ln \dfrac{x + 2y}{z^3}$; $P_0 = (5, -2, 3)$
21. In what direction from the point $(1, -1)$ is the instantaneous rate of change of $f(x, y) = 2x^2 + 2xy - 3y^2$ equal to 2? (There are two solutions.) In what direction from $(1, -1)$ does this function increase most rapidly? What is this most rapid rate of change?

22. In what direction from the point $(4, 1)$ is the function $f(x, y) = x/(y + 1)$ stationary? From the same point, in what direction is the instantaneous rate of change of this function equal to 1? (There are two solutions.) Can the rate of change from the point $(4, 1)$ in any direction ever equal 2? Explain.

23. Two adjacent edges of a flat rectangular metal plate coincide, respectively, with the positive x- and y-axes. For points other than the origin, the temperature $T(x, y)$ at an arbitrary point (x, y) is inversely proportional to the distance from P to the origin. At the point $P(8, 6)$ the temperature is 10°C. How rapidly is the temperature changing at P in the direction from P toward $Q(6, 10)$? In what direction from P does the temperature decrease most rapidly, and what is this rate of decrease?

24. Two adjacent edges of a large square metal plate are kept at temperatures $T = 0$ and $T = 100$, respectively, and the flat surfaces are well insulated. By taking the positive x- and y-axes along the edges held at $T = 0$ and $T = 100$, respectively, it can be shown that the temperature $T(x, y)$ at an arbitrary point in the plate is approximated by

$$T(x, y) = \frac{200}{\pi} \tan^{-1} \frac{y}{x}$$

How rapidly is the temperature changing at the point $(2, 4)$ in the direction of the vector $\mathbf{v} = 3\mathbf{i} - 4\mathbf{j}$? In what direction from $(2, 4)$ is the temperature increase most rapid, and what is this most rapid change?

25. A cross section of two long coaxial conducting cylindrical surfaces consists of the circles $x^2 + y^2 = 1$ and $x^2 + y^2 = 4$. If the smaller cylinder is held at electrostatic potential $V = 0$ and the larger at $V = 1$, then in the annular ring between the two it can be shown that the potential $V(x, y)$ is given by

$$V(x, y) = \frac{\ln(x^2 + y^2)}{\ln 4}$$

Find the rate of change in potential at $P(\frac{3}{2}, \frac{1}{2})$ in the direction from P toward $Q(\frac{3}{4}, 1)$. In what direction from P does V increase most rapidly, and what is this rate of change?

B

26. In Exercise 24 find the equation of the isotherm $T(x, y) = c$ for c between 0 and 100 and identify the graph. Show that $\nabla T(x_0, y_0)$ is orthogonal to the isotherm through (x_0, y_0).

27. In Exercise 25 find the equation of the equipotential $V(x, y) = c$ for c between 0 and 1 and identify the graph. Show that $\nabla V(x_0, y_0)$ is orthogonal to the equipotential curve through (x_0, y_0).

28. Prove Theorem 17.4 for the three-dimensional case, using an argument similar to that preceding the statement of the theorem.

29. Find a function $f(x, y)$ for which $\nabla f = \langle xe^x, e^{-y}\rangle$. Is this function unique? Explain.

30. Find the function f for which $\nabla f = (x \sin x)\mathbf{i} + (\cos y)\mathbf{j}$ and $f(\frac{\pi}{2}, 0) = 3$.

31. Show that if α, β, and γ are direction angles of the unit vector $\mathbf{u}$, then if f is differentiable at $P = (x, y, z)$,

$$D_{\mathbf{u}}f(P) = f_x(P) \cos \alpha + f_y(P) \cos \beta + f_z(P) \cos \gamma$$

In Exercises 32–37 $u = f(x, y)$ and $v = g(x, y)$ are differentiable functions, and c and α are arbitrary real numbers. Prove each statement.

32. $\nabla(cu) = c\nabla u$ **33.** $\nabla(u + v) = \nabla u + \nabla v$

34. $\nabla(uv) = u\nabla v + v\nabla u$

35. $\nabla\left(\dfrac{u}{v}\right) = \dfrac{v\nabla u - u\nabla v}{v^2}$ if $v \neq 0$

36. $\nabla u^\alpha = \alpha u^{\alpha-1}\nabla u$

37. If $w = h(u, v)$ and h is differentiable, then

$$\nabla w = \frac{\partial w}{\partial u}\nabla u + \frac{\partial w}{\partial v}\nabla v$$

17.4 TANGENT PLANES AND NORMAL LINES

Let S be a surface that is the graph of $F(x, y, z) = 0$. For example, S might be the ellipsoid whose equation is $x^2 + 2y^2 + 4z^2 = 16$. In this case $F(x, y, z) = x^2 + 2y^2 + 4z^2 - 16$. Let $P_0 = (x_0, y_0, z_0)$ be a point on S, and suppose F is differentiable at P_0 with $\nabla F(P_0) \neq 0$. Consider any curve C lying on the surface S and passing through P_0. Denote a vector equation of C by $\mathbf{r}(t) = \langle f(t), g(t), h(t)\rangle$, and denote by t_0 the value of t at which C passes through

P_0; that is, $P_0 = (f(t_0), g(t_0), h(t_0))$. Assume furthermore that $\mathbf{r}'(t_0)$ exists and is nonzero. Then, as we know from Chapter 15, $\mathbf{r}'(t_0)$ is a tangent vector to C at P_0. Because C lies on the surface S, all points $(f(t), g(t), h(t))$ on C satisfy the equation of the surface; that is, $F(f(t), g(t), h(t)) = 0$ and so $\frac{dF}{dt} = 0$ wherever this derivative exists. It does exist at t_0 and is found by the chain rule:

$$F_x(P_0)f'(t_0) + F_y(P_0)g'(t_0) + F_z(P_0)h'(t_0) = 0$$

or, in vector form,

$$\nabla F(P_0) \cdot \mathbf{r}'(t_0) = 0$$

Since $\mathbf{r}'(t_0)$ is tangent to C, it follows that $\nabla F(P_0)$ is orthogonal to the tangent line to C at P_0.

The argument just given applies to *every* curve C on S that passes through P_0 and has a tangent line there. Thus the plane through P_0 that is perpendicular to $\nabla F(P_0)$ must contain all of the tangent lines to such curves (see Figure 17.6). It is natural to call this plane the **tangent plane** to S at P_0.

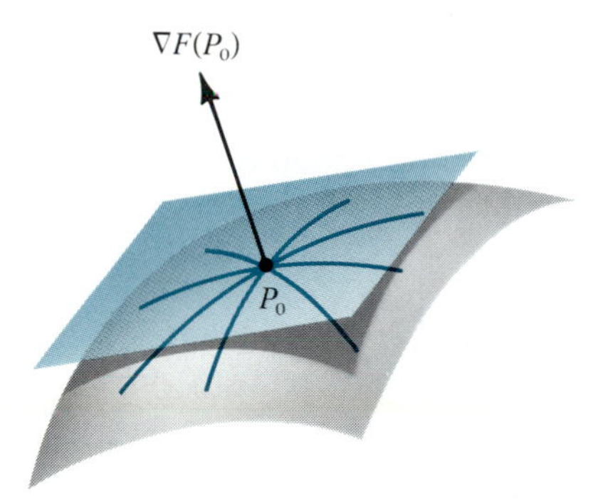

FIGURE 17.6

DEFINITON 17.4 Let S be a surface with the equation $F(x, y, z) = 0$. If $P_0 = (x_0, y_0, z_0)$ is a point on S at which F is differentiable, with $\nabla F(P_0) \neq \mathbf{0}$, then the plane through P_0 that has normal vector $\nabla F(P_0)$ is called the **tangent plane to S at P_0.** The line through P_0 with direction vector $\nabla F(P_0)$ is called the **normal line to S at P_0.** ■

Since $\nabla F(P_0) = \langle F_x(P_0), F_y(P_0), F_z(P_0)\rangle$ is a normal vector to the tangent plane, we know from Section 14.7 that the equation of the tangent plane can be written in the form:

THE TANGENT PLANE TO S AT P_0

$$F_x(P_0)(x - x_0) + F_y(P_0)(y - y_0) + F_z(P_0)(z - z_0) = 0 \qquad (17.26)$$

Also, the normal line has parametric equations:

THE NORMAL LINE TO S AT P_0

$$x = x_0 + F_x(P_0)t, \qquad y = y_0 + F_y(P_0)t, \qquad z = z_0 + F_z(P_0)t \qquad (17.27)$$

EXAMPLE 17.13 Find equations of the tangent plane and normal line to the ellipsoid $x^2 + 2y^2 + 4z^2 = 16$ at the point $(2, -2, 1)$.

Solution We let $F(x, y, z) = x^2 + 2y^2 + 4z^2 - 16$, so that $\nabla F = \langle 2x, 4y, 8z \rangle$. Since F_x, F_y, and F_z are continuous everywhere, F is differentiable and $\nabla F \neq \mathbf{0}$ for all points on the surface. Thus, a tangent plane exists everywhere. At $P_0 = (2, -2, 1)$ the vector $\nabla F(P_0) = \langle 4, -8, 8 \rangle$ is normal to the tangent plane. So the equation of the tangent plane is

$$4(x - 2) - 8(y + 2) + 8(z - 1) = 0$$

which, on simplification, becomes

$$x - 2y + 2z = 8$$

Parametric equations of the normal line at P_0 are $x = 2 + 4t$, $y = -2 - 8t$, and $z = 1 + 8t$.

We might note that since $\nabla F(P_0) = \langle 4, -8, 8 \rangle$ is normal to the surface, so is $\frac{1}{4} \nabla F(P_0) = \langle 1, -2, 2 \rangle$, and this could have been used to get the equations of both the tangent plane and the normal line. ■

An equation in the form $z = f(x, y)$ can be written in the form $F(x, y, z) = 0$, where $F(x, y, z) = f(x, y) - z$. For example, $z = x^2 + y^2$ would be written as $x^2 + y^2 - z = 0$. So equation (17.26) can be used to find the equation of the tangent plane. It is useful to obtain the general result for surfaces with equations in this form. We assume f is a differentiable function of two variables at (x_0, y_0), and we let $P_0 = (x_0, y_0, z_0)$, where $z_0 = f(x_0, y_0)$. It follows that $F(x, y, z) = f(x, y) - z$ is a differentiable function of three variables at P_0 (see Exercise 28). Furthermore, $\nabla F = \langle f_x, f_y, -1 \rangle \neq \mathbf{0}$. So the tangent plane at P_0 exists and has $\langle f_x(x_0, y_0), f_y(x_0, y_0), -1 \rangle$ as a normal vector. By equation (17.26) its equation is

$$f_x(x_0, y_0)(x - x_0) + f_y(x_0, y_0)(y - y_0) - (z - z_0) = 0 \qquad (17.28)$$

You may use this result to get the tangent plane when $z = f(x, y)$, or you may use equation (17.26) with $F(x, y, z) = f(x, y) - z$. The answers will be the same.

Equation (17.28) can be used to gain insight into the geometric interpretation of the differential $df = f_x \Delta x + f_y \Delta y$. If we let $\Delta x = x - x_0$ and $\Delta y = y - y_0$ and recall that $z_0 = f(x_0, y_0)$, then equation (17.28) becomes, on solving for z,

$$z = f(x_0, y_0) + f_x(x_0, y_0)\Delta x + f_y(x_0, y_0)\Delta y = f(x_0, y_0) + df$$

where we understand that df is evaluated at (x_0, y_0).

Now at (x_0, y_0) the vertical distances to the surface S and to its tangent plane T both equal $f(x_0, y_0)$. At $(x, y) = (x_0 + \Delta x, y_0 + \Delta y)$ the vertical distance to the surface S is $f(x_0, y_0) + \Delta f$, where $\Delta f = f(x_0 + \Delta x, y_0 + \Delta y) - f(x_0, y_0)$, whereas the vertical distance to T is $f(x_0, y_0) + df$. So the difference, $\Delta f - df$, represents the separation between the surface and the tangent plane. The smaller Δx and Δy are, the smaller this separation is, so that df becomes a better and better approximation to Δf. These ideas, which are analogous to those of the one-variable case, are illustrated in Figure 17.7.

Another result obtainable from equation (17.28) is stated in the following theorem. By a horizontal tangent plane, we mean one that is parallel to the xy-plane.

FIGURE 17.7

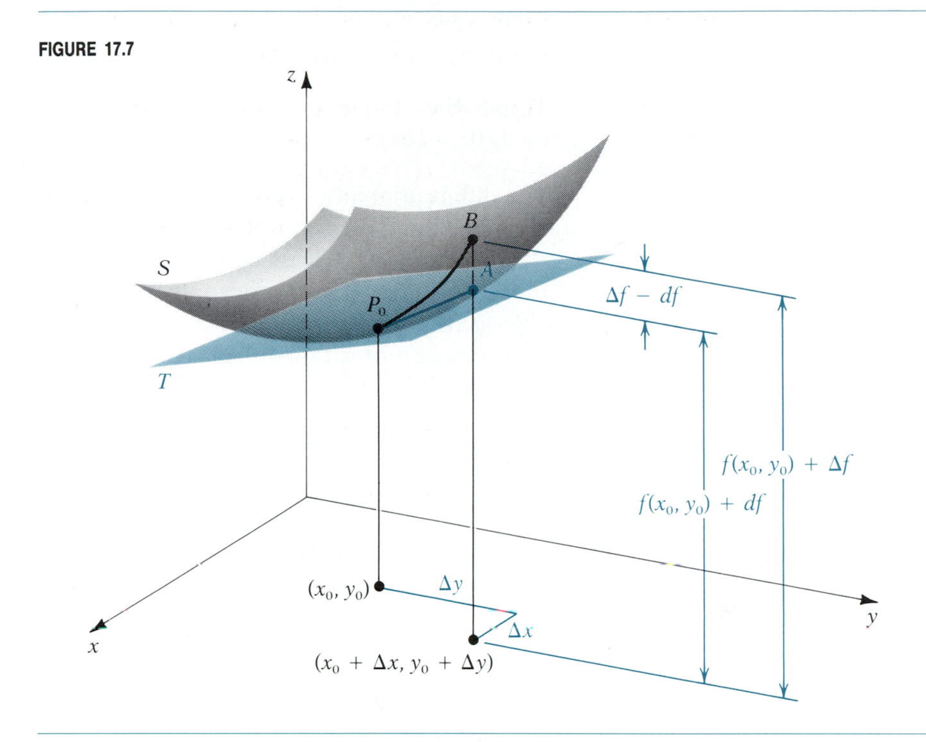

THEOREM 17.6 For a differentiable function f, the surface $z = f(x, y)$ has a horizontal tangent plane at (x_0, y_0) if and only if $\nabla f(x_0, y_0) = \mathbf{0}$.

Proof If $\nabla f(x_0, y_0) = \mathbf{0}$, then $f_x(x_0, y_0)$ and $f_y(x_0, y_0)$ are both 0. So equation (17.28) reduces to $z - z_0 = 0$, or $z = z_0$, which is the equation of a horizontal plane. Conversely, if the tangent plane is horizontal at (x_0, y_0), its equation must be of the form $z = z_0$, so that $z - z_0 = 0$. For equation (17.28) to reduce to this for all values of x and y, the coefficients of the x and y terms—namely, $f_x(x_0, y_0)$ and $f_y(x_0, y_0)$—must both be 0. So $\nabla f(x_0, y_0) = \mathbf{0}$. ■

EXAMPLE 17.14 Find all points on the surface $z = x^2 - 2xy^2 + 8x$ at which the tangent plane is horizontal.

Solution Let $f(x, y) = x^2 - 2xy^2 + 8x$. Then

$$\nabla f = \langle f_x, f_y \rangle = \langle 2x - 2y^2 + 8, -4xy \rangle$$

Both f_x and f_y are continuous and so by Theorem 17.2 f is differentiable everywhere. To find where $\nabla f = \mathbf{0}$ we solve the two equations $2x - 2y^2 + 8 = 0$ and $-4xy = 0$ simultaneously. From the second equation we must have $x = 0$ or $y = 0$. The corresponding values from the first equation are found as follows:

$$\underline{x = 0}: \quad -2y^2 + 8 = 0 \qquad\qquad \underline{y = 0}: \quad 2x + 8 = 0$$
$$y^2 = 4 \qquad\qquad x = -4$$
$$y = \pm 2$$

So the points in the domain at which $\nabla f = \mathbf{0}$ are $(0, 2)$, $(0, -2)$, and $(-4, 0)$. To find the corresponding points on the surface, we calculate $z = f(x, y)$ in each case:

$$f(0, 2) = 0 \qquad f(0, -2) = 0 \qquad f(-4, 0) = -16$$

Thus, the tangent plane is horizontal at $(0, 2, 0)$, $(0, -2, 0)$, and $(-4, 0, -16)$. ■

The differentiability hypothesis in Theorem 17.6 is essential, since otherwise a tangent plane does not even exist. For example, consider the function $f(x, y) = -\sqrt{|xy|}$, whose computer-generated graph is shown in Figure 17.8. It can be shown (see Exercise 41 in Exercise Set 17.1) that f is not differentiable at the origin, even though it is continuous there and $\nabla f(0, 0) = \mathbf{0}$. The nonexistence of a tangent plane at the origin can be seen clearly from the graph.

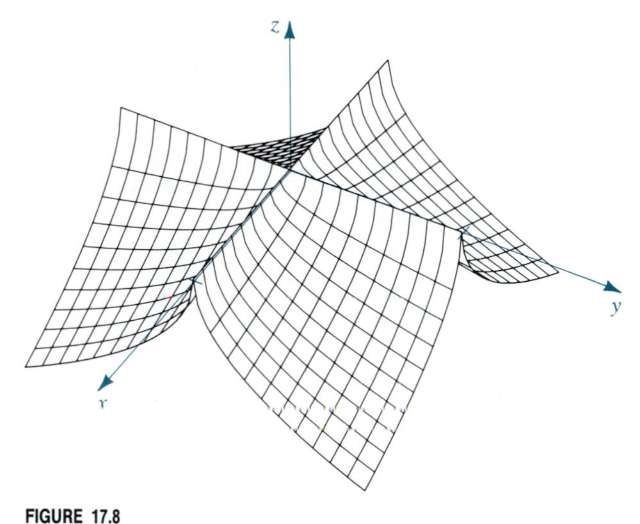

FIGURE 17.8

EXERCISE SET 17.4

A

In Exercises 1–14 find equations of the tangent plane and normal line to the given surface at the specified point.

1. $2x^2 + 3y^2 - z^2 = 5$; $(3, -2, 5)$
2. $z^2 = 3x^2 + 4y^2$; $(-2, 1, -4)$
3. $4x^2 + 3y^2 + 2z^2 = 12$; $(1, 0, -2)$
4. $xy + 2yz + 3xz = 16$; $(4, -2, 3)$
5. $(x + y)^2 + (y + z)^2 + (x + z)^2 = 10$; $(1, -1, 2)$
6. $xe^{2y-z} - 3 = 0$; $(3, 1, 2)$
7. $y = \ln\left(\dfrac{x + 2y}{y + 2z}\right) - 1$; $(3, -1, 1)$
8. $\sin(x/y) + \cos(y/z) = 0$; $(\pi, 1, \frac{2}{\pi})$
9. $z = x^2 - 2y^2$; $(5, 4, -7)$
10. $z = \ln\sqrt{x + y}$; $(3, -2, 0)$
11. $z = \tan^{-1}\frac{y}{x}$; $(1, -1, -\frac{\pi}{4})$
12. $z = \dfrac{x + y}{x - y}$; $(4, 3, 7)$
13. $z = e^{2x}\sin 3y$; $(0, \frac{\pi}{2}, -1)$
14. $z = \sqrt{\dfrac{x - 2y}{x + 2y}}$; $(5, -2, 3)$

In Exercises 15–20 find all points on the given surface at which the tangent plane is horizontal.

15. $z = x^2 - y^2 + 2x + 6y$
16. $z = 3x^2 + 4y^2 - 12x + 8y$
17. $z = x^2 - 2xy + 3y^2$
18. $z = xy + 4x - 2y$
19. $z = x^2 - 2xy + 4x - 2y$
20. $z = x^3 + 12xy - y^3$

21. Find the point on the hyperbolic paraboloid $z = 2x^2 - 3y^2$ at which the tangent plane is parallel to the plane $4x + 9y - 2z = 11$.

22. Find the point on the elliptic paraboloid $z = 3x^2 + 4y^2$ at which the tangent plane is perpendicular to the line through the points $(1, -2, 4)$ and $(-2, 0, 3)$.

B

In Exercises 23 and 24 assume F and G are differentiable functions at $P_0 = (x_0, y_0, z_0)$ and have nonzero gradients there.

23. The surfaces defined by $F(x, y, z) = 0$ and $G(x, y, z) = 0$ are said to be *tangent* at P_0 if they have the same tangent plane there.
   (a) Prove that the surfaces are tangent at P_0 if and only if $\nabla F(P_0) = k\nabla G(P_0)$ for some nonzero scalar k.
   (b) Find all points P_0 at which the surfaces $x^2 + 2y^2 - 2z^2 = 20$ and $xy - yz + 2xz = 5$ are tangent to each other.

24. The surfaces defined by $F(x, y, z) = 0$ and $G(x, y, z) = 0$ are said to be *orthogonal* at P_0 if their normal lines at P_0 are perpendicular to each other.
   (a) Prove that the surfaces are orthogonal at P_0 if and only if $\nabla F(P_0) \cdot \nabla G(P_0) = 0$.
   (b) Find two points P_0 at which the surfaces defined by $2x^2 - 3y^2 + 4z^2 = 10$ and $2x^2 + y^2 - 4z^2 + 2y - z = 21$ are orthogonal to each other.

25. Find all points on the surface $z = 2x^3 - 6x^2y + 9y^2 + 2y^3$ at which the tangent plane is horizontal.

26. The angle between a line l and a surface S is defined as the complement of the acute angle between l and the normal to S at the point where l pierces S. Find the angle between the line $\mathbf{r}(t) = \langle t, 2t, 2 - t\rangle$ and the elliptic cone $2z^2 = 4x^2 + y^2$ at each point of intersection.

27. The angle between a curve C and a surface S is defined as the angle between the tangent line to C and the surface at each point of intersection. (See Exercise 26.) Find the angle between the curve $\mathbf{r}(t) = \langle 1 - t, 2 + t, t^2\rangle$ and the paraboloid $9z = 4x^2 + y^2$ at each of their points of intersection.

28. Prove that if f is a function of two variables that is differentiable at (x_0, y_0), then $F(x, y, z) = f(x, y) - z$ is differentiable at $P_0 = (x_0, y_0, z_0)$, where $z_0 = f(x_0, y_0)$.

17.5 EXTREME VALUES

Just as in the case of one variable, some of the most important applications of multivariate differential calculus involve finding maximum and minimum values of functions. We will concentrate in this section on the two-variable case, but much of the theory can be extended to functions of three or more variables.

The definitions of local and absolute maxima and minima (which are referred to collectively as extreme values, or extrema) parallel those for functions of one variable.

DEFINITION 17.5 Let f be a function of two variables with domain D. Then f is said to have a **local maximum** at a point (x_0, y_0) in D if there exists some neighborhood N of (x_0, y_0) such that for all points (x, y) of D that lie in N, $f(x_0, y_0) \geq f(x, y)$. If this inequality holds true for all points (x, y) in D, then f is said to have an **absolute maximum** at (x_0, y_0). When the reverse inequality holds, f has a **local minimum** at (x_0, y_0) in the first instance and an **absolute minimum** in the second. ■

The terms *relative maximum* and *relative minimum* are often used instead of *local* maximum and minimum. If f has a local maximum at (x_0, y_0), then $f(x_0, y_0)$ is called a **local maximum value** of f, and the point $(x_0, y_0, f(x_0, y_0))$ is a **local maximum point** on the graph of f. This point is

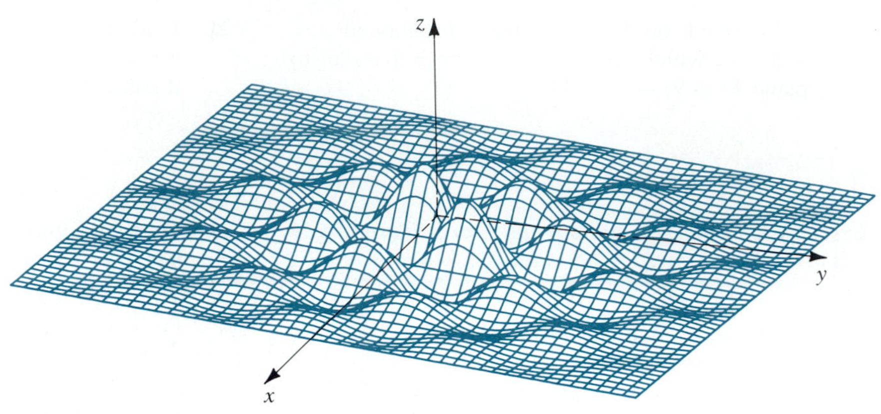

FIGURE 17.9

the highest point on the graph in its immediate vicinity. Similar remarks apply for a local minimum. The absolute maximum value of f, if it exists, is the largest of its local maximum values, and the absolute minimum value is the smallest of its local minimum values.

Figure 17.9 shows the computer-generated graph of the function

$$f(x, y) = 4e^{-\sqrt{x^2+y^2}/4} \sin x \sin y$$

having many local maxima and minima.

Not all functions have extreme values, but a fundamental theorem of analysis (proved in advanced calculus) states that *if f is continuous on a closed and bounded domain D, then f has both an absolute maximum and an absolute minimum in D.* Local (and absolute) extreme values can occur either at interior points of D or on its boundary. We will be concerned primarily with extrema that occur at interior points, but we illustrate by an example how to deal with boundary points when equations of the curves that make up the boundary are known.

Sometimes it is possible to determine maximum and minimum values without using calculus. For polynomial functions of degree 2, the technique of completing the square is especially useful in this regard, as illustrated in the following example.

EXAMPLE 17.15 Find the absolute extrema and where they occur for the function

$$f(x, y) = x^2 + y^2 - 2x + 4y + 10$$

Solution We complete the squares on x and y, getting

$$\begin{aligned} f(x, y) &= (x^2 - 2x + 1) + (y^2 + 4y + 4) + 10 - 1 - 4 \\ &= (x-1)^2 + (y+2)^2 + 5 \end{aligned}$$

The two squared terms are positive except when $x = 1$ and $y = -2$, when each is 0. So the absolute minimum value of f occurs at $(1, -2)$ and this minimum value is 5. There is no maximum value. ■

Since problems of this type are rather specialized, we need to develop methods for handling a wider class of problems. The first task is to find a systematic way of determining points in the domain at which maximum or minimum values *might* occur. Then we need a test to see the actual nature of the function at these points.

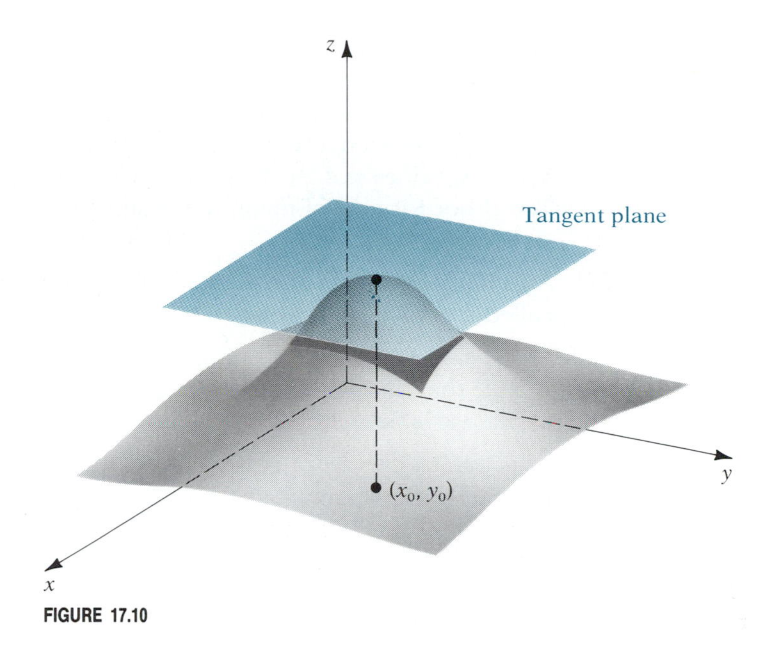

FIGURE 17.10

If (x_0, y_0) is an interior point of D at which f has a local maximum or minimum and if f is differentiable there, then it is geometrically evident that the tangent plane to the graph of f at $(x_0, y_0, f(x_0, y_0))$ must be horizontal. (See Figure 17.10 for the case of a maximum.) Thus, by Theorem 17.6 we know that $\nabla f(x_0, y_0) = \mathbf{0}$. The next theorem shows that this conclusion holds even if f is not differentiable, provided f_x and f_y exist at (x_0, y_0).

THEOREM 17.7 Let (x_0, y_0) be an interior point of the domain of f at which f has either a local maximum or a local minimum value. Then $\nabla f(x_0, y_0) = \mathbf{0}$ or else $\nabla f(x_0, y_0)$ does not exist.

Proof Suppose $\nabla f(x_0, y_0)$ does exist—that is, $f_x(x_0, y_0)$ and $f_y(x_0, y_0)$ both exist—and suppose for definiteness that f has a local maximum at (x_0, y_0). Then, as shown in Figure 17.11, the curves $z = f(x, y_0)$ and $z = f(x_0, y)$ formed by the

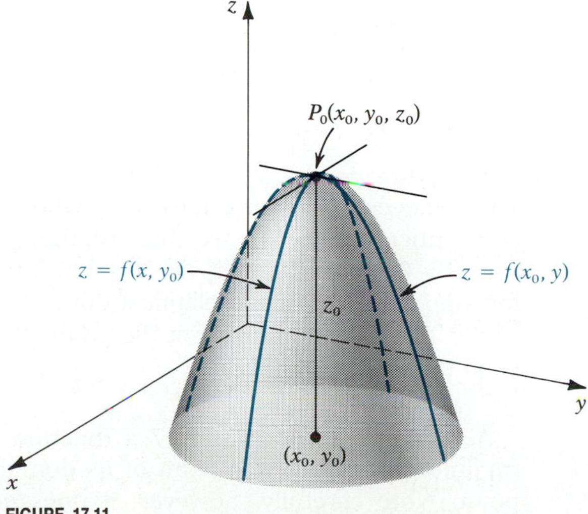

FIGURE 17.11

intersection of the surface $z = f(x, y)$ with the planes $y = y_0$ and $x = x_0$, respectively, have maximum points at $P_0(x_0, y_0, z_0)$, where $z_0 = f(x_0, y_0)$. Thus, their slopes at P_0 both equal 0. But these slopes are $f_x(x_0, y_0)$ and $f_y(x_0, y_0)$, respectively. So $\nabla f(x_0, y_0) = \mathbf{0}$. The only other possibility is that $\nabla f(x_0, y_0)$ does not exist. So the theorem is proved for a local maximum. The proof for a local minimum is similar. ∎

DEFINITION 17.6 An interior point in the domain of f at which $\nabla f = \mathbf{0}$ or ∇f fails to exist is called a **critical point** of f. ∎

Note that this definition is analogous to that of a critical point for a function of one variable, with ∇f replacing f'.

EXAMPLE 17.16 Find all critical points of

$$f(x, y) = x^3 - 3x^2y + 6y^2 + 24y$$

Solution We calculate $\nabla f = \langle 3x^2 - 6xy, -3x^2 + 12y + 24 \rangle$. Since this exists everywhere, the only critical points are those for which $\nabla f = \mathbf{0}$. We set $f_x = 0$ and $f_y = 0$ and solve simultaneously:

$$\begin{array}{c|c} 3x^2 - 6xy = 0 & -3x^2 + 12y + 24 = 0 \\ 3x(x - 2y) = 0 & x^2 - 4y - 8 = 0 \\ x = 0 \quad \text{or} \quad x = 2y & \end{array}$$

If $x = 0$, then the right-hand equation becomes $-4y - 8 = 0$, so that $y = -2$. Thus, $(0, -2)$ is a critical point. Substituting $x = 2y$ into the right-hand equation yields $4y^2 - 4y - 8 = 0$, whose solutions are readily found to be $y = 2$ and $y = -1$. The corresponding x values are 4 and -2. The critical points are therefore $(0, -2)$, $(4, 2)$, and $(-2, -1)$. ∎

EXAMPLE 17.17 Find all critical points of

$$f(x, y) = (3x^2 + 4y^2)^{1/2}$$

and determine the nature of the function at each.

Solution

$$\nabla f = \left\langle \frac{3x}{\sqrt{3x^2 + 4y^2}}, \frac{4y}{\sqrt{3x^2 + 4y^2}} \right\rangle$$

This is never $\mathbf{0}$. It is undefined only when $x = 0$ and $y = 0$. So $(0, 0)$ is the only critical point. We see that $f(0, 0) = 0$ and for all other points (x, y), $f(x, y) > 0$. Thus $f(0, 0)$ is the absolute minimum value. The graph of f is the upper nappe of the elliptical cone $z^2 = 3x^2 + 4y^2$, pictured in Figure 17.12. At the minimum point the graph comes to a sharp point, and there is no tangent plane there. ∎

According to Theorem 17.7 a function can have a local maximum or minimum at an interior point of its domain only if that point is a critical point. Note carefully, however, it *does not* say that f *will* have a local maximum or minimum at each critical point. We can indicate this briefly as follows:

$$\text{local extremum at } (x_0, y_0) \Longrightarrow (x_0, y_0) \text{ is a critical point}$$
$$(x_0, y_0) \text{ is a critical point } \not\Longrightarrow \text{ local extremum at } (x_0, y_0)$$

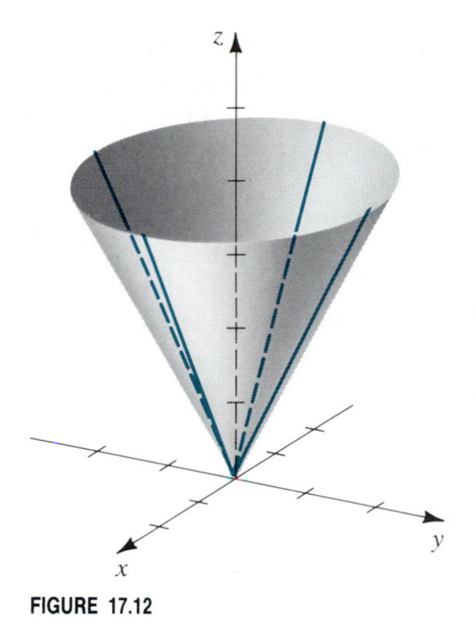

FIGURE 17.12

To verify that a point can be a critical point but not a point where the function has an extreme value, consider $f(x, y) = xy$. Since $\nabla f(x, y) = \langle y, x \rangle$, we see that $\nabla f(0, 0) = \mathbf{0}$. So the origin is a critical point. At this point the function value is 0, but for all other points (x, y) with x and y like in sign, $f(x, y) > 0$, whereas if x and y are unlike in sign, $f(x, y) < 0$. So there is no neighborhood of (0, 0) throughout which $f(0, 0)$ is either the largest or the smallest value. The graph of f in this case is the hyperbolic paraboloid shown in Figure 17.13. The origin in this case is a saddle point; the general definition of a saddle point is a critical point at which f assumes neither a local maximum nor a local minimum value.

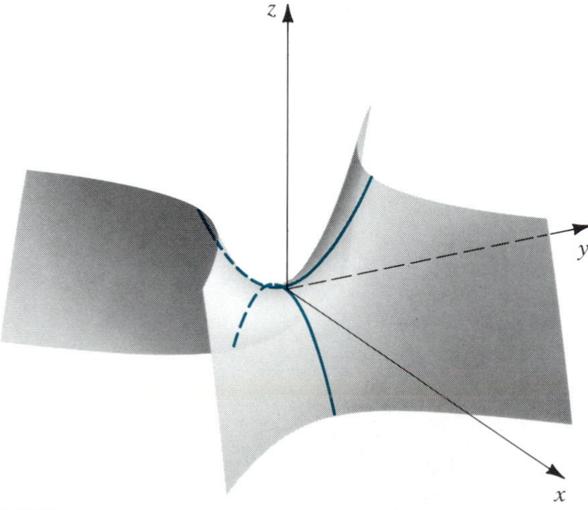

FIGURE 17.13

We now state a test that applies to critical points where $\nabla f = \mathbf{0}$. It is somewhat restrictive in that it assumes the continuity of all second-order partial derivatives, but fortunately in practice this condition is usually met. The validity of the test is shown in most advanced calculus textbooks.

A TEST FOR LOCAL EXTREMA

Let f have continuous second partial derivatives in some neighborhood of a critical point (x_0, y_0). Define

$$D(x, y) = f_{xx}(x, y)f_{yy}(x, y) - (f_{xy}(x, y))^2$$

1. If $D(x_0, y_0) > 0$, then f has a local maximum at (x_0, y_0) if $f_{xx}(x_0, y_0) < 0$ and f has a local minimum at (x_0, y_0) if $f_{xx}(x_0, y_0) > 0$.
2. If $D(x_0, y_0) < 0$, then f has a saddle point at (x_0, y_0).
3. If $D(x_0, y_0) = 0$, then the test is inconclusive.

Remarks D is called the **discriminant function** of f. When it is positive at (x_0, y_0), the test for maximum and minimum values parallels the second-derivative test for single-variable functions, with f_{xx} taking the place of f''.

If the test is inconclusive or not applicable, then it may be possible to determine the nature of f by examining its values near the critical point.

EXAMPLE 17.18 Find all local maximum and minimum values of the function f defined by

$$f(x, y) = x^2 - 2xy + 3y^2 + 6x - 10y + 5$$

Solution The critical points are those for which $\nabla f = \langle 2x - 2y + 6, -2x + 6y - 10\rangle = \langle 0, 0\rangle$. Setting each component equal to 0 and solving yields $(-2, 1)$ as the only critical point. (You should verify this.) To test this point we first calculate the second partials of f:

$$f_{xx} = 2 \qquad f_{xy} = -2 \qquad f_{yy} = 6$$

So $D(x, y) = 2(6) - (-2)^2 = 12 - 4 = 8 > 0$. Thus, $D(-2, 1) = 8$ also. Since $f_{xx}(-2, 1) = 2 > 0$, we conclude that f has a local minimum at $(-2, 1)$. To get the local minimum value, we calculate $f(-2, 1)$ and find it to be -6. There is no maximum value. A computer-generated graph of f is shown in Figure 17.14. ■

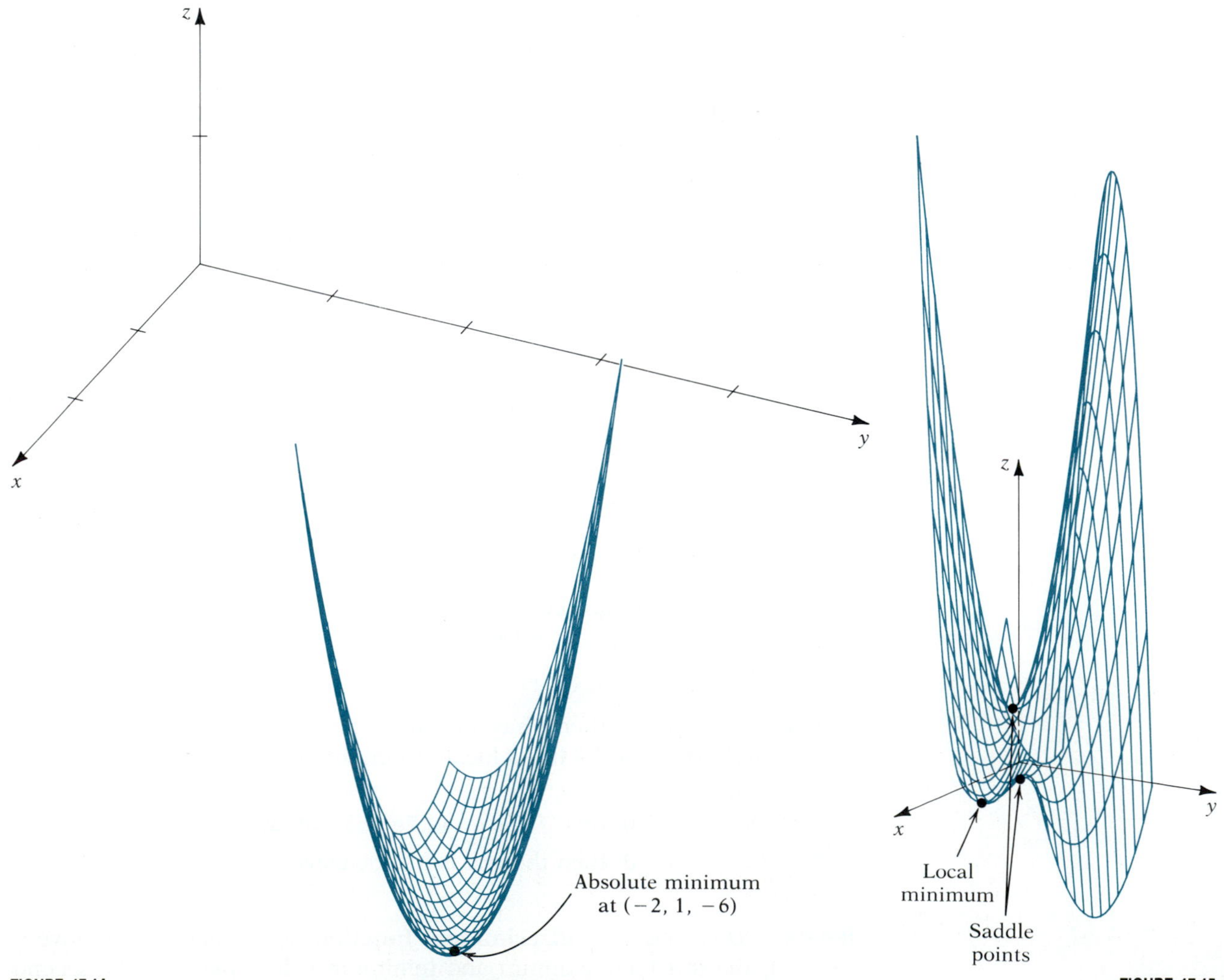

FIGURE 17.14

FIGURE 17.15

EXAMPLE 17.19 Find and classify all extrema of $f(x, y) = x^3 - 3x^2y + 6y^2 + 24y$.

Solution This is the function of Example 17.16, and we found the critical points to be $(0, -2)$, $(4, 2)$, and $(-2, -1)$. To test them we need the second partials:

$$f_x = 3x^2 - 6xy \qquad f_y = -3x^2 + 12y + 24$$
$$f_{xx} = 6x - 6y \qquad f_{xy} = -6x \qquad f_{yy} = 12$$

The following tabular arrangement helps to keep track of things.

(x_0, y_0)	$f_{xx}(x_0, y_0)$	$f_{xy}(x_0, y_0)$	$f_{yy}(x_0, y_0)$	$D(x_0, y_0)$	*Test result*
$(0, -2)$	12	0	12	144	Minimum
$(4, 2)$	12	-24	12	-432	Saddle point
$(-2, -1)$	-6	12	12	-216	Saddle point

So the only local extremum is $(0, -2)$, where f has a minimum value. The minimum value is $f(0, -2) = -24$. Figure 17.15 shows a computer-generated graph of f, where the vertical scale has been compressed. ■

EXAMPLE 17.20 A crate in the shape of a rectangular box is to be constructed so that its volume is 270 cu ft. The sides and top each cost \$1 per square foot to construct, and the bottom, which must be stronger, costs \$1.50 per square foot. What are the dimensions of the crate that will yield the minimum cost? What is the minimum cost?

Solution Let the dimensions x, y, and z be as shown in Figure 17.16. Then since there are two sides of area xz and two ends of area yz, the cost function C is given by

$$C = \overbrace{2xz}^{\text{sides}} + \overbrace{2yz}^{\text{ends}} + \overbrace{1.5xy}^{\text{bottom}} + \overbrace{xy}^{\text{top}} = 2(xz + yz) + 2.5xy$$

Also, we are given that the volume must be 270. So the variables are related by the equation

$$xyz = 270$$

We solve for z and substitute in the cost function, reducing C to a function of two variables only:

$$z = \frac{270}{xy}$$

$$C = 2\left(\frac{270}{y} + \frac{270}{x}\right) + 2.5xy$$

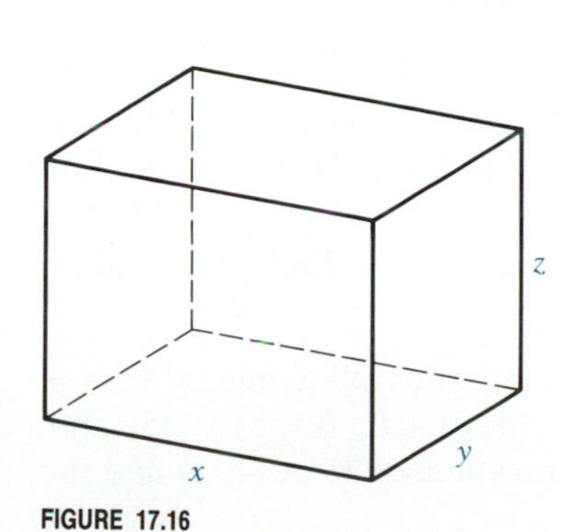

FIGURE 17.16

To find the minimum value, we proceed in the usual way:

$$\nabla C = \left\langle \frac{-540}{x^2} + \frac{5}{2}y, \frac{-540}{y^2} + \frac{5}{2}x \right\rangle$$

$$\frac{-540}{x^2} + \frac{5}{2}y = 0 \qquad\Bigg|\qquad \frac{-540}{y^2} + \frac{5}{2}x = 0$$

$$y = \frac{216}{x^2} \qquad\Bigg|\qquad x = \frac{216}{y^2}$$

Substituting for y from the bottom equation on the left in the one on the right and simplifying, we get

$$216x = x^4 \quad \text{or} \quad x(x^3 - 216) = 0$$

Clearly $x \neq 0$, so $x^3 = 216$ or $x = 6$. Thus $y = \frac{216}{36} = 6$. We can test to verify that this critical point results in a minimum value for C, but because it is the only critical point and it is clear from the nature of the problem that C has some minimum value, it is not essential that the test be carried out. For $x = 6$ and $y = 6$, we find $z = 7.5$. So the dimensions should be $6 \times 6 \times 7.5$ ft, and the cost is

$$C = 2\left(\tfrac{270}{6} + \tfrac{270}{6}\right) + 2.5(36) = \$270$$

■

The next example illustrates how to determine extreme values that occur on the boundary. The technique works whenever the boundary consists of a finite number of curves on each of which the given function can be expressed in terms of one variable.

EXAMPLE 17.21 Find the absolute maximum and minimum values of

$$f(x, y) = x^2 - 2xy + 3y^2 - 4x$$

on the closed trapezoidal region pictured in Figure 17.17.

Solution Name the boundary segments C_1, C_2, C_3, and C_4, as shown. First we look for points in the interior where extrema occur:

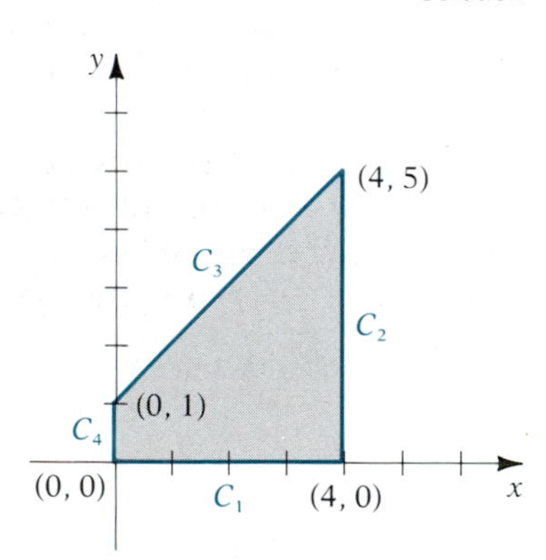

FIGURE 17.17

$$\nabla f = \langle 2x - 2y - 4, \ -2x + 6y \rangle$$

$$\begin{array}{r|r} 2x - 2y - 4 = 0 & -x + 3y = 0 \\ x = y + 2 & x = 3y \end{array}$$

Substituting, we get $3y = y + 2$, or $y = 1$. Thus $x = 3$. To test, we have $f_{xx} = 2$, $f_{xy} = -2$, and $f_{yy} = 6$. So $D(x, y) = 12 - (-2)^2 = 8 > 0$, and since $f_{xx} > 0$, we conclude that f has a local minimum at (3, 1). The value there is found to be $f(3, 1) = -6$.

Now we consider each boundary segment. On C_1 we have $y = 0$. So the function becomes $f(x, 0) = x^2 - 4x$. By setting the derivative equal to 0 and testing, we find that $x = 2$ yields a minimum value—namely, $f(2, 0) = -4$. The end point values are $f(0, 0) = 0$ and $f(4, 0) = 0$.

We sketch briefly the results along C_2, C_3, and C_4. You should verify these. In each case we are working with one variable only.

C_2: $\underline{x = 4}$. $f(4, y) = 3y^2 - 8y$. Minimum at $y = \frac{4}{3}$. $f(4, \frac{4}{3}) = -\frac{16}{3}$. End point values: $f(4, 0) = 0$, $f(4, 5) = 35$.

C_3: $\underline{y = x + 1}$. $f(x, x + 1) = 2x^2 + 3$. No interior critical values. End point values: $f(4, 5) = 35$, $f(0, 1) = 3$.

C_4: $\underline{x = 0}$. $f(0, y) = 3y^2$. No interior critical values. End point values already found.

The extreme values are to be found among the following: $f(3, 1) = -6$, $f(2, 0) = -4$, $f(4, \frac{4}{3}) = -\frac{16}{3}$, $f(0, 0) = 0$, $f(4, 0) = 0$, $f(4, 5) = 35$, and $f(0, 1) = 3$. So f assumes the absolute maximum value of 35 at (4, 5) and the absolute minimum value of -6 at (3, 1). ■

EXERCISE SET 17.5

A

In Exercises 1–4 find the extreme values of f and where they occur by completing the square.

1. $f(x, y) = x^2 + y^2 + 2x - 4y + 3$
2. $f(x, y) = 2x^2 + 3y^2 - 4x + 6y - 9$
3. $f(x, y) = x^4 + y^2 - 8y + 13$
4. $f(x, y) = x^2 + y^4 - 2y^2 + 4x + 1$

In Exercises 5–26 locate all critical points and at each such point determine whether f has a local maximum, a local minimum, or a saddle point.

5. $f(x, y) = x^2 - 6xy + 2y^3 - 8x - 16$
6. $f(x, y) = x^2 + y^3 - 4xy - 8x + 13y + 1$
7. $f(x, y) = 2x^3 - 6xy + y^2 + 30$
8. $f(x, y) = x^3 + 3xy^2 - 3x^2 - 3y^2 + 4$
9. $f(x, y) = 2x^2 + y^4 - 4xy + 2$
10. $f(x, y) = x^4 + 2y^2 + 8xy - 7$
11. $f(x, y) = 6xy - x^2 - y^3$
12. $f(x, y) = 4xy - 2x^2 - y^3 + 3$
13. $f(x, y) = x^4 - 2x^2y + y^3 - y$
14. $f(x, y) = x^4 - y^4 - 4x^2y^2 + 20y^2$
15. $f(x, y) = xy + \dfrac{1}{x} + \dfrac{2}{y}$
16. $f(x, y) = 4 - \dfrac{2}{x} - \dfrac{1}{y} - x^2y$
17. $f(x, y) = \dfrac{8}{x^2} - \dfrac{1}{y} + 2x - y$
18. $f(x, y) = \dfrac{8}{xy} + \dfrac{2}{x} - \dfrac{4}{y}$
19. $f(x, y) = x^3 + 2x^2y + y^3 + x$
20. $f(x, y) = 2x^3 + 3y^3 + xy^2 + 2y$
21. $f(x, y) = e^x(x^2 - y^2)$
22. $f(x, y) = e^{-y}(x^2 - 2x + 3y)$
23. $f(x, y) = \sin x \sin y,\ -\pi < x < \pi,\ -\pi < y < \pi$
24. $f(x, y) = \sin^2 x - 2\cos^2 y,\ -\dfrac{\pi}{4} < x < \dfrac{3\pi}{4},\ -\dfrac{\pi}{4} < y < \dfrac{3\pi}{4}$
25. $f(x, y) = x^2 - 2x\cos y + 1,\ 0 \le y < 2\pi$
26. $f(x, y) = y^2 - 4y(\sin x + \cos x),\ -\pi < x < \pi$
27. Show that $f(x, y) = 4 - x^{2/3} + 2x^{1/3}y^{1/3} - y^{2/3}$ has a critical point at (0, 0) for which ∇f does not exist and that f has a local (and absolute) maximum value there.

B

In Exercises 28–31 find the absolute maximum and absolute minimum values of f on the closed domain bounded by the given curves.

28. $f(x, y) = x^2 + 2y^3$; the line segments joining (0, 0), (2, 0), and (0, 1)
29. $f(x, y) = x^2y - xy^2 - y$; x-axis, y-axis, $x = 1$, $y = 1$
30. $f(x, y) = 2x^3 - 3x^2y + 2y^3 - 3y$; $x + y = \pm 1$, $x - y = \pm 1$
31. $f(x, y) = x^2 - xy - x + y$; $y = 5 - x^2$, $y = 0$
32. Find the absolute maximum and minimum values of $f(x, y) = x^3 - y^3 - 3x$ on the closed unit disk $x^2 + y^2 \le 1$. At what points do these extremes occur? (*Hint:* Use the parameterization $x = \cos t$ and $y = \sin t$ for $0 \le t \le 2\pi$.)
33. An open-top rectangular box is to have a volume of 256 cu ft. What dimensions will require the least amount of material?
34. Find the point on the plane $3x + 2y - z = 4$ that is nearest the origin. [*Hint:* Minimize the *square* of the distance of the point (x, y, z) from the origin, using the fact that z satisfies the given equation.]
35. The temperature on the surface of the hemisphere $z = \sqrt{1 - x^2 - y^2}$ is given by $T(x, y, z) = 400xyz^2$. Find the hottest and coldest temperatures on the hemisphere and the points where these extremes occur.
36. A company makes two types of automatic ice cream freezers, type A and type B. The cost C of producing x type A and y type B machines per day is
$C(x, y) = x^2 + xy + y^2 + 20x - 20y$
and the revenue from selling x type A and y type B machines per day is $R(x, y) = 100x + 80y$. How many machines of each type should be manufactured and sold each day to maximize profit? What is the maximum profit?

37. An open-top rectangular box is to be constructed with a divider in the middle. The unit cost of the divider is half that of the bottom and sides. If the volume is to be 320 cu in., find the dimensions that minimize the cost.

38. A common problem in experimental work is to find the line $y = mx + b$ that "fits" a set of data points $(x_1, y_1), (x_2, y_2), \ldots, (x_n, y_n)$ best in the sense that the sum of the squares of the vertical deviations of the data points from the line is minimum. This line is said to fit the data best in the sense of *least squares*. So the problem is to find m and b such that

$$F(m, b) = \sum_{k=1}^{n} (y_k - mx_k - b)^2$$

is a minimum. Determine the values of m and b.

39. Use the result of Exercise 38 to find the line that best fits the data points (0, 1), (1, 4), (2, 4), (3, 6), and (6, 5). Draw a graph showing the line and the data points.

17.6 CONSTRAINED EXTREMUM PROBLEMS

In extremum problems we often seek to maximize or minimize some function subject to a *constraint* on the variables. This was the case in Example 17.20. There we wanted to find the dimensions x, y, and z of a box that minimized the cost, subject to the constraint that the volume had to be constant at 270; that is, the constraint on the variables was that $xyz = 270$. In that example we solved the constraint equation for z and substituted into the cost function, thereby reducing it to a function of two variables. In this section we consider an alternate method of solving such problems that in many cases is easier. In fact, depending on the nature of the constraint equation, it may be difficult or impossible to use the substitution method.

The method we describe is called the **method of Lagrange multipliers,** after the French–Italian mathematician Joseph Louis Lagrange (1736–1813). We consider first the simplest case in which a function of two variables, say $f(x, y)$, is to be maximized or minimized subject to a constraint on x and y. We assume this constraint to be expressed as an equation, say $g(x, y) = 0$. Geometrically, $g(x, y) = 0$ represents a curve C in the xy-plane, and we say that f has a *constrained local maximum* at a point (x_0, y_0) on C provided there is a neighborhood N of (x_0, y_0) such that for all (x, y) *on* C that lie in N, $f(x_0, y_0) \geq f(x, y)$. (See Figure 17.18.) A similar definition applies for a *constrained local minimum*. The basis for the method of Lagrange multipliers is given in the following theorem.

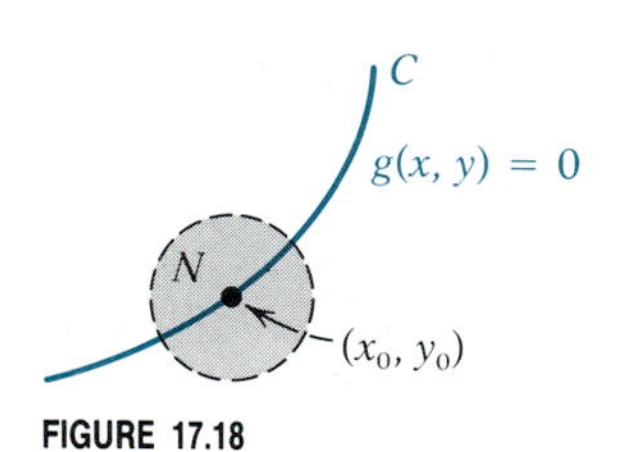

FIGURE 17.18

THEOREM 17.8 Let $f(x, y)$ have a constrained local maximum or minimum at (x_0, y_0), with the constraint curve C given by $g(x, y) = 0$. Then if f and g are differentiable in some neighborhood N of (x_0, y_0) and if $\nabla g(x_0, y_0) \neq \mathbf{0}$, there exists a constant λ such that

$$\nabla f(x_0, y_0) = \lambda \nabla g(x_0, y_0) \tag{17.29}$$

Proof It is shown in advanced calculus that under the given hypotheses, the curve C has a parametric representation

$$\begin{cases} x = x(t) \\ y = y(t) \end{cases}$$

with $x(t)$ and $y(t)$ differentiable, and such that the tangent vector $\langle x'(t_0), y'(t_0)\rangle \neq \mathbf{0}$, where t_0 is the parameter value that corresponds to (x_0, y_0); that is, $(x_0, y_0) = (x(t_0), y(t_0))$.

Write $F(t) = f(x(t), y(t))$. Then the single-variable function F has a local maximum or minimum at t_0, and so $F'(t_0) = 0$. Using the chain rule, we therefore have

$$\begin{aligned} 0 = F'(t_0) &= f_x(x_0, y_0)x'(t_0) + f_y(x_0, y_0)y'(t_0) \\ &= \nabla f(x_0, y_0) \cdot \langle x'(t_0), y'(t_0) \rangle \end{aligned}$$

It follows that $\nabla f(x_0, y_0)$ is orthogonal to C at (x_0, y_0).

Now $g(x, y) = 0$ can be interpreted as a level curve of $z = g(x, y)$, and as we saw in Section 17.3, ∇g is orthogonal to such a level curve at each point on the curve. In particular, $\nabla g(x_0, y_0)$ is orthogonal to C at (x_0, y_0). As we have just shown, $\nabla f(x_0, y_0)$ also has this property. Hence, $\nabla f(x_0, y_0)$ and $\nabla g(x_0, y_0)$ are parallel vectors. Thus,

$$\nabla f(x_0, y_0) = \lambda \nabla g(x_0, y_0)$$

for some constant λ. ■

The number λ asserted to exist by this theorem is called a **Lagrange multiplier.** To make use of the theorem to find where f can assume a constrained local extreme value, we write the component equations that arise from equation (17.29)—namely,

$$f_x(x_0, y_0) = \lambda g_x(x_0, y_0)$$
$$f_y(x_0, y_0) = \lambda g_y(x_0, y_0)$$

These, together with the constraint equation

$$g(x_0, y_0) = 0$$

constitute a system of three equations in the three unknowns x_0, y_0, and λ, which we must solve simultaneously. Our objective is to find x_0 and y_0, and the multiplier λ is just a means to an end. So we might attempt to eliminate λ from the three equations as a first step. However, there are times when this is not feasible. It might be best, in fact, in some cases to solve first for λ and then get x_0 and y_0. The examples that follow illustrate some possible strategies.

EXAMPLE 17.22 Find the points on the ellipse $x^2 + 2y^2 = 6$ at which the function $f(x, y) = x^2y$ assumes its largest and smallest values. What are these values?

Solution The constraint on points (x, y) is that they lie on the ellipse $x^2 + 2y^2 = 6$. We write this in the form $g(x, y) = 0$ by letting $g(x, y) = x^2 + 2y^2 - 6$. Now $\nabla f(x, y) = \langle 2xy, x^2 \rangle$ and $\nabla g(x, y) = \langle 2x, 4y \rangle$. So from $\nabla f = \lambda \nabla g$ and $g(x, y) = 0$, we get the three equations

$$\begin{cases} 2xy = 2\lambda x \\ x^2 = 4\lambda y \\ x^2 + 2y^2 - 6 = 0 \end{cases}$$

From the first of these we have $2x(y - \lambda) = 0$ so that either $x = 0$ or $\lambda = y$. If $x = 0$, then from the third equation $2y^2 = 6$, or $y = \pm\sqrt{3}$. When $\lambda = y$, we substitute for λ in the second equation and get $x^2 = 4y^2$, or $x = \pm 2y$. Then replacing x^2 by $4y^2$ in the third equation gives $6y^2 = 6$, or $y = \pm 1$, and so $x = \pm 2$.

Summarizing, we have found the points $(0, \pm\sqrt{3})$ and $(\pm 2, \pm 1)$ as candidates for places where f reaches extreme values. Next we calculate $f(x, y)$ at each point:

(x_0, y_0)	$(0, \sqrt{3})$	$(0, -\sqrt{3})$	$(2, 1)$	$(-2, 1)$	$(2, -1)$	$(-2, -1)$
$f(x_0, y_0)$	0	0	4	4	-4	-4

Clearly, f has an absolute maximum of 4 at $(2, 1)$ and $(-2, 1)$ and an absolute minimum of -4 at $(2, -1)$ and $(-2, -1)$. Now consider points (x, y) on the ellipse in a small neighborhood of $(0, \sqrt{3})$. Since $y > 0$ and $x^2 > 0$, $f(x, y) > 0$. So $f(0, \sqrt{3}) = 0$ is a local minimum value. By similar reasoning we see that $f(0, -\sqrt{3}) = 0$ is a local maximum. The results are summarized in Figure 17.19. ■

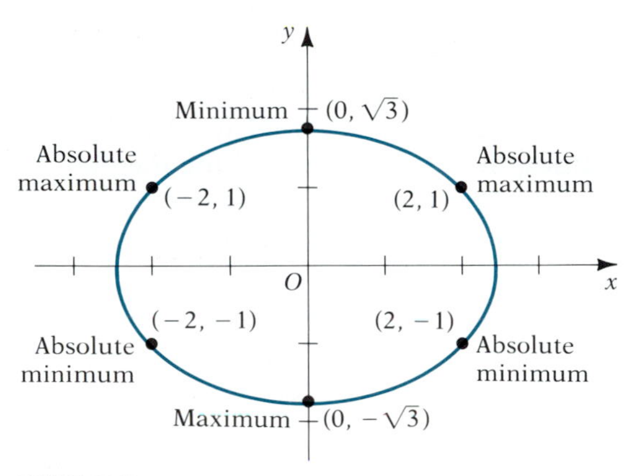

FIGURE 17.19

EXAMPLE 17.23 Find the point on the line $3x - 4y = 10$ that is nearest the origin.

Solution The distance from a point (x, y) to the origin is $\sqrt{x^2 + y^2}$, and this is a minimum if and only if its square is a minimum. So to simplify calculations, we use the square of the distance. We therefore want to minimize $f(x, y) = x^2 + y^2$ subject to the constraint that $3x - 4y = 10$. The constraint equation can be written as $g(x, y) = 0$, where $g(x, y) = 3x - 4y - 10$. Proceeding as in the preceding example, we have

$$\nabla f(x, y) = \lambda \nabla g(x, y)$$
$$\langle 2x, 2y \rangle = \lambda \langle 3, -4 \rangle$$

So we have three equations to solve:

$$\begin{cases} 2x = 3\lambda \\ 2y = -4\lambda \\ 3x - 4y - 10 = 0 \end{cases}$$

This time we solve for x and y in terms of λ from the first two equations and substitute into the third:

$$x = \frac{3\lambda}{2} \qquad y = -2\lambda$$

$$3\left(\frac{3\lambda}{2}\right) - 4(-2\lambda) = 10$$

Solving for λ, we get $\lambda = \frac{4}{5}$. So

$$x = \tfrac{3}{2}(\tfrac{4}{5}) = \tfrac{6}{5} \quad \text{and} \quad y = -2(\tfrac{4}{5}) = -\tfrac{8}{5}$$

Thus, the point to be tested is $(\frac{6}{5}, -\frac{8}{5})$. The geometry of the situation tells us that there is some minimum distance and no maximum distance (see Figure 17.20), and since there is only one critical point, it must be the point at which the distance is minimum. ■

The analogue of Theorem 17.8 for f and g functions of three (or more) variables also holds true, with the proof being virtually the same. We illustrate the procedure for the three-variable case in the next two examples.

EXAMPLE 17.24 Rework Example 17.20 using the method of Lagrange multipliers.

Solution The problem can be phrased as follows: minimize

$$C(x, y, z) = 2(xz + yz) + \tfrac{5}{2}xy$$

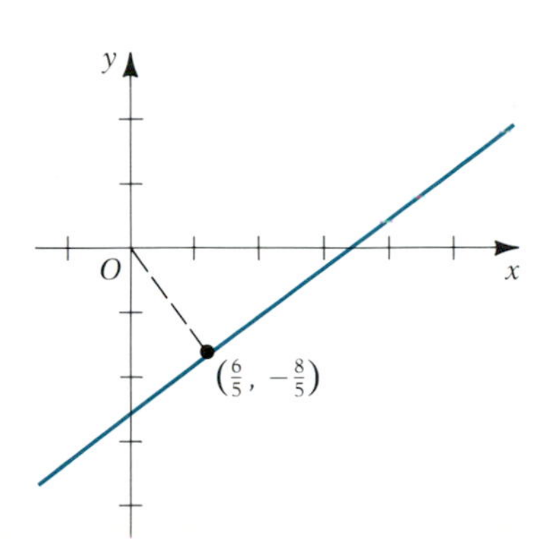

FIGURE 17.20

subject to the constraint $g(x, y, z) = 0$, where

$$g(x, y, z) = xyz - 270$$

We seek solutions to the system $\nabla C = \lambda \nabla g$ and $g(x, y, z) = 0$. In terms of components, these are

$$\begin{cases} 2z + \frac{5}{2}y = \lambda yz \\ 2z + \frac{5}{2}x = \lambda xz \\ 2(x + y) = \lambda xy \\ xyz - 270 = 0 \end{cases} \quad \text{or} \quad \begin{cases} 5y + 4z = 2\lambda yz \\ 5x + 4z = 2\lambda xz \\ 2x + 2y = \lambda xy \\ xyz = 270 \end{cases}$$

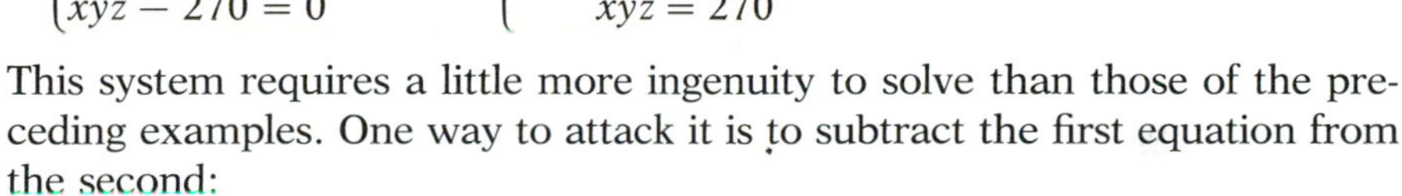

This system requires a little more ingenuity to solve than those of the preceding examples. One way to attack it is to subtract the first equation from the second:

$$5x - 5y = 2\lambda xz - 2\lambda yz$$
$$5(x - y) = 2\lambda z(x - y)$$
$$(x - y)(5 - 2\lambda z) = 0$$

So either $x = y$ or $2\lambda z = 5$. The constraint equation ensures that $z \neq 0$, so from $2\lambda z = 5$ we get $\lambda = \frac{5}{2z}$. When this is substituted into the first equation, we get

$$5y + 4z = 2(\tfrac{5}{2z})yz$$
$$5y + 4z = 5y$$
$$4z = 0$$

So $z = 0$, which is not possible. Thus, the only feasible solution is $x = y$. The third equation then gives $4x = \lambda x^2$, or $\lambda = \frac{4}{x}$, since $x \neq 0$. Putting this value of λ into the second equation of our system gives

$$5x + 4z = 2(\tfrac{4}{x})xz$$
$$5x - 4z = 0$$
$$z = \tfrac{5x}{4}$$

Now we have $y = x$ and $z = \frac{5x}{4}$, and so from the constraint equation $xyz = 270$, we get $5x^3/4 = 270$, or $x^3 = 216$. Finally, $x = 6$, $y = 6$, and $z = \frac{15}{2}$.

That a minimum value of C exists is evident from physical considerations, so $(6, 6, \frac{15}{2})$ must yield the minimum—namely,

$$C(6, 6, \tfrac{15}{2}) = 2(45 + 45) + \tfrac{5}{2}(36) = 270$$ ■

EXAMPLE 17.25 The largest box the United Parcel Service will accept is one for which the length plus the girth (distance around) is 108 in. What are the dimensions of the box of maximum volume that can be sent by UPS?

Solution Let the dimensions be x, y, and z as shown in Figure 17.21. Then we want to maximize $V = xyz$ subject to the constraint $x + 2(y + z) = 108$. Taking $g(x, y, z) = x + 2(y + z) - 108$, for the constrained maximum we must have $\nabla V = \lambda \nabla g$ and $g = 0$. These equations are

$$\begin{cases} yz = \lambda \\ xz = 2\lambda \\ xy = 2\lambda \\ x + 2(y + z) = 108 \end{cases}$$

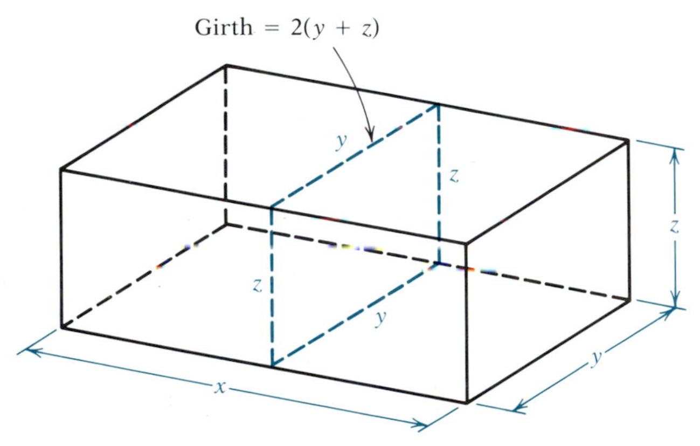

FIGURE 17.21

Eliminating λ from the first two equations yields $2yz = xz$. Since $z = 0$ is not a feasible solution, we get $x = 2y$. Again eliminating λ from the second and third equations of our system, we obtain $xz = xy$, or $z = y$ (since $x \neq 0$). Putting $x = 2y$ and $z = y$ into the constraint equation gives

$$\begin{aligned} 2y + 2(2y) &= 108 \\ 6y &= 108 \\ y &= 18 \end{aligned}$$

So the dimensions that give the maximum volume are $36 \times 18 \times 18$ in. ■

It can be proved that the method of Lagrange multipliers also works when there is more than one constraint equation. We then have a multiplier for each constraint. For example, all local extrema of a function $f(x, y, z)$ subject to the constraints $g(x, y, z) = 0$ and $h(x, y, z) = 0$ will occur at points for which

$$\nabla f = \lambda_1 \nabla g + \lambda_2 \nabla h \tag{17.30}$$

The next example illustrates this situation.

EXAMPLE 17.26 Find the points on the curve of intersection of the paraboloid of revolution $x^2 + y^2 + 2z = 4$ and the plane $x - y + 2z = 0$ that are closest to and farthest from the origin.

Solution As in Example 17.23 we find the minimum and maximum values of the square of the distance from the origin. Thus, we take

$$f(x, y, z) = x^2 + y^2 + z^2$$

where the points (x, y, z) are constrained to lie on the given paraboloid and the given plane. We write the constraints in the form $g(x, y, z) = 0$ and $h(x, y, z) = 0$ by letting

$$g(x, y, z) = x^2 + y^2 + 2z - 4$$

and

$$h(x, y, z) = x - y + 2z$$

From equation (17.30) we have

$$\langle 2x, 2y, 2z \rangle = \lambda_1 \langle 2x, 2y, 2 \rangle + \lambda_2 \langle 1, -1, 2 \rangle$$

or, in terms of components,

$$\begin{cases} 2x = 2x\lambda_1 + \lambda_2 \\ 2y = 2y\lambda_1 - \lambda_2 \\ 2z = 2\lambda_1 + 2\lambda_2 \end{cases}$$

and these three equations together with the two constraint equations constitute a system involving the five unknowns, x, y, z, λ_1, and λ_2 that we need to solve simultaneously. By eliminating λ_2 from the first two equations, we get

$$(x + y)(1 - \lambda_1) = 0$$

so that $y = -x$ or $\lambda_1 = 1$. We leave it as an exercise to show that $y = -x$ yields the points $(2, -2, -2)$ and $(-1, 1, 1)$, and that $\lambda_1 = 1$ also gives $(-1, 1, 1)$. Since $f(2, -2, -2) = 8$ and $f(-1, 1, 1) = 3$, we conclude that the maximum distance from the origin is $\sqrt{8} = 2\sqrt{2}$ and the minimum distance is $\sqrt{3}$. ■

EXERCISE SET 17.6

A

In all exercises use the method of Lagrange multipliers. In Exercises 1–8 find the local maxima and minima of f subject to the given constraint.

1. $f(x, y) = x^2 - y^2$; $x - 2y = 4$
2. $f(x, y) = 2x^2 + 3y^2$; $2x - 6y = 7$
3. $f(x, y) = x^2 + 2xy$; $y = x^2 - 2$
4. $f(x, y) = 2x - 4y$; $x^2 + y^2 = 5$
5. $f(x, y) = x^3 + 3y^2$; $xy + 4 = 0$
6. $f(x, y, z) = x^2 + 2y^2 - 2z^2$; $z = x^2y$
7. $f(x, y, z) = x - 2y - 3z$; $xyz = 36$
8. $f(x, y, z) = 6x^2 + 3y^2 + 4z^2$; $3x^2y + 2z^2 = 4$
9. Use Lagrange multipliers to find the distance from the line $2x + y = 3$ to the point $(1, -1)$.
10. Find the point on the plane $3x - y - 2z = 6$ that is nearest the origin.
11. A coordinate system is set up on a flat metal plate so that the temperature $T(x, y)$ in degrees Celsius at the point (x, y) is $T(x, y) = 10x^2y + 50$. Find the hottest

and coldest spots at points on the ellipse $2x^2 + 3y^2 = 9$. What are these extreme temperatures?

12. A company determines from experience that its monthly revenue from the sale of a certain product is

$$R(x, y) = y^2 + 5xy + 20x$$

where x is the amount spent on magazine ads and y is the amount spent on television commercials, both in thousands of dollars. If the company plans to spend a total of \$60,000 per month on advertising, how should it be divided to maximize R?

In Exercises 13–16 rework the specified exercises from Exercise Set 17.5 using Lagrange multipliers.

13. Exercise 33 **14.** Exercise 34

15. Exercise 35 **16.** Exercise 37

B

17. Supply the details for the solution of Example 17.26.

18. Prove that a function can have a constrained local extremum at a point but not have a local extremum there. [*Hint:* Consider $f(x, y) = xy$ with an appropriate constraint.]

19. Find the dimensions of the rectangular box of greatest volume that can be inscribed in the ellipsoid $2x^2 + y^2 + 4z^2 = 12$.

20. Find the dimensions of the cone of maximum volume that can be inscribed in a sphere of radius a.

21. Find the maximum and minimum values of $f(x, y, z) = x^2 + 2y^2 - 3z^2$ subject to the two constraints $2x^2 - 3y^2 = 8$ and $y^2 - 2z = 3$.

22. Use Lagrange multipliers to derive the formula

$$d = \frac{|Ax_0 + By_0 + Cz_0 + D|}{\sqrt{A^2 + B^2 + C^2}}$$

for the distance d from the plane $Ax + By + Cz + D = 0$ to the point (x_0, y_0, z_0).

23. Find the points on the curve of intersection of the ellipsoid $2x^2 + 3y^2 + 4z^2 = 6$ and the paraboloid $z = 4 - x^2 - 2y^2$ that are closest to the origin and farthest from the origin.

17.7 SUPPLEMENTARY EXERCISES

In Exercises 1 and 2 find df.

1. (a) $f(x, y) = \ln \dfrac{x^2}{\sqrt{1 - y^2}}$ (b) $f(x, y) = \dfrac{\sin x}{\cosh y}$

2. (a) $f(x, y) = \sin^{-1} \dfrac{x}{y}$, $y > 0$

(b) $f(x, y, z) = \dfrac{4x - 2z}{z + 3y}$

3. Approximate Δf using df.

(a) $f(x, y) = \dfrac{(x - y)^2}{x^2 + y^2}$ from $(2, -4)$ to $(2.02, -3.97)$

(b) $f(x, y) = \ln \sqrt{9 - x^2 - y^2}$ from $(-2, 1)$ to $(-1.99, 0.98)$

4. Use Definition 17.1 to show that $f(x, y) = x^2 - 2xy + 3y$ is differentiable throughout $\mathbf{R}^2$.

5. Suppose the electrostatic potential at a point in $\mathbf{R}^3$ is given by

$$V(x, y, z) = \frac{140z}{\sqrt{x^2 + y^2 + z^2}}$$

Find the approximate change in potential from $(3, -2, 6)$ to $(2.6, -1.8, 6.5)$.

6. A company makes two types of toasters, models A and B. It costs \$15 to produce each unit of model A and \$21 to produce each unit of model B. The revenue from producing and selling x model A units and y model B units is

$$R(x, y) = 42x + 56y - 0.02xy - 0.01x^2 - 0.03y^2$$

The current weekly production level is 150 model A and 100 model B units. Find the approximate increase in profit if 5 more model A units and 8 more model B units are produced each week. Approximately what is the profit at this new level?

7. Use Theorem 17.2 to show that f is differentiable except at $(0, 0)$ where

$$f(x, y) = \tan^{-1} \tfrac{x}{y}$$

8. Let

$$f(x, y) = \begin{cases} \dfrac{x^2 y}{x^4 + y^2} & \text{if } (x, y) \neq (0, 0) \\ 0 & \text{if } (x, y) = (0, 0) \end{cases}$$

Show that f_x and f_y both exist at $(0, 0)$ but that f is not differentiable there.

In Exercises 9–13 use an appropriate chain rule.

9. Find $\dfrac{dz}{dt}$ at $t = \dfrac{5\pi}{6}$ if $z = x^2 - 2xy - y^3$ and $x = \cos 2t$, $y = \sin 2t$.

10. Find $\dfrac{dz}{dt}$ at $t = 2$ if $z = \ln \sqrt{\dfrac{x+y}{x-y}}$ and $x = t + \dfrac{1}{t}$, $y = t - \dfrac{1}{t}$.

11. Find $\dfrac{dz}{dt}$ if $z = e^{-x^2/y}$ and $x = \sinh t$, $y = 1 + \cosh t$.

12. Find $\dfrac{\partial z}{\partial u}$ and $\dfrac{\partial z}{\partial v}$ if $z = \dfrac{x^2 - y^2}{xy}$ and $x = \dfrac{u}{v}$, $y = \dfrac{1}{u}$.

13. Find $\dfrac{\partial z}{\partial s}$ and $\dfrac{\partial z}{\partial t}$ at $(s, t) = \left(1, \dfrac{1}{2}\right)$ if $z = x \sin \pi y - y \cos \pi x$, $x = s^2 - t^2$, and $y = 2st$.

14. Find the equation of the tangent line to the graph of

$$2x^4 - 3x^2y + 4xy^2 + y^3 + 4 = 0$$

at $(-1, 1)$.

15. Find $\dfrac{\partial z}{\partial x}$ and $\dfrac{\partial z}{\partial y}$ if

$$y \ln(\cos xz) + xyz = 3$$

16. Let $z = f\left(\dfrac{x-y}{x+y}\right)$. Show that

$$x\frac{\partial z}{\partial x} + y\frac{\partial z}{\partial y} = 0$$

17. A water tank is in the form of a frustum of a cone as shown in the figure, with bottom radius 3 ft. Water is being drained from the tank at the constant rate of 10π cu ft/min. When 210π cu ft of water remain in the tank, the radius r of the upper surface of the water is 6 ft and is decreasing at the rate of 1 in./min. Find how fast the water level is falling at that instant.

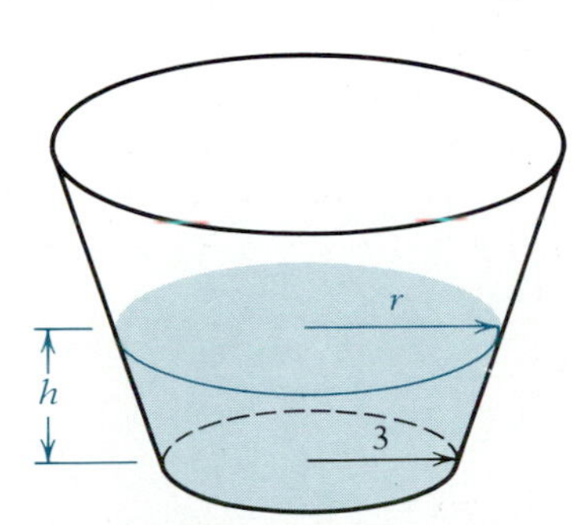

In Exercises 18 and 19 find the directional derivatives of f in the direction indicated.

18. $f(x, y) = \ln [x^2/\sqrt{x - y}]$ at $(1, -3)$ in the direction $6\mathbf{i} + 8\mathbf{j}$

19. $f(x, y, z) = x^2 y \cos \pi z$ at $(3, -1, 1)$ toward $(4, 1, -1)$

20. For the function

$$f(x, y, z) = \sqrt{\frac{x - 2y}{y - z}}$$

in what direction from $P_0 = (0, -2, -3)$ is $D_{\mathbf{u}}f(P_0)$ a maximum, and what is this maximum value?

21. Find $f(x, y)$ if $f(1, 1) = 1$ and

$$\nabla f = \left\langle \frac{1 - 2x^2}{x}, \frac{2y^2 - 1}{y} \right\rangle$$

In Exercises 22 and 23 find the equation of the tangent plane and normal line at the indicated point.

22. $z = x^2 - xy - 2y^2$ at $(2, -1, 4)$

23. $z = 2e^{(x-y)/z}$ at $(1, 1, 2)$

24. The accompanying figure shows a cross section of a long semicircular cylinder with a flat base. The curved surface is kept at electrostatic potential $V = 1$ and the base at $V = 0$. It can be shown that the potential $V(x, y)$ at points inside the region is

$$V(x, y) = \frac{2}{\pi} \tan^{-1}\left(\frac{2y}{1 - x^2 - y^2}\right)$$

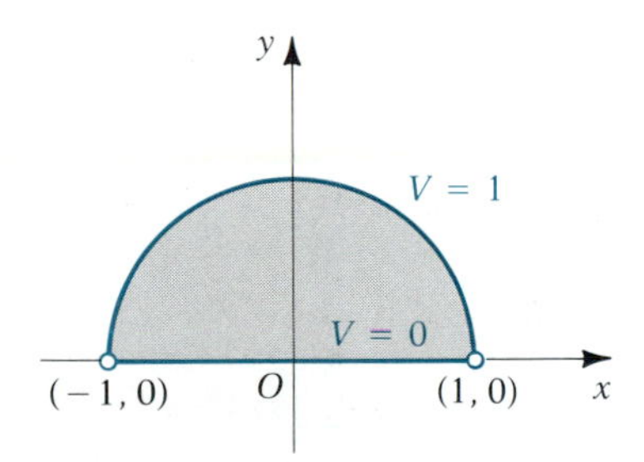

(a) Draw several equipotential curves $V(x, y) = c$ for $0 < c < 1$.
(b) Find the rate of change of V at $(\frac{1}{2}, \frac{1}{2})$ in the direction toward $(0, 1)$.
(c) Show that $\nabla V(\frac{1}{2}, \frac{1}{2})$ is orthogonal to the level curve through $(\frac{1}{2}, \frac{1}{2})$.

25. Find the angle between the hyperbolic paraboloid $z = x^2 - y^2$ and the line $\mathbf{r}(t) = \langle -t, 1 + 2t, 5(1 + t)\rangle$ at each of their points of intersection. (See Exercise 26 in Exercise Set 17.4.)

26. Find all maximum, minimum, and saddle points on the surface $z = x^3 - 2xy^2 - 9x + 4y^3$.

27. Find the absolute maximum and minimum values of the function

$$f(x, y) = \tfrac{4}{x} + \tfrac{1}{y} + 2xy$$

over the closed region bounded by $x = 1$, $x = 2y$, and $y = 2$, and give the points at which these extreme values occur.

28. Suppose the temperature in a three-dimensional region that contains the ellipsoid $x^2 + y^2 + 4z^2 = 12$ is given by $T(x, y, z) = xyz$. Find the hottest and coldest temperatures on the ellipsoid and where they occur.

29. Use the Lagrange multiplier method to find the dimensions of the right circular cylinder inscribed in a sphere of radius a that has a maximum (a) volume and (b) lateral surface area.

30. A company produces two types of electric brooms, the standard and deluxe models. Suppose the monthly profit in thousands of dollars from producing and selling x thousand standard models and y thousand deluxe models is given by

$$P(x, y) = 2x + 3y - 0.1x^2 - 0.2y^2 - 0.5xy - 4$$

If the combined total production of the two models is to be 6000 per month, how many of each model should be produced to maximize profit?

CHAPTER 18

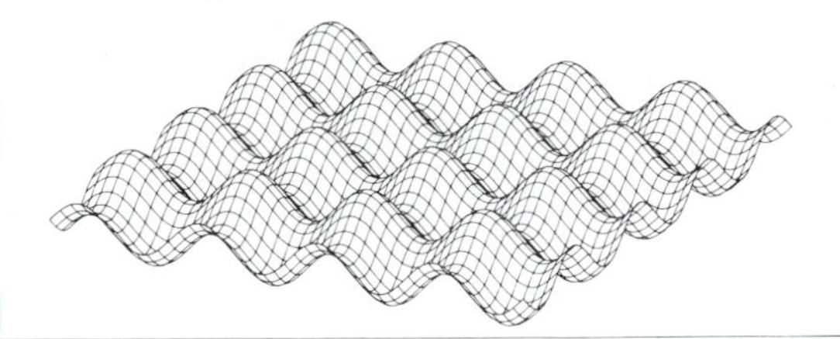

MULTIPLE INTEGRALS

18.1 DOUBLE INTEGRALS

We turn now to the integral calculus of functions of more than one variable, beginning with the definition of the integral of a function of two variables over a region in the xy-plane. The development will have much in common with that of the integral $\int_a^b f(x)\,dx$ of a single-variable function over an interval $[a, b]$ of the x-axis, so it would be useful for you to review Section 4.3 at this time.

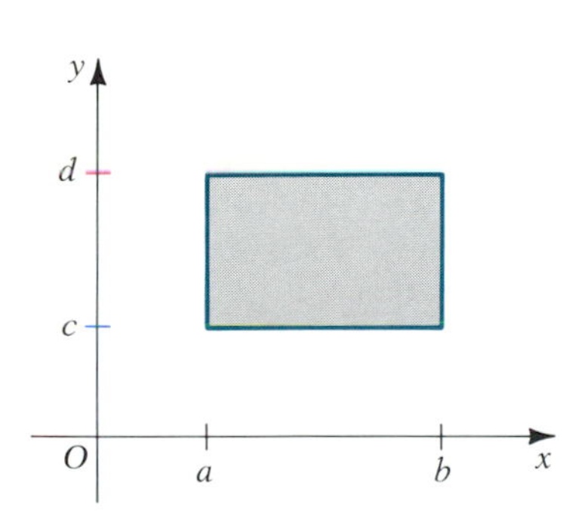

FIGURE 18.1 A closed interval in $\mathbf{R}^2$

We consider *bounded* functions only. To say that a function f of two variables is bounded means that there is a positive constant M such that $|f(x, y)| \leq M$ for all (x, y) in the domain of f. For functions of one variable we defined the integral over a closed interval $[a, b]$. The natural extension of such an interval to two dimensions is a rectangular region of the form $I = \{(x, y)\colon a \leq x \leq b,\ c \leq y \leq d\}$, an example of which is shown in Figure 18.1. We call this a *closed interval in* $\mathbf{R}^2$. This notion of a closed interval can easily be extended to $\mathbf{R}^3$, or to $\mathbf{R}^n$ in general. In $\mathbf{R}^3$, for example, we would have a rectangular box together with its interior. Although integrals over closed intervals in $\mathbf{R}^2$ are of interest, the usefulness of the integral can be greatly extended by considering more general regions as domains of integration. We call a subset R of $\mathbf{R}^2$ a **regular region** if it satisfies the following conditions:

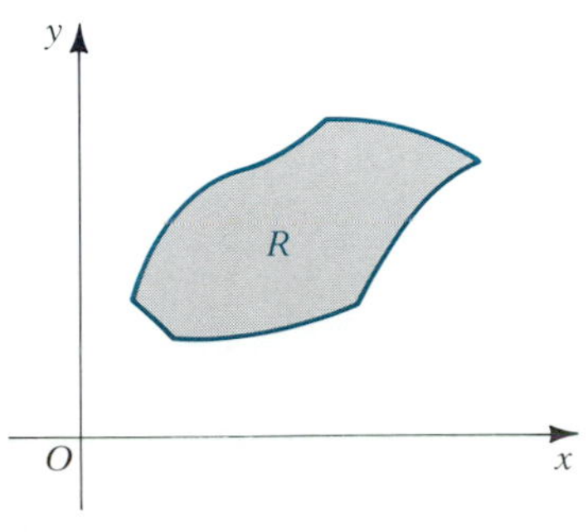

FIGURE 18.2

1. R is *closed*; that is, R contains its boundary.
2. R is *bounded*; that is, there is a closed interval I of $\mathbf{R}^2$ that contains R.
3. The boundary of R is a *sectionally smooth, simple closed curve*. This means that the boundary consists of a finite number of smooth arcs joined end to end to form a closed curve that does not cross itself.

An example of such a regular region is shown in Figure 18.2.

Now let f be a bounded function defined on a regular region R, and let I be any closed interval that contains R. We form a rectangular grid over I as in Figure 18.3 by means of horizontal and vertical lines drawn at arbitrary points on the boundary of I. The rectangles so formed constitute a *partition*

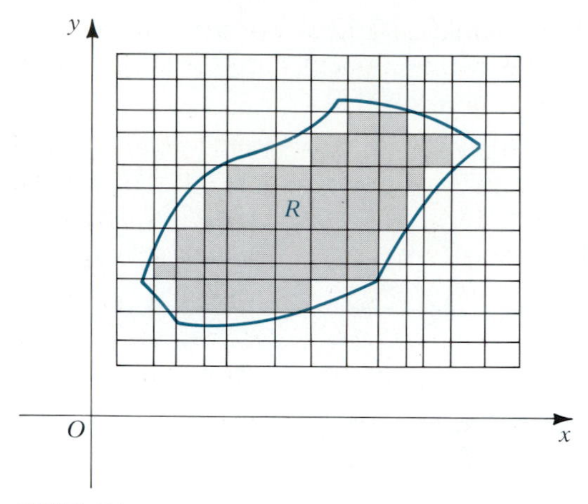

FIGURE 18.3

P of I. We denote by $\|P\|$ the length of the largest diagonal of the rectangles of P and call this the *norm* of P. The rectangles of P divide I into finitely many *subintervals*. We are interested only in those subintervals contained in R. These are shaded in Figure 18.3. Suppose there are n of these. We number them from 1 to n, and we denote the area of the ith one by ΔA_i. Let (x_i^*, y_i^*) be any point in the ith subinterval.

Now form the sum

$$\sum_{i=1}^{n} f(x_i^*, y_i^*)\,\Delta A_i \tag{18.1}$$

This is called a *Riemann sum* for f corresponding to the partition P. If we take finer and finer partitions (that is, partitions whose norms approach 0), the corresponding Riemann sums (18.1) may approach some limit L. What this means precisely is that, given $\varepsilon > 0$, there is a $\delta > 0$ such that for all partitions P with $\|P\| < \delta$,

$$\left|\sum_{i=1}^{n} f(x_i^*, y_i^*)\,\Delta A_i - L\right| < \varepsilon$$

independently of the choice of (x_i^*, y_i^*) in the ith subinterval. If this condition is satisfied, we call L the double integral of f over R, and denote it by

$$\iint_R f(x, y)\,dA$$

DEFINITION 18.1 THE DOUBLE INTEGRAL

If f is bounded on the regular region R, then

$$\iint_R f(x, y)\,dA = \lim_{\|P\|\to 0} \sum_{i=1}^{n} f(x_i^*, y_i^*)\,\Delta A_i \tag{18.2}$$

provided the limit exists. When the limit does exist, f is said to be *integrable* over R. ■

Although we will not prove it here, it can be shown that a *continuous* function is always integrable over any regular region. Most of the functions we work with will be continuous. In the next section we will see a way of evaluating double integrals that does not require direct calculation of the limit in equation (18.2). In fact, evaluating most integrals by calculating this limit would be a hopeless task. However, we can approximate the integral by means of Riemann sums for particular partitions. The next example illustrates this.

EXAMPLE 18.1 Approximate $\iint_R (x + 2y)\,dA$, where R is the region bounded by the lines $x = 0$, $x = 4$, $y = 0$, and $y = 1$. Use a partition formed by vertical lines at $x = 1$, 2, and 3, and a horizontal line at $y = \frac{1}{2}$. Take (x_i^*, y_i^*) as the lower left-hand corner of the ith subinterval.

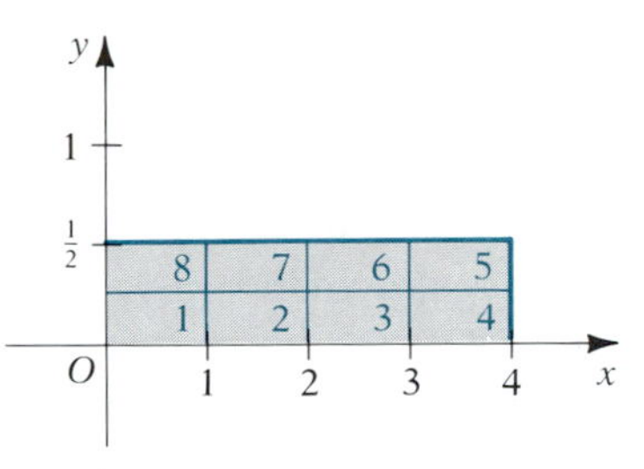

FIGURE 18.4

Solution Figure 18.4 shows the region R and the given partition of it. We have numbered the eight subintervals. In each case $\Delta A_i = (1)(\frac{1}{2}) = \frac{1}{2}$. The calculation of the Riemann sum is carried out using the following table:

i	(x_i^*, y_i^*)	$f(x_i^*, y_i^*)$	$f(x_i^*, y_i^*)\,\Delta A_i$
1	(0, 0)	0	0
2	(1, 0)	1	$\frac{1}{2}$
3	(2, 0)	2	1
4	(3, 0)	3	$\frac{3}{2}$
5	$(3, \frac{1}{2})$	4	2
6	$(2, \frac{1}{2})$	3	$\frac{3}{2}$
7	$(1, \frac{1}{2})$	2	1
8	$(0, \frac{1}{2})$	1	$\frac{1}{2}$

$$\sum_{i=1}^{8} f(x_i^*, y_i^*)\,\Delta A_i = 8$$

By the technique we will learn in the next section, we can show that the exact value of the integral is 12, so our approximation is not very good. We could improve it by using a partition with smaller norm. ■

If f is a nonnegative continuous function of R, we can see a geometric interpretation of the double integral as follows. As seen from Figure 18.5, the volume of the prism with base the ith subinterval formed by P and height of $f(x_i^*, y_i^*)$ is $f(x_i^*, y_i^*)\,\Delta A_i$. The Riemann sum $\sum_{i=1}^{n} f(x_i^*, y_i)A_i$, which adds the volumes of all such prisms, is therefore an approximation to the volume

FIGURE 18.5

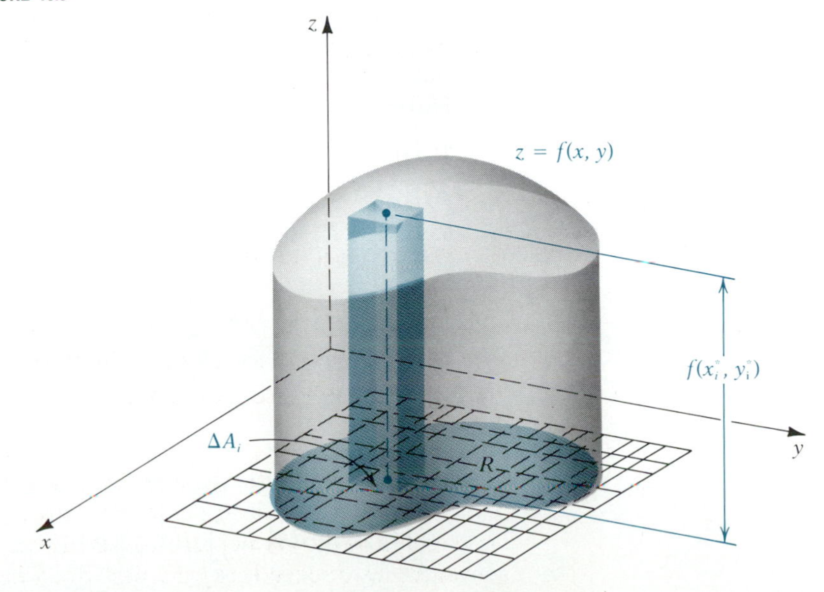

under the surface $z = f(x, y)$ and above the region R. For any given partition the Riemann sum may differ from the true volume for either of two reasons: (1) the subintervals that lie inside R may not fill out all of R, or (2) whereas the prisms have flat tops, the surface $z = f(x, y)$ may be curved. Nevertheless, it seems reasonable to assume that as we take finer and finer partitions, these deviations become negligible. The volume, in fact, is defined by the integral.

DEFINITION 18.2 Let f be a nonnegative continuous function on a regular region R. Then the volume V under the graph of f and above the region R is given by

$$V = \iint_R f(x, y)\, dA \tag{18.3}$$ ■

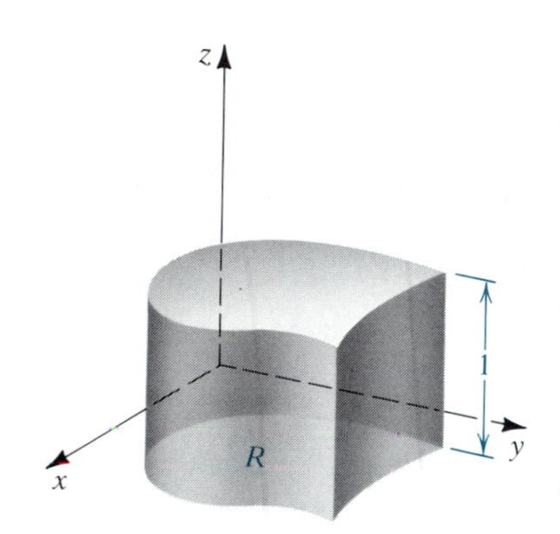

FIGURE 18.6

The special case of equation (18.3) in which $f(x, y) = 1$ for all (x, y) in R is of particular interest. A typical volume of this type is shown in Figure 18.6. By equation (18.3) the volume is

$$V = \iint_R 1\, dA = \iint_R dA$$

But the volume of such a slab can also be calculated by multiplying the area of the base by the altitude. So if we denote the area of R by A, we have $V = (A)(1) = A$. If we equate these two expressions for V, we get the following very useful formula for the area of R:

$$A = \iint_R dA \tag{18.4}$$

In fact, although we gave a motivation for this result, we can take equation (18.4) as the *definition* of the area of a regular region. In Exercise 54 of Exercise Set 18.2 you will be asked to show that when R is of a particular type, this definition is consistent with the area between two curves studied in Chapter 5.

A number of properties of the double integral can be proved based on Definition 18.1. Some of the more important ones are stated here. We assume all regions are regular and that each integral exists.

PROPERTIES OF THE DOUBLE INTEGRAL

1. $\iint_R cf(x, y)\, dA = c \iint_R f(x, y)\, dA$ for any constant c
2. $\iint_R [f(x, y) \pm g(x, y)]\, dA = \iint_R f(x, y)\, dA \pm \iint_R g(x, y)\, dA$
3. If $R = R_1 \cup R_2$, where R_1 and R_2 are nonoverlapping, then
$$\iint_R f(x, y)\, dA = \iint_{R_1} f(x, y)\, dA + \iint_{R_2} f(x, y)\, dA$$
4. If $f(x, y) \geq 0$ for all (x, y) in R, then
$$\iint_R f(x, y)\, dA \geq 0$$
5. If f is integrable over R, then so is $|f|$, and
$$\left| \iint_R f(x, y)\, dA \right| \leq \iint_R |f(x, y)|\, dA$$

EXERCISE SET 18.1

A

1. Redo Example 18.1, taking (x_i^*, y_i^*) as the midpoint of the ith subinterval.
2. Redo Example 18.1 with a partition that divides R into 16 squares $\frac{1}{2}$ unit on each side.

In Exercises 3–6 calculate the Riemann sum for f over R where P is the partition formed by vertical lines at the specified x values and horizontal lines at the specified y values. Take (x_i^*, y_i^*) as the lower right-hand corner of the ith subinterval.

3. $f(x, y) = 2xy$; $R = \{(x, y): 1 \le x \le 3,\ 0 \le y \le 4\}$; $x = 2$; $y = 1, 2, 3$
4. $f(x, y) = x^2 - y^2$; $R = \{(x, y): -1 \le x \le 2,\ 1 \le y \le 3\}$; $x = 0, 1$; $y = 2$
5. $f(x, y) = x^2 - 2xy$; $R = \{(x, y): 2 \le x \le 6,\ -2 \le y \le 4\}$; $x = 3, 4$; $y = 0, 2$
6. $f(x, y) = 2x^2 + 3y^2$; $R = \{(x, y): -2 \le x \le 2,\ 0 \le y \le 3\}$; $x = 0, 1$; $y = \frac{1}{2}, \frac{3}{2}$

In Exercises 7–10 approximate the volume under the surface $z = f(x, y)$ and above the region R, using a Riemann sum for the partition formed by the specified vertical and horizontal lines. Take (x_i^*, y_i^*) as the center of the ith subinterval.

7. $f(x, y) = 2x + 3y$; R is bounded by the triangle with vertices $(0, 0)$, $(4, 0)$, and $(0, 4)$; $x = 1, 2, 3$; $y = 1, 2, 3$
8. $f(x, y) = \sqrt{x^2 + y^2}$; $R = \{(x, y): 0 \le x \le 5,\ -x \le y \le x\}$; $x = 1, 2, 3, 4$; $y = \pm 1, \pm 2, \pm 3, \pm 4$
9. $f(x, y) = 2x^2 + 4y^2$; $R = \{(x, y): |x| + |y| \le 2\}$; $x = 0, \pm\frac{1}{2}, \pm 1, \pm\frac{3}{2}$; $y = 0, \pm\frac{1}{2}, \pm 1, \pm\frac{3}{2}$
10. $f(x, y) = x^2 + 2xy + 3y^2$; $R = \{(x, y): y - 4 \le x \le 4 - y,\ 0 \le y \le 2\}$; $x = \pm 1, \pm 2, \pm 3$; $y = \frac{1}{2}, 1, \frac{3}{2}$

B

11. Make use of one or more of the properties of the double integral to show that if f and g are integrable over R and $f(x, y) \le g(x, y)$ for all (x, y) in R, then
$$\iint_R f(x, y)\, dA \le \iint_R g(x, y)\, dA$$
12. If f is integrable over R and m and M are numbers such that $m \le f(x, y) \le M$ for all (x, y) in R, prove that
$$mA \le \iint_R f(x, y)\, dA \le MA$$
where A is the area of R.

C

Write a computer program that calculates the Riemann sum of f over R, where R is the closed interval whose boundary is the rectangle formed by the lines $x = a$, $x = b$, $y = c$, and $y = d$, where $a < b$ and $c < d$. Use a partition formed by n equally spaced vertical lines and m equally spaced horizontal lines. Take (x_i^*, y_i^*) as the center of the ith subinterval. Run this program with the values specified in Exercises 13 and 14.

13. $f(x, y) = e^x \sin y$; $a = 0$, $b = 2$, $c = 0$, $d = 4$; $m = 20$, $n = 40$
14. $f(x, y) = \sqrt[3]{2x^2y - 3y^2}$; $a = -4$, $b = 6$, $c = -1$, $d = 4$
    (a) $m = 5$, $n = 10$
    (b) $m = 10$, $n = 20$
    (c) $m = 20$, $n = 40$
    (d) $m = 50$, $n = 100$

18.2 EVALUATION OF DOUBLE INTEGRALS BY ITERATED INTEGRALS

As we have observed, evaluation of double integrals directly from Definition 18.1 is likely to be tedious, to say the least. Except in the simplest cases, in fact, direct calculation of the limit involved is not possible. Fortunately, integrals for a wide class of functions over suitably restricted regular regions

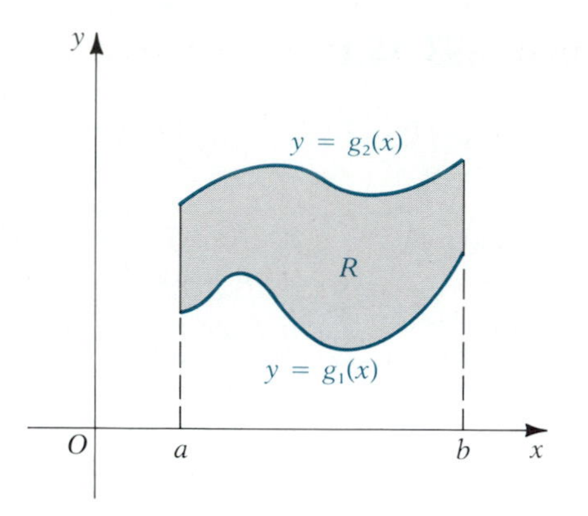

Type I region

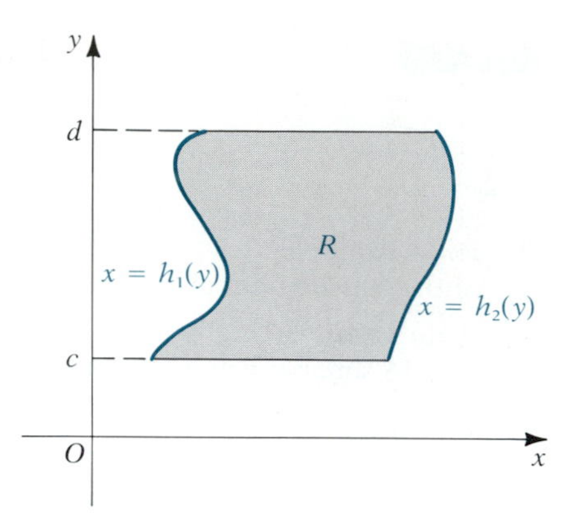

Type II region

FIGURE 18.7

can be evaluated by calculating two successive single integrals. The types of regions we refer to are shown in Figure 18.7. A region R is a type I region if R is bounded on the left and right by the vertical lines $x = a$ and $x = b$, respectively, and is bounded below and above by smooth curves $y = g_1(x)$ and $y = g_2(x)$, where $g_1(x) \le g_2(x)$. Thus,

$$R = \{(x, y): a \le x \le b,\ g_1(x) \le y \le g_2(x)\} \qquad \text{(type I)}$$

Similarly, a type II region is bounded on the left and right by smooth curves $x = h_1(y)$ and $x = h_2(y)$, with $h_1(y) \le h_2(y)$, and is bounded below and above by the horizontal lines $y = c$ and $y = d$. Thus,

$$R = \{(x, y): h_1(y) \le x \le h_2(y),\ c \le y \le d\} \qquad \text{(type II)}$$

Suppose R is a type I region and f is continuous on R. Then it can be shown that the value of the double integral of f over R is the same as the value of the **iterated integral**

$$\int_a^b \int_{g_1(x)}^{g_2(x)} f(x, y)\,dy\,dx \tag{18.5}$$

This iterated integral is defined to mean

$$\int_a^b \left[\int_{g_1(x)}^{g_2(x)} f(x, y)\,dy\right] dx$$

where the integral inside the bracket is evaluated first, *holding x fixed,* with the result being the integrand for the outer integral. The following example illustrates this.

EXAMPLE 18.2 Evaluate the iterated integral

$$\int_0^1 \int_0^x (x + 2y)\,dy\,dx$$

and show the region R determined by the limits.

Solution The integral is of the form of (18.5) with $g_1(x) = 0$ and $g_2(x) = x$. So R is the triangular region pictured in Figure 18.8. Notice that R is both a type I

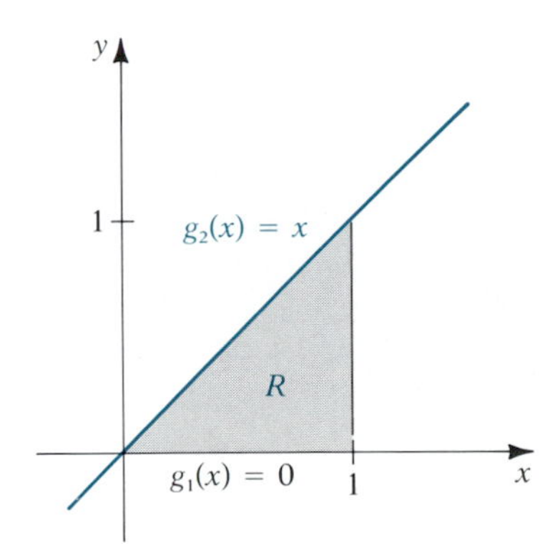

FIGURE 18.8

and a type II region, but we are treating it only as type I:

$$\int_0^1 \int_0^x (x + 2y)\,dy\,dx = \int_0^1 \left[\int_0^x (x + 2y)\,dy\right] dx$$ Hold x fixed and integrate with respect to y.

$$= \int_0^1 [xy + y^2]_0^x\,dx$$

$$= \int_0^1 2x^2\,dx$$ Now integrate with respect to x.

$$= \frac{2x^3}{3}\Big]_0^1 = \frac{2}{3}$$ ■

For a type II region the integral will be of the form

$$\int_c^d \int_{h_1(y)}^{h_2(y)} f(x, y)\,dx\,dy \tag{18.6}$$

which is understood to mean

$$\int_c^d \left[\int_{h_1(y)}^{h_2(y)} f(x, y)\,dx\right] dy$$

In evaluating the inner integral, y is held fixed. The result of this inner integral will be a function of y, which then is the integrand of the outer integral. Again, it can be shown that if f is continuous over R, the value of this iterated integral is the same as the value of the double integral of f over R.

We summarize the preceding discussion in the following theorem.

THEOREM 18.1 Let f be continuous on the region R. If R is of type I, then

$$\iint_R f(x, y)\,dA = \int_a^b \int_{g_1(x)}^{g_2(x)} f(x, y)\,dy\,dx$$

and if R is of type II, then

$$\iint_R f(x, y)\,dA = \int_c^d \int_{h_1(y)}^{h_2(y)} f(x, y)\,dx\,dy$$ ■

The importance of this theorem can hardly be overemphasized. Without it, evaluating most double integrals would be a hopeless task. With it, a wide class of double integrals can be evaluated by using our knowledge of single integrals.

Although we shall not give a formal proof of Theorem 18.1, we can show its plausibility, at least for $f(x, y) \geq 0$ on R. Consider such a nonnegative continuous function on a type I region. Then by Definition 18.2 the volume V under the surface $z = f(x, y)$ and above R is the double integral

$$V = \iint_R f(x, y)\,dA \tag{18.7}$$

But we can also calculate the volume another way. As indicated in Figure 18.9, we select an arbitrary x value between $x = a$ and $x = b$ and hold it fixed temporarily. To emphasize that it is held fixed, we designate it by x_0 but will soon remove the subscript. The plane $x = x_0$ intersects the given surface in a curve $z = f(x_0, y)$. Here y is the independent variable and z the dependent variable. We are interested in that portion of the curve from $y = g_1(x_0)$ to $y = g_2(x_0)$. Denote the area under its graph on this interval by

FIGURE 18.9

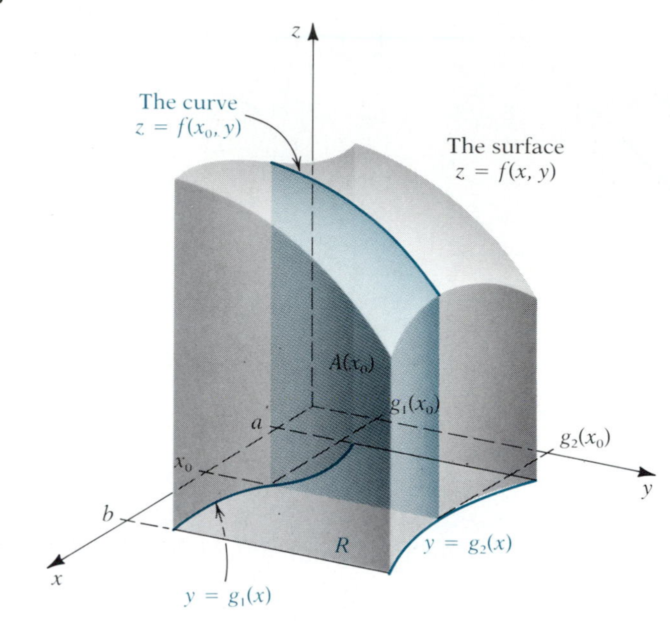

$A(x_0)$. Then from single-variable calculus we know that

$$A(x_0) = \int_{g_1(x_0)}^{g_2(x_0)} f(x_0, y)\,dy$$

Now $A(x_0)$ is the area of a typical cross section of the volume under the surface, taken perpendicular to the x-axis. So by the theory of Section 5.2 we get the total volume by integrating this cross-sectional area over the interval $[a, b]$. Since we now want x_0 to vary over this interval, we drop the subscript and write $V = \int_a^b A(x)\,dx$; that is,

$$V = \int_a^b \left[\int_{g_1(x)}^{g_2(x)} f(x, y)\,dy \right] dx \qquad (18.8)$$

Finally, we equate the two expressions for volume, equations (18.7) and (18.8), to get

$$\iint_R f(x, y)\,dA = \int_a^b \int_{g_1(x)}^{g_2(x)} f(x, y)\,dy\,dx \qquad (R \text{ is a type I region.})$$

A similar argument can be given when R is a type II region (see Exercise 53).

If R is a regular region not of type I or type II but one that can be divided into a finite number of nonoverlapping subregions, each of which is of type I or type II, then we can use property 3 of double integrals together with Theorem 18.1 to evaluate the double integral of f over R. Figure 18.10 illustrates this situation. In this case we could write

$$\iint_R f(x, y)\,dA = \iint_{R_1} f(x, y)\,dA + \iint_{R_2} f(x, y)\,dA$$

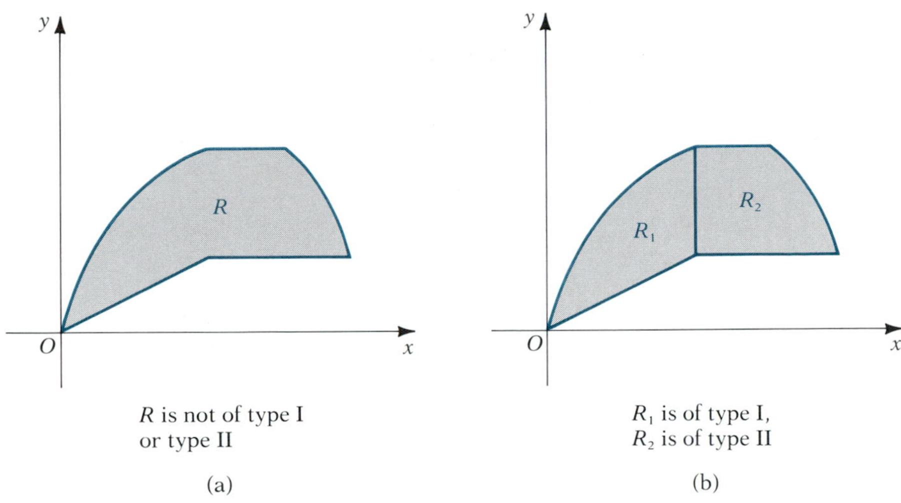

FIGURE 18.10

and then evaluate the two integrals on the right as iterated integrals. You might observe that the region shown in this case could be divided in different ways from that shown.

The next three examples illustrate the use of Theorem 18.1.

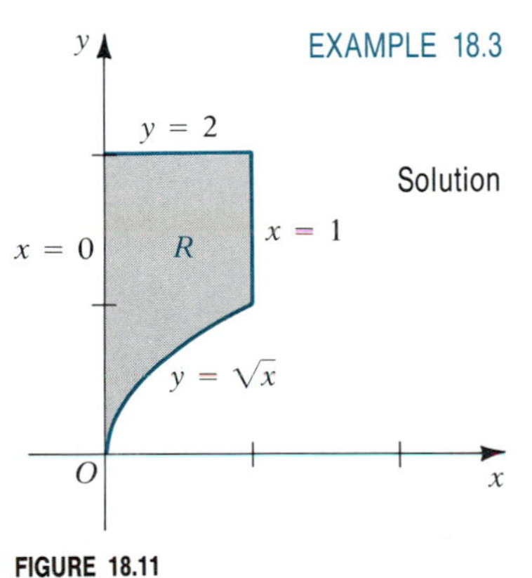

FIGURE 18.11

EXAMPLE 18.3 Evaluate the integral $\iint_R x^2y\,dA$, where R is bounded by $x = 0$, $x = 1$, $y = \sqrt{x}$, and $y = 2$. Draw the region.

Solution The region is pictured in Figure 18.11. It is a type I region, so by Theorem 18.1,

$$\iint_R x^2y\,dA = \int_0^1 \int_{\sqrt{x}}^2 x^2y\,dy\,dx$$

$$= \int_0^1 \left[\frac{x^2y^2}{2}\right]_{\sqrt{x}}^2 dx$$

$$= \int_0^1 \left(2x^2 - \frac{x^3}{2}\right) dx = \frac{2x^3}{3} - \frac{x^4}{8}\bigg]_0^1 = \frac{13}{24}$$ ■

EXAMPLE 18.4 Find the volume under the paraboloid $z = x^2 + 3y^2$ and above the region bounded by the lines $x = 0$, $y = 0$, and $x + y = 1$. Sketch the volume.

Solution The region R in the xy-plane is shown in Figure 18.12(a) and the three-dimensional volume is shown in Figure 18.12(b). The region R is both a type I and a type II region. Treating it as type I and using Definition 18.2 and Theorem 18.1, we have

$$V = \iint_R (x^2 + 3y^2)\,dA = \int_0^1 \int_0^{1-x} (x^2 + 3y^2)\,dy\,dx$$

$$= \int_0^1 [x^2y + y^3]_0^{1-x}\,dx$$

$$= \int_0^1 [x^2 - x^3 + (1-x)^3]\,dx$$

$$= \frac{x^3}{3} - \frac{x^4}{4} - \frac{(1-x)^4}{4}\bigg]_0^1 = \left(\frac{1}{3} - \frac{1}{4}\right) + \frac{1}{4} = \frac{1}{3}$$ ■

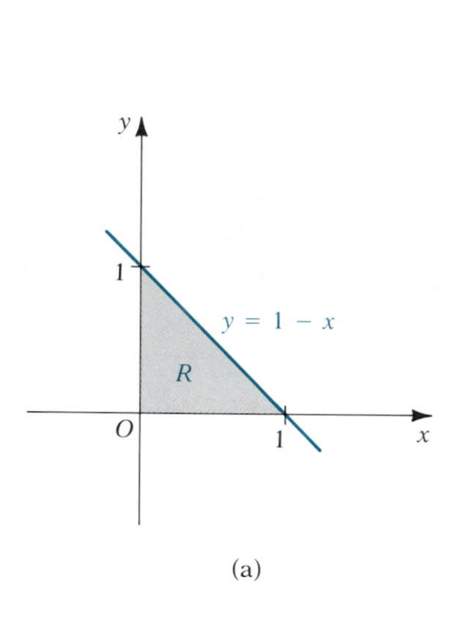

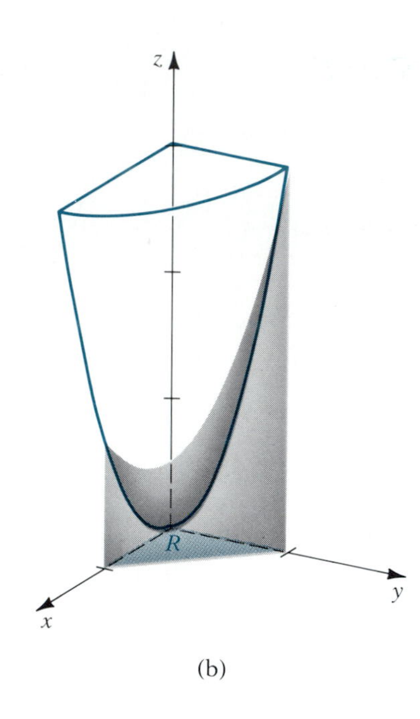

FIGURE 18.12

EXAMPLE 18.5 Using double integration, find the area of the region R bounded by the curves $y^2 = x$ and $x - y = 2$. Sketch the region.

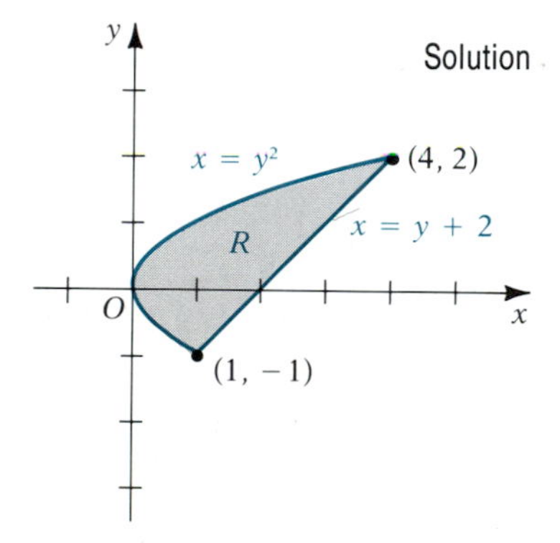

FIGURE 18.13

Solution The region is shown in Figure 18.13. Solving the two equations simultaneously, we find the points of intersection $(1, -1)$ and $(4, 2)$ as shown. (Verify.) The region is of type II. So we have from equation (18.4) and Theorem 18.1,

$$A = \iint_R dA = \int_{-1}^{2} \int_{y^2}^{y+2} dx\,dy = \int_{-1}^{2} [x]_{y^2}^{y+2}\,dy = \int_{-1}^{2} (y + 2 - y^2)\,dy$$

$$= \frac{y^2}{2} + 2y - \frac{y^3}{3}\Big]_{-1}^{2} = \left(2 + 4 - \frac{8}{3}\right) - \left(\frac{1}{2} - 2 + \frac{1}{3}\right) = \frac{9}{2} \quad ■$$

When a region is both a type I and a type II region, there sometimes can be an advantage in favoring one over the other, as the next example shows.

EXAMPLE 18.6 Evaluate the iterated integral

$$\int_0^1 \int_x^1 \sin y^2\,dy\,dx$$

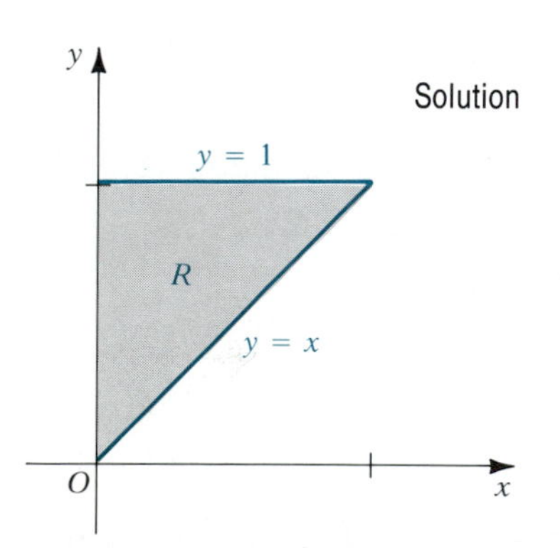

FIGURE 18.14

Solution Since $\sin y^2$ has no elementary antiderivative, we appear to be stymied. But the given iterated integral according to Theorem 18.1 equals the double integral $\iint_R \sin y^2\,dA$, where R is the type I region $\{(x, y): 0 \le x \le 1,\ x \le y \le 1\}$ pictured in Figure 18.14. But R can be equally well viewed as a type II region with left boundary $x = 0$ and right boundary $x = y$. Thus, the double integral, and hence the original iterated integral, equal

$$\int_0^1 \int_0^y \sin y^2\,dx\,dy = \int_0^1 [x \sin y^2]_0^y\,dy$$

Remember that y is constant at this stage.

$$= \int_0^1 y \sin y^2\,dy$$

$$= -\tfrac{1}{2} \cos y^2\Big]_0^1 = \tfrac{1}{2}(1 - \cos 1) \quad ■$$

EXERCISE SET 18.2

A

In Exercises 1–10 evaluate the iterated integral. Sketch the region R determined by the limits.

1. $\int_0^1 \int_1^2 (2x - 3y)\,dy\,dx$
2. $\int_0^2 \int_{-1}^1 (3x^2 - 2xy)\,dx\,dy$
3. $\int_1^2 \int_0^x xy\,dy\,dx$
4. $\int_0^1 \int_x^{4-x} (2x - 1)\,dy\,dx$
5. $\int_0^4 \int_{-2y}^{\sqrt{y}} y(2x - 1)\,dx\,dy$
6. $\int_{-2}^2 \int_{-y}^{\sqrt{2-y}} (x^3 + 2xy)\,dx\,dy$
7. $\int_0^{\sqrt{3}} \int_{2-\sqrt{4-x^2}}^{\sqrt{4-x^2}} x\,dy\,dx$
8. $\int_0^{\sqrt{5}} \int_{y^2-2}^{\sqrt{y^2+4}} y\,dx\,dy$
9. $\int_1^2 \int_{-y}^{4-y} \dfrac{y-4}{(x+4)^2}\,dx\,dy$
10. $\int_1^4 \int_{x-2}^{\sqrt{x}} (2xy + 3)\,dy\,dx$

In Exercises 11–21 evaluate the double integral of f over R, making use of Theorem 18.1. Sketch the region and identify its type.

11. $f(x, y) = 4x - 3y$; $R = \{(x, y): 0 \le x \le 4,\ 0 \le y \le 2\}$
12. $f(x, y) = x^2 - y^2$; $R = \{(x, y): 1 \le x \le 2,\ -1 \le y \le 3\}$
13. $f(x, y) = y/(1 + x^3)$; $R = \{(x, y): 0 \le x \le 1,\ 0 \le y \le 2x\}$
14. $f(x, y) = x + y$; $R = \{(x, y): y - 1 \le x \le \sqrt{1 - y^2},\ 0 \le y \le 1\}$
15. $f(x, y) = 2x - y$; $R = \{(x, y): y \le x \le \sqrt{4 - y^2},\ 0 \le y \le \sqrt{2}\}$
16. $f(x, y) = xe^y$; $R = \{(x, y): 0 \le x \le 1,\ x^2 \le y \le 2 - x^2\}$
17. $f(x, y) = \sqrt{(2x^2 + 7)/y}$; $R = \{(x, y): 1 \le x \le 3,\ x^2/4 \le y \le x^2\}$
18. $f(x, y) = e^{x+2y}$; R is bounded by the triangle with vertices (0, 0), (1, 1), and (3, 0).
19. $f(x, y) = 10xy^3$; R is bounded by the parallelogram with vertices (−1, 0), (0, 1), (1, 0), and (2, 1).
20. $f(x, y) = 4xy$; R is bounded by the triangle with vertices (−1, 1), (0, 2), and (1, 0).
21. $f(x, y) = y^2$; R is bounded by the triangle with vertices (0, 0), (2, 2), and (3, −1).

In Exercises 22–31 make use of equation (18.4) and Theorem 18.1 to find the area of the region bounded by the given curves. Sketch the region and identify its type.

22. $y = x^2$, $y = 2 - x^2$
23. $y = x$, $y = 3x - x^2$
24. $y^2 = 4x$, $y = 2x - 4$
25. $y = \sqrt{x}$, $x = 0$, $y = 4$
26. $y = e^x$, $y = e^{-x}$, $x = \ln 3$
27. $y = 1 - x^2$, $y = \ln x$, $y = 1$
28. $y = \cos \pi x$, $4x^2 + 4y = 1$, $-\frac{1}{2} \le x \le \frac{1}{2}$
29. $y = \cos^{-1} \frac{x}{2}$, $y = \tan^{-1} \frac{x}{3}$, $x = 0$
30. $y = \sqrt{x}$, $x + y = 0$, $x - y = 2$
31. Below $y = 3$ and $y = 3(x + 1)$, and above $y = x^2 - 1$

In Exercises 32–39 find the volume under the surface $z = f(x, y)$ that is above the region R bounded by the given curves.

32. $f(x, y) = 2 - x^2 - y^2$; $x = 0$, $x = 1$, $y = 0$, $y = 1$. Sketch the volume.
33. $f(x, y) = xy$; $y = \sqrt{8 - x^2}$, $y = x$, $y = 0$. Sketch the volume.
34. $f(x, y) = y\sqrt{1 + x^3}$; $y = x$, $x = 2$, $y = 0$
35. $f(x, y) = x + y$; $x = \sqrt{2 - y^2}$, $x = 0$, $x = y$. Sketch the volume.
36. $f(x, y) = x^2e^{xy}$; $xy = 1$, $x = 2$, $y = 2$
37. $f(x, y) = 1 + x^2 + y^2$; $x = 0$, $y = x$, $y = 1$, $y = 2$
38. $f(x, y) = e^{-y}$; $x = 0$, $x = 4$, $y = 0$, $y = \ln 2$. Sketch the volume.
39. $f(x, y) = 4 - x^2$; $y = 0$, $x = 2$, $y = x$. Sketch the volume.
40. Find the volume of the tetrahedron formed by the planes $3x + 4y + 2z = 12$, $x = 0$, $y = 0$, and $z = 0$.
41. Find the volume of the tetrahedron with vertices (0, 0, 0), (2, 0, 0), (0, 4, 0), and (0, 0, 4).
42. Find the volume inside the paraboloid $z = 4 - x^2 - y^2$ and above the xy-plane.

In Exercises 43–46 evaluate the iterated integral by changing the order of integration.

43. $\int_0^1 \int_x^1 e^{-y^2}\,dy\,dx$
44. $\int_0^1 \int_{\sqrt{y}}^1 \sqrt{1 + x^3}\,dx\,dy$
45. $\int_0^2 \int_y^2 \dfrac{y}{(1 + x^3)^2}\,dx\,dy$
46. $\int_0^2 \int_{x^2}^4 \dfrac{1}{1 + y^{3/2}}\,dy\,dx$

In Exercises 47–52 give an equivalent integral with the order of integration reversed.

47. $\int_0^4 \int_{\sqrt{x}}^2 f(x, y)\, dy\, dx$

48. $\int_0^{\pi/2} \int_0^{\sin x} f(x, y)\, dy\, dx$

49. $\int_1^3 \int_{y+1}^4 f(x, y)\, dx\, dy$

50. $\int_0^4 \int_{y^2/4}^{2\sqrt{y}} f(x, y)\, dx\, dy$

51. $\int_0^1 \int_{1-\sqrt{1-x^2}}^{\sqrt{2x-x^2}} f(x, y)\, dy\, dx$

52. $\int_1^2 \int_0^{\ln y} f(x, y)\, dx\, dy$

B

53. Use an argument similar to that used in the text for a type I region to show that if f is nonnegative and continuous on the type II region $R = \{(x, y): h_1(y) \leq x \leq h_2(y),\ c \leq y \leq d\}$, then

$$\iint_R f(x, y)\, dA = \int_c^d \int_{h_1(y)}^{h_2(y)} f(x, y)\, dx\, dy$$

54. Show that if R is a type I region, then equation (18.4) for the area of R is consistent with equation (5.1) for the area between the graphs of two functions.

In Exercises 55–57 evaluate the double integral.

55. $\iint_R x^2 e^y\, dA$; $R = \{(x, y): 0 \leq x \leq 1,\ -x \leq y \leq x\}$

56. $\iint_R (4xy + 3)\, dA$; R is bounded by $y = 0$ and $y = \sin x$ between $x = 0$ and $x = \pi$.

57. $\iint_R \sqrt{y}(x^2 + 1)\, dA$; R is the first-quadrant area bounded by $x^2 y = 4$, $x = 0$, $x = 2$, $y = 0$, and $y = 4$.

In Exercises 58–60 find the area of the region bounded by the given curves, using double integration.

58. Below $y = \sqrt{2x}$ and $x + 2y = 6$, and above $x - 4y = 0$

59. $y = x^2$, $y = x^3 - 2x$

60. $y^2 = x$, $y^2 = 8x$, $x + y = 6$, $5x + y = 48$ (first quadrant)

In Exercises 61–64 find the given volume.

61. Under the surface $z = 2x/y^2$ and above the region bounded by $xy = 6$ and $x^2 + y = 7$

62. Bounded by the surface $z = (1 - \sqrt{x} - \sqrt{y})^2$ and the coordinate planes

63. Bounded by the surface $z = (a^{2/3} - x^{2/3} - y^{2/3})^{3/2}$ and the xy-plane (*Hint:* To evaluate the inner integral use a trigonometric substitution.)

64. Common to the two cylinders $x^2 + y^2 = a^2$ and $x^2 + z^2 = a^2$

18.3 DOUBLE INTEGRALS IN POLAR COORDINATES

Frequently regions in the xy-plane are more conveniently described using polar coordinates rather than rectangular coordinates. When R is a regular region and f is bounded on R, we define the double integral of f over R using polar coordinates as follows. Enclose R in a *polar interval* $I = \{(r, \theta): a \leq r \leq b,\ \alpha \leq \theta \leq \beta\}$ as shown in Figure 18.15, and partition I as shown by rays and circular arcs, thus dividing I into finitely many polar subintervals. Denote the partition by P. The norm of P is again denoted by $\|P\|$ and is the length of the longest diagonal of the subintervals formed by P (just as with a rectangular interval, the diagonal of a polar interval is a line segment joining opposite vertices). Number those subintervals that lie entirely in R (shaded in Figure 18.15) from 1 to n. Let ΔA_i denote the area of the ith one of these, and let (r_i^*, θ_i^*) be polar coordinates of an arbitrary point in the ith subinterval. Then

$$\iint_R f(r, \theta)\, dA = \lim_{\|P\| \to 0} \sum_{i=1}^{n} f(r_i^*, \theta_i^*)\, \Delta A_i \tag{18.9}$$

provided this limit exists, independently of the choices of the points (r_i^*, θ_i^*). Again, it can be shown that the integral will exist when f is continuous on R.

To evaluate the integral of a continuous function by iterated integrals, we consider two types of regions only, analogous to type I and type II for

FIGURE 18.15

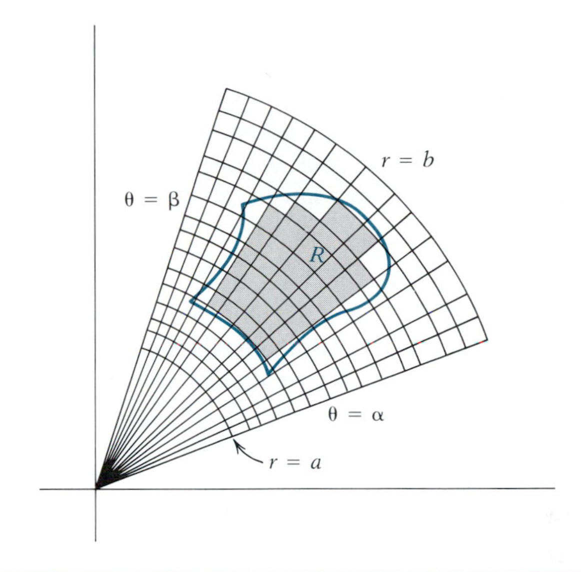

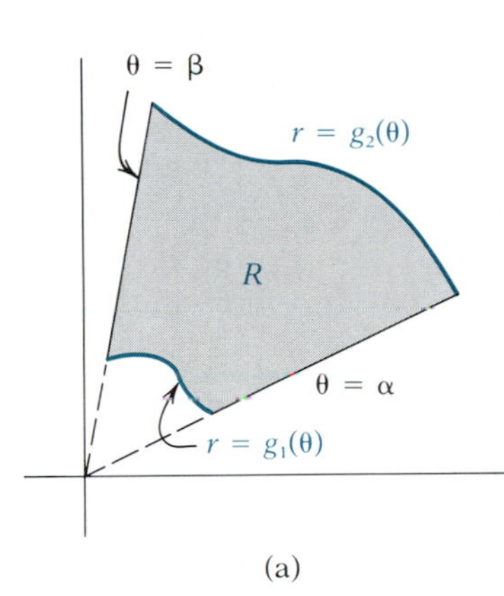

(a)

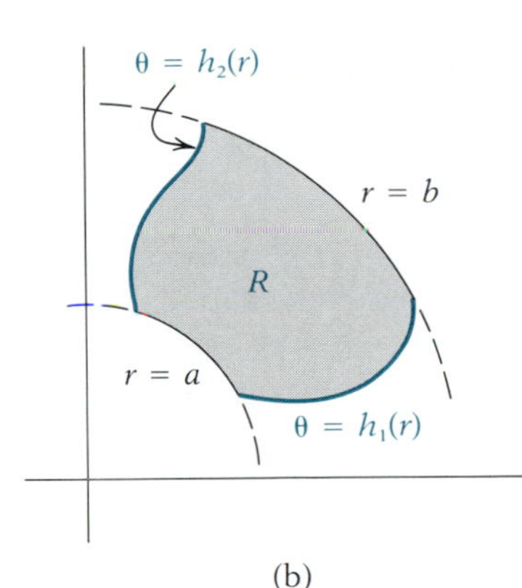

(b)

FIGURE 18.16

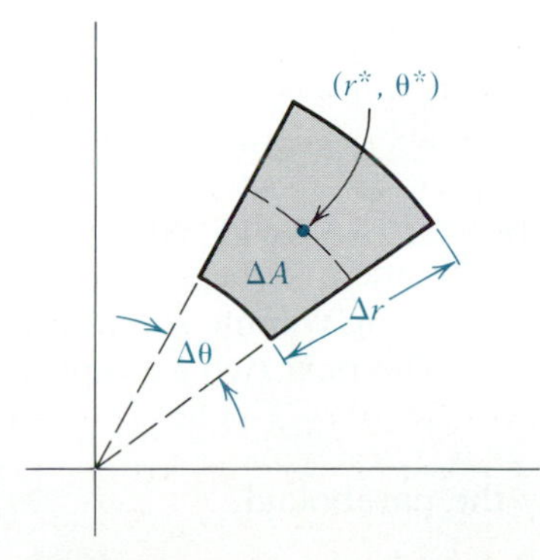

FIGURE 18.17

rectangular coordinates. Figure 18.16 illustrates these two types. In part a R is of the form $\{(r, \theta): g_1(\theta) \le r \le g_2(\theta), \alpha \le \theta \le \beta\}$, where g_1 and g_2 are smooth functions on $[\alpha, \beta]$, and in part b R is of the form $\{(r, \theta): a \le r \le b, h_1(r) \le \theta \le h_2(\theta)\}$, where h_1 and h_2 are smooth functions on $[a, b]$. Because the first type is the more common, we will concentrate our attention on it. To motivate the form of the iterated integral, consider a typical polar subinterval formed by a partition P. We have enlarged such a subinterval in Figure 18.17. To simplify notation we have deleted all subscripts. As shown, let $\Delta\theta$ be the angle between the two rays that form the subinterval, and let Δr be the change in radial distance between the two bounding arcs. In the definition of the double integral, we are free to choose an arbitrary point in each subinterval at which the function is evaluated. We now specify this point (r^*, θ^*) as the center of the subinterval. Then the inner radius is $r^* - \Delta r/2$ and the outer radius is $r^* + \Delta r/2$. The area ΔA is the difference in areas of two circular sectors. Recall that a circular sector of radius r and angle θ has area $\frac{1}{2}r^2\theta$. Thus,

$$\Delta A = \frac{1}{2}\left(r^* + \frac{\Delta r}{2}\right)^2 \Delta\theta - \frac{1}{2}\left(r^* - \frac{\Delta r}{2}\right)^2 \Delta\theta$$

Squaring and collecting terms (verify), we get

$$\Delta A = r^* \, \Delta r \, \Delta\theta \tag{18.10}$$

You might note that $r^* \Delta\theta$ is the length of the arc through the center, which is the average of the lengths of the inner and outer bounding arcs. So ΔA is the average arc length times the radial distance Δr. Notice the similarity with the formula for the area of a trapezoid. The Riemann sum in equation (18.9) can now be written

$$\sum_{i=1}^{n} f(r_i^*, \theta_i^*)\,\Delta A_i = \sum_{i=1}^{n} f(r_i^*, \theta_i^*) r_i^* (\Delta r\,\Delta\theta)_i$$

where by $(\Delta r\,\Delta\theta)_i$ we mean the product of Δr and $\Delta\theta$ for the ith subinterval.

This suggests the following result for a region of the type in Figure 18.16(a):

$$\iint_R f(r, \theta)\, dA = \int_\alpha^\beta \int_{g_1(\theta)}^{g_2(\theta)} f(r, \theta) r\, dr\, d\theta \tag{18.11}$$

Observe the factor r in the integrand of the iterated integral on the right. This discussion is intended to suggest where it comes from, but it is not a proof. A complete proof of equation (18.11) requires concepts studied in advanced calculus.

It is useful to write

$$dA = r\, dr\, d\theta$$

and to call this the *differential of area in polar coordinates*. For rectangular coordinates the analogous formula is $dA = dy\, dx$. So in going from a double integral in polar coordinates to an iterated integral, dA is replaced by $r\, dr\, d\theta$, whereas in rectangular coordinates dA is replaced by $dy\, dx$. (In each case the order of the differentials may be reversed, depending on the order of integration.)

When the region is of the type in Figure 18.16(b), the double integral is evaluated by

$$\iint_R f(r, \theta)\, dA = \int_a^b \int_{h_1(\theta)}^{h_2(\theta)} f(r, \theta) r\, d\theta\, dr \tag{18.12}$$

EXAMPLE 18.7 Evaluate the double integral $\iint_R r^2 \sin\theta\, dA$, where R is the region bounded by the polar axis and the upper half of the cardioid $r = 1 + \cos\theta$.

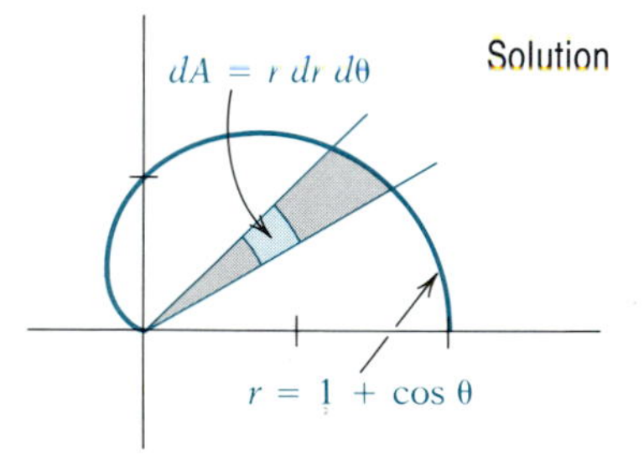

FIGURE 18.18

Solution The region R is pictured in Figure 18.18. We have shown a typical small sector of the type used in a partition along with a typical "differential element" that we imagine to have area $dA = r\, dr\, d\theta$. This is to aid in setting up the integral. From equation (18.11) we have

$$\begin{aligned}
\iint_R r^2 \sin\theta\, dA &= \int_0^\pi \int_0^{1+\cos\theta} (r^2 \sin\theta) \overbrace{r\, dr\, d\theta}^{dA} \\
&= \int_0^\pi \left[\int_0^{1+\cos\theta} r^3 \sin\theta\, dr \right] d\theta \\
&= \frac{1}{4} \int_0^\pi [r^4]_0^{1+\cos\theta} \sin\theta\, d\theta \\
&= \frac{1}{4} \int_0^\pi (1 + \cos\theta)^4 \sin\theta\, d\theta \\
&= -\frac{1}{4} \left[\frac{(1+\cos\theta)^5}{5} \right]_0^\pi = \frac{8}{5}
\end{aligned}$$

■

When f is continuous and nonnegative on R, $\iint_R f(r, \theta)\, dA$ again represents the volume above R and under the surface $z = f(r, \theta)$, and in the special case $f(r, \theta) = 1$ for all points (r, θ) in R, as before, the resulting volume is numerically the same as the area of R; that is, $A = \iint_R dA$. The next two examples illustrate these results.

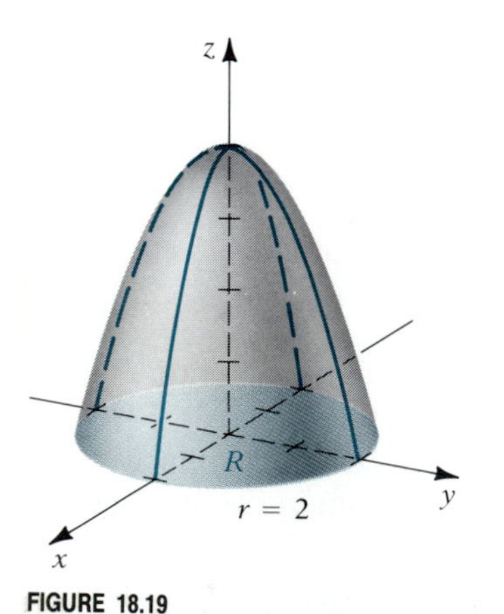

FIGURE 18.19

EXAMPLE 18.8 Use polar coordinates to find the volume below the paraboloid $z = 4 - x^2 - y^2$ and above the xy-plane.

Solution In polar coordinates $x^2 + y^2 = r^2$, so the equation of the paraboloid can be written as $z = 4 - r^2$. Thus we take $f(r, \theta) = 4 - r^2$. The boundary of the region R is found by putting $z = 0$, giving $r^2 = 4$ or $r = 2$ (see Figure 18.19). So we have

$$V = \iint_R (4 - r^2)\, dA = \int_0^{2\pi} \int_0^2 (4 - r^2) r\, dr\, d\theta$$

$$= \int_0^{2\pi} \left[2r^2 - \frac{r^4}{4} \right]_0^2 d\theta = 4 \int_0^{2\pi} d\theta = 8\pi$$

■

EXAMPLE 18.9 Find the area inside the circle $r = 5 \cos \theta$ and outside the limaçon $r = 2 + \cos \theta$.

Solution The curves are shown in Figure 18.20. The points of intersection are found by solving the equations simultaneously:

$$5 \cos \theta = 2 + \cos \theta$$

$$\cos \theta = \tfrac{1}{2}$$

$$\theta = \pm\tfrac{\pi}{3}, \qquad r = \tfrac{5}{2}$$

By symmetry, the total area is twice that for $0 \leq \theta \leq \frac{\pi}{3}$:

$$A = \iint_R dA = 2 \int_0^{\pi/3} \int_{2 + \cos\theta}^{5 \cos\theta} r\, dr\, d\theta = 2 \int_0^{\pi/3} \left[\frac{r^2}{2} \right]_{2 + \cos\theta}^{5 \cos\theta} d\theta$$

$$= \int_0^{\pi/3} [25 \cos^2 \theta - (4 + 4 \cos \theta + \cos^2 \theta)]\, d\theta$$

$$= 4 \int_0^{\pi/3} (6 \cos^2 \theta - \cos \theta - 1]\, d\theta$$

$$= 4 \int_0^{\pi/3} [3(1 + \cos 2\theta) - \cos \theta - 1]\, d\theta$$

$$= 4 \left[2\theta + \frac{3 \sin 2\theta}{2} - \sin \theta \right]_0^{\pi/3} = \frac{8\pi}{3} + \sqrt{3}$$

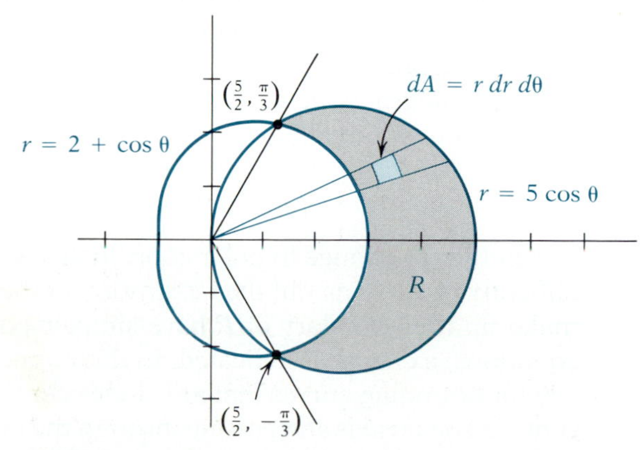

FIGURE 18.20

■

The next example illustrates how an iterated integral in rectangular coordinates can sometimes be evaluated more easily by changing to polar coordinates.

EXAMPLE 18.10 Evaluate the integral

$$\int_0^2 \int_0^{\sqrt{2x-x^2}} \sqrt{x^2+y^2}\, dy\, dx$$

by changing to polar coordinates.

Solution First we find the region R for which the given integral equals the double integral $\iint_R \sqrt{x^2+y^2}\, dA$. The limits tell us that R is of type I with lower bounding curve $y = 0$ and upper bounding curve $y = \sqrt{2x - x^2}$. Squaring and collecting terms, we find that the upper boundary is that part of the circle $x^2 + y^2 - 2x = 0$ for which $y \geq 0$. So the region R is that shown in Figure 18.21. In polar coordinates the circle $x^2 + y^2 - 2x = 0$ becomes $r = 2\cos\theta$. (Verify.) So in polar coordinates

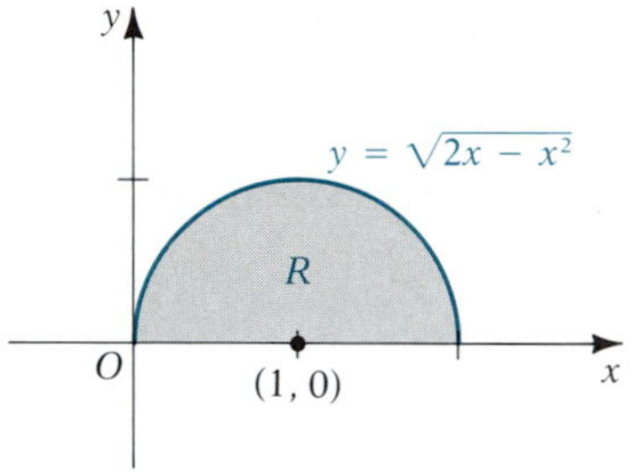

$$R = \{(r, \theta): 0 \leq r \leq 2\cos\theta,\ 0 \leq \theta \leq \tfrac{\pi}{2}\}$$

FIGURE 18.21

The integrand, $\sqrt{x^2+y^2}$, changes to r in polar coordinates, and so the double integral and hence the original iterated integral are equal to

$$\int_0^{\pi/2} \int_0^{2\cos\theta} r(r\,dr\,d\theta) = \int_0^{\pi/2} \left[\frac{r^3}{3}\right]_0^{2\cos\theta} d\theta$$

$$= \frac{8}{3}\int_0^{\pi/2} \cos^3\theta\, d\theta = \frac{8}{3}\int_0^{\pi/2} (1 - \sin^2\theta)\cos\theta\, d\theta$$

$$= \frac{8}{3}\left[\sin\theta - \frac{\sin^3\theta}{3}\right]_0^{\pi/2} = \frac{8}{3}\left[1 - \frac{1}{3}\right] = \frac{16}{9}$$ ■

The procedure suggested by this example for changing an iterated integral in rectangular coordinates to one in polar coordinates can be summarized as follows:

1. Sketch the region R determined by the limits of integration, and write the equations of the curves that form the boundary of R in polar coordinates.
2. Determine the new limits of integration as in equation (18.11) or (18.12), depending on whether R is of the type shown in Figure 18.16(a) or (b). (In some cases it may be necessary to express R as the union of two or more such regions.)
3. In the integrand replace x by $r\cos\theta$ and y by $r\sin\theta$. (Note that $x^2 + y^2$ can be changed immediately to r^2.)
4. Replace $dy\,dx$ or $dx\,dy$, whichever occurs, by $r\,dr\,d\theta$ or by $r\,d\theta\,dr$, depending on the order of integration to be used.

Whether to change to polar coordinates is determined by the comparative difficulty of carrying out the integration in the two systems. If the curves that make up the boundary of R have simpler polar equations than rectangular equations, a change is indicated. In this connection, be on the lookout especially for bounding curves that are circles centered at the origin ($r = a$), circles centered on an axis and passing through the origin ($r = a\cos\theta$ or $r = a\sin\theta$), and lines through the origin ($\theta = \alpha$). The nature of the integrand also is a factor to be considered. If the original integrand is $f(x, y)$, we see from applying steps 3 and 4 that the new integrand is $rf(r\cos\theta, r\sin\theta)$. Whether the antiderivatives needed in the iterated integration can be found clearly affects the decision. As you practice problems such as Exercises 20–31, you should begin to develop a feel for when a change is worthwhile.

EXERCISE SET 18.3

A

In Exercises 1–8 evaluate the double integral $\iint_R f(r, \theta)\, dA$, where R is the region described. Draw the region.

1. $f(r, \theta) = r\theta$; $|r| \le 2$
2. $f(r, \theta) = r(\sin\theta - \cos\theta)$; $\{(r, \theta): 0 \le r \le 1, \frac{\pi}{4} \le \theta \le \pi\}$
3. $f(r, \theta) = 2r + 1$; $1 \le |r| \le 4$
4. $f(r, \theta) = \sqrt{r}\sin\theta$; $\{(r, \theta): 0 \le r \le 1 - \cos\theta, 0 \le \theta \le \pi\}$
5. $f(r, \theta) = r^2\cos\theta$; $\{(r, \theta): 0 \le r \le 2\sin\theta, 0 \le \theta \le \frac{\pi}{2}\}$
6. $f(r, \theta) = r(\sin\frac{\theta}{2} - \cos\frac{\theta}{2})$; $1 \le |r| \le 2$
7. $f(r, \theta) = r^2\sin^2 2\theta$; $\{(r, \theta): 0 \le r \le 2\sqrt{\cos 2\theta}, 0 \le \theta \le \frac{\pi}{4}\}$
8. $f(r, \theta) = \sqrt{1 + \theta^3}$; $\{(r, \theta): 0 \le r \le \theta, 0 \le \theta \le 2\}$

In Exercises 9–19 use double integration and polar coordinates to find the area of the region described. Draw the region.

9. Inside $r = 2 + 2\cos\theta$
10. Inside $r^2 = 4\cos 2\theta$
11. Inside $r = 2\cos 2\theta$
12. Inside $r = 2 + \sin\theta$
13. Inside $r = 2(1 + \cos\theta)$ and outside $r = 1$
14. Inside $r = 8\sin\theta$ and outside $r = 3 + 2\sin\theta$
15. Inside $r^2 = 2\cos 2\theta$ and outside $r = 1$
16. Inside $r = 1 + 2\cos\theta$ and to the right of $r\cos\theta = 1$
17. Inside both $r = 2$ and $r = 3 - 2\cos\theta$
18. Inside the small loop of the limaçon $r = 1 + 2\cos\theta$
19. Inside both $r = 6\sin\theta$ and $r = 2(1 + \sin\theta)$

In Exercises 20–31 evaluate the iterated integral by changing to polar coordinates.

20. $\int_0^2 \int_0^{\sqrt{4-x^2}} x^2 y\, dy\, dx$
21. $\int_0^1 \int_x^{\sqrt{2-x^2}} (x^2 + y^2)^2\, dy\, dx$
22. $\int_{-1}^1 \int_{-\sqrt{1-x^2}}^{\sqrt{1-x^2}} (x^2 + y^2)^{3/2}\, dy\, dx$
23. $\int_0^2 \int_{-\sqrt{8-y^2}}^{-y} (x + y)\, dx\, dy$
24. $\int_0^4 \int_0^{\sqrt{4y-y^2}} y\, dx\, dy$
25. $\int_0^1 \int_x^{\sqrt{2x-x^2}} x\, dy\, dx$
26. $\int_{-\sqrt{3}}^{\sqrt{3}} \int_1^{\sqrt{4-y^2}} \frac{1}{(x^2 + y^2)^{3/2}}\, dx\, dy$
27. $\int_0^{\sqrt{3}} \int_1^{\sqrt{4-x^2}} \frac{x}{y}\, dy\, dx$
28. $\int_1^{\sqrt{3}} \int_1^x \frac{x^2 - y^2}{x^2 + y^2}\, dy\, dx$
29. $\int_0^1 \int_y^{1+\sqrt{1-y^2}} \sqrt{4 - x^2 - y^2}\, dx\, dy$
30. $\int_{-2}^2 \int_2^{2+\sqrt{4-x^2}} \frac{x + y}{y}\, dy\, dx$
31. $\int_{-a}^a \int_{-\sqrt{a^2-y^2}}^{\sqrt{a^2-y^2}} e^{-(x^2+y^2)}\, dx\, dy$

In Exercises 32–39 use polar coordinates to find the indicated volume.

32. Inside the ellipsoid $x^2 + y^2 + 4z^2 = 4$ and above the xy-plane
33. Inside the cone $z = 2 - \sqrt{x^2 + y^2}$ and above the xy-plane
34. A sphere of radius a
35. Inside the sphere $x^2 + y^2 + z^2 = a^2$ and outside the cylinder $x^2 + y^2 = b^2$, where $0 < b < a$
36. Inside the cylinder $x^2 + y^2 = a^2$ between the upper and lower sheets of the hyperboloid $z^2 - x^2 - y^2 = a^2$
37. Under the cone $z = \sqrt{x^2 + y^2}$ and above the region R in the xy-plane inside the circle $x^2 + y^2 = 2x$
38. Under the cylindrical surface $z = y^2$ and above the region $R = \{(x, y): 0 \le x \le 1, x \le y \le \sqrt{2 - x^2}\}$
39. Between the surfaces $z = 6 - x^2 - y^2$ and $z = 2$. (*Hint:* Use the difference of two volumes.)

B

40. Evaluate by changing to polar coordinates:
$$\int_{2-2\sqrt{2}}^{2+2\sqrt{2}} \int_{-y}^{(4-y^2)/4} x^2 y\, dx\, dy$$
41. Write as a single iterated integral in polar coordinates and evaluate:
$$\int_0^{3/2} \int_{\sqrt{2y-y^2}}^{\sqrt{4y-y^2}} xy\, dx\, dy + \int_{3/2}^3 \int_{y/\sqrt{3}}^{\sqrt{4y-y^2}} xy\, dx\, dy$$
42. Find the area bounded by the curves $r = 1$, $r = 2$, $r = \ln\theta$, and $r\theta = 1$.
43. Find the area between the y-axis and the parabola $y^2 = 4 - 4x$ that lies outside the circle $x^2 + y^2 - 4x = 0$.

44. Let $I = \int_0^\infty e^{-x^2}\,dx$. It can be shown that

$$I^2 = \left(\int_0^\infty e^{-x^2}\,dx\right)\left(\int_0^\infty e^{-y^2}\,dy\right) = \int_0^\infty \int_0^\infty e^{-(x^2+y^2)}\,dy\,dx$$

(a) Use this to find the value of I.

(b) Use the value of I to prove that the area under the normal probability density function

$$f(x) = \frac{1}{\sqrt{2\pi}\sigma} e^{-(x-\mu)^2/2\sigma^2}, \qquad -\infty < x < \infty$$

is 1.

18.4 MOMENTS, CENTROIDS, AND CENTER OF MASS

In this section we present some applications of the double integral that are especially important in physics and engineering. The ideas are also used in statistics.

Children playing on a seesaw quickly learn that the heavier child must sit closer to the pivot point, called the *fulcrum,* in order to bring the seesaw into balance. Denoting the weights of the children by w_1 and w_2 and the corresponding distances from the fulcrum by d_1 and d_2, as in Figure 18.22, we must have $w_1 d_1 = w_2 d_2$ for the seesaw to be balanced. This is called the *law of the lever* and was known to Archimedes.

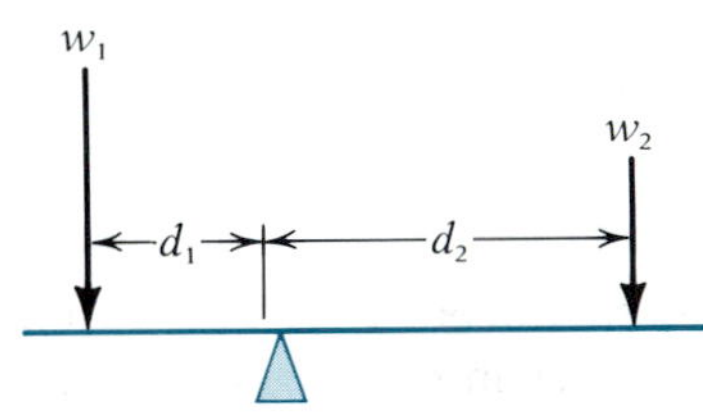

FIGURE 18.22

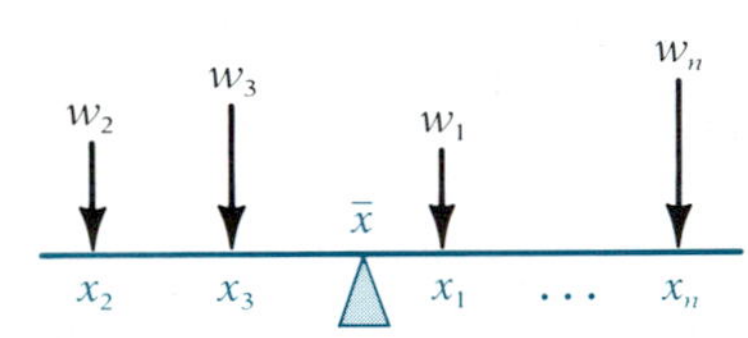

FIGURE 18.23

Now consider a finite number of weights, $w_1, w_2, \ldots, w_n$, placed at various points on an axis that has been coordinatized so that w_i has x-coordinate x_i, as in Figure 18.23. We wish to find the location of the fulcrum, also called the *center of gravity*, of the system. Denote its x-coordinate by $\bar{x}$. The product of a weight (or force) and its distance from the fulcrum is called the **moment** produced by that weight (or force). The moment thus is a measure of the tendency for the weight to cause a turning, or rotation, about the fulcrum. The directed distance of the weight from the fulcrum is called the **moment arm** of the weight. If the moment arm is positive, then the weight tends to produce a clockwise rotation, and if negative, a counterclockwise rotation. In Figure 18.23, for example, the moment arm of w_1 is the positive distance $x_1 - \bar{x}$, and the moment is $w_1(x_1 - \bar{x})$. The moment arm of w_3 is $x_3 - \bar{x}$, which is a negative number, and so w_3 produces the negative moment $w_3(x_3 - \bar{x})$. For balance, it must be true that the algebraic sum of all the moments is 0:

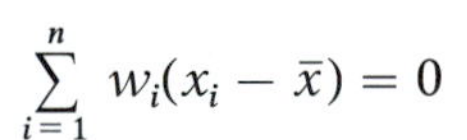

$$\sum_{i=1}^{n} w_i(x_i - \bar{x}) = 0$$

We can now solve for $\bar{x}$ from this equation:

$$\sum_{i=1}^{n} w_i x_i - \sum_{i=1}^{n} \bar{x} w_i = 0$$

Since $\bar{x}$ is a constant, we can factor it out of the second summation and then solve, getting

$$\bar{x} = \frac{\sum_{i=1}^{n} w_i x_i}{\sum_{i=1}^{n} w_i}$$

Since by Newton's second law the weight w of an object is its mass m times the acceleration of gravity g, we have for each i, $w_i = m_i g$, so that $\bar{x}$ can be rewritten in terms of masses as

$$\bar{x} = \frac{\sum_{i=1}^{n} m_i g x_i}{\sum_{i=1}^{n} m_i g} = \frac{\sum_{i=1}^{n} m_i x_i}{\sum_{i=1}^{n} m_i} \tag{18.13}$$

Because of the shift in emphasis from weight to mass as given by equation (18.13), it is customary to use the term **center of mass** instead of center of gravity for the balance point of the system. The product $m_i x_i$ is the moment of mass m_i about the origin, since x_i is the directed distance of m_i from the origin. Thus, we can say the result of equation (18.13) in words as

$$\bar{x} = \frac{\text{sum of moments about the origin}}{\text{sum of masses}}$$

If we let $m = \sum_{i=1}^{n} m_i$, then on clearing equation (18.13) of fractions, we see that

$$m\bar{x} = \sum_{i=1}^{n} m_i x_i$$

From this we conclude that if the masses were all concentrated at $\bar{x}$—that is, at the center of mass—the moment about the origin of m would equal the sum of the moments about the origin of the individual masses.

Now we extend these ideas to a system of masses in the xy-plane, as shown in Figure 18.24. We assume each mass to be concentrated at a point and refer to it as a *point mass*. This is, of course, an idealized concept. Let the point mass m_i be located at (x_i, y_i). If l is any line in the plane and d_i is the directed distance from l to (x_i, y_i), then the moment of m_i *about* l is $m_i d_i$, and the sum of the moments of all the masses about l is $\sum_{i=1}^{n} m_i d_i$. In particular, the sum of the moments about the x-axis is $\sum_{i=1}^{n} m_i y_i$, and about the y-axis is $\sum_{i=1}^{n} m_i x_i$. The point $(\bar{x}, \bar{y})$ for which the sum of the moments about the lines $x = \bar{x}$ and $y = \bar{y}$ each equals 0 is the center of mass of the system. By proceeding as in the one-dimensional case, we find that

$$\begin{aligned} \bar{x} &= \frac{\sum_{i=1}^{n} m_i x_i}{\sum_{i=1}^{n} m_i} = \frac{\text{sum of moments about } y\text{-axis}}{\text{sum of masses}} \\ \bar{y} &= \frac{\sum_{i=1}^{n} m_i y_i}{\sum_{i=1}^{n} m_i} = \frac{\text{sum of moments about } x\text{-axis}}{\text{sum of masses}} \end{aligned} \tag{18.14}$$

So far, what we have done involves systems of discrete masses on a line or in a plane. Next we extend the ideas to a continuously distributed mass over some plane region R that we assume to be regular. The thickness is considered negligible. We call such a two-dimensional body a *lamina*. Again, this is an idealized concept, since in any actual situation thickness does exist, however small it may be. Figure 18.25 illustrates such a lamina.

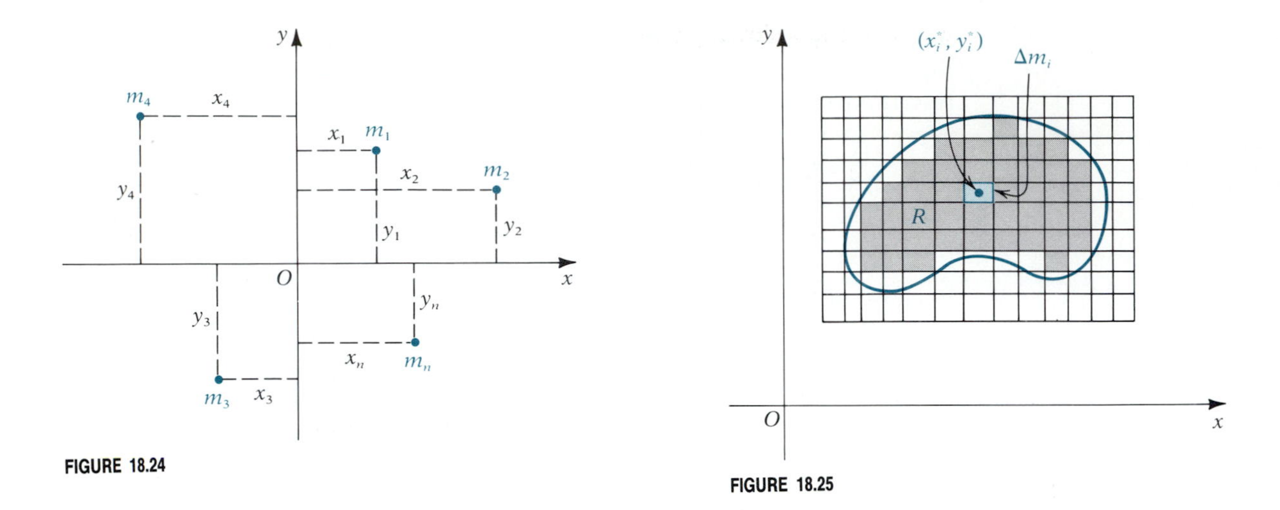

FIGURE 18.24

FIGURE 18.25

If the area of R is A and the density of the lamina is a constant ρ (mass per unit area), then the total mass of the lamina is $m = \rho A$. If the density is a variable, we consider the density function $\rho(x, y)$ defined as follows. If ΔA is the area of a rectangular region centered on the point (x, y) and m is the mass of this region, then

$$\rho(x, y) = \lim_{\delta \to 0} \frac{\Delta m}{\Delta A}$$

where δ is the length of the diagonal of the rectangle.

Now we proceed exactly as in defining the double integral of a function over R, partitioning a two-dimensional interval I that contains R as in Figure 18.25 and counting the subintervals inside R. In the ith such subinterval let (x_i^*, y_i^*) denote the center of the subinterval. If the norm $\|P\|$ of the partition is small, we would expect the density throughout that part of the lamina in this ith subinterval to be approximately equal to the constant value $\rho(x_i^*, y_i^*)$. Then its mass $\Delta m_i \approx \rho(x_i^*, y_i^*)\,\Delta A_i$. We think of the mass Δm_i as if it were a point mass concentrated at (x_i^*, y_i^*). From equation (18.14), the coordinates $(\bar{x}, \bar{y})$ of the center of mass of the lamina are approximated as follows:

$$\bar{x} \approx \frac{\sum_{i=1}^{n} (\Delta m_i) x_i}{\sum_{i=1}^{n} \Delta m_i} \quad \text{and} \quad \bar{y} \approx \frac{\sum_{i=1}^{n} (\Delta m_i) y_i}{\sum_{i=1}^{n} \Delta m_i}$$

The approximations become better and better as we take partitions with smaller and smaller norms. In the limit we obtain

$$\bar{x} = \frac{\iint_R x\, dm}{\iint_R dm} \text{ and } \bar{y} = \frac{\iint_R y\, dm}{\iint_R dm} \tag{18.15}$$

where $dm = \rho(x, y)\, dA$. Writing M_x and M_y for the moments about the x- and y-axes, respectively, and denoting the total mass of the lamina by m, we have the following equations:

$$\bar{x} = \frac{M_y}{m}, \qquad \bar{y} = \frac{M_x}{m}$$
$$M_y = \iint_R x\,dm$$
$$M_x = \iint_R y\,dm \tag{18.16}$$
$$m = \iint_R dm$$

Moments about the lines $x = a$ and $y = b$ are given by

$$M_{x=a} = \iint_R (x - a)\,dm$$
$$M_{y=b} = \iint_R (y - b)\,dm$$

Since $M_y = m\bar{x}$ and $M_x = m\bar{y}$, it follows that moments about the x- and y-axes would be unchanged if the entire mass m of the lamina were concentrated at the center of mass $(\bar{x}, \bar{y})$. This center of mass is the "balance point" of the lamina in the sense that moments taken about lines through it parallel to the axes equal 0. (See Exercise 32.)

EXAMPLE 18.11 A lamima in the shape of the triangular region in Figure 18.26 has density $\rho(x, y) = \sqrt{xy}$. Find the mass of the lamina and the center of mass.

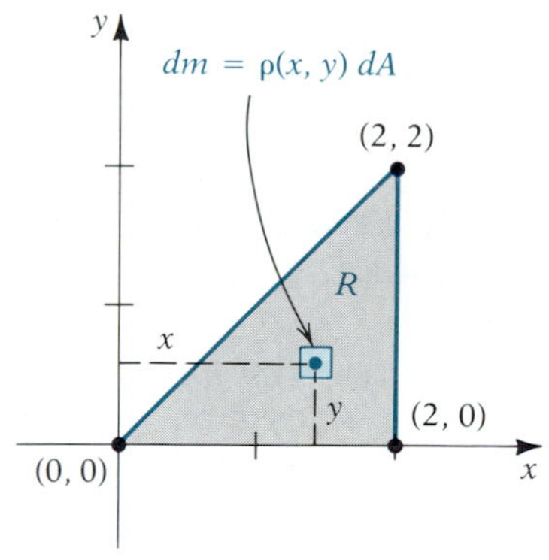

FIGURE 18.26

Solution The small rectangle shown in Figure 18.26 and labeled $dm = \rho(x, y)\,dA$ suggests a point mass. We sum up all such point masses to get the total mass:

$$m = \iint_R dm = \iint_R \rho(x, y)\,dA$$
$$= \int_0^2 \int_0^x \sqrt{xy}\,dy\,dx = \tfrac{16}{9}$$

(Fill in the details of the integration.) The moments M_y and M_x are similarly found by integrating the moments $x\,dm$ and $y\,dm$ of the "point mass":

$$M_y = \iint_R x\,dm = \iint_R x\rho(x, y)\,dA = \int_0^2 \int_0^x x\sqrt{xy}\,dy\,dx = \tfrac{8}{3}$$
$$M_x = \iint_R y\,dm = \iint_R y\rho(x, y)\,dA = \int_0^2 \int_0^2 y\sqrt{xy}\,dy\,dx = \tfrac{8}{5}$$

Thus,

$$\bar{x} = \frac{M_y}{m} = \frac{\frac{8}{3}}{\frac{16}{9}} = \frac{3}{2}$$

$$\bar{y} = \frac{M_x}{m} = \frac{\frac{8}{5}}{\frac{16}{9}} = \frac{9}{10}$$

■

If $\rho(x, y)$ is a constant, then the lamina is said to be **homogeneous,** and in this case $\bar{x}$ and $\bar{y}$ reduce to

$$\bar{x} = \frac{\iint\limits_R x\,dA}{\iint\limits_R dA} \text{ and } \bar{y} = \frac{\iint\limits_R y\,dA}{\iint\limits_R dA} \tag{18.17}$$

Since the integrals here no longer involve density, the physical nature of the lamina disappears from consideration; that is, $(\bar{x}, \bar{y})$ as given by equation (18.17) exists on a purely geometric basis for any regular region R. So we can think of equation (18.17) as defining the center of mass of a homogeneous lamina or, alternately, as defining the geometric center of a regular region in the xy-plane. To distinguish this latter interpretation from the former, we use the term **centroid.** Thus, centroid refers to a geometric object and center of mass to a physical object. They coincide when the lamina is homogeneous.

EXAMPLE 18.12 Find the centroid of the area bounded by the parabola $y = 4 - x^2$ and the x-axis.

Solution The region is shown in Figure 18.27. By symmetry the centroid lies on the y-axis, so that $\bar{x} = 0$. (Note that if we were working with a nonhomogeneous lamina, this conclusion would hold only if the mass were symmetrically distributed with respect to the y-axis). From equations (18.17) we have

$$\bar{y} = \frac{\iint\limits_R y\,dA}{\iint\limits_R dA} = \frac{\int_{-2}^{2}\int_0^{4-x^2} y\,dy\,dx}{\int_{-2}^{2}\int_0^{4-x^2} dy\,dx} = \frac{8}{5}$$

(Verify this result by supplying the details of the integration.) Thus the centroid is at $(0, \frac{8}{5})$. ■

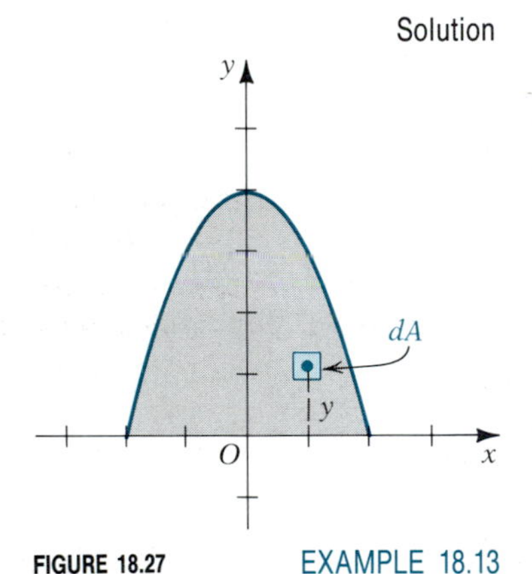

FIGURE 18.27

EXAMPLE 18.13 A homogeneous lamina occupies the region outside the circle $x^2 + y^2 = 1$ and inside the circle $x^2 + y^2 - 2x = 0$. Find its center of mass.

Solution The region R in this case is more easily described in polar coordinates (see Figure 18.28). The bounding curves are $r = 1$ and $r = 2\cos\theta$, as you can readily verify. These intersect at $(1, \pm\frac{\pi}{3})$. By symmetry, $\bar{y} = 0$. For $\bar{x}$ we have from equations (18.17)

$$\bar{x} = \frac{\iint\limits_R x\,dA}{\iint\limits_R dA} = \frac{\int_{-\pi/3}^{\pi/3}\int_1^{2\cos\theta} (r\cos\theta)(r\,dr\,d\theta)}{\int_{-\pi/3}^{\pi/3}\int_1^{2\cos\theta} r\,dr\,d\theta} = \frac{8\pi + 3\sqrt{3}}{4\pi + 6\sqrt{3}} \approx 1.321 \quad \text{(Verify.)}$$

■

Consider again a system of point masses in the xy-plane. Let m be such a point mass and l a line in the plane. Then, as we have seen, the moment of m about l is the product md, where d is the directed distance from l to m. To distinguish such a moment from those we are about to define, we sometimes refer to md as the **first moment.** Higher order moments of m about

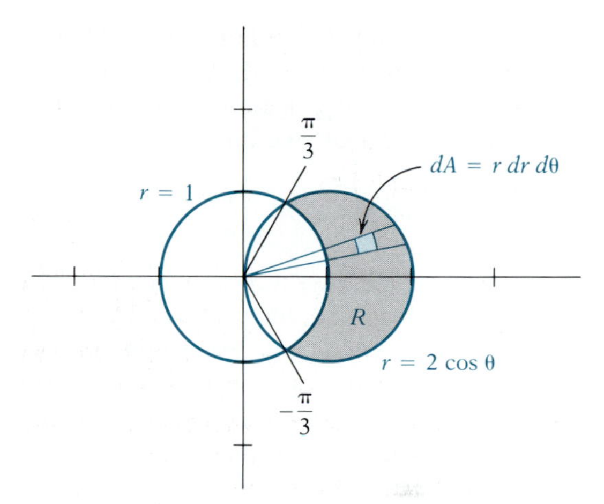

FIGURE 18.28

l are defined by

md^2, second moment
md^3, third moment
⋮
md^k, kth moment

We concentrate on the second moment because it is especially important in physics and engineering. Moments of orders 3 and 4 are also used, for example in structural design, but the primary use of higher order moments is in probability theory. We will say more about this later. In a physical system the second moment is also called the **moment of inertia** and the distance d is called the **radius of gyration**. This terminology suggests a rotation, and we will explain shortly in what sense the moment of inertia is related to a rotating mass.

For a lamina of the type we have been considering, using the same partition process as before, we define the moments of inertia about the x-axis and y-axis, respectively, as

$$I_x = \lim_{||P|| \to 0} \sum_{i=1}^{n} (y_i^*)^2 \, \Delta m_i = \iint_R y^2 \, dm$$
$$I_y = \lim_{||P|| \to 0} \sum_{i=1}^{n} (x_i^*)^2 \, \Delta m_i = \iint_R x^2 \, dm \tag{18.18}$$

where $dm = \rho(x, y)\, dA$. We also define the *polar moment of inertia*, I_0, by

$$I_0 = \iint_R (x^2 + y^2)\, dm \tag{18.19}$$

It follows that $I_0 = I_x + I_y$. This polar moment of inertia is sometimes said to be the moment of inertia about the origin. Perhaps a better way to describe it is as the moment of inertia about a line through the origin perpendicular to the xy-plane—that is, about the z-axis.

If a lamina of mass m has moment of inertia I with respect to a line, then the positive number r that satisfies $I = mr^2$ is called the *radius of gyration* of the lamina with respect to that line. This is consistent with the notion of radius of gyration for a point mass if we consider the entire mass of the lamina concentrated at a distance r from the line. In particular, if we denote by r_x the radius of gyration with respect to the y-axis and by r_y that with respect to the x-axis, we have

$$r_x = \sqrt{\frac{I_y}{m}} \quad \text{and} \quad r_y = \sqrt{\frac{I_x}{m}}$$

The point (r_x, r_y) plays a role for second moments analogous to that of $(\bar{x}, \bar{y})$ for first moments. In general, these points are not the same.

The polar moment of inertia is helpful in getting an intuitive feeling for what role moments of inertia play in a dynamic system. Consider a point mass m in the xy-plane located a distance r from the origin, and suppose it is rotating about the z-axis at the constant angular velocity ω (number of radians it turns through per unit of time). Its kinetic energy is $\frac{1}{2}mv^2$, where v is the linear velocity. Since $v = r\omega$, the kinetic energy becomes

$$KE = \tfrac{1}{2}m(r\omega)^2 = \tfrac{1}{2}(mr^2)\omega^2 = \tfrac{1}{2}I_0\omega^2$$

since $I_0 = m(x^2 + y^2) = mr^2$. The same result holds if we have a lamina in the xy-plane rotating about the z-axis, as the usual partitioning, summing, and passing to the limit would show. For example, the total kinetic energy of a rotating wheel is $\frac{1}{2}I_0\omega^2$. Now kinetic energy is equal to the work required to bring the object to rest. So for a constant angular velocity, the moment of inertia is a measure of the work required to bring the rotating wheel to a stop.

EXAMPLE 18.14 A lamina is bounded by the curves $y^2 = x$ and $x - y = 2$, and the density is $\rho(x, y) = 2x$. Find I_x, I_y, r_x, and r_y.

Solution By equations (18.18),

$$I_x = \iint_R y^2\,dm = \iint_R 2xy^2\,dA$$

$$I_y = \iint_R x^2\,dm = \iint_R 2x^3\,dA$$

The region occupied by the lamina is the type II region pictured in Figure 18.29. The points (4, 2) and (1, −1) of intersection are readily found. Using

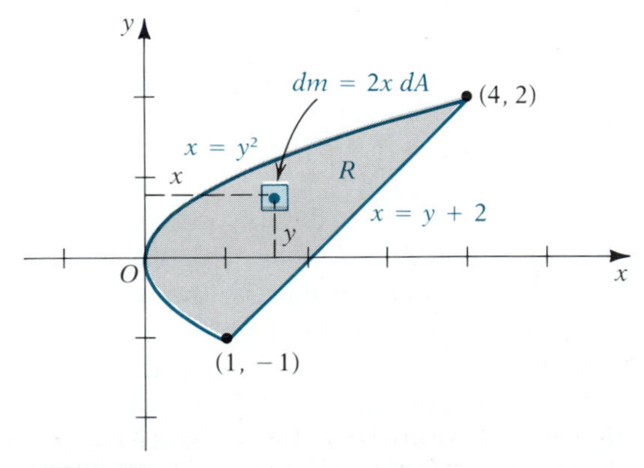

FIGURE 18.29

iterated integrals, we obtain

$$I_x = \int_{-1}^{2} \int_{y^2}^{y+2} 2xy^2 \, dx \, dy \quad \text{and} \quad I_y = \int_{-1}^{2} \int_{y^2}^{y+2} 2x^3 \, dx \, dy$$

The results, after some straightforward but tedious integration, are

$$I_x = \tfrac{531}{35} \quad \text{and} \quad I_y = \tfrac{369}{5}$$

Thus, $I_0 = I_x + I_y = \frac{3114}{35}$.

To find r_x and r_y we need the mass

$$m = \iint_R \rho \, dm = \int_{-1}^{2} \int_{y^2}^{y+2} 2x \, dx \, dy = \tfrac{72}{5} \qquad \text{(Verify.)}$$

So we get, after some simplification,

$$r_x = \sqrt{\frac{I_y}{m}} = \sqrt{\frac{41}{8}} \approx 2.26 \qquad r_y = \sqrt{\frac{I_x}{m}} = \sqrt{\frac{59}{56}} \approx 1.03$$

■

EXAMPLE 18.15 A lamina is the annular ring between the circles $x^2 + y^2 = 1$ and $x^2 + y^2 = 4$. The density at any point (x, y) is the reciprocal of its distance from the origin. Find the polar moment of inertia.

Solution The nature of the region (Figure 18.30) suggests that it would be convenient to use polar coordinates. In polar coordinates equation (18.19) becomes

$$I_0 = \iint_R r^2 \, dm$$

and $dm = \rho(r, \theta) \, dA = \frac{1}{r}(r \, dr \, d\theta) = dr \, d\theta$. Thus,

$$I_0 = \int_0^{2\pi} \int_1^2 r^2 \, dr \, d\theta = \frac{14\pi}{3}$$

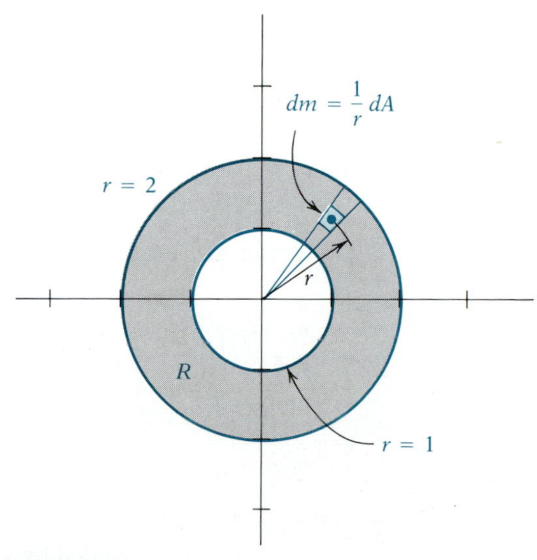

FIGURE 18.30

■

Now we will briefly relate the ideas studied in this section to probability theory. Consider first the one-dimensional case in which $f(x)$ is a continuous probability density function on the interval $[a, b]$. The *kth moment about the origin* M_k is

$$M_k = \int_a^b x^k f(x)\,dx$$

This is what we referred to in Section 5.8 as the expected value of x^k and denoted by $E[x^k]$. In particular, for $k = 1$, $M_1 = E[x] = \mu$, the mean of the distribution. Since f is a probability density function, $\int_a^b f(x)\,dx = 1$, and so we can write

$$\mu = \frac{\int_a^b xf(x)\,dx}{\int_a^b f(x)\,dx}$$

From this we see that μ is analogous to the center of mass. In fact, we can think of the probability as a continuously distributed "mass" on the interval $[a, b]$, where the "density" is $f(x)$. Indeed, the name *probability density* comes from this analogy.

The *kth moment about the mean* is defined by

$$M_k' = \int_a^b (x - \mu)^k f(x)\,dx$$

which is the same as the expected value $E[(x - \mu)^k]$. In particular, for $k = 2$, we get the variance σ^2:

$$\sigma^2 = E[(x - \mu)^2] = \int_a^b (x - \mu)^2 f(x)\,dx$$

These ideas can be extended to two dimensions. If $f(x, y) \geq 0$ and $\iint_R f(x, y)\,dA = 1$, then f can serve as a probability density function of two random variables. We define the x mean and y mean by

$$\mu_x = \iint_R xf(x, y)\,dA \quad \text{and} \quad \mu_y = \iint_R yf(x, y)\,dA$$

and these are analogous to $\bar{x}$ and $\bar{y}$ for a lamina with density $f(x, y)$. Similarly,

$$\sigma_x^2 = \iint_R (x - \mu_x)^2 f(x, y)\,dA \quad \text{and} \quad \sigma_y^2 = \iint_R (y - \mu_y)^2 f(x, y)\,dA$$

Moments of higher order are useful, but we will not consider them here.

EXERCISE SET 18.4

A

In Exercises 1–8 find the center of mass of the lamina that occupies the region R and has density $\rho(x, y)$.

1. $R = \{(x, y): 0 \leq x \leq 1,\ x \leq y \leq 1\}$; $\rho(x, y) = 2y$
2. $R = \{(x, y): 0 \leq x \leq 2,\ 0 \leq y \leq 1\}$; $\rho(x, y) = x + 2y$
3. $R = \{(x, y): -y \leq x \leq y,\ 0 \leq y \leq 1\}$; $\rho(x, y) = y^2$
4. $R = \{(x, y): 0 \leq x \leq 4,\ 0 \leq y \leq \sqrt{x}\}$; $\rho(x, y) = xy$
5. R is bounded by $y = x$ and $y = 2 - x^2$; $\rho(x, y) = 2$
6. R is bounded by $y = 2x - x^2$ and $y = 0$; $\rho(x, y) = 1$
7. R is bounded by $y = \sqrt{4 - x^2}$ and $y = 0$; $\rho(x, y) = \sqrt{x^2 + y^2}$

8. R is the region inside $x^2 + y^2 = 4x$ for which $x \geq 1$; $\rho(x, y) = y^2/x$

In Exercises 9–16 find the centroid of the region R.

9. $R = \{(x, y): -1 \leq x \leq 1,\ 0 \leq y \leq x^2\}$

10. $R = \{(x, y): -1 \leq x \leq 1,\ 0 \leq y \leq 2 - x^2\}$

11. $R = \{(r, \theta): 0 \leq r \leq a,\ 0 \leq \theta \leq \pi\}$

12. $R = \{(r, \theta): 0 \leq r \leq 2 \cos \theta,\ 0 \leq \theta \leq \frac{\pi}{4}\}$

13. R is the region bounded by $y = x^3 - x$ and $y = 7x$ for which $x \geq 0$.

14. $R = \{(x, y): -\frac{\pi}{2} \leq x \leq \frac{\pi}{2},\ 0 \leq y \leq \cos x\}$

15. $R = \{(x, y): 0 \leq x \leq 2,\ 0 \leq y \leq e^x\}$

16. R is the region above the x-axis and between $x^2 + y^2 = 1$ and $x^2 + y^2 = 16$.

In Exercises 17–24 find I_x, I_y, I_0, r_x, and r_y for the lamina that occupies the region R and has density $\rho(x, y)$.

17. Same as Exercise 1

18. Same as Exercise 2

19. Same as Exercise 3

20. Same as Exercise 4

21. $R = \{(x, y): 0 \leq x \leq 2,\ 0 \leq y \leq 2 - x\}$; $\rho(x, y) = x + y$

22. $R = \{(r, \theta): 0 \leq r \leq a,\ 0 \leq \theta \leq \frac{\pi}{2}\}$; $\rho(r, \theta) = \rho$ (constant)

23. R is the region under $y = \sin x$ and above the x-axis, from $x = 0$ to $x = \pi$; $\rho(x, y) = \rho$ (constant).

24. $R = \{(x, y): -\sqrt{3} \leq x \leq \sqrt{3},\ 1 \leq y \leq \sqrt{4 - x^2}\}$; $\rho(x, y) = \frac{x}{y}$

25. Prove that the moment of inertia of a homogeneous lamina of density ρ in the shape of a rectangle with base b and altitude h about its base is $I = (bh^3/3)\rho$.

26. Find the centroid of the region bounded by $y^2 = 4x$ and $2x - y = 4$.

27. Suppose a homogeneous lamina of density ρ occupies the region described in Exercise 26. Set up, but do not evaluate, iterated integrals for the moments of inertia about the lines $x + 2 = 0$ and $y - 4 = 0$.

28. Consider the limaçon $r = 3 + 2 \cos \theta$ and the circle $r = 3$. Set up, but do not evaluate, iterated integrals for $\bar{x}$ and I_0 for homogeneous laminas of density ρ that occupy each of the following regions:
(a) Outside the circle and inside the limaçon
(b) Outside the limaçon and inside the circle
(c) Inside both the limaçon and the circle

29. Let $f(x, y) = kxy$ on the region $R = \{(x, y): 0 \leq x \leq 1,\ 0 \leq y \leq 1\}$.
(a) Find k so that $f(x, y)$ will be a probability density function.
(b) Find μ_x and μ_y.
(c) Find σ_x^2 and σ_y^2.

30. Let $f(x, y) = k$ on the unit circular disk $\{(x, y): x^2 + y^2 \leq 1\}$.
(a) Find k so that $f(x, y)$ is a probability density function.
(b) Find σ_x^2 and σ_y^2.

31. Let $f(x, y) = k(1 - y)$ on the region $R = \{(x, y): 0 \leq x \leq y \leq 1\}$.
(a) Find k so that $f(x, y)$ is a probability density function.
(b) Find μ_x and μ_y.
(c) Find σ_x^2 and σ_y^2.

B

32. Prove that the first moment of a lamina about each of the lines $x = \bar{x}$ and $y = \bar{y}$ equals 0.

33. (a) Prove that the moment of inertia of a lamina about the line $x = \bar{x}$ is

$$I_{x=\bar{x}} = I_y - m\bar{x}^2$$

(b) State and prove an analogous result for the moment of inertia about the line $y = \bar{y}$.

34. Prove that the medians of a triangle intersect at the centroid of the region enclosed by the triangle.

35. If a homogeneous lamina of density ρ is bounded by a triangle of altitude h and base b, prove that the moment of inertia about its base is

$$I_b = \frac{bh^3\rho}{12}$$

36. A lamina of density $\rho(x, y) = y^2$ is in the shape of the circular disk $x^2 + y^2 \leq 1$. Find its moment of inertia with respect to the line $x = 2$.

37. Prove the following *first theorem of Pappus:* If R is a plane region and l is a line in the plane of R that does not intersect R, then the volume of the solid formed by revolving R about l is the area of R times the distance traveled by the centroid of R.

38. It can be shown that the function

$$f(x, y) = \tfrac{1}{8}xe^{-(x+y)/2}$$

defined on the infinite region $\{(x, y): 0 \leq x < \infty,\ 0 \leq y < \infty\}$ is a probability density function. Assuming that the double integrals over this unbounded region exist whenever the corresponding improper iterated integrals converge, find the following:
(a) μ_x (b) μ_y (c) σ_x^2 (d) σ_y^2

18.5 SURFACE AREA

We learned in Chapter 5 how to find areas of surfaces of revolution using single integration. We now show how double integration can be used to find areas of more general surfaces.

Let S be the surface that is the graph of a function f over a regular region R, and suppose f_x and f_y are continuous on R. We partition a two-dimensional interval containing R in the usual way and count those subintervals entirely contained in R. In the ith such subinterval, having area ΔA_i, select an arbitrary point (x_i^*, y_i^*). By the assumption of continuity of f_x and f_y, we know that f is differentiable in R, and so S has a tangent plane at each point. Let T_i denote its tangent plane at $P_i = (x_i^*, y_i^*, f(x_i^*, y_i^*))$. Now project the ith subinterval vertically. The resulting prism cuts out a patch of the surface S, whose area we denote by $\Delta\sigma_i$, and a corresponding patch of the tangent plane T_i, whose area we denote by ΔT_i. This is illustrated in Figure 18.31.

If the norm $\|P\|$ of the partition is small, we would expect $\Delta\sigma_i$ to be approximated by ΔT_i. Hence, denoting the area of S by $A(S)$, it is reasonable to suppose that

$$A(S) \approx \sum_{i=1}^{n} \Delta\sigma_i \approx \sum_{i=1}^{n} \Delta T_i$$

and that the approximation becomes better and better as we take partitions with norms shrinking toward 0. Our next task is to express ΔT_i in terms of the function f.

FIGURE 18.31

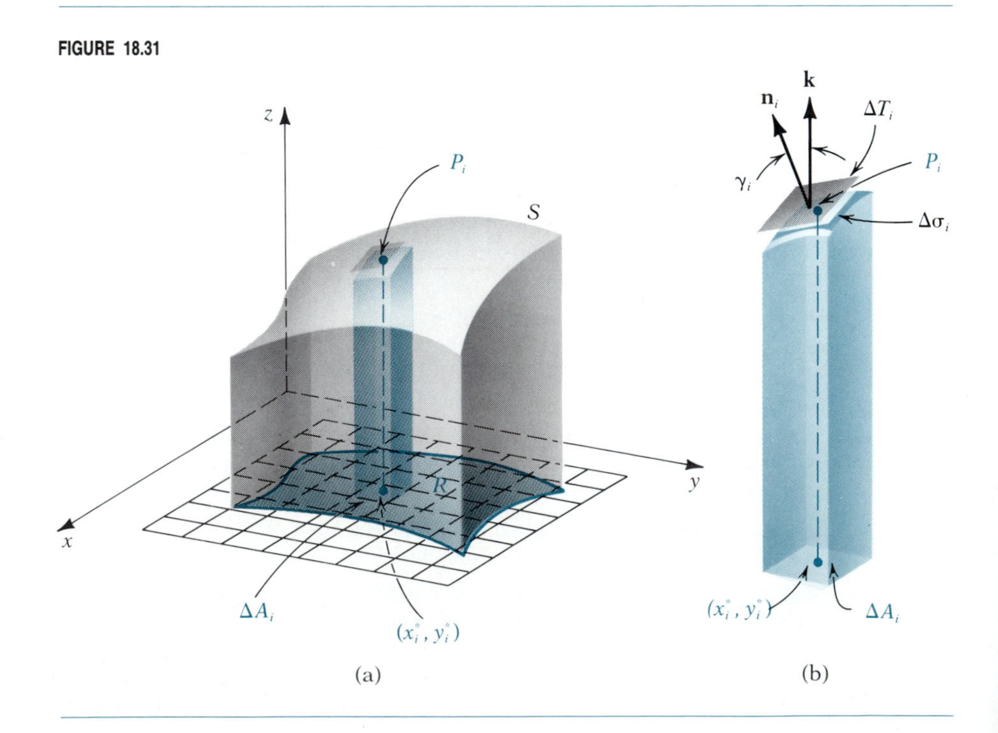

As shown in the enlarged view in Figure 18.31(b), let $\mathbf{n}_i$ be a normal vector to the tangent plane at P_i, and let γ_i be the angle between $\mathbf{n}_i$ and the unit vertical vector $\mathbf{k} = \langle 0, 0, 1\rangle$. Then it can be shown (see Exercise 20) that

$$\Delta T_i|\cos \gamma_i| = \Delta A_i \tag{18.20}$$

We learned in Section 17.4 that the vector $\langle f_x, f_y, -1\rangle$ is normal to the surface S, and so we may take

$$\mathbf{n}_i = \langle f_x(x_i^*, y_i^*), f_y(x_i^*, y_i^*), -1\rangle$$

Thus

$$\cos \gamma_i = \frac{\mathbf{n}_i \cdot \mathbf{k}}{|\mathbf{n}_i|\,|\mathbf{k}|} = \frac{-1}{\sqrt{f_x^2 + f_y^2 + 1}}$$

where f_x and f_y are evaluated at (x_i^*, y_i^*). Solving for ΔT_i from equation (18.20) and substituting for $\cos \gamma_i$, we have

$$\Delta T_i = \sqrt{1 + f_x^2 + f_y^2}\,\Delta A_i$$

Thus,

$$A(S) \approx \sum_{i=1}^{n} \sqrt{1 + [f_x(x_i^*, y_i^*)]^2 + [f_y(x_i^*, y_i^*)]^2}\,\Delta A_i$$

We *define* $A(S)$ by the limit of this as $\|P\| \to 0$:

$$A(S) = \iint_R \sqrt{1 + [f_x(x, y)]^2 + [f_y(x, y)]^2}\,dA \tag{18.21}$$

Remark Note the similarity between this formula for surface area and the formula for arc length developed in Section 5.4:

$$L = \int_a^b \sqrt{1 + [f'(x)]^2}\,dx$$

EXAMPLE 18.16 Find the area of that part of the cylinder $z = \sqrt{4 - x^2}$ above the region R bounded by $y = x$, $x = 0$, and $y = 1$. Sketch the surface.

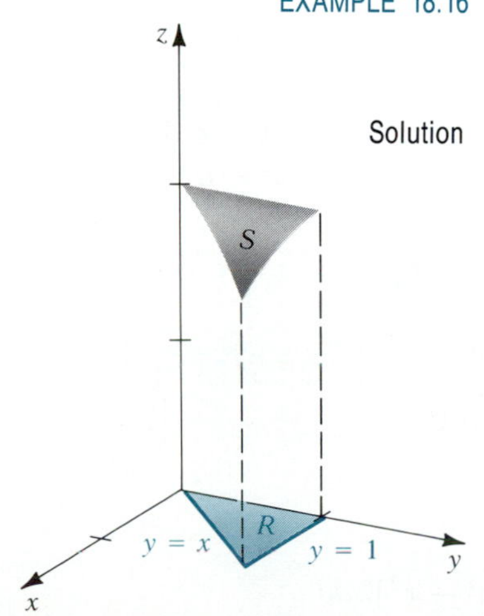

FIGURE 18.32

Solution Figure 18.32 shows the surface. Writing $f(x, y) = \sqrt{4 - x^2}$, we have

$$f_x(x, y) = \frac{-x}{\sqrt{4 - x^2}} \qquad f_y(x, y) = 0$$

So by equation (18.21),

$$\begin{aligned} A(S) &= \iint_R \sqrt{1 + \frac{x^2}{4 - x^2}}\,dA = \iint_R \frac{2}{\sqrt{4 - x^2}}\,dA \\ &= \int_0^1 \int_x^1 \frac{2}{\sqrt{4 - x^2}}\,dy\,dx = \int_0^1 \left(\frac{2}{\sqrt{4 - x^2}} - \frac{2x}{\sqrt{4 - x^2}}\right)dx \\ &= \left[2 \sin^{-1}\frac{x}{2} + 2\sqrt{4 - x^2}\right]_0^1 = \frac{\pi}{3} + 2\sqrt{3} - 4 \end{aligned}$$

■

EXAMPLE 18.17 Find the area of that portion of the paraboloid $z = 2 - x^2 - y^2$ that lies above the xy-plane.

Solution The surface S in question is shown in Figure 18.33. Writing $z = f(x, y)$, we have $f_x = -2x$ and $f_y = -2y$, so that by equation (18.21),

$$A(S) = \iint_R \sqrt{1 + 4x^2 + 4y^2}\, dA$$

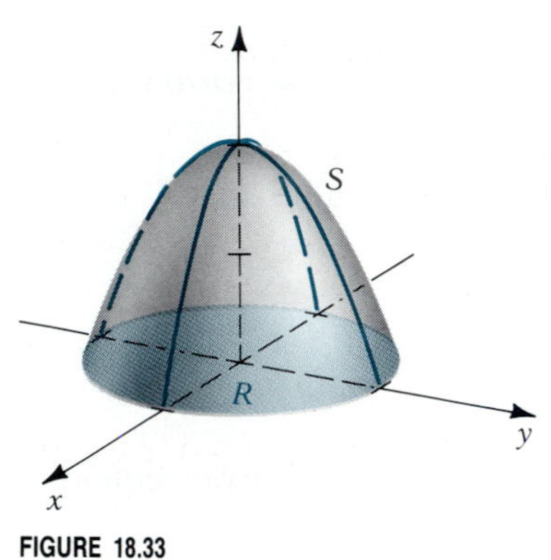

FIGURE 18.33

Both the nature of R and the integrand suggest using polar coordinates to evaluate the integral:

$$A(S) = \int_0^{2\pi} \int_0^{\sqrt{2}} \sqrt{1 + 4r^2}\, r\, dr\, d\theta = \int_0^{2\pi} \left[\tfrac{1}{8} \cdot \tfrac{2}{3}(1 + 4r^2)^{3/2}\right]_0^{\sqrt{2}} d\theta = \tfrac{13\pi}{3}$$ ■

If the equation for the surface S has the form $F(x, y, z) = 0$, where F_x, F_y, and F_z are continuous and $F_z \neq 0$, then we know from Section 17.2 that z is implicitly a function of x and y, say $z = f(x, y)$, and

$$f_x = \frac{\partial z}{\partial x} = -\frac{F_x}{F_z} \qquad f_y = \frac{\partial z}{\partial y} = -\frac{F_y}{F_z}$$

When these are substituted for f_x and f_y in equation (18.21), we get the alternate formula

$$A(S) = \iint_R \frac{\sqrt{F_x^2 + F_y^2 + F_z^2}}{|F_z|}\, dA \qquad (18.22)$$

By observing that the numerator is $|\nabla F|$ and that $\nabla F \cdot \mathbf{k} = F_z$, we can put equation (18.22) in the compact form

$$A(S) = \iint_R \frac{|\nabla F|}{|\nabla F \cdot \mathbf{k}|}\, dA \qquad (18.23)$$

In both equations (18.22) and (18.23) it must be understood that z is a function of x and y defined implicitly by $F(x, y, z) = 0$.

EXAMPLE 18.18 Find the area of that portion of the sphere $x^2 + y^2 + z^2 = a^2$ that is inside the cylinder $x^2 + y^2 = b^2$, where $0 < b < a$.

Solution From symmetry we can consider the upper hemisphere only and double the result (see Figure 18.34). Taking $F(x, y, z) = x^2 + y^2 + z^2 - a^2$, we have $\nabla F = \langle 2x, 2y, 2z \rangle$, so that

$$|\nabla F| = \sqrt{4x^2 + 4y^2 + 4z^2} = 2\sqrt{x^2 + y^2 + z^2} = 2a$$

Thus,

$$A(S) = 2 \iint_R \frac{2a}{2z}\, dA = 2 \iint_R \frac{a}{\sqrt{a^2 - (x^2 + y^2)}}\, dA$$

Using polar coordinates, we obtain

$$A(S) = 2a \int_0^{2\pi} \int_0^b \frac{r\, dr\, d\theta}{\sqrt{a^2 - r^2}} = 2a \int_0^{2\pi} \left[-\sqrt{a^2 - r^2}\right]_0^b d\theta$$

$$= 4\pi a\left[a - \sqrt{a^2 - b^2}\right]$$ ■

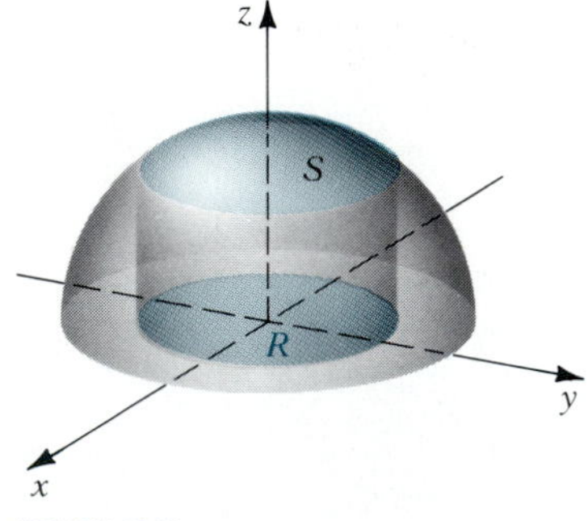

FIGURE 18.34

EXERCISE SET 18.5

A

1. Find the area of the portion of the plane $z + 2x + 3y = 6$ that lies above the rectangular region R bounded by $x = 0$, $y = 0$, $x = 2$, and $y = 1$.
2. Find the area of that portion of the plane $4x - 3y - 6z + 12 = 0$ that lies above the triangular region R with vertices $(3, 0)$, $(0, 0)$, and $(0, 4)$.
3. Find the area of the first-octant portion of the cylinder $z = 2 - y^2$ that is cut off by the planes $x = 0$, $y = x$, and $z = 0$.
4. Find the area of the portion of the cylinder $z = \sqrt{4 - y^2}$ that lies above region $R = \{(x, y): 0 \le x \le 3, 0 \le y \le 1\}$.
5. Find the area of the paraboloid $z = x^2 + y^2$ that lies inside the cylinder $x^2 + y^2 = 2$.
6. Find the area of the part of the sphere $x^2 + y^2 + z^2 = 4$ that is inside the cylinder $x^2 + y^2 - 2x = 0$.
7. Find the area of the part of the upper nappe of the cone $z^2 = x^2 + y^2$ that lies inside the cylinder $x^2 + y^2 - 4y = 0$.
8. Find the area of that portion of the cylinder $x^2 + z^2 = a^2$ that lies inside the cylinder $x^2 + y^2 = a^2$.
9. Find the area of the paraboloid $z = 1 + x^2 + y^2$ that is between the planes $z = 2$ and $z = 5$.
10. Find the area of the part of the sphere $x^2 + y^2 + z^2 = 25$ that is between the planes $z = 3$ and $z = 4$.

In Exercises 11–16 find $A(S)$, where S is the portion of the graph of $z = f(x, y)$ that lies above the region R.

11. $f(x, y) = 3x + y^2$; $R = \{(x, y): 0 \le x \le y, 0 \le y \le 2\}$
12. $f(x, y) = \frac{2}{3}(x^{3/2} + y^{3/2})$; $R = \{(x, y): 0 \le x \le 3, 0 \le y \le 3 - x\}$
13. $f(x, y) = \ln \cos x$; $R = \{(x, y): -\frac{\pi}{4} \le x \le \frac{\pi}{4}, 0 \le y \le \sec x\}$
14. $f(x, y) = (1 - y^{2/3})^{3/2}$; $R = \{(x, y): 0 \le x \le 1, 0 \le y \le (1 - x^{2/3})^{3/2}\}$
15. $f(x, y) = 2 - (x^2 + y^2)/2$; R is the region inside the lemniscate $r^2 = \cos 2\theta$.
16. $f(x, y) = xy$; R is the region outside $r = 1$ and inside $r = 2\cos\theta$.

B

17. Use the methods of this section to find the area of the surface of a sphere of radius a.
18. Use the methods of this section to find the lateral surface area of a right circular cone (one nappe) of base radius a and height h.
19. Using the accompanying figure, verify that when a nonnegative smooth function f of one variable, defined on $[a, b]$, is rotated about the x-axis, the equation of the resulting surface of revolution is $y^2 + z^2 = [f(x)]^2$. Using this and the methods of this section, show that its total surface area is

$$2\pi \int_a^b f(x)\sqrt{1 + [f'(x)]^2}\, dx$$

and so is in agreement with Definition 5.4.

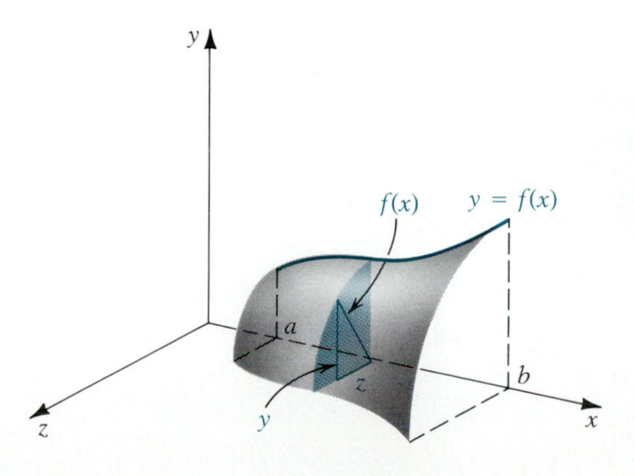

20. In the accompanying figure $PQRS$ is a parallelogram of area ΔT formed by intersecting a plane with the

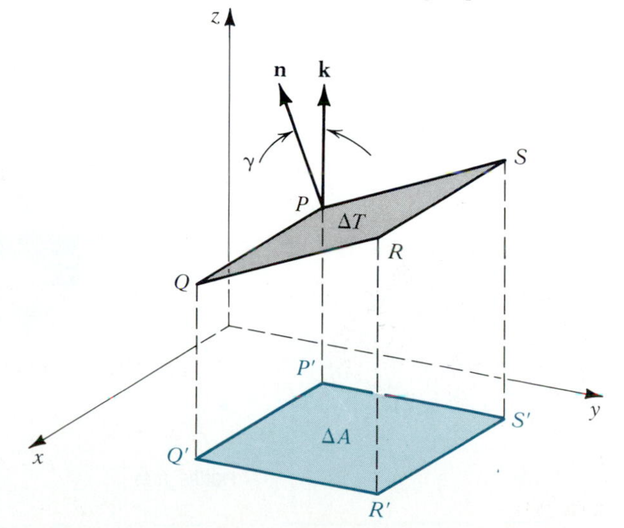

prism that has the rectangular base $P'Q'R'S'$. The normal $\mathbf{n}$ to the plane makes an angle γ with the unit vertical vector $\mathbf{k}$. If ΔA is the area of the rectangle $P'Q'R'S'$, show that

$$(\overrightarrow{PQ} \times \overrightarrow{PS}) \cdot \mathbf{k} = (\overrightarrow{P'Q'} \times \overrightarrow{P'S'}) \cdot \mathbf{k}$$

and then explain how it follows from this that

$$\Delta T|\cos \gamma| = \Delta A$$

(*Hint:* Write $\overrightarrow{PQ} = \overrightarrow{PP'} + \overrightarrow{P'Q'} + \overrightarrow{Q'Q}$ and write $\overrightarrow{PS}$ in a similar way.)

18.6 TRIPLE INTEGRALS

The ideas behind the triple integral of a function of three variables over a region in $\mathbf{R}^3$ are similar to those for a double integral, so we will be briefer in our treatment. Let f be a bounded function defined on a closed and bounded region Q of $\mathbf{R}^3$. We also want the boundary of Q to satisfy conditions analogous to those for a regular two-dimensional region, which in that case meant a sectionally smooth, simple closed curve. Without going more deeply into the theory of surfaces, it is difficult to describe precisely the corresponding conditions on the boundary of Q. Roughly speaking, its boundary consists of a finite number of ''simple, smooth surfaces'' that join together in smooth curves, and such that Q is completely enclosed by these surfaces. We again use the term *regular* to refer to such a region. We will shortly describe in detail certain special cases of regular regions that suffice for most applications.

To define the triple integral we begin by enclosing Q in a three-dimensional closed interval I of the form

$$I = \{(x, y, z): a_1 \leq x \leq a_2,\ b_1 \leq y \leq b_2,\ c_1 \leq z \leq c_2\}$$

We partition I by means of planes parallel to each of the coordinate planes, as indicated in Figure 18.35. Let P denote such a partition, and denote by $\|P\|$ (the norm of P) the length of the longest diagonal of all the subintervals of I formed by P. Number, say from 1 to n, the subintervals that are completely contained within Q, and in the ith such subinterval select a point (x_i^*, y_i^*, z_i^*) for $i = 1, 2, \ldots, n$. Let ΔV_i denote the volume of the ith

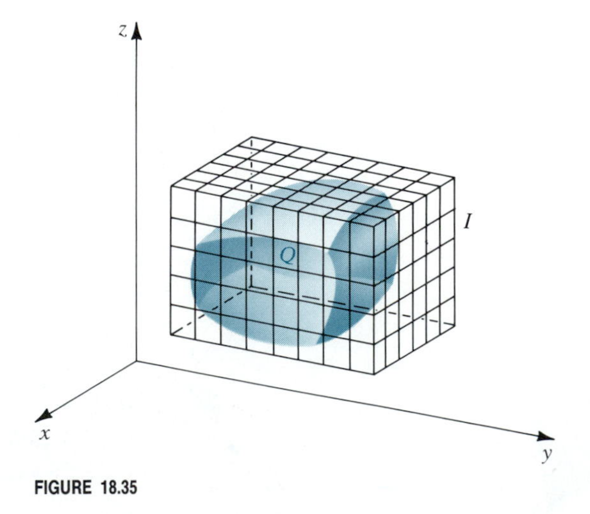

FIGURE 18.35

subinterval. Then

$$\iiint_Q f(x, y, z)\,dV = \lim_{\|P\|\to 0} \sum_{i=1}^{n} f(x_i^*, y_i^*, z_i^*)\,\Delta V_i \tag{18.24}$$

provided the limit on the right exists independently of the choices of the points (x_i^*, y_i^*, z_i^*). As with single and double integrals, the limit will exist if f is continuous on Q.

To evaluate the triple integral by an iterated integral, we consider first the special case of a region Q of the type

$$Q = \{(x, y, z): a \le x \le b,\ g_1(x) \le y \le g_2(x),\ k_1(x, y) \le z \le k_2(x, y)\}$$

where g_1 and g_2 have continuous derivatives, and k_1 and k_2 have continuous first partial derivatives. A region of this type is shown in Figure 18.36. The projection R of Q onto the xy-plane is a type I region. If f is continuous throughout Q, then it can be shown that

$$\iiint_Q f(x, y, z)\,dV = \int_a^b \int_{g_1(x)}^{g_2(x)} \int_{k_1(x,y)}^{k_2(x,y)} f(x, y, z)\,dz\,dy\,dx \tag{18.25}$$

As with iterated integrals in two variables, the order of the differentials indicates the order in which the integrations are to be performed: first z, then y, then x.

If the projection on Q is a type II region, so that Q is of the form

$$Q = \{(x, y, z): h_1(y) \le x \le h_2(y),\ c \le y \le d,\ k_1(x, y) \le z \le k_2(x, y)\}$$

FIGURE 18.36

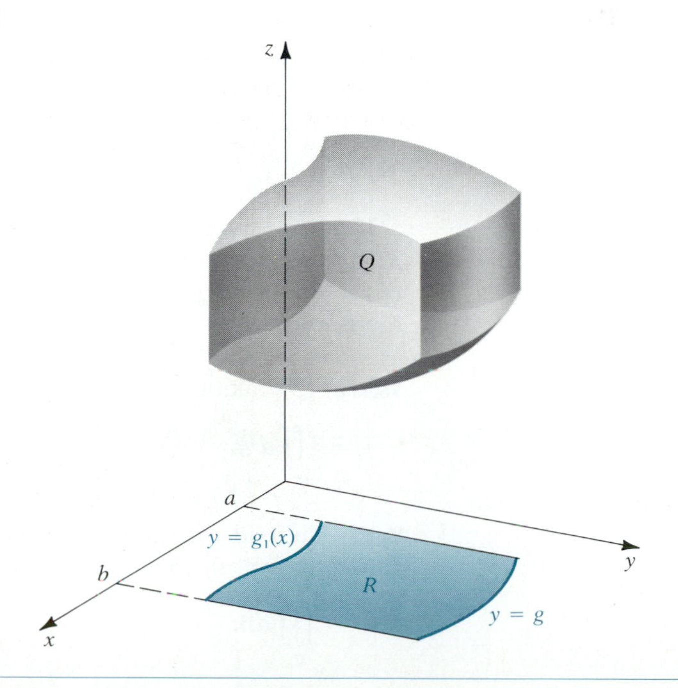

then the iterated integral becomes

$$\int_c^d \int_{h_1(y)}^{h_2(y)} \int_{k_1(x,y)}^{k_2(x,y)} f(x, y, z)\, dz\, dx\, dy$$

By various permutations on the variables of integration, we can get four other possibilities. For example, if

$$Q = \{(x, y, z): a < x < b,\ g_1(x) \le z \le g_2(x),\ k_1(x, z) \le y \le k_2(x, z)\}$$

then we get

$$\int_a^b \int_{g_1(x)}^{g_2(x)} \int_{k_1(x,z)}^{k_2(x,z)} f(x, y, z)\, dy\, dz\, dx$$

In this case the projection of Q onto the xz-plane is a type I region.

EXAMPLE 18.19 Evaluate the integral $\iiint_Q 2xz\, dV$, where Q is the region enclosed by the planes $x + y + z = 4$, $y = 3x$, $x = 0$, and $z = 0$.

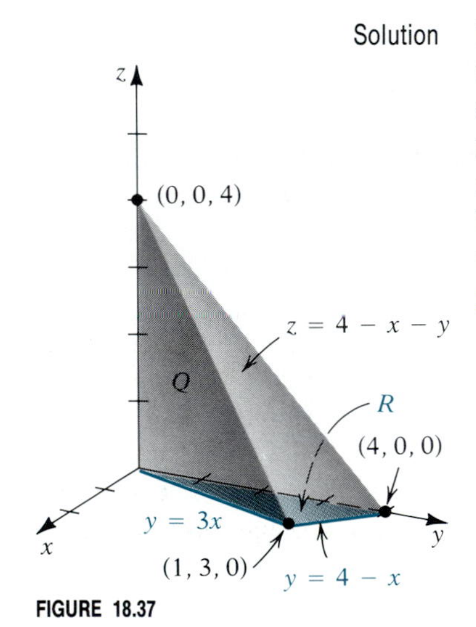

FIGURE 18.37

Solution The region Q is shown in Figure 18.37. Its projection onto the xy-plane is a type I region, so we will use equation (18.25). To find the limits on z, we consider a point (x, y, z) in Q and find the smallest and largest values of z so that the point remains in Q. In this case z varies from the horizontal plane $z = 0$ to the inclined plane $z = 4 - x - y$ that forms the top of the region. The limits on x and y are determined from the xy projection of Q just as with double integrals. In this case the limits on y are the traces $y = 3x$ and $y = 4 - x$. They intersect when $x = 1$, so that x varies from 0 to 1. Thus,

$$\iiint_Q 2xz\, dV = \int_0^1 \int_{3x}^{4-x} \int_0^{4-x-y} 2xz\, dz\, dy\, dx = \int_0^1 \int_{3x}^{4-x} x(4 - x - y)^2\, dy\, dx$$

$$= \int_0^1 \left[\frac{-x(4 - x - y)^3}{3}\right]_{3x}^{4-x} dx = \frac{1}{3}\int_0^1 (4 - 4x)^3\, dx$$

$$= -\frac{64}{12}(1 - x)^4\Big]_0^1 = \frac{8}{3}$$

■

The triple integral can be used to find volumes and centroids of regions in $\mathbf{R}^3$, mass, center of mass, and moments of inertia of solids. The ideas parallel those for the two-dimensional case, so we omit the details and simply give the results.

Volume By taking $f(x, y, z) = 1$, we get the volume of Q:

$$V = \iiint_Q dV \tag{18.26}$$

Mass If a solid is in the shape of a regular three-dimensional region Q and has density $\rho(x, y, z)$, then its mass m is

$$m = \iiint_Q dm \tag{18.27}$$

where $dm = \rho(x, y, z)\, dV$.

Moments The (first) moments with respect to the xy-plane, the xz-plane, and the yz-plane are

$$M_{xy} = \iiint_Q z\,dm, \quad M_{xz} = \iiint_Q y\,dm, \quad M_{yz} = \iiint_Q x\,dm \tag{18.28}$$

Center of Mass

$$\bar{x} = \frac{\iiint_Q x\,dm}{\iiint_Q dm}, \quad \bar{y} = \frac{\iiint_Q y\,dm}{\iiint_Q dm}, \quad \bar{z} = \frac{\iiint_Q z\,dm}{\iiint_Q dm} \tag{18.29}$$

Centroid The centroid of a region is the same as the center of mass of a homogeneous solid that occupies that region:

$$\bar{x} = \frac{\iiint_Q x\,dV}{V}, \quad \bar{y} = \frac{\iiint_Q y\,dV}{V}, \quad \bar{z} = \frac{\iiint_Q z\,dV}{V} \tag{18.30}$$

Moments of Inertia With respect to the coordinate axes:

$$I_x = \iiint_Q (y^2 + z^2)\,dm, \quad I_y = \iiint_Q (x^2 + z^2)\,dm, \quad I_z = \iiint_Q (x^2 + y^2)\,dm \tag{18.31}$$

EXAMPLE 18.20 A solid of density $\rho(x, y, z) = \sqrt{xyz}$ is in the shape of the region Q enclosed by the surfaces $z = 4 - x^2$, $z = x^2 + y^2$, $x = 0$, and $y = 0$. Set up, but do not evaluate, iterated integrals for (a) the mass m, (b) $\bar{x}$, and (c) I_z.

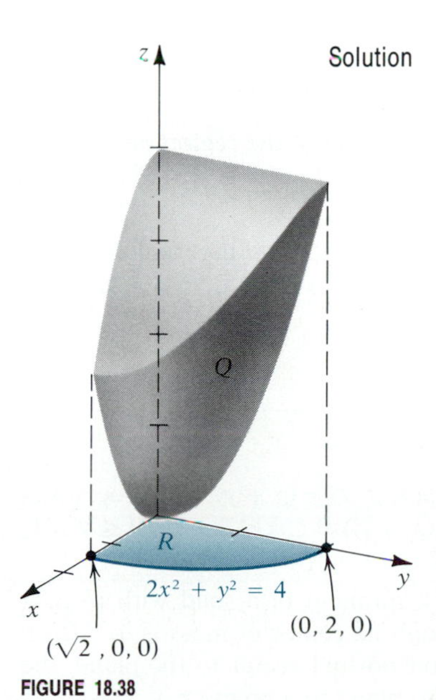

FIGURE 18.38

Solution The region Q is shown in Figure 18.38. The simultaneous solution of $z = 4 - x^2$ and $z = x^2 + y^2$ is $2x^2 + y^2 = 4$. The graph of this is the xy projection of the curve of intersection of these surfaces. Thus, the projection of Q onto the xy-plane is the region

$$R = \{(x, y): 0 \le x \le \sqrt{2},\ 0 \le y \le \sqrt{4 - 2x^2}\}$$

The z limits are from the lower bounding surface $z = x^2 + y^2$ to the upper bounding surface $z = 4 - x^2$.

(a) $m = \iiint_Q dm = \int_0^{\sqrt{2}} \int_0^{\sqrt{4-2x^2}} \int_{x^2+y^2}^{4-x^2} \sqrt{xyz}\,dz\,dy\,dx$

(b) $\bar{x} = \dfrac{M_{yz}}{m} = \dfrac{1}{m} \int_0^{\sqrt{2}} \int_0^{\sqrt{4-2x^2}} \int_{x^2+y^2}^{4-x^2} x\sqrt{xyz}\,dz\,dy\,dx$

(c) $I_z = \iiint_Q (x^2 + y^2)\,dm = \int_0^{\sqrt{2}} \int_0^{\sqrt{4-2x^2}} \int_{x^2+y^2}^{4-x^2} (x^2 + y^2)\sqrt{xyz}\,dz\,dy\,dx$

Note that we could have integrated first with respect to y, with limits from 0 to $\sqrt{z - x^2}$, since the xz projection of Q is the type I region $\{(x, z): 0 \le x \le \sqrt{2},\ x^2 \le z \le 4 - x^2\}$. Then we would have had for part a,

$$m = \int_0^{\sqrt{2}} \int_{x^2}^{4-x^2} \int_0^{\sqrt{z-x^2}} \sqrt{xyz}\,dy\,dz\,dx$$

and the limits for parts b and c would have been the same. ■

EXERCISE SET 18.6

A

In Exercises 1–6 evaluate the iterated integrals. Sketch the region Q determined by the limits of integration.

1. $\int_0^1 \int_1^2 \int_{-1}^1 (xy - 2yz)\, dy\, dz\, dx$
2. $\int_2^4 \int_0^1 \int_0^2 \frac{2x + y}{z^2}\, dx\, dy\, dz$
3. $\int_0^1 \int_0^{1-x} \int_0^2 x\, dz\, dy\, dx$
4. $\int_0^4 \int_0^{\sqrt{z}} \int_0^2 (2xy^2 - 1)\, dy\, dx\, dz$
5. $\int_0^4 \int_0^2 \int_{\sqrt{y}}^{6-y} 4xz\, dx\, dz\, dy$
6. $\int_0^1 \int_0^{2-2y} \int_2^{4-x-2y} z\, dz\, dx\, dy$

In Exercises 7–10 for the given region Q write an iterated integral whose value equals $\iiint_Q f(x, y, z)\, dV$.

7. Q is bounded by $z = 4 - x^2$, $y = x$, $y = 0$, and $z = 0$.
8. Q is the first-octant portion of the region inside the ellipsoid $4x^2 + y^2 + z^2 = 4$.
9. Q is bounded by $2x + 3y = 6$, $x + z = 3$, and the coordinate planes.
10. $Q = \{(x, y, z): \sqrt{3} \le x \le \sqrt{4 - z^2},\ 0 \le y \le 4,\ 1 \le z \le 2\}$
11. Evaluate the integral in Exercise 7 for $f(x, y, z) = \sqrt{z}$.
12. Evaluate the integral in Exercise 10 for $f(x, y, z) = x(3 - y)z^2$.

In Exercises 13–16 find the volume of the region Q bounded by the given surfaces.

13. $x + 2y + z = 4$ and the coordinate planes
14. $z = x^2 + y^2$, $z = 2$, $y = x$, in the first octant
15. $x^2 + z^2 = 4$, $y = x$, $y = 0$, $z = 0$, in the first octant
16. $x + z = 1$, $4x + y + z = 4$, and the coordinate planes

In Exercises 17 and 18 write five different iterated integrals equal to the given integral.

17. $\int_0^1 \int_0^x \int_0^{1-x} f(x, y, z)\, dz\, dy\, dx$
18. $\int_0^4 \int_0^{4-z} \int_0^{\sqrt{y}} f(x, y, z)\, dx\, dy\, dz$
19. Find the centroid of the region bounded by $y^2 = 2x$, $2x + z = 4$, and $z = 0$.
20. For the solid of density $\rho(x, y, z) = 2x$ bounded by the planes $x + y = 1$, $y + z = 1$, and the coordinate planes, find m and $\bar{x}$.

In Exercises 21–27 set up iterated integrals for the specified quantities for the solid that has density $\rho(x, y, z)$ and occupies the region Q.

21. Q is the region above $z = x^2 + y^2$ and below $z = 8 - x^2 - y^2$, $\rho(x, y, z) =$ distance from the xy-plane; m, $\bar{z}$, I_z.
22. The solid of Exercise 20; $\bar{z}$, I_y
23. Q is bounded by $x + 2y + 3z = 6$ and the coordinate planes, $\rho(x, y, z) = x^2yz$; center of mass.
24. Q is the region inside both $x^2 + y^2 = 4$ and $x^2 + y^2 + z^2 = 16$ above the xy-plane, $\rho(x, y, z) =$ distance from the z-axis; m, $\bar{z}$, I_z.
25. Q is the first-octant portion of the region bounded by $y^2 + z^2 = 4$, $z = x - 2y$, and the coordinate planes, $\rho(x, y, z) = \sqrt{x + y}$; $\bar{x}$, I_x.
26. Q is the first-octant portion of the region inside both $x^2 + y^2 = 1$ and $x^2 + z^2 = 1$, $\rho(x, y, z) = xyz$; center of mass.
27. Q is the first-octant portion of the region inside the ellipsoid $2x^2 + y^2 + z^2 = 4$, $\rho(x, y, z) =$ distance from the y-axis; $\bar{y}$, I_y.

B

28. Find the volume and the location of the centroid of a pyramid of height h and with a base that is a square of side a.
29. Evaluate the integral

$$\iiint_Q y^2 \sin xy\, dV$$

where Q is bounded by $z = x$, $2xy = \pi$, $y = \frac{\pi}{2}$, $y = \pi$, and $z = 0$. Give two possible physical interpretations of this integral.

30. A solid of density $\rho(x, y, z) = (x + y)^2 e^{(x+y)z}$ occupies the region $Q = \{(x, y, z): 0 \le x \le 2 - y,\ 0 \le y \le 1,\ 0 \le z \le 1\}$. Find $\bar{z}$.
31. Prove that the first moment of a solid with respect to any plane through its center of mass is 0. *Hint:* If $\langle a, b, c\rangle$ is a unit normal vector to the plane, the distance d from the plane to a point (x, y, z) is

$$d = a(x - \bar{x}) + b(y - \bar{y}) + c(z - \bar{z}) \qquad \text{(Verify.)}$$

32. The *parallel axis theorem* states that for a solid of mass m, the moment of inertia I_l about any line l is

$$I_l = I_{l'} + md^2$$

where l' is a line through the center of mass parallel to l and at a distance d from it. Prove this theorem.

18.7 CYLINDRICAL AND SPHERICAL COORDINATES

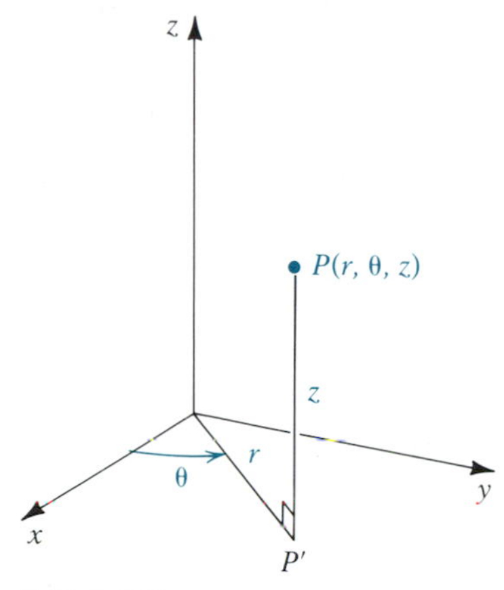

FIGURE 18.39

As we have seen, polar coordinates often can be used to simplify double integrals. In a similar way triple integrals can often be more readily evaluated by using one of two alternatives to rectangular coordinates, called **cylindrical coordinates** and **spherical coordinates.**

In a cylindrical coordinate system polar coordinates are used for two variables and the rectangular coordinate for the third variable. For example, in Figure 18.39 the point P has coordinates (r, θ, z). The projection P' of P onto the xy-plane has polar coordinates r and θ, and z is the usual rectangular z-coordinate. We could equally well use polar coordinates in the xz-or yz-plane, along with the appropriate third rectangular coordinate, but we will concentrate on the situation illustrated in Figure 18.39.

Rectangular and cylindrical coordinates are related by the equations

$$x = r\cos\theta \qquad y = r\sin\theta \qquad z = z$$

$$x^2 + y^2 = r^2 \qquad \tan\theta = \frac{y}{x},\ x \neq 0$$

Some common equations of surfaces in rectangular coordinates along with the corresponding cylindrical equations are:

Rectangular	$x^2 + y^2 = a^2$	$z^2 = a^2(x^2 + y^2)$	$x^2 + y^2 + z^2 = a^2$	$z = a(x^2 + y^2)$
Cylindrical	$r = a$	$z = ar$	$r^2 + z^2 = a^2$	$z = ar^2$
	Circular cylinder	Circular cone	Sphere	Paraboloid

The simplest type of closed region in $\mathbf{R}^3$ to describe in cylindrical coordinates is a set of the form

$$I = \{(r, \theta, z) : a \le r \le b,\ \alpha \le \theta \le \beta,\ c \le z \le d\}$$

Such a region is shown in Figure 18.40. We will call this is a *closed cylindrical interval.* Notice that its projection onto the xy-plane is a closed polar interval.

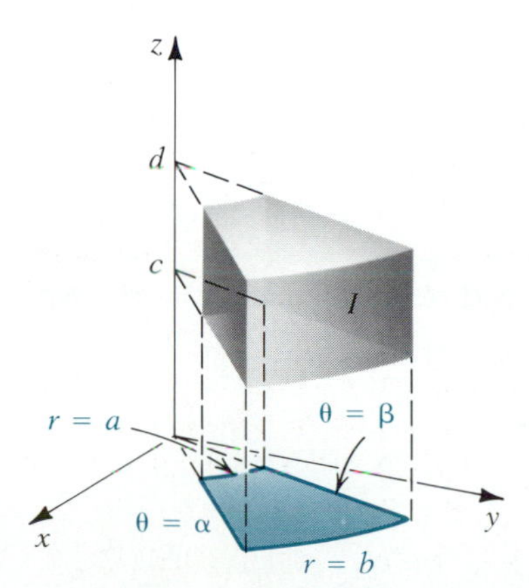

FIGURE 18.40

Now let Q be a closed and bounded region in $\mathbf{R}^3$ of the form

$$Q = \{(r, \theta, z) : (r, \theta) \text{ in } R,\ k_1(r, \theta) \le z \le k_2(r, \theta)\}$$

where R is a polar region of one of the types we studied in Section 18.5 and k_1 and k_2 have continuous first partial derivatives. Let f be a bounded function defined on Q. To define the triple integral of f over Q in cylindrical coordinates, we proceed in what has become a familiar way. First let I be any cylindrical interval that contains Q, and partition I by means of horizontal planes, planes that contain the z-axis, and cylinders centered on the z-axis. These are the surfaces that correspond, respectively, to the equations $z =$ constant, $\theta =$ constant, and $r =$ constant. They divide I into finitely many subintervals with shapes similar to I. As usual, call the partition P and its norm $\|P\|$ the length of the longest diagonal of all the subintervals. Number from 1 to n the subintervals contained entirely in Q, and denote their volumes by $\Delta V_1, \Delta V_2, \ldots, \Delta V_n$. Select an arbitrary point $(r_i^*, \theta_1^*, z_i^*)$ in the ith such subinterval. Then

$$\iiint_Q f(r, \theta, z)\, dV = \lim_{\|P\|\to 0} \sum_{i=1}^{n} f(r_i^*, \theta_1^*, z_i^*)\Delta V_i$$

provided the limit on the right side exists independently of the choices of $(r_i^*, \theta_i^*, z_i^*)$. It will exist when f is continuous on Q, and in that case the integral can be evaluated by

$$\iiint_Q f(r, \theta, z)\, dV = \iint_R \left[\int_{k_1(r,\theta)}^{k_2(r,\theta)} f(r, \theta, z)\, dz\right] dA \tag{18.32}$$

When the projection R of Q is of the form

$$R = \{(r, \theta): g_1(\theta) \le r \le g_2(\theta),\ \alpha \le \theta \le \beta\}$$

then equation (18.32) becomes

$$\iiint_Q f(r, \theta, z)\, dV = \int_\alpha^\beta \int_{g_1(\theta)}^{g_2(\theta)} \int_{k_1(r,\theta)}^{k_2(r,\theta)} f(r, \theta, z) r\, dz\, dr\, d\theta \tag{18.33}$$

since $dA = r\,dr\,d\theta$. We write

$$dV = r\,dz\,dr\,d\theta$$

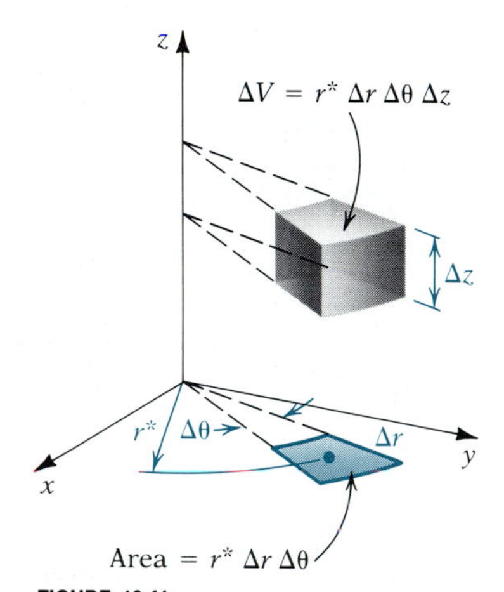

FIGURE 18.41

and call this the *differential of volume* in cylindrical coordinates. The form of this is plausible from a consideration of a typical volume element that results from a partition of the region Q, as shown in Figure 18.41. Its projection on the xy-plane has area $r^*\Delta r\Delta\theta$, where r^* is the radial distance to the midpoint of this projection. Since the thickness of the element is Δz, its volume is $r^*\Delta r\Delta\theta\Delta z$.

EXAMPLE 18.21 A homogeneous solid is bounded laterally by the circular cylinder $x^2 + y^2 = 3$, on the top by the sphere $x^2 + y^2 + z^2 = 4$, and on the bottom by the xy-plane. Find its center of mass and the moment of inertia with respect to the z-axis.

Solution The region occupied by the solid is shown in Figure 18.42. In cylindrical coordinates the bounding surfaces are $r = \sqrt{3}$, $z = \sqrt{4 - r^2}$, and $z = 0$. Denote the density by ρ. First we calculate the mass:

$$m = \iiint_Q dm = \iiint_Q \rho\, dV = \int_0^{2\pi}\int_0^{\sqrt{3}}\int_0^{\sqrt{4-r^2}} \rho r\, dz\, dr\, d\theta$$

$$= \rho \int_0^{2\pi}\int_0^{\sqrt{3}} r\sqrt{4 - r^2}\, dr\, d\theta = \rho \int_0^{2\pi} [-\tfrac{1}{2}\cdot\tfrac{2}{3}(4 - r^2)^{3/2}]_0^{\sqrt{3}}\, d\theta = \tfrac{14\pi}{3}\rho$$

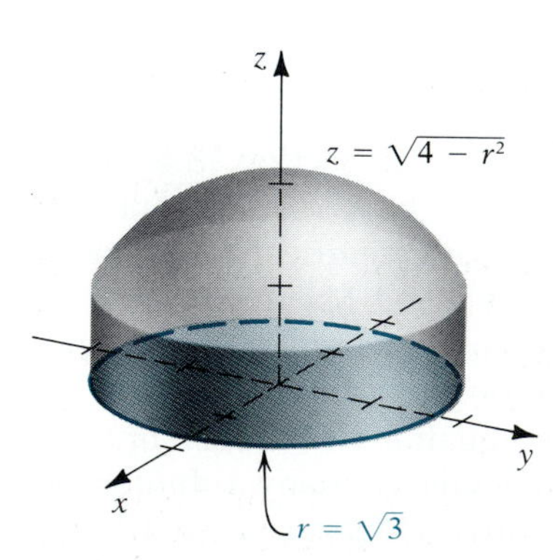

FIGURE 18.42

By symmetry we see that $\bar{x} = \bar{y} = 0$, so we need only calculate $\bar{z} = M_{xy}/m$:

$$M_{xy} = \iiint_Q z\, dm = \int_0^{2\pi}\int_0^{\sqrt{3}}\int_0^{\sqrt{4-r^2}} z\rho r\, dz\, dr\, d\theta$$

$$= \rho \int_0^{2\pi}\int_0^{\sqrt{3}} r\left(\frac{4 - r^2}{2}\right) dr\, d\theta$$

$$= \rho \int_0^{2\pi}\left[r^2 - \frac{r^4}{8}\right]_0^{\sqrt{3}} d\theta = \frac{15\pi}{4}\rho$$

So

$$\bar{z} = \frac{M_{xy}}{m} = \frac{15\pi\rho}{4} \cdot \frac{3}{14\pi\rho} = \frac{45}{56}$$

$$I_z = \iiint_Q (x^2 + y^2)\,dm = \iiint_Q r^2\,dm = \int_0^{2\pi} \int_0^{\sqrt{3}} \int_0^{\sqrt{4-r^2}} \rho r^3\,dz\,dr\,d\theta$$

$$= \rho \int_0^{2\pi} \int_0^{\sqrt{3}} r^3 \sqrt{4 - r^2}\,dr\,d\theta = \frac{94\pi}{15}\rho$$

(You should check the integration on r using trigonometric substitution, integration by parts, or the substitution $u = \sqrt{4 - r^2}$.) ■

EXAMPLE 18.22 Write an integral in cylindrical coordinates equivalent to

$$\int_0^1 \int_0^{\sqrt{4-x^2}} \int_{\sqrt{x^2+y^2}}^{6-x^2-y^2} f(x, y, z)\,dz\,dy\,dx$$

Solution The first thing to do is to sketch the region Q determined by the limits of integration. The lower and upper boundaries on z are, respectively,

$$z = \sqrt{x^2 + y^2} \quad \text{and} \quad z = 6 - x^2 - y^2$$

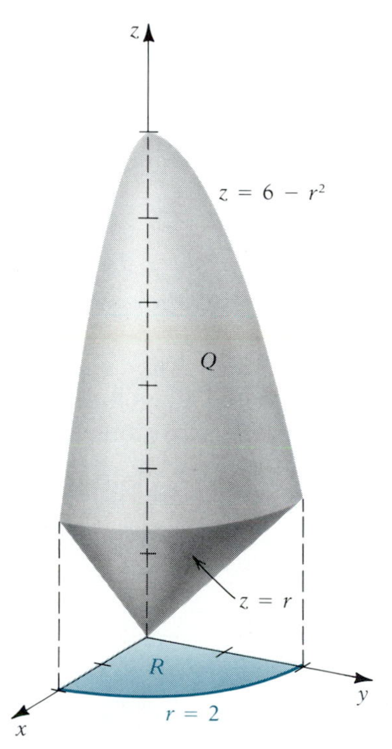

FIGURE 18.43

In cylindrical coordinates these are $z = r$ and $z = 6 - r^2$. They intersect when $r = 6 - r^2$, with the positive solution $r = 2$. The limits on x and y show that the projection R of Q onto the xy-plane is one-quarter of the circular disk bounded by $r = 2$, as shown in Figure 18.43. So an equivalent integral in cylindrical coordinates is

$$\int_0^{\pi/2} \int_0^2 \int_r^{6-r^2} f(r\cos\theta, r\sin\theta, z) r\,dz\,dr\,d\theta$$ ■

In the spherical coordinate system a point P in space is located by its distance ρ from the origin, the polar angle θ from the positive x-axis to the projection OP' of OP onto the xy-plane, and the angle ϕ from the z-axis to OP. This is illustrated in Figure 18.44. We restrict ρ and ϕ so that $\rho \geq 0$ and $0 \leq \phi \leq \pi$. From the right triangle $OP'P$ we see that

$$\overline{OP'} = \rho \sin\phi \quad \text{and} \quad \overline{P'P} = \rho\cos\phi$$

Thus, since $x = \overline{OP'}\cos\theta$, $y = \overline{OP'}\sin\theta$, $z = \overline{P'P}$, we have

$$\begin{cases} x = \rho\cos\theta\sin\phi \\ y = \rho\sin\theta\sin\phi \\ z = \rho\cos\phi \end{cases} \tag{18.34}$$

as the equations of transformation from rectangular to spherical coordinates. On squaring each of these and adding, we get (verify this)

$$x^2 + y^2 + z^2 = \rho^2 \tag{18.35}$$

The simplest surfaces to represent in spherical coordinates are those with equations of the form $\rho = a$, $\phi = \alpha$, and $\theta = \gamma$. These are, respectively, a sphere of radius a centered at the origin, a half-cone with vertex at the origin and axis along the z-axis, and a plane that contains the z-axis. These are illustrated in Figure 18.45.

To define the triple integral of a bounded function f over an appropriately restricted closed and bounded region Q, we proceed in a familiar way. Enclose Q in a closed *spherical interval* $I = \{(\rho, \theta, \phi): a \leq \rho \leq b, \alpha \leq \theta \leq \beta,$

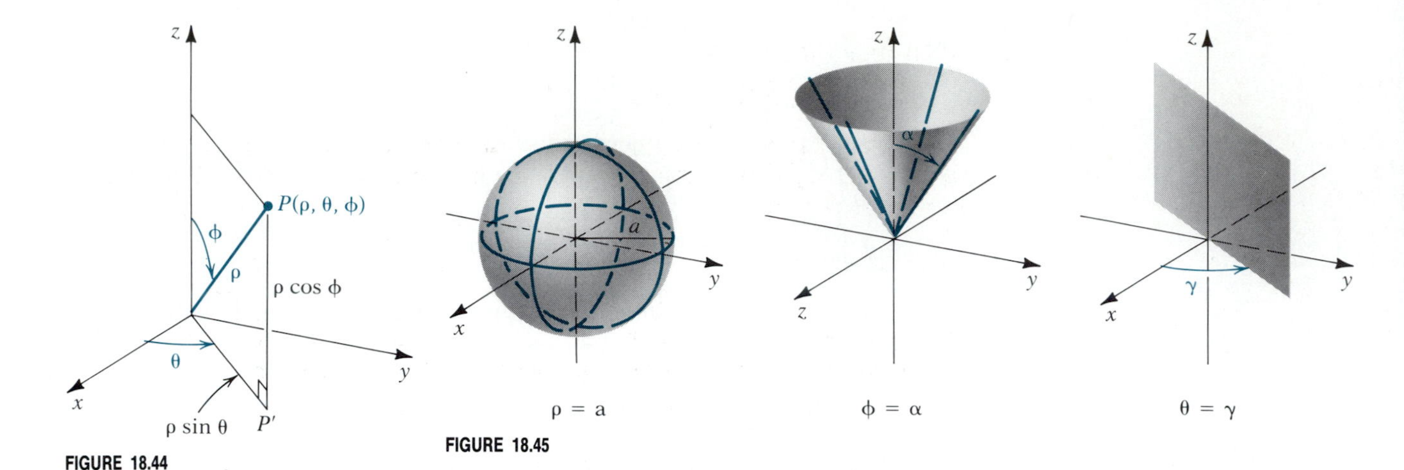

FIGURE 18.44

FIGURE 18.45

$\gamma \le \phi \le \delta\}$ and partition I by means of spheres $\rho =$ constant, half-cones $\phi =$ constant, and planes $\theta =$ constant. A typical subinterval into which this partition divides I is shown in Figure 18.46. In the usual way we count only those subintervals that lie entirely in Q. If these are numbered from 1 to n, we denote their volumes by $\Delta V_1, \Delta V_2, \ldots, \Delta V_n$. In the ith such subinterval we choose an arbitrary point $(\rho_i^*, \theta_i^*, \phi_i^*)$. Then we define

$$\iiint_Q f(\rho, \theta, \phi)\, dV = \lim_{\|P\| \to 0} \sum_{i=1}^{n} f(\rho_i^*, \theta_i^*, \phi_i^*)\, \Delta V_i$$

provided the limit on the right exists independently of the choices of the points $(\rho_i^*, \theta_i^*, \phi_i^*)$.

To motivate the appropriate form of dV when going to an iterated integral, consider the enlarged subinterval shown in Figure 18.46(b). Its volume ΔV is approximately the same as that of a rectangular box of dimensions $\Delta\rho$ by

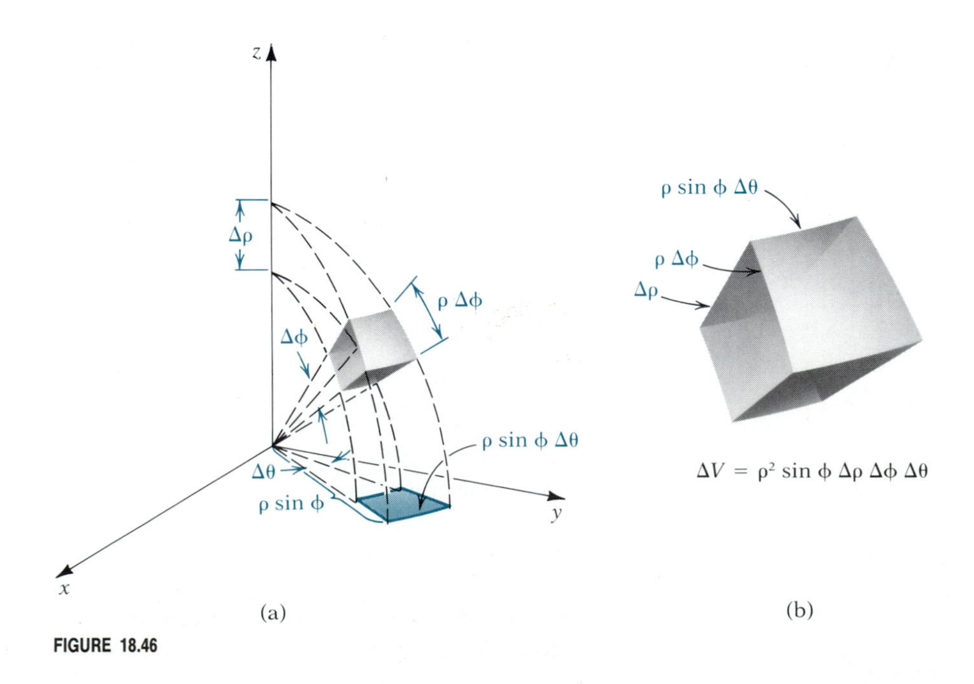

FIGURE 18.46

$\rho\,\Delta\phi$ by $\rho \sin\phi\,\Delta\theta$. So

$$\Delta V \approx \rho^2 \sin\phi\,\Delta\rho\,\Delta\phi\,\Delta\theta$$

The smaller the subinterval—that is, the smaller $\Delta\rho$, $\Delta\theta$, and $\Delta\phi$ are—the better the approximation. It can be proved that when f is continuous over a regular region Q,

$$\iiint_Q f(\rho, \theta, \phi)\, dV = \iiint_{\text{(appropriate limits)}} f(r, \theta, \phi)\,\rho^2 \sin\phi\, d\rho\, d\phi\, d\theta \tag{18.36}$$

The limits of integration depend on the nature of the bounding surfaces. We will illustrate some common types in the examples. We write

$$dV = \rho^2 \sin\phi\, d\rho\, d\phi\, d\theta$$

and call this the differential of volume in spherical coordinates. The factor $\rho^2 \sin\phi$ plays the analogous role to the factor r in the differential of volume in cylindrical coordinates.

Remark on Notation The radial variable ρ in spherical coordinates must not be confused with the density of a solid, which has been denoted by ρ also. With spherical coordinates, some other letter, such as δ, should be used for the density function.

EXAMPLE 18.23 Find the volume and the centroid of the "ice cream cone" shaped region (see Figure 18.47) bounded by the cone $z = \sqrt{3(x^2 + y^2)}$ and the hemisphere $z = \sqrt{4 - x^2 - y^2}$.

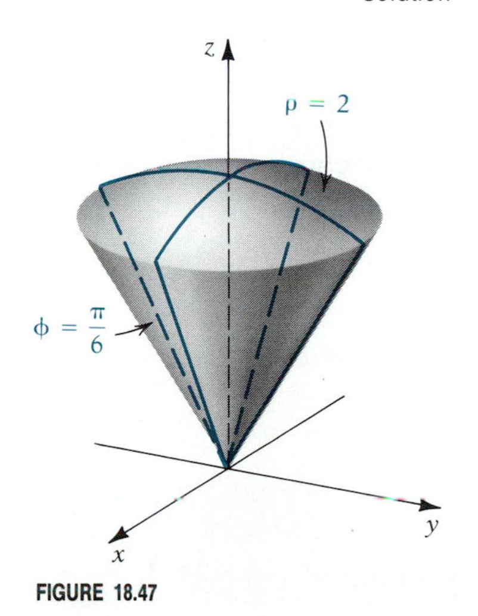

FIGURE 18.47

Solution For the cone the spherical equation is

$$\rho\cos\phi = \sqrt{3}\rho\sin\phi \quad \text{or} \quad \tan\phi = \frac{1}{\sqrt{3}}$$

so that $\phi = \frac{\pi}{6}$. The sphere has the equation $\rho = 2$. Thus,

$$\begin{aligned} V = \iiint_Q dV &= \int_0^{2\pi}\int_0^{\pi/6}\int_0^{2} \rho^2 \sin\phi\, d\rho\, d\phi\, d\theta \\ &= \int_0^{2\pi}\int_0^{\pi/6} \tfrac{8}{3}\sin\phi\, d\phi\, d\theta = \tfrac{8}{3}\int_0^{2\pi} [-\cos\phi]_0^{\pi/6}\, d\theta \\ &= \tfrac{8}{3}[-\tfrac{\sqrt{3}}{2} + 1]\cdot 2\pi = \tfrac{8\pi}{3}(2 - \sqrt{3}) \end{aligned}$$

By symmetry, $\bar{x} = \bar{y} = 0$, so we need only calculate $\bar{z}$. First we calculate M_{xy}:

$$\begin{aligned} M_{xy} &= \iiint_Q z\, dV = \iiint_Q (\rho\cos\phi)\, dV \\ &= \int_0^{2\pi}\int_0^{\pi/6}\int_0^{2} \rho^3 \sin\phi\cos\phi\, d\rho\, d\phi\, d\theta \\ &= \int_0^{2\pi}\int_0^{\pi/6} 4\sin\phi\cos\phi\, d\phi\, d\theta = 4\int_0^{2\pi}\left[\frac{\sin^2\phi}{2}\right]_0^{\pi/6} d\theta = \pi \end{aligned}$$

So

$$\bar{z} = \frac{M_{xy}}{V} = \frac{\pi}{\frac{8\pi}{3}(2 - \sqrt{3})} = \frac{3}{8(2 - \sqrt{3})}\cdot\frac{2 + \sqrt{3}}{2 + \sqrt{3}} = \frac{3(2 + \sqrt{3})}{8}$$ ■

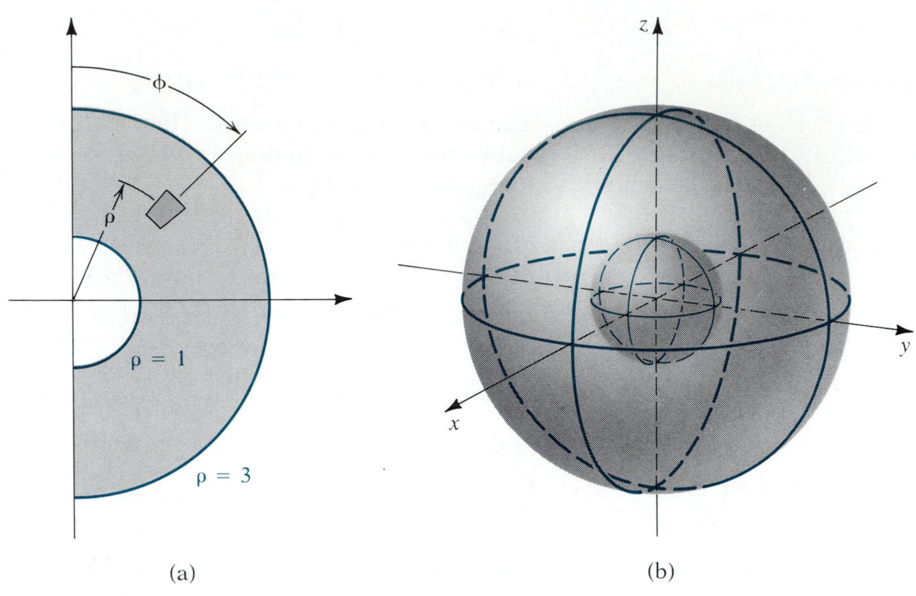

FIGURE 18.48

EXAMPLE 18.24 Find the mass of the solid between the two spheres $x^2 + y^2 + z^2 = 1$ and $x^2 + y^2 + z^2 = 9$ if the density is inversely proportional to the distance from the origin.

Solution The spheres have equations $\rho = 1$ and $\rho = 3$, and these are the limits of integration on ρ. If we next integrate with respect to ϕ, from $\phi = 0$ to $\phi = \pi$, we would then have integrated over the semiannular region shown in Figure 18.48(a). Then, allowing θ to vary from 0 to 2π in effect rotates this semiannular region around the z-axis to give the entire region shown in Figure 18.48(b). The density is of the form $\frac{k}{\rho}$ for some constant k, so we have

$$m = \iiint_Q dm = \int_0^{2\pi}\int_0^{\pi}\int_1^3 \frac{k}{\rho}(\rho^2 \sin\phi\, d\rho\, d\phi\, d\theta)$$

$$= k\int_0^{2\pi}\int_0^{\pi}\left[\frac{\rho^2}{2}\right]_1^3 \sin\phi\, d\phi\, d\theta = \frac{8k}{2}\int_0^{2\pi}\int_0^{\pi} \sin\phi\, d\phi\, d\theta = 16k\pi$$ ■

EXERCISE SET 18.7

A

In Exercises 1–6 evaluate the iterated integral. Sketch the region determined by the limits of integration.

1. $\int_0^{\pi}\int_0^1\int_0^4 rz\sin\theta\, dz\, dr\, d\theta$

2. $\int_0^{\pi/2}\int_1^2\int_0^{2-r^2} \frac{\cos\theta}{r^2}\, dz\, dr\, d\theta$

3. $\int_0^{2\pi}\int_0^1\int_0^r zr\, dz\, dr\, d\theta$

4. $\int_0^{\pi}\int_0^{2\sin\theta}\int_0^{\sqrt{4-r^2}} r\, dz\, dr\, d\theta$

5. $\int_0^{2\pi}\int_0^{\pi/2}\int_0^1 \rho^2\sin\phi\, d\rho\, d\phi\, d\theta$

6. $\int_0^{2\pi} \int_{\pi/6}^{\pi/4} \int_0^{1/\cos\phi} \rho^3 \sin\phi \cos\phi \, d\rho \, d\phi \, d\theta$

In Exercises 7–13 use cylindrical coordinates.

7. Find the volume of the region enclosed by the cylinder $x^2 + y^2 = 4$ and the paraboloid $z = 8 - x^2 - y^2$ that lies above the xy-plane.
8. Find the centroid of the region inside the hemisphere $z = \sqrt{4 - x^2 - y^2}$ and outside the cylinder $x^2 + y^2 = 1$.
9. A homogeneous solid of density δ is bounded by $z = \sqrt{x^2 + y^2}$, $x^2 + y^2 = 1$, and $z = 0$. Find I_z.
10. Find the mass of the solid of constant density δ bounded by the surfaces $z = r$, $z = 0$, and $r = 2\cos\theta$.
11. Find the center of mass of the solid in Exercise 10.
12. Find the center of mass of the homogeneous solid of density δ bounded above by $x^2 + y^2 + z^2 = 4$ and below by $3z = x^2 + y^2$.
13. A solid occupies the region

$$Q = \{(r, \theta, z): 1 \le r \le 3, \ 0 \le \theta \le \tfrac{\pi}{3}, \ 0 \le z \le 9 - r^2\}$$

Its density at any point is inversely proportional to the distance of the point from the z-axis. Find its center of mass.

In Exercises 14–18 use spherical coordinates.

14. Find the volume of a sphere of radius a.
15. Find the center of mass of a solid that occupies the first-octant region inside a sphere of radius a if the density is proportional to the distance from the z-axis.
16. Find the centroid of the region below the hemisphere $z = \sqrt{4 - x^2 - y^2}$ that lies between the upper nappes of the cones $z^2 = 3(x^2 + y^2)$ and $z^2 = x^2 + y^2$.
17. A solid that occupies the region above the xy-plane and between $z = \sqrt{9 - x^2 - y^2}$ and $z = \sqrt{1 - x^2 - y^2}$ has density proportional to the distance from the origin. Find its mass and its moment of inertia with respect to the z-axis.
18. A solid that occupies the region between the spheres $\rho = 1$ and $\rho = 2$ and inside the cone $\rho = \frac{\pi}{3}$ has density inversely proportional to the distance above the xy-plane. Find its center of mass.

In Exercises 19–23 evaluate the integral by changing to cylindrical or spherical coordinates.

19. $\int_0^2 \int_0^{\sqrt{4-x^2}} \int_0^{\sqrt{x^2+y^2}} (z + \sqrt{x^2 + y^2}) \, dz \, dy \, dx$
20. $\int_{-1}^1 \int_{-\sqrt{1-y^2}}^{\sqrt{1-y^2}} \int_1^{2-x^2-y^2} \frac{1}{z^2} \, dz \, dx \, dy$
21. $\int_0^{\sqrt{2}} \int_0^{\sqrt{2-x^2}} \int_{\sqrt{x^2+y^2}}^{\sqrt{4-x^2-y^2}} \sqrt{x^2+y^2+z^2} \, dz \, dy \, dx$
22. $\int_0^1 \int_{-\sqrt{1-z^2}}^{\sqrt{1-z^2}} \int_{-\sqrt{1-y^2-z^2}}^{\sqrt{1-y^2-z^2}} x^2 \, dx \, dy \, dz$
23. $\int_0^2 \int_{-\sqrt{2x-x^2}}^{\sqrt{2x-x^2}} \int_0^{\sqrt{4-x^2-y^2}} dz \, dy \, dx$

B

24. Find the centroid of the region that lies both inside the sphere $x^2 + y^2 + z^2 = 2az$ and outside the sphere $x^2 + y^2 + z^2 = a^2$.
25. A solid spherical ball $x^2 + y^2 + z^2 \le a^2$ has density equal to the distance from the xy-plane. The ball is cut by planes $z = h_1$ and $z = h_2$, where $0 < h_1 < h_2 < a$. Find the center of mass of the portion of the ball between these two planes. (*Hint:* Use cylindrical coordinates and integrate first with respect to r.)
26. Find the centroid of the region enclosed by the frustum of the cone shown in the accompanying figure.

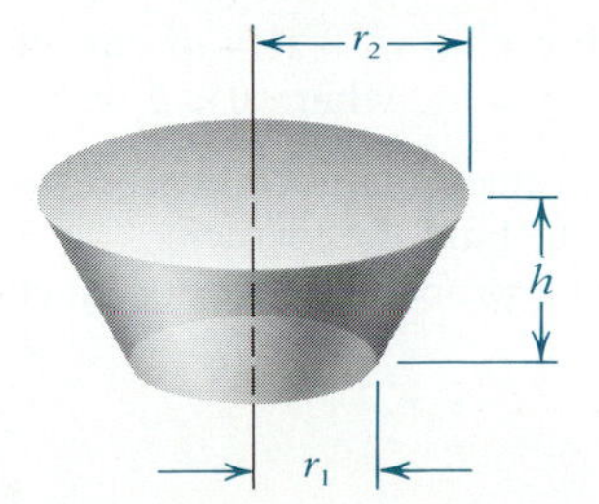

27. In a triple integral if a change from (x, y, z) coordinates to (u, v, w) coordinates is made by means of the transformation equations

$$x = x(u, v, w) \qquad y = y(u, v, w) \qquad z = z(u, v, w)$$

then it can be shown that the differential of volume $dV = dx\,dy\,dz$ is replaced by $|J(u, v, w)|\,du\,dv\,dw$, where the function J, called the **Jacobian** of the transformation, is defined by

$$J(u, v, w) = \begin{vmatrix} \frac{\partial x}{\partial u} & \frac{\partial x}{\partial v} & \frac{\partial x}{\partial w} \\ \frac{\partial y}{\partial u} & \frac{\partial y}{\partial v} & \frac{\partial y}{\partial w} \\ \frac{\partial z}{\partial u} & \frac{\partial z}{\partial v} & \frac{\partial z}{\partial w} \end{vmatrix}$$

Show that for cylindrical coordinates $|J(r, \theta, z)| = r$ and for spherical coordinates $|J(\rho, \theta, \phi)| = \rho^2 \sin\phi$.

18.8 SUPPLEMENTARY EXERCISES

In Exercises 1 and 2 evaluate the integrals. Sketch the region of integration.

1. $\int_1^5 \int_x^{2x+1} \frac{6x}{(x+y)^2}\, dy\, dx$
2. $\int_0^2 \int_{\sqrt{y}}^{2-y} xy\sqrt[3]{x^2+4y}\, dx\, dy$

In Exercises 3 and 4 evaluate $\iint_R f(x, y)\, dA$.

3. $f(x, y) = (x+y)/y^2$; $R = \{(x, y): y^2 \le x \le y+2, 1 \le y \le 2\}$
4. $f(x, y) = xe^y$; R is the region bounded by $y = x^2$ and $x - y + 2 = 0$.

In Exercises 5–9 use double integration.

5. Find the area of the region bounded by $y = \ln x$ and $y = (x-1)/(e-1)$.
6. Find the area outside the circle $r = 1$ and inside the limaçon $r = 3 + 4\cos\theta$.
7. Find the volume of the region under the graph of $f(x, y) = x + y$ that lies above $R = \{(x, y): 0 \le x \le 2, 0 \le y \le \sqrt{4 - x^2}\}$.
8. Find the volume of the region under the graph of

$$f(x, y) = \frac{xy}{x^2 + y^2}$$

above the region inside $r = 3\cos\theta$ that is outside $r = 1 + \cos\theta$.

9. Find the volume enclosed by the tetrahedron with vertices (0, 0, 0), (0, 2, 0), (0, 2, 4), and (1, 2, 0).
10. Give an equivalent integral with the order of integration reversed:

$$\int_0^4 \int_1^{\sqrt{2x+1}} f(x, y)\, dy\, dx$$

11. Evaluate the integral

$$\int_0^1 \int_{x^2}^1 x\cos^2 y^2\, dy\, dx$$

12. Change to polar coordinates and evaluate:

$$\int_0^3 \int_{y/3}^{\sqrt{10y - y^2}} (x + y)\, dx\, dy$$

13. A lamina occupies the region bounded by $2y = x^2$, $x = 2$, and $y = 0$. Its density is $\rho(x, y) = 5(x + y)$. Find $(\bar{x}, \bar{y})$ and (r_x, r_y).
14. Find the centroid of the leaf of the rose curve $r = 4\sin 2\theta$ that lies in the first quadrant.
15. A lamina occupies the region $R = \{(x, y): 0 \le x \le \pi, 0 \le y \le \sin x\}$ and has density $\rho(x, y) = y$. Find $(\bar{x}, \bar{y})$, I_x, and r_y.

In Exercises 16–21 find the volume described using triple integration in rectangular, cylindrical, or spherical coordinates, whichever seems most appropriate.

16. Inside $z = 4 - \sqrt{x^2 + y^2}$ and outside $x^2 + y^2 = 1$, in the first octant
17. Bounded by $z = \sqrt{x}$, $x + z = 2$, $y = 0$, and $y = 4$, in the first octant
18. Inside $x^2 + y^2 = 4$, above $z = 0$, and below $2x + y + z = 8$
19. Bounded by $z = 12 - x^2 - y^2$, $y = x^2$, and $x = 0$, in the first octant
20. Inside $x^2 + y^2 + z^2 = a^2$, between $z^2 = x^2 + y^2$ and $z^2 = 3(x^2 + y^2)$
21. Bounded by $y = x^2 + z^2$, $z = \sqrt{3}x$, $z = 0$, and $y = 4$, in the first octant (*Hint:* Use cylindrical coordinates having the polar variables in the xz-plane.)
22. A solid occupies the first-octant region bounded by the surfaces $y = x^2$, $y = z$, $y = 1$, $y = 4$, and $z = 0$. If its density is $\rho(x, y) = (x + z)/\sqrt{y}$, find its center of mass.
23. For the region common to the two half-cones $z = \sqrt{x^2 + y^2}$ and $z = 3 - 2\sqrt{x^2 + y^2}$, find (a) the centroid and (b) I_z.
24. For the solid described in Example 18.24 find the moment of inertia with respect to the z-axis.
25. Find the moment of inertia of the homogeneous solid inside $z = \sqrt{2 - x^2 - y^2}$, below $z = \sqrt{x^2 + y^2}$, and above $z = 0$, with respect to the line $\mathbf{r}(t) = 2\mathbf{j} + t\mathbf{k}$. (*Hint:* Use the law of cosines.)

In Exercises 26–28 find the surface area described.

26. The portion of the paraboloid $z = 4 - x^2 - y^2$ above the plane $2x + z = 4$
27. The portion of the surface $z = 2e^{x/2}\sin(y/2)$ that lies above the region $R = \{(x, y): 0 \le x \le \ln 3, 0 \le y \le e^x\}$
28. The band on the sphere $x^2 + y^2 + z^2 = a^2$ cut off by the planes $y = b_1$ and $y = b_2$, where $0 < b_1 < b_2 < a$
29. Let $f(x, y) = \frac{kx}{y}$ on the region bounded by $x - 2y = 0$, $y = 1$, $y = 3$, and $x = 0$. Find the following:
    (a) k so that f will be a probability density function
    (b) $P(x > 2)$
    (c) μ_x and μ_y
    (d) σ_x and σ_y

30. Rewrite each integral in either cylindrical or spherical coordinates, as seems most appropriate, but do not evaluate.

(a) $\int_0^2 \int_0^{\sqrt{2y-y^2}} \int_0^{2-\sqrt{x^2+y^2}} (xy - y^2)\, dz\, dx\, dy$

(b) $\int_{-1}^1 \int_{-\sqrt{1-x^2}}^{\sqrt{1-x^2}} \int_{x^2+y^2}^{\sqrt{2-x^2-y^2}} \frac{1}{(x^2+y^2+z^2)^2}\, dz\, dy\, dx$

(c) $\int_0^2 \int_1^{\sqrt{2x-x^2}} \int_0^{x^2+y^2} \frac{1}{1+x^2+y^2}\, dz\, dy\, dx$

(d) $\int_{-1}^1 \int_{-\sqrt{1-z^2}}^{\sqrt{1-z^2}} \int_{-\sqrt{5-x^2-z^2}}^{\sqrt{5-x^2-z^2}} \frac{z^2-x^2}{z^2+x^2}\, dy\, dx\, dz$

(*Hint:* Permute the variables in the transformation equations.)

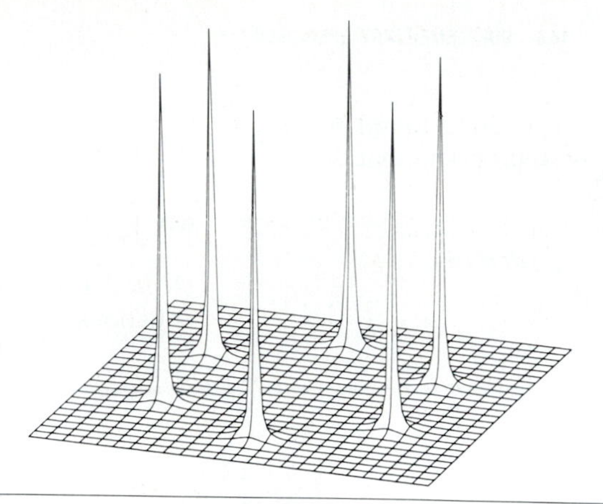

CHAPTER 19

VECTOR FIELD THEORY

19.1 LINE INTEGRALS

In Chapter 4 we defined the single integral $\int_a^b f(x)\,dx$ as the limit of Riemann sums formed from partitioning the interval $[a, b]$ on the x-axis. An important generalization is obtained by replacing the *linear interval* $[a, b]$ by what might be called a *curvilinear interval*—that is, by an arc of a curve. We suppose first that the curve is in $\mathbf{R}^2$. The extension to three dimensions is straightforward.

Let C denote an arc of a curve in the xy-plane with end points A and B, as in Figure 19.1(a). We assume that C is rectifiable (has finite length). Let f be a function of two variables that is defined and bounded on some domain that contains C in its interior. We partition C as shown in Figure 19.1(b) by points $P_0, P_1, P_2, \ldots, P_n$, where $P_0 = A$ and $P_n = B$. In each of the subarcs $\widehat{P_{i-1}P_i}$ so formed we select an arbitrary point $P_i^* = (x_i^*, y_i^*)$. Let Δs_i denote the arc length from P_{i-1} to P_i, and let $\|P\|$ denote the maxi-

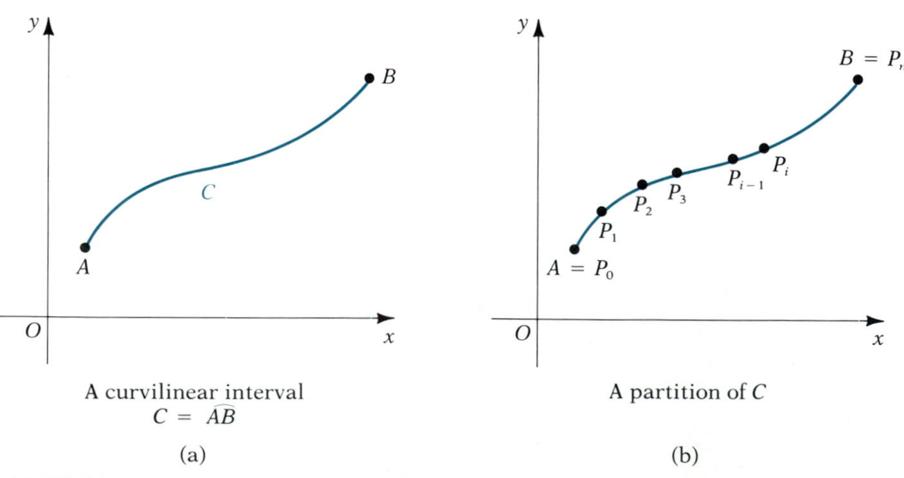

A curvilinear interval
$C = \widehat{AB}$

(a)

A partition of C

(b)

FIGURE 19.1

mum of these lengths for $i = 1, 2, \ldots, n$. We now define the integral of f *along C*.

DEFINITION 19.1 **LINE INTEGRAL OF f ALONG C**

$$\int_C f(x, y)\, ds = \lim_{||P|| \to 0} \sum_{i=1}^{n} f(x_i^*, y_i^*)\, \Delta s_i \tag{19.1}$$

provided the limit on the right exists independently of the choices of the points P_i^*. ■

Note that if f is the constant function $f(x, y) = 1$, then the integral along C gives the length of C:

$$\text{length of } C = \int_C ds$$

The integral $\int_C f(x, y)\, ds$ is customarily called the *line integral* of f along C. The name line integral is unfortunate, since we normally interpret "line" to mean "straight line." *Curvilinear integral* would be a better name. We will use line integral, however, since it occurs so widely in the literature. The limit in equation (19.1) can be shown to exist when f is continuous on C.

When f is a nonnegative continuous function on C we can give a geometric interpretation to $\int_C f(x, y)\, ds$ as the area of the cylindrical surface obtained by projecting C upward until it meets the surface $z = f(x, y)$. This is seen by observing that each term $f(x_i^*, y_i^*)\, \Delta s_i$ of the sum on the right-hand side of equation (19.1) gives the area of a rectangle that approximates the area of a vertical strip with curved base Δs_i. Figure 19.2(a) shows one such strip. In the limit we get the exact area, as illustrated in Figure 19.2(b). Note the

FIGURE 19.2

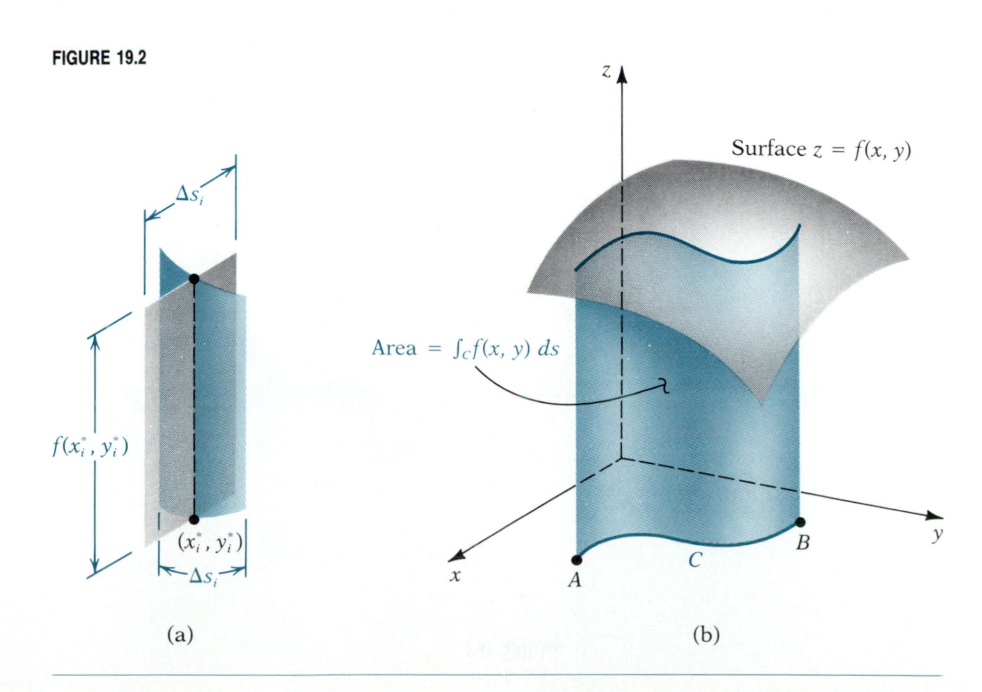

similarity of this to the interpretation of $\int_a^b f(x)\,dx$ as the area above the linear interval $[a, b]$ and below the graph of f, when f is nonnegative.

As is true for other integrals we have studied, Definition (19.1) does not provide a very useful means of calculating the values of line integrals. The following theorem, which we state without proof, provides a more efficient computational method. Recall that an arc C is *smooth* if it has a parameterization $x = x(t)$, $y = y(t)$, for $a \le t \le b$, where x' and y' are continuous on $[a, b]$ and not simultaneously 0 there. We saw in Chapter 13 that smooth curves are rectifiable.

THEOREM 19.1 If C is a smooth arc with parameterization $x = x(t)$, $y = y(t)$, where $a \le t \le b$, and if f is continuous on C, then

$$\int_C f(x, y)\,ds = \int_a^b f(x(t), y(t))s'(t)\,dt \tag{19.2}$$

where

$$s'(t) = \sqrt{(x'(t))^2 + (y'(t))^2}$$

■

Remarks

1. According to the theorem, line integrals are evaluated as ordinary integrals on the linear interval $[a, b]$ by writing both the integrand and the differential in terms of the parameter t.
2. It can be shown that the value of a line integral is unchanged when C is given a different parameterization. This is important, since there can be many ways of parameterizing the same curve.

EXAMPLE 19.1 Find the area of the portion of the parabolic cylinder $y = x^2/2$ from $x = 0$ to $x = 2$ that lies in the first octant and is bounded above by the cone $2z = \sqrt{x^2 + 4y^2}$.

Solution The surface in question is shown in Figure 19.3. The desired area is given by the line integral

$$\text{area} = \int_C f(x, y)\,ds$$

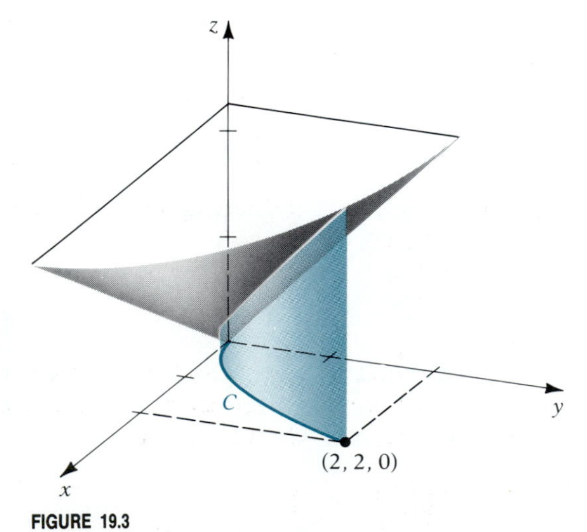

FIGURE 19.3

where C is the arc of the parabola $y = x^2/2$ from $(0, 0)$ to $(2, 2)$ and $f(x, y) = \sqrt{x^2 + 4y^2}/2$. We can get a parameterization of C by setting $x = t$:

$$\begin{cases} x = t \\ y = \dfrac{t^2}{2} \end{cases} \qquad 0 \le t \le 2$$

Thus, since $ds = \sqrt{(x')^2 + (y')^2}\, dt = \sqrt{1 + t^2}\, dt$, we get from equation (19.2)

$$\text{area} = \int_C \frac{\sqrt{x^2 + 4y^2}}{2}\, ds = \int_0^2 \frac{\sqrt{t^2 + t^4}}{2} \sqrt{1 + t^2}\, dt$$
$$= \frac{1}{2} \int_0^2 t(1 + t^2)\, dt = 3$$

■

If C is sectionally smooth, so that it is composed of a finite number of smooth arcs joined end to end, say $C_1, C_2, \ldots, C_n$, then we write $C = C_1 \cup C_2 \cup \cdots \cup C_n$ and define

$$\int_C f(x, y)\, ds = \int_{C_1} f(x, y)\, ds + \int_{C_2} f(x, y)\, ds + \cdots + \int_{C_n} f(x, y)\, ds$$

An important application of line integrals is in calculating the work done by a variable force $\mathbf{F}$ that acts on a particle as it moves along a curve. This is a natural extension of our consideration in Chapter 14 of the work done by $\mathbf{F}$ as the particle moves along a straight line. To motivate the definition, let $\mathbf{F}(x, y) = \langle f(x, y), g(x, y) \rangle$ be the variable force, and suppose $\mathbf{F}$ is continuous in some two-dimensional region that contains the smooth curve C: $\mathbf{r}(t) = \langle x(t), y(t) \rangle$, where $a \le t \le b$, in its interior. Let $A = (f(a), g(a))$ and $B = (f(b), g(b))$.

Partition C as in Definition 19.1. If the norm $\|P\|$ of the partition is small, the work ΔW_i done by the variable force $\mathbf{F}$ as the particle moves from P_{i-1} to P_i is approximated by the work done by the constant force $\mathbf{F}(x_i^*, y_i^*)$ in moving the particle along the straight line segment from P_{i-1} to P_i. (See Figure 19.4.) If we choose the intermediate point P_i^* properly, the unit tangent vector $\mathbf{T}$ at this point will be parallel to the line segment from P_{i-1} to P_i. (Why?) So the component of $\mathbf{F}(x_i^*, y_i^*)$ that acts in the direction of the line segment is $\text{Comp}_{\mathbf{T}}\mathbf{F} = \mathbf{F} \cdot \mathbf{T}$, evaluated at P_i^*, and since $|\overrightarrow{P_{i-1}P_i}|$ is approximately equal to Δs_i, we have

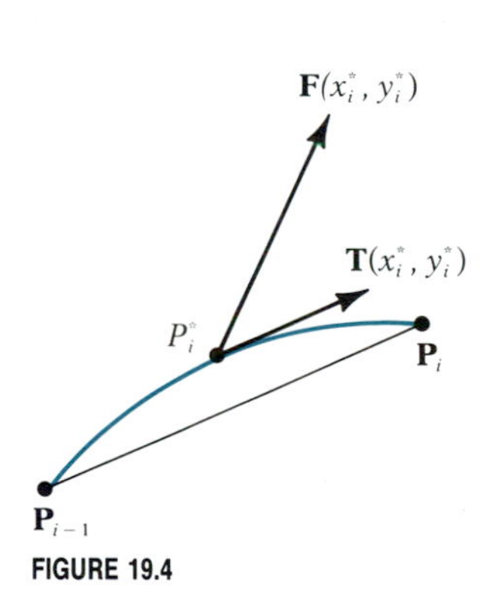

FIGURE 19.4

$$\Delta W_i \approx \mathbf{F}(x_i^*, y_i^*) \cdot \mathbf{T}(x_i^*, y_i^*)\, \Delta s_i$$

By summing over i and taking limits of such sums as $\|P\| \to 0$, we are led to the definition

$$W = \int_C \mathbf{F} \cdot \mathbf{T}\, ds \tag{19.3}$$

Before giving an application, we obtain a convenient alternate formulation of the integral in equation (19.3). We know from Chapter 15 that $\mathbf{T} = d\mathbf{r}/ds$, so formally we can write

$$\mathbf{T}\, ds = \left(\frac{d\mathbf{r}}{ds}\right) ds = d\mathbf{r}$$

Thus, we have

$$\int_C \mathbf{F} \cdot \mathbf{T}\, ds = \int_C \mathbf{F} \cdot d\mathbf{r} \tag{19.4}$$

We can evaluate this integral using Theorem 19.1, noting that, as a function of t, the unit tangent vector $\mathbf{T}$ can be written

$$\mathbf{T} = \frac{\mathbf{r}'(t)}{|\mathbf{r}'(t)|}$$

So by Theorem 19.1,

$$\int_C \mathbf{F} \cdot d\mathbf{r} = \int_C \mathbf{F} \cdot \mathbf{T}\, ds = \int_a^b \mathbf{F}(x(t), y(t)) \frac{\mathbf{r}'(t)}{|\mathbf{r}'(t)|} s'(t)\, dt$$

We know from Chapter 15 that $|\mathbf{r}'(t)| = s'(t)$. So the result can be written

$$\int_C \mathbf{F} \cdot d\mathbf{r} = \int_a^b \mathbf{F}(x(t), y(t)) \cdot \mathbf{r}'(t)\, dt \tag{19.5}$$

EXAMPLE 19.2 Find the work done by the force $\mathbf{F}(x, y) = x^2\mathbf{i} + 2xy\mathbf{j}$ on a particle as it moves along the curve C: $\mathbf{r}(t) = (\cos t)\mathbf{i} + (\sin t)\mathbf{j}$, where $0 \le t \le \pi$.

Solution Note that C is the upper half of the circle $x^2 + y^2 = 1$, with a counterclockwise orientation. Since $\mathbf{r}'(t) = (-\sin t)\mathbf{i} + (\cos t)\mathbf{j}$, we have by equation (19.5)

$$\begin{aligned} W = \int_C \mathbf{F} \cdot d\mathbf{r} &= \int_0^\pi [(\cos^2 t)\mathbf{i} + 2(\cos t \sin t)\mathbf{j}] \cdot [(-\sin t)\mathbf{i} + (\cos t)\mathbf{j}]\, dt \\ &= \int_0^\pi (-\cos^2 t \sin t + 2\cos^2 t \sin t)\, dt \\ &= \int_0^\pi \cos^2 t \sin t\, dt = -\frac{\cos^3 t}{3}\Big]_0^\pi = \frac{2}{3} \end{aligned}$$

The units depend on those for force and distance. ■

EXAMPLE 19.3 Find the work done by the force $\mathbf{F}(x, y) = \langle x - y, xy \rangle$ that acts on a particle in moving it from point A to point B along the curve $C = C_1 \cup C_2$ shown in Figure 19.5. Assume force is measured in dynes and distance in centimeters.

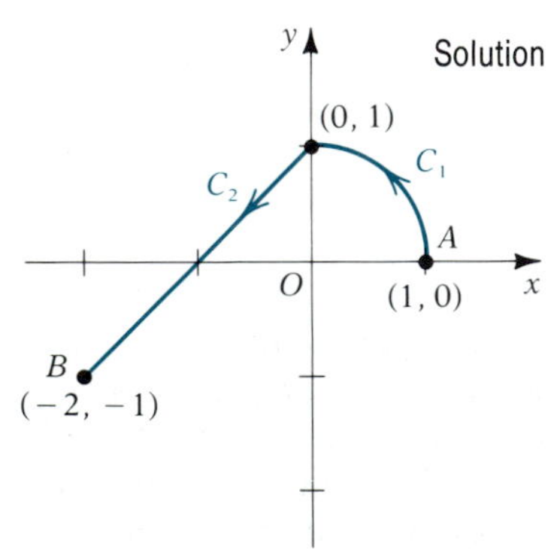

FIGURE 19.5

Solution First we obtain parameterizations for C_1 and C_2. For C_1 we have $\mathbf{r}(t) = \langle \cos t, \sin t \rangle$, where $0 \le t \le \frac{\pi}{2}$. For C_2 we could determine the rectangular equation of the line and then use x (or y) as a parameter. A more general approach that works on straight line segments in either two or three dimensions is to observe that a line segment from a point P to a point Q has the vector representation $\mathbf{r}(t) = \overrightarrow{OP} + t(\overrightarrow{PQ})$ for $0 \le t \le 1$. (Verify this.) Thus for C_2,

$$\mathbf{r}(t) = \langle 0, 1 \rangle + t\langle -2, -2 \rangle = \langle -2t, 1 - 2t \rangle, \qquad 0 \le t \le 1$$

The work done is given by

$$W = \int_C \mathbf{F} \cdot d\mathbf{r} = \int_{C_1} \mathbf{F} \cdot d\mathbf{r} + \int_{C_2} \mathbf{F} \cdot d\mathbf{r}$$

We evaluate each integral on the right separately and then add the results:

$$\int_{C_1} \mathbf{F} \cdot d\mathbf{r} = \int_0^{\pi/2} \langle \cos t - \sin t, \cos t \sin t \rangle \cdot \langle -\sin t, \cos t \rangle \, dt$$
$$= \int_0^{\pi/2} (-\sin t \cos t + \sin^2 t + \cos^2 t \sin t) \, dt = \tfrac{\pi}{4} - \tfrac{1}{6}$$

(Supply the missing steps.)

$$\int_{C_2} \mathbf{F} \cdot d\mathbf{r} = \int_0^1 \langle -2t - (1 - 2t), -2t(1 - 2t) \rangle \cdot \langle -2, -2 \rangle \, dt$$
$$= \int_0^1 (2 + 4t - 8t^2) \, dt = \tfrac{4}{3}$$

So $W = (\frac{\pi}{4} - \frac{1}{6}) + \frac{4}{3} = \frac{\pi}{4} + \frac{7}{6}$ ergs. ■

The integral $\int_C \mathbf{F} \cdot d\mathbf{r}$ can be expressed in yet another way by evaluating the dot product $\mathbf{F} \cdot d\mathbf{r}$, where $\mathbf{F}(x, y) = \langle f(x, y), g(x, y) \rangle$, if we write $d\mathbf{r} = \langle dx, dy \rangle$. This gives

$$\int_C \mathbf{F} \cdot d\mathbf{r} = \int_C f(x, y) \, dx + g(x, y) \, dy \tag{19.6}$$

The form on the right is sometimes called the *differential form* of the line integral. We illustrate it in the next example.

EXAMPLE 19.4 Evaluate the integral

$$\int_C (2x + y) \, dx + x^2 \, dy$$

where C is the curve

$$\begin{cases} x = t \\ y = 2t^2 \end{cases} \qquad 0 \leq t \leq 2$$

Solution We have $dx = dt$ and $dy = 4t \, dt$. So, on substituting for x, y, dx, and dy, we get

$$\int_C (2x + y) \, dx + x^2 \, dy = \int_0^2 [(2t + 2t^2) \, dt + t^2(4t \, dt)]$$
$$= 2 \int_0^2 (t + t^2 + 2t^3) \, dt = \tfrac{76}{3}$$ ■

Remark Although the line integral $\int_C f(x, y) \, ds$ is unchanged if C is given a new parameterization, the integral $\int_C \mathbf{F} \cdot \mathbf{T} \, ds$, or equivalently $\int_C \mathbf{F} \cdot d\mathbf{r}$, changes in sign if the orientation of C is reversed. This is because the unit tangent vector $\mathbf{T}$ reverses direction when the orientation is reversed. So the function $\mathbf{F} \cdot \mathbf{T}$ changes in sign. Thus, if $-C$ denotes the curve C with orientation reversed, we have

$$\int_{-C} \mathbf{F} \cdot d\mathbf{r} = -\int_C \mathbf{F} \cdot d\mathbf{r}$$

Vector functions of the form $\mathbf{F}(x, y) = \langle f(x, y), g(x, y) \rangle$, or in three dimensions $\mathbf{F}(x, y, z) = \langle f(x, y, z), g(x, y, z), h(x, y, z) \rangle$, play an important role in the physical sciences. We have seen one application in finding work, when $\mathbf{F}$ represents a force. Whatever the physical interpretation of $\mathbf{F}$, we call such a function a **vector field.** Corresponding to each point P in the domain of $\mathbf{F}$, there is a vector $\mathbf{F}(P)$. So the set of all such vectors is a "field of vectors." Figure 19.6 shows several vectors in the vector field $\mathbf{F}(x, y) = x\mathbf{i} + y\mathbf{j}$.

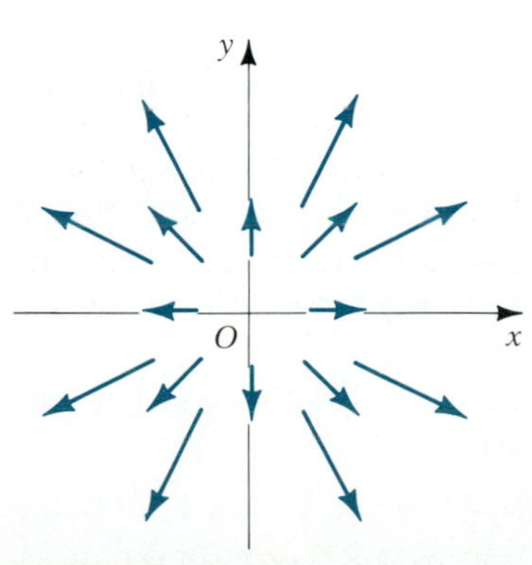

FIGURE 19.6 The vector field $x\mathbf{i} + y\mathbf{j}$

The two most important types of vector fields are those in which **F** represents a force and those in which **F** represents the velocity of a fluid (liquid or gas) at a point. We call these **force fields** and **velocity fields,** respectively. There can be many types of force fields, such as gravitational fields, magnetic fields, and electric fields. Functions that assign single real numbers to points in some region, such as those we studied in the last three chapters, are often called **scalar fields** to distinguish them from vector fields. As an example of this distinction, consider a flowing river. Its velocity field **F** gives the velocity vector at each point in the stream, and the scalar field T gives the temperature of the water at each point. We will use this terminology of vector and scalar fields extensively in the remainder of this chapter, and we will see how certain line integrals give important characteristics of velocity fields as well as force fields.

The concept of the line integral extends in a natural way to functions of three variables defined on space curves, and we will use such integrals freely from now on. For example, if the smooth curve C has vector representation $\mathbf{r}(t) = \langle x(t), y(t), z(t)\rangle$, and if $\mathbf{F}(x, y, z) = \langle f(x, y, z), g(x, y, z), h(x, y, z)\rangle$ is continuous on C, then

$$\int_C \mathbf{F}\cdot\mathbf{T}\,ds = \int_C \mathbf{F}\cdot d\mathbf{r} = \int_C \mathbf{F}(x(t), y(t), z(t))\cdot\mathbf{r}'(t)\,dt$$

and in component notation

$$\int_C \mathbf{F}\cdot d\mathbf{r} = \int_C f(x, y, z)\,dx + g(x, y, z)\,dy + h(x, y, z)\,dz$$

EXERCISE SET 19.1

A

In Exercises 1–6 evaluate the integral $\int_C f(x, y)\,ds$, where C has the given parameterization.

1. $f(x, y) = x^2y;\ x = 2t - 1,\ y = 3t,\ -1 \le t \le 2$
2. $f(x, y) = x - 2y;\ x = \cos 2t,\ y = \sin 2t,\ 0 \le t \le \frac{\pi}{2}$
3. $f(x, y) = x/y;\ x = 2t,\ y = t^2 + 1,\ 0 \le t \le 2$
4. $f(x, y) = ye^x;\ x = t,\ y = 2t - 3,\ 0 \le t \le 1$
5. $f(x, y) = x - y;\ x = 2\cos^2 t,\ y = \sin 2t,\ 0 \le t \le \frac{\pi}{2}$
6. $f(x, y) = (x^3 + 9x)/y;\ x = 3t,\ y = t^2 + 1,\ 0 \le t \le 2$

In Exercises 7–10 find the area of the vertical cylindrical surface that has xy trace C and is bounded above by $z = f(x, y)$ and below by the xy-plane.

7. C is the circle $x^2 + y^2 = 1$; $f(x, y) = 4 - x - 2y$.
8. C is the line segment from $(1, 0)$ to $(0, 2)$; $f(x, y) = x^2 + y^2$.
9. C is the curve $y = \sin x$ from $x = 0$ to $x = \frac{\pi}{2}$; $f(x, y) = \sin 2x$.
10. C is the parabolic arc $y = x^2$ from $(0, 0)$ to $(1, 1)$; $f(x, y) = \sqrt{1 + 4x^2}$.

*In Exercises 11–14 find the work done by the force **F** acting on a particle moving in the positive direction along C.*

11. $\mathbf{F}(x, y) = 16xy^2\mathbf{i} - (y/x^2)\mathbf{j}$; C is given by $x = \cosh t$, $y = \sinh t$, $0 \le t \le \ln 2$.
12. $\mathbf{F}(x, y) = x\mathbf{i} + y\mathbf{j}$; C is the arc of the parabola $y = 2x^2 - 1$ from $(0, -1)$ to $(1, 1)$.
13. $\mathbf{F}(x, y, z) = \left\langle \dfrac{x + y}{z}, xyz, \dfrac{1}{z^2}\right\rangle$; C has the vector equation $\mathbf{r}(t) = \left\langle t, t^2, \dfrac{1}{t}\right\rangle$, $1 \le t \le 3$.
14. $\mathbf{F}(x, y, z) = xy\mathbf{i} + xz\mathbf{j} + yz\mathbf{k}$; C is the arc of the circular helix, $x = \cos t$, $y = \sin t$, $z = t$ for $0 \le t \le \frac{\pi}{2}$.

In Exercises 15–20 evaluate the line integral $\int_C \mathbf{F}\cdot d\mathbf{r}$, where C is the graph of $\mathbf{r}(t)$.

15. $\mathbf{F}(x, y) = \langle xy, x - y\rangle$; $\mathbf{r}(t) = \langle 2t, 1 - t\rangle$, $1 \le t \le 2$
16. $\mathbf{F}(x, y) = \langle x^2 - y^2, 2xy\rangle$; $\mathbf{r}(t) = \langle \cos t, \sin t\rangle$, $0 \le t \le \pi$

17. $\mathbf{F}(x, y) = (x + y)\mathbf{i} + (3x - 2y)\mathbf{j}$; $\mathbf{r}(t) = t^2\mathbf{i} - 2t\mathbf{j}$, $-1 \le t \le 1$

18. $\mathbf{F}(x, y) = xy\mathbf{i} - x^2\mathbf{j}$; $\mathbf{r}(t) = e^t\mathbf{i} + e^{-t}\mathbf{j}$, $0 \le t \le 1$

19. $\mathbf{F}(x, y, z) = \langle x + y - z, x - 2z, y + z\rangle$; $\mathbf{r}(t) = \langle t + 1, t - 1, t\rangle$, $1 \le t \le 2$

20. $\mathbf{F}(x, y, z) = xyz^2\mathbf{i} + \frac{x}{y}\mathbf{j} + xz\mathbf{k}$; $\mathbf{r}(t) = t\mathbf{i} + (1/t^2)\mathbf{j} + \sqrt{t}\mathbf{k}$, $1 \le t \le 4$

In Exercises 21–26 evaluate the line integral along the given curve C.

21. $\int_C y\,dx - x\,dy$; C is the curve $y = e^x$, $0 \le x \le \ln 2$.

22. $\int_C 2xy\,dx + x^2\,dy$; C is the arc of the parabola $y = x^2$ from $(-1, 1)$ to $(2, 4)$.

23. $\int_C (x^2 - y^2)\,dx + xy\,dy$; C is the line segment from $(0, 1)$ to $(1, 2)$.

24. $\int_C (2x - 3y)\,dx + (4x + 2y)\,dy$; $C = C_1 \cup C_2$, where C_1 is the line segment from $(0, 0)$ to $(2, 0)$ and C_2 is the line segment from $(2, 0)$ to $(2, 4)$.

25. $\int_C x\,dx + y\,dy$; $C = C_1 \cup C_2 \cup C_3$, where C_1 is the arc of the parabola $y = x^2$ from $(0, 0)$ to $(1, 1)$, C_2 is the line segment from $(1, 1)$ to $(0, 1)$, and C_3 is the line segment from $(0, 1)$ to $(0, 0)$.

26. $\int_C (x^2 + y^2)\,dx + (1 - x^2)\,dy$; $C = C_1 \cup C_2$, where C_1 is the semicircle $y = \sqrt{1 - x^2}$ from $(1, 0)$ to $(-1, 0)$, and C_2 is the line segment from $(-1, 0)$ to $(1, 0)$.

In Exercises 27 and 28 show graphically the members of the vector field **F** *at the specified points.*

27. $\mathbf{F}(x, y) = y^2\mathbf{i} - xy\mathbf{j}$; $(\pm 1, 0)$, $(\pm 2, \pm 1)$, $(\pm 1, \pm 2)$, $(0, \pm 1)$

28. $\mathbf{F}(x, y) = \left\langle \dfrac{x}{\sqrt{x^2 + y^2}}, \dfrac{y}{\sqrt{x^2 + y^2}} \right\rangle$; $(\pm 2, 0)$, $(\pm 3, \pm 4)$, $(\pm 4, \pm 3)$, $(0, \pm 2)$

29. For a certain vector field **F**, defined on all of $\mathbf{R}^2$ except the origin, the vector at each point has length inversely proportional to the distance of that point from the origin, and it is directed toward the origin.
(a) Draw several typical vectors of the field **F**.
(b) Find, without integration, the work done by **F** on a particle moving on a circle $x^2 + y^2 = a^2$. Explain your reasoning.
(c) Find a formula for $\mathbf{F}(x, y)$.

30. Let $\mathbf{F}(x, y) = -y\mathbf{i} + x\mathbf{j}$. Find and show geometrically the vectors in this field at 45° intervals around the circle C: $x^2 + y^2 = a^2$ directed in a counterclockwise direction. Show that for this curve C, $\mathbf{F} \cdot \mathbf{T} = |\mathbf{F}| = a$ and use this to evaluate $\int_C \mathbf{F} \cdot \mathbf{T}\,ds$ by inspection.

31. Evaluate $\int_C xyz\,ds$ along the circular helix $x = 2\cos t$, $y = 2\sin t$, $z = t$, for $0 \le t \le \pi$.

32. Evaluate $\int_C (x + z)\,dx + (x - y)\,dy + (2y - z)\,dz$, where C is the line segment from $(2, -1, 3)$ to $(3, 0, 4)$.

33. Evaluate $\int_C (2x + y)\,dx + (x - y)\,dy$ where C is the path from A to B shown in the accompanying figure.

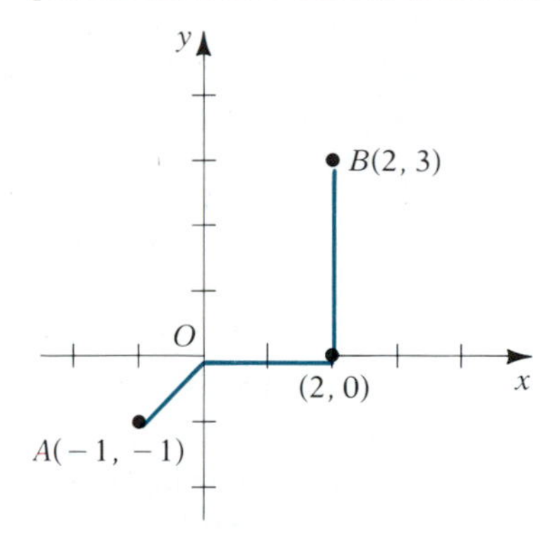

34. Evaluate $\int_C (x^2 + y^2)\,dx + 2xy\,dy$ where C is (a) the path shown in the figure for Exercise 33 and (b) the straight line segment from A to B.

35. Evaluate $\int_C x^2y\,dx + (x - 2y)\,dy$ where C is the closed curve shown in the accompanying figure.

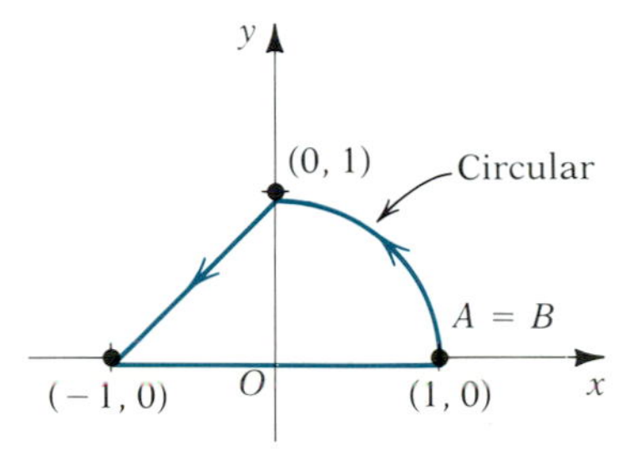

36. Show that

$$\int_C (2x + 3y)\,dx + (3x - 4y)\,dy = 0$$

for the curve C in the figure for Exercise 35.

37. Evaluate

$$\int_C (x + z)\,dx + (y - z)\,dy + (x - y)\,dz$$

where C is the path from A to B shown in the accompanying figure.

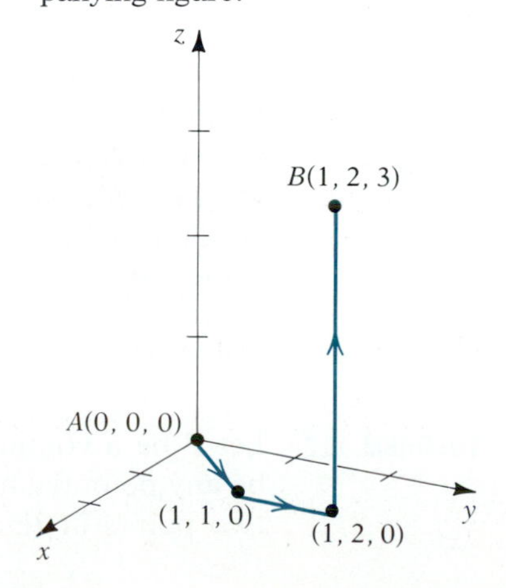

B

38. Evaluate $\int_C (x^2 - 4y)\,ds$, where C is defined by $x = 2t$, $y = t^2/2 - \ln t$, with $1 \le t \le e$.

39. Evaluate $\int_C \mathbf{F} \cdot d\mathbf{r}$ if C is given by $\mathbf{r}(t) = (e^t \cos t)\mathbf{i} + (e^t \sin t)\mathbf{j}$ for $0 \le t \le \pi$, and

$$\mathbf{F}(x, y) = \frac{-x}{\sqrt{x^2 + y^2}}\mathbf{i} + \frac{y}{\sqrt{x^2 + y^2}}\mathbf{j}$$

40. Evaluate $\int_C (x + y - z)\,ds$ where C is defined by $x = 4\cos^3 t$, $y = 4\sin^3 t$, $z = \sin 2t$, with $0 \le t \le \frac{\pi}{2}$.

41. Find the work done by the force field $\mathbf{F}(x, y, z) = \langle xz, xy^2, -yz \rangle$ on a unit mass moving along the elliptical helix $\mathbf{r}(t) = \langle 2\cos t, \sin t, 3t \rangle$, where $0 \le t \le 2\pi$.

Exercises 42 and 43 refer to a thin wire of density $\rho(x, y)$ (mass per unit length) in the shape of a plane smooth curve C.

42. By partitioning C, explain the plausibility of defining the mass m of the wire by $m = \int_C dm$, where $dm = \rho(x, y)\,ds$.

43. (a) Show that the natural definition of the center of mass $(\bar{x}, \bar{y})$ of the wire is

$$\bar{x} = \frac{\int_C x\,dm}{m} \qquad \bar{y} = \frac{\int_C y\,dm}{m}$$

(b) Similarly, show that the natural definition of the moments of inertia I_x and I_y are

$$I_x = \int_C y^2\,dm \qquad I_y = \int_C x^2\,dm$$

44. A wire is in the shape of the catenary $y = a\cosh\frac{x}{a}$, where $-a \le x \le a$, and its density at any point is inversely proportional to the distance from the x-axis to the point. Use the results of Exercises 42 and 43 to find (a) its center of mass and (b) I_x and I_y.

45. Using results analogous to those in Exercises 42 and 43 find the center of mass of a homogeneous wire of density ρ in the shape of the circular helix

$$\mathbf{r}(t) = (2\cos t)\mathbf{i} + (2\sin t)\mathbf{j} + (3t)\mathbf{k}$$

from $t = 0$ to $t = \frac{3\pi}{2}$. Also find I_x, I_y, and I_z.

46. An *inverse square force field* is of the form

$$\mathbf{F}(x, y, z) = \frac{k}{|\mathbf{r}|^2}\mathbf{u}$$

where $\mathbf{u} = \mathbf{r}/|\mathbf{r}|$. Find the work done by such a force field in moving a particle of unit mass from $(1, 1, 1)$ to $(2, 1, 3)$.

47. Use Newton's second law of motion, $\mathbf{F} = m\mathbf{a}$, to prove that the work done by a force $\mathbf{F}$ in moving a particle of mass m along a curve C: $\mathbf{r}(t) = \langle x(t), y(t), z(t) \rangle$, from $A = (x(a), y(a), z(a))$ to $B = (x(b), y(b), z(b))$, equals the change in kinetic energy $K(B) - K(A)$, where $K = \frac{1}{2}mv^2$. *Hint:* Show that

$$\begin{aligned}\int_0 \mathbf{F} \cdot d\mathbf{r} &= \int_a^b m\mathbf{r}'' \cdot \mathbf{r}'\,dt = \frac{m}{2}\int_a^b \frac{d}{dt}(\mathbf{r}' \cdot \mathbf{r}')\,dt \\ &= \frac{m}{2}\int_a^b \frac{d}{dt}|\mathbf{r}'|^2\,dt\end{aligned}$$

19.2 GRADIENT FIELDS AND PATH INDEPENDENCE

A special case of a vector field $\mathbf{F}$ occurs when $\mathbf{F} = \nabla\phi$ for some scalar function ϕ. When this is the case, $\mathbf{F}$ is called a **gradient field,** and ϕ is called a **potential function** for $\mathbf{F}$. If ϕ is such a potential function, then clearly $\phi + k$ for any constant k is also a potential function for $\mathbf{F}$. An example of a gradient field in two dimensions is $\mathbf{F}(x, y) = \langle 2x, 2y \rangle$, since $\mathbf{F} = \nabla\phi$ with $\phi(x, y) = x^2 + y^2$. We will see later a way of determining when a vector field is a gradient field and of finding its potential. For now we will concentrate on some special properties of gradient fields that are not shared by other vector fields. The most important is given in the following theorem, sometimes called the *fundamental theorem for line integrals.*

THEOREM 19.2 Let $\mathbf{F}$ be a continuous gradient field in an open region R of $\mathbf{R}^2$, and let ϕ be any potential function for $\mathbf{F}$. Then if $A = (x_1, y_1)$ and $B = (x_2, y_2)$ are any two points in R, and C is any piecewise smooth curve from A to B lying entirely in R,

$$\int_C \mathbf{F} \cdot d\mathbf{r} = \phi(x_2, y_2) - \phi(x_1, y_1) \tag{19.7}$$

Proof We prove the theorem when C is a smooth curve. The result for piecewise smooth curves can then be obtained by adding the results for each of the smooth components (see Exercise 38). Let $\mathbf{F}(x, y) = \langle f(x, y), g(x, y)\rangle$. Then, since $\mathbf{F} = \nabla\phi = \langle \phi_x, \phi_y \rangle$, we have $f(x, y) = \phi_x(x, y)$ and $g(x, y) = \phi_y(x, y)$. So

$$\begin{aligned}\int_C \mathbf{F} \cdot d\mathbf{r} &= \int_C f(x, y)\,dx + g(x, y)\,dy = \int_C \phi_x(x, y)\,dx + \phi_y(x, y)\,dy \\ &= \int_a^b [\phi_x(x(t), y(t))x'(t) + \phi_y(x(t), y(t))y'(t)]\,dt\end{aligned}$$

In Leibniz notation, this integral is

$$\int_a^b \left[\frac{\partial\phi}{\partial x}\frac{dx}{dt} + \frac{\partial\phi}{\partial y}\frac{dt}{dt}\right] dt$$

By the chain rule (17.15) the integrand is just $d\phi/dt$. So we have

$$\begin{aligned}\int_C \mathbf{F} \cdot d\mathbf{r} &= \int_a^b [\tfrac{d}{dt}\phi(x(t), y(t))]\,dt \\ &= \phi(x(b), y(b)) - \phi(x(a), y(a)) \quad \text{By the fundamental theorem of calculus} \\ &= \phi(x_2, y_2) - \phi(x_1, y_1)\end{aligned}$$

■

The striking feature of this theorem is that the value of the line integral $\int_C \mathbf{F} \cdot d\mathbf{r}$ depends only on the initial and final points of C and not on C itself. Thus, *any* piecewise smooth curve C from A to B in R will give the same value of the line integral. Because of this, the line integral $\int_C \mathbf{F} \cdot d\mathbf{r}$ is said to be *independent of the path*. The term *path* is used here to mean a piecewise smooth curve (which may, in particular, be simply a smooth curve). Our theorem can therefore be restated as follows:

If F is a continuous gradient field in an open region R, then $\int_C \mathbf{F} \cdot d\mathbf{r}$ is independent of the path C in R.

Because of this path independence, the integral $\int_C \mathbf{F} \cdot d\mathbf{r}$ is sometimes written as $\int_{(x_1, y_1)}^{(x_2, y_2)} \mathbf{F} \cdot d\mathbf{r}$ or even as $\int_A^B \mathbf{F} \cdot d\mathbf{r}$. Using the latter notation, the result of Theorem 19.2 assumes the form

$$\int_A^B \mathbf{F} \cdot d\mathbf{r} = \phi(B) - \phi(A)$$

which looks very much like the result of the fundamental theorem of calculus.

Another result is immediately apparent. If C is a *closed* curve, then the initial point $A = (x_1, y_1)$ and the final point $B = (x_2, y_2)$ coincide. Thus,

$$\int_C \mathbf{F} \cdot d\mathbf{r} = \phi(x_2, y_2) - \phi(x_1, y_1) = 0$$

So we have the following result.

If F is a continuous gradient field in an open region R and C is any piecewise smooth closed curve in R, then

$$\int_C \mathbf{F} \cdot d\mathbf{r} = 0$$

EXAMPLE 19.5 Evaluate the integral $\int_C 2x\,dx + 2y\,dy$, where C is a piecewise smooth curve from $A = (1, -2)$ to $B = (3, 6)$.

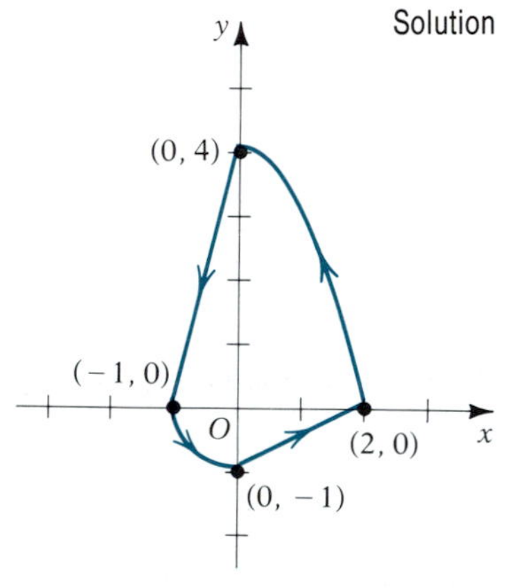

FIGURE 19.7

Solution As we observed earlier, the vector field $\mathbf{F} = \langle 2x, 2y \rangle$ is a gradient field with potential $\phi(x, y) = x^2 + y^2$, since $\nabla\phi = \langle 2x, 2y \rangle = \mathbf{F}$, and this is valid throughout all of $\mathbf{R}^2$. Thus, by Theorem 19.2 the integral

$$\int_C \mathbf{F} \cdot d\mathbf{r} = \int_C 2x\,dx + 2y\,dy$$

is independent of the path from A to B, and

$$\int_C 2x\,dx + 2y\,dy = \phi(3, 6) - \phi(1, -2)$$
$$= (9 + 36) - (1 + 4) = 40$$

■

In Exercise 29 you will be asked to evaluate the integral in Example 19.5 along various paths from A to B by parameterizing the curves to verify that the answer is always the same.

EXAMPLE 19.6 Evaluate the integral $\int_C 2x\,dx + 2y\,dy$, where C is the path shown in Figure 19.7.

Solution The hard way to do this problem is to parameterize each part of C and evaluate the integral on each part. The easy way is to use the fact that $\mathbf{F}(x, y) = \langle 2x, 2y \rangle$ is a gradient field, and so the integral $\int_C \mathbf{F} \cdot d\mathbf{r} = 0$ for every piecewise smooth closed path. Thus, $\int_C 2x\,dx + 2y\,dy = 0$. ■

The converse of Theorem 19.2 is also true, as we now show.

THEOREM 19.3 If $\mathbf{F}(x, y) = \langle f(x, y), g(x, y) \rangle$ is a continuous vector field in an open region R and $\int_C \mathbf{F} \cdot d\mathbf{r}$ is independent of the path C in R, then $\mathbf{F}$ is a gradient field.

Proof Let (a, b) denote any fixed point in R. We define a scalar function ϕ by

$$\phi(x, y) = \int_{(a,b)}^{(x,y)} \mathbf{F} \cdot d\mathbf{r}, \qquad (x, y) \text{ in } R$$

That this is a valid definition follows from the path independence, so that $\phi(x, y)$ depends only on x and y and not on the path of integration from (a, b) to (x, y). We now show that $\phi_x = f(x, y)$.

We fix (x, y) temporarily. By definition,

$$\phi_x(x, y) = \lim_{h \to 0} \frac{\phi(x + h, y) - \phi(x, y)}{h}$$

Since R is open, $|h|$ can be chosen small enough that the line segment from (x, y) to $(x + h, y)$ lies in R. Let C_1 be any path from (a, b) to (x, y), and let

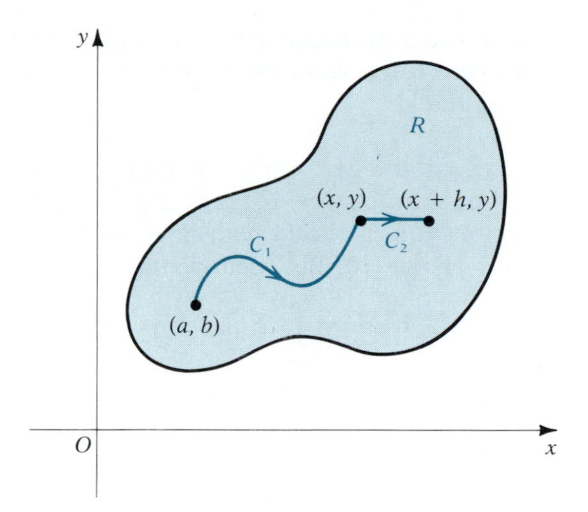

FIGURE 19.8

C_2 be the line segment from (x, y) to $(x + h, y)$ (see Figure 19.8). Then we have

$$\phi(x + h, y) - \phi(x, y) = \int_{C_1 \cup C_2} \mathbf{F} \cdot d\mathbf{r} - \int_{C_1} \mathbf{F} \cdot d\mathbf{r} = \int_{C_2} \mathbf{F} \cdot d\mathbf{r}$$

A parameterization of C_2 is $\mathbf{r}(t) = \langle t, y \rangle$, where $x \leq t \leq x + h$, so that $d\mathbf{r} = \langle dt, 0 \rangle$. So we have

$$\int_{C_2} \mathbf{F} \cdot d\mathbf{r} = \int_x^{x+h} f(t, y)\, dt = f(c, y)h \quad \text{By the mean value theorem for integrals}$$

where c is between x and $x + h$. Finally then

$$\phi_x(x, y) = \lim_{h \to 0} \frac{\phi(x + h, y) - \phi(x, y)}{h} = \lim_{h \to 0} \frac{f(c, y)h}{h} = f(x, y)$$

by the continuity of f.

A similar argument in which C_2 is the vertical segment from (x, y) to $(x, y + k)$ would show that $\phi_y(x, y) = g(x, y)$ (see Exercise 40). It follows that $\mathbf{F} = \nabla \phi$, and so $\mathbf{F}$ is a gradient field. ■

Remark The way the function ϕ was defined in the proof of this theorem provides one means of obtaining a potential function for $\mathbf{F}$, once we know that $\int_C \mathbf{F} \cdot d\mathbf{r}$ is independent of the path in a region R. We define ϕ by

$$\phi(x, y) = \int_{(a,b)}^{(x,y)} \mathbf{F} \cdot d\mathbf{r}$$

and evaluate the integral by selecting a convenient initial point (a, b) and a convenient path from (a, b) to (x, y). Often a path that consists of vertical and horizontal line segments is useful, or the line segment that goes from (a, b) to (x, y) might work well. The complete path, however, must lie in R. We will illustrate this technique after we have developed a test for determining path independence (see Example 19.8).

We have seen that when $\mathbf{F}$ is a continuous gradient field in an open region R and C is any piecewise smooth *closed* curve in R, $\int_C \mathbf{F} \cdot d\mathbf{r} = 0$. We now use Theorem 19.3 to show that the converse of this result is also true.

THEOREM 19.4 If $\mathbf{F}$ is a continuous vector field in an open region R with the property that $\int_C \mathbf{F} \cdot d\mathbf{r} = 0$ for *every* piecewise smooth closed curve C in R, then $\mathbf{F}$ is a gradient field.

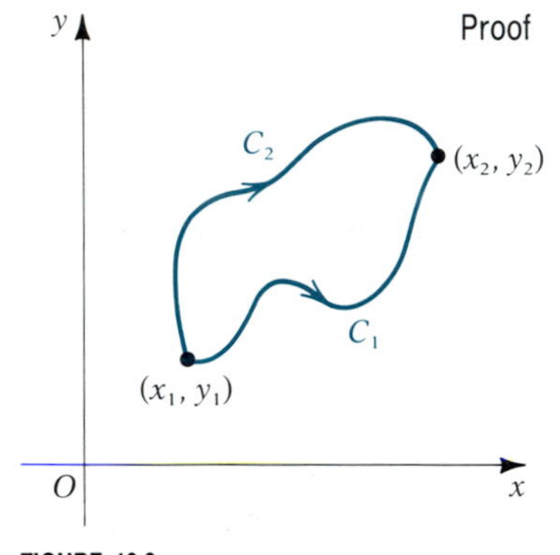

FIGURE 19.9

Proof Let (x_1, y_1) and (x_2, y_2) be any two points in R and consider any two piecewise smooth curves C_1 and C_2 in R from (x_1, y_1) to (x_2, y_2). Then $C_1 \cup (-C_2)$ in a closed path in R (see Figure 19.9), and so by hypothesis, $\int_{C_1 \cup (-C_2)} \mathbf{F} \cdot d\mathbf{r} = 0$. But then we have

$$\int_{C_1 \cup (-C_2)} \mathbf{F} \cdot d\mathbf{r} = \int_{C_1} \mathbf{F} \cdot d\mathbf{r} + \int_{-C_2} \mathbf{F} \cdot d\mathbf{r} = \int_{C_1} \mathbf{F} \cdot d\mathbf{r} - \int_{C_2} \mathbf{F} \cdot d\mathbf{r} = 0$$

so that

$$\int_{C_1} \mathbf{F} \cdot d\mathbf{r} = \int_{C_2} \mathbf{F} \cdot d\mathbf{r}$$

This proves that $\int_C \mathbf{F} \cdot d\mathbf{r}$ is independent of the path. Thus, by Theorem 19.3, $\mathbf{F}$ is a gradient field. ■

Recall from Section 17.1 that the differential of a differentiable function ϕ of two variables is given by

$$d\phi = \frac{\partial \phi}{\partial x}\,dx + \frac{\partial \phi}{\partial y}\,dy$$

When ϕ is a potential function for the continuous gradient field $\mathbf{F} = \langle f, g \rangle$, it follows that

$$d\phi = f(x, y)\,dx + g(x, y)\,dy$$

For this reason we call $f(x, y)\,dx + g(x, y)\,dy$ an *exact differential*, since it is the exact differential of ϕ. Using this terminology, we can say that *the continuous vector field* $\mathbf{F} = \langle f, g \rangle$ *in an open region R is a gradient field if and only if* $f\,dx + g\,dy$ *is an exact differential.*

In the study of gradient fields two crucial questions remain: (1) How can one tell whether a vector field is a gradient field? and (2) When it is a gradient field, how can a potential function for it be found? We discuss these now.

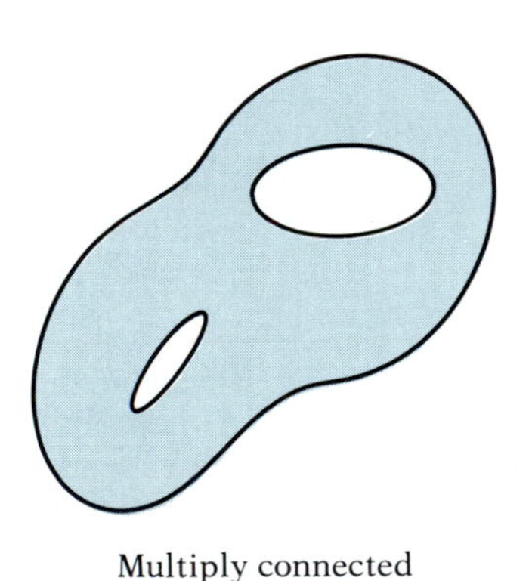
Simply connected
(a)

Multiply connected
(b)

FIGURE 19.10

A simple test for determining whether $\mathbf{F}$ is a gradient field is given in the following theorem. First, though, we must distinguish between two types of open regions. Recall that a region R is a *connected* set, in the sense that any two points in R can be joined by a polygonal path that lies entirely in R. We say that R is **simply connected** if, for every simple closed curve C in R, all points inside C are also in R. If R is not simply connected, it is said to be **multiply connected.** Figure 19.10 illustrates both types. Intuitively you can think of simple connectedness as meaning "no holes."

THEOREM 19.5 If f and g have continuous first partial derivatives in the simply connected open region R, then $\mathbf{F} = \langle f, g \rangle$ is a gradient field if and only if

$$\frac{\partial f}{\partial y} = \frac{\partial g}{\partial x} \tag{19.8}$$

Partial Proof We will prove only that when $\mathbf{F}$ is a gradient field, equation (19.8) is true. The converse is more difficult and will be omitted. Suppose then that $\mathbf{F}$ is a gradient field with potential function ϕ. Then $\mathbf{F} = \nabla\phi$ and so

$$f(x, y) = \frac{\partial\phi}{\partial x} \quad \text{and} \quad g(x, y) = \frac{\partial\phi}{\partial y}$$

Thus,

$$\frac{\partial f}{\partial y} = \frac{\partial^2\phi}{\partial y\,\partial x} \quad \text{and} \quad \frac{\partial g}{\partial x} = \frac{\partial^2\phi}{\partial x\,\partial y}$$

In view of the continuity of f_y and g_x, it follows from Theorem 16.1 that the two second-order mixed partials of ϕ are equal; that is,

$$\frac{\partial f}{\partial y} = \frac{\partial g}{\partial x}$$

It should be noted that this part of the proof did not require simple connectedness. It is only in proving the converse that this property comes into play. ■

This theorem provides the key for determining when a vector field is a gradient field, at least in a simply connected region. In the next example we use this test and then illustrate a general technique for finding a potential function.

EXAMPLE 19.7 Show that

$$\mathbf{F}(x, y) = \langle e^y - 2xy,\ xe^y - x^2 + 2y \rangle$$

is a gradient field throughout $\mathbf{R}^2$, and find a potential function for $\mathbf{F}$.

Solution The functions $f(x, y) = e^y - 2xy$ and $g(x, y) = xe^y - x^2 + 2y$ are each continuous and they have continuous partial derivatives throughout $\mathbf{R}^2$, which is a simply connected open region. Furthermore,

$$\frac{\partial f}{\partial y} = e^y - 2x \quad \text{and} \quad \frac{\partial g}{\partial x} = e^y - 2x$$

so that condition (19.8) is met. Thus $\mathbf{F}$ is a gradient field.

We know, then, that a function ϕ exists for which

$$\frac{\partial\phi}{\partial x} = e^y - 2xy \quad \text{and} \quad \frac{\partial\phi}{\partial y} = xe^y - x^2 + 2y$$

To find ϕ we can proceed in either of two ways: integrate $e^y - 2xy$ with respect to x while holding y fixed, or integrate $xe^y - x^2 + 2y$ with respect to y while holding x fixed. Let us choose the first:

$$\phi(x, y) = \int (e^y - 2xy)\,dx = xe^y - x^2y + C(y) \qquad y \text{ is held fixed.}$$

Note that we have written the "constant" of integration as $C(y)$ to allow for the fact that it may be a function of y because y was held fixed for this integration. To find $C(y)$, we force $\partial\phi/\partial y$ to equal $xe^y - x^2 + 2y$:

$$\frac{\partial\phi}{\partial y} = xe^y - x^2 + C'(y) = xe^y - x^2 + 2y$$

$$C'(y) = 2y$$

$$C(y) = y^2 + C_1$$

where C_1 is an arbitrary (numerical) constant. Since we are seeking any potential function for **F**, we might as well let $C_1 = 0$. Then

$$\phi(x, y) = xe^y - x^2y + y^2$$

You can verify that $\mathbf{F} = \nabla\phi$. ■

As we indicated in this example, once we have determined that the function $\mathbf{F} = \langle f, g\rangle$ is a gradient field, we know it has a potential function ϕ for which $\phi_x = f$ and $\phi_y = g$. So we can obtain $\phi(x, y)$ by integrating $f(x, y)$ with respect to x or $g(x, y)$ with respect to y. In the first case, the integration constant is a function $C(y)$ of y whose value can be determined by forcing $\phi_y = g$, as we did in the example. If the second approach is used, the integration constant is of the form $C(x)$, and its value is determined by forcing $\phi_x = f$. Although either way will work, sometimes one of the integrations may be simpler than the other, and so you should be on the lookout for this possibility.

An alternative way of finding a potential function when you know that **F** is a gradient field was suggested in the remark following Theorem 19.3. We illustrate this in the next example.

EXAMPLE 19.8 Find a potential function for the gradient field **F** of Example 19.7 using the method suggested in the remark following Theorem 19.3.

Solution The function **F** is

$$\mathbf{F}(x, y) = \langle e^y - 2xy, xe^y - x^2 + 2y\rangle$$

and this was shown in Example 19.7 to be a gradient field throughout $\mathbf{R}^2$. Define ϕ by $\phi(x, y) = \int_{(0,0)}^{(x,y)} \mathbf{F}\cdot d\mathbf{r}$, where the path taken from $(0, 0)$ to (x, y) is the polygonal path $C_1 \cup C_2$ shown in Figure 19.11. We can parameterize C_1 by $\mathbf{r}_1(t) = \langle t, 0\rangle$, $0 \le t \le x$, and C_2 by $\mathbf{r}_2(t) = \langle x, t\rangle$, $0 \le t \le y$. So

$$\begin{aligned}\phi &= \int_{C_1} \mathbf{F}\cdot d\mathbf{r}_1 + \int_{C_2} \mathbf{F}\cdot d\mathbf{r}_2 = \int_0^x \mathbf{F}(t, 0)\cdot\langle dt, 0\rangle + \int_0^y \mathbf{F}(x, t)\cdot\langle 0, dt\rangle \\ &= \int_0^x dt + \int_0^y (xe^t - x^2 + 2t)\,dt = x + [xe^t - x^2t + t^2]_0^y \\ &= x + [(xe^y - x^2y + y^2) - x] = xe^y - x^2y + y^2\end{aligned}$$

■

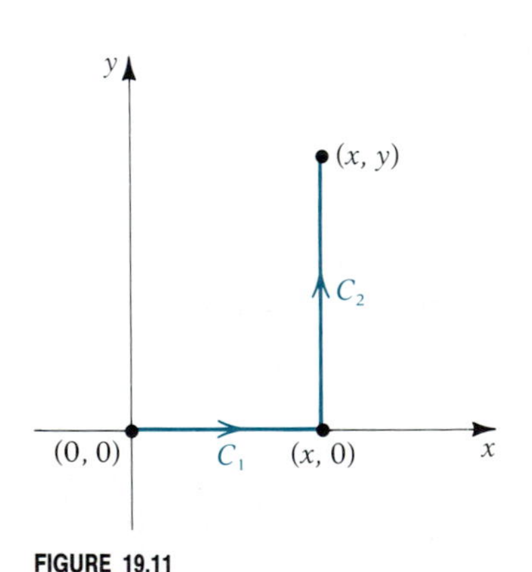

FIGURE 19.11

The main results of this section can be summarized by the following list of equivalent statements for a continuous vector field $\mathbf{F} = \langle f, g\rangle$ in an open region R. They are equivalent in the sense that if any one of the statements is true, each of the others is also true.

1. **F** is a gradient field.
2. $\int_C \mathbf{F}\cdot d\mathbf{r}$ is independent of the path.
3. $f\,dx + g\,dy$ is an exact differential.
4. $\int_C \mathbf{F}\cdot d\mathbf{r} = 0$ for every piecewise smooth closed curve C in R.

If, in addition, R is simply connected and f and g have continuous first partial derivatives in R, we can add a fifth statement equivalent to each of these four:

5. $\dfrac{\partial f}{\partial y} = \dfrac{\partial g}{\partial x}$

In physics a continuous force field **F** is said to be **conservative** if the work done by **F** on a particle as it moves from a point A to a point B is independent of the path from A to B. Since work is given by the integral $\int_C \mathbf{F} \cdot d\mathbf{r}$, it follows that *a force field is conservative if and only if it is a gradient field.* One consequence of this is the following *law of conservation of energy*:

> In a conservative force field the sum of the potential energy and kinetic energy of a particle is constant.

You will be asked to prove this in Exercise 43.

Most of the ideas of this section have natural extensions to three dimensions. In particular, the equivalence of the first four conditions given above remains true, where $\mathbf{F} = \langle f, g, h \rangle$. In condition 3 the differential form becomes $f\,dx + g\,dy + h\,dz$. We will show in Section 19.6 that the appropriate form of condition 5 in the three-dimensional case is the following:

$$\frac{\partial f}{\partial y} = \frac{\partial g}{\partial x} \qquad \frac{\partial g}{\partial z} = \frac{\partial h}{\partial y} \qquad \frac{\partial h}{\partial x} = \frac{\partial f}{\partial z} \tag{19.9}$$

EXERCISE SET 19.2

A

In Exercises 1–12 determine whether **F** *is a gradient field. If so, find a potential function for* **F** *using the method of Example 19.7. Assume the domain is* $\mathbf{R}^2$ *unless otherwise specified.*

1. $\mathbf{F}(x, y) = \langle 2x + 3y, 3x - 2y + 1 \rangle$
2. $\mathbf{F}(x, y) = (3x^2 - 4xy)\mathbf{i} + (6y^2 - 2x^2)\mathbf{j}$
3. $\mathbf{F}(x, y) = \langle x^2 - y^2, 2xy \rangle$
4. $\mathbf{F}(x, y) = \langle y \sin x, \cos x + y^2 \rangle$
5. $\mathbf{F}(x, y) = x^2(3y^2 + 2)\mathbf{i} + 2(x^3y - 1)\mathbf{j}$
6. $\mathbf{F}(x, y) = \left\langle \dfrac{xy}{\sqrt{x^2 + 1}}, \sqrt{x^2 + 1} \right\rangle$
7. $\mathbf{F}(x, y) = \langle xe^{xy}(xy + 2), x^3e^{xy} \rangle$
8. $\mathbf{F}(x, y) = \dfrac{x\mathbf{i} + y\mathbf{j}}{x^2 + y^2}$; in $\mathbf{R}^2 - \{(0, 0)\}$
9. $\mathbf{F}(x, y) = \left(\dfrac{y}{\sqrt{x^2 + y^2}}\right)\mathbf{i} + \left(\tan^{-1} \dfrac{y}{x}\right)\mathbf{j}$.
10. $\mathbf{F}(x, y) = \langle \sin^2 xy, 1 - \cos xy \rangle$
11. $\mathbf{F}(x, y) = \langle y^2 \cos xy, xy \cos xy + \sin xy - 2y \rangle$
12. $\mathbf{F}(x, y) = (2xy \tan^2 xy + 2 \tan xy + 1)\mathbf{i} + (2x^2 \sec^2 xy + 4y)\mathbf{j}$; in $\{(x, y): 0 < x < \infty, -\frac{\pi}{2x} < y < \frac{\pi}{2x}\}$

In Exercises 13–20 use results from Exercises 1–12 to evaluate the given integrals.

13. $\displaystyle\int_{(-1,-2)}^{(3,5)} (2x + 3y)\,dx + (3x - 2y + 1)\,dy$
14. $\displaystyle\int_{(0,2)}^{(1,4)} (3x^2 - 4xy)\,dx + (6y^2 - 2x^2)\,dy$
15. $\displaystyle\int_{(-1,2)}^{(3,1)} (3x^2y^2 + 2x^2)\,dx + (2x^3y - 2)\,dy$
16. $\displaystyle\int_{(0,0)}^{(2,5)} \frac{xy\,dx}{\sqrt{x^2 + 1}} + \sqrt{x^2 + 1}\,dy$
17. $\displaystyle\int_{(1,0)}^{(2,\ln 2)} xe^{xy}(xy + 2)\,dx + x^3e^{xy}\,dy$
18. $\displaystyle\int_{(1,1)}^{(3,-4)} \frac{x\,dx + y\,dy}{x^2 + y^2}$
19. $\displaystyle\int_{(1,0)}^{(5\pi/9,3/2)} \mathbf{F} \cdot d\mathbf{r}$, where **F** is given in Exercise 11
20. $\displaystyle\int_{(0,0)}^{(\pi/2,1/2)} \mathbf{F} \cdot d\mathbf{r}$, where **F** is given in Exercise 12

In Exercises 21–24 show that **F** *is independent of the path in the given region, and use the method shown in Example 19.8 to find a potential function for* **F**.

21. $\mathbf{F}(x, y) = \langle 2x - 2y^2, 3y^2 - 4xy \rangle$; all of $\mathbf{R}^2$
22. $\mathbf{F}(x, y) = \left\langle \dfrac{1}{x - y}, \dfrac{-1}{x - y} \right\rangle$; $\{(x, y): x > y\}$ [*Hint:* Use the straight-line path from $(1, 0)$ to (x, y).]
23. $\mathbf{F}(x, y) = \langle \sin y + y \sin x, x \cos y - \cos x \rangle$; all of $\mathbf{R}^2$
24. $\mathbf{F}(x, y) = \langle \ln \sqrt{x^2 + y^2}, \tan^{-1} \frac{x}{y} \rangle$; $\mathbf{R}^2 - \{(0, 0)\}$. [*Hint:* Use $C_1 \cup C_2$, where C_1 is the vertical line segment from $(1, 0)$ to $(1, y)$ and C_2 is the horizontal line segment from $(1, y)$ to (x, y).]

In Exercises 25–28 prove that the given integral is independent of the path and then evaluate it in two ways: (a) using Theorem 19.2, where C is any piecewise smooth curve from A to B, and (b) integrating along the specified curve.

25. $\int_C (2x - 2y)\,dx + (6y - 2x)\,dy$ from $A = (-1, 2)$ to $B = (3, -2)$; C is the straight line segment from A to B.

26. $\displaystyle\int_C \frac{x\,dx + y\,dy}{\sqrt{x^2 + y^2}}$ from $A = (-3, 4)$ to $B = (5, 12)$, where C does not pass through the origin; C is the horizontal line segment from A to $(5, 4)$ followed by the vertical line segment from $(5, 4)$ to B.

27. $\displaystyle\int_C \frac{-y\,dx + x\,dy}{x^2 + y^2}$ from $A = (\sqrt{6}, -\sqrt{6})$ to $B = (3, \sqrt{3})$, where C does not pass through the origin; C is the smaller arc of the circle $x^2 + y^2 = 12$ from A to B.

28. $\int_C e^{xy}(y \cos x - \sin x)\,dx + xe^{xy} \cos x\,dy$ from $A = (0, 4)$ to $B = (\frac{\pi}{3}, 0)$; C is the vertical line segment from A to $(0, 0)$ followed by the horizontal line segment from $(0, 0)$ to B.

29. Evaluate the integral in Example 19.5 by integrating along each of the following paths:
(a) C is the line segment from A to B.
(b) C is the horizontal line segment from A to $(3, -2)$ followed by the vertical line segment from $(3, -2)$ to B.
(c) C is the arc of the parabola $y = 2x^2 - 4x$ from A to B.

30. Evaluate the integral in Example 19.6 by integrating along each of the component curves of C.

31. Show that the force field

$$\mathbf{F}(x, y) = (2x - 3y)\mathbf{i} + 3(y^2 - x)\mathbf{j}$$

is conservative throughout $\mathbf{R}^2$, and find the work done by $\mathbf{F}$ on a particle moving from $(2, 1)$ to $(5, -3)$.

32. Show that the force field

$$\mathbf{F}(x, y) = \tfrac{y}{x}\mathbf{i} + (\ln x - \tfrac{2}{y})\mathbf{j}$$

is conservative in the region

$$R = \{(x, y): x > 0,\ y > 0\}$$

and find the work done by $\mathbf{F}$ on a particle as it goes once around the ellipse

$$4x^2 + y^2 - 16x - 6y + 9 = 0$$

33. Evaluate the integral

$$\int_C \cos x \cos y\,dx - \sin x \sin y\,dy$$

along each path in the accompanying figure. (*Hint:* There is an easy way and a hard way to do this.)

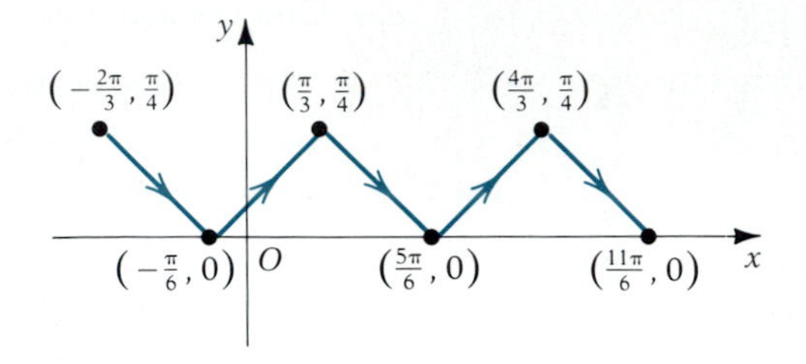

(a)

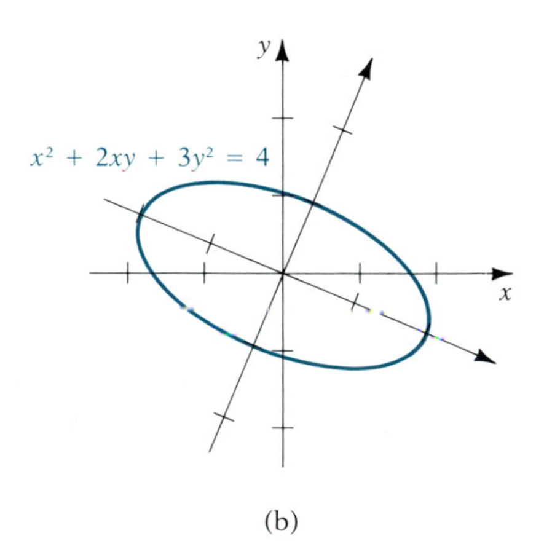

(b)

In Exercises 34–37 show that the given expression is an exact differential, and find ϕ so that it equals $d\phi$.

34. $(2x - 3y)\,dx + (8y - 3x)\,dy$

35. $(2x + y^2 \cos xy)\,dx + (\sin xy + xy \cos xy)\,dy$

36. $\displaystyle\frac{2xy}{(1 + x^2)^2}\,dx + \frac{x^2 y}{1 + x^2}\,dy$

37. $(\ln \cosh y)\,dx + (x \tanh y - 1)\,dy$

38. Using the proof of Theorem 19.2 for a smooth curve C, show that the theorem is also true for any piecewise smooth curve.

39. By inspection determine a potential function for $\mathbf{F}(x, y, z) = \langle 2x, 2y, 2z \rangle$ and use the result to evaluate

$$\int_{(1,-1,2)}^{(3,2,-1)} 2x\,dx + 2y\,dy + 2z\,dz$$

B

40. Complete the proof of Theorem 19.3 by showing that $\phi_y(x, y) = g(x, y)$, taking C_1 as any path from (a, b) to (x, y) and C_2 as the vertical line segment from (x, y) to $(x, y + k)$ for $|k|$ sufficiently small.

41. Using equations (19.9) show that

$$\mathbf{F}(x, y, z) = \langle 2xyz - 3z^2, x^2z + 8yz^3, x^2y - 6xz + 12y^2z^2\rangle$$

is a gradient field in $\mathbf{R}^3$, and follow a procedure similar to that used in Example 19.7 to find a potential function for $\mathbf{F}$.

42. Prove that the gravitational field

$$\mathbf{F} = -\frac{C}{|\mathbf{r}|^2}\mathbf{u}$$

of a point mass is conservative. Here $\mathbf{r}$ is the vector $\overrightarrow{OP}$ from the point located at O to a point P in space, and $\mathbf{u}$ is a unit vector in the direction of $\mathbf{r}$. Find the work done by $\mathbf{F}$ on an object as it moves from A to B, where $|\overrightarrow{OA}| = r_1$ and $|\overrightarrow{OB}| = r_2$.

43. Prove the law of conservation of energy as follows. Let $\mathbf{F}$ be a conservative force field in an open region R. The potential energy p due to $\mathbf{F}$ is defined by $\mathbf{F} = -\nabla p$. Show that for any piecewise smooth curve C in R from A to B, $\int_A^B \mathbf{F}\cdot d\mathbf{r} = p(A) - p(B)$. Use the result of Exercise 47 in Exercise Set 19.1 to conclude that $K(A) + p(A) = K(B) + p(B)$, where K is the kinetic energy.

19.3 GREEN'S THEOREM

In this section we take up a theorem that has far-reaching consequences in both pure and applied mathematics. The theorem is named in honor of the English mathematician and physicist George Green (1793–1841). It was contained in an 1828 essay by Green on electricity and magnetism that was privately circulated. The result was little noticed until it was found in 1846 by Lord Kelvin, who republished it for general circulation. Its great importance was then recognized by the scientific community. The Russian mathematician Michel Ostrogradski (1801–1861) discovered the same result independently, and so in Russia it is known as Ostrogradski's theorem.

The basic result is that under certain restrictions on the functions f and g and on the region R and the closed curve C that forms the boundary of R, the following equality holds true:

$$\int_C f\,dx + g\,dy = \iint_R \left(\frac{\partial g}{\partial x} - \frac{\partial f}{\partial y}\right) dA$$

Actually, there are several versions of the theorem, differing from one another in the restrictions placed on f, g, R, and C. We state the theorem for the case in which R is what we have called a regular region, but we prove it only for a special case. Later we will discuss how the theorem can be extended to certain multiply connected regions.

Recall that in $\mathbf{R}^2$ a regular region is a closed and bounded region with a boundary that is a simple closed curve made up of finitely many smooth arcs joined end to end. A regular region is necessarily simply connected.

THEOREM 19.6 **GREEN'S THEOREM IN THE PLANE**

Let R be a regular region in $\mathbf{R}^2$ with boundary C oriented in the counterclockwise direction. Then if f and g have continuous first partial derivatives in R,

$$\int_C f(x, y)\,dx + g(x, y)\,dy = \iint_R \left(\frac{\partial g}{\partial x} - \frac{\partial f}{\partial y}\right) dA \tag{19.10}$$ ■

We will prove Green's theorem only for a region R that is simultaneously of type I and type II, as defined in Section 18.2. We call a region of this type a **simple region.** In Exercise 37 you will be asked to show that when R can be divided by horizontal and vertical line segments into a finite number of simple regions, Green's theorem is true for R. Figure 19.12(a) shows an example of a simple region, and Figure 19.12(b) shows a region that has been divided into three simple regions.

Partial Proof of Green's Theorem

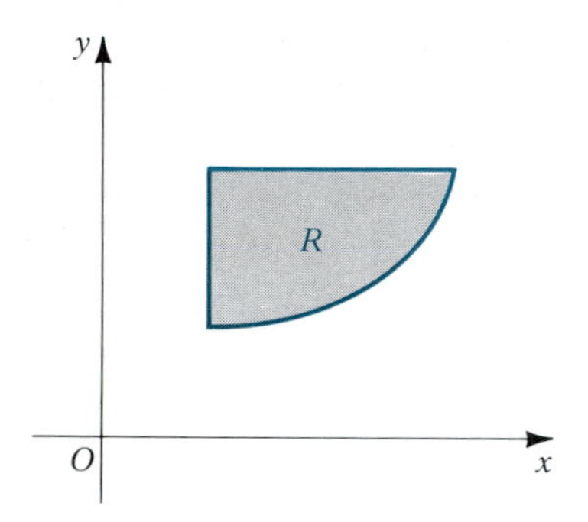

R is a simple region

(a)

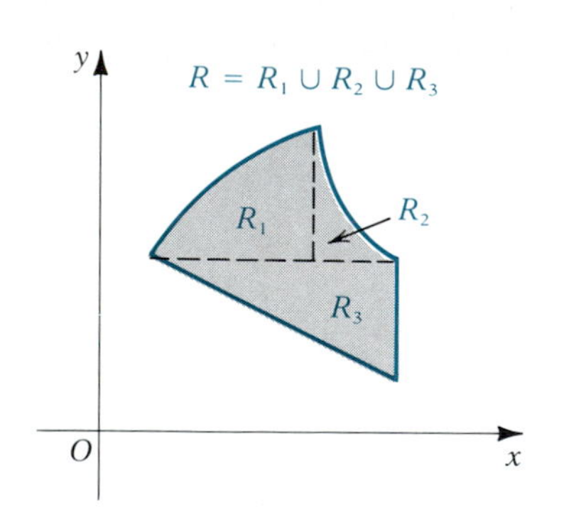

R_1, R_2, and R_3 are simple.

(b)

FIGURE 19.12

Let R be a simple region with boundary C oriented counterclockwise. Figure 19.13(a) shows such a region. In part b it is viewed as type I and in part c as type II. The proof of equation (19.10) is accomplished by showing that

$$\int_C f(x, y)\,dx = -\iint_R \frac{\partial f}{\partial y}\,dA \tag{19.11}$$

and

$$\int_C g(x, y)\,dy = \iint_R \frac{\partial g}{\partial x}\,dA \tag{19.12}$$

By adding the corresponding sides of these two equations, we get the desired result.

To prove equation (19.11) we use Figure 19.13(b). There we have indicated the lower bounding curve as C_1, with rectangular equation $y = y_1(x)$, and the upper bounding curve as C_2, with rectangular equation $y = y_2(x)$. Each of these can be parameterized with x as the parameter, where, for C_1, x varies from a to b and, for C_2, x varies from b to a, so that the total boundary $C = C_1 \cup C_2$ is oriented in the counterclockwise direction. We then have

$$-\iint_R \frac{\partial f}{\partial y}\,dA = -\int_a^b \left[\int_{y_1(x)}^{y_2(x)} \frac{\partial f}{\partial y}\,dy\right] dx$$

$$= -\int_a^b [f(x, y_2(x)) - f(x, y_1(x))]\,dx \qquad \text{By the fundamental theorem of calculus}$$

$$= \int_a^b f(x, y_1(x))\,dx - \int_a^b f(x, y_2(x))\,dx$$

$$= \int_a^b f(x, y_1(x))\,dx + \int_b^a f(x, y_2(x))\,dx$$

$$= \int_{C_1} f(x, y)\,dx + \int_{C_2} f(x, y)\,dx$$

$$= \int_C f(x, y)\,dx$$

This proves equation (19.11). The proof of equation (19.12) is carried out in a similar manner, using Figure 19.13(c), and you will be asked to do this in Exercise 36. Equation (19.10) follows, as previously indicated, by adding the results of equations (19.11) and (19.12). ■

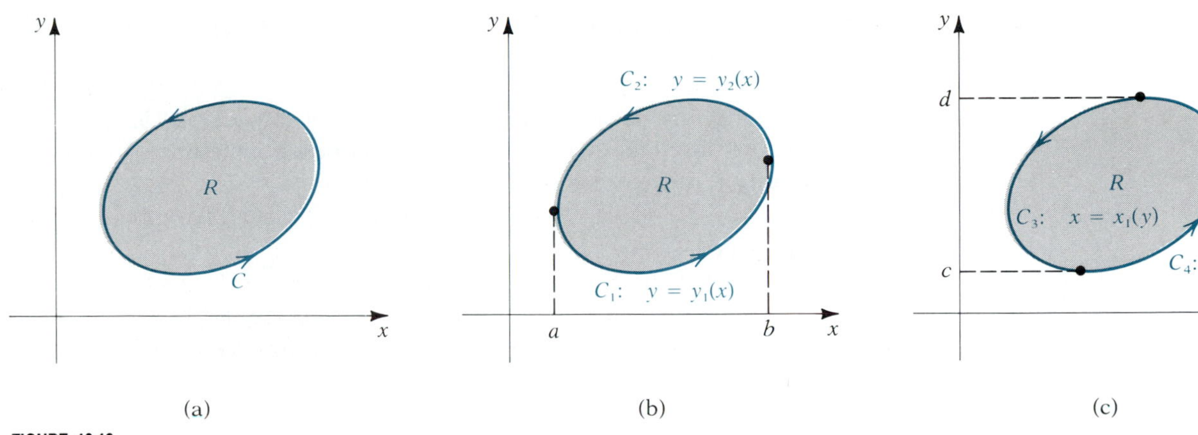

FIGURE 19.13

Remarks

1. There is a version of Green's theorem for $\mathbf{R}^3$ that we will consider in Section 19.6. This explains the phrase "in the plane" in the name of the theorem as stated here.
2. We can view equation (19.10) in either of two ways: (a) as a means of evaluating a line integral using a double integral, or (b) as a means of evaluating a double integral using a line integral. Both points of view are important, and we illustrate them in the examples that follow.

EXAMPLE 19.9 Use Green's theorem to evaluate the line integral

$$\int_C (e^{-x^2} + xy^2)\,dx + (x^3 + \sqrt{1 + y^3})\,dy$$

where C is the path shown in Figure 19.14.

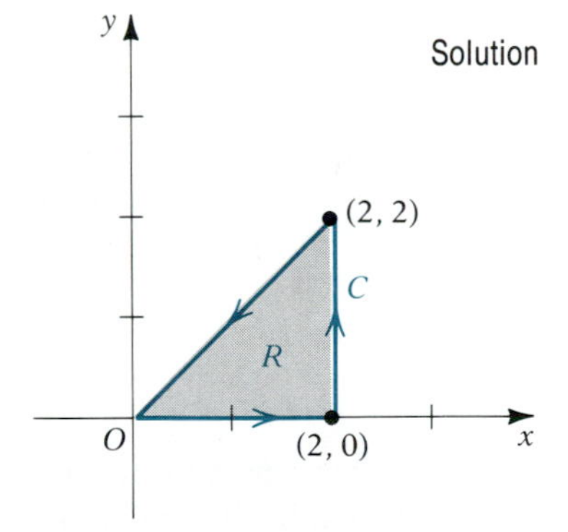

FIGURE 19.14

Solution Observe first that evaluating the integral by parameterizing the component curves of C would lead to integrals that cannot be evaluated by elementary means. (Try it.) To apply Green's theorem we take

$$f(x, y) = e^{-x^2} + xy^2 \quad \text{and} \quad g(x, y) = x^3 + \sqrt{1 + y^3}$$

The region R is the triangular region bounded by C. Since

$$\frac{\partial g}{\partial x} = 3x^2 \quad \text{and} \quad \frac{\partial f}{\partial y} = 2xy$$

we have, by Green's theorem,

$$\begin{aligned}
\int_C (e^{-x^2} + xy^2)\,dx + (x^3 + \sqrt{1 + y^3})\,dy &= \iint_R (3x^2 - 2xy)\,dA \\
&= \int_0^2 \int_0^x (3x^2 - 2xy)\,dy\,dx \\
&= \int_0^2 [3x^2y - xy^2]_0^x\,dx \\
&= \int_0^2 2x^3\,dx \\
&= 8
\end{aligned}$$

■

In this case Green's theorem was useful in evaluating a difficult line integral by means of a double integral. In the next example the situation is reversed.

EXAMPLE 19.10 Use Green's theorem to evaluate the integral $\iint_R y^2\,dA$, where R is the elliptical region shown in Figure 19.15.

Solution Direct evaluation by iterated integrals, though not impossible, is somewhat tedious. To use Green's theorem we must find functions f and g that satisfy the continuity requirements over R for which

$$\iint_R y^2\,dA = \iint_R \left(\frac{\partial g}{\partial x} - \frac{\partial f}{\partial y}\right) dA$$

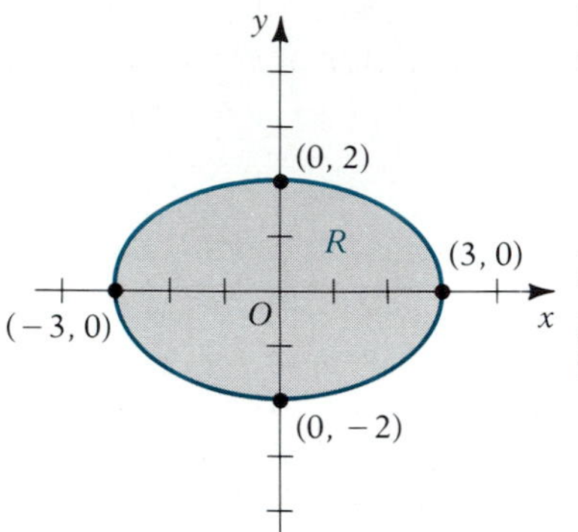

FIGURE 19.15

Many choices are possible, but a particularly simple one is $g(x, y) = xy^2$ and $f(x, y) = 0$. A parameterization of C that gives a counterclockwise orientation is $x = 3\cos t$, $y = 2\sin t$, where $0 \le t \le 2\pi$. So we have, by equation (19.10),

$$\begin{aligned}\iint_R y^2\,dA &= \iint_R \left(\frac{\partial g}{\partial x} - \frac{\partial f}{\partial y}\right) dA = \int_C f(x, y)\,dx + g(x, y)\,dy\\ &= \int_C xy^2\,dy = \int_0^{2\pi} (3\cos t)(4\sin^2 t)(2\cos t\,dt)\\ &= 6\int_0^{2\pi} \sin^2 2t\,dt = 3\int_0^{2\pi} (1 - \cos 4t)\,dt\\ &= 6\pi\end{aligned}$$

■

The idea used in this last example is the basis for the following theorem.

THEOREM 19.7 If R is a regular region of $\mathbf{R}^2$ with boundary C oriented in the counterclockwise direction, then the area A of R is given by each of the following:

$$\text{(a)}\quad A = \int_C x\,dy \qquad \text{(b)}\quad A = -\int_C y\,dx \qquad \text{(c)}\quad A = \tfrac{1}{2}\int_C x\,dy - y\,dx \tag{19.13}$$

Proof For form a take $g(x, y) = x$ and $f(x, y) = 0$. Green's theorem then gives $\int_C x\,dy = \iint_R 1\,dA = A$. For form b take $f(x, y) = y$ and $g(x, y) = 0$ and again apply Green's theorem. For form c add the results of forms a and b and divide by 2. ■

EXAMPLE 19.11 Find the area of the region R enclosed by the ellipse

$$\frac{x^2}{a^2} + \frac{y^2}{b^2} = 1$$

Solution Parametric equations of the ellipse that give it the correct orientation are $x = a\cos t$, $y = b\sin t$, where $0 \le t \le 2\pi$. Although each of equations (19.13) will give the same result, a little reflection should convince you that form c is the easiest to apply in this case. Using it, we get

$$\begin{aligned}A = \tfrac{1}{2}\int_C x\,dy - y\,dx &= \tfrac{1}{2}\int_0^{2\pi} [(a\cos t)(b\cos t\,dt) - (b\sin t)(-a\sin t\,dt)]\\ &= \tfrac{1}{2}\int_0^{2\pi} ab(\cos^2 t + \sin^2 t)\,dt\\ &= \pi ab\end{aligned}$$

■

We now discuss how to extend Green's theorem to certain multiply connected regions. In particular, suppose R is a closed and bounded region with a boundary that consists of finitely many simple closed curves that are sectionally smooth and do not intersect one another. Let each of these boundary curves be oriented so that when it is traversed in its positive di-

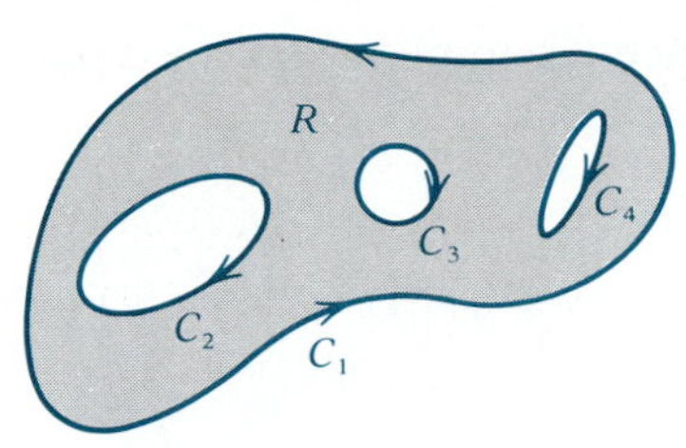

FIGURE 19.16 $C = C_1 \cup C_2 \cup C_3 \cup C_4$ is positively oriented with respect to R

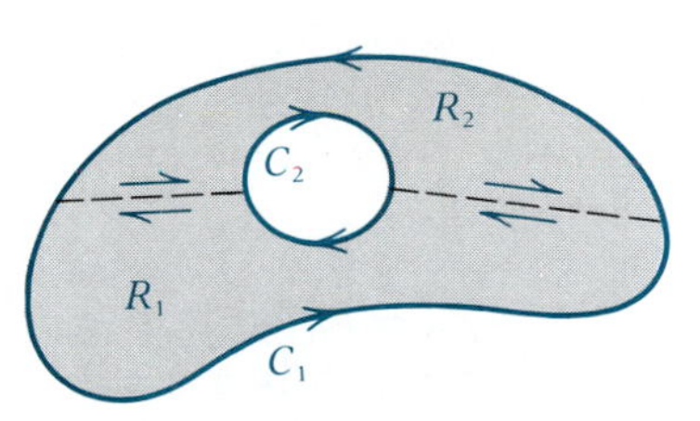

FIGURE 19.17

rection, the region R lies on the *left*, and let C be the union of all of these boundary curves. When the component curves of C are oriented according to this left-hand rule, we say that C is *positively oriented with respect to* R. We show such a region in Figure 19.16.

We now assert that Green's theorem holds for a region of the type described; that is, if f and g have continuous first partial derivatives in an open set that contains R, then

$$\int_C f\,dx + g\,dy = \iint_R \left(\frac{\partial g}{\partial x} - \frac{\partial f}{\partial y}\right) dA$$

We indicate the idea of the proof for the simple case indicated in Figure 19.17, in which R has outer boundary C_1 and inner boundary C_2. We divide R into two subregions R_1 and R_2 by line segments as shown. Now R_1 and R_2 are regular regions, and so Green's theorem applies for each of them. The double integrals over R_1 and R_2 add to give the double integrals over R. The line integral over the boundary of R_1 involves integrating over each of the line segments we have introduced from right to left, whereas the line integral over the boundary of R_2 involves integrating over these lines from left to right. Thus, when the two line integrals are added, the contributions from these line segments cancel one another, and we are left with the line integral over $C = C_1 \cup C_2$.

A very useful consequence of Green's theorem for multiply connected regions is the following.

THEOREM 19.8 Let C_1 and C_2 be any two nonintersecting piecewise smooth simple closed curves, and let R be the closed annular region bounded by C_1 and C_2. Then if f and g have continuous partial derivatives and

$$\frac{\partial f}{\partial y} = \frac{\partial g}{\partial x}$$

throughout some open set that contains R,

$$\int_{C_1} f\,dx + g\,dy = \int_{C_2} f\,dx + g\,dy$$

provided C_1 and C_2 are similarly oriented.

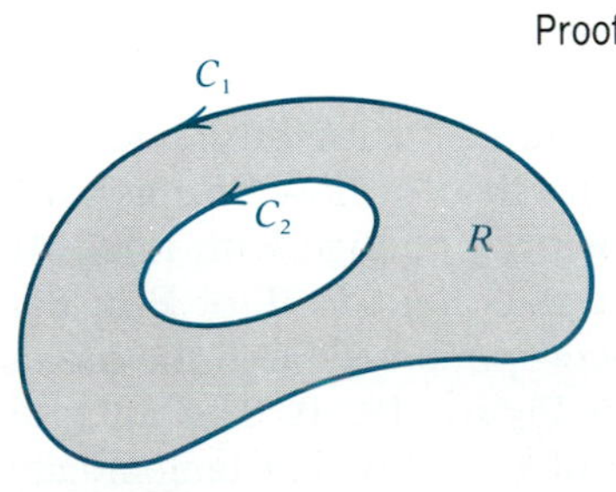

FIGURE 19.18

Proof Suppose C_1 and C_2 are both oriented in the counterclockwise direction and C_2 is interior to C_1 as shown in Figure 19.18. Then Green's theorem for the multiply connected region R applies, where the boundary of R is $C = C_1 \cup (-C_2)$, since this is positively oriented with respect to R. Thus,

$$\int_C f\,dx + g\,dy = \iint_R \left(\frac{\partial g}{\partial x} - \frac{\partial f}{\partial y}\right) dA = 0$$

But

$$\begin{aligned}\int_C f\,dx + g\,dy &= \int_{C_1} f\,dx + g\,dy + \int_{-C_2} f\,dx + g\,dy \\ &= \int_{C_1} f\,dx + g\,dy - \int_{C_2} f\,dx + g\,dy\end{aligned}$$

Since this equals 0, the result follows. If C_1 and C_2 are oriented in the clockwise direction, we let $C = (-C_1) \cup C_2$, and the reasoning follows the same lines. ■

This theorem sometimes enables us to replace a complicated closed path with a simpler one. The next example illustrates this.

EXAMPLE 19.12 Evaluate the integral

$$\int_C \frac{-y}{x^2+y^2}\,dx + \frac{x}{x^2+y^2}\,dy$$

where C is the ellipse $x^2 - 2xy + 3y^2 = 4$.

Solution The ellipse is pictured in Figure 19.19. Let

$$f(x, y) = \frac{-y}{x^2+y^2} \quad \text{and} \quad g(x, y) = \frac{x}{x^2+y^2}$$

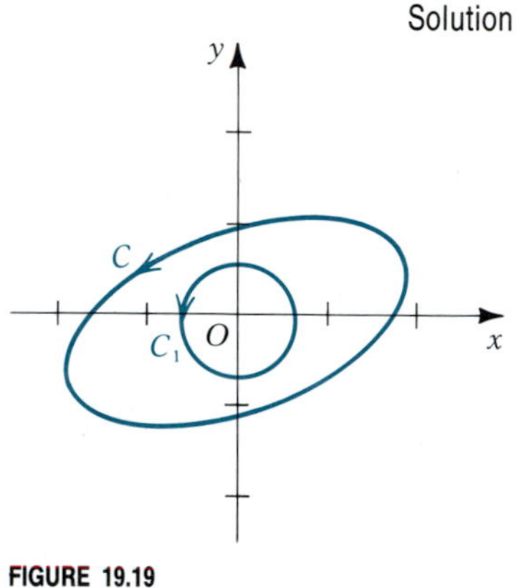

FIGURE 19.19

It is easily verified that for all points except $(0, 0)$, $\partial g/\partial x = \partial f/\partial y$. If it were not for the exceptional point, we could conclude that $\langle f, g\rangle$ is a gradient field (why?) and therefore that the integral around the closed path C is 0. But this reasoning breaks down, since f and g fail even to exist at the origin.

It is rather difficult to parameterize this ellipse, so we introduce a new curve C_1 inside C and with the origin in its interior. Because of its simplicity we might as well take C_1 to be a circle centered at the origin. Let its radius be a. It does not matter what a is so long as it is small enough that C_1 lies inside C. The conditions of Theorem 19.8 are now met, and so

$$\int_C \frac{-y}{x^2+y^2}\,dx + \frac{x}{x^2+y^2}\,dy = \int_{C_1} \frac{-y}{x^2+y^2}\,dx + \frac{x}{x^2+y^2}\,dy$$

A parameterization of C_1 is $x = a\cos t$, $y = a\sin t$, where $0 \le t \le 2\pi$. A straightforward calculation shows that with this parameterization, the integral on the right becomes $\int_0^{2\pi} dt = 2\pi$. The fact that the integral around the closed path is nonzero shows again that $\langle f, g\rangle$ is not a gradient field. ■

We conclude this section with a brief discussion of how Green's theorem can be used to determine certain characteristics of the flow of a fluid. We consider here a two-dimensional "laminar" flow (you can think of a thin sheet of water flowing over a flat surface), and in Sections 19.5 and 19.6 we will extend the ideas to three dimensions.

Let $\mathbf{F} = \langle f, g\rangle$ be the velocity field of the fluid, and let a region R bounded by the curve C, as in Green's theorem, be in the field of flow. We assume the flow is in *steady state*, meaning that $\mathbf{F}$ does not change with time. At each point P on C, we wish to consider the component $\mathbf{F}\cdot\mathbf{T}$ of $\mathbf{F}$ in the direction of the unit tangent vector $\mathbf{T}$ and the component $\mathbf{F}\cdot\mathbf{n}$ in the direction of the *outer* unit normal $\mathbf{n}$, as shown in Figure 19.20. The integral $\int_C \mathbf{F}\cdot\mathbf{T}\,ds$ gives the amount of fluid per unit of time flowing tangentially around C, whereas $\int_C \mathbf{F}\cdot\mathbf{n}\,ds$ gives the amount of fluid per unit of time flowing out of R orthogonally across C. These integrals are given names as follows:

FIGURE 19.20

$\int_C \mathbf{F}\cdot\mathbf{T}\,ds$, the **circulation** of F around C

$\int_C \mathbf{F}\cdot\mathbf{n}\,ds$, the **flux** of F through C

Although the circulation and flux can be evaluated directly as line integrals, we wish to show how Green's theorem can be used to find them. For the circulation this is immediate, since $\mathbf{F} \cdot \mathbf{T}\, ds = \mathbf{F} \cdot d\mathbf{r} = f\,dx + g\,dy$, so that Green's theorem gives

$$\text{circulation of } \mathbf{F} \text{ around } C = \int_C \mathbf{F} \cdot \mathbf{T}\, ds = \iint_R \left(\frac{\partial g}{\partial x} - \frac{\partial f}{\partial y}\right) dA \tag{19.14}$$

To obtain a similar representation of the flux we can show (see Exercise 33) that the outer unit normal vector $\mathbf{n}$ is given by

$$\mathbf{n} = \left\langle \frac{dy}{ds}, \frac{-dx}{ds} \right\rangle$$

It is easy to see, in fact, that $\mathbf{n}$ has length 1 and that it is orthogonal to $\mathbf{T}$. We can therefore write

$$\mathbf{F} \cdot \mathbf{n}\, ds = \langle f, g \rangle \cdot \left\langle \frac{dy}{ds}, \frac{-dx}{ds} \right\rangle ds = -g\,dx + f\,dy$$

Now we can apply Green's theorem, replacing f by $-g$ and g by f in equation (19.10), to get

$$\text{flux of } \mathbf{F} \text{ through } C = \int_C \mathbf{F} \cdot \mathbf{n}\, ds = \iint_R \left(\frac{\partial f}{\partial x} + \frac{\partial g}{\partial y}\right) dA \tag{19.15}$$

At each point P of R, the integrands of the double integrals in equations (19.14) and (19.15) are called, respectively, the **rotation of F at *P*** and the **divergence of F at *P*.** These are abbreviated *rot* $\mathbf{F}$ and *div* $\mathbf{F}$, respectively. So we have

$$\text{rot } \mathbf{F} = \frac{\partial g}{\partial x} - \frac{\partial f}{\partial y}$$

$$\text{div } \mathbf{F} = \frac{\partial f}{\partial x} + \frac{\partial g}{\partial y}$$

Each of these is a scalar function. Some authors use *scalar curl* or *two-dimensional curl* for rotation. (We will study the curl of a vector field in Section 19.6 and see how it is related to rotation.) With these definitions we can rewrite equations (19.14) and (19.15) as

$$\int_C \mathbf{F} \cdot \mathbf{T}\, ds = \iint_R (\text{rot } \mathbf{F})\, dA \tag{19.16}$$

$$\int_C \mathbf{F} \cdot \mathbf{n}\, ds = \iint_R (\text{div } \mathbf{F})\, dA \tag{19.17}$$

To get an intuitive understanding of the physical significance of the rotation and circulation of $\mathbf{F}$ at a point P in the field of flow, consider a small circle C_ε, of radius ε, centered at P, and denote the region it encloses by R_ε. Then by equation (19.16) the circulation around C_ε is

$$\int_{C_\varepsilon} \mathbf{F} \cdot \mathbf{T}\, ds = \iint_{R_\varepsilon} (\text{rot } \mathbf{F})\, dA$$

Now there is a mean value theorem for double integrals analogous to the one for single integrals that enables us to write (when rot **F** is continuous)

$$\iint_{R_\varepsilon} (\text{rot } \mathbf{F})\, dA = [\text{rot } \mathbf{F}(Q)](\text{area of } R_\varepsilon)$$

where Q is some point in R_ε. Solving for rot $\mathbf{F}(Q)$, we have

$$\text{rot } \mathbf{F}(Q) = \frac{\iint_{R_\varepsilon} (\text{rot } \mathbf{F})\, dA}{\text{area of } R_\varepsilon} = \frac{\int_{C_\varepsilon} \mathbf{F} \cdot \mathbf{T}\, ds}{\text{area of } R_\varepsilon}$$

Now if we let $\varepsilon \to 0$, then $Q \to P$, and we get

$$\text{rot } \mathbf{F}(P) = \lim_{\varepsilon \to 0} \frac{\text{circulation of } \mathbf{F} \text{ around } C_\varepsilon}{\text{area of } R_\varepsilon}$$

So rot **F** at a point P is the *circulation per unit area at P*. If rot $\mathbf{F} \neq 0$ at P, the fluid forms a whirlpool, or *vortex*, at P. If rot $\mathbf{F} = 0$ for all points of a region, then **F** is said to be **irrotational** in that region.

In an exactly analogous way we can show that

$$\text{div } \mathbf{F}(P) = \lim_{\varepsilon \to 0} \frac{\text{flux of } \mathbf{F} \text{ through } C_\varepsilon}{\text{area of } R_\varepsilon}$$

so that div **F** at a point P is the *flux per unit area* at that point. If div $\mathbf{F}(P) > 0$, fluid is emerging from P, and we say that P is a **source.** If div $\mathbf{F}(P) < 0$, fluid is flowing into P, and we say that P is a **sink.** If div $\mathbf{F} = 0$ for all points of a region, we say the fluid is **incompressible.**

Remark Although the concepts of circulation, flux, rotation, and divergence were introduced for fluid flow, the terms are frequently used for other types of vector fields as well.

EXAMPLE 19.13 Let $\mathbf{F} = -2xy\mathbf{i} + x^2\mathbf{j}$ be the velocity field of a two-dimensional fluid flow, and let R be the region enclosed by the triangle C that has vertices $(0, 0)$, $(2, 0)$, and $(2, 4)$ oriented counterclockwise. Use Green's theorem to find the circulation of **F** around C and the flux of **F** through C.

Solution The region R is shown in Figure 19.21. First we calculate rot **F** and div **F**. We write $f(x, y) = -2xy$ and $g(x, y) = x^2$, so that $\mathbf{F} = \langle f, g \rangle$. So

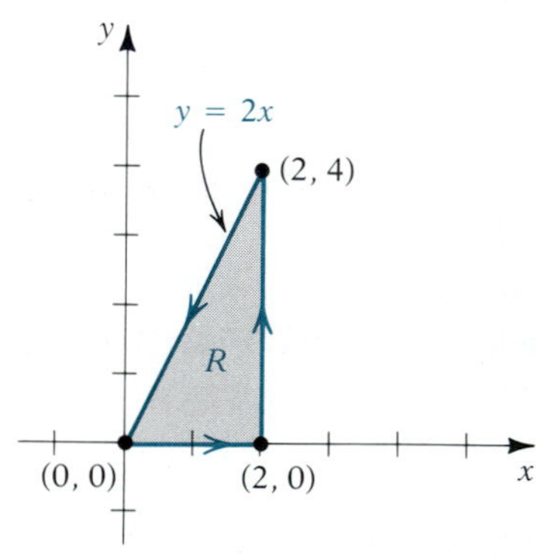

FIGURE 19.21

$$\text{rot } \mathbf{F} = \frac{\partial g}{\partial x} - \frac{\partial f}{\partial y} = 2x - (-2x) = 4x$$

and

$$\text{div } \mathbf{F} = \frac{\partial f}{\partial x} + \frac{\partial g}{\partial y} = -2y + 0 = -2y$$

Thus, by equations (19.16) and (19.17),

$$\begin{aligned}\text{circulation of } \mathbf{F} \text{ around } C &= \int_C \mathbf{F} \cdot \mathbf{T}\, ds = \iint_R (\text{rot } \mathbf{F})\, dA \\ &= \int_0^2 \int_0^{2x} 4x\, dy\, dx = \int_0^2 [4xy]_0^{2x}\, dx \\ &= \int_0^2 8x^2\, dx = \tfrac{64}{3}\end{aligned}$$

$$\text{flux of } \mathbf{F} \text{ through } C = \int_C \mathbf{F}\cdot\mathbf{n}\,ds = \iint_R (\operatorname{div}\mathbf{F})\,dA$$
$$= \int_0^2 \int_0^{2x} (-2y)\,dy\,dx = -\int_0^2 [y^2]_0^{2x}\,dx$$
$$= -\int_0^2 4x^2\,dx = -\tfrac{32}{3}$$

Since the circulation is positive, the net flow of fluid around C is in the counterclockwise direction, and since the flux is negative, there is a net inflow of fluid into R. ■

Note Since we are dealing with two-dimensional flow, the "amount" of fluid is given by area rather than volume. Thus, for example, if $|F|$ is in centimeters per second and distance is in centimeters, both circulation and flux will be in square centimeters per second.

EXERCISE SET 19.3

A

In Exercises 1–14 use Green's theorem to evaluate the line integral. In each case C has a counterclockwise orientation. For Exercises 1–4 take C to be the rectangle with vertices (0, 0), (2, 0), (2, 1), and (0, 1).

1. $\int_C (x^2y - 2y^2)\,dx + (x^3 - 2y^2)\,dy$
2. $\int_C (\ln\sqrt{x+1} + xy)\,dx + (x^2y + e^{y^2})\,dy$
3. $\int_C y\cos\dfrac{\pi x}{2}\,dx - x\sin\dfrac{\pi y}{2}\,dy$
4. $\int_C \dfrac{x}{1+y^2}\,dx + \dfrac{y}{1+x^2}\,dy$
5. $\int_C e^{x+2y}\,dx$, C is the triangle with vertices $(0, 0)$, $(1, 1)$, and $(0, 1)$.
6. $\int_C (\tan^{-1} x)\,dy$, C is the boundary of the region between $y = 2 - x^2$ and $y = x$.
7. $\int_C y\sin 2x\,dx + \sin^2 x\,dy$, C is the ellipse $2x^2 + 3y^2 = 6$.
8. $\int_C (x^2 + y^2)\,dx + (x^2 - y^2)\,dy$, C is the boundary of the region determined by $y = x^2$, $y = 0$, and $x = 1$.
9. $\int_C \left(x^2y + \dfrac{y^3}{3}\right)dx + (2x - y^5)\,dy$, C is the circle $x^2 + y^2 = 4$.
10. $\int_C (e^x + y^3)\,dx + (x^2 - \sqrt{y})\,dy$, C is the boundary of the region determined by $y = \sqrt{x}$, $y = 0$, and $x - y = 2$.
11. $\int_C y^2\,dx + x^2\,dy$, C is the boundary of the region determined by $x = -\sqrt{9 - y^2}$, $x + y = 3$, and $y = 0$.
12. $\int_C (2y^3 - 3x^2)\,dx + (2x^3 + 5y^2)\,dy$, C is the boundary of the region determined by $y = \sqrt{4 - x^2}$ and $y = 0$.
13. $\int_C y^2\,dx + 3xy\,dy$, C is the cardioid $r = 1 + \cos\theta$.
14. $\int_C \sqrt{x^2 + 1}\,dx + x(1 + y)\,dy$, C is the boundary of the region outside the circle $r = 1$ and inside the cardioid $r = 2(1 + \cos\theta)$.

In Exercises 15–18 make use of Green's theorem to evaluate each double integral by means of a line integral.

15. $\iint_R x\,dA$, R is the triangle with vertices $(0, 0)$, $(1, 1)$, and $(-1, 2)$.
16. $\iint_R [2(x - 1) - 2y]\,dA$, R is the circle $x^2 + y^2 = 2x$.
17. $\iint_R (x\sqrt{1 - y^2} - 4x^2y)\,dA$, R is the ellipse $x^2 + 4y^2 = 4$.
18. $\iint_R y\,dA$, R is the parallelogram with vertices $(0, 0)$, $(4, 0)$, $(5, 2)$, and $(1, 2)$.

In Exercises 19–25 find the area of the specified region using line integration.

19. Bounded by the parallelogram with vertices $(0, 0)$, $(3, 1)$, $(4, 3)$, and $(1, 2)$
20. Between $y = x^2$ and $y = 2x$
21. Bounded by the triangle with vertices $(0, 2)$, $(1, 1)$, and $(2, 3)$
22. Inside the loop of $x = t^3 - 1$, $y = t^3 - t$, where $-\infty < t < \infty$
23. Inside the four-cusp hypocycloid $x = a\cos^3 t$, $y = a\sin^3 t$, where $0 \le t \le 2\pi$

24. Bounded by $x^{1/2} + y^{1/2} = 1$, $x = 0$, and $y = 0$. (*Hint:* Take $t = x^{1/2}$.)

25. Above the x-axis and under one arch of the cycloid $x = a(t - \sin t)$, $y = a(1 - \cos t)$

26. Let R be the region inside the circle C: $x^2 + y^2 = a^2$, oriented in a counterclockwise direction. For each of the following vector fields find the circulation of $\mathbf{F}$ around C and the flux of $\mathbf{F}$ through C:
(a) $\mathbf{F}(x, y) = x\mathbf{i} + y\mathbf{j}$ (b) $\mathbf{F}(x, y) = -y\mathbf{i} + x\mathbf{j}$

In Exercises 27–30 $\mathbf{F}$ *is the velocity field of a two-dimensional fluid flow and R is the region enclosed by the curve C oriented in a counterclockwise direction. Find the circulation of* $\mathbf{F}$ *around C and the flux of* $\mathbf{F}$ *through C.*

27. $\mathbf{F}(x, y) = \langle x^2 - y^2, 2xy\rangle$; R is the region in the first quadrant bounded by $y = \sqrt{1 - x^2}$, $x = 0$, and $y = 0$.

28. $\mathbf{F}(x, y) = \langle xy^2 - 3y, 2x + x^2y\rangle$; C is the circle $x^2 + y^2 = 9$.

29. $\mathbf{F}(x, y) = -y^3\mathbf{i} + x^3\mathbf{j}$; C is the circle $x^2 + y^2 = 4$.

30. $\mathbf{F}(x, y) = (2x^2 - 3y^2)\mathbf{i} + (4y^2 - x^2)\mathbf{j}$; R is the region enclosed by the lines $y = x$, $y = 2 - x$, and $y = 0$.

B

In Exercises 31 and 32 C is the ellipse $x = 3\cos t$, $y = 2\sin t$, *where* $0 \le t \le 2\pi$.

31. Let $\mathbf{F} = \langle\sqrt{1 + x^4} - 4xy, x^3 - e^{y^2}\rangle$. Find the circulation of $\mathbf{F}$ around C.

32. Let $\mathbf{F} = \langle 2x + \cosh y^2, 4y - \sinh x^2\rangle$. Find the flux of $\mathbf{F}$ through C.

33. Let C be a smooth simple closed curve, oriented in a counterclockwise direction, and let C be parameterized by arc length. Prove that the outer unit normal $\mathbf{n}$ is given by

$$\mathbf{n} = \left\langle \frac{dy}{ds}, -\frac{dx}{ds} \right\rangle$$

[*Hint:* Let θ denote the smallest positive angle from the unit vector $\mathbf{i}$ to the unit tangent vector $\mathbf{T}$. Then

$$\mathbf{T} = \left\langle \frac{dx}{ds}, \frac{dy}{ds} \right\rangle = \langle \cos\theta, \sin\theta\rangle$$

Now show that $\mathbf{n} = \langle\cos(\theta - \frac{\pi}{2}), \sin(\theta - \frac{\pi}{2})\rangle$.]

34. Make use of Theorem 19.8 to evaluate the integral

$$\int_C \ln\sqrt{x^2 + y^2}\, dx - (\tan^{-1}\tfrac{y}{x})\, dy$$

where C is the limaçon $r = 4 + 2\cos\theta$. Explain why Green's theorem is not applicable in this case.

35. Prove that for

$$F(x, y) = \frac{x\mathbf{i} + y\mathbf{j}}{\sqrt{x^2 + y^2}}$$

and for any piecewise smooth simple closed curve C that does not pass through the origin,

$$\int_C \mathbf{F} \cdot d\mathbf{r} = 0$$

Consider the cases in which the origin is interior to C and exterior to C separately.

36. Complete the proof of Theorem 19.6 for a simple region by showing that equation (19.12) is true.

37. Prove Green's theorem for a region that can be divided by a horizontal or a vertical line segment into two simple regions. Extend this by mathematical induction to a region that can be so divided into finitely many simple regions.

19.4 SURFACE INTEGRALS

In defining the line integral we replaced the linear domain of integration of the ordinary Riemann integral by an arc of a curve. In a similar manner if we replace the plane region that forms the domain of integration of a double integral by a surface in space, we arrive at what is called a **surface integral.** Initially, we place only two restrictions on the surfaces we will use as domains of integration: they must be bounded and have finite area. When we say that a surface is bounded, we mean that it can be contained in some sufficiently large three-dimensional interval. For example, the surface could be a sphere, a portion of a paraboloid, or a part of a cylindrical surface.

Let S denote such a surface, and let g be a bounded function of three variables defined at all points on S. As in Figure 19.22 we partition S into

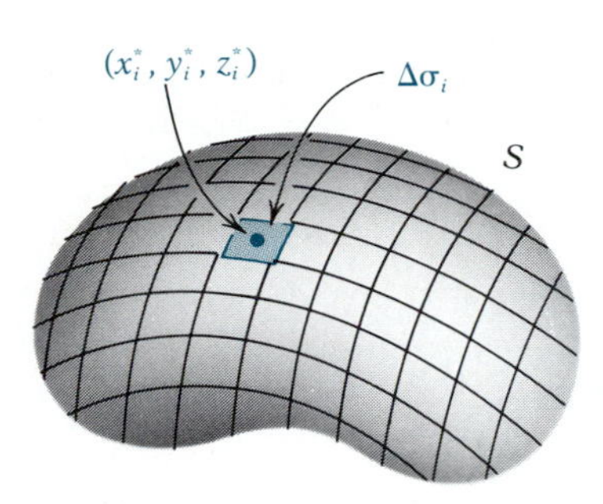

FIGURE 19.22

finitely many small surface elements that we will refer to as *cells*, having areas $\Delta\sigma_1, \Delta\sigma_2, \ldots, \Delta\sigma_n$. Denote the partition by P and define its norm $\|P\|$ as the length of the longest "diagonal" of all the cells. Choose a point (x_i^*, y_i^*, z_i^*) arbitrarily in the ith cell. Then we define the **surface integral of *g* over *S*** by

$$\iint_S g(x, y, z)\, d\sigma = \lim_{\|P\| \to 0} \sum_{i=1}^{n} g(x_i^*, y_i^*, z_i^*)\, \Delta\sigma_i \tag{19.18}$$

provided this limit exists independently of the manner of partitioning S and of the choices of the points (x_i^*, y_i^*, z_i^*) in the cells. It can be proved that the limit will exist when g is continuous on some open set that contains S.

It is readily apparent that if $g(x, y, z) = 1$ at all points of S, then the surface integral gives the area of S, since for all partitions, $\sum_{i=1}^{n} (1)\, \Delta\sigma_i = A(S)$; that is,

$$A(S) = \iint_S d\sigma \tag{19.19}$$

If we think of S as a lamina that has continuous density $\rho(x, y, z)$, then by the usual reasoning (see Exercise 31), we arrive at the mass, center of mass, and moments of inertia:

$$m = \iint_S \rho(x, y, z)\, d\sigma \tag{19.20}$$

and on writing $dm = \rho(x, y, z)\, d\sigma$,

$$\bar{x} = \frac{\iint_S x\, dm}{m} \qquad \bar{y} = \frac{\iint_S y\, dm}{m} \qquad \bar{z} = \frac{\iint_S z\, dm}{m} \tag{19.21}$$

$$I_x = \iint_S (y^2 + z^2)\, dm \qquad I_y = \iint_S (x^2 + z^2)\, dm \qquad I_z = \iint_S (x^2 + y^2)\, dm \tag{19.22}$$

There are many other important applications of surface integrals in fluid dynamics, heat transfer, and electric and magnetic field theory, to name a few. We will discuss one such application in fluid dynamics shortly. First we need to arrive at a workable computational method for evaluating surface integrals.

To obtain an equivalent formulation of a surface integral more suited for computation, we first impose a smoothness requirement on S, analogous to that for curves. In the case of a curve, smoothness means a continuously turning tangent line. For surfaces this continuity requirement is easier to state for a normal to the tangent plane than for the plane itself. We say that S is a **smooth surface** provided it has a tangent plane at each point with a normal vector that varies continuously on S. In particular, suppose S is the graph of an equation of the form $F(x, y, z) = 0$. Then S is smooth if ∇F is continuous and nonzero, since these conditions guarantee that S has a tangent plane everywhere with normal vector ∇F. We will deal only with smooth surfaces of this type.* If S is the graph of $z = f(x, y)$, where f_x and f_y are

* A more general representation of smooth surfaces involving parametric equations can be found in most advanced calculus texts.

FIGURE 19.23

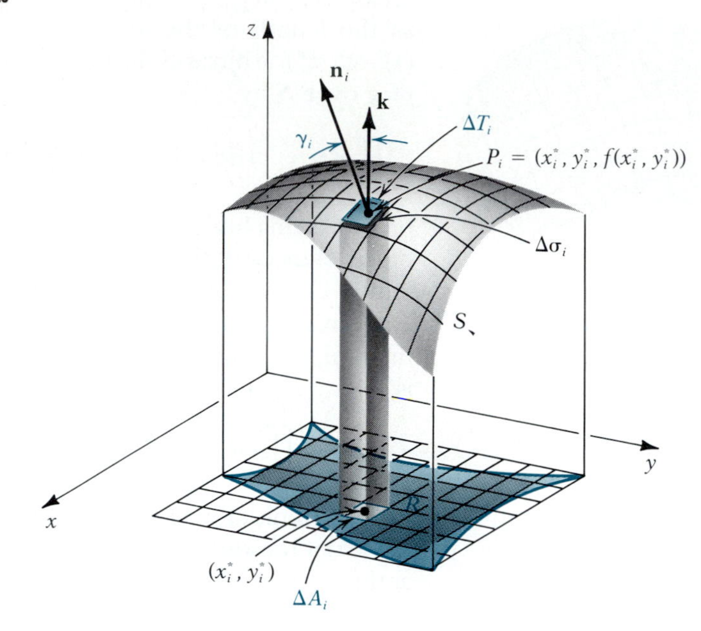

continuous, then S is smooth, since by setting $F(x, y, z) = f(x, y) - z$, we see that $\nabla F = \langle f_x, f_y, -1\rangle$ is continuous and nonzero. Similar remarks apply to equations of the form $x = f(y, z)$ and $y = f(x, z)$. We will concentrate on surfaces represented by $z = f(x, y)$, since the results can be extended to the other two cases by permuting the variables. If S is the union of finitely many smooth surfaces such that each pair is joined in a sectionally smooth curve, then we say S is a **sectionally smooth surface.**

Suppose that S is the smooth surface defined by $z = f(x, y)$, where the domain of f is a regular region R in the xy-plane. We will refer to R as the *xy projection* of S. From our definition of a smooth surface we know that f_x and f_y are continuous in R. Following exactly the same procedure as in Section 18.5, we partition an interval in the xy-plane that contains R and count those subintervals that lie altogether in R. Projecting the ith one of these vertically, we obtain a prism that cuts out a cell on S, with area denoted by $\Delta\sigma_i$ (see Figure 19.23). In this way the partition of R effectively induces a partition of S. As before, we choose (x_i^*, y_i^*) in the ith subinterval in R and let ΔT_i be the area of the tangent plane to S at $P_i = (x_i^*, y_i^*, f(x_i^*, y_i^*))$ that is cut out by this vertical prism. Then $\Delta\sigma_i \approx \Delta T_i$, and by equation (18.20),

$$\Delta T_i = |\sec \gamma_i|\,\Delta A_i$$

where, as shown in Figure 19.23, γ_i is the angle from the unit vector $\mathbf{k}$ to a unit normal $\mathbf{n}_i$ to S at P_i. Let g be a continuous function in some open three-dimensional region that contains S. By equation (19.18),

$$\iint_S g\,d\sigma = \lim_{\|P\|\to 0} \sum_{i=1}^{n} g(P_i)\,\Delta\sigma_i$$

It can be shown that under the smoothness hypothesis $\Delta\sigma_i$ can be replaced by ΔT_i and the limit on the right is unchanged, so that

$$\iint_S g\,d\sigma = \lim_{\|P\|\to 0}\sum_{i=1}^{n} g(P_i)\,\Delta T_i = \lim_{\|P\|\to 0}\sum_{i=1}^{n} g(P_i)|\sec\gamma_i|\,\Delta A_i$$

But this last expression is by definition the double integral of $g|\sec\gamma|$ over the region R. So we have

$$\iint_S g\,d\sigma = \iint_R g|\sec\gamma|\,dA \tag{19.23}$$

As we saw in Section 18.5, $|\sec\gamma_i| = \sqrt{1 + f_x^2(P_i) + f_y^2(P_i)}$. This substitution leads to the alternate formulation

$$\iint_S g\,d\sigma = \iint_R g(x, y, f(x, y))\sqrt{1 + f_x^2 + f_y^2}\,dA \tag{19.24}$$

If S is defined by $x = f(y, z)$ and has yz projection R, the corresponding results are

$$\iint_S g\,d\sigma = \iint_R g|\sec\alpha|\,dA \tag{19.25}$$

$$\iint_S g\,d\sigma = \iint_R g(f(y, z), y, z)\sqrt{1 + f_y^2 + f_z^2}\,dA \tag{19.26}$$

Similarly, when $y = f(x, z)$ and R is the xz projection of S,

$$\iint_S g\,d\sigma = \iint_R g|\sec\beta|\,dA \tag{19.27}$$

$$\iint_S g\,d\sigma = \iint_R g(x, f(x, z), z)\sqrt{1 + f_x^2 + f_z^2}\,dA \tag{19.28}$$

In equation (19.25) α is the angle between the unit normal vector $\mathbf{n}$ and the unit vector $\mathbf{i}$, and in equation (19.27) β is the angle between $\mathbf{n}$ and $\mathbf{j}$.

EXAMPLE 19.14 Evaluate the surface integral $\iint_S (2xy + xz)\,d\sigma$, where S is the first-octant portion of the plane $3x + 2y + z = 6$.

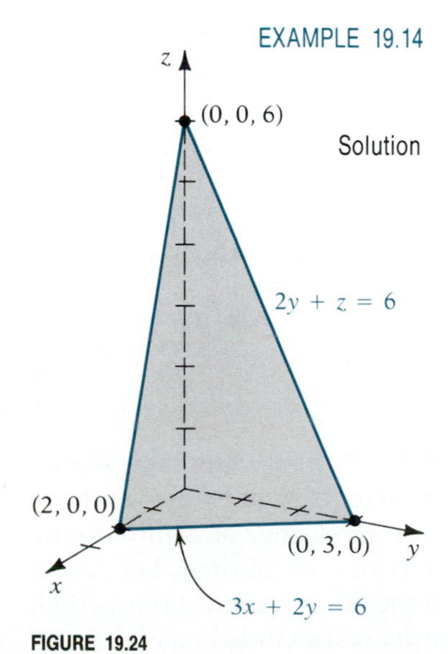

FIGURE 19.24

Solution We can treat S as the graph of an equation of any of the three forms we have discussed. We will do it in two ways, first using R as the xy projection of S and second as the yz projection.

In the first method, $z = f(x, y)$ (xy projection). Solving for z, we get $f(x, y) = 6 - 3x - 2y$. The region R is the triangular region in the xy-plane bounded by the x- and y-axes and the xy trace of S, $3x + 2y = 6$, as shown in Figure 19.24. Thus, $f_x = -3$ and $f_y = -2$, so that $\sec\gamma = \sqrt{1 + f_x^2 + f_y^2} = \sqrt{14}$. Using either equation (19.23) or (19.24) we get (omitting some details)

$$\begin{aligned}\iint_S (2xy + xz)\,d\sigma &= \iint_R [2xy + x(6 - 3x - 2y)]\sqrt{14}\,dA\\ &= \sqrt{14}\iint_R (6x - 3x^2)\,dA\\ &= 3\sqrt{14}\int_0^2\int_0^{(6-3x)/2} (2x - x^2)\,dy\,dx = 6\sqrt{14}\end{aligned}$$

In the second method, $x = f(y, z)$ (*yz* projection). Solving for x, we get $f(y, z) = (6 - 2y - z)/3$. This time the region R is the yz projection of S, bounded by the y- and z-axes and the line $2y + z = 6$. Since $f_y = -\frac{2}{3}$ and $f_z = -\frac{1}{3}$, we get $\sec\alpha = \sqrt{1 + f_y^2 + f_z^2} = \sqrt{14}/3$. Thus, by equation (19.25) or (19.26),

$$\iint_S (2xy + xz)\,d\sigma = \iint_R \left[2\left(\frac{6 - 2y - z}{3}\right)y + \left(\frac{6 - 2y - z}{3}\right)z\right]\frac{\sqrt{14}}{3}\,dA$$
$$= \frac{\sqrt{14}}{9}\int_0^3 \int_0^{6-2y} (12y - 4y^2 - 4yz + 6z - z^2)\,dz\,dy$$
$$= 6\sqrt{14}$$

Note that in this problem the integration is simpler using the first method. When there is a choice of methods, it pays to look ahead to anticipate which is simplest. ■

EXAMPLE 19.15 A homogeneous lamina of density ρ is in the shape of the portion of the paraboloid $z = x^2 + y^2$ between $z = 0$ and $z = 4$. Find the center of mass.

Solution The xy projection of S is the circular region bounded by $x^2 + y^2 = 4$, as shown in Figure 19.25. With $f(x, y) = x^2 + y^2$, we get

$$\sqrt{1 + f_x^2 + f_y^2} = \sqrt{1 + 4(x^2 + y^2)}$$

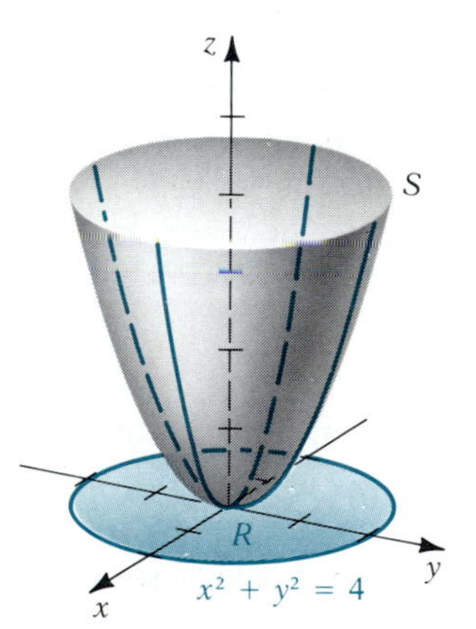

FIGURE 19.25

We first calculate the mass. By equations (19.20) and (19.24),

$$m = \iint_S \rho\,d\sigma = \iint_R \rho\sqrt{1 + 4(x^2 + y^2)}\,dA = \rho\int_0^{2\pi}\int_0^2 \sqrt{1 + 4r^2}\,r\,dr\,d\theta$$
$$= \frac{\pi\rho}{6}[(17)^{3/2} - 1]$$

where again we have omitted details of integration. By symmetry, $\bar{x} = \bar{y} = 0$. For $\bar{z}$ we have

$$m\bar{z} = \iint_S z\,dm = \iint_S z\rho\,d\sigma = \rho\iint_R (x^2 + y^2)\sqrt{1 + 4(x^2 + y^2)}\,dA$$
$$= \rho\int_0^{2\pi}\int_0^2 r^3\sqrt{1 + 4r^2}\,dr\,d\theta = \frac{\pi\rho}{60}[23(17)^{3/2} + 1]$$

(The integration with respect to r can be accomplished by the substitution $u = \sqrt{1 + 4r^2}$. You should supply the details.) Thus,

$$\bar{z} = \frac{1}{10}\,\frac{23(17)^{3/2} + 1}{(17)^{3/2} - 1} \approx 2.335$$ ■

Suppose now that $\mathbf{F}(x, y, z)$ is the velocity field of some fluid in steady state. The flow of water in a stream, ocean currents, wind flow, and the flow of blood in the vascular system can be assumed to have an approximate steady-state motion, at least over relatively short periods of time. Let S be a surface in the given vector field through which the fluid can flow unimpeded. You can think of S as being a screen or netting (or even an imaginary surface).

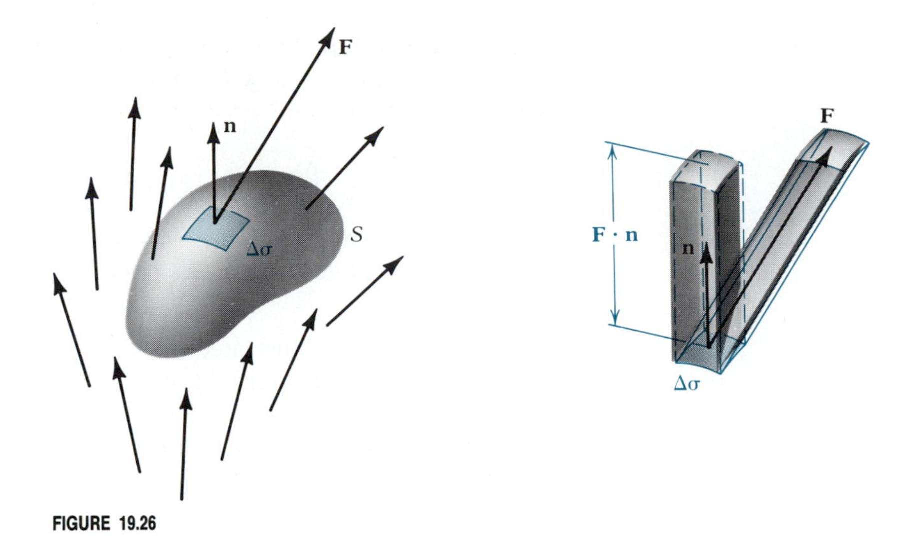

FIGURE 19.26

Let $\Delta\sigma$ be the area of one of the cells on S that results from a partition of S, and let $\mathbf{n}$ be a unit normal vector to S at an arbitrary point in this cell. If the cell is small, we can assume the velocity is approximately constant throughout the cell. The dot product $\mathbf{F} \cdot \mathbf{n}$ is the component of $\mathbf{F}$ in the direction of $\mathbf{n}$, and if we multiply this by $\Delta\sigma$ we get the approximate volume of fluid that flows orthogonally through this cell per unit of time. This is represented in Figure 19.26 by the prism in broken lines. By summing over all such cells and passing to the limit as the norms of partitions approach 0, we obtain the *flux of* $\mathbf{F}$ *through* S:

$$\text{flux of } \mathbf{F} \text{ through } S = \iint_S \mathbf{F} \cdot \mathbf{n}\, d\sigma \tag{19.29}$$

The flux is the total net volume of the fluid that flows through S per unit of time. This is analogous to flux in two dimensions. If S is a closed surface, we typically take $\mathbf{n}$ to be directed toward the exterior to S, called the *outer unit normal*. Then if the flux is positive, there is a net outflow of fluid through S, and we say there is a *source* inside S. If the flux is negative, there is a net inflow of fluid, and we say there is a *sink* inside S. If the flux is 0, the flow is said to be *incompressible*. As we have defined it, flux is the net volume that passes through S per unit of time. If the fluid has density $\rho = \rho(x, y, z)$, then the integral $\iint_S \rho\mathbf{F} \cdot \mathbf{n}\, d\sigma$ is the *mass flux*—that is, the net mass of fluid that passes through S per unit of time.

EXAMPLE 19.16 Find the flux of a fluid that has velocity field $\mathbf{F} = 4x\mathbf{i} + 4y\mathbf{j} + 3z\mathbf{k}$ through the parabolic surface S defined by $z = 4 - x^2 - y^2$ for $z \geq 0$ in the direction of outer unit normals.

Solution As shown in Figure 19.27, the outer normals are also the upward normals, so that

$$\mathbf{n} = \frac{-f_x\mathbf{i} - f_y\mathbf{j} + \mathbf{k}}{\sqrt{1 + f_x^2 + f_y^2}} = \frac{2x\mathbf{i} + 2y\mathbf{j} + \mathbf{k}}{\sqrt{1 + 4(x^2 + y^2)}}$$

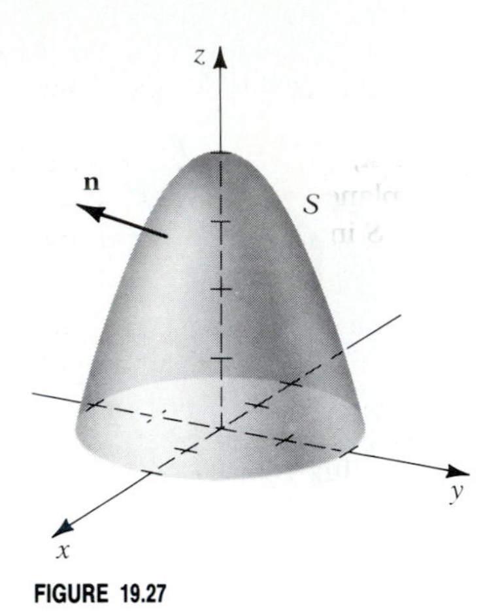

FIGURE 19.27

The flux is therefore

$$\iint_S \mathbf{F} \cdot \mathbf{n}\, d\sigma = \iint_R \frac{8x^2 + 8y^2 + 3(4 - x^2 - y^2)}{\sqrt{1 + 4(x^2 + y^2)}} \sqrt{1 + 4(x^2 + y^2)}\, dA$$

$$= \int_0^{2\pi} \int_0^2 (5r^2 + 12) r\, dr\, d\theta = 88\pi$$

If velocity is in feet per minute, for example, and area is in square feet, then the net amount of fluid that flows out of the surface each minute is 88π cu ft. ■

Although we have illustrated the idea of flux using fluid dynamics, we still call the integral $\iint_S \mathbf{F} \cdot \mathbf{n}\, d\sigma$ the flux of $\mathbf{F}$ through S for any vector field $\mathbf{F}$, whether or not it represents velocity. Some other areas in which this notion is useful are heat flow, electricity and magnetism, and the study of gravitational fields.

EXERCISE SET 19.4

A

In Exercises 1–8 evaluate the surface integral $\iint_S g(x, y, z)\, d\sigma$.

1. $g(x, y, z) = 2x - y + z$; S is the first-octant portion of the plane $x + y + z = 2$.
2. $g(x, y, z) = x^2y - 2z$; S is the first-octant portion of the plane $z = x + 2y$ that lies below $z = 4$.
3. $g(x, y, z) = xz$; S is the portion of the plane $z = 2x - 3y$ inside the cylinder $x^2 + y^2 = 9$.
4. $g(x, y, z) = xy$; S is the first-octant portion of the cylinder $x^2 + z^2 = 4$ between $y = 0$ and $y = 4$ and above $z = 1$.
5. $g(x, y, z) = x^2 + y^2 - z$; S is the portion of the paraboloid $z = 2 - x^2 - y^2$ above the xy-plane.
6. $g(x, y, z) = xyz$; S is the portion of the cone $z^2 = x^2 + y^2$ between $z = 1$ and $z = 2$.
7. $g(x, y, z) = 8/z^2$; S is the portion of the sphere $x^2 + y^2 + z^2 = 25$ above $z = 3$.
8. $g(x, y, z) = xz^2$; S is the portion of the parabolic cylinder $y = x^2$ in the first octant bounded by $y = 2$, $y = 6$, $z = 0$, and $z = 4$. (*Hint:* Use a yz projection.)

In Exercises 9–12 find the mass and center of mass of the lamina in the shape of the surface S with density $\rho(x, y, z)$.

9. S is the first-octant portion of the plane $x + 2y + 4z = 8$; $\rho(x, y, z) = z$.
10. S is the portion of the cylinder $3z = x^2$ lying above the region $R = \{(x, y): 0 \le x \le 2, 0 \le y \le 4\}$; $\rho =$ constant.
11. S is the portion of the paraboloid $z = 4 - x^2 - y^2$ that is inside the cylinder $x^2 + y^2 = 2$; $\rho =$ constant.
12. S is the upper portion of the sphere $x^2 + y^2 + z^2 = 16$ that is inside the cylinder $x^2 + y^2 = 8$; $\rho = 1/\sqrt{z}$.

In Exercises 13 and 14 set up, but do not evaluate, iterated integrals for I_x, I_y, and I_z for the specified laminas.

13. The lamina of Exercise 10
14. The lamina of Exercise 11

In Exercises 15 and 16 set up, but do not evaluate, iterated integrals for calculating the given surface integrals using (a) xy projections, (b) yz projections, and (c) xz projections.

15. $\iint_S x^2yz^3\, d\sigma$; S is the first-octant portion of the plane $2x + y - z = 0$ that is below the plane $z = 4$.
16. $\iint_S (x + 2yz)\, d\sigma$; S is the first-octant portion of the elliptic paraboloid $z = 12 - 3x^2 - 4y^2$.

In Exercises 17 and 18 set up, but do not evaluate, two iterated integrals for calculating the given surface integrals using projections on two different planes.

17. $\iint_S xyz\, d\sigma$; S is the portion of the cylinder $z = 2 - x^2$ in the first octant below $z = 1$ and between $y = 0$ and $y = 5$.
18. $\iint_S (xz/y)\, d\sigma$; S is the first-octant portion of the cylinder $y = e^x$ bounded by the planes $y = 2$ and $z = 3$.

*In Exercises 19–26 find the flux of **F** through S. Take **n** as the upward normal unless otherwise specified.*

19. $\mathbf{F} = \langle 3xy, yz, 2z \rangle$; S is the portion of the plane $x + 3y + z = 5$ lying above the region $R = \{(x, y): 0 \le x \le 2, 0 \le y \le 1\}$.

20. $\mathbf{F} = \langle x, y, z\rangle$; S is the upper portion of the sphere $x^2 + y^2 + z^2 = 25$ that is inside the cylinder $x^2 + y^2 = 16$.

21. $\mathbf{F} = 2x\mathbf{i} + 2y\mathbf{j} + 3z\mathbf{k}$; S is the portion of the paraboloid $z = 4 - x^2 - y^2$ above the xy-plane.

22. $\mathbf{F} = yz\mathbf{i} + (x^3/z)\mathbf{j} + 2z^2\mathbf{k}$; S is the first-octant portion of the cylinder $z = e^x$ bounded above by $z = 2$ and on the right by $y = 2$. Use downward directed normals.

23. $\mathbf{F} = \langle -x^3, y^3, -z\rangle$; S is the portion of the cone $z^2 = x^2 + y^2$ between $z = 1$ and $z = 2$. Use downward directed normals.

24. $\mathbf{F} = 3x\mathbf{i} + 3y\mathbf{j} - z\mathbf{k}$; S is the portion of the hemisphere $z = \sqrt{8 - x^2 - y^2}$ that is inside the cone $z^2 = x^2 + y^2$.

25. $\mathbf{F} = (x^3yz)\mathbf{i} + (x - y^2)\mathbf{j} + z^2\mathbf{k}$; S is the surface $z = 1 - |y|$ above the xy-plane and between $x = 0$ and $x = 4$. (*Hint:* Divide S into two parts and add the integrals over the separate parts.)

26. $\mathbf{F} = \langle xyz, x^2 - y^2, xz - yz\rangle$; S is the cube one unit on an edge that has one vertex at the origin and three of its edges on the positive coordinate axes. Use outward directed normals. Is there a source or a sink within S? (*Hint:* Find the flux through each face and add the results.)

B

27. Find the mass and center of mass of a homogeneous hemispherical shell of radius a and density $\rho = k$. (*Hint:* To evaluate the improper integral, first integrate over the circular region R_b with radius $b < a$, and after integration let $b \to a^-$.)

28. The velocity field for a certain liquid is $\mathbf{F} = x\mathbf{i} - y\mathbf{j} + z\mathbf{k}$. In it is submerged a closed surface that forms the boundary of the region below the hemisphere $z = \sqrt{2a^2 - x^2 - y^2}$ and above the paraboloid $z = x^2 + y^2$. Find the total flux through S, using outward directed normals.

29. Let $\mathbf{F}$ be the inverse square field

$$F = \frac{k\mathbf{u}}{|\mathbf{r}|^2}$$

where $\mathbf{u}$ is the unit vector in the direction of $\mathbf{r} = x\mathbf{i} + y\mathbf{j} + z\mathbf{k}$. Show that the flux of $\mathbf{F}$ through a sphere S centered at the origin is independent of the radius of S.

30. If $\mathbf{F} = \langle L, M, N\rangle$ and S is the graph of $z = f(x, y)$ that has projection R in the xy-plane, then

$$\iint_S \mathbf{F}\cdot\mathbf{n}\,d\sigma = \iint_R (-Lf_x - Mf_y + N)\,dA$$

assuming the appropriate hypotheses. Prove this result.

31. Give a derivation to justify equations (19.20), (19.21), and (19.22).

19.5 THE DIVERGENCE THEOREM

Green's theorem enabled us to express a line integral around a closed path C as a double integral over the region enclosed by C. The theorem we consider in this section enables us to express a surface integral over a closed surface S as a triple integral over the region enclosed by S. Before stating the theorem, we introduce the notion of the *divergence* of a vector field in $\mathbf{R}^3$, which is the natural extension of divergence in $\mathbf{R}^2$.

DEFINITION 19.2 Let $\mathbf{F} = \langle f, g, h\rangle$ be a vector field in $\mathbf{R}^3$ for which the partial derivatives of f, g, and h exist. The **divergence** of $\mathbf{F}$, written div $\mathbf{F}$, is defined by

$$\text{div } \mathbf{F} = \frac{\partial f}{\partial x} + \frac{\partial g}{\partial y} + \frac{\partial h}{\partial z}$$

■

Note that div $\mathbf{F}$ is a scalar. Another common notation for div $\mathbf{F}$ uses the so-called *del operator*, $\nabla = \langle \partial/\partial x, \partial/\partial y, \partial/\partial z\rangle$. Treating this as a vector, we can write div $\mathbf{F} = \nabla \cdot \mathbf{F}$.

The next theorem is known as both the *divergence theorem* and *Gauss' theorem*. A precise formulation of the hypotheses of the theorem would require a deeper background in three-dimensional regions and their boundaries than we have presented. Our formulation is sufficient for most applications.

THEOREM 19.9 THE DIVERGENCE THEOREM (GAUSS' THEOREM)

Let G be a closed and bounded three-dimensional region with boundary S that is a piecewise smooth closed surface. If $\mathbf{F}$ is a continuously differentiable vector field on some open set that contains G, then

$$\iint_S \mathbf{F} \cdot \mathbf{n}\, d\sigma = \iiint_G \operatorname{div} \mathbf{F}\, dV \tag{19.30}$$

where $\mathbf{n}$ is the outer unit normal to S. ■

Observe the similarity between this result and equation (19.17) for two dimensions.

If the direction angles of $\mathbf{n}$ are α, β, and γ, respectively, then $\mathbf{n} = \langle \cos \alpha, \cos \beta, \cos \gamma \rangle$. So if $\mathbf{F} = \langle f, g, h \rangle$, we can write equation (19.30) in the component form:

$$\iint_S (f \cos \alpha + g \cos \beta + h \cos \gamma)\, d\sigma = \iiint_G \left(\frac{\partial f}{\partial x} + \frac{\partial g}{\partial y} + \frac{\partial h}{\partial z} \right) dV \tag{19.31}$$

We will give a partial proof of the divergence theorem where the region G is of a particularly simple type. Suppose first that the xy projection of G is a regular region R and that the boundary of G consists of lower and upper smooth surfaces S_1 and S_2, respectively, and a lateral surface S_3 that is the portion of the vertical projection of the boundary of R that lies between S_1 and S_2. This is illustrated in Figure 19.28. We call a region of this type an

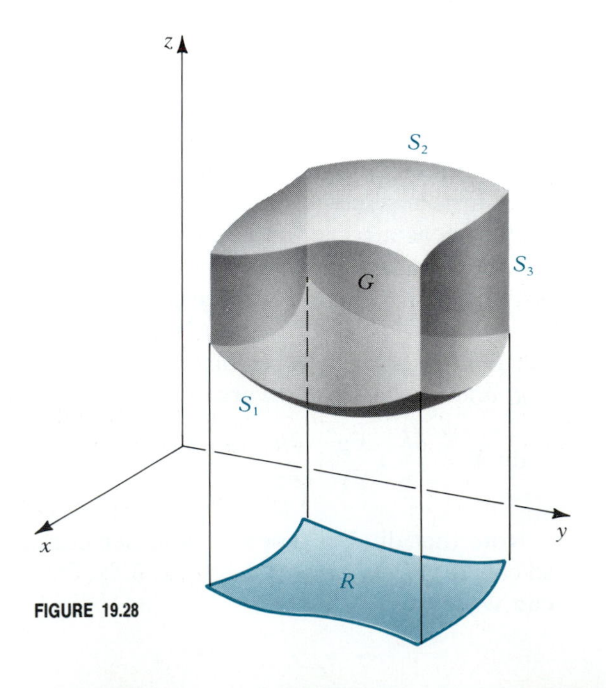

FIGURE 19.28

xy-simple region. We require that each vertical line through an interior point of R intersect G in a closed interval extending from S_1 to S_2 but allow the interval to shrink to a point for such lines through the boundary of R. This means that S_1 and S_2 may meet along some or all of the boundary. In the latter case there would be no surface S_3. Regions that are **yz-simple** and **xz-simple** are defined in an analogous way. Finally, we call G a **simple region** if it is simultaneously xy-simple, yz-simple, and xz-simple. It is for this type that we prove the divergence theorem.

Partial Proof of Theorem 19.9

Let G be a simple region. The theorem is proved by showing the validity of each of the equations

$$\iint_S f \cos \alpha \, d\sigma = \iiint_G \frac{\partial f}{\partial x} \, dV \tag{19.32}$$

$$\iint_S g \cos \beta \, d\sigma = \iiint_G \frac{\partial g}{\partial y} \, dV \tag{19.33}$$

$$\iint_S h \cos \gamma \, d\sigma = \iiint_G \frac{\partial h}{\partial z} \, dV \tag{19.34}$$

When these results are added, we get equation (19.31), which is equivalent to equation (19.30). We will prove equation (19.34) only and ask you to prove equations (19.32) and (19.33) in the exercises.

Since G is xy-simple, its boundary S can be decomposed into surfaces S_1, S_2, and S_3 as in Figure 19.28 (where S_3 may be empty, as mentioned). If S_3 is nonempty, on it the outer unit normal vectors are horizontal, so that $\gamma = \frac{\pi}{2}$ and $\cos \gamma = 0$. Thus,

$$\iint_{S_3} h \cos \gamma \, d\sigma = 0$$

It follows that whether or not S_3 is empty,

$$\begin{aligned}\iint_S h \cos \gamma \, d\sigma &= \iint_{S_1} h \cos \gamma \, d\sigma + \iint_{S_2} h \cos \gamma \, d\sigma + \iint_{S_3} h \cos \gamma \, d\sigma \\ &= \iint_{S_1} h \cos \gamma \, d\sigma + \iint_{S_2} h \cos \gamma \, d\sigma\end{aligned}$$

Let S_1 be the graph of $z = z_1(x, y)$ and S_2 the graph of $z = z_2(x, y)$. On S_2 the outer unit normal is directed upward, so that γ is acute. Thus, by equation (19.23),

$$\begin{aligned}\iint_{S_2} h \cos \gamma \, d\sigma &= \iint_R [h(x, y, z_2(x, y)) \cos \gamma] \sec \gamma \, dA \\ &= \iint_R h(x, y, z_2(x, y)) \, dA\end{aligned}$$

On S_1 the outer unit normal is directed downward, so that γ is obtuse, and $|\sec \gamma| = -\sec \gamma$. Equation (19.23) then gives

$$\begin{aligned}\iint_{S_1} h \cos \gamma \, d\sigma &= -\iint_R [h(x, y, z_1(x, y) \cos \gamma] \sec \sigma \, dA \\ &= -\iint_R h(x, y, z_1(x, y)) \, dA\end{aligned}$$

Combining these results, we get

$$\iint_S h \cos \gamma \, d\sigma = \iint_R [h(x, y, z_2(x, y)) - h(x, y, z_1(x, y))] \, dA \tag{19.35}$$

From the right-hand side of equation (19.34), we get

$$\begin{aligned}\iiint_G \frac{\partial h}{\partial z} dV &= \iint_R \left[\int_{z_1(x,y)}^{z_2(x,y)} \left(\frac{\partial h}{\partial z} \right) dz \right] dA \\ &= \iint_R [h(x, y, z_2(x, y)) - h(x, y, z_1(x, y))] \, dA\end{aligned}$$

and since this is the same as the right-hand side of equation (19.35), it follows that equation (19.34) is true. ■

The proof we have given can be extended to a region $G = G_1 \cup G_2$, where G_1 and G_2 are simple regions with an intersection that is a sectionally smooth surface, and by induction to any finite union of such simple regions. You will be asked to show this in Exercise 20.

EXAMPLE 19.17 Use the divergence theorem to evaluate $\iint_S \mathbf{F} \cdot \mathbf{n} \, d\sigma$ where S is the sphere $x^2 + y^2 + z^2 = a^2$ and $\mathbf{F} = xy^2\mathbf{i} + yz^2\mathbf{j} + x^2z\mathbf{k}$. Assume $\mathbf{n}$ is the outer unit normal.

Solution Let G be the sphere together with its interior. Then by equation (19.30),

$$\begin{aligned}\iint_S \mathbf{F} \cdot \mathbf{n} \, d\sigma &= \iiint_G \operatorname{div} \mathbf{F} \, dV \\ &= \iiint_G (y^2 + z^2 + x^2) \, dV\end{aligned}$$

It is convenient to change to spherical coordinates at this stage, getting the iterated integral

$$\iint_S \mathbf{F} \cdot \mathbf{n} \, d\sigma = \int_0^{\pi} \int_0^{2\pi} \int_0^{a} \rho^2(\rho^2 \sin \phi \, d\rho \, d\theta \, d\phi) = \frac{4\pi a^5}{5}$$

You may wish to try evaluating the surface integral in this problem without using the divergence theorem to compare the difficulty. ■

EXAMPLE 19.18 Evaluate the integral $\iint_S \mathbf{F} \cdot \mathbf{n} \, d\sigma$, where S is the boundary of the region G below the paraboloid $z = 4 - x^2 - y^2$, inside the cylinder $x^2 + y^2 = 1$, and above the xy-plane, and where

$$\mathbf{F} = \langle 2x + \sqrt{z^3}, 3y - e^{z^2}, (x^3 + y^3)^{4/3} \rangle$$

Use outer unit normals.

Solution Since $\operatorname{div} \mathbf{F} = 2 + 3 + 0 = 5$, we have by the divergence theorem

$$\begin{aligned}\iint_S \mathbf{F} \cdot \mathbf{n} \, d\sigma &= \iiint_G 5 \, dV \\ &= 5 \int_0^{2\pi} \int_0^{1} \int_0^{4-r^2} r \, dz \, dr \, d\theta \\ &= 5 \int_0^{2\pi} \int_0^{1} (4r - r^3) \, dr \, d\theta = \tfrac{35\pi}{2}\end{aligned}$$

where we used cylindrical coordinates in the iterated integral. Without the divergence theorem this problem would be virtually impossible. (Try it!)

■

If **F** is the velocity field of a fluid, we know that $\iint_S \mathbf{F} \cdot \mathbf{n}\, d\sigma$ is the flux of **F** through S. So, assuming appropriate conditions on **F**, S, and G, the divergence theorem says that

$$\text{flux of } \mathbf{F} \text{ through } S = \iiint_G \operatorname{div} \mathbf{F}\, dV$$

The physical interpretation of div **F** at a point P is analogous to that in two dimensions. We let S_ε be a sphere of radius ε centered at P and let G_ε be the region enclosed by S_ε. Then, using a mean value theorem for triple integrals, we can write

$$\iiint_{G_\varepsilon} \operatorname{div} \mathbf{F}\, dV = [\operatorname{div} \mathbf{F}(Q)]\ (\text{volume of } G_\varepsilon)$$

where Q is some point in G_ε. Thus,

$$\operatorname{div} \mathbf{F}(Q) = \frac{\text{flux of } \mathbf{F} \text{ through } S_\varepsilon}{\text{volume of } G_\varepsilon}$$

Now we let $\varepsilon \to 0$, so that $Q \to P$, and if div **F** is continuous, div $\mathbf{F}(Q) \to$ div $\mathbf{F}(P)$. Thus,

$$\operatorname{div} \mathbf{F}(P) = \lim_{\varepsilon \to 0} \frac{\text{flux of } \mathbf{F} \text{ through } S_\varepsilon}{\text{volume of } G_\varepsilon}$$

The limit on the right is the flux per unit volume at P, called the *flux density* of **F** at P. So the divergence of **F** at a point is the flux density at that point. Just as with two-dimensional flow, we call P a *source* if div $P > 0$ and a *sink* if div $P < 0$.

The divergence theorem says that we can obtain the flux of **F** through the closed surface S by integrating the flux density over the volume G enclosed by S:

$$\text{flux of } \mathbf{F} \text{ through } S = \iiint_G (\text{flux density of } \mathbf{F})\, dV$$

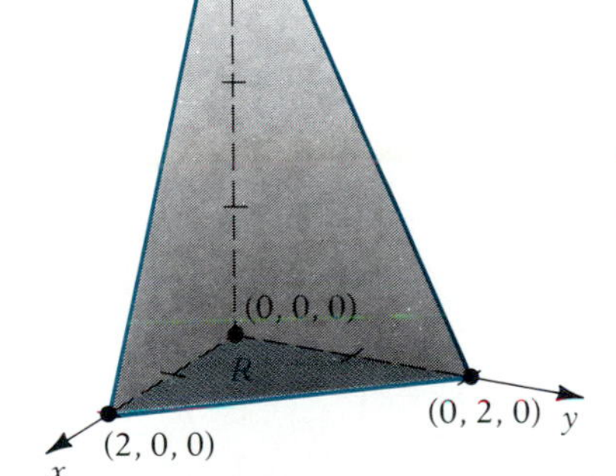

FIGURE 19.29

EXAMPLE 19.19 Use the divergence theorem to find the flux of the velocity field $\mathbf{F} = 2xy\mathbf{i} + 3yz\mathbf{j} + xz\mathbf{k}$ through the tetrahedron with vertices (0, 0, 0), (2, 0, 0), (0, 2, 0), and (0, 0, 4).

Solution Let S denote the surface of the tetrahedron. It is pictured in Figure 19.29. Let G denote the region enclosed by S. Its xy projection is the triangular region R as shown, bounded by the lines $x = 0$, $y = 0$, and $x + y = 2$. The inclined plane that forms the upper bounding surface of G has equation $2x + 2y + z = 4$ (verify), so that $z = 4 - 2x - 2y$. Applying the divergence theorem, we have

$$\begin{aligned} \text{flux of } \mathbf{F} \text{ through } S &= \iiint_G (\operatorname{div} \mathbf{F})\, dV \\ &= \int_0^2 \int_0^{2-x} \int_0^{4-2x-2y} (2y + 3z + x)\, dz\, dy\, dx = \tfrac{44}{3} \end{aligned}$$

The integration is straightforward (but tedious), and so we have omitted the details. ■

EXERCISE SET 19.5

A

In Exercises 1–6 use the divergence theorem to evaluate $\iint_S \mathbf{F} \cdot \mathbf{n}\, d\sigma$, where $\mathbf{n}$ is the outer unit normal.

1. $\mathbf{F} = x^2\mathbf{i} + y^2\mathbf{j} + z^2\mathbf{k}$; S is the rectangular parallelepiped formed by the planes $x = 1$, $x = 3$, $y = 2$, $y = 6$, $z = 0$, and $z = 4$.
2. $\mathbf{F} = (2x + y)\mathbf{i} + (x - y)\mathbf{j} + x^2y^3\mathbf{k}$; S is the tetrahedron formed by the planes $3x + 2y + z = 6$, $x = 0$, $y = 0$, and $z = 0$.
3. $\mathbf{F} = \langle x + z, y^2, yz \rangle$; S is the boundary of the first-octant region enclosed by the cylinder $z = 4 - x^2$ and the planes $y = 0$, $z = 0$, and $y = x$.
4. $\mathbf{F} = \langle xz, 2xy, 4yz \rangle$; S is the tetrahedron formed by the planes $x + y = 2$, $y + z = 2$, $x = 0$, $y = 0$, and $z = 0$.
5. $\mathbf{F} = (e^z \sin y)\mathbf{i} + (e^z \cos x)\mathbf{j} + z\mathbf{k}$; S is the boundary of the region inside the cylinder $x^2 + y^2 = 4$ between the planes $z = 0$ and $z = 6 + x + 2y$.
6. $\mathbf{F} = \langle x^3, y^3, \cosh x^3 \rangle$; S is the boundary of the region inside the cone $z = \sqrt{x^2 + y^2}$ between $z = 1$ and $z = 2$.

In Exercises 7–12 use the divergence theorem to find the flux of $\mathbf{F}$ through S in the direction of outer unit normals.

7. $\mathbf{F} = (2x - 3y)\mathbf{i} + (4y + 2z)\mathbf{j} + (x + z)\mathbf{k}$; S is the sphere $x^2 + y^2 + z^2 = 16$.
8. $\mathbf{F} = (e^y \cos z)\mathbf{i} + (e^z \sin x)\mathbf{j} + (e^{x^2+y^2})\mathbf{k}$; S is the ellipsoid $3x^2 + 7y^2 + 12z^2 = 84$.
9. $\mathbf{F} = \langle y/x, x/y, 1/z^2 \rangle$; S is the cube formed by the planes $x = 1$, $x = 2$, $y = 1$, $y = 2$, $z = 2$, and $z = 3$.
10. $\mathbf{F} = \langle xy^2, yz^2, zx^2 \rangle$; S is the "ice cream cone" formed by the cone $z = \sqrt{x^2 + y^2}$ and the hemisphere $z = \sqrt{4 - x^2 - y^2}$.
11. $\mathbf{F} = x^2y\mathbf{i} + xy^2\mathbf{j} + z^2\mathbf{k}$; S is the boundary of the region inside the paraboloid $z = x^2 + y^2$ and below the hemisphere $z = \sqrt{2 - x^2 - y^2}$.
12. $\mathbf{F} = xy\mathbf{i} + y^2\mathbf{j} + yz\mathbf{k}$; S is the boundary of the first-octant region inside the cylinder $x^2 + z^2 = 4$ between the planes $y = 0$ and $y = 2x$.

In Exercises 13–16 verify the divergence theorem by calculating $\iint_S \mathbf{F} \cdot \mathbf{n}\, d\sigma$ and $\iiint_G \operatorname{div} \mathbf{F}\, dV$.

13. $\mathbf{F} = 2x\mathbf{i} - 3y\mathbf{j} + 4z\mathbf{k}$; G is the region inside the paraboloid $z = x^2 + y^2$ and below the plane $z = 4$.
14. $\mathbf{F} = \langle x - y, x + y, 2x \rangle$; G is the region enclosed by the tetrahedron formed by the coordinate planes and the plane $x + y + z = 2$.
15. $\mathbf{F} = \langle 2x, 3y, z \rangle$; G is the first-octant region inside both of the cylinders $x^2 + y^2 = a^2$ and $x^2 + z^2 = a^2$.
16. $\mathbf{F} = x^{3/2}\mathbf{i} + y^{3/2}\mathbf{j} + z^{3/2}\mathbf{k}$; G is the region enclosed by the planes $x + y = 4$, $z = 4$, and the coordinate planes.
17. Show that if a region G and its boundary S satisfy the conditions of the divergence theorem, then for $\mathbf{F} = x\mathbf{i} + y\mathbf{j} + z\mathbf{k}$,
$$\iint_S \mathbf{F} \cdot \mathbf{r}\, d\sigma = 3V$$
where V is the volume of G.
18. Let G and S satisfy the divergence theorem hypotheses. Prove that the flux of any constant vector field through S is 0.

B

19. Verify equations (19.32) and (19.33) for a simple region.
20. Let $G = G_1 \cup G_2$, where G_1 and G_2 are simple regions with an intersection that is a sectionally smooth surface T. Prove that the divergence theorem holds true for G. Extend the result by induction to a finite union of simple regions. (*Hint:* The outer unit normals for G_1 and G_2 across their common boundary T are oppositely directed.)
21. Let S be the sphere $x^2 + y^2 + (z - a)^2 = a^2$ and $\mathbf{F} = \langle x^2, y^2, z^2 \rangle$. Show that $\iint_S \mathbf{F} \cdot \mathbf{n}\, d\sigma = 8\pi a^4/3$. (*Hint:* Use the divergence theorem and spherical coordinates.)
22. By Coulomb's law the force field $\mathbf{F}$ of a point charge of q coulombs located at the origin is
$$\mathbf{F} = \frac{cq\mathbf{u}}{|\mathbf{r}|^2}$$
where $\mathbf{r}$ is the position vector of a point P in space, $\mathbf{u}$ is a unit vector in the direction of $\mathbf{r}$, and c is a constant. Prove that the flux of $\mathbf{F}$ through any sectionally smooth surface S with the origin in its interior is $4\pi qc$. (*Hint:* Use the divergence theorem to show that if S_1 is a sphere centered at the origin lying inside S, then $\iint_S \mathbf{F} \cdot \mathbf{n}\, d\sigma = -\iint_{S_1} \mathbf{F} \cdot \mathbf{n}\, d\sigma$ with $\mathbf{n}$ directed toward the origin for S_1.)

Exercises 23–26 are to be done in sequence. The formulas in Exercises 23 and 24 are known as Green's identities.

23. If f is a scalar function whose second partials exist, the **Laplacian** of f, denoted by $\nabla^2 f$, is defined by

$$\nabla^2 f = \nabla \cdot \nabla f = \frac{\partial^2 f}{\partial x^2} + \frac{\partial^2 f}{\partial y^2} + \frac{\partial^2 f}{\partial z^2}$$

Prove that if u and v are scalar functions that satisfy appropriate continuity requirements and G and S are as in the divergence theorem, then

$$\iiint_G (u\nabla^2 v + \nabla u \cdot \nabla v)\, dV = \iint_S u\nabla v \cdot \mathbf{n}\, d\sigma$$

Hint: Take $\mathbf{F} = \left\langle u\frac{\partial v}{\partial x}, u\frac{\partial v}{\partial y}, u\frac{\partial v}{\partial z}\right\rangle$.

24. Using Exercise 23 prove that

$$\iiint_G (u\nabla^2 v - v\nabla^2 u)\, dV = \iint_S (u\nabla v - v\nabla u)\, d\sigma$$

(*Hint:* Make use of Exercise 23 twice, once as it stands and once with u and v interchanged.)

25. Prove that

$$\iiint_G \nabla^2 u\, dV = \iint_S \nabla u \cdot \mathbf{n}\, d\sigma$$

26. Let $\mathbf{F}$ be a gradient field with potential function ϕ. Prove that if $\operatorname{div} \mathbf{F} = 0$,

$$\iiint_G |\mathbf{F}|^2\, dV = \iint_S \phi\mathbf{F} \cdot \mathbf{n}\, d\sigma$$

19.6 STOKES' THEOREM

The main result of this section is Stokes' theorem, a generalization of Green's theorem to three dimensions, named for the English mathematical physicist George Stokes (1819–1903). Green's theorem relates a line integral around a closed curve C in the plane to the double integral over the region enclosed by C. In a similar way Stokes' theorem relates the line integral around a closed curve C in space to the surface integral over a surface that has C as its boundary. Before stating the theorem, we need to discuss how the orientation of C is related to the surface.

The theorem applies only to smooth surfaces that have two distinct sides, called **orientable surfaces.** It may surprise you to learn that some surfaces have only one side. The best known example is the *Möbius strip*, shown in Figure 19.30. You can easily construct such a surface by taking a strip of paper, giving one end a half twist, and pasting the ends together. To convince yourself that this is one-sided, take a crayon and start coloring the "top" side and continue until you return to the starting point. You will discover you have colored the entire surface!

FIGURE 19.30 Möbius strip

Another way to describe an orientable surface S is by means of its unit normal vectors. Suppose a direction for a unit normal can be chosen in such a way that, starting from any point on S, if any closed curve C on S is traversed, the unit normal will return to its original direction at the starting point. (You can convince yourself that this will not always happen on the Möbius strip.) Then S is orientable. We call whichever unit normal we have selected positive, and then we say that *S is oriented* with respect to that normal. In effect, we have designated a positive side to the surface. For example, a surface may be oriented by upward normals. Then we are calling the top side positive.

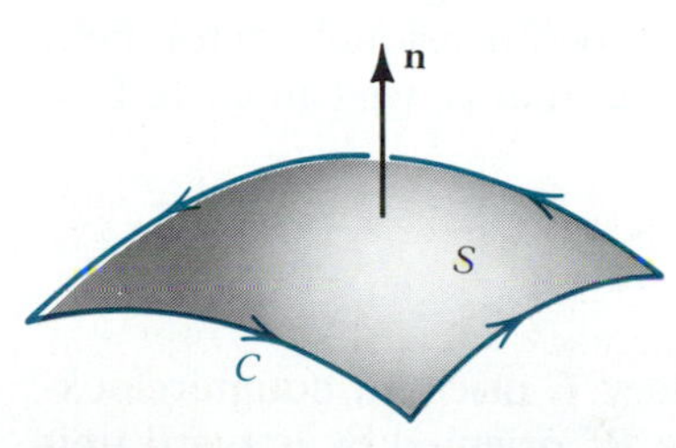

FIGURE 19.31

Now let S be a smooth oriented surface in the sense just described, and let its boundary be a sectionally smooth simple closed curve C. We will say that C is *positively oriented with respect to S* if, when viewed from the tip of a positive unit normal $\mathbf{n}$ to S, C is oriented in a counterclockwise direction. This is illustrated in Figure 19.31. The direction of C then is such that if you walked around C in its positive direction, with your head in the direction of $\mathbf{n}$, the surface S would always be on your left.

We give one more definition that will simplify the statement of Stokes' theorem.

DEFINITION 19.3 Let $\mathbf{F} = \langle f, g, h \rangle$ be a vector field for which the first partials of f, g, and h exist in some open region of $\mathbf{R}^3$. Then the **curl of F** is the vector

$$\text{curl } \mathbf{F} = \left\langle \frac{\partial h}{\partial y} - \frac{\partial g}{\partial z}, \frac{\partial f}{\partial z} - \frac{\partial h}{\partial x}, \frac{\partial g}{\partial x} - \frac{\partial f}{\partial y} \right\rangle \tag{19.36}$$ ■

Formally, curl $\mathbf{F}$ is the cross product $\nabla \times \mathbf{F}$, where $\nabla = \langle \partial/\partial x, \partial/\partial y, \partial/\partial z \rangle$. So we can compute curl $\mathbf{F}$ using the symbolic determinant

$$\text{curl } \mathbf{F} = \nabla \times \mathbf{F} = \begin{vmatrix} \mathbf{i} & \mathbf{j} & \mathbf{k} \\ \dfrac{\partial}{\partial x} & \dfrac{\partial}{\partial y} & \dfrac{\partial}{\partial z} \\ f & g & h \end{vmatrix} \tag{19.37}$$

EXAMPLE 19.20 Find curl $\mathbf{F}$, where

$$\mathbf{F}(x, y, z) = x^2y\mathbf{i} + (2y - z)\mathbf{j} + xyz\mathbf{k}$$

Solution Using equation (19.37) we get

$$\text{curl } \mathbf{F} = \begin{vmatrix} \mathbf{i} & \mathbf{j} & \mathbf{k} \\ \dfrac{\partial}{\partial x} & \dfrac{\partial}{\partial y} & \dfrac{\partial}{\partial z} \\ x^2y & 2y - z & xyz \end{vmatrix} = (xz + 1)\mathbf{i} - yz\mathbf{j} - x^2\mathbf{k}$$ ■

We are ready now for Stokes' theorem.

THEOREM 19.10 **STOKES' THEOREM**

Let C be a piecewise smooth simple closed curve that forms the boundary of a smooth oriented surface S, and let C be positively oriented with respect to S. Then if $\mathbf{F}$ is a continuously differentiable vector field on some open set that contains both S and C,

$$\int_C \mathbf{F} \cdot d\mathbf{r} = \iint_S (\text{curl } \mathbf{F}) \cdot \mathbf{n} \, d\sigma \tag{19.38}$$ ■

We will not give a proof, but we illustrate the result and see some of its consequences. First, let us show that the theorem does generalize Green's theorem in the plane. We can think of the two-dimensional vector field $\mathbf{F} = \langle f, g \rangle$ as being three dimensional, with $h = 0$; that is, we can write $\mathbf{F} = \langle f, g, 0 \rangle$. Then

$$\text{curl } \mathbf{F} = \left(\frac{\partial g}{\partial x} - \frac{\partial f}{\partial y} \right) \mathbf{k} \qquad \text{(Verify.)}$$

A regular region R in the plane with boundary C oriented counterclockwise can be thought of as a smooth surface in $\mathbf{R}^3$ oriented by upward unit normals; that is, $\mathbf{n} = \mathbf{k}$. Hence,

$$(\text{curl } \mathbf{F}) \cdot \mathbf{n} = \frac{\partial g}{\partial x} - \frac{\partial f}{\partial y}$$

and so equation (19.38) reduces to

$$\int_C f\,dx + g\,dy = \iint_R \left(\frac{\partial g}{\partial x} - \frac{\partial f}{\partial y}\right) dA$$

which is the conclusion in Green's theorem.

EXAMPLE 19.21 Use Stokes' theorem to evaluate the integral

$$\int_C (x^2 + y^2)\,dx + xy^2\,dy + xyz\,dz$$

where C is the boundary of the surface S consisting of the first-octant portion of the cylinder $z = 4 - x^2$ between the planes $y = 0$ and $y = 2x$. Orient S with upward unit normals and orient C positively with respect to S.

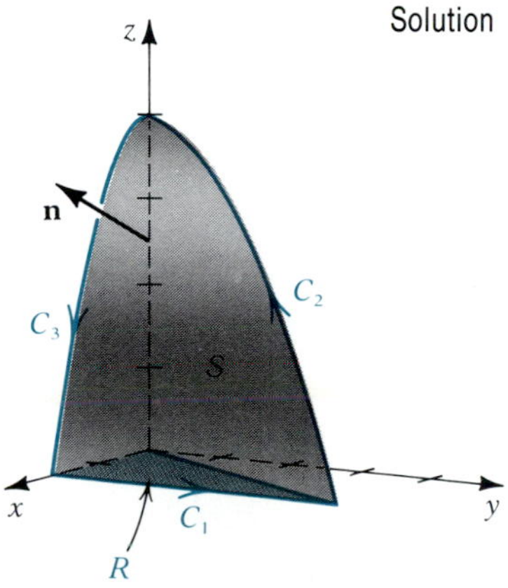

Solution In Figure 19.32 we show the surface S and its boundary $C = C_1 \cup C_2 \cup C_3$. We could parameterize each of these component curves and evaluate the integral directly, but it is easier to use Stokes' theorem. Let $\mathbf{F} = (x^2 + y^2)\mathbf{i} + xy^2\mathbf{j} + xyz\mathbf{k}$. A straightforward calculation gives curl $\mathbf{F} = xz\mathbf{i} - yz\mathbf{j} + (y^2 - 2y)\mathbf{k}$. The upward unit normal $\mathbf{n}$ to $z = 4 - x^2$ is

$$\mathbf{n} = \frac{-\dfrac{\partial z}{\partial x}\mathbf{i} - \dfrac{\partial z}{\partial y}\mathbf{j} + \mathbf{k}}{\sqrt{1 + \left(\dfrac{\partial z}{\partial x}\right)^2 + \left(\dfrac{\partial z}{\partial y}\right)^2}} = \frac{2x\mathbf{i} + \mathbf{k}}{\sqrt{1 + 4x^2}}$$

z
n
C_3
C_2
S
x
C_1
y
R

FIGURE 19.32

So we have

$$\int_C \mathbf{F} \cdot d\mathbf{r} = \iint_S (\text{curl } \mathbf{F}) \cdot \mathbf{n}\,d\sigma = \iint_S \frac{2x^2z + y^2 - 2y}{\sqrt{1 + 4x^2}}\,d\sigma$$

Using equation (19.24), this becomes

$$\iint_R \frac{2x^2(4 - x^2) + y^2 - 2y}{\sqrt{1 + 4x^2}}\sqrt{1 + 4x^2}\,dA$$

$$= \int_0^2 \int_0^{2x} (8x^2 - 2x^4 + y^2 - 2y)\,dy\,dx$$

$$= \frac{64}{3}$$

(You should supply the details.) ■

EXAMPLE 19.22 Verify Stokes' theorem for $\mathbf{F} = \langle x + y,\ y - x,\ z \rangle$, where S is the portion of the paraboloid $z = x^2 + y^2$ below $z = 4$, oriented by downward unit normals.

Solution The surface S and its boundary C are shown in Figure 19.33. Note that with the unit normal $\mathbf{n}$ directed downward, the positive orientation for C is clockwise as shown. We will verify equation (19.38) by calculating each side separately.

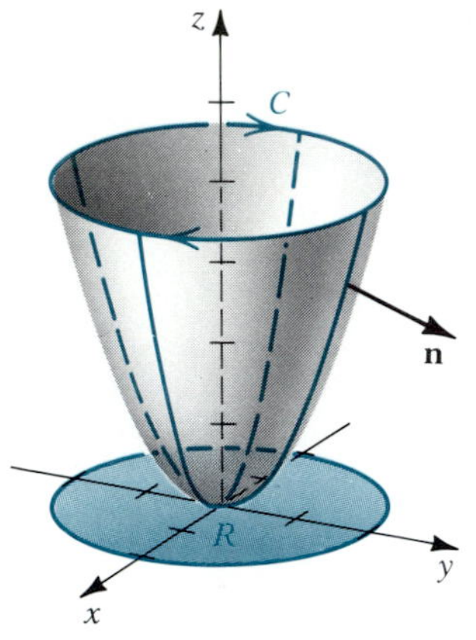

FIGURE 19.33

The curve C can be parameterized by

$$\begin{cases} x = 2\cos(-t) = 2\cos t \\ y = 2\sin(-t) = -2\sin t, \qquad 0 \le t \le 2\pi \\ z = 4 \end{cases}$$

So $dx = -2\sin t\,dt$, $dy = -2\cos t\,dt$, and $dz = 0$. Thus,

$$\begin{aligned}\int_C \mathbf{F}\cdot d\mathbf{r} &= \int_C (x+y)\,dx + (y-x)\,dy + z\,dz \\ &= \int_0^{2\pi} [(2\cos t - 2\sin t)(-2\sin t) \\ &\quad + (-2\sin t - 2\cos t)(-2\cos t)]\,dt \\ &= 4\int_0^{2\pi} (\sin^2 t + \cos^2 t)\,dt = 8\pi\end{aligned}$$

Next we calculate curl $\mathbf{F} = -2\mathbf{k}$, and for $z = x^2 + y^2$, the downward unit normal

$$\mathbf{n} = \frac{\dfrac{\partial z}{\partial x}\mathbf{i} + \dfrac{\partial z}{\partial y}\mathbf{j} - \mathbf{k}}{\sqrt{1 + \left(\dfrac{\partial z}{\partial x}\right)^2 + \left(\dfrac{\partial z}{\partial y}\right)^2}} = \frac{2x\mathbf{i} + 2y\mathbf{j} - \mathbf{k}}{\sqrt{1 + 4x^2 + 4y^2}}$$

So, by equation (19.24),

$$\begin{aligned}\iint_S (\text{curl }\mathbf{F})\cdot\mathbf{n}\,d\sigma &= \iint_S \frac{2}{\sqrt{1 + 4x^2 + 4y^2}}\,d\sigma \\ &= \iint_R \frac{2}{\sqrt{1 + 4x^2 + 4y^2}}\sqrt{1 + 4x^2 + 4y^2}\,dA \\ &= 2\iint_R dA = 2(\text{area of } R) = 2(4\pi) = 8\pi\end{aligned}$$

■

Comment Using ideas similar to those for extending Green's theorem to multiply connected regions, we can also extend Stokes' theorem to surfaces with holes, whose boundaries therefore consist of unions of two or more disjoint closed curves. We will not pursue these ideas here, however.

To gain some insight into the physical interpretation of Stokes' theorem, we return to the notion of fluid flow in which $\mathbf{F}$ is the velocity field of the fluid. For a closed curve C in the region of flow, then, just as in two-dimensional flow, the integral

$$\int_C \mathbf{F}\cdot d\mathbf{r} = \int_C \mathbf{F}\cdot\mathbf{T}\,ds$$

is called the *circulation of* $\mathbf{F}$ *around* C. Suppose now that S_ε is a small disk of radius ε, centered at a point P in the velocity field. Let $\mathbf{n}$ be the upward unit normal to S_ε, and let C_ε be the positively oriented boundary of S_ε, as in Figure 19.34. Then, by Stokes' theorem,

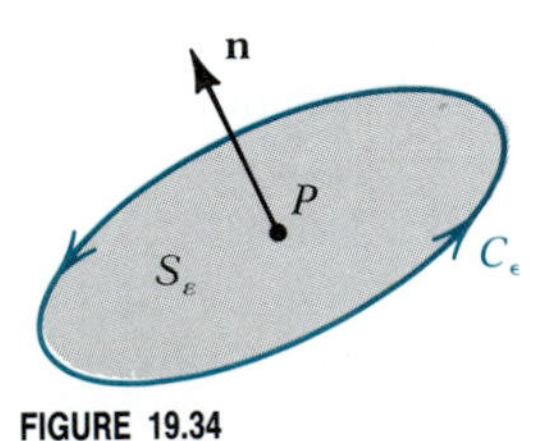

FIGURE 19.34

$$\int_{C_\varepsilon} \mathbf{F}\cdot d\mathbf{r} = \iint_{S_\varepsilon} (\text{curl }\mathbf{F})\cdot\mathbf{n}\,d\sigma$$

A mean value theorem for surface integrals enables us to write

$$\iint_{S_\varepsilon} (\text{curl }\mathbf{F}) \cdot \mathbf{n}\, d\sigma = [\text{curl }\mathbf{F}(Q) \cdot \mathbf{n}](\text{area of } S_\varepsilon)$$

where Q is some point in S_ε. Thus,

$$\text{curl }\mathbf{F}(Q) \cdot \mathbf{n} = \frac{\iint_{S_\varepsilon} (\text{curl }\mathbf{F}) \cdot \mathbf{n}\, d\sigma}{\text{area of } S_\varepsilon} = \frac{\int_{C_\varepsilon} \mathbf{F} \cdot d\mathbf{r}}{\text{area of } S_\varepsilon}$$

$$= \frac{\text{circulation of } \mathbf{F} \text{ around } C_\varepsilon}{\text{area of } S_\varepsilon}$$

Now we let $\varepsilon \to 0$, so that $Q \to P$, and assuming the continuity of curl $\mathbf{F}$, curl $\mathbf{F}(Q) \to$ curl $\mathbf{F}(P)$. So

$$\text{curl }\mathbf{F}(P) \cdot \mathbf{n} = \lim_{\varepsilon \to 0} \frac{\text{circulation of } \mathbf{F} \text{ around } C_\varepsilon}{\text{area of } S_\varepsilon}$$

The limit on the right is called the **rotation of F around n** at P. So we can write

$$\text{curl }\mathbf{F}(P) \cdot \mathbf{n} = \text{rotation of } \mathbf{F} \text{ around } \mathbf{n} \text{ at } P \tag{19.39}$$

As we saw earlier, when we interpret the two-dimensional vector field $\mathbf{F} = \langle f, g \rangle$ as being the same as the three-dimensional field $\mathbf{F} = \langle f, g, 0 \rangle$,

$$(\text{curl }\mathbf{F}) \cdot \mathbf{n} = \left(\frac{\partial g}{\partial x} - \frac{\partial f}{\partial y}\right)\mathbf{k} \cdot \mathbf{k} = \frac{\partial g}{\partial x} - \frac{\partial f}{\partial y}$$

and this is what we called the rotation of $\mathbf{F}$ at P in Section 19.3. So, in view of equation (19.39), we see that rot $\mathbf{F}$ for two dimensions is the same as the rotation of $\mathbf{F}$ about $\mathbf{k}$ in three dimensions.

From equation (19.39), if curl $\mathbf{F}(P) = 0$ or if $\mathbf{n}$ is perpendicular to curl $\mathbf{F}(P)$, then the rotation of $\mathbf{F}$ around $\mathbf{n}$ will be 0 at P. If curl $\mathbf{F} \neq 0$ at P, there is a vortex at P. The flow is irrotational in a region if curl $\mathbf{F} = 0$ for all points in that region.

The rotation at a point will be a maximum when $\mathbf{n}$ is in the direction of curl $\mathbf{F}$, the maximum value being $|\text{curl }\mathbf{F}|$ evaluated at the point. Suppose, for example, that curl $\mathbf{F}(P) \neq \mathbf{0}$ and a paddle wheel is submerged in the fluid at P, as in Figure 19.35. Then the paddle wheel will rotate so long as its axis is not perpendicular to curl $\mathbf{F}(P)$. It will rotate most rapidly when its axis is in the same direction as curl $\mathbf{F}(P)$.

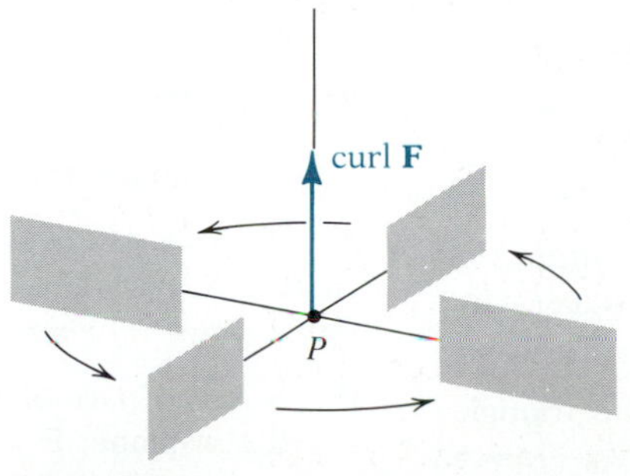

FIGURE 19.35

We close this section by giving a list of equivalent conditions for conservative vector fields, analogous to those in Section 19.2 for two-dimensional conservative fields. Proofs of the equivalences are similar to those for the two-dimensional case and will be omitted. We assume $\mathbf{F} = \langle f, g, h\rangle$ is a continuous vector field in an open region Q of $\mathbf{R}^3$.

1. $\mathbf{F}$ is a gradient field.
2. $\int_C \mathbf{F}\cdot d\mathbf{r}$ is independent of the path.
3. $f\,dx + g\,dy + h\,dz$ is an exact differential.
4. $\int_C \mathbf{F}\cdot d\mathbf{r} = 0$ for every piecewise smooth simple closed curve C in Q.

If Q is simply connected and $\mathbf{F}$ is continuously differentiable, we can add a fifth condition equivalent to these four:

5. curl $\mathbf{F} = \mathbf{0}$ in Q.

Without attempting a precise definition, to say that a three-dimensional region Q is simply connected means that every simple closed curve in Q can be continuously shrunk to a point without going outside Q. We will show that condition 5 is equivalent to the other four by showing that condition 5 implies 4 and condition 1 implies 5.

To prove that condition 5 implies 4, let C be a piecewise smooth, simple closed curve C in Q. Consider any smooth orientable surface S in Q that has C as its boundary, and orient S so that C has positive orientation with respect to S. Then by Stokes' theorem together with condition 5, we have

$$\int_C \mathbf{F}\cdot d\mathbf{r} = \iint_S (\text{curl}\,\mathbf{F})\cdot\mathbf{n}\,d\sigma = 0$$

To show that condition 1 implies 5, let $\mathbf{F} = \nabla\phi$. We ask you to show in Exercise 24 that for a scalar field ϕ that has continuous second partials, $\text{curl}(\nabla\phi) = \mathbf{0}$. So if $\mathbf{F}$ is continuously differentiable, $\text{curl}\,\mathbf{F} = \text{curl}(\nabla\phi) = \mathbf{0}$.

EXERCISE SET 19.6

A

In Exercises 1–6 find curl $\mathbf{F}$.

1. $\mathbf{F} = \langle 2xyz, x - y, y + 2z\rangle$
2. $\mathbf{F} = \langle x^2y^2, x^2z^2, y^2z^2\rangle$
3. $\mathbf{F} = e^xyz\mathbf{i} + e^yxz\mathbf{j} + e^zxy\mathbf{k}$
4. $\mathbf{F} = (y\ln xz)\mathbf{i} + (x\ln yz)\mathbf{j} + (z\ln xy)\mathbf{k}$
5. $\mathbf{F} = \langle \cos xy, \sin xy, e^{-z^2}\rangle$
6. $\mathbf{F} = \langle \ln\sqrt{x^2+y^2}, \tan^{-1}\frac{y}{x}, \frac{z}{xy}\rangle$

In Exercises 7–10 use Stokes' theorem to evaluate the line integral, where C is oriented in a counterclockwise direction when viewed from above.

7. $\int_C xy\,dx + (y+z)\,dy + (x - yz)\,dz$; C is the triangle formed by the traces of the plane $3x + 2y + z = 6$ on the coordinate planes.
8. $\int_C x^2yz\,dx + xy^2z^3\,dy + x^4y^3z^2\,dz$; C is the curve $\mathbf{r}(t) = \langle \cos t, \sin t, 2\rangle$, where $0 \le t \le 2\pi$.
9. $\int_C \mathbf{F}\cdot d\mathbf{r}$; $\mathbf{F} = \langle x + yz, 2yz, x - y\rangle$; C is the intersection of the cylinder $x^2 + y^2 = 4$ and the plane $x + y + z = 1$.
10. $\int_C \mathbf{F}\cdot d\mathbf{r}$; $\mathbf{F} = (x+y)\mathbf{i} + (x - z)\mathbf{j} + (y + z)\mathbf{k}$; C is the intersection of the hemisphere $z = \sqrt{1 - x^2 - y^2}$ and the cylinder $x^2 + y^2 = 2x$.

In Exercises 11–16 verify Stokes' theorem for the given surface S and vector field $\mathbf{F}$. *Assume S is oriented by upward unit normals unless otherwise specified.*

11. S is the part of the surface $z = 4 - x^2 - y^2$ above the xy-plane; $\mathbf{F} = \langle y - z, x - z, x - y\rangle$.

12. S is the part of the plane $x + z = 2$ in the first octant between $y = 0$ and $y = 4$; $\mathbf{F} = \langle x^2z, yz^2, x^2 + z^2\rangle$.
13. S is the triangular surface with vertices $(2, 0, 0)$, $(0, 1, 0)$, and $(0, 0, 3)$; $\mathbf{F} = (1 - xy)\mathbf{i} + (y + z)\mathbf{j} + (2x + 3z)\mathbf{k}$.
14. S is the hemisphere $z = \sqrt{4 - x^2 - y^2}$; $\mathbf{F} = xyz\mathbf{i} + (x + 1)\mathbf{j} + xz^2\mathbf{k}$.
15. S is the part of the cone $z = \sqrt{x^2 + y^2}$ below $z = 1$, oriented by downward unit normals; $\mathbf{F} = \langle 2x - 3z, xy + 2z, y - xz\rangle$.
16. S is the first-octant portion of the surface $z = e^x$ between $y = 0$ and $y = 3$, below $z = 2$, oriented by downward unit normals; $\mathbf{F} = \langle ye^x, ze^x, e^x\rangle$.

In Exercises 17 and 18 show that **F** *is a gradient field, and find a potential function for* **F**.

17. $\mathbf{F} = \langle 2xyz, x^2z, x^2y\rangle$
18. $\mathbf{F} = \langle 2y - 4z, 2x + 5z, 5y - 4x\rangle$

In Exercises 19 and 20 show that the given force field **F** *is conservative, and find the work done by* **F** *on a particle moving from A to B.*

19. $\mathbf{F} = (y - 2z)e^x\mathbf{i} + e^x\mathbf{j} - 2e^x\mathbf{k}$; $A = (0, 0, 0)$, $B = (0, -3, -5)$
20. $\mathbf{F} = 3xz\sqrt{x^2 + y^2}\mathbf{i} + 3yz\sqrt{x^2 + y^2}\mathbf{j} + [(x^2 + y^2)^{3/2} + 2]\mathbf{k}$; $A = (-2, 0, 3)$, $B = (3, 4, -1)$

In Exercises 21 and 22 show that the given expression is an exact differential and find a function for which it is the differential.

21. $e^z \cos x \cos y\, dx - e^z \sin x \sin y\, dy + e^z \sin x \cos y\, dz$
22. $(\ln y + \frac{z}{x})\, dx + (\frac{x}{y} - \ln z)\, dy + (\ln x - \frac{y}{z})\, dz$
23. Show that $\nabla \cdot (\nabla \times \mathbf{F}) = 0$; that is, div(curl **F**) = 0.
24. Show that $\nabla \times (\nabla\phi) = \mathbf{0}$; that is, curl(grad ϕ) = **0**.
25. Let S be a closed surface that satisfies the hypotheses of the divergence theorem and let **F** be a continuously differentiable vector field on some open region that contains S. Prove that

$$\iint_S (\text{curl } \mathbf{F}) \cdot \mathbf{n}\, d\sigma = 0$$

(*Hint:* Use the result of Exercise 23.)

B

26. A fluid has velocity field $\mathbf{F} = y^2\mathbf{i} + z^2\mathbf{j} + x^2\mathbf{k}$. Find its circulation around the curve of intersection of the surfaces $z = x^2 + y^2$ and $z = 2(x + y + 1)$, oriented counterclockwise when viewed from above.
27. Let $\mathbf{a} = \langle a_1, a_2, a_3\rangle$ be any constant vector and $\mathbf{r} = \langle x, y, z\rangle$ be the position vector of a point in $\mathbf{R}^3$.
    (a) Prove that $\text{curl}(\mathbf{a} \times \mathbf{r}) = 2\mathbf{a}$.
    (b) Show that $\int_C (\mathbf{a} \times \mathbf{r}) \cdot d\mathbf{r} = 2 \iint_S \mathbf{a} \cdot \mathbf{n}\, d\sigma$ where S and C satisfy Stokes' theorem.
28. Prove that if u has continuous first partials and v has continuous second partials in an open region Q of $\mathbf{R}^3$, then

$$\text{curl}(u\nabla v) = \nabla u \times \nabla v$$

29. Under the assumptions of Exercise 28 prove the following:
    (a) $\iint_S (\nabla u \times \nabla v) \cdot \mathbf{n}\, d\sigma = \int_C u\nabla v \cdot d\mathbf{r}$
    (b) $\int_C (u\nabla v - v\nabla u) \cdot d\mathbf{r} = 2 \iint_S (\nabla u \times \nabla v)\, d\sigma$
    where S and C are in Q and satisfy the hypotheses of Stokes' theorem.

19.7 SUPPLEMENTARY EXERCISES

1. Let C_1 be the quarter of the unit circle from $(1, 0)$ to $(0, 1)$, and let C_2 be the line segment from $(0, 1)$ to $(-1, 0)$, and let $C = C_1 \cup C_2$. Find $\int_C f(x, y)\, ds$ for $f(x, y) = (x + y)^2$.
2. Let $\mathbf{F} = \langle xy + z, x - y, z^2 - y^2\rangle$ and C be the arc of the helix $\mathbf{r}(t) = \langle 2\cos t, \sin t, 3t\rangle$ from $(2, 0, 0)$ to $(0, 1, \frac{3\pi}{2})$. Find $\int_C \mathbf{F} \cdot d\mathbf{r}$.
3. Find the work done by the force field $\mathbf{F}(x, y, z) = x^2\mathbf{i} - y^2\mathbf{j} + xyz\mathbf{k}$ acting on a unit mass along the curve $\mathbf{r}(t) = \sqrt{t}\mathbf{i} + t^{3/2}\mathbf{j} + (\ln t)\mathbf{k}$ from $(1, 1, 0)$ to $(\sqrt{e}, e^{3/2}, 1)$.
4. Evaluate $\int_C e^{-x} \cos y\, dx + e^{-x} \sin y\, dy$ for each path in the figure on the next page. (*Hint:* There is an easy way.)

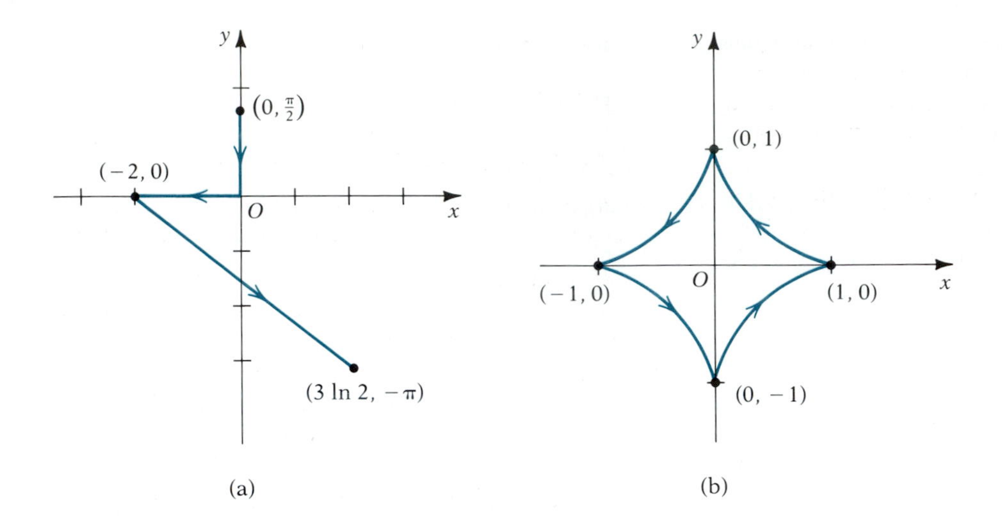

5. For each of the following determine whether $\mathbf{F}$ is a gradient field in the specified domain. If so, find a potential function for $\mathbf{F}$.

(a) $\mathbf{F} = \left\langle \dfrac{y^2}{(x^2 + y^2)^{3/2}} - 2, \dfrac{-xy}{(x^2 + y^2)^{3/2}} + 3 \right\rangle$; $\mathbf{R}^2 - \{(0, 0)\}$

(b) $\mathbf{F} = \left\langle \dfrac{y}{x(x + y)}, \dfrac{x}{y(x + y)} \right\rangle$; $x > 0$, $y > 0$

(c) $\mathbf{F} = y(2x + \tan xy)\mathbf{i} + x(x + \sec^2 xy)\mathbf{j}$; $\mathbf{R}$

(d) $\mathbf{F} = \langle 2x \tanh x^2 + \tanh y, x \operatorname{sech}^2 y - 2y \rangle$; $\mathbf{R}$

In Exercises 6 and 7 use Green's theorem to evaluate $\int_C \mathbf{F} \cdot d\mathbf{r}$, where C has a counterclockwise orientation.

6. $\mathbf{F} = \langle x^2y, x/y^2 \rangle$; C is the boundary of the region enclosed by $xy = 2$, $y = x + 1$, and $y = 1$.

7. $\mathbf{F} = (y^2 - \cos x^3)\mathbf{i} + (2xy + e^{y^2})\mathbf{j}$; $C = C_1 \cup C_2$, where C_1 is the arc of the parabola $y = x^2$ from $(-1, 1)$ to $(2, 4)$ and C_2 is the line segment from $(2, 4)$ to $(-1, 1)$.

In Exercises 8 and 9 evaluate the double integral using Green's theorem.

8. $\iint_R (3x^2 - 2y)\, dA$, where R is the region bounded by the triangle with vertices $(0, 0)$, $(1, 0)$, and $(1, 1)$.

9. $\iint_R (y^4/4 - x^3/3)\, dA$, where R is the region enclosed by the ellipse $4x^2 + 9y^2 = 36$.

In Exercises 10–12 use Green's theorem to find the area of R.

10. R is the region bounded by the quadrilateral with vertices $(0, 0)$, $(1, 2)$, $(3, 4)$, and $(-1, 3)$.

11. R is the region inside the loop of the *strophoid* $x = \cos 2t$, $y = \tan t \cos 2t$. (See the figure in the next column.)

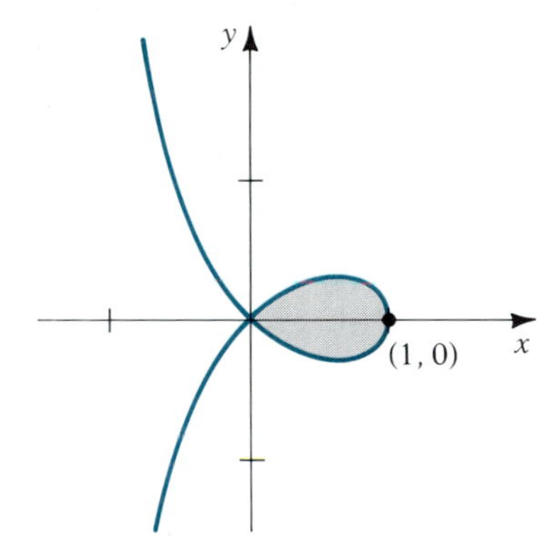

12. R is the region inside $\mathbf{r}(t) = \langle \cos t, \sin 2t \rangle$.

13. Evaluate the integral

$$\int_C \frac{xy^2\, dy - y^3\, dx}{x^2 + y^2}$$

where C is the star-shaped path that has vertices $(\pm 3, 0)$, $(\pm 1, 1)$, $(\pm 1, -1)$, and $(0, \pm 3)$ with a counterclockwise orientation. (*Hint:* Use Theorem 19.8.)

In Exercises 14 and 15 evaluate the surface integral $\iint_S g(x, y, z)\, d\sigma$.

14. $g(x, y, z) = 3y - x^2 - z$; S is the first-octant portion of the plane $z = 2x + y$ below $z = 4$.

15. $g(x, y, z) = 2x^2 + y^2 + z^2$; S is the portion of the sphere $x^2 + y^2 + z^2 = 5$ above $z = 1$.

16. A homogeneous lamina is in the shape of that portion of the hemisphere $z = \sqrt{4 - x^2 - y^2}$ that is inside the cylinder $(x - 1)^2 + y^2 = 1$.
(a) Find its mass and center of mass.
(b) Set up, but do not evaluate, iterated integrals to I_x, I_y, and I_z.

17. Set up, but do not evaluate, two iterated integrals for calculating $\iint_S x^2yz^3\,d\sigma$, using projections on two different planes, where S is the first-octant portion of the cylinder $z = 4 - x^2$ bounded by $y = 0$, $y = x$, and $z = 0$.

18. Let S be the portion of the upper half of the hyperboloid $x^2 + y^2 - z^2 = 3$ below $z = 2$, oriented by downward normals. Find the flux of

$$\mathbf{F} = \langle xy^2, -x^2y, -2\rangle$$

through S.

19. Use the divergence theorem to find the outward flux of $\mathbf{F} = 3x\mathbf{i} + 2y\mathbf{j} + z\mathbf{k}$ through the boundary of the region G between the paraboloids $z = 2 - x^2 - y^2$ and $z = x^2 + y^2$.

In Exercises 20 and 21 use the divergence theorem to evaluate $\iint_S \mathbf{F}\cdot\mathbf{n}\,d\sigma$, where $\mathbf{n}$ is the outer unit normal to S.

20. $\mathbf{F} = \langle x + \sqrt{1 + y^3},\, e^{x^3} + \ln(1 + z^2),\, xz\rangle$; S is the boundary of the region G between $z = 0$, $z = y$, and $y = 2x - x^2$.

21. $\mathbf{F} = 2xz\mathbf{i} - 2yz\mathbf{j} + z^2\mathbf{k}$; S is the boundary of the region G enclosed by the hemisphere

$$z = \sqrt{4 - x^2 - y^2}$$

the plane $z = 0$, and the cylinder $x^2 + y^2 = x + \sqrt{x^2 + y^2}$. (*Hint:* You will recognize the xy trace of the cylinder when you change to cylindrical coordinates.)

22. A function $f(x, y, z)$ is said to be **harmonic** in a region G if it satisfies the **Laplace equation**

$$\frac{\partial^2 f}{\partial x^2} + \frac{\partial^2 f}{\partial y^2} + \frac{\partial^2 f}{\partial z^2} = 0$$

for all (x, y, z) in G. Show that if u and v are harmonic in a bounded region G, with boundary S, where G and S satisfy the conditions of the divergence theorem, then

$$\iint_S u\nabla v\,d\sigma = \iint_S v\nabla u\,d\sigma$$

(See Exercise 24 in Exercise Set 19.5.)

In Exercises 23–25 use Stokes' theorem. In each case C is oriented counterclockwise when viewed from above.

23. Evaluate the integral $\int_C 2yz\,dx + 2xz\,dy + 3xy\,dz$, where C is the triangle with vertices $(3, 0, 0)$, $(0, 2, 0)$, and $(0, 0, 6)$.

24. Evaluate $\int_C \mathbf{F}\cdot d\mathbf{r}$ where $\mathbf{F} = \langle xz, y + 1, y^2\rangle$ and C is the union of the traces of the ellipsoid $2x^2 + 4y^2 + z^2 = 8$ on the first-quadrant parts of the coordinate planes.

25. Find the circulation of the velocity field $\mathbf{F} = yz\mathbf{i} + 8z\mathbf{j} + xy\mathbf{k}$ around the curve of intersection of the cylinder $x^2 + y^2 = 4y$ and the plane $3x + 2y + z = 12$.

26. Verify Stokes' theorem for $\mathbf{F} = \langle 2yz, y^2, xy\rangle$ and S the portion of the hemisphere $z = \sqrt{5 - x^2 - y^2}$ inside the cylinder $x^2 + y^2 = 1$, oriented by upward normals.

27. Make use of the divergence theorem to find the volume of the region bounded by the elliptic paraboloids $z = 2x^2 + 3y^2$ and $z = 4 - 2x^2 - y^2$. (*Hint:* Choose $\mathbf{F}$ so that div $\mathbf{F} = 1$.)

28. Show that the force field $\mathbf{F} = (2xy - 1)\mathbf{i} + (x^2 + 2z^2)\mathbf{j} + (4yz + 3z^2)\mathbf{k}$ is conservative in $\mathbf{R}$ and find the work done by $\mathbf{F}$ on a particle moving from $(0, 1, -1)$ to $(2, 1, 3)$.

29. Let u and v have continuous second partial derivatives in an open region G of $\mathbf{R}^3$, and let C be a sectionally smooth simple closed curve in G. Prove that

$$\int_C (u\nabla v + v\nabla u)\cdot d\mathbf{r} = 0$$

(*Hint:* Let S be a surface in G with C as its boundary such that S and C satisfy the conditions of Stokes' theorem. Then use Exercises 28 and 29 of Exercise Set 19.6.)

CHAPTER 20

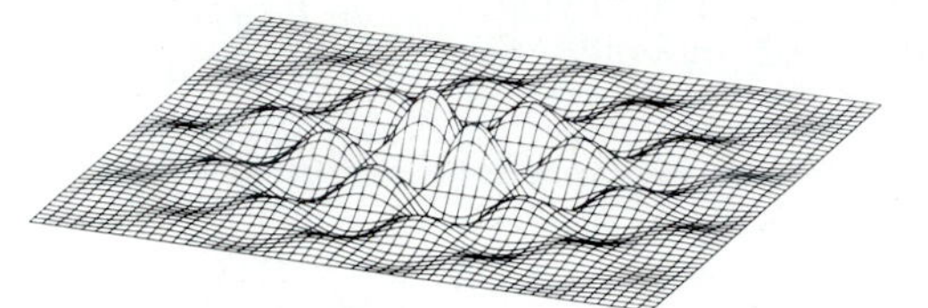

DIFFERENTIAL EQUATIONS

20.1 INTRODUCTION

We introduced the notion of a differential equation in Chapter 6 in our study of exponential growth and decay, and we described a technique of solution called separation of variables. In this chapter we will examine this technique further, along with certain other methods of solution applicable to a broad class of differential equations. We will also consider some additional important applications. In the process we will use many of the techniques of calculus we have studied, so it is appropriate to conclude our study of calculus with this chapter. The chapter can also be viewed as a prelude to what comes next, since a full course in differential equations often is taken as a follow-up to calculus.

The subject of differential equations is almost as old as calculus itself, with both Newton and Leibniz originating many of the concepts. A tremendous body of knowledge on the subject has been built up, but much is still not known, and the subject continues to be an area of intensive research. In one chapter we can only introduce this vast field, but the techniques we will present are nevertheless important. We leave to a later course the deeper theoretical aspects, in particular those having to do with the existence and uniqueness of solutions.

Recall that a differential equation is an equation that involves one or more derivatives of an unknown function. It might therefore be more appropriate to use the name "derivative equation," but this would go against long-established practice. We have seen, in fact, that certain equations containing derivatives can be written in terms of differentials. By a **solution** to a differential equation we mean a function that, when substituted for the dependent variable, causes the equation to be identically true on some interval.

When a differential equation involves derivatives of a function of one independent variable, it is called an **ordinary differential equation.** If there are two or more independent variables, then partial derivatives occur, and the equation is called a **partial differential equation.** For example, the equation of exponential growth and decay, $Q'(t) = kQ(t)$, is an ordinary differential equation, and

$$\frac{\partial u}{\partial t} = c\frac{\partial^2 u}{\partial x^2}$$

is a partial differential equation that arises in the study of heat transfer in physics. We will limit our consideration to ordinary differential equations.

The importance of differential equations in applications lies in their use as mathematical models to describe a wide variety of physical phenomena having to do with rates of change. These occur in biology, chemistry, economics, engineering, physics, and psychology, to name a few. For example, velocities, accelerations, chemical reaction rates, the rate of spread of a disease, learning rates, and growth rates of investments all enter into models that are expressed by differential equations.

One way of classifying differential equations is by their *order*. The order is defined to be the order of the highest derivative that occurs. For example, consider the following:

$$\frac{dy}{dx} = 3x^2$$

$$y'' - 2y' - 3y = e^x$$

$$\frac{d^4x}{dt^2} - (\sin t)\left(\frac{dx}{dt}\right)^2 = t^3 + 1$$

These are of orders 1, 2, and 4, respectively.

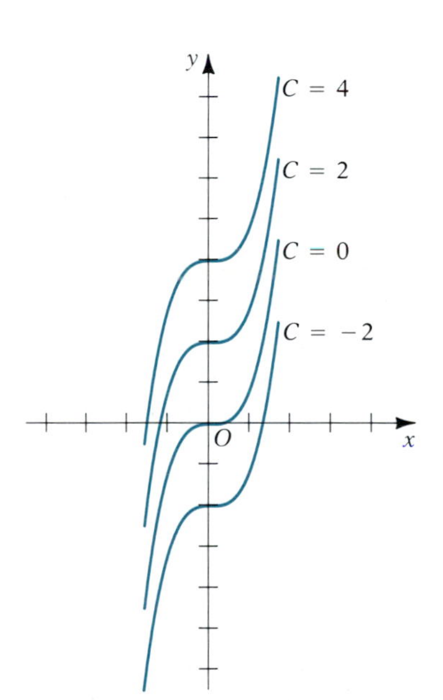

FIGURE 20.1 The family of solutions $y = x^3 + C$

Remark on Notation We will customarily use x as the independent variable and y as the dependent variable, so that when we write y', we mean $y'(x)$, or equivalently $\frac{dy}{dx}$, and similarly for higher derivatives. If some other independent variable is used, we will show it explicitly.

Let us consider the simple first-order differential equation $\frac{dy}{dx} = 3x^2$. As we know, the most general solution is $y = x^3 + C$, and we call this the **general solution.** All solutions to the equation can be obtained from the general solution by appropriate choices of the constant C. This general solution is really a **family** of solutions. We refer to it as a **one-parameter** family, since there is only one arbitrary constant (the parameter) involved. Several members of the family are sketched in Figure 20.1.

In solving differential equations, we often seek such a general solution from which all solutions can be obtained.* For first-order equations, the general solution will always be a one-parameter family. This seems plausible, since you would expect only one integration, giving rise to an arbitrary constant. Similarly, a second-order equation will have a two-parameter family as its general solution, and in general, an nth-order equation will have an n-parameter family as its general solution.

* Sometimes there are exceptional solutions not obtainable from the general solution. These are called **singular** solutions, but we will not consider them here.

Remark It is important to point out that, when speaking of the number of parameters in the general solution, we mean the number of *essential* constants. For example, the equation $y + C_1 = x^2 + C_2$ may appear to define a two-parameter family, but by writing it as $y = x^2 + (C_2 - C_1)$ and defining $C = C_2 - C_1$, we see that it is really the one-parameter family $y = x^2 + C$.

When specific values are given to the arbitrary constants in the general solution, we obtain a **particular** solution. In a first-order equation, specifying the value of y for one value of x is sufficient to determine the constant. For example, suppose we want the particular solution to $y' = 3x^2$ that satisfies $y(1) = 3$. Graphically, this means we want the curve of the family that passes through the point (1, 3). To find it, we substitute $x = 1$ and $y = 3$ into the general solution $y = x^3 + C$. This gives $3 = (1)^3 + C$, or $C = 2$. Thus, $y = x^3 + 2$ is the desired particular solution.

More generally, for a first-order differential equation, if we want the solution $y = y(x)$ that satisfies $y(x_0) = y_0$, we substitute $x = x_0$ and $y = y_0$ into the general solution to find C. The condition $y(x_0) = y_0$ is frequently called an **initial condition,** and the differential equation with such a specified initial condition is called an **initial value problem.** For a second-order equation, two initial conditions are needed to determine a particular solution, as the next example illustrates.

EXAMPLE 20.1 Verify that $y = C_1e^{3x} + C_2e^{-x}$ is a two-parameter family of solutions of the differential equation

$$y'' - 2y' - 3y = 0$$

and find the particular solution in this family that satisfies $y(0) = 2$ and $y'(0) = -1$. (*Note:* In Section 20.5 we will see that the given family is the general solution of this equation and see how it was obtained.)

Solution We first calculate y' and y'' for the given family:

$$y' = 3C_1e^{3x} - C_2e^{-x}$$
$$y'' = 9C_1e^{3x} + C_2e^{-x}$$

So

$$\begin{aligned} y'' - 2y' - 3y &= 9C_1e^{3x} + C_2e^{-x} - 2(3C_1e^{3x} - C_2e^{-x}) - 3(C_1e^{3x} + C_2e^{-x}) \\ &= (9C_1 - 6C_1 - 3C_1)e^{3x} + (C_2 + 2C_2 - 3C_2)e^{-x} = 0 \end{aligned}$$

Thus, the given function is a two-parameter family of solutions. To find C_1 and C_2 that satisfy the given initial conditions, we impose those conditions on y and y' to get

$$C_1 + C_2 = 2$$
$$3C_1 - C_2 = -1$$

The simultaneous solution is easily found to be $C_1 = \frac{1}{4}$, $C_2 = \frac{7}{4}$. So the particular solution in question is

$$y = \tfrac{1}{4}e^{3x} + \tfrac{7}{4}e^{-x}$$ ■

In the next example a particular solution of a second-order equation is obtained from its general solution by specifying values of y at two different values of x. Such values of y are called **boundary values** and are of the general form $y(x_0) = y_0$ and $y(x_1) = y_1$. The terminology reflects the fact that in applications the two known values of y occur at the end points (that is, on the boundary) of the interval under consideration. A differential equation to be solved, subject to such boundary conditions, is known as a **boundary value problem.**

EXAMPLE 20.2 Show that for all choices of the parameters C_1 and C_2, the function $y = C_1 \cos x + C_2 \sin x$ is a solution (it is the general solution) of the equation $y'' + y = 0$ and determine C_1 and C_2 such that the boundary conditions $y(0) = 2$ and $y(\frac{\pi}{2}) = 1$ are satisfied.

Solution Since for the given family $y' = -C_1 \sin x + C_2 \cos x$, and $y'' = -C_1 \cos x - C_2 \sin x$, we see that

$$y'' + y = (-C_1 \cos x - C_2 \sin x) + (C_1 \cos x + C_2 \sin x) = 0$$

So the equation is satisfied. From $y(0) = 2$, we get $C_1 = 2$, and from $y(\frac{\pi}{2}) = 1$, we get $C_2 = 1$. Thus, the solution to the boundary value problem is

$$y = 2 \cos x + \sin x$$

■

It is desirable to express a solution to a differential equation as an explicit function, $y = f(x)$, but sometimes this is difficult or even impossible and we settle for an **implicit solution** in the form $F(x, y) = 0$. The next example illustrates this.

EXAMPLE 20.3 Show that if y is defined implicitly as a function of x by the equation

$$x^3 - 2xy^2 + y^3 - 5 = C$$

then y is a solution to the differential equation

$$\frac{dy}{dx} = \frac{3x^2 - 2y^2}{4xy - 3y^2}$$

Solution Let $F(x, y) = x^3 - 2xy^2 + y^3 - 5 - C$. Then from Chapter 17 we know that

$$\frac{dy}{dx} = -\frac{F_x}{F_y} = -\frac{3x^2 - 2y^2}{-4xy + 3y^2} = \frac{3x^2 - 2y^2}{4xy - 3y^2}$$

so long as the denominator is nonzero. So the function y defined implicitly by $F(x, y) = 0$ satisfies the given differential equation. ■

In applications first- and second-order differential equations are especially useful, and we will concentrate on ways to solve these in the remainder of this chapter. Unfortunately, no single method will work for obtaining solutions to all first-order equations, much less second-order equations. In fact, there are relatively simple appearing first-order equations for which no known method exists for finding the general solution. In the next two sections, however, we will study four methods that will enable us to solve a large class of first-order equations. This will be followed by a section on applications, after which we will examine certain kinds of second-order equations.

It should be mentioned that there are various numerical methods for finding approximate solutions of initial value and boundary value problems. These are studied in courses on differential equations as well as numerical analysis.

EXERCISE SET 20.1

A

In Exercises 1–6 solve the given initial value problem.

1. $\dfrac{dy}{dx} = x - 2;\ y(1) = 4$

2. $\dfrac{dy}{dx} = xe^{-x};\ y(0) = -2$

3. $y' = \tan x$ on $[0, \frac{\pi}{2})$; $y(0) = 3$

4. $y' - x/\sqrt{9 + x^2} = 0;\ y(-4) = 3$

5. $y'' = 2;\ y(1) = 2,\ y'(1) = 3$

6. $d^2x/dt^2 = 3t;\ x(0) = 0,\ x'(0) = 2$

In Exercises 7–12 verify that for all choices of C_1 and C_2 the given function is a solution of the differential equation.

7. $y = C_1e^{-x} + C_2e^{-2x};\ d^2y/dx^2 + 3\,dy/dx + 2y = 0$

8. $y = C_1e^{-2x} + C_2e^{4x};\ y'' - 2y' - 8y = 0$

9. $y = C_1 \cos ax + C_2 \sin ax;\ y'' + a^2y = 0$

10. $y = C_1 \cosh ax + C_2 \sinh ax;\ y'' - a^2y = 0$

11. $x = e^{-t}(C_1 \cos t + C_2 \sin t);\ d^2x/dt^2 + 2\,dx/dt + 2x = 0$

12. $y = C_1x^{-1} + C_2x^{-2};\ x^2y'' + 4xy + 2y = 0 \quad (x > 0)$

In Exercises 13–18 find the particular solution of the problem specified that satisfies the given initial conditions or boundary conditions.

13. Exercise 7; $y(0) = 0,\ y'(0) = -2$

14. Exercise 8; $y(0) = 1,\ y'(0) = 0$

15. Exercise 9; $y(0) = 2,\ y(\frac{\pi}{2a}) = 1$

16. Exercise 10; $y(0) = 1,\ y'(0) = 4a$

17. Exercise 11; $x(0) = 0,\ x'(0) = 1$

18. Exercise 12; $y(1) = 2,\ y(2) = 3$

In Exercises 19–21 verify that any differentiable function $y = y(x)$ defined implicitly by the given equation is a solution of the differential equation.

19. $2x^3y - 3xy^2 + y^4 = 7;\ \dfrac{dy}{dx} = \dfrac{3y(y - 2x^2)}{2x^3 - 6xy + 4y^3}$

20. $y \tan xy - 2x = 1;\ \dfrac{dy}{dx} = \dfrac{2 - y^2 \tan^2 xy}{xy \sec^2 xy + \tan xy}$

21. $\tan^{-1}\dfrac{y}{x} + \ln\sqrt{x^2 + y^2} = 1;\ \dfrac{dy}{dx} = \dfrac{y - x}{y + x}$

B

In Exercises 22 and 23 solve the initial value problem by first showing that the given family is a solution and then determining the proper constants.

22. $y''' + 3y'' - 4y = e^{-2x};\ y(0) = 1,\ y'(0) = -2,\ y''(0) = 3;\ y = C_1e^x + C_2e^{-2x} + C_3xe^{-2x} - \frac{1}{6}x^2e^{-2x}$

23. $y''' + y' - 10y = 10x^2 - 2x + 5;\ y(0) = 2,\ y'(0) = 1,\ y''(0) = 0;\ y = e^{-x}(C_1 \cos 2x + C_2 \sin 2x) + C_3e^{2x} - x^2 - \frac{1}{2}$

In Exercises 24 and 25 use the given equation and those obtained for y' and y'' to eliminate the constants C_1 and C_2, obtaining a differential equation that has the given function as a solution.

24. $y = C_1x + C_2x^2$ **25.** $y = C_1x + C_2x \ln x$

26. The general solutions of the equations

$$\frac{dy}{dx} = F(x, y) \quad \text{and} \quad \frac{dy}{dx} = -\frac{1}{F(x, y)}$$

are called **orthogonal trajectories** of each other. Explain the graphical significance of such orthogonal trajectories. Find the family of orthogonal trajectories of the family $y = x^3 + C$ and draw several members of each family. (*Hint:* First find the differential equation that has $y = x^3 + C$ as its general solution.)

20.2 SOLUTION TECHNIQUES FOR FIRST-ORDER EQUATIONS (I): SEPARABLE EQUATIONS, HOMOGENEOUS EQUATIONS

Separable Equations

A first-order differential equation that can be written in the form

$$\frac{dy}{dx} = \frac{f(x)}{g(y)}, \qquad g(y) \neq 0$$

is said to be **separable.** We can write this in the equivalent *differential form*

$$g(y)\,dy = f(x)\,dx \tag{20.1}$$

and now we see that the variables are "separated"; that is, the terms involving y only appear on one side, and those involving x only appear on the other. The solution is obtained by observing that, since these differentials are equal, the antiderivatives differ by a constant:

$$\int g(y)\,dy = \int f(x)\,dx + C \tag{20.2}$$

Remark Equation (20.1) is really a shorthand notation for

$$g(y(x))y'(x)\,dx = f(x)\,dx$$

since if $y = y(x)$, $dy = y'(x)\,dx$. It therefore expresses the equality of the differentials of two functions of x. If G is an antiderivative of g and F is an antiderivative of F, then we can conclude that

$$G(y(x)) = F(x) + C$$

that is, the functions F and G differ by a constant. This is the meaning of equation (20.2). Formally, however, the procedure we have outlined works, and so we will use it with the understanding of its meaning as explained in this remark.

EXAMPLE 20.4 Find the general solution of the differential equation

$$\frac{dy}{dx} = \frac{2x}{y-1}, \qquad y > 1$$

Solution Separating variables, we get $(y-1)\,dy = 2x\,dx$. Thus, on taking antiderivatives,

$$\int (y-1)\,dy = \int 2x\,dx + C$$

$$\frac{(y-1)^2}{2} = x^2 + C$$

Since $y > 1$, we can solve explicitly for y by taking the positive square root:

$$y - 1 = \sqrt{2(x^2 + C)}$$

or, on writing $C_1 = 2C$,

$$y = 1 + \sqrt{2x^2 + C_1}$$

■

EXAMPLE 20.5 Find the general solution of the differential equation

$$(x^2 + 1)\,dy = xy\,dx$$

Solution Assume for the moment that $y \neq 0$. Then we can separate variables by dividing both sides of the equation by $(x^2 + 1)y$:

$$\frac{dy}{y} = \frac{x\,dx}{x^2 + 1}$$

Thus,

$$\ln|y| = \tfrac{1}{2}\ln(x^2 + 1) + C$$

We can solve for y more readily by writing the constant C in the form $\ln C_1$, where $C_1 > 0$. Then we have

$$\begin{aligned} \ln|y| &= \ln\sqrt{x^2 + 1} + \ln C_1 \\ &= \ln C_1\sqrt{x^2 + 1} \\ |y| &= C_1\sqrt{x^2 + 1} \end{aligned}$$

Thus,

$$y = \pm C_1\sqrt{x^2 + 1}$$

Finally, we write $C_2 = \pm C_1$ so that

$$y = C_2\sqrt{x^2 + 1}$$

Now let us return to the original equation and consider the case $y = 0$. The equation will be satisfied in this case if $dy = 0$ also—that is, if y is the zero function. But if we permit C_2 to be 0, the solution $y = C_2\sqrt{x^2 + 1}$ includes $y = 0$ as a particular case. Thus, we have obtained the general solution. ■

Remark The device of writing the constant of integration in the form $\ln C$ is frequently useful when other natural logarithm terms are involved. We will do this when convenient without further comment.

EXAMPLE 20.6 Solve the initial value problem

$$y' = xy - y + x - 1; \quad y(1) = 3$$

Solution We factor the right-hand side by grouping and then separate variables:

$$\begin{aligned} \frac{dy}{dx} &= y(x - 1) + (x - 1) \\ \frac{dy}{dx} &= (y + 1)(x - 1) \\ \frac{dy}{y + 1} &= (x - 1)\,dx \qquad (y \neq -1) \\ \ln|y + 1| &= \frac{(x - 1)^2}{2} + \ln C \end{aligned}$$

Since we are interested in the solution passing through (1, 3), we can assume $y > -1$, and so we can dispense with absolute values. This gives

$$\ln \frac{y+1}{C} = \frac{(x-1)^2}{2}$$
$$y = Ce^{(x-1)^2/2} - 1$$

Now we substitute $x = 1$ and $y = 3$ to get $3 = C - 1$, or $C = 4$. The desired particular solution is therefore

$$y = 4e^{(x-1)^2/2} - 1$$

■

Homogeneous Equations

Sometimes a differential equation in which the variables are not separable can be transformed into one in which they are. This is true in particular of so-called **homogeneous** first-order differential equations. To explain this we need some definitions.

DEFINITION 20.1 A function $f(x, y)$ is said to be **homogeneous of degree *n*** if for all $t > 0$ for which (tx, ty) is in the domain of f, the equation

$$f(tx, ty) = t^n f(x, y)$$

holds true. ■

EXAMPLE 20.7 Show that each of the following is homogeneous, and give the degree of homogeneity:

(a) $f(x, y) = x^3 - 2x^2y + xy^2$ (b) $g(x, y) = \dfrac{\sqrt{x^2 + y^2}}{x + y}$

Solution (a) $f(tx, ty) = t^3x^3 - 2(t^2x^2)(ty) + (tx)(t^2y^2)$

$$= t^3(x^3 - 2x^2y + xy^2) = t^3 f(x, y)$$

Thus f is homogeneous of degree 3.

(b) $g(tx, ty) = \dfrac{\sqrt{t^2x^2 + t^2y^2}}{tx + ty} = \dfrac{t\sqrt{x^2 + y^2}}{t(x + y)} = \dfrac{\sqrt{x^2 + y^2}}{x + y} = t^0 g(x, y)$

So g is homogeneous of degree 0. ■

DEFINITION 20.2 An equation that can be expressed in the form

$$\frac{dy}{dx} = F(x, y)$$

where F is homogeneous of degree 0 is said to be a first-order **homogeneous differential equation.** ■

EXAMPLE 20.8 Show that each of the following is homogeneous:

(a) $\dfrac{dy}{dx} = \dfrac{\sqrt{x^2 + y^2}}{x + y}$ (b) $\left(y + x\sin\dfrac{x}{y}\right)dx + \left(y\cos\dfrac{x}{y} - 2x\right)dy = 0$

Solution (a) We saw in Example 20.7, part b that $\sqrt{x^2 + y^2}/(x + y)$ is a homogeneous function of degree 0. So by Definition 20.2 the given differential equation is homogeneous.

(b) Solving for $\frac{dy}{dx}$, we get

$$\frac{dy}{dx} = \frac{y + x\sin\dfrac{x}{y}}{2x - y\cos\dfrac{x}{y}}$$

Denote the function on the right by $F(x, y)$. Then

$$F(tx, ty) = \frac{ty + tx\sin\dfrac{tx}{ty}}{2tx - ty\cos\dfrac{tx}{ty}} = \frac{t\left(y + x\sin\dfrac{x}{y}\right)}{t\left(2x - y\cos\dfrac{x}{y}\right)} = F(x, y)$$

So F is homogeneous of degree 0, which means that the differential equation is homogeneous. ■

As in part b of the preceding example, if a differential equation is in the differential form

$$f(x, y)\,dx + g(x, y)\,dy = 0$$

then the homogeneity requirement is satisfied *if f and g are homogeneous functions of the same degree.* This is seen by solving for $\frac{dy}{dx}$:

$$\frac{dy}{dx} = -\frac{f(x, y)}{g(x, y)}$$

and observing that $F = -\frac{f}{g}$ will be homogeneous of degree 0 when f and g are homogeneous of the same degree.

The important property common to all first-degree homogeneous differential equations is that either of the substitutions

$$v = \frac{y}{x} \quad \text{or} \quad v = \frac{x}{y}$$

will invariably transform the equation to one in which the variables are separable. We will illustrate this using $v = \frac{y}{x}$, but you should keep in mind that the second substitution $v = \frac{x}{y}$ also works and in some cases may result in simpler calculations.

Suppose, then, that $\frac{dy}{dx} = F(x, y)$ is homogeneous. Putting $v = \frac{y}{x}$, we obtain

$$y = vx \quad \text{and} \quad \frac{dy}{dx} = v + x\frac{dv}{dx} \tag{20.3}$$

and so we get

$$v + x\frac{dv}{dx} = F(x, vx)$$

Since F is homogeneous of degree 0,

$$F(x, vx) = x^0 F(1, v) = F(1, v)$$

Thus,

$$v + x\frac{dv}{dx} = F(1, v)$$

or

$$\frac{dv}{dx} = \frac{F(1, v) - v}{x}$$

The variables x and v can now be separated:

$$\frac{dv}{F(1, v) - v} = \frac{dx}{x}$$

If F were known, we could solve this equation and then replace v by $\frac{y}{x}$ to get the final result.

Remark The important thing to remember is that once you have identified a differential equation as being homogeneous, the substitution (20.3) should be made (or the similar substitution corresponding to $v = \frac{x}{y}$). Then in each individual case the variables can be separated and the resulting equation solved.

EXAMPLE 20.9 Find the general solution of the differential equation

$$\frac{dy}{dx} = \frac{x + y}{x - y}$$

Solution A quick check shows that the equation is homogeneous. By equations (20.3) we have, on substituting $v = \frac{y}{x}$,

$$v + \frac{dv}{dx} = \frac{1 + v}{1 - v}$$

After solving for $\frac{dv}{dx}$ and simplifying, we obtain

$$\frac{dv}{dx} = \frac{1 + v^2}{1 - v}$$

Now the variables are separable, and we get

$$\int \frac{1 - v}{1 + v^2}\, dv = \int dx + C$$

or

$$\int \frac{dv}{1 + v^2} - \int \frac{v\, dv}{1 + v^2} = x + C$$

$$\tan^{-1} v - \frac{1}{2}\ln(1 + v^2) = x + C$$

Finally, we replace v by $\frac{y}{x}$:

$$\tan^{-1}\frac{y}{x} - \frac{1}{2}\ln\left(\frac{x^2+y^2}{x^2}\right) = x + C$$

It would be difficult to solve for y, so we leave the answer in this implicit form. ■

EXAMPLE 20.10 Solve the initial value problem

$$(2x^2 - 3xy)\,dx + xy\,dy = 0; \qquad y(1) = 3$$

Solution The coefficients of dx and dy are both homogeneous functions of degree 2, so the differential equation is homogeneous. Putting $v = \frac{y}{x}$ gives $y = vx$ and $dy = v\,dx + x\,dv$. Thus, we get

$$(2x^2 - 3x^2v)\,dx + vx^2(v\,dx + x\,dv) = 0$$

At the given initial point $(1, 3)$, $x = 1$ and $y = 3$, so $v = \frac{y}{x} = 3$. Thus, we may limit our consideration to values of x and v near these values. In particular, we suppose $x > 0$ and $v > 2$. Then, on dividing the preceding equation by x^2 and simplifying, we get

$$(v^2 - 3v + 2)\,dx + xv\,dv = 0$$

We now separate variables and use partial fractions:

$$\frac{v\,dv}{(v-2)(v-1)} = -\frac{dx}{x}$$

$$\left(\frac{2}{v-2} - \frac{1}{v-1}\right)dv = -\frac{dx}{x}$$

Thus

$$2\ln(v-2) - \ln(v-1) = -\ln x + \ln C$$

$$\ln\frac{(v-2)^2}{v-1} = \ln\frac{C}{x}$$

$$(v-2)^2 = (v-1)\frac{C}{x}$$

Now we replace v by $\frac{y}{x}$ and simplify to get $(y - 2x)^2 = C(y - x)$. Imposing the initial condition $y(1) = 3$ gives $C = \frac{1}{2}$. Thus, the particular solution in question is given implicitly by

$$y = x + 2(y - 2x)^2$$

■

EXERCISE SET 20.2

A

In Exercises 1–5 use separation of variables to find the general solution.

1. $\dfrac{dy}{dx} = \dfrac{y}{x+1}$

2. $xy\,dy = (y^2 + 1)\,dx$

3. $y' = \dfrac{x+1}{y-2}$

4. $y' = \dfrac{y}{x^2y + x^2}$

5. $e^{-x}\,dy - (1 + y^2)\,dx = 0$

In Exercises 6–10 show that the equation is homogeneous, and find its general solution.

6. $y' = \dfrac{x+y}{x}$

7. $\dfrac{dy}{dx} = \dfrac{y(x+y)}{x(x-y)}$

8. $2x^2\,dx = (x^2 + y^2)\,dy$

9. $\dfrac{dy}{dx} = \dfrac{x}{x+2y}$

10. $xy\,dy = y\left(y + x\cos^2\dfrac{y}{x}\right)dx$

In Exercises 11–20 use whatever method is appropriate to find the general solution. If an initial value is given, find the particular solution that satisfies it.

11. $y' = x\sqrt{4-y^2}$

12. $y' = \dfrac{\ln x}{2xy - 3x}$; $y(1) = 1$

13. $x\,dy = (\sqrt{x^2 - y^2} + y)\,dx$

14. $\tan x \sec y\,dx = \sin^2 y\,dy$

15. $y' = \dfrac{2xy}{x^2 - y^2}$; $y(1) = 2$

16. $\sqrt{y^2 + 4}\,dx + (2y - xy)\,dy = 0$

17. $ye^{x/y}\,dx - (y + xe^{x/y})\,dy = 0$; $y(0) = e$

18. $\dfrac{dy}{dx} = \dfrac{3x - y}{x + y}$

19. $(xy - x + y - 1)\,dy = (xy + x)\,dx$

20. $xy' = y + x\tan\dfrac{y}{x}$; $y(2) = \dfrac{\pi}{3}$

B

21. Show that if $y' = F(x, y)$ is homogeneous, then

$$x = Ce^{\int dv/(F(1,v) - v)}$$

where $v = \frac{y}{x}$.

22. Show that with the substitution $v = \frac{x}{y}$, the homogeneous differential equation $\frac{dy}{dx} = F(x, y)$ is transformed into the separable equation

$$y\frac{dv}{dy} = \frac{1}{F(v, 1)} - v$$

23. Use the method of Exercise 22 to solve the differential equation

$$\frac{dy}{dx} = \frac{y^3 + x^2y}{x^3 + y^3}$$

24. Find the general solution of the equation

$$3(1 + y^2)\,dx + (y^2 - x^3y^2 - 2x^3y + 2y)\,dy = 0$$

25. Solve the initial value problem

$$y' = \frac{5xy + y - 5x - 1}{x^3 - 3x + 2}; \qquad y(0) = 2$$

26. Use the idea of Exercise 26 in Exercise Set 20.1 to find the family of orthogonal trajectories of the family $x^2 - xy + y^2 = C$.

27. Prove that $\frac{dy}{dx} = F(x, y)$ is homogeneous if and only if there is a function G of one variable such that

$$F(x, y) = G(\tfrac{y}{x})$$

20.3 SOLUTION TECHNIQUES FOR FIRST-ORDER EQUATIONS (II): EXACT EQUATIONS AND LINEAR EQUATIONS

Exact Equations

Consider a first-order differential equation in the differential form

$$f(x, y)\,dx + g(x, y)\,dy = 0 \tag{20.4}$$

where f and g have continuous first partial derivatives in some simply connected region R. If the expression on the left is the exact differential of some function ϕ in R, we call the equation an **exact differential equation.** From Section 19.2, therefore, we know the equation is exact if and only if

$$\frac{\partial g}{\partial x} = \frac{\partial f}{\partial y} \tag{20.5}$$

throughout R. In this case equation (20.4) is equivalent to

$$d\phi(x, y) = 0$$

whose solution is

$$\phi(x, y) = C \tag{20.6}$$

(See Exercise 35.) Equation (20.6) therefore is an implicit form of the solution to equation (20.4).

To find ϕ we use the method of Section 19.2. To review this, since

$$d\phi = f(x, y)\,dx + g(x, y)\,dy$$

we have

$$\frac{\partial \phi}{\partial x} = f(x, y) \quad \text{and} \quad \frac{\partial \phi}{\partial y} = g(x, y) \tag{20.7}$$

We can integrate the first of these, holding y fixed, and obtain a "constant" of integration that is a function of y. Then imposing the second condition of (20.7), we can determine this function of y. Alternately we can integrate the second equation of (20.7), holding x fixed, and determine the integration constant by imposing the first condition.

EXAMPLE 20.11 Show that the equation

$$(3x^2 - y)\,dx + (2y - x)\,dy = 0$$

is exact, and find its solution.

Solution Since

$$\frac{\partial}{\partial x}(2y - x) = -1 \quad \text{and} \quad \frac{\partial}{\partial y}(3x^2 - y) = -1$$

we see by condition (20.5) that the equation is exact throughout all of $\mathbf{R}^2$. We next find ϕ:

$$\phi(x, y) = \int (3x^2 - y)\,dx = x^3 - xy + C_1(y) \qquad \text{Holding } y \text{ fixed}$$

$$\frac{\partial \phi}{\partial y} = -x + C_1'(y) = 2y - x$$

$$C_1'(y) = 2y$$

$$C_1(y) = y^2$$

(At this point we can omit the constant of integration, since it can be combined with the constant in the final answer.) So

$$\phi(x, y) = x^3 - xy + y^2$$

and by equation (20.6) the solution (in implicit form) to the original problem is

$$x^3 - xy + y^2 = C$$

■

EXAMPLE 20.12 Solve the initial value problem

$$\frac{dy}{dx} = \frac{xe^{-x} + \cos y}{x \sin y - \cos y}; \qquad y(0) = \pi$$

Solution The equation can be written in the differential form

$$(xe^{-x} + \cos y)\, dx + (\cos y - x \sin y)\, dy = 0$$

The variables cannot be separated, and the equation is not homogeneous. Testing for exactness, we have

$$\frac{\partial}{\partial y}(xe^{-x} + \cos y) = -\sin y, \qquad \frac{\partial}{\partial x}(\cos y - x \sin y) = -\sin y$$

Thus, the equation is exact. To find ϕ we choose this time to begin by integrating the coefficient of dy with respect to y:

$$\phi(x, y) = \int (\cos y - x \sin y)\, dy = \sin y + x \cos y + C_1(x)$$

Now we must have $\partial\phi/\partial x = xe^{-x} + \cos y$. So

$$\cos y + C_1'(x) = xe^{-x} + \cos y$$
$$C_1'(x) = xe^{-x}$$
$$C_1(x) = -xe^{-x} - e^{-x} \qquad \text{Obtained by integrating by parts}$$

The general solution is therefore

$$\sin y + x \cos y - xe^{-x} - e^{-x} = C$$

Imposing the initial condition $y(0) = \pi$ gives $-1 = C$. The particular solution in question can therefore be written in the form

$$\sin y + x \cos y - e^{-x}(x + 1) + 1 = 0$$

■

Linear Equations

An nth-order differential equation is said to be **linear** if it can be written in the form

$$a_n(x)y^{(n)} + a_{n-1}(x)y^{(n-1)} + \cdots + a_1(x)y' + a_0(x)y = g(x) \tag{20.8}$$

with $a_n(x) \neq 0$. *Linear* is used because each term on the left involves y or one of its derivatives to the first degree only. To emphasize this reference to y and its derivatives, we sometimes say that the equation is *linear in y*. The coefficient functions $a_0(x), \ldots, a_n(x)$ may be highly nonlinear. For $n = 1$, we have the first-order linear equation

$$a_1(x)y' + a_0(x)y = g(x)$$

Under the assumption that $a_1(x) \neq 0$, we can divide both sides by this leading coefficient and obtain the equation in the form

$$y' + P(x)y = Q(x) \tag{20.9}$$

where $P(x) = a_0(x)/a_1(x)$ and $Q(x) = g(x)/a_1(x)$. We will refer to equation (20.9) as the **standard form** of the first-order linear differential equation. We will assume P and Q are continuous on some interval I.

To develop a technique for solving problems of this type, let us first rewrite equation (20.9) in the differential form

$$[P(x)y - Q(x)]\,dx + dy = 0$$

Our approach will be to see whether there is a function $\mu(x)$ that we can multiply by to make this an exact differential equation. Such a function is called an **integrating factor.** Let us assume for the moment that $\mu(x) > 0$ and $\mu'(x)$ exists on the interval I. Then, on multiplying by $\mu(x)$, we have

$$[\mu(x)P(x)y - \mu(x)Q(x)]\,dx + \mu(x)\,dy = 0 \tag{20.10}$$

The requirement for exactness is that

$$\frac{\partial}{\partial y}[\mu(x)P(x)y - \mu(x)Q(x)] = \frac{\partial}{\partial x}[\mu(x)]$$

that is,

$$\mu'(x) = \mu(x)P(x) \tag{20.11}$$

We can separate variables in equation (20.11) to get

$$\frac{d\mu}{\mu} = P(x)\,dx$$

Thus

$$\mu(x) = e^{\int P(x)dx} \tag{20.12}$$

is a suitable integrating factor. Note that $\mu(x)$ *is* positive. We can take the constant of integration as 0, since we want any integrating factor, so we might as well use the simplest one. We could proceed now to find a function ϕ for which the left-hand side of equation (20.10) is the exact differential, using the technique we have developed. However, in this case there is an easier way. If we multiply both sides of the standard form (20.9) by $\mu(x)$, we have

$$\mu(x)y' + \mu(x)P(x)y = \mu(x)Q(x)$$

But by equation (20.11) this is (suppressing the parentheses)

$$\mu y' + y\mu' = \mu Q$$

and the left-hand side is precisely the derivative of the product μy. Thus,

$$\frac{d}{dx}[\mu y] = \mu Q \tag{20.13}$$

So

$$\mu y = \int \mu Q\,dx + C$$

and

$$y = \frac{1}{\mu}\int \mu Q\,dx + \frac{C}{\mu}$$

Direct substitution will show that this is the solution we were seeking.

The key things to remember from this derivation are these:

1. If the equation is linear, write it in the standard form

   $y' + P(x)y = Q(x)$

2. Find the integrating factor

   $\mu(x) = e^{\int P(x)\,dx}$

3. Multiply both sides of the standard form by $\mu(x)$, thereby obtaining

   $\frac{d}{dx}[\mu y] = \mu Q$

   (It is a good idea to check at this stage that the left-hand side is indeed the derivative of the product of μ and y.)

4. Take antiderivatives and solve for y.

EXAMPLE 20.13 Solve the differential equation

$$xy' + 2y = 2x^2 + 3x, \qquad x > 0$$

Solution The equation is linear in y. To put it in standard form we divide by x:

$$y' + \tfrac{2}{x}y = 2x + 3$$

So $P(x) = \frac{2}{x}$ and $Q(x) = 2x + 3$. The integrating factor is therefore

$$\mu(x) = e^{\int P(x)\,dx} = e^{\int 2/x\,dx} = e^{2\ln x} = e^{\ln x^2} = x^2$$

(Note the use of the properties of logarithms and exponentials.) We next multiply both sides of the standard form by $\mu(x)$:

$$x^2y' + 2xy = 2x^3 + 3x^2$$

If our theory is correct, the left-hand side should be the derivative of μy—that is, of x^2y—and a quick check shows that this is correct. So we have

$$\frac{d}{dx}(x^2y) = 2x^3 + 3x^2$$

$$x^2y = \frac{x^4}{2} + x^3 + C$$

$$y = \frac{x^2}{2} + x + \frac{C}{x^2}$$

■

A Word of Caution A common mistake in this procedure is to forget the constant of integration in the next to last step. The constant is essential, however; otherwise we have not found the general solution.

EXAMPLE 20.14 Solve the initial value problem

$$\frac{dy}{dx} = \frac{x + y}{x - 1}; \qquad y(2) = 3$$

Solution We see the equation is linear in y, since both y and y' appear to the first degree. In standard form it is

$$y' + \left(\frac{1}{1 - x}\right)y = \frac{x}{x - 1}$$

Since we want the solution passing through (2, 3), we restrict x so that $x > 1$. Thus,

$$\mu(x) = e^{\int dx/(1-x)} = e^{-\ln|1-x|} = e^{\ln 1/|1-x|} = \frac{1}{|1-x|} = \frac{1}{x-1}$$

since $x > 1$. We multiply both sides by $\mu(x)$ to get

$$\frac{d}{dx}\left[\left(\frac{1}{x-1}\right)y\right] = \frac{x}{(x-1)^2} \qquad \text{(Verify.)}$$

Hence,

$$\frac{1}{x-1}\,y = \int \frac{x}{(x-1)^2}\,dx = \int \left[\frac{1}{x-1} + \frac{1}{(x-1)^2}\right] dx$$

(using partial fractions). So

$$\frac{1}{x-1}\,y = \ln(x-1) - \frac{1}{x-1} + C$$

$$y = (x-1)\ln(x-1) - 1 + C(x-1)$$

This is the general solution. Now we put $x = 2$ and $y = 3$, getting $C = 4$. Substituting this and simplifying give the particular solution

$$y = (x-1)\ln(x-1) + 4x - 5$$

■

EXERCISE SET 20.3

A

In Exercises 1–6 show that the given equation is exact, and find its general solution. If an initial condition is given, find the corresponding particular solution.

1. $\dfrac{dy}{dx} = \dfrac{x-y+1}{x-2}$

2. $(e^y + 2x)\,dx + (xe^y - 2xy)\,dy = 0$

3. $2xyy' = 3x^2 - y^2;\ y(1) = 2$

4. $\dfrac{dy}{dx} = \dfrac{y\sin x - \cos y}{\cos x - x\sin y}$

5. $\dfrac{y}{x}\,dx + (y + \ln x)\,dy = 0;\ y(1) = 4$

6. $y' = \dfrac{x\ln\cos^2 y}{x^2\tan y - 3y^2};\ y(0) = 2$

In Exercises 7–12 show that the given equation is linear, and find its general solution. If an initial condition is given, find the corresponding particular solution.

7. $y' + y = x$ **8.** $y' - 2xy = x$

9. $x\dfrac{dy}{dx} = x^2 - y;\ y(2) = 1$

10. $dy = (\sin x + y\tan x)\,dx \quad (-\frac{\pi}{2} < x < \frac{\pi}{2})$

11. $\dfrac{dy}{dx} + \dfrac{y}{1-x} = 1 - x^2;\ y(0) = 0$

12. $xy' = x^4e^x + 2y;\ y(1) = 0$

Solve Exercises 13–27 by any method.

13. $(x - 2y)\,dx + (3y - 2x)\,dy = 0$ **14.** $y' = x - xy$

15. $(y+1)\,dx + (x-1)\,dy = 0;\ y(0) = 2$

16. $y' + y\cot x = 0 \quad (0 < x < \pi)$

17. $y' - \dfrac{2xy}{x^2+1} = 1;\ y(1) = \pi$ **18.** $\dfrac{dy}{dx} = \dfrac{x^2 - y^2}{y(2x - y)}$

19. $\left(\dfrac{y}{x} - e^x + 2\right)dx + (\ln xy + 3y^2 - 1)\,dy = 0 \quad (x > 0)$

20. $(\cosh x)\,dy = (\cosh^3 x + y\sinh x)\,dx$

21. $\left(\dfrac{x}{\sqrt{x^2+y^2}} + 1\right)dx + \left(\dfrac{y}{\sqrt{x^2+y^2}} - 2\right)dy = 0;\ y(3) = 4$

22. $x\,dy + y(x+1)\,dx = x^2\,dx \quad (x > 0)$

23. $\left(\dfrac{x-y}{x}\right)dx + (y - \ln x)\,dy = 0 \quad (x > 0)$

24. $(e^x + 1)\,dy = e^x(x - y)\,dx$

B

25. $\frac{dy}{dx} = \frac{1+y^2}{y(1+x+y^2)}$ (*Hint:* Show it is linear in x.)

26. $y' + y\sin x = \sin 2x;\ y(\frac{\pi}{2}) = 3$

27. $[\ln(x^2+y^2)]\,dx = 2(\tan^{-1}\frac{y}{x})\,dy$

28. An equation of the form

$$y' + P(x)y = y^n Q(x) \qquad (n \neq 1)$$

is called a **Bernoulli** equation. Show that the substitution $v = y^{1-n}$ transforms it into an equation that is linear in v. (*Hint:* First divide both sides by y^n.)

In Exercises 29 and 30, use the method of Exercise 28 to find the general solution.

29. $y' + 2xy = xy^2$

30. $y' + \frac{2y}{x} = \sqrt{y}\sin x;\ y(\frac{\pi}{2}) = 4$

31. Show that a differential equation of the form

$$f(x, y)\,dx + g(x, y)\,dy = 0$$

if not already exact, can be made into an exact equation by multiplying both sides by an integrating factor of the form $\mu(x)$ provided $(f_y - g_x)/g$ is a function of x only. What is $\mu(x)$ in this case?

32. Find a condition analogous to that in Exercise 31 that will ensure the existence of an integrating factor of the form $\mu(y)$.

In Exercises 33 and 34 use the method of Exercise 31 or 32 to find an integrating factor, and solve the equation.

33. $(y^2 + 4x)\,dx + y\,dy = 0$

34. $2\sin x\cos^2 y\,dx + \cos x \sin 2y\,dy = 0$

35. Prove that if ϕ is differentiable in a region R of $\mathbf{R}^2$, then $y = y(x)$ is a solution to the differential equation $d\phi(x, y) = 0$ if and only if $\phi(x, y(x)) = C$.

20.4 APPLICATIONS OF FIRST-ORDER EQUATIONS

In this section we illustrate some of the many applications of first-order differential equations. The first example is familiar from Section 6.6.

EXAMPLE 20.15 The isotope plutonium-241 has a half-life of approximately 13.2 yr. Assume that the substance decays at a rate proportional to the amount present. Of a given initial amount, how much will remain after 5 yr? How long will it take for 90% of the original amount to decay?

Solution Let $Q(t)$ denote the quantity present at time t (in years). Then we assume that $\frac{dQ}{dt} = kQ$. Let $Q(0) = Q_0$. This initial value problem can be solved either by separating variables or by treating the equation as linear in Q. Choosing separation of variables, we get

$$\frac{dQ}{Q} = k\,dt$$
$$\ln Q = kt + \ln C$$
$$\ln \frac{Q}{C} = kt$$
$$Q(t) = Ce^{kt}$$

Letting $t = 0$ gives $Q_0 = C$. So the solution is

$$Q(t) = Q_0 e^{kt}$$

To find k we use the fact that

$$Q(13.2) = 0.5Q_0 \qquad \text{(Why?)}$$

Thus,

$$0.5Q_0 = Q_0 e^{13.2k}$$

So

$$k = \frac{\ln 0.5}{13.2} = -0.0525 \quad \text{(approximately)}$$

and the formula becomes

$$Q(t) = Q_0 e^{-0.0525t}$$

For $t = 5$, we have

$$Q(5) = Q_0 e^{(-0.0525)5} = 0.769 Q_0$$

So after 5 yr about 77% of the original amount remains.

To find how long it takes for 90% to decay, we set $Q(t) = 0.1 Q_0$ and solve for t, since if 90% has decayed, 10% remains:

$$0.1 Q_0 = Q_0 e^{-0.0525t}$$

$$t = \frac{\ln 0.1}{-0.0525} = 43.86 \text{ yr}$$

■

As our second example we solve a model for population growth that in general gives a more realistic estimate than the Malthus model. In most models the instantaneous rate of change of the size Q of the population is assumed to be some function of Q:

$$\frac{dQ}{dt} = f(Q) \tag{20.14}$$

For the Malthus model $f(Q)$ is taken to be kQ for $k > 0$, and just as in the previous example, the solution is $Q(t) = Q_0 e^{kt}$. In 1837 the Belgian mathematician Pierre-François Verhulst (1804–1849) proposed modifying the Malthus model by taking $f(Q) = \alpha Q - \beta Q^2$, where α and β are positive constants. Substracting the term βQ^2 was Verhulst's way of accounting for competition among members of the population. The next example shows one way to solve Verhulst's equation.

EXAMPLE 20.16 Solve the Verhulst equation for population growth

$$\frac{dQ}{dt} = \alpha Q - \beta Q^2 \qquad (\alpha > 0, \beta > 0)$$

if $Q(0) = Q_0$. Analyze the result.

Solution One way to solve the problem is by separating variables, making use of partial fractions in the integration. We choose to illustrate a method suggested in Exercise 28 of Exercise Set 20.3 whereby we make a substitution that transforms the equation into a linear one. Equations of this type are called **Bernoulli equations.** We first rewrite it as

$$\frac{dQ}{dt} - \alpha Q = -\beta Q^2$$

and then divide through by Q^2:

$$Q^{-2} \frac{dQ}{dt} - \alpha Q^{-1} = -\beta$$

Now let $v = Q^{-1}$:

$$\frac{dv}{dt} = -Q^{-2}\frac{dQ}{dt}$$

When we substitute and change all signs, we get

$$\frac{dv}{dt} + \alpha v = \beta$$

and now this is linear in v. The integrating factor is $\mu(t) = e^{\alpha t}$, and on multiplying by this, we get

$$\frac{d}{dt}[ve^{\alpha t}] = \beta e^{\alpha t}$$

Thus,

$$ve^{\alpha t} = \frac{\beta}{\alpha}e^{\alpha t} + C$$

The constant C can be found by putting $t = 0$, observing that $v(0) = \frac{1}{Q(0)} = \frac{1}{Q_0}$. Thus

$$C = \frac{1}{Q_0} - \frac{\beta}{\alpha} = \frac{\alpha - \beta Q_0}{\alpha Q_0}$$

Substituting this, we can solve for v and then invert to obtain Q, since $Q = \frac{1}{v}$. The details of the algebra are omitted, but the result is

$$Q(t) = \frac{\alpha Q_0}{\beta Q_0 + (\alpha - \beta Q_0)e^{-\alpha t}} \tag{20.15}$$

We make several observations about this solution. First, if β is small relative to α, the solution differs little from the Malthus solution $Q(t) = Q_0 e^{\alpha t}$, as you can see by neglecting the terms involving β. Second, if the denominator is written in the form

$$\beta Q_0(1 - e^{-\alpha t}) + \alpha e^{-\alpha t}$$

we see that for t near 0, the factor $1 - e^{-\alpha t} \approx 0$, so that again the solution is approximately $Q_0 e^{\alpha t}$. Finally, let us see what happens as $t \to \infty$:

$$\lim_{t\to\infty} Q(t) = \frac{\alpha Q_0}{\beta Q_0 + 0} = \frac{\alpha}{\beta}$$

Here, the result is strikingly different from that of Malthus. There is a limit to the size of the population, whereas in the Malthus model the growth is unlimited.

The Verhulst differential equation is sometimes referred to as a **logistic equation,** and its solution (20.15) is called the **law of logistic growth.** The graph of Q is shown in Figure 20.2. ■

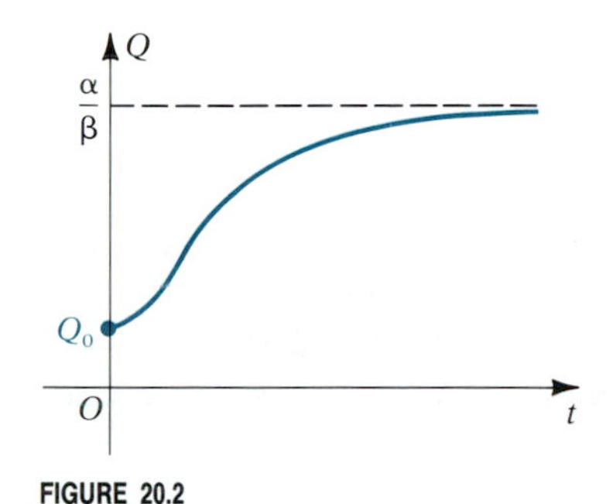

FIGURE 20.2

Remark If we put $m = \alpha/\beta$, the Verhulst equation can be written in the form

$$\frac{dQ}{dt} = \beta Q(m - Q)$$

and since m is the limiting maximum size of the population, this can be interpreted as saying that the rate of change of the population at any time is jointly proportional to the population at that time and the remaining capacity to expand. The solution (20.15) can then be written as

$$Q(t) = \frac{mQ_0}{Q_0 + (m - Q_0)e^{-m\beta t}}$$

(Compare Exercise 24 in Exercise Set 6.6.)

The third example is typical of *mixing problems.*

EXAMPLE 20.17 A tank with a capacity of 100 gal has 50 gal of pure water in it initially. Brine containing $\frac{1}{2}$ lb of salt per gallon is then fed into the tank at the rate of 4 gal/min, and the well-stirred mixture is allowed to drain out at the rate of 2 gal/min. Find how much salt is in the tank just as it begins to overflow.

Solution Let $Q(t)$ be the number of pounds of salt in the tank t minutes after the brine begins to enter. Because initially there was fresh water in the tank, we see that $Q(0) = 0$. The basic idea of this and other similar mixture problems is that the rate of change of Q is the rate at which salt enters the tank minus the rate at which it leaves:

$$\frac{dQ}{dt} = (\text{rate salt comes in}) - (\text{rate salt goes out})$$

Since 4 gal of brine come in each minute and each gallon contains $\frac{1}{2}$ lb of salt, it follows that salt enters the tank at the rate of 2 lb/min. To determine how much salt leaves the tank each minute, it is necessary to find the *concentration* (number of pounds per gallon) of salt in the tank at any given time. The net total increase in liquid in the tank is 2 gal/min (4 gal come in and 2 go out). So after t minutes it contains $50 + 2t$ gallons of solution (see Figure 20.3). Contained within this solution are $Q(t)$ pounds of salt. So we have at time t,

$$\text{concentration of salt in tank} = \frac{Q(t)}{50 + 2t} \text{ lb/gal}$$

Since 2 gal of solution leave the tank each minute, the amount of salt leaving is

$$(2 \text{ gal/min})\left(\frac{Q}{50 + 2t} \text{ lb/gal}\right) = \frac{Q}{25 + t} \text{ lb/min}$$

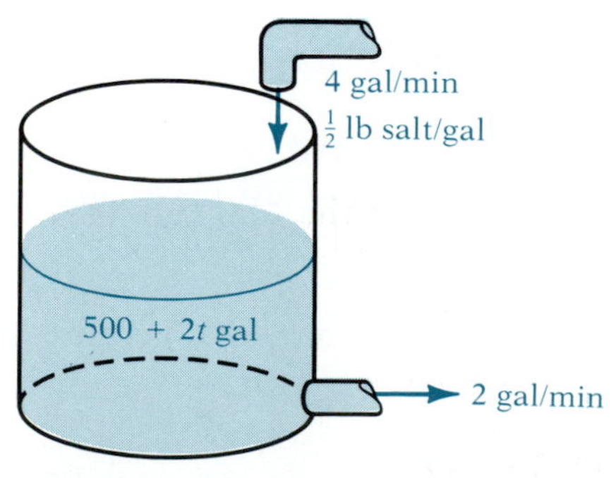

FIGURE 20.3

So the differential equation is

$$\frac{dQ}{dt} = 2 - \frac{Q}{25 + t} \quad \text{or} \quad \frac{dQ}{dt} + \left(\frac{1}{25 + t}\right)Q = 2$$

This is linear in Q with integrating factor

$$\mu(t) = e^{\int dt/(25+t)} = e^{\ln(25+t)} = 25 + t$$

Multiplying by this, we get

$$\frac{d}{dt}[(25 + t)Q] = 2(25 + t)$$

$$(25 + t)Q = (25 + t)^2 + C$$

$$Q(t) = 25 + t + \frac{C}{25 + t} \quad \text{(before the tank overflows)}$$

Since $Q(0) = 0$, we find that $C = -625$. The tank will overflow when $50 + 2t = 100$, or $t = 25$. The amount of salt in the tank at that instant is

$$Q(25) = 50 - \tfrac{625}{50} = 37.5 \text{ lb}$$ ■

In the next example we consider a falling body problem similar to ones we studied in Chapter 3 (see Example 3.26) but with the important difference that we now take air resistance into account.

EXAMPLE 20.18 From a height of s_0 feet above the ground an object of mass m is given an upward initial velocity of v_0 feet per second. Assume that air resistance has magnitude proportional to the speed. Find the height above the ground and the velocity of the object at time t. If the object could continue indefinitely without hitting the ground, what would be its limiting velocity?

Solution As shown in Figure 20.4, let the distance $s(t)$ be measured positively upward from the ground, and let the velocity be $v(t)$. Then $s(0) = s_0$ and $v(0) = v_0$. The forces that act on the body are the pull of gravity, with magnitude mg, acting downward, and air resistance acting opposite to the direction of motion. Let $k > 0$ be the proportionality constant for the air resistance. When the object is moving upward, $v > 0$, and the resisting force is downward and so equals $-kv$. When the object is moving downward $v < 0$, and the resisting force is upward, but again $-kv$ is the correct value, since this is positive when v is negative. Thus, in all cases the resisting force is $-kv$.

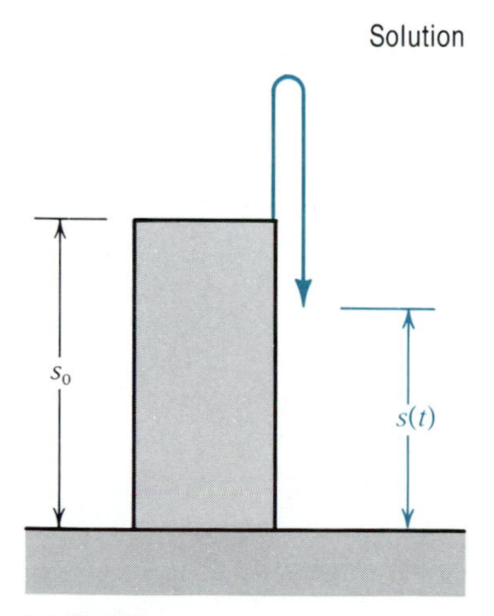

FIGURE 20.4

Now we use Newton's second law of motion, which for bodies of constant mass can be written as $m\frac{dv}{dt} = F$, where F is the net force acting on the body. In this case, then, we have

$$m\frac{dv}{dt} = -kv - mg$$

This can be written in the standard linear form

$$\frac{dv}{dt} + \frac{k}{m}v = -g$$

An integrating factor is $\mu(t) = e^{kt/m}$, and if we apply this to both sides we get

$$\frac{d}{dt}[ve^{kt/m}] = -ge^{kt/m}$$

Taking antiderivatives, we find

$$ve^{kt/m} = -\frac{gm}{k} e^{kt/m} + C_1$$

Putting $t = 0$ and $v = v_0$ gives

$$C_1 = v_0 + \frac{gm}{k}$$

Thus,

$$v(t) = \left(v_0 + \frac{gm}{k}\right) e^{-kt/m} - \frac{gm}{k}$$

Letting $t \to \infty$, we obtain

$$\lim_{t \to \infty} v(t) = -\frac{gm}{k}$$

So if the object could continue in motion indefinitely, it would approach a limiting velocity of $\frac{gm}{k}$ downward. Observe that this value is independent of the initial velocity.

To find $s(t)$ we integrate $v(t)$:

$$s(t) = -\frac{m}{k}\left(v_0 + \frac{gm}{k}\right) e^{-kt/m} - \frac{gmt}{k} + C_2$$

Since $s(0) = s_0$, we find that

$$C_2 = s_0 + \frac{m}{k}\left(v_0 + \frac{gm}{k}\right)$$

So we have

$$s(t) = s_0 + \frac{m}{k}\left(v_0 + \frac{gm}{k}\right)(1 - e^{-kt/m}) - \frac{gmt}{k}$$

■

In certain types of chemical reactions, called *second-order reactions,* two substances, say A and B, react with each other to form a third substance C. This is written symbolically as $A + B \to C$. Suppose that initially the concentration of substance A is a (usually given in moles per liter) and of substance B is b. If, after t seconds, x moles per liter of A and of B have decomposed, the concentrations of what is left of A and of B are $a - x$ and $b - x$, respectively, and the concentration of C is x. For a second-order reaction, the rate of change of x is jointly proportional to $a - x$ and $b - x$:

$$\frac{dx}{dt} = k(a - x)(b - x)$$

In the next example we solve this equation.

EXAMPLE 20.19 For the second-order chemical reaction described above, find the concentration $x(t)$ of C after time t.

Solution The initial value problem to be solved is

$$\frac{dx}{dt} = k(a - x)(b - x); \qquad x(0) = 0$$

We separate variables and make use of partial fractions to perform the integration:

$$\frac{dx}{(a-x)(b-x)} = k\,dt$$

$$\frac{1}{a-b}\int\left[\frac{1}{b-x} - \frac{1}{a-x}\right]dx = kt + C$$

$$\frac{1}{a-b}\ln\left(\frac{a-x}{b-x}\right) = kt + C$$

When $t = 0$, $x = 0$. So

$$C = \frac{1}{a-b}\ln\frac{a}{b}$$

and we obtain

$$\frac{1}{a-b}\left[\ln\frac{a-x}{b-x} - \ln\frac{a}{b}\right] = kt$$

$$\ln\frac{b(a-x)}{a(b-x)} = (a-b)kt$$

$$\frac{b(a-x)}{a(b-x)} = e^{(a-b)kt}$$

Now we solve for x. You should verify that the answer can be put in the form

$$x(t) = \frac{ab(e^{akt} - e^{bkt})}{ae^{akt} - be^{bkt}}$$ ■

In the theory of the flow of electricity through an electrical circuit, it is known that the algebraic sum of all voltage drops is 0. This result is known as **Kirchhoff's second law.** We will apply this to a circuit of the type shown in Figure 20.5, known as an **RL series circuit.** The customary units used are: $R =$ ohms, $L =$ henrys, $I(t) =$ amperes, and $E(t) =$ volts. It can be shown that the voltage drops are: across the resistor $= RI$ and across the inductor $= L\frac{dI}{dt}$. The electromotive force (for example, a generator) provides the only voltage increase. Thus, by Kirchhoff's second law, we have

$$L\frac{dI}{dt} + RI = E(t) \tag{20.16}$$

We make use of this equation in the next example.

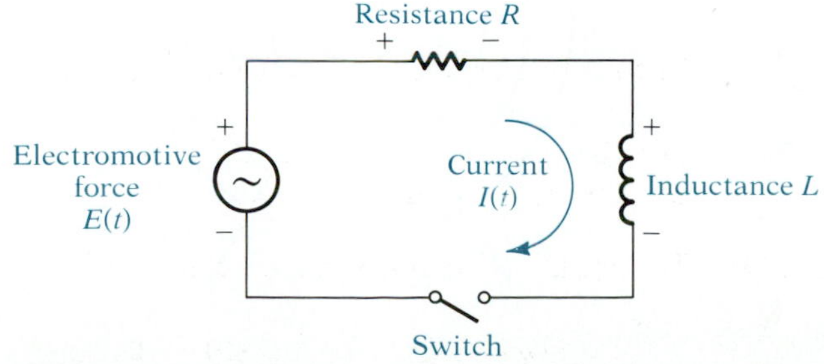

FIGURE 20.5

EXAMPLE 20.20 Find the current $I = I(t)$ in an RL circuit in which the electromotive force $E(t) = V$, a constant. Assume $I(0) = 0$. Draw the graph of I.

Solution From equation (20.16) we have

$$L\frac{dI}{dt} + RI = V$$

In standard form this linear equation is

$$\frac{dI}{dt} + \frac{R}{L}I = \frac{V}{L}$$

and an integrating factor is $e^{(R/L)t}$. So we have

$$\frac{d}{dt}[Ie^{(R/L)t}] = \frac{V}{L}e^{(R/L)t}$$

$$Ie^{(R/L)t} = \frac{V}{R}e^{(R/L)t} + C$$

From the initial condition $I(0) = 0$, we get $C = -\frac{V}{R}$. Thus,

$$I(t) = \frac{V}{R}(1 - e^{-(R/L)t})$$

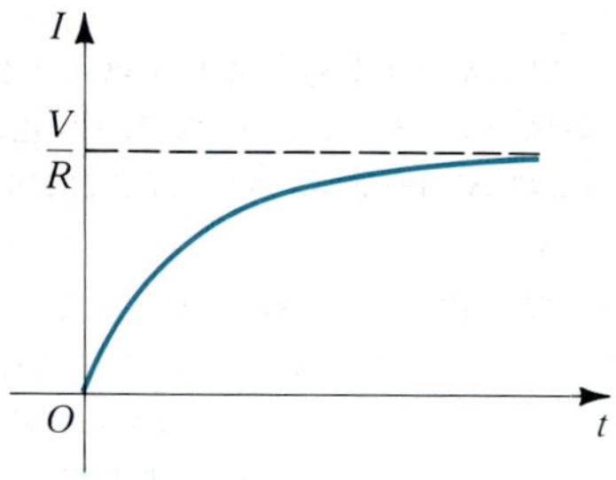

The graph is shown in Figure 20.6. ■

FIGURE 20.6

For our final example we give a geometric application. Suppose two one-parameter families of curves $F(x, y) = C_1$ and $G(x, y) = C_2$ have the property that whenever a curve of one family intersects a curve of the second family, it does so orthogonally (that is, the tangent lines are perpendicular). Then we say that the families are **orthogonal trajectories** of each other (see Exercise 26 in Exercise Set 20.1 and Exercise 26 in Exercise Set 20.2). For example, the family of circles $x^2 + y^2 = C_1$ and the family of lines $y = C_2x$ are easily seen to be orthogonal trajectories of each other (see Figure 20.7).

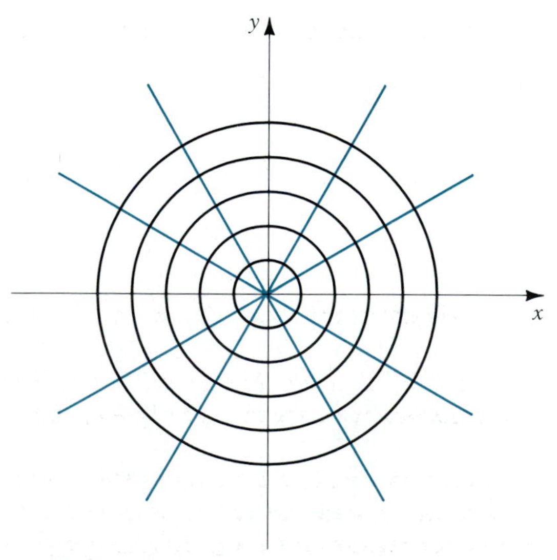

FIGURE 20.7 Orthogonal trajectories of circles $x^2 + y^2 = C_1$ are the lines $y = C_2x$

If the differential equation of one family is

$$\frac{dy}{dx} = \frac{f(x, y)}{g(x, y)}$$

then the differential equation of the family of orthogonal trajectories is

$$\frac{dy}{dx} = -\frac{g(x, y)}{f(x, y)}$$

This is the key to finding orthogonal trajectories of a given family, as we show in the example that follows.

EXAMPLE 20.21 Find the family of orthogonal trajectories of the family $3x^2 + y^2 = Cx$.

Solution To find the differential equation that has the given family as its general solution, we first differentiate both sides of the given equation:

$$6x + 2yy' = C$$

Now we substitute $C = 6x + 2yy'$ into the original equation:

$$3x^2 + y^2 = 6x^2 + 2xyy'$$

So, if $x \neq 0$ and $y \neq 0$,

$$y' = \frac{y^2 - 3x^2}{2xy}$$

The differential equation of the orthogonal trajectories is therefore

$$y' = \frac{2xy}{3x^2 - y^2}$$

To solve this, we observe that it is homogeneous, and so we substitute $y = vx$:

$$v + x\frac{dv}{dx} = \frac{2x^2v}{3x^2 - v^2x^2}$$

$$x\frac{dv}{dx} = \frac{2v}{3 - v^2} - v$$

$$x\frac{dv}{dx} = \frac{v^3 - v}{3 - v^2}$$

We now separate variables and make use of partial fractions to perform the integration:

$$\int \left[\frac{(3 - v^2)}{v(v + 1)(v - 1)}\right] dv = \int \frac{dx}{x}$$

$$\left(-\frac{3}{v} + \frac{1}{v + 1} + \frac{1}{v - 1}\right) dv = \frac{dx}{x}$$

$$-3 \ln|v| + \ln|v + 1| + \ln|v - 1| = \ln|x| + \ln C_1$$

$$\ln\left|\frac{v^2 - 1}{v^3}\right| = \ln C_1|x|$$

$$\left|\frac{v^2 - 1}{v^3}\right| = C_1|x|$$

FIGURE 20.8

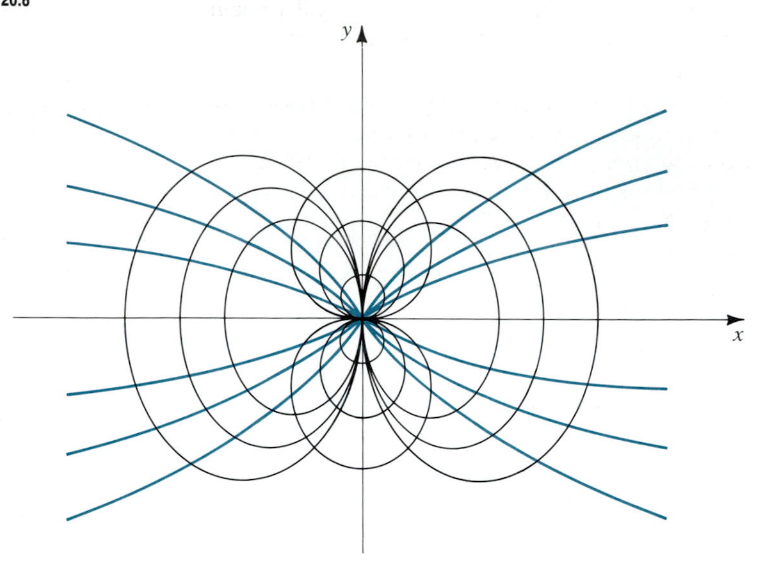

If we permit C_1 to be either positive or negative, we can remove the absolute values:

$$\frac{v^2 - 1}{v^3} = C_1 x$$

On replacing v by $\frac{y}{x}$ and simplifying, we get

$$x^2 = y^2(1 - C_1 y)$$

as the equation of the orthogonal trajectories. Figure 20.8 shows several curves of the two families. ■

EXERCISE SET 20.4

A

1. Assume that the rate of growth of a culture of bacteria is proportional to the number present. If a culture of 100 bacteria grows to a size of 150 after 2 hr, how long will it take for the size to double? How many will be present after 5 hr?
2. When interest on an investment is compounded continuously, the rate of growth of the amount of money in the account is proportional to the amount present, where the proportionality constant is the annual interest rate (expressed as a decimal). If an initial amount of P dollars is invested at r% compounded continuously, find a formula for the amount in the account at time t years after it is invested. What value of r would cause the amount to double after 7 yr?
3. Use the result of Exercise 2 to determine the amount after 30 yr that results from an initial amount of $1000 invested at 6% compounded continuously. How many years will it take for the amount in the account to be $5000?
4. If 10% of a radioactive substance decays after 33 yr, how much will remain after 200 yr? What is the half-life of the substance?
5. Radium-226 has a half-life of 1620 yr. How long will it take for 80% of a given amount to decay?
6. If 75% of a quantity of the radioactive isotope uranium-232 remains after 30 yr, how much will be present after 100 yr? What is its half-life?

7. The half-life of thorium-228 is approximately 1.913 yr. If 100 g are on hand , how much will remain after 3 yr? How long will it take for the amount left to be 10 g?

8. Newton's law of cooling states that the surface temperature of an object changes at a rate proportional to the difference between the temperature of the object and that of the surrounding medium. Let $T(t)$ be the temperature of the object at time t, and let T_m be the temperature of the surrounding medium. If $T(0) = T_0$ ($T_0 > T_m$), find a formula for $T(t)$.

9. Use the result of Exercise 8 to find the temperature of a body that was initially at 30°C, 30 min after it is placed in a medium of constant temperature 5°C if it cools to 20°C after 5 min. According to the model, will the object ever cool to 5°C? What does your answer tell about the model?

10. A thermometer registering 70°F is taken outside where the temperature is 28°F. After 5 min the thermometer registers 55°F. When will it register 30°F? (See Exercise 8.)

11. A vat contains 200 gal of a 20% dye solution. A 40% solution of the same dye is then fed into the tank at the rate of 10 gal/min, and the well-mixed solution is drained off at the same rate. Find an expression for the amount of pure dye in the tank at any time t. What is the concentration of dye in the tank after 30 min?

12. A tank with a 500-gal capacity initially contains 100 gal of brine with a salt concentration of 2 lb/gal. Fresh water is allowed to enter the tank at the rate of 3 gal/min, and the well-stirred mixture is drained from the tank at the rate of 1 gal/min. How much salt does the tank contain after 20 min? How long will it take for the concentration to be reduced to $\frac{1}{4}$ lb/gal? Will this happen before the tank overflows? What is the salt concentration just as the tank overflows?

13. Solve the Verhulst model (Example 20.16) by separating variables.

14. Supply all the missing details in the solution of the Verhulst model in Example 20.16. Show that for $Q_0 < \frac{\alpha}{\beta}$, $Q(t)$ is an increasing function, and for $Q_0 > \frac{\alpha}{\beta}$, it is decreasing.

15. A body of mass m is dropped from a balloon high above the earth. Assume air resistance is proportional to the square of the velocity. Use Newton's second law of motion to find a formula for the velocity at time t. What is the theoretical limiting velocity as $t \to \infty$? (Take the downward direction as positive.)

16. In Example 20.19 if the concentrations a and b are the same, then the solution given is not valid. (Why?) Solve the equation under this assumption, where $x(0) = 0$.

17. A 12-V battery is connected to an RL series circuit with a 6-ohm resistance and an inductance of 1 henry. If $I(0) = 0$, find $I(t)$.

18. Solve equation (20.16) if $E(t) = E_0 \cos \omega t$ and $I(0) = I_0$.

19. A reasonably accurate model for the rate of dissemination of a drug injected into the bloodstream is given by

$$\frac{dy}{dt} = a - by, \qquad a > 0,\ b > 0$$

where $y = y(t)$ is the concentration of the drug at time t. Find $y(t)$ if the initial concentration of the drug is y_0.

20. Find the orthogonal trajectories of the family $x^2 + y^2 = Cx$. Sketch several members of each family.

B

21. Third-order chemical reactions are rare in the gaseous state, and those that do occur are almost always of the form $2A + B \to C$. For example, $2NO + O_2 \to 2NO_2$. If x is the concentration of C at time t, and a and b are the initial concentrations of A and B, respectively, then the rate of change of x is given by

$$\frac{dx}{dt} = k(a - 2x)^2(b - x)$$

Solve this equation with the initial condition $x(0) = 0$.

22. One model for the spread of an infectious disease in a community of N individuals is that the rate of change of the number $x(t)$ infected is jointly proportional to the number infected and the number uninfected. Set up and solve the relevant differential equation. Suppose $x(0) = 1$. Determine the limiting value of $x(t)$ as $t \to \infty$. (Observe that this model is mathematically the same as the Verhulst population growth model.)

23. The air in a room with dimensions $12 \times 20 \times 8$ ft initially contains 1% carbon dioxide. Air that contains 0.02% carbon dioxide is then forced into the room at the rate of 200 cu ft/min, and the well-circulated air leaves the room at the same rate. What will be the concentration of carbon dioxide after 5 min? After how long a time will the concentration be reduced to 0.05%?

24. A problem that is important in ecology has to do with the rise and decline of two species, where one species is a predator and the other is its prey. This is called the predator–prey problem. The following model is known as the **Lotka–Volterra model,** proposed by the American mathematician and biologist A. J. Lotka (1880–1949) and the Italian mathematician Vito Volterra (1860–1940) in the study of the interaction between the Alaskan snowshoe hare and the lynx:

$$\begin{cases} \dfrac{dx}{dt} = x(a - by) \\ \dfrac{dy}{dt} = y(-c + dx) \end{cases}$$

where a, b, c, and d are positive constants.

Here $x(t)$ is the population of the predator at time t and $y(t)$ is the population of the prey. By dividing the second equation by the first, find y as a function of x.

In Exercises 25–27 different models for population growth are given by specifying $f(Q)$ in equation (20.14). Solve each model under the assumption that $Q(0) = Q_0$. The constants α and β are positive unless otherwise indicated.

25. $f(Q) = \alpha Q \cos \beta t$: this model is useful in describing seasonal growth, in which the population periodically increases and decreases. Sketch the graph of the solution.

26. $f(Q) = \alpha Q - \beta Q \ln Q$: this is called the **Gompertz model** and has applications in economic theory as well as population growth. For $\alpha > \beta \ln Q_0$ analyze and draw the graph of Q under each of the following circumstances:
(a) $\alpha > 0,\ \beta > 0$ (b) $\alpha > 0,\ \beta < 0$

27. $f(Q) = (Q - m)(\alpha - \beta Q)$: this is an appropriate model for the growth of a population that becomes extinct when its numbers are too small. Show that when $Q_0 < m$, $Q(t_1) = 0$ for some t_1. Find t_1.

20.5 SECOND-ORDER LINEAR EQUATIONS WITH CONSTANT COEFFICIENTS: THE HOMOGENEOUS CASE

The general second-order linear differential equation has the form

$$a_2(x)y'' + a_1(x)y' + a_0(x)y = g(x)$$

in which $a_2(x) \neq 0$. We limit our consideration to the case where the coefficient functions are constants, say $a_2(x) = a$, $a_1(x) = b$, and $a_0(x) = c$. When $g(x) = 0$ for all x on some interval, the equation is said to be **homogeneous.** Here we are using *homogeneous* in a different sense from that in Section 20.2. We will study the homogeneous case in this section and the nonhomogeneous case in the next section.

Consider, then, the second-order linear homogeneous differential equation with constant coefficients

$$ay'' + by' + cy = 0, \qquad a \neq 0 \tag{20.17}$$

It is convenient to introduce the notation $L[y]$ for the left-hand side of this equation. We refer to L as a **linear differential operator.** Saying it is linear means that it has the following two properties:

$$L[ky] = kL[y] \quad \text{for any constant } k$$
$$L[y_1 + y_2] = L[y_1] + L[y_2]$$

These can be readily verified using the corresponding properties of derivatives.

Suppose now that $y_1 = y_1(x)$ and $y_2 = y_2(x)$ are any two solutions of equation (20.17). In operator notation this means that $L[y_1] = 0$ and $L[y_2] = 0$.

Then, for any two constants C_1 and C_2, we have

$$L[C_1y_1] = C_1L[y_1] = 0 \quad \text{and} \quad L[C_2y_2] = C_2L[y_2] = 0$$

and so also

$$L[C_1y_1 + C_2y_2] = 0$$

The combination $C_1y_1 + C_2y_2$ is called a **linear combination** of y_1 and y_2. So we can say that any linear combination of two solutions of equation (20.17) is also a solution.

Clearly, if C_1 and C_2 are both 0, then $C_1y_1 + C_2y_2 = 0$. If these are the only values of C_1 and C_2 that cause the linear combination $C_1y_1 + C_2y_2$ to be 0, then y_1 and y_2 are said to be **linearly independent.** Otherwise, they are **linearly dependent.** For example, $y_1 = e^x$ and $y_2 = e^{-2x}$ are linearly independent, since no linear combination $C_1e^x + C_2e^{-2x}$ can be 0 except when both C_1 and C_2 are 0. On the other hand, $y_1 = 2x - 1$ and $y_2 = \frac{1}{2} - x$ are linearly dependent, since with $C_1 = 1$ and $C_2 = -2$,

$$C_1y_1 + C_2y_2 = 1(2x - 1) + (-2)(\tfrac{1}{2} - x) = 0$$

The importance of linear independence is that when y_1 and y_2 are solutions of equation (20.17) that have this property, $y = C_1y_1 + C_2y_2$ is the general solution of equation (20.17). We leave a proof of this to a course in differential equations.

Remark The definition we have given for the linear independence of two functions extends in a natural way to more than two. In the particular case of two functions, however, it is readily seen (see Exercise 38) that they are linearly independent if and only if *neither function is a multiple of the other.*

With this background, we can obtain the general solution of equation (20.17) if we can find two linearly independent solutions of it. For the remainder of this section we will concentrate on how to find two such solutions. It is natural to expect a solution of the form $y = e^{mx}$, since each derivative is a multiple of the function itself and hence, by choosing m appropriately, we can expect to have the terms in $L[y]$ add to give 0. We have $y' = me^{mx}$ and $y'' = m^2e^{mx}$, so that

$$\begin{aligned} L[y] = ay'' + by' + cy &= am^2e^{mx} + bme^{mx} + ce^{mx} \\ &= (am^2 + bm + c)e^{mx} \end{aligned}$$

This will be 0 if and only if

$$am^2 + bm + c = 0 \tag{20.18}$$

Equation (20.18) is called the **auxiliary equation** for equation (20.17). (It is sometimes referred to as the **characteristic equation** also.) Designate the roots of equation (20.18) by m_1 and m_2:

$$m_1 = \frac{-b + \sqrt{b^2 - 4ac}}{2a} \qquad m_2 = \frac{-b - \sqrt{b^2 - 4ac}}{2a}$$

We therefore have found that $y_1 = e^{m_1x}$ and $y_2 = e^{m_2x}$ are solutions of equation (20.17). Furthermore, if $m_1 \neq m_2$, they are linearly independent solutions, and so the general solution is

$$y = C_1e^{m_1x} + C_2e^{m_2x}$$

EXAMPLE 20.22 Find the general solution of the equation $y'' - 3y' - 4y = 0$.

Solution The auxiliary equation is $m^2 - 3m - 4 = 0$. (Notice how the pattern of the auxiliary equation resembles the pattern of the differential equation.) Its solutions are found to be $m_1 = 4$ and $m_2 = -1$. Thus, e^{4x} and e^{-x} are linearly independent solutions, and the general solution is

$$y = C_1 e^{4x} + C_2 e^{-x}$$

■

Let us now consider the case $m_1 = m_2$. This happens when the discriminant $b^2 - 4ac = 0$, so that

$$m_1 = \frac{-b + \sqrt{0}}{2a} \quad \text{and} \quad m_2 = \frac{-b - \sqrt{0}}{2a}$$

In this case we get only one solution:

$$y_1 = e^{-bx/2a}$$

By direct substitution into equation (20.17), it can be shown that a second solution, independent from y_1, is

$$y_2 = xe^{-bx/2a}$$

(See Exercise 39.) Thus, writing $m = -\frac{b}{2a}$, the general solution is

$$y = C_1 e^{mx} + C_2 x e^{mx}$$

EXAMPLE 20.23 Find the general solution of the equation $y'' - 4y' + 4y = 0$.

Solution The auxiliary equation is $m^2 - 4m + 4 = 0$, which has the double root $m = 2$. So the general solution is

$$y = C_1 e^{2x} + C_2 x e^{2x}$$

■

When the discriminant of the auxiliary equation, $b^2 - 4ac$, is negative, the roots are *complex conjugates*

$$m_1 = \alpha + i\beta \qquad m_2 = \alpha - i\beta$$

where i is the imaginary unit, satisfying $i^2 = -1$. In this case the general solution

$$y = C_1 e^{(\alpha + i\beta)x} + C_2 e^{(\alpha - i\beta)x} \tag{20.19}$$

though correct, is not very convenient to work with. A more suitable form can be obtained using the **Euler formula**:

$$e^{i\theta} = \cos\theta + i\sin\theta \tag{20.20}$$

To see why this is true, we begin by defining $e^{i\theta}$ as the result of replacing x by $i\theta$ in the Maclaurin series for e^x. (See Section 11.8.) When we do this and use $i^2 = -1$, so that $i^3 = -i$, $i^4 = 1, \ldots,$ we obtain

$$\begin{aligned} e^{i\theta} &= 1 + (i\theta) + \frac{(i\theta)^2}{2!} + \frac{(i\theta)^3}{3!} + \frac{(i\theta)^4}{4!} + \frac{(i\theta)^5}{5!} + \cdots \\ &= \left[1 - \frac{\theta^2}{2!} + \frac{\theta^4}{4!} - \cdots\right] + i\left[\theta - \frac{\theta^3}{3!} + \frac{\theta^5}{5!} - \cdots\right] \end{aligned}$$

where we have rearranged the terms into the so-called real and imaginary parts. But from Section 11.8 we recognize that the bracketed series are the Maclaurin series for $\cos\theta$ and $\sin\theta$, respectively. Thus, $e^{i\theta} = \cos\theta + i\sin\theta$.

Returning now to the solution in the form (20.19), we write

$$\begin{aligned} y &= C_1 e^{(\alpha+i\beta)x} + C_2 e^{(\alpha-i\beta x)} \\ &= e^{\alpha x}[C_1 e^{i\beta x} + C_2 e^{-i\beta x}] \\ &= e^{\alpha x}[C_1(\cos\beta x + i\sin\beta x) + C_2(\cos\beta x - i\sin\beta x)] \\ &= e^{\alpha x}[(C_1 + C_2)\cos\beta x + i(C_2 - C_1)\sin\beta x] \\ &= e^{\alpha x}[C_3\cos\beta x + C_4\sin\beta x] \end{aligned}$$

where C_3 and C_4 are new arbitrary constants. Note that in applying Euler's formula to $e^{-i\beta x}$, we used $\cos(-\beta x) = \cos\beta x$ and $\sin(-\beta x) = -\sin\beta x$. When the roots of the auxiliary equation are imaginary, we will use the result just obtained to write the general solution; that is, if the roots are $\alpha + i\beta$ and $\alpha - i\beta$, then the general solution is

$$y = e^{\alpha x}(C_1\cos\beta x + C_2\sin\beta x) \tag{20.21}$$

(There is no longer any need to call the constants C_3 and C_4.)

EXAMPLE 20.24 Find the general solution of each of the following:
(a) $y'' + 4y = 0$ (b) $y'' - 2y' + 3y = 0$

Solution (a) The auxiliary equation is $m^2 + 4 = 0$, with roots $m_1 = 2i$ and $m_2 = -2i$. So we have $\alpha = 0$ and $\beta = 2$. Thus, by equation (20.21) the general solution is

$$y = C_1\cos 2x + C_2\sin 2x$$

(b) The auxiliary equation is $m^2 - 2m + 3 = 0$ with the solution

$$m = \frac{2 \pm \sqrt{4 - 12}}{2} = 1 \pm i\sqrt{2}$$

The general solution is therefore

$$y = e^x(C_1\cos\sqrt{2}x + C_2\sin\sqrt{2}x)$$

■

The next two examples illustrate an initial value problem and a boundary value problem that involve second-order equations.

EXAMPLE 20.25 Solve the initial value problem

$$y'' - 2y' - 8y = 0; \qquad y(0) = 1,\ y'(0) = -4$$

Solution The auxiliary equation is

$$\begin{aligned} m^2 - 2m - 8 &= 0 \\ (m - 4)(m + 2) &= 0 \\ m &= 4,\ -2 \end{aligned}$$

Thus, the general solution is

$$y = C_1 e^{4x} + C_2 e^{-2x}$$

From $y(0) = 1$, we get $1 = C_1 + C_2$. Now we calculate y' and then apply the second initial condition:

$$y' = 4C_1e^{4x} - 2C_2e^{-2x}$$
$$-4 = 4C_1 - 2C_2 \quad \text{or} \quad 2C_1 - C_2 = -2$$

To find C_1 and C_2 we solve the system

$$\begin{cases} C_1 + C_2 = 1 \\ 2C_1 - C_2 = -2 \end{cases}$$

simultaneously. This gives $C_1 = -\frac{1}{3}$, $C_2 = \frac{4}{3}$. Thus, the desired particular solution is

$$y = -\tfrac{1}{3}e^{4x} + \tfrac{4}{3}e^{-2x}$$

■

EXAMPLE 20.26 Find the solution of the equation $x''(t) + x(t) = 0$ that satisfies the boundary conditions $x(0) = 3$ and $x(\frac{\pi}{2}) = 5$.

Solution The auxiliary equation $m^2 + 1 = 0$ has roots $\pm i$, and so by equation (10.21) the general solution is

$$x(t) = C_1 \cos t + C_2 \sin t$$

Since $x(0) = C_1$ and $x(\frac{\pi}{2}) = C_2$, we see immediately that $C_1 = 3$ and $C_2 = 5$. Thus, the particular solution in question is

$$x(t) = 3 \cos t + 5 \sin t$$

■

We summarize the results of this section.

To find the general solution of

$$ay'' + by' + cy = 0$$

first solve the auxiliary equation

$$am^2 + bm + c = 0$$

Denote the roots by m_1 and m_2.

Case 1. If m_1 and m_2 are real and unequal, the general solution is

$$y = C_1e^{m_1x} + C_2e^{m_2x}$$

Case 2. If m_1 and m_2 are real and equal, with common value m, say, then the general solution is

$$y = C_1e^{mx} + C_2xe^{mx}$$

Case 3. If m_1 and m_2 are the complex conjugates $m_1 = \alpha + i\beta$ and $m_2 = \alpha - i\beta$, with $\beta \neq 0$, then the general solution is

$$y = e^{\alpha x}(C_1 \cos \beta x + C_2 \sin \beta x)$$

EXERCISE SET 20.5

A

In Exercises 1–16 find the general solution. Unless otherwise indicated, the independent variable is x.

1. $y'' + 3y' + 2y = 0$

2. $y'' - y' - 2y = 0$

3. $y'' + 8y' + 16y = 0$

4. $y'' - 2y' + y = 0$

5. $y'' + 9y = 0$

6. $y'' + 9y' = 0$

7. $y'' - 2y' + 5y = 0$

8. $y'' + y' + 2y = 0$

9. $2y'' - 3y' - 5y = 0$

10. $4y'' + 5y' - 6y = 0$

11. $\dfrac{d^2y}{dt^2} - 3\dfrac{dy}{dt} - 4y = 0$

12. $2x''(t) + 6x'(t) + 5x(t) = 0$

13. $2\dfrac{d^2u}{dx^2} - 3\dfrac{du}{dx} = 0$

14. $\dfrac{d^2v}{dx^2} - 9v - 0$

15. $4s''(t) - 12s'(t) + 9s(t) = 0$

16. $\dfrac{d^2s}{dt^2} + 3\dfrac{ds}{dt} + 4s = 0$

In Exercises 17–24 solve the given initial value problem or boundary value problem.

17. $y'' + 2y' - 15y = 0;\ y(0) = 2,\ y'(0) = 3$

18. $2y'' - 3y' - 9y = 0;\ y(0) = -2,\ y'(0) = -6$

19. $y'' - 3y' = 0;\ y(0) = 2,\ y'(0) = 9$

20. $y'' - 9y = 0;\ y(0) = 0,\ y'(0) = 1$

21. $y'' - 4y' + 4y = 0,\ y(0) = 2,\ y(1) = e^2$

22. $y'' + 4y = 0;\ y(0) = 5,\ y(\frac{\pi}{4}) = 3$

23. $y'' + y = 0;\ y(\frac{\pi}{6}) = 0,\ y(\frac{\pi}{3}) = 1$

24. $y'' - 4y' + 5y = 0;\ y(0) = \frac{2}{3},\ y'(0) = -\frac{1}{3}$

B

25. Show that when the roots m_1 and m_2 of the auxiliary equation are real and unequal, the general solution of the differential equation (20.17) can be written in the form

$$y = e^{ux}(C_1 \cosh vx + C_2 \sinh vx)$$

where $u = -b/2a$ and $v = \sqrt{b^2 - 4ac}/2a$.

26. Find the general solution of $y'' + 3y' + y = 0$ in the form given in Exercise 25. Find the particular solution that satisfies $y(0) = 4,\ y'(0) = -3$.

27. An equation of the form

$$ax^2y'' + bxy' + cy = 0, \qquad a \neq 0,\ x > 0$$

is called an **Euler equation.** Show that the substitution $t = \ln x$ changes this into a linear equation with constant coefficients.

In Exercises 28–30 use the result of Exercise 27 to find the general solution.

28. $x^2y'' + 3xy' + 3y = 0,\ x > 0$

29. $\dfrac{d^2s}{dt^2} + \dfrac{1}{t}\dfrac{ds}{dt} = 0,\ t > 0$ (*Hint:* Multiply by t^2.)

30. $v\dfrac{d^2\psi}{dv^2} - \dfrac{1}{v}\psi = 0,\ v > 0$ (*Hint:* Multiply by v.)

In Exercises 31–36 assume that the ideas of this section extend to higher-order linear homogeneous equations with constant coefficients, and find the general solution.

31. $y''' - 2y'' - 3y' = 0$

32. $y^{(4)} - 16y = 0$

33. $y^{(4)} + 5y'' + 4y = 0$

34. $y''' - 3y' + 2y = 0$

35. $y''' - y'' + 4y' - 4y = 0$. Also find the particular solution that satisfies $y(0) = 0,\ y'(0) = 1,\ y''(0) = 2$.

36. $y''' - 3y'' + 3y' - y = 0$

37. Prove that $L[ky] = kL[y]$ and that $L[y_1 + y_2] = L[y_1] + L[y_2]$.

38. Prove that two functions $y_1(x)$ and $y_2(x)$ are linearly independent on an interval if and only if neither is a multiple of the other.

39. Prove that if the solutions of the auxiliary equation of $ay'' + by' + cy = 0$ are equal, then $y_2 = xe^{-bx/2a}$ is a solution of the differential equation.

20.6 SECOND-ORDER LINEAR EQUATIONS WITH CONSTANT COEFFICIENTS: THE NONHOMOGENEOUS CASE

In this section we consider equations of the form

$$ay'' + by' + cy = g(x), \qquad a \neq 0$$

where $g(x)$ is a continuous function on some interval I and is not identically 0. We refer to $g(x)$ as the **nonhomogeneous term.** Using the operator notation, we can write this equation as

$$L[y] = g(x) \tag{20.22}$$

The corresponding homogeneous equation

$$L[y] = 0 \tag{20.23}$$

plays a special role in obtaining the solution of the nonhomogeneous problem, and for this reason it is called the **complementary equation** to equation (20.22), and its solution is called the **complementary solution.** The next theorem shows the relationship between the solution of equation (20.22) and its complementary equation (20.23).

THEOREM 20.1 If y_p is any particular solution of the nonhomogeneous equation $L[y] = g(x)$, and if y_c is the complementary solution, then the general solution of $L[y] = g(x)$ is

$$y = y_c + y_p \tag{20.24}$$

Proof We must show that every solution of $L[y] = g(x)$ can be put in the form (20.24) for appropriate choices of the constants that occur in y_c. Now by hypothesis, $L[y_p] = g(x)$ and $L[y_c] = 0$. Suppose $y = Y(x)$ is any solution of equation (20.22); that is, $L[Y] = g(x)$. Then

$$L[Y - y_p] = L[Y] - L[y_p] = g(x) - g(x) = 0$$

Thus, $Y - y_p$ is a solution of the complementary equation $L[y] = 0$, whose general solution we know is y_c. This means, then, that for appropriate choices of the constants in y_c, $Y - y_p = y_c$ or $Y = y_c + y_p$. This completes the proof. ■

Comment The remarkable thing about this theorem is that y_p can be *any* particular solution. So, for example, two people might use two different particular solutions. Then, even though their answers $y_p + y_c$ would *look* different, they would, in fact, describe the same family.

EXAMPLE 20.27 Show that $y_p = x + 2$ is a particular solution of

$$y'' - 3y' + 2y = 2x + 1$$

and find the general solution.

Solution For $y_p = x + 2$, we have $y_p' = 1$ and $y_p'' = 0$. So

$$L[y_p] = y_p'' - 3y_p' + 2y_p = 0 - 3(1) + 2(x + 2) = 2x + 1$$

Thus, $y_p = x + 2$ is a particular solution. According to Theorem 20.1 the general solution is $y_c + y_p$, where y_c is the complementary solution; that is, y_c is the solution of the homogeneous equation $L[y] = 0$:

$$y'' - 3y' + 2y = 0$$

The auxiliary equation is $m^2 - 3m + 2 = 0$ and has roots $m_1 = 1$, $m_2 = 2$. Thus,

$$y_c = C_1e^x + C_2e^{2x}$$

The general solution of the nonhomogeneous equation is therefore

$$y = \underbrace{C_1e^x + C_2e^{2x}}_{y_c} + \underbrace{x + 2}_{y_p}$$ ■

The crucial question clearly is how to find y_p. Sometimes this can be done by inspection. For example, the equation $y'' + 2y' = 4$ has $y_p = 2x$ as a particular solution, since $y'_p = 2$ and $y''_p = 0$. Usually, however, more work is involved. We describe a procedure below, called the *method of undetermined coefficients*, that works when the nonhomogeneous term belongs to a certain class of functions. Although the restriction to this class limits the applicability of the method, it turns out that many of the functions that occur in applications fall in this class. A more general method is studied in differential equations courses.

Before describing the procedure in general, let us consider an example.

EXAMPLE 20.28 Find a particular solution to the equation

$$y'' + 2y' - 8y = g(x)$$

where:

(a) $g(x) = 5e^{3x}$ (b) $g(x) = 5e^{2x}$

Solution (a) It is reasonable to suppose that a solution of the form

$$y_p = Ae^{3x}$$

exists, where the coefficient A is yet to be determined (hence the name *undetermined* coefficients for this method). Trying this, we have $y'_p = 3Ae^{3x}$ and $y''_p = 9Ae^{3x}$. Substituting into the original equation gives

$$\begin{aligned} 9Ae^{3x} + 2(3Ae^{3x}) - 8(Ae^{3x}) &= 5e^{3x} \\ 7Ae^{3x} &= 5e^{3x} \\ A &= \tfrac{5}{7} \end{aligned}$$

So our trial solution works, with $A = \frac{5}{7}$; that is, a particular solution is

$$y_p = \tfrac{5}{7}e^{3x}$$

(b) Proceeding as above, we try

$$y_p = Ae^{2x}$$

Then $y'_p = 2Ae^{2x}$ and $y''_p = 4Ae^{2x}$. So on substitution we get

$$\begin{aligned} 4Ae^{2x} + 2(2Ae^{2x}) - 8(Ae^{2x}) &= 5e^{2x} \\ 0 &= 5e^{2x} \end{aligned}$$

Clearly, something has gone wrong, since we have arrived at an impossibility. A look at the complementary solution reveals the problem. You can easily verify that

$$y_c = C_1e^{2x} + C_2e^{-4x}$$

Taking $C_2 = 0$, we see that any function of the form C_1e^{2x} satisfies the homogeneous equation; that is, $L[C_1e^{2x}] = 0$. Since our trial solution $y_p = Ae^{2x}$ is of this form, it has no chance of satisfying the nonhomogeneous equation.

In this situation we multiply our initial trial solution by x. This is analogous to the repeated root situation for the auxiliary equation. Thus, we try

$$y_p = Axe^{2x}$$

Then

$$\begin{aligned} y_p' &= Ae^{2x} + 2Axe^{2x} \\ y_p'' &= 2Ae^{2x} + 2Ae^{2x} + 4Axe^{2x} \\ &= 4Ae^{2x} + 4Axe^{2x} \end{aligned}$$

Now substitute into the original equation:

$$4Ae^{2x} + 4Axe^{2x} + 2(Ae^{2x} + 2Axe^{2x}) - 8Axe^{2x} = 5e^{2x}$$

After collecting terms, this becomes $6Ae^{2x} = 5e^{2x}$, which is true if $A = \frac{5}{6}$. Our desired solution is therefore

$$y_p = \tfrac{5}{6}xe^{2x}$$

■

This example suggests the following algorithm for finding a particular solution:

1. Find y_c.
2. Determine a trial solution that has the same general form as the nonhomogeneous term $g(x)$.
3. If the trial solution has no term in common with y_c, substitute it into the original equation to find the unknown coefficients.
4. If the trial solution does have a term in common with y_c, multiply the trial solution by x. Use this as a new trial solution, and return to step 3.

Notice that when step 4 applies, the new trial solution may again have a term in common with y_c, in which case we have to multiply by x again. For example, in the problem

$$y'' - 4y' + 4y = 5e^{2x}$$

we would determine $y_c = C_1e^{2x} + C_2xe^{2x}$. (Verify.) Since both trial solutions Ae^{2x} and Axe^{2x} occur in y_c, we would use

$$y_p = Ax^2e^{2x}$$

The method of undetermined coefficients works when the nonhomogeneous term $g(x)$ is one of the following types: (1) an exponential function, (2) a polynomial function, or (3) a sine or cosine function, or else is a finite product of functions of one of these three types. In each case our initial trial for y_p is a generalized function of the same type as $g(x)$, where unknown coefficients are used and where all *derived* terms (that is, terms obtained by differentiation) are included. Here are some examples:

$g(x)$	Initial Trial for y_p
$3x^2$	$Ax^2 + Bx + C$
$5 \sin 2x$	$A \sin 2x + B \cos 2x$
$2e^{-x} \cos 3x$	$e^{-x}(A \cos 3x + B \sin 3x)$
$(3x+4)e^{2x}$	$(Ax + B)e^{2x}$
$x^2 \cos 3x$	$(Ax^2 + Bx + C) \cos 3x + (Dx^2 + Ex + F) \sin 3x$

Finally, suppose $g(x)$ is the sum of two functions of the type described above—say, $g(x) = g_1(x) + g_2(x)$. Then we find particular solutions y_{p_1} and y_{p_2} that satisfy

$$L[y_{p_1}] = g_1(x) \quad \text{and} \quad L[y_{p_2}] = g_2(x)$$

and set $y_p = y_{p_1} + y_{p_2}$. Since

$$L[y_p] = L[y_{p_1} + y_{p_2}] = L[y_{p_1}] + L[y_{p_2}] = g_1(x) + g_2(x)$$

it follows that y_p is a particular solution of the original equation. This can be extended in a natural way to any finite sum. The next example illustrates this additivity principle.

EXAMPLE 20.29 Solve the initial value problem

$$y'' - 2y' = 3x + 2e^{2x} \cos x; \qquad y(0) = 0,\ y'(0) = 1$$

Solution We find from the auxiliary equation $m^2 - 2m = 0$ that

$$y_c = C_1 + C_2 e^{2x}$$

Let $g_1(x) = 3x$ and $g_2(x) = 2e^{2x} \cos x$. Our initial trial solution for $L[y] = g_1(x)$ is $y_{p_1} = Ax + B$, but this contains a constant term (B) that duplicates a term (C_1) in y_c. Thus, we modify our initial trial and use

$$y_{p_1} = Ax + Bx^2$$

Substituting this into $L[y] = 3x$ enables us to find A and B. You should verify the results: $A = -\frac{3}{4}$, $B = -\frac{3}{4}$.

For the equation $L[y] = g_2(x)$, the initial trial is

$$y_{p_2} = e^{2x}(C \cos x + D \sin x)$$

and since this does not duplicate any term in y_c, we use it. (Even though e^{2x} occurs, it occurs only in combination with a sine or cosine and so does not duplicate C_2e^{2x} that occurs in y_c.) Calculating y'_{p_2} and y''_{p_2} and substituting into $L[y] = 2e^{2x} \cos x$ give

$$e^{2x}[(-C + 2D) \cos x + (-2C - D) \sin x] = 2e^{2x} \cos x$$

(Supply the missing steps.) For this to be an identity we must have

$$\begin{cases} -C + 2D = 2 \\ -2C - D = 0 \end{cases}$$

The simultaneous solution is $C = -\frac{2}{5}$, $D = \frac{4}{5}$.

A particular solution of the original differential equation is therefore

$$y_p = y_{p_1} + y_{p_2} = -\tfrac{3}{4}x - \tfrac{3}{4}x^2 + e^{2x}(-\tfrac{2}{5} \cos x + \tfrac{4}{5} \sin x)$$

and the general solution is $y = y_c + y_p$.

The determination of the constants C_1 and C_2 so that the two initial conditions are satisfied is straightforward (though a bit messy). The result is

$$y = -\frac{19}{40} + \frac{7}{8} e^{2x} - \frac{3}{4} x - \frac{3}{4} x^2 + e^{2x}\left(-\frac{2}{5} \cos x + \frac{4}{5} \sin x\right)$$ ■

EXERCISE SET 20.6

A

In Exercises 1–4 find a particular solution by inspection. Then give the general solution.

1. $y'' + y = 2x$ **2.** $y'' - 3y' = 1$

3. $y'' - y' - 2y = 3$ **4.** $y'' - 4y = 3x + 5$

In Exercises 5–18 find a particular solution using the method of undetermined coefficients.

5. $y'' + y' - 6y = 3e^x$

6. $2y'' - 3y' - 5y = 3x - 1$

7. $y'' + 4y = \sin x$ **8.** $y'' - 4y = 2x^2$

9. $y'' - y' - 2y = e^{2x}$ **10.** $y'' - 2y' = x - 2$

11. $y'' + y = 3 \cos x$ **12.** $y'' - y = 3e^x$

13. $y'' - 2y' + y = e^x$ **14.** $y'' + 4y' + 4y = 2e^{-2x}$

15. $y'' - 2y' = xe^{2x}$ **16.** $y'' + y = e^x \sin x$

17. $2y'' + y' + y = 2e^{-x} \cos x$

18. $3y'' - 2y' + 5y = (2x - 1)e^x$

In Exercises 19–26 determine an appropriate form for a particular solution, after finding the complementary solution. Do not calculate the coefficients.

19. $y'' + 3y' - 4y = x^3e^x$

20. $y'' - 4y' + 4y = 3e^{2x} + x \cos x$

21. $y'' - 2y' = 2x^2 - 1 + xe^{2x}$

22. $y'' - 2y' + 5y = 3e^x \sin 2x$

23. $y'' - 6y' + 25y = e^{3x} \cos 4x - 1$

24. $3y'' + 4y' - 7y = x^2 \sin 2x$

25. $2y'' + 3y' - 9y = xe^{-x} \sin x$

26. $y'' + y' + 2y = 2 \sin^2 x$

In Exercises 27–34 solve the initial value problem or boundary value problem.

27. $y'' - 4y = x + 3$; $y(0) = 0$, $y'(0) = 1$

28. $y'' + 2y' = 2e^{-x}$; $y(0) = 1$, $y'(0) = -1$

29. $y'' + y = \cos 2x$; $y(0) = 2$, $y(\frac{\pi}{2}) = 3$

30. $y'' + 3y' - 10y = x^2 + 2x$; $y(0) = 1$, $y'(0) = 0$

31. $y'' - 4y' + 5y = 2e^{-2x}$; $y(0) = -1$, $y'(0) = 2$

32. $y'' + 4y = 3 \sin x$; $y(0) = 4$, $y(\frac{\pi}{4}) = 0$

33. $y'' - y' - 2y = e^{2x}$; $y(0) = 2$, $y'(0) = 3$

34. $y'' + 4y' - 5y = xe^x$; $y(0) = 0$, $y'(0) = 1$

B

In Exercises 35–39 find the general solution.

35. $y'' - y = \cosh x$

36. $y'' + 4y = 2 \cos^2 x - 1$

37. $y'' + 4y = 2 \sin x \cos x$

38. $d^2x/dt^2 + \omega_0^2 x = k \cos \omega t$, where (a) $\omega \neq \omega_0$ and (b) $\omega = \omega_0$

39. $d^2x/dt^2 - 6(dx/dt) + 9x = 2e^{3t} + t^2$. Also find the particular solution that satisfies $x(0) = 4$, $x'(0) = -2$.

40. Prove that

$$ay'' + by' + cy = x^n \qquad (a \neq 0,\ n \geq 0)$$

has a particular solution of the form

$$y_p = A_n x^n + A_{n-1} x^{n-1} + \cdots + A_1 x + A_0$$

if and only if $c \neq 0$.

In Exercises 41–44 use the same procedure as for second-order equations to find a particular solution. Also give the general solution.

41. $y''' - 3y'' = x^2$

42. $y''' - y'' + 4y' - 4y = \sin 2x$

43. $y''' - 3y' + 2y = 2e^x + 3x$

44. $d^4y/dx^4 - 16y = \cosh 2x + \cos 2x$

20.7 THE VIBRATING SPRING

In this section we study an important application of second-order differential equations. The problem we discuss concerns a vibrating spring, but the ideas are applicable to a broad range of problems dealing with oscillatory motion.

Consider a spring of natural length l attached to a support, as shown in Figure 20.9(a). Suppose a weight W is attached, causing the spring to stretch a distance Δl as it comes to an equilibrium position, as shown in Figure 20.9(b). We assume the weight is not great enough to stretch the spring beyond its elastic limit. According to Hooke's law, the force exerted by the spring is proportional to the elongation. When the spring and weight are in equilibrium and the force exerted by the spring is W, we have $W = k\,\Delta l$, where k is the constant of proportionality (the *spring constant*). Since $W = mg$, where m is the mass of the spring, we have $mg = k\,\Delta l$, or

$$mg - k\,\Delta l = 0 \tag{20.25}$$

Now suppose the spring with attached weight is set in motion in some way. For example, it might be pulled below the equilibrium position and then released. As shown in Figure 20.9(c), we let $y = y(t)$ be the distance of the weight from the equilibrium position at time t. We will take y as positive downward. We want to determine y as a function of t. By Newton's second law of motion,

$$m\frac{d^2y}{dt^2} = F$$

where F is the summation of all the forces acting on the weight. The force of gravity is mg, acting downward. The force exerted by the spring is $k(\Delta l + y)$, since $\Delta l + y$ is the total displacement from the spring's natural length. When $\Delta l + y > 0$, the force of the spring is upward and so equals $-k(\Delta y + l)$. But this is also the correct sign when $\Delta l + y < 0$, since the

FIGURE 20.9

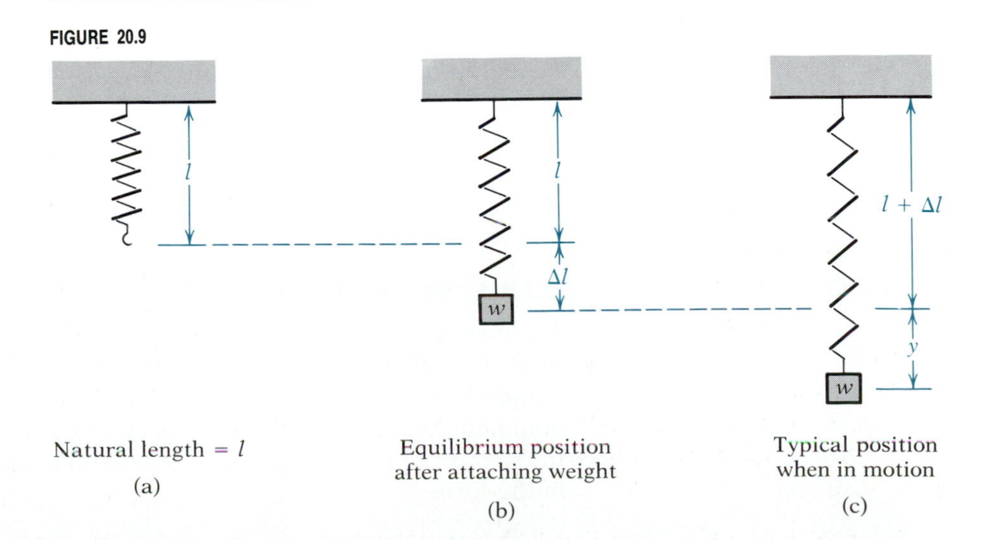

spring is then compressed and its force is downward, in agreement with $-k(\Delta l + y) > 0$.

Two other forces may need to be considered. If the action takes place in air, we might reasonably neglect air resistance, but often vibrations occur in some viscous medium, such as oil, and the resistance of the medium cannot be neglected. From experiments it can be shown that for relatively small velocities, it is reasonable to assume that the resisting force is proportional to the velocity. Since this force is always opposite to the direction of motion, it is of the form $-c\frac{dy}{dt}$ for $c > 0$. Finally, we allow for some external force $f(t)$. For example, $f(t)$ might be a force applied to the entire support system.

We can now sum all the forces and obtain from Newton's second law:

$$m\frac{d^2y}{dt^2} = \underbrace{mg}_{\substack{\text{force}\\\text{of}\\\text{gravity}}} - \underbrace{k(\Delta l + y)}_{\substack{\text{force of}\\\text{spring}}} - \underbrace{c\frac{dy}{dt}}_{\substack{\text{resisting}\\\text{force}}} + \underbrace{f(t)}_{\substack{\text{external}\\\text{force}}}$$

On applying equation (20.25), we can simplify this to obtain

$$m\frac{d^2y}{dt^2} + c\frac{dy}{dt} + ky = f(t) \tag{20.26}$$

This is the basic differential equation for the vibrating spring. We now consider certain special cases.

Undamped Free Vibrations

The simplest case is the one in which the resisting force is so small that it can be neglected, and for which there is no external force. We describe this situation by saying the motion is **undamped** (no resisting force) and **free** (no external force). Equation (20.26) then reduces to

$$m\frac{d^2y}{dt^2} + ky = 0$$

or, on dividing by m and writing $\omega^2 = \frac{k}{m}$,

$$\frac{d^2y}{dt^2} + \omega^2 y = 0 \tag{20.27}$$

This is a second-order linear homogeneous equation that has auxiliary equation $m^2 + \omega^2 = 0$, with roots $m = \pm i\omega$. So the general solution is

$$y = C_1 \cos \omega t + C_2 \sin \omega t \tag{20.28}$$

The motion described by equation (20.28) is called **simple harmonic motion.** Its **period** T is $\frac{2\pi}{\omega}$ and its **frequency** f is $\frac{1}{T} = \frac{\omega}{2\pi}$. The period is the time required to go through one complete cycle and return to the original position, and the frequency is the number of cycles completed per unit of time. In Exercise 15 you will be asked to show that equation (20.28) can be written in the form

$$y = C\sin(\omega t + \alpha) \tag{20.29}$$

FIGURE 20.10 Simple harmonic motion

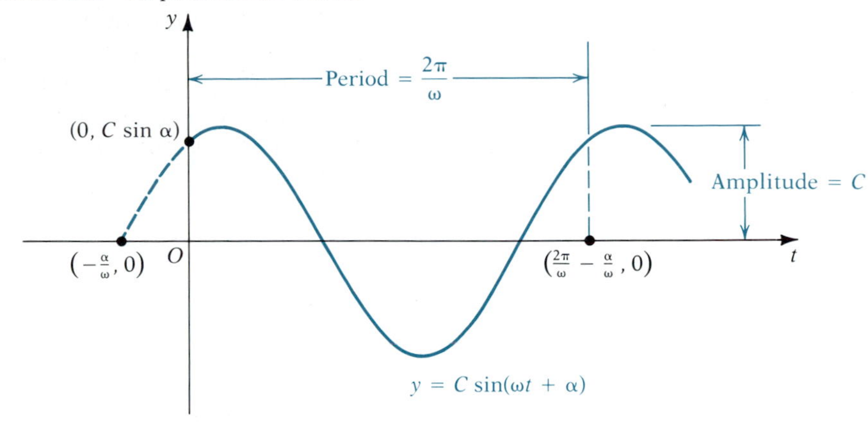

where $C = \sqrt{C_1^2 + C_2^2}$ and α satisfies $\sin\alpha = C_1/\sqrt{C_1^2 + C_2^2}$, $\cos\alpha = C_2/\sqrt{C_1^2 + C_2^2}$. In this form we see that the motion is sinusoidal, with amplitude C. Its graph is shown in Figure 20.10.

EXAMPLE 20.30 A spring is attached to a rigid overhead support. A 4-lb weight is attached to the end of the spring, causing it to stretch 6 in. Then the weight is pulled down 3 in. below the equilibrium position and released. Air resistance is negligible. Describe the resulting motion.

Solution For consistency we will use feet instead of inches as the unit of distance. Then we have $y(0) = \frac{1}{4}$ and $y'(0) = 0$, since the weight was released without imparting any velocity. Since the 4-lb weight stretched the spring $\frac{1}{2}$ ft, we get the spring constant k from Hooke's law:

$$4 = k \cdot \tfrac{1}{2}$$
$$k = 8 \quad \text{(pounds per foot)}$$

Approximating g as 32 ft/sec^2, we have

$$m = \frac{W}{g} = \frac{4}{32} = \frac{1}{8} \quad \text{(slugs)}$$

So using equation (20.27) we have the initial value problem

$$\frac{d^2y}{dt^2} + 64y = 0; \qquad y(0) = \frac{1}{4},\ y'(0) = 0$$

Here we have used the fact that $\omega^2 = \frac{k}{m} = 8 \cdot 8 = 64$. The general solution is

$$y = C_1 \cos 8t + C_2 \sin 8t$$

From the initial conditions we get $C_1 = \frac{1}{4}$ and $8C_2 = 0$, so that $C_2 = 0$. Thus, the equation of motion is

$$y = \tfrac{1}{4} \cos 8t$$

The period is $\frac{2\pi}{\omega} = \frac{\pi}{4}$, and the frequency is $\frac{4}{\pi}$ cycles per second. The amplitude is $\frac{1}{4}$. ■

Damped Free Vibrations

According to the undamped model, the motion of the spring continues forever without diminishing. This does not accurately describe reality, although it does give reasonable accuracy over a relatively short time span when the medium in which the motion takes place offers little resistance. When this resistance is taken into consideration and there is no external force, equation (20.26) takes the form

$$m\frac{d^2y}{dt^2} + c\frac{dy}{dt} + ky = 0$$

where it is important to remember that all of the constants m, c, and k are positive. The nature of the motion in this case depends on the relative sizes of the resistance constant c and the spring constant k. On solving the auxiliary equation, we get the roots

$$r_1 = \frac{-c + \sqrt{c^2 - 4km}}{2m} \quad \text{and} \quad r_2 = \frac{-c - \sqrt{c^2 - 4km}}{2m}$$

There are three cases to consider.

1. $c^2 > 4km$: the roots are real and unequal, and so the solution is of the form

   $$y = C_1e^{r_1t} + C_2e^{r_2t}$$

   Observe that since $\sqrt{c^2 - 4km} < c$, it follows that $r_1 < 0$ and $r_2 < 0$.
2. $c^2 = 4km$: the roots are real and equal. Denote the common value by r; that is, $r = -c/2m < 0$. The solution is therefore of the form

   $$y = (C_1 + C_2t)e^{rt}$$

3. $c^2 < 4km$: the roots are imaginary—say, $r_1 = \alpha + i\beta$ and $r_2 = \alpha - i\beta$, where $\alpha = -c/2m < 0$ and $\beta = \sqrt{4km - c^2}/2m > 0$. The solution then is of the form

   $$y = e^{\alpha t}(C_1 \cos \beta t + C_2 \sin \beta t)$$

We analyze the motion for each case. For case 1 the solution is of the form

$$y = C_1e^{r_1t} + C_2e^{r_2t}$$

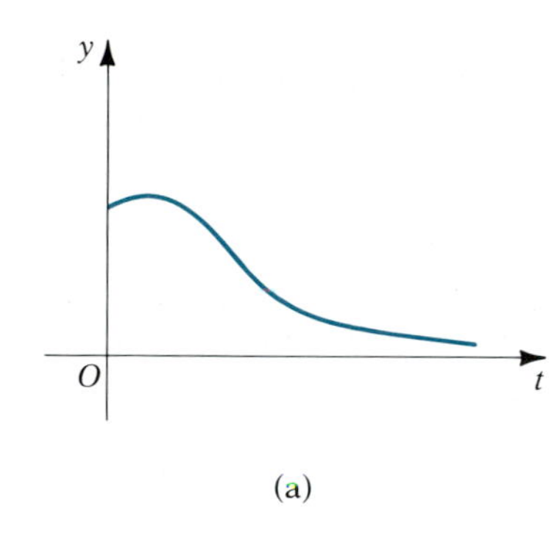

(a)

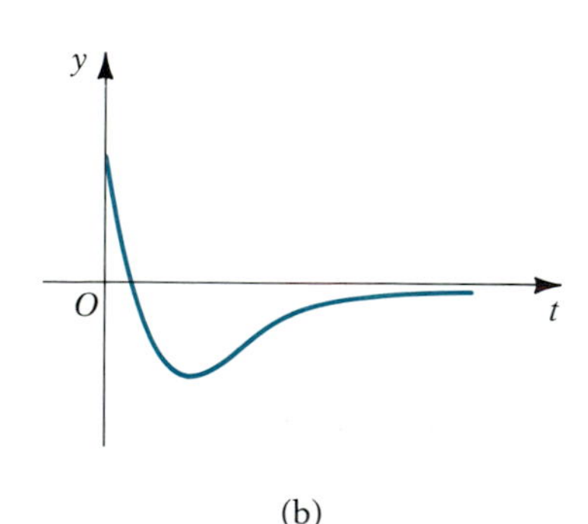

(b)

FIGURE 20.11 Overdamped motion

where both r_1 and r_2 are negative. As t increases, y approaches 0 and there is no oscillation. This type of motion is said to be **overdamped.** The resistance is so strong that the motion rapidly dies out. The form of the graph of y versus t depends on the initial conditions that determine C_1 and C_2. Figure 20.11 shows two possibilities.

In case 2 the solution is of the form

$$y = (C_1 + C_2t)e^{rt}, \qquad r < 0$$

Again there is no oscillation, and the motion tends to die out as t increases because of the factor e^{rt}. The slightest change in the resisting force changes the situation to either case 1 or case 3. We say this is **critically damped** motion. The graph is similar to that for overdamping, with the initial conditions determining the exact form. Figure 20.12 shows one possibility.

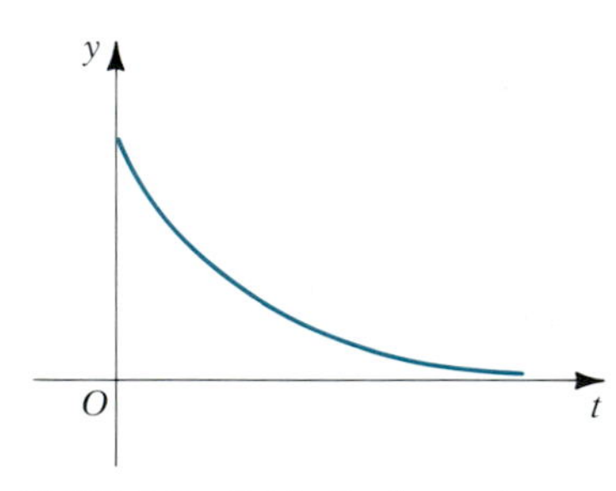

FIGURE 20.12 Critically damped motion

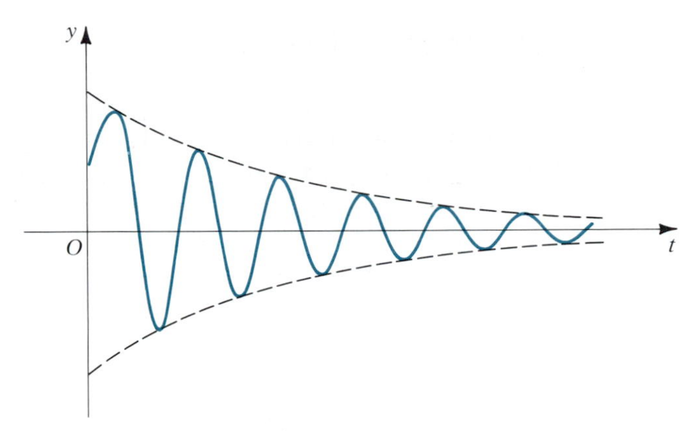

FIGURE 20.13 Underdamped motion

For case 3 the solution is of the form

$$y = e^{\alpha t}(C_1 \cos \beta t + C_1 \sin \beta t), \qquad \alpha < 0,\ \beta > 0$$

The motion in this case is oscillatory but with decreasing "amplitude" of the oscillations because of the $e^{\alpha t}$ factor. The motion is said to be **under-damped.** A typical situation is shown in Figure 20.13.

EXAMPLE 20.31 An 8-lb weight stretches a spring 1.6 ft beyond its natural length. A damping force equal in magnitude to that of the velocity is present. If the spring is pushed upward 4 in. above the equilibrium position and then given a downward velocity of 6 ft/sec, find the equation of motion.

Solution Using $g \approx 32$, we get $m = \frac{W}{g} = \frac{1}{4}$, and by Hooke's law $8 = k(1.6)$, so that $k = 5$. The resistance constant $c = 1$. So the differential equation is

$$\frac{1}{4}\frac{d^2y}{dt^2} + \frac{dy}{dt} + 5y = 0$$

or, equivalently,

$$\frac{d^2y}{dt^2} + 4\frac{dy}{dt} + 20y = 0$$

and the initial conditions are $y(0) = -\frac{1}{3}$, $y'(0) = 6$.

The auxiliary equation has roots $-2 \pm 4i$, so the general solution is

$$y = e^{-2t}(C_1 \cos 4t + C_2 \sin 4t)$$

To find C_1 and C_2 we impose the two initial conditions. From $y(0) = -\frac{1}{3}$, we have immediately that $C_1 = -\frac{1}{3}$. We find y' to be:

$$y' = e^{-2t}[(4C_2 - 2C_1) \cos 4t + (-4C_1 - 2C_2) \sin 4t] \qquad \text{(Verify.)}$$

and so from $y'(0) = 6$, we get

$$4C_2 - 2C_1 = 6$$

Substituting for C_1, we find $C_2 = \frac{4}{3}$. Thus, the solution is

$$y = -\frac{e^{-2t}}{3}(\cos 4t - 4 \sin 4t)$$

The motion is underdamped in this case. ■

Forced Vibrations

When there is an external force $f(t)$ present, we describe the motion as having **forced vibrations,** and $f(t)$ is called the **forcing function.** Frequently the forcing function is sinusoidal, as in the example that follows.

EXAMPLE 20.32 A 4-lb weight stretches a spring 2 ft beyond its natural length. Air resistance is negligible, but the system is subjected to an external force $f(t) = 2 \cos \lambda t$. The weight is pulled down 6 in. and released. Describe the motion for each of the following values of λ: (a) $\lambda = 3$ and (b) $\lambda = 4$.

Solution In the usual way we find $m = \frac{1}{8}$ and $k = 2$. We assume $c = 0$. Thus, equation (20.26) becomes

$$\frac{1}{8}\frac{d^2y}{dt^2} + 2y = 2 \cos \lambda t$$

or, equivalently,

$$\frac{d^2y}{dt^2} + 16y = 16 \cos \lambda t$$

and the initial conditions are $y(0) = \frac{1}{2}$, $y'(0) = 0$. The complementary solution y_c is

$$y_c = C_1 \cos 4t + C_2 \sin 4t$$

(a) For $\lambda = 3$, we expect a particular solution of the form

$$y_p = A \cos 3t + B \sin 3t$$

By substituting this into the equation and comparing coefficients, we get $A = \frac{16}{7}$, $B = 0$. The general solution is therefore

$$y = C_1 \cos 4t + C_2 \sin 4t + \tfrac{16}{7} \cos 3t$$

Imposing the initial conditions, we find $C_1 = -\frac{25}{14}$ and $C_2 = 0$, giving the equation of motion as

$$y = -\tfrac{25}{14} \cos 4t + \tfrac{16}{7} \cos 3t$$

(b) For $\lambda = 4$, the forcing function $f(t) = \cos 4t$ is a solution of the homogeneous equation, and so our trial solution y_p is of the form

$$y_p = t(A \cos 4t + B \sin 4t)$$

A straightforward calculation shows that $A = 0$, $B = 2$. Then applying the initial conditions to the general solution $y = y_c + y_p$, we obtain the equation of motion

$$y = \tfrac{1}{2} \cos 4t + 2\, t \sin 4t$$

■

It is instructive to analyze the motion in part b of the preceding example. The presence of the factor t in the term $2t \sin 4t$ means that as t increases, the values of t become unbounded. For example, when $t = (2n + 1)\pi/8$, we get

$$y\left(\frac{2n+1}{8}\pi\right) = \frac{(2n+1)\pi}{4}(-1)^n$$

so that as $n \to \infty$, $|y| \to \infty$, with y alternating between positive and negative values. This phenomenon is known as **resonance.** It occurs when the forcing function is periodic and has the same period as the complementary solution (called the natural period of the system). The consequences of resonance to a physical system can be catastrophic. The presence of the factor t indicates large oscillations. Physical structures usually break or the system changes so that the differential equation no longer describes it.

EXERCISE SET 20.7

A

In Exercises 1–10 find the equation of motion. Neglect the resisting force unless otherwise indicated. When the resisting force is considered, identify the motion as underdamped, overdamped, or critically damped.

1. A 4-lb weight stretches a spring 6 in. beyond its natural length. After the system comes to equilibrium, the weight is pulled down 6 more inches and released. Give the amplitude and period.
2. In Exercise 1 instead of the weight being pulled down, it is given an upward push from the equilibrium position of 4 ft/sec.
3. A 16-lb weight stretches a spring 2 ft beyond its natural length. The weight is then pulled down 8 in. below the equilibrium position and given an upward velocity of 12 ft/sec.
4. Repeat Exercise 3 under the assumption that there is a resisting force equal to 4 times the velocity.
5. A mass of 50 g stretches a spring 5 cm beyond its natural length. The mass is pushed upward 10 cm and released. [*Note:* In the cgs system $g = 980$ cm/sec^2, and so the weight is (50)(980) dynes.]
6. A 32-lb weight is attached to a spring with a natural length of 3 ft. When the system comes to equilibrium, the spring is 4 ft long. The system is in a viscous medium that produces a resisting force equal to the velocity and opposite in direction. The weight is started in motion with a downward velocity from the equilibrium position of 6 ft/sec.
7. A 20-g mass attached to a spring stretches it 10 cm beyond its natural length. The weight is attached to a dashpot mechanism that produces a resisting force of magnitude 420 $|v|$ dynes, where v is the velocity. The weight is pulled down 4 cm and also given an initial downward velocity of 7 cm/sec.
8. An 8-lb weight stretches a spring from its natural length of 20 in. to a length of 28 in. The weight is then pulled down an additional 4 in. and given a downward velocity of 4 ft/sec. Find (a) the time when the weight first passes through the equilibrium position and (b) its maximum displacement from the equilibrium position.
9. Repeat Exercise 1 if there is an external force $f(t) = 3 \cos 4t$ pounds acting on the system.
10. An 8-lb weight stretches a spring 1 ft beyond its natural length. The system is subjected to an external force of $f(t) = 2 \cos t + 3 \sin t$ pounds.
11. A weight of 8 lb is attached to a spring with a spring constant of 5 lb/ft. Assume a resisting force equal to $-c\frac{dy}{dt}$, where $c > 0$. Find c such that the motion is (a) overdamped, (b) critically damped, and (c) underdamped.
12. The angle θ from the vertical to a pendulum (see the figure) can be shown to satisfy the nonlinear second-order differential equation

$$\frac{d^2\theta}{dt^2} + \frac{g}{l}\sin\theta = 0$$

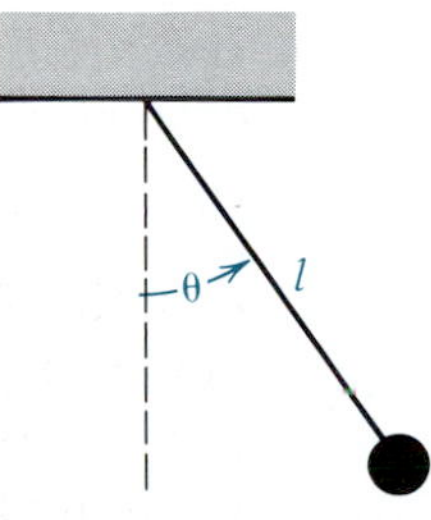

where l is the length of the pendulum rod. For small values of θ, $\sin\theta \approx \theta$. Using this approximation, determine the equation of motion of a pendulum 2 ft long, where θ is initially $\frac{1}{4}$ radian (with the positive direction for θ taken as counterclockwise, measured from the vertical), if the initial angular velocity is $d\theta/dt = \sqrt{3}$ radians per second.

13. For the pendulum of Exercise 12, find the following:
    (a) The maximum angle from the vertical through which the pendulum will swing
    (b) The time to complete one back and forth swing
    (c) The instant when the pendulum will be vertical for the first time
    (d) The magnitude of its velocity when it is in the vertical position

B

14. Prove that for free damped vibrations of a spring, in the overdamped and critically damped cases, the spring will pass through the equilibrium position at most one time.

15. Show that the equation

$$y = C_1 \cos \omega t + C_2 \sin \omega t$$

of simple harmonic motion can be expressed as either of the ways:
(a) $y = C \sin(\omega t + \alpha)$ (b) $y = C \cos(\omega t - \beta)$
where $C = \sqrt{C_1^2 + C_2^2}$. Describe the angles α and β.

16. Consider the following equation for damped forced vibrations:

$$m\frac{d^2y}{dt^2} + c\frac{dy}{dt} + ky = F_0 \cos \lambda t$$

Find the general solution. Show that if $c \neq 0$, there can be no resonance. Find $\lim_{t\to\infty} y(t)$. (This is called the *steady-state* solution.)

17. Consider a simple series circuit that contains a resistance of R ohms, an inductance of L henrys, a capacitance of C farads, and an electromotive force of $E(t)$ volts (see the figure). By one of Kirchhoff's laws, the current I, in amperes, satisfies

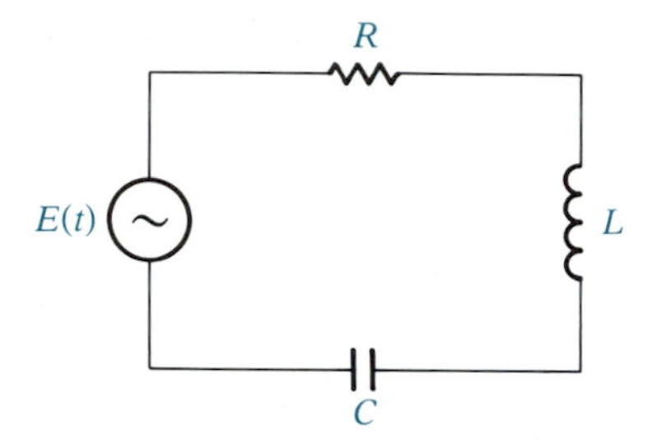

$$L\frac{dI}{dt} + RI + \frac{Q}{C} = E(t)$$

where Q is the electric charge, related to I by $I = \frac{dQ}{dt}$. By differentiating both sides of the differential equation, we obtain

$$L\frac{d^2I}{dt^2} + R\frac{dI}{dt} + \frac{I}{C} = E'(t)$$

Find I if $L = 10$, $R = 20$, $C = 0.02$, and $E(t) = 200 \sin t$. What is the steady-state current? (Let $t \to \infty$.)

20.8 SERIES SOLUTIONS

Using methods we have studied, we can solve a large number of differential equations, but there are many others for which none of the methods works. For example, we have no means yet of finding the general solution of $y'' + xy' + y = 0$. This is a homogeneous linear equation, but one of the coefficients is a variable, so the theory of Section 20.5 is not applicable. For this and other problems it is often possible to find a solution in the form of a power series, say

$$y = \sum_{n=0}^{\infty} a_n x^n \tag{20.30}$$

Recall from Chapter 11 that every such power series defines a function within its interval of convergence, $|x| < R$. We assume $R > 0$. The series can be differentiated term by term, and the resulting series converges to y' for $|x| < R$. Similarly, differentiating again, we obtain a series that converges to y'' if $|x| < R$, and so on and on. So if y is given by equation (20.30), then

$$y' = \sum_{n=1}^{\infty} n a_n x^{n-1}$$

and

$$y'' = \sum_{n=2}^{\infty} n(n-1)a_n x^{n-2}$$

Notice we started the index of summation with 1 for y' and with 2 for y''. This is because the constant term drops out on differentiation.

It is useful to shift the index of summation at times. For example, we can see by expanding the terms that

$$\sum_{n=0}^{\infty} a_n x^n = \sum_{n=1}^{\infty} a_{n-1} x^{n-1}$$

The key to handling such shifts is that when the initial value of n on the summation sign is increased, n in the summand must be decreased by the same amount, and vice versa. (More formally, in the example shown, we could make the change of index $n = m - 1$ and write the second sum in terms of m, but the way we have suggested is faster and is easy to apply.) Test yourself by seeing whether you agree that

$$\sum_{n=2}^{\infty} n(n-1)a_n x^{n-2} = \sum_{n=1}^{\infty} (n+1)(n)a_{n+1} x^{n-1} = \sum_{n=0}^{\infty} (n+2)(n+1)a_{n+2} x^n$$

We give two examples to illustrate solution by series. The first problem could be done more easily by separating variables, but we include it to illustrate the technique first on a simple problem and also to provide a means of comparing our answer with that obtained by separating variables.

EXAMPLE 20.33 Find the general solution of the equation $y' = 2xy$ using infinite series.

Solution We try a solution of the form (20.30): $y = \sum_{n=0}^{\infty} a_n x^n$. The equation can then be written as

$$\sum_{n=1}^{\infty} na_n x^{n-1} - 2\sum_{n=0}^{\infty} a_n x^{n+1} = 0$$

To have the same power of x appearing in the two summations, we shift the first index down by 1 and the second one up by 1:

$$\sum_{n=0}^{\infty} (n+1)a_{n+1} x^n - 2\sum_{n=1}^{\infty} a_{n-1} x^n = 0$$

After separating the $n = 0$ term of the first summation, both sums will begin with $n = 1$, and so they can be brought together. This gives

$$a_1 + \sum_{n=1}^{\infty} [(n+1)a_{n+1} - 2a_{n-1}]x^n = 0$$

We want this equation to be an identity in x. So we must have $a_1 = 0$, and every coefficient of x^n for $n = 1, 2, 3, \ldots$ equal to 0:

$$(n+1)a_{n+1} - 2a_{n-1} = 0, \qquad n \geq 1$$

The last equation can be written in the form

$$a_{n+1} = \frac{2a_{n-1}}{n+1}, \qquad n \geq 1$$

This is a **recursion formula** that enables us to find the coefficients $a_2, a_3, \ldots$ in sequence:

$$\underline{n=1}: \quad a_2 = \frac{2a_0}{2} = a_0$$

$$\underline{n=2}: \quad a_3 = \frac{2a_1}{3} = 0 \quad \text{since } a_1 = 0$$

$$\underline{n=3}: \quad a_4 = \frac{2a_2}{4} = \frac{a_2}{2} = \frac{a_0}{2} \quad \text{since } a_2 = a_0$$

$$\underline{n=4}: \quad a_5 = \frac{2a_3}{5} = 0 \quad \text{since } a_3 = 0$$

$$\underline{n=5}: \quad a_6 = \frac{2a_4}{6} = \frac{a_4}{3} = \frac{a_0}{3 \cdot 2} \quad \text{since } a_4 = \frac{a_0}{2}$$

$$\vdots \qquad\qquad \vdots$$

Continuing in this way, we see that

$$\begin{cases} a_{2k-1} = 0 & \text{for } k = 1, 2, 3, \ldots \\ a_{2k} = \dfrac{a_0}{k!} & \text{for } k = 1, 2, 3, \ldots \end{cases}$$

Since no condition is placed on a_0, it is arbitrary. Thus, our solution $y = \sum_{n=0}^{\infty} a_n x^n$ becomes

$$\begin{aligned} y &= a_0 + \sum_{k=1}^{\infty} a_{2k} x^{2k} \\ &= a_0 + a_0 \sum_{k=1}^{\infty} \frac{x^{2k}}{k!} \end{aligned}$$

Since $0! = 1$, we can combine the first term with the summation by starting k with 0:

$$y = a_0 \sum_{k=0}^{\infty} \frac{x^{2k}}{k!}$$

By the ratio test we can determine that the series converges for all values of x.

Now let us compare this solution with that obtained by separating variables. We have

$$\frac{dy}{y} = 2x\,dx$$

$$\ln y = x^2 + \ln C$$

$$\ln \frac{y}{C} = x^2$$

$$y = Ce^{x^2}$$

From Section 11.8 we see that the Maclaurin series for e^{x^2} is

$$e^{x^2} = \sum_{k=0}^{\infty} \frac{x^{2k}}{k!}$$

Thus, the two answers agree, where in our series solution a_0 is the arbitrary constant. ■

EXAMPLE 20.34 Find the general solution of the equation $y'' + xy' + y = 0$ using infinite series.

Solution This is the problem we posed at the beginning of this section. Substituting $y = \sum_{n=0}^{\infty} a_n x^n$, we get

$$\sum_{n=2}^{\infty} n(n-1)a_n x^{n-2} + \sum_{n=1}^{\infty} n a_n x^n + \sum_{n=0}^{\infty} a_n x^n = 0$$

We shift the index down by 2 on the first sum, and by observing that the second sum is unaltered by starting with $n = 0$, we obtain

$$\sum_{n=0}^{\infty} (n+2)(n+1)a_{n+2} x^n + \sum_{n=0}^{\infty} n a_n x^n + \sum_{n=0}^{\infty} a_n x^n = 0$$

The terms can be brought together now in one summation:

$$\sum_{n=0}^{\infty} [(n+2)(n+1)a_{n+2} + (n+1)a_n] x^n = 0$$

To be an identity in x, every coefficient must be 0. Thus,

$$(n+2)(n+1)a_{n+2} + (n+1)a_n = 0$$

or

$$a_{n+2} = -\frac{a_n}{n+2}, \qquad n \geq 0$$

We consider even subscripts and odd subscripts separately:

n Even	n Odd
$a_2 = -\frac{a_0}{2}$	$a_3 = -\frac{a_1}{3}$
$a_4 = -\frac{a_2}{4} = \frac{a_0}{2 \cdot 4}$	$a_5 = -\frac{a_3}{5} = \frac{a_1}{3 \cdot 5}$
$a_6 = -\frac{a_4}{6} = -\frac{a_0}{2 \cdot 4 \cdot 6}$	$a_7 = -\frac{a_5}{7} = -\frac{a_1}{3 \cdot 5 \cdot 7}$
$\vdots$	$\vdots$
$a_{2k} = \frac{(-1)^k a_0}{2 \cdot 4 \cdot 6 \cdot \cdots \cdot (2k)}$	$a_{2k+1} = \frac{(-1)^k a_1}{3 \cdot 5 \cdot 7 \cdot \cdots \cdot (2k+1)}$

Since a_0 and a_1 have no restrictions, they are arbitrary. The solution can now be written as

$$y = a_0\left[1 + \sum_{k=1}^{\infty} \frac{(-1)^k}{2 \cdot 4 \cdot 6 \cdot \cdots \cdot (2k)} x^{2k}\right] + a_1\left[x + \sum_{k=1}^{\infty} \frac{(-1)^k}{3 \cdot 5 \cdot 7 \cdot \cdots \cdot (2k+1)} x^{2k+1}\right]$$

This can be simplified somewhat by observing that $2 \cdot 4 \cdot 6 \cdot \cdots \cdot (2k) = 2^k \cdot k!$, and since $0! = 1$, we can combine the first term in each bracket with the sum by starting k with 0:

$$y = a_0 \sum_{k=0}^{\infty} \frac{(-1)^k x^{2k}}{2^k k!} + a_1 \sum_{k=0}^{\infty} \frac{(-1)^k}{1 \cdot 3 \cdot 5 \cdot \cdots \cdot (2k+1)} x^{2k+1}$$

Both series can be shown to converge for all x. If we let y_1 be the first sum and y_2 the second, and write $C_1 = a_0$ and $C_2 = a_1$, then the solution is seen to be in the familiar form $y = C_1y_1 + C_2y_2$. ∎

Remark In these two examples the recursion formulas were such that we could find explicit formulas for the coefficients. In some cases this is not possible, and the best we can do is calculate as many coefficients as we want in sequential order.

The procedure we have illustrated will work for nonhomogeneous equations also provided the nonhomogeneous term can be expanded in a power series. In particular, if it is a polynomial, it is already a (finite) power series. By comparing coefficients of like powers of x, the unknown coefficients a_n in the series $\sum a_n x^n$ can be found.

As you might expect, there is a good deal more to solution by power series than we have gone into here, but the deeper aspects will have to be deferred to a course on differential equations.

EXERCISE SET 20.8

A

In Exercises 1–13 find the general solution in terms of power series in x. Where possible, solve the equation by other means and compare answers. If initial values are given, find the particular solution.

1. $y' = x^2y$
2. $y' - 2xy = x$
3. $y'' = y$
4. $y'' - xy - y = 0$
5. $y'' = xy$
6. $y' - y = 2x;\ y(0) = 3$
7. $y'' + y = 0;\ y(0) = 1,\ y'(0) = 0$
8. $(1 - x)y'' = y' - xy$
9. $y'' + x^2y = 0$
10. $(1 + x^2)y'' + 2xy' = 0$
11. $(1 - x^2)y'' - 5xy' - 3y = 0$
12. $(1 + x^2)y'' - 3xy' - 5y = 0$

B

13. $y'' + xy' + 2y = x^2 + 1;\ y(0) = 2,\ y'(0) = 1$
14. Obtain the general solution of the equation

    $$y'' + (x - 2)y = 0$$

    as a power series about $x = 2$—that is, in the form $\sum a_n(x - 2)^n$.
15. Find two linearly independent solutions in the form of power series for the equation

    $$y'' + (1 - x)y = 0$$

    Then write the general solution. (*Hint:* For one solution take $a_0 = 1$ and $a_1 = 0$; then do the opposite.)
16. Obtain the first 10 nonzero terms of the Maclaurin series solution

    $$y(x) = y(0) + y'(0)x + \frac{y''(0)}{2!}x^2 + \frac{y'''(0)}{3!}x^3 + \cdots$$

    for the initial value problem

    $$y'' - xy' + 2y = 0; \qquad y(0) = 1,\ y'(0) = -1$$

    by carrying out the following steps:
    (a) Solve the equation for y''.
    (b) Put $x = 0$ and substitute the given values of $y(0)$ and $y'(0)$ to get $y''(0)$.
    (c) Differentiate the equation in step a to get y'''.
    (d) Put $x = 0$ and substitute known values for $y(0)$, $y'(0)$, and $y''(0)$ to get $y'''(0)$.
    (e) Differentiate the equation in step c and continue in this manner.
17. Follow the procedure in Exercise 16 to obtain the first 10 nonzero terms of the Maclaurin series solution of the initial value problem:

    $$y'' + (x - 1)\,y = 0; \qquad y(0) = 2,\ y'(0) = 1$$

20.9 SUPPLEMENTARY EXERCISES

In Exercises 1–26 find the general solution. If initial or boundary conditions are given, find the particular solution that satisfies them.

1. $(2 + e^{-x} \cos y)\, dx = (3 + e^{-x} \sin y)\, dy$
2. $xy' = y + \cot \frac{y}{x};\ y(1) = 0$
3. $y' = x\sqrt{1 - y^2};\ y(0) = 1$
4. $(\cosh x)y' + (\sinh x)y = \cosh^2 x$
5. $\dfrac{dy}{dx} = \dfrac{y(2x + y)}{x(x + y)}$
6. $xy' - 1 = e^{-y};\ y(0) = 0$
7. $(y - 3x)\, dy + (3y + 4x)\, dx = 0;\ y(1) = 0$
8. $\left(\dfrac{2x^2 + y^2}{\sqrt{x^2 + y^2}}\right) dx + \left(\dfrac{xy}{\sqrt{x^2 + y^2}} - 3\right) dy = 0;$ $y(3) = 4$
9. $\dfrac{dy}{dx} = 1 + \dfrac{y(1 - x)}{x(2 - x)};\ y(1) = 2$
10. $\dfrac{dy}{\tan y} = \dfrac{\sin 2x}{1 + \sin^2 x} dx$
11. $(\cos xy - xy \sin xy)\, dx = (x^2 \sin xy)\, dy$
12. $y' + \dfrac{3y}{x} = \dfrac{x}{y^2}$ (See Exercise 28 in Exercise Set 20.3.)
13. $e^{2x^2} dy + x(2ye^{2x^2} - 1)\, dx = 0;\ y(0) = 1$
14. $2y'' + 5y' - 3y = 0;\ y(0) = 1,\ y'(0) = 0$
15. $y'' + 6y' + 9y = 0;\ y(0) = 3,\ y'(0) = -4$
16. $y'' + 4y' + 5y = 0$
17. $\dfrac{d^2 s}{dt^2} + 2\dfrac{ds}{dt} = 0;\ s(0) = 0,\ s'(0) = 4$
18. $y'' + 9y = 0;\ y(0) = 3,\ y(\frac{\pi}{2}) = -2$
19. $\dfrac{d^2 x}{dt^2} - \dfrac{dx}{dt} - 6x = 2e^t$
20. $y'' + 2y' - 3y = 2e^{-x} - 1$
21. $y'' - y' = x + 3e^x$
22. $y'' + 4y = 2 \cos^2 x$
23. $2y'' + 3y' - 5y = e^x \cos x$
24. $y'' + y = xe^{-x}$
25. $y'' - y' = \sinh x;\ y(0) = 0,\ y'(0) = 1$
26. $y'' - 4y' + 4y = 8x^2;\ y(0) = 1,\ y'(0) = -1$

In Exercises 27 and 28 give the appropriate form for a particular solution. Do not calculate the coefficients.

27. $y'' - 3y' - 4y = xe^{-x} + \cos 2x$
28. $y'' + 2y' + 2y = x^2 e^{-x} \sin x$

29. If 10% of a certain radioactive substance decays in 2 yr, find its half-life. How long will it take for 90% to decay?
30. A 50% dye solution is fed at the rate of 8 L/min into a vat that originally contained 200 L of pure water, and the well-mixed solution is drained off at the same rate. How long will it take for the mixture to become a 25% dye solution?
31. A modification of the Malthus model that accounts for the harvesting of a population $Q(t)$ at a constant rate H (for example, the harvesting of a deer population by hunters or of some other animal species by predators) is given by

$$Q'(t) = \alpha Q(t) - H \qquad (\alpha > 0)$$

with $Q(0) = Q_0$. Solve this equation and show that the model predicts three possible outcomes, depending on the relative sizes of H and α: (1) the population dies out in a finite time, (2) the population grows without limit, or (3) the population size stays constant. Determine the relationship between H and α that produces each result.

32. In an *autocatalytic* chemical reaction a substance A is converted to a substance B in such a way that the reaction is stimulated by the substance being produced. If the original concentration of A is a and at time t the concentration of B is $x = x(t)$, then the reaction rate is modeled by

$$\frac{dx}{dt} = kx(a - x)$$

Find the solution if $x(0) = x_0$.

33. A 4-lb weight stretches a spring 0.64 ft beyond its natural length. The system is pushed up 4 in. from equilibrium and then given a downward velocity of 5 ft/sec. There is a damping force present of 0.25 v. Find the equation of motion.
34. In the equation

$$L\frac{d^2 I}{dt^2} + R\frac{dI}{dt} + \frac{I}{C} = E'(t)$$

obtained for the *RLC* circuit in Exercise 17 in Exercise Set 20.7, suppose $E(t) = E \sin \omega t$ and $R^2 C = 4\, L$. Find $I(t)$. To simplify notation, write $\alpha = \frac{R}{2L}$.

In Exercises 35 and 36 find the general solution using power series. Find the largest interval in which the solution is valid.

35. $y'' - 2xy' - 4y = 0$
36. $(1 - x^2)y'' - 6xy' - 4y = 0$

APPENDIX 1

MATHEMATICAL INDUCTION AND THE BINOMIAL FORMULA

A1.1 MATHEMATICAL INDUCTION

In this section we give a theorem called the **principle of mathematical induction** that provides a way of proving a broad class of statements, or *propositions,* about natural numbers. We denote by $P(n)$ such a proposition. P can be thought of as a function, called a *propositional function,* with domain the set $\mathbf{N}$ of natural numbers and range a set of propositions. For example, $P(n)$ might be the following proposition:

The product of any natural number n and its successor is no less than $3n - 1$.

(The *successor* of a natural number n is $n + 1$.) Of course, we could write $P(n)$ more succinctly as

$$n(n + 1) \geq 3n - 1$$

and usually propositions of the type we are discussing will be stated in symbols rather than words. Such a proposition may or may not be true, but if it satisfies the hypotheses of the following theorem, it *must* be true for all natural numbers.

THEOREM A1.1 THE PRINCIPLE OF MATHEMATICAL INDUCTION

A proposition $P(n)$ is true for all natural numbers n if the following two conditions are satisfied:

1. $P(1)$ is true.
2. If $P(k)$ is true for any natural number k, then $P(k + 1)$ is also true. ■

Note Condition 2 can be written symbolically:

$$P(k) \Rightarrow P(k + 1)$$

read "$P(k)$ implies $P(k + 1)$."

The proof is an immediate consequence of an axiom for **N** stating that any subset S of **N** that contains 1 and also contains the successor of each of its members must be the entire set **N**; that is, no proper subset of **N** has both of these properties. For the proof of the theorem we can take S to be the set of all numbers n for which $P(n)$ is true. The result then follows from the axiom stated.

The plausibility of the induction principle can be seen as follows. We know by condition 1 that $P(1)$ is true. Now we apply condition 2 with $k = 1$ to conclude that $P(2)$ is true. Applying condition 2 again with $k = 2$, we see that $P(3)$ is true. In a similar way, $P(4)$ follows from $P(3)$, and so on and on.

EXAMPLE A1.1 Use mathematical induction to prove that

$$n(n + 1) \geq 3n - 1$$

for all natural numbers n.

Solution Let $P(n)$ denote the inequality as stated. Then $P(1)$ is the statement $1(1 + 1) \geq 3(1) - 1$, which is true, since it simplifies to $2 \geq 2$. To prove that $P(k) \Rightarrow P(k + 1)$, we assume $P(k)$ is true—that is, that

$$P(k): \quad k(k + 1) \geq 3k - 1$$

This is called the **induction hypothesis.** We want to show that as a consequence of this assumption, $P(k + 1)$ also is true—that is,

$$P(k + 1): \quad (k + 1)(k + 2) \geq 3(k + 1) - 1$$

To show this, we begin by writing the product $(k + 1)(k + 2)$ in a way that enables us to use the induction hypothesis:

$$\begin{aligned} (k + 1)(k + 2) &= (k + 1) \cdot k + (k + 1) \cdot 2 && \text{Distributive law} \\ &\geq k(k + 1) + 4 && \text{Since } k \geq 1 \\ &\geq (3k - 1) + 4 && \text{By the induction hypothesis} \\ &= 3k + 3 \\ &> 3(k + 1) - 1 \end{aligned}$$

Thus, under the assumption that $P(k)$ is true, we have shown that $P(k + 1)$ is also true. Both conditions 1 and 2 of the induction principle are satisfied, and so the proof is complete. ■

A Word of Caution A fairly common mistake in an inductive proof is to prove $P(k) \Rightarrow P(k + 1)$ only and to think this is enough. This in itself does not assert the truth of the proposition for any natural number, but only that *if* it is true for some number k, it is true for its successor $k + 1$. So showing that $P(1)$ is true is an essential part of the proof.

EXAMPLE A1.2 Use mathematical induction to prove that the sum of the first n odd natural numbers is n^2.

Solution Let $P(n)$ be the given proposition. In symbols it can be written

$$1 + 3 + 5 + \cdots + (2n - 1) = n^2$$

For $n = 1$, this reads $1 = 1^2$. So $P(1)$ is true. Assume $P(k)$ is true:

$$P(k): \quad 1 + 3 + 5 + \cdots + (2k - 1) = k^2$$

We want to show that

$$P(k+1): \quad 1+3+5+\cdots+[2(k+1)-1]=(k+1)^2$$

follows as a consequence. The last bracketed term is the $(k+1)$st odd natural number, and it is seen to be $(2k+1)$ after simplifying. The odd number preceding it is $(2k-1)$, so we can write

$$1+3+5+\cdots+(2k+1)=[1+3+5+\cdots+(2k-1)]+(2k+1)$$

By the induction hypothesis we are assuming the bracketed terms add to give k^2, so the right-hand side becomes $k^2+2k+1=(k+1)^2$. Thus,

$$1+3+5+\cdots+(2k+1)=(k+1)^2$$

that is, $P(k) \Rightarrow P(k+1)$. So the proof is complete. ■

EXERCISE SET A1.1

A

Use mathematical induction to prove that the given proposition is true for all natural numbers n.

1. $2^n \geq 2n$ **2.** $1+2+3+\cdots+n=\dfrac{n(n+1)}{2}$

3. $\dfrac{1}{1\cdot 2}+\dfrac{1}{2\cdot 3}+\dfrac{1}{3\cdot 4}+\cdots+\dfrac{1}{n(n+1)}=\dfrac{n}{n+1}$

4. $n^2+1 \geq 2n$

5. (a) If $a>1$, $a^n>1$.
(b) If $0<a<1$, then $0<a^n<1$.

6. $(ab)^n=a^nb^n$

7. n^2+n is even.

8. 4^n-1 is a multiple of 3.

9. If $r \neq 1$, then $a+ar+ar^2+\cdots+ar^{n-1}=\dfrac{a(1-r^n)}{1-r}$.

B

10. $1^2+2^2+3^2+\cdots+n^2=\dfrac{n(n+1)(2n+1)}{6}$

11. $1^3+3^3+5^3+\cdots+(2n-1)^3=n^2(2n^2-1)$

12. $1^3+2^3+3^3+\cdots+n^3=(1+2+3+\cdots+n)^2$
(*Hint:* Use the result of Exercise 2.)

13. If $a \neq b$, $a-b$ is a factor of a^n-b^n. [*Hint:* In proving condition 2, write $a^{k+1}-b^{k+1}=a(a^k-b^k)+b^k(a-b)$.]

A1.2 THE BINOMIAL FORMULA

In this section we give the formula for the expansion of $(a+b)^n$, where n is a natural number. Before stating the formula in its full generality, let us consider the result for the first few values of n:

$$n=1: \quad (a+b)^1=a+b$$

$$n=2: \quad (a+b)^2=a^2+2ab+b^2$$

$$n=3: \quad (a+b)^3=a^3+3a^2b+3ab^2+b^3$$

This is as far as most students know from memory. (If you did not remember the last one, you can get it by multiplying the one before it by $a+b$.) If we multiply both sides again by $a + b$, we get

$$n = 4: \quad (a + b)^4 = a^4 + 4a^3b + 6a^2b^2 + 4ab^3 + b^4 \qquad \text{(Verify.)}$$

We could keep going, but it would quickly become tedious and we still would not have a general formula. We *can* anticipate the successive powers of a and b. For example, we would expect that in the expansion of $(a + b)^5$, the successive terms would involve, in order,

$$a^5,\ a^4b,\ a^3b^2,\ a^2b^3,\ ab^4,\ b^5$$

(Exponents on a decrease from 5 to 0 and exponents on b increase from 0 to 5.) But what are the coefficients? If you do not already know, perhaps you would like to conjecture. The brilliant French mathematician Blaise Pascal (1623–1662) discovered the following scheme for the coefficients in the expansions of $(a + b)^n$ beginning with $n = 0$:

$$\begin{array}{lccccccccccc}
n = 0: & & & & & & 1 & & & & & \\
n = 1: & & & & & 1 & & 1 & & & & \\
n = 2: & & & & 1 & & 2 & & 1 & & & \\
n = 3: & & & 1 & & 3 & & 3 & & 1 & & \\
n = 4: & & 1 & & 4 & & 6 & & 4 & & 1 & \\
n = 5: & 1 & & 5 & & 10 & & 10 & & 5 & & 1 \\
\vdots & & & & & & \vdots & & & & &
\end{array}$$

This is known as **Pascal's triangle.** Do you see how to get the next line? The following diagram should help:

$$\begin{array}{lccccccccccccc}
n = 5: & & 1 & & 5 & & 10 & & 10 & & 5 & & 1 & \\
n = 6: & 1 & & 6 & & 15 & & 20 & & 15 & & 6 & & 1
\end{array}$$

Pascal's triangle gives a simple method (an *algorithm*) for determining the coefficients in $(a + b)^n$ for any value of n, but we still do not have the general formula we are seeking. Besides, using Pascal's triangle to get $(a + b)^{20}$, say, would require a rather large triangle!

We now state the binomial formula in its full generality, but before proving it we will introduce some convenient notation and obtain some preliminary results.

THEOREM A1.2 THE BINOMIAL FORMULA

For all natural numbers n,

$$(a + b)^n = a^n + na^{n-1}b + \frac{n(n-1)}{1 \cdot 2}a^{n-2}b^2 + \frac{n(n-1)(n-2)}{1 \cdot 2 \cdot 3}a^{n-3}b^3 + \cdots + nab^{n-1} + b^n \qquad \text{(A1.1)}$$

■

The coefficient of the $(k + 1)$st term (that is, the term involving $a^{n-k}b^k$) is

$$\frac{n(n-1)(n-2)\cdots[n-(k-1)]}{1 \cdot 2 \cdot 3 \cdot \cdots \cdot k}$$

Observe that both numerator and denominator contain k factors. On the numerator the factors start with n and decrease consecutively, whereas on the denominator they start with 1 and increase consecutively. To simplify notation we represent this coefficient by the symbol $\binom{n}{k}$. So

$$\binom{n}{k} = \frac{n(n-1)(n-2)\cdots(n-k+1)}{1 \cdot 2 \cdot 3 \cdot \cdots \cdot k} \tag{A1.2}$$

For example,

$$\binom{7}{3} = \frac{7 \cdot 6 \cdot 5}{1 \cdot 2 \cdot 3} = 35$$

For certain purposes it is useful to put equation (A1.2) in another form. Recall that $1 \cdot 2 \cdot 3 \cdot \cdots \cdot k = k!$, and this is the denominator in equation (A1.2). The numerator in general is not a factorial, but it becomes $n!$ if it is multiplied by $(n-k)(n-k-1)\cdots(2)(1) = (n-k)!$ So, on multiplying numerator and denominator by $(n-k)!$, we get

$$\binom{n}{k} = \frac{n!}{k!(n-k)!} \tag{A1.3}$$

It is customary to define 0! to be 1, so that we always have

$$\binom{n}{0} = \binom{n}{n} = 1$$

obtained by putting $k = 0$ and $k = n$, in turn, in equation (A1.3).

In view of this notation, we can rewrite the binomial formula as

$$(a+b)^n = \binom{n}{0}a^n + \binom{n}{1}a^{n-1}b + \binom{n}{2}a^{n-2}b^2 + \cdots + \binom{n}{n-1}ab^{n-1} + \binom{n}{n}b^n \tag{A1.4}$$

The following lemma will be needed in the proof.

LEMMA A1.1 For any natural numbers k and r,

$$\binom{k}{r} + \binom{k}{r-1} = \binom{k+1}{r} \tag{A1.5}$$

Proof From equation (A1.3) we have

$$\begin{aligned}\binom{k}{r} + \binom{k}{r-1} &= \frac{k!}{r!(k-r)!} + \frac{k!}{(r-1)!(k-r+1)!} \\ &= \frac{k!}{(r-1)!(k-r)!}\left[\frac{1}{r} + \frac{1}{k-r+1}\right] \\ &= \frac{k!}{(r-1)!(k-r)!}\left[\frac{k+1}{r(k-r+1)}\right] \\ &= \frac{(k+1)!}{r!(k+1-r)!} \\ &= \binom{k+1}{r}\end{aligned}$$

Note that we made repeated use of the fact that for any natural number m, $(m + 1)! = (m + 1)m!$. You should follow through the steps of the following proof carefully. ■

Proof of the Binomial Formula Let $P(n)$ be the proposition stated in equation (A1.4). Then $P(1)$ says that

$$(a + b)^1 = \binom{1}{0} a^1 + \binom{1}{1} b^1$$

which is true, since both sides simplify to $a + b$. Now we assume the truth of $P(k)$:

$$(a + b)^k = \binom{k}{0} a^k + \binom{k}{1} a^{k-1}b + \binom{k}{2} a^{k-2}b^2 + \cdots + \binom{k}{k-1} ab^{k-1} + \binom{k}{k} b^k$$

In order to get $P(k + 1)$ we multiply both sides by $(a + b)$, obtaining (verify)

$$\begin{aligned}(a + b)^{k+1} &= \binom{k}{0} a^{k+1} + \binom{k}{1} a^k b + \binom{k}{2} a^{k-1}b^2 + \cdots + \binom{k}{k} ab^k \\ &\quad + \binom{k}{0} a^k b + \binom{k}{1} a^{k-1}b^2 + \cdots + \binom{k}{k-1} ab^k + \binom{k}{k} b^{k+1} \\ &= \binom{k}{0} a^{k+1} + \left[\binom{k}{1} + \binom{k}{0}\right] a^k b + \left[\binom{k}{2} + \binom{k}{1}\right] a^{k-1}b^2 \\ &\quad + \cdots + \left[\binom{k}{k} + \binom{k}{k-1}\right] ab^k + \binom{k}{k} b^{k+1}\end{aligned}$$

Now

$$\binom{k}{0} = \binom{k+1}{0} \quad \text{and} \quad \binom{k}{k} = \binom{k+1}{k+1}$$

each value being identically 1. Making these replacements and using equation (A1.5) from the lemma on each of the bracketed terms, we get

$$\begin{aligned}(a + b)^{k+1} &= \binom{k+1}{0} a^{k+1} + \binom{k+1}{1} a^k b + \binom{k+1}{2} a^{k-1}b^2 + \cdots \\ &\quad + \binom{k+1}{k} ab^k + \binom{k+1}{k+1} b^{k+1}\end{aligned}$$

This is precisely $P(k + 1)$. So $P(k) \Rightarrow P(k + 1)$, and the proof by induction is complete. ■

EXERCISE SET A1.2

A

1. Evaluate:

   (a) $\binom{8}{2}$ (b) $\binom{8}{6}$ (c) $\binom{12}{5}$ (d) $\binom{30}{2}$

2. Simplify:

   (a) $(n + 2)(n + 1)n!$ (b) $\dfrac{1}{k!} + \dfrac{1}{(k-1)!}$

(c) $\dfrac{n!}{(n-2)!}$ (d) $\dbinom{k}{3}+\dbinom{k}{4}$

In Exercises 3–6 expand and simplify.

3. $(x+y)^8$

4. $(a-b)^4$ [*Hint:* Write $a-b=a+(-b)$.]

5. $(3a+4b)^5$

6. $(x^2-\frac{2}{y})^6$

7. Write the first five terms in the expansion of $(2x-3y)^{10}$.

8. (a) Find the ninth term in the expansion of $(a+2b)^{12}$.
(b) Find the fourth term in the expansion of $(3a-b^2)^7$.

B

9. Show that

$$\binom{n}{0}+\binom{n}{1}+\binom{n}{2}+\cdots+\binom{n}{n}=2^n$$

[*Hint:* Consider $(1+1)^n$.]

10. Prove the identity

$$\frac{n-r+1}{r}\binom{n}{r-1}=\binom{n}{r}$$

APPENDIX 2

REVIEW OF TRIGONOMETRY

A2.1 ANGLES AND THEIR MEASURE

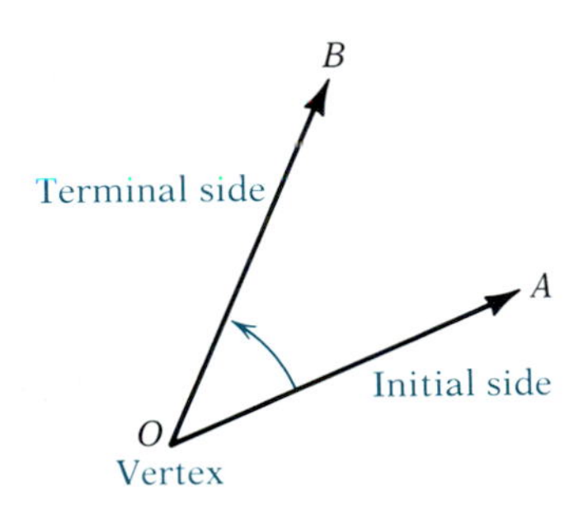

FIGURE A2.1

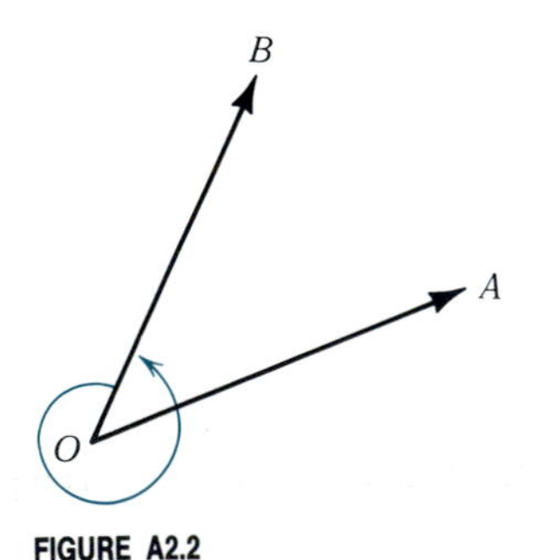

FIGURE A2.2

An **angle** is the union of two rays that have a common end point, called the **vertex.** It is useful to think of an angle as having been generated by rotating one of the rays while holding the other one fixed. In Figure A2.1, for example, we can consider that ray OB initially coincided with OA and then was rotated to its final position, as indicated by the small arc showing the direction. For the angle in the figure, the rotation is counterclockwise and the angle is said to be **positive.** If the rotation is clockwise, the angle is **negative.** The fixed ray is called the **initial side,** and the rotated ray is called the **terminal side.** The rotation may be through more than one complete revolution, as shown in Figure A2.2. A positive angle of less than one revolution is said to be a **primary angle.** Two angles with the same initial side and with terminal sides that also coincide are said to be **coterminal.** Every angle that is not primary is coterminal with one that is.

There are two principal ways of assigning a measure to angles, the **degree** and the **radian.** We assume you are familiar with degree measure and concentrate on radian measure because it is much more important in calculus.

Denote by θ an angle with initial side OA and terminal side OB. (Greek letters are frequently used to name angles.) Let C be a unit circle (circle of radius 1) centered at 0, as in Figure A2.3. Again think of θ as having been generated by rotating ray OB from ray OA. As it rotates, its point of intersection with the circle C sweeps out an arc on C, which we call the *arc subtended on C by θ.* For a positive angle θ let t be the length of the subtended arc, and for θ negative let t be the negative of the subtended arc length. Thus $t > 0$ when θ is positive, and $t < 0$ when θ is negative. We call t the **directed length** of the subtended arc. It is by definition the radian measure of θ.

DEFINITION A2.1 The **radian measure of an angle θ** is the directed length of the arc subtended by θ on the unit circle with center at the vertex of θ. ■

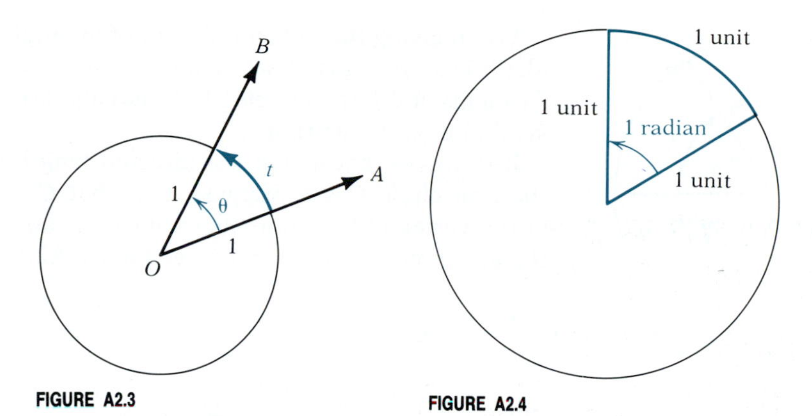

FIGURE A2.3

FIGURE A2.4

Remark on Notation It is customary to use the same letter to name an angle and to designate its measure. With this understanding, we can write $\theta = t$, where t is defined above. For example, if θ has radian measure 2, we write $\theta = 2$.

According to Definition A2.1, an angle of 1 radian subtends an arc of 1 unit on the unit circle. (see Figure A2.4). Since the circumference of the circle is 2π (circumference $= 2\pi r$, and $r = 1$), it follows that there are 2π (≈ 6.28) arcs of length 1 on the complete circle, and hence 2π radians in a complete revolution. Since there also are $360°$ in one revolution, we have $360° = 2\pi$ radians or

$$180° = \pi \text{ radians}$$

From this we get the conversion equations

$$1° = \left(\frac{\pi}{180}\right) \text{radians}$$
$$1 \text{ radian} = \left(\frac{180}{\pi}\right)^{\circ} \tag{A2.1}$$

The second of these shows that 1 radian is approximately $57.3°$.

EXAMPLE A2.1 (a) Change $240°$ to radians. (b) Change $\frac{7\pi}{4}$ radians to degrees.

Solution (a) We multiply by $\frac{\pi}{180}$: $(240)\left(\frac{\pi}{180}\right) = \frac{4\pi}{3}$ radians

(b) We multiply by $\frac{180}{\pi}$: $\left(\frac{7\pi}{4}\right)\left(\frac{180}{\pi}\right) = 315°$ ■

The following table shows the degree and radian measures of some frequently used angles. You should verify these.

Degrees	30°	45°	60°	90°	120°	135°	150°	180°
Radians	$\frac{\pi}{6}$	$\frac{\pi}{4}$	$\frac{\pi}{3}$	$\frac{\pi}{2}$	$\frac{2\pi}{3}$	$\frac{3\pi}{4}$	$\frac{5\pi}{4}$	π
Degrees	**210°**	**225°**	**240°**	**270°**	**300°**	**315°**	**330°**	**360°**
Radians	$\frac{7\pi}{6}$	$\frac{5\pi}{4}$	$\frac{4\pi}{3}$	$\frac{3\pi}{2}$	$\frac{5\pi}{3}$	$\frac{7\pi}{4}$	$\frac{11\pi}{6}$	2π

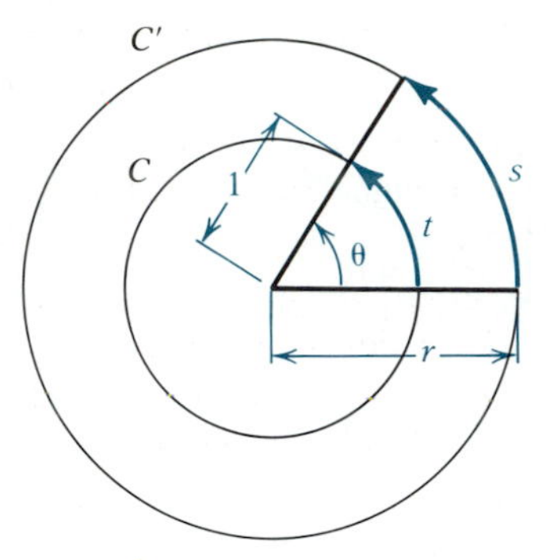

FIGURE A2.5

When giving the radian measure of an angle, it is customary to delete any identification. Thus, if we write $\alpha = \frac{\pi}{2}$ or $\beta = 10$, we mean α is an angle of $\frac{\pi}{2}$ radians and β is an angle of 10 radians. For degrees, however, the degree symbol must be written.

If $\theta = t$, we know that the directed length of the arc subtended by θ on the unit circle C is t. Suppose now that C' is any other circle with center at the vertex of θ. Denote its radius by r, and let s be the directed length of the arc subtended by θ on C' (see Figure A2.5). Then, by similarity, we have

$$\frac{s}{r} = \frac{t}{1} \quad \text{or} \quad s = rt$$

Since $\theta = t$, we can write this as

$$s = r\theta \tag{A2.2}$$

This formula provides a means of calculating the arc length s when we know the angle θ in radians. Alternatively, we can view it as another way of finding the radian measure of θ:

$$\theta = \frac{s}{r} \tag{A2.3}$$

EXAMPLE A2.2 Find the length of arc subtended by an angle of 72° on a circle of radius 10 in.

Solution First we express the angle θ in radians:

$$\theta = (72)\left(\frac{\pi}{180}\right) = \frac{2\pi}{5}$$

Then, by equation (A2.2),

$$s = (10)\left(\frac{2\pi}{5}\right) = 4\pi \text{ in.}$$

■

EXERCISE SET A2.1

A

In each part of Exercises 1 and 2 the degree measure of an angle is given. Find its radian measure.

1. (a) $50°$ (b) $450°$ (c) $-200°$ (d) $144°$

2. (a) $900°$ (b) $-25°$ (c) $(\frac{36}{\pi})°$ (d) $3.6°$

In each part of Exercises 3 and 4 the radian measure of an angle is given. Find its degree measure.

3. (a) $\frac{3\pi}{5}$ (b) $\frac{7\pi}{2}$ (c) $-\frac{13\pi}{12}$ (d) 3

4. (a) $\frac{5\pi}{9}$ (b) $\frac{3}{4}$ (c) $-\frac{8\pi}{3}$ (d) $-\frac{5}{2}$

5. Find the length of the arc subtended on a circle of radius 6 ft by an angle of 120°.

6. Find the arc length on the equator, in miles, subtended by an angle of 1° at the center of the earth. Take the earth's radius as 3960 mi.

7. What is the length of the arc swept out by the tip of the minute hand of a clock during the time interval from 6:00 to 6:20 if the minute hand is 4 in. long?

B

Suppose a point is moving around a circle of radius r at a constant rate. Its **linear speed** *is $v = s/t$ (distance divided by time), and its* **angular speed** *is $\omega = \theta/t$, giving the number of radians through which a radius to the point turns per unit of time. From $s = r\theta$, we get, on dividing by t, $v = r\omega$. Use this in Exercises 8–11.*

8. A flywheel 4 ft in diameter is revolving at the rate of 50 rpm (revolutions per minute). What is its angular speed in radians per second? What is the linear speed of a point on the rim in feet per second?

9. A train is traveling on a circular curve of radius $\frac{1}{2}$ mi at the rate of 30 mi/hr. Through what angle (in degrees) will it turn in 45 sec?

10. Assume the earth moves around the sun in a circular orbit with radius 93 million miles. A complete revolution takes 365 days. Find the approximate speed the earth moves in its orbit in miles per hour.

11. An automobile tire is 28 in. in diameter. If the car is traveling at 30 mi/hr, through how many revolutions per minute is the wheel turning?

A2.2 THE TRIGONOMETRIC FUNCTIONS

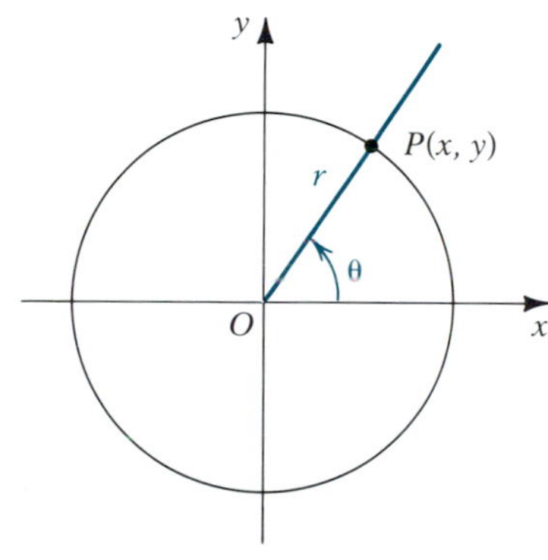

FIGURE A2.6

In this section we define the six **trigonometric functions** called the **sine, cosine, tangent, cotangent, secant,** and **cosecant.** We do this first in terms of angles and then in terms of real numbers. Although the latter is more important in calculus, the former approach is needed at times and has broad applications in what is referred to as numerical trigonometry, involving the study of triangles. If θ is an angle, we abbreviate the values of the six functions at θ by $\sin\theta$, $\cos\theta$, $\tan\theta$, $\cot\theta$, $\sec\theta$, and $\csc\theta$, respectively.

An angle is said to be in **standard position** with respect to a given coordinate system if its initial side coincides with the positive x-axis. As in Figure A2.6, let θ be such an angle, and let $P(x, y)$ be the point of intersection of its terminal side and the circle $x^2 + y^2 = r^2$, where r is an arbitrary positive number. Then we define the trigonometric functions of θ as follows.

DEFINITION A.2.2

$$\sin\theta = \frac{y}{r} \qquad \cot\theta = \frac{x}{y} \quad (y \neq 0)$$

$$\cos\theta = \frac{x}{r} \qquad \sec\theta = \frac{r}{x} \quad (x \neq 0)$$

$$\tan\theta = \frac{y}{x} \quad (x \neq 0) \qquad \csc\theta = \frac{r}{y} \quad (y \neq 0)$$

■

Remarks

1. That this definition is independent of the size of the radius r can be seen by looking at similar triangles.
2. The definition is valid whether θ is measured in radians or in degrees.
3. The sine and cosine are defined for all angles θ. For the tangent and secant, θ cannot terminate on the y-axis, and for the cotangent and cosecant, θ cannot terminate on the x-axis.

EXAMPLE A2.3 The terminal side of an angle θ in standard position passes through the point $(-1, 2)$. Find the six trigonometric functions of θ.

Solution Figure A2.7 illustrates the given information. For the circle $x^2 + y^2 = r^2$ to pass through $P(-1, 2)$, we must have $r = \sqrt{5}$. So by Definition A2.2,

$$\sin\theta = \frac{2}{\sqrt{5}} \qquad \tan\theta = -2 \qquad \sec\theta = -\sqrt{5}$$

$$\cos\theta = -\frac{1}{\sqrt{5}} \qquad \cot\theta = -\frac{1}{2} \qquad \csc\theta = \frac{\sqrt{5}}{2}$$

■

EXAMPLE A2.4 Given $\sec\theta = 3$ and $\sin\theta < 0$, find the other five trigonometric functions of θ.

Solution Since $\sec\theta = \frac{r}{x}$, we may take $r = 3$ and $x = 1$. Also $\sin\theta = \frac{y}{r}$, so that for $\sin\theta$ to be negative we must have $y < 0$. Substituting $x = 1$ and $r = 3$ into $x^2 + y^2 = r^2$ therefore yields $y = -\sqrt{3^2 - 1^2} = -2\sqrt{2}$. So we have

$$\sin\theta = \frac{-2\sqrt{2}}{3} \qquad \tan\theta = -2\sqrt{2} \qquad \csc\theta = -\frac{3}{2\sqrt{2}}$$

$$\cos\theta = \frac{1}{3} \qquad \cot\theta = -\frac{1}{2\sqrt{2}}$$

The (primary) angle θ is shown in Figure A2.8. ■

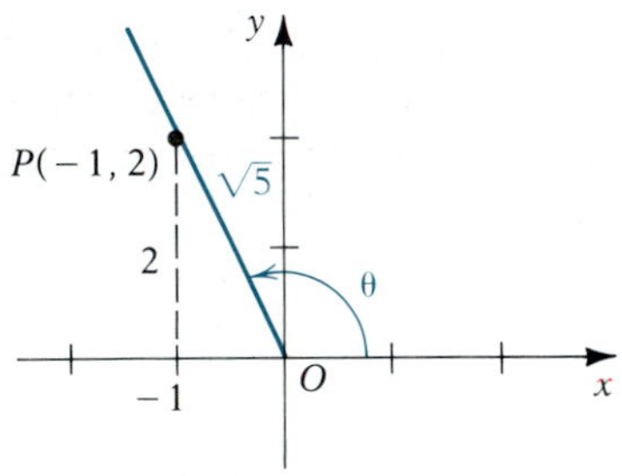

FIGURE A2.7

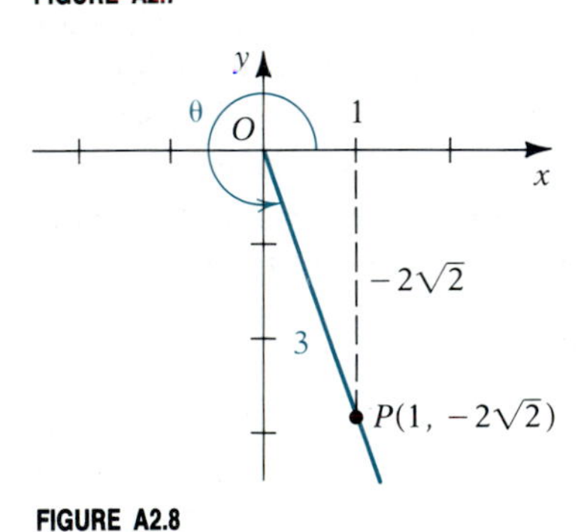

FIGURE A2.8

The trigonometric functions of an **acute** angle (one between 0° and 90°) can be expressed in terms of the sides of a right triangle. As illustrated in Figure A2.9, with the triangle placed so that the acute angle θ is in standard position, the vertex $P(x, y)$ is at a distance r from the origin, where r is the length of the hypotenuse. By the Pythagorean theorem, $x^2 + y^2 = r^2$, so that $P(x, y)$ is the intersection of the terminal side of θ and the circle of radius r centered at 0. The leg of the triangle of length x is called the **adjacent** side to θ, and the leg of length y is called the **opposite** side to θ. With the understanding that these designations stand for the lengths of the corresponding sides, and that the same is true for the hypotenuse, we have from Definition A2.2

$$\sin\theta = \frac{\text{opposite}}{\text{hypotenuse}} \qquad \cot\theta = \frac{\text{adjacent}}{\text{opposite}}$$

$$\cos\theta = \frac{\text{adjacent}}{\text{hypotenuse}} \qquad \sec\theta = \frac{\text{hypotenuse}}{\text{adjacent}}$$

$$\tan\theta = \frac{\text{opposite}}{\text{adjacent}} \qquad \csc\theta = \frac{\text{hypotenuse}}{\text{opposite}}$$

Now that we have these results, they can be applied to either acute angle of a right triangle without its being in standard position.

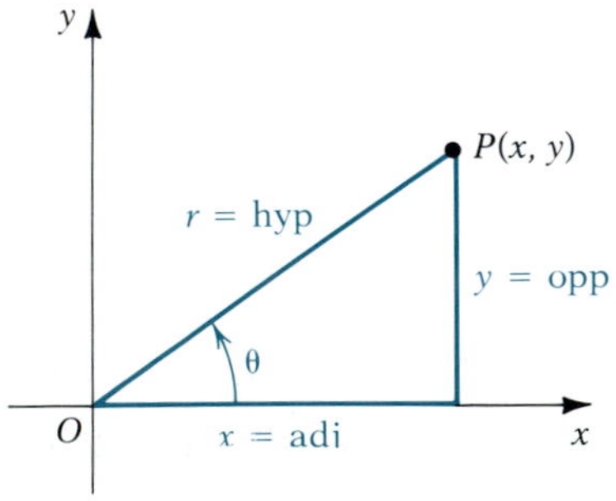

FIGURE A2.9

EXAMPLE A2.5 Find the six trigonometric functions of angle B in the right triangle in Figure A2.10.

Solution By the Pythagorean theorem the hypotenuse c is

$$c = \sqrt{3^2 + 2^2} = \sqrt{13}$$

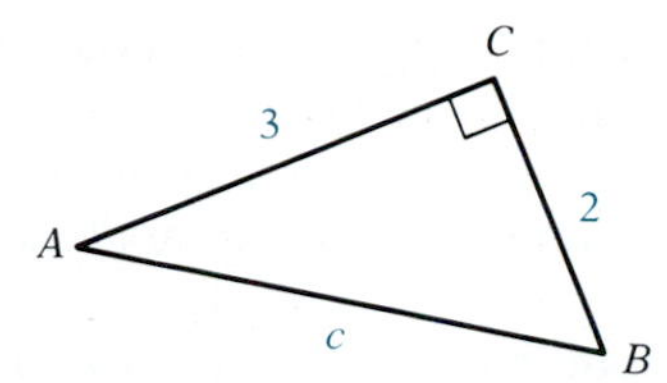

FIGURE A2.10

The side adjacent to B is of length 2 and the side opposite is of length 3. So we have

$$\sin B = \frac{3}{\sqrt{13}} \qquad \tan B = \frac{3}{2} \qquad \sec B = \frac{\sqrt{13}}{2}$$

$$\cos B = \frac{2}{\sqrt{13}} \qquad \cot B = \frac{2}{3} \qquad \csc B = \frac{\sqrt{13}}{3}$$

■

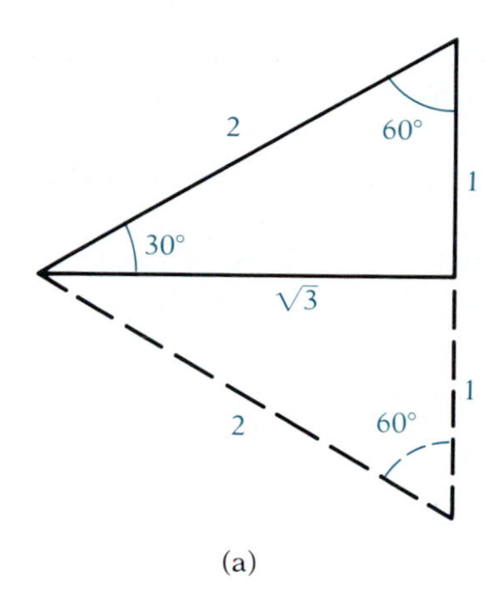

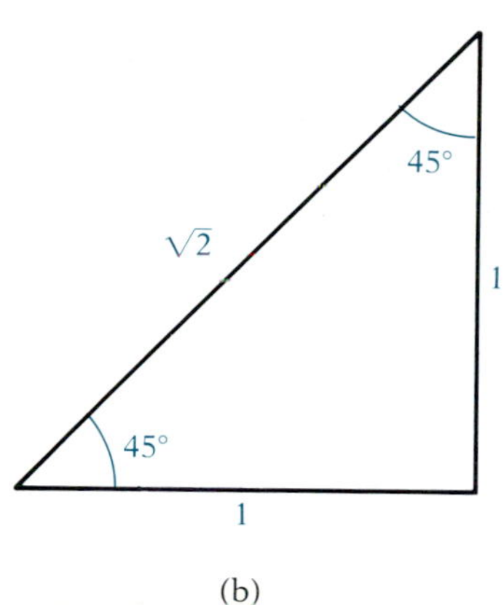

FIGURE A2.11

Two particular right triangles are of special importance, the *30°–60° right triangle* and the *45° right triangle,* shown in Figure A2.11. By reflecting the 30°–60° triangle as shown in part a, we see that it is half of an equilateral triangle. From this we deduce the following property.

In a 30°–60° right triangle, the side opposite the 30° angle is half the hypotenuse.

So if we make the hypotenuse 2, the side opposite the 30° angle is 1, and by the Pythagorean theorem the adjacent side is $\sqrt{3}$. All 30°–60° triangles are similar to this one. In the 45° right triangle the legs are equal in length. So if we make them 1 unit long, the hypotenuse is $\sqrt{2}$. From these two triangles we can write all the trigonometric functions of 30°, 45°, and 60° or, equivalently, of $\frac{\pi}{6}$, $\frac{\pi}{4}$, and $\frac{\pi}{3}$ radians. In fact, we can exploit these triangles to get the functions of many other related angles, as the next example illustrates.

EXAMPLE A2.6 Find all the trigonometric functions of (a) 240° and (b) $\frac{7\pi}{4}$.

Solution (a) Since 240° is 60° more than 180°, we can place our 30°–60° triangle in the third quadrant as shown in Figure A2.12 to find a convenient point on the terminal side. We must label the sides of the triangle with appropriate signs, in this case both negative, since x and y are negative in the third quadrant. So we have

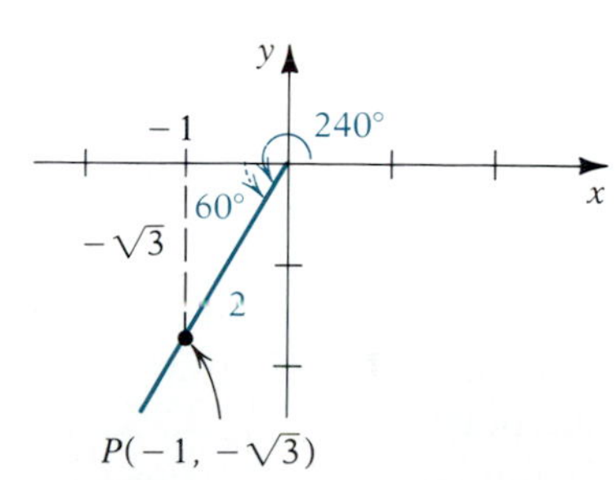

FIGURE A2.12

$$\sin 240° = -\frac{\sqrt{3}}{2} \qquad \tan 240° = \sqrt{3} \qquad \sec 240° = -2$$

$$\cos 240° = -\frac{1}{2} \qquad \cot 240° = \frac{1}{\sqrt{3}} \qquad \csc 240° = -\frac{2}{\sqrt{3}}$$

(b) The angle $\frac{7\pi}{4}$ lacks $\frac{\pi}{4}$ (or 45°) to be around to 2π. So we place our 45° triangle in the fourth quadrant as shown in Figure A2.13, giving the point $P(1, -1)$ on the terminal side. Thus,

$$\sin\frac{7\pi}{4} = -\frac{1}{\sqrt{2}} \qquad \tan\frac{7\pi}{4} = -1 \qquad \sec\frac{7\pi}{4} = \sqrt{2}$$

$$\cos\frac{7\pi}{4} = \frac{1}{\sqrt{2}} \qquad \cot\frac{7\pi}{4} = -1 \qquad \csc\frac{7\pi}{4} = -\sqrt{2}$$

■

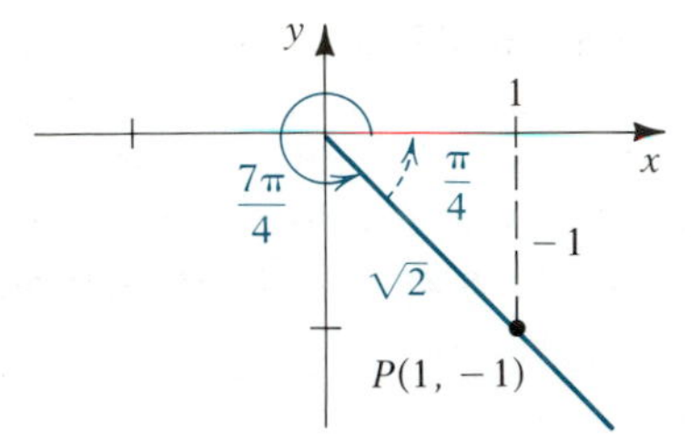

FIGURE A2.13

Note For an arbitrary angle θ, the acute angle between its terminal side and the x-axis is called the **reference angle** for θ. So, as seen in the preceding example, the reference angle for 240° is 60°, and the reference angle for $\frac{7\pi}{4}$ is $\frac{\pi}{4}$. Functions of θ and its reference angle either are the same or differ only in sign.

We now shift our emphasis from trigonometric functions of *angles* to trigonometric functions of *numbers*; that is, we define the functions with domains that are subsets of **R**. To do this we begin with an arbitrary real number t, and we consider the arc on the unit circle $x^2 + y^2 = 1$, of length $|t|$, measured from $(1, 0)$ in a counterclockwise direction if $t > 0$ and clockwise if $t < 0$. Let $P(x, y)$ denote the terminal point of the arc. Figure A2.14 illustrates this for the case $0 < t < \frac{\pi}{2}$. We define the six trigonometric functions of the real number t as follows.

DEFINITION A2.3

$$\sin t = y \qquad \cot t = \frac{x}{y} \quad (y \neq 0)$$

$$\cos t = x \qquad \sec t = \frac{1}{x} \quad (x \neq 0)$$

$$\tan t = \frac{y}{x} \quad (x \neq 0) \qquad \csc t = \frac{1}{y} \quad (y \neq 0)$$

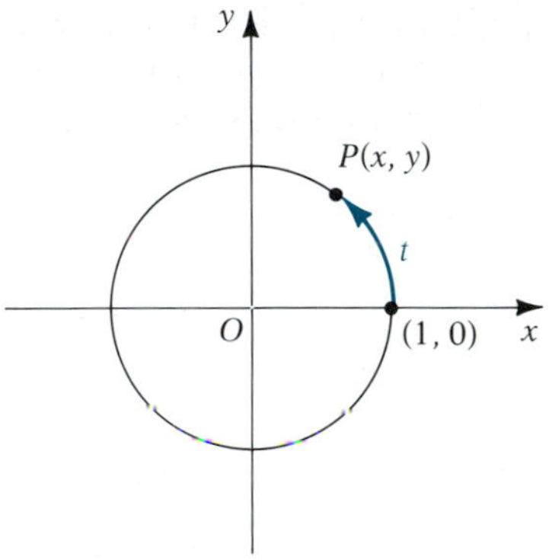

■

FIGURE A2.14

Since the angle with radian measure t, when placed in standard position, has the point $P(x, y)$ of Figure A2.14 on its terminal side, on comparing Definition A2.3 with Definition A2.2, we find the following relationship:

> Each trigonometric function of the *real number t* has the same value as that trigonometric function of the *angle whose radian measure is t*.

So when we see, for example, $\sin \frac{\pi}{6} = \frac{1}{2}$, we can interpret it either as meaning the sine of the real number $\frac{\pi}{6}$ is $\frac{1}{2}$ or as meaning the sine of the angle of $\frac{\pi}{6}$ radians is $\frac{1}{2}$. It makes no difference, since the value is the same in each case. We will shift from one point of view to the other whenever it is convenient to do so.

From Definition A2.3 (as well as from Definition A2.2) the following relationships can be seen:

$$\tan t = \frac{\sin t}{\cos t} \qquad \sec t = \frac{1}{\cos t}$$
$$\cot t = \frac{\cos t}{\sin t} \qquad \csc t = \frac{1}{\sin t} \tag{A2.4}$$

where t in each case is restricted so that the denominator is nonzero. These relationships point out the special importance of the sine and cosine. We concentrate our attention on these two basic functions.

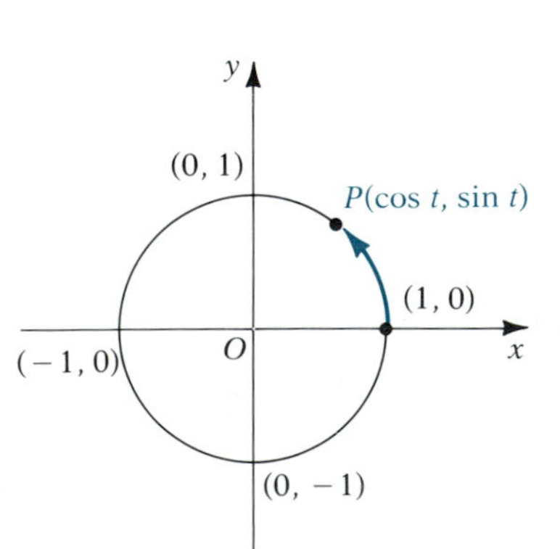

FIGURE A2.15

Figure A2.15 reveals the essential features of Definition A2.3, showing that $\cos t$ and $\sin t$ are the abscissa and ordinate, respectively, of the end point P of the arc of length $|t|$, starting from $(1, 0)$. Because P lies on the circle $x^2 + y^2 = 1$, we have for all real numbers t,

$$\cos^2 t + \sin^2 t = 1 \tag{A2.5}$$

This is of fundamental importance and will be used extensively in the next section.

Note By $\cos^2 t$ and $\sin^2 t$, we mean $(\cos t)^2$ and $(\sin t)^2$. Similar remarks apply to other exponents (except the exponent -1) and to the other functions.

To find the sine and cosine of any number, all we need to do is locate the point P on the unit circle. The x-coordinate is $\cos t$, and the y-coordinate is $\sin t$. We know that the circumference of the circle is 2π. So from Figure A2.15 we readily see the following special values:

$$\sin 0 = 0 \qquad \sin\frac{\pi}{2} = 1 \qquad \sin \pi = 0 \qquad \sin\frac{3\pi}{2} = -1$$

$$\cos 0 = 1 \qquad \cos\frac{\pi}{2} = 0 \qquad \cos \pi = -1 \qquad \cos\frac{3\pi}{2} = 0$$

Also, using the relationship with angles and appropriately placing the 30°–60° and 45° triangles, we can get functions of each of the following:

$$\frac{\pi}{6}, \frac{\pi}{4}, \frac{\pi}{3}, \frac{2\pi}{3}, \frac{3\pi}{4}, \frac{5\pi}{6}, \frac{7\pi}{6}, \frac{5\pi}{4}, \frac{4\pi}{3}, \frac{5\pi}{3}, \frac{7\pi}{4}, \frac{11\pi}{6}$$

(You will be asked to do this in the exercises.) We can get still more special values by observing that for any real number t, adding or subtracting an integral multiple of 2π (the length of the circumference) produces the same point P on the unit circle. Thus,

$$\sin(t + 2n\pi) = \sin t \quad \text{and} \quad \cos(t + 2n\pi) = \cos t$$

for $n = 0, \pm 1, \pm 2, \ldots$. When a function f satisfies $f(t + p) = f(t)$ for some constant p and for all t in its domain, it is said to be **periodic,** and the smallest positive number p for which this equation is true is called the (fundamental) **period** of f. So both the sine and cosine are periodic functions of period 2π.

By comparing the x- and y-coordinates of the points P and P_1 in Figure A2.16, corresponding to t and $-t$, respectively, we obtain the following further information about the nature of the sine and cosine:

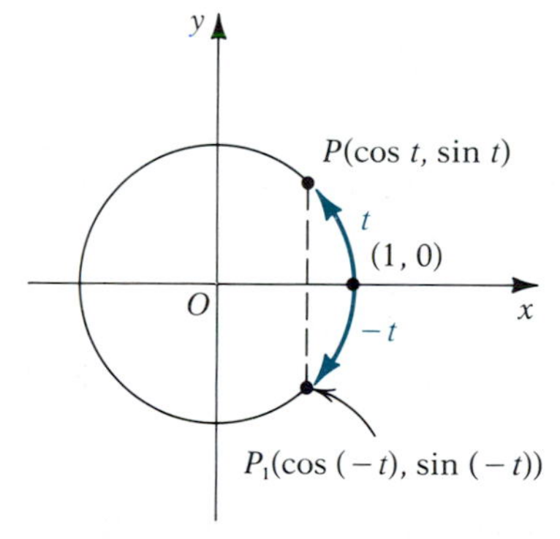

FIGURE A2.16

$$\cos(-t) = \cos t \qquad \sin(-t) = -\sin t$$

that is, the cosine is an even function and the sine is an odd function. You will be asked in the exercises to determine the evenness and oddness, as well as the period, of the other four functions.

We now have enough information to draw the graphs of $y = \sin t$ and $y = \cos t$; these are shown in Figures A2.17 and A2.18. For each function the domain is all of **R** and the range is the interval $[-1, 1]$.

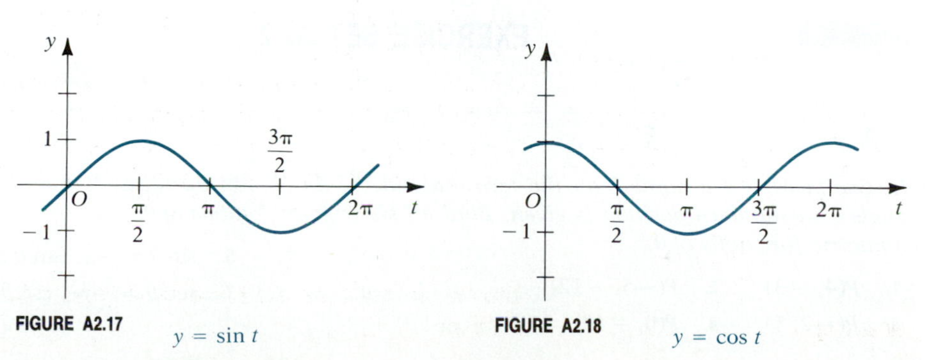

FIGURE A2.17 $y = \sin t$

FIGURE A2.18 $y = \cos t$

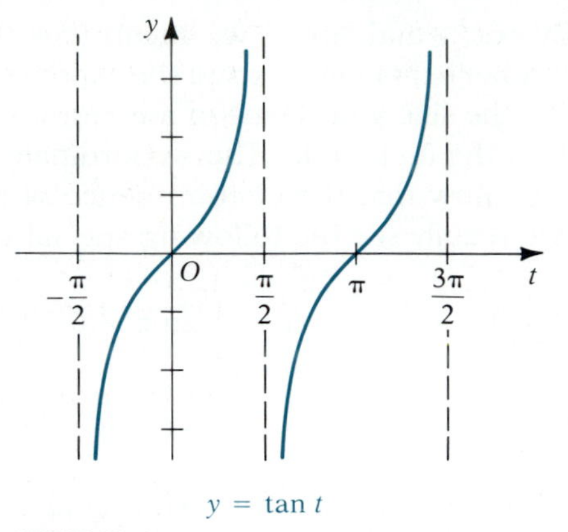

FIGURE A2.19

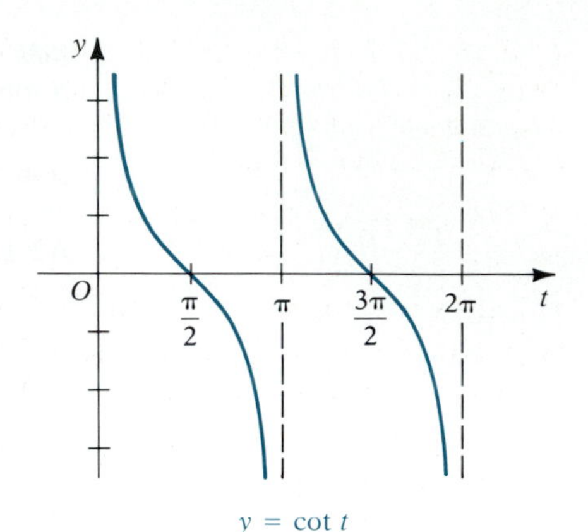

FIGURE A2.20

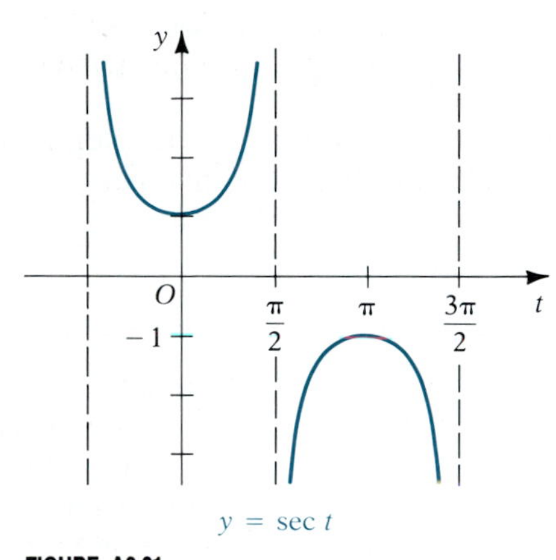

FIGURE A2.21

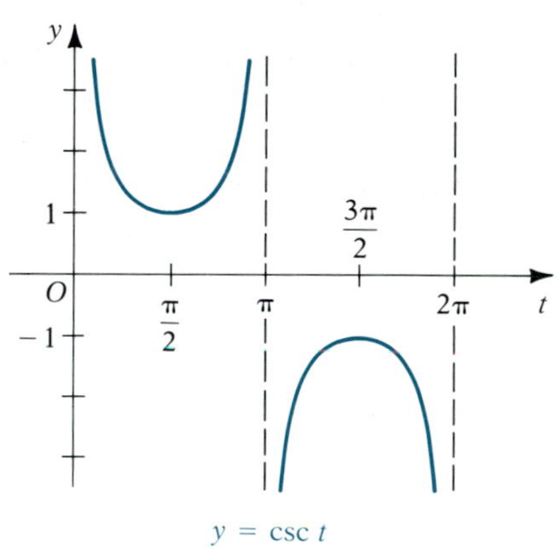

FIGURE A2.22

Graphs of the other four trigonometric functions are given in Figures A2.19–A2.22. In the exercises you will be asked to verify these.

Remark on Notation In Definition A2.3 the letter x was used as the abscissa in the given coordinate system, which is why we used a different letter, t, as the independent variable for the functions, and we have used this throughout. Now that we have obtained properties of the functions based on the definition, however, we are free to use any letter to name the independent variable, including x. In particular, for the graphs, we can rename the t-axis as the x-axis, if we prefer.

EXERCISE SET A2.2

A

In Exercises 1–4 a point P on the terminal side of an angle θ in standard position is given. Find all six trigonometric functions of θ.

1. $P(4, -3)$ **2.** $P(-5, -12)$

3. $P(-2, 5)$ **4.** $P(0, -10)$

In Exercises 5–8 find the other five trigonometric functions of θ.

5. $\sin\theta = -\frac{2}{3}$, $\tan\theta > 0$ **6.** $\cos\theta = \frac{1}{3}$, $\tan\theta < 0$

7. $\sec\theta = -\frac{25}{7}$, $\csc\theta > 0$ **8.** $\cot\theta = 2$, $\sec\theta < 0$

In Exercises 9–12 triangle ABC is a right triangle with right angle at C, and a, b, and c are the lengths of the sides opposite angles A, B, and C, respectively. Find the specified function values.

9. $a = 2, c = 5$; $\cot A$ **10.** $a = 1, b = 2$; $\sin B$

11. $b = 5, c = 7$; $\tan B$ **12.** $a = 8, c = 10$; $\sec B$

13. Show that $\cos n\pi = (-1)^n$ and $\sin(\frac{\pi}{2} + n\pi) = (-1)^n$.

14. Make a table showing the values of each trigonometric function at the following values of t:

$$0, \frac{\pi}{6}, \frac{\pi}{4}, \frac{\pi}{3}, \frac{\pi}{2}, \frac{2\pi}{3}, \frac{3\pi}{4}, \frac{5\pi}{6}, \pi, \frac{7\pi}{6}, \frac{5\pi}{4}, \frac{4\pi}{3}, \frac{3\pi}{2}, \frac{5\pi}{3}, \frac{7\pi}{4}, \frac{11\pi}{6}$$

(If a function is not defined at certain values of t, put NV for "no value.")

15. Use a geometric argument to show the following:
(a) $\sin(\pi - t) = \sin t$ (b) $\sin(\pi + t) = -\sin t$
(c) $\cos(\pi - t) = -\cos t$ (d) $\cos(\pi + t) = -\cos t$

16. Use the results of Exercise 15 to show the following:
(a) $\tan(\pi - t) = -\tan t$ (b) $\tan(\pi + t) = \tan t$
(c) $\cot(\pi - t) = -\cot t$ (d) $\cot(\pi + t) = \cot t$

17. Find the period of each of the following functions: tangent, cotangent, secant, cosecant. (*Hint:* For the tangent and cotangent, use the results of Exercise 16.)

18. Show that the tangent, cotangent, and cosecant are odd functions, and the secant is even.

19. Show that for $0 < t < \frac{\pi}{2}$, $\tan t$ and $\sec t$ become arbitrarily large as t approaches $\frac{\pi}{2}$, and that $\cot t$ and $\csc t$ become arbitrarily large as t approaches 0.

20. (a) For $-\frac{\pi}{2} < t < 0$, discuss the behavior of $\tan t$ and $\sec t$ as t approaches $-\frac{\pi}{2}$.
(b) For $\frac{\pi}{2} < t < \pi$, discuss the behavior of $\cot t$ and $\csc t$ as t approaches π.

21. Give the domain and range of $\tan t$, $\cot t$, $\sec t$, and $\csc t$.

22. Use the information in Exercises 14 and 16–21 to verify the graphs in Figures A2.19–A2.22.

B

23. Verify that the point Q shown in the accompanying figure has coordinates $(1, \tan t)$ for $0 < t < \frac{\pi}{2}$. Use this to confirm the fact that $\tan t$ becomes arbitrarily large as t approaches $\frac{\pi}{2}^-$. (*Hint:* Use similar triangles.)

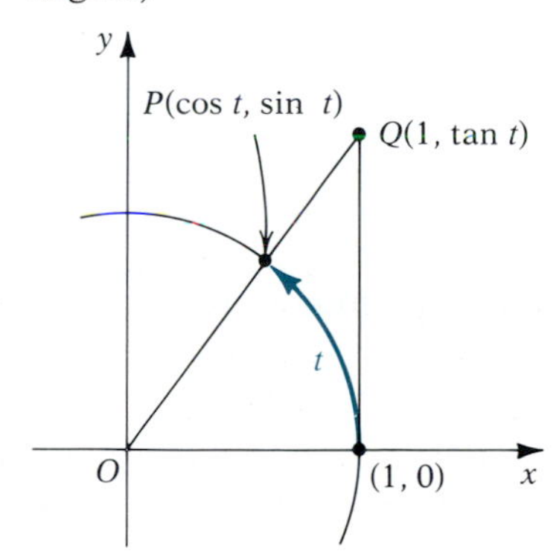

24. Let P_1, P_2, and P_3 be points on the unit circle corresponding to a directed arc from (1, 0) of $t = -\frac{\pi}{6}$, $t = \frac{\pi}{6}$, and $t = \frac{\pi}{2}$, respectively. Denote the coordinates of P_2 by (a, b). What are the coordinates of P_1? Show that the arcs $\widehat{P_1P_2}$ and $\widehat{P_2P_3}$ are equal in length, and conclude that the corresponding chords $\overline{P_1P_2}$ and $\overline{P_2P_3}$ are equal. Apply the distance formula to express the equality $(\overline{P_1P_2})^2 = (\overline{P_2P_3})^2$ in terms of a and b. What other equation must a and b satisfy? Find a and b. (This gives an alternate derivation of $\cos\frac{\pi}{6}$ and $\sin\frac{\pi}{6}$ without using angles.)

25. Use a similar procedure to that of Exercise 24 on the points P_1, P_2, and P_3 corresponding to $t = -\frac{\pi}{12}$, $t = \frac{\pi}{12}$, and $t = \frac{\pi}{4}$, respectively, to find $\cos\frac{\pi}{12}$ and $\sin\frac{\pi}{12}$.

26. Let $f(x) = a \sin bx$ and $g(x) = a \cos bx$, where $a > 0$ and $b > 0$. Show that f and g are periodic, and find the period of each. [*Hint:* $\sin(bx + 2\pi) = \sin b(x + \frac{2\pi}{b})$.]

27. Use the result of Exercise 26 to aid in drawing the graphs:
(a) $y = 2 \sin 3x$ (b) $y = 3 \cos \frac{x}{2}$

28. Let θ be any angle of an arbitrary triangle. Denote by a and b the lengths of the sides adjacent to θ and by c the length of the side opposite θ. Place θ in standard position as shown in the accompanying figure. Show that the coordinates of the vertex P are $x = b\cos\theta$, $y = b \sin\theta$. Apply the distance formula to compute c^2, thereby deriving the **law of cosines:**

$$c^2 = a^2 + b^2 - 2ab\cos\theta$$

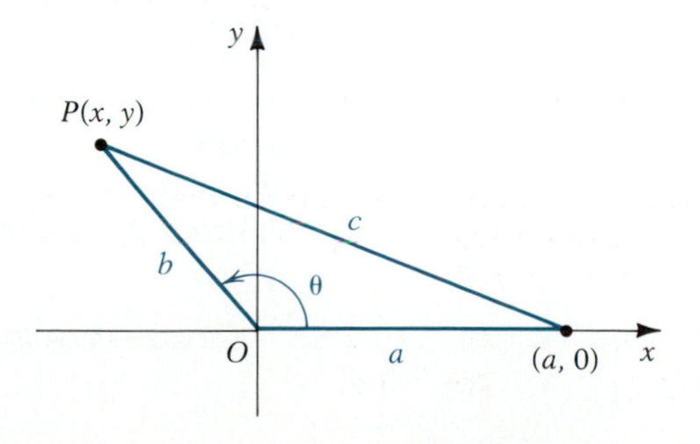

A2.3 TRIGONOMETRIC IDENTITIES

Relationships among the trigonometric functions greatly extend their usefulness. We have already seen some of these, as expressed by equations (A2.4) and (A2.5). The latter equation can be written as

$$\sin^2 t + \cos^2 t = 1$$

(or, equivalently, for any angle θ, $\sin^2 \theta + \cos^2 \theta = 1$). If we divide both sides of this first by $\cos^2 t$ and then by $\sin^2 t$, making use of the relationships given in equation (A2.4), we get

$$1 + \tan^2 t = \sec^2 t \tag{A2.6}$$

and

$$1 + \cot^2 t = \csc^2 t \tag{A2.7}$$

Of course, t must be restricted so that division by 0 does not occur. Equations (A2.4)–(A2.7) are examples of what are called **trigonometric identities,** since they are valid for all admissible numbers t. All can be expressed in terms of angles as well. We collect these together and refer to them as the *fundamental identities.* We understand that appropriate restrictions must be made on t without stating these explicitly.

THE FUNDAMENTAL TRIGONOMETRIC IDENTITIES

$$\tan t = \frac{\sin t}{\cos t} \qquad \sec t = \frac{1}{\cos t}$$

$$\cot t = \frac{\cos t}{\sin t} \qquad \csc t = \frac{1}{\sin t}$$

$$\sin^2 t + \cos^2 t = 1 \qquad 1 + \tan^2 t = \sec^2 t \qquad 1 + \cot^2 t = \csc^2 t$$

In the theorem that follows we prove another very important identity from which many others can be derived.

THEOREM A2.1 For any two real numbers α and β,

$$\cos(\alpha - \beta) = \cos \alpha \cos \beta + \sin \alpha \sin \beta \tag{A2.8}$$

Proof We prove the result for α and β that satisfy $0 \le \beta < \alpha < 2\pi$. Using periodicity and the fact that the cosine is even, so that $\cos(\beta - \alpha) = \cos(\alpha - \beta)$, the result for arbitrary α and β can be established (see Exercise 49). Let P_1, P_2, and P_3 be the end points of arcs on the unit circle, measured from (1, 0), of lengths $t = \alpha$, $t = \beta$, and $t = \alpha - \beta$, respectively. Then P_1, P_2, and P_3 have coordinates as shown in Figure A2.23. Because the arc length from P_2 to P_1 is the same as that from (1, 0) to P_3, it follows that the chord lengths labeled d_1 and d_2 are equal. Expressing $d_1^2 = d_2^2$ by the distance formula yields

$$(\cos \alpha - \cos \beta)^2 + (\sin \alpha - \sin \beta)^2 = [\cos(\alpha - \beta) - 1]^2 + [\sin(\alpha - \beta)]^2$$

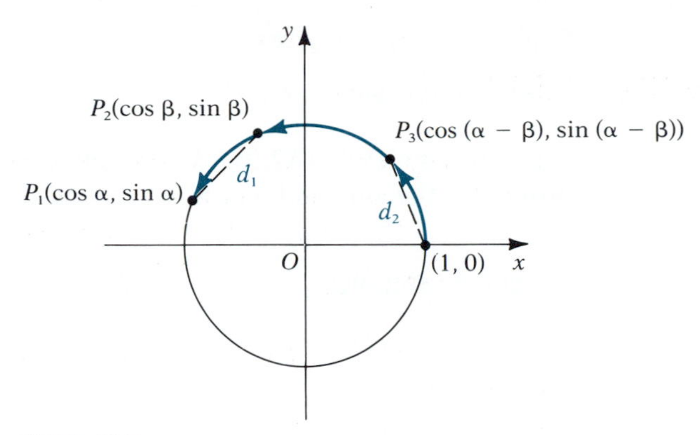

FIGURE A2.23

On expanding these and using $\sin^2 t + \cos^2 t = 1$ (three times), this simplifies to give equation (A2.8). ■

All the other identities we derive below can be thought of as corollaries to this basic result.

To obtain a formula for $\cos(\alpha + \beta)$, we write

$$\begin{aligned}\cos(\alpha+\beta) &= \cos[\alpha - (-\beta)] \\ &= \cos\alpha\cos(-\beta) + \sin\alpha\sin(-\beta) \quad \text{By equation (A2.8)}\end{aligned}$$

and since $\cos(-\beta) = \cos\beta$ and $\sin(-\beta) = -\sin\beta$, we get

$$\cos(\alpha+\beta) = \cos\alpha\cos\beta - \sin\alpha\sin\beta \tag{A2.9}$$

Corresponding identities for $\sin(\alpha+\beta)$ and $\sin(\alpha-\beta)$ can be obtained using the following two preliminary results. In equation (A2.8) we put $\alpha = \frac{\pi}{2}$, getting

$$\cos(\tfrac{\pi}{2} - \beta) = \cos\tfrac{\pi}{2}\cos\beta + \sin\tfrac{\pi}{2}\sin\beta$$

or, since $\cos\frac{\pi}{2} = 0$ and $\sin\frac{\pi}{2} = 1$,

$$\cos(\tfrac{\pi}{2} - \beta) = \sin\beta \tag{A2.10}$$

for all real numbers β. Now let $\beta = \frac{\pi}{2} - \gamma$. We get

$$\cos[\tfrac{\pi}{2} - (\tfrac{\pi}{2} - \gamma)] = \sin(\tfrac{\pi}{2} - \gamma)$$

that is,

$$\sin(\tfrac{\pi}{2} - \gamma) = \cos\gamma \tag{A2.11}$$

for all real numbers γ. Now in equation (A2.10) we replace β by $\alpha + \beta$ to get

$$\begin{aligned}\sin(\alpha+\beta) &= \cos[\tfrac{\pi}{2} - (\alpha+\beta)] \\ &= \cos[(\tfrac{\pi}{2} - \alpha) - \beta] \\ &= \cos(\tfrac{\pi}{2} - \alpha)\cos\beta + \sin(\tfrac{\pi}{2} - \alpha)\sin\beta \quad \text{By equation (A2.8)}\end{aligned}$$

So, by equations (A2.10) and (A2.11),

$$\sin(\alpha+\beta) = \sin\alpha\cos\beta + \cos\alpha\sin\beta \tag{A2.12}$$

Replacing β by $-\beta$ gives

$$\sin(\alpha - \beta) = \sin \alpha \cos \beta - \cos \alpha \sin \beta \tag{A2.13}$$

Equations (A2.8), (A2.9), (A2.12), and (A2.13) are known as **addition formulas** for the sine and cosine. We bring them together for emphasis.

ADDITION FORMULAS FOR THE SINE AND COSINE

$$\sin(\alpha + \beta) = \sin \alpha \cos \beta + \cos \alpha \sin \beta$$
$$\sin(\alpha - \beta) = \sin \alpha \cos \beta - \cos \alpha \sin \beta$$
$$\cos(\alpha + \beta) = \cos \alpha \cos \beta - \sin \alpha \sin \beta$$
$$\cos(\alpha - \beta) = \cos \alpha \cos \beta + \sin \alpha \sin \beta$$

We get the so-called **double angle formulas** by letting $\beta = \alpha$ in the formulas for $\sin(\alpha + \beta)$ and $\cos(\alpha + \beta)$:

$$\sin 2\alpha = \sin(\alpha + \alpha) = \sin \alpha \cos \alpha + \cos \alpha \sin \alpha$$

so that

$$\sin 2\alpha = 2 \sin \alpha \cos \alpha \tag{A2.14}$$

Also,

$$\cos 2\alpha = \cos(\alpha + \alpha) = \cos \alpha \cos \alpha - \sin \alpha \sin \alpha$$

that is,

$$\cos 2\alpha = \cos^2 \alpha - \sin^2 \alpha \tag{A2.15}$$

By combining this last result with the identity $\sin^2 \alpha + \cos^2 \alpha = 1$, we get two other useful forms. First, we replace $\sin^2 \alpha$ by $1 - \cos^2 \alpha$ to get

$$\cos 2\alpha = 2 \cos^2 \alpha - 1 \tag{A2.16}$$

Now replace $\cos^2 \alpha$ in equation (A2.15) by $1 - \sin^2 \alpha$:

$$\cos 2\alpha = 1 - 2 \sin^2 \alpha \tag{A2.17}$$

In calculus, equations (A2.16) and (A2.17) are often applied in the following equivalent forms:

$$\sin^2 \alpha = \tfrac{1}{2}(1 - \cos 2\alpha) \tag{A2.18}$$

$$\cos^2 \alpha = \tfrac{1}{2}(1 + \cos 2\alpha) \tag{A2.19}$$

By letting $\alpha = \frac{\theta}{2}$ and taking square roots, we get the **half angle formulas:**

$$\sin \frac{\theta}{2} = \pm\sqrt{\frac{1 - \cos \theta}{2}} \tag{A2.20}$$

$$\cos \frac{\theta}{2} = \pm\sqrt{\frac{1 + \cos \theta}{2}} \tag{A2.21}$$

Viewing θ as an angle, we can determine the quadrant in which $\frac{\theta}{2}$ lies, thus determining the correct sign for $\sin \frac{\theta}{2}$ and $\cos \frac{\theta}{2}$.

The double and half angle formulas for the sine and cosine are summarized here.

DOUBLE ANGLE AND HALF ANGLE FORMULAS

$$\sin 2\alpha = 2 \sin \alpha \cos \alpha \qquad \sin^2 \alpha = \frac{1}{2}(1 - \cos 2\alpha)$$

$$\cos 2\alpha = \cos^2 \alpha - \sin^2 \alpha \qquad \cos^2 \alpha = \frac{1}{2}(1 + \cos 2\alpha)$$

$$= 2 \cos^2 \alpha - 1 \qquad \sin \frac{\theta}{2} = \pm\sqrt{\frac{1 - \cos \theta}{2}}$$

$$= 1 - 2 \sin^2 \alpha \qquad \cos \frac{\theta}{2} = \pm\sqrt{\frac{1 + \cos \theta}{2}}$$

Still more identities can be derived from those we have so far. We give the most useful ones and give hints for deriving them in the exercises.

SUM FORMULAS

$$\sin x + \sin y = 2 \sin \frac{x + y}{2} \cos \frac{x - y}{2}$$

$$\sin x - \sin y = 2 \cos \frac{x + y}{2} \sin \frac{x - y}{2}$$

$$\cos x + \cos y = 2 \cos \frac{x + y}{2} \cos \frac{x - y}{2}$$

$$\cos x - \cos y = -2 \sin \frac{x + y}{2} \sin \frac{x - y}{2}$$

PRODUCT FORMULAS

$$\sin \alpha \cos \beta = \tfrac{1}{2}[\sin(\alpha + \beta) + \sin(\alpha - \beta)]$$

$$\sin \alpha \sin \beta = \tfrac{1}{2}[\cos(\alpha - \beta) - \cos(\alpha + \beta)]$$

$$\cos \alpha \cos \beta = \tfrac{1}{2}[\cos(\alpha + \beta) + \cos(\alpha - \beta)]$$

FORMULAS FOR THE TANGENT

$$\tan(\alpha + \beta) = \frac{\tan \alpha + \tan \beta}{1 - \tan \alpha \tan \beta} \qquad \tan(\alpha - \beta) = \frac{\tan \alpha - \tan \beta}{1 + \tan \alpha \tan \beta}$$

$$\tan 2\alpha = \frac{2 \tan \alpha}{1 - \tan^2 \alpha}$$

$$\tan \frac{1}{2}\theta = \pm\sqrt{\frac{1 - \cos \theta}{1 + \cos \theta}} = \frac{\sin \theta}{1 + \cos \theta} = \frac{1 - \cos \theta}{\sin \theta}$$

One of the most important applications of these identities in calculus occurs in the operation of integration, where it is often necessary to change the form of an expression involving trigonometric functions before the integration can be performed. One way to practice making such transformations is to verify the equality of two expressions by showing that one side can be transformed to another. This is called "proving an identity." We illustrate this with several examples. You should supply reasons for all steps, noting which of the basic identities we have used.

EXAMPLE A2.7 Prove the identity

$$\sec x - \sin x \tan x = \cos x$$

Solution

$$\sec x - \sin x \tan x = \frac{1}{\cos x} - \sin x\left(\frac{\sin x}{\cos x}\right) = \frac{1 - \sin^2 x}{\cos x} = \frac{\cos^2 x}{\cos x} = \cos x$$

■

EXAMPLE A2.8 Prove the identity

$$\frac{2}{\tan t + \cot t} = \sin 2t$$

Solution

$$\frac{2}{\tan t + \cot t} = \frac{2}{\dfrac{\sin t}{\cos t} + \dfrac{\cos t}{\sin t}} = \frac{2 \sin t \cos t}{\sin^2 t + \cos^2 t} = \sin 2t$$

■

EXAMPLE A2.9 Prove the identity

$$\frac{\tan^2 \theta}{1 + \tan^2 \theta} = \frac{1 - \cos 2\theta}{2}$$

Solution

$$\frac{\tan^2 \theta}{1 + \tan^2 \theta} = \frac{\tan^2 \theta}{\sec^2 \theta} = \frac{\sin^2 \theta}{\cos^2 \theta} \cdot \cos^2 \theta = \sin^2 \theta = \frac{1}{2}(1 - \cos 2\theta)$$

■

EXAMPLE A2.10 Prove the identity

$$\csc \alpha - \tan \frac{\alpha}{2} = \cot \alpha$$

Solution

$$\csc \alpha - \tan \frac{\alpha}{2} = \frac{1}{\sin \alpha} - \frac{1 - \cos \alpha}{\sin \alpha} = \frac{1 - (1 - \cos \alpha)}{\sin \alpha} = \frac{\cos \alpha}{\sin \alpha} = \cot \alpha$$

■

EXAMPLE A2.11 Prove the identity

$$\frac{\cos 3x + \cos x}{\sin 3x + \sin x} = \cot 2x$$

Solution

$$\frac{\cos 3x + \cos x}{\sin 3x + \sin x} = \frac{2 \cos \dfrac{3x + x}{2} \cos \dfrac{3x - x}{2}}{2 \sin \dfrac{3x + x}{2} \cos \dfrac{3x - x}{2}} = \frac{\cos 2x \cos x}{\sin 2x \cos x} = \cot 2x$$

■

EXAMPLE A2.12 Prove the identity

$$\frac{1 - \sec^2 x}{1 - \csc^2 x} = \tan^2 x \sec^2 x - \sec^2 x + 1$$

Solution

$$\begin{aligned}\frac{1 - \sec^2 x}{1 - \csc^2 x} = \frac{-\tan^2 x}{-\cot^2 x} &= (\tan^2 x)(\tan^2 x) = \tan^2 x(\sec^2 x - 1)\\ &= \tan^2 x \sec^2 x - \tan^2 x\\ &= \tan^2 x \sec^2 x - (\sec^2 x - 1)\\ &= \tan^2 x \sec^2 x - \sec^2 x + 1\end{aligned}$$

■

EXAMPLE A2.13 Prove the identity

$$\frac{\sin t}{1-\cos t} = \csc t + \cot t$$

Solution

$$\frac{\sin t}{1-\cos t} = \frac{\sin t}{1-\cos t}\cdot\frac{1+\cos t}{1+\cos t} = \frac{\sin t(1+\cos t)}{1-\cos^2 t}$$
$$= \frac{\sin t(1+\cos t)}{\sin^2 t} = \frac{1+\cos t}{\sin t}$$
$$= \frac{1}{\sin t} + \frac{\cos t}{\sin t} = \csc t + \cot t$$

■

EXERCISE SET A2.3

A

In Exercises 1–20 prove the identities.

1. $\dfrac{1+\sin x}{\tan x} = \cos x + \cot x$
2. $\dfrac{\sec t - \cos t}{\tan t} = \sin t$
3. $(\sin\alpha - \cos\alpha)^2 = 1 - \sin 2\alpha$
4. $\cos^4 x - \sin^4 x = \cos 2x$
5. $\dfrac{1-\cos t}{\sin t} + \dfrac{\sin t}{1-\cos t} = 2\csc t$
6. $\dfrac{1}{\sec x - \tan x} = \sec x + \tan x$
7. $\sec^2\theta + \csc^2\theta = \sec^2\theta\csc^2\theta$
8. $\dfrac{2\sin\theta - \sin 2\theta}{2\sin\theta + \sin 2\theta} = \tan^2\dfrac{\theta}{2}$
9. $\dfrac{\sin(\alpha+\beta)}{\sin\alpha\sin\beta} = \cot\alpha + \cot\beta$
10. $\dfrac{\cos(\alpha-\beta)}{\sin\alpha\sin\beta} = 1 + \cot\alpha\cot\beta$
11. $2\sin^2\dfrac{\theta}{2}(1+\cos\theta) = \sin^2\theta$
12. $2\cos\dfrac{3\theta}{2}\sin\dfrac{\theta}{2} = \sin 2\theta - \sin\theta$
13. $\dfrac{\sin 7x - \sin 5x}{\cos 7x + \cos 5x} = \tan x$
14. $\dfrac{\sin x - \sin y}{\cos x + \cos y} = \tan\dfrac{1}{2}(x-y)$
15. $\dfrac{1}{1-\sin t} + \dfrac{1}{1+\sin t} = 2\sec^2 t$
16. $\sec^4 x = \tan^2 x\sec^2 x + \sec^2 x$
17. $\tan x(1+\cos 2x) = \sin 2x$
18. $\dfrac{1+\tan^2\theta}{1-\tan^2\theta} = \sec 2\theta$
19. $\dfrac{\cos\theta + \cot\theta}{1+\csc\theta} = \cos\theta$
20. $1 + \tan\alpha\tan\dfrac{\alpha}{2} = \sec\alpha$
21. By writing
$$\frac{5\pi}{12} = \frac{3\pi}{12} + \frac{2\pi}{12} = \frac{\pi}{4} + \frac{\pi}{6}$$
find all trigonometric functions of $\frac{5\pi}{12}$.
22. Find all trigonometric functions of (a) $\frac{\pi}{12}$ and (b) $\frac{\pi}{8}$.
23. If α and β are primary angles in standard position with terminal sides that pass through $(-3, 4)$ and $(-12, -5)$, respectively, find:
(a) $\sin(\alpha+\beta)$ (b) $\cos(\alpha-\beta)$
24. Let t_1 and t_2 be numbers in the interval $[0, 2\pi)$ such that $\sin t_1 = -\frac{7}{25}$, $\sec t_1 > 0$, and $\cos t_2 = -\frac{4}{5}$, $\cot t_2 > 0$. Find:
(a) $\cos(t_1/2)$ (b) $\tan(t_1 - t_2)$
25. Let t_1 and t_2 be as defined in Exercise 24. Find:
(a) $\sin 2t_2$ (b) $\cos(t_1 + t_2)$
26. Write as a product:
(a) $\sin 5x + \sin 3x$ (b) $\cos 7x - \cos 5x$
27. Write as a sum or difference:
(a) $\sin 5x\cos 3x$ (b) $\sin 7x\sin 5x$

B

28. Derive the formula for $\tan(\alpha + \beta)$. [*Hint:* Write $\tan(\alpha + \beta) = \sin(\alpha + \beta)/\cos(\alpha + \beta)$ and after substituting identities, divide numerator and denominator by $\cos \alpha \cos \beta$.]

29. Derive the formulas for $\tan(\alpha - \beta)$ and $\tan 2\alpha$.

30. Derive the formula

$$\tan \frac{1}{2}\theta = \frac{\sin \theta}{1 + \cos \theta}$$

(*Hint:*

$$\frac{\sin \frac{1}{2}\theta}{\cos \frac{1}{2}\theta} = \frac{\sin \frac{1}{2}\theta}{\cos \frac{1}{2}\theta} \cdot \frac{2 \cos \frac{1}{2}\theta}{2 \cos \frac{1}{2}\theta} = \frac{2 \sin \frac{1}{2}\theta \cos \frac{1}{2}\theta}{2 \cos^2 \frac{1}{2}\theta}$$

Now use some identities.)

31. Derive the formula

$$\tan \frac{1}{2}\theta = \frac{1 - \cos \theta}{\sin \theta}$$

32. In equations (A2.12) and (A2.13) let $x = \alpha + \beta$ and $y = \alpha - \beta$ to obtain the formulas for $\sin x + \sin y$ and $\sin x - \sin y$.

33. In equations (A2.8) and (A2.9) let $x = \alpha + \beta$ and $y = \alpha - \beta$ to obtain the formulas for $\cos x + \cos y$ and $\cos x - \cos y$.

34. Add corresponding sides of equations (A2.12) and (A2.13) to obtain the formula for $\sin \alpha \cos \beta$.

35. Use the idea of Exercise 34 on equations (A2.8) and (A2.9) to obtain the formulas for $\cos \alpha \cos \beta$ and $\sin \alpha \sin \beta$.

36. Derive the formula

$$\cot(\alpha + \beta) = \frac{\cot \alpha \cot \beta - 1}{\cot \alpha + \cot \beta}$$

37. Prove the formulas:
(a) $\sin(n\pi + \theta) = (-1)^n \sin \theta$
(b) $\cos(n\pi + \theta) = (-1)^n \cos \theta$

38. Prove the formulas:

(a) $\sin\left(\frac{2n + 1}{2}\pi + \theta\right) = (-1)^n \cos \theta$

(b) $\cos\left(\frac{2n + 1}{2}\pi + \theta\right) = (-1)^{n+1} \sin \theta$

In Exercises 39–44 prove the identities.

39. $\frac{1}{1 - \sin \theta} = \sec^2 \theta + \sec \theta \tan \theta$

40. $\frac{\sin 3\theta}{\sin \theta} - \frac{\cos 3\theta}{\cos \theta} = 2$

41. $\cos^4 x = \frac{3}{8} + \frac{\cos 2x}{2} + \frac{\cos 4x}{8}$

42. $\frac{\sin^2 x \tan^2 x}{\tan x - \sin x} = \tan x + \sin x$

43. $\frac{\sin^3 t + \cos^3 t}{\sin t + \cos t} = 1 - \frac{\sin 2t}{2}$

44. $\sin 3x + \sin x = 4 \sin x \cos^2 x$

45. Find all values of x in $[0, 2\pi)$ that satisfy

$$\sin 5x \cos 4x = \cos 5x \sin 4x - 1$$

46. Find all values of x in $[0, 2\pi)$ that satisfy

$$2 \cos 2x \cos x = 1 - \sin 2x \sin x$$

47. By calculating $\sin(\frac{\pi}{12})$ in two ways, show that

$$\sqrt{2 - \sqrt{3}} = \tfrac{1}{2}(\sqrt{6} - \sqrt{2})$$

48. Prove the identity:

$$\frac{\sin(x + h) - \sin x}{h} = \frac{\sin \frac{1}{2}h}{\frac{1}{2}h} \cos(x + \tfrac{1}{2}h)$$

49. Complete the proof of Theorem A2.1 by considering α and β without any restrictions. Let α_1 and β_1 be numbers in $[0, 2\pi)$ such that $\alpha = \alpha_1 + 2m\pi$ and $\beta = \beta_1 + 2n\pi$ for integers m and n. Now show that equation (A2.8) holds by expressing $\cos(\alpha - \beta)$ in terms of α_1 and β_1.

50. Give an alternative proof of Theorem A2.1 as follows. In Figure A2.23, observe that the angle at O in triangle P_1OP_2 is $\alpha - \beta$. Now apply the law of cosines. (See Exercise 28 in Exercise Set A2.2.)

APPENDIX 3

TABLES

TABLE 1 Natural logarithms of numbers

x	$\ln x$	x	$\ln x$	x	$\ln x$
		4.5	1.5041	9.0	2.1972
0.1	−2.3026	4.6	1.5261	9.1	2.2083
0.2	−1.6094	4.7	1.5476	9.2	2.2192
0.3	−1.2040	4.8	1.5686	9.3	2.2300
0.4	−0.9163	4.9	1.5892	9.4	2.2407
0.5	−0.6931	5.0	1.6094	9.5	2.2513
0.6	−0.5108	5.1	1.6292	9.6	2.2618
0.7	−0.3567	5.2	1.6487	9.7	2.2721
0.8	−0.2231	5.3	1.6677	9.8	2.2824
0.9	−0.1054	5.4	1.6864	9.9	2.2925
1.0	0.0000	5.5	1.7047	10	2.3026
1.1	0.0953	5.6	1.7228	11	2.3979
1.2	0.1823	5.7	1.7405	12	2.4849
1.3	0.2624	5.8	1.7579	13	2.5649
1.4	0.3365	5.9	1.7750	14	2.6391
1.5	0.4055	6.0	1.7918	15	2.7081
1.6	0.4700	6.1	1.8083	16	2.7726
1.7	0.5306	6.2	1.8245	17	2.8332
1.8	0.5878	6.3	1.8405	18	2.8904
1.9	0.6419	6.4	1.8563	19	2.9444
2.0	0.6931	6.5	1.8718	20	2.9957
2.1	0.7419	6.6	1.8871	25	3.2189
2.2	0.7885	6.7	1.9021	30	3.4012
2.3	0.8329	6.8	1.9169	35	3.5553
2.4	0.8755	6.9	1.9315	40	3.6889
2.5	0.9163	7.0	1.9459	45	3.8067
2.6	0.9555	7.1	1.9601	50	3.9120
2.7	0.9933	7.2	1.9741	55	4.0073
2.8	1.0296	7.3	1.9879	60	4.0943
2.9	1.0647	7.4	2.0015	65	4.1744
3.0	1.0986	7.5	2.0149	70	4.2485
3.1	1.1314	7.6	2.0281	75	4.3175
3.2	1.1632	7.7	2.0412	80	4.3820
3.3	1.1939	7.8	2.0541	85	4.4427
3.4	1.2238	7.9	2.0669	90	4.4998
3.5	1.2528	8.0	2.0794	100	4.6052
3.6	1.2809	8.1	2.0919	110	4.7005
3.7	1.3083	8.2	2.1041	120	4.7875
3.8	1.3350	8.3	2.1163	130	4.8676
3.9	1.3610	8.4	2.1282	140	4.9416
4.0	1.3863	8.5	2.1401	150	5.0106
4.1	1.4110	8.6	2.1518	160	5.0752
4.2	1.4351	8.7	2.1633	170	5.1358
4.3	1.4586	8.8	2.1748	180	5.1930
4.4	1.4816	8.9	2.1861	190	5.2470

TABLE 2 Exponential Functions

x	e^x	e^{-x}	x	e^x	e^{-x}
0.00	1.0000	1.0000	1.5	4.4817	0.2231
0.01	1.0101	0.9901	1.6	4.9530	0.2019
0.02	1.0202	0.9802	1.7	5.4739	0.1827
0.03	1.0305	0.9705	1.8	6.0496	0.1653
0.04	1.0408	0.9608	1.9	6.6859	0.1496
0.05	1.0513	0.9512	2.0	7.3891	0.1353
0.06	1.0618	0.9418	2.1	8.1662	0.1225
0.07	1.0725	0.9324	2.2	9.0250	0.1108
0.08	1.0833	0.9231	2.3	9.9742	0.1003
0.09	1.0942	0.9139	2.4	11.023	0.0907
0.10	1.1052	0.9048	2.5	12.182	0.0821
0.11	1.1163	0.8958	2.6	13.464	0.0743
0.12	1.1275	0.8869	2.7	14.880	0.0672
0.13	1.1388	0.8781	2.8	16.445	0.0608
0.14	1.1503	0.8694	2.9	18.174	0.0550
0.15	1.1618	0.8607	3.0	20.086	0.0498
0.16	1.1735	0.8521	3.1	22.198	0.0450
0.17	1.1853	0.8437	3.2	24.533	0.0408
0.18	1.1972	0.8353	3.3	27.113	0.0369
0.19	1.2092	0.8270	3.4	29.964	0.0334
0.20	1.2214	0.8187	3.5	33.115	0.0302
0.21	1.2337	0.8106	3.6	36.598	0.0273
0.22	1.2461	0.8025	3.7	40.447	0.0247
0.23	1.2586	0.7945	3.8	44.701	0.0224
0.24	1.2712	0.7866	3.9	49.402	0.0202
0.25	1.2840	0.7788	4.0	54.598	0.0183
0.30	1.3499	0.7408	4.1	60.340	0.0166
0.35	1.4191	0.7047	4.2	66.686	0.0150
0.40	1.4918	0.6703	4.3	73.700	0.0136
0.45	1.5683	0.6376	4.4	81.451	0.0123
0.50	1.6487	0.6065	4.5	90.017	0.0111
0.55	1.7333	0.5769	4.6	99.484	0.0101
0.60	1.8221	0.5488	4.7	109.95	0.0091
0.65	1.9155	0.5220	4.8	121.51	0.0082
0.70	2.0138	0.4966	4.9	134.29	0.0074
0.75	2.1170	0.4724	5.0	148.41	0.0067
0.80	2.2255	0.4493	5.5	244.69	0.0041
0.85	2.3396	0.4274	6.0	403.43	0.0025
0.90	2.4596	0.4066	6.5	665.14	0.0015
0.95	2.5857	0.3867	7.0	1096.6	0.0009
1.0	2.7183	0.3679	7.5	1808.0	0.0006
1.1	3.0042	0.3329	8.0	2981.0	0.0003
1.2	3.3201	0.3012	8.5	4914.8	0.0002
1.3	3.6693	0.2725	9.0	8103.1	0.0001
1.4	4.0552	0.2466	10.0	22026	0.00005

TABLE 3 Trigonometric Functions

Degrees	*Radians*	sin	cos	tan	cot		
0	0.0000	0.0000	1.0000	0.0000		1.5708	90
1	0.0175	0.0175	0.9998	0.0175	57.290	1.5533	89
2	0.0349	0.0349	0.9994	0.0349	28.636	1.5359	88
3	0.0524	0.0523	0.9986	0.0524	19.081	1.5184	87
4	0.0698	0.0698	0.9976	0.0699	14.301	1.5010	86
5	0.0873	0.0872	0.9962	0.0875	11.430	1.4835	85
6	0.1047	0.1045	0.9945	0.1051	9.5144	1.4661	84
7	0.1222	0.1219	0.9925	0.1228	8.1443	1.4486	83
8	0.1396	0.1392	0.9903	0.1405	7.1154	1.4312	82
9	0.1571	0.1564	0.9877	0.1584	6.3138	1.4137	81
10	0.1745	0.1736	0.9848	0.1763	5.6713	1.3963	80
11	0.1920	0.1908	0.9816	0.1944	5.1446	1.3788	79
12	0.2094	0.2079	0.9781	0.2126	4.7046	1.3614	78
13	0.2269	0.2250	0.9744	0.2309	4.3315	1.3439	77
14	0.2443	0.2419	0.9703	0.2493	4.0108	1.3265	76
15	0.2618	0.2588	0.9659	0.2679	3.7321	1.3090	75
16	0.2793	0.2756	0.9613	0.2867	3.4874	1.2915	74
17	0.2967	0.2924	0.9563	0.3057	3.2709	1.2741	73
18	0.3142	0.3090	0.9511	0.3249	3.0777	1.2566	72
19	0.3316	0.3256	0.9455	0.3443	2.9042	1.2392	71
20	0.3491	0.3420	0.9397	0.3640	2.7475	1.2217	70
21	0.3665	0.3584	0.9336	0.3839	2.6051	1.2043	69
22	0.3840	0.3746	0.9272	0.4040	2.4751	1.1868	68
23	0.4014	0.3907	0.9205	0.4245	2.3559	1.1694	67
24	0.4189	0.4067	0.9135	0.4452	2.2460	1.1519	66
25	0.4363	0.4226	0.9063	0.4663	2.1445	1.1345	65
26	0.4538	0.4384	0.8988	0.4877	2.0503	1.1170	64
27	0.4712	0.4540	0.8910	0.5095	1.9626	1.0996	63
28	0.4887	0.4695	0.8829	0.5317	1.8807	1.0821	62
29	0.5061	0.4848	0.8746	0.5543	1.8040	1.0647	61
30	0.5236	0.5000	0.8660	0.5774	1.7321	1.0472	60
31	0.5411	0.5150	0.8572	0.6009	1.6643	1.0297	59
32	0.5585	0.5299	0.8480	0.6249	1.6003	1.0123	58
33	0.5760	0.5446	0.8387	0.6494	1.5399	0.9948	57
34	0.5934	0.5592	0.8290	0.6745	1.4826	0.9774	56
35	0.6109	0.5736	0.8192	0.7002	1.4281	0.9599	55
36	0.6283	0.5878	0.8090	0.7265	1.3764	0.9425	54
37	0.6458	0.6018	0.7986	0.7536	1.3270	0.9250	53
38	0.6632	0.6157	0.7880	0.7813	1.2799	0.9076	52
39	0.6807	0.6293	0.7771	0.8098	1.2349	0.8901	51
40	0.6981	0.6428	0.7660	0.8391	1.1918	0.8727	50
41	0.7156	0.6561	0.7547	0.8693	1.1504	0.8552	49
42	0.7330	0.6691	0.7431	0.9004	1.1106	0.8378	48
43	0.7505	0.6820	0.7314	0.9325	1.0724	0.8203	47
44	0.7679	0.6947	0.7193	0.9657	1.0355	0.8029	46
45	0.7854	0.7071	0.7071	1.0000	1.0000	0.7854	45
		cos	sin	cot	tan	*Radians*	*Degrees*

A NOTE ABOUT CHAPTER-OPENING ART

The design art at the beginning of each chapter was generated by computer by Professor Douglas Dunham of the University of Minnesota at Duluth. Each figure is a three-dimensional surface that is the graph of an equation in three variables. A list of these equations is given below.

Each surface is depicted by showing a series of curves of intersection of the surface with planes perpendicular to one or more of the coordinate planes. These curves are called *traces,* or *sections.* Hidden lines are not shown. That is, if a part of a trace falls behind a portion of the surface, as seen from the front, then that part is not shown. This requires a rather sophisticated computer program.

These drawings serve to illustrate the great variety and complexity of graphs of functions of two variables. As you might expect, some of the functions themselves are quite complicated, but observe also that some strange-looking graphs correspond to very simple functions:

Chapter 1 $z = \cos\sqrt{x^2 + y^2}$

Chapter 2 $z = -\frac{3}{2}[\cos x + \cos y - \cos(x + y)]$

Chapter 3 $z = 6\{e^{-[(x-4)^2+y^2]} + e^{-[(x+2)^2+(y-2\sqrt{3})^2]} + e^{-[(x+2)^2+(y+2\sqrt{3})^2}\}$

Chapter 4 $z = (x + y)\sin\left(\frac{1}{x + y}\right)$

Chapter 5 $z = -\sqrt{x^2 + y^2}\sin\frac{1}{\sqrt{x^2 + y^2}}$

Chapter 6 $z = 2e^{(-x^2+y^2)/16}\cos(x^2 + y^2)$

Chapter 7 $z = -\frac{x^2y}{x^4 + y^2}$

Chapter 8 $z = -2[\cos x + \cos y - \cos(x - y)]$

Chapter 9 $z = e^{-(\sqrt{x^2+y^2}-4)^2}\cos(x^2 + y^2)$

Chapter 10 $z = \frac{xy}{x^2 + y^2}$

Chapter 11 $z = xy\sin\left(\frac{1}{xy}\right)$

Chapter 12 $z = \sqrt{|xy|}$

Chapter 13 $z = -(x - y)\sin\left(\frac{1}{x - y}\right)$

Chapter 14 $z = |x^2 - y^2|$

Chapter 15 $z = 2\{e^{-[(x-2)^2+(y+2\sqrt{3})^2]} + 2e^{-[(x-4)^2+y^2]} + 3e^{-[(x-2)^2+(y-2\sqrt{3})]^2} + 4e^{-[(x+2)^2+(y-2\sqrt{3})^2]} + 5e^{-[(x+4)^2+y^2]} + 6e^{-[(x+2)^2+(y+2\sqrt{3})^2]}\}$

Chapter 16 $z^2 = 1 - \frac{x^2}{16} - \frac{y^2}{16}$

Chapter 17 $z = 5\{e^{-[x^2+(y+2)^2]/4} - e^{-[x^2+(y-2)^2]/4}\}$

Chapter 18 $z = \sin x + \sin y$

Chapter 19

$$z = \frac{1}{20}\left[\frac{1}{\sqrt{(x-4)^2 + y^2}} + \frac{1}{\sqrt{(x-2)^2 + (y - 2\sqrt{3})^2}} + \frac{1}{\sqrt{(x+2)^2 + (y - 2\sqrt{3})^2}} + \frac{1}{\sqrt{(x+4)^2 + y^2}} + \frac{1}{\sqrt{(x+2)^2 + (y + 2\sqrt{3})^2}} + \frac{1}{\sqrt{(x-2)^2 + (y + 2\sqrt{3})^2}}\right]$$

Chapter 20 $z = e^{-\sqrt{x^2+y^2}/4}\sin x \sin y$

A NOTE ABOUT THE CALCULUS TUTOR

If you are experiencing difficulty with any of the sections below, you may need help with the algebra, trigonometry, or geometry concepts underlying the calculus.

The Calculus Tutor is a study guide specifically designed to provide students with the crucial and specific precalculus support they need to succeed in calculus. All topics and problems in *The Calculus Tutor* are cross-referenced to Holder's *Calculus with Analytic Geometry*.

Section from Calculus	Corresponding Section from Calculus Tutor
1.2	3, 6, 12
1.3	4, 5, 9, 17
1.4	7, 10, 11, 18, 20
1.5	10, 21
1.6	3, 7, 10, 21
1.7	7, 15
1.8	6, 8, 15
1.9	1 4, 6, 7, 8, 9, 11, 15, 17, 21, 22
2.1	3, 10
2.3	1, 9, 11
2.4	1, 2
2.5	2
2.6	2
2.7	1, 2, 21, 22, 23
2.8	2, 5, 10, 14, 21
2.9	11, 21
2.10	10, 11, 18, 21
3.1	2, 3, 7, 12, 14, 15, 21, 22
3.2	2, 3, 7, 13, 14, 15, 21, 22
3.3	3, 13
3.4	3, 11, 16, 17, 18, 20, 21, 22
3.5	16, 17, 18
3.6	3, 13, 15, 23
3.7	2, 3, 21
3.8	11, 12, 14, 16, 18
4.2	1, 2
4.3	12, 13, 15
4.4	14, 21, 22
4.5	21
4.6	21

Section from Calculus	Corresponding Section from Calculus Tutor
5.1	9, 13, 23
5.2	9, 13, 16, 20, 23
5.3	9, 12, 19, 20
5.4	4
5.5	4, 19
5.6	16, 19, 20
5.7	9, 16, 17, 19
5.9	5, 11, 13, 20, 23
6.2	1, 3, 12, 21, 23, 24, 26
6.3	3, 13, 15, 25
6.4	1, 10, 15, 24, 26
6.5	15, 26
6.6	26
6.7	2, 11, 24, 26
7.1	4, 10, 11, 13, 17, 22, 25
7.2	11, 26
7.3	3
7.4	17, 22, 26
8.2	2, 21, 26
8.3	22, 26
8.4	4, 5, 21
8.5	9, 13, 14
8.6	9, 10, 21, 22
8.7	3
8.8	9, 11, 13
12.2	4, 9, 13
12.3	4, 11, 13
12.4	4, 5, 9
12.5	22
12.6	4, 9, 11, 17

ANSWERS TO ODD-NUMBERED EXERCISES

Note Answers to C exercises and complete program listings are contained in the Solutions Manual.

CHAPTER 1

Exercise Set 1.2, page 8

1. (a) 0.625 (b) 1.6 (c) 0.370370370 . . . (d) $-0.243243\ldots$ (e) 1.42857142857 . . . **3.** $\dfrac{41}{33}$

5. (a) $-10 < 2$ (b) $-7 > -9$ (c) $\dfrac{2}{3} > \dfrac{5}{9}$ (d) $3.6 < \dfrac{15}{4}$ (e) $0 > -100$

7. (a) $\{x: 2 \le x \le 5\}$ (b) $\{x: -2 < x \le 3\}$ (c) $\{x: 3 < x < 5\}$

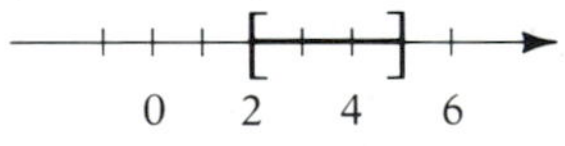

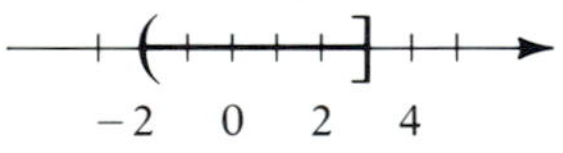

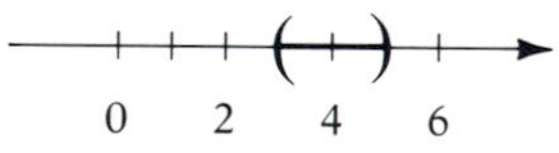

0 2 4 6   −2 0 2 4   0 2 4 6

(d) $\{x: x < 2\}$ (e) $\{x: x \ge 0\}$

−2 0 2 4   0 2 4 6

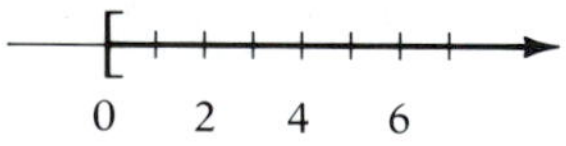

9. $(-\infty, 2)$ **11.** $\left(\dfrac{2}{3}, \dfrac{10}{3}\right]$ **13.** $\left(-\infty, \dfrac{1}{2}\right] \cup \left[\dfrac{5}{2}, \infty\right)$ **15.** $\left(-\dfrac{3}{2}, \dfrac{5}{2}\right)$ **17.** $[-1, 4]$ **19.** $\left(-1, \dfrac{5}{2}\right)$

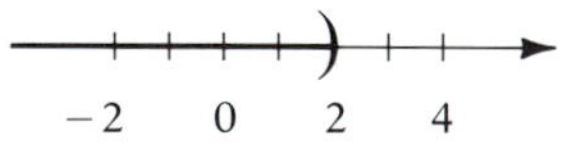

21. $(-\infty, -2) \cup (3, \infty)$ **23.** $\left[-4, \dfrac{1}{2}\right] \cup [3, \infty)$ **25.** $(-\infty, -5) \cup (2, \infty)$

27. $|x - 3| < 1$ **29.** $0 < |x - 5| < 2$ **31.** $|x - 4| < \varepsilon$ **33.** $a = -2, r = 3$

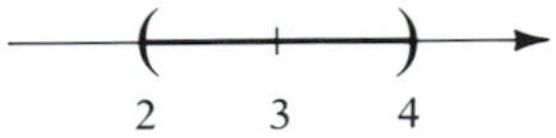

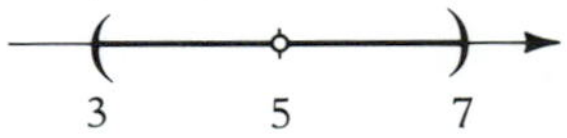

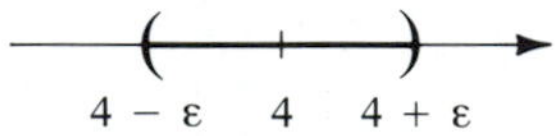

2 3 4   3 5 7 $4 - \varepsilon$ 4 $4 + \varepsilon$

35. Least upper bound = 3   −1 0 1 2 3 4

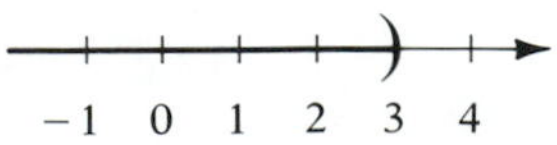

37. Least upper bound = 2   0 1 2

39. Since $a < b$, $b - a$ is positive. If $c > 0$, $c(b - a) = bc - ac$ is positive, so $ac < bc$. If $c < 0$, $-c > 0$, and $-c(b - a) = ac - bc$ is positive, so $ac > bc$.

41. Since $a \cdot \frac{1}{a} = 1 > 0$, $\frac{1}{a}$ cannot be negative, otherwise the product would be negative. Neither can it be 0, since this would make the product 0. So $\frac{1}{a}$ must be positive.

43. $|-a| = \begin{cases} -a & \text{if } -a > 0 \\ 0 & \text{if } -a = 0 \\ -(-a) & \text{if } -a < 0 \end{cases} = \begin{cases} -a & \text{if } a < 0 \\ 0 & \text{if } a = 0 \\ a & \text{if } a > 0 \end{cases} = |a|$

45. $|a - b| = |a + (-b)| \le |a| + |-b| = |a| + |b|$

47. By (i) M is an upper bound of S. Use (ii) to show no number less than M is an upper bound of S.

Exercise Set 1.3, page 17

1. $2x - 5y = 21$ **3.** $3x - 2y = 1$ **5.** $3x - 4y = 27$ **7.** $x + 3y = 5$ **9.** $x - 2y + 7 = 0$

11. Let $A = (2, 2)$, $B = (0, -1)$, $C = (-4, 1)$, $D = (-2, 4)$. Then $m_{AB} = m_{CD} = \dfrac{3}{2}$ and $m_{BC} = m_{AD} = -\dfrac{1}{2}$.

13. Center $(1, -3)$; radius 2 **15.** Center $(-2, 0)$; radius $\sqrt{3}$ **17.** No graph

19. Vertex $(-2, 1)$; axis $x = -2$

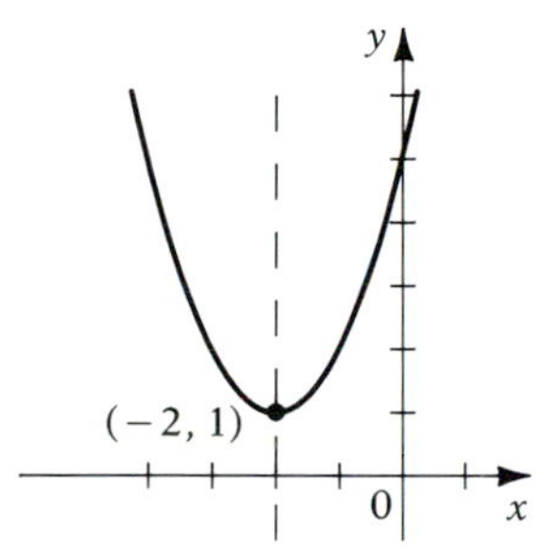

21. Vertex $\left(-\dfrac{5}{4}, \dfrac{57}{8}\right)$; axis $x = -\dfrac{5}{4}$

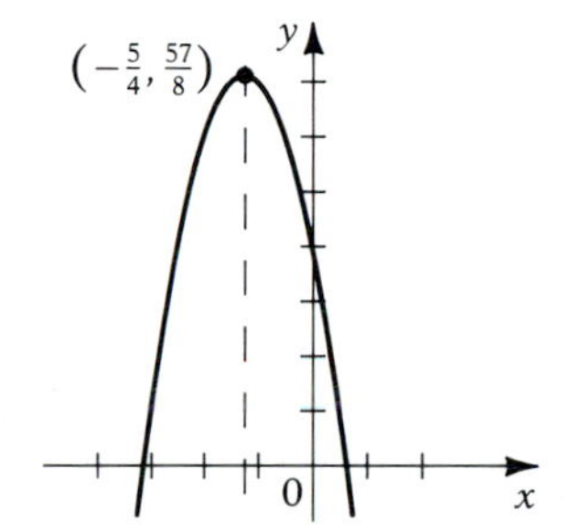

23. From figure $x - x_1 = x_2 - x$. So $2x = x_1 + x_2$ and $x = \dfrac{x_1 + x_2}{2}$. Similarly $y = \dfrac{y_1 + y_2}{2}$.

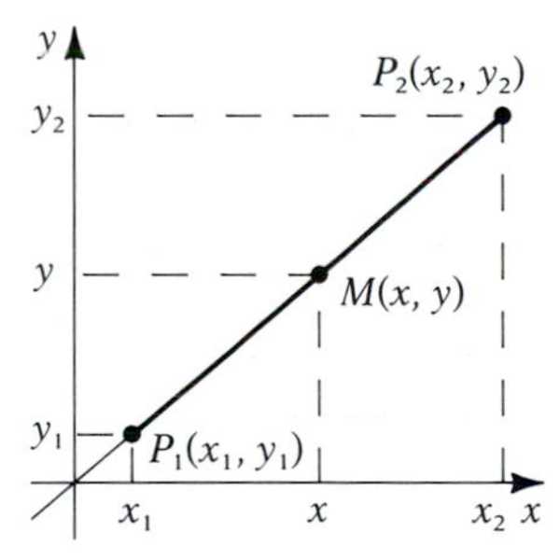

25. Point $(-1, 3)$ is on both circle and line. Radius to this point is perpendicular to the line.

27. $\left(\dfrac{5}{2}, \dfrac{1}{2}\right)$; $\left(x - \dfrac{5}{2}\right)^2 + \left(y - \dfrac{1}{2}\right)^2 = \dfrac{25}{2}$

29. $(1 \pm \sqrt{3}, -1)$

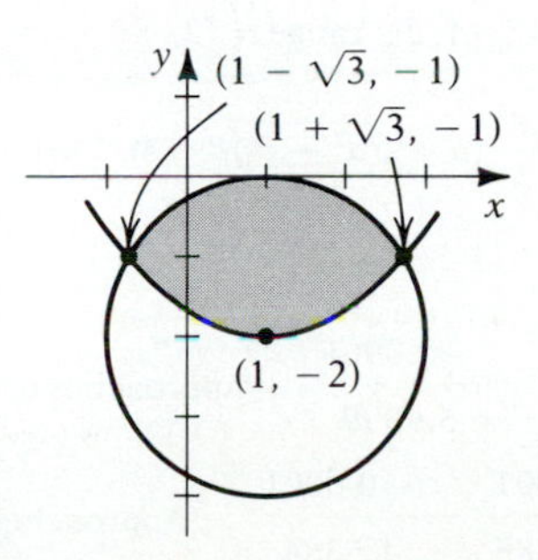

Exercise Set 1.4, page 23

1. (a) 0 (b) 0 (c) $-\dfrac{1}{2}$ (d) $\dfrac{a^2-1}{a^2+2}$ (e) $\dfrac{x^2+2x}{x+3}$ **3.** (a) 3 (b) 2 (c) 3 (d) $|t|$ (e) x^2

5. Domain $= \mathbf{R}$; range $= \{y: y \geq -4\}$ **7.** Domain $= \{t: t \leq 1\}$; range $= \{y: y \geq 0\}$ **9.** Domain $= \mathbf{R}$; range $= \mathbf{R}$

11.

x	0	± 1	± 2
y	1	2	5

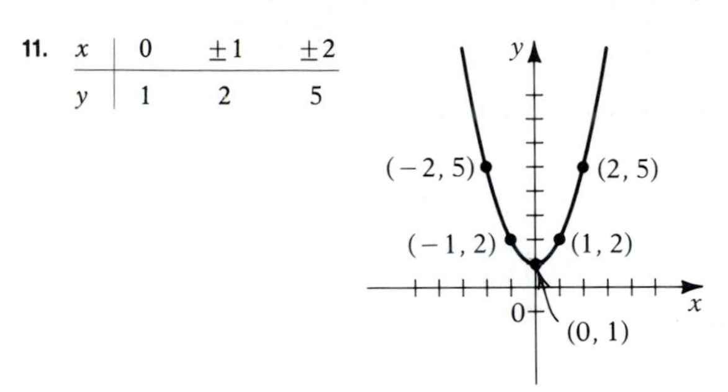

13.

x	0	1	2	3	4
y	0	3	4	3	0

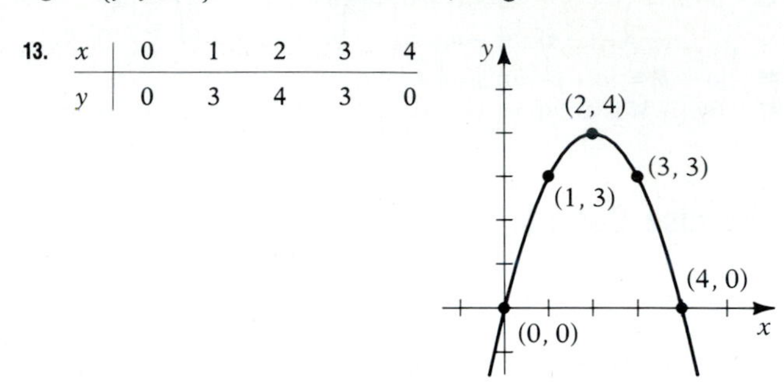

15.

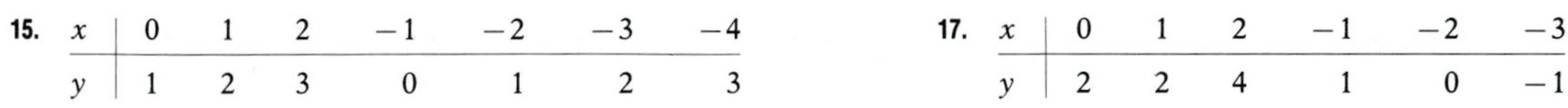

x	0	1	2	-1	-2	-3	-4
y	1	2	3	0	1	2	3

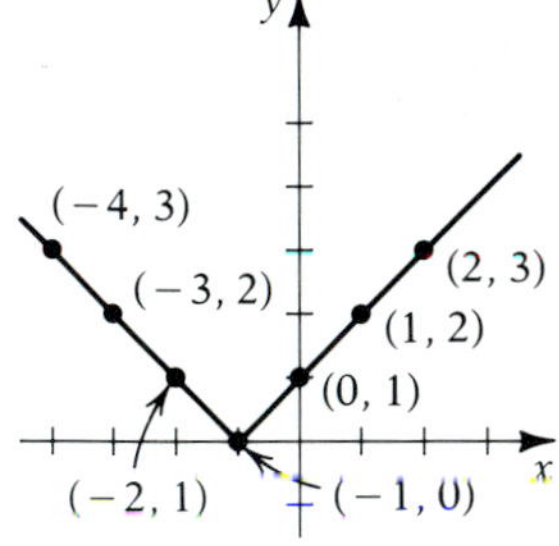

17.

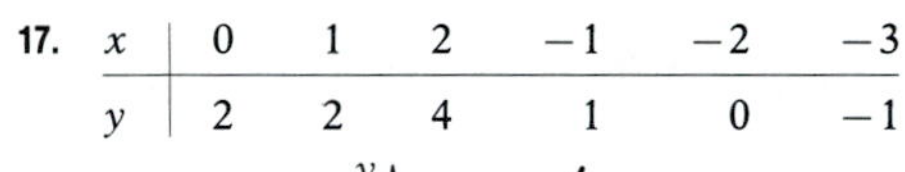

x	0	1	2	-1	-2	-3
y	2	2	4	1	0	-1

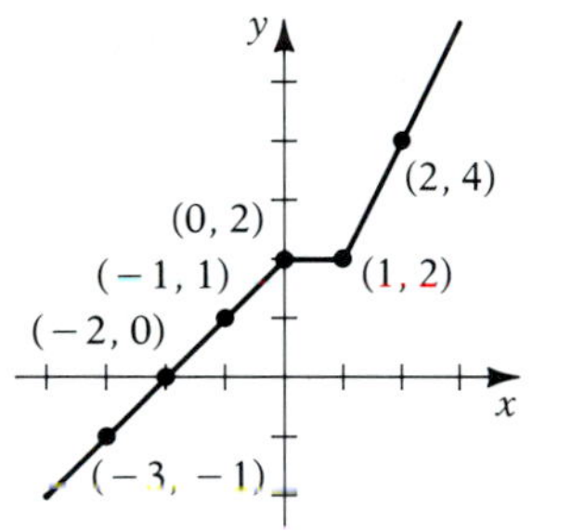

19.

x	0	± 2	± 3	± 4
y	4	0	5	12

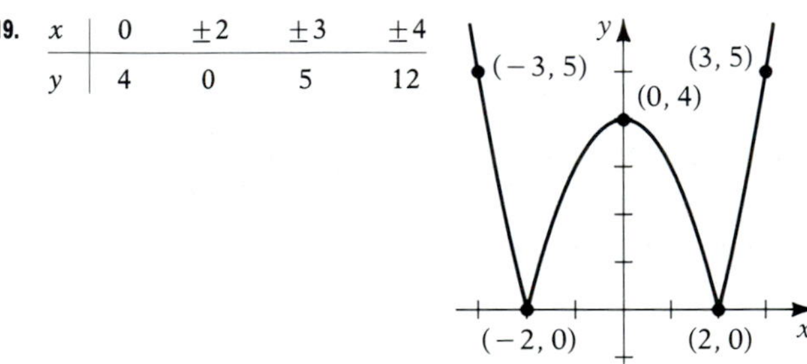

21. $\dfrac{g(x+h)-g(x)}{h} = \dfrac{\sqrt{x+h}-\sqrt{x}}{h} \cdot \dfrac{\sqrt{x+h}+\sqrt{x}}{\sqrt{x+h}+\sqrt{x}} = \dfrac{1}{\sqrt{x+h}+\sqrt{x}}$ **23.** $f\left(\dfrac{1}{x}\right) = \dfrac{\dfrac{1}{x}+2}{2\left(\dfrac{1}{x}\right)+1} = \dfrac{1+2x}{2+x} = \dfrac{1}{f(x)}$

25. (a) Odd (b) Odd (c) Neither (d) Even (e) Even

27. (a) $f(x) = 1000x - 25x^2$ (b) $\{x: 0 \leq x \leq 40\}$ **29.** (a) $V = 2\pi r^3$ (b) $S_L = 4\pi r^2$ (c) $S_T = 6\pi r^2$

31. $A = \dfrac{1}{3}(12b - 2b^2)$ **33.** Domain $= [-1, 3]$; range $= [0, 2]$ **35.** Domain $= (-1, 1)$; range $= [2, \infty)$

37. Domain $= (-\infty, -5) \cup (-5, -2) \cup [4, \infty)$; range $= \{y: y \neq \pm 1\}$ **39.** $V = \dfrac{\pi r^2}{3}(a + \sqrt{a^2 - r^2})$ **41.** (a) $V = \dfrac{\pi y^2}{12}(3x + 2y)$
(b) $S = \pi y(x + y)$ (c) $V = \dfrac{7\pi y^3}{6}$, $S = 5\pi y^2$ **43.** $f(x) = \begin{cases} 0.02x & \text{if } x \leq 20{,}000 \\ 400 & \text{if } x > 20{,}000 \end{cases}$

45.

h	0.1	-0.1	0.01	-0.01	0.001	-0.001	0.0001	-0.0001
$f(3+h)$	6.1	5.9	6.01	5.99	6.001	5.999	6.0001	5.9999

Approaches 6

47.

h	0.1	-0.1	0.01	-0.01	0.001	-0.001	0.0001	-0.0001
$f(2+h)$	1.61	1.47	1.546	1.531	1.5392	1.5378	1.5385	1.5384

Approaches $\dfrac{20}{13} \approx 1.5384615$

Exercise Set 1.5, page 29

1. $(f+g)(x) = 5x - 2$; domain $= \mathbf{R}$ $\quad (f-g)(x) = x - 8$; domain $= \mathbf{R}$

$(f \cdot g)(x) = 6x^2 - x - 15$; domain $= \mathbf{R}$ $\quad \left(\dfrac{f}{g}\right)(x) = \dfrac{3x-5}{2x+3}$; domain $= \left\{x : x \neq -\dfrac{3}{2}\right\}$

3. $(f+g)(x) = \dfrac{3x-1}{4}$; domain $= \mathbf{R}$ $\quad (f-g)(x) = \dfrac{x-3}{4}$; domain $= \mathbf{R}$

$(f \cdot g)(x) = \dfrac{x^2-1}{8}$; domain $= \mathbf{R}$ $\quad \left(\dfrac{f}{g}\right)(x) = 2\left(\dfrac{x-1}{x+1}\right)$; domain $= \{x : x \neq -1\}$

5. $(f+g)(x) = \sqrt{x+4} + \sqrt{4-x}$; domain $= [-4, 4]$ $\quad (f-g)(x) = \sqrt{x+4} - \sqrt{4-x}$; domain $= [-4, 4]$

$(f \cdot g)(x) = \sqrt{16 - x^2}$; domain $= [-4, 4]$ $\quad \left(\dfrac{f}{g}\right)(x) = \sqrt{\dfrac{4+x}{4-x}}$; domain $= [-4, 4)$

7. $(f+g)(x) = \dfrac{x^2+2}{x^2+x-2}$; domain $= \{x : x \neq 1, -2\}$ $\quad (f-g)(x) = \dfrac{2+2x-x^2}{x^2+x-2}$; domain $= \{x : x \neq 1, -2\}$

$(f \cdot g)(x) = \dfrac{x}{x^2+x-2}$; domain $= \{x : x \neq 1, -2\}$ $\quad \left(\dfrac{f}{g}\right)(x) = \dfrac{x+2}{x^2-x}$; domain $= \{x : x \neq 0, 1, -2\}$

9. $(f+g)(x) = 0$ $\quad (f-g)(x) = \begin{cases} -2 & \text{if } x < 0 \\ 2 & \text{if } x \geq 0 \end{cases}$ $\quad$ All have domain $\mathbf{R}$.

$(f \cdot g)(x) = -1$ $\quad \left(\dfrac{f}{g}\right)(x) = -1$

11. $(f+g)(x) = 1$; domain $= \mathbf{R}$ $\quad (f-g)(x) = \cos 2x$; domain $= \mathbf{R}$

$(f \cdot g)(x) = \dfrac{1}{4}\sin^2 2x$; domain $= \mathbf{R}$ $\quad \left(\dfrac{f}{g}\right)(x) = \cot^2 x$; domain $= \{x : x \neq n\pi, n = 0, \pm 1, \pm 2, \ldots\}$

13. Domain $= [1, \infty)$

$$(f-g)(x) = \sqrt{x} - \sqrt{x-1} \cdot \frac{\sqrt{x} + \sqrt{x+1}}{\sqrt{x} + \sqrt{x+1}} = \frac{1}{\sqrt{x} + \sqrt{x+1}} = \frac{1}{(f+g)(x)}$$

15. (a) All even (b) $f \pm g$ odd; $f \cdot g$ and f/g even (c) $f \pm g$ neither even nor odd; $f \cdot g$ and f/g odd

17. $(f \circ g)(x) = \cos(2x^2 - 1)$; domain $= \mathbf{R}$ $\quad (g \circ f)(x) = \cos 2x$; domain $= \mathbf{R}$

19. $\dfrac{x}{9-4x}$; domain $= \left\{x : x \neq \dfrac{3}{2}, \dfrac{9}{4}\right\}$

21. $f(x) = \sqrt{x}$; $g(x) = 2x + 3$ (Not unique)

23. $f(x) = x^{2/3}$; $g(x) = \dfrac{x}{x+2}$ (Not unique)

25. $f(x) = x^{3/2}$; $g(x) = x^2 - 1$ (Not unique)

27. $f(x) = \sin x$; $g(x) = \sqrt{x^2+1}$ (Not unique)

29. $[f \circ (g \circ h)](x) = [(f \circ g) \circ h](x) = \dfrac{8x^2 + 28x + 24}{x^2 + 2x + 1}$. In general, $f \circ (g \circ h) = (f \circ g) \circ h$.

31. Let $f(x) = P_1(x)/Q_1(x)$ and $g(x) = P_2(x)/Q_2(x)$, where P_1, Q_1, P_2, and Q_2 are polynomials. Show how each combination results in a quotient of polynomials. For $f + g$, $f - g$, and $f \cdot g$, domain $= \{x : Q_1(x) \neq 0, Q_2(x) \neq 0\}$. Domain of f/g is $\{x : P_2(x) \neq 0, Q_1(x) \neq 0, Q_2(x) \neq 0\}$.

33. $(f+g)(x) = \begin{cases} 2x - 1 & \text{if } x \geq 1 \\ 1 & \text{if } 0 \leq x < 1 \\ 1 - 2x & \text{if } x < 0 \end{cases}$ $\quad (f \cdot g)(x) = \begin{cases} x^2 - x & \text{if } x \in (-\infty, 0] \cup [1, \infty) \\ x - x^2 & \text{if } 0 < x < 1 \end{cases}$

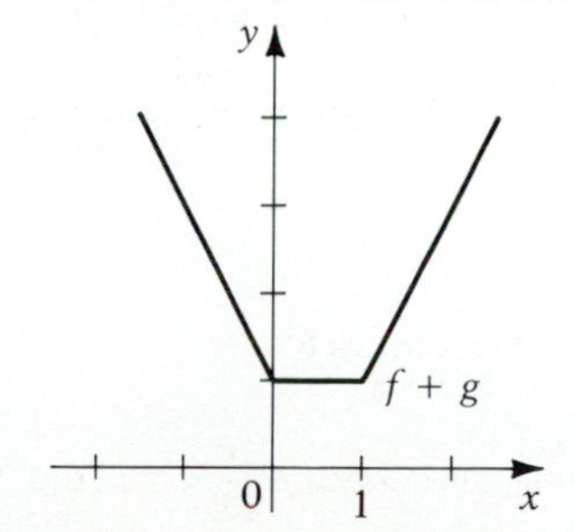

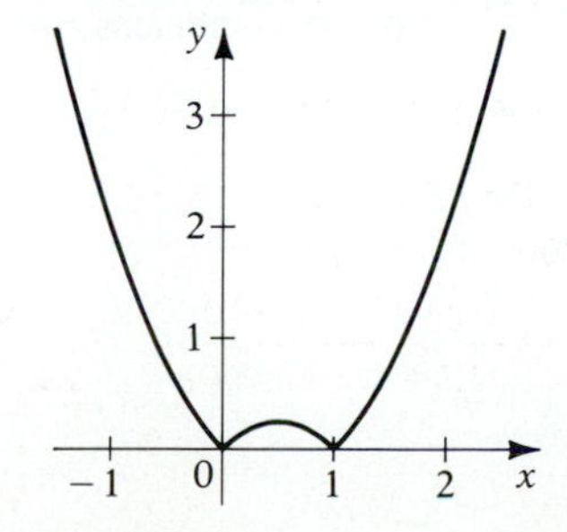

35. $\left(f \circ \dfrac{g}{h}\right)(x) = f\left(\dfrac{g(x)}{h(x)}\right) = \dfrac{1}{\dfrac{g(x)}{h(x)}} = \dfrac{\dfrac{1}{g(x)}}{\dfrac{1}{h(x)}} = \dfrac{(f \circ g)(x)}{(f \circ h)(x)}$ **37.** (a) $F(t) = \dfrac{gR^2 m}{\left(-\dfrac{1}{2}gt^2 + v_0 t + s_0\right)^2}$ (b) 19,360 newtons (approx.)

Exercise Set 1.6, page 36

1. 0 **3.** $\dfrac{9}{7}$ **5.** $\dfrac{34}{15}$ **7.** 1 **9.** $2 - \sqrt{3}$ **11.** 1 **13.** $\dfrac{2}{3}$ **15.** 8 **17.** -4 **19.** $-\dfrac{1}{10}$ **21.** 5 **23.** -7

25. $\dfrac{1}{2}$ **27.** $\dfrac{1}{3}$ **29.** $\dfrac{1}{2}$ **31.** 2 **33.** 1 **35.** 0 **37.** 12 **39.** $\dfrac{1}{4}$ **41.** -1 **43.** 6 **45.** 26 **47.** $\dfrac{1}{4}$

49. $\lim\limits_{x\to 2^+} f(x) = \lim\limits_{x\to 2^-} f(x) = 4$. So $\lim\limits_{x\to 2} f(x) = 4$. **51.** $\lim\limits_{x\to 0^+} f(x) = 1$; $\lim\limits_{x\to 0^-} f(x) = -1$. So $\lim\limits_{x\to 0} f(x)$ does not exist.

53. $\lim\limits_{x\to -1^+} f(x) = \lim\limits_{x\to -1^-} f(x) = -1$. So $\lim\limits_{x\to -1} f(x) = -1$. **55.** $\lim\limits_{x\to 0^+} f(x) = \lim\limits_{x\to 0^-} f(x) = 0$. So $\lim\limits_{x\to 0} f(x) = 0$

57. $\lim\limits_{x\to 0^+} f(x) = \lim\limits_{x\to 0^-} f(x) = 1$. So $\lim\limits_{x\to 0} f(x) = 1$ **59.** 0 **61.** 0 **63.** $-\dfrac{1}{2}$ **65.** $-\dfrac{1}{2}$ **67.** $\dfrac{3}{2}$ **69.** 0 **71.** 0

Exercise Set 1.7, page 41

1. Continuous **3.** Continuous **5.** Discontinuous; $\lim_{x\to -1} f(x)$ and $f(-1)$ both fail to exist. **7.** Continuous
9. Discontinuous; $\lim_{x\to 2} f(x)$ and $f(2)$ both fail to exist. **11.** Discontinuous; $f(-3)$ does not exist.
13. Discontinuous; $\lim_{x\to 0} f(x)$ does not exist.

15. Continuous on all of $\mathbf{R}$

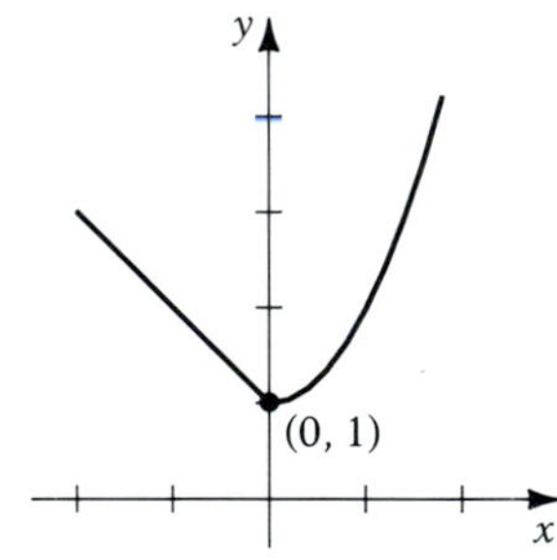

17. Continuous on $\{x : x \neq 0\}$

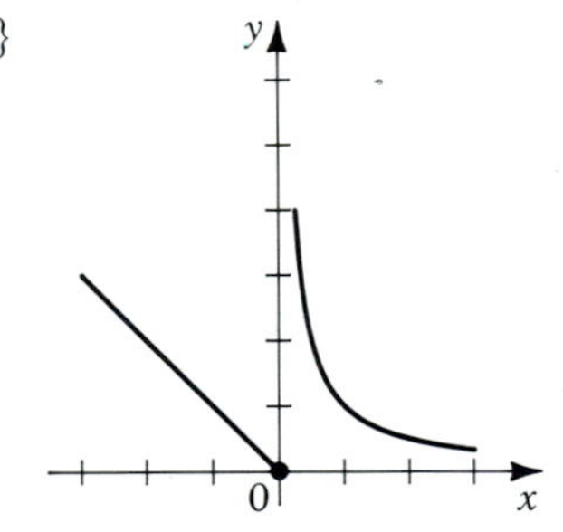

19. Continuous on $\{x : x \neq 0\}$

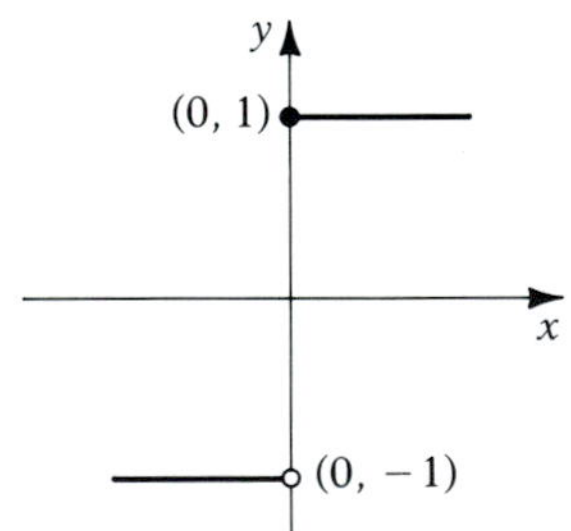

21. Continuous on all of $\mathbf{R}$

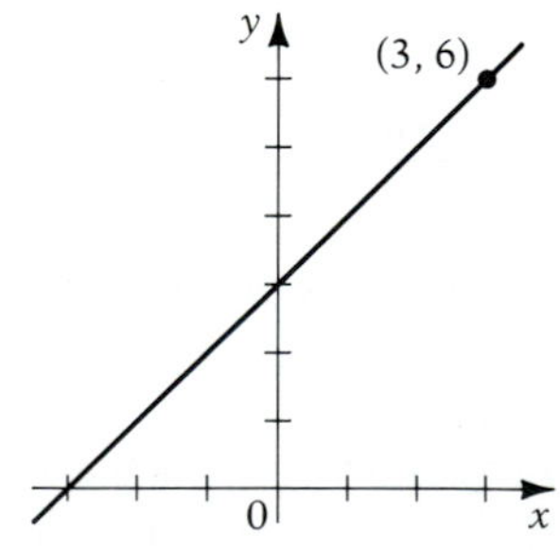

23. Continuous on $\{x : x \neq 2\}$

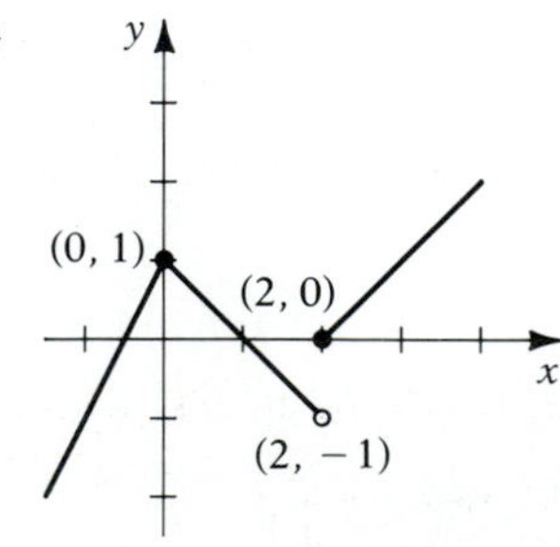

25. Discontinuous everywhere

27. $f(x) = \frac{1}{x}$ on $(0, 1)$ **29.** Yes, ϕ of Exercise 25; yes, ψ of Exercise 26.

31. (a) False, for example $f(x) = \begin{cases} 1 & \text{if } x \geq 0 \\ 0 & \text{if } x < 0 \end{cases}$, $g(x) = \begin{cases} 0 & \text{if } x \geq 0 \\ 1 & \text{if } x < 0 \end{cases}$ (b) False. Consider f and g of part a.

33. Show $P_n(a)$ and $P_n(-a)$ have opposite signs for a large. Then use Theorem 1.7.

Exercise Set 1.8, page 48

1. Take $\delta = \frac{\varepsilon}{2}$. **3.** Take $\delta = 3\varepsilon$. **5.** Take $\delta = \frac{\varepsilon}{4}$. **7.** Take $\delta = \frac{\varepsilon}{2}$. **9.** Take $\delta = 5\varepsilon$. **11.** Take $\delta = \varepsilon$.

13. Take $\delta = \frac{\varepsilon}{2}$. **15.** Take $\delta = \varepsilon$. **19.** Use $\big||f(x)| - |L|\big| \leq |f(x) - L|$.

21. Restrict x to be in $(a - 1, a + 1)$. Take $\delta = \min\left\{1, \frac{\varepsilon}{3(1 + |a|)^2}\right\}$. **23.** Restrict x to be in $\left(\frac{3}{2}, \frac{5}{2}\right)$. Take $\delta = \min\left\{\frac{1}{2}, \frac{\varepsilon}{2}\right\}$.

25. Take $\delta = \min\left\{\sqrt{\varepsilon}, \frac{\varepsilon}{2}\right\}$. **27.** Let $\varepsilon = \frac{1}{2}$ and show no δ works.

29. Let $\varepsilon = \frac{|f(a)|}{2}$ and for the corresponding δ, consider $x \in (a - \delta, a + \delta)$.

Supplementary Exercises 1.9, page 49

1. $\left(-\infty, \frac{1}{5}\right]$ **3.** $\left(-\frac{1}{2}, \frac{19}{6}\right)$ **5.** (a) $0 < |x + 2| < 1$ (b) The deleted neighborhood of 3 of radius δ

7. For (ii) use fact that if $L < M$, L is not an upper bound of S. **11.** $x^2 + y^2 - 6x + 5 = 0$

13. Parabola; vertex $\left(\frac{3}{2}, \frac{11}{4}\right)$; 1

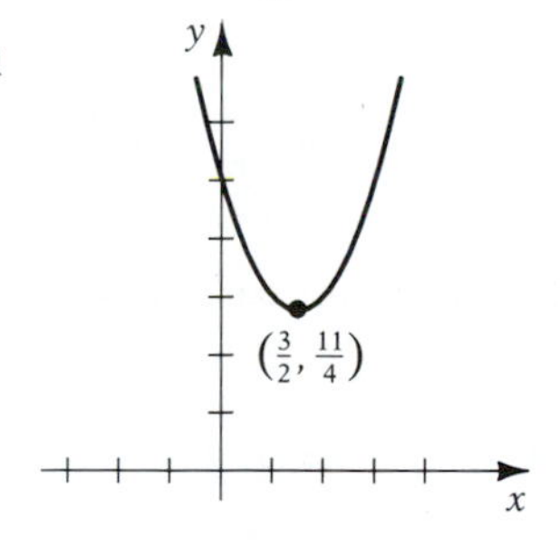

15. $A = \frac{\pi s^2}{12}$ **17.** $(f \circ g)(t) = -t - 2$, domain $= (-\infty, 1]$; $(g \circ f)(t) = \sqrt{4 - t^2}$, domain $= [-2, 2]$

19. $\frac{1}{(x + 1)^2}$ **21.** $(g \circ h)(t) = -\frac{t + 2}{2}$, domain $= \{t : t \neq -2\}$; $(h \circ g)(t) = \frac{1}{2t - 1}$, domain $= \left\{t : t \neq \frac{1}{2}, 1\right\}$

23. (a) $1 + x^2 - \tan\frac{\pi x}{4}$ (b) -2 (c) $\sec^2\frac{\pi x}{4}$ **25.** (a) 0 (b) $\frac{1}{4}$ **27.** (a) Does not exist (b) Does not exist

29. (a) $-\frac{1}{4}$ (b) -1 **31.** (a) 0 (b) 0 **33.** $\frac{1}{\sqrt{2x + 3}}$; $\frac{1}{\sqrt{x}}$

35. (a) $x = 2$; $f(2)$ and $\lim_{x \to 2} f(x)$ fail to exist. (b) $x = \pm\sqrt{2}$; both $f(x)$ and $\lim f(x)$ fail to exist at both points.

37. Take $\delta = \frac{\varepsilon}{2}$. **39.** Restrict x to be in $(2, 6)$; take $\delta = \min\{2, \varepsilon\}$. **41.** Take $\delta = \frac{\varepsilon}{|m|}$.

CHAPTER 2

Exercise Set 2.1, page 55

1. $y = 2x - 2$

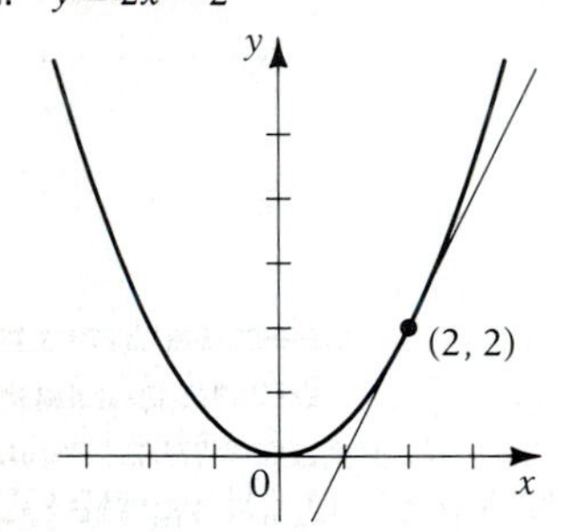

3. $y = 4 - x$

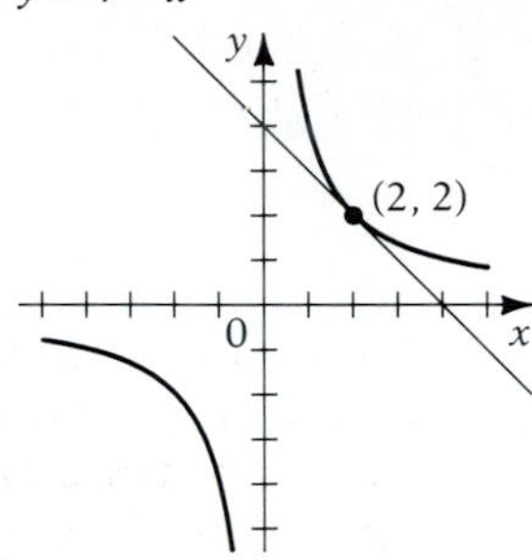

5. $x + 2y + 3 = 0$

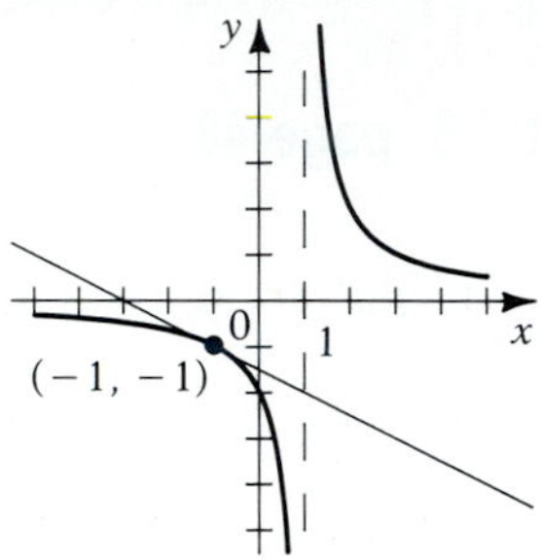

7. $y = 2x - 5$

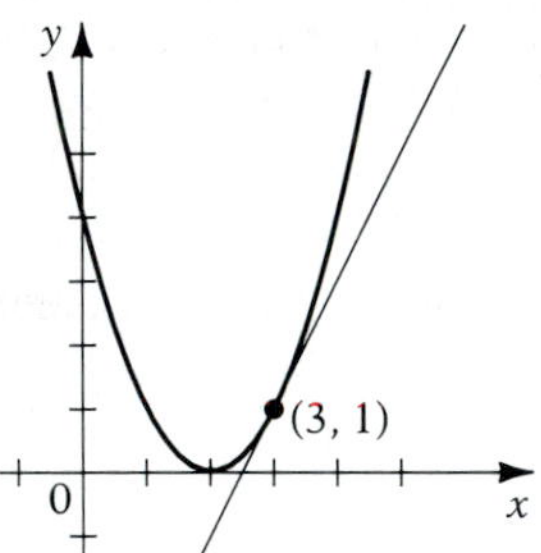

9. $4x + 2y + 1 = 0$

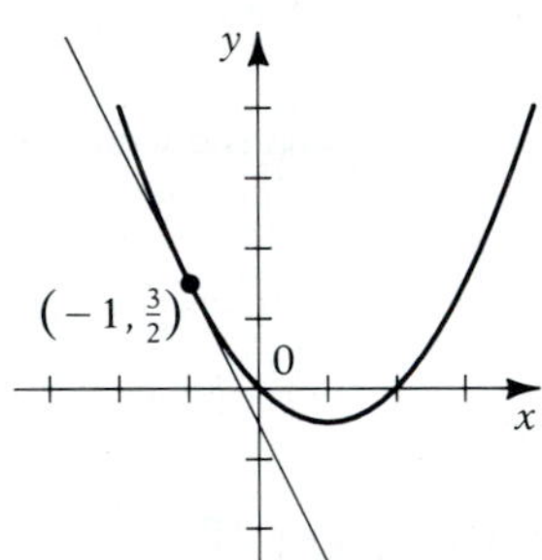

11. 2 **13.** $\frac{1}{2}$ **15.** 8 **17.** $\frac{1}{4}$ **19.** $-\frac{4}{9}$ **21.** $2(x_1 - 1)$ **23.** $\frac{1}{(x_1 + 1)^2}$ **25.** $x - 3y + 5 = 0$

Exercise Set 2.2, page 59

1. 3 **3.** -4 **5.** 1 **7.** $\frac{1}{6}$ **9.** -2 **11.** 3 **13.** -4 **15.** 1 **17.** $\frac{1}{6}$ **19.** -2 **21.** 2 **23.** $-\frac{2}{x^2}$

25. $\frac{1}{(x+1)^2}$ **27.** (a)

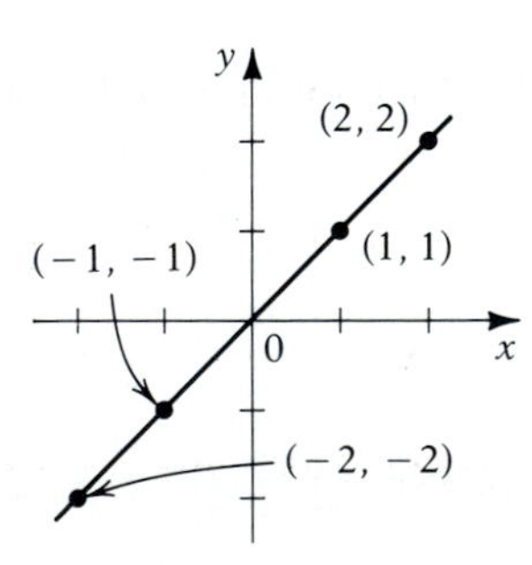

(b)

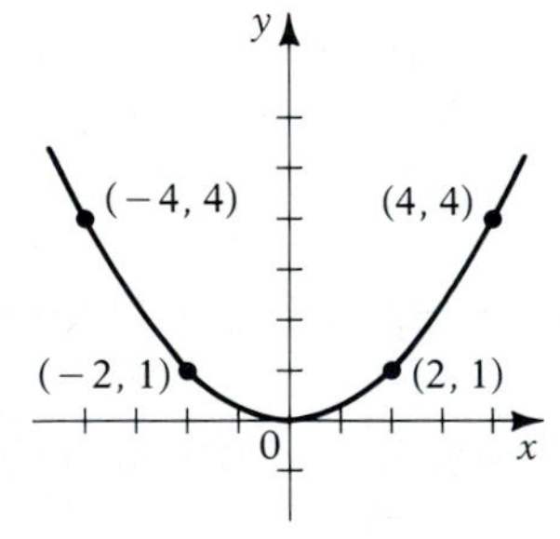

29. (a)

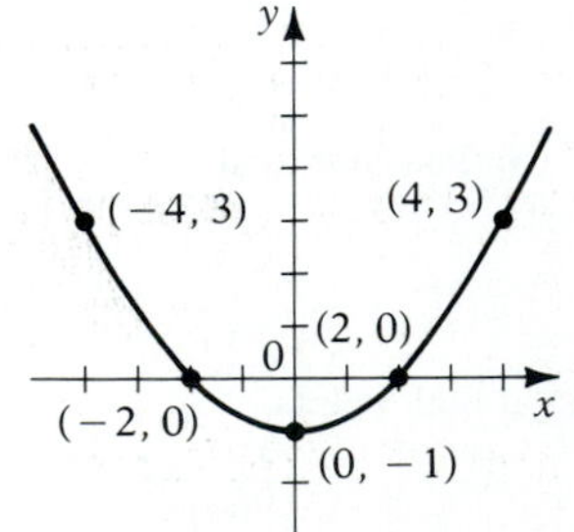

(b)

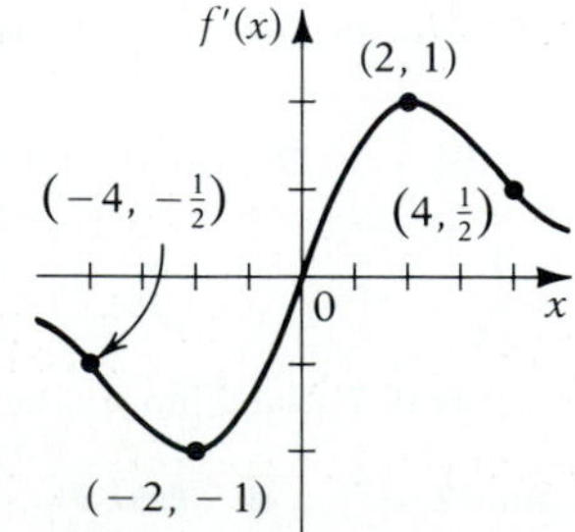

31. 1 **33.** $\dfrac{-a}{\sqrt{4-a^2}}, |a| < 2$ **35.** All have slope 2.

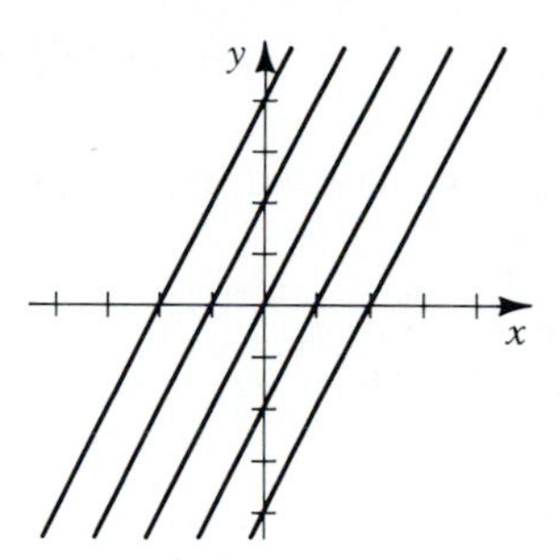

Exercise Set 2.3, page 64

1. 0 **3.** $4x^3$ **5.** -2 **7.** a **9.** $12x^2 - 14x + 8$ **11.** $5x^4 - 24x^3$ **13.** $2ax + b$ **15.** $4x + 5$ **17.** $18x + 24$

19. $15x^2 - 6x - 6$ **21.** $y = 5x - 4$ **23.** $\dfrac{4}{3}$ **25.** tangent: $y = 2x - 9$; normal: $x + 2y + 3 = 0$ **27.** 20 **29.** -10

31. $a = \dfrac{5}{4}, b = -4$ **33.** $(2, 4), (-2, 4)$; $y = 4x - 4$, $y = -4x - 4$ **35.** $v(t) = 48t^5 - 144t^3 + 108t$; $a(t) = 240t^4 - 432t^2 + 108$

Exercise Set 2.4, page 69

1. $4x^3 + 2x$ **3.** $6x - 20x^3$ **5.** $10x^4 - 24x^3 + 3x^2 - 6x$ **7.** $7x^6 - 15x^4 + 12x^3 + 9x^2 - 12x - 1$

9. $70x^6 + 72x^5 - 75x^4 - 52x^3 + 60x^2 - 21$ **11.** $\dfrac{-2}{(x-2)^2}$ **13.** $\dfrac{4x}{(x^2+1)^2}$ **15.** $\dfrac{-2x^2+6x+6}{(x^2+2x)^2}$ **17.** $\dfrac{-8x}{(x^2-4)^2}$

19. $\dfrac{-27x^2}{(x^3-1)^2}$ **21.** $-6x^{-4}$ **23.** $2x - 2x^{-3}$ **25.** $1 - \dfrac{4}{x^2}$ **27.** $\dfrac{8}{x^3}$ **29.** $y = 13 - 4x$ **31.** $\dfrac{27}{16}$

33. $\dfrac{-18x^4 + 36x^3 - 28x^2 + 10x - 10}{(3x^2+5)^2(x^2-2x)^2}$ **35.** $\dfrac{16x}{(x^2+4)^2}$ **37.** $f \cdot g \cdot h' + f \cdot g' \cdot h + f' \cdot g \cdot h$

39. $f_1' \cdot f_2 \cdots f_n + f_1 \cdot f_2' \cdot f_3 \cdots f_n + \cdots + f_1 \cdot f_2 \cdots f_{n-1} \cdot f_n'$

Exercise Set 2.5, page 72

1. $y' = 8x^3, y'' = 24x^2, y''' = 48x$ **3.** $y' = 12x^2 - 4x, y'' = 24x - 4, y''' = 24$ **5.** $y' = 2x + 2x^{-3}, y'' = 2 - 6x^{-4}, y''' = 24x^{-5}$

7. $\dfrac{dy}{dx} = 12x + 7, \dfrac{d^2y}{dx^2} = 12$ **9.** $\dfrac{dy}{dx} = 4x^3 - 6x - 5x^{-2}, \dfrac{d^2y}{dx^2} = 12x^2 - 6 + 10x^{-3}$ **11.** $\dfrac{dy}{dx} = 1 - 3x^{-2}, \dfrac{d^2y}{dx^2} = 6x^{-3}$

13. $3x^2 - 6x$ **15.** $\dfrac{5}{(x+3)^2}$ **17.** $\dfrac{3}{4}(x^{-1/2} - x^{-5/2})$ **19.** $2 + \dfrac{3}{2x^4}$ **21.** $n!a_n$; 0

23. $f'(x) = \dfrac{1}{2}x^{-1/2}, f''(x) = -\dfrac{1}{2^2}x^{-3/2}, f'''(x) = \dfrac{1 \cdot 3}{2^3}x^{-5/2}, f^{(4)}(x) = -\dfrac{1 \cdot 3 \cdot 5}{2^4}x^{-7/2}$; $f^{(n)}(x) = (-1)^{n-1}\dfrac{1 \cdot 3 \cdot 5 \cdots (2n-1)}{2^n}x^{-(2n+1)/2}$

25. $v(t) = 4t - \dfrac{t^2 + 6t + 1}{(t+3)^2}, a(t) = 4 - \dfrac{16}{(t+3)^3}$; $v(1) = \dfrac{7}{2}, a(1) = \dfrac{15}{4}$

Exercise Set 2.6, page 75

1. $8x(x^2+2)^3$ **3.** $-20x(1-2x^2)^4$ **5.** $120x^3(3x^4-2)^9$ **7.** $3(2x-3)(x^2-3x+4)^2$

9. $-8x(3x^2-1)(3x^4-2x^2+1)^{-3}$ **11.** $2(2x+1)(x^2+2)^2(8x^2+3x+4)$ **13.** $\dfrac{-12x}{(x^2+1)^4}$ **15.** $\dfrac{(3x-4)^2(6x+13)}{(x+1)^2}$

17. $-\dfrac{2x^3+6x}{(x^2-1)^3}$ **19.** $\dfrac{6(x-1)^2}{(x+1)^4}$ **21.** $6(2x+3)^3$ **23.** $-4x(x^2+4)^{-3}$ **25.** $18x^2(x^3-8)$ **27.** $12x(x^2-3)^2$

29. $\dfrac{-18x^2}{(x^3+4)^3}$ **31.** $4(3x^2-4x-1)(3x-2)$ **33.** $8(x^2+1)^2(7x^2+1)$ **35.** $\dfrac{2}{(1-x)^3}$

37. $(h\circ g)'(t)=\dfrac{-4t}{(t^2-1)^2}$, $(g\circ h)'(t)=\dfrac{-4(t+1)}{(t-1)^3}$ **39.** $3(y^2+3y-1)(y^2+3y-3)(2y+3)$ **41.** $\dfrac{4(6x^3-5x^2-6x+1)}{(x^2+1)^4}$

43. $36x^2(2x^3-3)[(2x^3-3)^2+4]^2$; $f(x)=x^3$, $g(x)=x^2+4$, $h(x)=2x^3-3$

Exercise Set 2.7, page 80

1. $3\cos 3x$ **3.** $2\sin(1-2x)$ **5.** $6x\sec^2(3x^2)$ **7.** $-6\csc 3x\cot 3x$ **9.** $5\sin 2x$ **11.** $4\tan(2x+3)\sec^2(2x+3)$

13. $-\dfrac{1}{x^2}\sec\left(\dfrac{1}{x}\right)\tan\left(\dfrac{1}{x}\right)$ **15.** $6x\sec^2 x^2\tan^2 x^2$ **17.** $\dfrac{-2\sec x\tan x}{(\sec x-1)^2}$ **19.** $4\cos 2x(1+\sin 2x)$ **21.** $\dfrac{1-\tan x}{\sec x}$

23. $3\csc 3x[1-(3x+2)\cot 3x]$ **25.** $-\sin x[\cos(\cos x)]$ **27.** $2\cos x - x\sin x$ **29.** $2\sec^2 x^2(4x^2\tan x^2+1)$

31. $\dfrac{(1-\cos x)^2}{\sin^3 x}$ **33.** $x\cos x+\sin x$ **37.** $g'(t)=-\dfrac{1}{t^2}\sec\dfrac{1}{t},\tan\dfrac{1}{t}$, $g''(t)=\dfrac{1}{t^4}\sec\dfrac{1}{t}\left(\sec^2\dfrac{1}{t}+\tan^2\dfrac{1}{t}+2t\tan\dfrac{1}{t}\right)$

Exercise Set 2.8, page 86

1. $\dfrac{x}{y}$ **3.** $\dfrac{2(1-xy^3)}{1+3x^2y^2}$ **5.** $\dfrac{3x^2}{1+3y^2}$ **7.** $\dfrac{2xy}{4y-x^2}$ **9.** $\dfrac{1}{2\sqrt{x+y}-1}$ **11.** $-\dfrac{y\cos x+\sin y}{x\cos y+\sin x}$ **13.** $\dfrac{3-x\sec(xy)\tan(xy)}{x\sec(xy)\tan(xy)}$

15. $\dfrac{\cos x\cos y-1}{\sin x\sin y-1}$ **17.** $\dfrac{x}{\sqrt{x^2+1}}$ **19.** $\dfrac{x}{(4-x^2)^{3/2}}$ **21.** $\dfrac{4x^2}{\sqrt[3]{2x^3+5}}$ **23.** $\dfrac{1}{(1-x^2)^{3/2}}$ **25.** $\dfrac{\cos x}{2\sqrt{\sin x}}$

27. $-\dfrac{1}{2x^{3/2}}\sec^2\left(\dfrac{1}{\sqrt{x}}\right)$ **29.** $-2\csc(2x+3)\cot(2x+3)$ **31.** $2\cos x$ **33.** $\dfrac{-4}{x^2\sqrt{4-x^2}}$ **35.** $-\dfrac{1}{y^3}$ **37.** $\dfrac{-4}{x^{1/2}y^2}$

39. $\dfrac{-1}{(1-x^2)^{3/2}}$ **41.** $-\dfrac{\cos\sqrt{x}+\sqrt{x}\sin\sqrt{x}}{4x^{3/2}}$ **43.** $\dfrac{dy}{dx}=\dfrac{x^{1/3}}{2y^{1/3}}$, $\dfrac{d^2y}{dx^2}=\dfrac{-a^{4/3}}{12x^{2/3}y^{5/3}}$ **45.** $\dfrac{8x^3t-3x^2t^2+10t^4}{2xt^3-12x^2t^2+6x}$

47. (a) $y_1=x$, $y_2=x^2$, $y_3=-x^2$; $y_1'=1$, $y_2'=2x$, $y_3'=-2x$ (b) $y'=\dfrac{y^2+4x^3y-5x^4}{3y^2-2xy-x^4}$

Exercise Set 2.9, page 91

1. $(2x+2)\,dx$ **3.** $-\dfrac{x\,dx}{\sqrt{4-x^2}}$ **5.** $\dfrac{(2-x)\,dx}{2(1-x)^{3/2}}$ **7.** $(x\cos x+\sin x)\,dx$ **9.** $(\sec x\tan x+\sec^2 x)\,dx$ **11.** $\dfrac{4x^5\,dx}{(x^6+1)^{1/3}}$

13. $\dfrac{1}{x^2}(x\sec^2 x-\tan x)\,dx$ **15.** $(2x\sin 2x^2)\,dx$ **17.** $\dfrac{3x^2-2y^2}{4xy}$ **19.** $\dfrac{3y-4x}{8y-3x}$ **21.** $-\dfrac{y^{1/3}}{x^{1/3}}$ **23.** $\dfrac{y\cos x-\sin y}{x\cos y-\sin x}$

25. (a) 1.8 cm^2 (b) 0.2 cm **27.** (a) 0.0056 (b) 0.0028 **29.** Take $x=9$, $\Delta x=1$. $\sqrt{10}\approx 3.167$; by calculator $\sqrt{10}\approx 3.16278$.

31. Take $x=-27$, $\Delta x=2$, $\sqrt[3]{-25}\approx -2.926$; by calculator $\sqrt[3]{-25}\approx -2.92402$.

33. max ≈ 19, min ≈ 13; by calculator max ≈ 18.627, min ≈ 13.623. **35.** -0.129 ft **37.** \$3

39. (a) $4\cos 2t\,dt$ (b) $\dfrac{4(2-x^2)}{\sqrt{4-x^2}}\cos t\,dt$, gives same results. **41.** $\dfrac{z(1+z^2)\sec^2 z+\tan z}{(1+z^2)^{3/2}}\,dz$ **43.** $\dfrac{8\pi}{3}$ cu ft ≈ 8.38 cu ft

45. (a) Take $x=4$, $\Delta x=0.2$; $\sqrt{4.2}\approx 2.05$. Error ≤ 0.000625. (b) Take $x=27$, $\Delta x=3$; $(30)^{2/3}\approx 9.7$. Error ≤ 0.01235.

Supplementary Exercises 2.10, page 93

1. 9 **3.** $\dfrac{1}{9}$ **5.** $60x(3x^2-2)^9$ **7.** $\dfrac{1-x}{(1-2x)^{3/2}}$ **9.** $\dfrac{1}{1+\cos x}$ **11.** $-\dfrac{(a^{2/3}-x^{2/3})^{1/2}}{x^{1/3}}$ **13.** $\sqrt{2}\sec^2 x$

15. tangent: $3x + y = 4$; normal: $x - 3y = 8$ **17.** $v(3) = 12$, $a(3) = -\dfrac{52}{5}$ **19.** $\dfrac{4x^3 + 2y^2}{3y^2 - 4xy}$ **21.** $\dfrac{\sec y - y\sec^2 x}{\tan x - x\sec y\tan y}$

23. $y' = -\dfrac{y^{1/2}}{x^{1/2}}$, $y'' = \dfrac{a^{1/2}}{2x^{3/2}}$ **25.** (a) 135.7 cm³ (b) 1.99 cm²

CHAPTER 3

Exercise Set 3.1, page 104

1. Increasing on $(-\infty, -1)$, decreasing on $(-1, \infty)$ **3.** Increasing on $(-\infty, -\frac{4}{3}) \cup (2, \infty)$, decreasing on $(-\frac{4}{3}, 2)$
5. Increasing on $(-\infty, -\frac{2}{3}) \cup (2, \infty)$, decreasing on $(-\frac{2}{3}, 2)$ **7.** Increasing on $(1, \infty)$, decreasing on $(-\infty, 0) \cup (0, 1)$
9. Increasing on $(-\infty, 2)$, decreasing on $(2, \infty)$ **11.** Increasing on $(-\infty, -1) \cup (3, \infty)$, decreasing on $(-1, 3)$
13. Local minimum point $(4, -9)$ **15.** Local maximum point $(-1, 8)$, local minimum point $(1, 4)$
17. Local maximum point $(-2, 32)$, local minimum point $(4, -76)$
19. Local maximum point $(0, 12)$, local minimum points $(-2, -4)$ and $(2, -4)$
21. Local minimum point $(3, -19)$ **23.** Local minimum point $(3, 1)$
25. Local and absolute minimum $= -1$, when $x = 2$. Absolute maximum $= 8$, when $x = 5$.
27. Local and absolute maximum $= \frac{256}{27}$, when $x = -\frac{2}{3}$. Local and absolute minimum $= 0$, when $x = 2$ or $x = -2$.
29. Local and absolute maximum $= 27$, when $x = 0$. Local and absolute minimum $= -27$, when $x = 3$.
31. Local and absolute maximum $= \sqrt{2}$, when $x = \dfrac{\pi}{4}$. Absolute minimum $= -1$, when $x = \pi$.
33. Local and absolute minimum $= 2 - \dfrac{\pi}{2} \approx 0.43$, when $x = \dfrac{\pi}{4}$. Absolute maximum $= 1 + \dfrac{\pi}{2} \approx 2.57$, when $x = -\dfrac{\pi}{4}$.
35. Local maximum $= \dfrac{5\sqrt{3}}{2} + \dfrac{\pi}{6} - 1 \approx 3.85$ when $x = \dfrac{\pi}{6}$. Local and absolute minimum $= -\dfrac{5\sqrt{3}}{2} + \dfrac{5\pi}{6} - 1 \approx -2.71$ when $x = \dfrac{5\pi}{6}$.
Absolute maximum $= 3 + 2\pi \approx 9.28$ when $x = 2\pi$.
37. Absolute maximum $= 24$, when $x = 3$. Absolute minimum $= -12$, when $x = -3$.
39. Absolute maximum $= 32$, when $x = -2$. Absolute minimum $= -76$, when $x = 4$.
41. Absolute maximum $= 21$, when $x = \pm 3$. Absolute minimum $= -4$, when $x = \pm 2$.
43. Absolute maximum $= 8$, when $x = \pm 4$. Absolute minimum $= -19$, when $x = 3$.
45. Absolute maximum $= 9$, when $x = -5$. Absolute minimum $= 1$, when $x = 3$. **47.** Absolute minimum $= 0$, when $x = 1$.
49. Absolute maximum $= 2$, when $x = \frac{1}{2}$. Local and absolute minimum $= -\frac{1}{4}$, when $x = -1$.
51. Absolute maximum $= \frac{1}{3}$, when $x = 3$ or 12. Local and absolute minimum $= 0$, when $x = 4$.
53. Absolute maximum $= 264$, when $x = 4$. Local and absolute minimum $= -177$, when $x = -3$.
55. Absolute maximum $= 13$, when $x = 3$. Local and absolute minimum $= -19$, when $x = -1$.
57. Local and absolute maximum ≈ 51.3, when $x \approx -1.92$. Local and absolute minimum ≈ 25.6, when $x \approx 0.97$.
59. Assume $\lim_{x \to a} f(x) = L < 0$. Take $\varepsilon = -\frac{L}{2}$ and use the definition of limit to show $f(x) < \frac{L}{2} < 0$ in a deleted neighborhood of a, contradicting hypothesis. For second part, take $f(x) = x^2$, for example, with $a = 0$.
61. $[f(0 + h) - f(0)]/h = \sin \dfrac{1}{h}$ does not approach a limit as $h \to 0$. **63.** One example is $f(x) = \begin{cases} x & \text{if } x \in (0, 1) \\ \frac{1}{2} & \text{if } x = 0 \text{ or } 1 \end{cases}$ on $[0, 1]$.
65. $\max(f + g) \le \max f + \max g$
If $\max f \ge 0$ and $\max g \ge 0$, then $\max(f \cdot g) \le (\max f)(\max g)$
If $\max f \ge 0$ and $\max g \le 0$, then $\max(f \cdot g) \le 0$
If $\max f \le 0$ and $\max g \le 0$, then $\max(f \cdot g) \ge 0$

Exercise Set 3.2, page 111

1. Concave upward on $(-\infty, \infty)$ **3.** Concave upward on $(-\infty, \infty)$ if $a > 0$ and downward if $a < 0$
5. Concave upward on $(-2, \infty)$, concave downward on $(-\infty, -2)$
7. Concave upward on $(-\infty, 0) \cup (1, \infty)$, concave downward on $(0, 1)$
9. Concave upward on $(-\infty, -1) \cup (1, \infty)$, concave downward on $(-1, 1)$
11. Local maximum point $(-1, 8)$, local minimum point $(1, 0)$ **13.** Local maximum point $(0, 7)$, local minimum point $(1, 6)$
15. Local maximum point $(0, 4)$, local minimum points $(1, 1)$ and $(-4, -124)$

17. Local maximum point $(-1, -2)$, local minimum point $(1, 2)$ **19.** Local minimum point $(0, 0)$

21. Increasing on $(-\infty, 0) \cup (2, \infty)$, decreasing on $(0, 2)$; local maximum point $(0, 0)$, local minimum point $(2, -4)$; concave downward on $(-\infty, 1)$, concave upward on $(1, \infty)$; inflection point $(1, -2)$.

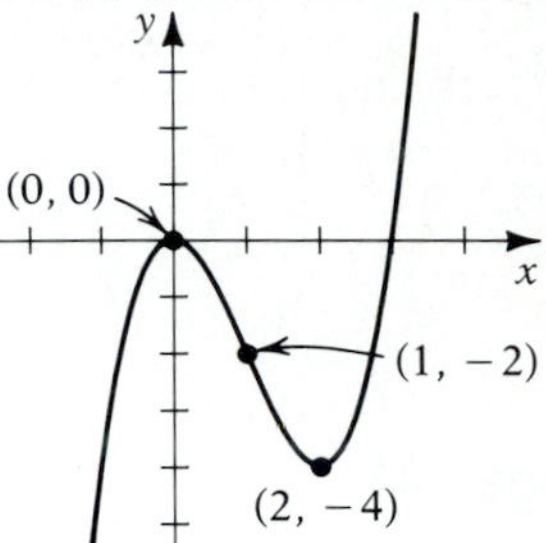

23. Increasing on $(-\infty, -3) \cup (1, \infty)$, decreasing on $(-3, 1)$; local maximum point $(-3, 12)$, local minimum point $(1, -20)$; concave downward on $(-\infty, -1)$, concave upward on $(-1, \infty)$; inflection point $(-1, -4)$.

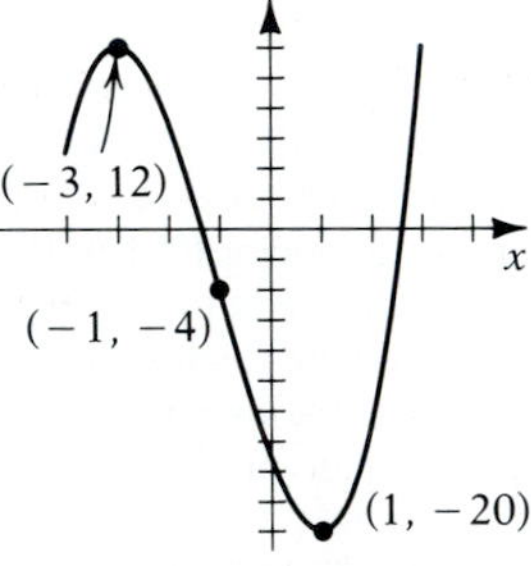

25. Increasing on $(-1, 1)$, decreasing on $(-\infty, -1) \cup (1, \infty)$; local maximum point $(1, 6)$, local minimum point $(-1, 2)$; concave downward on $(0, \infty)$, concave upward on $(-\infty, 0)$; inflection point $(0, 4)$.

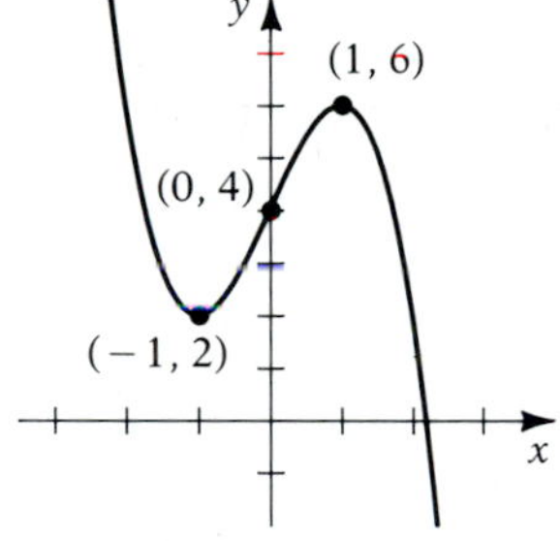

27. Increasing on $(3, \infty)$, decreasing on $(-\infty, 3)$; local minimum point $(3, -22)$; concave downward on $(0, 2)$, concave upward on $(-\infty, 0) \cup (2, \infty)$; inflection points $(0, 5)$, $(2, -11)$.

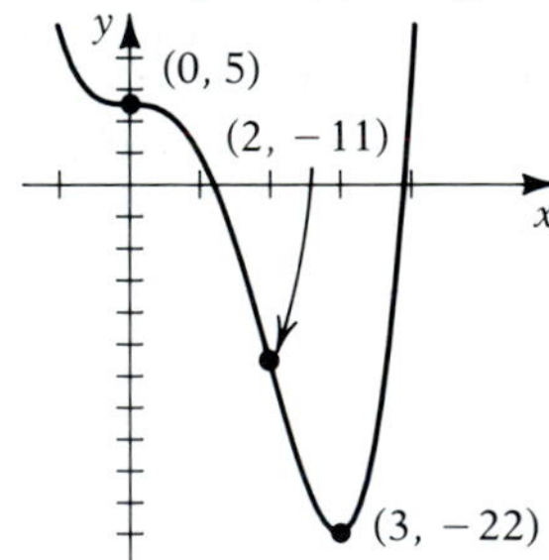

29. Increasing on $(-\sqrt{3}, 0) \cup (\sqrt{3}, \infty)$, decreasing on $(-\infty, -\sqrt{3}) \cup (0, \sqrt{3})$; local maximum point $(0, 10)$, local minimum points $(\pm\sqrt{3}, 1)$; concave downward on $(-1, 1)$, concave upward on $(-\infty, -1) \cup (1, \infty)$; inflection points $(\pm 1, 5)$.

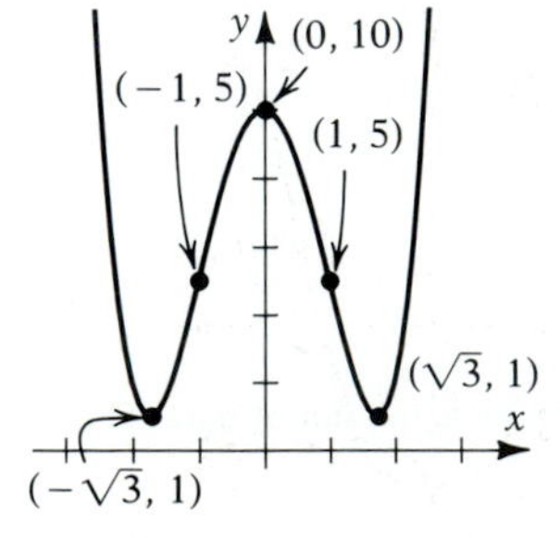

31. Increasing on $(-\infty, -3) \cup (-1, 1) \cup (3, \infty)$, decreasing on $(-3, -1) \cup (1, 3)$; local maximum points $(-3, \frac{72}{5})$, $(1, \frac{88}{15})$, local minimum points $(-1, -\frac{88}{15})$, $(3, -\frac{72}{5})$; concave downward on $(-\infty, -\sqrt{5}) \cup (0, \sqrt{5})$, concave upward on $(-\sqrt{5}, 0) \cup (\sqrt{5}, \infty)$; inflection points $\left(-\sqrt{5}, \dfrac{8\sqrt{5}}{3}\right)$, $(0, 0)$, $\left(\sqrt{5}, -\dfrac{8\sqrt{5}}{3}\right)$.

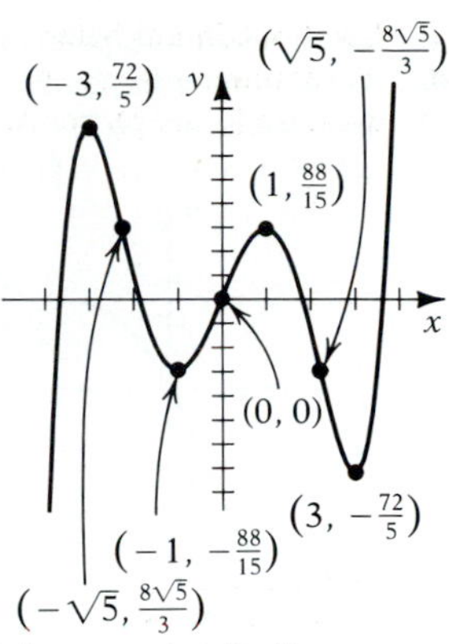

33. Increasing on $(-\infty, -1) \cup (3, \infty)$, decreasing on $(-1, 3)$; local maximum point $(-1, 32)$, local minimum point $(3, 0)$; concave downward on $(-\infty, 1)$, concave upward on $(1, \infty)$; inflection point $(1, 16)$.

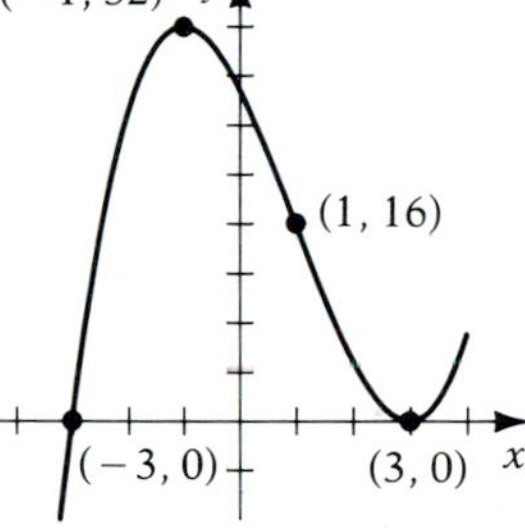

35. Increasing on $[0, \pi) \cup (3\pi, 4\pi]$, decreasing on $(\pi, 3\pi)$; local maximum point $(\pi, 3)$, local minimum point $(3\pi, -1)$; concave downward on $(0, 2\pi)$, concave upward on $(2\pi, 4\pi)$; inflection point $(2\pi, 1)$.

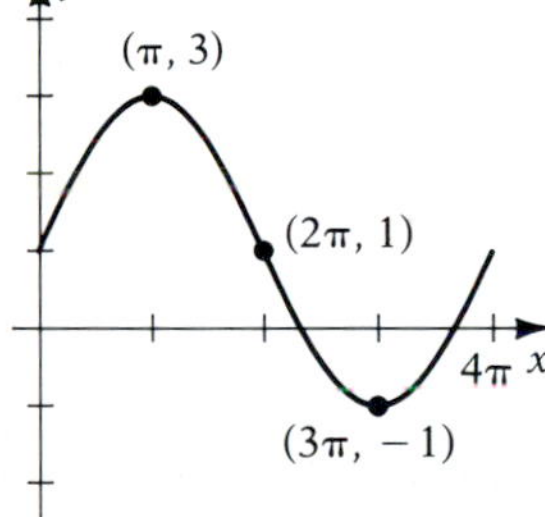

37. Increasing on $[0, 4\pi]$; no local maxima or minima; concave downward on $(0, \pi) \cup (2\pi, 3\pi)$, concave upward on $(\pi, 2\pi) \cup (3\pi, 4\pi)$; inflection points (π, π), $(2\pi, 2\pi)$, $(3\pi, 3\pi)$.

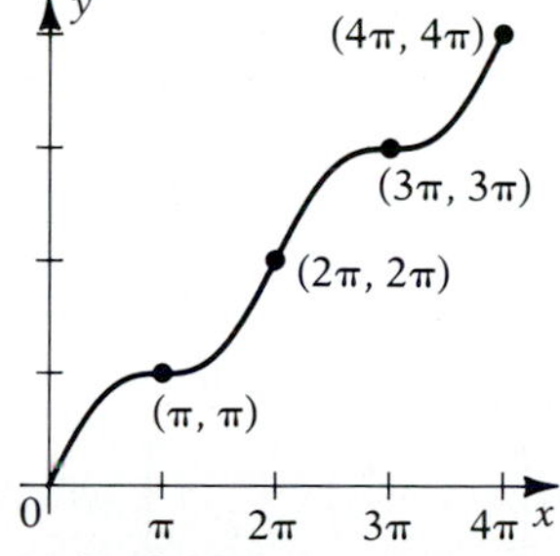

39. Increasing on $(0, \pi]$, decreasing on $[-\pi, 0)$; local minimum point $(0, -3)$, end point maxima $(\pm\pi, 5)$; concave downward on $\left[-\pi, -\dfrac{2\pi}{3}\right) \cup \left(\dfrac{2\pi}{3}, \pi\right]$, concave upward on $\left(-\dfrac{2\pi}{3}, \dfrac{2\pi}{3}\right)$; inflection points $\left(\pm\dfrac{2\pi}{3}, \dfrac{3}{2}\right)$.

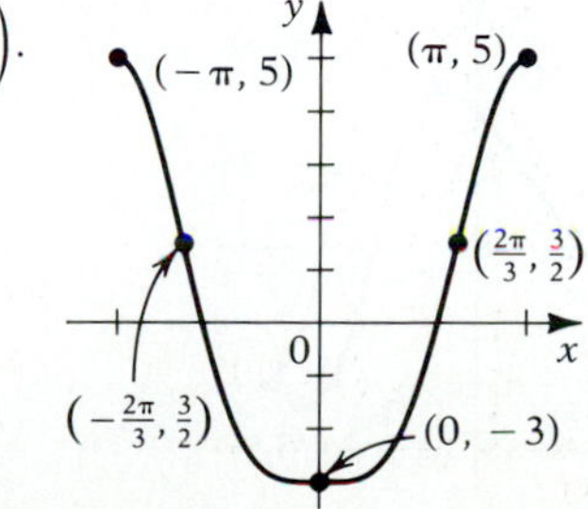

41. Local maximum points $(1, \frac{59}{12})$, $(2, \frac{13}{3})$, local minimum point $(-2, -\frac{19}{3})$.

43. Local minimum point $(-1, \frac{7}{2})$ **45.** Local minimum point $(3, 11)$

47. Increasing on $(-2, \infty)$, decreasing on $(-\infty, -2)$; local minimum point $(-2, -11)$; concave downward on $(-\frac{2}{3}, 2)$, concave upward on $(-\infty, -\frac{2}{3}) \cup (2, \infty)$; inflection points $(-\frac{2}{3}, -\frac{121}{27})$, $(2, 5)$.

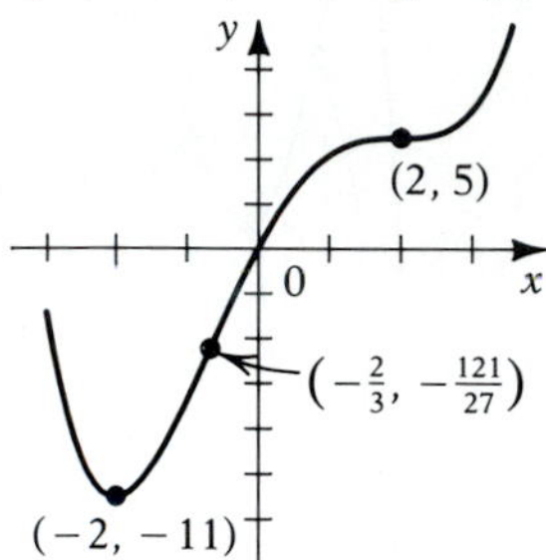

49. Increasing on $(-\infty, -\sqrt{2}) \cup (-1, 1) \cup (\sqrt{2}, \infty)$, decreasing on $(-\sqrt{2}, -1) \cup (1, \sqrt{2})$; local maximum points $(-\sqrt{2}, -4\sqrt{2} - 3)$, $(1, 3)$, local minimum points $(-1, -9)$, $(\sqrt{2}, 4\sqrt{2} - 3)$; concave downward on $(-\infty, -\sqrt{\frac{3}{2}}) \cup (0, \sqrt{\frac{3}{2}})$, concave upward on $(-\sqrt{\frac{3}{2}}, 0) \cup (\sqrt{\frac{3}{2}}, \infty)$; inflection points $(-\sqrt{\frac{3}{2}}, -\frac{19}{4}\sqrt{\frac{3}{2}} - 3)$, $(0, -3)$, and $(\sqrt{\frac{3}{2}}, \frac{19}{4}\sqrt{\frac{3}{2}} - 3)$.

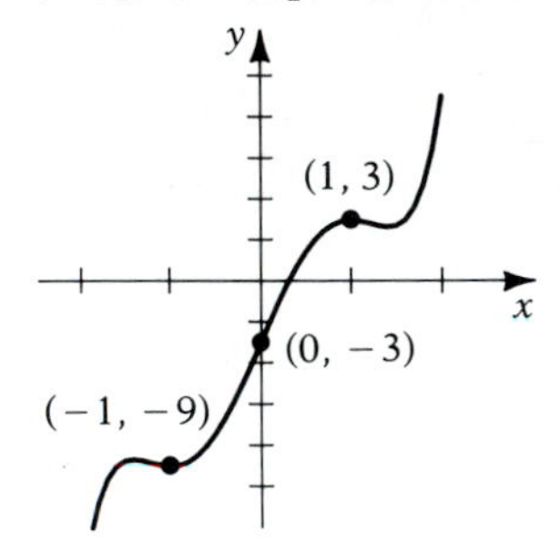

51. Increasing on $(-1, \infty)$, decreasing on $(-\infty, -1)$; local minimum $(-1, -15)$; concave downward on $(0, 2)$, concave upward on $(-\infty, 0) \cup (2, \infty)$; inflection points $(0, -4)$, $(2, 12)$.

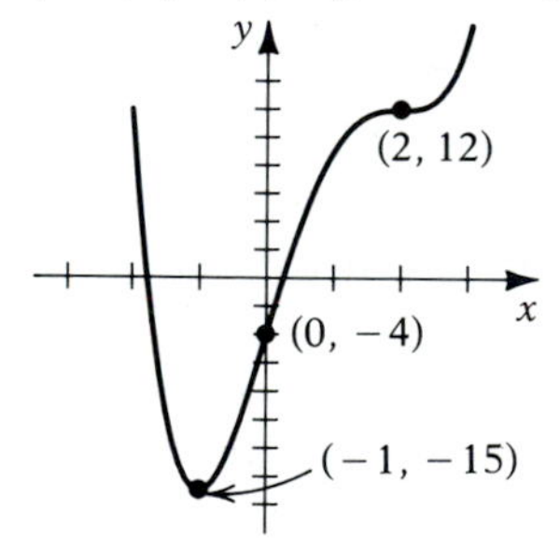

53. Increasing on $\left[-\pi, \frac{\pi}{6}\right) \cup \left(\frac{5\pi}{6}, \pi\right]$, decreasing on $\left(\frac{\pi}{6}, \frac{5\pi}{6}\right)$; local maximum point $\left(\frac{\pi}{6}, \sqrt{3}\right)$, local minimum point $\left(\frac{5\pi}{6}, -\sqrt{3}\right)$; concave downward on $\left(-\frac{\pi}{2}, \frac{\pi}{2}\right)$, concave upward on $\left[-\pi, -\frac{\pi}{2}\right) \cup \left(\frac{\pi}{2}, \pi\right]$; inflection points $\left(-\frac{\pi}{2}, 0\right)$, $\left(\frac{\pi}{2}, 0\right)$.

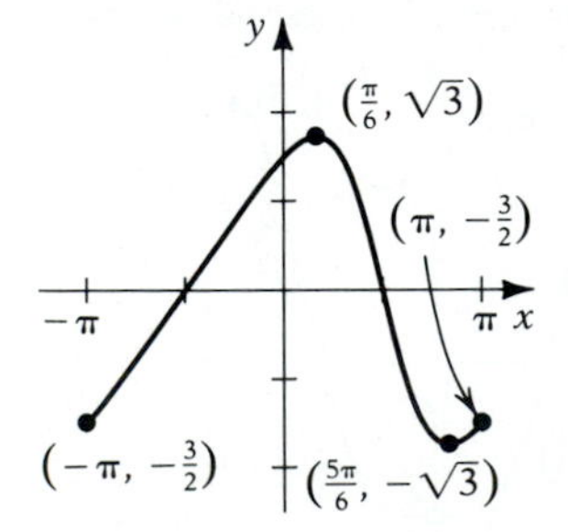

55. Increasing on $\left[-\frac{3\pi}{2}, \pi\right) \cup (0, \pi)$, decreasing on $(-\pi, 0) \cup \left(\pi, \frac{3\pi}{2}\right)$; local maximum points $(\pm\pi, 8)$, local minimum point $(0, 0)$; concave downward on $\left[-\frac{3\pi}{2}, -\frac{\pi}{2}\right) \cup \left(\frac{\pi}{2}, \frac{3\pi}{2}\right]$, concave upward on $\left(-\frac{\pi}{2}, \frac{\pi}{2}\right)$; inflection points $\left(\pm\frac{\pi}{2}, 3\right)$.

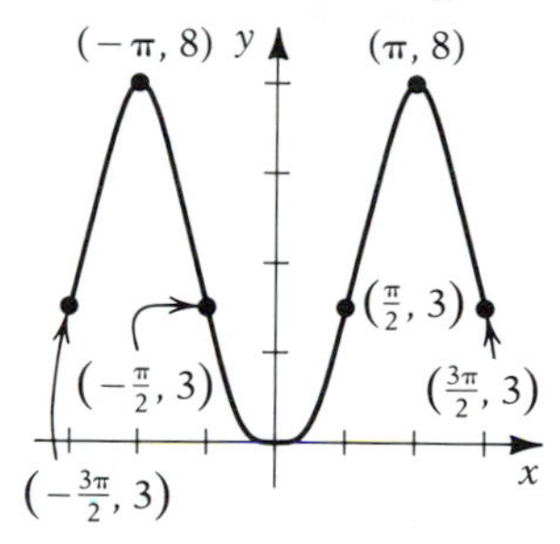

Exercise Set 3.3, page 121

1. (a) $-\infty$ (b) $+\infty$ **3.** (a) $+\infty$ (b) $-\infty$ **5.** (a) $-\infty$ (b) $-\infty$ **7.** 0 **9.** 1 **11.** 2
13. Does not exist **15.** vertical: $x = 3$; horizontal: $y = 1$ **17.** vertical: $x = \pm 3$; horizontal: $y = 1$
19. vertical: $x = -\frac{2}{3}$, $x = \frac{1}{2}$; horizontal: $y = 0$ **21.** vertical: $x = -3$; horizontal: none **23.** vertical: $x = -\frac{2}{3}$; horizontal: $y = \frac{1}{3}$
25. vertical: none; horizontal: $y = 0$ **27.** vertical: $x = 0$; horizontal: $y = 0$
29. vertical: $x = \frac{\pi}{2} + 2n\pi$, $x = -\frac{\pi}{6} + 2n\pi$, $x = -\frac{5\pi}{6} + 2n\pi$, $n = 0, \pm 1, \pm 2, \ldots$; horizontal: none

31. *Intercepts:* $(1, 0)$, $(0, -\frac{1}{2})$
Symmetry: none
Asymptotes: $x = -2$, $y = 1$
Extrema: none
Inflection points: none

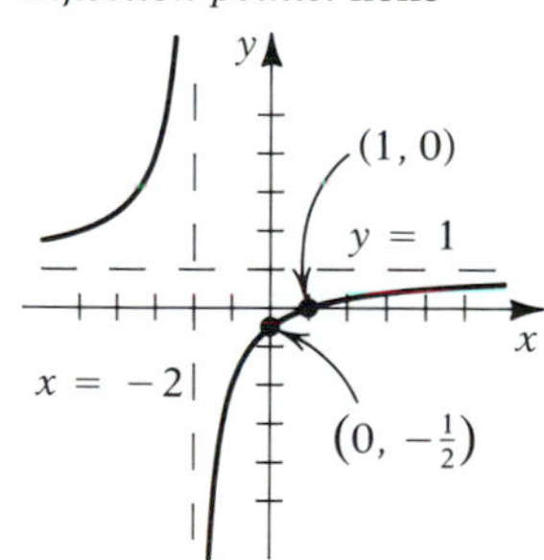

33. *Intercepts:* $(0, 2)$
Symmetry: y-axis
Asymptotes: $y = 0$
Extrema: local maximum point $(0, 2)$
Inflection points: $\left(\pm\frac{1}{\sqrt{3}}, \frac{3}{2}\right)$

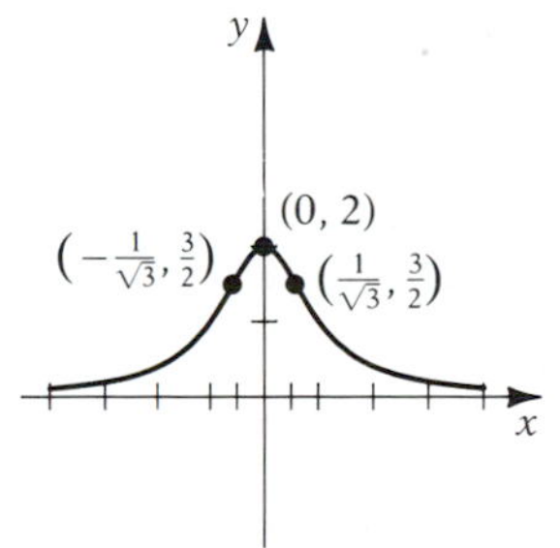

35. *Intercepts:* $(0, 0)$
Symmetry: y-axis
Asymptotes: $x = \pm 2$, $y = 1$
Extrema: local maximum point $(0, 0)$
Inflection points: none

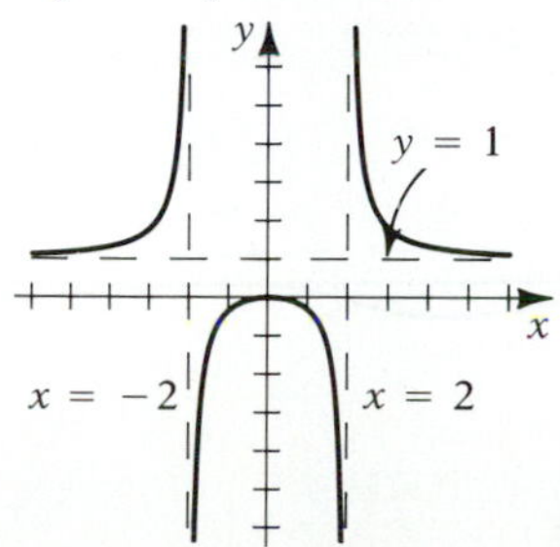

37. *Intercepts:* $(0, 0)$
Symmetry: y-axis
Asymptotes: $y = 1$
Extrema: local minimum point $(0, 0)$
Inflection points: $\left(\pm\frac{2}{\sqrt{3}}, \frac{1}{4}\right)$

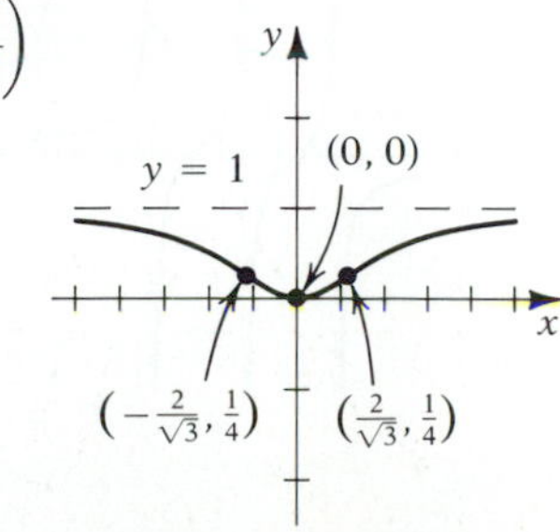

39. *Intercepts:* (0, 0)
Symmetry: origin
Asymptotes: $x = \pm 1$, $y = 0$
Extrema: none
Inflection points: (0, 0)

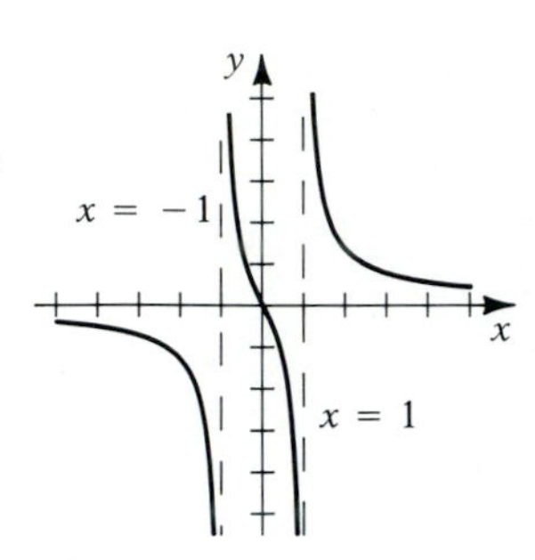

41. Cusp, local minimum

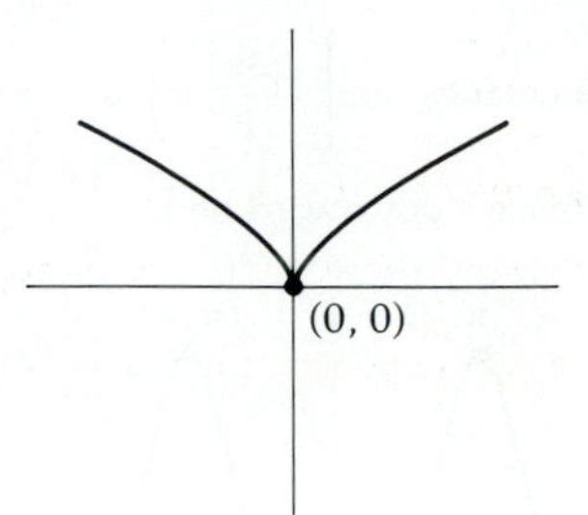

43. Inflection point

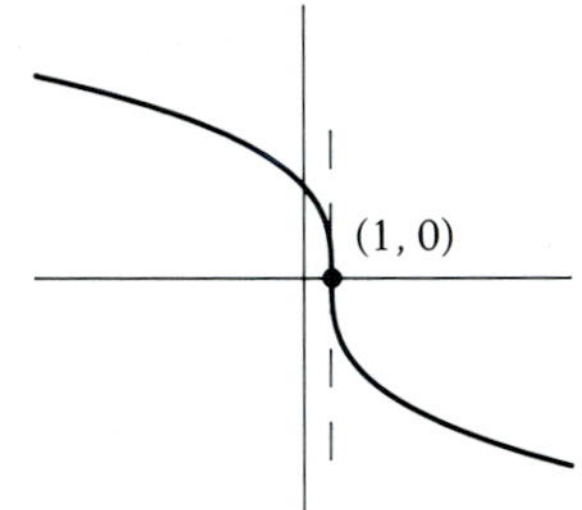

45. Cusp, local maximum

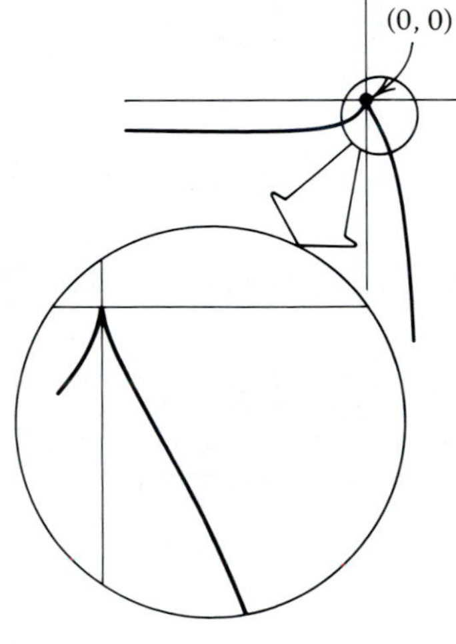

47. End point maximum

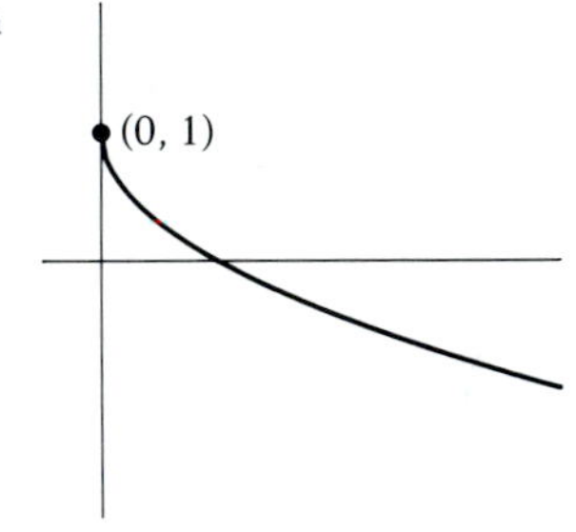

49. *Intercepts:* (0, 3), (−4.9, 0) approx.
Asymptotes: none
Symmetry: none
Extrema: local minimum (cusp) at (1, 2), local maximum at (0, 3)
Inflection points: none, concave down always

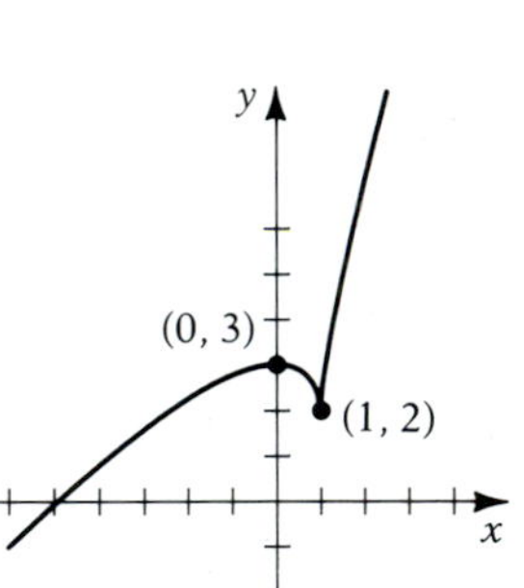

51. *Intercepts:* $(\pm 1, 0)$
Symmetry: origin
Asymptotes: $x = 0$, $x = \pm\sqrt{3}$, $y = 0$
Extrema: none
Inflection points: (By calculator)
(0.871, 0.124)
(−0.871, −0.124)

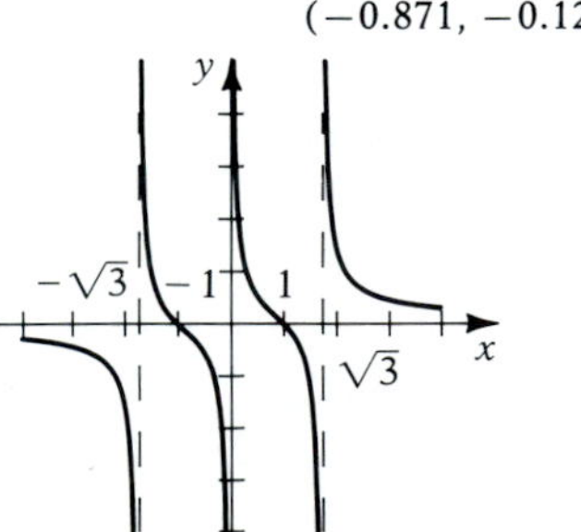

53. *Intercepts:* $\left(\frac{\pi}{2} + 2n\pi, 0\right)$, (0, 1)
Symmetry: none
Asymptotes: $x \approx 0.74$, $y = 0$
Extrema: local maxima at $x = \pm\frac{\pi}{2}, \pm\frac{5\pi}{2}, \pm\frac{9\pi}{2}, \ldots$

local minima at $x = 0, \pm\frac{3\pi}{2}, \pm\frac{7\pi}{2}, \ldots$

Inflection points: (Delete)

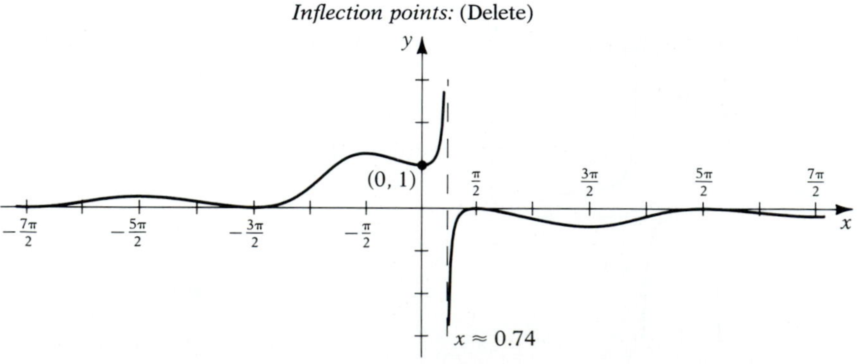

55. $y = x - 1$ **57.** $y = x$

59. *Intercepts:* $(-1, 0)$
Symmetry: none
Asymptotes: $x = 0$, $y = x$
Extrema: local minimum point $\left(\sqrt[3]{2}, \dfrac{3}{\sqrt[3]{4}}\right)$
Inflection points: none, always concave up

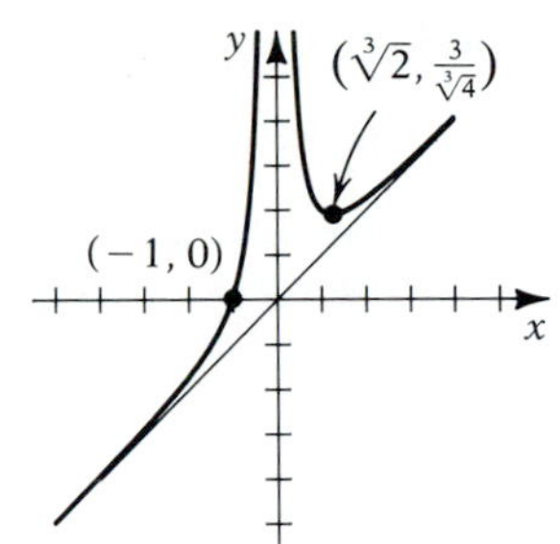

61. For arbitrary $\varepsilon > 0$, let $N = \left(\dfrac{1}{\varepsilon}\right)^{1/p}$. Show that if $x > N$ or if $x < -N$, $\left|\dfrac{1}{x^p}\right| < \varepsilon$ (assuming x^p is defined for $x < 0$).

Exercise Set 3.4, page 131

1. 12.5 ft by 25 ft **3.** 100 m by 120 m **5.** 3 ft by 2 ft by 1.5 ft **7.** $x = 2$, $y = 3$ **11.** $r = \dfrac{a\sqrt{6}}{3}$, $h = \dfrac{2a\sqrt{3}}{3}$ **13.** 45

15. 210 **17.** 70 **19.** 10 **21.** $\sqrt{3}$ ft by $\sqrt{6}$ ft **23.** 10 in. wide, 12 in. high **25.** 12.5 in. wide, 10 in. high

27. $r = \dfrac{2a\sqrt{2}}{3}$, $h = \dfrac{4a}{3}$ **29.** $x = 8\sqrt{5}$ **31.** rectangle width $= \dfrac{24}{\pi + 4}$, height $= \dfrac{12}{\pi + 4}$

33. 24 min; first car is 32 km from A, second car is 24 km from B, and they are 30 km apart.

35. $(3^{2/3} + 5^{2/3})^{3/2}$ ft ≈ 11.19 ft **37.** (a) 4 in. from B (b) 4.5 in. from B

Exercise Set 3.5, page 138

1. 5 m/sec **3.** 1 ft/sec **5.** 10π m²/sec **7.** $\dfrac{2000\pi}{3}$ m/sec **9.** $\dfrac{48}{125}$ m/sec **11.** $\dfrac{2}{5}$ ft/min **13.** $\dfrac{5}{2}$ m/sec

15. $-\dfrac{2}{13}$ rad/min **17.** $\dfrac{3}{4}$ cm/sec **19.** 20 N/sec **21.** $72{,}000\pi$ cm³/min **23.** $\dfrac{17}{4}$ m/sec **25.** $\dfrac{63\pi}{10}$ mm³/day

27. 10 cm/sec **29.** $\dfrac{8}{3}$ cm/min **31.** approx. 15.48 cm/min

Exercise Set 3.6, page 146

1. $c = \dfrac{3}{2}$ **3.** $c = \dfrac{3}{4}$ **5.** $c = \sqrt{3}$ **7.** $c = 1$ **9.** $c = \dfrac{3}{2}$ **11.** $c = \dfrac{7}{3}$ **13.** $c = \dfrac{2}{3}$ **15.** $c = \dfrac{3}{2}$

17. $f'(0)$ does not exist **19.** f is discontinuous at $\dfrac{\pi}{2}$ **21.** f is discontinuous at 0 **23.** $f'(1)$ does not exist

25. $f'(\frac{1}{3})$ does not exist, but $c = \frac{2}{3}$ satisfies conclusion. Theorem gives sufficient conditions only.

27. $\left(-\dfrac{1}{3}, \dfrac{37}{27}\right)$, $(2, -9)$

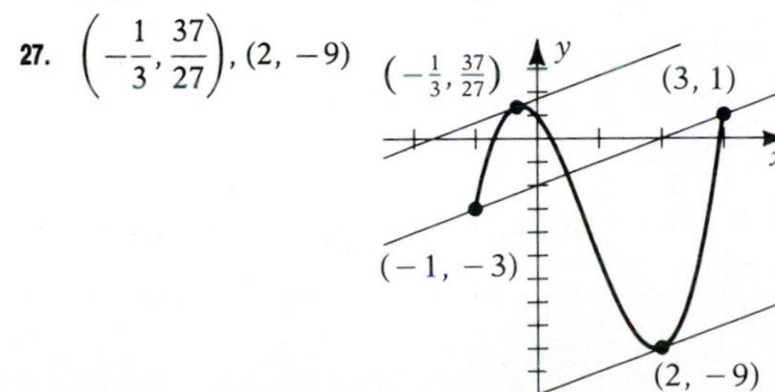

29. In the result of Exercise 28 let $h = x - a$ and $\theta = (c - a)/h$.

31. For any $x \in \left(0, \frac{\pi}{2}\right)$, there is a number $c \in (0, x)$ such that $(\tan x)/x = \sec^2 c > 1$.

35. Show that if $f'(a) > 0$ and $f'(b) < 0$, then f attains its maximum value at a number $c \in (a, b)$, and use Theorem 3.3. Use the fact that f is increasing in a right-neighborhood of a and decreasing in a left-neighborhood of b. Similarly, show f attains its minimum value in (a, b) if $f'(a) < 0$ and $f'(b) > 0$.

Exercise Set 3.7, page 152

1. $\frac{x^6}{6} + C$ **3.** $\frac{2x^3}{3} - 2x^2 + 5x + C$ **5.** $\frac{3x^4}{4} - \frac{5x^3}{3} + x^2 - 3x + C$ **7.** $-\frac{1}{2}\left(\frac{1}{x} + \frac{x^3}{3}\right)$ **9.** $\frac{4}{3}x^{3/2} - 6x^{1/2}$

11. $\frac{3x^2}{2} - \frac{8}{3}x^{3/2} + 7x + C$ **13.** $\frac{2}{5}x^{5/2} + \frac{3}{2}x^2 + C$ **15.** $-2\cos x - 3\sin x + C$ **17.** $\frac{x^2}{2} + \cot x + C$ **19.** $x + \csc x + C$

21. $f(x) = x^2 - 3x + 5$ **23.** $f(x) = x - \frac{x^2}{2} - \frac{4x^3}{3} + \frac{79}{2}$ **25.** $f(x) = \frac{x^3}{2} + 2$ **27.** $f(x) = \frac{x^4}{12} - \frac{x^2}{2} - \frac{2x}{3} + 2$

29. $y = \frac{x^2}{2} - \cos x + 4$ **31.** Rises 64 ft; hits ground when $t = 4$ sec; 64 ft/sec.

33. Rises 19.6 m above top of cliff; hits bottom when $t = 6$ sec. **35.** $\frac{x^3}{3} - x^2 - 4x + C,\ x \neq -2$

37. $\frac{2}{3}x^{3/2} - 2x^{1/2} + C$ **39.** $-\cot x + C$ **41.** $-\cot x - \tan x + C$ **43.** $s(0) = 0;\ s(3) = \frac{135}{4}$

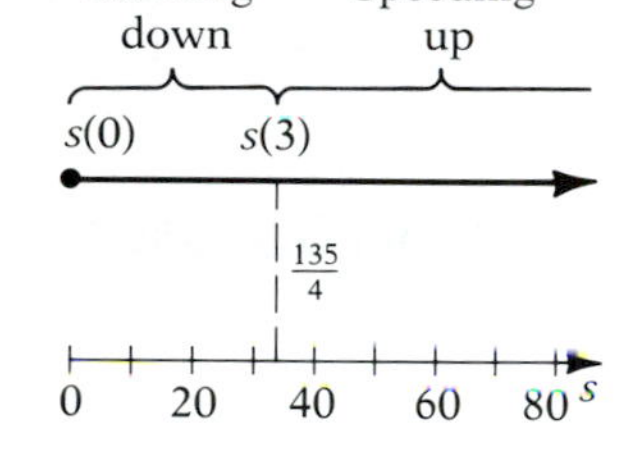

45. $s(0) = 2;\ s\left(\frac{2}{3}\right) = \frac{230}{81} \approx 2.84;\ s\left(\frac{3}{2}\right) = \frac{245}{64} \approx 3.83;\ s(3) = -\frac{1}{4};\ s(4) = -\frac{10}{3}$

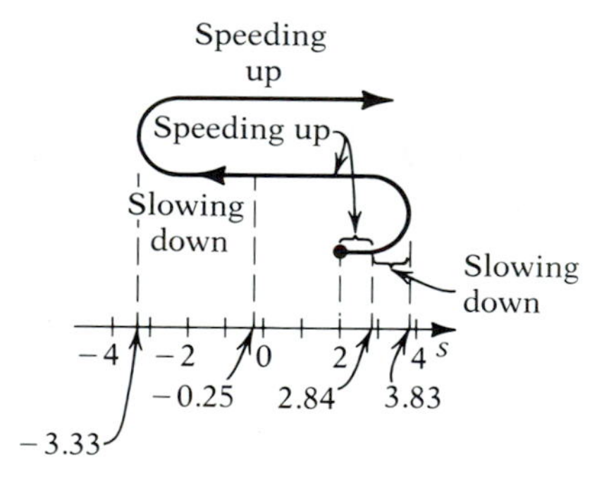

Supplementary Exercises 3.8, page 153

1. (a) Increasing on $(-1, 1) \cup (3, \infty)$; decreasing on $(-\infty, -1) \cup (1, 3)$. (b) Increasing on $(-\infty, -\frac{1}{2}) \cup (\frac{1}{2}, \infty)$.

3. Local maximum value $f(2) = 13$; local minimum value $f(-\frac{2}{3}) = \frac{95}{27}$.

5. Local maximum value $f(0) = 0$; local minimum value $f(2) = 4$.

7. Absolute maximum value $f(-2) = 13$; local and absolute minimum value $f(2) = -19$.

9. Local maximum value $f\left(\frac{\pi}{12}\right) = \frac{\sqrt{3}}{2} + \frac{\pi}{12} \approx 1.13$; local minimum value $f\left(\frac{5\pi}{12}\right) = -\frac{\sqrt{3}}{2} + \frac{5\pi}{12} \approx 0.44$; absolute maximum value $f\left(\frac{\pi}{2}\right) = \frac{\pi}{2} \approx 1.57$; absolute minimum value $f\left(-\frac{\pi}{2}\right) = -\frac{\pi}{2} \approx -1.57$.

11. (a) Concave upward on $(-\infty, -\frac{1}{2}) \cup (1, \infty)$; concave downward on $(-\frac{1}{2}, 1)$.
(b) Concave upward on $(-1, 1) \cup (3, \infty)$; concave downward on $(-\infty, -1) \cup (1, 3)$.

13. (a) $x = -1$, $y = 3$ (b) $x = \dfrac{5}{2}$, $x = -1$, $y = 0$

15. (a) $x = 0$, $x = (2n-1)\pi$ $(n = 0, \pm 1, \pm 2, \ldots)$, $y = 0$ (b) $x = 0$, $x = \dfrac{2n+1}{2}\pi$ $(n = 0, \pm 1, \pm 2, \ldots)$, $y = 0$

17. *Intercepts:* $(\pm 3.7, 0)$, $(\pm 2.1, 0)$ approx., $(0, 60)$
Symmetry: y-axis
Asymptotes: none
Extrema: local maximum point $(0, 60)$
local minimum points $(\pm 3, -21)$
Inflection points: $(\pm\sqrt{3}, 15)$

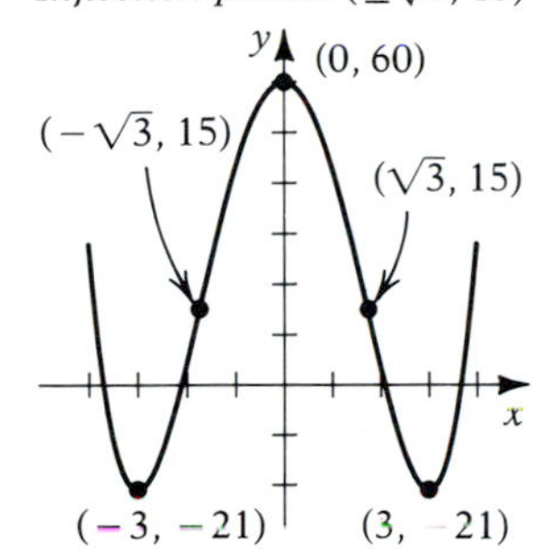

19. *Intercepts:* $(0, 0)$
Symmetry: origin
Asymptotes: $y = 0$
Extrema: local maximum point $(2, 2)$
local minimum point $(-2, -2)$
Inflection points: $(-2\sqrt{3}, -\sqrt{3})$, $(0, 0)$, $(2\sqrt{3}, \sqrt{3})$

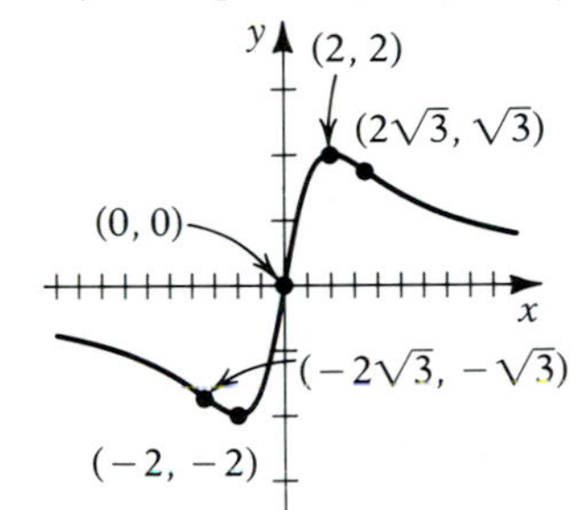

21. *Intercepts:* $(0, 15)$, x-intercepts between 1 and 2 and between 3 and 4
Symmetry: none
Asymptotes: none
Extrema: local minimum point $(3, -12)$
Inflection points: $(2, -1)$, $(0, 15)$
horizontal tangent at $(0, 15)$

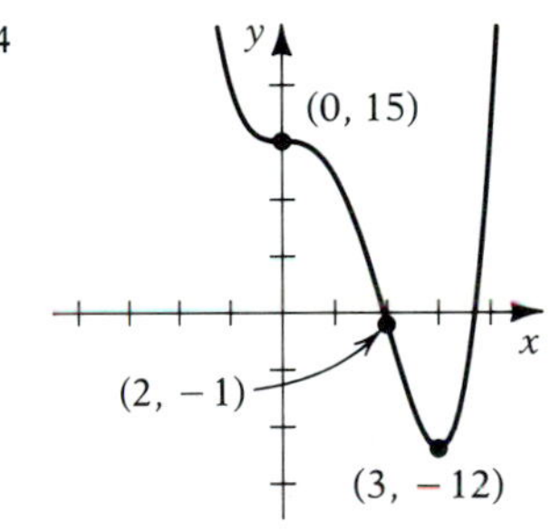

23. *Intercepts:* $(2, 0)$, $(-1, 0)$, $(0, 2)$
Symmetry: none
Asymptotes: $x = 1$, $y = x$
Extrema: none
Inflection points: none

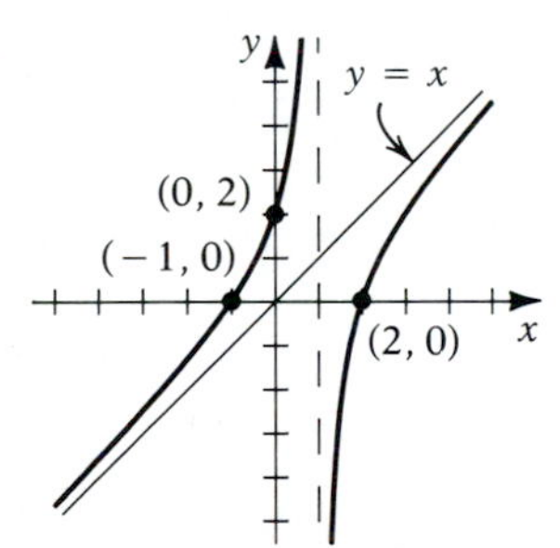

25. *Intercepts:* $(0, -1)$, x-intercepts between 0 and $\frac{1}{3}$, $\frac{1}{3}$ and 1, 8 and 9
Symmetry: none
Asymptotes: none
Extrema: local minimum point $(3, -1)$
local maximum point $(\frac{1}{3}, \frac{1}{3})$ (cusp)
Inflection points: none

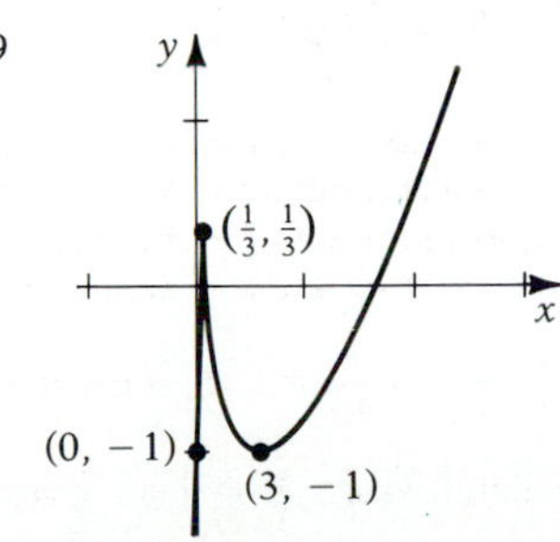

27. *Intercepts:* (0, 0)
Symmetry: none
Asymptotes: $x = 1$, $y = 0$
Extrema: local minimum point $(-1, -1)$
Inflection point: $(-2, -\frac{8}{9})$

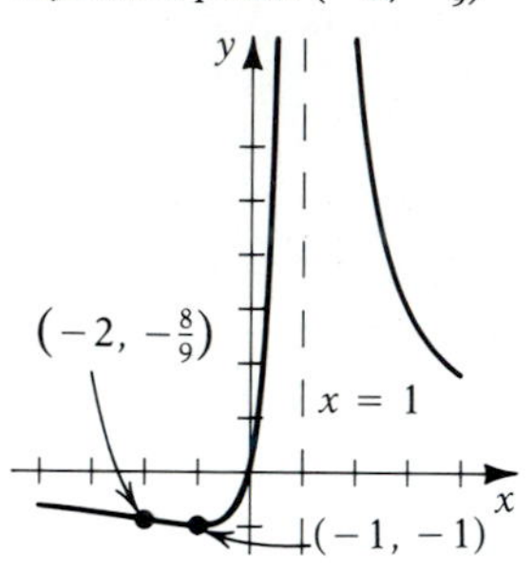

29. *Intercepts:* $\left(\pm\frac{1}{\sqrt{3}}, 0\right)$
Symmetry: origin
Asymptotes: $x = 0$, $y = 0$
Extrema: local maximum point (1, 2)
local minimum point $(-1, -2)$
Inflection points: $\left(\sqrt{2}, \frac{5}{2\sqrt{2}}\right), \left(-\sqrt{2}, -\frac{5}{2\sqrt{2}}\right)$

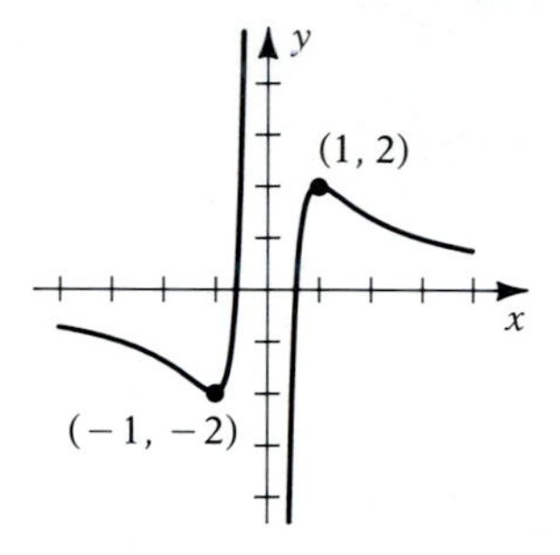

31. *Intercepts:* $(0, -1)$
Symmetry: y-axis
Asymptotes: $x = \pm\frac{\pi}{4}$ in $\left[-\frac{\pi}{2}, \frac{\pi}{2}\right]$
Extrema: local maximum points $(0, -1)$ in $\left[-\frac{\pi}{2}, \frac{\pi}{2}\right]$
local minimum points $\left(\pm\frac{\pi}{2}, 3\right)$ in $\left[-\frac{\pi}{2}, \frac{\pi}{2}\right]$
Inflection points: (Delete)
Periodicity: f is periodic with period π

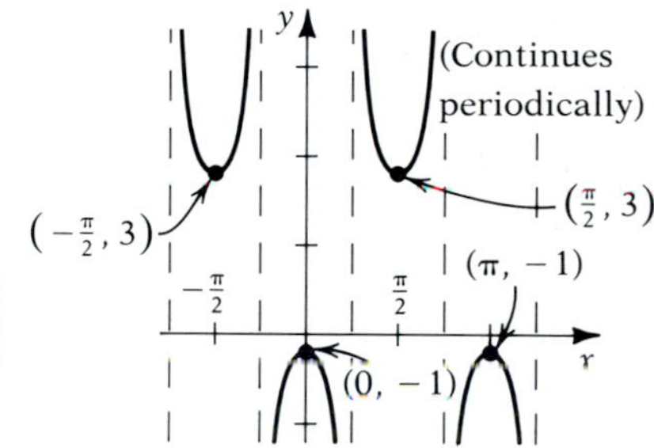

33. 6 ft by 6 ft by 9 ft **35.** 18 in. by 18 in. by 36 in. **37.** base radius = 5 in.; height = 10 in. **39.** $x = \frac{m}{2}$ **41.** $5\sqrt{5}$ ft

43. 17 mi/hr **45.** (a) $\frac{2}{\pi}$ ft/min (b) $\frac{1}{\pi\sqrt[3]{2}}$ ft/min **47.** (a) $c = 1$ (b) $c = 0$ **49.** (a) $c = \frac{1}{2}$ (b) $c = \frac{2 + \sqrt{19}}{3}$

51. (a) 7, because $f(x) = C$ and $f(2) = 7$. So $C = 7$. (b) $x^2 + 1$, because $f(x) = g(x) + C$, so $g(x) = x^2 + 4 - C$. From $g(1) = 2$, obtain $C = 3$.

53. (a) $-4 \cos x - 3 \sin x + C$ (b) $\tan x + 2 \sec x + C$ **55.** (a) $f(x) = x^3 + \frac{2}{x} + 1$ (b) $f(x) = -2 \cos x - \frac{x^2}{2} + 7$

57. (a) $f(x) = -\frac{1}{6x^3} - \frac{x^2}{2} - \frac{x}{2} + \frac{11}{6}$ (b) $f(x) = \cos x + x^2 - x + 1$

59. (a) $v(t) = 3t^2 - 12$, $s(t) = t^3 - 12t$
(b) 16 units to the left of the origin, when $t = 2$
(c) Slowing down from $t = 0$ to $t = 2$, speeding up for $t > 2$
(d)

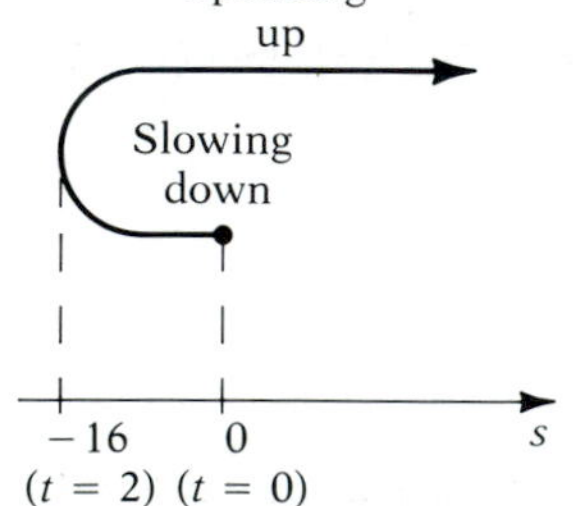

CHAPTER 4

Exercise Set 4.2, page 163

1. $1+\frac{1}{4}+\frac{1}{9}+\frac{1}{16}$ **3.** $1+\frac{2}{3}+\frac{3}{5}+\frac{4}{7}$ **5.** $\sqrt{n}$ **7.** $\frac{1}{2}-\frac{2}{5}+\frac{3}{10}-\cdots+\frac{(-1)^{m-1}m}{m^2+1}$

9. $\sin x+\frac{\sin 2x}{2}+\frac{\sin 3x}{3}+\cdots+\frac{\sin nx}{n}$ **11.** $\sum_{n=1}^{5}\frac{1}{n}$ **13.** $\sum_{n=1}^{4}\frac{1}{n(n+1)}$ **15.** $\sum_{k=1}^{n}\frac{2^k-1}{3^k}$ **17.** $\sum_{n=1}^{5}\frac{n+3}{2^{n+3}}$ or $\sum_{n=4}^{8}\frac{n}{2^n}$

19. $\sum_{n=1}^{5}(-1)^{n-1}\frac{2n}{2n+1}$ **21.** $\frac{n(n+1)(2n-5)}{6}$ **23.** $\frac{n(n+1)(n-1)(n+2)}{4}$ **27.** $L_5=\frac{12}{25}$, $U_5=\frac{22}{25}$

29. $L_5=\frac{1627}{1260}$, $U_5=\frac{1879}{1260}$ **31.** $L_n=4-\frac{2(n+1)}{n}$, $U_n=4-\frac{2(n-1)}{n}$; $\lim_{n\to\infty}L_n=\lim_{n\to\infty}U_n=2$

Figure is a triangle with base 2 and altitude 2. So area $=\frac{1}{2}bh=2$.

33. $L_n=1-\frac{2n^2+3n+1}{6n^2}$; area $=\frac{2}{3}$

35. (a) $\sum_{k=1}^{n}[f(k)-f(k-1)]=[f(1)-f(0)]+[f(2)-f(1)]+[f(3)-f(2)]+\cdots+[f(n)-f(n-1)]=f(n)-f(0)$

(b) $\sum_{k=1}^{n}[k^p-(k-1)^p]=n^p-0^p=n^p$, by part (a)

37. (b) From $\sum_{k=1}^{n}(4k^3-6k^2+4k-1)=\sum_{k=1}^{n}[k^4-(k-1)^4]=n^4$, obtain $\sum_{k=1}^{n}k^3=\frac{1}{4}\left[n^4+6\sum_{k=1}^{n}k^2-4\sum_{k=1}^{n}k+\sum_{k=1}^{n}1\right]$, and use equations (4.2) and (4.3) and property 1 to get the result.

39. $\frac{32}{3}$ **41.** 9

Exercise Set 4.3, page 175

1. -19 **3.** -5.0 **5.** 6 **7.** $-\frac{3}{2}$ **9.** 27 **11.** (a) 100; properties 2 and 4 (b) $\frac{20}{3}$; properties 2 and 4

13. $x(2-x)\geq 0$ on $[0, 2]$; property 5. **15.** $6x+x^2-x^3=x(3-x)(2+x)\geq 0$ on $[0, 3]$; property 5.

17. $x^2+4\geq 4x$ since $x^2-4x+4=(x-2)^2\geq 0$; property 6.

19. Absolute maximum of x^2-2x+3 on $[0, 3]$ is 6 and absolute minimum is 2; property 8.

21. Absolute maximum of $x/(x^2+1)$ on $[-2, 2]$ is $\frac{1}{2}$, and absolute minimum is $-\frac{1}{2}$; property 8. **23.** $\frac{27}{2}$ **25.** $\frac{9}{4}$

27. Absolute maximum and minimum values of $(x-2)/(x+2)$ on $[-1, 1]$ are $-\frac{1}{3}$ and -3, respectively. So by property 8,

$$-6\leq\int_{-1}^{1}\frac{x-2}{x+2}\,dx\leq-\frac{2}{3}<6.$$

Result now follows.

29. Absolute maximum and minimum values of $\sin x-\cos^2 x+2$ on $[-\pi, \pi]$ are 3 and $\frac{3}{4}$, respectively. Now use property 8.

31. $\frac{46}{3}$ **33.** $f_+(x)-f_-(x)=\begin{Bmatrix}f(x) & \text{if } x\geq 0\\ f(x) & \text{if } x<0\end{Bmatrix}=f(x)$; $f_+(x)+f_-(x)=\begin{Bmatrix}f(x) & \text{if } x\geq 0\\ -f(x) & \text{if } x<0\end{Bmatrix}=|f(x)|$

$$\left|\int_a^b f(x)\,dx\right|=\left|\int_a^b [f_+(x)-f_-(x)]\,dx\right|\leq\left|\int_a^b f_+(x)\,dx\right|+\left|\int_a^b f_-(x)\,dx\right|$$

$$=\int_a^b [f_+(x)+f_-(x)]\,dx=\int_a^b |f(x)|\,dx$$

35. For any partition P, if x_k^* is rational $\sum_{k=1}^{n}f(x_k^*)\,\Delta x_k=1$ and if x_k^* is irrational $\sum_{k=1}^{n}f(x_k^*)\,\Delta x_k=0$. So limit does not exist.

Exercise Set 4.4, page 181

1. $\frac{44}{3}$ **3.** 9 **5.** 16 **7.** $\frac{17}{6}$ **9.** $\frac{84}{5}$ **11.** -2 **13.** 4 **15.** $\sqrt{3}$ **17.** 2 **19.** $\frac{961}{15}$ **21.** $\frac{48}{5}$

23. $\frac{\pi^2}{3}+\sqrt{3}$ **25.** 1 **27.** $\sqrt{3}-1$ **29.** $\frac{28}{3}$ **31.** -1 **33.** $-\frac{15}{4}$ **35.** $c=4$ **37.** $c=\sqrt{3}$ **39.** $\frac{2}{\sqrt{1+x^2}}, x\in\mathbf{R}$

41. $\sqrt{x^3+8}, x\geq -2$ **43.** x^3-3x^2+1 **45.** $-\frac{2}{3}$ **47.** $-\frac{\sin x}{x}$ **51.** $\sqrt{3}$

53. $\frac{3-\sqrt{3}}{2}$ *Hint:* Use the fact that $1=\sin^2 x+\cos^2 x$.

57. Let $M>0$ be such that $|f(x)|\leq M$ on $[a, b]$. For $\varepsilon>0$, let $\delta=\varepsilon/M$. Then if $|x-x_0|<\delta$, $|F(x)-F(x_0)|=\left|\int_{x_0}^x f(t)\,dt\right|\leq M\left|\int_{x_0}^x dt\right|=M|x-x_0|<\varepsilon$

59. $\lim_{x\to b^-}\int_a^x f(t)\,dt=\lim_{x\to b^-}\int_a^x g(t)\,dt=\int_a^b g(t)\,d(t)$, and by Theorem 4.3 f is integrable on $[a, b]$. So by Exercise 57, $\int_a^x f(t)\,dt$ is continuous and thus $\lim_{x\to b^-}\int_a^x f(t)\,dt=\int_a^b f(t)\,dt$. Conclusion now follows.

61. Follow similar argument to that used in Exercise 59.

Exercise Set 4.5, page 187

1. $\frac{(1+x^2)^4}{8}+C$ **3.** $\frac{(x^3-1)^6}{18}+C$ **5.** $\frac{(3x^2+4)^{3/2}}{9}+C$ **7.** $\frac{2}{9}(1+3x)^{3/2}+C$ **9.** $\frac{1}{3}(x^2-2x+3)^{3/2}+C$

11. $-\frac{1}{2}\cos 2x+C$ **13.** $-\frac{1}{k}\cos kx+C$ **15.** $-\frac{1}{2}\cos 2x+C$ **17.** $\frac{1}{2}\tan x^2+C$ **19.** $\frac{2}{3}(\tan x)^{3/2}+C$

21. $-\frac{1}{1-\cos x}+C$ **23.** $-2\cot\sqrt{x}+C$ **25.** $\frac{1}{6}\left(\frac{x}{6}-\frac{1}{x}\right)^6+C$ **27.** $\frac{2}{5}(x^{3/2}-1)^{5/3}+C$ **29.** $\frac{121}{5}$ **31.** $\frac{33}{10}$

33. $\frac{254}{7}$ **35.** 60 **37.** $\frac{1}{3}$ **39.** $\frac{1}{8}$ **41.** -1 **43.** $\frac{7}{3}$ **45.** Use definition of $\int f(x)\,dx$ and properties of derivatives.

47. $-\frac{1}{\tan x+1}+C$ **49.** $-4\sqrt{1-\sqrt{x}}+C$ **51.** $\sin x-\frac{2}{3}\sin^3 x+C$ **53.** $\frac{d}{dx}\left[-2\left(\frac{1}{\sqrt{x}}\sin\frac{1}{\sqrt{x}}+\cos\frac{1}{\sqrt{x}}\right)\right]=\frac{1}{x^2}\cos\frac{1}{\sqrt{x}}$

Supplementary Exercises 4.6, page 188

1. (a) $\frac{1}{3}n(2+n)(5-4n)$ (b) $\frac{1}{4}(n)(1+n)(2+n)(1-n)$ **3.** $\sum_{k=1}^{4} f(x_k^*)\Delta x_k=3.06$; $\int_{-1}^{4} f(x)\,dx=\frac{25}{12}\approx 2.08$

5. (a) Absolute maximum of $(x+1)/(x-2)$ on $[-1, 1]$ is 0 and absolute minimum is -2, so $\left|\frac{x+1}{x-2}\right|\leq 2$. Now use integral properties 7 and 8.

(b) Absolute maximum and minimum of integrand on $\left[0, \frac{\pi}{4}\right]$ are $\sqrt{5}$ and 2, respectively. So

$2\leq\frac{\tan x+2}{\sec x}\leq\sqrt{5}<3.$

Result follows by integral property 8.

7. $-\frac{94}{3}$ **9.** $\frac{691}{20}$ **11.** $-\frac{57}{4}$ **13.** $\sqrt{3}$ **15.** 21 **17.** $\frac{3}{16}$ **19.** $-\frac{1}{2}\tan\frac{2}{x}+C$ **21.** $\sqrt{x^2-2x+4}+C$

23. $\frac{4}{3}(1+\sqrt{x})^{3/2}+C$ **25.** $2\csc\theta+C$ **27.** Show that $\frac{d}{dx}\left\{\frac{1}{27}[(9x^2-2)\sin 3x+6x\cos 3x]\right\}=x^2\cos 3x$ **29.** $\frac{t}{\sin^2 t+1}$

31. $-\frac{10}{3}$ **33.** $c=\frac{32}{25}$

CHAPTER 5

Exercise Set 5.1, page 196

1. Area $= \frac{14}{3}$

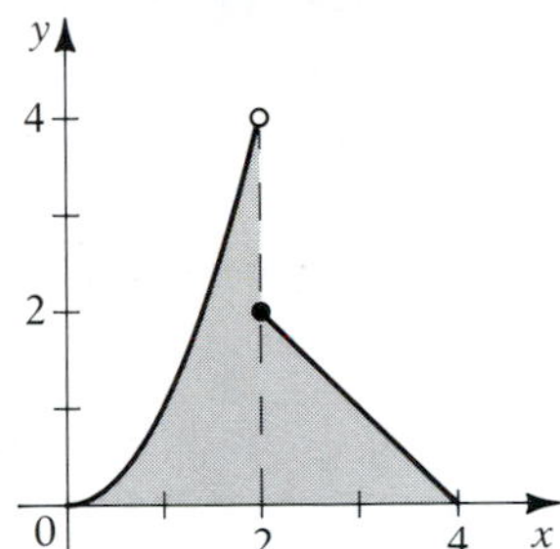

3. Area $= \frac{22}{3}$

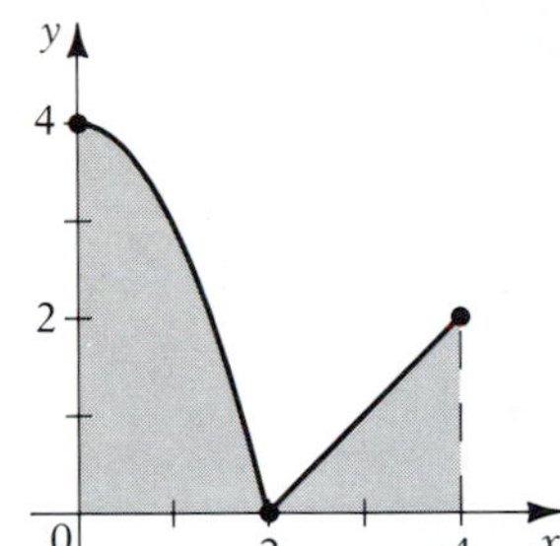

5. Area $= \frac{20}{3}$

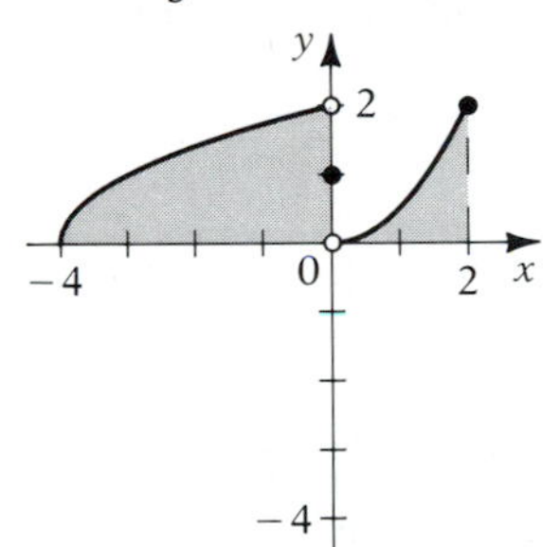

7. Area $= 3$

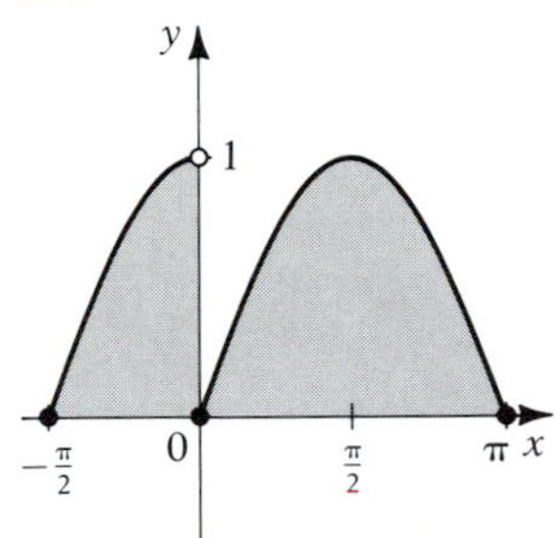

9. Area $= \frac{8\pi}{3}$

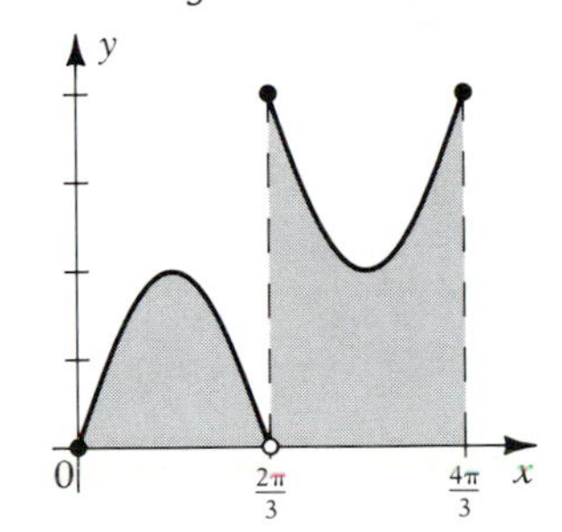

11. Area $= \frac{5}{6}$

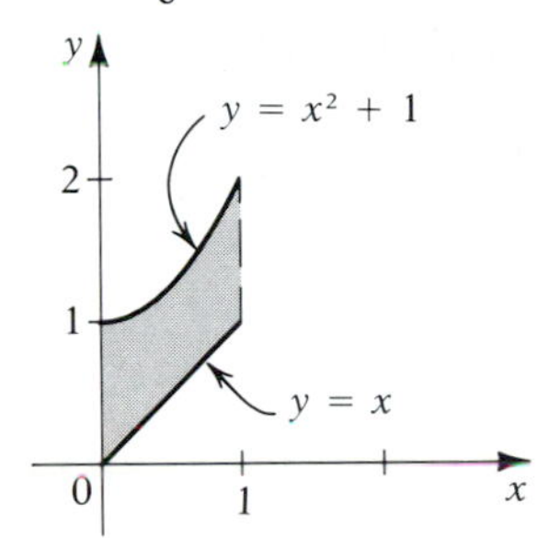

13. Area $= \frac{80}{3}$

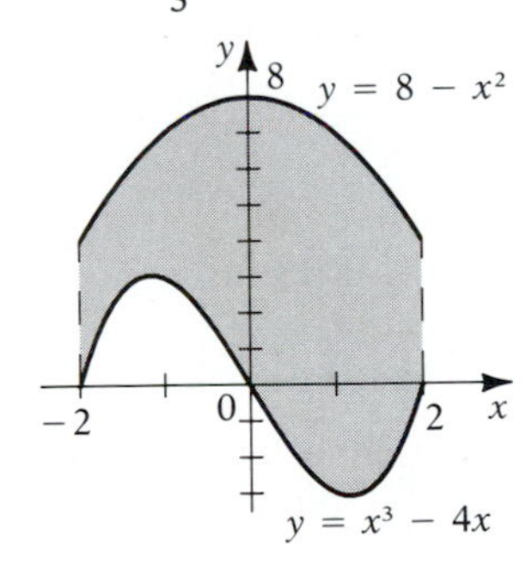

15. Area $= \frac{9}{2}$

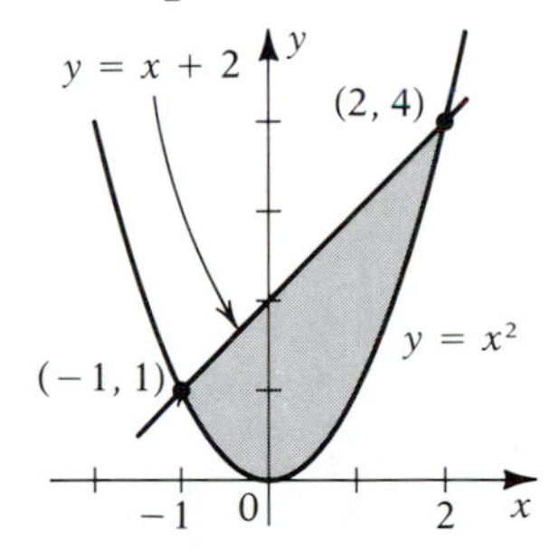

17. Area $= \frac{4}{3}$

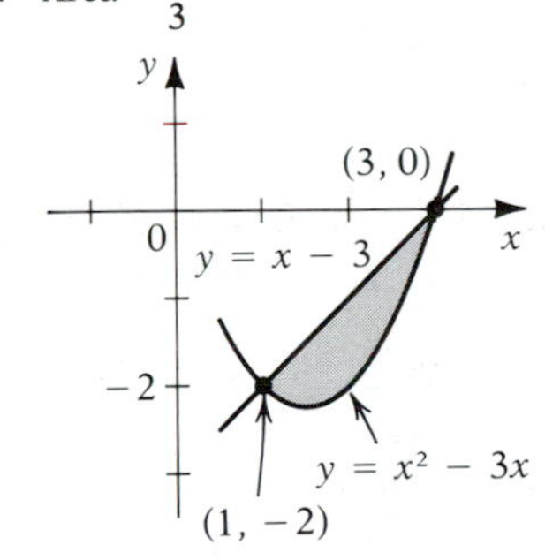

19. Area $= \frac{1}{3}$

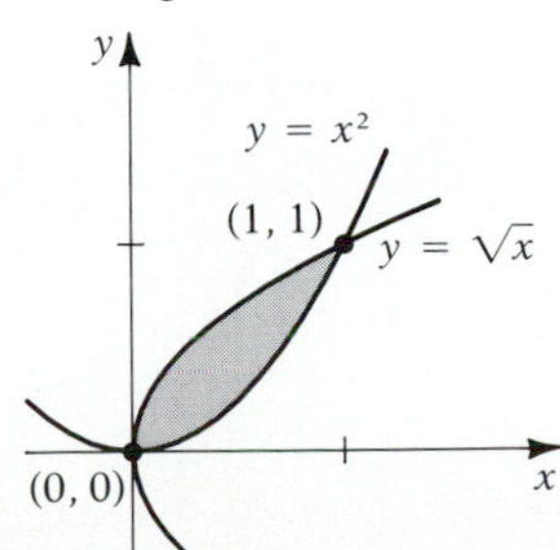

21. Area $= 8$

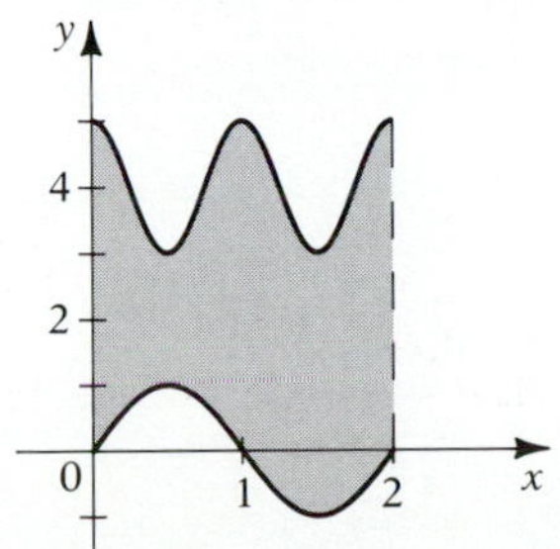

23. Area $= \frac{9}{4}$

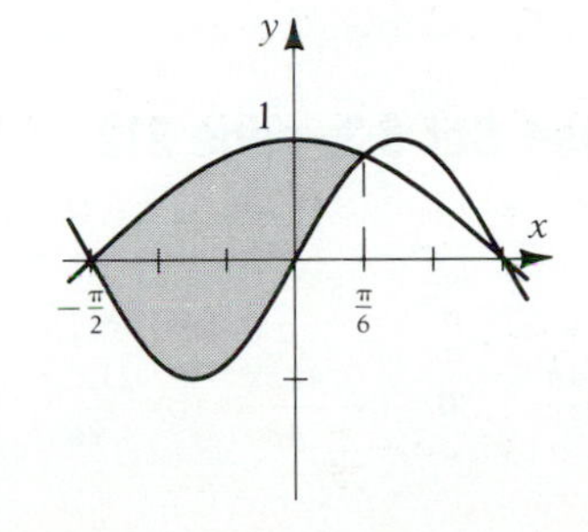

25. Area $= \frac{9}{2}$

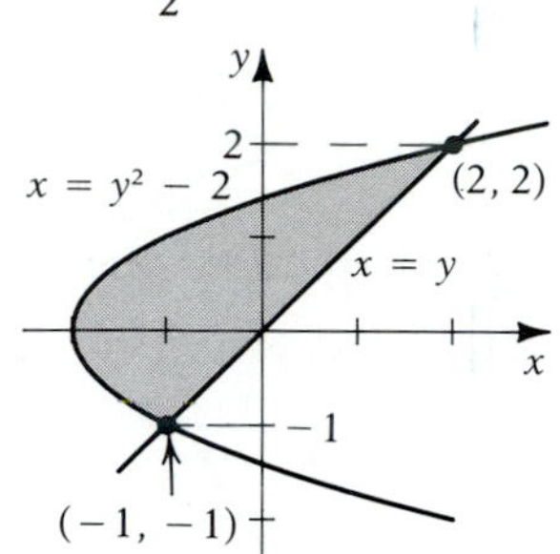

27. Area $= 8$

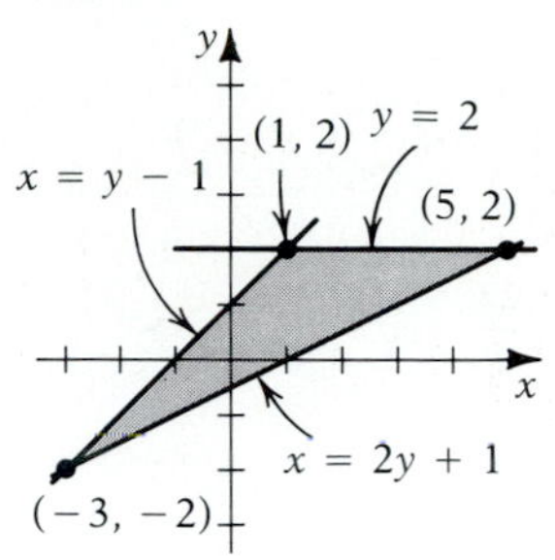

29. Area $= \frac{256}{27}$

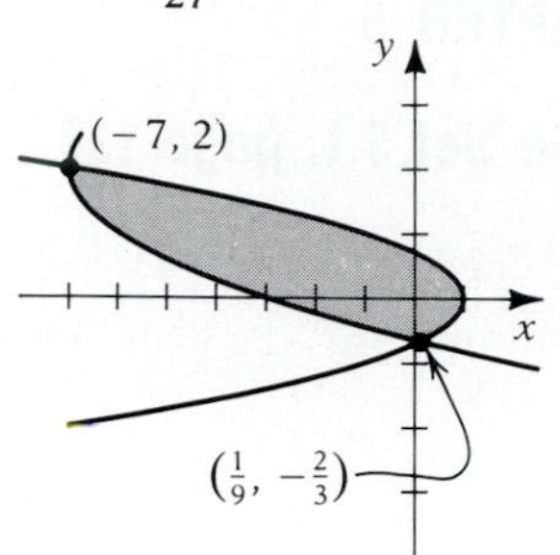

31. Area $= \frac{8}{3}$

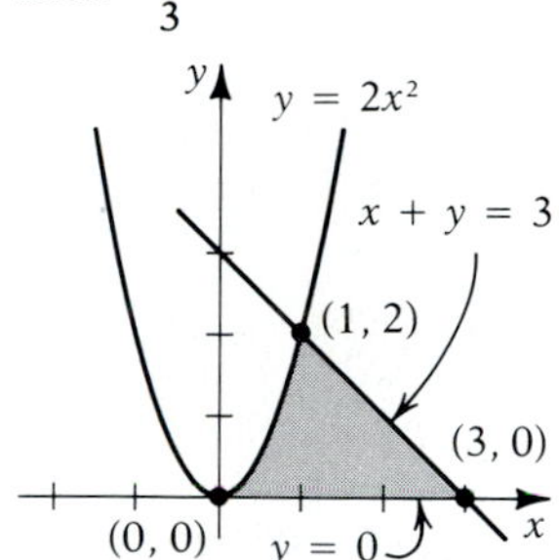

33. Area $= 3$

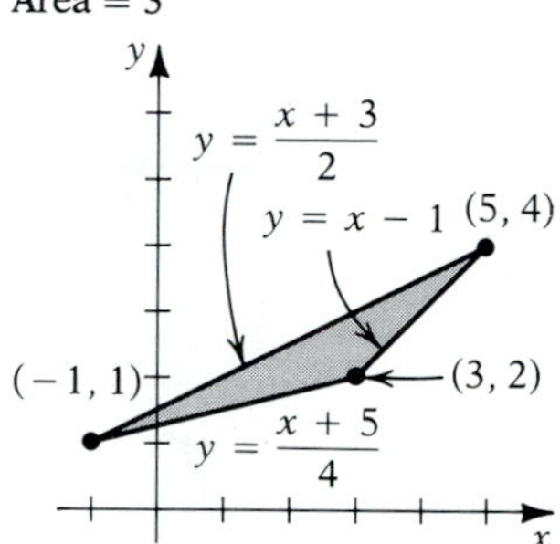

35. Area $= 8$

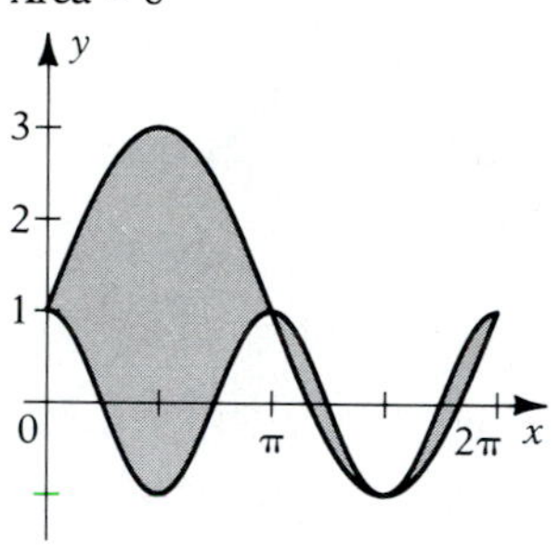

39. Area $= 3$

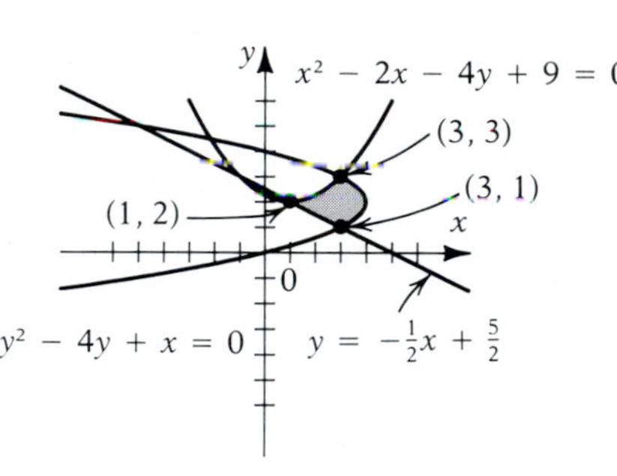

Exercise Set 5.2, page 206

1. $\frac{37\pi}{12}$ **3.** $\frac{256\pi}{15}$ **5.** 8π **7.** $\frac{16\pi}{7}$ **9.** $\frac{434\pi}{3}$ **11.** $\frac{92\pi}{15}$ **13.** $\frac{477\pi}{4}$ **15.** $\frac{72\pi}{5}$ **17.** $\frac{16\pi}{3}$ **19.** $\frac{32\pi}{3}$

21. $\frac{32\pi}{3}$ **23.** $\frac{99\pi}{5}$ **25.** 10π **27.** $\frac{250\pi}{3}$ **29.** $\frac{120\pi}{7}$ **31.** $V = \frac{4}{3}\pi a^3$ **33.** $\frac{28\pi}{3}$ **35.** (a) $\frac{88\pi}{3}$ (b) 24π

37. $\frac{5\pi^2}{6} - \frac{\pi}{4}$ **39.** 8π **41.** $\frac{16}{3}$ **43.** 2π **45.** $\frac{1612\pi}{15}$ **47.** $\frac{5000\pi}{3}L$ **49.** $8\sqrt{3}$ **51.** $\frac{256\pi}{3}$

Exercise Set 5.3, page 212

1. $\frac{16\pi}{3}$ **3.** 8π **5.** $\frac{32\pi}{3}$ **7.** $\frac{110\pi}{3}$ **9.** $\frac{381\pi}{14}$ **11.** $\frac{8\pi}{3}$ **13.** $\frac{12\pi}{5}$ **15.** $\frac{4\pi}{27}$ **17.** $\frac{16\pi}{3}$ **19.** $2\pi^2a^2b$

21. $\frac{1024\pi}{3}$ **23.** (a) $\frac{485\pi}{48}$ (b) $\frac{2125\pi}{96}$

Exercise Set 5.4, page 216

1. $3\sqrt{10}$ **3.** $\sqrt{13}$ **5.** $\frac{14}{3}$ **7.** $\frac{335}{27}$ **9.** $\frac{22}{3}$ **11.** $\frac{1}{54\sqrt{2}}[(77)^{3/2} - (32)^{3/2}]$ **13.** $\frac{56}{3}$ **15.** $6a$

17. On solving for x, there are two values: $x_1 = -\left(\frac{2y}{3}\right)^{3/2} - 1$ and $x_2 = \left(\frac{2y}{3}\right)^{3/2} - 1$. The point $(-2, \frac{3}{2})$ is on the graph of x_1, and $(7, 6)$ is on the graph of x_2. Total length $= L_1 + L_2$, where L_1 is the length of x_1 from $(-2, \frac{3}{2})$ to $(-1, 0)$ and L_2 is the length of x_2 from $(-1, 0)$ to $(7, 6)$. The result is $5\sqrt{5} + 2\sqrt{2} - 2$.

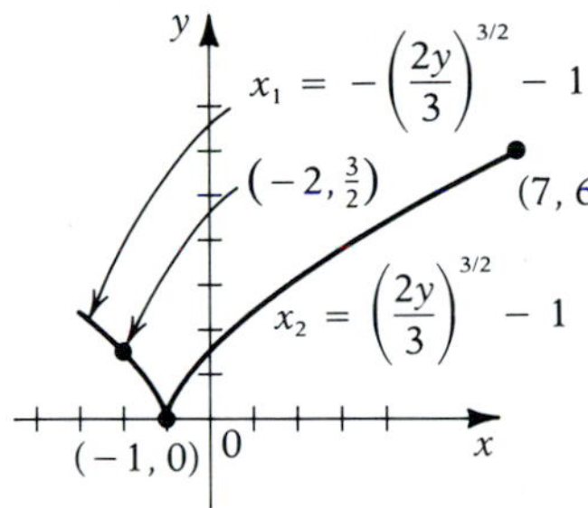

19. arc length $\approx ds = \sqrt{0.05} \approx 0.22$

Exercise Set 5.5, page 222

1. $24\pi\sqrt{5}$ **3.** $\frac{2\pi}{3}(5\sqrt{5} - 1)$ **5.** 2π **7.** $\frac{\pi}{27}[(145)^{3/2} - (10)^{3/2}]$ **9.** $\frac{\pi}{6}[(17)^{3/2} - 5^{3/2}]$ **11.** $S = 4\pi a^2$

13. (a) $5\pi\sqrt{13}$ (b) $6\pi\sqrt{13}$ **15.** $\frac{\pi}{36}(45\sqrt{5} - 16)$ **17.** $\frac{47\pi}{16}$ **19.** $2\pi ah$

Exercise Set 5.6, page 227

1. $\frac{4}{3}$ joules **3.** 24 in.-lb **5.** 30 erg **7.** $\frac{4}{5}$ dyne/cm **9.** 1350 ft-lb **11.** 1000 ft-lb

13. $14{,}400\pi\rho$ ft-lb $\approx$ 2,823,000 ft-lb **15.** $\frac{k}{3}$ erg **17.** $75\pi(10^4)$ in.-lb **19.** Approx. $6.87\ (10^6)$ joules

21. Approx. 108.3 ft-lb

Exercise Set 5.7, page 232

1. 56ρ **3.** 135ρ **5.** $\frac{184\rho}{3}$ **7.** 2ρ lb $\approx$ 124.8 lb when full; ρ lb $\approx$ 62.4 lb when water level is 1 ft.

9. (a) 5ρ N on shallow end; 80ρ N on deep end (b) $\frac{105\rho}{2}$ N on each side ($\rho \approx 999.5$ kg/cu m)

11. $F = \int_3^5 \rho x(2x - 6)\,dx = \frac{52\rho}{3}$ lb **13.** $\frac{304\rho}{15}$

15. As shown in the figure, corresponding to Δx_k in a partition of $[a, b]$, the length on the inclined plane is $\Delta x_k \sec \theta$. Obtain the Riemann sum $\sum_{k=1}^{n} \rho h(x^*)w(x^*)\Delta x_k \sec \theta$ and pass to the limit.

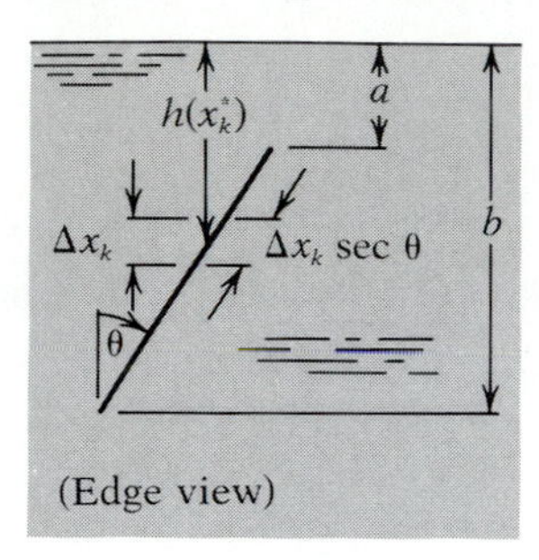

17. $300\pi\rho$ lb $\approx$ 60,600 lb

Exercise Set 5.8, page 238

1. $\frac{1}{28}$ **3.** $\frac{3}{38}$ **5.** $\frac{4}{3}$ **7.** 60 **9.** 1 **11.** $\frac{1}{2}$ **13.** $\frac{1}{4}$ **15.** $\frac{37}{125}$ **17.** $\mu = 1, \sigma^2 = \frac{1}{2}$ **19.** $\mu = \frac{1}{2}, \sigma^2 = \frac{1}{4}$

21. $\mu = \frac{4}{5}, \sigma^2 = \frac{4}{25}$ **23.** $\mu = \frac{1}{2}(a+b), \sigma^2 = \frac{1}{12}(b-a)^2$ **25.** $\frac{3}{4}$ **27.** (a) $\frac{1}{8}$ (b) $333\frac{1}{3}$ million barrels

29. 158 hours **31.** $\sigma^2 = E[(x-\mu)^2] = E[x^2 - 2\mu X + \mu^2]$. Now use Exercise 30 and definition of μ.

33. Find μ and then use $\sigma^2 = E[x^2] - \mu^2$. From this obtain $\int_0^1 \frac{x^2}{(1+x^2)^2}\,dx = \frac{\pi - 2}{8}$

Supplementary Exercises 5.9, page 239

1. Area $= \frac{46}{3}$

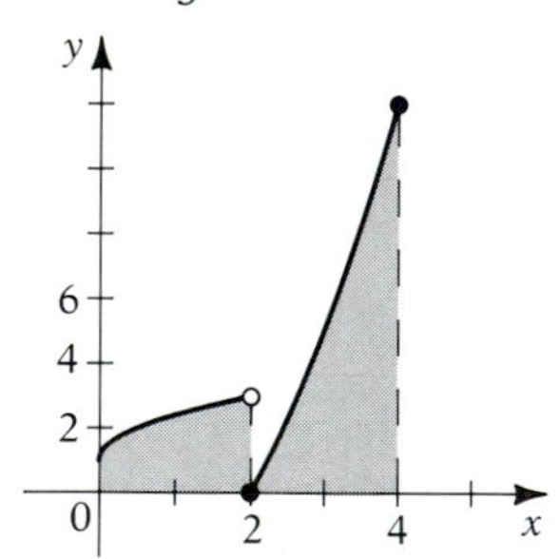

3. Area = 24

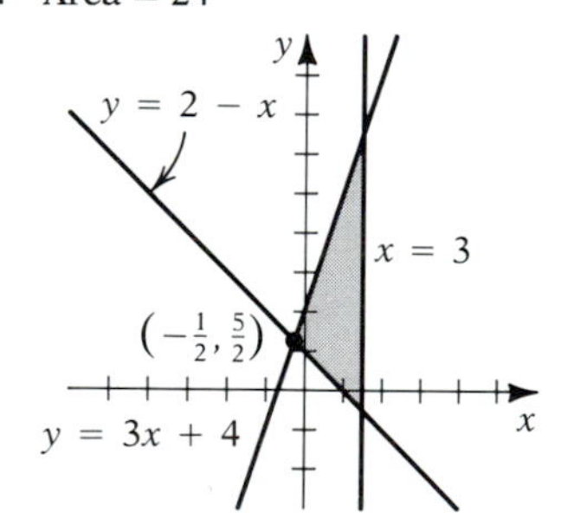

5. Area = 16

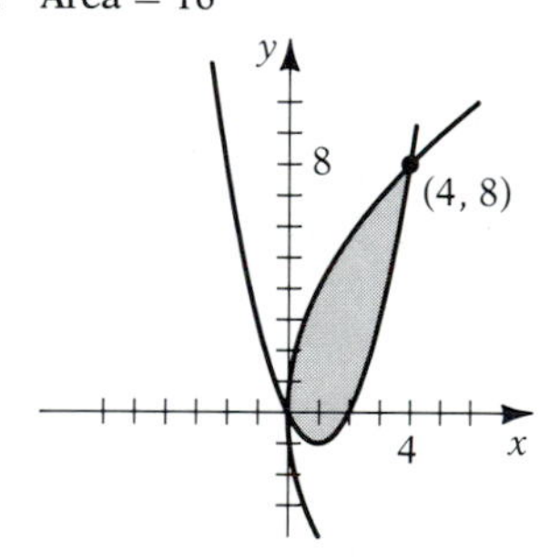

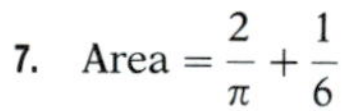

7. Area $= \frac{2}{\pi} + \frac{1}{6}$

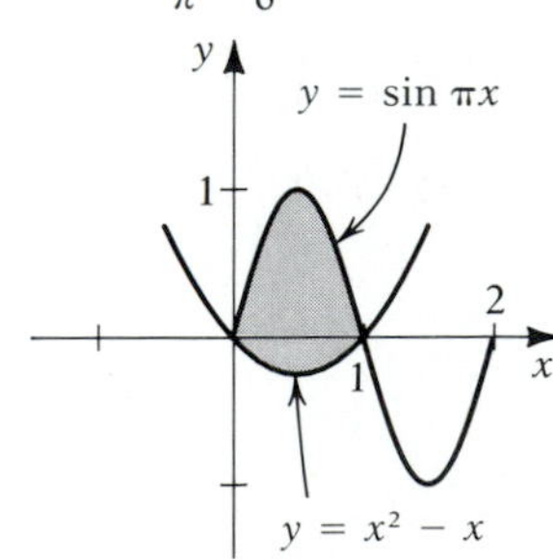

9. Area = 2

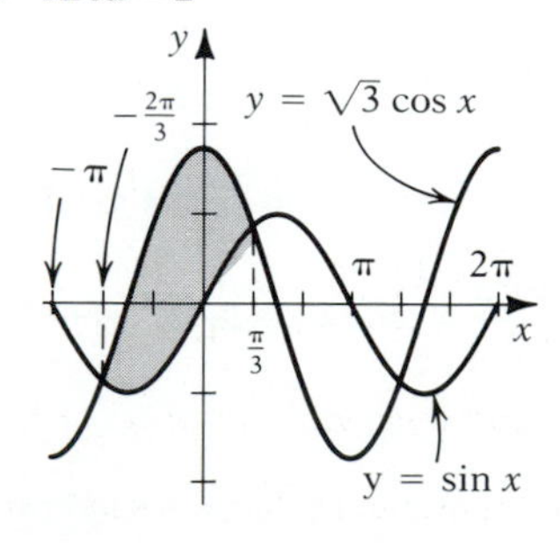

11. Volume $= \frac{\pi}{2}$

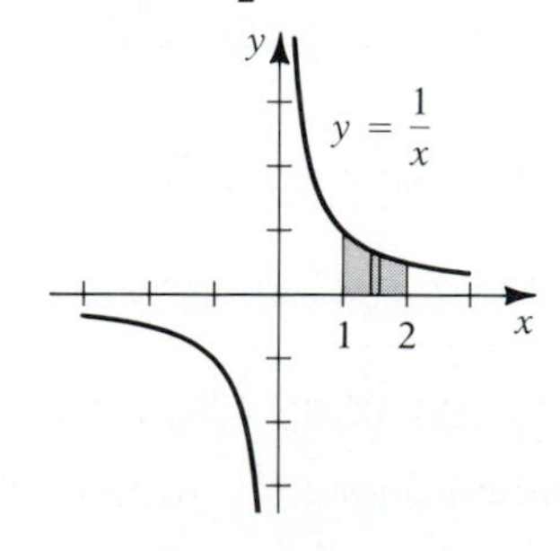

13. Volume $= \dfrac{176\pi}{3}$

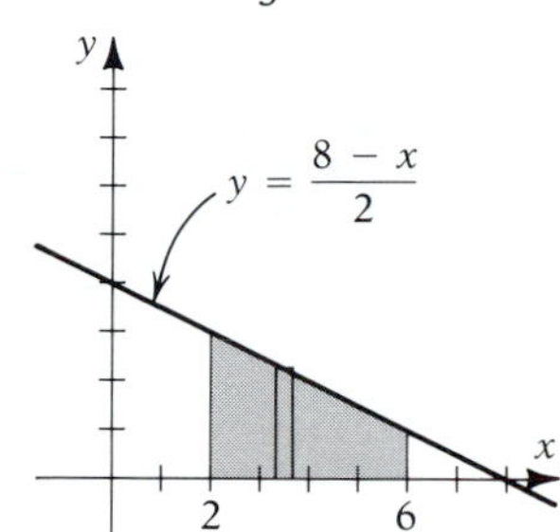

15. Volume $= \dfrac{500\pi}{3}$

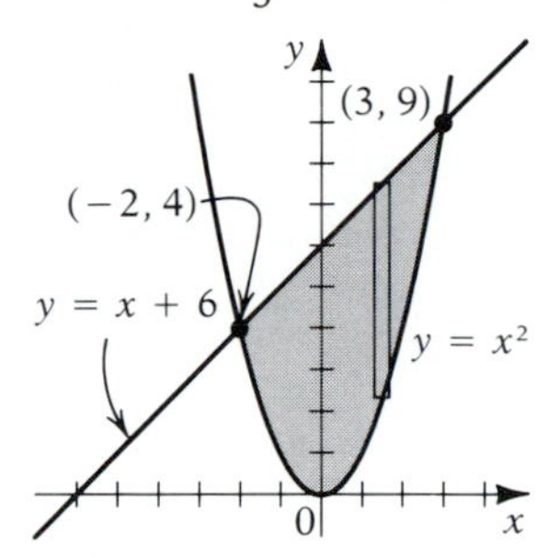

17. Volume $= \dfrac{28\pi}{3}$

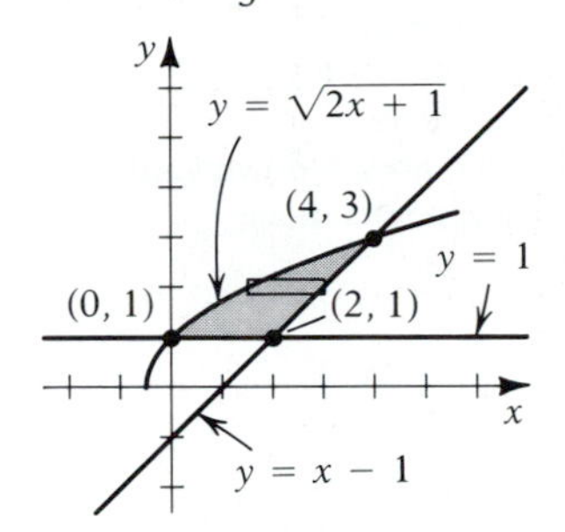

19. Volume $= \dfrac{625\pi}{6}$

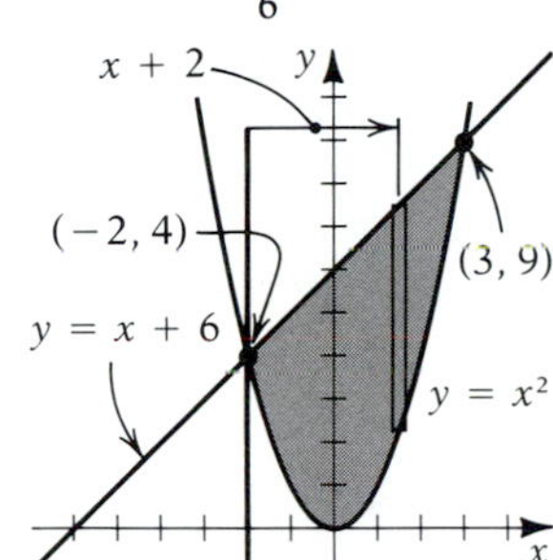

21. $\dfrac{32}{3}$ **23.** $\dfrac{1}{27}[(40)^{3/2} - 8]$ **25.** $\dfrac{22}{3}$ **27.** $\dfrac{49\pi}{9}$ **29.** (a) $\dfrac{1744\pi}{45}$ (b) $\dfrac{173\pi}{3}$

31. 37,500 ft-lb **33.** (a) $\dfrac{64\rho}{15}$ lb ≈ 266 lb (b) $\dfrac{28\rho\sqrt{2}}{15} \approx 165$ lb **35.** (a) $\dfrac{1}{2}$ (b) $\dfrac{17}{8}$ (c) $\dfrac{139}{320}$ (d) $\dfrac{2}{27}$ (e) $\dfrac{17}{27}$

CHAPTER 6

Exercise Set 6.2, page 250

1. Domain $= (0, \infty)$; $f'(x) = -\dfrac{1}{x}$ **3.** Domain $= (0, \infty)$; $f'(x) = \dfrac{3 \ln x}{x}$ **5.** Domain $= (-1, 1)$; $f'(x) = \dfrac{1}{x} - \dfrac{1}{2(1-x)}$

7. Domain $= (0, \infty)$; $f'(x) = 1 + \ln x$ **9.** Domain $= (-1, 1)$; $f'(x) = \dfrac{1}{x^2 - 1}$ **11.** Domain $= (1, \infty)$; $f'(x) = \dfrac{1}{x \ln x}$

13. Domain $(2n\pi, (2n+1)\pi)$, $n = 0, \pm 1, \pm 2, \cdots$; $f'(x) = \cot x$ **15.** Domain $= (0, \infty)$; $f'(x) = \dfrac{\ln x - 2}{x^2}$

17. Domain $= (-\infty, -1) \cup (1, 2)$; $f'(x) = \dfrac{x}{x^2 - 1} - \dfrac{1}{2 - x}$ **19.** Domain $= (-3, 0) \cup (2, \infty)$; $f'(x) = \dfrac{1}{x+3} - \dfrac{1}{x} - \dfrac{1}{x-2}$

21. $\ln(x^2 + 4) + C$ **23.** $-\dfrac{1}{6} \ln|2 - 3x^2| + C$, $x \neq \pm\sqrt{\dfrac{2}{3}}$ **25.** $\dfrac{1}{2}\ln(x^2 - 2x + 3) + C$ **27.** $2 \ln(\sqrt{x} + 1) + C$, $x > 0$

29. $\ln|x \sin x + \cos x| + C$, $x \neq -\cot x$ **31.** $\dfrac{(x-1)^2\sqrt{3x+4}}{x+2}\left[\dfrac{2}{x-1} - \dfrac{3}{2(3x+4)} - \dfrac{1}{x+2}\right]$

33. $\dfrac{1}{2}\sqrt{\dfrac{(x-1)(x+2)}{(x-3)^3}}\left[\dfrac{1}{x-1} + \dfrac{1}{x+2} - \dfrac{3}{x-3}\right]$ **35.** Local minimum value = 1, when $x = 1$

37. Local minimum value = 1, when $x = 1$ **39.** Local maximum value $= -\frac{3}{2}\ln 2$, when $x = \frac{1}{2}$

41. *Domain:* $(0,\infty)$
Intercepts: $(\frac{1}{2}, 0)$
Symmetry: none
Asymptotes: $x = 0$
Extrema: none
Concavity: always downward
Inflection points: none

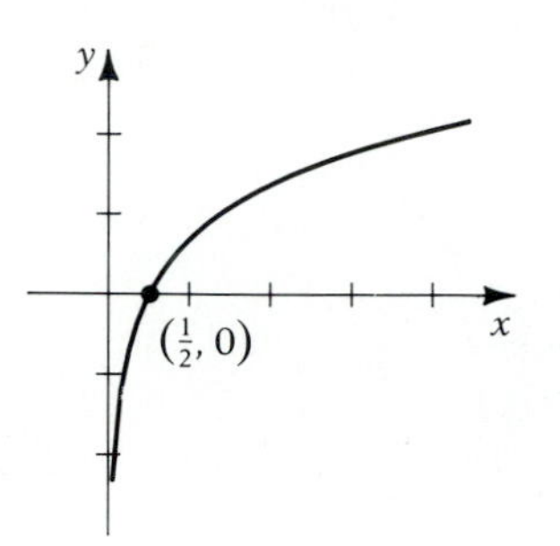

43. *Domain:* $(0, \infty)$
Intercepts: $(1, 0)$
Symmetry: none
Asymptotes: $x = 0$
Extrema: none
Concavity: always upward
Inflection points: none

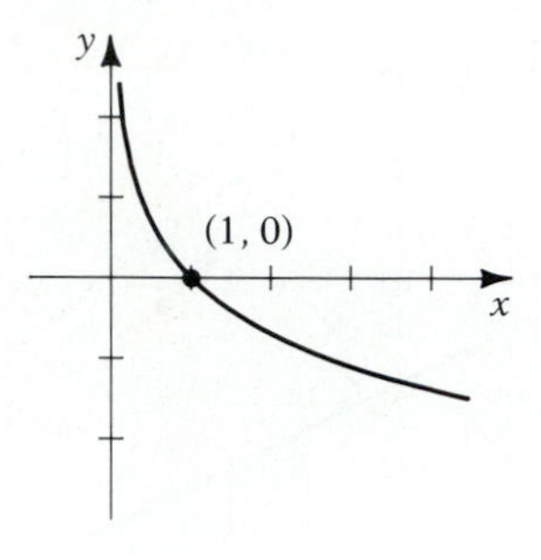

45. *Domain:* $(-\infty, -1) \cup (1, \infty)$
Intercepts: $(\pm\sqrt{2}, 0)$
Symmetry: y-axis
Asymptotes: $x = \pm 1$
Extrema: none
Concavity: always downward
Inflection points: none

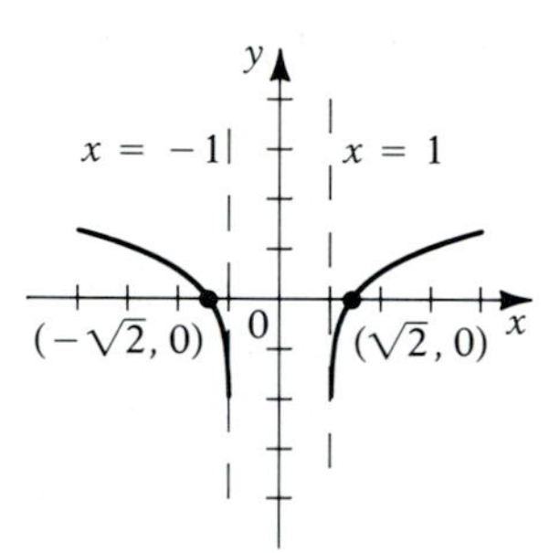

47. *Domain:* $\{x : x \neq 0\}$
Intercepts: $(\pm 1, 0)$
Symmetry: y-axis
Asymptotes: $x = 0$
Extrema: none
Concavity: always downward
Inflection points: none

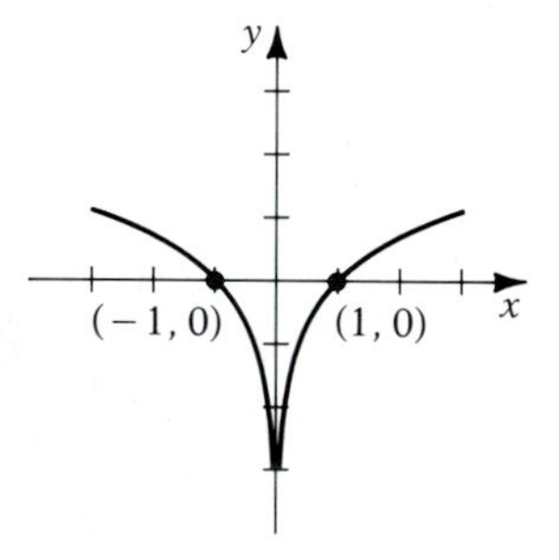

49. (a) 1.80 (b) 2.10 (c) 4.4 (d) 0.40 (e) 3.6 **51.** $f'(x) = 2\ln x + (\ln x)^2$, $f''(x) = \dfrac{2(1 + \ln x)}{x}$

53. $f'(x) = \dfrac{1}{x\sqrt{\ln x^2}}$, $f''(x) = -\dfrac{1 + \ln x^2}{x^2(\ln x^2)^{3/2}}$ **55.** $\dfrac{y(2x + \ln y)}{x(1 - x)}$ **57.** $-\ln(1 + \cos^2 x) + C$ **59.** $x - 2\ln|x - 1| - \dfrac{1}{x - 1} + C$

61. *Domain:* $\{x : x \neq \pm 1\}$
Intercepts: $(0, 0)$
Symmetry: origin
Asymptotes: $x = \pm 1$, $y = 0$
Extrema: none
Concavity: downward on $(0, 1) \cup (1, \infty)$, upward on $(-\infty, -1) \cup (-1, 0)$
Inflection points: $(0, 0)$

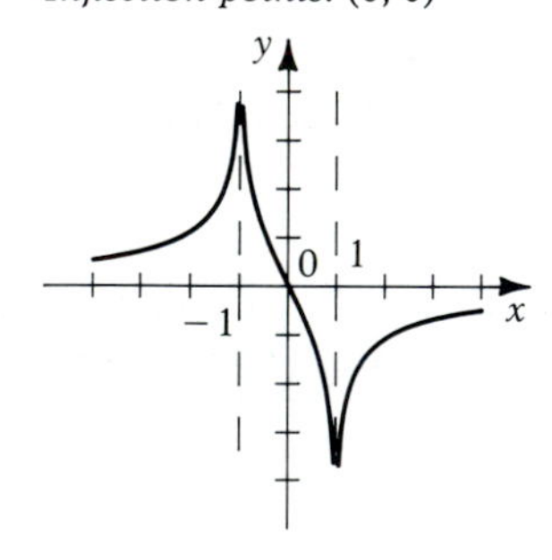

63. *Domain:* $(-\infty, 0) \cup (3, \infty)$
Intercepts: $(-1, 0)$
Symmetry: none
Asymptotes: $x = 0$, $x = 3$, $y = 0$
Extrema: local maximum point $(-3, \ln\frac{3}{2})$
Inflection points: between $x = -4$ and $x = -5$

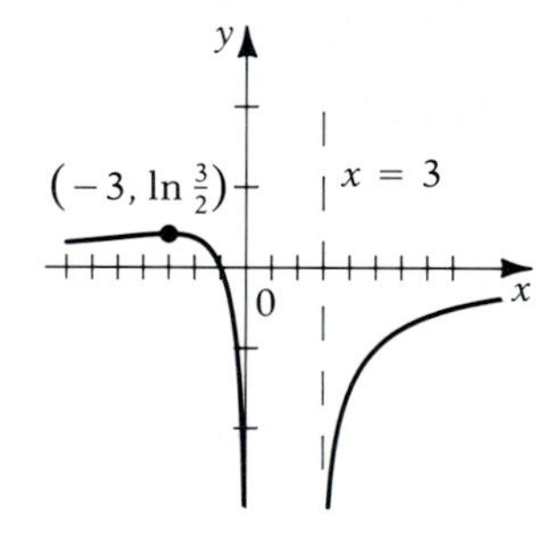

Exercise Set 6.3, page 258

1. $f^{-1}(x) = \dfrac{1 - x}{2}$ **3.** $f^{-1}(x) = \dfrac{4x - 3}{2}$ **5.** $f^{-1}(x) = \sqrt[3]{x + 1}$ **7.** $f^{-1}(x) = \sqrt{x}$; domain $= [0, \infty)$

9. $f^{-1}(x) = 1 + \sqrt{\dfrac{x - 3}{2}}$; domain $= [3, \infty)$ **11.** $f^{-1}(x) = \dfrac{1}{x}$; domain $= (0, \infty)$ **13.** $f^{-1}(x) = \dfrac{1 + x}{1 - x}$; domain $= (-\infty, 1)$

15. 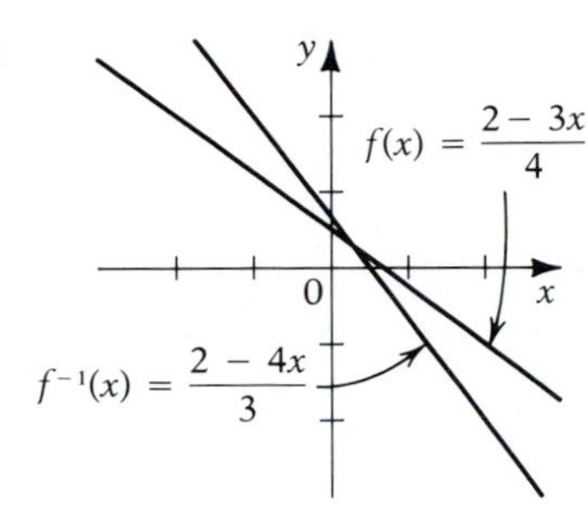

17. 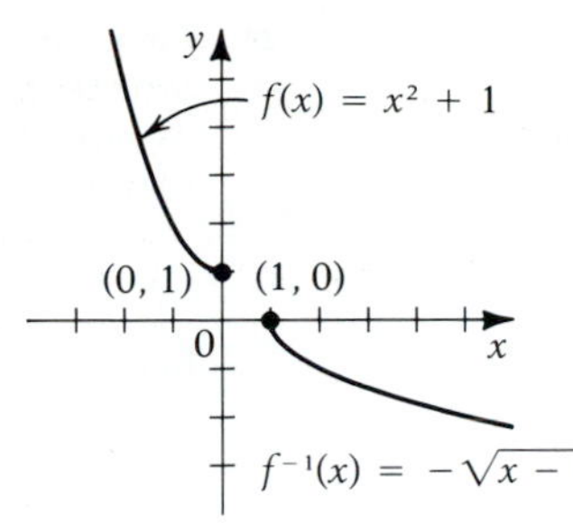

19. 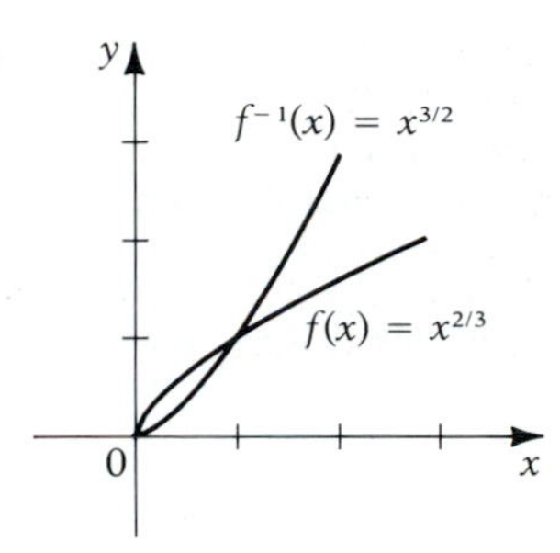

21. Take domain of $f = [1,\infty)$. Then $f^{-1}(x) = 1 + \sqrt{x-2}$ on $[2, \infty)$.

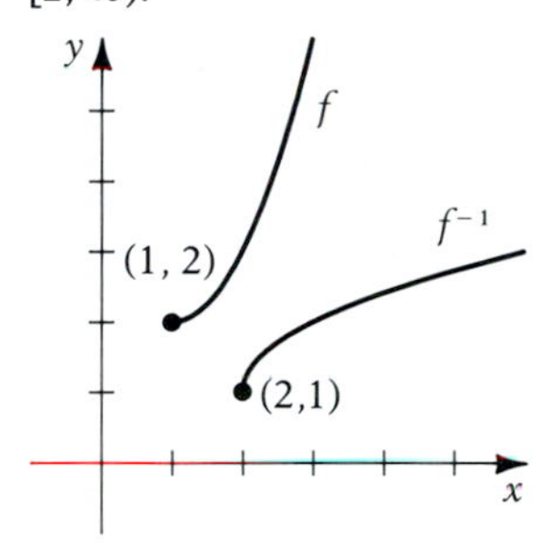

23. Take domain of $f = [1, \infty)$. Then $f^{-1}(x) = 1 + \sqrt{1-x}$ on $(-\infty, 1]$.

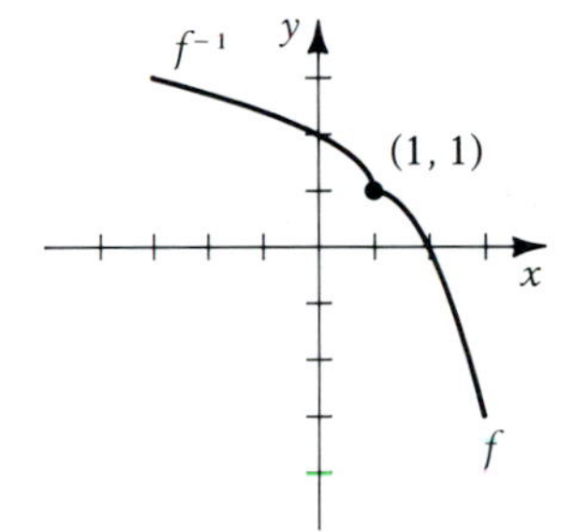

25. Take domain of $f = (1, \infty)$. Then $f^{-1}(x) = 1 + \dfrac{1}{\sqrt{x}}$ on $(0, \infty)$.

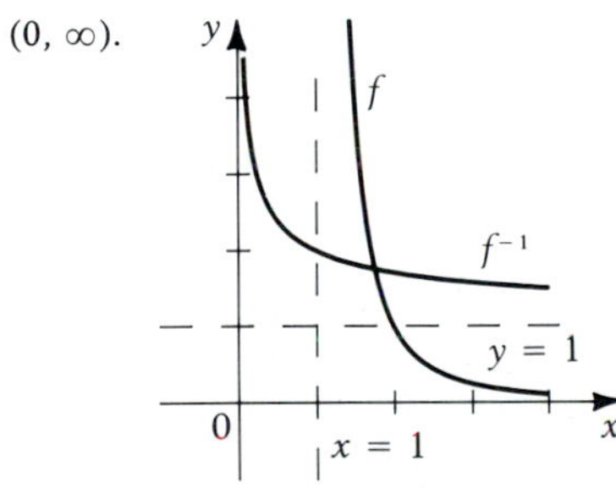

27. $f'(x) > 0$ on R. $(f^{-1})'(-3) = \dfrac{1}{5}$

29. $f'(x) > 0$ on $(0, \infty)$ $(f^{-1})'(y_0) = \dfrac{1}{8}$

31. $x_1^3 = x_2^3$ implies that $x_1 = x_2$. So $f(x) = x^3$ is $1-1$ on $\mathbf{R}$. But $f'(0) = 0$.

33. For $x_1, x_2 \in I$, $f'(x_1)$ and $f'(x_2)$ cannot be of opposite sign, since if they were, the intermediate value property for derivatives would say that $f'(c) = 0$ for some c in I, contrary to hypothesis. So either $f'(x) > 0$ for all $x \in I$ and f is increasing, or $f'(x) < 0$ for all $x \in I$ and f is decreasing.

Exercise Set 6.4, page 264

1. $2xe^{x^2}$ 3. $3x^2$ 5. $\dfrac{e^{\sqrt{x}}}{2\sqrt{x}}$ 7. $\dfrac{1-2x}{e^{2x}}$ 9. $e^x(x+1)$ 11. $e^{2x}(3\cos 3x + 2\sin 3x)$ 13. $\dfrac{e^x + e^{-x}}{2}$ 15. $\dfrac{4}{(e^x + e^{-x})^2}$

17. $e^x \sin(2e^x)$ 19. $\dfrac{-e^x}{(e^x - 1)^2}$ 21. $\dfrac{y - e^{x+y}}{e^{x+y} - x}$ 23. $\dfrac{e^{-x}\sin y - e^{-y}\sin x}{e^{-x}\cos y + e^{-y}\cos x}$ 25. $-\dfrac{1}{2}e^{1-x^2} + C$ 27. $e^{-\cos x} + C$

29. $\ln(e^x + e^{-x}) + C$ 31. $\dfrac{7}{3}$ 33. $\dfrac{3}{8}[1 - (1 - e^6)^{4/3}]$ 35. $\ln\dfrac{3}{2}$ 37. $3 - \sqrt{3}$ 39. $\dfrac{1}{1 - e^x} + C$

41. *Intercepts:* (0, 1)
Symmetry: y-axis
Asymptotes: $y = 0$
Extrema: local (and absolute) maximum point (0, 1)
Inflection points: $\left(\pm 2, \frac{1}{e}\right)$

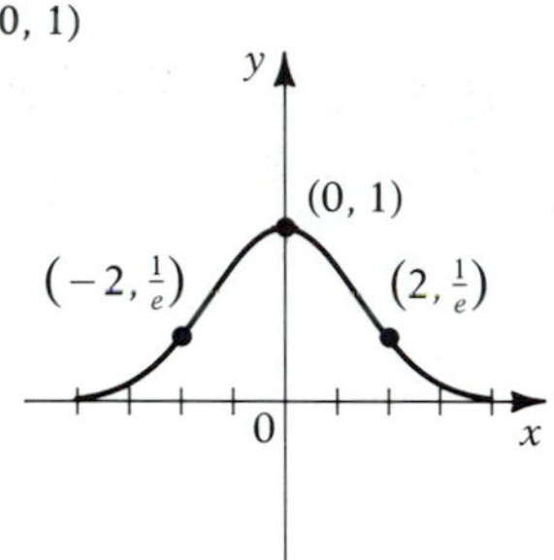

43. *Intercepts:* (0, 0)
Symmetry: none
Asymptotes: $y = 0$, on right
Extrema: local (and absolute)

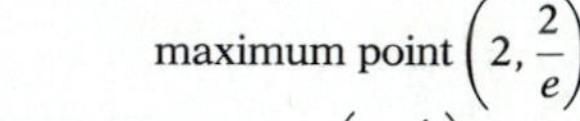
maximum point $\left(2, \frac{2}{e}\right)$

Inflection points: $\left(4, \frac{4}{e^2}\right)$

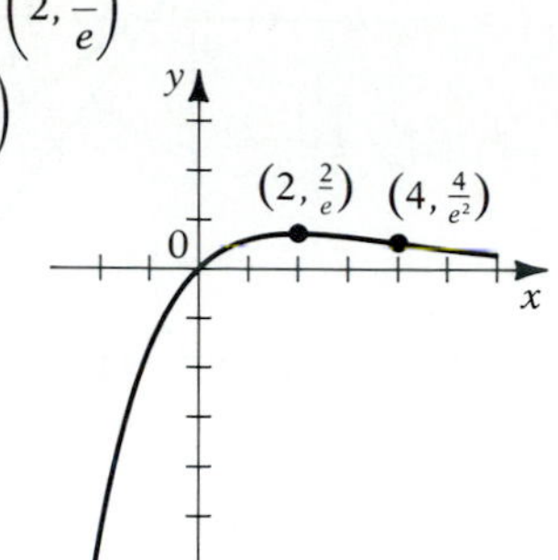

45. *Intercepts:* $(0, \frac{1}{2})$
Symmetry: not to x-axis, y-axis, or origin
Asymptotes: $y = 1$ on right
$y = 0$ on left
Extrema: none
Inflection points: $(0, \frac{1}{2})$

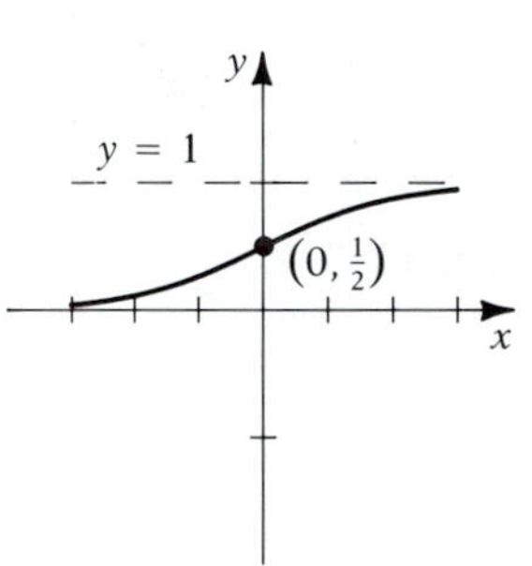

47. $\ln e^{x-y} = x - y$ and $\ln \frac{e^x}{e^y} = \ln e^x - \ln e^y = x - y$. So $\ln e^{x-y} = \ln \frac{e^x}{e^y}$, and since ln is 1 − 1, $e^{x-y} = \frac{e^x}{e^y}$.

49. (a) True, since $\ln 1 = 0$ (b) True, since $\ln e = 1$ **51.** πe **53.** $\frac{\pi}{2}\left(1 - \frac{1}{e^3}\right)$ **55.** $\frac{1}{\sqrt{x^2+1}}$

57. *Intercepts:* none
Symmetry: none
Asymptotes: $x = 0$, $y = 0$ on left
Extrema: local minimum point $(1, e^2)$
Inflection points: none

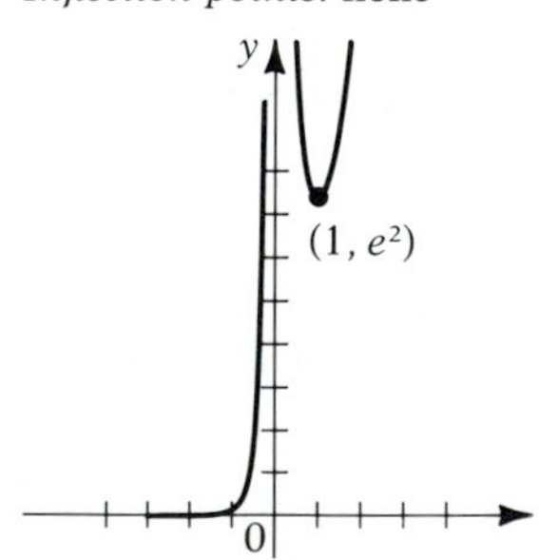

59. $v(\pi) = \frac{2}{e^\pi}$, $a(\pi) = -\frac{4}{e^\pi}$

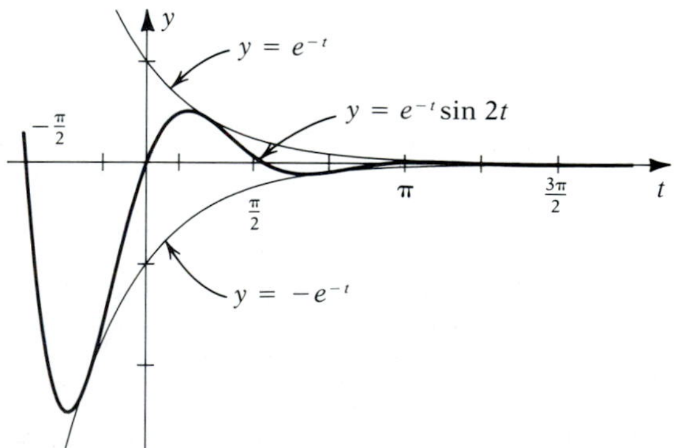

61. *Intercepts:* none
Symmetry: to line $x = \mu$
Asymptotes: $y = 0$
Extrema: local (and absolute) maximum point $\left(\mu, \frac{1}{\sqrt{2\pi}\,\sigma}\right)$
Inflection points: $\left(\mu \pm \sigma, \frac{1}{\sqrt{2\pi e}\sigma}\right)$

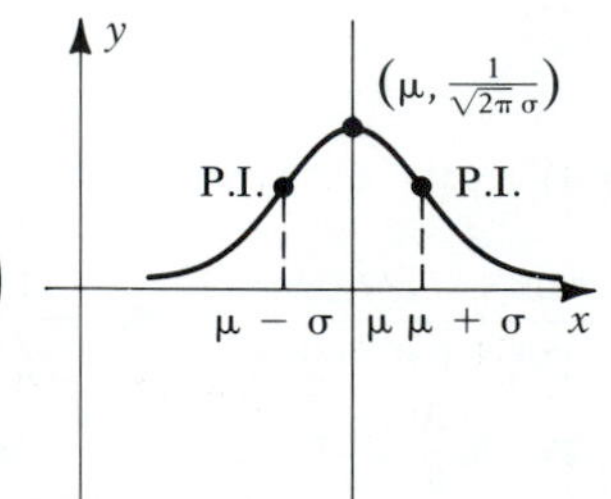

Exercise Set 6.5, page 274

1. (a) $e^{\sqrt{3}\ln\pi} \approx 7.26255$ (b) $e^{\pi\ln\pi} \approx 36.46216$ (c) $e^{e\ln\pi} \approx 22.45916$ (d) $e^{\sqrt{2}\ln\sqrt{2}} \approx 1.63253$ (e) $e^{\pi\ln(\sin 2)} \approx 0.74177$

3. $2^x \ln 2$ **5.** $x4^{-x}(2 - x\ln 4)$ **7.** $\dfrac{1}{\ln 3}\left[\dfrac{1}{x} - \dfrac{3x^2}{1+x^3}\right]$ **9.** $\dfrac{1}{x(x^2-1)\ln 10}$ **11.** $\dfrac{1}{x(\ln x)(\ln 2)}$ **13.** $x^x(1+\ln x)$

15. $x^{1/(1-x)}\left[\dfrac{1}{x(1-x)} - \dfrac{\ln x}{(1-x)^2}\right]$ **17.** $(1-x^2)^{1/x}\left[\dfrac{2}{x^2-1} - \dfrac{1}{x^2}\ln(1-x^2)\right]$ **19.** $\dfrac{5^x}{\ln 5} + C$ **21.** $\dfrac{1}{\ln 3}\ln(1+3^x) + C$

23. $\dfrac{2^{x^3-1}}{3\ln 2} + C$ **25.** $\dfrac{4}{\ln 2}$ **27.** $\dfrac{-1}{(\ln 3)(1-3^{-x})} + C$ **29.**

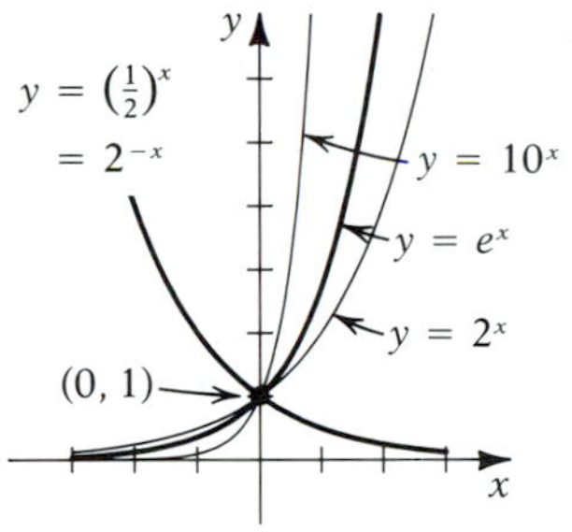

$y = (\frac{1}{2})^x = 2^{-x}$; $y = 10^x$; $y = e^x$; $y = 2^x$; $(0, 1)$

31. If $x > 0$, $\dfrac{d}{dx}\log_a|x| = \dfrac{d}{dx}\log_a x = \dfrac{1}{x}\ln a$

If $x < 0$, $\dfrac{d}{dx}\log_a|x| = \dfrac{d}{dx}\log_a(-x) = \dfrac{1}{(-x)}(-1)\ln a = \dfrac{1}{x}\ln a$

33. *Intercepts:* (0, 2)
Symmetry: y-axis
Asymptotes: $y = 0$
Extrema: local (and absolute) maximum point (0, 2)
Inflection points: $\left(\pm\dfrac{1}{\sqrt{2\ln 2}}, 2^{(1-1/\ln 4)}\right)$

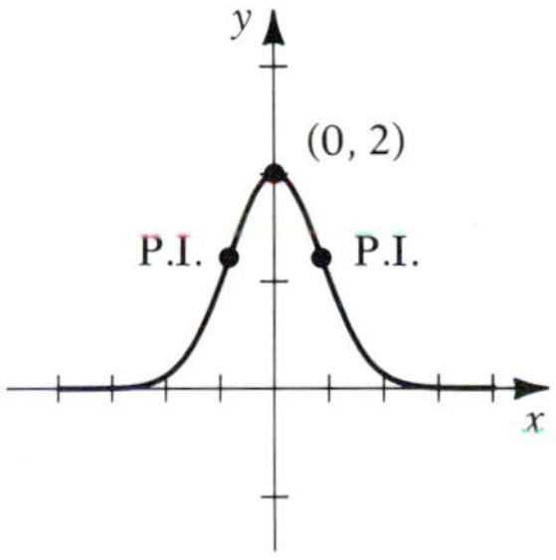

(0, 2); P.I.; P.I.

35. $x = \dfrac{\ln 3}{\ln \frac{9}{4}} \approx 1.355$ **37.** $x = \dfrac{2 + \ln\frac{3}{4}}{\ln 5 - 1} \approx 2.8097$

39. Property 2: $a^{x-y} = e^{(x-y)\ln a} = e^{x\ln a - y\ln a} = \dfrac{e^{x\ln a}}{e^{y\ln a}} = \dfrac{a^x}{a^y}$

Property 4: $(ab)^x = e^{x\ln(ab)} = e^{x(\ln a + \ln b)} = e^{x\ln a + x\ln b} = e^{x\ln a}\cdot e^{x\ln b} = a^x b^x$
Property 5 is done in a similar way.

41. $A_1(1) = P + Pr = P(1+r)$. True for $t = 1$
Assume $A_1(k) = P(1+r)^k$. During $(k+1)$-*st* year interest earned is $r(A_1(k))$. So

$A_1(k+1) = A_1(k) + r(A_1(k)) = P(1+r)^k + rP(1+r)^k = P(1+r)^{k+1}$

By mathematical induction, $A_1(t) = P(1+r)^t$ for $t = 1, 2, 3, \cdots$

43. $\dfrac{\pi}{\ln 2}$

Exercise Set 6.6, page 281

1. 13,500 **3.** 10 hr **5.** 16 kg **7.** $\dfrac{\ln 9}{\ln 2} \approx 3.17$ hr **9.** $\dfrac{20\ln 0.5}{\ln 0.7} \approx 38.87$ hr **11.** $1000(5)^{5/2} \approx 55{,}902$

13. $Q(12) = 200\left(\dfrac{1}{2}\right)^{3/2} \approx 70.7$ g. When $t = \dfrac{8\ln 10}{\ln 2} \approx 26.6$ yr, 10% remains **15.** $-6.25°$ **17.** $t = \dfrac{10\ln\frac{2}{15}}{\ln\frac{8}{15}} \approx 32.1$ min

19. 9.6 lb/sq in. **21.** $\dfrac{10 \ln 10}{\ln 2} \approx 33.2$ ft **23.** (a) $N(t) = N_1 e^{k(t-1)}$ (b) 5 **25.** $\dfrac{5730 \ln 10}{\ln 2} \approx 19{,}035$ yr

27. (a) $t = -\dfrac{1}{k} \ln(1 + \lambda)$ (b) approx. 5.67×10^9 yr

Supplementary Exercises 6.7, page 282

1. (a) $\dfrac{1}{x} + \dfrac{1}{2(x-2)}$, $x > 2$ (b) $\dfrac{1}{x} + \dfrac{1}{x-1} - \dfrac{1}{x-4} - \dfrac{1}{x+2}$
$x \in (-\infty, -2) \cup (0, 1) \cup (4, \infty)$ **3.** (a) $xe^{-x}(2 - x)$ (b) $-e^{-x}$

5. (a) $e^{-x}(\sin e^{-x} + \cos e^{-x})$ (b) $\dfrac{1}{2}\dfrac{e^x - e^{-x}}{e^x + e^{-x}} \ln(e^x + e^{-x})$ **7.** (a) $2(\ln x)x^{\ln x - 1}$ (b) $(\sec x)^{\tan x}[\tan^2 x + \sec^2 x \ln \sec x]$

9. (a) $\dfrac{y}{xye^y + x}$ (b) $\dfrac{x^2 - y^2 + 2y}{2x}$ **11.** (a) $-\dfrac{1}{x} + C$ (b) $\dfrac{1}{x - e^x} + C$ **13.** (a) $\dfrac{5}{72}$ (b) 2

15. (a) $\dfrac{1}{\ln 4} \ln(4^x + 1) + C$ (b) $\dfrac{\sqrt{3}}{4 \ln 2}$

17. *Intercepts:* (2, 0), (0, 0)
Symmetry: to $x = 1$
Asymptotes: $x = 1$
Extrema: none
Inflection points: none

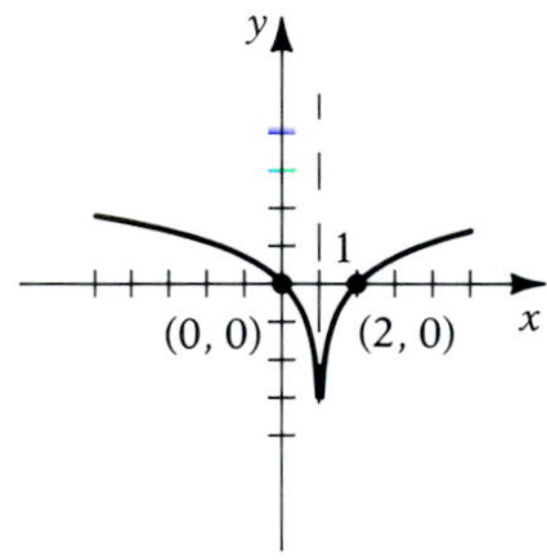

19. *Intercepts:* (0, 0)
Symmetry: origin
Asymptotes: $y = 0$
Extrema: local maximum point $(\sqrt{3}, 3\sqrt{3}e^{-3/2})$
local minimum point $(-\sqrt{3}, -3\sqrt{3}e^{-3/2})$
Inflection points: $(0, 0)$, $(1, e^{-1/2})$, $(\sqrt{6}, 6\sqrt{6}e^{-3})$
$(-1, -e^{-1/2})$, $(-\sqrt{6}, -6\sqrt{6}e^{-3})$

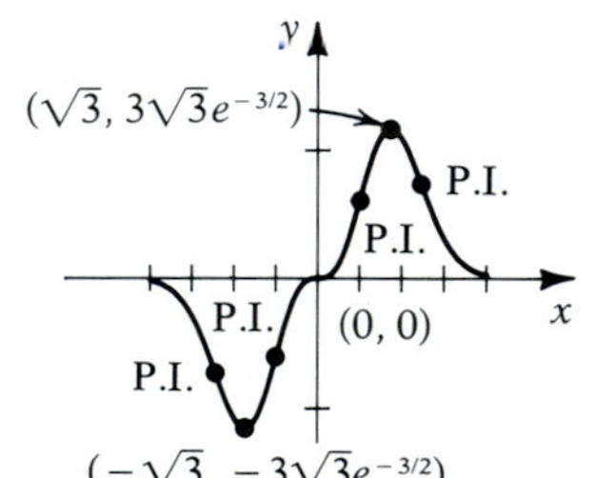

21. (a) $(f^{-1})'(y_0) = \dfrac{1}{4}$ (b) $(f^{-1})'(1) = \dfrac{1}{24}$ **23.** 2π **25.** (a) $\dfrac{\ln \frac{2}{5}}{\ln 48} \approx -0.2367$ (b) $\dfrac{\ln 3 - 1}{\ln \frac{3}{2} - 2} \approx -0.06184$

27. $Q(3) \approx 0.673Q_0$; after approx. 12.21 yr 20% will remain. **29.** $T(5) = (1 + (0.9)^{5/2}) \approx 35.37$°C; $T = 21$°C after approx. 56.87 min

CHAPTER 7

Exercise Set 7.1, page 292

1. (a) $\dfrac{\pi}{6}$ (b) $\dfrac{\pi}{3}$ (c) $\dfrac{\pi}{4}$ (d) $\dfrac{\pi}{3}$ (e) $\dfrac{\pi}{2}$ **3.** (a) $-\dfrac{\pi}{2}$ (b) π (c) 0 (d) $\dfrac{2\pi}{3}$ (e) $\dfrac{3\pi}{4}$ **5.** (a) $\dfrac{4}{3}$ (b) $\dfrac{3}{5}$

7. (a) $\dfrac{7}{9}$ (b) $-\dfrac{24}{25}$ **9.** (a) $\dfrac{\pi}{5}$ (b) $-\dfrac{\pi}{8}$ **11.** (a) $\dfrac{33}{65}$ (b) $\dfrac{2}{5\sqrt{5}}$ **13.** (a) 2.412 (b) 4.965 **15.** $\dfrac{6}{4x^2 + 9}$

17. $\dfrac{1}{2\sqrt{x-x^2}}$ **19.** $1+2x\tan^{-1}x$ **21.** $\dfrac{x\cos^{-1}x-\sqrt{1-x^2}}{(1-x^2)^{3/2}}$ **23.** $\dfrac{2x\sqrt{1-y^2}}{\sqrt{1-y^2}-1}$ **25.** $\dfrac{\pi}{2}$ **27.** $\dfrac{\pi}{12}$

29. $\dfrac{\pi}{16}$ **31.** $\dfrac{1}{3}\sin^{-1}\dfrac{3x}{4}+C$ **33.** $\dfrac{1}{3}\sec^{-1}\dfrac{4x}{3}+C$ **35.** $\dfrac{1}{3}\tan^{-1}(3x-2)+C$ **37.** $\dfrac{1}{2}\sin^{-1}x^2+C$ **39.** $-\dfrac{\pi}{6\sqrt{3}}$

41. $\dfrac{\pi}{3}$ **43.** $\dfrac{\pi^2}{2\sqrt{3}}$

55. Show that $\dfrac{d}{dx}(\sin^{-1}x+\sin^{-1}\sqrt{1-x^2})=0$, so that $\sin^{-1}x+\sin^{-1}\sqrt{1-x^2}=C$. Let $x=\frac{1}{2}$ to determine C. **57.** $\dfrac{\pi}{6}$

59. $\dfrac{\pi}{4}$ **61.** $\dfrac{\pi^2}{4}$ **63.** 2 m **65.** $-\dfrac{1}{65}$ rad/sec

Exercise Set 7.2, page 299

1. For sinh x, domain = **R**, range = **R**
For cosh x, domain = **R**, range = $[1, \infty)$
For tanh x, domain = **R**, range = $(-1, 1)$
For coth x, domain = $\{x: x \neq 0\}$, range = $(-\infty, -1) \cup (1, \infty)$
For sech x, domain = **R**, range = $(0, 1]$
For csch x, domain = $\{x: x \neq 0\}$, range = $\{y: y \neq 0\}$

3.

x	tanh x	coth x	sech x	csch x
0.5	0.462	2.16	0.887	1.92
1.0	0.762	1.31	0.648	0.851
2.0	0.964	1.04	0.266	0.276
4.0	0.999	1.00	0.037	0.037

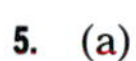
5. (a)

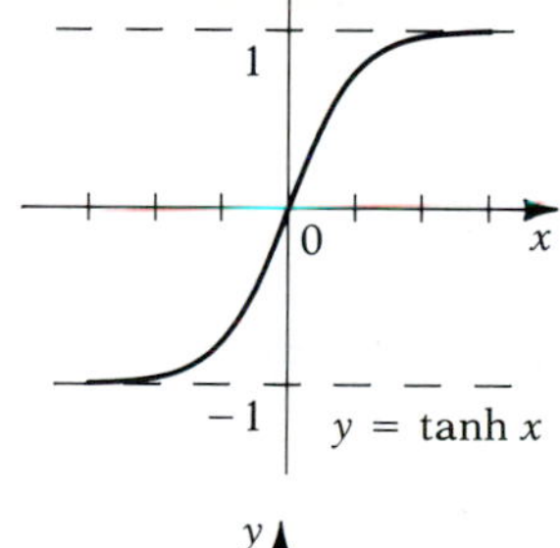

(b)

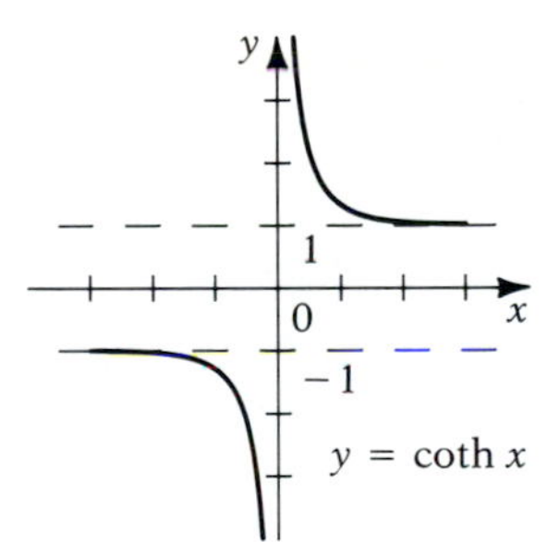

(c)

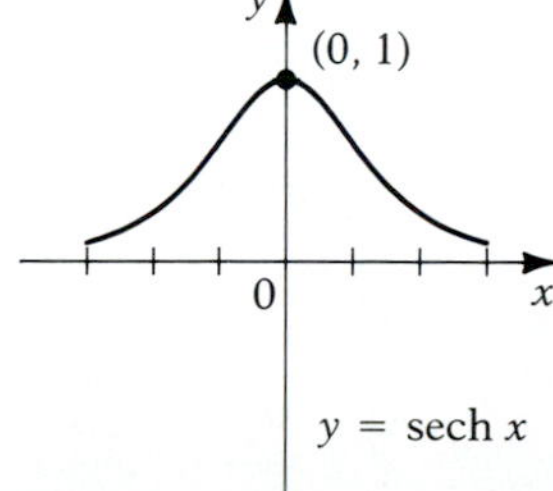

(d)

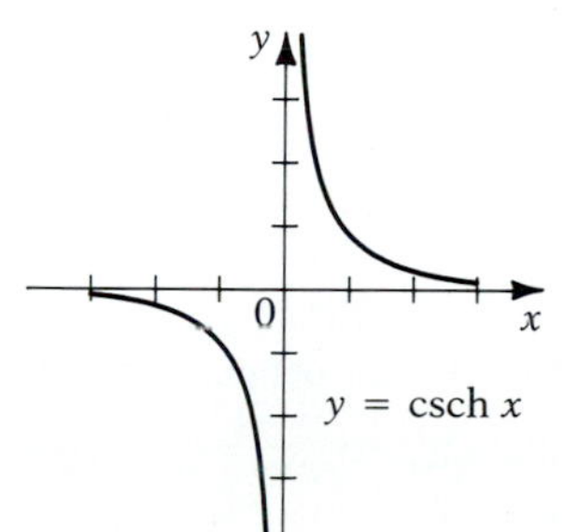

7. In identity 2 put $y=x$ **9.** Use identity 5 in the form $\cosh^2 x = 1 + 2\sinh^2 x$. Replace x by $\dfrac{x}{2}$, and solve for $\sinh\dfrac{x}{2}$.

11. Use identity 1. Divide both sides by $\cosh^2 x$.

13. From identities 4 and 5,

$$\tanh 2x = \frac{\sinh 2x}{\cosh 2x} = \frac{2\sinh x\cosh x}{\cosh^2 x + \sinh^2 x}$$

Now divide numerator and denominator by $\cosh^2 x$.

15. $\dfrac{1+\sinh x}{(1-\sinh x)^2}$ **17.** $\operatorname{sech} x$ **19.** $\coth^2 x$ **21.** $\dfrac{\operatorname{sech} x}{1+\operatorname{sech} x}$ **23.** $\dfrac{1}{x(2y-\tanh y)}$

25. $(\sinh x)^{\operatorname{sech} x}[\operatorname{csch} x - \operatorname{sech} x \tanh x \ln(\sinh x)]$ **27.** $\dfrac{(1+\tanh x)^2}{2}+C$ **29.** $\dfrac{\sinh 2x}{4}-\dfrac{x}{2}+C$ **31.** $e^{\sinh^2 x}+C$

33. $\dfrac{\cosh^3 x^2}{6}+C$ **35.** $x-\tanh x+C$ **37.** $\dfrac{-\operatorname{sech}^3 x}{3}+C$ **39.** $\dfrac{\cosh^5 x}{5}-\dfrac{\cosh^3 x}{3}+C$ **41.** $\sinh 2$ **43.** $2\tan^{-1}\dfrac{5}{3}$

45. $\pi(2-\tanh 2)$ **51.** $\tanh\dfrac{x}{2}=\dfrac{\sinh\frac{x}{2}}{\cosh\frac{x}{2}}\cdot\dfrac{2\cosh\frac{x}{2}}{2\cosh\frac{x}{2}}=\dfrac{2\sinh\frac{x}{2}\cosh\frac{x}{2}}{2\cosh^2\frac{x}{2}}=\dfrac{\sinh x}{1+\cosh x}$ by identities 4 and 7

55. $\ln\cosh x-\dfrac{\tanh^2 x}{2}-\dfrac{\tanh^4 x}{4}+C$ **57.** $\dfrac{3x}{2}-\coth x+\dfrac{\sinh 2x}{4}+C$

59. $\displaystyle\int \operatorname{csch} x\,dx=\int\frac{\sinh x}{\sinh^2 x}\,dx=\int\frac{\sinh x}{1-\cosh^2 x}\,dx$. Now let $u=\cosh x$, and use partial fractions.

63. (a) 30 ft (b) $20\cosh\dfrac{3}{2}+10\approx 57.05$ ft (c) $40\sinh\dfrac{3}{2}\approx 85.17$ ft

Exercise Set 7.3, page 305

1. Domain $=(-1,1)$
Range $=\mathbf{R}$

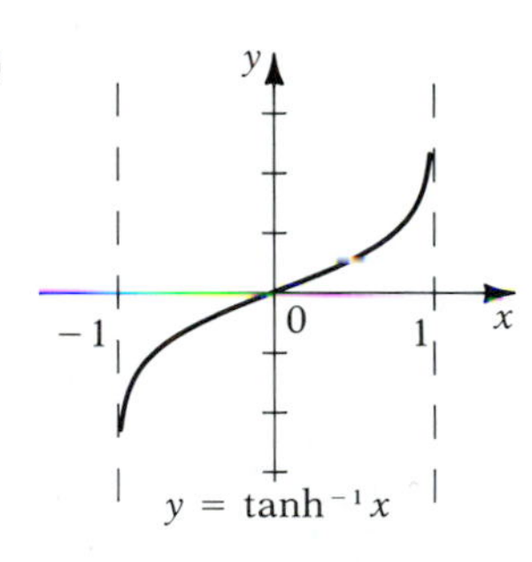

3. Domain $=(0,1]$
Range $=[0,\infty)$

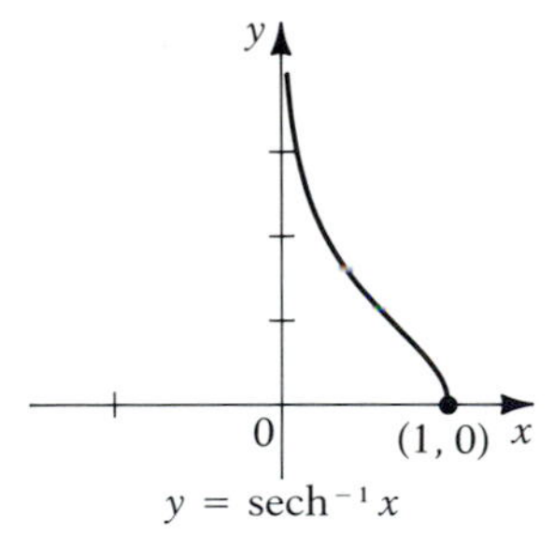

5. (a) Let $y=\operatorname{sech}^{-1} x$, $x>0$. Then $x=\operatorname{sech} y$. So $\cosh^{-1}\left(\dfrac{1}{x}\right)=\cosh^{-1}\left(\dfrac{1}{\operatorname{sech} y}\right)=\cosh^{-1}(\cosh y)=y=\operatorname{sech}^{-1} x$

(b) Use the idea of part a.

7. Let $y=\coth^{-1} x$, $|x|>1$. Then $x=\coth y$. So $1=(-\operatorname{csch}^2 y)y'$, and $y'=-\dfrac{1}{\operatorname{csch}^2 y}=-\dfrac{1}{\coth^2 y-1}=\dfrac{1}{1-x^2}$

9. Let $y=\operatorname{csch}^{-1} x$, $x\neq 0$. Then $x=\operatorname{csch} y$ and so $1=-(\operatorname{csch} y\coth y)\,y'$. Thus,

$$y'=-\frac{1}{\operatorname{csch} y\coth y}$$

From $\coth^2 y=1+\operatorname{csch}^2 y=1+x^2$, obtain

$$\coth y=\begin{cases}\sqrt{1+x^2} & \text{if } x>0\\ -\sqrt{1+x^2} & \text{if } x<0\end{cases}$$

Thus,

$$y'=-\frac{1}{|x|\sqrt{1+x^2}}$$

11. $\dfrac{2x}{\sqrt{x^4-1}}$ **13.** $-2\csc 2x$ **15.** $\operatorname{csch}^{-1} x-\dfrac{1}{\sqrt{1+x^2}}$ **17.** $-\dfrac{1}{\sqrt{-(x+x^2)}}$ **19.** $\dfrac{e^x}{1+2e^x}$

21. $\frac{1}{2}\cosh^{-1} 2x + C$ **23.** $-\operatorname{sech}^{-1} 2x + C$ **25.** $\cosh^{-1}(\tan x) + C$ **27.** $\frac{1}{2}\ln\frac{5}{3}$ **29.** $\sinh^{-1}(x+1) + C$

35. $\frac{1}{3}\sinh^{-1}\frac{3x}{2} + C$ **37.** $\frac{1}{12}\ln 5$ **39.** $-\frac{1}{2}\operatorname{csch}^{-1}\frac{e^x}{2} + C$

45. $\frac{d}{dx}\cosh^{-1} x = \frac{d}{dx}\ln(x + \sqrt{x^2-1}) = \frac{1}{x+\sqrt{x^2-1}}\left(1 + \frac{x}{\sqrt{x^2-1}}\right) = \frac{1}{\sqrt{x^2-1}},\ x > 1$

47. $\frac{d}{dx}\coth^{-1} x = \frac{d}{dx}\left(\frac{1}{2}\ln\frac{1+x}{1-x}\right) = \frac{1}{2}\frac{d}{dx}[\ln(1+x) - \ln(1-x)]$

$$= \frac{1}{2}\left(\frac{1}{1+x} + \frac{1}{1-x}\right) = \frac{1}{1-x^2}$$

49. For $x > 0$,

$$\frac{d}{dx}\operatorname{csch}^{-1} x = \frac{d}{dx}\ln\left(\frac{1+\sqrt{x^2+1}}{x}\right) = \frac{d}{dx}[\ln(1+\sqrt{x^2+1}) - \ln x]$$

$$= \frac{1}{1+\sqrt{x^2+1}}\cdot\frac{x}{\sqrt{x^2+1}} - \frac{1}{x} = -\frac{1}{x\sqrt{x^2+1}} = -\frac{1}{|x|\sqrt{x^2+1}}$$

For $x < 0$, use $|x| = -x$, and proceed in a similar way.

51. For $x > 0$, $\frac{d}{dx}\operatorname{csch}^{-1}|x| = -\frac{1}{x\sqrt{1+x^2}}$, and for $x < 0$

$$\frac{d}{dx}\operatorname{csch}^{-1}|x| = \frac{d}{dx}\operatorname{csch}(-x) = -\frac{1}{(-x)\sqrt{1+x^2}}(-1) = -\frac{1}{x\sqrt{1+x^2}}$$

So for all $u \neq 0$,

$$\int\frac{du}{u\sqrt{a^2+u^2}} = \frac{1}{a}\int\frac{\frac{1}{a}\,du}{\frac{u}{a}\sqrt{1+\left(\frac{u}{a}\right)^2}} = -\frac{1}{a}\operatorname{csch}^{-1}\frac{|u|}{a} + C \quad (a > 0)$$

53. (a) $v(t) = \sqrt{\frac{mg}{k}}\tanh\left(\sqrt{\frac{kg}{m}}\,t\right)$

(b) $s(t) = \frac{m}{k}\ln\cosh\left(\sqrt{\frac{kg}{m}}\,t\right)$

(c) $\lim_{t\to\infty} v(t) = \sqrt{\frac{mg}{k}}$

Supplementary Exercises 7.4, page 307

1. (a) $\frac{\pi}{2}$ (b) $\frac{5\pi}{6}$ (c) $-\frac{\pi}{6}$ (d) $\frac{5\pi}{4}$ (e) $\frac{\pi}{2}$ (f) $\frac{4\pi}{3}$ **3.** (a) $\frac{7}{25}$ (b) $\frac{\pi}{9}$ **5.** (a) 5.052 (b) 3.990

7. (a) $\frac{1}{\sqrt{e^{2x}-1}}$ (b) $\sqrt{1-x^2} + 2x\sec^{-1}\frac{1}{x}$ **9.** (a) $2\tanh x$ (b) $\operatorname{sech} x$ **11.** (a) $\frac{1}{\sqrt{x^2-1}}$ (b) $\sec x$

13. $\frac{1}{2}\sin\frac{2x}{3} + C$ **15.** $-\frac{\pi}{12\sqrt{2}}$ **17.** $\frac{\pi}{4}$ **19.** $-\operatorname{csch}^{-1} e^x + C$ **21.** $\frac{1}{2}\ln(1+\cosh 2x) + C$ **23.** $\frac{1}{2}\ln\cosh x^2 + C$

25. $\frac{9}{50}$ **27.** $\frac{1}{2}\ln 3$ **29.** $\frac{4\pi^2}{3}$ **31.** $5\sqrt{10}$ ft **33.** $\frac{\coth x\coth y + 1}{\coth x + \coth y}$ **39.** $A = -1,\ B = 0$

CHAPTER 8

Exercise Set 8.2, page 317

1. $-xe^{-x} - e^{-x} + C$ **3.** $\frac{\pi}{8} - \frac{1}{4}$ **5.** $e^x(x^4 - 4x^3 + 12x^2 - 24x + 24) + C$ **7.** $x \sec x - \ln|\sec x + \tan x| + C$

9. $\frac{3}{4}(\ln 2)^2 - \frac{5}{2}\ln 2 + \frac{3}{2}$ **11.** $-x \cot x + \ln|\sin x| + C$ **13.** $-\frac{x^2 \ln x}{2} + \frac{x^2}{4} + C$ **15.** $\sin x(\ln \sin x - 1) + C$

17. $\frac{1}{6} + \frac{\ln 3}{4}$ **19.** $-\frac{e^{-2x}}{8}(4x^3 + 6x^2 + 6x + 3) + C$ **21.** $x \tan x + \ln|\cos x| - \frac{x^2}{2} + C$ **25.** $\frac{1}{4}e^{2x} + \frac{1}{2}x + C$

27. $\frac{1}{2}[\operatorname{sech} x \tanh x + \tan^{-1}(\sinh x)] + C$ **29.** $\frac{x^3}{3}\cot^{-1} x + \frac{x^2}{6} - \frac{1}{6}\ln(1 + x^2) + C$ **31.** $4\pi(\pi - 2)$

33. $2(\sin\sqrt{x} - \sqrt{x}\cos\sqrt{x}) + C$ **35.** $\frac{x^2}{4} - \frac{x \sin 2x}{4} - \frac{\cos 2x}{8} + C$ **37.** $-\frac{\sqrt{1 - x^2}}{3}(x^2 + 2) + C$

39. $\frac{e^{ax}(a \sin bx - b \cos bx)}{a^2 + b^2} + C$ **43.** (a) $\frac{8}{15}$ (b) $\frac{5\pi}{32}$ (c) $\frac{63\pi}{512}$ (d) $\frac{16}{35}$

45. $\int u\,dv = u(v + C) - \int (v + C)\,du = uv + uC - \int v\,du - Cu = uv - \int v\,du$

Exercise Set 8.3, page 323

1. $-\cos x + \frac{2\cos^3 x}{3} - \frac{\cos^5 x}{5} + C$ **3.** $\frac{3x}{8} - \frac{\sin 2x}{4} + \frac{\sin 4x}{32} + C$ **5.** $\frac{16 - 7\sqrt{2}}{120}$ **7.** $\frac{x}{8} - \frac{\sin 4x}{32} + C$ **9.** $\frac{1}{4}$

11. $\frac{\sec^3 x}{3} - \sec x + C$ **13.** $\frac{1}{2}(\sec x \tan x - \ln|\sec x + \tan x|) + C$ **15.** $\ln 2 - \frac{3}{8}$ **17.** $\tan x - \cot x + C$ **19.** $\frac{3}{2} - \ln 2$

21. $\frac{\sin^3 3x}{9} - \frac{\sin^5 3x}{15} + C$ **23.** $\frac{10}{21}$ **25.** $\frac{1}{2}\left(\frac{\sin^3 x^2}{3} - \frac{2\sin^5 x^2}{5} - \frac{\sin^7 x^2}{7}\right) + C$ **27.** $-\frac{2}{3}\cot^3\frac{x}{2} + C$

29. $\frac{\sinh^2 x}{2} + \frac{\sinh^4 x}{4} + C$ **31.** $\frac{\tanh^4 x}{4} - \frac{\tanh^6 x}{6} + C$ **33.** $-\frac{\cos 5x}{10} - \frac{\cos x}{2} + C$ **35.** $\frac{\sin 5x}{10} + \frac{\sin x}{2} + C$

37. $\frac{1}{8}\left(\frac{5}{2}x - 2\sin 2x + \frac{3}{8}\sin 4x + \frac{1}{6}\sin^3 2x\right) + C$ **39.** $\frac{\sec^3 x}{3} - 2\ln|\sec x| - \cos x + C$

Exercise Set 8.4, page 328

1. $-\frac{1}{15}(1 - x^2)^{3/2}(3x^2 + 2) + C$ **3.** $\frac{1}{3}\sqrt{x^2 - 1}\,(x^2 + 2) + C$ **5.** $3\ln\left|\frac{3 - \sqrt{9 - x^2}}{x}\right| + \sqrt{9 - x^2} + C$

7. $\frac{8}{27\sqrt{3}}(3\sqrt{3} - 1)$ **9.** $\frac{\ln(2 + \sqrt{3})}{3\sqrt{3}} - \frac{1}{6}$ **11.** $\frac{1}{2}\left(\frac{x}{1 - x^2} + \ln\left|\frac{1 + x}{\sqrt{1 - x^2}}\right|\right) + C$ **13.** $-\frac{1}{4}\sqrt{9 - 4x^2} + C$ **15.** $\frac{1}{2}$

25. Let $x = \sinh t$ **27.** Let $x = 3\tanh t$, or let $x = 3\operatorname{sech} t$ if $0 < x \le 3$ and $x = -3\operatorname{sech} t$ if $-3 \le x < 0$.

29. πab **31.** $\frac{3}{8}\sin^{-1}(x - 1) + \frac{3}{8}(x - 1)\sqrt{2x - x^2} + \frac{1}{4}(x - 1)(2x - x^2)^{3/2} + C$ **33.** $\frac{1}{3}\left(\frac{3x + 2}{\sqrt{5 - 12x - 9x^2}} - \sin^{-1}\frac{3x + 2}{\sqrt{5}}\right) + C$

35. $\frac{1}{3}\sqrt{\tan^2 t - 1}\,(\tan^2 t + 2) + C$ **37.** $\frac{1}{2}(\sin^{-1} e^x - e^x\sqrt{1 - e^{2x}}) + C$

Exercise Set 8.5, page 336

1. $\frac{1}{2}\ln\left|\frac{x}{x+2}\right| + C$ **3.** $\frac{2}{5}\ln|x+2| + \frac{3}{5}\ln|x-3| + C$ **5.** $5\ln 6$ **7.** $\ln\frac{(x+2)^4}{|x+3|} + C$ **9.** $\ln\frac{\sqrt{|x^2-1|}}{|x|} + C$

11. $x + \ln(x+2)^2|x-1| + C$ **13.** $5\ln|x+3| - 6\ln|x+4| + 4\ln|x-2| + C$ **15.** $x^2 + x + \ln\left|\frac{x-1}{(x+1)^5}\right| + C$

17. $\ln 12 + 2$ **19.** $\ln\left|\frac{x^3}{x-2}\right| - \frac{4}{x-2} + C$ **21.** $\ln\frac{(x-2)^2}{x^2+4} + \frac{1}{2}\tan^{-1}\frac{x}{2} + C$ **23.** $\frac{\pi}{3\sqrt{3}} + \ln\sqrt{3} - \frac{4}{3}$

25. $\frac{1}{6}\ln\frac{(x-1)^2}{x^2+x+1} - \frac{1}{\sqrt{3}}\tan^{-1}\frac{2x+1}{\sqrt{3}} + C$ **27.** (a) $\ln\frac{4}{3}$ (b) $\pi\left(\frac{2}{3} + \ln\frac{9}{16}\right)$ **29.** $\frac{1}{b}\ln\left|\frac{u}{au+b}\right| + C$

31. $\frac{1}{b^2}\ln\left|\frac{u}{au+b}\right|$ **33.** $\frac{1}{3a^2}\left(\ln\frac{|u-a|}{\sqrt{u^2+au+a^2}} - \sqrt{3}\tan^{-1}\frac{2u+a}{\sqrt{3}a}\right) + C$ **35.** $\ln\left|\frac{x-1}{x+1}\right| - \frac{2}{x-1} + C$

37. $\frac{3}{2}\ln(x^2+4) - \ln(x^2+1) + \tan^{-1}x - 2\tan^{-1}\frac{x}{2} + C$ **39.** $Q(t) = \frac{mQ_0}{Q_0 + (m-Q_0)e^{-mkt}}$

Exercise Set 8.6, page 340

1. $2\left[\frac{x^{3/2}}{3} - \frac{x}{2} + \sqrt{x} - \ln(1+\sqrt{x})\right] + C$ **3.** $2(\sqrt{x-1} - \tan^{-1}\sqrt{x-1}) + C$ **5.** $-\frac{26}{27}$ **7.** $-4\left(\frac{1}{\sqrt[4]{x}} + \frac{1}{2\sqrt{x}}\right) + C$

9. $\frac{2x^{1/6}}{35}(15x - 21x^{2/3} + 35x^{1/3} - 105) + 6\tan^{-1}x^{1/6} + C$ **11.** $\sqrt{x^2+1} + \ln(\sqrt{x^2+1} - 1) + C$ **13.** $\frac{15}{4}$

15. $\frac{2\sin x}{\cos x - \sin x - 1} + C$ **17.** $\ln\left|\tan\frac{x}{2} - 1\right| + C$ **19.** $6\left[\ln\left|\frac{x^{1/6}-1}{x^{1/6}}\right| + \frac{1}{x^{1/6}-1} - \frac{1}{2(x^{1/6}-1)^2}\right] + C$

21. $\frac{1}{4}\left[\ln|\sqrt[3]{x^2+8} - 2| - \frac{1}{2}\ln|(x^2+8)^{2/3} + 2\sqrt[3]{x^2+8} + 4| + \sqrt{3}\tan^{-1}\frac{\sqrt[3]{x^2+8}+1}{\sqrt{3}}\right] + C$

23. $\frac{\pi}{6\sqrt{3}}(\sqrt{3} - 2)$ **25.** $2\sqrt{1+e^x} + \frac{1}{2}\ln\frac{\sqrt{1+e^x}-1}{\sqrt{1+e^x}+1} + C$

Exercise Set 8.7, page 343

1. $\frac{1}{2(2+3x)} - \frac{1}{4}\ln\left|\frac{2+3x}{x}\right| + C$ (#22) **3.** $\frac{4x^2+1}{8}\tan^{-1}2x - \frac{x}{4} + C$ (#97) **5.** $-\frac{\sqrt{4x-x^2}}{2x} + C$ (#66)

7. $\frac{1}{2x} + \frac{1}{4}\ln\left|\frac{x-2}{x}\right| + C$ (#23) **9.** $\frac{x}{9\sqrt{x^2+9}} + C$ (#49) **11.** $\frac{x-1}{2}\sqrt{2x-x^2} + \frac{1}{2}\cos^{-1}(1-x) + C$ (#59)

13. $e^{3x}\left(\frac{1}{3}x^4 - \frac{4}{9}x^3 + \frac{4}{9}x^2 - \frac{8}{27}x + \frac{8}{81}\right) + C$ (#102) **15.** $\frac{1}{5}\ln\left|\frac{x}{3x+5}\right| + C$ (#26) **17.** $\frac{3}{16(4x-3)} + \frac{1}{16}\ln|4x-3| + C$ (#20)

19. $\frac{e^{-2x}}{13}(3\sin 3x - 2\cos 3x) + C$ (#112) **21.** $-x + \frac{23}{20}\ln|x-4| - \frac{2}{5}\ln|x+1| + C$ (#26, #27)

23. $\frac{1}{3}\left[x^3\sin^{-1}x + \sqrt{1-x^2} - \frac{1}{3}(1-x^2)^{3/2}\right] + C$ (#98)

Supplementary Exercises 8.8, page 343

1. $-\frac{1}{4}e^{-2x}(2x^2 + 2x + 1) + C$ **3.** $\frac{\pi}{8} - \frac{\ln 2}{4}$ **5.** $2(\sqrt{x}\tan\sqrt{x} + \ln|\cos\sqrt{x}|) + C$ **7.** $\frac{\sin^6 x}{6} - \frac{\sin^8 x}{8} + C$ **9.** $\frac{1}{3}(1 + 3\sqrt{3})$

11. $-\frac{1}{4}x\cos^4 x + \frac{3x}{32} + \frac{\sin 2x}{16} + \frac{\sin 4x}{128} + C$ **13.** $\frac{1}{12}(3\cos 2x - 2\cos 6x) + C$ **15.** $\frac{2\sqrt{3}}{3} - \frac{\pi}{6}$ **17.** $\frac{\pi}{3}$

19. $\frac{1}{3}\sqrt{1+e^{2x}}(e^{2x}-2) + C$ **21.** $6\sin^{-1}\frac{x-2}{2} - 4\sqrt{4x-x^2} - \frac{1}{2}(x-2)\sqrt{4x-x^2} + C$ **23.** $1 + \ln\frac{3}{8}$

25. $\ln\frac{|x|}{\sqrt{x^2+1}} + \frac{1}{2}\left(\tan^{-1}x + \frac{x}{x^2+1}\right) + C$ **27.** $-\left(\frac{\pi}{2\sqrt{3}} + \frac{\ln 3}{2}\right)$ **29.** $\frac{1}{3}\sqrt{x^2-4}(x^2+8) + C$

31. (a) $\frac{\pi^2}{2}$ (b) $2\pi^2$ **33.** (a) $\frac{2\pi}{5}\ln 4$ (b) $10\pi\sin^{-1}\frac{3}{5}$ **35.** $y = \frac{1}{2}\left[1 - \frac{kt+C}{\sqrt{16+(kt+C)^2}}\right]$

CHAPTER 9

Exercise Set 9.2, page 355

1. (a) 0.509, $E_4 \le 0.03125$ (b) 0.50042, $E_4 \le 0.0026$. Exact value $= \frac{1}{2}$

3. (a) 0.869, $E_8 \le 0.0104$ (b) 0.8647, $E_8 \le 0.000034$. Exact value $= 1 - \frac{1}{e^2} \approx 0.8646647$

5. (a) 3.135, $E_{10} \le 0.0267$ (b) 3.14159, $E_{10} \le 0.0009$. Exact value $= \pi \approx 3.1415927$

7. (a) 6.94, $E_6 \le 0.75$ (b) 6.75, $E_6 = 0$. Exact value $= \frac{27}{4} = 6.75$ **9.** (a) $n = 30$ (b) $n = 5$ **11.** (a) $n = 50$ (b) $n = 8$

13. (a) $n = 116$ (b) $n = 11$ **15.** 1.76, with $n = 12$. Exact value $= 2\ln(1+\sqrt{2}) \approx 1.7627$ **17.** 53.8 ft-lb **19.** 1.53×10^7 lb

21. (a) 0.746, $E_{10} \le 0.00167$ (b) 0.7468, $E_{10} \le 0.0000467$ **23.** (a) 0.659, $E_{10} \le 0.00833$ (b) 0.6725, $E_{10} \le 0.0000933$

Exercise Set 9.3, page 362

1. 2.2247 **3.** −0.8850 **5.** 0.4429 **7.** 0.7140, 1.6449 **9.** −2.1245 **11.** $x^3 - 4 = 0$; 1.5874

13. $x^4 - 6 = 0$; 1.5651

15. $I = (2, 3)$, $x_0 = 2.9$ **17.** $I = (1, 1.5)$, $x_0 = 1.4$

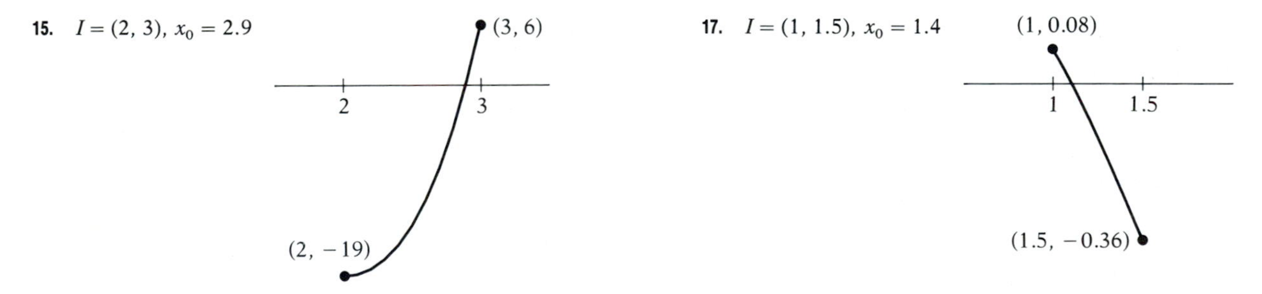

19. $f(-2) = 30$, $f(-1) = -2$; $f'(x) < 0$ and $f''(x) > 0$ in I; $f(-1.2) = 1.1296$; at least 3 places

21. $f(2) \approx -2.6$, $f(3) \approx 1.19$; $f'(x) > 0$ and $f''(x) > 0$ in I; $f(2.7) = 0.2965$; at least 8 places

23. Two; −1.4301 and 3.2480

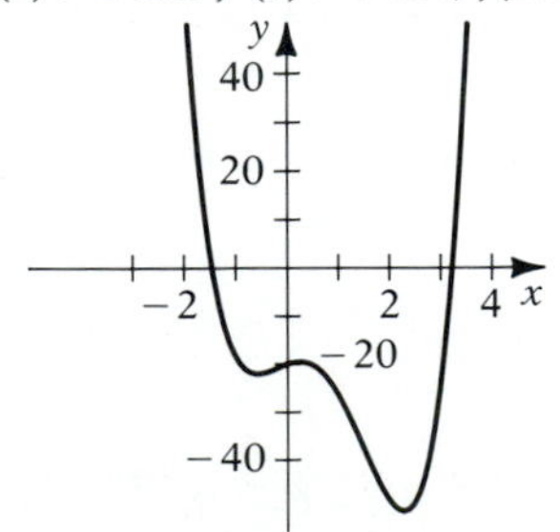

25. (a) $z_1, z_2, \ldots$ are on opposite sides of root and converge to root. Similarly for other cases.
(b) Equation of secant line joining $(x_0, f(x_0))$ and $(z_n, f(z_n))$ is $y - f(z_n) = m_n(x - z_n)$. The x-intercept is z_{n+1}.
(c) Tangent lines and secant lines have x-intercepts on opposite sides of r in each case. So $E_n = |x_n - r| \le |x_n - z_n|$.

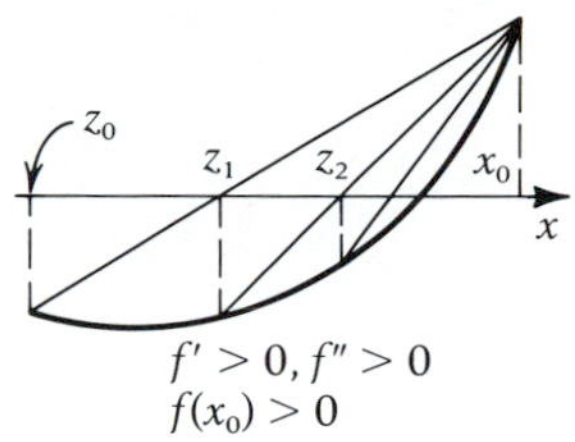

Exercise Set 9.4, page 373

1. $1 - x + \frac{x^2}{2!} - \frac{x^3}{3!} + \frac{x^4}{4!} - \frac{x^5}{5!}$ **3.** $1 - x + x^2 - x^3 + x^4 - x^5 + x^6$ **5.** $1 + \frac{x^2}{2!} + \frac{x^4}{4!} + \frac{x^6}{6!} + \frac{x^8}{8!}$

7. $1 - (x-1) + (x-1)^2 - (x-1)^3 + \cdots + (-1)^n(x-1)^n$ **9.** $x - \frac{x^3}{3!} + \frac{x^5}{5!} - \cdots + \frac{(-1)^k x^{2k-1}}{(2k-1)!}$, $n = 2k-1$ or $2k$, $k = 1, 2, 3, \ldots$

11. $-(x-1) - \frac{(x-1)^2}{2} - \frac{(x-1)^3}{3} - \cdots - \frac{(x-1)^n}{n}$ **13.** $R_n(x) = \frac{(-1)^{n+1}(x-1)^{n+1}}{c^{n+2}}$ $x \in (0, 2)$, c between 1 and x

15. For $n = 2k-1$ or $2k$, $R_n(x) = R_{2k}(x) = \frac{(-1)^k \cos c}{(2k+1)!} x^{2k+1}$, $k = 1, 2, 3, \ldots$; $x \in (-\infty, \infty)$, c between 0 and x

17. $R_n(x) = -\frac{(x-1)^{n+1}}{(2-c)^{n+1}(n+1)}$; $x \in (0, 2)$, c between 0 and x

19. $e^{-1} \approx P_5(1) \approx 0.36667$; $|R_5(1)| \le \frac{1}{6!} \approx 0.0013889$; 2 places, 0.37

21. $\sin 1.5 \approx P_7(1.5) = P_8(1.5) \approx 0.99739$; $|R_8(1.5)| \le \frac{(1.5)^9}{9!} \approx 0.0001059$; 3 places, 0.997

23. $\cos 2 \approx P_8(2) = P_9(2) \approx -0.415873$; $|R_9(2)| \le \frac{2^{10}}{(10)!} \approx 0.0002822$; 3 places, -0.416

25. $\sqrt{5} \approx P_5(5) \approx 2.236076$; $|R_5(5)| \le \frac{1 \cdot 3 \cdot 5 \cdot 7 \cdot 9}{2^{17} \cdot 6!} \approx 0.00001001$; 4 places, 2.2361

27. $e^3 \approx P_{12}(3) \approx 20.0852$; $|R_{12}(3)| \le \frac{3^{16}}{(13)!} \approx 0.0069$; 1 place, 20.1

29. $\cosh 1 \approx P_6(1) = P_7(1) \approx 1.54306$; $|R_7(1)| \le \frac{2}{8!} \approx 0.0000496$, 4 places, 1.5431

31. $\cos 58° = \cos\left(\frac{29\pi}{90}\right) \approx P_4\left(\frac{29\pi}{90}\right) \approx 0.5299193$; $\left|R_4\left(\frac{29\pi}{90}\right)\right| \le \frac{1}{5!}\left(\frac{\pi}{90}\right)^5 \approx 4.3 \times 10^{-10}$; 9 places, 0.5299193 (calculator limited to 7 places)

33. Calculate $P_n^{(k)}(x) = k!a_k + (k+1)!a_{k+1}(x-a) + (k+2)(k+1)\cdots 4 \cdot 3a_{k+2}(x-a)^2 + \cdots + n(n-1)(n-2)\cdots(n-k+1)a_n(x-a)^{n-k}$. Then $P_n^{(k)}(a) = k!a_k$, and put this equal to $f^{(k)}(a)$.

35. 0.98481

37. $P_n(x) = 2 + \frac{1}{2^2}(x-4) - \frac{1}{2^5 \cdot 2!}(x-4)^2 + \frac{1 \cdot 3}{2^8 \cdot 3!}(x-4)^3 + \cdots + (-1)^{n-1}\frac{1 \cdot 3 \cdot 5 \cdots (2n-3)}{2^{3n-1} \cdot n!}(x-4)^n$

$R_n(x) = (-1)^n \frac{1 \cdot 3 \cdot 5 \cdots (2n-1)}{2^{n+1}c^{(2n+1)/2}(n+1)!}(x-4)^{n+1}$, $x \in (0, 8)$, c between 4 and x; $\sqrt{3} \approx 1.7321$

39. (a) $P_n(x) = 1 + \alpha x + \dfrac{\alpha(\alpha-1)}{2!}x^2 + \dfrac{\alpha(\alpha-1)(\alpha-2)}{3!}x^3 + \cdots + \dfrac{\alpha(\alpha-1)(\alpha-2)\cdots(\alpha-n+1)}{n!}x^n$

$R_n(x) = \dfrac{\alpha(\alpha-1)(\alpha-2)\cdots(\alpha-n)}{(n+1)!(1+c)^{n+1-\alpha}}x^{n+1}$, c between 0 and x

(b) If $\alpha = m$, $R_m(x) = 0$. So $f(x) = P_m(x)$

(c) $f^{(n+1)}(x)$ fails to exist at $x = -1$ only. So the radius of the neighborhood of validity is $|0-(-1)| = 1$.

(d) $\sqrt[3]{1.5} = (1+0.5)^{1/3} \approx 1.14$; $|R_4(0.5)| \le 0.000943$

41. (a) $x = \dfrac{u-1}{u+1} = 1 - \dfrac{2}{u+1} < 1$ and $1 - \dfrac{2}{u+1} > 1 - \dfrac{2}{1} = -1$. So $x \in (-1, 1)$.

(b) For $n = 2k-1$ or $2k$,

$$P_n(x) = P_{2k-1}(x) = 2\left[x + \frac{x^3}{3} + \frac{x^5}{5} + \frac{x^7}{7} + \cdots + \frac{x^{2k-1}}{(2k-1)!}\right]$$

$$R_{2k-1}(x) = R_{2k}(x) = \left[\frac{1}{(1+c)^{2k+1}} - \frac{1}{(1-c)^{2k+1}}\right]\frac{x^{2k+1}}{2k+1}, \quad k = 1, 2, 3, \ldots,$$

c between 0 and x

(c) 0.693

Supplementary Exercises 9.5, page 374

1. Both trapezoidal and Simpson's rules give 1. For trapezoidal rule, $E_8 \le 0.0514$ and for Simpson's rule, $E_8 \le 0.002$. Exact value $= 1$.

3. Trapezoidal rule: -0.788; $E_4 \le 0.042$, 1 place accuracy; Simpson's rule: -0.7730; $E_4 \le 0.00208$, 2 place accuracy; Exact value $= 2 - 4\ln 2 \approx -0.7725887$.

5. $n = 130$; $n = 18$

7. Trapezoidal rule: 1.126; $E_{10} \le 0.000833$: 2 place accuracy, 1.13; Simpson's rule: 1.125387; $E_{10} \le 0.00004$; 4 place accuracy, 1.1254

9. $V \approx 19.09$, $E_{10} \le 0.00176$ **11.** 0.8437 **13.** -1.0580, 1.9646 **15.** $f_{\max} \approx 0.561096$ at $x \approx 0.860334$

17. $n = 4$ **19.** $f\left(\dfrac{\pi}{6}\right) \approx 0.142$, $f\left(\dfrac{\pi}{3}\right) \approx -0.523$; $f'(x) < 0$, $f''(x) < 0$ on I; $x_0 = \dfrac{\pi}{4}$; $f(x_0) < 0$

21. $P_5(x) = -x - \dfrac{x^2}{2} - \dfrac{x^3}{3} - \dfrac{x^4}{4} - \dfrac{x^5}{5}$; for $|x| \le 0.2$, $|R_5(x)| \le \dfrac{1}{4^6 \cdot 6} \approx 0.0000407$; 4 decimal place accuracy

23. $P_5(x) = 1 + 2\left(x - \dfrac{\pi}{4}\right) + 2\left(x - \dfrac{\pi}{4}\right)^2 + \dfrac{8}{3}\left(x - \dfrac{\pi}{4}\right)^3 + \dfrac{10}{3}\left(x - \dfrac{\pi}{4}\right)^4 + \dfrac{64}{15}\left(x - \dfrac{\pi}{4}\right)^5$

25. For $n = 2k-1$ or $2k$

$$P_n(x) = P_{2k-1}(x) = \sqrt{2}\left[\left(x - \frac{\pi}{4}\right) - \frac{\left(x - \frac{\pi}{4}\right)^3}{3!} + \frac{\left(x - \frac{\pi}{4}\right)^5}{5!} - \cdots + \frac{(-1)^k\left(x - \frac{\pi}{4}\right)^{2k-1}}{(2k-1)!}\right] \quad k = 1, 2, 3, \ldots$$

27. $P_3(x) = \dfrac{\pi}{4} + \dfrac{(x-1)}{2} - \dfrac{(x-1)^2}{4} - \dfrac{(x-1)^3}{12}$

$R_3(x) = \dfrac{c(1-c)(x-1)^4}{(1+c^2)^4}$, c between 1 and x

29.

x	$P_1(x)$	$P_2(x)$	$P_3(x)$	$P_4(x)$	e^x
0.25	1.25	1.28125	1.2838542	1.2840169	1.2840254
0.50	1.50	1.62500	1.6458333	1.6484375	1.6487213
0.75	1.75	2.03125	2.1015625	2.1147461	2.1170000
1.00	2.00	2.50000	2.6666667	2.7083333	2.7182818
−0.25	0.75	0.78125	0.7786458	0.7788086	0.7788080
−0.50	0.50	0.62500	0.6041667	0.6067708	0.6065307
−0.75	0.25	0.53125	0.4609375	0.4741211	0.4723666
−1.00	0.00	0.50000	0.3333333	0.3750000	0.3678794

CHAPTER 10

Exercise Set 10.1, page 381

1. 2 **3.** $\frac{1}{2}$ **5.** 1 **7.** 10 **9.** $\frac{1}{2}$ **11.** 6 **13.** $+\infty$ **15.** 4 **17.** 0 **19.** $-\infty$ **21.** $-\frac{1}{2}$ **23.** 1 **25.** 0
27. 0 **29.** $\frac{1}{2}$ **31.** $+\infty$ **33.** 0 **35.** Suppose $n - 1 < \alpha \leq n$. Then, apply L'Hôpital's rule n times to get

$$\lim_{x \to +\infty} \frac{\alpha(\alpha-1)(\alpha-2)\cdots(\alpha-n+1)x^{\alpha-n}}{e^x}$$

and use the fact that $\alpha - n \leq 0$.
37. 0 **39.** 0 **43.** Yes. In Exercise 42 both limits exist even though $\lim f'(x)/g'(x)$ does not.

Exercise Set 10.2, page 385

1. 0 **3.** 0 **5.** 0 **7.** $-\frac{1}{2}$ **9.** 1 **11.** e **13.** $+\infty$ **15.** 1 **17.** $\frac{1}{2}$ **19.** 1 **21.** $\frac{1}{e^2}$ **23.** 1 **25.** 0
27. e^2 **29.** $-\frac{1}{2}$ **31.** e^2 **33.** $\sqrt{e}$ **35.** 2 **37.** $\frac{1}{3}$ **39.** $-\infty$ **41.** 0 **43.** 0 **45.** 1 **47.** 1 **49.** 0

51. Let $y = f(x)^{g(x)}$, so that $\ln y = g(x)\ln(f(x))$. Now consider $\lim_{x\to a} g(x) = +\infty$ and $\lim_{x\to a} g(x) = -\infty$ separately.
53. For $f'(0)$ consider right-hand and left-hand limits of the difference quotient $[f(x) - f(0)]/x$. Use Exercise 51 in the first case.

Local maximum $e^{1/e}$ when $x = e$, local minimum 0 when $x \leq 0$; asymptote $y = 1$.

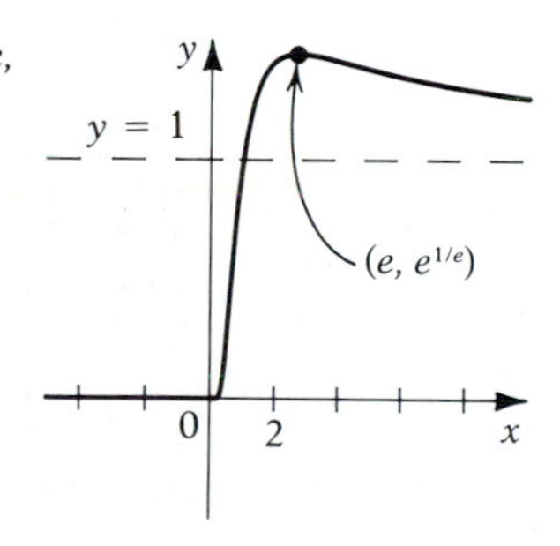

Exercise Set 10.3, page 394

1. $\frac{1}{2}$ **3.** 1 **5.** $\frac{\pi}{2}$ **7.** ln 2 **9.** $\frac{1}{2}$ **11.** $\frac{5}{e}$ **13.** Diverges **15.** 0 **17.** ln 3 **19.** ln 4
21. $\ln(\sqrt{2}+1)$ **23.** Diverges **25.** $\frac{1}{27}$ **27.** Converges **29.** Diverges **31.** Converges **33.** Converges
35. $\frac{1}{3}$; $e^{-1/3} \approx 0.72$ **37.** 2; $\frac{e^3-1}{e^4} \approx 0.35$ **39.** $\frac{\pi}{4}$ **41.** $\frac{1}{\ln 2}$ **45.** $e^{-3/2} \approx 0.223$ **47.** Converges if $p > 1$, diverges if $p \leq 1$
51. (a) $\frac{1}{s}$ $(s > 0)$ (b) $\frac{1}{s^2}$ $(s > 0)$ (c) $\frac{1}{s-1}$ $(s > 1)$ (d) $\frac{1}{1+s^2}$ $(s > 0)$ **53.** (a) π (b) $+\infty$ **55.** (a) $e^{-1} \approx 0.368$ (b) 4

Exercise Set 10.4, page 402

1. Diverges **3.** $\sqrt{3}$ **5.** $\frac{\pi}{3}$ **7.** Diverges **9.** Diverges **11.** Diverges **13.** $-2\sqrt{\ln 2}$ **15.** $\sqrt[3]{18}$
17. $\frac{3}{2}$ **19.** Diverges **21.** 0 **23.** Diverges **25.** Diverges **27.** Diverges **29.** Diverges **31.** Diverges **33.** 3 ln 2

35. 0 **37.** $\frac{\pi}{3}$ **39.** 2 **41.** π **43.** Diverges **45.** Diverges **47.** Diverges

49. Let $u = \frac{1}{x}$. Show that $\int_0^1 \cos\frac{1}{x}\,dx = \int_1^\infty \frac{\cos u}{u^2}\,du$. Now compare with $\int_1^\infty \frac{du}{u^2}$.

51. (a) Converges (b) Diverges (c) Converges (d) Diverges

53. (a) $\frac{\pi}{2}$ (b) $2\pi^2 - 4\pi$ (c) Compare with $C\int_0^2 \frac{dx}{\sqrt{2-x}}$, for some constant C.

Supplementary Exercises 10.5, page 404

1. 1 **3.** $-\frac{1}{2}$ **5.** $\frac{1}{2}$ **7.** 0 **9.** 0 **11.** $-\frac{1}{2}$ **13.** 1 **15.** 1 **17.** $1 - \frac{\pi}{4}$ **19.** $\frac{\pi}{4}$ **21.** $16\sqrt{3}$ **23.** $\frac{\pi}{2}$

25. Diverges **27.** Diverges **29.** Diverges **31.** $\frac{\pi}{4}$ **33.** $\frac{\pi}{4}$

37. For part b an equivalent inequality is $x^{\alpha+1}e^{-x} \le 1$ for x sufficiently large. Use L'Hôpital's rule to show that $\lim_{x\to\infty} x^{\alpha+1}e^{-x} = 0$, and from this draw the desired conclusion. For part c use comparison tests.

39. $\mu = \alpha\beta$, $\sigma^2 = \alpha\beta^2$

CHAPTER 11

Exercise Set 11.2, page 416

1. (a) $1, -\frac{2}{3}, \frac{3}{5}, -\frac{4}{7}, \frac{5}{9}$ (b) $2, \frac{4}{2!}, \frac{8}{3!}, \frac{16}{4!}, \frac{32}{5!}$ **3.** (a) $1, -2, 6, -24, 120$ (b) $1, 2, 2, 1, \frac{1}{2}$

5. (a) $\frac{2^n - 1}{2^n}$ (b) $\frac{n}{n^2+1}$ **7.** $\frac{1}{3}$ **9.** Diverges **11.** 0 **13.** Diverges **15.** 1 **17.** 0 **19.** 0 **21.** 0

23. Diverges **25.** 0 **27.** 0 **29.** 0 **31.** Increasing, since $a_{n+1} - a_n > 0$

33. Decreasing for $n \ge 2$, since $\frac{d}{dx}(x^2e^{-x}) \le 0$ for $x \ge 2$ **35.** Decreasing, since $a_{n+1}/a_n \le 1$ **37.** Decreasing, since $a_{n+1}/a_n < 1$

39. Increasing, since $\frac{d}{dx}\left[\ln x - \ln(x+1)\right] > 0$ for $x > 0$ **41.** Decreasing for $n \ge 2$, and bounded below by 0

43. Decreasing for $n \ge 3$, and bounded below by 0 **45.** Increasing, and bounded above by 2

47. Assume two limits, $L_1 \ne L_2$. Consider disjoint deleted neighborhoods of L_1 and L_2, respectively.

53. Use the fact that $|a_n - 0| = ||a_n| - 0|$.

57. Sequence is increasing, since $a_{n+1} - a_n = 1/(2n+1) - 1/(2n+2) > 0$, and it is bounded above, since $a_n < n/(n+1) < 1$.

Exercise Set 11.3, page 426

1. $\frac{1}{2} + \frac{2}{5} + \frac{3}{10} + \frac{4}{17} + \cdots + \frac{n}{n^2+1} + \cdots$; $S_1 = \frac{1}{2}$, $S_2 = \frac{9}{10}$, $S_3 = \frac{6}{5}$, $S_4 = \frac{122}{85}$

3. $\frac{\ln 2}{1!} + \frac{\ln 3}{2!} + \frac{\ln 4}{3!} + \frac{\ln 5}{4!} + \cdots + \frac{\ln(n+1)}{n!} + \cdots$

$S_1 = \ln 2$, $S_2 = \ln 2 + \frac{1}{2}\ln 3$, $S_3 = \ln 2 + \frac{1}{2}\ln 3 + \frac{1}{6}\ln 4$

$S_4 = \ln 2 + \frac{1}{2}\ln 3 + \frac{1}{6}\ln 4 + \frac{1}{24}\ln 5$

5. $\sum_{k=1}^{\infty} \frac{(-1)^{k-1}}{3^{k-1}}$; $S_n = \sum_{k=1}^{n} \frac{(-1)^{k-1}}{3^{k-1}}$ **7.** $\sum_{k=1}^{\infty} \frac{1}{(k+1)\ln(k+1)}$; $S_n = \sum_{k=1}^{n} \frac{1}{(k+1)\ln(k+1)}$

9. Converges to 1 **11.** Converges to 2 **13.** $a = 2$, $r = -\frac{1}{2}$; converges to $\frac{4}{3}$ **15.** $a = 3$, $r = \frac{1}{2}$; converges to 6

17. $a = \frac{2}{e}$, $r = -\frac{2}{e}$; converges to $\frac{2}{e+2}$ **19.** $a = 1$, $r = 0.99$; converges to 100 **21.** $\frac{5}{33}$ **23.** $\frac{4}{27}$ **25.** $\frac{1}{2}$ **27.** 1

29. 2 **31.** $-\frac{1}{2}$ **33.** 50 m **35.** (a) $\lim_{n\to\infty} \frac{n}{100n+1} = \frac{1}{100} \neq 0$ (b) $\lim_{n\to\infty} \frac{n}{(\ln n)^2} = +\infty \neq 0$

37. (a) False (b) False (c) True (d) False (e) True

39. Suppose $\sum(a_n + b_n)$ converges. Then $\sum b_n = \sum[(a_n + b_n) - a_n] = \sum(a_n + b_n) - \sum a_n$ also converges, contrary to hypothesis.

41. $|x| < \frac{3}{2}$; $\frac{6}{3+2x}$ **43.** $\frac{11\pi}{9}$ ft

Exercise Set 11.4, page 435

1. Converges **3.** Diverges **5.** Diverges **7.** Diverges **9.** Converges **11.** Converges **13.** Diverges
15. Converges **17.** Diverges **19.** Converges **21.** Diverges **23.** Diverges **25.** Converges **27.** Converges
29. Converges **31.** Converges **33.** Converges **35.** Diverges **37.** Converges **39.** Converges **41.** Converges
43. Converges **45.** Converges **47.** Converges **49.** Converges **51.** Converges **53.** Diverges **55.** Converges
57. Diverges **59.** Converges

61. Show that since $\lim_{n\to\infty} a_n = 0$, $0 \leq a_n < 1$ for $n \geq n_0$, for some n_0. So for $n \geq n_0$, $0 \leq a_n^2 \leq a_n$. Now use the comparison test.

63. Converges if and only if $p > 1$

Exercise Set 11.5, page 441

1. Converges **3.** Converges **5.** Converges **7.** Diverges **9.** Converges **11.** Error ≤ 0.001

13. Error $\leq \frac{1}{6^6} \approx 0.00002$ **15.** $n = 39{,}998$ **17.** $n = 101$ **19.** Absolutely convergent **21.** Conditionally convergent

23. Conditionally convergent **25.** Diverges **27.** Absolutely convergent **29.** Absolutely convergent

31. Series for $2S$ is conditionally convergent, and rearranging it changed its sum.

35. $\sum_{n=1}^{\infty} \frac{(-1)^{n-1}}{n} = \sum_{n=1}^{\infty} \left(\frac{1}{2n-1} - \frac{1}{2n}\right) = \sum_{n=1}^{\infty} \frac{1}{(2n-1)(2n)}$

37. $\sum p_n = \frac{1}{2}\sum[(p_n - q_n) + (p_n + q_n)] = \frac{1}{2}(\sum a_n + \sum|a_n|)$ and both $\sum a_n$ and $\sum|a_n|$ converge. Similarly, $\sum q_n = \frac{1}{2}\sum[(p_n + q_n) - (p_n - q_n)] = \frac{1}{2}(\sum|a_n| - \sum a_n)$.

39. The rearranged series converges to -1. At the nth stage $|S_n - (-1)| \leq p_k$, where p_k is the last term of $\sum p_n$ that was added, and $p_k \to 0$ as $k \to \infty$.

Exercise Set 11.6, page 447

1. $-1 \leq x < 1$ **3.** $-1 < x < 1$ **5.** $-\infty < x < \infty$ **7.** $-3 < x < 1$ **9.** $-\frac{3}{2} < x < \frac{3}{2}$ **11.** $-1 \leq x < 1$

13. $-1 < x < 1$ **15.** $x = 0$ only **17.** $-e < x < e$ **19.** $-3 < x \leq -1$ **21.** $-\frac{1}{2} < x \leq \frac{1}{2}$ **23.** $-e < x < e$

25. $-1 < x < 1$ **27.** $0 \leq x \leq 1$ **29.** $-1 < x < 3$ **31.** $-2 \leq x < 2$ **33.** $x = 0$ only **35.** $-\frac{3}{2} < x < \frac{3}{2}$

37. $R = \frac{1}{\lim_{n\to\infty} \sqrt{|a_n|}}$

Exercise Set 11.7, page 454

1. $\sum_{n=1}^{\infty} x^{n-1}, |x| < 1$ **3.** $\sum_{n=1}^{\infty} \frac{(-1)^n x^{2n-1}}{(2n-1)!}, |x| < \infty$ **5.** $\sum_{n=1}^{\infty} \frac{nx^{n-1}}{2^{n+1}}, |x| < 2$ **7.** $\sum_{n=0}^{\infty} \frac{x^{n+1}}{(n+1)^2} + C, |x| < 1$

9. $\sum_{n=0}^{\infty} \frac{(-1)^n x^{2n+1}}{(2n+1)!} + C, |x| < \infty$ **11.** $\sum_{n=0}^{\infty} \frac{(-1)^n x^{2n+1}}{2^n} + C, |x| < \sqrt{2}$ **13.** $\sum_{n=0}^{\infty} \left(\frac{1}{2}\right)^{n+1} = 1$ **15.** $\sum_{n=0}^{\infty} \frac{2^{n+1}-1}{(n+1)!} = e^2 - e$

17. $\sum_{n=0}^{\infty} \frac{(-1)^n}{2^n(2n+1)^2}$ **19.** $\sum_{n=1}^{\infty} (-1)^{n-1} nx^{n-1}, R = 1$ **21.** $-\sum_{n=0}^{\infty} \frac{x^{n+1}}{n+1}, R = 1$ **23.** $\sum_{n=1}^{\infty} n2^n x^n, R = \frac{1}{2}$

25. $\frac{1}{2} \sum_{n=0}^{\infty} \frac{(-1)^n x^{n+1}}{n+1}, R = 1$ **27.** $\sum_{n=0}^{\infty} \frac{(-1)^n x^{2n+1}}{2^{2n+1}(2n+1)}, R = \sqrt{2}$ **29.** $\frac{1}{2} \sum_{n=2}^{\infty} n(n-1)x^{n-2}, R = 1$ **33.** $\frac{x}{(1-x)^2}$

35. $2e$ **37.** (a) $\frac{1}{3} \sum_{n=0}^{\infty} \left[(-1)^n + \frac{1}{2^{n+1}}\right] x^n, R = 1$ (b) $\frac{1}{3} \sum_{n=1}^{\infty} n\left[(-1)^n + \frac{1}{2^{n+1}}\right] x^{n-1}, R = 1$

Exercise Set 11.8, page 464

1. $\sum_{n=0}^{\infty} \frac{(-1)^n}{(2n+1)} x^{2n+1}, -\infty < x < \infty$ **3.** $\sum_{n=0}^{\infty} \frac{x^{2n}}{(2n)!}, -\infty < x < \infty$ **5.** $\sum_{n=0}^{\infty} \frac{x^{2n+1}}{2n+1}, -1 < x < 1$

7. $\sum_{n=1}^{\infty} (-1)^{n-1} 2^{2n-1} \frac{x^{2n}}{(2n)!}, -\infty < x < \infty$ **9.** $\sum_{n=0}^{\infty} \binom{-3}{n} (-1)^n x^n, -1 < x < 1$ **11.** $\sum_{n=0}^{\infty} (\ln 2)^n \frac{x^n}{n!}, -\infty < x < \infty$

13. $\sum_{n=1}^{\infty} \frac{(-1)^{n+1} x^{2n-2}}{(2n)!}, -\infty < x < \infty$ **15.** $\sum_{n=0}^{\infty} (-1)^n \binom{\frac{1}{3}}{n} \frac{x^n}{2^{3n-1}}, -8 < x < 8$ **17.** $\sum_{n=0}^{\infty} (-1)^n \frac{x^n}{(2n)!}, -\infty < x < \infty$

19. $\sum_{n=0}^{\infty} \frac{e(x-1)^n}{n!}, -\infty < x < \infty$ **21.** $-\sum_{n=0}^{\infty} (x+1)^n, -2 < x < 0$

23. $\frac{1}{2}\left[1 + \frac{\sqrt{3}\pi}{6}(x-1) - \left(\frac{\pi}{6}\right)^2 \frac{(x-1)^2}{2!} - \sqrt{3}\left(\frac{\pi}{6}\right)^3 \frac{(x-1)^3}{3!} + \left(\frac{\pi}{6}\right)^4 \frac{(x-1)^4}{4!} + \cdots\right], -\infty < x < \infty$

29. $f^{(n)}(0) = 0$ for all n, so Maclaurin series is identically 0. **31.** 0.5077, error ≤ 0.000006 **33.** 0.19737, error ≤ 0.0000003

35. $1 + x^2 + \frac{2x^4}{3}$

Supplementary Exercises 11.9, page 465

1. (a) 2 (b) 0 **3.** (a) Increasing, bounded above by e; converges (b) Increasing, bounded above by 2; converges

5. Converse not true. For example, consider $a_n = \frac{(-1)^n n}{n+1}$. **7.** $S_n = 2 - \frac{2}{n+1}$; $\lim_{n\to\infty} S_n = 2$ **9.** (a) $\frac{3}{10}$ (b) $\frac{3}{4}$

11. Converges **13.** Converges **15.** Diverges **17.** Converges **19.** Conditionally convergent
21. Absolutely convergent **23.** Converges **25.** Converges **27.** Conditionally convergent **29.** Converges

31. Diverges **33.** Conditionally convergent **35.** $-\frac{3}{2} < x < \frac{3}{2}$ **37.** $1 \leq x < 3$ **39.** $-1 < x < 1$ **41.** $-\frac{1}{2} < x \leq \frac{3}{2}$

43. $-2 \sum_{n=0}^{\infty} \frac{x^{2n+2}}{2n+2}, R = 1$ **45.** $\sum_{n=0}^{\infty} \binom{-\frac{1}{2}}{n} \frac{x^{2n+1}}{2n+1}, R = 1$ **49.** $\frac{1}{2} \sum_{n=1}^{\infty} \frac{(-1)^{n-1}(x+1)^n}{n}, R = 1$

51. 0.444, error ≤ 0.0000521 **53.** $\sum_{n=0}^{\infty} \frac{(-1)^{n+1} 2^{2n+1} \left(x - \frac{\pi}{4}\right)^{2n+1}}{(2n+1)!}$

CHAPTER 12

Exercise Set 12.2, page 474

1. (a) $y^2 = 8x$ (b) $x^2 = -12y$ **3.** (a) $(x-1)^2 = 8(y+2)$ (b) $(x+3)^2 = -8(y-1)$

5. (a) $(x-2)^2 = 12(y+3)$ (b) $(y+1)^2 = 6\left(x+\dfrac{1}{2}\right)$ **7.** $(x-2)^2 = 4(y+4)$ **9.** $y^2 - 2x - 2y - 3 = 0$

11. (a) $y^2 = 8x$
Focus $(2, 0)$; directrix $x = -2$

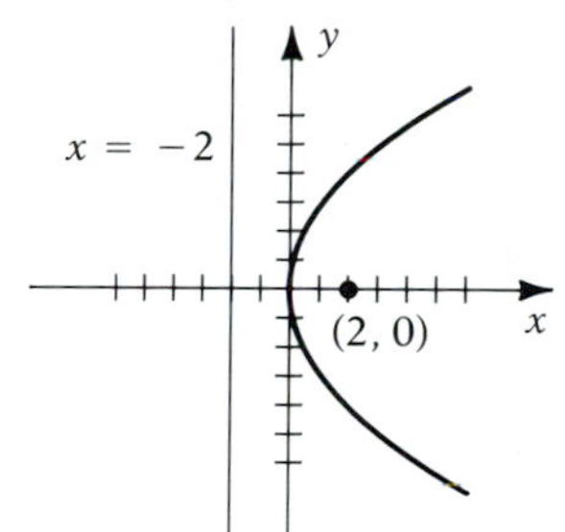

(b) $y^2 = -4x$
Focus $(-1, 0)$; directrix $x = 1$

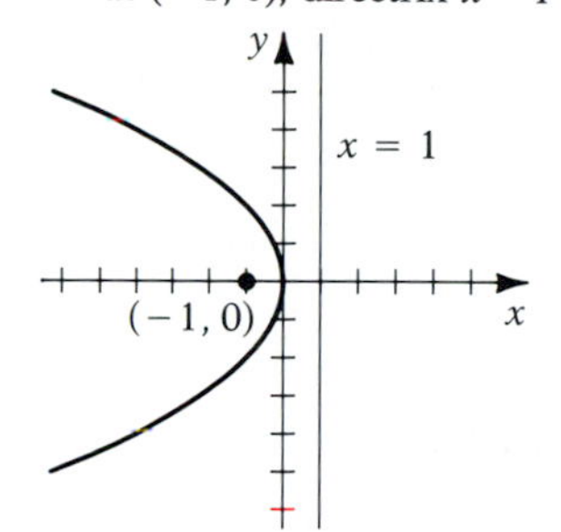

(c) $x^2 = -6y$
Focus $\left(0, -\dfrac{3}{2}\right)$; directrix $y = \dfrac{3}{2}$

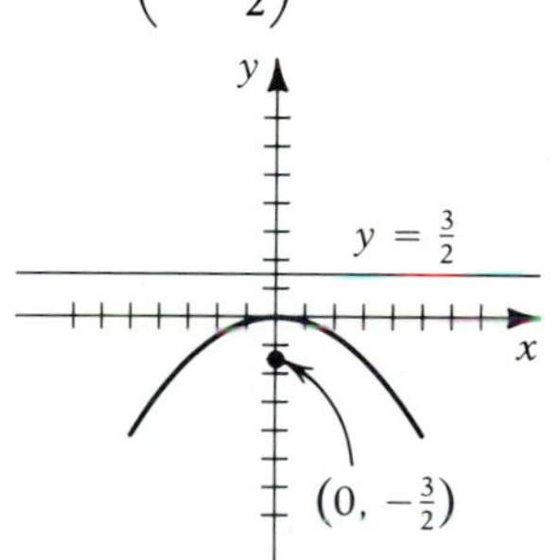

(d) $x^2 = 8y$
Focus $(0, 2)$; directrix $y = -2$

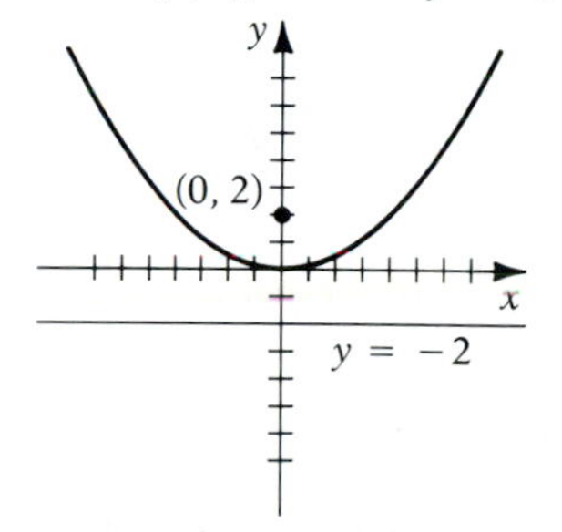

13. $(y-3)^2 = -8(x+4)$
Focus $(-6, 3)$; directrix $x = -2$

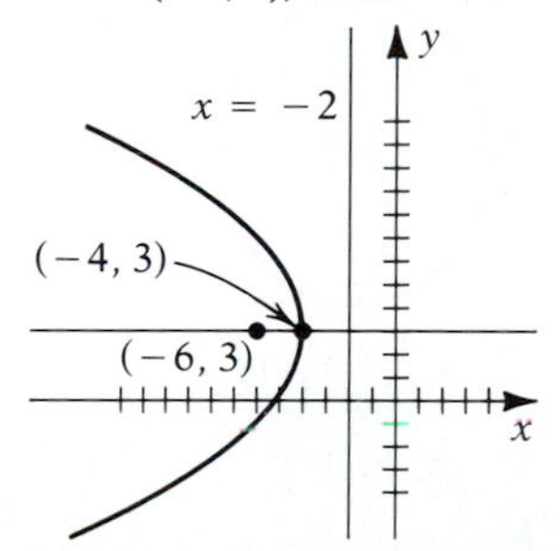

15. $\left(y-\dfrac{3}{2}\right)^2 = x - \dfrac{7}{4}$
Focus $\left(2, \dfrac{3}{2}\right)$; directrix $x = \dfrac{3}{2}$

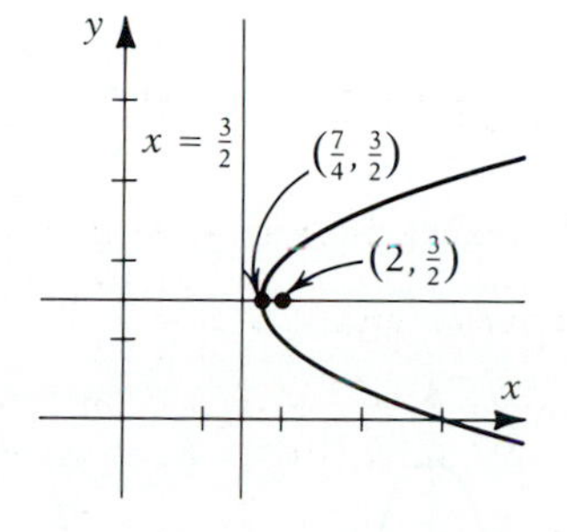

17. $\left(y+\dfrac{1}{2}\right)^2 = -6\left(x-\dfrac{3}{2}\right)$

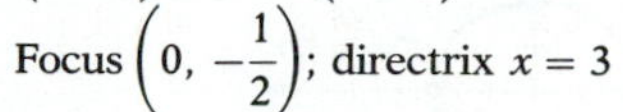

Focus $\left(0, -\dfrac{1}{2}\right)$; directrix $x = 3$

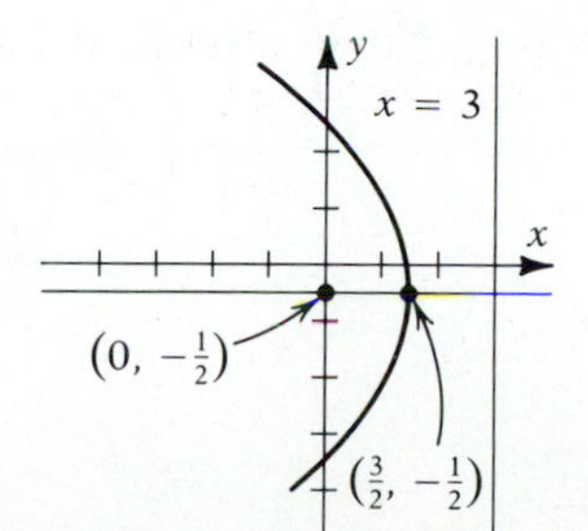

19. Area $= \dfrac{2}{3}$

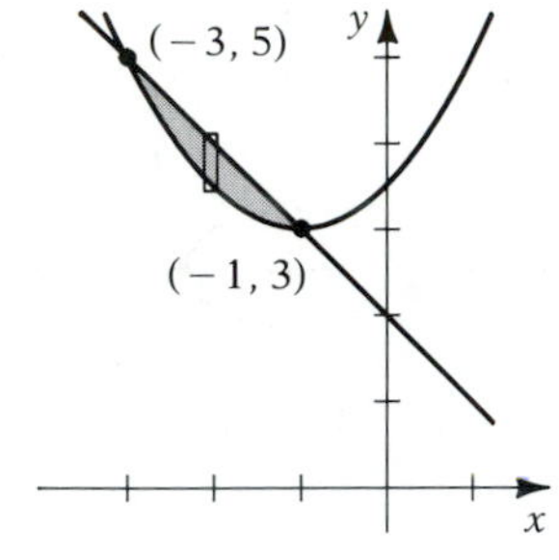

21. Area = 32

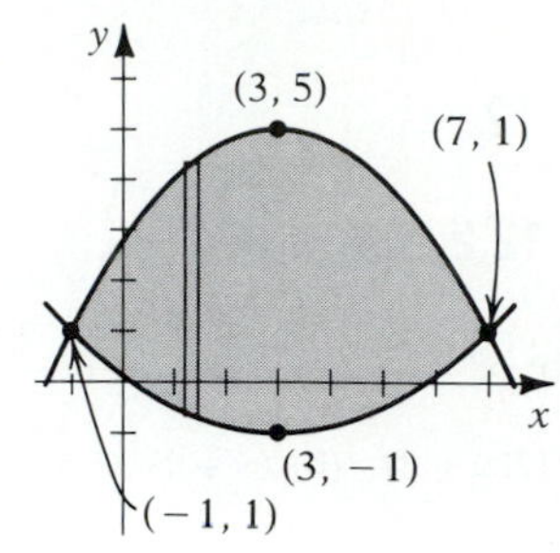

23. (a) $\sqrt{5} + \dfrac{1}{2}\ln(\sqrt{5} + 2)$ (b) $\dfrac{2\pi}{3}(5\sqrt{5} - 1)$

25. $\dfrac{50}{9}, \dfrac{80}{9}, 10, \dfrac{80}{9}, \dfrac{50}{9}$ **27.** $\dfrac{1}{3}(4\pi + \sqrt{3})$

29. Introduce axes so that the equation of the parabola is $y^2 = 4px$. Use the fact that $\tan\beta = dy/dx$ and $\tan\alpha = (m_2 - m_1)/(1 + m_1 m_2)$, where $m_1 = \tan\beta$ and $m_2 =$ slope of FP.

Exercise Set 12.3, page 481

1. (a) $\dfrac{x^2}{16} + \dfrac{y^2}{4} = 1$

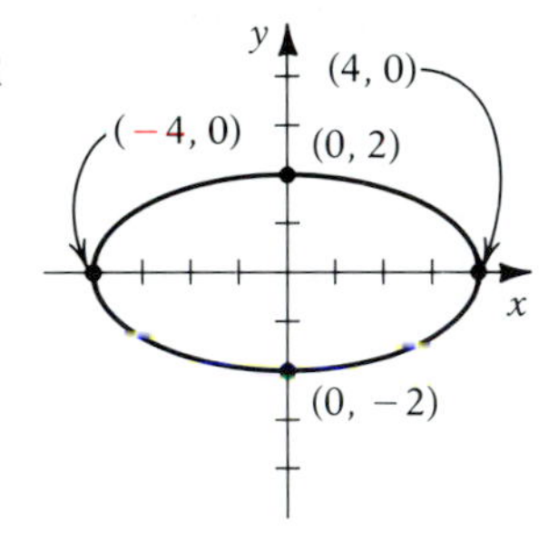

(b) $\dfrac{x^2}{25} + y^264 = 1$

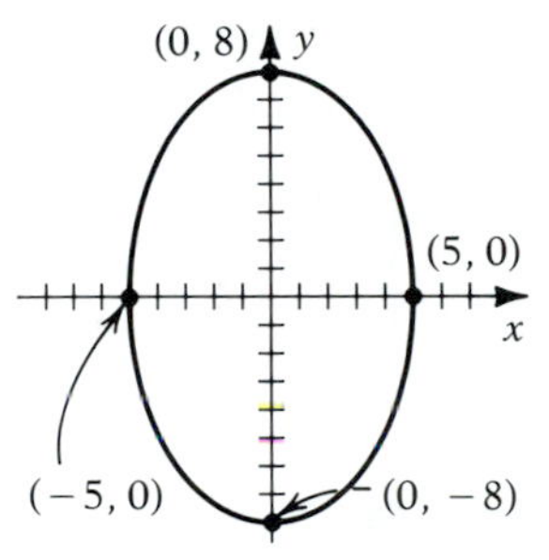

3. $\dfrac{(y-4)^2}{25} + \dfrac{(x+2)^2}{9} = 1$

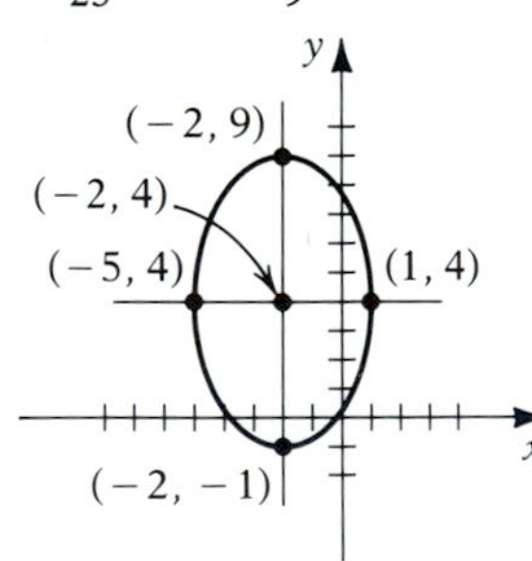

5. $\dfrac{(x-4)^2}{20} + \dfrac{(y+5)^2}{4} = 1$

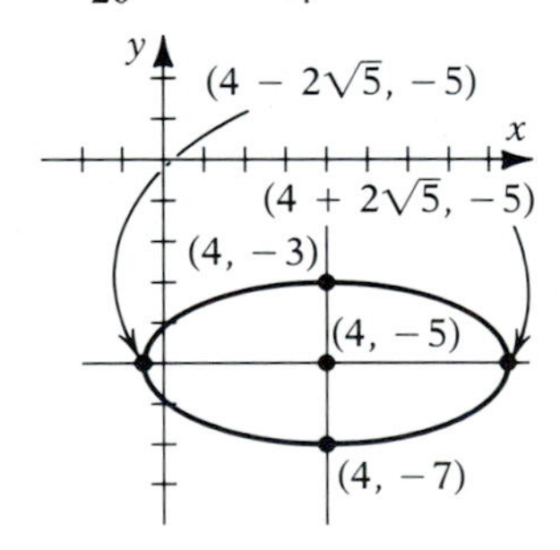

7. $\dfrac{(y+3)^2}{36} + \dfrac{(x-1)^2}{27} = 1$

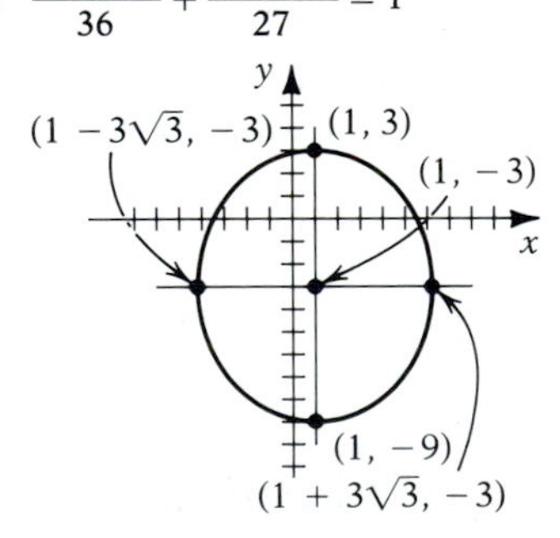

9. (a) $\dfrac{x^2}{4} + \dfrac{y^2}{25} = 1$

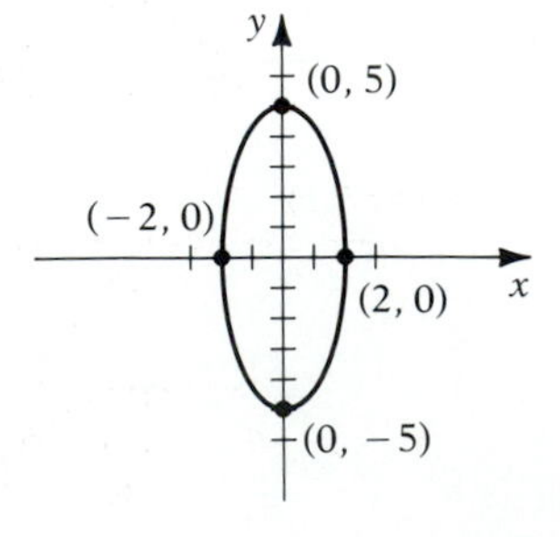

(b) $\dfrac{x^2}{5} + \dfrac{y^2}{4} = 1$

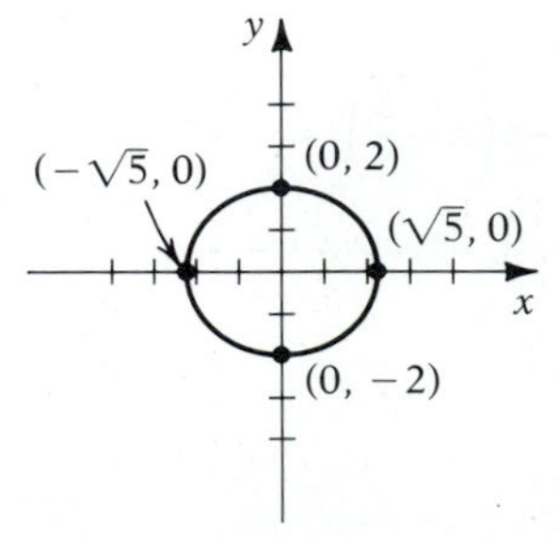

11. (a) $\dfrac{x^2}{4} + \dfrac{y^2}{4/9} = 1$

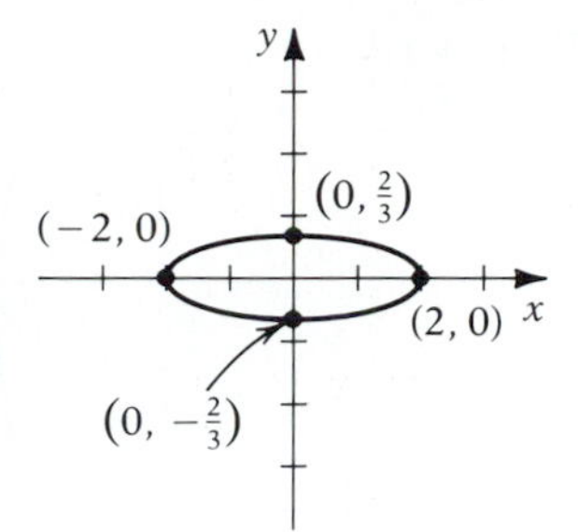

(b) $\dfrac{x^2}{3/5} + \dfrac{y^2}{1/5} = 1$

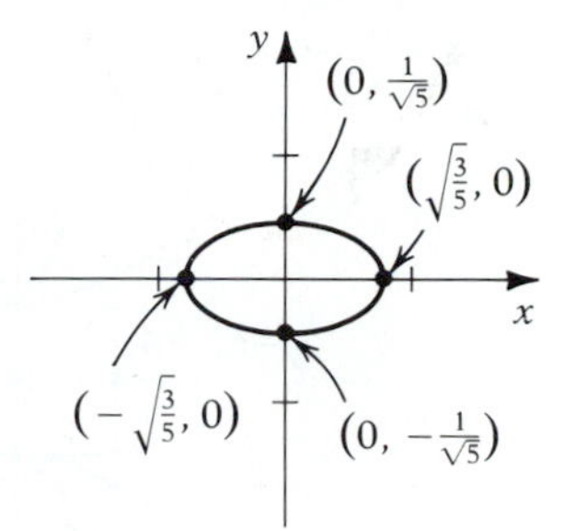

13. $\dfrac{(x-1)^2}{4} + \dfrac{(y+2)^2}{1} = 1$

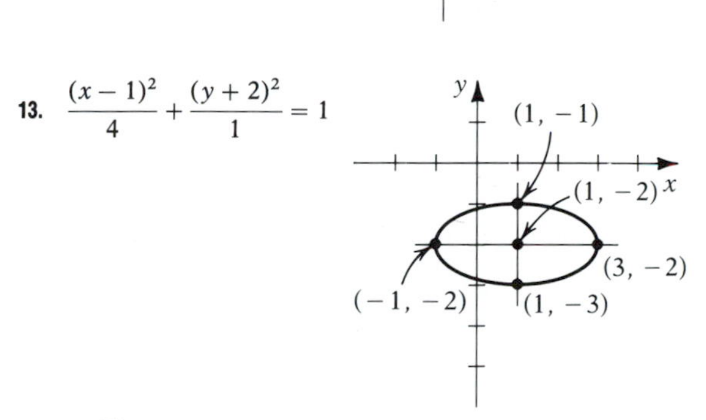

15. $\dfrac{\left(x - \dfrac{5}{4}\right)^2}{75/16} + \dfrac{\left(y + \dfrac{3}{4}\right)^2}{75/32} = 1$

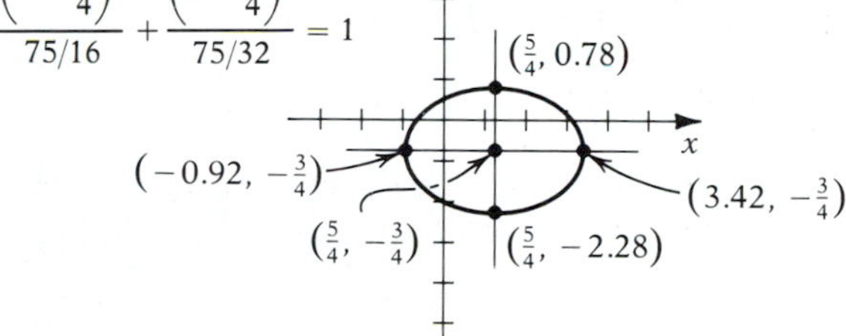

17. $2\sqrt{3}$ m

19. Tangent: $y = 2x - 8$
Normal: $x + 2y + 6 = 0$

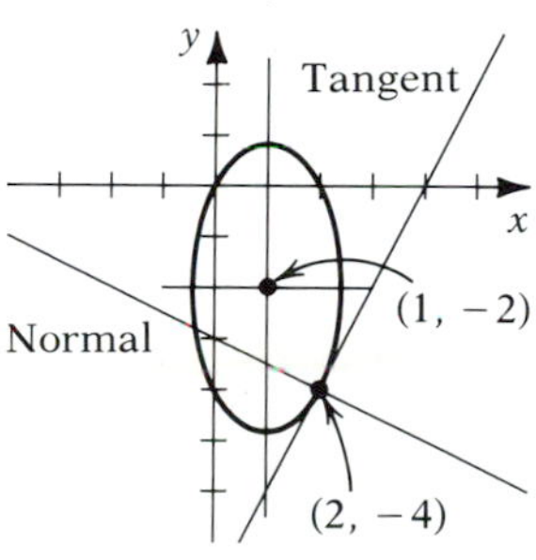

21. Length $= \sqrt{2}\,a$, height $= \sqrt{2}\,b$

23. Inscribed circle: $(x+2)^2 + (y+3)^2 = \dfrac{25}{2}$
Circumscribed circle: $(x+2)^2 + (y+3)^2 = 25$

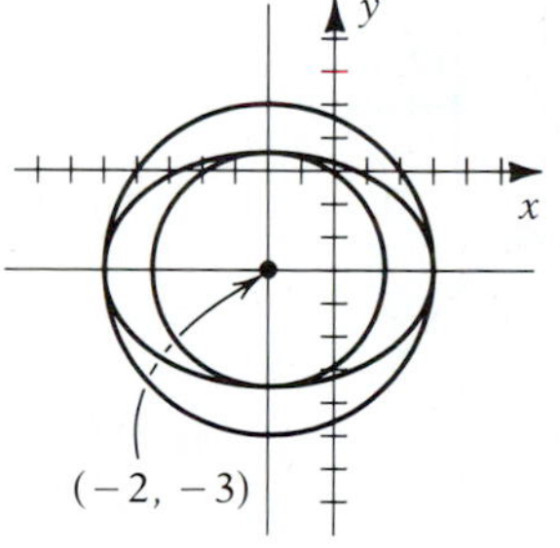
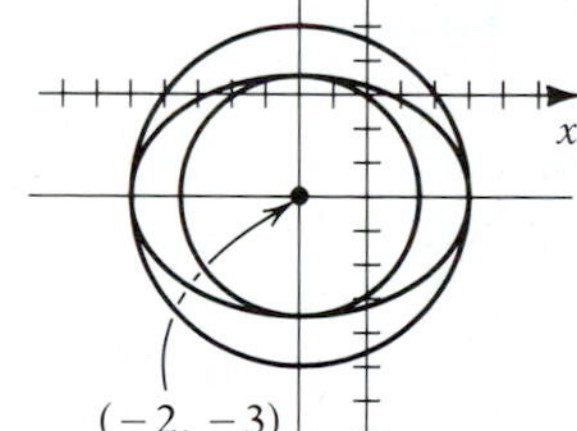

29. $\dfrac{\sqrt{3}}{2} + 2\pi$

Exercise Set 12.4, page 489

1. (a) $\dfrac{x^2}{4} - \dfrac{y^2}{5} = 1$

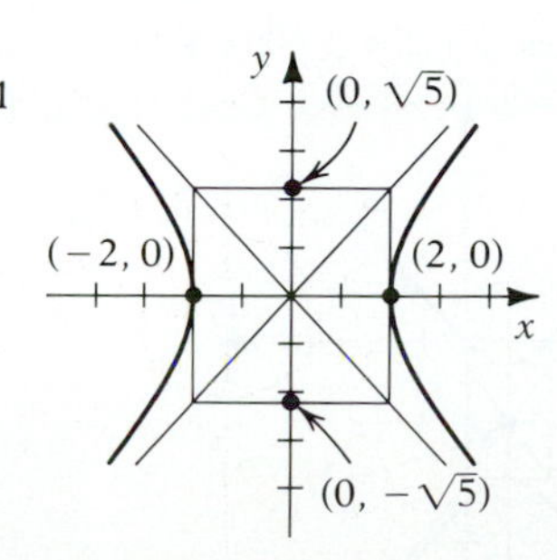

(b) $\dfrac{y^2}{9} - \dfrac{x^2}{7} = 1$

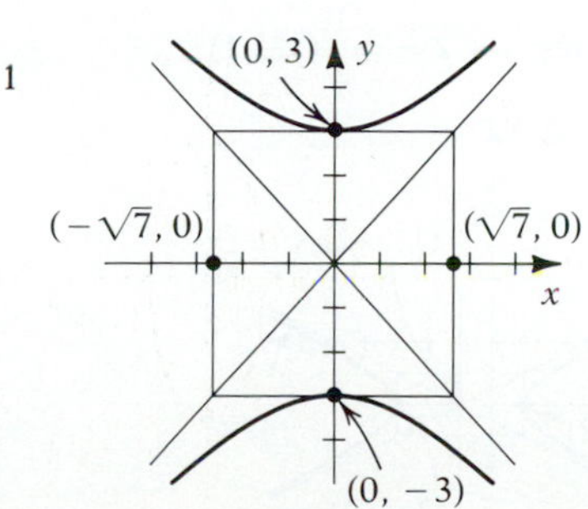

3. $\dfrac{(y+4)^2}{9} - \dfrac{(x+2)^2}{16} = 1$

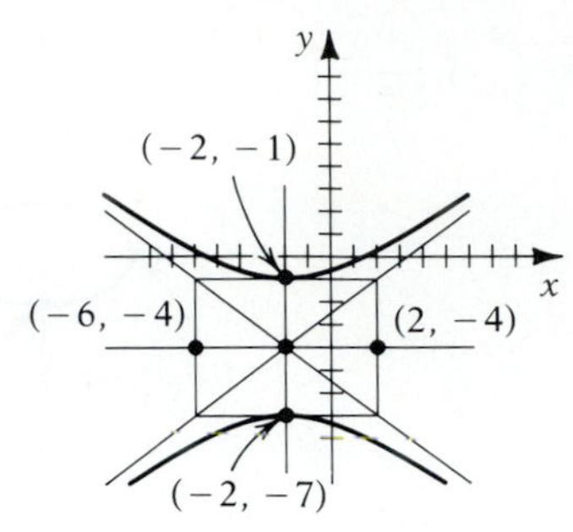

5. $\dfrac{(x-2)^2}{4} - \dfrac{(y+1)^2}{36} = 1$

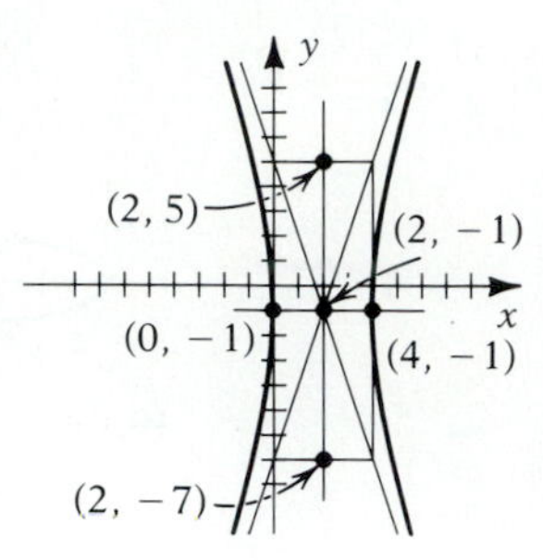

7. $\dfrac{(x-3)^2}{16} - \dfrac{(y-2)^2}{9} = 1$

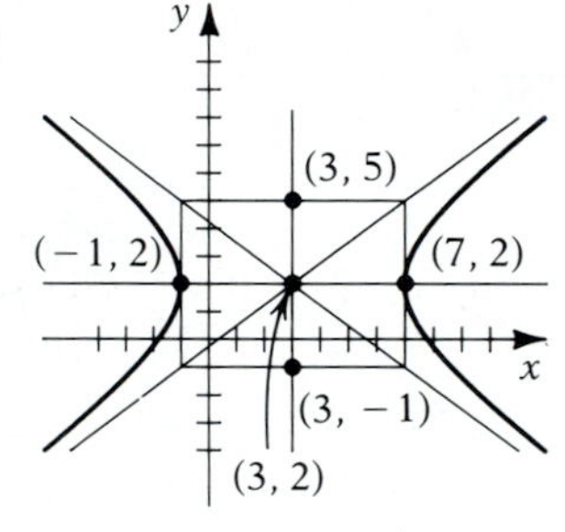

9. (a) $\dfrac{x^2}{9} - \dfrac{y^2}{4} = 1$

Vertices $(\pm 3, 0)$; $e = \dfrac{\sqrt{13}}{3}$

Asymptotes: $y = \pm\dfrac{2x}{3}$

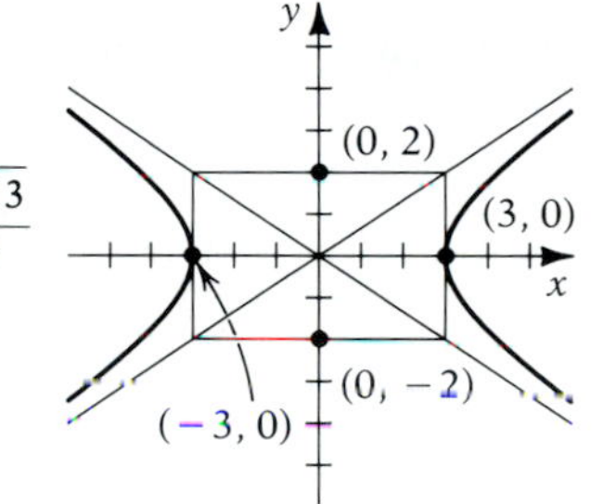

(b) $\dfrac{y^2}{4} - \dfrac{x^2}{9} = 1$

Vertices: $(0, \pm 2)$; $e = \dfrac{\sqrt{13}}{2}$

Asymptotes: $y = \pm\dfrac{2x}{3}$

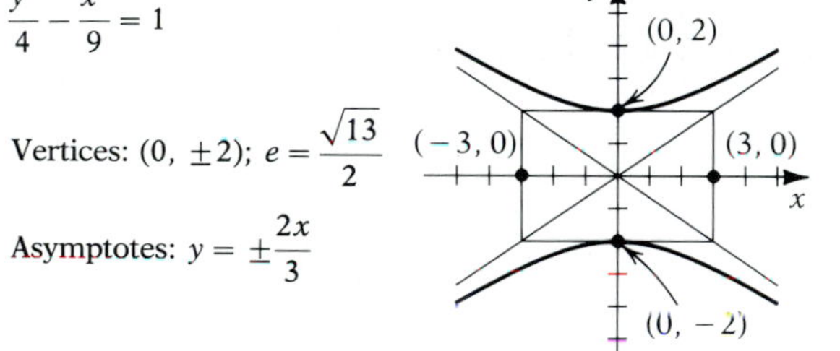

11. (a) $\dfrac{y^2}{9} - \dfrac{x^2}{4} = 1$

Vertices: $(0, \pm 3)$; $e = \dfrac{\sqrt{13}}{3}$

Asymptotes: $y = \pm\dfrac{3x}{2}$

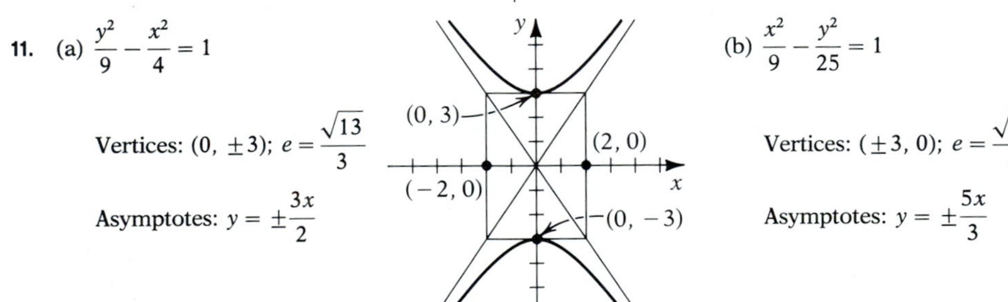

(b) $\dfrac{x^2}{9} - \dfrac{y^2}{25} = 1$

Vertices: $(\pm 3, 0)$; $e = \dfrac{\sqrt{34}}{3}$

Asymptotes: $y = \pm\dfrac{5x}{3}$

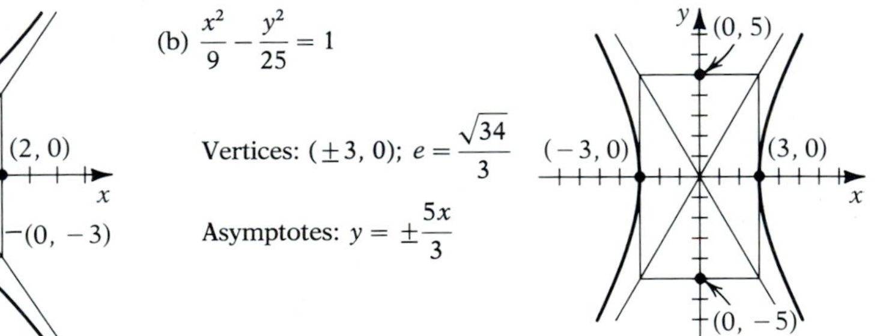

13. $\dfrac{(x-1)^2}{4} - \dfrac{(y+2)^2}{1} = 1$

Vertices: $(3, -2)$, $(-1, -2)$; $e = \dfrac{\sqrt{5}}{2}$

Asymptotes: $y + 2 = \pm\dfrac{1}{2}(x - 1)$

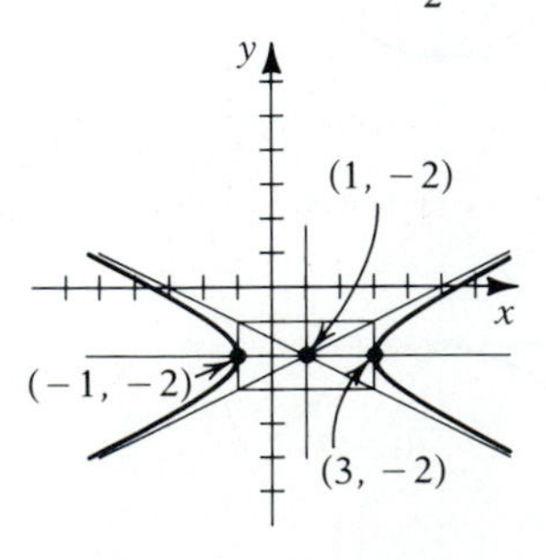

15. $\dfrac{(y+2)^2}{2} - \dfrac{(x+1)^2}{1} = 1$

Vertices: $(-1, -2 \pm \sqrt{2})$; $e = \dfrac{\sqrt{6}}{2}$

Asymptotes: $y + 2 = \pm\sqrt{2}(x + 1)$

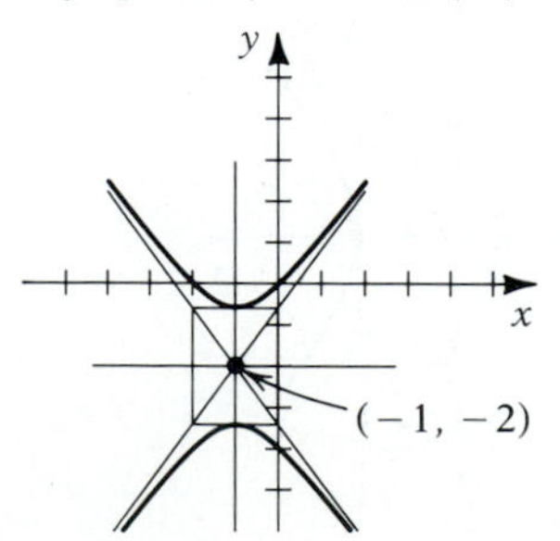

17. Area $= 6 - 9 \ln \sqrt{3}$

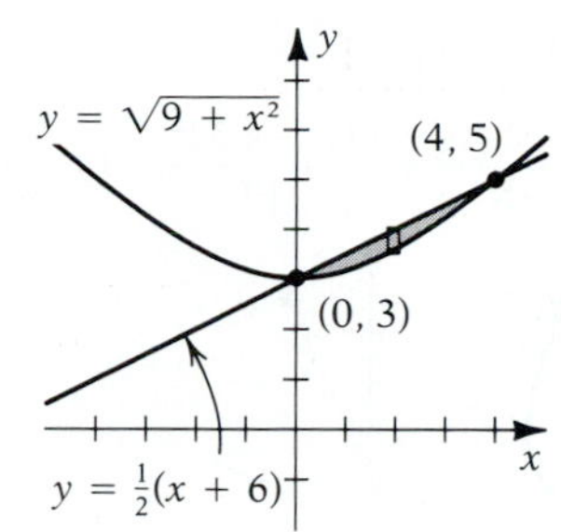

19. Area $= \dfrac{\pi}{\sqrt{2}} - \ln(1 + \sqrt{2})$

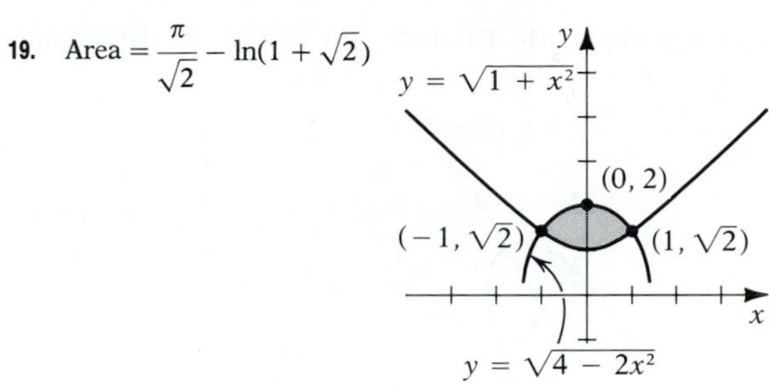

21. Put $x = c$ in $\dfrac{x^2}{a^2} - \dfrac{y^2}{b^2} = 1$, and solve for y. Use $c^2 = a^2 + b^2$.

23. $\pi\left(2\sqrt{7} - 1 + \dfrac{1}{\sqrt{2}} \ln \dfrac{\sqrt{2} + 1}{2\sqrt{2} + \sqrt{7}}\right)$ **25.** $(2, 3)$

27. Let (x_0, y_0) be a point on the hyperbola $\dfrac{x^2}{a^2} - \dfrac{y^2}{b^2} = 1$. Use the formula from Exercise 30 in Exercise Set 1.3 for the distance from a line to a point.

29. $3 \ln 3 + \dfrac{4}{3}$

Exercise Set 12.5, page 496

1. $xy = 2$

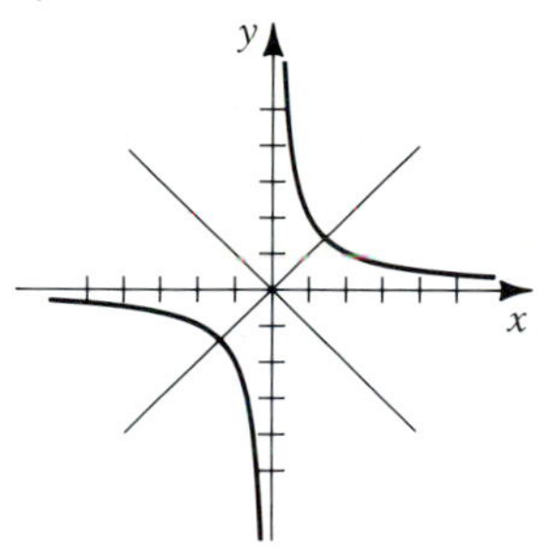

3. $(x - 3)(y + 2) = 8$

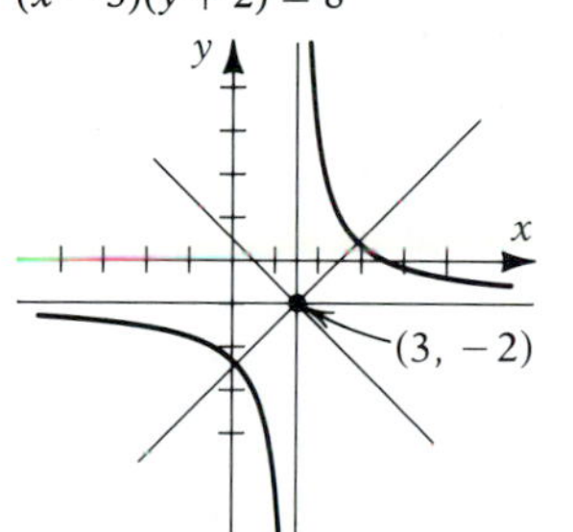

5. $(x + 1)(y - 2) = -2$

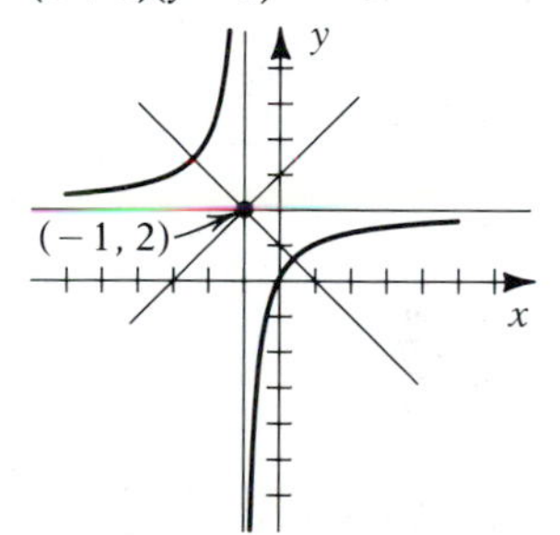

7. $\dfrac{x'^2}{44} + \dfrac{y'^2}{4} = 1$, ellipse

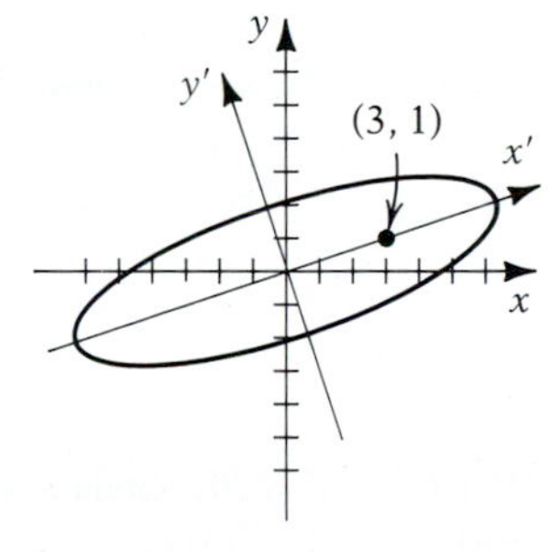

9. $\dfrac{y'^2}{4} - \dfrac{x'^2}{36} = 1$, hyperbola

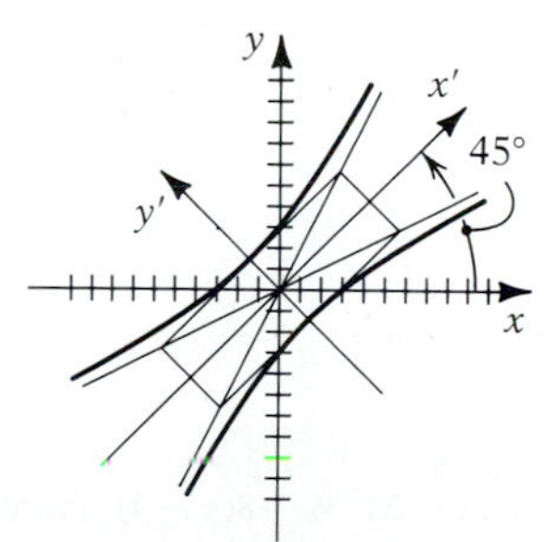

11. $\dfrac{x'^2}{21} + \dfrac{y'^2}{4} = 1$, ellipse

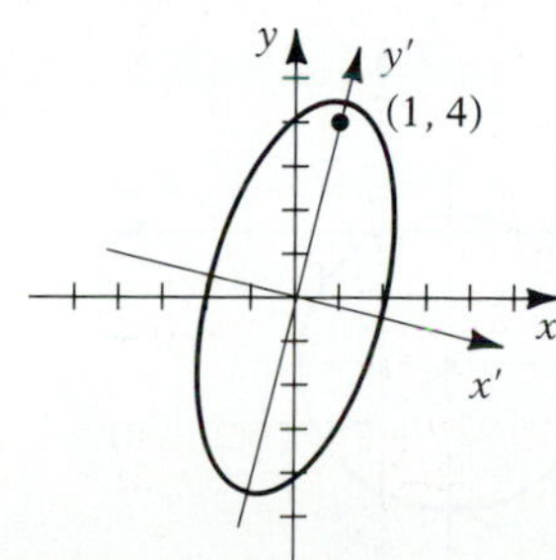

13. $y'^2 = -4(x' - 1)$, parabola

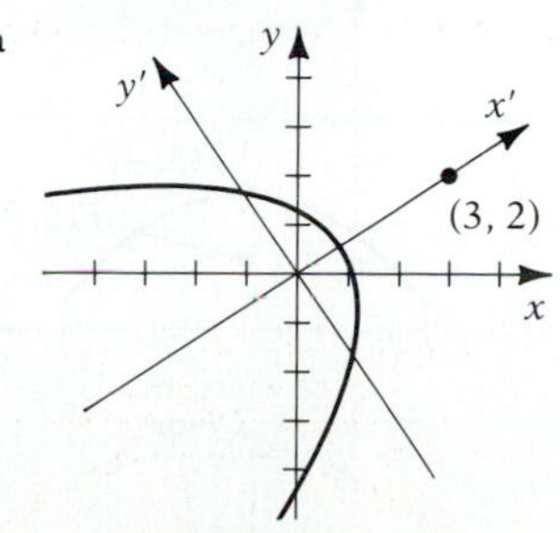

15. (a) Hyperbola (b) Parabola (c) Ellipse (d) Hyperbola (e) Hyperbola

17. $\dfrac{x'^2}{20} + \dfrac{(y'-2)^2}{4} = 1$, ellipse

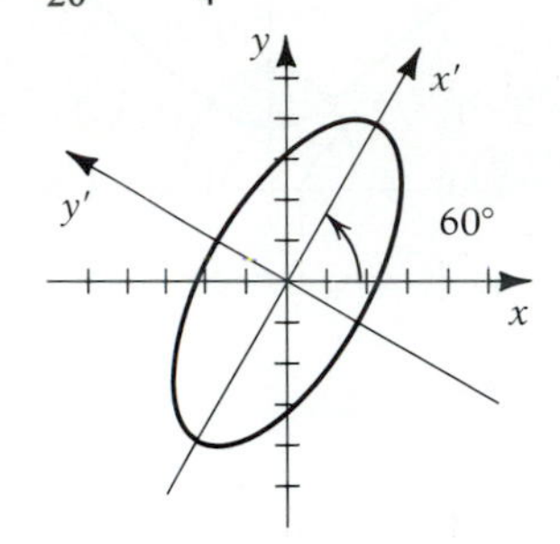

19. $\dfrac{(y'-2\sqrt{5})^2}{15} - \dfrac{(x'-\sqrt{5})^2}{10} = 1$, hyperbola

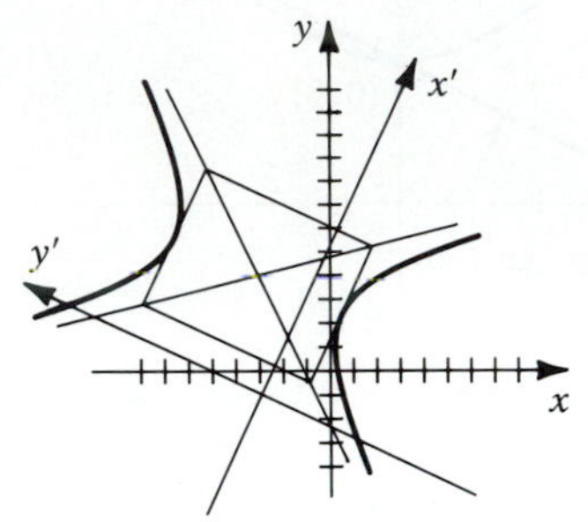

Supplementary Exercises 12.6, page 497

1. $\dfrac{(y-1)^2}{4} - \dfrac{(x-2)^2}{16} = 1$

Hyperbola, position II, center (2, 1)

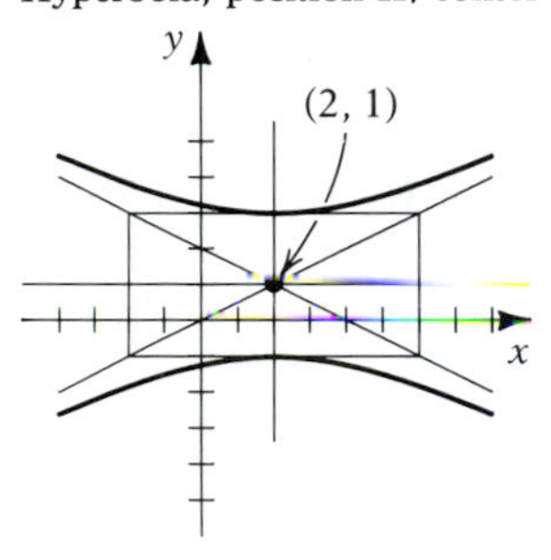

3. $(y+3)^2 = -2(x-1)$

Parabola, vertex $(1, -3)$, horizontal axis

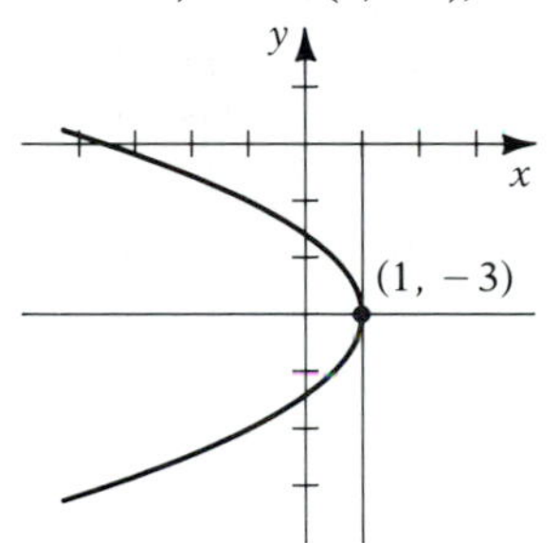

5. Degenerate circle; point $(3, -7)$

7. $y - 2 = \pm\dfrac{2}{3}(x+1)$

Degenerate hyperbola

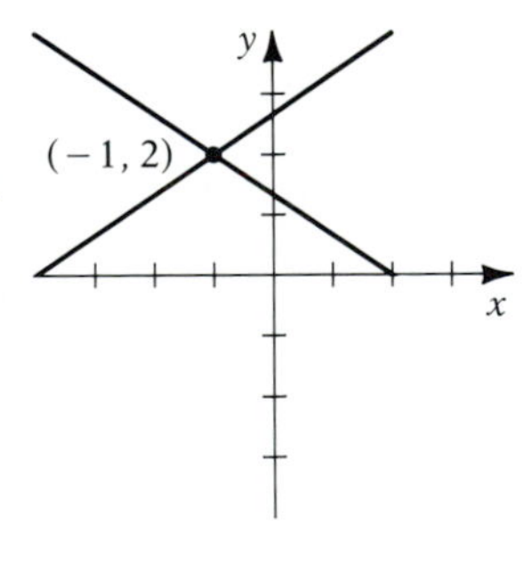

9. $\dfrac{x^2}{9} - \dfrac{y^2}{3} = 1$

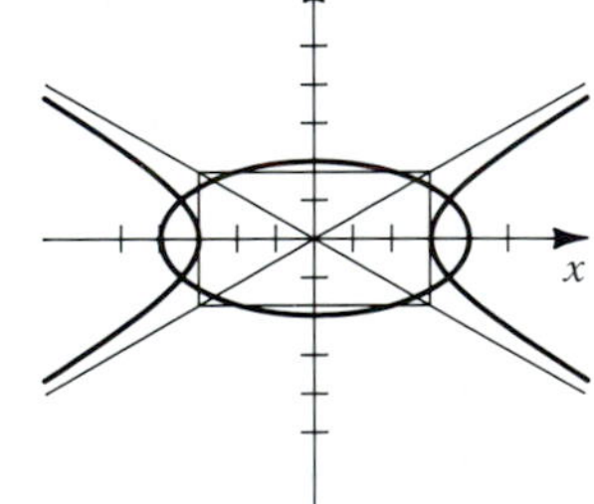

11. (a) $(x-3)^2 = -8(y-4)$, parabola

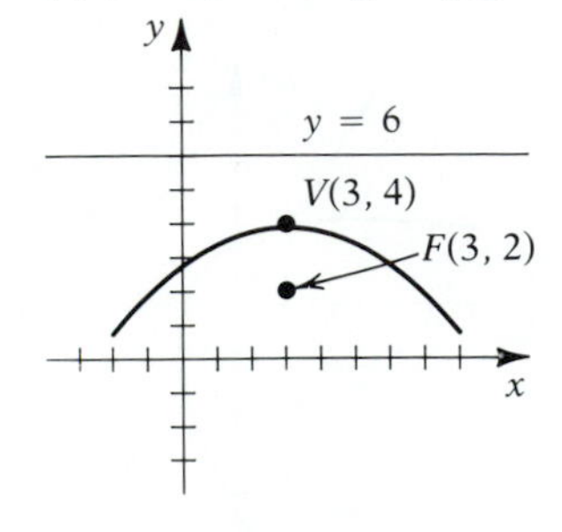

(b) $(x+2)^2 + (y+4)^2 = 20$, circle, radius $2\sqrt{5}$, center $(-2, -4)$

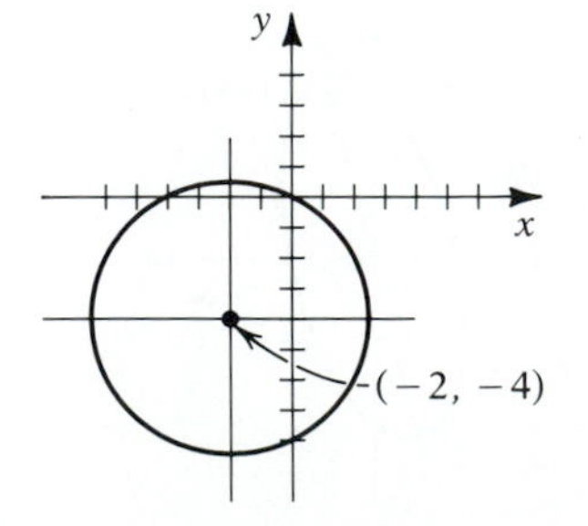

13. (a) $\frac{(x-2)^2}{4} + \frac{y^2}{3} = 1$, ellipse, center (2, 0), position I

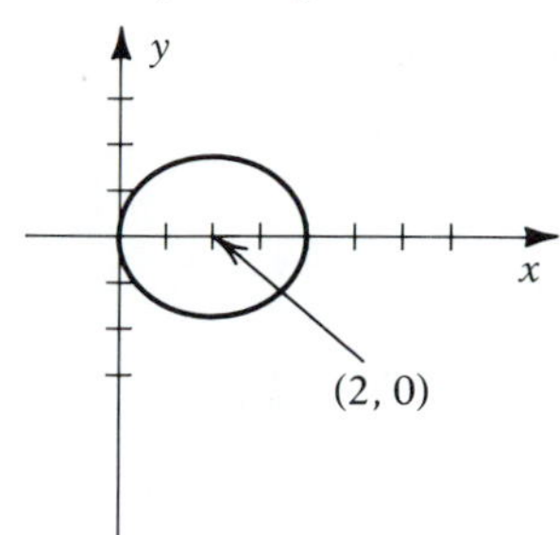

(b) $5x^2 - 6xy - 3y^2 - 8x = 4$, hyperbola, rotated through the angle $\theta = \sin^{-1} 3/\sqrt{10}$

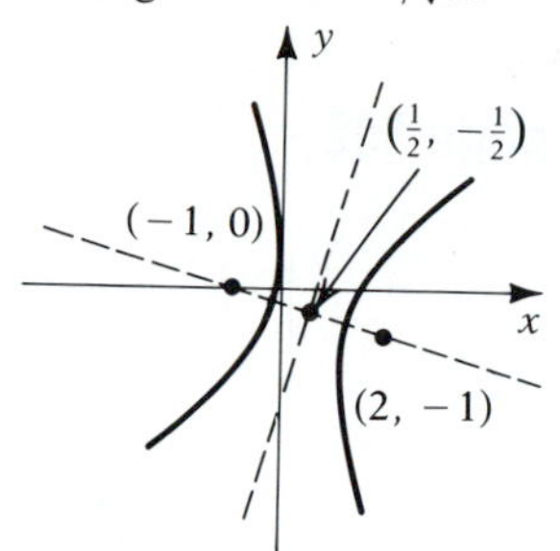

15. Inscribed circle: $x^2 + y^2 + 2x - 4 = 0$
Circumscribed circle: $x^2 + y^2 + 7x - 19 = 0$

17. Area $= \frac{1}{2} + \frac{2\pi}{3}$

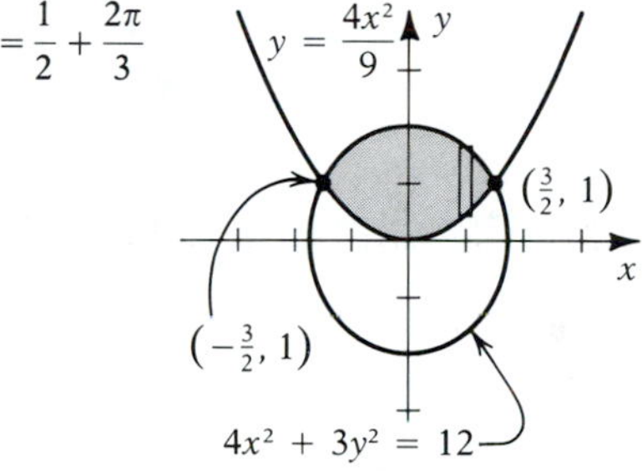

19. $\frac{512\pi}{9}$

21. Tangent: $x - y + 2 = 0$
Normal: $x + y = 8$

25. Hyperbola

$\frac{y'^2}{1} - \frac{x'^2}{3} = 1$

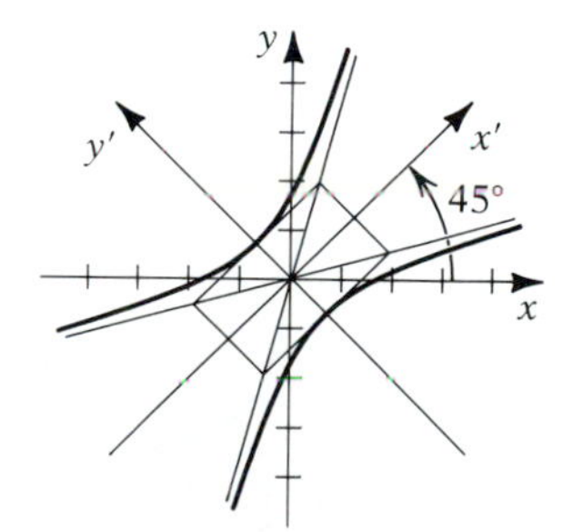

27. Ellipse

$\frac{x'^2}{26} + \frac{y'^2}{6} = 1$

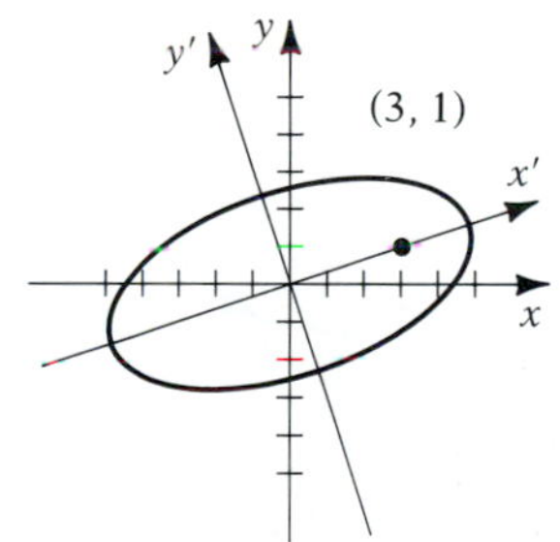

29. $\frac{PE}{PP'} = e$

$\sqrt{(x-p)^2 + y^2} = e|x - k|$

On squaring and rearranging terms, obtain

$x^2(1 - e^2) + y^2 + (2ke^2 - 2p)x + p^2 - e^2k^2 = 0$

If $0 < e < 1$, coefficients of x^2 and y^2 are positive—ellipse
If $e > 1$, coefficient of x^2 is negative—hyperbola
If $e = 1$, coefficient of x^2 is 0—parabola

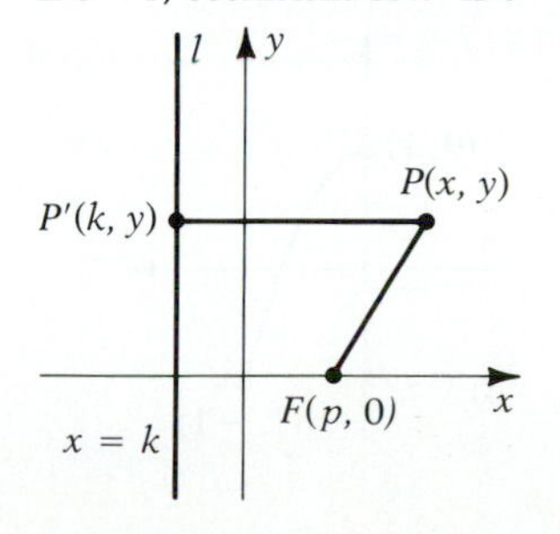

CHAPTER 13

Exercise Set 13.1, page 502

1.

t	x	y
0	3	-1
1	5	2
2	7	5
-1	1	-4

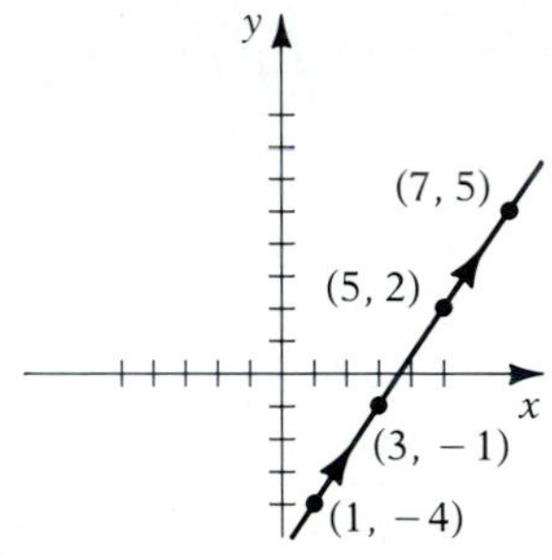

3.

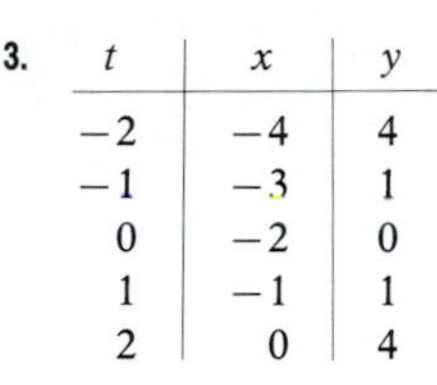

t	x	y
-2	-4	4
-1	-3	1
0	-2	0
1	-1	1
2	0	4

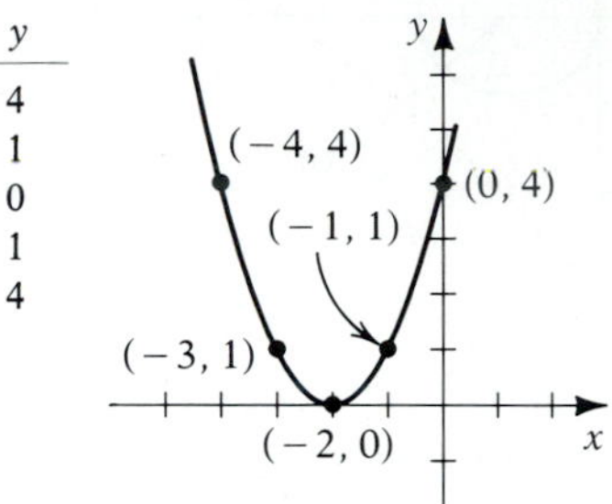

5.

v	x	y
$\frac{1}{2}$	2	$\frac{1}{2}$
1	1	1
2	$\frac{1}{2}$	2
3	$\frac{1}{3}$	3
$\frac{1}{3}$	3	$\frac{1}{3}$

7. $x + 3y = 1$
$(-5, 2)$ to $(10, -3)$

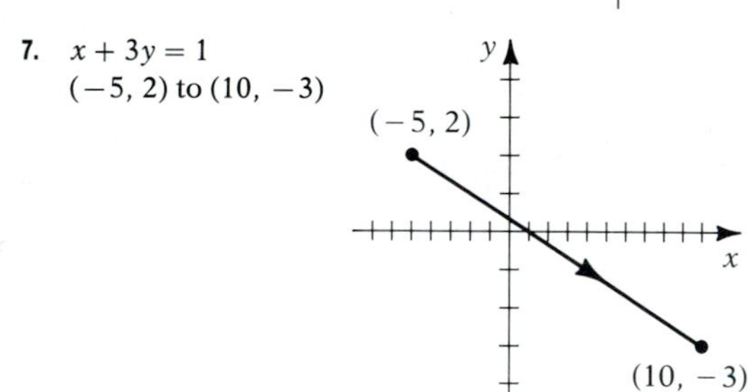

9. $x^2 + y^2 = 1$
$(1, 0)$ to $(-1, 0)$, $y < 0$

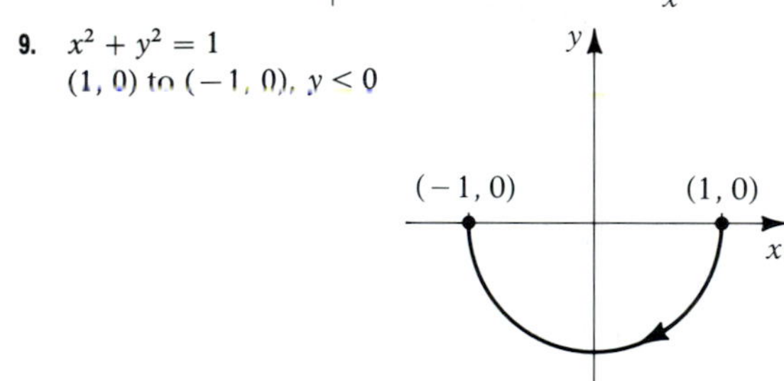

11. $xy = 1$, $x > 0$, $y > 0$

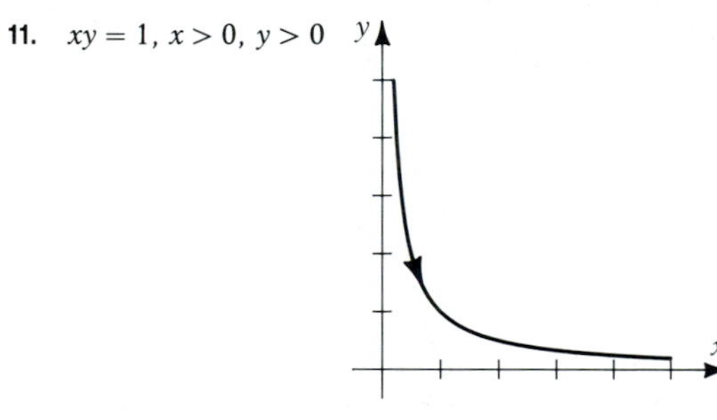

13. $y^2 = -(x - 2)$
$(1, 1)$ to $(1, -1)$

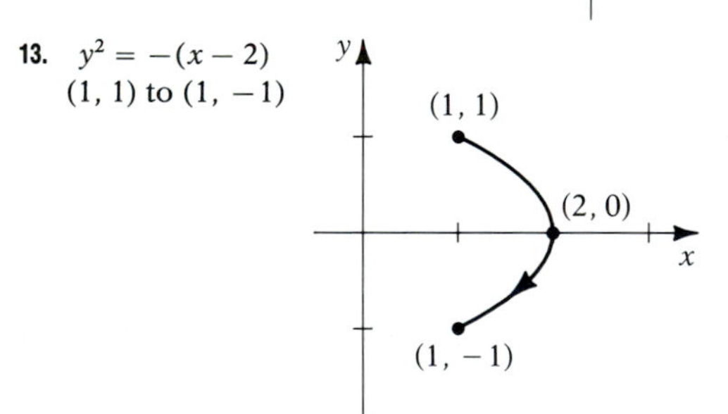

15. $x^2 - y^2 = 1$, $x \geq 1$

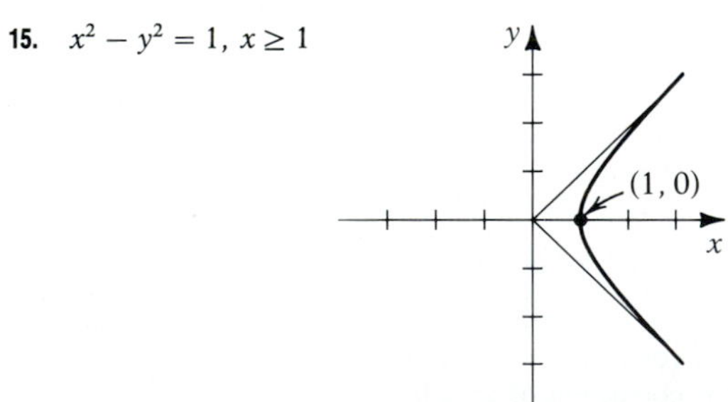

17. $\dfrac{(x-2)^2}{1} + \dfrac{(y-1)^2}{9} = 1$

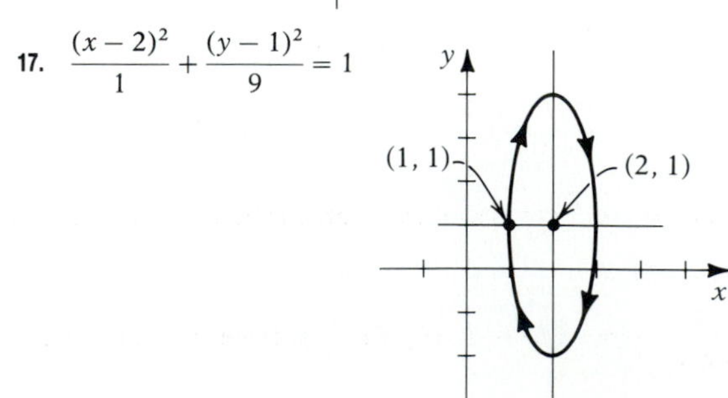

19. $y = 1 - 2x^2$, $x \geq 0$
$(0, 1)$ to $(1, -1)$

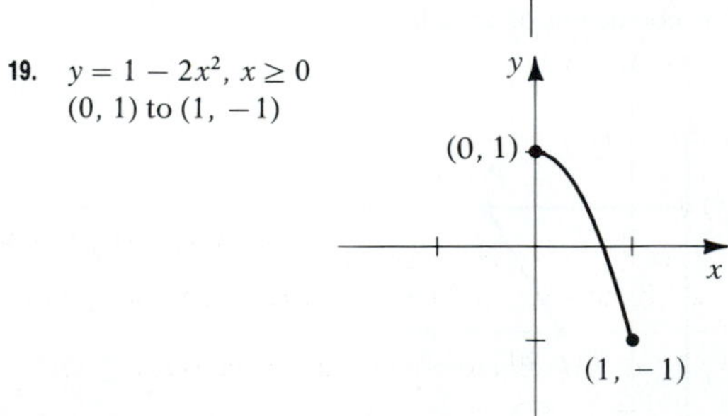

21. All four curves consist of the line segment with end points $(1, -2)$ and $(3, 2)$, with rectangular equation $y = 2x - 4$. The curves C_1, C_3, and C_4 are oriented from $(1, -2)$ to $(3, 2)$, and C_2 has the reverse orientation. If t represents time, then the point (x, y) covers the same path in each case but gets to positions along the way at different times.

23. Assume t_1 and t_2, both in $[0, 2]$, produce the same point, and show that $t_1 = t_2$. The rectangular equation is $y + 1 = (x - 1)^{2/3}$.

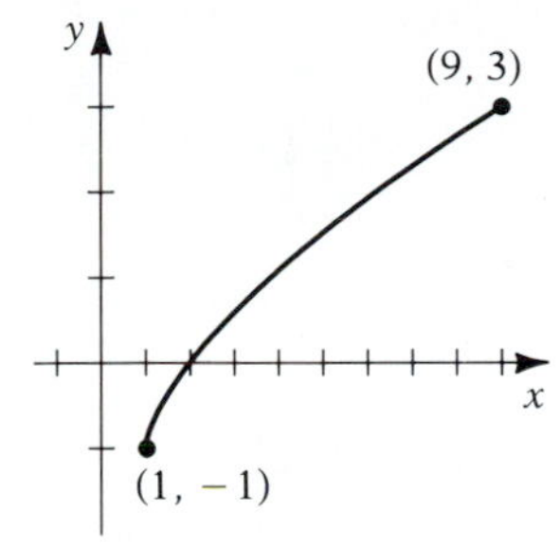

25.

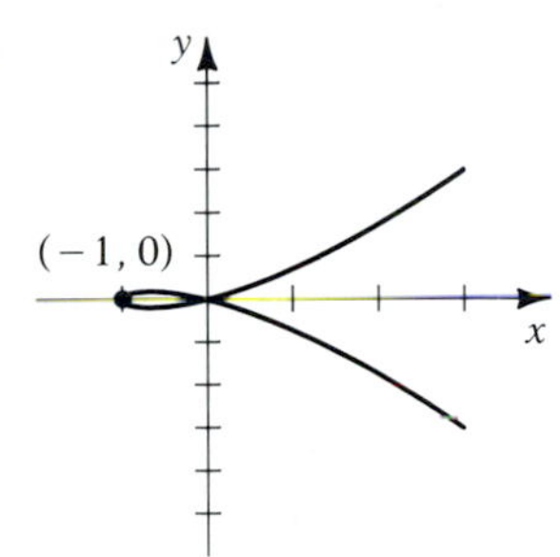

27. Circle, $\left(x - \frac{1}{2}\right)^2 + y^2 = \frac{1}{4}$

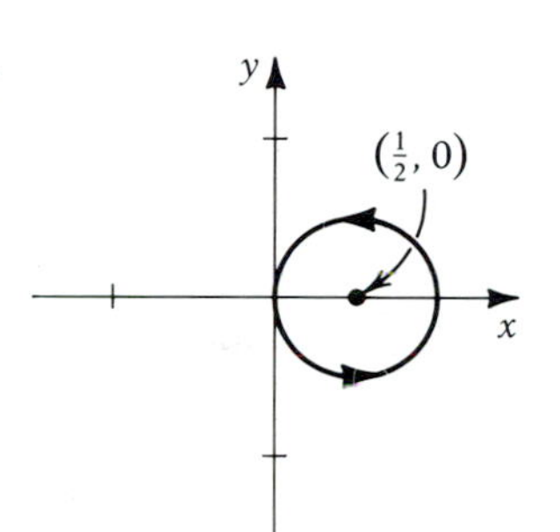

29. $x = r\left[\cos^{-1}\frac{r-y}{r} - \frac{\sqrt{2ry - y^2}}{r}\right]$ if $0 \le x \le \pi r$; $x = r\left[\left(2\pi - \cos^{-1}\frac{r-y}{r}\right) + \frac{\sqrt{2ry - y^2}}{r}\right]$ if $\pi r < x \le 2\pi r$.
Graph is periodic, of period $2\pi r$.

33. $$\begin{cases} x = (a + b)\cos\theta - b\cos\dfrac{a+b}{b}\theta \\ y = (a + b)\sin\theta - b\sin\dfrac{a+b}{b}\theta \end{cases} \qquad 0 \le \theta \le 2\pi$$

Exercise Set 13.2, page 507

1. 1 **3.** $-\frac{1}{3}$ **5.** $-\frac{1}{2}$ **7.** -4 **9.** $\frac{1}{20}$ **11.** $x - 8y = 13$ **13.** $4x - 3y = 11$

15. Horizontal tangent line at
$\left(-\frac{27}{64}, -\frac{49}{8}\right)$; vertical tangent lines at $\left(\frac{14}{27}, -\frac{52}{9}\right)$ and $(-18, 22)$.

17. For $t \ne 2$, C is smooth and the equations define y as a differentiable function of x, since $f'(t) \ne 0$, and both f' and g' are continuous.

$$y' = -\frac{1}{t^2 - 4t + 5}$$

19. $-\frac{3}{4}\csc^3 t$ **21.** -2

23. $y' = \frac{\sin t}{1 - \cos t}$, $y'' = -\frac{1}{a(1 - \cos t)^2}$; at $t = (2n + 1)\pi$, $y' = 0$ and $y'' < 0$; at $t = 2n\pi$, y' is undefined, with left-hand limit $-\infty$ and right-hand limit $+\infty$, so there is a local minimum; $y'' < 0$ everywhere it is defined.

25. For fixed ϕ in $(0, 2\pi)$, the normal line equation is $y - a(1 - \cos\phi) = \frac{\cos\phi - 1}{\sin\phi}[x - a(\phi - \sin\phi)]$. Its x-intercept is $a\phi$, and from Figure 13.6 $\overline{OA} = \widehat{AP} = a\phi$.

Exercise Set 13.3, page 512

1. $4\sqrt{13}$ **3.** $\dfrac{335}{27}$ **5.** $\dfrac{488}{27}$ **7.** 4 **9.** $2\sqrt{2}$ **11.** $\dfrac{3}{2}$ **13.** $\dfrac{9}{2}$ **15.** $\sqrt{3} + \dfrac{1}{2}\ln(\sqrt{3} + 2)$

17. $\dfrac{4088\pi}{27}$ **19.** $2\pi(\pi + 1)$ **21.** $\dfrac{8\pi}{3}(5\sqrt{5} - 2\sqrt{2})$ **23.** $70\,\pi$ **25.** (a) $8a$ (b) $\dfrac{64\pi a^2}{3}$

27. (a) $\dfrac{\sqrt{6} - 2}{\sqrt{3}} + \ln\dfrac{\sqrt{3} + 2}{\sqrt{2} + 1}$ (b) $\pi\left[\sqrt{2} - \dfrac{2}{3} - \ln(\sqrt{6} - \sqrt{3})\right]$ **29.** 1.497, error ≤ 0.0002

Exercise Set 13.4, page 517

1.

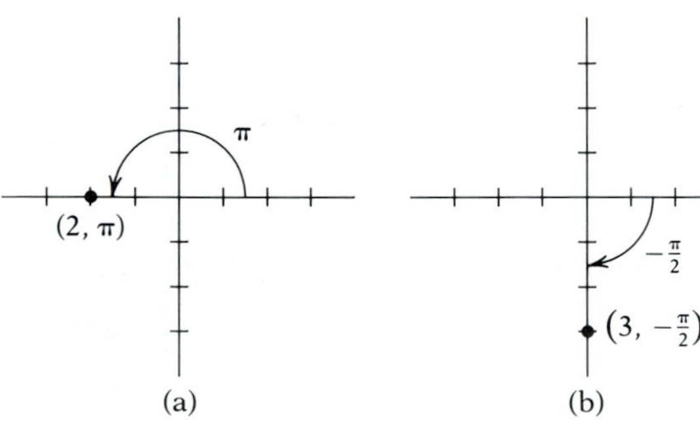

(a) (b)

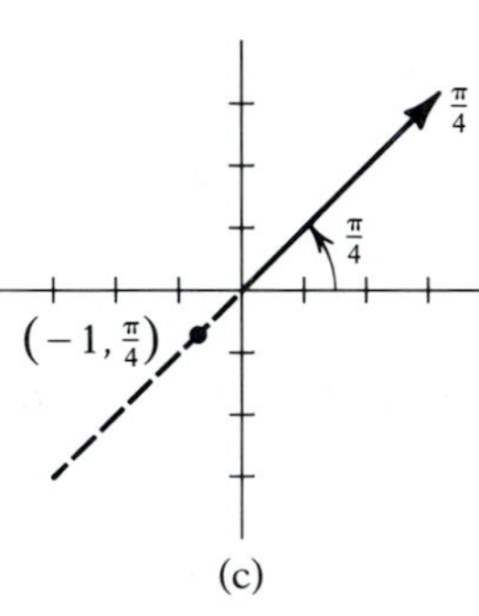

(c)

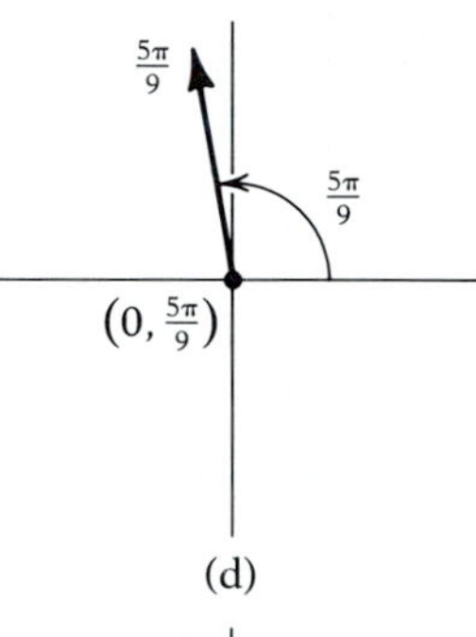

(d)

3.

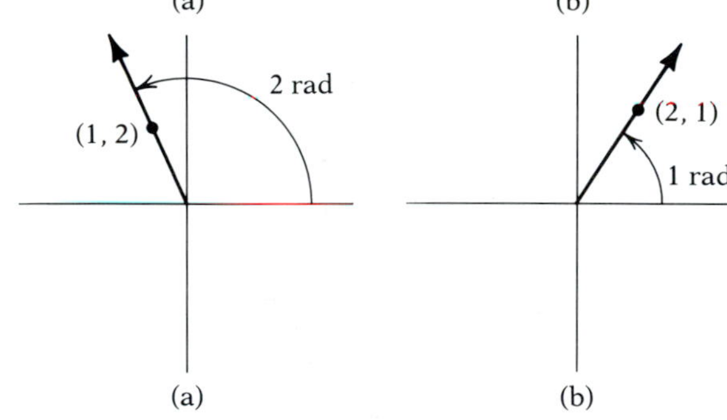

(a) (b)

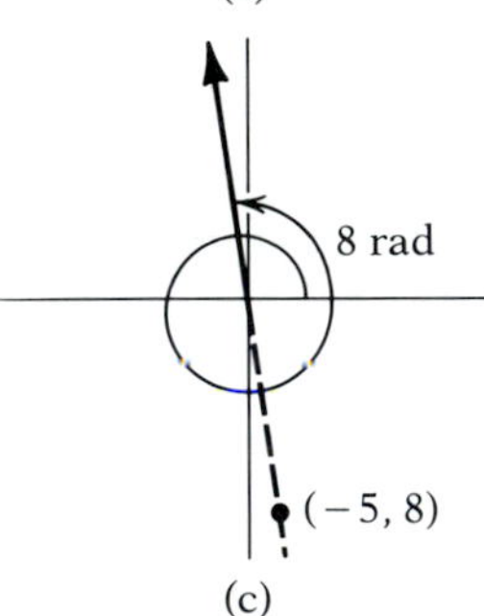

(c)

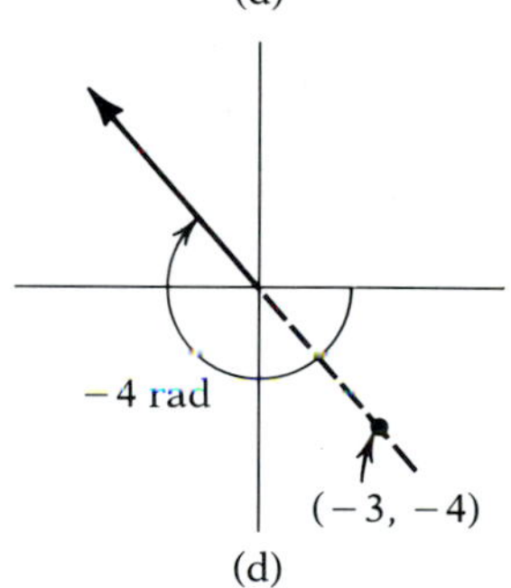

(d)

5. (a) $(-\sqrt{3}, 1)$ (b) $(0, 3)$ (c) $(0, 0)$ (d) $(\sqrt{2}, \sqrt{2})$

7. (a) $(4, 0)$ (b) $(4, \frac{\pi}{2})$ (c) $(4, \pi)$ (d) $\left(4, \dfrac{3\pi}{2}\right)$

9. (a) $x^2 + y^2 = 4$; circle

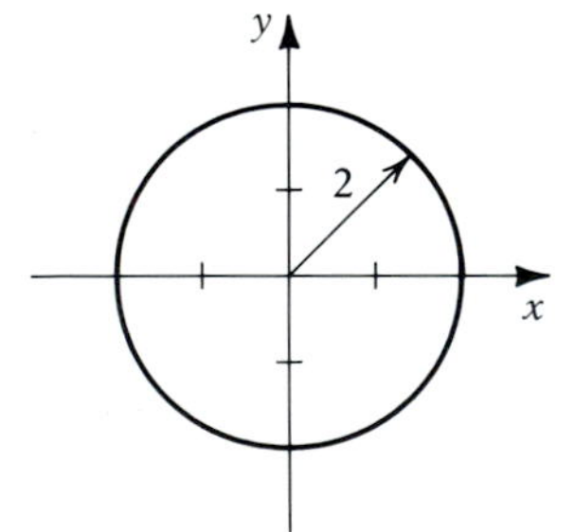

(b) $y = x$; line

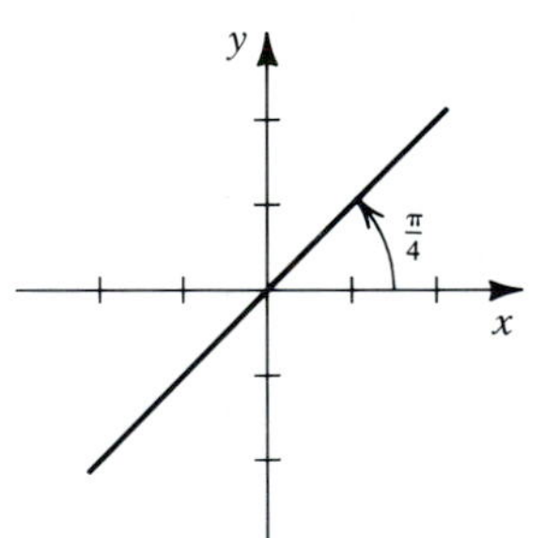

11. (a) $x = 3$; line

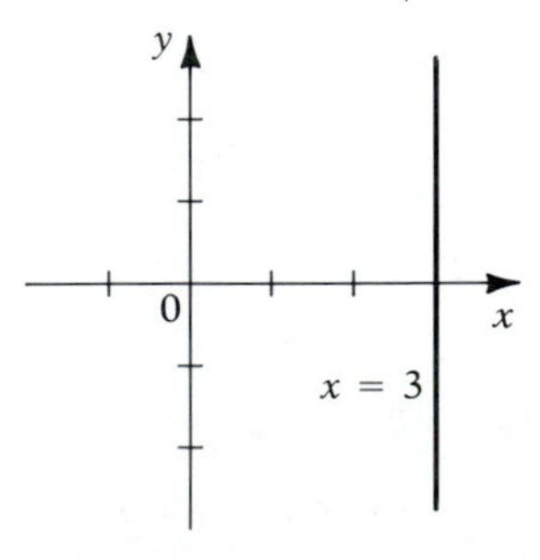

(b) $y = -1$; line

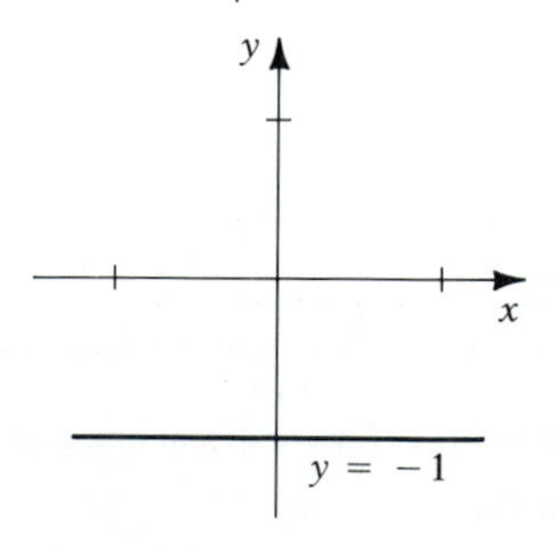

13. (a) $2x - 3y = 4$; line

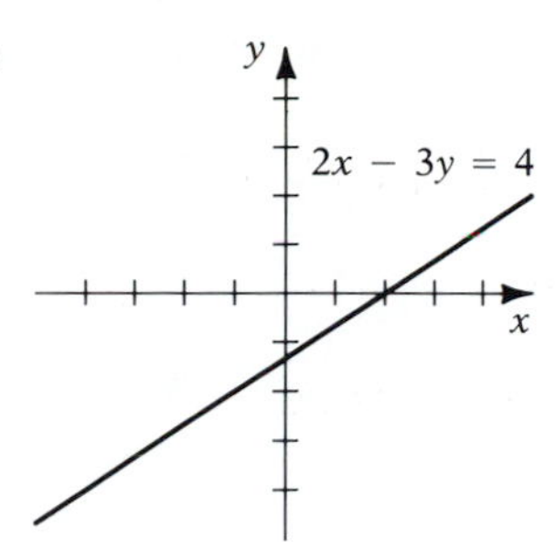

(b) $3x + 2y = 5$; line

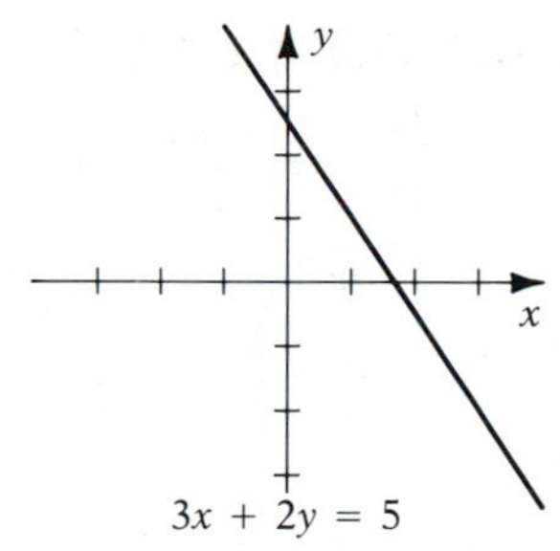

15. $(x^2 + y^2)^2 - 2ax^3 - ay^2(2x + 1) = 0$

17. $(x^2 + y^2)^2 + 4x^3 + y^2(4x - 1) + 3x^2 = 0$ **19.** $(x^2 + y^2)^3 = a^2(x^2 - y^2)^2$

21. $y^2 = 4(x + 1)$ **23.** (a) $r = a$ (b) $\theta = \dfrac{\pi}{3}$ **25.** (a) $r \sin\theta = -1$ (b) $r\cos\theta = 3$ **27.** $r = \tan\theta\sec\theta$ **29.** $r = \dfrac{4\cos\theta}{\cos 2\theta}$

31. $r^2 = \tan 2\theta$ **33.** $r = \dfrac{1}{1 + \cos\theta}$; $p = 1$, $e = 1$ **35.** $r = \dfrac{1}{1 + 2\cos\theta}$; $p = \dfrac{1}{2}$, $e = 2$

37. One example is $r = 2\cos\theta$. The pole is on the graph, since $(0, \pi/2)$ satisfies the equation. But $(0, 0)$ does not satisfy the equation.

Exercise Set 13.5, page 526

1. (a) Circle

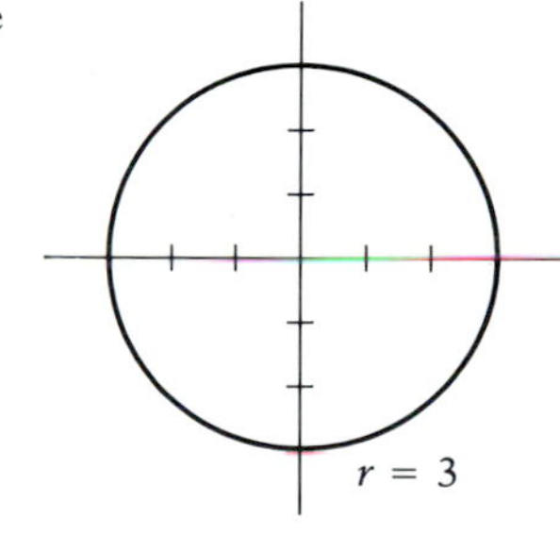

(b) Circle

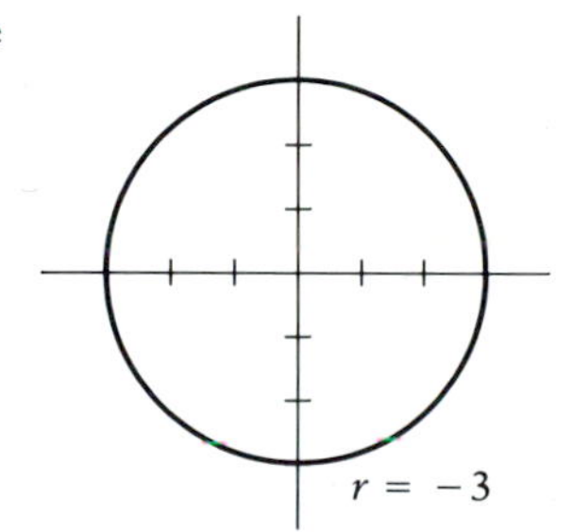

3. (a) Line

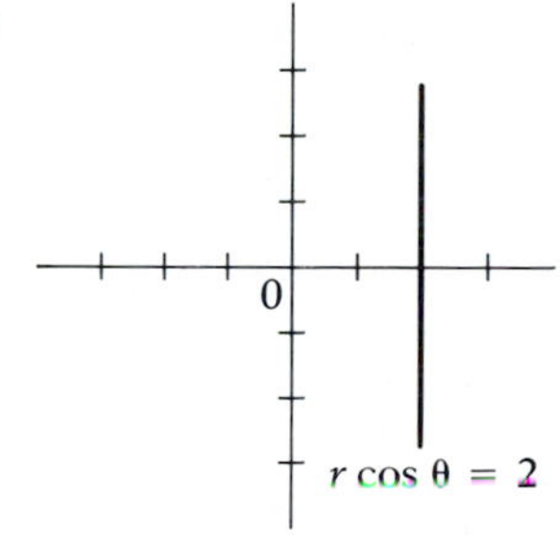

(b) Line

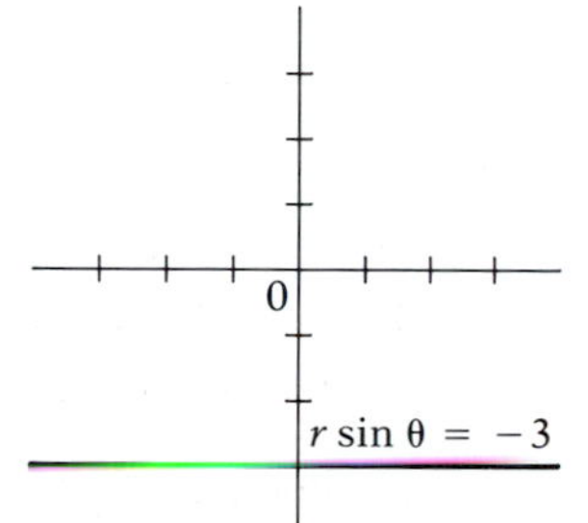

5. Circle

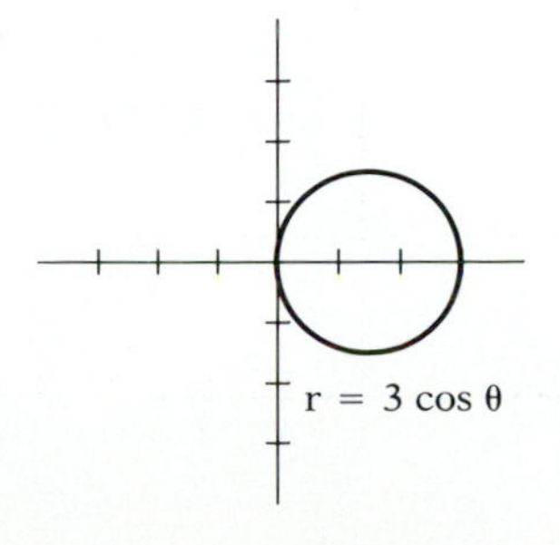

7. Cardioid

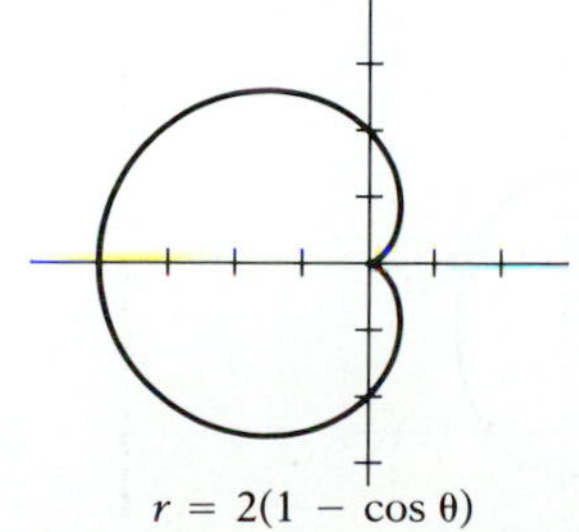

9. Cardioid

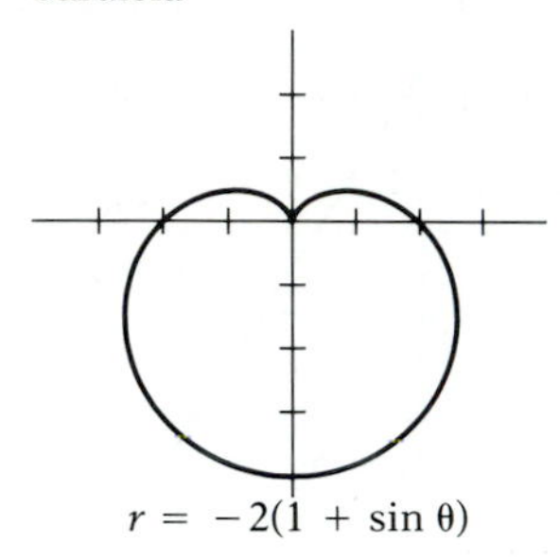

11. Limaçon without loop

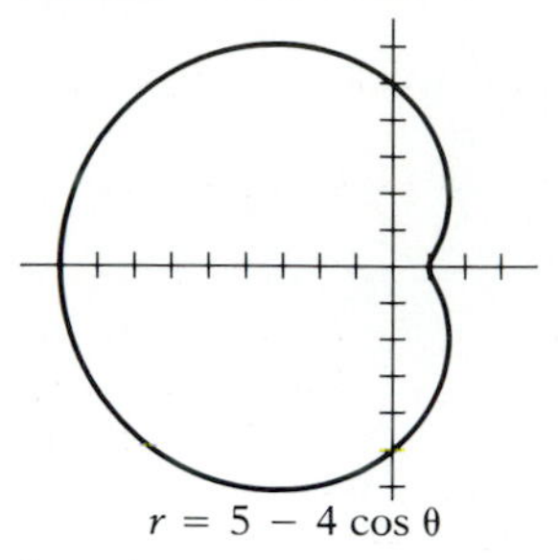

13. Limaçon with loop

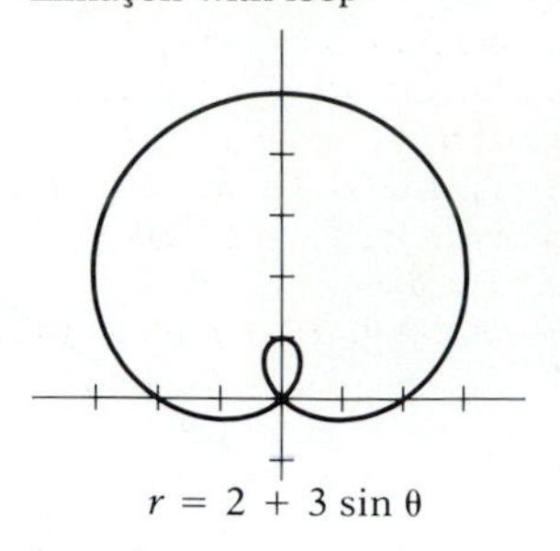

15. Limaçon without loop (no dimple)

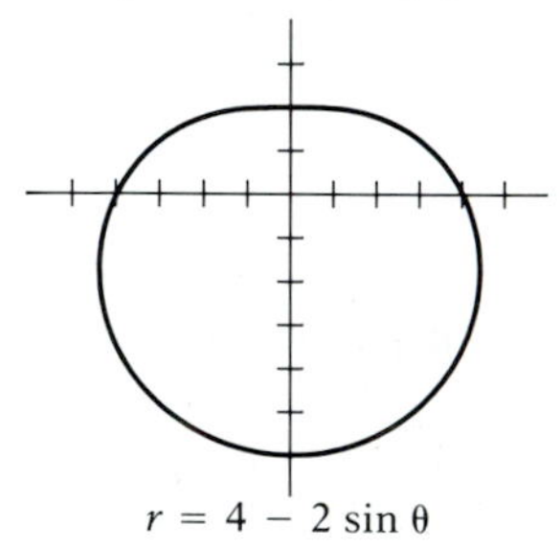

17. Lemniscate

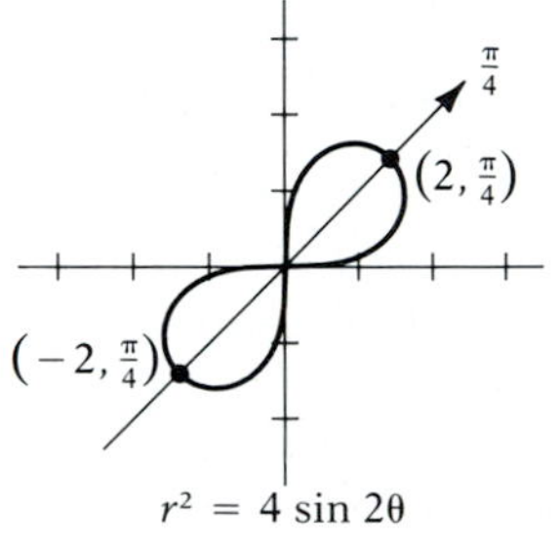

19. Lemniscate

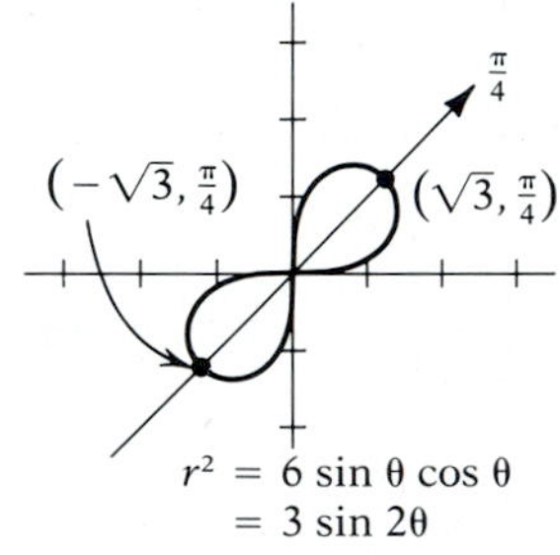

21. 3-leaf rose

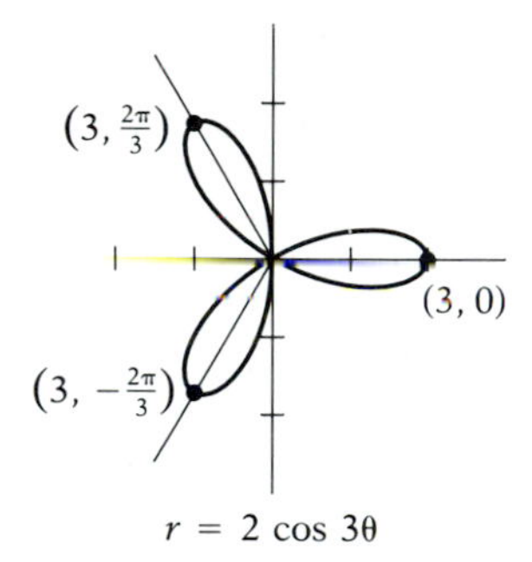

23. 4-leaf rose

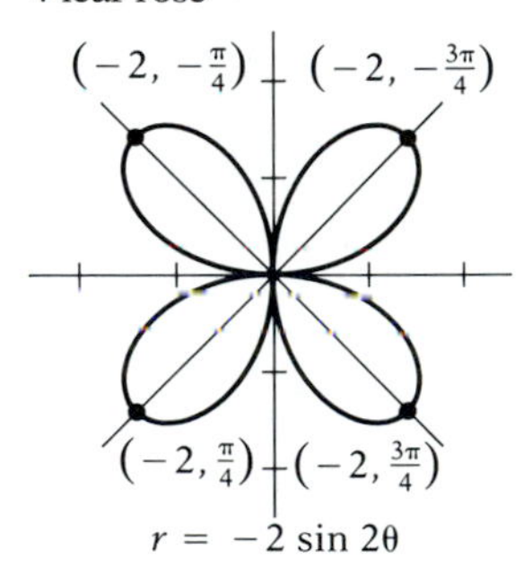

25. $r = \theta$

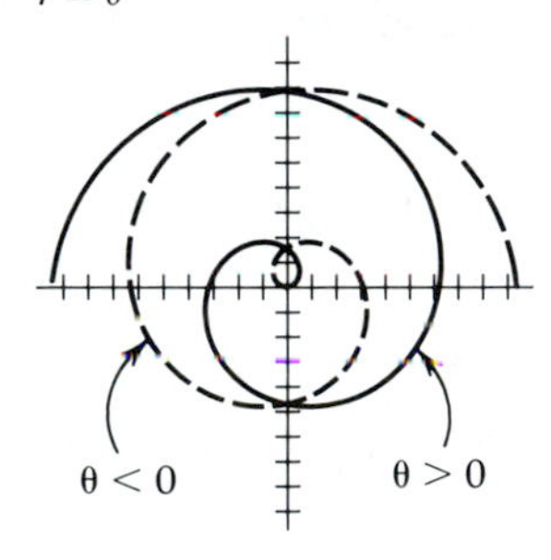

27. $r^2 = \theta$

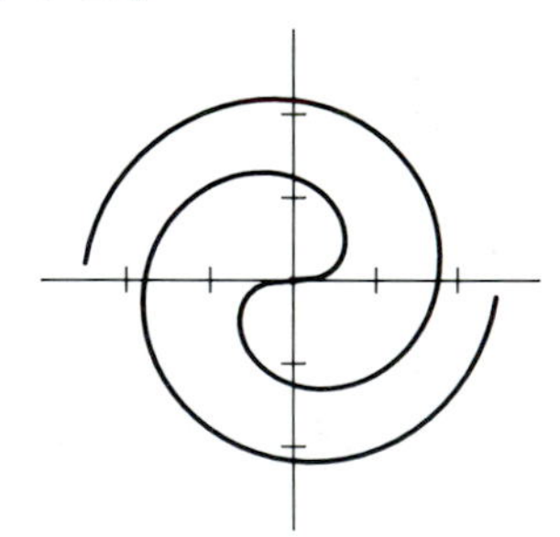

29. $2x - 3y = 6$, line

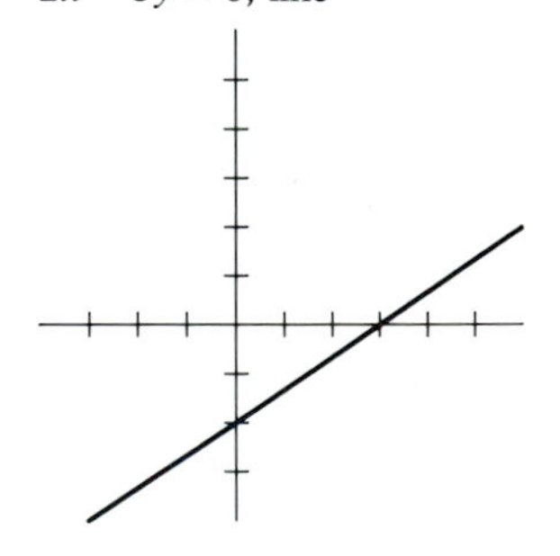

31. $y^2 = 4(x + 1)$, parabola

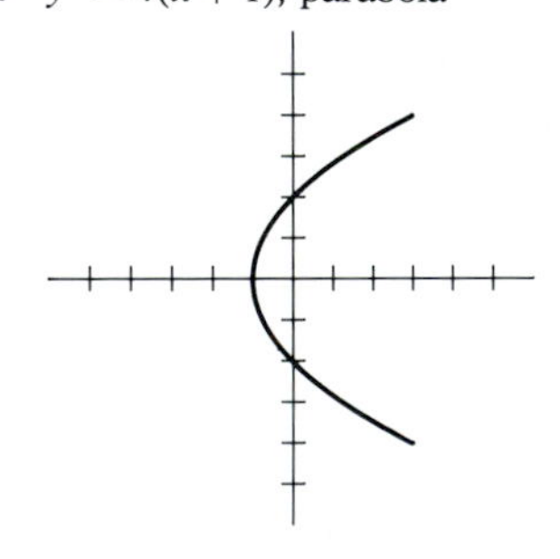

33. $\dfrac{x^2}{16} + \dfrac{(y-4)^2}{32} = 1$, ellipse

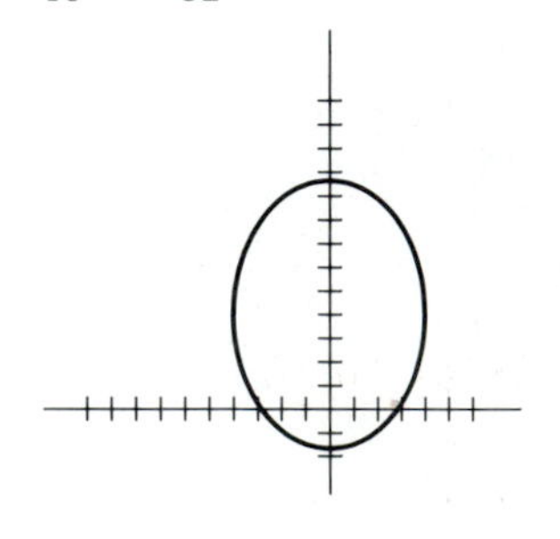

35. (a) Symmetric to polar axis if and only if whenever (r, θ) satisfies the equation, $(r, 2n\pi - \theta)$ or $(-r, (2n+1)\pi - \theta)$ satisfies the equation for some integer n.
(b) Symmetric to $\theta = \pi/2$ if and only if whenever (r, θ) satisfies the equation, $(r, (2n+1)\pi - \theta)$ or $(-r, 2n\pi - \theta)$ satisfies the equation for some integer n.
(c) Symmetric to the pole if and only if whenever (r, θ) satisfies the equation, $(-r, 2n\pi + \theta)$ or $(r, (2n+1)\pi + \theta)$ satisfies the equation for some integer n.

37. (a) Let $F(r, \theta) = r - \tan\theta$
$F(-r, \pi - \theta) = -F(r, \theta)$, so graph is symmetric to polar axis. $F(r, -\theta) \neq \pm F(r, \theta)$
(b) $F(-r, -\theta) = -F(r, \theta)$, so graph is symmetric to $\theta = \pi/2$. $F(r, \pi - \theta) \neq \pm F(r, \theta)$
(c) $F(r, \pi + \theta) = F(r, \theta)$, so graph is symmetric to the pole. $F(-r, \theta) \neq \pm F(r, \theta)$
(d) $F(-r, \pi - \theta) = -F(r, \theta)$, so graph is symmetric to the polar axis. $F(r, -\theta) \neq \pm F(r, \theta)$

39. (a) $F(r, \sin\theta) = F(r, \cos(\theta - \frac{\pi}{2}))$ (b) $F(r, -\cos\theta) = F(r, \cos(\theta - \pi))$ (c) $F(r, -\sin\theta) = F(r, \cos(\theta - \frac{3\pi}{2}))$

41. $$r = \sqrt{a^2 + b^2}\left(\frac{a}{\sqrt{a^2 + b^2}}\sin\theta + \frac{b}{\sqrt{a^2 + b^2}}\cos\theta\right)$$
$$= \sqrt{a^2 + b^2}\,(\cos\theta\cos\alpha + \sin\theta\sin\alpha) = \sqrt{a^2 + b^2}\,\cos(\theta - \alpha)$$

43.

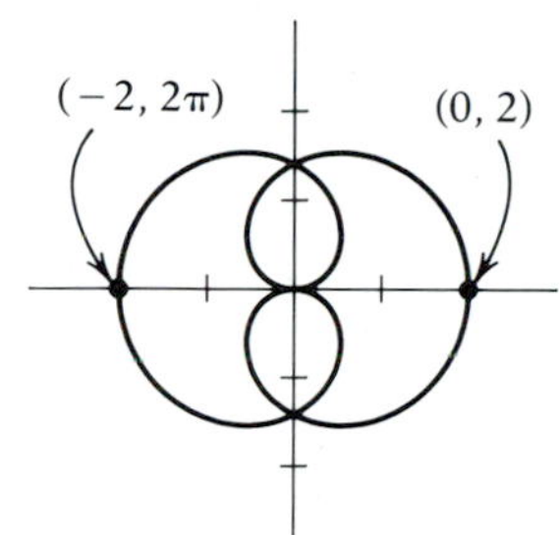

45.

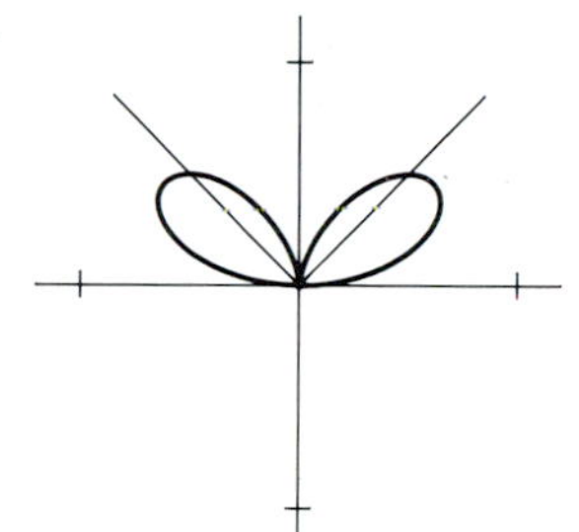

47.

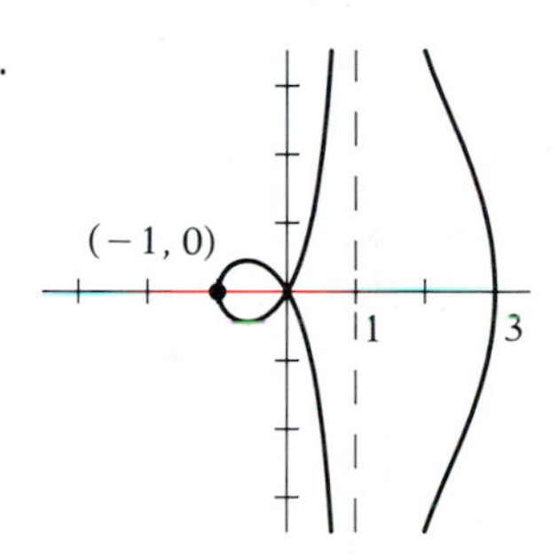

Exercise Set 13.6, page 531

1. 6π

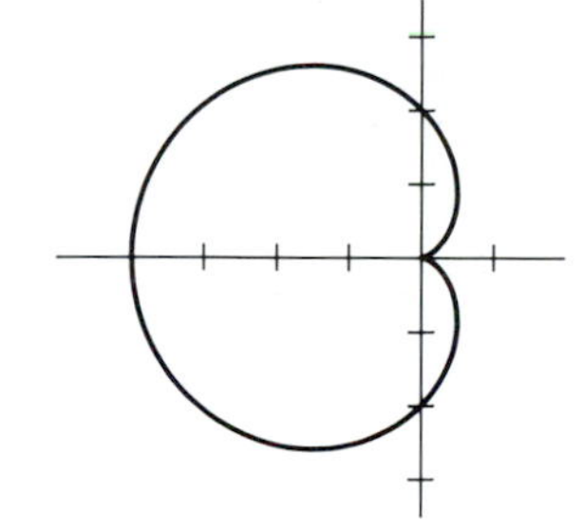

3. 9

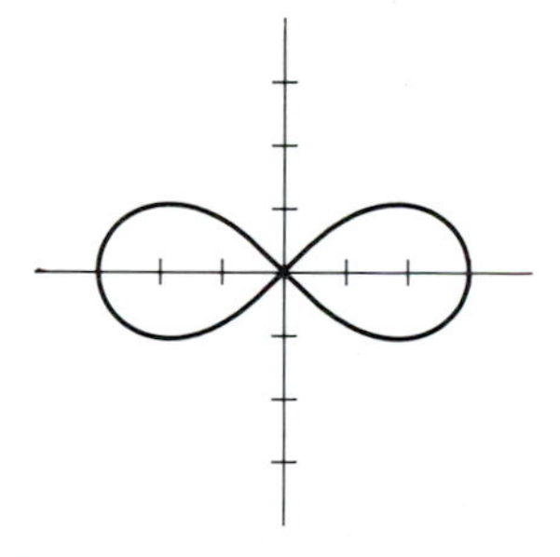

5. $\frac{3\pi}{2}$

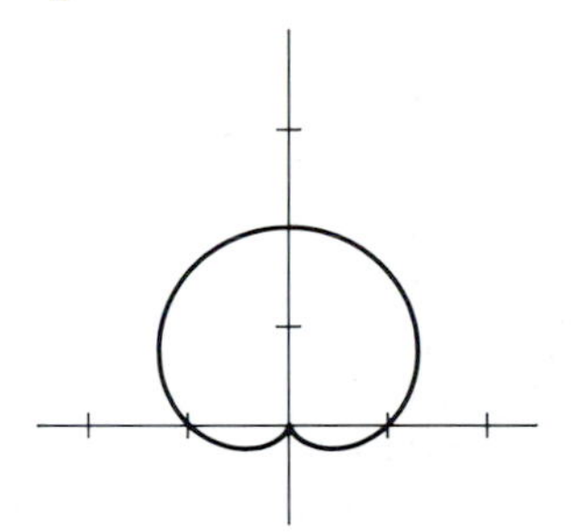

7. π

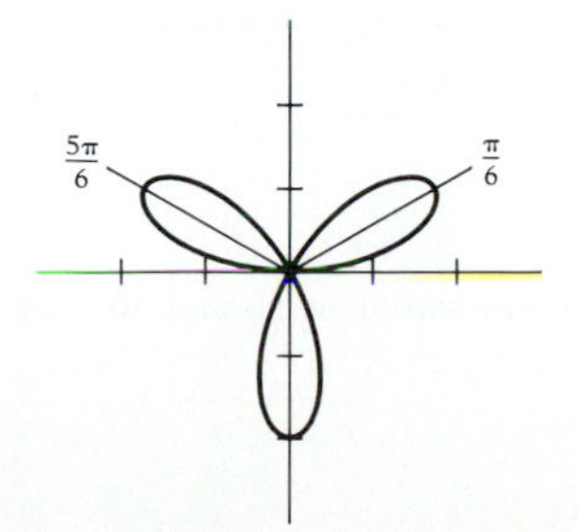

9. 4π

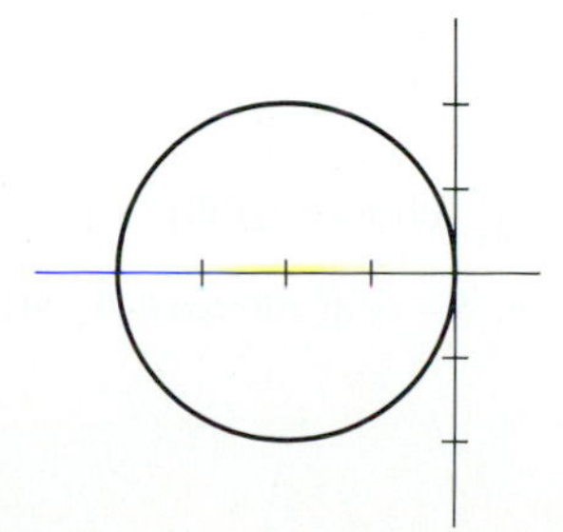

11. $8 + \pi$

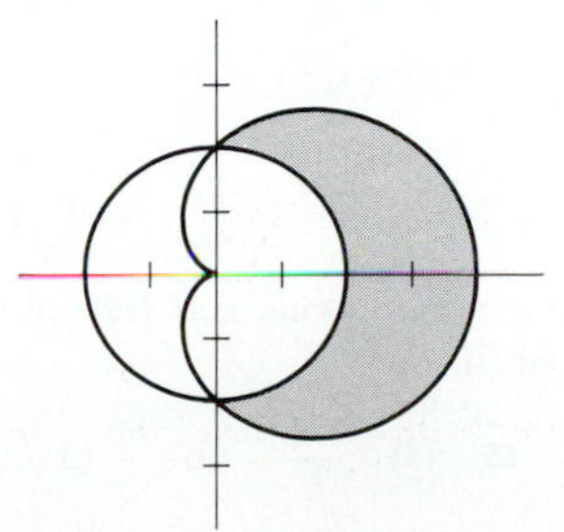

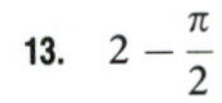

13. $2 - \frac{\pi}{2}$

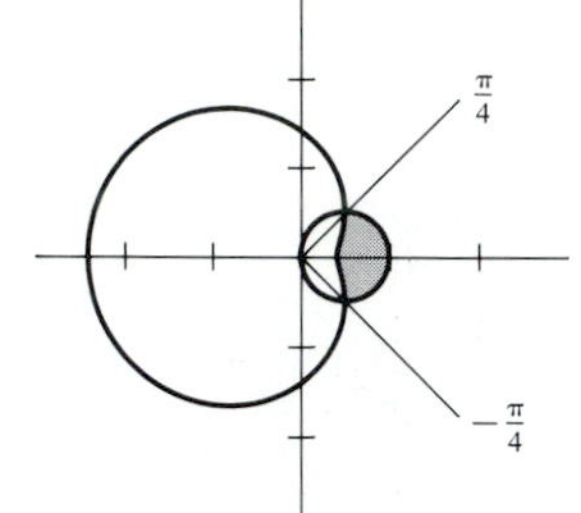

15. $\sqrt{3} - \frac{\pi}{3}$

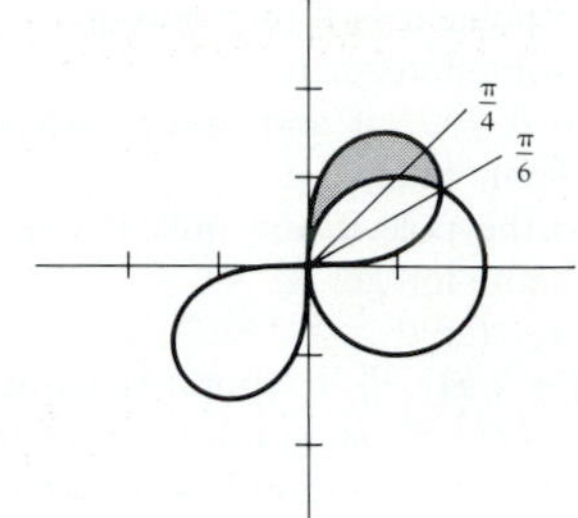

17. $\frac{-8\pi}{3} + 8\sqrt{3}$

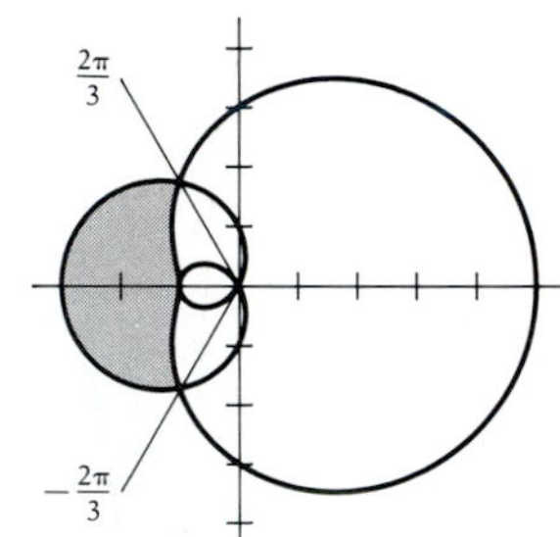

19. $9\sqrt{3} - 4\pi$

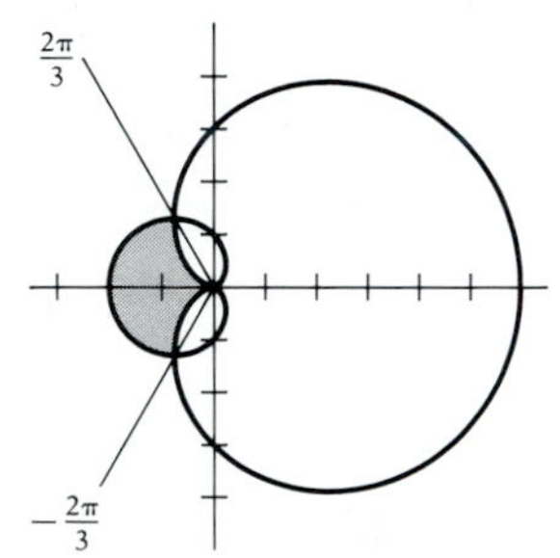

21. $\pi - \frac{3\sqrt{3}}{2}$ **23.** $\frac{11\pi}{3} + \frac{5\sqrt{3}}{2}$ **25.** $2\pi - 4$ **27.** $\pi + 3$ **29.** $1 + \frac{3\sqrt{3}}{2} - \pi$ **31.** $\frac{27}{2}\pi^3$

33. $\left(2, \pm\frac{\pi}{12}\right), \left(2, \pm\frac{5\pi}{12}\right), \left(2, \pm\frac{13\pi}{12}\right), \left(2, \pm\frac{17\pi}{12}\right)$

Exercise Set 13.7, page 537

1. $-\frac{1}{\sqrt{3}}$ **3.** $-\frac{1}{2}$ **5.** $\sqrt{3}$ **7.** $-\sqrt{3}$ **9.** $-\frac{\sqrt{3}}{2}$ **11.** 0 **13.** $\frac{1}{\sqrt{3}}$ **15.** 1 **17.** $\frac{1}{2}$ **19.** $-\frac{3}{2}$

21. $\tan\phi = 0$ **23.** $\cot\phi = 0$ **25.** $\tan\psi = \frac{1}{b}; \frac{\pi}{4}$ **27.** $\tan\alpha = \sqrt{3}$ at $\theta = -\frac{2\pi}{3}$, $\tan\alpha = -\sqrt{3}$ at $\theta = \frac{2\pi}{3}$.

29. $\tan\alpha = 0$ at $\theta = \pm\frac{\pi}{2}$; $\tan\alpha = \frac{-1}{3\sqrt{3}}$ at $\theta = \frac{\pi}{3}$; $\tan\alpha = \frac{1}{3\sqrt{3}}$ at $\theta = \frac{2\pi}{3}$.

31. $\cot\alpha = \frac{1 + \tan\psi_1 \tan\psi_2}{\tan\psi_2 - \tan\psi_1}$, so $\alpha = \frac{\pi}{2}$ if and only if $\tan\psi_1 \tan\psi_2 = -1$.

33. $\tan\psi_1 = \tan\theta$, $\tan\psi_2 = -\cot\theta$, so $\tan\psi_1 \tan\psi_2 = -1$ at all points of intersection.

35. $\tan\psi_1 = -\cot\theta$, $\tan\psi_2 = \tan\theta$. So $\tan\psi_1 \tan\psi_2 = -1$

37. $\tan\psi_1 = \frac{1 - \sin\theta}{-\cos\theta}$, $\tan\psi_2 = \frac{1 + \sin\theta}{\cos\theta}$; $\tan\psi_1 \tan\psi_2 = -\frac{1 - \sin^2\theta}{\cos^2\theta} = -1$. **39.** $\frac{7\sqrt{10}}{3}$ **41.** 4 **43.** 12

47. $\alpha = \phi_2 - \phi_1 = (\theta + \psi_2 + n\pi) - (\theta + \psi_1 + m\pi) = \psi_2 - \psi_1 + (n - m)\pi$ for integers n and m. So $\tan\alpha = \tan(\psi_2 - \psi_1)$

49. $1 + \frac{1}{2\sqrt{3}} \ln(2 + \sqrt{3})$

51. (a) $S = 2\pi \int y\, ds = 2\pi \int_\alpha^\beta r \sin\theta \sqrt{r^2 + \left(\frac{dr}{d\theta}\right)^2}\, d\theta = 2\pi \int_\alpha^\beta f(\theta) \sin\theta \sqrt{[f(\theta)]^2 + [f'(\theta)]^2}\, d\theta$

(b) For f continuous and $f(\theta)\cos\theta \geq 0$ on $\alpha \leq \theta \leq \beta$, $S = 2\pi \int_\alpha^\beta f(\theta) \cos\theta \sqrt{[f(\theta)]^2 + [f'(\theta)]^2}\, d\theta$. Derivation is similar to that in part a.

53. 2π **55.** (a) $\frac{4\pi\sqrt{2}}{5}$ (b) $\frac{8\pi}{5}(3\sqrt{2} - 4)$

Supplementary Exercises 13.8, page 538

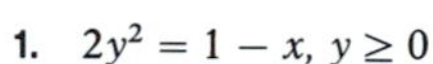

1. $2y^2 = 1 - x,\ y \geq 0$

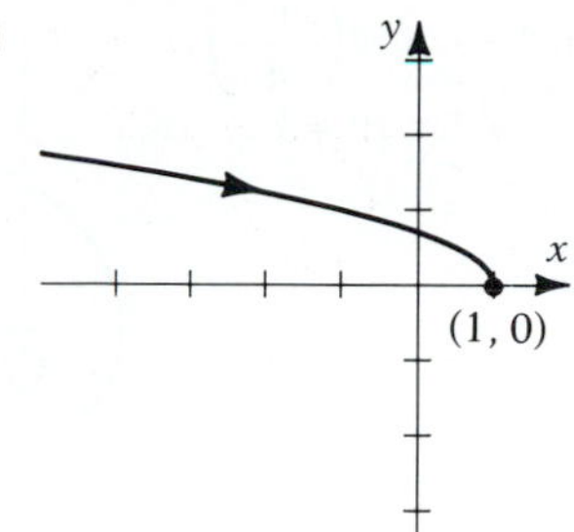

3. $y = 1 - e^{-x}$

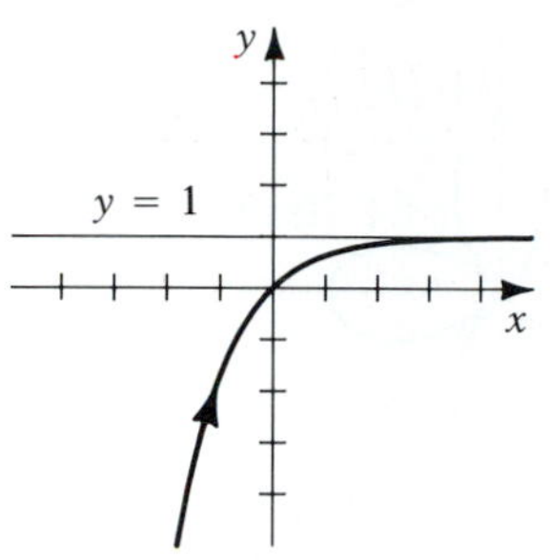

5. $(y-1)^2 = x + 2,\ y \geq 2,\ x \geq -1$

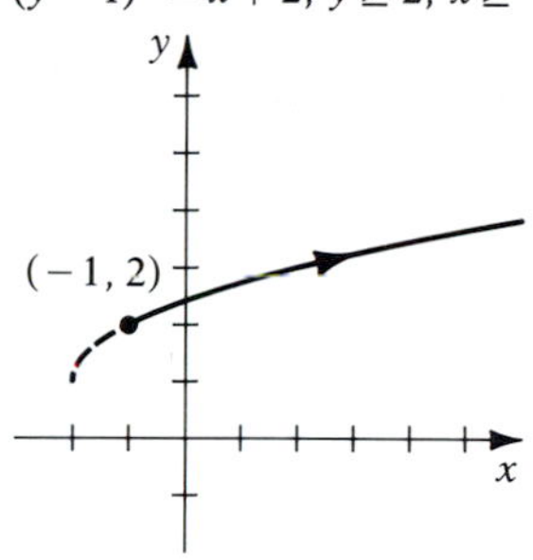

7. $\dfrac{dy}{dx} = 3t,\ \dfrac{d^2y}{dx^2} = \dfrac{3}{2t}$

9. $\dfrac{dy}{dx} = -\dfrac{1}{2}\sec^2 t,\ \dfrac{d^2y}{dx^2} = -\dfrac{1}{2}\sec^4 t$

11. Tangent: $3x + 4y = 8$; Normal: $4x - 3y + 6 = 0$

13. $\dfrac{20}{9} + \ln 3$

15. (a) $4 - 2\sqrt{2} - \ln(9 - 6\sqrt{2})$ (b) $8\pi\left(\sqrt{3} - \dfrac{\sqrt{2}}{2} + \dfrac{1}{2}\ln\dfrac{\sqrt{3}+2}{\sqrt{2}+1}\right)$

17. (a) $(x-1)^2 + (y-1)^2 = 2$, circle, center $(1, 1)$, radius $\sqrt{2}$

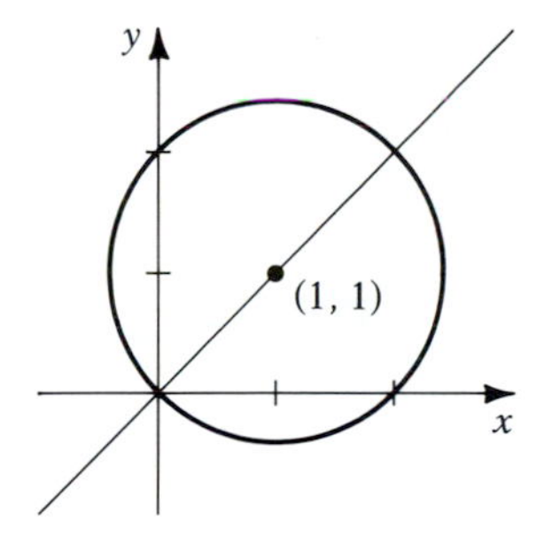

(b) $\dfrac{x^2}{\frac{1}{3}} + \dfrac{(y+\frac{1}{3})^2}{\frac{4}{9}} = 1$, ellipse, center $\left(0, -\dfrac{1}{3}\right)$, one focus at origin

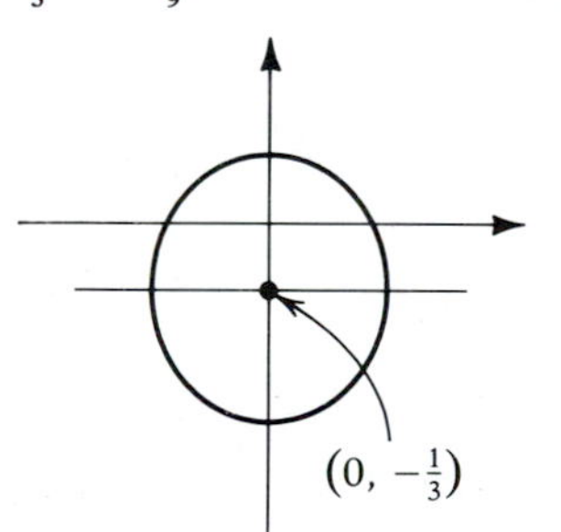

19. (a) $r = 2\sin^2\dfrac{\theta}{2} = 1 - \cos\theta$, cardioid, symmetric to the polar axis

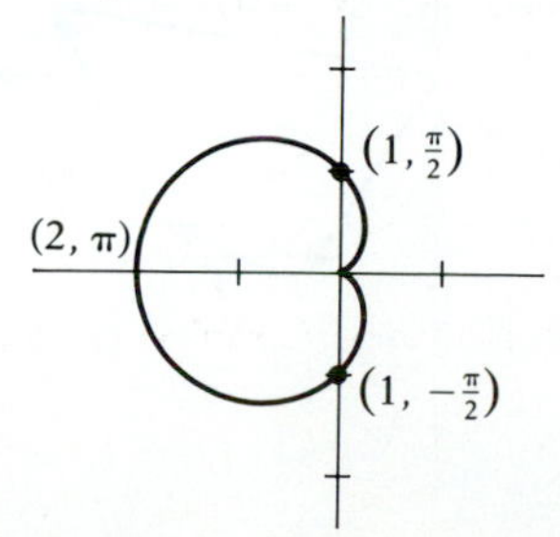

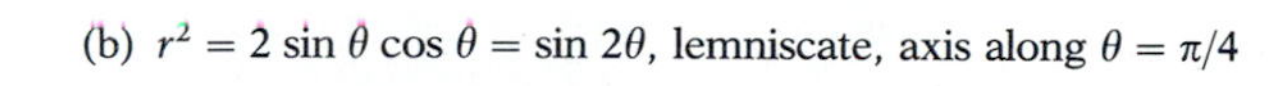

(b) $r^2 = 2\sin\theta\cos\theta = \sin 2\theta$, lemniscate, axis along $\theta = \pi/4$

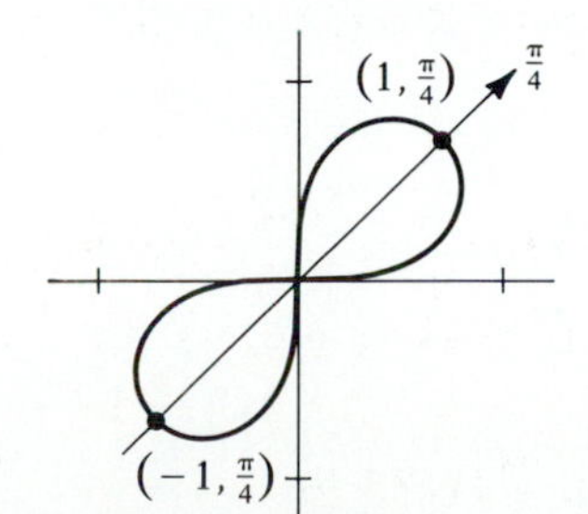

21. (a) $r = \sin^2\theta - \cos^2\theta = -\cos 2\theta$, 4-leaf rose, symmetric to $\theta = 0$, $\theta = \pi/2$, and the pole

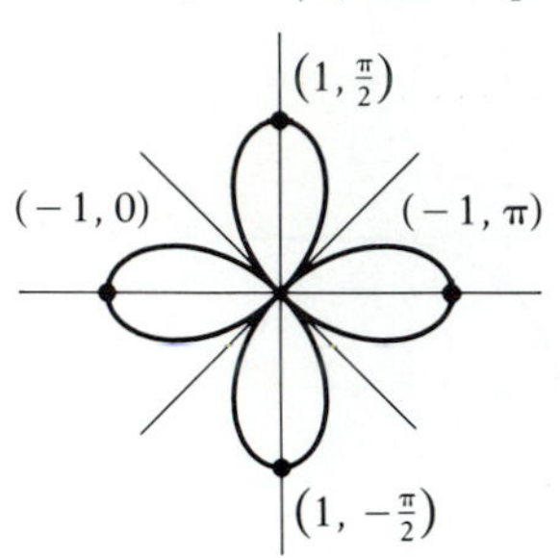

(b) $r = \dfrac{1}{\theta}$, spiral, asymptotic to $y = 1$, since

$$\lim_{\theta\to 0} r\sin\theta = \lim_{\theta\to 0} r\theta\left(\frac{\sin\theta}{\theta}\right) = \lim_{\theta\to 0} r\theta = 1$$

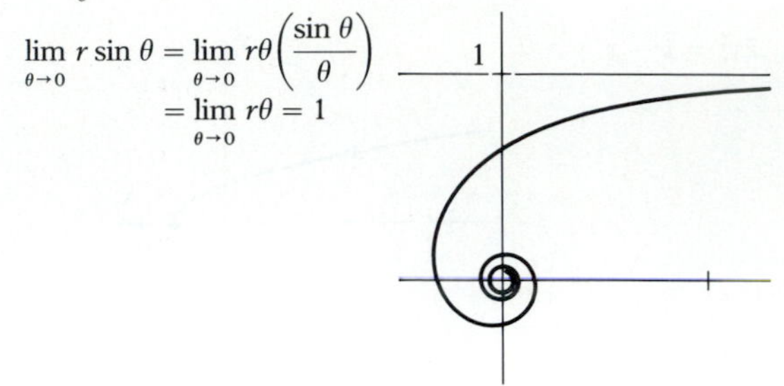

23. $\dfrac{5\pi}{2} - 3\sqrt{3}$ **25.** (a) $\dfrac{4\pi}{3} + 2\sqrt{3}$ (b) $\dfrac{8\pi}{3} - 2\sqrt{3}$ (c) $\dfrac{4\pi}{3} + 2\sqrt{3}$ **27.** $\alpha = \dfrac{\pi}{4}$ at $\theta = \pm\dfrac{\pi}{4}, \pm\dfrac{3\pi}{4}$

29. Curves intersect at $\theta = \pm\pi/3$, and at each point $\tan\psi_1 \tan\psi_2 = -1$ **31.** $2\sqrt{2} + 7\ln(3 + 2\sqrt{2})$

33. $\begin{cases} x = 2a\cot\theta \\ y = 2a\sin^2\theta \end{cases}$

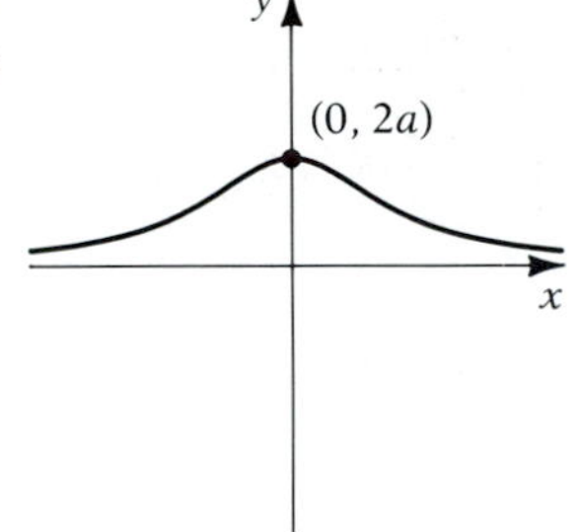

CHAPTER 14

Exercise Set 14.2, page 545

1. $\langle 2, 5\rangle$

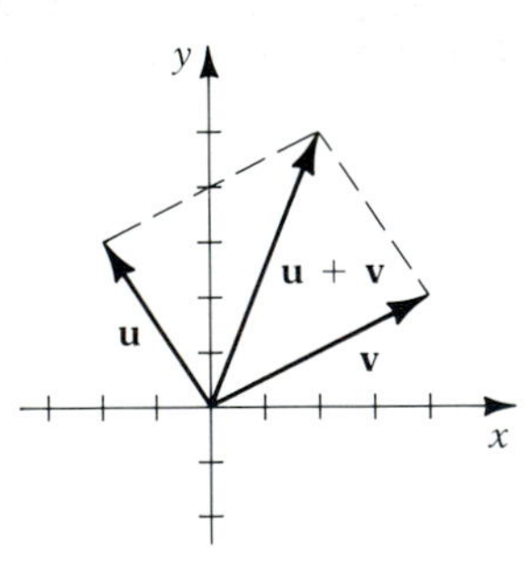

3. $\langle -3, 1\rangle$

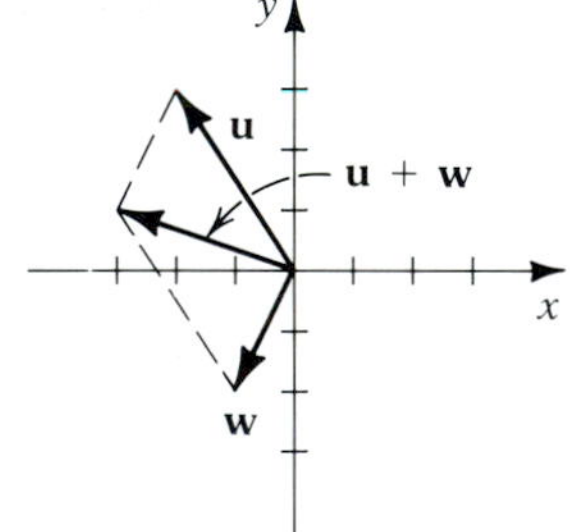

5. $\langle -6, 1\rangle$

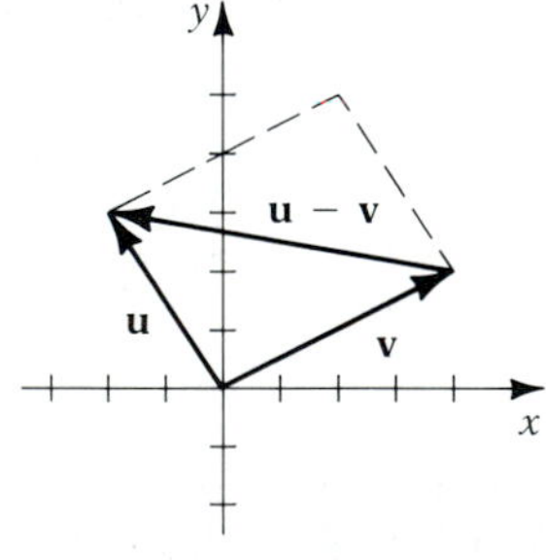

7. $\langle 1, -5\rangle$

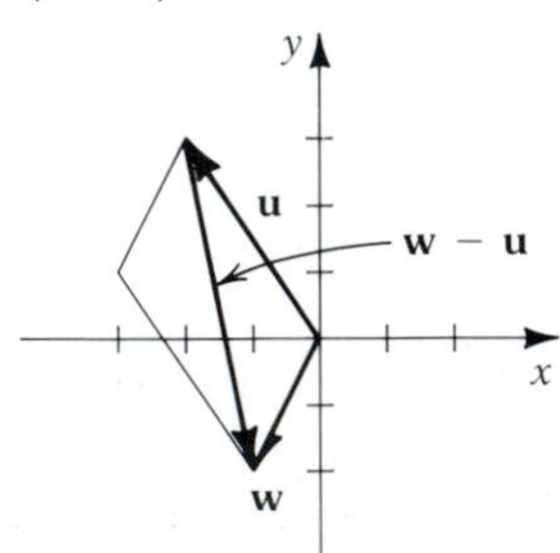

9. $\langle 0, 7\rangle$

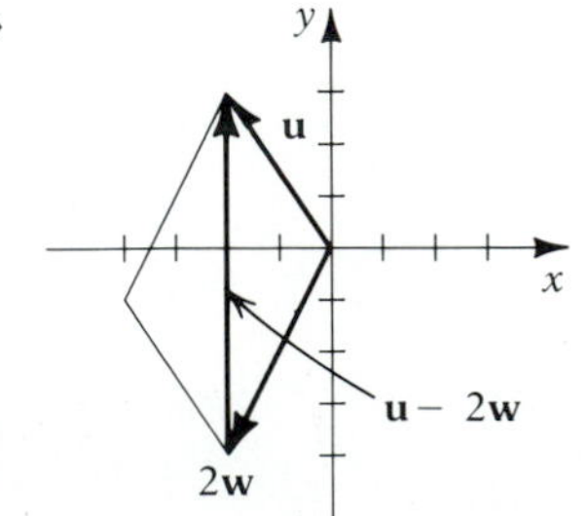

11. $\langle 11, 2\rangle$

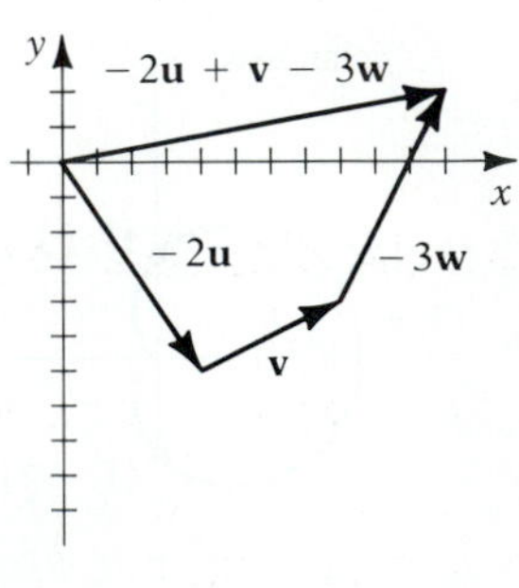

13. $\langle -4, -2\rangle$ **15.** $\langle -3, -4\rangle$ **17.**

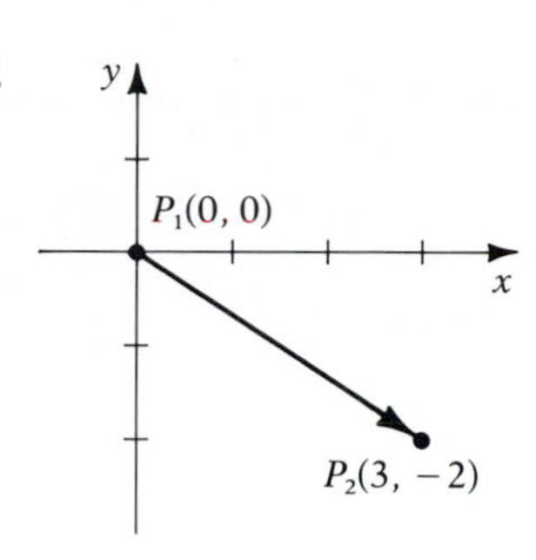

19.

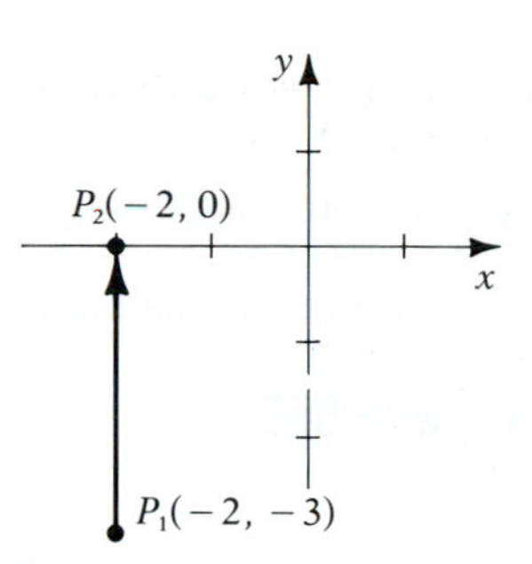

21. 5 **23.** 17 **25.** $\sqrt{5}$ **27.** (a) $\sqrt{5}$ (b) $\sqrt{41}$ (c) 7 (d) $5\sqrt{10}$

29. $|\mathbf{R}| = 10\sqrt{19} \approx 43.6$ N; $\theta \approx 46.6°$

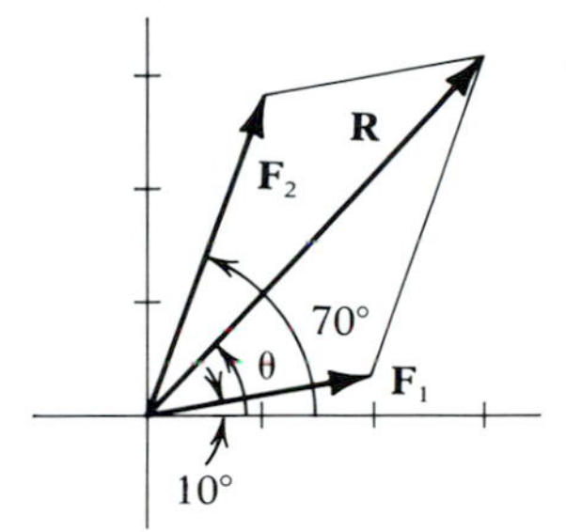

31. $|\mathbf{v}| = \sqrt{86{,}800} \approx 294.6$; actual heading $\approx 240.2°$.

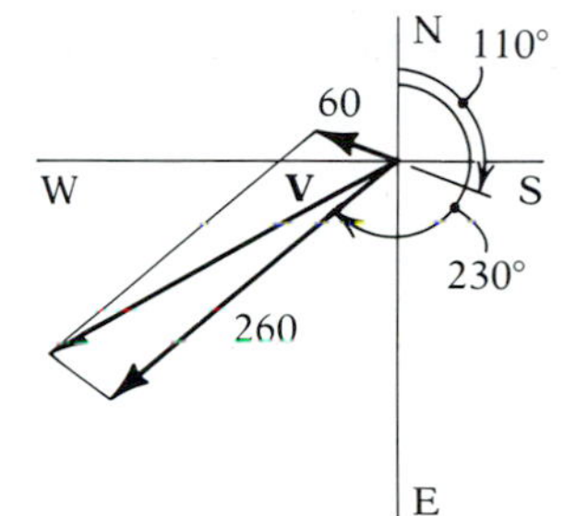

33. Use geometric vectors with initial points at the origin, and consider corresponding algebraic vectors.

37. $a = 2,\ b = -1$

39. If $\mathbf{u} \neq k\mathbf{v}$, the system

$$\begin{cases} au_1 + bu_1 = w_1 \\ au_2 + bu_2 = w_2 \end{cases}$$

has a unique solution for a and b.

41. Denote consecutive vertices by A, B, C, D. Let M_1 be the midpoint of AC and M_2 be the midpoint of BD. Show that $\overrightarrow{AM_1} = \frac{1}{2}(\overrightarrow{AB} + \overrightarrow{AD})$ and $\overrightarrow{AM_2} = \overrightarrow{AD} + \frac{1}{2}\overrightarrow{DB} = \overrightarrow{AD} + \frac{1}{2}(\overrightarrow{AB} - \overrightarrow{AD})$ and hence $\overrightarrow{AM_1} = \overrightarrow{AM_2}$. Conclude that M_1 and M_2 are the same point.

43. $\mathbf{a} = \dfrac{\mathbf{v}}{2} + \dfrac{\mathbf{w}}{2}$

$\mathbf{b} = \dfrac{\mathbf{u}}{2} + \dfrac{\mathbf{z}}{2}$

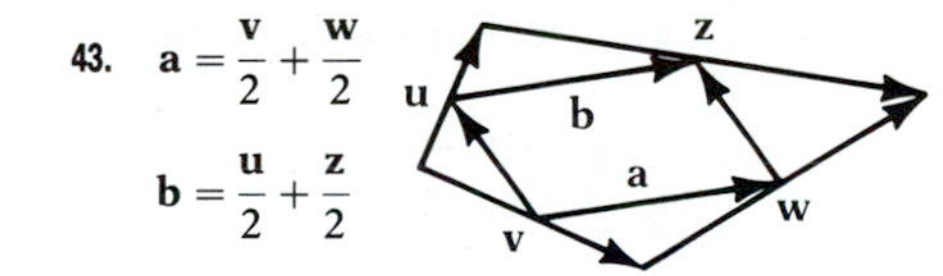

Express $\mathbf{z}$ in terms of $\mathbf{u}$, $\mathbf{v}$, and $\mathbf{w}$ to show that $\mathbf{a} = \mathbf{b}$.

45. Heading: 348.4° going, 191.6° returning; total time: 3.5 hr

Exercise Set 14.3, page 553

1. -6 **3.** -4 **5.** $\mathbf{u} \cdot \mathbf{v} = 0$ **7.** $\mathbf{u} \cdot \mathbf{v} = 0$ **9.** $-\dfrac{3}{5}$ **11.** $-\dfrac{1}{5\sqrt{2}}$ **13.** $\dfrac{7}{25}$ **15.** $\dfrac{1}{2};\ \theta = \dfrac{\pi}{3}$ **17.** $\dfrac{5}{13}\mathbf{i} + \dfrac{12}{13}\mathbf{j}$

19. $\left\langle -\dfrac{2}{\sqrt{13}}, -\dfrac{3}{\sqrt{13}} \right\rangle$ **21.** $\langle 2\sqrt{5}, -4\sqrt{5}\rangle$ **23.** $\dfrac{12}{5}\mathbf{i} + \dfrac{16}{5}\mathbf{j}$ **25.** $3\sqrt{5}\mathbf{i} - 6\sqrt{5}\mathbf{j}$ **27.** $7, -\dfrac{1}{7}$ **29.** $-\dfrac{5}{\sqrt{2}}$ **31.** $-\dfrac{5}{\sqrt{2}}$

33. 28 ft-lb **35.** 30 ft-lb **37.** $\text{Proj}_{\mathbf{v}}\mathbf{u} = \left\langle -\frac{1}{5}, -\frac{2}{5} \right\rangle$, $\text{Proj}_{\mathbf{v}}^{\perp}\mathbf{u} = \left\langle \frac{16}{5}, -\frac{8}{5} \right\rangle$ **39.** $\text{Proj}_{\mathbf{v}}\mathbf{u} = \frac{6}{5}\mathbf{i} - \frac{8}{5}\mathbf{j}$, $\text{Proj}_{\mathbf{v}}^{\perp}\mathbf{u} = \frac{24}{5}\mathbf{i} + \frac{18}{5}\mathbf{j}$

41. 500 lb **43.** Area = 20 **45.** Interior angles are $\cos^{-1}\frac{1}{5\sqrt{2}} \approx 81.9°$ and its supplement 98.1°.

49. Use equations (14.11) and (14.12) to show that $(\text{Proj}_{\mathbf{v}}\mathbf{u}) \cdot (\text{Proj}_{\mathbf{v}}^{\perp}\mathbf{u}) = 0$.

Exercise Set 14.4, page 560

1.

(3, 2, 4)

(a)

(4, −2, 1)

(b)

(−3, 2, 4)

(c)

(0, −5, −2)

(d)

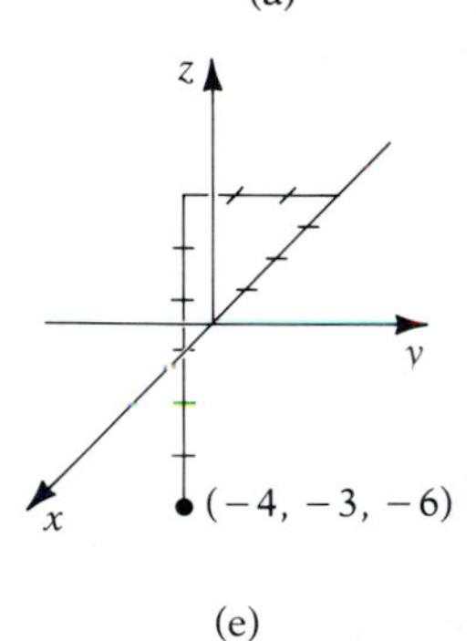

(e)

3. (a) xy-plane (b) yz-plane (c) xz-plane (d) x-axis (e) x-axis, y-axis, and z-axis

5. (a) $\langle -2, -4, 3 \rangle$ (b) $\langle -3, 1, 4 \rangle$ (c) $\langle 3, 2, -2 \rangle$ (d) $\langle 6, -2, -2 \rangle$ **7.** (a) -14 (b) -27 (c) $\sqrt{126}$ (d) $\langle 3, 2, -3 \rangle$

9. (a) $\left\langle \frac{2}{3}, -\frac{1}{3}, \frac{2}{3} \right\rangle$ (b) $\left\langle \frac{4}{5\sqrt{2}}, \frac{3}{5\sqrt{2}}, -\frac{1}{\sqrt{2}} \right\rangle$ (c) $\frac{1}{\sqrt{3}}\mathbf{i} - \frac{1}{\sqrt{3}}\mathbf{j} + \frac{1}{\sqrt{3}}\mathbf{k}$ (d) $\frac{2}{3\sqrt{5}}\mathbf{i} - \frac{4}{3\sqrt{5}}\mathbf{j} + \frac{5}{3\sqrt{5}}\mathbf{k}$ **11.** (a) $\mathbf{u} \cdot \mathbf{v} = 0$ (b) $\frac{1}{2}$

13. (a) $\cos\alpha = \frac{2}{3}$, $\cos\beta = -\frac{1}{3}$, $\cos\gamma = \frac{2}{3}$ (b) $\cos\alpha = \frac{3}{5\sqrt{2}}$, $\cos\beta = -\frac{1}{\sqrt{2}}$, $\cos\gamma = \frac{4}{5\sqrt{2}}$ **15.** $\left\langle -\frac{1}{2}, \frac{1}{2}, \frac{1}{\sqrt{2}} \right\rangle$

17. $\cos^2\gamma = 1 - (\cos^2\alpha + \cos^2\beta) \geq \frac{1}{2}$; so $\cos\gamma \geq \frac{1}{\sqrt{2}}$, and $\gamma \leq \frac{\pi}{4}$. **19.** $\frac{31}{5}$ **21.** $-\frac{2}{\sqrt{3}}$ **23.** 184 erg

25. $\text{Proj}_{\mathbf{v}}\mathbf{u} = \left\langle -\frac{2}{7}, -\frac{6}{7}, \frac{4}{7} \right\rangle$, $\text{Proj}_{\mathbf{v}}^{\perp}\mathbf{u} = \left\langle \frac{37}{7}, -\frac{1}{7}, \frac{17}{7} \right\rangle$ **27.** $\text{Proj}_{\mathbf{v}}\mathbf{u} = \frac{11}{14}\mathbf{i} + \frac{11}{7}\mathbf{j} - \frac{33}{14}\mathbf{k}$, $\text{Proj}_{\mathbf{v}}^{\perp}\mathbf{u} = \frac{17}{14}\mathbf{i} - \frac{32}{7}\mathbf{j} - \frac{37}{14}\mathbf{k}$

35. $\frac{2}{\sqrt{93}}\mathbf{i} - \frac{8}{\sqrt{93}}\mathbf{j} - \frac{5}{\sqrt{93}}\mathbf{k}$ **37.** $a = \frac{5}{3}$, $b = \frac{7}{3}$, $c = \frac{4}{3}$

Exercise Set 14.5, page 566

1. $\langle 3, -1, 4 \rangle$ **3.** $3\mathbf{i} + 9\mathbf{j} + 6\mathbf{k}$ **5.** $\langle -9, -1, -21 \rangle$ **7.** $19\mathbf{i} + 13\mathbf{j} + 5\mathbf{k}$ **9.** $\langle 11, -6, -2 \rangle$ **11.** $\mathbf{i} - \mathbf{j} + \mathbf{k}$

13. 12 **15.** $12\mathbf{i} + 12\mathbf{j} - 12\mathbf{k}$ **17.** 69 **19.** $(\mathbf{u} \times \mathbf{v}) \cdot \mathbf{v} = 0$ **21.** $\sqrt{146}$ **23.** $3\sqrt{78}$ **25.** $\frac{3\sqrt{5}}{2}$

27. $\left\langle \frac{4}{\sqrt{89}}, -\frac{3}{\sqrt{89}}, -\frac{8}{\sqrt{89}} \right\rangle$ **29.** 23 **31.** Show that the volume of the parallelepiped with edges $\overrightarrow{AB}$, $\overrightarrow{AC}$, and $\overrightarrow{AD}$ is 0.

Exercise Set 14.6, page 572

1. (a) $\mathbf{r} = \langle 2 - 3t, 5 + t, -1 + 2t \rangle$ (b) $x = 2 - 3t$, $y = 5 + t$, $z = -1 + 2t$
3. (a) $\mathbf{r} = (5 + 2t)\mathbf{i} + (8 - 3t)\mathbf{j} + (-6 + 4t)\mathbf{k}$ (b) $x = 5 + 2t$, $y = 8 - 3t$, $z = -6 + 4t$ **5.** $\mathbf{r} = \langle 4 - t, -1 + 3t, 8 - 3t \rangle$
7. $\mathbf{r} = \langle 7 - 4t, -2 + 3t, -4 + 2t \rangle$ **9.** $\cos^{-1} \dfrac{10}{3\sqrt{30}} \approx 52.5°$ **11.** $\cos^{-1} \dfrac{1}{\sqrt{29}\sqrt{14}} \approx 87.2°$
15. $x = 3 - 3t$, $y = -1 + t$, $z = 2 + 5t$ **17.** $\mathbf{r} = \langle 5 + 2t, 2 + t, -3 \rangle$
19. Lines intersect at $(-1, 2, 1)$; $x = -1 - 7t$, $y = 2 - 9t$, $z = 1 + 3t$ **21.** Skew **23.** Intersect at $(-1, 3, 2)$
27. (a) $\dfrac{x - 3}{-2} = \dfrac{y + 1}{4} = \dfrac{z - 4}{3}$ (b) $x = 3 - 2t$, $y = -1 + 4t$, $z = 4 + 3t$ (c) $\mathbf{r} = \langle 3 - 2t, -1 + 4t, 4 + 3t \rangle$
(d) xy-plane: $\left(\dfrac{17}{3}, \dfrac{-19}{3}, 0\right)$; xz-plane: $\left(-\dfrac{5}{2}, 0, \dfrac{19}{4}\right)$; yz-plane: $\left(0, 5, \dfrac{17}{2}\right)$
29. $\mathbf{r} = (-1 + 3t)\mathbf{i} + (2 + 4t)\mathbf{j} + (8 - 5t)\mathbf{k}$ or $\mathbf{r} = (-1 + 19t)\mathbf{i} + (2 - 8t)\mathbf{j} + (8 + 5t)\mathbf{k}$

Exercise Set 14.7, page 578

1. (a) $\langle x - 4, y - 2, z - 6 \rangle \cdot \langle 3, 2, -1 \rangle = 0$ (b) $3(x - 4) + 2(y - 2) - (z - 6) = 0$ (c) $3x + 2y - z - 10 = 0$
3. (a) $\langle x - 5, y + 3, z + 4 \rangle \cdot \langle 2, 3, -4 \rangle = 0$ (b) $2(x - 5) + 3(y + 3) - 4(z + 4) = 0$ (c) $2x + 3y - 4z - 17 = 0$
5. $x - 2y - 4z + 8 = 0$ **7.** $3x - 2y + 4z + 25 = 0$ **9.** $z = 4$ **11.** $3x - 4y + 5z = 0$ **13.** $x - 4y - 3z - 1 = 0$
15. $5x + y - 4z - 9 = 0$ **17.** $\cos^{-1} \dfrac{4}{\sqrt{29}\sqrt{14}} \approx 78.55°$ **19.** $\langle 3, -4, 2 \rangle \cdot \langle 2, 3, 3 \rangle = 0$ **21.** $3y - z = 2$
23. $17x + 5y + 19z = 55$

25.

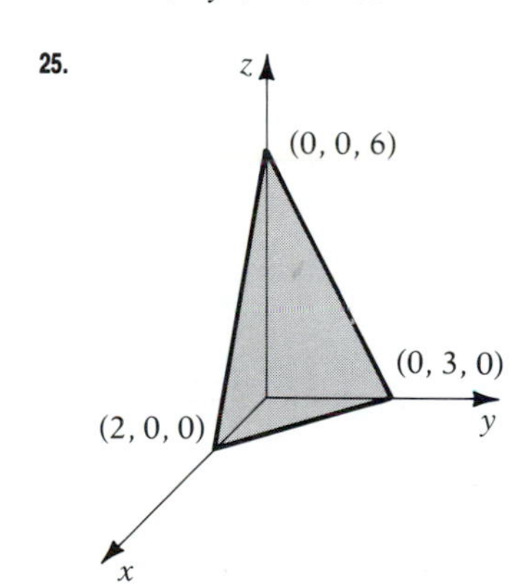

27.

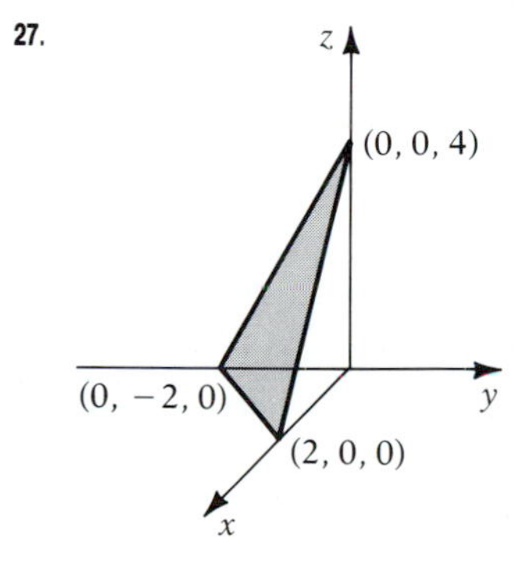

29.

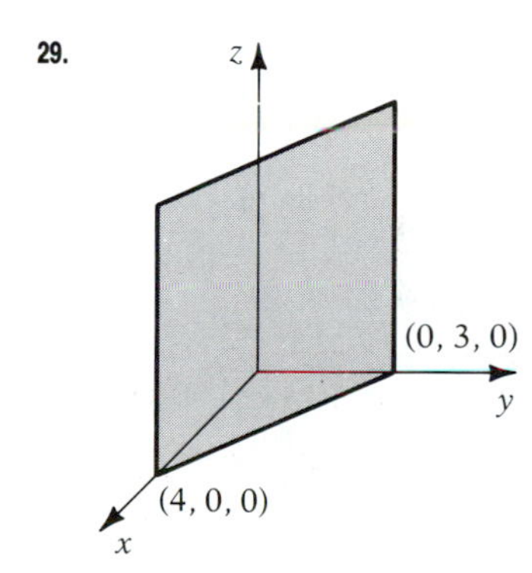

31.

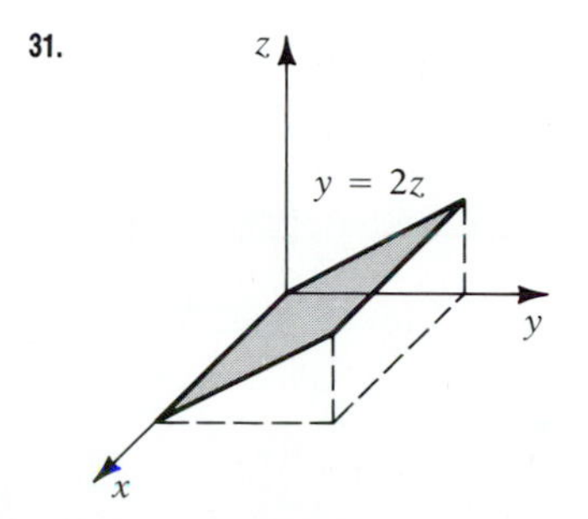

33.

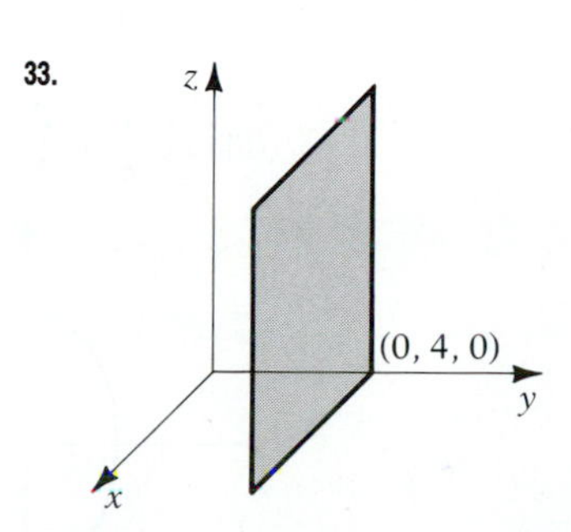

35. $\begin{cases} x = t \\ y = -13t \\ z = -2 + 8t \end{cases}$

37. $\dfrac{22}{5\sqrt{5}}$ **39.** Distance $= \dfrac{14}{9}$ **41.** Point of intersection is $(-1, -4, 3)$; $x - y - z = 0$. **43.** $(-1, 9, 4)$

Supplementary Exercises 14.8, page 579

1. (a) 201 (b) $\sqrt{5}\mathbf{i} + 2\sqrt{5}\mathbf{j}$ **3.** (a) 82 (b) 45° **5.** (a) $3\sqrt{30}$ (b) $\left\langle -\frac{20}{9}, \frac{10}{9}, \frac{20}{9} \right\rangle$

7. $\mathbf{z} = \frac{1}{2}\mathbf{v} + \frac{1}{2}\mathbf{w}$; express $\mathbf{w}$ in terms of $\mathbf{u}$ and $\mathbf{v}$ to get $\mathbf{z} = \frac{1}{2}\mathbf{u}$. **9.** Heading = 186.79°; speed = 253.74 mi/hr

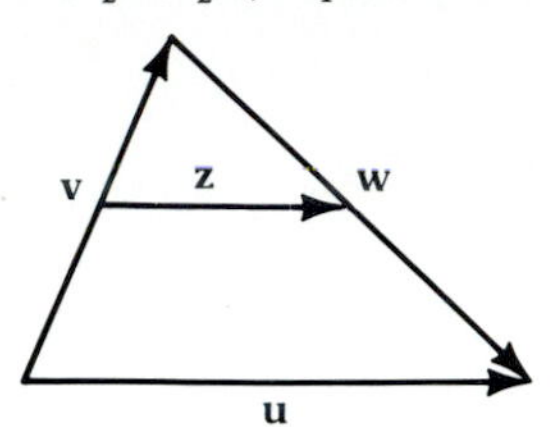

11. $\mathbf{r} = \langle 3 - t, 1 + 2t, -2 + 5t \rangle$; $x = 3 - t$, $y = 1 + 2t$, $z = -2 + 5t$

13. $\mathbf{r} = \langle 4 + 5t, 2 - 3t, -1 + 4t \rangle$; $x = 4 + 5t$, $y = 2 - 3t$, $z = -1 + 4t$ **15.** Skew **17.** $x - y + z = 1$

19. $\alpha = \cos^{-1}\frac{-2}{3\sqrt{5}} \approx 107.3°$, $\beta = \cos^{-1}\frac{4}{3\sqrt{5}} \approx 53.4°$, $\gamma = \cos^{-1}\frac{5}{3\sqrt{5}} \approx 41.8°$ **21.** (a) $\frac{16}{5}$ (b) $\frac{5}{3}$

CHAPTER 15

Exercise Set 15.1, page 586

1. $\{t : t \le 1, t \ne 0\}$ **3.** $\{t : t > 0, t \ne n\pi/2, n = 1, 2, 3, \ldots\}$ **5.** $\langle 1, 2, 3 \rangle$ **7.** $\mathbf{i} + 2\mathbf{j}$ **9.** $\{t : t < 1\}$ **11.** $\{t : -1 < t < 1\}$

13. (a) $\langle 3t - 1, t - 2, 3 - t \rangle$ (b) $\langle t + 3, 2t + 6, 3t + 6 \rangle$ **15.** (a) $\langle t^2 - 6, 3 - 3t - 2t^2, 3t + t^2 \rangle$ (b) $\langle t, 2, 1 - t \rangle$

17. (a) $(2t\cos t)\mathbf{i} + (2t\sin t)\mathbf{j} + (2t\sin t)\mathbf{k}$ (b) $\sin 2t + \sin t$ **19.** $\left\langle 2t, 2e^{2t}, -\frac{1}{t^2} \right\rangle$ **21.** $\langle t\cos t + \sin t, 1 + \ln t \rangle$

23. $e^{-t}\mathbf{i} + e^{2t}(2t + 1)\mathbf{j}$ **25.** $\frac{1}{(1 - t^2)^{3/2}}\mathbf{i} + \frac{1}{\sqrt{1 - t^2}}\mathbf{j} - \frac{t}{\sqrt{1 - t^2}}\mathbf{k}$ **27.** $\left\langle -\cos t + C_1, t - \sin t + C_2, \frac{t^2}{2} + C_3 \right\rangle$ **29.** $\left\langle \frac{\pi}{4}, \frac{2}{3}, \frac{1}{3} \right\rangle$

31. $\frac{\pi}{4}\mathbf{i} + \frac{1}{3}\mathbf{j}$ **33.** $-\frac{3\sqrt{3}}{8}\mathbf{i} + \left(\sqrt{3} - \frac{\pi}{3}\right)\mathbf{j} + \frac{\pi}{3}\mathbf{k}$ **35.** $6\mathbf{i} + \frac{23}{2}\mathbf{j} + 8\mathbf{k}$ **37.** $4\mathbf{i} - 2\mathbf{j} + 16\mathbf{k}$ **39.** $-194\mathbf{i} + 96\mathbf{j} + 80\mathbf{k}$

41. $\frac{26}{3}$

Exercise Set 15.2, page 592

1. Line through (2, 3, 5) with direction vector $\langle -1, 1, 2 \rangle$

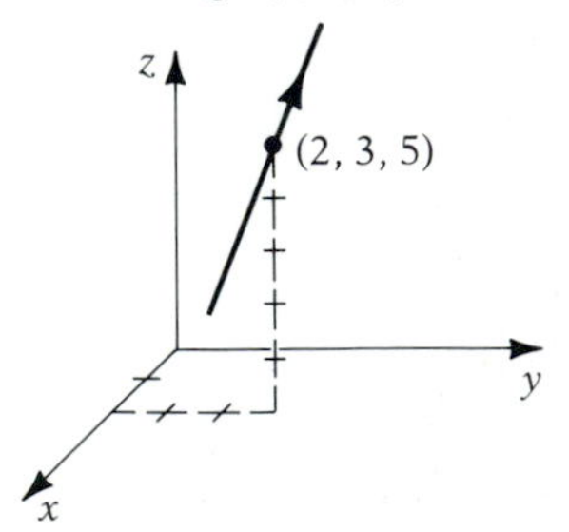

3. Ellipse $\frac{x^2}{4} + \frac{y^2}{9} = 1$

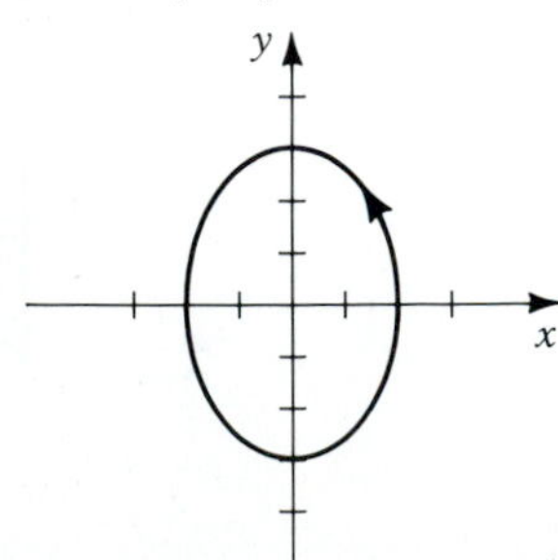

5. Circular helix

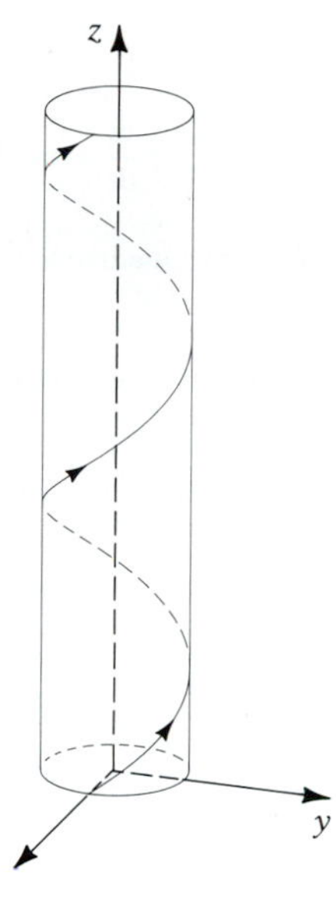

7. Elliptical helix

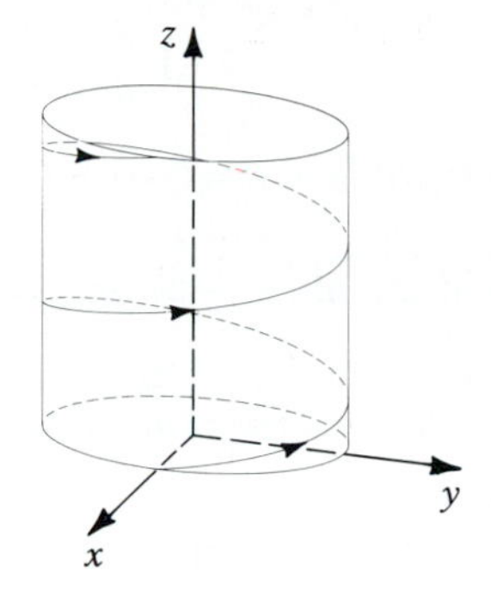

9. $\begin{cases} x = 2 + 2u \\ y = 3 + 6u \\ z = 1 + 3u \end{cases}$ **11.** $\begin{cases} x = 4 - 16u \\ y = \dfrac{1}{2} + u \\ z = \dfrac{1}{4} + 2u \end{cases}$

13. $\begin{cases} x = 2 - 2u \\ y = 4 - 5u \\ z = 1 - 2u \end{cases}$ **15.** $\mathbf{R}(u) = \left\langle u, 2\sqrt{u}, \dfrac{1}{\sqrt{u}} \right\rangle$, $u > 0$; direction unchanged

17. $\mathbf{R}(u) = (1 - u)\mathbf{i} + u\mathbf{j} + (1 + u)\mathbf{k}$, $-\dfrac{1}{2} \le u \le 1$; direction reversed **19.** $\langle 2, -3, 3 \rangle$ **21.** $-\mathbf{i} - 2\sqrt{3}\mathbf{j} \quad \mathbf{k}$

23. $\langle -6, 2, -3 \rangle$ **25.** $2\sqrt{14}$ **27.** $6(2\sqrt{6} - \sqrt{3})$ **29.** $\sqrt{5}\pi$ **31.** $\sqrt{65}(t + 1)$ **33.** $3(t^2 + 2t)$

35. $\dfrac{1}{27}[(9t^2 + 4)^{3/2} - 8]$ **39.** $\dfrac{2}{27}(343 - 80\sqrt{10})$ **41.** $\mathbf{R}(u) = \left\langle \dfrac{3u}{\sqrt{13}} + \dfrac{3}{2}, \dfrac{2u}{\sqrt{13}} + 5, 3 \right\rangle$, $\mathbf{R}'(u) = \left\langle \dfrac{3}{\sqrt{13}}, \dfrac{2}{\sqrt{13}}, 0 \right\rangle$, $|\mathbf{R}'(u)| = 1$

43. Make use of the chain rule.

Exercise Set 15.3, page 599

1. $\mathbf{v}\left(\dfrac{\pi}{4}\right) = \left\langle -\sqrt{2}, \dfrac{3}{2}\sqrt{2} \right\rangle$

$\mathbf{a}\left(\dfrac{\pi}{4}\right) = \left\langle -\sqrt{2}, -\dfrac{3}{2}\sqrt{2} \right\rangle$

$\left|\mathbf{v}\left(\dfrac{\pi}{4}\right)\right| = \dfrac{\sqrt{26}}{2}$

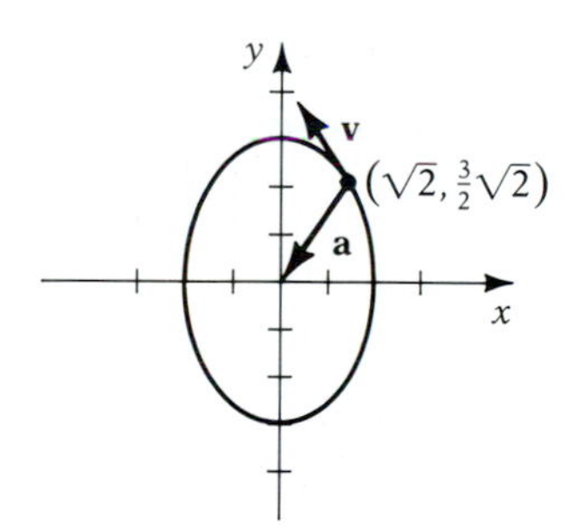

3.

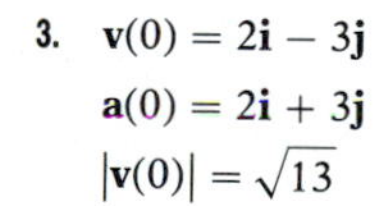

$\mathbf{v}(0) = 2\mathbf{i} - 3\mathbf{j}$

$\mathbf{a}(0) = 2\mathbf{i} + 3\mathbf{j}$

$|\mathbf{v}(0)| = \sqrt{13}$

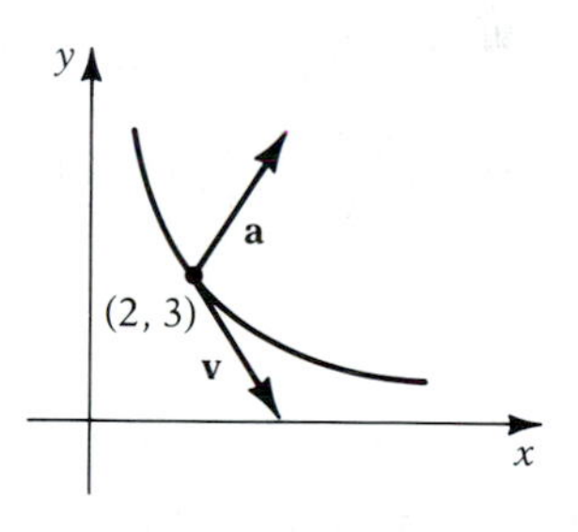

5. $\mathbf{v}(2) = \langle 2, 4 \rangle$

$\mathbf{a}(2) = \langle 0, 2 \rangle$

$|\mathbf{v}(2)| = 2\sqrt{5}$

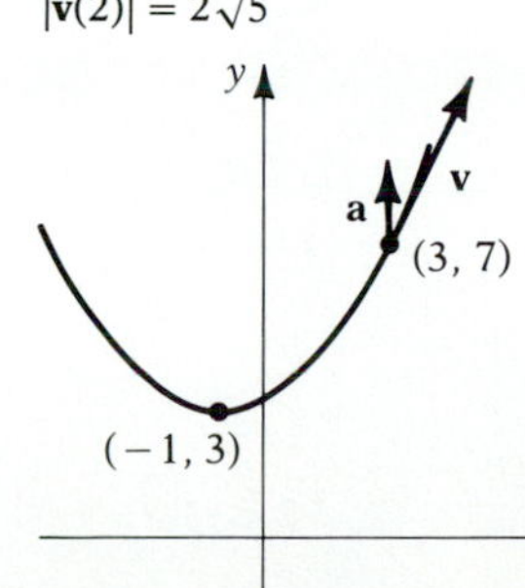

7. $\mathbf{v}(1) = 2\mathbf{i} + 2\mathbf{j} + 3\mathbf{k}$

$\mathbf{a}(1) = 2\mathbf{j} + 6\mathbf{k}$

$|\mathbf{v}(1)| = \sqrt{17}$

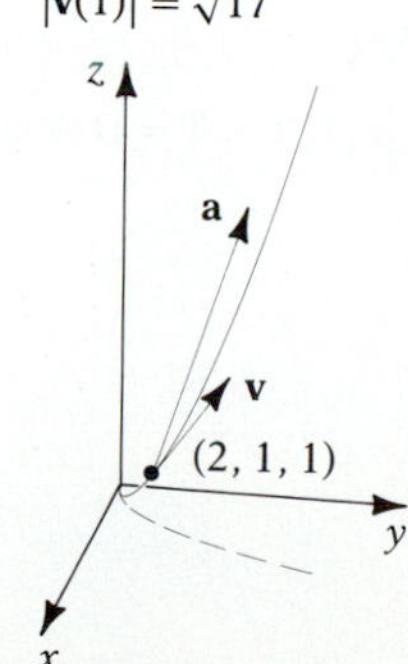

9. $\mathbf{v}(0) = \langle 1, 2, -1 \rangle$

$\mathbf{a}(0) = \langle 1, 0, 1 \rangle$

$|\mathbf{v}(0)| = \sqrt{6}$

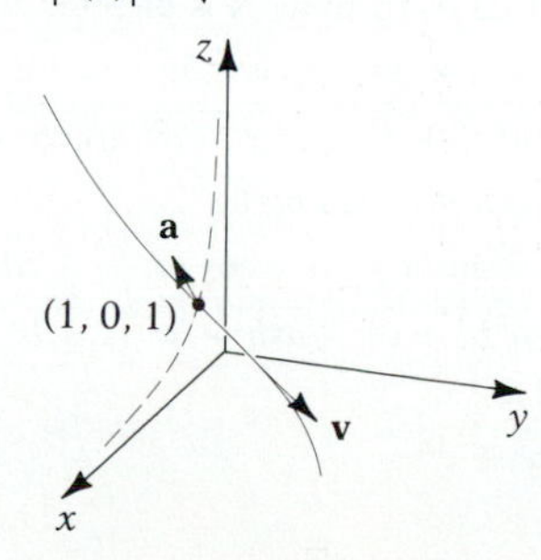

11. $\mathbf{v}(t) = -3(5-2t)^{1/2}\mathbf{i} + (t+2)\mathbf{j}$; $\mathbf{a}(t) = 3(5-2t)^{-1/2}\mathbf{i} + \mathbf{j}$; $\dfrac{ds}{dt} = 7 - t$

13. $\mathbf{v}(t) = \langle -t\sin t + \cos t,\ t\cos t + \sin t,\ (2t)^{1/2}\rangle$; $\mathbf{a}(t) = \langle -t\cos t - 2\sin t,\ -t\sin t + 2\cos t,\ (2t)^{-1/2}\rangle$; $\dfrac{ds}{dt} = t + 1$

15. $\mathbf{v}(t) = \langle e^t(\cos t - \sin t),\ e^t(\cos t + \sin t),\ e^t\rangle$; $\mathbf{a}(t) = \langle -2e^t\sin t,\ 2e^t\cos t,\ e^t\rangle$; $\dfrac{ds}{dt} = \sqrt{3}\,t$ **17.** (a) Quadrupled (b) Doubled

19. $h \approx 82.4$ mi **21.** $y_{max} \approx 2.94$ mi, $x_{max} \approx 20.4$ mi **23.** $v_0 \approx 848$ ft/sec

27. The motion is along a straight line through the point (b_1, b_2, b_3), where $\mathbf{B} = \langle b_1, b_2, b_3\rangle$, and has direction vector $\mathbf{A}$. The speed is $|\mathbf{A}|$.

29. Range $\approx 21{,}493$ m; impact speed ≈ 519.2 m/sec **31.** $\mathbf{v}(t) = 2\mathbf{i} - 3\mathbf{j} + \left(1 - \dfrac{t^2}{2}\right)\mathbf{k}$, $\mathbf{r}(t) = (4+2t)\mathbf{i} + (2-3t)\mathbf{j} + \left(t - \dfrac{t^3}{6}\right)\mathbf{k}$

Exercise Set 15.4, page 606

1. $\mathbf{T} = \left\langle \dfrac{1}{\sqrt{17}}, \dfrac{4}{\sqrt{17}}\right\rangle$, $\mathbf{N} = \left\langle -\dfrac{4}{\sqrt{17}}, \dfrac{1}{\sqrt{17}}\right\rangle$ **3.** $\mathbf{T} = -\dfrac{1}{\sqrt{2}}\mathbf{i} + \dfrac{1}{\sqrt{2}}\mathbf{j}$, $\mathbf{N} = -\dfrac{1}{\sqrt{2}}\mathbf{i} - \dfrac{1}{\sqrt{2}}\mathbf{j}$

5. $\mathbf{T} = \left\langle -\dfrac{1}{\sqrt{10}}, -\dfrac{1}{\sqrt{10}}, \dfrac{2}{\sqrt{5}}\right\rangle$, $\mathbf{N} = \left\langle \dfrac{1}{\sqrt{2}}, -\dfrac{1}{\sqrt{2}}, 0\right\rangle$ **7.** $\mathbf{T} = \dfrac{4}{5}\mathbf{i} + \dfrac{3}{5}\mathbf{j}$, $\mathbf{N} = -\dfrac{3}{5}\mathbf{i} + \dfrac{4}{5}\mathbf{j}$ **9.** $\mathbf{T} = \left\langle \dfrac{1}{3}, -\dfrac{2}{3}, \dfrac{2}{3}\right\rangle$, $\mathbf{N} = \left\langle \dfrac{2}{3}, \dfrac{2}{3}, \dfrac{1}{3}\right\rangle$

11. $\dfrac{6}{13\sqrt{13}}$ **13.** $\dfrac{3}{2\sqrt{2}}$ **15.** $\dfrac{32}{343}$ **17.** $\dfrac{2}{5\sqrt{5}}$ **19.** $\dfrac{4}{5\sqrt{5}}$ **21.** $\dfrac{14}{17\sqrt{17}}$ **23.** $\dfrac{2}{9}$ **25.** $\dfrac{\sqrt{2}}{3}$

27. $\left(\dfrac{7}{4}, 0\right)$

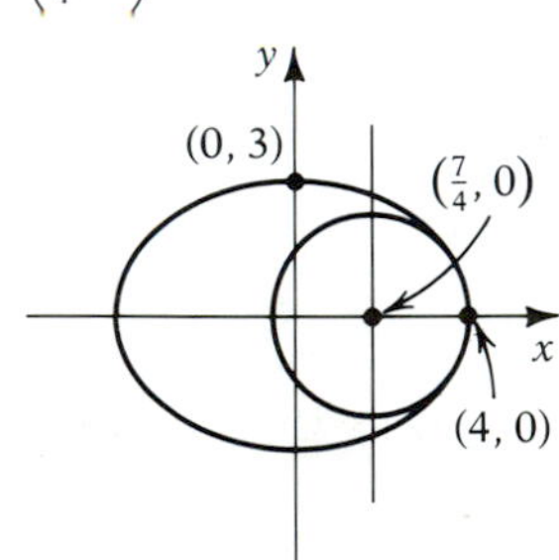

29. $(-2, 3)$

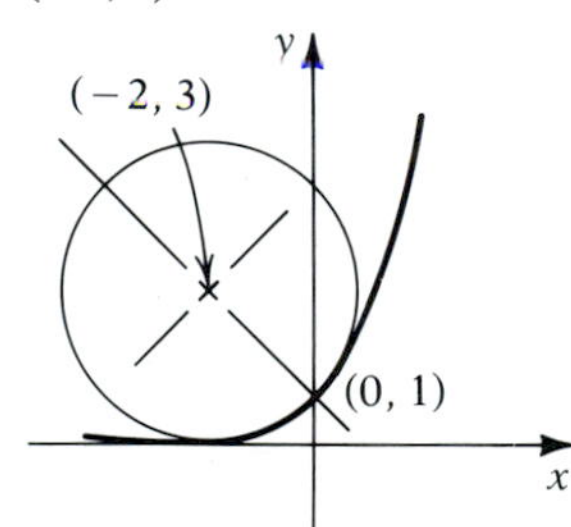

31. (a) $\mathbf{T} = \langle |\mathbf{T}|\cos\theta, |\mathbf{T}|\sin\theta\rangle = \langle \cos\theta, \sin\theta\rangle$ (b) $\dfrac{d\mathbf{T}}{ds} = \dfrac{d\mathbf{T}}{d\theta}\dfrac{d\theta}{ds} = \langle -\sin\theta, \cos\theta\rangle\dfrac{d\theta}{ds}$

$$\kappa = \left|\frac{d\mathbf{T}}{ds}\right| = \left|\frac{d\theta}{ds}\right|$$

(c) $\mathbf{N} = \dfrac{1}{\kappa}\dfrac{d\mathbf{T}}{ds} = \langle -\sin\theta, \cos\theta\rangle\dfrac{d\theta/ds}{|d\theta/ds|}$ and $\dfrac{d\theta}{ds}\Big/\left|\dfrac{d\theta}{ds}\right| = \begin{cases} 1 & \text{if } d\theta/ds > 0 \\ -1 & \text{if } d\theta/ds < 0 \end{cases}$

Consider cases to show $\mathbf{N}$ is directed toward concave side of C.

33. $\mathbf{T} = \left\langle \dfrac{-ab\sin bt}{\sqrt{a^2b^2+c^2}}, \dfrac{ab\cos bt}{\sqrt{a^2b^2+c^2}}, \dfrac{c}{\sqrt{a^2b^2+c^2}}\right\rangle$ **35.** Use $\mathbf{r}' = |\mathbf{r}'|\mathbf{T}$, $\mathbf{T}\times\mathbf{T} = \mathbf{O}$ and $\mathbf{B}\times\mathbf{T} = \mathbf{N}$

$\mathbf{N} = \langle -\cos bt, -\sin bt, 0\rangle$

$\mathbf{B} = \left\langle \dfrac{c\sin bt}{\sqrt{a^2b^2+c^2}}, \dfrac{-c\cos bt}{\sqrt{a^2b^2+c^2}}, \dfrac{ab}{\sqrt{a^2b^2+c^2}}\right\rangle$

$\kappa = \dfrac{|a|b^2}{a^2b^2+c^2}$

Exercise Set 15.5, page 610

1. $a_{\mathbf{T}} = \dfrac{4}{\sqrt{5}}$, $a_{\mathbf{N}} = \dfrac{2}{\sqrt{5}}$

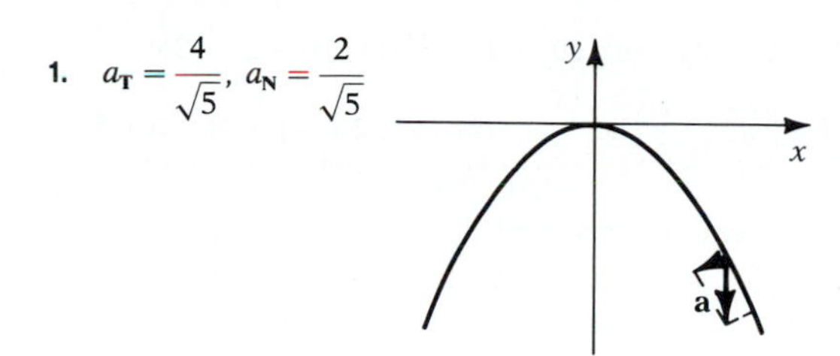

3. $a_{\mathbf{T}} = -\dfrac{3\sqrt{21}}{7}$, $a_{\mathbf{N}} = \dfrac{8}{\sqrt{7}}$

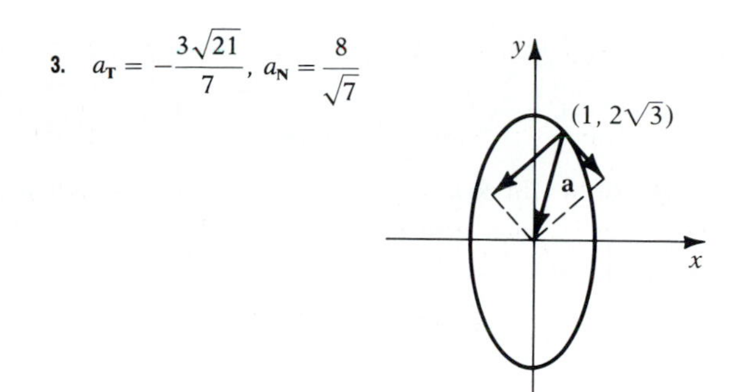

5. $a_{\mathbf{T}} = -\dfrac{2}{\sqrt{5}}$, $a_{\mathbf{N}} = \dfrac{1}{\sqrt{5}}$

y
(0, 4)
a
(−1, 0)
x

7. $\dfrac{1}{10\sqrt{5}}$ **9.** $\dfrac{8}{7\sqrt{7}}$ **11.** $\dfrac{1}{5\sqrt{5}}$

13. $a_{\mathbf{T}} = 0$, $a_{\mathbf{N}} = a\omega^2$ **15.** $a_{\mathbf{T}} = 4t$, $a_{\mathbf{N}} = 2$ **17.** $a_{\mathbf{T}} = 2e^t - e^{-t}$, $a_{\mathbf{N}} = 2$

21. $\mathbf{T} = \left\langle \dfrac{1}{2t^2+1}, \dfrac{2t}{2t^2+1}, \dfrac{2t^2}{2t^2+1} \right\rangle$

$\mathbf{N} = \left\langle \dfrac{-2t}{2t^2+1}, \dfrac{1-2t^2}{2t^2+1}, \dfrac{2t}{2t^2+1} \right\rangle$

$\kappa = \dfrac{2}{(2t^2+1)^2}$

23. $\mathbf{T} = \left\langle \dfrac{2}{2e^t+e^{-t}}, \dfrac{2e^t}{2e^t+e^{-t}}, \dfrac{-e^{-t}}{2e^t+e^{-t}} \right\rangle$

$\mathbf{N} = \left\langle \dfrac{-2e^t+e^{-t}}{2e^t+e^{-t}}, \dfrac{2}{2e^t+e^{-t}}, \dfrac{2}{2e^t+e^{-t}} \right\rangle$

$\kappa = \dfrac{2}{(2e^t+e^{-t})^2}$

27. Since $\mathbf{F} = m\mathbf{a}$ and is directed along the normal, it follows that $\mathbf{a} = a_{\mathbf{N}}\mathbf{N}$ and hence that $a_{\mathbf{T}} = 0$. Now use the fact that $a_{\mathbf{T}} = \dfrac{d^2s}{dt^2}$.

33. $\mathbf{v}(t) = (-2t\sin t^2)\mathbf{u}_r + 2t(1+\cos t^2)\mathbf{u}_\theta$

$\mathbf{a}(t) = -2(\sin t^2 + 2t^2 + 4t^2\cos t^2)\mathbf{u}_r + 2(1+\cos t^2 - 4t^2\sin t^2)\mathbf{u}_\theta$

Exercise Set 15.6, page 614

1. $\dfrac{(x-h)^2}{a^2} + \dfrac{(y-k)^2}{b^2} = 1$, where $h = -\dfrac{pe^2}{1-e^2}$

$a = \dfrac{pe}{1-e^2}$, $b = \dfrac{pe}{\sqrt{1-e^2}}$

3. Follow the same procedure as in Exercises 28 and 29 of Exercise Set 12.5.

7. (a) From part b of Exercise 6, show that

$$A(t) = \frac{c}{2}t + C$$

Let $t = 0$ to get C, and let $t = T$ to get the total area.

(b) Solve $\dfrac{c}{2}T = \pi ab$ for T, and use the values of a and b from Exercise 2.

Supplementary Exercises 15.7, page 615

1. (a) $t^3 \ln t - 2t^3 + 3 - 2t - t^2$ (b) $\langle e^{2t}, e^{3t}, 1 - e^t\rangle$ (c) $\langle t^2 - t\ln t, t^3 + 2, -2t - 2\rangle$ (d) $\langle 0, -2, 4\rangle$ (e) $\langle 3, -3, -3\rangle$

3. (a) $\left\langle \sin^{-1} t + C_1, -\sqrt{1 - t^2} + C_2, \ln\sqrt{\dfrac{1+t}{1-t}} + C_3\right\rangle$ (b) $\left(\dfrac{\pi}{6} + \dfrac{\sqrt{3}}{8}\right)\mathbf{i} - \dfrac{5}{24}\mathbf{j} + \left(\sqrt{3} - \dfrac{\pi}{3}\right)\mathbf{k}$ **5.** (a) $(4\ln 2)\mathbf{i} + \mathbf{j} + 4\mathbf{k}$ (b) 5

7. 30 **9.** (a) 6 (b) $\mathbf{R}(u) = \left\langle \ln\dfrac{u^2+1}{u}, 2\tan^{-1}\dfrac{1}{u}, -\sqrt{3}\ln u\right\rangle$; $e^{-2} \le u \le e$; orientation reversed

11.
$$\mathbf{T} = \left\langle \frac{2}{3}\left[\cos\left(\ln\frac{s+3}{3}\right) + \sin\left(\ln\frac{s+3}{3}\right)\right], \frac{2}{3}\left[\cos\left(\ln\frac{s+3}{3}\right) - \sin\left(\ln\frac{s+3}{3}\right)\right], \frac{1}{3}\right\rangle$$
$$\mathbf{N} = \frac{1}{\sqrt{2}}\left\langle \cos\left(\ln\frac{s+3}{3}\right) - \sin\left(\ln\frac{s+3}{3}\right), -\cos\left(\ln\frac{s+3}{3}\right) - \sin\left(\ln\frac{s+3}{3}\right), 0\right\rangle$$
$$\kappa = \frac{2\sqrt{2}}{3(s+3)}$$
$$\mathbf{T}(0) = \left\langle \frac{2}{3}, \frac{2}{3}, \frac{1}{3}\right\rangle, \mathbf{N}(0) = \left\langle \frac{1}{\sqrt{2}}, -\frac{1}{\sqrt{2}}, 0\right\rangle, \kappa(0) = \frac{2\sqrt{2}}{9}$$

13.
$$\mathbf{v}(t) = \langle e^{-t}(2\cos 2t - \sin 2t), -e^{-t}(2\sin 2t + \cos 2t), -2e^{-t}\rangle$$
$$\mathbf{a}(t) = \langle -e^{-t}(3\sin 2t + 4\cos 2t), e^{-t}(4\sin 2t - 3\cos 2t), 2e^{-t}\rangle$$
$$\frac{ds}{dt} = 3e^{-t}$$

15. (a) Speed $\approx$ 17,456 mi/hr (b) Height $\approx$ 1041 mi, speed $\approx$ 15,840 mi/hr **17.** (a) $\sin x$ (b) $\dfrac{8}{5\sqrt{10}}$

19. $\mathbf{a} = \left\langle -\dfrac{1}{(t+1)^2}, 0, 2\right\rangle$, $a_{\mathbf{T}} = -\dfrac{1}{(t+1)^2} + 2$, $a_{\mathbf{N}} = \dfrac{2}{t+1}$

21. The most distant point occurs when $\theta = \pi$. So $\mathbf{u}_r = -\mathbf{i}$, and since $\mathbf{d} = d\mathbf{i}$, from Equation (15.33), obtain

$$|\mathbf{v} \times \mathbf{c}| = |GM\mathbf{u}_r + \mathbf{d}| = GM - d$$

Now use the fact that $|\mathbf{v} \times \mathbf{c}| = |\mathbf{v}||\mathbf{c}|$. For the earth $v_m \approx$ 65,520 mi/hr.

CHAPTER 16

Exercise Set 16.1, page 619

1. (a) $\dfrac{3}{2}$ (b) 0 (c) $\dfrac{3}{22}$ (d) $\dfrac{1}{4}$ **3.** (a) 1 (b) 3 (c) 2 (d) 0

5. (a) 0 (b) a (c) $\dfrac{x^2 + 2x\Delta x + (\Delta x)^2 - y}{x + \Delta x - y^2}$ (d) $\dfrac{x^2 - y - \Delta y}{x - y^2 - 2y\Delta y - (\Delta y)^2}$ **7.** (a) $\dfrac{2}{3(h-3)}$ (b) $\dfrac{-4}{3(3+k)}$ **9.** $1 + \tan 2t$

11. (a) $\{(x, y): y \ne x^2\}$

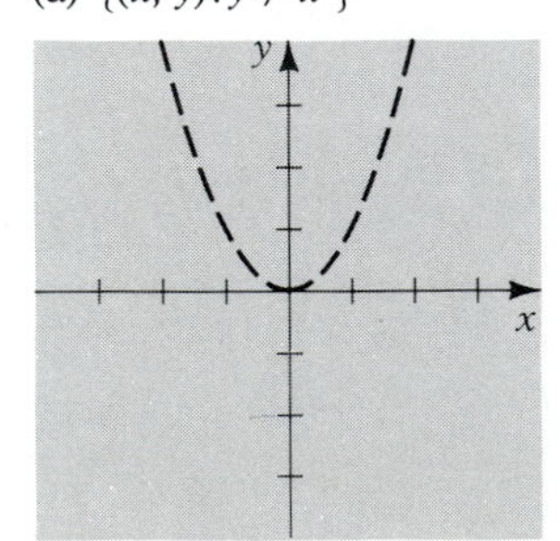

(b) $\{(x, y): y \ne \pm x\}$

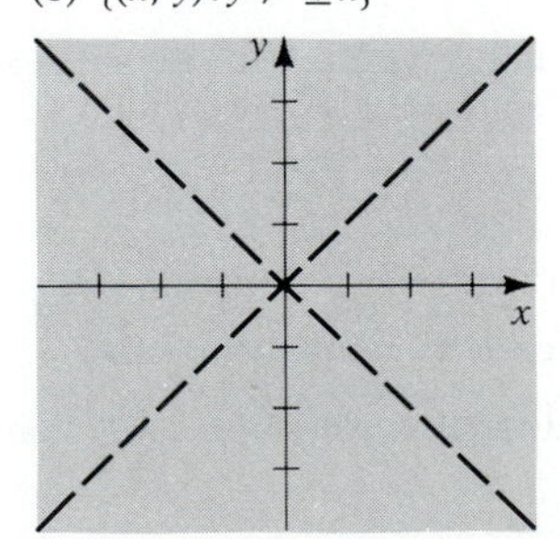

13. $\{(x, y): xy \geq 0\}$

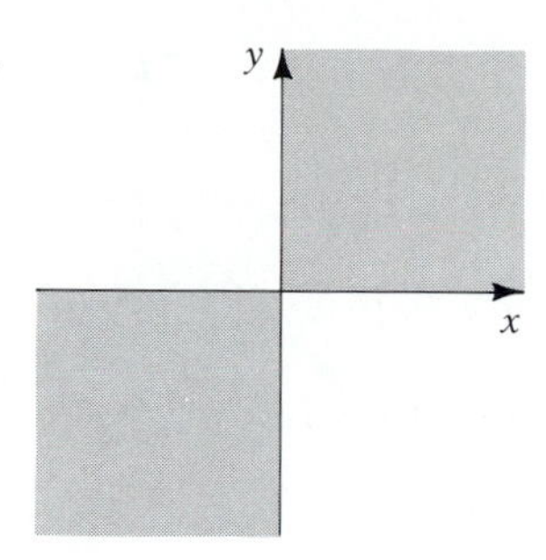

15. $\{(x, y): (x, y) \neq (0, 0)\}$

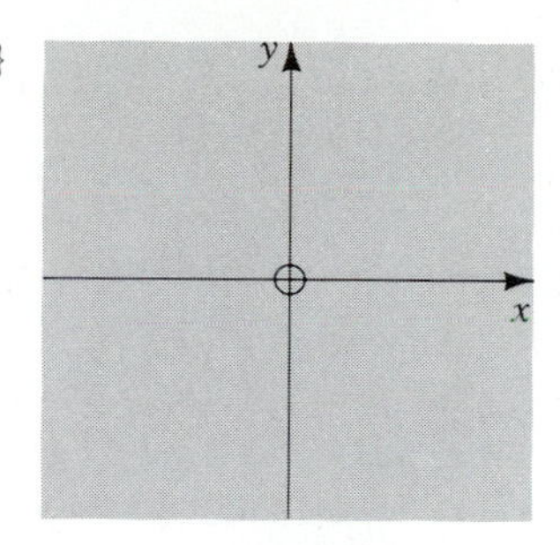

17. $\{(x, y): x^2 > y^2\}$

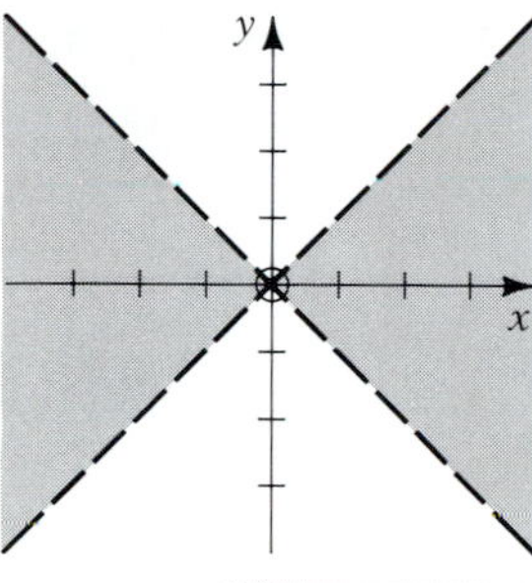

19. $\left\{(x, y): y < \dfrac{x^2}{2}\right\}$

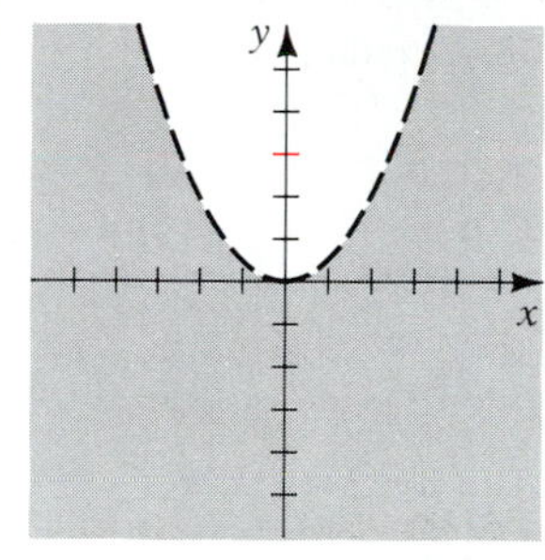

21. $\{(x, y): xy > 0\}$

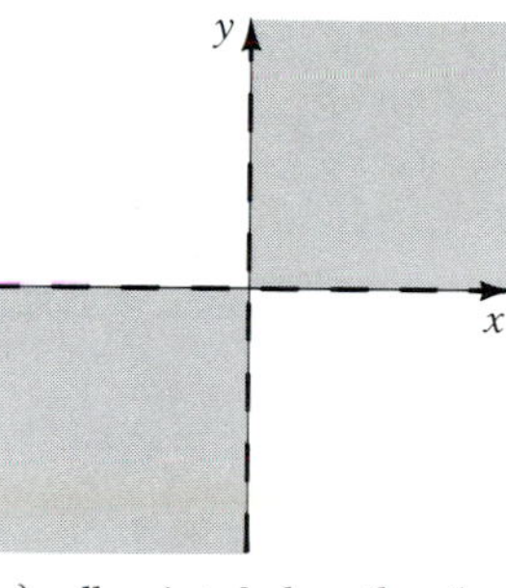

23. $\{(x, y, z): (x, y, z) \neq (1, -2, 3)\}$, all of $\mathbf{R}^3$ except $(1, -2, 3)$

25. $\{(x, y, z): z < x + y\}$, all points below the plane $z = x + y$

27. Can take $g(x, y) = x - 3y$ and $h(x, y) = 2y$

29. 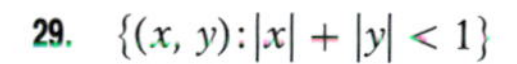
$\{(x, y): |x| + |y| < 1\}$

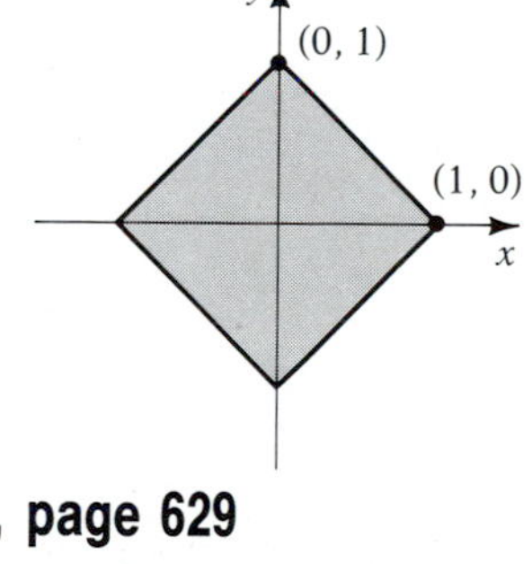

31. $C(r, h, p) = \pi rp(5r + 2h)$

Exercise Set 16.2, page 629

1.

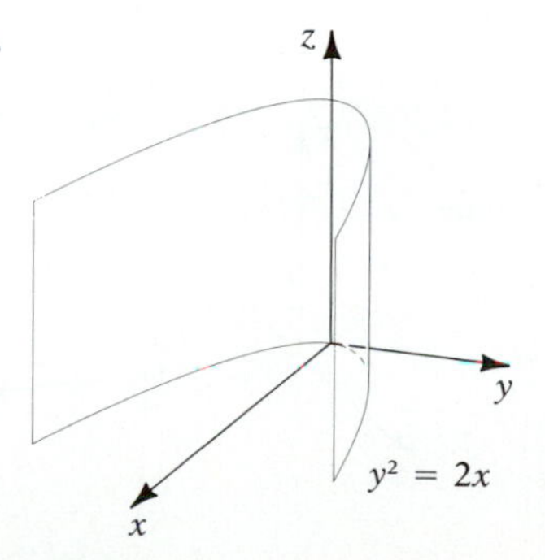

3.

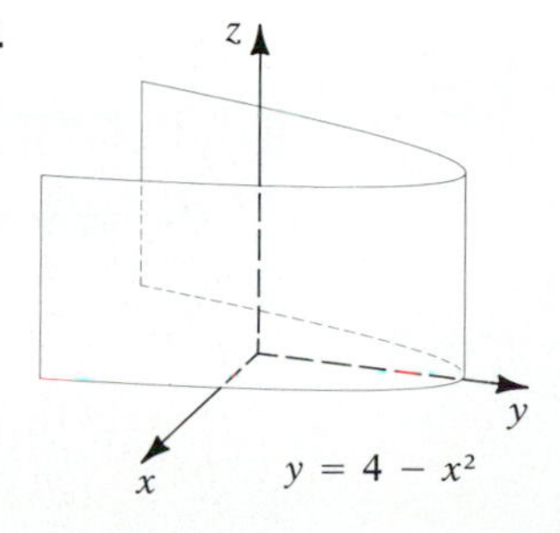

5.

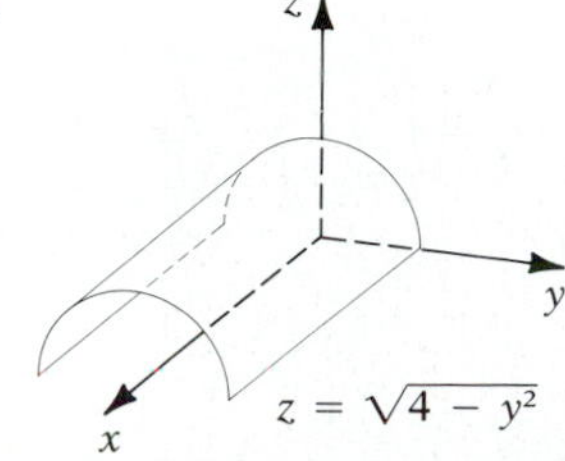

7.

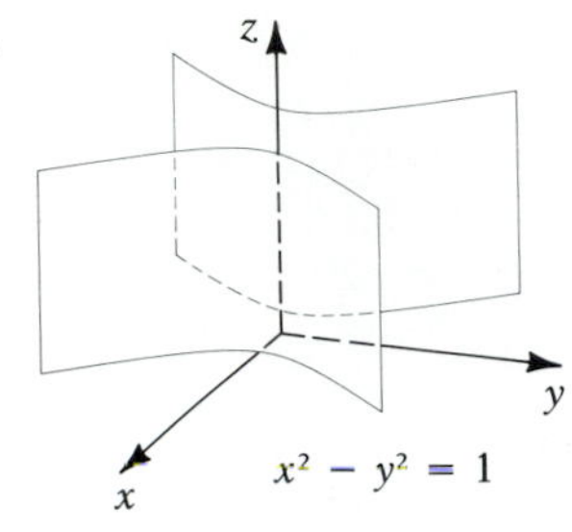

9.

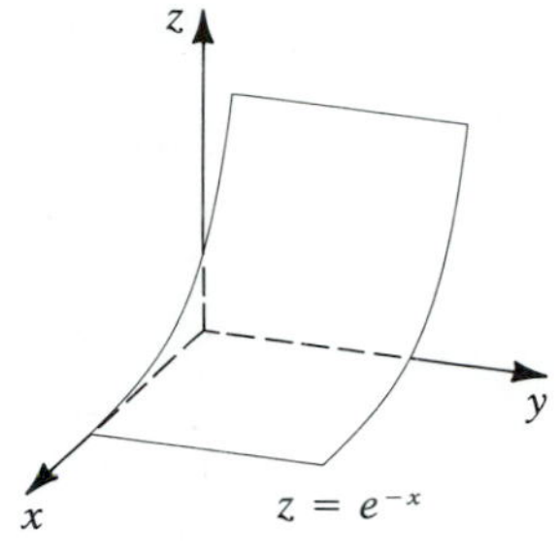

11.

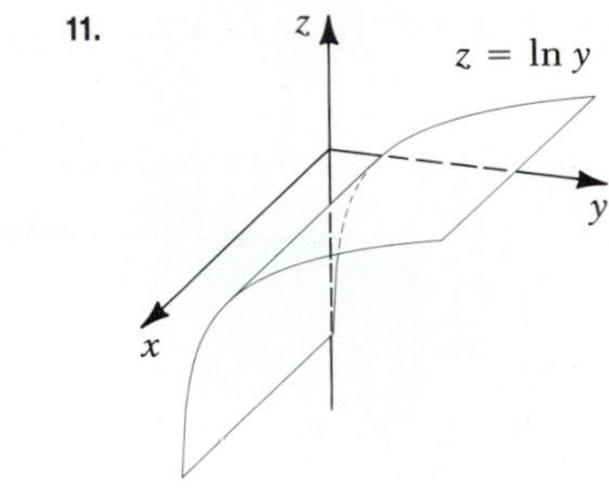

13. Center $(1, 3, 4)$; radius $= 4$

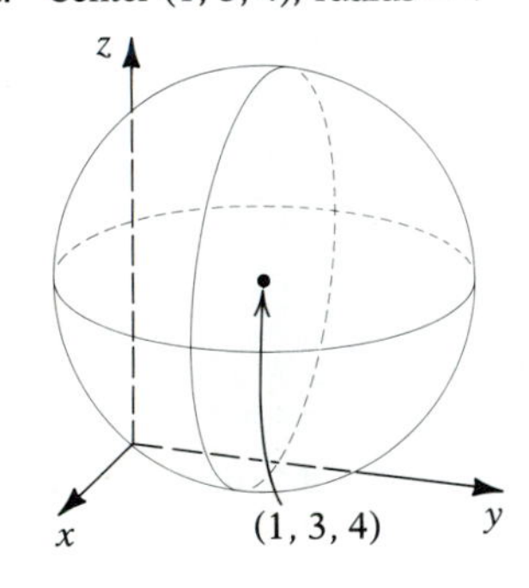

15. Degenerate sphere; point $(-5, 1, -3)$

17. Center $(3, -4, 4)$; radius $= 2\sqrt{2}$

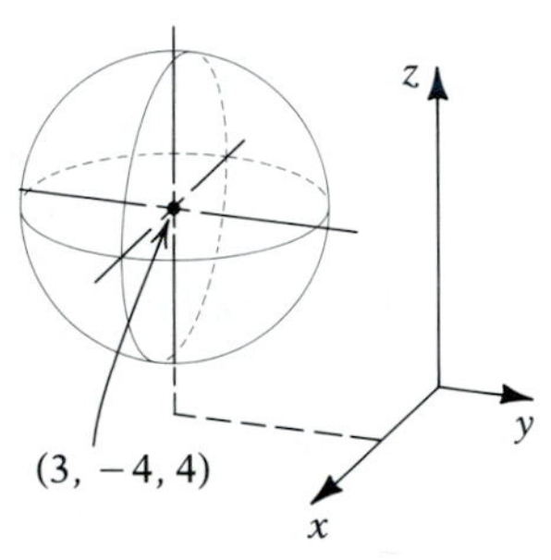

19. $\dfrac{x^2}{4} - \dfrac{y^2}{16} + \dfrac{z^2}{9} = 1$

Hyperboloid of one sheet

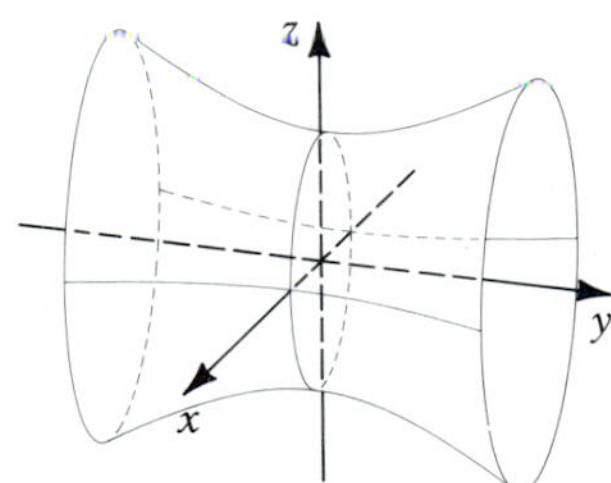

21. $9y^2 = 36x^2 + 16z^2$

Elliptic cone

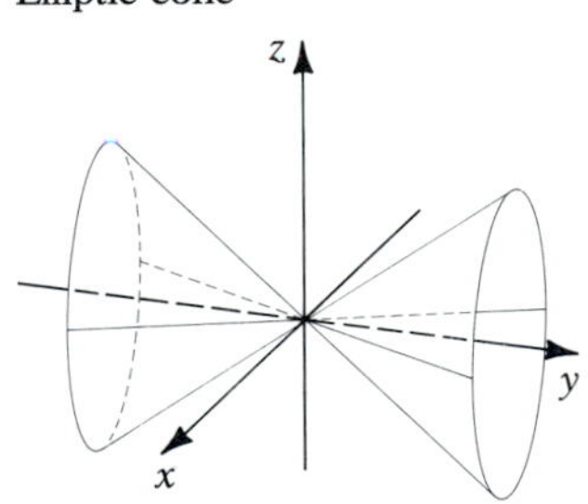

23. $16x = 4y^2 + 9z^2$

Elliptic paraboloid

25. $z = x^2 - y^2$

Hyperbolic paraboloid

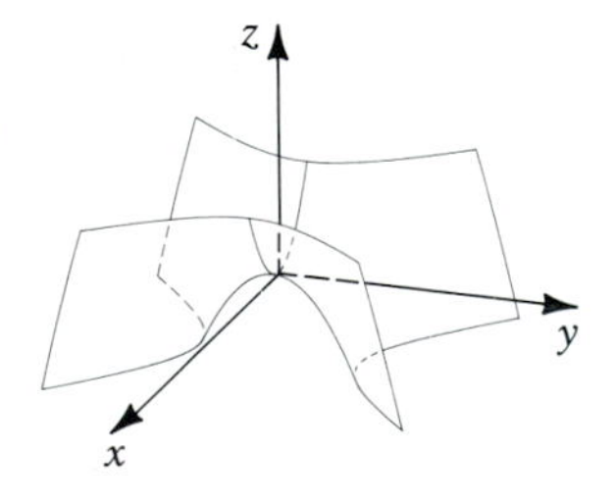

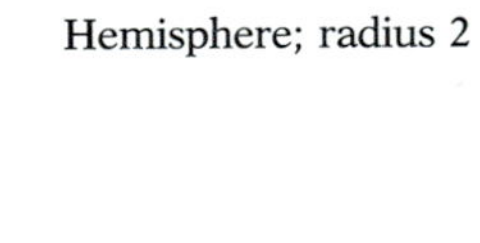

27. $z = \sqrt{4 - x^2 - y^2}$

Hemisphere; radius 2

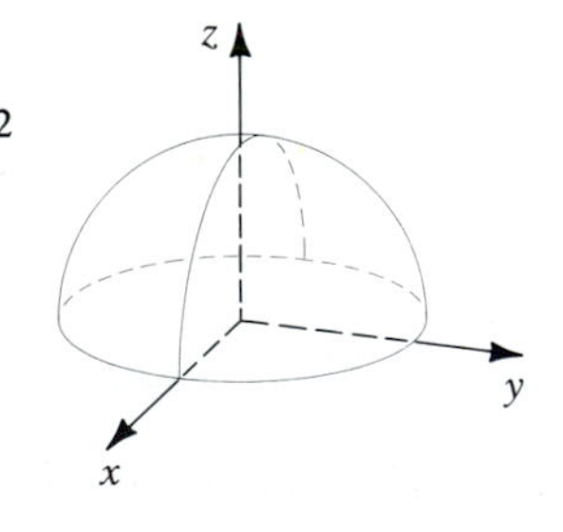

31. $\dfrac{x^2}{2} + \dfrac{y^2}{4} + \dfrac{z^2}{8} = 1$

Ellipsoid

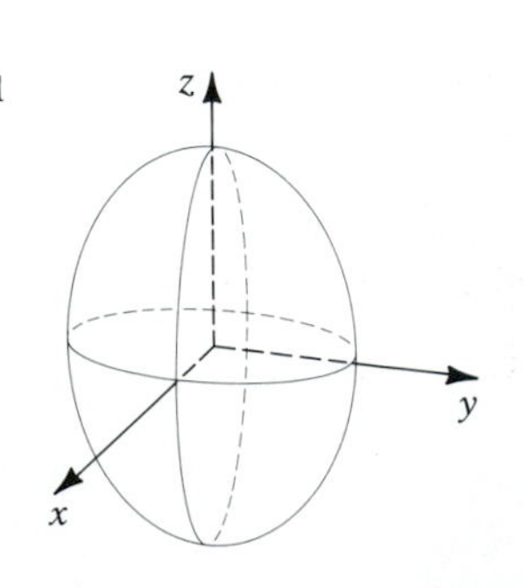

33. $y = \sqrt{4x^2 + z^2}$

One nappe of an elliptic cone

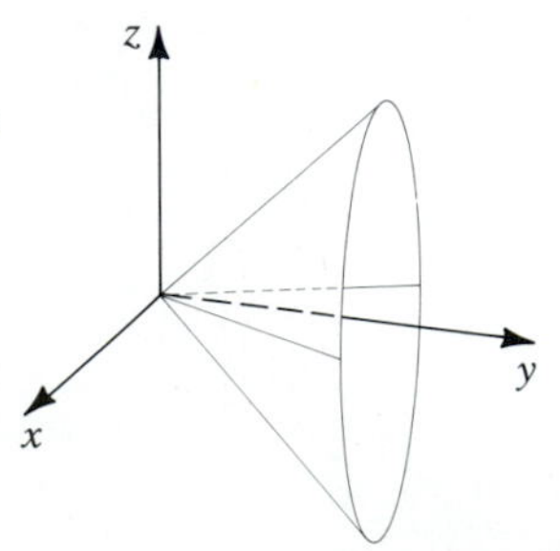

35. $\dfrac{(z+2)^2}{4} - \dfrac{(x-1)^2}{2} - \dfrac{(y-2)^2}{8} = 1$

Hyperboloid of two sheets; center $(1, 2, -2)$

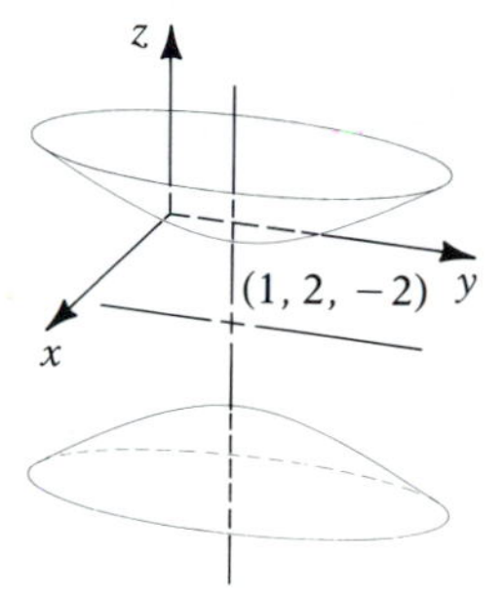

37. $4(z+5) = 4(x-3)^2 + (y-2)^2$

Elliptic paraboloid; vertex $(3, 2, -5)$

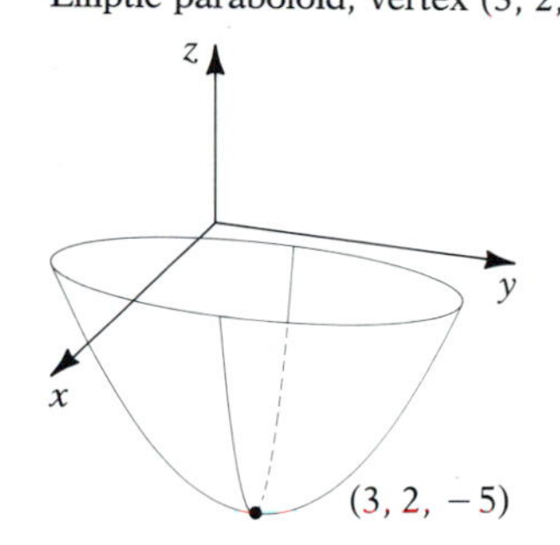

39. $z - 4 = \dfrac{(y+2)^2}{4} - \dfrac{(x-2)^2}{6}$

Hyperbolic paraboloid

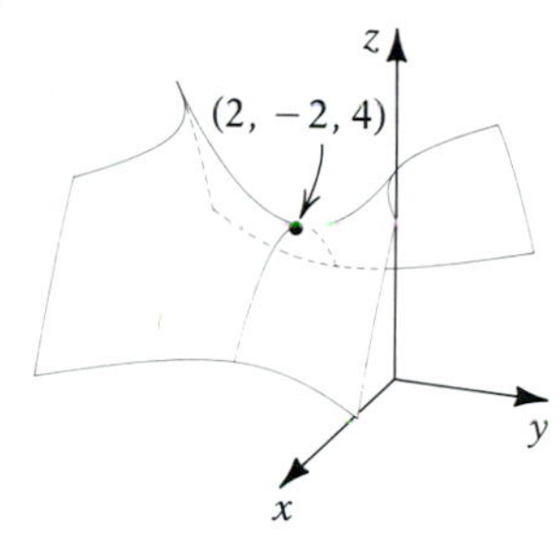

41.

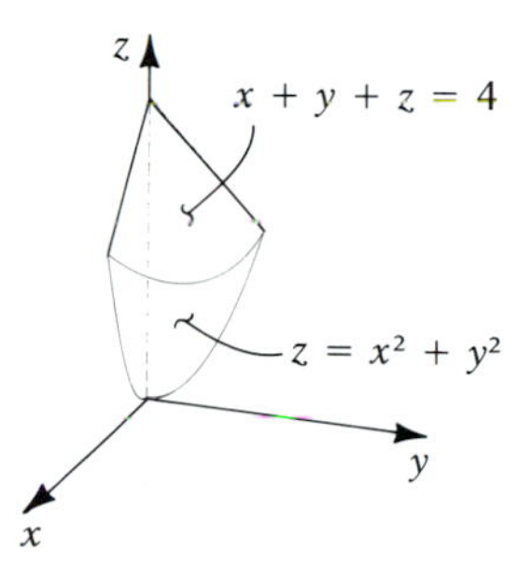

43.

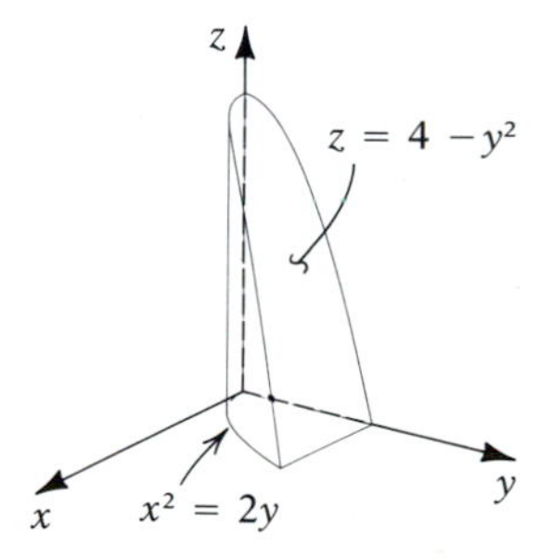

45.

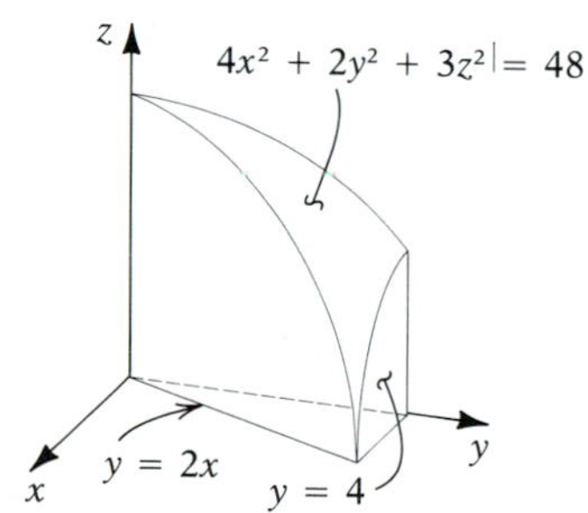

Exercise Set 16.3, page 633

1. Plane; domain $= \mathbf{R}^2$, range $= \mathbf{R}$

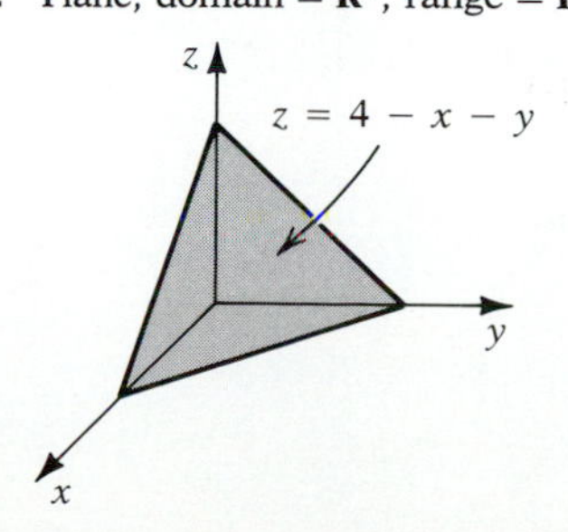

3. Plane; domain $= \mathbf{R}^2$, range $= \mathbf{R}$

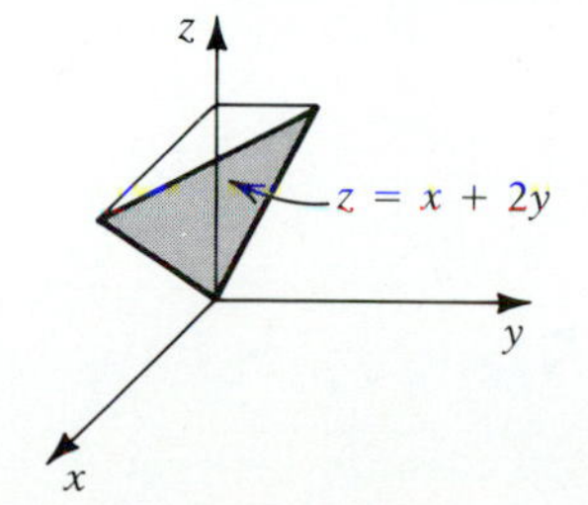

5. Upper half of a parabolic cylinder; domain = $\{(x, y): x \geq 0\}$, range = $\{z: z \geq 0\}$

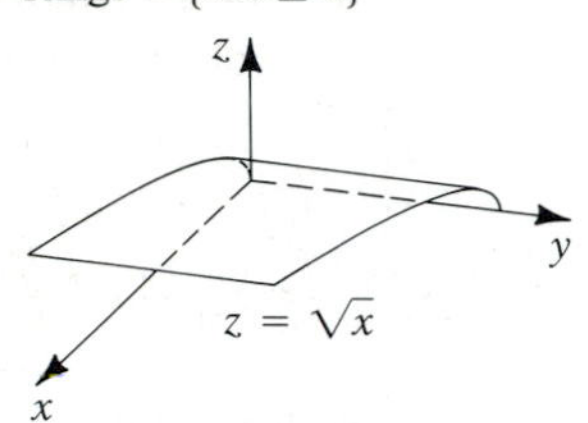

7. Parabolic cylinder; domain = $\mathbf{R}^2$, range = $\{z: z \leq 9\}$

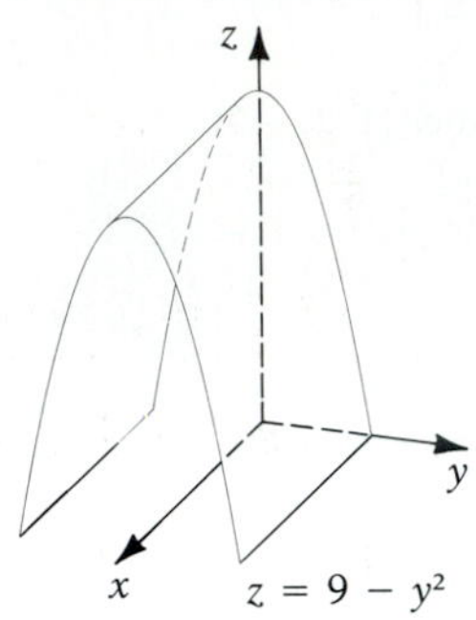

9. Elliptic paraboloid; domain = $\mathbf{R}^2$, range = $\{z: z \geq 0\}$

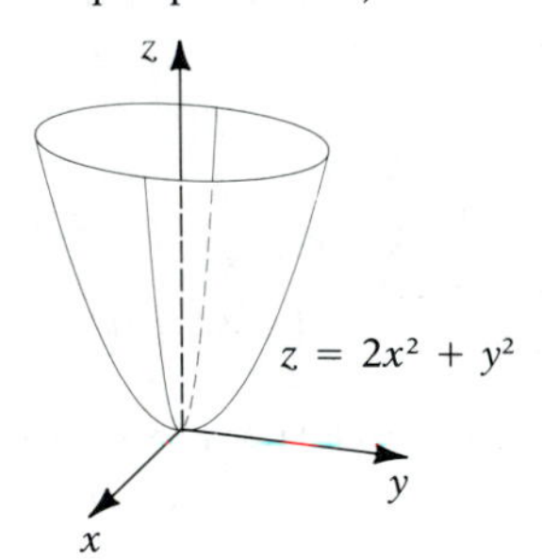

11. Upper half of a hyperboloid of 1 sheet; domain = $\{(x, y): 4x^2 - 3y^2 \leq 12\}$, range = $\{z: z \geq 0\}$

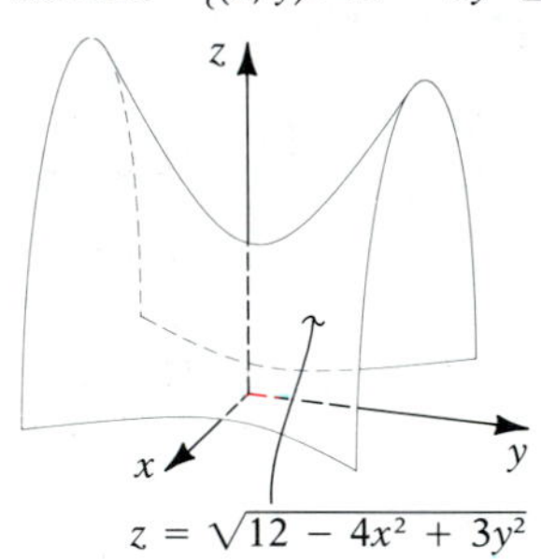

13. Paraboloid of revolution; domain = $\mathbf{R}^2$, range = $\{z: z \geq 1\}$

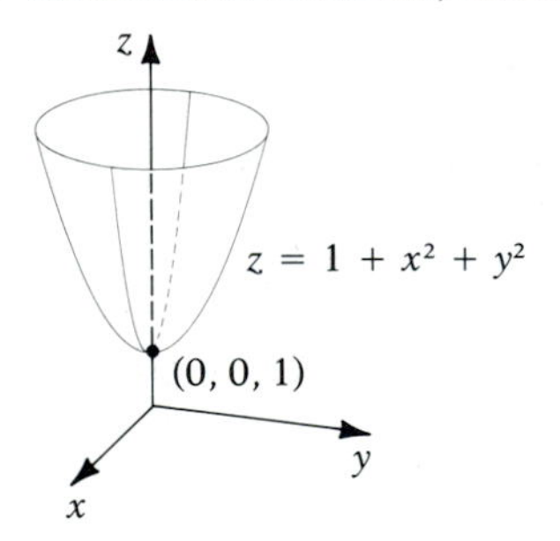

15. Upper half of a hyperboloid of 1 sheet; domain = $\{(x, y): x^2 + y^2 \geq 4\}$, range = $\{z: z \geq 0\}$

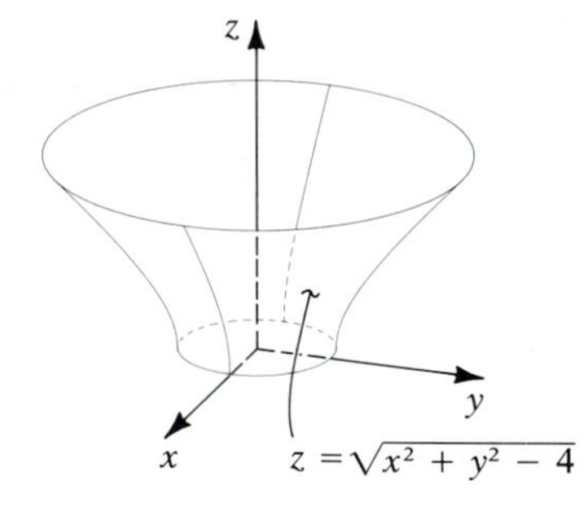

17. Upper half of a paraboloid of revolution; domain = $\{(x, y): x^2 \leq 4y\}$, range = $\{z: z \geq 0\}$

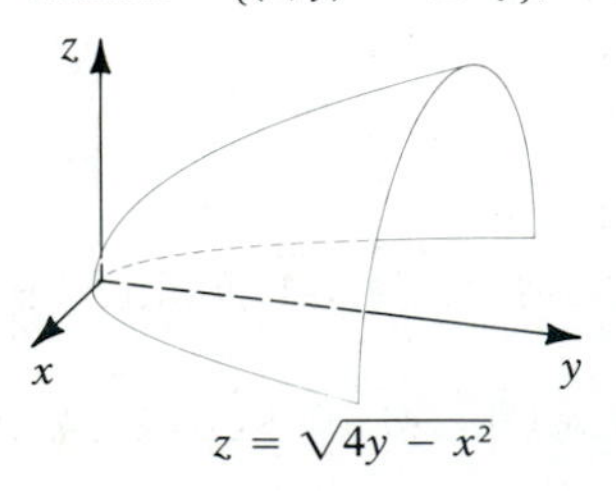

19. Hyperbolic paraboloid; domain = $\mathbf{R}^2$, range = $\mathbf{R}$

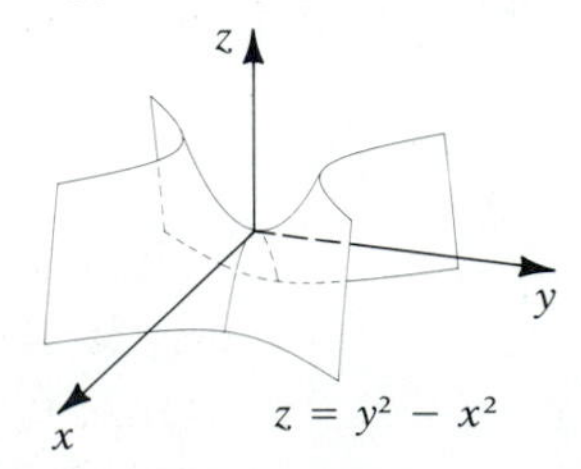

21. $3x - 5y = c$

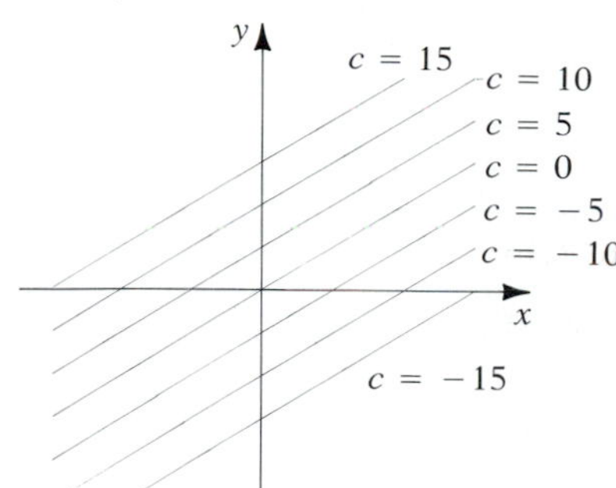

23. $x^2 - 2y = c$

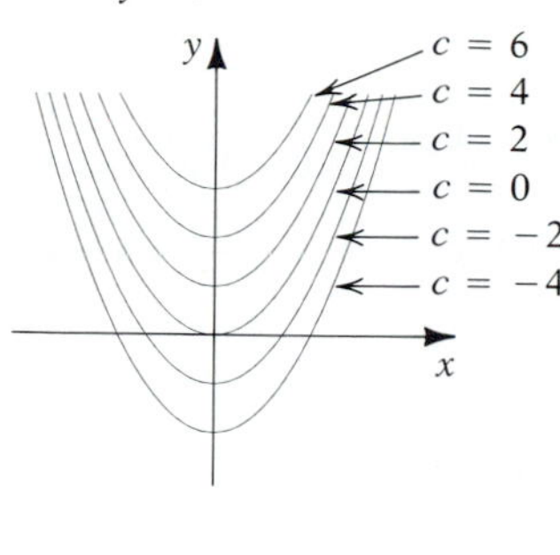

25. $x^2y = c$

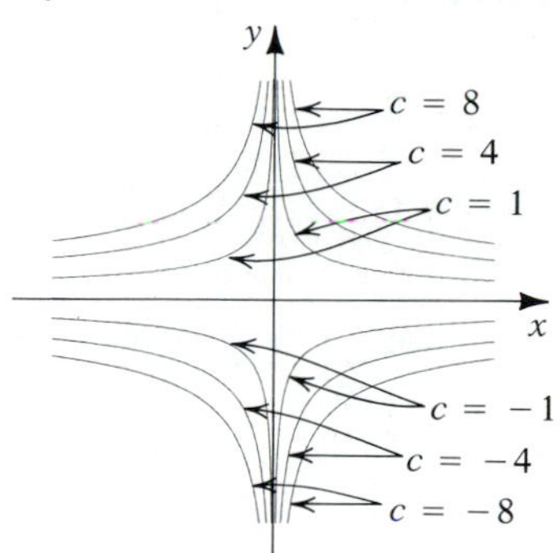

27. $x^2 = 1 + cy$

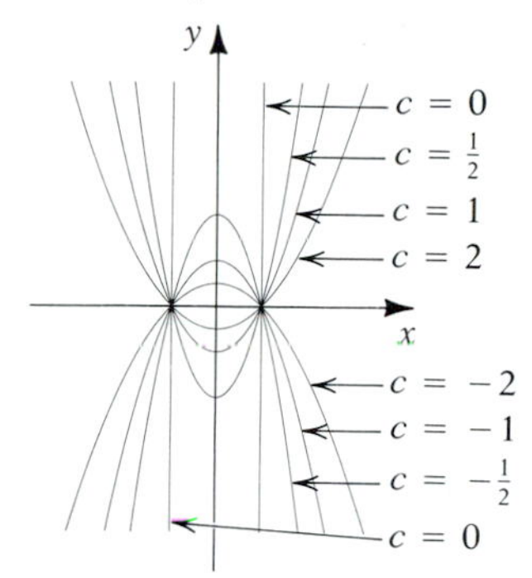

29. $x^2 - y^2 = c$

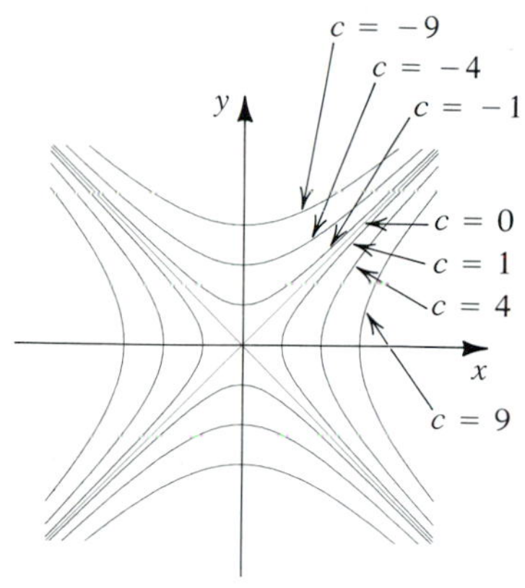

31. $y = \sin \pi x + c$

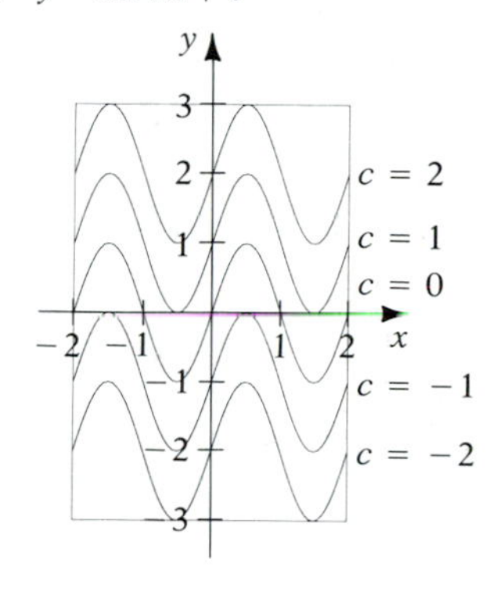

33. $x + y + z = c$

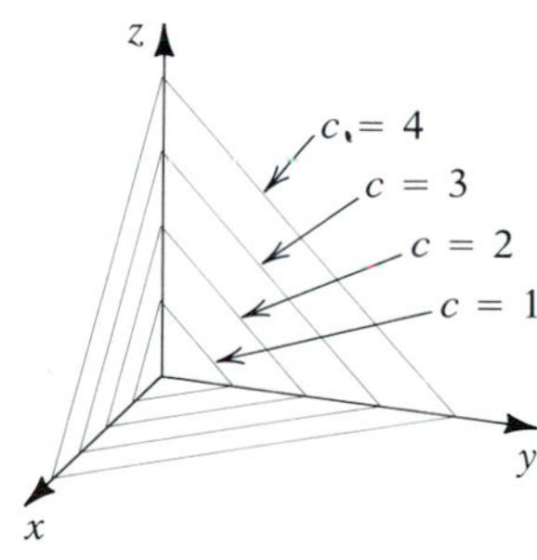

35. $2x^2 + 4y^2 + z^2 = c$

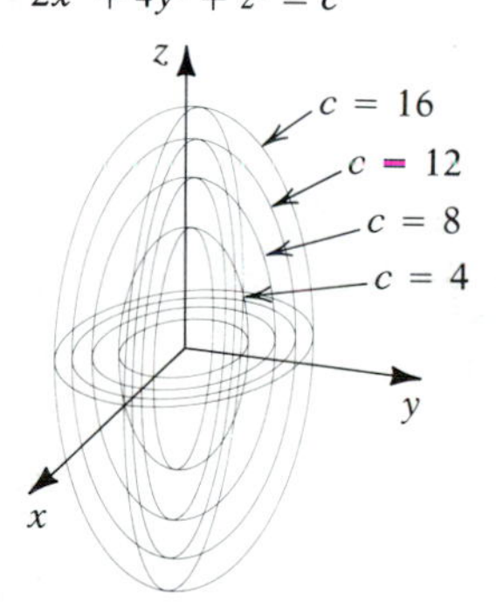

37. $y^2 = x - c$

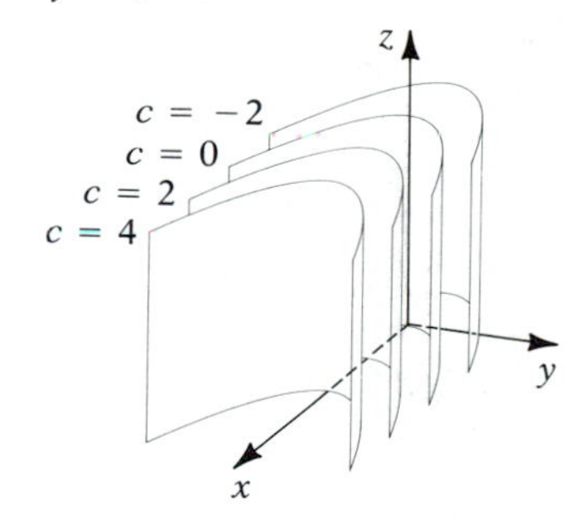

39. $T(x, y) = \dfrac{k}{\sqrt{x^2 + y^2}}$

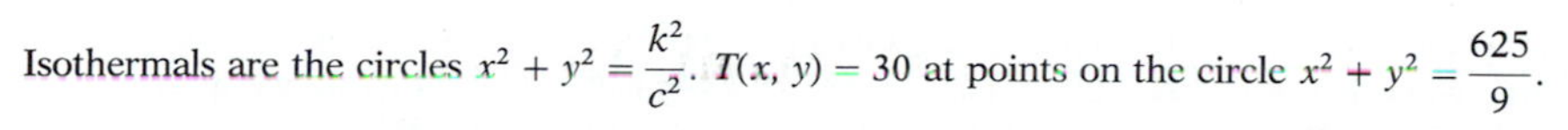

Isothermals are the circles $x^2 + y^2 = \dfrac{k^2}{c^2}$. $T(x, y) = 30$ at points on the circle $x^2 + y^2 = \dfrac{625}{9}$.

41. $v = \sqrt{\dfrac{kp}{d}} \qquad \dfrac{kp}{d} = c^2 \qquad\qquad p = \dfrac{v^2 d}{k} \qquad d = \dfrac{kc}{v^2}$

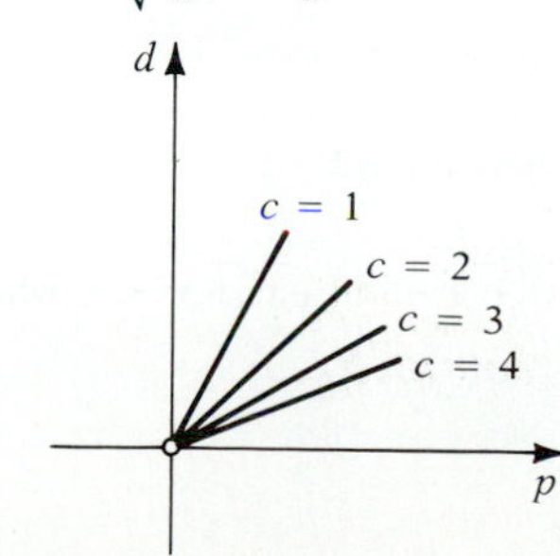

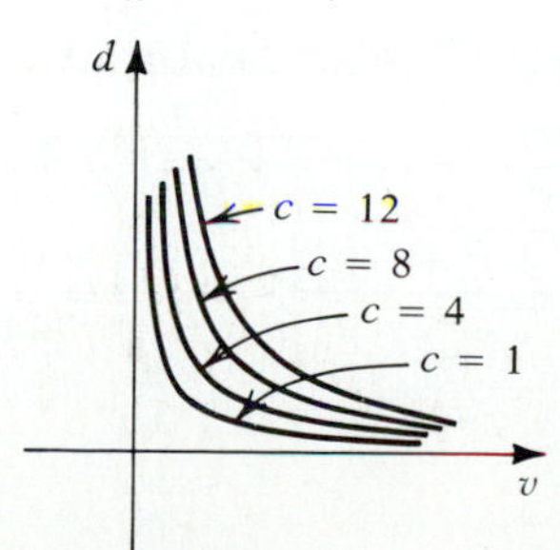

43. $y - \dfrac{y}{x^2 + y^2} = c$, or $\quad \sin\theta = \dfrac{cr}{r^2 - 1}$

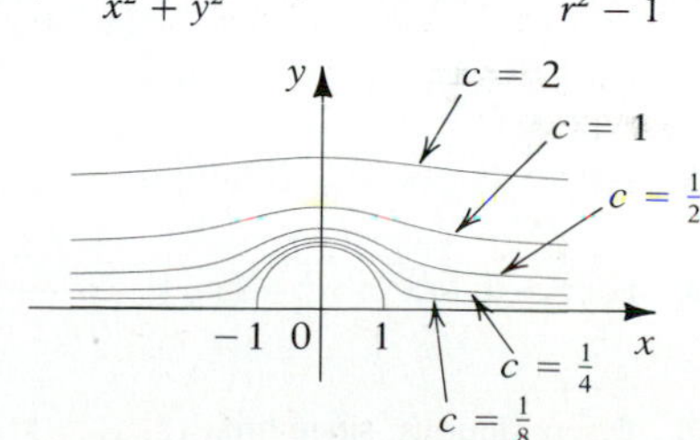

Exercise Set 16.4, page 641

1. $f_x = 2x,\ f_y = 2y$ **3.** $f_x = y\cos xy,\ f_y = x\cos xy$ **5.** $f_x = \dfrac{-2y}{(x-y)^2},\ f_y = \dfrac{2x}{(x-y)^2}$ **7.** $f_x = \ln y - \dfrac{y}{x},\ f_y = \dfrac{x}{y} - \ln x$

9. $f_x = \dfrac{-y^3}{(x^2-y^2)^{3/2}},\ f_y = \dfrac{x^3}{(x^2-y^2)^{3/2}}$ **11.** $\dfrac{\partial z}{\partial x} = x(1-x^2-y^2)^{-3/2},\ \dfrac{\partial z}{\partial y} = y(1-x^2-y^2)^{-3/2}$

13. $\dfrac{\partial z}{\partial x} = -\tan(x-y),\ \dfrac{\partial z}{\partial y} = \tan(x-y)$ **15.** $\dfrac{\partial z}{\partial x} = \dfrac{-y}{\sqrt{1-x^2y^2}},\ \dfrac{\partial z}{\partial y} = \dfrac{-x}{\sqrt{1-x^2y^2}}$ **17.** $f_x(3,-2) = -12,\ f_y(3,-2) = 17$

19. $f_r\left(0, \dfrac{\pi}{3}\right) = \dfrac{1}{2},\ f_\theta\left(0, \dfrac{\pi}{3}\right) = -\dfrac{\sqrt{3}}{2}$ **21.** $\left.\dfrac{\partial w}{\partial s}\right|_{(3,-2)} = -\dfrac{1}{2},\ \left.\dfrac{\partial w}{\partial t}\right|_{(3,-2)} = -1$ **25.** $f_x(3,2) = 0,\ f_y(3,2) = 0$

27. $f_{xx} = \dfrac{2(3y^2-x^2)}{(x^2+3y^2)^2},\ f_{xy} = f_{yx} = \dfrac{-12xy}{(x^2+3y^2)^2},\ f_{yy} = \dfrac{6(x^2-3y^2)}{(x^2+3y^2)^2}$ **29.** $f_{xx} = 2ye^{x^2y}(2x^2y+1),\ f_{xy} = f_{yx} = 2xe^{x^2y}(x^2y+1),\ f_{yy} = x^4e^{x^2y}$

31. $f_{xx} = e^{-x}\sin y - e^{-y}\cos x,\ f_{xy} = f_{yx} = e^{-y}\sin x - e^{-x}\cos y,\ f_{yy} = e^{-y}\cos x - e^{-x}\sin y$ **33.** $f_{xx} = -\cos x\cosh y = -f_{yy}$

35. $f_{xx} = \dfrac{2xy}{(x^2+y^2)^2} = -f_{yy}$ **37.** $g_x = -\cos x\sinh y = -f_y,\ g_y = -\sin x\cosh y = f_x,\ g_{xx} = \sin x\sinh y = -g_{yy}$

39. $g_x = \dfrac{-x}{x^2+y^2} = -f_y,\ g_y = \dfrac{-y}{x^2+y^2} = f_x,\ g_{xx} = \dfrac{x^2-y^2}{(x^2+y^2)^2} = -g_{yy}$

41. (c) Let m_1 = slope of $f(x,y) = c_1$ and m_2 = slope of $g(x,y) = c_2$. Then $m_1 = \dfrac{x+1}{y}$ $(y \neq 0)$ and $m_2 = \dfrac{-y}{x+1}$. So $m_1m_2 = -1$

(d)

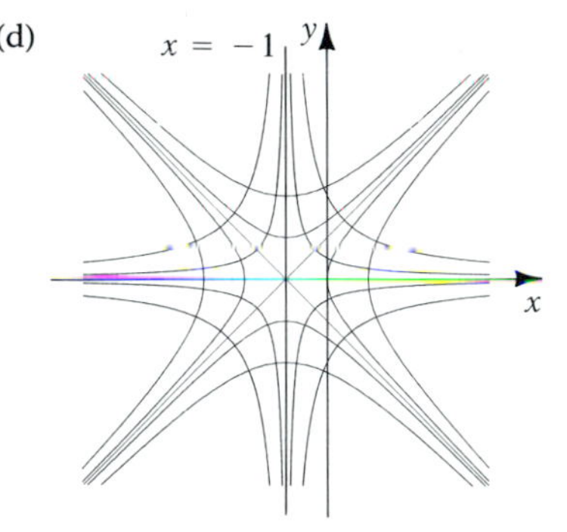

43. $f_x(0,0) = 1,\ f_y(0,0) = -1$

Exercise Set 16.5, page 650

1. 13 **3.** −8 **5.** 3 **7.** 4 **9.** 0 **11.** 4; for $\varepsilon > 0, |f(x,y) - 4| = |x^2+y^2| < \varepsilon$ if $0 < \sqrt{x^2+y^2} < \sqrt{\varepsilon}$. Take $\delta = \sqrt{\varepsilon}$.

13. −5; for $\varepsilon > 0$, $|f(x,y) - (-5)| = 2|x||y| \le 2(x^2+y^2) < \varepsilon$ if $0 < \sqrt{x^2+y^2} < \sqrt{\varepsilon/2}$. Take $\delta = \sqrt{\varepsilon/2}$.

15. 0; for $\varepsilon > 0$, $|f(x,y) - 0| \le x^2 + 3|x||y| + y^2 \le 4(x^2+y^2) < \varepsilon$ if $0 < \sqrt{x^2+y^2} < \sqrt{\varepsilon}/2$. Take $\delta = \sqrt{\varepsilon}/2$.

17. Limit along x-axis is 1, and limit along y-axis is −1. **19.** Limit along x-axis or y-axis is 0, and limit along $y = x$ is $\frac{1}{2}$.

21. Limit along x-axis is $\frac{3}{4}$, and limit along y-axis is −4. **23.** Limit along x-axis or y-axis is 0, and limit along $y = x^2$ is $\frac{1}{2}$.

25. $\mathbf{R}^2$ **27.** $\mathbf{R}^2$ **29.** $\{(x,y): x^2+y^2 \neq 1\}$ **31.** $\{(x,y): y \neq x^2+1\}$ **33.** $\{(x,y): y \neq 0\}$ **35.** $\mathbf{R}^2$

39. For $\varepsilon > 0$ $|f(x,y) - f(0,0)| < \dfrac{x^2+y^2}{\sqrt{x^2+y^2}} = \sqrt{x^2+y^2} < \varepsilon$ if $\sqrt{x^2+y^2} < \delta$, where $\delta = \varepsilon$.

41. For $\varepsilon > 0$, $|f(x,y) - f(0,0)| = |x^2-y^2| \le |x^2+y^2| < \varepsilon$ if $\sqrt{x^2+y^2} < \delta$, where $\delta = \sqrt{\varepsilon}$.

43. For $\varepsilon > 0$, $|f(x,y) - f(0,0)| \le \dfrac{|x||y|(|x|+|y|)}{x^2+y^2} \le \dfrac{2(x^2+y^2)^{3/2}}{x^2+y^2} = 2\sqrt{x^2+y^2} < \varepsilon$ if $\sqrt{x^2+y^2} < \delta$, where $\delta = \varepsilon/2$.

45. For $\varepsilon > 0$, $|\sqrt{x^2+y^2+1} - 1| = \left|\dfrac{\sqrt{x^2+y^2+1}-1}{1} \cdot \dfrac{\sqrt{x^2+y^2+1}+1}{\sqrt{x^2+y^2+1}+1}\right| = \dfrac{x^2+y^2}{\sqrt{x^2+y^2+1}+1} \le \sqrt{x^2+y^2} < \varepsilon$ if $\sqrt{x^2+y^2} < \delta$, where $\delta = \varepsilon$.

47. Discontinuous, since $\lim_{(x,x)\to(0,0)} f(x,x) = 2 \neq 1$

55. (a) $|f(x, y) - f(x_0, y_0)| = |x - x_0| \le \sqrt{(x - x_0)^2 + (y - y_0)^2}$. So for $\varepsilon > 0$, take $\delta = \varepsilon$. Similarly for $g(x, y)$.
(b) Use the results of part a and Exercise 54.

57. $f_x(0, 0) = f_y(0, 0) = 0$, but as $(x, y) \to (0, 0)$ along $y = x$, $f(x, y) \to \frac{1}{2} \neq 0$.

Supplementary Exercises 16.6, page 651

1. (a) $\{(x, y): x^2 + y^2 > 1\}$

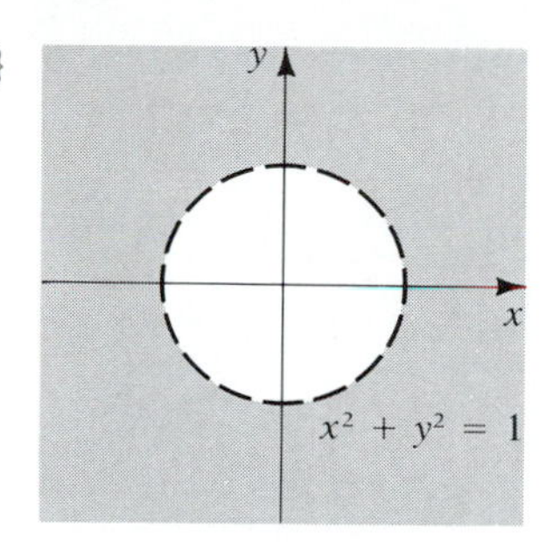

(b) $\{(x, y): x^2 > y^2\}$

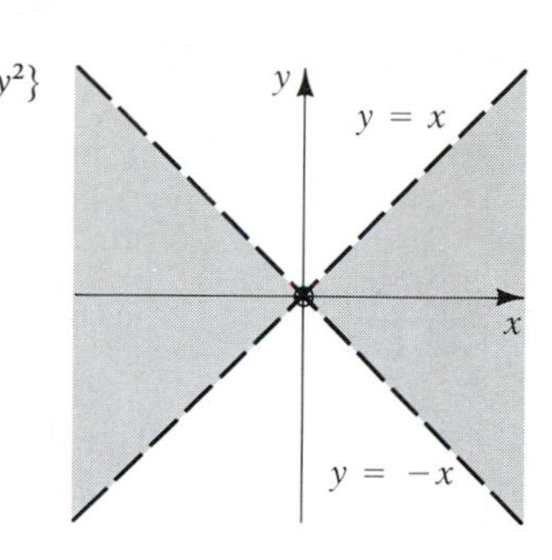

3. (a) Parabolic cylinder

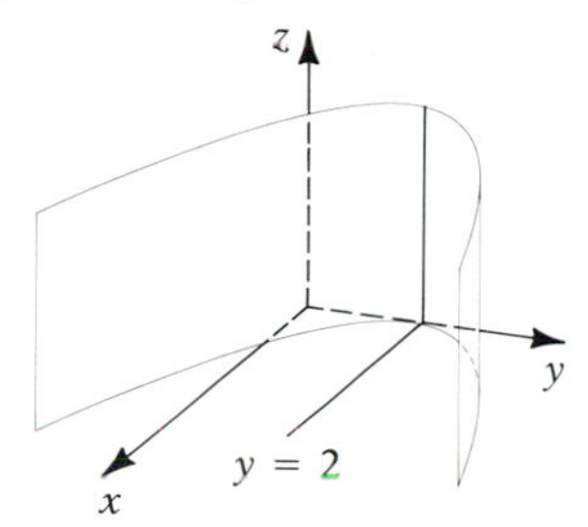

(b) Sphere, center (3, 2, 1), radius 2

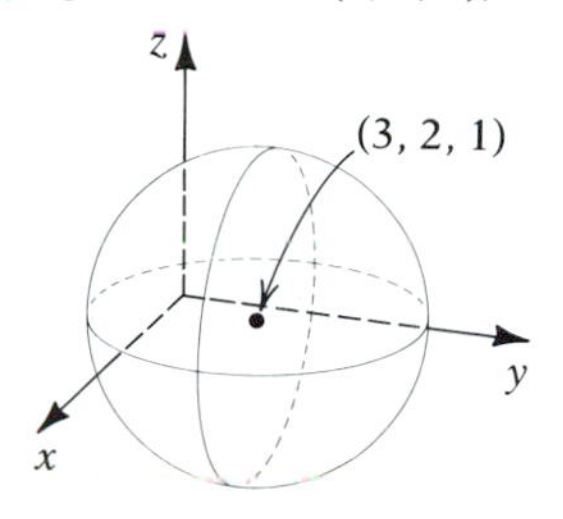

5. (a) Hyperbolic paraboloid

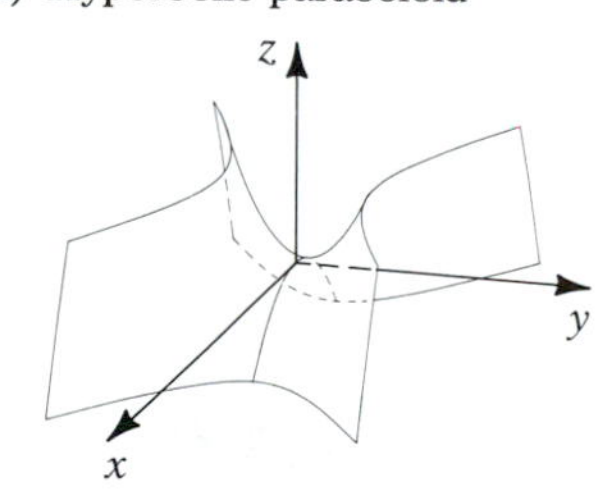

(b) Cylindrical surface

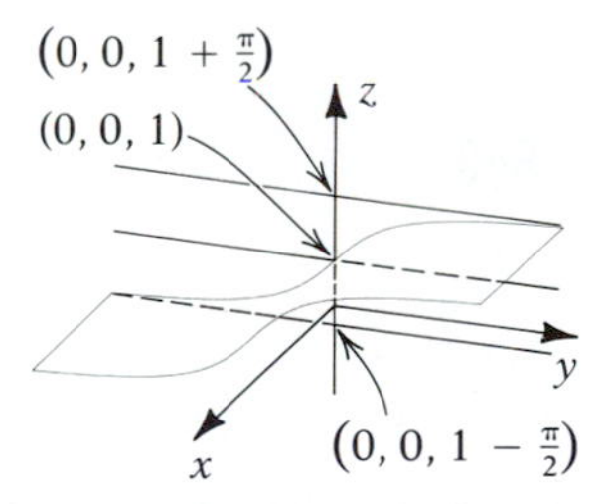

7. (a) Domain $= \mathbf{R}^2$, range $= \{z: z \ge 2\}$, upper half of elliptic hyperboloid of 2 sheets

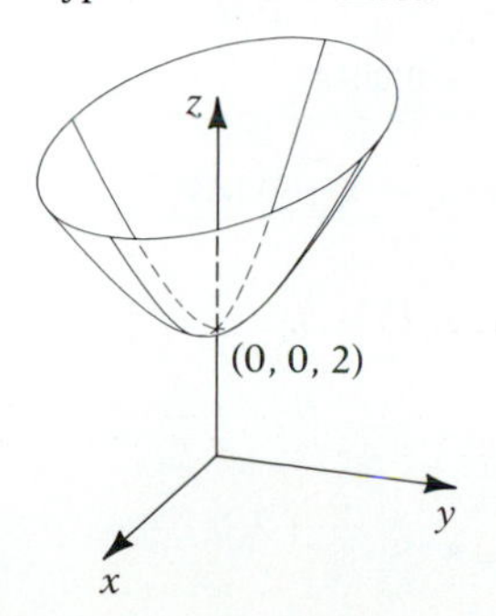

(b) Domain $= \mathbf{R}^2$, range $= \{z: z \le 4\}$, paraboloid of revolution

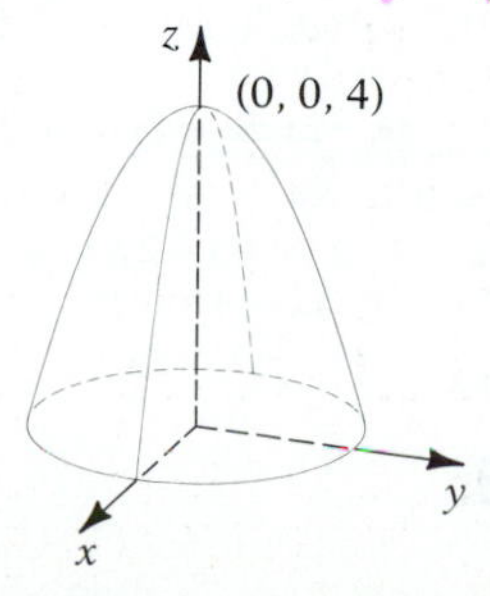

9. $T_y(1, 3) = -300e^{-5}$, decreasing

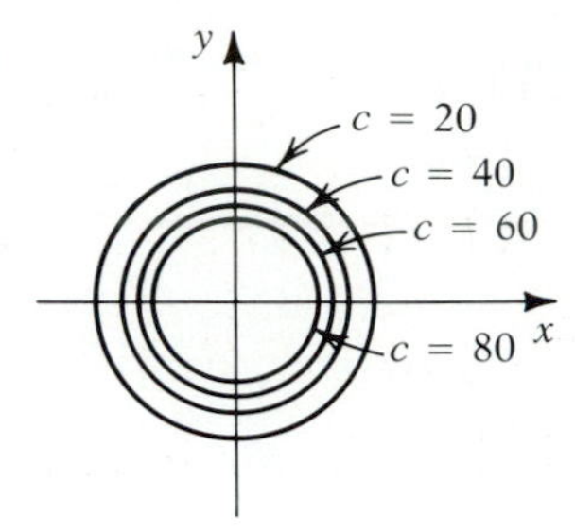

11. (a) $f_x = \frac{1}{x}\sinh\frac{2x}{y} - \frac{y}{x^2}\cosh^2\frac{x}{y}$, $f_y = -\frac{1}{y}\sinh\frac{2x}{y} + \frac{1}{x}\cosh^2\frac{x}{y}$ (b) $f_x = \frac{-1}{\sqrt{2x(y^2-2x)}}$, $f_y = \frac{2x}{y\sqrt{2x(y^2-2x)}}$

13. $f_x = \cos x\cosh y = g_y$, $f_y = \sin x\sinh y = -g_x$; $f_{xx} = -\sin x\cosh y = -f_{yy}$, $g_{xx} = -\cos x\sinh y = -g_{yy}$

15. (a) $\frac{-(x^2+y^2)}{(z^2-x^2-y^2)^{3/2}}$ (b) $\frac{x}{(z^2-x^2-y^2)^{3/2}}$ (c) $\frac{yz}{(z^2-x^2-y^2)^{3/2}}$ (d) $\frac{-xy}{(z^2-x^2-y^2)^{3/2}}$

17. $f(0, 0) = 0$. For $\varepsilon > 0$, $|f(x, y) - 0| \le \frac{|x^2|(|x| + 3|y|)}{x^2+y^2} \le 4\sqrt{x^2+y^2} < \varepsilon$ if $\sqrt{x^2+y^2} < \delta$, where $\delta = \varepsilon/4$.

19. (a) For $\varepsilon > 0$, $|(3x - 2y + 4) - 4| \le 3|x| + 2|y| \le 5\sqrt{x^2+y^2} < \varepsilon$ if $0 < \sqrt{x^2+y^2} < \delta$, where $\delta = \varepsilon/5$.

(b) For $\varepsilon > 0$, $\left|\frac{x^3-y^3}{x-y} - 0\right| = |x^2 + xy + y^2| \le x^2 + y^2 + 2|x||y| \le 3(x^2+y^2) < \varepsilon$ if $0 < \sqrt{x^2+y^2} < \delta$, where $\delta = \sqrt{\varepsilon/3}$.

21. (a) $f_x(0, 0) = 2$ (b) $f_x(x, y) = \frac{2x^4 + 12x^2y^2 + 2xy^3}{(x^2+2y^2)^2}$ if $(x, y) \ne (0, 0)$ (c) $f_x(x, y) \to \frac{16}{9} \ne f_x(0, 0)$ as $(x, y) \to (0, 0)$ along $y = x$

(d) $f_y(0, 0) = -\frac{1}{2}$; $f_y(x, y) = \frac{-2y^4 - 3x^2y^2 - 8x^3y}{(x^2+2y^2)^2} \to -\frac{13}{9} \ne f_y(0, 0)$ as $(x, y) \to (0, 0)$ along $y = x$.

23. Suppose S is open, and let $P \in S$. Then P is not a boundary point of S. So some neighborhood of P contains only points of S. Conversely, let S have the given property. Then no point of S is a boundary point of S. So S is open.

CHAPTER 17

Exercise Set 17.1, page 658

1. $\Delta f = 2\,\Delta x + 3\,\Delta y = f_x\,\Delta x + f_y\,\Delta y$; take $\varepsilon_1 = \varepsilon_2 = 0$

3. $\Delta f = 2x\,\Delta x + 6y\,\Delta y + (\Delta x)^2 + 3(\Delta y)^2 = f_x\,\Delta x + f_y\,\Delta y + \varepsilon_1\,\Delta x + \varepsilon_2\,\Delta y$, where $\varepsilon_1 = \Delta x$ and $\varepsilon_2 = 3\,\Delta y$

5. $\Delta f = 4xy\,\Delta x + 2x^2\,\Delta y + 2(y\,\Delta x + x\,\Delta y)\,\Delta x + 2\,\Delta x(x + \Delta x)\,\Delta y = f_x\,\Delta x + f_y\,\Delta y + \varepsilon_1\,\Delta x + \varepsilon_2\,\Delta y$, where $\varepsilon_1 = 2(y\,\Delta x + x\,\Delta y)$ and $\varepsilon_2 = 2\,\Delta x(x + \Delta x)$

7. $\Delta f = 2(x - 3y)\,\Delta x - 6(x - 3y)\,\Delta y + (\Delta x - 3\Delta y)\,\Delta x + (-3\,\Delta x + 9\,\Delta y)\,\Delta y = f_x\,\Delta x + f_y\,\Delta y + \varepsilon_1\,\Delta x + \varepsilon_2\,\Delta y$, where $\varepsilon_1 = \Delta x - 3\,\Delta y$ and $\varepsilon_2 = -3\,\Delta x + 9\,\Delta y$

9. $(4x - 3y)\,dx + (3y^2 - 3x)\,dy$ **11.** $ye^{xy}\,dx + xe^{xy}\,dy$ **13.** $\frac{2y}{(x+y)^2}\,dx - \frac{2x}{(x+y)^2}\,dy$ **15.** $\frac{y}{x^2+y^2}\,dx - \frac{x}{x^2+y^2}\,dy$

17. $-2y\tan 2xy\,dx - 2x\tan 2xy\,dy$ **19.** $2x\,dx + 2y\,dy + 2z\,dz$ **21.** 0.58 **23.** 0.012 **25.** -0.0044

27. 51.6 cu ft **29.** $dT = 0.2$, $\Delta T = 0.20594$ **31.** \$54.10

37. $\Delta f = (y + 3z - 4yz)\,\Delta x + (x - 2z - 4xz)\,\Delta y + (-2y + 3x - 4xy)\,\Delta z + \varepsilon_1\,\Delta x + \varepsilon_2\,\Delta y + \varepsilon_3\,\Delta z$, where $\varepsilon_1 = \Delta y - 4z\,\Delta y$, $\varepsilon_2 = -2\Delta z - 4x\,\Delta z$, and $\varepsilon_3 = 3\Delta x - 4y\,\Delta x - 4\Delta x\,\Delta y$

Exercise Set 17.2, page 665

1. 56 **3.** 2 **5.** $-\tan t - \cot t$ **7.** $1 - \cos t^2 + 2t^2\sin t^2$ **9.** $\left.\frac{\partial z}{\partial u}\right|_{(2,-1)} = -4$, $\left.\frac{\partial z}{\partial v}\right|_{(2,-1)} = 8$

11. $\left.\frac{\partial z}{\partial u}\right|_{(1,1/2)} = 2e^2, \left.\frac{\partial z}{\partial v}\right|_{(1,1/2)} = 0$ **13.** $\frac{\partial z}{\partial u} = 2v \tan 2uv, \frac{\partial z}{\partial v} = 2u \tan 2uv$ **15.** $\frac{\partial z}{\partial u} = -4u, \frac{\partial u}{\partial v} = 6v$

17. $\frac{2y^2 - 3x^2}{4y^3 - 4xy}$ **19.** $\frac{\sin y - y\cos x}{\sin x - x\cos y}$ **21.** $\frac{2x - x^2y + y^3}{2y + x^3 - xy^2}$ **23.** $\frac{\partial z}{\partial x} = -\frac{2xyz}{x^2y + 6z}, \frac{\partial z}{\partial y} = \frac{4y - x^2z}{x^2y + 6z}$

25. $\frac{\partial z}{\partial x} = \frac{2z(1 - 2xyz - 2y^2z)}{(x + y)(1 + 4xyz)}, \frac{\partial z}{\partial y} = \frac{2z(1 - 2xyz - 2x^2z)}{(x + y)(1 + 4xyz)}$ **27.** $\frac{\partial z}{\partial x} = -\frac{z\cos xz}{x\cos xz - y\tan yz}, \frac{\partial z}{\partial y} = \frac{z\tan yz}{x\cos xz - y\tan yz}$

29. $\frac{\partial w}{\partial x} = \frac{2yz(y^2 + z^2)}{(x^2 + y^2 + z^2)^{3/2}}, \frac{\partial w}{\partial y} = \frac{2xz(x^2 + z^2)}{(x^2 + y^2 + z^2)^{3/2}}, \frac{\partial w}{\partial z} = \frac{2xy(x^2 + y^2)}{(x^2 + y^2 + z^2)^{3/2}}$ **33.** 0.68 sq km/hr

35. Let $u = x/y$. Then

$$\frac{\partial z}{\partial x} = f'(u)\frac{\partial u}{\partial x} = f'(u)\frac{1}{y} \quad \text{and} \quad \frac{\partial z}{\partial y} = f'(u)\frac{\partial u}{\partial y} = f'(u)\left(-\frac{x}{y^2}\right).$$

So $x\frac{\partial z}{\partial x} + y\frac{\partial z}{\partial y} = 0$.

37. $\frac{\partial^2 z}{\partial u^2} = f_{xx}g_u^2 + 2f_{xy}g_uh_u + f_{yy}h_u^2 + f_xg_{uu} + f_yh_{uu}$

$\frac{\partial^2 z}{\partial v^2} = f_{xx}g_v^2 + 2f_{xy}g_vh_v + f_{yy}h_v^2 + f_xg_{vv} + f_yh_{vv}$ **41.** Approximately 0.91 sq ft/sec

Exercise Set 17.3, page 671

1. $\frac{27}{5}$ **3.** $-\frac{9}{10\sqrt{2}}$ **5.** $-\frac{7}{5}$ **7.** $\frac{1}{\sqrt{5}}$ **9.** $-\frac{3}{2\sqrt{2}}$ **11.** $\frac{16}{3}$ **13.** $\frac{1}{\sqrt{2}}$ **15.** $\frac{82}{\sqrt{10}}$

17. $\mathbf{u} = \left\langle \frac{4}{\sqrt{41}}, \frac{-5}{\sqrt{41}} \right\rangle$; $D_\mathbf{u}f(P_0) = \frac{\sqrt{41}}{27}$ **19.** $\mathbf{u} = \left\langle -\frac{1}{\sqrt{5}}, -\frac{2}{\sqrt{5}} \right\rangle$; $D_\mathbf{u}f(P_0) = \sqrt{5}$

21. $D_\mathbf{u}f(P_0) = 2$ if $\mathbf{u} = \langle 1, 0\rangle$ or $\left\langle -\frac{15}{17}, \frac{8}{17} \right\rangle$; max $D_\mathbf{u}f(P_0) = 2\sqrt{17}$ when $\mathbf{u} = \left\langle \frac{1}{\sqrt{17}}, \frac{4}{\sqrt{17}} \right\rangle$

23. $-\frac{2}{5\sqrt{5}}$; T decreases most rapidly in the direction $\mathbf{u} = \left\langle \frac{4}{5}, \frac{3}{5} \right\rangle$, and the rate of decrease is -1.

25. $-\frac{14}{5\sqrt{13}\ln 4}$; max $D_\mathbf{u}V(P) = \frac{2\sqrt{10}}{5\ln 4}$, when $\mathbf{u} = \left\langle \frac{3}{\sqrt{10}}, \frac{1}{\sqrt{10}} \right\rangle$

27. $x^2 + y^2 = 4^c$, circles centered at the origin with radii varying from 1 to 2. Using the vector equation $\mathbf{r}(t) = \langle 2^c\cos t, 2^c\sin t\rangle$, let t_0 correspond to $P_0 = (x_0, y_0)$. Then $\nabla V(P_0) = \frac{2}{4^c\ln 4}\langle x_0, y_0\rangle = \frac{2}{2^c\ln 4}\langle \cos t_0, \sin t_0\rangle$. Now show that $\nabla V(P_0)\cdot\mathbf{r}'(t_0) = 0$.

29. $f(x, y) = xe^x - e^x - e^{-y}$; not unique, since adding a constant does not change f_x or f_y.

Exercise Set 17.4, page 676

1. $6x - 6y - 5z = 5$; $\mathbf{r}(t) = \langle 3 + 6t, -2 - 6t, 5 - 5t\rangle$ **3.** $x - z = 3$; $\mathbf{r}(t) = \langle 1 + t, 0, -2 - t\rangle$
5. $3x + y + 4z = 10$; $\mathbf{r}(t) = \langle 1 + 3t, -1 + t, 2 + 4t\rangle$ **7.** $x - 2z = 1$, $\mathbf{r}(t) = \langle 3 + t, -1, 1 - 2t\rangle$
9. $10x - 16y - z + 7 = 0$; $\mathbf{r}(t) = \langle 5 + 10t, 4 - 16t, -7 - t\rangle$

11. $x + y - 2z = \frac{\pi}{2}$; $\mathbf{r}(t) = \left\langle 1 + t, -1 + t, -\frac{\pi}{4} - 2t \right\rangle$ **13.** $2x + z + 1 = 0$; $\mathbf{r}(t) = \left\langle 2t, \frac{\pi}{2}, -1 + t \right\rangle$

15. $(-1, 3)$ **17.** $(0, 0)$ **19.** $(-1, 1)$ **21.** $\left(\frac{1}{2}, -\frac{3}{4}, -\frac{19}{16}\right)$

23. (a) They have a common tangent plane at P_0 if and only if their normals are parallel at P_0.
(b) $(2, 3, -1)$ and $(-2, -3, 1)$, for $k = 4$

25. $(0, 0, 0)$ and $(2, 1, -3)$ **27.** $6.34°$ at $(3, 0, 4)$; $26.93°$ at $(0, 3, 1)$

Exercise Set 17.5, page 685

1. Absolute minimum $= -2$ at $(-1, 2)$ **3.** Absolute minimum $= -3$ at $(0, 4)$
5. Local minimum at $(16, 4)$; saddle point at $(1, -1)$ **7.** Local minimum at $(3, 9)$; saddle point at $(0, 0)$
9. Local minima at $(1, 1)$ and $(-1, -1)$; saddle point at $(0, 0)$ **11.** Local maximum at $(18, 6)$; saddle point at $(0, 0)$
13. Local minima at $(\pm 1, 1)$; saddle points at $\left(0, \pm\frac{1}{\sqrt{3}}\right)$ **15.** Local minimum at $\left(\frac{1}{\sqrt[3]{2}}, \sqrt[3]{4}\right)$
17. Local minimum at $(2, -1)$; saddle point at $(2, 1)$ **19.** No critical points
21. Local maximum at $(-2, 0)$; saddle point at $(0, 0)$
23. Local maxima at $\left(\frac{\pi}{2}, \frac{\pi}{2}\right)$ and $\left(-\frac{\pi}{2}, -\frac{\pi}{2}\right)$; local minima at $\left(\frac{\pi}{2}, -\frac{\pi}{2}\right)$ and $\left(-\frac{\pi}{2}, \frac{\pi}{2}\right)$; saddle point at $(0, 0)$
25. Local minima at $(1, 0)$ and $(-1, \pi)$; saddle points at $\left(0, \frac{\pi}{2}\right)$ and $\left(0, \frac{3\pi}{2}\right)$
27. Write f in the form $f(x, y) = 4 - (x^{1/3} - y^{1/3})^2$.
29. Absolute maximum $= 0$ at all points for which $y = 0$; absolute minimum $= -\frac{5}{4}$ at $\left(\frac{1}{2}, 1\right)$
31. Absolute maximum $= 5 + 4\sqrt{2}$ at $(-\sqrt{2}, 3)$; absolute minimum $= 5 - 4\sqrt{2}$ at $(\sqrt{2}, 3)$
33. Base 8 ft by 8 ft; height 4 ft
35. $T_{\max} = 50°$ at $\left(\frac{1}{2}, \frac{1}{2}, \frac{1}{\sqrt{2}}\right)$ and $\left(-\frac{1}{2}, -\frac{1}{2}, \frac{1}{\sqrt{2}}\right)$; $T_{\min} = -50°$ at $\left(\frac{1}{2}, -\frac{1}{2}, \frac{1}{\sqrt{2}}\right)$ and $\left(-\frac{1}{2}, \frac{1}{2}, \frac{1}{\sqrt{2}}\right)$
37. Length $= 10$ in.; width $= 8$ in.; height $= 4$ in. **39.** $y = \frac{30}{53}x + \frac{140}{53} \approx 0.57x + 2.64$

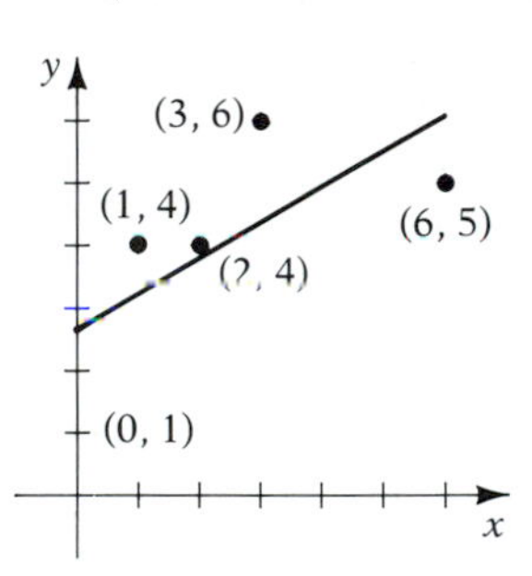

Exercise Set 17.6, page 691

1. Local minimum $= -\frac{16}{3}$ at $\left(-\frac{4}{3}, -\frac{8}{3}\right)$ **3.** Local maximum $= 3$ at $(-1, -1)$; local minimum $= -\frac{44}{27}$ at $\left(\frac{2}{3}, -\frac{14}{9}\right)$
5. Local minimum $= 20$ at $(2, -2)$ **7.** Local minimum $= 18$ at $(6, -3, -2)$ **9.** $\frac{2\sqrt{5}}{5}$
11. $T_{\max} = 80°$ at $(\pm\sqrt{3}, 1)$; $T_{\min} = 20°$ at $(\pm\sqrt{3}, -1)$ **19.** $2\sqrt{2}$ in the x-direction; 4 in the y-direction; 2 in the z-direction
21. $f_{\max} = \frac{223}{12}$ at $\left(\pm 2\sqrt{3}, \pm\frac{4}{\sqrt{3}}, \frac{7}{6}\right)$; $f_{\min} = -\frac{11}{4}$ at $\left(\pm 2, 0, -\frac{3}{2}\right)$
23. Nearest points: $\left(0, \pm\frac{\sqrt{29}}{4}, \frac{3}{8}\right)$; farthest points: $\left(\pm\frac{\sqrt{10}}{6}, \pm\frac{4}{3}, \frac{1}{6}\right)$

Supplementary Exercises, page 692

1. (a) $\frac{2}{x}\,dx + \frac{y}{1 - y^2}\,dy$ (b) $\frac{\cos x}{\cosh y}\,dx - \frac{\sin x \sinh y}{\cosh^2 y}\,dy$ **3.** (a) 0.0084 (b) 0.01 **5.** 6.57 **9.** $\sqrt{3} - \frac{1}{4}$
11. $-e^{1-\cosh t}\sinh t$ **13.** $\left.\frac{\partial z}{\partial s}\right|_{(1,1/2)} = \frac{\pi}{4}(4\sqrt{2} - 3) + \frac{1}{\sqrt{2}}$; $\left.\frac{\partial z}{\partial t}\right|_{(1,1/2)} = \sqrt{2} - \frac{\pi}{2}(3 + \sqrt{2})$ **15.** $\frac{\partial z}{\partial x} = -\frac{z}{x}$; $\frac{\partial z}{\partial y} = -\frac{\ln(\cos xz) + xz}{xy(1 - \tan xz)}$

17. $\frac{10}{3}$ in./min **19.** -4 **21.** $f(x, y) = y^2 - x^2 + \ln\frac{x}{y} + 1$ **23.** $2x - 2y - 3z + 6 = 0$; $\mathbf{r}(t) = \langle 1 + 2t, 1 - 2t, 2 - 3t\rangle$

25. $86.5°$ at $(1, -1, 0)$; $74.0°$ at $(-2, -3, -5)$ **27.** Absolute maximum $= \frac{35}{2}$ at $(4, 2)$; absolute minimum $= \frac{9}{\sqrt[3]{3}}$ at $\left(\sqrt[3]{3}, \frac{\sqrt[3]{3}}{2}\right)$

29. (a) $r = \frac{2a}{\sqrt{6}}$; $h = \frac{2a}{\sqrt{3}}$ (b) $r = \frac{a}{\sqrt{2}}$; $h = \sqrt{2}\,a$

CHAPTER 18

Exercise Set 18.1, page 699

1. 12 **3.** 60 **5.** 582 **7.** 35 **9.** $\frac{69}{4}$

Exercise Set 18.2, page 705

1. $-\frac{7}{2}$

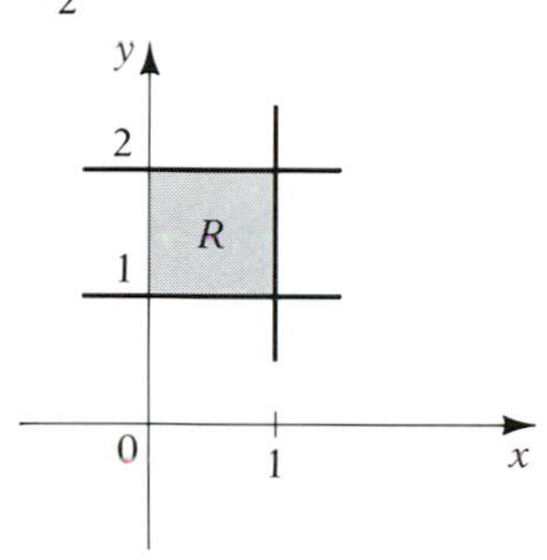

3. $\frac{15}{8}$

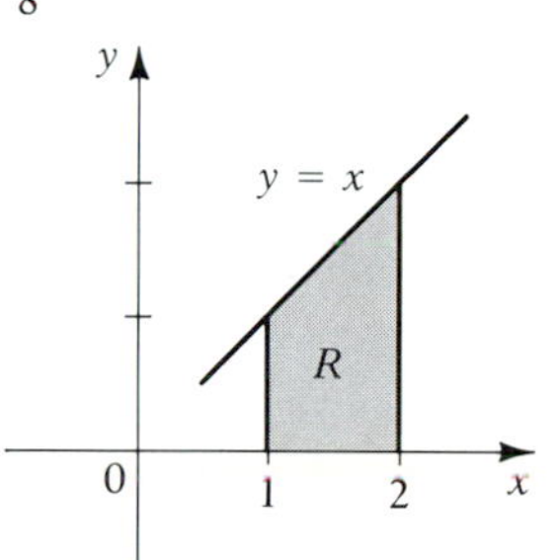

5. $-\frac{4352}{15}$

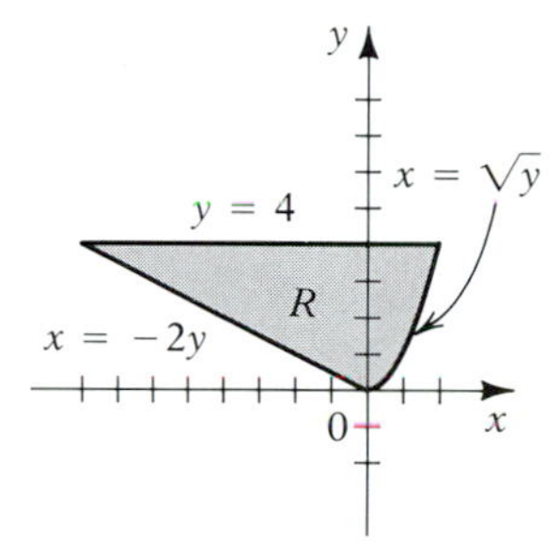

7. $\frac{5}{3}$

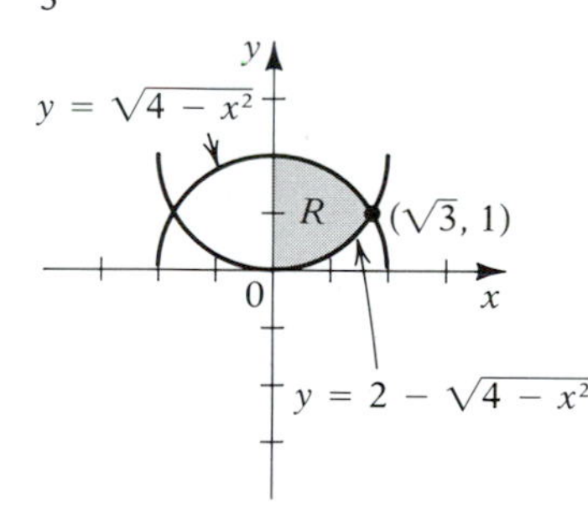

9. $4 \ln\frac{6}{7}$

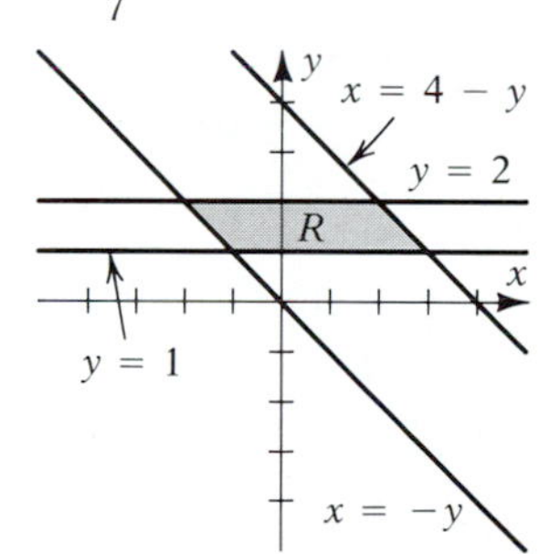

11. 40

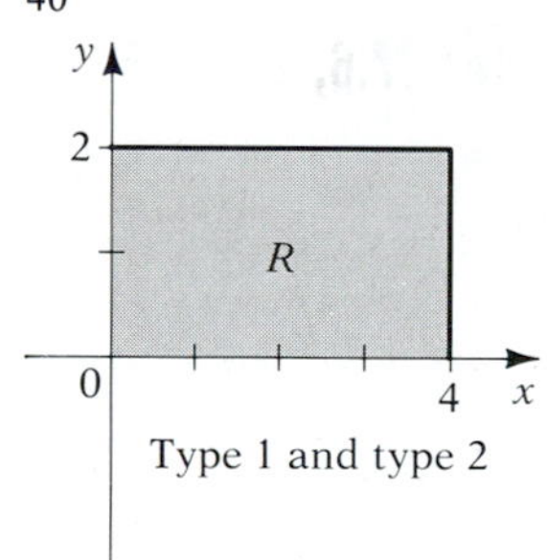

13. $\frac{2}{3}\ln 2$

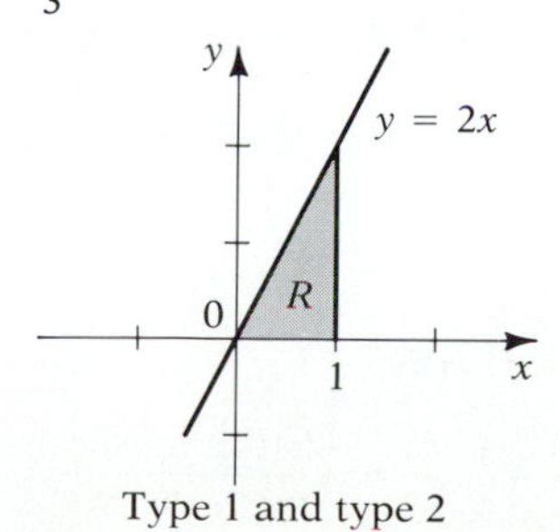

15. $4\sqrt{2} - \frac{8}{3}$

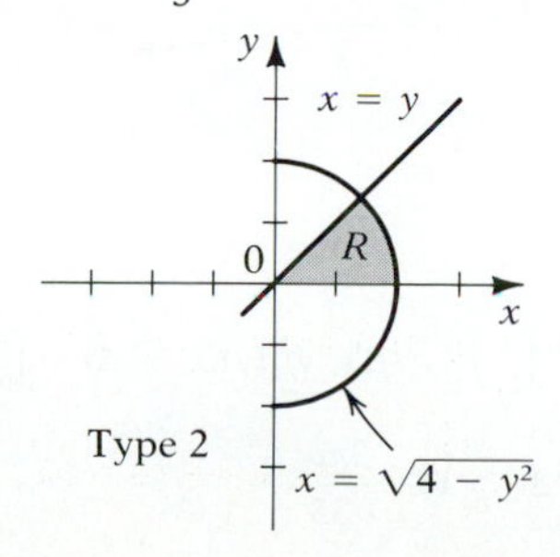

17. $\frac{49}{3}$

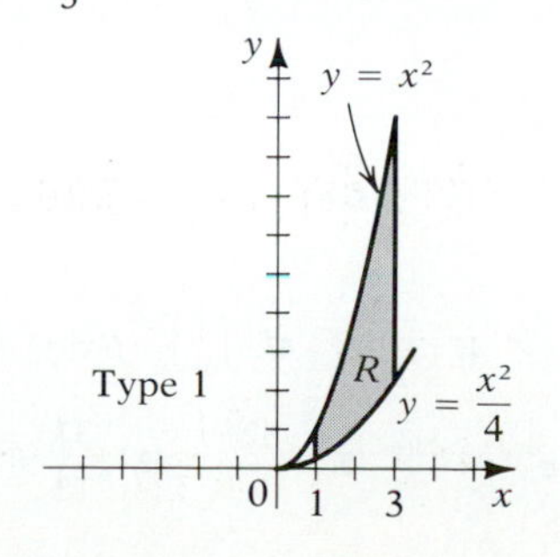

19. 4

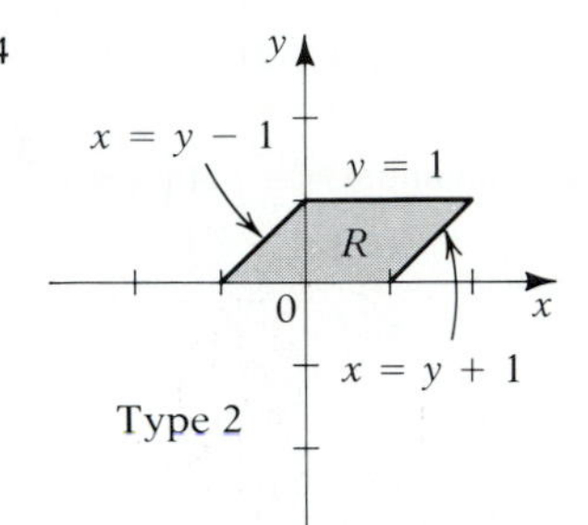

21. 2

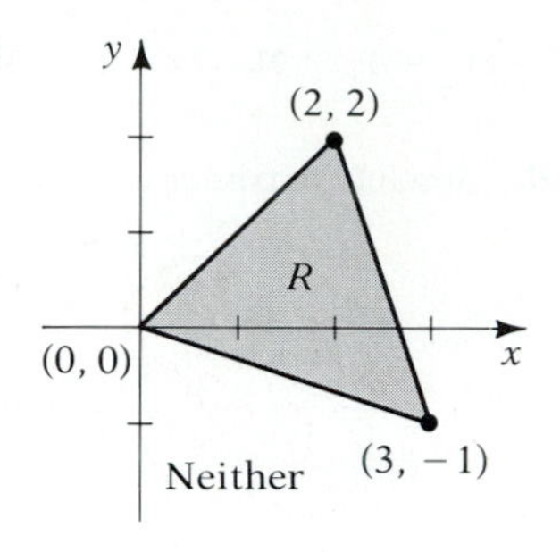

23. $\frac{4}{3}$

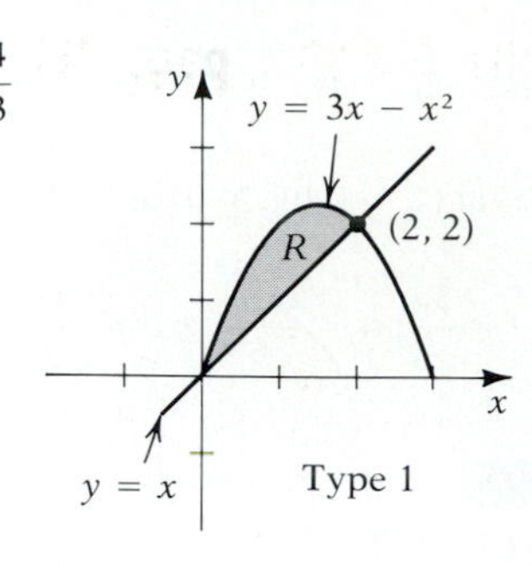

25. $\frac{64}{3}$

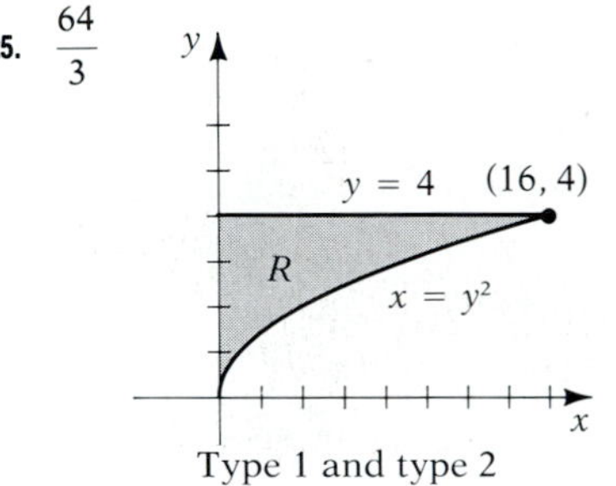

27. $e - \frac{5}{3}$

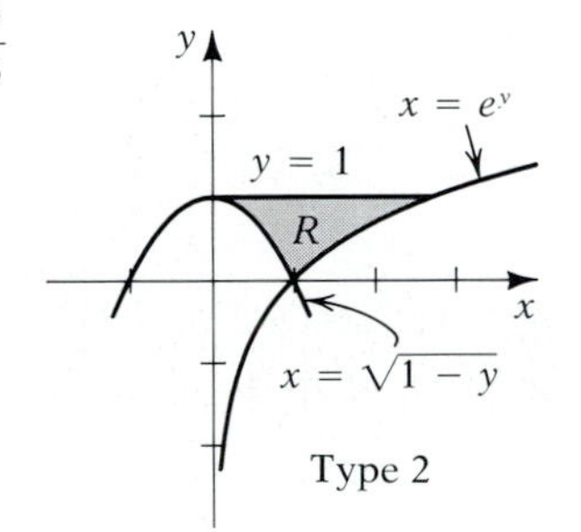

29. $1 + \frac{3}{2} \ln \frac{4}{3}$

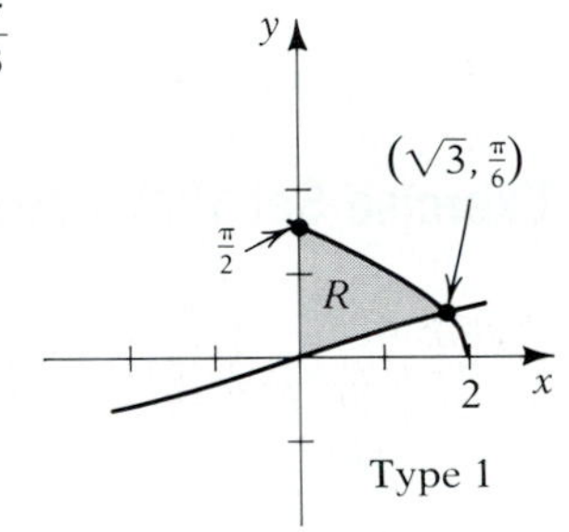

31. $\frac{15}{2}$

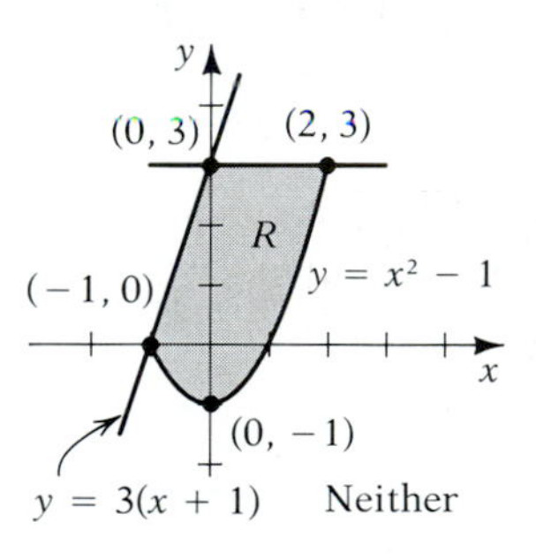

33. 2

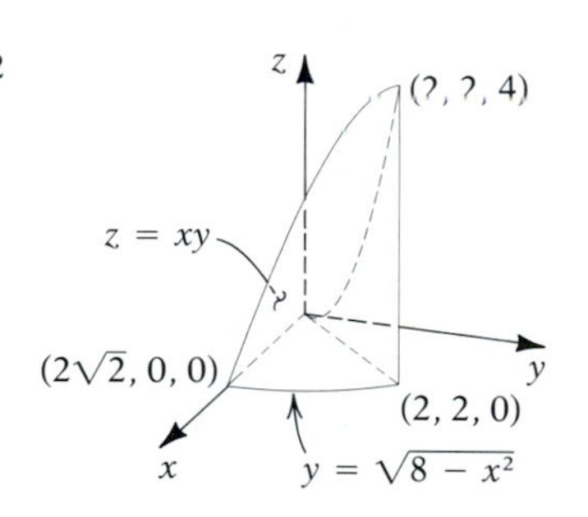

35. $\frac{2\sqrt{2}}{3}$

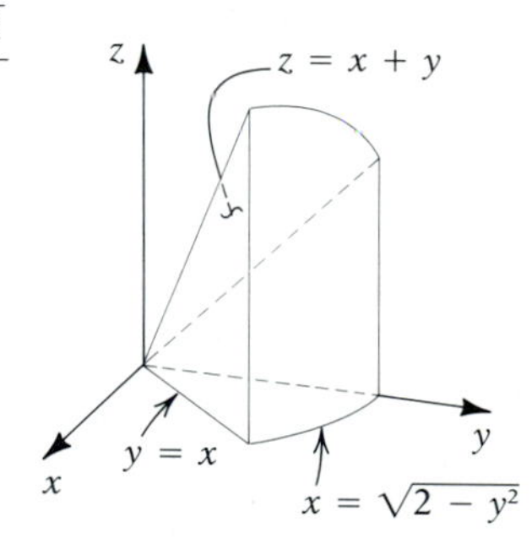

37. $\frac{13}{2}$

39. 4

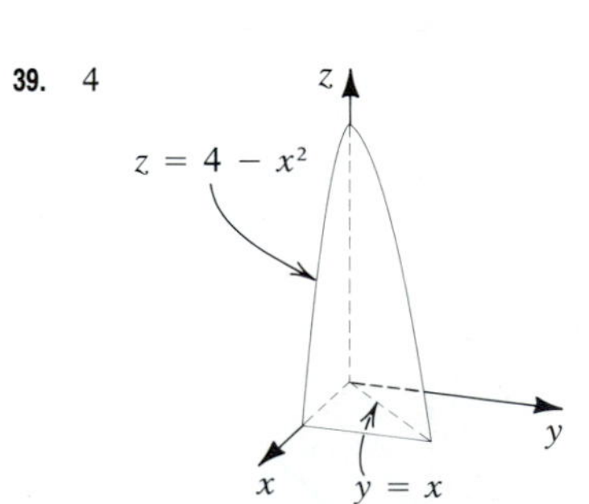

41. $\frac{16}{3}$

43. $\frac{e-1}{2e}$ **45.** $\frac{4}{27}$ **47.** $\int_0^2 \int_0^{y^2} f(x, y)\,dx\,dy$ **49.** $\int_2^4 \int_1^{x-1} f(x, y)\,dy\,dx$ **51.** $\int_0^1 \int_{1-\sqrt{1-y^2}}^{\sqrt{2y-y^2}} f(x, y)\,dx\,dy$

55. $e + 5e^{-1} + 4$ **57.** $\frac{82}{9}$ **59.** $\frac{37}{12}$ **61.** $\frac{7}{9} - \ln 2$ **63.** $\frac{2\pi a^3}{35}$

Exercise Set 18.3, page 711

1. $\frac{16\pi^2}{3}$

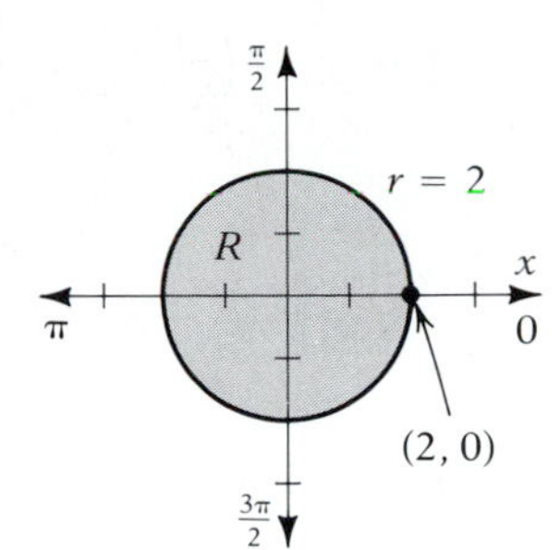

3. 99π

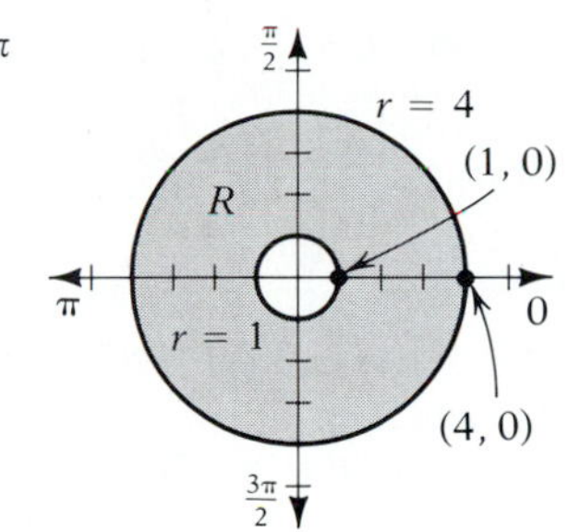

5. $\frac{4}{5}$

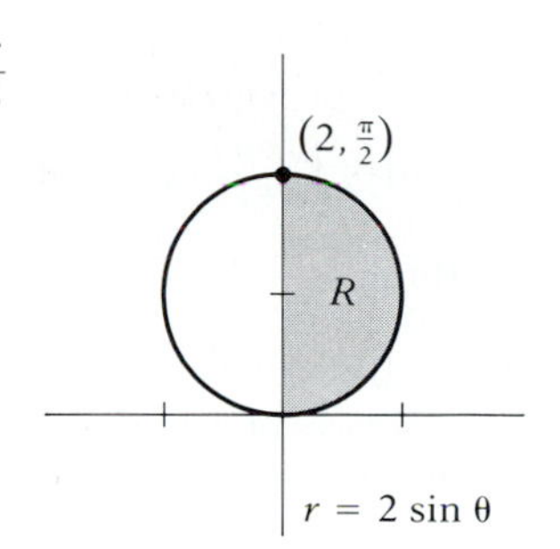

7. $\frac{\pi}{8}$

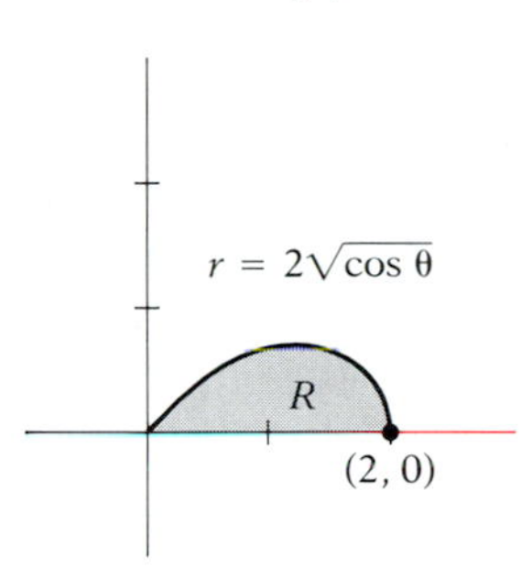

9. 6π

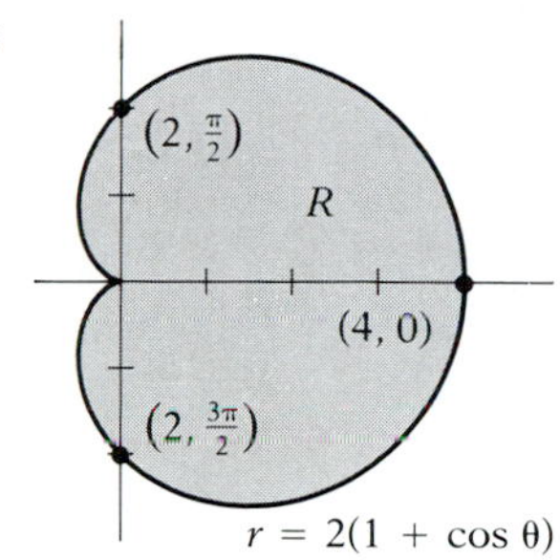

11. 2π

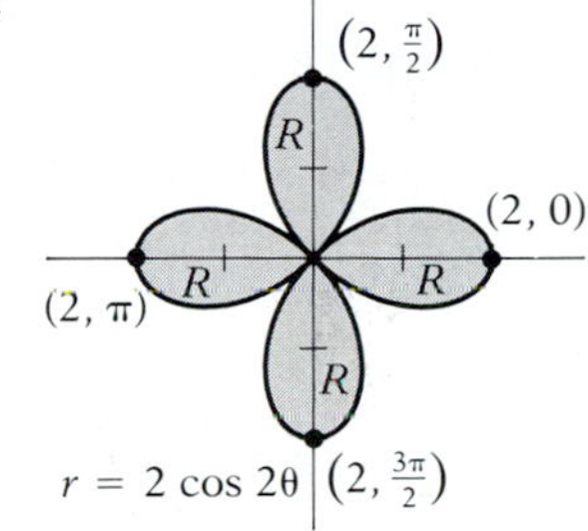

13. $\frac{10\pi}{3} + \frac{7\sqrt{3}}{2}$

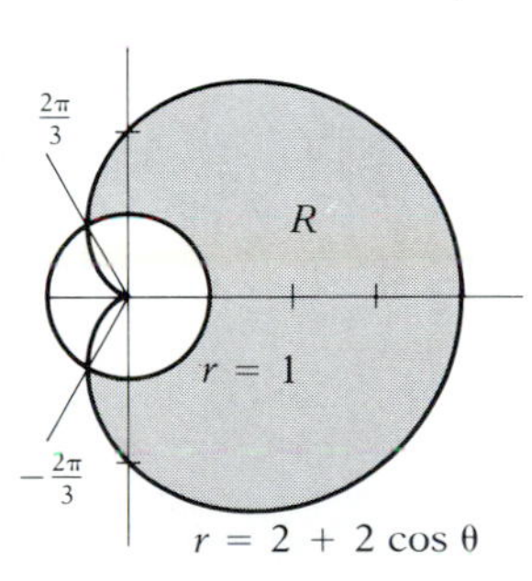

15. $\sqrt{3} - \frac{\pi}{3}$

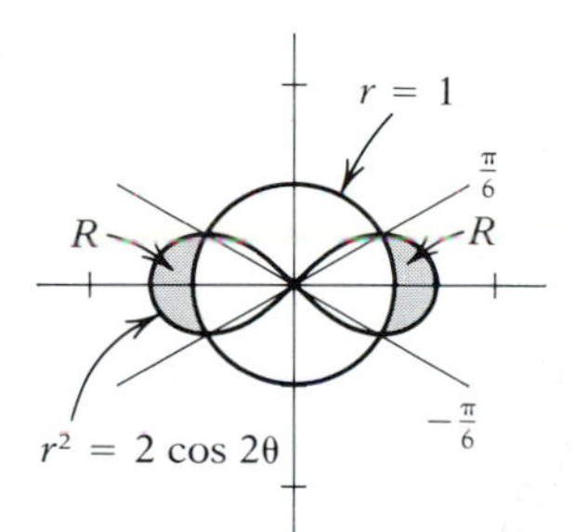

17. $\frac{19\pi}{3} - \frac{11\sqrt{3}}{2}$

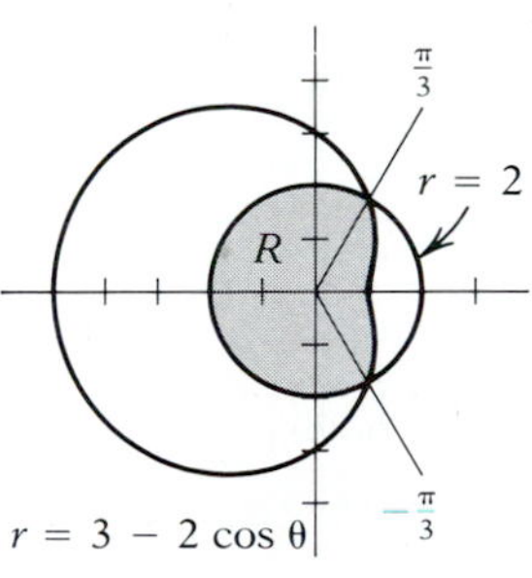

19. 5π

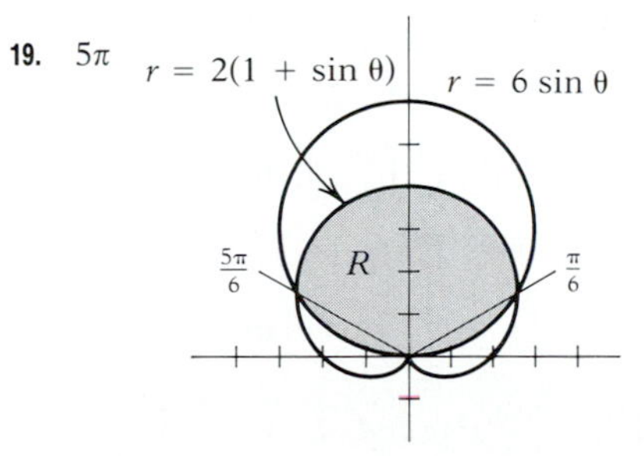

21. $\frac{\pi}{3}$ **23.** $\frac{16\sqrt{2}}{3}(1 - \sqrt{2})$ **25.** $\frac{\pi}{4} - \frac{2}{3}$ **27.** $\ln 4 - \frac{3}{4}$ **29.** $\frac{2\pi}{3} - \frac{20}{9\sqrt{2}}$ **31.** $\pi(1 - e^{-a^2})$ **33.** $\frac{8\pi}{3}$

35. $\frac{2\pi}{3}(a^2 - b^2)^{3/2}$ **37.** $\frac{32}{9}$ **39.** 8π **41.** $\int_0^{\pi/3} \int_{2 \sin \theta}^{4 \sin \theta} (r^2 \sin \theta \cos \theta)\, r\, dr\, d\theta = \frac{135}{32}$ **43.** $\frac{4}{3} - \frac{2\pi}{3} + \frac{17\sqrt{3}}{27}$

Exercise Set 18.4, page 720

1. $\left(\frac{1}{2}, \frac{2}{3}\right)$ **3.** $\left(0, \frac{4}{5}\right)$ **5.** $\left(-\frac{1}{2}, \frac{2}{5}\right)$ **7.** $\left(0, \frac{3}{\pi}\right)$ **9.** $\left(0, \frac{3}{10}\right)$ **11.** $\left(0, \frac{4a}{3\pi}\right)$ **13.** $\left(\frac{16\sqrt{2}}{15}, \frac{176\sqrt{2}}{35}\right)$

15. $\left(\frac{e^2+1}{e^2-1}, \frac{e^2+1}{4}\right)$ **17.** $I_x = \frac{1}{2}, I_y = \frac{1}{3}, I_0 = \frac{5}{6}, (r_x, r_y) = \left(\frac{1}{\sqrt{3}}, \frac{1}{\sqrt{2}}\right)$ **19.** $I_x = \frac{1}{3}, I_y = \frac{1}{9}, I_0 = \frac{4}{9}, (r_x, r_y) = \left(\frac{\sqrt{2}}{3}, \frac{\sqrt{6}}{3}\right)$

21. $I_x = \frac{32}{15}, I_y = \frac{32}{15}, I_0 = \frac{64}{15}, (r_x, r_y) = \left(\frac{2}{\sqrt{5}}, \frac{2}{\sqrt{5}}\right)$ **23.** $I_x = \frac{4\rho}{9}, I_y = \rho(\pi^2 - 4), I_0 = \rho\left(\pi^2 - \frac{32}{9}\right), (r_x, r_y) = \left(\sqrt{\frac{\pi^2 - 4}{2}}, \frac{2}{3\sqrt{2}}\right)$

27. $I_{x=-2} = \rho \int_{-2}^{4} \int_{y^2/4}^{(y+4)/2} (x+2)^2\, dx\, dy,\ I_{y=4} = \int_{-2}^{4} \int_{y^2/4}^{(y+4)/2} (4-y)^2\, dx\, dy$

29. (a) 4 (b) $\mu_x = \frac{2}{3}, \mu_y = \frac{2}{3}$ (c) $\sigma_x^2 = \frac{2}{9}, \sigma_y^2 = \frac{2}{9}$ **31.** (a) 6 (b) $\mu_x = \frac{1}{4}, \mu_y = \frac{1}{2}$ (c) $\sigma_x^2 = \frac{3}{80}, \sigma_y^2 = \frac{1}{20}$

33. (a) $I_{x=\bar{x}} = \iint_R (x-\bar{x})^2\, dm$. Expand and use the fact that $\iint_R x\, dm = m\bar{x}$. (b) $I_{y=\bar{y}} = I_x - m\bar{y}^2$

35. Take the vertices of the triangle as $(0, 0)$, $(b, 0)$, and (a, h).

37. Orient the axes so that l coincides with the y-axis. Show that $V = 2\pi \iint_R x\, dA$, and use the fact that $A\bar{x} = \iint_R x\, dA$.

Exercise Set 18.5, page 725

1. $2\sqrt{14}$ **3.** $\frac{13}{6}$ **5.** $\frac{13\pi}{3}$ **7.** $4\pi\sqrt{2}$ **9.** $\frac{\pi}{6}(17\sqrt{17} - 5\sqrt{5})$ **11.** $\frac{1}{6}(13\sqrt{26} - 5\sqrt{10})$ **13.** 2 **15.** $\frac{20}{9} - \frac{2\pi\sqrt{2}}{3}$

17. $4\pi a^2$

Exercise Set 18.6, page 730

1. 0

3. $\frac{1}{3}$

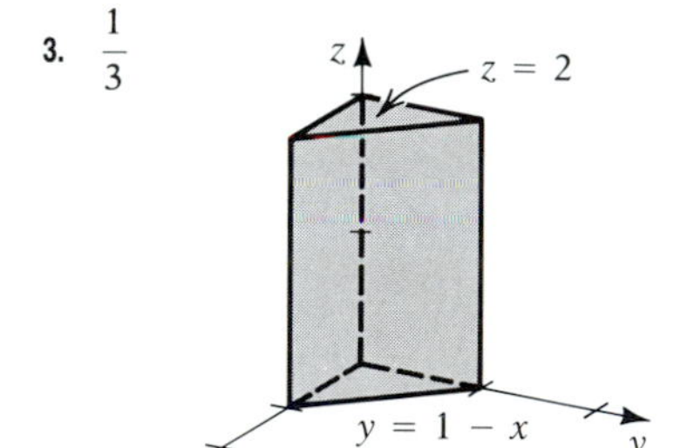

5. $\frac{736}{3}$

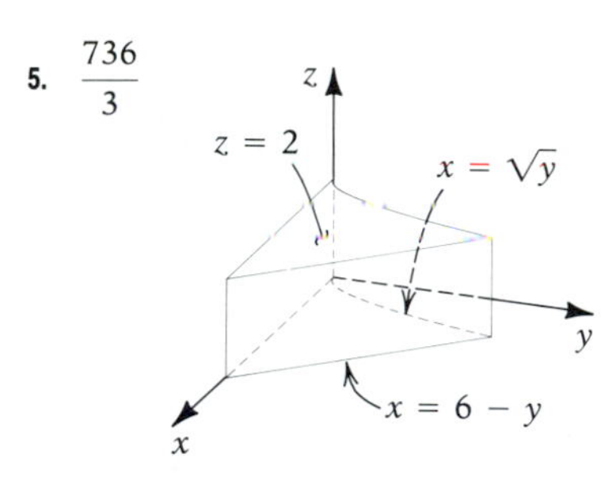

7. $\int_0^2 \int_0^x \int_0^{4-x^2} f(x, y, z)\, dz\, dy\, dx$ **9.** $\int_0^3 \int_0^{(6-2x)/3} \int_0^{3-x} f(x, y, z)\, dz\, dy\, dx$ **11.** $\frac{64}{15}$ **13.** $\frac{16}{3}$ **15.** $\frac{8}{3}$

17. $\int_0^1 \int_0^{1-z} \int_0^x f(x, y, z)\, dy\, dx\, dz$, $\int_0^1 \int_0^{1-x} \int_0^x f(x, y, z)\, dy\, dz\, dx$, $\int_0^1 \int_y^1 \int_0^{1-x} f(x, y, z)\, dz\, dx\, dy$, $\int_0^1 \int_0^{1-y} \int_y^{1-z} f(x, y, z)\, dx\, dz\, dy$, $\int_0^1 \int_0^{1-z} \int_y^{1-z} f(x, y, z)\, dx\, dy\, dz$

19. $\left(\frac{6}{7}, 0, \frac{8}{7}\right)$ **21.** $m = \int_0^{2\pi} \int_0^2 \int_{r^2}^{8-r^2} zr\, dz\, dr\, d\theta,\ \bar{z} = \frac{1}{m} \int_0^{2\pi} \int_0^2 \int_{r^2}^{8-r^2} z^2 r\, dz\, dr\, d\theta,\ I_z = \int_0^{2\pi} \int_0^2 \int_{r^2}^{8-r^2} zr^3\, dz\, dr\, d\theta$

23. $m = \int_0^6 \int_0^{(6-x)/2} \int_0^{(6-x-2y)/3} x^2 yz\, dz\, dy\, dx,\ \bar{x} = \frac{1}{m} \int_0^6 \int_0^{(6-x)/2} \int_0^{(6-x-2y)/3} x^3 yz\, dz\, dy\, dx$

$\bar{y} = \frac{1}{m} \int_0^6 \int_0^{(6-x)/2} \int_0^{(6-x-2y)/3} x^2 y^2 z\, dz\, dy\, dx,\ \bar{z} = \frac{1}{m} \int_0^6 \int_0^{(6-x)/2} \int_0^{(6-x-2y)/3} x^2 yz^2\, dz\, dy\, dx$

25. $m = \int_0^2 \int_0^{\sqrt{4-z^2}} \int_0^{2y+z} \sqrt{x+y}\, dx\, dy\, dz,\ \bar{x} = \frac{1}{m} \int_0^2 \int_0^{\sqrt{4-z^2}} \int_0^{2y+z} x\sqrt{x+y}\, dx\, dy\, dz,\ I_x = \int_0^2 \int_0^{\sqrt{4-z^2}} \int_0^{2y+z} (y^2 + z^2)\sqrt{x+y}\, dx\, dy\, dz$

27. $m = \int_0^{\sqrt{2}} \int_0^{\sqrt{4-2x^2}} \int_0^{\sqrt{4-2x^2-y^2}} \sqrt{x^2+z^2}\, dz\, dy\, dx,\ \bar{y} = \frac{1}{m} \int_0^{\sqrt{2}} \int_0^{\sqrt{4-2x^2}} \int_0^{\sqrt{4-2x^2-y^2}} y\sqrt{x^2+z^2}\, dz\, dy\, dx,$

$I_y = \int_0^{\sqrt{2}} \int_0^{\sqrt{4-2x^2}} \int_0^{\sqrt{4-2x^2-y^2}} (x^2+z^2)^{3/2}\, dz\, dy\, dx$

29. $\frac{\pi}{2}$; mass of solid occupying Q, with density $y^2 \sin xy$, or $m\bar{y}$, where the density is $y \sin xy$.

Exercise Set 18.7, page 736

1. 4π

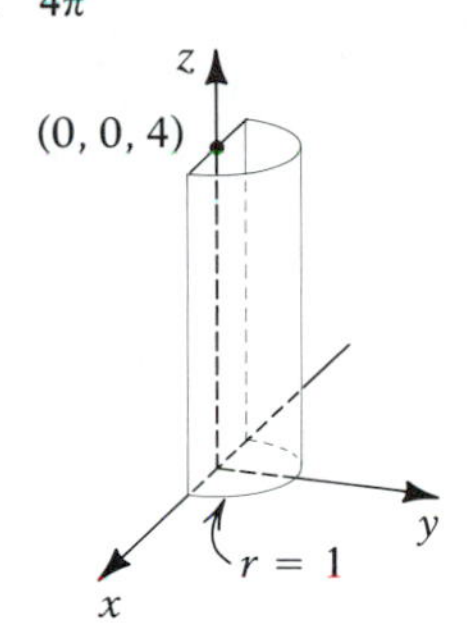

3. $\frac{\pi}{4}$

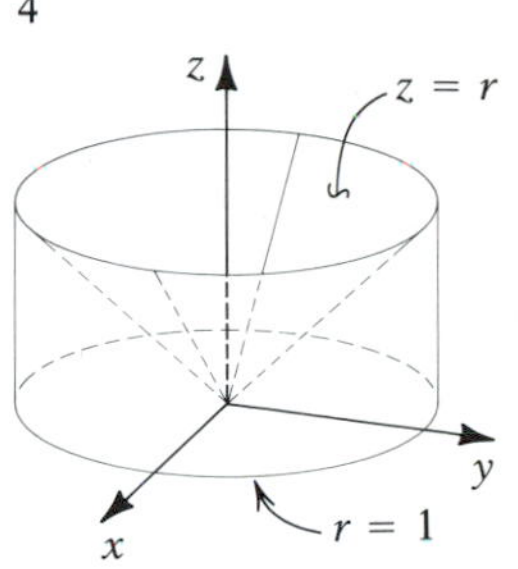

5. $\frac{2\pi}{3}$

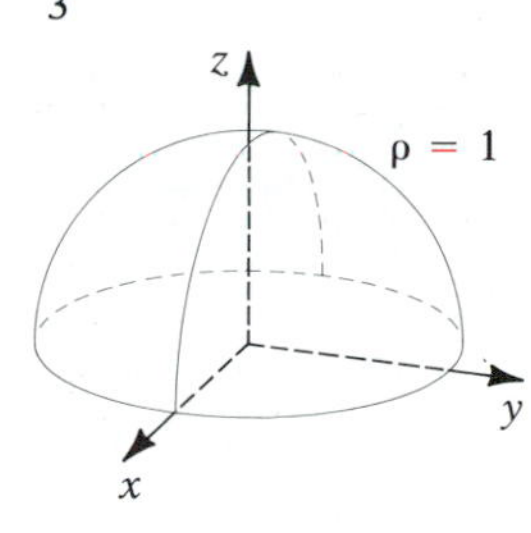

7. 24π **9.** $\frac{2\pi\delta}{5}$ **11.** $\left(\frac{6}{5}, 0, \frac{27\pi}{128}\right)$ **13.** $\left(\frac{18\sqrt{3}}{7\pi}, \frac{18}{7\pi}, \frac{102}{35}\right)$ **15.** $\left(\frac{64a}{15\pi^2}, \frac{64a}{15\pi^2}, \frac{16}{15\pi}\right)$

17. $m = 20\pi^2,\ I_z = \frac{1456\pi}{9}$ **19.** 4π **21.** $(2 - \sqrt{2})\pi$ **23.** $\frac{8}{9}(3\pi - 4)$ **25.** $\bar{x} = \bar{y} = 0,\ \bar{z} = \frac{4}{15}\left[\frac{5a^2(h_2^3 - h_1^3) - 3(h_2^5 - h_1^5)}{2a^2(h_2^2 - h_1^2) - (h_2^4 - h_1^4)}\right]$

Supplementary Exercises 18.8, page 738

1. $4 + \frac{2}{3}\ln 4$

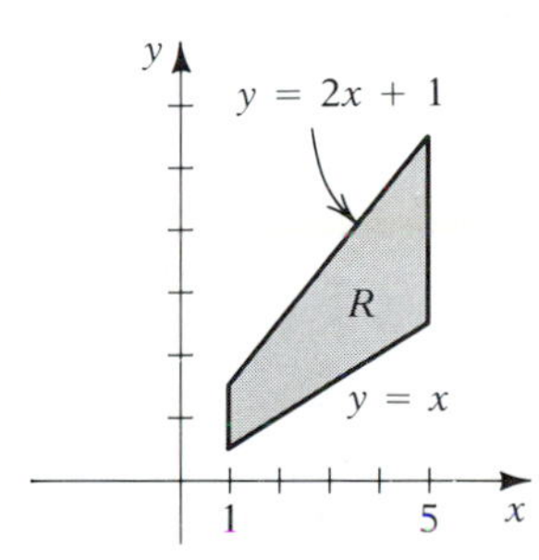

3. $4\ln 2 - \frac{1}{6}$ **5.** $\frac{3 - e}{2}$ **7.** $\frac{16}{3}$ **9.** $\frac{4}{3}$ **11.** $\frac{1}{16}(2 + \sin 2)$ **13.** $(\bar{x}, \bar{y}) = \left(\frac{34}{21}, \frac{110}{147}\right),\ (r_x, r_y) = \left(\frac{20}{7\sqrt{3}}, \frac{10}{3\sqrt{14}}\right)$

15. $(\bar{x}, \bar{y}) = \left(\frac{\pi}{2}, \frac{16}{9\pi}\right),\ I_x = \frac{3\pi}{32},\ r_y = \frac{\sqrt{6}}{4}$ **17.** $\frac{14}{3}$ **19.** $6\pi + \frac{111\sqrt{3}}{20}$ **21.** $\frac{16\pi}{9}$ **23.** (a) $\left(0, 0, \frac{5}{4}\right)$ (b) $\frac{3\pi}{10}$

25. $\frac{46\pi}{15}$ **27.** $\frac{2}{3}(8 - 2\sqrt{2})$ **29.** (a) $\frac{1}{8}$ (b) $1 - \frac{1}{4}\ln 3$ (c) $\mu_x = \frac{26}{9},\ \mu_y = \frac{13}{6}$ (d) $\sigma_x^2 = \frac{\sqrt{134}}{9},\ \sigma_y^2 = \frac{\sqrt{11}}{6}$

CHAPTER 19

Exercise Set 19.1, page 746

1. $\frac{27\sqrt{13}}{2}$ **3.** $4(\sqrt{5} - 1)$ **5.** $\pi - 2$ **7.** 8π **9.** $\frac{2}{3}(2\sqrt{2} - 1)$ **11.** $\frac{81}{64} - \ln\frac{5}{4}$ **13.** $\frac{200}{3}$ **15.** $-\frac{41}{6}$

17. $-\frac{20}{3}$ **19.** 3 **21.** $2 - \ln 4$ **23.** $-\frac{7}{6}$ **25.** 0

27. $\mathbf{F}(\pm 1, 0) = \langle 0, 0 \rangle$, $\mathbf{F}(2, 1) = \mathbf{F}(-2, -1) = \langle 1, -2 \rangle$, $\mathbf{F}(2, -1) = \mathbf{F}(-2, 1) = \langle 1, 2 \rangle$, $\mathbf{F}(1, 2) = \mathbf{F}(-1, -2) = \langle 4, -2 \rangle$, $\mathbf{F}(1, -2) = \mathbf{F}(-1, 2) = \langle 4, 2 \rangle$, $\mathbf{F}(0, \pm 1) = \langle 1, 0 \rangle$

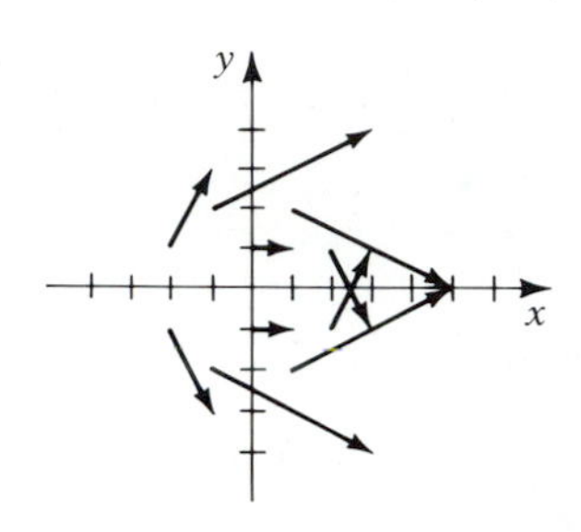

29. (a) (b) Work = 0 since **T** and **F** are orthogonal (c) $\mathbf{F}(x, y) = \left\langle -\frac{kx}{x^2 + y^2}, -\frac{ky}{x^2 + y^2} \right\rangle$

(a)

31. $-\pi\sqrt{5}$ **33.** 4 **35.** $\frac{5}{12} + \frac{3\pi}{16}$ **37.** $-\frac{1}{2}$ **39.** $\frac{3}{5}(1 - e^{\pi})$ **41.** $\frac{49\pi}{2}$

45. Center of mass $= \left(-\frac{4}{3\pi}, \frac{4}{3\pi}, \frac{9\pi}{4}\right)$; $I_x = I_y = \frac{3\rho\sqrt{13}}{8}(8 + 27\pi^3)$, $I_z = 6\pi\rho\sqrt{13}$

Exercise Set 19.2, page 755

1. $\phi(x, y) = x^3 + 3xy - y^2 + y$ **3.** Not a gradient field **5.** $\phi(x, y) = x^3y^2 - 2y + \frac{2}{3}x^3$ **7.** $\phi(x, y) = x^2e^{xy}$

9. Not a gradient field **11.** $\phi(x, y) = y \sin xy - y^2$ **13.** 33 **15.** $\frac{155}{3}$ **17.** 15 **19.** $-\frac{3}{2}$

21. $\phi(x, y) = x^2 + y^3 - 2xy^2$ **23.** $\phi(x, y) = x \sin y - y \cos x$ **25.** 16 **27.** $\frac{5\pi}{12}$ **29.** Result is 40 in each case

31. 44 **33.** (a) $\frac{\sqrt{6} - 2}{4}$ (b) 0 **35.** $\phi(x, y) = x^2 + y \sin xy$ **37.** $\phi(x, y) = x \ln \cosh y - y$ **39.** $\phi(x, y, z) = x^2 + y^2 + z^2$; 8

41. $\phi(x, y, z) = x^2yz - 3xz^2 + 4y^2z^3$

Exercise Set 19.3, page 765

1. $\frac{28}{3}$ **3.** $-\frac{4}{\pi}$ **5.** $\frac{1}{3}(3e^2 - 2e^3 - 1)$ **7.** 0 **9.** 0 **11.** -36 **13.** $\frac{5\pi}{3}$ **15.** 0 **17.** π **19.** 5 **21.** $\frac{3}{2}$

23. $\frac{3}{8}\pi a^2$ **25.** πa^2 **27.** Circulation $= \frac{4}{3}$; flux $= \frac{4}{3}$ **29.** Circulation $= 24\pi$; flux $= 0$ **31.** $\frac{81\pi}{4}$

33. If the origin is exterior to C, use Green's theorem. If the origin is interior to C, use Theorem 19.8 to replace C by a small circle lying inside C.

37. Show that integrals along common boundaries of subregions cancel.

Exercise Set 19.4, page 772

1. $\frac{8\sqrt{3}}{3}$ **3.** $\frac{1}{2}(81\pi\sqrt{14})$ **5.** $\frac{19\pi}{15}$ **7.** $\frac{32\pi}{3}$ **9.** $m = \frac{8\sqrt{21}}{3}$; $(\bar{x}, \bar{y}, \bar{z}) = (2, 1, 1)$ **11.** $m = \frac{13\pi k}{6}$; $(\bar{x}, \bar{y}, \bar{z}) = \left(0, 0, \frac{371}{30}\right)$

13. $I_x = \frac{k}{3}\int_0^2 \int_0^4 \left(y^2 + \frac{x^4}{9}\right)\sqrt{9 + 4x^2}\, dy\, dx$; $I_y = \frac{k}{3}\int_0^2 \int_0^4 \left(x^2 + \frac{x^4}{9}\right)\sqrt{9 + 4x^2}\, dy\, dx$; $I_z = \frac{k}{3}\int_0^2 \int_0^4 (x^2 + y^2)\sqrt{9 + 4x^2}\, dy\, dx$

15. (a) $\sqrt{6}\int_0^2 \int_0^{4-2x} x^2y(2x + y)^3\, dy\, dx$ (b) $\frac{\sqrt{6}}{8}\int_0^4 \int_0^z (z - y)^2yz^3\, dy\, dz$ (c) $\sqrt{6}\int_0^4 \int_0^{z/2} x^2(z - 2x)z^3\, dx\, dz$

17. xy-projection: $\int_1^{\sqrt{2}} \int_0^5 xy(2 - x^2)\sqrt{1 + 4x^2}\, dy\, dx$ yz-projection: $\frac{1}{2}\int_0^1 \int_0^5 yz\sqrt{9 - 4z}\, dy\, dz$ **19.** 19 **21.** 56π **23.** $\frac{14\pi}{3}$

25. $\frac{8}{3}$ **27.** $m = 2\pi a^2k$; $(\bar{x}, \bar{y}, \bar{z}) = \left(0, 0, \frac{a}{2}\right)$ **29.** Flux $= 4\pi k$

Exercise Set 19.5, page 778

1. 512 **3.** $\frac{52}{5}$ **5.** 24π **7.** $\frac{1792\pi}{3}$ **9.** $-\frac{59}{36}$ **11.** $\frac{11\pi}{10}$ **13.** $\iint_S \mathbf{F} \cdot \mathbf{n}\, d\sigma = \iiint_G \operatorname{div} \mathbf{F}\, dV = 24\pi$

15. $\iint_S \mathbf{F} \cdot \mathbf{n}\, d\sigma = \iiint_G \operatorname{div} \mathbf{F}\, dV = 4a^3$ **25.** Use Exercise 23 with appropriate choices of u and v.

Exercise Set 19.6, page 784

1. $\langle 1, 2xy, 1 - 2xz\rangle$ **3.** $x(e^z - e^y)\mathbf{i} + y(e^x - e^z)\mathbf{j} + z(e^y - e^x)\mathbf{k}$ **5.** $\langle 0, 0, y\cos xy + x\sin xy\rangle$ **7.** -35 **9.** -12π
17. $\phi(x, y, z) = x^2yz$ **19.** $\phi(x, y, z) = e^z \sin x \cos y$

Supplementary Exercises 19.7, page 785

1. $1 + \frac{\pi}{2} + \frac{\sqrt{2}}{3}$ **3.** $\frac{e^{3/2}(1 - e^3)}{3} + \frac{e^2 + 1}{4}$

5. (a) $\phi(x, y) = \frac{x}{\sqrt{x^2 + y^2}} - 2x + 3y$ (b) $\phi(x, y) = \ln\frac{xy}{x + y}$ (c) Not a gradient field (d) $\phi(x, y) = x\tanh y - y^2 + \ln\cosh x^2$

7. 0 **9.** 3π **11.** $2 - \frac{\pi}{2}$ **13.** π **15.** $\frac{4\pi}{3}(34 - 5\sqrt{5})$ **17.** xy-projection: $\int_0^2 \int_0^x x^2y(4 - x^2)^3\sqrt{1 + 4x^2}\, dy\, dx$
yz-projection: $\frac{1}{2}\int_0^2 \int_0^{4-y^2} yz^3\sqrt{(4 - z)(17 - 4z)}\, dz\, dy$

19. 6π **21.** $\frac{61\pi}{16}$ **23.** 0 **25.** -128π **27.** 2π

CHAPTER 20

Exercise Set 20.1, page 792

1. $y = \frac{x^2}{2} - 2x + C$ **3.** $y = \ln\sec x + 3$ **5.** $y = x^2 + x$ **13.** $y = -2e^{-x} + 2e^{-2x}$ **15.** $y = 2\cos ax + \sin ax$

17. $x = e^{-t}\sin t$ **23.** $y = \frac{2e^{-x}}{13}(8\cos 2x - \sin 2x) + \frac{33}{26}e^{2x} - x^2 - \frac{1}{2}$ **25.** $x^2y'' - xy' + y = 0$

Exercise Set 20.2, page 798

1. $y = C(x+1)$ **3.** $(y-2)^2 = (x+1)^2 + C$ **5.** $y = \tan(e^x + C)$ **7.** $xye^{x/y} = C$ **9.** $(x+y)(2y-x)^2 = C$

11. $y = 2\sin\left(\dfrac{x^2}{2} + C\right)$ **13.** $y = x\sin(\ln Cx)$ **15.** $y = \dfrac{2}{5}(x^2 + y^2)$ **17.** $e^{x/y} = \ln y$

19. $e^x(y+1)^2 = Ce^y(x+1)$ **23.** $(x-y)^2 = Ce^{-(x+y)^2/2y^2}$ **25.** $y = 1 + \dfrac{2-2x}{2+x}e^{-2x/(x-1)}$

27. If F is homogeneous of degree 0, then set $v = y/x$ to get $F(x, y) = F(x, vx) = x^{\circ}F(1, v) = F(1, v)$. Define G by $G(v) = F(1, v)$. Thus, $F(x, y) = G(y/x)$. Conversely, if it is known that $F(x, y) = G(y/x)$, then $F(tx, ty) = G(tx/ty) = G(y/x)$. So $F(tx, ty) = t^{\circ}F(x, y)$.

Exercise Set 20.3, page 804

1. $y = \dfrac{(x+1)^2}{2(x-2)} + \dfrac{C}{x-2}$ **3.** $xy^2 - x^3 = 3$ **5.** $y\ln x + \dfrac{y^2}{2} = 8$ **7.** $y = x - 1 + Ce^{-x}$ **9.** $y = \dfrac{x^2}{3} - \dfrac{2}{3x}$

11. $y = \dfrac{1}{2}(1-x)(1+x)^2 - \dfrac{1}{2}$ **13.** $x^2 - 4xy + 3y^2 = C$ **15.** $y = \dfrac{2+x}{1-x}$ **17.** $y = (x^2+1)\left(\dfrac{\pi}{4} + \tan^{-1}x\right)$

19. $y\ln xy - e^x + 2x + y^3 - 2y = C$ **21.** $\sqrt{x^2+y^2} + x - 2y = 0$ **23.** $y - y\ln x + \dfrac{y^2}{2} = C$ **25.** $x = 1 + y^2 + C\sqrt{1+y^2}$

27. $x\ln(x^2+y^2) - 2x - 2y\tan^{-1}\dfrac{y}{x} = C$ **29.** $y = \dfrac{2}{1+Ce^{x^2}}$ **31.** $\mu(x) = e^{\int[(f_y - g_x)/g]\,dx}$ **33.** $y^2 + 4x - 2 = Ce^{2x}$

Exercise Set 20.4, page 814

1. 3.42 hr; 276 **3.** \$6,049.65; 26.82 yr **5.** 3761.5 yr **7.** 33.7 yr; 6.35 yr

9. 6.17°; $\lim_{t\to\infty} T(t) = 5$, but $T(t) \neq 5$ for finite t, so model is only an approximation.

11. $Q(t) = 40(2 - e^{-t/20})$; concentration $\approx 35.5\%$ after 30 min **15.** $v(t) = \sqrt{\dfrac{mg}{k}}\tanh\sqrt{\dfrac{kg}{m}}\,t$; $\lim_{t\to\infty} v(t) = \sqrt{\dfrac{mg}{k}}$

17. $I(t) = 2(1 - e^{-6t})$ **19.** $y = \dfrac{a}{b}(1 - e^{-bt}) + y_0e^{-bt}$ **21.** $\ln\dfrac{b(a-2x)}{a(b-x)} + \dfrac{2x(2b-a)}{a(a-2x)} = (2b-a)^2kt$ **23.** 0.602%; 33.47 min

25. $Q(t) = Q_0e^{(\alpha/\beta)\sin\beta}$

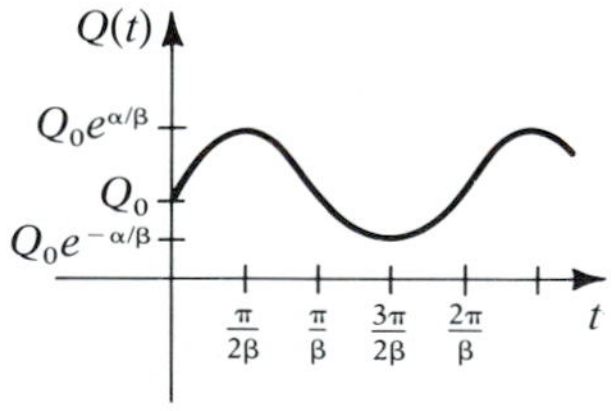

27. $Q(t) = \dfrac{\alpha(Q_0 - m) + m(\alpha - \beta Q_0)e^{-(\alpha - m\beta)t}}{\beta(Q_0 - m) + (\alpha - \beta Q_0)e^{-(\alpha - m\beta)t}}$

If $Q_0 < m$, $Q(t_1) = 0$ when $t_1 = \dfrac{1}{\alpha - m\beta}\ln\dfrac{m(\alpha - \beta Q_0)}{\alpha(m - Q_0)}$

Exercise Set 20.5, page 821

1. $y = C_1e^{-x} + C_2e^{-2x}$ **3.** $y = C_1e^{-4x} + C_2xe^{-4x}$ **5.** $y = C_1\cos 3x + C_2\sin 3x$ **7.** $y = e^x(C_1\cos 2x + C_2\sin 2x)$

9. $y = C_1e^{-x} + C_2e^{5x/2}$ **11.** $y = C_1e^{4t} + C_2e^{-t}$ **13.** $u = C_1 + C_2e^{3x/2}$ **15.** $s = e^{3t/2}(C_1 + C_2t)$ **17.** $y = \dfrac{3}{8}e^{-5x} + \dfrac{13}{8}e^{3x}$

19. $y = 3e^{3x} - 1$ **21.** $y = e^x(2 - x)$ **23.** $y = \sqrt{3}\sin x - \cos x$ **27.** Use the chain rule. **29.** $s = C_1 \ln t + C_2$

31. $y = C_1 + C_2 e^{3x} + C_3 e^{-x}$ **33.** $y = C_1 \cos x + C_2 \sin x + C_3 \cos 2x + C_4 \sin 2x$ **35.** $y = \frac{2}{5} e^x - \frac{2}{5} \cos 2x + \frac{3}{10} \sin 2x$

37. Use the fact that $b^2 - 4ac = 0$.

Exercise Set 20.6, page 826

1. $y_p = 2x$; $y = C_1 \cos x + C_2 \sin x + 2x$ **3.** $y_p = -\frac{3}{2}$; $y = C_1 e^{2x} + C_2 e^{-x} - \frac{3}{2}$ **5.** $y_p = -\frac{3}{4} e^x$ **7.** $y_p = \frac{1}{3} \sin x$

9. $y_p = \frac{1}{3} x e^{2x}$ **11.** $y_p = \frac{3}{2} x \sin x$ **13.** $y_p = \frac{1}{2} x^2 e^x$ **15.** $y_p = \frac{1}{4} x(x - 1)e^{2x}$ **17.** $y_p = -\frac{2}{3} e^{-x} \sin x$

19. $y_p = (Ax^4 + Bx^3 + Cx^2 + Dx)e^x$ **21.** $y_p = Ax^3 + Bx^2 + Cx + (Dx^2 + Ex)e^{2x}$ **23.** $y_p = e^{3x}(Ax \cos 4x + Bx \sin 4x) + C$

25. $y_p = e^{-x}[(Ax + B) \cos x + (Cx + D) \sin x]$ **27.** $y = \frac{11}{16} e^{2x} + \frac{1}{16} e^{-2x} - \frac{1}{4}(x + 3)$ **29.** $y = \frac{7}{3} \cos x + \frac{8}{3} \sin x - \frac{1}{3} \cos 2x$

31. $y = \frac{19e^{2x}}{17}(4 \sin x - \cos x) + \frac{2e^{-2x}}{17}$ **33.** $y = \frac{14}{9} e^{2x} + \frac{4}{9} e^{-x} + \frac{1}{3} x e^{2x}$ **35.** $y = C_1 \cosh x + C_2 \sinh x + \frac{1}{2} x \sinh x$

37. $y = C_1 \cos 2x + C_2 \sin 2x - \frac{1}{4} x \cos 2x$

39. $x = C_1 e^{3t} + C_2 t e^{3t} + t^2 e^{3t} + \frac{1}{9} t^2 + \frac{4}{27} t + \frac{2}{27}$; $x = \frac{106}{27} e^{3t} - \frac{376}{27} t e^{3t} + t^2 e^{3t} + \frac{1}{9} t^2 + \frac{4}{27} t + \frac{2}{27}$

41. $y_p = -\frac{x^4}{36} - \frac{x^3}{27} - \frac{x^2}{27}$; $y = C_1 + C_2 x + C_3 e^{3x} - \frac{x^4}{36} - \frac{x^3}{27} - \frac{x^2}{27}$ **43.** $y_p = \frac{1}{3} x^2 e^x + \frac{3x}{2} + \frac{9}{4}$; $y = C_1 e^x + C_2 x e^x + \frac{1}{3} x^2 e^x + \frac{3x}{2} + \frac{9}{4}$

Exercise Set 20.7, page 833

1. $y = \frac{1}{2} \cos 8t$; amplitude $= \frac{1}{2}$ ft, period $= \frac{\pi}{4}$ **3.** $y = \frac{2}{3} \cos 4t - 3 \sin 4t$ **5.** $y = -10 \cos 14t$

7. $y = 9e^{-7t} - 5e^{-4t}$; overdamped **9.** $y = \frac{7}{16} \cos 8t + \frac{1}{16} \cos 4t$ **11.** (a) $c > \sqrt{5}$ (b) $c = \sqrt{5}$ (c) $c < \sqrt{5}$

13. (a) $\frac{1}{2}$ radian (b) $t = \frac{\pi}{2}$ (c) $t = \frac{5\pi}{24}$ (d) 2 ft/sec

15. $C_1 \cos \omega t + C_2 \sin \omega t = \sqrt{C_1^2 + C_2^2}\left(\frac{C_1}{\sqrt{C_1^2 + C_2^2}} \cos \omega t + \frac{C_2}{\sqrt{C_1^2 + C_2^2}} \sin \omega t\right)$

Let α and β be defined by $\sin \alpha = C_1/\sqrt{C_1^2 + C_2^2} = \cos \beta$ and $\cos \alpha = C_2/\sqrt{C_1^2 + C_2^2} = \sin \beta$. Now use addition formulas for the sine and cosine.

17. $I(t) = e^{-t}(C_1 \cos 2t + C_2 \sin 2t) - 2 \cos t + 4 \sin t$; steady-state current is $I(t) = -2 \cos t + 4 \sin t$.

Exercise Set 20.8, page 838

1. $y = a_0 \sum_{k=0}^{\infty} \frac{x^{3k}}{3^k k!}$; by separating variables, $y = Ce^{x^3/3} = C \sum_{k=0}^{\infty} \frac{1}{k!}\left(\frac{x^3}{3}\right)^k$. Answers agree.

3. $y = a_0 \sum_{k=0}^{\infty} \frac{x^{2k}}{(2k)!} + a_1 \sum_{k=0}^{\infty} \frac{x^{2k+1}}{(2k+1)!} = a_0 \cosh x + a_1 \sinh x$; by solving as an exact equation obtain
$y = C_1 e^x + C_2 e^{-x} = C_3 \cosh x + C_4 \sinh x$. Answers agree.

5. $y = a_0\left[1 + \sum_{k=1}^{\infty} \frac{1 \cdot 4 \cdot 7 \cdot \cdots \cdot (3k-2)}{(3k)!} x^{3k}\right] + a_1\left[x + \sum_{k=1}^{\infty} \frac{2 \cdot 5 \cdot 8 \cdot \cdots \cdot (3k-1)}{(3k+1)!} x^{3k+1}\right]$

7. $y = \sum_{k=0}^{\infty} \frac{(-1)^k}{(2k)!} x^{2k} = \cos x$, agrees with solution as a linear homogeneous equation.

9. $y = a_0\left(1 - \frac{x^4}{3 \cdot 4} + \frac{x^8}{3 \cdot 4 \cdot 7 \cdot 8} - \frac{x^{12}}{3 \cdot 4 \cdot 7 \cdot 8 \cdot 11 \cdot 12} + \cdots\right) + a_1\left(x - \frac{x^5}{4 \cdot 5} + \frac{x^9}{4 \cdot 5 \cdot 8 \cdot 9} - \frac{x^{13}}{4 \cdot 5 \cdot 8 \cdot 9 \cdot 12 \cdot 13} + \cdots\right)$

11. $y = a_0 \sum_{k=0}^{\infty} \frac{1 \cdot 3 \cdot 5 \cdot \cdots \cdot (2k+1)}{2^k \cdot k!} x^{2k} + a_1 \sum_{k=0}^{\infty} \frac{2^k(k+1)!}{1 \cdot 3 \cdot 5 \cdot \cdots \cdot (2k+1)} x^{2k+1}$

13. $y = 2 + x - \frac{3}{2}x^2 + \frac{7}{4}\sum_{k=0}^{\infty} \frac{(-1)^k x^{2k+4}}{1 \cdot 3 \cdot 5 \cdot \cdots \cdot (2k+3)} + \sum_{k=1}^{\infty} \frac{(-1)^k x^{2k+1}}{2 \cdot 4 \cdot 6 \cdot \cdots \cdot (2k)}$

15. $y_1 = 1 - \frac{x^2}{2!} + \frac{x^3}{3!} + \frac{x^4}{4!} - \frac{4x^5}{5!} + \frac{3x^6}{6!} + \frac{9x^7}{7!} - \frac{27x^8}{8!} + \frac{12x^9}{9!} + \cdots$

$y_2 = x - \frac{x^3}{3!} + \frac{x^4}{4!} + \frac{x^5}{5!} - \frac{5x^6}{6!} + \frac{4x^7}{7!} + \frac{11x^8}{8!} - \frac{39x^9}{9!} + \frac{21x^{10}}{10!} + \cdots$

$y = C_1y_1 + C_2y_2$

17. $y = 2 + x + x^2 - \frac{x^3}{3!} - \frac{7x^5}{5!} + \frac{4x^6}{6!} - \frac{7x^7}{7!} + \frac{46x^8}{8!} - \frac{35x^9}{9!} + \frac{102x^{10}}{10!} + \cdots$

Supplementary Exercises 20.9, page 839

1. $2x + 3y - e^{-x}\cos y = C$ **3.** $\sin^{-1} y = \frac{1}{2}(x^2 + \pi)$ **5.** $Cx^2 = ye^{y/x}$ **7.** $\tan^{-1}\frac{y}{2x} = \frac{2}{3}\ln\frac{\sqrt{y^2 + 4x^2}}{2}$

9. $y = -\sqrt{2x - x^2}[\sin^{-1}(1 - x) + 2]$ **11.** $x\cos xy = C$ **13.** $y = -\frac{1}{2}e^{-2x^2} + \frac{3}{2}e^{-x^2}$ **15.** $y = 3e^{-3x} + 5xe^{-3x}$

17. $s = 2(1 - e^{-2t})$ **19.** $x = C_1e^{3t} + C_2e^{-2t} - \frac{1}{3}e^t$ **21.** $y = C_1 + C_2e^x - \frac{1}{2}x^2 - x + 3xe^x$

23. $y = C_1e^{-5x/2} + C_2e^x + \frac{e^x}{53}(7\sin x - 2\cos x)$ **25.** $y = \frac{1}{2}xe^x + \frac{1}{2}\sinh x$ **27.** $y_p = (Ax^2 + Bx)e^{-x} + C\cos 2x + D\sin 2x$

29. 13.16 yr; 43.71 yr **31.** $Q(t) = e^{\alpha t}\left(Q_0 - \frac{H}{\alpha}\right) + \frac{H}{\alpha}$

(1) If $H > \alpha Q_0$, $Q(t) = 0$ when $t = \frac{1}{\alpha}\ln\frac{H}{H - \alpha Q_0}$.

(2) If $H = \alpha Q_0$, $Q(t) = \frac{H}{\alpha}$ for all t.

(3) If $H < \alpha Q_0$, $\lim_{t\to\infty} Q(t) = +\infty$.

33. $y(t) = \frac{e^{-t}}{3}(2\sin 7t - \cos 7t)$ **35.** $y = a_0\sum_{k=0}^{\infty}\frac{2^{2k}k!}{(2k)!}x^{2k} + a_1\sum_{k=0}^{\infty}\frac{x^{2k+1}}{k!}$, $-\infty < x < \infty$

APPENDIX 1

Exercise Set A1.2, page A-6

1. (a) 28 (b) 28 (c) 792 (d) 435 **3.** $x^8 + 8x^7y + 28x^6y^2 + 56x^5y^3 + 70x^4y^4 + 56x^3y^5 + 28x^2y^6 + 8xy^7 + y^8$

5. $243a^5 + 1620a^4b + 4320a^3b^2 + 5760a^2b^3 + 3840ab^4 + 1024b^5$

7. $1024x^{10} - 15{,}360x^9y + 103{,}680x^8y^2 - 414{,}720x^7y^3 + 1{,}088{,}640x^6y^4$

APPENDIX 2

Exercise Set A2.1, page A-10

1. (a) $\frac{5\pi}{18}$ (b) $\frac{5\pi}{2}$ (c) $-\frac{10\pi}{9}$ (d) $\frac{4\pi}{5}$ **3.** (a) $108°$ (b) $630°$ (c) $-195°$ (d) $\left(\frac{540}{\pi}\right)^{\circ} \approx 171.89°$ **5.** 4π ft ≈ 12.57 ft

7. $\frac{8\pi}{3}$ in. ≈ 8.38 in. **9.** $\left(\frac{135}{\pi}\right)^{\circ} \approx 43.0°$ **11.** $\frac{7920}{7\pi}$ rev/min ≈ 360.14 rev/min

Exercise Set A2.2, page A-16

1. $\sin\theta = -\frac{3}{5}$, $\cos\theta = \frac{4}{5}$, $\tan\theta = -\frac{3}{4}$, $\cot\theta = -\frac{4}{3}$, $\sec\theta = \frac{5}{4}$, $\csc\theta = -\frac{5}{3}$

3. $\sin\theta = \frac{5}{\sqrt{29}}$, $\cos\theta = -\frac{2}{\sqrt{29}}$, $\tan\theta = -\frac{5}{2}$, $\cot\theta = -\frac{2}{5}$, $\sec\theta = -\frac{\sqrt{29}}{2}$, $\csc\theta = \frac{\sqrt{29}}{5}$

5. $\cos\theta = -\frac{\sqrt{5}}{3}$, $\tan\theta = \frac{2}{\sqrt{5}}$, $\cot\theta = \frac{2}{\sqrt{5}}$, $\sec\theta = -\frac{3}{\sqrt{5}}$, $\csc\theta = -\frac{3}{2}$

7. $\sin\theta = \frac{24}{25}$, $\cos\theta = -\frac{7}{25}$, $\tan\theta = -\frac{24}{7}$, $\cot\theta = -\frac{7}{24}$, $\csc\theta = \frac{25}{24}$ **9.** $\frac{\sqrt{21}}{2}$ **11.** $\frac{5}{2\sqrt{6}}$

13. When n is even, the angle $n\pi$ terminates on the positive x-axis, and $\frac{\pi}{2} + n\pi$ on the positive y-axis. When n is odd these angles terminate on the negative x- and y-axis, respectively.

17. The tangent and cotangent have period π; the secant and cosecant have period 2π. **19.** Make use of equations (A2.4).

21.

	Domain	Range
$\tan t$	$\left\{t : t \neq \frac{\pi}{2} + n\pi\right\}$	$\mathbf{R}$
$\cot t$	$\{t : t \neq n\pi\}$	$\mathbf{R}$
$\sec t$	$\left\{t : t \neq \frac{\pi}{2} + n\pi\right\}$	$\{y : \lvert y\rvert \geq 1\}$
$\csc t$	$\{t : t \neq n\pi\}$	$\{y : \lvert y\rvert \leq 1\}$

25. $\cos\frac{\pi}{12} = \frac{\sqrt{6}+\sqrt{2}}{4}$, $\sin\frac{\pi}{12} = \frac{\sqrt{6}-\sqrt{2}}{4}$

27. (a) $y = 2\sin 3x$ (b) $y = 3\cos\frac{x}{2}$

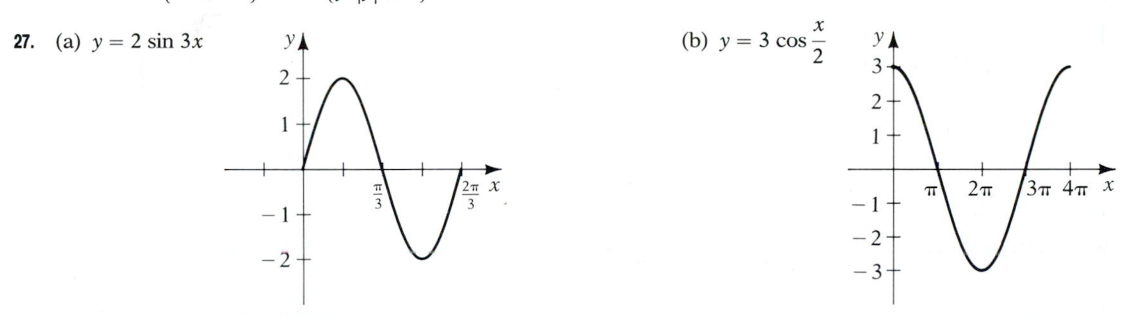

Exercise Set A2.3, page A-23

21. $\sin\frac{5\pi}{12} = \frac{\sqrt{6}+\sqrt{2}}{4}$, $\cos\frac{5\pi}{12} = \frac{\sqrt{6}-\sqrt{2}}{4}$, $\tan\frac{5\pi}{12} = 2+\sqrt{3}$, $\cot\frac{5\pi}{12} = 2-\sqrt{3}$, $\sec\frac{5\pi}{12} = \sqrt{6}+\sqrt{2}$, $\csc\frac{5\pi}{12} = \sqrt{6}-\sqrt{2}$

23. (a) $-\frac{33}{65}$ (b) $\frac{16}{65}$ **25.** (a) $\frac{24}{25}$ (b) $-\frac{117}{125}$ **27.** (a) $\frac{1}{2}(\sin 8x + \sin 2x)$ (b) $\frac{1}{2}(\cos 2x - \cos 12x)$

31. Make use of the result of Exercise 30. **45.** $\frac{3\pi}{2}$ **47.** Use $\frac{\pi}{12} = \frac{1}{2}\left(\frac{\pi}{6}\right)$ and $\frac{\pi}{12} = \frac{\pi}{3} - \frac{\pi}{4}$.

INDEX

Forms Involving $\sqrt{2au - u^2}$ *(continued)*

67. $$\int \frac{\sqrt{2au - u^2}}{u^n}\,du = \frac{(2au - u^2)^{3/2}}{(3 - 2n)au^n} + \frac{n-3}{(2n-3)a}\int \frac{\sqrt{2au - u^2}}{u^{n-1}}\,du, \quad n \neq \frac{3}{2}$$

68. $$\int \frac{u^n\,du}{\sqrt{2au - u^2}} = -\frac{u^{n-1}\sqrt{2au - u^2}}{n} + \frac{a(2n-1)}{n}\int \frac{u^{n-1}}{\sqrt{2au - u^2}}\,du$$

69. $$\int \frac{du}{u^n\sqrt{2au - u^2}} = \frac{\sqrt{2au - u^2}}{a(1 - 2n)u^n} + \frac{n-1}{(2n-1)a}\int \frac{du}{u^{n-1}\sqrt{2au - u^2}}$$

70. $$\int \frac{du}{(2au - u^2)^{3/2}} = \frac{u - a}{a^2\sqrt{2au - u^2}} + C$$

71. $$\int \frac{u\,du}{(2au - u^2)^{3/2}} = \frac{u}{a\sqrt{2au - u^2}} + C$$

Forms Containing Trigonometric Functions

72. $$\int \sin^2 u\,du = \frac{u}{2} - \frac{\sin 2u}{4} + C$$

73. $$\int \cos^2 u\,du = \frac{u}{2} + \frac{\sin 2u}{4} + C$$

74. $$\int \tan^2 u\,du = \tan u - u + C$$

75. $$\int \cot^2 u\,du = -\cot u - u + C$$

76. $$\int \sec^3 u\,du = \frac{1}{2}\sec u \tan u + \frac{1}{2}\ln|\sec u + \tan u| + C$$

77. $$\int \csc^3 u\,du = -\frac{1}{2}\csc u \cot u + \frac{1}{2}\ln|\csc u - \cot u| + C$$

78. $$\int \sin^n u\,du = -\frac{1}{n}\sin^{n-1} u \cos u + \frac{n-1}{n}\int \sin^{n-2} u\,du$$

79. $$\int \cos^n u\,du = \frac{1}{n}\cos^{n-1} u \sin u + \frac{n-1}{n}\int \cos^{n-2} u\,du$$

80. $$\int \tan^n u\,du = \frac{1}{n-1}\tan^{n-1} u - \int \tan^{n-2} u\,du$$

81. $$\int \cot^n u\,du = \frac{-1}{n-1}\cot^{n-1} u - \int \cot^{n-2} u\,du$$

82. $$\int \sec^n u\,du = \frac{1}{n-1}\tan u \sec^{n-2} u + \frac{n-2}{n-1}\int \sec^{n-2} u\,du$$

83. $$\int \csc^n u\,du = \frac{-1}{n-1}\cot u \csc^{n-2} u + \frac{n-2}{n-1}\int \csc^{n-2} u\,du$$

84. $$\int \sin mu \sin nu\,du = -\frac{\sin(m+n)u}{2(m+n)} + \frac{\sin(m-n)u}{2(m-n)} + C, \quad m^2 \neq n^2$$

85. $$\int \cos mu \cos nu\,du = \frac{\sin(m+n)u}{2(m+n)} + \frac{\sin(m-n)u}{2(m-n)} + C, \quad m^2 \neq n^2$$

86. $$\int \sin mu \cos nu\,du = -\frac{\cos(m+n)u}{2(m+n)} - \frac{\cos(m-n)u}{2(m-n)} + C, \quad m^2 \neq n^2$$

87. $$\int u \sin u\,du = \sin u - u \cos u + C$$

88. $$\int u \cos u\,du = \cos u + u \sin u + C$$

89. $$\int u^n \sin u\,du = -u^n \cos u + n\int u^{n-1} \cos u\,du$$

90. $$\int u^n \cos u\,du = u^n \sin u - n\int u^{n-1} \sin u\,du$$

91. $$\int \sin^m u \cos^n u\,du = -\frac{\sin^{m-1} u \cos^{n+1} u}{m+n} + \frac{m-1}{m+n}\int \sin^{m-2} u \cos^n u\,du$$
$$= -\frac{\sin^{m+1} u \cos^{n-1} u}{m+n} + \frac{n-1}{m+n}\int \sin^m u \cos^{n-2} u\,du$$

(If $m = -n$, use formula 80 or 81.)

Forms Containing Inverse Trigonometric Functions

92. $$\int \sin^{-1} u\, du = u \sin^{-1} u + \sqrt{1 - u^2} + C$$

93. $$\int \cos^{-1} u\, du = u \cos^{-1} u - \sqrt{1 - u^2} + C$$

94. $$\int \tan^{-1} u\, du = u \tan^{-1} u - \frac{1}{2} \ln(1 + u^2) + C$$

95. $$\int u \sin^{-1} u\, du = \frac{2u^2 - 1}{4} \sin^{-1} u + \frac{u\sqrt{1 - u^2}}{4} + C$$

96. $$\int u \cos^{-1} u\, du = \frac{2u^2 - 1}{4} \cos^{-1} u - \frac{u\sqrt{1 - u^2}}{4} + C$$

97. $$\int u \tan^{-1} u\, du = \frac{u^2 + 1}{2} \tan^{-1} u - \frac{u}{2} + C$$

98. $$\int u^n \sin^{-1} u\, du = \frac{1}{n+1}\left(u^{n+1} \sin^{-1} u - \int \frac{u^{n+1}\, du}{\sqrt{1 - u^2}}\right), \quad n \neq -1$$

99. $$\int u^n \cos^{-1} u\, du = \frac{1}{n+1}\left(u^{n+1} \cos^{-1} u + \int \frac{u^{n+1}\, du}{\sqrt{1 - u^2}}\right), \quad n \neq -1$$

100. $$\int u^n \tan^{-1} u\, du = \frac{1}{n+1}\left(u^{n+1} \tan^{-1} u - \int \frac{u^{n+1}\, du}{1 + u^2}\right), \quad n \neq -1$$

Exponential and Logarithmic Forms

101. $$\int u e^{au}\, du = \frac{1}{a^2}(au - 1)e^{au} + C$$

102. $$\int u^n e^{au}\, du = \frac{1}{a} u^n e^{au} - \frac{n}{a} \int u^{n-1} e^{au}\, du$$

103. $$\int \frac{e^u}{u^n}\, du = -\frac{e^u}{(n-1)u^{n-1}} + \frac{1}{n-1} \int \frac{e^u}{u^{n-1}}\, du$$

104. $$\int \ln u\, du = u \ln u - u + C$$

105. $$\int (\ln u)^n\, du = u(\ln u)^n - n \int (\ln u)^{n-1}\, du$$

106. $$\int u^n \ln u\, du = \frac{u^{n+1}}{n+1}\left(\ln u - \frac{1}{n+1}\right) + C$$

107. $$\int \frac{(\ln u)^n}{u}\, du = \frac{(\ln u)^{n+1}}{n+1} + C, \quad n \neq -1$$

108. $$\int \frac{du}{u \ln u} = \ln|\ln u| + C$$

109. $$\int u^m (\ln u)^n\, du = \frac{u^{m+1}(\ln u)^n}{m+1} - \frac{n}{m+1} \int u^m (\ln u)^{n-1}\, du$$

110. $$\int \frac{u^m}{(\ln u)^n}\, du = -\frac{u^{m+1}}{(n-1)(\ln u)^{n-1}} + \frac{m+1}{n-1} \int \frac{u^m}{(\ln u)^{n-1}}\, du$$

111. $$\int e^{au} \sin bu\, du = \frac{e^{au}}{a^2 + b^2}(a \sin bu - b \cos bu) + C$$

112. $$\int e^{au} \cos bu\, du = \frac{e^{au}}{a^2 + b^2}(a \cos bu + b \sin bu) + C$$

Forms Containing Hyperbolic Functions

113. $$\int \sinh u\, du = \cosh u + C$$

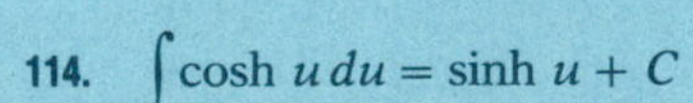

114. $$\int \cosh u\, du = \sinh u + C$$

115. $$\int \tanh u\, du = \ln \cosh u + C$$

116. $$\int \coth u\, du = \ln|\sinh u| + C$$

117. $$\int \operatorname{sech} u\, du = \tan^{-1}(\sinh u) + C$$

118. $$\int \operatorname{csch} u\, du = \ln\left|\tanh \frac{u}{2}\right| + C$$

119. $$\int \operatorname{sech}^2 u\, du = \tanh u + C$$

120. $$\int \operatorname{csch}^2 u\, du = -\coth u + C$$

121. $$\int \operatorname{sech} u \tanh u\, du = -\operatorname{sech} u + C$$

122. $$\int \operatorname{csch} u \coth u\, du = -\operatorname{csch} u + C$$

123. $$\int \sinh^2 u\, du = \frac{\sinh 2u}{4} - \frac{u}{2} + C$$

124. $$\int \cosh^2 u\, du = \frac{\sinh 2u}{4} + \frac{u}{2} + C$$

125. $$\int \tanh^2 u\, du = u - \tanh u + C$$

126. $$\int \coth^2 u\, du = u - \coth u + C$$

127. $$\int u^n \sinh u\, du = u^n \cosh u - n \int u^{n-1} \cosh u\, du$$

128. $$\int u^n \cosh u\, du = u^n \sinh u - n \int u^{n-1} \sinh u\, du$$